$136.95 per copy (in United States).
Price subject to change without prior notice.

0017

RSMeans
Construction Publishers & Consultants
63 Smiths Lane
Kingston, MA 02364-0800
(781) 422-5000

Copyright©2006 by Reed Construction Data, Inc.
All rights reserved.

Printed in the United States of America.
ISSN 0068-3531
ISBN 0-87629-851-X

The authors, editors, and engineers of RSMeans, a business unit of Reed Construction Data, apply diligence and judgement in locating and using reliable sources for the information published. **However, RSMeans makes no express or implied warranty or guarantee in connection with the content of the information contained herein, including the accuracy, correctness, value, sufficiency, or completeness of the data, methods, and other information contained herein. RSMeans makes no express or implied warranty of merchantability or fitness for a particular purpose.** RSMeans shall have no liability to any customer or third party for any loss, expense, or damage, including consequential, incidental, special, or punitive damage, including lost profits or lost revenue, caused directly or indirectly by any error or omission, or arising out of, or in connection with, the information contained herein.

No part of this publication may be reproduced, stored in a retrieval system, or transmitted in any form or by any means without prior written permission of Reed Construction Data.

RSMeans
Building Construction Cost Data
65th Annual Edition

Senior Editor
Phillip R. Waier, PE

Contributing Editors
Christopher Babbitt
Ted Baker
Barbara Balboni
Robert A. Bastoni
John H. Chiang, PE
Cheryl Elsmore
Robert J. Kuchta
Robert C. McNichols
Robert W. Mewis, CCC
Melville J. Mossman, PE
Jeannene D. Murphy
Stephen C. Plotner
Eugene R. Spencer
Marshall J. Stetson

Senior Engineering Operations Manager
John H. Ferguson, PE

Senior Vice President & General Manager
John Ware

Vice President of Direct Response
John M. Shea

Director of Product Development
Thomas J. Dion

Production Manager
Michael Kokernak

Production Coordinator
Wayne D. Anderson

Technical Support
Jonathan Forgit
Mary Lou Geary
Jill Goodman
Gary L. Hoitt
Genevieve Medeiros
Paula Reale-Camelio
Kathryn S. Rodriguez
Sheryl A. Rose

Book & Cover Design
Norman R. Forgit

Editorial Advisory Board

James E. Armstrong, CPE, CEM
Senior Energy Consultant
KEMA Services, Inc.

William R. Barry, CCC
Cost Consultant

Robert F. Cox, PhD
Department Head and Professor
of Building Construction Management
Purdue University

Roy F. Gilley, AIA
Principal
Gilley Design Associates

Kenneth K. Humphreys, PhD, PE, CCE

Patricia L. Jackson, PE
Jackson A&E Associates, Inc.

Martin F. Joyce
Executive Vice President
Bond Brothers, Inc.

 This book is recyclable.

 This book is printed on recycled stock.

 Reed Construction Data®

First Printing

Foreword

Reed Construction Data's portfolio of project information products and services includes national, regional, and local construction data, project leads, and project plans, specifications, and addenda available online or in print. Reed Bulletin (www.reedbulletin.com) and Reed CONNECT™ (www.reedconnect.com) deliver the most comprehensive, timely, and reliable project information to support contractors, distributers, and building product manufacturers in identifying, bidding, and tracking projects. The Reed First Source (www.reedfirstsource.com) suite of products used by design professionals for the search, selection, and specification of nationally available building projects consists of the First Source™ annual, SPEC-DATA™, MANU-SPEC™, First Source CAD, and manufacturer catalogs. Reed Design Registry, a database of more then 30,000 U.S. architectural firms, is also published by Reed Construction Data. RSMeans (www.rsmeans.com) provides construction cost data, training, and consulting services in print, CD-ROM, and online. Associated Construction Publications (www.reedpubs.com) reports on heavy, highway, and non-residential construction through a network of 14 regional construction magazines. Reed Construction Data (www.reedconstructiondata.com), headquartered in Atlanta, is a subsidiary of Reed Business Information (www.reedbusinessinformation.com), North America's largest business-to-business information provider. With more than 80 market-leading publications and 55 websites, Reed Business Information's wide range of services also includes research, business development, direct marketing lists, training and development programs, and technology solutions. Reed Business Information is a member of the Reed Elsevier plc group (NYSE: RUK and ENL)—a leading provider of global information-driven services and solutions in the science and medical, legal, education, and business-to-business industry sectors.

Our Mission

Since 1942, RSMeans has been actively engaged in construction cost publishing and consulting throughout North America.

Today, over 60 years after RSMeans began, our primary objective remains the same: to provide you, the construction and facilities professional, with the most current and comprehensive construction cost data possible.

Whether you are a contractor, an owner, an architect, an engineer, a facilities manager, or anyone else who needs a reliable construction cost estimate, you'll find this publication to be a highly useful and necessary tool.

Today, with the constant flow of new construction methods and materials, it's difficult to find the time to look at and evaluate all the different construction cost possibilities. In addition, because labor and material costs keep changing, last year's cost information is not a reliable basis for today's estimate or budget.

That's why so many construction professionals turn to RSMeans. We keep track of the costs for you, along with a wide range of other key information, from city cost indexes . . . to productivity rates . . . to crew composition . . . to contractor's overhead and profit rates.

RSMeans performs these functions by collecting data from all facets of the industry and organizing it in a format that is instantly accessible to you. From the preliminary budget to the detailed unit price estimate, you'll find the data in this book useful for all phases of construction cost determination.

The Staff, the Organization, and Our Services

When you purchase one of RSMeans' publications, you are, in effect, hiring the services of a full-time staff of construction and engineering professionals.

Our thoroughly experienced and highly qualified staff works daily at collecting, analyzing, and disseminating comprehensive cost information for your needs. These staff members have years of practical construction experience and engineering training prior to joining the firm. As a result, you can count on them not only for the cost figures, but also for additional background reference information that will help you create a realistic estimate.

The RSMeans organization is always prepared to help you solve construction problems through its five major divisions: Construction and Cost Data Publishing, Electronic Products and Services, Consulting and Research Services, Insurance Services, and Professional Development Services.

Besides a full array of construction cost estimating books, RSMeans also publishes a number of other reference works for the construction industry. Subjects include construction estimating and project and business management; special topics such as HVAC, roofing, plumbing, and hazardous waste remediation; and a library of facility management references.

In addition, you can access all of our construction cost data electronically using *Means CostWorks*® CD or on the Web.

What's more, you can increase your knowledge and improve your construction estimating and management performance with an RSMeans Construction Seminar or In-House Training Program. These two-day seminar programs offer unparalleled opportunities for everyone in your organization to get updated on a wide variety of construction-related issues.

RSMeans is also a worldwide provider of construction cost management and analysis services for commercial and government owners, and of claims and valuation services for insurers.

In short, RSMeans can provide you with the tools and expertise for constructing accurate and dependable construction estimates and budgets in a variety of ways.

Robert Snow Means Established a Tradition of Quality That Continues Today

Robert Snow Means spent years building RSMeans, making certain he always delivered a quality product.

Today, at RSMeans, we do more than talk about the quality of our data and the usefulness of our books. We stand behind all of our data, from historical cost indexes to construction materials and techniques to current costs.

If you have any questions about our products or services, please call us toll-free at 1-800-334-3509. Our customer service representatives will be happy to assist you. You can also visit our Web site at www.rsmeans.com.

Table of Contents

Foreword ii
How the Book Is Built: An Overview v
How To Use the Book: The Details vi
Unit Price Section 1
Reference Section 599
    Construction Equipment Rental Costs 601
    Crew Listings 613
    Historical Cost Indexes 642
    City Cost Indexes 643
    Location Factors 686
    Reference Tables 692
    Change Orders 752
    Square Foot Costs 757
    Abbreviations 767
Index 770
Other RSMeans Products and Services 797
Labor Trade Rates including Overhead & Profit Inside Back Cover

Related RSMeans Products and Services

Phillip R. Waier, P.E., Senior Editor of this cost data book, suggests the following RSMeans products and services as companion information resources to *RSMeans Building Construction Cost Data*:

Construction Cost Data Books
Assemblies Cost Data 2007
Square Foot Costs 2007

Reference Books
ADA Compliance Pricing Guide, 2nd Edition
Building Security: Strategies & Costs
Designing & Building with the IBC
Estimating Building Costs
Estimating Handbook, 2nd Edition
Green Building: Project Planning & Estimating, 2nd Ed.
How to Estimate with Means Data and CostWorks, 3rd Ed.
Plan Reading & Material Takeoff
Project Scheduling and Management for Construction
Unit Price Estimating Methods, 3rd Edition

Seminars and In-House Training
Unit Price Estimating
Means CostWorks® Training
Means Data for Job Order Contracting (JOC)
Plan Reading & Material Takeoff
Scheduling & Project Management
Mechanical & Electrical Estimating

RSMeans on the Internet
Visit RSMeans at **www.rsmeans.com.** The site contains useful interactive cost and reference material. Request or download **FREE** estimating software demos. Visit our bookstore for convenient ordering and to learn more about new publications and companion products.

RSMeans Electronic Data
Get the information found in RSMeans cost books electronically on *Means CostWorks*® CD or on the Web.

RSMeans Business Solutions
Engineers and Analysts offer research studies, benchmark analysis, predictive cost modeling, analytics, job order contracting, and real property management consultation, as well as custom-designed, web-based dashboards and calculators that apply RCD/RSMeans extensive databases. Clients include federal government agencies, architects, construction management firms, and institutional organizations such as school systems, health care facilities, associations, and corporations.

New! RSMeans for Job Order Contracting (JOC)
Best practice JOC tools for cost estimating and project management to help streamline delivery processes for renovation projects. Renovation is a $147 billion market in the U.S., and includes projects in school districts, municipalities, health care facilities, colleges and universities, and corporations.
- RSMeans Engineers consult in contracting methods and conduct JOC Facility Audits
- JOCWorks™ Software (Basic, Advanced, PRO levels)
- RSMeans Job Order Contracting Cost Data for the entire U.S.

Construction Costs for Software Applications
Over 25 unit price and assemblies cost databases are available through a number of leading estimating and facilities management software providers (listed below). For more information see the "Other RSMeans Products" pages at the back of this publication.
MeansData™ is also available to federal, state, and local government agencies as multi-year, multi-seat licenses.

- 3D International
- 4Clicks-Solutions, LLC
- Aepco, Inc.
- Applied Flow Technology
- ArenaSoft Estimating
- Ares Corporation
- Beck
- BSD – Building Systems Design, Inc.
- CMS – Construction Management Software
- Corecon Technologies, Inc.
- CorVet Systems
- Earth Tech
- Estimating Systems, Inc.
- HCSS
- Maximus Asset Solutions
- MC^2 – Management Computer Controls
- Sage Timberline Office
- Shaw Beneco Enterprises, Inc.
- US Cost, Inc.
- VFA – Vanderweil Facility Advisers
- WinEstimator, Inc.

How the Book Is Built: An Overview

NEW In 2007...

The Construction Specifications Institute (CSI) and Construction Specifications Canada (CSC) have produced the 2004 edition of MasterFormat, a new and updated system of titles and numbers used extensively to organize construction information.

All unit price data in the RSMeans cost data books is now arranged in the 50-division MasterFormat 2004 system.

A Powerful Construction Tool

You have in your hands one of the most powerful construction tools available today. A successful project is built on the foundation of an accurate and dependable estimate. This book will enable you to construct just such an estimate.

For the casual user the book is designed to be:

- quickly and easily understood so you can get right to your estimate.
- filled with valuable information so you can understand the necessary factors that go into the cost estimate.

For the regular user, the book is designed to be:

- a handy desk reference that can be quickly referred to for key costs.
- a comprehensive, fully reliable source of current construction costs and productivity rates so you'll be prepared to estimate any project.
- a source book for preliminary project cost, product selections, and alternate materials and methods.

To meet all of these requirements we have organized the book into the following clearly defined sections.

How To Use the Book: The Details

This section contains an in-depth explanation of how the book is arranged . . . and how you can use it to determine a reliable construction cost estimate. It includes information about how we develop our cost figures and how to completely prepare your estimate.

Unit Price Section

All unit price cost data has been divided into the 50 divisions according to the MasterFormat system of classification and numbering. For a listing of these divisions and an outline of their subdivisions, see the Unit Price Section Table of Contents.

Estimating tips are included at the beginning of each division.

Reference Section

This section includes information on Equipment Rental Costs, Crew Listings, Historical Cost Indexes, City Cost Indexes, Location Factors, Reference Tables, Change Orders, Square Foot Costs, and a listing of Abbreviations.

Equipment Rental Costs: This section contains the average costs to rent and operate hundreds of pieces of construction equipment.

Crew Listings: This section lists all the crews referenced in the book. For the purposes of this book, a crew is composed of more than one trade classification and/or the addition of power equipment to any trade classification. Power equipment is included in the cost of the crew. Costs are shown both with the bare labor rates and with the installing contractor's overhead and profit added. For each, the total crew cost per eight-hour day and the composite cost per labor-hour are listed.

Historical Cost Indexes: These indexes provide you with data to adjust construction costs over time.

City Cost Indexes: All costs in this book are U.S. national averages. Costs vary because of the regional economy. You can adjust costs by CSI Division to over 700 locations throughout the U.S. and Canada by using the data in this section.

Location Factors: You can adjust total project costs to over 900 locations throughout the U.S. and Canada by using the data in this section.

Reference Tables: At the beginning of selected major classifications in the Unit Price section are "reference numbers" shown in a shaded box. These numbers refer you to related information in the Reference Section. In this section, you'll find reference tables, explanations, estimating information that support how we develop the unit price data, technical data, and estimating procedures.

Change Orders: This section includes information on the factors that influence the pricing of change orders.

Square Foot Costs: This section contains costs for 59 different building types that allow you to make a rough estimate for the overall cost of a project or its major components.

Abbreviations: A listing of abbreviations used throughout this book, along with the terms they represent, is included in this section.

Index

A comprehensive listing of all terms and subjects in this book will help you quickly find what you need when you are not sure where it falls in MasterFormat.

The Scope of This Book

This book is designed to be as comprehensive and as easy to use as possible. To that end we have made certain assumptions and limited its scope in three key ways:

1. We have established material prices based on a national average.
2. We have computed labor costs based on a 30-city national average of union wage rates.
3. We have targeted the data for projects of a certain size range.

For a more detailed explanation of how the cost data is developed, see "How To Use the Book: The Details."

Project Size

This book is aimed primarily at commercial and industrial projects costing $1,000,000 and up, or large multi-family housing projects. Costs are primarily for new construction or major renovation of buildings rather than repairs or minor alterations.

With reasonable exercise of judgment the figures can be used for any building work. However, for civil engineering structures such as bridges, dams, highways, or the like, please refer to RSMeans Heavy Construction Cost Data.

How to Use the Book: The Details

What's Behind the Numbers? The Development of Cost Data

The staff at RSMeans continuously monitors developments in the construction industry in order to ensure reliable, thorough, and up-to-date cost information.

While *overall* construction costs may vary relative to general economic conditions, price fluctuations within the industry are dependent upon many factors. Individual price variations may, in fact, be opposite to overall economic trends. Therefore, costs are continually monitored and complete updates are published yearly. Also, new items are frequently added in response to changes in materials and methods.

Costs—$ (U.S.)

All costs represent U.S. national averages and are given in U.S. dollars. The RSMeans City Cost Indexes can be used to adjust costs to a particular location. The City Cost Indexes for Canada can be used to adjust U.S. national averages to local costs in Canadian dollars. No exchange rate conversion is necessary.

Material Costs

The RSMeans staff contacts manufacturers, dealers, distributors, and contractors all across the U.S. and Canada to determine national average material costs. If you have access to current material costs for your specific location, you may wish to make adjustments to reflect differences from the national average. Included within material costs are fasteners for a normal installation. RSMeans engineers use manufacturers' recommendations, written specifications, and/ or standard construction practice for size and spacing of fasteners. Adjustments to material costs may be required for your specific application or location. Material costs do not include sales tax.

Labor Costs

Labor costs are based on the average of wage rates from 30 major U.S. cities. Rates are determined from labor union agreements or prevailing wages for construction trades for the current year. Rates, along with overhead and profit markups, are listed on the inside back cover of this book.

- If wage rates in your area vary from those used in this book, or if rate increases are expected within a given year, labor costs should be adjusted accordingly.

Labor costs reflect productivity based on actual working conditions. These figures include time spent during a normal workday on tasks other than actual installation, such as material receiving and handling, mobilization at site, site movement, breaks, and cleanup.

Productivity data is developed over an extended period so as not to be influenced by abnormal variations and reflects a typical average.

Equipment Costs

Equipment costs include not only rental, but also operating costs for equipment under normal use. The operating costs include parts and labor for routine servicing such as repair and replacement of pumps, filters, and worn lines. Normal operating expendables, such as fuel, lubricants, tires, and electricity (where applicable), are also included. Extraordinary operating expendables with highly variable wear patterns, such as diamond bits and blades, are excluded. These costs are included under materials. Equipment rental rates are obtained from industry sources throughout North America—contractors, suppliers, dealers, manufacturers, and distributors.

Crew Equipment Cost/Day—The power equipment required for each crew is included in the crew cost. The daily cost for crew equipment is based on dividing the weekly bare rental rate by 5 (number of working days per week) and then adding the hourly operating cost times 8 (hours per day). This "Crew Equipment Cost/Day" is listed in the Reference Section.

Mobilization/Demobilization—The cost to move construction equipment from an equipment yard or rental company to the job site and back again is not included in equipment costs. Mobilization (to the site) and demobilization (from the site) costs can be found in the Unit Price section. If a piece of equipment is already at the job site, it is not appropriate to utilize mobil./demob. costs again in an estimate.

Overhead and Profit

Total Cost including O&P for the *Installing Contractor* is shown in the last column on the Unit Price pages of this book. This figure is the sum of the bare material cost plus 10% for overhead and profit, the bare labor cost plus total overhead and profit, and the bare equipment cost plus 10% for overhead and profit. Details for the calculation of Overhead and Profit on labor are shown on the inside back cover of this book. (See the "How To Use the Unit Price Pages" for an example of this calculation.)

General Conditions

General conditions, or general requirements of the contract, should also be added to the Total Cost including O&P when applicable. Costs for General Conditions are listed in Division 1 and the Reference Section of this book. General Conditions for the *Installing Contractor* may range from 0% to 10% of the Total Cost including O&P. For the *General or Prime Contractor*, costs for General Conditions may range from 5% to 15% of the Total Cost including O&P, with a figure of 10% as the most typical allowance.

Factors Affecting Costs

Costs can vary depending upon a number of variables. Here's how we have handled the main factors affecting costs.

Quality—The prices for materials and the workmanship upon which productivity is based represent sound construction work. They are also in line with U.S. government specifications.

Overtime—We have made no allowance for overtime. If you anticipate premium time or work beyond normal working hours, be sure to make an appropriate adjustment to your labor costs.

Productivity—The productivity, daily output, and labor-hour figures for each line item are based on working an eight-hour day in daylight hours in moderate temperatures. For work that extends beyond normal work hours or is performed under adverse conditions, productivity may decrease. (See the section in "How To Use the Unit Price Pages" for more on productivity.)

Size of Project—The size, scope of work, and type of construction project will have a significant impact on cost. Economies of scale can reduce costs for large projects. Unit costs can often run higher for small projects. Costs in this book are intended for the size and type of project as previously described in "How the Book Is Built: An Overview." Costs for projects of a significantly different size or type should be adjusted accordingly.

Location—Material prices in this book are for metropolitan areas. However, in dense urban areas, traffic and site storage limitations may increase costs. Beyond a 20-mile radius of large cities, extra trucking or transportation charges may also increase the material costs slightly. On the other hand, lower wage rates may be in effect. Be sure to consider both of these factors when preparing an estimate, particularly if the job site is located in a central city or remote rural location.

In addition, highly specialized subcontract items may require travel and per-diem expenses for mechanics.

Other Factors—
- season of year
- contractor management
- weather conditions
- local union restrictions
- building code requirements
- availability of:
  - adequate energy
  - skilled labor
  - building materials
- owner's special requirements/restrictions
- safety requirements
- environmental considerations

Unpredictable Factors—General business conditions influence "in-place" costs of all items. Substitute materials and construction methods may have to be employed. These may affect the installed cost and/or life cycle costs. Such factors may be difficult to evaluate and cannot necessarily be predicted on the basis of the job's location in a particular section of the country. Thus, where these factors apply, you may find significant but unavoidable cost variations for which you will have to apply a measure of judgment to your estimate.

Rounding of Costs

In general, all unit prices in excess of $5.00 have been rounded to make them easier to use and still maintain adequate precision of the results. The rounding rules we have chosen are in the following table.

Prices from . . .	Rounded to the nearest . . .
$.01 to $5.00	$.01
$5.01 to $20.00	$.05
$20.01 to $100.00	$.50
$100.01 to $300.00	$1.00
$300.01 to $1,000.00	$5.00
$1,000.01 to $10,000.00	$25.00
$10,000.01 to $50,000.00	$100.00
$50,000.01 and above	$500.00

Final Checklist

Estimating can be a straightforward process provided you remember the basics. Here's a checklist of some of the steps you should remember to complete before finalizing your estimate.

Did you remember to . . .

- factor in the City Cost Index for your locale
- take into consideration which items have been marked up and by how much?
- mark up the entire estimate sufficiently for your purposes?
- read the background information on techniques and technical matters that could impact your project time span and cost?
- include all components of your project in the final estimate?
- double check your figures for accuracy?
- call RSMeans if you have any questions about your estimate or the data you've found in our publications?

Remember, RSMeans stands behind its publications. If you have any questions about your estimate . . . about the costs you've used from our books . . . or even about the technical aspects of the job that may affect your estimate, feel free to call the RSMeans editors at 1-800-334-3509.

Unit Price Section

Table of Contents

Div. No.		Page
	General Requirements	**7**
01 11	Summary of Work	8
01 21	Allowances	9
01 31	Project Management & Coordination	10
01 32	Construction Progress Documentation	12
01 41	Regulatory Requirements	12
01 45	Quality Control	12
01 51	Temporary Utilities	14
01 52	Construction Facilities	15
01 54	Construction Aids	15
01 55	Vehicular Access & Parking	20
01 56	Temporary Barriers & Enclosures	20
01 58	Project Identification	21
01 71	Examination & Preparation	21
01 74	Cleaning & Waste Management	22
01 91	Commissioning	22
	Existing Conditions	**23**
02 21	Surveys	24
02 32	Geotechnical Investigations	24
02 41	Demolition	25
02 43	Structure Moving	30
02 58	Snow Control	30
02 65	Underground Storage Tank Removal	30
02 81	Transportation & Disposal of Hazardous Materials	31
02 82	Asbestos Remediation	31
02 83	Lead Remediation	35
02 85	Mold Remediation	36
	Concrete	**39**
03 01	Maintenance of Concrete	40
03 05	Common Work Results for Concrete	40
03 11	Concrete Forming	42
03 15	Concrete Accessories	50
03 21	Reinforcing Steel	55
03 22	Welded Wire Fabric Reinforcing	60
03 23	Stessing Tendons	61
03 24	Fibrous Reinforcing	61
03 30	Cast-In-Place Concrete	61
03 31	Structural Concrete	64
03 35	Concrete Finishing	66
03 37	Specialty Placed Concrete	67
03 39	Concrete Curing	68
03 41	Precast Structural Concrete	68
03 45	Precast Architectural Concrete	70
03 47	Site-Cast Concrete	71
03 48	Precast Concrete Specialties	71
03 51	Cast Roof Decks	71
03 52	Lightweight Concrete Roof Insulation	72
03 54	Cast Underlayment	72
03 62	Non-Shrink Grouting	72
03 63	Epoxy Grouting	73
03 81	Concrete Cutting	73
03 82	Concrete Boring	73
	Masonry	**75**
04 01	Maintenance of Masonry	76
04 05	Common Work Results for Masonry	77
04 21	Clay Unit Masonry	81
04 22	Concrete Unit Masonry	85
04 23	Glass Unit Masonry	91
04 24	Adobe Unit Masonry	91

Div. No.		Page
04 25	Unit Masonry Panels	92
04 27	Multiple-Wythe Unit Masonry	92
04 41	Dry-Placed Stone	93
04 43	Stone Masonry	93
04 51	Flue Liner Masonry	96
04 54	Refractory Brick Masonry	96
04 57	Masonry Fireplaces	96
04 71	Manufactured Brick Masonry	97
04 72	Cast Stone Masonry	97
04 73	Manufactured Stone Masonry	98
	Metals	**99**
05 01	Maintenance of Metals	101
05 05	Common Work Results for Metals	101
05 12	Structural Steel Framing	110
05 14	Structural Aluminum Framing	117
05 15	Wire Rope Assemblies	117
05 21	Steel Joist Framing	119
05 31	Steel Decking	121
05 35	Raceway Decking Assemblies	123
05 41	Structural Metal Stud Framing	123
05 42	Cold-Formed Metal Joist Framing	127
05 44	Cold-Formed Metal Trusses	132
05 51	Metal Stairs	133
05 52	Metal Railings	134
05 53	Metal Gratings	136
05 54	Metal Floor Plates	138
05 55	Metal Stair Treads & Nosings	138
05 56	Metal Castings	139
05 58	Formed Metal Fabrications	139
05 59	Metal Specialties	140
05 71	Decorative Metal Stairs	141
05 73	Decorative Metal Railings	141
05 75	Decorative Formed Metal	142
	Wood, Plastics, and Compos.	**143**
06 05	Common Work Results for Wood, Plastics & Composites	144
06 11	Wood Framing	153
06 12	Structural Panels	160
06 13	Heavy Timber	161
06 15	Wood Decking	161
06 16	Sheathing	162
06 17	Shop-Fabricated Structural Wood	164
06 18	Glued-Laminated Construction	164
06 22	Millwork	165
06 25	Prefinished Paneling	167
06 26	Board Paneling	169
06 43	Wood Stairs & Railings	169
06 44	Ornamental Woodwork	170
06 48	Wood Frames	171
06 49	Wood Screens & Exterior Wood Shutters	172
06 52	Plastic Structural Assemblies	173
06 65	Plastic Simulated Wood Trim	174
	Thermal and Moisture Protection	**177**
07 01	Operation & Maint. of Thermal and Moisture Protection	178
07 05	Common Work Results for Thermal and Moisture Protection	178
07 11	Dampproofing	179
07 12	Built-Up Bituminous Waterproofing	179
07 13	Sheet Waterproofing	180
07 16	Cementitious & Reactive Waterproofing	180

Div. No.		Page
07 17	Bentonite Waterproofing	180
07 19	Water Repellents	181
07 21	Thermal Insulation	181
07 22	Roof & Deck Insulation	184
07 24	Exterior Insulation & Finish Systems	185
07 26	Vapor Retarders	186
07 31	Shingles & Shakes	186
07 32	Roof Tiles	188
07 41	Roof Panels	189
07 42	Wall Panels	191
07 44	Faced Panels	192
07 46	Siding	193
07 51	Built-Up Bituminous Roofing	194
07 52	Modified Bituminous Membrane Roofing	196
07 53	Elastomeric Membrane Roofing	197
07 54	Thermoplastic Membrane Roofing	197
07 56	Fluid-Applied Roofing	198
07 57	Coated Foamed Roofing	198
07 58	Roll Roofing	199
07 61	Sheet Metal Roofing	199
07 65	Flexible Flashing	200
07 71	Roof Specialties	202
07 72	Roof Accessories	206
07 81	Applied Fireproofing	207
07 84	Firestopping	208
07 92	Joint Sealants	209
07 95	Expansion Control	210
	Openings	**211**
08 05	Common Work Results for Openings	212
08 11	Metal Doors & Frames	213
08 12	Metal Frames	214
08 13	Metal Doors	215
08 14	Wood Doors	218
08 16	Composite Doors	223
08 17	Integrated Door Opening Assemblies	224
08 31	Access Doors & Panels	225
08 32	Sliding Glass Doors	226
08 33	Coiling Doors & Grilles	226
08 34	Special Function Doors	228
08 36	Panel Doors	229
08 38	Traffic Doors	231
08 41	Entrances & Storefronts	232
08 42	Entrances	233
08 43	Storefronts	234
08 44	Curtain Wall & Glazed Assemblies	235
08 45	Translucent Wall & Roof Assemblies	235
08 51	Metal Windows	236
08 52	Wood Windows	237
08 53	Plastic Windows	242
08 56	Special Function Windows	244
08 62	Unit Skylights	244
08 63	Metal-Framed Skylights	245
08 71	Door Hardware	245
08 74	Access Control Hardware	253
08 75	Window Hardware	254
08 79	Hardware Accessories	254
08 81	Glass Glazing	255
08 83	Mirrors	257
08 84	Plastic Glazing	257
08 87	Glazing Surface Films	258
08 88	Special Function Glazing	259
08 91	Louvers	259
08 95	Vents	260

Div. No.		Page
	Finishes	**261**
09 01	Maintenance of Finishes	262
09 05	Common Work Results for Finishes	262
09 21	Plaster & Gypsum Board Assemblies	264
09 22	Supports for Plaster & Gypsum Board	265
09 23	Gypsum Plastering	269
09 24	Portland Cement Plastering	270
09 26	Veneer Plastering	270
09 28	Backing Boards & Underlayments	271
09 29	Gypsum Board	271
09 30	Tiling	275
09 51	Acoustical Ceilings	277
09 53	Acoustical Ceiling Suspension Assemblies	278
09 63	Masonry Flooring	278
09 64	Wood Flooring	279
09 65	Resilient Flooring	281
09 66	Terrazzo Flooring	283
09 67	Fluid-Applied Flooring	284
09 68	Carpeting	285
09 69	Access Flooring	287
09 72	Wall Coverings	287
09 77	Special Wall Surfacing	288
09 81	Acoustic Insulation	289
09 84	Acoustic Room Components	289
09 91	Painting	290
09 93	Staining & Transparent Finishing	302
09 96	High-Performance Coatings	302
09 97	Special Coatings	302
	Specialties	**305**
10 11	Visual Display Surfaces	306
10 13	Directories	308
10 14	Signage	309
10 17	Telephone Specialties	310
10 21	Compartments & Cubicles	310
10 22	Partitions	314
10 26	Wall & Door Protection	316
10 28	Toilet, Bath, & Laundry Accessories	317
10 31	Manufactured Fireplaces	319
10 32	Fireplace Specialties	320
10 35	Stoves	321
10 44	Fire Protection Specialties	321
10 51	Lockers	322
10 55	Postal Specialties	323
10 56	Storage Assemblies	323
10 57	Wardrobe & Closet Specialties	324
10 73	Protective Covers	325
10 74	Manufactured Exterior Specialties	326
10 75	Flagpoles	327
10 81	Pest Control Devices	328
10 88	Scales	328
	Equipment	**329**
11 05	Common Work Results for Equipment	330
11 11	Vehicle Service Equipment	330
11 12	Parking Control Equipment	330
11 13	Loading Dock Equipment	331
11 14	Pedestrian Control Equipment	332
11 16	Vault Equipment	332
11 17	Teller & Service Equipment	333
11 19	Detention Equipment	334
11 21	Mercantile & Service Equipment	334

Div. No.		Page
11 23	Commercial Laundry & Dry Cleaning Equipment	334
11 24	Maintenance Equipment	335
11 26	Unit Kitchens	336
11 27	Photographic Processing Equipment	336
11 31	Residential Appliances	336
11 33	Retractable Stairs	338
11 41	Food Storage Equipment	339
11 42	Food Preparation Equipment	341
11 43	Food Delivery Carts & Conveyors	341
11 44	Food Cooking Equipment	341
11 46	Food Dispensing Equipment	342
11 47	Ice Machines	343
11 48	Cleaning & Disposal Equipment	343
11 52	Audio-Visual Equipment	344
11 53	Laboratory Equipment	345
11 57	Vocational Shop Equipment	346
11 61	Theater & Stage Equipment	346
11 62	Musical Equipment	347
11 66	Athletic Equipment	347
11 67	Recreational Equipment	349
11 68	Play Field Equipment & Structures	349
11 71	Medical Sterilizing Equipment	351
11 72	Examination & Treatment Equipment	351
11 73	Patient Care Equipment	351
11 74	Dental Equipment	352
11 76	Operating Room Equipment	353
11 77	Radiology Equipment	353
11 78	Mortuary Equipment	353
11 82	Solid Waste Handling Equipment	353
11 91	Religious Equipment	354
	Furnishings	**357**
12 21	Window Blinds	358
12 22	Curtains & Drapes	358
12 23	Interior Shutters	359
12 24	Window Shades	359
12 32	Manufactured Wood Casework	360
12 35	Specialty Casework	362
12 36	Countertops	364
12 46	Furnishing Accessories	366
12 48	Rugs & Mats	366
12 51	Office Furniture	367
12 52	Seating	368
12 54	Hospitality Furniture	368
12 55	Detention Furniture	369
12 56	Institutional Furniture	369
12 61	Fixed Audience Seating	371
12 63	Stadium & Arena Seating	371
12 67	Pews & Benches	371
12 92	Interior Planters & Artificial Plants	372
12 93	Site Furnishings	372
	Special Construction	**375**
13 05	Common Work Results for Special Construction	376
13 11	Swimming Pools	379
13 17	Tubs & Pools	381
13 18	Ice Rinks	381
13 21	Controlled Environment Rooms	381
13 24	Special Activity Rooms	383
13 28	Athletic & Recreational Special Construction	384
13 31	Fabric Structures	384

Div. No.		Page
13 34	Fabricated Engineered Structures	386
13 36	Towers	392
13 42	Building Modules	393
13 48	Sound, Vibration, & Seismic Control	393
13 49	Radiation Protection	393
13 53	Meteorological Instrumentation	395
	Conveying Equipment	**397**
14 11	Manual Dumbwaiters	398
14 12	Electric Dumbwaiters	398
14 21	Electric Traction Elevators	398
14 24	Hydraulic Elevators	399
14 27	Custom Elevator Cabs	401
14 28	Elevator Equipment & Controls	401
14 31	Escalators	402
14 32	Moving Walks	403
14 42	Wheelchair Lifts	403
14 45	Vehicle Lifts	404
14 51	Correspondence & Parcel Lifts	404
14 91	Facility Chutes	404
14 92	Pneumatic Tube Systems	405
	Fire Suppression	**407**
21 05	Common Work Results for Fire Suppression	408
21 11	Facility Fire-Suppression Water-Service Piping	408
21 12	Fire-Suppression Standpipes	409
21 13	Fire-Suppression Sprinkler Systems	409
21 21	Carbon-Dioxide Fire-Extinguishing Systems	410
21 22	Clean-Agent Fire-Extinguishing Systems	411
21 31	Centrifugal Fire Pumps	411
	Plumbing	**413**
22 01	Operation & Maintenance of Plumbing	414
22 05	Common Work Results for Plumbing	414
22 07	Plumbing Insulation	419
22 11	Facility Water Distribution	420
22 13	Facility Sanitary Sewerage	427
22 14	Facility Storm Drainage	430
22 31	Domestic Water Softeners	431
22 33	Electric Domestic Water Heaters	431
22 34	Fuel-Fired Domestic Water Heaters	432
22 35	Domestic Water Heat Exchangers	433
22 41	Residential Plumbing Fixtures	433
22 42	Commercial Plumbing Fixtures	436
22 45	Emergency Plumbing Fixtures	439
22 47	Drinking Fountains & Water Coolers	440
22 51	Swimming Pool Plumbing Systems	441
22 52	Fountain Plumbing Systems	441
22 66	Chemical-Waste Systems for Lab. & Healthcare Facilities	442
	Heating, Ventilating, and Air Conditioning	**443**
23 05	Common Work Results for HVAC	444
23 07	HVAC Insulation	446

Div. No.		Page
23 09	Instrumentation & Control for HVAC	447
23 12	Facility Fuel Pumps	448
23 13	Facility Fuel-Storage Tanks	448
23 21	Hydronic Piping & Pumps	450
23 22	Steam & Condensate Piping & Pumps	452
23 31	HVAC Ducts & Casings	453
23 33	Air Duct Accessories	454
23 34	HVAC Fans	456
23 37	Air Outlets & Inlets	459
23 38	Ventilation Hoods	462
23 41	Particulate Air Filtration	462
23 42	Gas-Phase Air Filtration	462
23 43	Electronic Air Cleaners	463
23 51	Breechings, Chimneys, & Stacks	463
23 52	Heating Boilers	464
23 54	Furnaces	466
23 55	Fuel-Fired Heaters	467
23 56	Solar Energy Heating Equipment	467
23 57	Heat Exchangers for HVAC	469
23 62	Packaged Compressor & Condenser Units	469
23 63	Refrigerant Condensers	469
23 64	Packaged Water Chillers	470
23 65	Cooling Towers	471
23 73	Indoor Central-Station Air-Handling Units	472
23 74	Packaged Outdoor HVAC Equipment	472
23 81	Decentralized Unitary HVAC Equipment	473
23 82	Convection Heating & Cooling Units	474
23 83	Radiant Heating Units	476
23 84	Humidity Control Equipment	477
Electrical		**479**
26 05	Common Work Results for Electrical	480
26 09	Instrumentation & Control for Electrical Systems	496
26 12	Medium-Voltage Transformers	496
26 22	Low-Voltage Transformers	497
26 24	Switchboards & Panelboards	497
26 25	Enclosed Bus Assemblies	500
26 27	Low-Voltage Distribution Equipment	502
26 28	Low-Voltage Circuit Protective Devices	503
26 29	Low-Voltage Controllers	504
26 32	Packaged Generator Assemblies	504
26 35	Power Filters & Conditioners	504
26 42	Cathodic Protection	505
26 51	Interior Lighting	506
26 52	Emergency Lighting	507
26 53	Exit Signs	507
26 54	Classified Location Lighting	508
26 55	Special Purpose Lighting	508
26 56	Exterior Lighting	508
26 61	Lighting Sys. & Accessories	509
26 71	Motors	510

Div. No.		Page
Communications		**511**
27 13	Communications Backbone Cabling	512
27 41	Audio-Video Systems	512
27 51	Distributed Audio-Video Communications Systems	512
27 52	Healthcare Communications & Monitoring Systems	513
27 53	Distributed Systems	513
Electronic Safety and Secur.		**515**
28 13	Access Control	516
28 16	Intrusion Detection	516
28 23	Video Surveillance	516
28 31	Fire Detection & Alarm	517
28 33	Fuel-Gas Detection & Alarm	518
Earthwork		**519**
31 05	Common Work Results for Earthwork	520
31 06	Schedules for Earthwork	520
31 11	Clearing & Grubbing	521
31 13	Selective Tree & Shrub Removal and Trimming	522
31 14	Earth Stripping & Stockpiling	523
31 22	Grading	523
31 23	Excavation & Fill	523
31 25	Erosion & Sedimentation Controls	534
31 31	Soil Treatment	535
31 32	Soil Stabilization	535
31 33	Rock Stabilization	535
31 36	Gabions	535
31 37	Riprap	536
31 41	Shoring	536
31 43	Concrete Raising	537
31 45	Vibroflotation & Densification	537
31 46	Needle Beams	538
31 48	Underpinning	538
31 52	Cofferdams	538
31 56	Slurry Walls	539
31 62	Driven Piles	539
31 63	Bored Piles	541
Exterior Improvements		**545**
32 01	Operation & Maintenance of Exterior Improvements	546
32 06	Schedules for Exterior Improvements	546
32 11	Base Courses	547
32 12	Flexible Paving	547
32 13	Rigid Paving	548
32 14	Unit Paving	549
32 16	Curbs & Gutters	550
32 17	Paving Specialties	551
32 18	Athletic & Recreational Surfacing	553
32 31	Fences & Gates	553
32 32	Retaining Walls	557
32 34	Fabricated Bridges	559
32 35	Screening Devices	559
32 84	Planting Irrigation	559
32 91	Planting Preparation	561
32 92	Turf & Grasses	562

Div. No.		Page
32 93	Plants	563
32 94	Planting Accessories	564
32 96	Transplanting	565
Utilities		**567**
33 01	Operation & Maintenance of Utilities	569
33 05	Common Work Results for Utilities	569
33 11	Water Utility Distribution Piping	570
33 12	Water Utility Distribution Equipment	572
33 16	Water Utility Storage Tanks	573
33 21	Water Supply Wells	574
33 31	Sanitary Utility Sewerage Piping	574
33 36	Utility Septic Tanks	575
33 41	Storm Utility Drainage Piping	575
33 42	Culverts	577
33 44	Storm Utility Water Drains	578
33 46	Subdrainage	578
33 49	Storm Drainage Structures	579
33 51	Natural-Gas Distribution	580
33 52	Liquid Fuel Distribution	580
33 71	Electrical Utility Transmission & Distribution	581
33 81	Communications Structures	582
Transportation		**583**
34 01	Operation & Maintenance of Transportation	584
34 11	Rail Tracks	584
34 41	Roadway Signaling & Control Equipment	585
34 71	Roadway Construction	585
34 72	Railway Construction	587
Waterway and Marine		**589**
35 20	Waterway & Marine Construction and Equipment	590
35 51	Floating Construction	590
Material Processing and Handling Equipment		**591**
41 21	Conveyors	592
41 22	Cranes & Hoists	592
Pollution Control Equip.		**595**
44 11	Air Pollution Control Equipment	596
44 41	Packaged Water Treatment	597

How to Use the Unit Price Pages

The following is a detailed explanation of a sample entry in the Unit Price Section. Next to each bold number below is the described item with the appropriate component of the sample entry following in parentheses. Some prices are listed as bare costs; others as costs that include overhead and profit of the installing contractor. In most cases, if the work is to be subcontracted, the general contractor will need to add an additional markup (RSMeans suggests using 10%) to the figures in the column "Total Incl. O&P."

1. Division Number/Title
(03 30/Cast-In-Place Concrete)

Use the Unit Price Section Table of Contents to locate specific items. The sections are classified according to the CSI MasterFormat 2004 system.

2. Line Numbers
(03 30 53.40 3920)

Each unit price line item has been assigned a unique 12-digit code based on the CSI MasterFormat classification.

3. Description
(Concrete In Place, etc.)

Each line item is described in detail. Sub-items and additional sizes are indented beneath the appropriate line items. The first line or two after the main item (in boldface) may contain descriptive information that pertains to all line items beneath this boldface listing.

Items which include the symbol **CN** are updated in The *RSMeans Quarterly Update Service* online. To obtain access to this service contact RSMeans customer service at 1-800-334-3509.

4. Reference Number Information

R033053-50 You'll see reference numbers shown in shaded boxes at the beginning of some sections. These refer to related items in the Reference Section, visually identified by a vertical gray bar on the page edges.

The relation may be an estimating procedure that should be read before estimating, or technical information.

The "R" designates the Reference Section. The numbers refer to the MasterFormat 2004 classification system.

It is strongly recommended that you review all reference numbers that appear within the section in which you are working.

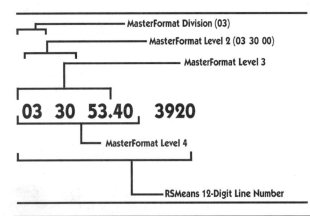

03 30 Cast-In-Place Concrete

03 30 53 – Miscellaneous Cast-In-Place Concrete

03 30 53.40 Concrete In Place		Crew	Daily Output	Labor-Hours	Unit	Material	2007 Bare Costs Labor	Equipment	Total	Total Incl O&P
3590	10' x 10' x 12" thick	C-14H	5	9.600	Ea.	565	350	4.32	919.32	1,175
3800	Footings, spread under 1 C.Y.	C-14C	38	1.942	C.Y.	195	103	.56	298.56	370
3850	Over 5 C.Y.		81.04	1.382		266	48.50	.26	314.76	370
3900	Footings, strip, 18" x 9", unreinforced		40	2.800		126	98.50	.53	225.03	293
3920	18" x 9", reinforced		35	3.200		147	112	.61	259.61	340
3925	20" x 10", unreinforced		45	2.489		122	87.50	.47	209.97	273
3930	20" x 10", reinforced		40	2.800		140	98.50	.53	239.03	310
3935	24" x 12", unreinforced		55	2.036		120	71.50	.39	191.89	244
3940	24" x 12", reinforced		48	2.333		138	82	.44	220.44	280
3945	36" x 12", unreinforced		70	1.600		117	56	.30	173.30	216
3950	36" x 12", reinforced		60	1.867		133	65.50	.35	198.85	249

Crew (C-14C)

The "Crew" column designates the typical trade or crew used to install the item. If an installation can be accomplished by one trade and requires no power equipment, that trade and the number of workers are listed (for example, "2 Carpenters"). If an installation requires a composite crew, a crew code designation is listed (for example, "C-14C"). You'll find full details on all composite crews in the Crew Listings.

- For a complete list of all trades utilized in this book and their abbreviations, see the inside back cover.

Crews

Crew No.	Bare Costs		Incl. Subs O & P		Cost Per Labor-Hour	
Crew C-14C	Hr.	Daily	Hr.	Daily	Bare Costs	Incl. O&P
1 Carpenter Foreman (out)	$38.70	$309.60	$60.25	$482.00	$35.15	$54.99
6 Carpenters	36.70	1761.60	57.15	2743.20		
2 Rodmen (reinf.)	41.30	660.80	67.75	1084.00		
4 Laborers	28.75	920.00	44.75	1432.00		
1 Cement Finisher	35.55	284.40	52.20	417.60		
1 Gas Engine Vibrator		21.40		23.54	.19	.21
112 L.H., Daily Totals		$3957.80		$6182.34	$35.34	$55.20

Productivity: Daily Output (35.0)/ Labor-Hours (3.20)

The "Daily Output" represents the typical number of units the designated crew will install in a normal 8-hour day. To find out the number of days the given crew would require to complete the installation, divide your quantity by the daily output. For example:

Quantity	÷	Daily Output	=	Duration
100 C.Y.	÷	35.0/ Crew Day	=	2.86 Crew Days

The "Labor-Hours" figure represents the number of labor-hours required to install one unit of work. To find out the number of labor-hours required for your particular task, multiply the quantity of the item times the number of labor-hours shown. For example:

Quantity	x	Productivity Rate	=	Duration
100 C.Y.	x	3.20 Labor-Hours/ C.Y.	=	320 Labor-Hours

Unit (C.Y.)

The abbreviated designation indicates the unit of measure upon which the price, production, and crew are based (C.Y. = Cubic Yard). For a complete listing of abbreviations, refer to the Abbreviations Listing in the Reference Section of this book.

Bare Costs:
Mat. (Bare Material Cost) (147)

The unit material cost is the "bare" material cost with no overhead or profit included. Costs shown reflect national average material prices for January of the current year and include delivery to the job site. No sales taxes are included.

Labor (112)

The unit labor cost is derived by multiplying bare labor-hour costs for Crew C-14C by labor-hour units. The bare labor-hour cost is found in the Crew Section under C-14C. (If a trade is listed, the hourly labor cost—the wage rate—is found on the inside back cover.)

Labor-Hour Cost Crew C-14C	x	Labor-Hour Units	=	Labor
$35.15	x	3.20	=	$112.00

Equip. (Equipment) (.61)

Equipment costs for each crew are listed in the description of each crew. Tools or equipment whose value justifies purchase or ownership by a contractor are considered overhead as shown on the inside back cover. The unit equipment cost is derived by multiplying the bare equipment hourly cost by the labor-hour units.

Equipment Cost Crew C-14C	x	Labor-Hour Units	=	Equip.
$.19	x	3.20	=	$.61

Total (259.61)

The total of the bare costs is the arithmetic total of the three previous columns: mat., labor, and equip.

Material	+	Labor	+	Equip.	=	Total
$147	+	$112	+	$.61	=	$259.61

Total Costs Including O & P

This figure is the sum of the bare material cost plus 10% for profit; the bare labor cost plus total overhead and profit (per the inside back cover or, if a crew is listed, from the crew listings); and the bare equipment cost plus 10% for profit.

Material is Bare Material Cost + 10% = 147 + 14.70	=	$161.70
Labor for Crew C-14C = Labor-Hour Cost (54.99) x Labor-Hour Units (3.20)	=	$175.97
Equip. is Bare Equip. Cost + 10% = .61 + .06	=	$ .67
Total (Rounded)	=	$340

Estimating Tips

01 20 00 Price & Payment Procedures

When estimating historic preservation projects (depending on the condition of the existing structure and the owner's requirements), a 15%–20% contingency or allowance is recommended, regardless of the stage of the drawings.

01 30 00 Administrative Requirements

- Before determining a final cost estimate, it is a good practice to review all the items listed in Subdivision 01 03 00 to make final adjustments for items that may need customizing to specific job conditions.
- Requirements for initial and periodic submittals can represent a significant cost to the General Requirements of a job. Thoroughly check the submittal specifications when estimating a project to determine any costs that should be included.

01 40 00 Quality Requirements

- All projects will require some degree of Quality Control. This cost is not included in the unit cost of construction listed in each division. Depending upon the terms of the contract, the various costs of inspection and testing can be the responsibility of either the owner or the contractor. Be sure to include the required costs in your estimate.

01 50 00 Temporary Facilities and Controls

- Barricades, access roads, safety nets, scaffolding, security, and many more requirements for the execution of a safe project are elements of direct cost. These costs can easily be overlooked when preparing an estimate. When looking through the major classifications of this subdivision, determine which items apply to each division in your estimate.

01 70 00 Execution and Closeout Requirements

- When preparing an estimate, thoroughly read the specifications to determine the requirements for Contract Closeout. Final cleaning, record documentation, operation and maintenance data, warranties and bonds, and spare parts and maintenance materials can all be elements of cost for the completion of a contract. Do not overlook these in your estimate.

Reference Numbers

Reference numbers are shown in shaded boxes at the beginning of some major classifications. These numbers refer to related items in the Reference Section. The reference information may be an estimating procedure, an alternate pricing method, or technical information.

Note: Not all subdivisions listed here necessarily appear in this publication.

No part of this publication may be reproduced, stored in a retrieval system, or transmitted in any form or by any means without prior written permission of Reed Construction Data.

01 11 Summary of Work

01 11 31 – Professional Consultants

01 11 31.10 Architectural Fees

		Crew	Daily Output	Labor-Hours	Unit	Material	2007 Bare Costs Labor	Equipment	Total	Total Incl O&P
0010	**ARCHITECTURAL FEES** R011110-10									
0020	For new construction									
0060	Minimum				Project					4.90%
0090	Maximum									16%
0100	For alteration work, to $500,000, add to fee									50%
0150	Over $500,000, add to fee				▼					25%

01 11 31.20 Construction Management Fees

0010	**CONSTRUCTION MANAGEMENT FEES**									
0020	$1,000,000 job, minimum				Project					4.50%
0050	Maximum									7.50%
0300	$5,000,000 job, minimum									2.50%
0350	Maximum				▼					4%

01 11 31.30 Engineering Fees

0010	**ENGINEERING FEES** R011110-30									
0020	Educational planning consultant, minimum				Project					.50%
0100	Maximum				"					2.50%
0200	Electrical, minimum				Contrct					4.10%
0300	Maximum									10.10%
0400	Elevator & conveying systems, minimum									2.50%
0500	Maximum									5%
0600	Food service & kitchen equipment, minimum									8%
0700	Maximum									12%
0800	Landscaping & site development, minimum									2.50%
0900	Maximum									6%
1000	Mechanical (plumbing & HVAC), minimum									4.10%
1100	Maximum				▼					10.10%
1200	Structural, minimum				Project					1%
1300	Maximum				"					2.50%

01 11 31.50 Models

0010	**MODELS**									
0020	Cardboard & paper, 1 building, minimum				Ea.	665			665	735
0050	Maximum					1,525			1,525	1,675
0100	2 buildings, minimum					890			890	980
0150	Maximum				▼	2,025			2,025	2,225
0200	Plexiglass and metal, basic layout				SF Flr.	.06			.06	.07
0210	Including equipment and personnel				"	.30			.30	.33
0300	Site plan layout, minimum				Ea.	1,275			1,275	1,400
0350	Maximum				"	2,125			2,125	2,350

01 11 31.75 Renderings

0010	**RENDERINGS** Color, matted, 20" x 30", eye level,									
0020	1 building, minimum				Ea.	1,850			1,850	2,050
0050	Average					2,650			2,650	2,925
0100	Maximum					4,250			4,250	4,675
1000	5 buildings, minimum					3,725			3,725	4,075
1100	Maximum					7,425			7,425	8,175
2000	Aerial perspective, color, 1 building, minimum					2,650			2,650	2,925
2100	Maximum					7,425			7,425	8,175
3000	5 buildings, minimum					5,300			5,300	5,825
3100	Maximum				▼	10,600			10,600	11,700

01 21 Allowances

01 21 16 – Contingency Allowances

01 21 16.50 Contingencies

		Crew	Daily Output	Labor-Hours	Unit	Material	2007 Bare Costs Labor	Equipment	Total	Total Incl O&P
0010	**CONTINGENCIES**									
0020	For estimate at conceptual stage				Project					20%
0050	Schematic stage									15%
0100	Preliminary working drawing stage (Design Dev.)									10%
0150	Final working drawing stage				↓					3%

01 21 55 – Job Conditions Allowance

01 21 55.50 Job Conditions

		Crew	Daily Output	Labor-Hours	Unit	Material	Labor	Equipment	Total	Total Incl O&P
0010	**JOB CONDITIONS** Modifications to total									
0020	project cost summaries									
0100	Economic conditions, favorable, deduct				Project				2%	2%
0200	Unfavorable, add								5%	5%
0300	Hoisting conditions, favorable, deduct								2%	2%
0400	Unfavorable, add								5%	5%
0500	General Contractor management, experienced, deduct								2%	2%
0600	Inexperienced, add								10%	10%
0700	Labor availability, surplus, deduct								1%	1%
0800	Shortage, add								10%	10%
0900	Material storage area, available, deduct								1%	1%
1000	Not available, add								2%	2%
1100	Subcontractor availability, surplus, deduct								5%	5%
1200	Shortage, add								12%	12%
1300	Work space, available, deduct								2%	2%
1400	Not available, add				↓				5%	5%

01 21 57 – Overtime Allowance

01 21 57.50 Overtime

		Crew	Daily Output	Labor-Hours	Unit	Material	Labor	Equipment	Total	Total Incl O&P
0010	**OVERTIME** for early completion of projects or where	R012909-90								
0020	labor shortages exist, add to usual labor, up to				Costs		100%			

01 21 61 – Cost Indexes

01 21 61.10 Construction Cost Index

		Crew	Daily Output	Labor-Hours	Unit	Material	Labor	Equipment	Total	Total Incl O&P
0010	**CONSTRUCTION COST INDEX** (Reference) over 930 zip code locations in									
0020	The U.S. and Canada, total bldg cost, min. (Clarksdale, MS)				%					67.20%
0050	Average									100%
0100	Maximum (New York, NY)				↓					130.90%

01 21 61.20 Historical Cost Indexes

0010	**HISTORICAL COST INDEXES** (See Reference section)									

01 21 61.30 Labor Index

		Crew	Daily Output	Labor-Hours	Unit	Material	Labor	Equipment	Total	Total Incl O&P
0010	**LABOR INDEX** (Reference) For over 930 zip code locations in									
0020	the U.S. and Canada, minimum (Clarksdale, MS)				%		29.90%			
0050	Average						100%			
0100	Maximum (New York, NY)				↓		164.50%			

01 21 61.50 Material Index

		Crew	Daily Output	Labor-Hours	Unit	Material	Labor	Equipment	Total	Total Incl O&P
0010	**MATERIAL INDEX** (Reference) For over 930 zip code locations in									
0020	the U.S. and Canada, minimum (Elizabethtown, KY)				%	90.70%				
0040	Average					100%				
0060	Maximum (Ketchikan, AK)				↓	141.60%				

01 21 63 – Taxes

01 21 63.10 Taxes

		Crew	Daily Output	Labor-Hours	Unit	Material	Labor	Equipment	Total	Total Incl O&P
0010	**TAXES**	R012909-80								
0020	Sales tax, State, average				%	4.84%				
0050	Maximum	R012909-85			↓	7.25%				

01 21 Allowances

01 21 63 – Taxes

01 21 63.10 Taxes		Crew	Daily Output	Labor-Hours	Unit	Material	2007 Bare Costs Labor	Equipment	Total	Total Incl O&P
0200	Social Security, on first $94,200 of wages				%		7.65%			
0300	Unemployment, combined Federal and State, minimum						.80%			
0350	Average						6.20%			
0400	Maximum						11.76%			

01 31 Project Management and Coordination

01 31 13 – Project Coordination

01 31 13.20 Field Personnel

		Crew	Daily Output	Labor-Hours	Unit	Material	Labor	Equipment	Total	Total Incl O&P
0010	**FIELD PERSONNEL**									
0020	Clerk, average				Week		350		350	545
0100	Field engineer, minimum						835		835	1,300
0120	Average						1,085		1,085	1,700
0140	Maximum						1,250		1,250	1,950
0160	General purpose laborer, average						1,150		1,150	1,800
0180	Project manager, minimum						1,550		1,550	2,400
0200	Average						1,775		1,775	2,750
0220	Maximum						2,025		2,025	3,150
0240	Superintendent, minimum						1,500		1,500	2,325
0260	Average						1,650		1,650	2,575
0280	Maximum						1,875		1,875	2,925
0290	Timekeeper, average						970		970	1,500

01 31 13.30 Insurance

					Unit		Labor			Total Incl O&P
0010	**INSURANCE**	R013113-40								
0020	Builders risk, standard, minimum				Job					.24%
0050	Maximum	R013113-50								.64%
0200	All-risk type, minimum									.25%
0250	Maximum	R013113-60								.62%
0400	Contractor's equipment floater, minimum				Value					.50%
0450	Maximum				"					1.50%
0600	Public liability, average				Job					2.02%
0800	Workers' compensation & employer's liability, average									
0850	by trade, carpentry, general				Payroll		18.39%			
0900	Clerical						.67%			
0950	Concrete						15.53%			
1000	Electrical						6.59%			
1050	Excavation						10.39%			
1100	Glazing						14.05%			
1150	Insulation						15.81%			
1200	Lathing						11.61%			
1250	Masonry						14.92%			
1300	Painting & decorating						13.19%			
1350	Pile driving						21.39%			
1400	Plastering						14.04%			
1450	Plumbing						8.10%			
1500	Roofing						32.30%			
1550	Sheet metal work (HVAC)						11.89%			
1600	Steel erection, structural						40.80%			
1650	Tile work, interior ceramic						9.53%			
1700	Waterproofing, brush or hand caulking						7.16%			
1800	Wrecking						39.10%			
2000	Range of 35 trades in 50 states, excl. wrecking, min.						2.04%			

01 31 Project Management and Coordination

01 31 13 – Project Coordination

01 31 13.30 Insurance

		Crew	Daily Output	Labor-Hours	Unit	Material	2007 Bare Costs Labor	2007 Bare Costs Equipment	Total	Total Incl O&P
2100	Average				Payroll		16.30%			
2200	Maximum						197.85%			

01 31 13.40 Main Office Expense

0010	**MAIN OFFICE EXPENSE** Average for General Contractors R013113-50									
0020	As a percentage of their annual volume									
0125	Annual volume under 1 million dollars				% Vol.				13.60%	
0145	Up to 2.5 million dollars								8%	
0150	Up to 4.0 million dollars								6.80%	
0200	Up to 7.0 million dollars								5.60%	
0250	Up to 10 million dollars								5.10%	
0300	Over 10 million dollars								3.90%	

01 31 13.50 Mark-Up

0010	**MARK-UP** For General Contractors for change									
0100	of scope of job as bid									
0200	Extra work, by subcontractors, add				%					10%
0250	By General Contractor, add									15%
0400	Omitted work, by subcontractors, deduct all but									5%
0450	By General Contractor, deduct all but									7.50%
0600	Overtime work, by subcontractors, add									15%
0650	By General Contractor, add									10%
1000	Installing contractors, on his own labor, minimum						46.90%			
1100	Maximum						87.30%			

01 31 13.60 Overhead

0010	**OVERHEAD** R013113-50									
0020	As percent of direct costs, minimum				%				5%	
0050	Average								13%	
0100	Maximum								30%	

01 31 13.80 Overhead and Profit

0010	**OVERHEAD & PROFIT** Allowance to add to items in this									
0020	book that do not include Subs O&P, average				%					25%
0100	Allowance to add to items in this book that									
0110	do include Subs O&P, minimum				%					5%
0150	Average									10%
0200	Maximum									15%
0300	Typical, by size of project, under $100,000								30%	
0350	$500,000 project								25%	
0400	$2,000,000 project								20%	
0450	Over $10,000,000 project								15%	

01 31 13.90 Performance Bond

0010	**PERFORMANCE BOND** R013113-80									
0020	For buildings, minimum				Job					.60%
0100	Maximum				"					2.50%

01 32 Construction Progress Documentation

01 32 13 – Scheduling of work

01 32 13.50 Scheduling

		Crew	Daily Output	Labor-Hours	Unit	Material	2007 Bare Costs Labor	Equipment	Total	Total Incl O&P
0010	**SCHEDULING**									
0020	Critical path, as % of architectural fee, minimum				%					.50%
0100	Maximum				"					1%
0300	Computer-update, micro, no plots, minimum				Ea.				455	500
0400	Including plots, maximum				"				1,455	1,600
0600	Rule of thumb, CPM scheduling, small job ($10 Million)				Job					.05%
0650	Large job ($50 Million +)									.03%
0700	Including cost control, small job									.08%
0750	Large job									.04%

01 32 33 – Photographic Documentation

01 32 33.50 Photographs

		Crew	Daily Output	Labor-Hours	Unit	Material	2007 Bare Costs Labor	Equipment	Total	Total Incl O&P
0010	**PHOTOGRAPHS**									
0020	8" x 10", 4 shots, 2 prints ea., std. mounting				Set	450			450	495
0100	Hinged linen mounts					530			530	580
0200	8" x 10", 4 shots, 2 prints each, in color					455			455	505
0300	For I.D. slugs, add to all above					5.30			5.30	5.85
0500	Aerial photos, initial fly-over, 6 shots, 1 print ea., 8" x 10"					790			790	870
0550	11" x 14" prints					920			920	1,025
0600	16" x 20" prints					1,150			1,150	1,250
0700	For full color prints, add					40%				40%
0750	Add for traffic control area					294			294	325
0900	For over 30 miles from airport, add per				Mile	5.30			5.30	5.85
1000	Vertical photography, 4 to 6 shots with									
1010	different scales, 1 print each				Set	1,100			1,100	1,200
1500	Time lapse equipment, camera and projector, buy					3,775			3,775	4,175
1550	Rent per month					565			565	620
1700	Cameraman and film, including processing, B.&W.				Day	1,375			1,375	1,525
1720	Color				"	1,375			1,375	1,525

01 41 Regulatory Requirements

01 41 26 – Permits

01 41 26.50 Permits

		Crew	Daily Output	Labor-Hours	Unit	Material	2007 Bare Costs Labor	Equipment	Total	Total Incl O&P
0010	**PERMITS**									
0020	Rule of thumb, most cities, minimum				Job					.50%
0100	Maximum				"					2%

01 45 Quality Control

01 45 23 – Testing and Inspecting Services

01 45 23.50 Testing

		Crew	Daily Output	Labor-Hours	Unit	Material	2007 Bare Costs Labor	Equipment	Total	Total Incl O&P
0010	**TESTING** and Inspecting Services									
0015	For concrete building costing $1,000,000, minimum				Project				4,725	5,200
0020	Maximum								38,000	41,800
0050	Steel building, minimum								4,727	5,200
0070	Maximum								14,818	16,300
0100	For building costing, $10,000,000, minimum								30,091	33,100
0150	Maximum								48,182	53,000
0200	Asphalt testing, compressive strength Marshall stability, set of 3				Ea.				145	165
0220	Density, set of 3								86	95
0250	Extraction, individual tests on sample								136	150

01 45 Quality Control

01 45 23 – Testing and Inspecting Services

01 45 23.50 Testing		Crew	Daily Output	Labor-Hours	Unit	Material	2007 Bare Costs Labor	2007 Bare Costs Equipment	Total	Total Incl O&P
0300	Penetration				Ea.				41	45
0350	Mix design, 5 specimens								182	200
0360	Additional specimen								36	40
0400	Specific gravity								41	45
0420	Swell test								64	70
0450	Water effect and cohesion, set of 6								182	200
0470	Water effect and plastic flow								64	70
0600	Concrete testing, aggregates, abrasion, ASTM C 131								136	150
0650	Absorption, ASTM C 127								42	46
0800	Petrographic analysis, ASTM C 295								773	850
0900	Specific gravity, ASTM C 127								50	55
1000	Sieve analysis, washed, ASTM C 136								59	65
1050	Unwashed								59	65
1200	Sulfate soundness								114	125
1300	Weight per cubic foot								36	40
1500	Cement, physical tests, ASTM C 150								318	350
1600	Chemical tests, ASTM C 150								245	270
1800	Compressive test, cylinder, delivered to lab, ASTM C 39								12	13
1900	Picked up by lab, minimum								14	15
1950	Average								18	20
2000	Maximum								27	30
2200	Compressive strength, cores (not incl. drilling), ASTM C 42								36	40
2250	Core drilling, 4" diameter (plus technician)				Inch				23	25
2260	Technician for core drilling				Hr.				45	50
2300	Patching core holes				Ea.				22	24
2400	Drying shrinkage at 28 days								236	260
2500	Flexural test beams, ASTM C 78								59	65
2600	Mix design, one batch mix								259	285
2650	Added trial batches								120	132
2800	Modulus of elasticity, ASTM C 469								164	180
2900	Tensile test, cylinders, ASTM C 496								45	50
3000	Water-Cement ratio curve, 3 batches								141	155
3100	4 batches								186	205
3300	Masonry testing, absorption, per 5 brick, ASTM C 67								45	50
3350	Chemical resistance, per 2 brick								50	55
3400	Compressive strength, per 5 brick, ASTM C 67								68	75
3420	Efflorescence, per 5 brick, ASTM C 67								68	75
3440	Imperviousness, per 5 brick								87	96
3470	Modulus of rupture, per 5 brick								86	95
3500	Moisture, block only								32	35
3550	Mortar, compressive strength, set of 3								23	25
4100	Reinforcing steel, bend test								55	61
4200	Tensile test, up to #8 bar								36	40
4220	#9 to #11 bar								41	45
4240	#14 bar and larger								64	70
4400	Soil testing, Atterberg limits, liquid and plastic limits								59	65
4510	Hydrometer analysis								109	120
4530	Specific gravity, ASTM D 354								44	48
4600	Sieve analysis, washed, ASTM D 422								55	60
4700	Unwashed, ASTM D 422								59	65
4710	Consolidation test (ASTM D2435), minimum								250	275
4715	Maximum								432	475
4720	Density and classification of undisturbed sample								73	80

01 45 Quality Control

01 45 23 – Testing and Inspecting Services

01 45 23.50 Testing

		Crew	Daily Output	Labor-Hours	Unit	Material	2007 Bare Costs Labor	2007 Bare Costs Equipment	Total	Total Incl O&P
4735	Soil density, nuclear method, ASTM D2922				Ea.				35	38.67
4740	Sand cone method ASTM D1556								27	30.17
4750	Moisture content, ASTM D 2216								9	10
4780	Permeability test, double ring infiltrometer								500	550
4800	Permeability, var. or constant head, undist., ASTM D 2434								227	250
4850	Recompacted								250	275
4900	Proctor compaction, 4" standard mold, ASTM D 698								123	135
4950	6" modified mold								68	75
5100	Shear tests, triaxial, minimum								409	450
5150	Maximum								545	600
5300	Direct shear, minimum, ASTM D 3080								318	350
5350	Maximum								409	450
5550	Technician for inspection, per day, earthwork								210	231
5650	Bolting								268	295
5750	Roofing								244	256
5790	Welding				▼				257	283
5820	Non-destructive testing, dye penetrant				Day				310	341
5840	Magnetic particle								310	341
5860	Radiography								450	495
5880	Ultrasonic				▼				309	340
6000	Welding certification, minimum				Ea.				91	100
6100	Maximum				"				250	275
7000	Underground storage tank									
7500	Volumetric tightness test, <=12,000 gal				Ea.				364	400
7510	<=30,000 gal				"				600	660
7600	Vadose zone (soil gas) sampling, 10-40 samples, min.				Day				1,364	1,500
7610	Maximum				"				2,273	2,500
7700	Ground water monitoring incl. drilling 3 wells, min.				Total				4,545	5,000
7710	Maximum				"				6,364	7,000
8000	X-ray concrete slabs				Ea.				182	200

01 51 Temporary Utilities

01 51 13 – Temporary Electricity

01 51 13.80 Temporary Utilities

		Crew	Daily Output	Labor-Hours	Unit	Material	2007 Bare Costs Labor	2007 Bare Costs Equipment	Total	Total Incl O&P
0010	**TEMPORARY UTILITIES**									
0100	Heat, incl. fuel and operation, per week, 12 hrs. per day	1 Skwk	100	.080	CSF Flr	10.35	3.04		13.39	16.10
0200	24 hrs. per day	"	60	.133		19.95	5.05		25	30
0350	Lighting, incl. service lamps, wiring & outlets, minimum	1 Elec	34	.235		2.63	10.35		12.98	18.30
0360	Maximum	"	17	.471		5.70	20.50		26.20	37.50
0400	Power for temp lighting only, per month, min/month 6.6 KWH								.75	.83
0450	Maximum/month 23.6 KWH								2.85	3.14
0600	Power for job duration incl. elevator, etc., minimum								47	51.70
0650	Maximum				▼				110	121
1000	Toilet, portable, see division 01 54 33.40									

01 52 Construction Facilities

01 52 13 – Field Offices and Sheds

01 52 13.20 Office and Storage Space

		Crew	Daily Output	Labor-Hours	Unit	Material	2007 Bare Costs Labor	Equipment	Total	Total Incl O&P
0010	**OFFICE AND STORAGE SPACE**									
0020	Trailer, furnished, no hookups, 20' x 8', buy	2 Skwk	1	16	Ea.	7,975	610		8,585	9,700
0250	Rent per month					201			201	221
0300	32' x 8', buy	2 Skwk	.70	22.857		11,900	870		12,770	14,400
0350	Rent per month					241			241	265
0400	50' x 10', buy	2 Skwk	.60	26.667		20,500	1,025		21,525	24,100
0450	Rent per month					330			330	365
0500	50' x 12', buy	2 Skwk	.50	32		25,600	1,225		26,825	30,100
0550	Rent per month					375			375	410
0700	For air conditioning, rent per month, add					41			41	45
0800	For delivery, add per mile				Mile	4.50			4.50	4.95
1000	Portable buildings, prefab, on skids, economy, 8' x 8'	2 Carp	265	.060	S.F.	85	2.22		87.22	97
1100	Deluxe, 8' x 12'	"	150	.107	"	95	3.91		98.91	111
1200	Storage boxes, 20' x 8', buy	2 Skwk	1.80	8.889	Ea.	4,175	340		4,515	5,100
1250	Rent per month					76			76	83.50
1300	40' x 8', buy	2 Skwk	1.40	11.429		6,175	435		6,610	7,450
1350	Rent per month					101			101	111
5000	Air supported structures, see division 13 31 13.13									

01 52 13.40 Field Office Expense

		Crew	Daily Output	Labor-Hours	Unit	Material	Labor	Equipment	Total	Total Incl O&P
0010	**FIELD OFFICE EXPENSE**									
0100	Field office expense, office equipment rental average				Month	150			150	165
0120	Office supplies, average				"	95			95	105
0125	Office trailer rental, see division 01 52 13.20									
0140	Telephone bill; avg. bill/month incl. long dist.				Month	210			210	231
0160	Field office lights & HVAC				"	110			110	121

01 54 Construction Aids

01 54 09 – Protection Equipment

01 54 09.50 Personnel Protective Equipment

		Crew	Daily Output	Labor-Hours	Unit	Material	Labor	Equipment	Total	Total Incl O&P
0010	**PERSONNEL PROTECTIVE EQUIPMENT**									
0015	Hazardous waste protection									
0020	Respirator mask only, full face, silicone				Ea.	207			207	228
0030	Half face, silicone					31.50			31.50	34.50
0040	Respirator cartridges, 2 req'd/mask, dust or asbestos					5.10			5.10	5.60
0050	Chemical vapor					4.30			4.30	4.73
0060	Combination vapor and dust					8.95			8.95	9.85
0100	Emergency escape breathing apparatus, 5 min					455			455	500
0110	10 min					525			525	575
0150	Self contained breathing apparatus with full face piece, 30 min					1,900			1,900	2,075
0160	60 min					3,025			3,025	3,325
0200	Encapsulating suits, limited use, level A					885			885	975
0210	Level B					178			178	196
0300	Over boots, latex				Pr.	4.11			4.11	4.52
0310	PVC					14.40			14.40	15.85
0320	Neoprene					51			51	56.50
0400	Gloves, nitrile/PVC					5.70			5.70	6.25
0410	Neoprene coated					26.50			26.50	29

01 54 09.60 Safety Nets

		Crew	Daily Output	Labor-Hours	Unit	Material	Labor	Equipment	Total	Total Incl O&P
0010	**SAFETY NETS**									
0020	No supports, stock sizes, nylon, 4" mesh				S.F.	1.10			1.10	1.21

01 54 Construction Aids

01 54 09 – Protection Equipment

01 54 09.60 Safety Nets

		Crew	Daily Output	Labor-Hours	Unit	Material	2007 Bare Costs Labor	2007 Bare Costs Equipment	Total	Total Incl O&P
0100	Polypropylene, 6" mesh				S.F.	1.59			1.59	1.75
0200	Small mesh debris nets, 1/4" & 3/4" mesh, stock sizes					.74			.74	.81
0220	Combined 4" mesh and 1/4" mesh, stock sizes					2.05			2.05	2.26
0300	Monthly rental, 4" mesh, stock sizes, 1st month					.50			.50	.55
0320	2nd month rental					.25			.25	.28
0340	Maximum rental/year					1.15			1.15	1.27

01 54 23 – Temporary Scaffolding and Platforms

01 54 23.60 Pump Staging

		Crew	Daily Output	Labor-Hours	Unit	Material	Labor	Equipment	Total	Total Incl O&P
0010	**PUMP STAGING**, Aluminum R015423-20									
0200	24' long pole section, buy				Ea.	370			370	405
0300	18' long pole section, buy					286			286	315
0400	12' long pole section, buy					193			193	212
0500	6' long pole section, buy					102			102	112
0600	6' long splice joint section, buy					76			76	83.50
0700	Pump jack					123			123	136
0900	Foldable brace					53.50			53.50	59
1000	Workbench/back safety rail support					65.50			65.50	72
1100	Scaffolding planks/workbench, 14" wide x 24' long					605			605	665
1200	Plank end safety rail					315			315	350
1250	Safety net, 22' long					291			291	320
1300	System in place, 50' working height, per use based on 50 uses	2 Carp	84.80	.189	C.S.F.	5.65	6.90		12.55	17.05
1400	100 uses		84.80	.189		2.83	6.90		9.73	13.90
1500	150 uses		84.80	.189		1.90	6.90		8.80	12.90

01 54 23.70 Scaffolding

		Crew	Daily Output	Labor-Hours	Unit	Material	Labor	Equipment	Total	Total Incl O&P
0010	**SCAFFOLDING** R015423-10									
0015	Steel tube, regular, no plank, labor only to erect & dismantle									
0090	Building exterior, wall face, 1 to 5 stories, 6'-4" x 5' frames	3 Carp	8	3	C.S.F.		110		110	171
0200	6 to 12 stories	4 Carp	8	4			147		147	229
0310	13 to 20 stories	5 Carp	8	5			184		184	286
0460	Building interior, wall face area, up to 16' high	3 Carp	12	2			73.50		73.50	114
0560	16' to 40' high		10	2.400			88		88	137
0800	Building interior floor area, up to 30' high		150	.160	C.C.F.		5.85		5.85	9.15
0900	Over 30' high	4 Carp	160	.200	"		7.35		7.35	11.45
0906	Complete system for face of walls, no plank, material only rent/mo				C.S.F.	34.50			34.50	37.50
0908	Interior spaces, no plank, material only rent/mo				"	3.30			3.30	3.63
0910	Steel tubular, heavy duty shoring, buy									
0920	Frames 5' high 2' wide				Ea.	82.50			82.50	91
0925	5' high 4' wide					93.50			93.50	103
0930	6' high 2' wide					94.50			94.50	104
0935	6' high 4' wide					110			110	121
0940	Accessories									
0945	Cross braces				Ea.	16			16	17.60
0950	U-head, 8" x 8"					19.30			19.30	21
0955	J-head, 4" x 8"					14.10			14.10	15.50
0960	Base plate, 8" x 8"					15.70			15.70	17.25
0965	Leveling jack					33.50			33.50	37
1000	Steel tubular, regular, buy									
1100	Frames 3' high 5' wide				Ea.	64			64	70
1150	5' high 5' wide					73.50			73.50	81
1200	6'-4" high 5' wide					92.50			92.50	102
1350	7'-6" high 6' wide					160			160	175
1500	Accessories cross braces					18.15			18.15	19.95

01 54 Construction Aids

01 54 23 – Temporary Scaffolding and Platforms

01 54 23.70 Scaffolding

		Crew	Daily Output	Labor-Hours	Unit	Material	2007 Bare Costs Labor	Equipment	Total	Total Incl O&P
1550	Guardrail post				Ea.	16.50			16.50	18.15
1600	Guardrail 7' section					8.80			8.80	9.70
1650	Screw jacks & plates					24			24	26.50
1700	Sidearm brackets					31			31	34
1750	8" casters					33			33	36.50
1800	Plank 2" x 10" x 16'-0"					44			44	48.50
1900	Stairway section					270			270	296
1910	Stairway starter bar					32			32	35
1920	Stairway inside handrail					58.50			58.50	64
1930	Stairway outside handrail					80.50			80.50	88.50
1940	Walk-thru frame guardrail				▼	40.50			40.50	44.50
2000	Steel tubular, regular, rent/mo.									
2100	Frames 3' high 5' wide				Ea.	5			5	5.50
2150	5' high 5' wide					5			5	5.50
2200	6'-4" high 5' wide					4.50			4.50	4.95
2250	7'-6" high 6' wide					7			7	7.70
2500	Accessories, cross braces					1			1	1.10
2550	Guardrail post					1			1	1.10
2600	Guardrail 7' section					1			1	1.10
2650	Screw jacks & plates					2			2	2.20
2700	Sidearm brackets					2			2	2.20
2750	8" casters					8			8	8.80
2800	Outrigger for rolling tower					3			3	3.30
2850	Plank 2" x 10" x 16'-0"					6			6	6.60
2900	Stairway section					40			40	44
2940	Walk-thru frame guardrail				▼	2.50			2.50	2.75
3000	Steel tubular, heavy duty shoring, rent/mo.									
3250	5' high 2' & 4' wide				Ea.	5			5	5.50
3300	6' high 2' & 4' wide					5			5	5.50
3500	Accessories, cross braces					1			1	1.10
3600	U - head, 8" x 8"					1			1	1.10
3650	J - head, 4" x 8"					1			1	1.10
3700	Base plate, 8" x 8"					1			1	1.10
3750	Leveling jack					2			2	2.20
5700	Planks, 2x10x16'-0", labor only to erect & remove to 50' H	3 Carp	72	.333			12.25		12.25	19.05
5800	Over 50' high	4 Carp	80	.400	▼		14.70		14.70	23
6000	Heavy duty shoring for elevated slab forms to 8'-2" high, floor area									
6100	Labor only to erect & dismantle	4 Carp	16	2	C.S.F.		73.50		73.50	114
6110	Materials only, rent.mo				"	29.50			29.50	32.50
6500	To 14'-8" high									
6600	Labor only to erect & dismantle	4 Carp	10	3.200	C.S.F.		117		117	183
6610	Materials only, rent/mo				"	43			43	47.50

01 54 23.75 Scaffolding Specialties

		Crew	Daily Output	Labor-Hours	Unit	Material	Labor	Equipment	Total	Total Incl O&P
0010	**SCAFFOLDING SPECIALTIES**									
1200	Sidewalk bridge, heavy duty steel posts & beams, including									
1210	parapet protection & waterproofing									
1220	8' to 10' wide, 2 posts	3 Carp	15	1.600	L.F.	31	58.50		89.50	126
1230	3 posts	"	10	2.400	"	47.50	88		135.50	189
1500	Sidewalk bridge using tubular steel									
1510	scaffold frames, including planking	3 Carp	45	.533	L.F.	5.60	19.55		25.15	36.50
1600	For 2 uses per month, deduct from all above					50%				
1700	For 1 use every 2 months, add to all above					100%				

01 54 Construction Aids

01 54 23 – Temporary Scaffolding and Platforms

01 54 23.75 Scaffolding Specialties

		Crew	Daily Output	Labor-Hours	Unit	Material	2007 Bare Costs Labor	Equipment	Total	Total Incl O&P
1900	Catwalks, 20" wide, no guardrails, 7' span, buy				Ea.	121			121	133
2000	10' span, buy					170			170	187
3720	Putlog, standard, 8' span, with hangers, buy					67			67	74
3730	Rent per month					10			10	11
3750	12' span, buy					101			101	111
3755	Rent per month					15			15	16.50
3760	Trussed type, 16' span, buy					232			232	255
3770	Rent per month					20			20	22
3790	22' span, buy					277			277	305
3795	Rent per month					30			30	33
3800	Rolling ladders with handrails, 30" wide, buy, 2 step					196			196	215
4000	7 step					590			590	645
4050	10 step					820			820	900
4100	Rolling towers, buy, 5' wide, 7' long, 10' high					1,175			1,175	1,275
4200	For 5' high added sections, to buy, add					184			184	202
4300	Complete incl. wheels, railings, outriggers,									
4350	21' high, to buy				Ea.	1,975			1,975	2,150
4400	Rent/month = 5% of purchase cost				"	98.75			98.75	108.50

01 54 23.80 Staging Aids

		Crew	Daily Output	Labor-Hours	Unit	Material	2007 Bare Costs Labor	Equipment	Total	Total Incl O&P
0010	**STAGING AIDS** and fall protection equipment									
0100	Sidewall staging bracket, tubular, buy				Ea.	35			35	38.50
0110	Cost each per day, based on 250 days use				Day	.14			.14	.15
0200	Guard post, buy				Ea.	18.20			18.20	20
0210	Cost each per day, based on 250 days use				Day	.07			.07	.08
0300	End guard chains, buy per pair				Pair	27.50			27.50	30.50
0310	Cost per set per day, based on 250 days use				Day	.14			.14	.15
1000	Roof shingling bracket, steel, buy				Ea.	7.15			7.15	7.85
1010	Cost each per day, based on 250 days use				Day	.03			.03	.03
1100	Wood bracket, buy				Ea.	14.85			14.85	16.35
1110	Cost each per day, based on 250 days use				Day	.06			.06	.07
2000	Ladder jack, aluminum, buy per pair				Pair	103			103	113
2010	Cost per pair per day, based on 250 days use				Day	.41			.41	.45
2100	Steel siderail jack, buy per pair				Pair	73			73	80.50
2110	Cost per pair per day, based on 250 days use				Day	.29			.29	.32
3000	Laminated wood plank, 2x10x16', buy				Ea.	44			44	48.50
3010	Cost each per day, based on 250 days use				Day	.18			.18	.19
3100	Aluminum scaffolding plank, 20" wide x 24' long, buy				Ea.	720			720	790
3110	Cost each per day, based on 250 days use				Day	2.87			2.87	3.16
4000	Nylon full body harness, lanyard and rope grab				Ea.	205			205	225
4010	Cost each per day, based on 250 days use				Day	.82			.82	.90
4100	Rope for safety line, 5/8" x 100' nylon, buy				Ea.	55			55	60.50
4110	Cost each per day, based on 250 days use				Day	.22			.22	.24
4200	Permanent U-Bolt roof anchor, buy				Ea.	32.50			32.50	35.50
4300	Temporary (one use) roof ridge anchor, buy				"	26			26	28.50
5000	Installation (setup and removal) of staging aids									
5010	Sidewall staging bracket	2 Carp	64	.250	Ea.		9.20		9.20	14.30
5020	Guard post with 2 wood rails	"	64	.250			9.20		9.20	14.30
5030	End guard chains, set	1 Carp	64	.125			4.59		4.59	7.15
5100	Roof shingling bracket		96	.083			3.06		3.06	4.76
5200	Ladder jack		64	.125			4.59		4.59	7.15
5300	Wood plank, 2x10x16'	2 Carp	80	.200			7.35		7.35	11.45
5310	Aluminum scaffold plank, 20" x 24'	"	40	.400			14.70		14.70	23

01 54 Construction Aids

01 54 23 – Temporary Scaffolding and Platforms

01 54 23.80 Staging Aids		Crew	Daily Output	Labor-Hours	Unit	Material	2007 Bare Costs Labor	Equipment	Total	Total Incl O&P
5410	Safety rope	1 Carp	40	.200	Ea.		7.35		7.35	11.45
5420	Permanent U-Bolt roof anchor (install only)	2 Carp	40	.400			14.70		14.70	23
5430	Temporary roof ridge anchor (install only)	1 Carp	64	.125	↓		4.59		4.59	7.15

01 54 26 – Temporary Swing Staging

01 54 26.50 Swing Staging

		Crew	Daily Output	Labor-Hours	Unit	Material	Labor	Equipment	Total	Total Incl O&P
0010	**SWING STAGING**, 500 lb cap., 2' wide to 24' long, hand operat									
0020	steel cable type, with 60' cables, buy				Ea.	5,100			5,100	5,625
0030	Rent per month				"	510			510	560
0600	Lightweight (not for masons) 24' long for 150' height,									
0610	manual type, buy				Ea.	5,500			5,500	6,050
0620	Rent per month					550			550	605
0700	Powered, electric or air, to 150' high, buy					18,800			18,800	20,700
0710	Rent per month					1,325			1,325	1,450
0780	To 300' high, buy					24,500			24,500	27,000
0800	Rent per month					1,725			1,725	1,900
1000	Bosun's chair or work basket 3' x 3.5', to 300' high, electric, buy					8,400			8,400	9,225
1010	Rent per month				↓	585			585	645
2200	Move swing staging (setup and remove)	E-4	2	16	Move		670	57.50	727.50	1,300

01 54 36 – Equipment Mobilization

01 54 36.50 Mobilization or Demob.

		Crew	Daily Output	Labor-Hours	Unit	Material	Labor	Equipment	Total	Total Incl O&P
0010	**MOBILIZATION OR DEMOB.** (One or the other, unless noted) R015433-10									
0015	Up to 25 mi haul dist (50 mi RT for mob/demob crew)									
0020	Dozer, loader, backhoe, excav., grader, paver, roller, 70 to 150 H.P.	B-34N	4	2	Ea.		59	124	183	228
0100	Above 150 HP	B-34K	3	2.667			79	192	271	335
0300	Scraper, towed type (incl. tractor), 6 C.Y. capacity		3	2.667			79	192	271	335
0400	10 C.Y.		2.50	3.200			94.50	231	325.50	400
0600	Self-propelled scraper, 15 C.Y.		2.50	3.200			94.50	231	325.50	400
0700	24 C.Y.		2	4			118	288	406	500
0900	Shovel or dragline, 3/4 C.Y.		3.60	2.222			65.50	160	225.50	277
1000	1-1/2 C.Y.		3	2.667			79	192	271	335
1100	Small equipment, placed in rear of, or towed by pickup truck	A-3A	8	1			28.50	11.40	39.90	56.50
1150	Equip up to 70 HP, on flatbed trailer behind pickup truck	A-3D	4	2			57	46	103	139
2000	Mob & demob truck-mounted crane up to 75 ton, driver only	1 Eqhv	3.60	2.222			88.50		88.50	133
2100	Crane, truck-mounted, over 75 ton	A-3E	2.50	6.400			222	36.50	258.50	380
2200	Crawler-mounted, up to 75 ton	A-3F	2	8			277	305	582	760
2300	Over 75 ton	A-3G	1.50	10.667	↓		370	445	815	1,050
2500	For each additional 5 miles haul distance, add						10%	10%		
3000	For large pieces of equipment, allow for assembly/knockdown									
3001	For mob/demob of vibroflotation equip, see section 31 45 13.10									
3100	For mob/demob of micro-tunneling equip, see section 33 05 23.19									
3200	For mob/demob of pile driving equip, see section 31 62 19.10									
3300	For mob/demob of caisson drilling equip, see section 31 63 26.13									

01 54 39 – Construction Equipment

01 54 39.70 Small Tools

		Crew	Daily Output	Labor-Hours	Unit	Material	Labor	Equipment	Total	Total Incl O&P
0010	**SMALL TOOLS** R013113-50									
0020	As % of contractor's work, minimum				Total				.45%	.50%
0100	Maximum				"				1.82%	2%

01 55 Vehicular Access and Parking

01 55 23 – Temporary Roads

01 55 23.50 Roads and Sidewalks

		Crew	Daily Output	Labor-Hours	Unit	Material	2007 Bare Costs Labor	2007 Bare Costs Equipment	Total	Total Incl O&P
0010	**ROADS AND SIDEWALKS** Temporary									
0050	Roads, gravel fill, no surfacing, 4" gravel depth	B-14	715	.067	S.Y.	3.87	2.04	.34	6.25	7.80
0100	8" gravel depth	"	615	.078	"	7.75	2.38	.40	10.53	12.60
1000	Ramp, 3/4" plywood on 2" x 6" joists, 16" O.C.	2 Carp	300	.053	S.F.	1.28	1.96		3.24	4.46
1100	On 2" x 10" joists, 16" O.C.	"	275	.058	"	1.89	2.14		4.03	5.40

01 56 Temporary Barriers and Enclosures

01 56 13 – Temporary Air Barriers

01 56 13.60 Tarpaulins

		Crew	Daily Output	Labor-Hours	Unit	Material	2007 Bare Costs Labor	2007 Bare Costs Equipment	Total	Total Incl O&P
0010	**TARPAULINS**									
0020	Cotton duck, 10 oz. to 13.13 oz. per S.Y., minimum				S.F.	.48			.48	.53
0050	Maximum					.62			.62	.68
0100	Polyvinyl coated nylon, 14 oz. to 18 oz., minimum					.48			.48	.53
0150	Maximum					.68			.68	.75
0200	Reinforced polyethylene 3 mils thick, white					.15			.15	.17
0300	4 mils thick, white, clear or black					.20			.20	.22
0400	5.5 mils thick, clear					.25			.25	.28
0500	White, fire retardant					.35			.35	.39
0600	7.5 mils, oil resistant, fire retardant					.40			.40	.44
0700	8.5 mils, black					.53			.53	.58
0710	Woven polyethylene, 6 mils thick					.35			.35	.39
0730	Polyester reinforced w/ integral fastening system 11 mils thick					1.07			1.07	1.18
0740	Mylar polyester, non-reinforced, 7 mils thick					1.17			1.17	1.29

01 56 13.90 Winter Protection

		Crew	Daily Output	Labor-Hours	Unit	Material	2007 Bare Costs Labor	2007 Bare Costs Equipment	Total	Total Incl O&P
0010	**WINTER PROTECTION**									
0100	Framing to close openings	2 Clab	750	.021	S.F.	.39	.61		1	1.38
0200	Tarpaulins hung over scaffolding, 8 uses, not incl. scaffolding		1500	.011		.25	.31		.56	.76
0250	Tarpaulin polyester reinf. w/ integral fastening system 11 mils thick		1600	.010		.80	.29		1.09	1.33
0300	Prefab fiberglass panels, steel frame, 8 uses		1200	.013		.85	.38		1.23	1.54

01 56 23 – Temporary Barricades

01 56 23.10 Barricades

		Crew	Daily Output	Labor-Hours	Unit	Material	2007 Bare Costs Labor	2007 Bare Costs Equipment	Total	Total Incl O&P
0010	**BARRICADES**									
0020	5' high, 3 rail @ 2" x 8", fixed	2 Carp	20	.800	L.F.	5.65	29.50		35.15	51.50
0150	Movable	"	30	.533	"	4.72	19.55		24.27	35.50
0300	Stock units, 6' high, 8' wide, plain, buy				Ea.	435			435	480
0350	With reflective tape, buy				"	525			525	580
0400	Break-a-way 3" PVC pipe barricade									
0410	with 3 ea. 1' x 4' reflectorized panels, buy				Ea.	305			305	335
0500	Plywood with steel legs, 32" wide					72			72	79
0600	Telescoping Christmas tree, 9' high, 5 flags, buy					122			122	134
0800	Traffic cones, PVC, 18" high					7.10			7.10	7.80
0850	28" high					14.20			14.20	15.65
1000	Guardrail, wooden, 3' high, 1" x 6", on 2" x 4" posts	2 Carp	200	.080	L.F.	1.14	2.94		4.08	5.85
1100	2" x 6", on 4" x 4" posts	"	165	.097		2.33	3.56		5.89	8.10
1200	Portable metal with base pads, buy					13.15			13.15	14.50
1250	Typical installation, assume 10 reuses	2 Carp	600	.027		1.31	.98		2.29	2.96
1300	Barricade tape, polyethelyne, 7 mil, 3" wide x 500' long roll				Ea.	25			25	27.50
5000	Barricades, see also division 01 54 33.40									

01 56 Temporary Barriers and Enclosures

01 56 26 – Temporary Fencing

01 56 26.50 Temporary Fencing

		Crew	Daily Output	Labor-Hours	Unit	Material	2007 Bare Costs Labor	Equipment	Total	Total Incl O&P
0010	**TEMPORARY FENCING**									
0020	Chain link, 11 ga, 5' high	2 Clab	400	.040	L.F.	6	1.15		7.15	8.40
0100	6' high		300	.053		6.50	1.53		8.03	9.55
0200	Rented chain link, 6' high, to 1000' (up to 12 mo.)		400	.040		2.85	1.15		4	4.93
0250	Over 1000' (up to 12 mo.)		300	.053		2.65	1.53		4.18	5.30
0350	Plywood, painted, 2" x 4" frame, 4' high	A-4	135	.178		5.20	6.30		11.50	15.45
0400	4" x 4" frame, 8' high	"	110	.218		10.35	7.70		18.05	23.50
0500	Wire mesh on 4" x 4" posts, 4' high	2 Carp	100	.160		10.05	5.85		15.90	20
0550	8' high	"	80	.200		15.15	7.35		22.50	28

01 56 29 – Temporary Protective Walkways

01 56 29.50 Protection

		Crew	Daily Output	Labor-Hours	Unit	Material	Labor	Equipment	Total	Total Incl O&P
0010	**PROTECTION**									
0020	Stair tread, 2" x 12" planks, 1 use	1 Carp	75	.107	Tread	4.29	3.91		8.20	10.80
0100	Exterior plywood, 1/2" thick, 1 use		65	.123		1.42	4.52		5.94	8.60
0200	3/4" thick, 1 use		60	.133		2.12	4.89		7.01	9.95
2200	Sidewalks, 2" x 12" planks, 2 uses		350	.023	S.F.	.72	.84		1.56	2.10
2300	Exterior plywood, 2 uses, 1/2" thick		750	.011		.24	.39		.63	.87
2400	5/8" thick		650	.012		.30	.45		.75	1.03
2500	3/4" thick		600	.013		.35	.49		.84	1.15

01 56 32 – Temporary Security

01 56 32.50 Watchman

		Crew	Daily Output	Labor-Hours	Unit	Material	Labor	Equipment	Total	Total Incl O&P
0010	**WATCHMAN**									
0020	Service, monthly basis, uniformed person, minimum				Hr.				18.18	20
0100	Maximum								45.45	50
0200	Person and command dog, minimum								25.45	28
0300	Maximum								54.55	60
0500	Sentry dog, leased, with job patrol (yard dog), 1 dog				Week				227	250
0600	2 dogs				"				291	320
0800	Purchase, trained sentry dog, minimum				Ea.				1,364	1,500
0900	Maximum				"				2,727	3,000

01 58 Project Identification

01 58 13 – Temporary Project Signage

01 58 13.50 Signs

		Crew	Daily Output	Labor-Hours	Unit	Material	Labor	Equipment	Total	Total Incl O&P
0010	**SIGNS**									
0020	High intensity reflectorized, no posts, buy				S.F.	16.55			16.55	18.20

01 71 Examination and Preparation

01 71 23 – Field Engineering

01 71 23.13 Construction Layout

		Crew	Daily Output	Labor-Hours	Unit	Material	Labor	Equipment	Total	Total Incl O&P
0010	**CONSTRUCTION LAYOUT**									
1100	Crew for layout of building, trenching or pipe laying, 2 person crew	A-6	1	16	Day		590	58	648	975
1200	3 person crew	A-7	1	24			970	58	1,028	1,575
1400	Crew for roadway layout, 4 person crew	A-8	1	32			1,250	58	1,308	2,000

01 71 23.19 Surveyor Stakes

		Crew	Daily Output	Labor-Hours	Unit	Material	Labor	Equipment	Total	Total Incl O&P
0010	**SURVEYOR STAKES**									
0020	Hardwood, 1" x 1" x 48" long				C	54			54	59.50

01 71 Examination and Preparation

01 71 23 – Field Engineering

01 71 23.19 Surveyor Stakes		Crew	Daily Output	Labor-Hours	Unit	Material	2007 Bare Costs Labor	Equipment	Total	Total Incl O&P
0100	2" x 2" x 18" long				C	68.50			68.50	75
0150	2" x 2" x 24" long					80			80	88
0200	2" x 2" x 30" long					74			74	81.50

01 74 Cleaning and Waste Management

01 74 13 – Progress Cleaning

01 74 13.20 Cleaning Up

		Crew	Daily Output	Labor-Hours	Unit	Material	Labor	Equipment	Total	Total Incl O&P
0010	**CLEANING UP**									
0020	After job completion, allow, minimum				Job					.30%
0040	Maximum				"					1%
0050	Cleanup of floor area, continuous, per day, during const.	A-5	24	.750	M.S.F.	1.70	21.50	1.47	24.67	37
0100	Final by GC at end of job	"	11.50	1.565	"	2.71	45	3.07	50.78	76.50
0200	Rubbish removal, see division 02 41 19.23									

01 91 Commissioning

01 91 13 – General Commissioning Requirements

01 91 13.50 Commissioning

		Crew	Daily Output	Labor-Hours	Unit	Material	Labor	Equipment	Total	Total Incl O&P
0010	**COMMISSIONING** Including documentation of design intent									
0100	performance verification, O&M, training, min				Project					.50%
0150	Maximum				"					.75%

Division 2 - Existing Conditions

Estimating Tips

02 20 00 Assessment

- If possible, visit the site and take an inventory of the type, quantity, and size of the trees. Certain trees may have a landscape resale value or firewood value. Stump disposal can be very expensive, particularly if they cannot be buried at the site. Consider using a bulldozer in lieu of hand-cutting trees.

- Estimators should visit the site to determine the need for haul road access, storage of materials, and security considerations. When estimating for access roads on unstable soil, consider using a geotextile stabilization fabric. It can greatly reduce the quantity of crushed stone or gravel. Sites of limited size and access can cause cost overruns due to lost productivity. Theft and damage is another consideration if the location is isolated. A temporary fence or security guards may be required. Investigate the site thoroughly.

02 30 00 Subsurface Investigation

In preparing estimates on structures involving earthwork or foundations, all information concerning soil characteristics should be obtained. Look particularly for hazardous waste, evidence of prior dumping of debris, and previous stream beds.

02 40 00 Demolition and Structure Moving

The costs shown for selective demolition do not include rubbish handling or disposal. These items should be estimated separately using RSMeans data or other sources.

- Historic preservation often requires that the contractor remove materials from the existing structure, rehab them, and replace them. The estimator must be aware of any related measures and precautions that must be taken when doing selective demolition and cutting and patching. Requirements may include special handling and storage, as well as security.

- In addition to Section 02 41 00, you can find selective demolition items in each division. Example: Roofing demolition is in Division 7.

Reference Numbers

Reference numbers are shown in shaded boxes at the beginning of some major classifications. These numbers refer to related items in the Reference Section. The reference information may be an estimating procedure, an alternate pricing method, or technical information.

Note: Not all subdivisions listed here necessarily appear in this publication.

No part of this publication may be reproduced, stored in a retrieval system, or transmitted in any form or by any means without prior written permission of Reed Construction Data.

02 21 Surveys

02 21 13 – Site Surveys

02 21 13.09 Topographical Surveys

		Crew	Daily Output	Labor-Hours	Unit	Material	2007 Bare Costs Labor	2007 Bare Costs Equipment	Total	Total Incl O&P
0010	**TOPOGRAPHICAL SURVEYS**									
0020	Minimum	A-7	3.30	7.273	Acre	17	294	17.60	328.60	495
0100	Maximum	A-8	.60	53.333	"	52	2,100	97	2,249	3,400

02 21 13.13 Boundary and Survey Markers

		Crew	Daily Output	Labor-Hours	Unit	Material	Labor	Equipment	Total	Total Incl O&P
0010	**BOUNDARY AND SURVEY MARKERS**									
0300	Lot location and lines, minimum, for large quantities	A-7	2	12	Acre	30	485	29	544	810
0320	Average	"	1.25	19.200		48	775	46.50	869.50	1,300
0400	Maximum, for small quantities	A-8	1	32	↓	64	1,250	58	1,372	2,050
0600	Monuments, 3' long	A-7	10	2.400	Ea.	30	97	5.80	132.80	188
0800	Property lines, perimeter, cleared land	"	1000	.024	L.F.	.03	.97	.06	1.06	1.58
0900	Wooded land	A-8	875	.037	"	.05	1.44	.07	1.56	2.34

02 21 13.16 Aerial Surveys

		Crew	Daily Output	Labor-Hours	Unit	Material	Labor	Equipment	Total	Total Incl O&P
0010	**AERIAL SURVEYS**									
1500	Aerial surveying, including ground control, minimum fee, 10 acres				Total					5,700
1510	100 acres									9,500
1550	From existing photography, deduct				↓					1,370
1600	2' contours, 10 acres				Acre					460
1650	20 acres									315
1800	50 acres									95
1850	100 acres									85
2000	1000 acres									17.85
2050	10,000 acres				↓					11.50
2150	For 1' contours and									
2160	dense urban areas, add to above				Acre					40%
3000	Inertial guidance system for									
3010	locating coordinates, rent per day				Ea.					4,000

02 32 Geotechnical Investigations

02 32 13 – Subsurface Drilling and Sampling

02 32 13.10 Boring and Exploratory Drilling

		Crew	Daily Output	Labor-Hours	Unit	Material	Labor	Equipment	Total	Total Incl O&P
0010	**BORING AND EXPLORATORY DRILLING**									
0020	Borings, initial field stake out & determination of elevations	A-6	1	16	Day		590	58	648	975
0100	Drawings showing boring details				Total		222		222	324
0200	Report and recommendations from P.E.						500		500	730
0300	Mobilization and demobilization, minimum	B-55	4	6	↓		172	215	387	505
0350	For over 100 miles, per added mile		450	.053	Mile		1.53	1.91	3.44	4.47
0600	Auger holes in earth, no samples, 2-1/2" diameter		78.60	.305	L.F.		8.75	10.95	19.70	25.50
0650	4" diameter		67.50	.356			10.20	12.75	22.95	30
0800	Cased borings in earth, with samples, 2-1/2" diameter		55.50	.432		16.55	12.40	15.50	44.45	54.50
0850	4" diameter	↓	32.60	.736		26	21	26.50	73.50	90.50
1000	Drilling in rock, "BX" core, no sampling	B-56	34.90	.458			15.05	34.50	49.55	61
1050	With casing & sampling		31.70	.505		16.55	16.55	38	71.10	85
1200	"NX" core, no sampling		25.92	.617			20.50	46.50	67	82
1250	With casing and sampling	↓	25	.640	↓	20	21	48	89	108
1400	Drill rig and crew with truck mounted auger	B-55	1	24	Day		690	860	1,550	2,025
1450	With crawler type drill	B-56	1	16	"		525	1,200	1,725	2,125
1500	For inner city borings add, minimum									10%
1510	Maximum									20%

02 32 Geotechnical Investigations

02 32 19 – Exploratory Excavations

02 32 19.10 Test Pits

		Crew	Daily Output	Labor-Hours	Unit	Material	2007 Bare Costs Labor	2007 Bare Costs Equipment	Total	Total Incl O&P
0010	**TEST PITS**									
0020	Hand digging, light soil	1 Clab	4.50	1.778	C.Y.		51		51	79.50
0100	Heavy soil	"	2.50	3.200			92		92	143
0120	Loader-backhoe, light soil	B-11M	28	.571			19.20	10.20	29.40	40.50
0130	Heavy soil	"	20	.800			27	14.25	41.25	56.50
1000	Subsurface exploration, mobilization				Mile				7.15	8.22
1010	Difficult access for rig, add				Hr.				143	164
1020	Auger borings, drill rig, incl. samples				L.F.				16.45	18.92
1030	Hand auger								24.31	27.96
1050	Drill and sample every 5', split spoon								21.45	24.67
1060	Extra samples				Ea.				29	33.35

02 41 Demolition

02 41 13 – Selective Site Demolition

02 41 13.15 Hydrodemolition

			Crew	Daily Output	Labor-Hours	Unit	Material	Labor	Equipment	Total	Total Incl O&P
0010	**HYDRODEMOLITION**	R024119-10									
0015	Hydrodemolition, concrete pavement, 4000 PSI, 2" depth		B-5	500	.112	S.F.		3.56	2.03	5.59	7.75
0120	4" depth			450	.124			3.96	2.25	6.21	8.60
0130	6" depth			400	.140			4.45	2.54	6.99	9.65
0410	6000 PSI, 2" depth			410	.137			4.34	2.47	6.81	9.40
0420	4" depth			350	.160			5.10	2.90	8	11.05
0430	6" depth			300	.187			5.95	3.38	9.33	12.85
0510	8000 PSI, 2" depth			330	.170			5.40	3.07	8.47	11.70
0520	4" depth			280	.200			6.35	3.62	9.97	13.80
0530	6" depth			240	.233			7.40	4.23	11.63	16.05

02 41 13.17 Demolish, Remove Pavement and Curb

			Crew	Daily Output	Labor-Hours	Unit	Material	Labor	Equipment	Total	Total Incl O&P
0010	**DEMOLISH, REMOVE PAVEMENT AND CURB**	R024119-10									
5010	Pavement removal, bituminous roads, 3" thick		B-38	690	.058	S.Y.		1.90	1.30	3.20	4.35
5050	4" to 6" thick			420	.095			3.11	2.14	5.25	7.15
5100	Bituminous driveways			640	.063			2.04	1.41	3.45	4.69
5200	Concrete to 6" thick, hydraulic hammer, mesh reinforced			255	.157			5.15	3.53	8.68	11.75
5300	Rod reinforced			200	.200			6.55	4.50	11.05	15
5400	Concrete, 7" to 24" thick, plain			33	1.212	C.Y.		39.50	27.50	67	91
5500	Reinforced			24	1.667	"		54.50	37.50	92	125
5600	With hand held air equipment, bituminous, to 6" thick		B-39	1900	.025	S.F.		.77	.09	.86	1.29
5700	Concrete to 6" thick, no reinforcing			1600	.030			.91	.11	1.02	1.53
5800	Mesh reinforced			1400	.034			1.04	.13	1.17	1.75
5900	Rod reinforced			765	.063			1.91	.24	2.15	3.21
6000	Curbs, concrete, plain		B-6	360	.067	L.F.		2.10	.68	2.78	3.96
6100	Reinforced			275	.087			2.74	.88	3.62	5.20
6200	Granite			360	.067			2.10	.68	2.78	3.96
6300	Bituminous			528	.045			1.43	.46	1.89	2.71

02 41 13.33 Minor Site Demolition

			Crew	Daily Output	Labor-Hours	Unit	Material	Labor	Equipment	Total	Total Incl O&P
0010	**MINOR SITE DEMOLITION**	R024119-10									
0015	No hauling, abandon catch basin or manhole		B-6	7	3.429	Ea.		108	35	143	205
0020	Remove existing catch basin or manhole, masonry			4	6			189	61	250	355
0030	Catch basin or manhole frames and covers, stored			13	1.846			58	18.70	76.70	110
0040	Remove and reset			7	3.429			108	35	143	205
0100	Roadside delineators, remove only		B-80	175	.183			5.70	3.20	8.90	12.30
0110	Remove and reset		"	100	.320			10	5.60	15.60	21.50

02 41 Demolition

02 41 13 – Selective Site Demolition

02 41 13.33 Minor Site Demolition

		Crew	Daily Output	Labor-Hours	Unit	Material	2007 Bare Costs Labor	2007 Bare Costs Equipment	Total	Total Incl O&P
0800	Guiderail, corrugated steel, remove only	B-80A	100	.240	L.F.		6.90	1.85	8.75	12.80
0850	Remove and reset	"	40	.600	"		17.25	4.63	21.88	32
0860	Guide posts, remove only	B-80B	120	.267	Ea.		8.20	2.26	10.46	15.15
0870	Remove and reset	B-55	50	.480			13.75	17.20	30.95	40.50
0900	Hydrants, fire, remove only	B-21A	5	8			288	115	403	565
0950	Remove and reset	"	2	20			720	288	1,008	1,425
1000	Masonry walls, block or tile, solid, remove	B-5	1800	.031	C.F.		.99	.56	1.55	2.14
1100	Cavity wall		2200	.025			.81	.46	1.27	1.76
1200	Brick, solid		900	.062			1.98	1.13	3.11	4.29
1300	With block back-up		1130	.050			1.58	.90	2.48	3.42
1400	Stone, with mortar		900	.062			1.98	1.13	3.11	4.29
1500	Dry set		1500	.037			1.19	.68	1.87	2.57
1600	Median barrier, precast concrete, remove and store	B-3	430	.112	L.F.		3.46	4.41	7.87	10.20
1610	Remove and reset	"	390	.123			3.81	4.86	8.67	11.25
2900	Pipe removal, sewer/water, no excavation, 12" diameter	B-6	175	.137			4.31	1.39	5.70	8.20
2930	15"-18" diameter		150	.160			5.05	1.62	6.67	9.55
2960	21"-24" diameter		120	.200			6.30	2.03	8.33	11.90
3000	27"-36" diameter		90	.267			8.40	2.70	11.10	15.90
3200	Steel, welded connections, 4" diameter		160	.150			4.72	1.52	6.24	8.90
3300	10" diameter		80	.300			9.45	3.04	12.49	17.85
3500	Railroad track removal, ties and track	B-13	330	.170			5.30	2.24	7.54	10.65
3600	Ballast	B-14	500	.096	C.Y.		2.92	.49	3.41	5.05
3700	Remove and re-install, ties & track using new bolts & spikes		50	.960	L.F.		29	4.87	33.87	50.50
3800	Turnouts using new bolts and spikes		1	48	Ea.		1,450	243	1,693	2,525
4000	Sidewalk removal, bituminous, 2-1/2" thick	B-6	325	.074	S.Y.		2.32	.75	3.07	4.39
4050	Brick, set in mortar		185	.130			4.08	1.32	5.40	7.70
4100	Concrete, plain, 4"		160	.150			4.72	1.52	6.24	8.90
4200	Mesh reinforced		150	.160			5.05	1.62	6.67	9.55
4300	Slab on grade removal, plain	B-5	45	1.244	C.Y.		39.50	22.50	62	86
4310	Mesh reinforced		33	1.697			54	30.50	84.50	117
4320	Rod reinforced		25	2.240			71	40.50	111.50	155
4400	For congested sites or small quantities, add up to								200%	200%
4450	For disposal on site, add	B-11A	232	.069			2.32	4.26	6.58	8.25
4500	To 5 miles, add	B-34D	76	.105			3.11	6	9.11	11.40

02 41 13.60 Fencing Demolition

			Crew	Daily Output	Labor-Hours	Unit	Material	Labor	Equipment	Total	Total Incl O&P
0010	**FENCING DEMOLITION**	R024119-10									
1600	Fencing, barbed wire, 3 strand		2 Clab	430	.037	L.F.		1.07		1.07	1.67
1650	5 strand		"	280	.057			1.64		1.64	2.56
1700	Chain link, posts & fabric, remove only, 8' to 10' high		B-6	445	.054			1.70	.55	2.25	3.21
1750	Remove and reset		"	70	.343			10.80	3.48	14.28	20.50

02 41 16 – Structure Demolition

02 41 16.13 Building Demolition

			Crew	Daily Output	Labor-Hours	Unit	Material	Labor	Equipment	Total	Total Incl O&P
0010	**BUILDING DEMOLITION** Large urban projects, incl. 20 mi. haul	R024119-10									
0011	No foundation or dump fees, C.F. is vol. of building standing										
0012	Steel		B-8	21500	.003	C.F.		.10	.12	.22	.28
0050	Concrete			15300	.004			.13	.17	.30	.40
0080	Masonry			20100	.003			.10	.13	.23	.30
0100	Mixture of types, average			20100	.003			.10	.13	.23	.30
0500	Small bldgs, or single bldgs, no salvage included, steel		B-3	14800	.003			.10	.13	.23	.29
0600	Concrete			11300	.004			.13	.17	.30	.38
0650	Masonry			14800	.003			.10	.13	.23	.29
0700	Wood			14800	.003			.10	.13	.23	.29

02 41 Demolition

02 41 16 – Structure Demolition

02 41 16.13 Building Demolition

		Crew	Daily Output	Labor-Hours	Unit	Material	2007 Bare Costs Labor	2007 Bare Costs Equipment	Total	Total Incl O&P
1000	Single family, one story house, wood, minimum				Ea.				3,200	3,520
1020	Maximum								5,500	6,050
1200	Two family, two story house, wood, minimum								4,200	4,620
1220	Maximum								8,000	8,800
1300	Three family, three story house, wood, minimum								5,500	6,050
1320	Maximum								9,700	10,670
5000	For buildings with no interior walls, deduct								50%	

02 41 16.15 Explosive/Implosive Demolition

		Crew	Daily Output	Labor-Hours	Unit	Material	Labor	Equipment	Total	Total Incl O&P
0010	**EXPLOSIVE/IMPLOSIVE DEMOLITION** Large projects, R024119-10									
0020	no disposal fee based on building volume, steel building	B-5B	16900	.003	C.F.		.10	.13	.23	.29
0100	Concrete building		16900	.003			.10	.13	.23	.29
0200	Masonry building		16900	.003			.10	.13	.23	.29
0400	Disposal of material, minimum	B-3	445	.108	C.Y.		3.34	4.26	7.60	9.85
0500	Maximum	"	365	.132	"		4.07	5.20	9.27	12

02 41 16.17 Bldg. Footings and Foundations Demolition

		Crew	Daily Output	Labor-Hours	Unit	Material	Labor	Equipment	Total	Total Incl O&P
0010	**BLDG. FOOTINGS AND FOUNDATIONS DEMOLITION** R024119-10									
0200	Floors, concrete slab on grade,									
0240	4" thick, plain concrete	B-9C	500	.080	S.F.		2.33	.36	2.69	4.03
0280	Reinforced, wire mesh		470	.085			2.48	.38	2.86	4.28
0300	Rods		400	.100			2.92	.45	3.37	5.05
0400	6" thick, plain concrete		375	.107			3.11	.48	3.59	5.35
0420	Reinforced, wire mesh		340	.118			3.43	.53	3.96	5.95
0440	Rods		300	.133			3.89	.60	4.49	6.70
1000	Footings, concrete, 1' thick, 2' wide	B-5	300	.187	L.F.		5.95	3.38	9.33	12.85
1080	1'-6" thick, 2' wide		250	.224			7.10	4.06	11.16	15.40
1120	3' wide		200	.280			8.90	5.05	13.95	19.30
1140	2' thick, 3' wide		175	.320			10.15	5.80	15.95	22
1200	Average reinforcing, add								10%	10%
1220	Heavy reinforcing, add								20%	20%
2000	Walls, block, 4" thick	1 Clab	180	.044	S.F.		1.28		1.28	1.99
2040	6" thick		170	.047			1.35		1.35	2.11
2080	8" thick		150	.053			1.53		1.53	2.39
2100	12" thick		150	.053			1.53		1.53	2.39
2200	For horizontal reinforcing, add								10%	10%
2220	For vertical reinforcing, add								20%	20%
2400	Concrete, plain concrete, 6" thick	B-9	160	.250			7.30	1.13	8.43	12.60
2420	8" thick		140	.286			8.35	1.29	9.64	14.35
2440	10" thick		120	.333			9.70	1.50	11.20	16.80
2500	12" thick		100	.400			11.65	1.80	13.45	20
2600	For average reinforcing, add								10%	10%
2620	For heavy reinforcing, add								20%	20%
4000	For congested sites or small quantities, add up to								200%	200%
4200	Add for disposal, on site	B-11A	232	.069	C.Y.		2.32	4.26	6.58	8.25
4250	To five miles	B-30	220	.109	"		3.55	8.35	11.90	14.60

02 41 19 – Selective Structure Demolition

02 41 19.13 Selective Building Demolition

0010	**SELECTIVE BUILDING DEMOLITION**
0020	Costs related to selective demolition of specific building components
0025	are included under Common Work Results (XX 05 00)
0030	in the component's appropriate division.

02 41 Demolition

02 41 19 – Selective Structure Demolition

02 41 19.16 Selective Demolition, Cutout

		Crew	Daily Output	Labor-Hours	Unit	Material	2007 Bare Costs Labor	Equipment	Total	Total Incl O&P
0010	**SELECTIVE DEMOLITION, CUTOUT** R024119-10									
0020	Concrete, elev. slab, light reinforcement, under 6 CF	B-9C	65	.615	C.F.		17.95	2.77	20.72	31
0050	Light reinforcing, over 6 C.F.	"	75	.533	"		15.55	2.40	17.95	26.50
0200	Slab on grade to 6" thick, not reinforced, under 8 S.F.	B-9	85	.471	S.F.		13.70	2.12	15.82	24
0250	8 - 16 S.F.		175	.229	"		6.65	1.03	7.68	11.50
0600	Walls, not reinforced, under 6 C.F.		60	.667	C.F.		19.45	3	22.45	34
0650	6 - 12 C.F.	↓	80	.500	↓		14.60	2.25	16.85	25
1000	Concrete, elevated slab, bar reinforced, under 6 C.F.	B-9C	45	.889			26	4	30	45
1050	Bar reinforced, over 6 C.F.	"	50	.800	↓		23.50	3.60	27.10	40.50
1200	Slab on grade to 6" thick, bar reinforced, under 8 S.F.	B-9	75	.533	S.F.		15.55	2.40	17.95	26.50
1250	8 - 16 S.F.	"	150	.267	"		7.75	1.20	8.95	13.40
1400	Walls, bar reinforced, under 6 C.F.	B-9C	50	.800	C.F.		23.50	3.60	27.10	40.50
1450	6 - 12 CF	"	70	.571	"		16.65	2.57	19.22	29
2000	Brick, to 4 S.F. opening, not including toothing									
2040	4" thick	B-9C	30	1.333	Ea.		39	6	45	67
2060	8" thick		18	2.222			65	10	75	112
2080	12" thick		10	4			117	18	135	202
2400	Concrete block, to 4 S.F. opening, 2" thick		35	1.143			33.50	5.15	38.65	57.50
2420	4" thick		30	1.333			39	6	45	67
2440	8" thick		27	1.481			43	6.65	49.65	74.50
2460	12" thick	↓	24	1.667			48.50	7.50	56	84
2600	Gypsum block, to 4 S.F. opening, 2" thick	B-9	80	.500			14.60	2.25	16.85	25
2620	4" thick		70	.571			16.65	2.57	19.22	29
2640	8" thick		55	.727			21	3.27	24.27	36.50
2800	Terra cotta, to 4 S.F. opening, 4" thick		70	.571			16.65	2.57	19.22	29
2840	8" thick		65	.615			17.95	2.77	20.72	31
2880	12" thick	↓	50	.800	↓		23.50	3.60	27.10	40.50
3000	Toothing masonry cutouts, brick, soft old mortar	1 Brhe	40	.200	V.L.F.		5.75		5.75	8.70
3100	Hard mortar		30	.267			7.65		7.65	11.65
3200	Block, soft old mortar		70	.114			3.27		3.27	4.98
3400	Hard mortar	↓	50	.160	↓		4.58		4.58	7
6000	Walls, interior, not including re-framing,									
6010	openings to 5 S.F.									
6100	Drywall to 5/8" thick	1 Clab	24	.333	Ea.		9.60		9.60	14.90
6200	Paneling to 3/4" thick		20	.400			11.50		11.50	17.90
6300	Plaster, on gypsum lath		20	.400			11.50		11.50	17.90
6340	On wire lath	↓	14	.571	↓		16.45		16.45	25.50
7000	Wood frame, not including re-framing, openings to 5 S.F.									
7200	Floors, sheathing and flooring to 2" thick	1 Clab	5	1.600	Ea.		46		46	71.50
7310	Roofs, sheathing to 1" thick, not including roofing		6	1.333			38.50		38.50	59.50
7410	Walls, sheathing to 1" thick, not including siding	↓	7	1.143	↓		33		33	51

02 41 19.18 Selective Demolition, Disposal Only

		Crew	Daily Output	Labor-Hours	Unit	Material	Labor	Equipment	Total	Total Incl O&P
0010	**SELECTIVE DEMOLITION, DISPOSAL ONLY** R024119-10									
0015	Urban bldg w/salvage value allowed									
0020	Including loading and 5 mile haul to dump									
0200	Steel frame	B-3	430	.112	C.Y.		3.46	4.41	7.87	10.20
0300	Concrete frame		365	.132			4.07	5.20	9.27	12
0400	Masonry construction		445	.108			3.34	4.26	7.60	9.85
0500	Wood frame	↓	247	.194	↓		6	7.65	13.65	17.75

02 41 19.19 Selective Demolition, Dump Charges

0010	**SELECTIVE DEMOLITION, DUMP CHARGES** R024119-10									
0020	Dump charges, typical urban city, tipping fees only									

02 41 Demolition

02 41 19 – Selective Structure Demolition

02 41 19.19 Selective Demolition, Dump Charges

		Crew	Daily Output	Labor-Hours	Unit	Material	2007 Bare Costs Labor	Equipment	Total	Total Incl O&P
0100	Building construction materials				Ton	90			90	90
0200	Trees, brush, lumber					65			65	65
0300	Rubbish only					78			78	78
0500	Reclamation station, usual charge					95			95	95

02 41 19.21 Selective Demolition, Gutting

		Crew	Daily Output	Labor-Hours	Unit	Material	Labor	Equipment	Total	Total Incl O&P
0010	**SELECTIVE DEMOLITION, GUTTING** R024119-10									
0020	Building interior, including disposal, dumpster fees not included									
0500	Residential building									
0560	Minimum	B-16	400	.080	SF Flr.		2.36	1.32	3.68	5.10
0580	Maximum	"	360	.089	"		2.62	1.47	4.09	5.70
0900	Commercial building									
1000	Minimum	B-16	350	.091	SF Flr.		2.69	1.51	4.20	5.85
1020	Maximum	"	250	.128	"		3.77	2.12	5.89	8.20

02 41 19.23 Selective Demolition, Rubbish Handling

		Crew	Daily Output	Labor-Hours	Unit	Material	Labor	Equipment	Total	Total Incl O&P
0010	**SELECTIVE DEMOLITION, RUBBISH HANDLING** R024119-10									
0020	The following are to be added to the demolition prices									
0400	Chute, circular, prefabricated steel, 18" diameter	B-1	40	.600	L.F.	37	17.65		54.65	68
0440	30" diameter	"	30	.800	"	44	23.50		67.50	85
0725	Dumpster, weekly rental, 1 dump/week, 20 C.Y. capacity (8 Tons)				Week	690			690	759
0800	30 C.Y. capacity (10 Tons)					900			900	990
0840	40 C.Y. capacity (13 Tons)					1,160			1,160	1,276
1000	Dust partition, 6 mil polyethylene, 1" x 3" frame	2 Carp	2000	.008	S.F.	.28	.29		.57	.77
1080	2" x 4" frame	"	2000	.008	"	.33	.29		.62	.82
2000	Load, haul, and dump, 50' haul	2 Clab	24	.667	C.Y.		19.15		19.15	30
2040	100' haul		16.50	.970			28		28	43.50
2080	Over 100' haul, add per 100 L.F.		35.50	.451			12.95		12.95	20
2120	In elevators, per 10 floors, add		140	.114			3.29		3.29	5.10
3000	Loading & trucking, including 2 mile haul, chute loaded	B-16	45	.711			21	11.80	32.80	45.50
3040	Hand loading truck, 50' haul	"	48	.667			19.65	11.05	30.70	42.50
3080	Machine loading truck	B-17	120	.267			8.25	5.10	13.35	18.30
5000	Haul, per mile, up to 8 C.Y. truck	B-34B	1165	.007			.20	.46	.66	.81
5100	Over 8 C.Y. truck	"	1550	.005			.15	.34	.49	.62

02 41 19.25 Selective Demolition, Saw Cutting

		Crew	Daily Output	Labor-Hours	Unit	Material	Labor	Equipment	Total	Total Incl O&P
0010	**SELECTIVE DEMOLITION, SAW CUTTING** R024119-10									
0015	Asphalt, up to 3" deep	B-89	1050	.015	L.F.	.30	.50	.30	1.10	1.43
0020	Each additional inch of depth	"	1800	.009		.07	.29	.18	.54	.72
1200	Masonry walls, hydraulic saw, brick, per inch of depth	B-89B	300	.053		.37	1.74	1.95	4.06	5.20
1220	Block walls, solid, per inch of depth	"	250	.064		.39	2.09	2.34	4.82	6.20
2000	Brick or masonry w/hand held saw, per inch of depth	A-1	125	.064		.31	1.84	.43	2.58	3.68
5000	Wood sheathing to 1" thick, on walls	1 Carp	200	.040			1.47		1.47	2.29
5020	On roof	"	250	.032			1.17		1.17	1.83

02 41 19.27 Selective Demolition, Torch Cutting

		Crew	Daily Output	Labor-Hours	Unit	Material	Labor	Equipment	Total	Total Incl O&P
0010	**SELECTIVE DEMOLITION, TORCH CUTTING** R024119-10									
0020	Steel, 1" thick plate	1 Clab	360	.022	L.F.	.20	.64		.84	1.21
0040	1" diameter bar	"	210	.038	Ea.		1.10		1.10	1.71
1000	Oxygen lance cutting, reinforced concrete walls									
1040	12" to 16" thick walls	1 Clab	10	.800	L.F.		23		23	36
1080	24" thick walls	"	6	1.333	"		38.50		38.50	59.50

02 43 Structure Moving

02 43 13 – Structure Relocation

02 43 13.13 Building Relocation

		Crew	Daily Output	Labor-Hours	Unit	Material	2007 Bare Costs Labor	2007 Bare Costs Equipment	Total	Total Incl O&P
0010	**BUILDING RELOCATION** One day move, up to 24' wide									
0020	Reset on new foundation, patch & hook-up, average move				Total					10,695
0040	Wood or steel frame bldg., based on ground floor area	B-4	185	.259	S.F.		7.60	1.95	9.55	13.95
0060	Masonry bldg., based on ground floor area	"	137	.350			10.25	2.64	12.89	18.80
0200	For 24' to 42' wide, add									15%
0220	For each additional day on road, add	B-4	1	48	Day		1,400	360	1,760	2,575
0240	Construct new basement, move building, 1 day									
0300	move, patch & hook-up, based on ground floor area	B-3	155	.310	S.F.	8.10	9.60	12.20	29.90	37

02 58 Snow Control

02 58 13 – Snow Fencing

02 58 13.10 Snow Fencing

		Crew	Daily Output	Labor-Hours	Unit	Material	Labor	Equipment	Total	Total Incl O&P
0010	**SNOW FENCING**									
7001	Snow fence on steel posts 10' O.C., 4' high	B-1	500	.048	L.F.	2.57	1.41		3.98	5.05

02 65 Underground Storage Tank Removal

02 65 10 – Underground Storage Tank Removal

02 65 10.30 Removal of Underground Storage Tanks

		Crew	Daily Output	Labor-Hours	Unit	Material	Labor	Equipment	Total	Total Incl O&P
0010	**REMOVAL OF UNDERGROUND STORAGE TANKS** R026510-20									
0011	Petroleum storage tanks, non-leaking									
0100	Excavate & load onto trailer									
0110	3000 gal. to 5000 gal. tank	B-14	4	12	Ea.		365	61	426	630
0120	6000 gal to 8000 gal tank	B-3A	3	13.333			410	259	669	915
0130	9000 gal to 12000 gal tank	"	2	20			615	390	1,005	1,375
0190	Known leaking tank add				%				100%	100%
0200	Remove sludge, water and remaining product from tank bottom									
0201	of tank with vacuum truck									
0300	3000 gal to 5000 gal tank	A-13	5	1.600	Ea.		59	133	192	235
0310	6000 gal to 8000 gal tank		4	2			73.50	166	239.50	293
0320	9000 gal to 12000 gal tank		3	2.667			98.50	221	319.50	390
0390	Dispose of sludge off-site, average				Gal.				5.15	5.72
0400	Insert inert solid CO_2 "dry ice" into tank									
0401	For cleaning/transporting tanks (1.5 lbs./100 gal. cap)	1 Clab	500	.016	Lb.	1.53	.46		1.99	2.40
1020	Haul tank to certified salvage dump, 100 miles round trip									
1023	3000 gal. to 5000 gal. tank				Ea.				660	726
1026	6000 gal. to 8000 gal. tank								780	858
1029	9,000 gal. to 12,000 gal. tank								1,050	1,155
1100	Disposal of contaminated soil to landfill									
1110	Minimum				C.Y.				118.80	132
1111	Maximum				"				356.40	396
1120	Disposal of contaminated soil to									
1121	bituminous concrete batch plant									
1130	Minimum				C.Y.				59.40	66
1131	Maximum				"				118.80	132
2010	Decontamination of soil on site incl poly tarp on top/bottom									
2011	Soil containment berm, and chemical treatment									
2020	Minimum	B-11C	100	.160	C.Y.	6.30	5.35	2.43	14.08	17.80
2021	Maximum	"	100	.160		8.15	5.35	2.43	15.93	19.85
2050	Disposal of decontaminated soil, minimum								71.10	79

02 65 Underground Storage Tank Removal

02 65 10 – Underground Storage Tank Removal

02 65 10.30 Removal of Underground Storage Tanks	Crew	Daily Output	Labor-Hours	Unit	Material	2007 Bare Costs Labor	Equipment	Total	Total Incl O&P
2055 Maximum				C.Y.				145.80	162

02 81 Transportation and Disposal of Hazardous Materials

02 81 20 – Hazardous Waste Handling

02 81 20.10 Hazardous Waste Cleanup/Pickup/Disposal

		Crew	Daily Output	Labor-Hours	Unit	Material	Labor	Equipment	Total	Total Incl O&P
0010	**HAZARDOUS WASTE CLEANUP/PICKUP/DISPOSAL**									
0100	For contractor rental equipment, i.e. Dozer,									
0110	Front end loader, Dump truck, etc., see Reference 01 54 33									
1000	Solid pickup									
1100	55 gal. drums				Ea.				217.80	242
1120	Bulk material, minimum				Ton				163.80	182
1130	Maximum				"				544.50	605
1200	Transportation to disposal site									
1220	Truckload = 80 drums or 25 C.Y. or 18 tons									
1260	Minimum				Mile				3.38	3.75
1270	Maximum				"				5.94	6.60
3000	Liquid pickup, vacuum truck, stainless steel tank									
3100	Minimum charge, 4 hours									
3110	1 compartment, 2200 gallon				Hr.				118.80	132
3120	2 compartment, 5000 gallon				"				118.80	132
3400	Transportation in 6900 gallon bulk truck				Mile				6.42	7.13
3410	In teflon lined truck				"				7.43	8.25
5000	Heavy sludge or dry vacuumable material				Hr.				118.80	132
6000	Dumpsite disposal charge, minimum				Ton				118.80	132
6020	Maximum				"				435.60	484

02 82 Asbestos Remediation

02 82 13 – Asbestos Abatement

02 82 13.41 Asbestos Abatement Equipment

		Crew	Daily Output	Labor-Hours	Unit	Material	Labor	Equipment	Total	Total Incl O&P
0010	**ASBESTOS ABATEMENT EQUIPMENT** and supplies, buy R028213-20									
0200	Air filtration device, 2000 C.F.M.				Ea.	2,500			2,500	2,750
0250	Large volume air sampling pump, minimum					310			310	340
0260	Maximum					475			475	520
0300	Airless sprayer unit, 2 gun					2,100			2,100	2,300
0350	Light stand, 500 watt					95			95	105
0400	Personal respirators									
0410	Negative pressure, 1/2 face, dual operation, min.				Ea.	22.50			22.50	25
0420	Maximum					24			24	26.50
0450	P.A.P.R., full face, minimum					450			450	495
0460	Maximum					840			840	925
0470	Supplied air, full face, incl. air line, minimum					500			500	550
0480	Maximum					720			720	790
0500	Personnel sampling pump, minimum					500			500	550
0510	Maximum					750			750	825
1500	Power panel, 20 unit, incl. G.F.I.					600			600	660
1600	Shower unit, including pump and filters					1,300			1,300	1,425
1700	Supplied air system (type C)					10,000			10,000	11,000
1750	Vacuum cleaner, HEPA, 16 gal., stainless steel, wet/dry					930			930	1,025
1760	55 gallon					1,700			1,700	1,875

02 82 Asbestos Remediation

02 82 13 – Asbestos Abatement

02 82 13.41 Asbestos Abatement Equipment

		Crew	Daily Output	Labor-Hours	Unit	Material	2007 Bare Costs Labor	Equipment	Total	Total Incl O&P
1800	Vacuum loader, 9-18 ton/hr				Ea.	90,000			90,000	99,000
1900	Water atomizer unit, including 55 gal. drum					230			230	253
2000	Worker protection, whole body, foot, head cover & gloves, plastic					5.60			5.60	6.15
2500	Respirator, single use					10.90			10.90	12
2550	Cartridge for respirator					11.10			11.10	12.20
2570	Glove bag, 7 mil, 50" x 64"					9.20			9.20	10.10
2580	10 mil, 44" x 60"					8.40			8.40	9.25
3000	HEPA vacuum for work area, minimun					1,050			1,050	1,150
3050	Maximum					3,100			3,100	3,425
6000	Disposable polyethelene bags, 6 mil, 3 C.F.					1.24			1.24	1.36
6300	Disposable fiber drums, 3 C.F.					6.50			6.50	7.15
6400	Pressure sensitive caution lables, 3" x 5"					1.19			1.19	1.31
6450	11" x 17"					7.05			7.05	7.75
6500	Negative air machine, 1800 C.F.M.					850			850	935

02 82 13.42 Preparation of Asbestos Containment Area

		Crew	Daily Output	Labor-Hours	Unit	Material	2007 Bare Costs Labor	Equipment	Total	Total Incl O&P
0010	**PREPARATION OF ASBESTOS CONTAINMENT AREA**									
0100	Pre-cleaning, HEPA vacuum and wet wipe, flat surfaces	A-10	12000	.005	S.F.	.01	.22		.23	.37
0200	Protect carpeted area, 2 layers 6 mil poly on 3/4" plywood	"	1000	.064		1.50	2.66		4.16	5.85
0300	Separation barrier, 2" x 4" @ 16", 1/2" plywood ea. side, 8' high	2 Carp	400	.040		1.25	1.47		2.72	3.67
0310	12' high		320	.050		1.40	1.84		3.24	4.40
0320	16' high		200	.080		1.50	2.94		4.44	6.20
0400	Personnel decontam. chamber, 2" x 4" @ 16", 3/4" ply ea. side		280	.057		2.50	2.10		4.60	6
0450	Waste decontam. chamber, 2" x 4" studs @ 16", 3/4" ply ea. side		360	.044		3	1.63		4.63	5.85
0500	Cover surfaces with polyethelene sheeting									
0501	Including glue and tape									
0550	Floors, each layer, 6 mil	A-10	8000	.008	S.F.	.05	.33		.38	.58
0551	4 mil		9000	.007		.03	.30		.33	.50
0560	Walls, each layer, 6 mil		6000	.011		.05	.44		.49	.75
0561	4 mil		7000	.009		.03	.38		.41	.63
0570	For heights above 12', add						20%			
0575	For heights above 20', add						30%			
0580	For fire retardant poly, add					100%				
0590	For large open areas, deduct					10%	20%			
0600	Seal floor penetrations with foam firestop to 36 Sq. In.	2 Carp	200	.080	Ea.	6.25	2.94		9.19	11.45
0610	36 Sq. In. to 72 Sq. In.		125	.128		12.50	4.70		17.20	21
0615	72 Sq. In. to 144 Sq. In.		80	.200		25	7.35		32.35	39
0620	Wall penetrations, to 36 square inches		180	.089		6.25	3.26		9.51	12
0630	36 Sq. In. to 72 Sq. In.		100	.160		12.50	5.85		18.35	23
0640	72 Sq. In. to 144 Sq. In.		60	.267		25	9.80		34.80	43
0800	Caulk seams with latex	1 Carp	230	.035	L.F.	.15	1.28		1.43	2.16
0900	Set up neg. air machine, 1-2k C.F.M. /25 M.C.F. volume	1 Asbe	4.30	1.860	Ea.		77		77	122

02 82 13.43 Bulk Asbestos Removal

		Crew	Daily Output	Labor-Hours	Unit	Material	2007 Bare Costs Labor	Equipment	Total	Total Incl O&P
0010	**BULK ASBESTOS REMOVAL**									
0020	Includes disposable tools and 2 suits and 1 respirator filter/day/worker									
0100	Beams, W 10 x 19	A-9	235	.272	L.F.	.76	11.30		12.06	18.75
0110	W 12 x 22		210	.305		.85	12.65		13.50	21
0120	W 14 x 26		180	.356		.99	14.80		15.79	24.50
0130	W 16 x 31		160	.400		1.12	16.60		17.72	27.50
0140	W 18 x 40		140	.457		1.27	19		20.27	31.50
0150	W 24 x 55		110	.582		1.62	24		25.62	40
0160	W 30 x 108		85	.753		2.10	31.50		33.60	52
0170	W 36 x 150		72	.889		2.48	37		39.48	61

02 82 Asbestos Remediation

02 82 13 – Asbestos Abatement

02 82 13.43 Bulk Asbestos Removal

		Crew	Daily Output	Labor-Hours	Unit	Material	2007 Bare Costs Labor	2007 Bare Costs Equipment	Total	Total Incl O&P
0200	Boiler insulation	A-9	480	.133	S.F.	.43	5.55		5.98	9.20
0210	With metal lath add				%				50%	
0300	Boiler breeching or flue insulation	A-9	520	.123	S.F.	.34	5.10		5.44	8.50
0310	For active boiler, add				%				100%	
0400	Duct or AHU insulation	A-10B	440	.073	S.F.	.20	3.03		3.23	5
0500	Duct vibration isolation joints, up to 24 Sq. In. duct	A-9	56	1.143	Ea.	3.19	47.50		50.69	78.50
0520	25 Sq. In. to 48 Sq. In. duct		48	1.333		3.72	55.50		59.22	91.50
0530	49 Sq. In. to 76 Sq. In. duct		40	1.600		4.46	66.50		70.96	110
0600	Pipe insulation, air cell type, up to 4" diameter pipe		900	.071	L.F.	.20	2.96		3.16	4.89
0610	4" to 8" diameter pipe		800	.080		.22	3.32		3.54	5.50
0620	10" to 12" diameter pipe		700	.091		.25	3.80		4.05	6.30
0630	14" to 16" diameter pipe		550	.116		.32	4.84		5.16	8
0650	Over 16" diameter pipe		650	.098	S.F.	.27	4.09		4.36	6.75
0700	With glove bag up to 3" diameter pipe		200	.320	L.F.	3.22	13.30		16.52	24.50
1000	Pipe fitting insulation up to 4" diameter pipe		320	.200	Ea.	.56	8.30		8.86	13.75
1100	6" to 8" diameter pipe		304	.211		.59	8.75		9.34	14.50
1110	10" to 12" diameter pipe		192	.333		.93	13.85		14.78	23
1120	14" to 16" diameter pipe		128	.500		1.39	21		22.39	34.50
1130	Over 16" diameter pipe		176	.364	S.F.	1.01	15.10		16.11	25
1200	With glove bag, up to 8" diameter pipe		75	.853	L.F.	6.55	35.50		42.05	63
2000	Scrape foam fireproofing from flat surface		2400	.027	S.F.	.07	1.11		1.18	1.83
2100	Irregular surfaces		1200	.053		.15	2.22		2.37	3.66
3000	Remove cementitious material from flat surface		1800	.036		.10	1.48		1.58	2.45
3100	Irregular surface		1000	.064		.13	2.66		2.79	4.34
4000	Scrape acoustical coating/fireproofing, from ceiling		3200	.020		.06	.83		.89	1.37
5000	Remove VAT from floor by hand		2400	.027		.07	1.11		1.18	1.83
5100	By machine	A-11	4800	.013		.04	.55	.01	.60	.93
5150	For 2 layers, add				%				50%	
6000	Remove contaminated soil from crawl space by hand	A-9	400	.160	C.F.	.45	6.65		7.10	11
6100	With large production vacuum loader	A-12	700	.091	"	.25	3.80	.95	5	7.30
7000	Radiator backing, not including radiator removal	A-9	1200	.053	S.F.	.15	2.22		2.37	3.66
8000	Cement-asbestos transite board	2 Asbe	1000	.016		.12	.66		.78	1.19
8100	Transite shingle siding	A-10B	750	.043		.20	1.78		1.98	3.03
8200	Shingle roofing	"	2000	.016		.07	.67		.74	1.13
8250	Built-up, no gravel, non-friable	B-2	1400	.029		.07	.83		.90	1.38
8300	Asbestos millboard	2 Asbe	1000	.016		.08	.66		.74	1.14
9000	For type C (supplied air) respirator equipment, add				%				10%	

02 82 13.44 Demolition In Asbestos Contaminated Area

		Crew	Daily Output	Labor-Hours	Unit	Material	2007 Bare Costs Labor	2007 Bare Costs Equipment	Total	Total Incl O&P
0010	**DEMOLITION IN ASBESTOS CONTAMINATED AREA**									
0200	Ceiling, including suspension system, plaster and lath	A-9	2100	.030	S.F.	.08	1.27		1.35	2.09
0210	Finished plaster, leaving wire lath		585	.109		.30	4.55		4.85	7.55
0220	Suspended acoustical tile		3500	.018		.05	.76		.81	1.26
0230	Concealed tile grid system		3000	.021		.06	.89		.95	1.47
0240	Metal pan grid system		1500	.043		.12	1.77		1.89	2.93
0250	Gypsum board		2500	.026		.07	1.06		1.13	1.76
0260	Lighting fixtures up to 2' x 4'		72	.889	Ea.	2.48	37		39.48	61
0400	Partitions, non load bearing									
0410	Plaster, lath, and studs	A-9	690	.093	S.F.	.81	3.85		4.66	7
0450	Gypsum board and studs	"	1390	.046	"	.13	1.91		2.04	3.16
9000	For type C (supplied air) respirator equipment, add				%				10%	

02 82 13.45 OSHA Testing

0010	**OSHA TESTING**									

02 82 Asbestos Remediation

02 82 13 – Asbestos Abatement

02 82 13.45 OSHA Testing

		Crew	Daily Output	Labor-Hours	Unit	Material	2007 Bare Costs Labor	2007 Bare Costs Equipment	Total	Total Incl O&P
0100	Certified technician, minimum				Day				272.73	300
0110	Maximum				"				454.55	500
0200	Personal sampling, PCM analysis, NIOSH 7400, minimum	1 Asbe	8	1	Ea.	2.75	41.50		44.25	68.50
0210	Maximum	"	4	2	"	3	83		86	134
0300	Industrial hygienist, minimum				Day				363.64	400
0310	Maximum				"				500	550
1000	Cleaned area samples	1 Asbe	8	1	Ea.	2.63	41.50		44.13	68.50
1100	PCM air sample analysis, NIOSH 7400, minimum		8	1		30	41.50		71.50	98.50
1110	Maximum		4	2		3.09	83		86.09	134
1200	TEM air sample analysis, NIOSH 7402, minimum								100	125
1210	Maximum								400	500

02 82 13.46 Decontamination of Asbestos Containment Area

		Crew	Daily Output	Labor-Hours	Unit	Material	Labor	Equipment	Total	Total Incl O&P
0010	**DECONTAMINATION OF ASBESTOS CONTAINMENT AREA**									
0100	Spray exposed substrate with surfactant (bridging)									
0200	Flat surfaces	A-9	6000	.011	S.F.	.36	.44		.80	1.10
0250	Irregular surfaces		4000	.016	"	.31	.67		.98	1.39
0300	Pipes, beams, and columns		2000	.032	L.F.	.56	1.33		1.89	2.72
1000	Spray encapsulate polyethelene sheeting		8000	.008	S.F.	.31	.33		.64	.87
1100	Roll down polyethelene sheeting		8000	.008	"		.33		.33	.53
1500	Bag polyethelene sheeting		400	.160	Ea.	.77	6.65		7.42	11.35
2000	Fine clean exposed substrate, with nylon brush		2400	.027	S.F.		1.11		1.11	1.75
2500	Wet wipe substrate		4800	.013			.55		.55	.88
2600	Vacuum surfaces, fine brush		6400	.010			.42		.42	.66
3000	Structural demolition									
3100	Wood stud walls	A-9	2800	.023	S.F.		.95		.95	1.50
3500	Window manifolds, not incl. window replacement		4200	.015			.63		.63	1
3600	Plywood carpet protection		2000	.032			1.33		1.33	2.10
4000	Remove custom decontamination facility	A-10A	8	3	Ea.	15	125		140	215
4100	Remove portable decontamination facility	3 Asbe	12	2	"	12.50	83		95.50	145
5000	HEPA vacuum, shampoo carpeting	A-9	4800	.013	S.F.	.05	.55		.60	.94
9000	Final cleaning of protected surfaces	A-10A	8000	.003	"		.13		.13	.20

02 82 13.47 Asbestos Waste Packaging, Handling, and Disposal

		Crew	Daily Output	Labor-Hours	Unit	Material	Labor	Equipment	Total	Total Incl O&P
0010	**ASBESTOS WASTE PACKAGING, HANDLING, AND DISPOSAL**									
0100	Collect and bag bulk material, 3 C.F. bags, by hand	A-9	400	.160	Ea.	1.24	6.65		7.89	11.85
0200	Large production vacuum loader	A-12	880	.073		.80	3.02	.75	4.57	6.50
1000	Double bag and decontaminate	A-9	960	.067		1.24	2.77		4.01	5.75
2000	Containerize bagged material in drums, per 3 C.F. drum	"	800	.080		6.50	3.32		9.82	12.40
3000	Cart bags 50' to dumpster	2 Asbe	400	.040			1.66		1.66	2.62
5000	Disposal charges, not including haul, minimum				C.Y.				45.45	50
5020	Maximum				"				159.09	175
5100	Remove refrigerant from system	1 Plum	40	.200	Lb.		8.95		8.95	13.50
9000	For type C (supplied air) respirator equipment, add				%					10%

02 82 13.48 Asbestos Encapsulation With Sealants

		Crew	Daily Output	Labor-Hours	Unit	Material	Labor	Equipment	Total	Total Incl O&P
0010	**ASBESTOS ENCAPSULATION WITH SEALANTS**									
0100	Ceilings and walls, minimum	A-9	21000	.003	S.F.	.27	.13		.40	.50
0110	Maximum		10600	.006		.42	.25		.67	.86
0200	Columns and beams, minimum		13300	.005		.27	.20		.47	.62
0210	Maximum		5325	.012		.47	.50		.97	1.31
0300	Pipes to 12" diameter including minor repairs, minimum		800	.080	L.F.	.37	3.32		3.69	5.65
0310	Maximum		400	.160	"	1.04	6.65		7.69	11.65

02 83 Lead Remediation

02 83 19 – Lead-Based Paint Remediation

02 83 19.23 Encapsulation of Lead-Based Paint

		Crew	Daily Output	Labor-Hours	Unit	Material	2007 Bare Costs Labor	Equipment	Total	Total Incl O&P
0010	**ENCAPSULATION OF LEAD-BASED PAINT**, water based polymer									
0020	Interior, brushwork, trim, under 6"	1 Pord	240	.033	L.F.	2.46	1.09		3.55	4.35
0030	6" to 12" wide		180	.044		3.28	1.45		4.73	5.80
0040	Balustrades		300	.027		1.98	.87		2.85	3.49
0050	Pipe to 4" diameter		500	.016		1.19	.52		1.71	2.10
0060	To 8" diameter		375	.021		1.57	.70		2.27	2.78
0070	To 12" diameter		250	.032		2.36	1.05		3.41	4.17
0080	To 16" diameter		170	.047		3.47	1.54		5.01	6.15
0090	Cabinets, ornate design		200	.040	S.F.	2.97	1.31		4.28	5.25
0100	Simple design		250	.032	"	2.36	1.05		3.41	4.17
0110	Doors, 3' x 7', both sides, incl. frame & trim									
0120	Flush	1 Pord	6	1.333	Ea.	30.50	43.50		74	99
0130	French, 10-15 lite		3	2.667		6.05	87		93.05	138
0140	Panel		4	2		36.50	65.50		102	139
0150	Louvered		2.75	2.909		33.50	95		128.50	180
0160	Windows, per interior side, per 15 S.F.									
0170	1 to 6 lite	1 Pord	14	.571	Ea.	21	18.70		39.70	51
0180	7 to 10 lite		7.50	1.067		23	35		58	78
0190	12 lite		5.75	1.391		31	45.50		76.50	103
0200	Radiators		8	1		73.50	32.50		106	130
0210	Grilles, vents		275	.029	S.F.	2.15	.95		3.10	3.80
0220	Walls, roller, drywall or plaster		1000	.008		.59	.26		.85	1.04
0230	With spunbonded reinforcing fabric		720	.011		.68	.36		1.04	1.30
0240	Wood		800	.010		.74	.33		1.07	1.30
0250	Ceilings, roller, drywall or plaster		900	.009		.68	.29		.97	1.19
0260	Wood		700	.011		.84	.37		1.21	1.48
0270	Exterior, brushwork, gutters and downspouts		300	.027	L.F.	1.98	.87		2.85	3.49
0280	Columns		400	.020	S.F.	1.47	.65		2.12	2.60
0290	Spray, siding		600	.013	"	.99	.44		1.43	1.75
0300	Miscellaneous									
0310	Electrical conduit, brushwork, to 2" diameter	1 Pord	500	.016	L.F.	1.19	.52		1.71	2.10
0320	Brick, block or concrete, spray		500	.016	S.F.	1.19	.52		1.71	2.10
0330	Steel, flat surfaces and tanks to 12"		500	.016		1.19	.52		1.71	2.10
0340	Beams, brushwork		400	.020		1.47	.65		2.12	2.60
0350	Trusses		400	.020		1.47	.65		2.12	2.60

02 83 19.26 Removal of Lead-Based Paint

		Crew	Daily Output	Labor-Hours	Unit	Material	Labor	Equipment	Total	Total Incl O&P
0010	**REMOVAL OF LEAD-BASED PAINT**, by chemicals, per application R028319-60									
0050	Baseboard, to 6" wide	1 Pord	64	.125	L.F.	1.62	4.09		5.71	7.95
0070	To 12" wide		32	.250	"	3.20	8.20		11.40	15.80
0200	Balustrades, one side		28	.286	S.F.	3.62	9.35		12.97	18.05
1400	Cabinets, simple design		32	.250		3.17	8.20		11.37	15.80
1420	Ornate design		25	.320		4.08	10.45		14.53	20
1600	Cornice, simple design		60	.133		1.71	4.36		6.07	8.45
1620	Ornate design		20	.400		5.05	13.10		18.15	25.50
2800	Doors, one side, flush		84	.095		1.23	3.11		4.34	6.05
2820	Two panel		80	.100		1.27	3.27		4.54	6.30
2840	Four panel		45	.178		2.25	5.80		8.05	11.25
2880	For trim, one side, add		64	.125	L.F.	1.62	4.09		5.71	7.95
3000	Fence, picket, one side		30	.267	S.F.	3.41	8.70		12.11	16.85
3200	Grilles, one side, simple design		30	.267		3.41	8.70		12.11	16.85
3220	Ornate design		25	.320		4.08	10.45		14.53	20
4400	Pipes, to 4" diameter		90	.089	L.F.	1.17	2.91		4.08	5.65

02 83 Lead Remediation

02 83 19 – Lead-Based Paint Remediation

02 83 19.26 Removal of Lead-Based Paint

		Crew	Daily Output	Labor-Hours	Unit	Material	2007 Bare Costs Labor	Equipment	Total	Total Incl O&P
4420	To 8" diameter	1 Pord	50	.160	L.F.	2.03	5.25		7.28	10.10
4440	To 12" diameter		36	.222		2.85	7.25		10.10	14.10
4460	To 16" diameter		20	.400	↓	5.05	13.10		18.15	25.50
4500	For hangers, add		40	.200	Ea.	2.54	6.55		9.09	12.65
4800	Siding		90	.089	S.F.	1.17	2.91		4.08	5.65
5000	Trusses, open		55	.145	SF Face	1.85	4.76		6.61	9.20
6200	Windows, one side only, double hung, 1/1 light, 24" x 48" high		4	2	Ea.	25.50	65.50		91	127
6220	30" x 60" high		3	2.667		34	87		121	169
6240	36" x 72" high		2.50	3.200		41	105		146	202
6280	40" x 80" high		2	4		51	131		182	253
6400	Colonial window, 6/6 light, 24" x 48" high		2	4		51	131		182	254
6420	30" x 60" high		1.50	5.333		68	174		242	335
6440	36" x 72" high		1	8		102	262		364	505
6480	40" x 80" high		1	8		102	262		364	505
6600	8/8 light, 24" x 48" high		2	4		51	131		182	254
6620	40" x 80" high		1	8		102	262		364	505
6800	12/12 light, 24" x 48" high		1	8		102	262		364	505
6820	40" x 80" high	↓	.75	10.667	↓	136	350		486	675
6840	Window frame & trim items, included in pricing above									

02 85 Mold Remediation

02 85 16 – Mold Remediation Preparation and Containment

02 85 16.40 Mold Remediation Plans and Methods

		Crew	Daily Output	Labor-Hours	Unit	Material	2007 Bare Costs Labor	Equipment	Total	Total Incl O&P
0010	**MOLD REMEDIATION PLANS AND METHODS**									
0020	Initial inspection, average 3 bedroom home				Total				236.36	260
0030	Initial inspection, average 5 bedroom home								327.27	360
0040	Air sample test, each								236.36	260
0050	Swab sample test, each								140.91	155
0060	Tape sample test, each								140.91	155
0070	After remediation air test, each								236.36	260
0080	Mold abatement plan, average 3 bedroom home								1,159.09	1,275
0090	Mold abatement plan, average 5 bedroom home								1,409.09	1,550
0100	Packup & removal of contents, average 3 bedroom home, excl storage								9,364	10,300
0110	Packup & removal of contents, average 5 bedroom home, excl storage				↓				23,455	25,800
0120	For demolition in mold contaminated areas, see Div. 08 82 13.44									
0130	For personal protection equipment, see Div. 02 82 13.41									

02 85 16.50 Preparation of Mold Containment Area

		Crew	Daily Output	Labor-Hours	Unit	Material	2007 Bare Costs Labor	Equipment	Total	Total Incl O&P
0010	**PREPARATION OF MOLD CONTAINMENT AREA**									
0100	Pre-cleaning, HEPA vacuum and wet wipe, flat surfaces	A-10	12000	.005	S.F.	.01	.22		.23	.37
0300	Separation barrier, 2" x 4" @ 16", 1/2" plywood ea. side, 8' high	2 Carp	400	.040		1.25	1.47		2.72	3.67
0310	12' high		320	.050		1.40	1.84		3.24	4.40
0320	16' high		200	.080		1.50	2.94		4.44	6.20
0400	Personnel decontam. chamber, 2" x 4" @ 16", 3/4" ply ea. side		280	.057		2.50	2.10		4.60	6
0450	Waste decontam. chamber, 2" x 4" studs @ 16", 3/4" ply each side	↓	360	.044	↓	3	1.63		4.63	5.85
0500	Cover surfaces with polyethelene sheeting									
0501	Including glue and tape									
0550	Floors, each layer, 6 mil	A-10	8000	.008	S.F.	.09	.33		.42	.63
0551	4 mil		9000	.007		.23	.30		.53	.72
0560	Walls, each layer, 6 mil		6000	.011		.09	.44		.53	.80
0561	4 mil	↓	7000	.009	↓	.07	.38		.45	.68
0570	For heights above 12', add						20%			

02 85 Mold Remediation

02 85 16 – Mold Remediation Preparation and Containment

02 85 16.50 Preparation of Mold Containment Area

		Crew	Daily Output	Labor-Hours	Unit	Material	2007 Bare Costs Labor	Equipment	Total	Total Incl O&P
0575	For heights above 20', add						30%			
0580	For fire retardant poly, add					100%				
0590	For large open areas, deduct					10%	20%			
0600	Seal floor penetrations with foam firestop to 36 sq in	2 Carp	200	.080	Ea.	6.25	2.94		9.19	11.45
0610	36 sq in to 72 sq in		125	.128		12.50	4.70		17.20	21
0615	72 sq in to 144 sq in		80	.200		25	7.35		32.35	39
0620	Wall penetrations, to 36 square inches		180	.089		6.25	3.26		9.51	12
0630	36 Sq. in. to 72 sq. in.		100	.160		12.50	5.85		18.35	23
0640	72 Sq. in. to 144 sq. in.	↓	60	.267	↓	25	9.80		34.80	43
0800	Caulk seams with latex	1 Carp	230	.035	L.F.	.15	1.28		1.43	2.16
0900	Set up neg. air machine, 1-2k C.F.M. /25 M.C.F. volume	1 Asbe	4.30	1.860	Ea.		77		77	122

02 85 33 – Removal and Disposal of Materials with Mold

02 85 33.50 Demolition In Mold Contaminated Area

		Crew	Daily Output	Labor-Hours	Unit	Material	2007 Bare Costs Labor	Equipment	Total	Total Incl O&P
0010	**DEMOLITION IN MOLD CONTAMINATED AREA**									
0200	Ceiling, including suspension system, plaster and lath	A-9	2100	.030	S.F.	.08	1.27		1.35	2.09
0210	Finished plaster, leaving wire lath		585	.109		.30	4.55		4.85	7.55
0220	Suspended acoustical tile		3500	.018		.05	.76		.81	1.26
0230	Concealed tile grid system		3000	.021		.06	.89		.95	1.47
0240	Metal pan grid system		1500	.043		.12	1.77		1.89	2.93
0250	Gypsum board		2500	.026		.07	1.06		1.13	1.76
0255	Plywood		2500	.026		.07	1.06		1.13	1.76
0260	Lighting fixtures up to 2' x 4'		72	.889	Ea.	2.48	37		39.48	61
0400	Partitions, non load bearing									
0410	Plaster, lath, and studs	A-9	690	.093	S.F.	.81	3.85		4.66	7
0450	Gypsum board and studs		1390	.046		.13	1.91		2.04	3.16
0465	Carpet & pad		1390	.046	↓	.13	1.91		2.04	3.16
0600	Pipe insulation, air cell type, up to 4" diameter pipe		900	.071	L.F.	.20	2.96		3.16	4.89
0610	4" to 8" diameter pipe		800	.080		.22	3.32		3.54	5.50
0620	10" to 12" diameter pipe		700	.091		.25	3.80		4.05	6.30
0630	14" to 16" diameter pipe		550	.116	↓	.32	4.84		5.16	8
0650	Over 16" diameter pipe	↓	650	.098	S.F.	.27	4.09		4.36	6.75
9000	For type C (supplied air) respirator equipment, add				%					10%

Division Notes

	CREW	DAILY OUTPUT	LABOR-HOURS	UNIT	2007 BARE COSTS				TOTAL INCL O&P
					MAT.	LABOR	EQUIP.	TOTAL	

Estimating Tips

General

- Carefully check all the plans and specifications. Concrete often appears on drawings other than structural drawings, including mechanical and electrical drawings for equipment pads. The cost of cutting and patching is often difficult to estimate. See Subdivisions 02 41 19 and 03 01 05 for demolition costs.
- Always obtain concrete prices from suppliers near the job site. A volume discount can often be negotiated depending upon competition in the area. Remember to add for waste, particularly for slabs and footings on grade.

03 10 00 Concrete Forming and Accessories

- A primary cost for concrete construction is forming. Most jobs today are constructed with prefabricated forms. The selection of the forms best suited for the job and the total square feet of forms required for efficient concrete forming and placing are key elements in estimating concrete construction. Enough forms must be available for erection to make efficient use of the concrete placing equipment and crew.
- Concrete accessories for forming and placing depend upon the systems used. Study the plans and specifications to assure that all special accessory requirements have been included in the cost estimate, such as anchor bolts, inserts, and hangers.

03 20 00 Concrete Reinforcing

- Ascertain that the reinforcing steel supplier has included all accessories, cutting, bending, and an allowance for lapping, splicing, and waste. A good rule of thumb is 10% for lapping, splicing, and waste. Also, 10% waste should be allowed for welded wire fabric.
- The unit price items in the section for Reinforcing In Place include the labor to install accessories such as beam and slab bolsters, high chairs, and bar ties and tie wire. The material cost for these accessories is not included; they may be obtained from the Accessories section.

03 30 00 Cast-in-Place Concrete

- When estimating structural concrete, pay particular attention to requirements for concrete additives, curing methods, and surface treatments. Special consideration for climate, hot or cold, must be included in your estimate. Be sure to include requirements for concrete placing equipment and concrete finishing.

03 40 00 Precast Concrete

03 50 00 Cast Decks and Underlayment

- The cost of hauling precast concrete structural members is often an important factor. For this reason, it is important to get a quote from the nearest supplier. It may become economically feasible to set up precasting beds on the site if the hauling costs are prohibitive.

Reference Numbers

Reference numbers are shown in shaded boxes at the beginning of some major classifications. These numbers refer to related items in the Reference Section. The reference information may be an estimating procedure, an alternate pricing method, or technical information.

Note: Not all subdivisions listed here necessarily appear in this publication.

03 01 Maintenance of Concrete

03 01 30 – Maintenance of Cast-In-Place Concrete

03 01 30.62 Concrete Patching

		Crew	Daily Output	Labor-Hours	Unit	Material	2007 Bare Costs Labor	2007 Bare Costs Equipment	Total	Total Incl O&P
0010	**CONCRETE PATCHING**									
0100	Floors, 1/4" thick, small areas, regular grout	1 Cefi	170	.047	S.F.	1	1.67		2.67	3.56
0150	Epoxy grout	"	100	.080	"	7.40	2.84		10.24	12.35
2000	Walls, including chipping, cleaning and epoxy grout									
2100	1/4" deep	1 Cefi	65	.123	S.F.	7.45	4.38		11.83	14.60
2150	1/2" deep	↓	50	.160	↓	14.90	5.70		20.60	25
2200	3/4" deep	↓	40	.200	↓	22.50	7.10		29.60	35

03 05 Common Work Results for Concrete

03 05 05 – Selective Concrete Demolition

03 05 05.10 Selective Concrete Demolition

		Crew	Daily Output	Labor-Hours	Unit	Material	2007 Bare Costs Labor	2007 Bare Costs Equipment	Total	Total Incl O&P
0010	**SELECTIVE CONCRETE DEMOLITION** R024119-10									
0012	Excludes saw cutting, torch cutting, loading or hauling									
0050	Break up into small pieces, minimum reinforcing	B-9	24	1.667	C.Y.		48.50	7.50	56	84
0060	Average reinforcing		16	2.500			73	11.25	84.25	125
0070	Maximum reinforcing	↓	8	5	↓		146	22.50	168.50	252
0150	Remove whole pieces, up to 2 tons per piece	E-18	36	1.111	Ea.		45.50	27	72.50	110
0160	2 - 5 tons per piece		30	1.333			55	32.50	87.50	133
0170	5 - 10 tons per piece		24	1.667			68.50	40.50	109	165
0180	10 - 15 tons per piece	↓	18	2.222			91.50	54.50	146	220
0250	Precast unit embedded in masonry, up to 1 CF	D-1	16	1			33.50		33.50	51
0260	1 - 2 CF		12	1.333			44.50		44.50	67.50
0270	2 - 5 CF		10	1.600			53.50		53.50	81
0280	5 - 10 CF	↓	8	2	↓		66.50		66.50	102

03 05 13 – Basic Concrete Materials

03 05 13.20 Concrete Admixtures and Surface Treatments

		Crew	Daily Output	Labor-Hours	Unit	Material	2007 Bare Costs Labor	2007 Bare Costs Equipment	Total	Total Incl O&P
0010	**CONCRETE ADMIXTURES AND SURFACE TREATMENTS**									
0040	Abrasives, aluminum oxide, over 20 tons				Lb.	1.29			1.29	1.42
0070	Under 1 ton					1.41			1.41	1.55
0100	Silicon carbide, black, over 20 tons					1.76			1.76	1.94
0120	Under 1 ton				↓	1.95			1.95	2.15
0200	Air entraining agent, .7 to 1.5 oz. per bag, 55 gallon lots				Gal.	9.50			9.50	10.45
0220	5 gallon lots					10			10	11
0300	Bonding agent, acrylic latex, 250 S.F. per gallon					18.60			18.60	20.50
0320	Epoxy resin, 80 S.F. per gallon				↓	40.50			40.50	44.50
0400	Calcium chloride, 50 lb. bags				Ton	420			420	460
0420	Less than truckload lots				Bag	14.30			14.30	15.75
0500	Carbon black, liquid, 2 to 8 lbs. per bag of cement				Lb.	4.70			4.70	5.15
0600	Colors, integral, 2 to 10 lb. per bag of cement, minimum					2.32			2.32	2.55
0610	Average					3.12			3.12	3.43
0620	Maximum				↓	4.87			4.87	5.35
0920	Dustproofing compound, 250 SF/gal, 5 gallon lots				Gal.	6.20			6.20	6.80
1010	Epoxy based, 200 SF/Gal, 5 gallon lots				"	24.50			24.50	27
1100	Hardeners, metallic, 55 lb. bags, natural (grey)				Lb.	.80			.80	.88
1200	Colors					1.17			1.17	1.29
1300	Non-metallic, 55 lb. bags, natural grey					.36			.36	.39
1320	Colors				↓	.63			.63	.69
1550	Release agent, for tilt slabs				Gal.	12.70			12.70	13.95
1570	For forms, average					6.60			6.60	7.25
1600	Sealer, hardener and dustproofer, epoxy-based, 200 SF/gal, min				↓	24.50			24.50	27

03 05 Common Work Results for Concrete

03 05 13 – Basic Concrete Materials

	03 05 13.20 Concrete Admixtures and Surface Treatments	Crew	Daily Output	Labor-Hours	Unit	Material	2007 Bare Costs Labor	Equipment	Total	Total Incl O&P
1620	Maximum				Gal.	27			27	29.50
1630	Solvent-based, 250 SF/gal, minimum					12.05			12.05	13.25
1640	Maximum					14.40			14.40	15.85
1650	Water based, 350 S.F., minimum					14.20			14.20	15.65
1660	Maximum					17.30			17.30	19
1900	Set retarder, 2 to 4 fl. oz. per bag of cement				↓	15.05			15.05	16.55
2000	Waterproofing, integral 1 lb. per bag of cement				Lb.	1.08			1.08	1.19
2100	Powdered metallic, 40 lbs. per 100 S.F., minimum					1.34			1.34	1.47
2120	Maximum				↓	2.68			2.68	2.95
3000	For integral colors, 2500 psi (5 bag mix)									
3100	Red, yellow or brown, 1.8 lb. per bag, add				C.Y.	21			21	23
3200	9.4 lb. per bag, add					109			109	120
3400	Black, 1.8 lb. per bag, add					28			28	31
3500	7.5 lb. per bag, add					117			117	129
3700	Green, 1.8 lb. per bag, add					44			44	48
3800	7.5 lb. per bag, add				↓	183			183	201

03 05 13.25 Aggregate

0010	**AGGREGATE** R033105-20									
0020	Expanded shale, C.L. lots, 52 lb. per C.F., minimum				Ton	36			36	39.50
0050	Maximum R033105-30				"	48			48	53
0100	Lightweight vermiculite or perlite, 4 C.F. bag, C.L. lots				Bag	6.30			6.30	6.95
0150	L.C.L. lots R033105-40				"	6.95			6.95	7.60
0250	Sand & stone, loaded at pit, crushed bank gravel				Ton	15.15			15.15	16.65
0350	Sand, washed, for concrete R033105-50					13.85			13.85	15.25
0400	For plaster or brick					13.85			13.85	15.25
0450	Stone, 3/4" to 1-1/2"					21			21	23.50
0500	3/8" roofing stone & 1/2" pea stone					15.95			15.95	17.50
0550	For trucking 10 miles (5 mile R/T), add to the above	B-34B	117	.068			2.02	4.53	6.55	8.10
0600	30 miles (15 mi R/T), add to the above	"	72	.111	↓		3.28	7.35	10.63	13.15
0850	Sand & stone, loaded at pit, crushed bank gravel				C.Y.	21			21	23.50
0950	Sand, washed, for concrete					19.25			19.25	21
1000	For plaster or brick					19.25			19.25	21
1050	Stone, 3/4" to 1-1/2"					23			23	25.50
1100	3/8" roofing stone & 1/2" pea stone					33			33	36.50
1150	For trucking 10 miles (5 mi R/T), add to the above	B-34B	78	.103			3.03	6.80	9.83	12.15
1200	30 miles (15 mi R/T), add to the above	"	48	.167	↓		4.93	11.05	15.98	19.75
1310	Quartz chips, 50 lb. bags				Cwt.	18.70			18.70	20.50
1330	Silica chips, 50 lb. bags					10.25			10.25	11.30
1410	White marble, 3/8" to 1/2", 50 lb. bags				↓	15.45			15.45	17
1430	3/4"				Ton	74.50			74.50	82

03 05 13.30 Cement

0010	**CEMENT** R033105-20									
0240	Portland, Type I/II, TL lots, 94 lb bags				Bag	8.80			8.80	9.70
0250	LTL/LCL lots R033105-30				"	10.60			10.60	11.65
0300	Trucked in bulk, per cwt **CN**				Cwt.	5.30			5.30	5.85
0400	Type III, high early strength, TL lots, 94 lb bags **CN** R033105-40				Bag	9.95			9.95	10.95
0420	L.T.L. or L.C.L. lots					11.15			11.15	12.25
0500	White, type III, high early strength, T.L. or C.L. lots, bags R033105-50					22			22	24.50
0520	L.T.L. or L.C.L. lots					23.50			23.50	25.50
0600	White, type I, T.L. or C.L. lots, bags					21			21	23
0620	L.T.L. or L.C.L. lots				↓	22			22	24.50

03 05 Common Work Results for Concrete

03 05 13 – Basic Concrete Materials

03 05 13.80 Waterproofing and Dampproofing

		Crew	Daily Output	Labor-Hours	Unit	Material	2007 Bare Costs Labor	2007 Bare Costs Equipment	Total	Total Incl O&P
0010	**WATERPROOFING AND DAMPPROOFING**									
0050	Integral waterproofing, add to cost of regular concrete				C.Y.	6.50			6.50	7.15

03 05 13.85 Winter Protection

		Crew	Daily Output	Labor-Hours	Unit	Material	Labor	Equipment	Total	Total Incl O&P
0010	**WINTER PROTECTION**									
0012	For heated ready mix, add, minimum				C.Y.	5.50			5.50	6.05
0050	Maximum				"	6.90			6.90	7.55
0100	Temporary heat to protect concrete, 24 hours, minimum	2 Clab	50	.320	M.S.F.	200	9.20		209.20	234
0150	Maximum	"	25	.640	"	260	18.40		278.40	315
0200	Temporary shelter for slab on grade, wood frame/polyethylene sheeting									
0201	Build or remove, minimum	2 Carp	10	1.600	M.S.F.	291	58.50		349.50	410
0210	Maximum	"	3	5.333	"	345	196		541	685
0500	Electrically heated pads, 110 volts, 15 watts per S.F., buy				S.F.	5.90			5.90	6.50
0600	20 watts per S.F., buy					7.85			7.85	8.65
0710	Electrically, heated pads, 15 watts/S.F., 20 uses, minimum					.21			.21	.23
0800	Maximum					.35			.35	.39

03 11 Concrete Forming

03 11 13 – Structural Cast-In-Place Concrete Forming

03 11 13.13 Concrete Slip Forming

			Crew	Daily Output	Labor-Hours	Unit	Material	Labor	Equipment	Total	Total Incl O&P
0010	**CONCRETE SLIP FORMING**	R031113-30									
0020	Silos, minimum		C-17E	3885	.021	SFCA	3.30	.79	.02	4.11	4.88
0050	Maximum			1095	.073		4.75	2.81	.08	7.64	9.70
1000	Buildings, minimum			3660	.022		2.10	.84	.02	2.96	3.64
1050	Maximum			875	.091		3.89	3.51	.09	7.49	9.85

03 11 13.20 Forms In Place, Beams and Girders

			Crew	Daily Output	Labor-Hours	Unit	Material	Labor	Equipment	Total	Total Incl O&P
0010	**FORMS IN PLACE, BEAMS AND GIRDERS**	R031113-40									
0500	Exterior spandrel, job-built plywood, 12" wide, 1 use		C-2	225	.213	SFCA	3.66	7.60		11.26	15.90
0550	2 use	R031113-60		275	.175		1.95	6.25		8.20	11.85
0600	3 use			295	.163		1.47	5.80		7.27	10.65
0650	4 use			310	.155		1.19	5.55		6.74	9.90
1000	18" wide, 1 use			250	.192		3.03	6.85		9.88	14.05
1050	2 use			275	.175		1.67	6.25		7.92	11.55
1100	3 use			305	.157		1.21	5.60		6.81	10.10
1150	4 use			315	.152		.99	5.45		6.44	9.55
1500	24" wide, 1 use			265	.181		2.76	6.45		9.21	13.10
1550	2 use			290	.166		1.55	5.90		7.45	10.90
1600	3 use			315	.152		1.11	5.45		6.56	9.65
1650	4 use			325	.148		.90	5.25		6.15	9.20
2000	Interior beam, job-built plywood, 12" wide, 1 use			300	.160		4.26	5.70		9.96	13.60
2050	2 use			340	.141		2.11	5.05		7.16	10.15
2100	3 use			364	.132		1.70	4.71		6.41	9.20
2150	4 use			377	.127		1.38	4.55		5.93	8.60
2500	24" wide, 1 use			320	.150		2.82	5.35		8.17	11.45
2550	2 use			365	.132		1.58	4.70		6.28	9.05
2600	3 use			385	.125		1.12	4.45		5.57	8.20
2650	4 use			395	.122		.91	4.34		5.25	7.75
3000	Encasing steel beam, hung, job-built plywood, 1 use			325	.148		2.74	5.25		7.99	11.20
3050	2 use			390	.123		1.51	4.40		5.91	8.50
3100	3 use			415	.116		1.09	4.13		5.22	7.65
3150	4 use			430	.112		.89	3.99		4.88	7.20

03 11 Concrete Forming

03 11 13 – Structural Cast-In-Place Concrete Forming

03 11 13.20 Forms In Place, Beams and Girders

		Crew	Daily Output	Labor-Hours	Unit	Material	2007 Bare Costs Labor	Equipment	Total	Total Incl O&P
3500	Bottoms only, to 30" wide, job-built plywood, 1 use	C-2	230	.209	SFCA	4.28	7.45		11.73	16.30
3550	2 use		265	.181		2.39	6.45		8.84	12.70
3600	3 use		280	.171		1.71	6.10		7.81	11.45
3650	4 use		290	.166		1.39	5.90		7.29	10.75
4000	Sides only, vertical, 36" high, job-built plywood, 1 use		335	.143		4.50	5.10		9.60	12.90
4050	2 use		405	.119		2.48	4.23		6.71	9.30
4100	3 use		430	.112		1.80	3.99		5.79	8.20
4150	4 use		445	.108		1.46	3.85		5.31	7.60
4500	Sloped sides, 36" high, 1 use		305	.157		4.34	5.60		9.94	13.55
4550	2 use		370	.130		2.42	4.63		7.05	9.85
4600	3 use		405	.119		1.74	4.23		5.97	8.50
4650	4 use		425	.113		1.41	4.03		5.44	7.85
5000	Upstanding beams, 36" high, 1 use		225	.213		5.50	7.60		13.10	17.90
5050	2 use		255	.188		3.07	6.70		9.77	13.85
5100	3 use		275	.175		2.23	6.25		8.48	12.15
5150	4 use		280	.171		1.81	6.10		7.91	11.55

03 11 13.25 Forms In Place, Columns

		Crew	Daily Output	Labor-Hours	Unit	Material	2007 Bare Costs Labor	Equipment	Total	Total Incl O&P
0010	**FORMS IN PLACE, COLUMNS** R031113-40									
0500	Round fiberglass, 4 use per mo., rent, 12" diameter	C-1	160	.200	L.F.	6.50	6.95		13.45	17.95
0550	16" diameter R031113-60		150	.213		7.80	7.40		15.20	20
0600	18" diameter		140	.229		8.65	7.95		16.60	22
0650	24" diameter		135	.237		10.80	8.25		19.05	24.50
0700	28" diameter		130	.246		12.05	8.55		20.60	26.50
0800	30" diameter		125	.256		12.60	8.90		21.50	27.50
0850	36" diameter		120	.267		16.75	9.25		26	33
1500	Round fiber tube, 1 use, 8" diameter		155	.206		1.56	7.15		8.71	12.85
1550	10" diameter		155	.206		2.04	7.15		9.19	13.40
1600	12" diameter		150	.213		2.35	7.40		9.75	14.15
1650	14" diameter		145	.221		3.31	7.65		10.96	15.60
1700	16" diameter		140	.229		3.98	7.95		11.93	16.75
1750	20" diameter		135	.237		6.25	8.25		14.50	19.65
1800	24" diameter		130	.246		8.05	8.55		16.60	22
1850	30" diameter		125	.256		11.45	8.90		20.35	26.50
1900	36" diameter		115	.278		15.25	9.65		24.90	32
1950	42" diameter		100	.320		30.50	11.10		41.60	51
2000	48" diameter		85	.376		39.50	13.05		52.55	63.50
2200	For seamless type, add					15%				
3000	Round, steel, 4 use per mo., rent, regular duty, 12" diameter	C-1	145	.221	L.F.	10.45	7.65		18.10	23.50
3050	16" diameter		125	.256		11.75	8.90		20.65	27
3100	Heavy duty, 20" diameter		105	.305		12.90	10.60		23.50	30.50
3150	24" diameter		85	.376		14.15	13.05		27.20	36
3200	30" diameter		70	.457		16.25	15.85		32.10	42.50
3250	36" diameter		60	.533		17.40	18.50		35.90	48
3300	48" diameter		50	.640		24	22		46	61
3350	60" diameter		45	.711		32	24.50		56.50	73.50
4000	Column capitals, steel, 4 uses/mo., 24" col, 4' cap diameter		12	2.667	Ea.	198	92.50		290.50	360
4050	5' cap diameter		11	2.909		226	101		327	405
4100	6' cap diameter		10	3.200		310	111		421	515
4150	7' cap diameter		9	3.556		380	123		503	605
4500	For second and succeeding months, deduct					50%				
5000	Job-built plywood, 8" x 8" columns, 1 use	C-1	165	.194	SFCA	2.24	6.75		8.99	12.95
5050	2 use		195	.164		1.29	5.70		6.99	10.25

03 11 Concrete Forming

03 11 13 – Structural Cast-In-Place Concrete Forming

03 11 13.25 Forms In Place, Columns

		Crew	Daily Output	Labor-Hours	Unit	Material	2007 Bare Costs Labor	Equipment	Total	Total Incl O&P
5100	3 use	C-1	210	.152	SFCA	.90	5.30		6.20	9.25
5150	4 use		215	.149		.74	5.15		5.89	8.85
5500	12" x 12" columns, 1 use		180	.178		2.23	6.15		8.38	12.05
5550	2 use		210	.152		1.22	5.30		6.52	9.60
5600	3 use		220	.145		.89	5.05		5.94	8.85
5650	4 use		225	.142		.72	4.94		5.66	8.50
6000	16" x 16" columns, 1 use		185	.173		2.26	6		8.26	11.85
6050	2 use		215	.149		1.19	5.15		6.34	9.35
6100	3 use		230	.139		.91	4.83		5.74	8.50
6150	4 use		235	.136		.74	4.73		5.47	8.15
6500	24" x 24" columns, 1 use		190	.168		2.57	5.85		8.42	11.95
6550	2 use		216	.148		1.41	5.15		6.56	9.55
6600	3 use		230	.139		1.03	4.83		5.86	8.65
6650	4 use		238	.134		.84	4.67		5.51	8.15
7000	36" x 36" columns, 1 use		200	.160		1.90	5.55		7.45	10.75
7050	2 use		230	.139		1.07	4.83		5.90	8.65
7100	3 use		245	.131		.76	4.53		5.29	7.90
7150	4 use		250	.128		.62	4.44		5.06	7.60
7500	Steel framed plywood, 4 use per mo., rent, 8" x 8"		340	.094		3.48	3.27		6.75	8.95
7550	10" x 10"		350	.091		2.45	3.17		5.62	7.65
7600	12" x 12"		370	.086		3.24	3		6.24	8.25
7650	16" x 16"		400	.080		3.60	2.78		6.38	8.30
7700	20" x 20"		420	.076		1.64	2.64		4.28	5.90
7750	24" x 24"		440	.073		1.54	2.52		4.06	5.60
7755	30" x 30"		440	.073		1.38	2.52		3.90	5.45

03 11 13.30 Forms In Place, Culvert

			Crew	Daily Output	Labor-Hours	Unit	Material	Labor	Equipment	Total	Total Incl O&P
0010	**FORMS IN PLACE, CULVERT**	R031113-40									
0015	5' to 8' square or rectangular, 1 use		C-1	170	.188	SFCA	5.15	6.55		11.70	15.80
0050	2 use	R031113-60		180	.178		3.02	6.15		9.17	12.95
0100	3 use			190	.168		2.32	5.85		8.17	11.65
0150	4 use			200	.160		1.97	5.55		7.52	10.80

03 11 13.35 Forms In Place, Elevated Slabs

			Crew	Daily Output	Labor-Hours	Unit	Material	Labor	Equipment	Total	Total Incl O&P
0010	**FORMS IN PLACE, ELEVATED SLABS**	R031113-40									
1000	Flat plate, job-built plywood, to 15' high, 1 use	R031113-60	C-2	470	.102	S.F.	4.40	3.65		8.05	10.55
1050	2 use			520	.092		2.42	3.30		5.72	7.80
1100	3 use			545	.088		1.76	3.15		4.91	6.85
1150	4 use			560	.086		1.43	3.06		4.49	6.35
1500	15' to 20' high ceilings, 4 use			495	.097		1.58	3.46		5.04	7.15
1600	21' to 35' high ceilings, 4 use			450	.107		2.33	3.81		6.14	8.50
2000	Flat slab, drop panels, job-built plywood, to 15' high, 1 use			449	.107		4.93	3.82		8.75	11.40
2050	2 use			509	.094		2.71	3.37		6.08	8.25
2100	3 use			532	.090		1.97	3.22		5.19	7.15
2150	4 use			544	.088		1.60	3.15		4.75	6.65
2250	15' to 20' high ceilings, 4 use			480	.100		4.54	3.57		8.11	10.55
2350	20' to 35' high ceilings, 4 use			435	.110		5.25	3.94		9.19	11.95
3000	Floor slab hung from steel beams, 1 use			485	.099		2.01	3.53		5.54	7.70
3050	2 use			535	.090		1.52	3.20		4.72	6.65
3100	3 use			550	.087		1.35	3.12		4.47	6.35
3150	4 use			565	.085		1.27	3.03		4.30	6.10
3500	Floor slab, with 20" metal pans, 1 use			415	.116		6.25	4.13		10.38	13.35
3550	2 use			445	.108		4.02	3.85		7.87	10.45
3600	3 use			475	.101		3.28	3.61		6.89	9.20

03 11 Concrete Forming

03 11 13 – Structural Cast-In-Place Concrete Forming

03 11 13.35 Forms In Place, Elevated Slabs

		Crew	Daily Output	Labor-Hours	Unit	Material	2007 Bare Costs Labor	Equipment	Total	Total Incl O&P
3650	4 use	C-2	500	.096	S.F.	2.91	3.43		6.34	8.55
3700	Floor slab with 30" pans, 1 use		418	.115		6.15	4.10		10.25	13.20
3720	2 use		455	.105		3.94	3.77		7.71	10.20
3740	3 use		470	.102		3.20	3.65		6.85	9.20
3760	4 use		480	.100		2.83	3.57		6.40	8.65
4000	Floor slab with 19" metal domes, 1 use		405	.119		6.60	4.23		10.83	13.85
4050	2 use		435	.110		4.37	3.94		8.31	10.95
4100	3 use		465	.103		3.63	3.69		7.32	9.75
4150	4 use		495	.097		3.26	3.46		6.72	9
4500	With 30" fiberglass domes, 1 use		405	.119		6.45	4.23		10.68	13.70
4520	2 use		450	.107		4.21	3.81		8.02	10.60
4530	3 use		460	.104		3.46	3.73		7.19	9.60
4550	4 use		470	.102		3.09	3.65		6.74	9.10
5000	Box out for slab openings, over 16" deep, 1 use		190	.253	SFCA	3.32	9		12.32	17.70
5050	2 use		240	.200	"	1.82	7.15		8.97	13.10
5500	Shallow slab box outs, to 10 S.F.		42	1.143	Ea.	10.55	41		51.55	75
5550	Over 10 S.F. (use perimeter)		600	.080	L.F.	1.41	2.86		4.27	6
6000	Bulkhead forms for slab, with keyway, 1 use, 2 piece		500	.096		1.54	3.43		4.97	7.05
6100	3 piece (see also edge forms)		460	.104		1.71	3.73		5.44	7.70
6200	Slab bulkhead form, 4-1/2" high, exp metal, w/ keyway & stakes	C-1	1200	.027		.84	.93		1.77	2.36
6210	5-1/2" high		1100	.029		.90	1.01		1.91	2.56
6215	7-1/2" high		960	.033		1.08	1.16		2.24	2.99
6220	9-1/2" high		840	.038		1.88	1.32		3.20	4.13
6500	Curb forms, wood, 6" to 12" high, on elevated slabs, 1 use		180	.178	SFCA	1.46	6.15		7.61	11.20
6550	2 use		205	.156		.80	5.40		6.20	9.35
6600	3 use		220	.145		.59	5.05		5.64	8.50
6650	4 use		225	.142		.48	4.94		5.42	8.20
7000	Edge forms to 6" high, on elevated slab, 4 use		500	.064	L.F.	.19	2.22		2.41	3.67
7500	Depressed area forms to 12" high, 4 use		300	.107		.74	3.70		4.44	6.55
7550	12" to 24" high, 4 use		175	.183		1	6.35		7.35	11
8000	Perimeter deck and rail for elevated slabs, straight		90	.356		13.55	12.35		25.90	34
8050	Curved		65	.492		18.60	17.10		35.70	47
8500	Void forms, round fiber, 3" diameter		450	.071		.67	2.47		3.14	4.58
8550	4" diameter		425	.075		.80	2.61		3.41	4.95
8600	6" diameter		400	.080		1.20	2.78		3.98	5.65
8650	8" diameter		375	.085		1.92	2.96		4.88	6.70
8700	10" diameter		350	.091		2.46	3.17		5.63	7.65
8750	12" diameter		300	.107		3.17	3.70		6.87	9.25

03 11 13.40 Forms In Place, Equipment Foundations

			Crew	Daily Output	Labor-Hours	Unit	Material	Labor	Equipment	Total	Total Incl O&P
0010	**FORMS IN PLACE, EQUIPMENT FOUNDATIONS**	R031113-40									
0020	1 use		C-2	160	.300	SFCA	2.59	10.70		13.29	19.55
0050	2 use	R031113-60		190	.253		1.42	9		10.42	15.60
0100	3 use			200	.240		1.04	8.55		9.59	14.50
0150	4 use			205	.234		.84	8.35		9.19	13.95

03 11 13.45 Forms In Place, Footings

			Crew	Daily Output	Labor-Hours	Unit	Material	Labor	Equipment	Total	Total Incl O&P
0010	**FORMS IN PLACE, FOOTINGS**	R031113-40									
0020	Continuous wall, plywood, 1 use		C-1	375	.085	SFCA	2.64	2.96		5.60	7.50
0050	2 use	R031113-60		440	.073		1.45	2.52		3.97	5.55
0100	3 use			470	.068		1.06	2.36		3.42	4.84
0150	4 use			485	.066		.86	2.29		3.15	4.52
0500	Dowel supports for footings or beams, 1 use			500	.064	L.F.	.84	2.22		3.06	4.39
1000	Integral starter wall, to 4" high, 1 use			400	.080		.86	2.78		3.64	5.25

03 11 Concrete Forming

03 11 13 – Structural Cast-In-Place Concrete Forming

03 11 13.45 Forms In Place, Footings

		Crew	Daily Output	Labor-Hours	Unit	Material	Labor	Equipment	Total	Total Incl O&P
1500	Keyway, 4 use, tapered wood, 2" x 4"	1 Carp	530	.015	L.F.	.20	.55		.75	1.08
1550	2" x 6"		500	.016		.29	.59		.88	1.23
2000	Tapered plastic, 2" x 3"		530	.015		.70	.55		1.25	1.63
2050	2" x 4"		500	.016		.90	.59		1.49	1.90
2250	For keyway hung from supports, add		150	.053		.84	1.96		2.80	3.98
3000	Pile cap, square or rectangular, job-built plywood, 1 use	C-1	290	.110	SFCA	2.46	3.83		6.29	8.65
3050	2 use		346	.092		1.35	3.21		4.56	6.50
3100	3 use		371	.086		.98	2.99		3.97	5.75
3150	4 use		383	.084		.80	2.90		3.70	5.40
4000	Triangular or hexagonal, 1 use		225	.142		2.90	4.94		7.84	10.90
4050	2 use		280	.114		1.60	3.97		5.57	7.95
4100	3 use		305	.105		1.16	3.64		4.80	6.95
4150	4 use		315	.102		.94	3.53		4.47	6.55
5000	Spread footings, job-built lumber, 1 use		305	.105		1.81	3.64		5.45	7.65
5050	2 use		371	.086		1	2.99		3.99	5.75
5100	3 use		401	.080		.72	2.77		3.49	5.10
5150	4 use		414	.077		.59	2.68		3.27	4.83
6000	Supports for dowels, plinths or templates, 2' x 2' footing		25	1.280	Ea.	5.45	44.50		49.95	75
6050	4' x 4' footing		22	1.455		10.95	50.50		61.45	90.50
6100	8' x 8' footing		20	1.600		22	55.50		77.50	111
6150	12' x 12' footing		17	1.882		26.50	65.50		92	131
7000	Plinths, job-built plywood, 1 use		250	.128	SFCA	2.76	4.44		7.20	9.95
7100	4 use		270	.119	"	.91	4.11		5.02	7.40

03 11 13.47 Forms In Place, Gas Station Forms

		Crew	Daily Output	Labor-Hours	Unit	Material	Labor	Equipment	Total	Total Incl O&P
0010	**FORMS IN PLACE, GAS STATION FORMS**									
0050	Curb fascia, with template, 12 ga. steel, left in place, 9" high	1 Carp	50	.160	L.F.	10.75	5.85		16.60	21
1000	Sign or light bases, 18" diameter, 9" high		9	.889	Ea.	67.50	32.50		100	126
1050	30" diameter, 13" high		8	1		108	36.50		144.50	175
2000	Island forms, 10' long, 9" high, 3'- 6" wide	C-1	10	3.200		300	111		411	505
2050	4' wide		9	3.556		310	123		433	530
2500	20' long, 9" high, 4' wide		6	5.333		500	185		685	840
2550	5' wide		5	6.400		520	222		742	915

03 11 13.50 Forms In Place, Grade Beam

		Crew	Daily Output	Labor-Hours	Unit	Material	Labor	Equipment	Total	Total Incl O&P
0010	**FORMS IN PLACE, GRADE BEAM** R031113-40									
0020	Job-built plywood, 1 use	C-2	530	.091	SFCA	1.73	3.23		4.96	6.95
0050	2 use R031113-60		580	.083		.95	2.96		3.91	5.65
0100	3 use		600	.080		.69	2.86		3.55	5.20
0150	4 use		605	.079		.56	2.83		3.39	5.05

03 11 13.55 Forms In Place, Mat Foundation

		Crew	Daily Output	Labor-Hours	Unit	Material	Labor	Equipment	Total	Total Incl O&P
0010	**FORMS IN PLACE, MAT FOUNDATION** R031113-40									
0020	Job-built plywood, 1 use	C-2	290	.166	SFCA	2.54	5.90		8.44	12
0050	2 use R031113-60		310	.155		.99	5.55		6.54	9.70
0100	3 use		330	.145		.70	5.20		5.90	8.85
0120	4 use		350	.137		.59	4.90		5.49	8.30

03 11 13.65 Forms In Place, Slab On Grade

		Crew	Daily Output	Labor-Hours	Unit	Material	Labor	Equipment	Total	Total Incl O&P
0010	**FORMS IN PLACE, SLAB ON GRADE** R031113-40									
1000	Bulkhead forms w/keyway, wood, 6" high, 1 use	C-1	510	.063	L.F.	.87	2.18		3.05	4.35
1050	2 uses R031113-60		400	.080		.48	2.78		3.26	4.85
1100	4 uses		350	.091		.28	3.17		3.45	5.25
1400	Bulkhead form for slab, 4-1/2" high, exp metal, incl keyway & stakes		1200	.027		.84	.93		1.77	2.36
1410	5-1/2" high		1100	.029		.90	1.01		1.91	2.56
1420	7-1/2" high		960	.033		1.08	1.16		2.24	2.99

03 11 Concrete Forming

03 11 13 – Structural Cast-In-Place Concrete Forming

03 11 13.65 Forms In Place, Slab On Grade

		Crew	Daily Output	Labor-Hours	Unit	Material	2007 Bare Costs Labor	2007 Bare Costs Equipment	Total	Total Incl O&P
1430	9-1/2" high	C-1	840	.038	L.F.	1.88	1.32		3.20	4.13
2000	Curb forms, wood, 6" to 12" high, on grade, 1 use		215	.149	SFCA	1.95	5.15		7.10	10.20
2050	2 use		250	.128		1.08	4.44		5.52	8.10
2100	3 use		265	.121		.78	4.19		4.97	7.40
2150	4 use		275	.116		.63	4.04		4.67	7
3000	Edge forms, wood, 4 use, on grade, to 6" high		600	.053	L.F.	.29	1.85		2.14	3.20
3050	7" to 12" high		435	.074	SFCA	.65	2.55		3.20	4.69
3500	For depressed slabs, 4 use, to 12" high		300	.107	L.F.	.66	3.70		4.36	6.45
3550	To 24" high		175	.183		.85	6.35		7.20	10.85
4000	For slab blockouts, to 12" high, 1 use		200	.160		.68	5.55		6.23	9.40
4050	To 24" high, 1 use		120	.267		.86	9.25		10.11	15.35
4100	Plastic (extruded), to 6" high, multiple use, on grade		800	.040		.40	1.39		1.79	2.60
5000	Screed, 24 ga. metal key joint, see Div 03 15 05.25									
5020	Wood, incl. wood stakes, 1" x 3"	C-1	900	.036	L.F.	.45	1.23		1.68	2.41
5050	2" x 4"		900	.036	"	.60	1.23		1.83	2.58
6000	Trench forms in floor, wood, 1 use		160	.200	SFCA	2.51	6.95		9.46	13.55
6050	2 use		175	.183		1.21	6.35		7.56	11.25
6100	3 use		180	.178		.88	6.15		7.03	10.55
6150	4 use		185	.173		.71	6		6.71	10.15
8760	Void form, corrugated fiberboard, 6" x 12", 10' long		240	.133	S.F.	.91	4.63		5.54	8.20

03 11 13.85 Forms In Place, Walls

		Crew	Daily Output	Labor-Hours	Unit	Material	2007 Bare Costs Labor	2007 Bare Costs Equipment	Total	Total Incl O&P
0010	**FORMS IN PLACE, WALLS** R031113-10									
0100	Box out for wall openings, to 16" thick, to 10 S.F.	C-2	24	2	Ea.	22	71.50		93.50	135
0150	Over 10 S.F. (use perimeter) R031113-40	"	280	.171	L.F.	1.93	6.10		8.03	11.70
0250	Brick shelf, 4" w, add to wall forms, use wall area abv shelf									
0260	1 use R031113-60	C-2	240	.200	SFCA	2.14	7.15		9.29	13.45
0300	2 use		275	.175		1.17	6.25		7.42	11
0350	4 use		300	.160		.85	5.70		6.55	9.85
0500	Bulkhead, with keyway, 1 use, 2 piece		265	.181	L.F.	3.49	6.45		9.94	13.90
0550	3 piece		175	.274		4.40	9.80		14.20	20
0600	Bulkhead forms with keyway, 1 piece expanded metal, 8" wall	C-1	1000	.032		1.08	1.11		2.19	2.92
0610	10" wall		800	.040		1.88	1.39		3.27	4.23
0620	12" wall		525	.061		2.29	2.12		4.41	5.80
0700	Buttress, to 8' high, 1 use	C-2	350	.137	SFCA	3.82	4.90		8.72	11.85
0750	2 use		430	.112		2.10	3.99		6.09	8.50
0800	3 use		460	.104		1.53	3.73		5.26	7.50
0850	4 use		480	.100		1.26	3.57		4.83	6.95
1000	Corbel or haunch, to 12" wide, add to wall forms, 1 use		150	.320	L.F.	2.14	11.45		13.59	20
1050	2 use		170	.282		1.18	10.10		11.28	17
1100	3 use		175	.274		.86	9.80		10.66	16.20
1150	4 use		180	.267		.70	9.50		10.20	15.60
2000	Wall, below grade, job-built plywood, to 8' high, 1 use		300	.160	SFCA	2.49	5.70		8.19	11.65
2050	2 use		365	.132		1.58	4.70		6.28	9.05
2100	3 use		425	.113		1.15	4.03		5.18	7.55
2150	4 use		435	.110		.94	3.94		4.88	7.20
2400	Over 8' to 16' high, 1 use		280	.171		5.20	6.10		11.30	15.25
2420	2 use		345	.139		1.98	4.97		6.95	9.95
2430	3 use		375	.128		1.68	4.57		6.25	8.95
2440	4 use		395	.122		1.53	4.34		5.87	8.45
2445	Exterior wall, 8' to 16' high, 1 use		280	.171		2.06	6.10		8.16	11.80
2450	2 use		345	.139		1.14	4.97		6.11	9
2500	3 use		375	.128		.81	4.57		5.38	8

03 11 Concrete Forming

03 11 13 – Structural Cast-In-Place Concrete Forming

03 11 13.85 Forms In Place, Walls		Crew	Daily Output	Labor-Hours	Unit	Material	2007 Bare Costs Labor	Equipment	Total	Total Incl O&P
2550	4 use	C-2	395	.122	SFCA	.66	4.34		5	7.50
2700	Over 16' high, 1 use		235	.204		2.33	7.30		9.63	13.90
2750	2 use		290	.166		1.28	5.90		7.18	10.60
2800	3 use		315	.152		.93	5.45		6.38	9.45
2850	4 use		330	.145		.76	5.20		5.96	8.95
3000	For architectural finish, add		1820	.026		5.45	.94		6.39	7.45
4000	Radial, smooth curved, job-built plywood, 1 use		245	.196		2.03	7		9.03	13.15
4050	2 use		300	.160		1.12	5.70		6.82	10.15
4100	3 use		325	.148		.81	5.25		6.06	9.10
4150	4 use		335	.143		.67	5.10		5.77	8.70
4200	Below grade, job-built plywood, 1 use		225	.213		3.04	7.60		10.64	15.20
4210	2 use		225	.213		1.69	7.60		9.29	13.70
4220	3 use		225	.213		1.36	7.60		8.96	13.35
4230	4 use		225	.213		.99	7.60		8.59	12.95
4300	Curved, 2' chords, job-built plywood, 1 use		290	.166		1.82	5.90		7.72	11.20
4350	2 use		355	.135		1	4.83		5.83	8.60
4400	3 use		385	.125		.73	4.45		5.18	7.75
4450	4 use		400	.120		.59	4.29		4.88	7.30
4500	Over 8' high, 1 use		290	.166		.71	5.90		6.61	10
4525	2 use		355	.135		.39	4.83		5.22	7.95
4550	3 use		385	.125		.28	4.45		4.73	7.25
4575	4 use		400	.120		.23	4.29		4.52	6.90
4600	Retaining wall, battered, job-built plyw'd, to 8' high, 1 use		300	.160		1.71	5.70		7.41	10.80
4650	2 use		355	.135		.94	4.83		5.77	8.55
4700	3 use		375	.128		.68	4.57		5.25	7.85
4750	4 use		390	.123		.50	4.40		4.90	7.40
4900	Over 8' to 16' high, 1 use		240	.200		1.87	7.15		9.02	13.15
4950	2 use		295	.163		1.03	5.80		6.83	10.20
5000	3 use		305	.157		.75	5.60		6.35	9.55
5050	4 use		320	.150		.61	5.35		5.96	9
5500	For gang wall forming, 192 S.F. sections, deduct					10%	10%			
5550	384 S.F. sections, deduct					20%	20%			
5750	Liners for forms (add to wall forms), A.B.S. plastic									
5800	Aged wood, 4" wide, 1 use	1 Carp	250	.032	SFCA	6.85	1.17		8.02	9.40
5820	2 use		400	.020		3.77	.73		4.50	5.30
5840	4 use		750	.011		2.23	.39		2.62	3.06
5900	Fractured rope rib, 1 use		250	.032		3.84	1.17		5.01	6.05
6000	4 use		750	.011		1.25	.39		1.64	1.98
6100	Ribbed look, 1/2" & 3/4" deep, 1 use		300	.027		5.45	.98		6.43	7.50
6200	4 use		800	.010		1.77	.37		2.14	2.52
6300	Rustic brick pattern, 1 use		250	.032		3.84	1.17		5.01	6.05
6400	4 use		750	.011		1.25	.39		1.64	1.98
6500	Striated, random, 3/8" x 3/8" deep, 1 use		300	.027		3.84	.98		4.82	5.75
6600	4 use		800	.010		1.25	.37		1.62	1.94
6800	Rustication strips, A.B.S. plastic, 2 piece snap-on									
6850	1" deep x 1-3/8" wide, 1 use	C-2	400	.120	L.F.	4.26	4.29		8.55	11.35
6900	2 use		600	.080		2.34	2.86		5.20	7.05
6950	4 use		800	.060		1.38	2.14		3.52	4.86
7050	Wood, beveled edge, 3/4" deep, 1 use		600	.080		.12	2.86		2.98	4.58
7100	1" deep, 1 use		450	.107		.17	3.81		3.98	6.15
7200	For solid board finish, uniform, 1 use, add to wall forms		300	.160	SFCA	.99	5.70		6.69	10
7300	Non-uniform finish		250	.192		.92	6.85		7.77	11.70
7500	Lintel or sill forms, 1 use	1 Carp	30	.267		2.92	9.80		12.72	18.45

03 11 Concrete Forming

03 11 13 – Structural Cast-In-Place Concrete Forming

03 11 13.85 Forms In Place, Walls

		Crew	Daily Output	Labor-Hours	Unit	Material	2007 Bare Costs Labor	2007 Bare Costs Equipment	Total	Total Incl O&P
7520	2 use	1 Carp	34	.235	SFCA	1.61	8.65		10.26	15.20
7540	3 use		36	.222		1.17	8.15		9.32	14
7560	4 use	↓	37	.216		.95	7.95		8.90	13.40
7800	Modular prefabricated plywood, to 8' high, 1 use	C-2	1180	.041		2.01	1.45		3.46	4.47
7820	2 use		1200	.040		1.11	1.43		2.54	3.44
7840	3 use		1240	.039		.80	1.38		2.18	3.03
7860	4 use		1260	.038		.66	1.36		2.02	2.85
8000	To 16' high, 1 use		715	.067		2.65	2.40		5.05	6.65
8020	2 use		740	.065		1.46	2.32		3.78	5.20
8040	3 use		770	.062		1.06	2.23		3.29	4.64
8060	4 use		790	.061		.87	2.17		3.04	4.34
8100	Over 16' high, 1 use		715	.067		3.18	2.40		5.58	7.25
8120	2 use		740	.065		1.75	2.32		4.07	5.55
8140	3 use		770	.062		1.27	2.23		3.50	4.87
8160	4 use		790	.061		1.06	2.17		3.23	4.55
8600	Pilasters, 1 use		270	.178		2.90	6.35		9.25	13.10
8620	2 use		330	.145		1.59	5.20		6.79	9.85
8640	3 use		370	.130		1.16	4.63		5.79	8.45
8660	4 use	↓	385	.125	↓	.94	4.45		5.39	8
9010	Steel framed plywood, based on 100 uses of purchased									
9020	forms, and 4 uses of bracing lumber									
9060	To 8' high	C-2	600	.080	SFCA	.42	2.86		3.28	4.91
9260	Over 8' to 16' high		450	.107		.42	3.81		4.23	6.40
9460	Over 16' to 20' high	↓	400	.120	↓	.42	4.29		4.71	7.10
9475	For elevated walls, add						10%			
9480	For battered walls, 1 side battered, add					10%	10%			
9485	For battered walls, 2 sides battered, add					15%	15%			

03 11 19 – Insulating Concrete Forming

03 11 19.10 Insulating Forms, Left In Place

		Crew	Daily Output	Labor-Hours	Unit	Material	2007 Bare Costs Labor	2007 Bare Costs Equipment	Total	Total Incl O&P
0010	**INSULATING FORMS, LEFT IN PLACE**									
0020	Forms left in place, S.F. is for one face, but incl. forms for both faces									
1000	Panel system, flat cavity, minimum	2 Carp	960	.017	S.F.	2.24	.61		2.85	3.41
1010	Maximum		960	.017		3.36	.61		3.97	4.65
1020	Grid cavity, minimum		960	.017		2.32	.61		2.93	3.50
1030	Maximum		960	.017		3.48	.61		4.09	4.78
1040	Post and beam cavity, minimum		960	.017		2.36	.61		2.97	3.55
1050	Maximum		960	.017		3.54	.61		4.15	4.84
1060	Plank system, flat cavity, minimum		1920	.008		2.60	.31		2.91	3.34
1070	Maximum		1920	.008		3.90	.31		4.21	4.77
1120	Block system, flat cavity, minimum		480	.033		2.48	1.22		3.70	4.63
1130	Maximum		480	.033		3.72	1.22		4.94	6
1140	Grid cavity, minimum		480	.033		2.48	1.22		3.70	4.63
1150	Maximum		480	.033		3.72	1.22		4.94	6
1160	Post and beam cavity, minimum		480	.033		2.48	1.22		3.70	4.63
1170	Maximum	↓	480	.033	↓	3.72	1.22		4.94	6

03 11 19.60 Roof Deck Form Boards

		Crew	Daily Output	Labor-Hours	Unit	Material	2007 Bare Costs Labor	2007 Bare Costs Equipment	Total	Total Incl O&P
0010	**ROOF DECK FORM BOARDS** R051223-50									
0050	Non-asbestos fiber cement, 1/8" thick	C-13	2950	.008	S.F.	2.25	.32	.04	2.61	3.08
0070	1/4" thick		2950	.008		2.59	.32	.04	2.95	3.45
0100	Fiberglass, 1" thick, economy	↓	2700	.009		2.80	.35	.04	3.19	3.74

03 11 Concrete Forming

03 11 23 – Permanent Stair Forming

03 11 23.75 Forms In Place, Stairs

		Crew	Daily Output	Labor-Hours	Unit	Material	2007 Bare Costs Labor	2007 Bare Costs Equipment	Total	Total Incl O&P
0010	**FORMS IN PLACE, STAIRS** R031113-40									
0015	(Slant length x width), 1 use	C-2	165	.291	S.F.	5.50	10.40		15.90	22
0050	2 use R031113-60		170	.282		3.29	10.10		13.39	19.30
0100	3 use		180	.267		2.55	9.50		12.05	17.65
0150	4 use		190	.253		2.18	9		11.18	16.45
1000	Alternate pricing method (1.0 L.F./S.F.), 1 use		100	.480	LF Rsr	5.50	17.15		22.65	32.50
1050	2 use		105	.457		3.29	16.30		19.59	29
1100	3 use		110	.436		2.55	15.60		18.15	27.50
1150	4 use		115	.417		2.18	14.90		17.08	25.50
2000	Stairs, cast on sloping ground (length x width), 1 use		220	.218	S.F.	2.11	7.80		9.91	14.45
2100	4 use		240	.200	"	.69	7.15		7.84	11.85

03 15 Concrete Accessories

03 15 05 – Concrete Forming Accessories

03 15 05.02 Anchor Bolts

		Crew	Daily Output	Labor-Hours	Unit	Material	2007 Bare Costs Labor	2007 Bare Costs Equipment	Total	Total Incl O&P
0010	**ANCHOR BOLTS**									
0015	J-type, plain, incl. nut and washer									
0020	1/2" diameter, 6" long	1 Carp	90	.089	Ea.	1.09	3.26		4.35	6.30
0050	10" long		85	.094		1.24	3.45		4.69	6.75
0100	12" long		85	.094		1.36	3.45		4.81	6.90
0200	5/8" diameter, 12" long		80	.100		1.37	3.67		5.04	7.20
0250	18" long		70	.114		1.61	4.19		5.80	8.30
0300	24" long		60	.133		1.85	4.89		6.74	9.65
0350	3/4" diameter, 8" long		80	.100		1.61	3.67		5.28	7.45
0400	12" long		70	.114		2.01	4.19		6.20	8.75
0450	18" long		60	.133		2.61	4.89		7.50	10.45
0500	24" long		50	.160		3.42	5.85		9.27	12.90
0600	7/8" diameter, 12" long		60	.133		2.66	4.89		7.55	10.50
0650	18" long		50	.160		3.56	5.85		9.41	13.05
0700	24" long		40	.200		3.67	7.35		11.02	15.50
0800	1" diameter, 12" long		55	.145		3.84	5.35		9.19	12.55
0850	18" long		45	.178		4.63	6.50		11.13	15.25
0900	24" long		35	.229		5.65	8.40		14.05	19.25
0950	36" long		25	.320		7.75	11.75		19.50	27
1200	1-1/2" diameter, 18" long		22	.364		13.70	13.35		27.05	36
1250	24" long		18	.444		16.30	16.30		32.60	43.50
1300	36" long		12	.667		20.50	24.50		45	60.50

03 15 05.12 Chamfer Strips

		Crew	Daily Output	Labor-Hours	Unit	Material	2007 Bare Costs Labor	2007 Bare Costs Equipment	Total	Total Incl O&P
0010	**CHAMFER STRIPS**									
2000	Polyvinyl chloride, 1/2" wide with leg	1 Carp	535	.015	L.F.	.26	.55		.81	1.14
2200	3/4" wide with leg		525	.015		.36	.56		.92	1.27
2400	1" radius with leg		515	.016		.74	.57		1.31	1.70
2800	1-1/2" radius with leg		500	.016		2.44	.59		3.03	3.59
5000	Wood, 1/2" wide		535	.015		.11	.55		.66	.97
5200	3/4" wide		525	.015		.12	.56		.68	1
5400	1" wide		515	.016		.17	.57		.74	1.08

03 15 05.15 Column Form Accessories

		Crew	Daily Output	Labor-Hours	Unit	Material	2007 Bare Costs Labor	2007 Bare Costs Equipment	Total	Total Incl O&P
0010	**COLUMN FORM ACCESSORIES**									
1000	Column clamps, adjustable to 24" x 24", buy				Set	97			97	107
1100	Rent per month					8.35			8.35	9.15

03 15 Concrete Accessories

03 15 05 – Concrete Forming Accessories

03 15 05.15 Column Form Accessories		Crew	Daily Output	Labor-Hours	Unit	Material	2007 Bare Costs Labor	Equipment	Total	Total Incl O&P
1300	For sizes to 30" x 30", buy				Set	137			137	150
1400	Rent per month					11.20			11.20	12.35
1600	For sizes to 36" x 36", buy					138			138	152
1700	Rent per month					14.35			14.35	15.80
2000	Bull winch (band iron) 36" x 36", buy					48.50			48.50	53.50
2100	Rent per month					4.44			4.44	4.88
2300	48" x 48", buy					50			50	55
2400	Rent per month					5.60			5.60	6.15
3000	Chain & wedge type 36" x 36", buy					68.50			68.50	75
3100	Rent per month					7.65			7.65	8.40
3300	60" x 60", buy					92			92	101
3400	Rent per month					10.70			10.70	11.80
4000	Friction collars 2'-6" dia., buy					695			695	765
4100	Rent per month					68.50			68.50	75
4300	4'-0" dia., buy					790			790	870
4400	Rent per month					86.50			86.50	95.50

03 15 05.20 Dovetail Anchor System		Crew	Daily Output	Labor-Hours	Unit	Material	Labor	Equipment	Total	Total Incl O&P
0010	**DOVETAIL ANCHOR SYSTEM**									
0500	Anchor slot, galv., filled, 24 ga.	1 Carp	425	.019	L.F.	.64	.69		1.33	1.78
0600	20 ga.		400	.020		.84	.73		1.57	2.06
0625	12 ga.		400	.020		1.28	.73		2.01	2.55
0800	16 oz. copper, foam filled		375	.021		2.47	.78		3.25	3.94
0900	26 ga. stainless steel, foam filled		375	.021		1.02	.78		1.80	2.35
1200	Brick anchor, corr., galv., 3-1/2" long, 16 ga.	1 Bric	10.50	.762	C	29.50	29		58.50	76.50
1300	12 ga.		10.50	.762		41	29		70	89
1500	Flat, galv., 3-1/2" long, 16 ga.		10.50	.762		29	29		58	76
1600	12 ga.		10.50	.762		42	29		71	90
2000	Cavity wall, corr., galv., 5" long, 16 ga.		10.50	.762		35	29		64	82.50
2100	12 ga.		10.50	.762		49.50	29		78.50	98.50
3000	Furring anchors, corr., galv., 1-1/2" long, 16 ga.		10.50	.762		13.90	29		42.90	59.50
3100	12 ga.		10.50	.762		23.50	29		52.50	69.50
6000	Stone anchors, 3-1/2" long, galv., 1/8" x 1" wide		10.50	.762		116	29		145	172
6100	1/4" x 1" wide		10.50	.762		146	29		175	204

03 15 05.25 Expansion Joints		Crew	Daily Output	Labor-Hours	Unit	Material	Labor	Equipment	Total	Total Incl O&P
0010	**EXPANSION JOINTS**									
0020	Keyed, cold, 24 ga, incl. stakes, 3-1/2" high	1 Carp	200	.040	L.F.	.69	1.47		2.16	3.05
0050	4-1/2" high		200	.040		.84	1.47		2.31	3.21
0100	5-1/2" high		195	.041		.90	1.51		2.41	3.33
0150	7-1/2" high		190	.042		1.08	1.55		2.63	3.60
0300	Poured asphalt, plain, 1/2" x 1"	1 Clab	450	.018		.44	.51		.95	1.28
0350	1" x 2"		400	.020		1.74	.58		2.32	2.81
0500	Neoprene, liquid, cold applied, 1/2" x 1"		450	.018		2.07	.51		2.58	3.08
0550	1" x 2"		400	.020		6.25	.58		6.83	7.80
0700	Polyurethane, poured, 2 part, 1/2" x 1"		400	.020		1.36	.58		1.94	2.40
0750	1" x 2"		350	.023		5.45	.66		6.11	7
0900	Rubberized asphalt, hot or cold applied, 1/2" x 1"		450	.018		.42	.51		.93	1.26
0950	1" x 2"		400	.020		1.50	.58		2.08	2.55
1100	Hot applied, fuel resistant, 1/2" x 1"		450	.018		1.16	.51		1.67	2.08
1150	1" x 2"		400	.020		2.24	.58		2.82	3.36
2000	Premolded, bituminous fiber, 1/2" x 6"	1 Carp	375	.021		.41	.78		1.19	1.67
2050	1" x 12"		300	.027		1.76	.98		2.74	3.46
2250	Cork with resin binder, 1/2" x 6"		375	.021		1.84	.78		2.62	3.24

03 15 Concrete Accessories

03 15 05 – Concrete Forming Accessories

03 15 05.25 Expansion Joints		Crew	Daily Output	Labor-Hours	Unit	Material	2007 Bare Costs Labor	Equipment	Total	Total Incl O&P
2300	1" x 12"	1 Carp	300	.027	L.F.	6.60	.98		7.58	8.75
2500	Neoprene sponge, closed cell, 1/2" x 6"		375	.021		1.56	.78		2.34	2.94
2550	1" x 12"		300	.027		7.15	.98		8.13	9.40
2750	Polyethylene foam, 1/2" x 6"		375	.021		.53	.78		1.31	1.80
2800	1" x 12"		300	.027		1.85	.98		2.83	3.56
3000	Polyethylene backer rod, 3/8" diameter		460	.017		.05	.64		.69	1.04
3050	3/4" diameter		460	.017		.09	.64		.73	1.08
3100	1" diameter		460	.017		.12	.64		.76	1.12
3500	Polyurethane foam, with polybutylene, 1/2" x 1/2"		475	.017		.95	.62		1.57	2.01
3550	1" x 1"		450	.018		1.95	.65		2.60	3.17
3750	Polyurethane foam, regular, closed cell, 1/2" x 6"		375	.021		.70	.78		1.48	1.99
3800	1" x 12"		300	.027		1.30	.98		2.28	2.95
4000	Polyvinyl chloride foam, closed cell, 1/2" x 6"		375	.021		2.73	.78		3.51	4.22
4050	1" x 12"		300	.027		6.95	.98		7.93	9.15
4250	Rubber, gray sponge, 1/2" x 6"		375	.021		3	.78		3.78	4.52
4300	1" x 12"		300	.027		11.25	.98		12.23	13.90
4500	Lead wool for joints, 1 ton lots				Lb.	2.02			2.02	2.22
5000	For installation in walls, add						75%			
5250	For installation in boxouts, add						25%			

03 15 05.30 Hangers		Crew	Daily Output	Labor-Hours	Unit	Material	Labor	Equipment	Total	Total Incl O&P
0010	**HANGERS**									
0020	Slab and beam form									
0500	Banding iron, 3/4" x 22 ga, 14 L.F. per lb or									
0550	1/2" x 14 ga, 7 L.F. per lb.				Lb.	1.25			1.25	1.38
1000	Fascia ties, coil type add to frame ties below				C	89.50			89.50	98.50
1500	Frame ties to 8-1/8"					224			224	246
1550	8-1/8" to 10-1/8"					239			239	263
5000	Snap tie hanger, to 30" overall length, 4000 #					530			530	585
5050	30" to 36" overall length					575			575	635
5100	42" to 48" overall length					660			660	725
5500	Steel beam hanger									
5600	Flange to 8-1/8"				C	435			435	475
5650	8-1/8" to 10-1/8"					365			365	400
6000	Tie hangers to 24" overall length, 6000 #					495			495	540
6100	30" to 36" overall length					580			580	640
6500	Tie back hanger, up to 12-1/8" flange					465			465	510
8500	Wire, black annealed, 9 ga				Cwt.	133			133	146
8600	16 ga				"	139			139	153

03 15 05.35 Inserts		Crew	Daily Output	Labor-Hours	Unit	Material	Labor	Equipment	Total	Total Incl O&P
0010	**INSERTS**									
1000	All size nut insert, 5/8" & 3/4", incl. nut	1 Carp	84	.095	Ea.	4.37	3.50		7.87	10.25
2000	Continuous slotted, 1-5/8" x 1-3/8"									
2100	12 ga., 3" long	1 Carp	65	.123	Ea.	3.96	4.52		8.48	11.40
2150	6" long		65	.123		5.15	4.52		9.67	12.70
2200	8 ga., 12" long		65	.123		12.30	4.52		16.82	20.50
2250	24" long		65	.123		20	4.52		24.52	29
2300	36" long		60	.133		28	4.89		32.89	38.50
2350	60" long		55	.145		43	5.35		48.35	56
7000	Threaded cast									
7100	1/4" diameter bolt	1 Carp	84	.095	Ea.	7.10	3.50		10.60	13.25
7350	7/8" diameter bolt	"	84	.095	"	10.60	3.50		14.10	17.10
9000	Wedge									

03 15 Concrete Accessories

03 15 05 – Concrete Forming Accessories

03 15 05.35 Inserts

		Crew	Daily Output	Labor-Hours	Unit	Material	2007 Bare Costs Labor	Equipment	Total	Total Incl O&P
9050	For 5/8" diameter bolt	1 Carp	60	.133	Ea.	4.78	4.89		9.67	12.85
9100	For 3/4" diameter bolt	"	60	.133	"	10.25	4.89		15.14	18.90
9800	Cut washers, "Black"									
9850	5/8" bolt				Ea.	.29			.29	.32
9950	For galvanized inserts, add					30%				

03 15 05.70 Shores

		Crew	Daily Output	Labor-Hours	Unit	Material	Labor	Equipment	Total	Total Incl O&P
0010	**SHORES**									
0020	Erect and strip, by hand, horizontal members									
0500	Aluminum joists and stringers	2 Carp	60	.267	Ea.		9.80		9.80	15.25
0600	Steel, adjustable beams		45	.356			13.05		13.05	20.50
0700	Wood joists		50	.320			11.75		11.75	18.30
0800	Wood stringers		30	.533			19.55		19.55	30.50
1000	Vertical members to 10' high		55	.291			10.70		10.70	16.65
1050	To 13' high		50	.320			11.75		11.75	18.30
1100	To 16' high		45	.356			13.05		13.05	20.50
1500	Reshoring		1400	.011	S.F.	.46	.42		.88	1.16
1600	Flying truss system	C-17D	9600	.009	SFCA		.34	.06	.40	.59
1760	Horizontal, aluminum joists, 6-1/4" high x 5' to 21' span, buy				L.F.	37			37	40.50
1770	Beams, 7-1/4" high x 4' to 30' span				"	51.50			51.50	56.50
1810	Horizontal, steel beam, adjustable, 4' to 7' span				Ea.	590			590	650
1830	6' to 10' span					760			760	835
1920	9' to 15' span					965			965	1,050
1940	12' to 20' span					985			985	1,075
1970	Steel stringer, W8x10, 4' to 16' span, buy				L.F.	21			21	23.50
3000	Rent for job duration, aluminum joist @ 2' O.C., per mo				SF Flr.	.93			.93	1.02
3050	Steel W8x10					.53			.53	.58
3060	Steel adjustable					1.90			1.90	2.09
3500	#1 post shore, steel, 5'-7" to 9'-6" high, 10000# cap., buy				Ea.	390			390	430
3550	#2 post shore, 7'-3" to 12'-10" high, 7800# capacity					420			420	460
3600	#3 post shore, 8'-10" to 16'-1" high, 3800# capacity					455			455	505
5010	Frame shoring systems, steel, 12000#/leg, buy									
5040	Frame, 2' wide x 6' high				Ea.	365			365	400
5250	X-brace					57.50			57.50	63.50
5550	Base plate					45.50			45.50	50
5600	Screw jack					145			145	160
5650	U-head, 8" x 8"					23			23	25.50

03 15 05.75 Sleeves and Chases

		Crew	Daily Output	Labor-Hours	Unit	Material	Labor	Equipment	Total	Total Incl O&P
0010	**SLEEVES AND CHASES**									
0100	Plastic, 1 use, 9" long, 2" diameter	1 Carp	100	.080	Ea.	1.55	2.94		4.49	6.30
0150	4" diameter		90	.089		4.14	3.26		7.40	9.65
0200	6" diameter		75	.107		7.30	3.91		11.21	14.15
0250	12" diameter		60	.133		21	4.89		25.89	30.50

03 15 05.80 Snap Ties, Flat Washer

		Crew	Daily Output	Labor-Hours	Unit	Material	Labor	Equipment	Total	Total Incl O&P
0010	**SNAP TIES, FLAT WASHER**									
0100	3000 lb., to 8"				C	121			121	133
0200	11" & 12"					145			145	160
0250	16"					154			154	170
0300	18"					153			153	169
0500	With plastic cone, to 8"					112			112	123
0600	11" & 12"					133			133	146
0650	16"					140			140	154
0700	18"					146			146	161

03 15 Concrete Accessories

03 15 05 – Concrete Forming Accessories

03 15 05.80 Snap Ties, Flat Washer

		Crew	Daily Output	Labor-Hours	Unit	Material	2007 Bare Costs Labor	Equipment	Total	Total Incl O&P
1000	5000 lb., to 8"				C	154			154	169
1150	11" & 12"					180			180	197
1200	16"					197			197	217
1250	18"					193			193	212
1500	With plastic cone, to 8"					192			192	211
1600	11" & 12"					228			228	250
1650	16"					252			252	277
1700	18"					261			261	287

03 15 05.85 Stair Tread Inserts

		Crew	Daily Output	Labor-Hours	Unit	Material	Labor	Equipment	Total	Total Incl O&P
0010	**STAIR TREAD INSERTS**									
0015	Cast iron, abrasive, 3" wide	1 Carp	90	.089	L.F.	8.45	3.26		11.71	14.40
0020	4" wide		80	.100		11.25	3.67		14.92	18.05
0040	6" wide		75	.107		16.90	3.91		20.81	24.50
0050	9" wide		70	.114		25.50	4.19		29.69	34.50
0100	12" wide		65	.123		34	4.52		38.52	44
0300	Cast aluminum, compared to cast iron, deduct					10%				
0500	Extruded aluminum safety tread, 3" wide	1 Carp	75	.107		9.60	3.91		13.51	16.65
0550	4" wide		75	.107		12.80	3.91		16.71	20
0600	6" wide		75	.107		19.25	3.91		23.16	27
0650	9" wide to resurface stairs		70	.114		29	4.19		33.19	38
1700	Cement fill for pan-type metal treads, plain	1 Cefi	115	.070	S.F.	1.93	2.47		4.40	5.75
1750	Non-slip	"	100	.080	"	2.12	2.84		4.96	6.50

03 15 05.95 Wall and Foundation Form Accessories

		Crew	Daily Output	Labor-Hours	Unit	Material	Labor	Equipment	Total	Total Incl O&P
0010	**WALL AND FOUNDATION FORM ACCESSORIES**									
2000	Footings, form braces, solid steel, adjustable				Ea.	15.85			15.85	17.40
2050	Spreaders for footer, adjustable				"	5.50			5.50	6.05
3000	Form oil, coverage varies greatly, minimum				Gal.	5.50			5.50	6.05
3050	Maximum				"	8.35			8.35	9.15
3500	Form patches, 1-3/4" diameter				C	82.50			82.50	91
3550	2-3/4" diameter				"	105			105	116
4000	Nail stakes, 3/4" diameter, 18" long				Ea.	2.56			2.56	2.82
4050	24" long					2.93			2.93	3.22
4200	30" long					3.47			3.47	3.82
4250	36" long					4.01			4.01	4.41

03 15 13 – Waterstops

03 15 13.50 Waterstops

		Crew	Daily Output	Labor-Hours	Unit	Material	Labor	Equipment	Total	Total Incl O&P
0010	**WATERSTOPS**									
0020	PVC, ribbed 3/16" thick, 4" wide	1 Carp	155	.052	L.F.	.91	1.89		2.80	3.95
0050	6" wide		145	.055		1.39	2.02		3.41	4.68
0500	Ribbed, PVC, with center bulb, 6" wide, 3/16" thick		135	.059		1.36	2.17		3.53	4.89
0550	3/8" thick		130	.062		2.12	2.26		4.38	5.85
0800	Dumbbell type, PVC, 6" wide, 3/16" thick		150	.053		1.31	1.96		3.27	4.49
0850	3/8" thick		145	.055		2.59	2.02		4.61	6
1000	9" wide, 3/8" thick, PVC, plain		130	.062		3.74	2.26		6	7.65
1050	Center bulb		130	.062		7.35	2.26		9.61	11.55
1250	Split PVC, 3/8" thick, 6" wide		145	.055		1.49	2.02		3.51	4.79
1300	9" wide		130	.062		2.86	2.26		5.12	6.65
2000	Rubber, flat dumbbell, 3/8" thick, 6" wide		145	.055		6.95	2.02		8.97	10.80
2050	9" wide		135	.059		10.45	2.17		12.62	14.90
2500	Flat dumbbell split, 3/8" thick, 6" wide		145	.055		1.49	2.02		3.51	4.79
2550	9" wide		135	.059		2.86	2.17		5.03	6.55
3000	Center bulb, 1/4" thick, 6" wide		145	.055		6.05	2.02		8.07	9.80

03 15 Concrete Accessories

03 15 13 – Waterstops

03 15 13.50 Waterstops

		Crew	Daily Output	Labor-Hours	Unit	Material	2007 Bare Costs Labor	Equipment	Total	Total Incl O&P
3050	9" wide	1 Carp	135	.059	L.F.	11.95	2.17		14.12	16.50
3500	Center bulb split, 3/8" thick, 6" wide		145	.055		7.35	2.02		9.37	11.25
3550	9" wide		135	.059		12.75	2.17		14.92	17.45
5000	Waterstop fittings, rubber, flat									
5010	Dumbbell or center bulb, 3/8" thick,									
5200	Field union, 6" wide	1 Carp	50	.160	Ea.	23	5.85		28.85	34.50
5250	9" wide		50	.160		27.50	5.85		33.35	39
5500	Flat cross, 6" wide		30	.267		48	9.80		57.80	68
5550	9" wide		30	.267		68.50	9.80		78.30	91
6000	Flat tee, 6" wide		30	.267		44.50	9.80		54.30	64
6050	9" wide		30	.267		58.50	9.80		68.30	79.50
6500	Flat ell, 6" wide		40	.200		40	7.35		47.35	55.50
6550	9" wide		40	.200		54	7.35		61.35	70.50
7000	Vertical tee, 6" wide		25	.320		38.50	11.75		50.25	61
7050	9" wide		25	.320		49.50	11.75		61.25	73
7500	Vertical ell, 6" wide		35	.229		34.50	8.40		42.90	51
7550	9" wide		35	.229		47.50	8.40		55.90	65

03 21 Reinforcing Steel

03 21 05 – Reinforcing Steel Accessories

03 21 05.10 Rebar Accessories

		Crew	Daily Output	Labor-Hours	Unit	Material	2007 Bare Costs Labor	Equipment	Total	Total Incl O&P
0010	**REBAR ACCESSORIES**									
0100	Beam bolsters, (BB) standard, lower, up to 1-1/2" high, plain				C.L.F.	66			66	72.50
0102	Galvanized					101			101	111
0104	Stainless					191			191	210
0106	Plastic					115			115	126
0108	Epoxy					157			157	173
0110	2-1/2" to 3" high, plain					86			86	94.50
0120	Galvanized					115			115	126
0140	Stainless					262			262	288
0160	Plastic					120			120	132
0162	Epoxy					167			167	184
0200	Upper, standard (BBU) to 1-1/2" high, plain					189			189	208
0210	2-1/2" to 3" high					340			340	375
0300	Beam bolster with plate (BBP) to 1-1/2" high, plain					218			218	239
0310	2-1/2" to 3" high					420			420	465
0500	Slab bolsters, continuous, plain (SB) 3/4" to 1" high, plain					54			54	59.50
0502	Galvanized					62.50			62.50	68.50
0504	Stainless					87.50			87.50	96.50
0506	Plastic					74			74	81.50
0510	1" to 2" high, plain					67.50			67.50	74
0515	Galvanized					79.50			79.50	87
0520	Stainless					150			150	165
0525	Plastic					89.50			89.50	98.50
0530	For bolsters with wire runners (SBR), add					76			76	83.50
0540	For bolsters with plates (SBP), add					184			184	202
0700	Clip or bar ties, 16 ga., plain, 3" long				C	15			15	16.50
0710	4" long					15.85			15.85	17.45
0720	6" long					18.55			18.55	20.50
0730	8" long					19.40			19.40	21.50
0900	Flange clips, expandable flanges, 10 ga., 12" O.C., continuous,									

03 21 Reinforcing Steel

03 21 05 – Reinforcing Steel Accessories

03 21 05.10 Rebar Accessories		Crew	Daily Output	Labor-Hours	Unit	Material	2007 Bare Costs Labor	Equipment	Total	Total Incl O&P
0910	galvanized, over 500 L.F., 4" to 8"				C.L.F.	55.50			55.50	61.50
0920	9" to 12"					76			76	83.50
0930	17" to 24"				▼	77.50			77.50	85.50
1200	High chairs, individual, no plates (1 HC), to 3" high, plain				C	81			81	89.50
1202	Galvanized					113			113	125
1204	Stainless					232			232	255
1206	Plastic					108			108	119
1210	5" high, plain					125			125	137
1212	Galvanized					163			163	179
1214	Stainless					355			355	390
1216	Plastic					151			151	167
1220	8" high, plain					270			270	297
1222	Galvanized					360			360	395
1224	Stainless					595			595	655
1226	Plastic					340			340	375
1230	12" high, plain					560			560	615
1232	Galvanized					715			715	790
1234	Stainless					1,075			1,075	1,175
1236	Plastic					630			630	690
1240	15" high, plain					1,050			1,050	1,150
1242	Galvanized					1,225			1,225	1,350
1244	Stainless					1,800			1,800	1,975
1246	Plastic					1,125			1,125	1,250
1250	For each added inch up to 24" high, plain, add					78			78	85.50
1252	Galvanized, add					90			90	99
1254	Stainless, add					101			101	111
1256	Plastic, add					90			90	99
1400	Individual high chairs, with plates, (HCP), to 5" high, add					355			355	395
1410	Over 5" high, add					410			410	450
1500	Bar chair (BC) for up to 1-3/4" high, plain					46			46	50.50
1520	Galvanized					53			53	58.50
1530	Stainless					142			142	156
1540	Plastic					94			94	103
1550	Joist chair (JC), joists up to 6", plain					55.50			55.50	61.50
1580	Galvanized					71			71	78
1600	Stainless					101			101	111
1620	Plastic					96			96	106
1630	Epoxy				▼	234			234	257
1700	Continuous high chairs, legs 8" O.C. (CHC) to 4" high, plain				C.L.F.	108			108	119
1705	Galvanized					133			133	146
1710	Stainless					283			283	310
1715	Plastic					133			133	147
1718	Epoxy					202			202	222
1720	6" high, plain					155			155	171
1725	Galvanized					223			223	245
1730	Stainless					330			330	360
1735	Plastic					213			213	234
1738	Epoxy					271			271	298
1740	8" high, plain					226			226	249
1745	Galvanized					315			315	345
1750	Stainless					390			390	430
1755	Plastic					295			295	325
1758	Epoxy				▼	410			410	450

03 21 Reinforcing Steel

03 21 05 – Reinforcing Steel Accessories

03 21 05.10 Rebar Accessories		Crew	Daily Output	Labor-Hours	Unit	Material	2007 Bare Costs Labor	Equipment	Total	Total Incl O&P
1760	12" high, plain				C.L.F.	555			555	610
1765	Galvanized					665			665	735
1770	Stainless					930			930	1,025
1775	Plastic					635			635	695
1778	Epoxy					970			970	1,075
1780	15" high, plain					625			625	685
1785	Galvanized					750			750	825
1790	Stainless					745			745	820
1795	Plastic					710			710	780
1798	Epoxy					1,275			1,275	1,400
1800	For each added 1" up to 24" high, plain, add					53			53	58.50
1820	Galvanized, add					65.50			65.50	72
1840	Stainless, add					58.50			58.50	64
1860	Plastic, add					58.50			58.50	64
1900	For continuous bottom plate, (CHCP), add					255			255	280
1940	For upper continuous high chairs, (CHCU), add					255			255	280
1960	For galvanized wire runners, add					239			239	263
2100	Paper tubing, 4' lengths, for #2 & #3 bar					95.50			95.50	105
2120	For #6 bar				↓	119			119	131
2200	Screed base, 1/2" diameter, 2-1/2" high, plain				C	247			247	272
2210	Galvanized					256			256	282
2220	5-1/2" high, plain					298			298	330
2250	Galvanized					315			315	345
2300	3/4" diameter, 2-1/2" high, plain					305			305	335
2310	Galvanized					325			325	355
2320	5-1/2" high, plain					380			380	415
2350	Galvanized					405			405	445
2400	Screed holder, 1/2" diam. for 1" I.D. pipe, plain, 6" long					256			256	282
2420	12" long					425			425	465
2500	3/4" diameter, for 1-1/2" I.D. pipe, 6" long					460			460	505
2520	12" long					760			760	835
2700	Screw anchor for bolts, plain, 1/2" diameter					168			168	185
2720	1" diameter					500			500	550
2740	1-1/2" diameter					830			830	915
2800	Screw eye bolts, 1/2" x 5" long					191			191	210
2820	1" x 9" long					760			760	835
2840	1-1/2" x 14" long					1,850			1,850	2,025
2900	Screw anchor bolts, 1/2" x up to 7" long					765			765	845
2920	1" x up to 12" long					2,475			2,475	2,725
3000	Slab lifting inserts, single, 3/4" dia., galv., 4" high					515			515	565
3010	6" high					630			630	695
3030	7" high					720			720	795
3100	1" diameter, 5" high					810			810	895
3120	7" high					855			855	945
3200	Double lifting inserts, 1" diameter, 5" high					1,600			1,600	1,775
3220	7" high					1,700			1,700	1,875
3330	1-1/4" diameter, 5" high				↓	1,750			1,750	1,925
3500	Sleeper clips for wood sleepers, 20 ga., galv., 2" wide				M	615			615	675
3520	4" wide					760			760	835
3600	Spacers, plastic for 1" bar clearance, average					94			94	103
3620	For 2" bar clearance, average				↓	114			114	125
3800	Subgrade chairs, 1/2" diameter, 3-1/2" high				C	505			505	555
3850	12" high				↓	1,400			1,400	1,550

03 21 Reinforcing Steel

03 21 05 – Reinforcing Steel Accessories

03 21 05.10 Rebar Accessories		Crew	Daily Output	Labor-Hours	Unit	Material	2007 Bare Costs Labor	Equipment	Total	Total Incl O&P
3900	3/4" diameter, 3-1/2" high				C	650			650	715
3950	12" high					1,525			1,525	1,675
4200	Subgrade stakes, 3/4" diameter, 12" long					525			525	575
4250	24" long					705			705	775
4300	1" diameter, 12" long					785			785	865
4350	24" long					1,150			1,150	1,275
4500	Tie wire, 16 ga. annealed steel, under 500 lbs.				Cwt.	152			152	167
4520	2,000 to 4,000 lbs.				"	143			143	157
4550	Tie wire holder, plastic case				Ea.	59.50			59.50	65.50
4600	Aluminum case				"	72			72	79.50

03 21 05.75 Splicing Reinforcing Bars

		Crew	Daily Output	Labor-Hours	Unit	Material	Labor	Equipment	Total	Total Incl O&P
0010	**SPLICING REINFORCING BARS** R032110-70									
0020	Including holding bars in place while splicing									
0100	Butt weld columns #4 bars	C-5	190	.295	Ea.	4.29	11.90	3.89	20.08	28
0110	#6 bars		150	.373		6.05	15.05	4.93	26.03	36
0130	#10 bars		95	.589		14.85	24	7.80	46.65	63
0150	#14 bars		65	.862		23.50	34.50	11.40	69.40	94.50
0280	Column splices, bar to bar end bearing									
0300	#7 or #8 bars	C-5	190	.295	Ea.	38.50	11.90	3.89	54.29	66
0310	#9 or #10 bars		170	.329		42.50	13.30	4.35	60.15	73
0320	#11 bars		160	.350		51.50	14.10	4.62	70.22	84
0330	#14 bars		150	.373		64.50	15.05	4.93	84.48	100
0340	#18 bars		140	.400		93	16.10	5.30	114.40	134
0500	Transition bar to bar end bearing, #14 to #18 bar					91.50			91.50	101
0520	#14 to #11 bar					70.50			70.50	77.50
0550	#10 to #9 bar					44.50			44.50	48.50
0560	#9 to #8 bar					41			41	45
0580	#8 to #7 bar					40			40	44
0600	For bolted speed sleeve type, deduct						15%			
0800	Mechanical butt splice, sleeve type w/ filler metal, compression									
0810	only, all grades, columns only #11 bars	C-5	68	.824	Ea.	45.50	33	10.90	89.40	115
0900	#14 bars		62	.903		73.50	36.50	11.95	121.95	153
0920	#18 bars		62	.903		134	36.50	11.95	182.45	220
1000	125% yield point, grade 60, columns only, #6 bars		68	.824		36.50	33	10.90	80.40	105
1020	#7 or #8 bars		68	.824		38.50	33	10.90	82.40	108
1030	#9 bars		68	.824		40	33	10.90	83.90	109
1040	#10 bars		68	.824		42.50	33	10.90	86.40	112
1050	#11 bars		68	.824		51.50	33	10.90	95.40	122
1060	#14 bars		62	.903		64.50	36.50	11.95	112.95	143
1070	#18 bars		62	.903		93	36.50	11.95	141.45	174
1200	Full tension, grade 60 steel, columns,									
1220	slabs or beams, #6, #7, #8 bars	C-5	68	.824	Ea.	12.20	33	10.90	56.10	79
1230	#9 bars		68	.824		13.35	33	10.90	57.25	80
1240	#10 bars		68	.824		14.85	33	10.90	58.75	82
1250	#11 bars		68	.824		15.75	33	10.90	59.65	83
1260	#14 bars		62	.903		23.50	36.50	11.95	71.95	97.50
1270	#18 bars		62	.903		36	36.50	11.95	84.45	111
1400	If equipment handling not required, deduct						50%			
1600	Mechanical threaded type, bar threading not included,									
1700	Straight bars, #10 & #11	C-5	140	.400	Ea.	15.75	16.10	5.30	37.15	49
1750	#14 bars		130	.431		23.50	17.35	5.70	46.55	60.50
1800	#18 bars		75	.747		36	30	9.85	75.85	99

03 21 Reinforcing Steel

03 21 05 – Reinforcing Steel Accessories

03 21 05.75 Splicing Reinforcing Bars		Crew	Daily Output	Labor-Hours	Unit	Material	2007 Bare Costs Labor	Equipment	Total	Total Incl O&P
2100	#11 to #18 & #14 to #18 transition	C-5	75	.747	Ea.	38.50	30	9.85	78.35	102
2400	Bent bars, #10 & #11		105	.533		15.75	21.50	7.05	44.30	59.50
2500	#14		90	.622		27	25	8.20	60.20	79
2600	#18		70	.800		44.50	32.50	10.55	87.55	112
2800	#11 to #14 transition		75	.747		25.50	30	9.85	65.35	87.50
2900	#11 to #18 & #14 to #18 transition	↓	70	.800	↓	42	32.50	10.55	85.05	110

03 21 10 – Uncoated Reinforcing Steel

03 21 10.60 Reinforcing In Place A615

			Crew	Daily Output	Labor-Hours	Unit	Material	Labor	Equipment	Total	Total Incl O&P
0015	**REINFORCING IN PLACE A615** Grade 60, incl. access. labor	R032110-10									
0100	Beams & Girders, #3 to #7	**CN**	4 Rodm	1.60	20	Ton	895	825		1,720	2,325
0150	#8 to #18	R032110-20		2.70	11.852		895	490		1,385	1,800
0200	Columns, #3 to #7			1.50	21.333		895	880		1,775	2,425
0250	#8 to #18			2.30	13.913		895	575		1,470	1,925
0300	Spirals, hot rolled, 8" to 15" diameter			2.20	14.545		1,175	600		1,775	2,275
0320	15" to 24" diameter	R032110-40		2.20	14.545		1,125	600		1,725	2,200
0330	24" to 36" diameter			2.30	13.913		1,075	575		1,650	2,125
0340	36" to 48" diameter	R032110-50		2.40	13.333		1,025	550		1,575	2,025
0360	48" to 64" diameter			2.50	12.800		1,125	530		1,655	2,100
0380	64" to 84" diameter	R032110-70		2.60	12.308		1,175	510		1,685	2,125
0390	84" to 96" diameter			2.70	11.852		1,225	490		1,715	2,150
0400	Elevated slabs, #4 to #7	R032110-80		2.90	11.034		950	455		1,405	1,800
0500	Footings, #4 to #7			2.10	15.238		850	630		1,480	1,950
0550	#8 to #18			3.60	8.889		805	365		1,170	1,475
0600	Slab on grade, #3 to #7			2.30	13.913		850	575		1,425	1,875
0700	Walls, #3 to #7			3	10.667		850	440		1,290	1,650
0750	#8 to #18		↓	4	8	↓	850	330		1,180	1,475
0900	Use the following for a rough estimate guide										
1000	Typical in place, average, under 10 ton job, #3 to #7		4 Rodm	1.80	17.778	Ton	925	735		1,660	2,225
1010	#8 to #18			2.70	11.852		945	490		1,435	1,850
1050	10 - 50 ton job, #3 to #7			2.10	15.238		910	630		1,540	2,025
1060	#8 to #18			3	10.667		925	440		1,365	1,750
1100	50 - 100 ton job, #3 - #7			2.20	14.545		885	600		1,485	1,950
1110	#8 to #18			3.10	10.323		905	425		1,330	1,700
1150	Over 100 ton job, #3 - #7			2.30	13.913		875	575		1,450	1,900
1160	#8 - #18		↓	3.20	10		895	415		1,310	1,675
1200	High strength steel, Grade 75, #14 bars only, add						65			65	71.50
2000	Unloading & sorting, add to above		C-5	100	.560			22.50	7.40	29.90	44.50
2200	Crane cost for handling, add to above, minimum			135	.415			16.70	5.50	22.20	33
2210	Average			92	.609			24.50	8.05	32.55	48.50
2220	Maximum		↓	35	1.600	↓		64.50	21	85.50	128
2400	Dowels, 2 feet long, deformed, #3		2 Rodm	520	.031	Ea.	.37	1.27		1.64	2.49
2410	#4			480	.033		.66	1.38		2.04	2.98
2420	#5			435	.037		1.03	1.52		2.55	3.62
2430	#6			360	.044	↓	1.48	1.84		3.32	4.64
2450	Longer and heavier dowels, add			725	.022	Lb.	.49	.91		1.40	2.04
2500	Smooth dowels, 12" long, 1/4" or 3/8" diameter			140	.114	Ea.	.74	4.72		5.46	8.55
2520	5/8" diameter			125	.128		1.29	5.30		6.59	10.05
2530	3/4" diameter		↓	110	.145	↓	1.60	6		7.60	11.60
2600	Dowel sleeves for CIP concrete, 2-part system										
2610	Sleeve base, plastic, for #5 bar, fasten to edge form		1 Rodm	200	.040	Ea.	.40	1.65		2.05	3.15
2615	Sleeve, plastic, for #5 bar x 9" long, snap onto base			400	.020		1.14	.83		1.97	2.61
2620	Sleeve base, for #6 bar			175	.046		.40	1.89		2.29	3.54

03 21 Reinforcing Steel

03 21 10 – Uncoated Reinforcing Steel

03 21 10.60 Reinforcing In Place A615

		Crew	Daily Output	Labor-Hours	Unit	Material	2007 Bare Costs Labor	Equipment	Total	Total Incl O&P
2625	Sleeve, for #6 bar	1 Rodm	350	.023	Ea.	1.17	.94		2.11	2.84
2630	Sleeve base, for #8 bar		150	.053		.49	2.20		2.69	4.15
2635	Sleeve, for #8 bar		300	.027		1.30	1.10		2.40	3.24
2700	Dowel caps, visual warning only, plastic, #3 to #8	2 Rodm	800	.020		.28	.83		1.11	1.67
2720	#7 to #14		750	.021		.50	.88		1.38	2
2750	Impalement protective, plastic, #3 to #7		800	.020		1.60	.83		2.43	3.12
2760	#7 to #11		775	.021		2.18	.85		3.03	3.80
2770	#11 to #16		750	.021		2.22	.88		3.10	3.89

03 21 13 – Galvanized Reinforcing Steel

03 21 13.10 Galvanized Reinforcing

		Crew	Daily Output	Labor-Hours	Unit	Material	2007 Bare Costs Labor	Equipment	Total	Total Incl O&P
0010	**GALVANIZED REINFORCING**									
0150	Galvanized, #3				Ton	740			740	815
0200	#4					740			740	815
0250	#5					725			725	795
0300	#6 or over					725			725	795
1000	For over 20 tons, #6 or larger, minimum					675			675	740
1500	Maximum					805			805	885

03 21 16 – Epoxy-Coated Reinforcing Steel

03 21 16.10 Epoxy-Coated Reinforcing

		Crew	Daily Output	Labor-Hours	Unit	Material	2007 Bare Costs Labor	Equipment	Total	Total Incl O&P
0010	**EPOXY-COATED REINFORCING**									
0100	Epoxy coated, A775				Ton	380			380	420

03 22 Welded Wire Fabric Reinforcing

03 22 05 – Uncoated Welded Wire Fabric

03 22 05.50 Welded Wire Fabric

		Crew	Daily Output	Labor-Hours	Unit	Material	2007 Bare Costs Labor	Equipment	Total	Total Incl O&P
0010	**WELDED WIRE FABRIC** R032205-30									
0050	Sheets									
0100	6 x 6 - W1.4 x W1.4 (10 x 10) 21 lb. per C.S.F. CN	2 Rodm	35	.457	C.S.F.	12.75	18.90		31.65	45
0200	6 x 6 - W2.1 x W2.1 (8 x 8) 30 lb. per C.S.F.		31	.516		15.40	21.50		36.90	52
0300	6 x 6 - W2.9 x W2.9 (6 x 6) 42 lb. per C.S.F.		29	.552		19.80	23		42.80	59.50
0400	6 x 6 - W4 x W4 (4 x 4) 58 lb. per C.S.F.		27	.593		28	24.50		52.50	70.50
0500	4 x 4 - W1.4 x W1.4 (10 x 10) 31 lb. per C.S.F.		31	.516		18.45	21.50		39.95	55.50
0600	4 x 4 - W2.1 x W2.1 (8 x 8) 44 lb. per C.S.F.		29	.552		23.50	23		46.50	63
0650	4 x 4 - W2.9 x W2.9 (6 x 6) 61 lb. per C.S.F.		27	.593		34	24.50		58.50	77
0700	4 x 4 - W4 x W4 (4 x 4) 85 lb. per C.S.F.		25	.640		44	26.50		70.50	92
0750	Rolls									
0800	2 x 2 - #14 galv., 21 lb/C.S.F., beam & column wrap	2 Rodm	6.50	2.462	C.S.F.	31.50	102		133.50	202
0900	2 x 2 - #12 galv. for gunite reinforcing	"	6.50	2.462	"	37.50	102		139.50	209

03 23 Stressing Tendons

03 23 05 – Stressing Tendons

03 23 05.50 Prestressing Steel

		Crew	Daily Output	Labor-Hours	Unit	Material	2007 Bare Costs Labor	Equipment	Total	Total Incl O&P
0010	**PRESTRESSING STEEL** R034136-90									
0100	Grouted strand, post-tensioned in field, 50' span, 100 kip	C-3	1200	.053	Lb.	1.98	2.02	.08	4.08	5.50
0150	300 kip		2700	.024		.87	.90	.04	1.81	2.43
0300	100' span, 100 kip		1700	.038		1.98	1.43	.06	3.47	4.53
0350	300 kip		3200	.020		1.71	.76	.03	2.50	3.13
0500	200' span, 100 kip		2700	.024		1.98	.90	.04	2.92	3.66
0550	300 kip		3500	.018		1.71	.69	.03	2.43	3.02
0800	Grouted bars, 50' span, 42 kip		2600	.025		.86	.93	.04	1.83	2.49
0850	143 kip		3200	.020		.83	.76	.03	1.62	2.16
1000	75' span, 42 kip		3200	.020		.88	.76	.03	1.67	2.21
1050	143 kip		4200	.015		.75	.58	.02	1.35	1.78
1200	Ungrouted strand, 50' span, 100 kip	C-4	1275	.025		.49	1.05	.02	1.56	2.28
1250	300 kip		1475	.022		.49	.91	.02	1.42	2.05
1400	100' span, 100 kip		1500	.021		.49	.89	.02	1.40	2.02
1450	300 kip		1650	.019		.49	.81	.01	1.31	1.89
1600	200' span, 100 kip		1500	.021		.49	.89	.02	1.40	2.02
1650	300 kip		1700	.019		.49	.79	.01	1.29	1.85
1800	Ungrouted bars, 50' span, 42 kip		1400	.023		.45	.96	.02	1.43	2.08
1850	143 kip		1700	.019		.45	.79	.01	1.25	1.80
2000	75' span, 42 kip		1800	.018		.45	.74	.01	1.20	1.73
2050	143 kip		2200	.015		.45	.61	.01	1.07	1.50
2220	Ungrouted single strand, 100' slab, 25 kip		1200	.027		.49	1.11	.02	1.62	2.39
2250	35 kip		1475	.022		.49	.91	.02	1.42	2.05

03 24 Fibrous Reinforcing

03 24 05 – Fibrous Reinforcing

03 24 05.30 Synthetic Fibers

		Crew	Daily Output	Labor-Hours	Unit	Material	2007 Bare Costs Labor	Equipment	Total	Total Incl O&P
0010	**SYNTHETIC FIBERS**									
0100	Synthetic fibers, add to concrete				Lb.	3.98			3.98	4.38
0110	1-1/2 lb. per C.Y.				C.Y.	6.15			6.15	6.75

03 24 05.70 Steel Fibers

		Crew	Daily Output	Labor-Hours	Unit	Material	2007 Bare Costs Labor	Equipment	Total	Total Incl O&P
0010	**STEEL FIBERS**									
0150	Steel fibers, add to concrete				Lb.	.46			.46	.51
0155	25 lb. per C.Y.				C.Y.	11.50			11.50	12.65
0160	50 lb. per C.Y.					23			23	25.50
0170	75 lb. per C.Y.					35.50			35.50	39
0180	100 lb. per C.Y.					46			46	50.50

03 30 Cast-In-Place Concrete

03 30 53 – Miscellaneous Cast-In-Place Concrete

03 30 53.40 Concrete In Place

		Crew	Daily Output	Labor-Hours	Unit	Material	2007 Bare Costs Labor	Equipment	Total	Total Incl O&P
0010	**CONCRETE IN PLACE**									
0020	Including forms (4 uses), concrete, placement, reinforcing									
0050	steel and finishing unless otherwise indicated R033053-50									
0300	Beams, 5 kip per L.F., 10' span	C-14A	15.62	12.804	C.Y.	310	470	46.50	826.50	1,125
0350	25' span	"	18.55	10.782		325	400	39.50	764.50	1,025
0500	Chimney foundations, industrial, minimum	C-14C	32.22	3.476		148	122	.66	270.66	355
0510	Maximum	"	23.71	4.724		172	166	.90	338.90	450
0700	Columns, square, 12" x 12", minimum reinforcing	C-14A	11.96	16.722		330	615	61	1,006	1,400

03 30 Cast-In-Place Concrete

03 30 53 – Miscellaneous Cast-In-Place Concrete

03 30 53.40 Concrete In Place		Crew	Daily Output	Labor-Hours	Unit	Material	2007 Bare Costs Labor	2007 Bare Costs Equipment	Total	Total Incl O&P
0720	Average reinforcing R033105-85	C-14A	10.13	19.743	C.Y.	515	730	72	1,317	1,800
0740	Maximum reinforcing		9.03	22.148		770	815	81	1,666	2,225
0800	16" x 16", minimum reinforcing		16.22	12.330		266	455	45	766	1,050
0820	Average reinforcing		12.57	15.911		445	585	58	1,088	1,475
0840	Maximum reinforcing		10.25	19.512		680	720	71	1,471	1,950
0900	24" x 24", minimum reinforcing		23.66	8.453		229	310	31	570	775
0920	Average reinforcing		17.71	11.293		400	415	41	856	1,150
0940	Maximum reinforcing		14.15	14.134		625	520	51.50	1,196.50	1,550
1000	36" x 36", minimum reinforcing		33.69	5.936		204	219	21.50	444.50	595
1020	Average reinforcing		23.32	8.576		350	315	31.50	696.50	915
1040	Maximum reinforcing		17.82	11.223		585	415	41	1,041	1,350
1100	Columns, round, tied, 12" diameter, minimum reinforcing		20.97	9.537		294	350	35	679	915
1120	Average reinforcing		15.27	13.098		480	485	48	1,013	1,325
1140	Maximum reinforcing		12.11	16.515		720	610	60.50	1,390.50	1,825
1200	16" diameter, minimum reinforcing		31.49	6.351		266	234	23	523	685
1220	Average reinforcing		19.12	10.460		455	385	38	878	1,150
1240	Maximum reinforcing		13.77	14.524		675	535	53	1,263	1,650
1300	20" diameter, minimum reinforcing		41.04	4.873		266	180	17.80	463.80	595
1320	Average reinforcing		24.05	8.316		435	305	30.50	770.50	995
1340	Maximum reinforcing		17.01	11.758		675	435	43	1,153	1,475
1400	24" diameter, minimum reinforcing		51.85	3.857		251	142	14.10	407.10	515
1420	Average reinforcing		27.06	7.391		440	273	27	740	935
1440	Maximum reinforcing		18.29	10.935		665	405	40	1,110	1,400
1500	36" diameter, minimum reinforcing		75.04	2.665		254	98.50	9.75	362.25	445
1520	Average reinforcing		37.49	5.335		420	197	19.45	636.45	790
1540	Maximum reinforcing		22.84	8.757		650	325	32	1,007	1,250
1900	Elevated slabs, flat slab with drops, 125 psf Sup. Load, 20' span	C-14B	38.45	5.410		265	199	19	483	620
1950	30' span		50.99	4.079		274	150	14.30	438.30	550
2100	Flat plate, 125 psf Sup. Load, 15' span		30.24	6.878		242	253	24	519	690
2150	25' span		49.60	4.194		248	155	14.70	417.70	530
2300	Waffle const., 30" domes, 125 psf Sup. Load, 20' span		37.07	5.611		350	207	19.70	576.70	730
2350	30' span		44.07	4.720		315	174	16.55	505.55	635
2500	One way joists, 30" pans, 125 psf Sup. Load, 15' span		27.38	7.597		425	280	26.50	731.50	940
2550	25' span		31.15	6.677		390	246	23.50	659.50	840
2700	One way beam & slab, 125 psf Sup. Load, 15' span		20.59	10.102		265	370	35.50	670.50	910
2750	25' span		28.36	7.334		246	270	25.50	541.50	720
2900	Two way beam & slab, 125 psf Sup. Load, 15' span		24.04	8.652		253	320	30.50	603.50	815
2950	25' span		35.87	5.799		216	214	20.50	450.50	595
3100	Elevated slabs including finish, not									
3110	including forms or reinforcing									
3150	Regular concrete, 4" slab	C-8	2613	.021	S.F.	1.39	.69	.27	2.35	2.87
3200	6" slab		2585	.022		2.06	.70	.27	3.03	3.63
3250	2-1/2" thick floor fill		2685	.021		.88	.68	.26	1.82	2.29
3300	Lightweight, 110# per C.F., 2-1/2" thick floor fill		2585	.022		1.12	.70	.27	2.09	2.60
3400	Cellular concrete, 1-5/8" fill, under 5000 S.F.		2000	.028		.74	.91	.35	2	2.59
3450	Over 10,000 S.F.		2200	.025		.71	.82	.32	1.85	2.38
3500	Add per floor for 3 to 6 stories high		31800	.002			.06	.02	.08	.11
3520	For 7 to 20 stories high		21200	.003			.09	.03	.12	.17
3540	Equipment pad, 3' x 3' x 6" thick	C-14H	45	1.067	Ea.	42.50	38.50	.48	81.48	108
3550	4' x 4' x 6" thick		30	1.600		64.50	58	.72	123.22	162
3560	5' x 5' x 8" thick		18	2.667		115	97	1.20	213.20	279
3570	6' x 6' x 8" thick		14	3.429		157	124	1.54	282.54	370
3580	8' x 8' x 10" thick		8	6		330	218	2.70	550.70	705

03 30 Cast-In-Place Concrete

03 30 53 – Miscellaneous Cast-In-Place Concrete

03 30 53.40 Concrete In Place		Crew	Daily Output	Labor-Hours	Unit	Material	2007 Bare Costs Labor	Equipment	Total	Total Incl O&P
3590	10' x 10' x 12" thick	C-14H	5	9.600	Ea.	565	350	4.32	919.32	1,175
3800	Footings, spread under 1 C.Y.	C-14C	38.07	2.942	C.Y.	195	103	.56	298.56	375
3850	Over 5 C.Y.		81.04	1.382		266	48.50	.26	314.76	370
3900	Footings, strip, 18" x 9", unreinforced		40	2.800		126	98.50	.53	225.03	293
3920	18" x 9", reinforced		35	3.200		147	112	.61	259.61	340
3925	20" x 10", unreinforced		45	2.489		122	87.50	.47	209.97	273
3930	20" x 10", reinforced		40	2.800		140	98.50	.53	239.03	310
3935	24" x 12", unreinforced		55	2.036		120	71.50	.39	191.89	244
3940	24" x 12", reinforced		48	2.333		138	82	.44	220.44	280
3945	36" x 12", unreinforced		70	1.600		117	56	.30	173.30	216
3950	36" x 12", reinforced		60	1.867		133	65.50	.35	198.85	249
4000	Foundation mat, under 10 C.Y.		38.67	2.896		197	102	.55	299.55	375
4050	Over 20 C.Y.		56.40	1.986		174	70	.38	244.38	300
4200	Grade walls, 8" thick, 8' high	C-14D	45.83	4.364		177	159	15.95	351.95	460
4250	14' high		27.26	7.337		226	268	27	521	700
4260	12" thick, 8' high		64.32	3.109		160	114	11.35	285.35	365
4270	14' high		40.01	4.999		180	183	18.25	381.25	505
4300	15" thick, 8' high		80.02	2.499		151	91.50	9.10	251.60	320
4350	12' high		51.26	3.902		160	143	14.25	317.25	415
4500	18' high		48.85	4.094		176	150	14.95	340.95	445
4520	Handicap access ramp, railing both sides, 3' wide	C-14H	14.58	3.292	L.F.	238	119	1.48	358.48	450
4525	5' wide		12.22	3.928		247	143	1.77	391.77	495
4530	With 6" curb and rails both sides, 3' wide		8.55	5.614		246	204	2.53	452.53	590
4535	5' wide		7.31	6.566		251	238	2.95	491.95	650
4650	Slab on grade, not including finish, 4" thick	C-14E	60.75	1.449	C.Y.	124	52.50	.35	176.85	219
4700	6" thick	"	92	.957	"	120	34.50	.23	154.73	187
4751	Slab on grade, incl. troweled finish, not incl. forms									
4760	or reinforcing, over 10,000 S.F., 4" thick	C-14F	3425	.021	S.F.	1.37	.70	.01	2.08	2.57
4820	6" thick		3350	.021		2.01	.72	.01	2.74	3.30
4840	8" thick		3184	.023		2.75	.76	.01	3.52	4.16
4900	12" thick		2734	.026		4.12	.88	.01	5.01	5.85
4950	15" thick		2505	.029		5.20	.96	.01	6.17	7.15
5000	Slab on grade, incl. textured finish, not incl. forms									
5001	or reinforcing, 4" thick	C-14G	2873	.019	S.F.	1.35	.64	.01	2	2.46
5010	6" thick		2590	.022		2.11	.71	.01	2.83	3.41
5020	8" thick		2320	.024		2.76	.79	.01	3.56	4.23
5200	Lift slab in place above the foundation, incl. forms,									
5210	reinforcing, concrete and columns, minimum	C-14B	2113	.098	S.F.	5.95	3.63	.35	9.93	12.60
5250	Average		1650	.126		6.45	4.65	.44	11.54	14.85
5300	Maximum		1500	.139		6.95	5.10	.49	12.54	16.20
5500	Lightweight, ready mix, including screed finish only,									
5510	not including forms or reinforcing									
5550	1:4 for structural roof decks	C-14B	260	.800	C.Y.	132	29.50	2.81	164.31	195
5600	1:6 for ground slab with radiant heat	C-14F	92	.783		134	26	.23	160.23	186
5650	1:3:2 with sand aggregate, roof deck	C-14B	260	.800		139	29.50	2.81	171.31	202
5700	Ground slab	C-14F	107	.673		139	22.50	.20	161.70	187
5900	Pile caps, incl. forms and reinf., sq. or rect., under 5 C.Y.	C-14C	54.14	2.069		162	72.50	.39	234.89	293
5950	Over 10 C.Y.		75	1.493		154	52.50	.28	206.78	251
6000	Triangular or hexagonal, under 5 C.Y.		53	2.113		123	74.50	.40	197.90	252
6050	Over 10 C.Y.		85	1.318		138	46.50	.25	184.75	225
6200	Retaining walls, gravity, 4' high see division 32 32 13.10	C-14D	66.20	3.021		143	110	11.05	264.05	340
6250	10' high		125	1.600		136	58.50	5.85	200.35	246
6300	Cantilever, level backfill loading, 8' high		70	2.857		156	104	10.45	270.45	345

03 30 Cast-In-Place Concrete

03 30 53 – Miscellaneous Cast-In-Place Concrete

03 30 53.40 Concrete In Place		Crew	Daily Output	Labor-Hours	Unit	Material	2007 Bare Costs Labor	Equipment	Total	Total Incl O&P
6350	16' high	C-14D	91	2.198	C.Y.	149	80.50	8	237.50	298
6800	Stairs, not including safety treads, free standing, 3'-6" wide	C-14H	83	.578	LF Nose	5.70	21	.26	26.96	39
6850	Cast on ground		125	.384	"	4.52	13.95	.17	18.64	26.50
7000	Stair landings, free standing		200	.240	S.F.	4.63	8.70	.11	13.44	18.75
7050	Cast on ground		475	.101	"	3.45	3.67	.05	7.17	9.55

03 31 Structural Concrete

03 31 05 – Normal Weight Structural Concrete

03 31 05.30 Concrete, Field Mix

0010	**CONCRETE, FIELD MIX**	R033105-65								
0015	FOB forms 2250 psi				C.Y.	92.50			92.50	102
0020	3000 psi				"	96.50			96.50	106

03 31 05.35 Normal Weight Concrete, Ready Mix

0010	**NORMAL WEIGHT CONCRETE, READY MIX**	R033105-10								
0012	Includes local aggregate, sand, portland cement, and water									
0015	Excludes all additives and treatments	R033105-20								
0020	2000 psi				C.Y.	99.50			99.50	110
0100	2500 psi	R033105-30				101			101	111
0150	3000 psi	**CN**				104			104	114
0200	3500 psi	R033105-40				106			106	116
0300	4000 psi					108			108	119
0350	4500 psi	R033105-50				110			110	121
0400	5000 psi	**CN**				114			114	125
0411	6000 psi					130			130	143
0412	8000 psi					212			212	233
0413	10,000 psi					300			300	330
0414	12,000 psi					365			365	400
1000	For high early strength cement, add					10%				
1010	For structural lightweight with regular sand, add					25%				
2000	For all lightweight aggregate, add					45%				

03 31 05.70 Placing Concrete

0010	**PLACING CONCRETE**	R033105-70								
0020	Includes labor and equipment to place and vibrate									
0050	Beams, elevated, small beams, pumped		C-20	60	1.067	C.Y.		33	12.50	45.50	65
0100	With crane and bucket		C-7	45	1.600			50	25	75	105
0200	Large beams, pumped		C-20	90	.711			22	8.35	30.35	43
0250	With crane and bucket		C-7	65	1.108			35	17.25	52.25	72.50
0400	Columns, square or round, 12" thick, pumped		C-20	60	1.067			33	12.50	45.50	65
0450	With crane and bucket		C-7	40	1.800			56.50	28	84.50	118
0600	18" thick, pumped		C-20	90	.711			22	8.35	30.35	43
0650	With crane and bucket		C-7	55	1.309			41	20.50	61.50	85.50
0800	24" thick, pumped		C-20	92	.696			21.50	8.15	29.65	42
0850	With crane and bucket		C-7	70	1.029			32.50	16	48.50	67
1000	36" thick, pumped		C-20	140	.457			14.20	5.35	19.55	28
1050	With crane and bucket		C-7	100	.720			22.50	11.20	33.70	47
1400	Elevated slabs, less than 6" thick, pumped		C-20	140	.457			14.20	5.35	19.55	28
1450	With crane and bucket		C-7	95	.758			24	11.80	35.80	49.50
1500	6" to 10" thick, pumped		C-20	160	.400			12.40	4.70	17.10	24.50
1550	With crane and bucket		C-7	110	.655			20.50	10.20	30.70	42.50
1600	Slabs over 10" thick, pumped		C-20	180	.356			11.05	4.17	15.22	21.50

03 31 Structural Concrete

03 31 05 – Normal Weight Structural Concrete

03 31 05.70 Placing Concrete		Crew	Daily Output	Labor-Hours	Unit	Material	2007 Bare Costs Labor	Equipment	Total	Total Incl O&P
1650	With crane and bucket	C-7	130	.554	C.Y.		17.35	8.60	25.95	36
1900	Footings, continuous, shallow, direct chute	C-6	120	.400			12.10	.36	12.46	19
1950	Pumped	C-20	150	.427			13.25	5	18.25	26
2000	With crane and bucket	C-7	90	.800			25	12.45	37.45	52
2100	Footings, continuous, deep, direct chute	C-6	140	.343			10.35	.31	10.66	16.30
2150	Pumped	C-20	160	.400			12.40	4.70	17.10	24.50
2200	With crane and bucket	C-7	110	.655			20.50	10.20	30.70	42.50
2400	Footings, spread, under 1 C.Y., direct chute	C-6	55	.873			26.50	.78	27.28	41.50
2450	Pumped	C-20	65	.985			30.50	11.55	42.05	59.50
2500	With crane and bucket	C-7	45	1.600			50	25	75	105
2600	Over 5 C.Y., direct chute	C-6	120	.400			12.10	.36	12.46	19
2650	Pumped	C-20	150	.427			13.25	5	18.25	26
2700	With crane and bucket	C-7	100	.720			22.50	11.20	33.70	47
2900	Foundation mats, over 20 C.Y., direct chute	C-6	350	.137			4.14	.12	4.26	6.55
2950	Pumped	C-20	400	.160			4.97	1.88	6.85	9.70
3000	With crane and bucket	C-7	300	.240			7.55	3.73	11.28	15.65
3200	Grade beams, direct chute	C-6	150	.320			9.65	.28	9.93	15.20
3250	Pumped	C-20	180	.356			11.05	4.17	15.22	21.50
3300	With crane and bucket	C-7	120	.600			18.80	9.35	28.15	39.50
3500	High rise, for more than 5 stories, pumped, add per story	C-20	2100	.030			.95	.36	1.31	1.84
3510	With crane and bucket, add per story	C-7	2100	.034			1.08	.53	1.61	2.24
3700	Pile caps, under 5 C.Y., direct chute	C-6	90	.533			16.10	.47	16.57	25.50
3750	Pumped	C-20	110	.582			18.05	6.85	24.90	35.50
3800	With crane and bucket	C-7	80	.900			28	14	42	59
3850	Pile cap, 5 C.Y. to 10 C.Y., direct chute	C-6	175	.274			8.30	.24	8.54	13
3900	Pumped	C-20	200	.320			9.95	3.76	13.71	19.40
3950	With crane and bucket	C-7	150	.480			15.05	7.45	22.50	31
4000	Over 10 C.Y., direct chute	C-6	215	.223			6.75	.20	6.95	10.60
4050	Pumped	C-20	240	.267			8.30	3.13	11.43	16.15
4100	With crane and bucket	C-7	185	.389			12.20	6.05	18.25	25.50
4300	Slab on grade, 4" thick, direct chute	C-6	110	.436			13.20	.39	13.59	21
4350	Pumped	C-20	130	.492			15.30	5.80	21.10	30
4400	With crane and bucket	C-7	110	.655			20.50	10.20	30.70	42.50
4600	Over 6" thick, direct chute	C-6	165	.291			8.80	.26	9.06	13.85
4650	Pumped	C-20	185	.346			10.75	4.06	14.81	21
4700	With crane and bucket	C-7	145	.497			15.60	7.75	23.35	32.50
4900	Walls, 8" thick, direct chute	C-6	90	.533			16.10	.47	16.57	25.50
4950	Pumped	C-20	100	.640			19.90	7.50	27.40	39
5000	With crane and bucket	C-7	80	.900			28	14	42	59
5050	12" thick, direct chute	C-6	100	.480			14.50	.43	14.93	23
5100	Pumped	C-20	110	.582			18.05	6.85	24.90	35.50
5200	With crane and bucket	C-7	90	.800			25	12.45	37.45	52
5300	15" thick, direct chute	C-6	105	.457			13.80	.41	14.21	22
5350	Pumped	C-20	120	.533			16.55	6.25	22.80	32.50
5400	With crane and bucket	C-7	95	.758			24	11.80	35.80	49.50
5600	Wheeled concrete dumping, add to placing costs above									
5610	Walking cart, 50' haul, add	C-18	32	.281	C.Y.		8.15	1.67	9.82	14.55
5620	150' haul, add	↓	24	.375			10.85	2.22	13.07	19.35
5700	250' haul, add	↓	18	.500			14.50	2.97	17.47	26
5800	Riding cart, 50' haul, add	C-19	80	.113			3.26	.99	4.25	6.15
5810	150' haul, add	↓	60	.150			4.35	1.31	5.66	8.20
5900	250' haul, add	↓	45	.200	↓		5.80	1.75	7.55	10.95

03 35 Concrete Finishing

03 35 29 – Tooled Concrete Finishing

03 35 29.30 Finishing Floors

		Crew	Daily Output	Labor-Hours	Unit	Material	2007 Bare Costs Labor	Equipment	Total	Total Incl O&P
0010	**FINISHING FLOORS**									
0020	Monolithic, screed finish	1 Cefi	900	.009	S.F.		.32		.32	.46
0100	Screed and bull float (darby) finish		725	.011			.39		.39	.58
0150	Screed, float, and broom finish		630	.013			.45		.45	.66
0200	Screed, float, and hand trowel		600	.013			.47		.47	.70
0250	Machine trowel		550	.015			.52		.52	.76
0400	Integral topping and finish, using 1:1:2 mix, 3/16" thick	C-10B	1000	.040		.07	1.26	.19	1.52	2.20
0450	1/2" thick		950	.042		.18	1.33	.20	1.71	2.43
0500	3/4" thick		850	.047		.28	1.48	.22	1.98	2.80
0600	1" thick		750	.053		.37	1.68	.25	2.30	3.23
0800	Granolithic topping, laid after, 1:1:1-1/2 mix, 1/2" thick		590	.068		.21	2.13	.32	2.66	3.82
0820	3/4" thick		580	.069		.31	2.17	.33	2.81	3.99
0850	1" thick		575	.070		.41	2.19	.33	2.93	4.13
0950	2" thick		500	.080		.82	2.52	.38	3.72	5.15
1200	Heavy duty, 1:1:2, 3/4" thick, preshrunk, gray, 20 MSF		320	.125		.28	3.93	.60	4.81	6.90
1300	100 MSF		380	.105		.28	3.31	.50	4.09	5.85
1600	Exposed local aggregate finish, minimum	1 Cefi	625	.013		.18	.46		.64	.87
1650	Maximum		465	.017		.26	.61		.87	1.19
1800	Floor abrasives, .25 psf, aluminum oxide		850	.009		.35	.33		.68	.88
1850	Silicon carbide		850	.009		.49	.33		.82	1.03
2000	Floor hardeners, metallic, light service, .50 psf, add		850	.009		.45	.33		.78	.99
2050	Medium service, .75 psf		750	.011		.68	.38		1.06	1.31
2100	Heavy service, 1.0 psf		650	.012		.91	.44		1.35	1.64
2150	Extra heavy, 1.5 psf		575	.014		1.36	.49		1.85	2.22
2300	Non-metallic, light service, .50 psf		850	.009		.17	.33		.50	.68
2350	Medium service, .75 psf		750	.011		.26	.38		.64	.85
2400	Heavy service, 1.00 psf		650	.012		.35	.44		.79	1.02
2450	Extra heavy, 1.50 psf		575	.014		.52	.49		1.01	1.30
2800	Trap rock wearing surface for monolithic floors									
2810	2.0 psf	C-10B	1250	.032	S.F.	.02	1.01	.15	1.18	1.72
3000	Floor coloring, dusted on, minimum (0.6 psf), add to above	1 Cefi	1300	.006		.39	.22		.61	.75
3050	Maximum (1.0 psf), add to above	"	625	.013		.65	.46		1.11	1.39
3100	Colored powder only				Lb.	.65			.65	.72
3600	1/2" topping using 0.6 psf powdered color	C-10B	590	.068	S.F.	4.95	2.13	.32	7.40	9.05
3650	1/2" topping using 1.0 psf powdered color	"	590	.068		5.20	2.13	.32	7.65	9.35
3800	Dustproofing, solvent-based, 1 coat	1 Cefi	1900	.004		.16	.15		.31	.39
3850	2 coats		1300	.006		.56	.22		.78	.94
4000	Epoxy-based, 1 coat		1500	.005		.12	.19		.31	.41
4050	2 coats		1500	.005		.25	.19		.44	.55
4400	Stair finish, float		275	.029			1.03		1.03	1.52
4500	Steel trowel finish		200	.040			1.42		1.42	2.09
4600	Silicon carbide finish, .25 psf		150	.053		.35	1.90		2.25	3.17

03 35 29.35 Control Joints, Saw Cut

		Crew	Daily Output	Labor-Hours	Unit	Material	Labor	Equipment	Total	Total Incl O&P
0010	**CONTROL JOINTS, SAW CUT**									
0100	Sawcut in green concrete									
0120	1" depth	C-27	2000	.008	L.F.	.09	.28	.06	.43	.59
0140	1-1/2" depth		1800	.009		.14	.32	.07	.53	.68
0160	2" depth		1600	.010		.18	.36	.07	.61	.80
0200	Clean out control joint of debris	C-28	6000	.001			.05		.05	.07
0300	Joint sealant									
0320	Backer rod, polyethylene, 1/4" diameter	1 Cefi	460	.017	L.F.	.04	.62		.66	.96
0340	Sealant, polyurethane									

03 35 Concrete Finishing

03 35 29 – Tooled Concrete Finishing

03 35 29.35 Control Joints, Saw Cut		Crew	Daily Output	Labor-Hours	Unit	Material	2007 Bare Costs Labor	Equipment	Total	Total Incl O&P
0360	1/4" x 1/4" (308 LF/Gal)	1 Cefi	270	.030	L.F.	.17	1.05		1.22	1.74
0380	1/4" x 1/2" (154 LF/Gal)	"	255	.031	"	.34	1.12		1.46	2.01

03 35 29.60 Finishing Walls

		Crew	Daily Output	Labor-Hours	Unit	Material	Labor	Equipment	Total	Total Incl O&P
0010	**FINISHING WALLS**									
0020	Break ties and patch voids	1 Cefi	540	.015	S.F.	.03	.53		.56	.80
0050	Burlap rub with grout		450	.018		.03	.63		.66	.96
0100	Carborundum rub, dry		270	.030			1.05		1.05	1.55
0150	Wet rub		175	.046			1.63		1.63	2.39
0300	Bush hammer, green concrete	B-39	1000	.048			1.46	.18	1.64	2.46
0350	Cured concrete	"	650	.074			2.25	.28	2.53	3.78
0600	Float finish, 1/16" thick	1 Cefi	300	.027		.25	.95		1.20	1.67
0700	Sandblast, light penetration	E-11	1100	.029		.25	.96	.16	1.37	2.08
0750	Heavy penetration	"	375	.085		.51	2.83	.48	3.82	5.85
0850	Grind form fins flush	1 Clab	700	.011	L.F.		.33		.33	.51

03 35 33 – Stamped Concrete Finishing

03 35 33.50 Slab Texture Stamping

		Crew	Daily Output	Labor-Hours	Unit	Material	Labor	Equipment	Total	Total Incl O&P
0010	**SLAB TEXTURE STAMPING**									
0020	Approx. 3 S.F.- 5 S.F. each, buy, minimum				Ea.	75			75	82.50
0030	Average					138			138	152
0120	Maximum					200			200	220
0200	Commonly used chemicals for texture systems									
0210	Hardener, colored powder				S.F.	.55			.55	.61
0220	Release agent, colored powder					.09			.09	.10
0230	Curing & sealing compound, solvent based					.05			.05	.06
0300	Broadcasting hardener & release agent, stamping	2 Cefi	1000	.016			.57		.57	.84

03 37 Specialty Placed Concrete

03 37 13 – Shotcrete

03 37 13.30 Gunite (Dry-Mix)

		Crew	Daily Output	Labor-Hours	Unit	Material	Labor	Equipment	Total	Total Incl O&P
0010	**GUNITE (DRY-MIX)**									
0020	Applied in 1" layers, no mesh included	C-8	2000	.028	S.F.	.29	.91	.35	1.55	2.09
0100	Mesh for gunite 2 x 2, #12, to 3" thick	2 Rodm	800	.020		.38	.83		1.21	1.77
0150	Over 3" thick	"	500	.032		.38	1.32		1.70	2.58
0300	Typical in place, including mesh, 2" thick, minimum	C-16	1000	.072		.95	2.47	.71	4.13	5.65
0350	Maximum		500	.144		.95	4.95	1.42	7.32	10.30
0500	4" thick, minimum		750	.096		1.53	3.30	.94	5.77	7.80
0550	Maximum		350	.206		1.53	7.05	2.02	10.60	14.85
0900	Prepare old walls, no scaffolding, minimum	C-10	1000	.024			.80		.80	1.19
0950	Maximum	"	275	.087			2.90		2.90	4.34
1100	For high finish requirement or close tolerance, add, minimum						50%			
1150	Maximum						110%			

03 39 Concrete Curing

03 39 13 – Water Concrete Curing

03 39 13.50 Water Curing

		Crew	Daily Output	Labor-Hours	Unit	Material	2007 Bare Costs Labor	Equipment	Total	Total Incl O&P
0010	**WATER CURING**									
0015	With burlap, 4 uses assumed, 7.5 oz.	2 Clab	55	.291	C.S.F.	7.25	8.35		15.60	21
0100	10 oz.	"	55	.291	"	13.10	8.35		21.45	27.50
0400	Curing blankets, 1" to 2" thick, buy, minimum				S.F.	.33			.33	.36
0450	Maximum				"	.50			.50	.54

03 39 23 – Membrane Concrete Curing

03 39 23.13 Chemical Compound Membrane Concrete Curing

		Crew	Daily Output	Labor-Hours	Unit	Material	Labor	Equipment	Total	Total Incl O&P
0010	**CHEMICAL COMPOUND MEMBRANE CONCRETE CURING**									
0300	Sprayed membrane curing compound	2 Clab	95	.168	C.S.F.	4.69	4.84		9.53	12.70
0700	Curing compound, solvent based, 400 S.F./gal, 55 gal. lots				Gal.	12.05			12.05	13.25
0720	5 gallon lots					14.40			14.40	15.85
0800	Water based, 250 S.F./gal, 55 gallon lots					14.20			14.20	15.65
0820	5 gallon lots					17.30			17.30	19

03 39 23.23 Sheet Membrane Concrete Curing

		Crew	Daily Output	Labor-Hours	Unit	Material	Labor	Equipment	Total	Total Incl O&P
0010	**SHEET MEMBRANE CONCRETE CURING**									
0200	Curing blanket, burlap/poly, 2-ply	2 Clab	70	.229	C.S.F.	16.70	6.55		23.25	28.50

03 41 Precast Structural Concrete

03 41 05 – Precast Concrete Members

03 41 05.10 Precast Beams

		Crew	Daily Output	Labor-Hours	Unit	Material	Labor	Equipment	Total	Total Incl O&P
0010	**PRECAST BEAMS** R034105-30									
0011	L-shaped, 20' span, 12" x 20"	C-11	32	2.250	Ea.	1,175	91.50	58	1,324.50	1,500
1000	Inverted tee beams, add to above, small beams					15%				
1050	Large beams					20%				
1200	Rectangular, 20' span, 12" x 20"	C-11	32	2.250		865	91.50	58	1,014.50	1,175
1250	18" x 36"		24	3		980	122	77	1,179	1,375
1300	24" x 44"		22	3.273		1,225	133	84	1,442	1,675
1400	30' span, 12" x 36"		24	3		1,325	122	77	1,524	1,750
1450	18" x 44"		20	3.600		1,650	146	92.50	1,888.50	2,175
1500	24" x 52"		16	4.500		2,150	183	116	2,449	2,800
1600	40' span, 12" x 52"		20	3.600		2,450	146	92.50	2,688.50	3,050
1650	18" x 52"		16	4.500		2,925	183	116	3,224	3,650
1700	24" x 52"		12	6		3,200	243	154	3,597	4,125
2000	"T" shaped, 20' span, 12" x 20"		32	2.250		1,475	91.50	58	1,624.50	1,850
2050	18" x 36"		24	3		1,675	122	77	1,874	2,125
2100	24" x 44"		22	3.273		2,075	133	84	2,292	2,625
2200	30' span, 12" x 36"		24	3		2,250	122	77	2,449	2,775
2250	18" x 44"		20	3.600		2,825	146	92.50	3,063.50	3,450
2300	24" x 52"		16	4.500		3,625	183	116	3,924	4,450
2500	40' span, 12" x 52"		20	3.600		4,175	146	92.50	4,413.50	4,925
2550	18" x 52"		16	4.500		4,950	183	116	5,249	5,900
2600	24" x 52"		12	6		5,425	243	154	5,822	6,575

03 41 05.15 Precast Columns

		Crew	Daily Output	Labor-Hours	Unit	Material	Labor	Equipment	Total	Total Incl O&P
0010	**PRECAST COLUMNS** R034105-30									
0020	Rectangular to 12' high, small columns	C-11	120	.600	L.F.	47	24.50	15.45	86.95	111
0050	Large columns		96	.750		82	30.50	19.30	131.80	165
0300	24' high, small columns		192	.375		47	15.20	9.65	71.85	88.50
0350	Large columns		144	.500		82	20.50	12.85	115.35	140

03 41 05.25 Precast Joists

		Crew	Daily Output	Labor-Hours	Unit	Material	Labor	Equipment	Total	Total Incl O&P
0010	**PRECAST JOISTS** R034105-30									

03 41 Precast Structural Concrete

03 41 05 – Precast Concrete Members

03 41 05.25 Precast Joists

		Crew	Daily Output	Labor-Hours	Unit	Material	2007 Bare Costs Labor	Equipment	Total	Total Incl O&P
0015	40 psf L.L., 6" deep for 12' spans	C-12	600	.080	L.F.	7.80	2.90	1.21	11.91	14.40
0050	8" deep for 16' spans		575	.083		13	3.02	1.26	17.28	20.50
0100	10" deep for 20' spans		550	.087		23	3.16	1.32	27.48	31.50
0150	12" deep for 24' spans	↓	525	.091	↓	31	3.31	1.38	35.69	41

03 41 13 – Precast Concrete Hollow Core Planks

03 41 13.50 Precast Slab Planks

			Crew	Daily Output	Labor-Hours	Unit	Material	Labor	Equipment	Total	Total Incl O&P
0010	**PRECAST SLAB PLANKS**	R034105-30									
0020	Prestressed roof/floor members, grouted, solid, 4" thick		C-11	2400	.030	S.F.	4.89	1.22	.77	6.88	8.40
0050	6" thick			2800	.026		6.30	1.04	.66	8	9.50
0100	Hollow, 8" thick	CN		3200	.023		6.70	.91	.58	8.19	9.60
0150	10" thick			3600	.020		7	.81	.51	8.32	9.70
0200	12" thick		↓	4000	.018	↓	7.95	.73	.46	9.14	10.55

03 41 16 – Precast Concrete Slabs

03 41 16.20 Precast Concrete Channel Slabs

		Crew	Daily Output	Labor-Hours	Unit	Material	Labor	Equipment	Total	Total Incl O&P
0010	**PRECAST CONCRETE CHANNEL SLABS**									
0020	Lightweight channel slab, 2.75" or 3.5" thk, straight	C-12	1575	.030	S.F.	5.15	1.10	.46	6.71	7.85
0050	Chopped up		785	.061		5.65	2.22	.92	8.79	10.65
0200	6" thick, span to 20'		1300	.037		5.95	1.34	.56	7.85	9.25
0300	8" thick, span to 24'	↓	1100	.044	↓	6.85	1.58	.66	9.09	10.70

03 41 16.50 Precast Lightweight Concrete Plank

		Crew	Daily Output	Labor-Hours	Unit	Material	Labor	Equipment	Total	Total Incl O&P
0010	**PRECAST LIGHTWEIGHT CONCRETE PLANK**									
0015	Lightweight plank, nailable, T&G, 2" thick	C-12	1800	.027	S.F.	3.46	.97	.40	4.83	5.75
0050	2-3/4" thick		1575	.030		3.48	1.10	.46	5.04	6.05
0100	3-3/4" thick	↓	1375	.035		4.77	1.26	.53	6.56	7.80
0150	For premium ceiling finish, add				↓	10%				
0200	For sloping roofs, slope over 4 in 12, add						25%			
0250	Slope over 6 in 12, add						150%			

03 41 23 – Precast Concrete Stairs

03 41 23.50 Precast Stairs

		Crew	Daily Output	Labor-Hours	Unit	Material	Labor	Equipment	Total	Total Incl O&P
0010	**PRECAST STAIRS**									
0020	Precast concrete treads on steel stringers, 3' wide	C-12	75	.640	Riser	107	23	9.65	139.65	164
0300	Front entrance, 5' wide with 48" platform, 2 risers		16	3	Flight	340	109	45	494	595
0350	5 risers		12	4		535	145	60.50	740.50	880
0500	6' wide, 2 risers		15	3.200		365	116	48.50	529.50	630
0550	5 risers		11	4.364		550	158	66	774	925
0700	7' wide, 2 risers		14	3.429		500	124	51.50	675.50	800
0750	5 risers	↓	10	4.800		855	174	72.50	1,101.50	1,300
1200	Basement entrance stairs, steel bulkhead doors, minimum	B-51	22	2.182		1,350	63.50	6.45	1,419.95	1,575
1250	Maximum	"	11	4.364	↓	2,175	127	12.85	2,314.85	2,600

03 41 33 – Precast Structural Pretensioned Concrete

03 41 33.60 Tees

			Crew	Daily Output	Labor-Hours	Unit	Material	Labor	Equipment	Total	Total Incl O&P
0010	**TEES**	R034105-30									
0020	Quad tee, short spans, roof		C-11	7200	.010	S.F.	6.25	.41	.26	6.92	7.90
0050	Floor			7200	.010		6.25	.41	.26	6.92	7.90
0200	Double tee, floor members, 60' span			8400	.009		7.80	.35	.22	8.37	9.45
0250	80' span			8000	.009		9.15	.37	.23	9.75	10.95
0300	Roof members, 30' span			4800	.015		6.60	.61	.39	7.60	8.75
0350	50' span			6400	.011		7.55	.46	.29	8.30	9.40
0400	Wall members, up to 55' high			3600	.020		10.55	.81	.51	11.87	13.60
0500	Single tee roof members, 40' span		↓	3200	.023	↓	8.35	.91	.58	9.84	11.45

03 41 Precast Structural Concrete

03 41 33 – Precast Structural Pretensioned Concrete

03 41 33.60 Tees

		Crew	Daily Output	Labor-Hours	Unit	Material	2007 Bare Costs Labor	Equipment	Total	Total Incl O&P
0550	80' span	C-11	5120	.014	S.F.	10.25	.57	.36	11.18	12.65
0600	100' span		6000	.012		15.40	.49	.31	16.20	18.15
0650	120' span		6000	.012		16.80	.49	.31	17.60	19.65
1000	Double tees, floor members									
1100	Lightweight, 20" x 8' wide, 45' span	C-11	20	3.600	Ea.	2,675	146	92.50	2,913.50	3,275
1150	24" x 8' wide, 50' span		18	4		2,900	162	103	3,165	3,600
1200	32" x 10' wide, 60' span		16	4.500		4,900	183	116	5,199	5,850
1250	Standard weight, 12" x 8' wide, 20' span		22	3.273		1,000	133	84	1,217	1,425
1300	16" x 8' wide, 25' span		20	3.600		1,325	146	92.50	1,563.50	1,825
1350	18" x 8' wide, 30' span **CN**		20	3.600		1,700	146	92.50	1,938.50	2,200
1400	20" x 8' wide, 45' span		18	4		1,900	162	103	2,165	2,500
1450	24" x 8' wide, 50' span		16	4.500		2,450	183	116	2,749	3,150
1500	32" x 10' wide, 60' span		14	5.143		4,500	209	132	4,841	5,450
2000	Roof members									
2050	Lightweight, 20" x 8' wide, 40' span	C-11	20	3.600	Ea.	2,250	146	92.50	2,488.50	2,825
2100	24" x 8' wide, 50' span		18	4		3,000	162	103	3,265	3,700
2150	32" x 10' wide, 60' span		16	4.500		4,950	183	116	5,249	5,900
2200	Standard weight, 12" x 8' wide, 30' span		22	3.273		1,500	133	84	1,717	1,975
2250	16" x 8' wide, 30' span		20	3.600		1,575	146	92.50	1,813.50	2,075
2300	18" x 8' wide, 30' span		20	3.600		1,750	146	92.50	1,988.50	2,275
2350	20" x 8' wide, 40' span		18	4		1,800	162	103	2,065	2,375
2400	24" x 8' wide, 50' span		16	4.500		2,400	183	116	2,699	3,100
2450	32" x 10' wide, 60' span		14	5.143		4,200	209	132	4,541	5,125

03 41 36 – Precast Structural Post-Tensioned Concrete

03 41 36.50 Post-Tensioned Jobs

			Crew	Daily Output	Labor-Hours	Unit	Material	Labor	Equipment	Total	Total Incl O&P
0010	**POST-TENSIONED JOBS**	R034105-30									
0100	Post-tensioned in place, small job	R034136-90	C-17B	8.50	9.647	C.Y.	630	370	36.50	1,036.50	1,300
0200	Large job		"	10	8.200	"	475	315	31	821	1,050

03 45 Precast Architectural Concrete

03 45 13 – Faced Architectural Precast Concrete

03 45 13.50 Precast Wall Panels

			Crew	Daily Output	Labor-Hours	Unit	Material	Labor	Equipment	Total	Total Incl O&P
0010	**PRECAST WALL PANELS**	R034513-10									
0050	Uninsulated 4" thick, smooth gray										
0150	Low rise, 4' x 8' x 4" thick		C-11	320	.225	S.F.	13.70	9.15	5.80	28.65	37.50
0200	8' x 8' x 4" thick			576	.125		13.55	5.05	3.22	21.82	27.50
0250	8' x 16' x 4" thick			1024	.070		13.45	2.85	1.81	18.11	22
0400	8' x 8', 4" thick, smooth gray			576	.125		13.55	5.05	3.22	21.82	27.50
0500	Exposed aggregate			576	.125		18.20	5.05	3.22	26.47	32.50
0600	High rise, 4' x 8' x 4" thick			288	.250		13.70	10.15	6.45	30.30	40
0650	8' x 8' x 4" thick			512	.141		13.55	5.70	3.62	22.87	29
0700	8' x 16' x 4" thick			768	.094		13.45	3.80	2.41	19.66	24
0800	Insulated panel, 2" polystyrene, add						.99			.99	1.09
0850	2" urethane, add						.87			.87	.96
1000	20' x 10', 6" thick, smooth gray **CN**		C-11	1400	.051		23	2.09	1.32	26.41	30.50
1100	Exposed aggregate		"	1800	.040		27.50	1.62	1.03	30.15	34
1200	Finishes, white, add						2.44			2.44	2.69
1250	Exposed aggregate, add						1.07			1.07	1.18
1300	Granite faced, domestic, add						26			26	28.50
2200	Fiberglass reinforced cement with urethane core										

03 45 Precast Architectural Concrete

03 45 13 – Faced Architectural Precast Concrete

03 45 13.50 Precast Wall Panels

		Crew	Daily Output	Labor-Hours	Unit	Material	2007 Bare Costs Labor	Equipment	Total	Total Incl O&P
2210	R20, 8' x 8', minimum	E-2	750	.075	S.F.	18.70	3.01	2.06	23.77	28
2220	Maximum	"	600	.093	"	21	3.77	2.58	27.35	32.50

03 47 Site-Cast Concrete

03 47 13 – Tilt-Up Concrete

03 47 13.50 Tilt-Up Wall Panels

			Crew	Daily Output	Labor-Hours	Unit	Material	Labor	Equipment	Total	Total Incl O&P
0010	**TILT-UP WALL PANELS**	R034713-20									
0015	Wall panel construction, walls only, 5-1/2" thick		C-14	1600	.090	S.F.	4.89	3.24	.76	8.89	11.25
0100	7-1/2" thick			1550	.093		6.15	3.34	.78	10.27	12.80
0500	Walls and columns, 5-1/2" thick walls, 12" x 12" columns			1565	.092		7.45	3.31	.78	11.54	14.20
0550	7-1/2" thick wall, 12" x 12" columns			1370	.105		9.20	3.78	.89	13.87	16.95
0800	Columns only, site precast, minimum			200	.720	L.F.	12.40	26	6.05	44.45	61
0850	Maximum			105	1.371	"	15.65	49.50	11.55	76.70	107

03 48 Precast Concrete Specialties

03 48 43 – Precast Concrete Trim

03 48 43.40 Precast Lintels

		Crew	Daily Output	Labor-Hours	Unit	Material	Labor	Equipment	Total	Total Incl O&P
0010	**PRECAST LINTELS**									
0800	Precast concrete, 4" wide, 8" high, to 5' long	D-10	28	1.429	Ea.	29	50	20.50	99.50	131
0850	5'-12' long		24	1.667		74	58.50	24	156.50	196
1000	6" wide, 8" high, to 5' long		26	1.538		40.50	54	22	116.50	152
1050	5'-12' long		22	1.818		98.50	63.50	26	188	234
1200	8" wide, 8" high, to 5' long		24	1.667		39	58.50	24	121.50	158
1250	5'-12' long		20	2		101	70	29	200	249
1400	10" wide, 8" high U-Shape, to 14' long		18	2.222		176	78	32	286	345
1450	12" wide, 8" high U-Shape, to 19' long		16	2.500		230	87.50	36	353.50	425

03 48 43.90 Precast Window Sills

		Crew	Daily Output	Labor-Hours	Unit	Material	Labor	Equipment	Total	Total Incl O&P
0010	**PRECAST WINDOW SILLS**									
0600	Precast concrete, 4" tapers to 3", 9" wide	D-1	70	.229	L.F.	11.75	7.60		19.35	24.50
0650	11" wide		60	.267		14.25	8.90		23.15	29
0700	13" wide, 3 1/2" tapers to 2 1/2", 12" wall		50	.320		15	10.65		25.65	33

03 51 Cast Roof Decks

03 51 13 – Cementitious Wood Fiber Decks

03 51 13.50 Cementitious/Wood Fiber Planks

			Crew	Daily Output	Labor-Hours	Unit	Material	Labor	Equipment	Total	Total Incl O&P
0010	**CEMENTITIOUS/WOOD FIBER PLANKS**	R051223-50									
0050	Plank, beveled, 1" thick		2 Carp	1000	.016	S.F.	2.02	.59		2.61	3.13
0100	Plank, T & G, 1-1/2" thick			975	.016		2.16	.60		2.76	3.32
0150	2" thick			950	.017		2.80	.62		3.42	4.04
0200	2-1/2" thick			925	.017		3.12	.63		3.75	4.42
0250	3" thick			900	.018		3.58	.65		4.23	4.96
0300	3-1/2" thick			875	.018		5.30	.67		5.97	6.90
0350	4" thick			850	.019		5.80	.69		6.49	7.50
1000	Bulb tee, sub-purlin and grout, 6' span, add		E-1	5000	.005		1.75	.19	.02	1.96	2.28
1100	8' span		"	4200	.006		1.98	.23	.03	2.24	2.61

03 52 Lightweight Concrete Roof Insulation

03 52 16 – Lightweight Insulating Concrete

03 52 16.13 Lightweight Cellular Insulating Concrete

		Crew	Daily Output	Labor-Hours	Unit	Material	2007 Bare Costs Labor	Equipment	Total	Total Incl O&P
0010	**LIGHTWEIGHT CELLULAR INSULATING CONCRETE** R035216-10									
0020	Portland cement and foaming agent	C-8	50	1.120	C.Y.	97.50	36	14.15	147.65	178

03 52 16.16 Lightweight Aggregate Insulating Concrete

		Crew	Daily Output	Labor-Hours	Unit	Material	Labor	Equipment	Total	Total Incl O&P
0010	**LIGHTWEIGHT AGGREGATE INSULATING CONCRETE** R035216-10									
0100	Poured vermiculite or perlite, field mix,									
0110	1:6 field mix	C-8	50	1.120	C.Y.	91.50	36	14.15	141.65	171
0200	Ready mix, 1:6 mix, roof fill, 2" thick		10000	.006	S.F.	.51	.18	.07	.76	.92
0250	3" thick		7700	.007		.76	.24	.09	1.09	1.30
0400	Expanded volcanic glass rock, with binder, minimum	2 Carp	1500	.011		.34	.39		.73	.98
0450	Maximum	"	1200	.013		1.01	.49		1.50	1.88

03 54 Cast Underlayment

03 54 13 – Gypsum Cement Underlayment

03 54 13.50 Gypsum Underlayment

		Crew	Daily Output	Labor-Hours	Unit	Material	Labor	Equipment	Total	Total Incl O&P
0010	**GYPSUM UNDERLAYMENT**									
1000	Poured gypsum, 2" thick, add to formboard above	C-8	6000	.009	S.F.	1.67	.30	.12	2.09	2.43
1100	3" thick	"	4800	.012	"	2.50	.38	.15	3.03	3.48

03 54 16 – Hydraulic Cement Underlayment

03 54 16.50 Cement Underlayment

		Crew	Daily Output	Labor-Hours	Unit	Material	Labor	Equipment	Total	Total Incl O&P
0010	**CEMENT UNDERLAYMENT**									
2510	Underlayment, P.C based self-leveling, 4100 psi, pumped, 1/4"	C-8	20000	.003	S.F.	1.46	.09	.04	1.59	1.79
2520	1/2"		19000	.003		2.92	.10	.04	3.06	3.40
2530	3/4"		18000	.003		4.38	.10	.04	4.52	5
2540	1"		17000	.003		5.85	.11	.04	6	6.60
2550	1-1/2"		15000	.004		8.75	.12	.05	8.92	9.90
2560	Hand mix, 1/2"	C-18	4000	.002		2.92	.07	.01	3	3.32
2610	Topping, P.C. based self-level/dry 6100 psi, pumped, 1/4"	C-8	20000	.003		2.28	.09	.04	2.41	2.69
2620	1/2"		19000	.003		4.56	.10	.04	4.70	5.20
2630	3/4"		18000	.003		6.85	.10	.04	6.99	7.70
2660	1"		17000	.003		9.10	.11	.04	9.25	10.25
2670	1-1/2"		15000	.004		13.70	.12	.05	13.87	15.30
2680	Hand mix, 1/2"	C-18	4000	.002		4.56	.07	.01	4.64	5.10

03 62 Non-Shrink Grouting

03 62 13 – Non-Metallic Non-Shrink Grouting

03 62 13.50 Grout, Non-Metallic Non-shrink

		Crew	Daily Output	Labor-Hours	Unit	Material	Labor	Equipment	Total	Total Incl O&P
0010	**GROUT, NON-METALLIC NON-SHRINK**									
0300	Non-shrink, non-metallic, 1" deep	1 Cefi	35	.229	S.F.	5.95	8.15		14.10	18.50
0350	2" deep	"	25	.320	"	11.90	11.40		23.30	30

03 62 16 – Metallic Non-Shrink Grouting

03 62 16.50 Grout, Metallic Non-Shrink

		Crew	Daily Output	Labor-Hours	Unit	Material	Labor	Equipment	Total	Total Incl O&P
0010	**GROUT, METALLIC NON-SHRINK**									
0020	Column & machine bases, non-shrink, metallic, 1" deep	1 Cefi	35	.229	S.F.	7.15	8.15		15.30	19.85
0050	2" deep	"	25	.320	"	14.35	11.40		25.75	32.50

03 63 Epoxy Grouting

03 63 05 – Grouting of Dowels and Fasteners

03 63 05.10 Epoxy Only

		Crew	Daily Output	Labor-Hours	Unit	Material	2007 Bare Costs Labor	2007 Bare Costs Equipment	Total	Total Incl O&P
0010	**EPOXY ONLY**									
1500	Chemical anchoring, epoxy cartridge, excludes layout, drilling, fastener									
1530	For fastener 3/4" dia x 6" embedment	B-89A	27	.593	Ea.	5.55	19.80	3.89	29.24	41.50
1535	1" dia x 8" embedment		24	.667		8.35	22.50	4.37	35.22	48.50
1540	1-1/4" dia x 10" embedment		21	.762		16.65	25.50	5	47.15	63.50
1545	1-3/4" dia x 12" embedment		20	.800		28	26.50	5.25	59.75	78
1550	14" embedment		17	.941		33.50	31.50	6.15	71.15	92.50
1555	2" dia x 12" embedment		16	1		44.50	33.50	6.55	84.55	108
1560	18" embedment		15	1.067		55.50	35.50	7	98	124

03 81 Concrete Cutting

03 81 13 – Flat Concrete Sawing

03 81 13.50 Concrete Floor/Slab Cutting

		Crew	Daily Output	Labor-Hours	Unit	Material	Labor	Equipment	Total	Incl O&P
0010	**CONCRETE FLOOR/SLAB CUTTING**									
0400	Concrete slabs, mesh reinforcing, up to 3" deep	B-89	980	.016	L.F.	.41	.53	.33	1.27	1.62
0420	Each additional inch of depth	"	1600	.010	"	.14	.33	.20	.67	.87

03 81 16 – Track Mounted Concrete Wall Sawing

03 81 16.50 Concrete Wall Cutting

		Crew	Daily Output	Labor-Hours	Unit	Material	Labor	Equipment	Total	Incl O&P
0010	**CONCRETE WALL CUTTING**									
0800	Concrete walls, hydraulic saw, plain, per inch of depth	B-89B	250	.064	L.F.	.37	2.09	2.34	4.80	6.20
0820	Rod reinforcing, per inch of depth	"	150	.107	"	.51	3.49	3.90	7.90	10.15

03 82 Concrete Boring

03 82 13 – Concrete Core Drilling

03 82 13.10 Core Drilling

		Crew	Daily Output	Labor-Hours	Unit	Material	Labor	Equipment	Total	Incl O&P
0010	**CORE DRILLING**									
0020	Reinf. conc slab, up to 6" thick, incl. bit, layout & set up									
0100	1" diameter core	B-89A	28	.571	Ea.	2.80	19.05	3.75	25.60	36.50
0150	Each added inch thick, add		300	.053		.50	1.78	.35	2.63	3.71
0300	3" diameter core		23	.696		6.20	23	4.56	33.76	48
0350	Each added inch thick, add		186	.086		1.12	2.87	.56	4.55	6.30
0500	4" diameter core		19	.842		6.20	28	5.50	39.70	57
0550	Each added inch thick, add		170	.094		1.41	3.14	.62	5.17	7.10
0700	6" diameter core		14	1.143		10.25	38	7.50	55.75	79
0750	Each added inch thick, add		140	.114		1.74	3.82	.75	6.31	8.70
0900	8" diameter core		11	1.455		14	48.50	9.55	72.05	101
0950	Each added inch thick, add		95	.168		2.35	5.60	1.10	9.05	12.55
1100	10" diameter core		10	1.600		18.70	53.50	10.50	82.70	115
1150	Each added inch thick, add		80	.200		3.09	6.70	1.31	11.10	15.25
1300	12" diameter core		9	1.778		22.50	59.50	11.65	93.65	130
1350	Each added inch thick, add		68	.235		3.70	7.85	1.54	13.09	17.95
1500	14" diameter core		7	2.286		27	76.50	15	118.50	166
1550	Each added inch thick, add		55	.291		4.71	9.70	1.91	16.32	22.50
1700	18" diameter core		4	4		35.50	134	26	195.50	276
1750	Each added inch thick, add		28	.571		6.20	19.05	3.75	29	40.50
1760	For horizontal holes, add to above								30%	30%
1770	Prestressed hollow core plank, 6" thick									
1780	1" diameter core	B-89A	52	.308	Ea.	1.86	10.25	2.02	14.13	20.50
1790	Each added inch thick, add		350	.046		.32	1.53	.30	2.15	3.05

03 82 Concrete Boring

03 82 13 – Concrete Core Drilling

03 82 13.10 Core Drilling

		Crew	Daily Output	Labor-Hours	Unit	Material	2007 Bare Costs Labor	2007 Bare Costs Equipment	Total	Total Incl O&P
1800	3" diameter core	B-89A	50	.320	Ea.	4.09	10.70	2.10	16.89	23.50
1810	Each added inch thick, add		240	.067		.68	2.23	.44	3.35	4.69
1820	4" diameter core		48	.333		5.45	11.15	2.19	18.79	25.50
1830	Each added inch thick, add		216	.074		.94	2.47	.49	3.90	5.40
1840	6" diameter core		44	.364		6.75	12.15	2.39	21.29	29
1850	Each added inch thick, add		175	.091		1.12	3.05	.60	4.77	6.65
1860	8" diameter core		32	.500		9.05	16.70	3.28	29.03	39.50
1870	Each added inch thick, add		118	.136		1.57	4.53	.89	6.99	9.75
1880	10" diameter core		28	.571		12.20	19.05	3.75	35	47
1890	Each added inch thick, add		99	.162		1.68	5.40	1.06	8.14	11.40
1900	12" diameter core		22	.727		14.85	24.50	4.77	44.12	59.50
1910	Each added inch thick, add		85	.188		2.47	6.30	1.23	10	13.90
1950	Minimum charge for above, 3" diameter core		7	2.286	Total		76.50	15	91.50	136
2000	4" diameter core		6.80	2.353			78.50	15.45	93.95	139
2050	6" diameter core		6	2.667			89	17.50	106.50	158
2100	8" diameter core		5.50	2.909			97	19.10	116.10	172
2150	10" diameter core		4.75	3.368			112	22	134	200
2200	12" diameter core		3.90	4.103			137	27	164	243
2250	14" diameter core		3.38	4.734			158	31	189	280
2300	18" diameter core		3.15	5.079			170	33.50	203.50	300
3010	Bits for core drill, diamond, premium, 1" diameter				Ea.	125			125	137
3020	3" diameter					310			310	340
3040	4" diameter					345			345	375
3050	6" diameter					550			550	605
3080	8" diameter					755			755	830
3120	12" diameter					1,175			1,175	1,300
3180	18" diameter					2,450			2,450	2,700
3240	24" diameter					3,275			3,275	3,600

03 82 16 – Concrete Drilling

03 82 16.10 Concrete Drilling

		Crew	Daily Output	Labor-Hours	Unit	Material	2007 Bare Costs Labor	2007 Bare Costs Equipment	Total	Total Incl O&P
0010	**CONCRETE DRILLING**									
0050	Up to 4" deep in conc/brick floor/wall, incl. bit & layout, no anchor									
0100	Holes, 1/4" diameter	1 Carp	75	.107	Ea.	.10	3.91		4.01	6.20
0150	For each additional inch of depth, add		430	.019		.02	.68		.70	1.09
0200	3/8" diameter		63	.127		.09	4.66		4.75	7.35
0250	For each additional inch of depth, add		340	.024		.02	.86		.88	1.36
0300	1/2" diameter		50	.160		.09	5.85		5.94	9.25
0350	For each additional inch of depth, add		250	.032		.02	1.17		1.19	1.86
0400	5/8" diameter		48	.167		.17	6.10		6.27	9.75
0450	For each additional inch of depth, add		240	.033		.04	1.22		1.26	1.95
0500	3/4" diameter		45	.178		.20	6.50		6.70	10.35
0550	For each additional inch of depth, add		220	.036		.05	1.33		1.38	2.14
0600	7/8" diameter		43	.186		.25	6.85		7.10	10.90
0650	For each additional inch of depth, add		210	.038		.06	1.40		1.46	2.25
0700	1" diameter		40	.200		.28	7.35		7.63	11.75
0750	For each additional inch of depth, add		190	.042		.07	1.55		1.62	2.49
0800	1-1/4" diameter		38	.211		.40	7.75		8.15	12.50
0850	For each additional inch of depth, add		180	.044		.10	1.63		1.73	2.65
0900	1-1/2" diameter		35	.229		.61	8.40		9.01	13.70
0950	For each additional inch of depth, add		165	.048		.15	1.78		1.93	2.94
1000	For ceiling installations, add						40%			

Division 4 - Masonry

Estimating Tips

04 05 00 Common Work Results for Masonry

- The terms *mortar* and *grout* are often used interchangeably, and incorrectly. Mortar is used to bed masonry units, seal the entry of air and moisture, provide architectural appearance, and allow for size variations in the units. Grout is used primarily in reinforced masonry construction and is used to bond the masonry to the reinforcing steel. Common mortar types are M(2500 psi), S(1800 psi), N(750 psi), and O(350 psi), and conform to ASTM C270. Grout is either fine or coarse and conforms to ASTM C476, and in-place strengths generally exceed 2500 psi. Mortar and grout are different components of masonry construction and are placed by entirely different methods. An estimator should be aware of their unique uses and costs.
- Waste, specifically the loss/droppings of mortar and the breakage of brick and block, is included in all masonry assemblies in this division. A factor of 25% is added for mortar and 3% for brick and concrete masonry units.
- Scaffolding or staging is not included in any of the Division 4 costs. Refer to section 01 54 23 for scaffolding and staging costs.

04 20 00 Unit Masonry

- The most common types of unit masonry are brick and concrete masonry. The major classifications of brick are building brick (ASTM C62), facing brick (ASTM C216), glazed brick, fire brick, and pavers. Many varieties of texture and appearance can exist within these classifications, and the estimator would be wise to check local custom and availability within the project area. For repair and remodeling jobs, matching the existing brick may be the most important criteria.
- Brick and concrete block are priced by the piece and then converted into a price per square foot of wall. Openings less than two square feet are generally ignored by the estimator because any savings in units used is offset by the cutting and trimming required.
- It is often difficult and expensive to find and purchase small lots of historic brick. Costs can vary widely. Many design issues affect costs, selection of mortar mix, and repairs or replacement of masonry materials. Cleaning techniques must be reflected in the estimate.
- All masonry walls, whether interior or exterior, require bracing. The cost of bracing walls during construction should be included by the estimator, and this bracing must remain in place until permanent bracing is complete. Permanent bracing of masonry walls is accomplished by masonry itself, in the form of pilasters or abutting wall corners, or by anchoring the walls to the structural frame. Accessories in the form of anchors, anchor slots, and ties are used, but their supply and installation can be by different trades. For instance, anchor slots on spandrel beams and columns are supplied and welded in place by the steel fabricator, but the ties from the slots into the masonry are installed by the bricklayer. Regardless of the installation method, the estimator must be certain that these accessories are accounted for in pricing.

Reference Numbers

Reference numbers are shown in shaded boxes at the beginning of some major classifications. These numbers refer to related items in the Reference Section. The reference information may be an estimating procedure, an alternate pricing method, or technical information.

Note: Not all subdivisions listed here necessarily appear in this publication.

04 01 Maintenance of Masonry

04 01 20 – Maintenance of Unit Masonry

04 01 20.20 Pointing Masonry

		Crew	Daily Output	Labor-Hours	Unit	Material	2007 Bare Costs Labor	Equipment	Total	Total Incl O&P
0010	**POINTING MASONRY**									
0300	Cut and repoint brick, hard mortar, running bond	1 Bric	80	.100	S.F.	.47	3.81		4.28	6.30
0320	Common bond		77	.104		.47	3.95		4.42	6.50
0360	Flemish bond		70	.114		.49	4.35		4.84	7.15
0400	English bond		65	.123		.49	4.68		5.17	7.70
0600	Soft old mortar, running bond		100	.080		.48	3.04		3.52	5.15
0620	Common bond		96	.083		.47	3.17		3.64	5.35
0640	Flemish bond		90	.089		.49	3.38		3.87	5.70
0680	English bond		82	.098		.49	3.71		4.20	6.20
0700	Stonework, hard mortar		140	.057	L.F.	.62	2.17		2.79	4
0720	Soft old mortar		160	.050	"	.62	1.90		2.52	3.59
1000	Repoint, mask and grout method, running bond		95	.084	S.F.	.62	3.20		3.82	5.55
1020	Common bond		90	.089		.62	3.38		4	5.85
1040	Flemish bond		86	.093		.62	3.54		4.16	6.10
1060	English bond		77	.104		.62	3.95		4.57	6.70
2000	Scrub coat, sand grout on walls, minimum		120	.067		3.05	2.54		5.59	7.20
2020	Maximum		98	.082		2.19	3.11		5.30	7.15

04 01 20.40 Sawing

		Crew	Daily Output	Labor-Hours	Unit	Material	Labor	Equipment	Total	Total Incl O&P
0010	**SAWING MASONRY**									
0050	Brick or block by hand, per inch depth	A-1	125	.064	L.F.		1.84	.43	2.27	3.34

04 01 30 – Unit Masonry Cleaning

04 01 30.20 Cleaning Masonry

		Crew	Daily Output	Labor-Hours	Unit	Material	Labor	Equipment	Total	Total Incl O&P
0010	**CLEANING MASONRY**									
0200	Chemical cleaning, new construction, brush and wash, minimum	D-1	1000	.016	S.F.	.05	.53		.58	.87
0220	Average		800	.020		.08	.67		.75	1.10
0240	Maximum		600	.027		.10	.89		.99	1.46
0260	Light restoration, minimum		800	.020		.07	.67		.74	1.10
0270	Average		400	.040		.11	1.33		1.44	2.15
0280	Maximum		330	.048		.14	1.62		1.76	2.62
0300	Heavy restoration, minimum		600	.027		.08	.89		.97	1.44
0310	Average		400	.040		.12	1.33		1.45	2.16
0320	Maximum		250	.064		.16	2.13		2.29	3.42
0400	High pressure water only, minimum	B-9	2000	.020			.58	.09	.67	1.01
0420	Average		1500	.027			.78	.12	.90	1.34
0440	Maximum		1000	.040			1.17	.18	1.35	2.02
2000	Steam cleaning, minimum		3000	.013			.39	.06	.45	.67
2020	Average		2500	.016			.47	.07	.54	.81
2040	Maximum		1500	.027			.78	.12	.90	1.34
4000	Add for masking doors and windows	1 Clab	800	.010		.06	.29		.35	.52
4200	Add for pedestrian protection				Job					10%

04 01 30.40 Masonry Building Cleaning

		Crew	Daily Output	Labor-Hours	Unit	Material	Labor	Equipment	Total	Total Incl O&P
0010	**MASONRY BUILDING CLEANING**									
0020	Minimum	B-9	3000	.013	S.F.		.39	.06	.45	.67
0100	Maximum		1500	.027			.78	.12	.90	1.34
0300	Common face brick		1750	.023			.67	.10	.77	1.15
0400	Wire cut face brick		1250	.032			.93	.14	1.07	1.61

04 01 30.60 Brick Washing

			Crew	Daily Output	Labor-Hours	Unit	Material	Labor	Equipment	Total	Total Incl O&P
0010	**BRICK WASHING**	R040130-10									
0012	Acid cleanser, smooth brick surface		1 Bric	560	.014	S.F.	.03	.54		.57	.86
0050	Rough brick			400	.020		.03	.76		.79	1.20
0060	Stone, acid wash			600	.013		.04	.51		.55	.81

04 01 Maintenance of Masonry

04 01 30 – Unit Masonry Cleaning

04 01 30.60 Brick Washing		Crew	Daily Output	Labor-Hours	Unit	Material	2007 Bare Costs Labor	Equipment	Total	Total Incl O&P
1000	Muriatic acid, price per gallon in 5 gallon lots				Gal.	5.10			5.10	5.60

04 05 Common Work Results for Masonry

04 05 05 – Selective Masonry Demolition

04 05 05.10 Selective Demolition

		Crew	Daily Output	Labor-Hours	Unit	Material	Labor	Equipment	Total	Total Incl O&P
0010	**SELECTIVE DEMOLITION** R024119-10									
0300	Concrete block walls, unreinforced, 2" thick	2 Clab	1200	.013	S.F.		.38		.38	.60
0310	4" thick		1150	.014			.40		.40	.62
0320	6" thick		1100	.015			.42		.42	.65
0330	8" thick		1050	.015			.44		.44	.68
0340	10" thick		1000	.016			.46		.46	.72
0360	12" thick		950	.017			.48		.48	.75
0380	Reinforced alternate courses, 2" thick		1130	.014			.41		.41	.63
0390	4" thick		1080	.015			.43		.43	.66
0400	6" thick		1035	.015			.44		.44	.69
0410	8" thick		990	.016			.46		.46	.72
0420	10" thick		940	.017			.49		.49	.76
0430	12" thick		890	.018			.52		.52	.80
0440	Reinforced alternate courses & vertically 48" OC, 4" thick		900	.018			.51		.51	.80
0450	6" thick		850	.019			.54		.54	.84
0460	8" thick		800	.020			.58		.58	.90
0480	10" thick		750	.021			.61		.61	.95
0490	12" thick		700	.023			.66		.66	1.02
1000	Chimney, 16" x 16", soft old mortar	1 Clab	55	.145	C.F.		4.18		4.18	6.50
1020	Hard mortar		40	.200			5.75		5.75	8.95
1030	16" x 20", soft old mortar		55	.145			4.18		4.18	6.50
1040	Hard mortar		40	.200			5.75		5.75	8.95
1050	16" x 24", soft old mortar		55	.145			4.18		4.18	6.50
1060	Hard mortar		40	.200			5.75		5.75	8.95
1080	20" x 20", soft old mortar		55	.145			4.18		4.18	6.50
1100	Hard mortar		40	.200			5.75		5.75	8.95
1110	20" x 24", soft old mortar		55	.145			4.18		4.18	6.50
1120	Hard mortar		40	.200			5.75		5.75	8.95
1140	20" x 32", soft old mortar		55	.145			4.18		4.18	6.50
1160	Hard mortar		40	.200			5.75		5.75	8.95
1200	48" x 48", soft old mortar		55	.145			4.18		4.18	6.50
1220	Hard mortar		40	.200			5.75		5.75	8.95
1250	Metal, high temp steel jacket, 24" diameter	E-2	130	.431	V.L.F.		17.40	11.90	29.30	43
1260	60" diameter	"	60	.933			37.50	26	63.50	93.50
1280	Flue lining, up to 12" x 12"	1 Clab	200	.040			1.15		1.15	1.79
1282	Up to 24" x 24"		150	.053			1.53		1.53	2.39
2000	Columns, 8" x 8", soft old mortar		48	.167			4.79		4.79	7.45
2020	Hard mortar		40	.200			5.75		5.75	8.95
2060	16" x 16", soft old mortar		16	.500			14.40		14.40	22.50
2100	Hard mortar		14	.571			16.45		16.45	25.50
2140	24" x 24", soft old mortar		8	1			29		29	45
2160	Hard mortar		6	1.333			38.50		38.50	59.50
2200	36" x 36", soft old mortar		4	2			57.50		57.50	89.50
2220	Hard mortar		3	2.667			76.50		76.50	119
2230	Alternate pricing method, soft old mortar		30	.267	C.F.		7.65		7.65	11.95
2240	Hard mortar		23	.348	"		10		10	15.55

04 05 Common Work Results for Masonry

04 05 05 – Selective Masonry Demolition

04 05 05.10 Selective Demolition	Crew	Daily Output	Labor-Hours	Unit	Material	2007 Bare Costs Labor	Equipment	Total	Total Incl O&P
3000 Copings, precast or masonry, to 8" wide									
3020 Soft old mortar	1 Clab	180	.044	L.F.		1.28		1.28	1.99
3040 Hard mortar	"	160	.050	"		1.44		1.44	2.24
3100 To 12" wide									
3120 Soft old mortar	1 Clab	160	.050	L.F.		1.44		1.44	2.24
3140 Hard mortar	"	140	.057	"		1.64		1.64	2.56
4000 Fireplace, brick, 30" x 24" opening									
4020 Soft old mortar	1 Clab	2	4	Ea.		115		115	179
4040 Hard mortar		1.25	6.400			184		184	286
4100 Stone, soft old mortar		1.50	5.333			153		153	239
4120 Hard mortar		1	8	▼		230		230	360
5000 Veneers, brick, soft old mortar		140	.057	S.F.		1.64		1.64	2.56
5020 Hard mortar		125	.064			1.84		1.84	2.86
5100 Granite and marble, 2" thick		180	.044			1.28		1.28	1.99
5120 4" thick		170	.047			1.35		1.35	2.11
5140 Stone, 4" thick		180	.044			1.28		1.28	1.99
5160 8" thick		175	.046	▼		1.31		1.31	2.05
5400 Alternate pricing method, stone, 4" thick		60	.133	C.F.		3.83		3.83	5.95
5420 8" thick	▼	85	.094	"		2.71		2.71	4.21

04 05 13 – Masonry Mortaring

04 05 13.10 Cement

0010 **CEMENT**									
0100 Masonry, 70 lb. bag, T.L. lots	**CN**			Bag	8.55			8.55	9.40
0150 L.T.L. lots					9.05			9.05	9.95
0200 White, 70 lb. bag, T.L. lots					15.40			15.40	16.90
0250 L.T.L. lots				▼	16.50			16.50	18.15

04 05 13.20 Lime

0010 **LIME**									
0020 Masons, hydrated, 50 lb. bag, T.L. lots	**CN**			Bag	7.55			7.55	8.30
0050 L.T.L. lots					8.70			8.70	9.60
0200 Finish, double hydrated, 50 lb. bag, T.L. lots					9.90			9.90	10.90
0250 L.T.L. lots				▼	11			11	12.10

04 05 13.23 Surface Bonding Masonry Mortaring

0010 **SURFACE BONDING MASONRY MORTARING**									
0020 Gray or white colors, not incl. block work	1 Bric	540	.015	S.F.	.12	.56		.68	.99

04 05 13.30 Mortar

0010 **MORTAR** R040513-10									
0020 With masonry cement									
0100 Type M, 1:1:6 mix	1 Brhe	143	.056	C.F.	4.30	1.60		5.90	7.15
0200 Type N, 1:3 mix	"	143	.056	"	4.24	1.60		5.84	7.10
2000 With portland cement and lime									
2100 Type M, 1:1/4:3 mix	1 Brhe	143	.056	C.F.	7.20	1.60		8.80	10.35
2200 Type N, 1:1:6 mix, 750 psi		143	.056		6	1.60		7.60	9.05
2300 Type O, 1:2:9 mix (Pointing Mortar)	▼	143	.056		6.60	1.60		8.20	9.70
2650 Pre-mixed, type S or N					4.81			4.81	5.30
2700 Mortar for glass block	1 Brhe	143	.056	▼	9.05	1.60		10.65	12.40
2900 Mortar for Fire Brick, 80 lb. bag, T.L. Lots				Bag	23.50			23.50	25.50

04 05 13.91 Masonry Restoration Mortaring

0010 **MASONRY RESTORATION MORTARING**									
0020 Masonry restoration mix				Lb.	1.79			1.79	1.97
0050 White				"	2.06			2.06	2.27

04 05 Common Work Results for Masonry

04 05 13 – Masonry Mortaring

04 05 13.93 Mortar Pigments

		Crew	Daily Output	Labor-Hours	Unit	Material	2007 Bare Costs Labor	Equipment	Total	Total Incl O&P
0010	**MORTAR PIGMENTS** R040513-10									
0020	range 2 to 10 lb. per bag of cement, minimum				Lb.	6.20			6.20	6.85
0050	Average					8.20			8.20	9.05
0100	Maximum					16.25			16.25	17.90

04 05 13.95 Sand

		Crew	Daily Output	Labor-Hours	Unit	Material	Labor	Equipment	Total	Total Incl O&P
0010	**SAND**									
0020	For mortar, per ton	CN			Ton	13.85			13.85	15.25
0050	With 10 mile haul					19.35			19.35	21.50
0100	With 30 mile haul					21.50			21.50	24
0200	Screened and washed, at the pit				C.Y.	19.25			19.25	21
0250	With 10 mile haul					27			27	29.50
0300	With 30 mile haul					30			30	33

04 05 13.98 Mortar Admixtures

		Crew	Daily Output	Labor-Hours	Unit	Material	Labor	Equipment	Total	Total Incl O&P
0010	**MORTAR ADMIXTURES**									
0020	Per quart (1 qt. to 2 bags of masonry cement)				Qt.	8.05			8.05	8.85

04 05 16 – Masonry Grouting

04 05 16.30 Grouting

		Crew	Daily Output	Labor-Hours	Unit	Material	Labor	Equipment	Total	Total Incl O&P
0010	**GROUTING** R040513-10									
0011	Bond beams & lintels, 8" deep, 6" thick, 0.15 C.F. per L.F.	D-4	1480	.022	L.F.	.63	.71	.10	1.44	1.89
0020	8" thick, 0.2 C.F. per L.F.		1400	.023		.94	.76	.11	1.81	2.31
0050	10" thick, 0.25 C.F. per L.F.		1200	.027		1.06	.88	.12	2.06	2.64
0060	12" thick, 0.3 C.F. per L.F.		1040	.031		1.27	1.02	.14	2.43	3.09
0200	Concrete block cores, solid, 4" thk., by hand, 0.067 C.F./S.F. of wall	D-8	1100	.036	S.F.	.28	1.25		1.53	2.21
0210	6" thick, pumped, 0.175 C.F. per S.F.	D-4	720	.044		.74	1.47	.21	2.42	3.27
0250	8" thick, pumped, 0.258 C.F. per S.F.		680	.047		1.09	1.56	.22	2.87	3.80
0300	10" thick, pumped, 0.340 C.F. per S.F.		660	.048		1.44	1.60	.23	3.27	4.26
0350	12" thick, pumped, 0.422 C.F. per S.F.		640	.050		1.78	1.65	.23	3.66	4.73
0500	Cavity walls, 2" space, pumped, 0.167 C.F./S.F. of wall		1700	.019		.71	.62	.09	1.42	1.82
0550	3" space, 0.250 C.F./S.F.		1200	.027		1.06	.88	.12	2.06	2.64
0600	4" space, 0.333 C.F. per S.F.		1150	.028		1.41	.92	.13	2.46	3.09
0700	6" space, 0.500 C.F. per S.F.		800	.040		2.11	1.32	.19	3.62	4.54
0800	Door frames, 3' x 7' opening, 2.5 C.F. per opening		60	.533	Opng.	10.55	17.65	2.50	30.70	41.50
0850	6' x 7' opening, 3.5 C.F. per opening		45	.711	"	14.80	23.50	3.33	41.63	55.50
2000	Grout, C476, for bond beams, lintels and CMU cores		350	.091	C.F.	4.23	3.02	.43	7.68	9.70

04 05 19 – Masonry Anchorage and Reinforcing

04 05 19.05 Anchor Bolts

		Crew	Daily Output	Labor-Hours	Unit	Material	Labor	Equipment	Total	Total Incl O&P
0010	**ANCHOR BOLTS**									
0020	Hooked, with nut and washer, 1/2" diam., 8" long	1 Bric	200	.040	Ea.	.74	1.52		2.26	3.13
0030	12" long		190	.042		1.36	1.60		2.96	3.94
0040	5/8" diameter, 8" long		180	.044		1.13	1.69		2.82	3.81
0050	12" long		170	.047		1.24	1.79		3.03	4.09
0060	3/4" diameter, 8" long		160	.050		1.61	1.90		3.51	4.67
0070	12" long		150	.053		2.01	2.03		4.04	5.30

04 05 19.16 Masonry Anchors

		Crew	Daily Output	Labor-Hours	Unit	Material	Labor	Equipment	Total	Total Incl O&P
0010	**MASONRY ANCHORS**									
0020	For brick veneer, galv., corrugated, 7/8" x 7", 22 Ga.	1 Bric	10.50	.762	C	6.40	29		35.40	51
0100	24 Ga.		10.50	.762		5.95	29		34.95	50.50
0150	16 Ga.		10.50	.762		22.50	29		51.50	69
0200	Buck anchors, galv., corrugated, 16 gauge, 2" bend, 8" x 2"		10.50	.762		124	29		153	181
0250	8" x 3"		10.50	.762		128	29		157	185

04 05 Common Work Results for Masonry

04 05 19 – Masonry Anchorage and Reinforcing

04 05 19.16 Masonry Anchors

		Crew	Daily Output	Labor-Hours	Unit	Material	2007 Bare Costs Labor	Equipment	Total	Total Incl O&P
0670	3/16" diameter	1 Bric	10.50	.762	C	18.10	29		47.10	64
0680	1/4" diameter		10.50	.762		36.50	29		65.50	84
0850	8" long, 3/16" diameter		10.50	.762		20	29		49	66
0855	1/4" diameter		10.50	.762		43	29		72	91.50
1000	Rectangular type, galvanized, 1/4" diameter, 2" x 6"		10.50	.762		41.50	29		70.50	89.50
1050	4" x 6"		10.50	.762		46.50	29		75.50	95
1100	3/16" diameter, 2" x 6"		10.50	.762		31	29		60	78
1150	4" x 6"		10.50	.762		27.50	29		56.50	74.50
1500	Rigid partition anchors, plain, 8" long, 1" x 1/8"		10.50	.762		79.50	29		108.50	132
1550	1" x 1/4"		10.50	.762		131	29		160	188
1580	1-1/2" x 1/8"		10.50	.762		106	29		135	160
1600	1-1/2" x 1/4"		10.50	.762		215	29		244	281
1650	2" x 1/8"		10.50	.762		137	29		166	195
1700	2" x 1/4"		10.50	.762		252	29		281	320

04 05 19.26 Masonry Reinforcing Bars

			Crew	Daily Output	Labor-Hours	Unit	Material	Labor	Equipment	Total	Total Incl O&P
0010	**MASONRY REINFORCING BARS**	R040519-50									
0015	Steel bars A615, placed horiz., #3 & #4 bars		1 Bric	450	.018	Lb.	.45	.68		1.13	1.52
0020	#5 & #6 bars			800	.010		.45	.38		.83	1.07
0050	Placed vertical, #3 & #4 bars			350	.023		.45	.87		1.32	1.81
0060	#5 & #6 bars			650	.012		.45	.47		.92	1.20
0200	Joint reinforcing, regular truss, to 6" wide, mill std galvanized			30	.267	C.L.F.	13.95	10.15		24.10	31
0250	12" wide			20	.400		15.35	15.20		30.55	40
0400	Cavity truss with drip section, to 6" wide			30	.267		13.45	10.15		23.60	30.50
0450	12" wide			20	.400		13.80	15.20		29	38

04 05 23 – Masonry Accessories

04 05 23.13 Control Joint

		Crew	Daily Output	Labor-Hours	Unit	Material	Labor	Equipment	Total	Total Incl O&P
0010	**CONTROL JOINT**									
0020	Rubber, 4" and wider wall	1 Bric	400	.020	L.F.	2.33	.76		3.09	3.72
0050	PVC, 4" wall		400	.020		1.20	.76		1.96	2.48
0100	Rubber, 6" wall		320	.025		3.21	.95		4.16	4.98
0120	PVC, 6" wall		320	.025		1.70	.95		2.65	3.32
0140	Rubber, 8" and wider wall		280	.029		3.29	1.09		4.38	5.25
0160	PVC, 8" wall		280	.029		2.16	1.09		3.25	4.03
0180	12" wall		240	.033		3.96	1.27		5.23	6.30

04 05 23.19 Vent Box

		Crew	Daily Output	Labor-Hours	Unit	Material	Labor	Equipment	Total	Total Incl O&P
0010	**VENT BOX**									
0020	Extruded aluminum, 4" deep, 2-3/8" x 8-1/8"	1 Bric	30	.267	Ea.	25	10.15		35.15	43
0050	5" x 8-1/8"		25	.320		31.50	12.20		43.70	53
0100	2-1/4" x 25"		25	.320		65	12.20		77.20	90
0150	5" x 16-1/2"		22	.364		55	13.85		68.85	81
0200	6" x 16-1/2"		22	.364		76	13.85		89.85	105
0250	7-3/4" x 16-1/2"		20	.400		74	15.20		89.20	105
0400	For baked enamel finish, add					35%				
0500	For cast aluminum, painted, add					60%				
1000	Stainless steel ventilators, 6" x 6"	1 Bric	25	.320		109	12.20		121.20	139
1050	8" x 8"		24	.333		115	12.70		127.70	146
1100	12" x 12"		23	.348		133	13.25		146.25	166
1150	12" x 6"		24	.333		116	12.70		128.70	147
1200	Foundation block vent, galv., 1-1/4" thk, 8" high, 16" long, no damper		30	.267		21.50	10.15		31.65	39
1250	For damper, add					7.15			7.15	7.85

04 05 Common Work Results for Masonry

04 05 23 – Masonry Accessories

04 05 23.95 Wall Plugs

		Crew	Daily Output	Labor-Hours	Unit	Material	2007 Bare Costs Labor	Equipment	Total	Total Incl O&P
0010	**WALL PLUGS**									
0020	26 ga., galvanized, plain	1 Bric	10.50	.762	C	29.50	29		58.50	76.50
0050	Wood filled	"	10.50	.762	"	100	29		129	154

04 21 Clay Unit Masonry

04 21 13 – Brick Masonry

04 21 13.13 Brick Veneer Masonry

			Crew	Daily Output	Labor-Hours	Unit	Material	Labor	Equipment	Total	Total Incl O&P
0010	**BRICK VENEER MASONRY**	R042110-20									
0015	Material costs incl. 3% brick and 25% mortar waste										
0020	Standard, select common, 4" x 2-2/3" x 8" (6.75/S.F.)		D-8	1.50	26.667	M	500	915		1,415	1,950
0050	Red, 4" x 2-2/3" x 8", running bond			1.50	26.667		570	915		1,485	2,025
0100	Full header every 6th course (7.88/S.F.)	R042110-50		1.45	27.586		570	945		1,515	2,075
0150	English, full header every 2nd course (10.13/S.F.)			1.40	28.571		570	980		1,550	2,125
0200	Flemish, alternate header every course (9.00/S.F.)			1.40	28.571		570	980		1,550	2,125
0250	Flemish, alt. header every 6th course (7.13/S.F.)			1.45	27.586		570	945		1,515	2,075
0300	Full headers throughout (13.50/S.F.)			1.40	28.571		565	980		1,545	2,125
0350	Rowlock course (13.50/S.F.)			1.35	29.630		565	1,025		1,590	2,175
0400	Rowlock stretcher (4.50/S.F.)			1.40	28.571		575	980		1,555	2,125
0450	Soldier course (6.75/S.F.)			1.40	28.571		570	980		1,550	2,125
0500	Sailor course (4.50/S.F.)			1.30	30.769		575	1,050		1,625	2,225
0601	Buff or gray face, running bond, (6.75/S.F.)			1.50	26.667		570	915		1,485	2,025
0700	Glazed face, 4" x 2-2/3" x 8", running bond			1.40	28.571		1,525	980		2,505	3,175
0750	Full header every 6th course (7.88/S.F.)			1.35	29.630		1,425	1,025		2,450	3,125
1000	Jumbo, 6" x 4" x 12", (3.00/S.F.)			1.30	30.769		1,575	1,050		2,625	3,325
1051	Norman, 4" x 2-2/3" x 12" (4.50/S.F.)			1.45	27.586		980	945		1,925	2,525
1100	Norwegian, 4" x 3-1/5" x 12" (3.75/S.F.)			1.40	28.571		945	980		1,925	2,550
1150	Economy, 4" x 4" x 8" (4.50 per S.F.)			1.40	28.571		920	980		1,900	2,500
1201	Engineer, 4" x 3-1/5" x 8", (5.63/S.F.)			1.45	27.586		580	945		1,525	2,100
1251	Roman, 4" x 2" x 12", (6.00/S.F.)			1.50	26.667		925	915		1,840	2,425
1300	S.C.R. 6" x 2-2/3" x 12" (4.50/S.F.)			1.40	28.571		1,100	980		2,080	2,725
1350	Utility, 4" x 4" x 12" (3.00/S.F.)			1.35	29.630		1,225	1,025		2,250	2,900
1360	For less than truck load lots, add						12			12	13.20
1400	For battered walls, add							30%			
1450	For corbels, add							75%			
1500	For curved walls, add							30%			
1550	For pits and trenches, deduct							20%			
1999	Alternate method of figuring by square foot										
2000	Standard, sel. common, 4" x 2-2/3" x 8", (6.75/S.F.)		D-8	230	.174	S.F.	3.85	5.95		9.80	13.30
2020	Standard, red, 4" x 2-2/3" x 8", running bond (6.75/SF)			220	.182		3.85	6.25		10.10	13.75
2050	Full header every 6th course (7.88/S.F.)			185	.216		4.49	7.40		11.89	16.25
2100	English, full header every 2nd course (10.13/S.F.)			140	.286		5.75	9.80		15.55	21.50
2150	Flemish, alternate header every course (9.00/S.F.)			150	.267		5.10	9.15		14.25	19.55
2200	Flemish, alt. header every 6th course (7.13/S.F.)			205	.195		4.07	6.70		10.77	14.65
2250	Full headers throughout (13.50/S.F.)			105	.381		7.65	13.05		20.70	28.50
2300	Rowlock course (13.50/S.F.)			100	.400		7.65	13.70		21.35	29.50
2350	Rowlock stretcher (4.50/S.F.)			310	.129		2.59	4.42		7.01	9.60
2400	Soldier course (6.75/S.F.)			200	.200		3.85	6.85		10.70	14.70
2450	Sailor course (4.50/S.F.)			290	.138		2.59	4.73		7.32	10.05
2600	Buff or gray face, running bond, (6.75/S.F.)			220	.182		4.08	6.25		10.33	14
2700	Glazed face brick, running bond			210	.190		9.70	6.55		16.25	20.50

04 21 Clay Unit Masonry

04 21 13 – Brick Masonry

04 21 13.13 Brick Veneer Masonry		Crew	Daily Output	Labor-Hours	Unit	Material	2007 Bare Costs Labor	Equipment	Total	Total Incl O&P
2750	Full header every 6th course (7.88/S.F.)	D-8	170	.235	S.F.	11.30	8.05		19.35	25
3000	Jumbo, 6" x 4" x 12" running bond (3.00/S.F.)		435	.092		4.28	3.15		7.43	9.50
3050	Norman, 4" x 2-2/3" x 12" running bond, (4.5/S.F.)		320	.125		4.65	4.29		8.94	11.60
3100	Norwegian, 4" x 3-1/5" x 12" (3.75/S.F.)		375	.107		3.44	3.66		7.10	9.35
3150	Economy, 4" x 4" x 8" (4.50/S.F.)		310	.129		4.09	4.42		8.51	11.25
3200	Engineer, 4" x 3-1/5" x 8" (5.63/S.F.)		260	.154		3.26	5.30		8.56	11.65
3250	Roman, 4" x 2" x 12" (6.00/S.F.)		250	.160		5.40	5.50		10.90	14.30
3300	SCR, 6" x 2-2/3" x 12" (4.50/S.F.)		310	.129		4.92	4.42		9.34	12.15
3350	Utility, 4" x 4" x 12" (3.00/S.F.)		450	.089		3.59	3.05		6.64	8.60
3400	For cavity wall construction, add					15%				
3450	For stacked bond, add					10%				
3500	For interior veneer construction, add					15%				
3550	For curved walls, add					30%				

04 21 13.15 Chimney

		Crew	Daily Output	Labor-Hours	Unit	Material	Labor	Equipment	Total	Total Incl O&P
0010	**CHIMNEY**									
0100	Brick, 16" x 16", 8" flue, scaff. not incl.	D-1	18.20	.879	V.L.F.	18.90	29.50		48.40	65.50
0150	16" x 20" with one 8" x 12" flue		16	1		29	33.50		62.50	83
0200	16" x 24" with two 8" x 8" flues		14	1.143		41.50	38		79.50	104
0250	20" x 20" with one 12" x 12" flue		13.70	1.168		34.50	39		73.50	97.50
0300	20" x 24" with two 8" x 12" flues		12	1.333		47.50	44.50		92	120
0350	20" x 32" with two 12" x 12" flues		10	1.600		60.50	53.50		114	148
1800	Metal, high temp. steel jacket, factory lining, 24" diam.	E-2	65	.862		198	35	24	257	305
1900	60" diameter	"	30	1.867		720	75.50	51.50	847	975
2100	Poured concrete, brick lining, 200' high x 10' diam.					6,525			6,525	7,175
2800	500' x 20' diameter					11,400			11,400	12,600

04 21 13.18 Columns

		Crew	Daily Output	Labor-Hours	Unit	Material	Labor	Equipment	Total	Total Incl O&P
0010	**COLUMNS** R042110-10									
0050	Brick, 8" x 8", 9 brick per course	D-1	56	.286	V.L.F.	4.94	9.55		14.49	19.95
0100	12" x 8", 13.5 brick		37	.432		7.40	14.40		21.80	30
0200	12" x 12", 20 brick		25	.640		11	21.50		32.50	44.50
0300	16" x 12", 27 brick		19	.842		14.80	28		42.80	59
0400	16" x 16", 36 brick		14	1.143		19.75	38		57.75	79.50
0500	20" x 16", 45 brick		11	1.455		24.50	48.50		73	101
0600	20" x 20", 56 brick		9	1.778		31	59.50		90.50	124
0700	24" x 20", 68 brick		7	2.286		37.50	76		113.50	157
0800	24" x 24", 81 brick		6	2.667		44.50	89		133.50	184
1000	36" x 36", 182 brick		3	5.333		100	178		278	380

04 21 13.30 Oversized Brick

		Crew	Daily Output	Labor-Hours	Unit	Material	Labor	Equipment	Total	Total Incl O&P
0010	**OVERSIZED BRICK**									
0100	Veneer, 4" x 2.25" x 16"	D-8	387	.103	S.F.	4.79	3.54		8.33	10.65
0105	4" x 2.75" x 16"		412	.097		4.33	3.33		7.66	9.80
0110	4" x 4" x 16"		460	.087		4.26	2.98		7.24	9.20
0120	4" x 8" x 16"		533	.075		5.20	2.57		7.77	9.65
0125	Loadbearing, 6" x 4" x 16", grouted and reinforced		387	.103		7.55	3.54		11.09	13.70
0130	8" x 4" x 16", grouted and reinforced		327	.122		8	4.19		12.19	15.20
0135	6" x 8" x 16", grouted and reinforced		440	.091		7.20	3.12		10.32	12.70
0140	8" x 8" x 16", grouted and reinforced		400	.100		8.10	3.43		11.53	14.10
0145	Curtainwall / reinforced veneer, 6" x 4" x 16"		387	.103		11.75	3.54		15.29	18.30
0150	8" x 4" x 16"		327	.122		14.25	4.19		18.44	22
0155	6" x 8" x 16"		440	.091		11.85	3.12		14.97	17.80
0160	8" x 8" x 16"		400	.100		14.35	3.43		17.78	21
0200	For 1 to 3 slots in face, add					25%				

04 21 Clay Unit Masonry

04 21 13 – Brick Masonry

04 21 13.30 Oversized Brick

		Crew	Daily Output	Labor-Hours	Unit	Material	2007 Bare Costs Labor	Equipment	Total	Total Incl O&P
0210	For 4 to 7 slots in face, add				S.F.	15%				
0220	For bond beams, add					20%				
0230	For bullnose shapes, add					20%				
0240	For open end knockout, add					10%				
0250	For white or gray color group, add					10%				
0260	For 135 degree corner, add					250%				

04 21 13.35 Common Building Brick

0010	**COMMON BUILDING BRICK**, Material Only	R042110-20								
0020	Standard, minimum				M	335			335	365
0050	Average (select)				"	405			405	450

04 21 13.40 Structural Brick

0010	**STRUCTURAL BRICK**									
0100	Standard unit, 4-5/8" x 2-3/4" x 9-5/8"	D-8	245	.163	S.F.	5.05	5.60		10.65	14.05
0120	Bond beam		225	.178		9.55	6.10		15.65	19.80
0140	V cut bond beam		225	.178		9.85	6.10		15.95	20
0160	Stretcher quoin, 5-5/8" x 2-3/4" x 9-5/8"		245	.163		9.50	5.60		15.10	18.95
0180	Corner quoin		245	.163		10.65	5.60		16.25	20
0200	Corner, 45 deg, 4-5/8" x 2-3/4" x 10-7/16"		235	.170		11.05	5.85		16.90	21

04 21 13.45 Face Brick

0010	**FACE BRICK** Material Only	R042110-20								
0300	Standard modular, 4" x 2-2/3" x 8", minimum				M	475			475	525
0350	Maximum	CN				545			545	600
0450	Economy, 4" x 4" x 8", minimum					800			800	880
0500	Maximum					1,050			1,050	1,150
0510	Economy, 4" x 4" x 12", minimum					1,100			1,100	1,200
0520	Maximum					1,425			1,425	1,575
0550	Jumbo, 6" x 4" x 12", minimum					1,300			1,300	1,425
0600	Maximum					1,650			1,650	1,825
0610	Jumbo, 8" x 4" x 12", minimum					1,300			1,300	1,425
0620	Maximum					1,650			1,650	1,825
0650	Norwegian, 4" x 3-1/5" x 12", minimum					805			805	885
0700	Maximum					1,050			1,050	1,150
0710	Norwegian, 6" x 3-1/5" x 12", minimum					1,175			1,175	1,300
0720	Maximum					1,525			1,525	1,675
0850	Standard glazed, plain colors, 4" x 2-2/3" x 8", minimum					1,325			1,325	1,450
0900	Maximum					1,700			1,700	1,875
1000	Deep trim shades, 4" x 2-2/3" x 8", minimum					1,325			1,325	1,450
1050	Maximum					1,475			1,475	1,625
1080	Jumbo utility, 4" x 4" x 12"					1,075			1,075	1,175
1120	4" x 8" x 8"					1,225			1,225	1,350
1140	4" x 8" x 16"					2,600			2,600	2,850
1260	Engineer, 4" x 3-1/5" x 8", minimum					480			480	530
1270	Maximum					560			560	615
1350	King, 4" x 2-3/4" x 10", minimum					495			495	545
1360	Maximum					530			530	585
1400	Norman, 4" x 2-3/4" x 12"					495			495	545
1450	Roman, 4" x 2" x 12"					495			495	545
1500	SCR, 6" x 2-2/3" x 12"					495			495	545
1550	Double, 4" x 5-1/3" x 8"					495			495	545
1600	Triple, 4" x 5-1/3" x 12"					495			495	545
1770	Standard modular, double glazed, 4" x 2-2/3" x 8"					1,125			1,125	1,225
1850	Jumbo, colored glazed ceramic, 6" x 4" x 12"					1,650			1,650	1,825

04 21 Clay Unit Masonry

04 21 13 – Brick Masonry

04 21 13.45 Face Brick

		Crew	Daily Output	Labor-Hours	Unit	Material	2007 Bare Costs Labor	Equipment	Total	Total Incl O&P
2050	Jumbo utility, glazed, 4" x 4" x 12"				M	1,225			1,225	1,350
2100	4" x 8" x 8"					1,700			1,700	1,875
2150	4" x 16" x 8"					2,950			2,950	3,250
2170	For less than truck load lots, add					12			12	13.20
2180	For buff or gray brick, add					16			16	17.60

04 21 26 – Glazed Structural Clay Tile Masonry

04 21 26.10 Structural Facing Tile

		Crew	Daily Output	Labor-Hours	Unit	Material	Labor	Equipment	Total	Total Incl O&P
0010	**STRUCTURAL FACING TILE**									
0020	6T series, 5-1/3" x 12", 2.3 pieces per S.F., glazed 1 side, 2" thick	D-8	225	.178	S.F.	6.50	6.10		12.60	16.45
0100	4" thick		220	.182		7.65	6.25		13.90	17.90
0150	Glazed 2 sides		195	.205		12.10	7.05		19.15	24
0250	6" thick		210	.190		11.70	6.55		18.25	23
0300	Glazed 2 sides		185	.216		13.90	7.40		21.30	26.50
0400	8" thick		180	.222		14.80	7.60		22.40	28
0500	Special shapes, group 1		400	.100	Ea.	5.85	3.43		9.28	11.60
0550	Group 2		375	.107		8.85	3.66		12.51	15.30
0600	Group 3		350	.114		11	3.92		14.92	18.05
0650	Group 4		325	.123		19	4.22		23.22	27.50
0700	Group 5		300	.133		25	4.57		29.57	34.50
0750	Group 6		275	.145		30	4.99		34.99	40.50
1000	Fire rated, 4" thick, 1 hr. rating		210	.190	S.F.	13.10	6.55		19.65	24.50
1300	Acoustic, 4" thick		210	.190	"	31	6.55		37.55	44
1400	For designer colors, add					25%				
2000	8W series, 8" x 16", 1.125 pieces per S.F.									
2050	2" thick, glazed 1 side	D-8	360	.111	S.F.	7.05	3.81		10.86	13.60
2100	4" thick, glazed 1 side		345	.116		7.70	3.98		11.68	14.50
2150	Glazed 2 sides		325	.123		11.50	4.22		15.72	19.05
2200	6" thick, glazed 1 side		330	.121		10.85	4.16		15.01	18.25
2250	8" thick, glazed 1 side		310	.129		12.60	4.42		17.02	20.50
2500	Special shapes, group 1		300	.133	Ea.	9.40	4.57		13.97	17.30
2550	Group 2		280	.143		12.40	4.90		17.30	21
2600	Group 3		260	.154		13	5.30		18.30	22.50
2650	Group 4		250	.160		29	5.50		34.50	40
2700	Group 5		240	.167		50.50	5.70		56.20	64
2750	Group 6		230	.174		55.50	5.95		61.45	70
3000	4" thick, glazed 1 side		345	.116	S.F.	9	3.98		12.98	15.95
3100	Acoustic, 4" thick		345	.116	"	11.85	3.98		15.83	19.10
3120	4W series, 8" x 8", 2.25 pieces per S.F.									
3125	2" thick, glazed 1 side	D-8	360	.111	S.F.	8.25	3.81		12.06	14.85
3130	4" thick, glazed 1 side		345	.116		8.25	3.98		12.23	15.10
3135	Glazed 2 sides		325	.123		11.80	4.22		16.02	19.40
3140	6" thick, glazed 1 side		330	.121		11.05	4.16		15.21	18.45
3150	8" thick, glazed 1 side		310	.129		15.90	4.42		20.32	24.50
3155	Special shapes, group I		300	.133	Ea.	5.80	4.57		10.37	13.35
3160	Group II		280	.143	"	7.05	4.90		11.95	15.25
3200	For designer colors, add					25%				
3300	For epoxy mortar joints, add				S.F.	1.38			1.38	1.52

04 21 29 – Terra Cotta Masonry

04 21 29.10 Terra Cotta Masonry

		Crew	Daily Output	Labor-Hours	Unit	Material	Labor	Equipment	Total	Total Incl O&P
0010	**TERRA COTTA MASONRY**									
0020	Split type, not glazed, 9" wide	D-1	90	.178	L.F.	8.75	5.95		14.70	18.65
0100	13" wide		80	.200		13.20	6.65		19.85	24.50

04 21 Clay Unit Masonry

04 21 29 – Terra Cotta Masonry

04 21 29.10 Terra Cotta Masonry

		Crew	Daily Output	Labor-Hours	Unit	Material	2007 Bare Costs Labor	Equipment	Total	Total Incl O&P
0200	Split type, glazed, 9" wide	D-1	90	.178	L.F.	14.85	5.95		20.80	25.50
0250	13" wide	↓	80	.200	↓	19.30	6.65		25.95	31
0500	Partition or back-up blocks, scored, in C.L. lots									
0700	Non-load bearing 12" x 12", 3" thick, special order	D-8	550	.073	S.F.	14.80	2.49		17.29	20
0750	4" thick, standard		500	.080		5.85	2.74		8.59	10.55
0800	6" thick		450	.089		6.90	3.05		9.95	12.25
0850	8" thick		400	.100		8.90	3.43		12.33	15
1000	Load bearing, 12" x 12", 4" thick, in walls		500	.080		6.10	2.74		8.84	10.85
1050	In floors		750	.053		6.10	1.83		7.93	9.50
1200	6" thick, in walls		450	.089		6.90	3.05		9.95	12.25
1250	In floors		675	.059		6.90	2.03		8.93	10.70
1400	8" thick, in walls		400	.100		8.90	3.43		12.33	15
1450	In floors		575	.070		8.90	2.39		11.29	13.45
1600	10" thick, in walls, special order		350	.114		11.10	3.92		15.02	18.15
1650	In floors, special order		500	.080		11.10	2.74		13.84	16.35
1800	12" thick, in walls, special order		300	.133		23.50	4.57		28.07	33
1850	In floors, special order	↓	450	.089		23.50	3.05		26.55	30.50
2000	For reinforcing with steel rods, add to above					15%	5%			
2100	For smooth tile instead of scored, add					3.01			3.01	3.31
2200	For L.C.L. quantities, add				↓	10%	10%			

04 21 29.20 Terra Cotta Tile

		Crew	Daily Output	Labor-Hours	Unit	Material	Labor	Equipment	Total	Total Incl O&P
0010	**TERRA COTTA TILE**									
0100	Square, hexagonal or lattice shapes, unglazed	1 Tilf	135	.059	S.F.	5.75	2.10		7.85	9.45
0300	Glazed, plain colors		130	.062		7.90	2.18		10.08	11.90
0400	Intense colors	↓	125	.064	↓	9.25	2.27		11.52	13.50

04 22 Concrete Unit Masonry

04 22 10 – Concrete Masonry Units

04 22 10.11 Autoclave Aerated Concrete Block

		Crew	Daily Output	Labor-Hours	Unit	Material	Labor	Equipment	Total	Total Incl O&P
0010	**AUTOCLAVE AERATED CONCRETE BLOCK**									
0050	Solid, 4" x 12" x 24", incl mortar	D-8	600	.067	S.F.	1.87	2.29		4.16	5.55
0060	6" x 12" x 24"		600	.067		2.52	2.29		4.81	6.25
0070	8" x 8" x 24"		575	.070		3.16	2.39		5.55	7.10
0080	10" x 12" x 24"		575	.070		4.12	2.39		6.51	8.15
0090	12" x 12" x 24"	↓	550	.073	↓	4.80	2.49		7.29	9.10

04 22 10.12 Chimney Block

		Crew	Daily Output	Labor-Hours	Unit	Material	Labor	Equipment	Total	Total Incl O&P
0010	**CHIMNEY BLOCK**									
0220	1 piece, with 8" x 8" flue, 16" x 16"	D-1	28	.571	V.L.F.	11.60	19.05		30.65	42
0230	2 piece, 16" x 16"		26	.615		13.35	20.50		33.85	45.50
0240	2 piece, with 8" x 12" flue, 16" x 20"	↓	24	.667	↓	24	22		46	60.50

04 22 10.14 Concrete Block, Back-Up

		Crew	Daily Output	Labor-Hours	Unit	Material	Labor	Equipment	Total	Total Incl O&P
0010	**CONCRETE BLOCK, BACK-UP** R042210-20									
0020	Normal weight, 8" x 16" units, tooled joint 1 side									
0050	Not-reinforced, 2000 psi, 2" thick	D-8	475	.084	S.F.	1.17	2.89		4.06	5.65
0200	4" thick		460	.087		1.29	2.98		4.27	5.95
0300	6" thick		440	.091		1.86	3.12		4.98	6.80
0350	8" thick		400	.100		2.02	3.43		5.45	7.40
0400	10" thick	↓	330	.121		2.74	4.16		6.90	9.30
0450	12" thick	D-9	310	.155		2.97	5.15		8.12	11.10
1000	Reinforced, alternate courses, 4" thick	D-8	450	.089	↓	1.41	3.05		4.46	6.20

04 22 Concrete Unit Masonry

04 22 10 – Concrete Masonry Units

04 22 10.14 Concrete Block, Back-Up		Crew	Daily Output	Labor-Hours	Unit	Material	2007 Bare Costs Labor	Equipment	Total	Total Incl O&P
1100	6" thick	D-8	430	.093	S.F.	1.98	3.19		5.17	7
1150	8" thick		395	.101		2.15	3.47		5.62	7.65
1200	10" thick		320	.125		2.87	4.29		7.16	9.65
1250	12" thick	D-9	300	.160		3.11	5.35		8.46	11.50
04 22 10.16 Concrete Block, Bond Beam										
0010	**CONCRETE BLOCK, BOND BEAM**									
0020	Not including grout or reinforcing									
0130	8" high, 8" thick	D-8	565	.071	L.F.	2.19	2.43		4.62	6.10
0150	12" thick	D-9	510	.094		3.05	3.14		6.19	8.15
0525	Lightweight, 6" thick	D-8	592	.068		2.20	2.32		4.52	5.95
0530	8" high, 8" thick	"	575	.070		2.19	2.39		4.58	6.05
0550	12" thick	D-9	520	.092		3.60	3.08		6.68	8.65
2000	Including grout and 2 #5 bars									
2100	Regular block, 8" high, 8" thick	D-8	300	.133	L.F.	4.14	4.57		8.71	11.50
2150	12" thick	D-9	250	.192		5.55	6.40		11.95	15.85
2500	Lightweight, 8" high, 8" thick	D-8	305	.131		4.60	4.50		9.10	11.90
2550	12" thick	D-9	255	.188		6.10	6.30		12.40	16.25
04 22 10.18 Concrete Block, Column										
0010	**CONCRETE BLOCK, COLUMN**									
0050	Including vertical reinforcing (4-#4 bars) and grout									
0160	1 piece unit, 16" x 16"	D-1	26	.615	V.L.F.	17.50	20.50		38	50.50
0170	2 piece units, 16" x 20"		24	.667		23	22		45	59.50
0180	20" x 20"		22	.727		33.50	24.50		58	74
0190	22" x 24"		18	.889		47.50	29.50		77	97
0200	20" x 32"		14	1.143		51.50	38		89.50	115
04 22 10.19 Concrete Block, Insulation Inserts										
0010	**CONCRETE BLOCK, INSULATION INSERTS**									
0100	Inserts, styrofoam, plant installed, add to block prices									
0200	8" x 16" units, 6" thick				S.F.	1.13			1.13	1.24
0250	8" thick					1.13			1.13	1.24
0300	10" thick					1.33			1.33	1.46
0350	12" thick					1.40			1.40	1.54
0500	8" x 8" units, 8" thick					.93			.93	1.02
0550	12" thick					1.13			1.13	1.24
04 22 10.23 Concrete Block, Decorative										
0010	**CONCRETE BLOCK, DECORATIVE**									
0020	Embossed, simulated brick face									
0100	8" x 16" units, 4" thick	D-8	400	.100	S.F.	3.02	3.43		6.45	8.55
0200	8" thick		340	.118		4.17	4.03		8.20	10.75
0250	12" thick		300	.133		5.50	4.57		10.07	13
0400	Embossed both sides									
0500	8" thick	D-8	300	.133	S.F.	4.68	4.57		9.25	12.10
0550	12" thick	"	275	.145	"	5.90	4.99		10.89	14.10
1000	Fluted high strength									
1100	8" x 16" x 4" thick, flutes 1 side,	D-8	345	.116	S.F.	3.59	3.98		7.57	10
1150	Flutes 2 sides		335	.119		4.36	4.09		8.45	11.05
1200	8" thick		300	.133		5.65	4.57		10.22	13.20
1250	For special colors, add					.35			.35	.39
1400	Deep grooved, smooth face									
1450	8" x 16" x 4" thick	D-8	345	.116	S.F.	2.34	3.98		6.32	8.65
1500	8" thick	"	300	.133	"	4.04	4.57		8.61	11.40
2000	Formblock, incl. inserts & reinforcing									

04 22 Concrete Unit Masonry

04 22 10 – Concrete Masonry Units

04 22 10.23 Concrete Block, Decorative		Crew	Daily Output	Labor-Hours	Unit	Material	2007 Bare Costs Labor	Equipment	Total	Total Incl O&P
2100	8" x 16" x 8" thick	D-8	345	.116	S.F.	4.10	3.98		8.08	10.55
2150	12" thick	"	310	.129	"	5.10	4.42		9.52	12.40
2500	Ground face									
2600	8" x 16" x 4" thick	D-8	345	.116	S.F.	3.82	3.98		7.80	10.25
2650	6" thick		325	.123		4.94	4.22		9.16	11.85
2700	8" thick		300	.133		5.75	4.57		10.32	13.25
2750	12" thick	D-9	265	.181		6.95	6.05		13	16.85
2900	For special colors, add, minimum					15%				
2950	For special colors, add, maximum					45%				
4000	Slump block									
4100	4" face height x 16" x 4" thick	D-1	165	.097	S.F.	3.69	3.23		6.92	8.95
4150	6" thick		160	.100		5.10	3.34		8.44	10.70
4200	8" thick		155	.103		5.65	3.44		9.09	11.45
4250	10" thick		140	.114		10.30	3.81		14.11	17.10
4300	12" thick		130	.123		10.75	4.10		14.85	18.10
4400	6" face height x 16" x 6" thick		155	.103		4.49	3.44		7.93	10.20
4450	8" thick		150	.107		6.20	3.56		9.76	12.20
4500	10" thick		130	.123		8.70	4.10		12.80	15.85
4550	12" thick		120	.133		9.85	4.45		14.30	17.55
5000	Split rib profile units, 1" deep ribs, 8 ribs									
5100	8" x 16" x 4" thick	D-8	345	.116	S.F.	2.68	3.98		6.66	9
5150	6" thick		325	.123		3.10	4.22		7.32	9.80
5200	8" thick		300	.133		3.56	4.57		8.13	10.85
5250	12" thick	D-9	275	.175		4.27	5.80		10.07	13.55
5400	For special deeper colors, 4" thick, add					1.02			1.02	1.12
5450	12" thick, add					.87			.87	.96
5600	For white, 4" thick, add					1.02			1.02	1.12
5650	6" thick, add					1.02			1.02	1.12
5700	8" thick, add					.95			.95	1.04
5750	12" thick, add					.87			.87	.96
6000	Split face									
6100	8" x 16" x 4" thick	D-8	350	.114	S.F.	2.55	3.92		6.47	8.75
6150	6" thick		325	.123		2.95	4.22		7.17	9.65
6200	8" thick		300	.133		3.46	4.57		8.03	10.75
6250	12" thick	D-9	270	.178		4.12	5.95		10.07	13.55
6300	For scored, add					.29			.29	.32
6400	For special deeper colors, 4" thick, add					.59			.59	.65
6450	6" thick, add					.59			.59	.65
6500	8" thick, add					.52			.52	.57
6550	12" thick, add					.44			.44	.48
6650	For white, 4" thick, add					1.02			1.02	1.12
6700	6" thick, add					1.02			1.02	1.12
6750	8" thick, add					.95			.95	1.04
6800	12" thick, add					.87			.87	.96
7000	Scored ground face, 2 to 5 scores									
7100	8" x 16" x 4" thick	D-8	340	.118	S.F.	4.57	4.03		8.60	11.20
7150	6" thick		310	.129		5.20	4.42		9.62	12.50
7200	8" thick		290	.138		6.15	4.73		10.88	13.95
7250	12" thick	D-9	265	.181		7.60	6.05		13.65	17.60
8000	Hexagonal face profile units, 8" x 16" units									
8100	4" thick, hollow	D-8	340	.118	S.F.	2.75	4.03		6.78	9.20
8200	Solid		340	.118		3.64	4.03		7.67	10.15
8300	6" thick, hollow		310	.129		3.08	4.42		7.50	10.15

04 22 Concrete Unit Masonry

04 22 10 – Concrete Masonry Units

04 22 10.23 Concrete Block, Decorative		Crew	Daily Output	Labor-Hours	Unit	Material	2007 Bare Costs Labor	Equipment	Total	Total Incl O&P
8350	8" thick, hollow	D-8	290	.138	S.F.	4.18	4.73		8.91	11.80
8500	For stacked bond, add						26%			
8550	For high rise construction, add per story	D-8	67.80	.590	M.S.F.		20		20	31
8600	For scored block, add					10%				
8650	For honed or ground face, per face, add				Ea.	.34			.34	.37
8700	For honed or ground end, per end, add				"	2.63			2.63	2.90
8750	For bullnose block, add					10%				
8800	For special color, add					13%				

04 22 10.24 Concrete Block, Exterior		Crew	Daily Output	Labor-Hours	Unit	Material	Labor	Equipment	Total	Total Incl O&P
0010	**CONCRETE BLOCK, EXTERIOR**									
0020	Reinforced alt courses, tooled joints 2 sides									
0100	Normal weight, 8" x 16" x 6" thick	D-8	395	.101	S.F.	2.20	3.47		5.67	7.70
0200	8" thick		360	.111		3.23	3.81		7.04	9.35
0250	10" thick		290	.138		3.86	4.73		8.59	11.45
0300	12" thick	D-9	250	.192		4	6.40		10.40	14.15
0500	Lightweight, 8" x 16" x 6" thick	D-8	450	.089		2.46	3.05		5.51	7.35
0600	8" thick		430	.093		3.27	3.19		6.46	8.45
0650	10" thick		395	.101		3.89	3.47		7.36	9.60
0700	12" thick	D-9	350	.137		5.60	4.57		10.17	13.10

04 22 10.26 Concrete Block Foundation Wall		Crew	Daily Output	Labor-Hours	Unit	Material	Labor	Equipment	Total	Total Incl O&P
0010	**CONCRETE BLOCK FOUNDATION WALL**									
0050	Normal-weight, cut joints, horiz joint reinf, no vert reinf									
0200	Hollow, 8" x 16" x 6" thick	D-8	455	.088	S.F.	2.28	3.01		5.29	7.10
0250	8" thick		425	.094		2.45	3.23		5.68	7.60
0300	10" thick		350	.114		3.19	3.92		7.11	9.45
0350	12" thick	D-9	300	.160		3.43	5.35		8.78	11.85
0500	Solid, 8" x 16" block, 6" thick	D-8	440	.091		2.30	3.12		5.42	7.25
0550	8" thick	"	415	.096		3.38	3.31		6.69	8.75
0600	12" thick	D-9	350	.137		5	4.57		9.57	12.45

04 22 10.28 Concrete Block, High Strength		Crew	Daily Output	Labor-Hours	Unit	Material	Labor	Equipment	Total	Total Incl O&P
0010	**CONCRETE BLOCK, HIGH STRENGTH**									
0050	Hollow, reinforced alternate courses, 8" x 16" units									
0200	3500 psi, 4" thick	D-8	440	.091	S.F.	1.73	3.12		4.85	6.65
0250	6" thick		395	.101		2.12	3.47		5.59	7.65
0300	8" thick		360	.111		3.16	3.81		6.97	9.30
0350	12" thick	D-9	250	.192		3.90	6.40		10.30	14.05
0500	5000 psi, 4" thick	D-8	440	.091		2.11	3.12		5.23	7.05
0550	6" thick		395	.101		2.65	3.47		6.12	8.20
0600	8" thick		360	.111		3.58	3.81		7.39	9.75
0650	12" thick	D-9	300	.160		5.30	5.35		10.65	13.95
1000	For 75% solid block, add					30%				
1050	For 100% solid block, add					50%				

04 22 10.30 Concrete Block, Interlocking		Crew	Daily Output	Labor-Hours	Unit	Material	Labor	Equipment	Total	Total Incl O&P
0010	**CONCRETE BLOCK, INTERLOCKING**									
0100	Not including grout or reinforcing									
0200	8" x 16" units, 2,000 psi, 8" thick	D-1	245	.065	S.F.	2.43	2.18		4.61	6
0300	12" thick		220	.073		3.60	2.43		6.03	7.65
0350	16" thick		185	.086		5.45	2.88		8.33	10.40
0400	Including grout & reinforcing, 8" thick	D-4	245	.131		7.05	4.32	.61	11.98	14.95
0450	12" thick		220	.145		8.40	4.81	.68	13.89	17.30
0500	16" thick		185	.173		10.40	5.70	.81	16.91	21

04 22 Concrete Unit Masonry

04 22 10 – Concrete Masonry Units

04 22 10.32 Concrete Block, Lintels

		Crew	Daily Output	Labor-Hours	Unit	Material	2007 Bare Costs Labor	Equipment	Total	Total Incl O&P
0010	**CONCRETE BLOCK, LINTELS**									
0100	Including grout and horizontal reinforcing									
0200	8" x 8" x 8", 1 #4 bar	D-4	300	.107	L.F.	4.29	3.53	.50	8.32	10.60
0250	2 #4 bars		295	.108		4.49	3.58	.51	8.58	10.95
0400	8" x 16" x 8", 1 #4 bar		275	.116		7.35	3.85	.54	11.74	14.55
0450	2 #4 bars		270	.119		7.55	3.92	.55	12.02	14.85
1000	12" x 8" x 8", 1 #4 bar		275	.116		6	3.85	.54	10.39	13.05
1100	2 #4 bars		270	.119		6.20	3.92	.55	10.67	13.40
1150	2 #5 bars		270	.119		6.45	3.92	.55	10.92	13.65
1200	2 #6 bars		265	.121		6.70	3.99	.57	11.26	14.05
1500	12" x 16" x 8", 1 #4 bar		250	.128		9.55	4.23	.60	14.38	17.55
1600	2 #3 bars		245	.131		9.60	4.32	.61	14.53	17.75
1650	2 #4 bars		245	.131		9.75	4.32	.61	14.68	17.95
1700	2 #5 bars		240	.133		10	4.41	.62	15.03	18.40

04 22 10.34 Concrete Block, Partitions

		Crew	Daily Output	Labor-Hours	Unit	Material	2007 Bare Costs Labor	Equipment	Total	Total Incl O&P
0010	**CONCRETE BLOCK, PARTITIONS** R042210-20									
0100	Acoustical slotted block									
0200	4" thick, type A-1	D-8	315	.127	S.F.	3.04	4.35		7.39	10
0210	8" thick		275	.145		4.48	4.99		9.47	12.55
0250	8" thick, type Q		275	.145		6.10	4.99		11.09	14.30
0260	4" thick, type RSC		315	.127		4.48	4.35		8.83	11.60
0270	6" thick		295	.136		4.48	4.65		9.13	12.05
0280	8" thick		275	.145		4.48	4.99		9.47	12.55
0290	12" thick		250	.160		4.48	5.50		9.98	13.30
0300	8" thick, type RSR		275	.145		4.48	4.99		9.47	12.55
0400	8" thick, type RSC/RF		275	.145		5.60	4.99		10.59	13.75
0410	10" thick		260	.154		6.10	5.30		11.40	14.75
0420	12" thick		250	.160		6.65	5.50		12.15	15.70
0430	12" thick, type RSC/RF-4		250	.160		7.70	5.50		13.20	16.80
0500	NRC .60 type R, 8" thick		265	.151		5.25	5.20		10.45	13.65
0600	NRC .65 type RR, 8" thick		265	.151		5.90	5.20		11.10	14.40
0700	NRC .65 type 4R-RF, 8" thick		265	.151		6.70	5.20		11.90	15.25
0710	NRC .70 type R, 12" thick		245	.163		7.50	5.60		13.10	16.70
1000	Lightweight block, tooled joints, 2 sides, hollow									
1100	Not reinforced, 8" x 16" x 4" thick	D-8	440	.091	S.F.	1.49	3.12		4.61	6.40
1150	6" thick		410	.098		2.03	3.35		5.38	7.35
1200	8" thick		385	.104		2.48	3.56		6.04	8.15
1250	10" thick		370	.108		3.21	3.71		6.92	9.20
1300	12" thick	D-9	350	.137		3.51	4.57		8.08	10.80
2000	Not reinforced, 8" x 24" x 4" thick, hollow		460	.104		1.09	3.48		4.57	6.50
2100	6" thick		440	.109		1.46	3.64		5.10	7.15
2150	8" thick		415	.116		1.83	3.86		5.69	7.85
2200	10" thick		385	.125		2.34	4.16		6.50	8.95
2250	12" thick		365	.132		2.58	4.39		6.97	9.50
2800	Solid, not reinforced, 8" x 16" x 2" thick	D-8	440	.091		1.30	3.12		4.42	6.15
2900	4" thick		420	.095		2.03	3.27		5.30	7.20
2950	6" thick		390	.103		2.37	3.52		5.89	7.95
3000	8" thick		365	.110		3.17	3.76		6.93	9.20
3050	10" thick		350	.114		3.79	3.92		7.71	10.10
3100	12" thick	D-9	330	.145		5.50	4.85		10.35	13.45
4000	Regular block, tooled joints, 2 sides, hollow									
4100	Not reinforced, 8" x 16" x 4" thick	D-8	430	.093	S.F.	1.22	3.19		4.41	6.20

04 22 Concrete Unit Masonry

04 22 10 – Concrete Masonry Units

04 22 10.34 Concrete Block, Partitions

		Crew	Daily Output	Labor-Hours	Unit	Material	2007 Bare Costs Labor	Equipment	Total	Total Incl O&P
4150	6" thick	D-8	400	.100	S.F.	1.78	3.43		5.21	7.15
4200	8" thick	CN	375	.107		1.94	3.66		5.60	7.70
4250	10" thick		360	.111		2.66	3.81		6.47	8.75
4300	12" thick	D-9	340	.141		2.89	4.71		7.60	10.35
4500	Reinforced alternate courses, 8" x 16" x 4" thick	D-8	425	.094		1.33	3.23		4.56	6.35
4550	6" thick		395	.101		1.90	3.47		5.37	7.40
4600	8" thick		370	.108		2.07	3.71		5.78	7.90
4650	10" thick		355	.113		2.91	3.86		6.77	9.10
4700	12" thick	D-9	335	.143		3.03	4.78		7.81	10.60
4900	Solid, not reinforced, 2" thick	D-8	435	.092		1.12	3.15		4.27	6.05
5000	3" thick		430	.093		1.10	3.19		4.29	6.05
5050	4" thick		415	.096		1.57	3.31		4.88	6.75
5100	6" thick		385	.104		1.80	3.56		5.36	7.40
5150	8" thick		360	.111		2.87	3.81		6.68	8.95
5200	12" thick	D-9	325	.148		4.48	4.93		9.41	12.45
5500	Solid, reinforced alternate courses, 4" thick	D-8	420	.095		1.64	3.27		4.91	6.80
5550	6" thick		380	.105		1.89	3.61		5.50	7.60
5600	8" thick		355	.113		2.96	3.86		6.82	9.15
5650	12" thick	D-9	320	.150		3.97	5		8.97	11.95

04 22 10.38 Concrete Brick

		Crew	Daily Output	Labor-Hours	Unit	Material	Labor	Equipment	Total	Total Incl O&P
0010	**CONCRETE BRICK**									
0100	Regular, 4 x 2-1/4 x 8	D-8	660	.061	Ea.	.38	2.08		2.46	3.58
0125	Rusticated, 4 x 2-1/4 x 8		660	.061		.43	2.08		2.51	3.63
0150	Frog, 4 x 2-1/4 x 8		660	.061		.41	2.08		2.49	3.61
0200	Double, 4 x 4-7/8 x 8		535	.075		.67	2.56		3.23	4.63

04 22 10.42 Concrete Block, Screen Block

		Crew	Daily Output	Labor-Hours	Unit	Material	Labor	Equipment	Total	Total Incl O&P
0010	**CONCRETE BLOCK, SCREEN BLOCK**									
0200	8" x 16", 4" thick	D-8	330	.121	S.F.	1.82	4.16		5.98	8.30
0300	8" thick		270	.148		2.76	5.10		7.86	10.80
0350	12" x 12", 4" thick		290	.138		2.05	4.73		6.78	9.45
0500	8" thick		250	.160		2.82	5.50		8.32	11.45

04 22 10.44 Glazed Concrete Block

		Crew	Daily Output	Labor-Hours	Unit	Material	Labor	Equipment	Total	Total Incl O&P
0010	**GLAZED CONCRETE BLOCK**									
0100	Single face, 8" x 16" units, 2" thick	D-8	360	.111	S.F.	7.15	3.81		10.96	13.65
0200	4" thick		345	.116		7.25	3.98		11.23	14.05
0250	6" thick		330	.121		7.80	4.16		11.96	14.85
0300	8" thick		310	.129		8.35	4.42		12.77	15.95
0350	10" thick		295	.136		9.35	4.65		14	17.40
0400	12" thick	D-9	280	.171		9.80	5.70		15.50	19.50
0700	Double face, 8" x 16" units, 4" thick	D-8	340	.118		11.05	4.03		15.08	18.30
0750	6" thick		320	.125		12.45	4.29		16.74	20
0800	8" thick		300	.133		13	4.57		17.57	21.50
1000	Jambs, bullnose or square, single face, 8" x 16", 2" thick		315	.127	Ea.	13.30	4.35		17.65	21.50
1050	4" thick		285	.140	"	14.20	4.81		19.01	23
1200	Caps, bullnose or square, 8" x 16", 2" thick		420	.095	L.F.	11.15	3.27		14.42	17.20
1250	4" thick		380	.105		14.50	3.61		18.11	21.50
1500	Cove base, 8" x 16", 2" thick		315	.127		7.15	4.35		11.50	14.55
1550	4" thick		285	.140		6.50	4.81		11.31	14.45
1600	6" thick		265	.151		7	5.20		12.20	15.60
1650	8" thick		245	.163		7.35	5.60		12.95	16.60

04 23 Glass Unit Masonry

04 23 13 – Vertical Glass Unit Masonry

04 23 13.10 Glass Block

		Crew	Daily Output	Labor-Hours	Unit	Material	2007 Bare Costs Labor	Equipment	Total	Total Incl O&P
0010	**GLASS BLOCK**									
0100	Plain, 4" thick, under 1,000 S.F., 6" x 6"	D-8	115	.348	S.F.	18.70	11.95		30.65	38.50
0150	8" x 8"		160	.250		10.60	8.55		19.15	24.50
0160	end block		160	.250		35	8.55		43.55	51.50
0170	90 deg corner		160	.250		34	8.55		42.55	50.50
0180	45 deg corner		160	.250		15.75	8.55		24.30	30.50
0200	12" x 12"		175	.229		13.55	7.85		21.40	27
0210	4" x 8"		160	.250		8.55	8.55		17.10	22.50
0220	6" x 8"		160	.250		10.10	8.55		18.65	24
0300	1,000 to 5,000 S.F., 6" x 6"		135	.296		18.35	10.15		28.50	35.50
0350	8" x 8"		190	.211		10.40	7.20		17.60	22.50
0400	12" x 12"		215	.186		13.25	6.40		19.65	24.50
0410	4" x 8"		215	.186		4.11	6.40		10.51	14.20
0420	6" x 8"		215	.186		4.48	6.40		10.88	14.65
0500	Over 5,000 S.F., 6" x 6"		145	.276		17.80	9.45		27.25	34
0550	8" x 8"		215	.186		10.05	6.40		16.45	21
0600	12" x 12"		240	.167		12.85	5.70		18.55	23
0610	4" x 8"		240	.167		4.36	5.70		10.06	13.50
0620	6" x 8"	▼	240	.167	▼	4.48	5.70		10.18	13.65
0700	For solar reflective blocks, add					100%				
1000	Thinline, plain, 3-1/8" thick, under 1,000 S.F., 6" x 6"	D-8	115	.348	S.F.	13.25	11.95		25.20	32.50
1050	8" x 8"		160	.250		10.60	8.55		19.15	24.50
1200	Over 5,000 S.F., 6" x 6"		145	.276		13.25	9.45		22.70	29
1250	8" x 8"		215	.186		7.60	6.40		14	18.05
1400	For cleaning block after installation (both sides), add	▼	1000	.040	▼	.10	1.37		1.47	2.20
4000	Accessories									
4100	Anchors, 20 ga. galv., 1-3/4" wide x 24" long				Ea.	3.07			3.07	3.38
4200	Emulsion asphalt				Gal.	10.60			10.60	11.70
4300	Expansion joint, fiberglass				L.F.	.61			.61	.67
4400	Steel mesh, double galvanized				"	.49			.49	.54

04 24 Adobe Unit Masonry

04 24 16 – Manufactured Adobe Unit Masonry

04 24 16.06 Adobe Brick

		Crew	Daily Output	Labor-Hours	Unit	Material	2007 Bare Costs Labor	Equipment	Total	Total Incl O&P
0010	**ADOBE BRICK**									
0060	Brick, 10" x 4" x 14", 2.6/S.F.	D-8	560	.071	S.F.	2.77	2.45		5.22	6.80
0080	12" x 4" x 16", 2.3/S.F.		580	.069		4.28	2.37		6.65	8.30
0100	10" x 4" x 16", 2.3/S.F.		590	.068		4.07	2.32		6.39	8
0120	8" x 4" x 16", 2.3/S.F.		560	.071		3.02	2.45		5.47	7.05
0140	4" x 4" x 16", 2.3/S.F.		540	.074		2.50	2.54		5.04	6.60
0160	6" x 4" x 16", 2.3/S.F.		540	.074		2.45	2.54		4.99	6.55
0180	4" x 4" x 12", 3.0/S.F.		520	.077		2.27	2.64		4.91	6.50
0200	8" x 4" x 12", 3.0/S.F.	▼	520	.077	▼	2.60	2.64		5.24	6.90

04 25 Unit Masonry Panels

04 25 10 – Unit Masonry Panels

04 25 10.10 Facing Panels	Crew	Daily Output	Labor-Hours	Unit	Material	2007 Bare Costs Labor	Equipment	Total	Total Incl O&P
0010 **FACING PANELS**									
0020 See division 07 44 13.10									

04 25 20 – Pre-Fabricated Masonry Panels

04 25 20.10 Pre-Fabricated Masonry Panels

0010 **PRE-FABRICATED MASONRY PANELS**									
0020 Prefabricated, 4" thick, minimum	C-11	775	.093	S.F.	8.65	3.77	2.39	14.81	18.80
0100 Maximum	"	500	.144		11.40	5.85	3.70	20.95	27
0200 4" brick & 2" concrete back-up, add					50%				
0300 4" brick & 1" urethane & 3" concrete back-up, add					70%				

04 27 Multiple-Wythe Unit Masonry

04 27 10 – Multiple-Wythe Unit Masonry

04 27 10.20 Cavity Walls

	Crew	Daily Output	Labor-Hours	Unit	Material	Labor	Equipment	Total	Total Incl O&P
0010 **CAVITY WALLS**									
0200 4" face brick, 4" block	D-8	165	.242	S.F.	5.20	8.30		13.50	18.35
0400 6" block		145	.276		5.65	9.45		15.10	20.50
0600 8" block		125	.320		5.70	10.95		16.65	23

04 27 10.30 Brick Walls

	Crew	Daily Output	Labor-Hours	Unit	Material	Labor	Equipment	Total	Total Incl O&P
0010 **BRICK WALLS** R042110-20									
0140 4" thick, facing, 4" x 2-2/3" x 8"	D-8	1.45	27.586	M	415	945		1,360	1,900
0150 4" thick, as back-up, 6.75 bricks per S.F.		1.60	25		415	855		1,270	1,750
0204 8" thick, 13.50 bricks per S.F.		1.80	22.222		435	760		1,195	1,625
0250 12" thick, 20.25 bricks per S.F.		1.90	21.053		435	720		1,155	1,575
0304 16" thick, 27.00 bricks per S.F.		2	20		440	685		1,125	1,525
0500 Reinforced, 4" wall, 4" x 2-2/3" x 8"		1.40	28.571		420	980		1,400	1,950
0550 8" thick, 13.50 bricks per S.F.		1.75	22.857		510	785		1,295	1,750
0600 12" thick, 20.25 bricks per S.F.		1.85	21.622		510	740		1,250	1,700
0650 16" thick, 27.00 bricks per S.F.		1.95	20.513		520	705		1,225	1,650
0660 4" thick, select common, face, 4" x 2-2/3" x 8"		1.45	27.586		495	945		1,440	2,000
0790 Alternate method of figuring by square foot									
0800 4" wall, face, 4" x 2-2/3" x 8"	D-8	215	.186	S.F.	3.79	6.40		10.19	13.85
0850 4" thick, as back up, 6.75 bricks per S.F.		240	.167		2.81	5.70		8.51	11.80
0900 8" thick wall, 13.50 brick per S.F.		135	.296		5.85	10.15		16	22
1000 12" thick wall, 20.25 bricks per S.F.		95	.421		8.80	14.45		23.25	31.50
1050 16" thick wall, 27.00 bricks per S.F.		75	.533		11.95	18.30		30.25	41
1200 Reinforced, 4" x 2-2/3" x 8", 4" wall		205	.195		2.81	6.70		9.51	13.30
1250 8" thick wall, 13.50 brick per S.F.		130	.308		5.85	10.55		16.40	22.50
1300 12" thick wall, 20.25 bricks per S.F.		90	.444		8.80	15.25		24.05	32.50
1350 16" thick wall, 27.00 bricks per S.F.		70	.571		11.95	19.60		31.55	43

04 27 10.40 Steps

	Crew	Daily Output	Labor-Hours	Unit	Material	Labor	Equipment	Total	Total Incl O&P
0010 **STEPS**									
0012 Entry steps, brick	D-1	.30	53.333	M	405	1,775		2,180	3,150

04 41 Dry-Placed Stone

04 41 10 – Dry Placed Stone

04 41 10.10 Rough Stone Wall

		Crew	Daily Output	Labor-Hours	Unit	Material	2007 Bare Costs Labor	Equipment	Total	Total Incl O&P
0011	**ROUGH STONE WALL**, Dry									
0100	Random fieldstone, under 18" thick	D-12	60	.533	C.F.	8.20	18.05		26.25	36.50
0150	Over 18" thick	"	63	.508	"	9.85	17.20		27.05	37

04 43 Stone Masonry

04 43 10 – Stone Masonry

04 43 10.05 Ashlar Veneer

		Crew	Daily Output	Labor-Hours	Unit	Material	Labor	Equipment	Total	Total Incl O&P
0011	**ASHLAR VENEER** 4" + or - thk, random or random rectangular									
0150	Sawn face, split joints, low priced stone	D-8	140	.286	S.F.	11.30	9.80		21.10	27.50
0200	Medium priced stone		130	.308		13.85	10.55		24.40	31.50
0300	High priced stone		120	.333		19.45	11.45		30.90	39
0600	Seam face, split joints, medium price stone		125	.320		14.55	10.95		25.50	32.50
0700	High price stone		120	.333		20.50	11.45		31.95	40
1000	Split or rock face, split joints, medium price stone		125	.320		14.30	10.95		25.25	32.50
1100	High price stone		120	.333		20.50	11.45		31.95	40

04 43 10.10 Bluestone

		Crew	Daily Output	Labor-Hours	Unit	Material	Labor	Equipment	Total	Total Incl O&P
0010	**BLUESTONE**									
0500	Sills, natural cleft, 10" wide to 6' long, 1-1/2" thick	D-11	70	.343	L.F.	12.85	12.20		25.05	32.50
0550	2" thick	"	63	.381		14.60	13.55		28.15	36.50
1000	Stair treads, natural cleft, 12" wide, 6' long, 1-1/2" thick	D-10	115	.348		18.90	12.20	5	36.10	45
1050	2" thick		105	.381		19.90	13.35	5.50	38.75	48.50
1100	Smooth finish, 1-1/2" thick		115	.348		23	12.20	5	40.20	49
1300	Thermal finish		115	.348		28.50	12.20	5	45.70	55.50
1350	2" thick		105	.381		32.50	13.35	5.50	51.35	62

04 43 10.45 Granite

			Crew	Daily Output	Labor-Hours	Unit	Material	Labor	Equipment	Total	Total Incl O&P
0010	**GRANITE**										
0050	Veneer, polished face, 3/4" to 1-1/2" thick										
0150	Low price, gray, light gray, etc.	CN	D-10	130	.308	S.F.	23.50	10.80	4.42	38.72	47
0220	High price, red, black, etc.	CN	"	130	.308	"	37	10.80	4.42	52.22	62
0300	1-1/2" to 2-1/2" thick, veneer										
0350	Low price, gray, light gray, etc.		D-10	130	.308	S.F.	25	10.80	4.42	40.22	48.50
0550	High price, red, black, etc.		"	130	.308	"	45.50	10.80	4.42	60.72	71.50
0700	2-1/2" to 4" thick, veneer										
0750	Low price, gray, light gray, etc.		D-10	110	.364	S.F.	33.50	12.75	5.25	51.50	62
0950	High price, red, black, etc.		"	110	.364		55	12.75	5.25	73	85.50
1000	For bush hammered finish, deduct						5%				
1050	Coarse rubbed finish, deduct						10%				
1100	Honed finish, deduct						5%				
1150	Thermal finish, deduct						18%				
2450	For radius under 5', add					L.F.	100%				
2500	Steps, copings, etc., finished on more than one surface										
2550	Minimum		D-10	50	.800	C.F.	87.50	28	11.50	127	152
2600	Maximum		"	50	.800	"	140	28	11.50	179.50	209
2800	Pavers, 4" x 4" x 4" blocks, split face and joints										
2850	Minimum		D-11	80	.300	S.F.	12.90	10.65		23.55	30.50
2900	Maximum		"	80	.300	"	25.50	10.65		36.15	44.50
3500	Curbing, city street type, See Division 32 16 13.13										
4000	Soffits, 2" thick, minimum		D-13	35	1.371	S.F.	36	48.50	16.45	100.95	132
4100	Maximum			35	1.371		80	48.50	16.45	144.95	180
4200	4" thick, minimum			35	1.371		55	48.50	16.45	119.95	153

04 43 Stone Masonry

04 43 10 – Stone Masonry

04 43 10.45 Granite		Crew	Daily Output	Labor-Hours	Unit	Material	2007 Bare Costs Labor	Equipment	Total	Total Incl O&P
4300	Maximum	D-13	35	1.371	S.F.	104	48.50	16.45	168.95	206
04 43 10.50 Lightweight Natural Stone										
0011	**LIGHTWEIGHT NATURAL STONE** Lava type									
0100	Veneer, rubble face, sawed back, irregular shapes	D-10	130	.308	S.F.	6.05	10.80	4.42	21.27	28
0200	Sawed face and back, irregular shapes	"	130	.308	"	6.05	10.80	4.42	21.27	28
04 43 10.55 Limestone										
0010	**LIMESTONE**									
0020	Veneer facing panels									
0500	Texture finish, light stick, 4-1/2" thick, 5' x 12'	D-4	300	.107	S.F.	46.50	3.53	.50	50.53	57
0750	5" thick, 5' x 14' panels	D-10	275	.145		50.50	5.10	2.09	57.69	65.50
1000	Sugarcube finish, 2" Thick, 3' x 5' panels		275	.145		25	5.10	2.09	32.19	37.50
1050	3" Thick, 4' x 9' panels		275	.145		37.50	5.10	2.09	44.69	51.50
1200	4" Thick, 5' x 11' panels		275	.145		39.50	5.10	2.09	46.69	53.50
1400	Sugarcube, textured finish, 4-1/2" thick, 5' x 12'		275	.145		43.50	5.10	2.09	50.69	58
1450	5" thick, 5' x 14' panels		275	.145		51.50	5.10	2.09	58.69	66.50
2000	Coping, sugarcube finish, top & 2 sides		30	1.333	C.F.	44.50	46.50	19.15	110.15	141
2100	Sills, lintels, jambs, trim, stops, sugarcube finish, average		20	2		66	70	29	165	210
2150	Detailed		20	2		66	70	29	165	210
2300	Steps, extra hard, 14" wide, 6" rise		50	.800	L.F.	22.50	28	11.50	62	80
3000	Quoins, plain finish, 6"x12"x12"	D-12	25	1.280	Ea.	60	43.50		103.50	132
3050	6"x16"x24"	"	25	1.280	"	80	43.50		123.50	154
04 43 10.60 Marble										
0011	**MARBLE**, ashlar, split face, 4" + or - thick, random									
0040	Lengths 1' to 4' & heights 2" to 7-1/2", average	D-8	175	.229	S.F.	14.55	7.85		22.40	28
0100	Base, polished, 3/4" or 7/8" thick, polished, 6" high	D-10	65	.615	L.F.	13.25	21.50	8.85	43.60	57
0300	Carvings or bas relief, from templates, average		80	.500	S.F.	118	17.50	7.20	142.70	164
0350	Maximum		80	.500	"	276	17.50	7.20	300.70	340
0600	Columns, cornices, mouldings, etc.									
0650	Hand or special machine cut, average	D-10	35	1.143	C.F.	118	40	16.45	174.45	209
0700	Maximum	"	35	1.143	"	253	40	16.45	309.45	360
1000	Facing, polished finish, cut to size, 3/4" to 7/8" thick									
1050	Average	D-10	130	.308	S.F.	19.45	10.80	4.42	34.67	42.50
1100	Maximum		130	.308		45	10.80	4.42	60.22	70.50
1300	1-1/4" thick, average		125	.320		28	11.20	4.60	43.80	53
1350	Maximum		125	.320		56.50	11.20	4.60	72.30	84
1500	2" thick, average		120	.333		32.50	11.70	4.79	48.99	59
1550	Maximum		120	.333		56.50	11.70	4.79	72.99	85
2200	Window sills, 6" x 3/4" thick	D-1	85	.188	L.F.	7.30	6.30		13.60	17.60
2500	Flooring, polished tiles, 12" x 12" x 3/8" thick									
2510	Thin set, average	D-11	90	.267	S.F.	9.75	9.50		19.25	25
2600	Maximum		90	.267		150	9.50		159.50	179
2700	Mortar bed, average		65	.369		9.90	13.15		23.05	31
2740	Maximum		65	.369		150	13.15		163.15	185
2780	Travertine, 3/8" thick, average	D-10	130	.308		12.40	10.80	4.42	27.62	35
2790	Maximum	"	130	.308		30.50	10.80	4.42	45.72	54.50
2800	Patio tile, non-slip, 1/2" thick, flame finish	D-11	75	.320		12.65	11.40		24.05	31.50
2900	Shower or toilet partitions, 7/8" thick partitions									
3050	3/4" or 1-1/4" thick stiles, polished 2 sides, average	D-11	75	.320	S.F.	37.50	11.40		48.90	59
3201	Soffits, add to above prices				"	20%	100%			
3210	Stairs, risers, 7/8" thick x 6" high	D-10	115	.348	L.F.	12.75	12.20	5	29.95	38
3360	Treads, 12" wide x 1-1/4" thick	"	115	.348	"	18.55	12.20	5	35.75	44.50
3500	Thresholds, 3' long, 7/8" thick, 4" to 5" wide, plain	D-12	24	1.333	Ea.	14.05	45		59.05	84

04 43 Stone Masonry

04 43 10 – Stone Masonry

04 43 10.60 Marble		Crew	Daily Output	Labor-Hours	Unit	Material	2007 Bare Costs Labor	Equipment	Total	Total Incl O&P
3550	Beveled	D-12	24	1.333	Ea.	16.30	45		61.30	86.50
3700	Window stools, polished, 7/8" thick, 5" wide	↓	85	.376	L.F.	13.10	12.75		25.85	34
04 43 10.75 Sandstone Or Brownstone										
0011	**SANDSTONE OR BROWNSTONE**									
0100	Sawed face veneer, 2-1/2" thick, to 2' x 4' panels	D-10	130	.308	S.F.	17.25	10.80	4.42	32.47	40
0150	4" thick, to 3'-6" x 8' panels		100	.400		17.25	14	5.75	37	47
0300	Split face, random sizes	↓	100	.400	↓	12.40	14	5.75	32.15	41.50
0350	Cut stone trim (limestone)									
0360	Ribbon stone, 4" thick, 5' pieces	D-8	120	.333	Ea.	152	11.45		163.45	184
0370	Cove stone, 4" thick, 5' pieces		105	.381		153	13.05		166.05	188
0380	Cornice stone, 10" to 12" wide		90	.444		189	15.25		204.25	230
0390	Band stone, 4" thick, 5' pieces		145	.276		97.50	9.45		106.95	121
0410	Window and door trim, 3" to 4" wide		160	.250		82.50	8.55		91.05	104
0420	Key stone, 18" long	↓	60	.667	↓	87	23		110	131
04 43 10.80 Slate										
0010	**SLATE**									
0050	Unfading green, mottled green & purple, gray & purple									
0100	Virginia, blue black									
0200	Exterior paving, natural cleft, 1" thick									
0500	24" x 24", Pennsylvania	D-12	120	.267	S.F.	11.50	9.05		20.55	26.50
0550	Vermont		120	.267		15.15	9.05		24.20	30.50
0600	Virginia	↓	120	.267	↓	18.05	9.05		27.10	33.50
1000	Interior flooring, natural cleft, 1/2" thick									
1300	24" x 24" Pennsylvania	D-12	120	.267	S.F.	7	9.05		16.05	21.50
1350	Vermont		120	.267		13.30	9.05		22.35	28.50
1400	Virginia	↓	120	.267	↓	13.05	9.05		22.10	28
2000	Facing panels, 1-1/4" thick, to 4' x 4' panels									
2100	Natural cleft finish, Pennsylvania	D-10	180	.222	S.F.	28.50	7.80	3.20	39.50	47
2110	Vermont		180	.222		25.50	7.80	3.20	36.50	43.50
2120	Virginia	↓	180	.222		29.50	7.80	3.20	40.50	48
2150	Sand rubbed finish, surface, add					9			9	9.90
2200	Honed finish, add					6.50			6.50	7.15
2500	Ribbon, natural cleft finish, 1" thick, to 9 S.F.	D-10	80	.500		11.60	17.50	7.20	36.30	47
2700	1-1/2" thick		78	.513		15.10	17.95	7.35	40.40	52
2850	2" thick	↓	76	.526	↓	18.10	18.45	7.55	44.10	56
3000	Roofing, see division 07 31 26.10									
3500	Stair treads, sand finish, 1" thick x 12" wide									
3600	3 L.F. to 6 L.F.	D-10	120	.333	L.F.	21.50	11.70	4.79	37.99	46.50
3700	Ribbon, sand finish, 1" thick x 12" wide									
3750	To 6 L.F.	D-10	120	.333	L.F.	15.65	11.70	4.79	32.14	40
4000	Stools or sills, sand finish, 1" thick, 6" wide	D-12	160	.200		10.50	6.75		17.25	22
4200	10" wide		90	.356		16	12.05		28.05	36
4400	2" thick, 6" wide		140	.229		17.75	7.75		25.50	31.50
4600	10" wide	↓	90	.356		26.50	12.05		38.55	48
4800	For lengths over 3', add				↓	25%				
04 43 10.85 Window Sill										
0010	**WINDOW SILL**									
0020	Bluestone, thermal top, 10" wide, 1-1/2" thick	D-1	85	.188	S.F.	15.45	6.30		21.75	26.50
0050	2" thick		75	.213	"	18	7.10		25.10	30.50
0100	Cut stone, 5" x 8" plain		48	.333	L.F.	11.40	11.10		22.50	29.50
0200	Face brick on edge, brick, 8" wide		80	.200		2.41	6.65		9.06	12.80
0400	Marble, 9" wide, 1" thick	↓	85	.188	↓	8.40	6.30		14.70	18.80

04 43 Stone Masonry

04 43 10 – Stone Masonry

04 43 10.85 Window Sill		Crew	Daily Output	Labor-Hours	Unit	Material	2007 Bare Costs Labor	Equipment	Total	Total Incl O&P
0900	Slate, colored, unfading, honed, 12" wide, 1" thick	D-1	85	.188	L.F.	17.10	6.30		23.40	28.50
0950	2" thick	↓	70	.229	↓	24	7.60		31.60	37.50

04 51 Flue Liner Masonry

04 51 10 – Flue Liner Masonry

04 51 10.10 Flue Lining

		Crew	Daily Output	Labor-Hours	Unit	Material	Labor	Equipment	Total	Total Incl O&P
0010	**FLUE LINING**									
0020	8" x 8"	D-1	125	.128	V.L.F.	3.92	4.27		8.19	10.80
0100	8" x 12"		103	.155		5.85	5.20		11.05	14.35
0200	12" x 12"		93	.172		7.65	5.75		13.40	17.15
0300	12" x 18"		84	.190		13.15	6.35		19.50	24
0400	18" x 18"		75	.213		18.80	7.10		25.90	31.50
0500	20" x 20"		66	.242		30.50	8.10		38.60	46
0600	24" x 24"		56	.286		39	9.55		48.55	57.50
1000	Round, 18" diameter		66	.242		28.50	8.10		36.60	43.50
1100	24" diameter	↓	47	.340	↓	59	11.35		70.35	82.50

04 54 Refractory Brick Masonry

04 54 10 – Refractory Brick Masonry

04 54 10.10 Fire Brick

		Crew	Daily Output	Labor-Hours	Unit	Material	Labor	Equipment	Total	Total Incl O&P
0010	**FIRE BRICK**									
0012	Low duty, 2000° F, 9" x 2-1/2" x 4-1/2"	D-1	.60	26.667	M	1,150	890		2,040	2,625
0050	High duty, 3000° F	"	.60	26.667	"	1,900	890		2,790	3,425

04 54 10.20 Fire Clay

		Crew	Daily Output	Labor-Hours	Unit	Material	Labor	Equipment	Total	Total Incl O&P
0010	**FIRE CLAY**									
0020	Gray, high duty, 100 lb. bag				Bag	53			53	58.50
0050	100 lb. drum, premixed (400 brick per drum)				Drum	66			66	72.50

04 57 Masonry Fireplaces

04 57 10 – Masonry Fireplaces

04 57 10.10 Fireplace

		Crew	Daily Output	Labor-Hours	Unit	Material	Labor	Equipment	Total	Total Incl O&P
0010	**FIREPLACE**									
0100	Brick fireplace, not incl. foundations or chimneys									
0110	30" x 29" opening, incl. chamber, plain brickwork	D-1	.40	40	Ea.	480	1,325		1,805	2,550
0200	Fireplace box only (110 brick)	"	2	8	"	158	267		425	580
0300	For elaborate brickwork and details, add					35%	35%			
0400	For hearth, brick & stone, add	D-1	2	8	Ea.	177	267		444	600
0410	For steel angle, damper, cleanouts, add		4	4		124	133		257	340
0600	Plain brickwork, incl. metal circulator		.50	32	↓	915	1,075		1,990	2,625
0800	Face brick only, standard size, 8" x 2-2/3" x 4"	↓	.30	53.333	M	480	1,775		2,255	3,225

04 71 Manufactured Brick Masonry

04 71 10 – Manufactured Brick Masonry

04 71 10.10 Simulated Brick		Crew	Daily Output	Labor-Hours	Unit	Material	2007 Bare Costs Labor	Equipment	Total	Total Incl O&P
0010	**SIMULATED BRICK**									
0020	Aluminum, baked on colors	1 Carp	200	.040	S.F.	3.17	1.47		4.64	5.80
0050	Fiberglass panels		200	.040		3.48	1.47		4.95	6.10
0100	Urethane pieces cemented in mastic		150	.053		6	1.96		7.96	9.65
0150	Vinyl siding panels		200	.040		2.40	1.47		3.87	4.93
0160	Cement base, brick, incl. mastic	D-1	100	.160		3.79	5.35		9.14	12.25
0170	Corner		50	.320	V.L.F.	9.50	10.65		20.15	26.50
0180	Stone face, incl. mastic		100	.160	S.F.	9.15	5.35		14.50	18.20
0190	Corner		50	.320	V.L.F.	10.10	10.65		20.75	27.50

04 72 Cast Stone Masonry

04 72 10 – Cast Stone Masonry Features

04 72 10.10 Coping

		Crew	Daily Output	Labor-Hours	Unit	Material	2007 Bare Costs Labor	Equipment	Total	Total Incl O&P
0010	**COPING**									
0050	Precast concrete, 10" wide, 4" tapers to 3-1/2", 8" wall	D-1	75	.213	L.F.	17.10	7.10		24.20	29.50
0100	12" wide, 3-1/2" tapers to 3", 10" wall		70	.229		11.15	7.60		18.75	24
0110	14" wide, 4" tapers to 3-1/2", 12" wall		65	.246		19.70	8.20		27.90	34
0150	16" wide, 4" tapers to 3-1/2", 14" wall		60	.267		15.35	8.90		24.25	30.50
0250	Precast concrete corners		40	.400	Ea.	25.50	13.35		38.85	48.50
0300	Limestone for 12" wall, 4" thick		90	.178	L.F.	13.50	5.95		19.45	24
0350	6" thick		80	.200		15.75	6.65		22.40	27.50
0500	Marble, to 4" thick, no wash, 9" wide		90	.178		20.50	5.95		26.45	31.50
0550	12" wide		80	.200		30.50	6.65		37.15	44
0700	Terra cotta, 9" wide		90	.178		5.05	5.95		11	14.55
0750	12" wide		80	.200		8.25	6.65		14.90	19.25
0800	Aluminum, for 12" wall		80	.200		11.95	6.65		18.60	23.50

04 72 20 – Cultured Stone Veneer

04 72 20.10 Cultured Stone Veneer

		Crew	Daily Output	Labor-Hours	Unit	Material	2007 Bare Costs Labor	Equipment	Total	Total Incl O&P
0010	**CULTURED STONE VENEER**									
0110	On wood frame and sheathing substrate, random sized cobbles, corner stones	D-8	70	.571	V.L.F.	13.05	19.60		32.65	44.50
0120	Field stones		140	.286	S.F.	9	9.80		18.80	25
0130	Random sized flats, corner stones		70	.571	V.L.F.	12.05	19.60		31.65	43.50
0140	Field stones		140	.286	S.F.	9.70	9.80		19.50	25.50
0150	Horizontal lined ledgestones, corner stones		75	.533	V.L.F.	13.05	18.30		31.35	42.50
0160	Field stones		150	.267	S.F.	9	9.15		18.15	24
0170	Random shaped flats, corner stones		65	.615	V.L.F.	13.05	21		34.05	46.50
0180	Field stones		150	.267	S.F.	9	9.15		18.15	24
0190	Random shaped / textured face, corner stones		65	.615	V.L.F.	13.05	21		34.05	46.50
0200	Field stones		130	.308	S.F.	9	10.55		19.55	26
0210	Random shaped river rock, corner stones		65	.615	V.L.F.	13.05	21		34.05	46.50
0220	Field stones		130	.308	S.F.	9	10.55		19.55	26
0240	On concrete or CMU substrate, random sized cobbles, corner stones		70	.571	V.L.F.	12.50	19.60		32.10	44
0250	Field stones		140	.286	S.F.	8.70	9.80		18.50	24.50
0260	Random sized flats, corner stones		70	.571	V.L.F.	11.45	19.60		31.05	42.50
0270	Field stones		140	.286	S.F.	9.45	9.80		19.25	25.50
0280	Horizontal lined ledgestones, corner stones		75	.533	V.L.F.	12.50	18.30		30.80	42
0290	Field stones		150	.267	S.F.	8.70	9.15		17.85	23.50
0300	Random shaped flats, corner stones		70	.571	V.L.F.	12.50	19.60		32.10	44
0310	Field stones		140	.286	S.F.	8.70	9.80		18.50	24.50
0320	Random shaped / textured face, corner stones		65	.615	V.L.F.	12.50	21		33.50	46

04 72 Cast Stone Masonry

04 72 20 – Cultured Stone Veneer

04 72 20.10 Cultured Stone Veneer		Crew	Daily Output	Labor-Hours	Unit	Material	2007 Bare Costs Labor	Equipment	Total	Total Incl O&P
0330	Field stones	D-8	130	.308	S.F.	8.70	10.55		19.25	25.50
0340	Random shaped river rock, corner stones	↓	65	.615	V.L.F.	12.50	21		33.50	46
0350	Field stones		130	.308	S.F.	8.70	10.55		19.25	25.50
0360	Cultured stone veneer, #15 felt weather resistant barrier	1 Clab	3700	.002	Sq.	4.23	.06		4.29	4.75
0370	#15 felt weather resistant barrier	1 Lath	85	.094	S.Y.	2.50	3.17		5.67	7.45
0390	Water table or window sill, 18" long	1 Bric	80	.100	Ea.	12.50	3.81		16.31	19.55

04 73 Manufactured Stone Masonry

04 73 20 – Manufactured Stone Masonry

04 73 20.10 Simulated Stone		Crew	Daily Output	Labor-Hours	Unit	Material	2007 Bare Costs Labor	Equipment	Total	Total Incl O&P
0010	**SIMULATED STONE**									
0100	Insulated fiberglass panels, 5/8" ply backer	L-4	200	.120	S.F.	11.40	4.13		15.53	18.95

Estimating Tips

05 05 00 Common Work Results for Metals

- Nuts, bolts, washers, connection angles, and plates can add a significant amount to both the tonnage of a structural steel job and the estimated cost. As a rule of thumb, add 10% to the total weight to account for these accessories.
- Type 2 steel construction, commonly referred to as "simple construction," consists generally of field-bolted connections with lateral bracing supplied by other elements of the building, such as masonry walls or x-bracing. The estimator should be aware, however, that shop connections may be accomplished by welding or bolting. The method may be particular to the fabrication shop and may have an impact on the estimated cost.

05 20 00 Metal Joists

- In any given project the total weight of open web steel joists is determined by the loads to be supported and the design. However, economies can be realized in minimizing the amount of labor used to place the joists. This is done by maximizing the joist spacing, and therefore minimizing the number of joists required to be installed on the job. Certain spacings and locations may be required by the design, but in other cases maximizing the spacing and keeping it as uniform as possible will keep the costs down.

05 30 00 Metal Decking

- The takeoff and estimating of metal deck involves more than simply the area of the floor or roof and the type of deck specified or shown on the drawings. Many different sizes and types of openings may exist. Small openings for individual pipes or conduits may be drilled after the floor/roof is installed, but larger openings may require special deck lengths as well as reinforcing or structural support. The estimator should determine who will be supplying this reinforcing. Additionally, some deck terminations are part of the deck package, such as screed angles and pour stops, and others will be part of the steel contract, such as angles attached to structural members and cast-in-place angles and plates. The estimator must ensure that all pieces are accounted for in the complete estimate.

05 50 00 Metal Fabrications

- The most economical steel stairs are those that use common materials, standard details, and most importantly, a uniform and relatively simple method of field assembly. Commonly available A36 channels and plates are very good choices for the main stringers of the stairs, as are angles and tees for the carrier members. Risers and treads are usually made by specialty shops, and it is most economical to use a typical detail in as many places as possible. The stairs should be pre-assembled and shipped directly to the site. The field connections should be simple and straightforward to be accomplished efficiently, and with minimum equipment and labor.

Reference Numbers

Reference numbers are shown in shaded boxes at the beginning of some major classifications. These numbers refer to related items in the Reference Section. The reference information may be an estimating procedure, an alternate pricing method, or technical information.

Note: Not all subdivisions listed here necessarily appear in this publication.

Division 5 - Metals

There's always a solution in steel.

In today's fast-paced, competitive construction industry, success is often a matter of access to the right information at the right time. The AISC Steel Solutions Center makes it easy for you to find, compare, select, and specify the right system for your project. From typical framing studies to total structural systems, **including project costs and schedules**, we can provide you with up-to-date information for your project—**for free!**

All you have to do is call 866.ASK.AISC or e-mail solutions@aisc.org.

Trademarks licensed from AISC

Your Connection to Ideas + Answers

312.670.2400 ■ solutions@aisc.org

The Ultimate Building Solution

VP Buildings is a recognized leader in the pre-engineered building system industry. We offer architects, engineers and building owners several advantages, including:

- VP Command, our proprietary software to design and estimate building costs to your exact project specifications.

- Over 1,000 Authorized VP Builders from Alaska to Argentina to help supply and build your project on time and on budget.

- VP offers the industry's best standard warranty.

To find a local authorized VP Builder or contact us, log on to *vp.com* or call 800-238-3246. If you are a general contractor interested in becoming an Authorized VP Builder, call 901-748-8000.

Manufacturing • Retail • Commercial • Institutional • Recreational

05 01 Maintenance of Metals

05 01 10 – Maintenance of Structural Metal Framing

05 01 10.51 Cleaning of Structural Metal Framing	Crew	Daily Output	Labor-Hours	Unit	Material	2007 Bare Costs Labor	Equipment	Total	Total Incl O&P
0010 **CLEANING OF STRUCTURAL METAL FRAMING**									
6125 Steel surface treatments, PDCA guidelines									
6170 Wire brush, hand (SSPC-SP2)	1 Psst	400	.020	S.F.	.02	.67		.69	1.25
6180 Power tool (SSPC-SP3)	"	700	.011		.05	.38		.43	.75
6215 Pressure washing, 2800-6000 S.F./day	1 Pord	10000	.001			.03		.03	.04
6220 Steam cleaning, 2800-4000 S.F./day	↓	2000	.004			.13		.13	.20
6225 Water blasting		2500	.003			.10		.10	.16
6230 Brush-off blast (SSPC-SP7)	E-11	1750	.018		.08	.61	.10	.79	1.22
6235 Com'l blast (SSPC-SP6), loose scale, fine pwder rust, 2.0 #/S.F. sand		1200	.027		.17	.88	.15	1.20	1.83
6240 Tight mill scale, little/no rust, 3.0 #/S.F. sand		1000	.032		.25	1.06	.18	1.49	2.26
6245 Exist coat blistered/pitted, 4.0 #/S.F. sand		875	.037		.34	1.21	.20	1.75	2.64
6250 Exist coat badly pitted/nodules, 6.7 #/S.F. sand		825	.039		.57	1.29	.22	2.08	3.03
6255 Near white blast (SSPC-SP10), loose scale, fine rust, 5.6 #/S.F. sand		450	.071		.48	2.36	.40	3.24	4.92
6260 Tight mill scale, little/no rust, 6.9 #/S.F. sand		325	.098		.59	3.26	.55	4.40	6.75
6265 Exist coat blistered/pitted, 9.0 #/S.F. sand		225	.142		.76	4.71	.80	6.27	9.60
6270 Exist coat badly pitted/nodules, 11.3 #/S.F. sand	↓	150	.213	↓	.96	7.05	1.19	9.20	14.20

05 05 Common Work Results for Metals

05 05 05 – Selective Metals Demolition

05 05 05.10 Selective Metals Demolition

	Crew	Daily Output	Labor-Hours	Unit	Material	Labor	Equipment	Total	Total Incl O&P
0010 **SELECTIVE METALS DEMOLITION** R024119-10									
0015 Excludes shores, bracing, cutting, loading, hauling, dumping									
0020 Remove nuts only up to 3/4" diameter	1 Sswk	480	.017	Ea.		.69		.69	1.25
0030 7/8" to 1-1/4" diameter		240	.033			1.38		1.38	2.50
0040 1-3/8" to 2" diameter		160	.050			2.07		2.07	3.75
0060 Unbolt and remove structural bolts up to 3/4" diameter		240	.033			1.38		1.38	2.50
0070 7/8" to 2" diameter		160	.050			2.07		2.07	3.75
0140 Light weight framing members, remove whole or cut up, up to 20 lb	↓	240	.033			1.38		1.38	2.50
0150 21 - 40 lb	2 Sswk	210	.076			3.15		3.15	5.70
0160 41 - 80 lb	3 Sswk	180	.133			5.50		5.50	10
0170 81 - 120 lb	4 Sswk	150	.213			8.80		8.80	16
0230 Structural members, remove whole or cut up, up to 500 lb	E-19	48	.500			20.50	20.50	41	57.50
0240 1/4 - 2 tons	E-18	36	1.111			45.50	27	72.50	110
0250 2 - 5 tons	E-24	30	1.067			43.50	24.50	68	103
0260 5 - 10 tons	E-20	24	2.667			108	53	161	247
0270 10 - 15 tons	E-2	18	3.111			126	86	212	310
0340 Fabricated item, remove whole or cut up, up to 20 lb	1 Sswk	96	.083			3.45		3.45	6.25
0350 21 - 40 lb	2 Sswk	84	.190			7.90		7.90	14.25
0360 41 - 80 lb	3 Sswk	72	.333			13.80		13.80	25
0370 81 - 120 lb	4 Sswk	60	.533			22		22	40
0380 121 - 500 lb	E-19	48	.500			20.50	20.50	41	57.50
0390 501 - 1000 lb	"	36	.667	↓		27	27.50	54.50	76.50

05 05 13 – Shop-Applied Coatings for Metal

05 05 13.50 Paints and Protective Coatings

	Crew	Daily Output	Labor-Hours	Unit	Material	Labor	Equipment	Total	Total Incl O&P
0010 **PAINTS AND PROTECTIVE COATINGS**									
5900 Galvanizing structural steel in shop, under 1 ton				Ton	595			595	655
5950 1 ton to 20 tons					490			490	540
6000 Over 20 tons				↓	435			435	480

05 05 Common Work Results for Metals

05 05 21 – Fastening Methods for Metal

05 05 21.10 Cutting Steel

		Crew	Daily Output	Labor-Hours	Unit	Material	2007 Bare Costs Labor	Equipment	Total	Total Incl O&P
0010	**CUTTING STEEL**									
0020	Hand burning, incl. preparation, torch cutting & grinding, no staging									
0100	Steel to 1/2" thick	E-25	320	.025	L.F.		1.08	.26	1.34	2.24
0150	3/4" thick		260	.031			1.33	.31	1.64	2.77
0200	1" thick	↓	200	.040	↓		1.73	.41	2.14	3.59

05 05 21.15 Drilling Steel

		Crew	Daily Output	Labor-Hours	Unit	Material	2007 Bare Costs Labor	Equipment	Total	Total Incl O&P
0010	**DRILLING STEEL**									
1910	Drilling & layout for steel, up to 1/4" deep, no anchor									
1920	Holes, 1/4" diameter	1 Sswk	112	.071	Ea.	.13	2.95		3.08	5.50
1925	For each additional 1/4" depth, add		336	.024		.13	.98		1.11	1.92
1930	3/8" diameter		104	.077		.15	3.18		3.33	5.90
1935	For each additional 1/4" depth, add		312	.026		.15	1.06		1.21	2.08
1940	1/2" diameter		96	.083		.17	3.45		3.62	6.45
1945	For each additional 1/4" depth, add		288	.028		.17	1.15		1.32	2.26
1950	5/8" diameter		88	.091		.27	3.76		4.03	7.10
1955	For each additional 1/4" depth, add		264	.030		.27	1.25		1.52	2.57
1960	3/4" diameter		80	.100		.30	4.14		4.44	7.85
1965	For each additional 1/4" depth, add		240	.033		.30	1.38		1.68	2.83
1970	7/8" diameter		72	.111		.36	4.59		4.95	8.70
1975	For each additional 1/4" depth, add		216	.037		.36	1.53		1.89	3.17
1980	1" diameter		64	.125		.41	5.15		5.56	9.80
1985	For each additional 1/4" depth, add	↓	192	.042	↓	.41	1.72		2.13	3.57
1990	For drilling up, add						40%			

05 05 21.90 Welding Steel

		Crew	Daily Output	Labor-Hours	Unit	Material	2007 Bare Costs Labor	Equipment	Total	Total Incl O&P
0010	**WELDING STEEL** R050521-20									
0020	Field welding, 1/8" E6011, cost per welder, no oper. engr	E-14	8	1	Hr.	3.97	43.50	14.40	61.87	98.50
0200	With 1/2 operating engineer	E-13	8	1.500		3.97	62	14.40	80.37	126
0300	With 1 operating engineer	E-12	8	2	↓	3.97	80	14.40	98.37	154
0500	With no operating engineer, 2# weld rod per ton	E-14	8	1	Ton	3.97	43.50	14.40	61.87	98.50
0600	8# E6011 per ton	"	2	4		15.90	173	57.50	246.40	395
0800	With one operating engineer per welder, 2# E6011 per ton	E-12	8	2		3.97	80	14.40	98.37	154
0900	8# E6011 per ton	"	2	8	↓	15.90	320	57.50	393.40	615
1200	Continuous fillet, stick welding, incl. equipment									
1300	Single pass, 1/8" thick, 0.1#/L.F.	E-14	150	.053	L.F.	.20	2.31	.77	3.28	5.25
1400	3/16" thick, 0.2#/L.F.		75	.107		.40	4.62	1.54	6.56	10.50
1500	1/4" thick, 0.3#/L.F.		50	.160		.60	6.95	2.30	9.85	15.75
1610	5/16" thick, 0.4#/L.F.		38	.211		.79	9.15	3.03	12.97	21
1800	3 passes, 3/8" thick, 0.5#/L.F.		30	.267		.99	11.55	3.84	16.38	26.50
2010	4 passes, 1/2" thick, 0.7#/L.F.		22	.364		1.39	15.75	5.25	22.39	36
2200	5 to 6 passes, 3/4" thick, 1.3#/L.F.		12	.667		2.58	29	9.60	41.18	66
2400	8 to 11 passes, 1" thick, 2.4#/L.F.	↓	6	1.333		4.77	58	19.20	81.97	131
2600	For all position welding, add, minimum						20%			
2700	Maximum						300%			
2900	For semi-automatic welding, deduct, minimum						5%			
3000	Maximum				↓		15%			
4000	Cleaning and welding plates, bars, or rods									
4010	to existing beams, columns, or trusses	E-14	12	.667	L.F.	.99	29	9.60	39.59	64

05 05 23 – Metal Fastenings

05 05 23.05 Anchor Bolts

		Crew	Daily Output	Labor-Hours	Unit	Material	2007 Bare Costs Labor	Equipment	Total	Total Incl O&P
0010	**ANCHOR BOLTS**									
0100	J-type, incl. hex nut & washer, 1/2" diameter x 6" long	2 Carp	70	.229	Ea.	1.09	8.40		9.49	14.25

05 05 Common Work Results for Metals

05 05 23 – Metal Fastenings

05 05 23.05 Anchor Bolts

		Crew	Daily Output	Labor-Hours	Unit	Material	2007 Bare Costs Labor	2007 Bare Costs Equipment	Total	Total Incl O&P
0110	12" long	2 Carp	65	.246	Ea.	1.36	9.05		10.41	15.55
0120	18" long		60	.267		1.77	9.80		11.57	17.20
0130	3/4" diameter x 8" long		50	.320		1.61	11.75		13.36	20
0140	12" long		45	.356		2.01	13.05		15.06	22.50
0150	18" long		40	.400		2.61	14.70		17.31	26
0160	1" diameter x 12" long		35	.457		3.84	16.80		20.64	30
0170	18" long		30	.533		4.63	19.55		24.18	35.50
0180	24" long		25	.640		5.65	23.50		29.15	42.50
0190	36" long		20	.800		7.75	29.50		37.25	54
0200	1-1/2" diameter x 18" long		22	.727		13.70	26.50		40.20	56.50
0210	24" long		16	1		16.30	36.50		52.80	75
0300	L-type, incl. hex nut & washer, 3/4" diameter x 12" long		45	.356		1.51	13.05		14.56	22
0310	18" long		40	.400		1.94	14.70		16.64	25
0320	24" long		35	.457		2.36	16.80		19.16	28.50
0330	30" long		30	.533		3	19.55		22.55	34
0340	36" long		25	.640		3.42	23.50		26.92	40.50
0350	1" diameter x 12" long		35	.457		2.63	16.80		19.43	29
0360	18" long		30	.533		3.27	19.55		22.82	34
0370	24" long		25	.640		4.04	23.50		27.54	41
0380	30" long		23	.696		4.77	25.50		30.27	45.50
0390	36" long		20	.800		5.45	29.50		34.95	51.50
0400	42" long		18	.889		6.65	32.50		39.15	58.50
0410	48" long		15	1.067		7.45	39		46.45	69
0420	1-1/4" diameter x 18" long		25	.640		5.90	23.50		29.40	43
0430	24" long		22	.727		7	26.50		33.50	49
0440	30" long		20	.800		8.10	29.50		37.60	54.50
0450	36" long		18	.889		9.20	32.50		41.70	61
0460	42" long		16	1		10.40	36.50		46.90	68.50
0470	48" long		14	1.143		11.90	42		53.90	78.50
0480	54" long		12	1.333		14	49		63	91.50
0490	60" long		10	1.600		15.40	58.50		73.90	108
0500	1-1/2" diameter x 18" long		22	.727		9.10	26.50		35.60	51.50
0510	24" long		19	.842		10.60	31		41.60	59.50
0520	30" long		17	.941		11.95	34.50		46.45	67
0530	36" long		16	1		13.75	36.50		50.25	72
0540	42" long		15	1.067		15.70	39		54.70	78.50
0550	48" long		13	1.231		17.60	45		62.60	90
0560	54" long		11	1.455		21.50	53.50		75	107
0570	60" long		9	1.778		23.50	65		88.50	128
0580	1-3/4" diameter x 18" long		20	.800		13.70	29.50		43.20	60.50
0590	24" long		18	.889		16.05	32.50		48.55	68.50
0600	30" long		17	.941		18.70	34.50		53.20	74.50
0610	36" long		16	1		21.50	36.50		58	80.50
0620	42" long		14	1.143		24	42		66	92
0630	48" long		12	1.333		26.50	49		75.50	105
0640	54" long		10	1.600		32.50	58.50		91	128
0650	60" long		8	2		35.50	73.50		109	153
0660	2" diameter x 24" long		17	.941		20.50	34.50		55	76.50
0670	30" long		15	1.067		23	39		62	86.50
0680	36" long		13	1.231		25.50	45		70.50	98.50
0690	42" long		11	1.455		28.50	53.50		82	114
0700	48" long		10	1.600		32.50	58.50		91	128
0710	54" long		9	1.778		38.50	65		103.50	145

05 05 Common Work Results for Metals

05 05 23 – Metal Fastenings

05 05 23.05 Anchor Bolts		Crew	Daily Output	Labor-Hours	Unit	Material	2007 Bare Costs Labor	Equipment	Total	Total Incl O&P
0720	60" long	2 Carp	8	2	Ea.	41.50	73.50		115	160
0730	66" long		7	2.286		44.50	84		128.50	180
0740	72" long		6	2.667		49	98		147	206
0990	For galvanized, add					75%				

05 05 23.10 Bolts and Hex Nuts

0010	**BOLTS & HEX NUTS**, Steel, A307									
0100	1/4" diameter, 1/2" long	1 Sswk	140	.057	Ea.	.07	2.36		2.43	4.36
0200	1" long		140	.057		.08	2.36		2.44	4.37
0300	2" long		130	.062		.11	2.54		2.65	4.73
0400	3" long		130	.062		.16	2.54		2.70	4.79
0500	4" long		120	.067		.18	2.76		2.94	5.20
0600	3/8" diameter, 1" long		130	.062		.12	2.54		2.66	4.74
0700	2" long		130	.062		.15	2.54		2.69	4.78
0800	3" long		120	.067		.20	2.76		2.96	5.20
0900	4" long		120	.067		.25	2.76		3.01	5.25
1000	5" long		115	.070		.31	2.88		3.19	5.55
1100	1/2" diameter, 1-1/2" long		120	.067		.24	2.76		3	5.25
1200	2" long		120	.067		.27	2.76		3.03	5.30
1300	4" long		115	.070		.41	2.88		3.29	5.65
1400	6" long		110	.073		.56	3.01		3.57	6.05
1500	8" long		105	.076		.73	3.15		3.88	6.50
1600	5/8" diameter, 1-1/2" long		120	.067		.47	2.76		3.23	5.50
1700	2" long		120	.067		.51	2.76		3.27	5.55
1800	4" long		115	.070		.71	2.88		3.59	6
1900	6" long		110	.073		.89	3.01		3.90	6.45
2000	8" long		105	.076		1.29	3.15		4.44	7.10
2100	10" long		100	.080		1.60	3.31		4.91	7.75
2200	3/4" diameter, 2" long		120	.067		.74	2.76		3.50	5.80
2300	4" long		110	.073		1.02	3.01		4.03	6.60
2400	6" long		105	.076		1.30	3.15		4.45	7.15
2500	8" long		95	.084		1.92	3.48		5.40	8.40
2600	10" long		85	.094		2.50	3.89		6.39	9.80
2700	12" long		80	.100		2.91	4.14		7.05	10.70
2800	1" diameter, 3" long		105	.076		1.90	3.15		5.05	7.80
2900	6" long		90	.089		2.93	3.68		6.61	9.85
3000	12" long		75	.107		5.50	4.41		9.91	14.05
3100	For galvanized, add					75%				
3200	For stainless, add					350%				

05 05 23.15 Chemical Anchors

0010	**CHEMICAL ANCHORS**									
0020	Includes layout & drilling									
1430	Chemical anchor, w/rod & epoxy cartridge, 3/4" diam. x 9-1/2" long	B-89A	27	.593	Ea.	13.70	19.80	3.89	37.39	50.50
1435	1" diameter x 11-3/4" long		24	.667		26.50	22.50	4.37	53.37	68.50
1440	1-1/4" diameter x 14" long		21	.762		50.50	25.50	5	81	101
1445	1-3/4" diameter x 15" long		20	.800		95	26.50	5.25	126.75	152
1450	18" long		17	.941		114	31.50	6.15	151.65	182
1455	2" diameter x 18" long		16	1		145	33.50	6.55	185.05	219
1460	24" long		15	1.067		190	35.50	7	232.50	272

05 05 23.20 Expansion Anchors

0010	**EXPANSION ANCHORS**									
0100	Anchors for concrete, brick or stone, no layout and drilling									
0200	Expansion shields, zinc, 1/4" diameter, 1-5/16" long, single	1 Carp	90	.089	Ea.	1.11	3.26		4.37	6.30

05 05 Common Work Results for Metals

05 05 23 – Metal Fastenings

05 05 23.20 Expansion Anchors		Crew	Daily Output	Labor-Hours	Unit	Material	2007 Bare Costs Labor	Equipment	Total	Total Incl O&P
0300	1-3/8" long, double	1 Carp	85	.094	Ea.	1.22	3.45		4.67	6.75
0400	3/8" diameter, 1-1/2" long, single		85	.094		1.83	3.45		5.28	7.40
0500	2" long, double		80	.100		2.26	3.67		5.93	8.20
0600	1/2" diameter, 2-1/16" long, single		80	.100		3.03	3.67		6.70	9.05
0700	2-1/2" long, double		75	.107		2.92	3.91		6.83	9.30
0800	5/8" diameter, 2-5/8" long, single		75	.107		4.33	3.91		8.24	10.85
0900	2-3/4" long, double		70	.114		4.33	4.19		8.52	11.30
1000	3/4" diameter, 2-3/4" long, single		70	.114		6.45	4.19		10.64	13.60
1100	3-15/16" long, double		65	.123		8.60	4.52		13.12	16.50
1500	Self drilling anchor, snap-off, for 1/4" diameter bolt		26	.308		.95	11.30		12.25	18.65
1600	3/8" diameter bolt		23	.348		1.39	12.75		14.14	21.50
1700	1/2" diameter bolt		20	.400		2.13	14.70		16.83	25.50
1800	5/8" diameter bolt		18	.444		3.56	16.30		19.86	29.50
1900	3/4" diameter bolt		16	.500		6	18.35		24.35	35
2100	Hollow wall anchors for gypsum wall board, plaster or tile									
2300	1/8" diameter, short	1 Carp	160	.050	Ea.	.30	1.84		2.14	3.19
2400	Long		150	.053		.36	1.96		2.32	3.45
2500	3/16" diameter, short		150	.053		.64	1.96		2.60	3.75
2600	Long		140	.057		.69	2.10		2.79	4.03
2700	1/4" diameter, short		140	.057		.78	2.10		2.88	4.13
2800	Long		130	.062		.88	2.26		3.14	4.49
3000	Toggle bolts, bright steel, 1/8" diameter, 2" long		85	.094		.28	3.45		3.73	5.70
3100	4" long		80	.100		.43	3.67		4.10	6.15
3200	3/16" diameter, 3" long		80	.100		.48	3.67		4.15	6.25
3300	6" long		75	.107		.67	3.91		4.58	6.85
3400	1/4" diameter, 3" long		75	.107		.54	3.91		4.45	6.70
3500	6" long		70	.114		.76	4.19		4.95	7.40
3600	3/8" diameter, 3" long		70	.114		1.04	4.19		5.23	7.70
3700	6" long		60	.133		1.82	4.89		6.71	9.60
3800	1/2" diameter, 4" long		60	.133		2.70	4.89		7.59	10.55
3900	6" long		50	.160		4.45	5.85		10.30	14.05
4000	Nailing anchors									
4100	Nylon nailing anchor, 1/4" diameter, 1" long	1 Carp	3.20	2.500	C	21	92		113	166
4200	1-1/2" long		2.80	2.857		26.50	105		131.50	193
4300	2" long		2.40	3.333		44.50	122		166.50	240
4400	Metal nailing anchor, 1/4" diameter, 1" long		3.20	2.500		32	92		124	178
4500	1-1/2" long		2.80	2.857		43.50	105		148.50	211
4600	2" long		2.40	3.333		55.50	122		177.50	252
5000	Screw anchors for concrete, masonry,									
5100	stone & tile, no layout or drilling included									
5200	Jute fiber, #6, #8, & #10, 1" long	1 Carp	240	.033	Ea.	.26	1.22		1.48	2.19
5300	#12, 1-1/2" long		200	.040		.38	1.47		1.85	2.71
5400	#14, 2" long		160	.050		.60	1.84		2.44	3.52
5500	#16, 2" long		150	.053		.63	1.96		2.59	3.74
5600	#20, 2" long		140	.057		1.01	2.10		3.11	4.38
5700	Lag screw shields, 1/4" diameter, short		90	.089		.48	3.26		3.74	5.65
5800	Long		85	.094		.57	3.45		4.02	6.05
5900	3/8" diameter, short		85	.094		.88	3.45		4.33	6.35
6000	Long		80	.100		1.04	3.67		4.71	6.85
6100	1/2" diameter, short		80	.100		1.22	3.67		4.89	7.05
6200	Long		75	.107		1.54	3.91		5.45	7.80
6300	3/4" diameter, short		70	.114		3.44	4.19		7.63	10.35
6400	Long		65	.123		4.16	4.52		8.68	11.65

05 05 Common Work Results for Metals

05 05 23 – Metal Fastenings

05 05 23.20 Expansion Anchors

		Crew	Daily Output	Labor-Hours	Unit	Material	2007 Bare Costs Labor	2007 Bare Costs Equipment	Total	Total Incl O&P
6600	Lead, #6 & #8, 3/4" long	1 Carp	260	.031	Ea.	.17	1.13		1.30	1.95
6700	#10 - #14, 1-1/2" long		200	.040		.25	1.47		1.72	2.57
6800	#16 & #18, 1-1/2" long		160	.050		.34	1.84		2.18	3.23
6900	Plastic, #6 & #8, 3/4" long		260	.031		.12	1.13		1.25	1.89
7000	#8 & #10, 7/8" long		240	.033		.05	1.22		1.27	1.96
7100	#10 & #12, 1" long		220	.036		.16	1.33		1.49	2.26
7200	#14 & #16, 1-1/2" long		160	.050		.09	1.84		1.93	2.96
8000	Wedge anchors, not including layout or drilling									
8050	Carbon steel, 1/4" diameter, 1-3/4" long	1 Carp	150	.053	Ea.	.48	1.96		2.44	3.58
8100	3 1/4" long		140	.057		.64	2.10		2.74	3.97
8150	3/8" diameter, 2-1/4" long		145	.055		.73	2.02		2.75	3.95
8200	5" long		140	.057		1.28	2.10		3.38	4.68
8250	1/2" diameter, 2-3/4" long		140	.057		1.11	2.10		3.21	4.49
8300	7" long		125	.064		1.89	2.35		4.24	5.75
8350	5/8" diameter, 3-1/2" long		130	.062		2.19	2.26		4.45	5.95
8400	8-1/2" long		115	.070		4.67	2.55		7.22	9.15
8450	3/4" diameter, 4-1/4" long		115	.070		2.65	2.55		5.20	6.90
8500	10" long		95	.084		6	3.09		9.09	11.40
8550	1" diameter, 6" long		100	.080		8.90	2.94		11.84	14.30
8575	9" long		85	.094		11.55	3.45		15	18.10
8600	12" long		75	.107		12.45	3.91		16.36	19.80
8650	1-1/4" diameter, 9" long		70	.114		16.15	4.19		20.34	24.50
8700	12" long		60	.133		20.50	4.89		25.39	30.50
8750	For type 303 stainless steel, add					350%				
8800	For type 316 stainless steel, add					450%				
8950	Self-drilling concrete screw, hex washer head, 3/16" dia x 1-3/4" long	1 Carp	300	.027	Ea.	.37	.98		1.35	1.93
8960	2-1/4" long		250	.032		.56	1.17		1.73	2.45
8970	Phillips flat head, 3/16" dia x 1-3/4" long		300	.027		.38	.98		1.36	1.94
8980	2-1/4" long		250	.032		.56	1.17		1.73	2.45

05 05 23.25 High Strength Bolts

		Crew	Daily Output	Labor-Hours	Unit	Material	2007 Bare Costs Labor	2007 Bare Costs Equipment	Total	Total Incl O&P
0010	**HIGH STRENGTH BOLTS** R050523-10									
0020	A325 Type 1, structural steel, bolt-nut-washer set									
0100	1/2" diameter x 1-1/2" long	1 Sswk	130	.062	Ea.	.37	2.54		2.91	5
0120	2" long		125	.064		.39	2.65		3.04	5.20
0150	3" long		120	.067		.50	2.76		3.26	5.55
0170	5/8" diameter x 1-1/2" long		125	.064		.59	2.65		3.24	5.45
0180	2" long		120	.067		.62	2.76		3.38	5.65
0190	3" long		115	.070		.73	2.88		3.61	6
0200	3/4" diameter x 2" long		120	.067		.85	2.76		3.61	5.95
0220	3" long		115	.070		.98	2.88		3.86	6.30
0250	4" long		110	.073		1.15	3.01		4.16	6.70
0300	6" long		105	.076		1.43	3.15		4.58	7.25
0350	8" long		95	.084		2.61	3.48		6.09	9.15
0360	7/8" diameter x 2" long		115	.070		1.28	2.88		4.16	6.60
0365	3" long		110	.073		1.46	3.01		4.47	7.05
0370	4" long		105	.076		1.70	3.15		4.85	7.55
0380	6" long		100	.080		2.08	3.31		5.39	8.30
0390	8" long		90	.089		3.14	3.68		6.82	10.10
0400	1" diameter x 2" long		105	.076		1.75	3.15		4.90	7.65
0420	3" long		100	.080		1.92	3.31		5.23	8.10
0450	4" long		95	.084		2.12	3.48		5.60	8.65
0500	6" long		90	.089		2.67	3.68		6.35	9.60

05 05 Common Work Results for Metals

05 05 23 – Metal Fastenings

05 05 23.25 High Strength Bolts

		Crew	Daily Output	Labor-Hours	Unit	Material	2007 Bare Costs Labor	Equipment	Total	Total Incl O&P
0550	8" long	1 Sswk	85	.094	Ea.	4.31	3.89		8.20	11.80
0600	1-1/4" diameter x 3" long		85	.094		3.99	3.89		7.88	11.45
0650	4" long		80	.100		1.81	4.14		5.95	9.50
0700	6" long		75	.107		5.20	4.41		9.61	13.70
0750	8" long		70	.114		6.35	4.73		11.08	15.50
1020	A490, bolt-nut-washer set									
1170	5/8" diameter x 1-1/2" long	1 Sswk	125	.064	Ea.	.83	2.65		3.48	5.70
1180	2" long		120	.067		.94	2.76		3.70	6
1190	3" long		115	.070		1.10	2.88		3.98	6.40
1200	3/4" diameter x 2" long		120	.067		1.10	2.76		3.86	6.20
1220	3" long		115	.070		1.25	2.88		4.13	6.60
1250	4" long		110	.073		1.41	3.01		4.42	7
1300	6" long		105	.076		1.95	3.15		5.10	7.85
1350	8" long		95	.084		3.13	3.48		6.61	9.75
1360	7/8" diameter x 2" long		115	.070		1.50	2.88		4.38	6.85
1365	3" long		110	.073		1.70	3.01		4.71	7.30
1370	4" long		105	.076		2.01	3.15		5.16	7.90
1380	6" long		100	.080		2.67	3.31		5.98	8.95
1390	8" long		90	.089		3.71	3.68		7.39	10.75
1400	1" diameter x 2" long		105	.076		2.04	3.15		5.19	7.95
1420	3" long		100	.080		2.33	3.31		5.64	8.55
1450	4" long		95	.084		2.59	3.48		6.07	9.15
1500	6" long		90	.089		3.26	3.68		6.94	10.25
1550	8" long		85	.094		4.79	3.89		8.68	12.30
1600	1-1/4" diameter x 3" long		85	.094		4.66	3.89		8.55	12.20
1650	4" long		80	.100		5.15	4.14		9.29	13.20
1700	6" long		75	.107		6.65	4.41		11.06	15.30
1750	8" long		70	.114		8.25	4.73		12.98	17.65

05 05 23.30 Lag Screws

		Crew	Daily Output	Labor-Hours	Unit	Material	Labor	Equipment	Total	Total Incl O&P
0010	**LAG SCREWS**									
0020	Steel, 1/4" diameter, 2" long	1 Carp	200	.040	Ea.	.10	1.47		1.57	2.40
0100	3/8" diameter, 3" long		150	.053		.27	1.96		2.23	3.35
0200	1/2" diameter, 3" long		130	.062		.44	2.26		2.70	4
0300	5/8" diameter, 3" long		120	.067		.86	2.45		3.31	4.76

05 05 23.35 Machine Screws

		Crew	Daily Output	Labor-Hours	Unit	Material	Labor	Equipment	Total	Total Incl O&P
0010	**MACHINE SCREWS**									
0020	Steel, round head, #8 x 1" long	1 Carp	4.80	1.667	C	2.63	61		63.63	98.50
0110	#8 x 2" long		2.40	3.333		5.75	122		127.75	197
0200	#10 x 1" long		4	2		3.75	73.50		77.25	118
0300	#10 x 2" long		2	4		7	147		154	237

05 05 23.40 Machinery Anchors

		Crew	Daily Output	Labor-Hours	Unit	Material	Labor	Equipment	Total	Total Incl O&P
0010	**MACHINERY ANCHORS**									
0020	Lower stud & coupling nut, fiber plug, connecting stud, washer & nut.									
0030	For flush mounted embedment in poured concrete heavy equip. pads.									
0200	1/2" diameter stud & bolt	E-16	40	.400	Ea.	61	16.95	2.88	80.83	101
0300	5/8" diameter		35	.457		67.50	19.35	3.29	90.14	113
0500	3/4" diameter		30	.533		78	22.50	3.84	104.34	131
0600	7/8" diameter		25	.640		85	27	4.61	116.61	148
0800	1" diameter		20	.800		89.50	34	5.75	129.25	166
0900	1-1/4" diameter		15	1.067		119	45	7.70	171.70	220

05 05 23.50 Powder Actuated Tools and Fasteners

0010	**POWDER ACTUATED TOOLS & FASTENERS**									

05 05 Common Work Results for Metals

05 05 23 – Metal Fastenings

05 05 23.50 Powder Actuated Tools and Fasteners		Crew	Daily Output	Labor-Hours	Unit	Material	2007 Bare Costs Labor	Equipment	Total	Total Incl O&P
0020	Stud driver, .22 caliber, buy, minimum				Ea.	355			355	390
0100	Maximum				"	575			575	630
0300	Powder charges for above, low velocity				C	18.30			18.30	20
0400	Standard velocity					26			26	28.50
0600	Drive pins & studs, 1/4" & 3/8" diam., to 3" long, minimum	1 Carp	4.80	1.667		13.70	61		74.70	111
0700	Maximum	"	4	2		53.50	73.50		127	173
0800	Pneumatic stud driver for 1/8" diameter studs				Ea.	2,475			2,475	2,725
0900	Drive pins for above, 1/2" to 3/4" long	1 Carp	1	8	M	565	294		859	1,075

05 05 23.55 Rivets

0010	**RIVETS**									
0100	Aluminum rivet & mandrel, 1/2" grip length x 1/8" diameter	1 Carp	4.80	1.667	C	5.50	61		66.50	102
0200	3/16" diameter		4	2		8.50	73.50		82	123
0300	Aluminum rivet, steel mandrel, 1/8" diameter		4.80	1.667		8.45	61		69.45	105
0400	3/16" diameter		4	2		7.75	73.50		81.25	123
0500	Copper rivet, steel mandrel, 1/8" diameter		4.80	1.667		9.50	61		70.50	106
0600	Monel rivet, steel mandrel, 1/8" diameter		4.80	1.667		27.50	61		88.50	126
0700	3/16" diameter		4	2		78.50	73.50		152	201
0800	Stainless rivet & mandrel, 1/8" diameter		4.80	1.667		13.70	61		74.70	111
0900	3/16" diameter		4	2		25.50	73.50		99	143
1000	Stainless rivet, steel mandrel, 1/8" diameter		4.80	1.667		10.70	61		71.70	107
1100	3/16" diameter		4	2		19.35	73.50		92.85	136
1200	Steel rivet and mandrel, 1/8" diameter		4.80	1.667		6.60	61		67.60	103
1300	3/16" diameter		4	2		9.95	73.50		83.45	125
1400	Hand riveting tool, minimum				Ea.	128			128	141
1500	Maximum					245			245	269
1600	Power riveting tool, minimum					925			925	1,025
1700	Maximum					2,325			2,325	2,575

05 05 23.80 Vibration Pads

0010	**VIBRATION PADS**									
0300	Laminated synthetic rubber impregnated cotton duck, 1/2" thick	2 Sswk	24	.667	S.F.	65	27.50		92.50	122
0400	1" thick		20	.800		131	33		164	204
0600	Neoprene bearing pads, 1/2" thick		24	.667		24.50	27.50		52	77
0700	1" thick		20	.800		50	33		83	116
0900	Fabric reinforced neoprene, 5000 psi, 1/2" thick		24	.667		11	27.50		38.50	62
1000	1" thick		20	.800		22	33		55	84
1200	Felt surfaced vinyl pads, cork and sisal, 5/8" thick		24	.667		28.50	27.50		56	81
1300	1" thick		20	.800		51.50	33		84.50	117
1500	Teflon bonded to 10 ga. carbon steel, 1/32" layer		24	.667		49	27.50		76.50	104
1600	3/32" layer		24	.667		73	27.50		100.50	131
1800	Bonded to 10 ga. stainless steel, 1/32" layer		24	.667		86.50	27.50		114	146
1900	3/32" layer		24	.667		112	27.50		139.50	174
2100	Circular machine leveling pad & stud				Kip	7.10			7.10	7.80

05 05 23.85 Welded Shear Connectors

0010	**WELDED SHEAR CONNECTORS**									
0020	3/4" diameter, 3-3/16" long	E-10	960	.017	Ea.	.43	.71	.31	1.45	2.10
0030	3-3/8" long		950	.017		.45	.71	.32	1.48	2.14
0200	3-7/8" long		945	.017		.49	.72	.32	1.53	2.19
0300	4-3/16" long		935	.017		.51	.72	.32	1.55	2.22
0500	4-7/8" long		930	.017		.57	.73	.32	1.62	2.31
0600	5-3/16" long		920	.017		.59	.74	.33	1.66	2.34
0800	5-3/8" long		910	.018		.60	.74	.33	1.67	2.37
0900	6-3/16" long		905	.018		.66	.75	.33	1.74	2.45

05 05 Common Work Results for Metals
05 05 23 – Metal Fastenings

05 05 23.85 Welded Shear Connectors		Crew	Daily Output	Labor-Hours	Unit	Material	2007 Bare Costs Labor	Equipment	Total	Total Incl O&P
1000	7-3/16" long	E-10	895	.018	Ea.	.82	.76	.34	1.92	2.64
1100	8-3/16" long		890	.018		.89	.76	.34	1.99	2.73
1500	7/8" diameter, 3-11/16" long		920	.017		.70	.74	.33	1.77	2.46
1600	4-3/16" long		910	.018		.75	.74	.33	1.82	2.53
1700	5-3/16" long		905	.018		.85	.75	.33	1.93	2.67
1800	6-3/16" long		895	.018		.95	.76	.34	2.05	2.79
1900	7-3/16" long		890	.018		1.06	.76	.34	2.16	2.91
2000	8-3/16" long		880	.018		1.15	.77	.34	2.26	3.04

05 05 23.87 Welded Studs		Crew	Daily Output	Labor-Hours	Unit	Material	Labor	Equipment	Total	Total Incl O&P
0010	**WELDED STUDS**									
0020	1/4" diameter, 2-11/16" long	E-10	1120	.014	Ea.	.28	.61	.27	1.16	1.70
0100	4-1/8" long		1080	.015		.26	.63	.28	1.17	1.74
0200	3/8" diameter, 4-1/8" long		1080	.015		.30	.63	.28	1.21	1.78
0300	6-1/8" long		1040	.015		.39	.65	.29	1.33	1.93
0400	1/2" diameter, 2-1/8" long		1040	.015		.29	.65	.29	1.23	1.81
0500	3-1/8" long		1025	.016		.35	.66	.29	1.30	1.90
0600	4-1/8" long		1010	.016		.41	.67	.30	1.38	1.99
0700	5-5/16" long		990	.016		.50	.68	.30	1.48	2.12
0800	6-1/8" long		975	.016		.54	.70	.31	1.55	2.20
0900	8-1/8" long		960	.017		.76	.71	.31	1.78	2.46
1000	5/8" diameter, 2-11/16" long		1000	.016		.50	.68	.30	1.48	2.11
1010	4-3/16" long		990	.016		.61	.68	.30	1.59	2.24
1100	6-9/16" long		975	.016		.80	.70	.31	1.81	2.48
1200	8-3/16" long		960	.017		1.07	.71	.31	2.09	2.80

05 05 23.90 Welding Rod		Crew	Daily Output	Labor-Hours	Unit	Material	Labor	Equipment	Total	Total Incl O&P
0010	**WELDING ROD**									
0020	Steel, type 6011, 1/8" dia, less than 500#				Lb.	1.99			1.99	2.19
0100	500# to 2,000#					1.79			1.79	1.97
0200	2,000# to 5,000#					1.68			1.68	1.85
0300	5/32" diameter, less than 500#					1.91			1.91	2.10
0310	500# to 2,000#					1.72			1.72	1.89
0320	2,000# to 5,000#					1.62			1.62	1.78
0400	3/16" dia, less than 500#					1.93			1.93	2.12
0500	500# to 2,000#					1.74			1.74	1.91
0600	2,000# to 5,000#					1.64			1.64	1.80
0620	Steel, type 6010, 1/8" dia, less than 500#					2.19			2.19	2.41
0630	500# to 2,000#					1.97			1.97	2.17
0640	2,000# to 5,000#					1.85			1.85	2.04
0650	Steel, type 7018 Low Hydrogen, 1/8" dia, less than 500#					1.89			1.89	2.08
0660	500# to 2,000#	CN				1.70			1.70	1.87
0670	2,000# to 5,000#					1.60			1.60	1.76
0700	Steel, type 7024 Jet Weld, 1/8" dia, less than 500#					1.99			1.99	2.19
0710	500# to 2,000#					1.79			1.79	1.97
0720	2,000# to 5,000#					1.68			1.68	1.85
1550	Aluminum, type 4043 TIG, 1/8" dia, less than 10#					6.85			6.85	7.55
1560	10# to 60#					6.20			6.20	6.80
1570	Over 60#					5.80			5.80	6.40
1600	Aluminum, type 5356 TIG, 1/8" dia, less than 10#					7.65			7.65	8.40
1610	10# to 60#					6.85			6.85	7.55
1620	Over 60#					6.45			6.45	7.10
1900	Cast iron, type 8 Nickel, 1/8" dia, less than 500#					23.50			23.50	26
1910	500# to 1,000#					21			21	23.50

05 05 Common Work Results for Metals

05 05 23 – Metal Fastenings

05 05 23.90 Welding Rod	Crew	Daily Output	Labor-Hours	Unit	Material	2007 Bare Costs Labor	Equipment	Total	Total Incl O&P
1920	Over 1,000#				Lb.	19.95			19.95	22
2000	Stainless steel, type 316/316L, 1/8" dia, less than 500#					10.85			10.85	11.90
2100	500# to 1000#					9.75			9.75	10.75
2220	Over 1000#					9.15			9.15	10.10

05 12 Structural Steel Framing

05 12 23 – Structural Steel for Buildings

05 12 23.05 Canopy Framing

		Crew	Daily Output	Labor-Hours	Unit	Material	Labor	Equipment	Total	Total Incl O&P
0010	**CANOPY FRAMING**									
0020	6" and 8" members	E-4	3000	.011	Lb.	1.23	.45	.04	1.72	2.20

05 12 23.10 Ceiling Supports

		Crew	Daily Output	Labor-Hours	Unit	Material	Labor	Equipment	Total	Total Incl O&P
0010	**CEILING SUPPORTS**									
1000	Entrance door/folding partition supports	E-4	60	.533	L.F.	20.50	22.50	1.92	44.92	65
1100	Linear accelerator door supports		14	2.286		93.50	95.50	8.25	197.25	285
1200	Lintels or shelf angles, hung, exterior hot dipped galv.		267	.120		14	5	.43	19.43	25
1250	Two coats primer paint instead of galv.		267	.120		12.15	5	.43	17.58	23
1400	Monitor support, ceiling hung, expansion bolted		4	8	Ea.	325	335	29	689	990
1450	Hung from pre-set inserts		6	5.333		350	223	19.20	592.20	810
1600	Motor supports for overhead doors		4	8		165	335	29	529	820
1700	Partition support for heavy folding partitions, without pocket		24	1.333	L.F.	46.50	56	4.80	107.30	158
1750	Supports at pocket only		12	2.667		93.50	112	9.60	215.10	315
2000	Rolling grilles & fire door supports		34	.941		40	39.50	3.39	82.89	119
2100	Spider-leg light supports, expansion bolted to ceiling slab		8	4	Ea.	133	167	14.40	314.40	470
2150	Hung from pre-set inserts		12	2.667	"	144	112	9.60	265.60	370
2400	Toilet partition support		36	.889	L.F.	46.50	37	3.20	86.70	123
2500	X-ray travel gantry support		12	2.667	"	160	112	9.60	281.60	390

05 12 23.15 Columns, Lightweight

		Crew	Daily Output	Labor-Hours	Unit	Material	Labor	Equipment	Total	Total Incl O&P
0010	**COLUMNS, LIGHTWEIGHT**									
1000	Lightweight units (lally), 3-1/2" diameter	E-2	780	.072	L.F.	3.14	2.90	1.98	8.02	10.65
1050	4" diameter	"	900	.062	"	4.62	2.51	1.72	8.85	11.35
5800	Adjustable jack post, 8' maximum height, 2-3/4" diameter				Ea.	25			25	27.50
5850	4" diameter				"	40			40	44

05 12 23.17 Columns, Structural

		Crew	Daily Output	Labor-Hours	Unit	Material	Labor	Equipment	Total	Total Incl O&P
0010	**COLUMNS, STRUCTURAL**	R051223-10								
0020	Shop fab'd for 100-ton, 1-2 story project, bolted conn's.									
0800	Steel, concrete filled, extra strong pipe, 3-1/2" diameter	E-2	660	.085	L.F.	34	3.42	2.34	39.76	45.50
0830	4" diameter		780	.072		37.50	2.90	1.98	42.38	48.50
0890	5" diameter		1020	.055		45	2.22	1.52	48.74	55
0930	6" diameter		1200	.047		59.50	1.88	1.29	62.67	70
0940	8" diameter		1100	.051		59.50	2.05	1.41	62.96	70.50
1100	For galvanizing, add				Lb.	.25			.25	.27
1300	For web ties, angles, etc., add per added lb.	1 Sswk	945	.008		1.03	.35		1.38	1.76
1500	Steel pipe, extra strong, no concrete, 3" to 5" diameter	E-2	16000	.004		1.03	.14	.10	1.27	1.48
1600	6" to 12" diameter		14000	.004		1.03	.16	.11	1.30	1.53
1700	Steel pipe, extra strong, no concrete, 3" diameter x 12'-0"		60	.933	Ea.	126	37.50	26	189.50	233
1750	4" diameter x 12'-0"		58	.966		184	39	26.50	249.50	300
1800	6" diameter x 12'-0"		54	1.037		350	42	28.50	420.50	490
1850	8" diameter x 14'-0"		50	1.120		625	45	31	701	800
1900	10" diameter x 16'-0"		48	1.167		900	47	32	979	1,100
1950	12" diameter x 18'-0"		45	1.244		1,200	50	34.50	1,284.50	1,450

05 12 Structural Steel Framing

05 12 23 – Structural Steel for Buildings

05 12 23.17 Columns, Structural

		Crew	Daily Output	Labor-Hours	Unit	Material	2007 Bare Costs Labor	Equipment	Total	Total Incl O&P
3300	Structural tubing, square, A500GrB, 4" to 6" square, light section	E-2	11270	.005	Lb.	1.03	.20	.14	1.37	1.63
3600	Heavy section		32000	.002	"	1.03	.07	.05	1.15	1.30
4000	Concrete filled, add				L.F.	4.02			4.02	4.42
4500	Structural tubing, sq, 4" x 4" x 1/4" x 12'-0"	E-2	58	.966	Ea.	169	39	26.50	234.50	283
4550	6" x 6" x 1/4" x 12'-0"		54	1.037		277	42	28.50	347.50	410
4600	8" x 8" x 3/8" x 14'-0"		50	1.120		600	45	31	676	775
4650	10" x 10" x 1/2" x 16'-0"		48	1.167		1,100	47	32	1,179	1,350
5100	Structural tubing, rect, 5" to 6" wide, light section		8000	.007	Lb.	1.03	.28	.19	1.50	1.83
5200	Heavy section		12000	.005		1.03	.19	.13	1.35	1.60
5300	7" to 10" wide, light section		15000	.004		1.03	.15	.10	1.28	1.50
5400	Heavy section		18000	.003		1.03	.13	.09	1.25	1.44
5500	Structural tubing, rect, 5" x 3" x 1/4" x 12'-0"		58	.966	Ea.	164	39	26.50	229.50	277
5550	6" x 4" x 5/16" x 12'-0"		54	1.037		256	42	28.50	326.50	385
5600	8" x 4" x 3/8" x 12'-0"		54	1.037		375	42	28.50	445.50	515
5650	10" x 6" x 3/8" x 14'-0"		50	1.120		600	45	31	676	775
5700	12" x 8" x 1/2" x 16'-0"		48	1.167		1,100	47	32	1,179	1,350
6800	W Shape, A992 steel, 2 tier, W8 x 24		1080	.052	L.F.	27	2.09	1.43	30.52	35
6850	W8 x 31		1080	.052		35	2.09	1.43	38.52	43.50
6900	W8 x 48		1032	.054		54	2.19	1.50	57.69	65
6950	W8 x 67		984	.057		75.50	2.30	1.57	79.37	88.50
7000	W10 x 45		1032	.054		50.50	2.19	1.50	54.19	61.50
7050	W10 x 68		984	.057		76.50	2.30	1.57	80.37	90
7100	W10 x 112		960	.058		126	2.35	1.61	129.96	145
7150	W12 x 50		1032	.054		56.50	2.19	1.50	60.19	67.50
7200	W12 x 87		984	.057		98	2.30	1.57	101.87	114
7250	W12 x 120		960	.058		135	2.35	1.61	138.96	155
7300	W12 x 190		912	.061		214	2.48	1.70	218.18	242
7350	W14 x 74		984	.057		83.50	2.30	1.57	87.37	97.50
7400	W14 x 120		960	.058		135	2.35	1.61	138.96	155
7450	W14 x 176		912	.061		198	2.48	1.70	202.18	224
8090	For projects 75 to 99 tons, add				All	10%				
8092	50 to 74 tons, add					20%				
8094	25 to 49 tons, add					30%	10%			
8096	10 to 24 tons, add					50%	25%			
8098	2 to 9 tons, add					75%	50%			
8099	Less than 2 tons, add					100%	100%			

05 12 23.20 Curb Edging

		Crew	Daily Output	Labor-Hours	Unit	Material	Labor	Equipment	Total	Total Incl O&P
0010	**CURB EDGING**									
0020	Steel angle w/anchors, on forms, 1" x 1", 0.8#/L.F.	E-4	350	.091	L.F.	1.72	3.83	.33	5.88	9.20
0100	2" x 2" angles, 3.92#/L.F.		330	.097		5.55	4.06	.35	9.96	13.85
0200	3" x 3" angles, 6.1#/L.F.		300	.107		8.80	4.46	.38	13.64	18.15
0300	4" x 4" angles, 8.2#/L.F.		275	.116		11.35	4.87	.42	16.64	22
1000	6" x 4" angles, 12.3#/L.F.		250	.128		16.40	5.35	.46	22.21	28.50
1050	Steel channels with anchors, on forms, 3" channel, 5#/L.F.		290	.110		6.90	4.62	.40	11.92	16.40
1100	4" channel, 5.4#/L.F.		270	.119		7.40	4.96	.43	12.79	17.55
1200	6" channel, 8.2#/L.F.		255	.125		11.35	5.25	.45	17.05	22.50
1300	8" channel, 11.5#/L.F.		225	.142		15.45	5.95	.51	21.91	28.50
1400	10" channel, 15.3#/L.F.		180	.178		20	7.45	.64	28.09	36
1500	12" channel, 20.7#/L.F.		140	.229		26.50	9.55	.82	36.87	48
2000	For curved edging, add					35%	10%			

05 12 23.40 Lightweight Framing

0010	**LIGHTWEIGHT FRAMING**	R051223-35								

05 12 Structural Steel Framing

05 12 23 – Structural Steel for Buildings

05 12 23.40 Lightweight Framing		Crew	Daily Output	Labor-Hours	Unit	Material	2007 Bare Costs Labor	Equipment	Total	Total Incl O&P
0200	For load-bearing steel studs see division 05 41 13.30									
0400	Angle framing, field fabricated, 4" and larger	E-3	440	.055	Lb.	.59	2.29	.26	3.14	5.10
	R051223-45									
0450	Less than 4" angles		265	.091	"	.62	3.81	.43	4.86	8.05
0460	1/2" x 1/2" x 1/8"		200	.120	L.F.	.12	5.05	.58	5.75	9.90
0462	3/4" x 3/4" x 1/8"		160	.150		.34	6.30	.72	7.36	12.55
0464	1" x 1" x 1/8"		135	.178		.49	7.45	.85	8.79	15.05
0466	1-1/4" x 1-1/4" x 3/16"		115	.209		.91	8.75	1	10.66	18
0468	1-1/2" x 1-1/2" x 3/16"		100	.240		1.11	10.10	1.15	12.36	20.50
0470	2" x 2" x 1/4"		90	.267		1.96	11.20	1.28	14.44	24
0472	2-1/2" x 2-1/2" x 1/4"		72	.333		2.52	14	1.60	18.12	30
0474	3" x 2" x 3/8"		65	.369		3.63	15.50	1.77	20.90	34
0476	3" x 3" x 3/8"		57	.421		4.43	17.70	2.02	24.15	39
0600	Channel framing, field fabricated, 8" and larger		500	.048	Lb.	.62	2.02	.23	2.87	4.58
0650	Less than 8" channels		335	.072	"	.62	3.01	.34	3.97	6.50
0660	C2 x 1.78		115	.209	L.F.	1.09	8.75	1	10.84	18.20
0662	C3 x 4.1		80	.300		2.52	12.60	1.44	16.56	27.50
0664	C4 x 5.4		66	.364		3.32	15.30	1.75	20.37	33
0666	C5 x 6.7		57	.421		4.12	17.70	2.02	23.84	39
0668	C6 x 8.2		55	.436		4.87	18.35	2.09	25.31	40.50
0670	C7 x 9.8		40	.600		6.05	25	2.88	33.93	55.50
0672	C8 x 11.5		36	.667		7.05	28	3.20	38.25	62
0710	Structural bar tee, field fabricated, 3/4" x 3/4" x 1/8"		160	.150		.34	6.30	.72	7.36	12.55
0712	1" x 1" x 1/8"		135	.178		.49	7.45	.85	8.79	15.05
0714	1-1/2" x 1-1/2" x 1/4"		114	.211		1.44	8.85	1.01	11.30	18.70
0716	2" x 2" x 1/4"		89	.270		1.96	11.35	1.29	14.60	24
0718	2-1/2" x 2-1/2" x 3/8"		72	.333		3.63	14	1.60	19.23	31.50
0720	3" x 3" x 3/8"		57	.421		4.43	17.70	2.02	24.15	39
0730	Structural zee, field fabricated, 1-1/4" x 1-3/4" x 1-3/4"		114	.211		.47	8.85	1.01	10.33	17.60
0732	2-11/16" x 3" x 2-11/16"		114	.211		1.09	8.85	1.01	10.95	18.30
0734	3-1/16" x 4" x 3-1/16"		133	.180		1.65	7.60	.87	10.12	16.50
0736	3-1/4" x 5" x 3-1/4"		133	.180		2.26	7.60	.87	10.73	17.20
0738	3-1/2" x 6" x 3-1/2"		160	.150		3.40	6.30	.72	10.42	15.95
0740	Junior beam, field fabricated, 3"		80	.300		3.51	12.60	1.44	17.55	28.50
0742	4"		72	.333		4.74	14	1.60	20.34	32.50
0744	5"		67	.358		6.15	15.05	1.72	22.92	36
0746	6"		62	.387		7.70	16.25	1.86	25.81	40
0748	7"		57	.421		9.40	17.70	2.02	29.12	44.50
0750	8"		53	.453		11.30	19.05	2.17	32.52	49.50
1000	Continuous slotted channel framing system, shop fab, min	2 Sswk	2400	.007	Lb.	3.18	.28		3.46	4
1200	Maximum	"	1600	.010		3.59	.41		4	4.70
1300	Cross bracing, rods, shop fabricated, 3/4" diameter	E-3	700	.034		1.23	1.44	.16	2.83	4.14
1310	7/8" diameter		850	.028		1.23	1.19	.14	2.56	3.65
1320	1" diameter		1000	.024		1.23	1.01	.12	2.36	3.31
1330	Angle, 5" x 5" x 3/8"		2800	.009		1.23	.36	.04	1.63	2.05
1350	Hanging lintels, shop fabricated, average		850	.028		1.23	1.19	.14	2.56	3.65
1380	Roof frames, shop fabricated, 3'-0" square, 5' span	E-2	4200	.013		1.23	.54	.37	2.14	2.69
1400	Tie rod, not upset, 1-1/2" to 4" diameter, with turnbuckle	2 Sswk	800	.020		1.33	.83		2.16	2.97
1420	No turnbuckle		700	.023		1.28	.95		2.23	3.12
1500	Upset, 1-3/4" to 4" diameter, with turnbuckle		800	.020		1.33	.83		2.16	2.97
1520	No turnbuckle		700	.023		1.28	.95		2.23	3.12

05 12 23.45 Lintels

| 0010 | LINTELS | | | | | | | | | |

05 12 Structural Steel Framing

05 12 23 – Structural Steel for Buildings

05 12 23.45 Lintels

		Crew	Daily Output	Labor-Hours	Unit	Material	2007 Bare Costs Labor	Equipment	Total	Total Incl O&P
0020	Plain steel angles, under 500 lb.	1 Bric	550	.015	Lb.	.79	.55		1.34	1.71
0100	500 to 1000 lb.	CN	640	.013		.77	.48		1.25	1.57
0200	1,000 to 2,000 lb.		640	.013		.75	.48		1.23	1.54
0300	2,000 to 4,000 lb.		640	.013		.73	.48		1.21	1.52
0500	For built-up angles and plates, add to above					.26			.26	.28
0700	For engineering, add to above					.10			.10	.11
0900	For galvanizing, add to above, under 500 lb.					.33			.33	.36
0950	500 to 2,000 lb.					.30			.30	.33
1000	Over 2,000 lb.					.25			.25	.27
2000	Steel angles, 3-1/2" x 3", 1/4" thick, 2'-6" long	1 Bric	47	.170	Ea.	11.05	6.50		17.55	22
2100	4'-6" long		26	.308		19.95	11.70		31.65	40
2600	4" x 3-1/2", 1/4" thick, 5'-0" long		21	.381		25.50	14.50		40	50
2700	9'-0" long		12	.667		46	25.50		71.50	89

05 12 23.60 Pipe Support Framing

		Crew	Daily Output	Labor-Hours	Unit	Material	Labor	Equipment	Total	Total Incl O&P
0010	**PIPE SUPPORT FRAMING**									
0020	Under 10#/L.F.	E-4	3900	.008	Lb.	1.37	.34	.03	1.74	2.16
0200	10.1 to 15#/L.F.		4300	.007		1.35	.31	.03	1.69	2.08
0400	15.1 to 20#/L.F.		4800	.007		1.33	.28	.02	1.63	2.01
0600	Over 20#/L.F.		5400	.006		1.31	.25	.02	1.58	1.91

05 12 23.65 Plates

		Crew	Daily Output	Labor-Hours	Unit	Material	Labor	Equipment	Total	Total Incl O&P
0010	**PLATES** R051223-80									
0020	For connections & stiffener plates, shop fabricated				S.F.					
0050	1/8" thick (5.1 Lb./S.F.)					5.25			5.25	5.75
0100	1/4" thick (10.2 Lb./S.F.)					10.45			10.45	11.50
0300	3/8" thick (15.3 Lb./S.F.)					15.70			15.70	17.25
0400	1/2" thick (20.4 Lb./S.F.)					21			21	23
0450	3/4" thick (30.6 Lb./S.F.)					31.50			31.50	34.50
0500	1" thick (40.8 Lb.S.F.)					42			42	46
2000	Steel plate, warehouse prices, no fabrication									
2100	1/4" thick (10.2 Lb./S.F.)				S.F.	4.49			4.49	4.94

05 12 23.70 Stressed Skin Roof and Ceiling System

		Crew	Daily Output	Labor-Hours	Unit	Material	Labor	Equipment	Total	Total Incl O&P
0010	**STRESSED SKIN ROOF & CEILING SYSTEM**									
0020	Double panel flat roof, spans to 100'	E-2	1150	.049	S.F.	8.20	1.97	1.35	11.52	13.90
0100	Double panel convex roof, spans to 200'		960	.058		13.35	2.35	1.61	17.31	20.50
0200	Double panel arched roof, spans to 300'		760	.074		20.50	2.97	2.04	25.51	30

05 12 23.75 Structural Steel Members

		Crew	Daily Output	Labor-Hours	Unit	Material	Labor	Equipment	Total	Total Incl O&P
0010	**STRUCTURAL STEEL MEMBERS** R051223-10									
0020	Shop fab'd for 100-ton, 1-2 story project, bolted conn's.									
0100	W 6 x 9	E-2	600	.093	L.F.	10.15	3.77	2.58	16.50	20.50
0120	x 16		600	.093		18.05	3.77	2.58	24.40	29
0140	x 20		600	.093		22.50	3.77	2.58	28.85	34.50
0300	W 8 x 10		600	.093		11.30	3.77	2.58	17.65	21.50
0320	x 15		600	.093		16.90	3.77	2.58	23.25	28
0350	x 21		600	.093		23.50	3.77	2.58	29.85	35.50
0360	x 24		550	.102		27	4.11	2.81	33.92	40
0370	x 28		550	.102		31.50	4.11	2.81	38.42	44.50
0500	x 31		550	.102		35	4.11	2.81	41.92	48.50
0520	x 35		550	.102		39.50	4.11	2.81	46.42	53.50
0540	x 48		550	.102		54	4.11	2.81	60.92	69.50
0600	W 10 x 12		600	.093		13.55	3.77	2.58	19.90	24
0620	x 15		600	.093		16.90	3.77	2.58	23.25	28
0700	x 22		600	.093		25	3.77	2.58	31.35	37

05 12 Structural Steel Framing

05 12 23 – Structural Steel for Buildings

05 12 23.75 Structural Steel Members		Crew	Daily Output	Labor-Hours	Unit	Material	2007 Bare Costs Labor	Equipment	Total	Total Incl O&P
0720	x 26	E-2	600	.093	L.F.	29.50	3.77	2.58	35.85	42
0740	x 33		550	.102		37	4.11	2.81	43.92	51
0900	x 49		550	.102		55.50	4.11	2.81	62.42	71
1100	W 12 x 14		880	.064		15.80	2.57	1.76	20.13	23.50
1300	x 22		880	.064		25	2.57	1.76	29.33	34
1500	x 26		880	.064		29.50	2.57	1.76	33.83	39
1520	x 35		810	.069		39.50	2.79	1.91	44.20	50.50
1560	x 50		750	.075		56.50	3.01	2.06	61.57	69.50
1580	x 58		750	.075		65.50	3.01	2.06	70.57	79.50
1700	x 72		640	.088		81	3.53	2.42	86.95	98.50
1740	x 87		640	.088		98	3.53	2.42	103.95	117
1900	W 14 x 26		990	.057		29.50	2.28	1.56	33.34	38
2100	x 30		900	.062		34	2.51	1.72	38.23	43
2300	x 34		810	.069		38.50	2.79	1.91	43.20	49
2320	x 43		810	.069		48.50	2.79	1.91	53.20	60.50
2340	x 53		800	.070		60	2.82	1.93	64.75	72.50
2360	x 74		760	.074		83.50	2.97	2.04	88.51	99.50
2380	x 90		740	.076		101	3.05	2.09	106.14	120
2500	x 120		720	.078		135	3.14	2.15	140.29	157
2700	W 16 x 26		1000	.056		29.50	2.26	1.55	33.31	38
2900	x 31		900	.062		35	2.51	1.72	39.23	44.50
3100	x 40		800	.070		45	2.82	1.93	49.75	56.50
3120	x 50		800	.070		56.50	2.82	1.93	61.25	69
3140	x 67		760	.074		75.50	2.97	2.04	80.51	90.50
3300	W 18 x 35	E-5	960	.083		39.50	3.40	1.73	44.63	51.50
3500	x 40		960	.083		45	3.40	1.73	50.13	57.50
3520	x 46		960	.083		52	3.40	1.73	57.13	65
3700	x 50		912	.088		56.50	3.58	1.82	61.90	70.50
3900	x 55		912	.088		62	3.58	1.82	67.40	76.50
3920	x 65		900	.089		73.50	3.63	1.85	78.98	89
3940	x 76		900	.089		85.50	3.63	1.85	90.98	103
3960	x 86		900	.089		97	3.63	1.85	102.48	115
3980	x 106		900	.089		120	3.63	1.85	125.48	139
4100	W 21 x 44		1064	.075		49.50	3.07	1.56	54.13	61.50
4300	x 50		1064	.075		56.50	3.07	1.56	61.13	69
4500	x 62		1036	.077		70	3.15	1.60	74.75	84.50
4700	x 68		1036	.077		76.50	3.15	1.60	81.25	92
4720	x 83		1000	.080		93.50	3.27	1.66	98.43	111
4740	x 93		1000	.080		105	3.27	1.66	109.93	123
4760	x 101		1000	.080		114	3.27	1.66	118.93	133
4780	x 122		1000	.080		138	3.27	1.66	142.93	159
4900	W 24 x 55		1110	.072		62	2.94	1.50	66.44	75
5100	x 62		1110	.072		70	2.94	1.50	74.44	84
5300	x 68		1110	.072		76.50	2.94	1.50	80.94	91.50
5500	x 76		1110	.072		85.50	2.94	1.50	89.94	101
5700	x 84		1080	.074		94.50	3.03	1.54	99.07	111
5720	x 94		1080	.074		106	3.03	1.54	110.57	124
5740	x 104		1050	.076		117	3.11	1.58	121.69	136
5760	x 117		1050	.076		132	3.11	1.58	136.69	152
5780	x 146		1050	.076		165	3.11	1.58	169.69	188
5800	W 27 x 84		1190	.067		94.50	2.75	1.40	98.65	110
5900	x 94		1190	.067		106	2.75	1.40	110.15	123
5920	x 114		1150	.070		129	2.84	1.45	133.29	148

05 12 Structural Steel Framing

05 12 23 – Structural Steel for Buildings

05 12 23.75 Structural Steel Members

		Crew	Daily Output	Labor-Hours	Unit	Material	2007 Bare Costs Labor	Equipment	Total	Total Incl O&P
5940	x 146	E-5	1150	.070	L.F.	165	2.84	1.45	169.29	188
5960	x 161		1150	.070		182	2.84	1.45	186.29	207
6100	W 30 x 99		1200	.067		112	2.72	1.39	116.11	129
6300	x 108		1200	.067		122	2.72	1.39	126.11	140
6500	x 116		1160	.069		131	2.82	1.43	135.25	151
6520	x 132		1160	.069		149	2.82	1.43	153.25	171
6540	x 148		1160	.069		167	2.82	1.43	171.25	191
6560	x 173		1120	.071		195	2.92	1.48	199.40	222
6580	x 191		1120	.071		215	2.92	1.48	219.40	244
6700	W 33 x 118		1176	.068		133	2.78	1.41	137.19	152
6900	x 130		1134	.071		147	2.88	1.47	151.35	168
7100	x 141		1134	.071		159	2.88	1.47	163.35	182
7120	x 169		1100	.073		191	2.97	1.51	195.48	217
7140	x 201		1100	.073		227	2.97	1.51	231.48	256
7300	W 36 x 135		1170	.068		152	2.79	1.42	156.21	173
7500	x 150		1170	.068		169	2.79	1.42	173.21	192
7600	x 170		1150	.070		192	2.84	1.45	196.29	218
7700	x 194		1125	.071		219	2.90	1.48	223.38	248
7900	x 230		1125	.071		259	2.90	1.48	263.38	292
7920	x 260		1035	.077		293	3.16	1.61	297.77	325
8100	x 300		1035	.077		340	3.16	1.61	344.77	375
8490	For projects 75 to 99 tons, add					10%				
8492	50 to 74 tons, add					20%				
8494	25 to 49 tons, add					30%	10%			
8496	10 to 24 tons, add					50%	25%			
8498	2 to 9 tons, add					75%	50%			
8499	Less than 2 tons, add					100%	100%			

05 12 23.77 Structural Steel Projects

			Crew	Daily Output	Labor-Hours	Unit	Material	Labor	Equipment	Total	Total Incl O&P
0010	**STRUCTURAL STEEL PROJECTS**	R050516-30									
0020	Shop fab'd for 100-ton, 1-2 story project, bolted conn's.										
0200	Apartments, nursing homes, etc., 1 to 2 stories	R050523-10	E-5	10.30	7.767	Ton	2,050	315	161	2,526	2,975
0300	3 to 6 stories		"	10.10	7.921		2,100	325	165	2,590	3,050
0400	7 to 15 stories	R051223-10	E-6	14.20	9.014		2,125	370	127	2,622	3,150
0500	Over 15 stories		"	13.90	9.209		2,225	375	130	2,730	3,225
0700	Offices, hospitals, etc., steel bearing, 1 to 2 stories **CN**	R051223-20	E-5	10.30	7.767		2,050	315	161	2,526	2,975
0800	3 to 6 stories		E-6	14.40	8.889		2,100	365	126	2,591	3,075
0900	7 to 15 stories	R051223-25		14.20	9.014		2,125	370	127	2,622	3,150
1000	Over 15 stories			13.90	9.209		2,225	375	130	2,730	3,225
1100	For multi-story masonry wall bearing construction, add	R051223-30						30%			
1300	Industrial bldgs., 1 story, beams & girders, steel bearing		E-5	12.90	6.202		2,050	253	129	2,432	2,825
1400	Masonry bearing		"	10	8		2,050	325	166	2,541	3,000
1500	Industrial bldgs., 1 story, under 10 tons,										
1510	steel from warehouse, trucked		E-2	7.50	7.467	Ton	2,450	300	206	2,956	3,450
1600	1 story with roof trusses, steel bearing		E-5	10.60	7.547		2,425	310	157	2,892	3,375
1700	Masonry bearing		"	8.30	9.639		2,425	395	200	3,020	3,550
1900	Monumental structures, banks, stores, etc., minimum		E-6	13	9.846		2,050	405	139	2,594	3,125
2000	Maximum		"	9	14.222		3,400	580	201	4,181	5,000
2200	Churches, minimum		E-5	11.60	6.897		1,900	282	143	2,325	2,750
2300	Maximum		"	5.20	15.385		2,550	630	320	3,500	4,250
2800	Power stations, fossil fuels, minimum		E-6	11	11.636		2,050	475	165	2,690	3,275
2900	Maximum			5.70	22.456		3,075	920	320	4,315	5,350
2950	Nuclear fuels, non-safety steel, minimum			7	18.286		2,050	750	259	3,059	3,850

05 12 Structural Steel Framing

05 12 23 – Structural Steel for Buildings

05 12 23.77 Structural Steel Projects

		Crew	Daily Output	Labor-Hours	Unit	Material	2007 Bare Costs Labor	Equipment	Total	Total Incl O&P
3000	Maximum	E-6	5.50	23.273	Ton	3,075	950	330	4,355	5,400
3040	Safety steel, minimum		2.50	51.200		3,000	2,100	725	5,825	7,775
3070	Maximum	↓	1.50	85.333		3,925	3,500	1,200	8,625	11,800
3100	Roof trusses, minimum	E-5	13	6.154		2,875	251	128	3,254	3,725
3200	Maximum		8.30	9.639		3,475	395	200	4,070	4,725
3210	Schools, minimum		14.50	5.517		2,050	225	115	2,390	2,775
3220	Maximum	↓	8.30	9.639		3,000	395	200	3,595	4,200
3400	Welded construction, simple commercial bldgs., 1 to 2 stories	E-7	7.60	10.526		2,100	430	234	2,764	3,300
3500	7 to 15 stories	E-9	8.30	15.422		2,425	630	256	3,311	4,025
3700	Welded rigid frame, 1 story, minimum	E-7	15.80	5.063		2,125	207	113	2,445	2,850
3800	Maximum	"	5.50	14.545	↓	2,775	595	325	3,695	4,450
3810	Fabrication shop costs (included in project material cost, above)									
3820	Mini mill base price, A992				Ton	590			590	650
3830	Mill extras and delivery to shop					210			210	231
3840	Shop extra, add for shop drawings and detailing					235			235	259
3850	Add for shop fabricating and handling					800			800	880
3860	Add for sandblasting and primer coat of paint					125			125	138
3870	Add for shop delivery to the job site					90			90	99
3880	Total material cost, shop fabricated, primed, delivered				↓	2,050			2,050	2,250
3900	High strength steel mill spec extras: A242, A441,									
3950	A529, A572 (42 ksi) and A992: same as A36 steel									
4000	Add to A992 price for A572 (50, 60, 65 ksi)				Ton	38			38	42
4100	A588 Weathering				"	45			45	49.50
4200	Mill size extras for W-Shapes: 0 to 30 plf: no extra charge									
4210	Member sizes 31 to 65 plf, add				Ton	.10			.10	.11
4220	Member sizes 66 to 100 plf, add					.50			.50	.55
4230	Member sizes 101 to 387 plf, add				↓	59.50			59.50	65.50
4300	Column base plates, light, up to 150 lb	2 Sswk	2000	.008	Lb.	1.13	.33		1.46	1.84
4400	Heavy, over 150 lb	E-2	7500	.007	"	1.18	.30	.21	1.69	2.05
4600	Castellated beams, light sections, to 50#/L.F., minimum		10.70	5.234	Ton	2,150	211	145	2,506	2,900
4700	Maximum		7	8		2,350	325	221	2,896	3,400
4900	Heavy sections, over 50# per L.F., minimum		11.70	4.786		2,250	193	132	2,575	2,950
5000	Maximum	↓	7.80	7.179		2,450	290	198	2,938	3,425
5390	For projects 75 to 99 tons, add					10%				
5392	50 to 74 tons, add					20%				
5394	25 to 49 tons, add					30%	10%			
5396	10 to 24 tons, add					50%	25%			
5398	2 to 9 tons, add					75%	50%			
5399	Less than 2 tons, add				↓	100%	100%			

05 12 23.80 Subpurlins

		Crew	Daily Output	Labor-Hours	Unit	Material	Labor	Equipment	Total	Total Incl O&P
0010	**SUBPURLINS** R051223-50									
0020	Bulb tees, painted, 32-5/8" O.C., 40 psf L.L.									
0100	Type 178, max 8'-9" span, 2.15 plf, 2" high x 1-5/8" wide	E-1	4200	.006	S.F.	1.45	.23	.03	1.71	2.03
0200	Type 218, max 10'-2" span, 3.19 plf, 2-1/8" high x 2-1/8" wide	"	3100	.008		1.68	.31	.04	2.03	2.43
1420	For 24-5/8" spacing, add					33%	33%			
1430	For 48-5/8" spacing, deduct				↓	50%	50%			

05 14 Structural Aluminum Framing

05 14 23 – Non-Exposed Structural Aluminum Framing

05 14 23.05 Aluminum Shapes

		Crew	Daily Output	Labor-Hours	Unit	Material	2007 Bare Costs Labor	Equipment	Total	Total Incl O&P
0010	**ALUMINUM SHAPES**									
0020	Structural shapes, 1" to 10" members, under 1 ton	E-2	1050	.053	Lb.	2.44	2.15	1.47	6.06	8.05
0050	1 to 5 tons		1330	.042		2.32	1.70	1.16	5.18	6.75
0100	Over 5 tons	CN	1330	.042		2.25	1.70	1.16	5.11	6.70
0300	Extrusions, over 5 tons, stock shapes		1330	.042		2.44	1.70	1.16	5.30	6.90
0400	Custom shapes		1330	.042		2.48	1.70	1.16	5.34	6.95

05 15 Wire Rope Assemblies

05 15 16 – Steel Wire Rope Assemblies

05 15 16.05 Accessories for Steel Wire Rope

		Crew	Daily Output	Labor-Hours	Unit	Material	2007 Bare Costs Labor	Equipment	Total	Total Incl O&P
0010	**ACCESSORIES FOR STEEL WIRE ROPE**									
1500	Thimbles, heavy duty, 1/4"	E-17	160	.100	Ea.	.57	4.24	.05	4.86	8.35
1510	1/2"		160	.100		2.51	4.24	.05	6.80	10.45
1520	3/4"		105	.152		5.70	6.45	.08	12.23	18.10
1530	1"		52	.308		11.40	13.05	.15	24.60	36
1540	1-1/4"		38	.421		17.55	17.85	.21	35.61	52
1550	1-1/2"		13	1.231		49.50	52	.62	102.12	149
1560	1-3/4"		8	2		102	84.50	1	187.50	266
1570	2"		6	2.667		148	113	1.33	262.33	370
1580	2-1/4"		4	4		201	169	2	372	530
1600	Clips, 1/4" diameter		160	.100		2.57	4.24	.05	6.86	10.55
1610	3/8" diameter		160	.100		2.82	4.24	.05	7.11	10.80
1620	1/2" diameter		160	.100		4.53	4.24	.05	8.82	12.70
1630	3/4" diameter		102	.157		7.35	6.65	.08	14.08	20
1640	1" diameter		64	.250		12.25	10.60	.13	22.98	33
1650	1-1/4" diameter		35	.457		20	19.35	.23	39.58	57.50
1670	1-1/2" diameter		26	.615		27	26	.31	53.31	77.50
1680	1-3/4" diameter		16	1		63	42.50	.50	106	147
1690	2" diameter		12	1.333		70	56.50	.67	127.17	180
1700	2-1/4" diameter		10	1.600		103	68	.80	171.80	238
2200	Jaw & jaw turnbuckles, 1/4" x 4"		160	.100		20.50	4.24	.05	24.79	30
2250	1/2" x 6"		96	.167		26	7.05	.08	33.13	41.50
2260	1/2" x 9"		77	.208		34.50	8.80	.10	43.40	54
2270	1/2" x 12"		66	.242		39	10.25	.12	49.37	61
2300	3/4" x 6"		38	.421		50.50	17.85	.21	68.56	88
2310	3/4" x 9"		30	.533		56	22.50	.27	78.77	103
2320	3/4" x 12"		28	.571		72	24	.29	96.29	124
2330	3/4" x 18"		23	.696		86	29.50	.35	115.85	148
2350	1" x 6"		17	.941		98	40	.47	138.47	181
2360	1" x 12"		13	1.231		108	52	.62	160.62	213
2370	1" x 18"		10	1.600		161	68	.80	229.80	300
2380	1" x 24"		9	1.778		178	75.50	.89	254.39	330
2400	1-1/4" x 12"		7	2.286		181	97	1.14	279.14	375
2410	1-1/4" x 18"		6.50	2.462		224	104	1.23	329.23	435
2420	1-1/4" x 24"		5.60	2.857		300	121	1.43	422.43	550
2450	1-1/2" x 12"		5.20	3.077		305	130	1.54	436.54	575
2460	1-1/2" x 18"		4	4		325	169	2	496	660
2470	1-1/2" x 24"		3.20	5		435	212	2.50	649.50	870
2500	1-3/4" x 18"		3.20	5		655	212	2.50	869.50	1,125
2510	1-3/4" x 24"		2.80	5.714		750	242	2.86	994.86	1,275
2550	2" x 24"		1.60	10		1,000	425	5	1,430	1,875

05 15 Wire Rope Assemblies

05 15 16 – Steel Wire Rope Assemblies

05 15 16.50 Steel Wire Rope		Crew	Daily Output	Labor-Hours	Unit	Material	2007 Bare Costs Labor	2007 Bare Costs Equipment	Total	Total Incl O&P
0010	**STEEL WIRE ROPE**									
0020	6 x 19, bright, fiber core, 5000' rolls, 1/2" diameter				L.F.	.89			.89	.97
0050	Steel core					1.17			1.17	1.28
0100	Fiber core, 1" diameter					2.98			2.98	3.28
0150	Steel core					3.40			3.40	3.74
0300	6 x 19, galvanized, fiber core, 1/2" diameter					1.30			1.30	1.44
0350	Steel core					1.49			1.49	1.64
0400	Fiber core, 1" diameter					3.82			3.82	4.20
0450	Steel core					4.01			4.01	4.41
0500	6 x 7, bright, IPS, fiber core, <500 L.F. w/acc., 1/4" diameter	E-17	6400	.003		.66	.11		.77	.92
0510	1/2" diameter		2100	.008		1.61	.32		1.93	2.35
0520	3/4" diameter		960	.017		2.91	.71	.01	3.63	4.50
0550	6 x 19, bright, IPS, IWRC, <500 L.F. w/acc., 1/4" diameter		5760	.003		.98	.12		1.10	1.29
0560	1/2" diameter		1730	.009		1.58	.39		1.97	2.46
0570	3/4" diameter		770	.021		2.75	.88	.01	3.64	4.62
0580	1" diameter		420	.038		4.66	1.61	.02	6.29	8.10
0590	1-1/4" diameter		290	.055		7.75	2.34	.03	10.12	12.75
0600	1-1/2" diameter		192	.083		9.50	3.53	.04	13.07	16.90
0610	1-3/4" diameter	E-18	240	.167		15.15	6.85	4.07	26.07	33
0620	2" diameter		160	.250		19.45	10.30	6.10	35.85	46.50
0630	2-1/4" diameter		160	.250		26	10.30	6.10	42.40	53.50
0650	6 x 37, bright, IPS, IWRC, <500 L.F. w/acc., 1/4" diameter	E-17	6400	.003		1.24	.11		1.35	1.56
0660	1/2" diameter		1730	.009		2.11	.39		2.50	3.04
0670	3/4" diameter		770	.021		3.40	.88	.01	4.29	5.35
0680	1" diameter		430	.037		5.40	1.58	.02	7	8.80
0690	1-1/4" diameter		290	.055		8.15	2.34	.03	10.52	13.20
0700	1-1/2" diameter		190	.084		11.65	3.57	.04	15.26	19.35
0710	1-3/4" diameter	E-18	260	.154		18.50	6.35	3.76	28.61	35.50
0720	2" diameter		200	.200		24	8.25	4.88	37.13	46.50
0730	2-1/4" diameter		160	.250		32	10.30	6.10	48.40	60
0800	6 x 19 & 6 x 37, swaged, 1/2" diameter	E-17	1220	.013		3.78	.56	.01	4.35	5.20
0810	9/16" diameter		1120	.014		4.39	.61	.01	5.01	5.95
0820	5/8" diameter		930	.017		5.20	.73	.01	5.94	7.10
0830	3/4" diameter		640	.025		6.65	1.06	.01	7.72	9.25
0840	7/8" diameter		480	.033		8.35	1.41	.02	9.78	11.80
0850	1" diameter		350	.046		10.20	1.94	.02	12.16	14.80
0860	1-1/8" diameter		288	.056		12.55	2.35	.03	14.93	18.10
0870	1-1/4" diameter		230	.070		15.20	2.95	.03	18.18	22
0880	1-3/8" diameter		192	.083		17.55	3.53	.04	21.12	26
0890	1-1/2" diameter	E-18	300	.133		21.50	5.50	3.26	30.26	36.50

05 15 16.70 Temporary Cable Safety Railing

		Crew	Daily Output	Labor-Hours	Unit	Material	Labor	Equipment	Total	Total Incl O&P
0010	**TEMPORARY CABLE SAFETY RAILING**, Each 100' strand incl.									
0020	2 eyebolts, 1 turnbuckle, 100' cable, 2 thimbles, 6 clips									
0100	One strand using 1/4" cable & accessories	2 Sswk	4	4	C.L.F.	129	165		294	440
0200	1/2" cable & accessories	"	2	8	"	240	330		570	865

05 21 Steel Joist Framing

05 21 13 – Deep Longspan Steel Joist Framing

05 21 13.50 Deep Longspan Joists		Crew	Daily Output	Labor-Hours	Unit	Material	2007 Bare Costs Labor	Equipment	Total	Total Incl O&P
0010	**DEEP LONGSPAN JOISTS**									
3010	DLH series, bolted cross bridging									
3020	Spans to 144' (shipped in 2 pieces), minimum	E-7	16	5	Ton	1,550	204	111	1,865	2,200
3040	Average		13	6.154		1,675	251	137	2,063	2,450
3100	Maximum		11	7.273	↓	2,025	297	162	2,484	2,925
3200	52DLH11, 26 Lb/LF		2000	.040	L.F.	21	1.63	.89	23.52	27
3220	52DLH16, 45 Lb/LF		2000	.040		38	1.63	.89	40.52	45.50
3240	56DLH11, 26 Lb/LF		2000	.040		22	1.63	.89	24.52	28
3260	56DLH16, 46 Lb/LF		2000	.040		38.50	1.63	.89	41.02	46.50
3280	60DLH12, 29 Lb/LF		2000	.040		24.50	1.63	.89	27.02	31
3300	60DLH17, 52 Lb/LF		2000	.040		43.50	1.63	.89	46.02	52
3320	64DLH12, 31 Lb/LF		2200	.036		26	1.49	.81	28.30	32
3340	64DLH17, 52 Lb/LF		2200	.036		43.50	1.49	.81	45.80	51.50
3360	68DLH13, 37 Lb/LF		2200	.036		31	1.49	.81	33.30	37.50
3380	68DLH18, 61 Lb/LF		2200	.036		51	1.49	.81	53.30	60
3400	72DLH14, 41 Lb/LF		2200	.036		34.50	1.49	.81	36.80	41.50
3420	72DLH19, 70 Lb/LF	↓	2200	.036	↓	59	1.49	.81	61.30	68
4010	SLH series, bolted cross bridging									
4020	Spans to 200' (shipped in 3 pieces), minimum	E-7	16	5	Ton	1,525	204	111	1,840	2,175
4040	Average		13	6.154		1,725	251	137	2,113	2,500
4060	Maximum		11	7.273	↓	2,050	297	162	2,509	2,950
4200	80SLH15, 40 Lb/LF		1500	.053	L.F.	34.50	2.18	1.19	37.87	43
4220	80SLH20, 75 Lb/LF		1500	.053		64.50	2.18	1.19	67.87	76
4240	88SLH16, 46 Lb/LF		1500	.053		39.50	2.18	1.19	42.87	48.50
4260	88SLH21, 89 Lb/LF		1500	.053		77	2.18	1.19	80.37	89.50
4280	96SLH17, 52 Lb/LF		1500	.053		45	2.18	1.19	48.37	54.50
4300	96SLH22, 102 Lb/LF		1500	.053		88	2.18	1.19	91.37	102
4320	104SLH18, 59 Lb/LF		1800	.044		51	1.82	.99	53.81	60.50
4340	104SLH23, 109 Lb/LF		1800	.044		94	1.82	.99	96.81	107
4360	112SLH19, 67 Lb/LF		1800	.044		58	1.82	.99	60.81	68
4380	112SLH24, 131 Lb/LF		1800	.044		113	1.82	.99	115.81	128
4400	120SLH20, 77 Lb/LF		1800	.044		66.50	1.82	.99	69.31	77.50
4420	120SLH25, 152 Lb/LF	↓	1800	.044	↓	131	1.82	.99	133.81	148

05 21 16 – Longspan Steel Joist Framing

05 21 16.50 Longspan Joists		Crew	Daily Output	Labor-Hours	Unit	Material	2007 Bare Costs Labor	Equipment	Total	Total Incl O&P
0010	**LONGSPAN JOISTS**									
2000	LH series, bolted cross bridging									
2020	Spans to 96', minimum	E-7	16	5	Ton	1,425	204	111	1,740	2,050
2040	Average		13	6.154		1,575	251	137	1,963	2,325
2080	Maximum		11	7.273	↓	1,850	297	162	2,309	2,750
2200	18LH04, 12 Lb/LF		1400	.057	L.F.	9.45	2.33	1.27	13.05	15.90
2220	18LH08, 19 Lb/LF		1400	.057		15	2.33	1.27	18.60	22
2240	20LH04, 12 Lb/LF		1400	.057		9.45	2.33	1.27	13.05	15.90
2260	20LH08, 19 Lb/LF		1400	.057		15	2.33	1.27	18.60	22
2280	24LH05, 13 Lb/LF		1400	.057		10.25	2.33	1.27	13.85	16.80
2300	24LH10, 23 Lb/LF		1400	.057		18.15	2.33	1.27	21.75	25.50
2320	28LH06, 16 Lb/LF		1800	.044		12.65	1.82	.99	15.46	18.20
2340	28LH11, 25 Lb/LF		1800	.044		19.75	1.82	.99	22.56	26
2360	32LH08, 17 Lb/LF		1800	.044		13.40	1.82	.99	16.21	19.05
2380	32LH13, 30 Lb/LF		1800	.044		23.50	1.82	.99	26.31	30.50
2400	36LH09, 21 Lb/LF		1800	.044		16.55	1.82	.99	19.36	22.50
2420	36LH14, 36 Lb/LF	↓	1800	.044	↓	28.50	1.82	.99	31.31	36

05 21 Steel Joist Framing

05 21 16 – Longspan Steel Joist Framing

05 21 16.50 Longspan Joists		Crew	Daily Output	Labor-Hours	Unit	Material	2007 Bare Costs Labor	Equipment	Total	Total Incl O&P
2440	40LH10, 21 Lb/LF	E-7	2200	.036	L.F.	16.55	1.49	.81	18.85	22
2460	40LH15, 36 Lb/LF		2200	.036		28.50	1.49	.81	30.80	35
2480	44LH11, 22 Lb/LF		2200	.036		17.35	1.49	.81	19.65	22.50
2500	44LH16, 42 Lb/LF		2200	.036		33	1.49	.81	35.30	40
2520	48LH11, 22 Lb/LF		2200	.036		17.35	1.49	.81	19.65	22.50
2540	48LH16, 42 Lb/LF		2200	.036		33	1.49	.81	35.30	40
6000	For welded cross bridging, add						30%			

05 21 19 – Open Web Steel Joist Framing

05 21 19.10 Open Web Joists

		Crew	Daily Output	Labor-Hours	Unit	Material	Labor	Equipment	Total	Total Incl O&P
0010	**OPEN WEB JOISTS**									
0020	K series, 40-ton job lots, horiz. bridging, spans to 30', minimum	E-7	15	5.333	Ton	1,275	218	119	1,612	1,925
0050	Average		12	6.667		1,425	272	148	1,845	2,225
0080	Maximum		9	8.889		1,700	365	198	2,263	2,725
0130	8K1, 5.1 Lb/LF		1200	.067	L.F.	3.63	2.72	1.48	7.83	10.40
0140	10K1, 5.0 Lb/LF		1200	.067		3.56	2.72	1.48	7.76	10.35
0160	12K3, 5.7 Lb/LF		1500	.053		4.06	2.18	1.19	7.43	9.60
0180	14K3, 6.0 Lb/LF		1500	.053		4.28	2.18	1.19	7.65	9.85
0200	16K3, 6.3 Lb/LF		1800	.044		4.49	1.82	.99	7.30	9.20
0220	16K6, 8.1 Lb/LF		1800	.044		5.75	1.82	.99	8.56	10.65
0240	18K5, 7.7 Lb/LF		2000	.040		5.50	1.63	.89	8.02	9.90
0260	18K9, 10.2 Lb/LF		2000	.040		7.25	1.63	.89	9.77	11.85
0410	Span 30' to 50', minimum		17	4.706	Ton	1,250	192	105	1,547	1,825
0440	Average **CN**		17	4.706		1,400	192	105	1,697	2,000
0460	Maximum		10	8		1,475	325	178	1,978	2,400
0500	20K5, 8.2 Lb/LF		2000	.040	L.F.	5.75	1.63	.89	8.27	10.15
0520	20K9, 10.8 Lb/LF		2000	.040		7.55	1.63	.89	10.07	12.15
0540	22K5, 8.8 Lb/LF		2000	.040		6.15	1.63	.89	8.67	10.65
0560	22K9, 11.3 Lb/LF		2000	.040		7.90	1.63	.89	10.42	12.55
0580	24K6, 9.7 Lb/LF		2200	.036		6.80	1.49	.81	9.10	10.95
0600	24K10, 13.1 Lb/LF		2200	.036		9.15	1.49	.81	11.45	13.60
0620	26K6, 10.6 Lb/LF		2200	.036		7.40	1.49	.81	9.70	11.65
0640	26K10, 13.8 Lb/LF		2200	.036		9.65	1.49	.81	11.95	14.15
0660	28K8, 12.7 Lb/LF		2400	.033		8.90	1.36	.74	11	13
0680	28K12, 17.1 Lb/LF		2400	.033		11.95	1.36	.74	14.05	16.35
0700	30K8, 13.2 Lb/LF		2400	.033		9.25	1.36	.74	11.35	13.35
0720	30K12, 17.6 Lb/LF		2400	.033		12.30	1.36	.74	14.40	16.75
1010	CS series, horizontal bridging									
1020	Spans to 30', minimum	E-7	15	5.333	Ton	1,325	218	119	1,662	1,975
1040	Average		12	6.667		1,475	272	148	1,895	2,250
1060	Maximum		9	8.889		1,725	365	198	2,288	2,750
1100	10CS2, 7.5 Lb/LF		1200	.067	L.F.	5.50	2.72	1.48	9.70	12.45
1120	12CS2, 8.0 Lb/LF		1500	.053		5.85	2.18	1.19	9.22	11.60
1140	14CS2, 8.0Lb/LF		1500	.053		5.85	2.18	1.19	9.22	11.60
1160	16CS2, 8.5 Lb/LF		1800	.044		6.20	1.82	.99	9.01	11.15
1180	16CS4, 14.5 Lb/LF		1800	.044		10.60	1.82	.99	13.41	15.95
1200	18CS2, 9.0 Lb/LF		2000	.040		6.60	1.63	.89	9.12	11.10
1220	18CS4, 15.0 Lb/LF		2000	.040		10.95	1.63	.89	13.47	15.90
1240	20CS2, 9.5 Lb/LF		2000	.040		6.95	1.63	.89	9.47	11.50
1260	20CS4, 16.5 Lb/LF		2000	.040		12.05	1.63	.89	14.57	17.10
1280	22CS2, 10.0 Lb/LF		2000	.040		7.30	1.63	.89	9.82	11.90
1300	22CS4, 16.5 Lb/LF		2000	.040		12.05	1.63	.89	14.57	17.10
1320	24CS2, 10.0 Lb/LF		2200	.036		7.30	1.49	.81	9.60	11.55

05 21 Steel Joist Framing

05 21 19 – Open Web Steel Joist Framing

05 21 19.10 Open Web Joists		Crew	Daily Output	Labor-Hours	Unit	Material	2007 Bare Costs Labor	Equipment	Total	Total Incl O&P
1340	24CS4, 16.5 Lb/LF	E-7	2200	.036	L.F.	12.05	1.49	.81	14.35	16.75
1360	26CS2, 10.0 Lb/LF		2200	.036		7.30	1.49	.81	9.60	11.55
1380	26CS4, 16.5 Lb/LF		2200	.036		12.05	1.49	.81	14.35	16.75
1400	28CS2, 10.5 Lb/LF		2400	.033		7.70	1.36	.74	9.80	11.65
1420	28CS4, 16.5 Lb/LF		2400	.033		12.05	1.36	.74	14.15	16.45
1440	30CS2, 11.0 Lb/LF		2400	.033		8.05	1.36	.74	10.15	12.05
1460	30CS4, 16.5 Lb/LF	↓	2400	.033	↓	12.05	1.36	.74	14.15	16.45
6100	For less than 40-ton job lots									
6102	For 30 to 39 tons, add					10%				
6104	20 to 29 tons, add					20%				
6106	10 to 19 tons, add					30%				
6107	5 to 9 tons, add					50%	25%			
6108	1 to 4 tons, add					75%	50%			
6109	Less than 1 ton, add					100%	100%			
6200	For shop prime paint other than mfrs. standard, add					20%				
6300	For bottom chord extensions, add per chord				Ea.	31			31	34
6400	Individual steel bearing plate, 6" x 6" x 1/4" with J-hook	1 Bric	160	.050	"	6.15	1.90		8.05	9.65

05 21 23 – Steel Joist Girder Framing

05 21 23.50 Joist Girders

		Crew	Daily Output	Labor-Hours	Unit	Material	Labor	Equipment	Total	Total Incl O&P
0010	**JOIST GIRDERS**									
7000	Joist girders, minimum	E-5	15	5.333	Ton	1,300	218	111	1,629	1,925
7020	Average		13	6.154		1,425	251	128	1,804	2,150
7040	Maximum		11	7.273		1,500	297	151	1,948	2,325
8000	Trusses, factory fabricated WT chords, average	↓	11	7.273	↓	4,675	297	151	5,123	5,800

05 31 Steel Decking

05 31 13 – Steel Floor Decking

05 31 13.50 Floor Decking

		Crew	Daily Output	Labor-Hours	Unit	Material	Labor	Equipment	Total	Total Incl O&P
0010	**FLOOR DECKING** R053100-10									
3200	3" deep, 22 gauge, under 50 squares	E-4	3600	.009	S.F.	2.14	.37	.03	2.54	3.06
3250	50-500 squares		3800	.008		1.71	.35	.03	2.09	2.55
3260	over 500 squares		4000	.008		1.54	.33	.03	1.90	2.33
3300	20 gauge, under 50 squares		3400	.009		2.47	.39	.03	2.89	3.47
3350	50-500 squares		3600	.009		1.98	.37	.03	2.38	2.89
3360	over 500 squares		3800	.008		1.78	.35	.03	2.16	2.63
3400	18 gauge, under 50 squares		3200	.010		3.19	.42	.04	3.65	4.31
3450	50-500 squares		3400	.009		2.55	.39	.03	2.97	3.56
3460	over 500 squares		3600	.009		2.30	.37	.03	2.70	3.24
3500	16 gauge, under 50 squares		3000	.011		4.22	.45	.04	4.71	5.50
3550	50-500 squares		3200	.010		3.38	.42	.04	3.84	4.52
3560	over 500 squares		3400	.009		3.04	.39	.03	3.46	4.09
3700	4-1/2" deep, long span roof, over 50 squares, 20 gauge		2700	.012		3.97	.50	.04	4.51	5.30
3800	18 gauge		2460	.013		5.10	.54	.05	5.69	6.70
3900	16 gauge		2350	.014		3.82	.57	.05	4.44	5.30
4100	6" deep, long span, 18 gauge		2000	.016		7.30	.67	.06	8.03	9.30
4200	16 gauge		1930	.017		5.45	.69	.06	6.20	7.35
4300	14 gauge		1860	.017		7.05	.72	.06	7.83	9.10
4500	7-1/2" deep, long span, 18 gauge		1690	.019		8.05	.79	.07	8.91	10.35
4600	16 gauge		1590	.020		6	.84	.07	6.91	8.20
4700	14 gauge	↓	1490	.021	↓	7.75	.90	.08	8.73	10.20

05 31 Steel Decking

05 31 13 – Steel Floor Decking

05 31 13.50 Floor Decking

		Crew	Daily Output	Labor-Hours	Unit	Material	2007 Bare Costs Labor	Equipment	Total	Total Incl O&P
4800	For painted instead of galvanized, deduct					2%				
5000	For acoustical perforated, with fiberglass, add				S.F.	1.05			1.05	1.16
5200	Non-cellular composite deck, galv., 2" deep, 22 gauge	E-4	3860	.008		1.47	.35	.03	1.85	2.28
5300	20 gauge		3600	.009		1.63	.37	.03	2.03	2.50
5400	18 gauge		3380	.009		2.07	.40	.03	2.50	3.03
5500	16 gauge		3200	.010		2.59	.42	.04	3.05	3.65
5700	3" deep, galv., 22 gauge		3200	.010		1.60	.42	.04	2.06	2.56
5800	20 gauge		3000	.011		1.79	.45	.04	2.28	2.82
5900	18 gauge **CN**		2850	.011		2.20	.47	.04	2.71	3.31
6000	16 gauge		2700	.012		2.94	.50	.04	3.48	4.18

05 31 23 – Steel Roof Decking

05 31 23.50 Roof Decking

		Crew	Daily Output	Labor-Hours	Unit	Material	Labor	Equipment	Total	Total Incl O&P
0010	**ROOF DECKING**									
2100	Open type, galv., 1-1/2" deep wide rib, 22 gauge, under 50 squares	E-4	4500	.007	S.F.	1.55	.30	.03	1.88	2.27
2200	50-500 squares **CN**		4900	.007		1.20	.27	.02	1.49	1.85
2400	Over 500 squares		5100	.006		1.11	.26	.02	1.39	1.72
2600	20 gauge, under 50 squares		3865	.008		1.82	.35	.03	2.20	2.66
2650	50-500 squares		4170	.008		1.45	.32	.03	1.80	2.21
2700	Over 500 squares		4300	.007		1.31	.31	.03	1.65	2.03
2900	18 gauge, under 50 squares		3800	.008		2.36	.35	.03	2.74	3.26
2950	50-500 squares		4100	.008		1.88	.33	.03	2.24	2.69
3000	Over 500 squares		4300	.007		1.70	.31	.03	2.04	2.46
3050	16 gauge, under 50 squares		3700	.009		3.17	.36	.03	3.56	4.17
3060	50-500 squares		4000	.008		2.53	.33	.03	2.89	3.43
3100	Over 500 squares		4200	.008		2.28	.32	.03	2.63	3.12
3150	For intermediate rib instead of wide rib, deduct					.03			.03	.03
3160	For narrow rib instead of wide rib, add					.47			.47	.52

05 31 33 – Steel Form Decking

05 31 33.50 Form Decking

		Crew	Daily Output	Labor-Hours	Unit	Material	Labor	Equipment	Total	Total Incl O&P
0010	**FORM DECKING**									
6100	Slab form, steel, 28 gauge, 9/16" deep, uncoated	E-4	4000	.008	S.F.	1.03	.33	.03	1.39	1.77
6200	Galvanized		4000	.008		.91	.33	.03	1.27	1.64
6220	24 gauge, 1" deep, uncoated		3900	.008		1.12	.34	.03	1.49	1.88
6240	Galvanized		3900	.008		1.32	.34	.03	1.69	2.10
6300	24 gauge, 1-5/16" deep, uncoated		3800	.008		1.20	.35	.03	1.58	1.99
6400	Galvanized		3800	.008		1.41	.35	.03	1.79	2.22
6500	22 gauge, 1-5/16" deep, uncoated		3700	.009		1.50	.36	.03	1.89	2.34
6600	Galvanized		3700	.009		1.53	.36	.03	1.92	2.37
6700	22 gauge, 2" deep uncoated		3600	.009		1.98	.37	.03	2.38	2.89
6800	Galvanized		3600	.009		1.94	.37	.03	2.34	2.84
7000	Sheet metal edge closure form, 12" wide with 2 bends, galv									
7100	18 gauge	E-14	360	.022	L.F.	3.19	.96	.32	4.47	5.60
7200	16 gauge	"	360	.022	"	4.33	.96	.32	5.61	6.85

05 35 Raceway Decking Assemblies

05 35 13 – Cellular Decking

05 35 13.50 Cellular Decking

		Crew	Daily Output	Labor-Hours	Unit	Material	2007 Bare Costs Labor	Equipment	Total	Total Incl O&P
0010	**CELLULAR DECKING**									
0200	Cellular units, galv, 2" deep, 20-20 gauge, over 15 squares	E-4	1460	.022	S.F.	6.15	.92	.08	7.15	8.50
0250	18-20 gauge		1420	.023		7	.94	.08	8.02	9.50
0300	18-18 gauge		1390	.023		7.15	.96	.08	8.19	9.75
0320	16-18 gauge		1360	.024		8.55	.98	.08	9.61	11.25
0340	16-16 gauge		1330	.024		9.50	1.01	.09	10.60	12.35
0400	3" deep, galvanized, 20-20 gauge		1375	.023		6.75	.97	.08	7.80	9.30
0500	18-20 gauge		1350	.024		8.15	.99	.09	9.23	10.90
0600	18-18 gauge		1290	.025		8.15	1.04	.09	9.28	10.95
0700	16-18 gauge		1230	.026		9.20	1.09	.09	10.38	12.15
0800	16-16 gauge		1150	.028		10	1.16	.10	11.26	13.20
1000	4-1/2" deep, galvanized, 20-18 gauge		1100	.029		9.40	1.22	.10	10.72	12.70
1100	18-18 gauge		1040	.031		9.35	1.29	.11	10.75	12.75
1200	16-18 gauge		980	.033		10.55	1.37	.12	12.04	14.20
1300	16-16 gauge		935	.034		11.50	1.43	.12	13.05	15.40
1500	For acoustical deck, add					15%				
1700	For cells used for ventilation, add					15%				
1900	For multi-story or congested site, add						50%			
8000	Metal deck and trench, 2" thick, 20 gauge, combination									
8010	60% cellular, 40% non-cellular, inserts and trench	R-4	1100	.036	S.F.	12.30	1.54	.10	13.94	16.35

05 41 Structural Metal Stud Framing

05 41 13 – Load-Bearing Metal Stud Framing

05 41 13.05 Bracing

		Crew	Daily Output	Labor-Hours	Unit	Material	Labor	Equipment	Total	Total Incl O&P
0010	**BRACING**, shear wall X-bracing, per 10' x 10' bay, one face									
0120	Metal strap, 20 ga x 4" wide	2 Carp	18	.889	Ea.	21.50	32.50		54	75
0130	6" wide		18	.889		34	32.50		66.50	88
0160	18 ga x 4" wide		16	1		31	36.50		67.50	91
0170	6" wide		16	1		46	36.50		82.50	108
0410	Continuous strap bracing, per horizontal row on both faces									
0420	Metal strap, 20 ga x 2" wide, studs 12" O.C.	1 Carp	7	1.143	C.L.F.	55	42		97	126
0430	16" O.C.		8	1		55	36.50		91.50	118
0440	24" O.C.		10	.800		55	29.50		84.50	106
0450	18 ga x 2" wide, studs 12" O.C.		6	1.333		75	49		124	159
0460	16" O.C.		7	1.143		75	42		117	148
0470	24" O.C.		8	1		75	36.50		111.50	140

05 41 13.10 Bridging

		Crew	Daily Output	Labor-Hours	Unit	Material	Labor	Equipment	Total	Total Incl O&P
0010	**BRIDGING**, solid between studs w/ 1-1/4" leg track, per stud bay									
0200	Studs 12" O.C., 18 ga x 2-1/2" wide	1 Carp	125	.064	Ea.	.93	2.35		3.28	4.68
0210	3-5/8" wide		120	.067		1.12	2.45		3.57	5.05
0220	4" wide		120	.067		1.18	2.45		3.63	5.10
0230	6" wide		115	.070		1.54	2.55		4.09	5.70
0240	8" wide		110	.073		1.96	2.67		4.63	6.30
0300	16 ga x 2-1/2" wide		115	.070		1.16	2.55		3.71	5.25
0310	3-5/8" wide		110	.073		1.42	2.67		4.09	5.70
0320	4" wide		110	.073		1.51	2.67		4.18	5.80
0330	6" wide		105	.076		1.93	2.80		4.73	6.45
0340	8" wide		100	.080		2.47	2.94		5.41	7.30
1200	Studs 16" O.C., 18 ga x 2-1/2" wide		125	.064		1.19	2.35		3.54	4.97
1210	3-5/8" wide		120	.067		1.44	2.45		3.89	5.40
1220	4" wide		120	.067		1.52	2.45		3.97	5.50

05 41 Structural Metal Stud Framing

05 41 13 – Load-Bearing Metal Stud Framing

05 41 13.10 Bridging		Crew	Daily Output	Labor-Hours	Unit	Material	2007 Bare Costs Labor	Equipment	Total	Total Incl O&P
1230	6" wide	1 Carp	115	.070	Ea.	1.98	2.55		4.53	6.15
1240	8" wide		110	.073		2.51	2.67		5.18	6.90
1300	16 ga x 2-1/2" wide		115	.070		1.49	2.55		4.04	5.60
1310	3-5/8" wide		110	.073		1.82	2.67		4.49	6.15
1320	4" wide		110	.073		1.93	2.67		4.60	6.30
1330	6" wide		105	.076		2.48	2.80		5.28	7.05
1340	8" wide		100	.080		3.17	2.94		6.11	8.05
2200	Studs 24" O.C., 18 ga x 2-1/2" wide		125	.064		1.72	2.35		4.07	5.55
2210	3-5/8" wide		120	.067		2.08	2.45		4.53	6.10
2220	4" wide		120	.067		2.20	2.45		4.65	6.25
2230	6" wide		115	.070		2.86	2.55		5.41	7.15
2240	8" wide		110	.073		3.63	2.67		6.30	8.15
2300	16 ga x 2-1/2" wide		115	.070		2.15	2.55		4.70	6.35
2310	3-5/8" wide		110	.073		2.63	2.67		5.30	7.05
2320	4" wide		110	.073		2.79	2.67		5.46	7.25
2330	6" wide		105	.076		3.58	2.80		6.38	8.30
2340	8" wide		100	.080		4.58	2.94		7.52	9.60
3000	Continuous bridging, per row									
3100	16 ga x 1-1/2" channel thru studs 12" O.C.	1 Carp	6	1.333	C.L.F.	51.50	49		100.50	133
3110	16" O.C.		7	1.143		51.50	42		93.50	123
3120	24" O.C.		8.80	.909		51.50	33.50		85	109
4100	2" x 2" angle x 18 ga, studs 12" O.C.		7	1.143		76	42		118	149
4110	16" O.C.		9	.889		76	32.50		108.50	135
4120	24" O.C.		12	.667		76	24.50		100.50	122
4200	16 ga, studs 12" O.C.		5	1.600		97	58.50		155.50	198
4210	16" O.C.		7	1.143		97	42		139	172
4220	24" O.C.		10	.800		97	29.50		126.50	152

05 41 13.25 Framing, Boxed Headers/Beams		Crew	Daily Output	Labor-Hours	Unit	Material	Labor	Equipment	Total	Total Incl O&P
0010	**FRAMING, BOXED HEADERS/BEAMS**									
0200	Double, 18 ga x 6" deep	2 Carp	220	.073	L.F.	5.35	2.67		8.02	10.05
0210	8" deep		210	.076		6	2.80		8.80	10.95
0220	10" deep		200	.080		7.25	2.94		10.19	12.50
0230	12" deep		190	.084		7.95	3.09		11.04	13.55
0300	16 ga x 8" deep		180	.089		6.85	3.26		10.11	12.65
0310	10" deep		170	.094		8.25	3.45		11.70	14.45
0320	12" deep		160	.100		8.95	3.67		12.62	15.55
0400	14 ga x 10" deep		140	.114		9.55	4.19		13.74	17.05
0410	12" deep		130	.123		10.50	4.52		15.02	18.60
1210	Triple, 18 ga x 8" deep		170	.094		8.65	3.45		12.10	14.95
1220	10" deep		165	.097		10.40	3.56		13.96	16.95
1230	12" deep		160	.100		11.45	3.67		15.12	18.30
1300	16 ga x 8" deep		145	.110		10	4.05		14.05	17.30
1310	10" deep		140	.114		11.90	4.19		16.09	19.60
1320	12" deep		135	.119		13	4.35		17.35	21
1400	14 ga x 10" deep		115	.139		13.05	5.10		18.15	22.50
1410	12" deep		110	.145		14.50	5.35		19.85	24

05 41 13.30 Framing, Stud Walls		Crew	Daily Output	Labor-Hours	Unit	Material	Labor	Equipment	Total	Total Incl O&P
0010	**FRAMING, STUD WALLS** w/ top & bottom track, no openings,									
0020	Headers, beams, bridging or bracing									
4100	8' high walls, 18 ga x 2-1/2" wide, studs 12" O.C.	2 Carp	54	.296	L.F.	8.95	10.85		19.80	27
4110	16" O.C.		77	.208		7.15	7.65		14.80	19.80
4120	24" O.C.		107	.150		5.35	5.50		10.85	14.45

05 41 Structural Metal Stud Framing

05 41 13 – Load-Bearing Metal Stud Framing

05 41 13.30 Framing, Stud Walls		Crew	Daily Output	Labor-Hours	Unit	Material	2007 Bare Costs Labor	Equipment	Total	Total Incl O&P
4130	3-5/8" wide, studs 12" O.C.	2 Carp	53	.302	L.F.	10.75	11.10		21.85	29
4140	16" O.C.		76	.211		8.60	7.75		16.35	21.50
4150	24" O.C.		105	.152		6.45	5.60		12.05	15.80
4160	4" wide, studs 12" O.C.		52	.308		11.30	11.30		22.60	30
4170	16" O.C.		74	.216		9.05	7.95		17	22.50
4180	24" O.C.		103	.155		6.75	5.70		12.45	16.35
4190	6" wide, studs 12" O.C.		51	.314		14.30	11.50		25.80	33.50
4200	16" O.C.		73	.219		11.45	8.05		19.50	25
4210	24" O.C.		101	.158		8.60	5.80		14.40	18.55
4220	8" wide, studs 12" O.C.		50	.320		17.50	11.75		29.25	37.50
4230	16" O.C.		72	.222		14.05	8.15		22.20	28
4240	24" O.C.		100	.160		10.60	5.85		16.45	21
4300	16 ga x 2-1/2" wide, studs 12" O.C.		47	.340		10.55	12.50		23.05	31
4310	16" O.C.		68	.235		8.35	8.65		17	22.50
4320	24" O.C.		94	.170		6.15	6.25		12.40	16.55
4330	3-5/8" wide, studs 12" O.C.		46	.348		12.70	12.75		25.45	34
4340	16" O.C.		66	.242		10.05	8.90		18.95	25
4350	24" O.C.		92	.174		7.40	6.40		13.80	18.10
4360	4" wide, studs 12" O.C.		45	.356		13.35	13.05		26.40	35
4370	16" O.C.		65	.246		10.60	9.05		19.65	25.50
4380	24" O.C.		90	.178		7.80	6.50		14.30	18.75
4390	6" wide, studs 12" O.C.		44	.364		16.85	13.35		30.20	39.50
4400	16" O.C.		64	.250		13.40	9.20		22.60	29
4410	24" O.C.		88	.182		9.90	6.65		16.55	21.50
4420	8" wide, studs 12" O.C.		43	.372		21	13.65		34.65	44.50
4430	16" O.C.		63	.254		16.50	9.30		25.80	32.50
4440	24" O.C.		86	.186		12.25	6.85		19.10	24
5100	10' high walls, 18 ga x 2-1/2" wide, studs 12" O.C.		54	.296		10.75	10.85		21.60	29
5110	16" O.C.		77	.208		8.50	7.65		16.15	21.50
5120	24" O.C.		107	.150		6.25	5.50		11.75	15.45
5130	3-5/8" wide, studs 12" O.C.		53	.302		12.95	11.10		24.05	31.50
5140	16" O.C.		76	.211		10.25	7.75		18	23.50
5150	24" O.C.		105	.152		7.55	5.60		13.15	17
5160	4" wide, studs 12" O.C.		52	.308		13.55	11.30		24.85	32.50
5170	16" O.C.		74	.216		10.75	7.95		18.70	24
5180	24" O.C.		103	.155		7.90	5.70		13.60	17.60
5190	6" wide, studs 12" O.C.		51	.314		17.15	11.50		28.65	37
5200	16" O.C.		73	.219		13.60	8.05		21.65	27.50
5210	24" O.C.		101	.158		10.05	5.80		15.85	20
5220	8" wide, studs 12" O.C.		50	.320		21	11.75		32.75	41.50
5230	16" O.C.		72	.222		16.60	8.15		24.75	31
5240	24" O.C.		100	.160		12.30	5.85		18.15	22.50
5300	16 ga x 2-1/2" wide, studs 12" O.C.		47	.340		12.75	12.50		25.25	33.50
5310	16" O.C.		68	.235		10	8.65		18.65	24.50
5320	24" O.C.		94	.170		7.25	6.25		13.50	17.75
5330	3-5/8" wide, studs 12" O.C.		46	.348		15.35	12.75		28.10	37
5340	16" O.C.		66	.242		12.05	8.90		20.95	27
5350	24" O.C.		92	.174		8.75	6.40		15.15	19.55
5360	4" wide, studs 12" O.C.		45	.356		16.15	13.05		29.20	38.50
5370	16" O.C.		65	.246		12.70	9.05		21.75	28
5380	24" O.C.		90	.178		9.20	6.50		15.70	20.50
5390	6" wide, studs 12" O.C.		44	.364		20.50	13.35		33.85	43.50
5400	16" O.C.		64	.250		16	9.20		25.20	32

05 41 Structural Metal Stud Framing

05 41 13 – Load-Bearing Metal Stud Framing

05 41 13.30 Framing, Stud Walls		Crew	Daily Output	Labor-Hours	Unit	Material	2007 Bare Costs Labor	Equipment	Total	Total Incl O&P
5410	24" O.C.	2 Carp	88	.182	L.F.	11.65	6.65		18.30	23
5420	8" wide, studs 12" O.C.		43	.372		25	13.65		38.65	49
5430	16" O.C.		63	.254		19.70	9.30		29	36
5440	24" O.C.		86	.186		14.35	6.85		21.20	26.50
6190	12' high walls, 18 ga x 6" wide, studs 12" O.C.		41	.390		20	14.30		34.30	44.50
6200	16" O.C.		58	.276		15.70	10.10		25.80	33
6210	24" O.C.		81	.198		11.45	7.25		18.70	24
6220	8" wide, studs 12" O.C.		40	.400		24.50	14.70		39.20	50
6230	16" O.C.		57	.281		19.20	10.30		29.50	37
6240	24" O.C.		80	.200		14.05	7.35		21.40	27
6390	16 ga x 6" wide, studs 12" O.C.		35	.457		24	16.80		40.80	52
6400	16" O.C.		51	.314		18.60	11.50		30.10	38.50
6410	24" O.C.		70	.229		13.40	8.40		21.80	28
6420	8" wide, studs 12" O.C.		34	.471		29.50	17.25		46.75	59
6430	16" O.C.		50	.320		23	11.75		34.75	43.50
6440	24" O.C.		69	.232		16.50	8.50		25	31.50
6530	14 ga x 3-5/8" wide, studs 12" O.C.		34	.471		22.50	17.25		39.75	52
6540	16" O.C.		48	.333		17.65	12.25		29.90	38.50
6550	24" O.C.		65	.246		12.65	9.05		21.70	28
6560	4" wide, studs 12" O.C.		33	.485		24	17.80		41.80	54
6570	16" O.C.		47	.340		18.60	12.50		31.10	40
6580	24" O.C.		64	.250		13.35	9.20		22.55	29
6730	12 ga x 3-5/8" wide, studs 12" O.C.		31	.516		31.50	18.95		50.45	64
6740	16" O.C.		43	.372		24	13.65		37.65	48
6750	24" O.C.		59	.271		17	9.95		26.95	34
6760	4" wide, studs 12" O.C.		30	.533		33.50	19.55		53.05	67.50
6770	16" O.C.		42	.381		26	14		40	50.50
6780	24" O.C.		58	.276		18.15	10.10		28.25	36
7390	16' high walls, 16 ga x 6" wide, studs 12" O.C.		33	.485		31	17.80		48.80	61.50
7400	16" O.C.		48	.333		24	12.25		36.25	45
7410	24" O.C.		67	.239		16.85	8.75		25.60	32
7420	8" wide, studs 12" O.C.		32	.500		38	18.35		56.35	70
7430	16" O.C.		47	.340		29.50	12.50		42	51.50
7440	24" O.C.		66	.242		21	8.90		29.90	37
7560	14 ga x 4" wide, studs 12" O.C.		31	.516		31	18.95		49.95	63.50
7570	16" O.C.		45	.356		24	13.05		37.05	47
7580	24" O.C.		61	.262		16.85	9.65		26.50	33.50
7590	6" wide, studs 12" O.C.		30	.533		39	19.55		58.55	73.50
7600	16" O.C.		44	.364		30	13.35		43.35	54
7610	24" O.C.		60	.267		21.50	9.80		31.30	39
7760	12 ga x 4" wide, studs 12" O.C.		29	.552		43.50	20.50		64	79.50
7770	16" O.C.		40	.400		33.50	14.70		48.20	60
7780	24" O.C.		55	.291		23.50	10.70		34.20	42
7790	6" wide, studs 12" O.C.		28	.571		55	21		76	93
7800	16" O.C.		39	.410		42.50	15.05		57.55	70
7810	24" O.C.		54	.296		29.50	10.85		40.35	49.50
8590	20' high walls, 14 ga x 6" wide, studs 12" O.C.		29	.552		47.50	20.50		68	84
8600	16" O.C.		42	.381		36.50	14		50.50	62.50
8610	24" O.C.		57	.281		25.50	10.30		35.80	44
8620	8" wide, studs 12" O.C.		28	.571		58.50	21		79.50	96.50
8630	16" O.C.		41	.390		45	14.30		59.30	72
8640	24" O.C.		56	.286		31.50	10.50		42	51
8790	12 ga x 6" wide, studs 12" O.C.		27	.593		68	22		90	109

05 41 Structural Metal Stud Framing

05 41 13 – Load-Bearing Metal Stud Framing

05 41 13.30 Framing, Stud Walls		Crew	Daily Output	Labor-Hours	Unit	Material	2007 Bare Costs Labor	Equipment	Total	Total Incl O&P
8800	16" O.C.	2 Carp	37	.432	L.F.	52	15.85		67.85	81.50
8810	24" O.C.		51	.314		36	11.50		47.50	57.50
8820	8" wide, studs 12" O.C.		26	.615		83	22.50		105.50	126
8830	16" O.C.		36	.444		63.50	16.30		79.80	95
8840	24" O.C.	↓	50	.320	↓	44	11.75		55.75	66.50

05 42 Cold-Formed Metal Joist Framing

05 42 13 – Cold-Formed Metal Floor Joist Framing

05 42 13.05 Bracing

		Crew	Daily Output	Labor-Hours	Unit	Material	Labor	Equipment	Total	Total Incl O&P
0010	**BRACING**, continuous, per row, top & bottom									
0120	Flat strap, 20 ga x 2" wide, joists at 12" O.C.	1 Carp	4.67	1.713	C.L.F.	57.50	63		120.50	162
0130	16" O.C.		5.33	1.501		55.50	55		110.50	147
0140	24" O.C.		6.66	1.201		53.50	44		97.50	128
0150	18 ga x 2" wide, joists at 12" O.C.		4	2		74	73.50		147.50	196
0160	16" O.C.		4.67	1.713		73	63		136	178
0170	24" O.C.	↓	5.33	1.501	↓	71.50	55		126.50	165

05 42 13.10 Bridging

		Crew	Daily Output	Labor-Hours	Unit	Material	Labor	Equipment	Total	Total Incl O&P
0010	**BRIDGING**, solid between joists w/ 1-1/4" leg track, per joist bay									
0230	Joists 12" O.C., 18 ga track x 6" wide	1 Carp	80	.100	Ea.	1.54	3.67		5.21	7.40
0240	8" wide		75	.107		1.96	3.91		5.87	8.25
0250	10" wide		70	.114		2.42	4.19		6.61	9.20
0260	12" wide		65	.123		2.81	4.52		7.33	10.15
0330	16 ga track x 6" wide		70	.114		1.93	4.19		6.12	8.65
0340	8" wide		65	.123		2.47	4.52		6.99	9.75
0350	10" wide		60	.133		3.04	4.89		7.93	10.95
0360	12" wide		55	.145		3.50	5.35		8.85	12.15
0440	14 ga track x 8" wide		60	.133		3.11	4.89		8	11.05
0450	10" wide		55	.145		3.82	5.35		9.17	12.50
0460	12" wide		50	.160		4.41	5.85		10.26	14
0550	12 ga track x 10" wide		45	.178		5.60	6.50		12.10	16.30
0560	12" wide		40	.200		6.35	7.35		13.70	18.45
1230	16" O.C., 18 ga track x 6" wide		80	.100		1.98	3.67		5.65	7.90
1240	8" wide		75	.107		2.51	3.91		6.42	8.85
1250	10" wide		70	.114		3.10	4.19		7.29	9.95
1260	12" wide		65	.123		3.60	4.52		8.12	11
1330	16 ga track x 6" wide		70	.114		2.48	4.19		6.67	9.25
1340	8" wide		65	.123		3.17	4.52		7.69	10.55
1350	10" wide		60	.133		3.89	4.89		8.78	11.90
1360	12" wide		55	.145		4.49	5.35		9.84	13.25
1440	14 ga track x 8" wide		60	.133		3.99	4.89		8.88	12
1450	10" wide		55	.145		4.90	5.35		10.25	13.70
1460	12" wide		50	.160		5.65	5.85		11.50	15.40
1550	12 ga track x 10" wide		45	.178		7.20	6.50		13.70	18.05
1560	12" wide		40	.200		8.15	7.35		15.50	20.50
2230	24" O.C., 18 ga track x 6" wide		80	.100		2.86	3.67		6.53	8.85
2240	8" wide		75	.107		3.63	3.91		7.54	10.10
2250	10" wide		70	.114		4.49	4.19		8.68	11.50
2260	12" wide		65	.123		5.20	4.52		9.72	12.75
2330	16 ga track x 6" wide		70	.114		3.58	4.19		7.77	10.50
2340	8" wide		65	.123		4.58	4.52		9.10	12.10
2350	10" wide	↓	60	.133	↓	5.65	4.89		10.54	13.80

05 42 Cold-Formed Metal Joist Framing

05 42 13 – Cold-Formed Metal Floor Joist Framing

05 42 13.10 Bridging		Crew	Daily Output	Labor-Hours	Unit	Material	2007 Bare Costs Labor	Equipment	Total	Total Incl O&P
2360	12" wide	1 Carp	55	.145	Ea.	6.50	5.35		11.85	15.45
2440	14 ga track x 8" wide		60	.133		5.80	4.89		10.69	13.95
2450	10" wide		55	.145		7.10	5.35		12.45	16.10
2460	12" wide		50	.160		8.20	5.85		14.05	18.15
2550	12 ga track x 10" wide		45	.178		10.40	6.50		16.90	21.50
2560	12" wide		40	.200		11.75	7.35		19.10	24.50

05 42 13.25 Framing, Band Joist

		Crew	Daily Output	Labor-Hours	Unit	Material	Labor	Equipment	Total	Total Incl O&P
0010	**FRAMING, BAND JOIST** (track) fastened to bearing wall									
0220	18 ga track x 6" deep	2 Carp	1000	.016	L.F.	1.26	.59		1.85	2.30
0230	8" deep		920	.017		1.60	.64		2.24	2.75
0240	10" deep		860	.019		1.97	.68		2.65	3.23
0320	16 ga track x 6" deep		900	.018		1.58	.65		2.23	2.75
0330	8" deep		840	.019		2.02	.70		2.72	3.31
0340	10" deep		780	.021		2.48	.75		3.23	3.90
0350	12" deep		740	.022		2.86	.79		3.65	4.38
0430	14 ga track x 8" deep		750	.021		2.54	.78		3.32	4.02
0440	10" deep		720	.022		3.12	.82		3.94	4.70
0450	12" deep		700	.023		3.60	.84		4.44	5.25
0540	12 ga track x 10" deep		670	.024		4.58	.88		5.46	6.40
0550	12" deep		650	.025		5.20	.90		6.10	7.10

05 42 13.30 Framing, Boxed Headers/Beams

		Crew	Daily Output	Labor-Hours	Unit	Material	Labor	Equipment	Total	Total Incl O&P
0010	**FRAMING, BOXED HEADERS/BEAMS**									
0200	Double, 18 ga x 6" deep	2 Carp	220	.073	L.F.	5.35	2.67		8.02	10.05
0210	8" deep		210	.076		6	2.80		8.80	10.95
0220	10" deep		200	.080		7.25	2.94		10.19	12.50
0230	12" deep		190	.084		7.95	3.09		11.04	13.55
0300	16 ga x 8" deep		180	.089		6.85	3.26		10.11	12.65
0310	10" deep		170	.094		8.25	3.45		11.70	14.45
0320	12" deep		160	.100		8.95	3.67		12.62	15.55
0400	14 ga x 10" deep		140	.114		9.55	4.19		13.74	17.05
0410	12" deep		130	.123		10.50	4.52		15.02	18.60
0500	12 ga x 10" deep		110	.145		12.65	5.35		18	22.50
0510	12" deep		100	.160		14.05	5.85		19.90	24.50
1210	Triple, 18 ga x 8" deep		170	.094		8.65	3.45		12.10	14.95
1220	10" deep		165	.097		10.40	3.56		13.96	16.95
1230	12" deep		160	.100		11.45	3.67		15.12	18.30
1300	16 ga x 8" deep		145	.110		10	4.05		14.05	17.30
1310	10" deep		140	.114		11.90	4.19		16.09	19.60
1320	12" deep		135	.119		13	4.35		17.35	21
1400	14 ga x 10" deep		115	.139		13.90	5.10		19	23.50
1410	12" deep		110	.145		15.30	5.35		20.65	25
1500	12 ga x 10" deep		90	.178		18.55	6.50		25.05	30.50
1510	12" deep		85	.188		20.50	6.90		27.40	33.50

05 42 13.40 Framing, Joists

		Crew	Daily Output	Labor-Hours	Unit	Material	Labor	Equipment	Total	Total Incl O&P
0010	**FRAMING, JOISTS**, no band joists (track), web stiffeners, headers,									
0020	Beams, bridging or bracing									
0030	Joists (2" flange) and fasteners, materials only									
0220	18 ga x 6" deep				L.F.	1.66			1.66	1.82
0230	8" deep					1.98			1.98	2.18
0240	10" deep					2.32			2.32	2.55
0320	16 ga x 6" deep					2.04			2.04	2.24
0330	8" deep					2.45			2.45	2.69

05 42 Cold-Formed Metal Joist Framing

05 42 13 – Cold-Formed Metal Floor Joist Framing

05 42 13.40 Framing, Joists

		Crew	Daily Output	Labor-Hours	Unit	Material	2007 Bare Costs Labor	Equipment	Total	Total Incl O&P
0340	10" deep				L.F.	2.85			2.85	3.13
0350	12" deep					3.23			3.23	3.56
0430	14 ga x 8" deep					3.09			3.09	3.40
0440	10" deep					3.55			3.55	3.90
0450	12" deep					4.04			4.04	4.45
0540	12 ga x 10" deep					5.20			5.20	5.70
0550	12" deep					5.90			5.90	6.50
1010	Installation of joists to band joists, beams & headers, labor only									
1220	18 ga x 6" deep	2 Carp	110	.145	Ea.		5.35		5.35	8.30
1230	8" deep		90	.178			6.50		6.50	10.15
1240	10" deep		80	.200			7.35		7.35	11.45
1320	16 ga x 6" deep		95	.168			6.20		6.20	9.65
1330	8" deep		70	.229			8.40		8.40	13.05
1340	10" deep		60	.267			9.80		9.80	15.25
1350	12" deep		55	.291			10.70		10.70	16.65
1430	14 ga x 8" deep		65	.246			9.05		9.05	14.05
1440	10" deep		45	.356			13.05		13.05	20.50
1450	12" deep		35	.457			16.80		16.80	26
1540	12 ga x 10" deep		40	.400			14.70		14.70	23
1550	12" deep		30	.533			19.55		19.55	30.50

05 42 13.45 Framing, Web Stiffeners

		Crew	Daily Output	Labor-Hours	Unit	Material	2007 Bare Costs Labor	Equipment	Total	Total Incl O&P
0010	**FRAMING, WEB STIFFENERS** at joist bearing, fabricated from									
0020	Stud piece (1-5/8" flange) to stiffen joist (2" flange)									
2120	For 6" deep joist, with 18 ga x 2-1/2" stud	1 Carp	120	.067	Ea.	1.98	2.45		4.43	6
2130	3-5/8" stud		110	.073		2.21	2.67		4.88	6.60
2140	4" stud		105	.076		2.15	2.80		4.95	6.70
2150	6" stud		100	.080		2.34	2.94		5.28	7.15
2160	8" stud		95	.084		2.41	3.09		5.50	7.45
2220	8" deep joist, with 2-1/2" stud		120	.067		2.17	2.45		4.62	6.20
2230	3-5/8" stud		110	.073		2.39	2.67		5.06	6.80
2240	4" stud		105	.076		2.35	2.80		5.15	6.95
2250	6" stud		100	.080		2.57	2.94		5.51	7.40
2260	8" stud		95	.084		2.77	3.09		5.86	7.85
2320	10" deep joist, with 2-1/2" stud		110	.073		3.06	2.67		5.73	7.55
2330	3-5/8" stud		100	.080		3.41	2.94		6.35	8.30
2340	4" stud		95	.084		3.38	3.09		6.47	8.50
2350	6" stud		90	.089		3.65	3.26		6.91	9.10
2360	8" stud		85	.094		3.71	3.45		7.16	9.50
2420	12" deep joist, with 2-1/2" stud		110	.073		3.24	2.67		5.91	7.70
2430	3-5/8" stud		100	.080		3.56	2.94		6.50	8.50
2440	4" stud		95	.084		3.50	3.09		6.59	8.65
2450	6" stud		90	.089		3.83	3.26		7.09	9.30
2460	8" stud		85	.094		4.13	3.45		7.58	9.95
3130	For 6" deep joist, with 16 ga x 3-5/8" stud		100	.080		2.31	2.94		5.25	7.10
3140	4" stud		95	.084		2.29	3.09		5.38	7.35
3150	6" stud		90	.089		2.52	3.26		5.78	7.90
3160	8" stud		85	.094		2.66	3.45		6.11	8.35
3230	8" deep joist, with 3-5/8" stud		100	.080		2.56	2.94		5.50	7.40
3240	4" stud		95	.084		2.51	3.09		5.60	7.60
3250	6" stud		90	.089		2.80	3.26		6.06	8.20
3260	8" stud		85	.094		3	3.45		6.45	8.70
3330	10" deep joist, with 3-5/8" stud		85	.094		3.51	3.45		6.96	9.25

05 42 Cold-Formed Metal Joist Framing

05 42 13 – Cold-Formed Metal Floor Joist Framing

05 42 13.45 Framing, Web Stiffeners

		Crew	Daily Output	Labor-Hours	Unit	Material	2007 Bare Costs Labor	Equipment	Total	Total Incl O&P
3340	4" stud	1 Carp	80	.100	Ea.	3.58	3.67		7.25	9.65
3350	6" stud		75	.107		3.90	3.91		7.81	10.40
3360	8" stud		70	.114		4.07	4.19		8.26	11
3430	12" deep joist, with 3-5/8" stud		85	.094		3.83	3.45		7.28	9.60
3440	4" stud		80	.100		3.75	3.67		7.42	9.85
3450	6" stud		75	.107		4.18	3.91		8.09	10.70
3460	8" stud		70	.114		4.47	4.19		8.66	11.45
4230	For 8" deep joist, with 14 ga x 3-5/8" stud		90	.089		3.34	3.26		6.60	8.75
4240	4" stud		85	.094		3.40	3.45		6.85	9.15
4250	6" stud		80	.100		3.69	3.67		7.36	9.75
4260	8" stud		75	.107		3.95	3.91		7.86	10.45
4330	10" deep joist, with 3-5/8" stud		75	.107		4.68	3.91		8.59	11.25
4340	4" stud		70	.114		4.65	4.19		8.84	11.65
4350	6" stud		65	.123		5.10	4.52		9.62	12.65
4360	8" stud		60	.133		5.35	4.89		10.24	13.45
4430	12" deep joist, with 3-5/8" stud		75	.107		4.98	3.91		8.89	11.60
4440	4" stud		70	.114		5.10	4.19		9.29	12.15
4450	6" stud		65	.123		5.50	4.52		10.02	13.10
4460	8" stud		60	.133		5.90	4.89		10.79	14.10
5330	For 10" deep joist, with 12 ga x 3-5/8" stud		65	.123		4.94	4.52		9.46	12.50
5340	4" stud		60	.133		5.10	4.89		9.99	13.20
5350	6" stud		55	.145		5.60	5.35		10.95	14.45
5360	8" stud		50	.160		6.15	5.85		12	15.95
5430	12" deep joist, with 3-5/8" stud		65	.123		5.45	4.52		9.97	13.05
5440	4" stud		60	.133		5.35	4.89		10.24	13.50
5450	6" stud		55	.145		6.10	5.35		11.45	15.05
5460	8" stud		50	.160		7.05	5.85		12.90	16.90

05 42 23 – Cold-Formed Metal Roof Joist Framing

05 42 23.05 Framing, Bracing

		Crew	Daily Output	Labor-Hours	Unit	Material	2007 Bare Costs Labor	Equipment	Total	Total Incl O&P
0010	**FRAMING, BRACING**									
0020	Continuous bracing, per row									
0100	16 ga x 1-1/2" channel thru rafters/trusses @ 16" O.C.	1 Carp	4.50	1.778	C.L.F.	51.50	65		116.50	159
0120	24" O.C.		6	1.333		51.50	49		100.50	133
0300	2" x 2" angle x 18 ga, rafters/trusses @ 16" O.C.		6	1.333		76	49		125	160
0320	24" O.C.		8	1		76	36.50		112.50	141
0400	16 ga, rafters/trusses @ 16" O.C.		4.50	1.778		97	65		162	208
0420	24" O.C.		6.50	1.231		97	45		142	177

05 42 23.10 Framing, Bridging

		Crew	Daily Output	Labor-Hours	Unit	Material	2007 Bare Costs Labor	Equipment	Total	Total Incl O&P
0010	**FRAMING, BRIDGING**									
0020	Solid, between rafters w/ 1-1/4" leg track, per rafter bay									
1200	Rafters 16" O.C., 18 ga x 4" deep	1 Carp	60	.133	Ea.	1.52	4.89		6.41	9.25
1210	6" deep		57	.140		1.98	5.15		7.13	10.20
1220	8" deep		55	.145		2.51	5.35		7.86	11.05
1230	10" deep		52	.154		3.10	5.65		8.75	12.20
1240	12" deep		50	.160		3.60	5.85		9.45	13.10
2200	24" O.C., 18 ga x 4" deep		60	.133		2.20	4.89		7.09	10
2210	6" deep		57	.140		2.86	5.15		8.01	11.15
2220	8" deep		55	.145		3.63	5.35		8.98	12.30
2230	10" deep		52	.154		4.49	5.65		10.14	13.75
2240	12" deep		50	.160		5.20	5.85		11.05	14.85

05 42 23.50 Framing, Parapets

0010	**FRAMING, PARAPETS**									

05 42 Cold-Formed Metal Joist Framing

05 42 23 – Cold-Formed Metal Roof Joist Framing

05 42 23.50 Framing, Parapets

		Crew	Daily Output	Labor-Hours	Unit	Material	2007 Bare Costs Labor	2007 Bare Costs Equipment	Total	Total Incl O&P
0100	3' high installed on 1st story, 18 ga x 4" wide studs, 12" O.C.	2 Carp	100	.160	L.F.	5.65	5.85		11.50	15.35
0110	16" O.C.		150	.107		4.80	3.91		8.71	11.40
0120	24" O.C.		200	.080		3.95	2.94		6.89	8.90
0200	6" wide studs, 12" O.C.		100	.160		7.20	5.85		13.05	17.05
0210	16" O.C.		150	.107		6.15	3.91		10.06	12.85
0220	24" O.C.		200	.080		5.05	2.94		7.99	10.15
1100	Installed on 2nd story, 18 ga x 4" wide studs, 12" O.C.		95	.168		5.65	6.20		11.85	15.85
1110	16" O.C.		145	.110		4.80	4.05		8.85	11.60
1120	24" O.C.		190	.084		3.95	3.09		7.04	9.15
1200	6" wide studs, 12" O.C.		95	.168		7.20	6.20		13.40	17.55
1210	16" O.C.		145	.110		6.15	4.05		10.20	13.05
1220	24" O.C.		190	.084		5.05	3.09		8.14	10.40
2100	Installed on gable, 18 ga x 4" wide studs, 12" O.C.		85	.188		5.65	6.90		12.55	16.95
2110	16" O.C.		130	.123		4.80	4.52		9.32	12.35
2120	24" O.C.		170	.094		3.95	3.45		7.40	9.75
2200	6" wide studs, 12" O.C.		85	.188		7.20	6.90		14.10	18.65
2210	16" O.C.		130	.123		6.15	4.52		10.67	13.80
2220	24" O.C.		170	.094		5.05	3.45		8.50	11

05 42 23.60 Framing, Roof Rafters

		Crew	Daily Output	Labor-Hours	Unit	Material	2007 Bare Costs Labor	2007 Bare Costs Equipment	Total	Total Incl O&P
0010	**FRAMING, ROOF RAFTERS**									
0100	Boxed ridge beam, double, 18 ga x 6" deep	2 Carp	160	.100	L.F.	5.35	3.67		9.02	11.60
0110	8" deep		150	.107		6	3.91		9.91	12.70
0120	10" deep		140	.114		7.25	4.19		11.44	14.50
0130	12" deep		130	.123		7.95	4.52		12.47	15.80
0200	16 ga x 6" deep		150	.107		6.10	3.91		10.01	12.80
0210	8" deep		140	.114		6.85	4.19		11.04	14.10
0220	10" deep		130	.123		8.25	4.52		12.77	16.10
0230	12" deep		120	.133		8.95	4.89		13.84	17.45
1100	Rafters, 2" flange, material only, 18 ga x 6" deep					1.66			1.66	1.82
1110	8" deep					1.98			1.98	2.18
1120	10" deep					2.32			2.32	2.55
1130	12" deep					2.70			2.70	2.97
1200	16 ga x 6" deep					2.04			2.04	2.24
1210	8" deep					2.45			2.45	2.69
1220	10" deep					2.85			2.85	3.13
1230	12" deep					3.23			3.23	3.56
2100	Installation only, ordinary rafter to 4:12 pitch, 18 ga x 6" deep	2 Carp	35	.457	Ea.		16.80		16.80	26
2110	8" deep		30	.533			19.55		19.55	30.50
2120	10" deep		25	.640			23.50		23.50	36.50
2130	12" deep		20	.800			29.50		29.50	45.50
2200	16 ga x 6" deep		30	.533			19.55		19.55	30.50
2210	8" deep		25	.640			23.50		23.50	36.50
2220	10" deep		20	.800			29.50		29.50	45.50
2230	12" deep		15	1.067			39		39	61
8100	Add to labor, ordinary rafters on steep roofs						25%			
8110	Dormers & complex roofs						50%			
8200	Hip & valley rafters to 4:12 pitch						25%			
8210	Steep roofs						50%			
8220	Dormers & complex roofs						75%			
8300	Hip & valley jack rafters to 4:12 pitch						50%			
8310	Steep roofs						75%			
8320	Dormers & complex roofs						100%			

05 42 Cold-Formed Metal Joist Framing

05 42 23 – Cold-Formed Metal Roof Joist Framing

05 42 23.70 Framing, Soffits and Canopies

		Crew	Daily Output	Labor-Hours	Unit	Material	2007 Bare Costs Labor	2007 Bare Costs Equipment	Total	Total Incl O&P
0010	**FRAMING, SOFFITS & CANOPIES**									
0130	Continuous ledger track @ wall, studs @ 16" O.C., 18 ga x 4" wide	2 Carp	535	.030	L.F.	1.01	1.10		2.11	2.82
0140	6" wide		500	.032		1.32	1.17		2.49	3.28
0150	8" wide		465	.034		1.67	1.26		2.93	3.81
0160	10" wide		430	.037		2.07	1.37		3.44	4.40
0230	Studs @ 24" O.C., 18 ga x 4" wide		800	.020		.97	.73		1.70	2.20
0240	6" wide		750	.021		1.26	.78		2.04	2.61
0250	8" wide		700	.023		1.60	.84		2.44	3.07
0260	10" wide		650	.025		1.97	.90		2.87	3.58
1000	Horizontal soffit and canopy members, material only									
1030	1-5/8" flange studs, 18 ga x 4" deep				L.F.	1.36			1.36	1.49
1040	6" deep					1.70			1.70	1.87
1050	8" deep					2.06			2.06	2.27
1140	2" flange joists, 18 ga x 6" deep					1.90			1.90	2.09
1150	8" deep					2.27			2.27	2.49
1160	10" deep					2.65			2.65	2.92
4030	Installation only, 18 ga, 1-5/8" flange x 4" deep	2 Carp	130	.123	Ea.		4.52		4.52	7.05
4040	6" deep		110	.145			5.35		5.35	8.30
4050	8" deep		90	.178			6.50		6.50	10.15
4140	2" flange, 18 ga x 6" deep		110	.145			5.35		5.35	8.30
4150	8" deep		90	.178			6.50		6.50	10.15
4160	10" deep		80	.200			7.35		7.35	11.45
6010	Clips to attach facia to rafter tails, 2" x 2" x 18 ga angle	1 Carp	120	.067		.90	2.45		3.35	4.80
6020	16 ga angle	"	100	.080		1.14	2.94		4.08	5.85

05 44 Cold-Formed Metal Trusses

05 44 13 – Cold-Formed Metal Roof Trusses

05 44 13.60 Framing, Roof Trusses

		Crew	Daily Output	Labor-Hours	Unit	Material	2007 Bare Costs Labor	2007 Bare Costs Equipment	Total	Total Incl O&P
0010	**FRAMING, ROOF TRUSSES**									
0020	Fabrication of trusses on ground, Fink (W) or King Post, to 4:12 pitch									
0120	18 ga x 4" chords, 16' span	2 Carp	12	1.333	Ea.	63.50	49		112.50	146
0130	20' span		11	1.455		79	53.50		132.50	170
0140	24' span		11	1.455		95	53.50		148.50	187
0150	28' span		10	1.600		111	58.50		169.50	214
0160	32' span		10	1.600		127	58.50		185.50	231
0250	6" chords, 28' span		9	1.778		139	65		204	255
0260	32' span		9	1.778		159	65		224	277
0270	36' span		8	2		179	73.50		252.50	310
0280	40' span		8	2		199	73.50		272.50	335
1120	5:12 to 8:12 pitch, 18 ga x 4" chords, 16' span		10	1.600		72.50	58.50		131	171
1130	20' span		9	1.778		90.50	65		155.50	202
1140	24' span		9	1.778		108	65		173	221
1150	28' span		8	2		127	73.50		200.50	253
1160	32' span		8	2		145	73.50		218.50	273
1250	6" chords, 28' span		7	2.286		159	84		243	305
1260	32' span		7	2.286		182	84		266	330
1270	36' span		6	2.667		204	98		302	375
1280	40' span		6	2.667		227	98		325	400
2120	9:12 to 12:12 pitch, 18 ga x 4" chords, 16' span		8	2		90.50	73.50		164	214
2130	20' span		7	2.286		113	84		197	255

05 44 Cold-Formed Metal Trusses

05 44 13 – Cold-Formed Metal Roof Trusses

05 44 13.60 Framing, Roof Trusses		Crew	Daily Output	Labor-Hours	Unit	Material	2007 Bare Costs Labor	Equipment	Total	Total Incl O&P
2140	24' span	2 Carp	7	2.286	Ea.	136	84		220	280
2150	28' span		6	2.667		158	98		256	325
2160	32' span		6	2.667		181	98		279	350
2250	6" chords, 28' span		5	3.200		199	117		316	400
2260	32' span		5	3.200		227	117		344	435
2270	36' span		4	4		256	147		403	510
2280	40' span		4	4		284	147		431	540
5120	Erection only of roof trusses, to 4:12 pitch, 16' span	F-6	48	.833			28.50	15.10	43.60	60.50
5130	20' span		46	.870			29.50	15.75	45.25	63.50
5140	24' span		44	.909			31	16.45	47.45	66
5150	28' span		42	.952			32.50	17.25	49.75	69.50
5160	32' span		40	1			34	18.10	52.10	73
5170	36' span		38	1.053			36	19.05	55.05	76.50
5180	40' span		36	1.111			38	20	58	80.50
5220	5:12 to 8:12 pitch, 16' span		42	.952			32.50	17.25	49.75	69.50
5230	20' span		40	1			34	18.10	52.10	73
5240	24' span		38	1.053			36	19.05	55.05	76.50
5250	28' span		36	1.111			38	20	58	80.50
5260	32' span		34	1.176			40	21.50	61.50	85.50
5270	36' span		32	1.250			42.50	22.50	65	91
5280	40' span		30	1.333			45.50	24	69.50	97
5320	9:12 to 12:12 pitch, 16' span		36	1.111			38	20	58	80.50
5330	20' span		34	1.176			40	21.50	61.50	85.50
5340	24' span		32	1.250			42.50	22.50	65	91
5350	28' span		30	1.333			45.50	24	69.50	97
5360	32' span		28	1.429			49	26	75	104
5370	36' span		26	1.538			52.50	28	80.50	112
5380	40' span		24	1.667			57	30	87	121

05 51 Metal Stairs

05 51 13 – Metal Pan Stairs

05 51 13.50 Pan Stairs

		Crew	Daily Output	Labor-Hours	Unit	Material	2007 Bare Costs Labor	Equipment	Total	Total Incl O&P
0010	**PAN STAIRS**, shop fabricated, steel stringers									
0200	Cement fill metal pan, picket rail, 3'-6" wide	E-4	35	.914	Riser	390	38.50	3.29	431.79	505
0300	4'-0" wide		30	1.067		445	44.50	3.84	493.34	575
0350	Wall rail, both sides, 3'-6" wide CN		53	.604		300	25.50	2.17	327.67	380
1500	Landing, steel pan, conventional		160	.200	S.F.	52.50	8.35	.72	61.57	73.50
1600	Pre-erected		255	.125	"	91.50	5.25	.45	97.20	111
1700	Pre-erected, steel pan tread, 3'-6" wide, 2 line pipe rail	E-2	87	.644	Riser	430	26	17.80	473.80	540

05 51 16 – Metal Floor Plate Stairs

05 51 16.50 Floor Plate Stairs

		Crew	Daily Output	Labor-Hours	Unit	Material	2007 Bare Costs Labor	Equipment	Total	Total Incl O&P
0010	**FLOOR PLATE STAIRS**, shop fabricated, steel stringers									
0400	Cast iron tread and pipe rail, 3'-6" wide	E-4	35	.914	Riser	420	38.50	3.29	461.79	535
0500	Checkered plate tread, industrial, 3'-6" wide		28	1.143		261	48	4.11	313.11	380
0550	Circular, for tanks, 3'-0" wide		33	.970		288	40.50	3.49	331.99	390
0600	For isolated stairs, add						100%			
0800	Custom steel stairs, 3'-6" wide, minimum	E-4	35	.914		390	38.50	3.29	431.79	505
0810	Average		30	1.067		525	44.50	3.84	573.34	660
0900	Maximum		20	1.600		655	67	5.75	727.75	845
1100	For 4' wide stairs, add					5%	5%			

05 51 Metal Stairs

05 51 16 – Metal Floor Plate Stairs

05 51 16.50 Floor Plate Stairs

		Daily	Labor-			2007 Bare Costs			Total
05 51 16.50 Floor Plate Stairs	Crew	Output	Hours	Unit	Material	Labor	Equipment	Total	Incl O&P
1300 For 5' wide stairs, add				Riser	10%	10%			

05 51 19 – Metal Grating Stairs

05 51 19.50 Grating Stairs

	Crew	Daily Output	Labor-Hours	Unit	Material	Labor	Equipment	Total	Total Incl O&P
0010 **GRATING STAIRS**, shop fabricated, steel stringers, safety nosing on treads									
0020 Grating tread and pipe railing, 3'-6" wide	E-4	35	.914	Riser	261	38.50	3.29	302.79	360
0100 4'-0" wide	"	30	1.067	"	340	44.50	3.84	388.34	460

05 51 23 – Metal Fire Escapes

05 51 23.25 Fire Escapes

	Crew	Daily Output	Labor-Hours	Unit	Material	Labor	Equipment	Total	Total Incl O&P
0010 **FIRE ESCAPES**, shop fabricated									
0200 2' wide balcony, 1" x 1/4" bars 1-1/2" O.C.	1 Sswk	5	1.600	L.F.	48	66		114	173
0400 1st story cantilevered stair, standard		.09	88.889	Ea.	2,000	3,675		5,675	8,850
0700 Platform & fixed stair, 36" x 40"	↓	.17	47.059	Flight	885	1,950		2,835	4,500
0900 For 3'-6" wide escapes, add to above					100%	150%			

05 51 23.50 Fire Escape Stairs

	Crew	Daily Output	Labor-Hours	Unit	Material	Labor	Equipment	Total	Total Incl O&P
0010 **FIRE ESCAPE STAIRS**, shop fabricated									
0020 One story, disappearing, stainless steel	2 Sswk	20	.800	V.L.F.	193	33		226	272
0100 Portable ladder				Ea.	58.50			58.50	64.50

05 51 33 – Metal Ladders

05 51 33.13 Vertical Metal Ladders

	Crew	Daily Output	Labor-Hours	Unit	Material	Labor	Equipment	Total	Total Incl O&P
0010 **VERTICAL METAL LADDERS**, shop fabricated									
0020 Steel, 20" wide, bolted to concrete, with cage	E-4	50	.640	V.L.F.	71	27	2.30	100.30	130
0100 Without cage		85	.376		33	15.75	1.36	50.11	66.50
0300 Aluminum, bolted to concrete, with cage		50	.640		96.50	27	2.30	125.80	157
0400 Without cage	↓	85	.376	↓	56	15.75	1.36	73.11	91.50

05 51 33.16 Inclined Metal Ladders

	Crew	Daily Output	Labor-Hours	Unit	Material	Labor	Equipment	Total	Total Incl O&P
0010 **INCLINED METAL LADDERS**, shop fabricated									
3900 Industrial ships ladder, 3' W, grating treads, 2 line pipe rail	E-4	30	1.067	Riser	149	44.50	3.84	197.34	249
4000 Aluminum	"	30	1.067	"	229	44.50	3.84	277.34	335

05 51 33.23 Alternating Tread Ladders

	Crew	Daily Output	Labor-Hours	Unit	Material	Labor	Equipment	Total	Total Incl O&P
0010 **ALTERNATING TREAD LADDERS**, shop fabricated									
1350 Alternating tread stair, 56/68°, steel, standard paint color	2 Sswk	50	.320	V.L.F.	155	13.25		168.25	195
1360 Non-standard paint color		50	.320		176	13.25		189.25	217
1370 Galvanized steel		50	.320		175	13.25		188.25	217
1380 Stainless steel		50	.320		261	13.25		274.25	310
1390 68°, aluminum	↓	50	.320	↓	191	13.25		204.25	234

05 52 Metal Railings

05 52 13 – Pipe and Tube Railings

05 52 13.50 Railings, Pipe

	Crew	Daily Output	Labor-Hours	Unit	Material	Labor	Equipment	Total	Total Incl O&P
0010 **RAILINGS, PIPE**, shop fabricated									
0020 Aluminum, 2 rail, satin finish, 1-1/4" diameter	E-4	160	.200	L.F.	23.50	8.35	.72	32.57	42
0030 Clear anodized		160	.200		29.50	8.35	.72	38.57	48
0040 Dark anodized		160	.200		33	8.35	.72	42.07	52.50
0080 1-1/2" diameter, satin finish		160	.200		28.50	8.35	.72	37.57	47
0090 Clear anodized		160	.200		31.50	8.35	.72	40.57	51
0100 Dark anodized		160	.200		35	8.35	.72	44.07	54.50
0140 Aluminum, 3 rail, 1-1/4" diam., satin finish		137	.234		36.50	9.80	.84	47.14	58.50
0150 Clear anodized	↓	137	.234	↓	45.50	9.80	.84	56.14	68.50

05 52 Metal Railings

05 52 13 – Pipe and Tube Railings

05 52 13.50 Railings, Pipe

		Crew	Daily Output	Labor-Hours	Unit	Material	2007 Bare Costs Labor	Equipment	Total	Total Incl O&P
0160	Dark anodized	E-4	137	.234	L.F.	50.50	9.80	.84	61.14	74
0200	1-1/2" diameter, satin finish		137	.234		43.50	9.80	.84	54.14	66.50
0210	Clear anodized		137	.234		49.50	9.80	.84	60.14	72.50
0220	Dark anodized		137	.234		54	9.80	.84	64.64	78
0500	Steel, 2 rail, on stairs, primed, 1-1/4" diameter		160	.200		19.45	8.35	.72	28.52	37.50
0520	1-1/2" diameter		160	.200		21.50	8.35	.72	30.57	39.50
0540	Galvanized, 1-1/4" diameter		160	.200		27	8.35	.72	36.07	45.50
0560	1-1/2" diameter		160	.200		30	8.35	.72	39.07	49
0580	Steel, 3 rail, primed, 1-1/4" diameter		137	.234		29	9.80	.84	39.64	50.50
0600	1-1/2" diameter		137	.234		31	9.80	.84	41.64	52.50
0620	Galvanized, 1-1/4" diameter		137	.234		40.50	9.80	.84	51.14	63
0640	1-1/2" diameter **CN**		137	.234		48	9.80	.84	58.64	71.50
0700	Stainless steel, 2 rail, 1-1/4" diam. #4 finish		137	.234		70	9.80	.84	80.64	95.50
0720	High polish		137	.234		113	9.80	.84	123.64	143
0740	Mirror polish		137	.234		141	9.80	.84	151.64	174
0760	Stainless steel, 3 rail, 1-1/2" diam., #4 finish		120	.267		106	11.15	.96	118.11	137
0770	High polish		120	.267		175	11.15	.96	187.11	213
0780	Mirror finish		120	.267		213	11.15	.96	225.11	255
0900	Wall rail, alum. pipe, 1-1/4" diam., satin finish		213	.150		13.55	6.30	.54	20.39	27
0905	Clear anodized		213	.150		16.50	6.30	.54	23.34	30
0910	Dark anodized		213	.150		20	6.30	.54	26.84	34
0915	1-1/2" diameter, satin finish		213	.150		15	6.30	.54	21.84	28.50
0920	Clear anodized		213	.150		18.85	6.30	.54	25.69	32.50
0925	Dark anodized		213	.150		23.50	6.30	.54	30.34	37.50
0930	Steel pipe, 1-1/4" diameter, primed		213	.150		11.80	6.30	.54	18.64	25
0935	Galvanized		213	.150		17.10	6.30	.54	23.94	31
0940	1-1/2" diameter		176	.182		12.15	7.60	.65	20.40	28
0945	Galvanized		213	.150		17.15	6.30	.54	23.99	31
0955	Stainless steel pipe, 1-1/2" diam., #4 finish		107	.299		56	12.50	1.08	69.58	85
0960	High polish		107	.299		114	12.50	1.08	127.58	149
0965	Mirror polish		107	.299		134	12.50	1.08	147.58	172
2000	Aluminum pipe & picket railing, double top rail, pickets @ 4-1/2" OC									
2010	36" high, straight & level	2 Sswk	80	.200	L.F.	73.50	8.25		81.75	95.50
2020	Curved & level		60	.267		103	11.05		114.05	134
2030	Straight & sloped		40	.400		84	16.55		100.55	123

05 52 16 – Industrial Railings

05 52 16.50 Railings, Industrial

		Crew	Daily Output	Labor-Hours	Unit	Material	2007 Bare Costs Labor	Equipment	Total	Total Incl O&P
0010	**RAILINGS, INDUSTRIAL**, shop fabricated									
0020	2 rail, 3'-6" high, 1-1/2" pipe	E-4	255	.125	L.F.	22.50	5.25	.45	28.20	35
0100	2" angle rail	"	255	.125		20.50	5.25	.45	26.20	32.50
0200	For 4" high kick plate, 10 gauge, add					4.70			4.70	5.15
0300	1/4" thick, add					6.05			6.05	6.65
0500	For curved rails, add					30%	30%			

05 53 Metal Gratings

05 53 13 – Floor Grating Frame

05 53 13.50 Grating Frame

		Crew	Daily Output	Labor-Hours	Unit	Material	2007 Bare Costs Labor	2007 Bare Costs Equipment	Total	Total Incl O&P
0010	**GRATING FRAME**, field fabricated									
0020	Aluminum, for gratings 1" to 1-1/2" deep	1 Sswk	70	.114	L.F.	7.80	4.73		12.53	17.10
0100	For each corner, add				Ea.	6.15			6.15	6.75

05 53 16 – Floor Grating Planks

05 53 16.50 Floor Grating Planks

		Crew	Daily Output	Labor-Hours	Unit	Material	2007 Bare Costs Labor	2007 Bare Costs Equipment	Total	Total Incl O&P
0010	**FLOOR GRATING PLANKS**, field fabricated from planks									
0020	Aluminum, 9-1/2" wide, 14 ga., 2" rib	E-4	950	.034	L.F.	14.80	1.41	.12	16.33	18.95
0200	Galvanized steel, 9-1/2" wide, 14 ga., 2-1/2" rib		950	.034		9.55	1.41	.12	11.08	13.20
0300	4" rib		950	.034		11.45	1.41	.12	12.98	15.30
0500	12 gauge, 2-1/2" rib		950	.034		12.70	1.41	.12	14.23	16.65
0600	3" rib		950	.034		14.85	1.41	.12	16.38	19.05
0800	Stainless steel, type 304, 16 ga., 2" rib		950	.034		25.50	1.41	.12	27.03	30.50
0900	Type 316		950	.034		31	1.41	.12	32.53	36.50

05 53 19 – Aluminum Floor Grating

05 53 19.50 Floor Grating, Aluminum

		Crew	Daily Output	Labor-Hours	Unit	Material	2007 Bare Costs Labor	2007 Bare Costs Equipment	Total	Total Incl O&P
0010	**FLOOR GRATING, ALUMINUM**, field fabricated from panels									
0110	Bearing bars @ 1-3/16" O.C., cross bars @ 4" O.C.,									
0111	Up to 300 S.F., 1" x 1/8" bar	E-4	900	.036	S.F.	10.20	1.49	.13	11.82	14.10
0112	Over 300 S.F.		850	.038		9.30	1.58	.14	11.02	13.20
0113	1-1/4" x 1/8" bar, up to 300 S.F.		800	.040		11.60	1.67	.14	13.41	15.95
0114	Over 300 S.F.		1000	.032		10.55	1.34	.12	12.01	14.15
0122	1-1/4" x 3/16" bar, up to 300 S.F.		750	.043		14.85	1.79	.15	16.79	19.75
0124	Over 300 S.F.		1000	.032		13.50	1.34	.12	14.96	17.40
0132	1-1/2" x 1/8" bar, up to 300 S.F.		700	.046		9.30	1.91	.16	11.37	13.85
0134	Over 300 S.F.		1000	.032		8.45	1.34	.12	9.91	11.85
0136	1-3/4" x 3/16" bar, up to 300 S.F.		500	.064		20	2.68	.23	22.91	27
0138	Over 300 S.F.		1000	.032		18.15	1.34	.12	19.61	22.50
0146	2-1/4" x 3/16" bar, up to 300 S.F.		600	.053		23.50	2.23	.19	25.92	30.50
0148	Over 300 S.F.		1000	.032		21.50	1.34	.12	22.96	26
0162	Cross bars @ 2" O.C., 1" x 1/8", up to 300 S.F.		600	.053		12.85	2.23	.19	15.27	18.40
0164	Over 300 S.F.		1000	.032		11.70	1.34	.12	13.16	15.40
0172	1-1/4" x 3/16" bar, up to 300 S.F.		600	.053		17.80	2.23	.19	20.22	24
0174	Over 300 S.F.		1000	.032		16.20	1.34	.12	17.66	20.50
0182	1-1/2" x 1/8" bar, up to 300 S.F.		600	.053		12.75	2.23	.19	15.17	18.30
0184	Over 300 S.F.		1000	.032		11.60	1.34	.12	13.06	15.30
0186	1-3/4" x 3/16" bar, up to 300 S.F.		600	.053		18.55	2.23	.19	20.97	25
0188	Over 300 S.F.		1000	.032		16.85	1.34	.12	18.31	21
0200	For straight cuts, add				L.F.	2.54			2.54	2.79
0300	For curved cuts, add					3.48			3.48	3.83
0400	For straight banding, add					3.09			3.09	3.40
0500	For curved banding, add					4.30			4.30	4.73
0600	For aluminum checkered plate nosings, add					4.23			4.23	4.65
0700	For straight toe plate, add					7.70			7.70	8.45
0800	For curved toe plate, add					9.60			9.60	10.55
1000	For cast aluminum abrasive nosings, add					5.15			5.15	5.65
1200	Expanded aluminum, .65# per S.F.	E-4	1050	.030	S.F.	4.48	1.28	.11	5.87	7.35
1400	Extruded I bars are 10% less than 3/16" bars									
1600	Heavy duty, all extruded plank, 3/4" deep, 1.8 # per S.F.	E-4	1100	.029	S.F.	12.55	1.22	.10	13.87	16.15
1700	1-1/4" deep, 2.9# per S.F.		1000	.032		13.40	1.34	.12	14.86	17.25
1800	1-3/4" deep, 4.2# per S.F.		925	.035		15.95	1.45	.12	17.52	20.50
1900	2-1/4" deep, 5.0# per S.F.		875	.037		18.30	1.53	.13	19.96	23
2100	For safety serrated surface, add					15%				

05 53 Metal Gratings

05 53 21 – Steel Floor Grating

05 53 21.50 Floor Grating, Steel

		Crew	Daily Output	Labor-Hours	Unit	Material	2007 Bare Costs Labor	Equipment	Total	Total Incl O&P
0010	**FLOOR GRATING, STEEL**, field fabricated from panels									
0050	Labor for installing, from ground/floor	E-4	845	.038	S.F.		1.58	.14	1.72	3.02
0100	Elevated		460	.070	"		2.91	.25	3.16	5.55
0300	Platforms, to 12' high, rectangular		3150	.010	Lb.	1.46	.43	.04	1.93	2.42
0400	Circular	▼	2300	.014	"	1.61	.58	.05	2.24	2.88
0410	Painted bearing bars @ 1-3/16"									
0412	Cross bars @ 4" O.C., 3/4" x 1/8" bar, up to 300 S.F.	E-2	500	.112	S.F.	6.15	4.52	3.09	13.76	18
0414	Over 300 S.F.		750	.075		5.60	3.01	2.06	10.67	13.60
0422	1-1/4" x 3/16", up to 300 S.F.		400	.140		8.30	5.65	3.87	17.82	23
0424	Over 300 S.F.		600	.093		7.55	3.77	2.58	13.90	17.65
0432	1-1/2" x 1/8", up to 300 S.F.		400	.140		7.45	5.65	3.87	16.97	22
0434	Over 300 S.F.		600	.093		6.75	3.77	2.58	13.10	16.80
0436	1-3/4" x 3/16", up to 300 S.F.		400	.140		10.80	5.65	3.87	20.32	26
0438	Over 300 S.F.		600	.093		9.80	3.77	2.58	16.15	20
0452	2-1/4" x 3/16", up to 300 S.F.		300	.187		12.80	7.55	5.15	25.50	33
0454	Over 300 S.F.		450	.124		11.65	5	3.44	20.09	25.50
0462	Cross bars @ 2" O.C., 3/4" x 1/8", up to 300 S.F.		500	.112		11.85	4.52	3.09	19.46	24.50
0464	Over 300 S.F.		750	.075		9.85	3.01	2.06	14.92	18.30
0472	1-1/4" x 3/16", up to 300 S.F.		400	.140		17.75	5.65	3.87	27.27	33.50
0474	Over 300 S.F.		600	.093		14.80	3.77	2.58	21.15	25.50
0482	1-1/2" x 1/8", up to 300 S.F.		400	.140		13.70	5.65	3.87	23.22	29
0484	Over 300 S.F.		600	.093		11.45	3.77	2.58	17.80	22
0486	1-3/4" x 3/16", up to 300 S.F.		400	.140		18.50	5.65	3.87	28.02	34.50
0488	Over 300 S.F.		600	.093		15.40	3.77	2.58	21.75	26.50
0502	2-1/4" x 3/16", up to 300 S.F.		300	.187		22	7.55	5.15	34.70	43
0504	Over 300 S.F.	▼	450	.124		18.40	5	3.44	26.84	32.50
0690	For galvanized grating, add				▼	25%				
0800	For straight cuts, add				L.F.	2.71			2.71	2.98
0900	For curved cuts, add					3.73			3.73	4.10
1000	For straight banding, add					3.13			3.13	3.44
1100	For curved banding, add					4.85			4.85	5.35
1200	For checkered plate nosings, add					5.50			5.50	6.05
1300	For straight toe or kick plate, add					7.80			7.80	8.55
1400	For curved toe or kick plate, add					9.50			9.50	10.45
1500	For abrasive nosings, add				▼	5.75			5.75	6.30
1510	For stair treads, see division 05 55 13.50									
1600	For safety serrated surface, minimum, add					15%				
1700	Maximum, add					25%				
2000	Stainless steel gratings, close spaced, 1" x 1/8" bars, up to 300 S.F.	E-4	450	.071	S.F.	35	2.98	.26	38.24	44
2100	Standard spacing, 3/4" x 1/8" bars		500	.064		27.50	2.68	.23	30.41	35
2200	1-1/4" x 3/16" bars		400	.080		43.50	3.35	.29	47.14	54
2400	Expanded steel grating, at ground, 3.0# per S.F.		900	.036		3.42	1.49	.13	5.04	6.60
2500	3.14# per S.F.		900	.036		3.63	1.49	.13	5.25	6.85
2600	4.0# per S.F.		850	.038		4.48	1.58	.14	6.20	7.95
2650	4.27# per S.F.		850	.038		4.84	1.58	.14	6.56	8.30
2700	5.0# per S.F.		800	.040		5.75	1.67	.14	7.56	9.55
2800	6.25# per S.F.		750	.043		7.30	1.79	.15	9.24	11.40
2900	7.0# per S.F.	▼	700	.046		8.20	1.91	.16	10.27	12.65
3100	For flattened expanded steel grating, add					8%				
3300	For elevated installation above 15', add				▼		15%			

05 54 Metal Floor Plates

05 54 13 – Floor Plates

05 54 13.20 Checkered Plates

		Crew	Daily Output	Labor-Hours	Unit	Material	2007 Bare Costs Labor	Equipment	Total	Total Incl O&P
0010	**CHECKERED PLATES**, field fabricated									
0020	1/4" & 3/8", 2000 to 5000 S.F., bolted	E-4	2900	.011	Lb.	.84	.46	.04	1.34	1.80
0100	Welded		4400	.007	"	.79	.30	.03	1.12	1.45
0300	Pit or trench cover and frame, 1/4" plate, 2' to 3' wide		100	.320	S.F.	22	13.40	1.15	36.55	50.50
0400	For galvanizing, add				Lb.	.45			.45	.50
0500	Platforms, 1/4" plate, no handrails included, rectangular	E-4	4200	.008		1.33	.32	.03	1.68	2.07
0600	Circular	"	2500	.013		1.86	.54	.05	2.45	3.07

05 54 13.70 Trench Covers

		Crew	Daily Output	Labor-Hours	Unit	Material	Labor	Equipment	Total	Total Incl O&P
0010	**TRENCH COVERS**, field fabricated									
0020	Cast iron grating with bar stops and angle frame, to 18" wide	1 Sswk	20	.400	L.F.	58.50	16.55		75.05	94.50
0100	Frame only (both sides of trench), 1" grating		45	.178		12.85	7.35		20.20	27.50
0150	2" grating		35	.229		19.95	9.45		29.40	39
0200	Aluminum, stock units, including frames and									
0210	3/8" plain cover plate, 4" opening	E-4	205	.156	L.F.	37	6.55	.56	44.11	53
0300	6" opening		185	.173		45.50	7.25	.62	53.37	64
0400	10" opening		170	.188		63	7.90	.68	71.58	84
0500	16" opening		155	.206		86	8.65	.74	95.39	111
0700	Add per inch for additional widths to 24"					3.43			3.43	3.77
0900	For custom fabrication, add					50%				
1100	For 1/4" plain cover plate, deduct					12%				
1500	For cover recessed for tile, 1/4" thick, deduct					12%				
1600	3/8" thick, add					5%				
1800	For checkered plate cover, 1/4" thick, deduct					12%				
1900	3/8" thick, add					2%				
2100	For slotted or round holes in cover, 1/4" thick, add					3%				
2200	3/8" thick, add					4%				
2300	For abrasive cover, add					12%				

05 55 Metal Stair Treads and Nosings

05 55 13 – Metal Stair Treads

05 55 13.50 Stair Treads

		Crew	Daily Output	Labor-Hours	Unit	Material	Labor	Equipment	Total	Total Incl O&P
0010	**STAIR TREADS**									
0020	Aluminum grating, 3' long, 1-1/2" x 3/16" rect. bars, 6" wide	1 Sswk	24	.333	Ea.	52	13.80		65.80	82
0100	12" wide		22	.364		76.50	15.05		91.55	111
0200	1-1/2" x 3/16" I-bars, 6" wide		24	.333		43.50	13.80		57.30	73
0300	12" wide		22	.364		70.50	15.05		85.55	105
0400	For abrasive nosings, add					13.60			13.60	14.95
0500	For narrow mesh, add					60%				
0600	Stair treads, not incl. stringers. See also div. 03150-660									
0700	Cast aluminum, abrasive, 3' long x 12" wide, 5/16" thick	1 Sswk	15	.533	Ea.	115	22		137	167
0800	3/8" thick		15	.533		122	22		144	175
0900	1/2" thick		15	.533		135	22		157	188
1000	Cast bronze, abrasive, 3/8" thick		8	1		525	41.50		566.50	650
1100	1/2" thick		8	1		690	41.50		731.50	830
1200	Cast iron, abrasive, 3' long x 12" wide, 3/8" thick		15	.533		101	22		123	151
1300	1/2" thick		15	.533		117	22		139	169
1400	Fiberglass reinforced plastic with safety nosing,									
1500	1-1/2" thick, 12" wide, 24" long	1 Sswk	22	.364	Ea.	115	15.05		130.05	153
1600	30" long		22	.364		120	15.05		135.05	159
1700	36" long		22	.364		131	15.05		146.05	171
2000	Steel grating, painted, 3' long, 1-1/4" x 3/16" bars, 6" wide		20	.400		44	16.55		60.55	78.50

05 55 Metal Stair Treads and Nosings

05 55 13 – Metal Stair Treads

05 55 13.50 Stair Treads

		Crew	Daily Output	Labor-Hours	Unit	Material	2007 Bare Costs Labor	Equipment	Total	Total Incl O&P
2010	12" wide	1 Sswk	18	.444	Ea.	51	18.40		69.40	89.50
2100	Add for abrasive nosing, 3' long, painted					13.75			13.75	15.15
2200	Galvanized					18.45			18.45	20.50
2300	Painting 3' treads, in shop, nonstandard paint					2.43			2.43	2.67
2400	Added coats, standard paint					1.04			1.04	1.14
2500	Expanded steel, 2-1/2" deep, 9" x 3' long, 18 gauge	1 Sswk	20	.400		24.50	16.55		41.05	57
2600	14 gauge	"	20	.400		30	16.55		46.55	63

05 56 Metal Castings

05 56 13 – Metal Castings

05 56 13.50 Construction Castings

		Crew	Daily Output	Labor-Hours	Unit	Material	2007 Bare Costs Labor	Equipment	Total	Total Incl O&P
0010	**CONSTRUCTION CASTINGS**									
0020	Manhole covers and frames see Division 33 44 13.13									
0100	Column bases, cast iron, 16" x 16", approx. 65 lbs.	E-4	46	.696	Ea.	113	29	2.50	144.50	179
0200	32" x 32", approx. 256 lbs.	"	23	1.391		420	58	5	483	575
0400	Cast aluminum for wood columns, 8" x 8"	1 Carp	32	.250		36.50	9.20		45.70	54.50
0500	12" x 12"	"	32	.250		78	9.20		87.20	100
0600	Miscellaneous C.I. castings, light sections, less than 150 lbs	E-4	3200	.010	Lb.	1.74	.42	.04	2.20	2.71
1100	Heavy sections, more than 150 lb		4200	.008		.85	.32	.03	1.20	1.55
1300	Special low volume items		3200	.010		3.03	.42	.04	3.49	4.13
1500	For ductile iron, add					100%				

05 58 Formed Metal Fabrications

05 58 21 – Formed Chain

05 58 21.05 Alloy Steel Chain

		Crew	Daily Output	Labor-Hours	Unit	Material	2007 Bare Costs Labor	Equipment	Total	Total Incl O&P
0010	**ALLOY STEEL CHAIN**, Grade 80									
0015	Self-colored, cut lengths, w/accessories, 1/4"	E-17	4	4	C.L.F.	495	169	2	666	845
0020	3/8"		2	8		635	340	4	979	1,325
0030	1/2"		1.20	13.333		1,000	565	6.65	1,571.65	2,125
0040	5/8"		.72	22.222		1,575	940	11.10	2,526.10	3,425
0050	3/4"	E-18	.48	83.333		2,300	3,425	2,025	7,750	10,800
0060	7/8"		.40	100		3,400	4,125	2,450	9,975	13,600
0070	1"		.35	114		5,575	4,700	2,800	13,075	17,500
0080	1-1/4"		.24	166		10,400	6,850	4,075	21,325	27,900
0110	Clevis slip hook, 1/4"				Ea.	14.30			14.30	15.70
0120	3/8"					21.50			21.50	23.50
0130	1/2"					36			36	39.50
0140	5/8"					54.50			54.50	60
0150	3/4"					114			114	125
0160	Eye/sling hook w/ hammerlock coupling, 7/8"					320			320	350
0170	1"					450			450	495
0180	1-1/4"					675			675	745

05 58 23 – Formed Columns

05 58 23.10 Aluminum Columns

		Crew	Daily Output	Labor-Hours	Unit	Material	2007 Bare Costs Labor	Equipment	Total	Total Incl O&P
0010	**ALUMINUM COLUMNS**									
0020	Aluminum, extruded, stock units, no cap or base, 6" diameter	E-4	240	.133	L.F.	10.45	5.60	.48	16.53	22
0100	8" diameter		170	.188		13.45	7.90	.68	22.03	30
0200	10" diameter		150	.213		18.30	8.95	.77	28.02	37
0300	12" diameter		140	.229		37	9.55	.82	47.37	59

05 58 Formed Metal Fabrications

05 58 23 – Formed Columns

05 58 23.10 Aluminum Columns

		Crew	Daily Output	Labor-Hours	Unit	Material	2007 Bare Costs Labor	Equipment	Total	Total Incl O&P
0400	15" diameter	E-4	120	.267	L.F.	44	11.15	.96	56.11	69.50
0410	Caps and bases, plain, 6" diameter				Set	19.55			19.55	21.50
0420	8" diameter					26			26	28.50
0430	10" diameter					33			33	36.50
0440	12" diameter					50			50	55
0450	15" diameter					90			90	99
0460	Caps, ornamental, minimum					242			242	266
0470	Maximum				▼	1,075			1,075	1,200
0500	For square columns, add to column prices above				L.F.	50%				
0700	Residential, flat, 8' high, plain	E-4	20	1.600	Ea.	88	67	5.75	160.75	224
0720	Fancy		20	1.600		167	67	5.75	239.75	310
0740	Corner type, plain		20	1.600		148	67	5.75	220.75	290
0760	Fancy	▼	20	1.600	▼	293	67	5.75	365.75	445

05 58 25 – Formed Lamp Posts

05 58 25.40 Lamp Posts

		Crew	Daily Output	Labor-Hours	Unit	Material	Labor	Equipment	Total	Total Incl O&P
0010	**LAMP POSTS**									
0020	Aluminum, 7' high, stock units, post only	1 Carp	16	.500	Ea.	28	18.35		46.35	59.50
0100	Mild steel, plain	"	16	.500	"	23	18.35		41.35	54

05 58 27 – Formed Guards

05 58 27.90 Window Guards

		Crew	Daily Output	Labor-Hours	Unit	Material	Labor	Equipment	Total	Total Incl O&P
0010	**WINDOW GUARDS**, shop fabricated									
0015	Expanded metal, steel angle frame, permanent	E-4	350	.091	S.F.	20.50	3.83	.33	24.66	30
0025	Steel bars, 1/2" x 1/2", spaced 5" O.C.	"	290	.110	"	14.30	4.62	.40	19.32	24.50
0030	Hinge mounted, add				Opng.	41.50			41.50	45.50
0040	Removable type, add				"	26.50			26.50	29
0050	For galvanized guards, add				S.F.	35%				
0070	For pivoted or projected type, add					105%	40%			
0100	Mild steel, stock units, economy	E-4	405	.079		5.60	3.31	.28	9.19	12.45
0200	Deluxe		405	.079	▼	11.50	3.31	.28	15.09	18.95
0400	Woven wire, stock units, 3/8" channel frame, 3' x 5' opening		40	.800	Opng.	151	33.50	2.88	187.38	230
0500	4' x 6' opening	▼	38	.842		242	35	3.03	280.03	335
0800	Basket guards for above, add					207			207	228
1000	Swinging guards for above, add				▼	71			71	78.50

05 59 Metal Specialties

05 59 63 – Detention Enclosures

05 59 63.13 Detention Enclosures

		Crew	Daily Output	Labor-Hours	Unit	Material	Labor	Equipment	Total	Total Incl O&P
0010	**DETENTION ENCLOSURES**									
0500	Bar front, rolling, 7/8" bars, 4" O.C., 7' high, 5' wide, with hardware	E-4	2	16	Ea.	5,075	670	57.50	5,802.50	6,875

05 71 Decorative Metal Stairs

05 71 13 – Fabricated Metal Spiral Stairs

05 71 13.50 Spiral Stairs	Crew	Daily Output	Labor-Hours	Unit	Material	2007 Bare Costs Labor	2007 Bare Costs Equipment	Total	Total Incl O&P
0010 **SPIRAL STAIRS**, shop fabricated									
1810 Spiral aluminum, 5'-0" diameter, stock units	E-4	45	.711	Riser	475	30	2.56	507.56	575
1820 Custom units		45	.711		895	30	2.56	927.56	1,050
1900 Spiral, cast iron, 4'-0" diameter, ornamental, minimum		45	.711		425	30	2.56	457.56	520
1920 Maximum		25	1.280		575	53.50	4.61	633.11	735
2000 Spiral, steel, industrial checkered plate, 4' diameter		45	.711		425	30	2.56	457.56	520
2200 Stock units, 6'-0" diameter		40	.800		515	33.50	2.88	551.38	630
3110 Spiral steel, stock units, primed, flat metal tread, 3'-6" dia	2 Carp	1.60	10	Flight	1,125	365		1,490	1,800
3120 4'-0" dia		1.45	11.034		1,275	405		1,680	2,025
3130 4'-6" dia		1.35	11.852		1,400	435		1,835	2,225
3140 5'-0" dia		1.25	12.800		1,525	470		1,995	2,400
3210 Galvanized, 3'-6" dia		1.60	10		1,950	365		2,315	2,725
3220 4'-0" dia		1.45	11.034		2,200	405		2,605	3,025
3230 4'-6" dia		1.35	11.852		2,375	435		2,810	3,300
3240 5'-0" dia		1.25	12.800		2,575	470		3,045	3,550
3310 Checkered plate tread, 3'-6" dia		1.45	11.034		1,375	405		1,780	2,125
3320 4'-0" dia		1.35	11.852		1,550	435		1,985	2,375
3330 4'-6" dia		1.25	12.800		1,700	470		2,170	2,600
3340 5'-0" dia		1.15	13.913		1,825	510		2,335	2,800
3410 Galvanized, 3'-6" dia		1.45	11.034		2,225	405		2,630	3,075
3420 4'-0" dia		1.35	11.852		2,500	435		2,935	3,425
3430 4'-6" dia		1.25	12.800		2,725	470		3,195	3,700
3440 5'-0" dia		1.15	13.913		2,925	510		3,435	4,025
3510 Red oak tread on flat metal, 3'-6" dia		1.35	11.852		1,850	435		2,285	2,725
3520 4'-0" dia		1.25	12.800		2,075	470		2,545	3,000
3530 4'-6" dia		1.15	13.913		2,250	510		2,760	3,275
3540 5'-0" dia		1.05	15.238		2,425	560		2,985	3,525

05 73 Decorative Metal Railings

05 73 16 – Wire Rope Decorative Metal Railings

05 73 16.10 Cable Railings

	Crew	Daily Output	Labor-Hours	Unit	Material	Labor	Equipment	Total	Total Incl O&P
0010 **CABLE RAILINGS**, with 316 stainless steel 1 x 19 cable, 3/16" diameter									
0100 1-3/4" diameter stainless steel posts x 42" high, cables 4" OC	2 Sswk	25	.640	L.F.	37	26.50		63.50	89

05 73 23 – Ornamental Railings

05 73 23.50 Railings, Ornamental

	Crew	Daily Output	Labor-Hours	Unit	Material	Labor	Equipment	Total	Total Incl O&P
0010 **RAILINGS, ORNAMENTAL**, shop fabricated									
0020 Aluminum, bronze or stainless, minimum	1 Sswk	24	.333	L.F.	29	13.80		42.80	57
0100 Maximum		9	.889		274	37		311	365
0200 Aluminum ornamental rail, minimum		15	.533		29	22		51	72
0300 Maximum		8	1		86.50	41.50		128	171
0400 Hand-forged wrought iron, minimum		12	.667		90	27.50		117.50	149
0500 Maximum		8	1		270	41.50		311.50	370
0550 Steel, minimum		12	.667		28	27.50		55.50	81
0560 Maximum		8	1		84	41.50		125.50	168
0600 Composite metal/wood/glass, minimum		6	1.333		173	55		228	290
0700 Maximum		5	1.600		345	66		411	500

05 75 Decorative Formed Metal

05 75 13 – Columns

05 75 13.20 Columns, Ornamental		Crew	Daily Output	Labor-Hours	Unit	Material	2007 Bare Costs Labor	Equipment	Total	Total Incl O&P
0010	**COLUMNS, ORNAMENTAL**, shop fabricated	R051223-10								
6400	Mild steel, flat, 9" wide, stock units, painted, plain		E-4	160	.200	L.F.	8.25	8.35	.72	17.32	25
6450	Fancy			160	.200		15.65	8.35	.72	24.72	33
6500	Corner columns, painted, plain			160	.200		13.90	8.35	.72	22.97	31
6550	Fancy		↓	160	.200	↓	27.50	8.35	.72	36.57	46

Estimating Tips

06 05 00 Common Work Results for Wood, Plastics, and Composites

- Common to any wood-framed structure are the accessory connector items such as screws, nails, adhesives, hangers, connector plates, straps, angles, and hold-downs. For typical wood-framed buildings, such as residential projects, the aggregate total for these items can be significant, especially in areas where seismic loading is a concern. For floor and wall framing, the material cost is based on 10 to 25 lbs. per MBF. Hold-downs, hangers, and other connectors should be taken off by the piece.

06 10 00 Carpentry

- Lumber is a traded commodity and therefore sensitive to supply and demand in the marketplace. Even in "budgetary" estimating of wood-framed projects, it is advisable to call local suppliers for the latest market pricing.
- Common quantity units for wood-framed projects are "thousand board feet" (MBF). A board foot is a volume of wood, 1" x 1' x 1', or 144 cubic inches. Board-foot quantities are generally calculated using nominal material dimensions—dressed sizes are ignored. Board foot per lineal foot of any stick of lumber can be calculated by dividing the nominal cross-sectional area by 12. As an example, 2,000 lineal feet of 2 x 12 equates to 4 MBF by dividing the nominal area, 2 x 12, by 12, which equals 2, and multiplying by 2,000 to give 4,000 board feet. This simple rule applies to all nominal dimensioned lumber.
- Waste is an issue of concern at the quantity takeoff for any area of construction. Framing lumber is sold in even foot lengths, i.e., 10', 12', 14', 16', and depending on spans, wall heights and the grade of lumber, waste is inevitable. A rule of thumb for lumber waste is 5% to 10% depending on material quality and the complexity of the framing.
- Wood in various forms and shapes is used in many projects, even where the main structural framing is steel, concrete, or masonry. Plywood as a back-up partition material and 2x boards used as blocking and cant strips around roof edges are two common examples. The estimator should ensure that the costs of all wood materials are included in the final estimate.

06 20 00 Finish Carpentry

- It is necessary to consider the grade of workmanship when estimating labor costs for erecting millwork and interior finish. In practice, there are three grades: premium, custom, and economy. The RSMeans daily output for base and case moldings is in the range of 200 to 250 L.F. per carpenter per day. This is appropriate for most average custom-grade projects. For premium projects, an adjustment to productivity of 25% to 50% should be made depending on the complexity of the job.

Reference Numbers

Reference numbers are shown in shaded boxes at the beginning of some major classifications. These numbers refer to related items in the Reference Section. The reference information may be an estimating procedure, an alternate pricing method, or technical information.

Note: Not all subdivisions listed here necessarily appear in this publication.

06 05 Common Work Results for Wood, Plastics and Composites

06 05 05 – Selective Wood and Plastics Demolition

06 05 05.10 Selective Demolition Wood Framing		Crew	Daily Output	Labor-Hours	Unit	Material	2007 Bare Costs Labor	2007 Bare Costs Equipment	Total	Total Incl O&P
0010	**SELECTIVE DEMOLITION WOOD FRAMING** R024119-10									
0100	Timber connector, nailed, small	1 Clab	96	.083	Ea.		2.40		2.40	3.73
0110	Medium		60	.133			3.83		3.83	5.95
0120	Large		48	.167			4.79		4.79	7.45
0130	Bolted, small		48	.167			4.79		4.79	7.45
0140	Medium		32	.250			7.20		7.20	11.20
0150	Large		24	.333			9.60		9.60	14.90
2958	Beams, 2" x 6"	2 Clab	1100	.015	L.F.		.42		.42	.65
2960	2" x 8"		825	.019			.56		.56	.87
2965	2" x 10"		665	.024			.69		.69	1.08
2970	2" x 12"		550	.029			.84		.84	1.30
2972	2" x 14"		470	.034			.98		.98	1.52
2975	4" x 8"	B-1	413	.058			1.71		1.71	2.66
2980	4" x 10"		330	.073			2.14		2.14	3.33
2985	4" x 12"		275	.087			2.57		2.57	4
3000	6" x 8"		275	.087			2.57		2.57	4
3040	6" x 10"		220	.109			3.21		3.21	5
3080	6" x 12"		185	.130			3.82		3.82	5.95
3120	8" x 12"		140	.171			5.05		5.05	7.85
3160	10" x 12"		110	.218			6.40		6.40	10
3162	Alternate pricing method		1.10	21.818	M.B.F.		640		640	1,000
3170	Blocking, in 16" OC wall framing, 2" x 4"	1 Clab	600	.013	L.F.		.38		.38	.60
3172	2" x 6"		400	.020			.58		.58	.90
3174	In 24" OC wall framing, 2" x 4"		600	.013			.38		.38	.60
3176	2" x 6"		400	.020			.58		.58	.90
3178	Alt method, wood blocking removal from wood framimg		.40	20	M.B.F.		575		575	895
3179	Wood blocking removal from steel framimg		.36	22.222	"		640		640	995
3180	Bracing, let in, 1" x 3", studs 16" OC		1050	.008	L.F.		.22		.22	.34
3181	Studs 24" OC		1080	.007			.21		.21	.33
3182	1" x 4", studs 16" OC		1050	.008			.22		.22	.34
3183	Studs 24" OC		1080	.007			.21		.21	.33
3184	1" x 6", studs 16" OC		1050	.008			.22		.22	.34
3185	Studs 24" OC		1080	.007			.21		.21	.33
3186	2" x 3", studs 16" OC		800	.010			.29		.29	.45
3187	Studs 24" OC		830	.010			.28		.28	.43
3188	2" x 4", studs 16" OC		800	.010			.29		.29	.45
3189	Studs 24" OC		830	.010			.28		.28	.43
3190	2" x 6", studs 16" OC		800	.010			.29		.29	.45
3191	Studs 24" OC		830	.010			.28		.28	.43
3192	2" x 8", studs 16" OC		800	.010			.29		.29	.45
3193	Studs 24" OC		830	.010			.28		.28	.43
3194	"T" shaped metal bracing, studs at 16" OC		1060	.008			.22		.22	.34
3195	Studs at 24" OC		1200	.007			.19		.19	.30
3196	Metal straps, studs at 16" OC		1200	.007			.19		.19	.30
3197	Studs at 24" OC		1240	.006			.19		.19	.29
3200	Columns, round, 8' to 14' tall		40	.200	Ea.		5.75		5.75	8.95
3202	Dimensional lumber sizes	2 Clab	1.10	14.545	M.B.F.		420		420	650
3250	Blocking, between joists	1 Clab	320	.025	Ea.		.72		.72	1.12
3252	Bridging, metal strap, between joists		320	.025	Pr.		.72		.72	1.12
3254	Wood, between joists		320	.025	"		.72		.72	1.12
3260	Door buck, studs, header & access, 8' high 2" x 4" wall, 3' wide		32	.250	Ea.		7.20		7.20	11.20
3261	4' wide		32	.250			7.20		7.20	11.20
3262	5' wide		32	.250			7.20		7.20	11.20

06 05 Common Work Results for Wood, Plastics and Composites

06 05 05 – Selective Wood and Plastics Demolition

06 05 05.10 Selective Demolition Wood Framing		Crew	Daily Output	Labor-Hours	Unit	Material	2007 Bare Costs Labor	Equipment	Total	Total Incl O&P
3263	6' wide	1 Clab	32	.250	Ea.		7.20		7.20	11.20
3264	8' wide		30	.267			7.65		7.65	11.95
3265	10' wide		30	.267			7.65		7.65	11.95
3266	12' wide		30	.267			7.65		7.65	11.95
3267	2" x 6" wall, 3' wide		32	.250			7.20		7.20	11.20
3268	4' wide		32	.250			7.20		7.20	11.20
3269	5' wide		32	.250			7.20		7.20	11.20
3270	6' wide		32	.250			7.20		7.20	11.20
3271	8' wide		30	.267			7.65		7.65	11.95
3272	10' wide		30	.267			7.65		7.65	11.95
3273	12' wide		30	.267			7.65		7.65	11.95
3274	Window buck, studs, header & access, 8' high 2" x 4" wall, 2' wide		24	.333			9.60		9.60	14.90
3275	3' wide		24	.333			9.60		9.60	14.90
3276	4' wide		24	.333			9.60		9.60	14.90
3277	5' wide		24	.333			9.60		9.60	14.90
3278	6' wide		24	.333			9.60		9.60	14.90
3279	7' wide		24	.333			9.60		9.60	14.90
3280	8' wide		22	.364			10.45		10.45	16.25
3281	10' wide		22	.364			10.45		10.45	16.25
3282	12' wide		22	.364			10.45		10.45	16.25
3283	2" x 6" wall, 2' wide		24	.333			9.60		9.60	14.90
3284	3' wide		24	.333			9.60		9.60	14.90
3285	4' wide		24	.333			9.60		9.60	14.90
3286	5' wide		24	.333			9.60		9.60	14.90
3287	6' wide		24	.333			9.60		9.60	14.90
3288	7' wide		24	.333			9.60		9.60	14.90
3289	8' wide		22	.364			10.45		10.45	16.25
3290	10' wide		22	.364			10.45		10.45	16.25
3291	12' wide		22	.364			10.45		10.45	16.25
3400	Fascia boards, 1" x 6"		500	.016	L.F.		.46		.46	.72
3440	1" x 8"		450	.018			.51		.51	.80
3480	1" x 10"		400	.020			.58		.58	.90
3490	2" x 6"		450	.018			.51		.51	.80
3500	2" x 8"		400	.020			.58		.58	.90
3510	2" x 10"		350	.023			.66		.66	1.02
3610	Furring, on wood walls or ceiling		4000	.002	S.F.		.06		.06	.09
3620	On masonry or concrete walls or ceiling		1200	.007	"		.19		.19	.30
3800	Headers over openings, 2 @ 2" x 6"		110	.073	L.F.		2.09		2.09	3.25
3840	2 @ 2" x 8"		100	.080			2.30		2.30	3.58
3880	2 @ 2" x 10"		90	.089			2.56		2.56	3.98
3885	Alternate pricing method		.26	30.651	M.B.F.		880		880	1,375
3920	Joists, 1" x 4"		1250	.006	L.F.		.18		.18	.29
3930	1" x 6"		1135	.007			.20		.20	.32
3940	1" x 8"		1000	.008			.23		.23	.36
3950	1" x 10"		895	.009			.26		.26	.40
3960	1" x 12"		765	.010			.30		.30	.47
4200	2" x 4"	2 Clab	1000	.016			.46		.46	.72
4230	2" x 6"		970	.016			.47		.47	.74
4240	2" x 8"		940	.017			.49		.49	.76
4250	2" x 10"		910	.018			.51		.51	.79
4280	2" x 12"		880	.018			.52		.52	.81
4281	2" x 14"		850	.019			.54		.54	.84
4282	Composite joists, 9-1/2"		960	.017			.48		.48	.75

06 05 Common Work Results for Wood, Plastics and Composites

06 05 05 – Selective Wood and Plastics Demolition

06 05 05.10 Selective Demolition Wood Framing		Crew	Daily Output	Labor-Hours	Unit	Material	2007 Bare Costs Labor	2007 Bare Costs Equipment	Total	Total Incl O&P
4283	11-7/8"	2 Clab	930	.017	L.F.		.49		.49	.77
4284	14"		897	.018			.51		.51	.80
4285	16"		865	.019			.53		.53	.83
4290	Wood joists, alternate pricing method		1.50	10.667	M.B.F.		305		305	475
4500	Open web joist, 12" deep		500	.032	L.F.		.92		.92	1.43
4505	14" deep		475	.034			.97		.97	1.51
4510	16" deep		450	.036			1.02		1.02	1.59
4520	18" deep		425	.038			1.08		1.08	1.68
4530	24" deep		400	.040			1.15		1.15	1.79
4550	Ledger strips, 1" x 2"	1 Clab	1200	.007			.19		.19	.30
4560	1" x 3"		1200	.007			.19		.19	.30
4570	1" x 4"		1200	.007			.19		.19	.30
4580	2" x 2"		1100	.007			.21		.21	.33
4590	2" x 4"		1000	.008			.23		.23	.36
4600	2" x 6"		1000	.008			.23		.23	.36
4601	2" x 8 or 2" x 10"		800	.010			.29		.29	.45
4602	4" x 6"		600	.013			.38		.38	.60
4604	4" x 8"		450	.018			.51		.51	.80
5400	Posts, 4" x 4"	2 Clab	800	.020			.58		.58	.90
5405	4" x 6"		550	.029			.84		.84	1.30
5410	4" x 8"		440	.036			1.05		1.05	1.63
5425	4" x 10"		390	.041			1.18		1.18	1.84
5430	4" x 12"		350	.046			1.31		1.31	2.05
5440	6" x 6"		400	.040			1.15		1.15	1.79
5445	6" x 8"		350	.046			1.31		1.31	2.05
5450	6" x 10"		320	.050			1.44		1.44	2.24
5455	6" x 12"		290	.055			1.59		1.59	2.47
5480	8" x 8"		300	.053			1.53		1.53	2.39
5500	10" x 10"		240	.067			1.92		1.92	2.98
5660	Tongue and groove floor planks		2	8	M.B.F.		230		230	360
5750	Rafters, ordinary, 16" OC, 2" x 4"		880	.018	S.F.		.52		.52	.81
5755	2" x 6"		840	.019			.55		.55	.85
5760	2" x 8"		820	.020			.56		.56	.87
5770	2" x 10"		820	.020			.56		.56	.87
5780	2" x 12"		810	.020			.57		.57	.88
5785	24" OC, 2" x 4"		1170	.014			.39		.39	.61
5786	2" x 6"		1117	.014			.41		.41	.64
5787	2" x 8"		1091	.015			.42		.42	.66
5788	2" x 10"		1091	.015			.42		.42	.66
5789	2" x 12"		1077	.015			.43		.43	.67
5795	Rafters, ordinary, 2" x 4" (alternate method)		862	.019	L.F.		.53		.53	.83
5800	2" x 6" (alternate method)		850	.019			.54		.54	.84
5840	2" x 8" (alternate method)		837	.019			.55		.55	.86
5855	2" x 10" (alternate method)		825	.019			.56		.56	.87
5865	2" x 12" (alternate method)		812	.020			.57		.57	.88
5870	Sill plate, 2" x 4"	1 Clab	1170	.007			.20		.20	.31
5871	2" x 6"		780	.010			.30		.30	.46
5872	2" x 8"		586	.014			.39		.39	.61
5873	Alternate pricing method		.78	10.256	M.B.F.		295		295	460
5885	Ridge board, 1" x 4"	2 Clab	900	.018	L.F.		.51		.51	.80
5886	1" x 6"		875	.018			.53		.53	.82
5887	1" x 8"		850	.019			.54		.54	.84
5888	1" x 10"		825	.019			.56		.56	.87

06 05 Common Work Results for Wood, Plastics and Composites

06 05 05 – Selective Wood and Plastics Demolition

06 05 05.10 Selective Demolition Wood Framing		Crew	Daily Output	Labor-Hours	Unit	Material	2007 Bare Costs Labor	Equipment	Total	Total Incl O&P
5889	1" x 12"	2 Clab	800	.020	L.F.		.58		.58	.90
5890	2" x 4"		900	.018			.51		.51	.80
5892	2" x 6"		875	.018			.53		.53	.82
5894	2" x 8"		850	.019			.54		.54	.84
5896	2" x 10"		825	.019			.56		.56	.87
5898	2" x 12"		800	.020			.58		.58	.90
6050	Rafter tie, 1" x 4"		1250	.013			.37		.37	.57
6052	1" x 6"		1135	.014			.41		.41	.63
6054	2" x 4"		1000	.016			.46		.46	.72
6056	2" x 6"	▼	970	.016			.47		.47	.74
6070	Sleepers, on concrete, 1" x 2"	1 Clab	4700	.002			.05		.05	.08
6075	1" x 3"		4000	.002			.06		.06	.09
6080	2" x 4"		3000	.003			.08		.08	.12
6085	2" x 6"	▼	2600	.003	▼		.09		.09	.14
6086	Sheathing from roof, 5/16"	2 Clab	1600	.010	S.F.		.29		.29	.45
6088	3/8"		1525	.010			.30		.30	.47
6090	1/2"		1400	.011			.33		.33	.51
6092	5/8"		1300	.012			.35		.35	.55
6094	3/4"		1200	.013			.38		.38	.60
6096	Board sheathing from roof		1400	.011			.33		.33	.51
6100	Sheathing, from walls, 1/4"		1200	.013			.38		.38	.60
6110	5/16"		1175	.014			.39		.39	.61
6120	3/8"		1150	.014			.40		.40	.62
6130	1/2"		1125	.014			.41		.41	.64
6140	5/8"		1100	.015			.42		.42	.65
6150	3/4"		1075	.015			.43		.43	.67
6152	Board sheathing from walls		1500	.011			.31		.31	.48
6158	Subfloor, with boards		1050	.015			.44		.44	.68
6160	Plywood, 1/2" thick		768	.021			.60		.60	.93
6162	5/8" thick		760	.021			.61		.61	.94
6164	3/4" thick		750	.021			.61		.61	.95
6165	1-1/8" thick	▼	720	.022			.64		.64	.99
6166	Underlayment, particle board, 3/8" thick	1 Clab	780	.010			.30		.30	.46
6168	1/2" thick		768	.010			.30		.30	.47
6170	5/8" thick		760	.011			.30		.30	.47
6172	3/4" thick	▼	750	.011	▼		.31		.31	.48
6200	Stairs and stringers, minimum	2 Clab	40	.400	Riser		11.50		11.50	17.90
6240	Maximum	"	26	.615	"		17.70		17.70	27.50
6300	Components, tread	1 Clab	110	.073	Ea.		2.09		2.09	3.25
6320	Riser		80	.100	"		2.88		2.88	4.48
6390	Stringer, 2" x 10"		260	.031	L.F.		.88		.88	1.38
6400	2" x 12"		260	.031			.88		.88	1.38
6410	3" x 10"		250	.032			.92		.92	1.43
6420	3" x 12"	▼	250	.032			.92		.92	1.43
6590	Wood studs, 2" x 3"	2 Clab	3076	.005			.15		.15	.23
6600	2" x 4"		2000	.008			.23		.23	.36
6640	2" x 6"	▼	1600	.010	▼		.29		.29	.45
6720	Wall framing, including studs plates and blocking, 2" x 4"	1 Clab	600	.013	S.F.		.38		.38	.60
6740	2" x 6"		480	.017	"		.48		.48	.75
6750	Headers, 2" x 4"		1125	.007	L.F.		.20		.20	.32
6755	2" x 6"		1125	.007			.20		.20	.32
6760	2" x 8"		1050	.008			.22		.22	.34
6765	2" x 10"	▼	1050	.008	▼		.22		.22	.34

06 05 Common Work Results for Wood, Plastics and Composites

06 05 05 – Selective Wood and Plastics Demolition

06 05 05.10 Selective Demolition Wood Framing

		Crew	Daily Output	Labor-Hours	Unit	Material	2007 Bare Costs Labor	Equipment	Total	Total Incl O&P
6770	2" x 12"	1 Clab	1000	.008	L.F.		.23		.23	.36
6780	4" x 10"		525	.015			.44		.44	.68
6785	4" x 12"		500	.016			.46		.46	.72
6790	6" x 8"		560	.014			.41		.41	.64
6795	6" x 10"		525	.015			.44		.44	.68
6797	6" x 12"		500	.016			.46		.46	.72
8000	Soffit, T & G wood		520	.015	S.F.		.44		.44	.69
8010	Hardboard, vinyl or aluminum		640	.013			.36		.36	.56
8030	Plywood	2 Carp	315	.051			1.86		1.86	2.90
9500	See Div. 02 41 19.23 for rubbish handling									

06 05 05.20 Selective Demolition Millwork and Trim

		Crew	Daily Output	Labor-Hours	Unit	Material	2007 Bare Costs Labor	Equipment	Total	Total Incl O&P
0010	**SELECTIVE DEMOLITION MILLWORK AND TRIM** R024119-10									
1000	Cabinets, wood, base cabinets, per L.F.	2 Clab	80	.200	L.F.		5.75		5.75	8.95
1020	Wall cabinets, per L.F.	"	80	.200	"		5.75		5.75	8.95
1060	Remove and reset, base cabinets	2 Carp	18	.889	Ea.		32.50		32.50	51
1070	Wall cabinets	"	20	.800	"		29.50		29.50	45.50
1100	Steel, painted, base cabinets	2 Clab	60	.267	L.F.		7.65		7.65	11.95
1120	Wall cabinets		60	.267	"		7.65		7.65	11.95
1200	Casework, large area		320	.050	S.F.		1.44		1.44	2.24
1220	Selective		200	.080	"		2.30		2.30	3.58
1500	Counter top, minimum		200	.080	L.F.		2.30		2.30	3.58
1510	Maximum		120	.133			3.83		3.83	5.95
1550	Remove and reset, minimum	2 Carp	50	.320			11.75		11.75	18.30
1560	Maximum	"	40	.400			14.70		14.70	23
2000	Paneling, 4' x 8' sheets	2 Clab	2000	.008	S.F.		.23		.23	.36
2100	Boards, 1" x 4"		700	.023			.66		.66	1.02
2120	1" x 6"		750	.021			.61		.61	.95
2140	1" x 8"		800	.020			.58		.58	.90
3000	Trim, baseboard, to 6" wide		1200	.013	L.F.		.38		.38	.60
3040	Greater than 6" and up to 12" wide		1000	.016			.46		.46	.72
3080	Remove and reset, minimum	2 Carp	400	.040			1.47		1.47	2.29
3090	Maximum	"	300	.053			1.96		1.96	3.05
3100	Ceiling trim	2 Clab	1000	.016			.46		.46	.72
3120	Chair rail		1200	.013			.38		.38	.60
3140	Railings with balusters		240	.067			1.92		1.92	2.98
3160	Wainscoting		700	.023	S.F.		.66		.66	1.02

06 05 23 – Wood, Plastic, and Composite Fastenings

06 05 23.10 Nails

		Crew	Daily Output	Labor-Hours	Unit	Material	2007 Bare Costs Labor	Equipment	Total	Total Incl O&P
0010	**NAILS**									
0020	Copper nails, plain				Lb.	6.60			6.60	7.25
0400	Stainless steel, plain					5.95			5.95	6.55
0500	Box, 3d to 20d, bright					.76			.76	.84
0520	Galvanized					.88			.88	.97
0600	Common, 3d to 60d, plain					.82			.82	.90
0700	Galvanized					1.06			1.06	1.17
0800	Aluminum					4.45			4.45	4.90
1000	Annular or spiral thread, 4d to 60d, plain					1.88			1.88	2.07
1200	Galvanized					1.95			1.95	2.15
1400	Drywall nails, plain					.79			.79	.87
1600	Galvanized					1.59			1.59	1.75
1800	Finish nails, 4d to 10d, plain					.91			.91	1
2000	Galvanized					1.51			1.51	1.66

06 05 Common Work Results for Wood, Plastics and Composites

06 05 23 – Wood, Plastic, and Composite Fastenings

06 05 23.10 Nails

		Crew	Daily Output	Labor-Hours	Unit	Material	2007 Bare Costs Labor	Equipment	Total	Total Incl O&P
2100	Aluminum				Lb.	4.10			4.10	4.51
2300	Flooring nails, hardened steel, 2d to 10d, plain					1.61			1.61	1.77
2400	Galvanized					2.25			2.25	2.48
2500	Gypsum lath nails, 1-1/8", 13 ga. flathead, blued					1.54			1.54	1.69
2600	Masonry nails, hardened steel, 3/4" to 3" long, plain					1.61			1.61	1.77
2700	Galvanized					2			2	2.20
2900	Roofing nails, threaded, galvanized					1.05			1.05	1.16
3100	Aluminum					4.80			4.80	5.30
3300	Compressed lead head, threaded, galvanized					1.50			1.50	1.65
3600	Siding nails, plain shank, galvanized					1.20			1.20	1.32
3800	Aluminum					4.11			4.11	4.52
5000	Add to prices above for cement coating					.10			.10	.11
5200	Zinc or tin plating					.13			.13	.14
5500	Vinyl coated sinkers, 8d to 16d					.57			.57	.63

06 05 23.20 Pneumatic Nails

		Crew	Daily Output	Labor-Hours	Unit	Material	2007 Bare Costs Labor	Equipment	Total	Total Incl O&P
0010	**PNEUMATIC NAILS**									
0020	Framing, per carton of 5000, 2"				Ea.	37			37	41
0100	2-3/8"					42.50			42.50	46.50
0200	Per carton of 4000, 3"					37.50			37.50	41.50
0300	3-1/4"					40			40	44
0400	Per carton of 5000, 2-3/8", galv.					57.50			57.50	63.50
0500	Per carton of 4000, 3", galv.					65			65	71.50
0600	3-1/4", galv.					80.50			80.50	88.50
0700	Roofing, per carton of 7200, 1"					34			34	37
0800	1-1/4"					31.50			31.50	34.50
0900	1-1/2"					36.50			36.50	40
1000	1-3/4"					45.50			45.50	50

06 05 23.40 Sheet Metal Screws

		Crew	Daily Output	Labor-Hours	Unit	Material	2007 Bare Costs Labor	Equipment	Total	Total Incl O&P
0010	**SHEET METAL SCREWS**									
0020	Steel, standard, #8 x 3/4", plain				C	3.29			3.29	3.62
0100	Galvanized					3.44			3.44	3.78
0300	#10 x 1", plain					4.49			4.49	4.94
0400	Galvanized					4.60			4.60	5.05
0600	With washers, #14 x 1", plain					10.20			10.20	11.20
0700	Galvanized					10.50			10.50	11.55
0900	#14 x 2", plain					18			18	19.80
1000	Galvanized					18.15			18.15	19.95
1500	Self-drilling, with washers, (pinch point) #8 x 3/4", plain					7.10			7.10	7.80
1600	Galvanized					7.20			7.20	7.90
1800	#10 x 3/4", plain					9.30			9.30	10.20
1900	Galvanized					9.40			9.40	10.35
3000	Stainless steel w/aluminum or neoprene washers, #14 x 1", plain					18.35			18.35	20
3100	#14 x 2", plain					25			25	27.50

06 05 23.50 Wood Screws

		Crew	Daily Output	Labor-Hours	Unit	Material	2007 Bare Costs Labor	Equipment	Total	Total Incl O&P
0010	**WOOD SCREWS**									
0020	Steel, #8 x 1" long				C	3.29			3.29	3.62
0100	Brass					11.50			11.50	12.65
0200	#8, 2" long, steel					5.60			5.60	6.15
0300	Brass					19.50			19.50	21.50
0400	#10, 1" long, steel					5.90			5.90	6.50
0500	Brass					14			14	15.40
0600	#10, 2" long, steel					6.30			6.30	6.90

06 05 Common Work Results for Wood, Plastics and Composites
06 05 23 – Wood, Plastic, and Composite Fastenings

06 05 23.50 Wood Screws		Crew	Daily Output	Labor-Hours	Unit	Material	2007 Bare Costs Labor	Equipment	Total	Total Incl O&P
0700	Brass				C	23			23	25.50
0800	#10, 3" long, steel					10.50			10.50	11.55
1000	#12, 2" long, steel					7.50			7.50	8.25
1100	Brass					30			30	33
1500	#12, 3" long, steel					11.50			11.50	12.65
2000	#12, 4" long, steel					43			43	47.50

06 05 23.60 Timber Connectors

		Crew	Daily Output	Labor-Hours	Unit	Material	Labor	Equipment	Total	Total Incl O&P
0010	**TIMBER CONNECTORS**									
0020	Add up cost of each part for total cost of connection									
0100	Connector plates, steel, with bolts, straight	2 Carp	75	.213	Ea.	22.50	7.85		30.35	37
0110	Tee		50	.320		33.50	11.75		45.25	55.50
0120	T-Strap, 14 gauge, 12" x 8" x 2"		50	.320		33.50	11.75		45.25	55.50
0150	Anchor plate, 7 gauge, 9" x 7"		75	.213		22.50	7.85		30.35	37
0200	Bolts, machine, sq. hd. with nut & washer, 1/2" diameter, 4" long	1 Carp	140	.057		.41	2.10		2.51	3.72
0300	7-1/2" long		130	.062		.72	2.26		2.98	4.31
0500	3/4" diameter, 7-1/2" long		130	.062		1.92	2.26		4.18	5.65
0610	Machine bolts, w/ nut, washer, 3/4" dia, 15" L, HD's & beam hangers		95	.084		3.52	3.09		6.61	8.70
0800	Drilling bolt holes in timber, 1/2" diameter		450	.018	Inch		.65		.65	1.02
0900	1" diameter		350	.023	"		.84		.84	1.31
1100	Framing anchors, 2 or 3 dimensional, 10 gauge, no nails incl.		175	.046	Ea.	.57	1.68		2.25	3.24
1150	Framing anchors, 18 gauge, 4 1/2" x 2 3/4"		175	.046		.57	1.68		2.25	3.24
1160	Framing anchors, 18 gauge, 4 1/2" x 3"		175	.046		.57	1.68		2.25	3.24
1170	Clip anchors plates, 18 gauge, 12" x 1 1/8"		175	.046		.57	1.68		2.25	3.24
1250	Holdowns, 3 gauge base, 10 gauge body		8	1		18.95	36.50		55.45	78
1260	Holdowns, 7 gauge 11 1/16" x 3 1/4"		8	1		18.95	36.50		55.45	78
1270	Holdowns, 7 gauge 14 3/8" x 3 1/8"		8	1		18.95	36.50		55.45	78
1275	Holdowns, 12 gauge 8" x 2 1/2"		8	1		18.95	36.50		55.45	78
1300	Joist and beam hangers, 18 ga. galv., for 2" x 4" joist		175	.046		.71	1.68		2.39	3.39
1400	2" x 6" to 2" x 10" joist		165	.048		.81	1.78		2.59	3.66
1600	16 ga. galv., 3" x 6" to 3" x 10" joist		160	.050		3.19	1.84		5.03	6.35
1700	3" x 10" to 3" x 14" joist		160	.050		3.71	1.84		5.55	6.95
1800	4" x 6" to 4" x 10" joist		155	.052		2.81	1.89		4.70	6.05
1900	4" x 10" to 4" x 14" joist		155	.052		3.81	1.89		5.70	7.15
2000	Two-2" x 6" to two-2" x 10" joists		150	.053		3.07	1.96		5.03	6.45
2100	Two-2" x 10" to two-2" x 14" joists		150	.053		3.07	1.96		5.03	6.45
2300	3/16" thick, 6" x 8" joist		145	.055		6.70	2.02		8.72	10.50
2400	6" x 10" joist		140	.057		7.85	2.10		9.95	11.90
2500	6" x 12" joist		135	.059		9.45	2.17		11.62	13.80
2700	1/4" thick, 6" x 14" joist		130	.062		11.75	2.26		14.01	16.45
2900	Plywood clips, extruded aluminum H clip, for 3/4" panels					.18			.18	.20
3000	Galvanized 18 ga. back-up clip					.17			.17	.19
3200	Post framing, 16 ga. galv. for 4" x 4" base, 2 piece	1 Carp	130	.062		6.60	2.26		8.86	10.75
3300	Cap		130	.062		3.24	2.26		5.50	7.10
3500	Rafter anchors, 18 ga. galv., 1-1/2" wide, 5-1/4" long		145	.055		.56	2.02		2.58	3.77
3600	10-3/4" long		145	.055		1.10	2.02		3.12	4.36
3800	Shear plates, 2-5/8" diameter		120	.067		1.98	2.45		4.43	6
3900	4" diameter		115	.070		4.55	2.55		7.10	9
4000	Sill anchors, embedded in concrete or block, 18-5/8" long		115	.070		1.33	2.55		3.88	5.45
4100	Spike grids, 4" x 4", flat or curved		120	.067		.49	2.45		2.94	4.35
4400	Split rings, 2-1/2" diameter		120	.067		1.60	2.45		4.05	5.55
4500	4" diameter		110	.073		2.47	2.67		5.14	6.90
4550	Tie plate, 20 gauge, 7" x 3 1/8"		110	.073		2.47	2.67		5.14	6.90

06 05 Common Work Results for Wood, Plastics and Composites

06 05 23 – Wood, Plastic, and Composite Fastenings

	06 05 23.60 Timber Connectors	Crew	Daily Output	Labor-Hours	Unit	Material	2007 Bare Costs Labor	Equipment	Total	Total Incl O&P
4560	Tie plate, 20 gauge, 5" x 4 1/8"	1 Carp	110	.073	Ea.	2.47	2.67		5.14	6.90
4575	Twist straps, 18 gauge, 12" x 1 1/4"		110	.073		2.47	2.67		5.14	6.90
4580	Twist straps, 18 gauge, 16" x 1 1/4"		110	.073		2.47	2.67		5.14	6.90
4600	Strap ties, 20 ga., 2-1/16" wide, 12 13/16" long		180	.044		1.44	1.63		3.07	4.12
4700	Strap ties, 16 ga., 1-3/8" wide, 12" long		180	.044		1.44	1.63		3.07	4.12
4800	24" long		160	.050		1.99	1.84		3.83	5.05
5000	Toothed rings, 2-5/8" or 4" diameter		90	.089		1.36	3.26		4.62	6.60
5200	Truss plates, nailed, 20 gauge, up to 32' span		17	.471	Truss	9.85	17.25		27.10	38
5400	Washers, 2" x 2" x 1/8"				Ea.	.31			.31	.34
5500	3" x 3" x 3/16"				"	.81			.81	.89
6101	Beam hangers, polymer painted									
6102	Bolted, 3 ga., (W x H x L)									
6104	3-1/4" x 9" x 12" top flange	1 Carp	1	8	C	7,750	294		8,044	8,950
6106	5-1/4" x 9" x 12" top flange		1	8		8,050	294		8,344	9,300
6108	5-1/4" x 11" x 11-3/4" top flange		1	8		18,100	294		18,394	20,400
6110	6-7/8" x 9" x 12" top flange		1	8		8,325	294		8,619	9,600
6112	6-7/8" x 11" x 13-1/2" top flange		1	8		19,000	294		19,294	21,400
6114	8-7/8" x 11" x 15-1/2" top flange		1	8		20,400	294		20,694	22,900
6116	Nailed, 3 ga., (W x H x L)									
6118	3-1/4" x 10-1/2" x 10" top flange	1 Carp	1.80	4.444	C	5,350	163		5,513	6,125
6120	3-1/4" x 10-1/2" x 12" top flange		1.80	4.444		6,200	163		6,363	7,050
6122	5-1/4" x 9-1/2" x 10" top flange		1.80	4.444		5,750	163		5,913	6,600
6124	5-1/4" x 9-1/2" x 12" top flange		1.80	4.444		6,525	163		6,688	7,450
6126	5-1/2" x 9-1/2" x 12" top flange		1.80	4.444		5,750	163		5,913	6,600
6128	6-7/8" x 8-1/2" x 12" top flange		1.80	4.444		5,950	163		6,113	6,800
6130	7-1/2" x 8-1/2" x 12" top flange		1.80	4.444		6,150	163		6,313	7,025
6132	8-7/8" x 7-1/2" x 14" top flange		1.80	4.444		6,800	163		6,963	7,725
6201	Beam and purlin hangers, galvanized, 12 ga.									
6202	Purlin or joist size, 3" x 8"	1 Carp	1.70	4.706	C	975	173		1,148	1,350
6204	3" x 10"		1.70	4.706		1,100	173		1,273	1,475
6206	3" x 12"		1.65	4.848		1,300	178		1,478	1,700
6208	3" x 14"		1.65	4.848		1,525	178		1,703	1,950
6210	3" x 16"		1.65	4.848		1,750	178		1,928	2,200
6212	4" x 8"		1.65	4.848		975	178		1,153	1,350
6214	4" x 10"		1.65	4.848		1,150	178		1,328	1,525
6216	4" x 12"		1.60	5		1,225	184		1,409	1,625
6218	4" x 14"		1.60	5		1,325	184		1,509	1,750
6220	4" x 16"		1.60	5		1,525	184		1,709	1,950
6224	6" x 10"		1.55	5.161		1,675	189		1,864	2,125
6226	6" x 12"		1.55	5.161		1,800	189		1,989	2,275
6228	6" x 14"		1.50	5.333		1,925	196		2,121	2,425
6230	6" x 16"		1.50	5.333		2,200	196		2,396	2,725
6300	Column bases									
6302	4 x 4, 16 ga.	1 Carp	1.80	4.444	C	1,350	163		1,513	1,750
6306	7 ga.		1.80	4.444		2,525	163		2,688	3,025
6314	6 x 6, 16 ga.		1.75	4.571		1,725	168		1,893	2,150
6318	7 ga.		1.75	4.571		3,550	168		3,718	4,150
6326	8 x 8, 7 ga.		1.65	4.848		5,900	178		6,078	6,775
6330	8 x 10, 7 ga.		1.65	4.848		6,300	178		6,478	7,225
6590	Joist hangers, heavy duty 12 ga., galvanized									
6592	2" x 4"	1 Carp	1.75	4.571	C	1,100	168		1,268	1,475
6594	2" x 6"		1.65	4.848		1,175	178		1,353	1,575
6595	2" x 6", 16 gauge		1.65	4.848		1,175	178		1,353	1,575

06 05 Common Work Results for Wood, Plastics and Composites

06 05 23 – Wood, Plastic, and Composite Fastenings

06 05 23.60 Timber Connectors		Crew	Daily Output	Labor-Hours	Unit	Material	2007 Bare Costs Labor	Equipment	Total	Total Incl O&P
6596	2" x 8"	1 Carp	1.65	4.848	C	1,250	178		1,428	1,675
6597	2" x 8", 16 gauge		1.65	4.848		1,250	178		1,428	1,675
6598	2" x 10"		1.65	4.848		1,350	178		1,528	1,775
6600	2" x 12"		1.65	4.848		1,525	178		1,703	1,950
6622	(2) 2" x 6"		1.60	5		1,475	184		1,659	1,900
6624	(2) 2" x 8"		1.60	5		1,600	184		1,784	2,050
6626	(2) 2" x 10"		1.55	5.161		1,800	189		1,989	2,275
6628	(2) 2" x 12"		1.55	5.161		2,175	189		2,364	2,700
6890	Purlin hangers, painted									
6892	12 ga., 2" x 6"	1 Carp	1.80	4.444	C	1,350	163		1,513	1,725
6894	2" x 8"		1.80	4.444		1,425	163		1,588	1,825
6896	2" x 10"		1.80	4.444		1,475	163		1,638	1,850
6898	2" x 12"		1.75	4.571		1,575	168		1,743	2,000
6934	(2) 2" x 6"		1.70	4.706		1,375	173		1,548	1,775
6936	(2) 2" x 8"		1.70	4.706		1,500	173		1,673	1,925
6938	(2) 2" x 10"		1.70	4.706		1,650	173		1,823	2,100
6940	(2) 2" x 12"		1.65	4.848		1,775	178		1,953	2,225

06 05 23.70 Rough Hardware

0010	**ROUGH HARDWARE**									
0020	Minimum					.50%				
0200	Maximum					1.50%				
0210	In seismic or hurricane areas, up to					10%				

06 05 23.80 Metal Bracing

0010	**METAL BRACING**									
0300	Let-in, "T" shaped, 22 ga. galv. steel, studs at 16" O.C.	1 Carp	5.80	1.379	C.L.F.	53	50.50		103.50	138
0400	Studs at 24" O.C.		6	1.333		53	49		102	135
0500	16 ga. galv. steel straps, studs at 16" O.C.		6	1.333		80	49		129	164
0600	Studs at 24" O.C.		6.20	1.290		80	47.50		127.50	162

06 05 73 – Wood Treatment

06 05 73.10 Lumber Treatment

0011	**LUMBER TREATMENT**									
0400	Fire retardant, wet				M.B.F.	330			330	365
0500	KDAT					315			315	345
0700	Salt treated, water borne, .40 lb. retention					150			150	166
0800	Oil borne, 8 lb. retention					176			176	194
1000	Kiln dried lumber, 1" & 2" thick, softwoods					101			101	111
1100	Hardwoods					107			107	118
1500	For small size 1" stock, add					13.60			13.60	14.95
1700	For full size rough lumber, add					20%				

06 05 73.20 Plywood Treatment

0010	**PLYWOOD TREATMENT**									
0020	Fire retardant, 1/4" thick				M.S.F.	252			252	277
0030	3/8" thick					276			276	305
0050	1/2" thick					296			296	325
0070	5/8" thick					315			315	345
0100	3/4" thick					345			345	380
0200	For KDAT, add					75			75	83
0500	Salt treated water borne, .25 lb., wet, 1/4" thick					138			138	152
0530	3/8" thick					144			144	159
0550	1/2" thick					150			150	166
0570	5/8" thick					164			164	180

06 05 Common Work Results for Wood, Plastics and Composites

06 05 73 – Wood Treatment

06 05 73.20 Plywood Treatment		Crew	Daily Output	Labor-Hours	Unit	Material	2007 Bare Costs Labor	Equipment	Total	Total Incl O&P
0600	3/4" thick				M.S.F.	170			170	187
0800	For KDAT add					75			75	83
0900	For .40 lb., per C.F. retention, add					63			63	69
1000	For certification stamp, add					37			37	40.50

06 11 Wood Framing

06 11 10 – Wood Framing

06 11 10.02 Blocking

		Crew	Daily Output	Labor-Hours	Unit	Material	Labor	Equipment	Total	Total Incl O&P
0010	**BLOCKING**									
2600	Miscellaneous, to wood construction									
2620	2" x 4"	1 Carp	.17	47.059	M.B.F.	560	1,725		2,285	3,325
2625	Pneumatic nailed		.21	38.095		560	1,400		1,960	2,800
2660	2" x 8"		.27	29.630		630	1,075		1,705	2,400
2665	Pneumatic nailed		.33	24.242		630	890		1,520	2,075
2720	To steel construction									
2740	2" x 4"	1 Carp	.14	57.143	M.B.F.	560	2,100		2,660	3,900
2780	2" x 8"	"	.21	38.095	"	630	1,400		2,030	2,875

06 11 10.04 Wood Bracing

		Crew	Daily Output	Labor-Hours	Unit	Material	Labor	Equipment	Total	Total Incl O&P
0010	**WOOD BRACING**									
0011	Let-in, with 1" x 6" boards, studs 16" OC	1 Carp	1.50	5.333	C.L.F.	65	196		261	375
0200	Studs @ 24" O.C.	"	2.30	3.478	"	65	128		193	271

06 11 10.06 Bridging

		Crew	Daily Output	Labor-Hours	Unit	Material	Labor	Equipment	Total	Total Incl O&P
0010	**BRIDGING**									
0011	Wood, for joists 16" OC, 1" x 3"	1 Carp	1.30	6.154	C.Pr.	33.50	226		259.50	385
0015	Pneumatic nailed		1.70	4.706		33.50	173		206.50	305
0100	2" x 3" bridging		1.30	6.154		44.50	226		270.50	400
0105	Pneumatic nailed		1.70	4.706		44.50	173		217.50	320
0300	Steel, galvanized, 18 ga., for 2" x 10" joists at 12" O.C.		1.30	6.154		128	226		354	490
0400	24" O.C.		1.40	5.714		155	210		365	495
0600	For 2" x 14" joists at 16" O.C.		1.30	6.154		152	226		378	515
0700	24" O.C.		1.40	5.714		250	210		460	600
0900	Compression type, 16" O.C., 2" x 8" joists		2	4		150	147		297	395
1000	2" x 12" joists		2	4		150	147		297	395

06 11 10.10 Beam and Girder Framing

		Crew	Daily Output	Labor-Hours	Unit	Material	Labor	Equipment	Total	Total Incl O&P
0010	**BEAM AND GIRDER FRAMING** R061110-30									
3500	Single, 2" x 6"	2 Carp	.70	22.857	M.B.F.	575	840		1,415	1,925
3505	Pneumatic nailed		.81	19.704		575	725		1,300	1,750
3520	2" x 8"		.86	18.605		630	685		1,315	1,775
3525	Pneumatic nailed		1	16.048		630	590		1,220	1,600
3540	2" x 10"		1	16		715	585		1,300	1,700
3545	Pneumatic nailed		1.16	13.793		715	505		1,220	1,575
3560	2" x 12"		1.10	14.545		715	535		1,250	1,625
3565	Pneumatic nailed		1.28	12.539		715	460		1,175	1,500
3580	2" x 14"		1.17	13.675		935	500		1,435	1,800
3585	Pneumatic nailed		1.36	11.791		935	435		1,370	1,700
3600	3" x 8"		1.10	14.545		1,300	535		1,835	2,250
3620	3" x 10"		1.25	12.800		1,300	470		1,770	2,150
3640	3" x 12"		1.35	11.852		1,300	435		1,735	2,125
3660	3" x 14"		1.40	11.429		1,625	420		2,045	2,450
3680	4" x 8"	F-3	2.66	15.038		1,400	560	272	2,232	2,725

06 11 Wood Framing

06 11 10 – Wood Framing

06 11 10.10 Beam and Girder Framing

		Crew	Daily Output	Labor-Hours	Unit	Material	2007 Bare Costs Labor	Equipment	Total	Total Incl O&P
3700	4" x 10"	F-3	3.16	12.658	M.B.F.	1,650	470	229	2,349	2,800
3720	4" x 12"		3.60	11.111		1,275	415	201	1,891	2,250
3740	4" x 14"		3.96	10.101		1,050	375	183	1,608	1,950
4000	Double, 2" x 6"	2 Carp	1.25	12.800		575	470		1,045	1,375
4005	Pneumatic nailed		1.45	11.034		575	405		980	1,275
4020	2" x 8"		1.60	10		630	365		995	1,275
4025	Pneumatic nailed		1.86	8.621		630	315		945	1,200
4040	2" x 10"		1.92	8.333		715	305		1,020	1,250
4045	Pneumatic nailed		2.23	7.185		715	264		979	1,200
4060	2" x 12"		2.20	7.273		715	267		982	1,200
4065	Pneumatic nailed		2.55	6.275		715	230		945	1,150
4080	2" x 14"		2.45	6.531		935	240		1,175	1,400
4085	Pneumatic nailed		2.84	5.634		935	207		1,142	1,350
5000	Triple, 2" x 6"		1.65	9.697		575	355		930	1,200
5005	Pneumatic nailed		1.91	8.377		575	305		880	1,125
5020	2" x 8"		2.10	7.619		630	280		910	1,125
5025	Pneumatic nailed		2.44	6.568		630	241		871	1,075
5040	2" x 10"		2.50	6.400		715	235		950	1,150
5045	Pneumatic nailed		2.90	5.517		715	202		917	1,100
5060	2" x 12"		2.85	5.614		715	206		921	1,100
5065	Pneumatic nailed		3.31	4.840		715	178		893	1,050
5080	2" x 14"		3.15	5.079		935	186		1,121	1,325
5085	Pneumatic nailed		3.35	4.770		935	175		1,110	1,300

06 11 10.12 Ceiling Framing

		Crew	Daily Output	Labor-Hours	Unit	Material	Labor	Equipment	Total	Total Incl O&P
0010	**CEILING FRAMING**									
6400	Suspended, 2" x 3"	2 Carp	.50	32	M.B.F.	595	1,175		1,770	2,475
6450	2" x 4"		.59	27.119		560	995		1,555	2,175
6500	2" x 6"		.80	20		575	735		1,310	1,775
6550	2" x 8"		.86	18.605		630	685		1,315	1,775

06 11 10.14 Column Framing

		Crew	Daily Output	Labor-Hours	Unit	Material	Labor	Equipment	Total	Total Incl O&P
0010	**COLUMN FRAMING**									
0400	4" x 4"	2 Carp	.52	30.769	M.B.F.	1,250	1,125		2,375	3,125
0420	4" x 6"		.55	29.091		1,400	1,075		2,475	3,200
0440	4" x 8"		.59	27.119		1,400	995		2,395	3,100
0460	6" x 6"		.65	24.615		1,600	905		2,505	3,175
0480	6" x 8"		.70	22.857		1,900	840		2,740	3,375
0500	6" x 10"		.75	21.333		2,175	785		2,960	3,625

06 11 10.18 Joist Framing

		Crew	Daily Output	Labor-Hours	Unit	Material	Labor	Equipment	Total	Total Incl O&P
0010	**JOIST FRAMING** R061110-30									
2650	Joists, 2" x 4"	2 Carp	.83	19.277	M.B.F.	560	705		1,265	1,725
2655	Pneumatic nailed		.96	16.667		560	610		1,170	1,575
2680	2" x 6"		1.25	12.800		575	470		1,045	1,375
2685	Pneumatic nailed		1.44	11.111		575	410		985	1,275
2700	2" x 8"		1.46	10.959		630	400		1,030	1,325
2705	Pneumatic nailed		1.68	9.524		630	350		980	1,250
2720	2" x 10"		1.49	10.738		715	395		1,110	1,400
2725	Pneumatic nailed		1.71	9.357		715	345		1,060	1,325
2740	2" x 12"		1.75	9.143		715	335		1,050	1,300
2745	Pneumatic nailed		2.01	7.960		715	292		1,007	1,250
2760	2" x 14"		1.79	8.939		935	330		1,265	1,525
2765	Pneumatic nailed		2.06	7.767		935	285		1,220	1,475
2780	3" x 6"		1.39	11.511		1,275	420		1,695	2,050

06 11 Wood Framing

06 11 10 – Wood Framing

06 11 10.18 Joist Framing

		Crew	Daily Output	Labor-Hours	Unit	Material	2007 Bare Costs Labor	Equipment	Total	Total Incl O&P
2790	3" x 8"	2 Carp	1.90	8.421	M.B.F.	1,300	310		1,610	1,900
2800	3" x 10"		1.95	8.205		1,300	300		1,600	1,900
2820	3" x 12"		1.80	8.889		1,300	325		1,625	1,950
2840	4" x 6"		1.60	10		1,400	365		1,765	2,100
2850	4" x 8"		4.15	3.855		1,400	141		1,541	1,775
2860	4" x 10"		2	8		1,650	294		1,944	2,275
2880	4" x 12"		1.80	8.889		1,275	325		1,600	1,900
3000	Composite wood joist 9-1/2" deep		.90	17.778	M.L.F.	1,850	650		2,500	3,075
3010	11-1/2" deep		.88	18.182		1,975	665		2,640	3,225
3020	14" deep		.82	19.512		2,325	715		3,040	3,675
3030	16" deep		.78	20.513		2,600	755		3,355	4,025
4000	Open web joist 12" deep		.88	18.182		1,825	665		2,490	3,050
4010	14" deep		.82	19.512		2,125	715		2,840	3,450
4020	16" deep		.78	20.513		2,200	755		2,955	3,600
4030	18" deep		.74	21.622		2,250	795		3,045	3,700
6000	Composite rim joist, 1-1/4" x 9-1/2"		90	.178		2,475	6.50		2,481.50	2,700
6010	1-1/4" x 11-1/2"		.88	18.182		2,725	665		3,390	4,050
6020	1-1/4" x 14-1/2"		.82	19.512		3,175	715		3,890	4,625
6030	1-1/4" x 16-1/2"		.78	20.513		3,200	755		3,955	4,675

06 11 10.24 Miscellaneous Framing

		Crew	Daily Output	Labor-Hours	Unit	Material	Labor	Equipment	Total	Total Incl O&P
0010	**MISCELLANEOUS FRAMING**									
8500	Firestops, 2" x 4"	2 Carp	.51	31.373	M.B.F.	560	1,150		1,710	2,425
8505	Pneumatic nailed		.62	25.806		560	945		1,505	2,100
8520	2" x 6"		.60	26.667		575	980		1,555	2,150
8525	Pneumatic nailed		.73	21.858		575	800		1,375	1,875
8540	2" x 8"		.60	26.667		630	980		1,610	2,225
8560	2" x 12"		.70	22.857		715	840		1,555	2,075
8600	Nailers, treated, wood construction, 2" x 4"		.53	30.189		785	1,100		1,885	2,600
8605	Pneumatic nailed		.64	25.157		785	925		1,710	2,325
8620	2" x 6"		.75	21.333		830	785		1,615	2,150
8625	Pneumatic nailed		.90	17.778		830	650		1,480	1,950
8640	2" x 8"		.93	17.204		860	630		1,490	1,925
8645	Pneumatic nailed		1.12	14.337		860	525		1,385	1,775
8660	Steel construction, 2" x 4"		.50	32		785	1,175		1,960	2,700
8680	2" x 6"		.70	22.857		830	840		1,670	2,225
8700	2" x 8"		.87	18.391		860	675		1,535	2,000
8760	Rough bucks, treated, for doors or windows, 2" x 6"		.40	40		830	1,475		2,305	3,200
8765	Pneumatic nailed		.48	33.333		830	1,225		2,055	2,825
8780	2" x 8"		.51	31.373		860	1,150		2,010	2,750
8785	Pneumatic nailed		.61	26.144		860	960		1,820	2,450
8800	Stair stringers, 2" x 10"		.22	72.727		715	2,675		3,390	4,925
8820	2" x 12"		.26	61.538		715	2,250		2,965	4,300
8840	3" x 10"		.31	51.613		1,300	1,900		3,200	4,375
8860	3" x 12"		.38	42.105		1,300	1,550		2,850	3,850
8870	Composite LSL, 1-1/4" x 11-1/2"		130	.123	L.F.	2.73	4.52		7.25	10.05
8880	1-1/4" x 14-1/2"		130	.123	"	3.17	4.52		7.69	10.55

06 11 10.26 Partitions

		Crew	Daily Output	Labor-Hours	Unit	Material	Labor	Equipment	Total	Total Incl O&P
0010	**PARTITIONS**									
0020	Single bottom and double top plate, no waste, std. & better lumber									
0180	2" x 4" studs, 8' high, studs 12" O.C.	2 Carp	80	.200	L.F.	4.12	7.35		11.47	16
0185	12" O.C., pneumatic nailed		96	.167		4.12	6.10		10.22	14.10
0200	16" O.C.		100	.160		3.37	5.85		9.22	12.85

06 11 Wood Framing

06 11 10 – Wood Framing

06 11 10.26 Partitions

		Crew	Daily Output	Labor-Hours	Unit	Material	2007 Bare Costs Labor	Equipment	Total	Total Incl O&P
0205	16" O.C., pneumatic nailed	2 Carp	120	.133	L.F.	3.37	4.89		8.26	11.30
0300	24" O.C.		125	.128		2.62	4.70		7.32	10.20
0305	24" O.C., pneumatic nailed		150	.107		2.62	3.91		6.53	9
0380	10' high, studs 12" O.C.		80	.200		4.87	7.35		12.22	16.80
0385	12" O.C., pneumatic nailed		96	.167		4.87	6.10		10.97	14.90
0400	16" O.C.		100	.160		3.93	5.85		9.78	13.45
0405	16" O.C., pneumatic nailed		120	.133		3.93	4.89		8.82	11.90
0500	24" O.C.		125	.128		2.99	4.70		7.69	10.60
0505	24" O.C., pneumatic nailed		150	.107		2.99	3.91		6.90	9.40
0580	12' high, studs 12" O.C.		65	.246		5.60	9.05		14.65	20.50
0585	12" O.C., pneumatic nailed		78	.205		5.60	7.55		13.15	17.90
0600	16" O.C.		80	.200		4.49	7.35		11.84	16.40
0605	16" O.C., pneumatic nailed		96	.167		4.49	6.10		10.59	14.50
0700	24" O.C.		100	.160		3.37	5.85		9.22	12.85
0705	24" O.C., pneumatic nailed		120	.133		3.37	4.89		8.26	11.30
0780	2" x 6" studs, 8' high, studs 12" O.C.		70	.229		6.35	8.40		14.75	20
0785	12" O.C., pneumatic nailed		84	.190		6.35	7		13.35	17.85
0800	16" O.C.		90	.178		5.20	6.50		11.70	15.85
0805	16" O.C., pneumatic nailed		108	.148		5.20	5.45		10.65	14.15
0900	24" O.C.		115	.139		4.03	5.10		9.13	12.40
0905	24" O.C., pneumatic nailed		138	.116		4.03	4.26		8.29	11.10
0980	10' high, studs 12" O.C.		70	.229		7.50	8.40		15.90	21.50
0985	12" O.C., pneumatic nailed		84	.190		7.50	7		14.50	19.10
1000	16" O.C.		90	.178		6.05	6.50		12.55	16.80
1005	16" O.C., pneumatic nailed		108	.148		6.05	5.45		11.50	15.10
1100	24" O.C.		115	.139		4.60	5.10		9.70	13
1105	24" O.C., pneumatic nailed		138	.116		4.60	4.26		8.86	11.70
1180	12' high, studs 12" O.C.		55	.291		8.65	10.70		19.35	26
1185	12" O.C., pneumatic nailed		66	.242		8.65	8.90		17.55	23.50
1200	16" O.C.		70	.229		6.90	8.40		15.30	20.50
1205	16" O.C., pneumatic nailed		84	.190		6.90	7		13.90	18.50
1300	24" O.C.		90	.178		5.20	6.50		11.70	15.85
1305	24" O.C., pneumatic nailed		108	.148		5.20	5.45		10.65	14.15
1400	For horizontal blocking, 2" x 4", add		600	.027		.37	.98		1.35	1.93
1500	2" x 6", add		600	.027		.58	.98		1.56	2.15
1600	For openings, add		250	.064			2.35		2.35	3.66
1700	Headers for above openings, material only, add				M.B.F.	630			630	695

06 11 10.28 Porch or Deck Framing

		Crew	Daily Output	Labor-Hours	Unit	Material	2007 Bare Costs Labor	Equipment	Total	Total Incl O&P
0010	**PORCH OR DECK FRAMING**									
0100	Treated lumber, posts or columns, 4" x 4"	2 Carp	390	.041	L.F.	1.41	1.51		2.92	3.89
0110	4" x 6"		275	.058		3.02	2.14		5.16	6.65
0120	4" x 8"		220	.073		3.33	2.67		6	7.80
0130	Girder, single, 4" x 4"		675	.024		1.41	.87		2.28	2.90
0140	4" x 6"		600	.027		3.02	.98		4	4.84
0150	4" x 8"		525	.030		3.33	1.12		4.45	5.40
0160	Double, 2" x 4"		625	.026		1.07	.94		2.01	2.64
0170	2" x 6"		600	.027		1.69	.98		2.67	3.38
0180	2" x 8"		575	.028		2.34	1.02		3.36	4.17
0190	2" x 10"		550	.029		3.17	1.07		4.24	5.15
0200	2" x 12"		525	.030		3.75	1.12		4.87	5.85
0210	Triple, 2" x 4"		575	.028		1.60	1.02		2.62	3.35
0220	2" x 6"		550	.029		2.54	1.07		3.61	4.45

06 11 Wood Framing

06 11 10 – Wood Framing

06 11 10.28 Porch or Deck Framing		Crew	Daily Output	Labor-Hours	Unit	Material	2007 Bare Costs Labor	Equipment	Total	Total Incl O&P
0230	2" x 8"	2 Carp	525	.030	L.F.	3.51	1.12		4.63	5.60
0240	2" x 10"		500	.032		4.76	1.17		5.93	7.10
0250	2" x 12"		475	.034		5.65	1.24		6.89	8.10
0260	Ledger, bolted 4' O.C., 2" x 4"		400	.040		.63	1.47		2.10	2.99
0270	2" x 6"		395	.041		.94	1.49		2.43	3.35
0280	2" x 8"		390	.041		1.26	1.51		2.77	3.73
0300	2" x 12"		380	.042		1.96	1.55		3.51	4.56
0310	Joists, 2" x 4"		1250	.013		.53	.47		1	1.32
0320	2" x 6"		1250	.013		.84	.47		1.31	1.66
0330	2" x 8"		1100	.015		1.17	.53		1.70	2.11
0340	2" x 10"		900	.018		1.58	.65		2.23	2.76
0350	2" x 12"		875	.018		1.90	.67		2.57	3.14
0360	Railings and trim, 1" x 4"	1 Carp	300	.027		.61	.98		1.59	2.19
0370	2" x 2"		300	.027		.30	.98		1.28	1.85
0380	2" x 4"		300	.027		.52	.98		1.50	2.10
0390	2" x 6"		300	.027		.83	.98		1.81	2.43
0400	Decking, 1" x 4"		275	.029	S.F.	1.91	1.07		2.98	3.76
0410	2" x 4"		300	.027		1.79	.98		2.77	3.49
0420	2" x 6"		320	.025		1.81	.92		2.73	3.42
0430	5/4" x 6"		320	.025		1.29	.92		2.21	2.85
0440	Balusters, square, 2" x 2"	2 Carp	660	.024	L.F.	.30	.89		1.19	1.72
0450	Turned, 2" x 2"		420	.038		.40	1.40		1.80	2.62
0460	Stair stringer, 2" x 10"		130	.123		1.58	4.52		6.10	8.80
0470	2" x 12"		130	.123		1.90	4.52		6.42	9.15
0480	Stair treads, 1" x 4"		140	.114		.64	4.19		4.83	7.25
0490	2" x 4"		140	.114		.53	4.19		4.72	7.15
0500	2" x 6"		160	.100		.80	3.67		4.47	6.60
0510	5/4" x 6"		160	.100		.60	3.67		4.27	6.35
0520	Turned handrail post, 4" x 4"		64	.250	Ea.	13.80	9.20		23	29.50
0530	Lattice panel, 4' x 8'		1600	.010	S.F.	.74	.37		1.11	1.38
0540	Cedar, posts or columns, 4" x 4"		390	.041	L.F.	3.02	1.51		4.53	5.65
0550	4" x 6"		275	.058		4.59	2.14		6.73	8.40
0560	4" x 8"		220	.073		6.90	2.67		9.57	11.70
0800	Decking, 1" x 4"		550	.029		1.43	1.07		2.50	3.24
0810	2" x 4"		600	.027		2.82	.98		3.80	4.62
0820	2" x 6"		640	.025		4.20	.92		5.12	6.05
0830	5/4" x 6"		640	.025		3.29	.92		4.21	5.05
0840	Railings and trim, 1" x 4"		600	.027		1.43	.98		2.41	3.10
0860	2" x 4"		600	.027		2.82	.98		3.80	4.62
0870	2" x 6"		600	.027		4.20	.98		5.18	6.15
0920	Stair treads, 1" x 4"		140	.114		1.43	4.19		5.62	8.15
0930	2" x 4"		140	.114		2.82	4.19		7.01	9.65
0940	2" x 6"		160	.100		4.20	3.67		7.87	10.30
0950	5/4" x 6"		160	.100		3.29	3.67		6.96	9.30
0980	Redwood, posts or columns, 4" x 4"		390	.041		6.50	1.51		8.01	9.50
0990	4" x 6"		275	.058		10.85	2.14		12.99	15.30
1000	4" x 8"		220	.073		19	2.67		21.67	25
1240	Redwood decking, 1" x 4"	1 Carp	275	.029	S.F.	5.10	1.07		6.17	7.25
1260	2" x 6"		340	.024		12.70	.86		13.56	15.30
1270	5/4" x 6"		320	.025		8.20	.92		9.12	10.45
1280	Railings and trim, 1" x 4"	2 Carp	600	.027	L.F.	1.50	.98		2.48	3.17
1310	2" x 6"		600	.027		5.85	.98		6.83	7.90
1420	Alternative decking, wood / plastic composite, 5/4" x 6"		640	.025		2.25	.92		3.17	3.90

06 11 Wood Framing

06 11 10 – Wood Framing

06 11 10.28 Porch or Deck Framing

		Crew	Daily Output	Labor-Hours	Unit	Material	2007 Bare Costs Labor	Equipment	Total	Total Incl O&P
1430	Vinyl, 1-1/2" x 5-1/2"	2 Carp	640	.025	L.F.	3.73	.92		4.65	5.55
1440	1" x 4" square edge fir		550	.029		1.37	1.07		2.44	3.16
1450	1" x 4" tongue and groove fir		450	.036		1.31	1.31		2.62	3.47
1460	1" x 4" mahogany		550	.029		1.31	1.07		2.38	3.10
1470	Accessories, joist hangers, 2" x 4"	1 Carp	160	.050	Ea.	.71	1.84		2.55	3.64
1480	2" x 6" through 2" x 12"	"	150	.053		.81	1.96		2.77	3.94
1530	Post footing, incl excav, backfill, tube form & concrete, 4' deep, 8" dia	F-7	12	2.667		11.45	87.50		98.95	149
1540	10" diameter		11	2.909		16.50	95		111.50	166
1550	12" diameter		10	3.200		22	105		127	187

06 11 10.30 Roof Framing

		Crew	Daily Output	Labor-Hours	Unit	Material	2007 Bare Costs Labor	Equipment	Total	Total Incl O&P
0010	**ROOF FRAMING**									
5250	Composite rafter, 9-1/2" deep	2 Carp	575	.028	L.F.	1.86	1.02		2.88	3.64
5260	11-1/2" deep		575	.028	"	1.98	1.02		3	3.76
6070	Fascia boards, 2" x 8"		.30	53.333	M.B.F.	630	1,950		2,580	3,750
6080	2" x 10"		.30	53.333		715	1,950		2,665	3,825
7000	Rafters, to 4 in 12 pitch, 2" x 6"		1	16		575	585		1,160	1,550
7060	2" x 8"		1.26	12.698		630	465		1,095	1,425
7300	Hip and valley rafters, 2" x 6"		.76	21.053		575	775		1,350	1,825
7360	2" x 8"		.96	16.667		630	610		1,240	1,650
7540	Hip and valley jacks, 2" x 6"		.60	26.667		575	980		1,555	2,150
7600	2" x 8"		.65	24.615		630	905		1,535	2,100
7780	For slopes steeper than 4 in 12, add						30%			
7790	For dormers or complex roofs, add						50%			
7800	Rafter tie, 1" x 4", #3	2 Carp	.27	59.259	M.B.F.	1,175	2,175		3,350	4,675
7820	Ridge board, #2 or better, 1" x 6" (CN)		.30	53.333		1,400	1,950		3,350	4,575
7840	1" x 8"		.37	43.243		1,325	1,575		2,900	3,925
7860	1" x 10"		.42	38.095		1,550	1,400		2,950	3,900
7880	2" x 6"		.50	32		575	1,175		1,750	2,450
7900	2" x 8"		.60	26.667		630	980		1,610	2,225
7920	2" x 10"		.66	24.242		715	890		1,605	2,150
7940	Roof cants, split, 4" x 4"		.86	18.605		1,250	685		1,935	2,450
7960	6" x 6"		1.80	8.889		1,600	325		1,925	2,275
7980	Roof curbs, untreated, 2" x 6"		.52	30.769		575	1,125		1,700	2,375
8000	2" x 12"		.80	20		715	735		1,450	1,925

06 11 10.32 Sill and Ledger Framing

		Crew	Daily Output	Labor-Hours	Unit	Material	2007 Bare Costs Labor	Equipment	Total	Total Incl O&P
0010	**SILL AND LEDGER FRAMING**									
4482	Ledgers, nailed, 2" x 4"	2 Carp	.50	32	M.B.F.	560	1,175		1,735	2,450
4484	2" x 6"		.60	26.667		575	980		1,555	2,150
4486	Bolted, not including bolts, 3" x 8"		.65	24.615		1,300	905		2,205	2,825
4488	3" x 12"		.70	22.857		1,300	840		2,140	2,750
4490	Mud sills, redwood, construction grade, 2" x 4"		.59	27.119		3,400	995		4,395	5,300
4492	2" x 6"		.78	20.513		3,400	755		4,155	4,925
4500	Sills, 2" x 4"		.40	40		560	1,475		2,035	2,900
4520	2" x 6"		.55	29.091		575	1,075		1,650	2,300
4540	2" x 8"		.67	23.881		630	875		1,505	2,075
4600	Treated, 2" x 4"		.36	44.444		785	1,625		2,410	3,425
4620	2" x 6"		.50	32		830	1,175		2,005	2,750
4640	2" x 8"		.60	26.667		860	980		1,840	2,475
4700	4" x 4"		.60	26.667		1,050	980		2,030	2,675
4720	4" x 6"		.70	22.857		1,500	840		2,340	2,950
4740	4" x 8"		.80	20		1,225	735		1,960	2,500
4760	4" x 10"		.87	18.391		1,225	675		1,900	2,400

06 11 Wood Framing

06 11 10 – Wood Framing

06 11 10.34 Sleepers		Crew	Daily Output	Labor-Hours	Unit	Material	2007 Bare Costs Labor	Equipment	Total	Total Incl O&P
0010	**SLEEPERS**									
0300	On concrete, treated, 1" x 2"	2 Carp	.39	41.026	M.B.F.	1,225	1,500		2,725	3,700
0320	1" x 3"		.50	32		1,250	1,175		2,425	3,200
0340	2" x 4"		.99	16.162		785	595		1,380	1,800
0360	2" x 6"		1.30	12.308		830	450		1,280	1,625

06 11 10.36 Soffit and Canopy Framing		Crew	Daily Output	Labor-Hours	Unit	Material	Labor	Equipment	Total	Total Incl O&P
0010	**SOFFIT AND CANOPY FRAMING**									
1300	Canopy or soffit framing, 1" x 4"	2 Carp	.30	53.333	M.B.F.	1,400	1,950		3,350	4,575
1340	1" x 8"		.50	32		1,325	1,175		2,500	3,275
1360	2" x 4"		.41	39.024		560	1,425		1,985	2,850
1400	2" x 8"		.67	23.881		630	875		1,505	2,075
1420	3" x 4"		.50	32		1,100	1,175		2,275	3,025
1460	3" x 8"		.60	26.667		1,300	980		2,280	2,950

06 11 10.38 Treated Lumber Framing Material		Crew	Daily Output	Labor-Hours	Unit	Material	Labor	Equipment	Total	Total Incl O&P
0010	**TREATED LUMBER FRAMING MATERIAL**									
0100	2" x 4"				M.B.F.	785			785	865
0110	2" x 6"					830			830	915
0120	2" x 8"					860			860	950
0130	2" x 10"					935			935	1,025
0140	2" x 12"					925			925	1,025
0200	4" x 4"					1,050			1,050	1,150
0210	4" x 6"					1,500			1,500	1,650
0220	4" x 8"					1,225			1,225	1,350

06 11 10.40 Wall Framing			Crew	Daily Output	Labor-Hours	Unit	Material	Labor	Equipment	Total	Total Incl O&P
0010	**WALL FRAMING**	R061110-30									
5860	Headers over openings, 2" x 6"		2 Carp	.36	44.444	M.B.F.	575	1,625		2,200	3,175
5865	2" x 6", pneumatic nailed			.43	37.209		575	1,375		1,950	2,750
5880	2" x 8"			.45	35.556		630	1,300		1,930	2,725
5885	2" x 8", pneumatic nailed			.54	29.630		630	1,075		1,705	2,400
5900	2" x 10"			.53	30.189		715	1,100		1,815	2,500
5905	2" x 10", pneumatic nailed			.67	23.881		715	875		1,590	2,150
5920	2" x 12"			.60	26.667		715	980		1,695	2,300
5925	2" x 12", pneumatic nailed			.72	22.222		715	815		1,530	2,050
5940	4" x 12"			.76	21.053		1,275	775		2,050	2,600
5945	4" x 12", pneumatic nailed			.92	17.391		1,275	640		1,915	2,400
5960	6" x 12"			.84	19.048		1,525	700		2,225	2,775
5965	6" x 12", pneumatic nailed			1.01	15.873		1,525	585		2,110	2,575
6000	Plates, untreated, 2" x 3"			.43	37.209		595	1,375		1,970	2,775
6005	2" x 3", pneumatic nailed			.52	30.769		595	1,125		1,720	2,400
6020	2" x 4"	CN		.53	30.189		560	1,100		1,660	2,350
6025	2" x 4", pneumatic nailed			.67	23.881		560	875		1,435	2,000
6040	2" x 6"			.75	21.333		575	785		1,360	1,850
6045	2" x 6", pneumatic nailed			.90	17.778		575	650		1,225	1,650
6120	Studs, 8' high wall, 2" x 3"			.60	26.667		595	980		1,575	2,175
6125	2" x 3", pneumatic nailed			.72	22.222		595	815		1,410	1,925
6140	2" x 4"			.92	17.391		560	640		1,200	1,625
6145	2" x 4", pneumatic nailed			1.10	14.493		560	530		1,090	1,450
6160	2" x 6"			1	16		575	585		1,160	1,550
6165	2" x 6", pneumatic nailed			1.20	13.333		575	490		1,065	1,400
6180	3" x 4"			.80	20		1,100	735		1,835	2,350
6185	3" x 4", pneumatic nailed			.96	16.667		1,100	610		1,710	2,150

06 11 Wood Framing

06 11 10 – Wood Framing

06 11 10.40 Wall Framing

		Daily Output	Labor-Hours	Unit	Material	2007 Bare Costs Labor	Equipment	Total	Total Incl O&P
		Crew							
8200	For 12' high walls, deduct					5%			
8220	For stub wall, 6' high, add					20%			
8240	3' high, add					40%			
8250	For second story & above, add					5%			
8300	For dormer & gable, add					15%			

06 11 10.42 Furring

		Crew	Daily Output	Labor-Hours	Unit	Material	Labor	Equipment	Total	Total Incl O&P
0010	**FURRING**									
0012	Wood strips, 1" x 2", on walls, on wood	1 Carp	550	.015	L.F.	.12	.53		.65	.96
0015	On wood, pneumatic nailed		710	.011		.12	.41		.53	.77
0300	On masonry		495	.016		.12	.59		.71	1.05
0400	On concrete		260	.031		.12	1.13		1.25	1.89
0600	1" x 3", on walls, on wood		550	.015		.22	.53		.75	1.08
0605	On wood, pneumatic nailed		710	.011		.22	.41		.63	.89
0700	On masonry		495	.016		.22	.59		.81	1.17
0800	On concrete		260	.031		.22	1.13		1.35	2.01
0850	On ceilings, on wood		350	.023		.22	.84		1.06	1.56
0855	On wood, pneumatic nailed		450	.018		.22	.65		.87	1.27
0900	On masonry		320	.025		.22	.92		1.14	1.68
0950	On concrete		210	.038		.22	1.40		1.62	2.43

06 11 10.44 Grounds

		Crew	Daily Output	Labor-Hours	Unit	Material	Labor	Equipment	Total	Total Incl O&P
0010	**GROUNDS**									
0020	For casework, 1" x 2" wood strips, on wood	1 Carp	330	.024	L.F.	.12	.89		1.01	1.52
0100	On masonry		285	.028		.12	1.03		1.15	1.73
0200	On concrete		250	.032		.12	1.17		1.29	1.96
0400	For plaster, 3/4" deep, on wood		450	.018		.12	.65		.77	1.15
0500	On masonry		225	.036		.12	1.31		1.43	2.16
0600	On concrete		175	.046		.12	1.68		1.80	2.74
0700	On metal lath		200	.040		.12	1.47		1.59	2.42

06 12 Structural Panels

06 12 10 – Structural Insulated Panels

06 12 10.10 Structural Insulated Panels

		Crew	Daily Output	Labor-Hours	Unit	Material	Labor	Equipment	Total	Total Incl O&P
0010	**STRUCTURAL INSULATED PANELS**									
0100	Structural insul. panels, 7/16" OSB both faces, EPS insul, 3-5/8" T	F-3	2075	.019	S.F.	3.71	.72	.35	4.78	5.55
0110	5-5/8" thick		1725	.023		4.25	.87	.42	5.54	6.50
0120	7-3/8" thick		1425	.028		4.75	1.05	.51	6.31	7.45
0130	9-3/8" thick		1125	.036		5.30	1.33	.64	7.27	8.55
0140	7/16" OSB one face, EPS insul, 3-5/8" thick		2175	.018		2.09	.69	.33	3.11	3.73
0150	5-5/8" thick		1825	.022		2.50	.82	.40	3.72	4.46
0160	7-3/8" thick		1525	.026		2.84	.98	.47	4.29	5.15
0170	9-3/8" thick		1225	.033		3.25	1.22	.59	5.06	6.10
0180	11-3/8" thick		925	.043		3.48	1.61	.78	5.87	7.20
0190	7/16" OSB - 1/2" GWB faces, EPS insul, 3-5/8" T		2075	.019		3.25	.72	.35	4.32	5.05
0200	5-5/8" thick		1725	.023		3.65	.87	.42	4.94	5.80
0210	7-3/8" thick		1425	.028		4	1.05	.51	5.56	6.60
0220	9-3/8" thick		1125	.036		4.40	1.33	.64	6.37	7.60
0230	11-3/8" thick		825	.048		4.68	1.81	.88	7.37	8.90
0240	7/16" OSB - 1/2" MRGWB faces, EPS insul, 3-5/8" T		2075	.019		3.25	.72	.35	4.32	5.05
0250	5-5/8" thick		1725	.023		3.65	.87	.42	4.94	5.80
0260	7-3/8" thick		1425	.028		4	1.05	.51	5.56	6.60

06 12 Structural Panels

06 12 10 – Structural Insulated Panels

06 12 10.10 Structural Insulated Panels		Crew	Daily Output	Labor-Hours	Unit	Material	2007 Bare Costs Labor	Equipment	Total	Total Incl O&P
0270	9-3/8" thick	F-3	1125	.036	S.F.	4.40	1.33	.64	6.37	7.60
0280	11-3/8" thick	▼	825	.048		4.68	1.81	.88	7.37	8.90
0300	For 1/2" GWB added to OSB skin, add					.60			.60	.66
0310	For 1/2" MRGWB added to OSB skin, add					.75			.75	.83
0320	For one T1-11 skin, add to OSB-OSB					.74			.74	.81
0330	For one 19/32" CDX skin, add to OSB-OSB				▼	.69			.69	.76

06 13 Heavy Timber

06 13 23 – Heavy Timber Construction

06 13 23.10 Heavy Framing

		Crew	Daily Output	Labor-Hours	Unit	Material	Labor	Equipment	Total	Total Incl O&P
0010	**HEAVY FRAMING**									
0020	Beams, single 6" x 10"	2 Carp	1.10	14.545	M.B.F.	2,150	535		2,685	3,200
0100	Single 8" x 16"		1.20	13.333		2,700	490		3,190	3,725
0200	Built from 2" lumber, multiple 2" x 14"		.90	17.778		935	650		1,585	2,050
0210	Built from 3" lumber, multiple 3" x 6"		.70	22.857		1,275	840		2,115	2,700
0220	Multiple 3" x 8"		.80	20		1,300	735		2,035	2,575
0230	Multiple 3" x 10"		.90	17.778		1,300	650		1,950	2,450
0240	Multiple 3" x 12"		1	16		1,300	585		1,885	2,375
0250	Built from 4" lumber, multiple 4" x 6"		.80	20		1,400	735		2,135	2,675
0260	Multiple 4" x 8"		.90	17.778		1,400	650		2,050	2,575
0270	Multiple 4" x 10"		1	16		1,650	585		2,235	2,750
0280	Multiple 4" x 12"		1.10	14.545		1,275	535		1,810	2,225
0290	Columns, structural grade, 1500f, 4" x 4"		.60	26.667		1,675	980		2,655	3,375
0300	6" x 6"		.65	24.615		1,450	905		2,355	3,000
0400	8" x 8"		.70	22.857		2,275	840		3,115	3,800
0500	10" x 10"		.75	21.333		2,150	785		2,935	3,600
0600	12" x 12"		.80	20		2,175	735		2,910	3,525
0800	Floor planks, 2" thick, T & G, 2" x 6"		1.05	15.238		1,875	560		2,435	2,950
0900	2" x 10"		1.10	14.545		1,875	535		2,410	2,900
1100	3" thick, 3" x 6"		1.05	15.238		1,350	560		1,910	2,350
1200	3" x 10"		1.10	14.545		1,350	535		1,885	2,300
1400	Girders, structural grade, 12" x 12"		.80	20		2,025	735		2,760	3,375
1500	10" x 16"	▼	1	16	▼	1,975	585		2,560	3,100
2050	Roof planks, see division 06 15 16.10									
2300	Roof purlins, 4" thick, structural grade	2 Carp	1.05	15.238	M.B.F.	1,375	560		1,935	2,400
2500	Roof trusses, add timber connectors, division 06090-800	"	.45	35.556	"	1,350	1,300		2,650	3,525

06 15 Wood Decking

06 15 16 – Wood Roof Decking

06 15 16.10 Wood Roof Decking

		Crew	Daily Output	Labor-Hours	Unit	Material	Labor	Equipment	Total	Total Incl O&P
0010	**WOOD ROOF DECKING**									
0020	For laminated decks, see division 06 15 23.10									
0200	For cementitious decks, see division 03 51 13.50									
0400	Cedar planks, 3" thick	2 Carp	320	.050	S.F.	6.40	1.84		8.24	9.90
0500	4" thick		250	.064		8.65	2.35		11	13.15
0700	Douglas fir, 3" thick		320	.050		2.40	1.84		4.24	5.50
0800	4" thick		250	.064		3.21	2.35		5.56	7.20
1000	Hemlock, 3" thick		320	.050		2.40	1.84		4.24	5.50
1100	4" thick	▼	250	.064	▼	3.20	2.35		5.55	7.20

06 15 Wood Decking

06 15 16 – Wood Roof Decking

06 15 16.10 Wood Roof Decking		Crew	Daily Output	Labor-Hours	Unit	Material	2007 Bare Costs Labor	Equipment	Total	Total Incl O&P
1300	Western white spruce, 3" thick	2 Carp	320	.050	S.F.	2.31	1.84		4.15	5.40
1400	4" thick	↓	250	.064	↓	3.08	2.35		5.43	7.05

06 15 23 – Laminated Wood Decking

06 15 23.10 Laminated Roof Deck

0010	**LAMINATED ROOF DECK**									
0020	Pine or hemlock, 3" thick	2 Carp	425	.038	S.F.	3.20	1.38		4.58	5.65
0100	4" thick		325	.049		4.26	1.81		6.07	7.50
0300	Cedar, 3" thick		425	.038		3.92	1.38		5.30	6.45
0400	4" thick		325	.049		4.99	1.81		6.80	8.30
0600	Fir, 3" thick		425	.038		3.26	1.38		4.64	5.75
0700	4" thick	↓	325	.049	↓	4.08	1.81		5.89	7.30

06 16 Sheathing

06 16 23 – Subflooring

06 16 23.10 Subfloor

0010	**SUBFLOOR** R061636-20									
0011	Plywood, CDX, 1/2" thick	2 Carp	1500	.011	SF Flr.	.47	.39		.86	1.13
0015	Pneumatic nailed		1860	.009		.47	.32		.79	1.01
0100	5/8" thick		1350	.012		.59	.43		1.02	1.33
0105	Pneumatic nailed		1674	.010		.59	.35		.94	1.20
0200	3/4" thick		1250	.013		.71	.47		1.18	1.51
0205	Pneumatic nailed		1550	.010		.71	.38		1.09	1.37
0300	1-1/8" thick, 2-4-1 including underlayment		1050	.015		1.54	.56		2.10	2.56
0450	1" x 8" S4S, laid regular		1000	.016		1.28	.59		1.87	2.32
0460	Laid diagonal		850	.019		1.28	.69		1.97	2.49
0500	1" x 10" S4S, laid regular		1100	.015		1.07	.53		1.60	2
0600	Laid diagonal	↓	900	.018	↓	1.07	.65		1.72	2.19

06 16 26 – Underlayment

06 16 26.10 Underlayment

0010	**UNDERLAYMENT** R061636-20									
0030	Plywood, underlayment grade, 3/8" thick	2 Carp	1500	.011	SF Flr.	.78	.39		1.17	1.47
0070	Pneumatic nailed		1860	.009		.78	.32		1.10	1.35
0100	1/2" thick		1450	.011		.90	.40		1.30	1.62
0105	Pneumatic nailed		1798	.009		.90	.33		1.23	1.50
0200	5/8" thick		1400	.011		1.09	.42		1.51	1.85
0205	Pneumatic nailed		1736	.009		1.09	.34		1.43	1.73
0300	3/4" thick		1300	.012		1.11	.45		1.56	1.92
0305	Pneumatic nailed		1612	.010		1.11	.36		1.47	1.79
0500	Particle board, 3/8" thick		1500	.011		.33	.39		.72	.97
0505	Pneumatic nailed		1860	.009		.33	.32		.65	.85
0600	1/2" thick		1450	.011		.37	.40		.77	1.04
0605	Pneumatic nailed		1798	.009		.37	.33		.70	.92
0800	5/8" thick		1400	.011		.41	.42		.83	1.10
0805	Pneumatic nailed		1736	.009		.41	.34		.75	.98
0900	3/4" thick		1300	.012		.64	.45		1.09	1.40
0905	Pneumatic nailed		1612	.010		.64	.36		1	1.27
1100	Hardboard, underlayment grade, 4' x 4', .215" thick	↓	1500	.011	↓	.39	.39		.78	1.04

06 16 Sheathing

06 16 36 – Wood Panel Product Sheathing

06 16 36.10 Sheathing		Crew	Daily Output	Labor-Hours	Unit	Material	2007 Bare Costs Labor	Equipment	Total	Total Incl O&P
0010	**SHEATHING**, plywood on roofs R061636-20									
0030	5/16" thick	2 Carp	1600	.010	S.F.	.62	.37		.99	1.25
0035	Pneumatic nailed R061110-30		1952	.008		.62	.30		.92	1.15
0050	3/8" thick		1525	.010		.44	.39		.83	1.08
0055	Pneumatic nailed		1860	.009		.44	.32		.76	.97
0100	1/2" thick **CN**		1400	.011		.47	.42		.89	1.17
0105	Pneumatic nailed		1708	.009		.47	.34		.81	1.06
0200	5/8" thick		1300	.012		.59	.45		1.04	1.35
0205	Pneumatic nailed		1586	.010		.59	.37		.96	1.23
0300	3/4" thick		1200	.013		.71	.49		1.20	1.54
0305	Pneumatic nailed		1464	.011		.71	.40		1.11	1.40
0500	Plywood on walls with exterior CDX, 3/8" thick		1200	.013		.44	.49		.93	1.24
0505	Pneumatic nailed		1488	.011		.44	.39		.83	1.09
0600	1/2" thick		1125	.014		.47	.52		.99	1.33
0605	Pneumatic nailed		1395	.011		.47	.42		.89	1.18
0700	5/8" thick		1050	.015		.59	.56		1.15	1.52
0705	Pneumatic nailed		1302	.012		.59	.45		1.04	1.35
0800	3/4" thick		975	.016		.71	.60		1.31	1.72
0805	Pneumatic nailed	↓	1209	.013	↓	.71	.49		1.20	1.54
1000	For shear wall construction, add						20%			
1200	For structural 1 exterior plywood, add				S.F.	10%				
1400	With boards, on roof 1" x 6" boards, laid horizontal	2 Carp	725	.022		1.42	.81		2.23	2.82
1500	Laid diagonal		650	.025		1.42	.90		2.32	2.97
1700	1" x 8" boards, laid horizontal		875	.018		1.28	.67		1.95	2.46
1800	Laid diagonal	↓	725	.022		1.28	.81		2.09	2.67
2000	For steep roofs, add						40%			
2200	For dormers, hips and valleys, add					5%	50%			
2400	Boards on walls, 1" x 6" boards, laid regular	2 Carp	650	.025		1.42	.90		2.32	2.97
2500	Laid diagonal		585	.027		1.42	1		2.42	3.12
2700	1" x 8" boards, laid regular		765	.021		1.28	.77		2.05	2.61
2800	Laid diagonal		650	.025		1.28	.90		2.18	2.82
2850	Gypsum, weatherproof, 1/2" thick		1125	.014		.62	.52		1.14	1.49
2900	With embedded glass mats		1100	.015		.52	.53		1.05	1.40
3000	Wood fiber, regular, no vapor barrier, 1/2" thick		1200	.013		.55	.49		1.04	1.37
3100	5/8" thick		1200	.013		.72	.49		1.21	1.55
3300	No vapor barrier, in colors, 1/2" thick		1200	.013		.78	.49		1.27	1.62
3400	5/8" thick		1200	.013		.96	.49		1.45	1.82
3600	With vapor barrier one side, white, 1/2" thick		1200	.013		.55	.49		1.04	1.37
3700	Vapor barrier 2 sides, 1/2" thick		1200	.013		.77	.49		1.26	1.61
3800	Asphalt impregnated, 25/32" thick		1200	.013		.28	.49		.77	1.07
3850	Intermediate, 1/2" thick	↓	1200	.013	↓	.18	.49		.67	.96

06 17 Shop-Fabricated Structural Wood

06 17 33 – Wood I-Joists

06 17 33.10 Wood I-Joists

		Crew	Daily Output	Labor-Hours	Unit	Material	2007 Bare Costs Labor	Equipment	Total	Total Incl O&P
0010	**WOOD I-JOISTS**									
0100	Plywood webs, incl. bridging & blocking, panels 24" O.C.									
1200	15' to 24' span, 50 psf live load	F-5	2400	.013	SF Flr.	2.12	.50		2.62	3.10
1300	55 psf live load		2250	.014		2.25	.53		2.78	3.30
1400	24' to 30' span, 45 psf live load		2600	.012		2.65	.46		3.11	3.63
1500	55 psf live load		2400	.013		2.96	.50		3.46	4.02
1600	Tubular steel open webs, 45 psf, 24" O.C., 40' span	F-3	6250	.006		2.13	.24	.12	2.49	2.84
1700	55' span		7750	.005		2.07	.19	.09	2.35	2.68
1800	70' span		9250	.004		2.69	.16	.08	2.93	3.30
1900	85 psf live load, 26' span		2300	.017		2.50	.65	.31	3.46	4.10

06 17 53 – Shop-Fabricated Wood Trusses

06 17 53.10 Roof Trusses

		Crew	Daily Output	Labor-Hours	Unit	Material	Labor	Equipment	Total	Total Incl O&P
0010	**ROOF TRUSSES**									
0020	For timber connectors, see div. 06 05 23.60									
0100	Fink (W) or King post type, 2'-0" O.C.									
0200	Metal plate connected, 4 in 12 slope									
0210	24' to 29' span	F-3	3000	.013	SF Flr.	1.84	.50	.24	2.58	3.06
0300	30' to 43' span		3000	.013		2.04	.50	.24	2.78	3.28
0400	44' to 60' span		3000	.013		2.25	.50	.24	2.99	3.52
0700	Glued and nailed, add					50%				

06 18 Glued-Laminated Construction

06 18 13 – Glued-Laminated Beams

06 18 13.20 Laminated Framing

		Crew	Daily Output	Labor-Hours	Unit	Material	Labor	Equipment	Total	Total Incl O&P
0010	**LAMINATED FRAMING**									
0020	30 lb., short term live load, 15 lb. dead load									
0200	Straight roof beams, 20' clear span, beams 8' O.C.	F-3	2560	.016	SF Flr.	1.89	.58	.28	2.75	3.29
0300	Beams 16' O.C.		3200	.013		1.36	.47	.23	2.06	2.47
0500	40' clear span, beams 8' O.C.		3200	.013		3.60	.47	.23	4.30	4.93
0600	Beams 16' O.C.		3840	.010		2.94	.39	.19	3.52	4.04
0800	60' clear span, beams 8' O.C.	F-4	2880	.017		6.20	.61	.37	7.18	8.15
0900	Beams 16' O.C.	"	3840	.013		4.61	.46	.28	5.35	6.05
1100	Tudor arches, 30' to 40' clear span, frames 8' O.C.	F-3	1680	.024		8.05	.89	.43	9.37	10.75
1200	Frames 16' O.C.	"	2240	.018		6.30	.67	.32	7.29	8.35
1400	50' to 60' clear span, frames 8' O.C.	F-4	2200	.022		8.70	.80	.48	9.98	11.35
1500	Frames 16' O.C.		2640	.018		7.40	.67	.40	8.47	9.60
1700	Radial arches, 60' clear span, frames 8' O.C.		1920	.025		8.15	.92	.55	9.62	11
1800	Frames 16' O.C.		2880	.017		6.25	.61	.37	7.23	8.25
2000	100' clear span, frames 8' O.C.		1600	.030		8.45	1.10	.66	10.21	11.70
2100	Frames 16' O.C.		2400	.020		7.40	.74	.44	8.58	9.75
2300	120' clear span, frames 8' O.C.		1440	.033		11.20	1.22	.74	13.16	15
2400	Frames 16' O.C.		1920	.025		10.25	.92	.55	11.72	13.30
2600	Bowstring trusses, 20' O.C., 40' clear span	F-3	2400	.017		5.05	.62	.30	5.97	6.85
2700	60' clear span	F-4	3600	.013		4.53	.49	.29	5.31	6.05
2800	100' clear span		4000	.012		6.40	.44	.27	7.11	8
2900	120' clear span		3600	.013		6.90	.49	.29	7.68	8.65
3000	For less than 1000 B.F., add					20%				
3050	For over 5000 B.F., deduct					10%				
3100	For premium appearance, add to S.F. prices					5%				
3300	For industrial type, deduct					15%				

06 18 Glued-Laminated Construction

06 18 13 – Glued-Laminated Beams

06 18 13.20 Laminated Framing		Crew	Daily Output	Labor-Hours	Unit	Material	2007 Bare Costs Labor	Equipment	Total	Total Incl O&P
3500	For stain and varnish, add				SF Flr.	5%				
3900	For 3/4" laminations, add to straight					25%				
4100	Add to curved					15%				
4300	Alternate pricing method: (use nominal footage of									
4310	components). Straight beams, camber less than 6"	F-3	3.50	11.429	M.B.F.	2,800	425	207	3,432	3,975
4400	Columns, including hardware		2	20		3,000	745	360	4,105	4,850
4600	Curved members, radius over 32'		2.50	16		3,075	595	290	3,960	4,625
4700	Radius 10' to 32'		3	13.333		3,050	500	241	3,791	4,375
4900	For complicated shapes, add maximum					100%				
5100	For pressure treating, add to straight					35%				
5200	Add to curved					45%				
6000	Laminated veneer members, southern pine or western species									
6050	1-3/4" wide x 5-1/2" deep	2 Carp	480	.033	L.F.	3.20	1.22		4.42	5.40
6100	9-1/2" deep		480	.033		4.56	1.22		5.78	6.90
6150	14" deep		450	.036		7.20	1.31		8.51	9.95
6200	18" deep		450	.036		9.70	1.31		11.01	12.70
6300	Parallel strand members, southern pine or western species									
6350	1-3/4" wide x 9-1/4" deep	2 Carp	480	.033	L.F.	4.25	1.22		5.47	6.60
6400	11-1/4" deep		450	.036		5.20	1.31		6.51	7.75
6450	14" deep		400	.040		6.25	1.47		7.72	9.20
6500	3-1/2" wide x 9-1/4" deep		480	.033		10	1.22		11.22	12.90
6550	11-1/4" deep		450	.036		12.50	1.31		13.81	15.80
6600	14" deep		400	.040		15.20	1.47		16.67	19
6650	7" wide x 9-1/4" deep		450	.036		21	1.31		22.31	25
6700	11-1/4" deep		420	.038		25	1.40		26.40	29.50
6750	14" deep		400	.040		27	1.47		28.47	32

06 22 Millwork

06 22 13 – Standard Pattern Wood Trim

06 22 13.15 Moldings, Base

		Crew	Daily Output	Labor-Hours	Unit	Material	Labor	Equipment	Total	Total Incl O&P
0010	**MOLDINGS, BASE**									
0500	Base, stock pine, 9/16" x 3-1/2"	1 Carp	240	.033	L.F.	2.21	1.22		3.43	4.33
0550	9/16" x 4-1/2"		200	.040		2.40	1.47		3.87	4.93
0561	Base shoe, oak, 3/4" x 1"		240	.033		1.13	1.22		2.35	3.14

06 22 13.30 Moldings, Casings

0010	**MOLDINGS, CASINGS**									
0090	Apron, stock pine, 5/8" x 2"	1 Carp	250	.032	L.F.	1.18	1.17		2.35	3.13
0110	5/8" x 3-1/2"		220	.036		1.83	1.33		3.16	4.09
0300	Band, stock pine, 11/16" x 1-1/8"		270	.030		.79	1.09		1.88	2.56
0350	11/16" x 1-3/4"		250	.032		1.12	1.17		2.29	3.06
0700	Casing, stock pine, 11/16" x 2-1/2"		240	.033		1.35	1.22		2.57	3.39
0750	11/16" x 3-1/2"		215	.037		1.83	1.37		3.20	4.14

06 22 13.35 Moldings, Ceilings

0010	**MOLDINGS, CEILINGS**									
0600	Bed, stock pine, 9/16" x 1-3/4"	1 Carp	270	.030	L.F.	.99	1.09		2.08	2.78
0650	9/16" x 2"		240	.033		1.19	1.22		2.41	3.21
1200	Cornice molding, stock pine, 9/16" x 1-3/4"		330	.024		.98	.89		1.87	2.47
1300	9/16" x 2-1/4"		300	.027		1.21	.98		2.19	2.85
2400	Cove scotia, stock pine, 9/16" x 1-3/4"		270	.030		.98	1.09		2.07	2.77
2500	11/16" x 2-3/4"		255	.031		1.30	1.15		2.45	3.22

06 22 Millwork

06 22 13 – Standard Pattern Wood Trim

06 22 13.35 Moldings, Ceilings		Crew	Daily Output	Labor-Hours	Unit	Material	2007 Bare Costs Labor	Equipment	Total	Total Incl O&P
2600	Crown, stock pine, 9/16" x 3-5/8"	1 Carp	250	.032	L.F.	1.86	1.17		3.03	3.88
2700	11/16" x 4-5/8"	↓	220	.036	↓	3.11	1.33		4.44	5.50

06 22 13.40 Moldings, Exterior

0010	**MOLDINGS, EXTERIOR**									
1500	Cornice, boards, pine, 1" x 2"	1 Carp	330	.024	L.F.	.32	.89		1.21	1.74
1700	1" x 6"		250	.032		1.07	1.17		2.24	3.01
2000	1" x 12"		180	.044		1.89	1.63		3.52	4.62
2200	Three piece, built-up, pine, minimum		80	.100		2.01	3.67		5.68	7.90
2300	Maximum		65	.123		5	4.52		9.52	12.55
3000	Corner board, sterling pine, 1" x 4"		200	.040		.73	1.47		2.20	3.09
3100	1" x 6"		200	.040		1.59	1.47		3.06	4.04
3350	Fascia, sterling pine, 1" x 6"		250	.032		1.59	1.17		2.76	3.58
3370	1" x 8"		225	.036		2.01	1.31		3.32	4.24
3400	Trim, exterior, sterling pine, back band		250	.032		.78	1.17		1.95	2.69
3500	Casing		250	.032		2.08	1.17		3.25	4.12
3600	Crown		250	.032		2	1.17		3.17	4.03
3700	Porch rail with balusters		22	.364		16.30	13.35		29.65	39
3800	Screen		395	.020		1.26	.74		2	2.55
4100	Verge board, sterling pine, 1" x 4"		200	.040		.83	1.47		2.30	3.20
4200	1" x 6"		200	.040		1.14	1.47		2.61	3.54
4300	2" x 6"		165	.048		1.86	1.78		3.64	4.82
4400	2" x 8"		165	.048	↓	2.46	1.78		4.24	5.50
4700	For redwood trim, add					200%				

06 22 13.45 Moldings, Trim

0010	**MOLDINGS, TRIM**									
0200	Astragal, stock pine, 11/16" x 1-3/4"	1 Carp	255	.031	L.F.	1.16	1.15		2.31	3.07
0250	1-5/16" x 2-3/16"		240	.033		2.71	1.22		3.93	4.88
0800	Chair rail, stock pine, 5/8" x 2-1/2"		270	.030		1.44	1.09		2.53	3.27
0900	5/8" x 3-1/2"		240	.033		2.20	1.22		3.42	4.32
1000	Closet pole, stock pine, 1-1/8" diameter		200	.040		.87	1.47		2.34	3.25
1100	Fir, 1-5/8" diameter		200	.040		1.60	1.47		3.07	4.05
3300	Half round, stock pine, 1/4" x 1/2"		270	.030		.23	1.09		1.32	1.94
3350	1/2" x 1"	↓	255	.031	↓	.60	1.15		1.75	2.45
3400	Handrail, fir, single piece, stock, hardware not included									
3450	1-1/2" x 1-3/4"	1 Carp	80	.100	L.F.	1.77	3.67		5.44	7.65
3470	Pine, 1-1/2" x 1-3/4"		80	.100		1.77	3.67		5.44	7.65
3500	1-1/2" x 2-1/2"		76	.105		1.63	3.86		5.49	7.80
3600	Lattice, stock pine, 1/4" x 1-1/8"		270	.030		.41	1.09		1.50	2.14
3700	1/4" x 1-3/4"		250	.032		.65	1.17		1.82	2.55
3800	Miscellaneous, custom, pine, 1" x 1"		270	.030		.34	1.09		1.43	2.06
3900	1" x 3"		240	.033		.69	1.22		1.91	2.66
4100	Birch or oak, nominal 1" x 1"		240	.033		.53	1.22		1.75	2.48
4200	Nominal 1" x 3"		215	.037		1.81	1.37		3.18	4.12
4400	Walnut, nominal 1" x 1"		215	.037		.87	1.37		2.24	3.09
4500	Nominal 1" x 3"		200	.040		2.61	1.47		4.08	5.15
4700	Teak, nominal 1" x 1"		215	.037		1.23	1.37		2.60	3.48
4800	Nominal 1" x 3"		200	.040		3.52	1.47		4.99	6.15
4900	Quarter round, stock pine, 1/4" x 1/4"		275	.029		.22	1.07		1.29	1.90
4950	3/4" x 3/4"		255	.031	↓	.64	1.15		1.79	2.49
5600	Wainscot moldings, 1-1/8" x 9/16", 2' high, minimum		76	.105	S.F.	10.75	3.86		14.61	17.80
5700	Maximum	↓	65	.123	"	19.75	4.52		24.27	28.50

06 22 Millwork

06 22 13 – Standard Pattern Wood Trim

06 22 13.50 Moldings, Window and Door

	06 22 13.50 Moldings, Window and Door	Crew	Daily Output	Labor-Hours	Unit	Material	2007 Bare Costs Labor	2007 Bare Costs Equipment	Total	Total Incl O&P
0010	**MOLDINGS, WINDOW AND DOOR**									
2800	Door moldings, stock, decorative, 1-1/8" wide, plain	1 Carp	17	.471	Set	46	17.25		63.25	77.50
2900	Detailed		17	.471	"	87.50	17.25		104.75	124
2960	Clear pine door jamb, no stops, 11/16" x 4-9/16"		240	.033	L.F.	3	1.22		4.22	5.20
3150	Door trim set, 1 head and 2 sides, pine, 2-1/2 wide		5.90	1.356	Opng.	23	50		73	103
3170	3-1/2" wide		5.30	1.509	"	31	55.50		86.50	121
3250	Glass beads, stock pine, 3/8" x 1/2"		275	.029	L.F.	.31	1.07		1.38	2
3270	3/8" x 7/8"		270	.030		.41	1.09		1.50	2.14
4850	Parting bead, stock pine, 3/8" x 3/4"		275	.029		.35	1.07		1.42	2.05
4870	1/2" x 3/4"		255	.031		.44	1.15		1.59	2.27
5000	Stool caps, stock pine, 11/16" x 3-1/2"		200	.040		2.20	1.47		3.67	4.71
5100	1-1/16" x 3-1/4"		150	.053		3.48	1.96		5.44	6.90
5300	Threshold, oak, 3' long, inside, 5/8" x 3-5/8"		32	.250	Ea.	9	9.20		18.20	24
5400	Outside, 1-1/2" x 7-5/8"		16	.500	"	37	18.35		55.35	69
5900	Window trim sets, including casings, header, stops,									
5910	stool and apron, 2-1/2" wide, minimum	1 Carp	13	.615	Opng.	30.50	22.50		53	69
5950	Average		10	.800		36	29.50		65.50	85
6000	Maximum		6	1.333		62.50	49		111.50	145

06 22 13.60 Moldings, Soffits

	06 22 13.60 Moldings, Soffits	Crew	Daily Output	Labor-Hours	Unit	Material	Labor	Equipment	Total	Total Incl O&P
0010	**MOLDINGS, SOFFITS**									
0200	Soffits, pine, 1" x 4"	2 Carp	420	.038	L.F.	.39	1.40		1.79	2.61
0210	1" x 6"		420	.038		.65	1.40		2.05	2.90
0220	1" x 8"		420	.038		.80	1.40		2.20	3.06
0230	1" x 10"		400	.040		.85	1.47		2.32	3.22
0240	1" x 12"		400	.040		1.10	1.47		2.57	3.49
0250	STK cedar, 1" x 4"		420	.038		.55	1.40		1.95	2.79
0260	1" x 6"		420	.038		.99	1.40		2.39	3.27
0270	1" x 8"		420	.038		1.29	1.40		2.69	3.60
0280	1" x 10"		400	.040		1.64	1.47		3.11	4.09
0290	1" x 12"		400	.040		1.98	1.47		3.45	4.47
1000	Exterior AC plywood, 1/4" thick		420	.038	S.F.	.70	1.40		2.10	2.95
1050	3/8" thick		420	.038		.78	1.40		2.18	3.04
1100	1/2" thick		420	.038		.90	1.40		2.30	3.17
1150	Polyvinyl chloride, white, solid	1 Carp	230	.035		.87	1.28		2.15	2.95
1160	Perforated	"	230	.035		.87	1.28		2.15	2.95

06 25 Prefinished Paneling

06 25 13 – Prefinished Hardboard Paneling

06 25 13.10 Paneling, Hardboard

	06 25 13.10 Paneling, Hardboard	Crew	Daily Output	Labor-Hours	Unit	Material	Labor	Equipment	Total	Total Incl O&P
0010	**PANELING, HARDBOARD**									
0050	Not incl. furring or trim, hardboard, tempered, 1/8" thick	2 Carp	500	.032	S.F.	.35	1.17		1.52	2.22
0100	1/4" thick		500	.032		.48	1.17		1.65	2.36
0300	Tempered pegboard, 1/8" thick		500	.032		.43	1.17		1.60	2.30
0400	1/4" thick		500	.032		.60	1.17		1.77	2.49
0600	Untempered hardboard, natural finish, 1/8" thick		500	.032		.36	1.17		1.53	2.23
0700	1/4" thick		500	.032		.38	1.17		1.55	2.25
0900	Untempered pegboard, 1/8" thick		500	.032		.35	1.17		1.52	2.22
1000	1/4" thick		500	.032		.39	1.17		1.56	2.26
1200	Plastic faced hardboard, 1/8" thick		500	.032		.59	1.17		1.76	2.48
1300	1/4" thick		500	.032		.78	1.17		1.95	2.69

06 25 Prefinished Paneling

06 25 13 – Prefinished Hardboard Paneling

06 25 13.10 Paneling, Hardboard

		Crew	Daily Output	Labor-Hours	Unit	Material	2007 Bare Costs Labor	Equipment	Total	Total Incl O&P
1500	Plastic faced pegboard, 1/8" thick	2 Carp	500	.032	S.F.	.56	1.17		1.73	2.45
1600	1/4" thick		500	.032		.69	1.17		1.86	2.59
1800	Wood grained, plain or grooved, 1/4" thick, minimum		500	.032		.52	1.17		1.69	2.40
1900	Maximum		425	.038		1.09	1.38		2.47	3.35
2100	Moldings for hardboard, wood or aluminum, minimum		500	.032	L.F.	.35	1.17		1.52	2.22
2200	Maximum		425	.038	"	.99	1.38		2.37	3.24

06 25 16 – Prefinished Plywood Paneling

06 25 16.10 Paneling, Plywood

		Crew	Daily Output	Labor-Hours	Unit	Material	Labor	Equipment	Total	Total Incl O&P
0010	**PANELING, PLYWOOD** R061636-20									
2400	Plywood, prefinished, 1/4" thick, 4' x 8' sheets									
2410	with vertical grooves. Birch faced, minimum	2 Carp	500	.032	S.F.	.85	1.17		2.02	2.77
2420	Average		420	.038		1.29	1.40		2.69	3.60
2430	Maximum		350	.046		1.89	1.68		3.57	4.69
2600	Mahogany, African		400	.040		2.41	1.47		3.88	4.94
2700	Philippine (Lauan)		500	.032		1.04	1.17		2.21	2.97
2900	Oak or Cherry, minimum		500	.032		2.02	1.17		3.19	4.05
3000	Maximum		400	.040		3.10	1.47		4.57	5.70
3200	Rosewood		320	.050		4.40	1.84		6.24	7.70
3400	Teak		400	.040		3.10	1.47		4.57	5.70
3600	Chestnut		375	.043		4.59	1.57		6.16	7.50
3800	Pecan		400	.040		1.98	1.47		3.45	4.47
3900	Walnut, minimum		500	.032		2.65	1.17		3.82	4.75
3950	Maximum		400	.040		5	1.47		6.47	7.80
4000	Plywood, prefinished, 3/4" thick, stock grades, minimum		320	.050		1.20	1.84		3.04	4.18
4100	Maximum		224	.071		5.15	2.62		7.77	9.80
4300	Architectural grade, minimum		224	.071		3.82	2.62		6.44	8.30
4400	Maximum		160	.100		5.85	3.67		9.52	12.10
4600	Plywood, "A" face, birch, V.C., 1/2" thick, natural		450	.036		1.81	1.31		3.12	4.02
4700	Select		450	.036		1.98	1.31		3.29	4.21
4900	Veneer core, 3/4" thick, natural		320	.050		1.91	1.84		3.75	4.96
5000	Select		320	.050		2.15	1.84		3.99	5.25
5200	Lumber core, 3/4" thick, natural		320	.050		2.87	1.84		4.71	6
5500	Plywood, knotty pine, 1/4" thick, A2 grade		450	.036		1.57	1.31		2.88	3.76
5600	A3 grade		450	.036		1.98	1.31		3.29	4.21
5800	3/4" thick, veneer core, A2 grade		320	.050		2.03	1.84		3.87	5.10
5900	A3 grade		320	.050		2.28	1.84		4.12	5.35
6100	Aromatic cedar, 1/4" thick, plywood		400	.040		2	1.47		3.47	4.49
6200	1/4" thick, particle board		400	.040		.97	1.47		2.44	3.36

06 25 26 – Panel Systems

06 25 26.10 Panel Systems

		Crew	Daily Output	Labor-Hours	Unit	Material	Labor	Equipment	Total	Total Incl O&P
0010	**PANEL SYSTEMS**									
0100	Raised panel, eng. wood core w/ wood veneer, std., paint grade	2 Carp	300	.053	S.F.	11.20	1.96		13.16	15.40
0110	Oak veneer		300	.053		18.60	1.96		20.56	23.50
0120	Maple veneer		300	.053		24	1.96		25.96	29
0130	Cherry veneer		300	.053		30	1.96		31.96	36
0300	Class I fire rated, paint grade		300	.053		13.40	1.96		15.36	17.80
0310	Oak veneer		300	.053		22.50	1.96		24.46	27.50
0320	Maple veneer		300	.053		28.50	1.96		30.46	34
0330	Cherry veneer		300	.053		36	1.96		37.96	42.50
0510	Beadboard, 5/8" MDF, standard, primed		300	.053		7.90	1.96		9.86	11.75
0520	Oak veneer, unfinished		300	.053		13.75	1.96		15.71	18.20
0530	Maple veneer, unfinished		300	.053		16.30	1.96		18.26	21

06 25 Prefinished Paneling

06 25 26 – Panel Systems

06 25 26.10 Panel Systems	Crew	Daily Output	Labor-Hours	Unit	Material	2007 Bare Costs Labor	Equipment	Total	Total Incl O&P
0610 Rustic paneling, 5/8" MDF, standard, maple veneer, unfinished	2 Carp	300	.053	S.F.	21	1.96		22.96	26
5000 For prefinished paneling, see division 06 25 16.10 & 06 25 13.10									

06 26 Board Paneling

06 26 13 – Profile Board Paneling

06 26 13.10 Paneling, Boards

		Crew	Daily Output	Labor-Hours	Unit	Material	Labor	Equipment	Total	Total Incl O&P
0010	**PANELING, BOARDS**									
6400	Wood board paneling, 3/4" thick, knotty pine	2 Carp	300	.053	S.F.	1.41	1.96		3.37	4.60
6500	Rough sawn cedar		300	.053		1.81	1.96		3.77	5.05
6700	Redwood, clear, 1" x 4" boards		300	.053		4.22	1.96		6.18	7.70
6900	Aromatic cedar, closet lining, boards	↓	275	.058	↓	3.25	2.14		5.39	6.90

06 43 Wood Stairs and Railings

06 43 13 – Wood Stairs

06 43 13.20 Prefabricated Wood Stairs

		Crew	Daily Output	Labor-Hours	Unit	Material	Labor	Equipment	Total	Total Incl O&P
0010	**PREFABRICATED WOOD STAIRS**									
0100	Box stairs, prefabricated, 3'-0" wide									
0110	Oak treads, up to 14 risers	2 Carp	39	.410	Riser	75.50	15.05		90.55	107
0600	With pine treads for carpet, up to 14 risers	"	39	.410	"	49	15.05		64.05	77.50
1100	For 4' wide stairs, add				Flight	25%				
1550	Stairs, prefabricated stair handrail with balusters	1 Carp	30	.267	L.F.	65	9.80		74.80	87
1700	Basement stairs, prefabricated, pine treads									
1710	Pine risers, 3' wide, up to 14 risers	2 Carp	52	.308	Riser	49	11.30		60.30	71.50
4000	Residential, wood, oak treads, prefabricated		1.50	10.667	Flight	985	390		1,375	1,675
4200	Built in place	↓	.44	36.364	"	1,475	1,325		2,800	3,700
4400	Spiral, oak, 4'-6" diameter, unfinished, prefabricated,									
4500	incl. railing, 9' high	2 Carp	1.50	10.667	Flight	4,675	390		5,065	5,725

06 43 13.30 Wood Stair Components

		Crew	Daily Output	Labor-Hours	Unit	Material	Labor	Equipment	Total	Total Incl O&P
0010	**WOOD STAIR COMPONENTS**									
0020	Balusters, turned, 3" high, pine, minimum	1 Carp	28	.286	Ea.	3.77	10.50		14.27	20.50
0100	Maximum		26	.308		19.40	11.30		30.70	39
0300	30" high birch balusters, minimum		28	.286		6.85	10.50		17.35	24
0400	Maximum		26	.308		27.50	11.30		38.80	47.50
0600	42" high, pine balusters, minimum		27	.296		5	10.85		15.85	22.50
0700	Maximum		25	.320		28	11.75		39.75	49
0900	42" high birch balusters, minimum		27	.296		10.90	10.85		21.75	29
1000	Maximum		25	.320	↓	39.50	11.75		51.25	61.50
1050	Baluster, stock pine, 1-1/4" x 1-1/4"		240	.033	L.F.	3.18	1.22		4.40	5.40
1100	1-3/4" x 1-3/4"		220	.036	"	9.10	1.33		10.43	12.10
1200	Newels, 3-1/4" wide, starting, minimum		7	1.143	Ea.	38	42		80	108
1300	Maximum		6	1.333		325	49		374	435
1500	Landing, minimum		5	1.600		106	58.50		164.50	209
1600	Maximum		4	2	↓	355	73.50		428.50	510
1800	Railings, oak, built-up, minimum		60	.133	L.F.	32.50	4.89		37.39	43
1900	Maximum		55	.145		47	5.35		52.35	60
2100	Add for sub rail		110	.073		5.40	2.67		8.07	10.10
2300	Risers, beech, 3/4" x 7-1/2" high		64	.125		5.90	4.59		10.49	13.65
2400	Fir, 3/4" x 7-1/2" high		64	.125		1.63	4.59		6.22	8.95
2600	Oak, 3/4" x 7-1/2" high	↓	64	.125	↓	6.20	4.59		10.79	14

06 43 Wood Stairs and Railings

06 43 13 – Wood Stairs

06 43 13.30 Wood Stair Components

		Crew	Daily Output	Labor-Hours	Unit	Material	2007 Bare Costs Labor	Equipment	Total	Total Incl O&P
2800	Pine, 3/4" x 7-1/2" high	1 Carp	66	.121	L.F.	3.11	4.45		7.56	10.35
2850	Skirt board, pine, 1" x 10"		55	.145		2.96	5.35		8.31	11.55
2900	1" x 12"		52	.154		3.52	5.65		9.17	12.65
3000	Treads, oak, 1-1/4" x 10" wide, 3' long		18	.444	Ea.	59	16.30		75.30	90.50
3100	4' long, oak		17	.471		79	17.25		96.25	114
3300	1-1/4" x 11-1/2" wide, 3' long, oak		18	.444		59	16.30		75.30	90.50
3400	6' long, oak		14	.571		118	21		139	163
3600	Beech treads, add					40%				
3800	For mitered return nosings, add				L.F.	3.98			3.98	4.38

06 43 16 – Wood Railings

06 43 16.10 Wood Railings

		Crew	Daily Output	Labor-Hours	Unit	Material	Labor	Equipment	Total	Total Incl O&P
0010	**WOOD RAILINGS**									
0020	Custom design, architectural grade, hardwood, minimum	1 Carp	38	.211	L.F.	5.60	7.75		13.35	18.20
0100	Maximum		30	.267		46.50	9.80		56.30	67
0300	Stock interior railing with spindles 6" O.C., 4' long		40	.200		21	7.35		28.35	35
0400	8' long		48	.167		21	6.10		27.10	33

06 44 Ornamental Woodwork

06 44 19 – Wood Grilles

06 44 19.10 Grilles

		Crew	Daily Output	Labor-Hours	Unit	Material	Labor	Equipment	Total	Total Incl O&P
0010	**GRILLES**									
0020	2' x 4' to 4' x 8', custom designs, unfinished, minimum	1 Carp	38	.211	S.F.	12.85	7.75		20.60	26
0050	Average		30	.267		28	9.80		37.80	46
0100	Maximum		19	.421		43	15.45		58.45	71
0300	As above, but prefinished, minimum		38	.211		12.85	7.75		20.60	26
0400	Maximum		19	.421		48	15.45		63.45	77

06 44 33 – Wood Mantels

06 44 33.10 Fireplace Mantels

		Crew	Daily Output	Labor-Hours	Unit	Material	Labor	Equipment	Total	Total Incl O&P
0010	**FIREPLACE MANTELS**									
0015	6" molding, 6' x 3'-6" opening, minimum	1 Carp	5	1.600	Opng.	147	58.50		205.50	254
0100	Maximum		5	1.600		183	58.50		241.50	294
0300	Prefabricated pine, colonial type, stock, deluxe		2	4		2,300	147		2,447	2,775
0400	Economy		3	2.667		395	98		493	585

06 44 33.20 Fireplace Mantel Beam

		Crew	Daily Output	Labor-Hours	Unit	Material	Labor	Equipment	Total	Total Incl O&P
0010	**FIREPLACE MANTEL BEAM**									
0020	Rough texture wood, 4" x 8"	1 Carp	36	.222	L.F.	5.45	8.15		13.60	18.70
0100	4" x 10"		35	.229	"	6.55	8.40		14.95	20.50
0300	Laminated hardwood, 2-1/4" x 10-1/2" wide, 6' long		5	1.600	Ea.	104	58.50		162.50	207
0400	8' long		5	1.600	"	145	58.50		203.50	251
0600	Brackets for above, rough sawn		12	.667	Pr.	9.60	24.50		34.10	48.50
0700	Laminated		12	.667	"	14.50	24.50		39	54

06 44 39 – Wood Posts and Columns

06 44 39.10 Decorative Beams

		Crew	Daily Output	Labor-Hours	Unit	Material	Labor	Equipment	Total	Total Incl O&P
0010	**DECORATIVE BEAMS**									
0020	Rough sawn cedar, non-load bearing, 4" x 4"	2 Carp	180	.089	L.F.	1.47	3.26		4.73	6.70
0100	4" x 6"		170	.094		2.21	3.45		5.66	7.85
0200	4" x 8"		160	.100		2.95	3.67		6.62	8.95
0300	4" x 10"		150	.107		3.68	3.91		7.59	10.15
0400	4" x 12"		140	.114		4.42	4.19		8.61	11.40

06 44 Ornamental Woodwork

06 44 39 - Wood Posts and Columns

06 44 39.10 Decorative Beams

		Crew	Daily Output	Labor-Hours	Unit	Material	2007 Bare Costs Labor	Equipment	Total	Total Incl O&P
0500	8" x 8"	2 Carp	130	.123	L.F.	7.50	4.52		12.02	15.30
1100	Beam connector plates see div. 06 05 23.60									

06 44 39.20 Columns

		Crew	Daily Output	Labor-Hours	Unit	Material	Labor	Equipment	Total	Total Incl O&P
0010	**COLUMNS**									
0050	Aluminum, round colonial, 6" diameter	2 Carp	80	.200	V.L.F.	16	7.35		23.35	29
0100	8" diameter		62.25	.257		17	9.45		26.45	33.50
0200	10" diameter		55	.291		19.10	10.70		29.80	37.50
0250	Fir, stock units, hollow round, 6" diameter		80	.200		17.20	7.35		24.55	30.50
0300	8" diameter		80	.200		20.50	7.35		27.85	34
0350	10" diameter		70	.229		25.50	8.40		33.90	41.50
0400	Solid turned, to 8' high, 3-1/2" diameter		80	.200		8.40	7.35		15.75	20.50
0500	4-1/2" diameter		75	.213		12.80	7.85		20.65	26.50
0600	5-1/2" diameter		70	.229		16.40	8.40		24.80	31
0800	Square columns, built-up, 5" x 5"		65	.246		10.20	9.05		19.25	25.50
0900	Solid, 3-1/2" x 3-1/2"		130	.123		6.70	4.52		11.22	14.40
1600	Hemlock, tapered, T & G, 12" diam, 10' high		100	.160		31.50	5.85		37.35	43.50
1700	16' high		65	.246		46.50	9.05		55.55	65
1900	10' high, 14" diameter		100	.160		78.50	5.85		84.35	95.50
2000	18' high		65	.246		63.50	9.05		72.55	84
2200	18" diameter, 12' high		65	.246		114	9.05		123.05	140
2300	20' high		50	.320		80	11.75		91.75	106
2500	20" diameter, 14' high		40	.400		129	14.70		143.70	165
2600	20' high		35	.457		120	16.80		136.80	158
2800	For flat pilasters, deduct					33%				
3000	For splitting into halves, add				Ea.	116			116	128
4000	Rough sawn cedar posts, 4" x 4"	2 Carp	250	.064	V.L.F.	3.70	2.35		6.05	7.75
4100	4" x 6"		235	.068		6.70	2.50		9.20	11.25
4200	6" x 6"		220	.073		12.50	2.67		15.17	17.90
4300	8" x 8"		200	.080		28.50	2.94		31.44	36

06 48 Wood Frames

06 48 13 - Exterior Wood Door Frames

06 48 13.10 Exterior Wood Door Frames

		Crew	Daily Output	Labor-Hours	Unit	Material	Labor	Equipment	Total	Total Incl O&P
0010	**EXTERIOR WOOD DOOR FRAMES**									
0400	Exterior frame, incl. ext. trim, pine, 5/4 x 4-9/16" deep	2 Carp	375	.043	L.F.	5.35	1.57		6.92	8.30
0420	5-3/16" deep		375	.043		10.90	1.57		12.47	14.45
0440	6-9/16" deep		375	.043		8.25	1.57		9.82	11.55
0600	Oak, 5/4 x 4-9/16" deep		350	.046		10.50	1.68		12.18	14.15
0620	5-3/16" deep		350	.046		12.60	1.68		14.28	16.45
0640	6-9/16" deep		350	.046		14.40	1.68		16.08	18.45
0800	Walnut, 5/4 x 4-9/16" deep		350	.046		12	1.68		13.68	15.80
0820	5-3/16" deep		350	.046		16.50	1.68		18.18	21
0840	6-9/16" deep		350	.046		19.20	1.68		20.88	23.50
1000	Sills, 8/4 x 8" deep, oak, no horns		100	.160		18	5.85		23.85	29
1020	2" horns		100	.160		18.25	5.85		24.10	29
1040	3" horns		100	.160		18.25	5.85		24.10	29
1100	8/4 x 10" deep, oak, no horns		90	.178		23	6.50		29.50	35.50
1120	2" horns		90	.178		21	6.50		27.50	33
1140	3" horns		90	.178		21	6.50		27.50	33
2000	Exterior, colonial, frame & trim, 3' opng., in-swing, minimum		22	.727	Ea.	310	26.50		336.50	380
2010	Average		21	.762		460	28		488	555

06 48 Wood Frames

06 48 13 – Exterior Wood Door Frames

06 48 13.10 Exterior Wood Door Frames

		Crew	Daily Output	Labor-Hours	Unit	Material	2007 Bare Costs Labor	Equipment	Total	Total Incl O&P
2020	Maximum	2 Carp	20	.800	Ea.	1,050	29.50		1,079.50	1,200
2100	5'-4" opening, in-swing, minimum		17	.941		355	34.50		389.50	445
2120	Maximum		15	1.067		1,050	39		1,089	1,200
2140	Out-swing, minimum		17	.941		360	34.50		394.50	450
2160	Maximum		15	1.067		1,100	39		1,139	1,250
2400	6'-0" opening, in-swing, minimum		16	1		340	36.50		376.50	425
2420	Maximum		10	1.600		1,100	58.50		1,158.50	1,300
2460	Out-swing, minimum		16	1		365	36.50		401.50	455
2480	Maximum		10	1.600		1,275	58.50		1,333.50	1,525
2600	For two sidelights, add, minimum		30	.533	Opng.	345	19.55		364.55	410
2620	Maximum		20	.800	"	1,125	29.50		1,154.50	1,275
2700	Custom birch frame, 3'-0" opening		16	1	Ea.	199	36.50		235.50	276
2750	6'-0" opening		16	1		300	36.50		336.50	385
2900	Exterior, modern, plain trim, 3' opng., in-swing, minimum		26	.615		33.50	22.50		56	72
2920	Average		24	.667		40	24.50		64.50	82
2940	Maximum		22	.727		48.50	26.50		75	95

06 48 16 – Interior Wood Door Frames

06 48 16.10 Interior Wood Door Frames

		Crew	Daily Output	Labor-Hours	Unit	Material	2007 Bare Costs Labor	Equipment	Total	Total Incl O&P
0010	**INTERIOR WOOD DOOR FRAMES**									
3000	Interior frame, pine, 11/16" x 3-5/8" deep	2 Carp	375	.043	L.F.	5.10	1.57		6.67	8.05
3020	4-9/16" deep		375	.043		6.25	1.57		7.82	9.35
3200	Oak, 11/16" x 3-5/8" deep		350	.046		4	1.68		5.68	7
3220	4-9/16" deep		350	.046		4.31	1.68		5.99	7.35
3240	5-3/16" deep		350	.046		4.47	1.68		6.15	7.55
3400	Walnut, 11/16" x 3-5/8" deep		350	.046		6.75	1.68		8.43	10.05
3420	4-9/16" deep		350	.046		7.10	1.68		8.78	10.40
3440	5-3/16" deep		350	.046		7.40	1.68		9.08	10.75
3600	Pocket door frame		16	1	Ea.	64	36.50		100.50	127
3800	Threshold, oak, 5/8" x 3-5/8" deep		200	.080	L.F.	3.75	2.94		6.69	8.70
3820	4-5/8" deep		190	.084		4.50	3.09		7.59	9.75
3840	5-5/8" deep		180	.089		7.40	3.26		10.66	13.25
4000	For casing see division 06 22 13.30 & 06 22 13.50									

06 49 Wood Screens and Exterior Wood Shutters

06 49 19 – Exterior Wood Shutters

06 49 19.10 Shutters, Exterior

		Crew	Daily Output	Labor-Hours	Unit	Material	2007 Bare Costs Labor	Equipment	Total	Total Incl O&P
0010	**SHUTTERS, EXTERIOR**									
0012	Aluminum, louvered, 1'-4" wide, 3'-0" long	1 Carp	10	.800	Pr.	48	29.50		77.50	98
0400	6'-8" long		9	.889		96.50	32.50		129	157
1000	Pine, louvered, primed, each 1'-2" wide, 3'-3" long		10	.800		91	29.50		120.50	146
1100	4'-7" long		10	.800		123	29.50		152.50	181
1500	Each 1'-6" wide, 3'-3" long		10	.800		96.50	29.50		126	152
1600	4'-7" long		10	.800		136	29.50		165.50	195
1620	Hemlock, louvered, 1'-2" wide, 5'-7" long		10	.800		149	29.50		178.50	210
1630	Each 1'-4" wide, 2'-2" long		10	.800		93	29.50		122.50	148
1670	4'-3" long		10	.800		111	29.50		140.50	168
1690	5'-11" long		10	.800		157	29.50		186.50	218
1700	Door blinds, 6'-9" long, each 1'-3" wide		9	.889		158	32.50		190.50	224
1710	1'-6" wide		9	.889		169	32.50		201.50	237
1720	Hemlock, solid raised panel, each 1'-4" wide, 3'-3" long		10	.800		150	29.50		179.50	211

06 49 Wood Screens and Exterior Wood Shutters

06 49 19 – Exterior Wood Shutters

06 49 19.10 Shutters, Exterior		Crew	Daily Output	Labor-Hours	Unit	Material	2007 Bare Costs Labor	Equipment	Total	Total Incl O&P
1740	4'-3" long	1 Carp	10	.800	Pr.	190	29.50		219.50	255
1770	5'-11" long		10	.800		254	29.50		283.50	325
1800	Door blinds, 6'-9" long, each 1'-3" wide		9	.889		286	32.50		318.50	365
1900	1'-6" wide		9	.889		310	32.50		342.50	395
2500	Polystyrene, solid raised panel, each 1'-4" wide, 3'-3" long		10	.800		40	29.50		69.50	89.50
2700	4'-7" long		10	.800		50.50	29.50		80	101
4500	Polystyrene, louvered, each 1'-2" wide, 3'-3" long		10	.800		27.50	29.50		57	75.50
4750	5'-3" long		10	.800		39	29.50		68.50	88
6000	Vinyl, louvered, each 1'-2" x 4'-7" long		10	.800		33.50	29.50		63	82.50
6200	Each 1'-4" x 6'-8" long		9	.889		47.50	32.50		80	103

06 52 Plastic Structural Assemblies

06 52 10 – Plastic Structural Assemblies

06 52 10.10 Castings, Fiberglass

		Crew	Daily Output	Labor-Hours	Unit	Material	Labor	Equipment	Total	Total Incl O&P
0010	**CASTINGS, FIBERGLASS**									
0100	Angle, 1" x 1" x 1/8" thick	2 Sswk	240	.067	L.F.	1.31	2.76		4.07	6.45
0120	3" x 3" x 1/4" thick		200	.080		5.45	3.31		8.76	12
0140	4" x 4" x 1/4" thick		200	.080		7	3.31		10.31	13.70
0160	4" x 4" x 3/8" thick		200	.080		10.80	3.31		14.11	17.85
0180	6" x 6" x 1/2" thick		160	.100		21.50	4.14		25.64	31
1000	Flat sheet, 1/8" thick		140	.114	S.F.	4.60	4.73		9.33	13.60
1020	1/4" thick		120	.133		9	5.50		14.50	19.90
1040	3/8" thick		100	.160		15.30	6.60		21.90	29
1060	1/2" thick		80	.200		20	8.25		28.25	37
2000	Handrail, 42" high, 2" diam. rails pickets 5' O.C.		32	.500	L.F.	46.50	20.50		67	88.50
3000	Round bar, 1/4" diam.		240	.067		.53	2.76		3.29	5.55
3020	1/2" diam.		200	.080		.92	3.31		4.23	7
3040	3/4" diam.		200	.080		1.87	3.31		5.18	8.05
3060	1" diam.		160	.100		3.29	4.14		7.43	11.10
3080	1-1/4" diam.		160	.100		3.75	4.14		7.89	11.65
3100	1-1/2" diam.		140	.114		4.59	4.73		9.32	13.60
3500	Round tube, 1" diam. x 1/8" thick		240	.067		1.79	2.76		4.55	6.95
3520	2" diam. x 1/4" thick		200	.080		5.75	3.31		9.06	12.35
3540	3" diam. x 1/4" thick		160	.100		9.20	4.14		13.34	17.60
4000	Square bar, 1/2" square		240	.067		3.62	2.76		6.38	8.95
4020	1" square		200	.080		4.64	3.31		7.95	11.10
4040	1-1/2" square		160	.100		9.45	4.14		13.59	17.90
4500	Square tube, 1" x 1" x 1/8" thick		240	.067		2.04	2.76		4.80	7.25
4520	2" x 2" x 1/8" thick		200	.080		3.95	3.31		7.26	10.35
4540	3" x 3" x 1/4" thick		160	.100		10.75	4.14		14.89	19.35
5000	Threaded rod, 3/8" diam.		320	.050		3.14	2.07		5.21	7.20
5020	1/2" diam.		320	.050		3.69	2.07		5.76	7.80
5040	5/8" diam.		280	.057		4.08	2.36		6.44	8.75
5060	3/4" diam.		280	.057		4.64	2.36		7	9.40
6000	Wide flange beam, 4" x 4" x 1/4" thick		120	.133		9.55	5.50		15.05	20.50
6020	6" x 6" x 1/4" thick		100	.160		18.15	6.60		24.75	32
6040	8" x 8" x 3/8" thick		80	.200		32	8.25		40.25	50

06 52 10.20 Fiberglass Stair Treads

		Crew	Daily Output	Labor-Hours	Unit	Material	Labor	Equipment	Total	Total Incl O&P
0010	**FIBERGLASS STAIR TREADS**									
0100	24" wide	2 Sswk	52	.308	Ea.	36	12.70		48.70	62.50
0140	30" wide		52	.308		43	12.70		55.70	70

06 52 Plastic Structural Assemblies

06 52 10 – Plastic Structural Assemblies

06 52 10.20 Fiberglass Stair Treads

	06 52 10.20 Fiberglass Stair Treads	Crew	Daily Output	Labor-Hours	Unit	Material	2007 Bare Costs Labor	Equipment	Total	Total Incl O&P
0180	36" wide	2 Sswk	52	.308	Ea.	50.50	12.70		63.20	78.50
0220	42" wide	↓	52	.308	↓	59.50	12.70		72.20	88.50

06 52 10.30 Fiberglass Grating

		Crew	Daily Output	Labor-Hours	Unit	Material	Labor	Equipment	Total	Total Incl O&P
0010	**FIBERGLASS GRATING**									
0100	Molded, green (for mod. corrosive environment)									
0140	1" x 4" mesh, 1" thick	2 Sswk	400	.040	S.F.	11.15	1.65		12.80	15.30
0180	1-1/2" square mesh, 1" thick		400	.040		16.60	1.65		18.25	21.50
0220	1-1/4" thick		400	.040		13.25	1.65		14.90	17.60
0260	1-1/2" thick		400	.040		24	1.65		25.65	29.50
0300	2" square mesh, 2" thick	↓	320	.050	↓	22.50	2.07		24.57	29
1000	Orange (for highly corrosive environment)									
1040	1" x 4" mesh, 1" thick	2 Sswk	400	.040	S.F.	14.05	1.65		15.70	18.45
1080	1-1/2" square mesh, 1" thick		400	.040		18.60	1.65		20.25	23.50
1120	1-1/4" thick		400	.040		19.30	1.65		20.95	24.50
1160	1-1/2" thick		400	.040		21	1.65		22.65	26
1200	2" square mesh, 2" thick	↓	320	.050		25	2.07		27.07	31.50
3000	Pultruded, green (for mod. corrosive environment)									
3040	1" O.C. bar spacing, 1" thick	2 Sswk	400	.040	S.F.	16.65	1.65		18.30	21.50
3080	1-1/2" thick		320	.050		18.50	2.07		20.57	24.50
3120	1-1/2" O.C. bar spacing, 1" thick		400	.040		13.50	1.65		15.15	17.85
3160	1-1/2" thick		400	.040	↓	15.15	1.65		16.80	19.65
4000	Grating support legs, fixed height, no base				Ea.	43.50			43.50	48
4040	With base					39			39	43
4080	Adjustable to 60"				↓	59			59	65

06 52 10.40 Fiberglass Floor Grating

		Crew	Daily Output	Labor-Hours	Unit	Material	Labor	Equipment	Total	Total Incl O&P
0010	**FIBERGLASS FLOOR GRATING**									
0100	Reinforced polyester, fire retardant, 1" x 4" grid, 1" thick	E-4	510	.063	S.F.	14.25	2.63	.23	17.11	20.50
0200	1-1/2" x 6" mesh, 1-1/2" thick		500	.064		16.65	2.68	.23	19.56	23.50
0300	With grit surface, 1-1/2" x 6" grid, 1-1/2" thick	↓	500	.064	↓	17	2.68	.23	19.91	24

06 65 Plastic Simulated Wood Trim

06 65 10 – Plastic Simulated Wood Trim

06 65 10.10 PVC Trim, Exterior

		Crew	Daily Output	Labor-Hours	Unit	Material	Labor	Equipment	Total	Total Incl O&P
0010	**PVC TRIM, EXTERIOR**									
0100	Cornerboards, 5/4" x 6" x 6"	1 Carp	240	.033	L.F.	6.60	1.22		7.82	9.15
0110	Door / window casing, 1" x 4"		200	.040		1.40	1.47		2.87	3.83
0120	1" x 6"		200	.040		2.18	1.47		3.65	4.69
0130	1" x 8"		195	.041		2.87	1.51		4.38	5.50
0140	1" x 10"		195	.041		3.65	1.51		5.16	6.35
0150	1" x 12"		190	.042		4.45	1.55		6	7.30
0160	5/4" x 4"		195	.041		1.78	1.51		3.29	4.30
0170	5/4" x 6"		195	.041		2.79	1.51		4.30	5.40
0180	5/4" x 8"		190	.042		3.69	1.55		5.24	6.45
0190	5/4" x 10"		190	.042		4.70	1.55		6.25	7.55
0200	5/4" x 12"		185	.043		5.70	1.59		7.29	8.75
0210	Fascia, 1" x 4"		250	.032		1.40	1.17		2.57	3.37
0220	1" x 6"		250	.032		2.18	1.17		3.35	4.23
0230	1" x 8"		225	.036		2.87	1.31		4.18	5.20
0240	1" x 10"		225	.036		3.65	1.31		4.96	6.05
0250	1" x 12"	↓	200	.040	↓	4.45	1.47		5.92	7.20

06 65 Plastic Simulated Wood Trim

06 65 10 – Plastic Simulated Wood Trim

06 65 10.10 PVC Trim, Exterior		Crew	Daily Output	Labor-Hours	Unit	Material	2007 Bare Costs Labor	Equipment	Total	Total Incl O&P
0260	5/4" x 4"	1 Carp	240	.033	L.F.	1.78	1.22		3	3.86
0270	5/4" x 6"		240	.033		2.79	1.22		4.01	4.97
0280	5/4" x 8"		215	.037		3.69	1.37		5.06	6.20
0290	5/4" x 10"		215	.037		4.70	1.37		6.07	7.30
0300	5/4" x 12"		190	.042		5.70	1.55		7.25	8.70
0310	Fascia, 1" x 4"		250	.032		1.40	1.17		2.57	3.37
0320	1" x 6"		250	.032		2.18	1.17		3.35	4.23
0330	1" x 8"		225	.036		2.87	1.31		4.18	5.20
0340	1" x 10"		225	.036		3.65	1.31		4.96	6.05
0350	1" x 12"		200	.040		4.45	1.47		5.92	7.20
0360	5/4" x 4"		240	.033		1.78	1.22		3	3.86
0370	5/4" x 6"		240	.033		2.79	1.22		4.01	4.97
0380	5/4" x 8"		215	.037		3.69	1.37		5.06	6.20
0390	5/4" x 10"		215	.037		4.70	1.37		6.07	7.30
0400	5/4" x 12"		190	.042		5.70	1.55		7.25	8.70
0410	Rake, 1" x 4"		200	.040		1.40	1.47		2.87	3.83
0420	1" x 6"		200	.040		2.18	1.47		3.65	4.69
0430	1" x 8"		190	.042		2.87	1.55		4.42	5.55
0440	1" x 10"		190	.042		3.65	1.55		5.20	6.45
0450	1" x 12"		180	.044		4.45	1.63		6.08	7.45
0460	5/4" x 4"		195	.041		1.78	1.51		3.29	4.30
0470	5/4" x 6"		195	.041		2.79	1.51		4.30	5.40
0480	5/4" x 8"		185	.043		3.69	1.59		5.28	6.55
0490	5/4" x 10"		185	.043		4.70	1.59		6.29	7.60
0500	5/4" x 12"		175	.046		5.70	1.68		7.38	8.90
0510	Rake trim, 1" x 4"		225	.036		1.40	1.31		2.71	3.57
0520	1" x 6"		225	.036		2.18	1.31		3.49	4.43
0560	5/4" x 4"		220	.036		1.78	1.33		3.11	4.04
0570	5/4" x 6"		220	.036		2.79	1.33		4.12	5.15
0610	Soffit, 1" x 4"	2 Carp	420	.038		1.40	1.40		2.80	3.72
0620	1" x 6"		420	.038		2.18	1.40		3.58	4.58
0630	1" x 8"		420	.038		2.87	1.40		4.27	5.35
0640	1" x 10"		400	.040		3.65	1.47		5.12	6.30
0650	1" x 12"		400	.040		4.45	1.47		5.92	7.20
0660	5/4" x 4"		410	.039		1.78	1.43		3.21	4.19
0670	5/4" x 6"		410	.039		2.79	1.43		4.22	5.30
0680	5/4" x 8"		410	.039		3.69	1.43		5.12	6.30
0690	5/4" x 10"		390	.041		4.70	1.51		6.21	7.50
0700	5/4" x 12"		390	.041		5.70	1.51		7.21	8.65

Division Notes

	CREW	DAILY OUTPUT	LABOR-HOURS	UNIT	2007 BARE COSTS				TOTAL INCL O&P
					MAT.	LABOR	EQUIP.	TOTAL	

Estimating Tips

07 10 00 Dampproofing and Waterproofing

- Be sure of the job specifications before pricing this subdivision. The difference in cost between waterproofing and dampproofing can be great. Waterproofing will hold back standing water. Dampproofing prevents the transmission of water vapor. Also included in this section are vapor retarding membranes.

07 20 00 Thermal Protection

- Insulation and fireproofing products are measured by area, thickness, volume or R value. Specifications may give only what the specific R value should be in a certain situation. The estimator may need to choose the type of insulation to meet that R value.

07 30 00 Steep Slope Roofing
07 40 00 Roofing and Siding Panels

- Many roofing and siding products are bought and sold by the square. One square is equal to an area that measures 100 square feet. This simple change in unit of measure could create a large error if the estimator is not observant. Accessories necessary for a complete installation must be figured into any calculations for both material and labor.

07 50 00 Membrane Roofing
07 60 00 Flashing and Sheet Metal
07 70 00 Roofing and Wall Specialties and Accessories

- The items in these subdivisions compose a roofing system. No one component completes the installation, and all must be estimated. Built-up or single-ply membrane roofing systems are made up of many products and installation trades. Wood blocking at roof perimeters or penetrations, parapet coverings, reglets, roof drains, gutters, downspouts, sheet metal flashing, skylights, smoke vents, and roof hatches all need to be considered along with the roofing material. Several different installation trades will need to work together on the roofing system. Inherent difficulties in the scheduling and coordination of various trades must be accounted for when estimating labor costs.

07 90 00 Joint Protection

- To complete the weather-tight shell, the sealants and caulkings must be estimated. Where different materials meet—at expansion joints, at flashing penetrations, and at hundreds of other locations throughout a construction project—they provide another line of defense against water penetration. Often, an entire system is based on the proper location and placement of caulking or sealants. The detailed drawings that are included as part of a set of architectural plans show typical locations for these materials. When caulking or sealants are shown at typical locations, this means the estimator must include them for all the locations where this detail is applicable. Be careful to keep different types of sealants separate, and remember to consider backer rods and primers if necessary.

Reference Numbers

Reference numbers are shown in shaded boxes at the beginning of some major classifications. These numbers refer to related items in the Reference Section. The reference information may be an estimating procedure, an alternate pricing method, or technical information.

Note: Not all subdivisions listed here necessarily appear in this publication.

07 01 Operation and Maint. of Thermal and Moisture Protection

07 01 50 – Maintenance of Membrane Roofing

07 01 50.10 Roof Coatings	Crew	Daily Output	Labor-Hours	Unit	Material	2007 Bare Costs Labor	Equipment	Total	Total Incl O&P
0010 **ROOF COATINGS**									
0012 Asphalt, brush grade, material only	CN			Gal.	7.45			7.45	8.15
0200 Asphalt base, fibered aluminum coating					8.95			8.95	9.85
0300 Asphalt primer, 5 gallon				↓	4.64			4.64	5.10
0600 Coal tar pitch, 200 lb. barrels				Ton	700			700	770
0700 Tar roof cement, 5 gal. lots				Gal.	7			7	7.70
0800 Glass fibered roof & patching cement, 5 gallon				"	4.96			4.96	5.45
0900 Reinforcing glass membrane, 450 S.F./roll				Ea.	46.50			46.50	51
1000 Neoprene roof coating, 5 gal, 2 gal/sq				Gal.	23.50			23.50	25.50
1100 Roof patch & flashing cement, 5 gallon					11.25			11.25	12.35
1200 Roof resaturant, glass fibered, 3 gal/sq					7.45			7.45	8.20
1300 Mineral rubber, 3 gal/sq					4.80			4.80	5.30

07 05 Common Work Results for Thermal and Moisture Protection

07 05 05 – Selective Demolition

07 05 05.10 Selective Demolition, Roofing and Siding

		Crew	Daily Output	Labor-Hours	Unit	Material	Labor	Equipment	Total	Total Incl O&P
0010	**SELECTIVE DEMOLITION, ROOFING AND SIDING** R024119-10									
0200	Waterproofing demo., scrape off vertical waterproofing, to 1/2" thick	2 Clab	2000	.008	S.F.		.23		.23	.36
0210	Over 1/2" thick		1750	.009	"		.26		.26	.41
0250	Protection / drain board		3900	.004	B.F.		.12		.12	.18
1000	Deck, roof, concrete plank	B-13	1680	.033	S.F.		1.04	.44	1.48	2.09
1100	Gypsum plank		3900	.014			.45	.19	.64	.90
1150	Metal decking	↓	3500	.016			.50	.21	.71	1
1200	Wood, boards, tongue and groove, 2" x 6"	2 Clab	960	.017			.48		.48	.75
1220	2" x 10"		1040	.015			.44		.44	.69
1280	Standard planks, 1" x 6"		1080	.015			.43		.43	.66
1320	1" x 8"		1160	.014			.40		.40	.62
1340	1" x 12"		1200	.013			.38		.38	.60
1350	Plywood, to 1" thick	↓	2000	.008	↓		.23		.23	.36
2000	Gutters, aluminum or wood, edge hung	1 Clab	240	.033	L.F.		.96		.96	1.49
2100	Built-in		100	.080	"		2.30		2.30	3.58
2200	Insulation removal, loose fitting		3000	.003	C.F.		.08		.08	.12
2250	Air barrier		3500	.002	S.F.		.07		.07	.10
2300	Batts or blankets		1400	.006	C.F.		.16		.16	.26
2350	Rigid board		3450	.002	B.F.		.07		.07	.10
2500	Roof accessories, plumbing vent flashing		14	.571	Ea.		16.45		16.45	25.50
2600	Adjustable metal chimney flashing		9	.889	"		25.50		25.50	40
2650	Coping, sheet metal, up to 12" wide	↓	240	.033	L.F.		.96		.96	1.49
2660	Concrete, up to 12" wide	2 Clab	160	.100	"		2.88		2.88	4.48
3000	Roofing, built-up, 5 ply roof, no gravel	B-2	1600	.025	S.F.		.73		.73	1.13
3001	Including gravel		890	.045			1.31		1.31	2.04
3100	Gravel removal, minimum		5000	.008			.23		.23	.36
3120	Maximum		2000	.020			.58		.58	.91
3400	Roof insulation board, up to 2" thick		3900	.010	↓		.30		.30	.47
3405	Over 2" thick	↓	7800	.005	B.F.		.15		.15	.23
3450	Roll roofing, cold adhesive	1 Clab	12	.667	Sq.		19.15		19.15	30
4000	Shingles, asphalt strip, 1 layer	B-2	3500	.011	S.F.		.33		.33	.52
4100	Slate		2500	.016			.47		.47	.73
4300	Wood	↓	2200	.018	↓		.53		.53	.83
4500	Skylight to 10 S.F.	1 Clab	8	1	Ea.		29		29	45
5000	Siding, metal, horizontal	↓	444	.018	S.F.		.52		.52	.81

07 05 Common Work Results for Thermal and Moisture Protection

07 05 05 – Selective Demolition

07 05 05.10 Selective Demolition, Roofing and Siding

		Crew	Daily Output	Labor-Hours	Unit	Material	2007 Bare Costs Labor	Equipment	Total	Total Incl O&P
5020	Vertical	1 Clab	400	.020	S.F.		.58		.58	.90
5200	Wood, boards, vertical		400	.020			.58		.58	.90
5220	Clapboards, horizontal		380	.021			.61		.61	.94
5240	Shingles		350	.023			.66		.66	1.02
5260	Textured plywood		725	.011			.32		.32	.49

07 11 Dampproofing

07 11 13 – Bituminous Damppoofing

07 11 13.10 Bituminous Asphalt Coating

		Crew	Daily Output	Labor-Hours	Unit	Material	Labor	Equipment	Total	Total Incl O&P
0010	**BITUMINOUS ASPHALT COATING**									
0030	Brushed on, below grade, 1 coat	1 Rofc	665	.012	S.F.	.13	.38		.51	.79
0100	2 coat		500	.016		.25	.51		.76	1.14
0300	Sprayed on, below grade, 1 coat, 25.6 S.F./gal.		830	.010		.13	.31		.44	.66
0400	2 coat, 20.5 S.F./gal.		500	.016		.25	.51		.76	1.13
0500	Asphalt coating, with fibers				Gal.	5.45			5.45	5.95
0600	Troweled on, asphalt with fibers, 1/16" thick	1 Rofc	500	.016	S.F.	.24	.51		.75	1.12
0700	1/8" thick		400	.020		.42	.64		1.06	1.54
1000	1/2" thick		350	.023		1.36	.73		2.09	2.72

07 11 16 – Cementitious Dampproofing

07 11 16.20 Cementitious Parging

		Crew	Daily Output	Labor-Hours	Unit	Material	Labor	Equipment	Total	Total Incl O&P
0010	**CEMENTITIOUS PARGING**									
0020	Portland cement, 2 coats, 1/2" thick	D-1	250	.064	S.F.	.20	2.13		2.33	3.47
0100	Waterproofed Portland cement	"	250	.064	"	2.03	2.13		4.16	5.50

07 12 Built-Up Bituminous Waterproofing

07 12 13 – Built-Up Asphalt Waterproofing

07 12 13.20 Membrane Waterproofing

		Crew	Daily Output	Labor-Hours	Unit	Material	Labor	Equipment	Total	Total Incl O&P
0010	**MEMBRANE WATERPROOFING**									
0012	On slabs, 1 ply, felt, mopped	G-1	3000	.019	S.F.	.25	.55	.11	.91	1.33
0300	On slabs, 2 ply, felt, mopped		2500	.022		.50	.66	.13	1.29	1.82
0400	On slabs, 2 ply, glass fiber fabric, mopped		1650	.034		.57	1.01	.20	1.78	2.56
0600	On slabs, 3 ply, felt, mopped		2100	.027		.74	.79	.16	1.69	2.33
0700	On slabs, 3 ply, glass fiber fabric, mopped		1550	.036		.77	1.07	.21	2.05	2.91
0710	Asphaltic hardboard protection board, 1/8" thick	2 Rofc	500	.032		.32	1.02		1.34	2.08
1000	For adhered 1/4" EPS protection board, add		3500	.005		.20	.15		.35	.47
1050	3/8" thick, add		3500	.005		.22	.15		.37	.49
1060	1/2" thick, add		3500	.005		.24	.15		.39	.51
1070	Fiberglass fabric, black, 20/10 mesh		116	.138	Sq.	11.95	4.39		16.34	20.50
1080	White, 20/10 mesh		116	.138	"	12.30	4.39		16.69	21
1100	1/16" urethane, troweled		200	.080	S.F.	.70	2.54		3.24	5.10
1200	Roller applied		120	.133	"	.70	4.24		4.94	7.95

07 13 Sheet Waterproofing

07 13 53 – Elastomeric Sheet Waterproofing

07 13 53.10 Elastomeric Sheet Waterproofing

		Crew	Daily Output	Labor-Hours	Unit	Material	2007 Bare Costs Labor	2007 Bare Costs Equipment	Total	Total Incl O&P
0010	**ELASTOMERIC SHEET WATERPROOFING**									
0090	EPDM, plain, 45 mils thick	2 Rofc	580	.028	S.F.	1.03	.88		1.91	2.62
0100	60 mils thick		570	.028		1.11	.89		2	2.73
0300	Nylon reinforced sheets, 45 mils thick		580	.028		1.01	.88		1.89	2.60
0400	60 mils thick	↓	570	.028	↓	1.28	.89		2.17	2.91
0600	Vulcanizing splicing tape for above, 2" wide				C.L.F.	40.50			40.50	44.50
0700	4" wide				"	63.50			63.50	69.50
0900	Adhesive, bonding, 60 SF per gal				Gal.	14			14	15.40
1000	Splicing, 75 SF per gal				"	25.50			25.50	28
1200	Neoprene sheets, plain, 45 mils thick	2 Rofc	580	.028	S.F.	1.49	.88		2.37	3.13
1300	60 mils thick		570	.028		2.13	.89		3.02	3.85
1500	Nylon reinforced, 45 mils thick		580	.028		1.86	.88		2.74	3.54
1600	60 mils thick		570	.028		2.11	.89		3	3.83
1800	120 mils thick	↓	500	.032	↓	3.40	1.02		4.42	5.45
1900	Adhesive, splicing, 150 S.F. per gal. per coat				Gal.	18.50			18.50	20.50
2100	Fiberglass reinforced, fluid applied, 1/8" thick	2 Rofc	500	.032	S.F.	1.65	1.02		2.67	3.55
2200	Polyethylene and rubberized asphalt sheets, 1/8" thick		550	.029		.55	.93		1.48	2.18
2210	Asphaltic hardboard protection board, 1/8" thick		500	.032		.32	1.02		1.34	2.08
2220	1/4" thick		450	.036		.56	1.13		1.69	2.54
2400	Polyvinyl chloride sheets, plain, 10 mils thick		580	.028		.16	.88		1.04	1.67
2500	20 mils thick		570	.028		.26	.89		1.15	1.80
2700	30 mils thick	↓	560	.029	↓	.36	.91		1.27	1.94
3000	Adhesives, trowel grade, 40-100 SF per gal				Gal.	23.50			23.50	26
3100	Brush grade, 100-250 SF per gal.				"	23.50			23.50	26
3300	Bitumen modified polyurethane, fluid applied, 55 mils thick	2 Rofc	665	.024	S.F.	.68	.77		1.45	2.05
3600	Vinyl plastic, sprayed on, 25 to 40 mils thick	"	475	.034	"	1.02	1.07		2.09	2.94

07 16 Cementitious and Reactive Waterproofing

07 16 16 – Crystalline Waterproofing

07 16 16.20 Cementitious Waterproofing

		Crew	Daily Output	Labor-Hours	Unit	Material	Labor	Equipment	Total	Total Incl O&P
0010	**CEMENTITIOUS WATERPROOFING**									
0020	1/8" application, sprayed on	G-2	1000	.024	S.F.	1.57	.73	.12	2.42	2.97
0030	2 coat, cementitious/metallic slurry, troweled, 1/4" thick	1 Cefi	2.48	3.226	C.S.F.	23.50	115		138.50	194
0040	3 coat, 3/8" thick		1.84	4.348		41	155		196	272
0050	4 coat, 1/2" thick	↓	1.20	6.667	↓	54.50	237		291.50	410

07 17 Bentonite Waterproofing

07 17 13 – Bentonite Panel Waterproofing

07 17 13.10 Bentonite Panels

		Crew	Daily Output	Labor-Hours	Unit	Material	Labor	Equipment	Total	Total Incl O&P
0010	**BENTONITE PANELS**									
0020	Panels, 4' x 4', 3/16" thick	1 Rofc	625	.013	S.F.	.81	.41		1.22	1.58
0100	Rolls, 3/8" thick, with geotextile fabric both sides	"	550	.015	"	.99	.46		1.45	1.88
0300	Granular bentonite, 50 lb. bags (.625 C.F.)				Bag	19.30			19.30	21
0400	3/8" thick, troweled on	1 Rofc	475	.017	S.F.	.97	.54		1.51	1.97
0500	Drain board, expanded polystyrene, binder encapsulated, 1-1/2" thick	1 Rohe	1600	.005		1.13	.12		1.25	1.44
0510	2" thick		1600	.005		1.51	.12		1.63	1.86
0520	3" thick		1600	.005		2.26	.12		2.38	2.69
0530	4" thick		1600	.005		3	.12		3.12	3.50
0600	With filter fabric, 1-1/2" thick	↓	1600	.005		1.37	.12		1.49	1.71

07 17 Bentonite Waterproofing

07 17 13 – Bentonite Panel Waterproofing

07 17 13.10 Bentonite Panels		Crew	Daily Output	Labor-Hours	Unit	Material	2007 Bare Costs Labor	Equipment	Total	Total Incl O&P
0625	2" thick	1 Rohe	1600	.005	S.F.	1.80	.12		1.92	2.18
0650	3" thick		1600	.005		2.55	.12		2.67	3.01
0675	4" thick		1600	.005		3.31	.12		3.43	3.84

07 19 Water Repellents

07 19 19 – Silicone Water Repellents

07 19 19.10 Silicone Water Repellents

		Crew	Daily Output	Labor-Hours	Unit	Material	Labor	Equipment	Total	Total Incl O&P
0010	**SILICONE WATER REPELLENTS**									
0020	Water base liquid, roller applied	2 Rofc	7000	.002	S.F.	.58	.07		.65	.76
0200	Silicone or stearate, sprayed on CMU, 1 coat	1 Rofc	4000	.002		.33	.06		.39	.47
0300	2 coats	"	3000	.003		.66	.08		.74	.86

07 21 Thermal Insulation

07 21 13 – Board Insulation

07 21 13.10 Rigid Insulation

		Crew	Daily Output	Labor-Hours	Unit	Material	Labor	Equipment	Total	Total Incl O&P
0010	**RIGID INSULATION**									
0040	Fiberglass, 1.5#/CF, unfaced, 1" thick, R4.1	1 Carp	1000	.008	S.F.	.44	.29		.73	.94
0060	1-1/2" thick, R6.2		1000	.008		.79	.29		1.08	1.33
0080	2" thick, R8.3		1000	.008		.68	.29		.97	1.21
0120	3" thick, R12.4		800	.010		.79	.37		1.16	1.44
0370	3#/CF, unfaced, 1" thick, R4.3		1000	.008		.49	.29		.78	1
0390	1-1/2" thick, R6.5		1000	.008		.94	.29		1.23	1.49
0400	2" thick, R8.7		890	.009		1.14	.33		1.47	1.76
0420	2-1/2" thick, R10.9		800	.010		1	.37		1.37	1.67
0440	3" thick, R13		800	.010		1	.37		1.37	1.67
0520	Foil faced, 1" thick, R4.3		1000	.008		.91	.29		1.20	1.46
0540	1-1/2" thick, R6.5		1000	.008		1.35	.29		1.64	1.95
0560	2" thick, R8.7		890	.009		1.69	.33		2.02	2.37
0580	2-1/2" thick, R10.9		800	.010		2	.37		2.37	2.77
0600	3" thick, R13		800	.010		2.17	.37		2.54	2.96
0670	6#/CF, unfaced, 1" thick, R4.3		1000	.008		.97	.29		1.26	1.53
0690	1-1/2" thick, R6.5		890	.009		1.49	.33		1.82	2.15
0700	2" thick, R8.7		800	.010		2.10	.37		2.47	2.88
0721	2-1/2" thick, R10.9		800	.010		2.30	.37		2.67	3.10
0741	3" thick, R13		730	.011		2.75	.40		3.15	3.66
0821	Foil faced, 1" thick, R4.3		1000	.008		1.37	.29		1.66	1.97
0840	1-1/2" thick, R6.5		890	.009		1.96	.33		2.29	2.67
0850	2" thick, R8.7		800	.010		2.56	.37		2.93	3.39
0880	2-1/2" thick, R10.9		800	.010		3.08	.37		3.45	3.96
0900	3" thick, R13		730	.011		3.68	.40		4.08	4.68
1500	Foamglass, 1-1/2" thick, R4.5		800	.010		1.31	.37		1.68	2.01
1550	3" thick, R9		730	.011		3.16	.40		3.56	4.11
1600	Isocyanurate, 4' x 8' sheet, foil faced, both sides									
1610	1/2" thick, R3.9	1 Carp	800	.010	S.F.	.30	.37		.67	.90
1620	5/8" thick, R4.5		800	.010		.50	.37		.87	1.12
1630	3/4" thick, R5.4		800	.010		.35	.37		.72	.96
1640	1" thick, R7.2		800	.010		.55	.37		.92	1.18
1650	1-1/2" thick, R10.8		730	.011		.64	.40		1.04	1.33
1660	2" thick, R14.4	CN	730	.011		.87	.40		1.27	1.59

07 21 Thermal Insulation

07 21 13 – Board Insulation

07 21 13.10 Rigid Insulation

		Crew	Daily Output	Labor-Hours	Unit	Material	2007 Bare Costs Labor	Equipment	Total	Total Incl O&P
1670	3" thick, R21.6	1 Carp	730	.011	S.F.	1.90	.40		2.30	2.72
1680	4" thick, R28.8		730	.011		2.14	.40		2.54	2.98
1700	Perlite, 1" thick, R2.77		800	.010		.30	.37		.67	.90
1750	2" thick, R5.55		730	.011		.59	.40		.99	1.28
1900	Extruded polystyrene, 25 PSI compressive strength, 1" thick, R5		800	.010		.47	.37		.84	1.09
1940	2" thick R10		730	.011		.99	.40		1.39	1.72
1960	3" thick, R15		730	.011		1.34	.40		1.74	2.10
2100	Expanded polystyrene, 1" thick, R3.85		800	.010		.22	.37		.59	.81
2120	2" thick, R7.69		730	.011		.59	.40		.99	1.28
2140	3" thick, R11.49		730	.011		.75	.40		1.15	1.46

07 21 13.13 Foam Board Insulation

0010	**FOAM BOARD INSULATION**									
0600	Polystyrene, expanded, 1" thick, R4	1 Carp	680	.012	S.F.	.29	.43		.72	.99
0700	2" thick, R8	"	675	.012	"	.59	.43		1.02	1.33

07 21 16 – Blanket Insulation

07 21 16.20 Blanket Insulation for Walls

0010	**BLANKET INSULATION FOR WALLS**									
0040	Fiberglass, kraft faced, batts or blankets									
0060	3-1/2" thick, R11, 11" wide	1 Carp	1150	.007	S.F.	.37	.26		.63	.81
0080	15" wide		1600	.005		.37	.18		.55	.70
0100	23" wide		1600	.005		.37	.18		.55	.70
0140	6" thick, R19, 11" wide		1000	.008		.43	.29		.72	.93
0160	15" wide		1350	.006		.43	.22		.65	.81
0180	23" wide		1600	.005		.43	.18		.61	.76
0200	9" thick, R30, 15" wide		1150	.007		.97	.26		1.23	1.47
0220	23" wide		1350	.006		.78	.22		1	1.20
0240	12" thick, R38, 15" wide		1000	.008		.90	.29		1.19	1.45
0260	23" wide		1350	.006		.90	.22		1.12	1.33
0400	Fiberglass, foil faced, batts or blankets									
0420	3-1/2" thick, R11, 15" wide	1 Carp	1600	.005	S.F.	.54	.18		.72	.88
0440	23" wide		1600	.005		.54	.18		.72	.88
0460	6" thick, R19, 15" wide		1350	.006		.53	.22		.75	.92
0480	23" wide		1600	.005		.53	.18		.71	.87
0500	9" thick, R30, 15" wide		1150	.007		.83	.26		1.09	1.31
0550	23" wide		1350	.006		.83	.22		1.05	1.25
0800	Fiberglass, unfaced, batts or blankets									
0820	3-1/2" thick, R11, 15" wide	1 Carp	1350	.006	S.F.	.36	.22		.58	.74
0830	23" wide		1600	.005		.36	.18		.54	.69
0860	6" thick, R19, 15" wide		1150	.007		.57	.26		.83	1.03
0880	23" wide		1350	.006		.57	.22		.79	.97
0900	9" thick, R30, 15" wide		1000	.008		.83	.29		1.12	1.37
0920	23" wide		1150	.007		.83	.26		1.09	1.31
0940	12" thick, R38, 15" wide		1000	.008		.89	.29		1.18	1.44
0960	23" wide		1150	.007		.89	.26		1.15	1.38
1300	Mineral fiber batts, kraft faced									
1320	3-1/2" thick, R12	1 Carp	1600	.005	S.F.	.38	.18		.56	.71
1340	6" thick, R19		1600	.005		.50	.18		.68	.84
1380	10" thick, R30		1350	.006		.74	.22		.96	1.15
1850	Friction fit wire insulation supports, 16" O.C.		960	.008	Ea.	.08	.31		.39	.57
1900	For foil backing, add				S.F.	.06			.06	.07

07 21 Thermal Insulation

07 21 23 – Loose-Fill Insulation

07 21 23.10 Loose-Fill Insulation	Crew	Daily Output	Labor-Hours	Unit	Material	2007 Bare Costs Labor	Equipment	Total	Total Incl O&P
0010 **LOOSE-FILL INSULATION**									
0020 R3.8 per inch	1 Carp	200	.040	C.F.	.63	1.47		2.10	2.98
0040 Ceramic type (perlite), R3.2 per inch		200	.040		1.70	1.47		3.17	4.16
0080 Fiberglass wool, R4 per inch		200	.040		.50	1.47		1.97	2.84
0100 Mineral wool, R3 per inch		200	.040		.39	1.47		1.86	2.72
0300 Polystyrene, R4 per inch		200	.040		3.06	1.47		4.53	5.65
0400 Vermiculite or perlite, R2.7 per inch		200	.040		1.70	1.47		3.17	4.16
0700 Wood fiber, R3.85 per inch		200	.040		.69	1.47		2.16	3.05
07 21 23.20 Masonry Loose-Fill Insulation									
0010 **MASONRY LOOSE-FILL INSULATION**									
0100 In cores of concrete block, 4" thick wall, .115 CF/SF	D-1	4800	.003	S.F.	.20	.11		.31	.39
0200 6" thick wall, .175 CF/SF		3000	.005		.30	.18		.48	.60
0300 8" thick wall, .258 CF/SF		2400	.007		.44	.22		.66	.82
0400 10" thick wall, .340 CF/SF		1850	.009		.58	.29		.87	1.08
0500 12" thick wall, .422 CF/SF		1200	.013		.72	.44		1.16	1.47
0550 For sand fill, deduct from above					70%				
0600 Poured cavity wall, vermiculite or perlite, water repellant	D-1	250	.064	C.F.	1.70	2.13		3.83	5.10
0700 Foamed in place, urethane in 2-5/8" cavity	G-2	1035	.023	S.F.	.41	.70	.12	1.23	1.65
0800 For each 1" added thickness, add	"	2372	.010	"	.12	.31	.05	.48	.66

07 21 26 – Blown Insulation

07 21 26.10 Blown Insulation	Crew	Daily Output	Labor-Hours	Unit	Material	2007 Bare Costs Labor	Equipment	Total	Total Incl O&P
0010 **BLOWN INSULATION**									
0020 Cellulose, 3-1/2" thick, R13	G-4	5000	.005	S.F.	.21	.14	.05	.40	.51
0030 5-3/16" thick, R19		3800	.006		.31	.19	.07	.57	.71
0050 6-1/2" thick, R22		3000	.008		.40	.24	.09	.73	.91
0100 8-11/16" thick, R30		2600	.009		.54	.27	.10	.91	1.12
0120 10-7/8" thick, R38		1800	.013		.69	.39	.15	1.23	1.53
1000 Fiberglass, 5" thick, R11		3800	.006		.21	.19	.07	.47	.60
1050 6" thick, R13		3000	.008		.25	.24	.09	.58	.75
1100 8-1/2" thick, R19		2200	.011		.36	.32	.12	.80	1.02
1200 10" thick, R22		1800	.013		.42	.39	.15	.96	1.23
1300 12" thick, R26		1500	.016		.50	.47	.18	1.15	1.47
2000 Mineral wool, 4" thick, R12		3500	.007		.23	.20	.08	.51	.64
2050 6" thick, R17		2500	.010		.26	.28	.11	.65	.85
2100 9" thick, R23		1750	.014		.34	.40	.15	.89	1.17
2500 Wall installation, incl. drilling & patching from outside, two 1"									
2510 diam. holes @ 16" O.C., top & mid-point of wall, add to above									
2700 For masonry	G-4	415	.058	S.F.	.06	1.70	.64	2.40	3.42
2800 For wood siding		840	.029		.06	.84	.31	1.21	1.73
2900 For stucco/plaster		665	.036		.06	1.06	.40	1.52	2.16

07 21 27 – Reflective Insulation

07 21 27.10 Reflective Insulation	Crew	Daily Output	Labor-Hours	Unit	Material	2007 Bare Costs Labor	Equipment	Total	Total Incl O&P
0010 **REFLECTIVE INSULATION**									
0020 Aluminum foil on reinforced scrim	1 Carp	19	.421	C.S.F.	13.90	15.45		29.35	39.50
0100 Reinforced with woven polyolefin		19	.421		17	15.45		32.45	42.50
0500 With single bubble air space, R8.8		15	.533		25	19.55		44.55	58
0600 With double bubble air space, R9.8		15	.533		26	19.55		45.55	59

07 21 29 – Sprayed Insulation

07 21 29.10 Sprayed Insulation

0010 **SPRAYED INSULATION**

07 21 Thermal Insulation

07 21 29 – Sprayed Insulation

07 21 29.10 Sprayed Insulation		Crew	Daily Output	Labor-Hours	Unit	Material	2007 Bare Costs Labor	Equipment	Total	Total Incl O&P
0020	Fibrous/cementitious, finished wall, 1" thick, R3.7	G-2	2050	.012	S.F.	.24	.36	.06	.66	.87
0100	Attic, 5.2" thick, R19	"	1550	.015	"	.39	.47	.08	.94	1.24
0300	Foam type, incl. preparation									
0600	3 #/CF, 1" thick, R3.8	G-2	770	.031	S.F.	.50	.95	.16	1.61	2.16
0700	2" thick, R7.5	"	475	.051	"	1.02	1.53	.26	2.81	3.74

07 22 Roof and Deck Insulation

07 22 16 – Roof Board Insulation

07 22 16.10 Roof Deck Insulation

		Crew	Daily Output	Labor-Hours	Unit	Material	Labor	Equipment	Total	Total Incl O&P
0010	**ROOF DECK INSULATION**									
0020	Fiberboard low density, 1/2" thick R1.39	1 Rofc	1000	.008	S.F.	.24	.25		.49	.69
0030	1" thick R2.78		800	.010		.42	.32		.74	1
0080	1 1/2" thick R4.17		800	.010		.63	.32		.95	1.23
0100	2" thick R5.56		800	.010		.84	.32		1.16	1.46
0110	Fiberboard high density, 1/2" thick R1.3		1000	.008		.22	.25		.47	.67
0120	1" thick R2.5		800	.010		.44	.32		.76	1.02
0130	1-1/2" thick R3.8		800	.010		.65	.32		.97	1.26
0200	Fiberglass, 3/4" thick R2.78		1000	.008		.55	.25		.80	1.04
0400	15/16" thick R3.70		1000	.008		.73	.25		.98	1.23
0460	1-1/16" thick R4.17		1000	.008		.92	.25		1.17	1.44
0600	1-5/16" thick R5.26		1000	.008		1.25	.25		1.50	1.81
0650	2-1/16" thick R8.33		800	.010		1.34	.32		1.66	2.01
0700	2-7/16" thick R10		800	.010		1.53	.32		1.85	2.22
1500	Foamglass, 1-1/2" thick R4.5		800	.010		1.30	.32		1.62	1.97
1530	3" thick R9		700	.011		2.65	.36		3.01	3.54
1600	Tapered for drainage		600	.013	B.F.	1.13	.42		1.55	1.96
1650	Perlite, 1/2" thick R1.32		1050	.008	S.F.	.33	.24		.57	.77
1655	3/4" thick R2.08		800	.010		.36	.32		.68	.94
1660	1" thick R2.78		800	.010		.45	.32		.77	1.04
1670	1-1/2" thick R4.17		800	.010		.47	.32		.79	1.06
1680	2" thick R5.56		700	.011		.78	.36		1.14	1.48
1685	2-1/2" thick R6.67		700	.011		.92	.36		1.28	1.63
1690	Tapered for drainage		800	.010	B.F.	.71	.32		1.03	1.32
1700	Polyisocyanurate, 2#/CF density, 3/4" thick, R5.1		1500	.005	S.F.	.44	.17		.61	.77
1705	1" thick R7.14		1400	.006		.50	.18		.68	.86
1715	1-1/2" thick R10.87		1250	.006		.63	.20		.83	1.04
1725	2" thick R14.29		1100	.007		.83	.23		1.06	1.30
1735	2-1/2" thick R16.67		1050	.008		1.01	.24		1.25	1.52
1745	3" thick R21.74		1000	.008		1.26	.25		1.51	1.82
1755	3-1/2" thick R25		1000	.008		1.93	.25		2.18	2.55
1765	Tapered for drainage		1400	.006	B.F.	1.93	.18		2.11	2.43
1900	Extruded Polystyrene									
1910	15 PSI compressive strength, 1" thick, R5	1 Rofc	1500	.005	S.F.	.45	.17		.62	.79
1920	2" thick, R10		1250	.006		.59	.20		.79	1
1930	3" thick R15		1000	.008		1.21	.25		1.46	1.76
1932	4" thick R20		1000	.008		1.56	.25		1.81	2.15
1934	Tapered for drainage		1500	.005	B.F.	.50	.17		.67	.84
1940	25 PSI compressive strength, 1" thick R5		1500	.005	S.F.	.63	.17		.80	.98
1942	2" thick R10		1250	.006		1.21	.20		1.41	1.68
1944	3" thick R15		1000	.008		1.84	.25		2.09	2.45
1946	4" thick R20		1000	.008		2.59	.25		2.84	3.28

07 22 Roof and Deck Insulation

07 22 16 – Roof Board Insulation

07 22 16.10 Roof Deck Insulation

		Crew	Daily Output	Labor-Hours	Unit	Material	2007 Bare Costs Labor	2007 Bare Costs Equipment	Total	Total Incl O&P
1948	Tapered for drainage	1 Rofc	1500	.005	B.F.	.53	.17		.70	.87
1950	40 psi compressive strength, 1" thick R5		1500	.005	S.F.	.47	.17		.64	.81
1952	2" thick R10		1250	.006		.91	.20		1.11	1.35
1954	3" thick R15		1000	.008		1.34	.25		1.59	1.90
1956	4" thick R20		1000	.008	↓	1.79	.25		2.04	2.40
1958	Tapered for drainage		1400	.006	B.F.	.67	.18		.85	1.05
1960	60 PSI compressive strength, 1" thick R5		1450	.006	S.F.	.56	.18		.74	.92
1962	2" thick R10		1200	.007		1	.21		1.21	1.46
1964	3" thick R15		975	.008		1.48	.26		1.74	2.07
1966	4" thick R20		950	.008	↓	2.07	.27		2.34	2.73
1968	Tapered for drainage		1400	.006	B.F.	.81	.18		.99	1.20
2010	Expanded polystyrene, 1#/CF density, 3/4" thick R2.89		1500	.005	S.F.	.29	.17		.46	.61
2020	1" thick R3.85		1500	.005		.29	.17		.46	.61
2100	2" thick R7.69		1250	.006		.59	.20		.79	1
2110	3" thick R11.49		1250	.006		.87	.20		1.07	1.31
2120	4" thick R15.38		1200	.007		.80	.21		1.01	1.24
2130	5" thick R19.23		1150	.007		1.01	.22		1.23	1.49
2140	6" thick R23.26		1150	.007	↓	1.18	.22		1.40	1.68
2150	Tapered for drainage	↓	1500	.005	B.F.	.47	.17		.64	.81
2400	Composites with 2" EPS									
2410	1" fiberboard	1 Rofc	950	.008	S.F.	1.04	.27		1.31	1.59
2420	7/16" oriented strand board		800	.010		1.23	.32		1.55	1.89
2430	1/2" plywood		800	.010		1.33	.32		1.65	2
2440	1" perlite	↓	800	.010	↓	1.09	.32		1.41	1.74
2450	Composites with 1-1/2" polyisocyanurate									
2460	1" fiberboard	1 Rofc	800	.010	S.F.	1.42	.32		1.74	2.10
2470	1" perlite		850	.009		1.49	.30		1.79	2.15
2480	7/16" oriented strand board	↓	800	.010	↓	1.72	.32		2.04	2.43

07 24 Exterior Insulation and Finish Systems

07 24 13 – Polymer Based Exterior Insulation and Finish Systems

07 24 13.10 Polymer Based Exterior Insulation and Finish Systems

		Crew	Daily Output	Labor-Hours	Unit	Material	2007 Bare Costs Labor	2007 Bare Costs Equipment	Total	Total Incl O&P
0010	**POLYMER BASED EXTERIOR INSULATION AND FINISH SYSTEMS**									
0095	Field applied, 1" EPS insulation	J-1	295	.136	S.F.	2.15	4.29	.39	6.83	9.30
0100	With 1/2" cement board sheathing		220	.182		2.87	5.75	.52	9.14	12.45
0105	2" EPS insulation		295	.136		2.52	4.29	.39	7.20	9.70
0110	With 1/2" cement board sheathing		220	.182		3.24	5.75	.52	9.51	12.85
0115	3" EPS insulation		295	.136		2.68	4.29	.39	7.36	9.85
0120	With 1/2" cement board sheathing		220	.182		3.40	5.75	.52	9.67	13
0125	4" EPS insulation		295	.136		3.11	4.29	.39	7.79	10.35
0130	With 1/2" cement board sheathing		220	.182		4.55	5.75	.52	10.82	14.25
0140	Premium finish add		1265	.032		.31	1	.09	1.40	1.95
0150	Heavy duty reinforcement add	↓	914	.044	↓	1.07	1.38	.12	2.57	3.41
0160	2.5#/S.Y. metal lath substrate add	1 Lath	75	.107	S.Y.	2.35	3.59		5.94	7.95
0170	3.4#/S.Y. metal lath substrate add	"	75	.107	"	2.55	3.59		6.14	8.15
0180	Color or texture change,	J-1	1265	.032	S.F.	.82	1	.09	1.91	2.51
0190	With substrate leveling base coat	1 Plas	530	.015		.82	.51		1.33	1.67
0210	With substrate sealing base coat	1 Pord	1224	.007	↓	.08	.21		.29	.41
0370	V groove shape in panel face				L.F.	.59			.59	.65
0380	U groove shape in panel face				"	.78			.78	.86
0440	For higher than one story, add						25%			

07 26 Vapor Retarders

07 26 10 – Vapor Retarders

07 26 10.10 Vapor Retarders

		Crew	Daily Output	Labor-Hours	Unit	Material	2007 Bare Costs Labor	2007 Bare Costs Equipment	Total	Total Incl O&P
0010	**VAPOR RETARDERS**									
0020	Aluminum and kraft laminated, foil 1 side	1 Carp	37	.216	Sq.	5.10	7.95		13.05	18
0100	Foil 2 sides		37	.216		8.05	7.95		16	21
0400	Asphalt felt sheathing paper, 15#		37	.216		4.23	7.95		12.18	17
0450	Housewrap, exterior, spun bonded polypropylene									
0470	Small roll	1 Carp	3800	.002	S.F.	.24	.08		.32	.38
0480	Large roll	"	4000	.002	"	.13	.07		.20	.25
0500	Material only, 3' x 111.1' roll				Ea.	80			80	88
0520	9' x 111.1' roll				"	130			130	143
0600	Polyethylene vapor barrier, standard, .002" thick	1 Carp	37	.216	Sq.	1.08	7.95		9.03	13.55
0700	.004" thick		37	.216		3.03	7.95		10.98	15.70
0900	.006" thick		37	.216		4.68	7.95		12.63	17.50
1200	.010" thick		37	.216		5.35	7.95		13.30	18.20
1300	Clear reinforced, fire retardant, .008" thick		37	.216		9.75	7.95		17.70	23
1350	Cross laminated type, .003" thick		37	.216		6.85	7.95		14.80	19.85
1400	.004" thick		37	.216		7.50	7.95		15.45	20.50
1500	Red rosin paper, 5 sq rolls, 4 lb per square		37	.216		2.09	7.95		10.04	14.65
1600	5 lbs. per square		37	.216		2.80	7.95		10.75	15.45
1800	Reinf. waterproof, .002" polyethylene backing, 1 side		37	.216		5.10	7.95		13.05	17.95
1900	2 sides		37	.216		6.70	7.95		14.65	19.70
2100	Roof deck vapor barrier, class 1 metal decks	1 Rofc	37	.216		14.75	6.90		21.65	28
2200	For all other decks	"	37	.216		10.50	6.90		17.40	23.50
2400	Waterproofed kraft with sisal or fiberglass fibers, minimum	1 Carp	37	.216		5.50	7.95		13.45	18.40
2500	Maximum	"	37	.216		13.60	7.95		21.55	27.50

07 31 Shingles and Shakes

07 31 13 – Asphalt Shingles

07 31 13.10 Asphalt Shingles

			Crew	Daily Output	Labor-Hours	Unit	Material	2007 Bare Costs Labor	2007 Bare Costs Equipment	Total	Total Incl O&P
0010	**ASPHALT SHINGLES**										
0100	Standard strip shingles										
0150	Inorganic, class A, 210-235 lb/sq	CN	1 Rofc	5.50	1.455	Sq.	41.50	46.50		88	125
0155	Pneumatic nailed			7	1.143		41.50	36.50		78	108
0200	Organic, class C, 235-240 lb/sq			5	1.600		44.50	51		95.50	136
0205	Pneumatic nailed			6.25	1.280		44.50	40.50		85	118
0250	Standard, laminated multi-layered shingles										
0300	Class A, 240-260 lb/sq		1 Rofc	4.50	1.778	Sq.	52	56.50		108.50	154
0305	Pneumatic nailed			5.63	1.422		52	45		97	134
0350	Class C, 260-300 lb/square, 4 bundles/square			4	2		51.50	63.50		115	165
0355	Pneumatic nailed			5	1.600		51.50	51		102.50	144
0400	Premium, laminated multi-layered shingles										
0450	Class A, 260-300 lb, 4 bundles/sq		1 Rofc	3.50	2.286	Sq.	67.50	72.50		140	197
0455	Pneumatic nailed			4.37	1.831		67.50	58		125.50	173
0500	Class C, 300-385 lb/square, 5 bundles/square			3	2.667		75.50	85		160.50	228
0505	Pneumatic nailed			3.75	2.133		75.50	68		143.50	199
0800	#15 felt underlayment			64	.125		4.23	3.98		8.21	11.40
0825	#30 felt underlayment			58	.138		7.05	4.39		11.44	15.20
0850	Self adhering polyethylene and rubberized asphalt underlayment			22	.364		49.50	11.55		61.05	74
0900	Ridge shingles			330	.024	L.F.	1.44	.77		2.21	2.89
0905	Pneumatic nailed			412.50	.019	"	1.44	.62		2.06	2.63
1000	For steep roofs (7 to 12 pitch or greater), add							50%			

07 31 Shingles and Shakes

07 31 16 – Metal Shingles

07 31 16.10 Aluminum Shingles		Crew	Daily Output	Labor-Hours	Unit	Material	2007 Bare Costs Labor	Equipment	Total	Total Incl O&P
0010	**ALUMINUM SHINGLES**									
0020	Mill finish, .019 thick	1 Carp	5	1.600	Sq.	178	58.50		236.50	287
0100	.020" thick	"	5	1.600		188	58.50		246.50	299
0300	For colors, add					16			16	17.60
0600	Ridge cap, .024" thick	1 Carp	170	.047	L.F.	1.99	1.73		3.72	4.88
0700	End wall flashing, .024" thick		170	.047		1.71	1.73		3.44	4.57
0900	Valley section, .024" thick		170	.047		2.65	1.73		4.38	5.60
1000	Starter strip, .024" thick		400	.020		1.48	.73		2.21	2.77
1200	Side wall flashing, .024" thick		170	.047		1.71	1.73		3.44	4.57
1500	Gable flashing, .024" thick		400	.020		1.61	.73		2.34	2.91

07 31 16.20 Steel Shingles

		Crew	Daily Output	Labor-Hours	Unit	Material	Labor	Equipment	Total	Total Incl O&P
0010	**STEEL SHINGLES**									
0012	Galvanized, 26 gauge	1 Rots	2.20	3.636	Sq.	194	116		310	410
0200	24 gauge	"	2.20	3.636		203	116		319	420
0300	For colored galvanized shingles, add					53			53	58.50
0500	For 1" factory applied polystyrene insulation, add					39			39	43

07 31 19 – Mineral-Fiber Cement Shingles

07 31 19.10 Fiber Cement Shingles

		Crew	Daily Output	Labor-Hours	Unit	Material	Labor	Equipment	Total	Total Incl O&P
0010	**FIBER CEMENT SHINGLES**									
0012	Field shingles, 16" x 9.35", 500 lb per square	1 Rofc	2.20	3.636	Sq.	315	116		431	540
0110	Starters, 16" x 9.35"		3	2.667	C.L.F.	106	85		191	261
0120	Hip & ridge, 4.75" x 14"		1	8	"	770	254		1,024	1,275
0200	Shakes, 16" x 9.35", 550 lb per square		2.20	3.636	Sq.	283	116		399	505
0300	Hip & ridge, 4.75 x 14"		1	8	C.L.F.	770	254		1,024	1,275
0400	Hexagonal, 16" x 16"		3	2.667	Sq.	212	85		297	375
0500	Square, 16" x 16"		3	2.667	"	190	85		275	355

07 31 26 – Slate Shingles

07 31 26.10 Slate Shingles

		Crew	Daily Output	Labor-Hours	Unit	Material	Labor	Equipment	Total	Total Incl O&P
0010	**SLATE SHINGLES** R073126-20									
0100	Buckingham Virginia black, 3/16" - 1/4" thick	1 Rots	1.75	4.571	Sq.	395	146		541	685
0200	1/4" thick		1.75	4.571		410	146		556	700
0900	Pennsylvania black, Bangor, #1 clear		1.75	4.571		475	146		621	775
1200	Vermont, unfading, green, mottled green		1.75	4.571		435	146		581	730
1300	Semi-weathering green & gray		1.75	4.571		345	146		491	630
1400	Purple		1.75	4.571		390	146		536	680
1500	Black or gray		1.75	4.571		445	146		591	740
1600	Red		1.75	4.571		1,125	146		1,271	1,475
1700	Variegated purple		1.75	4.571		410	146		556	700

07 31 29 – Wood Shingles and Shakes

07 31 29.13 Wood Shingles

		Crew	Daily Output	Labor-Hours	Unit	Material	Labor	Equipment	Total	Total Incl O&P
0010	**WOOD SHINGLES**									
0012	16" No. 1 red cedar shingles, 5" exposure, on roof	1 Carp	2.50	3.200	Sq.	189	117		306	390
0015	Pneumatic nailed		3.25	2.462		189	90.50		279.50	350
0200	7-1/2" exposure, on walls		2.05	3.902		126	143		269	360
0205	Pneumatic nailed		2.67	2.996		126	110		236	310
0300	18" No. 1 red cedar perfections, 5-1/2" exposure, on roof		2.75	2.909		167	107		274	350
0305	Pneumatic nailed		3.57	2.241		167	82		249	310
0500	7-1/2" exposure, on walls		2.25	3.556		123	130		253	340
0505	Pneumatic nailed		2.92	2.740		123	101		224	292
0600	Resquared, and rebutted, 5-1/2" exposure, on roof		3	2.667		240	98		338	415

07 31 Shingles and Shakes

07 31 29 – Wood Shingles and Shakes

07 31 29.13 Wood Shingles		Crew	Daily Output	Labor-Hours	Unit	Material	2007 Bare Costs Labor	Equipment	Total	Total Incl O&P
0605	Pneumatic nailed	1 Carp	3.90	2.051	Sq.	240	75.50		315.50	380
0900	7-1/2" exposure, on walls		2.45	3.265		176	120		296	380
0905	Pneumatic nailed		3.18	2.516		176	92.50		268.50	340
1000	Add to above for fire retardant shingles, 16" long					42			42	46
1050	18" long					42			42	46
1060	Preformed ridge shingles	1 Carp	400	.020	L.F.	1.75	.73		2.48	3.07
1100	Hand-split red cedar shakes, 1/2" thick x 24" long, 10" exp. on roof		2.50	3.200	Sq.	165	117		282	365
1105	Pneumatic nailed		3.25	2.462		165	90.50		255.50	320
1110	3/4" thick x 24" long, 10" exp. on roof		2.25	3.556		165	130		295	385
1115	Pneumatic nailed		2.92	2.740		165	101		266	340
1200	1/2" thick, 18" long, 8-1/2" exp. on roof		2	4		191	147		338	440
1205	Pneumatic nailed		2.60	3.077		191	113		304	385
1210	3/4" thick x 18" long, 8 1/2" exp. on roof		1.80	4.444		191	163		354	465
1215	Pneumatic nailed		2.34	3.419		191	125		316	405
1255	10" exp. on walls		2	4		216	147		363	465
1260	10" exposure on walls, pneumatic nailed		2.60	3.077		216	113		329	415
1700	Add to above for fire retardant shakes, 24" long					42			42	46
1800	18" long					42			42	46
1810	Ridge shakes	1 Carp	350	.023	L.F.	3	.84		3.84	4.61
2000	White cedar shingles, 16" long, extras, 5" exposure, on roof		2.40	3.333	Sq.	184	122		306	395
2005	Pneumatic nailed		3.12	2.564		184	94		278	350
2050	5" exposure on walls		2	4		184	147		331	430
2055	Pneumatic nailed		2.60	3.077		184	113		297	380
2100	7-1/2" exposure, on walls		2	4		132	147		279	375
2105	Pneumatic nailed		2.60	3.077		132	113		245	320
2150	"B" grade, 5" exposure on walls		2	4		170	147		317	415
2155	Pneumatic nailed		2.60	3.077		170	113		283	360
2300	For 15# organic felt underlayment on roof, 1 layer, add		64	.125		4.23	4.59		8.82	11.80
2400	2 layers, add		32	.250		8.45	9.20		17.65	23.50
2600	For steep roofs (7/12 pitch or greater), add to above						50%			
2700	Panelized systems, No.1 cedar shingles on 5/16" CDX plywood									
2800	On walls, 8' strips, 7" or 14" exposure	2 Carp	700	.023	S.F.	3.40	.84		4.24	5.05
3500	On roofs, 8' strips, 7" or 14" exposure	1 Carp	3	2.667	Sq.	320	98		418	500
3505	Pneumatic nailed	"	4	2	"	320	73.50		393.50	465

07 32 Roof Tiles

07 32 13 – Clay Roof Tiles

07 32 13.10 Clay Tiles		Crew	Daily Output	Labor-Hours	Unit	Material	2007 Bare Costs Labor	Equipment	Total	Total Incl O&P
0010	**CLAY TILES**									
0200	Lanai tile or Classic tile, 158 pc per sq	1 Rots	1.65	4.848	Sq.	470	155		625	785
0300	Americana, 158 pc per sq, most colors		1.65	4.848		605	155		760	935
0350	Green, gray or brown		1.65	4.848		590	155		745	910
0400	Blue		1.65	4.848		590	155		745	910
0600	Spanish tile, 171 pc per sq, red		1.80	4.444		298	142		440	565
0800	Blend		1.80	4.444		565	142		707	860
0900	Glazed white		1.80	4.444		625	142		767	930
1100	Mission tile, 192 pc per sq, machine scored finish, red		1.15	6.957		700	222		922	1,150
1700	French tile, 133 pc per sq, smooth finish, red		1.35	5.926		635	189		824	1,025
1750	Blue or green		1.35	5.926		820	189		1,009	1,225
1800	Norman black 317 pc per sq		1	8		930	256		1,186	1,450
2200	Williamsburg tile, 158 pc per sq, aged cedar		1.35	5.926		560	189		749	935

07 32 Roof Tiles

07 32 13 – Clay Roof Tiles

07 32 13.10 Clay Tiles		Crew	Daily Output	Labor-Hours	Unit	Material	2007 Bare Costs Labor	Equipment	Total	Total Incl O&P
2250	Gray or green	1 Rots	1.35	5.926	Sq.	560	189		749	935
2510	One piece mission tile, natural red, 75 pc per square	↓	1.65	4.848	↓	218	155		373	505
2530	Mission Tile, 134 pc per square	↓	1.15	6.957	↓	248	222		470	650

07 32 16 – Concrete Roof Tiles

07 32 16.10 Concrete Tiles

		Crew	Daily Output	Labor-Hours	Unit	Material	Labor	Equipment	Total	Total Incl O&P
0010	**CONCRETE TILES**									
0020	Corrugated, 13" x 16-1/2", 90 per sq, 950 lb per sq									
0050	Earthtone colors, nailed to wood deck	1 Rots	1.35	5.926	Sq.	93.50	189		282.50	425
0150	Blues		1.35	5.926		94.50	189		283.50	425
0200	Greens	↓	1.35	5.926	↓	94.50	189		283.50	425
0250	Premium colors	↓	1.35	5.926	↓	208	189		397	550
0500	Shakes, 13" x 16-1/2", 90 per sq, 950 lb per sq									
0600	All colors, nailed to wood deck	1 Rots	1.50	5.333	Sq.	245	170		415	560
1500	Accessory pieces, ridge & hip, 10" x 16-1/2", 8 lbs. each				Ea.	3.30			3.30	3.63
1700	Rake, 6-1/2" x 16-3/4", 9 lbs. each					3.30			3.30	3.63
1800	Mansard hip, 10" x 16-1/2", 9.2 lbs. each					3.30			3.30	3.63
1900	Hip starter, 10" x 16-1/2", 10.5 lbs. each					10.40			10.40	11.45
2000	3 or 4 way apex, 10" each side, 11.5 lbs. each				↓	12			12	13.20

07 32 19 – Metal Roof Tiles

07 32 19.10 Aluminum Roof Tiles

		Crew	Daily Output	Labor-Hours	Unit	Material	Labor	Equipment	Total	Total Incl O&P
0010	**ALUMINUM ROOF TILES**									
0020	Accessories included, .032" thick, mission tile	1 Carp	2.50	3.200	Sq.	605	117		722	850
0200	Spanish tiles	"	3	2.667	"	415	98		513	610

07 41 Roof Panels

07 41 13 – Metal Roof Panels

07 41 13.10 Aluminum Roof Panels

		Crew	Daily Output	Labor-Hours	Unit	Material	Labor	Equipment	Total	Total Incl O&P
0010	**ALUMINUM ROOF PANELS**									
0020	Corrugated or ribbed, .0155" thick, natural	G-3	1200	.027	S.F.	.72	.96		1.68	2.28
0300	Painted		1200	.027		1.05	.96		2.01	2.65
0400	Corrugated, .018" thick, on steel frame, natural finish		1200	.027		.95	.96		1.91	2.54
0600	Painted		1200	.027		1.17	.96		2.13	2.78
0700	Corrugated, on steel frame, natural, .024" thick		1200	.027		1.36	.96		2.32	2.99
0800	Painted		1200	.027		1.65	.96		2.61	3.31
0900	.032" thick, natural		1200	.027		1.75	.96		2.71	3.42
1200	Painted		1200	.027		2.24	.96		3.20	3.95
1300	V-Beam, on steel frame construction, .032" thick, natural		1200	.027		1.85	.96		2.81	3.53
1500	Painted		1200	.027		2.27	.96		3.23	3.99
1600	.040" thick, natural		1200	.027		2.25	.96		3.21	3.97
1800	Painted		1200	.027		2.71	.96		3.67	4.47
1900	.050" thick, natural		1200	.027		2.70	.96		3.66	4.46
2100	Painted		1200	.027		3.26	.96		4.22	5.10
2200	For roofing on wood frame, deduct		4600	.007	↓	.06	.25		.31	.46
2400	Ridge cap, .032" thick, natural	↓	800	.040	L.F.	2.13	1.45		3.58	4.58

07 41 13.20 Steel Roofing Panels

		Crew	Daily Output	Labor-Hours	Unit	Material	Labor	Equipment	Total	Total Incl O&P
0010	**STEEL ROOFING PANELS**									
0012	Corrugated or ribbed, on steel framing, 30 ga galv	G-3	1100	.029	S.F.	1.50	1.05		2.55	3.28
0100	28 ga		1050	.030		1.60	1.10		2.70	3.47
0300	26 ga		1000	.032		1.12	1.16		2.28	3.02
0400	24 ga	↓	950	.034	↓	2.60	1.22		3.82	4.74

07 41 Roof Panels

07 41 13 – Metal Roof Panels

07 41 13.20 Steel Roofing Panels

		Crew	Daily Output	Labor-Hours	Unit	Material	2007 Bare Costs Labor	Equipment	Total	Total Incl O&P
0600	Colored, 28 ga	G-3	1050	.030	S.F.	1.50	1.10		2.60	3.36
0700	26 ga		1000	.032		1.78	1.16		2.94	3.75
0710	Flat profile, 1-3/4" standing seams, 10" wide, standard finish, 26 ga		1000	.032		3.47	1.16		4.63	5.60
0715	24 ga		950	.034		4.03	1.22		5.25	6.30
0720	22 ga		900	.036		4.97	1.29		6.26	7.45
0725	Zinc aluminum alloy finish, 26 ga		1000	.032		2.72	1.16		3.88	4.78
0730	24 ga		950	.034		3.25	1.22		4.47	5.45
0735	22 ga		900	.036		3.72	1.29		5.01	6.10
0740	12" wide, standard finish, 26 ga		1000	.032		3.46	1.16		4.62	5.60
0745	24 ga		950	.034		4.55	1.22		5.77	6.90
0750	Zinc aluminum alloy finish, 26 ga		1000	.032		3.94	1.16		5.10	6.10
0755	24 ga		950	.034		3.24	1.22		4.46	5.45
0840	Flat profile, 1" x 3/8" batten, 12" wide, standard finish, 26 ga		1000	.032		3.05	1.16		4.21	5.15
0845	24 ga		950	.034		3.58	1.22		4.80	5.80
0850	22 ga		900	.036		4.29	1.29		5.58	6.70
0855	Zinc aluminum alloy finish, 26 ga		1000	.032		2.93	1.16		4.09	5
0860	24 ga		950	.034		3.27	1.22		4.49	5.50
0865	22 ga		900	.036		3.78	1.29		5.07	6.15
0870	16-1/2" wide, standard finish, 24 ga		950	.034		3.53	1.22		4.75	5.75
0875	22 ga		900	.036		3.95	1.29		5.24	6.35
0880	Zinc aluminum alloy finish, 24 ga		950	.034		3.08	1.22		4.30	5.25
0885	22 ga		900	.036		3.44	1.29		4.73	5.75
0890	Flat profile, 2" x 2" batten, 12" wide, standard finish, 26 ga		1000	.032		3.51	1.16		4.67	5.65
0895	24 ga		950	.034		4.19	1.22		5.41	6.50
0900	22 ga		900	.036		5.15	1.29		6.44	7.65
0905	Zinc aluminum alloy finish, 26 ga		1000	.032		3.27	1.16		4.43	5.40
0910	24 ga		950	.034		3.73	1.22		4.95	6
0915	22 ga		900	.036		4.35	1.29		5.64	6.80
0920	16-1/2" wide, standard finish, 24 ga		950	.034		3.86	1.22		5.08	6.15
0925	22 ga		900	.036		4.51	1.29		5.80	6.95
0930	Zinc aluminum alloy finish, 24 ga		950	.034		3.49	1.22		4.71	5.70
0935	22 ga		900	.036		3.97	1.29		5.26	6.35
1200	Ridge, galvanized, 10" wide		800	.040	L.F.	2.84	1.45		4.29	5.35
1210	20" wide		750	.043	"	4.40	1.54		5.94	7.25

07 41 33 – Plastic Roof Panels

07 41 33.10 Corrugated Fiberglass Panels

		Crew	Daily Output	Labor-Hours	Unit	Material	2007 Bare Costs Labor	Equipment	Total	Total Incl O&P
0010	**CORRUGATED FIBERGLASS PANELS**									
0012	Corrugated, 8 oz per SF	G-3	1000	.032	S.F.	1.53	1.16		2.69	3.47
0100	12 oz per SF		1000	.032		3.33	1.16		4.49	5.45
0300	Corrugated siding, 6 oz per SF		880	.036		1.53	1.31		2.84	3.71
0400	8 oz per SF		880	.036		2.30	1.31		3.61	4.56
0500	Fire retardant		880	.036		3.12	1.31		4.43	5.45
0600	12 oz. siding, textured		880	.036		3.22	1.31		4.53	5.55
0700	Fire retardant		880	.036		4.26	1.31		5.57	6.70
0900	Flat panels, 6 oz per SF, clear or colors		880	.036		1.78	1.31		3.09	3.99
1100	Fire retardant, class A		880	.036		3.09	1.31		4.40	5.45
1300	8 oz per SF, clear or colors		880	.036		2.31	1.31		3.62	4.57
1700	Sandwich panels, fiberglass, 1-9/16" thick, panels to 20 SF		180	.178		22.50	6.45		28.95	34.50
1900	As above, but 2-3/4" thick, panels to 100 SF		265	.121		16.55	4.37		20.92	25

07 42 Wall Panels

07 42 13 – Metal Wall Panels

07 42 13.10 Mansard Panels

		Crew	Daily Output	Labor-Hours	Unit	Material	2007 Bare Costs Labor	Equipment	Total	Total Incl O&P
0010	**MANSARD PANELS**									
0600	Stock units, straight surfaces	1 Shee	115	.070	S.F.	2.51	3.03		5.54	7.45
0700	Concave or convex surfaces		75	.107	"	2.76	4.65		7.41	10.20
0800	For framing, to 5' high, add		115	.070	L.F.	2.76	3.03		5.79	7.70
0900	Soffits, to 1' wide	↓	125	.064	S.F.	1.37	2.79		4.16	5.80

07 42 13.20 Aluminum Siding Panels

		Crew	Daily Output	Labor-Hours	Unit	Material	Labor	Equipment	Total	Total Incl O&P
0010	**ALUMINUM SIDING PANELS**									
0012	Corrugated, on steel framing, .019 thick, natural finish	G-3	775	.041	S.F.	1.11	1.49		2.60	3.53
0100	Painted		775	.041		1.20	1.49		2.69	3.63
0400	Farm type, .021" thick on steel frame, natural		775	.041		1.12	1.49		2.61	3.54
0600	Painted		775	.041		1.20	1.49		2.69	3.63
0700	Industrial type, corrugated, on steel, .024" thick, mill		775	.041		1.56	1.49		3.05	4.03
0900	Painted		775	.041		1.69	1.49		3.18	4.17
1000	.032" thick, mill		775	.041		1.80	1.49		3.29	4.29
1200	Painted		775	.041		2.20	1.49		3.69	4.73
1300	V-Beam, on steel frame, .032" thick, mill		775	.041		2.03	1.49		3.52	4.54
1500	Painted		775	.041		2.32	1.49		3.81	4.86
1600	.040" thick, mill		775	.041		2.46	1.49		3.95	5
1800	Painted		775	.041		2.89	1.49		4.38	5.50
1900	.050" thick, mill		775	.041		2.91	1.49		4.40	5.50
2100	Painted		775	.041		3.47	1.49		4.96	6.15
2200	Ribbed, 3" profile, on steel frame, .032" thick, natural		775	.041		1.75	1.49		3.24	4.24
2400	Painted		775	.041		2.22	1.49		3.71	4.75
2500	.040" thick, natural		775	.041		2	1.49		3.49	4.51
2700	Painted		775	.041		2.34	1.49		3.83	4.88
2750	.050" thick, natural		775	.041		2.31	1.49		3.80	4.85
2760	Painted		775	.041		2.65	1.49		4.14	5.25
3300	For siding on wood frame, deduct from above	↓	2800	.011	↓	.07	.41		.48	.72
3400	Screw fasteners, aluminum, self tapping, neoprene washer, 1"				M	152			152	167
3600	Stitch screws, self tapping, with neoprene washer, 5/8"				"	114			114	125
3630	Flashing, sidewall, .032" thick	G-3	800	.040	L.F.	2.10	1.45		3.55	4.55
3650	End wall, .040" thick		800	.040		2.46	1.45		3.91	4.95
3670	Closure strips, corrugated, .032" thick		800	.040		.62	1.45		2.07	2.92
3680	Ribbed, 4" or 8", .032" thick		800	.040		.62	1.45		2.07	2.92
3690	V-beam, .040" thick	↓	800	.040	↓	.85	1.45		2.30	3.18
3800	Horizontal, colored clapboard, 8" wide, plain	2 Carp	515	.031	S.F.	1.30	1.14		2.44	3.21
3900	Insulated		515	.031		1.64	1.14		2.78	3.58
4000	Vertical board & batten, colored, non-insulated	↓	515	.031		1.27	1.14		2.41	3.18
4200	For simulated wood design, add				↓	.09			.09	.10
4300	Corners for above, outside	2 Carp	515	.031	V.L.F.	1.95	1.14		3.09	3.93
4500	Inside corners	"	515	.031	"	1.15	1.14		2.29	3.05
4600	Sandwich panels, 1" insulation, single story	G-3	395	.081	S.F.	5.70	2.93		8.63	10.80
4900	Multi-story	"	345	.093		7.95	3.35		11.30	13.95
5100	For baked enamel finish 1 side, add				↓	.30			.30	.33
5200	See also Metal Facing Panels, division 07 44 73.10									

07 42 13.30 Steel Siding

		Crew	Daily Output	Labor-Hours	Unit	Material	Labor	Equipment	Total	Total Incl O&P
0010	**STEEL SIDING**									
0020	Beveled, vinyl coated, 8" wide, including fasteners	1 Carp	265	.030	S.F.	1.62	1.11		2.73	3.51
0050	10" wide	"	275	.029		1.73	1.07		2.80	3.56
0080	Galv, corrugated or ribbed, on steel frame, 30 gauge	G-3	800	.040		1.08	1.45		2.53	3.43
0100	28 gauge		795	.040		1.13	1.46		2.59	3.49
0300	26 gauge	↓	790	.041	↓	1.59	1.46		3.05	4.02

07 42 Wall Panels

07 42 13 – Metal Wall Panels

07 42 13.30 Steel Siding		Crew	Daily Output	Labor-Hours	Unit	Material	2007 Bare Costs Labor	Equipment	Total	Total Incl O&P
0400	24 gauge	G-3	785	.041	S.F.	1.60	1.47		3.07	4.04
0600	22 gauge		770	.042		1.84	1.50		3.34	4.35
0700	Colored, corrugated/ribbed, on steel frame, 10 yr fnsh, 28 ga.		800	.040		1.68	1.45		3.13	4.09
0900	26 gauge		795	.040		1.76	1.46		3.22	4.19
1000	24 gauge		790	.041		2.04	1.46		3.50	4.51
1020	20 gauge		785	.041		2.59	1.47		4.06	5.15
1200	Factory sandwich panel, 26 ga., 1" insulation, galvanized		380	.084		4.31	3.04		7.35	9.45
1300	Colored 1 side		380	.084		5.25	3.04		8.29	10.50
1500	Galvanized 2 sides		380	.084		6.55	3.04		9.59	11.90
1600	Colored 2 sides		380	.084		6.80	3.04		9.84	12.15
1800	Acrylic paint face, regular paint liner		380	.084		5.10	3.04		8.14	10.30
1900	For 2" thick polystyrene, add					.79			.79	.87
2000	22 ga, galv, 2" insulation, baked enamel exterior	G-3	360	.089		10	3.21		13.21	15.95
2100	P.V.F. exterior finish	"	360	.089		10.55	3.21		13.76	16.55

07 44 Faced Panels

07 44 13 – Aggregate Coated Panels

07 44 13.10 Exposed Aggregate Panels

		Crew	Daily Output	Labor-Hours	Unit	Material	Labor	Equipment	Total	Total Incl O&P
0010	**EXPOSED AGGREGATE PANELS**									
1400	Fiberglass polymer back-up,									
1500	Small size aggregate	F-3	445	.090	S.F.	5.05	3.35	1.63	10.03	12.55
1600	Medium size aggregate		445	.090		5.70	3.35	1.63	10.68	13.25
1700	Large size aggregate		445	.090		5.85	3.35	1.63	10.83	13.45

07 44 73 – Metal Faced Panels

07 44 73.10 Metal Faced Panels

		Crew	Daily Output	Labor-Hours	Unit	Material	Labor	Equipment	Total	Total Incl O&P
0010	**METAL FACED PANELS**									
0400	Textured aluminum, 4' x 8' x 5/16" plywood backing, single face	2 Shee	375	.043	S.F.	3.09	1.86		4.95	6.25
0600	Double face		375	.043		3.68	1.86		5.54	6.90
0700	4' x 10' x 5/16" plywood backing, single face		375	.043		3.45	1.86		5.31	6.65
0900	Double face		375	.043		4.86	1.86		6.72	8.20
1000	4' x 12' x 5/16" plywood backing, single face		375	.043		4.89	1.86		6.75	8.25
1300	Smooth aluminum, 1/4" panel, fluoropolymer finish, double face		375	.043		4.82	1.86		6.68	8.15
1350	Clear anodized finish, double face		375	.043		7.55	1.86		9.41	11.20
1400	Double face textured aluminum, structural panel, 1" EPS insulation		375	.043		4.19	1.86		6.05	7.50
1500	Accessories, outside corner	1 Shee	175	.046	L.F.	1.57	1.99		3.56	4.80
1600	Inside corner		175	.046		1.11	1.99		3.10	4.29
1800	Batten mounting clip		200	.040		.41	1.74		2.15	3.14
1900	Low profile batten		480	.017		.50	.73		1.23	1.67
2100	High profile batten		480	.017		1.17	.73		1.90	2.41
2200	Water table		200	.040		1.75	1.74		3.49	4.62
2400	Horizontal joint connector		200	.040		1.38	1.74		3.12	4.21
2500	Corner cap		200	.040		1.53	1.74		3.27	4.37
2700	H - moulding		480	.017		1.03	.73		1.76	2.25

07 46 Siding

07 46 23 – Wood Siding

07 46 23.10 Wood Board Siding

		Crew	Daily Output	Labor-Hours	Unit	Material	2007 Bare Costs Labor	Equipment	Total	Total Incl O&P
0010	**WOOD BOARD SIDING**									
3200	Wood, cedar bevel, A grade, 1/2" x 6"	1 Carp	250	.032	S.F.	3.05	1.17		4.22	5.20
3300	1/2" x 8"		275	.029		3.37	1.07		4.44	5.35
3500	3/4" x 10", clear grade		300	.027		4.76	.98		5.74	6.75
3600	"B" grade		300	.027		3.35	.98		4.33	5.20
3800	Cedar, rough sawn, 1" x 4", A grade, natural		240	.033		1.74	1.22		2.96	3.81
3900	Stained		240	.033		2.04	1.22		3.26	4.14
4100	1" x 12", board & batten, #3 & Btr., natural		260	.031		1.07	1.13		2.20	2.94
4200	Stained		260	.031		1.15	1.13		2.28	3.03
4400	1" x 8" channel siding, #3 & Btr., natural		250	.032		2.14	1.17		3.31	4.18
4500	Stained		250	.032		2.43	1.17		3.60	4.50
4700	Redwood, clear, beveled, vertical grain, 1/2" x 4"		200	.040		3.46	1.47		4.93	6.10
4750	1/2" x 6"		225	.036		2.86	1.31		4.17	5.20
4800	1/2" x 8"		250	.032		2.31	1.17		3.48	4.37
5000	3/4" x 10"		300	.027		3.80	.98		4.78	5.70
5200	Channel siding, 1" x 10", B grade		285	.028		2.50	1.03		3.53	4.35
5250	Redwood, T&G boards, B grade, 1" x 4"	2 Carp	300	.053		2.90	1.96		4.86	6.25
5270	1" x 8"	"	375	.043		2.60	1.57		4.17	5.30
5400	White pine, rough sawn, 1" x 8", natural	1 Carp	275	.029		1.07	1.07		2.14	2.84
5500	Stained	"	275	.029		1.15	1.07		2.22	2.93

07 46 26 – Hardboard Siding

07 46 26.10 Hardboard Siding

		Crew	Daily Output	Labor-Hours	Unit	Material	Labor	Equipment	Total	Total Incl O&P
0010	**HARDBOARD SIDING**									
0030	Lap siding, hardboard, 7/16" x 8", primed									
0051	Wood grain texture finish	L-2	552	.029	S.F.	1.13	.93		2.06	2.68
0100	Panels, 7/16" thick, smooth, textured or grooved, primed	2 Carp	700	.023		.92	.84		1.76	2.32
0200	Stained		700	.023		1	.84		1.84	2.41
0700	Particle board, overlaid, 3/8" thick		750	.021		.70	.78		1.48	1.99

07 46 29 – Plywood Siding

07 46 29.10 Plywood Siding

		Crew	Daily Output	Labor-Hours	Unit	Material	Labor	Equipment	Total	Total Incl O&P
0010	**PLYWOOD SIDING**									
0900	Plywood, medium density overlaid, 3/8" thick	2 Carp	750	.021	S.F.	.85	.78		1.63	2.16
1000	1/2" thick		700	.023		1.07	.84		1.91	2.49
1100	3/4" thick		650	.025		1.29	.90		2.19	2.83
1600	Texture 1-11, cedar, 5/8" thick, natural		675	.024		2.41	.87		3.28	4
1700	Factory stained		675	.024		1.92	.87		2.79	3.46
1900	Texture 1-11, fir, 5/8" thick, natural		675	.024		1.11	.87		1.98	2.57
2000	Factory stained		675	.024		1.70	.87		2.57	3.22
2050	Texture 1-11, S.Y.P., 5/8" thick, natural		675	.024		1.19	.87		2.06	2.66
2100	Factory stained		675	.024		1.23	.87		2.10	2.70
2200	Rough sawn cedar, 3/8" thick, natural		675	.024		1.19	.87		2.06	2.66
2300	Factory stained		675	.024		.95	.87		1.82	2.40
2500	Rough sawn fir, 3/8" thick, natural		675	.024		.75	.87		1.62	2.18
2600	Factory stained		675	.024		.95	.87		1.82	2.40
2800	Redwood, textured siding, 5/8" thick		675	.024		1.90	.87		2.77	3.44
3000	Polyvinyl chloride coated, 3/8" thick		750	.021		1.10	.78		1.88	2.43

07 46 33 – Plastic Siding

07 46 33.10 Vinyl Siding

		Crew	Daily Output	Labor-Hours	Unit	Material	Labor	Equipment	Total	Total Incl O&P
0010	**VINYL SIDING**									
0020	Clapboard profile, woodgrain texture, .048 thick, double 4	1 Carp	255	.031	S.F.	.78	1.15		1.93	2.65
0100	with 3/8" insulation		255	.031		.89	1.15		2.04	2.77

07 46 Siding

07 46 33 – Plastic Siding

07 46 33.10 Vinyl Siding

		Crew	Daily Output	Labor-Hours	Unit	Material	2007 Bare Costs Labor	Equipment	Total	Total Incl O&P
0200	Soffit and fascia	1 Carp	205	.039	S.F.	1.70	1.43		3.13	4.10
0300	Window and door trim moldings		185	.043	L.F.	.42	1.59		2.01	2.93
0500	Corner posts, outside corner		205	.039		1.53	1.43		2.96	3.91
0600	Inside corner		205	.039		.71	1.43		2.14	3.01

07 46 46 – Mineral-Fiber Cement Siding

07 46 46.10 Fiber Cement Siding

		Crew	Daily Output	Labor-Hours	Unit	Material	Labor	Equipment	Total	Total Incl O&P
0010	**FIBER CEMENT SIDING**									
0020	Lap siding, 5/16" thick, 6" wide, smooth texture	2 Carp	415	.039	S.F.	1.34	1.41		2.75	3.67
0025	Woodgrain texture		415	.039		1.34	1.41		2.75	3.67
0030	7-1/2" wide, smooth texture		425	.038		.94	1.38		2.32	3.18
0035	Woodgrain texture		425	.038		.94	1.38		2.32	3.18
0040	8" wide, smooth texture		425	.038		1.10	1.38		2.48	3.36
0045	Roughsawn texture		425	.038		1.10	1.38		2.48	3.36
0050	9-1/2" wide, smooth texture		440	.036		1.06	1.33		2.39	3.25
0055	Woodgrain texture		440	.036		1.06	1.33		2.39	3.25
0060	12" wide, smooth texture		455	.035		1.04	1.29		2.33	3.15
0065	Woodgrain texture		455	.035		1.04	1.29		2.33	3.15
0070	Panel siding, 5/16" thick, smooth texture		750	.021		.87	.78		1.65	2.18
0075	Stucco texture		750	.021		.87	.78		1.65	2.18
0080	Grooved woodgrain texture		750	.021		.87	.78		1.65	2.18
0085	V - grooved woodgrain texture		750	.021		.87	.78		1.65	2.18
0090	Wood starter strip		400	.040	L.F.	.31	1.47		1.78	2.63

07 46 73 – Soffit

07 46 73.10 Soffit

		Crew	Daily Output	Labor-Hours	Unit	Material	Labor	Equipment	Total	Total Incl O&P
0010	**SOFFIT**									
0012	Aluminum, residential, .020" thick	1 Carp	210	.038	S.F.	1.40	1.40		2.80	3.72
0100	Baked enamel on steel, 16 or 18 gauge		105	.076		4.97	2.80		7.77	9.80
0300	Polyvinyl chloride, white, solid		230	.035		.87	1.28		2.15	2.95
0400	Perforated		230	.035		.87	1.28		2.15	2.95
0500	For colors, add					.10			.10	.11

07 51 Built-Up Bituminous Roofing

07 51 13 – Built-Up Asphalt Roofing

07 51 13.10 Built-Up Roofing Components

		Crew	Daily Output	Labor-Hours	Unit	Material	Labor	Equipment	Total	Total Incl O&P
0010	**BUILT-UP ROOFING COMPONENTS**									
0012	Asphalt saturated felt, #30, 2 square per roll	1 Rofc	58	.138	Sq.	7.05	4.39		11.44	15.20
0200	#15, 4 sq per roll, plain or perforated, not mopped		58	.138		4.23	4.39		8.62	12.10
0300	Roll roofing, smooth, #65		15	.533		7.35	16.95		24.30	37
0500	#90		15	.533		23	16.95		39.95	54
0520	Mineralized		15	.533		18.05	16.95		35	49
0540	D.C. (Double coverage), 19" selvage edge		10	.800		39	25.50		64.50	86
0580	Adhesive (lap cement)				Gal.	4.37			4.37	4.81
0600	Steep, flat or dead level asphalt, 10 ton lots, bulk				Ton	310			310	340
0800	Packaged				"	525			525	575

07 51 13.13 Cold-Applied Built-Up Asphalt Roofing

		Crew	Daily Output	Labor-Hours	Unit	Material	Labor	Equipment	Total	Total Incl O&P
0010	**COLD-APPLIED BUILT-UP ASPHALT ROOFING**									
0020	3 ply system, installation only (components listed below)	G-5	50	.800	Sq.		23	3.06	26.06	42.50
0100	Spunbond poly. fabric, 1.35 oz/SY, 36"W, 10.8 Sq./roll				Ea.	152			152	167
0200	49" wide, 14.6 Sq./roll					210			210	231
0300	2.10 oz./S.Y., 36" wide, 10.8 Sq./roll					229			229	252

07 51 Built-Up Bituminous Roofing

07 51 13 – Built-Up Asphalt Roofing

07 51 13.13 Cold-Applied Built-Up Asphalt Roofing		Crew	Daily Output	Labor-Hours	Unit	Material	2007 Bare Costs Labor	Equipment	Total	Total Incl O&P
0400	49" wide, 14.6 Sq./roll				Ea.	310			310	340
0500	Base & finish coat, 3 gal./Sq., 5 gal./can				Gal.	4.04			4.04	4.44
0600	Coating, ceramic granules, 1/2 Sq./bag				Ea.	14.30			14.30	15.70
0700	Aluminum, 2 gal./Sq.				Gal.	11.50			11.50	12.65
0800	Emulsion, fibered or non-fibered, 4 gal./Sq.				"	5.30			5.30	5.80

07 51 13.20 Built-Up Roofing Systems

		Crew	Daily Output	Labor-Hours	Unit	Material	Labor	Equipment	Total	Total Incl O&P
0010	**BUILT-UP ROOFING SYSTEMS** R075113-20									
0120	Asphalt flood coat with gravel/slag surfacing, not including									
0140	Insulation, flashing or wood nailers									
0200	Asphalt base sheet, 3 plies #15 asphalt felt, mopped	G-1	22	2.545	Sq.	62.50	75.50	15.10	153.10	214
0350	On nailable decks		21	2.667		67.50	79	15.80	162.30	225
0500	4 plies #15 asphalt felt, mopped		20	2.800		88	83	16.60	187.60	256
0550	On nailable decks		19	2.947		79	87.50	17.50	184	254
0700	Coated glass base sheet, 2 plies glass (type IV), mopped		22	2.545		65	75.50	15.10	155.60	216
0850	3 plies glass, mopped		20	2.800		77	83	16.60	176.60	244
0950	On nailable decks		19	2.947		73	87.50	17.50	178	247
1100	4 plies glass fiber felt (type IV), mopped		20	2.800		94	83	16.60	193.60	263
1150	On nailable decks		19	2.947		85	87.50	17.50	190	261
1200	Coated & saturated base sheet, 3 plies #15 asph. felt, mopped		20	2.800		70.50	83	16.60	170.10	237
1250	On nailable decks		19	2.947		66	87.50	17.50	171	240
1300	4 plies #15 asphalt felt, mopped		22	2.545		82	75.50	15.10	172.60	235
2000	Asphalt flood coat, smooth surface									
2200	Asphalt base sheet & 3 plies #15 asphalt felt, mopped	G-1	24	2.333	Sq.	66.50	69	13.85	149.35	206
2400	On nailable decks		23	2.435		62.50	72	14.45	148.95	207
2600	4 plies #15 asphalt felt, mopped		24	2.333		78.50	69	13.85	161.35	218
2700	On nailable decks		23	2.435		74	72	14.45	160.45	220
2900	Coated glass fiber base sheet, mopped, and 2 plies of									
2910	glass fiber felt (type IV)	G-1	25	2.240	Sq.	59.50	66.50	13.30	139.30	193
3100	On nailable decks		24	2.333		56.50	69	13.85	139.35	194
3200	3 plies, mopped		23	2.435		72	72	14.45	158.45	218
3300	On nailable decks		22	2.545		67.50	75.50	15.10	158.10	219
3800	4 plies glass fiber felt (type IV), mopped		23	2.435		84.50	72	14.45	170.95	232
3900	On nailable decks		22	2.545		80	75.50	15.10	170.60	233
4000	Coated & saturated base sheet, 3 plies #15 asph. felt, mopped		24	2.333		65.50	69	13.85	148.35	204
4200	On nailable decks		23	2.435		61	72	14.45	147.45	206
4300	4 plies #15 organic felt, mopped		22	2.545		77	75.50	15.10	167.60	230
4500	Coal tar pitch with gravel/slag surfacing									
4600	4 plies #15 tarred felt, mopped	G-1	21	2.667	Sq.	131	79	15.80	225.80	296
4800	3 plies glass fiber felt (type IV), mopped	"	19	2.947	"	108	87.50	17.50	213	285
5000	Coated glass fiber base sheet, and 2 plies of									
5010	glass fiber felt, (type IV), mopped	G-1	19	2.947	Sq.	108	87.50	17.50	213	285
5300	On nailable decks		18	3.111		95.50	92.50	18.45	206.45	283
5600	4 plies glass fiber felt (type IV), mopped		21	2.667		149	79	15.80	243.80	315
5800	On nailable decks		20	2.800		137	83	16.60	236.60	310

07 51 13.30 Cants

		Crew	Daily Output	Labor-Hours	Unit	Material	Labor	Equipment	Total	Total Incl O&P
0010	**CANTS**									
0012	Lumber, treated, 4" x 4" cut diagonally	1 Rofc	325	.025	L.F.	1.38	.78		2.16	2.85
0100	Foamglass		325	.025		2.17	.78		2.95	3.72
0300	Mineral or fiber, trapezoidal, 1" x 4" x 48"		325	.025		.19	.78		.97	1.54
0400	1-1/2" x 5-5/8" x 48"		325	.025		.31	.78		1.09	1.67

07 51 13.40 Felts

0010	**FELTS**

07 51 Built-Up Bituminous Roofing

07 51 13 – Built-Up Asphalt Roofing

07 51 13.40 Felts

		Crew	Daily Output	Labor-Hours	Unit	Material	2007 Bare Costs Labor	Equipment	Total	Total Incl O&P
0012	Glass fibered roofing felt, #15, not mopped	1 Rofc	58	.138	Sq.	5.15	4.39		9.54	13.15
0300	Base sheet, #45, channel vented		58	.138		22	4.39		26.39	31.50
0400	#50, coated		58	.138		11.90	4.39		16.29	20.50
0500	Cap, mineral surfaced		58	.138		22	4.39		26.39	31.50
0600	Flashing membrane, #65		16	.500		33.50	15.90		49.40	64
0800	Coal tar fibered, #15, no mopping		58	.138		10	4.39		14.39	18.45
0900	Asphalt felt, #15, 4 sq per roll, no mopping CN		58	.138		4.23	4.39		8.62	12.10
1100	#30, 2 sq per roll		58	.138		7.05	4.39		11.44	15.20
1200	Double coated, #33		58	.138		8.15	4.39		12.54	16.45
1400	#40, base sheet		58	.138		9.30	4.39		13.69	17.70
1450	Coated and saturated		58	.138		8.15	4.39		12.54	16.40
1500	Tarred felt, organic, #15, 4 sq rolls		58	.138		11.80	4.39		16.19	20.50
1550	#30, 2 sq roll		58	.138		20	4.39		24.39	29.50
1700	Add for mopping above felts, per ply, asphalt, 24 lb per sq	G-1	192	.292		6.30	8.65	1.73	16.68	23.50
1800	Coal tar mopping, 30 lb per sq		186	.301		10.50	8.95	1.79	21.24	28.50
1900	Flood coat, with asphalt, 60 lb per sq		60	.933		15.75	27.50	5.55	48.80	70.50
2000	With coal tar, 75 lb per sq		56	1		26.50	29.50	5.95	61.95	86

07 51 13.50 Walkways for Built-Up Roofs

		Crew	Daily Output	Labor-Hours	Unit	Material	Labor	Equipment	Total	Total Incl O&P
0010	**WALKWAYS FOR BUILT-UP ROOFS**									
0020	Asphalt impregnated, 3' x 6' x 1/2" thick	1 Rofc	400	.020	S.F.	1.36	.64		2	2.58
0100	3' x 3' x 3/4" thick	"	400	.020		2.57	.64		3.21	3.91
0300	Concrete patio blocks, 2" thick, natural	1 Clab	115	.070		1.47	2		3.47	4.73
0400	Colors	"	115	.070		1.86	2		3.86	5.15

07 52 Modified Bituminous Membrane Roofing

07 52 13 – Atactic-Polypropylene-Modified Bituminous Membrane Roofing

07 52 13.10 APP Modified Bituminous Membrane

		Crew	Daily Output	Labor-Hours	Unit	Material	Labor	Equipment	Total	Total Incl O&P
0010	**APP MODIFIED BITUMINOUS MEMBRANE** R075213-30									
0020	Base sheet, #15 glass fiber felt, nailed to deck	1 Rofc	58	.138	Sq.	5.75	4.39		10.14	13.80
0030	Spot mopped to deck	G-1	295	.190		8.30	5.65	1.13	15.08	19.95
0040	Fully mopped to deck	"	192	.292		11.45	8.65	1.73	21.83	29
0050	#15 organic felt, nailed to deck	1 Rofc	58	.138		4.82	4.39		9.21	12.75
0060	Spot mopped to deck	G-1	295	.190		7.40	5.65	1.13	14.18	18.90
0070	Fully mopped to deck	"	192	.292		10.50	8.65	1.73	20.88	28
2100	APP mod., smooth surf. cap sheet, poly. reinf., torched, 160 mils	G-5	2100	.019	S.F.	.46	.55	.07	1.08	1.52
2150	170 mils		2100	.019		.53	.55	.07	1.15	1.59
2200	Granule surface cap sheet, poly. reinf., torched, 180 mils		2000	.020		.57	.58	.08	1.23	1.69
2250	Smooth surface flashing, torched, 160 mils		1260	.032		.46	.92	.12	1.50	2.19
2300	170 mils		1260	.032		.53	.92	.12	1.57	2.26
2350	Granule surface flashing, torched, 180 mils		1260	.032		.57	.92	.12	1.61	2.31
2400	Fibrated aluminum coating	1 Rofc	3800	.002		.09	.07		.16	.21

07 52 16 – Styrene-Butadiene-Styrene Modified Bituminous Membrane Roofing

07 52 16.10 SBS Modified Bituminous Membrane

		Crew	Daily Output	Labor-Hours	Unit	Material	Labor	Equipment	Total	Total Incl O&P
0010	**SBS MODIFIED BITUMINOUS MEMBRANE**									
0080	SBS modified, granule surf cap sheet, polyester rein., mopped									
0600	150 mils	G-1	2000	.028	S.F.	.51	.83	.17	1.51	2.15
1100	160 mils		2000	.028		.75	.83	.17	1.75	2.42
1500	Glass fiber reinforced, mopped, 160 mils		2000	.028		.47	.83	.17	1.47	2.11
1600	Smooth surface cap sheet, mopped, 145 mils		2100	.027		.47	.79	.16	1.42	2.03
1700	Smooth surface flashing, 145 mils		1260	.044		.47	1.32	.26	2.05	3.05

07 52 Modified Bituminous Membrane Roofing

07 52 16 – Styrene-Butadiene-Styrene Modified Bituminous Membrane Roofing

	07 52 16.10 SBS Modified Bituminous Membrane	Crew	Daily Output	Labor-Hours	Unit	Material	2007 Bare Costs Labor	Equipment	Total	Total Incl O&P
1800	150 mils	G-1	1260	.044	S.F.	.46	1.32	.26	2.04	3.04
1900	Granular surface flashing, 150 mils		1260	.044		.51	1.32	.26	2.09	3.09
2000	160 mils		1260	.044		.75	1.32	.26	2.33	3.36

07 53 Elastomeric Membrane Roofing

07 53 23 – Ethylene-Propylene-Diene-Monomer Roofing

07 53 23.20 Ethylene-Propylene-Diene-Monomer Roofing

		Crew	Daily Output	Labor-Hours	Unit	Material	2007 Bare Costs Labor	Equipment	Total	Total Incl O&P
0010	**ETHYLENE-PROPYLENE-DIENE-MONOMER ROOFING**									
0800	Chlorosulfonated polyethylene-hypalon (CSPE), 45 mils,									
0900	0.29 P.S.F., fully adhered	G-5	26	1.538	Sq.	129	44.50	5.90	179.40	224
1100	Loose-laid & ballasted with stone (10 P.S.F.)		51	.784		137	22.50	3	162.50	192
1200	Mechanically attached		35	1.143		132	33	4.38	169.38	206
1300	Plates with adhesive attachment		35	1.143		129	33	4.38	166.38	203
3500	Ethylene propylene diene monomer (EPDM), 45 mils, 0.28 P.S.F.									
3600	Loose-laid & ballasted with stone (10 P.S.F.)	G-5	51	.784	Sq.	63.50	22.50	3	89	112
3700	Mechanically attached		35	1.143		54	33	4.38	91.38	120
3800	Fully adhered with adhesive		26	1.538		78.50	44.50	5.90	128.90	168
4500	60 mils, 0.40 P.S.F.									
4600	Loose-laid & ballasted with stone (10 P.S.F.)	G-5	51	.784	Sq.	80	22.50	3	105.50	130
4700	Mechanically attached		35	1.143		69.50	33	4.38	106.88	137
4800	Fully adhered with adhesive		26	1.538		94	44.50	5.90	144.40	185
4810	45 mil, .28 PSF, membrane only CN					31.50			31.50	34.50
4820	60 mil, .40 PSF, membrane only					45			45	50
4850	Seam tape for membrane, 3" x 100' roll				Ea.	42.50			42.50	47
4900	Batten strips, 10' sections					3.10			3.10	3.41
4910	Cover tape for batten strips, 6" x 100' roll					158			158	174
4930	Plate anchors				M	56			56	61.50
4970	Adhesive for fully adhered systems, 60 S.F./gal.				Gal.	15.20			15.20	16.70
7500	Polyisobutylene (PIB), 100 mils, 0.57 P.S.F.									
7600	Loose-laid & ballasted with stone/gravel (10 P.S.F.)	G-5	51	.784	Sq.	133	22.50	3	158.50	188
7700	Partially adhered with adhesive		35	1.143		167	33	4.38	204.38	244
7800	Hot asphalt attachment		35	1.143		159	33	4.38	196.38	236
7900	Fully adhered with contact cement		26	1.538		172	44.50	5.90	222.40	271

07 54 Thermoplastic Membrane Roofing

07 54 19 – Polyvinyl-Chloride Roofing

07 54 19.10 Polyvinyl-Chloride Roofing

		Crew	Daily Output	Labor-Hours	Unit	Material	2007 Bare Costs Labor	Equipment	Total	Total Incl O&P
0010	**POLYVINYL-CHLORIDE ROOFING**									
8200	Heat welded seams									
8700	Reinforced, 48 mils, 0.33 P.S.F.									
8750	Loose-laid & ballasted with stone/gravel (12 P.S.F.)	G-5	51	.784	Sq.	108	22.50	3	133.50	160
8800	Mechanically attached		35	1.143		97.50	33	4.38	134.88	168
8850	Fully adhered with adhesive		26	1.538		143	44.50	5.90	193.40	239
8860	Reinforced, 60 mils, .40 P.S.F.									
8870	Loose-laid & ballasted with stone/gravel (12 P.S.F.)	G-5	51	.784	Sq.	110	22.50	3	135.50	163
8880	Mechanically attached		35	1.143		99.50	33	4.38	136.88	171
8890	Fully adhered with adhesive		26	1.538		145	44.50	5.90	195.40	241

07 54 Thermoplastic Membrane Roofing

07 54 23 – Thermoplastic Polyolefin Roofing

07 54 23.10 Thermoplastic Polyolefin Roofing		Crew	Daily Output	Labor-Hours	Unit	Material	2007 Bare Costs Labor	Equipment	Total	Total Incl O&P
0010	**THERMOPLASTIC POLYOLEFIN ROOFING**									
0100	45 mils, loose laid & ballasted with stone(1/2 ton/sq.)	G-5	51	.784	Sq.	96	22.50	3	121.50	148
0120	Fully adhered		25	1.600		149	46	6.15	201.15	249
0140	Mechanically attached		34	1.176		64.50	34	4.51	103.01	133
0160	Self adhered		35	1.143		76	33	4.38	113.38	145
0180	60 mil membrane, heat welded seams, ballasted		50	.800		112	23	3.06	138.06	165
0200	Fully adhered		25	1.600		165	46	6.15	217.15	266
0220	Mechanically attached		34	1.176		102	34	4.51	140.51	174
0240	Self adhered		35	1.143		90	33	4.38	127.38	160

07 56 Fluid-Applied Roofing

07 56 10 – Fluid-Applied Roofing

07 56 10.10 Elastomeric Roofing

		Crew	Daily Output	Labor-Hours	Unit	Material	Labor	Equipment	Total	Total Incl O&P
0010	**ELASTOMERIC ROOFING**									
0100	For Elastomeric waterproofing, see division 07 13 13.10									
0110	Acrylic rubber, fluid applied, 20 mils thick	G-5	2000	.020	S.F.	1.89	.58	.08	2.55	3.14
0120	50 mils, reinforced		1200	.033		2.94	.96	.13	4.03	5
0130	For walking surface, add		900	.044		.89	1.28	.17	2.34	3.34
0300	Hypalon neoprene, fluid applied, 20 mil thick, not-reinforced	G-1	1135	.049		2.15	1.46	.29	3.90	5.15
0600	Non-woven polyester, reinforced		960	.058		2.17	1.73	.35	4.25	5.70
0700	5 coat neoprene deck, 60 mil thick, under 10,000 SF		325	.172		4.79	5.10	1.02	10.91	15.05
0900	Over 10,000 SF		625	.090		4.46	2.66	.53	7.65	10
1300	Vinyl plastic traffic deck, sprayed, 2 to 4 mils thick		625	.090		1.40	2.66	.53	4.59	6.65
1500	Vinyl and neoprene membrane traffic deck		1550	.036		1.48	1.07	.21	2.76	3.69

07 57 Coated Foamed Roofing

07 57 13 – Sprayed Polyurethane Foam Roofing

07 57 13.10 Sprayed Polyurethane Foam Roofing

		Crew	Daily Output	Labor-Hours	Unit	Material	Labor	Equipment	Total	Total Incl O&P
0010	**SPRAYED POLYURETHANE FOAM ROOFING**									
1600	Polyurethane spray-on with 20 mil silicone rubber coating applied									
1700	1" thick, R7, minimum	G-2A	875	.027	S.F.	1.41	.77	.14	2.32	2.96
1800	Maximim		805	.030		1.89	.83	.15	2.87	3.62
1900	2" thick, R14, minimum		685	.035		2	.98	.18	3.16	4.02
2000	Maximum		575	.042		2.53	1.17	.21	3.91	4.93
2100	3" thick, R21, minimum		500	.048		3.41	1.34	.24	4.99	6.25
2200	Maximum		440	.055		4.37	1.53	.28	6.18	7.60

07 58 Roll Roofing

07 58 10 – Roll Roofing

07 58 10.10 Roll Roofing

	Crew	Daily Output	Labor-Hours	Unit	Material	2007 Bare Costs Labor	Equipment	Total	Total Incl O&P
0010 **ROLL ROOFING**									
0100 Asphalt, mineral surface									
0200 1 ply #15 organic felt, 1 ply mineral surfaced									
0300 Selvage roofing, lap 19", nailed & mopped	G-1	27	2.074	Sq.	52	61.50	12.30	125.80	175
0400 3 plies glass fiber felt (type IV), 1 ply mineral surfaced									
0500 Selvage roofing, lapped 19", mopped	G-1	25	2.240	Sq.	80	66.50	13.30	159.80	216
0600 Coated glass fiber base sheet, 2 plies of glass fiber									
0700 Felt (type IV), 1 ply mineral surfaced selvage									
0800 Roofing, lapped 19", mopped	G-1	25	2.240	Sq.	86.50	66.50	13.30	166.30	223
0900 On nailable decks	"	24	2.333	"	80.50	69	13.85	163.35	221
1000 3 plies glass fiber felt (type III), 1 ply mineral surfaced									
1100 Selvage roofing, lapped 19", mopped	G-1	25	2.240	Sq.	80	66.50	13.30	159.80	216

07 61 Sheet Metal Roofing

07 61 13 – Standing Seam Sheet Metal Roofing

07 61 13.10 Standing Seam Sheet Metal Roofing

	Crew	Daily Output	Labor-Hours	Unit	Material	2007 Bare Costs Labor	Equipment	Total	Total Incl O&P
0010 **STANDING SEAM SHEET METAL ROOFING**									
0400 Copper standing seam roofing, over 10 squares, 16 oz, 125 lb per sq	1 Shee	1.30	6.154	Sq.	845	268		1,113	1,350
0600 18 oz, 140 lb per sq		1.20	6.667		950	290		1,240	1,500
0700 20 oz, 150 lb per sq	↓	1.10	7.273		1,025	315		1,340	1,625
1200 For abnormal conditions or small areas, add					25%	100%			
1300 For lead-coated copper, add				↓	25%				

07 61 16 – Batten Seam Sheet Metal Roofing

07 61 16.10 Batten Seam Sheet Metal Roofing

	Crew	Daily Output	Labor-Hours	Unit	Material	2007 Bare Costs Labor	Equipment	Total	Total Incl O&P
0010 **BATTEN SEAM SHEET METAL ROOFING**									
0012 Copper batten seam roofing, 16 oz, 130 lb per sq	1 Shee	1.10	7.273	Sq.	880	315		1,195	1,450
0020 Lead batten seam roofing, 5 lb. per SF		1.20	6.667		395	290		685	885
0100 Zinc / copper alloy batten seam roofing, .020 thick		1.20	6.667		1,100	290		1,390	1,650
0200 18 oz, 145 lb per sq		1	8		980	350		1,330	1,600
0300 20 oz, 160 lb per sq		1	8		1,075	350		1,425	1,725
0500 Stainless steel batten seam roofing, type 304, 28 gauge		1.20	6.667		520	290		810	1,025
0600 26 gauge		1.15	6.957		645	305		950	1,175
0800 .027" thick		1.15	6.957		1,300	305		1,605	1,900
0900 .032" thick		1.10	7.273		1,500	315		1,815	2,150
1000 .040" thick	↓	1.05	7.619	↓	1,800	330		2,130	2,475

07 61 19 – Flat Seam Sheet Metal Roofing

07 61 19.10 Flat Seam Sheet Metal Roofing

	Crew	Daily Output	Labor-Hours	Unit	Material	2007 Bare Costs Labor	Equipment	Total	Total Incl O&P
0010 **FLAT SEAM SHEET METAL ROOFING**									
0900 Copper flat seam roofing, over 10 squares, 16 oz, 115 lb per sq	1 Shee	1.20	6.667	Sq.	780	290		1,070	1,300
1000 20 oz, 145 lb per sq		1.10	7.273		980	315		1,295	1,575
1100 Lead flat seam roofing, 5 lb per SF	↓	1.30	6.154	↓	395	268		663	850

07 65 Flexible Flashing

07 65 10 – Sheet Metal Flashing

07 65 10.10 Sheet Metal Flashing	Crew	Daily Output	Labor-Hours	Unit	Material	2007 Bare Costs Labor	2007 Bare Costs Equipment	Total	Total Incl O&P
0010 **SHEET METAL FLASHING**									
0011 Including up to 4 bends									
0020 Aluminum, mill finish, .013" thick	1 Rofc	145	.055	S.F.	.43	1.75		2.18	3.45
0030 .016" thick		145	.055		.63	1.75		2.38	3.67
0060 .019" thick		145	.055		.82	1.75		2.57	3.88
0100 .032" thick		145	.055		1.18	1.75		2.93	4.28
0200 .040" thick		145	.055		1.77	1.75		3.52	4.93
0300 .050" thick		145	.055		2.01	1.75		3.76	5.20
0325 Mill finish 5" x 7" step flashing, .016" thick		1920	.004	Ea.	.13	.13		.26	.37
0350 Mill finish 12" x 12" step flashing, .016" thick		1600	.005	"	.49	.16		.65	.81
0400 Painted finish, add				S.F.	.29			.29	.32
1000 .005" thick	1 Rofc	330	.024		1.33	.77		2.10	2.77
1100 .016" thick		330	.024		1.47	.77		2.24	2.93
1600 Copper, 16 oz, sheets, under 1000 lbs.		115	.070		5.05	2.21		7.26	9.30
1700 Over 4000 lbs.		155	.052		6.60	1.64		8.24	10.10
1900 20 oz sheets, under 1000 lbs.		110	.073		6.50	2.31		8.81	11.05
2000 Over 4000 lbs.		145	.055		6.60	1.75		8.35	10.30
2200 24 oz sheets, under 1000 lbs.		105	.076		7.95	2.42		10.37	12.85
2300 Over 4000 lbs.		135	.059		8.05	1.88		9.93	12.05
2500 32 oz sheets, under 1000 lbs.		100	.080		10.60	2.54		13.14	15.95
2600 Over 4000 lbs.		130	.062		16.70	1.96		18.66	21.50
2700 W shape for valleys, 16 oz, 24" wide		100	.080	L.F.	12.10	2.54		14.64	17.60
5800 Lead, 2.5 lb. per SF, up to 12" wide		135	.059	S.F.	3.03	1.88		4.91	6.55
5900 Over 12" wide		135	.059		3.63	1.88		5.51	7.20
8650 Copper, 16 oz		100	.080		4.33	2.54		6.87	9.10
8900 Stainless steel sheets, 32 ga, .010" thick		155	.052		3.13	1.64		4.77	6.20
9000 28 ga, .015" thick		155	.052		3.89	1.64		5.53	7.05
9100 26 ga, .018" thick		155	.052		4.71	1.64		6.35	8
9200 24 ga, .025" thick		155	.052		6.10	1.64		7.74	9.55
9290 For mechanically keyed flashing, add					40%				
9400 Terne coated stainless steel, .015" thick, 28 ga	1 Rofc	155	.052	S.F.	6.05	1.64		7.69	9.50
9500 .018" thick, 26 ga		155	.052		6.85	1.64		8.49	10.35
9600 Zinc and copper alloy (brass), .020" thick		155	.052		4.26	1.64		5.90	7.45
9700 .027" thick		155	.052		5.70	1.64		7.34	9.10
9800 .032" thick		155	.052		6.65	1.64		8.29	10.15
9900 .040" thick		155	.052		8.15	1.64		9.79	11.75

07 65 12 – Fabric and Mastic Flashings

07 65 12.10 Fabric and Mastic Flashings

	Crew	Daily Output	Labor-Hours	Unit	Material	Labor	Equipment	Total	Total Incl O&P
0010 **FABRIC AND MASTIC FLASHINGS**									
1300 Asphalt flashing cement, 5 gallon				Gal.	4.80			4.80	5.30
4900 Fabric, asphalt-saturated cotton, specification grade	1 Rofc	35	.229	S.Y.	2.12	7.25		9.37	14.70
5000 Utility grade		35	.229		1.34	7.25		8.59	13.80
5200 Open-mesh fabric, saturated, 40 oz per S.Y.		35	.229		1.49	7.25		8.74	14
5300 Close-mesh fabric, saturated, 17 oz per S.Y.		35	.229		1.56	7.25		8.81	14.05
5500 Fiberglass, resin-coated		35	.229		1.28	7.25		8.53	13.75
5600 Asphalt-coated, 40 oz per S.Y.		35	.229		8.70	7.25		15.95	22

07 65 13 – Laminated Sheet Flashing

07 65 13.10 Laminated Sheet Flashing

	Crew	Daily Output	Labor-Hours	Unit	Material	Labor	Equipment	Total	Total Incl O&P
0010 **LAMINATED SHEET FLASHING**, Including up to 4 bends									
0500 Fabric-backed 2 sides, .004" thick	1 Rofc	330	.024	S.F.	1.13	.77		1.90	2.55
0700 .005" thick		330	.024		1.33	.77		2.10	2.77
0750 Mastic-backed, self adhesive		460	.017		2.76	.55		3.31	3.98

07 65 Flexible Flashing

07 65 13 – Laminated Sheet Flashing

07 65 13.10 Laminated Sheet Flashing		Crew	Daily Output	Labor-Hours	Unit	Material	2007 Bare Costs Labor	Equipment	Total	Total Incl O&P
0800	Mastic-coated 2 sides, .004" thick	1 Rofc	330	.024	S.F.	1.13	.77		1.90	2.55
2800	Copper, paperbacked 1 side, 2 oz		330	.024		1.17	.77		1.94	2.60
2900	3 oz		330	.024		1.53	.77		2.30	2.99
3100	Paperbacked 2 sides, 2 oz		330	.024		1.18	.77		1.95	2.61
3150	3 oz		330	.024		1.52	.77		2.29	2.98
3200	5 oz		330	.024		2.28	.77		3.05	3.82
3250	7 oz		330	.024		3.72	.77		4.49	5.40
3400	Mastic-backed 2 sides, copper, 2 oz		330	.024		1.39	.77		2.16	2.84
3500	3 oz		330	.024		1.72	.77		2.49	3.20
3700	5 oz		330	.024		2.52	.77		3.29	4.08
3800	Fabric-backed 2 sides, copper, 2 oz		330	.024		1.48	.77		2.25	2.94
4000	3 oz		330	.024		1.91	.77		2.68	3.41
4100	5 oz		330	.024		2.60	.77		3.37	4.17
4300	Copper-clad stainless steel, .015" thick, under 500 lbs.		115	.070		4.32	2.21		6.53	8.50
4400	Over 2000 lbs.		155	.052		4.16	1.64		5.80	7.35
4600	.018" thick, under 500 lbs.		100	.080		5.70	2.54		8.24	10.60
4700	Over 2000 lbs.		145	.055		4.17	1.75		5.92	7.55
6100	Lead-coated copper, fabric-backed, 2 oz		330	.024		2	.77		2.77	3.51
6200	5 oz		330	.024		2.30	.77		3.07	3.84
6400	Mastic-backed 2 sides, 2 oz		330	.024		1.56	.77		2.33	3.03
6500	5 oz		330	.024		1.94	.77		2.71	3.44
6700	Paperbacked 1 side, 2 oz		330	.024		1.35	.77		2.12	2.80
6800	3 oz		330	.024		1.59	.77		2.36	3.06
7000	Paperbacked 2 sides, 2 oz		330	.024		1.39	.77		2.16	2.84
7100	5 oz		330	.024		2.27	.77		3.04	3.81
8550	3 ply copper and fabric, 3 oz		155	.052		2.26	1.64		3.90	5.25
8600	7 oz		155	.052		4.69	1.64		6.33	7.95
8700	Lead on copper and fabric, 5 oz		155	.052		2.30	1.64		3.94	5.30
8800	7 oz		155	.052		4.08	1.64		5.72	7.25
9300	Stainless steel, paperbacked 2 sides, .005" thick		330	.024		2.76	.77		3.53	4.35

07 65 19 – Plastic Sheet Flashing

07 65 19.10 Plastic Sheet Flashing

		Crew	Daily Output	Labor-Hours	Unit	Material	Labor	Equipment	Total	Total Incl O&P
0010	**PLASTIC SHEET FLASHING**									
7300	Polyvinyl chloride, black, .010" thick	1 Rofc	285	.028	S.F.	.21	.89		1.10	1.74
7400	.020" thick		285	.028		.29	.89		1.18	1.83
7600	.030" thick		285	.028		.38	.89		1.27	1.93
7700	.056" thick		285	.028		.91	.89		1.80	2.51
7900	Black or white for exposed roofs, .060" thick		285	.028		1.99	.89		2.88	3.70
8060	PVC tape, 5" x 45 mils, for joint covers, 100 L.F./roll				Ea.	97.50			97.50	107
8850	Polyvinyl chloride, .030" thick	1 Rofc	160	.050	S.F.	.35	1.59		1.94	3.09

07 65 23 – Rubber Sheet Flashing

07 65 23.10 Rubber Sheet Flashing

		Crew	Daily Output	Labor-Hours	Unit	Material	Labor	Equipment	Total	Total Incl O&P
0010	**RUBBER SHEET FLASHING**									
8100	Rubber, butyl, 1/32" thick	1 Rofc	285	.028	S.F.	.86	.89		1.75	2.46
8200	1/16" thick		285	.028		1.29	.89		2.18	2.93
8300	Neoprene, cured, 1/16" thick		285	.028		1.82	.89		2.71	3.51
8400	1/8" thick		285	.028		3.68	.89		4.57	5.55

07 65 26 – Self-Adhering Sheet Flashing

07 65 26.10 Self-Adhering Sheet Flashing

		Crew	Daily Output	Labor-Hours	Unit	Material	Labor	Equipment	Total	Total Incl O&P
0010	**SELF-ADHERING SHEET FLASHING**									
8500	Shower pan, bituminous membrane, 7 oz	1 Rofc	155	.052	S.F.	1.51	1.64		3.15	4.44

07 71 Roof Specialties

07 71 19 – Manufactured Gravel Stops and Fascias

07 71 19.10 Gravel Stop

		Crew	Daily Output	Labor-Hours	Unit	Material	2007 Bare Costs Labor	2007 Bare Costs Equipment	Total	Total Incl O&P
0010	**GRAVEL STOP**									
0020	Aluminum, .050" thick, 4" face height, mill finish	1 Shee	145	.055	L.F.	4.66	2.40		7.06	8.85
0080	Duranodic finish		145	.055		4.50	2.40		6.90	8.65
0100	Painted		145	.055		5.20	2.40		7.60	9.40
0300	6" face height		135	.059		5	2.58		7.58	9.50
0350	Duranodic finish		135	.059		5.30	2.58		7.88	9.85
0400	Painted		135	.059		6.15	2.58		8.73	10.80
0600	8" face height		125	.064		6.25	2.79		9.04	11.20
0650	Duranodic finish		125	.064		6.10	2.79		8.89	11
0700	Painted		125	.064		6.20	2.79		8.99	11.10
0900	12" face height, .080 thick, 2 piece		100	.080		8.35	3.48		11.83	14.50
0950	Duranodic finish		100	.080		7.65	3.48		11.13	13.80
1000	Painted		100	.080		9.05	3.48		12.53	15.30
1350	Galv steel, 24 ga., 4" leg, plain, with continuous cleat, 4" face		145	.055		1.83	2.40		4.23	5.70
1360	6" face height		145	.055		2.77	2.40		5.17	6.75
1500	Polyvinyl chloride, 6" face height		135	.059		4	2.58		6.58	8.40
1600	9" face height		125	.064		4.72	2.79		7.51	9.50
1800	Stainless steel, 24 ga., 6" face height		135	.059		8.75	2.58		11.33	13.60
1900	12" face height		100	.080		18.30	3.48		21.78	25.50
2100	20 ga., 6" face height		135	.059		9.90	2.58		12.48	14.90
2200	12" face height		100	.080		21	3.48		24.48	28.50

07 71 19.30 Fascia

		Crew	Daily Output	Labor-Hours	Unit	Material	Labor	Equipment	Total	Total Incl O&P
0010	**FASCIA**									
0100	Aluminum, reverse board and batten, .032" thick, colored, no furring incl	1 Shee	145	.055	S.F.	4.71	2.40		7.11	8.90
0300	Steel, galv and enameled, stock, no furring, long panels		145	.055		3.24	2.40		5.64	7.25
0600	Short panels		115	.070		4.32	3.03		7.35	9.40

07 71 23 – Manufactured Gutters and Downspouts

07 71 23.10 Downspouts

		Crew	Daily Output	Labor-Hours	Unit	Material	Labor	Equipment	Total	Total Incl O&P
0010	**DOWNSPOUTS**									
0020	Aluminum 2" x 3", .020" thick, embossed	1 Shee	190	.042	L.F.	.91	1.83		2.74	3.83
0100	Enameled		190	.042		1.51	1.83		3.34	4.49
0300	Enameled, .024" thick, 2" x 3"		180	.044		1.51	1.94		3.45	4.64
0400	3" x 4"		140	.057		3.49	2.49		5.98	7.70
0600	Round, corrugated aluminum, 3" diameter, .020" thick		190	.042		1.42	1.83		3.25	4.39
0700	4" diameter, .025" thick		140	.057		1.92	2.49		4.41	5.95
0900	Wire strainer, round, 2" diameter		155	.052	Ea.	1.89	2.25		4.14	5.55
1000	4" diameter		155	.052		2.04	2.25		4.29	5.70
1200	Rectangular, perforated, 2" x 3"		145	.055		2.30	2.40		4.70	6.25
1300	3" x 4"		145	.055		3.32	2.40		5.72	7.35
1500	Copper, round, 16 oz., stock, 2" diameter		190	.042	L.F.	9	1.83		10.83	12.75
1600	3" diameter		190	.042		7.35	1.83		9.18	10.95
1800	4" diameter		145	.055		7.85	2.40		10.25	12.35
1900	5" diameter		130	.062		13.85	2.68		16.53	19.40
2100	Rectangular, corrugated copper, stock, 2" x 3"		190	.042		6.50	1.83		8.33	10
2200	3" x 4"		145	.055		8.60	2.40		11	13.15
2400	Rectangular, plain copper, stock, 2" x 3"		190	.042		6.75	1.83		8.58	10.25
2500	3" x 4"		145	.055		8.40	2.40		10.80	12.95
2700	Wire strainers, rectangular, 2" x 3"		145	.055	Ea.	5	2.40		7.40	9.20
2800	3" x 4"		145	.055		6	2.40		8.40	10.30
3000	Round, 2" diameter		145	.055		4	2.40		6.40	8.10
3100	3" diameter		145	.055		6.10	2.40		8.50	10.45
3300	4" diameter		145	.055		11.30	2.40		13.70	16.10

07 71 Roof Specialties

07 71 23 – Manufactured Gutters and Downspouts

07 71 23.10 Downspouts

		Crew	Daily Output	Labor-Hours	Unit	Material	2007 Bare Costs Labor	Equipment	Total	Total Incl O&P
3400	5" diameter	1 Shee	115	.070	Ea.	16.60	3.03		19.63	23
3600	Lead-coated copper, round, stock, 2" diameter		190	.042	L.F.	6.90	1.83		8.73	10.45
3700	3" diameter		190	.042		6.90	1.83		8.73	10.45
3900	4" diameter		145	.055		9.60	2.40		12	14.25
4200	6" diameter, corrugated		105	.076		24	3.32		27.32	31.50
4300	Rectangular, corrugated, stock, 2" x 3"		190	.042		9.75	1.83		11.58	13.60
4500	Plain, stock, 2" x 3"		190	.042		9.75	1.83		11.58	13.60
4600	3" x 4"		145	.055		10	2.40		12.40	14.70
4800	Steel, galvanized, round, corrugated, 2" or 3" diam, 28 ga		190	.042		1.43	1.83		3.26	4.40
4900	4" diameter, 28 gauge		145	.055		2.15	2.40		4.55	6.05
5100	5" diameter, 28 gauge		130	.062		2.88	2.68		5.56	7.30
5200	26 gauge		130	.062		2.65	2.68		5.33	7.05
5400	6" diameter, 28 gauge		105	.076		4.31	3.32		7.63	9.85
5500	26 gauge		105	.076		5.25	3.32		8.57	10.90
5700	Rectangular, corrugated, 28 gauge, 2" x 3"		190	.042		1.43	1.83		3.26	4.40
5800	3" x 4"		145	.055		1.43	2.40		3.83	5.25
6000	Rectangular, plain, 28 gauge, galvanized, 2" x 3"		190	.042		1.55	1.83		3.38	4.54
6100	3" x 4"		145	.055		1.80	2.40		4.20	5.70
6300	Epoxy painted, 24 gauge, corrugated, 2" x 3"		190	.042		1.70	1.83		3.53	4.70
6400	3" x 4"		145	.055		1.89	2.40		4.29	5.80
6600	Wire strainers, rectangular, 2" x 3"		145	.055	Ea.	1.70	2.40		4.10	5.55
6700	3" x 4"		145	.055		3.38	2.40		5.78	7.40
6900	Round strainers, 2" or 3" diameter		145	.055		1.70	2.40		4.10	5.55
7000	4" diameter		145	.055		3.38	2.40		5.78	7.40
7200	5" diameter		145	.055		2.38	2.40		4.78	6.30
7300	6" diameter		115	.070		2.84	3.03		5.87	7.80
7500	Steel pipe, black, extra heavy, 4" diameter		20	.400	L.F.	16.35	17.40		33.75	45
7600	6" diameter		18	.444		30	19.35		49.35	63
7800	Stainless steel tubing, schedule 5, 2" x 3" or 3" diameter		190	.042		40.50	1.83		42.33	48
7900	3" x 4" or 4" diameter		145	.055		52	2.40		54.40	60.50
8100	4" x 5" or 5" diameter		135	.059		107	2.58		109.58	122
8200	Vinyl, rectangular, 2" x 3"		210	.038		.75	1.66		2.41	3.39
8300	Round, 2-1/2"		220	.036		.75	1.58		2.33	3.27

07 71 23.20 Elbows

		Crew	Daily Output	Labor-Hours	Unit	Material	2007 Bare Costs Labor	Equipment	Total	Total Incl O&P
0010	**ELBOWS**									
0020	Aluminum, 2" x 3", embossed	1 Shee	100	.080	Ea.	3.96	3.48		7.44	9.70
0100	Enameled		100	.080		2.95	3.48		6.43	8.60
0200	3" x 4", .025" thick, embossed		100	.080		3.96	3.48		7.44	9.70
0300	Enameled		100	.080		3.84	3.48		7.32	9.55
0400	Round corrugated, 3", embossed, .020" thick		100	.080		2.64	3.48		6.12	8.25
0500	4", .025" thick		100	.080		5.50	3.48		8.98	11.40
0600	Copper, 16 oz. round, 2" diameter		100	.080		13.40	3.48		16.88	20
0700	3" diameter		100	.080		11.20	3.48		14.68	17.65
0800	4" diameter		100	.080		17.50	3.48		20.98	24.50
1000	2" x 3" corrugated		100	.080		9.30	3.48		12.78	15.55
1100	3" x 4" corrugated		100	.080		15.90	3.48		19.38	23
1300	Vinyl, 2-1/2" diameter, 45° or 75°		100	.080		2.88	3.48		6.36	8.50
1400	Tee Y junction		75	.107		10.80	4.65		15.45	19.05

07 71 23.30 Gutters

		Crew	Daily Output	Labor-Hours	Unit	Material	2007 Bare Costs Labor	Equipment	Total	Total Incl O&P
0010	**GUTTERS**									
0012	Aluminum, stock units, 5" K type, .027" thick, plain	1 Shee	120	.067	L.F.	1.76	2.90		4.66	6.40
0100	Enameled		120	.067		1.85	2.90		4.75	6.50

07 71 Roof Specialties

07 71 23 – Manufactured Gutters and Downspouts

07 71 23.30 Gutters

		Crew	Daily Output	Labor-Hours	Unit	Material	2007 Bare Costs Labor	2007 Bare Costs Equipment	Total	Total Incl O&P
0300	5" K type type, .032" thick, plain	1 Shee	120	.067	L.F.	2.81	2.90		5.71	7.55
0400	Enameled		120	.067		2.36	2.90		5.26	7.10
0700	Copper, half round, 16 oz, stock units, 4" wide		120	.067		5.75	2.90		8.65	10.80
0900	5" wide		120	.067		7.30	2.90		10.20	12.55
1000	6" wide		115	.070		7.95	3.03		10.98	13.40
1200	K type, 16 oz, stock, 4" wide		120	.067		7.05	2.90		9.95	12.25
1300	5" wide		120	.067		7.70	2.90		10.60	12.95
1500	Lead coated copper, half round, stock, 4" wide		120	.067		10.50	2.90		13.40	16.05
1600	6" wide		115	.070		15.95	3.03		18.98	22
1800	K type, stock, 4" wide		120	.067		11.35	2.90		14.25	17
1900	5" wide		120	.067		12.45	2.90		15.35	18.15
2100	Stainless steel, half round or box, stock, 4" wide		120	.067		5.25	2.90		8.15	10.30
2200	5" wide		120	.067		5.65	2.90		8.55	10.70
2400	Steel, galv, half round or box, 28 ga, 5" wide, plain		120	.067		1.30	2.90		4.20	5.90
2500	Enameled		120	.067		1.35	2.90		4.25	5.95
2700	26 ga, stock, 5" wide		120	.067		1.42	2.90		4.32	6.05
2800	6" wide		120	.067		2.01	2.90		4.91	6.70
3000	Vinyl, O.G., 4" wide	1 Carp	110	.073		.92	2.67		3.59	5.15
3100	5" wide		110	.073		1.08	2.67		3.75	5.35
3200	4" half round, stock units		110	.073		.73	2.67		3.40	4.96
3250	Joint connectors				Ea.	1.47			1.47	1.62
3300	Wood, clear treated cedar, fir or hemlock, 3" x 4"	1 Carp	100	.080	L.F.	6.40	2.94		9.34	11.60
3400	4" x 5"	"	100	.080	"	7.40	2.94		10.34	12.70

07 71 23.35 Gutter Guard

		Crew	Daily Output	Labor-Hours	Unit	Material	Labor	Equipment	Total	Total Incl O&P
0010	**GUTTER GUARD**									
0020	6" wide strip, aluminum mesh	1 Carp	500	.016	L.F.	.39	.59		.98	1.34
0100	Vinyl mesh	"	500	.016	"	.40	.59		.99	1.35

07 71 26 – Reglets

07 71 26.10 Reglets

		Crew	Daily Output	Labor-Hours	Unit	Material	Labor	Equipment	Total	Total Incl O&P
0010	**REGLETS**									
0020	Aluminum, .025" thick, in concrete parapet	1 Carp	225	.036	L.F.	1.10	1.31		2.41	3.24
0100	Copper, 10 oz.		225	.036		2.25	1.31		3.56	4.51
0300	16 oz.		225	.036		4.40	1.31		5.71	6.85
0400	Galvanized steel, 24 gauge		225	.036		.80	1.31		2.11	2.91
0600	Stainless steel, .020" thick		225	.036		1.80	1.31		3.11	4.01
0700	Zinc and copper alloy, 20 oz.		225	.036		2.25	1.31		3.56	4.51
0900	Counter flashing for above, 12" wide, .032" aluminum	1 Shee	150	.053		1.40	2.32		3.72	5.10
1000	Copper, 10 oz.		150	.053		4.25	2.32		6.57	8.25
1200	16 oz.		150	.053		4.90	2.32		7.22	9
1300	Galvanized steel, .020" thick		150	.053		.79	2.32		3.11	4.45
1500	Stainless steel, .020" thick		150	.053		3.20	2.32		5.52	7.10
1600	Zinc and copper alloy, 20 oz.		150	.053		4.10	2.32		6.42	8.10

07 71 29 – Manufactured Roof Expansion Joints

07 71 29.10 Expansion Joints

		Crew	Daily Output	Labor-Hours	Unit	Material	Labor	Equipment	Total	Total Incl O&P
0010	**EXPANSION JOINTS**									
0300	Butyl or neoprene center with foam insulation, metal flanges									
0400	Aluminum, .032" thick for openings to 2-1/2"	1 Rofc	165	.048	L.F.	8.80	1.54		10.34	12.30
0600	For joint openings to 3-1/2"		165	.048		10.30	1.54		11.84	13.90
0610	For joint openings to 5"		165	.048		12.40	1.54		13.94	16.20
0620	For joint openings to 8"		165	.048		19.50	1.54		21.04	24
0700	Copper, 16 oz. for openings to 2-1/2"		165	.048		14.35	1.54		15.89	18.35

07 71 Roof Specialties

07 71 29 – Manufactured Roof Expansion Joints

07 71 29.10 Expansion Joints

		Crew	Daily Output	Labor-Hours	Unit	Material	2007 Bare Costs Labor	Equipment	Total	Total Incl O&P
0900	For joint openings to 3-1/2"	1 Rofc	165	.048	L.F.	16.45	1.54		17.99	20.50
0910	For joint openings to 5"		165	.048		19.30	1.54		20.84	23.50
0920	For joint openings to 8"		165	.048		29	1.54		30.54	34.50
1000	Galvanized steel, 26 ga. for openings to 2-1/2"		165	.048		7.25	1.54		8.79	10.55
1200	For joint openings to 3-1/2"		165	.048		8.55	1.54		10.09	12
1210	For joint openings to 5"		165	.048		10.70	1.54		12.24	14.40
1220	For joint openings to 8"		165	.048		18.25	1.54		19.79	22.50
1300	Lead-coated copper, 16 oz. for openings to 2-1/2"		165	.048		25.50	1.54		27.04	30.50
1500	For joint openings to 3-1/2"		165	.048		30	1.54		31.54	35.50
1600	Stainless steel, .018", for openings to 2-1/2"		165	.048		11	1.54		12.54	14.70
1800	For joint openings to 3-1/2"		165	.048		12.55	1.54		14.09	16.40
1810	For joint openings to 5"		165	.048		15.50	1.54		17.04	19.65
1820	For joint openings to 8"		165	.048		23.50	1.54		25.04	28.50
1900	Neoprene, double-seal type with thick center, 4-1/2" wide		125	.064		10.30	2.04		12.34	14.80
1950	Polyethylene bellows, with galv steel flat flanges		100	.080		3.99	2.54		6.53	8.70
1960	With galvanized angle flanges		100	.080		4.40	2.54		6.94	9.15
2000	Roof joint with extruded aluminum cover, 2"	1 Shee	115	.070		27.50	3.03		30.53	35
2100	Roof joint, plastic curbs, foam center, standard	1 Rofc	100	.080		10.05	2.54		12.59	15.35
2200	Large		100	.080		13.40	2.54		15.94	19.05
2300	Transitions, regular, minimum		10	.800	Ea.	87	25.50		112.50	139
2350	Maximum		4	2		110	63.50		173.50	230
2400	Large, minimum		9	.889		127	28.50		155.50	188
2450	Maximum		3	2.667		133	85		218	290
2500	Roof to wall joint with extruded aluminum cover	1 Shee	115	.070	L.F.	23.50	3.03		26.53	30.50
2600	See also divisions 03 15 05.25 & 07 95 13.50									
2700	Wall joint, closed cell foam on PVC cover, 9" wide	1 Rofc	125	.064	L.F.	3.58	2.04		5.62	7.40
2800	12" wide	"	115	.070	"	4.05	2.21		6.26	8.20

07 71 43 – Drip Edge

07 71 43.10 Drip Edge

		Crew	Daily Output	Labor-Hours	Unit	Material	2007 Bare Costs Labor	Equipment	Total	Total Incl O&P
0010	**DRIP EDGE**									
0020	Aluminum, .016" thick, 5" wide, mill finish	1 Carp	400	.020	L.F.	.42	.73		1.15	1.60
0100	White finish		400	.020		.39	.73		1.12	1.57
0200	8" wide, mill finish		400	.020		.62	.73		1.35	1.82
0300	Ice belt, 28" wide, mill finish		100	.080		3.80	2.94		6.74	8.75
0310	Vented, mill finish		400	.020		1.39	.73		2.12	2.67
0320	Painted finish		400	.020		1.51	.73		2.24	2.80
0400	Galvanized, 5" wide		400	.020		.43	.73		1.16	1.61
0500	8" wide, mill finish		400	.020		.52	.73		1.25	1.71
0510	Rake edge, aluminum, 1-1/2" x 1-1/2"		400	.020		.39	.73		1.12	1.57
0520	3-1/2" x 1-1/2"		400	.020		.44	.73		1.17	1.62

07 72 Roof Accessories

07 72 23 – Relief Vents

07 72 23.10 Roof Vents

		Crew	Daily Output	Labor-Hours	Unit	Material	2007 Bare Costs Labor	2007 Bare Costs Equipment	Total	Total Incl O&P
0010	**ROOF VENTS**									
0020	Mushroom shape, for built-up roofs, aluminum	1 Rofc	30	.267	Ea.	68.50	8.50		77	90
0100	PVC, 6" high	"	30	.267	"	28.50	8.50		37	46

07 72 26 – Ridge Vents

07 72 26.10 Ridge Vents

		Crew	Daily Output	Labor-Hours	Unit	Material	Labor	Equipment	Total	Total Incl O&P
0010	**RIDGE VENTS**									
0100	Aluminum strips, mill finish	1 Rofc	160	.050	L.F.	1.64	1.59		3.23	4.50
0150	Painted finish		160	.050	"	2.08	1.59		3.67	4.99
0200	Connectors		48	.167	Ea.	4.26	5.30		9.56	13.70
0300	End caps		48	.167	"	1.64	5.30		6.94	10.80
0400	Galvanized strips		160	.050	L.F.	2.14	1.59		3.73	5.05
0430	Molded polyethylene, shingles not included		160	.050	"	2.85	1.59		4.44	5.85
0440	End plugs		48	.167	Ea.	1.64	5.30		6.94	10.80
0450	Flexible roll, shingles not included		160	.050	L.F.	2.31	1.59		3.90	5.25
2300	Ridge vent strip, mill finish	1 Shee	155	.052	"	2.68	2.25		4.93	6.40

07 72 33 – Roof Hatches

07 72 33.10 Roof Hatches

		Crew	Daily Output	Labor-Hours	Unit	Material	Labor	Equipment	Total	Total Incl O&P
0010	**ROOF HATCHES**									
0500	2'-6" x 3', aluminum curb and cover	G-3	10	3.200	Ea.	715	116		831	965
0520	Galvanized steel curb and aluminum cover **CN**		10	3.200		695	116		811	945
0540	Galvanized steel curb and cover		10	3.200		555	116		671	790
0600	2'-6" x 4'-6", aluminum curb and cover		9	3.556		825	129		954	1,100
0800	Galvanized steel curb and aluminum cover		9	3.556		600	129		729	860
0900	Galvanized steel curb and cover		9	3.556		775	129		904	1,050
1100	4' x 4' aluminum curb and cover		8	4		1,475	145		1,620	1,825
1120	Galvanized steel curb and aluminum cover		8	4		600	145		745	885
1140	Galvanized steel curb and cover		8	4		1,425	145		1,570	1,775
1200	2'-6" x 8'-0", aluminum curb and cover		6.60	4.848		1,400	175		1,575	1,825
1400	Galvanized steel curb and aluminum cover		6.60	4.848		1,150	175		1,325	1,550
1500	Galvanized steel curb and cover		6.60	4.848		1,175	175		1,350	1,575
1800	For plexiglass panels, 2'-6" x 3'-0", add to above					400			400	440

07 72 36 – Smoke Vents

07 72 36.10 Smoke Hatches

		Crew	Daily Output	Labor-Hours	Unit	Material	Labor	Equipment	Total	Total Incl O&P
0010	**SMOKE HATCHES**									
0200	For 3'-0" long, add to roof hatches from division 07 72 33.10				Ea.	25%	5%			
0250	For 4'-0" long, add to roof hatches from division 07 72 33.10					20%	5%			
0300	For 8'-0" long, add to roof hatches from division 07 72 33.10					10%	5%			

07 72 36.20 Smoke Vents

		Crew	Daily Output	Labor-Hours	Unit	Material	Labor	Equipment	Total	Total Incl O&P
0010	**SMOKE VENTS**									
0100	4' x 4' aluminum cover and frame	G-3	13	2.462	Ea.	1,675	89		1,764	2,000
0200	Galvanized steel cover and frame		13	2.462		1,475	89		1,564	1,775
0300	4' x 8' aluminum cover and frame		8	4		2,300	145		2,445	2,750
0400	Galvanized steel cover and frame		8	4		1,950	145		2,095	2,350

07 72 53 – Snow Guards

07 72 53.10 Snow Guards

		Crew	Daily Output	Labor-Hours	Unit	Material	Labor	Equipment	Total	Total Incl O&P
0010	**SNOW GUARDS**									
0100	Slate & asphalt shingle roofs	1 Rofc	160	.050	Ea.	9	1.59		10.59	12.60
0200	Standing seam metal roofs		48	.167		17	5.30		22.30	27.50
0300	Surface mount for metal roofs		48	.167		7.25	5.30		12.55	17
0400	Double rail pipe type, including pipe		130	.062	L.F.	18.05	1.96		20.01	23

07 72 Roof Accessories

07 72 73 – Pitch Pockets

07 72 73.10 Pitch Pockets	Crew	Daily Output	Labor-Hours	Unit	Material	2007 Bare Costs Labor	Equipment	Total	Total Incl O&P
0010 **PITCH POCKETS**									
0100 Adjustable, 4" to 7", welded corners, 4" deep	1 Rofc	48	.167	Ea.	11	5.30		16.30	21
0200 Side extenders, 6"	"	240	.033	"	2.20	1.06		3.26	4.22

07 72 80 – Vents

07 72 80.30 Vents	Crew	Daily Output	Labor-Hours	Unit	Material	Labor	Equipment	Total	Total Incl O&P
0010 **VENTS**									
0020 Plastic, for insulated decks, 1 per M.S.F., minimum	1 Rofc	40	.200	Ea.	13	6.35		19.35	25
0100 Maximum		20	.400		30	12.70		42.70	54.50
0300 Aluminum		30	.267		13	8.50		21.50	28.50
0800 Polystyrene baffles, 12" wide for 16" O.C. rafter spacing	1 Carp	90	.089		.33	3.26		3.59	5.45
0900 For 24" O.C. rafter spacing	"	110	.073		.54	2.67		3.21	4.75

07 81 Applied Fireproofing

07 81 16 – Cementitious Fireproofing

07 81 16.10 Sprayed Cementitous Fireproofing	Crew	Daily Output	Labor-Hours	Unit	Material	Labor	Equipment	Total	Total Incl O&P
0010 **SPRAYED CEMENTITOUS FIREPROOFING**									
0050 Not incl tamping or canvas protection									
0100 1" thick, on flat plate steel	G-2	3000	.008	S.F.	.45	.24	.04	.73	.90
0200 Flat decking		2400	.010		.45	.30	.05	.80	1.01
0400 Beams		1500	.016		.45	.49	.08	1.02	1.32
0500 Corrugated or fluted decks		1250	.019		.67	.58	.10	1.35	1.74
0700 Columns, 1-1/8" thick		1100	.022		.50	.66	.11	1.27	1.68
0800 2-3/16" thick		700	.034		.96	1.04	.17	2.17	2.84
0850 For tamping, add						10%			
0900 For canvas protection, add	G-2	5000	.005	S.F.	.06	.15	.02	.23	.32
1000 Acoustical sprayed, 1" thick, finished, straight work, minimum		520	.046		.47	1.40	.23	2.10	2.92
1100 Maximum		200	.120		.50	3.64	.61	4.75	6.75
1300 Difficult access, minimum		225	.107		.50	3.24	.54	4.28	6.10
1400 Maximum		130	.185		.56	5.60	.94	7.10	10.20
1500 Intumescent epoxy fireproofing on wire mesh, 3/16" thick									
1550 1 hour rating, exterior use	G-2	136	.176	S.F.	5.85	5.35	.90	12.10	15.65
1600 Magnesium oxychloride, 35# to 40# density, 1/4" thick		3000	.008		1.23	.24	.04	1.51	1.76
1650 1/2" thick		2000	.012		2.47	.36	.06	2.89	3.35
1700 60# to 70# density, 1/4" thick		3000	.008		1.63	.24	.04	1.91	2.20
1750 1/2" thick		2000	.012		3.28	.36	.06	3.70	4.24
2000 Vermiculite cement, troweled or sprayed, 1/4" thick		3000	.008		1.11	.24	.04	1.39	1.63
2050 1/2" thick		2000	.012		2.21	.36	.06	2.63	3.06

07 84 Firestopping

07 84 13 – Penetration Firestopping

07 84 13.10 Firestopping		Crew	Daily Output	Labor-Hours	Unit	Material	2007 Bare Costs Labor	2007 Bare Costs Equipment	Total	Total Incl O&P
0010	**FIRESTOPPING** R078413-30									
0100	Metallic piping, non insulated									
0110	Through walls, 2" diameter	1 Carp	16	.500	Ea.	10.75	18.35		29.10	40.50
0120	4" diameter		14	.571		16.45	21		37.45	50.50
0130	6" diameter		12	.667		22	24.50		46.50	62.50
0140	12" diameter		10	.800		39	29.50		68.50	88.50
0150	Through floors, 2" diameter		32	.250		6.55	9.20		15.75	21.50
0160	4" diameter		28	.286		9.40	10.50		19.90	26.50
0170	6" diameter		24	.333		12.35	12.25		24.60	32.50
0180	12" diameter	↓	20	.400	↓	21	14.70		35.70	46
0190	Metallic piping, insulated									
0200	Through walls, 2" diameter	1 Carp	16	.500	Ea.	15.25	18.35		33.60	45.50
0210	4" diameter		14	.571		21	21		42	55.50
0220	6" diameter		12	.667		26.50	24.50		51	67.50
0230	12" diameter		10	.800		43.50	29.50		73	93.50
0240	Through floors, 2" diameter		32	.250		11.05	9.20		20.25	26.50
0250	4" diameter		28	.286		13.90	10.50		24.40	31.50
0260	6" diameter		24	.333		16.85	12.25		29.10	37.50
0270	12" diameter	↓	20	.400	↓	21	14.70		35.70	46
0280	Non metallic piping, non insulated									
0290	Through walls, 2" diameter	1 Carp	12	.667	Ea.	44.50	24.50		69	87
0300	4" diameter		10	.800		56	29.50		85.50	107
0310	6" diameter		8	1		77.50	36.50		114	143
0330	Through floors, 2" diameter		16	.500		34.50	18.35		52.85	66.50
0340	4" diameter		6	1.333		43	49		92	124
0350	6" diameter	↓	6	1.333	↓	51.50	49		100.50	133
0370	Ductwork, insulated & non insulated, round									
0380	Through walls, 6" diameter	1 Carp	12	.667	Ea.	22.50	24.50		47	62.50
0390	12" diameter		10	.800		45	29.50		74.50	95
0400	18" diameter		8	1		73	36.50		109.50	138
0410	Through floors, 6" diameter		16	.500		12.35	18.35		30.70	42
0420	12" diameter		14	.571		22.50	21		43.50	57
0430	18" diameter	↓	12	.667	↓	39.50	24.50		64	81.50
0440	Ductwork, insulated & non insulated, rectangular									
0450	With stiffener/closure angle, through walls, 6" x 12"	1 Carp	8	1	Ea.	18.75	36.50		55.25	77.50
0460	12" x 24"		6	1.333		25	49		74	104
0470	24" x 48"		4	2		71	73.50		144.50	192
0480	With stiffener/closure angle, through floors, 6" x 12"		10	.800		10.10	29.50		39.60	56.50
0490	12" x 24"		8	1		18.20	36.50		54.70	77
0500	24" x 48"	↓	6	1.333	↓	36	49		85	116
0510	Multi trade openings									
0520	Through walls, 6" x 12"	1 Carp	2	4	Ea.	39.50	147		186.50	273
0530	12" x 24"	"	1	8		158	294		452	630
0540	24" x 48"	2 Carp	1	16		635	585		1,220	1,600
0550	48" x 96"	"	.75	21.333		2,550	785		3,335	4,025
0560	Through floors, 6" x 12"	1 Carp	2	4		39.50	147		186.50	273
0570	12" x 24"	"	1	8		158	294		452	630
0580	24" x 48"	2 Carp	.75	21.333		635	785		1,420	1,925
0590	48" x 96"	"	.50	32	↓	2,550	1,175		3,725	4,625
0600	Structural penetrations, through walls									
0610	Steel beams, W8 x 10	1 Carp	8	1	Ea.	24.50	36.50		61	84
0620	W12 x 14		6	1.333		39.50	49		88.50	120
0630	W21 x 44	↓	5	1.600	↓	78.50	58.50		137	178

07 84 Firestopping

07 84 13 – Penetration Firestopping

07 84 13.10 Firestopping		Crew	Daily Output	Labor-Hours	Unit	Material	2007 Bare Costs Labor	Equipment	Total	Total Incl O&P
0640	W36 x 135	1 Carp	3	2.667	Ea.	191	98		289	360
0650	Bar joists, 18" deep		6	1.333		36	49		85	116
0660	24" deep		6	1.333		45	49		94	126
0670	36" deep		5	1.600		67.50	58.50		126	166
0680	48" deep	↓	4	2	↓	78.50	73.50		152	201
0690	Construction joints, floor slab at exterior wall									
0700	Precast, brick, block or drywall exterior									
0710	2" wide joint	1 Carp	125	.064	L.F.	5.60	2.35		7.95	9.85
0720	4" wide joint	"	75	.107	"	11.25	3.91		15.16	18.45
0730	Metal panel, glass or curtain wall exterior									
0740	2" wide joint	1 Carp	40	.200	L.F.	13.30	7.35		20.65	26
0750	4" wide joint	"	25	.320	"	18.15	11.75		29.90	38.50
0760	Floor slab to drywall partition									
0770	Flat joint	1 Carp	100	.080	L.F.	5.50	2.94		8.44	10.60
0780	Fluted joint		50	.160		11.25	5.85		17.10	21.50
0790	Etched fluted joint	↓	75	.107	↓	7.30	3.91		11.21	14.15
0800	Floor slab to concrete/masonry partition									
0810	Flat joint	1 Carp	75	.107	L.F.	12.35	3.91		16.26	19.70
0820	Fluted joint	"	50	.160	"	14.60	5.85		20.45	25
0830	Concrete/CMU wall joints									
0840	1" wide	1 Carp	100	.080	L.F.	6.75	2.94		9.69	11.95
0850	2" wide		75	.107		12.35	3.91		16.26	19.70
0860	4" wide	↓	50	.160	↓	23.50	5.85		29.35	35
0870	Concrete/CMU floor joints									
0880	1" wide	1 Carp	200	.040	L.F.	3.37	1.47		4.84	6
0890	2" wide		150	.053		6.20	1.96		8.16	9.85
0900	4" wide	↓	100	.080	↓	11.80	2.94		14.74	17.50

07 92 Joint Sealants

07 92 10 – Caulking and Sealants

07 92 10.10 Caulking and Sealants		Crew	Daily Output	Labor-Hours	Unit	Material	2007 Bare Costs Labor	Equipment	Total	Total Incl O&P
0010	**CAULKING AND SEALANTS**									
0020	Acoustical sealant, elastomeric, cartridges				Ea.	4.60			4.60	5.05
0030	Backer rod, polyethylene, 1/4" diameter	1 Bric	4.60	1.739	C.L.F.	4.44	66		70.44	106
0050	1/2" diameter		4.60	1.739		6.10	66		72.10	108
0070	3/4" diameter		4.60	1.739		8.60	66		74.60	110
0090	1" diameter	↓	4.60	1.739	↓	12.15	66		78.15	114
0100	Acrylic latex caulk, white									
0200	11 fl. oz cartridge				Ea.	2.40			2.40	2.64
0500	1/4" x 1/2"	1 Bric	248	.032	L.F.	.20	1.23		1.43	2.09
0600	1/2" x 1/2"		250	.032		.39	1.22		1.61	2.28
0800	3/4" x 3/4"		230	.035		.88	1.32		2.20	2.98
0900	3/4" x 1"		200	.040		1.18	1.52		2.70	3.61
1000	1" x 1"	↓	180	.044	↓	1.47	1.69		3.16	4.19
1400	Butyl based, bulk				Gal.	24			24	26.50
1500	Cartridges				"	29			29	32
1700	Bulk, in place 1/4" x 1/2", 154 L.F./gal.	1 Bric	230	.035	L.F.	.16	1.32		1.48	2.18
1800	1/2" x 1/2", 77 L.F./gal.	"	180	.044	"	.31	1.69		2	2.91
2000	Latex acrylic based, bulk				Gal.	25			25	27.50
2100	Cartridges					31			31	34
2300	Polysulfide compounds, 1 component, bulk				↓	47.50			47.50	52

07 92 Joint Sealants

07 92 10 – Caulking and Sealants

07 92 10.10 Caulking and Sealants

		Crew	Daily Output	Labor-Hours	Unit	Material	2007 Bare Costs Labor	Equipment	Total	Total Incl O&P
2600	1 or 2 component, in place, 1/4" x 1/4", 308 L.F./gal.	1 Bric	145	.055	L.F.	.15	2.10		2.25	3.36
2700	1/2" x 1/4", 154 L.F./gal.		135	.059		.31	2.25		2.56	3.77
2900	3/4" x 3/8", 68 L.F./gal.		130	.062		.70	2.34		3.04	4.33
3000	1" x 1/2", 38 L.F./gal.	↓	130	.062	↓	1.25	2.34		3.59	4.93
3200	Polyurethane, 1 or 2 component				Gal.	52.50			52.50	57.50
3500	Bulk, in place, 1/4" x 1/4"	1 Bric	150	.053	L.F.	.17	2.03		2.20	3.28
3600	1/2" x 1/4"		145	.055		.34	2.10		2.44	3.56
3800	3/4" x 3/8", 68 L.F./gal.		130	.062		.77	2.34		3.11	4.41
3900	1" x 1/2"	↓	110	.073	↓	1.36	2.77		4.13	5.70
4100	Silicone rubber, bulk				Gal.	40			40	44
4200	Cartridges				"	40			40	44
4400	Neoprene gaskets, closed cell, adhesive, 1/8" x 3/8"	1 Bric	240	.033	L.F.	.23	1.27		1.50	2.18
4500	1/4" x 3/4"		215	.037		.51	1.42		1.93	2.71
4700	1/2" x 1"		200	.040		1.14	1.52		2.66	3.57
4800	3/4" x 1-1/2"	↓	165	.048	↓	1.27	1.84		3.11	4.21
5500	Resin epoxy coating, 2 component, heavy duty				Gal.	27			27	29.50
5800	Tapes, sealant, P.V.C. foam adhesive, 1/16" x 1/4"				C.L.F.	4.98			4.98	5.50
5900	1/16" x 1/2"					7.35			7.35	8.10
5950	1/16" x 1"					12.20			12.20	13.45
6000	1/8" x 1/2"				↓	8.25			8.25	9.10
6200	Urethane foam, 2 component, handy pack, 1 C.F.				Ea.	30			30	33
6300	50.0 C.F. pack				C.F.	15.20			15.20	16.70

07 95 Expansion Control

07 95 13 – Expansion Joint Cover Assemblies

07 95 13.50 Expansion Joint Assemblies

		Crew	Daily Output	Labor-Hours	Unit	Material	2007 Bare Costs Labor	Equipment	Total	Total Incl O&P
0010	**EXPANSION JOINT ASSEMBLIES**									
0200	Floor cover assemblies, 1" space, aluminum	1 Sswk	38	.211	L.F.	17.05	8.70		25.75	34.50
0300	Bronze		38	.211		34.50	8.70		43.20	54
0500	2" space, aluminum		38	.211		20.50	8.70		29.20	38.50
0600	Bronze		38	.211		37	8.70		45.70	57
0800	Wall and ceiling assemblies, 1" space, aluminum		38	.211		10.15	8.70		18.85	27
0900	Bronze		38	.211		32.50	8.70		41.20	51.50
1100	2" space, aluminum		38	.211		17.25	8.70		25.95	34.50
1200	Bronze		38	.211		39.50	8.70		48.20	59.50
1400	Floor to wall assemblies, 1" space, aluminum		38	.211		15.50	8.70		24.20	33
1500	Bronze or stainless		38	.211		39	8.70		47.70	59
1700	Gym floor angle covers, aluminum, 3" x 3" angle		46	.174		13.40	7.20		20.60	28
1800	3" x 4" angle		46	.174		15.85	7.20		23.05	30.50
2000	Roof closures, aluminum, flat roof, low profile, 1" space		57	.140		31	5.80		36.80	44.50
2100	High profile		57	.140		38	5.80		43.80	52.50
2300	Roof to wall, low profile, 1" space		57	.140		17.15	5.80		22.95	29.50
2400	High profile	↓	57	.140	↓	22	5.80		27.80	35

Estimating Tips

08 10 00 Doors and Frames

- Most metal doors and frames look alike, but there may be significant differences among them. When estimating these items be sure to choose the line item that most closely compares to the specification or door schedule requirements regarding:
  - type of metal
  - metal gauge
  - door core material
  - fire rating
  - finish
- Wood and plastic doors vary considerably in price. The primary determinant is the veneer material. Lauan, birch, and oak are the most common veneers. Other variables include the following:
  - hollow or solid core
  - fire rating
  - flush or raised panel
  - finish

08 30 00 Specialty Doors and Frames

- There are many varieties of special doors, and they are usually priced per each. Add frames, hardware, or operators required for a complete installation.

08 40 00 Entrances, Storefronts, and Curtain Walls

- Glazed curtain walls consist of the metal tube framing and the glazing material. The cost data in this subdivision is presented for the metal tube framing alone or the composite wall. If your estimate requires a detailed takeoff of the framing, be sure to add the glazing cost.

08 50 00 Windows

- Most metal windows are delivered preglazed. However, some metal windows are priced without glass. Refer to 08 80 00 Glazing for glass pricing. The grade C indicates commercial grade windows, usually ASTM C-35.
- All wood windows are priced preglazed. The two glazing options priced are single pane float glass and insulating glass 1/2" thick. Add the cost of screens and grills if required.

08 70 00 Hardware

- Hardware costs add considerably to the cost of a door. The most efficient method to determine the hardware requirements for a project is to review the door schedule.
- Door hinges are priced by the pair, with most doors requiring 1-1/2 pairs per door. The hinge prices do not include installation labor because it is included in door installation. Hinges are classified according to the frequency of use.

08 80 00 Glazing

- Different openings require different types of glass. The three most common types are:
  - float
  - tempered
  - insulating
- Most exterior windows are glazed with insulating glass. Entrance doors and window walls, where the glass is less than 18" from the floor, are generally glazed with tempered glass. Interior windows and some residential windows are glazed with float glass.

Reference Numbers

Reference numbers are shown in shaded boxes at the beginning of some major classifications. These numbers refer to related items in the Reference Section. The reference information may be an estimating procedure, an alternate pricing method, or technical information.

Note: Not all subdivisions listed here necessarily appear in this publication.

08 05 Common Work Results for Openings

08 05 05 – Selective Windows and Doors Demolition

08 05 05.10 Selective Demolition Doors

		Crew	Daily Output	Labor-Hours	Unit	Material	2007 Bare Costs Labor	2007 Bare Costs Equipment	Total	Total Incl O&P
0010	**SELECTIVE DEMOLITION DOORS** R024119-10									
0200	Doors, exterior, 1-3/4" thick, single, 3' x 7' high	1 Clab	16	.500	Ea.		14.40		14.40	22.50
0220	Double, 6' x 7' high		12	.667			19.15		19.15	30
0500	Interior, 1-3/8" thick, single, 3' x 7' high		20	.400			11.50		11.50	17.90
0520	Double, 6' x 7' high		16	.500			14.40		14.40	22.50
0700	Bi-folding, 3' x 6'-8" high		20	.400			11.50		11.50	17.90
0720	6' x 6'-8" high		18	.444			12.80		12.80	19.90
0900	Bi-passing, 3' x 6'-8" high		16	.500			14.40		14.40	22.50
0940	6' x 6'-8" high		14	.571			16.45		16.45	25.50
1500	Remove and reset, minimum	1 Carp	8	1			36.50		36.50	57
1520	Maximum		6	1.333			49		49	76
2000	Frames, including trim, metal		8	1			36.50		36.50	57
2200	Wood	2 Carp	32	.500			18.35		18.35	28.50
3000	Special doors, counter doors		6	2.667			98		98	152
3100	Double acting		10	1.600			58.50		58.50	91.50
3200	Floor door (trap type), or access type		8	2			73.50		73.50	114
3300	Glass, sliding, including frames		12	1.333			49		49	76
3400	Overhead, commercial, 12' x 12' high		4	4			147		147	229
3440	up to 20' x 16' high		3	5.333			196		196	305
3445	up to 35' x 30' high		1	16			585		585	915
3500	Residential, 9' x 7' high		8	2			73.50		73.50	114
3540	16' x 7' high		7	2.286			84		84	131
3600	Remove and reset, minimum		4	4			147		147	229
3620	Maximum		2.50	6.400			235		235	365
3700	Roll-up grille		5	3.200			117		117	183
3800	Revolving door		2	8			294		294	455
3900	Storefront swing door		3	5.333			196		196	305
6600	Demo flexible transparent strip entrance	3 Shee	115	.209	SF Surf		9.10		9.10	14
7100	Remove double swing pneumatic doors, openers and sensors	2 Skwk	.50	32	Opng.		1,225		1,225	1,900
7110	Remove automatic operators, industrial, sliding doors, to 12' wide	"	.40	40	"		1,525		1,525	2,375
7550	Hangar door demo	2 Sswk	220	.073	S.F.		3.01		3.01	5.45
7570	Remove shock absorbing door	"	1.90	8.421	Opng.		350		350	630

08 05 05.20 Selective Demolition of Windows

		Crew	Daily Output	Labor-Hours	Unit	Material	2007 Bare Costs Labor	2007 Bare Costs Equipment	Total	Total Incl O&P
0010	**SELECTIVE DEMOLITION OF WINDOWS** R024119-10									
0200	Aluminum, including trim, to 12 S.F.	1 Clab	16	.500	Ea.		14.40		14.40	22.50
0240	To 25 S.F.		11	.727			21		21	32.50
0280	To 50 S.F.		5	1.600			46		46	71.50
0320	Storm windows/scree, to 12 S.F.		27	.296			8.50		8.50	13.25
0360	To 25 S.F.		21	.381			10.95		10.95	17.05
0400	To 50 S.F.		16	.500			14.40		14.40	22.50
0600	Glass, minimum		200	.040	S.F.		1.15		1.15	1.79
0620	Maximum		150	.053	"		1.53		1.53	2.39
1000	Steel, including trim, to 12 S.F.		13	.615	Ea.		17.70		17.70	27.50
1020	To 25 S.F.		9	.889			25.50		25.50	40
1040	To 50 S.F.		4	2			57.50		57.50	89.50
2000	Wood, including trim, to 12 S.F.		22	.364			10.45		10.45	16.25
2020	To 25 S.F.		18	.444			12.80		12.80	19.90
2060	To 50 S.F.		13	.615			17.70		17.70	27.50
2065	To 180 S.F.		8	1			29		29	45
4300	Remove bay/bow window	2 Carp	6	2.667			98		98	152
4410	Remove skylight, plstc domes, flush/curb mtd	G-3	395	.081	S.F.		2.93		2.93	4.53
5020	Remove and reset window, minimum	1 Carp	6	1.333	Ea.		49		49	76

08 05 Common Work Results for Openings

08 05 05 – Selective Windows and Doors Demolition

08 05 05.20 Selective Demolition of Windows	Crew	Daily Output	Labor-Hours	Unit	Material	2007 Bare Costs Labor	2007 Bare Costs Equipment	Total	Total Incl O&P
5040 Average	1 Carp	4	2	Ea.		73.50		73.50	114
5080 Maximum	↓	2	4	↓		147		147	229

08 11 Metal Doors and Frames

08 11 16 – Aluminum Doors and Frames

08 11 16.10 Aluminum Doors and Frames

		Crew	Daily Output	Labor-Hours	Unit	Material	Labor	Equipment	Total	Total Incl O&P
0010	**ALUMINUM DOORS AND FRAMES**, entrance, narrow stile									
0015	Standard hardware, clear finish, not incl. glass, 2'-6" x 7'-0" opng.	2 Sswk	2	8	Ea.	810	330		1,140	1,500
0020	3'-0" x 7'-0" opening		2	8		650	330		980	1,325
0030	3'-6" x 7'-0" opening		2	8		625	330		955	1,300
0100	3'-0" x 10'-0" opening, 3' high transom		1.80	8.889		945	370		1,315	1,725
0200	3'-6" x 10'-0" opening, 3' high transom		1.80	8.889		945	370		1,315	1,725
0280	5'-0" x 7'-0" opening		2	8		1,025	330		1,355	1,725
0300	6'-0" x 7'-0" opening		1.30	12.308	↓	1,000	510		1,510	2,025
0400	6'-0" x 10'-0" opening, 3' high transom		1.10	14.545	Pr.	1,225	600		1,825	2,450
0420	7'-0" x 7'-0" opening		1	16	"	1,025	660		1,685	2,325
0500	Wide stile, 2'-6" x 7'-0" opening		2	8	Ea.	855	330		1,185	1,550
0520	3'-0" x 7'-0" opening		2	8		845	330		1,175	1,525
0540	3'-6" x 7'-0" opening		2	8		915	330		1,245	1,600
0560	5'-0" x 7'-0" opening		2	8		1,350	330		1,680	2,075
0580	6'-0" x 7'-0" opening		1.30	12.308	Pr.	1,375	510		1,885	2,425
0600	7'-0" x 7'-0" opening	↓	1	16	"	1,475	660		2,135	2,825
1100	For full vision doors, with 1/2" glass, add				Leaf	55%				
1200	For non-standard size, add					67%				
1300	Light bronze finish, add					36%				
1400	Dark bronze finish, add					18%				
1500	For black finish, add					36%				
1600	Concealed panic device, add				↓	1,025			1,025	1,125
1700	Electric striker release, add				Opng.	262			262	288
1800	Floor check, add				Leaf	775			775	855
1900	Concealed closer, add				"	515			515	570
2000	Flush 3' x 7' Insulated, 12"x 12" lite, clear finish	2 Sswk	2	8	Ea.	1,050	330		1,380	1,750

08 11 63 – Metal Screen and Storm Doors and Frames

08 11 63.23 Aluminum Screen and Storm Doors and Frames

		Crew	Daily Output	Labor-Hours	Unit	Material	Labor	Equipment	Total	Total Incl O&P
0010	**ALUMINUM SCREEN AND STORM DOORS AND FRAMES**									
0020	Combination storm and screen									
0400	Clear anodic coating, 6'-8" x 2'-6" wide	2 Carp	15	1.067	Ea.	162	39		201	240
0420	2'-8" wide		14	1.143		191	42		233	276
0440	3'-0" wide	↓	14	1.143	↓	191	42		233	276
0500	For 7' door height, add					5%				
1000	Mill finish, 6'-8" x 2'-6" wide	2 Carp	15	1.067	Ea.	216	39		255	298
1020	2'-8" wide		14	1.143		216	42		258	305
1040	3'-0" wide	↓	14	1.143		234	42		276	325
1100	For 7'-0" door, add					5%				
1500	White painted, 6'-8" x 2'-6" wide	2 Carp	15	1.067	Ea.	262	39		301	350
1520	2'-8" wide		14	1.143		223	42		265	310
1540	3'-0" wide	↓	14	1.143		258	42		300	350
1600	For 7'-0" door, add				↓	5%				
2000	Wood door & screen, see division 08 14 33.20									

08 11 Metal Doors and Frames

08 11 74 – Sliding Metal Grilles

08 11 74.10 Rolling Grille Supports

		Crew	Daily Output	Labor-Hours	Unit	Material	2007 Bare Costs Labor	2007 Bare Costs Equipment	Total	Total Incl O&P
0010	**ROLLING GRILLE SUPPORTS**									
0020	Rolling grille supports, overhead framed	E-4	36	.889	L.F.	18.55	37	3.20	58.75	91.50

08 12 Metal Frames

08 12 13 – Hollow Metal Frames

08 12 13.13 Standard Hollow Metal Frames

		Crew	Daily Output	Labor-Hours	Unit	Material	2007 Bare Costs Labor	2007 Bare Costs Equipment	Total	Total Incl O&P
0010	**STANDARD HOLLOW METAL FRAMES**									
0020	16 ga., up to 5-3/4" jamb depth									
0025	6'-8" high, 3'-0" wide, single	2 Carp	16	1	Ea.	128	36.50		164.50	198
0028	3'-6" wide, single		16	1		109	36.50		145.50	177
0030	4'-0" wide, single		16	1		109	36.50		145.50	177
0040	6'-0" wide, double		14	1.143		143	42		185	224
0045	8'-0" wide, double		14	1.143		149	42		191	230
0100	7'-0" high, 3'-0" wide, single		16	1		132	36.50		168.50	202
0110	3'-6" wide, single		16	1		112	36.50		148.50	180
0112	4'-0" wide, single		16	1		143	36.50		179.50	214
0140	6'-0" wide, double		14	1.143		150	42		192	231
0145	8'-0" wide, double		14	1.143		140	42		182	220
1000	16 ga., up to 4-7/8" deep, 7'-0" H, 3'-0" W, single **CN**		16	1		123	36.50		159.50	192
1140	6'-0" wide, double		14	1.143		149	42		191	230
2800	14 ga., up to 3-7/8" deep, 7'-0" high, 3'-0" wide, single		16	1		127	36.50		163.50	196
2840	6'-0" wide, double		14	1.143		152	42		194	233
3000	14 ga., up to 5-3/4" deep, 6'-8" high, 3'-0" wide, single		16	1		133	36.50		169.50	204
3002	3'-6" wide, single		16	1		133	36.50		169.50	203
3005	4'-0" wide, single		16	1		133	36.50		169.50	203
3600	up to 5-3/4" jamb depth, 7'-0" high, 4'-0" wide, single		15	1.067		103	39		142	175
3620	6'-0" wide, double		12	1.333		162	49		211	254
3640	8'-0" wide, double		12	1.333		153	49		202	245
3700	8'-0" high, 4'-0" wide, single		15	1.067		136	39		175	211
3740	8'-0" wide, double		12	1.333		169	49		218	262
4000	6-3/4" deep, 7'-0" high, 4'-0" wide, single		15	1.067		162	39		201	239
4020	6'-0" wide, double		12	1.333		189	49		238	284
4040	8'-0", wide double		12	1.333		215	49		264	315
4100	8'-0" high, 4'-0" wide, single		15	1.067		162	39		201	239
4140	8'-0" wide, double		12	1.333		189	49		238	284
4400	8-3/4" deep, 7'-0" high, 4'-0" wide, single		15	1.067		152	39		191	228
4440	8'-0" wide, double		12	1.333		216	49		265	315
4500	8'-0" high, 4'-0" wide, single		15	1.067		163	39		202	240
4540	8'-0" wide, double		12	1.333		190	49		239	285
4900	For welded frames, add					44.50			44.50	49
5400	14 ga., "B" label, up to 5-3/4" deep, 7'-0" high, 4'-0" wide, single	2 Carp	15	1.067		161	39		200	238
5440	8'-0" wide, double		12	1.333		196	49		245	292
5800	6-3/4" deep, 7'-0" high, 4'-0" wide, single		15	1.067		140	39		179	215
5840	8'-0" wide, double		12	1.333		235	49		284	335
6200	8-3/4" deep, 7'-0" high, 4'-0" wide, single		15	1.067		201	39		240	282
6240	8'-0" wide, double		12	1.333		236	49		285	335
6300	For "A" label use same price as "B" label									
6400	For baked enamel finish, add					30%	15%			
6500	For galvanizing, add					15%				
6600	For hospital stop, add				Ea.	263			263	289

08 12 Metal Frames

08 12 13 – Hollow Metal Frames

08 12 13.13 Standard Hollow Metal Frames

		Crew	Daily Output	Labor-Hours	Unit	Material	2007 Bare Costs Labor	Equipment	Total	Total Incl O&P
7900	Transom lite frames, fixed, add	2 Carp	155	.103	S.F.	42.50	3.79		46.29	53
8000	Movable, add	"	130	.123	"	51.50	4.52		56.02	63.50

08 12 13.25 Channel Metal Frames

		Crew	Daily Output	Labor-Hours	Unit	Material	Labor	Equipment	Total	Total Incl O&P
0010	**CHANNEL METAL FRAMES**									
0020	Steel channels with anchors and bar stops									
0100	6" channel @ 8.2#/L.F., 3' x 7' door, weighs 150#	E-4	13	2.462	Ea.	185	103	8.85	296.85	400
0200	8" channel @ 11.5#/L.F., 6' x 8' door, weighs 275#		9	3.556		340	149	12.80	501.80	655
0300	8' x 12' door, weighs 400#		6.50	4.923		490	206	17.70	713.70	935
0400	10" channel @ 15.3#/L.F., 10' x 10' door, weighs 500# CN		6	5.333		615	223	19.20	857.20	1,100
0500	12' x 12' door, weighs 600#		5.50	5.818		740	243	21	1,004	1,275
0600	12" channel @ 20.7#/L.F., 12' x 12' door, weighs 825#		4.50	7.111		1,025	298	25.50	1,348.50	1,700
0700	12' x 16' door, weighs 1000#		4	8		1,225	335	29	1,589	1,975
0800	For frames without bar stops, light sections, deduct					15%				
0900	Heavy sections, deduct					10%				

08 13 Metal Doors

08 13 13 – Hollow Metal Doors

08 13 13.13 Standard Hollow Metal Doors

		Crew	Daily Output	Labor-Hours	Unit	Material	Labor	Equipment	Total	Total Incl O&P
0010	**STANDARD HOLLOW METAL DOORS** R081313-20									
0015	Flush, full panel, hollow core									
0020	1-3/8" thick, 20 ga., 2'-0" x 6'-8"	2 Carp	20	.800	Ea.	249	29.50		278.50	320
0040	2'-8" x 6'-8"		18	.889		256	32.50		288.50	335
0060	3'-0" x 6'-8"		17	.941		261	34.50		295.50	340
0100	3'-0" x 7'-0"		17	.941		279	34.50		313.50	360
0120	For vision lite, add					84.50			84.50	93
0140	For narrow lite, add					93			93	102
0320	Half glass, 20 ga., 2'-0" x 6'-8"	2 Carp	20	.800		370	29.50		399.50	455
0340	2'-8" x 6'-8"		18	.889		380	32.50		412.50	465
0360	3'-0" x 6'-8"		17	.941		385	34.50		419.50	480
0400	3'-0" x 7'-0"		17	.941		390	34.50		424.50	485
0410	1-3/8" thick, 18 ga., 2'-0" x 6'-8"		20	.800		298	29.50		327.50	375
0420	3'-0" x 6'-8"		17	.941		300	34.50		334.50	385
0425	3'-0" x 7'-0"		17	.941		315	34.50		349.50	400
0450	For vision lite, add					84.50			84.50	93
0452	For narrow lite, add					93			93	102
0460	Half glass, 18 ga., 2'-0" x 6'-8"	2 Carp	20	.800		420	29.50		449.50	510
0465	2'-8" x 6'-8"		18	.889		435	32.50		467.50	530
0470	3'-0" x 6'-8"		17	.941		425	34.50		459.50	520
0475	3'-0" x 7'-0"		17	.941		435	34.50		469.50	530
0500	Hollow core, 1-3/4" thick, full panel, 20 ga., 2'-8" x 6'-8"		18	.889		287	32.50		319.50	365
0520	3'-0" x 6'-8"		17	.941		261	34.50		295.50	340
0640	3'-0" x 7'-0"		17	.941		310	34.50		344.50	395
0680	4'-0" x 7'-0"		15	1.067		435	39		474	540
0700	4'-0" x 8'-0"		13	1.231		495	45		540	615
1000	18 ga., 2'-8" x 6'-8"		17	.941		288	34.50		322.50	370
1020	3'-0" x 6'-8"		16	1		280	36.50		316.50	360
1120	3'-0" x 7'-0" CN		17	.941		350	34.50		384.50	435
1180	4'-0" x 7'-0"		14	1.143		420	42		462	525
1200	4'-0" x 8'-0"		17	.941		500	34.50		534.50	605
1212	For vision lite, add					84.50			84.50	93
1214	For narrow lite, add					93			93	102

08 13 Metal Doors

08 13 13 – Hollow Metal Doors

08 13 13.13 Standard Hollow Metal Doors

		Crew	Daily Output	Labor-Hours	Unit	Material	2007 Bare Costs Labor	2007 Bare Costs Equipment	Total	Total Incl O&P
1230	Half glass, 20 ga., 2'-8" x 6'-8"	2 Carp	20	.800	Ea.	400	29.50		429.50	485
1240	3'-0" x 6'-8"		18	.889		395	32.50		427.50	485
1260	3'-0" x 7'-0"		18	.889		395	32.50		427.50	485
1320	18 ga., 2'-8" x 6'-8"		18	.889		430	32.50		462.50	525
1340	3'-0" x 6'-8"		17	.941		425	34.50		459.50	520
1360	3'-0" x 7'-0"		17	.941		435	34.50		469.50	535
1380	4'-0" x 7'-0"		15	1.067		540	39		579	655
1400	4'-0" x 8'-0"		14	1.143		620	42		662	745
1720	Insulated, 1-3/4" thick, full panel, 18 ga., 3'-0" x 6'-8"		15	1.067		375	39		414	475
1740	2'-8" x 7'-0"		16	1		390	36.50		426.50	485
1760	3'-0" x 7'-0"		15	1.067		385	39		424	485
1800	4'-0" x 8'-0"		13	1.231		565	45		610	690
1805	For vision lite, add					84.50			84.50	93
1810	For narrow lite, add					93			93	102
1820	Half glass, 18 ga., 3'-0" x 6'-8"	2 Carp	16	1		500	36.50		536.50	605
1840	2'-8" x 7'-0"		17	.941		510	34.50		544.50	615
1860	3'-0" x 7'-0"		16	1		550	36.50		586.50	660
1900	4'-0" x 8'-0"		14	1.143		610	42		652	740
2000	For bottom louver, add					149			149	164
2020	For baked enamel finish, add					30%	15%			
2040	For galvanizing, add					15%				

08 13 13.15 Metal Fire Doors

		Crew	Daily Output	Labor-Hours	Unit	Material	Labor	Equipment	Total	Total Incl O&P
0010	**METAL FIRE DOORS** R081313-20									
0015	Steel, flush, "B" label, 90 minute									
0020	Full panel, 20 ga., 2'-0" x 6'-8"	2 Carp	20	.800	Ea.	315	29.50		344.50	395
0040	2'-8" x 6'-8"		18	.889		320	32.50		352.50	405
0060	3'-0" x 6'-8"		17	.941		325	34.50		359.50	410
0080	3'-0" x 7'-0"		17	.941		335	34.50		369.50	425
0140	18 ga., 3'-0" x 6'-8"		16	1		370	36.50		406.50	460
0160	2'-8" x 7'-0"		17	.941		390	34.50		424.50	480
0180	3'-0" x 7'-0"		16	1		370	36.50		406.50	460
0200	4'-0" x 7'-0"		15	1.067		485	39		524	590
0220	For "A" label, 3 hour, 18 ga., use same price as "B" label									
0240	For vision lite, add				Ea.	105			105	116
0520	Flush, "B" label 90 min., composite, 20 ga., 2'-0" x 6'-8"	2 Carp	18	.889		405	32.50		437.50	495
0540	2'-8" x 6'-8"		17	.941		405	34.50		439.50	500
0560	3'-0" x 6'-8"		16	1		410	36.50		446.50	505
0580	3'-0" x 7'-0"		16	1		420	36.50		456.50	515
0640	Flush, "A" label 3 hour, composite, 18 ga., 3'-0" x 6'-8"		15	1.067		465	39		504	570
0660	2'-8" x 7'-0"		16	1		480	36.50		516.50	585
0680	3'-0" x 7'-0"		15	1.067		475	39		514	580
0700	4'-0" x 7'-0"		14	1.143		570	42		612	695

08 13 13.20 Residential Steel Doors

		Crew	Daily Output	Labor-Hours	Unit	Material	Labor	Equipment	Total	Total Incl O&P
0010	**RESIDENTIAL STEEL DOORS**									
0020	Prehung, insulated, exterior									
0030	Embossed, full panel, 2'-8" x 6'-8"	2 Carp	17	.941	Ea.	266	34.50		300.50	345
0040	3'-0" x 6'-8"		15	1.067		239	39		278	325
0060	3'-0" x 7'-0"		15	1.067		283	39		322	370
0070	5'-4" x 6'-8", double		8	2		500	73.50		573.50	665
0220	Half glass, 2'-8" x 6'-8"		17	.941		263	34.50		297.50	345
0240	3'-0" x 6'-8"		16	1		264	36.50		300.50	345
0260	3'-0" x 7'-0"		16	1		320	36.50		356.50	405

08 13 Metal Doors

08 13 13 – Hollow Metal Doors

08 13 13.20 Residential Steel Doors

		Crew	Daily Output	Labor-Hours	Unit	Material	2007 Bare Costs Labor	2007 Bare Costs Equipment	Total	Total Incl O&P
0270	5'-4" x 6'-8", double	2 Carp	8	2	Ea.	550	73.50		623.50	720
0720	Raised plastic face, full panel, 2'-8" x 6'-8"		16	1		266	36.50		302.50	350
0740	3'-0" x 6'-8"		15	1.067		269	39		308	355
0760	3'-0" x 7'-0"		15	1.067		271	39		310	360
0780	5'-4" x 6'-8", double		8	2		500	73.50		573.50	670
0820	Half glass, 2'-8" x 6'-8"		17	.941		298	34.50		332.50	385
0840	3'-0" x 6'-8"		16	1		300	36.50		336.50	385
0860	3'-0" x 7'-0"		16	1		330	36.50		366.50	420
0880	5'-4" x 6'-8", double		8	2		645	73.50		718.50	825
1320	Flush face, full panel, 2'-6" x 6'-8"		16	1		219	36.50		255.50	297
1340	3'-0" x 6'-8"		15	1.067		219	39		258	300
1360	3'-0" x 7'-0"		15	1.067		281	39		320	370
1380	5'-4" x 6'-8", double		8	2		455	73.50		528.50	615
1420	Half glass, 2'-8" x 6'-8"		17	.941		273	34.50		307.50	355
1440	3'-0" x 6'-8"		16	1		276	36.50		312.50	360
1460	3'-0" x 7'-0"		16	1		320	36.50		356.50	405
1480	5'-4" x 6'-8", double		8	2		535	73.50		608.50	705
1500	Sidelight, full lite, 1'-0" x 6'-8" with grille					213			213	235
1510	1'-0" x 6'-8", low e					245			245	269
1520	1'-0" x 6'-8", half lite					191			191	210
1530	1'-0" x 6'-8", half lite, low e					233			233	256
2300	Interior, residential, closet, bi-fold, 6'-8" x 2'-0" wide	2 Carp	16	1		141	36.50		177.50	213
2330	3'-0" wide		16	1		158	36.50		194.50	231
2360	4'-0" wide		15	1.067		241	39		280	325
2400	5'-0" wide		14	1.143		279	42		321	370
2420	6'-0" wide		13	1.231		315	45		360	415

08 13 16 – Aluminum Doors

08 13 16.10 Aluminum Doors

		Crew	Daily Output	Labor-Hours	Unit	Material	2007 Bare Costs Labor	2007 Bare Costs Equipment	Total	Total Incl O&P
0010	**ALUMINUM DOORS**, commercial, no glazing									
0020	Incl. hinges, push/pull, deadlock, cyl., threshold									
0800	Narrow stile, no glazing, standard hardware, pair of 2'-6" x 7'-0"				Pr.	910			910	1,000
1000	3'-0" x 7'-0", single	2 Carp	3	5.333	Ea.	530	196		726	885
1200	Pair of 3'-0" x 7'-0"		1.70	9.412	Pr.	1,075	345		1,420	1,725
1500	3'-6" x 7'-0", single		3	5.333	Ea.	555	196		751	915
2000	Medium stile, pair of 2'-6" x 7'-0"		1.70	9.412	Pr.	970	345		1,315	1,625
2100	3'-0" x 7'-0", single		3	5.333	Ea.	675	196		871	1,050
2200	Pair of 3'-0" x 7'-0"		1.70	9.412	Pr.	1,325	345		1,670	2,025
2300	3'-6" x 7'-0", single		5.33	3.002	Ea.	805	110		915	1,050
5000	Flush panel doors, pair of 2'-6" x 7'-0"	2 Sswk	2	8	Pr.	960	330		1,290	1,650
5050	3'-0" x 7'-0", single		2.50	6.400	Ea.	480	265		745	1,000
5100	Pair of 3'-0" x 7'-0"		2	8	Pr.	960	330		1,290	1,650
5150	3'-6" x 7'-0", single		2.50	6.400	Ea.	570	265		835	1,100

08 13 73 – Sliding Metal Doors

08 13 73.10 Steel Sliding Doors

		Crew	Daily Output	Labor-Hours	Unit	Material	2007 Bare Costs Labor	2007 Bare Costs Equipment	Total	Total Incl O&P
0010	**STEEL SLIDING DOORS**									
0020	Up to 50' x 18', electric, standard duty, minimum	L-5	360	.156	S.F.	23.50	6.45	2.05	32	39
0100	Maximum		340	.165		39	6.80	2.18	47.98	57.50
0500	Heavy duty, minimum		297	.189		32	7.80	2.49	42.29	51.50
0600	Maximum		277	.202		84	8.35	2.67	95.02	110

08 14 Wood Doors

08 14 13 – Carved Wood Doors

08 14 13.10 Carved Wood Doors

		Crew	Daily Output	Labor-Hours	Unit	Material	2007 Bare Costs Labor	Equipment	Total	Total Incl O&P
0010	**CARVED WOOD DOORS**									
3000	Solid wood, 1-3/4" thick stile and rail									
3020	Mahogany, 3'-0" x 7'-0", minimum	2 Carp	14	1.143	Ea.	790	42		832	935
3030	Maximum		10	1.600		1,325	58.50		1,383.50	1,550
3040	3'-6" x 8'-0", minimum		10	1.600		935	58.50		993.50	1,125
3050	Maximum		8	2		1,725	73.50		1,798.50	2,025
3100	Pine, 3'-0" x 7'-0", minimum		14	1.143		415	42		457	520
3110	Maximum		10	1.600		705	58.50		763.50	865
3120	3'-6" x 8'-0", minimum		10	1.600		675	58.50		733.50	830
3130	Maximum		8	2		1,125	73.50		1,198.50	1,350
3200	Red oak, 3'-0" x 7'-0", minimum		14	1.143		1,550	42		1,592	1,775
3210	Maximum		10	1.600		1,825	58.50		1,883.50	2,125
3220	3'-6" x 8'-0", minimum		10	1.600		1,700	58.50		1,758.50	1,950
3230	Maximum		8	2		3,000	73.50		3,073.50	3,425
4000	Hand carved door, mahogany									
4020	3'-0" x 7'-0", minimum	2 Carp	14	1.143	Ea.	1,525	42		1,567	1,750
4030	Maximum		11	1.455		2,975	53.50		3,028.50	3,350
4040	3'-6" x 8'-0", minimum		10	1.600		1,900	58.50		1,958.50	2,200
4050	Maximum		8	2		2,925	73.50		2,998.50	3,350
4200	Red oak, 3'-0" x 7'-0", minimum		14	1.143		4,625	42		4,667	5,175
4210	Maximum		11	1.455		12,700	53.50		12,753.50	14,000
4220	3'-6" x 8'-0", minimum		10	1.600		5,200	58.50		5,258.50	5,825
4280	For 6'-8" high door, deduct from 7'-0" door					33			33	36.50
4400	For custom finish, add					355			355	390
4600	Side light, mahogany, 7'-0" x 1'-6" wide, minimum	2 Carp	18	.889		850	32.50		882.50	985
4610	Maximum		14	1.143		2,500	42		2,542	2,825
4620	8'-0" x 1'-6" wide, minimum		14	1.143		1,600	42		1,642	1,850
4630	Maximum		10	1.600		1,825	58.50		1,883.50	2,100
4640	Side light, oak, 7'-0" x 1'-6" wide, minimum		18	.889		985	32.50		1,017.50	1,125
4650	Maximum		14	1.143		1,750	42		1,792	2,000
4660	8'-0" x 1-6" wide, minimum		14	1.143		925	42		967	1,100
4670	Maximum		10	1.600		1,750	58.50		1,808.50	2,025

08 14 16 – Flush Wood Doors

08 14 16.09 Flush Wood Doors

		Crew	Daily Output	Labor-Hours	Unit	Material	2007 Bare Costs Labor	Equipment	Total	Total Incl O&P
0010	**FLUSH WOOD DOORS**									
0015	Flush, int., 1-3/8", 7 ply, hollow core,									
0020	Lauan face, 2'-0" x 6'-8"	2 Carp	17	.941	Ea.	30	34.50		64.50	87
0080	3'-0" x 6'-8"		17	.941		47	34.50		81.50	106
0100	4'-0" x 6'-8"		16	1		72	36.50		108.50	136
0120	Birch face, 2'-0" x 6'-8"		17	.941		46.50	34.50		81	106
0140	2'-6" x 6'-8"		17	.941		78	34.50		112.50	140
0180	3'-0" x 6'-8"		17	.941		89.50	34.50		124	152
0200	4'-0" x 6'-8"		16	1		98	36.50		134.50	165
0220	Oak face, 2'-0" x 6'-8"		17	.941		74.50	34.50		109	136
0240	2'-6" x 6'-8"		17	.941		79.50	34.50		114	142
0280	3'-0" x 6'-8"		17	.941		85.50	34.50		120	148
0300	4'-0" x 6'-8"		16	1		108	36.50		144.50	176
0320	Walnut face, 2'-0" x 6'-8"		17	.941		151	34.50		185.50	220
0340	2'-6" x 6'-8"		17	.941		154	34.50		188.50	223
0380	3'-0" x 6'-8"		17	.941		160	34.50		194.50	230
0400	4'-0" x 6'-8"		16	1		181	36.50		217.50	256
0430	For 7'-0" high, add					14.90			14.90	16.40

08 14 Wood Doors

08 14 16 – Flush Wood Doors

08 14 16.09 Flush Wood Doors		Crew	Daily Output	Labor-Hours	Unit	Material	2007 Bare Costs Labor	Equipment	Total	Total Incl O&P
0440	For 8'-0" high, add				Ea.	21			21	23
0480	For prefinishing, clear, add					33			33	36
0500	For prefinishing, stain, add					44.50			44.50	49
1320	M.D. overlay on hardboard, 2'-0" x 6'-8"	2 Carp	17	.941		92.50	34.50		127	156
1340	2'-6" x 6'-8"		17	.941		92.50	34.50		127	156
1380	3'-0" x 6'-8"		17	.941		109	34.50		143.50	174
1400	4'-0" x 6'-8"		16	1		150	36.50		186.50	222
1420	For 7'-0" high, add					8.20			8.20	9
1440	For 8'-0" high, add					22			22	24
1720	H.P. plastic laminate, 2'-0" x 6'-8"	2 Carp	16	1		220	36.50		256.50	299
1740	2'-6" x 6'-8"		16	1		220	36.50		256.50	299
1780	3'-0" x 6'-8"		15	1.067		255	39		294	340
1800	4'-0" x 6'-8"		14	1.143		350	42		392	450
1820	For 7'-0" high, add					8.60			8.60	9.45
1840	For 8'-0" high, add					22			22	24
2020	5 ply particle core, lauan face, 2'-6" x 6'-8"	2 Carp	15	1.067		76	39		115	145
2040	3'-0" x 6'-8"		14	1.143		79	42		121	153
2080	3'-0" x 7'-0"		13	1.231		88	45		133	168
2100	4'-0" x 7'-0"		12	1.333		102	49		151	188
2120	Birch face, 2'-6" x 6'-8"		15	1.067		86.50	39		125.50	156
2140	3'-0" x 6'-8"		14	1.143		94.50	42		136.50	170
2180	3'-0" x 7'-0"		13	1.231		97	45		142	177
2200	4'-0" x 7'-0"		12	1.333		118	49		167	206
2220	Oak face, 2'-6" x 6'-8"		15	1.067		95.50	39		134.50	166
2240	3'-0" x 6'-8"		14	1.143		105	42		147	181
2280	3'-0" x 7'-0"		13	1.231		108	45		153	190
2300	4'-0" x 7'-0"		12	1.333		132	49		181	221
2320	Walnut face, 2'-0" x 6'-8"		15	1.067		106	39		145	177
2340	2'-6" x 6'-8"		14	1.143		121	42		163	199
2380	3'-0" x 6'-8"		13	1.231		136	45		181	220
2400	4'-0" x 6'-8"		12	1.333		178	49		227	272
2440	For 8'-0" high, add					26			26	28.50
2460	For 8'-0" high walnut, add					14.20			14.20	15.65
2480	For solid wood core, add					32			32	35
2720	For prefinishing, clear, add					20.50			20.50	23
2740	For prefinishing, stain, add					46			46	51
3320	M.D. overlay on hardboard, 2'-6" x 6'-8"	2 Carp	14	1.143		92	42		134	167
3340	3'-0" x 6'-8"		13	1.231		96.50	45		141.50	177
3380	3'-0" x 7'-0"		12	1.333		98.50	49		147.50	184
3400	4'-0" x 7'-0"		10	1.600		120	58.50		178.50	224
3440	For 8'-0" height, add					27			27	29.50
3460	For solid wood core, add					35			35	38.50
3720	H.P. plastic laminate, 2'-6" x 6'-8"	2 Carp	13	1.231		135	45		180	220
3740	3'-0" x 6'-8"		12	1.333		153	49		202	244
3780	3'-0" x 7'-0"		11	1.455		158	53.50		211.50	257
3800	4'-0" x 7'-0"		8	2		193	73.50		266.50	325
3840	For 8'-0" height, add					27			27	29.50
3860	For solid wood core, add					33			33	36
4000	Exterior, flush, solid wood stave core, birch, 1-3/4" x 7'-0" x 2'-6"	2 Carp	15	1.067		162	39		201	240
4020	2'-8" wide		15	1.067		170	39		209	248
4040	3'-0" wide		14	1.143		183	42		225	267
4100	Oak faced 1-3/4" x 7'-0" x 2'-6" wide		15	1.067		179	39		218	258
4120	2'-8" wide		15	1.067		191	39		230	271

08 14 Wood Doors

08 14 16 – Flush Wood Doors

08 14 16.09 Flush Wood Doors		Crew	Daily Output	Labor-Hours	Unit	Material	2007 Bare Costs Labor	Equipment	Total	Total Incl O&P
4140	3'-0" wide	2 Carp	14	1.143	Ea.	204	42		246	290
4200	Walnut faced, 1-3/4" x 7'-0" x 2'-6" wide		15	1.067		263	39		302	350
4220	2'-8" wide		15	1.067		275	39		314	360
4240	3'-0" wide	▼	14	1.143		286	42		328	380
4300	For 6'-8" high door, deduct from 7'-0" door					14.65			14.65	16.10

08 14 16.10 Wood Doors Decorator										
0010	**WOOD DOORS DECORATOR**									
0040	7 ply hollow core lauan face, 2'-6" x 6'-8"	2 Carp	17	.941	Ea.	34.50	34.50		69	92

08 14 16.20 Wood Fire Doors										
0010	**WOOD FIRE DOORS**									
0020	Particle core, 7 face plys, "B" label,									
0040	1 hour, birch face, 1-3/4" x 2'-6" x 6'-8"	2 Carp	14	1.143	Ea.	305	42		347	400
0080	3'-0" x 6'-8"		13	1.231		315	45		360	415
0090	3'-0" x 7'-0"		12	1.333		330	49		379	440
0100	4'-0" x 7'-0"		12	1.333		440	49		489	560
0140	Oak face, 2'-6" x 6'-8"		14	1.143		305	42		347	400
0180	3'-0" x 6'-8"		13	1.231		315	45		360	420
0190	3'-0" x 7'-0"		12	1.333		330	49		379	440
0200	4'-0" x 7'-0"		12	1.333		435	49		484	550
0240	Walnut face, 2'-6" x 6'-8"		14	1.143		400	42		442	505
0280	3'-0" x 6'-8"		13	1.231		410	45		455	520
0290	3'-0" x 7'-0"		12	1.333		430	49		479	545
0300	4'-0" x 7'-0"		12	1.333		580	49		629	710
0440	M.D. overlay on hardboard, 2'-6" x 6'-8"		15	1.067		268	39		307	355
0480	3'-0" x 6'-8"		14	1.143		279	42		321	370
0490	3'-0" x 7'-0"		13	1.231		295	45		340	395
0500	4'-0" x 7'-0"		12	1.333		360	49		409	470
0540	H.P. plastic laminate, 2'-6" x 6'-8"		13	1.231		360	45		405	465
0580	3'-0" x 6'-8"		12	1.333		375	49		424	490
0590	3'-0" x 7'-0"		11	1.455		380	53.50		433.50	505
0600	4'-0" x 7'-0"		10	1.600		485	58.50		543.50	625
0740	90 minutes, birch face, 1-3/4" x 2'-6" x 6'-8"		14	1.143		286	42		328	380
0780	3'-0" x 6'-8"		13	1.231		270	45		315	370
0790	3'-0" x 7'-0"		12	1.333		330	49		379	435
0800	4'-0" x 7'-0"		12	1.333		395	49		444	510
0840	Oak face, 2'-6" x 6'-8"		14	1.143		275	42		317	370
0880	3'-0" x 6'-8"		13	1.231		286	45		331	385
0890	3'-0" x 7'-0"		12	1.333		300	49		349	405
0900	4'-0" x 7'-0"		12	1.333		415	49		464	530
0940	Walnut face, 2'-6" x 6'-8"		14	1.143		380	42		422	480
0980	3'-0" x 6'-8"		13	1.231		385	45		430	495
0990	3'-0" x 7'-0"		12	1.333		405	49		454	520
1000	4'-0" x 7'-0"		12	1.333		585	49		634	715
1140	M.D. overlay on hardboard, 2'-6" x 6'-8"		15	1.067		300	39		339	390
1180	3'-0" x 6'-8"		14	1.143		310	42		352	405
1190	3'-0" x 7'-0"		13	1.231		320	45		365	420
1200	4'-0" x 7'-0"	▼	12	1.333		435	49		484	555
1240	For 8'-0" height, add					55			55	60
1260	For 8'-0" height walnut, add					72.50			72.50	80
1340	H.P. plastic laminate, 2'-6" x 6'-8"	2 Carp	13	1.231		375	45		420	480
1380	3'-0" x 6'-8"		12	1.333		390	49		439	505
1390	3'-0" x 7'-0"	▼	11	1.455		395	53.50		448.50	520

08 14 Wood Doors

08 14 16 – Flush Wood Doors

08 14 16.20 Wood Fire Doors

		Crew	Daily Output	Labor-Hours	Unit	Material	2007 Bare Costs Labor	2007 Bare Costs Equipment	Total	Total Incl O&P
1400	4'-0" x 7'-0"	2 Carp	10	1.600	Ea.	515	58.50		573.50	655
2200	Custom architectural "B" label, flush, 1-3/4" thick, birch,									
2210	Solid core									
2220	2'-6" x 7'-0"	2 Carp	15	1.067	Ea.	251	39		290	335
2260	3'-0" x 7'-0"		14	1.143		261	42		303	355
2300	4'-0" x 7'-0"		13	1.231		365	45		410	475
2420	4'-0" x 8'-0"		11	1.455		370	53.50		423.50	490
2480	For oak veneer, add					50%				
2500	For walnut veneer, add					75%				

08 14 23 – Clad Wood Doors

08 14 23.13 Metal-Faced Wood Doors

		Crew	Daily Output	Labor-Hours	Unit	Material	Labor	Equipment	Total	Total Incl O&P
0010	**METAL-FACED WOOD DOORS**									
0020	Interior, flush type, 3' x 7'	2 Carp	4.30	3.721	Opng.	185	137		322	415

08 14 23.20 Tin Clad Wood Doors

		Crew	Daily Output	Labor-Hours	Unit	Material	Labor	Equipment	Total	Total Incl O&P
0010	**TIN CLAD WOOD DOORS**									
0020	3 ply, 6' x 7', double sliding, doors only	2 Carp	1	16	Opng.	1,450	585		2,035	2,525
1000	For electric operator, add	1 Elec	2	4	"	2,575	176		2,751	3,075

08 14 33 – Stile and Rail Wood Doors

08 14 33.10 Wood Doors Paneled

		Crew	Daily Output	Labor-Hours	Unit	Material	Labor	Equipment	Total	Total Incl O&P
0010	**WOOD DOORS PANELED**									
0020	Interior, six panel, hollow core, 1-3/8" thick									
0040	Molded hardboard, 2'-0" x 6'-8"	2 Carp	17	.941	Ea.	49	34.50		83.50	108
0060	2'-6" x 6'-8"		17	.941		52.50	34.50		87	112
0070	2'-8" x 6'-8"		17	.941		55	34.50		89.50	115
0080	3'-0" x 6'-8"		17	.941		58	34.50		92.50	118
0140	Embossed print, molded hardboard, 2'-0" x 6'-8"		17	.941		52.50	34.50		87	112
0160	2'-6" x 6'-8"		17	.941		52.50	34.50		87	112
0180	3'-0" x 6'-8"		17	.941		58	34.50		92.50	118
0540	Six panel, solid, 1-3/8" thick, pine, 2'-0" x 6'-8"		15	1.067		128	39		167	201
0560	2'-6" x 6'-8"		14	1.143		143	42		185	224
0580	3'-0" x 6'-8"		13	1.231		165	45		210	252
1020	Two panel, bored rail, solid, 1-3/8" thick, pine, 1'-6" x 6'-8"		16	1		233	36.50		269.50	315
1040	2'-0" x 6'-8"		15	1.067		305	39		344	395
1060	2'-6" x 6'-8"		14	1.143		350	42		392	450
1340	Two panel, solid, 1-3/8" thick, fir, 2'-0" x 6'-8"		15	1.067		128	39		167	201
1360	2'-6" x 6'-8"		14	1.143		143	42		185	224
1380	3'-0" x 6'-8"		13	1.231		350	45		395	455
1740	Five panel, solid, 1-3/8" thick, fir, 2'-0" x 6'-8"		15	1.067		229	39		268	310
1760	2'-6" x 6'-8"		14	1.143		365	42		407	470
1780	3'-0" x 6'-8"		13	1.231		365	45		410	475

08 14 33.20 Wood Doors Residential

		Crew	Daily Output	Labor-Hours	Unit	Material	Labor	Equipment	Total	Total Incl O&P
0010	**WOOD DOORS RESIDENTIAL**									
0200	Exterior, combination storm & screen, pine									
0260	2'-8" wide	2 Carp	10	1.600	Ea.	272	58.50		330.50	390
0280	3'-0" wide		9	1.778		278	65		343	405
0300	7'-1" x 3'-0" wide		9	1.778		310	65		375	440
0400	Full lite, 6'-9" x 2'-6" wide		11	1.455		290	53.50		343.50	405
0420	2'-8" wide		10	1.600		290	58.50		348.50	410
0440	3'-0" wide		9	1.778		299	65		364	430
0500	7'-1" x 3'-0" wide		9	1.778		320	65		385	455
0700	Dutch door, pine, 1-3/4" x 6'-8" x 2'-8" wide, minimum		12	1.333		675	49		724	815

08 14 Wood Doors

08 14 33 – Stile and Rail Wood Doors

08 14 33.20 Wood Doors Residential		Crew	Daily Output	Labor-Hours	Unit	Material	2007 Bare Costs Labor	Equipment	Total	Total Incl O&P
0720	Maximum	2 Carp	10	1.600	Ea.	710	58.50		768.50	875
0800	3'-0" wide, minimum		12	1.333		685	49		734	830
0820	Maximum		10	1.600		750	58.50		808.50	915
1000	Entrance door, colonial, 1-3/4" x 6'-8" x 2'-8" wide		16	1		380	36.50		416.50	470
1020	6 panel pine, 3'-0" wide		15	1.067		400	39		439	500
1100	8 panel pine, 2'-8" wide		16	1		560	36.50		596.50	670
1120	3'-0" wide		15	1.067		540	39		579	650
1200	For tempered safety glass lites, (min of 2) add					65			65	71.50
1300	Flush, birch, solid core, 1-3/4" x 6'-8" x 2'-8" wide	2 Carp	16	1		97.50	36.50		134	164
1320	3'-0" wide		15	1.067		101	39		140	172
1350	7'-0" x 2'-8" wide		16	1		105	36.50		141.50	172
1360	3'-0" wide		15	1.067		114	39		153	187
1380	For tempered safety glass lites, add					97			97	107
2700	Interior, closet, bi-fold, w/hardware, no frame or trim incl.									
2720	Flush, birch, 6'-6" or 6'-8" x 2'-6" wide	2 Carp	13	1.231	Ea.	48.50	45		93.50	124
2740	3'-0" wide		13	1.231		52.50	45		97.50	129
2760	4'-0" wide		12	1.333		96	49		145	181
2780	5'-0" wide		11	1.455		97	53.50		150.50	190
2800	6'-0" wide		10	1.600		104	58.50		162.50	206
3000	Raised panel pine, 6'-6" or 6'-8" x 2'-6" wide		13	1.231		164	45		209	251
3020	3'-0" wide		13	1.231		209	45		254	300
3040	4'-0" wide		12	1.333		300	49		349	405
3060	5'-0" wide		11	1.455		360	53.50		413.50	480
3080	6'-0" wide		10	1.600		395	58.50		453.50	525
3200	Louvered, pine 6'-6" or 6'-8" x 2'-6" wide		13	1.231		104	45		149	186
3220	3'-0" wide		13	1.231		161	45		206	248
3240	4'-0" wide		12	1.333		195	49		244	291
3260	5'-0" wide		11	1.455		221	53.50		274.50	325
3280	6'-0" wide		10	1.600		245	58.50		303.50	360
4400	Bi-passing closet, incl. hardware and frame, no trim incl.									
4420	Flush, lauan, 6'-8" x 4'-0" wide	2 Carp	12	1.333	Opng.	165	49		214	258
4440	5'-0" wide		11	1.455		181	53.50		234.50	282
4460	6'-0" wide		10	1.600		194	58.50		252.50	305
4600	Flush, birch, 6'-8" x 4'-0" wide		12	1.333		207	49		256	305
4620	5'-0" wide		11	1.455		204	53.50		257.50	310
4640	6'-0" wide		10	1.600		248	58.50		306.50	365
4800	Louvered, pine, 6'-8" x 4'-0" wide		12	1.333		405	49		454	525
4820	5'-0" wide		11	1.455		385	53.50		438.50	505
4840	6'-0" wide		10	1.600		500	58.50		558.50	640
5000	Paneled, pine, 6'-8" x 4'-0" wide		12	1.333		485	49		534	605
5020	5'-0" wide		11	1.455		400	53.50		453.50	525
5040	6'-0" wide		10	1.600		575	58.50		633.50	720
6100	Folding accordion, closet, including track and frame									
6120	Vinyl, 2 layer, stock (see also division 10 22 26.13)	2 Carp	400	.040	Ea.	103	1.47		104.47	115
6140	Woven mahogany and vinyl, stock		400	.040		31	1.47		32.47	36.50
6160	Wood slats with vinyl overlay, stock		400	.040		126	1.47		127.47	141
6180	Economy vinyl, stock		400	.040	S.F.	1.83	1.47		3.30	4.30
6200	Rigid PVC		400	.040	"	4.88	1.47		6.35	7.65
6220	For custom partition, add					25%	10%			
7310	Passage doors, flush, no frame included									
7320	Hardboard, hollow core, 1-3/8" x 6'-8" x 1'-6" wide	2 Carp	18	.889	Ea.	40.50	32.50		73	96
7330	2'-0" wide		18	.889		40.50	32.50		73	95.50
7340	2'-6" wide		18	.889		44.50	32.50		77	100

08 14 Wood Doors

08 14 33 – Stile and Rail Wood Doors

08 14 33.20 Wood Doors Residential

		Crew	Daily Output	Labor-Hours	Unit	Material	2007 Bare Costs Labor	Equipment	Total	Total Incl O&P
7350	2'-8" wide	2 Carp	18	.889	Ea.	47	32.50		79.50	103
7360	3'-0" wide		17	.941		49.50	34.50		84	109
7420	Lauan, hollow core, 1-3/8" x 6'-8" x 1'-6" wide		18	.889		28.50	32.50		61	82.50
7440	2'-0" wide		18	.889		27	32.50		59.50	81
7450	2'-4" wide		18	.889		30.50	32.50		63	84.50
7460	2'-6" wide		18	.889		30.50	32.50		63	84.50
7480	2'-8" wide		18	.889		32.50	32.50		65	86.50
7500	3'-0" wide		17	.941		33.50	34.50		68	91
7700	Birch, hollow core, 1-3/8" x 6'-8" x 1'-6" wide		18	.889		36	32.50		68.50	90.50
7720	2'-0" wide		18	.889		40	32.50		72.50	95
7740	2'-6" wide		18	.889		44.50	32.50		77	100
7760	2'-8" wide		18	.889		46	32.50		78.50	102
7780	3'-0" wide CN		17	.941		50	34.50		84.50	109
8000	Pine louvered, 1-3/8" x 6'-8" x 1'-6" wide		19	.842		100	31		131	158
8020	2'-0" wide		18	.889		125	32.50		157.50	189
8040	2'-6" wide		18	.889		137	32.50		169.50	201
8060	2'-8" wide		18	.889		144	32.50		176.50	209
8080	3'-0" wide		17	.941		154	34.50		188.50	224
8300	Pine paneled, 1-3/8" x 6'-8" x 1'-6" wide		19	.842		109	31		140	168
8320	2'-0" wide		18	.889		125	32.50		157.50	189
8330	2'-4" wide		18	.889		137	32.50		169.50	202
8340	2'-6" wide		18	.889		140	32.50		172.50	205
8360	2'-8" wide		18	.889		152	32.50		184.50	218
8380	3'-0" wide		17	.941		161	34.50		195.50	231
8550	For over 20 doors, deduct					15%				

08 14 40 – Interior Cafe Doors

08 14 40.10 Interior Cafe Doors

		Crew	Daily Output	Labor-Hours	Unit	Material	Labor	Equipment	Total	Total Incl O&P
0010	**INTERIOR CAFE DOORS**									
6520	Interior cafe doors, 2'-6" opening, stock, panel pine	2 Carp	16	1	Ea.	188	36.50		224.50	263
6540	3'-0" opening	"	16	1	"	196	36.50		232.50	272
6550	Louvered pine									
6560	2'-6" opening	2 Carp	16	1	Ea.	164	36.50		200.50	237
8000	3'-0" opening		16	1		175	36.50		211.50	249
8010	2'-6" opening, hardwood		16	1		282	36.50		318.50	365
8020	3'-0" opening		16	1		310	36.50		346.50	400

08 16 Composite Doors

08 16 13 – Fiberglass Doors

08 16 13.10 Fiberglass Doors

		Crew	Daily Output	Labor-Hours	Unit	Material	Labor	Equipment	Total	Total Incl O&P
0010	**FIBERGLASS DOORS**									
0020	Exterior, fiberglass, door, 2'-8" wide x 6'-8" high	2 Carp	15	1.067	Ea.	305	39		344	395
0040	3'-0" wide x 6'-8" high		15	1.067		305	39		344	395
0060	3'-0" wide x 7'-0" high		15	1.067		375	39		414	475
0080	3'-0" wide x 6'-8" high, with two lites		15	1.067		370	39		409	465
0100	3'-0" wide x 7'-0" high, with two lites		15	1.067		440	39		479	545
0110	Half glass, 3'-0" wide x 6'-8" high		15	1.067		415	39		454	515
0120	3'-0" wide x 6'-8" high, low e		15	1.067		440	39		479	540
0130	3'-0" wide x 7'-0" high		15	1.067		485	39		524	590
0140	3'-0" wide x 7'-0" high, low e		15	1.067		510	39		549	620
0150	Side lights, 1'-0" wide x 6'-8" high,					281			281	310

08 16 Composite Doors

08 16 13 – Fiberglass Doors

08 16 13.10 Fiberglass Doors		Crew	Daily Output	Labor-Hours	Unit	Material	2007 Bare Costs Labor	Equipment	Total	Total Incl O&P
0160	1'-0" wide x 6'-8" high, low e				Ea.	293			293	325
0180	1'-0" wide x 6'-8" high, full glass					320			320	355
0190	1'-0" wide x 6'-8" high, low e					345			345	380

08 17 Integrated Door Opening Assemblies

08 17 13 – Integrated Metal Door Opening Assemblies

08 17 13.20 Tubular Steel Swing Doors

		Crew	Daily Output	Labor-Hours	Unit	Material	Labor	Equipment	Total	Incl O&P
0010	**TUBULAR STEEL SWING DOORS**									
0020	Tubular steel, 7' high, single, 3'-4" opening	2 Sswk	2.50	6.400	Ea.	640	265		905	1,175
0100	Double, 6'-0" opening	"	2	8	Pr.	810	330		1,140	1,500

08 17 23 – Integrated Wood Door Opening Assemblies

08 17 23.10 Pre-Hung Doors

		Crew	Daily Output	Labor-Hours	Unit	Material	Labor	Equipment	Total	Incl O&P
0010	**PRE-HUNG DOORS**									
0300	Exterior, wood, comb. storm & screen, 6'-9" x 2'-6" wide	2 Carp	15	1.067	Ea.	283	39		322	370
0320	2'-8" wide		15	1.067		283	39		322	370
0340	3'-0" wide		15	1.067		291	39		330	380
0360	For 7'-0" high door, add					26.50			26.50	29
1600	Entrance door, flush, birch, solid core									
1620	4-5/8" solid jamb, 1-3/4" x 6'-8" x 2'-8" wide	2 Carp	16	1	Ea.	285	36.50		321.50	370
1640	3'-0" wide		16	1		355	36.50		391.50	445
1642	5-5/8" jamb		16	1		320	36.50		356.50	405
1680	For 7'-0" high door, add					18.75			18.75	20.50
2000	Entrance door, colonial, 6 panel pine									
2020	4-5/8" solid jamb, 1-3/4" x 6'-8" x 2'-8" wide	2 Carp	16	1	Ea.	520	36.50		556.50	625
2040	3'-0" wide	"	16	1		520	36.50		556.50	625
2060	For 7'-0" high door, add					49			49	53.50
2200	For 5-5/8" solid jamb, add					38.50			38.50	42
4000	Interior, passage door, 4-5/8" solid jamb									
4400	Lauan, flush, solid core, 1-3/8" x 6'-8" x 2'-6" wide	2 Carp	20	.800	Ea.	178	29.50		207.50	241
4420	2'-8" wide		20	.800		178	29.50		207.50	241
4440	3'-0" wide		19	.842		191	31		222	258
4600	Hollow core, 1-3/8" x 6'-8" x 2'-6" wide		20	.800		120	29.50		149.50	178
4620	2'-8" wide		20	.800		120	29.50		149.50	178
4640	3'-0" wide		19	.842		121	31		152	181
4700	For 7'-0" high door, add					23			23	25.50
5000	Birch, flush, solid core, 1-3/8" x 6'-8" x 2'-6" wide	2 Carp	20	.800		166	29.50		195.50	228
5020	2'-8" wide		20	.800		190	29.50		219.50	255
5040	3'-0" wide		19	.842		200	31		231	268
5200	Hollow core, 1-3/8" x 6'-8" x 2'-6" wide		20	.800		137	29.50		166.50	197
5220	2'-8" wide		20	.800		144	29.50		173.50	204
5240	3'-0" wide		19	.842		144	31		175	206
5280	For 7'-0" high door, add					19.95			19.95	22
5500	Hardboard paneled, 1-3/8" x 6'-8" x 2'-6" wide	2 Carp	20	.800		138	29.50		167.50	198
5520	2'-8" wide		20	.800		145	29.50		174.50	205
5540	3'-0" wide		19	.842		143	31		174	205
6000	Pine paneled, 1-3/8" x 6'-8" x 2'-6" wide		20	.800		240	29.50		269.50	310
6020	2'-8" wide		20	.800		261	29.50		290.50	335
6040	3'-0" wide		19	.842		266	31		297	340
6500	For 5-5/8" solid jamb, add					13.95			13.95	15.35
6520	For split jamb, deduct					15.70			15.70	17.25

08 31 Access Doors and Panels

08 31 13 – Access Doors and Frames

08 31 13.10 Access Doors and Frames

	Crew	Daily Output	Labor-Hours	Unit	Material	2007 Bare Costs Labor	2007 Bare Costs Equipment	Total	Total Incl O&P
0010 **ACCESS DOORS AND FRAMES**									
1000 Fire rated door with lock									
1100 Metal, 12" x 12"	1 Carp	10	.800	Ea.	148	29.50		177.50	209
1150 18" x 18"		9	.889		194	32.50		226.50	264
1200 24" x 24"		9	.889		234	32.50		266.50	310
1250 24" x 36"		8	1		315	36.50		351.50	400
1300 24" x 48"		8	1		390	36.50		426.50	480
1350 36" x 36"		7.50	1.067		470	39		509	575
1400 48" x 48"		7.50	1.067		600	39		639	720
1600 Stainless steel, 12" x 12"		10	.800		265	29.50		294.50	340
1650 18" x 18"		9	.889		385	32.50		417.50	475
1700 24" x 24"		9	.889		475	32.50		507.50	570
1750 24" x 36"		8	1		605	36.50		641.50	720
2000 Flush door for finishing									
2100 Metal 8" x 8"	1 Carp	10	.800	Ea.	42.50	29.50		72	92.50
2150 12" x 12"	"	10	.800	"	48	29.50		77.50	98
3000 Recessed door for acoustic tile									
3100 Metal, 12" x 12"	1 Carp	4.50	1.778	Ea.	65.50	65		130.50	174
3150 12" x 24"		4.50	1.778		85.50	65		150.50	196
3200 24" x 24"		4	2		114	73.50		187.50	240
3250 24" x 36"		4	2		147	73.50		220.50	276
4000 Recessed door for drywall									
4100 Metal 12" x 12"	1 Carp	6	1.333	Ea.	73	49		122	156
4150 12" x 24"		5.50	1.455		108	53.50		161.50	202
4200 24" x 36"		5	1.600		172	58.50		230.50	281
6000 Standard door									
6100 Metal, 8" x 8"	1 Carp	10	.800	Ea.	37.50	29.50		67	87
6150 12" x 12"		10	.800		42.50	29.50		72	92.50
6200 18" x 18"		9	.889		59	32.50		91.50	116
6250 24" x 24"		9	.889		76.50	32.50		109	136
6300 24" x 36"		8	1		113	36.50		149.50	181
6350 36" x 36"		8	1		138	36.50		174.50	209
6500 Stainless steel, 8" x 8"		10	.800		75.50	29.50		105	129
6550 12" x 12"		10	.800		101	29.50		130.50	157
6600 18" x 18"		9	.889		186	32.50		218.50	256
6650 24" x 24"		9	.889		244	32.50		276.50	320

08 31 13.20 Bulkhead/Cellar Doors

	Crew	Daily Output	Labor-Hours	Unit	Material	2007 Bare Costs Labor	2007 Bare Costs Equipment	Total	Total Incl O&P
0010 **BULKHEAD/CELLAR DOORS**									
0020 Steel, not incl. sides, 44" x 62"	1 Carp	5.50	1.455	Ea.	300	53.50		353.50	415
0100 52" x 73"		5.10	1.569		320	57.50		377.50	440
0500 With sides and foundation plates, 57" x 45" x 24"		4.70	1.702		555	62.50		617.50	715
0600 42" x 49" x 51"		4.30	1.860		645	68.50		713.50	815

08 31 13.30 Commercial Floor Doors

	Crew	Daily Output	Labor-Hours	Unit	Material	2007 Bare Costs Labor	2007 Bare Costs Equipment	Total	Total Incl O&P
0010 **COMMERCIAL FLOOR DOORS**									
0020 Aluminum tile, steel frame, one leaf, 2' x 2' opng.	2 Sswk	3.50	4.571	Opng.	380	189		569	760
0050 3'-6" x 3'-6" opening		3.50	4.571		690	189		879	1,100
0500 Double leaf, 4' x 4' opening		3	5.333		1,000	221		1,221	1,500
0550 5' x 5' opening		3	5.333		1,475	221		1,696	2,025

08 31 13.35 Industrial Floor Doors

	Crew	Daily Output	Labor-Hours	Unit	Material	2007 Bare Costs Labor	2007 Bare Costs Equipment	Total	Total Incl O&P
0010 **INDUSTRIAL FLOOR DOORS**									
0020 Steel 300 psf L.L., single leaf, 2' x 2', 175#	2 Sswk	6	2.667	Opng.	610	110		720	870
0050 3' x 3' opening, 300#		5.50	2.909		840	120		960	1,150

08 31 Access Doors and Panels

08 31 13 – Access Doors and Frames

08 31 13.35 Industrial Floor Doors

		Crew	Daily Output	Labor-Hours	Unit	Material	2007 Bare Costs Labor	2007 Bare Costs Equipment	Total	Total Incl O&P
0300	Double leaf, 4' x 4' opening, 455#	2 Sswk	5	3.200	Opng.	1,275	132		1,407	1,650
0350	5' x 5' opening, 645#		4.50	3.556		1,650	147		1,797	2,100
1000	Aluminum, 300 psf L.L., single leaf, 2' x 2', 60#		6	2.667		600	110		710	860
1050	3' x 3' opening, 100#		5.50	2.909		935	120		1,055	1,250
1500	Double leaf, 4' x 4' opening, 160#		5	3.200		1,475	132		1,607	1,875
1550	5' x 5' opening, 235#		4.50	3.556		1,950	147		2,097	2,425
2000	Aluminum, 150 psf L.L., single leaf, 2' x 2', 60#		6	2.667		550	110		660	805
2050	3' x 3' opening, 95#		5.50	2.909		815	120		935	1,125
2500	Double leaf, 4' x 4' opening, 150#		5	3.200		1,300	132		1,432	1,675
2550	5' x 5' opening, 230#		4.50	3.556		1,750	147		1,897	2,200

08 31 13.40 Kennel Doors

		Crew	Daily Output	Labor-Hours	Unit	Material	Labor	Equipment	Total	Total Incl O&P
0010	**KENNEL DOORS**									
0020	2 way, swinging type, 13" x 19" opening	2 Carp	11	1.455	Opng.	207	53.50		260.50	310
0100	17" x 29" opening	"	11	1.455	"	223	53.50		276.50	330

08 32 Sliding Glass Doors

08 32 19 – Sliding Wood-Framed Glass Doors

08 32 19.15 Sliding Glass Vinyl-Clad Wood Doors

		Crew	Daily Output	Labor-Hours	Unit	Material	Labor	Equipment	Total	Total Incl O&P
0010	**SLIDING GLASS VINYL-CLAD WOOD DOORS**									
0012	Vinyl clad, 1" insul. glass, 6'-0" x 6'-10" high	2 Carp	4	4	Opng.	1,300	147		1,447	1,675
0030	6'-0" x 8'-0" high		4	4	Ea.	1,900	147		2,047	2,325
0100	8'-0" x 6'-10" high		4	4	Opng.	1,950	147		2,097	2,375
0500	3 leaf, 9'-0" x 6'-10" high		3	5.333		2,275	196		2,471	2,825
0600	12'-0" x 6'-10" high		3	5.333		3,100	196		3,296	3,700

08 33 Coiling Doors and Grilles

08 33 13 – Coiling Counter Doors

08 33 13.10 Coiling Counter Doors

		Crew	Daily Output	Labor-Hours	Unit	Material	Labor	Equipment	Total	Total Incl O&P
0010	**COILING COUNTER DOORS**									
0020	Manual, incl. frm and hdwe, galv. stl., 4' roll-up, 6' long	2 Carp	2	8	Opng.	980	294		1,274	1,525
0300	Galvanized steel, UL label		1.80	8.889		1,150	325		1,475	1,775
0600	Stainless steel, 4' high roll-up, 6' long		2	8		1,525	294		1,819	2,125
0700	10' long		1.80	8.889		2,300	325		2,625	3,050
2000	Aluminum, 4' high, 4' long		2.20	7.273		940	267		1,207	1,450
2020	6' long		2	8		1,050	294		1,344	1,600
2040	8' long		1.90	8.421		1,200	310		1,510	1,800
2060	10' long		1.80	8.889		1,450	325		1,775	2,075
2080	14' long		1.40	11.429		2,100	420		2,520	2,950
2100	6' high, 4' long		2	8		1,050	294		1,344	1,600
2120	6' long		1.60	10		1,225	365		1,590	1,925
2140	10' long		1.40	11.429		1,675	420		2,095	2,500

08 33 16 – Coiling Counter Grilles

08 33 16.10 Coiling Grilles

		Crew	Daily Output	Labor-Hours	Unit	Material	Labor	Equipment	Total	Total Incl O&P
0010	**COILING GRILLES**									
0015	Aluminum, manual, incl. frame, mill finish									
0020	Top coiling, 4' high, 4' long	2 Sswk	3.20	5	Opng.	1,200	207		1,407	1,700
0030	6' long		3.20	5		1,275	207		1,482	1,775
0040	8' long		2.40	6.667		1,450	276		1,726	2,100

08 33 Coiling Doors and Grilles

08 33 16 – Coiling Counter Grilles

08 33 16.10 Coiling Grilles		Crew	Daily Output	Labor-Hours	Unit	Material	2007 Bare Costs Labor	Equipment	Total	Total Incl O&P
0050	12' long	2 Sswk	2.40	6.667	Opng.	1,675	276		1,951	2,350
0060	16' long		1.60	10		2,100	415		2,515	3,050
0070	6' high, 4' long		3.20	5		1,300	207		1,507	1,800
0080	6' long		3.20	5		1,325	207		1,532	1,825
0090	8' long		2.40	6.667		1,475	276		1,751	2,125
0100	12' long		1.60	10		1,725	415		2,140	2,650
0110	16' long		1.20	13.333		2,225	550		2,775	3,450
0200	Side coiling, 8' high, 12' long		.60	26.667		1,950	1,100		3,050	4,150
0220	18' long		.50	32		3,000	1,325		4,325	5,700
0240	24' long		.40	40		3,125	1,650		4,775	6,450
0260	12' high, 12' long		.50	32		2,825	1,325		4,150	5,525
0280	18' long		.40	40		4,175	1,650		5,825	7,600
0300	24' long		.28	57.143		5,575	2,375		7,950	10,400

08 33 23 – Overhead Coiling Doors

08 33 23.10 Rolling Service Doors		Crew	Daily Output	Labor-Hours	Unit	Material	2007 Bare Costs Labor	Equipment	Total	Total Incl O&P
0010	**ROLLING SERVICE DOORS**									
0050	8' x 8' high	2 Sswk	1.60	10	Ea.	855	415		1,270	1,700
0100	10' x 10' high		1.40	11.429		1,475	475		1,950	2,450
0200	20' x 10' high		1	16		2,350	660		3,010	3,800
0300	12' x 12' high		1.20	13.333		1,475	550		2,025	2,625
0400	20' x 12' high		.90	17.778		2,425	735		3,160	4,000
0500	14' x 14' high		.80	20		2,750	825		3,575	4,525
0600	20' x 16' high		.60	26.667		2,725	1,100		3,825	5,000
0700	10' x 20' high		.50	32		1,650	1,325		2,975	4,200
1000	12' x 12', crank operated, crank on door side		.80	20		1,275	825		2,100	2,900
1100	Crank thru wall		.70	22.857		1,575	945		2,520	3,425
1300	For vision panel, add					234			234	257
1400	For 22 ga., deduct				S.F.	.83			.83	.91
1600	3' x 7' pass door within rolling steel door, new construction				Ea.	1,400			1,400	1,550
1700	Existing construction	2 Sswk	2	8		1,400	330		1,730	2,150
2000	Class A fire doors, manual, 20 ga., 8' x 8' high		1.40	11.429		1,100	475		1,575	2,075
2100	10' x 10' high		1.10	14.545		1,475	600		2,075	2,700
2200	20' x 10' high		.80	20		2,925	825		3,750	4,725
2300	12' x 12' high		1	16		1,950	660		2,610	3,350
2400	20' x 12' high		.80	20		3,375	825		4,200	5,200
2500	14' x 14' high		.60	26.667		2,525	1,100		3,625	4,775
2600	20' x 16' high		.50	32		4,225	1,325		5,550	7,025
2700	10' x 20' high		.40	40		2,675	1,650		4,325	5,925
3000	For 18 ga. doors, add				S.F.	.83			.83	.91
3300	For enamel finish, add				"	.99			.99	1.09
3600	For safety edge bottom bar, pneumatic, add				L.F.	13			13	14.30
3700	Electric, add					24.50			24.50	27
4000	For weatherstripping, extruded rubber, jambs, add					8.40			8.40	9.25
4100	Hood, add					5.90			5.90	6.50
4200	Sill, add					3.37			3.37	3.71
4500	Motor operators, to 14' x 14' opening	2 Sswk	5	3.200	Ea.	840	132		972	1,175
4600	Over 14' x 14', jack shaft type	"	5	3.200		830	132		962	1,150
4700	For fire door, additional fusible link, add					15.25			15.25	16.80

08 34 Special Function Doors

08 34 13 – Cold Storage Doors

08 34 13.10 Cold Storage Doors

	08 34 13.10 Cold Storage Doors	Crew	Daily Output	Labor-Hours	Unit	Material	2007 Bare Costs Labor	2007 Bare Costs Equipment	Total	Total Incl O&P
0010	**COLD STORAGE DOORS**									
0020	Single, 20 ga. galvanized steel									
0300	Horizontal sliding, 5' x 7', manual operation, 3.5" thick	2 Carp	2	8	Ea.	2,800	294		3,094	3,525
0400	4" thick		2	8		3,400	294		3,694	4,200
0500	6" thick		2	8		2,900	294		3,194	3,650
0800	5' x 7', power operation, 2" thick		1.90	8.421		4,825	310		5,135	5,775
0900	4" thick		1.90	8.421		4,900	310		5,210	5,850
1000	6" thick		1.90	8.421		5,575	310		5,885	6,600
1300	9' x 10', manual operation, 2" insulation		1.70	9.412		3,900	345		4,245	4,825
1400	4" insulation		1.70	9.412		4,025	345		4,370	4,975
1500	6" insulation		1.70	9.412		4,825	345		5,170	5,875
1800	Power operation, 2" insulation		1.60	10		6,700	365		7,065	7,950
1900	4" insulation		1.60	10		6,850	365		7,215	8,125
2000	6" insulation		1.70	9.412		7,775	345		8,120	9,100
2300	For stainless steel face, add					20%				
3000	Hinged, lightweight, 3' x 7'-0", galvanized 1 face, 2" thick	2 Carp	2	8	Ea.	1,250	294		1,544	1,825
3050	4" thick		1.90	8.421		1,475	310		1,785	2,100
3300	Aluminum doors, 3' x 7'-0", 4" thick		1.90	8.421		1,150	310		1,460	1,750
3350	6" thick		1.40	11.429		2,050	420		2,470	2,900
3600	Stainless steel, 3' x 7'-0", 4" thick		1.90	8.421		1,475	310		1,785	2,100
3650	6" thick		1.40	11.429		2,475	420		2,895	3,375
3900	Painted, 3' x 7'-0", 4" thick		1.90	8.421		1,050	310		1,360	1,625
3950	6" thick		1.40	11.429		2,000	420		2,420	2,875
5000	Bi-parting, electric operated									
5010	6' x 8' opening, galv. faces, 4" thick for cooler	2 Carp	.80	20	Opng.	6,300	735		7,035	8,075
5050	For freezer, 4" thick		.80	20		6,925	735		7,660	8,775
5300	For door buck framing and door protection, add		2.50	6.400		515	235		750	930
6000	Galvanized batten door, galvanized hinges, 4' x 7'		2	8		1,550	294		1,844	2,150
6050	6' x 8'		1.80	8.889		2,125	325		2,450	2,825
6500	Fire door, 3 hr., 6' x 8', single slide		.80	20		7,275	735		8,010	9,150
6550	Double, bi-parting		.70	22.857		11,000	840		11,840	13,400

08 34 16 – Hangar Doors

08 34 16.10 Hangar Doors

		Crew	Daily Output	Labor-Hours	Unit	Material	Labor	Equipment	Total	Incl O&P
0010	**HANGAR DOORS**									
0020	Bi-fold, ovhd., 20 PSF wind load, incl. elec. oper.									
0100	Without sheeting, 12' high x 40'	2 Sswk	240	.067	S.F.	13.70	2.76		16.46	20
0200	16' high x 60'		230	.070		17.45	2.88		20.33	24.50
0300	20' high x 80'		220	.073		17.75	3.01		20.76	25

08 34 36 – Darkroom Doors

08 34 36.10 Darkroom Doors

		Crew	Daily Output	Labor-Hours	Unit	Material	Labor	Equipment	Total	Incl O&P
0010	**DARKROOM DOORS**									
0015	Revolving, standard, 2 way, 36" diameter	2 Carp	3.10	5.161	Opng.	1,925	189		2,114	2,425
0020	41" diameter		3.10	5.161		2,075	189		2,264	2,600
0050	3 way, 51" diameter		1.40	11.429		2,600	420		3,020	3,500
1000	4 way, 49" diameter		1.40	11.429		3,075	420		3,495	4,025
2000	Hinged safety, 2 way, 41" diameter		2.30	6.957		2,475	255		2,730	3,125
2500	3 way, 51" diameter		1.40	11.429		3,100	420		3,520	4,075
3000	Pop out safety, 2 way, 41" diameter		3.10	5.161		3,025	189		3,214	3,625
4000	3 way, 51" diameter		1.40	11.429		3,025	420		3,445	3,975
5000	Wheelchair-type, pop out, 51" diameter		1.40	11.429		3,100	420		3,520	4,075
5020	72" diameter		.90	17.778		6,400	650		7,050	8,075
9300	For complete dark rooms, see division 13 21 53.50									

08 34 Special Function Doors

08 34 56 – Security Gates

08 34 56.10 Security Gates	Crew	Daily Output	Labor-Hours	Unit	Material	2007 Bare Costs Labor	Equipment	Total	Total Incl O&P
0010	SECURITY GATES - See division 10 22 16.10								

08 34 59 – Vault Doors and Day Gates

08 34 59.10 Vault Doors

		Crew	Daily Output	Labor-Hours	Unit	Material	Labor	Equipment	Total	Total Incl O&P
0010	**VAULT DOORS**									
0020	Door and frame, 32" x 78", clear opening									
0100	1 hour test, 32" door, weighs 750 lbs.	2 Sswk	1.50	10.667	Opng.	3,725	440		4,165	4,900
0200	2 hour test, 32" door, weighs 950 lbs.		1.30	12.308		3,925	510		4,435	5,250
0250	40" door, weighs 1130 lbs.		1	16		4,500	660		5,160	6,150
0300	4 hour test, 32" door, weighs 1025 lbs.		1.20	13.333		4,300	550		4,850	5,725
0350	40" door, weighs 1140 lbs.		.90	17.778		5,000	735		5,735	6,825
0500	For stainless steel front, including frame, add to above					1,800			1,800	1,975
0550	Back, add					1,800			1,800	1,975
0600	For time lock, two movement, add	1 Elec	2	4	Ea.	1,450	176		1,626	1,850
0650	Three movement, add	"	2	4		1,875	176		2,051	2,325
0800	Day gate, painted, steel, 32" wide	2 Sswk	1.50	10.667		1,675	440		2,115	2,650
0850	40" wide		1.40	11.429		1,875	475		2,350	2,900
0900	Aluminum, 32" wide		1.50	10.667		2,675	440		3,115	3,750
0950	40" wide		1.40	11.429		2,975	475		3,450	4,125
2050	Security vault door, class I, 3' wide, 3 1/2" thick	E-24	.19	166	Opng.	14,100	6,775	3,850	24,725	31,500
2100	Class II, 3' wide, 7" thick		.19	166		16,300	6,775	3,850	26,925	33,900
2150	Class III, 9R, 3' wide, 10" thick, minimum		.13	250		20,800	10,200	5,775	36,775	47,000
2160	Class V, type 1, 40" door		2.48	12.903	Ea.	5,825	525	298	6,648	7,675
2170	Class V, type 2, 40" door		2.48	12.903		5,800	525	298	6,623	7,625
2180	Day gate for class V vault	2 Sswk	2	8		1,200	330		1,530	1,925

08 34 63 – Detention Doors and Frames

08 34 63.13 Steel Detention Doors and Frames

		Crew	Daily Output	Labor-Hours	Unit	Material	Labor	Equipment	Total	Total Incl O&P
0010	**STEEL DETENTION DOORS AND FRAMES**									
1000	Doors & frames, 3' x 7', complete, with hardware, single plate	E-4	4	8	Ea.	3,825	335	29	4,189	4,825
1650	Double plate	"	4	8	"	4,700	335	29	5,064	5,800

08 34 73 – Sound Control Door Assemblies

08 34 73.10 Accoustical Doors

		Crew	Daily Output	Labor-Hours	Unit	Material	Labor	Equipment	Total	Total Incl O&P
0010	**ACCOUSTICAL DOORS**									
0020	Including framed seals, 3' x 7', wood, 27 STC rating	2 Carp	1.50	10.667	Ea.	400	390		790	1,050
0100	Steel, 40 STC rating		1.50	10.667		1,525	390		1,915	2,275
0200	45 STC rating		1.50	10.667		2,000	390		2,390	2,800
0300	48 STC rating		1.50	10.667		2,525	390		2,915	3,375
0400	52 STC rating		1.50	10.667		3,025	390		3,415	3,925

08 36 Panel Doors

08 36 13 – Sectional Doors

08 36 13.10 Overhead Commercial Doors

		Crew	Daily Output	Labor-Hours	Unit	Material	Labor	Equipment	Total	Total Incl O&P
0010	**OVERHEAD COMMERCIAL DOORS**									
1000	Stock, sectional, heavy duty, wood, 1-3/4" thick, 8' x 8' high	2 Carp	2	8	Ea.	625	294		919	1,150
1100	10' x 10' high		1.80	8.889		935	325		1,260	1,525
1200	12' x 12' high		1.50	10.667		1,375	390		1,765	2,100
1300	Chain hoist, 14' x 14' high		1.30	12.308		1,975	450		2,425	2,875
1400	12' x 16' high		1	16		1,750	585		2,335	2,850
1500	20' x 8' high		1.30	12.270		1,625	450		2,075	2,475
1600	20' x 16' high		.65	24.615		4,075	905		4,980	5,875

08 36 Panel Doors

08 36 13 – Sectional Doors

08 36 13.10 Overhead Commercial Doors		Crew	Daily Output	Labor-Hours	Unit	Material	2007 Bare Costs Labor	Equipment	Total	Total Incl O&P
1800	Center mullion openings, 8' high	2 Carp	4	4	Ea.	725	147		872	1,025
1900	20' high		2	8		1,350	294		1,644	1,950
2100	For medium duty custom door, deduct					5%	5%			
2150	For medium duty stock doors, deduct					10%	5%			
2300	Fiberglass and aluminum, heavy duty, sectional, 12' x 12' high	2 Carp	1.50	10.667	Ea.	2,000	390		2,390	2,800
2450	Chain hoist, 20' x 20' high		.50	32		5,025	1,175		6,200	7,350
2600	Steel, 24 ga. sectional, manual, 8' x 8' high		2	8		545	294		839	1,050
2650	10' x 10' high		1.80	8.889		750	325		1,075	1,325
2700	12' x 12' high		1.50	10.667		1,025	390		1,415	1,725
2800	Chain hoist, 20' x 14' high		.70	22.857		3,025	840		3,865	4,650
2850	For 1-1/4" rigid insulation and 26 ga. galv.									
2860	back panel, add				S.F.	2.64			2.64	2.90
2900	For electric trolley operator, 1/3 H.P., to 12' x 12', add	1 Carp	2	4	Ea.	715	147		862	1,025
2950	Over 12' x 12', 1/2 H.P., add	"	1	8	"	810	294		1,104	1,350

08 36 13.20 Residential Garage Doors		Crew	Daily Output	Labor-Hours	Unit	Material	Labor	Equipment	Total	Total Incl O&P
0010	**RESIDENTIAL GARAGE DOORS**									
0050	Hinged, wood, custom, double door, 9' x 7'	2 Carp	4	4	Ea.	400	147		547	670
0070	16' x 7'		3	5.333		680	196		876	1,050
0200	Overhead, sectional, incl. hardware, fiberglass, 9' x 7', standard		5.28	3.030		625	111		736	865
0220	Deluxe		5.28	3.030		800	111		911	1,050
0300	16' x 7', standard		6	2.667		1,150	98		1,248	1,400
0320	Deluxe		6	2.667		1,425	98		1,523	1,725
0500	Hardboard, 9' x 7', standard		8	2		420	73.50		493.50	580
0520	Deluxe		8	2		565	73.50		638.50	740
0600	16' x 7', standard		6	2.667		830	98		928	1,050
0620	Deluxe		6	2.667		965	98		1,063	1,225
0700	Metal, 9' x 7', standard		5.28	3.030		495	111		606	720
0720	Deluxe		8	2		670	73.50		743.50	850
0800	16' x 7', standard		3	5.333		630	196		826	1,000
0820	Deluxe		6	2.667		1,025	98		1,123	1,275
0900	Wood, 9' x 7', standard		8	2		525	73.50		598.50	690
0920	Deluxe		8	2		1,500	73.50		1,573.50	1,775
1000	16' x 7', standard		6	2.667		1,050	98		1,148	1,300
1020	Deluxe		6	2.667		2,200	98		2,298	2,575
1800	Door hardware, sectional	1 Carp	4	2		231	73.50		304.50	370
1810	Door tracks only		4	2		108	73.50		181.50	233
1820	One side only		7	1.143		74.50	42		116.50	148
3000	Swing-up, including hardware, fiberglass, 9' x 7', standard	2 Carp	8	2		725	73.50		798.50	910
3020	Deluxe		8	2		760	73.50		833.50	950
3100	16' x 7', standard		6	2.667		910	98		1,008	1,150
3120	Deluxe		6	2.667		940	98		1,038	1,175
3200	Hardboard, 9' x 7', standard		8	2		330	73.50		403.50	480
3220	Deluxe		8	2		440	73.50		513.50	595
3300	16' x 7', standard		6	2.667		460	98		558	660
3320	Deluxe		6	2.667		685	98		783	905
3400	Metal, 9' x 7', standard		8	2		365	73.50		438.50	515
3420	Deluxe		8	2		655	73.50		728.50	840
3500	16' x 7', standard		6	2.667		570	98		668	775
3520	Deluxe		6	2.667		915	98		1,013	1,150
3600	Wood, 9' x 7', standard		8	2		395	73.50		468.50	550
3620	Deluxe		8	2		715	73.50		788.50	900
3700	16' x 7', standard		6	2.667		690	98		788	910

08 36 Panel Doors

08 36 13 – Sectional Doors

08 36 13.20 Residential Garage Doors

	08 36 13.20 Residential Garage Doors	Crew	Daily Output	Labor-Hours	Unit	Material	2007 Bare Costs Labor	Equipment	Total	Total Incl O&P
3720	Deluxe	2 Carp	6	2.667	Ea.	970	98		1,068	1,225
3900	Door hardware only, swing up	1 Carp	4	2		117	73.50		190.50	243
3920	One side only		7	1.143		64.50	42		106.50	137
4000	For electric operator, economy, add		8	1		300	36.50		336.50	385
4100	Deluxe, including remote control		8	1		440	36.50		476.50	540
4500	For transmitter/receiver control, add to operator				Total	91			91	100
4600	Transmitters, additional				"	33			33	36.50

08 36 19 – Multi-Leaf Vertical Lift Doors

08 36 19.10 Multi-Leaf Vertical Lift Doors

		Crew	Daily Output	Labor-Hours	Unit	Material	Labor	Equipment	Total	Total Incl O&P
0010	**MULTI-LEAF VERTICAL LIFT DOORS**									
0020	Motorized, 14 ga. stl, incl., frm and ctrl pnl									
0050	16' x 16' high	L-10	.50	48	Ea.	18,500	2,000	1,450	21,950	25,400
0100	10' x 20' high		1.30	18.462		25,700	765	555	27,020	30,200
0120	15' x 20' high		1.30	18.462		31,500	765	555	32,820	36,500
0140	20' x 20' high		1	24		36,900	995	725	38,620	43,100
0160	25' x 20' high		1	24		40,800	995	725	42,520	47,400
0170	32' x 24' high		.75	32		38,600	1,325	965	40,890	45,800
0180	20' x 25' high		1	24		42,200	995	725	43,920	49,000
0200	25' x 25' high		.70	34.286		47,500	1,425	1,025	49,950	56,000
0220	25' x 30' high		.70	34.286		55,500	1,425	1,025	57,950	64,500
0240	30' x 30' high		.70	34.286		64,000	1,425	1,025	66,450	74,000
0260	35' x 30' high		.70	34.286		70,000	1,425	1,025	72,450	80,500

08 36 23 – Telescoping Vertical Lift Doors

08 36 23.10 Telescoping Steel Doors

		Crew	Daily Output	Labor-Hours	Unit	Material	Labor	Equipment	Total	Total Incl O&P
0010	**TELESCOPING STEEL DOORS**									
1000	Overhead, .03" thick, electric operated, 10' x 10'	E-3	.80	30	Ea.	10,800	1,250	144	12,194	14,200
2000	20' x 10'		.60	40		14,300	1,675	192	16,167	19,000
3000	20' x 16'		.40	60		17,200	2,525	288	20,013	23,800

08 38 Traffic Doors

08 38 13 – Flexible Strip Doors

08 38 13.10 Flexible Transparent Strip Doors

		Crew	Daily Output	Labor-Hours	Unit	Material	Labor	Equipment	Total	Total Incl O&P
0010	**FLEXIBLE TRANSPARENT STRIP DOORS**									
0100	12" strip width, 2/3 overlap	3 Shee	135	.178	SF Surf	5.45	7.75		13.20	17.95
0200	Full overlap		115	.209		5.70	9.10		14.80	20.50
0220	8" strip width, 1/2 overlap		140	.171		5.30	7.45		12.75	17.30
0240	Full overlap		120	.200		7	8.70		15.70	21
0300	Add for suspension system, header mount				L.F.	8.05			8.05	8.85
0400	Wall mount				"	8.15			8.15	8.95

08 38 19 – Rigid Traffic Doors

08 38 19.20 Double Acting Swing Doors

		Crew	Daily Output	Labor-Hours	Unit	Material	Labor	Equipment	Total	Total Incl O&P
0010	**DOUBLE ACTING SWING DOORS**									
1000	.063" aluminum, 7'-0" high, 4'-0" wide	2 Carp	4.20	3.810	Pr.	1,825	140		1,965	2,225
1025	6'-0" wide		4	4		2,200	147		2,347	2,650
1050	6'-8" wide		4	4		2,400	147		2,547	2,875
2000	Solid core wood, 3/4" thick, metal frame, stainless steel									
2010	base plate, 7' high opening, 4' wide	2 Carp	4	4	Pr.	2,175	147		2,322	2,625
2050	7' wide	"	3.80	4.211	"	2,400	155		2,555	2,900

08 38 Traffic Doors

08 38 19 – Rigid Traffic Doors

08 38 19.30 Shock Absorbing Doors

		Crew	Daily Output	Labor-Hours	Unit	Material	2007 Bare Costs Labor	Equipment	Total	Total Incl O&P
0010	**SHOCK ABSORBING DOORS**									
0020	Rigid, no frame, 1-1/2" thick, 5' x 7'	2 Sswk	1.90	8.421	Opng.	1,350	350		1,700	2,100
0100	8' x 8'		1.80	8.889		1,900	370		2,270	2,750
0500	Flexible, no frame, insulated, .16" thick, economy, 5' x 7'		2	8		1,650	330		1,980	2,400
0600	Deluxe		1.90	8.421		2,500	350		2,850	3,375
1000	8' x 8' opening, economy		2	8		2,575	330		2,905	3,425
1100	Deluxe		1.90	8.421		3,300	350		3,650	4,250

08 41 Entrances and Storefronts

08 41 13 – Aluminum-Framed Entrances and Storefronts

08 41 13.20 Tube Framing

		Crew	Daily Output	Labor-Hours	Unit	Material	2007 Bare Costs Labor	Equipment	Total	Total Incl O&P
0010	**TUBE FRAMING**, For window walls and store fronts, aluminum stock									
0050	Plain tube frame, mill finish, 1-3/4" x 1-3/4"	2 Glaz	103	.155	L.F.	7.30	5.60		12.90	16.55
0150	1-3/4" x 4"		98	.163		8.95	5.90		14.85	18.75
0200	1-3/4" x 4-1/2" CN		95	.168		10.40	6.05		16.45	20.50
0250	2" x 6"		89	.180		15.05	6.50		21.55	26.50
0350	4" x 4"		87	.184		14.70	6.65		21.35	26.50
0400	4-1/2" x 4-1/2"		85	.188		23.50	6.80		30.30	36
0450	Glass bead		240	.067		1.90	2.40		4.30	5.75
1000	Flush tube frame, mill finish, 1/4" glass, 1-3/4" x 4", open header		80	.200		8.80	7.20		16	20.50
1050	Open sill		82	.195		7.65	7.05		14.70	19.10
1100	Closed back header		83	.193		12.30	6.95		19.25	24
1150	Closed back sill		85	.188		11.70	6.80		18.50	23
1200	Vertical mullion, one piece		75	.213		13.05	7.70		20.75	26
1250	Two piece		73	.219		13.95	7.90		21.85	27.50
1300	90° or 180° vertical corner post		75	.213		22	7.70		29.70	35.50
1400	1-3/4" x 4-1/2", open header		80	.200		10.75	7.20		17.95	22.50
1450	Open sill		82	.195		8.85	7.05		15.90	20.50
1500	Closed back header		83	.193		12.95	6.95		19.90	25
1550	Closed back sill		85	.188		12.60	6.80		19.40	24
1600	Vertical mullion, one piece		75	.213		14.10	7.70		21.80	27
1650	Two piece		73	.219		14.90	7.90		22.80	28.50
1700	90° or 180° vertical corner post		75	.213		15.25	7.70		22.95	28.50
2000	Flush tube frame, mill fin. for ins. glass, 2" x 4-1/2", open header		75	.213		12.20	7.70		19.90	25
2050	Open sill		77	.208		9.90	7.50		17.40	22
2100	Closed back header		78	.205		11.40	7.40		18.80	24
2150	Closed back sill		80	.200		12.20	7.20		19.40	24.50
2200	Vertical mullion, one piece		70	.229		12.55	8.25		20.80	26.50
2250	Two piece		68	.235		13.35	8.50		21.85	27.50
2300	90° or 180° vertical corner post		70	.229		13.15	8.25		21.40	27
5000	Flush tube frame, mill fin., thermal brk., 2-1/4"x 4-1/2", open header		74	.216		12.30	7.80		20.10	25.50
5050	Open sill		75	.213		11.10	7.70		18.80	24
5100	Vertical mullion, one piece		69	.232		13.85	8.35		22.20	28
5150	Two piece		67	.239		16.05	8.60		24.65	30.50
5200	90° or 180° vertical corner post		69	.232		14.05	8.35		22.40	28
6980	Door stop (snap in)		380	.042		2.60	1.52		4.12	5.15
7000	For joints, 90°, clip type, add				Ea.	19.10			19.10	21
7050	Screw spline joint, add					15.80			15.80	17.40
7100	For joint other than 90°, add					33			33	36.50
8000	For bronze anodized aluminum, add					15%				

Since its founding in 1975, Reed Construction Data has developed an online and print portfolio of innovative products and services for the construction, design and manufacturing community. Our products and services are designed specifically to help construction industry professionals advance their businesses with timely, accurate and actionable project, product and cost data. Reed Construction Data is your all-inclusive source of construction information encompassing all phases of the construction process.

Reed Bulletin and Reed Connect™ deliver the most comprehensive, timely and reliable project information to support contractors, distributors and building product manufacturers in identifying, bidding and tracking projects – private and public, general building and civil. Reed Construction Data also offers in-depth construction activity statistics and forecasts covering major project categories, many at the county and metropolitan level.

Reed Bulletin
www.reedbulletin.com

- Project leads targeted by geographic region and formatted by construction stage – available online or in print.

- Locate those hard-to-find jobs that are more profitable to your business.

- Optional automatic e-mail updates sent whenever project details change.

- Download plans and specs online or order print copies.

Reed Connect
www.reedconnect.com

- Customized web-based project lead delivery service featuring advanced search capabilities.

- Manage and track actionable leads from planning to quote through winning the job.

- Competitive analysis tool to analyze lost sales opportunities.

- Potential integration with your CRM application.

Reed Research & Analytics
www.buildingteamforecast.com

Reed Construction Forecast

- Delivers timely construction industry activity combining historical data, current year projections and forecasts.

- Modeled at the individual MSA-level to capture changing local market conditions.

- Covers 21 major project categories.

Reed Construction Starts

- Available in a monthly report or as an interactive database.

- Data provided in square footage and dollar value.

- Highly effective and efficient business planning tool.

Market Fundamentals

- Metropolitan area-specific reporting.

- Five-year forecast of industry performance including major projects in development and underway.

- Property types include office, retail, hotel, warehouse and apartment.

For more information about Reed Construction Data, please call 877-REED411, visit our website at www.reedconstructiondata.com, or E-mail: marketing@reedbusiness.com

Reed Construction Data®

The design community utilizes the Reed First Source® suite of products to search, select and specify nationally available building products during the formative stages of project design, as well as during other stages of product selection and specification. Reed Design Registry is a detailed database of architecture firms. This tool features sophisticated search and sort capabilities to support the architect selection process and ensure locating the best manufacturers to partner with on your next project.

Reed First Source - The Leading Product Directory to "Find It, Choose It, Use It"

www.reedfirstsource.com

- Comprehensive directory of over 11,000 commercial building product manufacturers classified by MasterFormat™ 2004 categories.

- SPEC-DATA's 10-part format provides performance data along with technical and installation information.

- MANU-SPEC delivers manufacturer guide specifications in the CSI 3-part SectionFormat.

- Search for products, download CAD, research building codes and view catalogs online.

Reed Design Registry – The Premier Source of Architecture Firms

www.reedregistry.com

- Comprehensive website directory contains over 30,000 architect firms.

- Profiles list firm size, firm's specialties and more.

- Official database of AIA member-owned firms.

- Allows architects to share project information with the building community.

- Coming soon – Reed Registry to include engineers and landscape architects.

For more information about Reed Construction Data, please call 877-REED411, visit our website at www.reedconstructiondata.com, or E-mail: marketing@reedbusiness.com

08 41 Entrances and Storefronts

08 41 13 – Aluminum-Framed Entrances and Storefronts

08 41 13.20 Tube Framing

		Crew	Daily Output	Labor-Hours	Unit	Material	2007 Bare Costs Labor	Equipment	Total	Total Incl O&P
8020	For black finish, add					27%				
8050	For stainless steel materials, add					350%				
8100	For monumental grade, add					50%				
8150	For steel stiffener, add	2 Glaz	200	.080	L.F.	8.05	2.88		10.93	13.20
8200	For 2 to 5 stories, add per story				Story		5%			

08 41 19 – Stainless-Steel-Framed Entrances and Storefronts

08 41 19.10 Stainless-Steel and Glass Entrance Unit

		Crew	Daily Output	Labor-Hours	Unit	Material	Labor	Equipment	Total	Total Incl O&P
0010	**STAINLESS-STEEL AND GLASS ENTRANCE UNIT**, narrow stiles									
0020	3' x 7' opening, including hardware, minimum	2 Sswk	1.60	10	Opng.	5,150	415		5,565	6,425
0050	Average		1.40	11.429		5,600	475		6,075	7,000
0100	Maximum		1.20	13.333		6,000	550		6,550	7,600
1000	For solid bronze entrance units, statuary finish, add					60%				
1100	Without statuary finish, add					45%				
2000	Balanced doors, 3' x 7', economy	2 Sswk	.90	17.778	Ea.	7,000	735		7,735	9,025
2100	Premium	"	.70	22.857	"	12,100	945		13,045	15,000

08 41 26 – All-Glass Entrances and Storefronts

08 41 26.10 Window Walls Aluminum, Stock

		Crew	Daily Output	Labor-Hours	Unit	Material	Labor	Equipment	Total	Total Incl O&P
0010	**WINDOW WALLS ALUMINUM, STOCK,** including glazing									
0020	Minimum	H-2	160	.150	S.F.	29	5.05		34.05	39.50
0050	Average		140	.171		36	5.75		41.75	48.50
0100	Maximum		110	.218		115	7.35		122.35	138
0500	For translucent sandwich wall systems, see div. 07 41 33.10									
0850	Cost of the above walls depends on material,									
0860	finish, repetition, and size of units.									
0870	The larger the opening, the lower the S.F. cost									
1200	Double glazed acoustical window wall for airports,									
1220	including 1" thick glass with 2" x 4-1/2" tube frame	H-2	40	.600	S.F.	75	20		95	114

08 42 Entrances

08 42 26 – All-Glass Entrances

08 42 26.10 Swinging Glass Doors

		Crew	Daily Output	Labor-Hours	Unit	Material	Labor	Equipment	Total	Total Incl O&P
0010	**SWINGING GLASS DOORS**									
0020	Including hardware, 1/2" thick, tempered, 3' x 7' opening	2 Glaz	2	8	Opng.	1,850	288		2,138	2,450
0100	6' x 7' opening		1.40	11.429	"	3,650	410		4,060	4,625
9000	Minimum labor/equipment charge		2	8	Job		288		288	435

08 42 29 – Automatic Entrances

08 42 29.23 Sliding Automatic Entrances

		Crew	Daily Output	Labor-Hours	Unit	Material	Labor	Equipment	Total	Total Incl O&P
0010	**SLIDING AUTOMATIC ENTRANCES** 12' x 7'-6" opng., 5' x 7' door, 2 way traffic									
0020	Mat activated, panic pushout, incl. operator & hardware,									
0030	not including glass or glazing	2 Glaz	.70	22.857	Opng.	6,625	825		7,450	8,550
9000	Minimum labor/equipment charge	"	.70	22.857	Job		825		825	1,250

08 42 33 – Revolving Door Entrances

08 42 33.10 Revolving Door Entrances

		Crew	Daily Output	Labor-Hours	Unit	Material	Labor	Equipment	Total	Total Incl O&P
0010	**REVOLVING DOOR ENTRANCES**, Aluminum 6'-6" to 7'-0" in dia.									
0020	6'-10" to 7' high, stock units, minimum	4 Sswk	.75	42.667	Opng.	16,700	1,775		18,475	21,500
0050	Average		.60	53.333		20,500	2,200		22,700	26,600
0100	Maximum		.45	71.111		27,000	2,950		29,950	35,000
1000	Stainless steel		.30	105		32,300	4,375		36,675	43,600
1100	Solid bronze		.15	213		37,700	8,825		46,525	57,500

08 42 Entrances

08 42 33 – Revolving Door Entrances

08 42 33.10 Revolving Door Entrances

		Crew	Daily Output	Labor-Hours	Unit	Material	2007 Bare Costs Labor	2007 Bare Costs Equipment	Total	Total Incl O&P
1500	For automatic controls, add	2 Elec	2	8	Opng.	12,700	350		13,050	14,500

08 42 36 – Balanced Door Entrances

08 42 36.10 Balanced Entrance Doors

		Crew	Daily Output	Labor-Hours	Unit	Material	Labor	Equipment	Total	Incl O&P
0010	**BALANCED ENTRANCE DOORS**									
0020	Hardware & frame, alum. & glass, 3' x 7', econ.	2 Sswk	.90	17.778	Ea.	5,325	735		6,060	7,175
0150	Premium	"	.70	22.857	"	6,625	945		7,570	8,975

08 43 Storefronts

08 43 13 – Aluminum-Framed Storefronts

08 43 13.10 Aluminum-Framed Storefronts

		Crew	Daily Output	Labor-Hours	Unit	Material	Labor	Equipment	Total	Incl O&P
0010	**ALUMINUM-FRAMED STOREFRONTS**									
0020	Entrance, 3' x 7' opening, clear anodized finish	2 Sswk	7	2.286	Opng.	400	94.50		494.50	610
0040	Bronze finish		7	2.286		310	94.50		404.50	510
0060	Black finish		7	2.286		400	94.50		494.50	610
0500	6' x 7' opening, clear finish		6	2.667		320	110		430	555
0520	Bronze finish		6	2.667		535	110		645	790
0540	Black finish		6	2.667		415	110		525	655
1000	With 3' high transoms, 3' x 10' opening, clear finish		6.50	2.462		385	102		487	605
1050	Bronze finish		6.50	2.462		420	102		522	645
1100	Black finish		6.50	2.462		485	102		587	715
1500	With 3' high transoms, 6' x 10' opening, clear finish		5.50	2.909		470	120		590	735
1550	Bronze finish		5.50	2.909		505	120		625	775
1600	Black finish		5.50	2.909		585	120		705	860

08 43 13.20 Storefront Systems

		Crew	Daily Output	Labor-Hours	Unit	Material	Labor	Equipment	Total	Incl O&P
0010	**STOREFRONT SYSTEMS**, aluminum frame clear 3/8" plate glass									
0020	incl. 3' x 7' door with hardware (400 sq. ft. max. wall)									
0500	Wall height to 12' high, commercial grade	2 Glaz	150	.107	S.F.	14.25	3.85		18.10	21.50
0600	Institutional grade		130	.123		17.90	4.44		22.34	26.50
0700	Monumental grade		115	.139		27	5		32	37.50
1000	6' x 7' door with hardware, commercial grade		135	.119		14.55	4.27		18.82	22.50
1100	Institutional grade		115	.139		19.95	5		24.95	29.50
1200	Monumental grade		100	.160		37	5.75		42.75	49.50
1500	For bronze anodized finish, add					15%				
1600	For black anodized finish, add					30%				
1700	For stainless steel framing, add to monumental					75%				

08 43 29 – Sliding Storefronts

08 43 29.10 Sliding Panels

		Crew	Daily Output	Labor-Hours	Unit	Material	Labor	Equipment	Total	Incl O&P
0010	**SLIDING PANELS**									
0020	Mall fronts, aluminum & glass, 15' x 9' high	2 Glaz	1.30	12.308	Opng.	2,600	445		3,045	3,525
0100	24' x 9' high		.70	22.857		3,775	825		4,600	5,400
0200	48' x 9' high, with fixed panels		.90	17.778		7,025	640		7,665	8,700
0500	For bronze finish, add					17%				

08 44 Curtain Wall and Glazed Assemblies

08 44 13 – Glazed Aluminum Curtain Walls

08 44 13.10 Glazed Curtain Walls

		Crew	Daily Output	Labor-Hours	Unit	Material	2007 Bare Costs Labor	Equipment	Total	Total Incl O&P
0010	**GLAZED CURTAIN WALLS**, aluminum, stock, including glazing									
0020	Minimum	H-1	205	.156	S.F.	24	6.05		30.05	36.50
0050	Average, single glazed		195	.164		31	6.35		37.35	44.50
0150	Average, double glazed		180	.178		43.50	6.90		50.40	59.50
0200	Maximum	↓	160	.200	↓	119	7.75		126.75	144

08 45 Translucent Wall and Roof Assemblies

08 45 10 – Translucent Wall and Roof Assemblies

08 45 10.10 Skyroofs

		Crew	Daily Output	Labor-Hours	Unit	Material	2007 Bare Costs Labor	Equipment	Total	Total Incl O&P
0010	**SKYROOFS**, Translucent panels, 2-3/4" thick									
0020	Under 500 S.F.	G-3	395	.081	SF Hor.	29	2.93		31.93	36.50
0100	Over 5000 S.F.		465	.069		27.50	2.49		29.99	34
0300	Continuous vaulted, semi-circular, to 8' wide, double glazed		145	.221		46.50	8		54.50	64
0400	Single glazed		160	.200		31.50	7.25		38.75	45.50
0600	To 20' wide, single glazed		175	.183		35	6.60		41.60	49
0700	Over 20' wide, single glazed		200	.160		40.50	5.80		46.30	53.50
0900	Motorized opening type, single glazed, 1/3 opening		145	.221		42.50	8		50.50	59.50
1000	Full opening	↓	130	.246	↓	48.50	8.90		57.40	67.50
1200	Pyramid type units, self-supporting, to 30' clear opening,									
1300	square or circular, single glazed, minimum	G-3	200	.160	SF Hor.	23	5.80		28.80	34.50
1310	Average		165	.194		33	7		40	47
1400	Maximum		130	.246		47	8.90		55.90	66
1500	Grid type, 4' to 10' modules, single glass glazed, minimum		200	.160		30.50	5.80		36.30	42.50
1550	Maximum		128	.250		49.50	9.05		58.55	68.50
1600	Preformed acrylic, minimum		300	.107		36.50	3.86		40.36	46.50
1650	Maximum	↓	175	.183	↓	51	6.60		57.60	67
1800	Skyroofs, dome type, self-supporting, to 100' clear opening, circular									
1900	Rise to span ratio = 0.20									
1920	Minimum	G-3	197	.162	SF Hor.	14.60	5.85		20.45	25
1950	Maximum		113	.283		48.50	10.25		58.75	69.50
2100	Rise to span ratio = 0.33, minimum		169	.189		29	6.85		35.85	42.50
2150	Maximum		101	.317		57.50	11.45		68.95	81.50
2200	Rise to span ratio = 0.50, minimum		148	.216		44	7.80		51.80	60.50
2250	Maximum		87	.368		64	13.30		77.30	91
2400	Ridge units, continuous, to 8' wide, double		130	.246		112	8.90		120.90	137
2500	Single		200	.160		76.50	5.80		82.30	93.50
2700	Ridge and furrow units, over 4' O.C., double, minimum		200	.160		24	5.80		29.80	35.50
2750	Maximum		120	.267		44.50	9.65		54.15	64
2800	Single, minimum		214	.150		20.50	5.40		25.90	31
2850	Maximum		153	.209	↓	43.50	7.55		51.05	59.50
3000	Rolling roof, translucent panels, flat roof, minimum		253	.126	S.F.	18.55	4.57		23.12	27.50
3030	Maximum		160	.200		37	7.25		44.25	52
3100	Lean-to skyroof, long span, double, minimum		197	.162		25	5.85		30.85	36.50
3150	Maximum		101	.317		49.50	11.45		60.95	72.50
3300	Single, minimum		321	.100		18.65	3.60		22.25	26
3350	Maximum	↓	160	.200	↓	31	7.25		38.25	45

08 51 Metal Windows

08 51 13 – Aluminum Windows

08 51 13.10 Aluminum Sash

		Crew	Daily Output	Labor-Hours	Unit	Material	2007 Bare Costs Labor	Equipment	Total	Total Incl O&P
0010	**ALUMINUM SASH**									
0020	Stock, grade C, glaze & trim not incl., casement	2 Sswk	200	.080	S.F.	30	3.31		33.31	39
0050	Double hung		200	.080		30.50	3.31		33.81	39.50
0100	Fixed casement		200	.080		12.10	3.31		15.41	19.30
0150	Picture window		200	.080		13	3.31		16.31	20.50
0200	Projected window		200	.080		28	3.31		31.31	36.50
0250	Single hung		200	.080		14.10	3.31		17.41	21.50
0300	Sliding		200	.080		18.25	3.31		21.56	26
1000	Mullions for above, tubular		240	.067	L.F.	4.75	2.76		7.51	10.25
2000	Custom aluminum sash, grade HC, glazing not included, minimum		200	.080	S.F.	34	3.31		37.31	43.50
2100	Maximum		85	.188	"	44	7.80		51.80	62.50

08 51 13.20 Aluminum Windows

		Crew	Daily Output	Labor-Hours	Unit	Material	2007 Bare Costs Labor	Equipment	Total	Total Incl O&P
0010	**ALUMINUM WINDOWS**, incl. frame and glazing, commercial grade									
0020	See also division 05 12 23.40									
1000	Stock units, casement, 3'-1" x 3'-2" opening	2 Sswk	10	1.600	Ea.	315	66		381	470
1050	Add for storms					102			102	112
1600	Projected, with screen, 3'-1" x 3'-2" opening	2 Sswk	10	1.600		300	66		366	450
1700	Add for storms					100			100	110
2000	4'-5" x 5'-3" opening	2 Sswk	8	2		340	82.50		422.50	525
2100	Add for storms					110			110	121
2500	Enamel finish windows, 3'-1" x 3'-2"	2 Sswk	10	1.600		310	66		376	460
2600	4'-5" x 5'-3"		8	2		350	82.50		432.50	535
3000	Single hung, 2' x 3' opening, enameled, standard glazed		10	1.600		180	66		246	320
3100	Insulating glass		10	1.600		218	66		284	360
3300	2'-8" x 6'-8" opening, standard glazed		8	2		315	82.50		397.50	500
3400	Insulating glass		8	2		405	82.50		487.50	600
3700	3'-4" x 5'-0" opening, standard glazed		9	1.778		260	73.50		333.50	420
3800	Insulating glass		9	1.778		287	73.50		360.50	450
3890	Awning type, 3' x 3' opening standard glass		14	1.143		365	47.50		412.50	490
3900	Insulating glass		14	1.143		390	47.50		437.50	515
3910	3' x 4' opening, standard glass		10	1.600		430	66		496	590
3920	Insulating glass		10	1.600		495	66		561	660
3930	3' x 5'-4" opening, standard glass		10	1.600		515	66		581	690
3940	Insulating glass		10	1.600		610	66		676	790
3950	4' x 5'-4" opening, standard glass		9	1.778		565	73.50		638.50	760
3960	Insulating glass		9	1.778		680	73.50		753.50	880
4000	Sliding aluminum, 3' x 2' opening, standard glazed		10	1.600		190	66		256	330
4100	Insulating glass		10	1.600		204	66		270	345
4300	5' x 3' opening, standard glazed		9	1.778		290	73.50		363.50	455
4400	Insulating glass		9	1.778		340	73.50		413.50	505
4600	8' x 4' opening, standard glazed		6	2.667		305	110		415	535
4700	Insulating glass		6	2.667		495	110		605	740
5000	9' x 5' opening, standard glazed		4	4		465	165		630	810
5100	Insulating glass		4	4		740	165		905	1,125
5500	Sliding, with thermal barrier and screen, 6' x 4', 2 track		8	2		630	82.50		712.50	845
5700	4 track		8	2		800	82.50		882.50	1,025
6000	For above units with bronze finish, add					12%				
6200	For installation in concrete openings, add					5%				

08 51 23 – Steel Windows

08 51 23.10 Steel Sash

			Crew	Daily Output	Labor-Hours	Unit	Material	2007 Bare Costs Labor	Equipment	Total	Total Incl O&P
0010	**STEEL SASH** Custom units, glazing and trim not included	R085123-10									
0100	Casement, 100% vented		2 Sswk	200	.080	S.F.	42.50	3.31		45.81	53

08 51 Metal Windows

08 51 23 – Steel Windows

08 51 23.10 Steel Sash		Crew	Daily Output	Labor-Hours	Unit	Material	2007 Bare Costs Labor	Equipment	Total	Total Incl O&P
0200	50% vented	2 Sswk	200	.080	S.F.	39	3.31		42.31	48.50
0300	Fixed		200	.080		26.50	3.31		29.81	35.50
1000	Projected, commercial, 40% vented		200	.080		45.50	3.31		48.81	56
1100	Intermediate, 50% vented		200	.080		49.50	3.31		52.81	60.50
1500	Industrial, horizontally pivoted		200	.080		47	3.31		50.31	57.50
1600	Fixed		200	.080		26.50	3.31		29.81	35.50
2000	Industrial security sash, 50% vented		200	.080		50.50	3.31		53.81	61.50
2100	Fixed		200	.080		41	3.31		44.31	51
2500	Picture window		200	.080		25.50	3.31		28.81	34
3000	Double hung		200	.080		48	3.31		51.31	59
5000	Mullions for above, open interior face		240	.067	L.F.	8.60	2.76		11.36	14.45
5100	With interior cover		240	.067	"	14.20	2.76		16.96	20.50

08 51 23.20 Steel Windows

		Crew	Daily Output	Labor-Hours	Unit	Material	Labor	Equipment	Total	Total Incl O&P
0010	**STEEL WINDOWS** Stock, including frame, trim and insul. Glass									
0020	See also division 13 34 19.50									
1000	Custom units, double hung, 2'-8" x 4'-6" opening R085123-10	2 Sswk	12	1.333	Ea.	605	55		660	765
1100	2'-4" x 3'-9" opening		12	1.333		500	55		555	650
1500	Commercial projected, 3'-9" x 5'-5" opening		10	1.600		1,050	66		1,116	1,300
1600	6'-9" x 4'-1" opening		7	2.286		1,400	94.50		1,494.50	1,725
2000	Intermediate projected, 2'-9" x 4'-1" opening		12	1.333		595	55		650	755
2100	4'-1" x 5'-5" opening		10	1.600		1,200	66		1,266	1,450

08 51 66 – Metal Window Screens

08 51 66.10 Screens

		Crew	Daily Output	Labor-Hours	Unit	Material	Labor	Equipment	Total	Total Incl O&P
0010	**SCREENS**									
0020	For metal sash, aluminum or bronze mesh, flat screen	2 Sswk	1200	.013	S.F.	3.65	.55		4.20	5
0500	Wicket screen, inside window		1000	.016		5.55	.66		6.21	7.30
0800	Security screen, aluminum frame with stainless steel cloth		1200	.013		19.75	.55		20.30	22.50
0900	Steel grate, painted, on steel frame		1600	.010		11	.41		11.41	12.85
1000	For solar louvers, add		160	.100		20.50	4.14		24.64	30
4000	See also division 05 58 27.90									

08 52 Wood Windows

08 52 10 – Wood Windows

08 52 10.20 Awning Window

		Crew	Daily Output	Labor-Hours	Unit	Material	Labor	Equipment	Total	Total Incl O&P
0010	**AWNING WINDOW**, Including frame, screens and grills									
0100	Average quality, builders model, 34" x 22", double insulated glass	1 Carp	10	.800	Ea.	234	29.50		263.50	305
0200	Low E glass		10	.800		247	29.50		276.50	315
0300	40" x 28", double insulated glass		9	.889		295	32.50		327.50	375
0400	Low E Glass		9	.889		315	32.50		347.50	395
0500	48" x 36", double insulated glass		8	1		430	36.50		466.50	530
0600	Low E glass		8	1		455	36.50		491.50	555
1000	34" x 22"		10	.800		249	29.50		278.50	320
1100	40" x 22"		10	.800		271	29.50		300.50	345
1200	36" x 28"		9	.889		289	32.50		321.50	370
1300	36" x 36"		9	.889		320	32.50		352.50	405
1400	48" x 28"		8	1		345	36.50		381.50	435
1500	60" x 36"		8	1		500	36.50		536.50	605
2000	Bay, casement units, 8' x 5', 30° angle, w/screens, dbl Insul glass	2 Carp	2.50	6.400	Opng.	1,425	235		1,660	1,950
2200	36" x 25"	1 Carp	9	.889	Ea.	252	32.50		284.50	330
2300	40" x 30"		9	.889		315	32.50		347.50	395

08 52 Wood Windows

08 52 10 – Wood Windows

08 52 10.20 Awning Window

		Crew	Daily Output	Labor-Hours	Unit	Material	2007 Bare Costs Labor	Equipment	Total	Total Incl O&P
2400	48" x 28"	1 Carp	8	1	Ea.	325	36.50		361.50	410
2500	60" x 36"		8	1		345	36.50		381.50	435

08 52 10.30 Wood Windows

		Crew	Daily Output	Labor-Hours	Unit	Material	Labor	Equipment	Total	Total Incl O&P
0010	**WOOD WINDOWS**, double hung									
3200	20 S.F. and over	1 Carp	106	.075	S.F.	15.40	2.77		18.17	21.50
3800	Triple glazing for above, add					7.90			7.90	8.65
6000	Replacement sash, double hung, double glazing, to 12 S.F.	1 Carp	64	.125		17.65	4.59		22.24	26.50
6100	12 S.F. to 20 S.F.		94	.085		17.70	3.12		20.82	24.50
7000	Sash, single lite, 2'-0" x 2'-0" high		20	.400	Ea.	34	14.70		48.70	60
7050	2'-6" x 2'-0" high		19	.421		39	15.45		54.45	67
7100	2'-6" x 2'-6" high		18	.444		45	16.30		61.30	75
7150	3'-0" x 2'-0" high		17	.471		48	17.25		65.25	80

08 52 10.40 Casement Window

		Crew	Daily Output	Labor-Hours	Unit	Material	Labor	Equipment	Total	Total Incl O&P
0010	**CASEMENT WINDOW**, including frame, screen and grills									
0100	Avg. quality, bldrs. model, 2'-0" x 3'-0" H, dbl. insulated glass	1 Carp	10	.800	Ea.	263	29.50		292.50	335
0150	Low E glass		10	.800		281	29.50		310.50	355
0200	2'-0" x 4'-6" high, double insulated glass		9	.889		330	32.50		362.50	410
0250	Low E glass		9	.889		350	32.50		382.50	435
0300	2'-4" x 6'-0" high, double insulated glass		8	1		405	36.50		441.50	500
0350	Low E glass		8	1		425	36.50		461.50	520
0522	Vinyl clad, premium, double insulated glass, 2'-0" x 3'-0"		10	.800		264	29.50		293.50	335
0524	2'-0" x 4'-0"		9	.889		310	32.50		342.50	390
0525	2'-0" x 5'-0"		8	1		355	36.50		391.50	445
0528	2'-0" x 6'-0"		8	1		400	36.50		436.50	495
8100	Metal clad, deluxe, dbl. insul. glass, 2'-0" x 3'-0" high		10	.800		206	29.50		235.50	273
8120	2'-0" x 4'-0" high		9	.889		248	32.50		280.50	325
8140	2'-0" x 5'-0" high		8	1		282	36.50		318.50	365
8160	2'-0" x 6'-0" high		8	1		325	36.50		361.50	410
8200	For multiple leaf units, deduct for stationary sash									
8220	2' high				Ea.	21.50			21.50	23.50
8240	4'-6" high					24.50			24.50	27
8260	6' high					33			33	36.50
8300	For installation, add per leaf						15%			

08 52 10.50 Double Hung

		Crew	Daily Output	Labor-Hours	Unit	Material	Labor	Equipment	Total	Total Incl O&P
0010	**DOUBLE HUNG**, Including frame, screens and grills R085216-10									
0100	Avg. quality, bldrs. model, 2'-0" x 3'-0" high, dbl insul. glass	1 Carp	10	.800	Ea.	181	29.50		210.50	245
0150	Low E glass		10	.800		190	29.50		219.50	255
0200	3'-0" x 4'-0" high, double insulated glass		9	.889		244	32.50		276.50	320
0250	Low E glass		9	.889		256	32.50		288.50	335
0300	4'-0" x 4'-6" high, double insulated glass		8	1		310	36.50		346.50	395
0350	Low E glass		8	1		335	36.50		371.50	420
1000	Vinyl clad, premium, double insulated glass, 2'-6" x 3'-0"		10	.800		218	29.50		247.50	286
1100	3'-0" x 3'-6"		10	.800		256	29.50		285.50	330
1200	3'-0" x 4'-0"		9	.889		365	32.50		397.50	450
1300	3'-0" x 4'-6"		9	.889		340	32.50		372.50	425
1400	3'-0" x 5'-0"		8	1		310	36.50		346.50	395
1500	3'-6" x 6'-0"		8	1		360	36.50		396.50	455
2000	Metal clad, deluxe, dbl. insul. glass, 2'-6" x 3'-0" high		10	.800		238	29.50		267.50	310
2100	3'-0" x 3'-6" high		10	.800		272	29.50		301.50	345
2200	3'-0" x 4'-0" high		9	.889		286	32.50		318.50	365
2300	3'-0" x 4'-6" high		9	.889		300	32.50		332.50	380
2400	3'-0" x 5'-0" high		8	1		325	36.50		361.50	415

08 52 Wood Windows

08 52 10 – Wood Windows

08 52 10.50 Double Hung		Crew	Daily Output	Labor-Hours	Unit	Material	2007 Bare Costs Labor	Equipment	Total	Total Incl O&P
2500	3'-6" x 6'-0" high	1 Carp	8	1	Ea.	395	36.50		431.50	490

08 52 10.55 Picture Window

0010	**PICTURE WINDOW**, Including frame and grills									
0100	Average quality, bldrs. model, 3'-6" x 4'-0" high, dbl insulated glass	2 Carp	12	1.333	Ea.	350	49		399	460
0150	Low E glass		12	1.333		385	49		434	500
0200	4'-0" x 4'-6" high, double insulated glass		11	1.455		430	53.50		483.50	560
0250	Low E glass		11	1.455		450	53.50		503.50	580
0300	5'-0" x 4'-0" high, double insulated glass		11	1.455		505	53.50		558.50	640
0350	Low E glass		11	1.455		525	53.50		578.50	660
0400	6'-0" x 4'-6" high, double insulated glass		10	1.600		545	58.50		603.50	690
0450	Low E glass		10	1.600		565	58.50		623.50	710

08 52 10.65 Wood Sash

0010	**WOOD SASH**, Including glazing but not trim									
0050	Custom, 5'-0" x 4'-0", 1" dbl. glazed, 3/16" thick lites	2 Carp	3.20	5	Ea.	164	184		348	465
0100	1/4" thick lites		5	3.200		169	117		286	370
0200	1" thick, triple glazed		5	3.200		385	117		502	610
0300	7'-0" x 4'-6" high, 1" double glazed, 3/16" thick lites		4.30	3.721		395	137		532	645
0400	1/4" thick lites		4.30	3.721		445	137		582	700
0500	1" thick, triple glazed		4.30	3.721		505	137		642	770
0600	8'-6" x 5'-0" high, 1" double glazed, 3/16" thick lites		3.50	4.571		530	168		698	845
0700	1/4" thick lites		3.50	4.571		580	168		748	900
0800	1" thick, triple glazed		3.50	4.571		585	168		753	905
0900	Window frames only, based on perimeter length				L.F.	3.60			3.60	3.96
1200	Window sill, stock, per lineal foot					7.50			7.50	8.25
1250	Casing, stock					3			3	3.30

08 52 10.70 Sliding Windows

0010	**SLIDING WINDOWS**									
0100	Average quality, bldrs. model, 3'-0" x 3'-0" high, double insulated	1 Carp	10	.800	Ea.	233	29.50		262.50	300
0120	Low E glass		10	.800		254	29.50		283.50	325
0200	4'-0" x 3'-6" high, double insulated		9	.889		248	32.50		280.50	325
0220	Low E glass		9	.889		283	32.50		315.50	360
0300	6'-0" x 5'-0" high, double insulated		8	1		400	36.50		436.50	495
0320	Low E glass		8	1		440	36.50		476.50	540

08 52 13 – Metal-Clad Wood Windows

08 52 13.10 Metal-Clad Wood Windows

0010	**METAL-CLAD WOOD WINDOWS**									
2000	Metal clad, deluxe, double insulated glass, 34" x 22"	1 Carp	10	.800	Ea.	232	29.50		261.50	300
2100	40" x 22"	"	10	.800	"	273	29.50		302.50	345

08 52 13.35 Metal-Clad Wood Windows

0010	**METAL-CLAD WOOD WINDOWS**									
2000	Metal clad, deluxe, dbl. insul. glass, 4'-0" x 4'-0" high	2 Carp	12	1.333	Ea.	330	49		379	435
2100	4'-0" x 6'-0" high		11	1.455		485	53.50		538.50	620
2200	5'-0" x 6'-0" high		10	1.600		535	58.50		593.50	675
2300	6'-0" x 6'-0" high		10	1.600		615	58.50		673.50	765
2400	Metal clad, deluxe, double insulated glass, 3'-0" x 3'-0" high	1 Carp	10	.800		305	29.50		334.50	385
2420	4'-0" x 3'-6" high		9	.889		375	32.50		407.50	465
2440	5'-0" x 4'-0" high		9	.889		455	32.50		487.50	550
2460	6'-0" x 5'-0" high		8	1		670	36.50		706.50	795

08 52 Wood Windows

08 52 16 – Plastic-Clad Wood Windows

08 52 16.10 Bow Window

		Crew	Daily Output	Labor-Hours	Unit	Material	2007 Bare Costs Labor	2007 Bare Costs Equipment	Total	Total Incl O&P
0010	**BOW WINDOW**									
0020	End panels operable									
1000	Bow type, casement, wood, bldrs mdl, 8' x 5' dbl insltd glass, 4 panel	2 Carp	10	1.600	Ea.	1,200	58.50		1,258.50	1,425
1050	Low E glass		10	1.600		1,275	58.50		1,333.50	1,500
1100	10'-0" x 5'-0", double insulated glass, 6 panels		6	2.667		1,325	98		1,423	1,625
1200	Low E glass, 6 panels		6	2.667		1,400	98		1,498	1,700
1300	Vinyl clad, bldrs model, double insulated glass, 6'-0" x 4'-0", 3 panel		10	1.600		1,225	58.50		1,283.50	1,450
1340	9'-0" x 4'-0", 4 panel		8	2		1,450	73.50		1,523.50	1,725
1380	10'-0" x 6'-0", 5 panels		7	2.286		2,375	84		2,459	2,725
1420	12'-0" x 6'-0", 6 panels		6	2.667		3,000	98		3,098	3,450
1600	Metal clad, casement, bldrs mdl, 6'-0" x 4'-0", dbl insltd gls, 3 panels		10	1.600		855	58.50		913.50	1,025
1640	9'-0" x 4'-0", 4 panels		8	2		1,200	73.50		1,273.50	1,450
1680	10'-0" x 5'-0", 5 panels		7	2.286		1,650	84		1,734	1,950
1720	12'-0" x 6'-0", 6 panels		6	2.667		2,300	98		2,398	2,675
2000	Bay window, casement, builders model, 8' x 5' dbl insul glass, 4 panels		10	1.600		1,675	58.50		1,733.50	1,925
2050	Low E glass,		10	1.600		2,025	58.50		2,083.50	2,325
2100	12'-0" x 6'-0", double insulated glass, 6 panels		6	2.667		2,075	98		2,173	2,450
2200	Low E glass		6	2.667		2,150	98		2,248	2,500
2280	6'-0" x 4'-0"		11	1.455		1,075	53.50		1,128.50	1,275
2300	Vinyl clad, premium, double insulated glass, 8'-0" x 5'-0"		10	1.600		1,275	58.50		1,333.50	1,500
2340	10'-0" x 5'-0"		8	2		1,800	73.50		1,873.50	2,125
2380	10'-0" x 6'-0"		7	2.286		1,900	84		1,984	2,200
2420	12'-0" x 6'-0"		6	2.667		2,250	98		2,348	2,625
2600	Metal clad, deluxe, dbl insul. glass, 8'-0" x 5'-0" high, 4 panels		10	1.600		1,475	58.50		1,533.50	1,725
2640	10'-0" x 5'-0" high, 5 panels		8	2		1,575	73.50		1,648.50	1,875
2680	10'-0" x 6'-0" high, 5 panels		7	2.286		1,875	84		1,959	2,175
2720	12'-0" x 6'-0" high, 6 panels		6	2.667		2,600	98		2,698	3,000
3000	Double hung, bldrs. model, bay, 8' x 4' high, dbl insulated glass		10	1.600		1,175	58.50		1,233.50	1,375
3050	Low E glass		10	1.600		1,250	58.50		1,308.50	1,475
3100	9'-0" x 5'-0" high, doublel insulated glass		6	2.667		1,250	98		1,348	1,550
3200	Low E glass		6	2.667		1,325	98		1,423	1,625
3300	Vinyl clad, premium, double insulated glass, 7'-0" x 4'-6"		10	1.600		1,225	58.50		1,283.50	1,425
3340	8'-0" x 4'-6"		8	2		1,250	73.50		1,323.50	1,500
3380	8'-0" x 5'-0"		7	2.286		1,300	84		1,384	1,550
3420	9'-0" x 5'-0"		6	2.667		1,325	98		1,423	1,625
3600	Metal clad, deluxe, dbl insul. glass, 7'-0" x 4'-0" high		10	1.600		1,125	58.50		1,183.50	1,325
3640	8'-0" x 4'-0" high		8	2		1,175	73.50		1,248.50	1,400
3680	8'-0" x 5'-0" high		7	2.286		1,200	84		1,284	1,450
3720	9'-0" x 5'-0" high		6	2.667		1,275	98		1,373	1,550

08 52 16.30 Palladian Windows

		Crew	Daily Output	Labor-Hours	Unit	Material	2007 Bare Costs Labor	2007 Bare Costs Equipment	Total	Total Incl O&P
0010	**PALLADIAN WINDOWS**									
0020	Vinyl clad, double insulated glass, including frame and grills									
0040	3'-2" x 2'-6" high	2 Carp	11	1.455	Ea.	1,200	53.50		1,253.50	1,400
0060	3'-2" x 4'-10"		11	1.455		1,475	53.50		1,528.50	1,700
0080	3'-2" x 6'-4"		10	1.600		1,675	58.50		1,733.50	1,950
0100	4'-0" x 4'-0"		10	1.600		1,450	58.50		1,508.50	1,700
0120	4'-0" x 5'-4"	3 Carp	10	2.400		1,750	88		1,838	2,050
0140	4'-0" x 6'-0"		9	2.667		1,950	98		2,048	2,300
0160	4'-0" x 7'-4"		9	2.667		1,950	98		2,048	2,300
0180	5'-5" x 4'-10"		9	2.667		2,050	98		2,148	2,425
0200	5'-5" x 6'-10"		9	2.667		2,375	98		2,473	2,750
0220	5'-5" x 7'-9"		9	2.667		2,575	98		2,673	2,975

08 52 Wood Windows

08 52 16 – Plastic-Clad Wood Windows

08 52 16.30 Palladian Windows

		Crew	Daily Output	Labor-Hours	Unit	Material	2007 Bare Costs Labor	Equipment	Total	Total Incl O&P
0240	6'-0" x 7'-11"	3 Carp	8	3	Ea.	3,225	110		3,335	3,725
0260	8'-0" x 6'-0"	↓	8	3	↓	2,850	110		2,960	3,325

08 52 16.40 Transom Windows

		Crew	Daily Output	Labor-Hours	Unit	Material	Labor	Equipment	Total	Total Incl O&P
0010	**TRANSOM WINDOWS**									
0200	44" x 48"	1 Carp	12	.667	Ea.	515	24.50		539.50	610
1000	Vinyl clad, premium, dbl. insul. glass, 4'-0" x 4'-0"	2 Carp	12	1.333		470	49		519	590
1100	4'-0" x 6'-0"	↓	11	1.455		875	53.50		928.50	1,050
1200	5'-0" x 6'-0"		10	1.600		970	58.50		1,028.50	1,175
1300	6'-0" x 6'-0"		10	1.600	↓	990	58.50		1,048.50	1,175

08 52 16.70 Vinyl Clad, Premium, DBL. Insulated Glass

		Crew	Daily Output	Labor-Hours	Unit	Material	Labor	Equipment	Total	Total Incl O&P
0010	**VINYL CLAD, PREMIUM, DBL. INSULATED GLASS**									
1000	3'-0" x 3'-0"	1 Carp	10	.800	Ea.	510	29.50		539.50	605
1050	4'-0" x 3'-6"		9	.889		630	32.50		662.50	740
1100	5'-0" x 4'-0"		9	.889		760	32.50		792.50	885
1150	6'-0" x 5'-0"	↓	8	1	↓	965	36.50		1,001.50	1,100

08 52 50 – Window Accessories

08 52 50.10 Window Grille or Muntin

		Crew	Daily Output	Labor-Hours	Unit	Material	Labor	Equipment	Total	Total Incl O&P
0010	**WINDOW GRILLE OR MUNTIN,** snap in type									
0020	Standard pattern interior grills									
2000	Wood, awning window, glass size 28" x 16" high	1 Carp	30	.267	Ea.	21.50	9.80		31.30	39
2060	44" x 24" high		32	.250		31	9.20		40.20	49
2100	Casement, glass size, 20" x 36" high		30	.267		26.50	9.80		36.30	44.50
2180	20" x 56" high		32	.250	↓	38	9.20		47.20	56.50
2200	Double hung, glass size, 16" x 24" high		24	.333	Set	46	12.25		58.25	70
2280	32" x 32" high		34	.235	"	123	8.65		131.65	149
2500	Picture, glass size, 48" x 48" high		30	.267	Ea.	141	9.80		150.80	170
2580	60" x 68" high		28	.286	"	108	10.50		118.50	135
2600	Sliding, glass size, 14" x 36" high		24	.333	Set	25	12.25		37.25	46.50
2680	36" x 36" high	↓	22	.364	"	38.50	13.35		51.85	63

08 52 66 – Wood Window Screens

08 52 66.10 Wood Screens

		Crew	Daily Output	Labor-Hours	Unit	Material	Labor	Equipment	Total	Total Incl O&P
0010	**WOOD SCREENS**									
0020	Over 3 S.F., 3/4" frames	2 Carp	375	.043	S.F.	4.18	1.57		5.75	7.05
0100	1-1/8" frames	"	375	.043	"	6.60	1.57		8.17	9.70

08 52 69 – Wood Storm Windows

08 52 69.10 Storm Windows

		Crew	Daily Output	Labor-Hours	Unit	Material	Labor	Equipment	Total	Total Incl O&P
0010	**STORM WINDOWS,** aluminum residential									
0300	Basement, mill finish, incl. fiberglass screen									
0320	1'-10" x 1'-0" high	2 Carp	30	.533	Ea.	29.50	19.55		49.05	63
0340	2'-9" x 1'-6" high	"	30	.533	"	32	19.55		51.55	66
1600	Double-hung, combination, storm & screen									
2000	Average quality, clear anodic coating, 2'-0" x 3'-5" high	2 Carp	30	.533	Ea.	77.50	19.55		97.05	116
2020	2'-6" x 5'-0" high		28	.571		98	21		119	141
2040	4'-0" x 6'-0" high		25	.640		115	23.50		138.50	164
2400	White painted, 2'-0" x 3'-5" high		30	.533		76.50	19.55		96.05	115
2420	2'-6" x 5'-0" high		28	.571		84.50	21		105.50	126
2440	4'-0" x 6'-0" high		25	.640		92.50	23.50		116	139
2600	Mill finish, 2'-0" x 3'-5" high		30	.533		69.50	19.55		89.05	107
2620	2'-6" x 5'-0" high		28	.571		77.50	21		98.50	118
2640	4'-0" x 6-8" high	↓	25	.640	↓	87	23.50		110.50	133

08 53 Plastic Windows

08 53 13 – Vinyl Windows

08 53 13.20 Vinyl Single Hung Windows		Crew	Daily Output	Labor-Hours	Unit	Material	2007 Bare Costs Labor	Equipment	Total	Total Incl O&P
0010	**VINYL SINGLE HUNG WINDOWS**									
0100	Grids, low E, J fin, ext. jambs, 21" x 53"	2 Carp	18	.889	Ea.	146	32.50		178.50	212
0110	21" x 57"		17	.941		150	34.50		184.50	219
0120	21" x 65"		16	1		156	36.50		192.50	228
0130	25" x 41"		20	.800		138	29.50		167.50	198
0140	25" x 49"		18	.889		152	32.50		184.50	219
0150	25" x 57"		17	.941		156	34.50		190.50	225
0160	25" x 65"		16	1		162	36.50		198.50	235
0170	29" x 41"		18	.889		147	32.50		179.50	212
0180	29" x 53"		18	.889		157	32.50		189.50	224
0190	29" x 57"		17	.941		161	34.50		195.50	231
0200	29" x 65"		16	1		167	36.50		203.50	241
0210	33" x 41"		20	.800		152	29.50		181.50	213
0220	33" x 53"		18	.889		163	32.50		195.50	231
0230	33" x 57"		17	.941		167	34.50		201.50	238
0240	33" x 65"		16	1		174	36.50		210.50	248
0250	37" x 41"		20	.800		160	29.50		189.50	222
0260	37" x 53"		18	.889		172	32.50		204.50	240
0270	37" x 57"		17	.941		175	34.50		209.50	247
0280	37" x 65"		16	1		182	36.50		218.50	258
08 53 13.30 Vinyl Double Hung Windows										
0010	**VINYL DOUBLE HUNG WINDOWS**									
0100	Grids, low E, J fin, ext. jambs, 21" x 53"	2 Carp	18	.889	Ea.	167	32.50		199.50	235
0102	21" x 37"		18	.889		149	32.50		181.50	215
0104	21" x 41"		18	.889		153	32.50		185.50	219
0106	21" x 49"		18	.889		160	32.50		192.50	227
0110	21" x 57"		17	.941		171	34.50		205.50	242
0120	21" x 65"		16	1		177	36.50		213.50	252
0128	25" x 37"		20	.800		157	29.50		186.50	219
0130	25" x 41"		20	.800		161	29.50		190.50	223
0140	25" x 49"		18	.889		166	32.50		198.50	233
0145	25" x 53"		18	.889		172	32.50		204.50	241
0150	25" x 57"		17	.941		173	34.50		207.50	244
0160	25" x 65"		16	1		184	36.50		220.50	259
0162	25" x 69"		16	1		191	36.50		227.50	267
0164	25" x 77"		16	1		202	36.50		238.50	279
0168	29" x 37"		18	.889		162	32.50		194.50	229
0170	29" x 41"		18	.889		166	32.50		198.50	233
0172	29" x 49"		18	.889		174	32.50		206.50	242
0180	29" x 53"		18	.889		178	32.50		210.50	246
0190	29" x 57"		17	.941		181	34.50		215.50	253
0200	29" x 65"		16	1		188	36.50		224.50	264
0202	29" x 69"		16	1		195	36.50		231.50	272
0205	29" x 77"		16	1		207	36.50		243.50	284
0208	33" x 37"		20	.800		167	29.50		196.50	229
0210	33" x 41"		20	.800		170	29.50		199.50	233
0215	33" x 49"		20	.800		179	29.50		208.50	243
0220	33" x 53"		18	.889		183	32.50		215.50	252
0230	33" x 57"		17	.941		187	34.50		221.50	260
0240	33" x 65"		16	1		192	36.50		228.50	268
0242	33" x 69"		16	1		204	36.50		240.50	281
0246	33" x 77"		16	1		214	36.50		250.50	292

08 53 Plastic Windows

08 53 13 – Vinyl Windows

08 53 13.30 Vinyl Double Hung Windows		Crew	Daily Output	Labor-Hours	Unit	Material	2007 Bare Costs Labor	Equipment	Total	Total Incl O&P
0250	37" x 41"	2 Carp	20	.800	Ea.	174	29.50		203.50	238
0255	37" x 49"		20	.800		183	29.50		212.50	248
0260	37" x 53"		18	.889		191	32.50		223.50	261
0270	37" x 57"		17	.941		195	34.50		229.50	268
0280	37" x 65"		16	1		200	36.50		236.50	277
0282	37" x 69"		16	1		262	36.50		298.50	345
0286	37" x 77"		16	1		274	36.50		310.50	355
0300	Solid vinyl, average quality, double insulated glass, 2'-0" x 3'-0"	1 Carp	10	.800		253	29.50		282.50	325
0310	3'-0" x 4'-0"		9	.889		165	32.50		197.50	232
0320	4'-0" x 4'-6"		8	1		264	36.50		300.50	345
0330	Premium, double insulated glass, 2'-6" x 3'-0"		10	.800		180	29.50		209.50	244
0340	3'-0" x 3'-6"		9	.889		210	32.50		242.50	282
0350	3'-0" x 4'-0"		9	.889		222	32.50		254.50	295
0360	3'-0" x 4'-6"		9	.889		226	32.50		258.50	299
0370	3'-0" x 5'-0"		8	1		242	36.50		278.50	325
0380	3'-6" x 6'-0"		8	1		279	36.50		315.50	360

08 53 13.40 Vinyl Casement Windows		Crew	Daily Output	Labor-Hours	Unit	Material	2007 Bare Costs Labor	Equipment	Total	Total Incl O&P
0010	**VINYL CASEMENT WINDOWS**									
0100	Grids, low E, J fin, ext. jambs, 1 lt, 21" x 41"	2 Carp	20	.800	Ea.	216	29.50		245.50	284
0110	21" x 47"		20	.800		236	29.50		265.50	305
0120	21" x 53"		20	.800		255	29.50		284.50	325
0128	24" x 35"		19	.842		208	31		239	276
0130	24" x 41"		19	.842		226	31		257	296
0140	24" x 47"		19	.842		244	31		275	315
0150	24" x 53"		19	.842		263	31		294	340
0158	28" x 35"		19	.842		221	31		252	292
0160	28" x 41"		19	.842		239	31		270	310
0170	28" x 47"		19	.842		258	31		289	330
0180	28" x 53"		19	.842		284	31		315	365
0184	28" x 59"		19	.842		290	31		321	370
0188	Two lites, 33" x 35"		18	.889		355	32.50		387.50	440
0190	33" x 41"		18	.889		380	32.50		412.50	470
0200	33" x 47"		18	.889		410	32.50		442.50	500
0210	33" x 53"		18	.889		440	32.50		472.50	530
0212	33" x 59"		18	.889		465	32.50		497.50	560
0215	33" x 72"		18	.889		480	32.50		512.50	580
0220	41" x 41"		18	.889		415	32.50		447.50	510
0230	41" x 47"		18	.889		445	32.50		477.50	540
0240	41" x 53"		17	.941		475	34.50		509.50	575
0242	41" x 59"		17	.941		500	34.50		534.50	605
0246	41" x 72"		17	.941		520	34.50		554.50	630
0250	47" x 41"		17	.941		420	34.50		454.50	520
0260	47" x 47"		17	.941		450	34.50		484.50	550
0270	47" x 53"		17	.941		475	34.50		509.50	575
0272	47" x 59"		17	.941		520	34.50		554.50	625
0280	56" x 41"		15	1.067		450	39		489	555
0290	56" x 47"		15	1.067		475	39		514	580
0300	56" x 53"		15	1.067		520	39		559	630
0302	56" x 59"		15	1.067		540	39		579	655
0310	56" x 72"		15	1.067		590	39		629	705
0340	Solid vinyl, premium, double insulated glass, 2'-0" x 3'-0" high	1 Carp	10	.800		233	29.50		262.50	300
0360	2'-0" x 4'-0" high		9	.889		267	32.50		299.50	345

08 53 Plastic Windows

08 53 13 – Vinyl Windows

08 53 13.40 Vinyl Casement Windows	Crew	Daily Output	Labor-Hours	Unit	Material	2007 Bare Costs Labor	Equipment	Total	Total Incl O&P
0380 2'-0" x 5'-0" high	1 Carp	8	1	Ea.	266	36.50		302.50	350
08 53 13.50 Vinyl Picture Windows									
0010 **VINYL PICTURE WINDOWS**									
0100 Grids, low E, J fin, ext. jambs, 33" x 47"	2 Carp	12	1.333	Ea.	217	49		266	315
0110 35" x 71"		12	1.333		230	49		279	330
0120 41" x 47"		12	1.333		252	49		301	355
0130 41" x 71"		12	1.333		273	49		322	375
0140 47" x 47"		12	1.333		285	49		334	390
0150 47" x 71"		11	1.455		299	53.50		352.50	415
0160 53" x 47"		11	1.455		280	53.50		333.50	395
0170 53" x 71"		11	1.455		293	53.50		346.50	410
0180 59" x 47"		11	1.455		320	53.50		373.50	435
0190 59" x 71"		11	1.455		340	53.50		393.50	460
0200 71" x 47"		10	1.600		350	58.50		408.50	480
0210 71" x 71"		10	1.600		370	58.50		428.50	500

08 56 Special Function Windows

08 56 63 – Detention Windows

08 56 63.13 Visitor Cubicle Windows	Crew	Daily Output	Labor-Hours	Unit	Material	2007 Bare Costs Labor	Equipment	Total	Total Incl O&P
0010 **VISITOR CUBICLE WINDOWS**									
4000 Visitor cubicle, vision panel, no intercom	E-4	2	16	Ea.	2,750	670	57.50	3,477.50	4,325

08 62 Unit Skylights

08 62 13 – Domed Unit Skylights

08 62 13.20 Skylights	Crew	Daily Output	Labor-Hours	Unit	Material	2007 Bare Costs Labor	Equipment	Total	Total Incl O&P
0010 **SKYLIGHTS**, Plastic domes, flush or curb mounted ten or									
0100 more units, curb not included									
0300 Nominal size under 10 S.F., double	G-3	130	.246	S.F.	24.50	8.90		33.40	40.50
0400 Single		160	.200		19.20	7.25		26.45	32
0600 10 S.F. to 20 S.F., double		315	.102		18.55	3.67		22.22	26
0700 Single		395	.081		17.25	2.93		20.18	23.50
0900 20 S.F. to 30 S.F., double		395	.081		16.45	2.93		19.38	22.50
1000 Single		465	.069		15	2.49		17.49	20.50
1200 30 S.F. to 65 S.F., double		465	.069		14.50	2.49		16.99	19.80
1300 Single		610	.052		16	1.90		17.90	20.50
1500 For insulated 4" curbs, double, add					25%				
1600 Single, add					30%				
1800 For integral insulated 9" curbs, double, add					30%				
1900 Single, add					40%				
2120 Ventilating insulated plexiglass dome with									
2130 curb mounting, 36" x 36"	G-3	12	2.667	Ea.	380	96.50		476.50	570
2150 52" x 52"		12	2.667		570	96.50		666.50	775
2160 28" x 52"		10	3.200		445	116		561	670
2170 36" x 52"		10	3.200		480	116		596	710
2180 For electric opening system, add					285			285	315
2200 Field fabricated, factory type, aluminum and wire glass	G-3	120	.267	S.F.	14.70	9.65		24.35	31
2300 Insulated safety glass with aluminum frame		160	.200		85.50	7.25		92.75	106
2400 Sandwich panels, fiberglass, for walls, 1-9/16" thick, to 250 SF		200	.160		15.55	5.80		21.35	26
2500 250 SF and up		265	.121		13.95	4.37		18.32	22

08 62 Unit Skylights

08 62 13 – Domed Unit Skylights

08 62 13.20 Skylights

		Crew	Daily Output	Labor-Hours	Unit	Material	2007 Bare Costs Labor	Equipment	Total	Total Incl O&P
2700	As above, but for roofs, 2-3/4" thick, to 250 SF	G-3	295	.108	S.F.	22.50	3.92		26.42	30.50
2800	250 SF and up	↓	330	.097	↓	18.35	3.51		21.86	25.50

08 63 Metal-Framed Skylights

08 63 23 – Ridge Metal-Framed Skylights

08 63 23.10 Prefabricated

		Crew	Daily Output	Labor-Hours	Unit	Material	Labor	Equipment	Total	Total Incl O&P
0010	**PREFABRICATED** glass block with metal frame									
0020	Minimum	G-3	265	.121	S.F.	43.50	4.37		47.87	54.50
0100	Maximum	"	160	.200	"	86.50	7.25		93.75	107

08 71 Door Hardware

08 71 13 – Automatic Door Operators

08 71 13.10 Automatic Openers Commercial

		Crew	Daily Output	Labor-Hours	Unit	Material	Labor	Equipment	Total	Total Incl O&P
0010	**AUTOMATIC OPENERS COMMERCIAL**									
0020	Pneumatic, incl opener, motion sens, control box, tubing, compressor									
0050	For single swing door, per opening	2 Skwk	.80	20	Ea.	3,775	760		4,535	5,325
0100	Pair, per opening		.50	32	Opng.	6,300	1,225		7,525	8,825
1000	For single sliding door, per opening		.60	26.667		4,175	1,025		5,200	6,175
1300	Bi-parting pair	↓	.50	32	↓	6,325	1,225		7,550	8,850
1420	Electronic door opener incl motion sens, 12V control box, motor									
1450	For single swing door, per opening	2 Skwk	.80	20	Opng.	3,250	760		4,010	4,750
1500	Pair, per opening		.50	32		5,325	1,225		6,550	7,750
1600	For single sliding door, per opening		.60	26.667		3,525	1,025		4,550	5,450
1700	Bi-parting pair	↓	.50	32		5,400	1,225		6,625	7,850
1750	Handicap actuator buttons, 2, including 12V DC wiring, add	1 Carp	1.50	5.333	Pr.	380	196		576	725

08 71 13.20 Automatic Openers Industrial

		Crew	Daily Output	Labor-Hours	Unit	Material	Labor	Equipment	Total	Total Incl O&P
0010	**AUTOMATIC OPENERS INDUSTRIAL**									
0015	Sliding doors up to 6' wide	2 Skwk	.60	26.667	Opng.	4,950	1,025		5,975	7,000
0200	To 12' wide	"	.40	40	"	6,025	1,525		7,550	9,000
0400	Over 12' wide, add per L.F. of excess				L.F.	660			660	725
1000	Swing doors, to 5' wide	2 Skwk	.80	20	Ea.	2,875	760		3,635	4,350
1860	Add for controls, wall pushbutton, 3 button		4	4		176	152		328	430
1870	Ceiling pull cord	↓	4.30	3.721	↓	151	141		292	385

08 71 20 – Hardware

08 71 20.10 Bolts, Flush

		Crew	Daily Output	Labor-Hours	Unit	Material	Labor	Equipment	Total	Total Incl O&P
0010	**BOLTS, FLUSH**									
0020	Standard, concealed	1 Carp	7	1.143	Ea.	20	42		62	87.50
0800	Automatic fire exit	"	5	1.600		273	58.50		331.50	390
1600	For electric release, add	1 Elec	3	2.667		111	117		228	296
3000	Barrel, brass, 2" long	1 Carp	40	.200		4.37	7.35		11.72	16.25
3020	4" long		40	.200		3.22	7.35		10.57	15
3060	6" long	↓	40	.200	↓	8.85	7.35		16.20	21

08 71 20.15 Hardware

		Crew	Daily Output	Labor-Hours	Unit	Material	Labor	Equipment	Total	Total Incl O&P
0009	**HARDWARE**									
0010	Average hardware percentage for hardware, total job cost									
0025	Minimum									.75%
0050	Maximum									3.50%
0500	Total hardware for building, average distribution					85%	15%			

08 71 Door Hardware

08 71 20 – Hardware

08 71 20.15 Hardware

		Crew	Daily Output	Labor-Hours	Unit	Material	2007 Bare Costs Labor	Equipment	Total	Total Incl O&P
1000	Door hardware, apartment, interior				Door	129			129	142
1500	Hospital bedroom, minimum					288			288	315
2000	Maximum					630			630	695
2100	Pocket door				Ea.	129			129	142
2250	School, single exterior, incl. lever, not incl. panic device				Door	425			425	470
2500	Single interior, regular use, no lever included					284			284	315
2550	Including handicap lever					385			385	425
2600	Heavy use, incl. lever and closer					495			495	545
2850	Stairway, single interior					710			710	780
3100	Double exterior, with panic device				Pr.	1,000			1,000	1,100
3600	Toilet, public, single interior				Door	156			156	172

08 71 20.20 Door Protectors

		Crew	Daily Output	Labor-Hours	Unit	Material	Labor	Equipment	Total	Total Incl O&P
0010	**DOOR PROTECTORS**									
0020	1-3/4" x 3/4" U channel	2 Carp	80	.200	L.F.	18.75	7.35		26.10	32
0021	1-3/4" x 1-1/4" U channel		80	.200	"	8.75	7.35		16.10	21
1000	Tear drop, spring-stl, 8" high x 19" long		15	1.067	Ea.	81	39		120	151
1010	Tear drop, spring-stl, 8" high x 32" long		15	1.067		101	39		140	172
1100	8" high x19" long		15	1.067		207	39		246	289
1200	8" high x 32" long		15	1.067		258	39		297	345

08 71 20.30 Door Closers

		Crew	Daily Output	Labor-Hours	Unit	Material	Labor	Equipment	Total	Total Incl O&P
0010	**DOOR CLOSERS**									
0020	Adjustable backcheck, 3 way mount, all sizes, regular arm	1 Carp	6	1.333	Ea.	141	49		190	231
0040	Hold open arm		6	1.333		159	49		208	251
0100	Fusible link		6.50	1.231		124	45		169	207
0200	Non sized, regular arm		6	1.333		138	49		187	228
0240	Hold open arm		6	1.333		172	49		221	265
0400	4 way mount, non sized, regular arm		6	1.333		190	49		239	285
0440	Hold open arm		6	1.333		204	49		253	300
2000	Backcheck and adjustable power, hinge face mount									
2010	All sizes, regular arm	1 Carp	6.50	1.231	Ea.	175	45		220	263
2040	Hold open arm		6.50	1.231		188	45		233	278
2400	Top jamb mount, all sizes, regular arm		6	1.333		175	49		224	268
2440	Hold open arm		6	1.333		188	49		237	283
2800	Top face mount, all sizes, regular arm		6.50	1.231		175	45		220	263
2840	Hold open arm		6.50	1.231		187	45		232	277
4000	Backcheck, overhead concealed, all sizes, regular arm		5.50	1.455		185	53.50		238.50	286
4040	Concealed arm		5	1.600		197	58.50		255.50	310
4400	Compact overhead, concealed, all sizes, regular arm		5.50	1.455		335	53.50		388.50	455
4440	Concealed arm		5	1.600		350	58.50		408.50	475
4800	Concealed in door, all sizes, regular arm		5.50	1.455		124	53.50		177.50	220
4840	Concealed arm		5	1.600		134	58.50		192.50	240
4900	Floor concealed, all sizes, single acting		2.20	3.636		158	133		291	380
4940	Double acting		2.20	3.636		204	133		337	430
5000	For cast aluminum cylinder, deduct					16.75			16.75	18.45
5040	For delayed action, add					29.50			29.50	32.50
5080	For fusible link arm, add					12.15			12.15	13.35
5120	For shock absorbing arm, add					36.50			36.50	40
5160	For spring power adjustment, add					28			28	30.50
6000	Closer-holder, hinge face mount, all sizes, exposed arm	1 Carp	6.50	1.231		128	45		173	212
7000	Electronic closer-holder, hinge facemount, concealed arm		5	1.600		195	58.50		253.50	305
7400	With built-in detector		5	1.600		590	58.50		648.50	740
8000	Surface mounted, stand. duty, parallel arm, primed, traditional		6	1.333		135	49		184	225

08 71 Door Hardware

08 71 20 – Hardware

08 71 20.30 Door Closers

		Crew	Daily Output	Labor-Hours	Unit	Material	2007 Bare Costs Labor	Equipment	Total	Total Incl O&P
8030	Light duty	1 Carp	6	1.333	Ea.	80	49		129	164
8050	Heavy duty		6	1.333		161	49		210	254
8100	Standard duty, parallel arm, modern		6	1.333		151	49		200	243
8150	Heavy duty		6	1.333		171	49		220	264

08 71 20.31 Door Closers

		Crew	Daily Output	Labor-Hours	Unit	Material	Labor	Equipment	Total	Total Incl O&P
0010	**DOOR CLOSERS**									
0015	Door closer, rack and pinion	1 Carp	6.50	1.231	Ea.	134	45		179	219
1520	Door, single acting, standard arm		1	8		184	294		478	660
1526	Frame, single acting, standard arm		1	8		320	294		614	810
1530	Hold open arm		1	8		335	294		629	825
1534	Double acting, standard arm		1	8		430	294		724	930
1536	Hold open arm		1	8		440	294		734	940
1540	Hold open arm		1	8		198	294		492	675
1554	Floor, center hung, single acting, bottom arm		1	8		270	294		564	750
1558	Double acting		1	8		295	294		589	780
1560	Offset hung, single acting, bottom arm		1	8		310	294		604	800
6500	Electro magnetic closer/holder									
6510	Single point, no detector	1 Carp	1	8	Ea.	350	294		644	840
6515	Including detector		1	8		640	294		934	1,150
6520	Multi-point, no detector		1	8		640	294		934	1,150
6524	Including detector		1	8		740	294		1,034	1,275
6550	Electric automatic operators									
6555	Operator	1 Carp	1	8	Ea.	2,300	294		2,594	2,975
6570	Wall plate actuator		1	8		183	294		477	655
8010	Light duty, regular arm		1	8		101	294		395	565
8032	Parallel arm		1	8		104	294		398	570
8034	Hold open arm		1	8		112	294		406	580
8036	Fusible link arm		1	8		164	294		458	635
8040	Medium duty, regular arm		1	8		128	294		422	595
8042	Extra duty, parallel arm		1	8		138	294		432	605
8044	Hold open arm		1	8		147	294		441	615
8046	Positive stop arm		1	8		151	294		445	620
8052	Heavy duty, regular arm		1	8		155	294		449	625
8054	Top jamb mount		1	8		155	294		449	625
8056	Extra duty parallel arm		1	8		155	294		449	625
8058	Hold open arm		1	8		170	294		464	640
8060	Positive stop arm		1	8		169	294		463	640
8062	Fusible link arm		1	8		202	294		496	675
8080	Universal heavy duty, regular arm		1	8		167	294		461	640
8084	Parallel arm		1	8		167	294		461	640
8088	Extra duty, parallel arm		1	8		174	294		468	645
8090	Hold open arm		1	8		180	294		474	655
8094	Positive stop arm		1	8		184	294		478	660
8098	For delayed action add		1	8		22.50	294		316.50	480

08 71 20.35 Panic Devices

		Crew	Daily Output	Labor-Hours	Unit	Material	Labor	Equipment	Total	Total Incl O&P
0010	**PANIC DEVICES**									
0015	For rim locks, single door exit only	1 Carp	6	1.333	Ea.	375	49		424	490
0020	Outside key and pull		5	1.600		425	58.50		483.50	560
0200	Bar and vertical rod, exit only		5	1.600		545	58.50		603.50	690
0210	Outside key and pull		4	2		650	73.50		723.50	830
0400	Bar and concealed rod		4	2		550	73.50		623.50	720
0600	Touch bar, exit only		6	1.333		435	49		484	555

08 71 Door Hardware

08 71 20 – Hardware

08 71 20.35 Panic Devices

		Crew	Daily Output	Labor-Hours	Unit	Material	2007 Bare Costs Labor	2007 Bare Costs Equipment	Total	Total Incl O&P
0610	Outside key and pull	1 Carp	5	1.600	Ea.	525	58.50		583.50	665
0700	Touch bar and vertical rod, exit only		5	1.600		595	58.50		653.50	745
0710	Outside key and pull		4	2		695	73.50		768.50	875
1000	Mortise, bar, exit only		4	2		490	73.50		563.50	650
1600	Touch bar, exit only		4	2		560	73.50		633.50	730
2000	Narrow stile, rim mounted, bar, exit only		6	1.333		590	49		639	720
2010	Outside key and pull		5	1.600		635	58.50		693.50	790
2200	Bar and vertical rod, exit only		5	1.600		605	58.50		663.50	755
2210	Outside key and pull		4	2		605	73.50		678.50	780
2400	Bar and concealed rod, exit only		3	2.667		705	98		803	925
3000	Mortise, bar, exit only		4	2		515	73.50		588.50	680
3600	Touch bar, exit only		4	2		745	73.50		818.50	935

08 71 20.40 Lockset

		Crew	Daily Output	Labor-Hours	Unit	Material	Labor	Equipment	Total	Total Incl O&P
0010	**LOCKSET**, Standard duty									
0020	Non-keyed, passage	1 Carp	12	.667	Ea.	43	24.50		67.50	85.50
0100	Privacy		12	.667		54	24.50		78.50	97.50
0400	Keyed, single cylinder function **CN**		10	.800		75.50	29.50		105	129
0420	Hotel		8	1		107	36.50		143.50	175
0500	Lever handled, keyed, single cylinder function		10	.800		134	29.50		163.50	193
1000	Heavy duty with sectional trim, non-keyed, passages		12	.667		125	24.50		149.50	176
1100	Privacy		12	.667		158	24.50		182.50	212
1400	Keyed, single cylinder function		10	.800		187	29.50		216.50	252
1420	Hotel		8	1		279	36.50		315.50	360
1600	Communicating		10	.800		221	29.50		250.50	289
1690	For re-core cylinder, add					31.50			31.50	34.50
1700	Residential, interior door, minimum	1 Carp	16	.500		15.10	18.35		33.45	45
1720	Maximum		8	1		40	36.50		76.50	101
1800	Exterior, minimum		14	.571		33.50	21		54.50	69.50
1810	Average		8	1		67	36.50		103.50	131
1820	Maximum		8	1		141	36.50		177.50	213

08 71 20.41 Dead Locks

		Crew	Daily Output	Labor-Hours	Unit	Material	Labor	Equipment	Total	Total Incl O&P
0010	**DEAD LOCKS**									
0011	Mortise heavy duty outside key (security item)	1 Carp	9	.889	Ea.	133	32.50		165.50	197
0020	Double cylinder		9	.889		147	32.50		179.50	213
0100	Medium duty, outside key		10	.800		103	29.50		132.50	159
0110	Double cylinder		10	.800		129	29.50		158.50	188
1000	Tubular, standard duty, outside key		10	.800		55.50	29.50		85	107
1010	Double cylinder		10	.800		71.50	29.50		101	124
1200	Night latch, outside key		10	.800		70	29.50		99.50	123

08 71 20.42 Mortise Locksets

		Crew	Daily Output	Labor-Hours	Unit	Material	Labor	Equipment	Total	Total Incl O&P
0010	**MORTISE LOCKSETS**, Comm., wrought knobs & full escutcheon trim									
0020	Non-keyed, passage, minimum	1 Carp	9	.889	Ea.	162	32.50		194.50	229
0030	Maximum		8	1		261	36.50		297.50	345
0040	Privacy, minimum		9	.889		172	32.50		204.50	240
0050	Maximum		8	1		282	36.50		318.50	365
0100	Keyed, office/entrance/apartment, minimum		8	1		198	36.50		234.50	274
0110	Maximum		7	1.143		340	42		382	440
0120	Single cylinder, typical, minimum		8	1		169	36.50		205.50	243
0130	Maximum		7	1.143		315	42		357	410
0200	Hotel, minimum		7	1.143		204	42		246	290
0210	Maximum		6	1.333		330	49		379	440
0300	Communication, double cylinder, minimum		8	1		204	36.50		240.50	281

08 71 Door Hardware

08 71 20 – Hardware

08 71 20.42 Mortise Locksets		Crew	Daily Output	Labor-Hours	Unit	Material	2007 Bare Costs Labor	Equipment	Total	Total Incl O&P
0310	Maximum	1 Carp	7	1.143	Ea.	264	42		306	355
1000	Wrought knobs and sectional trim, non-keyed, passage, minimum		10	.800		107	29.50		136.50	164
1010	Maximum		9	.889		212	32.50		244.50	284
1040	Privacy, minimum		10	.800		125	29.50		154.50	184
1050	Maximum		9	.889		226	32.50		258.50	300
1100	Keyed, entrance, office/apartment, minimum		9	.889		186	32.50		218.50	256
1110	Maximum		8	1		269	36.50		305.50	355
1120	Single cylinder, typical, minimum		9	.889		179	32.50		211.50	248
1130	Maximum	↓	8	1	↓	261	36.50		297.50	345
2000	Cast knobs and full escutcheon trim									
2010	Non-keyed, passage, minimum	1 Carp	9	.889	Ea.	228	32.50		260.50	300
2020	Maximum		8	1		370	36.50		406.50	460
2040	Privacy, minimum		9	.889		274	32.50		306.50	350
2050	Maximum		8	1		400	36.50		436.50	495
2120	Keyed, single cylinder, typical, minimum		8	1		274	36.50		310.50	355
2130	Maximum		7	1.143		430	42		472	540
2200	Hotel, minimum		7	1.143		305	42		347	400
2210	Maximum		6	1.333		545	49		594	675
3000	Cast knob and sectional trim, non-keyed, passage, minimum		10	.800		177	29.50		206.50	241
3010	Maximum		10	.800		355	29.50		384.50	435
3040	Privacy, minimum		10	.800		202	29.50		231.50	268
3050	Maximum		10	.800		355	29.50		384.50	435
3100	Keyed, office/entrance/apartment, minimum		9	.889		229	32.50		261.50	305
3110	Maximum		9	.889		360	32.50		392.50	445
3120	Single cylinder, typical, minimum		9	.889		229	32.50		261.50	305
3130	Maximum	↓	9	.889		445	32.50		477.50	540
3190	For re-core cylinder, add					31			31	34
3800	Cipher lockset (security item)	1 Carp	13	.615	↓	720	22.50		742.50	830
3900	Keyless, pushbutton type									
4000	Residential/light commercial, deadbolt, standard	1 Carp	9	.889	Ea.	103	32.50		135.50	165
4010	Heavy duty		9	.889		123	32.50		155.50	186
4020	Industrial, heavy duty, with deadbolt		9	.889		243	32.50		275.50	320
4030	Key override		9	.889		270	32.50		302.50	350
4040	Lever activated handle		9	.889		295	32.50		327.50	375
4050	Key override		9	.889		325	32.50		357.50	410
4060	Double sided pushbutton type		8	1		540	36.50		576.50	650
4070	Key override	↓	8	1	↓	580	36.50		616.50	695

08 71 20.45 Peepholes		Crew	Daily Output	Labor-Hours	Unit	Material	Labor	Equipment	Total	Total Incl O&P
0010	**PEEPHOLES**									
2010	Peephole	1 Carp	32	.250	Ea.	14.30	9.20		23.50	30

08 71 20.50 Door Stops		Crew	Daily Output	Labor-Hours	Unit	Material	Labor	Equipment	Total	Total Incl O&P
0010	**DOOR STOPS**									
0020	Holder & bumper, floor or wall	1 Carp	32	.250	Ea.	31.50	9.20		40.70	49.50
1300	Wall bumper, 4" diameter, with rubber pad, aluminum		32	.250		9.50	9.20		18.70	25
1600	Door bumper, floor type, aluminum		32	.250		4.94	9.20		14.14	19.75
1900	Plunger type, door mounted	↓	32	.250	↓	26	9.20		35.20	43.50

08 71 20.55 Push-Pull Plates		Crew	Daily Output	Labor-Hours	Unit	Material	Labor	Equipment	Total	Total Incl O&P
0010	**PUSH-PULL PLATES**									
0100	Push plate, .050 thick, 4" x 16", aluminum	1 Carp	12	.667	Ea.	6.95	24.50		31.45	45.50
0500	Bronze		12	.667		17.65	24.50		42.15	57.50
1500	Pull handle and push bar, aluminum		11	.727		123	26.50		149.50	177
2000	Bronze	↓	10	.800	↓	159	29.50		188.50	221

08 71 Door Hardware

08 71 20 – Hardware

08 71 20.55 Push-Pull Plates		Crew	Daily Output	Labor-Hours	Unit	Material	2007 Bare Costs Labor	Equipment	Total	Total Incl O&P
3000	Push plate both sides, aluminum	1 Carp	14	.571	Ea.	15.05	21		36.05	49
3500	Bronze		13	.615		37.50	22.50		60	76
4000	Door pull, designer style, cast aluminum, minimum		12	.667		65.50	24.50		90	110
5000	Maximum		8	1		330	36.50		366.50	420
6000	Cast bronze, minimum		12	.667		77	24.50		101.50	123
7000	Maximum		8	1		360	36.50		396.50	450
8000	Walnut, minimum		12	.667		59	24.50		83.50	103
9000	Maximum		8	1		330	36.50		366.50	415

08 71 20.60 Entrance Locks										
0010	**ENTRANCE LOCKS**									
0015	Cylinder, grip handle deadlocking latch	1 Carp	9	.889	Ea.	122	32.50		154.50	185
0020	Deadbolt		8	1		148	36.50		184.50	220
0100	Push and pull plate, dead bolt		8	1		141	36.50		177.50	212
0900	For handicapped lever, add					154			154	169

08 71 20.65 Thresholds										
0010	**THRESHOLDS**									
0011	Threshold 3' long saddles aluminum	1 Carp	48	.167	L.F.	3.70	6.10		9.80	13.60
0100	Aluminum, 8" wide, 1/2" thick		12	.667	Ea.	35	24.50		59.50	76.50
0500	Bronze		60	.133	L.F.	32	4.89		36.89	43
0600	Bronze, panic threshold, 5" wide, 1/2" thick		12	.667	Ea.	60	24.50		84.50	104
0700	Rubber, 1/2" thick, 5-1/2" wide		20	.400		34	14.70		48.70	60.50
0800	2-3/4" wide		20	.400		14.75	14.70		29.45	39.50

08 71 20.70 Floor Checks										
0010	**FLOOR CHECKS**, for over 3' wide									
0020	For over 3' wide doors single acting	1 Carp	2.50	3.200	Ea.	465	117		582	700
0500	Double acting	"	2.50	3.200	"	595	117		712	840

08 71 20.75 Door Hardware Accessories										
0010	**DOOR HARDWARE ACCESSORIES**									
0050	Door closing coordinator, 36" (for paired openings up to 56")	1 Carp	8	1	Ea.	89	36.50		125.50	155
0060	48" (for paired openings up to 84")		8	1		95	36.50		131.50	161
0070	56" (for paired openings up to 96")		8	1		104	36.50		140.50	172

08 71 20.80 Hasps										
0010	**HASPS**, steel assembly									
0015	3"	1 Carp	26	.308	Ea.	2.75	11.30		14.05	20.50
0020	4-1/2"		13	.615		3.51	22.50		26.01	39
0040	6"		12.50	.640		5.30	23.50		28.80	42.50

08 71 20.90 Hinges										
0010	**HINGES** R087120-10									
0012	Full mortise, avg. freq., steel base, USP, 4-1/2" x 4-1/2"				Pr.	21.50			21.50	23.50
0100	5" x 5", USP					36			36	39.50
0200	6" x 6", USP					76.50			76.50	84.50
0400	Brass base, 4-1/2" x 4-1/2", US10					44.50			44.50	49
0500	5" x 5", US10					65			65	71.50
0600	6" x 6", US10					110			110	121
0800	Stainless steel base, 4-1/2" x 4-1/2", US32					66.50			66.50	73
0900	For non removable pin, add (security item)				Ea.	2.42			2.42	2.66
0910	For floating pin, driven tips, add					2.77			2.77	3.05
0930	For hospital type tip on pin, add					11.95			11.95	13.15
0940	For steeple type tip on pin, add					10.45			10.45	11.50
0950	Full mortise, high frequency, steel base, 3-1/2" x 3-1/2", US26D				Pr.	23			23	25.50
1000	4-1/2" x 4-1/2", USP					52.50			52.50	57.50

08 71 Door Hardware

08 71 20 – Hardware

08 71 20.90 Hinges

		Crew	Daily Output	Labor-Hours	Unit	Material	2007 Bare Costs Labor	Equipment	Total	Total Incl O&P
1100	5" x 5", USP				Pr.	49			49	54
1200	6" x 6", USP					119			119	131
1400	Brass base, 3-1/2" x 3-1/2", US4					41			41	45
1430	4-1/2" x 4-1/2", US10					70			70	77.50
1500	5" x 5", US10					105			105	115
1600	6" x 6", US10					152			152	167
1800	Stainless steel base, 4-1/2" x 4-1/2", US32					113			113	124
1810	5" x 4-1/2", US32				↓	157			157	172
1930	For hospital type tip on pin, add				Ea.	7.10			7.10	7.80
1950	Full mortise, low frequency, steel base, 3-1/2" x 3-1/2", US26D				Pr.	9.40			9.40	10.30
2000	4-1/2" x 4-1/2", USP					10.65			10.65	11.75
2100	5" x 5", USP					26.50			26.50	29
2200	6" x 6", USP					53			53	58
2300	4-1/2" x 4-1/2", US3					15.45			15.45	17
2310	5" x 5", US3					38			38	42
2400	Brass bass, 4-1/2" x 4-1/2", US10					37			37	41
2500	5" x 5", US10					56.50			56.50	62
2800	Stainless steel base, 4-1/2" x 4-1/2", US32				↓	64			64	70.50

08 71 20.91 Special Hinges

		Crew	Daily Output	Labor-Hours	Unit	Material	Labor	Equipment	Total	Total Incl O&P
0010	**SPECIAL HINGES**									
0015	Paumelle, high frequency									
0020	Steel base, 6" x 4-1/2", US10				Pr.	136			136	150
0100	Bronze base, 5" x 4-1/2", US10					172			172	189
0200	Paumelle, average frequency, steel base, 4-1/2" x 3-1/2", US10					92.50			92.50	102
0400	Olive knuckle, low frequency, brass base, 6" x 4-1/2", US10				↓	157			157	173
1000	Electric hinge with concealed conductor, average frequency									
1010	Steel base, 4-1/2" x 4-1/2", US26D				Pr.	291			291	320
1100	Bronze base, 4-1/2" x 4-1/2", US26D				"	305			305	335
1200	Electric hinge with concealed conductor, high frequency									
1210	Steel base, 4-1/2" x 4-1/2", US26D				Pr.	217			217	239
1600	Double weight, 800 lb., steel base, removable pin, 5" x 6", USP					125			125	138
1700	Steel base-welded pin, 5" x 6", USP					142			142	156
1800	Triple weight, 2000 lb., steel base, welded pin, 5" x 6", USP					144			144	158
2000	Pivot reinf., high frequency, steel base, 7-3/4" door plate, USP					169			169	186
2200	Bronze base, 7-3/4" door plate, US10				↓	199			199	219
3000	Swing clear, full mortise, full or half surface, high frequency,									
3010	Steel base, 5" high, USP				Pr.	145			145	160
3200	Swing clear, full mortise, average frequency									
3210	Steel base, 4-1/2" high, USP				Pr.	115			115	127
4000	Wide throw, average frequency, steel base, 4-1/2" x 6", USP					88			88	97
4200	High frequency, steel base, 4-1/2" x 6", USP				↓	135			135	148
4600	Spring hinge, single acting, 6" flange, steel				Ea.	49.50			49.50	54.50
4700	Brass					87			87	95.50
4900	Double acting, 6" flange, steel					86.50			86.50	95.50
4950	Brass				↓	142			142	156
9000	Continuous hinge, steel, full mortise, heavy duty	2 Carp	64	.250	L.F.	11.25	9.20		20.45	26.50

08 71 20.95 Kick Plates

		Crew	Daily Output	Labor-Hours	Unit	Material	Labor	Equipment	Total	Total Incl O&P
0010	**KICK PLATES**									
0020	Stainless steel	1 Carp	15	.533	Ea.	28	19.55		47.55	61.50
0500	Bronze		15	.533		35.50	19.55		55.05	69.50
2000	Aluminum, .050, with 3 beveled edges, 10" x 28"		15	.533		20	19.55		39.55	52.50
2010	10" x 30"	↓	15	.533	↓	22	19.55		41.55	54.50

08 71 Door Hardware

08 71 20 – Hardware

08 71 20.95 Kick Plates		Crew	Daily Output	Labor-Hours	Unit	Material	2007 Bare Costs Labor	Equipment	Total	Total Incl O&P
2020	10" x 34"	1 Carp	15	.533	Ea.	23	19.55		42.55	55.50
2040	10" x 38"	↓	15	.533	↓	25.50	19.55		45.05	58.50

08 71 21 – Astragals

08 71 21.10 Astragals

		Crew	Daily Output	Labor-Hours	Unit	Material	Labor	Equipment	Total	Total Incl O&P
0010	**ASTRAGALS**									
0400	One piece, overlapping cadmium plated steel, flat, 3/16" x 2"	1 Carp	90	.089	L.F.	3.10	3.26		6.36	8.50
0600	Prime coated steel, flat, 1/8" x 3"		90	.089		4.31	3.26		7.57	9.85
0800	Stainless steel, flat, 3/32" x 1-5/8"		90	.089		16.15	3.26		19.41	23
1000	Aluminum, flat, 1/8" x 2"		90	.089		3.11	3.26		6.37	8.50
1200	Nail on, "T" extrusion		120	.067		.69	2.45		3.14	4.57
1300	Vinyl bulb insert		105	.076		1.12	2.80		3.92	5.60
1600	Screw on, "T" extrusion		90	.089		4.53	3.26		7.79	10.10
1700	Vinyl insert		75	.107		3.06	3.91		6.97	9.45
2000	"L" extrusion, neoprene bulbs		75	.107		1.76	3.91		5.67	8.05
2100	Neoprene sponge insert		75	.107		5.35	3.91		9.26	12
2200	Magnetic		75	.107		8.85	3.91		12.76	15.80
2400	Spring hinged security seal, with cam		75	.107		5.70	3.91		9.61	12.35
2600	Spring loaded locking bolt, vinyl insert		45	.178		7.75	6.50		14.25	18.70
2800	Neoprene sponge strip, "Z" shaped, aluminum		60	.133		3.68	4.89		8.57	11.65
2900	Solid neoprene strip, nail on aluminum strip	↓	90	.089	↓	3.09	3.26		6.35	8.50
3000	One piece stile protection									
3020	Neoprene fabric loop, nail on aluminum strips	1 Carp	60	.133	L.F.	.55	4.89		5.44	8.20
3110	Flush mounted aluminum extrusion, 1/2" x 1-1/4"		60	.133		2.93	4.89		7.82	10.80
3140	3/4" x 1-3/8"		60	.133		3.72	4.89		8.61	11.70
3160	1-1/8" x 1-3/4"		60	.133		5.95	4.89		10.84	14.15
3300	Mortise, 9/16" x 3/4"		60	.133		3.15	4.89		8.04	11.05
3320	13/16" x 1-3/8"		60	.133		3.42	4.89		8.31	11.35
3600	Spring bronze strip, nail on type		105	.076		2.69	2.80		5.49	7.30
3620	Screw on, with retainer		75	.107		2.20	3.91		6.11	8.50
3800	Flexible stainless steel housing, pile insert, 1/2" door		105	.076		6.10	2.80		8.90	11.05
3820	3/4" door		105	.076		6.85	2.80		9.65	11.90
4000	Extruded aluminum retainer, flush mount, pile insert		105	.076		2.13	2.80		4.93	6.70
4080	Mortise, felt insert		90	.089		3.79	3.26		7.05	9.25
4160	Mortise with spring, pile insert		90	.089		2.96	3.26		6.22	8.35
4400	Rigid vinyl retainer, mortise, pile insert		105	.076		2.12	2.80		4.92	6.70
4600	Wool pile filler strip, aluminum backing	↓	105	.076	↓	2.13	2.80		4.93	6.70
5000	Two piece overlapping astragal, extruded aluminum retainer									
5010	Pile insert	1 Carp	60	.133	L.F.	2.76	4.89		7.65	10.65
5020	Vinyl bulb insert		60	.133		1.78	4.89		6.67	9.55
5040	Vinyl flap insert		60	.133		5.55	4.89		10.44	13.70
5060	Solid neoprene flap insert		60	.133		5.50	4.89		10.39	13.65
5080	Hypalon rubber flap insert		60	.133		5.60	4.89		10.49	13.75
5090	Snap on cover, pile insert		60	.133		6.40	4.89		11.29	14.65
5400	Magnetic aluminum, surface mounted		60	.133		21.50	4.89		26.39	31.50
5500	Interlocking aluminum, 5/8" x 1" neoprene bulb insert		45	.178		3.44	6.50		9.94	13.95
5600	Adjustable aluminum, 9/16" x 21/32", pile insert	↓	45	.178		16.30	6.50		22.80	28
5790	For vinyl bulb, deduct					.42			.42	.46
5800	Magnetic, adjustable, 9/16" x 21/32"	1 Carp	45	.178	↓	21	6.50		27.50	33
6000	Two piece stile protection									
6010	Cloth backed rubber loop, 1" gap, nail on aluminum strips	1 Carp	45	.178	L.F.	3.52	6.50		10.02	14
6040	Screw on aluminum strips		45	.178		5.50	6.50		12	16.20
6100	1-1/2" gap, screw on aluminum extrusion	↓	45	.178	↓	4.93	6.50		11.43	15.55

08 71 Door Hardware

08 71 21 - Astragals

	08 71 21.10 Astragals	Crew	Daily Output	Labor-Hours	Unit	Material	2007 Bare Costs Labor	Equipment	Total	Total Incl O&P
6240	Vinyl fabric loop, slotted aluminum extrusion, 1" gap	1 Carp	45	.178	L.F.	1.74	6.50		8.24	12.05
6300	1-1/4" gap		45	.178		5.20	6.50		11.70	15.85

08 71 25 - Weatherstripping

08 71 25.10 Weatherstripping

		Crew	Daily Output	Labor-Hours	Unit	Material	2007 Bare Costs Labor	Equipment	Total	Total Incl O&P
0010	**WEATHERSTRIPPING**									
1000	Doors, wood frame, interlocking, for 3' x 7' door, zinc	1 Carp	3	2.667	Opng.	13.95	98		111.95	167
1100	Bronze		3	2.667		22	98		120	176
1300	6' x 7' opening, zinc		2	4		15.25	147		162.25	246
1400	Bronze		2	4		29	147		176	261
1700	Wood frame, spring type, bronze									
1800	3' x 7' door	1 Carp	7.60	1.053	Opng.	17.90	38.50		56.40	79.50
1900	6' x 7' door	"	7	1.143	"	21.50	42		63.50	89
2200	Metal frame, spring type, bronze									
2300	3' x 7' door	1 Carp	3	2.667	Opng.	30	98		128	185
2400	6' x 7' door	"	2.50	3.200	"	41.50	117		158.50	229
2500	For stainless steel, spring type, add					133%				
2700	Metal frame, extruded sections, 3' x 7' door, aluminum	1 Carp	2	4	Opng.	40.50	147		187.50	274
2800	Bronze		2	4		102	147		249	340
3100	6' x 7' door, aluminum		1.20	6.667		51.50	245		296.50	435
3200	Bronze		1.20	6.667		121	245		366	515
3500	Threshold weatherstripping									
3650	Door sweep, flush mounted, aluminum	1 Carp	25	.320	Ea.	12	11.75		23.75	31.50
3700	Vinyl		25	.320		14.20	11.75		25.95	34
5000	Garage door bottom weatherstrip, 12' aluminum, clear		14	.571		19.20	21		40.20	53.50
5010	Bronze		14	.571		73	21		94	113
5050	Bottom protection, 12' aluminum, clear		14	.571		22	21		43	56.50
5100	Bronze		14	.571		90.50	21		111.50	132

08 74 Access Control Hardware

08 74 13 - Card Key Access Control Hardware

08 74 13.50 Card Key Access

		Crew	Daily Output	Labor-Hours	Unit	Material	2007 Bare Costs Labor	Equipment	Total	Total Incl O&P
0010	**CARD KEY ACCESS**									
0020	Card type, 1 time zone, minimum				Ea.	330			330	360
0040	Maximum					1,075			1,075	1,175
0060	3 time zones, minimum					785			785	865
0080	Maximum					1,850			1,850	2,050
0100	System with printer, and control console, 3 zones				Total	9,075			9,075	9,975
0120	6 zones				"	11,900			11,900	13,100
0140	For each door, minimum, add				Ea.	1,325			1,325	1,450
0160	Maximum, add				"	1,975			1,975	2,175

08 74 19 - Biometric Identity Access Control Hardware

08 74 19.50 Biometric Identity Access

		Crew	Daily Output	Labor-Hours	Unit	Material	2007 Bare Costs Labor	Equipment	Total	Total Incl O&P
0010	**BIOMETRIC IDENTITY ACCESS**									
0220	Hand geometry scanner, mem of 512 users, excl striker/powr	1 Elec	3	2.667	Ea.	1,725	117		1,842	2,075
0230	Memory upgrade for, adds 9,700 user profiles		8	1		225	44		269	315
0240	Adds 32,500 user profiles		8	1		525	44		569	645
0250	Prison type, memory of 256 users, excl striker, power		3	2.667		2,175	117		2,292	2,550
0260	Memory upgrade for, adds 3,300 user profiles		8	1		180	44		224	264
0270	Adds 9,700 user profiles		8	1		360	44		404	460
0280	Adds 27,900 user profiles		8	1		505	44		549	620

08 74 Access Control Hardware

08 74 19 – Biometric Identity Access Control Hardware

08 74 19.50 Biometric Identity Access		Crew	Daily Output	Labor-Hours	Unit	Material	2007 Bare Costs Labor	Equipment	Total	Total Incl O&P
0290	All weather, mem of 512 users, excl striker/pwr	1 Elec	3	2.667	Ea.	3,225	117		3,342	3,725
0300	Facial & fingerprint scanner, combination unit, excl striker/power	↓	3	2.667		4,200	117		4,317	4,800
0310	Access for, for initial setup, excl striker/power		3	2.667	↓	1,000	117		1,117	1,275

08 75 Window Hardware

08 75 30 – Weatherstripping

08 75 30.10 Weatherstripping

		Crew	Daily Output	Labor-Hours	Unit	Material	Labor	Equipment	Total	Total Incl O&P
0010	**WEATHERSTRIPPING**, Window, double hung, 3' X 5'									
0020	Zinc	1 Carp	7.20	1.111	Opng.	12.05	41		53.05	77
0100	Bronze		7.20	1.111		24	41		65	90
0500	As above but heavy duty, zinc	↓	4.60	1.739		15.55	64		79.55	117
0600	Bronze	↓	4.60	1.739	↓	27	64		91	130
9000	Minimum labor/equipment charge	1 Clab	4.60	1.739	Job		50		50	78

08 79 Hardware Accessories

08 79 13 – Key Storage Equipment

08 79 13.10 Key Cabinets

		Crew	Daily Output	Labor-Hours	Unit	Material	Labor	Equipment	Total	Total Incl O&P
0010	**KEY CABINETS**									
0020	Wall mounted, 60 key capacity	1 Carp	20	.400	Ea.	67.50	14.70		82.20	97.50
0200	Drawer type, 600 key capacity	1 Clab	15	.533		725	15.35		740.35	825
0300	2,400 key capacity		20	.400		2,600	11.50		2,611.50	2,875
0400	Tray type, 20 key capacity		50	.160		37.50	4.60		42.10	48.50
0500	50 key capacity	↓	40	.200	↓	75	5.75		80.75	91.50

08 79 20 – Door Accessories

08 79 20.10 Door Accessories

		Crew	Daily Output	Labor-Hours	Unit	Material	Labor	Equipment	Total	Total Incl O&P
0010	**DOOR ACCESSORIES**									
0140	Door bolt, surface, 4"	1 Carp	32	.250	Ea.	8.45	9.20		17.65	23.50
0160	Door latch	"	12	.667	"	6.50	24.50		31	45
0200	Sliding closet door									
0220	Track and hanger, single	1 Carp	10	.800	Ea.	46	29.50		75.50	96
0240	Double		8	1		65	36.50		101.50	129
0260	Door guide, single		48	.167		21.50	6.10		27.60	33
0280	Double		48	.167		29	6.10		35.10	41.50
0600	Deadbolt and lock cover plate, brass or stainless steel		30	.267		24	9.80		33.80	42
0620	Hole cover plate, brass or chrome		35	.229		6.20	8.40		14.60	19.90
2240	Mortise lockset, passage, lever handle		9	.889		174	32.50		206.50	242
4000	Security chain, standard	↓	18	.444	↓	6.85	16.30		23.15	33

08 81 Glass Glazing

08 81 10 – Float Glass

08 81 10.10 Float Glass

		Crew	Daily Output	Labor-Hours	Unit	Material	2007 Bare Costs Labor	Equipment	Total	Total Incl O&P
0010	**FLOAT GLASS**, 3/16" thick R088110-10									
0020	3/16" Plain	2 Glaz	130	.123	S.F.	4.87	4.44		9.31	12.05
0200	Tempered, clear		130	.123		5.70	4.44		10.14	13
0300	Tinted		130	.123		7.25	4.44		11.69	14.70
0600	1/4" thick, clear, plain **CN**		120	.133		5.95	4.81		10.76	13.85
0700	Tinted		120	.133		5.60	4.81		10.41	13.45
0800	Tempered, clear		120	.133		7.10	4.81		11.91	15.10
0900	Tinted		120	.133		10.10	4.81		14.91	18.40
1600	3/8" thick, clear, plain		75	.213		9.60	7.70		17.30	22
1700	Tinted		75	.213		11.20	7.70		18.90	24
1800	Tempered, clear		75	.213		14.30	7.70		22	27.50
1900	Tinted		75	.213		18.10	7.70		25.80	31.50
2200	1/2" thick, clear, plain		55	.291		18.95	10.50		29.45	37
2300	Tinted		55	.291		19.65	10.50		30.15	37.50
2400	Tempered, clear		55	.291		21.50	10.50		32	39.50
2500	Tinted		55	.291		27.50	10.50		38	46
2800	5/8" thick, clear, plain		45	.356		19.65	12.80		32.45	41
2900	Tempered, clear		45	.356		22.50	12.80		35.30	44
3200	3/4" thick, clear, plain		35	.457		25.50	16.50		42	53
3300	Tempered, clear		35	.457		29.50	16.50		46	57.50
3600	1" thick, clear, plain		30	.533		42	19.25		61.25	75.50
8900	For low emissivity coating for 3/16" & 1/4" only, add to above					15%				

08 81 13 – Decorative Glass Glazing

08 81 13.10 Beveled Glass

		Crew	Daily Output	Labor-Hours	Unit	Material	Labor	Equipment	Total	Total Incl O&P
0010	**BEVELED GLASS**, with design patterns									
0020	Minimum	2 Glaz	150	.107	S.F.	41.50	3.85		45.35	52
0050	Average		125	.128		98	4.61		102.61	115
0100	Maximum		100	.160		173	5.75		178.75	199

08 81 13.20 Faceted Glass

		Crew	Daily Output	Labor-Hours	Unit	Material	Labor	Equipment	Total	Total Incl O&P
0010	**FACETED GLASS**, Color tinted 3/4" thick									
0020	Minimum	2 Glaz	95	.168	S.F.	35.50	6.05		41.55	48
0100	Maximum	"	75	.213	"	61.50	7.70		69.20	79.50

08 81 13.30 Sandblasted Glass

		Crew	Daily Output	Labor-Hours	Unit	Material	Labor	Equipment	Total	Total Incl O&P
0010	**SANDBLASTED GLASS**, float glass									
0020	1/8" thick	2 Glaz	160	.100	S.F.	7.80	3.61		11.41	14
0100	3/16" thick		130	.123		8.60	4.44		13.04	16.20
0500	1/4" thick		120	.133		8.95	4.81		13.76	17.15
0600	3/8" thick		75	.213		9.75	7.70		17.45	22.50

08 81 20 – Vision Panels

08 81 20.10 Full Vision

		Crew	Daily Output	Labor-Hours	Unit	Material	Labor	Equipment	Total	Total Incl O&P
0010	**FULL VISION**, window system with 3/4" glass mullions									
0020	Up to 10' high	H-2	130	.185	S.F.	57	6.20		63.20	72.50
0100	10' to 20' high, minimum		110	.218		60.50	7.35		67.85	77.50
0150	Average		100	.240		65.50	8.05		73.55	84.50
0200	Maximum		80	.300		73.50	10.10		83.60	96.50

08 81 25 – Glazing Variables

08 81 25.10 Glazing Variables

		Crew	Daily Output	Labor-Hours	Unit	Material	Labor	Equipment	Total	Total Incl O&P
0010	**GLAZING VARIABLES** R088110-10									
0500	For high rise glazing, exterior, add per S.F. per story				S.F.					.12
0600	For glass replacement, add				"		100%			

08 81 Glass Glazing

08 81 25 – Glazing Variables

08 81 25.10 Glazing Variables

		Crew	Daily Output	Labor-Hours	Unit	Material	2007 Bare Costs Labor	Equipment	Total	Total Incl O&P
0700	For gasket settings, add				L.F.	4.20			4.20	4.62
0900	For sloped glazing, add				S.F.		25%			
2000	Fabrication, polished edges, 1/4" thick				Inch	.36			.36	.40
2100	1/2" thick					.92			.92	1.01
2500	Mitered edges, 1/4" thick					.92			.92	1.01
2600	1/2" thick					1.48			1.48	1.63

08 81 30 – Insulating Glass

08 81 30.10 Insulating Glass

			Crew	Daily Output	Labor-Hours	Unit	Material	Labor	Equipment	Total	Total Incl O&P
0010	**INSULATING GLASS**, 2 lites 1/8" float, 1/2" thk under 15 S.F.	R088110-10									
0020	Clear		2 Glaz	95	.168	S.F.	8.60	6.05		14.65	18.65
0100	Tinted			95	.168		12.70	6.05		18.75	23
0200	2 lites 3/16" float, for 5/8" thk unit, 15 to 30 S.F., clear			90	.178		9.80	6.40		16.20	20.50
0300	Tinted			90	.178		10.10	6.40		16.50	21
0400	1" thk, dbl. glazed, 1/4" float, 30-70 S.F., clear	CN		75	.213		14.45	7.70		22.15	27.50
0500	Tinted			75	.213		16.85	7.70		24.55	30
0600	1" thick double glazed, 1/4" float, 1/4" wire			75	.213		22	7.70		29.70	35.50
0700	1/4" float, 1/4" tempered			75	.213		21	7.70		28.70	35
0800	1/4" wire, 1/4" tempered			75	.213		29.50	7.70		37.20	44
0900	Both lites, 1/4" wire			75	.213		26.50	7.70		34.20	41
2000	Both lites, light & heat reflective			85	.188		22.50	6.80		29.30	35
2500	Heat reflective, film inside, 1" thick unit, clear			85	.188		19.65	6.80		26.45	32
2600	Tinted			85	.188		21	6.80		27.80	34
3000	Film on weatherside, clear, 1/2" thick unit			95	.168		14.05	6.05		20.10	24.50
3100	5/8" thick unit			90	.178		16.70	6.40		23.10	28
3200	1" thick unit			85	.188		19.35	6.80		26.15	32

08 81 35 – Translucent Glass

08 81 35.10 Obscure Glass

		Crew	Daily Output	Labor-Hours	Unit	Material	Labor	Equipment	Total	Total Incl O&P
0010	**OBSCURE GLASS**, 1/8" thick									
0020	Minimum	2 Glaz	140	.114	S.F.	7.90	4.12		12.02	14.95
0100	Maximum		125	.128		9.50	4.61		14.11	17.45
0300	7/32" thick, minimum		120	.133		8.80	4.81		13.61	17
0400	Maximum		105	.152		11.20	5.50		16.70	20.50

08 81 35.20 Patterned Glass

		Crew	Daily Output	Labor-Hours	Unit	Material	Labor	Equipment	Total	Total Incl O&P
0010	**PATTERNED GLASS**, colored, 1/8" thick									
0020	Minimum	2 Glaz	140	.114	S.F.	8.90	4.12		13.02	16
0100	Maximum		125	.128		10.90	4.61		15.51	19
0300	7/32" thick, minimum		120	.133		10.55	4.81		15.36	18.90
0400	Maximum		105	.152		11.70	5.50		17.20	21

08 81 45 – Sheet Glass

08 81 45.10 Sheet Glass

		Crew	Daily Output	Labor-Hours	Unit	Material	Labor	Equipment	Total	Total Incl O&P
0010	**SHEET GLASS**, gray									
0020	1/8" thick	2 Glaz	160	.100	S.F.	4.26	3.61		7.87	10.15
0200	1/4" thick	"	130	.123	"	5.55	4.44		9.99	12.85

08 81 50 – Spandrel Glass

08 81 50.10 Spandrel Glass

		Crew	Daily Output	Labor-Hours	Unit	Material	Labor	Equipment	Total	Total Incl O&P
0010	**SPANDREL GLASS**, 1/4" thick standard colors									
0020	Up to 100 S.F.	2 Glaz	110	.145	S.F.	13.65	5.25		18.90	23
0200	1,000 to 2,000 S.F.	"	120	.133	"	12.80	4.81		17.61	21.50
0300	For custom colors, add				Total	10%				
0500	For 3/8" thick, add				S.F.	8.05			8.05	8.85

08 81 Glass Glazing

08 81 50 – Spandrel Glass

08 81 50.10 Spandrel Glass		Crew	Daily Output	Labor-Hours	Unit	Material	2007 Bare Costs Labor	Equipment	Total	Total Incl O&P
1000	For double coated, 1/4" thick, add				S.F.	3.01			3.01	3.31
1200	For insulation on panels, add					4.97			4.97	5.45
2000	Panels, insulated, with aluminum backed fiberglass, 1" thick	2 Glaz	120	.133		12	4.81		16.81	20.50
2100	2" thick	"	120	.133		13.95	4.81		18.76	22.50
2500	With galvanized steel backing, add					4.15			4.15	4.57

08 81 55 – Window Glass

08 81 55.10 Window Glass

		Crew	Daily Output	Labor-Hours	Unit	Material	Labor	Equipment	Total	Total Incl O&P
0010	**WINDOW GLASS**, clear float, stops, putty bed									
0015	1/8" thick, clear float	2 Glaz	480	.033	S.F.	3.94	1.20		5.14	6.15
0500	3/16" thick, clear		480	.033		4.86	1.20		6.06	7.15
0600	Tinted		480	.033		5.40	1.20		6.60	7.75
0700	Tempered		480	.033		6.55	1.20		7.75	9

08 81 65 – Wire Glass

08 81 65.10 Wire Glass

		Crew	Daily Output	Labor-Hours	Unit	Material	Labor	Equipment	Total	Total Incl O&P
0010	**WIRE GLASS**, Chicken wire									
0012	1/4" thick rough obscure	2 Glaz	135	.119	S.F.	13	4.27		17.27	21
1000	Polished wire, 1/4" thick, diamond, clear		135	.119		16.95	4.27		21.22	25
1500	Pinstripe, obscure		135	.119		15.95	4.27		20.22	24

08 83 Mirrors

08 83 13 – Mirrored Glass Glazing

08 83 13.10 Mirrors

		Crew	Daily Output	Labor-Hours	Unit	Material	Labor	Equipment	Total	Total Incl O&P
0010	**MIRRORS**, No frames, wall type, 1/4" plate glass, polished edge									
0100	Up to 5 S.F.	2 Glaz	125	.128	S.F.	6.90	4.61		11.51	14.55
0200	Over 5 S.F.		160	.100		6.65	3.61		10.26	12.80
0500	Door type, 1/4" plate glass, up to 12 S.F.		160	.100		7.30	3.61		10.91	13.45
1000	Float glass, up to 10 S.F., 1/8" thick		160	.100		4.41	3.61		8.02	10.30
1100	3/16" thick		150	.107		5.15	3.85		9	11.45
1500	12" x 12" wall tiles, square edge, clear		195	.082		1.66	2.96		4.62	6.30
1600	Veined		195	.082		4.49	2.96		7.45	9.40
2000	1/4" thick, stock sizes, one way transparent		125	.128		16.15	4.61		20.76	25
2010	Bathroom, unframed, laminated		160	.100		12	3.61		15.61	18.65

08 83 13.15 Reflective Glass

		Crew	Daily Output	Labor-Hours	Unit	Material	Labor	Equipment	Total	Total Incl O&P
0010	**REFLECTIVE GLASS**									
0100	1/4" float with fused metallic oxide fixed	2 Glaz	115	.139	S.F.	11.70	5		16.70	20.50
0500	1/4" float glass with reflective applied coating	"	115	.139	"	9.75	5		14.75	18.30

08 84 Plastic Glazing

08 84 10 – Plastic Glazing

08 84 10.10 Plexiglass Acrylic

		Crew	Daily Output	Labor-Hours	Unit	Material	Labor	Equipment	Total	Total Incl O&P
0010	**PLEXIGLASS ACRYLIC**, clear, masked,									
0020	1/8" thick, cut sheets	2 Glaz	170	.094	S.F.	3.55	3.39		6.94	9.05
0200	Full sheets		195	.082		1.85	2.96		4.81	6.50
0500	1/4" thick, cut sheets		165	.097		6.25	3.50		9.75	12.20
0600	Full sheets		185	.086		3.41	3.12		6.53	8.45
0900	3/8" thick, cut sheets		155	.103		11.50	3.72		15.22	18.30
1000	Full sheets		180	.089		6.20	3.20		9.40	11.65
1300	1/2" thick, cut sheets		135	.119		13.25	4.27		17.52	21

08 84 Plastic Glazing

08 84 10 – Plastic Glazing

08 84 10.10 Plexiglass Acrylic

		Crew	Daily Output	Labor-Hours	Unit	Material	2007 Bare Costs Labor	Equipment	Total	Total Incl O&P
1400	Full sheets	2 Glaz	150	.107	S.F.	12.80	3.85		16.65	19.90
1700	3/4" thick, cut sheets		115	.139		47	5		52	59
1800	Full sheets		130	.123		27	4.44		31.44	36.50
2100	1" thick, cut sheets		105	.152		53	5.50		58.50	66.50
2200	Full sheets		125	.128		32.50	4.61		37.11	43
3000	Colored, 1/8" thick, cut sheets		170	.094		10.90	3.39		14.29	17.15
3200	Full sheets		195	.082		7.05	2.96		10.01	12.25
3500	1/4" thick, cut sheets		165	.097		12.20	3.50		15.70	18.75
3600	Full sheets		185	.086		8.35	3.12		11.47	13.85
4000	Mirrors, untinted, cut sheets, 1/8" thick		185	.086		5.15	3.12		8.27	10.35
4200	1/4" thick		180	.089		8.25	3.20		11.45	13.95

08 84 20 – Polycarbonate

08 84 20.10 Polycarbonate

0010	**POLYCARBONATE**, clear, masked, cut sheets									
0020	1/8" thick	2 Glaz	170	.094	S.F.	6.15	3.39		9.54	11.95
0500	3/16" thick		165	.097		7.45	3.50		10.95	13.50
1000	1/4" thick		155	.103		8.25	3.72		11.97	14.70
1500	3/8" thick		150	.107		15.15	3.85		19	22.50

08 84 30 – Vinyl Glass

08 84 30.10 Vinyl Glass

0010	**VINYL GLASS**, Steel mesh reinforced, stock sizes									
0020	.090" thick	2 Glaz	170	.094	S.F.	8.35	3.39		11.74	14.35
0500	.120" thick		170	.094		10.55	3.39		13.94	16.80
1000	.250" thick		155	.103		14.75	3.72		18.47	22
1500	For non-standard sizes, add					15%				

08 87 Glazing Surface Films

08 87 13 – Solar Control Films

08 87 13.10 Solar Films On Glass

0010	**SOLAR FILMS ON GLASS**									
2000	Solar film on glass, not including glass, minimum	2 Glaz	180	.089	S.F.	4.71	3.20		7.91	10.05
2050	Solar film on glass, not including glass, maximum	"	225	.071	"	10.90	2.56		13.46	15.90

08 87 16 – Safety Films

08 87 16.10 Window Protection Film

0010	**WINDOW PROTECTION FILM**									
2010	Window protection film, controls blast damage	2 Glaz	80	.200	S.F.	4.71	7.20		11.91	16.10

08 87 53 – Security Films

08 87 53.10 Security Film

0010	**SECURITY FILM**, clear, 32000psi tensile strength, adhered to glass									
0100	.002" thick, daylight installation	H-2	950	.025	S.F.	.85	.85		1.70	2.24
0150	.004" thick, daylight installation	R088110-10	800	.030		1.50	1.01		2.51	3.19
0200	.006" thick, daylight installation		700	.034		1.60	1.15		2.75	3.52
0210	Install for anchorage		600	.040		1.78	1.34		3.12	4.01
0400	.007" thick, daylight istallation		600	.040		1.70	1.34		3.04	3.92
0410	Install for anchorage		500	.048		1.89	1.61		3.50	4.54
0500	.008" thick, daylight installation		500	.048		2	1.61		3.61	4.66
0510	Install for anchorage		500	.048		2.22	1.61		3.83	4.90
0600	.015" thick, daylight installation		400	.060		3.10	2.02		5.12	6.50
0610	Install for anchorage		400	.060		2.22	2.02		4.24	5.50

08 87 Glazing Surface Films

08 87 53 – Security Films

08 87 53.10 Security Film		Crew	Daily Output	Labor-Hours	Unit	Material	2007 Bare Costs Labor	Equipment	Total	Total Incl O&P
0900	Security Film Anchorage, mechanical attachment and cover plate	H-3	370	.043	L.F.	7.30	1.37		8.67	10.15
0950	Security film anchorage, wet glaze structural caulking	1 Glaz	225	.036	"	.80	1.28		2.08	2.82
1000	Adhered security film removal	1 Clab	275	.029	S.F.		.84		.84	1.30

08 88 Special Function Glazing

08 88 40 – Acoustical Glass Units

08 88 40.10 Acoustical Glass Units

		Crew	Daily Output	Labor-Hours	Unit	Material	Labor	Equipment	Total	Total Incl O&P
0010	**ACOUSTICAL GLASS UNITS**, 1 lite at 3/8", 1 lite at 3/16"									
0020	For 1" thick	2 Glaz	100	.160	S.F.	24.50	5.75		30.25	36
0100	For 4" thick	"	80	.200	"	41.50	7.20		48.70	57

08 88 56 – Ballistics-Resistant Glazing

08 88 56.10 Laminated Glass

		Crew	Daily Output	Labor-Hours	Unit	Material	Labor	Equipment	Total	Total Incl O&P
0010	**LAMINATED GLASS**									
0020	Clear float .03" vinyl 1/4"	2 Glaz	90	.178	S.F.	10	6.40		16.40	20.50
0100	3/8" thick		78	.205		15.35	7.40		22.75	28
0200	.06" vinyl, 1/2" thick		65	.246		17.95	8.85		26.80	33
1000	5/8" thick		90	.178		21	6.40		27.40	32.50
2000	Bullet-resisting, 1-3/16" thick, to 15 S.F.		16	1		65	36		101	126
2100	Over 15 S.F.		16	1		61	36		97	122
2500	2-1/4" thick, to 15 S.F.		12	1.333		69	48		117	149
2600	Over 15 S.F.		12	1.333		61	48		109	141

08 91 Louvers

08 91 19 – Fixed Louvers

08 91 19.10 Louvers

		Crew	Daily Output	Labor-Hours	Unit	Material	Labor	Equipment	Total	Total Incl O&P
0010	**LOUVERS**									
0020	Aluminum with screen, residential, 8" x 8"	1 Carp	38	.211	Ea.	10.20	7.75		17.95	23.50
0100	12" x 12"		38	.211		11.50	7.75		19.25	24.50
0200	12" x 18"		35	.229		13	8.40		21.40	27.50
0250	14" x 24"		30	.267		19.50	9.80		29.30	37
0300	18" x 24"		27	.296		24	10.85		34.85	43.50
0500	24" x 30"		24	.333		33	12.25		45.25	55.50
0700	Triangle, adjustable, small		20	.400		28.50	14.70		43.20	54
0800	Large		15	.533		49	19.55		68.55	84
1200	Extruded aluminum, see division 23 37 15.40									
2100	Midget, aluminum, 3/4" deep, 1" diameter	1 Carp	85	.094	Ea.	.76	3.45		4.21	6.25
2150	3" diameter		60	.133		1.62	4.89		6.51	9.40
2200	4" diameter		50	.160		2.98	5.85		8.83	12.45
2250	6" diameter		30	.267		3.52	9.80		13.32	19.10

08 95 Vents

08 95 13 – Soffit Vents

08 95 13.10 Wall Louvers

		Crew	Daily Output	Labor-Hours	Unit	Material	2007 Bare Costs Labor	Equipment	Total	Total Incl O&P
0010	**WALL LOUVERS**									
2330	Soffit vent, continuous, 3" wide, aluminum, mill finish	1 Carp	200	.040	L.F.	.51	1.47		1.98	2.85
2340	Baked enamel finish		200	.040	"	.46	1.47		1.93	2.80
2400	Under eaves vent, aluminum, mill finish, 16" x 4"		48	.167	Ea.	1.94	6.10		8.04	11.70
2500	16" x 8"		48	.167	"	2.04	6.10		8.14	11.80

08 95 16 – Wall Vents

08 95 16.10 Louvers

		Crew	Daily Output	Labor-Hours	Unit	Material	2007 Bare Costs Labor	Equipment	Total	Total Incl O&P
0010	**LOUVERS**									
0020	Redwood, 2'-0" diameter, full circle	1 Carp	16	.500	Ea.	144	18.35		162.35	188
0100	Half circle		16	.500		138	18.35		156.35	181
0200	Octagonal		16	.500		110	18.35		128.35	150
0300	Triangular, 5/12 pitch, 5'-0" at base		16	.500		233	18.35		251.35	286
7000	Vinyl gable vent, 8" x 8"		38	.211		10.90	7.75		18.65	24
7020	12" x 12"		38	.211		22.50	7.75		30.25	37
7080	12" x 18"		35	.229		29	8.40		37.40	45
7200	18" x 24"		30	.267		35	9.80		44.80	54

Division 9 - Finishes

Estimating Tips

General

- Room Finish Schedule: A complete set of plans should contain a room finish schedule. If one is not available, it would be well worth the time and effort to obtain one.

09 20 00 Plaster and Gypsum Board

- Lath is estimated by the square yard plus a 5% allowance for waste. Furring, channels, and accessories are measured by the linear foot. An extra foot should be allowed for each accessory miter or stop.
- Plaster is also estimated by the square yard. Deductions for openings vary by preference, from zero deduction to 50% of all openings over 2 feet in width. The estimator should allow one extra square foot for each linear foot of horizontal interior or exterior angle located below the ceiling level. Also, double the areas of small radius work.
- Drywall accessories, studs, track, and acoustical caulking are all measured by the linear foot. Drywall taping is figured by the square foot. Gypsum wallboard is estimated by the square foot. No material deductions should be made for door or window openings under 32 S.F.

09 60 00 Flooring

- Tile and terrazzo areas are taken off on a square foot basis. Trim and base materials are measured by the linear foot. Accent tiles are listed per each. Two basic methods of installation are used. Mud set is approximately 30% more expensive than thin set. In terrazzo work, be sure to include the linear footage of embedded decorative strips, grounds, machine rubbing, and power cleanup.
- Wood flooring is available in strip, parquet, or block configuration. The latter two types are set in adhesives with quantities estimated by the square foot. The laying pattern will influence labor costs and material waste. In addition to the material and labor for laying wood floors, the estimator must make allowances for sanding and finishing these areas unless the flooring is prefinished.
- Sheet flooring is measured by the square yard. Roll widths vary, so consideration should be given to use the most economical width, as waste must be figured into the total quantity. Consider also the installation methods available, direct glue down or stretched.

09 70 00 Wall Finishes

- Wall coverings are estimated by the square foot. The area to be covered is measured, length by height of wall above baseboards, to calculate the square footage of each wall. This figure is divided by the number of square feet in the single roll which is being used. Deduct, in full, the areas of openings such as doors and windows. Where a pattern match is required allow 25%-30% waste.

09 80 00 Acoustic Treatment

- Acoustical systems fall into several categories. The takeoff of these materials should be by the square foot of area with a 5% allowance for waste. Do not forget about scaffolding, if applicable, when estimating these systems.

09 90 00 Painting and Coating

- A major portion of the work in painting involves surface preparation. Be sure to include cleaning, sanding, filling, and masking costs in the estimate.
- Protection of adjacent surfaces is not included in painting costs. When considering the method of paint application, an important factor is the amount of protection and masking required. These must be estimated separately and may be the determining factor in choosing the method of application.

Reference Numbers

Reference numbers are shown in shaded boxes at the beginning of some major classifications. These numbers refer to related items in the Reference Section. The reference information may be an estimating procedure, an alternate pricing method, or technical information.

Note: Not all subdivisions listed here necessarily appear in this publication.

No part of this publication may be reproduced, stored in a retrieval system, or transmitted in any form or by any means without prior written permission of Reed Construction Data.

09 01 Maintenance of Finishes

09 01 60 – Maintenance of Flooring

09 01 60.10 Carpet Maintenance

		Crew	Daily Output	Labor-Hours	Unit	Material	2007 Bare Costs Labor	2007 Bare Costs Equipment	Total	Total Incl O&P
0010	**CARPET MAINTENANCE**									
0020	Steam clean, per cleaning, minimum	1 Clab	3000	.003	S.F.	.07	.08		.15	.20
0500	Maximum	"	2000	.004	"	.12	.12		.24	.31

09 01 70 – Maintenance of Wall Finishes

09 01 70.10 Gypsum Wallboard Repairs

		Crew	Daily Output	Labor-Hours	Unit	Material	Labor	Equipment	Total	Total Incl O&P
0010	**GYPSUM WALLBOARD REPAIRS**									
0100	Fill and sand, pin / nail holes	1 Carp	960	.008	Ea.		.31		.31	.48
0110	Screw head pops		480	.017			.61		.61	.95
0120	Dents, up to 2" square		48	.167		.01	6.10		6.11	9.55
0130	2" to 4" square		24	.333		.03	12.25		12.28	19.10
0140	Cut square, patch, sand and finish, holes, up to 2" square		12	.667		.04	24.50		24.54	38
0150	2" to 4" square		11	.727		.09	26.50		26.59	41.50
0160	4" to 8" square		10	.800		.24	29.50		29.74	46
0170	8" to 12" square		8	1		.46	36.50		36.96	57.50

09 05 Common Work Results for Finishes

09 05 05 – Selective Finishes Demolition

09 05 05.10 Selective Demolition, Ceilings

		Crew	Daily Output	Labor-Hours	Unit	Material	Labor	Equipment	Total	Total Incl O&P
0010	**SELECTIVE DEMOLITION, CEILINGS** R024119-10									
0200	Ceiling, drywall, furred and nailed or screwed	2 Clab	800	.020	S.F.		.58		.58	.90
0220	On metal frame		760	.021			.61		.61	.94
0240	On suspension system, including system		720	.022			.64		.64	.99
1000	Plaster, lime and horse hair, on wood lath, incl. lath		700	.023			.66		.66	1.02
1020	On metal lath		570	.028			.81		.81	1.26
1100	Gypsum, on gypsum lath		720	.022			.64		.64	.99
1120	On metal lath		500	.032			.92		.92	1.43
1200	Suspended ceiling, mineral fiber, 2' x 2' or 2' x 4'		1500	.011			.31		.31	.48
1250	On suspension system, incl. system		1200	.013			.38		.38	.60
1500	Tile, wood fiber, 12" x 12", glued		900	.018			.51		.51	.80
1540	Stapled		1500	.011			.31		.31	.48
1580	On suspension system, incl. system		760	.021			.61		.61	.94
2000	Wood, tongue and groove, 1" x 4"		1000	.016			.46		.46	.72
2040	1" x 8"		1100	.015			.42		.42	.65
2400	Plywood or wood fiberboard, 4' x 8' sheets		1200	.013			.38		.38	.60

09 05 05.20 Selective Demolition, Flooring

		Crew	Daily Output	Labor-Hours	Unit	Material	Labor	Equipment	Total	Total Incl O&P
0010	**SELECTIVE DEMOLITION, FLOORING** R024119-10									
0200	Brick with mortar	2 Clab	475	.034	S.F.		.97		.97	1.51
0400	Carpet, bonded, including surface scraping		2000	.008			.23		.23	.36
0440	Scrim applied		8000	.002			.06		.06	.09
0480	Tackless		9000	.002			.05		.05	.08
0600	Composition, acrylic or epoxy		400	.040			1.15		1.15	1.79
0700	Concrete, scarify skin	A-1A	225	.036			1.35	1.34	2.69	3.58
0800	Resilient, sheet goods	2 Clab	1400	.011			.33		.33	.51
0820	For gym floors	"	900	.018			.51		.51	.80
0850	Vinyl or rubber cove base	1 Clab	1000	.008	L.F.		.23		.23	.36
0860	Vinyl or rubber cove base, molded corner	"	1000	.008	Ea.		.23		.23	.36
0900	Vinyl composition tile, 12" x 12"	2 Clab	1000	.016	S.F.		.46		.46	.72
2000	Tile, ceramic, thin set		675	.024			.68		.68	1.06
2020	Mud set		625	.026			.74		.74	1.15
2200	Marble, slate, thin set		675	.024			.68		.68	1.06

09 05 Common Work Results for Finishes

09 05 05 – Selective Finishes Demolition

09 05 05.20 Selective Demolition, Flooring

		Crew	Daily Output	Labor-Hours	Unit	Material	2007 Bare Costs Labor	2007 Bare Costs Equipment	Total	Total Incl O&P
2220	Mud set	2 Clab	625	.026	S.F.		.74		.74	1.15
2600	Terrazzo, thin set		450	.036			1.02		1.02	1.59
2620	Mud set		425	.038			1.08		1.08	1.68
2640	Cast in place	↓	300	.053			1.53		1.53	2.39
3000	Wood, block, on end	1 Carp	400	.020			.73		.73	1.14
3200	Parquet		450	.018			.65		.65	1.02
3400	Strip flooring, interior, 2-1/4" x 25/32" thick		325	.025			.90		.90	1.41
3500	Exterior, porch flooring, 1" x 4"		220	.036			1.33		1.33	2.08
3800	Subfloor, tongue and groove, 1" x 6"		325	.025			.90		.90	1.41
3820	1" x 8"		430	.019			.68		.68	1.06
3840	1" x 10"		520	.015			.56		.56	.88
4000	Plywood, nailed		600	.013			.49		.49	.76
4100	Glued and nailed	↓	400	.020			.73		.73	1.14
8000	Remove flooring, bead blast, minimum	A-1A	1000	.008			.30	.30	.60	.80
8100	Maximum		400	.020			.76	.76	1.52	2.01
8150	Mastic only	↓	1500	.005	↓		.20	.20	.40	.54

09 05 05.30 Selective Demolition, Walls and Partitions

		Crew	Daily Output	Labor-Hours	Unit	Material	2007 Bare Costs Labor	2007 Bare Costs Equipment	Total	Total Incl O&P
0010	**SELECTIVE DEMOLITION, WALLS AND PARTITIONS** R024119-10									
0100	Brick, 4" to 12" thick	B-9C	220	.182	C.F.		5.30	.82	6.12	9.15
0200	Concrete block, 4" thick		1000	.040	S.F.		1.17	.18	1.35	2.02
0280	8" thick	↓	810	.049			1.44	.22	1.66	2.48
0300	Exterior stucco 1" thick over mesh	B-9	3200	.013			.36	.06	.42	.63
1000	Drywall, nailed or screwed	1 Clab	1000	.008			.23		.23	.36
1020	Glued and nailed		900	.009			.26		.26	.40
1500	Fiberboard, nailed		900	.009			.26		.26	.40
1520	Glued and nailed		800	.010			.29		.29	.45
1568	Plenum barrier, sheet lead		300	.027			.77		.77	1.19
2000	Movable walls, metal, 5' high		300	.027			.77		.77	1.19
2020	8' high	↓	400	.020			.58		.58	.90
2200	Metal or wood studs, finish 2 sides, fiberboard	B-1	520	.046			1.36		1.36	2.11
2250	Lath and plaster		260	.092			2.72		2.72	4.23
2300	Plasterboard (drywall)		520	.046			1.36		1.36	2.11
2350	Plywood	↓	450	.053			1.57		1.57	2.44
3000	Plaster, lime and horsehair, on wood lath	1 Clab	400	.020			.58		.58	.90
3020	On metal lath		335	.024			.69		.69	1.07
3400	Gypsum or perlite, on gypsum lath		410	.020			.56		.56	.87
3420	On metal lath	↓	300	.027			.77		.77	1.19
3600	Plywood, one side	B-1	1500	.016			.47		.47	.73
3750	Terra cotta block and plaster, to 6" thick	"	175	.137	↓		4.03		4.03	6.30
3800	Toilet partitions, slate or marble	1 Clab	5	1.600	Ea.		46		46	71.50
3820	Hollow metal	"	8	1	"		29		29	45
5000	Wallcovering, vinyl	1 Pape	700	.011	S.F.		.38		.38	.57
5040	Designer	"	480	.017	"		.55		.55	.83

09 21 Plaster and Gypsum Board Assemblies

09 21 13 – Plaster Assemblies

09 21 13.10 Plaster Partition Wall

		Crew	Daily Output	Labor-Hours	Unit	Material	2007 Bare Costs Labor	Equipment	Total	Total Incl O&P
0010	**PLASTER PARTITION WALL**									
0400	Stud walls, 3.4 lb. metal lath, 3 coat gypsum plaster, 2 sides									
0600	2" x 4" wood studs, 16" O.C.	J-2	315	.152	S.F.	3.21	4.87	.36	8.44	11.30
0700	2-1/2" metal studs, 25 ga., 12" O.C.		325	.148		2.97	4.72	.35	8.04	10.75
0800	3-5/8" metal studs, 25 ga., 16" O.C.	↓	320	.150	↓	2.99	4.80	.36	8.15	10.95
0900	Gypsum lath, 2 coat vermiculite plaster, 2 sides									
1000	2" x 4" wood studs, 16" O.C.	J-2	355	.135	S.F.	3.58	4.32	.32	8.22	10.80
1200	2-1/2" metal studs, 25 ga., 12" O.C.		365	.132		3.17	4.21	.31	7.69	10.20
1300	3-5/8" metal studs, 25 ga., 16" O.C.	↓	360	.133	↓	3.26	4.26	.32	7.84	10.40

09 21 16 – Gypsum Board Assemblies

09 21 16.23 Gypsum Board Shaft Wall Assemblies

		Crew	Daily Output	Labor-Hours	Unit	Material	Labor	Equipment	Total	Total Incl O&P
0010	**GYPSUM BOARD SHAFT WALL ASSEMBLIES**									
0030	1" thick coreboard wall liner on shaft side									
0040	2-hour assembly with double layer									
0060	5/8" fire rated gypsum board on room side	2 Carp	220	.073	S.F.	1.23	2.67		3.90	5.50
0100	3-hour assembly with triple layer									
0300	5/8" fire rated gypsum board on room side	2 Carp	180	.089	S.F.	1.62	3.26		4.88	6.90
0400	4-hour assembly, 1" coreboard, 5/8" fire rated gypsum board									
0600	and 3/4" galv. metal furring channels, 24" O.C., with									
0700	Double layer 5/8" fire rated gypsum board on room side	2 Carp	110	.145	S.F.	1.38	5.35		6.73	9.80
0900	For taping & finishing, add per side	1 Carp	1050	.008	"	.04	.28		.32	.48
1000	For insulation, see div. 07 21 26.00									

09 21 16.33 Partition Wall

		Crew	Daily Output	Labor-Hours	Unit	Material	Labor	Equipment	Total	Total Incl O&P
0010	**PARTITION WALL** Stud wall, 8' to 12' high									
0050	1/2", interior, gypsum board, std, tape & finish 2 sides									
0500	Installed on and incl., 2" x 4" wood studs, 16" O.C.	2 Carp	310	.052	S.F.	1.14	1.89		3.03	4.21
1000	Metal studs, NLB, 25 ga., 16" O.C., 3-5/8" wide		350	.046		1.11	1.68		2.79	3.83
1200	6" wide		330	.048		1.30	1.78		3.08	4.20
1400	Water resistant, on 2" x 4" wood studs, 16" O.C.		310	.052		1.28	1.89		3.17	4.36
1600	Metal studs, NLB, 25 ga., 16" O.C., 3-5/8" wide		350	.046		1.25	1.68		2.93	3.98
1800	6" wide		330	.048		1.44	1.78		3.22	4.36
2000	Fire res., 2 layers, 1-1/2 hr., on 2" x 4" wood studs, 16" O.C.		210	.076		2.02	2.80		4.82	6.55
2200	Metal studs, NLB, 25 ga., 16" O.C., 3-5/8" wide		250	.064		1.99	2.35		4.34	5.85
2400	6" wide		230	.070		2.18	2.55		4.73	6.40
2600	Fire & water res., 2 layers, 1-1/2 hr., 2" x 4" studs, 16" O.C.		210	.076		2.02	2.80		4.82	6.55
2800	Metal studs, NLB, 25 ga., 16" O.C., 3-5/8" wide		250	.064		1.99	2.35		4.34	5.85
3000	6" wide	↓	230	.070	↓	2.18	2.55		4.73	6.40
3200	5/8", interior, gypsum board, std, tape & finish 2 sides									
3400	Installed on and including 2" x 4" wood studs, 16" O.C.	2 Carp	300	.053	S.F.	1.28	1.96		3.24	4.46
3600	24" O.C.		330	.048		1.19	1.78		2.97	4.08
3800	Metal studs, NLB, 25 ga., 16" O.C., 3-5/8" wide		340	.047		1.25	1.73		2.98	4.06
4000	6" wide		320	.050		1.44	1.84		3.28	4.45
4200	24" O.C., 3-5/8" wide		360	.044		1.15	1.63		2.78	3.80
4400	6" wide		340	.047		1.29	1.73		3.02	4.11
4800	Water resistant, on 2" x 4" wood studs, 16" O.C.		300	.053		1.28	1.96		3.24	4.46
5000	24" O.C.		330	.048		1.19	1.78		2.97	4.08
5200	Metal studs, NLB, 25 ga. 16" O.C., 3-5/8" wide		340	.047		1.25	1.73		2.98	4.06
5400	6" wide		320	.050		1.44	1.84		3.28	4.45
5600	24" O.C., 3-5/8" wide		360	.044		1.15	1.63		2.78	3.80
5800	6" wide		340	.047		1.29	1.73		3.02	4.11
6000	Fire res., 2 layers, 2 hr., on 2" x 4" wood studs, 16" O.C.		205	.078		1.93	2.86		4.79	6.60
6200	24" O.C.	↓	235	.068	↓	1.93	2.50		4.43	6

09 21 Plaster and Gypsum Board Assemblies

09 21 16 – Gypsum Board Assemblies

09 21 16.33 Partition Wall		Crew	Daily Output	Labor-Hours	Unit	Material	2007 Bare Costs Labor	Equipment	Total	Total Incl O&P
6400	Metal studs, NLB, 25 ga., 16" O.C., 3-5/8" wide	2 Carp	245	.065	S.F.	2.05	2.40		4.45	6
6600	6" wide		225	.071		2.18	2.61		4.79	6.45
6800	24" O.C., 3-5/8" wide		265	.060		1.89	2.22		4.11	5.55
7000	6" wide		245	.065		2.03	2.40		4.43	5.95
7200	Fire & water res., 2 layers, 2 hr., 2" x 4" studs, 16" O.C.		205	.078		2.02	2.86		4.88	6.70
7400	24" O.C.		235	.068		1.93	2.50		4.43	6
7600	Metal studs, NLB, 25 ga., 16" O.C., 3-5/8" wide		245	.065		1.99	2.40		4.39	5.90
7800	6" wide		225	.071		2.18	2.61		4.79	6.45
8000	24" O.C., 3-5/8" wide		265	.060		1.89	2.22		4.11	5.55
8200	6" wide		245	.065		2.03	2.40		4.43	5.95
8600	1/2" blueboard, mesh tape both sides									
8620	Installed on and including 2" x 4" wood studs, 16" O.C.	2 Carp	300	.053	S.F.	1.28	1.96		3.24	4.46
8640	Metal studs, NLB, 25 ga., 16" O.C., 3-5/8" wide		340	.047		1.25	1.73		2.98	4.06
8660	6" wide		320	.050		1.44	1.84		3.28	4.45
9000	Exterior, 1/2" gypsum sheathing, 1/2" gypsum finished, interior,									
9100	including foil faced insulation, metal studs, 20 ga.									
9200	16" O.C., 3-5/8" wide	2 Carp	290	.055	S.F.	2.14	2.02		4.16	5.50
9400	6" wide	"	270	.059	"	2.38	2.17		4.55	6

09 22 Supports for Plaster and Gypsum Board

09 22 03 – Fastening Methods for Finishes

09 22 03.20 Drilling Plaster/Drywall

		Crew	Daily Output	Labor-Hours	Unit	Material	Labor	Equipment	Total	Total Incl O&P
0010	**DRILLING PLASTER/DRYWALL**									
1100	Drilling & layout for drywall/plaster walls, up to 1" deep, no anchor									
1200	Holes, 1/4" diameter	1 Carp	150	.053	Ea.	.01	1.96		1.97	3.06
1300	3/8" diameter		140	.057		.01	2.10		2.11	3.28
1400	1/2" diameter		130	.062		.01	2.26		2.27	3.53
1500	3/4" diameter		120	.067		.03	2.45		2.48	3.84
1600	1" diameter		110	.073		.04	2.67		2.71	4.20
1700	1-1/4" diameter		100	.080		.05	2.94		2.99	4.62
1800	1-1/2" diameter		90	.089		.08	3.26		3.34	5.20
1900	For ceiling installations, add						40%			

09 22 13 – Metal Furring

09 22 13.13 Metal Channel Furring

		Crew	Daily Output	Labor-Hours	Unit	Material	Labor	Equipment	Total	Total Incl O&P
0010	**METAL CHANNEL FURRING**									
0030	Beams and columns, 7/8" galv. channels, 12" O.C.	1 Lath	155	.052	S.F.	.27	1.74		2.01	2.89
0050	16" O.C.		170	.047		.22	1.59		1.81	2.60
0070	24" O.C.		185	.043		.15	1.46		1.61	2.33
0100	Ceilings, on steel, 7/8" channels, galvanized, 12" O.C.		210	.038		.24	1.28		1.52	2.18
0300	16" O.C.		290	.028		.22	.93		1.15	1.63
0400	24" O.C.		420	.019		.15	.64		.79	1.12
0600	1-5/8" channels, galvanized, 12" O.C.		190	.042		.37	1.42		1.79	2.52
0700	16" O.C.		260	.031		.33	1.04		1.37	1.91
0900	24" O.C.		390	.021		.22	.69		.91	1.27
0930	7/8" channels with sound isolation clips, 12" O.C.		120	.067		1.27	2.25		3.52	4.75
0940	16" O.C.		100	.080		1.79	2.70		4.49	6
0950	24" O.C.		165	.048		1.18	1.63		2.81	3.72
0960	1-5/8" channels, galvanized, 12" O.C.		110	.073		1.40	2.45		3.85	5.20
0970	16" O.C.		100	.080		1.91	2.70		4.61	6.10
0980	24" O.C.		155	.052		1.25	1.74		2.99	3.97

09 22 Supports for Plaster and Gypsum Board

09 22 13 – Metal Furring

09 22 13.13 Metal Channel Furring

		Crew	Daily Output	Labor-Hours	Unit	Material	2007 Bare Costs Labor	2007 Bare Costs Equipment	Total	Total Incl O&P
1000	Walls, 7/8" channels, galvanized, 12" O.C.	1 Lath	235	.034	S.F.	.24	1.15		1.39	1.98
1200	16" O.C.		265	.030		.22	1.02		1.24	1.76
1300	24" O.C.		350	.023		.15	.77		.92	1.31
1500	1-5/8" channels, galvanized, 12" O.C.		210	.038		.37	1.28		1.65	2.32
1600	16" O.C.		240	.033		.33	1.12		1.45	2.04
1800	24" O.C.		305	.026		.22	.88		1.10	1.56
1920	7/8" channels with sound isolation clips, 12" O.C.		125	.064		1.27	2.16		3.43	4.61
1940	16" O.C.		100	.080		1.79	2.70		4.49	6
1950	24" O.C.		150	.053		1.18	1.80		2.98	3.97
1960	1-5/8" channels, galvanized, 12" O.C.		115	.070		1.40	2.34		3.74	5.05
1970	16" O.C.		95	.084		1.91	2.84		4.75	6.35
1980	24" O.C.		140	.057		1.25	1.93		3.18	4.25

09 22 16 – Non-Structural Metal Framing

09 22 16.13 Metal Studs and Track

		Crew	Daily Output	Labor-Hours	Unit	Material	Labor	Equipment	Total	Total Incl O&P
0010	**METAL STUDS AND TRACK**									
1600	Non-load bearing, galv. 8' high, 25 ga. 1-5/8" wide, 16" O.C.	1 Carp	619	.013	S.F.	.30	.47		.77	1.07
1610	24" O.C.		950	.008		.23	.31		.54	.73
1620	2-1/2" wide, 16" O.C.		613	.013		.36	.48		.84	1.14
1630	24" O.C.		938	.009		.27	.31		.58	.79
1640	3-5/8" wide, 16" O.C.		600	.013		.41	.49		.90	1.21
1650	24" O.C.		925	.009		.31	.32		.63	.83
1660	4" wide, 16" O.C.		594	.013		.50	.49		.99	1.32
1670	24" O.C.		925	.009		.37	.32		.69	.90
1680	6" wide, 16" O.C.		588	.014		.62	.50		1.12	1.46
1690	24" O.C.		906	.009		.46	.32		.78	1.01
1700	20 ga. studs, 1-5/8" wide, 16" O.C.		494	.016		.50	.59		1.09	1.48
1710	24" O.C.		763	.010		.37	.38		.75	1.01
1720	2-1/2" wide, 16" O.C.		488	.016		.58	.60		1.18	1.58
1730	24" O.C.		750	.011		.44	.39		.83	1.09
1740	3-5/8" wide, 16" O.C.		481	.017		.65	.61		1.26	1.67
1750	24" O.C.		738	.011		.49	.40		.89	1.16
1760	4" wide, 16" O.C.		475	.017		.78	.62		1.40	1.82
1770	24" O.C.		738	.011		.58	.40		.98	1.26
1780	6" wide, 16" O.C.		469	.017		.90	.63		1.53	1.97
1790	24" O.C.		725	.011		.68	.40		1.08	1.38
2000	Non-load bearing, galv. 10' high, 25 ga. 1-5/8" wide, 16" O.C.		495	.016		.28	.59		.87	1.23
2100	24" O.C.		760	.011		.21	.39		.60	.83
2200	2-1/2" wide, 16" O.C.		490	.016		.34	.60		.94	1.30
2250	24" O.C.		750	.011		.25	.39		.64	.89
2300	3-5/8" wide, 16" O.C.		480	.017		.39	.61		1	1.38
2350	24" O.C.		740	.011		.29	.40		.69	.94
2400	4" wide, 16" O.C.		475	.017		.47	.62		1.09	1.48
2450	24" O.C.		740	.011		.35	.40		.75	1
2500	6" wide, 16" O.C.		470	.017		.58	.62		1.20	1.61
2550	24" O.C.		725	.011		.43	.40		.83	1.10
2600	20 ga. studs, 1-5/8" wide, 16" O.C.		395	.020		.47	.74		1.21	1.68
2650	24" O.C.		610	.013		.35	.48		.83	1.13
2700	2-1/2" wide, 16" O.C.		390	.021		.55	.75		1.30	1.77
2750	24" O.C.		600	.013		.41	.49		.90	1.21
2800	3-5/8" wide, 16" O.C. CN		385	.021		.62	.76		1.38	1.87
2850	24" O.C.		590	.014		.46	.50		.96	1.28
2900	4" wide, 16" O.C.		380	.021		.74	.77		1.51	2.01

09 22 Supports for Plaster and Gypsum Board

09 22 16 – Non-Structural Metal Framing

09 22 16.13 Metal Studs and Track

		Crew	Daily Output	Labor-Hours	Unit	Material	2007 Bare Costs Labor	Equipment	Total	Total Incl O&P
2950	24" O.C.	1 Carp	590	.014	S.F.	.54	.50		1.04	1.38
3000	6" wide, 16" O.C.		375	.021		.86	.78		1.64	2.16
3050	24" O.C.		580	.014		.63	.51		1.14	1.49
3060	Non-load bearing, galv, 12' high, 25 ga. 1-5/8" wide, 16" O.C.		413	.019		.27	.71		.98	1.41
3070	24" O.C.		633	.013		.20	.46		.66	.94
3080	2-1/2" wide, 16" O.C.		408	.020		.32	.72		1.04	1.48
3090	24" O.C.		625	.013		.24	.47		.71	.99
3100	3-5/8" wide, 16" O.C.		400	.020		.37	.73		1.10	1.55
3110	24" O.C.		617	.013		.27	.48		.75	1.04
3120	4" wide, 16" O.C.		396	.020		.45	.74		1.19	1.64
3130	24" O.C.		617	.013		.33	.48		.81	1.10
3140	6" wide, 16" O.C.		392	.020		.56	.75		1.31	1.78
3150	24" O.C.		604	.013		.41	.49		.90	1.21
3160	20 ga. studs, 1-5/8" wide, 16" O.C.		329	.024		.45	.89		1.34	1.89
3170	24" O.C.		508	.016		.33	.58		.91	1.26
3180	2-1/2" wide, 16" O.C.		325	.025		.52	.90		1.42	1.99
3190	24" O.C.		500	.016		.38	.59		.97	1.33
3200	3-5/8" wide, 16" O.C.		321	.025		.59	.91		1.50	2.07
3210	24" O.C.		492	.016		.43	.60		1.03	1.40
3220	4" wide, 16" O.C.		317	.025		.71	.93		1.64	2.22
3230	24" O.C.		492	.016		.51	.60		1.11	1.49
3240	6" wide, 16" O.C.		313	.026		.82	.94		1.76	2.36
3250	24" O.C.		483	.017		.60	.61		1.21	1.61
5000	Load bearing studs, see division 05 41 13.30									

09 22 26 – Suspension Systems

09 22 26.13 Ceiling Suspension Systems

		Crew	Daily Output	Labor-Hours	Unit	Material	2007 Bare Costs Labor	Equipment	Total	Total Incl O&P
0010	**CEILING SUSPENSION SYSTEMS** For gypsum board or plaster									
8000	Suspended ceilings, including carriers									
8200	1-1/2" carriers, 24" O.C. with:									
8300	7/8" channels, 16" O.C.	1 Lath	165	.048	S.F.	.44	1.63		2.07	2.91
8320	24" O.C.		200	.040		.37	1.35		1.72	2.41
8400	1-5/8" channels, 16" O.C.		155	.052		.55	1.74		2.29	3.20
8420	24" O.C.		190	.042		.44	1.42		1.86	2.60
8600	2" carriers, 24" O.C. with:									
8700	7/8" channels, 16" O.C.	1 Lath	155	.052	S.F.	.46	1.74		2.20	3.10
8720	24" O.C.		190	.042		.39	1.42		1.81	2.54
8800	1-5/8" channels, 16" O.C.		145	.055		.57	1.86		2.43	3.40
8820	24" O.C.		180	.044		.46	1.50		1.96	2.74

09 22 36 – Lath

09 22 36.13 Gypsum Lath

			Crew	Daily Output	Labor-Hours	Unit	Material	2007 Bare Costs Labor	Equipment	Total	Total Incl O&P
0010	**GYPSUM LATH**	R092000-50									
0020	Plain or perforated, nailed, 3/8" thick		1 Lath	85	.094	S.Y.	4.50	3.17		7.67	9.65
0100	1/2" thick, nailed			80	.100		4.23	3.37		7.60	9.65
0300	Clipped to steel studs, 3/8" thick			75	.107		4.50	3.59		8.09	10.30
0400	1/2" thick			70	.114		4.23	3.85		8.08	10.40
0600	Firestop gypsum base, to steel studs, 3/8" thick			70	.114		3.51	3.85		7.36	9.60
0700	1/2" thick			65	.123		4.59	4.15		8.74	11.25
0900	Foil back, to steel studs, 3/8" thick			75	.107		3.87	3.59		7.46	9.60
1000	1/2" thick			70	.114		4.05	3.85		7.90	10.20
1500	For ceiling installations, add			216	.037			1.25		1.25	1.86
1600	For columns and beams, add			170	.047			1.59		1.59	2.36

09 22 Supports for Plaster and Gypsum Board

09 22 36 – Lath

09 22 36.23 Metal Lath

		Crew	Daily Output	Labor-Hours	Unit	Material	2007 Bare Costs Labor	2007 Bare Costs Equipment	Total	Total Incl O&P
0010	**METAL LATH** R092000-50									
0020	Diamond, expanded, 2.5 lb. per S.Y., painted				S.Y.	2.92			2.92	3.21
0100	Galvanized, 2.5 lb. per S.Y.					2.50			2.50	2.75
0300	3.4 lb. per S.Y., painted					4.03			4.03	4.43
0400	Galvanized					3.72			3.72	4.09
0600	For 15# asphalt sheathing paper, add					.38			.38	.42
0900	Flat rib, 1/8" high, 2.75 lb., painted					3.59			3.59	3.95
1000	Foil backed					3.98			3.98	4.38
1200	3.4 lb. per S.Y., painted					4.78			4.78	5.25
1300	Galvanized					4.44			4.44	4.88
1500	For 15# asphalt sheating paper, add					.38			.38	.42
1800	High rib, 3/8" high, 3.4 lb. per S.Y., painted					5.05			5.05	5.55
1900	Galvanized				↓	4.22			4.22	4.64
2400	High rib, 3/4" high, painted, .60 lb. per S.F.				S.F.	.48			.48	.53
2500	.75 lb. per S.F.				"	1.04			1.04	1.14
2800	Stucco mesh, painted, 3.6 lb.				S.Y.	3.50			3.50	3.85
3000	K-lath, perforated, absorbent paper, regular					3.69			3.69	4.06
3100	Heavy duty					4.36			4.36	4.80
3300	Waterproof, heavy duty, grade B backing					4.27			4.27	4.70
3400	Fire resistant backing					4.72			4.72	5.20
3600	2.5 lb. diamond painted, on wood framing, on walls	1 Lath	85	.094		2.92	3.17		6.09	7.95
3700	On ceilings		75	.107		2.92	3.59		6.51	8.55
3900	3.4 lb. diamond painted, on wood framing, on walls		80	.100		4.78	3.37		8.15	10.25
4000	On ceilings		70	.114		4.78	3.85		8.63	11
4200	3.4 lb. diamond painted, wired to steel framing		75	.107		4.78	3.59		8.37	10.60
4300	On ceilings		60	.133		4.78	4.49		9.27	11.95
4500	Columns and beams, wired to steel		40	.200		4.78	6.75		11.53	15.30
4600	Cornices, wired to steel		35	.229		4.78	7.70		12.48	16.70
4800	Screwed to steel studs, 2.5 lb.		80	.100		2.92	3.37		6.29	8.20
4900	3.4 lb.		75	.107		4.03	3.59		7.62	9.80
5100	Rib lath, painted, wired to steel, on walls, 2.5 lb.		75	.107		3.59	3.59		7.18	9.30
5200	3.4 lb.		70	.114		5.05	3.85		8.90	11.30
5400	4.0 lb.	↓	65	.123		5.20	4.15		9.35	11.90
5500	For self-furring lath, add					.09			.09	.10
5700	Suspended ceiling system, incl. 3.4 lb. diamond lath, painted	1 Lath	15	.533		12.75	17.95		30.70	41
5800	Galvanized	"	15	.533	↓	13.20	17.95		31.15	41.50
6000	Hollow metal stud partitions, 3.4 lb. painted lath both sides									
6010	Non-load bearing, 25 ga., w/rib lath 2-1/2" studs, 12" O.C.	1 Lath	20.30	.394	S.Y.	13.95	13.30		27.25	35
6300	16" O.C.		21.10	.379		13.10	12.80		25.90	33.50
6350	24" O.C.		22.70	.352		12.40	11.90		24.30	31.50
6400	3-5/8" studs, 16" O.C.		19.50	.410		13.60	13.85		27.45	35.50
6600	24" O.C.		20.40	.392		12.70	13.20		25.90	33.50
6700	4" studs, 16" O.C.		20.40	.392		14.35	13.20		27.55	35.50
6900	24" O.C.		21.60	.370		13.25	12.50		25.75	33
7000	6" studs, 16" O.C.		19.50	.410		15.35	13.85		29.20	37.50
7100	24" O.C.		21.10	.379		14	12.80		26.80	34.50
7200	L.B. partitions, 16 ga., w/rib lath, 2-1/2" studs, 16" O.C.		20	.400		13.35	13.50		26.85	34.50
7300	3-5/8" studs, 16 ga.		19.70	.406		15.20	13.70		28.90	37.50
7500	4" studs, 16 ga.		19.50	.410		15.80	13.85		29.65	38
7600	6" studs, 16 ga.	↓	18.70	.428	↓	18.75	14.40		33.15	42

09 22 36.43 Security Mesh

0010	**SECURITY MESH**, expanded metal, flat, screwed to framing									

09 22 Supports for Plaster and Gypsum Board

09 22 36 – Lath

09 22 36.43 Security Mesh

		Crew	Daily Output	Labor-Hours	Unit	Material	2007 Bare Costs Labor	Equipment	Total	Total Incl O&P
0100	On walls, 3/4", 1.76 lb/S.F.	2 Carp	1500	.011	S.F.	1.37	.39		1.76	2.12
0110	1-1/2", 1.14 lb/S.F.		1600	.010		1.08	.37		1.45	1.76
0200	On ceilings, 3/4", 1.76 lb/S.F.		1350	.012		1.37	.43		1.80	2.19
0210	1-1/2", 1.14 lb/S.F.	↓	1450	.011	↓	1.08	.40		1.48	1.82

09 22 36.83 Accessories, Plaster

		Crew	Daily Output	Labor-Hours	Unit	Material	Labor	Equipment	Total	Total Incl O&P
0010	**ACCESSORIES, PLASTER**									
0020	Casing bead, expanded flange, galvanized	1 Lath	2.70	2.963	C.L.F.	40.50	100		140.50	194
0900	Channels, cold rolled, 16 ga., 3/4" deep, galvanized					24.50			24.50	27
1200	1-1/2" deep, 16 ga., galvanized					37			37	41
1620	Corner bead, expanded bullnose, 3/4" radius, #10, galvanized	1 Lath	2.60	3.077		31.50	104		135.50	189
1650	#1, galvanized		2.55	3.137		48.50	106		154.50	211
1670	Expanded wing, 2-3/4" wide, galv. #1		2.65	3.019		32	102		134	188
1700	Inside corner, (corner rite) 3" x 3", painted		2.60	3.077		24.50	104		128.50	181
1750	Strip-ex, 4" wide, painted		2.55	3.137		25.50	106		131.50	185
1800	Expansion joint, 3/4" grounds, limited expansion, galv., 1 piece		2.70	2.963		96	100		196	255
2100	Extreme expansion, galvanized, 2 piece	↓	2.60	3.077	↓	150	104		254	320

09 23 Gypsum Plastering

09 23 13 – Acoustical Gypsum Plastering

09 23 13.10 Perlite or Vermiculite Plaster

		Crew	Daily Output	Labor-Hours	Unit	Material	Labor	Equipment	Total	Total Incl O&P
0010	**PERLITE OR VERMICULITE PLASTER** R092000-50									
0020	In 100 lb. bags, under 200 bags				Bag	13.65			13.65	15.05
0100	Over 200 bags				"	12.60			12.60	13.85
0300	2 coats, no lath included, on walls	J-1	92	.435	S.Y.	3.60	13.75	1.23	18.58	26.50
0400	On ceilings	"	79	.506		3.60	16	1.44	21.04	29.50
0600	2 coats, on and incl. 3/8" gypsum lath, on metal studs	J-2	84	.571		7.80	18.25	1.35	27.40	37.50
0700	On ceilings	"	70	.686		7.80	22	1.63	31.43	43.50
0900	3 coats, no lath included, on walls	J-1	74	.541		6.10	17.10	1.54	24.74	34.50
1000	On ceilings	"	63	.635		6.10	20	1.80	27.90	39
1200	3 coats, on and incl. painted metal lath, on metal studs	J-2	72	.667		11.20	21.50	1.58	34.28	46
1300	On ceilings		61	.787		11.20	25	1.86	38.06	52.50
1500	3 coats, on and incl. suspended metal lath ceiling	↓	37	1.297		18.90	41.50	3.07	63.47	87
1700	For irregular or curved surfaces, add to above						30%			
1800	For columns and beams, add to above						50%			
1900	For soffits, add to ceiling prices				↓		40%			

09 23 20 – Gypsum Plaster

09 23 20.10 Gypsum Plaster

		Crew	Daily Output	Labor-Hours	Unit	Material	Labor	Equipment	Total	Total Incl O&P
0010	**GYPSUM PLASTER** R092000-50									
0020	80# bag, less than 1 ton				Bag	13.60			13.60	14.95
0100	Over 1 ton				"	12.95			12.95	14.25
0300	2 coats, no lath included, on walls	J-1	105	.381	S.Y.	3.36	12.05	1.08	16.49	23
0400	On ceilings	"	92	.435		3.36	13.75	1.23	18.34	26
0600	2 coats on and incl. 3/8" gypsum lath on steel, on walls	J-2	97	.495		7.85	15.85	1.17	24.87	34
0700	On ceilings	"	83	.578		7.85	18.50	1.37	27.72	38
0900	3 coats, no lath included, on walls	J-1	87	.460		4.72	14.55	1.31	20.58	28.50
1000	On ceilings	"	78	.513		4.72	16.20	1.46	22.38	31.50
1200	3 coats on and including painted metal lath, on wood studs	J-2	86	.558		9.20	17.85	1.32	28.37	38.50
1300	On ceilings	"	76.50	.627	↓	9.20	20	1.49	30.69	42.50
1600	For irregular or curved surfaces, add						30%			
1800	For columns & beams, add						50%			

09 23 Gypsum Plastering

09 23 20 – Gypsum Plaster

09 23 20.20 Gauging Plaster

		Crew	Daily Output	Labor-Hours	Unit	Material	2007 Bare Costs Labor	Equipment	Total	Total Incl O&P
0010	**GAUGING PLASTER** R092000-50									
0020	100 lb. bags, less than 1 ton				Bag	22.50			22.50	25
0100	Over 1 ton				"	17.60			17.60	19.35

09 23 20.30 Keenes Cement

		Crew	Daily Output	Labor-Hours	Unit	Material	Labor	Equipment	Total	Total Incl O&P
0010	**KEENES CEMENT** R092000-50									
0020	In 100 lb. bags, less than 1 ton				Bag	22			22	24
0100	Over 1 ton				"	21.50			21.50	23.50
0300	Finish only, add to plaster prices, standard	J-1	215	.186	S.Y.	2.07	5.90	.53	8.50	11.75
0400	High quality	"	144	.278	"	2.09	8.80	.79	11.68	16.45

09 24 Portland Cement Plastering

09 24 23 – Portland Cement Stucco

09 24 23.40 Stucco

		Crew	Daily Output	Labor-Hours	Unit	Material	Labor	Equipment	Total	Total Incl O&P
0010	**STUCCO** R092000-50									
0015	3 coats 1" thick, float finish, with mesh, on wood frame	J-2	63	.762	S.Y.	5.65	24.50	1.81	31.96	45
0100	On masonry construction, no mesh incl.	J-1	67	.597		2.15	18.90	1.70	22.75	32.50
0300	For trowel finish, add	1 Plas	170	.047			1.58		1.58	2.39
0400	For 3/4" thick, on masonry, deduct	J-1	880	.045		.60	1.44	.13	2.17	2.97
0600	For coloring and special finish, add, minimum		685	.058		.39	1.85	.17	2.41	3.40
0700	Maximum		200	.200		1.35	6.35	.57	8.27	11.65
0900	For soffits, add	J-2	155	.310		2.09	9.90	.73	12.72	18.05
1000	Exterior stucco, with bonding agent, 3 coats, on walls, no mesh incl.	J-1	200	.200		3.50	6.35	.57	10.42	14
1200	Ceilings		180	.222		3.50	7.05	.63	11.18	15.20
1300	Beams		80	.500		3.50	15.80	1.42	20.72	29.50
1500	Columns		100	.400		3.50	12.65	1.14	17.29	24.50
1600	Mesh, painted, nailed to wood, 1.8 lb.	1 Lath	60	.133		4.80	4.49		9.29	12
1800	3.6 lb.		55	.145		3.50	4.90		8.40	11.15
1900	Wired to steel, painted, 1.8 lb.		53	.151		4.80	5.10		9.90	12.90
2100	3.6 lb.		50	.160		3.50	5.40		8.90	11.90

09 26 Veneer Plastering

09 26 13 – Gypsum Veneer Plastering

09 26 13.20 Blueboard

		Crew	Daily Output	Labor-Hours	Unit	Material	Labor	Equipment	Total	Total Incl O&P
0010	**BLUEBOARD** For use with thin coat									
0100	plaster application (see division 09 26 13.80)									
1000	3/8" thick, on walls or ceilings, standard, no finish included	2 Carp	1900	.008	S.F.	.30	.31		.61	.81
1100	With thin coat plaster finish		875	.018		.38	.67		1.05	1.47
1400	On beams, columns, or soffits, standard, no finish included		675	.024		.35	.87		1.22	1.73
1450	With thin coat plaster finish		475	.034		.42	1.24		1.66	2.38
3000	1/2" thick, on walls or ceilings, standard, no finish included		1900	.008		.32	.31		.63	.83
3100	With thin coat plaster finish		875	.018		.40	.67		1.07	1.49
3300	Fire resistant, no finish included		1900	.008		.32	.31		.63	.83
3400	With thin coat plaster finish		875	.018		.40	.67		1.07	1.49
3450	On beams, columns, or soffits, standard, no finish included		675	.024		.37	.87		1.24	1.75
3500	With thin coat plaster finish		475	.034		.45	1.24		1.69	2.41
3700	Fire resistant, no finish included		675	.024		.37	.87		1.24	1.75
3800	With thin coat plaster finish		475	.034		.45	1.24		1.69	2.41
5000	5/8" thick, on walls or ceilings, fire resistant, no finish included		1900	.008		.45	.31		.76	.98

09 26 Veneer Plastering

09 26 13 – Gypsum Veneer Plastering

09 26 13.20 Blueboard

		Crew	Daily Output	Labor-Hours	Unit	Material	2007 Bare Costs Labor	Equipment	Total	Total Incl O&P
5100	With thin coat plaster finish	2 Carp	875	.018	S.F.	.53	.67		1.20	1.63
5500	On beams, columns, or soffits, no finish included		675	.024		.52	.87		1.39	1.92
5600	With thin coat plaster finish		475	.034		.60	1.24		1.84	2.57
6000	For high ceilings, over 8' high, add		3060	.005			.19		.19	.30
6500	For over 3 stories high, add per story		6100	.003			.10		.10	.15

09 26 13.80 Thin Coat Plaster

		Crew	Daily Output	Labor-Hours	Unit	Material	Labor	Equipment	Total	Total Incl O&P
0010	**THIN COAT PLASTER**									
0012	1 coat veneer, not incl. lath	J-1	3600	.011	S.F.	.08	.35	.03	.46	.65
1000	In 50 lb. bags				Bag	10.50			10.50	11.55

09 28 Backing Boards and Underlayments

09 28 13 – Cementitious Backing Boards

09 28 13.10 Cementitious Backerboard

		Crew	Daily Output	Labor-Hours	Unit	Material	Labor	Equipment	Total	Total Incl O&P
0010	**CEMENTITIOUS BACKERBOARD**									
0070	Cementitious backerboard, on floor, 3' x 4' x 1/2" sheets	2 Carp	525	.030	S.F.	.96	1.12		2.08	2.80
0080	3' x 5' x 1/2" sheets		525	.030		.93	1.12		2.05	2.76
0090	3' x 6' x 1/2" sheets		525	.030		.72	1.12		1.84	2.54
0100	3' x 4' x 5/8" sheets		525	.030		1.08	1.12		2.20	2.93
0110	3' x 5' x 5/8" sheets		525	.030		1.06	1.12		2.18	2.91
0120	3' x 6' x 5/8" sheets		525	.030		1.12	1.12		2.24	2.97
0150	On wall, 3' x 4' x 1/2" sheets		350	.046		.96	1.68		2.64	3.67
0160	3' x 5' x 1/2" sheets		350	.046		.93	1.68		2.61	3.63
0170	3' x 6' x 1/2" sheets		350	.046		.72	1.68		2.40	3.41
0180	3' x 4' x 5/8" sheets		350	.046		1.08	1.68		2.76	3.80
0190	3' x 5' x 5/8" sheets		350	.046		1.06	1.68		2.74	3.78
0200	3' x 6' x 5/8" sheets		350	.046		1.12	1.68		2.80	3.84
0250	On counter, 3' x 4' x 1/2" sheets		180	.089		.96	3.26		4.22	6.15
0260	3' x 5' x 1/2" sheets		180	.089		.93	3.26		4.19	6.10
0270	3' x 6' x 1/2" sheets		180	.089		.72	3.26		3.98	5.90
0300	3' x 4' x 5/8" sheets		180	.089		1.08	3.26		4.34	6.30
0310	3' x 5' x 5/8" sheets		180	.089		1.06	3.26		4.32	6.25
0320	3' x 6' x 5/8" sheets		180	.089		1.12	3.26		4.38	6.35

09 29 Gypsum Board

09 29 10 – Gypsum Board

09 29 10.10 Gypsum Board Ceilings

		Crew	Daily Output	Labor-Hours	Unit	Material	Labor	Equipment	Total	Total Incl O&P
0010	**GYPSUM BOARD CEILINGS**, fire rated, finished									
0100	Screwed to grid, channel or joists, 1/2" thick	2 Carp	765	.021	S.F.	.38	.77		1.15	1.62
0150	Mold resistant		765	.021		.47	.77		1.24	1.72
0200	5/8" thick		765	.021		.38	.77		1.15	1.62
0250	Mold resistant		765	.021		.54	.77		1.31	1.79
0300	Over 8' high, 1/2" thick		615	.026		.38	.95		1.33	1.91
0350	Mold resistant		615	.026		.47	.95		1.42	2.01
0400	5/8" thick		615	.026		.38	.95		1.33	1.91
0450	Mold resistant		615	.026		.54	.95		1.49	2.08
0600	Grid suspension system, direct hung									
0700	1-1/2" C.R.C., with 7/8" hi hat furring channel, 16" O.C.	2 Carp	600	.027	S.F.	1.15	.98		2.13	2.79
0800	24" O.C.		900	.018		1.05	.65		1.70	2.18
0900	3-5/8" C.R.C., with 7/8" hi hat furring channel, 16" O.C.		600	.027		1.22	.98		2.20	2.86

09 29 Gypsum Board

09 29 10 – Gypsum Board

09 29 10.10 Gypsum Board Ceilings		Crew	Daily Output	Labor-Hours	Unit	Material	Labor	Equipment	Total	Total Incl O&P
1000	24" O.C.	2 Carp	900	.018	S.F.	1.06	.65		1.71	2.19

09 29 10.30 Gypsum Board			Crew	Daily Output	Labor-Hours	Unit	Material	Labor	Equipment	Total	Total Incl O&P
0010	**GYPSUM BOARD** on walls & ceilings	R092910-10									
0100	Nailed or screwed to studs unless otherwise noted										
0150	3/8" thick, on walls, standard, no finish included		2 Carp	2000	.008	S.F.	.33	.29		.62	.82
0200	On ceilings, standard, no finish included			1800	.009		.33	.33		.66	.87
0250	On beams, columns, or soffits, no finish included			675	.024		.33	.87		1.20	1.71
0300	1/2" thick, on walls, standard, no finish included	CN		2000	.008		.32	.29		.61	.81
0350	Taped and finished (level 4 finish)			965	.017		.36	.61		.97	1.35
0390	With compound skim coat (level 5 finish)			775	.021		.40	.76		1.16	1.62
0400	Fire resistant, no finish included			2000	.008		.38	.29		.67	.88
0450	Taped and finished (level 4 finish)			965	.017		.42	.61		1.03	1.41
0490	With compound skim coat (level 5 finish)			775	.021		.46	.76		1.22	1.69
0500	Water resistant, no finish included			2000	.008		.39	.29		.68	.89
0550	Taped and finished (level 4 finish)			965	.017		.43	.61		1.04	1.42
0590	With compound skim coat (level 5 finish)			775	.021		.47	.76		1.23	1.70
0600	Prefinished, vinyl, clipped to studs			900	.018		.60	.65		1.25	1.68
0700	Mold resistant, no finish included			2000	.008		.43	.29		.72	.93
0710	Taped and finished (level 4 finish)			965	.017		.47	.61		1.08	1.47
0720	With compound skim coat (level 5 finish)			775	.021		.51	.76		1.27	1.74
1000	On ceilings, standard, no finish included			1800	.009		.32	.33		.65	.86
1050	Taped and finished (level 4 finish)			765	.021		.36	.77		1.13	1.60
1090	With compound skim coat (level 5 finish)			610	.026		.40	.96		1.36	1.94
1100	Fire resistant, no finish included			1800	.009		.38	.33		.71	.93
1150	Taped and finished (level 4 finish)			765	.021		.42	.77		1.19	1.66
1195	With compound skim coat (level 5 finish)			610	.026		.46	.96		1.42	2.01
1200	Water resistant, no finish included			1800	.009		.39	.33		.72	.94
1250	Taped and finished (level 4 finish)			765	.021		.43	.77		1.20	1.67
1290	With compound skim coat (level 5 finish)			610	.026		.47	.96		1.43	2.02
1310	Mold resistant, no finish included			1800	.009		.43	.33		.76	.98
1320	Taped and finished (level 4 finish)			765	.021		.47	.77		1.24	1.72
1330	With compound skim coat (level 5 finish)			610	.026		.51	.96		1.47	2.06
1500	On beams, columns, or soffits, standard, no finish included			675	.024		.37	.87		1.24	1.75
1550	Taped and finished (level 4 finish)			475	.034		.36	1.24		1.60	2.32
1590	With compound skim coat (level 5 finish)			540	.030		.40	1.09		1.49	2.13
1600	Fire resistant, no finish included			675	.024		.38	.87		1.25	1.77
1650	Taped and finished (level 4 finish)			475	.034		.42	1.24		1.66	2.38
1690	With compound skim coat (level 5 finish)			540	.030		.46	1.09		1.55	2.20
1700	Water resistant, no finish included			675	.024		.45	.87		1.32	1.84
1750	Taped and finished (level 4 finish)			475	.034		.43	1.24		1.67	2.39
1790	With compound skim coat (level 5 finish)			540	.030		.47	1.09		1.56	2.21
1800	Mold resistant, no finish included			675	.024		.49	.87		1.36	1.89
1810	Taped and finished (level 4 finish)			475	.034		.47	1.24		1.71	2.44
1820	With compound skim coat (level 5 finish)			540	.030		.51	1.09		1.60	2.25
2000	5/8" thick, on walls, standard, no finish included			2000	.008		.39	.29		.68	.89
2050	Taped and finished (level 4 finish)			965	.017		.43	.61		1.04	1.42
2090	With compound skim coat (level 5 finish)			775	.021		.47	.76		1.23	1.70
2100	Fire resistant, no finish included			2000	.008		.38	.29		.67	.88
2150	Taped and finished (level 4 finish)			965	.017		.42	.61		1.03	1.41
2195	With compound skim coat (level 5 finish)			775	.021		.46	.76		1.22	1.69
2200	Water resistant, no finish included			2000	.008		.39	.29		.68	.89
2250	Taped and finished (level 4 finish)			965	.017		.43	.61		1.04	1.42

09 29 Gypsum Board

09 29 10 – Gypsum Board

09 29 10.30 Gypsum Board

		Crew	Daily Output	Labor-Hours	Unit	Material	2007 Bare Costs Labor	2007 Bare Costs Equipment	Total	Total Incl O&P
2290	With compound skim coat (level 5 finish)	2 Carp	775	.021	S.F.	.47	.76		1.23	1.70
2300	Prefinished, vinyl, clipped to studs		900	.018		.70	.65		1.35	1.79
2510	Mold resistant, no finish included		2000	.008		.49	.29		.78	1
2520	Taped and finished (level 4 finish)		965	.017		.53	.61		1.14	1.53
2530	With compound skim coat (level 5 finish)		775	.021		.57	.76		1.33	1.81
3000	On ceilings, standard, no finish included		1800	.009		.39	.33		.72	.94
3050	Taped and finished (level 4 finish)		765	.021		.43	.77		1.20	1.67
3090	With compound skim coat (level 5 finish)		615	.026		.47	.95		1.42	2.01
3100	Fire resistant, no finish included		1800	.009		.38	.33		.71	.93
3150	Taped and finished (level 4 finish)		765	.021		.42	.77		1.19	1.66
3190	With compound skim coat (level 5 finish)		615	.026		.46	.95		1.41	2
3200	Water resistant, no finish included		1800	.009		.39	.33		.72	.94
3250	Taped and finished (level 4 finish)		765	.021		.43	.77		1.20	1.67
3290	With compound skim coat (level 5 finish)		615	.026		.47	.95		1.42	2.01
3300	Mold resistant, no finish included		1800	.009		.49	.33		.82	1.05
3310	Taped and finished (level 4 finish)		765	.021		.53	.77		1.30	1.78
3320	With compound skim coat (level 5 finish)		615	.026		.57	.95		1.52	2.12
3500	On beams, columns, or soffits, no finish included		675	.024		.45	.87		1.32	1.84
3550	Taped and finished (level 4 finish)		475	.034		.49	1.24		1.73	2.46
3590	With compound skim coat (level 5 finish)		380	.042		.54	1.55		2.09	3.01
3600	Fire resistant, no finish included		675	.024		.44	.87		1.31	1.83
3650	Taped and finished (level 4 finish)		475	.034		.48	1.24		1.72	2.45
3690	With compound skim coat (level 5 finish)		380	.042		.46	1.55		2.01	2.92
3700	Water resistant, no finish included		675	.024		.45	.87		1.32	1.84
3750	Taped and finished (level 4 finish)		475	.034		.49	1.24		1.73	2.46
3790	With compound skim coat (level 5 finish)		380	.042		.47	1.55		2.02	2.93
3800	Mold resistant, no finish included		675	.024		.56	.87		1.43	1.97
3810	Taped and finished (level 4 finish)		475	.034		.61	1.24		1.85	2.59
3820	With compound skim coat (level 5 finish)		380	.042		.57	1.55		2.12	3.04
4000	Fireproofing, beams or columns, 2 layers, 1/2" thick, incl finish		330	.048		.80	1.78		2.58	3.65
4010	Mold resistant		330	.048		.90	1.78		2.68	3.76
4050	5/8" thick		300	.053		.84	1.96		2.80	3.97
4060	Mold resistant		300	.053		1.06	1.96		3.02	4.22
4100	3 layers, 1/2" thick		225	.071		1.18	2.61		3.79	5.35
4110	Mold resistant		225	.071		1.33	2.61		3.94	5.50
4150	5/8" thick		210	.076		1.26	2.80		4.06	5.75
4160	Mold resistant		210	.076		1.59	2.80		4.39	6.10
5050	For 1" thick coreboard on columns		480	.033		.45	1.22		1.67	2.40
5100	For foil-backed board, add					.10			.10	.11
5200	For work over 8' high, add	2 Carp	3060	.005			.19		.19	.30
5270	For textured spray, add	2 Lath	1600	.010		.05	.34		.39	.56
5300	For over 3 stories high, add per story	2 Carp	6100	.003			.10		.10	.15
5350	For finishing inner corners, add		950	.017	L.F.	.09	.62		.71	1.06
5355	For finishing outer corners, add		1250	.013		.20	.47		.67	.95
5500	For acoustical sealant, add per bead	1 Carp	500	.016		.03	.59		.62	.94
5550	Sealant, 1 quart tube				Ea.	5.15			5.15	5.65
5600	Sound deadening board, 1/4" gypsum	2 Carp	1800	.009	S.F.	.31	.33		.64	.85
5650	1/2" wood fiber	"	1800	.009	"	.41	.33		.74	.96

09 29 10.50 High Abuse Gypsum Board

		Crew	Daily Output	Labor-Hours	Unit	Material	Labor	Equipment	Total	Total Incl O&P
0010	**HIGH ABUSE GYPSUM BOARD**, fiber reinforced, nailed or									
0100	screwed to studs unless otherwise noted									
0110	1/2" thick, on walls, no finish included	2 Carp	1800	.009	S.F.	.56	.33		.89	1.13

09 29 Gypsum Board

09 29 10 – Gypsum Board

09 29 10.50 High Abuse Gypsum Board

		Crew	Daily Output	Labor-Hours	Unit	Material	2007 Bare Costs Labor	Equipment	Total	Total Incl O&P
0120	Taped and finished (level 4 finish)	2 Carp	870	.018	S.F.	.60	.67		1.27	1.71
0130	With compound skim coat (level 5 finish)		700	.023		.64	.84		1.48	2.02
0150	On ceilings, no finish included		1620	.010		.56	.36		.92	1.18
0160	Taped and finished (level 4 finish)		690	.023		.60	.85		1.45	1.99
0170	With compound skim coat (level 5 finish)		550	.029		.64	1.07		1.71	2.37
0210	5/8" thick, on walls, no finish included		1800	.009		.69	.33		1.02	1.27
0220	Taped and finished (level 4 finish)		870	.018		.73	.67		1.40	1.85
0230	With compound skim coat (level 5 finish)		700	.023		.77	.84		1.61	2.16
0250	On ceilings, no finish included		1620	.010		.69	.36		1.05	1.32
0260	Taped and finished (level 4 finish)		690	.023		.73	.85		1.58	2.13
0270	With compound skim coat (level 5 finish)		550	.029		.77	1.07		1.84	2.51
0310	5/8" thick, on walls, very high impact, no finish included		1800	.009		.69	.33		1.02	1.27
0320	Taped and finished (level 4 finish)		870	.018		.73	.67		1.40	1.85
0330	With compound skim coat (level 5 finish)		700	.023		.77	.84		1.61	2.16
0350	On ceilings, no finish included		1620	.010		.69	.36		1.05	1.32
0360	Taped and finished (level 4 finish)		690	.023		.73	.85		1.58	2.13
0370	With compound skim coat (level 5 finish)		550	.029		.77	1.07		1.84	2.51
0400	High abuse, gypsum core, paper face									
0410	1/2" thick, on walls, no finish included	2 Carp	1800	.009	S.F.	.63	.33		.96	1.20
0420	Taped and finished (level 4 finish)		870	.018		.67	.67		1.34	1.79
0430	With compound skim coat (level 5 finish)		700	.023		.71	.84		1.55	2.09
0450	On ceilings, no finish included		1620	.010		.63	.36		.99	1.25
0460	Taped and finished (level 4 finish)		690	.023		.67	.85		1.52	2.07
0470	With compound skim coat (level 5 finish)		550	.029		.71	1.07		1.78	2.44
0510	5/8" thick, on walls, no finish included		1800	.009		.68	.33		1.01	1.26
0520	Taped and finished (level 4 finish)		870	.018		.72	.67		1.39	1.84
0530	With compound skim coat (level 5 finish)		700	.023		.76	.84		1.60	2.15
0550	On ceilings, no finish included		1620	.010		.68	.36		1.04	1.31
0560	Taped and finished (level 4 finish)		690	.023		.72	.85		1.57	2.12
0570	With compound skim coat (level 5 finish)		550	.029		.76	1.07		1.83	2.50
1000	For high ceilings, over 8' high, add		2750	.006			.21		.21	.33
1010	For over 3 stories high, add per story		5500	.003			.11		.11	.17

09 29 15 – Gypsum Board Accessories

09 29 15.10 Accessories, Gypsum Board

		Crew	Daily Output	Labor-Hours	Unit	Material	2007 Bare Costs Labor	Equipment	Total	Total Incl O&P
0010	**ACCESSORIES, GYPSUM BOARD**									
0020	Casing bead, galvanized steel	1 Carp	2.90	2.759	C.L.F.	20.50	101		121.50	181
0100	Vinyl		3	2.667		21.50	98		119.50	176
0300	Corner bead, galvanized steel, 1" x 1"		4	2		13.50	73.50		87	129
0400	Corner bead, galvanized steel, 1-1/4" x 1-1/4"		3.50	2.286		21	84		105	154
0600	Vinyl corner bead		4	2		21	73.50		94.50	138
0900	Furring channel, galv. steel, 7/8" deep, standard		2.60	3.077		26.50	113		139.50	205
1000	Resilient		2.55	3.137		27	115		142	209
1100	J trim, galvanized steel, 1/2" wide		3	2.667		22	98		120	177
1120	5/8" wide		2.95	2.712		22	99.50		121.50	179
1140	L trim, galvanized		3	2.667		16.90	98		114.90	171
1150	U trim, galvanized		2.95	2.712		18.90	99.50		118.40	176
1160	Screws #6 x 1" A				M	5.80			5.80	6.40
1170	#6 x 1-5/8" A				"	9.20			9.20	10.10
1200	For stud partitions, see Divisions 05 41 13.30 and 09 22 16.13									
1500	Z stud, galvanized steel, 1-1/2" wide	1 Carp	2.60	3.077	C.L.F.	36.50	113		149.50	216
1600	2" wide	"	2.55	3.137	"	51	115		166	235

09 30 Tiling

09 30 13 – Ceramic Tiling

09 30 13.10 Ceramic Tile		Crew	Daily Output	Labor-Hours	Unit	Material	2007 Bare Costs Labor	Equipment	Total	Total Incl O&P
0010	**CERAMIC TILE**									
0050	Base, using 1' x 4" high pc. with 1" x 1" tiles, mud set	D-7	82	.195	L.F.	4.48	6.10		10.58	13.95
0100	Thin set	"	128	.125		4.26	3.92		8.18	10.45
0300	For 6" high base, 1" x 1" tile face, add					.70			.70	.77
0400	For 2" x 2" tile face, add to above					.37			.37	.41
0600	Cove base, 4-1/4" x 4-1/4" high, mud set	D-7	91	.176		3.48	5.50		8.98	11.95
0700	Thin set		128	.125		3.50	3.92		7.42	9.60
0900	6" x 4-1/4" high, mud set		100	.160		3.23	5		8.23	10.90
1000	Thin set		137	.117		3.23	3.66		6.89	8.95
1200	Sanitary cove base, 6" x 4-1/4" high, mud set		93	.172		3.55	5.40		8.95	11.80
1300	Thin set		124	.129		4.04	4.05		8.09	10.40
1500	6" x 6" high, mud set		84	.190		4.47	6		10.47	13.65
1600	Thin set		117	.137		4.47	4.29		8.76	11.20
1800	Bathroom accessories, average		82	.195	Ea.	10.35	6.10		16.45	20.50
1900	Bathtub, 5', rec. 4-1/4" x 4-1/4" tile wainscot, adhesive set 6' high		2.90	5.517		152	173		325	420
2100	7' high wainscot		2.50	6.400		174	201		375	485
2200	8' high wainscot		2.20	7.273		184	228		412	540
2400	Bullnose trim, 4-1/4" x 4-1/4", mud set		82	.195	L.F.	3.08	6.10		9.18	12.40
2500	Thin set		128	.125		2.85	3.92		6.77	8.90
2700	6" x 4-1/4" bullnose trim, mud set		84	.190		2.54	6		8.54	11.55
2800	Thin set		124	.129		2.54	4.05		6.59	8.75
3000	Floors, natural clay, random or uniform, thin set, color group 1		183	.087	S.F.	3.95	2.74		6.69	8.40
3100	Color group 2		183	.087		4.26	2.74		7	8.70
3255	Floors, glazed, thin set, 6" x 6", color group 1		200	.080		3.36	2.51		5.87	7.40
3260	8" x 8" tile		250	.064		3.36	2.01		5.37	6.65
3270	12" x 12" tile		325	.049		4.22	1.54		5.76	6.90
3280	16" x 16" tile		550	.029		5.85	.91		6.76	7.80
3285	Border, 6" x 12" tile		275	.058		11.10	1.83		12.93	14.90
3290	3" x 12" tile		200	.080		32.50	2.51		35.01	39.50
3300	Porcelain type, 1 color, color group 2, 1" x 1"		183	.087		4.57	2.74		7.31	9.10
3310	2" x 2" or 2" x 1", thin set		190	.084		4.90	2.64		7.54	9.30
3350	For random blend, 2 colors, add					.85			.85	.94
3360	4 colors, add					1.20			1.20	1.32
3370	For color group 3, add					.49			.49	.54
3380	For abrasive non-slip tile, add					.48			.48	.53
4300	Specialty tile, 4-1/4" x 4-1/4" x 1/2", decorator finish	D-7	183	.087		9.70	2.74		12.44	14.75
4500	Add for epoxy grout, 1/16" joint, 1" x 1" tile		800	.020		.60	.63		1.23	1.58
4600	2" x 2" tile		820	.020		.54	.61		1.15	1.49
4800	Pregrouted sheets, walls, 4-1/4" x 4-1/4", 6" x 4-1/4"									
4810	and 8-1/2" x 4-1/4", 4 S.F. sheets, silicone grout	D-7	240	.067	S.F.	4.59	2.09		6.68	8.10
5100	Floors, unglazed, 2 S.F. sheets,									
5110	urethane adhesive	D-7	180	.089	S.F.	4.57	2.79		7.36	9.15
5400	Walls, interior, thin set, 4-1/4" x 4-1/4" tile	**CN**	190	.084		2.16	2.64		4.80	6.25
5500	6" x 4-1/4" tile		190	.084		2.51	2.64		5.15	6.65
5700	8-1/2" x 4-1/4" tile		190	.084		3.55	2.64		6.19	7.80
5800	6" x 6" tile		200	.080		2.98	2.51		5.49	6.95
5810	8" x 8" tile		225	.071		3.97	2.23		6.20	7.65
5820	12" x 12" tile		300	.053		3.25	1.67		4.92	6.05
5830	16" x 16" tile		500	.032		3.52	1		4.52	5.35
6000	Decorated wall tile, 4-1/4" x 4-1/4", minimum		270	.059		3.95	1.86		5.81	7.10
6100	Maximum		180	.089		41.50	2.79		44.29	50
6300	Exterior walls, frostproof, mud set, 4-1/4" x 4-1/4"		102	.157		4.30	4.92		9.22	11.95
6400	1-3/8" x 1-3/8"		93	.172		4.13	5.40		9.53	12.45

09 30 Tiling

09 30 13 – Ceramic Tiling

09 30 13.10 Ceramic Tile

		Crew	Daily Output	Labor-Hours	Unit	Material	2007 Bare Costs Labor	Equipment	Total	Total Incl O&P
6600	Crystalline glazed, 4-1/4" x 4-1/4", mud set, plain	D-7	100	.160	S.F.	3.63	5		8.63	11.35
6700	4-1/4" x 4-1/4", scored tile		100	.160		4.42	5		9.42	12.20
6900	6" x 6" plain		93	.172		4.82	5.40		10.22	13.20
7000	For epoxy grout, 1/16" joints, 4-1/4" tile, add		800	.020		.35	.63		.98	1.31
7200	For tile set in dry mortar, add		1735	.009			.29		.29	.42
7300	For tile set in portland cement mortar, add	↓	290	.055			1.73		1.73	2.54
9000	Regrout tile 4-1/2 x 4-1/2, or larger, wall	1 Tilf	100	.080		.14	2.84		2.98	4.32
9220	Floor	"	125	.064	↓	.15	2.27		2.42	3.50

09 30 13.30 Ceramic Tile Panels

0010	**CERAMIC TILE PANELS**									
0020	Insulated, over 1000 S.F., 1-1/2" thick	D-7	220	.073	S.F.	9.30	2.28		11.58	13.55
0100	2-1/2" thick	"	220	.073	"	9.95	2.28		12.23	14.30

09 30 16 – Quarry Tiling

09 30 16.10 Quarry Tile

0010	**QUARRY TILE** Base, cove or sanitary, 2" or 5" high, mud set									
0100	1/2" thick	D-7	110	.145	L.F.	4.79	4.56		9.35	11.95
0300	Bullnose trim, red, mud set, 6" x 6" x 1/2" thick		120	.133		4.02	4.18		8.20	10.55
0400	4" x 4" x 1/2" thick		110	.145		4.53	4.56		9.09	11.70
0600	4" x 8" x 1/2" thick, using 8" as edge		130	.123	↓	3.99	3.86		7.85	10.05
0700	Floors, mud set, 1,000 S.F. lots, red, 4" x 4" x 1/2" thick		120	.133	S.F.	4.08	4.18		8.26	10.65
0900	6" x 6" x 1/2" thick		140	.114		3.36	3.59		6.95	8.95
1000	4" x 8" x 1/2" thick	↓	130	.123		4.08	3.86		7.94	10.15
1300	For waxed coating, add					.65			.65	.72
1500	For non-standard colors, add					.39			.39	.43
1600	For abrasive surface, add					.46			.46	.51
1800	Brown tile, imported, 6" x 6" x 3/4"	D-7	120	.133		4.85	4.18		9.03	11.50
1900	8" x 8" x 1"		110	.145		5.50	4.56		10.06	12.75
2100	For thin set mortar application, deduct		700	.023			.72		.72	1.05
2200	For epoxy grout & mortar, 6" x 6" x 1/2", add		350	.046		1.78	1.43		3.21	4.06
2700	Stair tread, 6" x 6" x 3/4", plain		50	.320		5.10	10.05		15.15	20.50
2800	Abrasive		47	.340		5.15	10.70		15.85	21.50
3000	Wainscot, 6" x 6" x 1/2", thin set, red		105	.152		3.85	4.78		8.63	11.25
3100	Non-standard colors		105	.152	↓	4.29	4.78		9.07	11.70
3300	Window sill, 6" wide, 3/4" thick		90	.178	L.F.	4.94	5.60		10.54	13.65
3400	Corners	↓	80	.200	Ea.	5.40	6.30		11.70	15.15

09 30 23 – Glass Mosaic Tiling

09 30 23.10 Glass Mosaics

0010	**GLASS MOSAICS** 3/4" tile on 12" sheets, standard grout									
0300	Color group 1 & 2	D-7	73	.219	S.F.	16.50	6.90		23.40	28.50
0350	Color group 3		73	.219		17.35	6.90		24.25	29
0400	Color group 4		73	.219		23.50	6.90		30.40	36
0450	Color group 5		73	.219		26.50	6.90		33.40	39.50
0500	Color group 6		73	.219		32	6.90		38.90	45.50
0600	Color group 7		73	.219		37.50	6.90		44.40	51
0700	Color group 8, golds, silvers & specialties	↓	64	.250	↓	54	7.85		61.85	71

09 30 29 – Metal Tiling

09 30 29.10 Metal Tile

0010	**METAL TILE** 4' x 4' sheet, 24 ga., tile pattern, nailed									
0200	Stainless steel	2 Carp	512	.031	S.F.	24.50	1.15		25.65	29
0400	Aluminized steel	"	512	.031	"	13.15	1.15		14.30	16.25

09 51 Acoustical Ceilings

09 51 23 – Acoustical Tile Ceilings

09 51 23.10 Suspended Acoustic Ceiling Tiles

		Crew	Daily Output	Labor-Hours	Unit	Material	2007 Bare Costs Labor	Equipment	Total	Total Incl O&P
0010	**SUSPENDED ACOUSTIC CEILING TILES**, Not including									
0100	suspension system									
0300	Fiberglass boards, film faced, 2' x 2' or 2' x 4', 5/8" thick	1 Carp	625	.013	S.F.	.59	.47		1.06	1.38
0400	3/4" thick		600	.013		1.36	.49		1.85	2.26
0500	3" thick, thermal, R11		450	.018		1.43	.65		2.08	2.59
0600	Glass cloth faced fiberglass, 3/4" thick		500	.016		1.74	.59		2.33	2.82
0700	1" thick		485	.016		1.93	.61		2.54	3.06
0820	1-1/2" thick, nubby face		475	.017		2.48	.62		3.10	3.69
1110	Mineral fiber tile, lay-in, 2' x 2' or 2' x 4', 5/8" thick, fine texture		625	.013		.47	.47		.94	1.25
1115	Rough textured		625	.013		1.14	.47		1.61	1.98
1125	3/4" thick, fine textured		600	.013		1.32	.49		1.81	2.21
1130	Rough textured		600	.013		1.65	.49		2.14	2.58
1135	Fissured		600	.013		2.09	.49		2.58	3.06
1150	Tegular, 5/8" thick, fine textured		470	.017		1.24	.62		1.86	2.33
1155	Rough textured		470	.017		1.62	.62		2.24	2.75
1165	3/4" thick, fine textured		450	.018		1.77	.65		2.42	2.97
1170	Rough textured		450	.018		2	.65		2.65	3.22
1175	Fissured		450	.018		3.10	.65		3.75	4.43
1180	For aluminum face, add					5.15			5.15	5.70
1185	For plastic film face, add					.82			.82	.90
1190	For fire rating, add					.38			.38	.42
1300	Mirror faced panels, 15/16" thick, 2' x 2'	1 Carp	500	.016		10.70	.59		11.29	12.70
1900	Eggcrate, acrylic, 1/2" x 1/2" x 1/2" cubes		500	.016		1.55	.59		2.14	2.62
2100	Polystyrene eggcrate, 3/8" x 3/8" x 1/2" cubes		510	.016		1.30	.58		1.88	2.33
2200	1/2" x 1/2" x 1/2" cubes		500	.016		1.74	.59		2.33	2.82
2400	Luminous panels, prismatic, acrylic		400	.020		1.89	.73		2.62	3.22
2500	Polystyrene		400	.020		.97	.73		1.70	2.21
2700	Flat white acrylic		400	.020		3.29	.73		4.02	4.76
2800	Polystyrene		400	.020		2.25	.73		2.98	3.62
3000	Drop pan, white, acrylic		400	.020		4.82	.73		5.55	6.45
3100	Polystyrene		400	.020		4.03	.73		4.76	5.55
3600	Perforated aluminum sheets, .024" thick, corrugated, painted		490	.016		1.92	.60		2.52	3.04
3700	Plain		500	.016		3.25	.59		3.84	4.49
3720	Mineral fiber, 24" x 24" or 48", reveal edge, painted, 5/8" thick		600	.013		1.18	.49		1.67	2.06
3740	3/4" thick		575	.014		1.91	.51		2.42	2.90

09 51 23.30 Suspended Ceilings, Complete

		Crew	Daily Output	Labor-Hours	Unit	Material	Labor	Equipment	Total	Total Incl O&P
0010	**SUSPENDED CEILINGS, COMPLETE** Including standard									
0100	suspension system but not incl. 1-1/2" carrier channels									
0600	Fiberglass ceiling board, 2' x 4' x 5/8", plain faced,	1 Carp	500	.016	S.F.	1.14	.59		1.73	2.16
0700	Offices, 2' x 4' x 3/4"		380	.021		1.91	.77		2.68	3.30
0800	Mineral fiber, on 15/16" T bar susp. 2' x 2' x 3/4" lay-in board		345	.023		2	.85		2.85	3.53
0810	2' x 4' x 5/8" tile		380	.021		1.02	.77		1.79	2.32
0820	Tegular, 2' x 2' x 5/8" tile on 9/16" grid		250	.032		2.08	1.17		3.25	4.12
0830	2' x 4' x 3/4" tile		275	.029		2.48	1.07		3.55	4.39
0900	Luminous panels, prismatic, acrylic		255	.031		2.44	1.15		3.59	4.47
1200	Metal pan with acoustic pad, steel		75	.107		3.62	3.91		7.53	10.10
1300	Painted aluminum		75	.107		2.47	3.91		6.38	8.80
1500	Aluminum, degreased finish		75	.107		4.18	3.91		8.09	10.70
1600	Stainless steel		75	.107		7.85	3.91		11.76	14.75
1800	Tile, Z bar suspension, 5/8" mineral fiber tile		150	.053		1.88	1.96		3.84	5.10
1900	3/4" mineral fiber tile		150	.053		2.01	1.96		3.97	5.25
2400	For strip lighting, see division 26 51 13.50									

09 51 Acoustical Ceilings

09 51 23 – Acoustical Tile Ceilings

09 51 23.30 Suspended Ceilings, Complete	Crew	Daily Output	Labor-Hours	Unit	Material	2007 Bare Costs Labor	Equipment	Total	Total Incl O&P
2500 For rooms under 500 S.F., add				S.F.		25%			

09 51 53 – Direct-Applied Acoustical Ceilings

09 51 53.10 Ceiling Tile

		Crew	Daily Output	Labor-Hours	Unit	Material	Labor	Equipment	Total	Total Incl O&P
0010	**CEILING TILE**, Stapled or cemented									
0100	12" x 12" or 12" x 24", not including furring									
0600	Mineral fiber, vinyl coated, 5/8" thick	CN 1 Carp	1000	.008	S.F.	1.65	.29		1.94	2.28
0700	3/4" thick		1000	.008		1.52	.29		1.81	2.13
0900	Fire rated, 3/4" thick, plain faced		1000	.008		1.39	.29		1.68	1.99
1000	Plastic coated face		1000	.008		1.18	.29		1.47	1.76
1200	Aluminum faced, 5/8" thick, plain		1000	.008		1.31	.29		1.60	1.90
3700	Wall application of above, add		3100	.003			.09		.09	.15
3900	For ceiling primer, add					.13			.13	.14
4000	For ceiling cement, add					.36			.36	.40

09 53 Acoustical Ceiling Suspension Assemblies

09 53 23 – Metal Acoustical Ceiling Suspension Assemblies

09 53 23.30 Ceiling Suspension Systems

		Crew	Daily Output	Labor-Hours	Unit	Material	Labor	Equipment	Total	Total Incl O&P
0010	**CEILING SUSPENSION SYSTEMS** for boards and tile									
0050	Class A suspension system, 15/16" T bar, 2' x 4' grid	CN 1 Carp	800	.010	S.F.	.55	.37		.92	1.17
0300	2' x 2' grid	"	650	.012		.68	.45		1.13	1.45
0350	For 9/16" grid, add					.16			.16	.18
0360	For fire rated grid, add					.09			.09	.10
0370	For colored grid, add					.20			.20	.22
0400	Concealed Z bar suspension system, 12" module	1 Carp	520	.015		.65	.56		1.21	1.60
0600	1-1/2" carrier channels, 4' O.C., add	"	470	.017		.12	.62		.74	1.10
0700	Carrier channels for ceilings with									
0900	recessed lighting fixtures, add	1 Carp	460	.017	S.F.	.22	.64		.86	1.23
1040	Hanging wire, 12 ga., 4' long		65	.123	C.S.F.	7.20	4.52		11.72	14.95
1080	8' long		65	.123	"	14.40	4.52		18.92	23
1200	Seismic bracing, 2-1/2" compression post with four bracing wires		50	.160	Ea.	2.74	5.85		8.59	12.15

09 63 Masonry Flooring

09 63 13 – Brick Flooring

09 63 13.10 Brick Flooring

		Crew	Daily Output	Labor-Hours	Unit	Material	Labor	Equipment	Total	Total Incl O&P
0010	**BRICK FLOORING**									
0020	Acid proof shales, red, 8" x 3-3/4" x 1-1/4" thick	D-7	.43	37.209	M	780	1,175		1,955	2,575
0050	2-1/4" thick	D-1	.40	40		865	1,325		2,190	2,975
0200	Acid proof clay brick, 8" x 3-3/4" x 2-1/4" thick	"	.40	40		805	1,325		2,130	2,925
0260	Cast ceramic, pressed, 4" x 8" x 1/2", unglazed	D-7	100	.160	S.F.	5.40	5		10.40	13.30
0270	Glazed		100	.160		7.20	5		12.20	15.25
0280	Hand molded flooring, 4" x 8" x 3/4", unglazed		95	.168		7.15	5.30		12.45	15.60
0290	Glazed		95	.168		8.95	5.30		14.25	17.60
0300	8" hexagonal, 3/4" thick, unglazed		85	.188		7.80	5.90		13.70	17.25
0310	Glazed		85	.188		14.10	5.90		20	24
0400	Heavy duty industrial, cement mortar bed, 2" thick, not incl. brick	D-1	80	.200		.74	6.65		7.39	10.95
0450	Acid proof joints, 1/4" wide	"	65	.246		1.24	8.20		9.44	13.85
0500	Pavers, 8" x 4", 1" to 1-1/4" thick, red	D-7	95	.168		3.15	5.30		8.45	11.20
0510	Ironspot	"	95	.168		4.44	5.30		9.74	12.65
0540	1-3/8" to 1-3/4" thick, red	D-1	95	.168		3.03	5.60		8.63	11.90

09 63 Masonry Flooring

09 63 13 – Brick Flooring

	09 63 13.10 Brick Flooring	Crew	Daily Output	Labor-Hours	Unit	Material	2007 Bare Costs Labor	Equipment	Total	Total Incl O&P
0560	Ironspot	D-1	95	.168	S.F.	4.39	5.60		9.99	13.40
0580	2-1/4" thick, red		90	.178		3.09	5.95		9.04	12.40
0590	Ironspot		90	.178		4.79	5.95		10.74	14.25
0870	For epoxy joints, add		600	.027		2.35	.89		3.24	3.94
0880	For Furan underlayment, add		600	.027		1.94	.89		2.83	3.48
0890	For waxed surface, steam cleaned, add	A-1H	1000	.008		.17	.23	.06	.46	.62

09 63 40 – Stone Flooring

09 63 40.10 Marble

		Crew	Daily Output	Labor-Hours	Unit	Material	Labor	Equipment	Total	Total Incl O&P
0010	**MARBLE**									
0020	Thin gauge tile, 12" x 6", 3/8", White Carara	D-7	60	.267	S.F.	9.10	8.35		17.45	22.50
0100	Travertine		60	.267		10.70	8.35		19.05	24
0200	12" x 12" x 3/8", thin set, floors		60	.267		6.45	8.35		14.80	19.40
0300	On walls		52	.308		9.35	9.65		19	24.50
1000	Marble Threshold, 4" Wide x 36" Long x 5/8" Thick, White		60	.267	Ea.	6.45	8.35		14.80	19.40

09 63 40.20 Slate Tile

		Crew	Daily Output	Labor-Hours	Unit	Material	Labor	Equipment	Total	Total Incl O&P
0010	**SLATE TILE**									
0020	Vermont, 6" x 6" x 1/4" thick, thin set	D-7	180	.089	S.F.	5	2.79		7.79	9.60
0200	See also division 32 14 40.10									

09 64 Wood Flooring

09 64 16 – Wood Block Flooring

09 64 16.10 End Grain Block Flooring

		Crew	Daily Output	Labor-Hours	Unit	Material	Labor	Equipment	Total	Total Incl O&P
0010	**END GRAIN BLOCK FLOORING**									
0020	End grain flooring, coated, 2" thick	1 Carp	295	.027	S.F.	3.19	1		4.19	5.05
0400	Natural finish, 1" thick, fir		125	.064		3.30	2.35		5.65	7.30
0600	1-1/2" thick, pine		125	.064		3.24	2.35		5.59	7.20
0700	2" thick, pine		125	.064		4.45	2.35		6.80	8.55

09 64 19 – Wood Composition Flooring

09 64 19.10 Wood Composition

		Crew	Daily Output	Labor-Hours	Unit	Material	Labor	Equipment	Total	Total Incl O&P
0010	**WOOD COMPOSITION** Gym floors									
0100	2-1/4" x 6-7/8" x 3/8", on 2" grout setting bed	D-7	150	.107	S.F.	5.60	3.35		8.95	11.05
0200	Thin set, on concrete	"	250	.064		5.10	2.01		7.11	8.55
0300	Sanding and finishing, add	1 Carp	200	.040		.75	1.47		2.22	3.12

09 64 23 – Wood Parquet Flooring

09 64 23.10 Wood Parquet

		Crew	Daily Output	Labor-Hours	Unit	Material	Labor	Equipment	Total	Total Incl O&P
0010	**WOOD PARQUET** flooring									
5200	Parquetry, standard, 5/16" thick, not incl. finish, oak, minimum	1 Carp	160	.050	S.F.	4.16	1.84		6	7.45
5300	Maximum		100	.080		4.91	2.94		7.85	9.95
5500	Teak, minimum		160	.050		4.69	1.84		6.53	8
5600	Maximum		100	.080		8.20	2.94		11.14	13.55
5650	13/16" thick, select grade oak, minimum		160	.050		9.05	1.84		10.89	12.80
5700	Maximum		100	.080		13.75	2.94		16.69	19.65
5800	Custom parquetry, including finish, minimum		100	.080		15.15	2.94		18.09	21
5900	Maximum		50	.160		20	5.85		25.85	31
6700	Parquetry, prefinished white oak, 5/16" thick, minimum		160	.050		3.61	1.84		5.45	6.85
6800	Maximum		100	.080		7.30	2.94		10.24	12.60
7000	Walnut or teak, parquetry, minimum		160	.050		5.25	1.84		7.09	8.65
7100	Maximum		100	.080		9.15	2.94		12.09	14.65
7200	Acrylic wood parquet blocks, 12" x 12" x 5/16",									

09 64 Wood Flooring

09 64 23 – Wood Parquet Flooring

09 64 23.10 Wood Parquet		Crew	Daily Output	Labor-Hours	Unit	Material	2007 Bare Costs Labor	Equipment	Total	Total Incl O&P
7210	irradiated, set in epoxy	1 Carp	160	.050	S.F.	7.65	1.84		9.49	11.25

09 64 29 – Wood Strip and Plank Flooring

09 64 29.10 Wood

		Crew	Daily Output	Labor-Hours	Unit	Material	Labor	Equipment	Total	Total Incl O&P
0010	**WOOD**									
0020	Fir, vertical grain, 1" x 4", not incl. finish, B & better	1 Carp	255	.031	S.F.	2.69	1.15		3.84	4.75
0100	C grade & better		255	.031		2.53	1.15		3.68	4.57
4000	Maple, strip, 25/32" x 2-1/4", not incl. finish, select		170	.047		5.15	1.73		6.88	8.35
4100	#2 & better		170	.047		3.22	1.73		4.95	6.25
4300	33/32" x 3-1/4", not incl. finish, #1 grade		170	.047		3.99	1.73		5.72	7.10
4400	#2 & better		170	.047		3.55	1.73		5.28	6.60
4600	Oak, white or red, 25/32" x 2-1/4", not incl. finish									
4700	#1 common	1 Carp	170	.047	S.F.	3.09	1.73		4.82	6.10
4900	Select quartered, 2-1/4" wide		170	.047		2.82	1.73		4.55	5.80
5000	Clear		170	.047		3.89	1.73		5.62	6.95
6100	Prefinished, white oak, prime grade, 2-1/4" wide		170	.047		6.40	1.73		8.13	9.75
6200	3-1/4" wide		185	.043		8.20	1.59		9.79	11.45
6400	Ranch plank		145	.055		7.90	2.02		9.92	11.85
6500	Hardwood blocks, 9" x 9", 25/32" thick		160	.050		5.65	1.84		7.49	9.10
7400	Yellow pine, 3/4" x 3-1/8", T & G, C & better, not incl. finish		200	.040		2.35	1.47		3.82	4.88
7500	Refinish wood floor, sand, 2 cts poly, wax, soft wood, min.	1 Clab	400	.020		.76	.58		1.34	1.74
7600	Hard wood, max		130	.062		1.14	1.77		2.91	4
7800	Sanding and finishing, 2 coats polyurethane		295	.027		.76	.78		1.54	2.05
7900	Subfloor and underlayment, see division 06 16 36.00									
8015	Transition molding, 2 1/4" wide, 5' long	1 Carp	19.20	.417	Ea.	15	15.30		30.30	40.50
8300	Floating floor, wood composition strip, complete.	1 Clab	133	.060	S.F.	4.24	1.73		5.97	7.35
8310	Floating floor components, T & G wood composite strips					3.64			3.64	4.01
8320	Film					.15			.15	.16
8330	Foam					.26			.26	.29
8340	Adhesive					.24			.24	.27
8350	Installation kit					.17			.17	.19
8360	Trim, 2" wide x 3' long				L.F.	2.44			2.44	2.68
8370	Reducer moulding				"	4.21			4.21	4.63

09 64 66 – Wood Athletic Flooring

09 64 66.10 Wood Athletic Flooring

		Crew	Daily Output	Labor-Hours	Unit	Material	Labor	Equipment	Total	Total Incl O&P
0010	**WOOD ATHLETIC FLOORING**									
0600	Gym floor, in mastic, over 2 ply felt, #2 & better									
0700	25/32" thick maple	1 Carp	100	.080	S.F.	3.80	2.94		6.74	8.75
0900	33/32" thick maple		98	.082		4.39	3		7.39	9.50
1000	For 1/2" corkboard underlayment, add		750	.011		.85	.39		1.24	1.55
1300	For #1 grade maple, add					.45			.45	.50
1600	Maple flooring, over sleepers, #2 & better									
1700	25/32" thick	1 Carp	85	.094	S.F.	4.11	3.45		7.56	9.90
1900	33/32" thick	"	83	.096		4.78	3.54		8.32	10.75
2000	For #1 grade, add					.48			.48	.53
2200	For 3/4" subfloor, add	1 Carp	350	.023		1.07	.84		1.91	2.49
2300	With two 1/2" subfloors, 25/32" thick	"	69	.116		5.15	4.26		9.41	12.30
2500	Maple, incl. finish, #2 & btr., 25/32" thick, on rubber									
2600	Sleepers, with two 1/2" subfloors	1 Carp	76	.105	S.F.	5.50	3.86		9.36	12.05
2800	With steel spline, double connection to channels	"	73	.110		5.90	4.02		9.92	12.75
2900	For 33/32" maple, add					.64			.64	.70
3100	For #1 grade maple, add					.48			.48	.53
3500	For termite proofing all of the above, add					.25			.25	.28

09 64 Wood Flooring

09 64 66 – Wood Athletic Flooring

09 64 66.10 Wood Athletic Flooring		Crew	Daily Output	Labor-Hours	Unit	Material	2007 Bare Costs Labor	Equipment	Total	Total Incl O&P
3700	Portable hardwood, prefinished panels	1 Carp	83	.096	S.F.	7.35	3.54		10.89	13.60
3720	Insulated with polystyrene, 1" thick, add		165	.048		.64	1.78		2.42	3.47
3750	Running tracks, Sitka spruce surface, 25/32" x 2-1/4"		62	.129		13.50	4.74		18.24	22
3770	3/4" plywood surface, finished	↓	100	.080	↓	3.20	2.94		6.14	8.10

09 65 Resilient Flooring

09 65 10 – Resilient Tile Underlayment

09 65 10.10 Resilient Tile Underlayment

		Crew	Daily Output	Labor-Hours	Unit	Material	Labor	Equipment	Total	Total Incl O&P
0010	**RESILIENT TILE UNDERLAYMENT**									
3600	Latex underlayment, 1/8" thk., cementitious for resilient flooring	1 Tilf	160	.050	S.F.	1.56	1.78		3.34	4.33
4000	Latex underlayment, liquid, fortified				Gal.	35			35	38.50

09 65 13 – Resilient Base and Accessories

09 65 13.13 Resilient Base

		Crew	Daily Output	Labor-Hours	Unit	Material	Labor	Equipment	Total	Total Incl O&P
0010	**RESILIENT BASE**									
0800	Base, cove, rubber or vinyl, .080" thick									
1100	Standard colors, 2-1/2" high	1 Tilf	315	.025	L.F.	.65	.90		1.55	2.04
1150	4" high		315	.025		.70	.90		1.60	2.09
1200	6" high		315	.025		1.10	.90		2	2.53
1450	1/8" thick, standard colors, 2-1/2" high		315	.025		.75	.90		1.65	2.15
1500	4" high	CN	315	.025		.64	.90		1.54	2.02
1600	Corners, 2-1/2" high		315	.025	Ea.	1.35	.90		2.25	2.81
1630	4" high		315	.025		2.25	.90		3.15	3.80
1660	6" high	↓	315	.025	↓	2.35	.90		3.25	3.91

09 65 13.23 Resilient Stair Treads and Risers

		Crew	Daily Output	Labor-Hours	Unit	Material	Labor	Equipment	Total	Total Incl O&P
0010	**RESILIENT STAIR TREADS AND RISERS**									
0300	Rubber, molded tread, 12" wide, 5/16" thick, black	1 Tilf	115	.070	L.F.	10.40	2.47		12.87	15.05
0400	Colors		115	.070		10.95	2.47		13.42	15.65
0600	1/4" thick, black		115	.070		10.25	2.47		12.72	14.90
0700	Colors		115	.070		10.75	2.47		13.22	15.45
0900	Grip strip safety tread, colors, 5/16" thick		115	.070		13.90	2.47		16.37	18.90
1000	3/16" thick		120	.067	↓	9.55	2.37		11.92	13.95
1200	Landings, smooth sheet rubber, 1/8" thick		120	.067	S.F.	5	2.37		7.37	8.95
1300	3/16" thick		120	.067	"	6.40	2.37		8.77	10.50
1500	Nosings, 3" wide, 3/16" thick, black		140	.057	L.F.	3.07	2.03		5.10	6.35
1600	Colors		140	.057		3.02	2.03		5.05	6.30
1800	Risers, 7" high, 1/8" thick, flat		250	.032		3.61	1.14		4.75	5.65
1900	Coved		250	.032		3.23	1.14		4.37	5.20
2100	Vinyl, molded tread, 12" wide, colors, 1/8" thick		115	.070		5	2.47		7.47	9.10
2200	1/4" thick		115	.070	↓	6.75	2.47		9.22	11.05
2300	Landing material, 1/8" thick		200	.040	S.F.	5	1.42		6.42	7.60
2400	Riser, 7" high, 1/8" thick, coved		175	.046	L.F.	2.50	1.62		4.12	5.15
2500	Tread and riser combined, 1/8" thick	↓	80	.100	"	7.85	3.55		11.40	13.85

09 65 16 – Resilient Sheet Flooring

09 65 16.10 Resilient Sheet Flooring

		Crew	Daily Output	Labor-Hours	Unit	Material	Labor	Equipment	Total	Total Incl O&P
0010	**RESILIENT SHEET FLOORING**									
5900	Rubber, sheet goods, 36" wide, 1/8" thick	1 Tilf	120	.067	S.F.	5.50	2.37		7.87	9.50
5950	3/16" thick		100	.080		7.50	2.84		10.34	12.40
6000	1/4" thick		90	.089		9	3.16		12.16	14.55
8000	Vinyl sheet goods, backed, .065" thick, minimum		250	.032		2.75	1.14		3.89	4.70
8050	Maximum	↓	200	.040	↓	3.03	1.42		4.45	5.40

09 65 Resilient Flooring

09 65 16 – Resilient Sheet Flooring

09 65 16.10 Resilient Sheet Flooring

		Crew	Daily Output	Labor-Hours	Unit	Material	2007 Bare Costs Labor	Equipment	Total	Total Incl O&P
8100	.080" thick, minimum	1 Tilf	230	.035	S.F.	3	1.23		4.23	5.10
8150	Maximum		200	.040		4.25	1.42		5.67	6.75
8200	.125" thick, minimum		230	.035		3.35	1.23		4.58	5.50
8250	Maximum		200	.040		5	1.42		6.42	7.60
8700	Adhesive cement, 1 gallon does 200 to 300 S.F.				Gal.	18.05			18.05	19.85
8800	Asphalt primer, 1 gallon per 300 S.F.					10.80			10.80	11.90
8900	Emulsion, 1 gallon per 140 S.F.					13.75			13.75	15.15

09 65 19 – Resilient Tile Flooring

09 65 19.10 Resilient Tile Flooring

		Crew	Daily Output	Labor-Hours	Unit	Material	Labor	Equipment	Total	Total Incl O&P
0010	**RESILIENT TILE FLOORING**									
2200	Cork tile, standard finish, 1/8" thick	1 Tilf	315	.025	S.F.	4.74	.90		5.64	6.50
2250	3/16" thick		315	.025		5.30	.90		6.20	7.15
2300	5/16" thick		315	.025		6.15	.90		7.05	8.10
2350	1/2" thick		315	.025		7.10	.90		8	9.10
2500	Urethane finish, 1/8" thick		315	.025		5.65	.90		6.55	7.55
2550	3/16" thick		315	.025		6.25	.90		7.15	8.15
2600	5/16" thick		315	.025		7.85	.90		8.75	9.90
2650	1/2" thick		315	.025		10.85	.90		11.75	13.25
6050	Tile, marbleized colors, 12" x 12", 1/8" thick		400	.020		5.60	.71		6.31	7.20
6100	3/16" thick		400	.020		8.25	.71		8.96	10.15
6300	Special tile, plain colors, 1/8" thick		400	.020		6.50	.71		7.21	8.20
6350	3/16" thick		400	.020		8.85	.71		9.56	10.80
6410	Raised, radial or square, minimum		400	.020		10	.71		10.71	12.05
6430	Maximum		400	.020		10	.71		10.71	12.05
6450	For golf course, skating rink, etc., 1/4" thick		275	.029		10	1.03		11.03	12.50
6700	Synthetic turf, 3/8" thick		90	.089		4.70	3.16		7.86	9.80
6750	Interlocking 2' x 2' squares, 1/2" thick, not									
6810	cemented, for playgrounds, minimum	1 Tilf	210	.038	S.F.	3.90	1.35		5.25	6.30
6850	Maximum		190	.042		10	1.49		11.49	13.20
7000	Vinyl composition tile, 12" x 12", 1/16" thick		500	.016		.97	.57		1.54	1.90
7050	Embossed		500	.016		1.75	.57		2.32	2.76
7100	Marbleized		500	.016		1.75	.57		2.32	2.76
7150	Solid		500	.016		2.25	.57		2.82	3.31
7200	3/32" thick, embossed		500	.016		1.17	.57		1.74	2.12
7250	Marbleized		500	.016		2	.57		2.57	3.03
7300	Solid		500	.016		2	.57		2.57	3.03
7350	1/8" thick, marbleized	CN	500	.016		1.39	.57		1.96	2.36
7400	Solid		500	.016		2.07	.57		2.64	3.11
7450	Conductive		500	.016		4.82	.57		5.39	6.15
7500	Vinyl tile, 12" x 12", .050" thick, minimum		500	.016		2.20	.57		2.77	3.25
7550	Maximum		500	.016		5	.57		5.57	6.35
7600	1/8" thick, minimum		500	.016		3.60	.57		4.17	4.79
7650	Solid colors		500	.016		5.35	.57		5.92	6.70
7700	Marbleized or Travertine pattern		500	.016		5.15	.57		5.72	6.50
7750	Florentine pattern		500	.016		5.20	.57		5.77	6.55
7800	Maximum		500	.016		11	.57		11.57	12.95

09 65 33 – Conductive Resilient Flooring

09 65 33.10 Conductive Resilient Flooring

		Crew	Daily Output	Labor-Hours	Unit	Material	Labor	Equipment	Total	Total Incl O&P
0010	**CONDUCTIVE RESILIENT FLOORING**									
1700	Conductive flooring, rubber tile, 1/8" thick	1 Tilf	315	.025	S.F.	4.50	.90		5.40	6.25
1800	Homogeneous vinyl tile, 1/8" thick	"	315	.025	"	5.70	.90		6.60	7.55

09 66 Terrazzo Flooring

09 66 13 – Portland Cement Terrazzo Flooring

09 66 13.10 Portland Cement Terrazzo		Crew	Daily Output	Labor-Hours	Unit	Material	2007 Bare Costs Labor	Equipment	Total	Total Incl O&P
0010	**PORTLAND CEMENT TERRAZZO**, cast-in-place R096613-10									
0020	Cove base, 6" high, 16ga. zinc	1 Mstz	20	.400	L.F.	3.26	14.05		17.31	24
0100	Curb, 6" high and 6" wide		6	1.333		5.25	47		52.25	75
0300	Divider strip for floors, 14 ga., 1-1/4" deep, zinc		375	.021		1.19	.75		1.94	2.41
0400	Brass		375	.021		2.11	.75		2.86	3.42
0600	Heavy top strip 1/4" thick, 1-1/4" deep, zinc		300	.027		1.65	.94		2.59	3.20
0900	Galv. bottoms, brass		300	.027		2.85	.94		3.79	4.52
1200	For thin set floors, 16 ga., 1/2" x 1/2", zinc		350	.023		.72	.80		1.52	1.97
1300	Brass		350	.023		1.43	.80		2.23	2.75
1500	Floor, bonded to concrete, 1-3/4" thick, gray cement	J-3	130	.123	S.F.	2.80	3.91	1.96	8.67	11
1600	White cement, mud set		130	.123		3.19	3.91	1.96	9.06	11.40
1800	Not bonded, 3" total thickness, gray cement		115	.139		3.50	4.41	2.21	10.12	12.80
1900	White cement, mud set		115	.139		3.82	4.41	2.21	10.44	13.15
2100	For Venetian terrazzo, 1" topping, add					50%	50%			
2200	For heavy duty abrasive terrazzo, add					50%	50%			
2700	Monolithic terrazzo, 1/2" thick									
2710	10' panels	J-3	125	.128	S.F.	2.55	4.06	2.04	8.65	11
3000	Stairs, cast in place, pan filled treads		30	.533	L.F.	2.53	16.90	8.50	27.93	37
3100	Treads and risers		14	1.143	"	5.25	36.50	18.20	59.95	79
3300	For stair landings, add to floor prices						50%			
3400	Stair stringers and fascia	J-3	30	.533	S.F.	4.26	16.90	8.50	29.66	39
3600	For abrasive metal nosings on stairs, add		150	.107	L.F.	7.25	3.38	1.70	12.33	14.85
3700	For abrasive surface finish, add		600	.027	S.F.	1	.85	.42	2.27	2.81
3900	For raised abrasive strips, add		150	.107	L.F.	.95	3.38	1.70	6.03	7.90
4000	Wainscot, bonded, 1-1/2" thick		30	.533	S.F.	3.22	16.90	8.50	28.62	38
4200	Epoxy terrazzo, 1/4" thick		40	.400	"	4.76	12.70	6.35	23.81	31
4300	Stone chips, onyx gemstone, per 50 lb. bag				Bag	15.55			15.55	17.10

09 66 13.30 Terrazzo, Precast

		Crew	Daily Output	Labor-Hours	Unit	Material	Labor	Equipment	Total	Total Incl O&P
0010	**TERRAZZO, PRECAST**									
0020	Base, 6" high, straight	1 Mstz	35	.229	L.F.	10.15	8.05		18.20	23
0100	Cove		30	.267		11.55	9.35		20.90	26.50
0300	8" high base, straight		30	.267		10.30	9.35		19.65	25
0400	Cove		25	.320		15.20	11.25		26.45	33
0600	For white cement, add					.41			.41	.45
0700	For 16 ga. zinc toe strip, add					1.52			1.52	1.67
0900	Curbs, 4" x 4" high	1 Mstz	19	.421		29	14.80		43.80	53.50
1000	8" x 8" high	"	15	.533		33	18.75		51.75	64
1200	Floor tiles, non-slip, 1" thick, 12" x 12"	D-1	29	.552	S.F.	16.95	18.40		35.35	46.50
1300	1-1/4" thick, 12" x 12"		29	.552		19	18.40		37.40	49
1500	16" x 16"		23	.696		20.50	23		43.50	58
1600	1-1/2" thick, 16" x 16"		21	.762		18.90	25.50		44.40	59.50
1800	For Venetian terrazzo, add					5.65			5.65	6.25
1900	For white cement, add					.53			.53	.58
2400	Stair treads, 1-1/2" thick, non-slip, three line pattern	2 Mstz	70	.229	L.F.	38	8.05		46.05	54
2500	Nosing and two lines		70	.229		38	8.05		46.05	54
2700	2" thick treads, straight		60	.267		41	9.35		50.35	59
2800	Curved		50	.320		53.50	11.25		64.75	75
3000	Stair risers, 1" thick, to 6" high, straight sections		60	.267		9.30	9.35		18.65	24
3100	Cove		50	.320		13.45	11.25		24.70	31.50
3300	Curved, 1" thick, to 6" high, vertical		48	.333		18.40	11.70		30.10	37
3400	Cove		38	.421		35	14.80		49.80	60
3600	Stair tread and riser, single piece, straight, minimum		60	.267		48.50	9.35		57.85	67.50

09 66 Terrazzo Flooring

09 66 13 – Portland Cement Terrazzo Flooring

09 66 13.30 Terrazzo, Precast

		Crew	Daily Output	Labor-Hours	Unit	Material	2007 Bare Costs Labor	Equipment	Total	Total Incl O&P
3700	Maximum	2 Mstz	40	.400	L.F.	63	14.05		77.05	89.50
3900	Curved tread and riser, minimum		40	.400		68	14.05		82.05	95.50
4000	Maximum		32	.500		85.50	17.60		103.10	121
4200	Stair stringers, notched, 1" thick		25	.640		28	22.50		50.50	63.50
4300	2" thick		22	.727		33	25.50		58.50	73.50
4500	Stair landings, structural, non-slip, 1-1/2" thick		85	.188	S.F.	30.50	6.60		37.10	43
4600	3" thick		75	.213		43	7.50		50.50	58.50
4800	Wainscot, 12" x 12" x 1" tiles	1 Mstz	12	.667		6.20	23.50		29.70	41.50
4900	16" x 16" x 1-1/2" tiles	"	8	1		13.30	35		48.30	66

09 66 16 – Terrazzo Floor Tile

09 66 16.10 Tile or Terrazzo Base

		Crew	Daily Output	Labor-Hours	Unit	Material	2007 Bare Costs Labor	Equipment	Total	Total Incl O&P
0010	**TILE OR TERRAZZO BASE**									
0020	Scratch coat only	1 Mstz	150	.053	S.F.	.40	1.87		2.27	3.19
0500	Scratch and brown coat only	"	75	.107	"	.76	3.75		4.51	6.35

09 66 33 – Conductive Terrazzo Flooring

09 66 33.10 Conductive Terrazzo

		Crew	Daily Output	Labor-Hours	Unit	Material	2007 Bare Costs Labor	Equipment	Total	Total Incl O&P
0010	**CONDUCTIVE TERRAZZO**									
2400	Bonded conductive floor for hospitals	J-3	90	.178	S.F.	3.82	5.65	2.83	12.30	15.60
2500	Epoxy terrazzo, 1/4" thick, minimum		100	.160		4.33	5.10	2.55	11.98	15
2550	Average		75	.213		4.77	6.75	3.40	14.92	18.95
2600	Maximum		60	.267		5.25	8.45	4.25	17.95	23

09 67 Fluid-Applied Flooring

09 67 20 – Epoxy-Marble Chip Flooring

09 67 20.13 Elastomeric Liquid Flooring

		Crew	Daily Output	Labor-Hours	Unit	Material	2007 Bare Costs Labor	Equipment	Total	Total Incl O&P
0010	**ELASTOMERIC LIQUID FLOORING**									
0020	Cementitous acrylic, 1/4" thick	C-6	520	.092	S.F.	1.43	2.79	.08	4.30	5.95
0100	3/8" thick	"	450	.107		1.80	3.22	.09	5.11	7.05
0200	Methyl methacrylate, 1/4" thick	C-8A	3000	.016		5.15	.50		5.65	6.40
0210	1/8" thick	"	3000	.016		2.83	.50		3.33	3.87
0300	Cupric oxychloride, on bond coat, minimum	C-6	480	.100		3.06	3.02	.09	6.17	8.10
0400	Maximum		420	.114		5.10	3.45	.10	8.65	11
2400	Mastic, hot laid, 2 coat, 1-1/2" thick, standard, minimum		690	.070		3.54	2.10	.06	5.70	7.20
2500	Maximum		520	.092		4.55	2.79	.08	7.42	9.40
2700	Acidproof, minimum		605	.079		4.55	2.40	.07	7.02	8.75
2800	Maximum		350	.137		6.30	4.14	.12	10.56	13.50
3000	Neoprene, trowelled on, 1/4" thick, minimum		545	.088		3.49	2.66	.08	6.23	8.05
3100	Maximum		430	.112		4.80	3.37	.10	8.27	10.60
4300	Polyurethane, with suspended vinyl chips, minimum		1065	.045		6.80	1.36	.04	8.20	9.60
4500	Maximum		860	.056		9.80	1.69	.05	11.54	13.40

09 67 20.16 Epoxy Terrazzo

		Crew	Daily Output	Labor-Hours	Unit	Material	2007 Bare Costs Labor	Equipment	Total	Total Incl O&P
0010	**EPOXY TERRAZZO**									
1800	Epoxy terrazzo, 1/4" thick, chemical resistant, minimum	J-3	200	.080	S.F.	5.15	2.54	1.27	8.96	10.80
1900	Maximum		150	.107		7.50	3.38	1.70	12.58	15.10
2100	Conductive, minimum		210	.076		6.75	2.42	1.21	10.38	12.30
2200	Maximum		160	.100		8.75	3.17	1.59	13.51	16

09 67 20.19 Polyacrylate Terrazzo

		Crew	Daily Output	Labor-Hours	Unit	Material	2007 Bare Costs Labor	Equipment	Total	Total Incl O&P
0010	**POLYACRYLATE TERRAZZO**									
3150	Polyacrylate terrazzo, 1/4" thick, minimum	C-6	735	.065	S.F.	2.91	1.97	.06	4.94	6.30

09 67 Fluid-Applied Flooring

09 67 20 – Epoxy-Marble Chip Flooring

09 67 20.19 Polyacrylate Terrazzo		Crew	Daily Output	Labor-Hours	Unit	Material	2007 Bare Costs Labor	Equipment	Total	Total Incl O&P
3170	Maximum	C-6	480	.100	S.F.	3.59	3.02	.09	6.70	8.70
3200	3/8" thick, minimum		620	.077		3.80	2.34	.07	6.21	7.85
3220	Maximum		480	.100		5.05	3.02	.09	8.16	10.30
3300	Conductive terrazzo, 1/4" thick, minimum		450	.107		6.10	3.22	.09	9.41	11.75
3330	Maximum		305	.157		7.55	4.76	.14	12.45	15.75
3350	3/8" thick, minimum		365	.132		8	3.97	.12	12.09	15.05
3370	Maximum		255	.188		9.45	5.70	.17	15.32	19.35
3450	Granite, conductive, 1/4" thick, minimum		695	.069		7.80	2.09	.06	9.95	11.85
3470	Maximum		420	.114		10.05	3.45	.10	13.60	16.45
3500	3/8" thick, minimum		695	.069		11.40	2.09	.06	13.55	15.80
3520	Maximum		380	.126		13.75	3.82	.11	17.68	21

09 67 20.26 Quartz Flooring

		Crew	Daily Output	Labor-Hours	Unit	Material	Labor	Equipment	Total	Total Incl O&P
0010	**QUARTZ FLOORING**									
0600	Epoxy, with colored quartz chips, broadcast, minimum	C-6	675	.071	S.F.	2.42	2.15	.06	4.63	6.05
0700	Maximum		490	.098		2.93	2.96	.09	5.98	7.90
0900	Trowelled, minimum		560	.086		3.12	2.59	.08	5.79	7.50
1000	Maximum		480	.100		4.55	3.02	.09	7.66	9.75
1200	Heavy duty epoxy topping, 1/4" thick,									
1300	500 to 1,000 S.F.	C-6	420	.114	S.F.	4.59	3.45	.10	8.14	10.45
1500	1,000 to 2,000 S.F.		450	.107		3.80	3.22	.09	7.11	9.25
1600	Over 10,000 S.F.		480	.100		3.57	3.02	.09	6.68	8.70
3600	Polyester, with colored quartz chips, 1/16" thick, minimum		1065	.045		2.63	1.36	.04	4.03	5.05
3700	Maximum		560	.086		3.58	2.59	.08	6.25	8
3900	1/8" thick, minimum		810	.059		3.07	1.79	.05	4.91	6.20
4000	Maximum		675	.071		4.15	2.15	.06	6.36	7.95
4200	Polyester, heavy duty, compared to epoxy, add		2590	.019		1.27	.56	.02	1.85	2.28

09 68 Carpeting

09 68 05 – Carpet Accessories

09 68 05.11 Flooring Transition Strip

		Crew	Daily Output	Labor-Hours	Unit	Material	Labor	Equipment	Total	Total Incl O&P
0010	**FLOORING TRANSITION STRIP**									
0107	Clamp down brass divider, 12' strip, vinyl to carpet	1 Tilf	31.25	.256	Ea.	25	9.10		34.10	41
0117	Vinyl to hard surface	"	31.25	.256	"	25	9.10		34.10	41

09 68 10 – Carpet Pad

09 68 10.10 Carpet Pad

		Crew	Daily Output	Labor-Hours	Unit	Material	Labor	Equipment	Total	Total Incl O&P
0010	**CARPET PAD**, commercial grade									
9000	Sponge rubber pad, minimum	1 Tilf	150	.053	S.Y.	3.89	1.89		5.78	7.05
9100	Maximum		150	.053		8.65	1.89		10.54	12.35
9200	Felt pad, minimum		150	.053		3.95	1.89		5.84	7.15
9300	Maximum		150	.053		7.05	1.89		8.94	10.55
9400	Bonded urethane pad, minimum		150	.053		4.15	1.89		6.04	7.35
9500	Maximum		150	.053		7.10	1.89		8.99	10.60
9600	Prime urethane pad, minimum		150	.053		2.38	1.89		4.27	5.40
9700	Maximum		150	.053		4.40	1.89		6.29	7.60

09 68 13 – Tile Carpeting

09 68 13.10 Carpet Tile

		Crew	Daily Output	Labor-Hours	Unit	Material	Labor	Equipment	Total	Total Incl O&P
0010	**CARPET TILE**									
0100	Tufted nylon, 18" x 18", hard back, 20 oz.	1 Tilf	150	.053	S.Y.	21.50	1.89		23.39	27
0110	26 oz.		150	.053		37	1.89		38.89	43.50
0200	Cushion back, 20 oz.		150	.053		27	1.89		28.89	33

09 68 Carpeting

09 68 13 – Tile Carpeting

09 68 13.10 Carpet Tile		Crew	Daily Output	Labor-Hours	Unit	Material	2007 Bare Costs Labor	Equipment	Total	Total Incl O&P
0210	26 oz.	1 Tilf	150	.053	S.Y.	42.50	1.89		44.39	50

09 68 16 – Sheet Carpeting

09 68 16.10 Sheet Carpet

			Crew	Daily Output	Labor-Hours	Unit	Material	Labor	Equipment	Total	Total Incl O&P
0010	**SHEET CARPET**										
0700	Nylon, level loop, 26 oz., light to medium traffic	CN	1 Tilf	75	.107	S.Y.	22.50	3.79		26.29	30.50
0720	28 oz., light to medium traffic			75	.107		23.50	3.79		27.29	31.50
0900	32 oz., medium traffic			75	.107		28.50	3.79		32.29	37
1100	40 oz., medium to heavy traffic			75	.107		42	3.79		45.79	51.50
2920	Nylon plush, 30 oz., medium traffic			57	.140		21	4.98		25.98	30.50
3000	36 oz., medium traffic			75	.107		27.50	3.79		31.29	36
3100	42 oz., medium to heavy traffic			70	.114		32	4.06		36.06	41
3200	46 oz., medium to heavy traffic			70	.114		36.50	4.06		40.56	46
3300	54 oz., heavy traffic			70	.114		40	4.06		44.06	50
3340	60 oz., heavy traffic			70	.114		56	4.06		60.06	67.50
3665	Olefin, 24 oz., light to medium traffic			75	.107		12.75	3.79		16.54	19.60
3670	26 oz., medium traffic			75	.107		13.50	3.79		17.29	20.50
3680	28 oz., medium to heavy traffic			75	.107		15	3.79		18.79	22
3700	32 oz., medium to heavy traffic			75	.107		19	3.79		22.79	26.50
3730	42 oz., heavy traffic			70	.114		25.50	4.06		29.56	34
4110	Wool, level loop, 40 oz., medium traffic			75	.107		95	3.79		98.79	111
4500	50 oz., medium to heavy traffic			75	.107		97.50	3.79		101.29	113
4700	Patterned, 32 oz., medium to heavy traffic			70	.114		96	4.06		100.06	112
4900	48 oz., heavy traffic			70	.114		98	4.06		102.06	114
5000	For less than full roll, add						25%				
5100	For small rooms, less than 12' wide, add							25%			
5200	For large open areas (no cuts), deduct							25%			
5600	For bound carpet baseboard, add		1 Tilf	300	.027	L.F.	1.65	.95		2.60	3.21
5610	For stairs, not incl. price of carpet, add		"	30	.267	Riser		9.45		9.45	13.90
5620	For borders and patterns, add to labor							18%			
8950	For tackless, stretched installation, add padding to above										
9850	For "branded" fiber, add					S.Y.	25%				

09 68 20 – Athletic Carpet

09 68 20.10 Indoor Athletic Carpet

		Crew	Daily Output	Labor-Hours	Unit	Material	Labor	Equipment	Total	Total Incl O&P
0010	**INDOOR ATHLETIC CARPET**									
3700	Polyethylene, in rolls, no base incl., landscape surfaces	1 Tilf	275	.029	S.F.	2.95	1.03		3.98	4.77
3800	Nylon action surface, 1/8" thick		275	.029		2.90	1.03		3.93	4.71
3900	1/4" thick		275	.029		4.18	1.03		5.21	6.10
4000	3/8" thick		275	.029		5.25	1.03		6.28	7.30
4100	Golf tee surface with foam back		235	.034		5.20	1.21		6.41	7.45
4200	Practice putting, knitted nylon surface		235	.034		4.41	1.21		5.62	6.60
4400	Polyurethane, thermoset, prefabricated in place, indoor									
4500	3/8" thick for basketball, gyms, etc.	1 Tilf	100	.080	S.F.	4.15	2.84		6.99	8.75
4600	1/2" thick for professional sports		95	.084		5.55	2.99		8.54	10.55
4700	Outdoor, 1/4" thick, smooth, for tennis		100	.080		4.31	2.84		7.15	8.90
4800	Rough, for track, 3/8" thick		95	.084		4.99	2.99		7.98	9.90
5000	Poured in place, indoor, with finish, 1/4" thick		80	.100		3.07	3.55		6.62	8.60
5050	3/8" thick		65	.123		3.72	4.37		8.09	10.50
5100	1/2" thick		50	.160		4.92	5.70		10.62	13.75
5500	Polyvinyl chloride, sheet goods for gyms, 1/4" thick		80	.100		6.50	3.55		10.05	12.35
5600	3/8" thick		60	.133		7.35	4.73		12.08	15.05

09 69 Access Flooring

09 69 13 – Rigid-Grid Access Flooring

09 69 13.10 Access Floors

		Crew	Daily Output	Labor-Hours	Unit	Material	2007 Bare Costs Labor	2007 Bare Costs Equipment	Total	Total Incl O&P
0010	**ACCESS FLOORS**									
0015	System pricing including panels, pedestals, stringers, and laminate cover									
0100	Computer room, greater than 6,000 S.F.	4 Carp	750	.043	S.F.	8.15	1.57		9.72	11.40
0110	Less than 6,000 S.F.	2 Carp	375	.043		9.40	1.57		10.97	12.80
0120	Office greater than 6,000 S.F.	4 Carp	1050	.030		4.30	1.12		5.42	6.45
0250	Panels, particle board or steel, 1250# load, no covering, under 6,000 S.F.	2 Carp	600	.027		4	.98		4.98	5.90
0300	Over 6,000 S.F.		640	.025		3.50	.92		4.42	5.30
0400	Aluminum, 24" panels		500	.032		30	1.17		31.17	35.50
0600	For carpet covering, add					8.30			8.30	9.10
0700	For vinyl floor covering, add					6.45			6.45	7.10
0900	For high pressure laminate covering, add					5.25			5.25	5.80
0910	For snap on stringer system, add	2 Carp	1	16		1.40	585		586.40	915
0950	Office applications, steel or concrete panels,									
0960	no covering, over 6,000 S.F.	2 Carp	960	.017	S.F.	9.75	.61		10.36	11.70
1000	Machine cutouts after initial installation	1 Carp	50	.160	Ea.	4.87	5.85		10.72	14.50
1050	Pedestals, 6" to 12"	2 Carp	85	.188		7.70	6.90		14.60	19.20
1100	Air conditioning grilles, 4" x 12"	1 Carp	17	.471		64	17.25		81.25	97.50
1150	4" x 18"	"	14	.571		87.50	21		108.50	129
1200	Approach ramps, minimum	2 Carp	60	.267	S.F.	24.50	9.80		34.30	42.50
1300	Maximum	"	40	.400	"	32.50	14.70		47.20	58.50
1500	Handrail, 2 rail, aluminum	1 Carp	15	.533	L.F.	96.50	19.55		116.05	137

09 72 Wall Coverings

09 72 23 – Wallpapering

09 72 23.10 Wallpaper

		Crew	Daily Output	Labor-Hours	Unit	Material	2007 Bare Costs Labor	2007 Bare Costs Equipment	Total	Total Incl O&P
0010	**WALLPAPER** including sizing; add 10-30% waste @ takeoff R097223-10									
0050	Aluminum foil	1 Pape	275	.029	S.F.	.91	.96		1.87	2.44
0100	Copper sheets, .025" thick, vinyl backing		240	.033		4.89	1.10		5.99	7.05
0300	Phenolic backing		240	.033		6.35	1.10		7.45	8.60
0600	Cork tiles, light or dark, 12" x 12" x 3/16"		240	.033		3.93	1.10		5.03	5.95
0700	5/16" thick		235	.034		3.35	1.12		4.47	5.40
0900	1/4" basketweave		240	.033		5.15	1.10		6.25	7.30
1000	1/2" natural, non-directional pattern		240	.033		6.45	1.10		7.55	8.75
1100	3/4" natural, non-directional pattern		240	.033		9.90	1.10		11	12.50
1200	Granular surface, 12" x 36", 1/2" thick		385	.021		1.11	.68		1.79	2.25
1300	1" thick		370	.022		1.44	.71		2.15	2.65
1500	Polyurethane coated, 12" x 12" x 3/16" thick		240	.033		3.48	1.10		4.58	5.50
1600	5/16" thick		235	.034		4.95	1.12		6.07	7.15
1800	Cork wallpaper, paperbacked, natural		480	.017		1.86	.55		2.41	2.88
1900	Colors		480	.017		2.45	.55		3	3.53
2100	Flexible wood veneer, 1/32" thick, plain woods		100	.080		2.08	2.63		4.71	6.25
2200	Exotic woods		95	.084		3.16	2.77		5.93	7.65
2400	Gypsum-based, fabric-backed, fire									
2500	resistant for masonry walls, minimum, 21 oz./S.Y.	1 Pape	800	.010	S.F.	.72	.33		1.05	1.29
2700	Maximum, (small quantities)	"	640	.013		1.20	.41		1.61	1.94
2750	Acrylic, modified, semi-rigid PVC, .028" thick	2 Carp	330	.048		1	1.78		2.78	3.87
2800	.040" thick	"	320	.050		1.31	1.84		3.15	4.30
3000	Vinyl wall covering, fabric-backed, lightweight, (12-15 oz./S.Y.)	1 Pape	640	.013		.60	.41		1.01	1.28
3300	Medium weight, type 2, (20-24 oz./S.Y.)		480	.017		.72	.55		1.27	1.62
3400	Heavy weight, type 3, (28 oz./S.Y.)		435	.018		1.46	.61		2.07	2.52
3600	Adhesive, 5 gal. lots, (18SY/Gal.)				Gal.	9.40			9.40	10.30

09 72 Wall Coverings

09 72 23 – Wallpapering

09 72 23.10 Wallpaper

		Crew	Daily Output	Labor-Hours	Unit	Material	2007 Bare Costs Labor	2007 Bare Costs Equipment	Total	Total Incl O&P
3700	Wallpaper, average workmanship, solid pattern, low cost paper	1 Pape	640	.013	S.F.	.32	.41		.73	.97
3900	basic patterns (matching required), avg. cost paper		535	.015		.65	.49		1.14	1.46
4000	Paper at $85 per double roll, quality workmanship	↓	435	.018	↓	2.17	.61		2.78	3.30
4100	Linen wall covering, paper backed									
4150	Flame treatment, minimum				S.F.	.79			.79	.87
4180	Maximum					1.37			1.37	1.51
4200	Grass cloths with lining paper, minimum	1 Pape	400	.020		.68	.66		1.34	1.74
4300	Maximum	"	350	.023	↓	2.19	.75		2.94	3.54

09 77 Special Wall Surfacing

09 77 33 – Fiberglass Reinforced Panels

09 77 33.10 Fiberglass Reinforced Plastic

		Crew	Daily Output	Labor-Hours	Unit	Material	Labor	Equipment	Total	Total Incl O&P
0010	**FIBERGLASS REINFORCED PLASTIC** panels, .090" thick, on walls									
0020	Adhesive mounted, embossed surface	2 Carp	640	.025	S.F.	1.52	.92		2.44	3.10
0030	Smooth surface		640	.025		1.78	.92		2.70	3.39
0040	Fire rated, embossed surface		640	.025		1.98	.92		2.90	3.61
0050	Nylon rivet mounted, on drywall, embossed surface		480	.033		1.41	1.22		2.63	3.45
0060	Smooth surface		480	.033		1.66	1.22		2.88	3.73
0070	Fire rated, embossed surface		480	.033		1.85	1.22		3.07	3.94
0080	On masonry, embossed surface		320	.050		1.41	1.84		3.25	4.41
0090	Smooth surface		320	.050		1.78	1.84		3.62	4.82
0100	Fire rated, embossed surface		320	.050		1.88	1.84		3.72	4.93
0110	Nylon rivet and adhesive mounted, on drywall, embossed surface		240	.067		1.66	2.45		4.11	5.65
0120	Smooth surface		240	.067		1.88	2.45		4.33	5.90
0130	Fire rated, embossed surface		240	.067		2.09	2.45		4.54	6.10
0140	On masonry, embossed surface		190	.084		1.66	3.09		4.75	6.65
0150	Smooth surface		190	.084		1.88	3.09		4.97	6.90
0160	Fire rated, embossed surface	↓	190	.084	↓	2.09	3.09		5.18	7.10
0170	For moldings add	1 Carp	250	.032	L.F.	.28	1.17		1.45	2.14
0180	On ceilings, for lay in grid system, embossed surface		400	.020	S.F.	1.52	.73		2.25	2.81
0190	Smooth surface		400	.020		1.78	.73		2.51	3.10
0200	Fire rated, embossed surface	↓	400	.020	↓	1.98	.73		2.71	3.32

09 77 43 – Panel Systems

09 77 43.20 Slatwall Panels and Accessories

		Crew	Daily Output	Labor-Hours	Unit	Material	Labor	Equipment	Total	Total Incl O&P
0010	**SLATWALL PANELS AND ACCESSORIES**									
0100	Slatwall panel, 4' x 8' x 3/4" T, MDF, paint grade	1 Carp	500	.016	S.F.	1.53	.59		2.12	2.59
0110	Melamine finish		500	.016		2.20	.59		2.79	3.33
0120	High pressure plastic laminate finish	↓	500	.016		3.51	.59		4.10	4.77
0130	Aluminum channel inserts, add				↓	3.38			3.38	3.72
0200	Accessories, corner forms, 8' L				L.F.	4.01			4.01	4.41
0210	T-connector, 8' L					5.55			5.55	6.10
0220	J-mold, 8' L					1.42			1.42	1.56
0230	Edge cap, 8' L					.93			.93	1.02
0240	Finish end cap, 8' L				↓	3.38			3.38	3.72
0300	Display hook, 4" L				Ea.	1.06			1.06	1.17
0310	6" L					1.15			1.15	1.27
0320	8" L					1.22			1.22	1.34
0330	10" L					1.38			1.38	1.52
0340	12" L					1.53			1.53	1.68
0350	Acrylic, 4" L				↓	.86			.86	.95

09 77 Special Wall Surfacing

09 77 43 – Panel Systems

09 77 43.20 Slatwall Panels and Accessories

		Crew	Daily Output	Labor-Hours	Unit	Material	2007 Bare Costs Labor	Equipment	Total	Total Incl O&P
0360	6" L				Ea.	1			1	1.10
0370	8" L					1.05			1.05	1.16
0380	10" L					1.15			1.15	1.27
0400	Waterfall hanger, metal, 12" - 16"					6.05			6.05	6.70
0410	Acrylic					9.95			9.95	10.95
0500	Shelf bracket, metal, 8"					4.25			4.25	4.68
0510	10"					4.54			4.54	4.99
0520	12"					4.89			4.89	5.40
0530	14"					5.50			5.50	6.05
0540	16"					6.10			6.10	6.70
0550	Acrylic, 8"					3.23			3.23	3.55
0560	10"					3.62			3.62	3.98
0570	12"					4.01			4.01	4.41
0580	14"					4.53			4.53	4.98
0600	Shelf, acrylic, 12" x 16" x 1/4"					26			26	28.50
0610	12" x 24" x 1/4"					43.50			43.50	48

09 81 Acoustic Insulation

09 81 16 – Acoustic Blanket Insulation

09 81 16.10 Sound Attenuation Blanket

		Crew	Daily Output	Labor-Hours	Unit	Material	Labor	Equipment	Total	Total Incl O&P
0010	**SOUND ATTENUATION BLANKET**									
0020	Blanket, 1" thick	1 Carp	925	.009	S.F.	.26	.32		.58	.78
0500	1-1/2" thick		920	.009		.27	.32		.59	.80
1000	2" thick		915	.009		.32	.32		.64	.85
1500	3" thick		910	.009		.50	.32		.82	1.05
3000	Thermal or acoustical batt above ceiling, 2" thick		900	.009		.48	.33		.81	1.04
3100	3" thick		900	.009		.72	.33		1.05	1.30
3200	4" thick		900	.009		.89	.33		1.22	1.49
3400	Urethane plastic foam, open cell, on wall, 2" thick	2 Carp	2050	.008		2.96	.29		3.25	3.71
3500	3" thick		1550	.010		3.93	.38		4.31	4.91
3600	4" thick		1050	.015		5.50	.56		6.06	6.90
3700	On ceiling, 2" thick		1700	.009		2.95	.35		3.30	3.79
3800	3" thick		1300	.012		3.93	.45		4.38	5
3900	4" thick		900	.018		5.50	.65		6.15	7.05
4000	Nylon matting 0.4" thick, with carbon black spinerette									
4010	plus polyester fabric, on floor	J-4	4000	.004	S.F.	2.19	.13		2.32	2.59
4200	Fiberglass reinf. backer board underlayment, 7/16" thick, on floor	"	800	.020	"	1.84	.63		2.47	2.94

09 84 Acoustic Room Components

09 84 13 – Fixed Sound-Absorptive Panels

09 84 13.10 Fixed Panels

		Crew	Daily Output	Labor-Hours	Unit	Material	Labor	Equipment	Total	Total Incl O&P
0010	**FIXED PANELS** Perforated steel facing, painted with									
0100	Fiberglass or mineral filler, no backs, 2-1/4" thick, modular									
0200	space units, ceiling or wall hung, white or colored	1 Carp	100	.080	S.F.	9.90	2.94		12.84	15.40
0300	Fiberboard sound deadening panels, 1/2" thick	"	600	.013	"	.32	.49		.81	1.11
0500	Fiberglass panels, 4' x 8' x 1" thick, with									
0600	glass cloth face for walls, cemented	1 Carp	155	.052	S.F.	6.50	1.89		8.39	10.10
0700	1-1/2" thick, dacron covered, inner aluminum frame,									
0710	wall mounted	1 Carp	300	.027	S.F.	7.95	.98		8.93	10.25

09 84 Acoustic Room Components

09 84 13 – Fixed Sound-Absorptive Panels

09 84 13.10 Fixed Panels

		Crew	Daily Output	Labor-Hours	Unit	Material	2007 Bare Costs Labor	2007 Bare Costs Equipment	Total	Total Incl O&P
0900	Mineral fiberboard panels, fabric covered, 30"x 108",									
1000	3/4" thick, concealed spline, wall mounted	1 Carp	150	.053	S.F.	5.75	1.96		7.71	9.35

09 84 36 – Sound-Absorbing Ceiling Units

09 84 36.10 Barriers

		Crew	Daily Output	Labor-Hours	Unit	Material	Labor	Equipment	Total	Total Incl O&P
0010	**BARRIERS** Plenum									
0600	Aluminum foil, fiberglass reinf., parallel with joists	1 Carp	275	.029	S.F.	.76	1.07		1.83	2.50
0700	Perpendicular to joists		180	.044		.94	1.63		2.57	3.57
0900	Aluminum mesh, kraft paperbacked		275	.029		.75	1.07		1.82	2.49
0970	Fiberglass batts, kraft faced, 3-1/2" thick		1400	.006		.27	.21		.48	.63
0980	6" thick		1300	.006		.45	.23		.68	.85
1000	Sheet lead, 1 lb., 1/64" thick, perpendicular to joists		150	.053		2.89	1.96		4.85	6.25
1100	Vinyl foam reinforced, 1/8" thick, 1.0 lb. per S.F.		150	.053		4.18	1.96		6.14	7.65

09 91 Painting

09 91 03 – Paint Restoration

09 91 03.20 Sanding

		Crew	Daily Output	Labor-Hours	Unit	Material	Labor	Equipment	Total	Total Incl O&P
0010	**SANDING** and puttying interior trim, compared to									
0100	Painting 1 coat, on quality work				L.F.		100%			
0300	Medium work						50%			
0400	Industrial grade						25%			
0500	Surface protection, placement and removal									
0510	Basic drop cloths	1 Pord	6400	.001	S.F.		.04		.04	.06
0520	Masking with paper		800	.010		.06	.33		.39	.56
0530	Volume cover up (using plastic sheathing, or building paper)		16000	.001			.02		.02	.02

09 91 03.30 Exterior Surface Preparation

		Crew	Daily Output	Labor-Hours	Unit	Material	Labor	Equipment	Total	Total Incl O&P
0010	**EXTERIOR SURFACE PREPARATION**									
0015	Doors, per side, not incl. frames or trim									
0020	Scrape & sand									
0030	Wood, flush	1 Pord	616	.013	S.F.		.42		.42	.64
0040	Wood, detail		496	.016			.53		.53	.79
0050	Wood, louvered		280	.029			.93		.93	1.41
0060	Wood, overhead		616	.013			.42		.42	.64
0070	Wire brush									
0080	Metal, flush	1 Pord	640	.013	S.F.		.41		.41	.62
0090	Metal, detail		520	.015			.50		.50	.76
0100	Metal, louvered		360	.022			.73		.73	1.09
0110	Metal or fibr., overhead		640	.013			.41		.41	.62
0120	Metal, roll up		560	.014			.47		.47	.70
0130	Metal, bulkhead		640	.013			.41		.41	.62
0140	Power wash, based on 2500 lb. operating pressure									
0150	Metal, flush	B-9	2240	.018	S.F.		.52	.08	.60	.90
0160	Metal, detail		2120	.019			.55	.08	.63	.95
0170	Metal, louvered		2000	.020			.58	.09	.67	1.01
0180	Metal or fibr., overhead		2400	.017			.49	.08	.57	.84
0190	Metal, roll up		2400	.017			.49	.08	.57	.84
0200	Metal, bulkhead		2200	.018			.53	.08	.61	.92
0400	Windows, per side, not incl. trim									
0410	Scrape & sand									
0420	Wood, 1-2 lite	1 Pord	320	.025	S.F.		.82		.82	1.23
0430	Wood, 3-6 lite		280	.029			.93		.93	1.41

09 91 Painting

09 91 03 – Paint Restoration

09 91 03.30 Exterior Surface Preparation

		Crew	Daily Output	Labor-Hours	Unit	Material	2007 Bare Costs Labor	Equipment	Total	Total Incl O&P
0440	Wood, 7-10 lite	1 Pord	240	.033	S.F.		1.09		1.09	1.64
0450	Wood, 12 lite		200	.040			1.31		1.31	1.97
0460	Wood, Bay / Bow	↓	320	.025	↓		.82		.82	1.23
0470	Wire brush									
0480	Metal, 1-2 lite	1 Pord	480	.017	S.F.		.55		.55	.82
0490	Metal, 3-6 lite		400	.020			.65		.65	.98
0500	Metal, Bay / Bow	↓	480	.017	↓		.55		.55	.82
0510	Power wash, based on 2500 lb. operating pressure									
0520	1-2 lite	B-9	4400	.009	S.F.		.27	.04	.31	.45
0530	3-6 lite		4320	.009			.27	.04	.31	.47
0540	7-10 lite		4240	.009			.27	.04	.31	.48
0550	12 lite		4160	.010			.28	.04	.32	.49
0560	Bay / Bow	↓	4400	.009	↓		.27	.04	.31	.45
0600	Siding, scrape and sand, light=10-30%, med.=30-70%									
0610	Heavy=70-100%, % of surface to sand									
0650	Texture 1-11, light	1 Pord	480	.017	S.F.		.55		.55	.82
0660	Med.		440	.018			.59		.59	.89
0670	Heavy		360	.022			.73		.73	1.09
0680	Wood shingles, shakes, light		440	.018			.59		.59	.89
0690	Med.		360	.022			.73		.73	1.09
0700	Heavy		280	.029			.93		.93	1.41
0710	Clapboard, light		520	.015			.50		.50	.76
0720	Med.		480	.017			.55		.55	.82
0730	Heavy	↓	400	.020			.65		.65	.98
0740	Wire brush									
0750	Aluminum, light	1 Pord	600	.013	S.F.		.44		.44	.66
0760	Med.		520	.015			.50		.50	.76
0770	Heavy	↓	440	.018	↓		.59		.59	.89
0780	Pressure wash, based on 2500 lb.. operating pressure									
0790	Stucco	B-9	3080	.013	S.F.		.38	.06	.44	.65
0800	Aluminum or vinyl		3200	.013			.36	.06	.42	.63
0810	Siding, masonry, brick & block	↓	2400	.017	↓		.49	.08	.57	.84
1300	Miscellaneous, wire brush									
1310	Metal, pedestrian gate	1 Pord	100	.080	S.F.		2.62		2.62	3.94
8000	For Chemical Washing, see Division 04 01 30.00									
8010	For Steam Cleaning, see Division 04 01 30.00									
8020	For Sand Blasting, see Division 05 01 10.51 and 03 35 29.60									

09 91 03.40 Interior Surface Preparation

		Crew	Daily Output	Labor-Hours	Unit	Material	Labor	Equipment	Total	Total Incl O&P
0010	**INTERIOR SURFACE PREPARATION**									
0020	Doors									
0030	Scrape & sand									
0040	Wood, flush	1 Pord	616	.013	S.F.		.42		.42	.64
0050	Wood, detail		496	.016			.53		.53	.79
0060	Wood, louvered	↓	280	.029	↓		.93		.93	1.41
0070	Wire brush									
0080	Metal, flush	1 Pord	640	.013	S.F.		.41		.41	.62
0090	Metal, detail		520	.015			.50		.50	.76
0100	Metal, louvered	↓	360	.022	↓		.73		.73	1.09
0110	Hand wash									
0120	Wood, flush	1 Pord	2160	.004	S.F.		.12		.12	.18
0130	Wood, detailed		2000	.004			.13		.13	.20
0140	Wood, louvered	↓	1360	.006	↓		.19		.19	.29

09 91 Painting

09 91 03 – Paint Restoration

09 91 03.40 Interior Surface Preparation

		Crew	Daily Output	Labor-Hours	Unit	Material	2007 Bare Costs Labor	Equipment	Total	Total Incl O&P
0150	Metal, flush	1 Pord	2160	.004	S.F.		.12		.12	.18
0160	Metal, detail		2000	.004			.13		.13	.20
0170	Metal, louvered	↓	1360	.006	↓		.19		.19	.29
0400	Windows, per side, not incl. trim									
0410	Scrape & sand									
0420	Wood, 1-2 lite	1 Pord	360	.022	S.F.		.73		.73	1.09
0430	Wood, 3-6 lite		320	.025			.82		.82	1.23
0440	Wood, 7-10 lite		280	.029			.93		.93	1.41
0450	Wood, 12 lite		240	.033			1.09		1.09	1.64
0460	Wood, Bay / Bow	↓	360	.022	↓		.73		.73	1.09
0470	Wire brush									
0480	Metal, 1-2 lite	1 Pord	520	.015	S.F.		.50		.50	.76
0490	Metal, 3-6 lite		440	.018			.59		.59	.89
0500	Metal, Bay / Bow	↓	520	.015	↓		.50		.50	.76
0600	Walls, sanding, light=10-30%									
0610	Med.=30-70%, heavy=70-100%, % of surface to sand									
0650	Walls, sand									
0660	Drywall, gypsum, plaster, light	1 Pord	3077	.003	S.F.		.09		.09	.13
0670	Drywall, gypsum, plaster, med.		2160	.004			.12		.12	.18
0680	Drywall, gypsum, plaster, heavy		923	.009			.28		.28	.43
0690	Wood, T&G, light		2400	.003			.11		.11	.16
0700	Wood, T&G, med.		1600	.005			.16		.16	.25
0710	Wood, T&G, heavy	↓	800	.010	↓		.33		.33	.49
0720	Walls, wash									
0730	Drywall, gypsum, plaster	1 Pord	3200	.003	S.F.		.08		.08	.12
0740	Wood, T&G		3200	.003			.08		.08	.12
0750	Masonry, brick & block, smooth		2800	.003			.09		.09	.14
0760	Masonry, brick & block, coarse	↓	2000	.004	↓		.13		.13	.20
8000	For Chemical Washing, see Division 04 01 30.00									
8010	For Steam Cleaning, see Division 04 01 30.00									
8020	For Sand Blasting, see Division 03 35 29.60 and 05 01 10.51									.

09 91 13 – Exterior Painting

09 91 13.30 Fences

			Crew	Daily Output	Labor-Hours	Unit	Material	Labor	Equipment	Total	Total Incl O&P
0010	**FENCES**	R099100-20									
0100	Chain link or wire metal, one side, water base										
0110	Roll & brush, first coat		1 Pord	960	.008	S.F.	.06	.27		.33	.48
0120	Second coat			1280	.006		.06	.20		.26	.37
0130	Spray, first coat			2275	.004		.06	.12		.18	.24
0140	Second coat		↓	2600	.003	↓	.06	.10		.16	.22
0150	Picket, water base										
0160	Roll & brush, first coat		1 Pord	865	.009	S.F.	.06	.30		.36	.53
0170	Second coat			1050	.008		.06	.25		.31	.44
0180	Spray, first coat			2275	.004		.06	.12		.18	.24
0190	Second coat		↓	2600	.003	↓	.06	.10		.16	.22
0200	Stockade, water base										
0210	Roll & brush, first coat		1 Pord	1040	.008	S.F.	.06	.25		.31	.45
0220	Second coat			1200	.007		.06	.22		.28	.40
0230	Spray, first coat			2275	.004		.06	.12		.18	.24
0240	Second coat		↓	2600	.003	↓	.06	.10		.16	.22

09 91 13.42 Miscellaneous, Exterior

0010	**MISCELLANEOUS, EXTERIOR**	R099100-20	
0015	For painting metals, see Div. 09 97 13.23		

09 91 Painting

09 91 13 – Exterior Painting

09 91 13.42 Miscellaneous, Exterior

		Crew	Daily Output	Labor-Hours	Unit	Material	2007 Bare Costs Labor	2007 Bare Costs Equipment	Total	Total Incl O&P
0100	Railing, ext., decorative wood, incl. cap & baluster									
0110	newels & spindles @ 12" O.C.									
0120	Brushwork, stain, sand, seal & varnish									
0130	First coat	1 Pord	90	.089	L.F.	.50	2.91		3.41	4.92
0140	Second coat	"	120	.067	"	.50	2.18		2.68	3.83
0150	Rough sawn wood, 42" high, 2" x 2" verticals, 6" O.C.									
0160	Brushwork, stain, each coat	1 Pord	90	.089	L.F.	.16	2.91		3.07	4.55
0170	Wrought iron, 1" rail, 1/2" sq. verticals									
0180	Brushwork, zinc chromate, 60" high, bars 6" O.C.									
0190	Primer	1 Pord	130	.062	L.F.	.52	2.01		2.53	3.60
0200	Finish coat		130	.062		.16	2.01		2.17	3.21
0210	Additional coat		190	.042		.19	1.38		1.57	2.28
0220	Shutters or blinds, single panel, 2' x 4', paint all sides									
0230	Brushwork, primer	1 Pord	20	.400	Ea.	.55	13.10		13.65	20.50
0240	Finish coat, exterior latex		20	.400		.49	13.10		13.59	20
0250	Primer & 1 coat, exterior latex		13	.615		.92	20		20.92	31.50
0260	Spray, primer		35	.229		.81	7.45		8.26	12.15
0270	Finish coat, exterior latex		35	.229		1.03	7.45		8.48	12.40
0280	Primer & 1 coat, exterior latex		20	.400		.87	13.10		13.97	20.50
0290	For louvered shutters, add				S.F.	10%				
0300	Stair stringers, exterior, metal									
0310	Roll & brush, zinc chromate, to 14", each coat	1 Pord	320	.025	L.F.	.05	.82		.87	1.29
0320	Rough sawn wood, 4" x 12"									
0330	Roll & brush, exterior latex, each coat	1 Pord	215	.037	L.F.	.07	1.22		1.29	1.91
0340	Trellis/lattice, 2" x 2" @ 3" O.C. with 2" x 8" supports									
0350	Spray, latex, per side, each coat	1 Pord	475	.017	S.F.	.07	.55		.62	.91
0450	Decking, Ext., sealer, alkyd, brushwork, sealer coat		1140	.007		.06	.23		.29	.42
0460	1st coat		1140	.007		.06	.23		.29	.42
0470	2nd coat		1300	.006		.04	.20		.24	.35
0500	Paint, alkyd, brushwork, primer coat		1140	.007		.07	.23		.30	.42
0510	1st coat		1140	.007		.07	.23		.30	.43
0520	2nd coat		1300	.006		.05	.20		.25	.36
0600	Sand paint, alkyd, brushwork, 1 coat		150	.053		.10	1.74		1.84	2.73

09 91 13.60 Siding Exterior

		Crew	Daily Output	Labor-Hours	Unit	Material	2007 Bare Costs Labor	2007 Bare Costs Equipment	Total	Total Incl O&P
0010	**SIDING EXTERIOR**, Alkyd (oil base) R099100-20									
0450	Steel siding, oil base, paint 1 coat, brushwork	2 Pord	2015	.008	S.F.	.06	.26		.32	.45
0500	Spray		4550	.004		.09	.12		.21	.27
0800	Paint 2 coats, brushwork		1300	.012		.11	.40		.51	.74
1000	Spray		4550	.004		.15	.12		.27	.34
1200	Stucco, rough, oil base, paint 2 coats, brushwork		1300	.012		.11	.40		.51	.74
1400	Roller		1625	.010		.12	.32		.44	.61
1600	Spray		2925	.005		.13	.18		.31	.41
1800	Texture 1-11 or clapboard, oil base, primer coat, brushwork		1300	.012		.09	.40		.49	.71
2000	Spray		4550	.004		.09	.12		.21	.27
2400	Paint 2 coats, brushwork		810	.020		.17	.65		.82	1.15
2600	Spray		2600	.006		.19	.20		.39	.50
3400	Stain 2 coats, brushwork		950	.017		.11	.55		.66	.95
4000	Spray		3050	.005		.12	.17		.29	.39
4200	Wood shingles, oil base primer coat, brushwork		1300	.012		.08	.40		.48	.70
4400	Spray		3900	.004		.07	.13		.20	.28
5000	Paint 2 coats, brushwork		810	.020		.14	.65		.79	1.12
5200	Spray		2275	.007		.13	.23		.36	.50

09 91 Painting

09 91 13 – Exterior Painting

09 91 13.60 Siding Exterior		Crew	Daily Output	Labor-Hours	Unit	Material	2007 Bare Costs Labor	Equipment	Total	Total Incl O&P
6500	Stain 2 coats, brushwork	2 Pord	950	.017	S.F.	.11	.55		.66	.95
7000	Spray	↓	2660	.006		.15	.20		.35	.46
8000	For latex paint, deduct					10%				
8100	For work over 12' H, from pipe scaffolding, add						15%			
8200	For work over 12' H, from extension ladder, add						25%			
8300	For work over 12' H, from swing staging, add				↓		35%			

09 91 13.62 Siding, Misc.

		Crew	Daily Output	Labor-Hours	Unit	Material	Labor	Equipment	Total	Total Incl O&P
0010	**SIDING, MISC.** R099100-20									
0100	Aluminum siding									
0110	Brushwork, primer	2 Pord	2275	.007	S.F.	.05	.23		.28	.41
0120	Finish coat, exterior latex		2275	.007		.05	.23		.28	.40
0130	Primer & 1 coat exterior latex		1300	.012		.11	.40		.51	.73
0140	Primer & 2 coats exterior latex	↓	975	.016	↓	.15	.54		.69	.98
0150	Mineral Fiber shingles									
0160	Brushwork, primer	2 Pord	1495	.011	S.F.	.09	.35		.44	.63
0170	Finish coat, industrial enamel		1495	.011		.10	.35		.45	.64
0180	Primer & 1 coat enamel		810	.020		.19	.65		.84	1.18
0190	Primer & 2 coats enamel		540	.030		.29	.97		1.26	1.78
0200	Roll, primer		1625	.010		.10	.32		.42	.59
0210	Finish coat, industrial enamel		1625	.010		.11	.32		.43	.60
0220	Primer & 1 coat enamel		975	.016		.21	.54		.75	1.04
0230	Primer & 2 coats enamel		650	.025		.32	.81		1.13	1.56
0240	Spray, primer		3900	.004		.07	.13		.20	.28
0250	Finish coat, industrial enamel		3900	.004		.09	.13		.22	.30
0260	Primer & 1 coat enamel		2275	.007		.17	.23		.40	.53
0270	Primer & 2 coats enamel		1625	.010		.26	.32		.58	.77
0280	Waterproof sealer, first coat		4485	.004		.07	.12		.19	.26
0290	Second coat	↓	5235	.003	↓	.07	.10		.17	.22
0300	Rough wood incl. shingles, shakes or rough sawn siding									
0310	Brushwork, primer	2 Pord	1280	.013	S.F.	.11	.41		.52	.74
0320	Finish coat, exterior latex		1280	.013		.08	.41		.49	.71
0330	Primer & 1 coat exterior latex		960	.017		.20	.55		.75	1.04
0340	Primer & 2 coats exterior latex		700	.023		.28	.75		1.03	1.43
0350	Roll, primer		2925	.005		.15	.18		.33	.43
0360	Finish coat, exterior latex		2925	.005		.10	.18		.28	.38
0370	Primer & 1 coat exterior latex		1790	.009		.25	.29		.54	.72
0380	Primer & 2 coats exterior latex		1300	.012		.35	.40		.75	1
0390	Spray, primer		3900	.004		.13	.13		.26	.34
0400	Finish coat, exterior latex		3900	.004		.08	.13		.21	.29
0410	Primer & 1 coat exterior latex		2600	.006		.20	.20		.40	.52
0420	Primer & 2 coats exterior latex		2080	.008		.28	.25		.53	.69
0430	Waterproof sealer, first coat		4485	.004		.12	.12		.24	.32
0440	Second coat	↓	4485	.004	↓	.07	.12		.19	.26
0450	Smooth wood incl. butt, T&G, beveled, drop or B&B siding									
0460	Brushwork, primer	2 Pord	2325	.007	S.F.	.08	.23		.31	.43
0470	Finish coat, exterior latex		1280	.013		.08	.41		.49	.71
0480	Primer & 1 coat exterior latex		800	.020		.16	.65		.81	1.16
0490	Primer & 2 coats exterior latex		630	.025		.25	.83		1.08	1.52
0500	Roll, primer		2275	.007		.09	.23		.32	.45
0510	Finish coat, exterior latex		2275	.007		.09	.23		.32	.45
0520	Primer & 1 coat exterior latex		1300	.012		.18	.40		.58	.81
0530	Primer & 2 coats exterior latex	↓	975	.016		.27	.54		.81	1.11

09 91 Painting

09 91 13 – Exterior Painting

09 91 13.62 Siding, Misc.

		Crew	Daily Output	Labor-Hours	Unit	Material	2007 Bare Costs Labor	2007 Bare Costs Equipment	Total	Total Incl O&P
0540	Spray, primer	2 Pord	4550	.004	S.F.	.07	.12		.19	.25
0550	Finish coat, exterior latex		4550	.004		.08	.12		.20	.26
0560	Primer & 1 coat exterior latex		2600	.006		.15	.20		.35	.46
0570	Primer & 2 coats exterior latex		1950	.008		.23	.27		.50	.65
0580	Waterproof sealer, first coat		5230	.003		.07	.10		.17	.23
0590	Second coat	↓	5980	.003		.07	.09		.16	.21
0600	For oil base paint, add					10%				

09 91 13.70 Doors and Windows, Exterior

		Crew	Daily Output	Labor-Hours	Unit	Material	Labor	Equipment	Total	Total Incl O&P
0010	**DOORS AND WINDOWS, EXTERIOR** R099100-20									
0100	Door frames & trim, only									
0110	Brushwork, primer	1 Pord	512	.016	L.F.	.05	.51		.56	.83
0120	Finish coat, exterior latex		512	.016		.06	.51		.57	.84
0130	Primer & 1 coat, exterior latex		300	.027		.11	.87		.98	1.43
0140	Primer & 2 coats, exterior latex	↓	265	.030	↓	.17	.99		1.16	1.68
0150	Doors, flush, both sides, incl. frame & trim									
0160	Roll & brush, primer	1 Pord	10	.800	Ea.	3.83	26		29.83	43.50
0170	Finish coat, exterior latex		10	.800		4.63	26		30.63	44.50
0180	Primer & 1 coat, exterior latex		7	1.143		8.45	37.50		45.95	65.50
0190	Primer & 2 coats, exterior latex		5	1.600		13.10	52.50		65.60	93
0200	Brushwork, stain, sealer & 2 coats polyurethane		4	2		17.50	65.50		83	118
0210	Doors, French, both sides, 10-15 lite, incl. frame & trim									
0220	Brushwork, primer	1 Pord	6	1.333	Ea.	1.92	43.50		45.42	67.50
0230	Finish coat, exterior latex		6	1.333		2.31	43.50		45.81	68
0240	Primer & 1 coat, exterior latex		3	2.667		4.23	87		91.23	136
0250	Primer & 2 coats, exterior latex		2	4		6.40	131		137.40	204
0260	Brushwork, stain, sealer & 2 coats polyurethane		2.50	3.200		6.35	105		111.35	164
0270	Doors, louvered, both sides, incl. frame & trim									
0280	Brushwork, primer	1 Pord	7	1.143	Ea.	3.83	37.50		41.33	60
0290	Finish coat, exterior latex		7	1.143		4.63	37.50		42.13	61
0300	Primer & 1 coat, exterior latex		4	2		8.45	65.50		73.95	108
0310	Primer & 2 coats, exterior latex		3	2.667		12.80	87		99.80	145
0320	Brushwork, stain, sealer & 2 coats polyurethane		4.50	1.778		17.50	58		75.50	107
0330	Doors, panel, both sides, incl. frame & trim									
0340	Roll & brush, primer	1 Pord	6	1.333	Ea.	3.83	43.50		47.33	69.50
0350	Finish coat, exterior latex		6	1.333		4.63	43.50		48.13	70.50
0360	Primer & 1 coat, exterior latex		3	2.667		8.45	87		95.45	140
0370	Primer & 2 coats, exterior latex		2.50	3.200		12.80	105		117.80	171
0380	Brushwork, stain, sealer & 2 coats polyurethane		3	2.667		17.50	87		104.50	150
0400	Windows, per ext. side, based on 15 SF									
0410	1 to 6 lite									
0420	Brushwork, primer	1 Pord	13	.615	Ea.	.76	20		20.76	31.50
0430	Finish coat, exterior latex		13	.615		.91	20		20.91	31.50
0440	Primer & 1 coat, exterior latex		8	1		1.67	32.50		34.17	51
0450	Primer & 2 coats, exterior latex		6	1.333		2.53	43.50		46.03	68.50
0460	Stain, sealer & 1 coat varnish	↓	7	1.143	↓	2.50	37.50		40	59
0470	7 to 10 lite									
0480	Brushwork, primer	1 Pord	11	.727	Ea.	.76	24		24.76	37
0490	Finish coat, exterior latex		11	.727		.91	24		24.91	37
0500	Primer & 1 coat, exterior latex		7	1.143		1.67	37.50		39.17	58
0510	Primer & 2 coats, exterior latex		5	1.600		2.53	52.50		55.03	81.50
0520	Stain, sealer & 1 coat varnish		6	1.333		2.50	43.50		46	68.50
0530	12 lite									

09 91 Painting

09 91 13 – Exterior Painting

09 91 13.70 Doors and Windows, Exterior

		Crew	Daily Output	Labor-Hours	Unit	Material	2007 Bare Costs Labor	Equipment	Total	Total Incl O&P
0540	Brushwork, primer	1 Pord	10	.800	Ea.	.76	26		26.76	40.50
0550	Finish coat, exterior latex		10	.800		.91	26		26.91	40.50
0560	Primer & 1 coat, exterior latex		6	1.333		1.67	43.50		45.17	67.50
0570	Primer & 2 coats, exterior latex		5	1.600		2.53	52.50		55.03	81.50
0580	Stain, sealer & 1 coat varnish	↓	6	1.333	↓	2.49	43.50		45.99	68
0590	For oil base paint, add					10%				

09 91 13.80 Trim, Exterior

		Crew	Daily Output	Labor-Hours	Unit	Material	Labor	Equipment	Total	Total Incl O&P
0010	**TRIM, EXTERIOR** R099100-20									
0100	Door frames & trim (see Doors, interior or exterior)									
0110	Fascia, latex paint, one coat coverage									
0120	1" x 4", brushwork	1 Pord	640	.013	L.F.	.02	.41		.43	.64
0130	Roll		1280	.006		.02	.20		.22	.33
0140	Spray		2080	.004		.02	.13		.15	.21
0150	1" x 6" to 1" x 10", brushwork		640	.013		.06	.41		.47	.69
0160	Roll		1230	.007		.07	.21		.28	.39
0170	Spray		2100	.004		.05	.12		.17	.25
0180	1" x 12", brushwork		640	.013		.06	.41		.47	.69
0190	Roll		1050	.008		.07	.25		.32	.44
0200	Spray	↓	2200	.004	↓	.05	.12		.17	.24
0210	Gutters & downspouts, metal, zinc chromate paint									
0220	Brushwork, gutters, 5", first coat	1 Pord	640	.013	L.F.	.06	.41		.47	.68
0230	Second coat		960	.008		.05	.27		.32	.47
0240	Third coat		1280	.006		.04	.20		.24	.36
0250	Downspouts, 4", first coat		640	.013		.06	.41		.47	.68
0260	Second coat		960	.008		.05	.27		.32	.47
0270	Third coat	↓	1280	.006	↓	.04	.20		.24	.36
0280	Gutters & downspouts, wood									
0290	Brushwork, gutters, 5", primer	1 Pord	640	.013	L.F.	.05	.41		.46	.68
0300	Finish coat, exterior latex		640	.013		.05	.41		.46	.68
0310	Primer & 1 coat exterior latex		400	.020		.11	.65		.76	1.10
0320	Primer & 2 coats exterior latex		325	.025		.17	.81		.98	1.40
0330	Downspouts, 4", primer		640	.013		.05	.41		.46	.68
0340	Finish coat, exterior latex		640	.013		.05	.41		.46	.68
0350	Primer & 1 coat exterior latex		400	.020		.11	.65		.76	1.10
0360	Primer & 2 coats exterior latex	↓	325	.025	↓	.09	.81		.90	1.30
0370	Molding, exterior, up to 14" wide									
0380	Brushwork, primer	1 Pord	640	.013	L.F.	.06	.41		.47	.69
0390	Finish coat, exterior latex		640	.013		.07	.41		.48	.69
0400	Primer & 1 coat exterior latex		400	.020		.13	.65		.78	1.13
0410	Primer & 2 coats exterior latex		315	.025		.13	.83		.96	1.40
0420	Stain & fill		1050	.008		.06	.25		.31	.44
0430	Shellac		1850	.004		.06	.14		.20	.28
0440	Varnish	↓	1275	.006	↓	.07	.21		.28	.39

09 91 13.90 Walls, Masonry (CMU), Exterior

		Crew	Daily Output	Labor-Hours	Unit	Material	Labor	Equipment	Total	Total Incl O&P
0350	**WALLS, MASONRY (CMU), EXTERIOR**									
0360	Concrete masonry units (CMU), smooth surface									
0370	Brushwork, latex, first coat	1 Pord	640	.013	S.F.	.03	.41		.44	.65
0380	Second coat		960	.008		.04	.27		.31	.46
0390	Waterproof sealer, first coat		736	.011		.24	.36		.60	.79
0400	Second coat		1104	.007		.24	.24		.48	.62
0410	Roll, latex, paint, first coat		1465	.005		.06	.18		.24	.34
0420	Second coat	↓	1790	.004	↓	.05	.15		.20	.27

09 91 Painting

09 91 13 – Exterior Painting

09 91 13.90 Walls, Masonry (CMU), Exterior

		Crew	Daily Output	Labor-Hours	Unit	Material	2007 Bare Costs Labor	Equipment	Total	Total Incl O&P
0430	Waterproof sealer, first coat	1 Pord	1680	.005	S.F.	.24	.16		.40	.49
0440	Second coat		2060	.004		.24	.13		.37	.45
0450	Spray, latex, paint, first coat		1950	.004		.05	.13		.18	.26
0460	Second coat		2600	.003		.04	.10		.14	.19
0470	Waterproof sealer, first coat		2245	.004		.24	.12		.36	.44
0480	Second coat		2990	.003		.24	.09		.33	.39
0490	Concrete masonry unit (CMU), porous									
0500	Brushwork, latex, first coat	1 Pord	640	.013	S.F.	.11	.41		.52	.74
0510	Second coat		960	.008		.05	.27		.32	.47
0520	Waterproof sealer, first coat		736	.011		.24	.36		.60	.79
0530	Second coat		1104	.007		.24	.24		.48	.62
0540	Roll latex, first coat		1465	.005		.08	.18		.26	.36
0550	Second coat		1790	.004		.05	.15		.20	.28
0560	Waterproof sealer, first coat		1680	.005		.24	.16		.40	.49
0570	Second coat		2060	.004		.24	.13		.37	.45
0580	Spray latex, first coat		1950	.004		.06	.13		.19	.26
0590	Second coat		2600	.003		.04	.10		.14	.19
0600	Waterproof sealer, first coat		2245	.004		.24	.12		.36	.44
0610	Second coat		2990	.003		.24	.09		.33	.39

09 91 23 – Interior Painting

09 91 23.20 Cabinets and Casework

		Crew	Daily Output	Labor-Hours	Unit	Material	Labor	Equipment	Total	Total Incl O&P
0010	**CABINETS AND CASEWORK** R099100-20									
1000	Primer coat, oil base, brushwork	1 Pord	650	.012	S.F.	.05	.40		.45	.67
2000	Paint, oil base, brushwork, 1 coat		650	.012		.07	.40		.47	.68
3000	Stain, brushwork, wipe off		650	.012		.05	.40		.45	.67
4000	Shellac, 1 coat, brushwork		650	.012		.05	.40		.45	.67
4500	Varnish, 3 coats, brushwork, sand after 1st coat		325	.025		.18	.81		.99	1.40
5000	For latex paint, deduct					10%				

09 91 23.33 Doors and Windows, Interior Alkyd (Oil Base)

		Crew	Daily Output	Labor-Hours	Unit	Material	Labor	Equipment	Total	Total Incl O&P
0010	**DOORS AND WINDOWS, INTERIOR ALKYD (OIL BASE)**									
0500	Flush door & frame, 3' x 7', oil, primer, brushwork	1 Pord	10	.800	Ea.	2.16	26		28.16	42
1000	Paint, 1 coat		10	.800		2.38	26		28.38	42
1400	Stain, brushwork, wipe off		18	.444		1.12	14.55		15.67	23
1600	Shellac, 1 coat, brushwork		25	.320		1.13	10.45		11.58	17
1800	Varnish, 3 coats, brushwork, sand after 1st coat		9	.889		3.69	29		32.69	47.50
2000	Panel door & frame, 3' x 7', oil, primer, brushwork		6	1.333		2.05	43.50		45.55	68
2200	Paint, 1 coat		6	1.333		2.38	43.50		45.88	68
2600	Stain, brushwork, panel door, 3' x 7', not incl. frame		16	.500		1.12	16.35		17.47	25.50
2800	Shellac, 1 coat, brushwork		22	.364		1.13	11.90		13.03	19.15
3000	Varnish, 3 coats, brushwork, sand after 1st coat		7.50	1.067		3.69	35		38.69	56.50
4400	Windows, including frame and trim, per side									
4600	Colonial type, 6/6 lites, 2' x 3', oil, primer, brushwork	1 Pord	14	.571	Ea.	.32	18.70		19.02	28.50
5800	Paint, 1 coat		14	.571		.38	18.70		19.08	28.50
6200	3' x 5' opening, 6/6 lites, primer coat, brushwork		12	.667		.81	22		22.81	34
6400	Paint, 1 coat		12	.667		.94	22		22.94	34
6800	4' x 8' opening, 6/6 lites, primer coat, brushwork		8	1		1.73	32.50		34.23	51
7000	Paint, 1 coat		8	1		2.01	32.50		34.51	51
8000	Single lite type, 2' x 3', oil base, primer coat, brushwork		33	.242		.32	7.95		8.27	12.30
8200	Paint, 1 coat		33	.242		.38	7.95		8.33	12.35
8600	3' x 5' opening, primer coat, brushwork		20	.400		.81	13.10		13.91	20.50
8800	Paint, 1 coat		20	.400		.94	13.10		14.04	20.50
9200	4' x 8' opening, primer coat, brushwork		14	.571		1.73	18.70		20.43	30

09 91 Painting

09 91 23 – Interior Painting

09 91 23.33 Doors and Windows, Interior Alkyd (Oil Base)		Crew	Daily Output	Labor-Hours	Unit	Material	2007 Bare Costs Labor	Equipment	Total	Total Incl O&P
9400	Paint, 1 coat	1 Pord	14	.571	Ea.	2.01	18.70		20.71	30

09 91 23.35 Doors and Windows, Interior Latex

0010	**DOORS & WINDOWS, INTERIOR LATEX** R099100-20									
0100	Doors flush, both sides, incl. frame & trim									
0110	Roll & brush, primer	1 Pord	10	.800	Ea.	3.95	26		29.95	44
0120	Finish coat, latex		10	.800		4.43	26		30.43	44.50
0130	Primer & 1 coat latex		7	1.143		8.40	37.50		45.90	65
0140	Primer & 2 coats latex		5	1.600		12.55	52.50		65.05	92.50
0160	Spray, both sides, primer		20	.400		4.16	13.10		17.26	24.50
0170	Finish coat, latex		20	.400		4.65	13.10		17.75	25
0180	Primer & 1 coat latex		11	.727		8.85	24		32.85	46
0190	Primer & 2 coats latex		8	1		13.30	32.50		45.80	63.50
0200	Doors, French, both sides, 10-15 lite, incl. frame & trim									
0210	Roll & brush, primer	1 Pord	6	1.333	Ea.	1.98	43.50		45.48	67.50
0220	Finish coat, latex		6	1.333		2.22	43.50		45.72	68
0230	Primer & 1 coat latex		3	2.667		4.19	87		91.19	136
0240	Primer & 2 coats latex		2	4		6.30	131		137.30	204
0260	Doors, louvered, both sides, incl. frame & trim									
0270	Roll & brush, primer	1 Pord	7	1.143	Ea.	3.95	37.50		41.45	60.50
0280	Finish coat, latex		7	1.143		4.43	37.50		41.93	61
0290	Primer & 1 coat latex		4	2		8.15	65.50		73.65	107
0300	Primer & 2 coats, latex		3	2.667		12.80	87		99.80	145
0320	Spray, both sides, primer		20	.400		4.16	13.10		17.26	24.50
0330	Finish coat, latex		20	.400		4.65	13.10		17.75	25
0340	Primer & 1 coat, latex		11	.727		8.85	24		32.85	46
0350	Primer & 2 coats, latex		8	1		13.55	32.50		46.05	64
0360	Doors, panel, both sides, incl. frame & trim									
0370	Roll & brush, primer	1 Pord	6	1.333	Ea.	4.16	43.50		47.66	70
0380	Finish coat, latex		6	1.333		4.43	43.50		47.93	70.50
0390	Primer & 1 coat, latex		3	2.667		8.40	87		95.40	140
0400	Primer & 2 coats, latex		2.50	3.200		12.80	105		117.80	171
0420	Spray, both sides, primer		10	.800		4.16	26		30.16	44
0430	Finish coat, latex		10	.800		4.65	26		30.65	44.50
0440	Primer & 1 coat, latex		5	1.600		8.85	52.50		61.35	88.50
0450	Primer & 2 coats, latex		4	2		13.55	65.50		79.05	113
0460	Windows, per interior side, based on 15 SF									
0470	1 to 6 lite									
0480	Brushwork, primer	1 Pord	13	.615	Ea.	.78	20		20.78	31.50
0490	Finish coat, enamel		13	.615		.87	20		20.87	31.50
0500	Primer & 1 coat enamel		8	1		1.65	32.50		34.15	51
0510	Primer & 2 coats enamel		6	1.333		2.53	43.50		46.03	68.50
0530	7 to 10 lite									
0540	Brushwork, primer	1 Pord	11	.727	Ea.	.78	24		24.78	37
0550	Finish coat, enamel		11	.727		.87	24		24.87	37
0560	Primer & 1 coat enamel		7	1.143		1.65	37.50		39.15	58
0570	Primer & 2 coats enamel		5	1.600		2.53	52.50		55.03	81.50
0590	12 lite									
0600	Brushwork, primer	1 Pord	10	.800	Ea.	.78	26		26.78	40.50
0610	Finish coat, enamel		10	.800		.87	26		26.87	40.50
0620	Primer & 1 coat enamel		6	1.333		1.65	43.50		45.15	67.50
0630	Primer & 2 coats enamel		5	1.600		2.53	52.50		55.03	81.50
0650	For oil base paint, add					10%				

09 91 Painting

09 91 23 – Interior Painting

09 91 23.40 Floors, Interior

		Crew	Daily Output	Labor-Hours	Unit	Material	2007 Bare Costs Labor	Equipment	Total	Total Incl O&P
0010	**FLOORS, INTERIOR**									
0100	Concrete paint									
0110	Brushwork, latex, block filler									
0120	1st coat	1 Pord	975	.008	S.F.	.13	.27		.40	.55
0130	2nd coat		1150	.007		.09	.23		.32	.44
0140	3rd coat	↓	1300	.006	↓	.07	.20		.27	.38
0150	Roll, latex, block filler									
0160	1st coat	1 Pord	2600	.003	S.F.	.18	.10		.28	.35
0170	2nd coat		3250	.002		.11	.08		.19	.24
0180	3rd coat	↓	3900	.002	↓	.08	.07		.15	.19
0190	Spray, latex, block filler									
0200	1st coat	1 Pord	2600	.003	S.F.	.15	.10		.25	.32
0210	2nd coat		3250	.002		.08	.08		.16	.21
0220	3rd coat	↓	3900	.002	↓	.07	.07		.14	.17
0300	Acid stain and sealer									
0310	Stain, one coat	1 Pord	650	.012	S.F.	.11	.40		.51	.73
0320	Two coats		570	.014		.22	.46		.68	.93
0330	Acrylic sealer, one coat		2600	.003		.15	.10		.25	.32
0340	Two coats	↓	1400	.006	↓	.30	.19		.49	.61

09 91 23.52 Miscellaneous, Interior

		Crew	Daily Output	Labor-Hours	Unit	Material	2007 Bare Costs Labor	Equipment	Total	Total Incl O&P
0010	**MISCELLANEOUS, INTERIOR** R099100-20									
2400	Floors, conc./wood, oil base, primer/sealer coat, brushwork	2 Pord	1950	.008	S.F.	.05	.27		.32	.46
2450	Roller		5200	.003		.06	.10		.16	.21
2600	Spray		6000	.003		.06	.09		.15	.19
2650	Paint 1 coat, brushwork		1950	.008		.06	.27		.33	.47
2800	Roller		5200	.003		.06	.10		.16	.22
2850	Spray		6000	.003		.07	.09		.16	.20
3000	Stain, wood floor, brushwork, 1 coat		4550	.004		.05	.12		.17	.23
3200	Roller		5200	.003		.06	.10		.16	.21
3250	Spray		6000	.003		.06	.09		.15	.19
3400	Varnish, wood floor, brushwork		4550	.004		.06	.12		.18	.23
3450	Roller		5200	.003		.06	.10		.16	.22
3600	Spray	↓	6000	.003		.07	.09		.16	.20
3650	For dust proofing or anti skid, see division 03 35 29.30									
3800	Grilles, per side, oil base, primer coat, brushwork	1 Pord	520	.015	S.F.	.11	.50		.61	.88
3850	Spray		1140	.007		.11	.23		.34	.47
3920	Paint 2 coats, brushwork		325	.025		.24	.81		1.05	1.48
3940	Spray	↓	650	.012	↓	.28	.40		.68	.92
5000	Pipe, thru 4" diameter, primer or sealer coat, oil base, brushwork	2 Pord	1250	.013	L.F.	.06	.42		.48	.69
5100	Spray		2165	.007		.05	.24		.29	.42
5350	Paint 2 coats, brushwork		775	.021		.11	.68		.79	1.14
5400	Spray		1240	.013		.12	.42		.54	.76
6300	Thru 16" diameter, primer or sealer coat, brushwork		310	.052		.23	1.69		1.92	2.79
6350	Spray		540	.030		.25	.97		1.22	1.74
6500	Paint 2 coats, brushwork		195	.082		.43	2.68		3.11	4.52
6550	Spray	↓	310	.052	↓	.48	1.69		2.17	3.07
7000	Trim, wood, incl. puttying, under 6" wide									
7200	Primer coat, oil base, brushwork	1 Pord	650	.012	L.F.	.03	.40		.43	.64
7250	Paint, 1 coat, brushwork		650	.012		.03	.40		.43	.64
7450	3 coats		325	.025		.09	.81		.90	1.31
7500	Over 6" wide, primer coat, brushwork		650	.012		.05	.40		.45	.67
7550	Paint, 1 coat, brushwork	↓	650	.012	↓	.06	.40		.46	.68

09 91 Painting

09 91 23 – Interior Painting

09 91 23.52 Miscellaneous, Interior

		Crew	Daily Output	Labor-Hours	Unit	Material	2007 Bare Costs Labor	2007 Bare Costs Equipment	Total	Total Incl O&P
7650	3 coats	1 Pord	325	.025	L.F.	.18	.81		.99	1.41
8000	Cornice, simple design, primer coat, oil base, brushwork		650	.012	S.F.	.05	.40		.45	.67
8250	Paint, 1 coat		650	.012		.06	.40		.46	.68
8350	Ornate design, primer coat		350	.023		.05	.75		.80	1.18
8400	Paint, 1 coat		350	.023		.06	.75		.81	1.19
8600	Balustrades, primer coat, oil base, brushwork		520	.015		.05	.50		.55	.82
8650	Paint, 1 coat		520	.015		.06	.50		.56	.83
8900	Trusses and wood frames, primer coat, oil base, brushwork		800	.010		.05	.33		.38	.55
8950	Spray		1200	.007		.06	.22		.28	.39
9220	Paint 2 coats, brushwork		500	.016		.12	.52		.64	.92
9240	Spray		600	.013		.14	.44		.58	.81
9260	Stain, brushwork, wipe off		600	.013		.05	.44		.49	.72
9280	Varnish, 3 coats, brushwork		275	.029		.18	.95		1.13	1.62
9350	For latex paint, deduct					10%				

09 91 23.72 Walls and Ceilings, Interior

		Crew	Daily Output	Labor-Hours	Unit	Material	2007 Bare Costs Labor	2007 Bare Costs Equipment	Total	Total Incl O&P
0010	**WALLS AND CEILINGS, INTERIOR** R099100-20									
0100	Concrete, dry wall or plaster, oil base, primer or sealer coat									
0200	Smooth finish, brushwork	1 Pord	1150	.007	S.F.	.05	.23		.28	.40
0240	Roller		1350	.006		.05	.19		.24	.35
0280	Spray		2750	.003		.04	.10		.14	.18
0300	Sand finish, brushwork		975	.008		.05	.27		.32	.46
0340	Roller		1150	.007		.06	.23		.29	.40
0380	Spray		2275	.004		.04	.12		.16	.22
0800	Paint 2 coats, smooth finish, brushwork		680	.012		.11	.38		.49	.70
0840	Roller		800	.010		.11	.33		.44	.61
0880	Spray		1625	.005		.09	.16		.25	.34
0900	Sand finish, brushwork		605	.013		.10	.43		.53	.76
0940	Roller		1020	.008		.11	.26		.37	.51
0980	Spray		1700	.005		.09	.15		.24	.33
1200	Paint 3 coats, smooth finish, brushwork		510	.016		.16	.51		.67	.94
1240	Roller		650	.012		.16	.40		.56	.79
1280	Spray		1625	.005		.14	.16		.30	.40
1600	Glaze coating, 2 coats, spray, clear		1200	.007		.42	.22		.64	.79
1640	Multicolor		1200	.007		.87	.22		1.09	1.28
1700	For latex paint, deduct					10%				
1800	For ceiling installations, add						25%			
2000	Masonry or concrete block, oil base, primer or sealer coat									
2100	Smooth finish, brushwork	1 Pord	1224	.007	S.F.	.07	.21		.28	.40
2180	Spray		2400	.003		.08	.11		.19	.24
2200	Sand finish, brushwork		1089	.007		.08	.24		.32	.45
2280	Spray		2400	.003		.08	.11		.19	.24
2800	Paint 2 coats, smooth finish, brushwork		756	.011		.16	.35		.51	.70
2880	Spray		1360	.006		.15	.19		.34	.46
2900	Sand finish, brushwork		672	.012		.16	.39		.55	.77
2980	Spray		1360	.006		.15	.19		.34	.46
3600	Glaze coating, 3 coats, spray, clear		900	.009		.60	.29		.89	1.10
3620	Multicolor		900	.009		1	.29		1.29	1.54
4000	Block filler, 1 coat, brushwork		425	.019		.13	.62		.75	1.08
4100	Silicone, water repellent, 2 coats, spray		2000	.004		.27	.13		.40	.49
4120	For latex paint, deduct					10%				
8200	For work 8 - 15' H, add						10%			
8300	For work over 15' H, add						20%			

09 91 Painting

09 91 23 – Interior Painting

09 91 23.75 Dry Fall Painting

		Crew	Daily Output	Labor-Hours	Unit	Material	2007 Bare Costs Labor	2007 Bare Costs Equipment	Total	Total Incl O&P
0010	**DRY FALL PAINTING**									
0100	Walls									
0200	Wallboard and smooth plaster, one coat, brush	1 Pord	910	.009	S.F.	.04	.29		.33	.48
0210	Roll		1560	.005		.04	.17		.21	.30
0220	Spray		2600	.003		.04	.10		.14	.20
0230	Two coats, brush		520	.015		.09	.50		.59	.86
0240	Roll		877	.009		.09	.30		.39	.55
0250	Spray		1560	.005		.09	.17		.26	.35
0260	Concrete or textured plaster, one coat, brush		747	.011		.04	.35		.39	.58
0270	Roll		1300	.006		.04	.20		.24	.35
0280	Spray		1560	.005		.04	.17		.21	.30
0290	Two coats, brush		422	.019		.09	.62		.71	1.03
0300	Roll		747	.011		.09	.35		.44	.63
0310	Spray		1300	.006		.09	.20		.29	.40
0320	Concrete block, one coat, brush		747	.011		.04	.35		.39	.58
0330	Roll		1300	.006		.04	.20		.24	.35
0340	Spray		1560	.005		.04	.17		.21	.30
0350	Two coats, brush		422	.019		.09	.62		.71	1.03
0360	Roll		747	.011		.09	.35		.44	.63
0370	Spray		1300	.006		.09	.20		.29	.40
0380	Wood, one coat, brush		747	.011		.04	.35		.39	.58
0390	Roll		1300	.006		.04	.20		.24	.35
0400	Spray		877	.009		.04	.30		.34	.50
0410	Two coats, brush		487	.016		.09	.54		.63	.91
0420	Roll		747	.011		.09	.35		.44	.63
0430	Spray		650	.012		.09	.40		.49	.71
0440	Ceilings									
0450	Wallboard and smooth plaster, one coat, brush	1 Pord	600	.013	S.F.	.04	.44		.48	.71
0460	Roll		1040	.008		.04	.25		.29	.43
0470	Spray		1560	.005		.04	.17		.21	.30
0480	Two coats, brush		341	.023		.09	.77		.86	1.25
0490	Roll		650	.012		.09	.40		.49	.71
0500	Spray		1300	.006		.09	.20		.29	.40
0510	Concrete or textured plaster, one coat, brush		487	.016		.04	.54		.58	.86
0520	Roll		877	.009		.04	.30		.34	.50
0530	Spray		1560	.005		.04	.17		.21	.30
0540	Two coats, brush		276	.029		.09	.95		1.04	1.53
0550	Roll		520	.015		.09	.50		.59	.86
0560	Spray		1300	.006		.09	.20		.29	.40
0570	Structural steel, bar joists or metal deck, one coat, spray		1560	.005		.04	.17		.21	.30
0580	Two coats, spray		1040	.008		.09	.25		.34	.48

09 93 Staining and Transparent Finishing

09 93 23 – Interior Staining and Finishing

09 93 23.10 Varnish		Crew	Daily Output	Labor-Hours	Unit	Material	2007 Bare Costs Labor	2007 Bare Costs Equipment	Total	Total Incl O&P
0010	**VARNISH**									
0012	1 coat + sealer, on wood trim, no sanding included	1 Pord	400	.020	S.F.	.07	.65		.72	1.06
0100	Hardwood floors, 2 coats, no sanding included, roller	"	1890	.004	"	.14	.14		.28	.36

09 96 High-Performance Coatings

09 96 56 – Epoxy Coatings

09 96 56.20 Wall Coatings

		Crew	Daily Output	Labor-Hours	Unit	Material	Labor	Equipment	Total	Total Incl O&P
0010	**WALL COATINGS**									
0100	Acrylic glazed coatings, minimum	1 Pord	525	.015	S.F.	.27	.50		.77	1.05
0200	Maximum		305	.026		.57	.86		1.43	1.92
0300	Epoxy coatings, minimum		525	.015		.35	.50		.85	1.14
0400	Maximum		170	.047		1.07	1.54		2.61	3.50
0600	Exposed aggregate, troweled on, 1/16" to 1/4", minimum		235	.034		.53	1.11		1.64	2.25
0700	Maximum (epoxy or polyacrylate)		130	.062		1.14	2.01		3.15	4.28
0900	1/2" to 5/8" aggregate, minimum		130	.062		1.06	2.01		3.07	4.20
1000	Maximum		80	.100		1.81	3.27		5.08	6.90
1500	Exposed aggregate, sprayed on, 1/8" aggregate, minimum		295	.027		.49	.89		1.38	1.87
1600	Maximum		145	.055		.91	1.80		2.71	3.71
1800	High build epoxy, 50 mil, minimum		390	.021		.59	.67		1.26	1.66
1900	Maximum		95	.084		1.01	2.75		3.76	5.25
2100	Laminated epoxy with fiberglass, minimum		295	.027		.64	.89		1.53	2.03
2200	Maximum		145	.055		1.13	1.80		2.93	3.95
2400	Sprayed perlite or vermiculite, 1/16" thick, minimum		2935	.003		.23	.09		.32	.38
2500	Maximum		640	.013		.65	.41		1.06	1.34
2700	Vinyl plastic wall coating, minimum		735	.011		.29	.36		.65	.86
2800	Maximum		240	.033		.71	1.09		1.80	2.42
3000	Urethane on smooth surface, 2 coats, minimum		1135	.007		.22	.23		.45	.59
3100	Maximum		665	.012		.48	.39		.87	1.12
3600	Ceramic-like glazed coating, cementitious, minimum		440	.018		.42	.59		1.01	1.35
3700	Maximum		345	.023		.70	.76		1.46	1.91
3900	Resin base, minimum		640	.013		.29	.41		.70	.94
4000	Maximum		330	.024		.47	.79		1.26	1.71

09 97 Special Coatings

09 97 13 – Steel Coatings

09 97 13.23 Exterior Steel Coatings

		Crew	Daily Output	Labor-Hours	Unit	Material	Labor	Equipment	Total	Total Incl O&P
0010	**EXTERIOR STEEL COATINGS** R050516-30									
6100	Cold galvanizing, brush in field	1 Psst	1100	.007	S.F.	.07	.24		.31	.51
6510	Paints & protective coatings, sprayed in field									
6520	Alkyds, primer	2 Psst	3600	.004	S.F.	.05	.15		.20	.33
6540	Gloss topcoats		3200	.005		.05	.17		.22	.36
6560	Silicone alkyd		3200	.005		.10	.17		.27	.42
6610	Epoxy, primer		3000	.005		.17	.18		.35	.51
6630	Intermediate or topcoat		2800	.006		.13	.19		.32	.50
6650	Enamel coat		2800	.006		.17	.19		.36	.53
6700	Epoxy ester, primer		2800	.006		.47	.19		.66	.86
6720	Topcoats		2800	.006		.11	.19		.30	.47
6810	Latex primer		3600	.004		.05	.15		.20	.33
6830	Topcoats		3200	.005		.06	.17		.23	.38

09 97 Special Coatings

09 97 13 – Steel Coatings

09 97 13.23 Exterior Steel Coatings

		Crew	Daily Output	Labor-Hours	Unit	Material	2007 Bare Costs Labor	2007 Bare Costs Equipment	Total	Total Incl O&P
6910	Universal primers, one part, phenolic, modified alkyd	2 Psst	2000	.008	S.F.	.09	.27		.36	.58
6940	Two part, epoxy spray		2000	.008		.26	.27		.53	.77
7000	Zinc rich primers, self cure, spray, inorganic		1800	.009		.52	.30		.82	1.11
7010	Epoxy, spray, organic	↓	1800	.009	↓	.13	.30		.43	.69
7020	Above one story, spray painting simple structures, add						25%			
7030	Intricate structures, add						50%			

Division Notes

	CREW	DAILY OUTPUT	LABOR-HOURS	UNIT	2007 BARE COSTS				TOTAL INCL O&P
					MAT.	LABOR	EQUIP.	TOTAL	

Estimating Tips

General

- The items in this division are usually priced per square foot or each.
- Many items in Division 10 require some type of support system or special anchors that are not usually furnished with the item. The required anchors must be added to the estimate in the appropriate division.
- Some items in Division 10, such as lockers, may require assembly before installation. Verify the amount of assembly required. Assembly can often exceed installation time.

10 20 00 Interior Specialties

- Support angles and blocking are not included in the installation of toilet compartments, shower/dressing compartments, or cubicles. Appropriate line items from Divisions 5 or 6 may need to be added to support the installations.
- Toilet partitions are priced by the stall. A stall consists of a side wall, pilaster, and door with hardware. Toilet tissue holders and grab bars are extra.
- The required acoustical rating of a folding partition can have a significant impact on costs. Verify the sound transmission coefficient rating of the panel priced to the specification requirements.
- Grab bar installation does not include supplemental blocking or backing to support the required load. When grab bars are installed at an existing facility, provisions must be made to attach the grab bars to solid structure.

Reference Numbers

Reference numbers are shown in shaded boxes at the beginning of some major classifications. These numbers refer to related items in the Reference Section. The reference information may be an estimating procedure, an alternate pricing method, or technical information.

Note: Not all subdivisions listed here necessarily appear in this publication.

No part of this publication may be reproduced, stored in a retrieval system, or transmitted in any form or by any means without prior written permission of Reed Construction Data.

10 11 Visual Display Surfaces

10 11 13 – Chalkboards

10 11 13.13 Fixed Chalkboards

		Crew	Daily Output	Labor-Hours	Unit	Material	2007 Bare Costs Labor	Equipment	Total	Total Incl O&P
0010	**FIXED CHALKBOARDS** Porcelain enamel steel									
3900	Wall hung									
4000	Aluminum frame and chalktrough									
4200	3' x 4'	2 Carp	16	1	Ea.	158	36.50		194.50	230
4300	3' x 5'		15	1.067		230	39		269	315
4500	4' x 8'		14	1.143		350	42		392	450
4600	4' x 12'	↓	13	1.231	↓	510	45		555	630
4700	Wood frame and chalktrough									
4800	3' x 4'	2 Carp	16	1	Ea.	164	36.50		200.50	237
5000	3' x 5'		15	1.067		228	39		267	310
5100	4' x 5'		14	1.143		230	42		272	320
5300	4' x 8'	↓	13	1.231		325	45		370	430
5400	Liquid chalk, white porcelain enamel, wall hung									
5420	Deluxe units, aluminum trim and chalktrough									
5450	4' x 4'	2 Carp	16	1	Ea.	210	36.50		246.50	288
5500	4' x 8'		14	1.143		355	42		397	455
5550	4' x 12'	↓	12	1.333	↓	485	49		534	610
5700	Wood trim and chalktrough									
5900	4' x 4'	2 Carp	16	1	Ea.	450	36.50		486.50	550
6000	4' x 6'		15	1.067		545	39		584	660
6200	4' x 8'	↓	14	1.143		635	42		677	765
6300	Liquid chalk, felt tip markers					1.50			1.50	1.65
6500	Erasers					1.53			1.53	1.68
6600	Board cleaner, 8 oz. bottle				↓	4.13			4.13	4.54

10 11 13.23 Modular-Support-Mounted Chalkboards

		Crew	Daily Output	Labor-Hours	Unit	Material	Labor	Equipment	Total	Total Incl O&P
0010	**MODULAR-SUPPORT-MOUNTED CHALKBOARDS**									
0400	Sliding chalkboards									
0450	Vertical, one sliding board with back panel, wall mounted									
0500	8' x 4'	2 Carp	8	2	Ea.	1,575	73.50		1,648.50	1,850
0520	8' x 8'		7.50	2.133		2,225	78.50		2,303.50	2,575
0540	8' x 12'	↓	7	2.286	↓	2,775	84		2,859	3,175
0600	Two sliding boards, with back panel									
0620	8' x 4'	2 Carp	8	2	Ea.	2,375	73.50		2,448.50	2,750
0640	8' x 8'		7.50	2.133		3,350	78.50		3,428.50	3,800
0660	8' x 12'	↓	7	2.286		4,175	84		4,259	4,700
0700	Horizontal, two track									
0800	4' x 8', 2 sliding panels	2 Carp	8	2	Ea.	1,625	73.50		1,698.50	1,925
0820	4' x 12', 2 sliding panels		7.50	2.133		2,075	78.50		2,153.50	2,400
0840	4' x 16', 4 sliding panels	↓	7	2.286	↓	2,800	84		2,884	3,200
0900	Four track, four sliding panels									
0920	4' x 8'	2 Carp	8	2	Ea.	2,600	73.50		2,673.50	3,000
0940	4' x 12'		7.50	2.133		2,075	78.50		2,153.50	2,400
0960	4' x 16'	↓	7	2.286	↓	4,225	84		4,309	4,775
1200	Vertical, motor operated									
1400	One sliding panel with back panel									
1450	10' x 4'	2 Carp	4	4	Ea.	5,850	147		5,997	6,650
1500	10' x 10'		3.75	4.267	↓	6,825	157		6,982	7,750
1550	10' x 16'	↓	3.50	4.571		7,825	168		7,993	8,875
1700	Two sliding panels with back panel									
1750	10' x 4'	2 Carp	4	4	Ea.	9,350	147		9,497	10,500
1800	10' x 10'		3.75	4.267		10,700	157		10,857	11,900
1850	10' x 16'	↓	3.50	4.571		12,100	168		12,268	13,600

10 11 Visual Display Surfaces

10 11 13 – Chalkboards

10 11 13.23 Modular-Support-Mounted Chalkboards

		Crew	Daily Output	Labor-Hours	Unit	Material	2007 Bare Costs Labor	Equipment	Total	Total Incl O&P
2000	Three sliding panels with back panel									
2100	10' x 4'	2 Carp	4	4	Ea.	12,200	147		12,347	13,600
2150	10' x 10'		3.75	4.267		14,200	157		14,357	15,800
2200	10' x 16'		3.50	4.571		16,200	168		16,368	18,100
2400	For projection screen, glass beaded, add				S.F.	9.85			9.85	10.85
2500	For remote control, 1 panel control, add				Ea.	455			455	500
2600	2 panel control, add				"	785			785	865
2800	For units without back panels, deduct				S.F.	5.65			5.65	6.25
2850	For liquid chalk porcelain panels, add				"	6.20			6.20	6.80
3000	Swing leaf, any comb. of chalkboard & cork, aluminum frame									
3100	Floor style, 6 panels									
3150	30" x 40" panels				Ea.	910			910	1,000
3200	48" x 40" panels				"	1,950			1,950	2,125
3300	Wall mounted, 6 panels									
3400	30" x 40" panels	2 Carp	16	1	Ea.	810	36.50		846.50	945
3450	48" x 40" panels	"	16	1	"	1,625	36.50		1,661.50	1,825
3600	Extra panels for swing leaf units									
3700	30" x 40" panels				Ea.	263			263	290
3750	48" x 40" panels				"	315			315	345

10 11 13.43 Portable Chalkboards

		Crew	Daily Output	Labor-Hours	Unit	Material	Labor	Equipment	Total	Total Incl O&P
0010	**PORTABLE CHALKBOARDS**									
0100	Freestanding, reversible									
0120	Economy, wood frame, 4' x 6'									
0140	Chalkboard both sides				Ea.	575			575	630
0160	Chalkboard one side, cork other side				"	470			470	520
0200	Standard, lightweight satin finished aluminum, 4' x 6'									
0220	Chalkboard both sides				Ea.	590			590	650
0240	Chalkboard one side, cork other side				"	500			500	555
0300	Deluxe, heavy duty extruded aluminum, 4' x 6'									
0320	Chalkboard both sides				Ea.	1,200			1,200	1,325
0340	Chalkboard one side, cork other side				"	1,100			1,100	1,200

10 11 23 – Tackboards

10 11 23.10 Fixed Tackboards

		Crew	Daily Output	Labor-Hours	Unit	Material	Labor	Equipment	Total	Total Incl O&P
0010	**FIXED TACKBOARDS**									
0020	Cork sheets, unbacked, no frame, 1/4" thick	2 Carp	290	.055	S.F.	3.60	2.02		5.62	7.10
0100	1/2" thick		290	.055		6.15	2.02		8.17	9.90
0300	Fabric-face, no frame, on 7/32" cork underlay		290	.055		5.10	2.02		7.12	8.75
0400	On 1/4" cork on 1/4" hardboard		290	.055		6.60	2.02		8.62	10.45
0600	With edges wrapped		290	.055		7.50	2.02		9.52	11.40
0700	On 7/16" fire retardant core		290	.055		5.15	2.02		7.17	8.80
0900	With edges wrapped		290	.055		7.20	2.02		9.22	11.05
1000	Designer fabric only, cut to size					1.90			1.90	2.09
1200	1/4" vinyl cork, on 1/4" hardboard, no frame	2 Carp	290	.055		6.70	2.02		8.72	10.50
1300	On 1/4" coreboard		290	.055		6.50	2.02		8.52	10.30
2000	For map and display rail, economy, add		385	.042	L.F.	2.10	1.53		3.63	4.69
2100	Deluxe, add		350	.046	"	3.06	1.68		4.74	6
2120	Prefabricated, 1/4" cork, 3' x 5' with aluminum frame		16	1	Ea.	116	36.50		152.50	185
2140	Wood frame		16	1		144	36.50		180.50	215
2160	4' x 4' with aluminum frame		16	1		94.50	36.50		131	161
2180	Wood frame		16	1		132	36.50		168.50	202
2200	4' x 8' with aluminum frame		14	1.143		182	42		224	266
2210	With wood frame		14	1.143		98	42		140	174

10 11 Visual Display Surfaces

10 11 23 – Tackboards

10 11 23.10 Fixed Tackboards

		Crew	Daily Output	Labor-Hours	Unit	Material	2007 Bare Costs Labor	Equipment	Total	Total Incl O&P
2220	4' x 12' with aluminum frame	2 Carp	12	1.333	Ea.	262	49		311	365
2230	Bulletin board case, single glass door, with lock									
2240	36" x 24", economy	2 Carp	12	1.333	Ea.	214	49		263	310
2250	Deluxe		12	1.333		440	49		489	560
2260	42" x 30", economy		12	1.333		245	49		294	345
2270	Deluxe		12	1.333		470	49		519	595
2300	Glass enclosed cabinets, alum., cork panel, hinged doors									
2400	3' x 3', 1 door	2 Carp	12	1.333	Ea.	510	49		559	635
2500	4' x 4', 2 door		11	1.455		850	53.50		903.50	1,025
2600	4' x 7', 3 door		10	1.600		1,425	58.50		1,483.50	1,675
2800	4' x 10', 4 door		8	2		1,875	73.50		1,948.50	2,175
2900	For lights, add per door opening	1 Elec	13	.615		156	27		183	212
3100	Horizontal sliding units, 4 doors, 4' x 8', 8' x 4'	2 Carp	9	1.778		1,625	65		1,690	1,900
3200	4' x 12'		7	2.286		2,075	84		2,159	2,400
3400	8 doors, 4' x 16'		5	3.200		2,675	117		2,792	3,125
3500	4' x 24'		4	4		3,600	147		3,747	4,200

10 11 23.20 Control Boards

		Crew	Daily Output	Labor-Hours	Unit	Material	Labor	Equipment	Total	Total Incl O&P
0010	**CONTROL BOARDS**									
0020	Magnetic, porcelain finish, 18" x 24", framed	2 Carp	8	2	Ea.	199	73.50		272.50	335
0100	24" x 36"		7.50	2.133		295	78.50		373.50	445
0200	36" x 48"		7	2.286		390	84		474	560
0300	48" x 72"		6	2.667		860	98		958	1,100
0400	48" x 96"		5	3.200		1,075	117		1,192	1,350

10 13 Directories

10 13 10 – Building Directories

10 13 10.10 Directory Boards

		Crew	Daily Output	Labor-Hours	Unit	Material	Labor	Equipment	Total	Total Incl O&P
0010	**DIRECTORY BOARDS**									
0050	Plastic, glass covered, 30" x 20"	2 Carp	3	5.333	Ea.	242	196		438	570
0100	36" x 48"		2	8		835	294		1,129	1,375
0300	Grooved cork, 30" x 20"		3	5.333		355	196		551	695
0400	36" x 48"		2	8		300	294		594	785
0600	Black felt, 30" x 20"		3	5.333		198	196		394	520
0700	36" x 48"		2	8		355	294		649	845
0900	Outdoor, weatherproof, black plastic, 36" x 24"		2	8		685	294		979	1,200
1000	36" x 36"		1.50	10.667		790	390		1,180	1,475
1800	Indoor, economy, open face, 18" x 24"		7	2.286		113	84		197	256
1900	24" x 36"		7	2.286		137	84		221	281
2000	36" x 24"		6	2.667		156	98		254	325
2100	36" x 48"		6	2.667		234	98		332	410
2400	Building directory, alum., black felt panels, 1 door, 24" x 18"		4	4		296	147		443	555
2500	36" x 24"		3.50	4.571		340	168		508	635
2600	48" x 32"		3	5.333		525	196		721	885
2700	36" x 48", 2 door		2.50	6.400		580	235		815	1,000
2800	36" x 60"		2	8		880	294		1,174	1,425
2900	48" x 60"		1	16		855	585		1,440	1,850
3100	For bronze enamel finish, add					15%				
3200	For bronze anodized finish, add					25%				
3400	For illuminated directory, single door unit, add					156			156	172
3500	For 6" header panel, 6 letters per foot, add				L.F.	27			27	30

10 14 Signage

10 14 19 – Dimensional Letter Signage

10 14 19.10 Exterior Signs		Crew	Daily Output	Labor-Hours	Unit	Material	2007 Bare Costs Labor	2007 Bare Costs Equipment	Total	Total Incl O&P
0010	**EXTERIOR SIGNS**									
0020	Letters, 2" high, 3/8" deep, cast bronze	1 Carp	24	.333	Ea.	21.50	12.25		33.75	42.50
0140	1/2" deep, cast aluminum		18	.444		21.50	16.30		37.80	49
0160	Cast bronze		32	.250		28	9.20		37.20	45
0300	6" high, 5/8" deep, cast aluminum		24	.333		26	12.25		38.25	47.50
0400	Cast bronze		24	.333		42.50	12.25		54.75	66
0600	8" high, 3/4" deep, cast aluminum		14	.571		32	21		53	67.50
0700	Cast bronze		20	.400		81.50	14.70		96.20	113
0900	10" high, 1" deep, cast aluminum		18	.444		38.50	16.30		54.80	68
1000	Bronze		18	.444		90.50	16.30		106.80	125
1200	12" high, 1-1/4" deep, cast aluminum		12	.667		48	24.50		72.50	91
1500	Cast bronze		18	.444		120	16.30		136.30	158
1600	14" high, 2-5/16" deep, cast aluminum		12	.667		76.50	24.50		101	123
1800	Fabricated stainless steel, 6" high, 2" deep		20	.400		119	14.70		133.70	153
1900	12" high, 3" deep		18	.444		140	16.30		156.30	180
2100	18" high, 3" deep		12	.667		206	24.50		230.50	264
2200	24" high, 4" deep		10	.800		287	29.50		316.50	360
2700	Acrylic, on high density foam, 12" high, 2" deep		20	.400		19.75	14.70		34.45	44.50
2800	18" high, 2" deep		18	.444		49.50	16.30		65.80	79.50
3900	Plaques, custom, 20" x 30", for up to 450 letters, cast aluminum	2 Carp	4	4		895	147		1,042	1,225
4000	Cast bronze		4	4		1,200	147		1,347	1,525
4200	30" x 36", up to 900 letters cast aluminum		3	5.333		1,825	196		2,021	2,300
4300	Cast bronze		3	5.333		2,350	196		2,546	2,875
4500	36" x 48", for up to 1300 letters, cast bronze		2	8		3,525	294		3,819	4,325
4800	Signs, reflective alum. directional signs, dbl. face, 2-way, w/bracket		30	.533		71	19.55		90.55	109
4900	4-way		30	.533		114	19.55		133.55	156
5100	Exit signs, 24 ga. alum., 14" x 12" surface mounted	1 Carp	30	.267		35.50	9.80		45.30	54.50
5200	10" x 7"		20	.400		25	14.70		39.70	50.50
5400	Bracket mounted, double face, 12" x 10"		30	.267		33	9.80		42.80	52
5500	Sticky back, stock decals, 14" x 10"	1 Clab	50	.160		8.60	4.60		13.20	16.60
6000	Interior elec., wall mount, fiberglass panels, 2 lamps, 6"	1 Elec	8	1		104	44		148	180
6100	8"	"	8	1		48	44		92	118
6400	Replacement sign faces, 6" or 8"	1 Clab	50	.160		26.50	4.60		31.10	36

10 14 23 – Panel Signage

10 14 23.13 Engraved Interior Panel Signage

		Crew	Daily Output	Labor-Hours	Unit	Material	Labor	Equipment	Total	Total Incl O&P
0010	**ENGRAVED INTERIOR PANEL SIGNAGE**									
1010	Flexible door sign, adhesive back, w/Braille, 5/8" letters, 4" x 4"	1 Clab	32	.250	Ea.	25.50	7.20		32.70	39
1050	6" x 6"		32	.250		37	7.20		44.20	51.50
1100	8" x 2"		32	.250		26.50	7.20		33.70	40
1150	8" x 4"		32	.250		33	7.20		40.20	47.50
1200	8" x 8"		32	.250		48	7.20		55.20	64
1250	12" x 2"		32	.250		35	7.20		42.20	49.50
1300	12" x 6"		32	.250		49.50	7.20		56.70	65.50
1350	12" x 12"		32	.250		96.50	7.20		103.70	117
1500	Graphic symbols, 2" x 2"		32	.250		12.20	7.20		19.40	24.50
1550	6" x 6"		32	.250		31	7.20		38.20	45
1600	8" x 8"		32	.250		31	7.20		38.20	45

10 14 53 – Traffic Signage

10 14 53.20 Traffic Signs

		Crew	Daily Output	Labor-Hours	Unit	Material	Labor	Equipment	Total	Total Incl O&P
0010	**TRAFFIC SIGNS**									
0012	Stock, 24" x 24", no posts, .080" alum. reflectorized	B-80	70	.457	Ea.	59.50	14.30	8	81.80	96.50
0100	High intensity		70	.457		59.50	14.30	8	81.80	96.50

10 14 Signage

10 14 53 – Traffic Signage

10 14 53.20 Traffic Signs	Crew	Daily Output	Labor-Hours	Unit	Material	2007 Bare Costs Labor	Equipment	Total	Total Incl O&P
0300	30" x 30", reflectorized	B-80	70	.457	Ea.	121	14.30	8	143.30	164
0400	High intensity		70	.457		121	14.30	8	143.30	164
0600	Guide and directional signs, 12" x 18", reflectorized		70	.457		38.50	14.30	8	60.80	73.50
0700	High intensity		70	.457		37	14.30	8	59.30	71.50
0900	18" x 24", stock signs, reflectorized		70	.457		43	14.30	8	65.30	78.50
1000	High intensity		70	.457		43	14.30	8	65.30	78.50
1200	24" x 24", stock signs, reflectorized		70	.457		53.50	14.30	8	75.80	90
1300	High intensity		70	.457		53.50	14.30	8	75.80	90
1500	Add to above for steel posts, galvanized, 10'-0" upright, bolted		200	.160		19.30	5	2.80	27.10	32
1600	12'-0" upright, bolted		140	.229		25.50	7.15	4	36.65	43.50
1800	Highway road signs, aluminum, over 20 S. F., reflectorized		350	.091	S.F.	24.50	2.86	1.60	28.96	33
2000	High intensity		350	.091		24.50	2.86	1.60	28.96	33
2200	Highway, suspended over road, 80 S.F. min., reflectorized		165	.194		24.50	6.05	3.40	33.95	40
2300	High intensity		165	.194		24.50	6.05	3.40	33.95	40

10 17 Telephone Specialties

10 17 16 – Telephone Enclosures

10 17 16.10 Telephone Enclosures

		Crew	Daily Output	Labor-Hours	Unit	Material	Labor	Equipment	Total	Total Incl O&P
0010	**TELEPHONE ENCLOSURES**									
0300	Shelf type, wall hung, minimum	2 Carp	5	3.200	Ea.	1,075	117		1,192	1,350
0400	Maximum		5	3.200		2,750	117		2,867	3,200
0600	Booth type, painted steel, indoor or outdoor, minimum		1.50	10.667		3,375	390		3,765	4,300
0700	Maximum (stainless steel)		1.50	10.667		11,200	390		11,590	13,000
1300	Outdoor, acoustical, on post		3	5.333		1,475	196		1,671	1,925
1400	Phone carousel, pedestal mounted with dividers		.60	26.667		5,625	980		6,605	7,725
1900	Outdoor, drive-up type, wall mounted		4	4		905	147		1,052	1,225
2000	Post mounted, stainless steel posts		3	5.333		1,400	196		1,596	1,825
2200	Directory shelf, wall mounted, stainless steel									
2300	3 binders	2 Carp	8	2	Ea.	1,075	73.50		1,148.50	1,300
2500	4 binders		7	2.286		1,225	84		1,309	1,475
2600	5 binders		6	2.667		1,575	98		1,673	1,875
2800	Table type, stainless steel, 4 binders		8	2		1,275	73.50		1,348.50	1,525
2900	7 binders		7	2.286		1,575	84		1,659	1,850

10 21 Compartments and Cubicles

10 21 13 – Toilet Compartments

10 21 13.13 Metal Toilet Compartments

		Crew	Daily Output	Labor-Hours	Unit	Material	Labor	Equipment	Total	Total Incl O&P
0010	**METAL TOILET COMPARTMENTS**									
0110	Cubicles, ceiling hung									
0200	Painted metal	2 Carp	4	4	Ea.	430	147		577	705
0500	Stainless steel	"	4	4		1,250	147		1,397	1,600
0600	For handicap units, incl. 52" grab bars, add					420			420	460
0900	Floor and ceiling anchored									
1000	Painted metal	2 Carp	5	3.200	Ea.	470	117		587	700
1300	Stainless steel	"	5	3.200		1,450	117		1,567	1,775
1400	For handicap units, incl. 52" grab bars, add					300			300	330
1610	Floor mounted									
1700	Painted metal	2 Carp	7	2.286	Ea.	530	84		614	710
2000	Stainless steel	"	7	2.286		1,550	84		1,634	1,825

10 21 Compartments and Cubicles

10 21 13 – Toilet Compartments

10 21 13.13 Metal Toilet Compartments

		Crew	Daily Output	Labor-Hours	Unit	Material	2007 Bare Costs Labor	Equipment	Total	Total Incl O&P
2100	For handicap units, incl. 52" grab bars, add				Ea.	300			300	330
2200	For juvenile units, deduct				↓	39.50			39.50	43.50
2450	Floor mounted, headrail braced									
2500	Painted metal	2 Carp	6	2.667	Ea.	470	98		568	665
2800	Stainless steel	"	6	2.667		1,450	98		1,548	1,725
2900	For handicap units, incl. 52" grab bars, add					320			320	355
3000	Wall hung partitions, painted metal	2 Carp	7	2.286		595	84		679	785
3300	Stainless steel	"	7	2.286		1,425	84		1,509	1,675
3400	For handicap units, incl. 52" grab bars, add				↓	320			320	355
4000	Screens, entrance, floor mounted, 58" high, 48" wide									
4200	Painted metal	2 Carp	15	1.067	Ea.	232	39		271	315
4500	Stainless steel	"	15	1.067	"	765	39		804	905
4650	Urinal screen, 18" wide									
4700	Painted metal	2 Carp	8	2	Ea.	208	73.50		281.50	345
5000	Stainless steel	"	8	2	"	550	73.50		623.50	720
5100	Floor mounted, head rail braced									
5300	Painted metal	2 Carp	8	2	Ea.	198	73.50		271.50	330
5600	Stainless steel	"	8	2	"	640	73.50		713.50	820
5750	Pilaster, flush									
5800	Painted metal	2 Carp	10	1.600	Ea.	269	58.50		327.50	390
6100	Stainless steel		10	1.600		635	58.50		693.50	785
6300	Urinal screen, post braced, painted metal		10	1.600		294	58.50		352.50	415
6600	Stainless steel	↓	10	1.600	↓	480	58.50		538.50	620
6700	Wall hung, bracket supported									
6800	Painted metal	2 Carp	10	1.600	Ea.	281	58.50		339.50	400
7100	Stainless steel		10	1.600		440	58.50		498.50	575
7400	Flange supported, painted metal		10	1.600		218	58.50		276.50	330
7700	Stainless steel		10	1.600		510	58.50		568.50	650
7800	Wedge type, painted metal		10	1.600		257	58.50		315.50	375
8100	Stainless steel	↓	10	1.600	↓	525	58.50		583.50	665

10 21 13.16 Plastic-Laminate-Clad Toilet Compartments

		Crew	Daily Output	Labor-Hours	Unit	Material	Labor	Equipment	Total	Total Incl O&P
0010	**PLASTIC-LAMINATE-CLAD TOILET COMPARTMENTS**									
0110	Cubicles, ceiling hung									
0300	Plastic laminate on particle board	2 Carp	4	4	Ea.	585	147		732	875
0600	For handicap units, incl. 52" grab bars, add				"	420			420	460
0900	Floor and ceiling anchored									
1100	Plastic laminate on particle board	2 Carp	5	3.200	Ea.	735	117		852	995
1400	For handicap units, incl. 52" grab bars, add				"	300			300	330
1610	Floor mounted									
1800	Plastic laminate on particle board	2 Carp	7	2.286	Ea.	580	84		664	770
2450	Floor mounted, headrail braced									
2600	Plastic laminate on particle board	2 Carp	6	2.667	Ea.	720	98		818	945
3400	For handicap units, incl. 52" grab bars, add					320			320	355
4300	Entrance screen, floor mtd., plas. lam., 58" high, 48" wide	2 Carp	15	1.067		465	39		504	575
4800	Urinal screen, 18" wide, ceiling braced, plastic laminate		8	2		310	73.50		383.50	455
5400	Floor mounted, headrail braced		8	2		355	73.50		428.50	505
5900	Pilaster, flush, plastic laminate		10	1.600		390	58.50		448.50	520
6400	Post braced, plastic laminate	↓	10	1.600	↓	350	58.50		408.50	475
6700	Wall hung, bracket supported									
6900	Plastic laminate on particle board	2 Carp	10	1.600	Ea.	167	58.50		225.50	275
7450	Flange supported									
7500	Plastic laminate on particle board	2 Carp	10	1.600	Ea.	425	58.50		483.50	560

10 21 Compartments and Cubicles

10 21 13 – Toilet Compartments

10 21 13.19 Plastic Toilet Compartments

		Crew	Daily Output	Labor-Hours	Unit	Material	2007 Bare Costs Labor	Equipment	Total	Total Incl O&P
0010	**PLASTIC TOILET COMPARTMENTS**									
0110	Cubicles, ceiling hung									
0250	Phenolic	2 Carp	4	4	Ea.	825	147		972	1,150
0600	For handicap units, incl. 52" grab bars, add				"	420			420	460
0900	Floor and ceiling anchored									
1050	Phenolic	2 Carp	5	3.200	Ea.	985	117		1,102	1,250
1400	For handicap units, incl. 52" grab bars, add				"	300			300	330
1610	Floor mounted									
1750	Phenolic	2 Carp	7	2.286	Ea.	1,200	84		1,284	1,425
2100	For handicap units, incl. 52" grab bars, add					300			300	330
2200	For juvenile units, deduct					39.50			39.50	43.50
2450	Floor mounted, headrail braced									
2550	Phenolic	2 Carp	6	2.667	Ea.	950	98		1,048	1,200

10 21 13.40 Stone Toilet Compartments

		Crew	Daily Output	Labor-Hours	Unit	Material	Labor	Equipment	Total	Total Incl O&P
0010	**STONE TOILET COMPARTMENTS**									
0100	Cubicles, ceiling hung, marble	2 Marb	2	8	Ea.	1,625	290		1,915	2,225
0600	For handicap units, incl. 52" grab bars, add					420			420	460
0800	Floor & ceiling anchored, marble	2 Marb	2.50	6.400		1,775	232		2,007	2,300
1400	For handicap units, incl. 52" grab bars, add					300			300	330
1600	Floor mounted, marble	2 Marb	3	5.333		1,050	193		1,243	1,450
2400	Floor mounted, headrail braced, marble	"	3	5.333		995	193		1,188	1,400
2900	For handicap units, incl. 52" grab bars, add					320			320	355
4100	Entrance screen, floor mounted marble, 58" high, 48" wide	2 Marb	9	1.778		655	64.50		719.50	820
4600	Urinal screen, 18" wide, ceiling braced, marble	D-1	6	2.667		655	89		744	855
5100	Floor mounted, head rail braced									
5200	Marble	D-1	6	2.667	Ea.	575	89		664	765
5700	Pilaster, flush, marble		9	1.778		730	59.50		789.50	895
6200	Post braced, marble		9	1.778		725	59.50		784.50	885

10 21 16 – Shower and Dressing Compartments

10 21 16.10 Partitions, Shower

		Crew	Daily Output	Labor-Hours	Unit	Material	Labor	Equipment	Total	Total Incl O&P
0010	**PARTITIONS, SHOWER** Floor mounted, no plumbing									
0100	Cabinet, incl. base, no door, painted steel, 1" thick walls	2 Shee	5	3.200	Ea.	825	139		964	1,125
0300	With door, fiberglass		4.50	3.556		665	155		820	975
0600	Galvanized and painted steel, 1" thick walls		5	3.200		865	139		1,004	1,175
0800	Stall, 1" thick wall, no base, enameled steel		5	3.200		940	139		1,079	1,250
1200	Stainless steel		5	3.200		1,525	139		1,664	1,900
1400	For double entry type, no doors, deduct					10%				
1500	Circular fiberglass, cabinet 36" diameter,	2 Shee	4	4		685	174		859	1,025
1700	One piece, 36" diameter, less door		4	4		575	174		749	905
1800	With door		3.50	4.571		945	199		1,144	1,350
2000	Curved shell shower, no door needed		3	5.333		825	232		1,057	1,275
2300	For fiberglass seat, add to both above					117			117	129
2400	Glass stalls, with doors, no receptors, chrome on brass	2 Shee	3	5.333		1,350	232		1,582	1,850
2700	Anodized aluminum	"	4	4		935	174		1,109	1,300
2900	Marble shower stall, stock design, with shower door	2 Marb	1.20	13.333		2,075	485		2,560	3,000
3000	With curtain		1.30	12.308		1,825	445		2,270	2,700
3200	Receptors, precast terrazzo, 32" x 32"		14	1.143		240	41.50		281.50	325
3300	48" x 34"		9.50	1.684		390	61		451	525
3500	Plastic, simulated terrazzo receptor, 32" x 32"		14	1.143		115	41.50		156.50	190
3600	32" x 48"		12	1.333		162	48.50		210.50	252
3800	Precast concrete, colors, 32" x 32"		14	1.143		180	41.50		221.50	261

10 21 Compartments and Cubicles

10 21 16 – Shower and Dressing Compartments

10 21 16.10 Partitions, Shower

		Crew	Daily Output	Labor-Hours	Unit	Material	2007 Bare Costs Labor	2007 Bare Costs Equipment	Total	Total Incl O&P
3900	48" x 48"	2 Marb	8	2	Ea.	310	72.50		382.50	450
4100	Shower doors, economy plastic, 24" wide	1 Shee	9	.889		110	38.50		148.50	181
4200	Tempered glass door, economy		8	1		160	43.50		203.50	243
4400	Folding, tempered glass, aluminum frame		6	1.333		355	58		413	480
4500	Sliding, tempered glass, 48" opening		6	1.333		234	58		292	345
4700	Deluxe, tempered glass, chrome on brass frame, minimum		8	1		204	43.50		247.50	292
4800	Maximum		1	8		805	350		1,155	1,425
4850	On anodized aluminum frame, minimum		2	4		132	174		306	415
4900	Maximum		1	8		475	350		825	1,050
5100	Shower enclosure, tempered glass, anodized alum. frame									
5120	2 panel & door, corner unit, 32" x 32"	1 Shee	2	4	Ea.	465	174		639	780
5140	Neo-angle corner unit, 16" x 24" x 16"	"	2	4		855	174		1,029	1,200
5200	Shower surround, 3 wall, polypropylene, 32" x 32"	1 Carp	4	2		249	73.50		322.50	390
5220	PVC, 32" x 32"		4	2		292	73.50		365.50	435
5240	Fiberglass		4	2		340	73.50		413.50	485
5250	2 wall, polypropylene, 32" x 32"		4	2		242	73.50		315.50	380
5270	PVC		4	2		300	73.50		373.50	445
5290	Fiberglass		4	2		335	73.50		408.50	485
5300	Tub doors, tempered glass & frame, minimum	1 Shee	8	1		187	43.50		230.50	272
5400	Maximum		6	1.333		435	58		493	570
5600	Chrome plated, brass frame, minimum		8	1		247	43.50		290.50	340
5700	Maximum		6	1.333		605	58		663	755
5900	Tub/shower enclosure, temp. glass, alum. frame, minimum		2	4		335	174		509	640
6200	Maximum		1.50	5.333		695	232		927	1,125
6500	On chrome-plated brass frame, minimum		2	4		460	174		634	780
6600	Maximum		1.50	5.333		985	232		1,217	1,425
6800	Tub surround, 3 wall, polypropylene	1 Carp	4	2		196	73.50		269.50	330
6900	PVC		4	2		298	73.50		371.50	445
7000	Fiberglass, minimum		4	2		335	73.50		408.50	485
7100	Maximum		3	2.667		570	98		668	780

10 21 23 – Cubicles

10 21 23.16 Cubicle Track and Hardware

		Crew	Daily Output	Labor-Hours	Unit	Material	2007 Bare Costs Labor	2007 Bare Costs Equipment	Total	Total Incl O&P
0010	**CUBICLE TRACK AND HARDWARE**									
0020	Curtain track, box channel, ceiling mounted	1 Carp	135	.059	L.F.	4.96	2.17		7.13	8.85
0100	Suspended	"	100	.080	"	6.75	2.94		9.69	12
0300	Curtains, nylon mesh tops, fire resistant, 11 oz. per lineal yard									
0310	Polyester oxford cloth, 9' ceiling height	1 Carp	425	.019	L.F.	10.95	.69		11.64	13.15
0500	8' ceiling height		425	.019		9.30	.69		9.99	11.30
0700	Designer oxford cloth		425	.019		23.50	.69		24.19	27
0800	I.V. track systems									
0820	I.V. track, oval	1 Carp	135	.059	L.F.	4.82	2.17		6.99	8.70
0830	I.V. trolley		32	.250	Ea.	33	9.20		42.20	51
0840	I.V. pendent, (tree, 5 hook)		32	.250	"	118	9.20		127.20	144

10 22 Partitions

10 22 13 – Wire Mesh Partitions

10 22 13.10 Partitions, Woven Wire

		Crew	Daily Output	Labor-Hours	Unit	Material	2007 Bare Costs Labor	2007 Bare Costs Equipment	Total	Total Incl O&P
0010	**PARTITIONS, WOVEN WIRE** For tool or stockroom enclosures									
0100	Channel frame, 1-1/2" diamond mesh, 10 ga. wire, painted									
0300	Wall panels, 4'-0" wide, 7' high	2 Carp	25	.640	Ea.	120	23.50		143.50	169
0400	8' high		23	.696		131	25.50		156.50	184
0600	10' high		18	.889		151	32.50		183.50	217
0700	For 5' wide panels, add					5%				
0900	Ceiling panels, 10' long, 2' wide	2 Carp	25	.640		91.50	23.50		115	137
1000	4' wide		15	1.067		131	39		170	205
1200	Panel with service window & shelf, 5' wide, 7' high		20	.800		282	29.50		311.50	355
1300	8' high		15	1.067		290	39		329	380
1500	Sliding doors, full height, 3' wide, 7' high		6	2.667		300	98		398	480
1600	10' high		5	3.200		405	117		522	630
1800	6' wide sliding door, 7' full height		5	3.200		475	117		592	705
1900	10' high		4	4		560	147		707	845
2100	Swinging doors, 3' wide, 7' high, no transom		6	2.667		245	98		343	420
2200	7' high, 3' transom		5	3.200		288	117		405	500

10 22 16 – Folding Gates

10 22 16.10 Security Gates

		Crew	Daily Output	Labor-Hours	Unit	Material	Labor	Equipment	Total	Total Incl O&P
0010	**SECURITY GATES** For roll up type, see division 08 33 13.10									
0300	Scissors type folding gate, ptd. steel, single, 6-1/2' high, 5-1/2' wide	2 Sswk	4	4	Opng.	157	165		322	475
0350	6-1/2' wide		4	4		158	165		323	475
0400	7-1/2' wide		4	4		178	165		343	495
0600	Double gate, 8' high, 8' wide		2.50	6.400		258	265		523	765
0650	10' wide		2.50	6.400		277	265		542	785
0700	12' wide		2	8		400	330		730	1,050
0750	14' wide		2	8		460	330		790	1,100
0900	Door gate, folding steel, 4' wide, 61" high		4	4		79.50	165		244.50	390
1000	71" high		4	4		82.50	165		247.50	390
1200	81" high		4	4		89.50	165		254.50	400
1300	Window gates, 2' to 4' wide, 31" high		4	4		52.50	165		217.50	360
1500	55" high		3.75	4.267		86	176		262	415
1600	79" high		3.50	4.571		99.50	189		288.50	450

10 22 19 – Demountable Partitions

10 22 19.43 Demountable Composite Partitions

			Crew	Daily Output	Labor-Hours	Unit	Material	Labor	Equipment	Total	Total Incl O&P
0010	**DEMOUNTABLE COMPOSITE PARTITIONS**, add for doors										
0100	Do not deduct door openings from total L.F.										
0900	Demountable gypsum system on 2" to 2-1/2"										
1000	steel studs, 9' high, 3" to 3-3/4" thick										
1200	Vinyl clad gypsum	CN	2 Carp	48	.333	L.F.	55	12.25		67.25	79.50
1300	Fabric clad gypsum			44	.364		137	13.35		150.35	172
1500	Steel clad gypsum			40	.400		148	14.70		162.70	186
1600	1.75 system, aluminum framing, vinyl clad hardboard,										
1800	paper honeycomb core panel, 1-3/4" to 2-1/2" thick										
1900	9' high		2 Carp	48	.333	L.F.	92.50	12.25		104.75	121
2100	7' high			60	.267		83	9.80		92.80	106
2200	5' high			80	.200		70	7.35		77.35	88.50
2250	Unitized gypsum system										
2300	Unitized panel, 9' high, 2" to 2-1/2" thick										
2350	Vinyl clad gypsum		2 Carp	48	.333	L.F.	119	12.25		131.25	150
2400	Fabric clad gypsum		"	44	.364	"	196	13.35		209.35	236
2500	Unitized mineral fiber system										
2510	Unitized panel, 9' high, 2-1/4" thick, aluminum frame										

10 22 Partitions

10 22 19 – Demountable Partitions

10 22 19.43 Demountable Composite Partitions

		Crew	Daily Output	Labor-Hours	Unit	Material	2007 Bare Costs Labor	Equipment	Total	Total Incl O&P
2550	Vinyl clad mineral fiber	2 Carp	48	.333	L.F.	118	12.25		130.25	149
2600	Fabric clad mineral fiber	"	44	.364	"	176	13.35		189.35	215
2800	Movable steel walls, modular system									
2900	Unitized panels, 9' high, 48" wide									
3100	Baked enamel, pre-finished	2 Carp	60	.267	L.F.	134	9.80		143.80	162
3200	Fabric clad steel		56	.286	"	194	10.50		204.50	229
5310	Trackless wall, cork finish, semi-acoustic, 1-5/8" thick, minimum		325	.049	S.F.	37.50	1.81		39.31	44
5320	Maximum		190	.084		35	3.09		38.09	43.50
5330	Acoustic, 2" thick, minimum		305	.052		30	1.93		31.93	36
5340	Maximum		225	.071		52	2.61		54.61	61.50
5500	For acoustical partitions, add, minimum					2.16			2.16	2.38
5550	Maximum					10.05			10.05	11.10
5700	For doors, see Div. 08 11 00.00 & 08 16 00.00									
5800	For door hardware, see div. 08 71 00.00									
6100	In-plant modular office system, w/prehung hollow core door									
6200	3" thick polystyrene core panels									
6250	12' x 12', 2 wall	2 Clab	3.80	4.211	Ea.	2,875	121		2,996	3,350
6300	4 wall		1.90	8.421		4,100	242		4,342	4,875
6350	16' x 16', 2 wall		3.60	4.444		4,475	128		4,603	5,100
6400	4 wall		1.80	8.889		5,875	256		6,131	6,875

10 22 23 – Portable Partitions, Screens, and Panels

10 22 23.13 Wall Screens

		Crew	Daily Output	Labor-Hours	Unit	Material	2007 Bare Costs Labor	Equipment	Total	Total Incl O&P
0010	**WALL SCREENS**, divider panels, free standing, fiber core									
0020	Fabric face straight									
0100	3'-0" long, 4'-0" high	2 Carp	100	.160	L.F.	106	5.85		111.85	125
0200	5'-0" high		90	.178		111	6.50		117.50	132
0500	6'-0" high		75	.213		114	7.85		121.85	137
0900	5'-0" long, 4'-0" high		175	.091		76.50	3.36		79.86	90
1000	5'-0" high		150	.107		88	3.91		91.91	103
1500	6'-0" high		125	.128		91	4.70		95.70	107
1600	6'-0" long, 5'-0" high		162	.099		76	3.62		79.62	89
3100	Curved, 3'-0" long, 5'-0" high		90	.178		89.50	6.50		96	109
3150	6'-0" high		75	.213		103	7.85		110.85	125
3200	Economical panels, fabric face, 4'-0" long, 5'-0" high		132	.121		47	4.45		51.45	59
3250	6'-0" high		112	.143		45.50	5.25		50.75	58.50
3300	5'-0" long, 5'-0" high		150	.107		36.50	3.91		40.41	46
3350	6'-0" high		125	.128		45	4.70		49.70	57
3380	3'-0" curved, 5'-0" high		90	.178		89.50	6.50		96	109
3390	6'-0" high		75	.213		103	7.85		110.85	125
3450	Acoustical panels, 60 to 90 NRC, 3'-0" long, 5'-0" high		90	.178		105	6.50		111.50	126
3550	6'-0" high		75	.213		99.50	7.85		107.35	122
3600	5'-0" long, 5'-0" high		150	.107		73	3.91		76.91	86.50
3650	6'-0" high		125	.128		81	4.70		85.70	97
3700	6'-0" long, 5'-0" high		162	.099		71	3.62		74.62	83.50
3750	6'-0" high		138	.116		68	4.26		72.26	81
3800	Economy acoustical panels, 40 NRC, 4'-0" long, 5'-0" high		132	.121		47	4.45		51.45	59
3850	6'-0" high		112	.143		52.50	5.25		57.75	66
3900	5'-0" long, 6'-0" high		125	.128		45	4.70		49.70	57
3950	6'-0" long, 5'-0" high		162	.099		29.50	3.62		33.12	38
4000	Metal chalkboard, 6'-6" high, chalkboard, 1 side		125	.128		98	4.70		102.70	115
4100	Metal chalkboard, 2 sides		120	.133		112	4.89		116.89	131
4300	Tackboard, both sides		123	.130		88.50	4.77		93.27	104

10 22 Partitions

10 22 26 – Operable Partitions

10 22 26.13 Accordion Folding Partitions

		Crew	Daily Output	Labor-Hours	Unit	Material	2007 Bare Costs Labor	Equipment	Total	Total Incl O&P
0010	**ACCORDION FOLDING PARTITIONS**									
0100	Vinyl covered, over 150 S.F., frame not included									
0300	Residential, 1.25 lb. per S.F., 8' maximum height	2 Carp	300	.053	S.F.	18.25	1.96		20.21	23
0400	Commercial, 1.75 lb. per S.F., 8' maximum height		225	.071		21	2.61		23.61	27
0600	2 lb. per S.F., 17' maximum height		150	.107		21.50	3.91		25.41	29.50
0700	Industrial, 4 lb. per S.F., 20' maximum height		75	.213		31	7.85		38.85	46
0900	Acoustical, 3 lb. per S.F., 17' maximum height		100	.160		24	5.85		29.85	35.50
1200	5 lb. per S.F., 20' maximum height		95	.168		33.50	6.20		39.70	46.50
1300	5.5 lb. per S.F., 17' maximum height		90	.178		39.50	6.50		46	53
1400	Fire rated, 4.5 psf, 20' maximum height		160	.100		39.50	3.67		43.17	48.50
1500	Vinyl clad wood or steel, electric operation, 5.0 psf		160	.100		45.50	3.67		49.17	55.50
1900	Wood, non-acoustic, birch or mahogany, to 10' high		300	.053		24.50	1.96		26.46	30

10 22 26.33 Folding Panel Partitions

		Crew	Daily Output	Labor-Hours	Unit	Material	Labor	Equipment	Total	Total Incl O&P
0010	**FOLDING PANEL PARTITIONS**, acoustic, wood									
0100	Vinyl faced, to 18' high, 6 psf, minimum	2 Carp	60	.267	S.F.	46.50	9.80		56.30	66.50
0150	Average		45	.356		55.50	13.05		68.55	81.50
0200	Maximum		30	.533		71.50	19.55		91.05	109
0400	Formica or hardwood finish, minimum		60	.267		47.50	9.80		57.30	68
0500	Maximum		30	.533		51	19.55		70.55	86.50
0600	Wood, low acoustical type, 4.5 psf, to 14' high		50	.320		34.50	11.75		46.25	56.50
1100	Steel, acoustical, 9 to 12 lb. per S.F., vinyl faced, minimum		60	.267		49.50	9.80		59.30	69.50
1200	Maximum		30	.533		60	19.55		79.55	96.50
1700	Aluminum framed, acoustical, to 12' high, 5.5 psf, minimum		60	.267		33	9.80		42.80	52
1800	Maximum		30	.533		40	19.55		59.55	74.50
2000	6.5 lb. per S.F., minimum		60	.267		35	9.80		44.80	54
2100	Maximum		30	.533		43	19.55		62.55	78

10 22 26.43 Sliding Partitions

		Crew	Daily Output	Labor-Hours	Unit	Material	Labor	Equipment	Total	Total Incl O&P
0010	**SLIDING PARTITIONS**									
0020	Acoustic air wall, 1-5/8" thick, minimum	2 Carp	375	.043	S.F.	28	1.57		29.57	33.50
0100	Maximum		365	.044		48.50	1.61		50.11	55.50
0300	2-1/4" thick, minimum		360	.044		31.50	1.63		33.13	37.50
0400	Maximum		330	.048		55.50	1.78		57.28	64
0600	For track type, add to above				L.F.	103			103	114
0700	Overhead track type, acoustical, 3" thick, 11 psf, minimum	2 Carp	350	.046	S.F.	71	1.68		72.68	81
0800	Maximum	"	300	.053	"	86	1.96		87.96	97.50

10 26 Wall and Door Protection

10 26 13 – Corner Guards

10 26 13.10 Corner Guards

		Crew	Daily Output	Labor-Hours	Unit	Material	Labor	Equipment	Total	Total Incl O&P
0010	**CORNER GUARDS**									
0020	Steel angle w/anchors, 1" x 1" x 1/4", 1.5#/L.F.	2 Carp	160	.100	L.F.	5.75	3.67		9.42	12.05
0100	2" x 2" x 1/4" angles, 3.2#/L.F.		150	.107		10.45	3.91		14.36	17.60
0200	3" x 3" x 5/16" angles, 6.1#/L.F.		140	.114		15.85	4.19		20.04	24
0300	4" x 4" x 5/16" angles, 8.2#/L.F.		120	.133		18.95	4.89		23.84	28.50
0350	For angles drilled and anchored to masonry, add					15%	120%			
0370	Drilled and anchored to concrete, add					20%	170%			
0400	For galvanized angles, add					35%				
0450	For stainless steel angles, add					100%				
0500	Steel door track/wheel guards, 4'-0" high	E-4	22	1.455	Ea.	128	61	5.25	194.25	257
0800	Pipe bumper for truck doors, 8' long, 6" diameter, filled		20	1.600		350	67	5.75	422.75	510

10 26 Wall and Door Protection

10 26 13 – Corner Guards

10 26 13.10 Corner Guards		Crew	Daily Output	Labor-Hours	Unit	Material	2007 Bare Costs Labor	Equipment	Total	Total Incl O&P
0900	8" diameter	E-4	20	1.600	Ea.	535	67	5.75	607.75	715

10 26 13.20 Corner Protection

		Crew	Daily Output	Labor-Hours	Unit	Material	Labor	Equipment	Total	Total Incl O&P
0010	**CORNER PROTECTION**									
0100	Stainless steel, 16 ga., adhesive mount, 3-1/2" leg	1 Sswk	80	.100	L.F.	19.95	4.14		24.09	29.50
0200	12 ga. stainless, adhesive mount	"	80	.100		34	4.14		38.14	45
0300	For screw mount, add						10%			
0500	Vinyl acrylic, adhesive mount, 3" leg	1 Carp	128	.063		6.85	2.29		9.14	11.05
0550	1-1/2" leg		160	.050		4.72	1.84		6.56	8.05
0600	Screw mounted, 3" leg		80	.100		7.15	3.67		10.82	13.55
0650	1-1/2" leg		100	.080		4.62	2.94		7.56	9.65
0700	Clear plastic, screw mounted, 2-1/2"		60	.133		3.12	4.89		8.01	11.05
1000	Vinyl cover, alum. retainer, surface mount, 3" x 3"		48	.167		6.60	6.10		12.70	16.80
1050	2" x 2"		48	.167		5.20	6.10		11.30	15.30
1100	Flush mounted, 3" x 3"		32	.250		14.45	9.20		23.65	30
1150	2" x 2"		32	.250		10.65	9.20		19.85	26

10 26 16 – Bumper Guards

10 26 16.10 Wallguard

		Crew	Daily Output	Labor-Hours	Unit	Material	Labor	Equipment	Total	Total Incl O&P
0010	**WALLGUARD**									
0400	Rub rail, vinyl, adhesive mounted	1 Carp	185	.043	L.F.	4.20	1.59		5.79	7.10
0500	Neoprene, aluminum backing, 1-1/2" x 2"		110	.073		7.20	2.67		9.87	12.05
1000	Trolley rail, PVC, clipped to wall, 5" high		185	.043		8.25	1.59		9.84	11.55
1050	8" high		180	.044		10.25	1.63		11.88	13.80
1200	Bed bumper, vinyl acrylic, alum. retainer, 21" long		10	.800	Ea.	39.50	29.50		69	89
1300	53" long with aligner		9	.889	"	117	32.50		149.50	179
1400	Bumper, vinyl cover, alum. retain., cush. mnt., 1-1/2" x 2-3/4"		80	.100	L.F.	11.90	3.67		15.57	18.80
1500	2" x 4-1/4"		80	.100		16.70	3.67		20.37	24
1600	Surface mounted, 1-3/4" x 3-5/8"		80	.100		11.30	3.67		14.97	18.10
2000	Crash rail, vinyl cover, alum. retainer, 1" x 4"		110	.073		12.40	2.67		15.07	17.80
2100	1" x 8"		90	.089		15.50	3.26		18.76	22
2150	Vinyl inserts, aluminum plate, 1" x 2-1/2"		110	.073		12.50	2.67		15.17	17.90
2200	1" x 5"		90	.089		17.15	3.26		20.41	24
3000	Handrail/bumper, vinyl cover, alum. retainer									
3010	Bracket mounted, flat rail, 5-1/2"	1 Carp	80	.100	L.F.	15	3.67		18.67	22
3100	6-1/2"		80	.100		23.50	3.67		27.17	31.50
3200	Bronze bracket, 1-3/4" diam. rail		80	.100		19.25	3.67		22.92	26.50

10 28 Toilet, Bath, and Laundry Accessories

10 28 13 – Toilet Accessories

10 28 13.13 Commercial Toilet Accessories

		Crew	Daily Output	Labor-Hours	Unit	Material	Labor	Equipment	Total	Total Incl O&P
0010	**COMMERCIAL TOILET ACCESSORIES**									
0200	Curtain rod, stainless steel, 5' long, 1" diameter	1 Carp	13	.615	Ea.	33	22.50		55.50	71.50
0300	1-1/4" diameter		13	.615		32	22.50		54.50	70
0400	Diaper changing station, horizontal, wall mounted, plastic		10	.800		224	29.50		253.50	292
0500	Dispenser units, combined soap & towel dispensers,									
0510	mirror and shelf, flush mounted	1 Carp	10	.800	Ea.	365	29.50		394.50	450
0600	Towel dispenser and waste receptacle,									
0610	18 gallon capacity	1 Carp	10	.800	Ea.	380	29.50		409.50	465
0800	Grab bar, straight, 1-1/4" diameter, stainless steel, 18" long		24	.333		23.50	12.25		35.75	44.50
0900	24" long		23	.348		22.50	12.75		35.25	44.50
1000	30" long		22	.364		25	13.35		38.35	48.50

10 28 Toilet, Bath, and Laundry Accessories

10 28 13 – Toilet Accessories

10 28 13.13 Commercial Toilet Accessories	Crew	Daily Output	Labor-Hours	Unit	Material	2007 Bare Costs Labor	Equipment	Total	Total Incl O&P
1100 36" long	1 Carp	20	.400	Ea.	24	14.70		38.70	49
1105 42" long		20	.400		27.50	14.70		42.20	53
1200 1-1/2" diameter, 24" long		23	.348		24	12.75		36.75	46.50
1300 36" long		20	.400		26	14.70		40.70	51.50
1310 42" long		18	.444		30	16.30		46.30	58.50
1500 Tub bar, 1-1/4" diameter, 24" x 36"		14	.571		92.50	21		113.50	135
1600 Plus vertical arm		12	.667		95	24.50		119.50	143
1900 End tub bar, 1" diameter, 90° angle, 16" x 32"		12	.667		130	24.50		154.50	181
2010 Tub/shower/toilet, 2-wall, 36" x 24"		12	.667		84	24.50		108.50	131
2300 Hand dryer, surface mounted, electric, 115 volt, 20 amp		4	2		515	73.50		588.50	680
2400 230 volt, 10 amp		4	2		535	73.50		608.50	700
2600 Hat and coat strip, stainless steel, 4 hook, 36" long		24	.333		53	12.25		65.25	77
2700 6 hook, 60" long		20	.400		92.50	14.70		107.20	125
3000 Mirror, with stainless steel 3/4" square frame, 18" x 24"		20	.400		53.50	14.70		68.20	82
3100 36" x 24"		15	.533		135	19.55		154.55	179
3200 48" x 24"		10	.800		180	29.50		209.50	244
3300 72" x 24"		6	1.333		300	49		349	405
3500 With 5" stainless steel shelf, 18" x 24"		20	.400		195	14.70		209.70	238
3600 36" x 24"		15	.533		279	19.55		298.55	335
3700 48" x 24"		10	.800		176	29.50		205.50	240
3800 72" x 24"		6	1.333		335	49		384	445
4100 Mop holder strip, stainless steel, 5 holders, 48" long		20	.400		84	14.70		98.70	115
4200 Napkin/tampon dispenser, recessed		15	.533		660	19.55		679.55	760
4300 Robe hook, single, regular		36	.222		12.60	8.15		20.75	26.50
4400 Heavy duty, concealed mounting		36	.222		13.85	8.15		22	28
4600 Soap dispenser, chrome, surface mounted, liquid		20	.400		51.50	14.70		66.20	79.50
4700 Powder		20	.400		48	14.70		62.70	75.50
5000 Recessed stainless steel, liquid		10	.800		175	29.50		204.50	238
5100 Powder		10	.800		213	29.50		242.50	280
5300 Soap tank, stainless steel, 1 gallon		10	.800		176	29.50		205.50	239
5400 5 gallon		5	1.600		242	58.50		300.50	360
5600 Shelf, stainless steel, 5" wide, 18 ga., 24" long		24	.333		55	12.25		67.25	79.50
5700 48" long		16	.500		98.50	18.35		116.85	137
5800 8" wide shelf, 18 ga., 24" long		22	.364		63	13.35		76.35	90.50
5900 48" long		14	.571		125	21		146	171
6000 Toilet seat cover dispenser, stainless steel, recessed		20	.400		132	14.70		146.70	169
6050 Surface mounted		15	.533		29.50	19.55		49.05	63
6100 Toilet tissue dispenser, surface mounted, SS, single roll		30	.267		13.50	9.80		23.30	30
6200 Double roll		24	.333		18.90	12.25		31.15	40
6400 Towel bar, stainless steel, 18" long		23	.348		39	12.75		51.75	63
6500 30" long		21	.381		87.50	14		101.50	118
6700 Towel dispenser, stainless steel, surface mounted		16	.500		39	18.35		57.35	71.50
6800 Flush mounted, recessed		10	.800		257	29.50		286.50	330
7000 Towel holder, hotel type, 2 guest size		20	.400		46	14.70		60.70	74
7200 Towel shelf, stainless steel, 24" long, 8" wide		20	.400		91	14.70		105.70	123
7400 Tumbler holder, tumbler only		30	.267		35	9.80		44.80	54
7500 Soap, tumbler & toothbrush		30	.267		25	9.80		34.80	43
7700 Wall urn ash receiver, surface mount, 11" long		12	.667		132	24.50		156.50	184
7800 7-1/2", long		18	.444		127	16.30		143.30	166
8000 Waste receptacles, stainless steel, with top, 13 gallon		10	.800		282	29.50		311.50	355
8100 36 gallon		8	1		505	36.50		541.50	610

10 28 Toilet, Bath, and Laundry Accessories

10 28 16 – Bath Accessories

10 28 16.20 Medicine Cabinets		Crew	Daily Output	Labor-Hours	Unit	Material	2007 Bare Costs Labor	Equipment	Total	Total Incl O&P
0010	**MEDICINE CABINETS**									
0020	With mirror, st. st. frame, 16" x 22", unlighted	1 Carp	14	.571	Ea.	76.50	21		97.50	117
0100	Wood frame		14	.571		106	21		127	150
0300	Sliding mirror doors, 20" x 16" x 4-3/4", unlighted		7	1.143		95	42		137	171
0400	24" x 19" x 8-1/2", lighted		5	1.600		150	58.50		208.50	257
0600	Triple door, 30" x 32", unlighted, plywood body		7	1.143		225	42		267	315
0700	Steel body		7	1.143		296	42		338	390
0900	Oak door, wood body, beveled mirror, single door		7	1.143		145	42		187	225
1000	Double door		6	1.333		350	49		399	460
1200	Hotel cabinets, stainless, with lower shelf, unlighted		10	.800		190	29.50		219.50	255
1300	Lighted		5	1.600		282	58.50		340.50	400

10 28 23 – Laundry Accessories

10 28 23.13 Built-In Ironing Boards

		Crew	Daily Output	Labor-Hours	Unit	Material	Labor	Equipment	Total	Total Incl O&P
0010	**BUILT-IN IRONING BOARDS**									
0020	Including cabinet, board & light, minimum	1 Carp	2	4	Ea.	297	147		444	555
0100	Maximum, see also Div. 11 23 13.13	"	1.50	5.333	"	530	196		726	890

10 31 Manufactured Fireplaces

10 31 13 – Manufactured Fireplace Chimneys

10 31 13.10 Fireplace Chimneys

		Crew	Daily Output	Labor-Hours	Unit	Material	Labor	Equipment	Total	Total Incl O&P
0010	**FIREPLACE CHIMNEYS**									
0500	Chimney dbl. wall, all stainless, over 8'-6", 7" diam., add	1 Carp	33	.242	V.L.F.	60.50	8.90		69.40	80.50
0600	10" diameter, add		32	.250		61.50	9.20		70.70	82.50
0700	12" diameter, add		31	.258		84	9.45		93.45	107
0800	14" diameter, add		30	.267		111	9.80		120.80	137
1000	Simulated brick chimney top, 4' high, 16" x 16"		10	.800	Ea.	226	29.50		255.50	295
1100	24" x 24"		7	1.143	"	420	42		462	530

10 31 13.20 Chimney Accessories

		Crew	Daily Output	Labor-Hours	Unit	Material	Labor	Equipment	Total	Total Incl O&P
0010	**CHIMNEY ACCESSORIES**									
0020	Chimney screens, galv., 13" x 13" flue	1 Bric	8	1	Ea.	48.50	38		86.50	111
0050	24" x 24" flue		5	1.600		112	61		173	216
0200	Stainless steel, 13" x 13" flue		8	1		296	38		334	385
0250	20" x 20" flue		5	1.600		400	61		461	540
2400	Squirrel and bird screens, galvanized, 8" x 8" flue		16	.500		42	19.05		61.05	75
2450	13" x 13" flue		12	.667		45.50	25.50		71	88.50

10 31 16 – Manufactured Fireplace Forms

10 31 16.10 Fireplace Forms

		Crew	Daily Output	Labor-Hours	Unit	Material	Labor	Equipment	Total	Total Incl O&P
0010	**FIREPLACE FORMS**									
1800	Fireplace forms, no accessories, 32" opening	1 Bric	3	2.667	Ea.	570	101		671	785
1900	36" opening		2.50	3.200		725	122		847	985
2000	40" opening		2	4		965	152		1,117	1,275
2100	78" opening		1.50	5.333		1,400	203		1,603	1,850

10 31 23 – Manufactured Fireplaces

10 31 23.10 Fireplace, Prefabricated

		Crew	Daily Output	Labor-Hours	Unit	Material	Labor	Equipment	Total	Total Incl O&P
0010	**FIREPLACE, PREFABRICATED**, free standing or wall hung									
0100	With hood & screen, minimum	1 Carp	1.30	6.154	Ea.	1,075	226		1,301	1,525
0150	Average		1	8		1,550	294		1,844	2,150
0200	Maximum		.90	8.889		3,825	325		4,150	4,700

10 31 Manufactured Fireplaces

10 31 23 – Manufactured Fireplaces

10 31 23.10 Fireplace, Prefabricated

		Crew	Daily Output	Labor-Hours	Unit	Material	2007 Bare Costs Labor	Equipment	Total	Total Incl O&P
1500	Simulated logs, gas fired, 40,000 BTU, 2' long, minimum	1 Carp	7	1.143	Set	560	42		602	680
1600	Maximum		6	1.333		775	49		824	930
1700	Electric, 1,500 BTU, 1'-6" long, minimum		7	1.143		154	42		196	235
1800	11,500 BTU, maximum		6	1.333		345	49		394	455
2000	Fireplace, built-in, 36" hearth, radiant		1.30	6.154	Ea.	605	226		831	1,025
2100	Recirculating, small fan		1	8		965	294		1,259	1,500
2150	Large fan		.90	8.889		1,675	325		2,000	2,350
2200	42" hearth, radiant		1.20	6.667		815	245		1,060	1,275
2300	Recirculating, small fan		.90	8.889		1,025	325		1,350	1,650
2350	Large fan		.80	10		1,675	365		2,040	2,425
2400	48" hearth, radiant		1.10	7.273		1,575	267		1,842	2,150
2500	Recirculating, small fan		.80	10		1,950	365		2,315	2,725
2550	Large fan		.70	11.429		3,025	420		3,445	3,975
3000	See through, including doors		.80	10		2,575	365		2,940	3,400
3200	Corner (2 wall)		1	8		2,575	294		2,869	3,275

10 32 Fireplace Specialties

10 32 13 – Fireplace Dampers

10 32 13.10 Dampers

		Crew	Daily Output	Labor-Hours	Unit	Material	2007 Bare Costs Labor	Equipment	Total	Total Incl O&P
0010	**DAMPERS**									
0800	Damper, rotary control, steel, 30" opening	1 Bric	6	1.333	Ea.	71.50	50.50		122	156
0850	Cast iron, 30" opening		6	1.333		79.50	50.50		130	165
0880	36" opening		6	1.333		86.50	50.50		137	172
0900	48" opening		6	1.333		123	50.50		173.50	213
0920	60" opening		6	1.333		276	50.50		326.50	380
0950	72" opening		5	1.600		330	61		391	460
1000	84" opening, special order		5	1.600		710	61		771	880
1050	96" opening, special order		4	2		720	76		796	910
1200	Steel plate, poker control, 60" opening		8	1		252	38		290	335
1250	84" opening, special opening		5	1.600		460	61		521	600
1400	"Universal" type, chain operated, 32" x 20" opening		8	1		196	38		234	274
1450	48" x 24" opening		5	1.600		292	61		353	415

10 32 23 – Fireplace Doors

10 32 23.10 Doors

		Crew	Daily Output	Labor-Hours	Unit	Material	2007 Bare Costs Labor	Equipment	Total	Total Incl O&P
0010	**DOORS**									
0400	Cleanout doors and frames, cast iron, 8" x 8"	1 Bric	12	.667	Ea.	33.50	25.50		59	75.50
0450	12" x 12"		10	.800		41	30.50		71.50	91.50
0500	18" x 24"		8	1		118	38		156	187
0550	Cast iron frame, steel door, 24" x 30"		5	1.600		254	61		315	370
1600	Dutch Oven door and frame, cast iron, 12" x 15" opening		13	.615		103	23.50		126.50	149
1650	Copper plated, 12" x 15" opening		13	.615		201	23.50		224.50	257

10 35 Stoves

10 35 13 – Heating Stoves

10 35 13.10 Woodburning Stoves		Crew	Daily Output	Labor-Hours	Unit	Material	2007 Bare Costs Labor	Equipment	Total	Total Incl O&P
0010	**WOODBURNING STOVES**									
0015	Cast iron, minimum	2 Carp	1.30	12.308	Ea.	1,100	450		1,550	1,900
0020	Average		1	16		1,600	585		2,185	2,675
0030	Maximum		.80	20		2,650	735		3,385	4,075
0050	For gas log lighter, add					45.50			45.50	50

10 44 Fire Protection Specialties

10 44 13 – Fire Extinguisher Cabinets

10 44 13.53 Fire Extinguisher Cabinets

		Crew	Daily Output	Labor-Hours	Unit	Material	Labor	Equipment	Total	Total Incl O&P
0010	**FIRE EXTINGUISHER CABINETS**, not equipped, 20 ga. steel box									
0040	recessed, D.S. glass in door, box size given									
1000	Portable extinguisher, single, 8" x 12" x 27", alum. door & frame	Q-12	8	2	Ea.	102	80		182	232
1100	Steel door and frame	"	8	2	"	81	80		161	209
3000	Hose rack assy., 1-1/2" valve & 100' hose, 24" x 40" x 5-1/2"									
3100	Aluminum door and frame	Q-12	6	2.667	Ea.	248	106		354	435
3200	Steel door and frame		6	2.667		177	106		283	355
3300	Stainless steel door and frame		6	2.667		355	106		461	555
4000	Hose rack assy., 2-1/2" x 1-1/2" valve, 100' hose, 24" x 40" x 8"									
4100	Aluminum door and frame	Q-12	6	2.667	Ea.	251	106		357	435
4200	Steel door and frame		6	2.667		169	106		275	345
4300	Stainless steel door and frame		6	2.667		335	106		441	525
5000	Hose rack assy., 2-1/2" x 1-1/2" valve, 100' hose									
5010	and extinguisher, 30" x 40" x 8"									
5100	Aluminum door and frame	Q-12	5	3.200	Ea.	320	128		448	540
5200	Steel door and frame		5	3.200		201	128		329	415
5300	Stainless steel door and frame		5	3.200		385	128		513	615
8000	Valve cabinet for 2-1/2" FD angle valve, 18" x 18" x 8"									
8100	Aluminum door and frame	Q-12	12	1.333	Ea.	110	53		163	201
8200	Steel door and frame		12	1.333		91	53		144	180
8300	Stainless steel door and frame		12	1.333		148	53		201	243

10 44 16 – Fire Extinguishers

10 44 16.13 Portable Fire Extinguishers

		Crew	Daily Output	Labor-Hours	Unit	Material	Labor	Equipment	Total	Total Incl O&P
0010	**PORTABLE FIRE EXTINGUISHERS**									
0120	CO_2, portable with swivel horn, 5 lb.				Ea.	130			130	143
0140	With hose and "H" horn, 10 lb.					160			160	176
0160	15 lb.					190			190	209
0180	20 lb.					220			220	242
1000	Dry chemical, pressurized									
1040	Standard type, portable, painted, 2-1/2 lb.				Ea.	27.50			27.50	30.50
1060	5 lb.					60			60	66
1080	10 lb.					80			80	88
1100	20 lb.					110			110	121
1120	30 lb.					209			209	230
1300	Standard type, wheeled, 150 lb.					1,600			1,600	1,750
2000	ABC all purpose type, portable, 2-1/2 lb.					30			30	33
2060	5 lb.					45			45	49.50
2080	9-1/2 lb.					68			68	75
2100	20 lb.					98			98	108
3000	Dry chemical, outside cartridge to -65°F, painted, 9 lb.					200			200	220
3060	26 lb.					250			250	275

10 44 Fire Protection Specialties

10 44 16 – Fire Extinguishers

10 44 16.13 Portable Fire Extinguishers	Crew	Daily Output	Labor-Hours	Unit	Material	2007 Bare Costs Labor	Equipment	Total	Total Incl O&P
5000 Pressurized water, 2-1/2 gallon, stainless steel				Ea.	85			85	93.50
5060 With anti-freeze					135			135	149
9400 Installation of extinguishers, 12 or more, on wood	1 Carp	30	.267			9.80		9.80	15.25
9420 On masonry or concrete	"	15	.533	↓		19.55		19.55	30.50

10 44 16.16 Wheeled Fire Extinguisher Units

0010 **WHEELED FIRE EXTINGUISHER UNITS**									
0350 CO_2, portable, with swivel horn									
0360 Wheeled type, cart mounted, 50 lb.				Ea.	1,200			1,200	1,325
0400 100 lb.				"	2,525			2,525	2,775
2200 ABC all purpose type									
2300 Wheeled, 45 lb.				Ea.	650			650	715
2360 150 lb.				"	1,600			1,600	1,750

10 51 Lockers

10 51 13 – Metal Lockers

10 51 13.10 Lockers

		Crew	Daily Output	Labor-Hours	Unit	Material	Labor	Equipment	Total	Total Incl O&P
0011	**LOCKERS** Steel, baked enamel									
0110	Single tier box locker, 12" x 15" x 72"	1 Shee	8	1	Ea.	176	43.50		219.50	260
0120	18" x 15" x 72"		8	1		191	43.50		234.50	277
0130	12" x 18" x 72"		8	1		186	43.50		229.50	272
0140	18" x 18" x 72"		8	1		203	43.50		246.50	290
0410	Double tier, 12" x 15" x 36"		21	.381		197	16.60		213.60	243
0420	18" x 15" x 36"		21	.381		231	16.60		247.60	280
0430	12" x 18" x 36"		21	.381		196	16.60		212.60	241
0440	18" x 18" x 36"		21	.381		223	16.60		239.60	271
0500	Two person, 18" x 15" x 72"		8	1		248	43.50		291.50	340
0510	18" x 18" x 72"		8	1		268	43.50		311.50	360
0520	Duplex, 15" x 15" x 72"		8	1		268	43.50		311.50	360
0530	15" x 21" x 72"		8	1	↓	315	43.50		358.50	410
0600	5 tier box lockers, minimum		30	.267	Opng.	34	11.60		45.60	55.50
0700	Maximum		24	.333		50.50	14.50		65	78.50
0900	6 tier box lockers, minimum		36	.222		35	9.70		44.70	53.50
1000	Maximum		30	.267	↓	40.50	11.60		52.10	62.50
1100	Wire meshed wardrobe, floor. mtd., open front varsity type	↓	7.50	1.067	Ea.	199	46.50		245.50	291
2400	16-person locker unit with clothing rack									
2500	72 wide x 15" deep x 72" high	1 Shee	8	1	Ea.	495	43.50		538.50	610
2550	18" deep	"	8	1	"	540	43.50		583.50	660
3000	Wall mounted lockers, 4 person, with coat bar									
3100	48" wide x 18" deep x 12" high	1 Shee	8	1	Ea.	252	43.50		295.50	345
3250	Rack w/ 24 wire mesh baskets		1.50	5.333	Set	335	232		567	730
3260	30 baskets		1.25	6.400		266	279		545	725
3270	36 baskets		.95	8.421		335	365		700	935
3280	42 baskets	↓	.80	10	↓	585	435		1,020	1,300
3300	For built-in lock with 2 keys, add				Ea.	6.90			6.90	7.55
3600	For hanger rods, add				"	1.86			1.86	2.05

10 51 53 – Locker Room Benches

10 51 53.10 Benches

		Crew	Daily Output	Labor-Hours	Unit	Material	Labor	Equipment	Total	Total Incl O&P
0010	**BENCHES**									
2100	Locker bench, laminated maple, top only	1 Shee	100	.080	L.F.	13.35	3.48		16.83	20
2200	Pedestals, steel pipe	"	25	.320	Ea.	36	13.95		49.95	61

10 55 Postal Specialties

10 55 23 – Mail Boxes

10 55 23.10 Mail Boxes

		Crew	Daily Output	Labor-Hours	Unit	Material	2007 Bare Costs Labor	2007 Bare Costs Equipment	Total	Total Incl O&P
0010	**MAIL BOXES**									
0020	Horiz., key lock, 5"H x 6"W x 15"D, alum., rear load	1 Carp	34	.235	Ea.	35	8.65		43.65	52
0100	Front loading		34	.235		39.50	8.65		48.15	57
0200	Double, 5"H x 12"W x 15"D, rear loading		26	.308		64.50	11.30		75.80	88.50
0300	Front loading		26	.308		67.50	11.30		78.80	91.50
0500	Quadruple, 10"H x 12"W x 15"D, rear loading		20	.400		119	14.70		133.70	154
0600	Front loading		20	.400		122	14.70		136.70	157
0800	Vertical, front load, 15"H x 5"W x 6"D, alum., per compartment		34	.235		27.50	8.65		36.15	44
0900	Bronze, duranodic finish		34	.235		47	8.65		55.65	65
1000	Steel, enameled		34	.235		34	8.65		42.65	51
1700	Alphabetical directories, 120 names		10	.800		122	29.50		151.50	180
1800	Letter collection box		6	1.333		640	49		689	780
1900	Letter slot, residential		20	.400		64.50	14.70		79.20	94
2000	Post office type		8	1		235	36.50		271.50	315
2200	Post office counter window, with grille		2	4		550	147		697	835
2250	Key keeper, single key, aluminum		26	.308		39.50	11.30		50.80	61
2300	Steel, enameled		26	.308		107	11.30		118.30	136

10 55 91 – Mail Chutes

10 55 91.10 Mail Chutes

		Crew	Daily Output	Labor-Hours	Unit	Material	2007 Bare Costs Labor	2007 Bare Costs Equipment	Total	Total Incl O&P
0010	**MAIL CHUTES**									
0020	Aluminum & glass, 14-1/4" wide, 4-5/8" deep	2 Shee	4	4	Floor	680	174		854	1,025
0100	8-5/8" deep		3.80	4.211		830	183		1,013	1,200
0300	8-3/4" x 3-1/2", aluminum		5	3.200		630	139		769	910
0400	Bronze or stainless		4.50	3.556		975	155		1,130	1,325
0600	Lobby collection boxes, aluminum		5	3.200	Ea.	1,875	139		2,014	2,300
0700	Bronze or stainless		4.50	3.556	"	2,425	155		2,580	2,900

10 56 Storage Assemblies

10 56 13 – Metal Storage Shelving

10 56 13.10 Shelving

		Crew	Daily Output	Labor-Hours	Unit	Material	2007 Bare Costs Labor	2007 Bare Costs Equipment	Total	Total Incl O&P
0010	**SHELVING**									
0020	Metal, industrial, cross-braced, 3' wide, 12" deep	1 Sswk	175	.046	SF Shlf	9	1.89		10.89	13.30
0100	24" deep		330	.024		6.20	1		7.20	8.60
0300	4' wide, 12" deep		185	.043		9.80	1.79		11.59	14.05
0400	24" deep		380	.021		6.30	.87		7.17	8.50
1200	Enclosed sides, cross-braced back, 3' wide, 12" deep		175	.046		14.05	1.89		15.94	18.85
1300	24" deep		290	.028		11.95	1.14		13.09	15.20
1500	Fully enclosed, sides and back, 3' wide, 12" deep		150	.053		11.60	2.21		13.81	16.75
1600	24" deep		255	.031		8.70	1.30		10	11.90
1800	4' wide, 12" deep		150	.053		13.25	2.21		15.46	18.60
1900	24" deep		290	.028		7.90	1.14		9.04	10.75
2200	Wide span, 1600 lb. capacity per shelf, 6' wide, 24" deep		380	.021		9	.87		9.87	11.50
2400	36" deep		440	.018		7.90	.75		8.65	10
2600	8' wide, 24" deep		440	.018		7.65	.75		8.40	9.80
2800	36" deep		520	.015		6.25	.64		6.89	8.05
4000	Pallet racks, steel frame 5,000 lb. capacity, 8' long, 36" deep	2 Sswk	450	.036		12.10	1.47		13.57	16
4200	42" deep		500	.032		11	1.32		12.32	14.50
4400	48" deep		520	.031		10.20	1.27		11.47	13.55

10 56 13.20 Parts Bins

0010	**PARTS BINS** Metal, gray baked enamel finish									

10 56 Storage Assemblies

10 56 13 – Metal Storage Shelving

10 56 13.20 Parts Bins

		Crew	Daily Output	Labor-Hours	Unit	Material	2007 Bare Costs Labor	Equipment	Total	Total Incl O&P
0100	6'-3" high, 3' wide									
0300	12 bins, 18" wide x 12" high, 12" deep	2 Clab	10	1.600	Ea.	271	46		317	370
0400	24" deep		10	1.600		335	46		381	440
0600	72 bins, 6" wide x 6" high, 12" deep		8	2		380	57.50		437.50	510
0700	18" deep		8	2		615	57.50		672.50	765
1000	7'-3" high, 3' wide									
1200	14 bins, 18" wide x 12" high, 12" deep	2 Clab	10	1.600	Ea.	263	46		309	360
1300	24" deep		10	1.600		345	46		391	450
1500	84 bins, 6" wide x 6" high, 12" deep		8	2		615	57.50		672.50	765
1600	24" deep		8	2		770	57.50		827.50	935

10 57 Wardrobe and Closet Specialties

10 57 13 – Hat and Coat Racks

10 57 13.10 Coat Racks and Wardrobes

		Crew	Daily Output	Labor-Hours	Unit	Material	2007 Bare Costs Labor	Equipment	Total	Total Incl O&P
0010	**COAT RACKS AND WARDROBES**									
0020	Floor model hat & coat racks, 6 hangers									
0050	Standing, beech wood, 21" x 21" x 72", chrome				Ea.	223			223	245
0100	18 gauge tubular steel, 21" x 21" x 69", wood walnut				"	237			237	261
0500	16 gauge steel frame, 22 gauge steel shelves									
0650	Single pedestal, 30" x 18" x 63"				Ea.	170			170	187
0800	Single face rack, 29" x 18-1/2" x 62"					183			183	201
0900	51" x 18-1/2" x 70"					330			330	365
0910	Double face rack, 39" x 26" x 70"					220			220	242
0920	63" x 26" x 70"					375			375	415
0940	For 2" ball casters, add				Set	80.50			80.50	89
1400	Utility hook strips, 3/8" x 2-1/2" x 18", 6 hooks	1 Carp	48	.167	Ea.	47.50	6.10		53.60	62
1500	34" long, 12 hooks	"	48	.167	"	51	6.10		57.10	65.50
1650	Wall mounted racks, 16 gauge steel frame, 22 gauge steel shelves									
1850	12" x 15" x 26", 6 hangers	1 Carp	32	.250	Ea.	135	9.20		144.20	162
2000	12" x 15" x 50", 12 hangers	"	32	.250	"	158	9.20		167.20	187
2150	Wardrobe cabinet, steel, baked enamel finish									
2300	36" x 21" x 78", incl. top shelf & hanger rod				Ea.	272			272	299

10 57 23 – Closet and Utility Shelving

10 57 23.19 Wood Closet and Utility Shelving

		Crew	Daily Output	Labor-Hours	Unit	Material	2007 Bare Costs Labor	Equipment	Total	Total Incl O&P
0010	**WOOD CLOSET AND UTILITY SHELVING**									
0020	Pine, clear grade, no edge band, 1" x 8"	1 Carp	115	.070	L.F.	2	2.55		4.55	6.20
0100	1" x 10"		110	.073		2.24	2.67		4.91	6.60
0200	1" x 12"		105	.076		2.67	2.80		5.47	7.30
0600	Plywood, 3/4" thick with lumber edge, 12" wide		75	.107		1.45	3.91		5.36	7.70
0700	24" wide		70	.114		2.56	4.19		6.75	9.35
0900	Bookcase, clear grade pine, shelves 12" O.C., 8" deep, /SF shelf		70	.114	S.F.	6.50	4.19		10.69	13.70
1000	12" deep shelves		65	.123	"	8.70	4.52		13.22	16.60
1200	Adjustable closet rod and shelf, 12" wide, 3' long		20	.400	Ea.	8.70	14.70		23.40	32.50
1300	8' long		15	.533	"	24	19.55		43.55	57
1500	Prefinished shelves with supports, stock, 8" wide		75	.107	L.F.	3.83	3.91		7.74	10.30
1600	10" wide		70	.114	"	4.26	4.19		8.45	11.25

10 73 Protective Covers

10 73 13 – Awnings

10 73 13.10 Awnings, Fabric

		Crew	Daily Output	Labor-Hours	Unit	Material	2007 Bare Costs Labor	Equipment	Total	Total Incl O&P
0010	**AWNINGS, FABRIC**									
0020	Including acrylic canvas and frame, standard design									
0100	Door and window, slope, 3' high, 4' wide	1 Carp	4.50	1.778	Ea.	615	65		680	780
0110	6' wide		3.50	2.286		795	84		879	1,000
0120	8' wide		3	2.667		970	98		1,068	1,225
0200	Quarter round convex, 4' wide		3	2.667		955	98		1,053	1,200
0210	6' wide		2.25	3.556		1,250	130		1,380	1,575
0220	8' wide		1.80	4.444		1,525	163		1,688	1,925
0300	Dome, 4' wide		7.50	1.067		370	39		409	465
0310	6' wide		3.50	2.286		830	84		914	1,050
0320	8' wide		2	4		1,475	147		1,622	1,850
0350	Elongated dome, 4' wide		1.33	6.015		1,375	221		1,596	1,875
0360	6' wide		1.11	7.207		1,650	265		1,915	2,225
0370	8' wide		1	8		1,950	294		2,244	2,575
1000	Entry or walkway, peak, 12' long, 4' wide	2 Carp	.90	17.778		4,375	650		5,025	5,850
1010	6' wide		.60	26.667		6,750	980		7,730	8,950
1020	8' wide		.40	40		9,325	1,475		10,800	12,500
1100	Radius with dome end, 4' wide		1.10	14.545		3,325	535		3,860	4,475
1110	6' wide		.70	22.857		5,350	840		6,190	7,175
1120	8' wide		.50	32		7,600	1,175		8,775	10,200
2000	Retractable lateral arm awning, manual									
2010	To 12' wide, 8' - 6" projection	2 Carp	1.70	9.412	Ea.	990	345		1,335	1,650
2020	To 14' wide, 8' - 6" projection		1.10	14.545		1,150	535		1,685	2,100
2030	To 19' wide, 8' - 6" projection		.85	18.824		1,575	690		2,265	2,800
2040	To 24' wide, 8' - 6" projection		.67	23.881		1,975	875		2,850	3,550
2050	Motor for above, add	1 Carp	2.67	3		850	110		960	1,100
3000	Patio/deck canopy with frame									
3010	12' wide, 12' projection	2 Carp	2	8	Ea.	1,400	294		1,694	1,975
3020	16' wide, 14' projection	"	1.20	13.333		2,175	490		2,665	3,150
9000	For fire retardant canvas, add					7%				
9010	For lettering or graphics, add					35%				
9020	For painted or coated acrylic canvas, deduct					8%				
9030	For translucent or opaque vinyl canvas, add					10%				
9040	For 6 or more units, deduct					20%	15%			

10 73 16 – Canopies

10 73 16.20 Canopies

		Crew	Daily Output	Labor-Hours	Unit	Material	2007 Bare Costs Labor	Equipment	Total	Total Incl O&P
0010	**CANOPIES**									
0020	Wall hung, .032", aluminum, prefinished, 8' x 10'	K-2	1.30	18.462	Ea.	1,900	695	143	2,738	3,450
0300	8' x 20'		1.10	21.818		3,775	825	168	4,768	5,750
0500	10' x 10'		1.30	18.462		2,125	695	143	2,963	3,700
0700	10' x 20'		1.10	21.818		3,925	825	168	4,918	5,925
1000	12' x 20'		1	24		4,825	905	185	5,915	7,100
1360	12' x 30'		.80	30		7,100	1,125	232	8,457	10,000
1700	12' x 40'		.60	40		8,650	1,500	310	10,460	12,500
1900	For free standing units, add					20%	10%			
2300	Aluminum entrance canopies, flat soffit, .032"									
2500	3'-6" x 4'-0", clear anodized	2 Carp	4	4	Ea.	795	147		942	1,100
2700	Bronze anodized		4	4		1,400	147		1,547	1,750
3000	Polyurethane painted		4	4		1,125	147		1,272	1,475
3300	4'-6" x 10'-0", clear anodized		2	8		2,200	294		2,494	2,850
3500	Bronze anodized		2	8		2,800	294		3,094	3,525
3700	Polyurethane painted		2	8		2,350	294		2,644	3,025

10 73 Protective Covers

10 73 16 – Canopies

10 73 16.20 Canopies		Crew	Daily Output	Labor-Hours	Unit	Material	2007 Bare Costs Labor	Equipment	Total	Total Incl O&P
4000	Wall downspout, 10 L.F., clear anodized	1 Carp	7	1.143	Ea.	136	42		178	216
4300	Bronze anodized		7	1.143		237	42		279	325
4500	Polyurethane painted		7	1.143		204	42		246	290
7000	Carport, baked vinyl finish, .032", 20' x 10', no foundations, min.	K-2	4	6	Car	3,400	227	46.50	3,673.50	4,200
7250	Maximum		2	12	"	6,800	455	92.50	7,347.50	8,375
7500	Walkway cover, to 12' wide, stl., vinyl finish, .032",no fndtns., min.		250	.096	S.F.	19.95	3.62	.74	24.31	29
7750	Maximum		200	.120	"	21.50	4.53	.93	26.96	33

10 74 Manufactured Exterior Specialties

10 74 23 – Cupolas

10 74 23.10 Cupolas

		Crew	Daily Output	Labor-Hours	Unit	Material	Labor	Equipment	Total	Total Incl O&P
0010	**CUPOLAS**									
0020	Stock units, pine, painted, 18" sq., 28" high, alum. roof	1 Carp	4.10	1.951	Ea.	158	71.50		229.50	285
0100	Copper roof		3.80	2.105		184	77.50		261.50	320
0300	23" square, 33" high, aluminum roof		3.70	2.162		345	79.50		424.50	505
0400	Copper roof		3.30	2.424		460	89		549	650
0600	30" square, 37" high, aluminum roof		3.70	2.162		485	79.50		564.50	655
0700	Copper roof		3.30	2.424		520	89		609	710
0900	Hexagonal, 31" wide, 46" high, copper roof		4	2		650	73.50		723.50	830
1000	36" wide, 50" high, copper roof		3.50	2.286		715	84		799	915
1200	For deluxe stock units, add to above					25%				
1400	For custom built units, add to above					50%	50%			

10 74 29 – Steeples

10 74 29.10 Steeples

		Crew	Daily Output	Labor-Hours	Unit	Material	Labor	Equipment	Total	Total Incl O&P
0010	**STEEPLES**									
4000	Steeples, translucent fiberglass, 30" square, 15' high	F-3	2	20	Ea.	4,550	745	360	5,655	6,550
4150	25' high		1.80	22.222		5,300	830	400	6,530	7,550
4350	Opaque fiberglass, 24" square, 14' high		2	20		3,775	745	360	4,880	5,700
4500	28' high		1.80	22.222		4,250	830	400	5,480	6,400
4600	Aluminum, baked finish, 14' high, 16" square					3,700			3,700	4,075
4620	20' high, 3'-6" base					6,250			6,250	6,875
4640	35' high, 8' base					24,300			24,300	26,700
4660	60' high, 14' base					53,000			53,000	58,000
4680	152' high, custom					446,000			446,000	490,500
4700	Porcelain enamel steeples, custom, 40' high	F-3	.50	80		13,700	2,975	1,450	18,125	21,300
4800	60' high	"	.30	133		23,800	4,975	2,425	31,200	36,600

10 74 46 – Window Wells

10 74 46.10 Area Window Wells

		Crew	Daily Output	Labor-Hours	Unit	Material	Labor	Equipment	Total	Total Incl O&P
0010	**AREA WINDOW WELLS**, Galvanized steel									
0020	20 ga., 3'-2" wide, 1' deep	1 Sswk	29	.276	Ea.	13.60	11.40		25	35.50
0100	2' deep		23	.348		24	14.40		38.40	52.50
0300	16 ga., galv., 3'-2" wide, 1' deep		29	.276		19.15	11.40		30.55	41.50
0400	3' deep		23	.348		39	14.40		53.40	68.50
0600	Welded grating for above, 15 lbs., painted		45	.178		43	7.35		50.35	61
0700	Galvanized		45	.178		84.50	7.35		91.85	106
0900	Translucent plastic cap for above		60	.133		14.10	5.50		19.60	25.50

10 75 Flagpoles

10 75 16 – Ground-Set Flagpoles

10 75 16.10 Flagpoles

		Crew	Daily Output	Labor-Hours	Unit	Material	2007 Bare Costs Labor	2007 Bare Costs Equipment	Total	Total Incl O&P
0010	**FLAGPOLES**, Ground set									
0050	Not including base or foundation									
0100	Aluminum, tapered, ground set 20' high	K-1	2	8	Ea.	790	261	92.50	1,143.50	1,375
0200	25' high		1.70	9.412		1,025	305	109	1,439	1,725
0300	30' high		1.50	10.667		1,050	350	124	1,524	1,825
0400	35' high		1.40	11.429		1,475	375	132	1,982	2,350
0500	40' high		1.20	13.333		2,075	435	154	2,664	3,125
0600	50' high		1	16		2,425	520	185	3,130	3,700
0700	60' high		.90	17.778		4,425	580	206	5,211	5,975
0800	70' high		.80	20		7,675	655	232	8,562	9,725
1100	Counterbalanced, internal halyard, 20' high		1.80	8.889		2,050	290	103	2,443	2,825
1200	30' high		1.50	10.667		2,275	350	124	2,749	3,175
1300	40' high		1.30	12.308		3,800	400	143	4,343	4,950
1400	50' high		1	16		7,000	520	185	7,705	8,725
2820	Aluminum, electronically operated, 30' high		1.40	11.429		4,175	375	132	4,682	5,325
2840	35' high		1.30	12.308		4,500	400	143	5,043	5,725
2860	39' high		1.10	14.545		4,900	475	168	5,543	6,325
2880	45' high		1	16		6,275	520	185	6,980	7,925
2900	50' high		.90	17.778		6,675	580	206	7,461	8,450
3000	Fiberglass, tapered, ground set, 23' high		2	8		1,050	261	92.50	1,403.50	1,675
3100	29'-7" high		1.50	10.667		1,275	350	124	1,749	2,075
3200	36'-1" high		1.40	11.429		1,650	375	132	2,157	2,550
3300	39'-5" high		1.20	13.333		2,150	435	154	2,739	3,225
3400	49'-2" high		1	16		4,375	520	185	5,080	5,825
3500	59' high		.90	17.778		6,275	580	206	7,061	8,025
4300	Steel, direct imbedded installation									
4400	Internal halyard, 20' high	K-1	2.50	6.400	Ea.	1,425	209	74	1,708	1,950
4500	25' high		2.50	6.400		1,575	209	74	1,858	2,150
4600	30' high		2.30	6.957		1,875	227	80.50	2,182.50	2,500
4700	40' high		2.10	7.619		2,775	249	88	3,112	3,525
4800	50' high		1.90	8.421		4,075	275	97.50	4,447.50	5,025
5000	60' high		1.80	8.889		6,175	290	103	6,568	7,375
5100	70' high		1.60	10		7,550	325	116	7,991	8,925
5200	80' high		1.40	11.429		9,950	375	132	10,457	11,600
5300	90' high		1.20	13.333		13,600	435	154	14,189	15,800
5500	100' high		1	16		14,100	520	185	14,805	16,500
6400	Wood poles, tapered, clear vertical grain fir with tilting									
6410	base, not incl. foundation, 4" butt, 25' high	K-1	1.90	8.421	Ea.	655	275	97.50	1,027.50	1,250
6800	6" butt, 30' high	"	1.30	12.308	"	815	400	143	1,358	1,675
7300	Foundations for flagpoles, including									
7400	excavation and concrete, to 35' high poles	C-1	10	3.200	Ea.	525	111		636	755
7600	40' to 50' high		3.50	9.143		955	315		1,270	1,550
7700	Over 60' high		2	16		1,075	555		1,630	2,050

10 75 23 – Wall-Mounted Flagpoles

10 75 23.10 Flagpoles

		Crew	Daily Output	Labor-Hours	Unit	Material	Labor	Equipment	Total	Total Incl O&P
0010	**FLAGPOLES**, Structure mounted									
0100	Fiberglass, vertical wall set, 19'-8" long	K-1	1.50	10.667	Ea.	1,225	350	124	1,699	2,025
0200	23' long		1.40	11.429		1,300	375	132	1,807	2,150
0300	26'-3" long		1.30	12.308		1,900	400	143	2,443	2,850
0800	19'-8" long outrigger		1.30	12.308		1,250	400	143	1,793	2,150
1300	Aluminum, vertical wall set, tapered, with base, 20' high		1.20	13.333		960	435	154	1,549	1,900
1400	29'-6" high		1	16		2,175	520	185	2,880	3,425

10 75 Flagpoles

10 75 23 – Wall-Mounted Flagpoles

10 75 23.10 Flagpoles

		Crew	Daily Output	Labor-Hours	Unit	Material	2007 Bare Costs Labor	2007 Bare Costs Equipment	Total	Total Incl O&P
2400	Outrigger poles with base, 12' long	K-1	1.30	12.308	Ea.	860	400	143	1,403	1,725
2500	14' long	↓	1	16	↓	1,375	520	185	2,080	2,550

10 81 Pest Control Devices

10 81 13 – Bird Control Devices

10 81 13.10 Bird Control Netting

		Crew	Daily Output	Labor-Hours	Unit	Material	Labor	Equipment	Total	Total Incl O&P
0010	**BIRD CONTROL NETTING**									
0020	1/8" square mesh	4 Clab	4000	.008	S.F.	.04	.23		.27	.40
0100	1/4" square mesh		4000	.008		.05	.23		.28	.42
0120	1/2" square mesh		4000	.008		.05	.23		.28	.42
0140	5/8" x 3/4" mesh		4000	.008		.05	.23		.28	.42
0160	1-1/4" x 1-1/2" mesh		4000	.008		.06	.23		.29	.43
0200	4" square mesh	↓	4000	.008	↓	.07	.23		.30	.44
1000	Poly clips				Ea.	.01			.01	.01

10 88 Scales

10 88 05 – Commercial Scales

10 88 05.10 Scales

		Crew	Daily Output	Labor-Hours	Unit	Material	Labor	Equipment	Total	Total Incl O&P
0010	**SCALES** Built-in floor scale, not incl. foundations									
0100	Dial type, 5 ton capacity, 8' x 6' platform	3 Carp	.50	48	Ea.	6,850	1,750		8,600	10,300
0300	9' x 7' platform		.40	60		8,900	2,200		11,100	13,200
0400	10 ton capacity, steel platform, 8' x 6' platform		.40	60		10,800	2,200		13,000	15,300
0600	9' x 7' platform	↓	.35	68.571	↓	10,100	2,525		12,625	15,000
0700	Truck scales, incl. steel weigh bridge,									
0800	not including foundation, pits									
1550	Digital, electronic, 100 ton capacity, steel deck 12' x 10' platform	3 Carp	.20	120	Ea.	11,400	4,400		15,800	19,400
1600	40' x 10' platform		.14	171		23,200	6,300		29,500	35,300
1640	60' x 10' platform		.13	184		28,700	6,775		35,475	42,200
1680	70' x 10' platform	↓	.12	200		32,100	7,350		39,450	46,700
2000	For standard automatic printing device, add					2,300			2,300	2,525
2100	For remote reading electronic system, add					2,100			2,100	2,300
2300	Concrete foundation pits for above, 8' x 6', 5 C.Y. required	C-1	.50	64		875	2,225		3,100	4,400
2400	14' x 6' platform, 10 C.Y. required		.35	91.429		1,275	3,175		4,450	6,350
2600	50' x 10' platform, 30 C.Y. required		.25	128		1,725	4,450		6,175	8,825
2700	70' x 10' platform, 40 C.Y. required	↓	.15	213		3,775	7,400		11,175	15,700
2750	Crane scales, dial, 1 ton capacity					930			930	1,025
2780	5 ton capacity					1,075			1,075	1,200
2800	Digital, 1 ton capacity					1,950			1,950	2,125
2850	10 ton capacity				↓	3,900			3,900	4,300
2900	Low profile electronic warehouse scale,									
3000	not incl. printer, 4' x 4' platform, 10,000 lb. capacity	2 Carp	.30	53.333	Ea.	1,900	1,950		3,850	5,150
3300	5' x 7' platform, 10,000 lb. capacity		.25	64		3,775	2,350		6,125	7,800
3400	20,000 lb. capacity	↓	.20	80		4,850	2,925		7,775	9,925
3500	For printers, incl. time, date & numbering, add					1,200			1,200	1,325
3800	Portable, beam type, capacity 1000#, platform 18" x 24"					705			705	775
3900	Dial type, capacity 2000#, platform 24" x 24"					1,200			1,200	1,325
4000	Digital type, capacity 1000#, platform 24" x 30"					1,950			1,950	2,125
4100	Portable contractor truck scales, 50 ton cap., 40' x 10' platform					30,200			30,200	33,200
4200	60' x 10' platform				↓	27,000			27,000	29,700

Estimating Tips

General

- The items in this division are usually priced per square foot or each. Many of these items are purchased by the owner for installation by the contractor. Check the specifications for responsibilities and include time for receiving, storage, installation, and mechanical and electrical hook-ups in the appropriate divisions.
- Many items in Division 11 require some type of support system that is not usually furnished with the item. Examples of these systems include blocking for the attachment of casework and support angles for ceiling-hung projection screens. The required blocking or supports must be added to the estimate in the appropriate division.
- Some items in Division 11 may require assembly or electrical hookups. Verify the amount of assembly required or the need for a hard electrical connection and add the appropriate costs.

Reference Numbers

Reference numbers are shown in shaded boxes at the beginning of some major classifications. These numbers refer to related items in the Reference Section. The reference information may be an estimating procedure, an alternate pricing method, or technical information.

Note: Not all subdivisions listed here necessarily appear in this publication.

11 05 Common Work Results for Equipment

11 05 10 – Equipment Installation

11 05 10.10 Equipment Installation	Crew	Daily Output	Labor-Hours	Unit	Material	2007 Bare Costs Labor	2007 Bare Costs Equipment	Total	Total Incl O&P
0010 **EQUIPMENT INSTALLATION**									
0020 Industrial equipment, minimum	E-2	12	4.667	Ton	188	129		317	465
0200 Maximum	"	2	28	"	1,125	775		1,900	2,800

11 11 Vehicle Service Equipment

11 11 13 – Compressed-Air Vehicle Service Equipment

11 11 13.10 Compressed Air Equipment	Crew	Daily Output	Labor-Hours	Unit	Material	Labor	Equipment	Total	Total Incl O&P
0010 **COMPRESSED AIR EQUIPMENT**									
0030 Compressors, electric, 1-1/2 H.P., standard controls	L-4	1.50	16	Ea.	295	550		845	1,175
0550 Dual controls		1.50	16		500	550		1,050	1,400
0600 5 H.P., 115/230 volt, standard controls		1	24		1,800	825		2,625	3,250
0650 Dual controls		1	24		1,925	825		2,750	3,375

11 11 19 – Vehicle Lubrication Equipment

11 11 19.10 Lubrication Equipment	Crew	Daily Output	Labor-Hours	Unit	Material	Labor	Equipment	Total	Total Incl O&P
0010 **LUBRICATION EQUIPMENT**									
3000 Lube equipment, 3 reel type, with pumps, not including piping	L-4	.50	48	Set	7,525	1,650		9,175	10,900

11 11 33 – Vehicle Spray Painting Equipment

11 11 33.10 Spray Painting Equipment	Crew	Daily Output	Labor-Hours	Unit	Material	Labor	Equipment	Total	Total Incl O&P
0010 **SPRAY PAINTING EQUIPMENT**									
4000 Spray painting booth, 26' long, complete	L-4	.40	60	Ea.	14,000	2,075		16,075	18,600

11 12 Parking Control Equipment

11 12 13 – Parking Key and Card Control Units

11 12 13.10 Parking Control Units	Crew	Daily Output	Labor-Hours	Unit	Material	Labor	Equipment	Total	Total Incl O&P
0010 **PARKING CONTROL UNITS**									
5100 Card reader	1 Elec	2	4	Ea.	1,800	176		1,976	2,225
5120 Proximity with customer display	2 Elec	1	16		5,500	700		6,200	7,100
6000 Parking control software, minimum	1 Elec	.50	16		21,900	700		22,600	25,200
6020 Maximum	"	.20	40		91,000	1,750		92,750	103,000

11 12 16 – Parking Ticket Dispensers

11 12 16.10 Ticket Dispensers	Crew	Daily Output	Labor-Hours	Unit	Material	Labor	Equipment	Total	Total Incl O&P
0010 **TICKET DISPENSERS**									
5900 Ticket spitter with time/date stamp, standard	2 Elec	2	8	Ea.	6,275	350		6,625	7,425
5920 Mag stripe encoding	"	2	8	"	18,600	350		18,950	20,900

11 12 26 – Parking Fee Collection Equipment

11 12 26.13 Parking Fee Coin Collection Equipment

	Crew	Daily Output	Labor-Hours	Unit	Material	Labor	Equipment	Total	Total Incl O&P
0010 **PARKING FEE COIN COLLECTION EQUIPMENT**									
5200 Cashier booth, average	B-22	1	30	Ea.	9,450	1,025	245	10,720	12,200
5300 Collector station, pay on foot	2 Elec	.20	80		109,500	3,500		113,000	125,500
5320 Credit card only	"	.50	32		20,200	1,400		21,600	24,400

11 12 26.23 Fee Equipment

	Crew	Daily Output	Labor-Hours	Unit	Material	Labor	Equipment	Total	Total Incl O&P
0010 **FEE EQUIPMENT**									
5600 Fee computer	1 Elec	1.50	5.333	Ea.	13,200	234		13,434	14,900

11 12 33 – Parking Gates

11 12 33.13 Parking Gates

	Crew	Daily Output	Labor-Hours	Unit	Material	Labor	Equipment	Total	Total Incl O&P
0010 **PARKING GATES**									

11 12 Parking Control Equipment

11 12 33 – Parking Gates

11 12 33.13 Parking Gates		Crew	Daily Output	Labor-Hours	Unit	Material	2007 Bare Costs Labor	Equipment	Total	Total Incl O&P
5000	Barrier gate with programmable controller	2 Elec	3	5.333	Ea.	3,275	234		3,509	3,950
5020	Industrial		3	5.333		4,425	234		4,659	5,225
5500	Exit verifier		1	16		17,400	700		18,100	20,300
5700	Full sign, 4" letters	1 Elec	2	4		1,225	176		1,401	1,575
5800	Inductive loop	2 Elec	4	4		169	176		345	445
5950	Vehicle detector, microprocessor based	1 Elec	3	2.667		385	117		502	600

11 13 Loading Dock Equipment

11 13 13 – Loading Dock Bumpers

11 13 13.10 Dock Bumpers		Crew	Daily Output	Labor-Hours	Unit	Material	2007 Bare Costs Labor	Equipment	Total	Total Incl O&P
0010	**DOCK BUMPERS** Bolts not included									
0020	2" x 6" to 4" x 8", average	1 Carp	.30	26.667	M.B.F.	1,150	980		2,130	2,800
0050	Bumpers, rubber blocks 4-1/2" thk, 10" H, 14" long		26	.308	Ea.	44	11.30		55.30	66
0200	24" long		22	.364		86.50	13.35		99.85	116
0300	36" long		17	.471		79.50	17.25		96.75	115
0500	12" high, 14" long		25	.320		91	11.75		102.75	118
0550	24" long		20	.400		101	14.70		115.70	134
0600	36" long		15	.533		112	19.55		131.55	154
0800	Rubber blocks 6" thick, 10" high, 14" long		22	.364		82.50	13.35		95.85	112
0850	24" long		18	.444		110	16.30		126.30	147
0900	36" long		13	.615		131	22.50		153.50	179
0910	20" high, 11" long		13	.615		133	22.50		155.50	182
0920	Extruded rubber bumpers, T section, 22" x 22" x 3" thick		41	.195		49	7.15		56.15	65
0940	Molded rubber bumpers, 24" x 12" x 3" thick		20	.400		44.50	14.70		59.20	72
1000	Welded installation of above bumpers	E-14	8	1		3.40	43.50	14.40	61.30	98
1100	For drilled anchors, add per anchor	1 Carp	36	.222		6.10	8.15		14.25	19.40
1300	Steel bumpers, see Div. 10 26 13.10									
1350	Wood bumpers, see Div. 11 13 13.10									
2200	Dock boards, heavy duty, 60" x 60", aluminum, 5,000 lb. cap.				Ea.	1,100			1,100	1,225
2700	9,000 lb. capacity					1,275			1,275	1,400
3200	15,000 lb. capacity					1,450			1,450	1,575

11 13 16 – Loading Dock Seals and Shelters

11 13 16.10 Loading Docks		Crew	Daily Output	Labor-Hours	Unit	Material	2007 Bare Costs Labor	Equipment	Total	Total Incl O&P
0010	**LOADING DOCKS**									
3600	Door seal for door perimeter, 12" x 12", vinyl covered	1 Carp	26	.308	L.F.	26	11.30		37.30	46
3900	Folding gates, see Div. 10 22 16.10									
4200	Platform lifter, 6' x 6', portable, 3,000 lb. capacity				Ea.	8,300			8,300	9,150
4250	4,000 lb. capacity					10,200			10,200	11,300
4400	Fixed, 6' x 8', 5,000 lb. capacity	E-16	.70	22.857		8,950	970	165	10,085	11,800
4500	Levelers, hinged for trucks, 10 ton capacity, 6' x 8'		1.08	14.815		4,525	625	107	5,257	6,225
4650	7' x 8'		1.08	14.815		4,525	625	107	5,257	6,225
4670	Air bag power operated, 10 ton cap., 6'x8'		1.08	14.815		5,250	625	107	5,982	7,025
4680	7' x 8'		1.08	14.815		5,275	625	107	6,007	7,050
4700	Hydraulic, 10 ton capacity, 6' x 8' **CN**		1.08	14.815		7,875	625	107	8,607	9,900
4800	7' x 8'		1.08	14.815		8,475	625	107	9,207	10,600
5000	Lights for loading docks, single arm, 24" long	1 Elec	3.80	2.105		133	92.50		225.50	285
5700	Double arm, 60" long	"	3.80	2.105		160	92.50		252.50	315
5800	Loading dock safety restraints, manual style	E-16	1.08	14.815		2,875	625	107	3,607	4,425
5900	Automatic style	"	1.08	14.815		4,850	625	107	5,582	6,575
6200	Shelters, fabric, for truck or train, scissor arms, minimum	1 Carp	1	8		1,400	294		1,694	2,000

11 13 Loading Dock Equipment

11 13 16 – Loading Dock Seals and Shelters

11 13 16.10 Loading Docks		Crew	Daily Output	Labor-Hours	Unit	Material	2007 Bare Costs Labor	2007 Bare Costs Equipment	Total	Total Incl O&P
6300	Maximum	1 Carp	.50	16	Ea.	1,825	585		2,410	2,925

11 14 Pedestrian Control Equipment

11 14 13 – Pedestrian Gates

11 14 13.13 Portable Posts and Railings

		Crew	Daily Output	Labor-Hours	Unit	Material	Labor	Equipment	Total	Total Incl O&P
0010	**PORTABLE POSTS AND RAILINGS**									
0020	Portable for pedestrian traffic control, standard, minimum				Ea.	103			103	113
0100	Maximum					164			164	180
0300	Deluxe posts, minimum					152			152	167
0400	Maximum					299			299	330
0600	Ropes for above posts, plastic covered, 1-1/2" diameter				L.F.	9.85			9.85	10.80
0700	Chain core				"	10.05			10.05	11.10
1500	Portable security or safety barrier, black with 7' yellow strap				Ea.	215			215	236
1510	12' yellow strap					250			250	275
1550	Sign holder, standard design					54			54	59.50

11 14 13.19 Turnstiles

		Crew	Daily Output	Labor-Hours	Unit	Material	Labor	Equipment	Total	Total Incl O&P
0010	**TURNSTILES**									
0020	One way, 4 arm, 46" diameter, economy, manual	2 Carp	5	3.200	Ea.	740	117		857	1,000
0100	Electric		1.20	13.333		1,200	490		1,690	2,075
0300	High security, galv., 5'-5" diameter, 7' high, manual		1	16		3,675	585		4,260	4,975
0350	Electric		.60	26.667		4,150	980		5,130	6,100
0420	Three arm, 24" opening, light duty, manual		2	8		1,175	294		1,469	1,725
0450	Heavy duty		1.50	10.667		2,900	390		3,290	3,775
0460	Manual, with registering & controls, light duty		2	8		2,525	294		2,819	3,225
0470	Heavy duty		1.50	10.667		3,050	390		3,440	3,950
0480	Electric, heavy duty		1.10	14.545		3,525	535		4,060	4,700
0500	For coin or token operating, add					565			565	620
1200	One way gate with horizontal bars, 5'-5" diameter									
1300	7' high, recreation or transit type	2 Carp	.80	20	Ea.	4,075	735		4,810	5,650
1500	For electronic counter, add				"	203			203	223

11 16 Vault Equipment

11 16 16 – Safes

11 16 16.10 Safes

		Crew	Daily Output	Labor-Hours	Unit	Material	Labor	Equipment	Total	Total Incl O&P
0010	**SAFES**									
0015	Office, 4 hr. rating, 30" x 18" x 18" inside				Ea.	3,700			3,700	4,075
0100	60" x 36" x 18"					8,050			8,050	8,850
0200	1 hr. rating, 30" x 18" x 18"					1,825			1,825	2,000
0250	40" x 18" x 18"					3,500			3,500	3,850
0300	60" x 36" x 18", double door					4,550			4,550	5,000
0400	Data, 4 hr. rating, 23-1/2" x 19-1/2" x 17" inside					3,225			3,225	3,550
0450	52" x 19" x 17"					8,000			8,000	8,800
0500	63" x 34" x 16", double door					12,200			12,200	13,500
0550	52" x 34" x 16", inside					11,500			11,500	12,600
0600	1 hr. rating, 27" x 19" x 16"					4,350			4,350	4,775
0700	63" x 34" x 16"					8,075			8,075	8,900
0750	Diskette, 1 hr., 14" x 12" x 11", inside					3,075			3,075	3,375
0800	Money, "B" label, 9" x 14" x 14"					435			435	475
0900	Tool resistive, 24" x 24" x 20"					2,950			2,950	3,225

11 16 Vault Equipment

11 16 16 – Safes

11 16 16.10 Safes		Crew	Daily Output	Labor-Hours	Unit	Material	2007 Bare Costs Labor	Equipment	Total	Total Incl O&P
1050	Tool and torch resistive, 24" x 24" x 20"				Ea.	7,200			7,200	7,925
1150	Jewelers, 23" x 20" x 18"					9,200			9,200	10,100
1200	63" x 25" x 18"					16,100			16,100	17,700
1300	For handling into building, add, minimum	A-2	8.50	2.824			81	16.65	97.65	144
1400	Maximum	"	.78	30.769			880	181	1,061	1,575

11 17 Teller and Service Equipment

11 17 13 – Teller Equipment Systems

11 17 13.10 Bank Equipment

		Crew	Daily Output	Labor-Hours	Unit	Material	Labor	Equipment	Total	Total Incl O&P
0010	**BANK EQUIPMENT**									
0020	Alarm system, police	2 Elec	1.60	10	Ea.	4,450	440		4,890	5,550
0100	With vault alarm	"	.40	40		17,700	1,750		19,450	22,000
0400	Bullet resistant teller window, 44" x 60"	1 Glaz	.60	13.333		2,975	480		3,455	4,000
0500	48" x 60"	"	.60	13.333		3,675	480		4,155	4,775
3000	Counters for banks, frontal only	2 Carp	1	16	Station	1,650	585		2,235	2,750
3100	Complete with steel undercounter	"	.50	32	"	3,250	1,175		4,425	5,400
4600	Door and frame, bullet-resistant, with vision panel, minimum	2 Sswk	1.10	14.545	Ea.	3,675	600		4,275	5,150
4700	Maximum		1.10	14.545		5,000	600		5,600	6,600
4800	Drive-up window, drawer & mike, not incl. glass, minimum		1	16		4,950	660		5,610	6,650
4900	Maximum		.50	32		8,325	1,325		9,650	11,600
5000	Night depository, with chest, minimum		1	16		6,825	660		7,485	8,700
5100	Maximum		.50	32		9,675	1,325		11,000	13,100
5200	Package receiver, painted		3.20	5		1,250	207		1,457	1,750
5300	Stainless steel		3.20	5		2,100	207		2,307	2,700
5400	Partitions, bullet-resistant, 1-3/16" glass, 8' high	2 Carp	10	1.600	L.F.	180	58.50		238.50	290
5450	Acrylic	"	10	1.600	"	340	58.50		398.50	465
5500	Pneumatic tube systems, 2 lane drive-up, complete	L-3	.25	64	Total	23,400	2,575		25,975	29,800
5550	With T.V. viewer	"	.20	80	"	45,400	3,225		48,625	55,000
5570	Safety deposit boxes, minimum	1 Sswk	44	.182	Opng.	52	7.50		59.50	71
5580	Maximum, 10" x 15" opening		19	.421		110	17.40		127.40	153
5590	Teller locker, average		15	.533		1,425	22		1,447	1,625
5600	Pass thru, bullet-res. window, painted steel, 24" x 36"	2 Sswk	1.60	10	Ea.	2,025	415		2,440	2,975
5700	48" x 48"		1.20	13.333		2,425	550		2,975	3,650
5800	72" x 40"		.80	20		3,025	825		3,850	4,825
5900	For stainless steel frames, add					20%				
6100	Surveillance system, video camera, complete	2 Elec	1	16	Ea.	13,800	700		14,500	16,300
6110	For each additional camera, add				"	900			900	990
6120	CCTV system, see Div. 27 41 19.10									
6200	Twenty-four hour teller, single unit,									
6300	automated deposit, cash and memo	L-3	.25	64	Ea.	41,100	2,575		43,675	49,200
7000	Vault front, see Div. 08 34 59.10									

11 19 Detention Equipment

11 19 30 – Prison Equipment

11 19 30.10 Detention Equipment	Crew	Daily Output	Labor-Hours	Unit	Material	2007 Bare Costs Labor	2007 Bare Costs Equipment	Total	Total Incl O&P
0010 **DETENTION EQUIPMENT**									
3000 Toilet apparatus including wash basin, average	L-8	1.50	13.333	Ea.	3,150	510		3,660	4,250

11 21 Mercantile and Service Equipment

11 21 13 – Cash Registers and Checking Equipment

11 21 13.10 Checkout Counter

	Crew	Daily Output	Labor-Hours	Unit	Material	Labor	Equipment	Total	Total Incl O&P
0010 **CHECKOUT COUNTER**									
0020 Supermarket conveyor, single belt	2 Clab	10	1.600	Ea.	2,425	46		2,471	2,725
0100 Double belt, power take-away		9	1.778		4,050	51		4,101	4,525
0400 Double belt, power take-away, incl. side scanning		7	2.286		5,050	65.50		5,115.50	5,650
0800 Warehouse or bulk type		6	2.667		5,900	76.50		5,976.50	6,625
1000 Scanning system, 2 lanes, w/registers, scan gun & memory				System	14,700			14,700	16,200
1100 10 lanes, single processor, full scan, with scales				"	140,000			140,000	154,000
2000 Register, restaurant, minimum				Ea.	595			595	655
2100 Maximum					2,625			2,625	2,875
2150 Store, minimum					595			595	655
2200 Maximum					2,625			2,625	2,875

11 21 53 – Barber and Beauty Shop Equipment

11 21 53.10 Barber Equipment

	Crew	Daily Output	Labor-Hours	Unit	Material	Labor	Equipment	Total	Total Incl O&P
0010 **BARBER EQUIPMENT**									
0020 Chair, hydraulic, movable, minimum	1 Carp	24	.333	Ea.	440	12.25		452.25	505
0050 Maximum	"	16	.500		2,750	18.35		2,768.35	3,050
0200 Wall hung styling station with mirrors, minimum	L-2	8	2		420	64		484	565
0300 Maximum	"	4	4		1,825	128		1,953	2,225
0500 Sink, hair washing basin, rough plumbing not incl.	1 Plum	8	1		286	45		331	385
1000 Sterilizer, liquid solution for tools					128			128	140
1100 Total equipment, rule of thumb, per chair, minimum	L-8	1	20		1,500	765		2,265	2,825
1150 Maximum	"	1	20		4,075	765		4,840	5,675

11 23 Commercial Laundry and Dry Cleaning Equipment

11 23 13 – Dry Cleaning Equipment

11 23 13.13 Dry Cleaning Equipment

	Crew	Daily Output	Labor-Hours	Unit	Material	Labor	Equipment	Total	Total Incl O&P
0010 **DRY CLEANING EQUIPMENT** Not incl. rough-in									
2000 Dry cleaners, electric, 20 lb. capacity	L-1	.20	80	Ea.	32,800	3,550		36,350	41,400
2050 25 lb. capacity		.17	94.118		44,200	4,175		48,375	55,000
2100 30 lb. capacity		.15	106		46,600	4,725		51,325	58,500
2150 60 lb. capacity		.09	177		72,000	7,875		79,875	91,000

11 23 16 – Drying and Conditioning Equipment

11 23 16.13 Dryers

	Crew	Daily Output	Labor-Hours	Unit	Material	Labor	Equipment	Total	Total Incl O&P
0010 **DRYERS**, not including rough-in									
1500 Industrial, 30 lb. capacity	1 Plum	2	4	Ea.	2,925	179		3,104	3,500
1600 50 lb. capacity	"	1.70	4.706		3,125	211		3,336	3,775
4700 Lint collector, ductwork not included, 8,000 to 10,000 C.F.M.	Q-10	.30	80		8,400	3,250		11,650	14,300

11 23 19 – Finishing Equipment

11 23 19.13 Folders and Spreaders

	Crew	Daily Output	Labor-Hours	Unit	Material	Labor	Equipment	Total	Total Incl O&P
0010 **FOLDERS AND SPREADERS**									
3500 Folders, blankets & sheets, minimum	1 Elec	.17	47.059	Ea.	29,500	2,075		31,575	35,500
3700 King size with automatic stacker		.10	80		53,500	3,500		57,000	64,000

11 23 Commercial Laundry and Dry Cleaning Equipment

11 23 19 – Finishing Equipment

11 23 19.13 Folders and Spreaders	Crew	Daily Output	Labor-Hours	Unit	Material	2007 Bare Costs Labor	Equipment	Total	Total Incl O&P
3800 For conveyor delivery, add	1 Elec	.45	17.778	Ea.	6,000	780		6,780	7,750
4900 Spreader feeders, 240V, 2 station	L-6	.70	17.143		49,300	765		50,065	55,000
4920 4 station	"	.35	34.286		60,000	1,525		61,525	68,500

11 23 23 – Commercial Ironing Equipment

11 23 23.13 Irons and Pressers

	Crew	Daily Output	Labor-Hours	Unit	Material	Labor	Equipment	Total	Incl O&P
0010 **IRONS AND PRESSERS**									
4500 Ironers, institutional, 110", single roll	1 Elec	.20	40	Ea.	29,100	1,750		30,850	34,600
4800 Pressers, low capacity air operated	L-6	1.75	6.857		8,600	305		8,905	9,925
4820 Hand operated		1.75	6.857		7,875	305		8,180	9,100
4840 Ironer 48", 240V		3.50	3.429		101,000	153		101,153	111,500
6600 Hand operated presser		.70	17.143		5,425	765		6,190	7,100
6620 Mushroom press 115V		.70	17.143		6,775	765		7,540	8,600

11 23 26 – Commercial Washers and Extractors

11 23 26.13 Washers and Extractors

	Crew	Daily Output	Labor-Hours	Unit	Material	Labor	Equipment	Total	Incl O&P
0010 **WASHERS AND EXTRACTORS**, not including rough-in									
6000 Combination washer/extractor, 20 lb. capacity	L-6	1.50	8	Ea.	5,100	355		5,455	6,125
6100 30 lb. capacity		.80	15		8,275	670		8,945	10,100
6200 50 lb. capacity		.68	17.647		9,700	785		10,485	11,900
6300 75 lb. capacity		.30	40		18,200	1,775		19,975	22,700
6350 125 lb. capacity		.16	75		24,400	3,350		27,750	31,800
6380 Washer extractor/dryer, 110 lb., 240V		1	12		8,025	535		8,560	9,650
6400 Washer extractor, 135 lb, 240V		1	12		22,400	535		22,935	25,500
6450 Pass through		1	12		63,000	535		63,535	70,500
6500 200 lb.		1	12		63,000	535		63,535	70,500
6550 Pass through		1	12		69,000	535		69,535	76,500
6600 Extractor, low capacity		1.75	6.857		6,475	305		6,780	7,575

11 23 33 – Coin-Operated Laundry Equipment

11 23 33.13 Coin Operated Laundry Equipment

	Crew	Daily Output	Labor-Hours	Unit	Material	Labor	Equipment	Total	Incl O&P
0010 **COIN OPERATED LAUNDRY EQUIPMENT**									
0990 Dryer, gas fired									
1000 Commercial, 30 lb. capacity, coin operated, single	1 Plum	3	2.667	Ea.	3,050	119		3,169	3,525
1100 Double stacked	"	2	4		6,050	179		6,229	6,950
4860 Coin dry cleaner 20 lb.	L-6	1.75	6.857		25,400	305		25,705	28,400
5290 Clothes washer									
5300 Commercial, coin operated, average	1 Plum	3	2.667	Ea.	1,100	119		1,219	1,400

11 24 Maintenance Equipment

11 24 19 – Vacuum Cleaning Systems

11 24 19.10 Vacuum Cleaning

	Crew	Daily Output	Labor-Hours	Unit	Material	Labor	Equipment	Total	Incl O&P
0010 **VACUUM CLEANING**									
0020 Central, 3 inlet, residential	1 Skwk	.90	8.889	Total	680	340		1,020	1,275
0200 Commercial		.70	11.429		1,275	435		1,710	2,075
0400 5 inlet system, residential		.50	16		1,025	610		1,635	2,075
0600 7 inlet system, commercial		.40	20		1,150	760		1,910	2,450
0800 9 inlet system, residential		.30	26.667		1,450	1,025		2,475	3,175
4010 Rule of thumb: First 1200 S.F., installed								1,125	1,240
4020 For each additional S.F., add				S.F.					.20

11 26 Unit Kitchens

11 26 13 – Metal Unit Kitchens

11 26 13.10 Unit Kitchens		Crew	Daily Output	Labor-Hours	Unit	Material	2007 Bare Costs Labor	2007 Bare Costs Equipment	Total	Total Incl O&P
0010	**UNIT KITCHENS**									
1500	Combination range, refrigerator and sink, 30" wide, minimum	L-1	2	8	Ea.	840	355		1,195	1,450
1550	Maximum		1	16		2,675	710		3,385	4,000
1570	60" wide, average		1.40	11.429		2,225	505		2,730	3,175
1590	72" wide, average		1.20	13.333		3,100	590		3,690	4,300
1600	Office model, 48" wide	CN	2	8		2,250	355		2,605	3,000
1620	Refrigerator and sink only		2.40	6.667		2,950	296		3,246	3,700
1640	Combination range, refrigerator, sink, microwave									
1660	oven and ice maker	L-1	.80	20	Ea.	3,450	885		4,335	5,125

11 27 Photographic Processing Equipment

11 27 13 – Darkroom Processing Equipment

11 27 13.10 Darkroom Equipment		Crew	Daily Output	Labor-Hours	Unit	Material	Labor	Equipment	Total	Total Incl O&P
0010	**DARKROOM EQUIPMENT**									
0020	Developing sink, 5" deep, 24" x 48"	Q-1	2	8	Ea.	4,700	325		5,025	5,625
0050	48" x 52"		1.70	9.412		4,750	380		5,130	5,800
0200	10" deep, 24" x 48"		1.70	9.412		5,875	380		6,255	7,025
0250	24" x 108"		1.50	10.667		2,125	430		2,555	3,000
0500	Dryers, dehumidified filtered air, 36" x 25" x 68" high	L-7	6	4.667		4,550	165		4,715	5,275
0550	48" x 25" x 68" high		5	5.600		8,425	199		8,624	9,575
2000	Processors, automatic, color print, minimum		4	7		11,900	248		12,148	13,500
2050	Maximum		.60	46.667		12,200	1,650		13,850	16,000
2300	Black and white print, minimum		2	14		9,025	495		9,520	10,700
2350	Maximum		.80	35		58,000	1,250		59,250	65,500
2600	Manual processor, 16" x 20" maximum print size		2	14		8,900	495		9,395	10,600
2650	20" x 24" maximum print size		1	28		8,475	995		9,470	10,900
3000	Viewing lites, 20" x 24"		6	4.667		365	165		530	660
3100	20" x 24" with color correction		6	4.667		520	165		685	825
3500	Washers, round, minimum sheet 11" x 14"	Q-1	2	8		2,875	325		3,200	3,650
3550	Maximum sheet 20" x 24"		1	16		3,375	645		4,020	4,700
3800	Square, minimum sheet 20" x 24"		1	16		3,025	645		3,670	4,300
3900	Maximum sheet 50" x 56"		.80	20		4,700	805		5,505	6,400
4500	Combination tank sink, tray sink, washers, with									
4510	dry side tables, average	Q-1	.45	35.556	Ea.	8,975	1,425		10,400	12,000

11 31 Residential Appliances

11 31 13 – Residential Kitchen Appliances

11 31 13.13 Cooking Equipment		Crew	Daily Output	Labor-Hours	Unit	Material	Labor	Equipment	Total	Total Incl O&P
0010	**COOKING EQUIPMENT**									
0020	Cooking range, 30" free standing, 1 oven, minimum	2 Clab	10	1.600	Ea.	270	46		316	370
0050	Maximum		4	4		1,550	115		1,665	1,875
0150	2 oven, minimum		10	1.600		1,575	46		1,621	1,825
0200	Maximum		10	1.600		1,725	46		1,771	1,975
0350	Built-in, 30" wide, 1 oven, minimum	1 Elec	6	1.333		470	58.50		528.50	600
0400	Maximum	2 Carp	2	8		1,450	294		1,744	2,050
0500	2 oven, conventional, minimum		4	4		1,300	147		1,447	1,650
0550	1 conventional, 1 microwave, maximum		2	8		1,650	294		1,944	2,250
0700	Free-standing, 1 oven, 21" wide range, minimum	2 Clab	10	1.600		310	46		356	410
0750	21" wide, maximum	"	4	4		325	115		440	540

11 31 Residential Appliances

11 31 13 – Residential Kitchen Appliances

11 31 13.13 Cooking Equipment

		Crew	Daily Output	Labor-Hours	Unit	Material	2007 Bare Costs Labor	Equipment	Total	Total Incl O&P
0900	Counter top cook tops, 4 burner, standard, minimum	1 Elec	6	1.333	Ea.	215	58.50		273.50	325
0950	Maximum		3	2.667		585	117		702	820
1050	As above, but with grille and griddle attachment, minimum		6	1.333		520	58.50		578.50	655
1100	Maximum		3	2.667		870	117		987	1,125
1200	Induction cooktop, 30" wide		3	2.667		590	117		707	820
1250	Microwave oven, minimum		4	2		85.50	88		173.50	225
1300	Maximum		2	4		420	176		596	720

11 31 13.23 Refrigeration Equipment

		Crew	Daily Output	Labor-Hours	Unit	Material	Labor	Equipment	Total	Total Incl O&P
0010	**REFRIGERATION EQUIPMENT**									
2000	Deep freeze, 15 to 23 C.F., minimum	2 Clab	10	1.600	Ea.	475	46		521	595
2050	Maximum		5	3.200		600	92		692	805
2200	30 C.F., minimum		8	2		805	57.50		862.50	975
2250	Maximum		3	5.333		915	153		1,068	1,250
5200	Icemaker, automatic, 20 lb. per day	1 Plum	7	1.143		850	51		901	1,000
5350	51 lb. per day	"	2	4		1,075	179		1,254	1,450
5500	Refrigerator, no frost, 10 C.F. to 12 C.F. minimum	2 Clab	10	1.600		410	46		456	520
5600	Maximum		6	2.667		500	76.50		576.50	670
5750	14 C.F. to 16 C.F., minimum		9	1.778		445	51		496	570
5800	Maximum		5	3.200		435	92		527	625
5950	18 C.F. to 20 C.F., minimum		8	2		530	57.50		587.50	670
6000	Maximum		4	4		795	115		910	1,050
6150	21 C.F. to 29 C.F., minimum		7	2.286		705	65.50		770.50	875
6200	Maximum		3	5.333		2,250	153		2,403	2,725

11 31 13.33 Kitchen Cleaning Equipment

		Crew	Daily Output	Labor-Hours	Unit	Material	Labor	Equipment	Total	Total Incl O&P
0010	**KITCHEN CLEANING EQUIPMENT**									
2750	Dishwasher, built-in, 2 cycles, minimum	L-1	4	4	Ea.	268	177		445	560
2800	Maximum		2	8		310	355		665	870
2950	4 or more cycles, minimum		4	4		285	177		462	580
2960	Average		4	4		380	177		557	680
3000	Maximum		2	8		530	355		885	1,125

11 31 13.43 Waste Disposal Equipment

		Crew	Daily Output	Labor-Hours	Unit	Material	Labor	Equipment	Total	Total Incl O&P
0010	**WASTE DISPOSAL EQUIPMENT**									
1750	Compactor, residential size, 4 to 1 compaction, minimum	1 Carp	5	1.600	Ea.	440	58.50		498.50	575
1800	Maximum	"	3	2.667		520	98		618	725
3300	Garbage disposal, sink type, minimum	L-1	10	1.600		50	71		121	161
3350	Maximum	"	10	1.600		159	71		230	281

11 31 13.53 Kitchen Ventilation Equipment

		Crew	Daily Output	Labor-Hours	Unit	Material	Labor	Equipment	Total	Total Incl O&P
0010	**KITCHEN VENTILATION EQUIPMENT**									
4150	Hood for range, 2 speed, vented, 30" wide, minimum	L-3	5	3.200	Ea.	40.50	129		169.50	242
4200	Maximum		3	5.333		595	214		809	980
4300	42" wide, minimum		5	3.200		242	129		371	465
4330	Custom		5	3.200		670	129		799	930
4350	Maximum		3	5.333		820	214		1,034	1,225
4500	For ventless hood, 2 speed, add					16.15			16.15	17.75
4650	For vented 1 speed, deduct from maximum					42			42	46

11 31 23 – Residential Laundry Appliances

11 31 23.13 Washers

		Crew	Daily Output	Labor-Hours	Unit	Material	Labor	Equipment	Total	Total Incl O&P
0010	**WASHERS**									
5000	Washers, residential, 4 cycle, average	1 Plum	3	2.667	Ea.	740	119		859	995
6650	Washing machine, automatic, minimum		3	2.667		310	119		429	520
6700	Maximum		1	8		1,100	360		1,460	1,775

11 31 Residential Appliances

11 31 23 – Residential Laundry Appliances

11 31 23.23 Dryers

		Crew	Daily Output	Labor-Hours	Unit	Material	2007 Bare Costs Labor	Equipment	Total	Total Incl O&P
0010	**DRYERS**									
0500	Dryers, gas fired residential, 16 lb. capacity, average	1 Plum	3	2.667	Ea.	580	119		699	820
3200	Dryer, automatic, minimum	L-2	3	5.333		269	171		440	560
3250	Maximum	"	2	8		665	256		921	1,125
7450	Vent kits for dryers	1 Carp	10	.800		14.65	29.50		44.15	61.50

11 31 33 – Miscellaneous Residential Appliances

11 31 33.13 Sump Pumps

		Crew	Daily Output	Labor-Hours	Unit	Material	Labor	Equipment	Total	Total Incl O&P
0010	**SUMP PUMPS**									
6400	Sump pump cellar drainer, pedestal, 1/3 H.P., molded PVC base	1 Plum	3	2.667	Ea.	96	119		215	286
6450	Solid brass	"	2	4	"	201	179		380	490
6460	Sump pump, see also division 22 14 29.16									

11 31 33.23 Water Heaters

		Crew	Daily Output	Labor-Hours	Unit	Material	Labor	Equipment	Total	Total Incl O&P
0010	**WATER HEATERS**									
6900	Water heater, electric, glass lined, 30 gallon, minimum	L-1	5	3.200	Ea.	345	142		487	590
6950	Maximum		3	5.333		480	237		717	885
7100	80 gallon, minimum		2	8		630	355		985	1,225
7150	Maximum		1	16		875	710		1,585	2,000
7180	Water heater, gas, glass lined, 30 gallon, minimum	2 Plum	5	3.200		565	143		708	835
7220	Maximum		3	5.333		785	239		1,024	1,225
7260	50 gallon, minimum		2.50	6.400		595	287		882	1,075
7300	Maximum		1.50	10.667		830	480		1,310	1,625
7310	Water heater, see also division 22 33 30.13									

11 31 33.43 Air Quality

		Crew	Daily Output	Labor-Hours	Unit	Material	Labor	Equipment	Total	Total Incl O&P
0010	**AIR QUALITY**									
2450	Dehumidifier, portable, automatic, 15 pint				Ea.	159			159	175
2550	40 pint					194			194	213
3550	Heater, electric, built-in, 1250 watt, ceiling type, minimum	1 Elec	4	2		74.50	88		162.50	213
3600	Maximum		3	2.667		122	117		239	310
3700	Wall type, minimum		4	2		106	88		194	247
3750	Maximum		3	2.667		140	117		257	330
3900	1500 watt wall type, with blower		4	2		131	88		219	275
3950	3000 watt		3	2.667		267	117		384	470
4850	Humidifier, portable, 8 gallons per day					160			160	176
5000	15 gallons per day					193			193	212

11 33 Retractable Stairs

11 33 10 – Disappearing Stairs

11 33 10.10 Disappearing Stairway

		Crew	Daily Output	Labor-Hours	Unit	Material	Labor	Equipment	Total	Total Incl O&P
0010	**DISAPPEARING STAIRWAY** No trim included									
0100	Custom grade, pine, 8'-6" ceiling, minimum	1 Carp	4	2	Ea.	122	73.50		195.50	248
0150	Average		3.50	2.286		123	84		207	266
0200	Maximum		3	2.667		204	98		302	375
0500	Heavy duty, pivoted, from 7'-7" to 12'-10" floor to floor		3	2.667		400	98		498	590
0600	16'-0" ceiling		2	4		1,325	147		1,472	1,700
0800	Economy folding, pine, 8'-6" ceiling		4	2		111	73.50		184.50	236
0900	9'-6" ceiling		4	2		121	73.50		194.50	247
1000	Fire escape, galvanized steel, 8'-0" to 10'-4" ceiling	2 Carp	1	16		1,400	585		1,985	2,475
1010	10'-6" to 13'-6" ceiling		1	16		1,775	585		2,360	2,875
1100	Automatic electric, aluminum, floor to floor height, 8' to 9'		1	16		7,300	585		7,885	8,950

11 33 Retractable Stairs

11 33 10 – Disappearing Stairs

11 33 10.10 Disappearing Stairway		Crew	Daily Output	Labor-Hours	Unit	Material	2007 Bare Costs Labor	Equipment	Total	Total Incl O&P
1400	11' to 12'	2 Carp	.90	17.778	Ea.	7,850	650		8,500	9,650
1700	14' to 15'	↓	.70	22.857	↓	8,350	840		9,190	10,500

11 41 Food Storage Equipment

11 41 13 – Refrigerated Food Storage Cases

11 41 13.10 Refrigerated Food Cases

		Crew	Daily Output	Labor-Hours	Unit	Material	Labor	Equipment	Total	Total Incl O&P
0010	**REFRIGERATED FOOD CASES**									
0030	Dairy, multi-deck, 12' long	Q-5	3	5.333	Ea.	9,025	217		9,242	10,300
0100	For rear sliding doors, add					1,300			1,300	1,425
0200	Delicatessen case, service deli, 12' long, single deck	Q-5	3.90	4.103		6,075	167		6,242	6,925
0300	Multi-deck, 18 S.F. shelf display		3	5.333		7,425	217		7,642	8,500
0400	Freezer, self-contained, chest-type, 30 C.F.		3.90	4.103		4,425	167		4,592	5,125
0500	Glass door, upright, 78 C.F.		3.30	4.848		8,450	197		8,647	9,575
0600	Frozen food, chest type, 12' long		3.30	4.848		6,100	197		6,297	7,000
0700	Glass door, reach-in, 5 door		3	5.333		11,700	217		11,917	13,200
0800	Island case, 12' long, single deck		3.30	4.848		6,900	197		7,097	7,900
0900	Multi-deck		3	5.333		14,600	217		14,817	16,400
1000	Meat case, 12' long, single deck		3.30	4.848		5,025	197		5,222	5,825
1050	Multi-deck		3.10	5.161		8,625	210		8,835	9,825
1100	Produce, 12' long, single deck		3.30	4.848		6,650	197		6,847	7,600
1200	Multi-deck	↓	3.10	5.161	↓	7,250	210		7,460	8,300

11 41 13.20 Refrigerated Food Storage Equipment

		Crew	Daily Output	Labor-Hours	Unit	Material	Labor	Equipment	Total	Total Incl O&P
0010	**REFRIGERATED FOOD STORAGE EQUIPMENT**									
2350	Cooler, reach-in, beverage, 6' long	Q-1	6	2.667	Ea.	3,775	108		3,883	4,300
4300	Freezers, reach-in, 44 C.F.		4	4		3,700	161		3,861	4,325
4500	68 C.F.	↓	3	5.333		6,075	215		6,290	7,000
4600	Freezer, pre-fab, 8' x 8' w/refrigeration	2 Carp	.45	35.556		8,500	1,300		9,800	11,400
4620	8' x 12'		.35	45.714		10,400	1,675		12,075	14,000
4640	8' x 16'		.25	64		13,000	2,350		15,350	18,000
4660	8' x 20'	↓	.17	94.118		15,600	3,450		19,050	22,600
4680	Reach-in, 1 compartment	Q-1	4	4		1,325	161		1,486	1,700
4700	2 compartment		3	5.333		3,475	215		3,690	4,150
8300	Refrigerators, reach-in type, 44 C.F.		5	3.200		5,450	129		5,579	6,200
8310	With glass doors, 68 C.F.	↓	4	4		5,775	161		5,936	6,600
8320	Refrigerator, reach-in, 1 compartment	R-18	7.80	3.333		1,725	113		1,838	2,075
8330	2 compartment		6.20	4.194		2,325	142		2,467	2,775
8340	3 compartment	↓	5.60	4.643		3,500	157		3,657	4,100
8350	Pre-fab, with refrigeration, 8' x 8'	2 Carp	.45	35.556		5,625	1,300		6,925	8,200
8360	8' x 12'		.35	45.714		7,150	1,675		8,825	10,500
8370	8' x 16'		.25	64		9,175	2,350		11,525	13,800
8380	8' x 20'	↓	.17	94.118		11,200	3,450		14,650	17,800
8390	Pass-thru/roll-in, 1 compartment	R-18	7.80	3.333		3,450	113		3,563	3,950
8400	2 compartment		6.24	4.167		4,975	141		5,116	5,700
8410	3 compartment	↓	5.60	4.643		6,550	157		6,707	7,475
8420	Walk-in, alum, door & floor only, no refrig, 6' x 6' x 7'-6"	2 Carp	1.40	11.429		7,075	420		7,495	8,425
8430	10' x 6' x 7'-6"		.55	29.091		10,100	1,075		11,175	12,900
8440	12' x 14' x 7'-6"		.25	64		14,100	2,350		16,450	19,200
8450	12' x 20' x 7'-6"	↓	.17	94.118		17,300	3,450		20,750	24,400
8460	Refrigerated cabinets, mobile					2,975			2,975	3,250
8470	Refrigerator/freezer, reach-in, 1 compartment	R-18	5.60	4.643		3,725	157		3,882	4,350
8480	2 compartment	"	4.80	5.417	↓	5,250	183		5,433	6,050

11 41 Food Storage Equipment

11 41 13 – Refrigerated Food Storage Cases

11 41 13.30 Wine Cellar

		Crew	Daily Output	Labor-Hours	Unit	Material	2007 Bare Costs Labor	Equipment	Total	Total Incl O&P
0010	**WINE CELLAR**, refrigerated, Redwood interior, carpeted, walk-in type									
0020	6'-8" high, including racks									
0200	80"W x 48"D for 900 bottles	2 Carp	1.50	10.667	Ea.	3,125	390		3,515	4,025
0250	80"W x 72" D for 1300 bottles		1.33	12.030		4,125	440		4,565	5,250
0300	80"W x 94" D for 1900 bottles		1.17	13.675		5,350	500		5,850	6,675
0400	80"W x 124" D for 2500 bottles	↓	1	16	↓	6,475	585		7,060	8,050
0600	Portable cabinets, red oak, reach-in temp. & humidity controlled									
0650	26-5/8"W x 26-1/2"D x 68"H for 235 bottles				Ea.	2,825			2,825	3,100
0660	32"W x 21-1/2"D x 73-1/2"H for 144 bottles					3,275			3,275	3,600
0670	32"W x 29-1/2"D x 73-1/2"H for 288 bottles					3,300			3,300	3,625
0680	39-1/2"W x 29-1/2"D x 86-1/2"H for 440 bottles					3,750			3,750	4,125
0690	52-1/2"W x 29-1/2"D x 73-1/2"H for 468 bottles					4,300			4,300	4,725
0700	52-1/2"W x 29-1/2"D x 86-1/2"H for 572 bottles				↓	4,525			4,525	4,975
0730	Portable, red oak, can be built-in with glass door									
0750	23-7/8"W x 24"D x 34-1/2"H for 50 bottles				Ea.	1,025			1,025	1,125

11 41 33 – Food Storage Shelving

11 41 33.20 Food Storage Shelving

		Crew	Daily Output	Labor-Hours	Unit	Material	Labor	Equipment	Total	Total Incl O&P
0010	**FOOD STORAGE SHELVING**									
8600	Stainless steel shelving, louvered 4-tier, 20" x 3'	1 Clab	6	1.333	Ea.	1,025	38.50		1,063.50	1,175
8605	20" x 4'		6	1.333		1,450	38.50		1,488.50	1,650
8610	20" x 6'		6	1.333		2,100	38.50		2,138.50	2,375
8615	24" x 3'		6	1.333		1,325	38.50		1,363.50	1,500
8620	24" x 4'		6	1.333		1,550	38.50		1,588.50	1,750
8625	24" x 6'		6	1.333		2,250	38.50		2,288.50	2,525
8630	Flat 4-tier, 20" x 3'		6	1.333		840	38.50		878.50	985
8635	20" x 4'		6	1.333		1,050	38.50		1,088.50	1,200
8640	20" x 5'		6	1.333		1,175	38.50		1,213.50	1,325
8645	24" x 3'		6	1.333		915	38.50		953.50	1,050
8650	24" x 4'		6	1.333		610	38.50		648.50	735
8655	24" x 6'		6	1.333		1,925	38.50		1,963.50	2,150
8700	Galvanized shelving, louvered 4-tier, 20" x 3'		6	1.333		520	38.50		558.50	630
8705	20" x 4'		6	1.333		575	38.50		613.50	695
8710	20" x 6'		6	1.333		605	38.50		643.50	725
8715	24" x 3'		6	1.333		460	38.50		498.50	565
8720	24" x 4'		6	1.333		620	38.50		658.50	740
8725	24" x 6'		6	1.333		885	38.50		923.50	1,025
8730	Flat 4-tier, 20" x 3'		6	1.333		370	38.50		408.50	465
8735	20" x 4'		6	1.333		415	38.50		453.50	520
8740	20" x 6'		6	1.333		730	38.50		768.50	860
8745	24" x 3'		6	1.333		360	38.50		398.50	460
8750	24" x 4'		6	1.333		560	38.50		598.50	675
8755	24" x 6'		6	1.333		755	38.50		793.50	890
8760	Stainless steel dunnage rack, 24" x 3'		8	1		410	29		439	495
8765	24" x 4'		8	1		705	29		734	820
8770	Galvanized dunnage rack, 24" x 3'		8	1		141	29		170	200
8775	24" x 4'	↓	8	1	↓	133	29		162	192

11 42 Food Preparation Equipment

11 42 10 – Food Preparation Equipment

11 42 10.10 Food Preparation Equipment	Crew	Daily Output	Labor-Hours	Unit	Material	2007 Bare Costs Labor	2007 Bare Costs Equipment	Total	Total Incl O&P
0010 **FOOD PREPARATION EQUIPMENT**									
1700 Choppers, 5 pounds	R-18	7	3.714	Ea.	1,375	125		1,500	1,725
1720 16 pounds		5	5.200		1,700	176		1,876	2,150
1740 35 to 40 pounds	↓	4	6.500		2,525	219		2,744	3,100
1840 Coffee brewer, 5 burners	1 Plum	3	2.667		1,025	119		1,144	1,300
1850 Coffee urn, twin 6 gallon urns		2	4		2,650	179		2,829	3,175
1860 Single, 3 gallon	↓	3	2.667		1,675	119		1,794	2,025
3000 Fast food equipment, total package, minimum	6 Skwk	.08	600		145,000	22,800		167,800	195,000
3100 Maximum	"	.07	685		197,500	26,100		223,600	257,500
3800 Food mixers, 20 quarts	L-7	7	4		3,350	142		3,492	3,925
3850 40 quarts		5.40	5.185		7,250	184		7,434	8,250
3900 60 quarts		5	5.600		11,000	199		11,199	12,400
4040 80 quarts		3.90	7.179		11,500	255		11,755	13,000
4080 130 quarts		2.20	12.727		17,900	450		18,350	20,400
4100 Floor type, 20 quarts		15	1.867		2,650	66		2,716	3,000
4120 60 quarts		14	2		7,925	71		7,996	8,825
4140 80 quarts		12	2.333		9,825	82.50		9,907.50	10,900
4160 140 quarts	↓	8.60	3.256		23,100	115		23,215	25,600
6700 Peelers, small	R-18	8	3.250		1,200	110		1,310	1,500
6720 Large	"	6	4.333		3,800	146		3,946	4,425
6800 Pulper/extractor, close coupled, 5 HP	1 Plum	1.90	4.211		3,375	189		3,564	3,975
8580 Slicer with table	R-18	9	2.889	↓	4,275	97.50		4,372.50	4,850

11 43 Food Delivery Carts and Conveyors

11 43 13 – Food Delivery Carts

11 43 13.10 Food Delivery Carts	Crew	Daily Output	Labor-Hours	Unit	Material	2007 Bare Costs Labor	2007 Bare Costs Equipment	Total	Total Incl O&P
0010 **FOOD DELIVERY CARTS**									
1650 Cabinet, heated, 1 compartment, reach-in	R-18	5.60	4.643	Ea.	2,450	157		2,607	2,950
1655 Pass-thru roll-in		5.60	4.643		4,125	157		4,282	4,775
1660 2 compartment, reach-in	↓	4.80	5.417		6,150	183		6,333	7,050
1670 Mobile					2,350			2,350	2,575
6850 Mobile rack w/pan slide					1,050			1,050	1,150
9180 Tray and silver dispenser, mobile	1 Clab	16	.500	↓	680	14.40		694.40	775

11 44 Food Cooking Equipment

11 44 13 – Commercial Ranges

11 44 13.10 Cooking Equipment	Crew	Daily Output	Labor-Hours	Unit	Material	2007 Bare Costs Labor	2007 Bare Costs Equipment	Total	Total Incl O&P
0010 **COOKING EQUIPMENT**									
0020 Bake oven, gas, one section	Q-1	8	2	Ea.	4,425	80.50		4,505.50	5,000
0300 Two sections		7	2.286		9,100	92		9,192	10,100
0600 Three sections	↓	6	2.667		14,500	108		14,608	16,200
0900 Electric convection, single deck	L-7	4	7		4,450	248		4,698	5,275
1300 Broiler, without oven, standard	Q-1	8	2		2,950	80.50		3,030.50	3,375
1550 Infra-red	L-7	4	7		6,750	248		6,998	7,800
4750 Fryer, with twin baskets, modular model	Q-1	7	2.286		1,725	92		1,817	2,050
5000 Floor model, on 6" legs	"	5	3.200		2,175	129		2,304	2,600
5100 Extra single basket, large					62			62	68
5300 Griddle, SS, 24" plate, w/4" legs, elec, 208V, 3 phase, 3' long	Q-1	7	2.286		915	92		1,007	1,150
5550 4' long	"	6	2.667	↓	1,250	108		1,358	1,525

11 44 Food Cooking Equipment

11 44 13 – Commercial Ranges

11 44 13.10 Cooking Equipment		Crew	Daily Output	Labor-Hours	Unit	Material	2007 Bare Costs Labor	Equipment	Total	Total Incl O&P
6200	Iced tea brewer	1 Plum	3.44	2.326	Ea.	595	104		699	810
6350	Kettle, w/steam jacket, tilting, w/positive lock, SS, 20 gallons	L-7	7	4		5,350	142		5,492	6,125
6600	60 gallons	"	6	4.667		7,550	165		7,715	8,550
6900	Range, restaurant type, 6 burners and 1 standard oven, 36" wide	Q-1	7	2.286		1,875	92		1,967	2,225
6950	Convection		7	2.286		4,600	92		4,692	5,200
7150	2 standard ovens, 24" griddle, 60" wide		6	2.667		3,825	108		3,933	4,375
7200	1 standard, 1 convection oven		6	2.667		5,975	108		6,083	6,725
7450	Heavy duty, single 34" standard oven, open top		5	3.200		4,300	129		4,429	4,925
7500	Convection oven		5	3.200		5,900	129		6,029	6,700
7700	Griddle top		6	2.667		1,900	108		2,008	2,225
7750	Convection oven		6	2.667		6,750	108		6,858	7,575
8850	Steamer, electric 27 KW	L-7	7	4		9,425	142		9,567	10,600
9100	Electric, 10 KW or gas 100,000 BTU	"	5	5.600		4,500	199		4,699	5,250
9150	Toaster, conveyor type, 16-22 slices per minute					1,125			1,125	1,225
9160	Pop-up, 2 slot					590			590	650
9200	For deluxe models of above equipment, add					75%				
9400	Rule of thumb: Equipment cost based									
9410	on kitchen work area									
9420	Office buildings, minimum	L-7	77	.364	S.F.	64.50	12.90		77.40	91
9450	Maximum		58	.483		109	17.10		126.10	147
9550	Public eating facilities, minimum		77	.364		84.50	12.90		97.40	113
9600	Maximum		46	.609		137	21.50		158.50	185
9750	Hospitals, minimum		58	.483		87	17.10		104.10	122
9800	Maximum		39	.718		145	25.50		170.50	200

11 46 Food Dispensing Equipment

11 46 16 – Service Line Equipment

11 46 16.10 Food Dispensing Equipment

		Crew	Daily Output	Labor-Hours	Unit	Material	Labor	Equipment	Total	Total Incl O&P
0010	**FOOD DISPENSING EQUIPMENT**									
1050	Butter pat dispenser	1 Clab	13	.615	Ea.	810	17.70		827.70	920
1100	Bread dispenser, counter top		13	.615		770	17.70		787.70	875
1900	Cup and glass dispenser, drop in		4	2		1,100	57.50		1,157.50	1,300
1920	Disposable cup, drop in		16	.500		335	14.40		349.40	390
2650	Dish dispenser, drop in, 12"		11	.727		1,250	21		1,271	1,400
2660	Mobile		10	.800		2,100	23		2,123	2,325
3300	Food warmer, counter, 1.2 KW					545			545	600
3550	1.6 KW					1,600			1,600	1,775
3600	Well, hot food, built-in, rectangular, 12" x 20"	R-30	10	2.600		680	90.50		770.50	890
3610	Circular, 7 qt		10	2.600		276	90.50		366.50	445
3620	Refrigerated, 2 compartments		10	2.600		1,900	90.50		1,990.50	2,225
3630	3 compartments		9	2.889		1,850	100		1,950	2,200
3640	4 compartments		8	3.250		2,650	113		2,763	3,100
4720	Frost cold plate		9	2.889		7,375	100		7,475	8,250
5700	Hot chocolate dispenser	1 Plum	4	2		880	89.50		969.50	1,100
6250	Jet spray dispenser	R-18	4.50	5.778		2,750	195		2,945	3,325
6300	Juice dispenser, concentrate	"	4.50	5.778		1,275	195		1,470	1,725
6690	Milk dispenser, bulk, 2 flavor	R-30	8	3.250		1,250	113		1,363	1,550
6695	3 flavor	"	8	3.250		1,700	113		1,813	2,050
8800	Serving counter, straight	1 Carp	40	.200	L.F.	650	7.35		657.35	725
8820	Curved section	"	30	.267	"	815	9.80		824.80	910
8825	Solid surface, see section 12 36 61.16									

11 46 Food Dispensing Equipment

11 46 16 – Service Line Equipment

11 46 16.10 Food Dispensing Equipment		Crew	Daily Output	Labor-Hours	Unit	Material	2007 Bare Costs Labor	Equipment	Total	Total Incl O&P
8830	Soft serve ice cream machine, medium	R-18	11	2.364	Ea.	7,350	80		7,430	8,200
8840	Large	"	9	2.889	"	13,500	97.50		13,597.50	14,900

11 47 Ice Machines

11 47 10 – Ice Machines

11 47 10.10 Ice Machines

		Crew	Daily Output	Labor-Hours	Unit	Material	Labor	Equipment	Total	Total Incl O&P
0010	**ICE MACHINES**									
5800	Ice cube maker, 50 pounds per day	Q-1	6	2.667	Ea.	1,250	108		1,358	1,525
5900	250 pounds per day		1.20	13.333		2,000	540		2,540	3,000
6050	500 pounds per day		4	4		2,775	161		2,936	3,300
6060	With bin		1.20	13.333		2,350	540		2,890	3,400
6090	1000 pounds per day, with bin		1	16		3,950	645		4,595	5,325
6100	Ice flakers, 300 pounds per day		1.60	10		2,850	405		3,255	3,750
6120	600 pounds per day		.95	16.842		3,450	680		4,130	4,825
6130	1000 pounds per day		.75	21.333		5,150	860		6,010	6,950
6140	2000 pounds per day		.65	24.615		15,400	995		16,395	18,400
6160	Ice storage bin, 500 pound capacity	Q-5	1	16		1,925	650		2,575	3,075
6180	1000 pound	"	.56	28.571		2,600	1,150		3,750	4,625

11 48 Cleaning and Disposal Equipment

11 48 13 – Commercial Dishwashers

11 48 13.10 Dishwashers

		Crew	Daily Output	Labor-Hours	Unit	Material	Labor	Equipment	Total	Total Incl O&P
0010	**DISHWASHERS**									
2700	Dishwasher, commercial, rack type									
2720	10 to 12 racks per hour	Q-1	3.20	5	Ea.	3,625	202		3,827	4,275
2750	Semi-automatic 38 to 50 racks per hour	"	1.30	12.308		7,150	495		7,645	8,625
2800	Automatic, 190 to 230 racks per hour	L-6	.35	34.286		10,300	1,525		11,825	13,600
2820	235 to 275 racks per hour		.25	48		24,700	2,125		26,825	30,400
2840	8,750 to 12,500 dishes per hour		.10	120		44,900	5,350		50,250	57,500
2950	Dishwasher hood, canopy type	L-3A	10	1.200	L.F.	485	48.50		533.50	605
2960	Pant leg type	"	2.50	4.800	Ea.	6,775	194		6,969	7,775
5200	Garbage disposal 1.5 HP, 100 GPH	L-1	4.80	3.333		1,425	148		1,573	1,800
5210	3 HP, 120 GPH		4.60	3.478		2,250	154		2,404	2,700
5220	5 HP, 250 GPH		4.50	3.556		3,225	158		3,383	3,775
6750	Pot sink, 3 compartment	1 Plum	7.25	1.103	L.F.	690	49.50		739.50	830
6760	Pot washer, small		1.60	5	Ea.	17,100	224		17,324	19,100
6770	Large		1.20	6.667		38,400	299		38,699	42,800
9170	Trash compactor, small, up to 125 lb. compacted weight	L-4	4	6		18,200	207		18,407	20,300
9175	Large, up to 175 lb. compacted weight	"	3	8		22,200	276		22,476	24,800

11 52 Audio-Visual Equipment

11 52 13 – Projection Screens

11 52 13.10 Projection Screens

		Crew	Daily Output	Labor-Hours	Unit	Material	2007 Bare Costs Labor	Equipment	Total	Total Incl O&P
0010	**PROJECTION SCREENS** Wall or ceiling hung, matte white									
0100	Manually operated, economy	2 Carp	500	.032	S.F.	5.40	1.17		6.57	7.80
0300	Intermediate		450	.036		6.30	1.31		7.61	9
0400	Deluxe		400	.040	↓	8.70	1.47		10.17	11.90
0600	Electric operated, matte white, 25 S.F., economy		5	3.200	Ea.	790	117		907	1,050
0700	Deluxe		4	4		1,650	147		1,797	2,050
0900	50 S.F., economy		3	5.333		1,025	196		1,221	1,450
1000	Deluxe		2	8		1,850	294		2,144	2,475
1200	Heavy duty, electric operated, 200 S.F.		1.50	10.667		3,600	390		3,990	4,575
1300	400 S.F.	↓	1	16	↓	4,450	585		5,035	5,825
1500	Rigid acrylic in wall, for rear projection, 1/4" thick	2 Glaz	30	.533	S.F.	43.50	19.25		62.75	77
1600	1/2" thick (maximum size 10' x 20')	"	25	.640	"	77	23		100	120

11 52 16 – Projectors

11 52 16.10 Movie Equipment

		Crew	Daily Output	Labor-Hours	Unit	Material	2007 Bare Costs Labor	Equipment	Total	Total Incl O&P
0010	**MOVIE EQUIPMENT**									
0020	Changeover, minimum				Ea.	430			430	475
0100	Maximum					835			835	920
0400	Film transport, incl. platters and autowind, minimum					4,650			4,650	5,100
0500	Maximum					13,200			13,200	14,500
0800	Lamphouses, incl. rectifiers, xenon, 1,000 watt	1 Elec	2	4		6,150	176		6,326	7,000
0900	1,600 watt		2	4		6,575	176		6,751	7,475
1000	2,000 watt		1.50	5.333		7,050	234		7,284	8,100
1100	4,000 watt	↓	1.50	5.333		8,725	234		8,959	9,925
1400	Lenses, anamorphic, minimum					1,175			1,175	1,300
1500	Maximum					2,650			2,650	2,925
1800	Flat 35 mm, minimum					1,025			1,025	1,125
1900	Maximum					1,600			1,600	1,750
2200	Pedestals, for projectors					1,375			1,375	1,500
2300	Console type					9,900			9,900	10,900
2600	Projector mechanisms, incl. soundhead, 35 mm, minimum					10,200			10,200	11,300
2700	Maximum				↓	14,100			14,100	15,500
3000	Projection screens, rigid, in wall, acrylic, 1/4" thick	2 Glaz	195	.082	S.F.	38.50	2.96		41.46	47
3100	1/2" thick	"	130	.123	"	45	4.44		49.44	56
3300	Electric operated, heavy duty, 400 S.F.	2 Carp	1	16	Ea.	2,725	585		3,310	3,925
3320	Theatre projection screens, matte white, including frames	"	200	.080	S.F.	6.15	2.94		9.09	11.30
3400	Also see division 11 52 13.10									
3700	Sound systems, incl. amplifier, mono, minimum	1 Elec	.90	8.889	Ea.	3,050	390		3,440	3,925
3800	Dolby/Super Sound, maximum		.40	20		16,700	880		17,580	19,700
4100	Dual system, 2 channel, front surround, minimum		.70	11.429		4,275	500		4,775	5,450
4200	Dolby/Super Sound, 4 channel, maximum	↓	.40	20		15,300	880		16,180	18,100
4500	Sound heads, 35 mm					4,900			4,900	5,375
4900	Splicer, tape system, minimum					685			685	755
5000	Tape type, maximum					1,225			1,225	1,350
5300	Speakers, recessed behind screen, minimum	1 Elec	2	4		980	176		1,156	1,325
5400	Maximum	"	1	8		2,850	350		3,200	3,675
5700	Seating, painted steel, upholstered, minimum	2 Carp	35	.457		122	16.80		138.80	160
5800	Maximum	"	28	.571		390	21		411	465
6100	Rewind tables, minimum					2,450			2,450	2,700
6200	Maximum				↓	4,350			4,350	4,800
7000	For automation, varying sophistication, minimum	1 Elec	1	8	System	2,200	350		2,550	2,950
7100	Maximum	2 Elec	.30	53.333	"	5,150	2,350		7,500	9,125

11 53 Laboratory Equipment

11 53 03 – Laboratory Test Equipment

11 53 03.13 Test Equipment

		Crew	Daily Output	Labor-Hours	Unit	Material	2007 Bare Costs Labor	Equipment	Total	Total Incl O&P
0010	**TEST EQUIPMENT**									
1700	Thermometer, electric, portable				Ea.	345			345	380
1800	Titration unit, four 2000 ml reservoirs				"	9,950			9,950	10,900

11 53 13 – Laboratory Fume Hoods

11 53 13.13 Recirculating Laboratory Fume Hoods

		Crew	Daily Output	Labor-Hours	Unit	Material	Labor	Equipment	Total	Total Incl O&P
0010	**RECIRCULATING LABORATORY FUME HOODS**									
0600	Fume hood, with countertop & base, not including HVAC									
0610	Simple, minimum	2 Carp	5.40	2.963	L.F.	875	109		984	1,125
0620	Complex, including fixtures		2.40	6.667		1,450	245		1,695	1,975
0630	Special, maximum	↓	1.70	9.412	↓	1,775	345		2,120	2,500
0670	Service fixtures, average				Ea.	102			102	112
0680	For sink assembly with hot and cold water, add	1 Plum	1.40	5.714		685	256		941	1,150
0750	Glove box, fiberglass, bacteriological					14,700			14,700	16,200
0760	Controlled atmosphere					15,200			15,200	16,800
0770	Radioisotope					14,700			14,700	16,200
0780	Carcinogenic				↓	14,700			14,700	16,200

11 53 13.23 Exhaust Hoods

		Crew	Daily Output	Labor-Hours	Unit	Material	Labor	Equipment	Total	Total Incl O&P
0010	**EXHAUST HOODS**									
0650	Ductwork, minimum	2 Shee	1	16	Hood	2,750	695		3,445	4,100
0660	Maximum	"	.50	32	"	4,875	1,400		6,275	7,525

11 53 16 – Laboratory Incubators

11 53 16.13 Incubators

		Crew	Daily Output	Labor-Hours	Unit	Material	Labor	Equipment	Total	Total Incl O&P
0010	**INCUBATORS**									
1000	Incubators, minimum				Ea.	3,275			3,275	3,600
1010	Maximum				"	18,600			18,600	20,500

11 53 19 – Laboratory Sterilizers

11 53 19.13 Sterilizers

		Crew	Daily Output	Labor-Hours	Unit	Material	Labor	Equipment	Total	Total Incl O&P
0010	**STERILIZERS**									
0700	Glassware washer, undercounter, minimum	L-1	1.80	8.889	Ea.	6,100	395		6,495	7,300
0710	Maximum	"	1	16		8,950	710		9,660	10,900
1850	Utensil washer-sanitizer	1 Plum	2	4	↓	10,100	179		10,279	11,400

11 53 23 – Laboratory Refrigerators

11 53 23.13 Refrigerators

		Crew	Daily Output	Labor-Hours	Unit	Material	Labor	Equipment	Total	Total Incl O&P
0010	**REFRIGERATORS**									
1200	Refrigerator, blood bank, 28.6 C.F. emergency signal				Ea.	6,450			6,450	7,100
1210	Reach-in, 16.9 C.F.				"	6,600			6,600	7,250

11 53 33 – Emergency Safety Appliances

11 53 33.13 Emergency Equipment

		Crew	Daily Output	Labor-Hours	Unit	Material	Labor	Equipment	Total	Total Incl O&P
0010	**EMERGENCY EQUIPMENT**									
1400	Safety equipment, eye wash, hand held				Ea.	405			405	445
1450	Deluge shower				"	640			640	705

11 53 43 – Service Fittings and Accessories

11 53 43.13 Fittings

		Crew	Daily Output	Labor-Hours	Unit	Material	Labor	Equipment	Total	Total Incl O&P
0010	**FITTINGS**									
1600	Sink, one piece plastic, flask wash, hose, free standing	1 Plum	1.60	5	Ea.	1,725	224		1,949	2,225
1610	Epoxy resin sink, 25" x 16" x 10"	"	2	4	"	192	179		371	480
1950	Utility table, acid resistant top with drawers	2 Carp	30	.533	L.F.	142	19.55		161.55	187
8000	Alternate pricing method: as percent of lab furniture									
8050	Installation, not incl. plumbing & duct work				% Furn.				20%	22%

11 53 Laboratory Equipment

11 53 43 – Service Fittings and Accessories

11 53 43.13 Fittings	Crew	Daily Output	Labor-Hours	Unit	Material	2007 Bare Costs Labor	Equipment	Total	Total Incl O&P
8100 Plumbing, final connections, simple system				% Furn.				9.09%	10%
8110 Moderately complex system								13.64%	15%
8120 Complex system								18.18%	20%
8150 Electrical, simple system								9.09%	10%
8160 Moderately complex system								18.18%	20%
8170 Complex system								31.80%	35%

11 57 Vocational Shop Equipment

11 57 10 – Shop Equipment

11 57 10.10 Vocational Shop Equipment

		Crew	Daily Output	Labor-Hours	Unit	Material	Labor	Equipment	Total	Total Incl O&P
0010	VOCATIONAL SHOP EQUIPMENT									
0020	Benches, work, wood, average	2 Carp	5	3.200	Ea.	475	117		592	705
0100	Metal, average		5	3.200		405	117		522	630
0400	Combination belt & disc sander, 6"		4	4		810	147		957	1,125
0700	Drill press, floor mounted, 12", 1/2 H.P.		4	4		375	147		522	640
0800	Dust collector, not incl. ductwork, 6" diameter	1 Shee	1.10	7.273		2,850	315		3,165	3,650
1000	Grinders, double wheel, 1/2 H.P.	2 Carp	5	3.200		201	117		318	405
1300	Jointer, 4", 3/4 H.P.		4	4		985	147		1,132	1,300
1600	Kilns, 16 C.F., to 2000°		4	4		2,400	147		2,547	2,875
1900	Lathe, woodworking, 10", 1/2 H.P.		4	4		685	147		832	980
2200	Planer, 13" x 6"		4	4		1,200	147		1,347	1,550
2500	Potter's wheel, motorized		4	4		1,000	147		1,147	1,325
2800	Saws, band, 14", 3/4 H.P.		4	4		780	147		927	1,100
3100	Metal cutting band saw, 14"		4	4		2,050	147		2,197	2,500
3400	Radial arm saw, 10", 2 H.P.		4	4		745	147		892	1,050
3700	Scroll saw, 24"		4	4		1,525	147		1,672	1,900
4000	Table saw, 10", 3 H.P.		4	4		1,775	147		1,922	2,175
4300	Welder AC arc, 30 amp capacity		4	4		2,225	147		2,372	2,675

11 61 Theater and Stage Equipment

11 61 23 – Folding and Portable Stages

11 61 23.10 Folding and Portable Stages

		Crew	Daily Output	Labor-Hours	Unit	Material	Labor	Equipment	Total	Total Incl O&P
0010	FOLDING AND PORTABLE STAGES									
1500	Flooring, portable oak parquet, 3' x 3' sections				S.F.	12.15			12.15	13.35
1600	Cart to carry 225 S.F. of flooring				Ea.	365			365	400
5000	Stages, portable with steps, folding legs, stock, 8" high				SF Stg.	24			24	26
5100	16" high					22			22	24
5200	32" high					32.50			32.50	36
5300	40" high					54			54	59
6000	Telescoping platforms, extruded alum., straight, minimum	4 Carp	157	.204		26	7.50		33.50	40
6100	Maximum		77	.416		35	15.25		50.25	62.50
6500	Pie-shaped, minimum		150	.213		56	7.85		63.85	73.50
6600	Maximum		70	.457		62.50	16.80		79.30	94.50
6800	For 3/4" plywood covered deck, deduct					3.53			3.53	3.88
7000	Band risers, steel frame, plywood deck, minimum	4 Carp	275	.116		27.50	4.27		31.77	37
7100	Maximum	"	138	.232		58	8.50		66.50	77.50
7500	Chairs for above, self-storing, minimum	2 Carp	43	.372	Ea.	91	13.65		104.65	122
7600	Maximum	"	40	.400	"	158	14.70		172.70	196

11 61 Theater and Stage Equipment

11 61 33 – Rigging Systems and Controls

11 61 33.10 Controls

		Crew	Daily Output	Labor-Hours	Unit	Material	2007 Bare Costs Labor	Equipment	Total	Total Incl O&P
0010	**CONTROLS**									
0050	Control boards with dimmers and breakers, minimum	1 Elec	1	8	Ea.	12,100	350		12,450	13,800
0100	Average		.50	16		33,600	700		34,300	38,100
0150	Maximum	↓	.20	40	↓	111,000	1,750		112,750	125,000
8000	Rule of thumb: total stage equipment, minimum	4 Carp	100	.320	SF Stg.	88	11.75		99.75	115
8100	Maximum	"	25	1.280	"	495	47		542	620

11 61 43 – Stage Curtains

11 61 43.10 Curtains

		Crew	Daily Output	Labor-Hours	Unit	Material	Labor	Equipment	Total	Total Incl O&P
0010	**CURTAINS**									
0500	Curtain track, straight, light duty	2 Carp	20	.800	L.F.	22.50	29.50		52	70
0600	Heavy duty		18	.889		48	32.50		80.50	104
0700	Curved sections		12	1.333	↓	148	49		197	238
1000	Curtains, velour, medium weight		600	.027	S.F.	7.15	.98		8.13	9.40
1150	Silica based yarn, inherently fire retardant	↓	50	.320	"	13.80	11.75		25.55	33.50

11 62 Musical Equipment

11 62 16 – Carillons

11 62 16.10 Carillons

		Crew	Daily Output	Labor-Hours	Unit	Material	Labor	Equipment	Total	Total Incl O&P
0010	**CARILLONS**									
0300	Carillon, 4 octave (48 bells), with keyboard				System	694,500			694,500	764,000
0320	2 octave (24 bells)					289,500			289,500	318,500
0340	3 to 4 bell peal, minimum					63,500			63,500	70,000
0360	Maximum				↓	463,000			463,000	509,500
0380	Cast bronze bell, average				Ea.	81,000			81,000	89,000
0400	Electronic, digital, minimum					14,500			14,500	15,900
0410	With keyboard, maximum				↓	69,500			69,500	76,500

11 66 Athletic Equipment

11 66 13 – Exercise Equipment

11 66 13.10 Exercise Equipment

		Crew	Daily Output	Labor-Hours	Unit	Material	Labor	Equipment	Total	Total Incl O&P
0010	**EXERCISE EQUIPMENT**									
0020	Abdominal rack, 2 board capacity				Ea.	455			455	500
0050	Abdominal board, upholstered					510			510	560
0200	Bicycle trainer, minimum					790			790	870
0300	Deluxe, electric					4,025			4,025	4,425
0400	Bar bell set, chrome plated steel, 25 lbs.					237			237	261
0420	100 lbs.					360			360	395
0450	200 lbs.				↓	695			695	765
0500	Weight plates, cast iron, per lb.				Lb.	4.86			4.86	5.35
0520	Storage rack, 10 station				Ea.	765			765	845
0600	Circuit training apparatus, 12 machines minimum	2 Clab	1.25	12.800	Set	26,800	370		27,170	30,000
0700	Average		1	16		32,900	460		33,360	36,800
0800	Maximum	↓	.75	21.333		38,900	615		39,515	43,800
0820	Dumbbell set, cast iron, with rack and 5 pair				↓	570			570	630
0900	Squat racks	2 Clab	5	3.200	Ea.	740	92		832	960
1200	Multi-station gym machine, 5 station					7,725			7,725	8,500
1250	9 station					10,800			10,800	11,900
1280	Rowing machine, hydraulic					1,425			1,425	1,575

11 66 Athletic Equipment

11 66 13 – Exercise Equipment

11 66 13.10 Exercise Equipment

		Crew	Daily Output	Labor-Hours	Unit	Material	2007 Bare Costs Labor	2007 Bare Costs Equipment	Total	Total Incl O&P
1300	Treadmill, manual				Ea.	1,400			1,400	1,550
1320	Motorized					3,500			3,500	3,850
1340	Electronic					5,300			5,300	5,825
1360	Cardio-testing					7,550			7,550	8,300
1400	Treatment/massage tables, minimum					535			535	590
1420	Deluxe, with accessories					665			665	735
4150	Exercise equipment, bicycle trainer					700			700	770
4180	Chinning bar, adjustable, wall mounted	1 Carp	5	1.600		295	58.50		353.50	415
4200	Exercise ladder, 16' x 1'-7", suspended	L-2	3	5.333		1,275	171		1,446	1,675
4210	High bar, floor plate attached	1 Carp	4	2		1,100	73.50		1,173.50	1,350
4240	Parallel bars, adjustable		4	2		2,975	73.50		3,048.50	3,375
4270	Uneven parallel bars, adjustable	↓	4	2		2,875	73.50		2,948.50	3,300
4280	Wall mounted, adjustable	L-2	1.50	10.667	Set	835	340		1,175	1,450
4300	Rope, ceiling mounted, 18' long	1 Carp	3.66	2.186	Ea.	220	80		300	365
4330	Side horse, vaulting		5	1.600		1,700	58.50		1,758.50	1,975
4360	Treadmill, motorized, deluxe, training type	↓	5	1.600		3,500	58.50		3,558.50	3,950
4390	Weight lifting multi-station, minimum	2 Clab	1	16		395	460		855	1,150
4450	Maximum	"	.50	32	↓	14,700	920		15,620	17,600

11 66 23 – Gymnasium Equipment

11 66 23.13 Basketball Equipment

		Crew	Daily Output	Labor-Hours	Unit	Material	Labor	Equipment	Total	Total Incl O&P
0010	**BASKETBALL EQUIPMENT**									
1000	Basketball backstops, wall mtd., 6' extended, fixed, minimum	L-2	1	16	Ea.	1,100	510		1,610	2,025
1100	Maximum		1	16		1,625	510		2,135	2,600
1200	Swing up, minimum		1	16		1,325	510		1,835	2,250
1250	Maximum		1	16		5,200	510		5,710	6,525
1300	Portable, manual, heavy duty, spring operated		1.90	8.421		11,000	270		11,270	12,600
1400	Ceiling suspended, stationary, minimum		.78	20.513		1,975	655		2,630	3,200
1450	Fold up, with accessories, maximum	↓	.40	40		5,925	1,275		7,200	8,525
1600	For electrically operated, add	1 Elec	1	8	↓	1,950	350		2,300	2,650
5800	Wall pads, 1-1/2" thick	2 Carp	640	.025	S.F.	6.80	.92		7.72	8.95

11 66 23.19 Boxing Ring

		Crew	Daily Output	Labor-Hours	Unit	Material	Labor	Equipment	Total	Total Incl O&P
0010	**BOXING RING**									
4100	Boxing ring, elevated, 22' x 22'	L-4	.10	240	Ea.	6,825	8,275		15,100	20,400
4110	For cellular plastic foam padding, add		.10	240		2,650	8,275		10,925	15,800
4120	Floor level, including posts and ropes only, 20' x 20'		.80	30		2,100	1,025		3,125	3,900
4130	Canvas, 30' x 30'	↓	5	4.800	↓	1,050	165		1,215	1,425

11 66 23.47 Gym Mats

		Crew	Daily Output	Labor-Hours	Unit	Material	Labor	Equipment	Total	Total Incl O&P
0010	**GYM MATS**									
5500	Gym mats, 2" thick, naugahyde covered				S.F.	3.59			3.59	3.95
5600	Vinyl/nylon covered					6.20			6.20	6.85
6000	Wrestling mats, 1" thick, heavy duty				↓	5.50			5.50	6.05

11 66 43 – Interior Scoreboards

11 66 43.10 Scoreboards

		Crew	Daily Output	Labor-Hours	Unit	Material	Labor	Equipment	Total	Total Incl O&P
0010	**SCOREBOARDS**									
7000	Scoreboards, baseball, minimum	R-3	1.30	15.385	Ea.	3,050	665	126	3,841	4,475
7200	Maximum		.05	400		14,800	17,300	3,275	35,375	45,700
7300	Football, minimum		.86	23.256		3,775	1,000	190	4,965	5,850
7400	Maximum		.20	100		12,000	4,325	815	17,140	20,600
7500	Basketball (one side), minimum		2.07	9.662		2,125	420	79	2,624	3,025
7600	Maximum		.30	66.667		5,275	2,875	545	8,695	10,700
7700	Hockey-basketball (four sides), minimum	↓	.25	80	↓	5,300	3,450	655	9,405	11,700

11 66 Athletic Equipment

11 66 43 – Interior Scoreboards

11 66 43.10 Scoreboards		Crew	Daily Output	Labor-Hours	Unit	Material	2007 Bare Costs Labor	Equipment	Total	Total Incl O&P
7800	Maximum	R-3	.15	133	Ea.	5,375	5,775	1,100	12,250	15,700

11 66 53 – Gymnasium Dividers

11 66 53.10 Divider Curtains

		Crew	Daily Output	Labor-Hours	Unit	Material	Labor	Equipment	Total	Total Incl O&P
0010	**DIVIDER CURTAINS**									
4500	Gym divider curtain, mesh top, vinyl bottom, manual	L-4	500	.048	S.F.	8.05	1.65		9.70	11.40
4700	Electric roll up	L-7	400	.070	"	11.75	2.48		14.23	16.80

11 67 Recreational Equipment

11 67 13 – Bowling Alley Equipment

11 67 13.10 Bowling Alleys

		Crew	Daily Output	Labor-Hours	Unit	Material	Labor	Equipment	Total	Total Incl O&P
0010	**BOWLING ALLEYS** Including alley, pinsetter, scorer,									
0020	counters and misc. supplies, minimum	4 Carp	.20	160	Lane	35,100	5,875		40,975	47,800
0150	Average		.19	168		40,000	6,175		46,175	53,500
0300	Maximum		.18	177		46,000	6,525		52,525	60,500
0400	Combo table ball rack, add					995			995	1,100
0600	For automatic scorer, add, minimum					6,350			6,350	7,000
0700	Maximum					8,800			8,800	9,675

11 67 23 – Shooting Range Equipment

11 67 23.10 Shooting Range

		Crew	Daily Output	Labor-Hours	Unit	Material	Labor	Equipment	Total	Total Incl O&P
0010	**SHOOTING RANGE** Incl. bullet traps, target provisions, controls,									
0100	separators, ceiling system, etc. Not incl. structural shell									
0200	Commercial	L-9	.64	56.250	Point	17,000	1,875		18,875	21,700
0300	Law enforcement		.28	128		22,000	4,275		26,275	31,100
0400	National Guard armories		.71	50.704		14,000	1,700		15,700	18,100
0500	Reserve training centers		.71	50.704		7,000	1,700		8,700	10,400
0600	Schools and colleges		.32	112		24,000	3,750		27,750	32,500
0700	Major acadamies		.19	189		34,000	6,325		40,325	47,600
0800	For acoustical treatment, add					10%	10%			
0900	For lighting, add					28%	25%			
1000	For plumbing, add					5%	5%			
1100	For ventilating system, add, minimum					40%	40%			
1200	Add, average					25%	25%			
1300	Add, maximum					35%	35%			

11 68 Play Field Equipment and Structures

11 68 13 – Playground Equipment

11 68 13.10 Playground Equipment

		Crew	Daily Output	Labor-Hours	Unit	Material	Labor	Equipment	Total	Total Incl O&P
0010	**PLAYGROUND EQUIPMENT** See also individual items									
0200	Bike rack, 10' long, permanent	B-1	12	2	Ea.	590	59		649	735
0400	Horizontal monkey ladder, 14' long, 6' high		4	6		900	177		1,077	1,275
0590	Parallel bars, 10' long		4	6		300	177		477	605
0600	Posts, tether ball set, 2-3/8" O.D.		12	2		250	59		309	365
0800	Poles, multiple purpose, 10'-6" long		12	2	Pr.	150	59		209	257
1000	Ground socket for movable posts, 2-3/8" post		10	2.400		107	70.50		177.50	228
1100	3-1/2" post		10	2.400		150	70.50		220.50	275
1300	See-saw, spring, steel, 2 units		6	4	Ea.	900	118		1,018	1,175
1400	4 units		4	6		1,200	177		1,377	1,600
1500	6 units		3	8		1,500	235		1,735	2,025

11 68 Play Field Equipment and Structures

11 68 13 - Playground Equipment

11 68 13.10 Playground Equipment

		Crew	Daily Output	Labor-Hours	Unit	Material	2007 Bare Costs Labor	Equipment	Total	Total Incl O&P
1700	Shelter, fiberglass golf tee, 3 person	B-1	4.60	5.217	Ea.	3,175	154		3,329	3,725
1900	Slides, stainless steel bed, 12' long, 6' high		3	8		4,200	235		4,435	5,000
2000	20' long, 10' high		2	12		2,400	355		2,755	3,200
2200	Swings, plain seats, 8' high, 4 seats		2	12		1,200	355		1,555	1,875
2300	8 seats		1.30	18.462		2,000	545		2,545	3,050
2500	12' high, 4 seats		2	12		1,700	355		2,055	2,425
2600	8 seats		1.30	18.462		2,275	545		2,820	3,350
2800	Whirlers, 8' diameter		3	8		3,000	235		3,235	3,675
2900	10' diameter		3	8		3,500	235		3,735	4,225

11 68 13.20 Modular Playground

		Crew	Daily Output	Labor-Hours	Unit	Material	Labor	Equipment	Total	Total Incl O&P
0010	**MODULAR PLAYGROUND** Basic components									
0100	Deck, square, steel, 48" x 48"	B-1	1	24	Ea.	850	705		1,555	2,025
0110	Recycled polyurethane		1	24		850	705		1,555	2,025
0120	Triangular, steel, 48" side		1	24		600	705		1,305	1,750
0130	Post, steel, 5" square		18	1.333	L.F.	25	39		64	88.50
0140	Aluminum, 2-3/8" square		20	1.200		21	35.50		56.50	78
0150	5" square		18	1.333		30	39		69	94
0160	Roof, square poly, 54" side		18	1.333	Ea.	1,200	39		1,239	1,375
0170	Wheelchair transfer module, for 3' high deck		3	8	"	2,600	235		2,835	3,225
0180	Guardrail, pipe, 36" high		60	.400	L.F.	120	11.75		131.75	150
0190	Steps, deck-to-deck, 3 - 8" steps		8	3	Ea.	1,200	88.50		1,288.50	1,450
0200	Activity panel, crawl through panel		2	12		400	355		755	990
0210	Alphabet/spelling panel		2	12		600	355		955	1,200
0360	With guardrails		3	8		2,200	235		2,435	2,800
0370	Crawl tunnel, straight, 56" long		4	6		1,200	177		1,377	1,600
0380	90°, 4' long		4	6		1,200	177		1,377	1,600
1200	Slide, tunnel, for 56" high deck		8	3		1,800	88.50		1,888.50	2,100
1210	Straight, poly		8	3		1,800	88.50		1,888.50	2,100
1220	Stainless steel, 54" high deck		6	4		250	118		368	460
1230	Curved, poly, 40" high deck		6	4		550	118		668	790
1240	Spiral slide, 56" - 72" high		5	4.800		2,700	141		2,841	3,200
1300	Ladder, vertical, for 24" - 72" high deck		5	4.800		600	141		741	880
1310	Horizontal, 8' long		5	4.800		700	141		841	990
1320	Corkscrew climber, 6' high		3	8		500	235		735	915
1330	Fire pole for 72" high deck		6	4		550	118		668	790
1340	Bridge, ring climber, 8' long		4	6		1,100	177		1,277	1,475
1350	Suspension		4	6	L.F.	500	177		677	825

11 68 16 - Play Structures

11 68 16.30 Platform/Paddle Tennis Court

		Crew	Daily Output	Labor-Hours	Unit	Material	Labor	Equipment	Total	Total Incl O&P
0010	**PLATFORM/PADDLE TENNIS COURT** Complete with lighting, etc.									
0100	Aluminum slat deck with aluminum frame	B-1	.08	300	Court	48,800	8,825		57,625	67,000
0500	Aluminum slat deck and wood frame	C-1	.12	266		48,800	9,250		58,050	68,000
0800	Aluminum deck heater, add	B-1	1.18	20.339		4,875	600		5,475	6,300
0900	Douglas fir planking and wood frame 2" x 6" x 30'	C-1	.12	266		46,200	9,250		55,450	65,500
1000	Plywood deck with steel frame		.12	266		46,200	9,250		55,450	65,500
1100	Steel slat deck with wood frame		.12	266		32,500	9,250		41,750	50,000

11 68 33 - Athletic Field Equipment

11 68 33.13 Football Field Equipment

		Crew	Daily Output	Labor-Hours	Unit	Material	Labor	Equipment	Total	Total Incl O&P
0010	**FOOTBALL FIELD EQUIPMENT**									
0020	Goal posts, steel, football, double post	B-1	1.50	16	Pr.	1,800	470		2,270	2,700
0100	Deluxe, single post		1.50	16		2,900	470		3,370	3,925

11 68 Play Field Equipment and Structures

11 68 33 – Athletic Field Equipment

11 68 33.13 Football Field Equipment	Crew	Daily Output	Labor-Hours	Unit	Material	2007 Bare Costs Labor	2007 Bare Costs Equipment	Total	Total Incl O&P
0300 Football, convertible to soccer	B-1	1.50	16	Pr.	2,900	470		3,370	3,900
0500 Soccer, regulation	↓	2	12	↓	2,275	355		2,630	3,075

11 71 Medical Sterilizing Equipment

11 71 10 – Medical Sterilizing Equipment

11 71 10.10 Medical Sterilizing Equipment

		Crew	Daily Output	Labor-Hours	Unit	Material	Labor	Equipment	Total	Total Incl O&P
0010	MEDICAL STERILIZING EQUIPMENT									
0700	Distiller, water, steam heated, 50 gal. capacity	1 Plum	1.40	5.714	Ea.	17,100	256		17,356	19,200
5600	Sterilizers, floor loading, 26" x 62" x 42", single door, steam					146,500			146,500	161,500
5650	Double door, steam					188,500			188,500	207,500
5800	General purpose, 20" x 20" x 38", single door					14,100			14,100	15,500
6000	Portable, counter top, steam, minimum					3,525			3,525	3,875
6020	Maximum					5,500			5,500	6,050
6050	Portable, counter top, gas, 17" x 15" x 32-1/2"					36,300			36,300	40,000
6150	Manual washer/sterilizer, 16" x 16" x 26"	1 Plum	2	4	↓	49,800	179		49,979	55,500
6200	Steam generators, electric 10 KW to 180 KW, freestanding									
6250	Minimum	1 Elec	3	2.667	Ea.	7,400	117		7,517	8,325
6300	Maximum	"	.70	11.429	↓	26,600	500		27,100	30,000
8200	Bed pan washer-sanitizer	1 Plum	2	4	↓	6,600	179		6,779	7,525

11 72 Examination and Treatment Equipment

11 72 13 – Examination Equipment

11 72 13.13 Examination Equipment

		Crew	Daily Output	Labor-Hours	Unit	Material	Labor	Equipment	Total	Total Incl O&P
0010	EXAMINATION EQUIPMENT									
0300	Blood pressure unit, mercurial, wall				Ea.	118			118	130
0400	Diagnostic set, wall					600			600	660
4400	Scale, physician's, with height rod				↓	288			288	315

11 72 53 – Treatment Equipment

11 72 53.13 Treatment Equipment

		Crew	Daily Output	Labor-Hours	Unit	Material	Labor	Equipment	Total	Total Incl O&P
0010	TREATMENT EQUIPMENT									
6500	Surgery table, minor minimum	1 Sswk	.70	11.429	Ea.	12,800	475		13,275	15,000
6520	Maximum	"	.50	16		24,100	660		24,760	27,700
6700	Surgical lights, doctor's office, single arm	2 Elec	2	8		1,200	350		1,550	1,825
6750	Dual arm	"	1	16	↓	5,375	700		6,075	6,950

11 73 Patient Care Equipment

11 73 10 – Patient Care Equipment

11 73 10.10 Patient Care Equipment

		Crew	Daily Output	Labor-Hours	Unit	Material	Labor	Equipment	Total	Total Incl O&P
0010	PATIENT CARE EQUIPMENT									
0750	Exam room furnishings, average per room				Ea.	5,975			5,975	6,575
1800	Heat therapy unit, humidified, 26" x 78" x 28"				"	3,150			3,150	3,450
2100	Hubbard tank with accessories, stainless steel,									
2110	125 GPM at 45 psi water pressure				Ea.	24,300			24,300	26,800
2150	For electric overhead hoist, add					2,650			2,650	2,925
2900	K-Module for heat therapy, 20 oz. capacity, 75° to 110°F					375			375	415
3600	Paraffin bath, 126°F, auto controlled					1,350			1,350	1,500
3900	Parallel bars for walking training, 12'-0"				↓	1,125			1,125	1,225

11 73 Patient Care Equipment

11 73 10 – Patient Care Equipment

11 73 10.10 Patient Care Equipment

		Crew	Daily Output	Labor-Hours	Unit	Material	2007 Bare Costs Labor	2007 Bare Costs Equipment	Total	Total Incl O&P
4600	Station, dietary, medium, with ice				Ea.	14,800			14,800	16,300
4700	Medicine					6,700			6,700	7,375
7000	Tables, physical therapy, walk off, electric	2 Carp	3	5.333		2,950	196		3,146	3,550
7150	Standard, vinyl top with base cabinets, minimum		3	5.333		1,200	196		1,396	1,625
7200	Maximum	↓	2	8		3,575	294		3,869	4,375
8400	Whirlpool bath, mobile, sst, 18" x 24" x 60"					4,150			4,150	4,575
8450	Fixed, incl. mixing valves	1 Plum	2	4	↓	8,500	179		8,679	9,650

11 74 Dental Equipment

11 74 10 – Dental Equipment

11 74 10.10 Dental Equipment

		Crew	Daily Output	Labor-Hours	Unit	Material	Labor	Equipment	Total	Total Incl O&P
0010	**DENTAL EQUIPMENT**									
0020	Central suction system, minimum	1 Plum	1.20	6.667	Ea.	1,025	299		1,324	1,600
0100	Maximum	"	.90	8.889		4,575	400		4,975	5,625
0300	Air compressor, minimum	1 Skwk	.80	10		2,100	380		2,480	2,900
0400	Maximum		.50	16		6,650	610		7,260	8,275
0600	Chair, electric or hydraulic, minimum		.50	16		3,700	610		4,310	5,025
0700	Maximum	↓	.25	32		13,300	1,225		14,525	16,600
0800	Doctor's/assistant's stool, minimum					200			200	220
0850	Maximum					750			750	825
1000	Drill console with accessories, minimum	1 Skwk	1.60	5		1,450	190		1,640	1,900
1100	Maximum		1.60	5		4,200	190		4,390	4,925
2000	Light, ceiling mounted, minimum		8	1		800	38		838	940
2100	Maximum	↓	8	1		2,200	38		2,238	2,475
2200	Unit light, minimum	2 Skwk	5.33	3.002		560	114		674	795
2210	Maximum		5.33	3.002		1,050	114		1,164	1,350
2220	Track light, minimum		3.20	5		2,000	190		2,190	2,500
2230	Maximum	↓	3.20	5		3,425	190		3,615	4,075
2300	Sterilizers, steam portable, minimum					1,050			1,050	1,150
2350	Maximum					9,600			9,600	10,600
2600	Steam, institutional					4,900			4,900	5,400
2650	Dry heat, electric, portable, 3 trays					1,125			1,125	1,225
2700	Ultra-sonic cleaner, portable, minimum					299			299	330
2750	Maximum (institutional)					1,400			1,400	1,550
3000	X-ray unit, wall, minimum	1 Skwk	4	2		1,900	76		1,976	2,225
3010	Maximum		4	2		4,750	76		4,826	5,350
3100	Panoramic unit	↓	.60	13.333		13,600	505		14,105	15,700
3105	Deluxe, minimum	2 Skwk	1.60	10		16,000	380		16,380	18,200
3110	Maximum	"	1.60	10		50,000	380		50,380	55,500
3500	Developers, X-ray, average	1 Plum	5.33	1.501		4,575	67		4,642	5,125
3600	Maximum	"	5.33	1.501	↓	7,475	67		7,542	8,325

11 76 Operating Room Equipment

11 76 10 - Operating Room Equipment

11 76 10.10 Operating Room Equipment

		Crew	Daily Output	Labor-Hours	Unit	Material	2007 Bare Costs Labor	2007 Bare Costs Equipment	Total	Total Incl O&P
0010	**OPERATING ROOM EQUIPMENT**									
5000	Scrub, surgical, stainless steel, single station, minimum	1 Plum	3	2.667	Ea.	2,200	119		2,319	2,600
5100	Maximum					7,575			7,575	8,350
6550	Major surgery table, minimum	1 Sswk	.50	16		29,100	660		29,760	33,200
6570	Maximum	"	.50	16		96,000	660		96,660	106,500
6800	Surgical lights, major operating room, dual head, minimum	2 Elec	1	16		16,100	700		16,800	18,900
6850	Maximum	"	1	16		26,900	700		27,600	30,700

11 77 Radiology Equipment

11 77 10 - Radiology Equipment

11 77 10.10 Radiology Equipment

		Crew	Daily Output	Labor-Hours	Unit	Material	Labor	Equipment	Total	Total Incl O&P
0010	**RADIOLOGY EQUIPMENT**									
8700	X-ray, mobile, minimum				Ea.	12,400			12,400	13,600
8750	Maximum					69,500			69,500	76,500
8900	Stationary, minimum					38,700			38,700	42,600
8950	Maximum					201,000			201,000	221,500
9150	Developing processors, minimum					10,800			10,800	11,800
9200	Maximum					28,300			28,300	31,100

11 78 Mortuary Equipment

11 78 13 - Mortuary Refrigerators

11 78 13.10 Mortuary Equipment

		Crew	Daily Output	Labor-Hours	Unit	Material	Labor	Equipment	Total	Total Incl O&P
0010	**MORTUARY EQUIPMENT**									
0015	Autopsy table, standard	1 Plum	1	8	Ea.	8,200	360		8,560	9,575
0020	Deluxe	"	.60	13.333		13,300	595		13,895	15,600
3200	Mortuary refrigerator, end operated, 2 capacity					11,400			11,400	12,500
3300	6 capacity					20,500			20,500	22,500

11 78 16 - Crematorium Equipment

11 78 16.10 Crematory

		Crew	Daily Output	Labor-Hours	Unit	Material	Labor	Equipment	Total	Total Incl O&P
0010	**CREMATORY**									
1500	Crematory, not including building, 1 place	Q-3	.20	160	Ea.	55,000	6,825		61,825	71,000
1750	2 place	"	.10	320	"	78,500	13,700		92,200	107,000

11 82 Solid Waste Handling Equipment

11 82 19 - Packaged Incinerators

11 82 19.10 Packaged Incinerators

		Crew	Daily Output	Labor-Hours	Unit	Material	Labor	Equipment	Total	Total Incl O&P
0010	**PACKAGED INCINERATORS**									
4400	Incinerator, gas, not incl. chimney, elec. or pipe, 50#/hr., minimum	Q-3	.80	40	Ea.	21,500	1,700		23,200	26,200
4420	Maximum		.70	45.714		28,000	1,950		29,950	33,700
4440	200 lb. per hr., minimum (batch type)		.60	53.333		28,000	2,275		30,275	34,200
4460	Maximum (with feeder)		.50	64		54,500	2,725		57,225	64,000
4480	400 lb. per hr., minimum (batch type)		.30	106		33,100	4,550		37,650	43,400
4500	Maximum (with feeder)		.25	128		62,500	5,475		67,975	77,000
4520	800 lb. per hr., with feeder, minimum		.20	160		81,500	6,825		88,325	100,000
4540	Maximum		.17	188		111,500	8,025		119,525	134,500
4560	1,200 lb. per hr., with feeder, minimum		.15	213		118,000	9,100		127,100	143,500
4580	Maximum		.11	290		142,000	12,400		154,400	174,500

11 82 Solid Waste Handling Equipment

11 82 19 – Packaged Incinerators

11 82 19.10 Packaged Incinerators

		Crew	Daily Output	Labor-Hours	Unit	Material	2007 Bare Costs Labor	Equipment	Total	Total Incl O&P
4600	2,000 lb. per hr., with feeder, minimum	Q-3	.10	320	Ea.	206,000	13,700		219,700	247,000
4620	Maximum		.05	640		346,500	27,300		373,800	422,500
4700	For heat recovery system, add, minimum		.25	128		67,000	5,475		72,475	81,500
4710	Add, maximum		.11	290		214,000	12,400		226,400	254,000
4720	For automatic ash conveyor, add		.50	64		28,100	2,725		30,825	35,000
4750	Large municipal incinerators, incl. stack, minimum		.25	128	Ton/day	17,100	5,475		22,575	27,000
4850	Maximum		.10	320	"	45,500	13,700		59,200	70,500

11 82 26 – Waste Compactors and Destructors

11 82 26.10 Compactors

		Crew	Daily Output	Labor-Hours	Unit	Material	Labor	Equipment	Total	Total Incl O&P
0010	**COMPACTORS**									
0020	Compactors, 115 volt, 250#/hr., chute fed	L-4	1	24	Ea.	10,100	825		10,925	12,400
0100	Hand fed		2.40	10		7,375	345		7,720	8,650
0300	Multi-bag, 230 volt, 600#/hr, chute fed		1	24		9,150	825		9,975	11,400
0400	Hand fed		1	24		7,850	825		8,675	9,925
0500	Containerized, hand fed, 2 to 6 C.Y. containers, 250#/hr.		1	24		9,725	825		10,550	12,000
0550	For chute fed, add per floor		1	24		1,125	825		1,950	2,500
1000	Heavy duty industrial compactor, 0.5 C.Y. capacity		1	24		6,575	825		7,400	8,500
1050	1.0 C.Y. capacity		1	24		10,000	825		10,825	12,300
1100	3 C.Y. capacity		.50	48		13,700	1,650		15,350	17,600
1150	5.0 C.Y. capacity		.50	48		16,900	1,650		18,550	21,200
1200	Combination shredder/compactor (5,000 lbs./hr.)		.50	48		33,400	1,650		35,050	39,300
1400	For handling hazardous waste materials, 55 gallon drum packer, std.					15,500			15,500	17,100
1410	55 gallon drum packer w/HEPA filter					19,400			19,400	21,300
1420	55 gallon drum packer w/charcoal & HEPA filter					25,900			25,900	28,400
1430	All of the above made explosion proof, add					11,800			11,800	13,000
5500	Shredder, municipal use, 35 ton per hour					254,500			254,500	280,000
5600	60 ton per hour					542,000			542,000	596,500
5750	Shredder & baler, 50 ton per day					508,500			508,500	559,000
5800	Shredder, industrial, minimum					20,000			20,000	22,000
5850	Maximum					107,500			107,500	118,000
5900	Baler, industrial, minimum					8,025			8,025	8,825
5950	Maximum					468,500			468,500	515,500
6000	Transfer station compactor, with power unit									
6050	and pedestal, not including pit, 50 ton per hour				Ea.	160,500			160,500	176,500

11 91 Religious Equipment

11 91 13 – Baptistries

11 91 13.10 Baptistry

		Crew	Daily Output	Labor-Hours	Unit	Material	Labor	Equipment	Total	Total Incl O&P
0010	**BAPTISTRY**									
0150	Baptistry, fiberglass, 3'-6" deep, x 13'-7" long,									
0160	steps at both ends, incl. plumbing, minimum	L-8	1	20	Ea.	3,475	765		4,240	5,000
0200	Maximum	"	.70	28.571		5,675	1,100		6,775	7,950
0250	Add for filter, heater and lights					1,250			1,250	1,375

11 91 23 – Sanctuary Equipment

11 91 23.10 Sanctuary Furnishings

		Crew	Daily Output	Labor-Hours	Unit	Material	Labor	Equipment	Total	Total Incl O&P
0010	**SANCTUARY FURNISHINGS**									
0020	Altar, wood, custom design, plain	1 Carp	1.40	5.714	Ea.	1,925	210		2,135	2,425
0050	Deluxe	"	.20	40		9,225	1,475		10,700	12,500
0070	Granite or marble, average	2 Marb	.50	32		8,525	1,150		9,675	11,200
0090	Deluxe	"	.20	80		24,300	2,900		27,200	31,100

11 91 Religious Equipment

11 91 23 – Sanctuary Equipment

11 91 23.10 Sanctuary Furnishings		Crew	Daily Output	Labor-Hours	Unit	Material	2007 Bare Costs Labor	Equipment	Total	Total Incl O&P
0100	Arks, prefabricated, plain	2 Carp	.80	20	Ea.	7,300	735		8,035	9,175
0130	Deluxe, maximum	"	.20	80		91,000	2,925		93,925	104,500
0500	Reconciliation room, wood, prefabricated, single, plain	1 Carp	.60	13.333		2,425	490		2,915	3,425
0550	Deluxe		.40	20		6,675	735		7,410	8,500
0650	Double, plain		.40	20		4,850	735		5,585	6,500
0700	Deluxe		.20	40		15,200	1,475		16,675	19,000
1000	Lecterns, wood, plain		5	1.600		690	58.50		748.50	850
1100	Deluxe		2	4		4,975	147		5,122	5,675
2000	Pulpits, hardwood, prefabricated, plain		2	4		1,225	147		1,372	1,575
2100	Deluxe		1.60	5	↓	8,325	184		8,509	9,425
2500	Railing, hardwood, average		25	.320	L.F.	151	11.75		162.75	184
3000	Seating, individual, oak, contour, laminated		21	.381	Person	133	14		147	169
3100	Cushion seat		21	.381		121	14		135	155
3200	Fully upholstered		21	.381		108	14		122	141
3300	Combination, self-rising	↓	21	.381		335	14		349	385
3500	For cherry, add				↓	30%				
5000	Wall cross, aluminum, extruded, 2" x 2" section	1 Carp	34	.235	L.F.	110	8.65		118.65	134
5150	4" x 4" section		29	.276		158	10.10		168.10	190
5300	Bronze, extruded, 1" x 2" section		31	.258		217	9.45		226.45	253
5350	2-1/2" x 2-1/2" section		34	.235		330	8.65		338.65	375
5450	Solid bar stock, 1/2" x 3" section		29	.276		430	10.10		440.10	490
5600	Fiberglass, stock		34	.235		89	8.65		97.65	111
5700	Stainless steel, 4" deep, channel section		29	.276		350	10.10		360.10	400
5800	4" deep box section	↓	29	.276	↓	475	10.10		485.10	535

Division Notes

	CREW	DAILY OUTPUT	LABOR-HOURS	UNIT	MAT.	LABOR	EQUIP.	TOTAL	TOTAL INCL O&P

Division 12 - Furnishings

Estimating Tips

General

- The items in this division are usually priced per square foot or each. Most of these items are purchased by the owner and placed by the supplier. Do not assume the items in Division 12 will be purchased and installed by the supplier. Check the specifications for responsibilities and include receiving, storage, installation, and mechanical and electrical hookups in the appropriate divisions.

- Some items in this division require some type of support system that is not usually furnished with the item. Examples of these systems include blocking for the attachment of casework and heavy drapery rods. The required blocking must be added to the estimate in the appropriate division.

Reference Numbers

Reference numbers are shown in shaded boxes at the beginning of some major classifications. These numbers refer to related items in the Reference Section. The reference information may be an estimating procedure, an alternate pricing method, or technical information.

Note: Not all subdivisions listed here necessarily appear in this publication.

No part of this publication may be reproduced, stored in a retrieval system, or transmitted in any form or by any means without prior written permission of Reed Construction Data.

12 21 Window Blinds

12 21 13 – Horizontal Louver Blinds

12 21 13.13 Metal Horizontal Louver Blinds

		Crew	Daily Output	Labor-Hours	Unit	Material	2007 Bare Costs Labor	2007 Bare Costs Equipment	Total	Total Incl O&P
0010	**METAL HORIZONTAL LOUVER BLINDS**									
0020	Horizontal, 1" aluminum slats, solid color, stock	1 Carp	590	.014	S.F.	3.52	.50		4.02	4.65
0090	Custom, minimum		590	.014		3.10	.50		3.60	4.19
0100	Maximum		440	.018		7.50	.67		8.17	9.30
0250	2" aluminum slats, custom, minimum		590	.014		4.50	.50		5	5.75
0350	Maximum		440	.018		8.65	.67		9.32	10.55
0450	Stock, minimum		590	.014		4.72	.50		5.22	6
0500	Maximum		440	.018		7.70	.67		8.37	9.50
0600	2" steel slats, stock, minimum		590	.014		1.79	.50		2.29	2.75
0630	Maximum		440	.018		5	.67		5.67	6.55
0750	Custom, minimum		590	.014		1.62	.50		2.12	2.56
0850	Maximum		400	.020		8	.73		8.73	9.95

12 21 16 – Vertical Louver Blinds

12 21 16.13 Metal Vertical Louver Blinds

		Crew	Daily Output	Labor-Hours	Unit	Material	Labor	Equipment	Total	Total Incl O&P
0010	**METAL VERTICAL LOUVER BLINDS**									
1500	Vertical, 3" to 5" PVC or cloth strips, minimum	CN 1 Carp	460	.017	S.F.	7.20	.64		7.84	8.90
1600	Maximum		400	.020		9.40	.73		10.13	11.45
1800	4" aluminum slats, minimum		460	.017		3.50	.64		4.14	4.84
1900	Maximum		400	.020		8.60	.73		9.33	10.60
1950	Mylar mirror-finish strips, to 8" wide, minimum		460	.017		14.30	.64		14.94	16.75
1970	Maximum		400	.020		21.50	.73		22.23	24.50

12 22 Curtains and Drapes

12 22 16 – Drapery Track and Accessories

12 22 16.10 Drapery Hardware

		Crew	Daily Output	Labor-Hours	Unit	Material	Labor	Equipment	Total	Total Incl O&P
0010	**DRAPERY HARDWARE**									
0030	Standard traverse, per foot, minimum	1 Carp	59	.136	L.F.	2.66	4.98		7.64	10.70
0100	Maximum		51	.157	"	10.90	5.75		16.65	21
4000	Traverse rods, adjustable, 28" to 48"		22	.364	Ea.	20.50	13.35		33.85	43.50
4020	48" to 84"		20	.400		25.50	14.70		40.20	51
4040	66" to 120"		18	.444		29.50	16.30		45.80	57.50
4060	84" to 156"		16	.500		32.50	18.35		50.85	64
4080	100" to 180"		14	.571		41.50	21		62.50	78
4090	156" to 228"		13	.615		48.50	22.50		71	88.50
4100	228" to 312"		13	.615		58.50	22.50		81	99
4200	Double rods, adjustable, 30" to 48"		9	.889		46.50	32.50		79	102
4220	48" to 86"		9	.889		61.50	32.50		94	119
4240	86" to 150"		8	1		55.50	36.50		92	119
4260	100" to 180"		7	1.143		57	42		99	129
4300	Tray and curtain rod, adjustable, 30" to 48"		9	.889		22.50	32.50		55	76
4320	48" to 86"		9	.889		34	32.50		66.50	88.50
4340	86" to 150"		8	1		44	36.50		80.50	106
4360	100" to 180"		7	1.143		48.50	42		90.50	119
4600	Valance, pinch pleated fabric, 12" deep, up to 54" long, minimum					36			36	39.50
4610	Maximum					90			90	99
4620	Up to 77" long, minimum					55			55	60.50
4630	Maximum					145			145	160
5000	Stationary rods, first 2 feet					9.70			9.70	10.70
5020	Each additional foot, add				L.F.	3.57			3.57	3.93

12 22 Curtains and Drapes

12 22 16 – Drapery Track and Accessories

12 22 16.20 Blast Curtains	Crew	Daily Output	Labor-Hours	Unit	Material	2007 Bare Costs Labor	Equipment	Total	Total Incl O&P
0010 **BLAST CURTAINS**, per LF horizontal opening width, off-white or gray fabric									
0100 Blast curtains, drapery system, complete, including hardware, minimum				L.F.				119	131
0120 Average								129	142
0140 Maximum				↓				138	152

12 23 Interior Shutters

12 23 10 – Interior Shutters

12 23 10.10 Interior Shutters

	Crew	Daily Output	Labor-Hours	Unit	Material	Labor	Equipment	Total	Total Incl O&P
0010 **INTERIOR SHUTTERS**, wood, louvered									
0200 Two panel, 27" wide, 36" high	1 Carp	5	1.600	Set	114	58.50		172.50	217
0300 33" wide, 36" high		5	1.600		149	58.50		207.50	256
0500 47" wide, 36" high		5	1.600		199	58.50		257.50	310
1000 Four panel, 27" wide, 36" high		5	1.600		134	58.50		192.50	240
1100 33" wide, 36" high		5	1.600		165	58.50		223.50	273
1300 47" wide, 36" high	↓	5	1.600	↓	234	58.50		292.50	350

12 23 13 – Wood Interior Shutters

12 23 13.13 Wood Panels

	Crew	Daily Output	Labor-Hours	Unit	Material	Labor	Equipment	Total	Total Incl O&P
0010 **WOOD PANELS**									
3000 Wood folding panels with movable louvers, 7" x 20" each	1 Carp	17	.471	Pr.	48	17.25		65.25	80
3300 8" x 28" each		17	.471		69.50	17.25		86.75	104
3450 9" x 36" each		17	.471		83	17.25		100.25	118
3600 10" x 40" each		17	.471		93.50	17.25		110.75	130
4000 Fixed louver type, stock units, 8" x 20" each		17	.471		72	17.25		89.25	107
4150 10" x 28" each		17	.471		80	17.25		97.25	115
4300 12" x 36" each		17	.471		96	17.25		113.25	133
4450 18" x 40" each		17	.471		156	17.25		173.25	198
5000 Insert panel type, stock, 7" x 20" each		17	.471		17.80	17.25		35.05	46.50
5150 8" x 28" each		17	.471		32.50	17.25		49.75	63
5300 9" x 36" each		17	.471		41.50	17.25		58.75	72.50
5450 10" x 40" each		17	.471		44.50	17.25		61.75	76
5600 Raised panel type, stock, 10" x 24" each		17	.471		191	17.25		208.25	238
5650 12" x 26" each		17	.471		144	17.25		161.25	186
5700 14" x 30" each		17	.471		163	17.25		180.25	206
5750 16" x 36" each	↓	17	.471		183	17.25		200.25	228
6000 For custom built pine, add					22%				
6500 For custom built hardwood blinds, add				↓	42%				

12 24 Window Shades

12 24 13 – Roller Window Shades

12 24 13.10 Shades

	Crew	Daily Output	Labor-Hours	Unit	Material	Labor	Equipment	Total	Total Incl O&P
0010 **SHADES**									
0020 Basswood, roll-up, stain finish, 3/8" slats	1 Carp	300	.027	S.F.	10.80	.98		11.78	13.40
0200 7/8" slats		300	.027		10.20	.98		11.18	12.75
0300 Vertical side slide, stain finish, 3/8" slats		300	.027		16.35	.98		17.33	19.50
0400 7/8" slats	↓	300	.027		16.35	.98		17.33	19.50
0500 For fire retardant finishes, add					16%				
0600 For "B" rated finishes, add					20%				
0900 Mylar, single layer, non-heat reflective	1 Carp	685	.012	↓	6.60	.43		7.03	7.90

12 24 Window Shades

12 24 13 – Roller Window Shades

12 24 13.10 Shades		Crew	Daily Output	Labor-Hours	Unit	Material	2007 Bare Costs Labor	Equipment	Total	Total Incl O&P
1000	Double layered, heat reflective	1 Carp	685	.012	S.F.	9.10	.43		9.53	10.65
1100	Triple layered, heat reflective	↓	685	.012		10.55	.43		10.98	12.30
1200	For metal roller instead of wood, add per				Shade	3.92			3.92	4.31
1300	Vinyl coated cotton, standard	1 Carp	685	.012	S.F.	2.21	.43		2.64	3.10
1400	Lightproof decorator shades		685	.012		2	.43		2.43	2.87
1500	Vinyl, lightweight, 4 gauge		685	.012		.50	.43		.93	1.22
1600	Heavyweight, 6 gauge		685	.012		1.54	.43		1.97	2.36
1700	Vinyl laminated fiberglass, 6 ga., translucent		685	.012		2.22	.43		2.65	3.11
1800	Lightproof		685	.012		3.54	.43		3.97	4.56
3000	Woven aluminum, 3/8" thick, lightproof and fireproof	↓	350	.023	↓	4.97	.84		5.81	6.75

12 32 Manufactured Wood Casework

12 32 13 – Manufactured Wood-Veneer-Faced Casework

12 32 13.10 Manufactured Wood Casework		Crew	Daily Output	Labor-Hours	Unit	Material	Labor	Equipment	Total	Total Incl O&P
0010	**MANUFACTURED WOOD CASEWORK**									
0300	Built-in drawer units, pine, 18" deep, 32" high, unfinished									
0400	Minimum	2 Carp	53	.302	L.F.	120	11.10		131.10	149
0500	Maximum	"	40	.400	"	146	14.70		160.70	184
0700	Kitchen base cabinets, hardwood, not incl. counter tops,									
0710	24" deep, 35" high, prefinished									
0800	One top drawer, one door below, 12" wide	2 Carp	24.80	.645	Ea.	195	23.50		218.50	251
0840	18" wide		23.30	.687		223	25		248	284
0880	24" wide		22.30	.717		265	26.50		291.50	335
1000	Four drawers, 12" wide		24.80	.645		350	23.50		373.50	420
1040	18" wide		23.30	.687		320	25		345	390
1060	24" wide		22.30	.717		345	26.50		371.50	420
1200	Two top drawers, two doors below, 27" wide		22	.727		295	26.50		321.50	365
1260	36" wide		20.30	.788		365	29		394	450
1300	48" wide		18.90	.847		410	31		441	500
1500	Range or sink base, two doors below, 30" wide		21.40	.748		289	27.50		316.50	365
1540	36" wide		20.30	.788		320	29		349	395
1580	48" wide	↓	18.90	.847		355	31		386	440
1800	For sink front units, deduct					56			56	61.50
2000	Corner base cabinets, 36" wide, standard	2 Carp	18	.889		420	32.50		452.50	510
2100	Lazy Susan with revolving door	"	16.50	.970	↓	415	35.50		450.50	510
4000	Kitchen wall cabinets, hardwood, 12" deep with two doors									
4050	12" high, 30" wide	2 Carp	24.80	.645	Ea.	157	23.50		180.50	209
4100	36" wide		24	.667		183	24.50		207.50	239
4400	15" high, 30" wide		24	.667		167	24.50		191.50	222
4440	36" wide		22.70	.705		189	26		215	249
4700	24" high, 30" wide		23.30	.687		207	25		232	267
4720	36" wide		22.70	.705		229	26		255	293
5000	30" high, one door, 12" wide		22	.727		140	26.50		166.50	196
5040	18" wide		20.90	.766		173	28		201	235
5060	24" wide		20.30	.788		195	29		224	260
5300	Two doors, 27" wide		19.80	.808		234	29.50		263.50	305
5340	36" wide		18.80	.851		267	31		298	345
5380	48" wide		18.40	.870		330	32		362	410
6000	Corner wall, 30" high, 24" wide		18	.889		159	32.50		191.50	226
6050	30" wide		17.20	.930		193	34		227	265
6100	36" wide	↓	16.50	.970	↓	209	35.50		244.50	286

12 32 Manufactured Wood Casework

12 32 13 – Manufactured Wood-Veneer-Faced Casework

12 32 13.10 Manufactured Wood Casework		Crew	Daily Output	Labor-Hours	Unit	Material	2007 Bare Costs Labor	Equipment	Total	Total Incl O&P
6500	Revolving Lazy Susan	2 Carp	15.20	1.053	Ea.	355	38.50		393.50	450
7000	Broom cabinet, 84" high, 24" deep, 18" wide		10	1.600		430	58.50		488.50	560
7500	Oven cabinets, 84" high, 24" deep, 27" wide		8	2		630	73.50		703.50	810
7750	Valance board trim		396	.040	L.F.	9	1.48		10.48	12.20
9000	For deluxe models of all cabinets, add					40%				
9500	For custom built in place, add					25%	10%			
9550	Rule of thumb, kitchen cabinets not including									
9560	appliances & counter top, minimum	2 Carp	30	.533	L.F.	104	19.55		123.55	146
9600	Maximum	"	25	.640	"	250	23.50		273.50	310
9610	For metal cabinets, see division 12 35 70.13									

12 32 13.15 Manufactured Wood Casework Frames

		Crew	Daily Output	Labor-Hours	Unit	Material	Labor	Equipment	Total	Total Incl O&P
0010	**MANUFACTURED WOOD CASEWORK FRAMES**									
0050	Base cabinets, counter storage, 36" high, one bay									
0100	18" wide	1 Carp	2.70	2.963	Ea.	115	109		224	295
0400	Two bay, 36" wide		2.20	3.636		175	133		308	400
1100	Three bay, 54" wide		1.50	5.333		208	196		404	535
2800	Book cases, one bay, 7' high, 18" wide		2.40	3.333		135	122		257	340
3500	Two bay, 36" wide		1.60	5		195	184		379	500
4100	Three bay, 54" wide		1.20	6.667		325	245		570	735
5100	Coat racks, one bay, 7' high, 24" wide		4.50	1.778		135	65		200	250
5300	Two bay, 48" wide		2.75	2.909		187	107		294	370
5800	Three bay, 72" wide		2.10	3.810		276	140		416	525
6100	Wall mounted cabinet, one bay, 24" high, 18" wide		3.60	2.222		74	81.50		155.50	209
6800	Two bay, 36" wide		2.20	3.636		108	133		241	325
7400	Three bay, 54" wide		1.70	4.706		135	173		308	415
8400	30" high, one bay, 18" wide		3.60	2.222		80.50	81.50		162	216
9000	Two bay, 36" wide		2.15	3.721		107	137		244	330
9400	Three bay, 54" wide		1.60	5		133	184		317	435
9800	Wardrobe, 7' high, single, 24" wide		2.70	2.963		148	109		257	330
9880	Partition & adjustable shelves, 48" wide		1.70	4.706		189	173		362	475
9950	Partition, adjustable shelves & drawers, 48" wide		1.40	5.714		283	210		493	635

12 32 13.20 Manufactured Wood Casework Doors

		Crew	Daily Output	Labor-Hours	Unit	Material	Labor	Equipment	Total	Total Incl O&P
0010	**MANUFACTURED WOOD CASEWORK DOORS**									
2000	Glass panel, hardwood frame									
2200	12" wide, 18" high	1 Carp	34	.235	Ea.	14.30	8.65		22.95	29
2600	30" high		32	.250		24	9.20		33.20	41
4450	18" wide, 18" high		32	.250		21.50	9.20		30.70	38
4550	30" high		29	.276		26.50	10.10		36.60	45
5000	Hardwood, raised panel									
5100	12" wide, 18" high	1 Carp	16	.500	Ea.	24	18.35		42.35	55
5200	30" high		15	.533		40	19.55		59.55	74.50
5500	18" wide, 18" high		15	.533		36	19.55		55.55	70
5600	30" high		14	.571		60	21		81	98.50
6000	Plastic laminate on particle board									
6100	12" wide, 18" high	1 Carp	25	.320	Ea.	10.35	11.75		22.10	29.50
6140	30" high		23	.348		17.25	12.75		30	39
6500	18" wide, 18" high		24	.333		15.50	12.25		27.75	36
6600	30" high		22	.364		26	13.35		39.35	49.50

12 32 13.25 Manufactured Wood Casework Drawer Fronts

		Crew	Daily Output	Labor-Hours	Unit	Material	Labor	Equipment	Total	Total Incl O&P
0010	**MANUFACTURED WOOD CASEWORK DRAWER FRONTS**									
0100	Solid hardwood front									
1000	4" high, 12" wide	1 Carp	17	.471	Ea.	3.50	17.25		20.75	31

12 32 Manufactured Wood Casework

12 32 13 – Manufactured Wood-Veneer-Faced Casework

12 32 13.25 Manufactured Wood Casework Drawer Fronts		Crew	Daily Output	Labor-Hours	Unit	Material	2007 Bare Costs Labor	Equipment	Total	Total Incl O&P
1200	18" wide	1 Carp	16	.500	Ea.	5.05	18.35		23.40	34
2800	Plastic laminate on particle board front									
3000	4" high, 12" wide	1 Carp	17	.471	Ea.	4.39	17.25		21.64	32
3200	18" wide	"	16	.500	"	6.60	18.35		24.95	36

12 32 13.30 Manufactured Wood Casework Vanities

0010	**MANUFACTURED WOOD CASEWORK VANITIES**									
8000	Vanity bases, 2 doors, 30" high, 21" deep, 24" wide	2 Carp	20	.800	Ea.	197	29.50		226.50	262
8050	30" wide		16	1		226	36.50		262.50	305
8100	36" wide		13.33	1.200		300	44		344	400
8150	48" wide		11.43	1.400		360	51.50		411.50	475
9000	For deluxe models of all vanities, add to above					40%				
9500	For custom built in place, add to above					25%	10%			

12 32 13.35 Manufactured Wood Casework Hardware

0010	**MANUFACTURED WOOD CASEWORK HARDWARE**									
1000	Catches, minimum	1 Carp	235	.034	Ea.	.84	1.25		2.09	2.87
1040	Maximum	"	80	.100	"	5.20	3.67		8.87	11.45
2000	Door/drawer pulls, handles									
2200	Handles and pulls, projecting, metal, minimum	1 Carp	160	.050	Ea.	3.70	1.84		5.54	6.95
2240	Maximum		68	.118		7.85	4.32		12.17	15.35
2300	Wood, minimum		160	.050		3.90	1.84		5.74	7.15
2340	Maximum		68	.118		7.15	4.32		11.47	14.60
2600	Flush, metal, minimum		160	.050		3.90	1.84		5.74	7.15
2640	Maximum		68	.118		7.15	4.32		11.47	14.60
3000	Drawer tracks/glides, minimum		48	.167	Pr.	6.60	6.10		12.70	16.85
3040	Maximum		24	.333		19.25	12.25		31.50	40
4000	Cabinet hinges, minimum		160	.050		2.25	1.84		4.09	5.35
4040	Maximum		68	.118		8.50	4.32		12.82	16.05

12 35 Specialty Casework

12 35 50 – Educational/Library Casework

12 35 50.13 Educational Casework

0010	**EDUCATIONAL CASEWORK**									
5000	School, 24" deep, metal, 84" high units	2 Carp	15	1.067	L.F.	345	39		384	440
5150	Counter height units		20	.800		232	29.50		261.50	300
5450	Wood, custom fabricated, 32" high counter		20	.800		193	29.50		222.50	258
5600	Add for counter top		56	.286		20.50	10.50		31	39
5800	84" high wall units		15	1.067		375	39		414	475
6000	Laminated plastic finish is same price as wood									

12 35 53 – Laboratory Casework

12 35 53.13 Metal Laboratory Casework

0010	**METAL LABORATORY CASEWORK**									
0020	Cabinets, base, door units, metal	2 Carp	18	.889	L.F.	167	32.50		199.50	234
0300	Drawer units		18	.889		375	32.50		407.50	460
0700	Tall storage cabinets, open, 7' high		20	.800		355	29.50		384.50	435
0900	With glazed doors		20	.800		450	29.50		479.50	540
1300	Wall cabinets, metal, 12-1/2" deep, open		20	.800		114	29.50		143.50	171
1500	With doors		20	.800		236	29.50		265.50	305
6300	Rule of thumb: lab furniture including installation & connection									
6320	High school				S.F.				30.55	33.60
6340	College								45	49.50

12 35 Specialty Casework

12 35 53 – Laboratory Casework

12 35 53.13 Metal Laboratory Casework

		Crew	Daily Output	Labor-Hours	Unit	Material	2007 Bare Costs Labor	2007 Bare Costs Equipment	Total	Total Incl O&P
6360	Clinical, health care				S.F.				38.65	42.50
6380	Industrial								62.50	68.50

12 35 59 – Display Casework

12 35 59.10 Display Cases

		Crew	Daily Output	Labor-Hours	Unit	Material	Labor	Equipment	Total	Total Incl O&P
0010	**DISPLAY CASES** Free standing, all glass									
0020	Aluminum frame, 42" high x 36" x 12" deep	2 Carp	8	2	Ea.	1,475	73.50		1,548.50	1,750
0100	70" high x 48" x 18" deep	"	6	2.667		2,600	98		2,698	3,000
0500	For wood bases, add					9%				
0600	For hardwood frames, deduct					8%				
0700	For bronze, baked enamel finish, add					10%				
2000	Wall mounted, glass front, aluminum frame									
2010	Non-illuminated, one section 3' x 4' x 1'-4"	2 Carp	5	3.200	Ea.	2,125	117		2,242	2,525
2100	5' x 4' x 1'-4"		5	3.200		2,300	117		2,417	2,725
2200	6' x 4' x 1'-4"		4	4		2,725	147		2,872	3,225
2500	Two sections, 8' x 4' x 1'-4"		2	8		1,850	294		2,144	2,500
2600	10' x 4' x 1'-4"		2	8		2,025	294		2,319	2,675
3000	Three sections, 16' x 4' x 1'-4"		1.50	10.667		3,500	390		3,890	4,450
3500	For fluorescent lights, add				Section	269			269	296
4000	Table exhibit cases, 2' wide, 3' high, 4' long, flat top	2 Carp	5	3.200	Ea.	790	117		907	1,050
4100	3' wide, 3' high, 4' long, sloping top	"	3	5.333	"	1,200	196		1,396	1,625

12 35 70 – Healthcare Casework

12 35 70.13 Hospital Casework

		Crew	Daily Output	Labor-Hours	Unit	Material	Labor	Equipment	Total	Total Incl O&P
0010	**HOSPITAL CASEWORK**									
0500	Hospital, base cabinets, laminated plastic	2 Carp	10	1.600	L.F.	221	58.50		279.50	335
1000	Stainless steel	"	10	1.600		405	58.50		463.50	540
1200	For all drawers, add					23			23	25.50
1300	Cabinet base trim, 4" high, enameled steel	2 Carp	200	.080		37	2.94		39.94	45.50
1400	Stainless steel		200	.080		74	2.94		76.94	86
1450	Counter top, laminated plastic, no backsplash		40	.400		38	14.70		52.70	65
1650	With backsplash		40	.400		47.50	14.70		62.20	75
1800	For sink cutout, add		12.20	1.311	Ea.		48		48	75
1900	Stainless steel counter top		40	.400	L.F.	124	14.70		138.70	159
2000	For drop-in stainless 43" x 21" sink, add				Ea.	815			815	895
2500	Wall cabinets, laminated plastic	2 Carp	15	1.067	L.F.	166	39		205	243
2600	Enameled steel		15	1.067		203	39		242	285
2700	Stainless steel		15	1.067		405	39		444	505
2800	For glass doors, add					27.50			27.50	30.50

12 35 70.16 Nurse Station Casework

		Crew	Daily Output	Labor-Hours	Unit	Material	Labor	Equipment	Total	Total Incl O&P
0010	**NURSE STATION CASEWORK**									
2100	Nurses station, door type, laminated plastic	2 Carp	10	1.600	L.F.	256	58.50		314.50	375
2200	Enameled steel		10	1.600		245	58.50		303.50	360
2300	Stainless steel		10	1.600		490	58.50		548.50	630
2400	For drawer type, add					208			208	229

12 35 80 – Commercial Kitchen Casework

12 35 80.13 Metal Kitchen

		Crew	Daily Output	Labor-Hours	Unit	Material	Labor	Equipment	Total	Total Incl O&P
0010	**METAL KITCHEN** Casework									
3500	Kitchen, base cabinets, metal, minimum	2 Carp	30	.533	L.F.	59.50	19.55		79.05	96
3600	Maximum		25	.640		152	23.50		175.50	204
3700	Wall cabinets, metal, minimum		30	.533		59.50	19.55		79.05	96
3800	Maximum		25	.640		137	23.50		160.50	188

12 36 Countertops

12 36 23 – Plastic Countertops

12 36 23.10 Countertops		Crew	Daily Output	Labor-Hours	Unit	Material	2007 Bare Costs Labor	Equipment	Total	Total Incl O&P
0010	**COUNTERTOPS**									
0020	Stock plastic laminate, 24" wide w/ backsplash, minimum	1 Carp	30	.267	L.F.	9.55	9.80		19.35	26
0100	Maximum		25	.320		17.30	11.75		29.05	37.50
0300	Custom plastic, 7/8" thick, aluminum molding, no splash		30	.267		19.10	9.80		28.90	36.50
0400	Cove splash		30	.267		25	9.80		34.80	43
0600	1-1/4" thick, no splash		28	.286		28.50	10.50		39	48
0700	Square splash		28	.286		27.50	10.50		38	46.50
0900	Square edge, plastic face, 7/8" thick, no splash		30	.267		24	9.80		33.80	42
1000	With splash		30	.267		30.50	9.80		40.30	49
1200	For stainless channel edge, 7/8" thick, add					2.51			2.51	2.76
1300	1-1/4" thick, add					2.95			2.95	3.25
1500	For solid color suede finish, add					2.45			2.45	2.70
1700	For end splash, add				Ea.	16.35			16.35	18
1900	For cut outs, standard, add, minimum	1 Carp	32	.250		3.27	9.20		12.47	17.90
2000	Maximum		8	1		5.45	36.50		41.95	63
2100	Postformed, including backsplash and front edge		30	.267	L.F.	9.80	9.80		19.60	26
2110	Mitred, add		12	.667	Ea.		24.50		24.50	38
2200	Built-in place, 25" wide, plastic laminate		25	.320	L.F.	13.10	11.75		24.85	32.50
2300	Ceramic tile mosaic		25	.320		28.50	11.75		40.25	49.50
2500	Marble, stock, with splash, 1/2" thick, minimum	1 Bric	17	.471		35	17.90		52.90	66
2700	3/4" thick, maximum	"	13	.615		87.50	23.50		111	132
2900	Maple, solid, laminated, 1-1/2" thick, no splash	1 Carp	28	.286		57	10.50		67.50	79.50
3000	With square splash		28	.286		68	10.50		78.50	91
3200	Stainless steel		24	.333	S.F.	131	12.25		143.25	163
3400	Recessed cutting block with trim, 16" x 20" x 1"		8	1	Ea.	66.50	36.50		103	130

12 36 53 – Laboratory Countertops

12 36 53.10 Laboratory Countertops		Crew	Daily Output	Labor-Hours	Unit	Material	Labor	Equipment	Total	Total Incl O&P
0010	**LABORATORY COUNTERTOPS**									
0020	Counter tops, not incl. base cabinets, acidproof, minimum	2 Carp	82	.195	S.F.	27.50	7.15		34.65	41.50
0030	Maximum		70	.229		37.50	8.40		45.90	54
0040	Stainless steel		82	.195		80.50	7.15		87.65	99.50

12 36 61 – Simulated Stone Countertops

12 36 61.16 Solid Surface Countertops

		Crew	Daily Output	Labor-Hours	Unit	Material	Labor	Equipment	Total	Total Incl O&P
0010	**SOLID SURFACE COUNTERTOPS**, Acrylic polymer									
0020	Pricing for orders of 100 L.F. or greater									
0100	25" wide, solid colors	2 Carp	28	.571	L.F.	44.50	21		65.50	81.50
0200	Patterned colors		28	.571		56.50	21		77.50	94.50
0300	Premium patterned colors		28	.571		70.50	21		91.50	111
0400	With silicone attached 4" backsplash, solid colors		27	.593		49	22		71	88
0500	Patterned colors		27	.593		62	22		84	102
0600	Premium patterned colors		27	.593		77	22		99	119
0700	With hard seam attached 4" backsplash, solid colors		23	.696		49	25.50		74.50	94
0800	Patterned colors		23	.696		62	25.50		87.50	108
0900	Premium patterned colors		23	.696		77	25.50		102.50	125
1000	Pricing for order of 51 - 99 L.F.									
1100	25" wide, solid colors	2 Carp	24	.667	L.F.	51.50	24.50		76	94.50
1200	Patterned colors		24	.667		65	24.50		89.50	110
1300	Premium patterned colors		24	.667		81.50	24.50		106	128
1400	With silicone attached 4" backsplash, solid colors		23	.696		56.50	25.50		82	102
1500	Patterned colors		23	.696		71.50	25.50		97	119
1600	Premium patterned colors		23	.696		89	25.50		114.50	138
1700	With hard seam attached 4" backsplash, solid colors		20	.800		56.50	29.50		86	108

12 36 Countertops

12 36 61 – Simulated Stone Countertops

12 36 61.16 Solid Surface Countertops

		Crew	Daily Output	Labor-Hours	Unit	Material	2007 Bare Costs Labor	Equipment	Total	Total Incl O&P
1800	Patterned colors	2 Carp	20	.800	L.F.	71.50	29.50		101	124
1900	Premium patterned colors	↓	20	.800	↓	89	29.50		118.50	143
2000	Pricing for order of 1 - 50 L.F.									
2100	25" wide, solid colors	2 Carp	20	.800	L.F.	60	29.50		89.50	112
2200	Patterned colors		20	.800		76.50	29.50		106	130
2300	Premium patterned colors		20	.800		95.50	29.50		125	151
2400	With silicone attached 4" backsplash, solid colors		19	.842		66	31		97	121
2500	Patterned colors		19	.842		83.50	31		114.50	140
2600	Premium patterned colors		19	.842		104	31		135	163
2700	With hard seam attached 4" backsplash, solid colors		15	1.067		66	39		105	134
2800	Patterned colors		15	1.067		83.50	39		122.50	153
2900	Premium patterned colors	↓	15	1.067	↓	104	39		143	176
3000	Sinks, pricing for order of 100 or greater units									
3100	Single bowl, hard seamed, solid colors, 13" x 17"	1 Carp	3	2.667	Ea.	300	98		398	480
3200	10" x 15"		7	1.143		139	42		181	219
3300	Cutouts for sinks	↓	8	1	↓		36.50		36.50	57
3400	Sinks, pricing for order of 51 - 99 units									
3500	Single bowl, hard seamed, solid colors, 13" x 17"	1 Carp	2.55	3.137	Ea.	345	115		460	560
3600	10" x 15"		6	1.333		160	49		209	252
3700	Cutouts for sinks	↓	7	1.143	↓		42		42	65.50
3800	Sinks, pricing for order of 1 - 50 units									
3900	Single bowl, hard seamed, solid colors, 13" x 17"	1 Carp	2	4	Ea.	405	147		552	675
4000	10" x 15"		4.55	1.758		188	64.50		252.50	305
4100	Cutouts for sinks		5.25	1.524			56		56	87
4200	Cooktop cutouts, pricing for 100 or greater units		4	2		22	73.50		95.50	139
4300	51 - 99 units		3.40	2.353		25.50	86.50		112	162
4400	1 - 50 units	↓	3	2.667	↓	30	98		128	185

12 36 61.17 Solid Surface Vanity Tops

		Crew	Daily Output	Labor-Hours	Unit	Material	2007 Bare Costs Labor	Equipment	Total	Total Incl O&P
0010	**SOLID SURFACE VANITY TOPS**									
0015	Solid surface, center bowl, 17" x 19"	1 Carp	12	.667	Ea.	184	24.50		208.50	240
0020	19" x 25"		12	.667		223	24.50		247.50	283
0030	19" x 31"		12	.667		270	24.50		294.50	335
0040	19" x 37"		12	.667		315	24.50		339.50	385
0050	22" x 25"		10	.800		196	29.50		225.50	262
0060	22" x 31"		10	.800		229	29.50		258.50	298
0070	22" x 37"		10	.800		266	29.50		295.50	340
0080	22" x 43"		10	.800		305	29.50		334.50	380
0090	22" x 49"		10	.800		335	29.50		364.50	415
0110	22" x 55"		8	1		380	36.50		416.50	475
0120	22" x 61"		8	1		435	36.50		471.50	535
0220	Double bowl, 22" x 61"		8	1		495	36.50		531.50	595
0230	Double bowl, 22" x 73"	↓	8	1	↓	685	36.50		721.50	805
0240	For aggregate colors, add					35%				
0250	For faucets and fittings see 22 41 39.10									

12 36 61.19 Engineered Stone Countertops

		Crew	Daily Output	Labor-Hours	Unit	Material	2007 Bare Costs Labor	Equipment	Total	Total Incl O&P
0010	**ENGINEERED STONE COUNTERTOPS**									
0100	25" wide, 4" backsplash, color group A, minimum	2 Carp	15	1.067	L.F.	16.15	39		55.15	79
0110	Maximum		15	1.067		41.50	39		80.50	107
0120	Color group B, minimum		15	1.067		21	39		60	84
0130	Maximum		15	1.067		48.50	39		87.50	114
0140	Color group C, minimum		15	1.067		29	39		68	93
0150	Maximum	↓	15	1.067	↓	59.50	39		98.50	127

12 36 Countertops

12 36 61 – Simulated Stone Countertops

12 36 61.19 Engineered Stone Countertops		Crew	Daily Output	Labor-Hours	Unit	Material	2007 Bare Costs Labor	Equipment	Total	Total Incl O&P
0160	Color group D, minimum	2 Carp	15	1.067	L.F.	36	39		75	101
0170	Maximum	↓	15	1.067	↓	69.50	39		108.50	137

12 46 Furnishing Accessories

12 46 13 – Ash Receptacles

12 46 13.10 Ash/Trash Receivers

		Crew	Daily Output	Labor-Hours	Unit	Material	Labor	Equipment	Total	Total Incl O&P
0010	**ASH/TRASH RECEIVERS**									
1000	Ash urn, cylindrical metal									
1020	8" diameter, 20" high	1 Clab	60	.133	Ea.	100	3.83		103.83	116
1060	10" diameter, 26" high	"	60	.133	"	110	3.83		113.83	127
2000	Combination ash/trash urn, metal									
2020	8" diameter, 20" high	1 Clab	60	.133	Ea.	73	3.83		76.83	86.50
2050	10" diameter, 26" high	"	60	.133	"	99	3.83		102.83	115

12 46 19 – Clocks

12 46 19.50 Clocks

0010	**CLOCKS**									
0080	12" diameter, single face	1 Elec	8	1	Ea.	75.50	44		119.50	149
0100	Double face	"	6.20	1.290	"	144	56.50		200.50	243

12 46 33 – Waste Receptacles

12 46 33.13 Trash Receptacles

0010	**TRASH RECEPTACLES**									
4000	Trash receptacle, metal									
4020	8" diameter, 15" high	1 Clab	60	.133	Ea.	67	3.83		70.83	79.50
4040	10" diameter, 18" high		60	.133		108	3.83		111.83	125
5040	16" x 8" x 14" high	↓	60	.133	↓	17	3.83		20.83	24.50
5500	Trash receptacle, plastic, with lid									
5520	35 gallon	1 Clab	60	.133	Ea.	211	3.83		214.83	238
5540	45 gallon		60	.133		250	3.83		253.83	281
5550	Plastic recycling barrel, w/lid & wheels, 32 gal		60	.133		56	3.83		59.83	67.50
5560	65 gal		60	.133		107	3.83		110.83	124
5570	95 gal	↓	60	.133	↓	310	3.83		313.83	345

12 48 Rugs and Mats

12 48 13 – Entrance Floor Mats and Frames

12 48 13.13 Entrance Floor Mats

0010	**ENTRANCE FLOOR MATS**									
0020	Recessed, in-laid black rubber, 3/8" thick, solid	1 Clab	155	.052	S.F.	17.80	1.48		19.28	22
0050	Perforated		155	.052		12.40	1.48		13.88	15.90
0100	1/2" thick, solid		155	.052		14.90	1.48		16.38	18.65
0150	Perforated		155	.052		20.50	1.48		21.98	25
0200	In colors, 3/8" thick, solid		155	.052		16.05	1.48		17.53	19.95
0250	Perforated		155	.052		16.55	1.48		18.03	20.50
0300	1/2" thick, solid		155	.052		21	1.48		22.48	25.50
0350	Perforated		155	.052		21.50	1.48		22.98	26
0500	Link mats, including nosings, aluminum, 3/8" thick CN		155	.052		19.25	1.48		20.73	23.50
0550	Black rubber with galvanized tie rods		155	.052		15	1.48		16.48	18.80
0600	Steel, galvanized, 3/8" thick		155	.052		8.30	1.48		9.78	11.45
0650	Vinyl, in colors	↓	155	.052	↓	19.30	1.48		20.78	23.50

12 48 Rugs and Mats

12 48 13 – Entrance Floor Mats and Frames

12 48 13.13 Entrance Floor Mats		Crew	Daily Output	Labor-Hours	Unit	Material	2007 Bare Costs Labor	Equipment	Total	Total Incl O&P
0750	Add for nosings, rubber				L.F.	3.25			3.25	3.58
0850	Recess frames for above mats, aluminum	1 Carp	100	.080		3.62	2.94		6.56	8.55
0870	Bronze	"	100	.080	↓	5.25	2.94		8.19	10.30
0900	Skate lock tile, 24" x 24" x 1/2" thick, rubber, black	1 Clab	125	.064	S.F.	15.05	1.84		16.89	19.40
0950	Color		125	.064	"	19.20	1.84		21.04	24
1000	12" x 24" border, black		75	.107	L.F.	16.55	3.07		19.62	23
1100	Color		75	.107	"	22.50	3.07		25.57	29.50
1150	12" x 12" outside corner, black		100	.080	S.F.	14.20	2.30		16.50	19.25
1200	Color		100	.080		18.40	2.30		20.70	24
1500	Duckboard, aluminum slats		155	.052		15.90	1.48		17.38	19.80
1700	Hardwood strips on rubber base, to 54" wide		155	.052		13	1.48		14.48	16.60
1800	Assembled with brass rods and vinyl spacers, to 48" wide		155	.052		17.35	1.48		18.83	21.50
1850	Tire fabric, 3/4" thick		155	.052		11.25	1.48		12.73	14.70
1900	Vinyl, 36" wide, in colors, hollow top & bottoms		155	.052		9.10	1.48		10.58	12.30
1950	Solid top & bottom members	↓	155	.052	↓	9.10	1.48		10.58	12.30

12 51 Office Furniture

12 51 16 – Case Goods

12 51 16.13 Metal Case Goods

0010	**METAL CASE GOODS**									
0020	Desks, 29" high, double pedestal, 30" x 60", metal, minimum				Ea.	395			395	435
0030	Maximum					1,050			1,050	1,150
0600	Desks, single pedestal, 30" x 60", metal, minimum					355			355	390
0620	Maximum					855			855	940
0720	Desks, secretarial, 30" x 60", metal, minimum					315			315	345
0730	Maximum					570			570	630
0740	Return, 20" x 42", minimum					242			242	266
0750	Maximum					380			380	415
0940	59" x 12" x 23" high, steel, minimum					176			176	194
0960	Maximum				↓	247			247	272

12 51 16.16 Wood Case Goods

0010	**WOOD CASE GOODS**									
0150	Desk, 29" high, double pedestal, 30" x 60"									
0160	Wood, minimum				Ea.	380			380	415
0180	Maximum				"	2,050			2,050	2,250
0630	Single pedestal, 30" x 60"									
0640	Wood, minimum				Ea.	455			455	500
0650	Maximum					640			640	705
0670	Executive return, 24" x 42", with box, file, wood, minimum					320			320	355
0680	Maximum				↓	640			640	705
0790	Desk, 29" high, secretarial, 30" x 60"									
0800	Wood, minimum				Ea.	410			410	450
0810	Maximum					2,075			2,075	2,275
0820	Return, 20" x 42", minimum					365			365	400
0830	Maximum					1,125			1,125	1,225
0900	Desktop organizer, 72" x 14" x 36" high, wood, minimum					158			158	174
0920	Maximum				↓	300			300	330
1110	Furniture, credenza, 29" high, 18" to 22" x 60" to 72"									
1120	Wood, minimum				Ea.	560			560	615
1140	Maximum				"	2,625			2,625	2,875

12 51 Office Furniture

12 51 23 – Office Tables

12 51 23.33 Conference Tables

	Crew	Daily Output	Labor-Hours	Unit	Material	2007 Bare Costs Labor	Equipment	Total	Total Incl O&P
0010 **CONFERENCE TABLES**									
6000 Table, conference									
6050 Boat, 96" x 42", minimum				Ea.	700			700	770
6150 Maximum					3,125			3,125	3,450
6720 Rectangle, 96" x 42", minimum					885			885	975
6740 Maximum				↓	2,250			2,250	2,475

12 52 Seating

12 52 23 – Office Seating

12 52 23.13 Office Chairs

	Crew	Daily Output	Labor-Hours	Unit	Material	2007 Bare Costs Labor	Equipment	Total	Total Incl O&P
0010 **OFFICE CHAIRS**									
2000 Chairs, office type, executive, minimum				Ea.	225			225	247
2150 Maximum					1,850			1,850	2,025
2200 Management, minimum					171			171	188
2250 Maximum					2,000			2,000	2,200
2280 Task, minimum					109			109	120
2290 Maximum					615			615	675
2300 Arm kit, minimum					50.50			50.50	55.50
2320 Maximum				↓	77			77	84.50

12 54 Hospitality Furniture

12 54 13 – Hotel and Motel Furniture

12 54 13.10 Furniture, Hotel

	Crew	Daily Output	Labor-Hours	Unit	Material	2007 Bare Costs Labor	Equipment	Total	Total Incl O&P
0010 **FURNITURE, HOTEL**									
0020 Standard quality set, minimum				Room	2,225			2,225	2,450
0200 Maximum				"	7,900			7,900	8,675

12 54 16 – Restaurant Furniture

12 54 16.10 Tables, Folding

	Crew	Daily Output	Labor-Hours	Unit	Material	2007 Bare Costs Labor	Equipment	Total	Total Incl O&P
0010 **TABLES, FOLDING** Laminated plastic tops									
1000 Tubular steel legs with glides									
1020 18" x 60", minimum				Ea.	194			194	213
1040 Maximum					950			950	1,050
1840 36" x 96", minimum					231			231	254
1860 Maximum					1,700			1,700	1,875
2000 Round, wood stained, plywood top, 60" diameter, minimum					330			330	360
2020 Maximum				↓	2,050			2,050	2,250

12 54 16.20 Furniture, Restaurant

	Crew	Daily Output	Labor-Hours	Unit	Material	2007 Bare Costs Labor	Equipment	Total	Total Incl O&P
0010 **FURNITURE, RESTAURANT**									
0020 Bars, built-in, front bar	1 Carp	5	1.600	L.F.	231	58.50		289.50	345
0200 Back bar	"	5	1.600	"	168	58.50		226.50	277
0300 Booth seating see Div. 12 54 16.70									
2000 Chair, bentwood side chair, metal, minimum				Ea.	78.50			78.50	86.50
2020 Maximum					92			92	101
2600 Upholstered seat & back, arms, minimum					158			158	174
2620 Maximum				↓	350			350	385

12 54 16.70 Booths

	Crew	Daily Output	Labor-Hours	Unit	Material	2007 Bare Costs Labor	Equipment	Total	Total Incl O&P
0010 **BOOTHS**									
1000 Banquet, upholstered seat and back, custom									

12 54 Hospitality Furniture

12 54 16 – Restaurant Furniture

12 54 16.70 Booths

		Crew	Daily Output	Labor-Hours	Unit	Material	2007 Bare Costs Labor	Equipment	Total	Total Incl O&P
1500	Straight, minimum	2 Carp	40	.400	L.F.	145	14.70		159.70	182
1520	Maximum		36	.444		281	16.30		297.30	335
1600	"L" or "U" shape, minimum		35	.457		147	16.80		163.80	188
1620	Maximum	↓	30	.533	↓	262	19.55		281.55	320
1800	Upholstered outside finished backs for									
1810	single booths and custom banquets									
1820	Minimum	2 Carp	44	.364	L.F.	16.70	13.35		30.05	39.50
1840	Maximum	"	40	.400	"	50	14.70		64.70	78
3000	Fixed seating, one piece plastic chair and									
3010	plastic laminate table top									
3100	Two seat, 24" x 24" table, minimum	F-7	30	1.067	Ea.	585	35		620	695
3120	Maximum		26	1.231		830	40.50		870.50	980
3200	Four seat, 24" x 48" table, minimum		28	1.143		580	37.50		617.50	695
3220	Maximum	↓	24	1.333		985	43.50		1,028.50	1,150
5000	Mount in floor, wood fiber core with									
5010	plastic laminate face, single booth									
5050	24" wide	F-7	30	1.067	Ea.	224	35		259	300
5100	48" wide	"	28	1.143	"	284	37.50		321.50	370

12 55 Detention Furniture

12 55 13 – Detention Bunks

12 55 13.13 Cots

		Crew	Daily Output	Labor-Hours	Unit	Material	2007 Bare Costs Labor	Equipment	Total	Total Incl O&P
0010	**COTS**									
2500	Cot, bolted, single, painted steel	E-4	20	1.600	Ea.	315	67	5.75	387.75	475
2700	Stainless steel	"	20	1.600	"	875	67	5.75	947.75	1,075

12 56 Institutional Furniture

12 56 13 – Bank Furniture

12 56 13.10 Bank Furniture

0010	**BANK FURNITURE** See Div. 12 51 16.13									

12 56 33 – Classroom Furniture

12 56 33.10 Furniture, School

		Crew	Daily Output	Labor-Hours	Unit	Material	Labor	Equipment	Total	Total Incl O&P
0010	**FURNITURE, SCHOOL**									
0500	Classroom, movable chair & desk type, minimum				Set				70	77
0600	Maximum				"				128	141
1000	Chair, molded plastic,									
1100	Integral tablet arm, minimum				Ea.	64			64	70.50
1150	Maximum					190			190	209
2000	Desk, single pedestal, top book compartment, minimum					56			56	61.50
2020	Maximum					76.50			76.50	84
2200	Flip top, minimum					87.50			87.50	96.50
2220	Maximum				↓	104			104	115

12 56 43 – Dormitory Furniture

12 56 43.10 Furniture, Dormitory

		Crew	Daily Output	Labor-Hours	Unit	Material	Labor	Equipment	Total	Total Incl O&P
0010	**FURNITURE, DORMITORY**									
1000	Chest, four drawer, minimum				Ea.	420			420	460
1020	Maximum				"	550			550	605
1050	Built-in, minimum	2 Carp	13	1.231	L.F.	120	45		165	203

12 56 Institutional Furniture

12 56 43 – Dormitory Furniture

12 56 43.10 Furniture, Dormitory

		Crew	Daily Output	Labor-Hours	Unit	Material	2007 Bare Costs Labor	Equipment	Total	Total Incl O&P
1150	Maximum	2 Carp	10	1.600	L.F.	222	58.50		280.50	335
1200	Desk top, built-in, laminated plastic, 24" deep, minimum		50	.320		35.50	11.75		47.25	57.50
1300	Maximum		40	.400		107	14.70		121.70	140
1450	30" deep, minimum		50	.320		45.50	11.75		57.25	68.50
1550	Maximum		40	.400		199	14.70		213.70	242
1750	Dressing unit, built-in, minimum		12	1.333		180	49		229	274
1850	Maximum	↓	8	2	↓	540	73.50		613.50	710
8000	Rule of thumb: total cost for furniture, minimum				Student				2,025	2,225
8050	Maximum				"				3,900	4,250

12 56 51 – Library Furniture

12 56 51.10 Library Furniture

		Crew	Daily Output	Labor-Hours	Unit	Material	2007 Bare Costs Labor	Equipment	Total	Total Incl O&P
0010	**LIBRARY FURNITURE**									
0100	Attendant desk, 36" x 62" x 29" high	1 Carp	16	.500	Ea.	2,500	18.35		2,518.35	2,775
0200	Book display, "A" frame display, both sides, 42" x 42" x 60" high		16	.500		2,125	18.35		2,143.35	2,375
0220	Table with bulletin board, 42" x 24" x 49" high		16	.500		1,550	18.35		1,568.35	1,725
0800	Card catalogue, 30 tray unit		16	.500		2,825	18.35		2,843.35	3,150
0840	60 tray unit	↓	16	.500		4,975	18.35		4,993.35	5,500
0880	72 tray unit	2 Carp	16	1		5,650	36.50		5,686.50	6,275
1000	Carrels, single face, initial unit	1 Carp	16	.500		695	18.35		713.35	795
1500	Double face, initial unit	2 Carp	16	1		1,050	36.50		1,086.50	1,200
1710	Carrels, hardwood, 36" x 24", minimum	1 Carp	5	1.600		715	58.50		773.50	875
1720	Maximum	"	4	2		1,100	73.50		1,173.50	1,325
2700	Card catalog file, 60 trays, complete					4,525			4,525	4,975
2720	Alternate method: each tray					75.50			75.50	83
3100	Chairs, wood				↓	156			156	171
3500	Charging desk, built-in, with counter, plastic laminated top	1 Carp	7	1.143	L.F.	435	42		477	545
4000	Dictionary stand, stationary		16	.500	Ea.	745	18.35		763.35	850
4020	Revolving		16	.500		295	18.35		313.35	355
4200	Exhibit case, table style, 60" x 28" x 36"		11	.727	↓	4,100	26.50		4,126.50	4,550
6010	Bookshelf, mtl, 90" high, 10" shelf, dbl face		11.50	.696	L.F.	153	25.50		178.50	208
6020	Single face	↓	12	.667	"	145	24.50		169.50	198
6050	For 8" shelving, subtract from above					10%				
6060	For 12" shelving, add to above					10%				
6070	For 42" high with countertop, subtract from above					20%				
6100	Mobile compacted shelving, hand crank, 9'-0" high									
6110	Double face, including track, 3' section				Ea.	1,075			1,075	1,175
6150	For electrical operation, add					25%				
6200	Magazine shelving, 82" high, 12" deep, single face	1 Carp	11.50	.696	L.F.	224	25.50		249.50	286
6210	Double face		11.50	.696	"	203	25.50		228.50	263
7000	Tables, card catalog reference, 24" x 60" x 42"	↓	16	.500	Ea.	930	18.35		948.35	1,050
7200	Reading table, laminated top, 60" x 36"				"	500			500	550

12 56 70 – Healthcare Furniture

12 56 70.10 Furniture, Hospital

		Crew	Daily Output	Labor-Hours	Unit	Material	2007 Bare Costs Labor	Equipment	Total	Total Incl O&P
0010	**FURNITURE, HOSPITAL**									
0020	Beds, manual, minimum				Ea.	795			795	875
0100	Maximum					2,475			2,475	2,725
0300	Manual and electric beds, minimum					1,025			1,025	1,125
0400	Maximum					2,750			2,750	3,025
0600	All electric hospital beds, minimum					1,275			1,275	1,400
0700	Maximum					3,975			3,975	4,375
0900	Manual, nursing home beds, minimum					785			785	865
1000	Maximum				↓	1,625			1,625	1,800

12 56 Institutional Furniture

12 56 70 – Healthcare Furniture

12 56 70.10 Furniture, Hospital		Crew	Daily Output	Labor-Hours	Unit	Material	2007 Bare Costs Labor	Equipment	Total	Total Incl O&P
1020	Overbed table, laminated top, minimum				Ea.	335			335	370
1040	Maximum				"	835			835	915
1100	Patient wall systems, not incl. plumbing, minimum				Room	1,000			1,000	1,100
1200	Maximum				"	1,850			1,850	2,025
2000	Geriatric chairs, minimum				Ea.	480			480	525
2020	Maximum				"	805			805	885

12 61 Fixed Audience Seating

12 61 13 – Upholstered Audience Seating

12 61 13.13 Auditorium Chairs

		Crew	Daily Output	Labor-Hours	Unit	Material	Labor	Equipment	Total	Total Incl O&P
0010	**AUDITORIUM CHAIRS**									
2000	Auditorium chair, all veneer construction	2 Carp	22	.727	Ea.	155	26.50		181.50	212
2200	Veneer back, padded seat		22	.727		195	26.50		221.50	257
2350	Fully upholstered, spring seat		22	.727		195	26.50		221.50	257
2450	For tablet arms, add					51			51	56
2500	For fire retardancy, CATB-133, add					22			22	24

12 61 13.23 Lecture Hall

		Crew	Daily Output	Labor-Hours	Unit	Material	Labor	Equipment	Total	Total Incl O&P
0010	**LECTURE HALL** seating									
1000	Lecture hall, pedestal type, minimum	2 Carp	22	.727	Ea.	146	26.50		172.50	202
1200	Maximum	"	14.50	1.103	"	490	40.50		530.50	605

12 63 Stadium and Arena Seating

12 63 13 – Stadium and Arena Bench Seating

12 63 13.13 Bleachers

		Crew	Daily Output	Labor-Hours	Unit	Material	Labor	Equipment	Total	Total Incl O&P
0010	**BLEACHERS**									
3000	Bleachers, telescoping, manual to 15 tier, minimum	F-5	65	.492	Seat	74	18.30		92.30	110
3100	Maximum		60	.533		111	19.85		130.85	153
3300	16 to 20 tier, minimum		60	.533		177	19.85		196.85	226
3400	Maximum		55	.582		222	21.50		243.50	278
3600	21 to 30 tier, minimum		50	.640		185	24		209	240
3700	Maximum		40	.800		222	30		252	291
3900	For integral power operation, add, minimum	2 Elec	300	.053		37	2.34		39.34	44
4000	Maximum	"	250	.064		59	2.81		61.81	69
5000	Benches, folding, in wall, 14' table, 2 benches	L-4	2	12	Set	625	415		1,040	1,325

12 67 Pews and Benches

12 67 13 – Pews

12 67 13.13 Sanctuary Pews

		Crew	Daily Output	Labor-Hours	Unit	Material	Labor	Equipment	Total	Total Incl O&P
0010	**SANCTUARY PEWS**									
1500	Pews, bench type, hardwood, minimum	1 Carp	20	.400	L.F.	67	14.70		81.70	97
1550	Maximum	"	15	.533		133	19.55		152.55	178
1570	For kneeler, add					16			16	17.60

12 92 Interior Planters and Artificial Plants

12 92 33 – Interior Planters

12 92 33.10 Planters

		Crew	Daily Output	Labor-Hours	Unit	Material	2007 Bare Costs Labor	Equipment	Total	Total Incl O&P
0010	**PLANTERS**									
1000	Fiberglass, hanging, 12" diameter, 7" high				Ea.	89			89	98
1500	Rectangular, 48" long, 16" high x 15" wide					455			455	500
1650	60" long, 30" high, 28" wide					885			885	975
2000	Round, 12" diameter, 13" high					108			108	118
2050	25" high					118			118	130
5000	Square, 10" side, 20" high					115			115	127
5100	14" side, 15" high					181			181	199
6000	Metal bowl, 32" diameter, 8" high, minimum					360			360	395
6050	Maximum					455			455	500
8750	Wood, fiberglass liner, square									
8780	14" square, 15" high, minimum				Ea.	221			221	243
8800	Maximum					272			272	299
9400	Plastic cylinder, molded, 10" diameter, 10" high					8.95			8.95	9.85
9500	11" diameter, 11" high					28.50			28.50	31

12 93 Site Furnishings

12 93 23 – Trash and Litter Receptors

12 93 23.10 Trash Receptacles

		Crew	Daily Output	Labor-Hours	Unit	Material	Labor	Equipment	Total	Total Incl O&P
0010	**TRASH RECEPTACLES**									
0020	Trash receptacle, fiberglass, 2' square, 18" high	2 Clab	30	.533	Ea.	279	15.35		294.35	330
0100	2' square, 2'-6" high		30	.533		390	15.35		405.35	455
0300	Circular, 2' diameter, 18" high		30	.533		251	15.35		266.35	300
0400	2' diameter, 2'-6" high		30	.533		335	15.35		350.35	395

12 93 23.20 Trash Closure

		Crew	Daily Output	Labor-Hours	Unit	Material	Labor	Equipment	Total	Total Incl O&P
0010	**TRASH CLOSURE** Steel with pullover cover									
0020	2'-3" wide, 4'-7" high, 6'-2" long	2 Clab	5	3.200	Ea.	1,025	92		1,117	1,275
0100	10'-1" long		4	4		1,350	115		1,465	1,675
0300	Wood, 10' wide, 6' high, 10' long		1.20	13.333		1,075	385		1,460	1,775

12 93 33 – Manufactured Planters

12 93 33.10 Planters

		Crew	Daily Output	Labor-Hours	Unit	Material	Labor	Equipment	Total	Total Incl O&P
0010	**PLANTERS**									
0012	Concrete, sandblasted, precast, 48" diameter, 24" high	2 Clab	15	1.067	Ea.	615	30.50		645.50	725
0100	Fluted, precast, 7' diameter, 36" high		10	1.600		1,025	46		1,071	1,225
0300	Fiberglass, circular, 36" diameter, 24" high		15	1.067		450	30.50		480.50	545
0320	36" diameter, 27" high		12	1.333		360	38.50		398.50	455
0330	33" high		15	1.067		470	30.50		500.50	565
0335	24" diameter, 36" high		15	1.067		400	30.50		430.50	490
0340	60" diameter, 39" high		8	2		1,075	57.50		1,132.50	1,275
0400	60" diameter, 24" high		10	1.600		760	46		806	905
0600	Square, 24" side, 36" high		15	1.067		435	30.50		465.50	530
0610	24" side, 27" high		12	1.333		286	38.50		324.50	375
0620	24" side, 16" high		20	.800		315	23		338	385
0700	48" side, 36" high		15	1.067		975	30.50		1,005.50	1,125
0900	Planter/bench, 72" square, 36" high		5	3.200		1,200	92		1,292	1,475
1000	96" square, 27" high		5	3.200		1,850	92		1,942	2,200
1200	Wood, square, 48" side, 24" high		15	1.067		950	30.50		980.50	1,100
1300	Circular, 48" diameter, 30" high		10	1.600		780	46		826	930
1500	72" diameter, 30" high		10	1.600		1,400	46		1,446	1,600
1600	Planter/bench, 72"		5	3.200		3,075	92		3,167	3,525

12 93 Site Furnishings

12 93 43 – Site Seating and Tables

12 93 43.13 Site Seating		Crew	Daily Output	Labor-Hours	Unit	Material	2007 Bare Costs Labor	Equipment	Total	Total Incl O&P
0010	**SITE SEATING**									
0012	Seating, benches, park, precast conc, w/backs, wood rails, 4' long	2 Clab	5	3.200	Ea.	350	92		442	530
0100	8' long		4	4		735	115		850	990
0300	Fiberglass, without back, one piece, 4' long		10	1.600		505	46		551	625
0400	8' long		7	2.286		1,050	65.50		1,115.50	1,250
0500	Steel barstock pedestals w/backs, 2" x 3" wood rails, 4' long		10	1.600		875	46		921	1,025
0510	8' long		7	2.286		1,025	65.50		1,090.50	1,225
0520	3" x 8" wood plank, 4' long		10	1.600		880	46		926	1,050
0530	8' long		7	2.286		920	65.50		985.50	1,100
0540	Backless, 4" x 4" wood plank, 4' square		10	1.600		855	46		901	1,000
0550	8' long		7	2.286		810	65.50		875.50	995
0600	Aluminum pedestals, with backs, aluminum slats, 8' long		8	2		266	57.50		323.50	385
0610	15' long		5	3.200		420	92		512	605
0620	Portable, aluminum slats, 8' long		8	2		370	57.50		427.50	500
0630	15' long		5	3.200		545	92		637	745
0800	Cast iron pedestals, back & arms, wood slats, 4' long		8	2		325	57.50		382.50	445
0820	8' long		5	3.200		940	92		1,032	1,175
0840	Backless, wood slats, 4' long		8	2		560	57.50		617.50	705
0860	8' long		5	3.200		710	92		802	925
1700	Steel frame, fir seat, 10' long		10	1.600		200	46		246	291

Division Notes

		CREW	DAILY OUTPUT	LABOR-HOURS	UNIT	\multicolumn{4}{c}{2007 BARE COSTS}	TOTAL INCL O&P
						MAT.	LABOR	EQUIP.	TOTAL	

Estimating Tips

General

- The items and systems in this division are usually estimated, purchased, supplied, and installed as a unit by one or more subcontractors. The estimator must ensure that all parties are operating from the same set of specifications and assumptions, and that all necessary items are estimated and will be provided. Many times the complex items and systems are covered, but the more common ones, such as excavation or a crane, are overlooked for the very reason that everyone assumes nobody could miss them. The estimator should be the central focus and be able to ensure that all systems are complete.
- Another area where problems can develop in this division is at the interface between systems. The estimator must ensure, for instance, that anchor bolts, nuts, and washers are estimated and included for the air-supported structures and pre-engineered buildings to be bolted to their foundations. Utility supply is a common area where essential items or pieces of equipment can be missed or overlooked due to the fact that each subcontractor may feel it is another's responsibility. The estimator should also be aware of certain items which may be supplied as part of a package but installed by others, and ensure that the installing contractor's estimate includes the cost of installation. Conversely, the estimator must also ensure that items are not costed by two different subcontractors, resulting in an inflated overall estimate.

13 30 00 Special Structures

- The foundations and floor slab, as well as rough mechanical and electrical, should be estimated, as this work is required for the assembly and erection of the structure. Generally, as noted in the book, the pre-engineered building comes as a shell and additional features, such as windows and doors, must be included by the estimator. Here again, the estimator must have a clear understanding of the scope of each portion of the work and all the necessary interfaces.

Reference Numbers

Reference numbers are shown in shaded boxes at the beginning of some major classifications. These numbers refer to related items in the Reference Section. The reference information may be an estimating procedure, an alternate pricing method, or technical information.

Note: Not all subdivisions listed here necessarily appear in this publication.

13 05 Common Work Results for Special Construction

13 05 05 – Selective Special Construction Demolition

13 05 05.10 Selective Demolition, Air Supported Structures	Crew	Daily Output	Labor-Hours	Unit	Material	2007 Bare Costs Labor	Equipment	Total	Total Incl O&P
0010 **SELECTIVE DEMOLITION, AIR SUPPORTED STRUCTURES**									
0020 Tank covers, scrim, dbl layer, vinyl poly w/ hdw, blower & controls									
0050 Round and rectangular R024119-10	B-2	9000	.004	S.F.		.13		.13	.20
0100 Warehouse structures									
0120 Poly/vinyl fabric, 28 oz, incl tension cables & inflation system	4 Clab	9000	.004	SF Flr.		.10		.10	.16
0150 Warehouse, reinforced vinyl, 12 oz., 3000 S.F.	"	5000	.006			.18		.18	.29
0200 12,000 to 24,000 S.F.	8 Clab	20000	.003			.09		.09	.14
0250 Warehouse, tedlar vinyl fabric, 28 oz. w/liner, to 3000 S.F.	4 Clab	5000	.006			.18		.18	.29
0300 12,000 to 24,000 S.F.	8 Clab	20000	.003			.09		.09	.14
0350 Greenhouse/shelter, woven polyethylene with liner									
0400 3000 S.F.	4 Clab	5000	.006	SF Flr.		.18		.18	.29
0450 12,000 to 24,000 S.F.	8 Clab	20000	.003			.09		.09	.14
0500 Tennis/gymnasium, poly/vinyl fabric, 28 oz., incl thermal liner	4 Clab	9000	.004			.10		.10	.16
0600 Stadium/convention center, teflon coated fiberglass, incl thermal liner	9 Clab	40000	.002			.05		.05	.08
0700 Doors, air lock, 15' long, 10' x 10'	2 Carp	1.50	10.667	Ea.		390		390	610
0720 15' x 15'		.80	20			735		735	1,150
0750 Revolving personnel door, 6' dia. x 6'-6" high		1.50	10.667			390		390	610

13 05 05.20 Selective Demolition, Garden Houses	Crew	Daily Output	Labor-Hours	Unit	Material	Labor	Equipment	Total	Total Incl O&P
0010 **SELECTIVE DEMOLITION, GARDEN HOUSES** R024119-10									
0020 Garden house, prefab, wood, excl foundation, average	2 Clab	400	.040	SF Flr.		1.15		1.15	1.79

13 05 05.25 Selective Demolition, Geodesic Domes	Crew	Daily Output	Labor-Hours	Unit	Material	Labor	Equipment	Total	Total Incl O&P
0010 **SELECTIVE DEMOLITION, GEODESIC DOMES**									
0050 Geodesic dome, shell only, interlocking plywood panels, 30' diameter	F-5	3.20	10	Ea.		370		370	580
0060 34' diameter		2.30	13.913			520		520	805
0070 39' diameter		2	16			595		595	925
0080 45' diameter		2.20	14.545			540		540	845
0090 55' diameter		2	16			595		595	925
0100 60' diameter		2	16			595		595	925
0110 65' diameter		1.60	20			745		745	1,150

13 05 05.30 Selective Demolition, Greenhouses	Crew	Daily Output	Labor-Hours	Unit	Material	Labor	Equipment	Total	Total Incl O&P
0010 **SELECTIVE DEMOLITION, GREENHOUSES** R024119-10									
0020 Greenhouse, resi-type, free standing, excl foundations, 9' long x 8' wide	2 Clab	160	.100	SF Flr.		2.88		2.88	4.48
0030 9' long x 11' wide		170	.094			2.71		2.71	4.21
0040 9' long x 14' wide		220	.073			2.09		2.09	3.25
0050 9' long x 17' wide		320	.050			1.44		1.44	2.24
0060 Lean-to type, 4' wide		64	.250			7.20		7.20	11.20
0070 7' wide		120	.133			3.83		3.83	5.95
0080 Geodesic hemishere, 1/8" plexiglass glazing, 8' dia		4	4	Ea.		115		115	179
0090 24' dia		.80	20			575		575	895
0100 48' dia		.40	40			1,150		1,150	1,800

13 05 05.35 Selective Demolition, Hangars	Crew	Daily Output	Labor-Hours	Unit	Material	Labor	Equipment	Total	Total Incl O&P
0010 **SELECTIVE DEMOLITION, HANGARS**									
0020 Hangars, prefab, steel T type, galv roof & walls, incl doors, excl fndtn	E-2	2550	.022	SF Flr.		.89	.61	1.50	2.20
0030 Circular type, prefab, steel frame, plastic skin, incl foundation, 80' dia	"	.50	112	Total		4,525	3,100	7,625	11,200

13 05 05.45 Selective Demolition, Lightning Protection	Crew	Daily Output	Labor-Hours	Unit	Material	Labor	Equipment	Total	Total Incl O&P
0010 **SELECTIVE DEMOLITION, LIGHTNING PROTECTION**									
0020 Air terminal & base, copper, 3/8" dia x 10", to 75' h	1 Clab	16	.500	Ea.		14.40		14.40	22.50
0030 1/2" dia x 12", over 75' h		16	.500			14.40		14.40	22.50
0050 Aluminum, 1/2" dia x 12", to 75' h		16	.500			14.40		14.40	22.50
0060 5/8" dia x 12", over 75' h		16	.500			14.40		14.40	22.50
0070 Cable, copper, 220 lb per thousand feet, to 75' high		640	.013	L.F.		.36		.36	.56

Means CostWorks®

Maximize Your Estimating & Budgeting Efforts with the RSMeans Online Construction Cost Estimator... or Means CostWorks 2007 CD-Rom

RSMeans Online Construction Cost Estimator: *Estimates that are efficient, customized & online*

- Create database-driven, customized estimates based on your specific building needs in a 24/7 secure online environment.
- Have immediate access to the entire RSMeans unit cost (70,000+ line items) & assemblies (19,000) databases, as well as referencing both union & open shop labor rates.
- Streamline your process with an optional "read-write" capability.

Means CostWorks 2007 CD-ROM *The cost data titles you know and trust – delivered electronically in a quick & easy CD format.*

- Purchase your titles separately or in value-priced packages & turn your desktop or laptop PC into a comprehensive cost library!
- Easily search & organize detailed cost data.
- Localize costs to your geographic area.
- Change & calculate results.

RSMeans

Get started with Means CostWorks today!

Fax: 1-800-632-6732, phone: 1-800-334-3509, or at www.rsmeans.com/cwcw.asp
For information on network pricing call 1-800-334-3509.

Catalog No.	Means CostWorks 2007 CD Subscription (12 Months)	Price	Cost
65067	Assemblies Cost Data	$241.95	
65017	Building Construction Cost Data	$169.95	
65317	Metric Construction Cost Data	$201.95	
65157	Open Shop Building Construction Cost Data	$169.95	
65117	Concrete & Masonry Cost Data	$169.95	
65037	Electrical Cost Data	$169.95	
65167	Heavy Construction Cost Data	$169.95	
65097	Interior Cost Data	$169.95	
65187	Light Commercial Cost Data	$210.95	
65027	Mechanical Cost Data	$169.95	
65217	Plumbing Cost Data	$169.95	
65287	Site Work & Landscape Cost Data	$169.95	
65057	Square Foot Costs	$239.95	
65607*	CostWorks CD Estimator	$299.95	
65207	Facilities Construction Cost Data	$392.95	
65047	Repair & Remodeling Cost Data	$210.95	
65177	Residential Cost Data	$195.95	
65307	Facilities Maintenance & Repair Cost Data	$392.95	

CD Package Subscription* (12 Months)

Catalog No.	Title	Price	Cost
65517	Builder's Package	$469.95	
65527	Building Professional's Package	$549.95	
65537	Facility Manager's Package	$712.95	
65547	Design Professional's Package	$539.95	

*(Packages include FREE Means CD Estimator)
All items shipped on a single CD with software code to unlock specific selections purchased.
Buy additional keys any time you need to unlock additional cost data books!

RSMeans Online Construction Cost Estimator Subscription

Catalog No.	Title	Price	Cost
66100	1 Yr. Subscription – Standard Edition	Call for Pricing	
66300	1 Yr. Subscription – Customized Data Base Utility	Call for Pricing	

Name: _____

Company: _____

Street: _____

City: _____ State: ____ Zip: _____

Phone: (___) _____ Fax: (___) _____

E-mail: _____

☐ Bill me   ☐ Please charge my credit card

☐ Visa   ☐ MasterCard   ☐ American Express   ☐ Discover

Card # _____ Expiration Date: _____

Cardholder Signature: _____

Subtotal	
MA residents please add 5% sales tax	
Shipping & Handling (CD)	$6.00
Total	

Reed Construction Data®

CWCW-2007

13 05 Common Work Results for Special Construction

13 05 05 – Selective Special Construction Demolition

13 05 05.45 Selective Demolition, Lightning Protection

		Crew	Daily Output	Labor-Hours	Unit	Material	2007 Bare Costs Labor	2007 Bare Costs Equipment	Total	Total Incl O&P
0080	375 lb per thousand feet, over 75' high	1 Clab	460	.017	L.F.		.50		.50	.78
0090	Aluminum, 101 lb per thousand feet, to 75' high		560	.014			.41		.41	.64
0100	199 lb per thousand feet, over 75' high		480	.017			.48		.48	.75
0110	Arrester, 175 V AC, to ground		16	.500	Ea.		14.40		14.40	22.50
0120	650 V AC, to ground		13	.615	"		17.70		17.70	27.50

13 05 05.50 Selective Demolition, Pre-Engineered Steel Buildings

		Crew	Daily Output	Labor-Hours	Unit	Material	Labor	Equipment	Total	Total Incl O&P
0010	**SELECTIVE DEMOLITION, PRE-ENGINEERED STEEL BUILDINGS**									
0500	Pre-engd steel bldgs, rigid frame, clear span & multi post, excl salvage									
0550	3,500 to 7,500 S.F	L-11	1350	.024	SF Flr.		.79	1.15	1.94	2.56
0600	7,501 to 12,500 S.F		2000	.016			.54	.78	1.32	1.73
0650	12,500 S.F or greater		2200	.015			.49	.71	1.20	1.57
0700	Pre-engd steel building components									
0710	Entrance canopy, including frame 4' x 4'	E-24	8	4	Ea.		162	92.50	254.50	385
0720	4' x 8'	"	7	4.571			186	106	292	440
0730	H.M doors, self framing, single leaf	2 Skwk	8	2			76		76	118
0740	Double leaf		5	3.200			122		122	189
0760	Gutter, eave type		600	.027	L.F.		1.01		1.01	1.58
0770	Sash, single slide, double slide or fixed		24	.667	Ea.		25.50		25.50	39.50
0780	Skylight, fiberglass, to 30 S.F.		16	1			38		38	59
0785	Roof vents, circular, 12" to 24" diameter		12	1.333			50.50		50.50	79
0790	Continuous, 10' long		8	2			76		76	118
0900	Shelters, aluminum frame									
0910	Aluminum frame, acrylic glazing, 3' x 9' x 8' high	2 Skwk	2	8	Ea.		305		305	475
0920	9' x 12' x 8' high	"	1.50	10.667	"		405		405	630

13 05 05.60 Selective Demolition, Silos

		Crew	Daily Output	Labor-Hours	Unit	Material	Labor	Equipment	Total	Total Incl O&P
0010	**SELECTIVE DEMOLITION, SILOS**									
0020	Silo, conc stave, indstrl, conical/sloping bott, excl fndtn, 12' dia, 35' h	D-8	.22	181	Ea.		6,225		6,225	9,475
0030	16' dia, 45' h		.16	250			8,575		8,575	13,000
0040	25' dia, 75' h		.10	400			13,700		13,700	20,900
0050	Steel, factory fabricated, 30,000 gal cap, painted or epoxy lined	L-5	2	28			1,150	370	1,520	2,450

13 05 05.65 Selective Demolition, Sound Control

		Crew	Daily Output	Labor-Hours	Unit	Material	Labor	Equipment	Total	Total Incl O&P
0010	**SELECTIVE DEMOLITION, SOUND CONTROL** R024119-10									
0120	Acoustical enclosure, 4" thick walls & ceiling panels, 8 lbs/SF	3 Carp	144	.167	SF Surf		6.10		6.10	9.55
0130	10.5 lbs/SF		128	.188			6.90		6.90	10.70
0140	Reverb chamber, parallel walls, 4" thick		120	.200			7.35		7.35	11.45
0150	Skewed walls, parallel roof, 4" thick		110	.218			8		8	12.45
0160	Skewed walls/roof, 4" layer/air space		96	.250			9.20		9.20	14.30
0170	Sound-absorbing panels, painted metal, 2'-6" x 8', under 1,000 SF		430	.056			2.05		2.05	3.19
0180	Over 1,000 SF		480	.050			1.84		1.84	2.86
0190	Flexible transparent curtain, clear	3 Shee	430	.056			2.43		2.43	3.75
0192	50% clear, 50% foam		430	.056			2.43		2.43	3.75
0194	25% clear, 75% foam		430	.056			2.43		2.43	3.75
0196	100% foam		430	.056			2.43		2.43	3.75
0200	Audio-masking sys, incl speakers, amplfr, signal gnrtr, clng mntd, 5,000 SF									
0205	Ceiling mounted, 5,000 SF	2 Elec	4800	.003	S.F.		.15		.15	.22
0210	10,000 SF		5600	.003			.13		.13	.19
0220	Plenum mounted, 5,000 SF		7600	.002			.09		.09	.14
0230	10,000 SF		8800	.002			.08		.08	.12

13 05 05.70 Selective Demolition, Special Purpose Rooms

		Crew	Daily Output	Labor-Hours	Unit	Material	Labor	Equipment	Total	Total Incl O&P
0010	**SELECTIVE DEMOLITION, SPECIAL PURPOSE ROOMS** R024119-10									
0100	Audiometric rooms, under 500 S.F. surface	4 Carp	200	.160	SF Surf		5.85		5.85	9.15
0110	Over 500 S.F. surface	"	240	.133	"		4.89		4.89	7.60

13 05 Common Work Results for Special Construction

13 05 05 – Selective Special Construction Demolition

13 05 05.70 Selective Demolition, Special Purpose Rooms

		Crew	Daily Output	Labor-Hours	Unit	Material	2007 Bare Costs Labor	2007 Bare Costs Equipment	Total	Total Incl O&P
0200	Clean rooms, 12' x 12' soft wall, class 100	1 Carp	.30	26.667	Ea.		980		980	1,525
0210	Class 1000		.30	26.667			980		980	1,525
0220	Class 10,000		.35	22.857			840		840	1,300
0230	Class 100,000		.35	22.857			840		840	1,300
0300	Darkrooms, shell complete, 8' high	2 Carp	220	.073	SF Flr.		2.67		2.67	4.16
0310	12' high		110	.145	"		5.35		5.35	8.30
0350	Darkrooms doors, mini-cylindrical, revolving		4	4	Ea.		147		147	229
0400	Music room, practice modular		140	.114	SF Surf		4.19		4.19	6.55
0500	Refrigeration structures and finishes									
0510	Wall finish, 2 coat portland cement plaster, 1/2" thick	1 Clab	200	.040	S.F.		1.15		1.15	1.79
0520	Fiberglass panels, 1/8" thick		400	.020			.58		.58	.90
0530	Ceiling finish, polystyrene plastic, 1" to 2" thick		500	.016			.46		.46	.72
0540	4" thick		450	.018			.51		.51	.80
0550	Refrigerator, prefab aluminum walk-in, 7'-6" high, 6' x 6' OD	2 Carp	100	.160	SF Flr.		5.85		5.85	9.15
0560	10' x 10' OD		160	.100			3.67		3.67	5.70
0570	Over 150 S.F.		200	.080			2.94		2.94	4.57
0600	Sauna, prefabricated, including heater & controls, 7' high, to 30 S.F.		120	.133			4.89		4.89	7.60
0610	To 40 S.F.		140	.114			4.19		4.19	6.55
0620	To 60 S.F.		175	.091			3.36		3.36	5.25
0630	To 100 S.F.		220	.073			2.67		2.67	4.16
0640	To 130 S.F.		250	.064			2.35		2.35	3.66
0650	Steam bath, heater, timer, head, single, to 140 CF	1 Plum	2.20	3.636	Ea.		163		163	245
0660	To 300 CF		2.20	3.636			163		163	245
0670	Steam bath, comm. size, w/blow-down assembly, to 800 CF		1.80	4.444			199		199	300
0680	To 2500 CF		1.60	5			224		224	335
0690	Steam bath, comm. size, multiple, for motels, apts, 500 CF, 2 baths		2	4			179		179	270
0700	1,000 CF, 4 baths		1.40	5.714			256		256	385

13 05 05.75 Selective Demolition, Storage Tanks

			Crew	Daily Output	Labor-Hours	Unit	Material	2007 Bare Costs Labor	2007 Bare Costs Equipment	Total	Total Incl O&P
0010	**SELECTIVE DEMOLITION, STORAGE TANKS**										
0500	Steel tank, single wall, above ground, not incl fdn, pumps or piping										
0510	Single wall, 275 gallon	R024119-10	Q-1	3	5.333	Ea.		215		215	325
0520	550 thru 2,000 gallon		B-34P	2	12			450	214	664	915
0530	5,000 thru 10,000 gallon		B-34Q	2	12			455	600	1,055	1,350
0540	15,000 thru 30,000 gallon		B-34N	2	4			118	247	365	455
0600	Steel tank, double wall, above ground not incl fdn, pumps & piping										
0620	500 thru 2,000 gallon		B-34P	2	12	Ea.		450	214	664	915

13 05 05.85 Selective Demolition, Swimming Pool Equip

		Crew	Daily Output	Labor-Hours	Unit	Material	2007 Bare Costs Labor	2007 Bare Costs Equipment	Total	Total Incl O&P
0010	**SELECTIVE DEMOLITION, SWIMMING POOL EQUIP**									
0020	Diving stand, stainless steel, 3 meter	2 Clab	3	5.333	Ea.		153		153	239
0030	1 meter		5	3.200			92		92	143
0040	Diving board, 16' long, aluminum		5.40	2.963			85		85	133
0050	Fiberglass		5.40	2.963			85		85	133
0070	Ladders, heavy duty, stainless steel, 2 tread		14	1.143			33		33	51
0080	4 tread		12	1.333			38.50		38.50	59.50
0090	Lifeguard chair, stainless steel, fixed		5	3.200			92		92	143
0100	Slide, tubular, fiberglass, aluminum handrails & ladder, 5', straight		4	4			115		115	179
0110	8', curved		6	2.667			76.50		76.50	119
0120	10', curved		3	5.333			153		153	239
0130	12' straight, with platform		2.50	6.400			184		184	286
0140	Removable access ramp, stainless steel		4	4			115		115	179
0150	Removable stairs, stainless steel, collapsible		4	4			115		115	179

13 05 Common Work Results for Special Construction

13 05 05 – Selective Special Construction Demolition

13 05 05.90 Selective Demolition, Tension Structures

		Crew	Daily Output	Labor-Hours	Unit	Material	2007 Bare Costs Labor	Equipment	Total	Total Incl O&P
0010	**SELECTIVE DEMOLITION, TENSION STRUCTURES**									
0020	Tension structure, steel/alum frame, fabric shell, 60' clear span, 6,000 SF	B-41	2000	.022	SF Flr.		.66	.12	.78	1.16
0030	12,000 SF		2200	.020			.60	.11	.71	1.05
0040	80' clear span, 20,800 SF		2440	.018			.54	.10	.64	.94
0050	100' clear span, 10,000 SF	L-5	4350	.013			.53	.17	.70	1.13
0060	26,000 SF	"	4600	.012			.50	.16	.66	1.07

13 05 05.95 Selective Demolition, X-Ray/Radio Freq Protection

		Crew	Daily Output	Labor-Hours	Unit	Material	2007 Bare Costs Labor	Equipment	Total	Total Incl O&P
0010	**SELECTIVE DEMOLITION, X-RAY/RADIO FREQ PROTECTION**									
0020	Shielding lead, lined door frame, excl hdwe, 1/16" thick	1 Clab	4.80	1.667	Ea.		48		48	74.50
0030	Lead sheets, 1/16" thick R024119-10	2 Clab	270	.059	S.F.		1.70		1.70	2.65
0040	1/8" thick		240	.067			1.92		1.92	2.98
0050	Lead shielding, 1/4" thick		270	.059			1.70		1.70	2.65
0060	1/2" thick		240	.067			1.92		1.92	2.98
0070	Lead glass, 1/4" thick, 2.0 mm LE, 12" x 16"	2 Glaz	26	.615	Ea.		22		22	33.50
0080	24" x 36"		16	1			36		36	54.50
0090	36" x 60"		4	4			144		144	218
0100	Lead glass window frame, with 1/16" lead & voice passage, 36" x 60"		4	4			144		144	218
0110	Lead glass window frame, 24" x 36"		16	1			36		36	54.50
0120	Lead gypsum board, 5/8" thick with 1/16" lead	2 Clab	320	.050	S.F.		1.44		1.44	2.24
0130	1/8" lead		280	.057			1.64		1.64	2.56
0140	1/32" lead		400	.040			1.15		1.15	1.79
0150	Butt joints, 1/8" lead or thicker, 2" x 7' long batten strip		480	.033	Ea.		.96		.96	1.49
0160	X-ray protection, average radiography room, up to 300 SF, 1/16" lead, min		.50	32	Total		920		920	1,425
0170	Maximum		.30	53.333			1,525		1,525	2,375
0180	Deep therapy X-ray room, 250 KV cap, up to 300 SF, 1/4" lead, min		.20	80			2,300		2,300	3,575
0190	Max		.12	133			3,825		3,825	5,975
0880	Radio frequency shielding, prefab or screen-type copper or steel, minimum		360	.044	SF Surf		1.28		1.28	1.99
0890	Average		310	.052			1.48		1.48	2.31
0895	Maximum		290	.055			1.59		1.59	2.47

13 11 Swimming Pools

13 11 13 – Below-Grade Swimming Pools

13 11 13.50 Swimming Pools

		Crew	Daily Output	Labor-Hours	Unit	Material	2007 Bare Costs Labor	Equipment	Total	Total Incl O&P
0010	**SWIMMING POOLS** Residential in-ground, vinyl lined, concrete									
0020	Sides including equipment, sand bottom	B-52	300	.187	SF Surf	12.25	6.30	1.39	19.94	25
0100	Metal or polystyrene sides R131113-20	B-14	410	.117		10.25	3.56	.59	14.40	17.40
0200	Add for vermiculite bottom					.79			.79	.87
0500	Gunite bottom and sides, white plaster finish									
0600	12' x 30' pool	B-52	145	.386	SF Surf	23	13.10	2.88	38.98	48
0720	16' x 32' pool		155	.361		20.50	12.25	2.69	35.44	44.50
0750	20' x 40' pool		250	.224		18.35	7.60	1.67	27.62	33.50
0810	Concrete bottom and sides, tile finish									
0820	12' x 30' pool	B-52	80	.700	SF Surf	23	23.50	5.20	51.70	68
0830	16' x 32' pool		95	.589		19.05	19.95	4.39	43.39	57
0840	20' x 40' pool		130	.431		15.15	14.60	3.21	32.96	42.50
1100	Motel, gunite with plaster finish, incl. medium									
1150	capacity filtration & chlorination	B-52	115	.487	SF Surf	28	16.50	3.63	48.13	60.50
1200	Municipal, gunite with plaster finish, incl. high									
1250	capacity filtration & chlorination	B-52	100	.560	SF Surf	36.50	18.95	4.17	59.62	74
1350	Add for formed gutters				L.F.	53.50			53.50	59

13 11 Swimming Pools

13 11 13 – Below-Grade Swimming Pools

13 11 13.50 Swimming Pools

		Crew	Daily Output	Labor-Hours	Unit	Material	2007 Bare Costs Labor	2007 Bare Costs Equipment	Total	Total Incl O&P
1360	Add for stainless steel gutters				L.F.	158			158	174
1600	For water heating system, see division 23 52 28.10									
1700	Filtration and deck equipment only, as % of total				Total				20%	20%
1800	Deck equipment, rule of thumb, 20' x 40' pool				SF Pool				1.18	1.30
1900	5000 S.F. pool				"				1.73	1.90
3000	Painting pools, preparation + 3 coats, 20' x 40' pool, epoxy	2 Pord	.33	48.485	Total	700	1,575		2,275	3,150
3100	Rubber base paint, 18 gallons	"	.33	48.485		530	1,575		2,105	2,950
3500	42' x 82' pool, 75 gallons, epoxy paint	3 Pord	.14	171		2,950	5,600		8,550	11,700
3600	Rubber base paint	"	.14	171		2,275	5,600		7,875	10,900

13 11 46 – Swimming Pool Accessories

13 11 46.50 Swimming Pool Equipment

		Crew	Daily Output	Labor-Hours	Unit	Material	2007 Bare Costs Labor	2007 Bare Costs Equipment	Total	Total Incl O&P
0010	**SWIMMING POOL EQUIPMENT**									
0020	Diving stand, stainless steel, 3 meter	2 Carp	.40	40	Ea.	6,150	1,475		7,625	9,050
0300	1 meter		2.70	5.926		5,475	217		5,692	6,375
0600	Diving boards, 16' long, aluminum		2.70	5.926		3,050	217		3,267	3,700
0700	Fiberglass		2.70	5.926		2,450	217		2,667	3,050
1100	Gutter system, stainless steel, with grating, stock,									
1110	contains supply and drainage system	E-1	20	1.200	L.F.	205	48.50	5.75	259.25	315
1120	Integral gutter and 5' high wall system, stainless steel	"	10	2.400	"	310	97.50	11.50	419	520
1200	Ladders, heavy duty, stainless steel, 2 tread	2 Carp	7	2.286	Ea.	535	84		619	720
1500	4 tread		6	2.667		620	98		718	830
1800	Lifeguard chair, stainless steel, fixed		2.70	5.926		1,375	217		1,592	1,875
1900	Portable					1,150			1,150	1,250
2100	Lights, underwater, 12 volt, with transformer, 300 watt	1 Elec	.40	20		175	880		1,055	1,500
2200	110 volt, 500 watt, standard		.40	20		122	880		1,002	1,425
2400	Low water cutoff type		.40	20		156	880		1,036	1,475
2800	Heaters, see division 23 52 28.10									
3000	Pool covers, reinforced vinyl	3 Clab	1800	.013	S.F.	.33	.38		.71	.96
3050	Automatic, electric								8.77	9.65
3100	Vinyl water tube	3 Clab	3200	.008		.46	.22		.68	.85
3200	Maximum	"	3000	.008		.58	.23		.81	1
3250	Sealed air bubble polyethylene solar blanket					.24			.24	.26
3300	Slides, tubular, fiberglass, aluminum handrails & ladder, 5'-0", straight	2 Carp	1.60	10	Ea.	2,325	365		2,690	3,125
3320	8'-0", curved		3	5.333		8,875	196		9,071	10,100
3400	10'-0", curved		1	16		18,000	585		18,585	20,700
3420	12'-0", straight with platform		1.20	13.333		4,950	490		5,440	6,200
4500	Hydraulic lift, movable pool bottom, single ram									
4520	Under 1,000 S.F. area	L-9	.03	1200	Ea.	80,000	40,000		120,000	152,500
4600	Four ram lift, over 1,000 S.F.	"	.02	1800		96,500	60,000		156,500	203,000
5000	Removable access ramp, stainless steel	2 Clab	2	8		7,925	230		8,155	9,075
5500	Removable stairs, stainless steel, collapsible	"	2	8		3,850	230		4,080	4,575

13 17 Tubs and Pools

13 17 23 – Therapeutic Pools

13 17 23.50 Therapeutic Pools	Crew	Daily Output	Labor-Hours	Unit	Material	2007 Bare Costs Labor	Equipment	Total	Total Incl O&P
0010 **THERAPEUTIC POOLS** See division 22 41 19.10									

13 18 Ice Rinks

13 18 13 – Ice Rink Floor Systems

13 18 13.50 Ice Skating

		Crew	Daily Output	Labor-Hours	Unit	Material	Labor	Equipment	Total	Total Incl O&P
0010	**ICE SKATING** Equipment incl. refrigeration, plumbing & cooling									
0020	coils & concrete slab, 85′ x 200′ rink									
0300	55° system, 5 mos., 100 ton				Total	496,000			496,000	545,500
0700	90° system, 12 mos., 135 ton				"	550,500			550,500	605,500
1200	Subsoil heating system (recycled from compressor), 85′ x 200′	Q-7	.27	118	Ea.	25,000	5,100		30,100	35,200
1300	Subsoil insulation, 2 lb. polystyrene with vapor barrier, 85′ x 200′	2 Carp	.14	114	"	30,000	4,200		34,200	39,500

13 18 16 – Ice Rink Dasher Boards

13 18 16.50 Ice Rink Dasher Boards

		Crew	Daily Output	Labor-Hours	Unit	Material	Labor	Equipment	Total	Total Incl O&P
0010	**ICE RINK DASHER BOARDS**									
1000	Dasher boards, 1/2" H.D. polyethylene faced steel frame, 3′ acrylic									
1020	screen at sides, 5′ acrylic ends, 85′ x 200′	F-5	.06	533	Ea.	115,000	19,800		134,800	157,500
1100	Fiberglass & aluminum construction, same sides and ends	"	.06	533	"	160,000	19,800		179,800	207,000

13 21 Controlled Environment Rooms

13 21 13 – Clean Rooms

13 21 13.50 Clean Rooms

		Crew	Daily Output	Labor-Hours	Unit	Material	Labor	Equipment	Total	Total Incl O&P
0010	**CLEAN ROOMS**									
1100	Clean room, soft wall, 12′ x 12′, Class 100	1 Carp	.18	44.444	Ea.	13,600	1,625		15,225	17,600
1110	Class 1,000		.18	44.444		10,600	1,625		12,225	14,200
1120	Class 10,000		.21	38.095		9,000	1,400		10,400	12,100
1130	Class 100,000		.21	38.095		8,325	1,400		9,725	11,300
2800	Ceiling grid support, slotted channel struts 4′-0" O.C., ea. way				S.F.				5.91	6.50
3000	Ceiling panel, vinyl coated foil on mineral substrate									
3020	Sealed, non-perforated				S.F.				1.27	1.40
4000	Ceiling panel seal, silicone sealant, 150 L.F./gal.	1 Carp	150	.053	L.F.	.24	1.96		2.20	3.31
4100	Two sided adhesive tape	"	240	.033	"	.11	1.22		1.33	2.02
4200	Clips, one per panel				Ea.	.95			.95	1.05
6000	HEPA filter, 2′x4′, 99.97% eff., 3" dp beveled frame (silicone seal)					290			290	320
6040	6" deep skirted frame (channel seal)					335			335	370
6100	99.99% efficient, 3" deep beveled frame (silicone seal)					315			315	345
6140	6" deep skirted frame (channel seal)					355			355	395
6200	99.999% efficient, 3" deep beveled frame (silicone seal)					355			355	395
6240	6" deep skirted frame (channel seal)					400			400	440
7000	Wall panel systems, including channel strut framing									
7020	Polyester coated aluminum, particle board				S.F.				18.18	20
7100	Porcelain coated aluminum, particle board								31.82	35
7400	Wall panel support, slotted channel struts, to 12′ high								16.36	18

13 21 26 – Cold Storage Rooms

13 21 26.50 Refrigeration

		Crew	Daily Output	Labor-Hours	Unit	Material	Labor	Equipment	Total	Total Incl O&P
0010	**REFRIGERATION**									
0020	Curbs, 12" high, 4" thick, concrete	2 Carp	58	.276	L.F.	3.70	10.10		13.80	19.80
1000	Doors, see division 08 34 13.10									
2400	Finishes, 2 coat portland cement plaster, 1/2" thick	1 Plas	48	.167	S.F.	1.02	5.60		6.62	9.55

13 21 Controlled Environment Rooms

13 21 26 – Cold Storage Rooms

13 21 26.50 Refrigeration		Crew	Daily Output	Labor-Hours	Unit	Material	2007 Bare Costs Labor	Equipment	Total	Total Incl O&P
2500	For galvanized reinforcing mesh, add	1 Lath	335	.024	S.F.	.68	.80		1.48	1.95
2700	3/16" thick latex cement	1 Plas	88	.091		1.74	3.05		4.79	6.50
2900	For glass cloth reinforced ceilings, add	"	450	.018		.42	.60		1.02	1.36
3100	Fiberglass panels, 1/8" thick	1 Carp	149.45	.054		2.30	1.96		4.26	5.60
3200	Polystyrene, plastic finish ceiling, 1" thick		274	.029		2.12	1.07		3.19	4
3400	2" thick		274	.029		2.42	1.07		3.49	4.33
3500	4" thick		219	.037		2.69	1.34		4.03	5.05
3800	Floors, concrete, 4" thick	1 Cefi	93	.086		1.01	3.06		4.07	5.60
3900	6" thick	"	85	.094		1.51	3.35		4.86	6.55
4000	Insulation, 1" to 6" thick, cork				B.F.	.99			.99	1.09
4100	Urethane					.97			.97	1.07
4300	Polystyrene, regular					.66			.66	.73
4400	Bead board					.49			.49	.54
4600	Installation of above, add per layer	2 Carp	657.60	.024	S.F.	.32	.89		1.21	1.74
4700	Wall and ceiling juncture		298.90	.054	L.F.	1.58	1.96		3.54	4.80
4900	Partitions, galvanized sandwich panels, 4" thick, stock		219.20	.073	S.F.	6.65	2.68		9.33	11.45
5000	Aluminum or fiberglass		219.20	.073	"	7.25	2.68		9.93	12.15
5200	Prefab walk-in, 7'-6" high, aluminum, incl. door & floors,									
5210	not incl. partitions or refrigeration, 6' x 6' O.D. nominal	2 Carp	54.80	.292	SF Flr.	119	10.70		129.70	148
5500	10' x 10' O.D. nominal		82.20	.195		95.50	7.15		102.65	116
5700	12' x 14' O.D. nominal		109.60	.146		85.50	5.35		90.85	103
5800	12' x 20' O.D. nominal		109.60	.146		74.50	5.35		79.85	90.50
6100	For 8'-6" high, add					5%				
6300	Rule of thumb for complete units, w/o doors & refrigeration, cooler	2 Carp	146	.110		108	4.02		112.02	125
6400	Freezer		109.60	.146		127	5.35		132.35	148
6600	Shelving, plated or galvanized, steel wire type		360	.044	SF Hor.	9.45	1.63		11.08	12.90
6700	Slat shelf type		375	.043		11.65	1.57		13.22	15.25
6900	For stainless steel shelving, add					300%				
7000	Vapor barrier, on wood walls	2 Carp	1644	.010	S.F.	.14	.36		.50	.71
7200	On masonry walls	"	1315	.012	"	.36	.45		.81	1.10
7500	For air curtain doors, see division 23 34 33.10									

13 21 48 – Sound-Conditioned Rooms

13 21 48.10 Anechoic Chambers

		Crew	Daily Output	Labor-Hours	Unit	Material	Labor	Equipment	Total	Total Incl O&P
0010	**ANECHOIC CHAMBERS** Standard units, 7' ceiling heights									
0100	Area for pricing is net inside dimensions									
0300	200 cycles per second cutoff, 25 S.F. floor area				SF Flr.	1,425			1,425	1,575
0400	50 S.F.								1,045	1,150
0600	75 S.F.								1,000	1,100
0700	100 S.F.					1,075			1,075	1,200
0900	For 150 cycles per second cutoff, add to 100 S.F. room									30%
1000	For 100 cycles per second cutoff, add to 100 S.F. room									45%

13 21 48.15 Audiometric Rooms

		Crew	Daily Output	Labor-Hours	Unit	Material	Labor	Equipment	Total	Total Incl O&P
0010	**AUDIOMETRIC ROOMS**									
0020	Under 500 S.F. surface	4 Carp	98	.327	SF Surf	46	12		58	69.50
0100	Over 500 S.F. surface	"	120	.267	"	44	9.80		53.80	64

13 21 53 – Darkrooms

13 21 53.50 Darkrooms

		Crew	Daily Output	Labor-Hours	Unit	Material	Labor	Equipment	Total	Total Incl O&P
0010	**DARKROOMS**									
0020	Shell, complete except for door, 64 S.F., 8' high	2 Carp	128	.125	SF Flr.	52	4.59		56.59	64
0100	12' high		64	.250		65.50	9.20		74.70	86.50
0500	120 S.F. floor, 8' high		120	.133		39	4.89		43.89	50.50

13 21 Controlled Environment Rooms

13 21 53 – Darkrooms

13 21 53.50 Darkrooms

		Crew	Daily Output	Labor-Hours	Unit	Material	2007 Bare Costs Labor	Equipment	Total	Total Incl O&P
0600	12' high	2 Carp	60	.267	SF Flr.	50.50	9.80		60.30	71
0800	240 S.F. floor, 8' high		120	.133		28.50	4.89		33.39	38.50
0900	12' high		60	.267		39	9.80		48.80	58.50
1200	Mini-cylindrical, revolving, unlined, 4' diameter		3.50	4.571	Ea.	2,900	168		3,068	3,425
1400	5'-6" diameter		2.50	6.400		4,100	235		4,335	4,900
1600	Add for lead lining, inner cylinder, 1/32" thick					1,425			1,425	1,550
1700	1/16" thick					1,650			1,650	1,800
1800	Add for lead lining, inner and outer cylinder, 1/32" thick					2,600			2,600	2,875
1900	1/16" thick					2,725			2,725	3,000
2000	For darkroom door, see division 08 34 36.10									

13 21 56 – Music Rooms

13 21 56.50 Music Rooms

		Crew	Daily Output	Labor-Hours	Unit	Material	Labor	Equipment	Total	Total Incl O&P
0010	**MUSIC ROOMS**									
0020	Practice room, modular, perforated steel, under 500 S.F.	2 Carp	70	.229	SF Surf	28.50	8.40		36.90	44
0100	Over 500 S.F.	"	80	.200	"	24	7.35		31.35	38

13 24 Special Activity Rooms

13 24 16 – Saunas

13 24 16.50 Saunas

		Crew	Daily Output	Labor-Hours	Unit	Material	Labor	Equipment	Total	Total Incl O&P
0010	**SAUNAS**									
0020	Prefabricated, incl. heater & controls, 7' high, 6' x 4', C/C	L-7	2.20	12.727	Ea.	3,875	450		4,325	4,950
0050	6' x 4', C/P		2	14		3,575	495		4,070	4,725
0400	6' x 5', C/C		2	14		4,325	495		4,820	5,525
0450	6' x 5', C/P		2	14		4,050	495		4,545	5,225
0600	6' x 6', C/C		1.80	15.556		4,625	550		5,175	5,925
0650	6' x 6', C/P		1.80	15.556		4,300	550		4,850	5,575
0800	6' x 9', C/C		1.60	17.500		5,775	620		6,395	7,300
0850	6' x 9', C/P		1.60	17.500		5,475	620		6,095	6,975
1000	8' x 12', C/C		1.10	25.455		8,950	905		9,855	11,200
1050	8' x 12', C/P		1.10	25.455		8,200	905		9,105	10,400
1200	8' x 8', C/C		1.40	20		6,800	710		7,510	8,600
1250	8' x 8', C/P		1.40	20		6,400	710		7,110	8,125
1400	8' x 10', C/C		1.20	23.333		7,575	825		8,400	9,600
1450	8' x 10', C/P		1.20	23.333		7,025	825		7,850	9,000
1600	10' x 12', C/C		1	28		9,475	995		10,470	11,900
1650	10' x 12', C/P		1	28		8,575	995		9,570	11,000
1700	Door only, cedar, 2'x6', with tempered insulated glass window	2 Carp	3.40	4.706		505	173		678	825
1800	Prehung, incl. jambs, pulls & hardware	"	12	1.333		515	49		564	645
2500	Heaters only (incl. above), wall mounted, to 200 C.F.					500			500	550
2750	To 300 C.F.					615			615	675
3000	Floor standing, to 720 C.F., 10,000 watts, w/controls	1 Elec	3	2.667		1,575	117		1,692	1,900
3250	To 1,000 C.F., 16,000 watts	"	3	2.667		1,750	117		1,867	2,100

13 24 26 – Steam Baths

13 24 26.50 Steam Baths

		Crew	Daily Output	Labor-Hours	Unit	Material	Labor	Equipment	Total	Total Incl O&P
0010	**STEAM BATHS**									
0020	Heater, timer & head, single, to 140 C.F.	1 Plum	1.20	6.667	Ea.	1,125	299		1,424	1,700
0500	To 300 C.F.		1.10	7.273		1,275	325		1,600	1,900
1000	Commercial size, with blow-down assembly, to 800 C.F.		.90	8.889		4,375	400		4,775	5,425
1500	To 2500 C.F.		.80	10		6,900	450		7,350	8,275
2000	Multiple, motels, apts., 2 baths, w/ blow-down assm., 500 C.F.	Q-1	1.30	12.308		4,325	495		4,820	5,500

13 24 Special Activity Rooms

13 24 26 – Steam Baths

13 24 26.50 Steam Baths		Crew	Daily Output	Labor-Hours	Unit	Material	2007 Bare Costs Labor	Equipment	Total	Total Incl O&P
2500	4 baths	Q-1	.70	22.857	Ea.	4,725	920		5,645	6,575
2700	Conversion unit for residential tub, including door				↓	3,550			3,550	3,900

13 28 Athletic and Recreational Special Construction

13 28 33 – Athletic and Recreational Court Walls

13 28 33.50 Sport Court

		Crew	Daily Output	Labor-Hours	Unit	Material	Labor	Equipment	Total	Total Incl O&P
0010	**SPORT COURT**									
0020	Floors, No. 2 & better maple, 25/32" thick				SF Flr.				6.05	6.65
0100	Walls, laminated plastic bonded to galv. steel studs				SF Wall				7.68	8.45
0150	Laminated fiberglass surfacing, minimum								1.91	2.10
0180	Maximum				↓				2.15	2.37
0300	Squash, regulation court in existing building, minimum				Court	30,000			30,000	33,000
0400	Maximum				"	37,000			37,000	40,700
0450	Rule of thumb for components:									
0470	Walls	3 Carp	.15	160	Court	8,000	5,875		13,875	18,000
0500	Floor	"	.25	96		7,000	3,525		10,525	13,200
0550	Lighting	2 Elec	.60	26.667		2,000	1,175		3,175	3,950
0600	Handball, racquetball court in existing building, minimum	C-1	.20	160		33,000	5,550		38,550	45,000
0800	Maximum	"	.10	320		40,000	11,100		51,100	61,500
0900	Rule of thumb for components: walls	3 Carp	.12	200		10,000	7,350		17,350	22,400
1000	Floor		.25	96		7,500	3,525		11,025	13,700
1100	Ceiling	↓	.33	72.727		2,300	2,675		4,975	6,675
1200	Lighting	2 Elec	.60	26.667	↓	2,000	1,175		3,175	3,950

13 31 Fabric Structures

13 31 13 – Air-Supported Fabric Structures

13 31 13.09 Air Supported Tank Covers

		Crew	Daily Output	Labor-Hours	Unit	Material	Labor	Equipment	Total	Total Incl O&P
0010	**AIR SUPPORTED TANK COVERS**, vinyl polyester									
0100	scrim, double layer, with hardware, blower, standby & controls									
0200	Round, 75' diameter	B-2	4500	.009	S.F.	7.75	.26		8.01	8.95
0300	100' diameter		5000	.008		7.05	.23		7.28	8.10
0400	150' diameter		5000	.008		5.60	.23		5.83	6.50
0500	Rectangular, 20' x 20'		4500	.009		17.65	.26		17.91	19.80
0600	30' x 40'		4500	.009		17.65	.26		17.91	19.80
0700	50' x 60'	↓	4500	.009		17.65	.26		17.91	19.80
0800	For single wall construction, deduct, minimum					.60			.60	.66
0900	Maximum					1.77			1.77	1.95
1000	For maximum resistance to atmosphere or cold, add				↓	.87			.87	.96
1100	For average shipping charges, add				Total	1,500			1,500	1,650

13 31 13.13 Single-Walled Air-Supported Structures

		Crew	Daily Output	Labor-Hours	Unit	Material	Labor	Equipment	Total	Total Incl O&P
0010	**SINGLE-WALLED AIR-SUPPORTED STRUCTURES** R133113-10									
0020	Site preparation, incl. anchor placement and utilities	B-11B	1000	.016	SF Flr.	.88	.52	.23	1.63	2.03
0030	For concrete curb, see division 03 30 53.40									
0050	Warehouse, polyester/vinyl fabric, 28 oz., over 10 yr. life, welded									
0060	Seams, tension cables, primary & auxiliary inflation system,									
0070	airlock, personnel doors and liner									
0100	5,000 S.F.	4 Clab	5000	.006	SF Flr.	19	.18		19.18	21.50
0250	12,000 S.F.	"	6000	.005		13.65	.15		13.80	15.25
0400	24,000 S.F.	8 Clab	12000	.005	↓	9.60	.15		9.75	10.80

13 31 Fabric Structures

13 31 13 – Air-Supported Fabric Structures

13 31 13.13 Single-Walled Air-Supported Structures

		Crew	Daily Output	Labor-Hours	Unit	Material	2007 Bare Costs Labor	Equipment	Total	Total Incl O&P
0500	50,000 S.F.	8 Clab	12500	.005	SF Flr.	8.90	.15		9.05	10.05
0700	12 oz. reinforced vinyl fabric, 5 yr. life, sewn seams,									
0710	accordian door, including liner									
0750	3000 S.F.	4 Clab	3000	.011	SF Flr.	9.10	.31		9.41	10.50
0800	12,000 S.F.	"	6000	.005		7.70	.15		7.85	8.75
0850	24,000 S.F.	8 Clab	12000	.005		6.55	.15		6.70	7.45
0950	Deduct for single layer					.78			.78	.86
1000	Add for welded seams					1.01			1.01	1.11
1050	Add for double layer, welded seams included					2.08			2.08	2.29
1250	Tedlar/vinyl fabric, 28 oz., with liner, over 10 yr. life,									
1260	incl. overhead and personnel doors									
1300	3000 S.F.	4 Clab	3000	.011	SF Flr.	18.50	.31		18.81	21
1450	12,000 S.F.	"	6000	.005		13.05	.15		13.20	14.60
1550	24,000 S.F.	8 Clab	12000	.005		10.10	.15		10.25	11.35
1700	Deduct for single layer					1.44			1.44	1.58
2250	Greenhouse/shelter, woven polyethylene with liner, 2 yr. life,									
2260	sewn seams, including doors									
2300	3000 S.F.	4 Clab	3000	.011	SF Flr.	6.60	.31		6.91	7.75
2350	12,000 S.F.	"	6000	.005		6.70	.15		6.85	7.65
2450	24,000 S.F.	8 Clab	12000	.005		5.60	.15		5.75	6.40
2550	Deduct for single layer					.63			.63	.69
2600	Tennis/gymnasium, polyester/vinyl fabric, 28 oz., over 10 yr. life,									
2610	including thermal liner, heat and lights									
2650	7,200 S.F.	4 Clab	6000	.005	SF Flr.	17.80	.15		17.95	19.85
2750	13,000 S.F.	"	6500	.005		13.65	.14		13.79	15.20
2850	Over 24,000 S.F.	8 Clab	12000	.005		12.45	.15		12.60	13.95
2860	For low temperature conditions, add					.87			.87	.96
2870	For average shipping charges, add				Total	4,225			4,225	4,650
2900	Thermal liner, translucent reinforced vinyl				SF Flr.	.87			.87	.96
2950	Metalized mylar fabric and mesh, double liner				"	1.77			1.77	1.95
3050	Stadium/convention center, teflon coated fiberglass, heavy weight,									
3060	over 20 yr. life, incl. thermal liner and heating system									
3100	Minimum	9 Clab	26000	.003	SF Flr.	43.50	.08		43.58	48
3110	Maximum	"	19000	.004	"	51.50	.11		51.61	57
3400	Doors, air lock, 15' long, 10' x 10'	2 Carp	.80	20	Ea.	16,300	735		17,035	19,200
3600	15' x 15'	"	.50	32		23,000	1,175		24,175	27,100
3700	For each added 5' length, add					4,125			4,125	4,525
3900	Revolving personnel door, 6' diameter, 6'-6" high	2 Carp	.80	20		11,500	735		12,235	13,900
4200	Double wall, self supporting, shell only, minimum				SF Flr.				19.09	21
4300	Maximum				"				35.45	39

13 31 23 – Tensioned Fabric Structures

13 31 23.50 Tension Structures

		Crew	Daily Output	Labor-Hours	Unit	Material	Labor	Equipment	Total	Total Incl O&P
0010	**TENSION STRUCTURES** Rigid steel/alum. frame, vyl. coated poly									
0100	fabric shell, 60' clear span, not incl. foundations or floors									
0200	6,000 S.F.	B-41	1000	.044	SF Flr.	11.55	1.31	.25	13.11	15
0300	12,000 S.F.		1100	.040		10.10	1.19	.23	11.52	13.25
0400	80' clear span, 20,800 S.F.		1220	.036		10.30	1.08	.20	11.58	13.25
0410	100' clear span, 10,000 S.F.	L-5	2175	.026		11.50	1.07	.34	12.91	14.90
0430	26,000 S.F.		2300	.024		10.50	1.01	.32	11.83	13.70
0450	36,000 S.F.		2500	.022		10.25	.93	.30	11.48	13.25
0460	120' clear span, 24,000 S.F.		3000	.019		12.40	.77	.25	13.42	15.30
0470	150' clear span, 30,000 S.F.		6000	.009		12.85	.39	.12	13.36	14.90

13 31 Fabric Structures

13 31 23 – Tensioned Fabric Structures

13 31 23.50 Tension Structures		Crew	Daily Output	Labor-Hours	Unit	Material	2007 Bare Costs Labor	Equipment	Total	Total Incl O&P
0480	200' clear span, 40,000 S.F.	E-6	8000	.016	SF Flr.	15.65	.65	.23	16.53	18.60
0500	For roll-up door, 12' x 14', add	L-2	1	16	Ea.	4,450	510		4,960	5,700
0600	For personnel doors, add, minimum				SF Flr.	5%				
0700	Add, maximum					15%				
0800	For site work, simple foundation, etc., add, minimum								1.25	1.95
0900	Add, maximum								2.75	3.05

13 34 Fabricated Engineered Structures

13 34 13 – Glazed Structures

13 34 13.13 Greenhouses

		Crew	Daily Output	Labor-Hours	Unit	Material	Labor	Equipment	Total	Total Incl O&P
0010	**GREENHOUSES**, Shell only, stock units, not incl. 2' stub walls,									
0020	foundation, floors, heat or compartments									
0300	Residential type, free standing, 8'-6" long x 7'-6" wide	2 Carp	59	.271	SF Flr.	46.50	9.95		56.45	67
0400	10'-6" wide		85	.188		36	6.90		42.90	50.50
0600	13'-6" wide		108	.148		32	5.45		37.45	43.50
0700	17'-0" wide		160	.100		36	3.67		39.67	45
0900	Lean-to type, 3'-10" wide		34	.471		41.50	17.25		58.75	72.50
1000	6'-10" wide		58	.276		32	10.10		42.10	51
1500	Commercial, custom, truss frame, incl. equip., plumbing, elec.,									
1510	benches and controls, under 2,000 S.F., minimum				SF Flr.				27.27	30
1550	Maximum					46			46	50.50
1700	Over 5,000 S.F., minimum					28			28	30.50
1750	Maximum								27.27	30
2000	Institutional, custom, rigid frame, including compartments and									
2010	multi-controls, under 500 S.F., minimum				SF Flr.				70	77
2050	Maximum					120			120	132
2150	Over 2,000 S.F., minimum					47.50			47.50	52
2200	Maximum								59.09	65
2400	Concealed rigid frame, under 500 S.F., minimum								81.82	90
2450	Maximum								100	110
2550	Over 2,000 S.F., minimum								61.82	68
2600	Maximum								71.82	79
2800	Lean-to type, under 500 S.F., minimum								72.73	80
2850	Maximum								109.09	120
3000	Over 2,000 S.F., minimum								40	44
3050	Maximum								66.36	73
3600	For 1/4" clear plate glass, add				SF Surf	1.86			1.86	2.05
3700	For 1/4" tempered glass, add				"	4.21			4.21	4.63
3900	For cooling, add, minimum				SF Flr.	2.81			2.81	3.09
4000	Maximum					6.95			6.95	7.65
4200	For heaters, 13.6 MBH, add					5.35			5.35	5.85
4300	60 MBH, add					2			2	2.20
4500	For benches, 2' x 3'-6", add				SF Hor.	24			24	26.50
4600	3' x 10', add				S.F.	12.95			12.95	14.25
4800	For controls, add, minimum				Total	2,400			2,400	2,650
4900	Maximum				"	14,300			14,300	15,700
5100	For humidification equipment, add				M.C.F.	6.15			6.15	6.75
5200	For vinyl shading, add				S.F.	1.29			1.29	1.42
6000	Geodesic hemisphere, 1/8" plexiglass glazing									
6050	8' diameter	2 Carp	2	8	Ea.	2,700	294		2,994	3,425
6150	24' diameter		.35	45.714		13,900	1,675		15,575	17,900

13 34 Fabricated Engineered Structures

13 34 13 – Glazed Structures

13 34 13.13 Greenhouses

	Crew	Daily Output	Labor-Hours	Unit	Material	2007 Bare Costs Labor	Equipment	Total	Total Incl O&P
6250 48' diameter	2 Carp	.20	80	Ea.	37,600	2,925		40,525	46,000

13 34 13.19 Swimming Pool Enclosures

	Crew	Daily Output	Labor-Hours	Unit	Material	Labor	Equipment	Total	Total Incl O&P
0010 **SWIMMING POOL ENCLOSURES** Translucent, free standing									
0020 not including foundations, heat or light									
0200 Economy, minimum	2 Carp	200	.080	SF Hor.	12.85	2.94		15.79	18.70
0300 Maximum		100	.160		36	5.85		41.85	48.50
0400 Deluxe, minimum		100	.160		42.50	5.85		48.35	55.50
0600 Maximum		70	.229		196	8.40		204.40	228
0700 For motorized roof, 40% opening, solid roof, add					6.95			6.95	7.60
0800 Skylight type roof, add					7.40			7.40	8.15
0900 Air-inflated, including blowers and heaters, minimum								3.18	3.50
1000 Maximum								6.09	6.70

13 34 16 – Grandstands and Bleachers

13 34 16.13 Grandstands

	Crew	Daily Output	Labor-Hours	Unit	Material	Labor	Equipment	Total	Total Incl O&P
0010 **GRANDSTANDS** Permanent, municipal, including foundation									
0050 Steel understructure w/aluminum closed deck, minimum				Seat				118.18	130
0100 Maximum								205	225
0300 Steel, minimum					50			50	55
0400 Maximum					100			100	110
0600 Aluminum, extruded, stock design, minimum								63.64	70
0700 Maximum								104.55	115
0900 Composite, steel, wood and plastic, stock design, minimum					125			125	138
1000 Maximum					250			250	275

13 34 16.53 Bleachers

	Crew	Daily Output	Labor-Hours	Unit	Material	Labor	Equipment	Total	Total Incl O&P
0010 **BLEACHERS**									
0020 Bleachers, outdoor, portable, 3 to 5 tiers, to 300' long, min	2 Sswk	120	.133	Seat	39.50	5.50		45	53
0100 Maximum, less than 15' long, prefabricated		80	.200		51.50	8.25		59.75	71.50
0200 6 to 20 tiers, minimum, up to 300' long		120	.133		47	5.50		52.50	62
0300 Max., under 15', (highly prefabricated, on wheels)		80	.200		68.50	8.25		76.75	90
0500 Permanent grandstands, wood seat, steel frame, 24" row									
0600 3 to 15 tiers, minimum	2 Sswk	60	.267	Seat	117	11.05		128.05	149
0700 Maximum		48	.333		128	13.80		141.80	166
0900 16 to 30 tiers, minimum		60	.267		135	11.05		146.05	168
0950 Average		55	.291		169	12.05		181.05	207
1000 Maximum		48	.333		202	13.80		215.80	248
1200 Seat backs only, 30" row, fiberglass		160	.100		25.50	4.14		29.64	35.50
1300 Steel and wood		160	.100		29.50	4.14		33.64	40
1400 NOTE: average seating is 1.5' in width									

13 34 19 – Metal Building Systems

13 34 19.50 Pre-Engineered Steel Buildings

	Crew	Daily Output	Labor-Hours	Unit	Material	Labor	Equipment	Total	Total Incl O&P
0010 **PRE-ENGINEERED STEEL BUILDINGS** R133419-10									
0100 Clear span rigid frame, 26 ga. colored roofing and siding									
0150 20' wide, 10' eave height	E-2	425	.132	SF Flr.	7.85	5.30	3.64	16.79	22
0160 14' eave height		350	.160		8.30	6.45	4.42	19.17	25
0170 16' eave height		320	.175		8.75	7.05	4.84	20.64	27
0180 20' eave height		275	.204		9.65	8.20	5.65	23.50	31
0190 24' eave height		240	.233		11	9.40	6.45	26.85	35.50
0200 30' to 40' wide, 10' eave height		535	.105		6.50	4.22	2.89	13.61	17.65
0300 14' eave height		450	.124		6.85	5	3.44	15.29	20
0400 16' eave height		415	.135		7.25	5.45	3.73	16.43	21.50
0500 20' eave height		360	.156		7.90	6.30	4.30	18.50	24.50

13 34 Fabricated Engineered Structures

13 34 19 – Metal Building Systems

13 34 19.50 Pre-Engineered Steel Buildings	Crew	Daily Output	Labor-Hours	Unit	Material	2007 Bare Costs Labor	Equipment	Total	Total Incl O&P
0600　24' eave height	E-2	320	.175	SF Flr.	8.90	7.05	4.84	20.79	27.50
0700　50' to 100' wide, 10' eave height		865	.065		5.60	2.61	1.79	10	12.65
0800　14' eave height		770	.073		5.95	2.93	2.01	10.89	13.85
0900　16' eave height		730	.077		6.30	3.10	2.12	11.52	14.65
1000　20' eave height		660	.085		6.80	3.42	2.34	12.56	16
1100　24' eave height	▼	605	.093	▼	7.45	3.73	2.56	13.74	17.45
1200　Clear span tapered beam frame, 26 ga. colored roofing/siding									
1300　30' wide, 10' eave height	E-2	535	.105	SF Flr.	7.20	4.22	2.89	14.31	18.40
1400　14' eave height		450	.124		7.90	5	3.44	16.34	21
1500　16' eave height		415	.135		8.50	5.45	3.73	17.68	23
1600　20' eave height		360	.156		9.35	6.30	4.30	19.95	26
1700　40' wide, 10' eave height		600	.093		6.60	3.77	2.58	12.95	16.60
1800　14' eave height		510	.110		7.20	4.43	3.03	14.66	18.90
1900　16' eave height		475	.118		7.50	4.76	3.26	15.52	20
2000　20' eave height		415	.135		8.20	5.45	3.73	17.38	22.50
2100　50' to 80' wide, 10' eave height		770	.073		6.25	2.93	2.01	11.19	14.20
2200　14' eave height		675	.083		6.70	3.35	2.29	12.34	15.65
2300　16' eave height		635	.088		6.95	3.56	2.44	12.95	16.50
2400　20' eave height	▼	565	.099	▼	7.75	4	2.74	14.49	18.50
2500　Single post 2-span frame, 26 ga. colored roofing and siding									
2600　80' wide, 14' eave height	E-2	740	.076	SF Flr.	5.35	3.05	2.09	10.49	13.50
2700　16' eave height		695	.081		5.70	3.25	2.23	11.18	14.40
2800　20' eave height		625	.090		6.25	3.62	2.48	12.35	15.80
2900　24' eave height		570	.098		6.75	3.96	2.71	13.42	17.25
3000　100' wide, 14' eave height		835	.067		5.20	2.71	1.85	9.76	12.50
3100　16' eave height		795	.070		5.15	2.84	1.95	9.94	12.70
3200　20' eave height		730	.077		6	3.10	2.12	11.22	14.30
3300　24' eave height		670	.084		6.50	3.37	2.31	12.18	15.55
3400　120' wide, 14' eave height		870	.064		5.15	2.60	1.78	9.53	12.15
3500　16' eave height		830	.067		5.40	2.72	1.86	9.98	12.65
3600　20' eave height		765	.073		5.85	2.95	2.02	10.82	13.70
3700　24' eave height	▼	705	.079	▼	6.35	3.21	2.19	11.75	14.95
3800　Double post 3-span frame, 26 ga. colored roofing and siding									
3900　150' wide, 14' eave height	E-2	925	.061	SF Flr.	4.46	2.44	1.67	8.57	11
4000　16' eave height		890	.063		4.60	2.54	1.74	8.88	11.35
4100　20' eave height		820	.068		4.89	2.76	1.89	9.54	12.25
4200　24' eave height	▼	765	.073		5.75	2.95	2.02	10.72	13.65
4300　Triple post 4-span frame, 26 ga. colored roofing and siding									
4400　160' wide, 14' eave height	E-2	970	.058	SF Flr.	4.32	2.33	1.60	8.25	10.55
4500　16' eave height		930	.060		4.52	2.43	1.66	8.61	11
4600　20' eave height		870	.064		4.54	2.60	1.78	8.92	11.45
4700　24' eave height		815	.069		5.40	2.77	1.90	10.07	12.85
4800　200' wide, 14' eave height		1030	.054		4.16	2.19	1.50	7.85	10.05
4900　16' eave height		995	.056		4.27	2.27	1.56	8.10	10.35
5000　20' eave height		935	.060		4.63	2.42	1.65	8.70	11.10
5100　24' eave height	▼	885	.063	▼	5.40	2.55	1.75	9.70	12.30
5200　Accessory items: add to the basic building cost above									
5250　Eave overhang, 2' wide, 26 ga., with soffit	E-2	360	.156	L.F.	24	6.30	4.30	34.60	41.50
5300　4' wide, without soffit		300	.187		23	7.55	5.15	35.70	43.50
5350　With soffit		250	.224		32	9.05	6.20	47.25	57.50
5400　6' wide, without soffit		250	.224		30.50	9.05	6.20	45.75	56.50
5450　With soffit		200	.280	▼	40	11.30	7.75	59.05	72
5500　Entrance canopy, incl. frame, 4' x 4'	▼	25	2.240	Ea.	305	90.50	62	457.50	560

13 34 Fabricated Engineered Structures

13 34 19 – Metal Building Systems

13 34 19.50 Pre-Engineered Steel Buildings		Crew	Daily Output	Labor-Hours	Unit	Material	2007 Bare Costs Labor	Equipment	Total	Total Incl O&P
5550	4' x 8'	E-2	19	2.947	Ea.	415	119	81.50	615.50	750
5600	End wall roof overhang, 4' wide, without soffit	↓	850	.066	L.F.	15.25	2.66	1.82	19.73	23.50
5650	With soffit		500	.112	"	22.50	4.52	3.09	30.11	36
5700	Doors, H.M. self-framing, incl. butts, lockset and trim									
5750	Single leaf, 3070 (3' x 7'), economy	2 Sswk	5	3.200	Opng.	440	132		572	725
5800	Deluxe		4	4		515	165		680	870
5825	Glazed		4	4		530	165		695	880
5850	3670 (3'-6" x 7')		4	4		540	165		705	890
5900	4070 (4' x 7')		3	5.333		570	221		791	1,025
5950	Double leaf, 6070 (6' x 7')		2	8		770	330		1,100	1,450
6000	Glazed	↓	2	8		945	330		1,275	1,650
6050	Framing only, for openings, 3' x 7'	E-2	25	2.240		143	90.50	62	295.50	380
6100	10' x 10'		21	2.667		470	108	73.50	651.50	785
6150	For windows below, 2020 (2' x 2')		25	2.240		151	90.50	62	303.50	390
6200	4030 (4' x 3')	↓	22	2.545	↓	185	103	70.50	358.50	460
6250	Flashings, 26 ga., corner or eave, painted	2 Sswk	240	.067	L.F.	3.71	2.76		6.47	9.05
6300	Galvanized		240	.067		3.02	2.76		5.78	8.30
6350	Rake flashing, painted		240	.067		3.98	2.76		6.74	9.35
6400	Galvanized		240	.067		3.31	2.76		6.07	8.65
6450	Ridge flashing, 18" wide, painted		240	.067		4.94	2.76		7.70	10.45
6500	Galvanized		240	.067		4.32	2.76		7.08	9.75
6550	Gutter, eave type, 26 ga., painted		320	.050		5.15	2.07		7.22	9.40
6600	Galvanized		320	.050		2.91	2.07		4.98	6.95
6650	Valley type, between buildings, painted		120	.133		8.15	5.50		13.65	18.95
6700	Galvanized	↓	120	.133	↓	8.75	5.50		14.25	19.65
6750	Insulation, rated .6 lb density, vinyl faced									
6800	1-1/2" thick, R5	2 Carp	2300	.007	S.F.	.24	.26		.50	.66
6850	3" thick, R10		2300	.007		.32	.26		.58	.75
6900	4" thick, R13		2300	.007		.45	.26		.71	.90
6950	Foil/scrim/kraft (FSK) faced, 1-1/2" thick, R5		2300	.007		.25	.26		.51	.68
7000	2" thick, R6		2300	.007		.32	.26		.58	.75
7050	3" thick, R10		2300	.007		.40	.26		.66	.84
7100	4" thick, R13		2300	.007		.51	.26		.77	.96
7150	Met polyester/scrim/kraft (PSK) facing, 1-1/2" thk, R5		2300	.007		.41	.26		.67	.85
7200	2" thick, R6		2300	.007		.47	.26		.73	.92
7250	3" thick, R11		2300	.007		.51	.26		.77	.96
7300	4" thick, R13		2300	.007		.62	.26		.88	1.08
7350	Vinyl/scrim/foil (VSF), 1-1/2" thick, R5		2300	.007		.36	.26		.62	.80
7400	2" thick, R6		2300	.007		.45	.26		.71	.90
7450	3" thick, R10		2300	.007		.58	.26		.84	1.04
7500	4" thick, R13	↓	2300	.007	↓	.67	.26		.93	1.14
7650	Sash, single slide, glazed, with screens, 2020 (2'x 2')	E-1	22	1.091	Opng.	83	44	5.25	132.25	173
7700	3030 (3' x 3')		14	1.714		187	69.50	8.25	264.75	335
7750	4030 (4' x 3')		13	1.846		250	75	8.85	333.85	415
7800	6040 (6' x 4')		12	2		500	81	9.60	590.60	700
7850	Double slide sash, 3030 (3' x 3')		14	1.714		157	69.50	8.25	234.75	300
7900	6040 (6' x 4')		12	2		415	81	9.60	505.60	610
7950	Fixed glass, no screens, 3030 (3' x 3')		14	1.714		201	69.50	8.25	278.75	350
8000	6040 (6' x 4')		12	2		535	81	9.60	625.60	740
8050	Prefinished storm sash, 3030 (3' x 3')	↓	70	.343	↓	65	13.90	1.65	80.55	97.50
8100	Siding and roofing, see division 07 31 00.00 & 07 41 00.00									
8200	Skylight, fiberglass panels, to 30 S.F.	E-1	10	2.400	Ea.	100	97.50	11.50	209	290
8250	Larger sizes, add for excess over 30 S.F.	"	300	.080	S.F.	3.33	3.24	.38	6.95	9.65

13 34 Fabricated Engineered Structures

13 34 19 – Metal Building Systems

13 34 19.50 Pre-Engineered Steel Buildings

		Crew	Daily Output	Labor-Hours	Unit	Material	2007 Bare Costs Labor	Equipment	Total	Total Incl O&P
8300	Roof vents, circular with damper, birdscreen									
8350	and operator hardware, painted									
8400	26 ga., 12" diameter	1 Sswk	4	2	Ea.	65.50	82.50		148	222
8450	20" diameter		3	2.667		135	110		245	350
8500	24 ga., 24" diameter		2	4		267	165		432	595
8550	Galvanized	↓	2	4		224	165		389	545
8600	Continuous, 26 ga., 10' long, 9" wide	2 Sswk	4	4		21.50	165		186.50	325
8650	12" wide	"	4	4	↓	25.50	165		190.50	330

13 34 23 – Fabricated Structures

13 34 23.10 Comfort Stations

		Crew	Daily Output	Labor-Hours	Unit	Material	Labor	Equipment	Total	Total Incl O&P
0010	**COMFORT STATIONS** Prefab., stock, w/doors, windows & fixt.									
0100	Not incl. interior finish or electrical									
0300	Mobile, on steel frame, minimum				S.F.	36.50			36.50	40
0350	Maximum					57.50			57.50	63.50
0400	Permanent, including concrete slab, minimum	B-12J	50	.320		142	10.95	18.10	171.05	193
0500	Maximum	"	43	.372	↓	206	12.75	21	239.75	270
0600	Alternate pricing method, mobile, minimum				Fixture	1,775			1,775	1,950
0650	Maximum					2,650			2,650	2,925
0700	Permanent, minimum	B-12J	.70	22.857		10,200	785	1,300	12,285	13,800
0750	Maximum	"	.50	32	↓	17,000	1,100	1,800	19,900	22,400

13 34 23.15 Domes

		Crew	Daily Output	Labor-Hours	Unit	Material	Labor	Equipment	Total	Total Incl O&P
0010	**DOMES**									
0020	Revolv alum, elec drv for astronomy observation, shell only, stk units									
0600	10'-0" diameter, 800#, dome	2 Carp	.25	64	Ea.	11,000	2,350		13,350	15,800
0700	Base		.67	23.881		3,800	875		4,675	5,550
0900	18'-0" diameter, 2,500#, dome		.17	94.118		32,100	3,450		35,550	40,700
1000	Base		.33	48.485		10,600	1,775		12,375	14,500
1200	24'-0" diameter, 4,500#, dome		.08	200		59,500	7,350		66,850	77,000
1300	Base	↓	.25	64	↓	19,200	2,350		21,550	24,800
1500	Bulk storage, shell only, dual radius hemispher. arch, steel									
1600	framing, corrugated steel covering, 150' diameter	E-2	550	.102	SF Flr.	28	4.11	2.81	34.92	40.50
1700	400' diameter	"	720	.078		22.50	3.14	2.15	27.79	33
1800	Wood framing, wood decking, to 400' diameter	F-4	400	.120	↓	20.50	4.41	2.65	27.56	32
1900	Radial framed wood (2" x 6"), 1/2" thick									
2000	plywood, asphalt shingles, 50' diameter	F-3	2000	.020	SF Flr.	38.50	.75	.36	39.61	44
2100	60' diameter		1900	.021		29.50	.79	.38	30.67	34
2200	72' diameter		1800	.022		26.50	.83	.40	27.73	30.50
2300	116' diameter		1730	.023		23	.86	.42	24.28	27.50
2400	150' diameter	↓	1500	.027	↓	20	1	.48	21.48	24

13 34 23.16 Fabricated Control Booths

		Crew	Daily Output	Labor-Hours	Unit	Material	Labor	Equipment	Total	Total Incl O&P
0010	**FABRICATED CONTROL BOOTHS**									
0100	Guard House, prefab conc w/bullet resistant doors & windows, roof and wiring									
0110	8' x 8', Level III	L-10	1	24	Ea.	20,000	995	725	21,720	24,500
0120	8' x 8', Level IV	"	1	24	"	25,000	995	725	26,720	30,000

13 34 23.25 Garage Costs

		Crew	Daily Output	Labor-Hours	Unit	Material	Labor	Equipment	Total	Total Incl O&P
0010	**GARAGE COSTS**									
0020	Public parking, average				Car				14,000	15,400
0100	See also Square Foot Costs in Reference Section									
0300	Residential, prefab shell, stock, wood, single car, minimum	2 Carp	1	16	Total	3,175	585		3,760	4,425
0350	Maximum		.67	23.881		6,525	875		7,400	8,550
0400	Two car, minimum	↓	.67	23.881	↓	7,275	875		8,150	9,400

13 34 Fabricated Engineered Structures

13 34 23 – Fabricated Structures

13 34 23.25 Garage Costs		Crew	Daily Output	Labor-Hours	Unit	Material	2007 Bare Costs Labor	Equipment	Total	Total Incl O&P
0450	Maximum	2 Carp	.50	32	Total	10,400	1,175		11,575	13,200
13 34 23.30 Garden House										
0010	**GARDEN HOUSE** Prefab wood, no floors or foundations									
0100	32 to 200 S.F., minimum	2 Carp	200	.080	SF Flr.	24	2.94		26.94	31
0300	Maximum	"	48	.333	"	40	12.25		52.25	63
13 34 23.35 Geodesic Domes										
0010	**GEODESIC DOMES** Shell only, interlocking plywood panels R133423-30									
0400	30' diameter	F-5	1.60	20	Ea.	19,300	745		20,045	22,400
0500	34' diameter		1.14	28.070		22,800	1,050		23,850	26,600
0600	39' diameter		1	32		25,600	1,200		26,800	30,100
0700	45' diameter	F-3	1.13	35.556		31,500	1,325	645	33,470	37,500
0750	55' diameter		1	40		45,200	1,500	725	47,425	53,000
0800	60' diameter		1	40		58,500	1,500	725	60,725	67,500
0850	65' diameter		.80	50		71,500	1,875	905	74,280	83,000
1100	Aluminum panel, with 6" insulation									
1200	100' diameter				SF Flr.	38			38	41.50
1300	500' diameter				"	32.50			32.50	35.50
1600	Aluminum framed, plexiglass closure panels									
1700	40' diameter				SF Flr.	96			96	105
1800	200' diameter				"	76.50			76.50	84
2100	Aluminum framed, aluminum closure panels									
2200	40' diameter				SF Flr.	30			30	33.50
2300	100' diameter					22.50			22.50	25
2400	200' diameter					21.50			21.50	24
2500	For VRP faced bonded fiberglass insulation, add									10
2700	Aluminum framed, fiberglass sandwich panel closure									
2800	6' diameter	2 Carp	150	.107	SF Flr.	33.50	3.91		37.41	43
2900	28' diameter	"	350	.046	"	30.50	1.68		32.18	36
13 34 23.45 Kiosks										
0010	**KIOSKS**									
0020	Round, 5' diameter, 8' high, 1/4" fiberglass wall				Ea.	5,600			5,600	6,175
0100	1" insulated double wall, fiberglass					6,375			6,375	7,025
0500	Rectangular, 5' x 9', 7'-6" high, 1/4" fiberglass wall					8,150			8,150	8,975
0600	1" insulated double wall, fiberglass					9,700			9,700	10,700
13 34 23.60 Portable Booths										
0010	**PORTABLE BOOTHS** Prefab aluminum with doors, windows, ext. roof									
0100	lights wiring & insulation, 15 S.F. building, O.D., painted, minimum				S.F.	287			287	315
0300	30 S.F. building, minimum					197			197	216
0400	50 S.F. building, minimum					146			146	161
0600	80 S.F. building, minimum					118			118	130
0700	100 S.F. building, minimum					109			109	120
0900	Acoustical booth, 27 Db @ 1,000 Hz, 15 S.F. floor				Ea.	3,375			3,375	3,725
1000	7' x 7'-6", including light & ventilation					6,950			6,950	7,625
1200	Ticket booth, galv. steel, not incl. foundations., 4' x 4'					5,300			5,300	5,825
1300	4' x 6'					6,175			6,175	6,800
13 34 23.70 Shelters										
0010	**SHELTERS**									
0020	Aluminum frame, acrylic glazing, 3' x 9' x 8' high	2 Sswk	1.14	14.035	Ea.	2,750	580		3,330	4,075
0100	9' x 12' x 8' high	"	.73	21.918	"	4,850	905		5,755	7,000

13 34 Fabricated Engineered Structures

13 34 43 – Aircraft Hangars

13 34 43.50 Hangars

		Crew	Daily Output	Labor-Hours	Unit	Material	2007 Bare Costs Labor	Equipment	Total	Total Incl O&P
0010	**HANGARS** Prefabricated steel T hangars, Galv. steel roof &									
0100	walls, incl. electric bi-folding doors, 4 or more units,									
0110	not including floors or foundations, minimum	E-2	1275	.044	SF Flr.	10.70	1.77	1.21	13.68	16.15
0130	Maximum		1063	.053		11.65	2.13	1.46	15.24	18.10
0900	With bottom rolling doors, minimum		1386	.040		10.50	1.63	1.12	13.25	15.60
1000	Maximum		966	.058		11.80	2.34	1.60	15.74	18.80
1200	Alternate pricing method:									
1300	Galv. roof and walls, electric bi-folding doors, minimum	E-2	1.06	52.830	Plane	14,000	2,125	1,450	17,575	20,700
1500	Maximum		.91	61.538		16,700	2,475	1,700	20,875	24,500
1600	With bottom rolling doors, minimum		1.25	44.800		10,300	1,800	1,250	13,350	15,900
1800	Maximum		.97	57.732		12,300	2,325	1,600	16,225	19,400
2000	Circular type, prefab., steel frame, plastic skin, electric									
2010	door, including foundations, 80' diameter,									
2020	for up to 5 light planes, minimum	E-2	.50	112	Total	70,500	4,525	3,100	78,125	88,500
2200	Maximum	"	.25	224	"	79,000	9,050	6,200	94,250	109,500

13 34 53 – Agricultural Structures

13 34 53.50 Silos

		Crew	Daily Output	Labor-Hours	Unit	Material	Labor	Equipment	Total	Total Incl O&P
0010	**SILOS** Concrete stave industrial, not incl. foundations, conical or									
0100	sloping bottoms, 12' diameter, 35' high	D-8	.11	363	Ea.	20,200	12,500		32,700	41,300
0200	16' diameter, 45' high		.08	500		27,000	17,100		44,100	56,000
0400	25' diameter, 75' high		.05	800		66,500	27,400		93,900	114,500
0500	Steel, factory fab., 30,000 gallon cap., painted, minimum	L-5	1	56		17,200	2,325	740	20,265	23,900
0700	Maximum		.50	112		27,400	4,650	1,475	33,525	40,000
0800	Epoxy lined, minimum		1	56		28,200	2,325	740	31,265	35,900
1000	Maximum		.50	112		35,700	4,650	1,475	41,825	49,100

13 36 Towers

13 36 13 – Metal Towers

13 36 13.50 Control Towers

		Crew	Daily Output	Labor-Hours	Unit	Material	Labor	Equipment	Total	Total Incl O&P
0010	**CONTROL TOWERS**									
0020	12' x 10', incl. instruments, min.				Ea.	609,000			609,000	670,000
0100	Maximum								927,273	1,020,000
0500	With standard 40' tower, average					850,000			850,000	935,000
1000	Temporary portable control towers, 8' x 12',									
1010	complete with one position communications, minimum				Ea.				266,000	293,000
2000	For fixed facilities, depending on height, minimum								57,200	63,000
2010	Maximum								115,000	127,000

13 42 Building Modules

13 42 63 – Detention Cell Modules

13 42 63.16 Steel Detention Cell Modules	Crew	Daily Output	Labor-Hours	Unit	Material	2007 Bare Costs Labor	Equipment	Total	Total Incl O&P
0010 **STEEL DETEENTION CELL MODULES**									
2000 Cells, prefab., 5' to 6' wide, 7' to 8' high, 7' to 8' deep,									
2010 bar front, cot, not incl. plumbing	E-4	1.50	21.333	Ea.	8,400	895	77	9,372	10,900

13 48 Sound, Vibration, and Seismic Control

13 48 13 – Manufactured Sound and Vibration Control Components

13 48 13.50 Audio Masking

	Crew	Daily Output	Labor-Hours	Unit	Material	Labor	Equipment	Total	Total Incl O&P
0010 **AUDIO MASKING**, acoustical enclosure, 4" thick wall and ceili									
0020 8# per S.F., up to 12' span	3 Carp	72	.333	SF Surf	29	12.25		41.25	51
0300 Better quality panels, 10.5# per S.F.		64	.375		33	13.75		46.75	57.50
0400 Reverb-chamber, 4" thick, parallel walls		60	.400		41	14.70		55.70	68
0600 Skewed wall, parallel roof, 4" thick panels		55	.436		46.50	16		62.50	76.50
0700 Skewed walls, skewed roof, 4" layers, 4" air space		48	.500		52.50	18.35		70.85	86.50
0900 Sound-absorbing panels, pntd mtl, 2'-6" x 8', under 1,000 S.F.		215	.112		10.85	4.10		14.95	18.35
1100 Over 1000 S.F.		240	.100		10.45	3.67		14.12	17.20
1200 Fabric faced		240	.100		8.45	3.67		12.12	15
1500 Flexible transparent curtain, clear	3 Shee	215	.112		6.55	4.86		11.41	14.70
1600 50% foam		215	.112		9.15	4.86		14.01	17.55
1700 75% foam		215	.112		9.15	4.86		14.01	17.55
1800 100% foam		215	.112		9.15	4.86		14.01	17.55
3100 Audio masking system, including speakers, amplification									
3110 and signal generator									
3200 Ceiling mounted, 5,000 S.F.	2 Elec	2400	.007	S.F.	1.19	.29		1.48	1.75
3300 10,000 S.F.		2800	.006		.96	.25		1.21	1.43
3400 Plenum mounted, 5,000 S.F.		3800	.004		.95	.18		1.13	1.33
3500 10,000 S.F.		4400	.004		.64	.16		.80	.94

13 49 Radiation Protection

13 49 13 – Lead Sheet

13 49 13.50 Lead Sheets

	Crew	Daily Output	Labor-Hours	Unit	Material	Labor	Equipment	Total	Total Incl O&P
0010 **LEAD SHEETS**									
0300 Lead sheets, 1/16" thick	2 Lath	135	.119	S.F.	5.85	3.99		9.84	12.35
0400 1/8" thick		120	.133		11.90	4.49		16.39	19.75
0500 Lead shielding, 1/4" thick		135	.119		25	3.99		28.99	33.50
0550 1/2" thick		120	.133		44.50	4.49		48.99	55.50
0950 Lead headed nails (average 1 lb. per sheet)				Lb.	6.20			6.20	6.80
1000 Butt joints in 1/8" lead or thicker, 2" batten strip x 7' long	2 Lath	240	.067	Ea.	15.35	2.25		17.60	20.50
1200 X-ray protection, average radiography or fluoroscopy									
1210 room, up to 300 S.F. floor, 1/16" lead, minimum	2 Lath	.25	64	Total	5,125	2,150		7,275	8,850
1500 Maximum, 7'-0" walls	"	.15	106	"	6,225	3,600		9,825	12,200
1600 Deep therapy X-ray room, 250 KV capacity,									
1800 up to 300 S.F. floor, 1/4" lead, minimum	2 Lath	.08	200	Total	18,800	6,750		25,550	30,700
1900 Maximum, 7'-0" walls	"	.06	266	"	24,300	8,975		33,275	40,100

13 49 19 – Lead-Lined Materials

13 49 19.50 Shielding Lead

	Crew	Daily Output	Labor-Hours	Unit	Material	Labor	Equipment	Total	Total Incl O&P
0010 **SHIELDING LEAD**									
0100 Laminated lead in wood doors, 1/16" thick, no hardware				S.F.	37			37	41
0200 Lead lined door frame, not incl. hardware,									
0210 1/16" thick lead, butt prepared for hardware	1 Lath	2.40	3.333	Ea.	495	112		607	710

13 49 Radiation Protection

13 49 19 – Lead-Lined Materials

13 49 19.50 Shielding Lead

		Crew	Daily Output	Labor-Hours	Unit	Material	2007 Bare Costs Labor	Equipment	Total	Total Incl O&P
0850	Window frame with 1/16" lead and voice passage, 36" x 60"	2 Glaz	2	8	Ea.	925	288		1,213	1,450
0870	24" x 36" frame		8	2		750	72		822	935
0900	Lead gypsum board, 5/8" thick with 1/16" lead		160	.100	S.F.	5.60	3.61		9.21	11.60
0910	1/8" lead	CN	140	.114		11.35	4.12		15.47	18.75
0930	1/32" lead	2 Lath	200	.080		3.64	2.70		6.34	8

13 49 21 – Lead Glazing

13 49 21.50 Lead Glazing

		Crew	Daily Output	Labor-Hours	Unit	Material	Labor	Equipment	Total	Total Incl O&P
0010	**LEAD GLAZING**									
0600	Lead glass, 1/4" thick, 2.0 mm LE, 12" x 16"	2 Glaz	13	1.231	Ea.	240	44.50		284.50	330
0700	24" x 36"		8	2		950	72		1,022	1,150
0800	36" x 60"		2	8		3,225	288		3,513	3,975
2000	X-ray viewing panels, clear lead plastic									
2010	7 mm thick, 0.3 mm LE, 2.3 lbs/S.F.	H-3	139	.115	S.F.	119	3.65		122.65	137
2020	12 mm thick, 0.5 mm LE, 3.9 lbs/S.F.		82	.195		161	6.20		167.20	187
2030	18 mm thick, 0.8mm LE, 5.9 lbs/S.F.		54	.296		175	9.40		184.40	207
2040	22 mm thick, 1.0 mm LE, 7.2 lbs/S.F.		44	.364		179	11.55		190.55	215
2050	35 mm thick, 1.5 mm LE, 11.5 lbs/S.F.		28	.571		201	18.10		219.10	249
2060	46 mm thick, 2.0 mm LE, 15.0 lbs/S.F.		21	.762		264	24		288	330
2090	For panels 12 S.F. to 48 S.F., add crating charge				Ea.					50

13 49 23 – Modular Shielding Partitions

13 49 23.50 Modular Shielding Partitions

		Crew	Daily Output	Labor-Hours	Unit	Material	Labor	Equipment	Total	Total Incl O&P
0010	**MODULAR SHIELDING PARTITIONS**									
4000	X-ray barriers, modular, panels mounted within framework for									
4002	attaching to floor, wall or ceiling, upper portion is clear lead									
4005	plastic window panels 48"H, lower portion is opaque leaded									
4008	steel panels 36"H, structural supports not incl.									
4010	1-section barrier, 36"W x 84"H overall									
4020	0.5 mm LE panels	H-3	6.40	2.500	Ea.	2,700	79.50		2,779.50	3,075
4030	0.8 mm LE panels		6.40	2.500		2,875	79.50		2,954.50	3,275
4040	1.0 mm LE panels		5.33	3.002		2,925	95		3,020	3,375
4050	1.5 mm LE panels		5.33	3.002		3,125	95		3,220	3,575
4060	2-section barrier, 72"W x 84"H overall									
4070	0.5 mm LE panels	H-3	4	4	Ea.	5,500	127		5,627	6,250
4080	0.8 mm LE panels		4	4		5,825	127		5,952	6,600
4090	1.0 mm LE panels		3.56	4.494		5,950	142		6,092	6,750
5000	1.5 mm LE panels		3.20	5		6,750	159		6,909	7,675
5010	3-section barrier, 108"W x 84"H overall									
5020	0.5 mm LE panels	H-3	3.20	5	Ea.	8,275	159		8,434	9,350
5030	0.8 mm LE panels		3.20	5		8,725	159		8,884	9,850
5040	1.0 mm LE panels		2.67	5.993		8,875	190		9,065	10,100
5050	1.5 mm LE panels		2.46	6.504		9,650	206		9,856	10,900
7000	X-ray barriers, mobile, mounted within framework w/casters on									
7005	bottom, clear lead plastic window panels on upper portion,									
7010	opaque on lower, 30"W x 75"H overall, incl. framework									
7020	24"H upper w/0.5 mm LE, 48"H lower w/0.8 mm LE	1 Carp	16	.500	Ea.	2,125	18.35		2,143.35	2,350
7030	48"W x 75"H overall, incl. framework									
7040	36"H upper w/0.5 mm LE, 36"H lower w/0.8 mm LE	1 Carp	16	.500	Ea.	3,750	18.35		3,768.35	4,150
7050	36"H upper w/1.0 mm LE, 36"H lower w/1.5 mm LE	"	16	.500	"	4,575	18.35		4,593.35	5,050
7060	72"W x 75"H overall, incl. framework									
7070	36"H upper w/0.5 mm LE, 36"H lower w/0.8 mm LE	1 Carp	16	.500	Ea.	4,450	18.35		4,468.35	4,925
7080	36"H upper w/1.0 mm LE, 36"H lower w/1.5 mm LE	"	16	.500	"	5,625	18.35		5,643.35	6,200

13 49 Radiation Protection

13 49 33 – Radio Frequency Shielding

13 49 33.50 Shielding, Radio Frequency		Crew	Daily Output	Labor-Hours	Unit	Material	2007 Bare Costs Labor	Equipment	Total	Total Incl O&P
0010	**SHIELDING, RADIO FREQUENCY**									
0020	Prefabricated or screen-type copper or steel, minimum	2 Carp	180	.089	SF Surf	25	3.26		28.26	32.50
0100	Average	↓	155	.103	↓	27.50	3.79		31.29	36
0150	Maximum	↓	145	.110	↓	32.50	4.05		36.55	42.50

13 53 Meteorological Instrumentation

13 53 09 – Weather Instrumentation

13 53 09.50 Weather Station

		Crew	Daily Output	Labor-Hours	Unit	Material	Labor	Equipment	Total	Total Incl O&P
0010	**WEATHER STATION**									
0020	Remote recording, minimum				Ea.	3,000			3,000	3,300
0100	Maximum				"	27,000			27,000	29,700

Division Notes

		CREW	DAILY OUTPUT	LABOR-HOURS	UNIT	2007 BARE COSTS				TOTAL INCL O&P
						MAT.	LABOR	EQUIP.	TOTAL	

Estimating Tips

General

- Many products in Division 14 will require some type of support or blocking for installation not included with the item itself. Examples are supports for conveyors or tube systems, attachment points for lifts, and footings for hoists or cranes. Add these supports in the appropriate division.

14 10 00 Dumbwaiters
14 20 00 Elevators

- Dumbwaiters and elevators are estimated and purchased in a method similar to buying a car. The manufacturer has a base unit with standard features. Added to this base unit price will be whatever options the owner or specifications require. Increased load capacity, additional vertical travel, additional stops, higher speed, and cab finish options are items to be considered. When developing an estimate for dumbwaiters and elevators, remember that some items needed by the installers may have to be included as part of the general contract. Examples are:
  - shaftway
  - rail support brackets
  - machine room
  - electrical supply
  - sill angles
  - electrical connections
  - pits
  - roof penthouses
  - pit ladders

  Check the job specifications and drawings before pricing.
- Installation of elevators and handicapped lifts in historic structures can require significant additional costs. The associated structural requirements may involve cutting into and repairing finishes, mouldings, flooring, etc. The estimator must account for these special conditions.

14 30 00 Escalators and Moving Walks

- Escalators and moving walks are specialty items installed by specialty contractors. There are numerous options associated with these items. For specific options contact a manufacturer or contractor. In a method similar to estimating dumbwaiters and elevators, you should verify the extent of general contract work and add items as necessary.

14 40 00 Lifts
14 90 00 Other Conveying Equipment

- Products such as correspondence lifts, conveyors, chutes, pneumatic tube systems, material handling cranes, and hoists, as well as other items specified in this subdivision, may require trained installers. The general contractor might not have any choice as to who will perform the installation or when it will be performed. Long lead times are often required for these products, making early decisions in scheduling necessary.

Reference Numbers

Reference numbers are shown in shaded boxes at the beginning of some major classifications. These numbers refer to related items in the Reference Section. The reference information may be an estimating procedure, an alternate pricing method, or technical information.

Note: Not all subdivisions listed here necessarily appear in this publication.

14 11 Manual Dumbwaiters

14 11 10 – Manual Dumbwaiters

14 11 10.20 Manual Dumbwaiters		Crew	Daily Output	Labor-Hours	Unit	Material	2007 Bare Costs Labor	Equipment	Total	Total Incl O&P
0010	**MANUAL DUMBWAITERS**									
0020	2 stop, minimum	2 Elev	.75	21.333	Ea.	2,575	1,150		3,725	4,550
0100	Maximum	↓	.50	32	"	5,875	1,700		7,575	9,000
0300	For each additional stop, add		.75	21.333	Stop	935	1,150		2,085	2,725

14 12 Electric Dumbwaiters

14 12 10 – Electric Dumbwaiters

14 12 10.10 Electric Dumbwaiters

		Crew	Daily Output	Labor-Hours	Unit	Material	Labor	Equipment	Total	Total Incl O&P
0010	**ELECTRIC DUMBWAITERS**									
0020	2 stop, minimum	2 Elev	.13	123	Ea.	6,525	6,575		13,100	17,000
0100	Maximum	↓	.11	145	"	19,600	7,775		27,375	33,100
0600	For each additional stop, add		.54	29.630	Stop	2,875	1,575		4,450	5,500
0750	Correspondence lift, 1 floor, 2 stop, 45 lb capacity	2 Elec	.20	80	Ea.	7,375	3,500		10,875	13,400

14 21 Electric Traction Elevators

14 21 13 – Electric Traction Freight Elevators

14 21 13.10 Electric Traction Freight Elevators

			Crew	Daily Output	Labor-Hours	Unit	Material	Labor	Equipment	Total	Total Incl O&P
0010	**ELECTRIC TRACTION FREIGHT ELEVATORS**	R142000-10									
0425	Electric freight, base unit, 4000 lb, 200 fpm, 4 stop, std. fin.		2 Elev	.05	320	Ea.	78,500	17,100		95,600	112,000
0450	For 5000 lb capacity, add						5,625			5,625	6,200
0500	For 6000 lb capacity, add						9,925			9,925	10,900
0525	For 7000 lb capacity, add						13,300			13,300	14,600
0550	For 8000 lb capacity, add						18,200			18,200	20,100
0575	For 10000 lb capacity, add						21,700			21,700	23,900
0600	For 12000 lb capacity, add						26,500			26,500	29,200
0625	For 16000 lb capacity, add						31,900			31,900	35,100
0650	For 20000 lb capacity, add						35,200			35,200	38,700
0675	For increased speed, 250 fpm, add						10,600			10,600	11,600
0700	300 fpm, geared electric, add						13,200			13,200	14,500
0725	350 fpm, geared electric, add						15,500			15,500	17,100
0750	400 fpm, geared electric, add						17,600			17,600	19,300
0775	500 fpm, gearless electric, add						22,400			22,400	24,700
0800	600 fpm, gearless electric, add						24,800			24,800	27,300
0825	700 fpm, gearless electric, add						28,900			28,900	31,800
0850	800 fpm, gearless electric, add						32,100			32,100	35,300
0875	For class "B" loading, add						1,900			1,900	2,100
0900	For class "C-1" loading, add						4,700			4,700	5,175
0925	For class "C-2" loading, add						5,625			5,625	6,175
0950	For class "C-3" loading, add						7,725			7,725	8,500
0975	For travel over 40 V.L.F., add		2 Elev	7.25	2.207	V.L.F.	112	118		230	299
1000	For number of stops over 4, add		"	.27	59.259	Stop	2,000	3,175		5,175	6,925

14 21 13.20 Elevator Systems

0010	**ELEVATOR SYSTEMS**										

14 21 23 – Electric Traction Passenger Elevators

14 21 23.10 Electric Traction Passenger Elevators

		Crew	Daily Output	Labor-Hours	Unit	Material	Labor	Equipment	Total	Total Incl O&P
0010	**ELECTRIC TRACTION PASSENGER ELEVATORS**									
1625	Electric pass., base unit, 2000 lb, 200 fpm, 4 stop, std. fin.	2 Elev	.05	320	Ea.	73,500	17,100		90,600	106,500
1650	For 2500 lb capacity, add	↓				3,000			3,000	3,300

14 21 Electric Traction Elevators

14 21 23 – Electric Traction Passenger Elevators

14 21 23.10 Electric Traction Passenger Elevators		Crew	Daily Output	Labor-Hours	Unit	Material	2007 Bare Costs Labor	Equipment	Total	Total Incl O&P
1675	For 3000 lb capacity, add				Ea.	4,575			4,575	5,025
1700	For 3500 lb capacity, add					6,500			6,500	7,150
1725	For 4000 lb capacity, add					6,800			6,800	7,475
1750	For 4500 lb capacity, add					8,950			8,950	9,850
1775	For 5000 lb capacity, add					11,400			11,400	12,500
1800	For increased speed, 250 fpm, geared electric, add					2,525			2,525	2,775
1825	300 fpm, geared electric, add					5,175			5,175	5,700
1850	350 fpm, geared electric, add					6,150			6,150	6,750
1875	400 fpm, geared electric, add					8,725			8,725	9,600
1900	500 fpm, gearless electric, add					41,000			41,000	45,100
1925	600 fpm, gearless electric, add					43,300			43,300	47,600
1950	700 fpm, gearless electric, add					47,400			47,400	52,000
1975	800 fpm, gearless electric, add					52,000			52,000	57,500
2000	For travel over 40 V.L.F., add	2 Elev	7.25	2.207	V.L.F.	114	118		232	300
2025	For number of stops over 4, add		.27	59.259	Stop	2,675	3,175		5,850	7,675
2400	Electric hospital, base unit, 4000 lb, 200 fpm, 4 stop, std fin.		.05	320	Ea.	80,500	17,100		97,600	114,000
2425	For 4500 lb capacity, add					5,550			5,550	6,100
2450	For 5000 lb capacity, add					7,250			7,250	7,950
2475	For increased speed, 250 fpm, geared electric, add					2,700			2,700	2,975
2500	300 fpm, geared electric, add					5,150			5,150	5,650
2525	350 fpm, geared electric, add					6,275			6,275	6,900
2550	400 fpm, geared electric, add					8,725			8,725	9,600
2575	500 fpm, gearless electric, add					39,200			39,200	43,100
2600	600 fpm, gearless electric, add					43,300			43,300	47,600
2625	700 fpm, gearless electric, add					47,300			47,300	52,000
2650	800 fpm, gearless electric, add					52,000			52,000	57,500
2675	For travel over 40 V.L.F., add	2 Elev	7.25	2.207	V.L.F.	114	118		232	300
2700	For number of stops over 4, add	"	.27	59.259	Stop	3,175	3,175		6,350	8,225

14 21 33 – Electric Traction Residential Elevators

14 21 33.20 Electric Traction Residential Elevators		Crew	Daily Output	Labor-Hours	Unit	Material	Labor	Equipment	Total	Total Incl O&P
0010	**ELECTRIC TRACTION RESIDENTIAL ELEVATORS**									
7000	Residential, cab type, 1 floor, 2 stop, minimum	2 Elev	.20	80	Ea.	8,925	4,275		13,200	16,200
7100	Maximum		.10	160		15,100	8,550		23,650	29,300
7200	2 floor, 3 stop, minimum		.12	133		13,200	7,125		20,325	25,200
7300	Maximum		.06	266		21,600	14,200		35,800	45,000

14 24 Hydraulic Elevators

14 24 13 – Hydraulic Freight Elevators

14 24 13.10 Hydraulic Freight Elevators		Crew	Daily Output	Labor-Hours	Unit	Material	Labor	Equipment	Total	Total Incl O&P
0010	**HYDRAULIC FREIGHT ELEVATORS**									
1025	Hydraulic freight, base unit, 2000 lb, 50 fpm, 2 stop, std. fin.	2 Elev	.10	160	Ea.	38,600	8,550		47,150	55,000
1050	For 2500 lb capacity, add					3,000			3,000	3,300
1075	For 3000 lb capacity, add					4,725			4,725	5,200
1100	For 3500 lb capacity, add					7,150			7,150	7,875
1125	For 4000 lb capacity, add					7,650			7,650	8,425
1150	For 4500 lb capacity, add					8,250			8,250	9,100
1175	For 5000 lb capacity, add					11,300			11,300	12,400
1200	For 6000 lb capacity, add					11,900			11,900	13,100
1225	For 7000 lb capacity, add					18,100			18,100	20,000
1250	For 8000 lb capacity, add					20,400			20,400	22,400

14 24 Hydraulic Elevators

14 24 13 – Hydraulic Freight Elevators

14 24 13.10 Hydraulic Freight Elevators		Crew	Daily Output	Labor-Hours	Unit	Material	2007 Bare Costs Labor	Equipment	Total	Total Incl O&P
1275	For 10000 lb capacity, add				Ea.	21,600			21,600	23,800
1300	For 12000 lb capacity, add					26,100			26,100	28,700
1325	For 16000 lb capacity, add					34,000			34,000	37,500
1350	For 20000 lb capacity, add					37,800			37,800	41,600
1375	For increased speed, 100 fpm, add					810			810	890
1400	125 fpm, add					1,650			1,650	1,825
1425	150 fpm, add					3,125			3,125	3,425
1450	175 fpm, add					4,450			4,450	4,900
1475	For class "B" loading, add					1,850			1,850	2,050
1500	For class "C-1" loading, add					4,650			4,650	5,125
1525	For class "C-2" loading, add					5,575			5,575	6,150
1550	For class "C-3" loading, add					7,675			7,675	8,450
1575	For travel over 20 V.L.F., add		2 Elev	7.25	2.207	V.L.F.	405	118		523	625
1600	For number of stops over 2, add		"	.27	59.259	Stop	3,325	3,175		6,500	8,375

14 24 23 – Hydraulic Passenger Elevators

14 24 23.10 Hydraulic Passenger Elevators

			Crew	Daily Output	Labor-Hours	Unit	Material	Labor	Equipment	Total	Total Incl O&P
0010	**HYDRAULIC PASSENGER ELEVATORS**										
2050	Hyd. pass., base unit, 1500 lb, 100 fpm, 2 stop, std. fin.	CN	2 Elev	.10	160	Ea.	32,800	8,550		41,350	48,800
2075	For 2000 lb capacity, add						1,150			1,150	1,250
2100	For 2500 lb capacity, add						2,350			2,350	2,575
2125	For 3000 lb capacity, add						4,175			4,175	4,600
2150	For 3500 lb capacity, add						6,150			6,150	6,775
2175	For 4000 lb capacity, add						7,425			7,425	8,175
2200	For 4500 lb capacity, add						8,950			8,950	9,825
2225	For 5000 lb capacity, add						13,600			13,600	14,900
2250	For increased speed, 125 fpm, add						1,350			1,350	1,475
2275	150 fpm, add						2,650			2,650	2,925
2300	175 fpm, add						4,300			4,300	4,725
2325	200 fpm, add						6,225			6,225	6,850
2350	For travel over 12 V.L.F., add		2 Elev	7.25	2.207	V.L.F.	410	118		528	625
2375	For number of stops over 2, add			.27	59.259	Stop	3,650	3,175		6,825	8,750
2725	Hydraulic hospital, base unit, 4000 lb, 100 fpm, 2 stop, std. fin.			.10	160	Ea.	45,400	8,550		53,950	62,500
2750	For 4000 lb capacity, add						5,475			5,475	6,025
2775	For 4500 lb capacity, add						6,450			6,450	7,075
2800	For 5000 lb capacity, add						9,375			9,375	10,300
2825	For increased speed, 125 fpm, add						1,525			1,525	1,675
2850	150 fpm, add						2,575			2,575	2,825
2875	175 fpm, add						4,150			4,150	4,550
2900	200 fpm, add						6,025			6,025	6,650
2925	For travel over 12 V.L.F., add		2 Elev	7.25	2.207	V.L.F.	251	118		369	450
2950	For number of stops over 2, add		"	.27	59.259	Stop	3,175	3,175		6,350	8,200

14 27 Custom Elevator Cabs

14 27 13 – Custom Elevator Cab Finishes

14 27 13.10 Cab Finishes

		Crew	Daily Output	Labor-Hours	Unit	Material	2007 Bare Costs Labor	Equipment	Total	Total Incl O&P
0010	**CAB FINISHES**									
3325	Passenger elevator cab finishes (based on 3500 lb cab size)									
3350	Acrylic panel ceiling				Ea.	475			475	520
3375	Aluminum eggcrate ceiling					545			545	600
3400	Stainless steel doors					2,875			2,875	3,150
3425	Carpet flooring					395			395	435
3450	Epoxy flooring					325			325	355
3475	Quarry tile flooring					500			500	550
3500	Slate flooring					700			700	770
3525	Textured rubber flooring					128			128	140
3550	Stainless steel walls					2,850			2,850	3,125
3575	Stainless steel returns at door					600			600	660
4450	Hospital elevator cab finishes (based on 3500 lb cab size)									
4475	Aluminum eggcrate ceiling				Ea.	545			545	600
4500	Stainless steel doors					1,750			1,750	1,925
4525	Epoxy flooring					325			325	355
4550	Quarry tile flooring					500			500	550
4575	Textured rubber flooring					128			128	140
4600	Stainless steel walls					2,850			2,850	3,125
4625	Stainless steel returns at door					600			600	660

14 28 Elevator Equipment and Controls

14 28 10 – Elevator Equipment and Controls

14 28 10.10 Elevator Controls and Doors

		Crew	Daily Output	Labor-Hours	Unit	Material	2007 Bare Costs Labor	Equipment	Total	Total Incl O&P
0010	**ELEVATOR CONTROLS AND DOORS**									
2975	Passenger elevator options									
3000	2 car group automatic controls	2 Elev	.66	24.242	Ea.	2,600	1,300		3,900	4,800
3025	3 car group automatic controls		.44	36.364		3,925	1,950		5,875	7,225
3050	4 car group automatic controls		.33	48.485		6,500	2,600		9,100	11,000
3075	5 car group automatic controls		.26	61.538		8,850	3,275		12,125	14,600
3100	6 car group automatic controls		.22	72.727		13,400	3,875		17,275	20,500
3125	Intercom service		3	5.333		345	285		630	805
3150	Duplex car selective collective		.66	24.242		2,875	1,300		4,175	5,100
3175	Center opening 1 speed doors		2	8		1,500	425		1,925	2,275
3200	Center opening 2 speed doors		2	8		1,775	425		2,200	2,575
3225	Rear opening doors (opposite front)		2	8		4,150	425		4,575	5,200
3250	Side opening 2 speed doors		2	8		3,950	425		4,375	4,975
3275	Automatic emergency power switching		.66	24.242		935	1,300		2,235	2,950
3300	Manual emergency power switching		8	2		360	107		467	555
3625	Hall finishes, stainless steel doors					990			990	1,100
3650	Stainless steel frames					990			990	1,100
3675	12 month maintenance contract								2,590	2,850
3700	Signal devices, hall lanterns	2 Elev	8	2		365	107		472	565
3725	Position indicators, up to 3		9.40	1.702		255	91		346	415
3750	Position indicators, per each over 3		32	.500		70.50	26.50		97	118
3775	High speed heavy duty door opener					1,800			1,800	1,975
3800	Variable voltage, O.H. gearless machine, min.	2 Elev	.16	100		26,400	5,350		31,750	37,000
3815	Maximum		.07	228		58,000	12,200		70,200	82,000
3825	Basement installed geared machine		.33	48.485		13,000	2,600		15,600	18,200
3850	Freight elevator options									
3875	Doors, bi-parting	2 Elev	.66	24.242	Ea.	4,475	1,300		5,775	6,850

14 28 Elevator Equipment and Controls

14 28 10 – Elevator Equipment and Controls

14 28 10.10 Elevator Controls and Doors

		Crew	Daily Output	Labor-Hours	Unit	Material	2007 Bare Costs Labor	Equipment	Total	Total Incl O&P
3900	Power operated door and gate	2 Elev	.66	24.242	Ea.	18,100	1,300		19,400	21,800
3925	Finishes, steel plate floor					785			785	865
3950	14 ga. 1/4" x 4' steel plate walls					1,925			1,925	2,100
3975	12 month maintenance contract								1,955	2,150
4000	Signal devices, hall lanterns	2 Elev	8	2		365	107		472	565
4025	Position indicators, up to 3		9.40	1.702		255	91		346	415
4050	Position indicators, per each over 3		32	.500		70.50	26.50		97	118
4075	Variable voltage basement installed geared machine		.66	24.242		14,500	1,300		15,800	17,900
4100	Hospital elevator options									
4125	2 car group automatic controls	2 Elev	.66	24.242	Ea.	2,600	1,300		3,900	4,800
4150	3 car group automatic controls		.44	36.364		3,925	1,950		5,875	7,200
4175	4 car group automatic controls		.33	48.485		6,500	2,600		9,100	11,000
4200	5 car group automatic controls		.26	61.538		8,875	3,275		12,150	14,700
4225	6 car group automatic controls		.22	72.727		13,400	3,875		17,275	20,600
4250	Intercom service		3	5.333		345	285		630	805
4275	Duplex car selective collective		.66	24.242		2,875	1,300		4,175	5,100
4300	Center opening 1 speed doors		2	8		1,500	425		1,925	2,275
4325	Center opening 2 speed doors		2	8		1,975	425		2,400	2,800
4350	Rear opening doors (opposite front)		2	8		4,150	425		4,575	5,200
4375	Side opening 2 speed doors		2	8		6,325	425		6,750	7,600
4400	Automatic emergency power switching		.66	24.242		935	1,300		2,235	2,950
4425	Manual emergency power switching		8	2		360	107		467	555
4675	Hall finishes, stainless steel doors					990			990	1,100
4700	Stainless steel frames					990			990	1,100
4725	12 month maintenance contract								3,910	4,300
4750	Signal devices, hall lanterns	2 Elev	8	2		365	107		472	565
4775	Position indicators, up to 3		9.40	1.702		255	91		346	415
4800	Position indicators, per each over 3		32	.500		70.50	26.50		97	118
4825	High speed heavy duty door opener					1,800			1,800	1,975
4850	Variable voltage, O.H. gearless machine, min.	2 Elev	.16	100		26,400	5,350		31,750	37,000
4865	Maximum		.07	228		58,000	12,200		70,200	82,000
4875	Basement installed geared machine		.33	48.485		13,000	2,600		15,600	18,200
5000	Drilling for piston, casing included, 18" diameter	B-48	80	.700	V.L.F.	41	23	47.50	111.50	133

14 31 Escalators

14 31 10 – Escalators

14 31 10.10 Escalators

			Crew	Daily Output	Labor-Hours	Unit	Material	Labor	Equipment	Total	Total Incl O&P
0010	**ESCALATORS**	R143110-10									
1000	Glass, 32" wide x 10' floor to floor height		M-1	.07	457	Ea.	68,500	23,200	685	92,385	111,000
1010	48" wide x 10' floor to floor height			.07	457		75,500	23,200	685	99,385	118,500
1020	32" wide x 15' floor to floor height	CN		.06	533		72,000	27,100	800	99,900	120,500
1030	48" wide x 15' floor to floor height			.06	533		76,000	27,100	800	103,900	125,000
1040	32" wide x 20' floor to floor height			.05	653		74,000	33,100	980	108,080	132,000
1050	48" wide x 20' floor to floor height			.05	653		79,000	33,100	980	113,080	137,500
1060	32" wide x 25' floor to floor height			.04	800		79,500	40,600	1,200	121,300	149,500
1070	48" wide x 25' floor to floor height			.04	800		85,500	40,600	1,200	127,300	156,000
1080	Enameled steel, 32" wide x 10' floor to floor height			.07	457		70,000	23,200	685	93,885	112,500
1090	48" wide x 10' floor to floor height			.07	457		75,500	23,200	685	99,385	118,500
1110	32" wide x 15' floor to floor height			.06	533		72,000	27,100	800	99,900	120,500
1120	48" wide x 15' floor to floor height			.06	533		76,000	27,100	800	103,900	125,000
1130	32" wide x 20' floor to floor height			.05	653		74,000	33,100	980	108,080	132,000

14 31 Escalators

14 31 10 – Escalators

14 31 10.10 Escalators

		Crew	Daily Output	Labor-Hours	Unit	Material	2007 Bare Costs Labor	Equipment	Total	Total Incl O&P
1140	48" wide x 20' floor to floor height	M-1	.05	653	Ea.	79,000	33,100	980	113,080	137,500
1150	32" wide x 25' floor to floor height		.04	800		79,500	40,600	1,200	121,300	149,500
1160	48" wide x 25' floor to floor height		.04	800		85,500	40,600	1,200	127,300	156,000
1170	Stainless steel, 32" wide x 10' floor to floor height		.07	457		77,000	23,200	685	100,885	120,500
1180	48" wide x 10' floor to floor height		.07	457		80,500	23,200	685	104,385	124,000
1500	32" wide x 15' floor to floor height		.06	533		78,500	27,100	800	106,400	127,500
1700	48" wide x 15' floor to floor height		.06	533		82,000	27,100	800	109,900	132,000
2300	32" wide x 25' floor to floor height		.04	800		89,000	40,600	1,200	130,800	159,500
2500	48" wide x 25' floor to floor height	↓	.04	800	↓	94,000	40,600	1,200	135,800	165,500

14 32 Moving Walks

14 32 10 – Moving Walks

14 32 10.10 Moving Walks

			Crew	Daily Output	Labor-Hours	Unit	Material	Labor	Equipment	Total	Total Incl O&P
0010	**MOVING WALKS**	R143210-20									
0020	Walk, 27" tread width, minimum		M-1	6.50	4.923	L.F.	750	250	7.40	1,007.40	1,200
0100	300' to 500', maximum			4.43	7.223		1,050	365	10.85	1,425.85	1,700
0300	48" tread width walk, minimum			4.43	7.223		1,700	365	10.85	2,075.85	2,425
0400	100' to 350', maximum			3.82	8.377		2,000	425	12.55	2,437.55	2,850
0600	Ramp, 12° incline, 36" tread width, minimum			5.27	6.072		1,400	310	9.10	1,719.10	2,000
0700	70' to 90' maximum			3.82	8.377		1,975	425	12.55	2,412.55	2,825
0900	48" tread width, minimum			3.57	8.964		2,025	455	13.45	2,493.45	2,925
1000	40' to 70', maximum		↓	2.91	10.997	↓	2,575	560	16.50	3,151.50	3,675

14 42 Wheelchair Lifts

14 42 13 – Inclined Wheelchair Lifts

14 42 13.10 Inclined Wheelchair Lifts

		Crew	Daily Output	Labor-Hours	Unit	Material	Labor	Equipment	Total	Total Incl O&P
0010	**INCLINED WHEELCHAIR LIFTS**									
7700	Stair climber (chair lift), single seat, minimum	2 Elev	1	16	Ea.	4,325	855		5,180	6,025
7800	Maximum		.20	80		5,925	4,275		10,200	12,900
8700	Stair lift, minimum		1	16		11,700	855		12,555	14,200
8900	Maximum	↓	.20	80	↓	18,500	4,275		22,775	26,800

14 42 16 – Vertical Wheelchair Lifts

14 42 16.10 Vertical Wheelchair Lifts

		Crew	Daily Output	Labor-Hours	Unit	Material	Labor	Equipment	Total	Total Incl O&P
0010	**VERTICAL WHEELCHAIR LIFTS**									
8000	Wheelchair lift, minimum	2 Elev	1	16	Ea.	5,925	855		6,780	7,775
8500	Maximum	"	.50	32	"	14,000	1,700		15,700	18,000

14 45 Vehicle Lifts

14 45 10 – Vehicle Lifts

14 45 10.10 Hydraulic Lifts

		Crew	Daily Output	Labor-Hours	Unit	Material	2007 Bare Costs Labor	2007 Bare Costs Equipment	Total	Total Incl O&P
0010	**HYDRAULIC LIFTS**									
0200	Double post, 9000 lb frame	L-4	1.15	20.870	Ea.	5,150	720		5,870	6,800
0400	26000 lb frame		.22	109		24,900	3,750		28,650	33,300
2200	Hoists, single post, 8,000# capacity, swivel arms		.40	60		4,750	2,075		6,825	8,450
2400	Two posts, adjustable frames, 11,000# capacity		.25	96		6,100	3,300		9,400	11,900
2500	24,000# capacity		.15	160		8,125	5,500		13,625	17,500
2700	7,500# capacity, frame supports		.50	48		6,775	1,650		8,425	10,000
2800	Four post, roll on ramp		.50	48		6,100	1,650		7,750	9,300
2810	Hydraulic lifts, above ground, 2 post, clear floor, 6000 lb cap		2.67	8.989		6,700	310		7,010	7,825
2815	9000 lb capacity		2.29	10.480		15,900	360		16,260	18,100
2820	15,000 lb capacity		2	12		17,900	415		18,315	20,300
2825	30,000 lb capacity		1.60	15		39,700	515		40,215	44,400
2830	4 post, ramp style, 25,000 lb capacity		2	12		15,500	415		15,915	17,700
2835	35,000 lb capacity		1	24		72,500	825		73,325	81,000
2840	50,000 lb capacity		1	24		81,000	825		81,825	90,500
2845	75,000 lb capacity		1	24		94,000	825		94,825	105,000
2850	For drive thru tracks, add, minimum					995			995	1,100
2855	Maximum					1,700			1,700	1,875
2860	Ramp extensions, 3' (set of 2)					820			820	900
2865	Rolling jack platform					2,850			2,850	3,125
2870	Elec/hyd jacking beam					7,625			7,625	8,375
2880	Scissor lift, portable, 6000 lb capacity					7,475			7,475	8,225

14 51 Correspondence and Parcel Lifts

14 51 10 – Correspondence and Parcel Lifts

14 51 10.10 Correspondence Lifts

		Crew	Daily Output	Labor-Hours	Unit	Material	Labor	Equipment	Total	Total Incl O&P
0010	**CORRESPONDENCE LIFTS**									
0020	1 floor, 2 stop, 25 lb capacity, electric	2 Elev	.20	80	Ea.	5,275	4,275		9,550	12,200
0100	Hand, 5 lb capacity	"	.20	80	"	2,000	4,275		6,275	8,575

14 51 10.20 Parcel Lifts

		Crew	Daily Output	Labor-Hours	Unit	Material	Labor	Equipment	Total	Total Incl O&P
0010	**PARCEL LIFTS**									
0020	20" x 20", 100 lb capacity, electric, per floor	2 Mill	.25	64	Ea.	8,250	2,450		10,700	12,700

14 91 Facility Chutes

14 91 33 – Laundry and Linen Chutes

14 91 33.10 Chutes

		Crew	Daily Output	Labor-Hours	Unit	Material	Labor	Equipment	Total	Total Incl O&P
0010	**CHUTES**									
0050	Aluminized steel, 16 ga., 18" diameter	2 Shee	3.50	4.571	Floor	840	199		1,039	1,225
0100	24" diameter		3.20	5		885	218		1,103	1,300
0200	30" diameter		3	5.333		1,025	232		1,257	1,475
0300	36" diameter		2.80	5.714		1,150	249		1,399	1,625
0400	Galvanized steel, 16 ga., 18" diameter		3.50	4.571		810	199		1,009	1,200
0500	24" diameter		3.20	5		910	218		1,128	1,325
0600	30" diameter		3	5.333		1,025	232		1,257	1,475
0700	36" diameter		2.80	5.714		1,200	249		1,449	1,700
0800	Stainless steel, 18" diameter		3.50	4.571		1,850	199		2,049	2,325
0900	24" diameter		3.20	5		2,000	218		2,218	2,525
1000	30" diameter		3	5.333		2,350	232		2,582	2,950
1005	36" diameter		2.80	5.714		2,600	249		2,849	3,250

14 91 Facility Chutes

14 91 33 – Laundry and Linen Chutes

	14 91 33.10 Chutes	Crew	Daily Output	Labor-Hours	Unit	Material	2007 Bare Costs Labor	2007 Bare Costs Equipment	Total	Total Incl O&P
1200	Linen chute bottom collector, aluminized steel	2 Shee	4	4	Ea.	1,125	174		1,299	1,525
1300	Stainless steel		4	4		1,375	174		1,549	1,800
1500	Refuse, bottom hopper, aluminized steel, 18" diameter		3	5.333		675	232		907	1,100
1600	24" diameter		3	5.333		820	232		1,052	1,250
1800	36" diameter		3	5.333		1,375	232		1,607	1,850

14 91 82 – Trash Chutes

	14 91 82.10 Trash Chutes	Crew	Daily Output	Labor-Hours	Unit	Material	Labor	Equipment	Total	Total Incl O&P
0010	**TRASH CHUTES**									
2900	Package chutes, spiral type, minimum	2 Shee	4.50	3.556	Floor	1,975	155		2,130	2,400
3000	Maximum	"	1.50	10.667	"	5,125	465		5,590	6,375

14 92 Pneumatic Tube Systems

14 92 10 – Pneumatic Tube Systems

	14 92 10.10 Pneumatic Tube Systems	Crew	Daily Output	Labor-Hours	Unit	Material	Labor	Equipment	Total	Total Incl O&P
0010	**PNEUMATIC TUBE SYSTEMS**									
0020	100' long, stock									
0100	3" diameter	2 Stpi	.12	133	Total	5,325	6,025		11,350	15,000
0300	4" diameter	"	.09	177	"	6,025	8,025		14,050	18,700
0400	Twin tube, two stations or more, conventional system									
0600	2-1/2" round	2 Stpi	62.50	.256	L.F.	10.75	11.55		22.30	29.50
0700	3" round		46	.348		12.35	15.70		28.05	37
0900	4" round		49.60	.323		13.65	14.60		28.25	37
1000	4" x 7" oval		37.60	.426		19.95	19.25		39.20	51
1050	Add for blower		2	8	System	4,225	360		4,585	5,200
1110	Plus for each round station, add		7.50	2.133	Ea.	475	96.50		571.50	665
1150	Plus for each oval station, add		7.50	2.133	"	475	96.50		571.50	665
1200	Alternate pricing method: base cost, minimum		.75	21.333	Total	5,175	965		6,140	7,125
1300	Maximum		.25	64	"	10,300	2,900		13,200	15,700
1500	Plus total system length, add, minimum		93.40	.171	L.F.	6.65	7.75		14.40	18.95
1600	Maximum		37.60	.426	"	19.65	19.25		38.90	50.50
1800	Completely automatic system, 4" round, 15 to 50 stations		.29	55.172	Station	16,300	2,500		18,800	21,700
2200	51 to 144 stations		.32	50		12,600	2,250		14,850	17,300
2400	6" round or 4" x 7" oval, 15 to 50 stations		.24	66.667		20,400	3,025		23,425	26,900
2800	51 to 144 stations		.23	69.565		17,100	3,150		20,250	23,500

Division Notes

		CREW	DAILY OUTPUT	LABOR-HOURS	UNIT	2007 BARE COSTS				TOTAL INCL O&P
						MAT.	LABOR	EQUIP.	TOTAL	

Estimating Tips

Pipe for fire protection and all uses is located in Section 22 22 13.

When installing piping above 10' review Section 22 02 02.20 for percentage adds due to the elevated work.

Many, but not all, areas require backflow protection in the fire system. It is advisable to check local building codes for specific requirements.

For your reference, the following is a list of the most applicable Fire Codes and Standards which may be purchased from the NFPA, 1 Batterymarch Park, Quincy, MA 02169-7471.

NFPA 1: Uniform Fire Code
NFPA 10: Portable Fire Extinguishers
NFPA 11: Low-, Medium-, and High-Expansion Foam
NFPA 12: Carbon Dioxide Extinguishing Systems (Also companion 12A)
NFPA 13: Installation of Sprinkler Systems (Also companion 13D, 13E, and 13R)
NFPA 14: Installation of Standpipe and Hose Systems
NFPA 15: Water Spray Fixed Systems for Fire Protection
NFPA 16: Installation of Foam-Water Sprinkler and Foam-Water Spray Systems
NFPA 17: Dry Chemical Extinguishing Systems (Also companion 17A)
NFPA 18: Wetting Agents
NFPA 20: Installation of Stationary Pumps for Fire Protection
NFPA 22: Water Tanks for Private Fire Protection
NFPA 24: Installation of Private Fire Service Mains and their Appurtenances
NFPA 25: Inspection, Testing and Maintenance of Water-Based Fire Protection

Reference Numbers

Reference numbers are shown in shaded boxes at the beginning of some major classifications. These numbers refer to related items in the Reference Section. The reference information may be an estimating procedure, an alternate pricing method, or technical information.

Note: Not all subdivisions listed here necessarily appear in this publication.

Note: **i2 Trade Service**, *in part, has been used as a reference source for some of the material prices used in Division 21.*

No part of this publication may be reproduced, stored in a retrieval system, or transmitted in any form or by any means without prior written permission of Reed Construction Data.

21 05 Common Work Results for Fire Suppression

21 05 23 – General-Duty Valves for Water-Based Fire-Suppression Piping

21 05 23.50 General-Duty Valves for Water-Based Fire Supp.	Crew	Daily Output	Labor-Hours	Unit	Material	2007 Bare Costs Labor	Equipment	Total	Total Incl O&P
0010 **GENERAL-DUTY VALVES FOR WATER-BASED FIRE SUPPRESSION PIPING**									
6200 Valves and components									
6500 Check, swing, C.I. body, brass fittings, auto. ball drip									
6520 4" size	Q-12	3	5.333	Ea.	196	213		409	535
6800 Check, wafer, butterfly type, C.I. body, bronze fittings									
6820 4" size	Q-12	4	4	Ea.	235	160		395	500

21 11 Facility Fire-Suppression Water-Service Piping

21 11 16 – Facility Fire Hydrants

21 11 16.50 Facility Fire Hydrants	Crew	Daily Output	Labor-Hours	Unit	Material	2007 Bare Costs Labor	Equipment	Total	Total Incl O&P
0010 **FACILITY FIRE HYDRANTS**									
3750 Hydrants, wall, w/caps, single, flush, polished brass									
3800 2-1/2" x 2-1/2"	Q-12	5	3.200	Ea.	135	128		263	340
3840 2-1/2" x 3"	"	5	3.200	"	180	128		308	390
3950 Double, flush, polished brass									
4000 2-1/2" x 2-1/2" x 4"	Q-12	5	3.200	Ea.	360	128		488	590
4040 2-1/2" x 2-1/2" x 6"	"	4.60	3.478		605	139		744	875
4200 For polished chrome, add					10%				
4350 Double, projecting, polished brass									
4400 2-1/2" x 2-1/2" x 4"	Q-12	5	3.200	Ea.	148	128		276	355
4450 2-1/2" x 2-1/2" x 6"	"	4.60	3.478	"	286	139		425	525
4460 Valve control, dbl. flush/projecting hydrant, cap &									
4470 chain, ext. rod & cplg., escutcheon, polished brass	Q-12	8	2	Ea.	203	80		283	345

21 11 19 – Fire-Department Connections

21 11 19.50 Fire-Department Connections	Crew	Daily Output	Labor-Hours	Unit	Material	2007 Bare Costs Labor	Equipment	Total	Total Incl O&P
0010 **FIRE-DEPARTMENT CONNECTIONS**									
7140 Standpipe connections, wall, w/plugs & chains									
7160 Single, flush, brass, 2-1/2" x 2-1/2"	Q-12	5	3.200	Ea.	107	128		235	310
7180 2-1/2" x 3"	"	5	3.200	"	110	128		238	315
7240 For polished chrome, add					15%				
7280 Double, flush, polished brass									
7300 2-1/2" x 2-1/2" x 4"	Q-12	5	3.200	Ea.	345	128		473	570
7330 2-1/2" x 2-1/2" x 6"	"	4.60	3.478	"	730	139		869	1,000
7400 For polished chrome, add					15%				
7440 For sill cock combination, add				Ea.	60.50			60.50	66.50
7900 Three way, flush, polished brass									
7920 2-1/2" (3) x 4"	Q-12	4.80	3.333	Ea.	1,600	133		1,733	1,950
7930 2-1/2" (3) x 6"	"	4.80	3.333		1,325	133		1,458	1,675
8000 For polished chrome, add					9%				
8020 Three way, projecting, polished brass									
8040 2-1/2"(3) x 4"	Q-12	4.80	3.333	Ea.	1,050	133		1,183	1,375

21 12 Fire-Suppression Standpipes

21 12 13 – Fire-Suppression Hoses and Nozzles

21 12 13.50 Fire-Suppression Hoses and Nozzles	Crew	Daily Output	Labor-Hours	Unit	Material	2007 Bare Costs Labor	Equipment	Total	Total Incl O&P
0010 **FIRE-SUPPRESSION HOSES AND NOZZLES**									
0200 Adapters, rough brass, straight hose threads									
0220 One piece, female to male, rocker lugs									
0240 1" x 1"				Ea.	32.50			32.50	35.50
2200 Hose, less couplings									
2260 Synthetic jacket, lined, 300 lb. test, 1-1/2" diameter	Q-12	2600	.006	L.F.	1.89	.25		2.14	2.45
2280 2-1/2" diameter		2200	.007		3.25	.29		3.54	4.02
2360 High strength, 500 lb. test, 1-1/2" diameter		2600	.006		2.01	.25		2.26	2.58
2380 2-1/2" diameter		2200	.007		3.50	.29		3.79	4.29
5600 Nozzles, brass									
5620 Adjustable fog, 3/4" booster line				Ea.	83.50			83.50	91.50
5630 1" booster line					96.50			96.50	106
5640 1-1/2" leader line					59			59	64.50
5660 2-1/2" direct connection					198			198	218
5780 For chrome plated, add					8%				
5850 Electrical fire, adjustable fog, no shock									
5900 1-1/2"				Ea.	315			315	350
5920 2-1/2"					425			425	470
5980 For polished chrome, add					6%				
6200 Heavy duty, comb. adj. fog and str. stream, with handle									
6210 1" booster line				Ea.	305			305	335

21 12 19 – Fire-Suppression Hose Racks

21 12 19.50 Fire-Suppression Hose Racks

	Crew	Daily Output	Labor-Hours	Unit	Material	Labor	Equipment	Total	Total Incl O&P
0010 **FIRE-SUPPRESSION HOSE RACKS**									
2600 Hose rack, swinging, for 1-1/2" diameter hose,									
2620 Enameled steel, 50' & 75' lengths of hose	Q-12	20	.800	Ea.	26	32		58	76.50
2640 100' and 125' lengths of hose	"	20	.800	"	26	32		58	76.50

21 12 23 – Fire-Suppression Hose Valves

21 12 23.70 Fire-Suppression Hose Valves

	Crew	Daily Output	Labor-Hours	Unit	Material	Labor	Equipment	Total	Total Incl O&P
0010 **FIRE-SUPPRESSION HOSE VALVES**									
0080 Wheel handle, 300 lb., 1-1/2"	1 Spri	12	.667	Ea.	35.50	29.50		65	83.50
0090 2-1/2"	"	7	1.143	"	60	50.50		110.50	142
0100 For polished brass, add					35%				
0110 For polished chrome, add					50%				

21 13 Fire-Suppression Sprinkler Systems

21 13 13 – Wet-Pipe Sprinkler Systems

21 13 13.50 Wet-Pipe Sprinkler Systems

	Crew	Daily Output	Labor-Hours	Unit	Material	Labor	Equipment	Total	Total Incl O&P
0010 **WET-PIPE SPRINKLER SYSTEMS**									
2600 Sprinkler heads, not including supply piping									
3700 Standard spray, pendent or upright, brass, 135° to 286°F									
3730 1/2" NPT, 7/16" orifice	1 Spri	16	.500	Ea.	8.65	22		30.65	43
3740 1/2" NPT, 1/2" orifice **CN**	"	16	.500		12.60	22		34.60	47.50
3860 For wax and lead coating, add					17.65			17.65	19.40
3880 For wax coating, add					10.65			10.65	11.70
3900 For lead coating, add					11.40			11.40	12.50
3920 For 360°F, same cost									
3930 For 400°F, add				Ea.	29.50			29.50	32
3940 For 500°F, add				"	29.50			29.50	32
4500 Sidewall, horizontal, brass, 135° to 286°F									

21 13 Fire-Suppression Sprinkler Systems

21 13 13 – Wet-Pipe Sprinkler Systems

21 13 13.50 Wet-Pipe Sprinkler Systems		Crew	Daily Output	Labor-Hours	Unit	Material	2007 Bare Costs Labor	Equipment	Total	Total Incl O&P
4520	1/2" NPT, 1/2" orifice	1 Spri	16	.500	Ea.	12.65	22		34.65	47.50
4540	For 360°F, same cost									
4800	Recessed pendent, brass, 135° to 286°F									
4820	1/2" NPT, 3/8" orifice	1 Spri	10	.800	Ea.	44	35.50		79.50	102
4830	1/2" NPT, 7/16" orifice	↓	10	.800		44	35.50		79.50	102
4840	1/2" NPT, 1/2" orifice	↓	10	.800	↓	58	35.50		93.50	118

21 13 16 – Dry-Pipe Sprinkler Systems

21 13 16.50 Dry-Pipe Sprinkler Systems

		Crew	Daily Output	Labor-Hours	Unit	Material	Labor	Equipment	Total	Total Incl O&P
0010	**DRY-PIPE SPRINKLER SYSTEMS**									
0600	Accelerator	1 Spri	8	1	Ea.	410	44.50		454.50	515
2600	Sprinkler heads, not including supply piping									
2640	Dry, pendent, 1/2" orifice, 3/4" or 1" NPT									
2700	15-1/4" to 18" length	1 Spri	14	.571	Ea.	81	25.50		106.50	127
2710	18-1/4" to 21" length		13	.615		83.50	27.50		111	133
2720	21-1/4" to 24" length		13	.615		87	27.50		114.50	137
2730	24-1/4" to 27" length	↓	13	.615	↓	90	27.50		117.50	140

21 13 26 – Deluge Fire-Suppression Sprinkler Systems

21 13 26.50 Deluge Fire-Suppression Sprinkler Systems

		Crew	Daily Output	Labor-Hours	Unit	Material	Labor	Equipment	Total	Total Incl O&P
0010	**DELUGE FIRE-SUPPRESSION SPRINKLER SYSTEMS**									
6200	Valves and components									
7000	Deluge, assembly, incl. trim, pressure									
7020	operated relief, emergency release, gauges									
7040	2" size	Q-12	2	8	Ea.	1,600	320		1,920	2,225
7060	3" size	"	1.50	10.667	"	1,850	425		2,275	2,700

21 13 39 – Foam-Water Systems

21 13 39.50 Foam-Water Systems

		Crew	Daily Output	Labor-Hours	Unit	Material	Labor	Equipment	Total	Total Incl O&P
0010	**FOAM-WATER SYSTEMS**									
2600	Sprinkler heads, not including supply piping									
3600	Foam-water, pendent or upright, 1/2" NPT	1 Spri	12	.667	Ea.	94	29.50		123.50	148

21 21 Carbon-Dioxide Fire-Extinguishing Systems

21 21 16 – Carbon-Dioxide Fire-Extinguishing Equipment

21 21 16.50 CO2 Fire Extinguishing System

		Crew	Daily Output	Labor-Hours	Unit	Material	Labor	Equipment	Total	Total Incl O&P
0010	**CO_2 FIRE EXTINGUISHING SYSTEM**									
0042	For detectors and control stations, see division 28 32 33									
1000	Dispersion nozzle, CO_2, 3" x 5"	1 Plum	18	.444	Ea.	52.50	19.90		72.40	88
2000	Extinguisher, CO_2 system, high pressure, 75 lb. cylinder	Q-1	6	2.667		1,000	108		1,108	1,250
3000	Electro/mechanical release	L-1	4	4		131	177		308	410
3400	Manual pull station	1 Plum	6	1.333	↓	47.50	59.50		107	142

21 22 Clean-Agent Fire-Extinguishing Systems

21 22 16 – Clean-Agent Fire-Extinguishing Equipment

21 22 16.50 FM200 Fire Extinguishing System	Crew	Daily Output	Labor-Hours	Unit	Material	2007 Bare Costs Labor	Equipment	Total	Total Incl O&P
0010 **FM200 FIRE EXTINGUISHING SYSTEM**									
1100 Dispersion nozzle FM200, 1-1/2"	1 Plum	14	.571	Ea.	52.50	25.50		78	96.50
2400 Extinguisher, FM 200 system, filled, with mounting bracket									
2540 196 lb. container	Q-1	4	4	Ea.	6,100	161		6,261	6,950
6000 Average FM200 system, minimum				C.F.	1.38			1.38	1.52
6020 Maximum				"	2.75			2.75	3.03

21 31 Centrifugal Fire Pumps

21 31 13 – Electric-Drive, Centrifugal Fire Pumps

21 31 13.50 Electric-Drive Fire Pumps

	Crew	Daily Output	Labor-Hours	Unit	Material	Labor	Equipment	Total	Total Incl O&P
0010 **ELECTRIC-DRIVE FIRE PUMPS** Including controller, fittings and relief valve									
3100 250 GPM, 55 psi, 15 HP, 3,550 RPM, 2" pump	Q-13	.70	45.714	Ea.	14,500	1,925		16,425	18,900
3200 500 GPM, 50 psi, 27 HP, 1770 RPM, 4" pump		.68	47.059		18,000	1,975		19,975	22,800
3350 750 GPM, 50 psi, 44 HP, 1,770 RPM, 5" pump		.64	50		26,300	2,100		28,400	32,100
3400 750 GPM, 100 psi, 66 HP, 3550 RPM, 4" pump	↓	.58	55.172		23,300	2,325		25,625	29,100
5000 For jockey pump 1", 3 HP, with control, add	Q-12	2	8	↓	3,225	320		3,545	4,025

21 31 16 – Diesel-Drive, Centrifugal Fire Pumps

21 31 16.50 Diesel-Drive Fire Pumps

	Crew	Daily Output	Labor-Hours	Unit	Material	Labor	Equipment	Total	Total Incl O&P
0010 **DIESEL-DRIVE FIRE PUMPS** Including controller, fittings and relief valve									
0050 500 GPM, 50 psi, 27 HP, 4" pump	Q-13	.64	50	Ea.	62,000	2,100		64,100	71,000
0200 750 GPM, 50 psi, 44 HP, 5" pump		.60	53.333		66,000	2,250		68,250	76,000
0400 1000 GPM, 100 psi, 89 HP, 4" pump		.56	57.143		71,000	2,400		73,400	81,500
0700 2,000 GPM, 100 psi, 167 HP, 6" pump		.34	94.118		75,000	3,975		78,975	88,500
0950 3500 GPM, 100 psi, 300 HP, 10" pump	↓	.24	133	↓	132,000	5,625		137,625	153,500

Division Notes

	CREW	DAILY OUTPUT	LABOR-HOURS	UNIT	2007 BARE COSTS				TOTAL INCL O&P
					MAT.	LABOR	EQUIP.	TOTAL	

Estimating Tips

22 10 00 Plumbing Piping and Pumps

This subdivision is primarily basic pipe and related materials. The pipe may be used by any of the mechanical disciplines, i.e., plumbing, fire protection, heating, and air conditioning.

- The piping section lists the add to labor for elevated pipe installation. These adds apply to all elevated pipe, fittings, valves, insulation, etc., that are placed above 10' high. CAUTION: the correct percentage may vary for the same pipe. For example, the percentage add for the basic pipe installation should be based on the maximum height that the craftsman must install for that particular section. If the pipe is to be located 14' above the floor but it is suspended on threaded rod from beams, the bottom flange of which is 18' high (4' rods), then the height is actually 18' and the add is 20%. The pipe coverer, however, does not have to go above the 14', and so his or her add should be 10%.
- Most pipe is priced first as straight pipe with a joint (coupling, weld, etc.) every 10' and a hanger usually every 10'. There are exceptions with hanger spacing such as for cast iron pipe (5') and plastic pipe (3 per 10'). Following each type of pipe there are several lines listing sizes and the amount to be subtracted to delete couplings and hangers. This is for pipe that is to be buried or supported together on trapeze hangers. The reason that the couplings are deleted is that these runs are usually long, and frequently longer lengths of pipe are used. By deleting the couplings, the estimator is expected to look up and add back the correct reduced number of couplings.
- When preparing an estimate it may be necessary to approximate the fittings. Fittings usually run between 25% and 50% of the cost of the pipe. The lower percentage is for simpler runs, and the higher number is for complex areas such as mechanical rooms.
- For historic restoration projects, the systems must be as invisible as possible, and pathways must be sought for pipes, conduit, and ductwork. While installations in accessible spaces (such as basements and attics) are relatively straightforward to estimate, labor costs may be more difficult to determine when delivery systems must be concealed.

22 40 00 Plumbing Fixtures

- Plumbing fixture costs usually require two lines: the fixture itself and its "rough-in, supply and waste."
- In the Assemblies Section (Plumbing D2010) for the desired fixture, the System Components Group at the center of the page shows the fixture on the first line. The rest of the list (fittings, pipe, tubing, etc.) will total up to what we refer to in the Unit Price section as "Rough-in, supply, waste and vent." Note that for most fixtures we allow a nominal 5' of tubing to reach from the fixture to a main or riser.
- Remember that gas- and oil-fired units need venting.

Reference Numbers

Reference numbers are shown in shaded boxes at the beginning of some major classifications. These numbers refer to related items in the Reference Section. The reference information may be an estimating procedure, an alternate pricing method, or technical information.

Note: Not all subdivisions listed here necessarily appear in this publication.

Note: **i2 Trade Service**, *in part, has been used as a reference source for some of the material prices used in Division 22.*

22 01 Operation and Maintenance of Plumbing

22 01 02 – General Plumbing

22 01 02.10 Boilers, General

		Crew	Daily Output	Labor-Hours	Unit	Material	2007 Bare Costs Labor	Equipment	Total	Total Incl O&P
0010	**BOILERS, GENERAL**, Prices do not include flue piping, elec. wiring,									
0020	gas or oil piping, boiler base, pad, or tankless unless noted									
0100	Boiler H.P.: 10 KW = 34 lbs/steam/hr = 33,475 BTU/hr. R235000-50									
0120										
0150	To convert SFR to BTU rating: Hot water, 150 x SFR;									
0160	Forced hot water, 180 x SFR; steam, 240 x SFR									

22 01 02.20 Piping

		Crew	Daily Output	Labor-Hours	Unit	Material	Labor	Equipment	Total	Total Incl O&P
0010	**PIPING**									
1000	Add to labor for elevated installation									
1080	10' to 15' high						10%			
1100	15' to 20' high						20%			
1120	20' to 25' high						25%			
1140	25' to 30' high						35%			
1160	30' to 35' high						40%			
1180	35' to 40' high						50%			
1200	Over 40' high						55%			

22 05 Common Work Results for Plumbing

22 05 05 – Selective Plumbing Demolition

22 05 05.10 Plumbing Demolition

		Crew	Daily Output	Labor-Hours	Unit	Material	Labor	Equipment	Total	Total Incl O&P
0010	**PLUMBING DEMOLITION**									
1020	Fixtures, including 10' piping									
1100	Bath tubs, cast iron	1 Plum	4	2	Ea.		89.50		89.50	135
1120	Fiberglass		6	1.333			59.50		59.50	90
1140	Steel		5	1.600			71.50		71.50	108
1200	Lavatory, wall hung		10	.800			36		36	54
1220	Counter top		8	1			45		45	67.50
1300	Sink, single compartment		8	1			45		45	67.50
1320	Double compartment		7	1.143			51		51	77
1400	Water closet, floor mounted		8	1			45		45	67.50
1420	Wall mounted		7	1.143			51		51	77
1500	Urinal, floor mounted		4	2			89.50		89.50	135
1520	Wall mounted		7	1.143			51		51	77
1600	Water fountains, free standing		8	1			45		45	67.50
1620	Wall or deck mounted		6	1.333			59.50		59.50	90
2000	Piping, metal, up thru 1-1/2" diameter		200	.040	L.F.		1.79		1.79	2.70
2050	2" thru 3-1/2" diameter		150	.053			2.39		2.39	3.59
2100	4" thru 6" diameter	2 Plum	100	.160			7.15		7.15	10.80
2150	8" thru 14" diameter	"	60	.267			11.95		11.95	17.95
2153	16" thru 20" diameter	Q-18	70	.343			14.45	.88	15.33	23
2155	24" thru 26" diameter		55	.436			18.40	1.12	19.52	28.50
2156	30" thru 36" diameter		40	.600			25.50	1.54	27.04	39.50
2212	Deduct for salvage, aluminum scrap				Ton				550	605
2214	Brass scrap								460	505
2216	Copper scrap								1,092	1,200
2218	Lead scrap								259	285
2220	Steel scrap								77	85
2250	Water heater, 40 gal.	1 Plum	6	1.333	Ea.		59.50		59.50	90

22 05 Common Work Results for Plumbing

22 05 23 – General-Duty Valves for Plumbing Piping

22 05 23.10 Valves, Brass		Crew	Daily Output	Labor-Hours	Unit	Material	2007 Bare Costs Labor	Equipment	Total	Total Incl O&P
0010	**VALVES, BRASS**									
0500	Gas cocks, threaded									
0530	1/2"	1 Plum	24	.333	Ea.	9.75	14.95		24.70	33.50
0540	3/4"		22	.364		14.25	16.30		30.55	40
0550	1"		19	.421		12.45	18.85		31.30	42
0560	1-1/4"		15	.533		24	24		48	62.50

22 05 23.20 Valves, Bronze		Crew	Daily Output	Labor-Hours	Unit	Material	2007 Bare Costs Labor	Equipment	Total	Total Incl O&P
0010	**VALVES, BRONZE**									
1020	Angle, 150 lb., rising stem, threaded									
1030	1/8"	1 Plum	24	.333	Ea.	62.50	14.95		77.45	91
1040	1/4"		24	.333		64.50	14.95		79.45	93.50
1050	3/8"		24	.333		64.50	14.95		79.45	93.50
1060	1/2"		22	.364		64.50	16.30		80.80	95.50
1070	3/4"		20	.400		87	17.90		104.90	123
1080	1"		19	.421		125	18.85		143.85	167
1100	1-1/2"		13	.615		210	27.50		237.50	273
1110	2"		11	.727		340	32.50		372.50	425
1380	Ball, 150 psi, threaded									
1400	1/4"	1 Plum	24	.333	Ea.	9.55	14.95		24.50	33
1430	3/8"		24	.333		9.55	14.95		24.50	33
1450	1/2"		22	.364		9.55	16.30		25.85	35
1460	3/4"		20	.400		15.75	17.90		33.65	44.50
1470	1"		19	.421		19.85	18.85		38.70	50.50
1480	1-1/4"		15	.533		21	24		45	59
1490	1-1/2"		13	.615		27	27.50		54.50	71.50
1500	2"		11	.727		33	32.50		65.50	85
1750	Check, swing, class 150, regrinding disc, threaded									
1800	1/8"	1 Plum	24	.333	Ea.	30	14.95		44.95	55.50
1830	1/4"		24	.333		30	14.95		44.95	55.50
1840	3/8"		24	.333		31	14.95		45.95	56.50
1850	1/2"		24	.333		31	14.95		45.95	56.50
1860	3/4"		20	.400		45	17.90		62.90	76.50
1870	1"		19	.421		65.50	18.85		84.35	101
1880	1-1/4"		15	.533		94	24		118	139
1890	1-1/2"		13	.615		161	27.50		188.50	219
1900	2"		11	.727		161	32.50		193.50	226
1910	2-1/2"	Q-1	15	1.067		325	43		368	420
2000	For 200 lb, add					5%	10%			
2040	For 300 lb, add					15%	15%			
2850	Gate, N.R.S., soldered, 125 psi									
2900	3/8"	1 Plum	24	.333	Ea.	25	14.95		39.95	50
2920	1/2"		24	.333		22	14.95		36.95	46.50
2940	3/4"		20	.400		25	17.90		42.90	54.50
2950	1"		19	.421		36	18.85		54.85	68
2960	1-1/4"		15	.533		57	24		81	98.50
2970	1-1/2"		13	.615		61	27.50		88.50	109
2980	2"		11	.727		86.50	32.50		119	144
2990	2-1/2"	Q-1	15	1.067		216	43		259	305
3000	3"	"	13	1.231		282	49.50		331.50	385
3850	Rising stem, soldered, 300 psi									
3950	1"	1 Plum	19	.421	Ea.	92.50	18.85		111.35	131
3980	2"	"	11	.727		246	32.50		278.50	320

22 05 Common Work Results for Plumbing

22 05 23 – General-Duty Valves for Plumbing Piping

22 05 23.20 Valves, Bronze		Crew	Daily Output	Labor-Hours	Unit	Material	2007 Bare Costs Labor	Equipment	Total	Total Incl O&P
4000	3"	Q-1	13	1.231	Ea.	775	49.50		824.50	930
4250	Threaded, class 150									
4310	1/4"	1 Plum	24	.333	Ea.	33	14.95		47.95	59
4320	3/8"		24	.333		33	14.95		47.95	59
4330	1/2"		24	.333		31	14.95		45.95	56.50
4340	3/4"		20	.400		36.50	17.90		54.40	67.50
4350	1"		19	.421		49	18.85		67.85	82.50
4360	1-1/4"		15	.533		65	24		89	108
4370	1-1/2"		13	.615		82	27.50		109.50	132
4380	2"	↓	11	.727		112	32.50		144.50	172
4390	2-1/2"	Q-1	15	1.067		260	43		303	350
4400	3"	"	13	1.231		390	49.50		439.50	505
4500	For 300 psi, threaded, add					100%	15%			
4540	For chain operated type, add				↓	15%				
4850	Globe, class 150, rising stem, threaded									
4920	1/4"	1 Plum	24	.333	Ea.	47.50	14.95		62.45	74.50
4940	3/8" size		24	.333		47.50	14.95		62.45	74.50
4950	1/2"		24	.333		47.50	14.95		62.45	74.50
4960	3/4"		20	.400		64.50	17.90		82.40	98
4970	1"		19	.421		101	18.85		119.85	140
4980	1-1/4"		15	.533		160	24		184	212
4990	1-1/2"		13	.615		194	27.50		221.50	256
5000	2"	↓	11	.727		291	32.50		323.50	370
5010	2-1/2"	Q-1	15	1.067		585	43		628	710
5020	3"	"	13	1.231	↓	835	49.50		884.50	990
5120	For 300 lb threaded, add					50%	15%			
5600	Relief, pressure & temperature, self-closing, ASME, threaded									
5640	3/4"	1 Plum	28	.286	Ea.	90	12.80		102.80	118
5650	1"		24	.333		131	14.95		145.95	167
5660	1-1/4"		20	.400		262	17.90		279.90	315
5670	1-1/2"		18	.444		500	19.90		519.90	580
5680	2"	↓	16	.500	↓	595	22.50		617.50	690
5950	Pressure, poppet type, threaded									
6000	1/2"	1 Plum	30	.267	Ea.	28	11.95		39.95	49
6040	3/4"	"	28	.286	"	30	12.80		42.80	52.50
6400	Pressure, water, ASME, threaded									
6440	3/4"	1 Plum	28	.286	Ea.	58	12.80		70.80	83.50
6450	1"		24	.333		124	14.95		138.95	159
6460	1-1/4"		20	.400		196	17.90		213.90	242
6470	1-1/2"		18	.444		232	19.90		251.90	286
6480	2"		16	.500		360	22.50		382.50	430
6490	2-1/2"	↓	15	.533	↓	1,725	24		1,749	1,925
6900	Reducing, water pressure									
6920	300 psi to 25-75 psi, threaded or sweat									
6940	1/2"	1 Plum	24	.333	Ea.	165	14.95		179.95	205
6950	3/4"		20	.400		165	17.90		182.90	209
6960	1"		19	.421		256	18.85		274.85	310
6970	1-1/4"		15	.533		460	24		484	540
6980	1-1/2"	↓	13	.615	↓	690	27.50		717.50	800
8350	Tempering, water, sweat connections									
8400	1/2"	1 Plum	24	.333	Ea.	57	14.95		71.95	85.50
8440	3/4"	"	20	.400	"	70	17.90		87.90	104
8650	Threaded connections									

22 05 Common Work Results for Plumbing

22 05 23 – General-Duty Valves for Plumbing Piping

22 05 23.20 Valves, Bronze		Crew	Daily Output	Labor-Hours	Unit	Material	2007 Bare Costs Labor	Equipment	Total	Total Incl O&P
8700	1/2"	1 Plum	24	.333	Ea.	70	14.95		84.95	99.50
8740	3/4"		20	.400		279	17.90		296.90	330
8750	1"		19	.421		310	18.85		328.85	370
8760	1-1/4"		15	.533		490	24		514	570
8770	1-1/2"		13	.615		535	27.50		562.50	625
8780	2"		11	.727		800	32.50		832.50	930
22 05 23.60 Valves, Plastic										
0010	**VALVES, PLASTIC**									
1100	Angle, PVC, threaded									
1110	1/4"	1 Plum	26	.308	Ea.	57	13.80		70.80	83
1120	1/2"		26	.308		57	13.80		70.80	83
1130	3/4"		25	.320		67.50	14.35		81.85	95.50
1140	1"		23	.348		80.50	15.60		96.10	112
1150	Ball, PVC, socket or threaded, single union									
1230	1/2"	1 Plum	26	.308	Ea.	20.50	13.80		34.30	43
1240	3/4"		25	.320		23.50	14.35		37.85	47.50
1250	1"		23	.348		29.50	15.60		45.10	56
1260	1-1/4"		21	.381		39.50	17.05		56.55	69
1270	1-1/2"		20	.400		47	17.90		64.90	79
1280	2"		17	.471		68.50	21		89.50	107
1360	For PVC, flanged, add					100%	15%			
1650	CPVC, socket or threaded, single union									
1700	1/2"	1 Plum	26	.308	Ea.	37	13.80		50.80	61
1720	3/4"		25	.320		46.50	14.35		60.85	72.50
1730	1"		23	.348		55.50	15.60		71.10	84.50
1750	1-1/4"		21	.381		93	17.05		110.05	128
1760	1-1/2"		20	.400		93	17.90		110.90	129
1840	For CPVC, flanged, add					65%	15%			
1880	For true union, socket or threaded, add					50%	5%			
2050	Polypropylene, threaded									
2100	1/4"	1 Plum	26	.308	Ea.	36	13.80		49.80	60
2120	3/8"		26	.308		36	13.80		49.80	60
2130	1/2"		26	.308		36	13.80		49.80	60
2140	3/4"		25	.320		45.50	14.35		59.85	71.50
2150	1"		23	.348		53.50	15.60		69.10	82.50
2160	1-1/4"		21	.381		77.50	17.05		94.55	111
2170	1-1/2"		20	.400		89.50	17.90		107.40	126
2180	2"		17	.471		122	21		143	167
4850	Foot valve, PVC, socket or threaded									
4900	1/2"	1 Plum	34	.235	Ea.	53	10.55		63.55	74
4930	3/4"		32	.250		60	11.20		71.20	83
4940	1"		28	.286		77.50	12.80		90.30	105
4950	1-1/4"		27	.296		150	13.25		163.25	185
4960	1-1/2"		26	.308		150	13.80		163.80	186
6350	Y sediment strainer, PVC, socket or threaded									
6400	1/2"	1 Plum	26	.308	Ea.	36	13.80		49.80	60
6440	3/4"		24	.333		39	14.95		53.95	65.50
6450	1"		23	.348		47	15.60		62.60	75.50
6460	1-1/4"		21	.381		77.50	17.05		94.55	111
6470	1-1/2"		20	.400		77.50	17.90		95.40	113

22 05 Common Work Results for Plumbing

22 05 76 – Facility Drainage Piping Cleanouts

22 05 76.10 Cleanouts

		Crew	Daily Output	Labor-Hours	Unit	Material	2007 Bare Costs Labor	2007 Bare Costs Equipment	Total	Total Incl O&P
0010	**CLEANOUTS**									
0060	Floor type									
0080	Round or square, scoriated nickel bronze top									
0100	2" pipe size	1 Plum	10	.800	Ea.	112	36		148	177
0120	3" pipe size		8	1		168	45		213	253
0140	4" pipe size	↓	6	1.333	↓	168	59.50		227.50	275
0980	Round top, recessed for terrazzo									
1000	2" pipe size	1 Plum	9	.889	Ea.	112	40		152	183
1080	3" pipe size		6	1.333		168	59.50		227.50	275
1100	4" pipe size		4	2		168	89.50		257.50	320
1120	5" pipe size	Q-1	6	2.667	↓	214	108		322	395

22 05 76.20 Cleanout Tees

		Crew	Daily Output	Labor-Hours	Unit	Material	2007 Bare Costs Labor	2007 Bare Costs Equipment	Total	Total Incl O&P
0010	**CLEANOUT TEES**									
0100	Cast iron, B&S, with countersunk plug									
0200	2" pipe size	1 Plum	4	2	Ea.	153	89.50		242.50	305
0220	3" pipe size		3.60	2.222		167	99.50		266.50	335
0240	4" pipe size	↓	3.30	2.424		207	109		316	390
0280	6" pipe size	Q-1	5	3.200	↓	560	129		689	810
0500	For round smooth access cover, same price									
4000	Plastic, tees and adapters. Add plugs									
4010	ABS, DWV									
4020	Cleanout tee, 1-1/2" pipe size	1 Plum	15	.533	Ea.	7.75	24		31.75	44.50
4030	2" pipe size	Q-1	27	.593		9.10	24		33.10	46
4040	3" pipe size		21	.762		18.25	30.50		48.75	66
4050	4" pipe size	↓	16	1		32	40.50		72.50	95.50
4100	Cleanout plug, 1-1/2" pipe size	1 Plum	32	.250		1.42	11.20		12.62	18.40
4110	2" pipe size	Q-1	56	.286		1.86	11.50		13.36	19.40
4120	3" pipe size		36	.444		3.01	17.90		20.91	30.50
4130	4" pipe size	↓	30	.533		5.85	21.50		27.35	39
4180	Cleanout adapter fitting, 1-1/2" pipe size	1 Plum	32	.250		2.24	11.20		13.44	19.30
4190	2" pipe size	Q-1	56	.286		3.14	11.50		14.64	21
4200	3" pipe size		36	.444		8.85	17.90		26.75	36.50
4210	4" pipe size	↓	30	.533	↓	14.20	21.50		35.70	48
5000	PVC, DWV									
5010	Cleanout tee, 1-1/2" pipe size	1 Plum	15	.533	Ea.	8.10	24		32.10	45
5020	2" pipe size	Q-1	27	.593		9.60	24		33.60	46.50
5030	3" pipe size		21	.762		21.50	30.50		52	69.50
5040	4" pipe size	↓	16	1		28.50	40.50		69	91.50
5090	Cleanout plug, 1-1/2" pipe size	1 Plum	32	.250		1.53	11.20		12.73	18.55
5100	2" pipe size	Q-1	56	.286		1.58	11.50		13.08	19.10
5110	3" pipe size		36	.444		3.03	17.90		20.93	30.50
5120	4" pipe size		30	.533		4.48	21.50		25.98	37.50
5130	6" pipe size	↓	24	.667		14.90	27		41.90	57
5170	Cleanout adapter fitting, 1-1/2" pipe size	1 Plum	32	.250		1.86	11.20		13.06	18.90
5180	2" pipe size	Q-1	56	.286		2.63	11.50		14.13	20
5190	3" pipe size		36	.444		7.40	17.90		25.30	35
5200	4" pipe size		30	.533		12.20	21.50		33.70	46
5210	6" pipe size	↓	24	.667	↓	32.50	27		59.50	76.50

22 07 Plumbing Insulation

22 07 19 – Plumbing Piping Insulation

22 07 19.10 Piping Insulation	Crew	Daily Output	Labor-Hours	Unit	Material	2007 Bare Costs Labor	Equipment	Total	Total Incl O&P
0010 **PIPING INSULATION**									
0100 Rule of thumb, as a percentage of total mechanical costs				Job				10%	
0110 Insulation req'd is based on the surface size/area to be covered									
4000 Pipe covering (price copper tube one size less than IPS)									
6600 Fiberglass, with all service jacket									
6840 1" wall, 1/2" iron pipe size	Q-14	240	.067	L.F.	.89	2.49		3.38	4.92
6870 1" iron pipe size		220	.073		1.04	2.72		3.76	5.45
6900 2" iron pipe size		200	.080		1.31	2.99		4.30	6.15
6920 3" iron pipe size		180	.089		1.59	3.32		4.91	7
6940 4" iron pipe size		150	.107		2.11	3.98		6.09	8.60
7320 2" wall, 1/2" iron pipe size		220	.073		2.60	2.72		5.32	7.15
7440 6" iron pipe size		100	.160		5.20	6		11.20	15.20
7460 8" iron pipe size		80	.200		6.35	7.45		13.80	18.75
7480 10" iron pipe size		70	.229		7.60	8.55		16.15	22
7490 12" iron pipe size	↓	65	.246	↓	8.50	9.20		17.70	24
7800 For fiberglass with standard canvas jacket, deduct					5%				
7802 For fittings, add 3 L.F. for each fitting									
7804 plus 4 L.F. for each flange of the fitting									
7808 Finishes, for .010" aluminum jacket, add				S.F.		20%			
7810 Finishes, .010" aluminum jacket									
7811 Finishes, for .010" aluminum jacket, add	Q-14	200	.080	S.F.	.59	2.99		3.58	5.35
7812 For .016" aluminum jacket, add	↓	200	.080	↓	.86	2.99		3.85	5.65
7813 For .010" stainless steel, add		160	.100		2.59	3.74		6.33	8.75
7814 For single layer of felt, add					10%	10%			
7879 Rubber tubing, flexible closed cell foam									
7880 3/8" wall, 1/4" iron pipe size	1 Asbe	120	.067	L.F.	.37	2.77		3.14	4.78
7910 1/2" iron pipe size		115	.070		.47	2.89		3.36	5.10
7920 3/4" iron pipe size		115	.070		.57	2.89		3.46	5.20
7930 1" iron pipe size		110	.073		.66	3.02		3.68	5.50
7950 1-1/2" iron pipe size		110	.073		.88	3.02		3.90	5.75
8100 1/2" wall, 1/4" iron pipe size		90	.089		.36	3.69		4.05	6.25
8130 1/2" iron pipe size		89	.090		.44	3.73		4.17	6.40
8140 3/4" iron pipe size		89	.090		.49	3.73		4.22	6.45
8150 1" iron pipe size		88	.091		.54	3.77		4.31	6.55
8170 1-1/2" iron pipe size		87	.092		.76	3.82		4.58	6.90
8180 2" iron pipe size		86	.093		.97	3.86		4.83	7.15
8200 3" iron pipe size		85	.094		1.37	3.91		5.28	7.65
8300 3/4" wall, 1/4" iron pipe size		90	.089		.56	3.69		4.25	6.45
8330 1/2" iron pipe size		89	.090		.73	3.73		4.46	6.70
8340 3/4" iron pipe size		89	.090		.89	3.73		4.62	6.90
8350 1" iron pipe size		88	.091		1.01	3.77		4.78	7.05
8370 1-1/2" iron pipe size		87	.092		1.54	3.82		5.36	7.75
8380 2" iron pipe size		86	.093		1.81	3.86		5.67	8.10
8400 3" iron pipe size		85	.094		2.76	3.91		6.67	9.20
8444 1" wall, 1/2" iron pipe size		86	.093		1.41	3.86		5.27	7.65
8445 3/4" iron pipe size		84	.095		1.71	3.95		5.66	8.15
8446 1" iron pipe size		84	.095		1.99	3.95		5.94	8.45
8447 1-1/4" iron pipe size		82	.098		2.25	4.05		6.30	8.90
8448 1-1/2" iron pipe size		82	.098		2.61	4.05		6.66	9.25
8449 2" iron pipe size		80	.100		3.49	4.15		7.64	10.40
8450 2-1/2" iron pipe size	↓	80	.100	↓	4.55	4.15		8.70	11.55
8456 Rubber insulation tape, 1/8" x 2" x 30'				Ea.	11.40			11.40	12.50

22 11 Facility Water Distribution

22 11 13 – Facility Water Distribution Piping

22 11 13.14 Pipe, Brass

		Crew	Daily Output	Labor-Hours	Unit	Material	2007 Bare Costs Labor	2007 Bare Costs Equipment	Total	Total Incl O&P
0010	**PIPE, BRASS**, Plain end									
0900	Field threaded, coupling & clevis hanger 10' O.C.									
0920	Regular weight									
1120	1/2" diameter	1 Plum	48	.167	L.F.	4.38	7.45		11.83	16.05
1140	3/4" diameter		46	.174		6	7.80		13.80	18.30
1160	1" diameter		43	.186		8.80	8.35		17.15	22.50
1180	1-1/4" diameter	Q-1	72	.222		13.50	8.95		22.45	28.50
1200	1-1/2" diameter		65	.246		16.10	9.95		26.05	32.50
1220	2" diameter		53	.302		22.50	12.20		34.70	43

22 11 13.23 Pipe, Copper

		Crew	Daily Output	Labor-Hours	Unit	Material	Labor	Equipment	Total	Total Incl O&P
0010	**PIPE, COPPER**, Solder joints									
1000	Type K tubing, couplings & clevis hangers 10' O.C.									
1100	1/4" diameter	1 Plum	84	.095	L.F.	2.71	4.27		6.98	9.40
1200	1" diameter		66	.121		9.80	5.45		15.25	18.90
1260	2" diameter		40	.200		25	8.95		33.95	41
2000	Type L tubing, couplings & hangers 10' O.C.									
2100	1/4" diameter	1 Plum	88	.091	L.F.	1.93	4.07		6	8.30
2120	3/8" diameter		84	.095		2.81	4.27		7.08	9.50
2140	1/2" diameter CN		81	.099		3.41	4.42		7.83	10.40
2160	5/8" diameter		79	.101		4.89	4.54		9.43	12.25
2180	3/4" diameter		76	.105		5.20	4.72		9.92	12.80
2200	1" diameter		68	.118		7.40	5.25		12.65	16.10
2220	1-1/4" diameter		58	.138		10.45	6.20		16.65	21
2240	1-1/2" diameter		52	.154		13.40	6.90		20.30	25
2260	2" diameter		42	.190		21	8.55		29.55	36
2280	2-1/2" diameter	Q-1	62	.258		32.50	10.40		42.90	51
2300	3" diameter		56	.286		43.50	11.50		55	65.50
2320	3-1/2" diameter		43	.372		58.50	15		73.50	87
2340	4" diameter		39	.410		74	16.55		90.55	106
2360	5" diameter		34	.471		184	19		203	231
2380	6" diameter	Q-2	40	.600		225	25		250	285
2400	8" diameter	"	36	.667		405	28		433	485
2410	For other than full hard temper, add					21%				
2590	For silver solder, add						15%			
4000	Type DWV tubing, couplings & hangers 10' O.C.									
4100	1-1/4" diameter	1 Plum	60	.133	L.F.	9.20	5.95		15.15	19.10
4120	1-1/2" diameter		54	.148		11.55	6.65		18.20	22.50
4140	2" diameter		44	.182		15.45	8.15		23.60	29.50
4160	3" diameter	Q-1	58	.276		27	11.15		38.15	46.50
4180	4" diameter		40	.400		47	16.15		63.15	76
4200	5" diameter		36	.444		135	17.90		152.90	176
4220	6" diameter	Q-2	42	.571		191	24		215	246

22 11 13.44 Pipe, Steel

		Crew	Daily Output	Labor-Hours	Unit	Material	Labor	Equipment	Total	Total Incl O&P
0010	**PIPE, STEEL**									
0012	The steel pipe in this section does not include fittings such as ells, tees									
0014	For fittings either add a % (usually 25 to 35%) or see									
0015	the Mechanical or Plumbing Cost Data									
0020	All pipe sizes are to Spec. A-53 unless noted otherwise R221113-50									
0050	Schedule 40, threaded, with couplings, and clevis type									
0060	hangers sized for covering, 10' O.C.									
0540	Black, 1/4" diameter	1 Plum	66	.121	L.F.	1.95	5.45		7.40	10.30
0550	3/8" diameter		65	.123		1.77	5.50		7.27	10.25

22 11 Facility Water Distribution

22 11 13 – Facility Water Distribution Piping

22 11 13.44 Pipe, Steel

		Crew	Daily Output	Labor-Hours	Unit	Material	2007 Bare Costs Labor	Equipment	Total	Total Incl O&P
0560	1/2" diameter	1 Plum	63	.127	L.F.	1.78	5.70		7.48	10.50
0570	3/4" diameter		61	.131		2.10	5.90		8	11.15
0580	1" diameter		53	.151		3.08	6.75		9.83	13.55
0590	1-1/4" diameter	Q-1	89	.180		4.06	7.25		11.31	15.35
0600	1-1/2" diameter		80	.200		4.78	8.05		12.83	17.40
0610	2" diameter	CN	64	.250		6.40	10.10		16.50	22
0620	2-1/2" diameter		50	.320		10	12.90		22.90	30.50
0630	3" diameter		43	.372		13	15		28	37
0640	3-1/2" diameter		40	.400		17.65	16.15		33.80	44
0650	4" diameter		36	.444		19	17.90		36.90	48
0660	5" diameter		26	.615		28	25		53	68
0670	6" diameter	Q-2	31	.774		36	32.50		68.50	88
1290	Galvanized, 1/4" diameter	1 Plum	66	.121		2.12	5.45		7.57	10.50
1300	3/8" diameter		65	.123		2.34	5.50		7.84	10.90
1310	1/2" diameter		63	.127		2.44	5.70		8.14	11.25
1320	3/4" diameter		61	.131		2.94	5.90		8.84	12.10
1330	1" diameter		53	.151		4.02	6.75		10.77	14.55
1340	1-1/4" diameter	Q-1	89	.180		5.40	7.25		12.65	16.80
1350	1-1/2" diameter		80	.200		6.35	8.05		14.40	19.15
1360	2" diameter		64	.250		8.50	10.10		18.60	24.50
1370	2-1/2" diameter		50	.320		14.15	12.90		27.05	35
1380	3" diameter		43	.372		18.10	15		33.10	42.50
1390	3-1/2" diameter		40	.400		23	16.15		39.15	50
1400	4" diameter		36	.444		26	17.90		43.90	56
1410	5" diameter		26	.615		37.50	25		62.50	78.50
1420	6" diameter	Q-2	31	.774		48.50	32.50		81	102
1430	8" diameter		27	.889		79.50	37		116.50	143
1440	10" diameter		23	1.043		103	43.50		146.50	180
1450	12" diameter		18	1.333		129	56		185	226
2000	Welded, sch. 40, on yoke & roll hangers, sized for covering, 10' O.C.									
2040	Black, 1" diameter	Q-15	93	.172	L.F.	3.42	6.95	.66	11.03	14.95
2070	2" diameter		61	.262		6.55	10.60	1.01	18.16	24
2090	3" diameter		43	.372		11.15	15	1.43	27.58	36.50
2110	4" diameter		37	.432		15.25	17.45	1.66	34.36	44.50
2120	5" diameter		32	.500		21.50	20	1.92	43.42	56
2130	6" diameter	Q-16	36	.667		27.50	28	1.71	57.21	74.50
2140	8" diameter		29	.828		44	34.50	2.12	80.62	103
2150	10" diameter		24	1		63	42	2.56	107.56	135
2160	12" diameter		19	1.263		76.50	53	3.23	132.73	168

22 11 13.48 Pipe, Grooved-Joint, Steel Fittings and Valves

		Crew	Daily Output	Labor-Hours	Unit	Material	2007 Bare Costs Labor	Equipment	Total	Total Incl O&P
0010	**PIPE, GROOVED-JOINT, STEEL FITTINGS AND VALVES**									
0012	Fittings are ductile iron. Steel fittings noted.									
0020	Pipe includes coupling & clevis type hanger 10' O.C.									
1000	Schedule 40, black									
1040	3/4" diameter	1 Plum	71	.113	L.F.	2.71	5.05		7.76	10.60
1050	1" diameter		63	.127		3.28	5.70		8.98	12.15
1060	1-1/4" diameter		58	.138		4.38	6.20		10.58	14.10
1070	1-1/2" diameter		51	.157		5.05	7.05		12.10	16.10
1080	2" diameter		40	.200		6.55	8.95		15.50	20.50
1090	2-1/2" diameter	Q-1	57	.281		9.30	11.30		20.60	27
1100	3" diameter		50	.320		11.60	12.90		24.50	32
1110	4" diameter		45	.356		15.80	14.35		30.15	39

22 11 Facility Water Distribution

22 11 13 – Facility Water Distribution Piping

22 11 13.48 Pipe, Grooved-Joint, Steel Fittings and Valves		Crew	Daily Output	Labor-Hours	Unit	Material	2007 Bare Costs Labor	Equipment	Total	Total Incl O&P
1120	5" diameter	Q-1	37	.432	L.F.	22.50	17.45		39.95	50.50
1130	6" diameter	Q-2	42	.571	↓	28.50	24		52.50	67.50
1800	Galvanized									
1840	3/4" diameter	1 Plum	71	.113	L.F.	3.53	5.05		8.58	11.50
1850	1" diameter		63	.127		4.08	5.70		9.78	13.05
1860	1-1/4" diameter		58	.138		5.45	6.20		11.65	15.30
1870	1-1/2" diameter		51	.157		6.40	7.05		13.45	17.55
1880	2" diameter	↓	40	.200		8.20	8.95		17.15	22.50
1890	2-1/2" diameter	Q-1	57	.281		12.40	11.30		23.70	30.50
1900	3" diameter		50	.320		15.90	12.90		28.80	37
1910	4" diameter		45	.356		22	14.35		36.35	46
1920	5" diameter	↓	37	.432		29	17.45		46.45	58
1930	6" diameter	Q-2	42	.571	↓	37	24		61	77
3990	Fittings: cplg. & labor required at joints not incl. in fitting									
3994	price. Add 1 per joint for installed price.									
4000	Elbow, 90° or 45°, painted									
4030	3/4" diameter	1 Plum	50	.160	Ea.	26	7.15		33.15	39.50
4040	1" diameter		50	.160		13.90	7.15		21.05	26
4050	1-1/4" diameter		40	.200		13.90	8.95		22.85	29
4060	1-1/2" diameter		33	.242		13.90	10.85		24.75	31.50
4070	2" diameter	↓	25	.320		13.90	14.35		28.25	37
4080	2-1/2" diameter	Q-1	40	.400		13.90	16.15		30.05	40
4090	3" diameter		33	.485		24.50	19.55		44.05	56.50
4100	4" diameter		25	.640		27	26		53	68.50
4110	5" diameter	↓	20	.800		64.50	32.50		97	120
4120	6" diameter	Q-2	25	.960		76	40		116	144
4250	For galvanized elbows, add				↓	26%				
4690	Tee, painted									
4700	3/4" diameter	1 Plum	38	.211	Ea.	28	9.45		37.45	45
4740	1" diameter		33	.242		21.50	10.85		32.35	40
4750	1-1/4" diameter		27	.296		21.50	13.25		34.75	43.50
4760	1-1/2" diameter		22	.364		21.50	16.30		37.80	48
4770	2" diameter	↓	17	.471		21.50	21		42.50	55
4780	2-1/2" diameter	Q-1	27	.593		21.50	24		45.50	59.50
4790	3" diameter		22	.727		30	29.50		59.50	77
4800	4" diameter		17	.941		45.50	38		83.50	107
4810	5" diameter	↓	13	1.231		107	49.50		156.50	192
4820	6" diameter	Q-2	17	1.412		123	59		182	225
4900	For galvanized tees, add				↓	24%				
4906	Couplings, rigid style, painted									
4908	1" diameter	1 Plum	100	.080	Ea.	10.30	3.58		13.88	16.75
4909	1-1/4" diameter		100	.080		10.30	3.58		13.88	16.75
4910	1-1/2" diameter		67	.119		10.30	5.35		15.65	19.40
4912	2" diameter	↓	50	.160		10.50	7.15		17.65	22.50
4914	2-1/2" diameter	Q-1	80	.200		12.10	8.05		20.15	25.50
4916	3" diameter		67	.239		14.10	9.65		23.75	30
4918	4" diameter		50	.320		16	12.90		28.90	37
4920	5" diameter	↓	40	.400		19.60	16.15		35.75	46
4922	6" diameter	Q-2	50	.480	↓	34.50	20		54.50	68
4940	Flexible, standard, painted									
4950	3/4" diameter	1 Plum	100	.080	Ea.	7.40	3.58		10.98	13.55
4960	1" diameter		100	.080		7.40	3.58		10.98	13.55
4970	1-1/4" diameter	↓	80	.100	↓	9.80	4.48		14.28	17.55

22 11 Facility Water Distribution

22 11 13 – Facility Water Distribution Piping

22 11 13.48 Pipe, Grooved-Joint, Steel Fittings and Valves		Crew	Daily Output	Labor-Hours	Unit	Material	2007 Bare Costs Labor	Equipment	Total	Total Incl O&P
4980	1-1/2" diameter	1 Plum	67	.119	Ea.	10.70	5.35		16.05	19.80
4990	2" diameter	↓	50	.160		11.40	7.15		18.55	23.50
5000	2-1/2" diameter	Q-1	80	.200		13.50	8.05		21.55	27
5010	3" diameter		67	.239		14.90	9.65		24.55	31
5020	3-1/2" diameter		57	.281		21.50	11.30		32.80	41
5030	4" diameter		50	.320		21.50	12.90		34.40	43.50
5040	5" diameter	↓	40	.400		33	16.15		49.15	61
5050	6" diameter	Q-2	50	.480	↓	39.50	20		59.50	73

22 11 13.64 Pipe, Stainless Steel

		Crew	Daily Output	Labor-Hours	Unit	Material	Labor	Equipment	Total	Total Incl O&P
0010	**PIPE, STAINLESS STEEL**									
3500	Threaded, couplings and hangers 10' O.C.									
3520	Schedule 40, type 304									
3540	1/4" diameter	1 Plum	54	.148	L.F.	6.30	6.65		12.95	16.90
3550	3/8" diameter		53	.151		7.55	6.75		14.30	18.45
3560	1/2" diameter		52	.154		9.50	6.90		16.40	21
3580	1" diameter	↓	45	.178		15.30	7.95		23.25	29
3610	2" diameter	Q-1	57	.281		27	11.30		38.30	46.50
3640	4" diameter	Q-2	51	.471		85.50	19.70		105.20	124
3740	For small quantities, add				↓	10%				
4250	Schedule 40, type 316									
4290	1/4" diameter	1 Plum	54	.148	L.F.	5.75	6.65		12.40	16.35
4300	3/8" diameter		53	.151		7	6.75		13.75	17.85
4310	1/2" diameter		52	.154		8.30	6.90		15.20	19.50
4320	3/4" diameter		51	.157		10.25	7.05		17.30	22
4330	1" diameter	↓	45	.178		13.50	7.95		21.45	27
4360	2" diameter	Q-1	57	.281		28	11.30		39.30	47.50
4390	4" diameter	Q-2	51	.471		91.50	19.70		111.20	130
4490	For small quantities, add				↓	10%				

22 11 13.74 Pipe, Plastic

		Crew	Daily Output	Labor-Hours	Unit	Material	Labor	Equipment	Total	Total Incl O&P
0010	**PIPE, PLASTIC**									
0020	Fiberglass reinforced, couplings 10' O.C., hangers 3 per 10'									
1800	PVC, couplings 10' O.C., hangers 3 per 10'									
1820	Schedule 40									
1860	1/2" diameter	1 Plum	54	.148	L.F.	1.04	6.65		7.69	11.15
1870	3/4" diameter		51	.157		1.21	7.05		8.26	11.90
1880	1" diameter		46	.174		1.47	7.80		9.27	13.30
1890	1-1/4" diameter		42	.190		1.92	8.55		10.47	14.95
1900	1-1/2" diameter	↓	36	.222		2.07	9.95		12.02	17.30
1910	2" diameter CN	Q-1	59	.271		2.75	10.95		13.70	19.50
1920	2-1/2" diameter		56	.286		3.88	11.50		15.38	21.50
1930	3" diameter		53	.302		5.15	12.20		17.35	24
1940	4" diameter		48	.333		6.85	13.45		20.30	27.50
1950	5" diameter	↓	43	.372		9.40	15		24.40	33
1960	6" diameter	↓	39	.410	↓	12.15	16.55		28.70	38.50
4100	DWV type, schedule 40, couplings 10' O.C., hangers 3 per 10'									
4120	ABS									
4140	1-1/4" diameter	1 Plum	42	.190	L.F.	1.52	8.55		10.07	14.50
4150	1-1/2" diameter	"	36	.222		1.53	9.95		11.48	16.70
4160	2" diameter	Q-1	59	.271	↓	1.91	10.95		12.86	18.55
4400	PVC									
4410	1-1/4" diameter	1 Plum	42	.190	L.F.	1.64	8.55		10.19	14.65
4420	1-1/2" diameter	"	36	.222	↓	1.43	9.95		11.38	16.60

22 11 Facility Water Distribution

22 11 13 – Facility Water Distribution Piping

22 11 13.74 Pipe, Plastic

		Crew	Daily Output	Labor-Hours	Unit	Material	2007 Bare Costs Labor	2007 Bare Costs Equipment	Total	Total Incl O&P
4460	2" diameter	Q-1	59	.271	L.F.	1.75	10.95		12.70	18.35
4470	3" diameter		53	.302		3.42	12.20		15.62	22
4480	4" diameter		48	.333		4.58	13.45		18.03	25
4490	6" diameter		39	.410		8.60	16.55		25.15	34.50
5360	CPVC, couplings 10' O.C., hangers 3 per 10'									
5380	Schedule 40									
5460	1/2" diameter	1 Plum	54	.148	L.F.	2.34	6.65		8.99	12.55
5470	3/4" diameter		51	.157		3.21	7.05		10.26	14.10
5480	1" diameter		46	.174		4	7.80		11.80	16.10
5490	1-1/4" diameter		42	.190		4.73	8.55		13.28	18.05
5500	1-1/2" diameter		36	.222		5.30	9.95		15.25	21
5510	2" diameter	Q-1	59	.271		6.60	10.95		17.55	24
5520	2-1/2" diameter		56	.286		10.85	11.50		22.35	29.50
5530	3" diameter		53	.302		13.35	12.20		25.55	33

22 11 19 – Domestic Water Piping Specialties

22 11 19.10 Flexible Connectors

		Crew	Daily Output	Labor-Hours	Unit	Material	Labor	Equipment	Total	Total Incl O&P
0010	**FLEXIBLE CONNECTORS**, Corrugated, 7/8" O.D., 1/2" I.D.									
0050	Gas, seamless brass, steel fittings									
0200	12" long	1 Plum	36	.222	Ea.	13.70	9.95		23.65	30
0220	18" long		36	.222		17	9.95		26.95	33.50
0240	24" long		34	.235		20	10.55		30.55	38
0280	36" long		32	.250		24	11.20		35.20	43.50
0340	60" long		30	.267		36	11.95		47.95	58
2000	Water, copper tubing, dielectric separators									
2100	12" long	1 Plum	36	.222	Ea.	10.30	9.95		20.25	26.50
2260	24" long	"	34	.235	"	15.50	10.55		26.05	33

22 11 19.14 Flexible Metal Hose

		Crew	Daily Output	Labor-Hours	Unit	Material	Labor	Equipment	Total	Total Incl O&P
0010	**FLEXIBLE METAL HOSE**, Connectors, standard lengths									
0100	Bronze braided, bronze ends									
0120	3/8" diameter x 12"	1 Stpi	26	.308	Ea.	16.95	13.90		30.85	39.50
0160	3/4" diameter x 12"		20	.400		23.50	18.10		41.60	53
0180	1" diameter x 18"		19	.421		30.50	19.05		49.55	62
0200	1-1/2" diameter x 18"		13	.615		48.50	28		76.50	95
0220	2" diameter x 18"		11	.727		58	33		91	114

22 11 19.26 Pressure Regulators

		Crew	Daily Output	Labor-Hours	Unit	Material	Labor	Equipment	Total	Total Incl O&P
0010	**PRESSURE REGULATORS**									
3000	Steam, high capacity, bronze body, stainless steel trim									
3020	Threaded, 1/2" diameter	1 Stpi	24	.333	Ea.	1,100	15.05		1,115.05	1,225
3030	3/4" diameter		24	.333		1,100	15.05		1,115.05	1,225
3040	1" diameter		19	.421		1,225	19.05		1,244.05	1,375
3060	1-1/4" diameter		15	.533		1,350	24		1,374	1,525
3080	1-1/2" diameter		13	.615		1,550	28		1,578	1,750
3100	2" diameter		11	.727		1,900	33		1,933	2,150
3120	2-1/2" diameter	Q-5	12	1.333		2,375	54		2,429	2,700
3140	3" diameter	"	11	1.455		2,725	59		2,784	3,075
3500	Flanged connection, iron body, 125 lb. W.S.P.									
3520	3" diameter	Q-5	11	1.455	Ea.	2,975	59		3,034	3,375
3540	4" diameter	"	5	3.200	"	3,750	130		3,880	4,325

22 11 19.38 Water Supply Meters

0010	**WATER SUPPLY METERS**									
2000	Domestic/commercial, bronze									

22 11 Facility Water Distribution

22 11 19 – Domestic Water Piping Specialties

22 11 19.38 Water Supply Meters

		Crew	Daily Output	Labor-Hours	Unit	Material	2007 Bare Costs Labor	2007 Bare Costs Equipment	Total	Total Incl O&P
2020	Threaded									
2060	5/8" diameter, to 20 GPM	1 Plum	16	.500	Ea.	40	22.50		62.50	77.50
2080	3/4" diameter, to 30 GPM		14	.571		67.50	25.50		93	113
2100	1" diameter, to 50 GPM		12	.667		94	30		124	148
2300	Threaded/flanged									
2340	1-1/2" diameter, to 100 GPM	1 Plum	8	1	Ea.	330	45		375	435
2360	2" diameter, to 160 GPM	"	6	1.333	"	415	59.50		474.50	545
2600	Flanged, compound									
2640	3" diameter, 320 GPM	Q-1	3	5.333	Ea.	1,950	215		2,165	2,450
2660	4" diameter, to 500 GPM		1.50	10.667		3,025	430		3,455	3,975
2680	6" diameter, to 1,000 GPM		1	16		4,350	645		4,995	5,750
2700	8" diameter, to 1,800 GPM		.80	20		8,600	805		9,405	10,700

22 11 19.42 Backflow Preventers

		Crew	Daily Output	Labor-Hours	Unit	Material	2007 Bare Costs Labor	2007 Bare Costs Equipment	Total	Total Incl O&P
0010	**BACKFLOW PREVENTERS**, Includes valves									
0020	and four test cocks, corrosion resistant, automatic operation									
4000	Reduced pressure principle									
4100	Threaded, bronze, valves are ball									
4120	3/4" pipe size	1 Plum	16	.500	Ea.	230	22.50		252.50	287
4140	1" pipe size		14	.571		248	25.50		273.50	310
4150	1-1/4" pipe size		12	.667		425	30		455	515
4160	1-1/2" pipe size		10	.800		465	36		501	570
4180	2" pipe size		7	1.143		525	51		576	650
5000	Flanged, valves are OS&Y									
5060	2-1/2" pipe size	Q-1	5	3.200	Ea.	2,825	129		2,954	3,300
5080	3" pipe size		4.50	3.556		2,950	143		3,093	3,475
5100	4" pipe size		3	5.333		3,725	215		3,940	4,425
5120	6" pipe size	Q-2	3	8		5,375	335		5,710	6,425
5600	Flanged, iron, valves are OS&Y									
5660	2-1/2" pipe size	Q-1	5	3.200	Ea.	2,250	129		2,379	2,675
5680	3" pipe size		4.50	3.556		2,375	143		2,518	2,825
5700	4" pipe size		3	5.333		2,975	215		3,190	3,600
5720	6" pipe size	Q-2	3	8		4,300	335		4,635	5,225
5740	8" pipe size		2	12		7,575	500		8,075	9,075
5760	10" pipe size		1	24		10,100	1,000		11,100	12,600

22 11 19.46 Vacuum Breakers

		Crew	Daily Output	Labor-Hours	Unit	Material	2007 Bare Costs Labor	2007 Bare Costs Equipment	Total	Total Incl O&P
0010	**VACUUM BREAKERS**, Hot or cold water									
1030	Anti-siphon, brass									
1040	1/4" size	1 Plum	24	.333	Ea.	27.50	14.95		42.45	52.50
1050	3/8" size		24	.333		27.50	14.95		42.45	52.50
1060	1/2" size		24	.333		31	14.95		45.95	56.50
1080	3/4" size		20	.400		36.50	17.90		54.40	67
1100	1" size		19	.421		57	18.85		75.85	91.50
1120	1-1/4" size		15	.533		100	24		124	146
1140	1-1/2" size		13	.615		118	27.50		145.50	171
1160	2" size		11	.727		183	32.50		215.50	251
1300	For polished chrome, (1/4" thru 1"), add					50%				
1900	Vacuum relief, water service, bronze									
2000	1/2" size	1 Plum	30	.267	Ea.	26	11.95		37.95	47

22 11 19.54 Water Hammer Arresters/Shock Absorbers

		Crew	Daily Output	Labor-Hours	Unit	Material	2007 Bare Costs Labor	2007 Bare Costs Equipment	Total	Total Incl O&P
0010	**WATER HAMMER ARRESTERS/SHOCK ABSORBERS**									
0490	Copper									
0500	3/4" male I.P.S. For 1 to 11 fixtures	1 Plum	12	.667	Ea.	15.55	30		45.55	62

22 11 Facility Water Distribution

22 11 19 – Domestic Water Piping Specialties

22 11 19.54 Water Hammer Arresters/Shock Absorbers

		Crew	Daily Output	Labor-Hours	Unit	Material	2007 Bare Costs Labor	Equipment	Total	Total Incl O&P
0600	1" male I.P.S., For 12 to 32 fixtures	1 Plum	8	1	Ea.	39.50	45		84.50	111
0700	1-1/4" male I.P.S. For 33 to 60 fixtures		8	1		46	45		91	118
0800	1-1/2" male I.P.S. For 61 to 113 fixtures		8	1		62.50	45		107.50	137
0900	2" male I.P.S. For 114 to 154 fixtures		8	1		100	45		145	178
1000	2-1/2" male I.P.S. For 155 to 330 fixtures		4	2		285	89.50		374.50	450

22 11 19.64 Hydrants

		Crew	Daily Output	Labor-Hours	Unit	Material	2007 Bare Costs Labor	Equipment	Total	Total Incl O&P
0010	**HYDRANTS**									
0050	Wall type, moderate climate, bronze, encased									
0200	3/4" IPS connection	1 Plum	16	.500	Ea.	435	22.50		457.50	515
0300	1" IPS connection	"	14	.571		435	25.50		460.50	520
0500	Anti-siphon type					410			410	450
1000	Non-freeze, bronze, exposed									
1100	3/4" IPS connection, 4" to 9" thick wall	1 Plum	14	.571	Ea.	295	25.50		320.50	365
1120	10" to 14" thick wall		12	.667		320	30		350	395
1140	15" to 19" thick wall		12	.667		355	30		385	435
1160	20" to 24" thick wall		10	.800		385	36		421	480
1200	For 1" IPS connection, add					15%	10%			
1240	For 3/4" adapter type vacuum breaker, add				Ea.	37			37	41
1280	For anti-siphon type, add				"	79			79	87
2000	Non-freeze bronze, encased, anti-siphon type									
2100	3/4" IPS connection, 5" to 9" thick wall	1 Plum	14	.571	Ea.	745	25.50		770.50	860
2140	15" to 19" thick wall	"	12	.667	"	805	30		835	930
3000	Ground box type, bronze frame, 3/4" IPS connection									
3080	Non-freeze, all bronze, polished face, set flush									
3100	2 feet depth of bury	1 Plum	8	1	Ea.	540	45		585	665
3180	6 feet depth of bury		7	1.143		710	51		761	855
3220	8 feet depth of bury		5	1.600		790	71.50		861.50	980
3400	For 1" IPS connection, add					15%	10%			
3550	For 2" connection, add					445%	24%			
3600	For tapped drain port in box, add					50.50			50.50	55.50
5000	Moderate climate, all bronze, polished face									
5020	and scoriated cover, set flush									
5100	3/4" IPS connection	1 Plum	16	.500	Ea.	385	22.50		407.50	460
5120	1" IPS connection	"	14	.571		385	25.50		410.50	465
5200	For tapped drain port in box, add					50.50			50.50	55.50

22 11 23 – Domestic Water Pumps

22 11 23.10 General Utility Pumps

		Crew	Daily Output	Labor-Hours	Unit	Material	2007 Bare Costs Labor	Equipment	Total	Total Incl O&P
0010	**GENERAL UTILITY PUMPS**									
2000	Single stage									
3000	Double suction,									
3190	75 HP, to 2500 GPM	Q-3	.28	114	Ea.	13,600	4,875		18,475	22,400
3220	100 HP, to 3000 GPM		.26	123		17,300	5,250		22,550	26,900
3240	150 HP, to 4000 GPM		.24	133		23,200	5,700		28,900	34,200

22 13 Facility Sanitary Sewerage

22 13 16 – Sanitary Waste and Vent Piping

22 13 16.20 Pipe, Cast Iron		Crew	Daily Output	Labor-Hours	Unit	Material	2007 Bare Costs Labor	Equipment	Total	Total Incl O&P
0010	**PIPE, CAST IRON**, Soil, on hangers 5' O.C.									
0020	Single hub, service wt., lead & oakum joints 10' O.C.									
2120	2" diameter	Q-1	63	.254	L.F.	5.20	10.25		15.45	21
2140	3" diameter		60	.267		7.25	10.75		18	24
2160	4" diameter	CN	55	.291		9.35	11.75		21.10	28
2180	5" diameter	Q-2	76	.316		13	13.20		26.20	34
2200	6" diameter	"	73	.329		15.90	13.75		29.65	38
2220	8" diameter	Q-3	59	.542		24.50	23		47.50	62
2240	10" diameter		54	.593		41	25.50		66.50	83
2260	12" diameter		48	.667		58	28.50		86.50	107
2320	For service weight, double hub, add					10%				
2340	For extra heavy, single hub, add					48%	4%			
2360	For extra heavy, double hub, add					71%	4%			
2400	Lead for caulking, (1#/dia in)	Q-1	160	.100	Lb.	1.04	4.03		5.07	7.20
2420	Oakum for caulking, (1/8#/dia in)	"	40	.400	"	3.58	16.15		19.73	28.50
4000	No hub, couplings 10' O.C.									
4100	1-1/2" diameter	Q-1	71	.225	L.F.	5.45	9.10		14.55	19.65
4120	2" diameter		67	.239		5.60	9.65		15.25	20.50
4160	4" diameter		58	.276		9.65	11.15		20.80	27.50

22 13 16.50 Shower Drains

		Crew	Daily Output	Labor-Hours	Unit	Material	Labor	Equipment	Total	Total Incl O&P
0010	**SHOWER DRAINS**									
2780	Shower, with strainer, uniform diam. trap, bronze top									
2800	2" and 3" pipe size	Q-1	8	2	Ea.	242	80.50		322.50	385
2820	4" pipe size	"	7	2.286		267	92		359	435
2840	For galvanized body, add					94.50			94.50	104

22 13 16.60 Traps

		Crew	Daily Output	Labor-Hours	Unit	Material	Labor	Equipment	Total	Total Incl O&P
0010	**TRAPS**									
0030	Cast iron, service weight									
0050	Running P trap, without vent									
1100	2"	Q-1	16	1	Ea.	31	40.50		71.50	94.50
1140	3"		14	1.143		52	46		98	127
1150	4"		13	1.231		90	49.50		139.50	174
1160	6"	Q-2	17	1.412		390	59		449	520
1180	Running trap, single hub, with vent									
2080	3" pipe size, 3" vent	Q-1	14	1.143	Ea.	71	46		117	148
2120	4" pipe size, 4" vent	"	13	1.231		97	49.50		146.50	182
2300	For double hub, vent, add					10%	20%			
3000	P trap, B&S, 2" pipe size	Q-1	16	1		21.50	40.50		62	84
3040	3" pipe size	"	14	1.143		32	46		78	105
3350	Deep seal trap, B&S									
3400	1-1/4" pipe size	Q-1	14	1.143	Ea.	38.50	46		84.50	112
3410	1-1/2" pipe size		14	1.143		38.50	46		84.50	112
3420	2" pipe size		14	1.143		25.50	46		71.50	97.50
3440	3" pipe size		12	1.333		39	54		93	124
4700	Copper, drainage, drum trap									
4800	3" x 5" solid, 1-1/2" pipe size	1 Plum	16	.500	Ea.	62	22.50		84.50	102
4840	3" x 6" swivel, 1-1/2" pipe size	"	16	.500	"	98	22.50		120.50	142
5100	P trap, standard pattern									
5200	1-1/4" pipe size	1 Plum	18	.444	Ea.	45.50	19.90		65.40	80
5240	1-1/2" pipe size		17	.471		44	21		65	80
5260	2" pipe size		15	.533		68	24		92	111
5280	3" pipe size		11	.727		163	32.50		195.50	229

22 13 Facility Sanitary Sewerage

22 13 16 – Sanitary Waste and Vent Piping

22 13 16.60 Traps		Crew	Daily Output	Labor-Hours	Unit	Material	2007 Bare Costs Labor	Equipment	Total	Total Incl O&P
5340	With cleanout and slip joint									
5360	1-1/4" pipe size	1 Plum	18	.444	Ea.	43.50	19.90		63.40	78
5400	1-1/2" pipe size	"	17	.471	"	72	21		93	111

22 13 16.80 Vent Flashing Caps

		Crew	Daily Output	Labor-Hours	Unit	Material	Labor	Equipment	Total	Total Incl O&P
0010	**VENT FLASHING CAPS**									
0120	Vent caps									
0140	Cast iron									
0180	2-1/2" - 3-5/8" pipe	1 Plum	21	.381	Ea.	36.50	17.05		53.55	65.50
0190	4" - 4-1/8" pipe	"	19	.421	"	44	18.85		62.85	76.50
0900	Vent flashing									
1000	Aluminum with lead ring									
1020	1-1/4" pipe	1 Plum	20	.400	Ea.	8.15	17.90		26.05	36
1030	1-1/2" pipe		20	.400		8.80	17.90		26.70	36.50
1040	2" pipe		18	.444		8.75	19.90		28.65	39.50
1050	3" pipe		17	.471		9.70	21		30.70	42
1060	4" pipe		16	.500		11.70	22.50		34.20	46.50
1350	Copper with neoprene ring									
1400	1-1/4" pipe	1 Plum	20	.400	Ea.	19.95	17.90		37.85	49
1430	1-1/2" pipe		20	.400		19.95	17.90		37.85	49
1440	2" pipe		18	.444		21	19.90		40.90	53
1450	3" pipe		17	.471		24.50	21		45.50	58.50
1460	4" pipe		16	.500		27	22.50		49.50	63.50

22 13 19 – Sanitary Waste Piping Specialties

22 13 19.13 Sanitary Drains

		Crew	Daily Output	Labor-Hours	Unit	Material	Labor	Equipment	Total	Total Incl O&P
0010	**SANITARY DRAINS**									
0400	Deck, auto park, C.I., 13" top									
0440	3", 4", 5", and 6" pipe size	Q-1	8	2	Ea.	820	80.50		900.50	1,025
0480	For galvanized body, add				"	385			385	425
2000	Floor, medium duty, C.I., deep flange, 7" dia top									
2040	2" and 3" pipe size	Q-1	12	1.333	Ea.	115	54		169	207
2080	For galvanized body, add					48.50			48.50	53
2120	For polished bronze top, add					63.50			63.50	70
2400	Heavy duty, with sediment bucket, C.I., 12" dia. loose grate									
2420	2", 3", 4", 5", and 6" pipe size	Q-1	9	1.778	Ea.	395	71.50		466.50	545
2460	For polished bronze top, add				"	162			162	179
2500	Heavy duty, cleanout & trap w/bucket, C.I., 15" top									
2540	2", 3", and 4" pipe size	Q-1	6	2.667	Ea.	3,650	108		3,758	4,175
2560	For galvanized body, add					845			845	930
2580	For polished bronze top, add					400			400	440

22 13 23 – Sanitary Waste Interceptors

22 13 23.10 Interceptors

		Crew	Daily Output	Labor-Hours	Unit	Material	Labor	Equipment	Total	Total Incl O&P
0010	**INTERCEPTORS**									
0150	Grease, cast iron, 4 GPM, 8 lb. fat capacity	1 Plum	4	2	Ea.	680	89.50		769.50	880
0200	7 GPM, 14 lb. fat capacity		4	2		940	89.50		1,029.50	1,150
1000	10 GPM, 20 lb. fat capacity		4	2		1,100	89.50		1,189.50	1,350
1040	15 GPM, 30 lb. fat capacity		4	2		1,650	89.50		1,739.50	1,925
1060	20 GPM, 40 lb. fat capacity		3	2.667		2,000	119		2,119	2,375
1120	Fabricated steel, 50 GPM, 100 lb. fat capacity	Q-1	2	8		3,675	325		4,000	4,525
1160	100 GPM, 200 lb. fat capacity	"	2	8		8,350	325		8,675	9,675
1580	For seepage pan, add					7%				
3000	Hair, cast iron, 1-1/4" and 1-1/2" pipe connection	1 Plum	8	1	Ea.	254	45		299	350

22 13 Facility Sanitary Sewerage

22 13 23 – Sanitary Waste Interceptors

22 13 23.10 Interceptors		Crew	Daily Output	Labor-Hours	Unit	Material	2007 Bare Costs Labor	Equipment	Total	Total Incl O&P
3100	For chrome-plated cast iron, add				Ea.	156			156	172
4000	Oil, fabricated steel, 10 GPM, 2" pipe size	1 Plum	4	2		1,500	89.50		1,589.50	1,775
4100	15 GPM, 2" or 3" pipe size		4	2		2,075	89.50		2,164.50	2,425
4120	20 GPM, 2" or 3" pipe size		3	2.667		2,500	119		2,619	2,925
4220	100 GPM, 3" pipe size	Q-1	2	8		8,350	325		8,675	9,675
6000	Solids, precious metals recovery, C.I., 1-1/4" to 2" pipe	1 Plum	4	2		385	89.50		474.50	555
6100	Dental Lab., large, C.I., 1-1/2" to 2" pipe	"	3	2.667		1,350	119		1,469	1,650

22 13 29 – Sanitary Sewerage Pumps

22 13 29.13 Wet-Pit-Mounted, Vertical Sewerage Pumps

		Crew	Daily Output	Labor-Hours	Unit	Material	Labor	Equipment	Total	Total Incl O&P
0010	**WET-PIT-MOUNTED, VERTICAL SEWERAGE PUMPS**									
0020	Controls incl. alarm/disconnect panel w/wire. Excavation not included									
0260	Simplex, 9 GPM at 60 PSIG, 70 gal. tank				Ea.	2,650			2,650	2,925
0300	For manway, 26" I.D., 18" high, add					335			335	370
0340	26" I.D., 36" high, add					465			465	510
0380	43" I.D., 4' high, add					520			520	575

22 13 29.14 Sewage Ejector Pumps

		Crew	Daily Output	Labor-Hours	Unit	Material	Labor	Equipment	Total	Total Incl O&P
0010	**SEWAGE EJECTOR PUMPS**, With operating and level controls									
0100	Simplex system incl. tank, cover, pump 15' head									
0500	37 gal PE tank, 12 GPM, 1/2 HP, 2" discharge	Q-1	3.20	5	Ea.	410	202		612	755
0510	3" discharge		3.10	5.161		445	208		653	805
0530	87 GPM, .7 HP, 2" discharge		3.20	5		625	202		827	990
0540	3" discharge		3.10	5.161		675	208		883	1,050
0600	45 gal. coated stl tank, 12 GPM, 1/2 HP, 2" discharge		3	5.333		725	215		940	1,125
0610	3" discharge		2.90	5.517		755	223		978	1,175
0630	87 GPM, .7 HP, 2" discharge		3	5.333		930	215		1,145	1,350
0640	3" discharge		2.90	5.517		985	223		1,208	1,400
0660	134 GPM, 1 HP, 2" discharge		2.80	5.714		1,000	230		1,230	1,450
0680	3" discharge		2.70	5.926		1,050	239		1,289	1,525
0700	70 gal. PE tank, 12 GPM, 1/2 HP, 2" discharge		2.60	6.154		800	248		1,048	1,250
0710	3" discharge		2.40	6.667		855	269		1,124	1,350
0730	87 GPM, 0.7 HP, 2" discharge		2.50	6.400		1,025	258		1,283	1,525
0740	3" discharge		2.30	6.957		1,100	281		1,381	1,625
0760	134 GPM, 1 HP, 2" discharge		2.20	7.273		1,125	293		1,418	1,675
0770	3" discharge		2	8		1,200	325		1,525	1,775

22 13 43 – Facility Packaged Sewage Pumping Stations

22 13 43.10 Sewage Pumping Stations

		Crew	Daily Output	Labor-Hours	Unit	Material	Labor	Equipment	Total	Total Incl O&P
0010	**SEWAGE PUMPING STATIONS** Prefabricated steel, concrete									
0020	or fiberglass, 200 GPM	C-17D	.17	494	Total	39,400	19,000	3,575	61,975	76,500
0200	1,000 GPM		.07	1200		55,500	46,200	8,675	110,375	142,000
0500	Add for generator unit, 200 GPM, steel		.34	247		28,500	9,500	1,775	39,775	48,100
0600	Concrete		.51	164		18,000	6,325	1,200	25,525	31,100
1000	Add for generator unit, 1,000 GPM, steel		.30	280		30,300	10,800	2,025	43,125	52,000
1200	Concrete		.38	221		25,500	8,500	1,600	35,600	43,100
1500	For drilled water well, if required, add	B-23	.50	80		7,350	2,325	7,375	17,050	19,800

22 14 Facility Storm Drainage

22 14 23 – Storm Drainage Piping Specialties

22 14 23.33 Backwater Valves		Crew	Daily Output	Labor-Hours	Unit	Material	2007 Bare Costs Labor	Equipment	Total	Total Incl O&P
0010	**BACKWATER VALVES**, C.I. Body.									
6980	Bronze gate and automatic flapper valves									
7000	3" and 4" pipe size	Q-1	13	1.231	Ea.	1,175	49.50		1,224.50	1,350
7100	5" and 6" pipe size	"	13	1.231	"	1,800	49.50		1,849.50	2,050
7240	Bronze flapper valve, bolted cover									
7260	2" pipe size	Q-1	16	1	Ea.	340	40.50		380.50	435
7300	4" pipe size	"	13	1.231	↓	655	49.50		704.50	795
7340	6" pipe size	Q-2	17	1.412		940	59		999	1,125

22 14 26 – Facility Storm Drains

22 14 26.13 Roof Drains

		Crew	Daily Output	Labor-Hours	Unit	Material	Labor	Equipment	Total	Total Incl O&P
0010	**ROOF DRAINS**									
0140	Cornice, C.I., 45° or 90° outlet									
0200	3" and 4" pipe size	Q-1	12	1.333	Ea.	188	54		242	287
0260	For galvanized body, add					38			38	41.50
0280	For polished bronze dome, add				↓	32			32	35.50
3860	Roof, flat metal deck, C.I. body, 12" C.I. dome									
3890	3" pipe size	Q-1	14	1.143	Ea.	225	46		271	315
3920	6" pipe size	"	10	1.600	"	390	64.50		454.50	525
4620	Main, all aluminum, 12" low profile dome									
4640	2", 3" and 4" pipe size	Q-1	14	1.143	Ea.	249	46		295	345

22 14 26.16 Facility Area Drains

		Crew	Daily Output	Labor-Hours	Unit	Material	Labor	Equipment	Total	Total Incl O&P
0010	**FACILITY AREA DRAINS**									
4980	Scupper floor, oblique strainer, C.I.									
5000	6" x 7" top, 2", 3" and 4" pipe size	Q-1	16	1	Ea.	162	40.50		202.50	239
5100	8" x 12" top, 5" and 6" pipe size	"	14	1.143		315	46		361	415
5160	For galvanized body, add					40%				
5200	For polished bronze strainer, add				↓	85%				

22 14 26.19 Facility Trench Drains

		Crew	Daily Output	Labor-Hours	Unit	Material	Labor	Equipment	Total	Total Incl O&P
0010	**FACILITY TRENCH DRAINS**									
5980	Trench, floor, hvy duty, modular, C.I., 12" x 12" top									
6000	2", 3", 4", 5", & 6" pipe size	Q-1	8	2	Ea.	510	80.50		590.50	680
6100	For polished bronze top, add				"	760			760	840
6600	Trench, floor, for cement concrete encasement									
6610	Not including trenching or concrete									
6640	Polyester polymer concrete									
6650	4" internal width, with grate									
6660	Light duty galvanized grate	Q-1	120	.133	L.F.	28	5.40		33.40	38.50
6670	Medium duty iron grate		115	.139		39	5.60		44.60	51.50
6680	Heavy duty iron grate	↓	110	.145	↓	49.50	5.85		55.35	63.50
6700	12" internal width, with grate									
6770	Heavy duty galvanized grate	Q-1	80	.200	L.F.	121	8.05		129.05	145
6800	Fiberglass									
6810	8" internal width, with grate									
6820	Medium duty galvanized grate	Q-1	115	.139	L.F.	82	5.60		87.60	98.50
6830	Heavy duty iron grate	"	110	.145	"	85	5.85		90.85	102

22 14 29 – Sump Pumps

22 14 29.13 Wet-Pit-Mounted, Vertical Sump Pumps

		Crew	Daily Output	Labor-Hours	Unit	Material	Labor	Equipment	Total	Total Incl O&P
0010	**WET-PIT-MOUNTED, VERTICAL SUMP PUMPS**									
0400	Molded PVC base, 21 GPM at 15' head, 1/3 HP	1 Plum	5	1.600	Ea.	96	71.50		167.50	214
0800	Iron base, 21 GPM at 15' head, 1/3 HP		5	1.600		123	71.50		194.50	243
1200	Solid brass, 21 GPM at 15' head, 1/3 HP	↓	5	1.600	↓	201	71.50		272.50	330

22 14 Facility Storm Drainage

22 14 29 – Sump Pumps

22 14 29.16 Submersible Sump Pumps

		Crew	Daily Output	Labor-Hours	Unit	Material	2007 Bare Costs Labor	Equipment	Total	Total Incl O&P
0010	**SUBMERSIBLE SUMP PUMPS**									
7000	Sump pump, automatic									
7100	Plastic, 1-1/4" discharge, 1/4 HP	1 Plum	6	1.333	Ea.	122	59.50		181.50	224
7140	1/3 HP		5	1.600		146	71.50		217.50	269
7160	1/2 HP		5	1.600		180	71.50		251.50	305
7180	1-1/2" discharge, 1/2 HP		4	2		200	89.50		289.50	355
7500	Cast iron, 1-1/4" discharge, 1/4 HP		6	1.333		140	59.50		199.50	244
7540	1/3 HP		6	1.333		165	59.50		224.50	271
7560	1/2 HP	↓	5	1.600	↓	199	71.50		270.50	325

22 31 Domestic Water Softeners

22 31 13 – Domestic Water Softeners

22 31 13.10 Residential Water Softeners

		Crew	Daily Output	Labor-Hours	Unit	Material	Labor	Equipment	Total	Total Incl O&P
0010	**RESIDENTIAL WATER SOFTENERS**									
7350	Water softener, automatic, to 30 grains per gallon	2 Plum	5	3.200	Ea.	465	143		608	725
7400	To 100 grains per gallon	"	4	4	"	645	179		824	980

22 31 16 – Commercial Domestic Water Softeners

22 31 16.10 Water Softeners

		Crew	Daily Output	Labor-Hours	Unit	Material	Labor	Equipment	Total	Total Incl O&P
0010	**WATER SOFTENERS**									
5800	Softener systems, automatic, intermediate sizes									
5820	available, may be used in multiples.									
6000	Hardness capacity between regenerations and flow									
6100	150,000 grains, 37 GPM cont., 51 GPM peak	Q-1	1.20	13.333	Ea.	4,500	540		5,040	5,750
6200	300,000 grains, 81 GPM cont., 113 GPM peak		1	16		6,675	645		7,320	8,325
6300	750,000 grains, 160 GPM cont., 230 GPM peak		.80	20		8,700	805		9,505	10,800
6400	900,000 grains, 185 GPM cont., 270 GPM peak	↓	.70	22.857	↓	14,000	920		14,920	16,800

22 33 Electric Domestic Water Heaters

22 33 13 – Instantaneous Electric Domestic Water Heaters

22 33 13.10 Hot Water Dispensers

		Crew	Daily Output	Labor-Hours	Unit	Material	Labor	Equipment	Total	Total Incl O&P
0010	**HOT WATER DISPENSERS**									
0160	Commercial, 100 cup, 11.3 amp	1 Plum	14	.571	Ea.	345	25.50		370.50	420
3180	Household, 60 cup	"	14	.571	"	173	25.50		198.50	230

22 33 30 – Residential, Electric Domestic Water Heaters

22 33 30.13 Residential, Small-Capacity Electric Domestic Water Heaters

		Crew	Daily Output	Labor-Hours	Unit	Material	Labor	Equipment	Total	Total Incl O&P
0010	**RESIDENTIAL, SMALL-CAPACITY ELECTRIC DOMESTIC WATER HEATERS**									
1000	Residential, electric, glass lined tank, 5 yr, 10 gal., single element	1 Plum	2.30	3.478	Ea.	273	156		429	535
1040	20 gallon, single element		2.20	3.636		345	163		508	625
1060	30 gallon, double element		2.20	3.636		385	163		548	670
1080	40 gallon, double element		2	4		410	179		589	720
1100	52 gallon, double element		2	4		460	179		639	775
1180	120 gallon, double element	↓	1.40	5.714	↓	975	256		1,231	1,450

22 33 33 – Light-Commercial Electric Domestic Water Heaters

22 33 33.10 Commercial Electric Water Heaters

0010	**COMMERCIAL ELECTRIC WATER HEATERS**
4000	Commercial, 100° rise. NOTE: for each size tank, a range of
4010	heaters between the ones shown are available

22 33 Electric Domestic Water Heaters

22 33 33 – Light-Commercial Electric Domestic Water Heaters

22 33 33.10 Commercial Electric Water Heaters		Crew	Daily Output	Labor-Hours	Unit	Material	2007 Bare Costs Labor	2007 Bare Costs Equipment	Total	Total Incl O&P
4020	Electric									
4100	5 gal., 3 kW, 12 GPH, 208V	1 Plum	2	4	Ea.	2,550	179		2,729	3,075
4120	10 gal., 6 kW, 25 GPH, 208V		2	4		2,825	179		3,004	3,375
4140	50 gal., 9 kW, 37 GPH, 208V		1.80	4.444		3,850	199		4,049	4,550
4160	50 gal., 36 kW, 148 GPH, 208V	↓	1.80	4.444		5,900	199		6,099	6,800
4300	200 gal., 15 kW, 61 GPH, 480V	Q-1	1.70	9.412		18,500	380		18,880	21,000
4320	200 gal., 120 kW , 490 GPH, 480V		1.70	9.412		25,300	380		25,680	28,400
4460	400 gal., 30 kW, 123 GPH, 480V	↓	1	16		25,500	645		26,145	29,100
5400	Modulating step control, 2-5 steps	1 Elec	5.30	1.509		1,850	66.50		1,916.50	2,125
5440	6-10 steps		3.20	2.500		2,375	110		2,485	2,800
5460	11-15 steps		2.70	2.963		2,875	130		3,005	3,350
5480	16-20 steps	↓	1.60	5	↓	3,350	220		3,570	4,000

22 34 Fuel-Fired Domestic Water Heaters

22 34 30 – Residential Gas Domestic Water Heaters

22 34 30.13 Residential, Atmospheric, Gas Domestic Water Heaters

		Crew	Daily Output	Labor-Hours	Unit	Material	Labor	Equipment	Total	Total Incl O&P
0010	**RESIDENTIAL, ATMOSPHERIC, GAS DOMESTIC WATER HEATERS**									
2000	Gas fired, foam lined tank, 10 yr, vent not incl.,									
2040	30 gallon	1 Plum	2	4	Ea.	625	179		804	960
2100	75 gallon	↓	1.50	5.333		880	239		1,119	1,325
2120	100 gallon	↓	1.30	6.154	↓	1,350	276		1,626	1,925

22 34 36 – Commercial Gas Domestic Water Heaters

22 34 36.13 Commercial, Atmospheric, Gas Domestic Water Heaters

		Crew	Daily Output	Labor-Hours	Unit	Material	Labor	Equipment	Total	Total Incl O&P
0010	**COMMERCIAL, ATMOSPHERIC, GAS DOMESTIC WATER HEATERS**									
6000	Gas fired, flush jacket, std. controls, vent not incl.									
6040	75 MBH input, 73 GPH	1 Plum	1.40	5.714	Ea.	1,575	256		1,831	2,100
6060	98 MBH input, 95 GPH		1.40	5.714		3,600	256		3,856	4,325
6080	120 MBH input, 110 GPH CN		1.20	6.667		3,900	299		4,199	4,725
6180	200 MBH input, 192 GPH		.60	13.333		4,300	595		4,895	5,650
6200	250 MBH input, 245 GPH		.50	16		4,675	715		5,390	6,225
6900	For low water cutoff, add		8	1		257	45		302	350
6960	For bronze body hot water circulator, add	↓	4	2	↓	1,300	89.50		1,389.50	1,550

22 34 46 – Oil-Fired Domestic Water Heaters

22 34 46.10 Residential Oil-Fired Water Heaters

		Crew	Daily Output	Labor-Hours	Unit	Material	Labor	Equipment	Total	Total Incl O&P
0010	**RESIDENTIAL OIL-FIRED WATER HEATERS**									
3000	Oil fired, glass lined tank, 5 yr, vent not included, 30 gallon	1 Plum	2	4	Ea.	665	179		844	1,000
3040	50 gallon		1.80	4.444		1,375	199		1,574	1,800
3060	70 gallon		1.50	5.333		1,725	239		1,964	2,250
3080	85 gallon	↓	1.40	5.714	↓	4,750	256		5,006	5,600

22 34 46.20 Commercial Oil-Fired Water Heaters

		Crew	Daily Output	Labor-Hours	Unit	Material	Labor	Equipment	Total	Total Incl O&P
0010	**COMMERCIAL OIL-FIRED WATER HEATERS**									
8000	Oil fired, glass lined, UL listed, std. controls, vent not incl.									
8060	140 gal., 140 MBH input, 134 GPH	Q-1	2.13	7.512	Ea.	13,800	305		14,105	15,700
8080	140 gal., 199 MBH input, 191 GPH		2	8		14,300	325		14,625	16,200
8100	140 gal., 255 MBH input, 247 GPH		1.60	10		14,700	405		15,105	16,800
8160	140 gal., 540 MBH input, 519 GPH		.96	16.667		19,500	670		20,170	22,500
8180	140 gal., 720 MBH input, 691 GPH	↓	.92	17.391		19,900	700		20,600	23,000
8280	201 gal., 1250 MBH input, 1200 GPH	Q-2	1.22	19.672		30,500	825		31,325	34,800
8300	201 gal., 1500 MBH input, 1441 GPH	"	1.16	20.690		33,200	865		34,065	37,800
8900	For low water cutoff, add	1 Plum	8	1		257	45		302	350

22 34 Fuel-Fired Domestic Water Heaters

22 34 46 – Oil-Fired Domestic Water Heaters

22 34 46.20 Commercial Oil-Fired Water Heaters		Crew	Daily Output	Labor-Hours	Unit	Material	2007 Bare Costs Labor	Equipment	Total	Total Incl O&P
8960	For bronze body hot water circulator, add	1 Plum	4	2	Ea.	1,000	89.50		1,089.50	1,250

22 35 Domestic Water Heat Exchangers

22 35 30 – Water Heating by Steam

22 35 30.10 Water Heating Transfer Package

0010	**WATER HEATING TRANSFER PACKAGE**, Complete controls									
0020	expansion tank, converter, air separator									
1000	Hot water, 180°F enter, 200°F leaving, 15# steam									
1010	One pump system, 28 GPM	Q-6	.75	32	Ea.	14,200	1,350		15,550	17,600
1020	35 GPM		.70	34.286		15,600	1,450		17,050	19,400
1040	55 GPM		.65	36.923		17,700	1,550		19,250	21,900
1060	130 GPM		.55	43.636		22,000	1,850		23,850	27,000
1080	255 GPM		.40	60		29,800	2,525		32,325	36,600
1100	550 GPM		.30	80		37,700	3,375		41,075	46,600

22 41 Residential Plumbing Fixtures

22 41 06 – Plumbing Fixtures General

22 41 06.10 Plumbing Fixture Notes

0010	**PLUMBING FIXTURE NOTES**, Incl. trim fittings unless otherwise noted R224000-40									
0080	For rough-in, supply, waste, and vent, see add for each type									
0122	For electric water coolers, see division 22 47 16.10									
0160	For color, unless otherwise noted, add					Ea.	20%				

22 41 13 – Residential Water Closets, Urinals, and Bidets

22 41 13.40 Water Closets

0010	**WATER CLOSETS** R224000-40									
0032	For automatic flush, see 22 41 39.10 0972									
0150	Tank type, vitreous china, incl. seat, supply pipe w/stop									
0200	Wall hung									
0400	Two piece, close coupled	Q-1	5.30	3.019	Ea.	550	122		672	790
0960	For rough-in, supply, waste, vent and carrier		2.73	5.861		490	236		726	895
1000	Floor mounted, one piece		5.30	3.019		530	122		652	770
1100	Two piece, close coupled **CN**		5.30	3.019		183	122		305	385
1960	For color, add					30%				
1980	For rough-in, supply, waste and vent	Q-1	3.05	5.246	Ea.	218	212		430	560

22 41 16 – Residential Lavatories and Sinks

22 41 16.10 Lavatories

0010	**LAVATORIES**, With trim, white unless noted otherwise R224000-40									
0500	Vanity top, porcelain enamel on cast iron									
0600	20" x 18"	Q-1	6.40	2.500	Ea.	220	101		321	395
0640	33" x 19" oval		6.40	2.500		460	101		561	655
0720	19" round		6.40	2.500		207	101		308	380
0860	For color, add					25%				
1000	Cultured marble, 19" x 17", single bowl	Q-1	6.40	2.500	Ea.	169	101		270	340
1040	25" x 19", single bowl	"	6.40	2.500	"	181	101		282	350
1580	For color, same price									
1900	Stainless steel, self-rimming, 25" x 22", single bowl, ledge	Q-1	6.40	2.500	Ea.	253	101		354	430
1960	17" x 22", single bowl		6.40	2.500		246	101		347	425
2600	Steel, enameled, 20" x 17", single bowl		5.80	2.759		145	111		256	325

22 41 Residential Plumbing Fixtures

22 41 16 – Residential Lavatories and Sinks

22 41 16.10 Lavatories

		Crew	Daily Output	Labor-Hours	Unit	Material	2007 Bare Costs Labor	Equipment	Total	Total Incl O&P
2660	19" round	Q-1	5.80	2.759	Ea.	142	111		253	325
2900	Vitreous china, 20" x 16", single bowl		5.40	2.963		265	120		385	470
2960	20" x 17", single bowl		5.40	2.963		157	120		277	355
3020	19" round, single bowl		5.40	2.963		156	120		276	350
3200	22" x 13", single bowl		5.40	2.963		247	120		367	450
3560	For color, add					50%				
3580	Rough-in, supply, waste and vent for all above lavatories	Q-1	2.30	6.957	Ea.	169	281		450	605
4000	Wall hung									
4040	Porcelain enamel on cast iron, 16" x 14", single bowl	Q-1	8	2	Ea.	360	80.50		440.50	515
4180	20" x 18", single bowl		8	2		279	80.50		359.50	425
4240	22" x 19", single bowl		8	2		475	80.50		555.50	645
4580	For color, add					30%				
6000	Vitreous china, 18" x 15", single bowl with backsplash	Q-1	7	2.286	Ea.	213	92		305	375
6500	For color, add					30%				
6960	Rough-in, supply, waste and vent for above lavatories	Q-1	1.66	9.639	Ea.	365	390		755	985

22 41 16.30 Sinks

		Crew	Daily Output	Labor-Hours	Unit	Material	Labor	Equipment	Total	Total Incl O&P
0010	**SINKS**, With faucets and drain. R224000-40									
2000	Kitchen, counter top style, P.E. on C.I., 24" x 21" single bowl	Q-1	5.60	2.857	Ea.	251	115		366	450
2100	31" x 22" single bowl		5.60	2.857		237	115		352	435
2200	32" x 21" double bowl		4.80	3.333		325	134		459	560
2300	42" x 21" double bowl		4.80	3.333		500	134		634	750
3000	Stainless steel, self rimming, 19" x 18" single bowl		5.60	2.857		380	115		495	595
3100	25" x 22" single bowl		5.60	2.857		425	115		540	640
4000	Steel, enameled, with ledge, 24" x 21" single bowl		5.60	2.857		145	115		260	330
4100	32" x 21" double bowl		4.80	3.333		166	134		300	385
4960	For color sinks except stainless steel, add					10%				
4980	For rough-in, supply, waste and vent, counter top sinks	Q-1	2.14	7.477		203	300		503	680
5000	Kitchen, raised deck, P.E. on C.I.									
5100	32" x 21", dual level, double bowl	Q-1	2.60	6.154	Ea.	268	248		516	670
5700	For color, add					20%				
5790	For rough-in, supply, waste & vent, sinks	Q-1	1.85	8.649		203	350		553	750

22 41 19 – Residential Bathtubs

22 41 19.10 Baths

		Crew	Daily Output	Labor-Hours	Unit	Material	Labor	Equipment	Total	Total Incl O&P
0010	**BATHS**									
0100	Tubs, recessed porcelain enamel on cast iron, with trim									
0180	48" x 42"	Q-1	4	4	Ea.	1,675	161		1,836	2,075
0220	72" x 36"		3	5.333		1,725	215		1,940	2,225
2000	Enameled formed steel, 4'-6" long		5.80	2.759		345	111		456	545
4000	Soaking, acrylic with pop-up drain, 60" x 32" x 21" deep		5.50	2.909		895	117		1,012	1,150
4100	60" x 48" x 18-1/2" deep		5	3.200		750	129		879	1,025
9600	Rough-in, supply, waste and vent, for all above tubs, add		2.07	7.729		219	310		529	710

22 41 23 – Residential Shower Receptors and Basins

22 41 23.20 Showers

		Crew	Daily Output	Labor-Hours	Unit	Material	Labor	Equipment	Total	Total Incl O&P
0010	**SHOWERS**									
1500	Stall, with drain only. Add for valve and door/curtain									
3000	Fiberglass, one piece, with 3 walls, 32" x 32" square	Q-1	5.50	2.909	Ea.	455	117		572	675
3100	36" x 36" square		5.50	2.909		525	117		642	750
3250	64" x 65-3/4" x 81-1/2" fold. seat, whlchr.		3.80	4.211		3,075	170		3,245	3,625
4000	Polypropylene, stall only, w/ molded-stone floor, 30" x 30"		2	8		340	325		665	855

22 41 Residential Plumbing Fixtures

22 41 23 – Residential Shower Receptors and Basins

22 41 23.20 Showers

		Crew	Daily Output	Labor-Hours	Unit	Material	2007 Bare Costs Labor	Equipment	Total	Total Incl O&P
4200	Rough-in, supply, waste and vent for above showers	Q-1	2.05	7.805	Ea.	310	315		625	820

22 41 23.40 Shower System Components

		Crew	Daily Output	Labor-Hours	Unit	Material	Labor	Equipment	Total	Total Incl O&P
0010	**SHOWER SYSTEM COMPONENTS**									
4520	Showers, fiberglass receptor only	1 Plum	8	1	Ea.	187	45		232	273
5000	Built-in, head, arm, 4 GPM valve		4	2		123	89.50		212.50	271
5200	Head, arm, by-pass, integral stops, handles	↓	3.60	2.222	↓	206	99.50		305.50	375

22 41 36 – Residential Laundry Trays

22 41 36.10 Laundry Sinks

		Crew	Daily Output	Labor-Hours	Unit	Material	Labor	Equipment	Total	Total Incl O&P
0010	**LAUNDRY SINKS**, With trim									
0020	Porcelain enamel on cast iron, black iron frame									
0050	24" x 21", single compartment	Q-1	6	2.667	Ea.	365	108		473	560
0100	26" x 21", single compartment	"	6	2.667	"	375	108		483	570
2000	Molded stone, on wall hanger or legs									
2020	22" x 23", single compartment	Q-1	6	2.667	Ea.	122	108		230	296
2100	45" x 21", double compartment	"	5	3.200	"	213	129		342	430
3000	Plastic, on wall hanger or legs									
3020	18" x 23", single compartment	Q-1	6.50	2.462	Ea.	97	99.50		196.50	256
3300	40" x 24", double compartment		5.50	2.909		227	117		344	425
5000	Stainless steel, counter top, 22" x 17" single compartment		6	2.667		57.50	108		165.50	225
5200	33" x 22", double compartment		5	3.200		69.50	129		198.50	271
9600	Rough-in, supply, waste and vent, for all laundry sinks	↓	2.14	7.477	↓	203	300		503	680

22 41 39 – Residential Faucets, Supplies, and Trim

22 41 39.10 Faucets and Fittings

		Crew	Daily Output	Labor-Hours	Unit	Material	Labor	Equipment	Total	Total Incl O&P
0010	**FAUCETS AND FITTINGS**									
0150	Bath, faucets, diverter spout combination, sweat	1 Plum	8	1	Ea.	104	45		149	182
0200	For integral stops, IPS unions, add					109			109	120
0420	Bath, press-bal mix valve w/diverter, spout, shower hd, arm/flange	1 Plum	8	1	↓	119	45		164	199
0810	Bidet									
0812	Fitting, over the rim, swivel spray/pop-up drain	1 Plum	8	1	Ea.	154	45		199	237
0840	Flush valves, with vacuum breaker									
0850	Water closet									
0860	Exposed, rear spud	1 Plum	8	1	Ea.	118	45		163	197
0920	Urinal									
0930	Exposed, stall	1 Plum	8	1	Ea.	108	45		153	187
0940	Wall, (washout)		8	1		108	45		153	187
0950	Pedestal, top spud		8	1		110	45		155	189
0960	Concealed, stall		8	1		124	45		169	204
0970	Wall (washout)	↓	8	1	↓	139	45		184	221
0971	Automatic flush sensor and operator for									
0972	urinals or water closets	1 Plum	5.33	1.501	Ea.	355	67		422	490
1000	Kitchen sink faucets, top mount, cast spout		10	.800		55	36		91	115
1100	For spray, add	↓	24	.333	↓	16.15	14.95		31.10	40.50
1300	Single control lever handle									
1310	With pull out spray									
1320	Polished chrome	1 Plum	10	.800	Ea.	167	36		203	237
1330	Polished brass		10	.800		200	36		236	274
1340	White	↓	10	.800	↓	184	36		220	256
1348	With spray thru escutcheon									
1350	Polished chrome	1 Plum	10	.800	Ea.	117	36		153	182
1360	Polished brass		10	.800		140	36		176	208
1370	White	↓	10	.800	↓	128	36		164	195

22 41 Residential Plumbing Fixtures

22 41 39 – Residential Faucets, Supplies, and Trim

22 41 39.10 Faucets and Fittings		Crew	Daily Output	Labor-Hours	Unit	Material	2007 Bare Costs Labor	Equipment	Total	Total Incl O&P
2000	Laundry faucets, shelf type, IPS or copper unions	1 Plum	12	.667	Ea.	45	30		75	94.50
2100	Lavatory faucet, centerset, without drain		10	.800		40	36		76	98
2210	Porcelain cross handles and pop-up drain									
2220	Polished chrome	1 Plum	6.66	1.201	Ea.	138	54		192	233
2230	Polished brass	"	6.66	1.201	"	208	54		262	310
2260	Single lever handle and pop-up drain									
2270	Black nickel	1 Plum	6.66	1.201	Ea.	220	54		274	325
2280	Polished brass		6.66	1.201		220	54		274	325
2290	Polished chrome		6.66	1.201		161	54		215	258
2810	Automatic sensor and operator, with faucet head		6.15	1.301		330	58.50		388.50	455
3000	Service sink faucet, cast spout, pail hook, hose end		14	.571		76.50	25.50		102	123
4000	Shower by-pass valve with union		18	.444		54	19.90		73.90	89.50
4200	Shower thermostatic mixing valve, concealed		8	1		246	45		291	340
4220	Shower pressure balancing mixing valve,									
4230	With shower head, arm, flange and diverter tub spout									
4240	Chrome	1 Plum	6.14	1.303	Ea.	141	58.50		199.50	243
4250	Polished brass		6.14	1.303		198	58.50		256.50	305
4260	Satin		6.14	1.303		198	58.50		256.50	305
4270	Polished chrome/brass		6.14	1.303		162	58.50		220.50	266
5000	Sillcock, compact, brass, IPS or copper to hose		24	.333		6.65	14.95		21.60	30

22 42 Commercial Plumbing Fixtures

22 42 13 – Commercial Water Closets, Urinals, and Bidets

22 42 13.30 Urinals

		Crew	Daily Output	Labor-Hours	Unit	Material	Labor	Equipment	Total	Total Incl O&P
0010	**URINALS**									
3000	Wall hung, vitreous china, with hanger & self-closing valve									
3100	Siphon jet type	Q-1	3	5.333	Ea.	325	215		540	685
3120	Blowout type		3	5.333		360	215		575	720
3300	Rough-in, supply, waste & vent		2.83	5.654		176	228		404	540
5000	Stall type, vitreous china, includes valve		2.50	6.400		540	258		798	985
6980	Rough-in, supply, waste and vent		1.99	8.040		244	325		569	760

22 42 13.40 Water Closets

		Crew	Daily Output	Labor-Hours	Unit	Material	Labor	Equipment	Total	Total Incl O&P
0010	**WATER CLOSETS**									
3000	Bowl only, with flush valve, seat									
3100	Wall hung	Q-1	5.80	2.759	Ea.	315	111		426	510
3200	For rough-in, supply, waste and vent, single WC		2.56	6.250		520	252		772	950
3300	Floor mounted		5.80	2.759		375	111		486	580
3350	With wall outlet		5.80	2.759		535	111		646	755
3400	For rough-in, supply, waste and vent, single WC		2.84	5.634		248	227		475	615

22 42 16 – Commercial Lavatories and Sinks

22 42 16.10 Handwasher-Dryer Module

		Crew	Daily Output	Labor-Hours	Unit	Material	Labor	Equipment	Total	Total Incl O&P
0010	**HANDWASHER-DRYER MODULE**									
0030	Wall mounted									
0040	With electric dryer									
0050	Sensor operated	Q-1	8	2	Ea.	2,300	80.50		2,380.50	2,650
0110	Sensor operated (ADA)		8	2		2,175	80.50		2,255.50	2,525
0140	Sensor operated (ADA), surface mounted		8	2		2,825	80.50		2,905.50	3,225
0150	With paper towels									
0180	Sensor operated	Q-1	8	2	Ea.	2,250	80.50		2,330.50	2,600

22 42 Commercial Plumbing Fixtures

22 42 16 – Commercial Lavatories and Sinks

22 42 16.40 Service Sinks

		Crew	Daily Output	Labor-Hours	Unit	Material	2007 Bare Costs Labor	2007 Bare Costs Equipment	Total	Total Incl O&P
0010	**SERVICE SINKS**									
6650	Service, floor, corner, P.E. on C.I., 28" x 28"	Q-1	4.40	3.636	Ea.	605	147		752	885
6790	For rough-in, supply, waste & vent, floor service sinks		1.64	9.756		500	395		895	1,150
7000	Service, wall, P.E. on C.I., roll rim, 22" x 18"		4	4		500	161		661	795
7100	24" x 20"		4	4		750	161		911	1,075
8600	Vitreous china, 22" x 20"		4	4		415	161		576	705
8960	For stainless steel rim guard, front or side, add					38.50			38.50	42.50
8980	For rough-in, supply, waste & vent, wall service sinks	Q-1	1.30	12.308		690	495		1,185	1,500

22 42 23 – Commercial Shower Receptors and Basins

22 42 23.30 Group Showers

		Crew	Daily Output	Labor-Hours	Unit	Material	Labor	Equipment	Total	Total Incl O&P
0010	**GROUP SHOWERS**									
6000	Group, w/pressure balancing valve, rough-in and rigging not included									
6800	Column, 6 heads, no receptors, less partitions	Q-1	3	5.333	Ea.	2,300	215		2,515	2,850
6900	With stainless steel partitions		1	16		4,900	645		5,545	6,375
7600	5 heads, no receptors, less partitions		3	5.333		1,750	215		1,965	2,250
7620	4 heads (1 handicap) no receptors, less partitions		3	5.333		2,450	215		2,665	3,025
7700	With stainless steel partitions		1	16		1,750	645		2,395	2,900
8000	Wall, 2 heads, no receptors, less partitions		4	4		1,000	161		1,161	1,375
8100	With stainless steel partitions		2	8		2,450	325		2,775	3,175

22 42 33 – Wash Fountains

22 42 33.20 Wash Fountains

		Crew	Daily Output	Labor-Hours	Unit	Material	Labor	Equipment	Total	Total Incl O&P
0010	**WASH FOUNTAINS**									
1900	Group, foot control									
2000	Precast terrazzo, circular, 36" diam., 5 or 6 persons	Q-2	3	8	Ea.	3,250	335		3,585	4,075
2100	54" diameter for 8 or 10 persons		2.50	9.600		3,950	400		4,350	4,950
2400	Semi-circular, 36" diam. for 3 persons		3	8		2,925	335		3,260	3,700
2500	54" diam. for 4 or 5 persons		2.50	9.600		3,600	400		4,000	4,575
2700	Quarter circle (corner), 54" for 3 persons		3.50	6.857		3,600	287		3,887	4,400
3000	Stainless steel, circular, 36" diameter		3.50	6.857		3,725	287		4,012	4,525
3100	54" diameter		2.80	8.571		4,925	360		5,285	5,950
3400	Semi-circular, 36" diameter		3.50	6.857		3,225	287		3,512	3,975
3500	54" diameter		2.80	8.571		4,250	360		4,610	5,225
5610	Group, infrared control, barrier free									
5614	Precast terrazzo									
5620	Semi-circular 36" diam. for 3 persons	Q-2	3	8	Ea.	4,300	335		4,635	5,225
5630	46" diam. for 4 persons		2.80	8.571		4,800	360		5,160	5,825
5640	Circular, 54" diam. for 8 persons, button control		2.50	9.600		7,025	400		7,425	8,325
5700	Rough-in, supply, waste and vent for above wash fountains	Q-1	1.82	8.791		315	355		670	880
6200	Duo for small washrooms, stainless steel		2	8		2,200	325		2,525	2,900
6500	Rough-in, supply, waste & vent for duo fountains		2.02	7.921		169	320		489	665

22 42 39 – Commercial Faucets, Supplies, and Trim

22 42 39.30 Carriers and Supports

		Crew	Daily Output	Labor-Hours	Unit	Material	Labor	Equipment	Total	Total Incl O&P
0010	**CARRIERS AND SUPPORTS**, For plumbing fixtures.									
0500	Drinking fountain, wall mounted									
0600	Plate type with studs, top back plate	1 Plum	7	1.143	Ea.	51	51		102	133
0700	Top front and back plate		7	1.143		63	51		114	147
0800	Top & bottom, front & back plates, w/bearing jacks		7	1.143		91.50	51		142.50	178
3000	Lavatory, concealed arm									
3050	Floor mounted, single									
3100	High back fixture	1 Plum	6	1.333	Ea.	285	59.50		344.50	405

22 42 Commercial Plumbing Fixtures

22 42 39 – Commercial Faucets, Supplies, and Trim

22 42 39.30 Carriers and Supports		Crew	Daily Output	Labor-Hours	Unit	Material	2007 Bare Costs Labor	Equipment	Total	Total Incl O&P
3200	Flat slab fixture	1 Plum	6	1.333	Ea.	246	59.50		305.50	360
3220	Paraplegic	↓	6	1.333	↓	335	59.50		394.50	460
3250	Floor mounted, back to back									
3300	High back fixtures	1 Plum	5	1.600	Ea.	405	71.50		476.50	555
3400	Flat slab fixtures		5	1.600		650	71.50		721.50	825
3430	Paraplegic	↓	5	1.600	↓	455	71.50		526.50	610
3500	Wall mounted, in stud or masonry									
3600	High back fixture	1 Plum	6	1.333	Ea.	168	59.50		227.50	275
3700	Flat slab fixture	"	6	1.333	"	216	59.50		275.50	330
4600	Sink, floor mounted									
4650	Exposed arm system									
4700	Single heavy fixture	1 Plum	5	1.600	Ea.	475	71.50		546.50	635
4750	Single heavy sink with slab		5	1.600		385	71.50		456.50	535
4800	Back to back, standard fixtures		5	1.600		350	71.50		421.50	495
4850	Back to back, heavy fixtures		5	1.600		425	71.50		496.50	575
4900	Back to back, heavy sink with slab	↓	5	1.600	↓	425	71.50		496.50	575
4950	Exposed offset arm system									
5000	Single heavy deep fixture	1 Plum	5	1.600	Ea.	455	71.50		526.50	610
5100	Plate type system									
5200	With bearing jacks, single fixture	1 Plum	5	1.600	Ea.	520	71.50		591.50	685
5300	With exposed arms, single heavy fixture		5	1.600		670	71.50		741.50	845
5400	Wall mounted, exposed arms, single heavy fixture		5	1.600		244	71.50		315.50	375
6000	Urinal, floor mounted, 2" or 3" coupling, blowout type		6	1.333		305	59.50		364.50	425
6100	With fixture or hanger bolts, blowout or washout		6	1.333		215	59.50		274.50	325
6200	With bearing plate		6	1.333		216	59.50		275.50	330
6300	Wall mounted, plate type system	↓	6	1.333	↓	186	59.50		245.50	295
6980	Water closet, siphon jet									
7000	Horizontal, adjustable, caulk									
7040	Single, 4" pipe size	1 Plum	5.33	1.501	Ea.	395	67		462	535
7050	4" pipe size, paraplegic		5.33	1.501		395	67		462	530
7060	5" pipe size		5.33	1.501		705	67		772	875
7100	Double, 4" pipe size		5	1.600		925	71.50		996.50	1,125
7110	4" pipe size, paraplegic		5	1.600		925	71.50		996.50	1,125
7120	5" pipe size	↓	5	1.600	↓	1,175	71.50		1,246.50	1,400
7160	Horizontal, adjustable, extended, caulk									
7180	Single, 4" pipe size	1 Plum	5.33	1.501	Ea.	530	67		597	685
7200	5" pipe size		5.33	1.501		820	67		887	1,000
7240	Double, 4" pipe size		5	1.600		925	71.50		996.50	1,125
7260	5" pipe size	↓	5	1.600	↓	1,175	71.50		1,246.50	1,400
7400	Vertical, adjustable, caulk or thread									
7440	Single, 4" pipe size	1 Plum	5.33	1.501	Ea.	540	67		607	695
7460	5" pipe size		5.33	1.501		670	67		737	835
7480	6" pipe size		5	1.600		765	71.50		836.50	950
7520	Double, 4" pipe size		5	1.600		810	71.50		881.50	1,000
7540	5" pipe size		5	1.600		935	71.50		1,006.50	1,125
7560	6" pipe size	↓	4	2	↓	1,025	89.50		1,114.50	1,250
7600	Vertical, adjustable, extended, caulk									
7620	Single, 4" pipe size	1 Plum	5.33	1.501	Ea.	540	67		607	695
7640	5" pipe size		5.33	1.501		670	67		737	835
7680	6" pipe size		5	1.600		765	71.50		836.50	950
7720	Double, 4" pipe size		5	1.600		810	71.50		881.50	1,000
7740	5" pipe size		5	1.600		935	71.50		1,006.50	1,125
7760	6" pipe size	↓	4	2	↓	1,025	89.50		1,114.50	1,250

22 42 Commercial Plumbing Fixtures

22 42 39 – Commercial Faucets, Supplies, and Trim

22 42 39.30 Carriers and Supports

		Crew	Daily Output	Labor-Hours	Unit	Material	2007 Bare Costs Labor	Equipment	Total	Total Incl O&P
7780	Water closet, blow out									
7800	Vertical offset, caulk or thread									
7820	Single, 4" pipe size	1 Plum	5.33	1.501	Ea.	405	67		472	545
7840	Double, 4" pipe size	"	5	1.600	"	555	71.50		626.50	720
7880	Vertical offset, extended, caulk									
7900	Single, 4" pipe size	1 Plum	5.33	1.501	Ea.	515	67		582	665
7920	Double, 4" pipe size	"	5	1.600	"	795	71.50		866.50	980
7960	Vertical, for floor mounted back-outlet									
7980	Single, 4" thread, 2" vent	1 Plum	5.33	1.501	Ea.	360	67		427	500
8000	Double, 4" thread, 2" vent	"	6	1.333	"	1,050	59.50		1,109.50	1,275
8040	Vertical, for floor mounted back-outlet, extended									
8060	Single, 4" caulk, 2" vent	1 Plum	6	1.333	Ea.	360	59.50		419.50	490
8080	Double, 4" caulk, 2" vent	"	6	1.333	"	1,050	59.50		1,109.50	1,275
8200	Water closet, residential									
8220	Vertical centerline, floor mount									
8240	Single, 3" caulk, 2" or 3" vent	1 Plum	6	1.333	Ea.	330	59.50		389.50	450
8260	4" caulk, 2" or 4" vent		6	1.333		425	59.50		484.50	555
8280	3" copper sweat, 3" vent		6	1.333		295	59.50		354.50	415
8300	4" copper sweat, 4" vent		6	1.333		355	59.50		414.50	485
8400	Vertical offset, floor mount									
8420	Single, 3" or 4" caulk, vent	1 Plum	4	2	Ea.	410	89.50		499.50	585
8440	3" or 4" copper sweat, vent		5	1.600		410	71.50		481.50	560
8460	Double, 3" or 4" caulk, vent		4	2		705	89.50		794.50	910
8480	3" or 4" copper sweat, vent		5	1.600		705	71.50		776.50	885
9000	Water cooler (electric), floor mounted									
9100	Plate type with bearing plate, single	1 Plum	6	1.333	Ea.	207	59.50		266.50	320

22 45 Emergency Plumbing Fixtures

22 45 13 – Emergency Showers

22 45 13.10 Emergency Showers

		Crew	Daily Output	Labor-Hours	Unit	Material	Labor	Equipment	Total	Total Incl O&P
0010	**EMERGENCY SHOWERS**, Rough-in not included									
5000	Shower, single head, drench, ball valve, pull, freestanding	Q-1	4	4	Ea.	287	161		448	560
5200	Horizontal or vertical supply		4	4		264	161		425	535
6000	Multi-nozzle, eye/face wash combination		4	4		535	161		696	835
6400	Multi-nozzle, 12 spray, shower only		4	4		1,425	161		1,586	1,825
6600	For freeze-proof, add		6	2.667		305	108		413	495

22 45 16 – Eyewash Equipment

22 45 16.10 Eyewash Equipment

		Crew	Daily Output	Labor-Hours	Unit	Material	Labor	Equipment	Total	Total Incl O&P
0010	**EYEWASH EQUIPMENT**, Rough-in not included									
1000	Eye wash fountain									
1400	Plastic bowl, pedestal mounted	Q-1	4	4	Ea.	205	161		366	470
1600	Unmounted		4	4		145	161		306	405
1800	Wall mounted		4	4		148	161		309	405
2000	Stainless steel, pedestal mounted		4	4		315	161		476	590
2200	Unmounted		4	4		200	161		361	465
2400	Wall mounted		4	4		198	161		359	460

22 45 19 – Self-Contained Eyewash Equipment

22 45 19.10 Self-Contained Eyewash Equipment

		Crew	Daily Output	Labor-Hours	Unit	Material	Labor	Equipment	Total	Total Incl O&P
0010	**SELF-CONTAINED EYEWASH EQUIPMENT**									
3000	Eye wash, portable, self-contained				Ea.	415			415	460

22 45 Emergency Plumbing Fixtures

22 45 26 – Eye/Face Wash Equipment

	22 45 26.10 Eye/Face Wash Equipment	Crew	Daily Output	Labor-Hours	Unit	Material	2007 Bare Costs Labor	Equipment	Total	Total Incl O&P
0010	**EYE/FACE WASH EQUIPMENT**, Rough-in not included									
4000	Eye and face wash, combination fountain									
4200	Stainless steel, pedestal mounted	Q-1	4	4	Ea.	440	161		601	725
4400	Unmounted		4	4		330	161		491	610
4600	Wall mounted		4	4		335	161		496	615

22 47 Drinking Fountains and Water Coolers

22 47 13 – Drinking Fountains

22 47 13.10 Drinking Fountains

		Crew	Daily Output	Labor-Hours	Unit	Material	Labor	Equipment	Total	Total Incl O&P
0010	**DRINKING FOUNTAINS**, For connection to cold water supply R224000-40									
1000	Wall mounted, non-recessed									
1400	Bronze, with no back	1 Plum	4	2	Ea.	1,125	89.50		1,214.50	1,375
1800	Cast aluminum, enameled, for correctional institutions		4	2		700	89.50		789.50	900
2000	Fiberglass, 12" back, single bubbler unit		4	2		625	89.50		714.50	825
2040	Dual bubbler		3.20	2.500		905	112		1,017	1,175
2400	Precast stone, no back		4	2		565	89.50		654.50	755
2700	Stainless steel, single bubbler, no back		4	2		1,175	89.50		1,264.50	1,400
2740	With back		4	2		1,025	89.50		1,114.50	1,250
2780	Dual handle & wheelchair projection type		4	2		605	89.50		694.50	800
2820	Dual level for handicapped type		3.20	2.500		1,150	112		1,262	1,450
3300	Vitreous china									
3340	7" back	1 Plum	4	2	Ea.	460	89.50		549.50	640
3940	For vandal-resistant bottom plate, add					55			55	60.50
3960	For freeze-proof valve system, add	1 Plum	2	4		450	179		629	765
3980	For rough-in, supply and waste, add	"	2.21	3.620		121	162		283	375
4000	Wall mounted, semi-recessed									
4200	Poly-marble, single bubbler	1 Plum	4	2	Ea.	715	89.50		804.50	920
4600	Stainless steel, satin finish, single bubbler		4	2		830	89.50		919.50	1,050
4900	Vitreous china, single bubbler		4	2		610	89.50		699.50	805
5980	For rough-in, supply and waste, add		1.83	4.372		121	196		317	430
6000	Wall mounted, fully recessed									
6400	Poly-marble, single bubbler	1 Plum	4	2	Ea.	815	89.50		904.50	1,025
6800	Stainless steel, single bubbler		4	2		910	89.50		999.50	1,125
7560	For freeze-proof valve system, add		2	4		550	179		729	875
7580	For rough-in, supply and waste, add		1.83	4.372		121	196		317	430
7600	Floor mounted, pedestal type									
7700	Aluminum, architectural style, C.I. base	1 Plum	2	4	Ea.	485	179		664	805
7780	Wheelchair handicap unit		2	4		1,150	179		1,329	1,550
8400	Stainless steel, architectural style		2	4		1,250	179		1,429	1,650
8600	Enameled iron, heavy duty service, 2 bubblers		2	4		1,150	179		1,329	1,550
8660	4 bubblers		2	4		1,225	179		1,404	1,625
8880	For freeze-proof valve system, add		2	4		355	179		534	660
8900	For rough-in, supply and waste, add		1.83	4.372		121	196		317	430
9100	Deck mounted									
9500	Stainless steel, circular receptor	1 Plum	4	2	Ea.	315	89.50		404.50	485
9760	White enameled steel, 14" x 9" receptor		4	2		276	89.50		365.50	440
9860	White enameled cast iron, 24" x 16" receptor		3	2.667		275	119		394	485
9980	For rough-in, supply and waste, add		1.83	4.372		121	196		317	430

22 47 Drinking Fountains and Water Coolers

22 47 16 – Pressure Water Coolers

22 47 16.10 Electric Water Coolers

	22 47 16.10 Electric Water Coolers	Crew	Daily Output	Labor-Hours	Unit	Material	2007 Bare Costs Labor	Equipment	Total	Total Incl O&P
0010	**ELECTRIC WATER COOLERS** R224000-50									
0100	Wall mounted, non-recessed									
0140	4 GPH	Q-1	4	4	Ea.	525	161		686	825
0160	8 GPH, barrier free, sensor operated		4	4		665	161		826	975
0180	8.2 GPH	↓	4	4		590	161		751	895
0600	For hot and cold water, add					152			152	167
0640	For stainless steel cabinet, add					77.50			77.50	85
1000	Dual height, 8.2 GPH	Q-1	3.80	4.211		850	170		1,020	1,200
1040	14.3 GPH	"	3.80	4.211		900	170		1,070	1,250
1240	For stainless steel cabinet, add					118			118	130
2600	Wheelchair type, 8 GPH	Q-1	4	4		1,500	161		1,661	1,900
3300	Semi-recessed, 8.1 GPH		4	4		885	161		1,046	1,225
3320	12 GPH	↓	4	4	↓	935	161		1,096	1,275
4600	Floor mounted, flush-to-wall									
4640	4 GPH	1 Plum	3	2.667	Ea.	550	119		669	785
4680	8.2 GPH		3	2.667		605	119		724	845
4720	14.3 GPH		3	2.667		655	119		774	900
4960	14 GPH hot and cold water	↓	3	2.667		915	119		1,034	1,175
4980	For stainless steel cabinet, add					118			118	130
5000	Dual height, 8.2 GPH	1 Plum	2	4		920	179		1,099	1,275
5040	14.3 GPH	"	2	4		950	179		1,129	1,325
5120	For stainless steel cabinet, add					155			155	170
9800	For supply, waste & vent, all coolers	1 Plum	2.21	3.620	↓	121	162		283	375

22 51 Swimming Pool Plumbing Systems

22 51 19 – Swimming Pool Water Treatment Equipment

22 51 19.50 Swimming Pool Filtration Equipment

		Crew	Daily Output	Labor-Hours	Unit	Material	Labor	Equipment	Total	Total Incl O&P
0010	**SWIMMING POOL FILTRATION EQUIPMENT**									
0900	Filter system, sand or diatomite type, incl. pump, 6,000 gal./hr.	2 Plum	1.80	8.889	Total	1,225	400		1,625	1,950
1020	Add for chlorination system, 800 S.F. pool		3	5.333	Ea.	211	239		450	590
1040	5,000 S.F. pool	↓	3	5.333	"	1,450	239		1,689	1,950

22 52 Fountain Plumbing Systems

22 52 16 – Fountain Pumps

22 52 16.10 Fountain Pumps

		Crew	Daily Output	Labor-Hours	Unit	Material	Labor	Equipment	Total	Total Incl O&P
0010	**FOUNTAIN PUMPS**									
0100	Pump w/controls									
0200	Single phase, 100' cord, 1/2 H.P. pump	2 Skwk	4.40	3.636	Ea.	3,325	138		3,463	3,900
0300	3/4 H.P. pump		4.30	3.721		3,800	141		3,941	4,400
0400	1 H.P. pump		4.20	3.810		3,850	145		3,995	4,450
0500	1-1/2 H.P. pump		4.10	3.902		3,975	148		4,123	4,600
0600	2 H.P. pump		4	4		4,025	152		4,177	4,650
0700	Three phase, 200' cord, 5 H.P. pump		3.90	4.103		9,775	156		9,931	10,900
0800	7-1/2 H.P. pump		3.80	4.211		11,000	160		11,160	12,300
0900	10 H.P. pump		3.70	4.324		12,300	164		12,464	13,800
1000	15 H.P. pump	↓	3.60	4.444	↓	14,600	169		14,769	16,300

22 52 Fountain Plumbing Systems

22 52 33 – Fountain Ancillary

22 52 33.10 Fountain Ancillary

22 52 33.10 Fountain Ancillary		Crew	Daily Output	Labor-Hours	Unit	Material	2007 Bare Costs Labor	2007 Bare Costs Equipment	Total	Total Incl O&P
0010	**FOUNTAIN ANCILLARY**									
1100	Nozzles, minimum	2 Skwk	8	2	Ea.	244	76		320	385
1200	Maximum		8	2		320	76		396	475
1300	Lights w/mounting kits, 200 watt		18	.889		405	34		439	500
1400	300 watt		18	.889		445	34		479	545
1500	500 watt		18	.889		490	34		524	595
1600	Color blender		12	1.333		380	50.50		430.50	500

22 66 Chemical-Waste Systems for Lab. and Healthcare Facilities

22 66 53 – Laboratory Chemical-Waste and Vent Piping

22 66 53.30 Glass Pipe

22 66 53.30 Glass Pipe		Crew	Daily Output	Labor-Hours	Unit	Material	2007 Bare Costs Labor	2007 Bare Costs Equipment	Total	Total Incl O&P
0010	**GLASS PIPE**, Borosilicate, couplings & hangers 10' O.C.									
0020	Drainage									
1100	1-1/2" diameter	Q-1	52	.308	L.F.	9.35	12.40		21.75	29
1120	2" diameter		44	.364		12.50	14.65		27.15	36
1140	3" diameter		39	.410		16.75	16.55		33.30	43.50
1160	4" diameter		30	.533		30.50	21.50		52	66
1180	6" diameter		26	.615		52	25		77	94.50

22 66 53.60 Corrosion Resistant Pipe

22 66 53.60 Corrosion Resistant Pipe		Crew	Daily Output	Labor-Hours	Unit	Material	2007 Bare Costs Labor	2007 Bare Costs Equipment	Total	Total Incl O&P
0010	**CORROSION RESISTANT PIPE**, No couplings or hangers									
0020	Iron alloy, drain, mechanical joint									
1000	1-1/2" diameter	Q-1	70	.229	L.F.	34.50	9.20		43.70	52
1100	2" diameter		66	.242		35.50	9.80		45.30	53.50
1120	3" diameter		60	.267		45.50	10.75		56.25	66
1140	4" diameter		52	.308		58.50	12.40		70.90	83
2980	Plastic, epoxy, fiberglass filament wound, B&S joint									
3000	2" diameter	Q-1	62	.258	L.F.	7.65	10.40		18.05	24
3100	3" diameter		51	.314		8.80	12.65		21.45	28.50
3120	4" diameter		45	.356		12.90	14.35		27.25	35.50
3140	6" diameter		32	.500		18.15	20		38.15	50.50
3980	Polyester, fiberglass filament wound, B&S joint									
4000	2" diameter	Q-1	62	.258	L.F.	10.55	10.40		20.95	27.50
4100	3" diameter		51	.314		14.55	12.65		27.20	35
4120	4" diameter		45	.356		22.50	14.35		36.85	46
4140	6" diameter		32	.500		37.50	20		57.50	71.50
4980	Polypropylene, acid resistant, fire retardant, schedule 40									
5000	1-1/2" diameter	Q-1	68	.235	L.F.	3.96	9.50		13.46	18.60
5100	2" diameter		62	.258		5.40	10.40		15.80	21.50
5120	3" diameter		51	.314		11	12.65		23.65	31
5140	4" diameter		45	.356		14.05	14.35		28.40	37
5980	Proxylene, fire retardant, Schedule 40									
6000	1-1/2" diameter	Q-1	68	.235	L.F.	6.15	9.50		15.65	21
6100	2" diameter		62	.258		8.45	10.40		18.85	25
6120	3" diameter		51	.314		15.30	12.65		27.95	36
6140	4" diameter		45	.356		21.50	14.35		35.85	45.50

Estimating Tips

23 10 00 Facility Fuel Systems
- The prices in this subdivision for above- and below-ground storage tanks do not include foundations or hold-down slabs. The estimator should refer to Divisions 3 and 31 for foundation system pricing. In addition to the foundations, required tank accessories, such as tank gauges, leak detection devices, and additional manholes and piping, must be added to the tank prices.

23 50 00 Central Heating Equipment
- When estimating the cost of an HVAC system, check to see who is responsible for providing and installing the temperature control system. It is possible to overlook controls, assuming that they would be included in the electrical estimate.
- When looking up a boiler, be careful on specified capacity. Some manufacturers rate their products on output while others use input.
- Include HVAC insulation for pipe, boiler, and duct (wrap and liner).
- Be careful when looking up mechanical items to get the correct pressure rating and connection type (thread, weld, flange).

23 70 00 Central HVAC Equipment
- Combination heating and cooling units are sized by the air conditioning requirements. (See Reference No. R236000-20 for preliminary sizing guide.)
- A ton of air conditioning is nominally 400 CFM.
- Rectangular duct is taken off by the linear foot for each size, but its cost is usually estimated by the pound. Remember that SMACNA standards now base duct on internal pressure.
- Prefabricated duct is estimated and purchased like pipe: straight sections and fittings.
- Note that cranes or other lifting equipment are not included on any lines in Division 15. For example, if a crane is required to lift a heavy piece of pipe into place high above a gym floor, or to put a rooftop unit on the roof of a four-story building, etc., it must be added. Due to the potential for extreme variation—from nothing additional required to a major crane or helicopter—we feel that including a nominal amount for "lifting contingency" would be useless and detract from the accuracy of the estimate. When using equipment rental from RSMeans do not forget to include the cost of the operator(s).

Reference Numbers
Reference numbers are shown in shaded boxes at the beginning of some major classifications. These numbers refer to related items in the Reference Section. The reference information may be an estimating procedure, an alternate pricing method, or technical information.

Note: Not all subdivisions listed here necessarily appear in this publication.

Note: **i2 Trade Service**, *in part, has been used as a reference source for some of the material prices used in Division 23.*

No part of this publication may be reproduced, stored in a retrieval system, or transmitted in any form or by any means without prior written permission of Reed Construction Data.

23 05 Common Work Results for HVAC

23 05 05 – Selective HVAC Demolition

23 05 05.10 HVAC Demolition

		Crew	Daily Output	Labor-Hours	Unit	Material	2007 Bare Costs Labor	Equipment	Total	Total Incl O&P
0010	**HVAC DEMOLITION**									
0100	Air conditioner, split unit, 3 ton	Q-5	2	8	Ea.		325		325	490
0150	Package unit, 3 ton	Q-6	3	8	"		335		335	510
0298	Boilers									
0300	Electric, up thru 148 kW	Q-19	2	12	Ea.		500		500	750
0310	150 thru 518 kW	"	1	24			1,000		1,000	1,500
0320	550 thru 2000 kW	Q-21	.40	80			3,400		3,400	5,125
0330	2070 kW and up	"	.30	106			4,550		4,550	6,825
0340	Gas and/or oil, up thru 150 MBH	Q-7	2.20	14.545			625		625	940
0350	160 thru 2000 MBH		.80	40			1,725		1,725	2,600
0360	2100 thru 4500 MBH		.50	64			2,750		2,750	4,150
0370	4600 thru 7000 MBH		.30	106			4,600		4,600	6,900
0380	7100 thru 12,000 MBH		.16	200			8,600		8,600	13,000
0390	12,200 thru 25,000 MBH		.12	266			11,500		11,500	17,300
1000	Ductwork, 4" high, 8" wide	1 Clab	200	.040	L.F.		1.15		1.15	1.79
1100	6" high, 8" wide		165	.048			1.39		1.39	2.17
1200	10" high, 12" wide		125	.064			1.84		1.84	2.86
1300	12"-14" high, 16"-18" wide		85	.094			2.71		2.71	4.21
1400	18" high, 24" wide		67	.119			3.43		3.43	5.35
1500	30" high, 36" wide		56	.143			4.11		4.11	6.40
1540	72" wide		50	.160			4.60		4.60	7.15
3000	Mechanical equipment, light items. Unit is weight, not cooling.	Q-5	.90	17.778	Ton		725		725	1,100
3600	Heavy items	"	1.10	14.545	"		590		590	890
3700	Deduct for salvage (when applicable), minimum				Job				59	65
3710	Maximum				"				455	490

23 05 23 – General-Duty Valves for HVAC Piping

23 05 23.30 Valves, Iron Body

		Crew	Daily Output	Labor-Hours	Unit	Material	2007 Bare Costs Labor	Equipment	Total	Total Incl O&P
0010	**VALVES, IRON BODY**									
1020	Butterfly, wafer type, gear actuator, 200 lb.									
1030	2"	1 Plum	14	.571	Ea.	153	25.50		178.50	207
1040	2-1/2"	Q-1	9	1.778		129	71.50		200.50	250
1050	3"		8	2		134	80.50		214.50	269
1060	4"		5	3.200		162	129		291	370
1070	5"	Q-2	5	4.800		191	201		392	510
1080	6"	"	5	4.800		212	201		413	535
1650	Gate, 125 lb., N.R.S.									
2150	Flanged									
2200	2"	1 Plum	5	1.600	Ea.	380	71.50		451.50	530
2240	2-1/2"	Q-1	5	3.200		390	129		519	625
2260	3"		4.50	3.556		435	143		578	695
2280	4"		3	5.333		625	215		840	1,025
2300	6"	Q-2	3	8		1,075	335		1,410	1,675
3550	OS&Y, 125 lb., flanged									
3600	2"	1 Plum	5	1.600	Ea.	269	71.50		340.50	405
3660	3"	Q-1	4.50	3.556		310	143		453	560
3680	4"	"	3	5.333		445	215		660	815
3700	6"	Q-2	3	8		735	335		1,070	1,325
3900	For 175 lb, flanged, add					200%	10%			
5450	Swing check, 125 lb., threaded									
5470	1"	1 Plum	13	.615	Ea.	430	27.50		457.50	510
5500	2"	"	11	.727		490	32.50		522.50	585
5540	2-1/2"	Q-1	15	1.067		520	43		563	640

23 05 Common Work Results for HVAC

23 05 23 – General-Duty Valves for HVAC Piping

23 05 23.30 Valves, Iron Body

		Crew	Daily Output	Labor-Hours	Unit	Material	2007 Bare Costs Labor	Equipment	Total	Total Incl O&P
5550	3"	Q-1	13	1.231	Ea.	580	49.50		629.50	710
5560	4"	↓	10	1.600	↓	990	64.50		1,054.50	1,200
5950	Flanged									
6000	2"	1 Plum	5	1.600	Ea.	185	71.50		256.50	310
6040	2-1/2"	Q-1	5	3.200		225	129		354	440
6050	3"		4.50	3.556		241	143		384	480
6060	4"	↓	3	5.333		241	215		456	590
6070	6"	Q-2	3	8	↓	650	335		985	1,225

23 05 23.80 Valves, Steel

		Crew	Daily Output	Labor-Hours	Unit	Material	2007 Bare Costs Labor	Equipment	Total	Total Incl O&P
0010	**VALVES, STEEL**									
0800	Cast									
1350	Check valve, swing type, 150 lb., flanged									
1370	1"	1 Plum	10	.800	Ea.	435	36		471	535
1400	2"	"	8	1		705	45		750	845
1440	2-1/2"	Q-1	5	3.200		705	129		834	970
1450	3"		4.50	3.556		720	143		863	1,000
1460	4"	↓	3	5.333		1,025	215		1,240	1,450
1540	For 300 lb., flanged, add					50%	15%			
1548	For 600 lb., flanged, add				↓	110%	20%			
1950	Gate valve, 150 lb., flanged									
2000	2"	1 Plum	8	1	Ea.	650	45		695	785
2040	2-1/2"	Q-1	5	3.200		920	129		1,049	1,225
2050	3"		4.50	3.556		920	143		1,063	1,250
2060	4"	↓	3	5.333		1,150	215		1,365	1,575
2070	6"	Q-2	3	8		1,775	335		2,110	2,475
3650	Globe valve, 150 lb., flanged									
3700	2"	1 Plum	8	1	Ea.	820	45		865	970
3740	2-1/2"	Q-1	5	3.200		1,050	129		1,179	1,350
3750	3"		4.50	3.556		1,050	143		1,193	1,375
3760	4"	↓	3	5.333		1,525	215		1,740	2,000
3770	6"	Q-2	3	8	↓	2,400	335		2,735	3,125
5150	Forged									
5650	Check valve, class 800, horizontal, socket									
5698	Threaded									
5700	1/4"	1 Plum	24	.333	Ea.	67.50	14.95		82.45	97
5720	3/8"		24	.333		67.50	14.95		82.45	97
5730	1/2"		24	.333		67.50	14.95		82.45	97
5740	3/4"		20	.400		73.50	17.90		91.40	108
5750	1"		19	.421		85.50	18.85		104.35	123
5760	1-1/4"	↓	15	.533	↓	168	24		192	221

23 05 93 – Testing, Adjusting, and Balancing for HVAC

23 05 93.10 Balancing, Air

		Crew	Daily Output	Labor-Hours	Unit	Material	2007 Bare Costs Labor	Equipment	Total	Total Incl O&P
0010	**BALANCING, AIR**, (Subcontractor's quote incl. material and labor.)									
0900	Heating and ventilating equipment									
1000	Centrifugal fans, utility sets				Ea.				209	322.32
1100	Heating and ventilating unit								315	483.48
1200	In-line fan								315	483.48
1300	Propeller and wall fan								83	91.32
1400	Roof exhaust fan								196	214.88
2000	Air conditioning equipment, central station								635	698.36
2100	Built-up low pressure unit								587	644.64
2200	Built-up high pressure unit				↓				684	752.08

23 05 Common Work Results for HVAC

23 05 93 – Testing, Adjusting, and Balancing for HVAC

23 05 93.10 Balancing, Air		Crew	Daily Output	Labor-Hours	Unit	Material	2007 Bare Costs Labor	Equipment	Total	Total Incl O&P
2500	Multi-zone A.C. and heating unit				Ea.				440	483.48
2600	For each zone over one, add								97.77	107.44
2700	Package A.C. unit								244	268.60
2800	Rooftop heating and cooling unit								342	376.04
3000	Supply, return, exhaust, registers & diffusers, avg. height ceiling								58.65	64.46
3100	High ceiling								88	96.70
3200	Floor height								48.88	53.72

23 05 93.20 Balancing, Water

0010	**BALANCING, WATER**, (Subcontractor's quote incl. material and labor.)									
0050	Air cooled condenser				Ea.				173.26	190.40
0080	Boiler								348.50	382.97
0100	Cabinet unit heater								59.40	65.28
0200	Chiller								421	462.40
0300	Convector								49.50	54.40
0500	Cooling tower								322	353.60
0600	Fan coil unit, unit ventilator								89.10	97.92
0700	Fin tube and radiant panels								99	108.80
0800	Main and duct re-heat coils								91.60	100.64
0810	Heat exchanger								91.60	100.64
1000	Pumps								218	239.36
1100	Unit heater								69.30	76.16

23 07 HVAC Insulation

23 07 13 – Duct Insulation

23 07 13.10 Duct Insulation

		Crew	Daily Output	Labor-Hours	Unit	Material	2007 Bare Costs Labor	Equipment	Total	Total Incl O&P
0010	**DUCT INSULATION**									
0100	Rule of thumb, as a percentage of total mechanical costs				Job				10%	
0110	Insulation req'd is based on the surface size/area to be covered									
3000	Ductwork									
3020	Blanket type, fiberglass, flexible									
3030	Fire resistant liner, black coating one side									
3050	1/2" thick, 2 lb. density	Q-14	380	.042	S.F.	.47	1.57		2.04	3.01
3060	1" thick, 1-1/2 lb. density		350	.046		.62	1.71		2.33	3.38
3070	1-1/2" thick, 1-1/2 lb. density		320	.050		1.09	1.87		2.96	4.15
3080	2" thick, 1-1/2 lb. density		300	.053		1.24	1.99		3.23	4.51
3140	FRK vapor barrier wrap, .75 lb. density									
3160	1" thick	Q-14	350	.046	S.F.	.30	1.71		2.01	3.03
3170	1-1/2" thick		320	.050		.32	1.87		2.19	3.30
3180	2" thick		300	.053		.47	1.99		2.46	3.67
3190	3" thick		260	.062		.58	2.30		2.88	4.27
3200	4" thick		242	.066		.96	2.47		3.43	4.96
3210	Vinyl jacket, same as FRK									
3280	Unfaced, 1 lb. density									
3310	1" thick	Q-14	360	.044	S.F.	.40	1.66		2.06	3.06
3320	1-1/2" thick		330	.048		.55	1.81		2.36	3.47
3330	2" thick		310	.052		.62	1.93		2.55	3.73
3795	Finishes									
3800	1/2" cement over 1" wire mesh, incl. corner bead	Q-14	100	.160	S.F.	1.75	6		7.75	11.40
3820	Glass cloth, pasted on		170	.094		2.95	3.52		6.47	8.80
3900	Weatherproof, non-metallic, 2 lb. per S.F.		180	.089		2.89	3.32		6.21	8.45
9600	Minimum labor/equipment charge	1 Stpi	4	2	Job		90.50		90.50	136

23 07 HVAC Insulation

23 07 16 – HVAC Equipment Insulation

23 07 16.10 HVAC Equipment Insulation

	23 07 16.10 HVAC Equipment Insulation	Crew	Daily Output	Labor-Hours	Unit	Material	2007 Bare Costs Labor	2007 Bare Costs Equipment	Total	Total Incl O&P
0010	**HVAC EQUIPMENT INSULATION**									
0100	Rule of thumb, as a percentage of total mechanical costs				Job				10%	
0110	Insulation req'd is based on the surface size/area to be covered									
1000	Boiler, 1-1/2" calcium silicate, 1/2" cement finish	Q-14	50	.320	S.F.	3.98	11.95		15.93	23.50
1020	2" fiberglass	"	80	.200	"	2.74	7.45		10.19	14.80
2000	Breeching, 2" calcium silicate with 1/2" cement finish, no lath									
2020	Rectangular	Q-14	42	.381	S.F.	5.20	14.25		19.45	28
2040	Round	"	38.70	.413	"	5.45	15.45		20.90	30.50

23 09 Instrumentation and Control for HVAC

23 09 33 – Electric and Electronic Control System for HVAC

23 09 33.10 Electronic Control Systems

		Crew	Daily Output	Labor-Hours	Unit	Material	2007 Bare Costs Labor	2007 Bare Costs Equipment	Total	Total Incl O&P
0010	**ELECTRONIC CONTROL SYSTEMS**									
0021	For electronic costs, add to division 23 09 43.10				Ea.					15%

23 09 43 – Pneumatic Control System for HVAC

23 09 43.10 Pneumatic Control Systems

		Crew	Daily Output	Labor-Hours	Unit	Material	2007 Bare Costs Labor	2007 Bare Costs Equipment	Total	Total Incl O&P
0010	**PNEUMATIC CONTROL SYSTEMS**									
0011	Including a nominal 50 Ft. of tubing. Add control panelboard if req'd.									
0100	Heating and Ventilating, split system									
0200	Mixed air control, economizer cycle, panel readout, tubing									
0220	Up to 10 tons	Q-19	.68	35.294	Ea.	3,150	1,475		4,625	5,675
0240	For 10 to 20 tons		.63	37.915		3,375	1,575		4,950	6,075
0260	For over 20 tons		.58	41.096		3,650	1,725		5,375	6,600
0300	Heating coil, hot water, 3 way valve,									
0320	Freezestat, limit control on discharge, readout	Q-5	.69	23.088	Ea.	2,350	940		3,290	4,000
0500	Cooling coil, chilled water, room									
0520	Thermostat, 3 way valve	Q-5	2	8	Ea.	1,050	325		1,375	1,650
0600	Cooling tower, fan cycle, damper control,									
0620	Control system including water readout in/out at panel	Q-19	.67	35.821	Ea.	4,150	1,500		5,650	6,800
1000	Unit ventilator, day/night operation,									
1100	freezestat, ASHRAE, cycle 2	Q-19	.91	26.374	Ea.	2,300	1,100		3,400	4,175
2000	Compensated hot water from boiler, valve control,									
2100	readout and reset at panel, up to 60 GPM	Q-19	.55	43.956	Ea.	4,300	1,825		6,125	7,475
2120	For 120 GPM		.51	47.059		4,600	1,975		6,575	8,000
2140	For 240 GPM		.49	49.180		4,800	2,050		6,850	8,375
3000	Boiler room combustion air, damper to 5 SF, controls		1.37	17.582		2,075	735		2,810	3,375
3500	Fan coil, heating and cooling valves, 4 pipe control system		3	8		935	335		1,270	1,525
3600	Heat exchanger system controls		.86	27.907		2,000	1,175		3,175	3,975
4000	Pneumatic thermostat, including controlling room radiator valve	Q-5	2.43	6.593		625	268		893	1,100
4060	Pump control system	Q-19	3	8		960	335		1,295	1,550
4500	Air supply for pneumatic control system									
4600	Tank mounted duplex compressor, starter, alternator,									
4620	piping, dryer, PRV station and filter									
4630	1/2 HP	Q-19	.68	35.139	Ea.	7,700	1,475		9,175	10,700
4660	1-1/2 HP		.58	41.739		9,400	1,750		11,150	12,900
4690	5 HP		.42	57.143		22,400	2,375		24,775	28,200

23 12 Facility Fuel Pumps

23 12 16 – Facility Gasoline Dispensing Pumps

23 12 16.20 Fuel Dispensing Equipment	Crew	Daily Output	Labor-Hours	Unit	Material	2007 Bare Costs Labor	Equipment	Total	Total Incl O&P
0010 **FUEL DISPENSING EQUIPMENT**									
1100 Product dispenser with vapor recovery for 6 nozzles, installed, not									
1110 including piping to storage tanks				Ea.	19,800			19,800	21,800

23 13 Facility Fuel-Storage Tanks

23 13 13 – Facility Underground Fuel-Oil, Storage Tanks

23 13 13.09 Single-Wall Steel Fuel-Oil Tanks

		Crew	Daily Output	Labor-Hours	Unit	Material	Labor	Equipment	Total	Total Incl O&P
0010	**SINGLE-WALL STEEL FUEL-OIL TANKS**									
5000	Steel underground, sti-P3, set in place, not incl. hold-down bars.									
5500	Excavation, pad, pumps and piping not included									
5510	Single wall, 500 gallon capacity, 7 gauge shell	Q-5	2.70	5.926	Ea.	1,000	241		1,241	1,475
5520	1,000 gallon capacity, 7 gauge shell	"	2.50	6.400		2,300	260		2,560	2,925
5530	2,000 gallon capacity, 1/4" thick shell	Q-7	4.60	6.957		3,700	300		4,000	4,525
5535	2,500 gallon capacity, 7 gauge shell	Q-5	3	5.333		4,025	217		4,242	4,750
5540	5,000 gallon capacity, 1/4" thick shell	Q-7	3.20	10		6,300	430		6,730	7,575
5580	15,000 gallon capacity, 5/16" thick shell		1.70	18.824		13,200	810		14,010	15,800
5600	20,000 gallon capacity, 5/16" thick shell		1.50	21.333		19,800	920		20,720	23,100
5610	25,000 gallon capacity, 3/8" thick shell		1.30	24.615		22,500	1,050		23,550	26,400
5620	30,000 gallon capacity, 3/8" thick shell		1.10	29.091		31,800	1,250		33,050	36,800
5630	40,000 gallon capacity, 3/8" thick shell		.90	35.556		37,400	1,525		38,925	43,400
5640	50,000 gallon capacity, 3/8" thick shell		.80	40		46,800	1,725		48,525	54,000

23 13 13.23 Glass-Fbr-Reinfcd-Plas., Undergrnd Fuel-Oil, Stor.

		Crew	Daily Output	Labor-Hours	Unit	Material	Labor	Equipment	Total	Total Incl O&P
0010	**GLASS-FIBER-REINFCD-PLASTIC, UNDERGRND FUEL-OIL, STORAGE**									
0210	Fiberglass, underground, single wall, U.L. listed, not including									
0220	manway or hold-down strap									
0240	2,000 gallon capacity	Q-7	4.57	7.002	Ea.	4,925	300		5,225	5,875
0250	4,000 gallon capacity		3.55	9.014		7,025	390		7,415	8,300
0260	6,000 gallon capacity		2.67	11.985		7,775	515		8,290	9,350
0280	10,000 gallon capacity		2	16		10,500	690		11,190	12,500
0284	15,000 gallon capacity		1.68	19.048		16,400	820		17,220	19,300
0290	20,000 gallon capacity		1.45	22.069		20,900	950		21,850	24,400
0500	For manway, fittings and hold-downs, add					20%	15%			
1020	Fiberglass, underground, double wall, U.L. listed									
1030	includes manways, not incl. hold-down straps									
1040	600 gallon capacity	Q-5	2.42	6.612	Ea.	5,600	269		5,869	6,550
1050	1,000 gallon capacity	"	2.25	7.111		7,375	289		7,664	8,525
1060	2,500 gallon capacity	Q-7	4.16	7.692		10,100	330		10,430	11,600
1070	3,000 gallon capacity		3.90	8.205		10,900	355		11,255	12,500
1080	4,000 gallon capacity		3.64	8.791		12,900	380		13,280	14,800
1090	6,000 gallon capacity		2.42	13.223		15,200	570		15,770	17,700
1100	8,000 gallon capacity CN		2.08	15.385		17,400	660		18,060	20,100
1110	10,000 gallon capacity		1.82	17.582		19,400	755		20,155	22,500
1120	12,000 gallon capacity		1.70	18.824		23,800	810		24,610	27,400
2210	Fiberglass, underground, single wall, U.L. listed, including									
2220	hold-down straps, no manways									
2240	2,000 gallon capacity	Q-7	3.55	9.014	Ea.	5,250	390		5,640	6,350
2250	4,000 gallon capacity		2.90	11.034		7,325	475		7,800	8,775
2260	6,000 gallon capacity		2	16		8,400	690		9,090	10,300
2280	10,000 gallon capacity		1.60	20		11,100	860		11,960	13,500
2284	15,000 gallon capacity		1.39	23.022		17,000	990		17,990	20,200
2290	20,000 gallon capacity		1.14	28.070		21,900	1,200		23,100	25,800

23 13 Facility Fuel-Storage Tanks

23 13 13 – Facility Underground Fuel-Oil, Storage Tanks

23 13 13.23 Glass-Fbr-Reinfcd-Plas., Undergrnd Fuel-Oil, Stor.

		Crew	Daily Output	Labor-Hours	Unit	Material	2007 Bare Costs Labor	2007 Bare Costs Equipment	Total	Total Incl O&P
3020	Fiberglass, underground, double wall, U.L. listed									
3030	includes manways and hold-down straps									
3040	600 gallon capacity	Q-5	1.86	8.602	Ea.	5,900	350		6,250	7,025
3050	1,000 gallon capacity	"	1.70	9.412		7,675	385		8,060	9,025
3060	2,500 gallon capacity	Q-7	3.29	9.726		10,400	420		10,820	12,100
3070	3,000 gallon capacity		3.13	10.224		11,200	440		11,640	13,100
3080	4,000 gallon capacity		2.93	10.922		13,200	470		13,670	15,300
3090	6,000 gallon capacity		1.86	17.204		15,900	740		16,640	18,500
3100	8,000 gallon capacity		1.65	19.394		18,000	835		18,835	21,100
3110	10,000 gallon capacity		1.48	21.622		20,000	930		20,930	23,400
3120	12,000 gallon capacity		1.40	22.857		24,500	985		25,485	28,400

23 13 23 – Facility Aboveground Fuel-Oil, Storage Tanks

23 13 23.13 Vertical, Steel, Aboveground Fuel-Oil, Storage Tanks

		Crew	Daily Output	Labor-Hours	Unit	Material	2007 Bare Costs Labor	2007 Bare Costs Equipment	Total	Total Incl O&P
0010	**VERTICAL, STEEL, ABOVEGROUND FUEL-OIL, STORAGE TANKS**									
4000	Fixed roof oil storage tanks, steel, (1 BBL=42 GAL w/ fdn 3'D x 1'W)									
4200	5,000 barrels				Ea.				157,273	173,000
4300	24,000 barrels								269,091	296,000
4500	56,000 barrels								535,000	588,500
4600	110,000 barrels								777,273	855,000
4800	143,000 barrels								996,818	1,096,500
4900	224,000 barrels								1,495,909	1,645,500
5100	Floating roof gasoline tanks, steel, 5,000 barrels								172,727	190,000
5200	25,000 barrels								295,455	325,000
5400	55,000 barrels								590,909	650,000
5500	100,000 barrels								854,545	940,000
5700	150,000 barrels								1,095,455	1,205,000
5800	225,000 barrels								1,645,455	1,810,000

23 13 23.16 Horizontal, Steel, Aboveground Fuel-Oil, Storage Tanks

		Crew	Daily Output	Labor-Hours	Unit	Material	2007 Bare Costs Labor	2007 Bare Costs Equipment	Total	Total Incl O&P
0010	**HORIZONTAL, STEEL, ABOVEGROUND FUEL-OIL, STORAGE TANKS**									
3000	Steel, storage, above ground, including cradles, coating,									
3020	fittings, not including fdn, pumps or piping									
3040	Single wall, interior, 275 gallon	Q-5	5	3.200	Ea.	330	130		460	560
3060	550 gallon	"	2.70	5.926		1,575	241		1,816	2,100
3080	1,000 gallon	Q-7	5	6.400		2,525	276		2,801	3,225
3100	1,500 gallon		4.75	6.737		3,750	290		4,040	4,550
3120	2,000 gallon		4.60	6.957		4,300	300		4,600	5,200
3140	5,000 gallon		3.20	10		5,825	430		6,255	7,050
3150	10,000 gallon		2	16		16,800	690		17,490	19,400
3160	15,000 gallon		1.70	18.824		19,500	810		20,310	22,600
3170	20,000 gallon		1.45	22.069		24,400	950		25,350	28,300
3180	25,000 gallon		1.30	24.615		27,300	1,050		28,350	31,600
3190	30,000 gallon		1.10	29.091		33,500	1,250		34,750	38,700
3320	Double wall, 500 gallon capacity	Q-5	2.40	6.667		2,575	271		2,846	3,250
3330	2000 gallon capacity	Q-7	4.15	7.711		6,450	330		6,780	7,600
3340	4000 gallon capacity		3.60	8.889		11,500	385		11,885	13,200
3350	6000 gallon capacity		2.40	13.333		13,600	575		14,175	15,800
3360	8000 gallon capacity		2	16		17,400	690		18,090	20,100
3370	10000 gallon capacity		1.80	17.778		19,200	765		19,965	22,300
3380	15000 gallon capacity		1.50	21.333		29,100	920		30,020	33,500
3390	20000 gallon capacity		1.30	24.615		33,200	1,050		34,250	38,100
3400	25000 gallon capacity		1.15	27.826		40,300	1,200		41,500	46,200
3410	30000 gallon capacity		1	32		44,200	1,375		45,575	50,500

23 21 Hydronic Piping and Pumps

23 21 20 – Hydronic HVAC Piping Specialties

23 21 20.10 Air Control

		Crew	Daily Output	Labor-Hours	Unit	Material	2007 Bare Costs Labor	2007 Bare Costs Equipment	Total	Total Incl O&P
0010	**AIR CONTROL**									
0030	Air separator, with strainer									
0040	2" diameter	Q-5	6	2.667	Ea.	750	108		858	990
0080	2-1/2" diameter		5	3.200		855	130		985	1,125
0100	3" diameter		4	4		1,300	163		1,463	1,675
0120	4" diameter		3	5.333		1,875	217		2,092	2,400
0130	5" diameter	Q-6	3.60	6.667		2,400	281		2,681	3,050
0140	6" diameter	"	3.40	7.059		2,850	298		3,148	3,600

23 21 20.18 Automatic Air Vent

		Crew	Daily Output	Labor-Hours	Unit	Material	Labor	Equipment	Total	Total Incl O&P
0010	**AUTOMATIC AIR VENT**									
0020	Cast iron body, stainless steel internals, float type									
0060	1/2" NPT inlet, 300 psi	1 Stpi	12	.667	Ea.	80.50	30		110.50	134
0220	3/4" NPT inlet, 250 psi	"	10	.800		258	36		294	340
0340	1-1/2" NPT inlet, 250 psi	Q-5	12	1.333		800	54		854	960

23 21 20.42 Expansion Joints

		Crew	Daily Output	Labor-Hours	Unit	Material	Labor	Equipment	Total	Total Incl O&P
0010	**EXPANSION JOINTS**									
0100	Bellows type, neoprene cover, flanged spool									
0140	6" face to face, 1-1/4" diameter	1 Stpi	11	.727	Ea.	231	33		264	305
0160	1-1/2" diameter	"	10.60	.755		231	34		265	305
0180	2" diameter	Q-5	13.30	1.203		235	49		284	335
0190	2-1/2" diameter		12.40	1.290		244	52.50		296.50	350
0200	3" diameter		11.40	1.404		277	57		334	390
0480	10" face to face, 2" diameter		13	1.231		345	50		395	455
0500	2-1/2" diameter		12	1.333		365	54		419	480
0520	3" diameter		11	1.455		370	59		429	495
0540	4" diameter		8	2		415	81.50		496.50	575
0560	5" diameter		7	2.286		500	93		593	690
0580	6" diameter		6	2.667		515	108		623	730

23 21 20.46 Expansion Tanks

		Crew	Daily Output	Labor-Hours	Unit	Material	Labor	Equipment	Total	Total Incl O&P
0010	**EXPANSION TANKS**									
1507	Fiberglass and steel single / double wall storage, see Div 23 13 23.19									
1512	Tank leak detection systems, see Div 28 33 33.50									
2000	Steel, liquid expansion, ASME, painted, 15 gallon capacity	Q-5	17	.941	Ea.	410	38.50		448.50	515
2020	24 gallon capacity		14	1.143		440	46.50		486.50	555
2040	30 gallon capacity		12	1.333		460	54		514	585
2060	40 gallon capacity		10	1.600		540	65		605	690
2080	60 gallon capacity		8	2		645	81.50		726.50	830
2100	80 gallon capacity		7	2.286		690	93		783	900
2120	100 gallon capacity		6	2.667		935	108		1,043	1,200
3000	Steel ASME expansion, rubber diaphragm, 19 gal. cap. accept.		12	1.333		1,950	54		2,004	2,200
3020	31 gallon capacity		8	2		2,150	81.50		2,231.50	2,500
3040	61 gallon capacity		6	2.667		3,025	108		3,133	3,500
3080	119 gallon capacity		4	4		3,275	163		3,438	3,850
3100	158 gallon capacity		3.80	4.211		4,550	171		4,721	5,250
3140	317 gallon capacity		2.80	5.714		6,850	232		7,082	7,875
3180	528 gallon capacity		2.40	6.667		11,100	271		11,371	12,600
5950										

23 21 20.58 Hydronic Heating Control Valves

		Crew	Daily Output	Labor-Hours	Unit	Material	Labor	Equipment	Total	Total Incl O&P
0010	**HYDRONIC HEATING CONTROL VALVES**									
0100	Radiator supply, 1/2" diameter	1 Stpi	24	.333	Ea.	65	15.05		80.05	94
0120	3/4" diameter		20	.400		66	18.10		84.10	99.50
0140	1" diameter		19	.421		76.50	19.05		95.55	113

23 21 Hydronic Piping and Pumps

23 21 20 – Hydronic HVAC Piping Specialties

23 21 20.58 Hydronic Heating Control Valves

		Crew	Daily Output	Labor-Hours	Unit	Material	2007 Bare Costs Labor	2007 Bare Costs Equipment	Total	Total Incl O&P
0160	1-1/4" diameter	1 Stpi	15	.533	Ea.	98	24		122	145
0500	For low pressure steam, add					25%				

23 21 20.70 Steam Traps

		Crew	Daily Output	Labor-Hours	Unit	Material	Labor	Equipment	Total	Total Incl O&P
0010	**STEAM TRAPS**									
0030	Cast iron body, threaded									
0040	Inverted bucket									
0050	1/2" pipe size	1 Stpi	12	.667	Ea.	117	30		147	175
0070	3/4" pipe size		10	.800		204	36		240	279
0100	1" pipe size		9	.889		315	40		355	405
0120	1-1/4" pipe size		8	1		475	45		520	590
1000	Float & thermostatic, 15 psi									
1010	3/4" pipe size	1 Stpi	16	.500	Ea.	95	22.50		117.50	138
1020	1" pipe size		15	.533		114	24		138	162
1040	1-1/2" pipe size		9	.889		201	40		241	282
1060	2" pipe size		6	1.333		370	60.50		430.50	495

23 21 20.76 Strainers, Y Type, Bronze Body

		Crew	Daily Output	Labor-Hours	Unit	Material	Labor	Equipment	Total	Total Incl O&P
0010	**STRAINERS, Y TYPE, BRONZE BODY**									
0050	Screwed, 150 lb., 1/4" pipe size	1 Stpi	24	.333	Ea.	12.80	15.05		27.85	36.50
0070	3/8" pipe size		24	.333		16.85	15.05		31.90	41
0100	1/2" pipe size		20	.400		16.85	18.10		34.95	45.50
0140	1" pipe size		17	.471		19.85	21.50		41.35	54
0160	1-1/2" pipe size		14	.571		42.50	26		68.50	86
0180	2" pipe size		13	.615		57	28		85	105
0182	3" pipe size		12	.667		405	30		435	490
0200	300 lb., 2-1/2" pipe size	Q-5	17	.941		305	38.50		343.50	395
0220	3" pipe size		16	1		600	40.50		640.50	720
0240	4" pipe size		15	1.067		1,375	43.50		1,418.50	1,575
0500	For 300 lb rating 1/4" thru 2", add					15%				
1000	Flanged, 150 lb., 1-1/2" pipe size	1 Stpi	11	.727	Ea.	315	33		348	395
1020	2" pipe size	"	8	1		380	45		425	485
1030	2-1/2" pipe size	Q-5	5	3.200		560	130		690	815
1040	3" pipe size		4.50	3.556		695	145		840	980
1060	4" pipe size		3	5.333		1,050	217		1,267	1,475
1100	6" pipe size	Q-6	3	8		2,000	335		2,335	2,700
1106	8" pipe size	"	2.60	9.231		2,200	390		2,590	3,000
1500	For 300 lb rating, add					40%				

23 21 20.78 Strainers, Y Type, Iron Body

		Crew	Daily Output	Labor-Hours	Unit	Material	Labor	Equipment	Total	Total Incl O&P
0010	**STRAINERS, Y TYPE, IRON BODY**									
0050	Screwed, 250 lb., 1/4" pipe size	1 Stpi	20	.400	Ea.	7.05	18.10		25.15	35
0070	3/8" pipe size		20	.400		7.05	18.10		25.15	35
0100	1/2" pipe size		20	.400		7.05	18.10		25.15	35
0140	1" pipe size		16	.500		11.40	22.50		33.90	46.50
0160	1-1/2" pipe size		12	.667		18.80	30		48.80	66
0180	2" pipe size		8	1		29	45		74	100
0220	3" pipe size	Q-5	11	1.455		160	59		219	264
0240	4" pipe size	"	5	3.200		270	130		400	495
0500	For galvanized body, add					50%				
1000	Flanged, 125 lb., 1-1/2" pipe size	1 Stpi	11	.727	Ea.	91	33		124	150
1020	2" pipe size	"	8	1		68	45		113	143
1040	3" pipe size	Q-5	4.50	3.556		90	145		235	315
1060	4" pipe size	"	3	5.333		164	217		381	505
1080	5" pipe size	Q-6	3.40	7.059		256	298		554	730

23 21 Hydronic Piping and Pumps

23 21 20 – Hydronic HVAC Piping Specialties

23 21 20.78 Strainers, Y Type, Iron Body

		Crew	Daily Output	Labor-Hours	Unit	Material	2007 Bare Costs Labor	Equipment	Total	Total Incl O&P
1100	6" pipe size	Q-6	3	8	Ea.	315	335		650	855
1500	For 250 lb rating, add					20%				
2000	For galvanized body, add					50%				
2500	For steel body, add					40%				

23 21 20.88 Venturi Flow

		Crew	Daily Output	Labor-Hours	Unit	Material	Labor	Equipment	Total	Total Incl O&P
0010	**VENTURI FLOW**, Measuring device									
0050	1/2" diameter	1 Stpi	24	.333	Ea.	217	15.05		232.05	262
0120	1" diameter		19	.421		214	19.05		233.05	264
0140	1-1/4" diameter		15	.533		266	24		290	330
0160	1-1/2" diameter		13	.615		278	28		306	345
0180	2" diameter		11	.727		284	33		317	360
0220	3" diameter	Q-5	14	1.143		405	46.50		451.50	515
0240	4" diameter	"	11	1.455		600	59		659	750
0280	6" diameter	Q-6	3.50	6.857		880	289		1,169	1,400
0500	For meter, add					1,600			1,600	1,750

23 21 23 – Hydronic Pumps

23 21 23.13 In-Line Centrifugal Hydronic Pumps

		Crew	Daily Output	Labor-Hours	Unit	Material	Labor	Equipment	Total	Total Incl O&P
0010	**IN-LINE CENTRIFUGAL HYDRONIC PUMPS**									
0600	Bronze, sweat connections, 1/40 HP, in line									
0640	3/4" size	Q-1	16	1	Ea.	151	40.50		191.50	227
1000	Flange connection, 3/4" to 1-1/2" size									
1040	1/12 HP	Q-1	6	2.667	Ea.	410	108		518	615
1060	1/8 HP		6	2.667		710	108		818	940
1100	1/3 HP		6	2.667		795	108		903	1,025
1140	2" size, 1/6 HP		5	3.200		1,025	129		1,154	1,325
1180	2-1/2" size, 1/4 HP		5	3.200		1,325	129		1,454	1,650
2000	Cast iron, flange connection									
2040	3/4" to 1-1/2" size, in line, 1/12 HP	Q-1	6	2.667	Ea.	269	108		377	460
2100	1/3 HP		6	2.667		500	108		608	710
2140	2" size, 1/6 HP		5	3.200		550	129		679	800
2180	2-1/2" size, 1/4 HP		5	3.200		720	129		849	990
2220	3" size, 1/4 HP		4	4		735	161		896	1,050
2600	For non-ferrous impeller, add					3%				

23 22 Steam and Condensate Piping and Pumps

23 22 13 – Steam and Condensate Heating Piping

23 22 13.23 Aboveground Steam and Condensate Piping

		Crew	Daily Output	Labor-Hours	Unit	Material	Labor	Equipment	Total	Total Incl O&P
0010	**ABOVEGROUND STEAM AND CONDENSATE HEATING PIPING**									
0100	500 lb. per hour	1 Stpi	14	.571	Ea.	2,300	26		2,326	2,575
0140	1500 lb. per hour	"	7	1.143	"	2,475	51.50		2,526.50	2,800

23 22 23 – Steam Condensate Pumps

23 22 23.10 Condensate Return System

		Crew	Daily Output	Labor-Hours	Unit	Material	Labor	Equipment	Total	Total Incl O&P
0010	**CONDENSATE RETURN SYSTEM**									
0200	Simplex, 3/4 H.P. mtr, float switch, controls, 10 Gal. C.I. rcvr, 6-15GPM	Q-1	1	16	Ea.	2,175	645		2,820	3,350
1000	Duplex, 2 pumps, 3/4 H.P. motors, float switch,									
1060	alternator asssembly, 15 Gal. C.I. receiver	Q-1	.50	32	Ea.	3,425	1,300		4,725	5,725

23 31 HVAC Ducts and Casings

23 31 13 – Metal Ducts

23 31 13.13 Rectangular Metal Ducts

		Crew	Daily Output	Labor-Hours	Unit	Material	2007 Bare Costs Labor	2007 Bare Costs Equipment	Total	Total Incl O&P
0010	**RECTANGULAR METAL DUCTS** R233100-20									
0020	Fabricated rectangular, includes fittings, joints, supports,									
0030	allowance for flexible connections, no insulation									
0031	NOTE: Fabrication and installation are combined									
0040	as LABOR cost. Approx. 25% fittings assumed.									
0050	Add to labor for elevated installation									
0051	of fabricated ductwork									
0052	10' to 15' high						6%			
0053	15' to 20' high						12%			
0054	20' to 25' high						15%			
0055	25' to 30' high						21%			
0056	30' to 35' high						24%			
0057	35' to 40' high						30%			
0058	Over 40' high						33%			
0072	For duct insulation and lining see 23 07 13.10 3000									
0100	Aluminum, alloy 3003-H14, under 100 lb.	Q-10	75	.320	Lb.	3.37	13		16.37	23.50
0110	100 to 500 lb.		80	.300		2.25	12.20		14.45	21.50
0120	500 to 1,000 lb.		95	.253		2.13	10.25		12.38	18.20
0140	1,000 to 2,000 lb.		120	.200		2.05	8.15		10.20	14.80
0150	2,000 to 5,000 lb. CN		130	.185		1.98	7.50		9.48	13.75
0160	Over 5,000 lb.		145	.166		1.98	6.75		8.73	12.55
0500	Galvanized steel, under 200 lb.		235	.102		.95	4.15		5.10	7.45
0520	200 to 500 lb.		245	.098		.81	3.98		4.79	7.05
0540	500 to 1,000 lb.		255	.094		.78	3.83		4.61	6.75
0560	1,000 to 2,000 lb.		265	.091		.73	3.68		4.41	6.50
0570	2,000 to 5,000 lb.		275	.087		.69	3.55		4.24	6.20
0580	Over 5,000 lb. CN		285	.084		.69	3.42		4.11	6.05
1000	Stainless steel, type 304, under 100 lb.		165	.145		3	5.90		8.90	12.40
1020	100 to 500 lb.		175	.137		2.61	5.55		8.16	11.45
1030	500 to 1,000 lb.		190	.126		2.52	5.15		7.67	10.65
1040	1,000 to 2,000 lb.		200	.120		2.34	4.88		7.22	10.05
1050	2,000 to 5,000 lb.		225	.107		2.25	4.34		6.59	9.20
1060	Over 5,000 lb.		235	.102		2.16	4.15		6.31	8.80
1100	For medium pressure ductwork, add						15%			
1200	For high pressure ductwork, add						40%			
1210	For welded ductwork, add						85%			
1220	For 30% fittings, add						11%			
1224	For 40% fittings, add						34%			
1228	For 50% fittings, add						56%			
1232	For 60% fittings, add						79%			
1236	For 70% fittings, add						101%			
1240	For 80% fittings, add						124%			
1244	For 90% fittings, add						147%			
1248	For 100% fittings, add						169%			

23 31 13.19 Metal Duct Fittings

		Crew	Daily Output	Labor-Hours	Unit	Material	Labor	Equipment	Total	Total Incl O&P
0010	**METAL DUCT FITTINGS**									
0050	Air extractors, 12" x 4"	1 Shee	24	.333	Ea.	16.50	14.50		31	40.50
0100	8" x 6"		22	.364		16.50	15.85		32.35	42.50
0200	20" x 8"		16	.500		37.50	22		59.50	74.50
0280	24" x 12"		10	.800		50.50	35		85.50	109
2000	Fabrics for flexible connections, with metal edge		100	.080	L.F.	3.33	3.48		6.81	9
2100	Without metal edge		160	.050	"	1.84	2.18		4.02	5.40

23 31 HVAC Ducts and Casings

23 31 16 – Nonmetal Ducts

23 31 16.13 Fibrous-Glass Ducts		Crew	Daily Output	Labor-Hours	Unit	Material	2007 Bare Costs Labor	2007 Bare Costs Equipment	Total	Total Incl O&P
0010	**FIBROUS-GLASS DUCTS** R233100-20									
3490	Rigid fiberglass duct board, foil reinf. kraft facing									
3500	Rectangular, 1" thick, alum. faced, (FRK), std. weight	Q-10	350	.069	SF Surf	.62	2.79		3.41	4.98

23 33 Air Duct Accessories

23 33 13 – Dampers

23 33 13.13 Volume-Control Dampers

		Crew	Daily Output	Labor-Hours	Unit	Material	Labor	Equipment	Total	Total Incl O&P
0010	**VOLUME-CONTROL DAMPERS**									
5990	Multi-blade dampers, opposed blade, 8" x 6"	1 Shee	24	.333	Ea.	20	14.50		34.50	44.50
5994	8" x 8"		22	.364		21	15.85		36.85	47.50
5996	10" x 10"		21	.381		24	16.60		40.60	52
6000	12" x 12"		21	.381		27	16.60		43.60	55.50
6020	12" x 18"		18	.444		37	19.35		56.35	70.50
6030	14" x 10"		20	.400		26.50	17.40		43.90	56
6031	14" x 14"		17	.471		33	20.50		53.50	67.50
6033	16" x 12"		17	.471		33	20.50		53.50	67.50
6035	16" x 16"		16	.500		41	22		63	78.50
6037	18" x 16"		15	.533		45	23		68	85.50
6038	18" x 18"		15	.533		49	23		72	89.50
6070	20" x 16"		14	.571		49	25		74	92
6072	20" x 20"		13	.615		58.50	27		85.50	106
6074	22" x 18"		14	.571		58.50	25		83.50	103
6076	24" x 16"		11	.727		57.50	31.50		89	113
6078	24" x 20"		8	1		68	43.50		111.50	142
6080	24" x 24"		8	1		79	43.50		122.50	154
6110	26" x 26"		6	1.333		85.50	58		143.50	184
6133	30" x 30"	Q-9	6.60	2.424		124	95		219	283
6135	32" x 32"		6.40	2.500		144	98		242	310
6180	48" x 36"		5.60	2.857		236	112		348	435
8000	Multi-blade dampers, parallel blade									
8100	8" x 8"	1 Shee	24	.333	Ea.	59	14.50		73.50	87.50
8140	16" x 10"		20	.400		78	17.40		95.40	113
8200	24" x 16"		11	.727		100	31.50		131.50	159
8260	30" x 18"		7	1.143		127	50		177	217

23 33 13.16 Fire Dampers

		Crew	Daily Output	Labor-Hours	Unit	Material	Labor	Equipment	Total	Total Incl O&P
0010	**FIRE DAMPERS**									
3000	Fire damper, curtain type, 1-1/2 hr rated, vertical, 6" x 6"	1 Shee	24	.333	Ea.	14.75	14.50		29.25	39
3020	8" x 6"		22	.364		15.10	15.85		30.95	41
3240	16" x 14"		18	.444		32.50	19.35		51.85	66
3400	24" x 20"		8	1		45.50	43.50		89	117

23 33 13.28 Splitter Damper Assembly

		Crew	Daily Output	Labor-Hours	Unit	Material	Labor	Equipment	Total	Total Incl O&P
0009	**SPLITTER DAMPER ASSEMBLY**									
0010	Self locking, 1' rod	1 Shee	24	.333	Ea.	17.40	14.50		31.90	41.50
7020	3' rod		22	.364		22.50	15.85		38.35	49.50
7040	4' rod		20	.400		25	17.40		42.40	54.50
7060	6' rod		18	.444		30.50	19.35		49.85	63.50

23 33 19 – Duct Silencers

23 33 19.10 Duct Silencers

		Crew	Daily Output	Labor-Hours	Unit	Material	Labor	Equipment	Total	Total Incl O&P
0009	**DUCT SILENCERS**									

23 33 Air Duct Accessories

23 33 19 − Duct Silencers

23 33 19.10 Duct Silencers

		Crew	Daily Output	Labor-Hours	Unit	Material	2007 Bare Costs Labor	Equipment	Total	Total Incl O&P
0010	Silencers, noise control for air flow, duct				MCFM	49			49	54

23 33 33 − Duct-Mounting Access Doors

23 33 33.13 Duct Access Doors

		Crew	Daily Output	Labor-Hours	Unit	Material	Labor	Equipment	Total	Total Incl O&P
0010	**DUCT ACCESS DOORS**									
1000	Duct access door, insulated, 6" x 6"	1 Shee	14	.571	Ea.	12.90	25		37.90	52.50
1020	10" x 10"		11	.727		19.80	31.50		51.30	71
1040	12" x 12"		10	.800		21	35		56	77
1050	12" x 18"		9	.889		33	38.50		71.50	95.50
1070	18" x 18"		8	1		37.50	43.50		81	109
1074	24" x 18"		8	1		45	43.50		88.50	117

23 33 46 − Flexible Ducts

23 33 46.10 Flexible Ducts

		Crew	Daily Output	Labor-Hours	Unit	Material	Labor	Equipment	Total	Total Incl O&P
0010	**FLEXIBLE DUCTS** R233100-20									
1280	Add to labor for elevated installation									
1282	of prefabricated (purchased) ductwork									
1283	10' to 15' high						10%			
1284	15' to 20' high						20%			
1285	20' to 25' high						25%			
1286	25' to 30' high						35%			
1287	30' to 35' high						40%			
1288	35' to 40' high						50%			
1289	Over 40' high						55%			
1300	Flexible, coated fiberglass fabric on corr. resist. metal helix									
1400	pressure to 12" (WG) UL-181									
1500	Non-insulated, 3" diameter	Q-9	400	.040	L.F.	1.08	1.57		2.65	3.61
1540	5" diameter		320	.050		1.35	1.96		3.31	4.51
1560	6" diameter		280	.057		1.60	2.24		3.84	5.20
1580	7" diameter		240	.067		1.90	2.61		4.51	6.10
1600	8" diameter		200	.080		2.20	3.14		5.34	7.25
1640	10" diameter		160	.100		2.76	3.92		6.68	9.10
1660	12" diameter		120	.133		3.34	5.25		8.59	11.70
1900	Insulated, 1" thick, PE jacket, 3" diameter		380	.042		2.22	1.65		3.87	4.99
1910	4" diameter		340	.047		2.22	1.84		4.06	5.30
1920	5" diameter		300	.053		2.39	2.09		4.48	5.85
1940	6" diameter		260	.062		2.54	2.41		4.95	6.50
1960	7" diameter		220	.073		2.95	2.85		5.80	7.65
1980	8" diameter		180	.089		3.15	3.48		6.63	8.80
2020	10" diameter		140	.114		3.71	4.48		8.19	11
2040	12" diameter		100	.160		4.62	6.25		10.87	14.75

23 33 53 − Duct Liners

23 33 53.10 Duct Liners

		Crew	Daily Output	Labor-Hours	Unit	Material	Labor	Equipment	Total	Total Incl O&P
0010	**DUCT LINERS**									
3490	Board type, fiberglass liner, 3 lb. density									
3500	Fire resistant, black pigmented, 1 side									
3520	1" thick	Q-14	150	.107	S.F.	1.77	3.98		5.75	8.25
3540	1-1/2" thick		130	.123		2.17	4.60		6.77	9.65
3560	2" thick		120	.133		2.61	4.98		7.59	10.70
3600	FSK vapor barrier									
3620	1" thick	Q-14	150	.107	S.F.	1.94	3.98		5.92	8.45
3630	1-1/2" thick		130	.123		2.50	4.60		7.10	10
3640	2" thick		120	.133		3.04	4.98		8.02	11.20

23 33 Air Duct Accessories

23 33 53 – Duct Liners

23 33 53.10 Duct Liners	Crew	Daily Output	Labor-Hours	Unit	Material	2007 Bare Costs Labor	Equipment	Total	Total Incl O&P
3680 No finish									
3700 1" thick	Q-14	170	.094	S.F.	1.03	3.52		4.55	6.70
3710 1-1/2" thick	↓	140	.114	↓	1.54	4.27		5.81	8.45
3720 2" thick		130	.123		1.94	4.60		6.54	9.40

23 34 HVAC Fans

23 34 13 – Axial HVAC Fans

23 34 13.10 Axial HVAC Fans

	Crew	Daily Output	Labor-Hours	Unit	Material	2007 Bare Costs Labor	Equipment	Total	Total Incl O&P
0010 **AXIAL HVAC FANS** R233400-10									
0020 Air conditioning and process air handling									
0030 Axial flow, compact, low sound, 2.5" S.P.									
0050 3,800 CFM, 5 HP	Q-20	3.40	5.882	Ea.	4,025	236		4,261	4,800
0080 6,400 CFM, 5 HP		2.80	7.143		4,500	287		4,787	5,400
0100 10,500 CFM, 7-1/2 HP		2.40	8.333		5,625	335		5,960	6,675
0120 15,600 CFM, 10 HP		1.60	12.500		7,075	500		7,575	8,550
1500 Vaneaxial, low pressure, 2000 CFM, 1/2 HP		3.60	5.556		1,375	223		1,598	1,875
1520 4,000 CFM, 1 HP		3.20	6.250		1,575	251		1,826	2,125
1540 8,000 CFM, 2 HP	↓	2.80	7.143	↓	2,050	287		2,337	2,700

23 34 14 – Blower HVAC Fans

23 34 14.10 Blower HVAC Fans

	Crew	Daily Output	Labor-Hours	Unit	Material	2007 Bare Costs Labor	Equipment	Total	Total Incl O&P
0010 **BLOWER HVAC FANS**									
2500 Ceiling fan, right angle, extra quiet, 0.10" S.P.									
2520 95 CFM	Q-20	20	1	Ea.	173	40		213	252
2540 210 CFM		19	1.053		204	42.50		246.50	289
2560 385 CFM		18	1.111		259	44.50		303.50	355
2580 885 CFM		16	1.250		510	50		560	635
2600 1,650 CFM		13	1.538		705	62		767	870
2620 2,960 CFM	↓	11	1.818		940	73		1,013	1,125
2640 For wall or roof cap, add	1 Shee	16	.500		173	22		195	224
2660 For straight thru fan, add					10%				
2680 For speed control switch, add	1 Elec	16	.500	↓	94	22		116	137
7500 Utility set, steel construction, pedestal, 1/4" S.P.									
7520 Direct drive, 150 CFM, 1/8 HP	Q-20	6.40	3.125	Ea.	680	125		805	935
7540 485 CFM, 1/6 HP		5.80	3.448		855	138		993	1,150
7560 1950 CFM, 1/2 HP		4.80	4.167		1,000	167		1,167	1,350
7580 2410 CFM, 3/4 HP		4.40	4.545		1,850	182		2,032	2,300
7600 3328 CFM, 1-1/2 HP	↓	3	6.667	↓	2,050	268		2,318	2,650
7680 V-belt drive, drive cover, 3 phase									
7700 800 CFM, 1/4 HP	Q-20	6	3.333	Ea.	560	134		694	820
7720 1,300 CFM, 1/3 HP		5	4		585	161		746	890
7740 2,000 CFM, 1 HP		4.60	4.348		695	175		870	1,025
7760 2,900 CFM, 3/4 HP	↓	4.20	4.762	↓	935	191		1,126	1,325

23 34 16 – Centrifugal HVAC Fans

23 34 16.10 Centrifugal HVAC Fans

	Crew	Daily Output	Labor-Hours	Unit	Material	2007 Bare Costs Labor	Equipment	Total	Total Incl O&P
0010 **CENTRIFUGAL HVAC FANS**									
0200 In-line centrifugal, supply/exhaust booster									
0220 aluminum wheel/hub, disconnect switch, 1/4" S.P.									
0240 500 CFM, 10" diameter connection	Q-20	3	6.667	Ea.	985	268		1,253	1,475
0260 1,380 CFM, 12" diameter connection		2	10		1,050	400		1,450	1,775
0280 1,520 CFM, 16" diameter connection	↓	2	10	↓	1,150	400		1,550	1,875

23 34 HVAC Fans

23 34 16 – Centrifugal HVAC Fans

23 34 16.10 Centrifugal HVAC Fans		Crew	Daily Output	Labor-Hours	Unit	Material	2007 Bare Costs Labor	Equipment	Total	Total Incl O&P
0300	2,560 CFM, 18" diameter connection	Q-20	1	20	Ea.	1,225	805		2,030	2,575
0320	3,480 CFM, 20" diameter connection		.80	25		1,475	1,000		2,475	3,150
0326	5,080 CFM, 20" diameter connection	▼	.75	26.667	▼	1,600	1,075		2,675	3,400
3500	Centrifugal, airfoil, motor and drive, complete									
3520	1000 CFM, 1/2 HP	Q-20	2.50	8	Ea.	1,175	320		1,495	1,775
3540	2,000 CFM, 1 HP		2	10		1,300	400		1,700	2,050
3560	4,000 CFM, 3 HP		1.80	11.111		1,575	445		2,020	2,400
3580	8,000 CFM, 7-1/2 HP		1.40	14.286		2,475	575		3,050	3,600
3600	12,000 CFM, 10 HP	▼	1	20	▼	3,025	805		3,830	4,575
4500	Corrosive fume resistant, plastic									
4600	roof ventilators, centrifugal, V belt drive, motor									
4620	1/4" S.P., 250 CFM, 1/4 HP	Q-20	6	3.333	Ea.	2,525	134		2,659	3,000
4640	895 CFM, 1/3 HP		5	4		2,750	161		2,911	3,275
4660	1630 CFM, 1/2 HP		4	5		3,250	201		3,451	3,875
4680	2240 CFM, 1 HP	▼	3	6.667	▼	3,400	268		3,668	4,125
5000	Utility set, centrifugal, V belt drive, motor									
5020	1/4" S.P., 1200 CFM, 1/4 HP	Q-20	6	3.333	Ea.	3,075	134		3,209	3,600
5040	1520 CFM, 1/3 HP		5	4		3,075	161		3,236	3,650
5060	1850 CFM, 1/2 HP		4	5		3,100	201		3,301	3,700
5080	2180 CFM, 3/4 HP		3	6.667		3,125	268		3,393	3,850
5100	1/2" S.P., 3600 CFM, 1 HP		2	10		4,525	400		4,925	5,625
5120	4250 CFM, 1-1/2 HP		1.60	12.500		4,600	500		5,100	5,825
5140	4800 CFM, 2 HP	▼	1.40	14.286	▼	4,650	575		5,225	6,000
7000	Roof exhauster, centrifugal, aluminum housing, 12" galvanized									
7020	curb, bird screen, back draft damper, 1/4" S.P.									
7100	Direct drive, 320 CFM, 11" sq. damper	Q-20	7	2.857	Ea.	385	115		500	600
7120	600 CFM, 11" sq. damper		6	3.333		390	134		524	635
7140	815 CFM, 13" sq. damper		5	4		390	161		551	675
7160	1450 CFM, 13" sq. damper		4.20	4.762		505	191		696	850
7180	2050 CFM, 16" sq. damper		4	5		500	201		701	860
7200	V-belt drive, 1650 CFM, 12" sq. damper		6	3.333		835	134		969	1,125
7220	2750 CFM, 21" sq. damper		5	4		965	161		1,126	1,300
7230	3500 CFM, 21" sq. damper		4.50	4.444		1,075	178		1,253	1,450
7240	4910 CFM, 23" sq. damper		4	5		1,325	201		1,526	1,750
7260	8525 CFM, 28" sq. damper		3	6.667		1,650	268		1,918	2,225
7280	13,760 CFM, 35" sq. damper		2	10		2,275	400		2,675	3,125
7300	20,558 CFM, 43" sq. damper	▼	1	20		4,825	805		5,630	6,550
7320	For 2 speed winding, add					15%				
7340	For explosionproof motor, add					340			340	375
7360	For belt driven, top discharge, add				▼	15%				
8500	Wall exhausters, centrifugal, auto damper, 1/8" S.P.									
8520	Direct drive, 610 CFM, 1/20 HP	Q-20	14	1.429	Ea.	251	57.50		308.50	365
8540	796 CFM, 1/12 HP		13	1.538		259	62		321	380
8560	822 CFM, 1/6 HP		12	1.667		410	67		477	550
8580	1,320 CFM, 1/4 HP	▼	12	1.667	▼	415	67		482	555
9500	V-belt drive, 3 phase									
9520	2,800 CFM, 1/4 HP	Q-20	9	2.222	Ea.	1,150	89		1,239	1,375
9540	3,740 CFM, 1/2 HP	"	8	2.500	"	1,175	100		1,275	1,450

23 34 23 – HVAC Power Ventilators

23 34 23.10 HVAC Power Ventilators

0010	**HVAC POWER VENTILATORS**									
3000	Paddle blade air circulator, 3 speed switch.									

23 34 HVAC Fans

23 34 23 – HVAC Power Ventilators

23 34 23.10 HVAC Power Ventilators	Crew	Daily Output	Labor-Hours	Unit	Material	2007 Bare Costs Labor	Equipment	Total	Total Incl O&P
3020 42", 5,000 CFM high, 3000 CFM low	1 Elec	2.40	3.333	Ea.	80	146		226	305
3040 52", 6,500 CFM high, 4000 CFM low	"	2.20	3.636	"	85.50	160		245.50	330
3100 For antique white motor, same cost									
3200 For brass plated motor, same cost									
3300 For light adaptor kit, add				Ea.	28.50			28.50	31.50
6000 Propeller exhaust, wall shutter, 1/4" S.P.									
6020 Direct drive, two speed									
6100 375 CFM, 1/10 HP	Q-20	10	2	Ea.	310	80.50		390.50	465
6120 730 CFM, 1/7 HP		9	2.222		345	89		434	515
6140 1000 CFM, 1/8 HP		8	2.500		480	100		580	685
6200 4720 CFM, 1 HP	↓	5	4	↓	770	161		931	1,100
6300 V-belt drive, 3 phase									
6320 6175 CFM, 3/4 HP	Q-20	5	4	Ea.	645	161		806	955
6340 7500 CFM, 3/4 HP		5	4		680	161		841	995
6360 10,100 CFM, 1 HP		4.50	4.444		830	178		1,008	1,200
6380 14,300 CFM, 1-1/2 HP	↓	4	5	↓	975	201		1,176	1,375
6650 Residential, bath exhaust, grille, back draft damper									
6660 50 CFM	Q-20	24	.833	Ea.	39.50	33.50		73	94
6670 110 CFM		22	.909		54.50	36.50		91	116
6680 Light combination, squirrel cage, 100 watt, 70 CFM	↓	24	.833	↓	63	33.50		96.50	120
6700 Light/heater combination, ceiling mounted									
6710 70 CFM, 1450 watt	Q-20	24	.833	Ea.	76	33.50		109.50	135
6800 Heater combination, recessed, 70 CFM		24	.833		36	33.50		69.50	90.50
6820 With 2 infrared bulbs		23	.870		54	35		89	113
6900 Kitchen exhaust, grille, complete, 160 CFM		22	.909		63.50	36.50		100	126
6910 180 CFM		20	1		54	40		94	121
6920 270 CFM		18	1.111		97.50	44.50		142	176
6930 350 CFM	↓	16	1.250	↓	76	50		126	161
6940 Residential roof jacks and wall caps									
6944 Wall cap with back draft damper									
6946 3" & 4" dia. round duct	1 Shee	11	.727	Ea.	13.80	31.50		45.30	64
6948 6" dia. round duct	"	11	.727	"	33.50	31.50		65	85.50
6958 Roof jack with bird screen and back draft damper									
6960 3" & 4" dia. round duct	1 Shee	11	.727	Ea.	13.30	31.50		44.80	63.50
6962 3-1/4" x 10" rectangular duct	"	10	.800	"	24.50	35		59.50	80.50
6980 Transition									
6982 3-1/4" x 10" to 6" dia. round	1 Shee	20	.400	Ea.	14.15	17.40		31.55	42.50

23 34 33 – Air Curtains

23 34 33.10 Air Curtains	Crew	Daily Output	Labor-Hours	Unit	Material	2007 Bare Costs Labor	Equipment	Total	Total Incl O&P
0010 **AIR CURTAINS**, Incl. motor starters, transformers,									
0050 door switches & temperature controls									
0100 Shipping and receiving doors, unheated, minimal wind stoppage									
0150 8' high, multiples of 3' wide	2 Shee	6	2.667	L.F.	330	116		446	540
0160 5' wide		10	1.600		320	69.50		389.50	460
0210 10' high, multiples of 4' wide		8	2		315	87		402	480
0250 12' high, 3'-6" wide		7	2.286		345	99.50		444.50	535
0260 12' wide		6	2.667		315	116		431	525
0350 16' high, 3'-6" wide		7	2.286		330	99.50		429.50	520
0360 12' wide	↓	6	2.667	↓	365	116		481	580
0500 Maximum wind stoppage									
0550 10' high, multiples of 4' wide	2 Shee	8	2	L.F.	253	87		340	410
0650 14' high, multiples of 4' wide	↓	8	2	↓	250	87		337	410

23 34 HVAC Fans

23 34 33 – Air Curtains

23 34 33.10 Air Curtains

		Crew	Daily Output	Labor-Hours	Unit	Material	2007 Bare Costs Labor	Equipment	Total	Total Incl O&P
0750	20' high, multiples of 8' wide	2 Shee	6	2.667	L.F.	865	116		981	1,125
1100	Heated, maximum wind stoppage, steam heat									
1150	10' high, multiples of 4' wide	2 Shee	6	2.667	L.F.	435	116		551	660
1250	14' high, multiples of 4' wide		6	2.667		475	116		591	705
1350	20' high, multiples of 8' wide	↓	4	4	↓	1,325	174		1,499	1,725
1500	Customer entrance doors, unheated, minimal wind stoppage									
1550	10' high, multiples of 3' wide	2 Shee	6	2.667	L.F.	290	116		406	500
1560	5' wide	↓	10	1.600		315	69.50		384.50	450
1650	Maximum wind stoppage, 12' high, multiples of 4' wide		8	2	↓	233	87		320	390
1700	Heated, minimal wind stoppage, electric heat									
1750	8' high, multiples of 3' wide	2 Shee	6	2.667	L.F.	580	116		696	815
1760	Multiples of 5' wide		10	1.600		415	69.50		484.50	560
1850	10' high, multiples of 3' wide		6	2.667		490	116		606	720
1860	Multiples of 5' wide	↓	10	1.600		535	69.50		604.50	690
1950	Maximum wind stoppage, steam heat									
1960	12' high, multiples of 4' wide	2 Shee	8	2	L.F.	525	87		612	715
2000	Walk-in coolers and freezers, ambient air, minimal wind stoppage									
2050	8' high, multiples of 3' wide	2 Shee	6	2.667	L.F.	215	116		331	415
2060	Multiples of 5' wide		10	1.600		234	69.50		303.50	365
2250	Maximum wind stoppage, 12' high, multiples of 3' wide		6	2.667		249	116		365	455
2450	Conveyor openings or service windows, unheated, 5' high		5	3.200		276	139		415	520
2460	Heated, electric, 5' high, 2'-6" wide	↓	5	3.200	↓	260	139		399	500

23 37 Air Outlets and Inlets

23 37 13 – Diffusers, Registers, and Grilles

23 37 13.10 Diffusers

		Crew	Daily Output	Labor-Hours	Unit	Material	2007 Bare Costs Labor	Equipment	Total	Total Incl O&P
0010	**DIFFUSERS**, Aluminum, opposed blade damper unless noted									
0100	Ceiling, linear, also for sidewall									
0500	Perforated, 24" x 24" lay-in panel size, 6" x 6"	1 Shee	16	.500	Ea.	75	22		97	116
0520	8" x 8"		15	.533		77	23		100	121
0530	9" x 9"		14	.571		80.50	25		105.50	127
0540	10" x 10"		14	.571		83	25		108	130
0560	12" x 12"		12	.667		85.50	29		114.50	139
0590	16" x 16"		11	.727		116	31.50		147.50	177
0600	18" x 18"		10	.800		125	35		160	191
0610	20" x 20"		10	.800		142	35		177	211
0620	24" x 24"		9	.889		165	38.50		203.50	241
1000	Rectangular, 1 to 4 way blow, 6" x 6"		16	.500		45.50	22		67.50	83.50
1010	8" x 8"		15	.533		54	23		77	95.50
1014	9" x 9"		15	.533		58	23		81	100
1016	10" x 10"		15	.533		68.50	23		91.50	111
1020	12" x 6"		15	.533		77.50	23		100.50	121
1040	12" x 9"		14	.571		84.50	25		109.50	132
1060	12" x 12"		12	.667		80	29		109	133
1070	14" x 6"		13	.615		84	27		111	134
1074	14" x 14"		12	.667		105	29		134	160
1150	18" x 18"		9	.889		147	38.50		185.50	222
1160	21" x 21"		8	1		186	43.50		229.50	271
1170	24" x 12"		10	.800		139	35		174	206
1500	Round, butterfly damper, steel, 6" diameter		18	.444		18.70	19.35		38.05	50.50
1520	8" diameter	↓	16	.500		20	22		42	55.50

23 37 Air Outlets and Inlets

23 37 13 – Diffusers, Registers, and Grilles

23 37 13.10 Diffusers		Crew	Daily Output	Labor-Hours	Unit	Material	2007 Bare Costs Labor	Equipment	Total	Total Incl O&P
1540	10" diameter	1 Shee	14	.571	Ea.	25	25		50	66
1560	12" diameter		12	.667		33	29		62	81
1580	14" diameter		10	.800		42.50	35		77.50	101
2000	T bar mounting, 24" x 24" lay-in frame, 6" x 6"		16	.500		94.50	22		116.50	138
2020	9" x 9"		14	.571		105	25		130	154
2040	12" x 12"		12	.667		135	29		164	193
2060	15" x 15"		11	.727		173	31.50		204.50	239
2080	18" x 18"		10	.800		190	35		225	264
6000	For steel diffusers instead of aluminum, deduct					10%				

23 37 13.30 Grilles

		Crew	Daily Output	Labor-Hours	Unit	Material	Labor	Equipment	Total	Total Incl O&P
0010	**GRILLES**									
0020	Aluminum									
1000	Air return, 6" x 6"	1 Shee	26	.308	Ea.	14.30	13.40		27.70	36.50
1020	10" x 6"		24	.333		16.90	14.50		31.40	41
1080	16" x 8"		22	.364		25.50	15.85		41.35	52.50
1100	12" x 12"		22	.364		25.50	15.85		41.35	52.50
1120	24" x 12"		18	.444		45	19.35		64.35	79.50
1220	24" x 18"		16	.500		49.50	22		71.50	88
1280	36" x 24"		14	.571		99	25		124	148
3000	Filter grille with filter, 12" x 12"		24	.333		42.50	14.50		57	69
3020	18" x 12"		20	.400		62	17.40		79.40	95
3040	24" x 18"		18	.444		75.50	19.35		94.85	113
3060	24" x 24"		16	.500		118	22		140	163
6000	For steel grilles instead of aluminum in above, deduct					10%				

23 37 13.60 Registers

		Crew	Daily Output	Labor-Hours	Unit	Material	Labor	Equipment	Total	Total Incl O&P
0010	**REGISTERS**									
0980	Air supply									
1000	Ceiling/wall, O.B. damper, anodized aluminum									
1010	One or two way deflection, adj. curved face bars									
1020	8" x 4"	1 Shee	26	.308	Ea.	23.50	13.40		36.90	46.50
1120	12" x 12"		18	.444		38	19.35		57.35	72
1240	20" x 6"		18	.444		38.50	19.35		57.85	72
1340	24" x 8"		13	.615		52	27		79	98.50
1350	24" x 18"		12	.667		96	29		125	151
2700	Above registers in steel instead of aluminum, deduct					10%				
4000	Floor, toe operated damper, enameled steel									
4020	4" x 8"	1 Shee	32	.250	Ea.	20	10.90		30.90	39
4100	8" x 10"		22	.364		24.50	15.85		40.35	51.50
4140	10" x 10"		20	.400		29	17.40		46.40	59
4220	14" x 14"		16	.500		103	22		125	147
4240	14" x 20"		15	.533		137	23		160	187
4980	Air return									
5000	Ceiling or wall, fixed 45° face blades									
5010	Adjustable O.B. damper, anodized aluminum									
5020	4" x 8"	1 Shee	26	.308	Ea.	25.50	13.40		38.90	48.50
5060	6" x 10"		19	.421		28	18.35		46.35	59.50
5280	24" x 24"		11	.727		120	31.50		151.50	181
5300	24" x 36"		8	1		187	43.50		230.50	273
6000	For steel construction instead of aluminum, deduct					10%				

23 37 15 – Louvers

23 37 15.40 Louvers

| 0010 | **LOUVERS** | | | | | | | | | |

23 37 Air Outlets and Inlets

23 37 15 – Louvers

23 37 15.40 Louvers

		Crew	Daily Output	Labor-Hours	Unit	Material	2007 Bare Costs Labor	Equipment	Total	Total Incl O&P
0100	Aluminum, extruded, with screen, mill finish									
1002	Brick vent, (see also division 04 05 23.19)									
1100	Standard, 4" deep, 8" wide, 5" high	1 Shee	24	.333	Ea.	30	14.50		44.50	55
1200	Modular, 4" deep, 7-3/4" wide, 5" high		24	.333		31	14.50		45.50	56.50
1300	Speed brick, 4" deep, 11-5/8" wide, 3-7/8" high		24	.333		31	14.50		45.50	56.50
1400	Fuel oil brick, 4" deep, 8" wide, 5" high		24	.333		53.50	14.50		68	81.50
2000	Cooling tower and mechanical equip., screens, light weight		40	.200	S.F.	13.40	8.70		22.10	28
2020	Standard weight		35	.229		35.50	9.95		45.45	54.50
2500	Dual combination, automatic, intake or exhaust		20	.400		48.50	17.40		65.90	80.50
2520	Manual operation		20	.400		36	17.40		53.40	67
2540	Electric or pneumatic operation		20	.400		36	17.40		53.40	67
2560	Motor, for electric or pneumatic		14	.571	Ea.	420	25		445	500
3000	Fixed blade, continuous line									
3100	Mullion type, stormproof	1 Shee	28	.286	S.F.	36	12.45		48.45	59
3200	Stormproof		28	.286		36	12.45		48.45	59
3300	Vertical line		28	.286		43	12.45		55.45	66
3500	For damper to use with above, add					50%	30%			
3520	Motor, for damper, electric or pneumatic	1 Shee	14	.571	Ea.	420	25		445	500
4000	Operating, 45°, manual, electric or pneumatic		24	.333	S.F.	36	14.50		50.50	62.50
4100	Motor, for electric or pneumatic		14	.571	Ea.	420	25		445	500
4200	Penthouse, roof		56	.143	S.F.	21.50	6.20		27.70	33
4300	Walls		40	.200		50.50	8.70		59.20	69
5000	Thinline, under 4" thick, fixed blade		40	.200		21	8.70		29.70	36.50
5010	Finishes, applied by mfr. at additional cost, available in colors									
5020	Prime coat only, add				S.F.	2.90			2.90	3.19
5040	Baked enamel finish coating, add					5.35			5.35	5.85
5060	Anodized finish, add					5.80			5.80	6.35
5080	Duranodic finish, add					10.45			10.45	11.50
5100	Fluoropolymer finish coating, add					16.40			16.40	18.05
9980	For small orders (under 10 pieces), add					25%				

23 37 23 – HVAC Gravity Ventilators

23 37 23.10 HVAC Gravity Ventilators

		Crew	Daily Output	Labor-Hours	Unit	Material	2007 Bare Costs Labor	Equipment	Total	Total Incl O&P
0010	**HVAC GRAVITY VENTILATORS**, Incl. base and damper									
1280	Rotary ventilators, wind driven, galvanized									
1300	4" neck diameter	Q-9	20	.800	Ea.	47.50	31.50		79	101
1340	6" neck diameter		16	1		53	39		92	119
1400	12" neck diameter		10	1.600		77	62.50		139.50	181
1500	24" neck diameter, 3,100 CFM		8	2		224	78.50		302.50	370
1540	36" neck diameter, 5,500 CFM		6	2.667		660	105		765	885
2000	Stationary, gravity, syphon, galvanized									
2160	6" neck diameter, 66 CFM	Q-9	16	1	Ea.	41	39		80	106
2240	12" neck diameter, 160 CFM		10	1.600		86	62.50		148.50	192
2340	24" neck diameter, 900 CFM		8	2		235	78.50		313.50	380
2380	36" neck diameter, 2,000 CFM		6	2.667		735	105		840	970
4200	Stationary mushroom, aluminum, 16" orifice diameter		10	1.600		385	62.50		447.50	520
4220	26" orifice diameter		6.15	2.602		570	102		672	780
4230	30" orifice diameter		5.71	2.802		835	110		945	1,100
4240	38" orifice diameter		5	3.200		1,200	125		1,325	1,525
4250	42" orifice diameter		4.70	3.404		1,575	133		1,708	1,950
4260	50" orifice diameter		4.44	3.604		1,875	141		2,016	2,300
5000	Relief vent									
5500	Rectangular, aluminum, galvanized curb									

23 37 Air Outlets and Inlets

23 37 23 – HVAC Gravity Ventilators

23 37 23.10 HVAC Gravity Ventilators	Crew	Daily Output	Labor-Hours	Unit	Material	2007 Bare Costs Labor	Equipment	Total	Total Incl O&P
5510 intake/exhaust, 0.05" SP									
5600 500 CFM, 12" x 16"	Q-9	8	2	Ea.	410	78.50		488.50	570
5640 1000 CFM, 12" x 24"		6.60	2.424		485	95		580	680
5680 3000 CFM, 20" x 42"		4	4		845	157		1,002	1,175

23 38 Ventilation Hoods

23 38 13 – Commercial-Kitchen Hoods

23 38 13.10 Hood and Ventilation Equipment	Crew	Daily Output	Labor-Hours	Unit	Material	2007 Bare Costs Labor	Equipment	Total	Total Incl O&P
0010 **HOOD AND VENTILATION EQUIPMENT**									
2970 Exhaust hood, sst, gutter on all sides, 4' x 4' x 2'	1 Carp	1.80	4.444	Ea.	3,375	163		3,538	3,975
2980 4' x 4' x 7'	"	1.60	5		5,250	184		5,434	6,050
7950 Hood fire protection system, minimum	Q-1	3	5.333		3,575	215		3,790	4,250
8050 Maximum	"	1	16		26,600	645		27,245	30,300

23 41 Particulate Air Filtration

23 41 13 – Panel Air Filters

23 41 13.10 Panel Air Filters	Crew	Daily Output	Labor-Hours	Unit	Material	2007 Bare Costs Labor	Equipment	Total	Total Incl O&P
0010 **PANEL AIR FILTERS**									
2950 Mechanical media filtration units									
3000 High efficiency type, with frame, non-supported				MCFM	45			45	49.50
3100 Supported type				"	60			60	66
5500 Throwaway glass or paper media type				Ea.	7.20			7.20	7.90

23 41 16 – Renewable-Media Air Filters

23 41 16.10 Renewable-Media Air Filters	Crew	Daily Output	Labor-Hours	Unit	Material	2007 Bare Costs Labor	Equipment	Total	Total Incl O&P
0010 **RENEWABLE-MEDIA AIR FILTERS**									
5000 Renewable disposable roll				MCFM	200			200	220

23 41 19 – Washable Air Filters

23 41 19.10 Washable Air Filters	Crew	Daily Output	Labor-Hours	Unit	Material	2007 Bare Costs Labor	Equipment	Total	Total Incl O&P
0010 **WASHABLE AIR FILTERS**									
4500 Permanent washable				MCFM	20			20	22

23 41 23 – Extended Surface Filters

23 41 23.10 Extended Surface Filters	Crew	Daily Output	Labor-Hours	Unit	Material	2007 Bare Costs Labor	Equipment	Total	Total Incl O&P
0010 **EXTENDED SURFACE FILTERS**									
4000 Medium efficiency, extended surface				MCFM	5.50			5.50	6.05

23 42 Gas-Phase Air Filtration

23 42 13 – Activated-Carbon Air Filtration

23 42 13.10 Activated-Carbon Air Filtration	Crew	Daily Output	Labor-Hours	Unit	Material	2007 Bare Costs Labor	Equipment	Total	Total Incl O&P
0010 **ACTIVATED-CARBON AIR FILTRATION**									
0050 Activated charcoal type, full flow				MCFM	600			600	660
0060 Activated charcoal type, full flow, impregnated media 12" deep					200			200	220
0070 Activated charcoal type, HEPA filter & frame for field erection					200			200	220
0080 Activated charcoal type, HEPA filter-diffuser, ceiling install.					275			275	305

23 43 Electronic Air Cleaners

23 43 13 – Washable Electronic Air Cleaners

23 43 13.10 Washable Electronic Air Cleaners	Crew	Daily Output	Labor-Hours	Unit	Material	2007 Bare Costs Labor	Equipment	Total	Total Incl O&P
0010 **WASHABLE ELECTRONIC AIR CLEANERS**									
2000 Electronic air cleaner, duct mounted									
2150 400 - 1000 CFM	1 Shee	2.30	3.478	Ea.	695	151		846	1,000
2200 1000 - 1400 CFM		2.20	3.636		725	158		883	1,050
2250 1400 - 2000 CFM	↓	2.10	3.810	↓	800	166		966	1,125

23 51 Breechings, Chimneys, and Stacks

23 51 13 – Draft Control Devices

23 51 13.13 Draft-Induction Fans

0010 **DRAFT-INDUCTION FANS**									
1000 Breeching installation									
1800 Hot gas, 600°F, variable pitch pulley and motor									
1860 8" diam. inlet, 1/4 H.P., 1phase, 1120 CFM	Q-9	4	4	Ea.	1,750	157		1,907	2,175
1900 12" diam. inlet, 3/4 H.P., 3 phase, 2960 CFM		3	5.333		2,400	209		2,609	2,975
1940 18" diam. inlet, 3 H.P., 3 phase, 9120 CFM		2	8		3,725	315		4,040	4,575
1980 24" diam. inlet, 7-1/2 H.P., 3 phase, 17,760 CFM	↓	.80	20	↓	5,850	785		6,635	7,625
2300 For multi-blade damper at fan inlet, add					20%				

23 51 23 – Gas Vents

23 51 23.10 Gas Vents

0010 **GAS VENTS**, Prefab metal, U.L. listed									
0020 Gas, double wall, galvanized steel									
0080 3" diameter	Q-9	72	.222	V.L.F.	3.96	8.70		12.66	17.80
0100 4" diameter		68	.235		4.95	9.20		14.15	19.65
0120 5" diameter		64	.250		5.75	9.80		15.55	21.50
0140 6" diameter		60	.267		6.70	10.45		17.15	23.50
0160 7" diameter		56	.286		9.85	11.20		21.05	28
0180 8" diameter		52	.308		10.95	12.05		23	30.50
0200 10" diameter		48	.333		23	13.05		36.05	45.50
0220 12" diameter		44	.364		31	14.25		45.25	56
0260 16" diameter	↓	40	.400		70.50	15.70		86.20	102
0300 20" diameter	Q-10	36	.667		107	27		134	160
0480 38" diameter	"	24	1	↓	340	40.50		380.50	440

23 51 26 – All-Fuel Vent Chimneys

23 51 26.30 All-Fuel Vent Chimneys, Double Wall, Stainless Steel

0010 **ALL-FUEL VENT CHIMNEYS, DOUBLE WALL, STAINLESS STEEL**									
7800 All fuel, double wall, stainless steel, 6" diameter	Q-9	60	.267	V.L.F.	32	10.45		42.45	51.50
7802 7" diameter		56	.286		41.50	11.20		52.70	63
7804 8" diameter		52	.308		48.50	12.05		60.55	72
7806 10" diameter		48	.333		71	13.05		84.05	98
7808 12" diameter		44	.364		95	14.25		109.25	127
7810 14" diameter	↓	42	.381	↓	125	14.95		139.95	160

23 51 33 – Insulated Sectional Chimneys

23 51 33.10 Insulated Sectional Chimneys

0010 **INSULATED SECTIONAL CHIMNEYS**									
9000 High temp. (2000° F), steel jacket, acid resist. refractory lining									
9010 11 ga. galvanized jacket, U.L. listed									
9020 Straight section, 48" long, 10" diameter	Q-10	13.30	1.805	Ea.	380	73.50		453.50	530
9040 18" diameter		7.40	3.243		580	132		712	845
9070 36" diameter	↓	2.70	8.889		1,250	360		1,610	1,900

23 51 Breechings, Chimneys, and Stacks

23 51 33 – Insulated Sectional Chimneys

23 51 33.10 Insulated Sectional Chimneys

		Crew	Daily Output	Labor-Hours	Unit	Material	2007 Bare Costs Labor	Equipment	Total	Total Incl O&P
9090	48" diameter	Q-11	2.70	11.852	Ea.	1,800	490		2,290	2,750
9120	Tee section, 10" diameter	Q-10	4.40	5.455		805	222		1,027	1,225
9140	18" diameter		2.40	10		1,150	405		1,555	1,900
9170	36" diameter	↓	.80	30		3,750	1,225		4,975	6,000
9190	48" diameter	Q-11	.90	35.556		6,400	1,475		7,875	9,300
9220	Cleanout pier section, 10" diameter	Q-10	3.50	6.857		620	279		899	1,100
9240	18" diameter		1.90	12.632		835	515		1,350	1,700
9270	36" diameter	↓	.75	32		1,525	1,300		2,825	3,675
9290	48" diameter	Q-11	.75	42.667		2,325	1,775		4,100	5,300
9320	For drain, add					53%				
9330	Elbow, 30° and 45°, 10" diameter	Q-10	6.60	3.636		525	148		673	810
9350	18" diameter		3.70	6.486		845	264		1,109	1,325
9380	36" diameter	↓	1.30	18.462		2,100	750		2,850	3,475
9400	48" diameter	Q-11	1.35	23.704		3,675	985		4,660	5,575
9430	For 60° and 90° elbow, add					112%				
9440	End cap, 10" diameter	Q-10	26	.923		900	37.50		937.50	1,050
9460	18" diameter		15	1.600		1,275	65		1,340	1,500
9490	36" diameter	↓	5	4.800		2,225	195		2,420	2,750
9510	48" diameter	Q-11	5.30	6.038		3,275	251		3,526	3,975
9540	Increaser (1 diameter), 10" diameter	Q-10	6.60	3.636		445	148		593	720
9560	18" diameter		3.70	6.486		695	264		959	1,175
9590	36" diameter	↓	1.30	18.462		1,375	750		2,125	2,650
9592	42" diameter	Q-11	1.60	20		1,500	830		2,330	2,925
9594	48" diameter		1.35	23.704		1,800	985		2,785	3,525
9596	54" diameter	↓	1.10	29.091		2,200	1,200		3,400	4,250
9600	For expansion joints, add to straight section					7%				
9610	For 1/4" hot rolled steel jacket, add					157%				
9620	For 2950°F very high temperature, add				↓	89%				
9630	26 ga. aluminized jacket, straight section, 48" long									
9640	10" diameter	Q-10	15.30	1.569	V.L.F.	224	64		288	345
9660	18" diameter		8.50	2.824		345	115		460	555
9690	36" diameter		3.10	7.742	↓	795	315		1,110	1,350
9810	Draw band, galv. stl., 11 gauge, 10" diameter		32	.750	Ea.	60.50	30.50		91	114
9820	12" diameter		30	.800		65	32.50		97.50	122
9830	18" diameter		26	.923		81	37.50		118.50	148
9860	36" diameter	↓	18	1.333		138	54		192	235
9880	48" diameter	Q-11	20	1.600	↓	236	66.50		302.50	360

23 52 Heating Boilers

23 52 13 – Electric Boilers

23 52 13.10 Electric Boilers, ASME

0010	**ELECTRIC BOILERS, ASME**, Standard controls and trim.									
1000	Steam, 6 KW, 20.5 MBH	Q-19	1.20	20	Ea.	3,175	835		4,010	4,725
1160	60 KW, 205 MBH		1	24		5,150	1,000		6,150	7,175
1220	112 KW, 382 MBH		.75	32		7,075	1,325		8,400	9,775
1280	222 KW, 758 MBH	↓	.55	43.636		11,700	1,825		13,525	15,600
1380	518 KW, 1768 MBH	Q-21	.36	88.889		19,800	3,800		23,600	27,400
1480	814 KW, 2778 MBH		.25	128		27,700	5,450		33,150	38,600
1600	2,340 KW, 7984 MBH	↓	.16	200		56,500	8,525		65,025	75,000
2000	Hot water, 7.5 KW, 25.6 MBH	Q-19	1.30	18.462		3,225	770		3,995	4,675
2100	90 KW, 307 MBH	↓	1.10	21.818	↓	4,700	910		5,610	6,550

23 52 Heating Boilers

23 52 13 – Electric Boilers

23 52 13.10 Electric Boilers, ASME

		Crew	Daily Output	Labor-Hours	Unit	Material	2007 Bare Costs Labor	Equipment	Total	Total Incl O&P
2220	296 KW, 1010 MBH	Q-19	.55	43.636	Ea.	10,300	1,825		12,125	14,100
2500	1,036 KW, 3536 MBH	Q-21	.34	94.118		24,000	4,000		28,000	32,400
2680	2,400 KW, 8191 MBH	↓	.25	128		50,000	5,450		55,450	63,000
2820	3,600 KW, 12,283 MBH		.16	200	↓	70,500	8,525		79,025	90,500

23 52 23 – Cast-Iron Boilers

23 52 23.20 Gas-Fired Boilers

		Crew	Daily Output	Labor-Hours	Unit	Material	Labor	Equipment	Total	Total Incl O&P
0010	**GAS-FIRED BOILERS,** Natural or propane, standard controls.									
1000	Cast iron, with insulated jacket									
2000	Steam, gross output, 81 MBH	Q-7	1.40	22.857	Ea.	1,650	985		2,635	3,300
2080	203 MBH		.90	35.556		2,675	1,525		4,200	5,225
2180	400 MBH		.56	56.838		4,125	2,450		6,575	8,200
2240	765 MBH		.43	74.419		8,950	3,200		12,150	14,700
2320	1,875 MBH		.30	106		13,600	4,600		18,200	21,800
2440	4,720 MBH		.15	207		26,000	8,950		34,950	42,100
2480	6,100 MBH		.13	246		64,000	10,600		74,600	86,500
2540	6,970 MBH		.10	320		72,000	13,800		85,800	99,500
3000	Hot water, gross output, 80 MBH		1.46	21.918		1,575	945		2,520	3,150
3140	320 MBH CN		.80	40		3,200	1,725		4,925	6,125
3260	1,088 MBH		.40	80		10,800	3,450		14,250	17,100
3380	3,264 MBH		.18	179		20,300	7,750		28,050	34,000
3480	6,100 MBH		.13	250		64,500	10,800		75,300	87,000
3540	6,970 MBH	↓	.09	359		72,500	15,500		88,000	103,000
7000	For tankless water heater, add				↓	10%				

23 52 23.30 Gas/Oil Fired Boilers

		Crew	Daily Output	Labor-Hours	Unit	Material	Labor	Equipment	Total	Total Incl O&P
0010	**GAS/OIL FIRED BOILERS,** Combination with burners and controls.									
1000	Cast iron with insulated jacket									
2000	Steam, gross output, 720 MBH	Q-7	.43	74.074	Ea.	7,975	3,200		11,175	13,600
2080	1,600 MBH		.30	107		13,100	4,600		17,700	21,300
2140	2,700 MBH		.19	165		18,700	7,150		25,850	31,200
2280	5,520 MBH		.14	235		71,000	10,100		81,100	93,000
2340	6,390 MBH		.11	296		76,000	12,800		88,800	102,500
2380	6,970 MBH	↓	.09	372	↓	80,500	16,000		96,500	112,500
2900	Hot water, gross output									
2910	200 MBH	Q-6	.62	39.024	Ea.	6,525	1,650		8,175	9,650
2920	300 MBH		.49	49.080		6,525	2,075		8,600	10,300
2930	400 MBH		.41	57.971		7,625	2,450		10,075	12,100
2940	500 MBH	↓	.36	67.039		8,225	2,825		11,050	13,300
3000	584 MBH	Q-7	.44	72.072		10,700	3,100		13,800	16,500
3060	1,460 MBH		.28	113		16,000	4,875		20,875	24,900
3160	4,088 MBH		.16	195		38,200	8,400		46,600	54,500
3300	13,500 MBH, 403.3 BHP	↓	.04	727		118,000	31,300		149,300	177,000

23 52 23.40 Oil-Fired Boilers

		Crew	Daily Output	Labor-Hours	Unit	Material	Labor	Equipment	Total	Total Incl O&P
0010	**OIL-FIRED BOILERS,** Standard controls, flame retention burner.									
1000	Cast iron, with insulated flush jacket									
2000	Steam, gross output, 109 MBH	Q-7	1.20	26.667	Ea.	2,350	1,150		3,500	4,325
2060	207 MBH		.90	35.556		3,225	1,525		4,750	5,825
2180	1,084 MBH		.38	85.106		8,125	3,675		11,800	14,500
2280	3,000 MBH		.19	170		18,200	7,325		25,525	31,000
2380	5,520 MBH		.14	235		60,000	10,100		70,100	81,000
2460	6,970 MBH	↓	.09	363		77,000	15,700		92,700	108,000
3000	Hot water, same price as steam									
4000	For tankless coil in smaller sizes, add				Ea.	15%				

23 52 Heating Boilers

23 52 26 – Steel Boilers

23 52 26.40 Oil-Fired Boilers

		Crew	Daily Output	Labor-Hours	Unit	Material	2007 Bare Costs Labor	Equipment	Total	Total Incl O&P
0010	**OIL-FIRED BOILERS**, Standard controls, flame retention burner									
5000	Steel, with insulated flush jacket									
7000	Hot water, gross output, 103 MBH	Q-6	1.60	15	Ea.	1,450	635		2,085	2,550
7120	420 MBH		.70	34.483		5,475	1,450		6,925	8,225
7320	3,150 MBH	↓	.13	184	↓	22,200	7,775		29,975	36,100
7340	For tankless coil in steam or hot water, add					7%				

23 52 28 – Swimming Pool Boilers

23 52 28.10 Swimming Pool Heaters

		Crew	Daily Output	Labor-Hours	Unit	Material	Labor	Equipment	Total	Total Incl O&P
0010	**SWIMMING POOL HEATERS**, Not including wiring, external									
0020	piping, base or pad,									
0160	Gas fired, input, 155 MBH	Q-6	1.50	16	Ea.	1,500	675		2,175	2,675
0200	199 MBH		1	24		1,575	1,000		2,575	3,275
0280	500 MBH		.40	60		6,400	2,525		8,925	10,900
0400	1,800 MBH	↓	.14	171		13,900	7,225		21,125	26,200
2000	Electric, 12 KW, 4,800 gallon pool	Q-19	3	8		2,050	335		2,385	2,750
2020	15 KW, 7,200 gallon pool		2.80	8.571		2,075	360		2,435	2,800
2040	24 KW, 9,600 gallon pool		2.40	10		2,775	420		3,195	3,675
2100	55 KW, 24,000 gallon pool	↓	1.20	20	↓	3,950	835		4,785	5,600

23 54 Furnaces

23 54 13 – Electric-Resistance Furnaces

23 54 13.10 Electric-Resistance Furnaces

		Crew	Daily Output	Labor-Hours	Unit	Material	Labor	Equipment	Total	Total Incl O&P
0010	**ELECTRIC-RESISTANCE FURNACES**, Hot air, blowers, std. controls.									
0011	not including gas, oil or flue piping									
1000	Electric, UL listed									
1020	10.2 MBH	Q-20	5	4	Ea.	325	161		486	605
1040	17.1 MBH		4.80	4.167		340	167		507	630
1060	27.3 MBH		4.60	4.348		410	175		585	720
1100	34.1 MBH	↓	4.40	4.545	↓	425	182		607	745

23 54 16 – Fuel-Fired Furnaces

23 54 16.13 Gas-Fired Furnaces

		Crew	Daily Output	Labor-Hours	Unit	Material	Labor	Equipment	Total	Total Incl O&P
0010	**GAS-FIRED FURNACES**									
3000	Gas, AGA certified, upflow, direct drive models									
3020	45 MBH input	Q-9	4	4	Ea.	455	157		612	740
3040	60 MBH input		3.80	4.211		605	165		770	920
3060	75 MBH input		3.60	4.444		640	174		814	975
3100	100 MBH input	↓	3.20	5	↓	680	196		876	1,050

23 54 16.16 Oil-Fired Furnaces

		Crew	Daily Output	Labor-Hours	Unit	Material	Labor	Equipment	Total	Total Incl O&P
0010	**OIL-FIRED FURNACES**									
6000	Oil, UL listed, atomizing gun type burner									
6020	56 MBH output	Q-9	3.60	4.444	Ea.	1,550	174		1,724	1,975
6030	84 MBH output		3.50	4.571		1,575	179		1,754	2,000
6040	95 MBH output		3.40	4.706		1,575	184		1,759	2,025
6060	134 MBH output		3.20	5		1,625	196		1,821	2,100
6080	151 MBH output	↓	3	5.333	↓	1,900	209		2,109	2,425

23 55 Fuel-Fired Heaters

23 55 13 – Fuel-Fired Duct Heaters

23 55 13.16 Gas-Fired Duct Heaters

		Crew	Daily Output	Labor-Hours	Unit	Material	2007 Bare Costs Labor	Equipment	Total	Total Incl O&P
0010	**GAS-FIRED DUCT HEATERS**, Includes burner, controls, stainless steel									
0020	heat exchanger. Gas fired, electric ignition									
0030	Indoor installation									
0100	120 MBH output	Q-5	4	4	Ea.	1,675	163		1,838	2,075
0130	200 MBH output		2.70	5.926		1,750	241		1,991	2,300
0140	240 MBH output		2.30	6.957		2,325	283		2,608	2,975
0180	320 MBH output		1.60	10		2,800	405		3,205	3,675
0300	For powered venter and adapter, add					370			370	410
0502	For required flue pipe, see 23 51 23.10									
1000	Outdoor installation, with vent cap									
1020	75 MBH output	Q-5	4	4	Ea.	2,175	163		2,338	2,650
1060	120 MBH output		4	4		2,775	163		2,938	3,300
1100	187 MBH output		3	5.333		3,975	217		4,192	4,700
1140	300 MBH output		1.80	8.889		5,175	360		5,535	6,250
1180	450 MBH output		1.40	11.429		5,725	465		6,190	7,000
1400	For powered venter, add					10%				

23 55 33 – Fuel-Fired Unit Heaters

23 55 33.13 Oil-Fired Unit Heaters

		Crew	Daily Output	Labor-Hours	Unit	Material	Labor	Equipment	Total	Total Incl O&P
0010	**OIL-FIRED UNIT HEATERS**, Cabinet, grilles, fan, ctrl, burner, no piping									
6000	Oil fired, suspension mounted, 94 MBH output	Q-5	4	4	Ea.	2,400	163		2,563	2,900
6040	140 MBH output		3	5.333		2,575	217		2,792	3,150
6060	184 MBH output		3	5.333		2,775	217		2,992	3,375

23 55 33.16 Gas-Fired Unit Heaters

		Crew	Daily Output	Labor-Hours	Unit	Material	Labor	Equipment	Total	Total Incl O&P
0010	**GAS-FIRED UNIT HEATERS**, Cabinet, grilles, fan, ctrls., burner, no piping									
0022	thermostat, no piping. For flue see 23 51 23.10									
1000	Gas fired, floor mounted									
1100	60 MBH output	Q-5	10	1.600	Ea.	635	65		700	795
1140	100 MBH output		8	2		700	81.50		781.50	890
1180	180 MBH output		6	2.667		1,000	108		1,108	1,275
2000	Suspension mounted, propeller fan, 20 MBH output		8.50	1.882		495	76.50		571.50	655
2040	60 MBH output		7	2.286		605	93		698	805
2060	80 MBH output		6	2.667		695	108		803	930
2100	130 MBH output		5	3.200		900	130		1,030	1,175
2240	320 MBH output		2	8		1,850	325		2,175	2,550
2500	For powered venter and adapter, add					315			315	350
5000	Wall furnace, 17.5 MBH output	Q-5	6	2.667		585	108		693	810
5020	24 MBH output		5	3.200		595	130		725	850
5040	35 MBH output		4	4		770	163		933	1,100

23 56 Solar Energy Heating Equipment

23 56 16 – Packaged Solar Heating Equipment

23 56 16.40 Packaged Solar Heating Equipment

		Crew	Daily Output	Labor-Hours	Unit	Material	Labor	Equipment	Total	Total Incl O&P
0010	**PACKAGED SOLAR HEATING EQUIPMENT** R235616-60									
0020	System/Package prices, not including connecting									
0030	pipe, insulation, or special heating/plumbing fixtures									
0500	Hot water, standard package, low temperature									
0540	1 collector, circulator, fittings, 65 gal. tank	Q-1	.50	32	Ea.	2,100	1,300		3,400	4,250
0580	2 collectors, circulator, fittings, 120 gal. tank		.40	40		3,125	1,625		4,750	5,850
0620	3 collectors, circulator, fittings, 120 gal. tank		.34	47.059		4,050	1,900		5,950	7,300
0700	Medium temperature package									

23 56 Solar Energy Heating Equipment

23 56 16 – Packaged Solar Heating Equipment

23 56 16.40 Packaged Solar Heating Equipment	Crew	Daily Output	Labor-Hours	Unit	Material	2007 Bare Costs Labor	Equipment	Total	Total Incl O&P
0720	1 collector, circulator, fittings, 80 gal. tank	Q-1	.50	32	Ea.	1,825	1,300		3,125	3,975
0740	2 collectors, circulator, fittings, 120 gal. tank		.40	40		2,675	1,625		4,300	5,375
0780	3 collectors, circulator, fittings, 120 gal. tank		.30	53.333		3,625	2,150		5,775	7,200
0980	For each additional 120 gal. tank, add					920			920	1,000

23 56 19 – Solar Heating Components

23 56 19.50 Solar Heating Components

		Crew	Daily Output	Labor-Hours	Unit	Material	Labor	Equipment	Total	Total Incl O&P
0010	**SOLAR HEATING COMPONENTS**									
2300	Circulators, air									
2310	Blowers									
2400	Reversible fan, 20" diameter, 2 speed	Q-9	18	.889	Ea.	108	35		143	173
2520	Space & DHW system, less duct work	"	.50	32		1,475	1,250		2,725	3,550
2870	1/12 HP, 30 GPM	Q-1	10	1.600		330	64.50		394.50	460
3000	Collector panels, air with aluminum absorber plate									
3010	Wall or roof mount									
3040	Flat black, plastic glazing									
3080	4' x 8'	Q-9	6	2.667	Ea.	635	105		740	855
3200	Flush roof mount, 10' to 16' x 22" wide	"	96	.167	L.F.	282	6.55		288.55	320
3300	Collector panels, liquid with copper absorber plate									
3330	Alum. frame, 4' x 8', 5/32" single glazing	Q-1	9.50	1.684	Ea.	670	68		738	840
3390	Alum. frame, 4' x 10', 5/32" single glazing		6	2.667		800	108		908	1,050
3450	Flat black, alum. frame, 3.5' x 7.5'		9	1.778		560	71.50		631.50	730
3500	4' x 8'		5.50	2.909		690	117		807	930
3520	4' x 10'		10	1.600		830	64.50		894.50	1,000
3540	4' x 12.5'		5	3.200		1,025	129		1,154	1,325
3600	Liquid, full wetted, plastic, alum. frame, 3' x 10'		5	3.200		195	129		324	410
3650	Collector panel mounting, flat roof or ground rack		7	2.286		198	92		290	355
3670	Roof clamps		70	.229	Set	2.35	9.20		11.55	16.45
3700	Roof strap, teflon	1 Plum	205	.039	L.F.	9.50	1.75		11.25	13.10
3900	Differential controller with two sensors									
3930	Thermostat, hard wired	1 Plum	8	1	Ea.	79.50	45		124.50	155
4100	Five station with digital read-out	"	3	2.667	"	218	119		337	420
4300	Heat exchanger									
4580	Fluid to fluid package includes two circulating pumps									
4590	expansion tank, check valve, relief valve									
4600	controller, high temperature cutoff and sensors	Q-1	2.50	6.400	Ea.	695	258		953	1,150
4650	Heat transfer fluid									
4700	Propylene glycol, inhibited anti-freeze	1 Plum	28	.286	Gal.	8.80	12.80		21.60	29
8250	Water storage tank with heat exchanger and electric element									
8300	80 gal. with 2" x 2 lb. density insulation	1 Plum	1.60	5	Ea.	940	224		1,164	1,350
8380	120 gal. with 2" x 2 lb. density insulation		1.40	5.714		1,050	256		1,306	1,525
8400	120 gal. with 2" x 2 lb. density insul., 40 S.F. heat coil		1.40	5.714		1,200	256		1,456	1,700

23 57 Heat Exchangers for HVAC

23 57 16 – Steam-to-Water Heat Exchangers

23 57 16.10 Steam-to-Water Heat Exchangers	Crew	Daily Output	Labor-Hours	Unit	Material	2007 Bare Costs Labor	Equipment	Total	Total Incl O&P
0010 **STEAM-TO-WATER HEAT EXCHANGERS**									
0016 Shell & tube type, 2 or 4 pass, 3/4" O.D. copper tubes,									
0020 C.I. heads, C.I. tube sheet, steel shell									
0100 Hot water 40°F to 180°F, by steam at 10 PSI									
0120 8 GPM	Q-5	6	2.667	Ea.	1,375	108		1,483	1,700
0140 10 GPM		5	3.200		2,075	130		2,205	2,500
0160 40 GPM		4	4		3,250	163		3,413	3,825
0180 64 GPM		2	8		4,975	325		5,300	5,950
0200 96 GPM	↓	1	16		6,650	650		7,300	8,300
0220 120 GPM	Q-6	1.50	16	↓	8,725	675		9,400	10,600

23 57 19 – Liquid-to-Liquid Heat Exchangers

23 57 19.13 Plate-Type, Liquid-to-Liquid Heat Exchangers

	Crew	Daily Output	Labor-Hours	Unit	Material	Labor	Equipment	Total	Total Incl O&P
0010 **PLATE-TYPE, LIQUID-TO-LIQUID HEAT EXCHANGERS**									
3000 Plate type,									
3100 400 GPM	Q-6	.80	30	Ea.	26,900	1,275		28,175	31,500
3120 800 GPM	"	.50	48		46,300	2,025		48,325	54,000
3140 1200 GPM	Q-7	.34	94.118		69,000	4,050		73,050	82,000
3160 1800 GPM	"	.24	133	↓	91,500	5,750		97,250	109,000

23 57 19.16 Shell-Type, Liquid-to-Liquid Heat Exchangers

	Crew	Daily Output	Labor-Hours	Unit	Material	Labor	Equipment	Total	Total Incl O&P
0010 **SHELL-TYPE, LIQUID-TO-LIQUID HEAT EXCHANGERS**									
1000 Hot water 40°F to 140°F, by water at 200°F									
1020 7 GPM	Q-5	6	2.667	Ea.	1,700	108		1,808	2,050
1040 16 GPM		5	3.200		2,425	130		2,555	2,850
1060 34 GPM		4	4		3,650	163		3,813	4,275
1100 74 GPM	↓	1.50	10.667	↓	6,625	435		7,060	7,925

23 62 Packaged Compressor and Condenser Units

23 62 13 – Packaged Air-Cooled Refrigerant Compressor and Condenser Units

23 62 13.10 Packaged Air-Cooled Refrigerant Condensing Units

	Crew	Daily Output	Labor-Hours	Unit	Material	Labor	Equipment	Total	Total Incl O&P
0010 **PACKAGED AIR-COOLED REFRIGERANT CONDENSING UNITS**									
0020 Condensing unit									
0030 Air cooled, compressor, standard controls									
0050 1.5 ton	Q-5	2.50	6.400	Ea.	805	260		1,065	1,275
0500 5 ton		.60	26.667		1,575	1,075		2,650	3,350
0600 10 ton	↓	.50	32		4,125	1,300		5,425	6,500
0700 20 ton	Q-6	.40	60	↓	7,975	2,525		10,500	12,600

23 63 Refrigerant Condensers

23 63 13 – Air-Cooled Refrigerant Condensers

23 63 13.10 Air-Cooled Refrig. Condensers

	Crew	Daily Output	Labor-Hours	Unit	Material	Labor	Equipment	Total	Total Incl O&P
0010 **AIR-COOLED REFRIG. CONDENSERS**, Ratings are for 30° F TD, R-22.									
0080 Air cooled, belt drive, propeller fan									
0240 50 ton	Q-6	.69	34.985	Ea.	11,500	1,475		12,975	14,800
0280 59 ton		.58	41.308		12,700	1,750		14,450	16,600
0320 73 ton		.47	51.173		15,000	2,150		17,150	19,800
0360 86 ton		.40	60.302		17,800	2,550		20,350	23,400
0380 88 ton	↓	.39	61.697	↓	19,100	2,600		21,700	24,900
1550 Air cooled, direct drive, propeller fan									

23 63 Refrigerant Condensers

23 63 13 – Air-Cooled Refrigerant Condensers

23 63 13.10 Air-Cooled Refrig. Condensers		Crew	Daily Output	Labor-Hours	Unit	Material	2007 Bare Costs Labor	Equipment	Total	Total Incl O&P
1590	1 ton	Q-5	3.80	4.211	Ea.	585	171		756	905
1600	1-1/2 ton		3.60	4.444		700	181		881	1,025
1620	2 ton		3.20	5		770	203		973	1,150
1640	5 ton		2	8		1,575	325		1,900	2,225
1660	10 ton		1.40	11.429		2,425	465		2,890	3,350
1690	16 ton		1.10	14.545		3,400	590		3,990	4,650
1720	26 ton		.84	19.002		4,400	775		5,175	6,025
1760	41 ton	Q-6	.77	31.008		8,075	1,300		9,375	10,900
1800	63 ton	"	.55	44.037		12,700	1,850		14,550	16,800

23 64 Packaged Water Chillers

23 64 13 – Absorption Water Chillers

23 64 13.16 Indirect-Fired Absorption Water Chillers

		Crew	Daily Output	Labor-Hours	Unit	Material	Labor	Equipment	Total	Total Incl O&P
0010	**INDIRECT-FIRED ABSORPTION WATER CHILLERS**									
0020	Steam or hot water, water cooled									
0050	100 ton	Q-7	.13	240	Ea.	125,000	10,300		135,300	153,000
0400	420 ton	"	.10	323	"	231,500	13,900		245,400	275,500

23 64 16 – Centrifugal Water Chillers

23 64 16.10 Centrifugal Water Chillers

		Crew	Daily Output	Labor-Hours	Unit	Material	Labor	Equipment	Total	Total Incl O&P
0010	**CENTRIFUGAL WATER CHILLERS**, With standard controls.									
0020	Centrifugal liquid chiller, water cooled									
0030	not including water tower									
0100	Open drive, 2000 ton	Q-7	.07	477	Ea.	617,500	20,600		638,100	710,000

23 64 19 – Reciprocating Water Chillers

23 64 19.10 Reciprocating Water Chillers

		Crew	Daily Output	Labor-Hours	Unit	Material	Labor	Equipment	Total	Total Incl O&P
0010	**RECIPROCATING WATER CHILLERS**, With standard controls.									
0494	Water chillers, integral air cooled condenser									
0600	100 ton cooling	Q-7	.25	129	Ea.	65,000	5,550		70,550	80,000
0980	Water cooled, multiple compr., semi-hermetic, tower not incl.									
1000	15 ton cooling	Q-6	.36	65.934	Ea.	13,200	2,775		15,975	18,700
1020	20 ton cooling	Q-7	.41	78.049		14,500	3,350		17,850	21,100
1060	30 ton cooling		.31	101		17,100	4,400		21,500	25,400
1100	50 ton cooling		.28	113		30,700	4,900		35,600	41,200
1160	100 ton cooling		.18	179		55,000	7,750		62,750	72,000
1180	120 ton cooling		.16	196		62,000	8,450		70,450	80,500
1200	140 ton cooling		.16	202		73,000	8,725		81,725	93,500
1451	Water cooled, dual compressors, semi-hermetic, tower not incl.									
1500	80 ton cooling	Q-7	.14	222	Ea.	27,800	9,575		37,375	44,900
1520	100 ton cooling		.14	228		34,100	9,850		43,950	52,500
1540	120 ton cooling		.14	231		41,300	9,975		51,275	60,500

23 64 23 – Scroll Water Chillers

23 64 23.10 Scroll Water Chillers

		Crew	Daily Output	Labor-Hours	Unit	Material	Labor	Equipment	Total	Total Incl O&P
0010	**SCROLL WATER CHILLERS**, With standard controls.									
0490	Packaged w/integral air cooled condenser, 15 ton cool	Q-7	.37	86.486	Ea.	16,100	3,725		19,825	23,300
0500	20 ton cooling		.34	94.118		21,300	4,050		25,350	29,500
0520	40 ton cooling		.30	108		28,900	4,650		33,550	38,800
0680	Scroll water cooled, single compressor, hermetic, tower not incl.									
0700	2 to 5 ton cooling	Q-5	.57	28.070	Ea.	3,425	1,150		4,575	5,500
0740	8 ton cooling	"	.31	52.117		4,950	2,125		7,075	8,650

23 64 Packaged Water Chillers

23 64 23 – Scroll Water Chillers

23 64 23.10 Scroll Water Chillers		Crew	Daily Output	Labor-Hours	Unit	Material	2007 Bare Costs Labor	Equipment	Total	Total Incl O&P
0760	10 ton cooling	Q-6	.36	67.039	Ea.	6,050	2,825		8,875	10,900
0800	20 ton cooling	Q-7	.38	83.990		10,200	3,625		13,825	16,700
0820	30 ton cooling	"	.33	96.096	↓	11,500	4,150		15,650	18,800

23 64 26 – Rotary-Screw Water Chillers

23 64 26.10 Rotary-Screw Water Chillers

		Crew	Daily Output	Labor-Hours	Unit	Material	Labor	Equipment	Total	Total Incl O&P
0010	**ROTARY-SCREW WATER CHILLERS**, With standard controls.									
0110	Screw, liquid chiller, air cooled, insulated evaporator									
0120	130 ton	Q-7	.14	228	Ea.	69,000	9,850		78,850	91,000
0124	160 ton		.13	246		85,000	10,600		95,600	109,500
0128	180 ton		.13	250		95,500	10,800		106,300	121,500
0132	210 ton		.12	258		105,500	11,100		116,600	132,500
0136	270 ton		.12	266		120,500	11,500		132,000	150,000
0140	320 ton	↓	.12	275	↓	151,000	11,900		162,900	184,500
1450	Water cooled, tower not included									
1580	150 ton cooling, screw compressors	Q-7	.13	240	Ea.	57,500	10,400		67,900	78,500
1620	200 ton cooling, screw compressors		.13	250		74,500	10,800		85,300	98,000
1660	291 ton cooling, screw compressors	↓	.12	260	↓	83,500	11,200		94,700	109,000

23 65 Cooling Towers

23 65 13 – Forced-Draft Cooling Towers

23 65 13.10 Forced-Draft Cooling Towers

		Crew	Daily Output	Labor-Hours	Unit	Material	Labor	Equipment	Total	Total Incl O&P
0010	**FORCED-DRAFT COOLING TOWERS**, Packaged units.									
0070	Galvanized steel									
0080	Induced draft, crossflow									
0100	Vertical, belt drive, 61 tons	Q-6	90	.267	TonAC	91	11.25		102.25	117
0150	100 ton		100	.240		72.50	10.10		82.60	95
0200	115 ton		109	.220		70.50	9.30		79.80	91.50
0250	131 ton		120	.200		62.50	8.45		70.95	81
0260	162 ton	↓	132	.182	↓	52.50	7.65		60.15	69
1500	Induced air, double flow									
1900	Vertical, gear drive, 167 ton	Q-6	126	.190	TonAC	97	8.05		105.05	119
2000	297 ton		129	.186		69.50	7.85		77.35	88.50
2100	582 ton		132	.182		57.50	7.65		65.15	74.50
2150	849 ton		142	.169		56.50	7.15		63.65	73
2200	1016 ton	↓	150	.160	↓	54	6.75		60.75	69.50
3000	For higher capacities, use multiples									
3500	For pumps and piping, add	Q-6	38	.632	TonAC	45.50	26.50		72	90
4000	For absorption systems, add				"	75%	75%			
5000	Fiberglass									
5010	Draw thru									
5100	60 ton	Q-6	1.50	16	Ea.	3,275	675		3,950	4,625
5120	125 ton		.99	24.242		6,700	1,025		7,725	8,925
5140	300 ton		.43	55.814		15,700	2,350		18,050	20,900
5160	600 ton		.22	109		28,200	4,600		32,800	37,900
5180	1000 ton	↓	.15	160	↓	48,400	6,750		55,150	63,000
6000	Stainless steel									
6010	Induced draft, crossflow, horizontal, belt drive									
6100	57 ton	Q-6	1.50	16	Ea.	9,225	675		9,900	11,100
6120	91 ton		.99	24.242		13,300	1,025		14,325	16,200
6140	111 ton	↓	.43	55.814	↓	15,600	2,350		17,950	20,800

23 65 Cooling Towers

23 65 13 – Forced-Draft Cooling Towers

23 65 13.10 Forced-Draft Cooling Towers	Crew	Daily Output	Labor-Hours	Unit	Material	2007 Bare Costs Labor	Equipment	Total	Total Incl O&P
6160 126 ton	Q-6	.22	109	Ea.	16,900	4,600		21,500	25,500

23 73 Indoor Central-Station Air-Handling Units

23 73 39 – Indoor, Direct Gas-Fired Heating and Ventilating Units

23 73 39.10 Make-Up Air Unit

		Crew	Daily Output	Labor-Hours	Unit	Material	Labor	Equipment	Total	Total Incl O&P
0010	**MAKE-UP AIR UNIT**									
0020	Indoor suspension, natural/LP gas, direct fired,									
0032	standard control. For flue see division 23 51 23.10									
0040	70°F temperature rise, MBH is input									
0100	2000 CFM, 168 MBH	Q-6	3	8	Ea.	7,525	335		7,860	8,775
0160	6000 CFM, 502 MBH		1.50	16		9,500	675		10,175	11,400
0220	12,000 CFM, 1005 MBH	↓	1	24		11,800	1,000		12,800	14,500
0300	24,000 CFM, 2007 MBH	Q-7	1	32		15,400	1,375		16,775	19,000
0400	50,000 CFM, 4180 MBH	"	.80	40		21,100	1,725		22,825	25,800
0600	For discharge louver assembly, add					10%				
0700	For filters, add					20%				
0800	For air shut-off damper section, add					10%				
0900	For vertical unit, add				↓	10%				

23 74 Packaged Outdoor HVAC Equipment

23 74 33 – Packaged, Outdoor, Heating and Cooling Makeup Air-Conditioners

23 74 33.10 Roof Top Air Conditioners

			Crew	Daily Output	Labor-Hours	Unit	Material	Labor	Equipment	Total	Total Incl O&P
0010	**ROOF TOP AIR CONDITIONERS**, Standard controls, curb, economizer.										
1000	Single zone, electric cool, gas heat										
1090	2 ton cooling, 55 MBH heating	R236000-20	Q-5	.93	17.204	Ea.	2,900	700		3,600	4,250
1100	3 ton cooling, 60 MBH heating			.70	22.857		3,075	930		4,005	4,775
1120	4 ton cooling, 95 MBH heating			.61	26.403		4,475	1,075		5,550	6,550
1140	5 ton cooling, 112 MBH heating			.56	28.521		4,900	1,150		6,050	7,150
1145	6 ton cooling, 140 MBH heating			.52	30.769		5,725	1,250		6,975	8,175
1150	7.5 ton cooling, 170 MBH heating		↓	.50	32.258		7,150	1,300		8,450	9,850
1160	10 ton cooling, 200 MBH heating		Q-6	.67	35.982		9,175	1,525		10,700	12,400
1170	12.5 ton cooling, 230 MBH heating			.63	37.975		10,500	1,600		12,100	13,900
1190	18 ton cooling, 330 MBH heating		↓	.52	45.889		17,100	1,925		19,025	21,700
1200	20 ton cooling, 360 MBH heating		Q-7	.67	47.976		19,200	2,075		21,275	24,200
1210	25 ton cooling, 450 MBH heating			.56	57.554		23,500	2,475		25,975	29,600
1220	30 ton cooling, 540 MBH heating		↓	.47	68.376	↓	28,300	2,950		31,250	35,500
2000	Multizone, electric cool, gas heat, economizer										
2100	15 ton cooling, 360 MBH heating		Q-7	.61	52.545	Ea.	48,600	2,275		50,875	57,000
2120	20 ton cooling, 360 MBH heating	CN		.53	60.038		65,000	2,575		67,575	75,500
2200	40 ton cooling, 540 MBH heating			.28	113		112,500	4,900		117,400	131,000
2210	50 ton cooling, 540 MBH heating			.23	142		139,500	6,125		145,625	162,500
2220	70 ton cooling, 1500 MBH heating			.16	198		151,000	8,550		159,550	179,500
2240	80 ton cooling, 1500 MBH heating			.14	228		173,000	9,850		182,850	205,000
2260	90 ton cooling, 1500 MBH heating			.13	256		184,000	11,000		195,000	219,000
2280	105 ton cooling, 1500 MBH heating		↓	.11	290		199,500	12,500		212,000	238,500
2400	For hot water heat coil, deduct						5%				
2500	For steam heat coil, deduct						2%				
2600	For electric heat, deduct					↓	3%	5%			

23 81 Decentralized Unitary HVAC Equipment

23 81 13 – Packaged Terminal Air-Conditioners

23 81 13.10 Packaged Terminal Air-Conditioners

	23 81 13.10 Packaged Terminal Air-Conditioners	Crew	Daily Output	Labor-Hours	Unit	Material	2007 Bare Costs Labor	Equipment	Total	Total Incl O&P
0010	**PACKAGED TERMINAL AIR-CONDITIONERS**, Cabinet, wall sleeve,									
0100	louver, electric heat, thermostat, manual changeover, 208 V									
0200	6,000 BTUH cooling, 8800 BTU heat	Q-5	6	2.667	Ea.	1,000	108		1,108	1,275
0220	9,000 BTUH cooling, 13,900 BTU heat		5	3.200		1,050	130		1,180	1,375
0240	12,000 BTUH cooling, 13,900 BTU heat		4	4		1,150	163		1,313	1,525
0260	15,000 BTUH cooling, 13,900 BTU heat		3	5.333		1,375	217		1,592	1,850
0500	For hot water coil, increase heat by 10%, add					5%	10%			
1000	For steam, increase heat output by 30%, add					8%	10%			

23 81 19 – Self-Contained Air-Conditioners

23 81 19.20 Self-Contained Single Package

		Crew	Daily Output	Labor-Hours	Unit	Material	Labor	Equipment	Total	Total Incl O&P
0010	**SELF-CONTAINED SINGLE PACKAGE**									
0100	Air cooled, for free blow or duct, not incl. remote condenser									
0200	3 ton cooling	Q-5	1	16	Ea.	3,125	650		3,775	4,400
0220	5 ton cooling	Q-6	1.20	20		3,700	845		4,545	5,350
0240	10 ton cooling	Q-7	1	32		6,325	1,375		7,700	9,025
0260	20 ton cooling		.90	35.556		12,700	1,525		14,225	16,200
0280	30 ton cooling		.80	40		18,500	1,725		20,225	23,000
0340	60 ton cooling	Q-8	.40	80		45,800	3,450	154	49,404	56,000
0490	For duct mounting no price change									
0500	For steam heating coils, add				Ea.	10%	10%			
1000	Water cooled for free blow or duct, not including tower									
1010	Constant volume									
1100	3 ton cooling	Q-6	1	24	Ea.	2,975	1,000		3,975	4,800
1120	5 ton cooling	"	1	24		3,850	1,000		4,850	5,750
1140	10 ton cooling	Q-7	.90	35.556		7,600	1,525		9,125	10,700
1160	20 ton cooling		.80	40		25,000	1,725		26,725	30,100
1180	30 ton cooling		.70	45.714		33,100	1,975		35,075	39,400

23 81 23 – Computer-Room Air-Conditioners

23 81 23.10 Computer Room Units

		Crew	Daily Output	Labor-Hours	Unit	Material	Labor	Equipment	Total	Total Incl O&P
0010	**COMPUTER ROOM UNITS**									
1000	Air cooled, includes remote condenser but not									
1020	interconnecting tubing or refrigerant									
1080	3 ton	Q-5	.50	32	Ea.	14,800	1,300		16,100	18,300
1120	5 ton		.45	35.556		15,800	1,450		17,250	19,600
1160	6 ton		.30	53.333		26,300	2,175		28,475	32,200
1200	8 ton		.27	59.259		27,600	2,400		30,000	34,000
1240	10 ton		.25	64		28,900	2,600		31,500	35,700
1280	15 ton		.22	72.727		31,700	2,950		34,650	39,400
1290	18 ton		.20	80		37,900	3,250		41,150	46,600
1320	20 ton	Q-6	.29	82.759		38,700	3,500		42,200	47,900
1360	23 ton	"	.28	85.714		47,400	3,625		51,025	57,500

23 81 43 – Air-Source Unitary Heat Pumps

23 81 43.10 Air-Source Heat Pumps

		Crew	Daily Output	Labor-Hours	Unit	Material	Labor	Equipment	Total	Total Incl O&P
0010	**AIR-SOURCE HEAT PUMPS**, Not including interconnecting tubing.									
1000	Air to air, split system, not including curbs, pads, or ductwork									
1020	2 ton cooling, 8.5 MBH heat @ 0°F	Q-5	1.20	13.333	Ea.	1,600	540		2,140	2,575
1060	5 ton cooling, 27 MBH heat @ 0°F		.50	32		2,500	1,300		3,800	4,700
1080	7.5 ton cooling, 33 MBH heat @ 0°F		.30	53.333		4,800	2,175		6,975	8,550
1100	10 ton cooling, 50 MBH heat @ 0°F	Q-6	.38	63.158		6,700	2,675		9,375	11,400
1120	15 ton cooling, 64 MBH heat @ 0°F		.26	92.308		9,725	3,900		13,625	16,600
1130	20 ton cooling, 85 MBH heat @ 0°F		.20	120		14,600	5,050		19,650	23,600

23 81 Decentralized Unitary HVAC Equipment

23 81 43 – Air-Source Unitary Heat Pumps

23 81 43.10 Air-Source Heat Pumps

		Crew	Daily Output	Labor-Hours	Unit	Material	2007 Bare Costs Labor	2007 Bare Costs Equipment	Total	Total Incl O&P
1140	25 ton cooling, 119 MBH heat @ 0°F	Q-6	.20	120	Ea.	17,600	5,050		22,650	26,900
1300	Supplementary electric heat coil, included									
1500	Single package, not including curbs, pads, or plenums									
1520	2 ton cooling, 6.5 MBH heat @ 0°F	Q-5	1.50	10.667	Ea.	2,650	435		3,085	3,575
1580	4 ton cooling, 13 MBH heat @ 0°F		.96	16.667		3,725	680		4,405	5,125
1640	7.5 ton cooling, 35 MBH heat @ 0°F		.40	40		6,150	1,625		7,775	9,225

23 81 46 – Water-Source Unitary Heat Pumps

23 81 46.10 Water Source Heat Pumps

		Crew	Daily Output	Labor-Hours	Unit	Material	Labor	Equipment	Total	Total Incl O&P
0010	**WATER SOURCE HEAT PUMPS**, Not incl. conn. tubing or wtr. source									
2000	Water source to air, single package									
2100	1 ton cooling, 13 MBH heat @ 75°F	Q-5	2	8	Ea.	1,200	325		1,525	1,825
2140	2 ton cooling, 19 MBH heat @ 75°F		1.70	9.412		1,375	385		1,760	2,100
2220	5 ton cooling, 29 MBH heat @ 75°F		.90	17.778		2,100	725		2,825	3,400
3960	For supplementary heat coil, add					10%				
4000	For increase in capacity thru use									
4020	of solar collector, size boiler at 60%									

23 82 Convection Heating and Cooling Units

23 82 16 – Air Coils

23 82 16.10 Flanged Coils

		Crew	Daily Output	Labor-Hours	Unit	Material	Labor	Equipment	Total	Total Incl O&P
0010	**FLANGED COILS**									
0500	Chilled water cooling, 6 rows, 24" x 48"	Q-5	3.20	5	Ea.	2,975	203		3,178	3,575
1000	Direct expansion cooling, 6 rows, 24" x 48"		2.80	5.714		3,250	232		3,482	3,925
1500	Hot water heating, 1 row, 24" x 48"		4	4		1,200	163		1,363	1,550
2000	Steam heating, 1 row, 24" x 48"		3.06	5.229		1,675	213		1,888	2,175

23 82 16.20 Duct Heaters

		Crew	Daily Output	Labor-Hours	Unit	Material	Labor	Equipment	Total	Total Incl O&P
0010	**DUCT HEATERS**, Electric, 480 V, 3 Ph.									
0020	Finned tubular insert, 500°F									
0100	8" wide x 6" high, 4.0 kW	Q-20	16	1.250	Ea.	640	50		690	775
0120	12" high, 8.0 kW		15	1.333		1,050	53.50		1,103.50	1,250
0140	18" high, 12.0 kW		14	1.429		1,475	57.50		1,532.50	1,725
0160	24" high, 16.0 kW		13	1.538		1,900	62		1,962	2,200
0180	30" high, 20.0 kW		12	1.667		2,325	67		2,392	2,675
0300	12" wide x 6" high, 6.7 kW		15	1.333		675	53.50		728.50	825
0360	24" high, 26.7 kW		12	1.667		1,975	67		2,042	2,275
0700	24" wide x 6" high, 17.8 kW		13	1.538		770	62		832	940
0760	24" high, 71.1 kW		10	2		2,450	80.50		2,530.50	2,825
8000	To obtain BTU multiply kW by 3413									

23 82 19 – Fan Coil Units

23 82 19.10 Fan Coil Air Conditioning

		Crew	Daily Output	Labor-Hours	Unit	Material	Labor	Equipment	Total	Total Incl O&P
0010	**FAN COIL AIR CONDITIONING** Cabinet mounted, filters, controls.									
0100	Chilled water, 1/2 ton cooling	Q-5	8	2	Ea.	625	81.50		706.50	810
0120	1 ton cooling		6	2.667		750	108		858	990
0140	1.5 ton cooling		5.50	2.909		860	118		978	1,125
0150	2 ton cooling		5.25	3.048		1,100	124		1,224	1,375
0180	3 ton cooling		4	4		1,775	163		1,938	2,200
0190	7.5 ton cooling		2.70	5.926		1,900	241		2,141	2,450
0262	For hot water coil, add					40%	10%			
0940	Direct expansion, for use w/air cooled condensing, 1.5 ton cooling	Q-5	5	3.200	Ea.	515	130		645	760
1000	5 ton cooling	"	3	5.333		955	217		1,172	1,375

23 82 Convection Heating and Cooling Units

23 82 19 – Fan Coil Units

23 82 19.10 Fan Coil Air Conditioning

		Crew	Daily Output	Labor-Hours	Unit	Material	2007 Bare Costs Labor	Equipment	Total	Total Incl O&P
1040	10 ton cooling	Q-6	2.60	9.231	Ea.	2,625	390		3,015	3,475
1060	20 ton cooling	"	.70	34.286	↓	4,600	1,450		6,050	7,225

23 82 19.20 Heating and Ventilating Units

		Crew	Daily Output	Labor-Hours	Unit	Material	Labor	Equipment	Total	Total Incl O&P
0010	**HEATING AND VENTILATING UNITS**, Classroom units									
0020	Includes filter, heating/cooling coils, standard controls									
0080	750 CFM, 2 tons cooling	Q-6	2	12	Ea.	3,375	505		3,880	4,475
0120	1250 CFM, 3 tons cooling		1.40	17.143		4,125	725		4,850	5,650
0140	1500 CFM, 4 tons cooling	↓	.80	30		4,425	1,275		5,700	6,750
0500	For electric heat, add					35%				
1000	For no cooling, deduct				↓	25%	10%			

23 82 27 – Infrared Units

23 82 27.10 Infrared Units

		Crew	Daily Output	Labor-Hours	Unit	Material	Labor	Equipment	Total	Total Incl O&P
0010	**INFRARED UNITS**									
0020	Gas fired, unvented, electric ignition, 100% shutoff.									
0030	Piping and wiring not included									
0060	Input, 15 MBH	Q-5	7	2.286	Ea.	425	93		518	610
0120	45 MBH		5	3.200		435	130		565	675
0160	60 MBH		4	4		485	163		648	775
0240	120 MBH	↓	2	8	↓	895	325		1,220	1,475

23 82 29 – Radiators

23 82 29.10 Hydronic Heating

		Crew	Daily Output	Labor-Hours	Unit	Material	Labor	Equipment	Total	Total Incl O&P
0010	**HYDRONIC HEATING**, Terminal units, not incl. main supply pipe									
1000	Radiation									
1100	Panel, baseboard, C.I., including supports, no covers	Q-5	46	.348	L.F.	33.50	14.15		47.65	58.50
3000	Radiators, cast iron									
3100	Free standing or wall hung, 6 tube, 25" high	Q-5	96	.167	Section	24	6.80		30.80	36.50
3200	4 tube, 19" high	"	96	.167	"	16.65	6.80		23.45	28.50
3250	Adj. brackets, 2 per wall radiator up to 30 sections	1 Stpi	32	.250	Ea.	19.25	11.30		30.55	38
9500	To convert SFR to BTU rating: Hot water, 150 x SFR									
9510	Forced hot water, 180 x SFR; steam, 240 x SFR									

23 82 33 – Convectors

23 82 33.10 Convectors

		Crew	Daily Output	Labor-Hours	Unit	Material	Labor	Equipment	Total	Total Incl O&P
0010	**CONVECTORS**, Terminal units, not incl. main supply pipe									
1990	Convector unit, floor recessed, flush, with trim									
2000	for under large glass wall areas, no damper	Q-5	20	.800	L.F.	29	32.50		61.50	81
2100	For unit with damper				"	41.50			41.50	45.50
2210	17" H x 24" L	Q-5	10	1.600	Ea.	81	65		146	187
2214	17" H x 36" L		8.60	1.860		122	75.50		197.50	248
2218	17" H x 48" L		7.40	2.162		162	88		250	310
2222	21" H x 24" L		9	1.778		81	72.50		153.50	198
2226	21" H x 36" L		8.20	1.951		122	79.50		201.50	253
2228	21" H x 48" L	↓	6.80	2.353	↓	162	95.50		257.50	320
2240	For knob operated damper, add					140%				
2241	For metal trim strips, add	Q-5	64	.250	Ea.	4.40	10.15		14.55	20
2243	For snap-on inlet grille, add					10%	10%			
2245	For hinged access door, add	Q-5	64	.250	Ea.	15.70	10.15		25.85	32.50
2246	For air chamber, auto-venting, add	"	58	.276	"	5.10	11.20		16.30	22.50

23 82 36 – Finned-Tube Radiation Heaters

23 82 36.10 Finned Tube Radiation

		Crew	Daily Output	Labor-Hours	Unit	Material	Labor	Equipment	Total	Total Incl O&P
0010	**FINNED TUBE RADIATION**, Terminal units, not incl. main supply pipe									

23 82 Convection Heating and Cooling Units

23 82 36 – Finned-Tube Radiation Heaters

23 82 36.10 Finned Tube Radiation

		Crew	Daily Output	Labor-Hours	Unit	Material	2007 Bare Costs Labor	Equipment	Total	Total Incl O&P
1150	Fin tube, wall hung, 14" slope top cover, with damper									
1200	1-1/4" copper tube, 4-1/4" alum. fin	Q-5	38	.421	L.F.	34	17.15		51.15	63.50
1250	1-1/4" steel tube, 4-1/4" steel fin	"	36	.444	"	32.50	18.10		50.60	62.50
1500	Note: fin tube may also require corners, caps, etc.									

23 82 39 – Unit Heaters

23 82 39.16 Propeller Unit Heaters

		Crew	Daily Output	Labor-Hours	Unit	Material	Labor	Equipment	Total	Total Incl O&P
0010	**PROPELLER UNIT HEATERS**									
3950	Unit heaters, propeller, 115 V 2 psi steam, 60°F entering air									
4000	Horizontal, 12 MBH	Q-5	12	1.333	Ea.	279	54		333	385
4060	43.9 MBH		8	2		430	81.50		511.50	595
4140	96.8 MBH		6	2.667		640	108		748	865
4180	157.6 MBH		4	4		850	163		1,013	1,175
4240	286.9 MBH		2	8		1,350	325		1,675	2,000
4250	326.0 MBH		1.90	8.421		1,650	345		1,995	2,350
4260	364 MBH		1.80	8.889		1,875	360		2,235	2,625
4270	404 MBH		1.60	10		2,300	405		2,705	3,125
4300	For vertical diffuser, add					160			160	176
4310	Vertical flow, 40 MBH	Q-5	11	1.455		430	59		489	565
4314	58.5 MBH		8	2		485	81.50		566.50	650
4326	131.0 MBH		4	4		675	163		838	990
4346	297.0 MBH		1.80	8.889		1,150	360		1,510	1,825
4354	420 MBH, (460 V)	Q-6	1.80	13.333		1,625	560		2,185	2,625
4358	500 MBH, (460 V)		1.71	14.035		2,300	590		2,890	3,425
4362	570 MBH, (460 V)		1.40	17.143		3,425	725		4,150	4,850
4366	620 MBH, (460 V)		1.30	18.462		3,925	780		4,705	5,475
4370	960 MBH, (460 V)		1.10	21.818		6,775	920		7,695	8,850

23 83 Radiant Heating Units

23 83 33 – Electric Radiant Heaters

23 83 33.10 Electric Heating

		Crew	Daily Output	Labor-Hours	Unit	Material	Labor	Equipment	Total	Total Incl O&P
0010	**ELECTRIC HEATING**, Not incl. Conduit or feed wiring.									
1100	Rule of thumb: Baseboard units, including control	1 Elec	4.40	1.818	kW	84	80		164	212
1300	Baseboard heaters, 2' long, 375 watt		8	1	Ea.	35	44		79	104
1400	3' long, 500 watt		8	1		39	44		83	109
1600	4' long, 750 watt		6.70	1.194		49	52.50		101.50	132
1800	5' long, 935 watt		5.70	1.404		58	61.50		119.50	156
2000	6' long, 1125 watt		5	1.600		64.50	70		134.50	176
2400	8' long, 1500 watt		4	2		81.50	88		169.50	221
2950	Wall heaters with fan, 120 to 277 volt									
3600	Thermostats, integral	1 Elec	16	.500	Ea.	19.50	22		41.50	54
3800	Line voltage, 1 pole	"	8	1	"	23.50	44		67.50	91.50

23 84 Humidity Control Equipment

23 84 13 – Humidifiers

23 84 13.10 Humidifiers		Crew	Daily Output	Labor-Hours	Unit	Material	2007 Bare Costs Labor	Equipment	Total	Total Incl O&P
0010	**HUMIDIFIERS**									
0520	Steam, room or duct, filter, regulators, auto. controls, 220 V									
0540	11 lb. per hour	Q-5	6	2.667	Ea.	2,175	108		2,283	2,575
0560	22 lb. per hour		5	3.200		2,400	130		2,530	2,825
0580	33 lb. per hour		4	4		2,450	163		2,613	2,950
0600	50 lb. per hour		4	4		3,025	163		3,188	3,575
0620	100 lb. per hour		3	5.333		3,600	217		3,817	4,300

Division Notes

Estimating Tips

26 05 00 Common Work Results for Electrical

- Conduit should be taken off in three main categories–power distribution, branch power, and branch lighting–so the estimator can concentrate on systems and components, therefore making it easier to ensure all items have been accounted for.
- For cost modifications for elevated conduit installation, add the percentages to labor according to the height of installation, and only to the quantities exceeding the different height levels, not to the total conduit quantities.
- Remember that aluminum wiring of equal ampacity is larger in diameter than copper and may require larger conduit.
- If more than three wires at a time are being pulled, deduct percentages from the labor hours of that grouping of wires.
- The estimator should take the weights of materials into consideration when completing a takeoff. Topics to consider include: How will the materials be supported? What methods of support are available? How high will the support structure have to reach? Will the final support structure be able to withstand the total burden? Is the support material included or separate from the fixture, equipment, and material specified?
- Do not overlook the costs for equipment used in the installation. If scaffolding or highlifts are available in the field, contractors may use them in lieu of the proposed ladders and rolling staging.

26 20 00 Low-Voltage Electrical Transmission

- Supports and concrete pads may be shown on drawings for the larger equipment, or the support system may be only a piece of plywood for the back of a panelboard. In either case, it must be included in the costs.

26 40 00 Electrical and Cathodic Protection

- When taking off grounding system, identify separately the type and size of wire, and list each unique type of ground connection.

26 50 00 Lighting

- Fixtures should be taken off room by room, using the fixture schedule, specifications, and the ceiling plan. For large concentrations of lighting fixtures in the same area, deduct the percentages from labor hours.

Reference Numbers

Reference numbers are shown in shaded boxes at the beginning of some major classifications. These numbers refer to related items in the Reference Section. The reference information may be an estimating procedure, an alternate pricing method, or technical information.

Note: Not all subdivisions listed here necessarily appear in this publication.

Note: **i2 Trade Service,** *in part, has been used as a reference source for some of the material prices used in Division 26.*

No part of this publication may be reproduced, stored in a retrieval system, or transmitted in any form or by any means without prior written permission of Reed Construction Data.

26 05 Common Work Results for Electrical

26 05 05 – Selective Electrical Demolition

26 05 05.10 Electrical Demolition	Crew	Daily Output	Labor-Hours	Unit	Material	2007 Bare Costs Labor	2007 Bare Costs Equipment	Total	Total Incl O&P
0010 **ELECTRICAL DEMOLITION**									
0020 Conduit to 15' high, including fittings & hangers									
0100 Rigid galvanized steel, 1/2" to 1" diameter	1 Elec	242	.033	L.F.		1.45		1.45	2.16
0120 1-1/4" to 2"	"	200	.040			1.76		1.76	2.61
0140 2-1/2" to 3-1/2"	2 Elec	302	.053			2.33		2.33	3.46
0160 4" to 6"	"	160	.100			4.39		4.39	6.55
0200 Electric metallic tubing (EMT), 1/2" to 1"	1 Elec	394	.020			.89		.89	1.33
0220 1-1/4" to 1-1/2"	"	326	.025			1.08		1.08	1.60
0260 3-1/2" to 4"	2 Elec	310	.052			2.27		2.27	3.37
0270 Armored cable, (BX) avg. 50' runs									
0280 #14, 2 wire	1 Elec	690	.012	L.F.		.51		.51	.76
0290 #14, 3 wire		571	.014			.62		.62	.92
0300 #12, 2 wire		605	.013			.58		.58	.86
0310 #12, 3 wire		514	.016			.68		.68	1.02
0320 #10, 2 wire		514	.016			.68		.68	1.02
0330 #10, 3 wire		425	.019			.83		.83	1.23
0340 #8, 3 wire		342	.023			1.03		1.03	1.53
0350 Non metallic sheathed cable (Romex)									
0360 #14, 2 wire	1 Elec	720	.011	L.F.		.49		.49	.73
0370 #14, 3 wire		657	.012			.53		.53	.80
0380 #12, 2 wire		629	.013			.56		.56	.83
0390 #10, 3 wire		450	.018			.78		.78	1.16
0400 Wiremold raceway, including fittings & hangers									
0420 No. 3000	1 Elec	250	.032	L.F.		1.40		1.40	2.09
0440 No. 4000		217	.037			1.62		1.62	2.41
0460 No. 6000		166	.048			2.12		2.12	3.15
0465 Telephone/power pole		12	.667	Ea.		29.50		29.50	43.50
0470 Non-metallic, straight section		480	.017	L.F.		.73		.73	1.09
0500 Channels, steel, including fittings & hangers									
0520 3/4" x 1-1/2"	1 Elec	308	.026	L.F.		1.14		1.14	1.70
0540 1-1/2" x 1-1/2"		269	.030			1.31		1.31	1.94
0560 1-1/2" x 1-7/8"		229	.035			1.53		1.53	2.28
0600 Copper bus duct, indoor, 3 phase									
0610 Including hangers & supports									
0620 225 amp	2 Elec	135	.119	L.F.		5.20		5.20	7.75
0640 400 amp		106	.151			6.65		6.65	9.85
0660 600 amp		86	.186			8.15		8.15	12.15
0680 1000 amp		60	.267			11.70		11.70	17.45
0700 1600 amp		40	.400			17.55		17.55	26
0720 3000 amp		10	1.600			70		70	105
1300 Transformer, dry type, 1 ph, incl. removal of									
1320 supports, wire & conduit terminations									
1340 1 kVA	1 Elec	7.70	1.039	Ea.		45.50		45.50	68
1420 75 kVA	2 Elec	2.50	6.400	"		281		281	420
1440 3 Phase to 600V, primary									
1460 3 kVA	1 Elec	3.85	2.078	Ea.		91		91	136
1520 75 kVA	2 Elec	2.70	5.926			260		260	385
1550 300 kVA	R-3	1.80	11.111			480	91	571	820
1570 750 kVA	"	1.10	18.182			785	149	934	1,350
1800 Wire, THW-THWN-THHN, removed from									
1810 in place conduit, to 15' high									
1830 #14	1 Elec	65	.123	C.L.F.		5.40		5.40	8.05
1840 #12		55	.145			6.40		6.40	9.50

26 05 Common Work Results for Electrical

26 05 05 – Selective Electrical Demolition

26 05 05.10 Electrical Demolition		Crew	Daily Output	Labor-Hours	Unit	Material	2007 Bare Costs Labor	Equipment	Total	Total Incl O&P
1850	#10	1 Elec	45.50	.176	C.L.F.		7.70		7.70	11.50
1860	#8		40.40	.198			8.70		8.70	12.95
1870	#6	↓	32.60	.245			10.75		10.75	16.05
1880	#4	2 Elec	53	.302			13.25		13.25	19.75
1890	#3		50	.320			14.05		14.05	21
1900	#2		44.60	.359			15.75		15.75	23.50
1910	1/0		33.20	.482			21		21	31.50
1920	2/0		29.20	.548			24		24	36
1930	3/0		25	.640			28		28	42
1940	4/0		22	.727			32		32	47.50
1950	250 kcmil		20	.800			35		35	52.50
1960	300 kcmil		19	.842			37		37	55
1970	350 kcmil		18	.889			39		39	58
1980	400 kcmil		17	.941			41.50		41.50	61.50
1990	500 kcmil	↓	16.20	.988	↓		43.50		43.50	64.50
2000	Interior fluorescent fixtures, incl. supports									
2010	& whips, to 15' high									
2100	Recessed drop-in 2' x 2', 2 lamp	2 Elec	35	.457	Ea.		20		20	30
2120	2' x 4', 2 lamp		33	.485			21.50		21.50	31.50
2140	2' x 4', 4 lamp		30	.533			23.50		23.50	35
2160	4' x 4', 4 lamp	↓	20	.800	↓		35		35	52.50
2180	Surface mount, acrylic lens & hinged frame									
2200	1' x 4', 2 lamp	2 Elec	44	.364	Ea.		15.95		15.95	24
2220	2' x 2', 2 lamp		44	.364			15.95		15.95	24
2260	2' x 4', 4 lamp		33	.485			21.50		21.50	31.50
2280	4' x 4', 4 lamp	↓	23	.696	↓		30.50		30.50	45.50
2300	Strip fixtures, surface mount									
2320	4' long, 1 lamp	2 Elec	53	.302	Ea.		13.25		13.25	19.75
2340	4' long, 2 lamp		50	.320			14.05		14.05	21
2360	8' long, 1 lamp		42	.381			16.70		16.70	25
2380	8' long, 2 lamp	↓	40	.400	↓		17.55		17.55	26
2400	Pendant mount, industrial, incl. removal									
2410	of chain or rod hangers, to 15' high									
2420	4' long, 2 lamp	2 Elec	35	.457	Ea.		20		20	30
2440	8' long, 2 lamp	"	27	.593	"		26		26	38.50

26 05 13 – Medium-Voltage Cables

26 05 13.16 Medium-Voltage, Single Cable

		Crew	Daily Output	Labor-Hours	Unit	Material	Labor	Equipment	Total	Total Incl O&P
0010	**MEDIUM-VOLTAGE, SINGLE CABLE** Splicing & terminations not included									
0040	Copper, XLP shielding, 5 kV, #6	2 Elec	4.40	3.636	C.L.F.	206	160		366	465
0050	#4		4.40	3.636		267	160		427	530
0100	#2		4	4		315	176		491	610
0200	#1		4	4		365	176		541	660
0400	1/0		3.80	4.211		400	185		585	715
0600	2/0		3.60	4.444		500	195		695	840
0800	4/0		3.20	5		655	220		875	1,050
1000	250 kcmil	3 Elec	4.50	5.333		810	234		1,044	1,250
1200	350 kcmil		3.90	6.154		1,050	270		1,320	1,550
1400	500 kcmil	↓	3.60	6.667		1,275	293		1,568	1,825
1600	15 kV, ungrounded neutral, #1	2 Elec	4	4		440	176		616	745
1800	1/0		3.80	4.211		525	185		710	855
2000	2/0		3.60	4.444		600	195		795	950
2200	4/0	↓	3.20	5	↓	795	220		1,015	1,200

26 05 Common Work Results for Electrical

26 05 13 – Medium-Voltage Cables

26 05 13.16 Medium-Voltage, Single Cable

		Crew	Daily Output	Labor-Hours	Unit	Material	2007 Bare Costs Labor	Equipment	Total	Total Incl O&P
2400	250 kcmil	3 Elec	4.50	5.333	C.L.F.	880	234		1,114	1,325
2600	350 kcmil		3.90	6.154		1,125	270		1,395	1,625
2800	500 kcmil	↓	3.60	6.667	↓	1,375	293		1,668	1,925

26 05 19 – Low-Voltage Electrical Power Conductors and Cables

26 05 19.20 Armored Cable

		Crew	Daily Output	Labor-Hours	Unit	Material	Labor	Equipment	Total	Total Incl O&P
0010	**ARMORED CABLE**									
0050	600 volt, copper (BX), #14, 2 conductor, solid	1 Elec	2.40	3.333	C.L.F.	74.50	146		220.50	300
0100	3 conductor, solid		2.20	3.636		118	160		278	370
0150	#12, 2 conductor, solid		2.30	3.478		75.50	153		228.50	310
0200	3 conductor, solid		2	4		121	176		297	395
0250	#10, 2 conductor, solid		2	4		138	176		314	410
0300	3 conductor, solid		1.60	5		190	220		410	535
0350	#8, 3 conductor, solid		1.30	6.154		355	270		625	790
0400	3 conductor with PVC jacket, in cable tray, #6	↓	3.10	2.581		405	113		518	615
0450	#4	2 Elec	5.40	2.963		520	130		650	770
0500	#2		4.60	3.478		695	153		848	985
0550	#1		4	4		940	176		1,116	1,275
0600	1/0		3.60	4.444		1,100	195		1,295	1,525
0650	2/0		3.40	4.706		1,350	207		1,557	1,775
0700	3/0		3.20	5		1,575	220		1,795	2,050
0750	4/0	↓	3	5.333		1,800	234		2,034	2,325
0800	250 kcmil	3 Elec	3.60	6.667		2,025	293		2,318	2,650
0850	350 kcmil		3.30	7.273		2,700	320		3,020	3,450
0900	500 kcmil	↓	3	8	↓	3,500	350		3,850	4,375
1050	5 kV, copper, 3 conductor with PVC jacket,									
1060	non-shielded, in cable tray, #4	2 Elec	380	.042	L.F.	8	1.85		9.85	11.55
1100	#2		360	.044		10.35	1.95		12.30	14.30
1200	#1		300	.053		13.20	2.34		15.54	18
1400	1/0		290	.055		15.25	2.42		17.67	20.50
1600	2/0		260	.062		17.60	2.70		20.30	23.50
2000	4/0	↓	240	.067		23.50	2.93		26.43	30.50
2100	250 kcmil	3 Elec	330	.073		32	3.19		35.19	40.50
2150	350 kcmil		315	.076		39.50	3.34		42.84	48.50
2200	500 kcmil	↓	270	.089	↓	53	3.90		56.90	64
2400	15 kV, copper, 3 conductor with PVC jacket galv steel armored									
2500	grounded neutral, in cable tray, #2	2 Elec	300	.053	L.F.	16.75	2.34		19.09	22
2600	#1		280	.057		17.85	2.51		20.36	23.50
2800	1/0		260	.062		20.50	2.70		23.20	26.50
2900	2/0		220	.073		27	3.19		30.19	35
3000	4/0	↓	190	.084		30.50	3.70		34.20	39.50
3100	250 kcmil	3 Elec	270	.089		34.50	3.90		38.40	43.50
3150	350 kcmil		240	.100		40.50	4.39		44.89	51
3200	500 kcmil	↓	210	.114	↓	54	5		59	67
3400	15 kV, copper, 3 conductor with PVC jacket,									
3450	ungrounded neutral, in cable tray, #2	2 Elec	260	.062	L.F.	18.05	2.70		20.75	24
3500	#1		230	.070		20	3.05		23.05	26.50
3600	1/0		200	.080		23	3.51		26.51	30.50
3700	2/0		190	.084		28	3.70		31.70	36.50
3800	4/0	↓	160	.100		34	4.39		38.39	44
4000	250 kcmil	3 Elec	210	.114		39.50	5		44.50	51
4050	350 kcmil		195	.123		52	5.40		57.40	65.50
4100	500 kcmil	↓	180	.133	↓	63.50	5.85		69.35	78

26 05 Common Work Results for Electrical

26 05 19 – Low-Voltage Electrical Power Conductors and Cables

26 05 19.20 Armored Cable		Crew	Daily Output	Labor-Hours	Unit	Material	2007 Bare Costs Labor	Equipment	Total	Total Incl O&P
9010	600 volt, copper (MC) steel clad, #14, 2 wire	1 Elec	2.40	3.333	C.L.F.	75.50	146		221.50	300
9020	3 wire		2.20	3.636		117	160		277	365
9040	#12, 2 wire		2.30	3.478		77.50	153		230.50	310
9050	3 wire		2	4		119	176		295	390
9070	#10, 2 wire		2	4		139	176		315	415
9080	3 wire		1.60	5		216	220		436	565
9100	#8, 2 wire, stranded		1.80	4.444		269	195		464	585
9110	3 wire, stranded		1.30	6.154		415	270		685	855
9200	600 volt, copper (MC) aluminum clad, #14, 2 wire		2.65	3.019		75.50	133		208.50	280
9210	3 wire		2.45	3.265		117	143		260	340
9220	4 wire		2.20	3.636		158	160		318	410
9230	#12, 2 wire		2.55	3.137		77.50	138		215.50	290
9240	3 wire		2.20	3.636		119	160		279	370
9250	4 wire		2	4		161	176		337	440
9260	#10, 2 wire		2.20	3.636		139	160		299	390
9270	3 wire		1.80	4.444		216	195		411	530
9280	4 wire		1.55	5.161		340	227		567	710

26 05 19.35 Cable Terminations		Crew	Daily Output	Labor-Hours	Unit	Material	2007 Bare Costs Labor	Equipment	Total	Total Incl O&P
0010	**CABLE TERMINATIONS**									
0015	Wire connectors, screw type, #22 to #14	1 Elec	260	.031	Ea.	.08	1.35		1.43	2.10
0020	#18 to #12		240	.033		.09	1.46		1.55	2.28
0025	#18 to #10		240	.033		.14	1.46		1.60	2.33
0030	Screw-on connectors, insulated, #18 to #12		240	.033		.09	1.46		1.55	2.28
0035	#16 to #10		230	.035		.12	1.53		1.65	2.40
0040	#14 to #8		210	.038		.26	1.67		1.93	2.78
0045	#12 to #6		180	.044		.44	1.95		2.39	3.38
0050	Terminal lugs, solderless, #16 to #10		50	.160		.48	7		7.48	11
0100	#8 to #4		30	.267		.74	11.70		12.44	18.25
0150	#2 to #1		22	.364		1	15.95		16.95	25
0200	1/0 to 2/0		16	.500		2.22	22		24.22	35
0250	3/0		12	.667		3.40	29.50		32.90	47
0300	4/0		11	.727		3.40	32		35.40	51
0350	250 kcmil		9	.889		3.40	39		42.40	61.50
0400	350 kcmil		7	1.143		4.40	50		54.40	79.50
0450	500 kcmil		6	1.333		8.60	58.50		67.10	96.50
1600	Crimp 1 hole lugs, copper or aluminum, 600 volt									
1620	#14	1 Elec	60	.133	Ea.	.34	5.85		6.19	9.05
1630	#12		50	.160		.47	7		7.47	10.95
1640	#10		45	.178		.47	7.80		8.27	12.10
1780	#8		36	.222		1.81	9.75		11.56	16.50
1800	#6		30	.267		2.92	11.70		14.62	20.50
2000	#4		27	.296		3.15	13		16.15	23
2200	#2		24	.333		4.30	14.65		18.95	26.50
2400	#1		20	.400		5.15	17.55		22.70	31.50
2600	2/0		15	.533		6.95	23.50		30.45	42.50
2800	3/0		12	.667		8.60	29.50		38.10	53
3000	4/0		11	.727		9.35	32		41.35	58
3200	250 kcmil		9	.889		11.15	39		50.15	70.50
3400	300 kcmil		8	1		12.50	44		56.50	79.50
3500	350 kcmil		7	1.143		12.95	50		62.95	89
3600	400 kcmil		6.50	1.231		16.10	54		70.10	98
3800	500 kcmil		6	1.333		19.50	58.50		78	109

26 05 Common Work Results for Electrical

26 05 19 – Low-Voltage Electrical Power Conductors and Cables

26 05 19.50 Mineral Insulated Cable		Crew	Daily Output	Labor-Hours	Unit	Material	2007 Bare Costs Labor	Equipment	Total	Total Incl O&P
0010	**MINERAL INSULATED CABLE** 600 volt									
0100	1 conductor, #12	1 Elec	1.60	5	C.L.F.	253	220		473	605
0200	#10		1.60	5		330	220		550	690
0400	#8		1.50	5.333		390	234		624	775
0500	#6		1.40	5.714		435	251		686	855
0600	#4	2 Elec	2.40	6.667		605	293		898	1,100
0800	#2		2.20	7.273		805	320		1,125	1,350
0900	#1		2.10	7.619		920	335		1,255	1,500
1000	1/0		2	8		1,050	350		1,400	1,700
1100	2/0		1.90	8.421		1,275	370		1,645	1,950
1200	3/0		1.80	8.889		1,500	390		1,890	2,225
1400	4/0		1.60	10		1,725	440		2,165	2,550
1410	250 kcmil	3 Elec	2.40	10		1,950	440		2,390	2,800
1420	350 kcmil		1.95	12.308		2,200	540		2,740	3,225
1430	500 kcmil		1.95	12.308		2,775	540		3,315	3,850

26 05 19.55 Non-Metallic Sheathed Cable										
0010	**NON-METALLIC SHEATHED CABLE** 600 volt									
0100	Copper with ground wire, (Romex)									
0150	#14, 2 conductor	1 Elec	2.70	2.963	C.L.F.	38.50	130		168.50	236
0200	3 conductor		2.40	3.333		53.50	146		199.50	277
0250	#12, 2 conductor		2.50	3.200		58	140		198	273
0300	3 conductor		2.20	3.636		87.50	160		247.50	335
0350	#10, 2 conductor		2.20	3.636		96.50	160		256.50	345
0400	3 conductor		1.80	4.444		138	195		333	440
0450	#8, 3 conductor		1.50	5.333		224	234		458	595
0500	#6, 3 conductor		1.40	5.714		355	251		606	770
0550	SE type SER aluminum cable, 3 RHW and									
0600	1 bare neutral, 3 #8 & 1 #8	1 Elec	1.60	5	C.L.F.	133	220		353	470
0650	3 #6 & 1 #6	"	1.40	5.714		150	251		401	540
0700	3 #4 & 1 #6	2 Elec	2.40	6.667		168	293		461	620
0750	3 #2 & 1 #4		2.20	7.273		248	320		568	745
0800	3 #1/0 & 1 #2		2	8		375	350		725	935
0850	3 #2/0 & 1 #1		1.80	8.889		440	390		830	1,075
0900	3 #4/0 & 1 #2/0		1.60	10		630	440		1,070	1,350

26 05 19.90 Wire										
0010	**WIRE** R260519-92									
0020	600 volt type THW, copper, solid, #14	1 Elec	13	.615	C.L.F.	10	27		37	51
0030	#12		11	.727		15.20	32		47.20	64
0040	#10		10	.800		23	35		58	78
0050	Stranded, #14 R260533-22		13	.615		11.60	27		38.60	53
0100	#12		11	.727		17.60	32		49.60	67
0120	#10		10	.800		27	35		62	82
0140	#8		8	1		45	44		89	115
0160	#6		6.50	1.231		75	54		129	163
0180	#4	2 Elec	10.60	1.509		118	66.50		184.50	229
0200	#3		10	1.600		148	70		218	267
0220	#2		9	1.778		186	78		264	320
0240	#1		8	2		235	88		323	390
0260	1/0		6.60	2.424		266	106		372	450
0280	2/0		5.80	2.759		335	121		456	545
0300	3/0		5	3.200		415	140		555	670
0350	4/0		4.40	3.636		520	160		680	815

26 05 Common Work Results for Electrical

26 05 19 – Low-Voltage Electrical Power Conductors and Cables

26 05 19.90 Wire

		Crew	Daily Output	Labor-Hours	Unit	Material	2007 Bare Costs Labor	Equipment	Total	Total Incl O&P
0400	250 kcmil	3 Elec	6	4	C.L.F.	600	176		776	925
0420	300 kcmil		5.70	4.211		715	185		900	1,075
0450	350 kcmil		5.40	4.444		840	195		1,035	1,225
0480	400 kcmil		5.10	4.706		960	207		1,167	1,350
0490	500 kcmil		4.80	5		1,200	220		1,420	1,625
0540	Aluminum, stranded, #6	1 Elec	8	1		25.50	44		69.50	93.50
0560	#4	2 Elec	13	1.231		31.50	54		85.50	115
0580	#2		10.60	1.509		42.50	66.50		109	145
0600	#1		9	1.778		62	78		140	185
0620	1/0		8	2		75	88		163	214
0640	2/0		7.20	2.222		88	97.50		185.50	242
0680	3/0		6.60	2.424		109	106		215	278
0700	4/0		6.20	2.581		122	113		235	305
0720	250 kcmil	3 Elec	8.70	2.759		149	121		270	345
0740	300 kcmil		8.10	2.963		205	130		335	420
0760	350 kcmil		7.50	3.200		208	140		348	440
0780	400 kcmil		6.90	3.478		244	153		397	495
0800	500 kcmil		6	4		269	176		445	555
0850	600 kcmil		5.70	4.211		340	185		525	650
0880	700 kcmil		5.10	4.706		395	207		602	745
0900	750 kcmil		4.80	5		400	220		620	760
0920	Type THWN-THHN, copper, solid, #14	1 Elec	13	.615		10	27		37	51
0940	#12		11	.727		15.20	32		47.20	64
0960	#10		10	.800		23	35		58	78
1000	Stranded, #14		13	.615		11.60	27		38.60	53
1200	#12		11	.727		17.60	32		49.60	67
1250	#10	CN	10	.800		27	35		62	82
1300	#8		8	1		45	44		89	115
1350	#6		6.50	1.231		75	54		129	163
1400	#4	2 Elec	10.60	1.509		118	66.50		184.50	229

26 05 23 – Control-Voltage Electrical Power Cables

26 05 23.10 Control Cable

		Crew	Daily Output	Labor-Hours	Unit	Material	Labor	Equipment	Total	Total Incl O&P
0010	**CONTROL CABLE**									
0020	600 volt, copper, #14 THWN wire with PVC jacket, 2 wires	1 Elec	9	.889	C.L.F.	26.50	39		65.50	87
0100	4 wires		7	1.143		46.50	50		96.50	126
0200	6 wires		6	1.333		83	58.50		141.50	179
0300	8 wires		5.30	1.509		102	66.50		168.50	211
0400	10 wires		4.80	1.667		121	73		194	242
0500	12 wires		4.30	1.860		139	81.50		220.50	275
0600	14 wires		3.80	2.105		167	92.50		259.50	320
0700	16 wires		3.50	2.286		170	100		270	335
0800	18 wires		3.30	2.424		191	106		297	370
0900	20 wires		3	2.667		221	117		338	415
1000	22 wires		2.80	2.857		230	125		355	440

26 05 26 – Grounding and Bonding for Electrical Systems

26 05 26.80 Grounding

		Crew	Daily Output	Labor-Hours	Unit	Material	Labor	Equipment	Total	Total Incl O&P
0010	**GROUNDING**									
0030	Rod, copper clad, 8' long, 1/2" diameter	1 Elec	5.50	1.455	Ea.	14.05	64		78.05	110
0050	3/4" diameter		5.30	1.509		29.50	66.50		96	131
0080	10' long, 1/2" diameter		4.80	1.667		18.10	73		91.10	129
0100	3/4" diameter		4.40	1.818		32.50	80		112.50	155
0130	15' long, 3/4" diameter		4	2		87	88		175	227

26 05 Common Work Results for Electrical

26 05 26 – Grounding and Bonding for Electrical Systems

26 05 26.80 Grounding

		Crew	Daily Output	Labor-Hours	Unit	Material	2007 Bare Costs Labor	Equipment	Total	Total Incl O&P
0390	Bare copper wire, stranded, #8	1 Elec	11	.727	C.L.F.	40	32		72	91.50
0400	#6		10	.800		73	35		108	133
0600	#2	2 Elec	10	1.600		167	70		237	289
0800	3/0		6.60	2.424		375	106		481	570
1000	4/0		5.70	2.807		470	123		593	700
1200	250 kcmil	3 Elec	7.20	3.333		555	146		701	830
1800	Water pipe ground clamps, heavy duty									
2000	Bronze, 1/2" to 1" diameter	1 Elec	8	1	Ea.	21	44		65	88.50
2100	1-1/4" to 2" diameter		8	1		27.50	44		71.50	96
2200	2-1/2" to 3" diameter		6	1.333		46.50	58.50		105	138
2800	Brazed connections, #6 wire		12	.667		12.90	29.50		42.40	57.50
3000	#2 wire		10	.800		17.30	35		52.30	71.50
3100	3/0 wire		8	1		26	44		70	94
3200	4/0 wire		7	1.143		29.50	50		79.50	107
3400	250 kcmil wire		5	1.600		34.50	70		104.50	143
3600	500 kcmil wire		4	2		42.50	88		130.50	178

26 05 33 – Raceway and Boxes for Electrical Systems

26 05 33.05 Conduit

		Crew	Daily Output	Labor-Hours	Unit	Material	2007 Bare Costs Labor	Equipment	Total	Total Incl O&P
0010	**CONDUIT** To 15' high, includes 2 terminations, 2 elbows,									
0020	11 beam clamps, and 11 couplings per 100 L.F.									
0300	Aluminum, 1/2" diameter	1 Elec	100	.080	L.F.	2.08	3.51		5.59	7.55
0500	3/4" diameter		90	.089		2.81	3.90		6.71	8.90
0700	1" diameter		80	.100		3.72	4.39		8.11	10.65
1000	1-1/4" diameter		70	.114		5	5		10	12.95
1030	1-1/2" diameter		65	.123		6	5.40		11.40	14.65
1050	2" diameter		60	.133		8.20	5.85		14.05	17.70
1070	2-1/2" diameter		50	.160		13.30	7		20.30	25
1100	3" diameter	2 Elec	90	.178		18.15	7.80		25.95	31.50
1130	3-1/2" diameter		80	.200		24	8.80		32.80	39.50
1140	4" diameter		70	.229		29	10.05		39.05	47
1750	Rigid galvanized steel, 1/2" diameter	1 Elec	90	.089		2.48	3.90		6.38	8.55
1770	3/4" diameter		80	.100		2.83	4.39		7.22	9.65
1800	1" diameter		65	.123		3.89	5.40		9.29	12.35
1830	1-1/4" diameter		60	.133		5.40	5.85		11.25	14.65
1850	1-1/2" diameter		55	.145		6.25	6.40		12.65	16.35
1870	2" diameter		45	.178		8	7.80		15.80	20.50
1900	2-1/2" diameter		35	.229		15.05	10.05		25.10	31.50
1930	3" diameter	2 Elec	50	.320		18.40	14.05		32.45	41
1950	3-1/2" diameter		44	.364		23	15.95		38.95	49
1970	4" diameter		40	.400		26	17.55		43.55	54.50
2500	Steel, intermediate conduit (IMC), 1/2" diameter	1 Elec	100	.080		1.92	3.51		5.43	7.35
2530	3/4" diameter		90	.089		2.37	3.90		6.27	8.40
2550	1" diameter		70	.114		3.32	5		8.32	11.10
2570	1-1/4" diameter		65	.123		4.42	5.40		9.82	12.90
2600	1-1/2" diameter		60	.133		5.20	5.85		11.05	14.40
2630	2" diameter		50	.160		6.75	7		13.75	17.85
2650	2-1/2" diameter		40	.200		12.80	8.80		21.60	27
2670	3" diameter	2 Elec	60	.267		16.80	11.70		28.50	36
2700	3-1/2" diameter		54	.296		21	13		34	42.50
2730	4" diameter		50	.320		24	14.05		38.05	47.50
5000	Electric metallic tubing (EMT), 1/2" diameter	1 Elec	170	.047		.60	2.07		2.67	3.74
5020	3/4" diameter	CN	130	.062		.99	2.70		3.69	5.10

26 05 Common Work Results for Electrical

26 05 33 – Raceway and Boxes for Electrical Systems

26 05 33.05 Conduit

		Crew	Daily Output	Labor-Hours	Unit	Material	2007 Bare Costs Labor	2007 Bare Costs Equipment	Total	Total Incl O&P
5040	1" diameter	1 Elec	115	.070	L.F.	1.74	3.05		4.79	6.45
5060	1-1/4" diameter		100	.080		2.76	3.51		6.27	8.30
5080	1-1/2" diameter		90	.089		3.50	3.90		7.40	9.65
5100	2" diameter		80	.100		4.41	4.39		8.80	11.40
5120	2-1/2" diameter		60	.133		9	5.85		14.85	18.55
5140	3" diameter	2 Elec	100	.160		11.75	7		18.75	23.50
5160	3-1/2" diameter		90	.178		15.20	7.80		23	28.50
5180	4" diameter		80	.200		16.85	8.80		25.65	31.50
9900	Add to labor for higher elevated installation									
9910	15' to 20' high, add						10%			
9920	20' to 25' high, add						20%			
9930	25' to 30' high, add						25%			
9940	30' to 35' high, add						30%			
9950	35' to 40' high, add						35%			
9960	Over 40' high, add						40%			
9980	Allow. for cond. ftngs., 5% min.-20% max.									

26 05 33.45 Wireway

		Crew	Daily Output	Labor-Hours	Unit	Material	Labor	Equipment	Total	Total Incl O&P
0010	**WIREWAY** to 15' high									
0100	Screw cover, NEMA 1 w/ fittings and supports, 2-1/2" x 2-1/2"	1 Elec	45	.178	L.F.	11.45	7.80		19.25	24
0200	4" x 4"	"	40	.200		12.65	8.80		21.45	27
0400	6" x 6"	2 Elec	60	.267		19.70	11.70		31.40	39
0600	8" x 8"	"	40	.400		33	17.55		50.55	62.50
4475	Screw cover, NEMA 3R w/ fittings and supports, 4" x 4"	1 Elec	36	.222		27	9.75		36.75	44
4480	6" x 6"	2 Elec	55	.291		34	12.75		46.75	56.50
4485	8" x 8"		36	.444		55	19.50		74.50	89.50
4490	12" x 12"		18	.889		77	39		116	143
9980	Allow. for wireway ftngs., 5% min.-20% max.									

26 05 33.50 Outlet Boxes

		Crew	Daily Output	Labor-Hours	Unit	Material	Labor	Equipment	Total	Total Incl O&P
0010	**OUTLET BOXES**									
0020	Pressed steel, octagon, 4"	1 Elec	20	.400	Ea.	2.73	17.55		20.28	29
0060	Covers, blank		64	.125		1.14	5.50		6.64	9.40
0100	Extension rings		40	.200		4.53	8.80		13.33	18.05
0150	Square, 4"		20	.400		2.35	17.55		19.90	28.50
0200	Extension rings		40	.200		4.58	8.80		13.38	18.10
0250	Covers, blank		64	.125		1.29	5.50		6.79	9.55
0300	Plaster rings		64	.125		2.51	5.50		8.01	10.90
0650	Switchbox		27	.296		4.38	13		17.38	24
1100	Concrete, floor, 1 gang		5.30	1.509		73	66.50		139.50	179
2000	Poke-thru fitting, fire rated, for 3-3/4" floor		6.80	1.176		100	51.50		151.50	187
2040	For 7" floor		6.80	1.176		100	51.50		151.50	187
2100	Pedestal, 15 amp, duplex receptacle & blank plate		5.25	1.524		103	67		170	214
2120	Duplex receptacle and telephone plate		5.25	1.524		104	67		171	214
2140	Pedestal, 20 amp, duplex recept. & phone plate		5	1.600		105	70		175	220
2200	Abandonment plate		32	.250		30.50	11		41.50	50

26 05 33.65 Pull Boxes

		Crew	Daily Output	Labor-Hours	Unit	Material	Labor	Equipment	Total	Total Incl O&P
0010	**PULL BOXES**									
0100	Sheet metal, pull box, NEMA 1, type SC, 6" W x 6" H x 4" D	1 Elec	8	1	Ea.	11.65	44		55.65	78.50
0200	8" W x 8" H x 4" D		8	1		16	44		60	83
0300	10" W x 12" H x 6" D		5.30	1.509		28	66.50		94.50	130
0400	16" W x 20" H x 8" D		4	2		106	88		194	248
0500	20" W x 24" H x 8" D		3.20	2.500		125	110		235	300
0600	24" W x 36" H x 8" D		2.70	2.963		176	130		306	390

26 05 Common Work Results for Electrical

26 05 33 – Raceway and Boxes for Electrical Systems

26 05 33.65 Pull Boxes		Crew	Daily Output	Labor-Hours	Unit	Material	2007 Bare Costs Labor	Equipment	Total	Total Incl O&P
0650	Pull box, hinged, NEMA 1, 6" W x 6" H x 4" D	1 Elec	8	1	Ea.	11.90	44		55.90	78.50
0800	12" W x 16" H x 6" D		4.70	1.702		39	74.50		113.50	154
1000	20" W x 20" H x 6" D		3.60	2.222		78.50	97.50		176	232
1200	20" W x 20" H x 8" D		3.20	2.500		157	110		267	335
1400	24" W x 36" H x 8" D		2.70	2.963		252	130		382	470
1600	24" W x 42" H x 8" D	↓	2	4	↓	380	176		556	680
2100	Pull box, NEMA 3R, type SC, raintight & weatherproof									
2150	6" L x 6" W x 6" D	1 Elec	10	.800	Ea.	20	35		55	74.50
2200	8" L x 6" W x 6" D		8	1		25	44		69	93
2250	10" L x 6" W x 6" D		7	1.143		33	50		83	111
2300	12" L x 12" W x 6" D		5	1.600		47	70		117	157
2350	16" L x 16" W x 6" D		4.50	1.778		93.50	78		171.50	219
2400	20" L x 20" W x 6" D		4	2		128	88		216	272
2450	24" L x 18" W x 8" D		3	2.667		140	117		257	330
2500	24" L x 24" W x 10" D		2.50	3.200		188	140		328	415
2550	30" L x 24" W x 12" D		2	4		340	176		516	635
2600	36" L x 36" W x 12" D	↓	1.50	5.333	↓	450	234		684	845
2800	Cast iron, pull boxes for surface mounting									
3000	NEMA 4, watertight & dust tight									
3050	6" L x 6" W x 6" D	1 Elec	4	2	Ea.	171	88		259	320
3100	8" L x 6" W x 6" D		3.20	2.500		232	110		342	420
3150	10" L x 6" W x 6" D		2.50	3.200		289	140		429	530
3200	12" L x 12" W x 6" D		2.30	3.478		490	153		643	760
3250	16" L x 16" W x 6" D		1.30	6.154		1,000	270		1,270	1,500
3300	20" L x 20" W x 6" D		.80	10		1,850	440		2,290	2,700
3350	24" L x 18" W x 8" D		.70	11.429		1,950	500		2,450	2,900
3400	24" L x 24" W x 10" D		.50	16		3,000	700		3,700	4,350
3450	30" L x 24" W x 12" D		.40	20		5,025	880		5,905	6,825
3500	36" L x 36" W x 12" D	↓	.20	40	↓	6,200	1,750		7,950	9,450
6000	J.I.C. wiring boxes, NEMA 12, dust tight & drip tight									
6050	6" L x 8" W x 4" D	1 Elec	10	.800	Ea.	46.50	35		81.50	104
6100	8" L x 10" W x 4" D		8	1		58.50	44		102.50	130
6150	12" L x 14" W x 6" D		5.30	1.509		95	66.50		161.50	203
6200	14" L x 16" W x 6" D		4.70	1.702		113	74.50		187.50	235
6250	16" L x 20" W x 6" D		4.40	1.818		241	80		321	385
6300	24" L x 30" W x 6" D		3.20	2.500		355	110		465	555
6350	24" L x 30" W x 8" D		2.90	2.759		380	121		501	595
6400	24" L x 36" W x 8" D		2.70	2.963		420	130		550	655
6450	24" L x 42" W x 8" D		2.30	3.478		460	153		613	730
6500	24" L x 48" W x 8" D	↓	2	4	↓	500	176		676	810

26 05 36 – Cable Trays for Electrical Systems

26 05 36.10 Cable Tray Ladder Type

		Crew	Daily Output	Labor-Hours	Unit	Material	Labor	Equipment	Total	Total Incl O&P
0010	**CABLE TRAY LADDER TYPE** w/ ftngs. & supports, 4" dp., to 15' elev.									
0160	Galvanized steel tray									
0170	4" rung spacing, 6" wide	2 Elec	98	.163	L.F.	10.85	7.15		18	22.50
0200	12" wide		86	.186		13.05	8.15		21.20	26.50
0400	18" wide		82	.195		15.15	8.55		23.70	29.50
0600	24" wide		78	.205		17.45	9		26.45	32.50
3200	Aluminum tray, 4" deep, 6" rung spacing, 6" wide		134	.119		13.65	5.25		18.90	23
3220	12" wide		124	.129		15.30	5.65		20.95	25.50
3230	18" wide		114	.140		17	6.15		23.15	28
3240	24" wide	↓	106	.151	↓	19.75	6.65		26.40	32

26 05 Common Work Results for Electrical

26 05 36 – Cable Trays for Electrical Systems

26 05 36.10 Cable Tray Ladder Type		Crew	Daily Output	Labor-Hours	Unit	Material	2007 Bare Costs Labor	2007 Bare Costs Equipment	Total	Total Incl O&P
9980	Allow. for tray ftngs., 5% min.-20% max.									

26 05 39 – Underfloor Raceways for Electrical Systems

26 05 39.30 Conduit In Concrete Slab

		Crew	Daily Output	Labor-Hours	Unit	Material	Labor	Equipment	Total	Total Incl O&P
0010	**CONDUIT IN CONCRETE SLAB** Including terminations,									
0020	fittings and supports									
3230	PVC, schedule 40, 1/2" diameter	1 Elec	270	.030	L.F.	.73	1.30		2.03	2.75
3250	3/4" diameter		230	.035		.88	1.53		2.41	3.24
3270	1" diameter		200	.040		1.20	1.76		2.96	3.93
3300	1-1/4" diameter		170	.047		1.68	2.07		3.75	4.93
3330	1-1/2" diameter		140	.057		2.02	2.51		4.53	5.95
3350	2" diameter		120	.067		2.60	2.93		5.53	7.20
4350	Rigid galvanized steel, 1/2" diameter		200	.040		2.26	1.76		4.02	5.10
4400	3/4" diameter		170	.047		2.61	2.07		4.68	5.95
4450	1" diameter		130	.062		3.67	2.70		6.37	8.05
4500	1-1/4" diameter		110	.073		4.96	3.19		8.15	10.20
4600	1-1/2" diameter		100	.080		5.80	3.51		9.31	11.60
4800	2" diameter		90	.089		7.35	3.90		11.25	13.90

26 05 39.40 Conduit In Trench

		Crew	Daily Output	Labor-Hours	Unit	Material	Labor	Equipment	Total	Total Incl O&P
0010	**CONDUIT IN TRENCH** Includes terminations and fittings									
0020	Does not include excavation or backfill, see div. 31 23 16.00									
0200	Rigid galvanized steel, 2" diameter	1 Elec	150	.053	L.F.	7.05	2.34		9.39	11.25
0400	2-1/2" diameter	"	100	.080		13.75	3.51		17.26	20.50
0600	3" diameter	2 Elec	160	.100		17.05	4.39		21.44	25.50
0800	3-1/2" diameter		140	.114		22	5		27	31.50
1000	4" diameter		100	.160		24	7		31	37
1200	5" diameter		80	.200		52	8.80		60.80	70
1400	6" diameter		60	.267		76	11.70		87.70	101

26 05 43 – Underground Ducts and Raceways for Electrical Systems

26 05 43.10 Trench Duct

		Crew	Daily Output	Labor-Hours	Unit	Material	Labor	Equipment	Total	Total Incl O&P
0010	**TRENCH DUCT** Steel with cover									
0020	Standard adjustable, depths to 4"									
0100	Straight, single compartment, 9" wide	2 Elec	40	.400	L.F.	82	17.55		99.55	116
0200	12" wide		32	.500		93	22		115	135
0400	18" wide		26	.615		124	27		151	176
0600	24" wide		22	.727		159	32		191	223
0800	30" wide		20	.800		192	35		227	264
1000	36" wide		16	1		231	44		275	320
1200	Horizontal elbow, 9" wide		5.40	2.963	Ea.	305	130		435	530
1400	12" wide		4.60	3.478		350	153		503	610
1600	18" wide		4	4		450	176		626	755
1800	24" wide		3.20	5		630	220		850	1,025
2000	30" wide		2.60	6.154		840	270		1,110	1,325
2200	36" wide		2.40	6.667		1,100	293		1,393	1,625
2400	Vertical elbow, 9" wide		5.40	2.963		106	130		236	310
2600	12" wide		4.60	3.478		115	153		268	355
2800	18" wide		4	4		132	176		308	405
3000	24" wide		3.20	5		164	220		384	505
3200	30" wide		2.60	6.154		181	270		451	600
3400	36" wide		2.40	6.667		199	293		492	655
3600	Cross, 9" wide		4	4		495	176		671	805
3800	12" wide		3.20	5		525	220		745	905

26 05 Common Work Results for Electrical

26 05 43 – Underground Ducts and Raceways for Electrical Systems

26 05 43.10 Trench Duct

		Crew	Daily Output	Labor-Hours	Unit	Material	2007 Bare Costs Labor	Equipment	Total	Total Incl O&P
4000	18" wide	2 Elec	2.60	6.154	Ea.	630	270		900	1,100
4200	24" wide		2.20	7.273		800	320		1,120	1,350
4400	30" wide		2	8		1,025	350		1,375	1,675
4600	36" wide		1.80	8.889		1,300	390		1,690	2,000
4800	End closure, 9" wide		14.40	1.111		31	49		80	107
5000	12" wide		12	1.333		35.50	58.50		94	126
5200	18" wide		10	1.600		54.50	70		124.50	165
5400	24" wide		8	2		72	88		160	210
5600	30" wide		6.60	2.424		90	106		196	257
5800	36" wide		5.80	2.759		107	121		228	298
6000	Tees, 9" wide		4	4		300	176		476	590
6200	12" wide		3.60	4.444		350	195		545	675
6400	18" wide		3.20	5		450	220		670	820
6600	24" wide		3	5.333		645	234		879	1,050
6800	30" wide		2.60	6.154		840	270		1,110	1,325
7000	36" wide		2	8		1,100	350		1,450	1,725
7200	Riser, and cabinet connector, 9" wide		5.40	2.963		132	130		262	340
7400	12" wide		4.60	3.478		154	153		307	395
7600	18" wide		4	4		189	176		365	470
7800	24" wide		3.20	5		229	220		449	575
8000	30" wide		2.60	6.154		263	270		533	690
8200	36" wide		2	8		305	350		655	860
8400	Insert assembly, cell to conduit adapter, 1-1/4"	1 Elec	16	.500		52.50	22		74.50	90

26 05 43.20 Underfloor Duct

		Crew	Daily Output	Labor-Hours	Unit	Material	2007 Bare Costs Labor	Equipment	Total	Total Incl O&P
0010	**UNDERFLOOR DUCT**									
0100	Duct, 1-3/8" x 3-1/8" blank, standard	2 Elec	160	.100	L.F.	10	4.39		14.39	17.55
0200	1-3/8" x 7-1/4" blank, super duct		120	.133		20	5.85		25.85	30.50
0400	7/8" or 1-3/8" insert type, 24" O.C., 1-3/8" x 3-1/8", std.		140	.114		13.40	5		18.40	22
0600	1-3/8" x 7-1/4", super duct		100	.160		23.50	7		30.50	36
0800	Junction box, single duct, 1 level, 3-1/8"	1 Elec	4	2	Ea.	300	88		388	460
1000	Junction box, single duct, 1 level, 7-1/4"		2.70	2.963		350	130		480	580
1200	1 level, 2 duct, 3-1/8"		3.20	2.500		400	110		510	605
1400	Junction box, 1 level, 2 duct, 7-1/4"		2.30	3.478		1,025	153		1,178	1,350
1600	Triple duct, 3-1/8"		2.30	3.478		680	153		833	975
1800	Insert to conduit adapter, 3/4" & 1"		32	.250		24.50	11		35.50	43.50
2000	Support, single cell		27	.296		37	13		50	60
2200	Super duct		16	.500		37	22		59	73
2400	Double cell		16	.500		37	22		59	73
2600	Triple cell		11	.727		37	32		69	88
2800	Vertical elbow, standard duct		10	.800		66.50	35		101.50	126
3000	Super duct		8	1		66.50	44		110.50	139
3200	Cabinet connector, standard duct		32	.250		50	11		61	71.50
3400	Super duct		27	.296		50	13		63	74.50
3600	Conduit adapter, 1" to 1-1/4"		32	.250		50	11		61	71.50
3800	2" to 1-1/4"		27	.296		60	13		73	85.50
4000	Outlet, low tension (tele, computer, etc.)		8	1		70	44		114	143
4200	High tension, receptacle (120 volt)		8	1		70	44		114	143

26 05 80 – Wiring Connections

26 05 80.10 Motor Connections

		Crew	Daily Output	Labor-Hours	Unit	Material	2007 Bare Costs Labor	Equipment	Total	Total Incl O&P
0010	**MOTOR CONNECTIONS**									
0020	Flexible conduit and fittings, 115 volt, 1 phase, up to 1 HP motor	1 Elec	8	1	Ea.	8.75	44		52.75	75
0120	230 volt, 10 HP motor, 3 phase		4.20	1.905		14.40	83.50		97.90	140

26 05 Common Work Results for Electrical

26 05 80 – Wiring Connections

26 05 80.10 Motor Connections

		Crew	Daily Output	Labor-Hours	Unit	Material	2007 Bare Costs Labor	Equipment	Total	Total Incl O&P
0200	25 HP motor	1 Elec	2.70	2.963	Ea.	29.50	130		159.50	227
0400	50 HP motor	↓	2.20	3.636	↓	76	160		236	320
0600	100 HP motor	↓	1.50	5.333	↓	183	234		417	550

26 05 90 – Residential Wiring

26 05 90.10 Residential Wiring

		Crew	Daily Output	Labor-Hours	Unit	Material	Labor	Equipment	Total	Total Incl O&P
0010	**RESIDENTIAL WIRING**									
0020	20' avg. runs and #14/2 wiring incl. unless otherwise noted									
1000	Service & panel, includes 24' SE-AL cable, service eye, meter,									
1010	Socket, panel board, main bkr., ground rod, 15 or 20 amp									
1020	1-pole circuit breakers, and misc. hardware									
1100	100 amp, with 10 branch breakers	1 Elec	1.19	6.723	Ea.	515	295		810	1,000
1110	With PVC conduit and wire		.92	8.696		575	380		955	1,200
1120	With RGS conduit and wire		.73	10.959		735	480		1,215	1,525
1150	150 amp, with 14 branch breakers		1.03	7.767		800	340		1,140	1,400
1170	With PVC conduit and wire		.82	9.756		925	430		1,355	1,675
1180	With RGS conduit and wire	↓	.67	11.940		1,250	525		1,775	2,125
1200	200 amp, with 18 branch breakers	2 Elec	1.80	8.889		1,050	390		1,440	1,725
1220	With PVC conduit and wire		1.46	10.959		1,175	480		1,655	2,000
1230	With RGS conduit and wire	↓	1.24	12.903		1,600	565		2,165	2,600
1800	Lightning surge suppressor for above services, add	1 Elec	32	.250	↓	46	11		57	67
2000	Switch devices									
2100	Single pole, 15 amp, Ivory, with a 1-gang box, cover plate,									
2110	Type NM (Romex) cable	1 Elec	17.10	.468	Ea.	12.65	20.50		33.15	44.50
2120	Type MC (BX) cable		14.30	.559		26.50	24.50		51	65.50
2130	EMT & wire		5.71	1.401		32	61.50		93.50	127
2150	3-way, #14/3, type NM cable		14.55	.550		17.35	24		41.35	55
2170	Type MC cable		12.31	.650		37	28.50		65.50	83
2180	EMT & wire		5	1.600		35.50	70		105.50	144
2200	4-way, #14/3, type NM cable		14.55	.550		31.50	24		55.50	70.50
2220	Type MC cable		12.31	.650		51	28.50		79.50	98.50
2230	EMT & wire		5	1.600		49.50	70		119.50	160
2250	S.P., 20 amp, #12/2, type NM cable		13.33	.600		21	26.50		47.50	62
2270	Type MC cable		11.43	.700		31	30.50		61.50	80
2280	EMT & wire		4.85	1.649		41.50	72.50		114	154
2290	S.P. rotary dimmer, 600W, no wiring		17	.471		17.45	20.50		37.95	50
2300	S.P. rotary dimmer, 600W, type NM cable		14.55	.550		25	24		49	63.50
2320	Type MC cable		12.31	.650		39	28.50		67.50	85.50
2330	EMT & wire		5	1.600		46.50	70		116.50	156
2350	3-way rotary dimmer, type NM cable		13.33	.600		23	26.50		49.50	64.50
2370	Type MC cable		11.43	.700		37	30.50		67.50	86
2380	EMT & wire	↓	4.85	1.649	↓	44.50	72.50		117	157
2400	Interval timer wall switch, 20 amp, 1-30 min., #12/2									
2410	Type NM cable	1 Elec	14.55	.550	Ea.	46	24		70	86.50
2420	Type MC cable		12.31	.650		51	28.50		79.50	98.50
2430	EMT & wire	↓	5	1.600	↓	66.50	70		136.50	178
2500	Decorator style									
2510	S.P., 15 amp, type NM cable	1 Elec	17.10	.468	Ea.	16.75	20.50		37.25	49
2520	Type MC cable		14.30	.559		30.50	24.50		55	70
2530	EMT & wire		5.71	1.401		36	61.50		97.50	131
2550	3-way, #14/3, type NM cable		14.55	.550		21.50	24		45.50	59.50
2570	Type MC cable		12.31	.650		41	28.50		69.50	87.50
2580	EMT & wire	↓	5	1.600		39.50	70		109.50	149

26 05 Common Work Results for Electrical

26 05 90 – Residential Wiring

26 05 90.10 Residential Wiring		Crew	Daily Output	Labor-Hours	Unit	Material	2007 Bare Costs Labor	Equipment	Total	Total Incl O&P
2600	4-way, #14/3, type NM cable	1 Elec	14.55	.550	Ea.	35.50	24		59.50	75
2620	Type MC cable		12.31	.650		55	28.50		83.50	103
2630	EMT & wire		5	1.600		53.50	70		123.50	164
2650	S.P., 20 amp, #12/2, type NM cable		13.33	.600		25	26.50		51.50	66.50
2670	Type MC cable		11.43	.700		35.50	30.50		66	84.50
2680	EMT & wire		4.85	1.649		45.50	72.50		118	158
2700	S.P., slide dimmer, type NM cable		17.10	.468		31	20.50		51.50	65
2720	Type MC cable		14.30	.559		45	24.50		69.50	86
2730	EMT & wire		5.71	1.401		52.50	61.50		114	150
2750	S.P., touch dimmer, type NM cable		17.10	.468		28	20.50		48.50	61.50
2770	Type MC cable		14.30	.559		42	24.50		66.50	82.50
2780	EMT & wire		5.71	1.401		49.50	61.50		111	146
2800	3-way touch dimmer, type NM cable		13.33	.600		48	26.50		74.50	92
2820	Type MC cable		11.43	.700		62	30.50		92.50	114
2830	EMT & wire		4.85	1.649		69.50	72.50		142	185
3000	Combination devices									
3100	S.P. switch/15 amp recpt., Ivory, 1-gang box, plate									
3110	Type NM cable	1 Elec	11.43	.700	Ea.	23	30.50		53.50	71
3120	Type MC cable		10	.800		37	35		72	93
3130	EMT & wire		4.40	1.818		44	80		124	168
3150	S.P. switch/pilot light, type NM cable		11.43	.700		23.50	30.50		54	71.50
3170	Type MC cable		10	.800		37.50	35		72.50	94
3180	EMT & wire		4.43	1.806		45	79.50		124.50	168
3190	2-S.P. switches, 2-#14/2, no wiring		14	.571		7.35	25		32.35	45.50
3200	2-S.P. switches, 2-#14/2, type NM cables		10	.800		29.50	35		64.50	85
3220	Type MC cable		8.89	.900		51	39.50		90.50	115
3230	EMT & wire		4.10	1.951		49.50	85.50		135	183
3250	3-way switch/15 amp recpt., #14/3, type NM cable		10	.800		31.50	35		66.50	87
3270	Type MC cable		8.89	.900		51	39.50		90.50	115
3280	EMT & wire		4.10	1.951		49.50	85.50		135	183
3300	2-3 way switches, 2-#14/3, type NM cables		8.89	.900		44.50	39.50		84	108
3320	Type MC cable		8	1		77	44		121	150
3330	EMT & wire		4	2		58	88		146	195
3350	S.P. switch/20 amp recpt., #12/2, type NM cable		10	.800		35.50	35		70.50	91.50
3370	Type MC cable		8.89	.900		40	39.50		79.50	103
3380	EMT & wire		4.10	1.951		55.50	85.50		141	190
3400	Decorator style									
3410	S.P. switch/15 amp recpt., type NM cable	1 Elec	11.43	.700	Ea.	27	30.50		57.50	75
3420	Type MC cable		10	.800		41	35		76	97.50
3430	EMT & wire		4.40	1.818		48.50	80		128.50	172
3450	S.P. switch/pilot light, type NM cable		11.43	.700		28	30.50		58.50	76
3470	Type MC cable		10	.800		41.50	35		76.50	98.50
3480	EMT & wire		4.40	1.818		49	80		129	173
3500	2-S.P. switches, 2-#14/2, type NM cables		10	.800		34	35		69	89.50
3520	Type MC cable		8.89	.900		55	39.50		94.50	120
3530	EMT & wire		4.10	1.951		53.50	85.50		139	187
3550	3-way/15 amp recpt., #14/3, type NM cable		10	.800		35.50	35		70.50	91.50
3570	Type MC cable		8.89	.900		55	39.50		94.50	120
3580	EMT & wire		4.10	1.951		53.50	85.50		139	187
3650	2-3 way switches, 2-#14/3, type NM cables		8.89	.900		48.50	39.50		88	113
3670	Type MC cable		8	1		81	44		125	155
3680	EMT & wire		4	2		62.50	88		150.50	200
3700	S.P. switch/20 amp recpt., #12/2, type NM cable		10	.800		39.50	35		74.50	96

26 05 Common Work Results for Electrical

26 05 90 – Residential Wiring

26 05 90.10 Residential Wiring		Crew	Daily Output	Labor-Hours	Unit	Material	2007 Bare Costs Labor	Equipment	Total	Total Incl O&P
3720	Type MC cable	1 Elec	8.89	.900	Ea.	44	39.50		83.50	108
3730	EMT & wire	↓	4.10	1.951	↓	60	85.50		145.50	194
4000	Receptacle devices									
4010	Duplex outlet, 15 amp recpt., Ivory, 1-gang box, plate									
4015	Type NM cable	1 Elec	14.55	.550	Ea.	11	24		35	48
4020	Type MC cable		12.31	.650		25	28.50		53.50	70
4030	EMT & wire		5.33	1.501		30.50	66		96.50	132
4050	With #12/2, type NM cable		12.31	.650		14.95	28.50		43.45	59
4070	Type MC cable		10.67	.750		25	33		58	76.50
4080	EMT & wire		4.71	1.699		35.50	74.50		110	150
4100	20 amp recpt., #12/2, type NM cable		12.31	.650		23	28.50		51.50	68
4120	Type MC cable		10.67	.750		33	33		66	85.50
4130	EMT & wire	↓	4.71	1.699	↓	43.50	74.50		118	159
4140	For GFI see line 4300 below									
4150	Decorator style, 15 amp recpt., type NM cable	1 Elec	14.55	.550	Ea.	15.10	24		39.10	52.50
4170	Type MC cable		12.31	.650		29	28.50		57.50	74.50
4180	EMT & wire		5.33	1.501		34.50	66		100.50	136
4200	With #12/2, type NM cable		12.31	.650		19.05	28.50		47.55	63.50
4220	Type MC cable		10.67	.750		29	33		62	81
4230	EMT & wire		4.71	1.699		39.50	74.50		114	155
4250	20 amp recpt. #12/2, type NM cable		12.31	.650		27	28.50		55.50	72.50
4270	Type MC cable		10.67	.750		37.50	33		70.50	90
4280	EMT & wire		4.71	1.699		47.50	74.50		122	164
4300	GFI, 15 amp recpt., type NM cable		12.31	.650		39.50	28.50		68	86
4320	Type MC cable		10.67	.750		53.50	33		86.50	108
4330	EMT & wire		4.71	1.699		58.50	74.50		133	176
4350	GFI with #12/2, type NM cable		10.67	.750		43.50	33		76.50	96.50
4370	Type MC cable		9.20	.870		53.50	38		91.50	116
4380	EMT & wire		4.21	1.900		64	83.50		147.50	194
4400	20 amp recpt., #12/2 type NM cable		10.67	.750		45	33		78	98.50
4420	Type MC cable		9.20	.870		55	38		93	118
4430	EMT & wire		4.21	1.900		65.50	83.50		149	196
4500	Weather-proof cover for above receptacles, add	↓	32	.250	↓	4.55	11		15.55	21.50
4550	Air conditioner outlet, 20 amp-240 volt recpt.									
4560	30' of #12/2, 2 pole circuit breaker									
4570	Type NM cable	1 Elec	10	.800	Ea.	59.50	35		94.50	118
4580	Type MC cable		9	.889		71	39		110	137
4590	EMT & wire		4	2		78.50	88		166.50	218
4600	Decorator style, type NM cable		10	.800		63.50	35		98.50	123
4620	Type MC cable		9	.889		75.50	39		114.50	141
4630	EMT & wire	↓	4	2	↓	83	88		171	222
4650	Dryer outlet, 30 amp-240 volt recpt., 20' of #10/3									
4660	2 pole circuit breaker									
4670	Type NM cable	1 Elec	6.41	1.248	Ea.	78.50	55		133.50	168
4680	Type MC cable		5.71	1.401		76	61.50		137.50	175
4690	EMT & wire	↓	3.48	2.299	↓	81.50	101		182.50	240
4700	Range outlet, 50 amp-240 volt recpt., 30' of #8/3									
4710	Type NM cable	1 Elec	4.21	1.900	Ea.	116	83.50		199.50	251
4720	Type MC cable		4	2		160	88		248	305
4730	EMT & wire		2.96	2.703		114	119		233	300
4750	Central vacuum outlet, Type NM cable		6.40	1.250		72	55		127	161
4770	Type MC cable		5.71	1.401		86.50	61.50		148	187
4780	EMT & wire	↓	3.48	2.299		88.50	101		189.50	248

26 05 Common Work Results for Electrical

26 05 90 – Residential Wiring

26 05 90.10 Residential Wiring	Crew	Daily Output	Labor-Hours	Unit	Material	2007 Bare Costs Labor	Equipment	Total	Total Incl O&P
4800 30 amp-110 volt locking recpt., #10/2 circ. bkr.									
4810 Type NM cable	1 Elec	6.20	1.290	Ea.	83	56.50		139.50	176
4820 Type MC cable		5.40	1.481		101	65		166	208
4830 EMT & wire	↓	3.20	2.500	↓	101	110		211	274
4900 Low voltage outlets									
4910 Telephone recpt., 20' of 4/C phone wire	1 Elec	26	.308	Ea.	9.35	13.50		22.85	30.50
4920 TV recpt., 20' of RG59U coax wire, F type connector	"	16	.500	"	16.25	22		38.25	50.50
4950 Door bell chime, transformer, 2 buttons, 60' of bellwire									
4970 Economy model	1 Elec	11.50	.696	Ea.	60	30.50		90.50	112
4980 Custom model		11.50	.696		93.50	30.50		124	149
4990 Luxury model, 3 buttons	↓	9.50	.842	↓	243	37		280	320
6000 Lighting outlets									
6050 Wire only (for fixture), type NM cable	1 Elec	32	.250	Ea.	9.65	11		20.65	27
6070 Type MC cable		24	.333		17.85	14.65		32.50	41.50
6080 EMT & wire		10	.800		22	35		57	76.50
6100 Box (4"), and wire (for fixture), type NM cable		25	.320		17.60	14.05		31.65	40.50
6120 Type MC cable		20	.400		26	17.55		43.55	54.50
6130 EMT & wire	↓	11	.727	↓	29.50	32		61.50	80
6200 Fixtures (use with lines 6050 or 6100 above)									
6210 Canopy style, economy grade	1 Elec	40	.200	Ea.	27	8.80		35.80	42.50
6220 Custom grade		40	.200		50	8.80		58.80	68
6250 Dining room chandelier, economy grade		19	.421		81	18.50		99.50	117
6260 Custom grade		19	.421		239	18.50		257.50	291
6270 Luxury grade		15	.533		530	23.50		553.50	620
6310 Kitchen fixture (fluorescent), economy grade		30	.267		57	11.70		68.70	80
6320 Custom grade		25	.320		165	14.05		179.05	203
6350 Outdoor, wall mounted, economy grade		30	.267		28	11.70		39.70	48.50
6360 Custom grade		30	.267		106	11.70		117.70	134
6370 Luxury grade		25	.320		240	14.05		254.05	285
6410 Outdoor PAR floodlights, 1 lamp, 150 watt		20	.400		27	17.55		44.55	55.50
6420 2 lamp, 150 watt each		20	.400		44	17.55		61.55	74.50
6430 For infrared security sensor, add		32	.250		92	11		103	117
6450 Outdoor, quartz-halogen, 300 watt flood		20	.400		40	17.55		57.55	70
6600 Recessed downlight, round, pre-wired, 50 or 75 watt trim		30	.267		37	11.70		48.70	58
6610 With shower light trim		30	.267		46	11.70		57.70	68
6620 With wall washer trim		28	.286		55	12.55		67.55	79
6630 With eye-ball trim	↓	28	.286		55	12.55		67.55	79
6640 For direct contact with insulation, add					1.85			1.85	2.04
6700 Porcelain lamp holder	1 Elec	40	.200		3.80	8.80		12.60	17.25
6710 With pull switch		40	.200		4.13	8.80		12.93	17.60
6750 Fluorescent strip, 1-20 watt tube, wrap around diffuser, 24"		24	.333		53	14.65		67.65	80.50
6760 1-40 watt tube, 48"		24	.333		65	14.65		79.65	93.50
6770 2-40 watt tubes, 48"		20	.400		79	17.55		96.55	113
6780 With residential ballast		20	.400		89.50	17.55		107.05	125
6800 Bathroom heat lamp, 1-250 watt		28	.286		41	12.55		53.55	63.50
6810 2-250 watt lamps	↓	28	.286	↓	67	12.55		79.55	92
6820 For timer switch, see line 2400									
6900 Outdoor post lamp, incl. post, fixture, 35' of #14/2									
6910 Type NMC cable	1 Elec	3.50	2.286	Ea.	195	100		295	365
6920 Photo-eye, add		27	.296		32	13		45	54.50
6950 Clock dial time switch, 24 hr., w/enclosure, type NM cable		11.43	.700		60	30.50		90.50	112
6970 Type MC cable		11	.727		74	32		106	129
6980 EMT & wire	↓	4.85	1.649		79.50	72.50		152	195

26 05 Common Work Results for Electrical

26 05 90 – Residential Wiring

26 05 90.10 Residential Wiring	Crew	Daily Output	Labor-Hours	Unit	Material	2007 Bare Costs Labor	Equipment	Total	Total Incl O&P
7000 Alarm systems									
7050 Smoke detectors, box, #14/3, type NM cable	1 Elec	14.55	.550	Ea.	35.50	24		59.50	75
7070 Type MC cable	↓	12.31	.650		49.50	28.50		78	97
7080 EMT & wire	↓	5	1.600	↓	48.50	70		118.50	158
7090 For relay output to security system, add					12.90			12.90	14.20
8000 Residential equipment									
8050 Disposal hook-up, incl. switch, outlet box, 3' of flex									
8060 20 amp-1 pole circ. bkr., and 25' of #12/2									
8070 Type NM cable	1 Elec	10	.800	Ea.	33.50	35		68.50	89
8080 Type MC cable		8	1		44.50	44		88.50	115
8090 EMT & wire	↓	5	1.600		56	70		126	167
8100 Trash compactor or dishwasher hook-up, incl. outlet box,									
8110 3' of flex, 15 amp-1 pole circ. bkr., and 25' of #14/2									
8120 Type NM cable	1 Elec	10	.800	Ea.	23.50	35		58.50	78.50
8130 Type MC cable		8	1		39.50	44		83.50	109
8140 EMT & wire	↓	5	1.600	↓	47.50	70		117.50	157
8150 Hot water sink dispensor hook-up, use line 8100									
8200 Vent/exhaust fan hook-up, type NM cable	1 Elec	32	.250	Ea.	9.65	11		20.65	27
8220 Type MC cable		24	.333		17.85	14.65		32.50	41.50
8230 EMT & wire	↓	10	.800		22	35		57	76.50
8250 Bathroom vent fan, 50 CFM (use with above hook-up)									
8260 Economy model	1 Elec	15	.533	Ea.	22.50	23.50		46	60
8270 Low noise model		15	.533		32	23.50		55.50	70
8280 Custom model	↓	12	.667	↓	117	29.50		146.50	173
8300 Bathroom or kitchen vent fan, 110 CFM									
8310 Economy model	1 Elec	15	.533	Ea.	60.50	23.50		84	102
8320 Low noise model	"	15	.533	"	79	23.50		102.50	122
8350 Paddle fan, variable speed (w/o lights)									
8360 Economy model (AC motor)	1 Elec	10	.800	Ea.	105	35		140	169
8370 Custom model (AC motor)		10	.800		182	35		217	253
8380 Luxury model (DC motor)		8	1		360	44		404	460
8390 Remote speed switch for above, add	↓	12	.667	↓	26	29.50		55.50	72
8500 Whole house exhaust fan, ceiling mount, 36", variable speed									
8510 Remote switch, incl. shutters, 20 amp-1 pole circ. bkr.									
8520 30' of #12/2, type NM cable	1 Elec	4	2	Ea.	750	88		838	955
8530 Type MC cable		3.50	2.286		765	100		865	990
8540 EMT & wire	↓	3	2.667		780	117		897	1,025
8600 Whirlpool tub hook-up, incl. timer switch, outlet box									
8610 3' of flex, 20 amp-1 pole GFI circ. bkr.									
8620 30' of #12/2, type NM cable	1 Elec	5	1.600	Ea.	116	70		186	233
8630 Type MC cable		4.20	1.905		121	83.50		204.50	257
8640 EMT & wire	↓	3.40	2.353		131	103		234	298
8650 Hot water heater hook-up, incl. 1-2 pole circ. bkr., box;									
8660 3' of flex, 20' of #10/2, type NM cable	1 Elec	5	1.600	Ea.	37	70		107	146
8670 Type MC cable		4.20	1.905		52	83.50		135.50	181
8680 EMT & wire	↓	3.40	2.353	↓	50	103		153	209
9000 Heating/air conditioning									
9050 Furnace/boiler hook-up, incl. firestat, local on-off switch									
9060 Emergency switch, and 40' of type NM cable	1 Elec	4	2	Ea.	55.50	88		143.50	193
9070 Type MC cable		3.50	2.286		77	100		177	234
9080 EMT & wire	↓	1.50	5.333		87	234		321	445
9100 Air conditioner hook-up, incl. local 60 amp disc. switch									
9110 3' sealtite, 40 amp, 2 pole circuit breaker									

26 05 Common Work Results for Electrical

26 05 90 – Residential Wiring

26 05 90.10 Residential Wiring		Crew	Daily Output	Labor-Hours	Unit	Material	2007 Bare Costs Labor	Equipment	Total	Total Incl O&P
9130	40' of #8/2, type NM cable	1 Elec	3.50	2.286	Ea.	210	100		310	380
9140	Type MC cable		3	2.667		291	117		408	495
9150	EMT & wire	↓	1.30	6.154	↓	236	270		506	660
9200	Heat pump hook-up, 1-40 & 1-100 amp 2 pole circ. bkr.									
9210	Local disconnect switch, 3' sealtite									
9220	40' of #8/2 & 30' of #3/2									
9230	Type NM cable	1 Elec	1.30	6.154	Ea.	610	270		880	1,075
9240	Type MC cable		1.08	7.407		720	325		1,045	1,275
9250	EMT & wire	↓	.94	8.511	↓	645	375		1,020	1,275
9500	Thermostat hook-up, using low voltage wire									
9520	Heating only	1 Elec	24	.333	Ea.	7.10	14.65		21.75	30
9530	Heating/cooling	"	20	.400	"	8.60	17.55		26.15	35.50

26 09 Instrumentation and Control for Electrical Systems

26 09 13 – Electrical Power Monitoring and Control

26 09 13.10 Switchboard Instruments

		Crew	Daily Output	Labor-Hours	Unit	Material	Labor	Equipment	Total	Total Incl O&P
0010	**SWITCHBOARD INSTRUMENTS** 3 phase, 4 wire									
0100	AC indicating, ammeter & switch	1 Elec	8	1	Ea.	1,775	44		1,819	2,050
0200	Voltmeter & switch		8	1		1,775	44		1,819	2,050
0300	Wattmeter		8	1		3,525	44		3,569	3,975
0400	AC recording, ammeter		4	2		6,300	88		6,388	7,050
0500	Voltmeter		4	2		6,300	88		6,388	7,050
0600	Ground fault protection, zero sequence		2.70	2.963		5,550	130		5,680	6,325
0700	Ground return path		2.70	2.963		5,550	130		5,680	6,325
0800	3 current transformers, 5 to 800 amp		2	4		2,600	176		2,776	3,100
0900	1000 to 1500 amp		1.30	6.154		3,725	270		3,995	4,500
1200	2000 to 4000 amp		1	8		4,400	350		4,750	5,350
1300	Fused potential transformer, maximum 600 volt	↓	8	1	↓	975	44		1,019	1,150

26 12 Medium-Voltage Transformers

26 12 19 – Pad-Mounted, Liquid-Filled, Medium-Voltage Transformers

26 12 19.10 Oil Filled Transformer

		Crew	Daily Output	Labor-Hours	Unit	Material	Labor	Equipment	Total	Total Incl O&P
0010	**OIL FILLED TRANSFORMER** primary delta or Y,									
0050	Pad mounted 5 kV or 15 kV, with taps, 277/480 V secondary, 3 phase									
0100	150 kVA	R-3	.65	30.769	Ea.	8,125	1,325	251	9,701	11,200
0200	300 kVA		.45	44.444		11,200	1,925	365	13,490	15,600
0300	500 kVA		.40	50		16,200	2,175	410	18,785	21,500
0400	750 kVA		.38	52.632		20,100	2,275	430	22,805	26,000
0500	1000 kVA		.26	76.923		23,800	3,325	630	27,755	31,900
0600	1500 kVA		.23	86.957		28,300	3,775	710	32,785	37,600
0700	2000 kVA		.20	100		35,800	4,325	815	40,940	46,800
0800	3750 kVA	↓	.16	125	↓	67,000	5,400	1,025	73,425	83,000

26 22 Low-Voltage Transformers

26 22 13 – Low-Voltage Distribution Transformers

26 22 13.10 Dry Type Transformer	Crew	Daily Output	Labor-Hours	Unit	2007 Bare Costs Material	2007 Bare Costs Labor	2007 Bare Costs Equipment	Total	Total Incl O&P
0010 **DRY TYPE TRANSFORMER**									
0050 Single phase, 240/480 volt primary, 120/240 volt secondary									
0100 1 kVA	1 Elec	2	4	Ea.	235	176		411	520
0300 2 kVA		1.60	5		355	220		575	715
0500 3 kVA		1.40	5.714		435	251		686	855
0700 5 kVA	↓	1.20	6.667		600	293		893	1,100
0900 7.5 kVA	2 Elec	2.20	7.273		835	320		1,155	1,400
1100 10 kVA		1.60	10		1,025	440		1,465	1,775
1300 15 kVA		1.20	13.333		1,400	585		1,985	2,425
1500 25 kVA	CN	1	16		1,900	700		2,600	3,150
1700 37.5 kVA		.80	20		2,475	880		3,355	4,025
1900 50 kVA		.70	22.857		2,950	1,000		3,950	4,725
2100 75 kVA		.65	24.615		3,900	1,075		4,975	5,875
2190 480V primary 120/240V secondary, nonvent., 15 kVA		1.20	13.333		1,475	585		2,060	2,500
2200 25 kVA		.90	17.778		2,175	780		2,955	3,550
2210 37 kVA		.75	21.333		2,575	935		3,510	4,250
2220 50 kVA	↓	.65	24.615	↓	3,075	1,075		4,150	4,975
2300 3 phase, 480 volt primary 120/208 volt secondary									
2310 Ventilated, 3 kVA	1 Elec	1	8	Ea.	805	350		1,155	1,400
2700 6 kVA		.80	10		1,100	440		1,540	1,875
2900 9 kVA	↓	.70	11.429		1,250	500		1,750	2,125
3100 15 kVA	2 Elec	1.10	14.545		1,675	640		2,315	2,800
3300 30 kVA		.90	17.778		1,975	780		2,755	3,300
3500 45 kVA		.80	20		2,350	880		3,230	3,900
3700 75 kVA	↓	.70	22.857		3,550	1,000		4,550	5,400
3900 112.5 kVA	R-3	.90	22.222		4,725	960	182	5,867	6,825
4100 150 kVA		.85	23.529		6,150	1,025	192	7,367	8,500
4300 225 kVA		.65	30.769		8,350	1,325	251	9,926	11,500
4500 300 kVA		.55	36.364		10,600	1,575	297	12,472	14,300
4700 500 kVA		.45	44.444		17,500	1,925	365	19,790	22,500
4800 750 kVA	↓	.35	57.143	↓	30,600	2,475	465	33,540	37,900

26 24 Switchboards and Panelboards

26 24 13 – Switchboards

26 24 13.10 Switchboards

	Crew	Daily Output	Labor-Hours	Unit	Material	Labor	Equipment	Total	Total Incl O&P
0010 **SWITCHBOARDS** Incoming main service section									
0100 Aluminum bus bars, not including CT's or PT's									
0200 No main disconnect, includes CT compartment									
0300 120/208 volt, 4 wire, 600 amp	2 Elec	1	16	Ea.	3,650	700		4,350	5,075
0400 800 amp		.88	18.182		3,650	800		4,450	5,225
0500 1000 amp		.80	20		4,400	880		5,280	6,125
0600 1200 amp		.72	22.222		4,400	975		5,375	6,275
0700 1600 amp		.66	24.242		4,400	1,075		5,475	6,400
0800 2000 amp		.62	25.806		4,725	1,125		5,850	6,875
1000 3000 amp	↓	.56	28.571	↓	6,250	1,250		7,500	8,750
2000 Fused switch & CT compartment									
2100 120/208 volt, 4 wire, 400 amp	2 Elec	1.12	14.286	Ea.	3,775	625		4,400	5,075
2200 600 amp		.94	17.021		4,475	745		5,220	6,000
2300 800 amp		.84	19.048		9,900	835		10,735	12,200
2400 1200 amp	↓	.68	23.529	↓	12,900	1,025		13,925	15,800
2900 Pressure switch & CT compartment									

26 24 Switchboards and Panelboards

26 24 13 – Switchboards

26 24 13.10 Switchboards

		Crew	Daily Output	Labor-Hours	Unit	Material	2007 Bare Costs Labor	Equipment	Total	Total Incl O&P
3000	120/208 volt, 4 wire, 800 amp	2 Elec	.80	20	Ea.	8,875	880		9,755	11,100
3100	1200 amp		.66	24.242		17,200	1,075		18,275	20,500
3200	1600 amp		.62	25.806		18,300	1,125		19,425	21,800
3300	2000 amp		.56	28.571		19,400	1,250		20,650	23,300
4400	Circuit breaker, molded case & CT compartment									
4600	3 pole, 4 wire, 600 amp	2 Elec	.94	17.021	Ea.	7,650	745		8,395	9,525
4800	800 amp		.84	19.048		9,175	835		10,010	11,400
5000	1200 amp		.68	23.529		12,500	1,025		13,525	15,300
5100	Copper bus bars, not incl. CT's or PT's, add, minimum					15%				

26 24 13.30 Switchboards

		Crew	Daily Output	Labor-Hours	Unit	Material	Labor	Equipment	Total	Total Incl O&P
0010	**SWITCHBOARDS** distribution section									
0100	Aluminum bus bars, not including breakers									
0200	120/208 or 277/480 volt, 4 wire, 600 amp	2 Elec	1	16	Ea.	1,625	700		2,325	2,825
0300	800 amp		.88	18.182		2,075	800		2,875	3,500
0400	1000 amp		.80	20		2,075	880		2,955	3,600
0500	1200 amp		.72	22.222		2,975	975		3,950	4,725
0600	1600 amp		.66	24.242		3,475	1,075		4,550	5,400
0700	2000 amp		.62	25.806		4,075	1,125		5,200	6,150
0800	2500 amp		.60	26.667		4,600	1,175		5,775	6,800
0900	3000 amp		.56	28.571		5,575	1,250		6,825	8,000
0950	4000 amp		.52	30.769		8,150	1,350		9,500	11,000

26 24 13.40 Switchboards

		Crew	Daily Output	Labor-Hours	Unit	Material	Labor	Equipment	Total	Total Incl O&P
0010	**SWITCHBOARDS** feeder section group mounted devices									
0030	Circuit breakers									
0160	FA frame, 15 to 60 amp, 240 volt, 1 pole	1 Elec	8	1	Ea.	91	44		135	166
0280	FA frame, 70 to 100 amp, 240 volt, 1 pole		7	1.143		120	50		170	207
0420	KA frame, 70 to 225 amp		3.20	2.500		1,000	110		1,110	1,275
0430	LA frame, 125 to 400 amp		2.30	3.478		2,275	153		2,428	2,725
0460	MA frame, 450 to 600 amp		1.60	5		3,775	220		3,995	4,475
0470	700 to 800 amp		1.30	6.154		4,875	270		5,145	5,775
0480	MAL frame, 1000 amp		1	8		5,075	350		5,425	6,100
0490	PA frame, 1200 amp		.80	10		10,300	440		10,740	12,000
0500	Branch circuit, fusible switch, 600 volt, double 30/30 amp		4	2		1,025	88		1,113	1,250
0550	60/60 amp		3.20	2.500		1,025	110		1,135	1,300
0600	100/100 amp		2.70	2.963		1,425	130		1,555	1,775
0650	Single, 30 amp		5.30	1.509		670	66.50		736.50	840
0700	60 amp		4.70	1.702		670	74.50		744.50	850
0750	100 amp		4	2		965	88		1,053	1,175
0800	200 amp		2.70	2.963		1,425	130		1,555	1,775
0850	400 amp		2.30	3.478		2,700	153		2,853	3,200
0900	600 amp		1.80	4.444		3,200	195		3,395	3,825
0950	800 amp		1.30	6.154		5,250	270		5,520	6,175
1000	1200 amp		.80	10		6,250	440		6,690	7,525

26 24 16 – Panelboards

26 24 16.20 Panelboard and Load Center Circuit Breakers

		Crew	Daily Output	Labor-Hours	Unit	Material	Labor	Equipment	Total	Total Incl O&P
0010	**PANELBOARD AND LOAD CENTER CIRCUIT BREAKERS**									
0050	Bolt-on, 10,000 amp IC, 120 volt, 1 pole									
0100	15 to 50 amp	1 Elec	10	.800	Ea.	13.80	35		48.80	67.50
0200	60 amp		8	1		13.80	44		57.80	80.50
0300	70 amp		8	1		26	44		70	94.50
0350	240 volt, 2 pole									
0400	15 to 50 amp	1 Elec	8	1	Ea.	30.50	44		74.50	99

26 24 Switchboards and Panelboards

26 24 16 – Panelboards

26 24 16.20 Panelboard and Load Center Circuit Breakers		Crew	Daily Output	Labor-Hours	Unit	Material	2007 Bare Costs Labor	Equipment	Total	Total Incl O&P
0500	60 amp	1 Elec	7.50	1.067	Ea.	30.50	47		77.50	103
0600	80 to 100 amp		5	1.600		78	70		148	191
0700	3 pole, 15 to 60 amp		6.20	1.290		96	56.50		152.50	191
0800	70 amp		5	1.600		121	70		191	238
0900	80 to 100 amp		3.60	2.222		137	97.50		234.50	296
1000	22,000 amp I.C., 240 volt, 2 pole, 70 - 225 amp		2.70	2.963		590	130		720	845
1100	3 pole, 70 - 225 amp		2.30	3.478		655	153		808	945
1200	14,000 amp I.C., 277 volts, 1 pole, 15 - 30 amp		8	1		36.50	44		80.50	106
1300	22,000 amp I.C., 480 volts, 2 pole, 70 - 225 amp		2.70	2.963		590	130		720	845
1400	3 pole, 70 - 225 amp		2.30	3.478		730	153		883	1,025

26 24 16.30 Panelboards

		Crew	Daily Output	Labor-Hours	Unit	Material	Labor	Equipment	Total	Total Incl O&P
0010	**PANELBOARDS** (Commercial use)									
0050	NQOD, w/20 amp 1 pole bolt-on circuit breakers									
0100	3 wire, 120/240 volts, 100 amp main lugs									
0150	10 circuits	1 Elec	1	8	Ea.	455	350		805	1,025
0200	14 circuits		.88	9.091		535	400		935	1,175
0250	18 circuits		.75	10.667		585	470		1,055	1,325
0300	20 circuits		.65	12.308		655	540		1,195	1,525
0350	225 amp main lugs, 24 circuits	2 Elec	1.20	13.333		745	585		1,330	1,700
0400	30 circuits		.90	17.778		870	780		1,650	2,100
0450	36 circuits		.80	20		990	880		1,870	2,400
0500	38 circuits		.72	22.222		1,075	975		2,050	2,625
0550	42 circuits		.66	24.242		1,125	1,075		2,200	2,800
0600	4 wire, 120/208 volts, 100 amp main lugs, 12 circuits	1 Elec	1	8		515	350		865	1,100
0650	16 circuits		.75	10.667		595	470		1,065	1,350
0700	20 circuits		.65	12.308		690	540		1,230	1,575
0750	24 circuits CN		.60	13.333		750	585		1,335	1,700
0800	30 circuits		.53	15.094		865	665		1,530	1,925
0850	225 amp main lugs, 32 circuits	2 Elec	.90	17.778		975	780		1,755	2,225
0900	34 circuits		.84	19.048		995	835		1,830	2,350
0950	36 circuits		.80	20		1,025	880		1,905	2,425
1000	42 circuits		.68	23.529		1,150	1,025		2,175	2,800
1200	NEHB, w/20 amp, 1 pole bolt-on circuit breakers									
1250	4 wire, 277/480 volts, 100 amp main lugs, 12 circuits	1 Elec	.88	9.091	Ea.	985	400		1,385	1,675
1300	20 circuits	"	.60	13.333		1,450	585		2,035	2,475
1350	225 amp main lugs, 24 circuits	2 Elec	.90	17.778		1,675	780		2,455	3,000
1400	30 circuits		.80	20		2,025	880		2,905	3,525
1450	36 circuits		.72	22.222		2,350	975		3,325	4,025
1600	NQOD panel, w/20 amp, 1 pole, circuit breakers									
1650	3 wire, 120/240 volt with main circuit breaker									
1700	100 amp main, 12 circuits	1 Elec	.80	10	Ea.	630	440		1,070	1,350
1750	20 circuits	"	.60	13.333		810	585		1,395	1,750
1800	225 amp main, 30 circuits	2 Elec	.68	23.529		1,550	1,025		2,575	3,250
1850	42 circuits		.52	30.769		1,800	1,350		3,150	3,975
1900	400 amp main, 30 circuits		.54	29.630		2,150	1,300		3,450	4,275
1950	42 circuits		.50	32		2,400	1,400		3,800	4,725
2000	4 wire, 120/208 volts with main circuit breaker									
2050	100 amp main, 24 circuits	1 Elec	.47	17.021	Ea.	945	745		1,690	2,150
2100	30 circuits	"	.40	20		1,075	880		1,955	2,475
2200	225 amp main, 32 circuits	2 Elec	.72	22.222		1,800	975		2,775	3,425
2250	42 circuits		.56	28.571		1,975	1,250		3,225	4,050
2300	400 amp main, 42 circuits		.48	33.333		2,675	1,475		4,150	5,100

26 24 Switchboards and Panelboards

26 24 16 – Panelboards

26 24 16.30 Panelboards

		Crew	Daily Output	Labor-Hours	Unit	Material	2007 Bare Costs Labor	Equipment	Total	Total Incl O&P
2350	600 amp main, 42 circuits	2 Elec	.40	40	Ea.	3,950	1,750		5,700	6,975
2400	NEHB, with 20 amp, 1 pole circuit breaker									
2450	4 wire, 277/480 volts with main circuit breaker									
2500	100 amp main, 24 circuits	1 Elec	.42	19.048	Ea.	1,950	835		2,785	3,375
2550	30 circuits	"	.38	21.053		2,275	925		3,200	3,875
2600	225 amp main, 30 circuits	2 Elec	.72	22.222		2,875	975		3,850	4,600
2650	42 circuits	"	.56	28.571		3,525	1,250		4,775	5,775

26 24 19 – Motor-Control Centers

26 24 19.40 Motor Starters and Controls

		Crew	Daily Output	Labor-Hours	Unit	Material	Labor	Equipment	Total	Total Incl O&P
0010	**MOTOR STARTERS AND CONTROLS**									
0050	Magnetic, FVNR, with enclosure and heaters, 480 volt									
0100	5 HP, size 0	1 Elec	2.30	3.478	Ea.	231	153		384	480
0200	10 HP, size 1	"	1.60	5		259	220		479	610
0300	25 HP, size 2	2 Elec	2.20	7.273		485	320		805	1,000
0400	50 HP, size 3		1.80	8.889		795	390		1,185	1,450
0500	100 HP, size 4		1.20	13.333		1,775	585		2,360	2,825
0600	200 HP, size 5		.90	17.778		4,125	780		4,905	5,700
0700	Combination, with motor circuit protectors, 5 HP, size 0	1 Elec	1.80	4.444		750	195		945	1,125
0800	10 HP, size 1	"	1.30	6.154		780	270		1,050	1,250
0900	25 HP, size 2	2 Elec	2	8		1,100	350		1,450	1,725
1000	50 HP, size 3		1.32	12.121		1,575	530		2,105	2,525
1200	100 HP, size 4		.80	20		3,425	880		4,305	5,075
1400	Combination, with fused switch, 5 HP, size 0	1 Elec	1.80	4.444		575	195		770	920
1600	10 HP, size 1	"	1.30	6.154		615	270		885	1,075
1800	25 HP, size 2	2 Elec	2	8		995	350		1,345	1,625
2000	50 HP, size 3		1.32	12.121		1,675	530		2,205	2,650
2200	100 HP, size 4		.80	20		2,925	880		3,805	4,525

26 25 Enclosed Bus Assemblies

26 25 13 – Enclosed Bus Assemblies

26 25 13.40 Copper Bus Duct

		Crew	Daily Output	Labor-Hours	Unit	Material	Labor	Equipment	Total	Total Incl O&P
0010	**COPPER BUS DUCT** 10 ft. long									
0050	3 pole 4 wire, plug-in/indoor, straight section, 225 amp	2 Elec	40	.400	L.F.	159	17.55		176.55	201
1000	400 amp		32	.500		159	22		181	208
1500	600 amp		26	.615		159	27		186	215
2400	800 amp		20	.800		189	35		224	261
2450	1000 amp		18	.889		209	39		248	288
2500	1350 amp		16	1		299	44		343	395
2510	1600 amp		12	1.333		340	58.50		398.50	455
2520	2000 amp		10	1.600		430	70		500	575
2550	Feeder, 600 amp		28	.571		139	25		164	191
2600	800 amp		22	.727		169	32		201	234
2700	1000 amp		20	.800		189	35		224	261
2800	1350 amp		18	.889		279	39		318	365
2900	1600 amp		14	1.143		320	50		370	425
3000	2000 amp		12	1.333		410	58.50		468.50	535
3100	Elbows, 225 amp		4	4	Ea.	945	176		1,121	1,300
3200	400 amp		3.60	4.444		945	195		1,140	1,350
3300	600 amp		3.20	5		945	220		1,165	1,375
3400	800 amp		2.80	5.714		1,025	251		1,276	1,500

26 25 Enclosed Bus Assemblies

26 25 13 – Enclosed Bus Assemblies

26 25 13.40 Copper Bus Duct

		Crew	Daily Output	Labor-Hours	Unit	Material	2007 Bare Costs Labor	Equipment	Total	Total Incl O&P
3500	1000 amp	2 Elec	2.60	6.154	Ea.	1,150	270		1,420	1,650
3600	1350 amp		2.40	6.667		1,350	293		1,643	1,900
3700	1600 amp		2.20	7.273		1,450	320		1,770	2,075
3800	2000 amp		1.80	8.889		1,800	390		2,190	2,550
4000	End box, 225 amp		34	.471		127	20.50		147.50	171
4100	400 amp		32	.500		127	22		149	173
4200	600 amp		28	.571		127	25		152	178
4300	800 amp		26	.615		127	27		154	180
4400	1000 amp		24	.667		127	29.50		156.50	184
4500	1350 amp		22	.727		127	32		159	188
4600	1600 amp		20	.800		127	35		162	193
4700	2000 amp		18	.889		156	39		195	230
4800	Cable tap box end, 225 amp		3.20	5		950	220		1,170	1,375
5000	400 amp		2.60	6.154		950	270		1,220	1,450
5100	600 amp		2.20	7.273		950	320		1,270	1,525
5200	800 amp		2	8		1,075	350		1,425	1,700
5300	1000 amp		1.60	10		1,175	440		1,615	1,950
5400	1350 amp		1.40	11.429		1,400	500		1,900	2,300
5500	1600 amp		1.20	13.333		1,575	585		2,160	2,600
5600	2000 amp		1	16		1,800	700		2,500	3,025
5700	Switchboard stub, 225 amp		5.40	2.963		955	130		1,085	1,250
5800	400 amp		4.60	3.478		955	153		1,108	1,275
5900	600 amp		4	4		955	176		1,131	1,300
6000	800 amp		3.20	5		1,150	220		1,370	1,600
6100	1000 amp		3	5.333		1,350	234		1,584	1,825
6200	1350 amp		2.60	6.154		1,725	270		1,995	2,300
6300	1600 amp		2.40	6.667		1,950	293		2,243	2,575
6400	2000 amp		2	8		2,350	350		2,700	3,125
6490	Tee fittings, 225 amp		2.40	6.667		1,300	293		1,593	1,850
6500	400 amp		2	8		1,300	350		1,650	1,950
6600	600 amp		1.80	8.889		1,300	390		1,690	2,000
6700	800 amp		1.60	10		1,500	440		1,940	2,300
6800	1350 amp		1.20	13.333		2,150	585		2,735	3,250
7000	1600 amp		1	16		2,450	700		3,150	3,750
7100	2000 amp		.80	20		2,900	880		3,780	4,500
7200	Plug-in fusible switches w/3 fuses, 600 volt, 3 pole, 30 amp	1 Elec	4	2		565	88		653	755
7300	60 amp		3.60	2.222		635	97.50		732.50	845
7400	100 amp		2.70	2.963		975	130		1,105	1,275
7500	200 amp	2 Elec	3.20	5		1,750	220		1,970	2,225
7600	400 amp		1.40	11.429		5,075	500		5,575	6,350
7700	600 amp		.90	17.778		5,775	780		6,555	7,500
7800	800 amp		.66	24.242		7,775	1,075		8,850	10,100
7900	1200 amp		.50	32		14,600	1,400		16,000	18,100
8000	Plug-in circuit breakers, molded case, 15 to 50 amp	1 Elec	4.40	1.818		535	80		615	710
8100	70 to 100 amp	"	3.10	2.581		595	113		708	825
8200	150 to 225 amp	2 Elec	3.40	4.706		1,625	207		1,832	2,075
8300	250 to 400 amp		1.40	11.429		2,825	500		3,325	3,875
8400	500 to 600 amp		1	16		3,825	700		4,525	5,275
8500	700 to 800 amp		.64	25		4,725	1,100		5,825	6,825
8600	900 to 1000 amp		.56	28.571		6,750	1,250		8,000	9,300
8700	1200 amp		.44	36.364		8,125	1,600		9,725	11,300

26 27 Low-Voltage Distribution Equipment

26 27 16 – Electrical Cabinets and Enclosures

26 27 16.10 Cabinets		Crew	Daily Output	Labor-Hours	Unit	Material	2007 Bare Costs Labor	Equipment	Total	Total Incl O&P
0010	CABINETS									
7000	Cabinets, current transformer									
7050	Single door, 24" H x 24" W x 10" D	1 Elec	1.60	5	Ea.	160	220		380	500
7100	30" H x 24" W x 10" D		1.30	6.154		175	270		445	595
7150	36" H x 24" W x 10" D		1.10	7.273		202	320		522	695
7200	30" H x 30" W x 10" D		1	8		210	350		560	755
7250	36" H x 30" W x 10" D		.90	8.889		287	390		677	895
7300	36" H x 36" W x 10" D		.80	10		305	440		745	990
7500	Double door, 48" H x 36" W x 10" D		.60	13.333		570	585		1,155	1,500
7550	24" H x 24" W x 12" D		1	8		248	350		598	800

26 27 23 – Indoor Service Poles

26 27 23.40 Surface Raceway

26 27 23.40 Surface Raceway		Crew	Daily Output	Labor-Hours	Unit	Material	2007 Bare Costs Labor	Equipment	Total	Total Incl O&P
0010	SURFACE RACEWAY									
0090	Metal, straight section									
0100	No. 500	1 Elec	100	.080	L.F.	.95	3.51		4.46	6.30
0110	No. 700		100	.080		1.07	3.51		4.58	6.45
0400	No. 1500, small pancake		90	.089		1.93	3.90		5.83	7.90
0600	No. 2000, base & cover, blank		90	.089		1.88	3.90		5.78	7.85
0800	No. 3000, base & cover, blank		75	.107		3.80	4.68		8.48	11.15
1000	No. 4000, base & cover, blank		65	.123		6.15	5.40		11.55	14.80
1200	No. 6000, base & cover, blank		50	.160		10.30	7		17.30	22
2400	Fittings, elbows, No. 500		40	.200	Ea.	1.72	8.80		10.52	14.95
2800	Elbow cover, No. 2000		40	.200		3.28	8.80		12.08	16.65
2880	Tee, No. 500		42	.190		3.30	8.35		11.65	16.10
2900	No. 2000		27	.296		10.30	13		23.30	30.50
3000	Switch box, No. 500		16	.500		11.20	22		33.20	45
3400	Telephone outlet, No. 1500		16	.500		12.40	22		34.40	46
3600	Junction box, No. 1500		16	.500		8.70	22		30.70	42
3800	Plugmold wired sections, No. 2000									
4000	1 circuit, 6 outlets, 3 ft. long	1 Elec	8	1	Ea.	31.50	44		75.50	101
4100	2 circuits, 8 outlets, 6 ft. long	"	5.30	1.509	"	52.50	66.50		119	157

26 27 26 – Wiring Devices

26 27 26.10 Low Voltage Switching

26 27 26.10 Low Voltage Switching		Crew	Daily Output	Labor-Hours	Unit	Material	2007 Bare Costs Labor	Equipment	Total	Total Incl O&P
0010	LOW VOLTAGE SWITCHING									
3600	Relays, 120 V or 277 V standard	1 Elec	12	.667	Ea.	28	29.50		57.50	74.50
3800	Flush switch, standard		40	.200		9.75	8.80		18.55	24
4000	Interchangeable		40	.200		12.75	8.80		21.55	27
4100	Surface switch, standard		40	.200		7.15	8.80		15.95	21
4200	Transformer 115 V to 25 V		12	.667		100	29.50		129.50	154
4400	Master control, 12 circuit, manual		4	2		102	88		190	243
4500	25 circuit, motorized		4	2		110	88		198	252
4600	Rectifier, silicon		12	.667		33	29.50		62.50	79.50
4800	Switchplates, 1 gang, 1, 2 or 3 switch, plastic		80	.100		3.24	4.39		7.63	10.10
5000	Stainless steel		80	.100		8.75	4.39		13.14	16.20
5400	2 gang, 3 switch, stainless steel		53	.151		16.90	6.65		23.55	28.50
5500	4 switch, plastic		53	.151		7.25	6.65		13.90	17.80
5800	3 gang, 9 switch, stainless steel		32	.250		54	11		65	76

26 27 26.20 Wiring Devices

26 27 26.20 Wiring Devices			Crew	Daily Output	Labor-Hours	Unit	Material	2007 Bare Costs Labor	Equipment	Total	Total Incl O&P
0010	WIRING DEVICES										
0200	Toggle switch, quiet type, single pole, 15 amp	CN	1 Elec	40	.200	Ea.	5.10	8.80		13.90	18.70
0600	3 way, 15 amp			23	.348		7.65	15.25		22.90	31

26 27 Low-Voltage Distribution Equipment

26 27 26 – Wiring Devices

	26 27 26.20 Wiring Devices	Crew	Daily Output	Labor-Hours	Unit	Material	2007 Bare Costs Labor	Equipment	Total	Total Incl O&P
0900	4 way, 15 amp	1 Elec	15	.533	Ea.	21.50	23.50		45	59
1650	Dimmer switch, 120 volt, incandescent, 600 watt, 1 pole		16	.500		11.25	22		33.25	45
2460	Receptacle, duplex, 120 volt, grounded, 15 amp		40	.200		1.23	8.80		10.03	14.40
2470	20 amp		27	.296		9.40	13		22.40	29.50
2490	Dryer, 30 amp		15	.533		5.85	23.50		29.35	41.50
2500	Range, 50 amp		11	.727		11.95	32		43.95	60.50
2600	Wall plates, stainless steel, 1 gang		80	.100		2.06	4.39		6.45	8.80
2800	2 gang		53	.151		3.80	6.65		10.45	14.05
3200	Lampholder, keyless		26	.308		11.20	13.50		24.70	32.50
3400	Pullchain with receptacle		22	.364		12.20	15.95		28.15	37.50

26 27 73 – Door Chimes

26 27 73.10 Doorbell System

		Crew	Daily Output	Labor-Hours	Unit	Material	Labor	Equipment	Total	Total Incl O&P
0010	**DOORBELL SYSTEM** Incl. transformer, button & signal									
0100	6" bell	1 Elec	4	2	Ea.	89	88		177	229
0200	Buzzer	"	4	2	"	70.50	88		158.50	209

26 28 Low-Voltage Circuit Protective Devices

26 28 16 – Enclosed Switches and Circuit Breakers

26 28 16.10 Circuit Breakers

		Crew	Daily Output	Labor-Hours	Unit	Material	Labor	Equipment	Total	Total Incl O&P
0010	**CIRCUIT BREAKERS** (in enclosure)									
0100	Enclosed (NEMA 1), 600 volt, 3 pole, 30 amp	1 Elec	3.20	2.500	Ea.	530	110		640	750
0200	60 amp		2.80	2.857		530	125		655	770
0400	100 amp		2.30	3.478		605	153		758	895
0600	225 amp		1.50	5.333		1,400	234		1,634	1,900
0700	400 amp	2 Elec	1.60	10		2,400	440		2,840	3,300
0800	600 amp		1.20	13.333		3,475	585		4,060	4,700
1000	800 amp		.94	17.021		4,525	745		5,270	6,075

26 28 16.20 Safety Switches

		Crew	Daily Output	Labor-Hours	Unit	Material	Labor	Equipment	Total	Total Incl O&P
0010	**SAFETY SWITCHES**									
0100	General duty 240 volt, 3 pole NEMA 1, fusible, 30 amp	1 Elec	3.20	2.500	Ea.	92	110		202	264
0200	60 amp		2.30	3.478		155	153		308	400
0300	100 amp		1.90	4.211		267	185		452	570
0400	200 amp		1.30	6.154		575	270		845	1,025
0500	400 amp	2 Elec	1.80	8.889		1,450	390		1,840	2,175
0600	600 amp	"	1.20	13.333		2,700	585		3,285	3,850
2900	Heavy duty, 240 volt, 3 pole NEMA 1 fusible									
2910	30 amp	1 Elec	3.20	2.500	Ea.	148	110		258	325
3000	60 amp		2.30	3.478		251	153		404	505
3300	100 amp		1.90	4.211		395	185		580	710
3500	200 amp		1.30	6.154		680	270		950	1,150
3700	400 amp	2 Elec	1.80	8.889		1,750	390		2,140	2,500
3900	600 amp	"	1.20	13.333		3,000	585		3,585	4,175

26 29 Low-Voltage Controllers

26 29 13 – Enclosed Controllers

26 29 13.20 Control Stations

26 29 13.20 Control Stations	Crew	Daily Output	Labor-Hours	Unit	Material	2007 Bare Costs Labor	Equipment	Total	Total Incl O&P
0010 **CONTROL STATIONS**									
0050 NEMA 1, heavy duty, stop/start	1 Elec	8	1	Ea.	131	44		175	210
0100 Stop/start, pilot light		6.20	1.290		179	56.50		235.50	281
0200 Hand/off/automatic		6.20	1.290		97	56.50		153.50	192
0400 Stop/start/reverse	↓	5.30	1.509	↓	177	66.50		243.50	293

26 32 Packaged Generator Assemblies

26 32 13 – Engine Generators

26 32 13.13 Diesel-Engine-Driven Generator Sets

		Crew	Daily Output	Labor-Hours	Unit	Material	Labor	Equipment	Total	Total Incl O&P
0010	**DIESEL-ENGINE-DRIVEN GENERATOR SETS**									
2000	Diesel engine, including battery, charger,									
2010	muffler, automatic transfer switch & day tank, 30 kW	R-3	.55	36.364	Ea.	17,000	1,575	297	18,872	21,400
2100	50 kW		.42	47.619		20,900	2,050	390	23,340	26,500
2200	75 kW		.35	57.143		27,200	2,475	465	30,140	34,200
2300	100 kW		.31	64.516		30,200	2,800	525	33,525	38,000
2400	125 kW		.29	68.966		31,800	2,975	565	35,340	40,100
2500	150 kW		.26	76.923		36,500	3,325	630	40,455	45,800
2600	175 kW		.25	80		39,800	3,450	655	43,905	49,600
2700	200 kW		.24	83.333		41,000	3,600	680	45,280	51,000
2800	250 kW		.23	86.957		48,300	3,775	710	52,785	59,500
2900	300 kW		.22	90.909		52,000	3,925	745	56,670	64,000
3000	350 kW		.20	100		59,000	4,325	815	64,140	72,500
3100	400 kW		.19	105		73,000	4,550	860	78,410	87,500
3200	500 kW	↓	.18	111	↓	91,500	4,800	910	97,210	109,000

26 32 13.16 Gas-Engine-Driven Generator Sets

		Crew	Daily Output	Labor-Hours	Unit	Material	Labor	Equipment	Total	Total Incl O&P
0010	**GAS-ENGINE-DRIVEN GENERATOR SETS**									
0020	Gas or gasoline operated, includes battery,									
0050	charger, muffler & transfer switch									
0200	3 phase 4 wire, 277/480 volt, 7.5 kW	R-3	.83	24.096	Ea.	6,350	1,050	197	7,597	8,775
0300	11.5 kW		.71	28.169		9,000	1,225	230	10,455	12,000
0400	20 kW		.63	31.746		10,600	1,375	259	12,234	14,000
0500	35 kW		.55	36.364		12,700	1,575	297	14,572	16,700
0600	80 kW		.40	50		21,700	2,175	410	24,285	27,600
0700	100 kW		.33	60.606		23,900	2,625	495	27,020	30,700
0800	125 kW		.28	71.429		48,800	3,100	585	52,485	59,000
0900	185 kW	↓	.25	80	↓	64,500	3,450	655	68,605	77,000

26 35 Power Filters and Conditioners

26 35 13 – Capacitors

26 35 13.10 Capacitors

		Crew	Daily Output	Labor-Hours	Unit	Material	Labor	Equipment	Total	Total Incl O&P
0010	**CAPACITORS** Indoor									
0020	240 volts, single & 3 phase, 0.5 kVAR	1 Elec	2.70	2.963	Ea.	315	130		445	545
0100	1.0 kVAR		2.70	2.963		380	130		510	615
0150	2.5 kVAR		2	4		430	176		606	735
0200	5.0 kVAR		1.80	4.444		515	195		710	855
0250	7.5 kVAR		1.60	5		600	220		820	985
0300	10 kVAR		1.50	5.333		720	234		954	1,150
0350	15 kVAR		1.30	6.154		995	270		1,265	1,500
0400	20 kVAR	↓	1.10	7.273	↓	1,225	320		1,545	1,825

26 35 Power Filters and Conditioners

26 35 13 − Capacitors

26 35 13.10 Capacitors

		Crew	Daily Output	Labor-Hours	Unit	Material	2007 Bare Costs Labor	Equipment	Total	Total Incl O&P
0450	25 kVAR	1 Elec	1	8	Ea.	1,450	350		1,800	2,125
1000	480 volts, single & 3 phase, 1 kVAR		2.70	2.963		289	130		419	515
1050	2 kVAR		2.70	2.963		330	130		460	560
1100	5 kVAR		2	4		420	176		596	720
1150	7.5 kVAR		2	4		450	176		626	755
1200	10 kVAR		2	4		505	176		681	815
1250	15 kVAR		2	4		625	176		801	945
1300	20 kVAR		1.60	5		695	220		915	1,100
1350	30 kVAR		1.50	5.333		870	234		1,104	1,300
1400	40 kVAR		1.20	6.667		1,100	293		1,393	1,650
1450	50 kVAR	↓	1.10	7.273	↓	1,250	320		1,570	1,850

26 42 Cathodic Protection

26 42 16 − Passive Cathodic Protection for Underground Storage Tank

26 42 16.50 Cathodic Protection

		Crew	Daily Output	Labor-Hours	Unit	Material	2007 Bare Costs Labor	Equipment	Total	Total Incl O&P
0010	**CATHODIC PROTECTION**									
1000	Anodes, magnesium type, 9 #	R-15	18.50	2.595	Ea.	24	111	15.65	150.65	210
1010	17 #		13	3.692		41	158	22.50	221.50	305
1020	32 #		10	4.800		72	205	29	306	415
1030	48 #	↓	7.20	6.667		110	285	40.50	435.50	590
1100	Graphite type w/ epoxy cap, 3" x 60" (32 #)	R-22	8.40	4.438		120	164		284	380
1110	4" x 80" (68 #)		6	6.213		225	229		454	595
1120	6" x 72" (80 #)		5.20	7.169		1,125	264		1,389	1,650
1130	6" x 36" (45 #)	↓	9.60	3.883		570	143		713	840
2000	Rectifiers, silicon type, air cooled, 28 V/10 A	R-19	3.50	5.714		1,950	251		2,201	2,525
2010	20 V/20 A		3.50	5.714		2,100	251		2,351	2,675
2100	Oil immersed, 28 V/10 A		3	6.667		2,750	293		3,043	3,450
2110	20 V/20 A	↓	3	6.667	↓	2,950	293		3,243	3,675
3000	Anode backfill, coke breeze	R-22	3850	.010	Lb.	.15	.36		.51	.71
4000	Cable, OR2, No. 8		2.40	15.533	M.L.F.	233	575		808	1,125
4010	No. 6		2.40	15.533		330	575		905	1,225
4020	No. 4		2.40	15.533		445	575		1,020	1,350
4030	No. 2		2.40	15.533		730	575		1,305	1,675
4040	No. 1		2.20	16.945		910	625		1,535	1,950
4050	No. 1/0		2.20	16.945		1,175	625		1,800	2,250
4060	No. 2/0		2.20	16.945		1,400	625		2,025	2,475
4070	No. 4/0	↓	2	18.640	↓	4,075	685		4,760	5,500
5000	Test station, 7 terminal box, flush curb type w/lockable cover	R-19	12	1.667	Ea.	62.50	73.50		136	178
5010	Reference cell, 2" dia PVC conduit, cplg, plug, set flush	"	4.80	4.167	"	34	183		217	310

26 51 Interior Lighting

26 51 13 – Interior Lighting Fixtures, Lamps, and Ballasts

26 51 13.50 Interior Lighting Fixtures		Crew	Daily Output	Labor-Hours	Unit	Material	2007 Bare Costs Labor	2007 Bare Costs Equipment	Total	Total Incl O&P
0010	**INTERIOR LIGHTING FIXTURES** Including lamps, mounting									
0030	hardware and connections									
0100	Fluorescent, C.W. lamps, troffer, recess mounted in grid, RS									
0200	Acrylic lens, 1'W x 4'L, two 40 watt	1 Elec	5.70	1.404	Ea.	50.50	61.50		112	147
0210	1'W x 4'L, three 40 watt		5.40	1.481		59	65		124	162
0300	2'W x 2'L, two U40 watt		5.70	1.404		54	61.50		115.50	151
0400	2'W x 4'L, two 40 watt		5.30	1.509		54	66.50		120.50	158
0500	2'W x 4'L, three 40 watt		5	1.600		57.50	70		127.50	169
0600	2'W x 4'L, four 40 watt	CN	4.70	1.702		61	74.50		135.50	178
0700	4'W x 4'L, four 40 watt	2 Elec	6.40	2.500		325	110		435	520
0800	4'W x 4'L, six 40 watt		6.20	2.581		335	113		448	535
0900	4'W x 4'L, eight 40 watt		5.80	2.759		345	121		466	560
0910	Acrylic lens, 1'W x 4'L, two 32 watt	1 Elec	5.70	1.404		60	61.50		121.50	158
0930	2'W x 2'L, two U32 watt		5.70	1.404		80	61.50		141.50	180
0940	2'W x 4'L, two 32 watt		5.30	1.509		72	66.50		138.50	178
0950	2'W x 4'L, three 32 watt		5	1.600		76.50	70		146.50	189
0960	2'W x 4'L, four 32 watt		4.70	1.702		78.50	74.50		153	198
1000	Surface mounted, RS									
1030	Acrylic lens with hinged & latched door frame									
1100	1'W x 4'L, two 40 watt	1 Elec	7	1.143	Ea.	73.50	50		123.50	156
1110	1'W x 4'L, three 40 watt		6.70	1.194		75.50	52.50		128	161
1200	2'W x 2'L, two U40 watt		7	1.143		79	50		129	161
1300	2'W x 4'L, two 40 watt		6.20	1.290		90.50	56.50		147	184
1400	2'W x 4'L, three 40 watt		5.70	1.404		91.50	61.50		153	192
1500	2'W x 4'L, four 40 watt		5.30	1.509		93.50	66.50		160	202
1600	4'W x 4'L, four 40 watt	2 Elec	7.20	2.222		445	97.50		542.50	635
1700	4'W x 4'L, six 40 watt		6.60	2.424		485	106		591	695
1800	4'W x 4'L, eight 40 watt		6.20	2.581		505	113		618	725
1900	2'W x 8'L, four 40 watt		6.40	2.500		163	110		273	345
2000	2'W x 8'L, eight 40 watt		6.20	2.581		187	113		300	375
2100	Strip fixture									
2130	Surface mounted									
2200	4' long, one 40 watt RS	1 Elec	8.50	.941	Ea.	29.50	41.50		71	94
2300	4' long, two 40 watt RS		8	1		31.50	44		75.50	101
2400	4' long, one 40 watt, SL		8	1		43	44		87	113
2500	4' long, two 40 watt, SL		7	1.143		59	50		109	139
2600	8' long, one 75 watt, SL	2 Elec	13.40	1.194		44	52.50		96.50	127
2700	8' long, two 75 watt, SL	"	12.40	1.290		53	56.50		109.50	143
2800	4' long, two 60 watt, HO	1 Elec	6.70	1.194		86	52.50		138.50	173
2900	8' long, two 110 watt, HO	2 Elec	10.60	1.509		90.50	66.50		157	199
3000	Pendent mounted, industrial, white porcelain enamel									
3100	4' long, two 40 watt, RS	1 Elec	5.70	1.404	Ea.	48.50	61.50		110	145
3200	4' long, two 60 watt, HO	"	5	1.600		77.50	70		147.50	190
3300	8' long, two 75 watt, SL	2 Elec	8.80	1.818		92	80		172	220
3400	8' long, two 110 watt, HO	"	8	2		118	88		206	260
3470	Troffer, air handling, 2'W x 4'L with four 40 watt, RS	1 Elec	4	2		88.50	88		176.50	229
3480	2'W x 2'L with two U40 watt RS		5.50	1.455		77	64		141	180
3490	Air connector insulated, 5" diameter		20	.400		60.50	17.55		78.05	92.50
3500	6" diameter		20	.400		61.50	17.55		79.05	93.50
4450	Incandescent, high hat can, round alzak reflector, prewired									
4470	100 watt	1 Elec	8	1	Ea.	62.50	44		106.50	135
4480	150 watt		8	1		93	44		137	168
4500	300 watt		6.70	1.194		215	52.50		267.50	315

26 51 Interior Lighting

26 51 13 – Interior Lighting Fixtures, Lamps, and Ballasts

26 51 13.50 Interior Lighting Fixtures	Crew	Daily Output	Labor-Hours	Unit	Material	2007 Bare Costs Labor	Equipment	Total	Total Incl O&P
4600 Square glass lens with metal trim, prewired									
4630 100 watt	1 Elec	6.70	1.194	Ea.	47	52.50		99.50	130
4700 200 watt		6.70	1.194		86.50	52.50		139	173
4800 300 watt		5.70	1.404		125	61.50		186.50	230
4900 Ceiling/wall, surface mounted, metal cylinder, 75 watt		10	.800		48	35		83	106
4920 150 watt		10	.800		69	35		104	129
5200 Ceiling, surface mounted, opal glass drum									
5300 8", one 60 watt lamp	1 Elec	10	.800	Ea.	40	35		75	96.50
5400 10", two 60 watt lamps		8	1		44.50	44		88.50	115
5500 12", four 60 watt lamps		6.70	1.194		60	52.50		112.50	144
6010 Vapor tight, incandescent, ceiling mounted, 200 watt		6.20	1.290		58.50	56.50		115	149
6100 Fluorescent, surface mounted, 2 lamps, 4'L, RS, 40 watt		3.20	2.500		104	110		214	277
6850 Vandalproof, surface mounted, fluorescent, two 40 watt		3.20	2.500		220	110		330	405
6860 Incandescent, one 150 watt		8	1		54.50	44		98.50	126
7500 Ballast replacement, by weight of ballast, to 15' high									
7520 Indoor fluorescent, less than 2 lbs.	1 Elec	10	.800	Ea.		35		35	52.50
7540 Two 40W, watt reducer, 2 to 5 lbs.		9.40	.851		30	37.50		67.50	88.50
7560 Two F96 slimline, over 5 lbs.		8	1		56.50	44		100.50	128
7580 Vaportite ballast, less than 2 lbs.		9.40	.851			37.50		37.50	55.50
7600 2 lbs. to 5 lbs.		8.90	.899			39.50		39.50	58.50
7620 Over 5 lbs.		7.60	1.053			46		46	69
7630 Electronic ballast for two tubes		8	1		35	44		79	104
7640 Dimmable ballast one lamp		8	1		62.50	44		106.50	135
7650 Dimmable ballast two-lamp		7.60	1.053		102	46		148	181

26 52 Emergency Lighting

26 52 13 – Emergency Lighting

26 52 13.10 Emergency Lighting and Battery Units

	Crew	Daily Output	Labor-Hours	Unit	Material	2007 Bare Costs Labor	Equipment	Total	Total Incl O&P
0010 **EMERGENCY LIGHTING AND BATTERY UNITS**									
0300 Emergency light units, battery operated									
0350 Twin sealed beam light, 25 watt, 6 volt each									
0500 Lead battery operated	1 Elec	4	2	Ea.	122	88		210	265
0700 Nickel cadmium battery operated		4	2		580	88		668	770
0900 Self-contained fluorescent lamp pack		10	.800		135	35		170	202

26 53 Exit Signs

26 53 13 – Exit Lighting

26 53 13.10 Exit Lighting

	Crew	Daily Output	Labor-Hours	Unit	Material	2007 Bare Costs Labor	Equipment	Total	Total Incl O&P
0010 **EXIT LIGHTING**									
0080 Exit light ceiling or wall mount, incandescent, single face	1 Elec	8	1	Ea.	42	44		86	112
0100 Double face		6.70	1.194		48	52.50		100.50	131
0200 L.E.D. standard, single face		8	1		65	44		109	137
0220 Double face		6.70	1.194		69	52.50		121.50	154
0240 L.E.D. w/battery unit, single face		4.40	1.818		125	80		205	257
0260 Double face		4	2		127	88		215	271

26 54 Classified Location Lighting

26 54 13 – Classified Lighting

26 54 13.20 Explosionproof

		Crew	Daily Output	Labor-Hours	Unit	Material	2007 Bare Costs Labor	Equipment	Total	Total Incl O&P
0010	**EXPLOSIONPROOF**									
6510	Incandescent, ceiling mounted, 200 watt	1 Elec	4	2	Ea.	730	88		818	935
6600	Fluorescent, RS, 4' long, ceiling mounted, two 40 watt	"	2.70	2.963	"	1,850	130		1,980	2,225

26 55 Special Purpose Lighting

26 55 61 – Theatrical Lighting

26 55 61.10 Lights

		Crew	Daily Output	Labor-Hours	Unit	Material	Labor	Equipment	Total	Total Incl O&P
0010	**LIGHTS**									
2000	Lights, border, quartz, reflector, vented,									
2100	colored or white	1 Elec	20	.400	L.F.	157	17.55		174.55	198
2500	Spotlight, follow spot, with transformer, 2,100 watt	"	4	2	Ea.	1,250	88		1,338	1,500
2600	For no transformer, deduct					825			825	910
3000	Stationary spot, fresnel quartz, 6" lens	1 Elec	4	2		99	88		187	240
3100	8" lens		4	2		198	88		286	350
3500	Ellipsoidal quartz, 1,000W, 6" lens		4	2		299	88		387	460
3600	12" lens		4	2		495	88		583	675
4000	Strobe light, 1 to 15 flashes per second, quartz		3	2.667		680	117		797	925
4500	Color wheel, portable, five hole, motorized		4	2		158	88		246	305

26 56 Exterior Lighting

26 56 13 – Lighting Poles and Standards

26 56 13.10 Lighting Poles

		Crew	Daily Output	Labor-Hours	Unit	Material	Labor	Equipment	Total	Total Incl O&P
0010	**LIGHTING POLES**									
2800	Light poles, anchor base									
2820	not including concrete bases									
2840	Aluminum pole, 8' high	1 Elec	4	2	Ea.	610	88		698	800
3000	20' high	R-3	2.90	6.897		790	298	56.50	1,144.50	1,375
3200	30' high		2.60	7.692		1,550	335	63	1,948	2,275
3400	35' high		2.30	8.696		1,675	375	71	2,121	2,475
3600	40' high		2	10		1,900	435	81.50	2,416.50	2,825
3800	Bracket arms, 1 arm	1 Elec	8	1		104	44		148	180
4000	2 arms		8	1		209	44		253	296
4200	3 arms		5.30	1.509		315	66.50		381.50	445
4400	4 arms		5.30	1.509		420	66.50		486.50	560
4500	Steel pole, galvanized, 8' high		3.80	2.105		530	92.50		622.50	720
4600	20' high	R-3	2.60	7.692		940	335	63	1,338	1,600
4800	30' high		2.30	8.696		1,100	375	71	1,546	1,875
5000	35' high		2.20	9.091		1,200	395	74.50	1,669.50	2,000
5200	40' high		1.70	11.765		1,500	510	96	2,106	2,525
5400	Bracket arms, 1 arm	1 Elec	8	1		154	44		198	235
5600	2 arms		8	1		238	44		282	330
5800	3 arms		5.30	1.509		258	66.50		324.50	385
6000	4 arms		5.30	1.509		360	66.50		426.50	495

26 56 19 – Roadway Lighting

26 56 19.20 Roadway Area Luminaire

		Crew	Daily Output	Labor-Hours	Unit	Material	Labor	Equipment	Total	Total Incl O&P
0010	**ROADWAY AREA LUMINAIRE**									
2650	Roadway area luminaire, low pressure sodium, 135 watt	1 Elec	2	4	Ea.	550	176		726	865
2700	180 watt	"	2	4		580	176		756	900
2750	Metal halide, 400 watt	2 Elec	4.40	3.636		460	160		620	745

26 56 Exterior Lighting

26 56 19 – Roadway Lighting

26 56 19.20 Roadway Area Luminaire		Crew	Daily Output	Labor-Hours	Unit	Material	2007 Bare Costs Labor	Equipment	Total	Total Incl O&P
2760	1000 watt	2 Elec	4	4	Ea.	520	176		696	830
2780	High pressure sodium, 400 watt		4.40	3.636		475	160		635	765
2790	1000 watt		4	4		545	176		721	860

26 56 23 – Area Lighting

26 56 23.10 Exterior Fixtures

0010	**EXTERIOR FIXTURES** With lamps									
0200	Wall mounted, incandescent, 100 watt	1 Elec	8	1	Ea.	29	44		73	97
0400	Quartz, 500 watt		5.30	1.509		54	66.50		120.50	158
1100	Wall pack, low pressure sodium, 35 watt		4	2		220	88		308	375
1150	55 watt		4	2		263	88		351	420

26 56 36 – Flood Lighting

26 56 36.20 Floodlights

0010	**FLOODLIGHTS** with ballast and lamp,									
1400	pole mounted, pole not included									
1950	Metal halide, 175 watt	1 Elec	2.70	2.963	Ea.	300	130		430	525
2000	400 watt	2 Elec	4.40	3.636		380	160		540	655
2200	1000 watt	"	4	4		555	176		731	870
2340	High pressure sodium, 70 watt	1 Elec	2.70	2.963		216	130		346	430
2400	400 watt	2 Elec	4.40	3.636		365	160		525	640
2600	1000 watt	"	4	4		625	176		801	950

26 61 Lighting Systems and Accessories

26 61 23 – Lamps

26 61 23.10 Lamps

		Crew	Daily Output	Labor-Hours	Unit	Material	Labor	Equipment	Total	Total Incl O&P
0010	**LAMPS**									
0080	Fluorescent, rapid start, cool white, 2' long, 20 watt	1 Elec	1	8	C	310	350		660	865
0100	4' long, 40 watt		.90	8.889		345	390		735	960
0200	Slimline, 4' long, 40 watt		.90	8.889		735	390		1,125	1,375
0210	4' long, 30 watt energy saver		.90	8.889		735	390		1,125	1,375
0400	High output, 4' long, 60 watt		.90	8.889		890	390		1,280	1,550
0410	8' long, 95 watt energy saver		.80	10		885	440		1,325	1,625
0500	8' long, 110 watt		.80	10		885	440		1,325	1,625
0512	2' long, T5, 14 watt energy saver		1	8		1,225	350		1,575	1,875
0514	3' long, T5, 21 watt energy saver		.90	8.889		1,225	390		1,615	1,925
0516	4' long, T5, 28 watt energy saver		.90	8.889		1,050	390		1,440	1,725
0560	Twin tube compact lamp		.90	8.889		555	390		945	1,200
0570	Double twin tube compact lamp		.80	10		1,350	440		1,790	2,125
0600	Mercury vapor, mogul base, deluxe white, 100 watt		.30	26.667		2,825	1,175		4,000	4,850
0700	250 watt		.30	26.667		3,725	1,175		4,900	5,850
0800	400 watt		.30	26.667		3,000	1,175		4,175	5,050
0900	1000 watt		.20	40		7,000	1,750		8,750	10,300
1000	Metal halide, mogul base, 175 watt		.30	26.667		3,525	1,175		4,700	5,625
1200	400 watt		.30	26.667		3,800	1,175		4,975	5,925
1300	1000 watt		.20	40		9,150	1,750		10,900	12,700
1350	High pressure sodium, 70 watt		.30	26.667		4,100	1,175		5,275	6,275
1380	250 watt		.30	26.667		4,675	1,175		5,850	6,900
1400	400 watt		.30	26.667		4,800	1,175		5,975	7,025
1450	1000 watt		.20	40		13,300	1,750		15,050	17,200
3000	Guards, fluorescent lamp, 4' long		1	8		960	350		1,310	1,575
3200	8' long		.90	8.889		1,925	390		2,315	2,675

26 71 Motors

26 71 13 – Motors

26 71 13.20 Motors		Crew	Daily Output	Labor-Hours	Unit	Material	2007 Bare Costs Labor	2007 Bare Costs Equipment	Total	Total Incl O&P
0010	**MOTORS** 230/460 volts, 60 HZ									
0050	Dripproof, premium efficiency, 1.15 service factor									
0060	1800 RPM, 1/4 HP	1 Elec	5.33	1.501	Ea.	150	66		216	263
0070	1/3 HP		5.33	1.501		164	66		230	278
0080	1/2 HP		5.33	1.501		193	66		259	310
0090	3/4 HP		5.33	1.501		215	66		281	335
0100	1 HP		4.50	1.778		226	78		304	365
0250	5 HP		4.50	1.778		345	78		423	495
0350	10 HP		4	2		590	88		678	775
0450	20 HP CN	2 Elec	5.20	3.077		980	135		1,115	1,275

Estimating Tips

27 20 00 Data Communications
27 30 00 Voice Communications
27 40 00 Audio-Video Communications

- When estimating material costs for special systems, it is always prudent to obtain manufacturers' quotations for equipment prices and special installation requirements which will affect the total costs.

Reference Numbers

Reference numbers are shown in shaded boxes at the beginning of some major classifications. These numbers refer to related items in the Reference Section. The reference information may be an estimating procedure, an alternate pricing method, or technical information.

Note: Not all subdivisions listed here necessarily appear in this publication.

Note: **i2 Trade Service**, *in part, has been used as a reference source for some of the material prices used in Division 27.*

No part of this publication may be reproduced, stored in a retrieval system, or transmitted in any form or by any means without prior written permission of Reed Construction Data.

27 13 Communications Backbone Cabling

27 13 23 – Communications Optical Fiber Backbone Cabling

27 13 23.13 Communications Optical Fiber		Crew	Daily Output	Labor-Hours	Unit	Material	2007 Bare Costs Labor	2007 Bare Costs Equipment	Total	Total Incl O&P
0010	**COMMUNICATIONS OPTICAL FIBER**									
0020	Fiber optics cable only. Added costs depend on the type of fiber									
0030	special connectors, optical modems, and networking parts.									
0040	Specialized tools & techniques cause installation costs to vary.									
0070	Cable, minimum, bulk simplex	1 Elec	8	1	C.L.F.	23	44		67	91
0080	Cable, maximum, bulk plenum quad	"	2.29	3.493	"	66	153		219	300
0150	Fiber optic jumper				Ea.	55.50			55.50	61
0200	Fiber optic pigtail					30			30	33
0300	Fiber optic connector	1 Elec	24	.333		18.45	14.65		33.10	42.50
0350	Fiber optic finger splice		32	.250		32	11		43	51.50
0400	Transceiver (low cost bi-directional)		8	1		295	44		339	390
0450	Rack housing, 4 rack spaces (12 panels)		2	4		450	176		626	755
0500	Fiber optic patch panel (12 ports)		6	1.333		178	58.50		236.50	283

27 41 Audio-Video Systems

27 41 19 – Portable Audio-Video Equipment

27 41 19.10 T.V. Systems

		Crew	Daily Output	Labor-Hours	Unit	Material	Labor	Equipment	Total	Total Incl O&P
0010	**T.V. SYSTEMS** not including rough-in wires, cables & conduits									
0100	Master TV antenna system									
0200	VHF reception & distribution, 12 outlets	1 Elec	6	1.333	Outlet	183	58.50		241.50	288
0400	30 outlets		10	.800		120	35		155	185
0600	100 outlets		13	.615		121	27		148	173
0800	VHF & UHF reception & distribution, 12 outlets		6	1.333		182	58.50		240.50	287
1000	30 outlets		10	.800		120	35		155	185
1200	100 outlets		13	.615		123	27		150	175
1400	School and deluxe systems, 12 outlets		2.40	3.333		240	146		386	480
1600	30 outlets		4	2		210	88		298	360
1800	80 outlets		5.30	1.509		202	66.50		268.50	320

27 51 Distributed Audio-Video Communications Systems

27 51 16 – Public Address and Mass Notification Systems

27 51 16.10 Public Address System

		Crew	Daily Output	Labor-Hours	Unit	Material	Labor	Equipment	Total	Total Incl O&P
0010	**PUBLIC ADDRESS SYSTEM**									
0100	Conventional, office	1 Elec	5.33	1.501	Speaker	113	66		179	222
0200	Industrial	"	2.70	2.963	"	218	130		348	435
0400	Explosionproof system is 3 times cost of central control									
0600	Installation costs run about 120% of material cost									

27 51 19 – Sound Masking Systems

27 51 19.10 Sound System

		Crew	Daily Output	Labor-Hours	Unit	Material	Labor	Equipment	Total	Total Incl O&P
0010	**SOUND SYSTEM** not including rough-in wires, cables & conduits									
0100	Components, outlet, projector	1 Elec	8	1	Ea.	52	44		96	123
0200	Microphone		4	2		58	88		146	195
0400	Speakers, ceiling or wall		8	1		98.50	44		142.50	174
0600	Trumpets		4	2		183	88		271	335
0800	Privacy switch		8	1		73	44		117	146
1000	Monitor panel		4	2		325	88		413	490
1200	Antenna, AM/FM		4	2		182	88		270	330
1400	Volume control		8	1		73	44		117	146
1600	Amplifier, 250 watts		1	8		1,450	350		1,800	2,125

27 51 Distributed Audio-Video Communications Systems

27 51 19 – Sound Masking Systems

27 51 19.10 Sound System

		Crew	Daily Output	Labor-Hours	Unit	Material	2007 Bare Costs Labor	Equipment	Total	Total Incl O&P
1800	Cabinets	1 Elec	1	8	Ea.	710	350		1,060	1,300
2000	Intercom, 25 station capacity, master station	2 Elec	2	8		1,700	350		2,050	2,400
2200	Remote station	1 Elec	8	1		137	44		181	217
2400	Intercom outlets		8	1		80.50	44		124.50	154
2600	Handset		4	2		266	88		354	425
2800	Emergency call system, 12 zones, annunciator		1.30	6.154		800	270		1,070	1,275
3000	Bell		5.30	1.509		83	66.50		149.50	190
3200	Light or relay		8	1		41.50	44		85.50	111
3400	Transformer		4	2		182	88		270	330
3600	House telephone, talking station		1.60	5		390	220		610	755
3800	Press to talk, release to listen	▼	5.30	1.509		91	66.50		157.50	199
4000	System-on button					54.50			54.50	60
4200	Door release	1 Elec	4	2		97.50	88		185.50	238
4400	Combination speaker and microphone		8	1		166	44		210	248
4600	Termination box		3.20	2.500		52	110		162	221
4800	Amplifier or power supply		5.30	1.509	▼	600	66.50		666.50	760
5000	Vestibule door unit		16	.500	Name	110	22		132	154
5200	Strip cabinet		27	.296	Ea.	208	13		221	248
5400	Directory	▼	16	.500	"	98	22		120	141

27 52 Healthcare Communications and Monitoring Systems

27 52 23 – Nurse Call/Code Blue Systems

27 52 23.10 Nurse Call Systems

		Crew	Daily Output	Labor-Hours	Unit	Material	2007 Bare Costs Labor	Equipment	Total	Total Incl O&P
0010	**NURSE CALL SYSTEMS**									
0100	Single bedside call station	1 Elec	8	1	Ea.	202	44		246	288
0200	Ceiling speaker station		8	1		59	44		103	130
0400	Emergency call station		8	1		99.50	44		143.50	175
0600	Pillow speaker		8	1		186	44		230	270
0800	Double bedside call station		4	2		191	88		279	340
1000	Duty station		4	2		154	88		242	300
1200	Standard call button		8	1		80.50	44		124.50	154
1400	Lights, corridor, dome or zone indicator	▼	8	1	▼	45	44		89	115
1600	Master control station for 20 stations	2 Elec	.65	24.615	Total	3,575	1,075		4,650	5,550

27 53 Distributed Systems

27 53 13 – Clock Systems

27 53 13.50 Clock Systems

		Crew	Daily Output	Labor-Hours	Unit	Material	2007 Bare Costs Labor	Equipment	Total	Total Incl O&P
0010	**CLOCK SYSTEMS**, not including wires & conduits									
0100	Time system components, master controller	1 Elec	.33	24.242	Ea.	1,625	1,075		2,700	3,375
0200	Program bell		8	1		53.50	44		97.50	125
0400	Combination clock & speaker		3.20	2.500		189	110		299	370
0600	Frequency generator		2	4		6,350	176		6,526	7,225
0800	Job time automatic stamp recorder, minimum		4	2		445	88		533	620
1000	Maximum	▼	4	2		680	88		768	880
1200	Time stamp for correspondence, hand operated					340			340	375
1400	Fully automatic					495			495	545
1600	Master time clock system, clocks & bells, 20 room	4 Elec	.20	160		4,075	7,025		11,100	15,000
1800	50 room	"	.08	400		9,400	17,600		27,000	36,400
2000	Time clock, 100 cards in & out, 1 color	1 Elec	3.20	2.500	▼	950	110		1,060	1,225

27 53 Distributed Systems

27 53 13 – Clock Systems

27 53 13.50 Clock Systems	Crew	Daily Output	Labor-Hours	Unit	Material	2007 Bare Costs Labor	Equipment	Total	Total Incl O&P
2200 2 colors	1 Elec	3.20	2.500	Ea.	1,025	110		1,135	1,300
2400 With 3 circuit program device, minimum		2	4		375	176		551	675
2600 Maximum		2	4		545	176		721	860
2800 Metal rack for 25 cards		7	1.143		60	50		110	141
3000 Watchman's tour station		8	1		63	44		107	135
3200 Annunciator with zone indication		1	8		230	350		580	780
3400 Time clock with tape		1	8		610	350		960	1,200

Estimating Tips

- When estimating material costs for electronic safety and security systems, it is always prudent to obtain manufacturers' quotations for equipment prices and special installation requirements, which affect the total cost.
- Fire alarm systems consist of control panels, annunciator panels, battery with rack, charger, and fire alarm actuating and indicating devices. Some fire alarm systems include speakers, telephone lines, door closer controls, and other components. Be careful not to overlook the costs related to installation for these items.

Also be aware of costs for integrated automation instrumentation and terminal devices, control equipment, control wiring, and programming.

- Security equipment includes items such as CCTV, access control, and other detection and identification systems to perform alert and alarm functions. Be sure to consider the costs related to installation for this security equipment, such as for integrated automation instrumentation and terminal devices, control equipment, control wiring, and programming.

Reference Numbers

Reference numbers are shown in shaded boxes at the beginning of some major classifications. These numbers refer to related items in the Reference Section. The reference information may be an estimating procedure, an alternate pricing method, or technical information.

Note: Not all subdivisions listed here necessarily appear in this publication.

No part of this publication may be reproduced, stored in a retrieval system, or transmitted in any form or by any means without prior written permission of Reed Construction Data.

28 13 Access Control

28 13 53 – Security Access Detection

28 13 53.13 Security Access Metal Detectors

		Crew	Daily Output	Labor-Hours	Unit	Material	2007 Bare Costs Labor	2007 Bare Costs Equipment	Total	Total Incl O&P
0010	**SECURITY ACCESS METAL DETECTORS**									
0240	Metal detector, hand-held, wand type, unit only				Ea.				81.82	90
0250	Metal detector, walk through portal type, single zone	1 Elec	2	4		4,000	176		4,176	4,650
0260	Multi zone	"	2	4		5,000	176		5,176	5,750

28 13 53.16 Security Access X-Ray Equipment

		Crew	Daily Output	Labor-Hours	Unit	Material	Labor	Equipment	Total	Total Incl O&P
0010	**SECURITY ACCESS X-RAY EQUIPMENT**									
0290	X-ray machine, desk top, for mail/small packages/letters	1 Elec	4	2	Ea.	3,000	88		3,088	3,425
0300	Conveyor type, incl monitor, minimum		2	4		14,000	176		14,176	15,700
0310	Maximum		2	4		25,000	176		25,176	27,800
0320	X-ray machine, large unit, for airports, incl monitor, min	2 Elec	1	16		35,000	700		35,700	39,600
0330	Maximum	"	.50	32		60,000	1,400		61,400	68,000

28 13 53.23 Security Access Explosive Detection Equipment

		Crew	Daily Output	Labor-Hours	Unit	Material	Labor	Equipment	Total	Total Incl O&P
0010	**SECURITY ACCESS EXPLOSIVE DETECTION EQUIPMENT**									
0270	Explosives detector, walk through portal type	1 Elec	2	4	Ea.	3,500	176		3,676	4,100
0280	Hand-held, battery operated				"				25,455	28,000

28 16 Intrusion Detection

28 16 16 – Intrusion Detection Systems Infrastructure

28 16 16.50 Intrusion Detection

		Crew	Daily Output	Labor-Hours	Unit	Material	Labor	Equipment	Total	Total Incl O&P
0010	**INTRUSION DETECTION**, not including wires & conduits									
0100	Burglar alarm, battery operated, mechanical trigger	1 Elec	4	2	Ea.	254	88		342	410
0200	Electrical trigger		4	2		305	88		393	465
0400	For outside key control, add		8	1		72	44		116	145
0600	For remote signaling circuitry, add		8	1		114	44		158	192
0800	Card reader, flush type, standard		2.70	2.963		850	130		980	1,125
1000	Multi-code		2.70	2.963		1,100	130		1,230	1,400
1200	Door switches, hinge switch		5.30	1.509		53.50	66.50		120	158
1400	Magnetic switch		5.30	1.509		63	66.50		129.50	168
1600	Exit control locks, horn alarm		4	2		315	88		403	480
1800	Flashing light alarm		4	2		355	88		443	525
2000	Indicating panels, 1 channel		2.70	2.963		335	130		465	565
2200	10 channel	2 Elec	3.20	5		1,150	220		1,370	1,575
2400	20 channel		2	8		2,250	350		2,600	3,000
2600	40 channel		1.14	14.035		4,075	615		4,690	5,425
2800	Ultrasonic motion detector, 12 volt	1 Elec	2.30	3.478		210	153		363	460
3000	Infrared photoelectric detector	"	2.30	3.478		173	153		326	420

28 23 Video Surveillance

28 23 13 – Video Surveillance Control and Management Systems

28 23 13.10 Closed Circuit Television System

		Crew	Daily Output	Labor-Hours	Unit	Material	Labor	Equipment	Total	Total Incl O&P
0010	**CLOSED CIRCUIT TELEVISION SYSTEM**									
2000	Closed circuit, surveillance, one station (camera & monitor)	2 Elec	2.60	6.154	Total	1,150	270		1,420	1,675
2200	For additional camera stations, add	1 Elec	2.70	2.963	Ea.	650	130		780	910
2400	Industrial quality, one station (camera & monitor)	2 Elec	2.60	6.154	Total	2,400	270		2,670	3,050
2600	For additional camera stations, add	1 Elec	2.70	2.963	Ea.	1,475	130		1,605	1,825
2610	For low light, add		2.70	2.963		1,175	130		1,305	1,500
2620	For very low light, add		2.70	2.963		8,725	130		8,855	9,800
2800	For weatherproof camera station, add		1.30	6.154		910	270		1,180	1,400
3000	For pan and tilt, add		1.30	6.154		2,350	270		2,620	2,975

28 23 Video Surveillance

28 23 13 – Video Surveillance Control and Management Systems

28 23 13.10 Closed Circuit Television System		Crew	Daily Output	Labor-Hours	Unit	Material	2007 Bare Costs Labor	Equipment	Total	Total Incl O&P
3200	For zoom lens - remote control, add, minimum	1 Elec	2	4	Ea.	2,175	176		2,351	2,625
3400	Maximum		2	4		7,925	176		8,101	8,950
3410	For automatic iris for low light, add		2	4		1,900	176		2,076	2,325
3600	Educational T.V. studio, basic 3 camera system, black & white,									
3800	electrical & electronic equip. only, minimum	4 Elec	.80	40	Total	11,300	1,750		13,050	15,000
4000	Maximum (full console)		.28	114		47,900	5,025		52,925	60,000
4100	As above, but color system, minimum		.28	114		63,500	5,025		68,525	77,500
4120	Maximum		.12	266		275,500	11,700		287,200	320,500
4200	For film chain, black & white, add	1 Elec	1	8	Ea.	12,900	350		13,250	14,700
4250	Color, add		.25	32		15,600	1,400		17,000	19,300
4400	For video tape recorders, add, minimum		1	8		2,700	350		3,050	3,500
4600	Maximum	4 Elec	.40	80		22,500	3,500		26,000	30,000

28 23 23 – Video Surveillance Systems Infrastructure

28 23 23.50 Video Surveillance

		Crew	Daily Output	Labor-Hours	Unit	Material	Labor	Equipment	Total	Total Incl O&P
0010	**VIDEO SURVEILLANCE**									
0200	Video cameras, wireless, hidden in exit signs, clocks, etc, incl receiver	1 Elec	3	2.667	Ea.	300	117		417	505
0210	Accessories for, VCR, single camera		3	2.667		500	117		617	725
0220	For multiple cameras		3	2.667		1,500	117		1,617	1,825
0230	Video cameras, wireless, for under vehicle searching, complete		2	4		3,500	176		3,676	4,100

28 31 Fire Detection and Alarm

28 31 23 – Fire Detection and Alarm Annunciation Panels and Fire Stations

28 31 23.50 Alarm Panels and Devices

		Crew	Daily Output	Labor-Hours	Unit	Material	Labor	Equipment	Total	Total Incl O&P
0010	**ALARM PANELS AND DEVICES**									
3594	Fire, alarm control panel									
3600	4 zone	2 Elec	2	8	Ea.	940	350		1,290	1,550
3800	8 zone		1	16		1,400	700		2,100	2,600
4000	12 zone		.67	23.988		1,825	1,050		2,875	3,575
4020	Alarm device	1 Elec	8	1		122	44		166	200
4050	Actuating device		8	1		292	44		336	385
4200	Battery and rack		4	2		725	88		813	930
4400	Automatic charger		8	1		445	44		489	555
4600	Signal bell		8	1		52	44		96	123
4800	Trouble buzzer or manual station		8	1		37	44		81	106
5600	Strobe and horn		5.30	1.509		95	66.50		161.50	204
5800	Fire alarm horn		6.70	1.194		36.50	52.50		89	118
6000	Door holder, electro-magnetic		4	2		77.50	88		165.50	217
6200	Combination holder and closer		3.20	2.500		430	110		540	640
6400	Code transmitter		4	2		690	88		778	890
6600	Drill switch		8	1		86.50	44		130.50	161
6800	Master box		2.70	2.963		3,100	130		3,230	3,600
7000	Break glass station		8	1		50	44		94	121
7800	Remote annunciator, 8 zone lamp		1.80	4.444		201	195		396	510
8000	12 zone lamp	2 Elec	2.60	6.154		345	270		615	780
8200	16 zone lamp	"	2.20	7.273		345	320		665	855

28 31 43 – Fire Detection Sensors

28 31 43.50 Fire and Heat Detectors

		Crew	Daily Output	Labor-Hours	Unit	Material	Labor	Equipment	Total	Total Incl O&P
0010	**FIRE & HEAT DETECTORS**									
5000	Detector, rate of rise	1 Elec	8	1	Ea.	35	44		79	104

28 31 Fire Detection and Alarm

28 31 46 – Smoke Detection Sensors

28 31 46.50 Smoke Detectors		Crew	Daily Output	Labor-Hours	Unit	Material	2007 Bare Costs Labor	Equipment	Total	Total Incl O&P
0010	**SMOKE DETECTORS**									
5200	Smoke detector, ceiling type	1 Elec	6.20	1.290	Ea.	79	56.50		135.50	171
5400	Duct type	"	3.20	2.500	"	250	110		360	440

28 33 Fuel-Gas Detection and Alarm

28 33 33 – Fuel-Gas Detection Sensors

28 33 33.50 Tank Leak Detection Systems

		Crew	Daily Output	Labor-Hours	Unit	Material	2007 Bare Costs Labor	Equipment	Total	Total Incl O&P
0010	**TANK LEAK DETECTION SYSTEMS** Liquid and vapor									
0100	For hydrocarbons and hazardous liquids/vapors									
0120	Controller, data acquisition, incl. printer, modem, RS232 port									
0140	24 channel, for use with all probes				Ea.	2,175			2,175	2,400
0160	9 channel, for external monitoring				"	1,700			1,700	1,850
0200	Probes									
0210	Well monitoring									
0220	Liquid phase detection				Ea.	760			760	835
0230	Hydrocarbon vapor, fixed position					700			700	770
0240	Hydrocarbon vapor, float mounted					630			630	695
0250	Both liquid and vapor hydrocarbon					940			940	1,025
0300	Secondary containment, liquid phase									
0310	Pipe trench/manway sump				Ea.	340			340	375
0320	Double wall pipe and manual sump					330			330	365
0330	Double wall fiberglass annular space					305			305	335
0340	Double wall steel tank annular space					305			305	335
0500	Accessories									
0510	Modem, non-dedicated phone line				Ea.	273			273	300
0600	Monitoring, internal									
0610	Automatic tank gauge, incl. overfill				Ea.	1,100			1,100	1,200
0620	Product line				"	1,175			1,175	1,300
0700	Monitoring, special									
0710	Cathodic protection				Ea.	650			650	715
0720	Annular space chemical monitor				"	885			885	975

Estimating Tips

31 05 00 Common Work Results for Earthwork

- Estimating the actual cost of performing earthwork requires careful consideration of the variables involved. This includes items such as type of soil, whether water will be encountered, dewatering, whether banks need bracing, disposal of excavated earth, and length of haul to fill or spoil sites, etc. If the project has large quantities of cut or fill, consider raising or lowering the site to reduce costs, while paying close attention to the effect on site drainage and utilities.
- If the project has large quantities of fill, creating a borrow pit on the site can significantly lower the costs.
- It is very important to consider what time of year the project is scheduled for completion. Bad weather can create large cost overruns from dewatering, site repair, and lost productivity from cold weather.

Reference Numbers

Reference numbers are shown in shaded boxes at the beginning of some major classifications. These numbers refer to related items in the Reference Section. The reference information may be an estimating procedure, an alternate pricing method, or technical information.

Note: Not all subdivisions listed here necessarily appear in this publication.

No part of this publication may be reproduced, stored in a retrieval system, or transmitted in any form or by any means without prior written permission of Reed Construction Data.

31 05 Common Work Results for Earthwork

31 05 13 – Soils for Earthwork

31 05 13.10 Borrow

		Crew	Daily Output	Labor-Hours	Unit	Material	2007 Bare Costs Labor	Equipment	Total	Total Incl O&P
0010	**BORROW**	R312316-40								
0020	Spread, 200 H.P. dozer, no compaction, 2 mi. RT haul									
0200	Common borrow	B-15	600	.047	C.Y.	6	1.49	3.41	10.90	12.65
0700	Screened loam	**CN**	600	.047		21	1.49	3.41	25.90	29
0800	Topsoil, weed free	↓	600	.047		22	1.49	3.41	26.90	30
0900	For 5 mile haul, add	B-34B	200	.040	↓		1.18	2.65	3.83	4.75

31 05 16 – Aggregates for Earthwork

31 05 16.10 Borrow

		Crew	Daily Output	Labor-Hours	Unit	Material	2007 Bare Costs Labor	Equipment	Total	Total Incl O&P
0010	**BORROW**	R312316-40								
0020	Spread, with 200 H.P. dozer, no compaction, 2 mi RT haul									
0100	Bank run gravel	**CN** B-15	600	.047	C.Y.	21.50	1.49	3.41	26.40	29.50
0300	Crushed stone (1.40 tons per CY), 1-1/2"		600	.047		32	1.49	3.41	36.90	41
0320	3/4"	**CN**	600	.047		32	1.49	3.41	36.90	41
0340	1/2"		600	.047		36	1.49	3.41	40.90	45.50
0360	3/8"		600	.047		27	1.49	3.41	31.90	35.50
0400	Sand, washed, concrete		600	.047		29.50	1.49	3.41	34.40	38.50
0500	Dead or bank sand		600	.047		6.25	1.49	3.41	11.15	12.95
0600	Select structural fill	↓	600	.047		9.45	1.49	3.41	14.35	16.45
0900	For 5 mile haul, add	B-34B	200	.040	↓		1.18	2.65	3.83	4.75

31 05 23 – Cement and Concrete for Earthwork

31 05 23.30 Plant Mixed Bituminous Concrete

		Crew	Daily Output	Labor-Hours	Unit	Material	2007 Bare Costs Labor	Equipment	Total	Total Incl O&P
0010	**PLANT MIXED BITUMINOUS CONCRETE**									
0020	Asphaltic concrete plant mix (145 LB per c.f.)	**CN**			Ton	48			48	53
0040	Asphaltic concrete less than 300 tons add trucking costs									
0050	See Div. 31 23 16 for hauling costs									
0200	All weather patching mix, hot					48			48	53
0250	Cold patch					53			53	58.50
0300	Berm mix				↓	48			48	53

31 06 Schedules for Earthwork

31 06 60 – Schedules for Special Foundations and Load Bearing Elements

31 06 60.14 Piling Special Costs

		Crew	Daily Output	Labor-Hours	Unit	Material	2007 Bare Costs Labor	Equipment	Total	Total Incl O&P
0010	**PILING SPECIAL COSTS**									
0011	Pile caps, see Division 03 30 53.40									
0500	Cutoffs, concrete piles, plain	1 Pile	5.50	1.455	Ea.		52		52	85.50
0600	With steel thin shell, add		38	.211			7.55		7.55	12.35
0700	Steel pile or "H" piles		19	.421			15.10		15.10	24.50
0800	Wood piles	↓	38	.211	↓		7.55		7.55	12.35
0900	Pre-augering up to 30' deep, average soil, 24" diameter	B-43	180	.267	L.F.		8.45	19.45	27.90	34.50
0920	36" diameter		115	.417			13.25	30.50	43.75	54
0960	48" diameter		70	.686			22	50	72	88.50
0980	60" diameter	↓	50	.960	↓		30.50	70	100.50	124
1000	Testing, any type piles, test load is twice the design load									
1050	50 ton design load, 100 ton test				Ea.				15,900	17,013
1100	100 ton design load, 200 ton test								20,405	22,330
1150	150 ton design load, 300 ton test								26,510	28,270
1200	200 ton design load, 400 ton test								28,820	31,680
1250	400 ton design load, 800 ton test				↓				33,330	37,070
1500	Wet conditions, soft damp ground									
1600	Requiring mats for crane, add								40%	40%

31 06 Schedules for Earthwork

31 06 60 – Schedules for Special Foundations and Load Bearing Elements

31 06 60.14 Piling Special Costs		Crew	Daily Output	Labor-Hours	Unit	Material	2007 Bare Costs Labor	Equipment	Total	Total Incl O&P
1700	Barge mounted driving rig, add								30%	30%

31 06 60.15 Mobilization

		Crew	Daily Output	Labor-Hours	Unit	Material	Labor	Equipment	Total	Total Incl O&P
0010	**MOBILIZATION**									
0020	Set up & remove, air compressor, 600 C.F.M.	A-5	3.30	5.455	Ea.		157	10.70	167.70	256
0100	1200 C.F.M.	"	2.20	8.182			235	16.05	251.05	385
0200	Crane, with pile leads and pile hammer, 75 ton	B-19	.60	106			3,925	2,750	6,675	9,275
0300	150 ton	"	.36	177			6,550	4,575	11,125	15,500
0500	Drill rig, for caissons, to 36", minimum	B-43	2	24			765	1,750	2,515	3,100
0600	Up to 84"	"	1	48			1,525	3,500	5,025	6,200
0800	Auxiliary boiler, for steam small	A-5	1.66	10.843			310	21.50	331.50	510
0900	Large	"	.83	21.687			625	42.50	667.50	1,025
1100	Rule of thumb: complete pile driving set up, small	B-19	.45	142			5,250	3,675	8,925	12,400
1200	Large	"	.27	237			8,750	6,100	14,850	20,600

31 11 Clearing and Grubbing

31 11 10 – Clearing and Grubbing

31 11 10.10 Clear and Grub

		Crew	Daily Output	Labor-Hours	Unit	Material	Labor	Equipment	Total	Total Incl O&P
0010	**CLEAR AND GRUB**									
0020	Cut & chip light trees to 6" diam.	B-7	1	48	Acre		1,475	1,075	2,550	3,475
0150	Grub stumps and remove	B-30	2	12			390	920	1,310	1,600
0200	Cut & chip medium, trees to 12" diam.	B-7	.70	68.571			2,100	1,550	3,650	4,950
0250	Grub stumps and remove	B-30	1	24			780	1,825	2,605	3,225
0300	Cut & chip heavy, trees to 24" diam.	B-7	.30	160			4,900	3,625	8,525	11,600
0350	Grub stumps and remove	B-30	.50	48			1,550	3,675	5,225	6,425
0400	If burning is allowed, reduce cut & chip									40%
3000	Chipping stumps, to 18" deep, 12" diam.	B-86	20	.400	Ea.		15.35	5.30	20.65	29
3040	18" diameter		16	.500			19.20	6.65	25.85	36.50
3080	24" diameter		14	.571			22	7.60	29.60	41.50
3100	30" diameter		12	.667			25.50	8.85	34.35	48.50
3120	36" diameter		10	.800			30.50	10.65	41.15	58
3160	48" diameter		8	1			38.50	13.30	51.80	72.50
5000	Tree thinning, feller buncher, conifer									
5080	Up to 8" diameter	B-93	240	.033	Ea.		1.28	2.14	3.42	4.28
5120	12" diameter		160	.050			1.92	3.21	5.13	6.40
5240	Hardwood, up to 4" diameter		240	.033			1.28	2.14	3.42	4.28
5280	8" diameter		180	.044			1.71	2.85	4.56	5.70
5320	12" diameter		120	.067			2.56	4.28	6.84	8.55
7000	Tree removal, congested area, aerial lift truck									
7040	8" diameter	B-85	7	5.714	Ea.		176	105	281	390
7080	12" diameter		6	6.667			206	123	329	450
7120	18" diameter		5	8			247	148	395	540
7160	24" diameter		4	10			310	185	495	680
7240	36" diameter		3	13.333			410	246	656	905
7280	48" diameter		2	20			615	370	985	1,350

31 13 Selective Tree and Shrub Removal and Trimming

31 13 13 – Selective Tree and Shrub Removal

31 13 13.10 Selective Clearing

		Crew	Daily Output	Labor-Hours	Unit	Material	2007 Bare Costs Labor	2007 Bare Costs Equipment	Total	Total Incl O&P
0010	**SELECTIVE CLEARING**									
0020	Clearing brush with brush saw	A-1C	.25	32	Acre		920	92	1,012	1,525
0100	By hand	1 Clab	.12	66.667			1,925		1,925	2,975
0300	With dozer, ball and chain, light clearing	B-11A	2	8			269	495	764	955
0400	Medium clearing		1.50	10.667			360	660	1,020	1,275
0500	With dozer and brush rake, light		10	1.600			53.50	99	152.50	191
0550	Medium brush to 4" diameter		8	2			67	124	191	239
0600	Heavy brush to 4" diameter		6.40	2.500			84	154	238	298
1000	Brush mowing, tractor w/rotary mower, no removal									
1020	Light density	B-84	2	4	Acre		154	118	272	360
1040	Medium density		1.50	5.333			205	157	362	480
1080	Heavy density		1	8			305	235	540	725

31 13 13.20 Selective Tree Removal

		Crew	Daily Output	Labor-Hours	Unit	Material	2007 Bare Costs Labor	2007 Bare Costs Equipment	Total	Total Incl O&P
0010	**SELECTIVE TREE REMOVAL** With tractor, large tract, firm									
0020	level terrain, no boulders, less than 12" diam. trees									
0300	300 HP dozer, up to 400 trees/acre, 0 to 25% hardwoods	B-10M	.75	16	Acre		565	1,725	2,290	2,750
0340	25% to 50% hardwoods		.60	20			705	2,175	2,880	3,450
0370	75% to 100% hardwoods		.45	26.667			940	2,900	3,840	4,600
0400	500 trees/acre, 0% to 25% hardwoods		.60	20			705	2,175	2,880	3,450
0440	25% to 50% hardwoods		.48	25			880	2,700	3,580	4,300
0470	75% to 100% hardwoods		.36	33.333			1,175	3,625	4,800	5,750
0500	More than 600 trees/acre, 0 to 25% hardwoods		.52	23.077			810	2,500	3,310	3,975
0540	25% to 50% hardwoods		.42	28.571			1,000	3,100	4,100	4,925
0570	75% to 100% hardwoods		.31	38.710			1,350	4,200	5,550	6,700
0900	Large tract clearing per tree									
1500	300 HP dozer, to 12" diameter, softwood	B-10M	320	.038	Ea.		1.32	4.07	5.39	6.50
1550	Hardwood		100	.120			4.22	13	17.22	20.50
1600	12" to 24" diameter, softwood		200	.060			2.11	6.50	8.61	10.35
1650	Hardwood		80	.150			5.30	16.25	21.55	26
1700	24" to 36" diameter, softwood		100	.120			4.22	13	17.22	20.50
1750	Hardwood		50	.240			8.45	26	34.45	41.50
1800	36" to 48" diameter, softwood		70	.171			6.05	18.60	24.65	29.50
1850	Hardwood		35	.343			12.05	37	49.05	59.50
2000	Stump removal on site by hydraulic backhoe, 1-1/2 C.Y.									
2040	4" to 6" diameter	B-17	60	.533	Ea.		16.50	10.20	26.70	36.50
2050	8" to 12" diameter	B-30	33	.727			23.50	55.50	79	97
2100	14" to 24" diameter		25	.960			31	73.50	104.50	129
2150	26" to 36" diameter		16	1.500			49	115	164	201
3000	Remove selective trees, on site using chain saws and chipper,									
3050	not incl. stumps, up to 6" diameter	B-7	18	2.667	Ea.		82	60.50	142.50	194
3100	8" to 12" diameter		12	4			123	90.50	213.50	290
3150	14" to 24" diameter		10	4.800			147	109	256	350
3200	26" to 36" diameter		8	6			184	136	320	435
3300	Machine load, 2 mile haul to dump, 12" diam. tree, add								208	312

31 14 Earth Stripping and Stockpiling

31 14 13 – Soil Stripping and Stockpiling

31 14 13.23 Topsoil Stripping and Stockpiling

		Crew	Daily Output	Labor-Hours	Unit	Material	2007 Bare Costs Labor	Equipment	Total	Total Incl O&P
0010	**TOPSOIL STRIPPING AND STOCKPILING**									
0020	200 H.P. dozer, ideal conditions	B-10B	2300	.005	C.Y.		.18	.43	.61	.75
0100	Adverse conditions	"	1150	.010			.37	.86	1.23	1.51
0200	300 HP dozer, ideal conditions	B-10M	3000	.004			.14	.43	.57	.69
0300	Adverse conditions	"	1650	.007			.26	.79	1.05	1.26
0400	400 HP dozer, ideal conditions	B-10X	3900	.003			.11	.43	.54	.63
0500	Adverse conditions	"	2000	.006			.21	.84	1.05	1.24
0600	Clay, dry and soft, 200 HP dozer, ideal conditions	B-10B	1600	.008			.26	.62	.88	1.08
0700	Adverse conditions	"	800	.015			.53	1.24	1.77	2.16
1000	Medium hard, 300 HP dozer, ideal conditions	B-10M	2000	.006			.21	.65	.86	1.04
1100	Adverse conditions	"	1100	.011			.38	1.18	1.56	1.88
1200	Very hard, 400 HP dozer, ideal conditions	B-10X	2600	.005			.16	.65	.81	.96
1300	Adverse conditions	"	1340	.009			.32	1.25	1.57	1.86
1400	Loam or topsoil, remove and stockpile on site									
1420	6" deep, 200' haul	B-10B	865	.014	C.Y.		.49	1.14	1.63	2
1430	300' haul		520	.023			.81	1.90	2.71	3.32
1440	500' haul		225	.053			1.88	4.39	6.27	7.70
1450	Alternate method: 6" deep, 200' haul		5090	.002	S.Y.		.08	.19	.27	.34
1460	500' haul		1325	.009	"		.32	.75	1.07	1.30

31 22 Grading

31 22 16 – Fine Grading

31 22 16.10 Finish Grading

		Crew	Daily Output	Labor-Hours	Unit	Material	2007 Bare Costs Labor	Equipment	Total	Total Incl O&P
0010	**FINISH GRADING**									
0012	Finish grading area to be paved with grader, small area	B-11L	400	.040	S.Y.		1.34	1.26	2.60	3.44
0100	Large area		2000	.008			.27	.25	.52	.69
1100	Fine grade for slab on grade, machine		1040	.015			.52	.49	1.01	1.32
1150	Hand grading	B-18	700	.034			1.01	.05	1.06	1.63

31 23 Excavation and Fill

31 23 16 – Excavation

31 23 16.13 Excavating, Trench

		Crew	Daily Output	Labor-Hours	Unit	Material	2007 Bare Costs Labor	Equipment	Total	Total Incl O&P
0010	**EXCAVATING, TRENCH** or continuous footing									
0020	Common earth with no sheeting or dewatering included									
0050	1' to 4' deep, 3/8 C.Y. excavator	B-11C	150	.107	B.C.Y.		3.58	1.62	5.20	7.25
0060	1/2 C.Y. excavator	B-11M	200	.080			2.69	1.43	4.12	5.65
0090	4' to 6' deep, 1/2 C.Y. excavator	"	200	.080			2.69	1.43	4.12	5.65
0100	5/8 C.Y. excavator	B-12Q	250	.064			2.19	1.85	4.04	5.40
0110	3/4 C.Y. excavator	B-12F	300	.053			1.83	1.72	3.55	4.68
0300	1/2 C.Y. excavator, truck mounted	B-12J	200	.080			2.74	4.53	7.27	9.15
0500	6' to 10' deep, 3/4 C.Y. excavator	B-12F	225	.071			2.44	2.30	4.74	6.25
0510	1 C.Y. excavator	B-12A	400	.040			1.37	1.50	2.87	3.75
0600	1 C.Y. excavator, truck mounted	B-12K	400	.040			1.37	2.57	3.94	4.92
0610	1-1/2 C.Y. excavator	B-12B	600	.027			.91	1.29	2.20	2.82
0900	10' to 14' deep, 3/4 C.Y. excavator	B-12F	200	.080			2.74	2.58	5.32	7.05
0910	1 C.Y. excavator	B-12A	360	.044			1.52	1.67	3.19	4.17
1000	1-1/2 C.Y. excavator	B-12B	540	.030			1.02	1.44	2.46	3.13
1300	14' to 20' deep, 1 C.Y. excavator	B-12A	320	.050			1.71	1.88	3.59	4.69
1310	1-1/2 C.Y. excavator	B-12B	480	.033			1.14	1.62	2.76	3.53

31 23 Excavation and Fill

31 23 16 – Excavation

31 23 16.13 Excavating, Trench

		Crew	Daily Output	Labor-Hours	Unit	Material	2007 Bare Costs Labor	Equipment	Total	Total Incl O&P
1320	2-1/2 C.Y. excavator	B-12S	850	.019	B.C.Y.		.65	1.57	2.22	2.71
1340	20' to 24' deep, 1 C.Y. excavator	B-12A	288	.056			1.90	2.09	3.99	5.20
1342	1-1/2 C.Y. excavator	B-12B	432	.037			1.27	1.80	3.07	3.92
1344	2-1/2 C.Y. excavator	B-12S	765	.021			.72	1.74	2.46	3.02
1352	4' to 6' deep, 1/2 C.Y. excavator w/ trench box	B-13H	188	.085			2.92	5.25	8.17	10.20
1354	5/8 C.Y. excavator	"	235	.068			2.33	4.19	6.52	8.20
1356	3/4 C.Y. excavator	B-13G	282	.057			1.95	2.11	4.06	5.30
1362	6' to 10' deep, 3/4 C.Y. excavator w/trench box	"	212	.075			2.59	2.81	5.40	7.05
1370	1 C.Y. excavator	B-13D	376	.043			1.46	1.81	3.27	4.22
1371	1-1/2 C.Y. excavator	B-13E	564	.028			.97	1.52	2.49	3.16
1374	10' to 14' deep, 3/4 C.Y. excavator w/trench box	B-13G	188	.085			2.92	3.17	6.09	7.95
1375	1 C.Y. excavator	B-13D	338	.047			1.62	2.02	3.64	4.70
1376	1-1/2 C.Y. excavator	B-13E	508	.032			1.08	1.68	2.76	3.50
1381	14' to 20' deep, 1 C.Y. excavator w/trench box	B-13D	301	.053			1.82	2.26	4.08	5.25
1382	1-1/2 C.Y. excavator	B-13E	451	.035			1.22	1.90	3.12	3.95
1383	2-1/2 C.Y. excavator	B-13J	799	.020			.69	1.77	2.46	2.99
1386	20' to 24' deep, 1 C.Y. excavator w/trench box	B-13D	271	.059			2.02	2.51	4.53	5.85
1387	1-1/2 C.Y. excavator	B-13E	406	.039			1.35	2.11	3.46	4.38
1388	2-1/2 C.Y. excavator	B-13J	719	.022			.76	1.96	2.72	3.33
1395	Hydraulic shoring, SF trench wall protected stable matl, 4' W	2 Clab	2700	.006	SF Wall	.09	.17		.26	.37
1397	semi-stable material, 4' W	"	2400	.007		.16	.19		.35	.47
1398	Rent hyraulic shoring per day per SF protected area stable matl					.09			.09	.10
1399	semi-stable material					.16			.16	.17
1400	By hand with pick and shovel 2' to 6' deep, light soil	1 Clab	8	1	B.C.Y.		29		29	45
1500	Heavy soil	"	4	2	"		57.50		57.50	89.50
1700	For tamping backfilled trenches, air tamp, add	A-1G	100	.080	E.C.Y.		2.30	.38	2.68	4
1900	Vibrating plate, add	B-18	180	.133	"		3.92	.20	4.12	6.30
2100	Trim sides and bottom for concrete pours, common earth		1500	.016	S.F.		.47	.02	.49	.76
2300	Hardpan		600	.040	"		1.18	.06	1.24	1.90
2400	Pier and spread footing excavation, add to above				B.C.Y.				30%	30%
3000	Backfill trench, F.E. loader, wheel mtd., 1 C.Y. bucket									
3020	Minimal haul	B-10R	400	.030	L.C.Y.		1.06	.52	1.58	2.17
3040	100' haul	"	200	.060			2.11	1.04	3.15	4.35
3080	2-1/4 C.Y. bucket, minimum haul	B-10T	600	.020			.70	.58	1.28	1.71
3090	100' haul	"	300	.040			1.41	1.16	2.57	3.42
5020	Loam & Sandy clay with no sheeting or dewatering included									
5050	1' to 4' deep, 3/8 C.Y. excavator	B-11C	162	.099	B.C.Y.		3.32	1.50	4.82	6.70
5060	1/2 C.Y. excavator	B-11M	216	.074			2.49	1.32	3.81	5.25
5080	4' to 6' deep, 1/2 C.Y. excavator	"	216	.074			2.49	1.32	3.81	5.25
5090	5/8 C.Y. excavator	B-12Q	276	.058			1.99	1.67	3.66	4.88
5100	3/4 C.Y. excavator	B-12F	324	.049			1.69	1.59	3.28	4.34
5130	1/2 C.Y. excavator, truck mounted	B-12J	216	.074			2.54	4.19	6.73	8.50
5140	6' to 10' deep, 3/4 C.Y. excavator	B-12F	243	.066			2.26	2.13	4.39	5.80
5150	1 C.Y. excavator	B-12A	432	.037			1.27	1.39	2.66	3.47
5160	1 C.Y. excavator, truck mounted	B-12K	432	.037			1.27	2.38	3.65	4.55
5170	1-1/2 C.Y. excavator	B-12B	648	.025			.85	1.20	2.05	2.61
5190	10' to 14' deep, 3/4 C.Y. excavator	B-12F	216	.074			2.54	2.39	4.93	6.50
5200	1 C.Y. excavator	B-12A	389	.041			1.41	1.55	2.96	3.85
5210	1-1/2 C.Y. excavator	B-12B	583	.027			.94	1.33	2.27	2.90
5250	14' to 20' deep, 1 C.Y. excavator	B-12A	346	.046			1.59	1.74	3.33	4.33
5260	1-1/2 C.Y. excavator	B-12B	518	.031			1.06	1.50	2.56	3.27
5270	2-1/2 C.Y. excavator	B-12S	918	.017			.60	1.45	2.05	2.51
5300	20' to 24' deep, 1 C.Y. excavator	B-12A	311	.051			1.76	1.94	3.70	4.83

31 23 Excavation and Fill

31 23 16 – Excavation

31 23 16.13 Excavating, Trench

		Crew	Daily Output	Labor-Hours	Unit	Material	2007 Bare Costs Labor	2007 Bare Costs Equipment	Total	Total Incl O&P
5310	1-1/2 C.Y. excavator	B-12B	467	.034	B.C.Y.		1.17	1.66	2.83	3.62
5320	2-1/2 C.Y. excavator	B-12S	826	.019			.66	1.61	2.27	2.79
5352	4' to 6' deep, 1/2 C.Y. excavator w/trench box	B-13H	205	.078			2.68	4.81	7.49	9.40
5354	5/8 C.Y. excavator	"	257	.062			2.13	3.83	5.96	7.50
5356	3/4 C.Y. excavator	B-13G	308	.052			1.78	1.93	3.71	4.85
5362	6' to 10' deep, 3/4 C.Y. excavator w/ trench box	"	231	.069			2.37	2.58	4.95	6.45
5364	1 C.Y. excavator	B-13D	410	.039			1.34	1.66	3	3.87
5366	1-1/2 C.Y. excavator	B-13E	616	.026			.89	1.39	2.28	2.89
5370	10' to 14' deep, 3/4 C.Y. excavator w/ trench box	B-13G	205	.078			2.68	2.91	5.59	7.30
5372	1 C.Y. excavator	B-13D	370	.043			1.48	1.84	3.32	4.28
5374	1-1/2 C.Y. excavator	B-13E	554	.029			.99	1.54	2.53	3.21
5382	14' to 20' deep, 1 C.Y. excavator w/ trench box	B-13D	329	.049			1.67	2.07	3.74	4.83
5384	1-1/2 C.Y. excavator	B-13E	492	.033			1.11	1.74	2.85	3.61
5386	2-1/2 C.Y. excavator	B-13J	872	.018			.63	1.62	2.25	2.74
5392	20' to 24' deep, 1 C.Y. excavator w/ trench box	B-13D	295	.054			1.86	2.31	4.17	5.40
5394	1-1/2 C.Y. excavator	B-13E	444	.036			1.24	1.93	3.17	4.01
5396	2-1/2 C.Y. excavator	B-13J	785	.020			.70	1.80	2.50	3.05
6020	Sand & gravel with no sheeting or dewatering included									
6050	1' to 4' deep, 3/8 C.Y. excavator	B-11C	165	.097	B.C.Y.		3.26	1.47	4.73	6.60
6060	1/2 C.Y. excavator	B-11M	220	.073			2.44	1.30	3.74	5.15
6080	4' to 6' deep, 1/2 C.Y. excavator	"	220	.073			2.44	1.30	3.74	5.15
6090	5/8 C.Y. excavator	B-12Q	275	.058			1.99	1.68	3.67	4.90
6100	3/4 C.Y. excavator	B-12F	330	.048			1.66	1.56	3.22	4.26
6130	1/2 C.Y. excavator, truck mounted	B-12J	220	.073			2.49	4.12	6.61	8.35
6140	6' to 10' deep, 3/4 C.Y. excavator	B-12F	248	.065			2.21	2.08	4.29	5.65
6150	1 C.Y. excavator	B-12A	440	.036			1.25	1.37	2.62	3.40
6160	1 C.Y. excavator, truck mounted	B-12K	440	.036			1.25	2.33	3.58	4.46
6170	1-1/2 C.Y. excavator	B-12B	660	.024			.83	1.18	2.01	2.56
6190	10' to 14' deep, 3/4 C.Y. excavator	B-12F	220	.073			2.49	2.35	4.84	6.40
6200	1 C.Y. excavator	B-12A	396	.040			1.38	1.52	2.90	3.79
6210	1-1/2 C.Y. excavator	B-12B	594	.027			.92	1.31	2.23	2.85
6250	14' to 20' deep, 1 C.Y. excavator	B-12A	352	.045			1.56	1.71	3.27	4.26
6260	1-1/2 C.Y. excavator	B-12B	528	.030			1.04	1.47	2.51	3.21
6270	2-1/2 C.Y. excavator	B-12S	935	.017			.59	1.43	2.02	2.47
6300	20' to 24' deep, 1 C.Y. excavator	B-12A	317	.050			1.73	1.90	3.63	4.73
6310	1-1/2 C.Y. excavator	B-12B	475	.034			1.15	1.63	2.78	3.56
6320	2-1/2 C.Y. excavator	B-12S	842	.019			.65	1.58	2.23	2.74
6352	4' to 6' deep, 1/2 C.Y. excavator w/ trench box	B-13H	209	.077			2.62	4.71	7.33	9.20
6354	5/8 C.Y. excavator	"	261	.061			2.10	3.77	5.87	7.35
6356	3/4 C.Y. excavator	B-13G	314	.051			1.75	1.90	3.65	4.76
6362	6' to 10' deep, 3/4 C.Y. excavator w/ trench box	"	236	.068			2.32	2.52	4.84	6.35
6364	1 C.Y. excavator	B-13D	418	.038			1.31	1.63	2.94	3.80
6366	1-1/2 C.Y. excavator	B-13E	627	.026			.87	1.36	2.23	2.84
6370	10' to 14' deep, 3/4 C.Y. excavator w/ trench box	B-13G	209	.077			2.62	2.85	5.47	7.15
6372	1 C.Y. excavator	B-13D	376	.043			1.46	1.81	3.27	4.22
6374	1-1/2 C.Y. excavator	B-13E	564	.028			.97	1.52	2.49	3.16
6382	14' to 20' deep, 1 C.Y. excavator w/ trench box	B-13D	334	.048			1.64	2.04	3.68	4.75
6384	1-1/2 C.Y. excavator	B-13E	502	.032			1.09	1.70	2.79	3.54
6386	2-1/2 C.Y. excavator	B-13J	888	.018			.62	1.59	2.21	2.69
6392	20' to 24' deep, 1 C.Y. excavator w/ trench box	B-13D	301	.053			1.82	2.26	4.08	5.25
6394	1-1/2 C.Y. excavator	B-13E	452	.035			1.21	1.89	3.10	3.93
6396	2-1/2 C.Y. excavator	B-13J	800	.020			.69	1.77	2.46	2.99
7020	Dense hard clay with no sheeting or dewatering included									

31 23 Excavation and Fill

31 23 16 – Excavation

31 23 16.13 Excavating, Trench

		Crew	Daily Output	Labor-Hours	Unit	Material	2007 Bare Costs Labor	2007 Bare Costs Equipment	Total	Total Incl O&P
7050	1' to 4' deep, 3/8 C.Y. excavator	B-11C	132	.121	B.C.Y.		4.07	1.84	5.91	8.25
7060	1/2 C.Y. excavator	B-11M	176	.091			3.05	1.62	4.67	6.45
7080	4' to 6' deep, 1/2 C.Y. excavator	"	176	.091			3.05	1.62	4.67	6.45
7090	5/8 C.Y. excavator	B-12Q	220	.073			2.49	2.10	4.59	6.10
7100	3/4 C.Y. excavator	B-12F	264	.061			2.08	1.96	4.04	5.30
7130	1/2 C.Y. excavator, truck mounted	B-12J	176	.091			3.12	5.15	8.27	10.40
7140	6' to 10' deep, 3/4 C.Y. excavator	B-12F	198	.081			2.77	2.61	5.38	7.10
7150	1 C.Y. excavator	B-12A	352	.045			1.56	1.71	3.27	4.26
7160	1 C.Y. excavator, truck mounted	B-12K	352	.045			1.56	2.91	4.47	5.60
7170	1-1/2 C.Y. excavator	B-12B	528	.030			1.04	1.47	2.51	3.21
7190	10' to 14' deep, 3/4 C.Y. excavator	B-12F	176	.091			3.12	2.93	6.05	8
7200	1 C.Y. excavator	B-12A	317	.050			1.73	1.90	3.63	4.73
7210	1-1/2 C.Y. excavator	B-12B	475	.034			1.15	1.63	2.78	3.56
7250	14' to 20' deep, 1 C.Y. excavator	B-12A	282	.057			1.95	2.13	4.08	5.30
7260	1-1/2 C.Y. excavator	B-12B	422	.038			1.30	1.84	3.14	4.01
7270	2-1/2 C.Y. excavator	B-12S	748	.021			.73	1.78	2.51	3.08
7300	20' to 24' deep, 1 C.Y. excavator	B-12A	254	.063			2.16	2.37	4.53	5.90
7310	1-1/2 C.Y. excavator	B-12B	380	.042			1.44	2.04	3.48	4.46
7320	2-1/2 C.Y. excavator	B-12S	673	.024			.81	1.98	2.79	3.43

31 23 16.14 Excavating, Utility Trench

		Crew	Daily Output	Labor-Hours	Unit	Material	Labor	Equipment	Total	Total Incl O&P
0010	**EXCAVATING, UTILITY TRENCH** Common earth									
0050	Trenching with chain trencher, 12 H.P., operator walking									
0100	4" wide trench, 12" deep	B-53	800	.010	L.F.		.37	.06	.43	.63
0150	18" deep		750	.011			.39	.07	.46	.66
0200	24" deep		700	.011			.42	.07	.49	.71
0300	6" wide trench, 12" deep		650	.012			.45	.08	.53	.77
0350	18" deep		600	.013			.49	.09	.58	.83
0400	24" deep		550	.015			.54	.09	.63	.91
0450	36" deep		450	.018			.66	.11	.77	1.11
0600	8" wide trench, 12" deep		475	.017			.62	.11	.73	1.06
0650	18" deep		400	.020			.74	.13	.87	1.25
0700	24" deep		350	.023			.84	.15	.99	1.43
0750	36" deep		300	.027			.98	.17	1.15	1.67
1000	Backfill by hand including compaction, add									
1050	4" wide trench, 12" deep	A-1G	800	.010	L.F.		.29	.05	.34	.50
1100	18" deep		530	.015			.43	.07	.50	.76
1150	24" deep		400	.020			.58	.10	.68	1.01
1300	6" wide trench, 12" deep		540	.015			.43	.07	.50	.74
1350	18" deep		405	.020			.57	.09	.66	.98
1400	24" deep		270	.030			.85	.14	.99	1.49
1450	36" deep		180	.044			1.28	.21	1.49	2.22
1600	8" wide trench, 12" deep		400	.020			.58	.10	.68	1.01
1650	18" deep		265	.030			.87	.14	1.01	1.51
1700	24" deep		200	.040			1.15	.19	1.34	2
1750	36" deep		135	.059			1.70	.28	1.98	2.96
2000	Chain trencher, 40 H.P. operator riding									
2050	6" wide trench and backfill, 12" deep	B-54	1200	.007	L.F.		.25	.19	.44	.58
2100	18" deep		1000	.008			.29	.23	.52	.69
2150	24" deep		975	.008			.30	.23	.53	.72
2200	36" deep		900	.009			.33	.25	.58	.77
2250	48" deep		750	.011			.39	.30	.69	.92
2300	60" deep		650	.012			.45	.35	.80	1.06

31 23 Excavation and Fill

31 23 16 – Excavation

31 23 16.14 Excavating, Utility Trench

		Crew	Daily Output	Labor-Hours	Unit	Material	2007 Bare Costs Labor	2007 Bare Costs Equipment	Total	Total Incl O&P
2400	8" wide trench and backfill, 12" deep	B-54	1000	.008	L.F.		.29	.23	.52	.69
2450	18" deep		950	.008			.31	.24	.55	.73
2500	24" deep		900	.009			.33	.25	.58	.77
2550	36" deep		800	.010			.37	.28	.65	.87
2600	48" deep		650	.012			.45	.35	.80	1.06
2700	12" wide trench and backfill, 12" deep		975	.008			.30	.23	.53	.72
2750	18" deep		860	.009			.34	.26	.60	.81
2800	24" deep		800	.010			.37	.28	.65	.87
2850	36" deep		725	.011			.41	.31	.72	.95
3000	16" wide trench and backfill, 12" deep		835	.010			.35	.27	.62	.83
3050	18" deep		750	.011			.39	.30	.69	.92
3100	24" deep	▼	700	.011	▼		.42	.32	.74	.99
3200	Compaction with vibratory plate, add								50%	50%
5100	Hand excavate and trim for pipe bells after trench excavation									
5200	8" pipe	1 Clab	155	.052	L.F.		1.48		1.48	2.31
5300	18" pipe	"	130	.062	"		1.77		1.77	2.75

31 23 16.16 Structural Excavation for Minor Structures

		Crew	Daily Output	Labor-Hours	Unit	Material	2007 Bare Costs Labor	2007 Bare Costs Equipment	Total	Total Incl O&P
0010	**STRUCTURAL EXCAVATION FOR MINOR STRUCTURES** R312316-40									
0015	Hand, pits to 6' deep, sandy soil	1 Clab	8	1	B.C.Y.		29		29	45
0100	Heavy soil or clay		4	2			57.50		57.50	89.50
0300	Pits 6' to 12' deep, sandy soil		5	1.600			46		46	71.50
0500	Heavy soil or clay		3	2.667			76.50		76.50	119
0700	Pits 12' to 18' deep, sandy soil		4	2			57.50		57.50	89.50
0900	Heavy soil or clay		2	4			115		115	179
1100	Hand loading trucks from stock pile, sandy soil		12	.667			19.15		19.15	30
1300	Heavy soil or clay	▼	8	1	▼		29		29	45
1500	For wet or muck hand excavation, add to above				%				50%	50%
6000	Machine excavation, for spread and mat footings, elevator pits,									
6001	and small building foundations									
6030	Common earth, hydraulic backhoe, 1/2 C.Y. bucket	B-12E	55	.291	B.C.Y.		9.95	6.50	16.45	22.50
6035	3/4 C.Y. bucket	B-12F	90	.178			6.10	5.75	11.85	15.60
6040	1 C.Y. bucket	B-12A	108	.148			5.10	5.55	10.65	13.90
6050	1-1/2 C.Y. bucket	B-12B	144	.111			3.81	5.40	9.21	11.70
6060	2 C.Y. bucket	B-12C	200	.080			2.74	4.97	7.71	9.65
6070	Sand and gravel, 3/4 C.Y. bucket	B-12F	100	.160			5.50	5.15	10.65	14.10
6080	1 C.Y. bucket	B-12A	120	.133			4.57	5	9.57	12.50
6090	1-1/2 C.Y. bucket	B-12B	160	.100			3.43	4.85	8.28	10.60
6100	2 C.Y. bucket	B-12C	220	.073			2.49	4.52	7.01	8.80
6110	Clay, till, or blasted rock, 3/4 C.Y. bucket	B-12F	80	.200			6.85	6.45	13.30	17.60
6120	1 C.Y. bucket	B-12A	95	.168			5.75	6.35	12.10	15.75
6130	1-1/2 C.Y. bucket	B-12B	130	.123			4.22	5.95	10.17	13
6140	2 C.Y. bucket	B-12C	175	.091	▼		3.13	5.70	8.83	11.05
9010	For mobilization or demobilization, see Div. 01 54 36.50									
9020	For dewatering, see Div. 31 23 19.20									
9022	For larger structures, see Bulk Excavation, Div. 31 23 16.42									
9024	For loading onto trucks, add								15%	
9026	For hauling, see Div. 31 23 23.18									
9030	For sheeting or soldier bms/lagging, see 31 52 & 31 62									
9040	For trench excavation of strip ftgs, see 31 23 16.13									

31 23 16.26 Rock Removal

		Crew	Daily Output	Labor-Hours	Unit	Material	2007 Bare Costs Labor	2007 Bare Costs Equipment	Total	Total Incl O&P
0010	**ROCK REMOVAL** R312316-40									
0015	Drilling only rock, 2" hole for rock bolts	B-47	316	.076	L.F.		2.44	3.87	6.31	8

31 23 Excavation and Fill

31 23 16 – Excavation

31 23 16.26 Rock Removal

		Crew	Daily Output	Labor-Hours	Unit	Material	2007 Bare Costs Labor	2007 Bare Costs Equipment	Total	Total Incl O&P
0800	2-1/2" hole for pre-splitting	B-47	600	.040	L.F.		1.28	2.04	3.32	4.22
4600	Quarry operations, 2-1/2" to 3-1/2" diameter	↓	715	.034	↓		1.08	1.71	2.79	3.54

31 23 16.30 Drilling and Blasting Rock

		Crew	Daily Output	Labor-Hours	Unit	Material	Labor	Equipment	Total	Total Incl O&P
0010	**DRILLING AND BLASTING ROCK**									
0020	Rock, open face, under 1500 CY	B-47	225	.107	B.C.Y.	2.45	3.43	5.45	11.33	13.90
0100	Over 1500 C.Y.		300	.080		2.45	2.57	4.07	9.09	11.15
0200	Areas where blasting mats are required, under 1500 C.Y.		175	.137		2.45	4.40	7	13.85	17.15
0250	Over 1500 C.Y.	↓	250	.096		2.45	3.08	4.89	10.42	12.85
0300	Bulk drilling and blasting, can vary greatly, average								9.90	11
0500	Pits, average								39.60	44
1300	Deep hole method, up to 1500 C.Y.	B-47	50	.480		2.45	15.40	24.50	42.35	53
1400	Over 1500 C.Y.		66	.364		2.45	11.70	18.50	32.65	41
1900	Restricted areas, up to 1500 C.Y.		13	1.846		2.45	59.50	94	155.95	197
2000	Over 1500 C.Y.		20	1.200		2.45	38.50	61	101.95	129
2200	Trenches, up to 1500 C.Y.		22	1.091		7.10	35	55.50	97.60	123
2300	Over 1500 C.Y.		26	.923		7.10	29.50	47	83.60	105
2500	Pier holes, up to 1500 C.Y.		22	1.091		2.45	35	55.50	92.95	118
2600	Over 1500 C.Y.	↓	31	.774		2.45	25	39.50	66.95	84.50
2800	Boulders under 1/2 C.Y., loaded on truck, no hauling	B-100	80	.150			5.30	7.35	12.65	16.10
2900	Boulders, drilled, blasted	B-47	100	.240	↓	2.45	7.70	12.20	22.35	28
3100	Jackhammer operators with foreman compressor, air tools	B-9	1	40	Day		1,175	180	1,355	2,025
3300	Track drill, compressor, operator and foreman	B-47	1	24	"		770	1,225	1,995	2,525
3500	Blasting caps				Ea.	4.56			4.56	5
3700	Explosives					.34			.34	.37
3900	Blasting mats, rent, for first day					107			107	117
4000	Per added day					35.50			35.50	39
4200	Preblast survey for 6 room house, individual lot, minimum	A-6	2.40	6.667			247	24	271	405
4300	Maximum	"	1.35	11.852	↓		440	43	483	725
4500	City block within zone of influence, minimum	A-8	25200	.001	S.F.		.05		.05	.08
4600	Maximum	"	15100	.002	"		.08		.08	.13

31 23 16.42 Excavating, Bulk Bank Measure

		Crew	Daily Output	Labor-Hours	Unit	Material	Labor	Equipment	Total	Total Incl O&P
0010	**EXCAVATING, BULK BANK MEASURE** Common earth piled R312316-40									
0020	For loading onto trucks, add								15%	15%
0050	For mobilization and demobilization, see division 01 54 36.50 R312316-45									
0100	For hauling, see division 31 23 23.18									
0200	Backhoe, hydraulic, crawler mtd., 1 C.Y. cap. = 75 C.Y./hr.	B-12A	600	.027	B.C.Y.		.91	1	1.91	2.50
0250	1-1/2 C.Y. cap. = 100 C.Y./hr.	B-12B	800	.020			.69	.97	1.66	2.12
0260	2 C.Y. cap. = 130 C.Y./hr.	B-12C	1040	.015			.53	.96	1.49	1.86
0300	3 C.Y. cap. = 160 C.Y./hr.	B-12D	1280	.013			.43	1.69	2.12	2.51
0310	Wheel mounted, 1/2 C.Y. cap. = 30 C.Y./hr.	B-12E	240	.067			2.29	1.49	3.78	5.15
0360	3/4 C.Y. cap. = 45 C.Y./hr.	B-12F	360	.044			1.52	1.43	2.95	3.91
0500	Clamshell, 1/2 C.Y. cap. = 20 C.Y./hr.	B-12G	160	.100			3.43	3.82	7.25	9.45
0550	1 C.Y. cap. = 35 C.Y./hr.	B-12H	280	.057			1.96	3.67	5.63	7.05
0950	Dragline, 1/2 C.Y. cap. = 30 C.Y./hr.	B-12I	240	.067			2.29	3.15	5.44	6.95
1000	3/4 C.Y. cap. = 35 C.Y./hr.	"	280	.057			1.96	2.70	4.66	5.95
1050	1-1/2 C.Y. cap. = 65 C.Y./hr.	B-12P	520	.031			1.05	1.97	3.02	3.78
1200	Front end loader, track mtd., 1-1/2 C.Y. cap. = 70 C.Y./hr.	B-10N	560	.021			.75	.60	1.35	1.81
1250	2-1/2 C.Y. cap. = 95 C.Y./hr.	B-100	760	.016			.56	.78	1.34	1.69
1300	3 C.Y. cap. = 130 C.Y./hr.	B-10P	1040	.012			.41	.80	1.21	1.50
1350	5 C.Y. cap. = 160 C.Y./hr.	B-10Q	1280	.009			.33	.90	1.23	1.48
1500	Wheel mounted, 3/4 C.Y. cap. = 45 C.Y./hr.	B-10R	360	.033			1.17	.58	1.75	2.41
1550	1-1/2 C.Y. cap. = 80 C.Y./hr.	B-10S	640	.019	↓		.66	.43	1.09	1.48

31 23 Excavation and Fill

31 23 16 – Excavation

31 23 16.42 Excavating, Bulk Bank Measure

		Crew	Daily Output	Labor-Hours	Unit	Material	2007 Bare Costs Labor	2007 Bare Costs Equipment	Total	Total Incl O&P
1600	2-1/4 C.Y. cap. = 100 C.Y./hr.	B-10T	800	.015	B.C.Y.		.53	.44	.97	1.28
1650	5 C.Y. cap. = 185 C.Y./hr.	B-10U	1480	.008			.29	.51	.80	.99
1800	Hydraulic excavator, truck mtd, 1/2 C.Y. = 30 C.Y./hr.	B-12J	240	.067			2.29	3.77	6.06	7.65
1850	48 inch bucket, 1 C.Y. = 45 C.Y./hr.	B-12K	360	.044			1.52	2.85	4.37	5.45
3700	Shovel, 1/2 C.Y. capacity = 55 C.Y./hr.	B-12L	440	.036			1.25	1.42	2.67	3.46
3750	3/4 C.Y. capacity = 85 C.Y./hr.	B-12M	680	.024			.81	1.16	1.97	2.51
3800	1 C.Y. capacity = 120 C.Y./hr.	B-12N	960	.017			.57	1.09	1.66	2.07
3850	1-1/2 C.Y. capacity = 160 C.Y./hr.	B-12O	1280	.013			.43	.83	1.26	1.56
3900	3 C.Y. cap. = 250 C.Y./hr.	B-12T	2000	.008			.27	.74	1.01	1.23
4000	For soft soil or sand, deduct								15%	15%
4100	For heavy soil or stiff clay, add								60%	60%
4200	For wet excavation with clamshell or dragline, add								100%	100%
4250	All other equipment, add								50%	50%
4400	Clamshell in sheeting or cofferdam, minimum	B-12H	160	.100			3.43	6.40	9.83	12.30
4450	Maximum	"	60	.267			9.15	17.15	26.30	33
8000	For hauling excavated material, see div. 31 23 23.18									

31 23 16.46 Excavating, Bulk, Dozer

		Crew	Daily Output	Labor-Hours	Unit	Material	2007 Bare Costs Labor	2007 Bare Costs Equipment	Total	Total Incl O&P
0010	**EXCAVATING, BULK, DOZER** Open site									
2000	80 H.P., 50' haul, sand & gravel	B-10L	460	.026	B.C.Y.		.92	.77	1.69	2.25
2020	Common earth		400	.030			1.06	.89	1.95	2.58
2040	Clay		250	.048			1.69	1.42	3.11	4.13
2200	150' haul, sand & gravel		230	.052			1.84	1.54	3.38	4.49
2220	Common earth		200	.060			2.11	1.77	3.88	5.15
2240	Clay		125	.096			3.38	2.84	6.22	8.25
2400	300' haul, sand & gravel		120	.100			3.52	2.96	6.48	8.60
2420	Common earth		100	.120			4.22	3.55	7.77	10.30
2440	Clay		65	.185			6.50	5.45	11.95	15.85
3000	105 H.P., 50' haul, sand & gravel	B-10W	700	.017			.60	.71	1.31	1.70
3020	Common earth		610	.020			.69	.81	1.50	1.94
3040	Clay		385	.031			1.10	1.29	2.39	3.08
3200	150' haul, sand & gravel		310	.039			1.36	1.60	2.96	3.83
3220	Common earth		270	.044			1.56	1.83	3.39	4.40
3240	Clay		170	.071			2.48	2.91	5.39	7
3300	300' haul, sand & gravel		140	.086			3.02	3.53	6.55	8.45
3320	Common earth		120	.100			3.52	4.12	7.64	9.90
3340	Clay		100	.120			4.22	4.95	9.17	11.85
4000	200 H.P., 50' haul, sand & gravel	B-10B	1400	.009			.30	.71	1.01	1.24
4020	Common earth		1230	.010			.34	.80	1.14	1.40
4040	Clay		770	.016			.55	1.28	1.83	2.24
4200	150' haul, sand & gravel		595	.020			.71	1.66	2.37	2.91
4220	Common earth		516	.023			.82	1.92	2.74	3.35
4240	Clay		325	.037			1.30	3.04	4.34	5.30
4400	300' haul, sand & gravel		310	.039			1.36	3.19	4.55	5.60
4420	Common earth		270	.044			1.56	3.66	5.22	6.40
4440	Clay		170	.071			2.48	5.80	8.28	10.20
5000	300 H.P., 50' haul, sand & gravel	B-10M	1900	.006			.22	.69	.91	1.09
5020	Common earth		1650	.007			.26	.79	1.05	1.26
5040	Clay		1025	.012			.41	1.27	1.68	2.03
5200	150' haul, sand & gravel		920	.013			.46	1.41	1.87	2.26
5220	Common earth		800	.015			.53	1.63	2.16	2.59
5240	Clay		500	.024			.84	2.60	3.44	4.14
5400	300' haul, sand & gravel		470	.026			.90	2.77	3.67	4.41

31 23 Excavation and Fill

31 23 16 – Excavation

31 23 16.46 Excavating, Bulk, Dozer

		Crew	Daily Output	Labor-Hours	Unit	Material	2007 Bare Costs Labor	2007 Bare Costs Equipment	Total	Total Incl O&P
5420	Common earth	B-10M	410	.029	B.C.Y.		1.03	3.17	4.20	5.05
5440	Clay	↓	250	.048			1.69	5.20	6.89	8.25
5500	460 H.P., 50' haul, sand & gravel	B-10X	1930	.006			.22	.87	1.09	1.29
5510	Common earth		1680	.007			.25	1	1.25	1.48
5520	Clay		1050	.011			.40	1.60	2	2.37
5530	150' haul, sand & gravel		1290	.009			.33	1.30	1.63	1.93
5540	Common earth		1120	.011			.38	1.50	1.88	2.22
5550	Clay		700	.017			.60	2.40	3	3.55
5560	300' haul, sand & gravel		660	.018			.64	2.54	3.18	3.76
5570	Common earth		575	.021			.73	2.92	3.65	4.33
5580	Clay	↓	350	.034			1.21	4.79	6	7.10
6000	700 H.P., 50' haul, sand & gravel	B-10V	3500	.003			.12	.97	1.09	1.25
6010	Common earth		3035	.004			.14	1.12	1.26	1.44
6020	Clay		1925	.006			.22	1.77	1.99	2.28
6030	150' haul, sand & gravel		2025	.006			.21	1.68	1.89	2.17
6040	Common earth		1750	.007			.24	1.95	2.19	2.51
6050	Clay		1100	.011			.38	3.10	3.48	3.99
6060	300' haul, sand & gravel		1030	.012			.41	3.31	3.72	4.26
6070	Common earth		900	.013			.47	3.78	4.25	4.87
6080	Clay	↓	550	.022	↓		.77	6.20	6.97	7.95

31 23 16.50 Excavation, Bulk, Scrapers

		Crew	Daily Output	Labor-Hours	Unit	Material	Labor	Equipment	Total	Total Incl O&P
0010	**EXCAVATION, BULK, SCRAPERS** R312316-40									
0100	Elev. scraper 11 C.Y., sand & gravel 1500' haul, 1/4 dozer	B-33F	690	.020	B.C.Y.		.72	1.72	2.44	2.99
0150	3000' haul		610	.023			.82	1.94	2.76	3.38
0200	5000' haul		505	.028			.99	2.35	3.34	4.08
0300	Common earth, 1500' haul		600	.023			.83	1.97	2.80	3.43
0350	3000' haul		530	.026			.94	2.24	3.18	3.89
0400	5000' haul		440	.032			1.13	2.69	3.82	4.68
0500	Clay, 1500' haul		375	.037			1.33	3.16	4.49	5.50
0550	3000' haul		330	.042			1.51	3.59	5.10	6.25
0600	5000' haul	↓	275	.051	↓		1.81	4.31	6.12	7.50
1000	Self propelled scraper, 14 C.Y. 1/4 push dozer, sand									
1050	Sand and gravel, 1500' haul	B-33D	920	.015	B.C.Y.		.54	2.22	2.76	3.26
1100	3000' haul		805	.017			.62	2.54	3.16	3.73
1200	5000' haul		645	.022			.77	3.17	3.94	4.65
1300	Common earth, 1500' haul		800	.018			.62	2.55	3.17	3.76
1350	3000' haul		700	.020			.71	2.92	3.63	4.29
1400	5000' haul		560	.025			.89	3.65	4.54	5.35
1500	Clay, 1500' haul		500	.028			1	4.08	5.08	6
1550	3000' haul		440	.032			1.13	4.64	5.77	6.80
1600	5000' haul	↓	350	.040			1.43	5.85	7.28	8.55
2000	21 C.Y., 1/4 push dozer, sand & gravel, 1500' haul	B-33E	1180	.012			.42	2.43	2.85	3.32
2100	3000' haul		910	.015			.55	3.16	3.71	4.30
2200	5000' haul		750	.019			.67	3.83	4.50	5.25
2300	Common earth, 1500' haul		1030	.014			.48	2.79	3.27	3.81
2350	3000' haul		790	.018			.63	3.64	4.27	4.96
2400	5000' haul		650	.022			.77	4.42	5.19	6.05
2500	Clay, 1500' haul		645	.022			.77	4.46	5.23	6.05
2550	3000' haul		495	.028			1.01	5.80	6.81	7.95
2600	5000' haul	↓	405	.035			1.23	7.10	8.33	9.65
2700	Towed, 10 C.Y., 1/4 push dozer, sand & gravel, 1500' haul	B-33B	560	.025			.89	3.18	4.07	4.85
2720	3000' haul	↓	450	.031	↓		1.11	3.96	5.07	6.05

31 23 Excavation and Fill

31 23 16 – Excavation

31 23 16.50 Excavation, Bulk, Scrapers

		Crew	Daily Output	Labor-Hours	Unit	Material	2007 Bare Costs Labor	2007 Bare Costs Equipment	Total	Total Incl O&P
2730	5000' haul	B-33B	365	.038	B.C.Y.		1.37	4.89	6.26	7.45
2750	Common earth, 1500' haul		420	.033			1.19	4.25	5.44	6.45
2770	3000' haul		400	.035			1.25	4.46	5.71	6.80
2780	5000' haul		310	.045			1.61	5.75	7.36	8.80
2800	Clay, 1500' haul		315	.044			1.58	5.65	7.23	8.65
2820	3000' haul		300	.047			1.66	5.95	7.61	9.10
2840	5000' haul		225	.062			2.22	7.95	10.17	12.05
2900	15 C.Y., 1/4 push dozer, sand & gravel, 1500' haul	B-33C	800	.018			.62	2.23	2.85	3.40
2920	3000' haul		640	.022			.78	2.79	3.57	4.25
2940	5000' haul		520	.027			.96	3.43	4.39	5.25
2960	Common earth, 1500' haul		600	.023			.83	2.97	3.80	4.53
2980	3000' haul		560	.025			.89	3.18	4.07	4.85
3000	5000' haul		440	.032			1.13	4.05	5.18	6.20
3020	Clay, 1500' haul		450	.031			1.11	3.96	5.07	6.05
3040	3000' haul		420	.033			1.19	4.25	5.44	6.45
3060	5000' haul		320	.044			1.56	5.55	7.11	8.50

31 23 19 – Dewatering

31 23 19.20 Dewatering

		Crew	Daily Output	Labor-Hours	Unit	Material	Labor	Equipment	Total	Total Incl O&P
0010	**DEWATERING**									
0020	Excavate drainage trench, 2' wide, 2' deep	B-11C	90	.178	C.Y.		5.95	2.70	8.65	12.05
0100	2' wide, 3' deep, with backhoe loader	"	135	.119			3.98	1.80	5.78	8.10
0200	Excavate sump pits by hand, light soil	1 Clab	7.10	1.127			32.50		32.50	50.50
0300	Heavy soil	"	3.50	2.286			65.50		65.50	102
0500	Pumping 8 hr., attended 2 hrs. per day, including 20 L.F.									
0550	of suction hose & 100 L.F. discharge hose									
0600	2" diaphragm pump used for 8 hours	B-10H	4	3	Day		106	15.30	121.30	177
0650	4" diaphragm pump used for 8 hours	B-10I	4	3			106	21.50	127.50	184
0800	8 hrs. attended, 2" diaphragm pump	B-10H	1	12			420	61	481	710
0900	3" centrifugal pump	B-10J	1	12			420	70	490	715
1000	4" diaphragm pump	B-10I	1	12			420	86	506	735
1100	6" centrifugal pump	B-10K	1	12			420	300	720	975
1300	CMP, incl. excavation 3' deep, 12" diameter	B-6	115	.209	L.F.	10.85	6.55	2.12	19.52	24.50
1400	18" diameter		100	.240	"	13.50	7.55	2.43	23.48	29
1600	Sump hole construction, incl. excavation and gravel, pit		1250	.019	C.F.	.86	.60	.19	1.65	2.09
1700	With 12" gravel collar, 12" pipe, corrugated, 16 ga.		70	.343	L.F.	18.75	10.80	3.48	33.03	41
1800	15" pipe, corrugated, 16 ga.		55	.436		24	13.70	4.42	42.12	52.50
1900	18" pipe, corrugated, 16 ga.		50	.480		28	15.10	4.87	47.97	59.50
2000	24" pipe, corrugated, 14 ga.		40	.600		33.50	18.85	6.10	58.45	72.50
2200	Wood lining, up to 4' x 4', add		300	.080	SFCA	15.90	2.52	.81	19.23	22
9950	See Div. 31 23 19.40 for wellpoints									
9960	See div. 31 23 19.30 for deep well systems									

31 23 19.30 Wells

		Crew	Daily Output	Labor-Hours	Unit	Material	Labor	Equipment	Total	Total Incl O&P
0010	**WELLS** For dewatering 10' to 20' deep, 2' diameter									
0020	with steel casing, minimum	B-6	165	.145	V.L.F.	39	4.57	1.47	45.04	51.50
0050	Average		98	.245		39	7.70	2.48	49.18	57.50
0100	Maximum		49	.490		39	15.40	4.97	59.37	72
0300	For dewatering pumps see Reference 01 54 33									
0500	For domestic water wells, see Division 33 21 13.10									

31 23 19.40 Wellpoints

		Crew	Daily Output	Labor-Hours	Unit	Material	Labor	Equipment	Total	Total Incl O&P
0010	**WELLPOINTS** For equipment rental, see 01 54 33 R312319-90									
0100	Installation and removal of single stage system									
0110	Labor only, .75 labor-hours per L.F., minimum	1 Clab	10.70	.748	LF Hdr		21.50		21.50	33.50

31 23 Excavation and Fill

31 23 19 – Dewatering

31 23 19.40 Wellpoints

		Crew	Daily Output	Labor-Hours	Unit	Material	2007 Bare Costs Labor	Equipment	Total	Total Incl O&P
0200	2.0 labor-hours per L.F., maximum	1 Clab	4	2	LF Hdr		57.50		57.50	89.50
0400	Pump operation, 4 @ 6 hr. shifts									
0410	Per 24 hour day	4 Eqlt	1.27	25.197	Day		930		930	1,400
0500	Per 168 hour week, 160 hr. straight, 8 hr. double time		.18	177	Week		6,550		6,550	9,875
0550	Per 4.3 week month		.04	800	Month		29,500		29,500	44,400
0600	Complete installation, operation, equipment rental, fuel &									
0610	removal of system with 2" wellpoints 5' O.C.									
0700	100' long header, 6" diameter, first month	4 Eqlt	3.23	9.907	LF Hdr	148	365		513	715
0800	Thereafter, per month		4.13	7.748		118	286		404	560
1000	200' long header, 8" diameter, first month		6	5.333		131	197		328	440
1100	Thereafter, per month		8.39	3.814		66.50	141		207.50	285
1300	500' long header, 8" diameter, first month		10.63	3.010		52	111		163	224
1400	Thereafter, per month		20.91	1.530		37	56.50		93.50	126
1600	1,000' long header, 10" diameter, first month		11.62	2.754		44.50	101		145.50	202
1700	Thereafter, per month		41.81	.765		22	28		50	67
1900	Note: above figures include pumping 168 hrs. per week									
1910	and include the pump operator and one stand-by pump.									

31 23 23 – Fill

31 23 23.13 Backfill

		Crew	Daily Output	Labor-Hours	Unit	Material	Labor	Equipment	Total	Total Incl O&P
0010	**BACKFILL** R312323-30									
0015	By hand, no compaction, light soil	1 Clab	14	.571	L.C.Y.		16.45		16.45	25.50
0100	Heavy soil		11	.727	"		21		21	32.50
0300	Compaction in 6" layers, hand tamp, add to above		20.60	.388	E.C.Y.		11.15		11.15	17.40
0400	Roller compaction operator walking, add	B-10A	100	.120			4.22	1.34	5.56	7.85
0500	Air tamp, add	B-9D	190	.211			6.15	1.13	7.28	10.80
0600	Vibrating plate, add	A-1D	60	.133			3.83	.49	4.32	6.50
0800	Compaction in 12" layers, hand tamp, add to above	1 Clab	34	.235			6.75		6.75	10.55
0900	Roller compaction operator walking, add	B-10A	150	.080			2.81	.89	3.70	5.25
1000	Air tamp, add	B-9	285	.140			4.09	.63	4.72	7.05
1100	Vibrating plate, add	A-1E	90	.089			2.56	.40	2.96	4.42
1300	Dozer backfilling, bulk, up to 300' haul, no compaction	B-10B	1200	.010	L.C.Y.		.35	.82	1.17	1.44
1400	Air tamped, add	B-11B	80	.200	E.C.Y.		6.55	2.92	9.47	13.25
1600	Compacting backfill, 6" to 12" lifts, vibrating roller	B-10C	800	.015			.53	1.67	2.20	2.64
1700	Sheepsfoot roller	B-10D	750	.016			.56	1.82	2.38	2.86
1900	Dozer backfilling, trench, up to 300' haul, no compaction	B-10B	900	.013	L.C.Y.		.47	1.10	1.57	1.92
2000	Air tamped, add	B-11B	80	.200	E.C.Y.		6.55	2.92	9.47	13.25
2200	Compacting backfill, 6" to 12" lifts, vibrating roller	B-10C	700	.017			.60	1.91	2.51	3.02
2300	Sheepsfoot roller	B-10D	650	.018			.65	2.10	2.75	3.30
2350	Spreading in 8" layers, small dozer	B-10B	1060	.011	L.C.Y.		.40	.93	1.33	1.64

31 23 23.14 Backfill, Structural

		Crew	Daily Output	Labor-Hours	Unit	Material	Labor	Equipment	Total	Total Incl O&P
0010	**BACKFILL, STRUCTURAL** Dozer or F.E. loader									
0020	From existing stockpile, no compaction									
2000	80 H.P., 50' haul, sand & gravel	B-10L	1100	.011	L.C.Y.		.38	.32	.70	.93
2020	Common earth		975	.012			.43	.36	.79	1.06
2040	Clay		850	.014			.50	.42	.92	1.22
2400	300' haul, sand & gravel		370	.032			1.14	.96	2.10	2.78
2420	Common earth		330	.036			1.28	1.08	2.36	3.12
2440	Clay		290	.041			1.46	1.22	2.68	3.56
3000	105 H.P., 50' haul, sand & gravel	B-10W	1350	.009			.31	.37	.68	.88
3020	Common earth		1225	.010			.34	.40	.74	.96
3040	Clay		1100	.011			.38	.45	.83	1.07
3300	300' haul, sand & gravel		465	.026			.91	1.06	1.97	2.55

31 23 Excavation and Fill

31 23 23 – Fill

31 23 23.14 Backfill, Structural

		Crew	Daily Output	Labor-Hours	Unit	Material	2007 Bare Costs Labor	2007 Bare Costs Equipment	Total	Total Incl O&P
3320	Common earth	B-10W	415	.029	L.C.Y.		1.02	1.19	2.21	2.86
3340	Clay	↓	370	.032			1.14	1.34	2.48	3.20
4000	200 H.P., 50' haul, sand & gravel	B-10B	2500	.005			.17	.40	.57	.69
4020	Common earth		2200	.005			.19	.45	.64	.78
4040	Clay		1950	.006			.22	.51	.73	.89
4400	300' haul, sand & gravel		805	.015			.52	1.23	1.75	2.15
4420	Common earth		735	.016			.57	1.35	1.92	2.35
4440	Clay	↓	660	.018			.64	1.50	2.14	2.62
5000	300 H.P., 50' haul, sand & gravel	B-10M	3170	.004			.13	.41	.54	.65
5020	Common earth		2900	.004			.15	.45	.60	.71
5040	Clay		2700	.004			.16	.48	.64	.77
5400	300' haul, sand & gravel		1500	.008			.28	.87	1.15	1.38
5420	Common earth		1350	.009			.31	.96	1.27	1.54
5440	Clay	↓	1225	.010	↓		.34	1.06	1.40	1.69
6010	For trench backfill, see div. 31 23 16.13 & 31 23 16.14									
6100	For compaction, see div. 31 23 23.24									

31 23 23.16 Fill By Borrow and Utility Bedding

		Crew	Daily Output	Labor-Hours	Unit	Material	Labor	Equipment	Total	Total Incl O&P
0010	**FILL BY BORROW AND UTILITY BEDDING**									
0015	FILL BY BORROW, load, 1 mile haul, spread with dozer									
0020	for embankments	B-15	1200	.023	L.C.Y.	6	.75	1.71	8.46	9.60
0035	Select fill for shoulders & embankments	"	1200	.023	"	9.45	.75	1.71	11.91	13.40
0040	For hauling over 1 mile, add to above per C.Y., see Div 31 23 23.18				Mile				1.50	1.50
0049	Utility bedding, for pipe & conduit, not incl. compaction									
0050	Crushed or screened bank run gravel	B-6	150	.160	L.C.Y.	24	5.05	1.62	30.67	36
0100	Crushed stone 3/4" to 1/2"		150	.160		32	5.05	1.62	38.67	44.50
0200	Sand, dead or bank	↓	150	.160	↓	6.25	5.05	1.62	12.92	16.45
0500	Compacting bedding in trench	A-1D	90	.089	E.C.Y.		2.56	.32	2.88	4.34
0600	If material source exceeds 2 miles, add for extra mileage.									
0610	See 31 23 23.18 for hauling kilometer add.									

31 23 23.17 Fill

		Crew	Daily Output	Labor-Hours	Unit	Material	Labor	Equipment	Total	Total Incl O&P
0010	**FILL**, spread dumped material, no compaction									
0020	By dozer, no compaction	B-10B	1000	.012	L.C.Y.		.42	.99	1.41	1.73
0100	By hand	1 Clab	12	.667	"		19.15		19.15	30
0500	Gravel fill, compacted, under floor slabs, 4" deep	B-37	10000	.005	S.F.	.19	.15	.01	.35	.45
0600	6" deep		8600	.006		.28	.17	.01	.46	.59
0700	9" deep		7200	.007		.47	.20	.02	.69	.85
0800	12" deep		6000	.008	↓	.66	.24	.02	.92	1.13
1000	Alternate pricing method, 4" deep		120	.400	E.C.Y.	14.20	12.15	1.02	27.37	35.50
1100	6" deep		160	.300		14.20	9.15	.77	24.12	30.50
1200	9" deep		200	.240		14.20	7.30	.61	22.11	27.50
1300	12" deep	↓	220	.218	↓	14.20	6.65	.56	21.41	26.50
1500	For fill under exterior paving, see Div. 32 11 23.23									

31 23 23.18 Hauling

		Crew	Daily Output	Labor-Hours	Unit	Material	Labor	Equipment	Total	Total Incl O&P
0010	**HAULING**, excavated or borrow, loose cubic yards R312316-40									
0012	no loading included, highway haulers									
0020	6 C.Y. dump truck, 1/4 mile round trip, 5.0 loads/hr.	B-34A	195	.041	L.C.Y.		1.21	1.89	3.10	3.95
0030	1/2 mile round trip, 4.1 loads/hr.		160	.050			1.48	2.30	3.78	4.81
0040	1 mile round trip, 3.3 loads/hr.		130	.062			1.82	2.83	4.65	5.95
0100	2 mile round trip, 2.6 loads/hr.		100	.080			2.36	3.68	6.04	7.70
0150	3 mile round trip, 2.1 loads/hr.		80	.100			2.96	4.61	7.57	9.60
0200	4 mile round trip, 1.8 loads/hr.	↓	70	.114			3.38	5.25	8.63	11
0310	12 C.Y. dump truck, 1/4 mile round trip 3.7 loads/hr.	B-34B	288	.028	↓		.82	1.84	2.66	3.29

31 23 Excavation and Fill

31 23 23 – Fill

31 23 23.18 Hauling

		Crew	Daily Output	Labor-Hours	Unit	Material	2007 Bare Costs Labor	Equipment	Total	Total Incl O&P
0320	1/2 mile round trip, 3.2 loads/hr.	B-34B	250	.032	L.C.Y.		.95	2.12	3.07	3.79
0330	1 mile round trip 2.7, loads/hr.		210	.038			1.13	2.52	3.65	4.52
0400	2 mile round trip, 2.2 loads/hr.		180	.044			1.31	2.94	4.25	5.25
0450	3 mile round trip, 1.9 loads/hr.		170	.047			1.39	3.12	4.51	5.60
0500	4 mile round trip, 1.6 loads/hr.		125	.064			1.89	4.24	6.13	7.60
0540	5 mile round trip, 1 load/hr.		78	.103			3.03	6.80	9.83	12.15
0550	10 mile round trip, 0.60 load/hr.		58	.138			4.08	9.15	13.23	16.35
0560	20 mile round trip, 0.4 load/hr.	↓	39	.205			6.05	13.60	19.65	24.50
0600	16.5 C.Y. dump trailer, 1 mile round trip, 2.6 loads/hr.	B-34C	280	.029			.84	1.59	2.43	3.04
0700	2 mile round trip, 2.1 loads/hr.		225	.036			1.05	1.97	3.02	3.79
1000	3 mile round trip, 1.8 loads/hr.		193	.041			1.22	2.30	3.52	4.42
1100	4 mile round trip, 1.6 loads/hr.		172	.047			1.37	2.58	3.95	4.96
1110	5 mile round trip, 1 load/hr.		108	.074			2.19	4.11	6.30	7.90
1120	10 mile round trip, .60 load/hr.		80	.100			2.96	5.55	8.51	10.65
1130	20 mile round trip, .4 load/hr.	↓	54	.148			4.38	8.20	12.58	15.80
1150	20 C.Y. dump trailer, 1 mile round trip, 2.5 loads/hr.	B-34D	325	.025			.73	1.40	2.13	2.66
1200	2 mile round trip, 2 loads/hr.		260	.031			.91	1.75	2.66	3.33
1220	3 mile round trip, 1.7 loads/hr.		221	.036			1.07	2.06	3.13	3.92
1240	4 mile round trip, 1.5 loads/hr.		195	.041			1.21	2.34	3.55	4.44
1245	5 mile round trip, 1.1 load/hr.		143	.056			1.65	3.19	4.84	6.05
1250	10 mile round trip, .75 load/hr.		110	.073			2.15	4.15	6.30	7.90
1255	20 mile round trip, .5 load/hr.	↓	78	.103			3.03	5.85	8.88	11.15
1300	Hauling in medium traffic, add								20%	20%
1400	Heavy traffic, add								30%	30%
1600	Grading at dump, or embankment if required, by dozer	B-10B	1000	.012	↓		.42	.99	1.41	1.73
1800	Spotter at fill or cut, if required	1 Clab	8	1	Hr.		29		29	45

31 23 23.24 Compaction, Structural

			Crew	Daily Output	Labor-Hours	Unit	Material	Labor	Equipment	Total	Total Incl O&P
0010	**COMPACTION, STRUCTURAL**	R312323-30									
0020	Steel wheel tandem roller, 5 tons		B-10E	8	1.500	Hr.		53	15.25	68.25	97
0100	10 tons		B-10F	8	1.500	"		53	26.50	79.50	109
0300	Sheepsfoot or wobbly wheel roller, 8" lifts, common fill		B-10G	1300	.009	E.C.Y.		.32	.74	1.06	1.30
0400	Select fill		"	1500	.008			.28	.64	.92	1.13
0600	Vibratory plate, 8" lifts, common fill		A-1D	200	.040			1.15	.15	1.30	1.95
0700	Select fill		"	216	.037	↓		1.06	.14	1.20	1.81

31 25 Erosion and Sedimentation Controls

31 25 13 – Erosion Controls

31 25 13.10 Synthetic Erosion Control

		Crew	Daily Output	Labor-Hours	Unit	Material	Labor	Equipment	Total	Total Incl O&P
0010	**SYNTHETIC EROSION CONTROL**									
0020	Jute mesh, 100 SY per roll, 4' wide, stapled	B-80A	2400	.010	S.Y.	.94	.29	.08	1.31	1.56
0100	Plastic netting, stapled, 2" x 1" mesh, 20 mil	B-1	2500	.010		.71	.28		.99	1.22
0200	Polypropylene mesh, stapled, 6.5 oz./S.Y.		2500	.010		1.47	.28		1.75	2.06
0300	Tobacco netting, or jute mesh #2, stapled	↓	2500	.010	↓	.08	.28		.36	.53
1000	Silt fence, polypropylene, 3' high, ideal conditions	2 Clab	1600	.010	L.F.	.34	.29		.63	.82
1100	Adverse conditions	"	950	.017	"	.34	.48		.82	1.12
1200	Place and remove hay bales	A-2	3	8	Ton	151	229	47	427	575
1250	Hay bales, staked	"	2500	.010	L.F.	6.05	.28	.06	6.39	7.15

31 31 Soil Treatment

31 31 16 – Termite Control

31 31 16.13 Chemical Termite Control

		Crew	Daily Output	Labor-Hours	Unit	Material	2007 Bare Costs Labor	Equipment	Total	Total Incl O&P
0010	**CHEMICAL TERMITE CONTROL**									
0020	Slab and walls, residential	1 Skwk	1200	.007	SF Flr.	.29	.25		.54	.71
0100	Commercial, minimum		2496	.003		.31	.12		.43	.53
0200	Maximum		1645	.005	↓	.47	.18		.65	.80
0400	Insecticides for termite control, minimum		14.20	.563	Gal.	11.85	21.50		33.35	46.50
0500	Maximum	↓	11	.727	"	20.50	27.50		48	65.50

31 32 Soil Stabilization

31 32 13 – Soil Mixing Stabilization

31 32 13.30 Calcium Chloride

		Crew	Daily Output	Labor-Hours	Unit	Material	Labor	Equipment	Total	Total Incl O&P
0010	**CALCIUM CHLORIDE**									
0020	Calcium chloride delivered, 100 LB bags, truckload lots				Ton	535			535	590
0030	Solution, 4 lb. flake per gallon, tank truck delivery				Gal.	1.14			1.14	1.25

31 33 Rock Stabilization

31 33 13 – Rock Bolting and Grouting

31 33 13.10 Rock Bolting

		Crew	Daily Output	Labor-Hours	Unit	Material	Labor	Equipment	Total	Total Incl O&P
0010	**ROCK BOLTING**									
2020	Hollow core, prestressable anchor, 1" diameter, 5' long	2 Skwk	32	.500	Ea.	87.50	19		106.50	126
2025	10' long		24	.667		175	25.50		200.50	232
2060	2" diameter, 5' long		32	.500		325	19		344	390
2065	10' long		24	.667		655	25.50		680.50	760
2100	Super high-tensile, 3/4" diameter, 5' long		32	.500		19.85	19		38.85	51.50
2105	10' long		24	.667		39.50	25.50		65	83
2160	2" diameter, 5' long		32	.500		177	19		196	224
2165	10' long	↓	24	.667		355	25.50		380.50	430
4400	Drill hole for rock bolt, 1-3/4" diam., 5' long (for 3/4" bolt)	B-56	17	.941			31	71	102	125
4405	10' long		9	1.778			58.50	134	192.50	236
4420	2" diameter, 5' long (for 1" bolt)		13	1.231			40.50	92.50	133	164
4425	10' long		7	2.286			75	172	247	305
4460	3-1/2" diameter, 5' long (for 2" bolt)		10	1.600			52.50	120	172.50	212
4465	10' long	↓	5	3.200	↓		105	241	346	425

31 36 Gabions

31 36 13 – Gabion Boxes

31 36 13.10 Gabion Boxes

		Crew	Daily Output	Labor-Hours	Unit	Material	Labor	Equipment	Total	Total Incl O&P
0010	**GABION BOXES**									
0400	Gabions, galvanized steel mesh mats or boxes, stone filled, 6" deep	B-13	200	.280	S.Y.	20.50	8.80	3.70	33	40
0500	9" deep		163	.344		31.50	10.75	4.54	46.79	56.50
0600	12" deep		153	.366		33	11.45	4.84	49.29	59.50
0700	18" deep		102	.549		41.50	17.20	7.25	65.95	80.50
0800	36" deep	↓	60	.933	↓	70	29.50	12.35	111.85	136

31 37 Riprap

31 37 13 – Machined Riprap

31 37 13.10 Rip-Rap and Rock Lining

		Crew	Daily Output	Labor-Hours	Unit	Material	2007 Bare Costs Labor	Equipment	Total	Total Incl O&P
0010	**RIP-RAP AND ROCK LINING**, Random, broken stone									
0100	Machine placed for slope protection	B-12G	62	.258	L.C.Y.	25.50	8.85	9.85	44.20	52.50
0110	3/8 to 1/4 C.Y. pieces, grouted	B-13	80	.700	S.Y.	60.50	22	9.25	91.75	111
0200	18" minimum thickness, not grouted	"	53	1.057	"	15.85	33	13.95	62.80	84
0300	Dumped, 50 lb. average	B-11A	800	.020	Ton	23	.67	1.24	24.91	28
0350	100 lb. average		700	.023		33	.77	1.41	35.18	38.50
0370	300 lb. average		600	.027		38.50	.90	1.65	41.05	45

31 41 Shoring

31 41 13 – Timber Shoring

31 41 13.10 Shoring

		Crew	Daily Output	Labor-Hours	Unit	Material	Labor	Equipment	Total	Total Incl O&P
0010	**SHORING**									
0020	Shoring, existing building, with timber, no salvage allowance	B-51	2.20	21.818	M.B.F.	715	635	64.50	1,414.50	1,850
1000	On cribbing with 35 ton screw jacks, per box and jack	"	3.60	13.333	Jack	52	385	39.50	476.50	700

31 41 16 – Sheet Piling

31 41 16.10 Sheet Piling

		Crew	Daily Output	Labor-Hours	Unit	Material	Labor	Equipment	Total	Total Incl O&P
0010	**SHEET PILING**									
0020	Sheet piling steel, not incl. wales, 22 psf, 15' excav., left in place	B-40	10.81	5.920	Ton	1,000	218	275	1,493	1,750
0100	Drive, extract & salvage		6	10.667		450	395	495	1,340	1,675
0300	20' deep excavation, 27 psf, left in place		12.95	4.942		1,000	182	230	1,412	1,650
0400	Drive, extract & salvage		6.55	9.771		450	360	455	1,265	1,575
0600	25' deep excavation, 38 psf, left in place		19	3.368		1,000	124	157	1,281	1,475
0700	Drive, extract & salvage		10.50	6.095		450	225	284	959	1,150
0900	40' deep excavation, 38 psf, left in place		21.20	3.019		1,000	111	140	1,251	1,425
1000	Drive, extract & salvage		12.25	5.224		450	193	243	886	1,075
1200	15' deep excavation, 22 psf, left in place		983	.065	S.F.	11.75	2.40	3.03	17.18	20
1300	Drive, extract & salvage		545	.117		5.05	4.33	5.45	14.83	18.40
1500	20' deep excavation, 27 psf, left in place		960	.067		14.75	2.46	3.10	20.31	23.50
1600	Drive, extract & salvage		485	.132		6.55	4.87	6.15	17.57	21.50
1800	25' deep excavation, 38 psf, left in place		1000	.064		21.50	2.36	2.98	26.84	31
1900	Drive, extract & salvage		553	.116		8.95	4.27	5.40	18.62	22.50
2100	Rent steel sheet piling and wales, first month				Ton	240			240	264
2200	Per added month					24			24	26.50
2300	Rental piling left in place, add to rental					800			800	880
2500	Wales, connections & struts, 2/3 salvage					245			245	269
2700	High strength piling, 50,000 psi, add					54.50			54.50	60
2800	55,000 psi, add					57.50			57.50	63.50
3000	Tie rod, not upset, 1-1/2" to 4" diameter with turnbuckle					1,775			1,775	1,975
3100	No turnbuckle					1,375			1,375	1,500
3300	Upset, 1-3/4" to 4" diameter with turnbuckle					2,000			2,000	2,200
3400	No turnbuckle					1,750			1,750	1,925
3600	Lightweight, 18" to 28" wide, 7 ga., 9.22 psf, and									
3610	9 ga., 8.6 psf, minimum				Lb.	.74			.74	.81
3700	Average					.80			.80	.88
3750	Maximum					.92			.92	1.01
3900	Wood, solid sheeting, incl. wales, braces and spacers,									
3910	drive, extract & salvage, 8' deep excavation	B-31	330	.121	S.F.	1.75	3.73	.51	5.99	8.30
4000	10' deep, 50 S.F./hr. in & 150 S.F./hr. out		300	.133		1.80	4.10	.56	6.46	9
4100	12' deep, 45 S.F./hr. in & 135 S.F./hr. out		270	.148		1.85	4.55	.63	7.03	9.85
4200	14' deep, 42 S.F./hr. in & 126 S.F./hr. out		250	.160		1.91	4.92	.68	7.51	10.50

31 41 Shoring

31 41 16 – Sheet Piling

31 41 16.10 Sheet Piling

		Crew	Daily Output	Labor-Hours	Unit	Material	2007 Bare Costs Labor	Equipment	Total	Total Incl O&P
4300	16' deep, 40 S.F./hr. in & 120 S.F./hr. out	B-31	240	.167	S.F.	1.97	5.10	.70	7.77	10.95
4400	18' deep, 38 S.F./hr. in & 114 S.F./hr. out		230	.174		2.03	5.35	.73	8.11	11.35
4500	20' deep, 35 S.F./hr. in & 105 S.F./hr. out		210	.190		2.10	5.85	.80	8.75	12.30
4520	Left in place, 8' deep, 55 S.F./hr.		440	.091		3.15	2.79	.38	6.32	8.25
4540	10' deep, 50 S.F./hr.		400	.100		3.31	3.07	.42	6.80	8.90
4560	12' deep, 45 S.F./hr.		360	.111		3.50	3.42	.47	7.39	9.65
4565	14' deep, 42 S.F./hr.		335	.119		3.70	3.67	.50	7.87	10.30
4570	16' deep, 40 S.F./hr.		320	.125		3.94	3.84	.53	8.31	10.90
4580	18' deep, 38 S.F./hr.		305	.131		4.20	4.03	.55	8.78	11.55
4590	20' deep, 35 S.F./hr.		280	.143		4.50	4.39	.60	9.49	12.45
4700	Alternate pricing, left in place, 8' deep		1.76	22.727	M.B.F.	710	700	96	1,506	1,975
4800	Drive, extract and salvage, 8' deep		1.32	30.303	"	630	930	128	1,688	2,275
5000	For treated lumber add cost of treatment to lumber									
5010	See Division 06 05 73.10									

31 43 Concrete Raising

31 43 13 – Pressure Grouting

31 43 13.13 Concrete Pressure Grouting

		Crew	Daily Output	Labor-Hours	Unit	Material	2007 Bare Costs Labor	Equipment	Total	Total Incl O&P
0010	**CONCRETE PRESSURE GROUTING**									
0020	Grouting, pressure, cement & sand, 1:1 mix, minimum	B-61	124	.323	Bag	9.95	9.95	2.44	22.34	29
0100	Maximum		51	.784	"	9.95	24	5.95	39.90	55
0200	Cement and sand, 1:1 mix, minimum		250	.160	C.F.	19.90	4.92	1.21	26.03	31
0300	Maximum		100	.400		30	12.30	3.03	45.33	55.50
0400	Epoxy cement grout, minimum		137	.292		104	9	2.21	115.21	130
0500	Maximum		57	.702		104	21.50	5.30	130.80	153
0600	Structural epoxy grout				Gal.	64			64	70
0700	Alternate pricing method: (Add for materials)									
0710	5 person crew and equipment	B-61	1	40	Day		1,225	305	1,530	2,225

31 45 Vibroflotation and Densification

31 45 13 – Vibroflotation

31 45 13.10 Vibroflotation

		Crew	Daily Output	Labor-Hours	Unit	Material	2007 Bare Costs Labor	Equipment	Total	Total Incl O&P
0010	**VIBROFLOTATION** R314513-90									
0900	Vibroflotation compacted sand cylinder, minimum	B-60	750	.075	V.L.F.		2.51	1.71	4.22	5.70
0950	Maximum		325	.172			5.80	3.95	9.75	13.20
1100	Vibro replacement compacted stone cylinder, minimum		500	.112			3.77	2.57	6.34	8.55
1150	Maximum		250	.224			7.55	5.15	12.70	17.15
1300	Mobilization and demobilization, minimum		.47	119	Total		4,000	2,725	6,725	9,125
1400	Maximum		.14	400	"		13,500	9,175	22,675	30,600

31 46 Needle Beams

31 46 13 – Cantilever Needle Beams

31 46 13.10 Cantilever Needle Beams		Crew	Daily Output	Labor-Hours	Unit	Material	2007 Bare Costs Labor	Equipment	Total	Total Incl O&P
0010	**CANTILEVER NEEDLE BEAMS**									
0400	Block, concrete, 8" thick	B-9	7.10	5.634	Ea.	48.50	164	25.50	238	340
0420	12" thick		6.70	5.970		58.50	174	27	259.50	365
0800	Brick, 4" thick with 8" backup block		5.70	7.018		58.50	205	31.50	295	420
1000	Brick, solid, 8" thick		6.20	6.452		48.50	188	29	265.50	380
1040	12" thick		4.90	8.163		58.50	238	36.50	333	475
1080	16" thick	↓	4.50	8.889		79	259	40	378	535
2000	Add for additional floors of shoring	B-1	6	4	↓	48.50	118		166.50	237

31 48 Underpinning

31 48 13 – Underpinning Piers

31 48 13.10 Underpinning Foundations

		Crew	Daily Output	Labor-Hours	Unit	Material	Labor	Equipment	Total	Total Incl O&P
0010	**UNDERPINNING FOUNDATIONS** Including excavation,									
0020	forming, reinforcing, concrete and equipment									
0100	5' to 16' below grade, 100 to 500 C.Y.	B-52	2.30	24.348	C.Y.	285	825	181	1,291	1,800
0200	Over 500 C.Y.		2.50	22.400		257	760	167	1,184	1,650
0400	16' to 25' below grade, 100 to 500 C.Y.		2	28		315	950	209	1,474	2,050
0500	Over 500 C.Y.		2.10	26.667		296	905	199	1,400	1,950
0700	26' to 40' below grade, 100 to 500 C.Y.		1.60	35		340	1,175	261	1,776	2,475
0800	Over 500 C.Y.	↓	1.80	31.111		315	1,050	232	1,597	2,225
0900	For under 50 C.Y., add					10%	40%			
1000	For 50 C.Y. to 100 C.Y., add					5%	20%			

31 52 Cofferdams

31 52 16 – Timber Cofferdams

31 52 16.10 Cofferdams

		Crew	Daily Output	Labor-Hours	Unit	Material	Labor	Equipment	Total	Total Incl O&P
0010	**COFFERDAMS**, incl. mobilization and temporary sheeting									
0080	Soldier beams & lagging H piles with 3" wood sheeting									
0090	horizontal between piles, including removal of wales & braces									
0100	No hydrostatic head, 15' deep, 1 line of braces, minimum	B-50	545	.206	S.F.	7.15	7.20	3.46	17.81	23
0200	Maximum		495	.226		7.95	7.95	3.81	19.71	25.50
0400	15' to 22' deep with 2 lines of braces, 10" H, minimum		360	.311		8.40	10.90	5.25	24.55	32.50
0500	Maximum		330	.339		9.55	11.90	5.70	27.15	36
0700	23' to 35' deep with 3 lines of braces, 12" H, minimum		325	.345		11	12.10	5.80	28.90	38
0800	Maximum		295	.380		11.90	13.30	6.40	31.60	41
1000	36' to 45' deep with 4 lines of braces, 14" H, minimum		290	.386		12.35	13.55	6.50	32.40	42
1100	Maximum		265	.423		13	14.80	7.10	34.90	45.50
1300	No hydrostatic head, left in place, 15' dp., 1 line of braces, min.		635	.176		9.55	6.20	2.97	18.72	23.50
1400	Maximum		575	.195		10.20	6.85	3.28	20.33	26
1600	15' to 22' deep with 2 lines of braces, minimum		455	.246		14.30	8.65	4.14	27.09	34
1700	Maximum		415	.270		15.90	9.45	4.54	29.89	37.50
1900	23' to 35' deep with 3 lines of braces, minimum		420	.267		17	9.35	4.49	30.84	38.50
2000	Maximum		380	.295		18.80	10.35	4.96	34.11	42.50
2200	36' to 45' deep with 4 lines of braces, minimum		385	.291		20.50	10.20	4.89	35.59	44
2300	Maximum	↓	350	.320		24	11.20	5.40	40.60	50
2350	Lagging only, 3" thick wood between piles 8' O.C., minimum	B-46	400	.120		1.59	3.92	.09	5.60	8.15
2370	Maximum		250	.192		2.38	6.25	.14	8.77	12.80
2400	Open sheeting no bracing, for trenches to 10' deep, min.		1736	.028		.72	.90	.02	1.64	2.26
2450	Maximum	↓	1510	.032	↓	.79	1.04	.02	1.85	2.56

31 52 Cofferdams

31 52 16 – Timber Cofferdams

31 52 16.10 Cofferdams

		Crew	Daily Output	Labor-Hours	Unit	Material	2007 Bare Costs Labor	Equipment	Total	Total Incl O&P
2500	Tie-back method, add to open sheeting, add, minimum				S.F.				20%	20%
2550	Maximum				↓				60%	60%
2700	Tie-backs only, based on tie-backs total length, minimum	B-46	86.80	.553	L.F.	14.25	18.05	.40	32.70	45
2750	Maximum		38.50	1.247	"	25	40.50	.90	66.40	93.50
3500	Tie-backs only, typical average, 25' long		2	24	Ea.	625	785	17.30	1,427.30	1,950
3600	35' long	↓	1.58	30.380	"	835	990	22	1,847	2,550

31 56 Slurry Walls

31 56 23 – Lean Concrete Slurry Walls

31 56 23.20 Slurry Trench

		Crew	Daily Output	Labor-Hours	Unit	Material	2007 Bare Costs Labor	Equipment	Total	Total Incl O&P
0010	**SLURRY TRENCH** Excavated slurry trench in wet soils									
0020	backfilled with 3000 PSI concrete, no reinforcing steel									
0050	Minimum	C-7	333	.216	C.F.	7.70	6.80	3.36	17.86	22.50
0100	Maximum		200	.360	"	12.90	11.30	5.60	29.80	37.50
0200	Alternate pricing method, minimum		150	.480	S.F.	15.40	15.05	7.45	37.90	48
0300	Maximum	↓	120	.600	↓	23	18.80	9.35	51.15	65
0500	Reinforced slurry trench, minimum	B-48	177	.316		11.55	10.30	21.50	43.35	52
0600	Maximum	"	69	.812		38.50	26.50	55	120	144
0800	Haul for disposal, 2 mile haul, excavated material, add	B-34B	99	.081	C.Y.		2.39	5.35	7.74	9.60
0900	Haul bentonite castings for disposal, add	"	40	.200	"		5.90	13.25	19.15	24

31 62 Driven Piles

31 62 13 – Concrete Piles

31 62 13.23 Prestressed Concrete Piles

		Crew	Daily Output	Labor-Hours	Unit	Material	2007 Bare Costs Labor	Equipment	Total	Total Incl O&P
0010	**PRESTRESSED CONCRETE PILES**, 200 piles									
0020	unless specified otherwise, not incl. pile caps or mobilization									
2200	Precast, prestressed, 50' long, 12" diam., 2-3/8" wall	B-19	720	.089	V.L.F.	14.55	3.28	2.29	20.12	23.50
2300	14" diameter, 2-1/2" wall		680	.094		19.15	3.47	2.43	25.05	29
2500	16" diameter, 3" wall	↓	640	.100		26.50	3.69	2.58	32.77	37.50
2600	18" diameter, 3" wall	B-19A	600	.107		33.50	3.93	3.68	41.11	47.50
2800	20" diameter, 3-1/2" wall		560	.114		38.50	4.22	3.94	46.66	53.50
2900	24" diameter, 3-1/2" wall	↓	520	.123		47	4.54	4.25	55.79	64
3100	Precast, prestressed, 40' long, 10" thick, square	B-19	700	.091		10.55	3.37	2.36	16.28	19.55
3200	12" thick, square		680	.094		13.25	3.47	2.43	19.15	23
3400	14" thick, square		600	.107		15.70	3.93	2.75	22.38	26.50
3500	Octagonal		640	.100		21.50	3.69	2.58	27.77	32.50
3700	16" thick, square		560	.114		25	4.22	2.95	32.17	37.50
3800	Octagonal	↓	600	.107		26	3.93	2.75	32.68	38
4000	18" thick, square	B-19A	520	.123		30	4.54	4.25	38.79	45
4100	Octagonal	B-19	560	.114		31	4.22	2.95	38.17	44
4300	20" thick, square	B-19A	480	.133		37	4.92	4.60	46.52	53.50
4400	Octagonal	B-19	520	.123		34	4.54	3.17	41.71	48
4600	24" thick, square	B-19A	440	.145		52.50	5.35	5	62.85	72
4700	Octagonal	B-19	480	.133		49	4.92	3.44	57.36	65.50
4730	Precast, prestressed, 60' long, 10" thick, square		700	.091		11.10	3.37	2.36	16.83	20
4740	12" thick, square (60' long)		680	.094		13.80	3.47	2.43	19.70	23.50
4750	Mobilization for 10,000 L.F. pile job, add		3300	.019			.72	.50	1.22	1.68
4800	25,000 L.F. pile job, add	↓	8500	.008	↓		.28	.19	.47	.65

31 62 Driven Piles

31 62 16 – Steel Piles

31 62 16.13 Sheet Steel Piles

	Crew	Daily Output	Labor-Hours	Unit	Material	2007 Bare Costs Labor	Equipment	Total	Total Incl O&P
0010 **SHEET STEEL PILES**									
0100 Step tapered, round, concrete filled									
0110 8" tip, 60 ton capacity, 30' depth	B-19	760	.084	V.L.F.	8.20	3.11	2.17	13.48	16.30
0120 60' depth		740	.086		9.25	3.19	2.23	14.67	17.70
0130 80' depth		700	.091		9.60	3.37	2.36	15.33	18.50
0150 10" tip, 90 ton capacity, 30' depth		700	.091		10.10	3.37	2.36	15.83	19.05
0160 60' depth		690	.093		10.40	3.42	2.39	16.21	19.50
0170 80' depth		670	.096		11.20	3.52	2.46	17.18	20.50
0190 12" tip, 120 ton capacity, 30' depth		660	.097		14.10	3.58	2.50	20.18	24
0200 60' depth, 12" diameter		630	.102		14.15	3.75	2.62	20.52	24.50
0210 80' depth		590	.108		12.30	4	2.80	19.10	23
0250 "H" Sections, 50' long, HP8 x 36		640	.100		13.85	3.69	2.58	20.12	24
0400 HP10 X 42		610	.105		16.15	3.87	2.70	22.72	27
0500 HP10 X 57		610	.105		22	3.87	2.70	28.57	33
0700 HP12 X 53		590	.108		20.50	4	2.80	27.30	32
0800 HP12 X 74	B-19A	590	.108		29	4	3.74	36.74	42
1000 HP14 X 73		540	.119		28.50	4.37	4.09	36.96	43
1100 HP14 X 89		540	.119		35	4.37	4.09	43.46	50
1300 HP14 X 102		510	.125		40	4.63	4.33	48.96	56
1400 HP14 X 117		510	.125		45.50	4.63	4.33	54.46	62.50
1600 Splice on standard points, not in leads, 8" or 10"	1 Sswl	5	1.600	Ea.	90.50	66		156.50	220
1700 12" or 14"		4	2		132	82.50		214.50	295
1900 Heavy duty points, not in leads, 10" wide		4	2		140	82.50		222.50	305
2100 14" wide		3.50	2.286		181	94.50		275.50	370
2600 Pipe piles, 50' lg. 8" diam., 29 lb. per L.F., no concrete	B-19	500	.128	V.L.F.	14.95	4.72	3.30	22.97	27.50
2700 Concrete filled		460	.139		15.95	5.15	3.59	24.69	29.50
2900 10" diameter, 34 lb. per L.F., no concrete		500	.128		18.60	4.72	3.30	26.62	31.50
3000 Concrete filled		450	.142		20.50	5.25	3.67	29.42	35.50
3200 12" diameter, 44 lb. per L.F., no concrete		475	.135		23	4.97	3.47	31.44	37
3300 Concrete filled		415	.154		24.50	5.70	3.98	34.18	40
3500 14" diameter, 46 lb. per L.F., no concrete		430	.149		24.50	5.50	3.84	33.84	40
3600 Concrete filled		355	.180		27	6.65	4.65	38.30	45
3800 16" diameter, 52 lb. per L.F., no concrete		385	.166		27.50	6.15	4.29	37.94	44.50
3900 Concrete filled		335	.191		31	7.05	4.93	42.98	50.50
4100 18" diameter, 59 lb. per L.F., no concrete		355	.180		36	6.65	4.65	47.30	55
4200 Concrete filled		310	.206		36.50	7.60	5.30	49.40	58
4400 Splices for pipe piles, not in leads, 8" diameter	1 Sswl	4.67	1.713	Ea.	67	71		138	202
4500 14" diameter		3.79	2.111		88	87.50		175.50	255
4600 16" diameter		3.03	2.640		109	109		218	315
4800 Points, standard, 8" diameter		4.61	1.735		75	72		147	213
4900 14" diameter		4.05	1.975		104	81.50		185.50	263
5000 16" diameter		3.37	2.374		127	98		225	320
5200 Points, heavy duty, 10" diameter		2.89	2.768		52	114		166	265
5300 14" or 16" diameter		2.02	3.960		83	164		247	390
5500 For reinforcing steel, add		1150	.007	Lb.	.62	.29		.91	1.20
5700 For thick wall sections, add				"	.70			.70	.77

31 62 19 – Timber Piles

31 62 19.10 Timber Piles

	Crew	Daily Output	Labor-Hours	Unit	Material	2007 Bare Costs Labor	Equipment	Total	Total Incl O&P
0010 **TIMBER PILES**, Friction or end bearing, not including									
0050 mobilization or demobilization									
0100 Untreated piles, up to 30' long, 12" butts, 8" points	B-19	625	.102	V.L.F.	6.90	3.78	2.64	13.32	16.50
0200 30' to 39' long, 12" butts, 8" points		700	.091		6.90	3.37	2.36	12.63	15.55

31 62 Driven Piles

31 62 19 – Timber Piles

31 62 19.10 Timber Piles

		Crew	Daily Output	Labor-Hours	Unit	Material	2007 Bare Costs Labor	Equipment	Total	Total Incl O&P
0300	40' to 49' long, 12" butts, 7" points	B-19	720	.089	V.L.F.	6.90	3.28	2.29	12.47	15.30
0400	50' to 59' long, 13" butts, 7" points		800	.080		6.95	2.95	2.06	11.96	14.60
0500	60' to 69' long, 13" butts, 7" points		840	.076		7.85	2.81	1.96	12.62	15.25
0600	70' to 80' long, 13" butts, 6" points		840	.076		8.70	2.81	1.96	13.47	16.20
0800	Treated piles, 12 lb. per C.F.,									
0810	friction or end bearing, ASTM class B									
1000	Up to 30' long, 12" butts, 8" points	B-19	625	.102	V.L.F.	10.15	3.78	2.64	16.57	20
1100	30' to 39' long, 12" butts, 8" points		700	.091		9.95	3.37	2.36	15.68	18.90
1200	40' to 49' long, 12" butts, 7" points		720	.089		10.65	3.28	2.29	16.22	19.40
1300	50' to 59' long, 13" butts, 7" points		800	.080		11.75	2.95	2.06	16.76	19.85
1400	60' to 69' long, 13" butts, 6" points	B-19A	840	.076		16.05	2.81	2.63	21.49	25
1500	70' to 80' long, 13" butts, 6" points	"	840	.076		20	2.81	2.63	25.44	29.50
1600	Treated piles, C.C.A., 2.5# per C.F.									
1610	8" butts, 10' long	B-19	400	.160	V.L.F.	7.40	5.90	4.12	17.42	22
1620	11' to 16' long		500	.128		7.40	4.72	3.30	15.42	19.30
1630	17' to 20' long		575	.111		7.40	4.10	2.87	14.37	17.80
1640	10" butts, 10' to 16' long		500	.128		8.45	4.72	3.30	16.47	20.50
1650	17' to 20' long		575	.111		8.45	4.10	2.87	15.42	18.95
1660	21' to 40' long		700	.091		8.45	3.37	2.36	14.18	17.25
1670	12" butts, 10' to 20' long		575	.111		9.10	4.10	2.87	16.07	19.70
1680	21' to 35' long		650	.098		9.10	3.63	2.54	15.27	18.60
1690	36' to 40' long		700	.091		9.10	3.37	2.36	14.83	18
1695	14" butts, to 40' long		700	.091		13.05	3.37	2.36	18.78	22.50
1700	Boot for pile tip, minimum	1 Pile	27	.296	Ea.	21	10.65		31.65	40.50
1800	Maximum		21	.381		62.50	13.70		76.20	91
2000	Point for pile tip, minimum		20	.400		21	14.35		35.35	46.50
2100	Maximum		15	.533		75	19.15		94.15	114
2300	Splice for piles over 50' long, minimum	B-46	35	1.371		52	45	.99	97.99	130
2400	Maximum		20	2.400		62.50	78.50	1.73	142.73	196
2600	Concrete encasement with wire mesh and tube		331	.145	V.L.F.	9.80	4.74	.10	14.64	18.45
2700	Mobilization for 10,000 L.F. pile job, add	B-19	3300	.019			.72	.50	1.22	1.68
2800	25,000 L.F. pile job, add	"	8500	.008			.28	.19	.47	.65

31 63 Bored Piles

31 63 26 – Drilled Caissons

31 63 26.13 Fixed End Cassion Piles

		Crew	Daily Output	Labor-Hours	Unit	Material	2007 Bare Costs Labor	Equipment	Total	Total Incl O&P
0010	**FIXED END CASSION PILES** R316326-60									
0020	per C.Y., not incl. mobilization, boulder removal, disposal									
0100	Open style, machine drilled, to 50' deep, in stable ground, no									
0110	casings or ground water, 18" diam., 0.065 C.Y./L.F.	B-43	200	.240	V.L.F.	8.20	7.65	17.50	33.35	40
0200	24" diameter, 0.116 C.Y./L.F.		190	.253		14.65	8.05	18.40	41.10	49
0300	30" diameter, 0.182 C.Y./L.F.		150	.320		23	10.15	23.50	56.65	66.50
0400	36" diameter, 0.262 C.Y./L.F.		125	.384		33	12.20	28	73.20	86.50
0500	48" diameter, 0.465 C.Y./L.F.		100	.480		58.50	15.25	35	108.75	127
0600	60" diameter, 0.727 C.Y./L.F.		90	.533		92	16.95	39	147.95	170
0700	72" diameter, 1.05 C.Y./L.F.		80	.600		133	19.05	43.50	195.55	224
0800	84" diameter, 1.43 C.Y./L.F.		75	.640		181	20.50	46.50	248	282
1000	For bell excavation and concrete, add									
1020	4' bell diameter, 24" shaft, 0.444 C.Y.	B-43	20	2.400	Ea.	46	76.50	175	297.50	360
1040	6' bell diameter, 30" shaft, 1.57 C.Y.		5.70	8.421		163	268	615	1,046	1,275
1060	8' bell diameter, 36" shaft, 3.72 C.Y.		2.40	20		385	635	1,450	2,470	3,000

31 63 Bored Piles

31 63 26 – Drilled Caissons

31 63 26.13 Fixed End Cassion Piles		Crew	Daily Output	Labor-Hours	Unit	Material	2007 Bare Costs Labor	Equipment	Total	Total Incl O&P
1080	9' bell diameter, 48" shaft, 4.48 C.Y.	B-43	2	24	Ea.	465	765	1,750	2,980	3,625
1100	10' bell diameter, 60" shaft, 5.24 C.Y.		1.70	28.235		545	895	2,050	3,490	4,250
1120	12' bell diameter, 72" shaft, 8.74 C.Y.		1	48		910	1,525	3,500	5,935	7,200
1140	14' bell diameter, 84" shaft, 13.6 C.Y.	↓	.70	68.571	↓	1,425	2,175	5,000	8,600	10,400
1200	Open style, machine drilled, to 50' deep, in wet ground, pulled									
1300	casing and pumping, 18" diameter, 0.065 C.Y./L.F.	B-48	160	.350	V.L.F.	8.20	11.40	23.50	43.10	52.50
1400	24" diameter, 0.116 C.Y./L.F.		125	.448		14.65	14.55	30.50	59.70	72
1500	30" diameter, 0.182 C.Y./L.F.		85	.659		23	21.50	44.50	89	108
1600	36" diameter, 0.262 C.Y./L.F.	↓	60	.933		33	30.50	63.50	127	153
1700	48" diameter, 0.465 C.Y./L.F.	B-49	55	1.600		58.50	54.50	82.50	195.50	239
1800	60" diameter, 0.727 C.Y./L.F.		35	2.514		92	85.50	130	307.50	375
1900	72" diameter, 1.05 C.Y./L.F.		30	2.933		133	99.50	151	383.50	465
2000	84" diameter, 1.43 C.Y./L.F.	↓	25	3.520	↓	181	119	181	481	585
2100	For bell excavation and concrete, add									
2120	4' bell diameter, 24" shaft, 0.444 C.Y.	B-48	19.80	2.828	Ea.	46	92	192	330	405
2140	6' bell diameter, 30" shaft, 1.57 C.Y.		5.70	9.825		163	320	665	1,148	1,400
2160	8' bell diameter, 36" shaft, 3.72 C.Y.		2.40	23.333		385	760	1,575	2,720	3,350
2180	9' bell diameter, 48" shaft, 4.48 C.Y.	B-49	3.30	26.667		465	905	1,375	2,745	3,425
2200	10' bell diameter, 60" shaft, 5.24 C.Y.		2.80	31.429		545	1,075	1,625	3,245	4,025
2220	12' bell diameter, 72" shaft, 8.74 C.Y.		1.60	55		910	1,875	2,825	5,610	7,000
2240	14' bell diameter, 84" shaft, 13.6 C.Y.	↓	1	88	↓	1,425	2,975	4,525	8,925	11,200
2300	Open style, machine drilled, to 50' deep, in soft rocks and									
2400	medium hard shales, 18" diameter, 0.065 C.Y./L.F.	B-49	50	1.760	V.L.F.	8.20	59.50	90.50	158.20	202
2500	24" diameter, 0.116 C.Y./L.F.		30	2.933		14.65	99.50	151	265.15	335
2600	30" diameter, 0.182 C.Y./L.F.		20	4.400		23	149	227	399	505
2700	36" diameter, 0.262 C.Y./L.F.		15	5.867		33	199	300	532	680
2800	48" diameter, 0.465 C.Y./L.F.		10	8.800		58.50	298	455	811.50	1,025
2900	60" diameter, 0.727 C.Y./L.F.		7	12.571		92	425	650	1,167	1,475
3000	72" diameter, 1.05 C.Y./L.F.		6	14.667		133	495	755	1,383	1,750
3100	84" diameter, 1.43 C.Y./L.F.	↓	5	17.600	↓	181	595	905	1,681	2,125
3200	For bell excavation and concrete, add									
3220	4' bell diameter, 24" shaft, 0.444 C.Y.	B-49	10.90	8.073	Ea.	46	274	415	735	935
3240	6' bell diameter, 30" shaft, 1.57 C.Y.		3.10	28.387		163	965	1,475	2,603	3,275
3260	8' bell diameter, 36" shaft, 3.72 C.Y.		1.30	67.692		385	2,300	3,500	6,185	7,800
3280	9' bell diameter, 48" shaft, 4.48 C.Y.		1.10	80		465	2,725	4,125	7,315	9,250
3300	10' bell diameter, 60" shaft, 5.24 C.Y.		.90	97.778		545	3,325	5,050	8,920	11,300
3320	12' bell diameter, 72" shaft, 8.74 C.Y.		.60	146		910	4,975	7,550	13,435	17,000
3340	14' bell diameter, 84" shaft, 13.6 C.Y.		.40	220	↓	1,425	7,450	11,300	20,175	25,600
3600	For rock excavation, sockets, add, minimum		120	.733	C.F.		25	38	63	80
3650	Average		95	.926			31.50	47.50	79	101
3700	Maximum	↓	48	1.833	↓		62	94.50	156.50	200
3900	For 50' to 100' deep, add				V.L.F.				7%	7%
4000	For 100' to 150' deep, add								25%	25%
4100	For 150' to 200' deep, add				↓				30%	30%
4200	For casings left in place, add				Lb.	.87			.87	.96
4300	For other than 50 lb. reinf. per C.Y., add or deduct				"	.92			.92	1.01
4400	For steel "I" beam cores, add	B-49	8.30	10.602	Ton	1,650	360	545	2,555	2,950
4500	Load and haul excess excavation, 2 miles	B-34B	178	.045	L.C.Y.		1.33	2.98	4.31	5.35
4600	For mobilization, 50 mile radius, rig to 36"	B-43	2	24	Ea.		765	1,750	2,515	3,100
4650	Rig to 84"	B-48	1.75	32			1,050	2,175	3,225	3,975
4700	For low headroom, add								50%	
5000	Bottom inspection	1 Skwk	1.20	6.667	↓		253		253	395

31 63 Bored Piles

31 63 26 – Drilled Caissons

31 63 26.16 Concrete Caissons for Marine Construction

	31 63 26.16 Concrete Caissons for Marine Construction	Crew	Daily Output	Labor-Hours	Unit	Material	2007 Bare Costs Labor	Equipment	Total	Total Incl O&P
0010	**CONCRETE CAISSONS FOR MARINE CONSTRUCTION**									
0100	Caissons, incl. mobilization and demobilization, up to 50 miles									
0200	Uncased shafts, 30 to 80 tons cap., 17" diam., 10' depth	B-44	88	.727	V.L.F.	21	26.50	11.80	59.30	78
0300	25' depth		165	.388		14.85	14.05	6.30	35.20	46
0400	80-150 ton capacity, 22" diameter, 10' depth		80	.800		26	29	13	68	89
0500	20' depth		130	.492		21	17.85	8	46.85	60.50
0700	Cased shafts, 10 to 30 ton capacity, 10-5/8" diam., 20' depth		175	.366		14.85	13.25	5.95	34.05	44
0800	30' depth		240	.267		13.85	9.65	4.33	27.83	35.50
0850	30 to 60 ton capacity, 12" diameter, 20' depth		160	.400		21	14.50	6.50	42	53
0900	40' depth		230	.278		16	10.10	4.52	30.62	38.50
1000	80 to 100 ton capacity, 16" diameter, 20' depth		160	.400		29.50	14.50	6.50	50.50	62.50
1100	40' depth		230	.278		27.50	10.10	4.52	42.12	51.50
1200	110 to 140 ton capacity, 17-5/8" diameter, 20' depth		160	.400		32	14.50	6.50	53	65
1300	40' depth		230	.278		29.50	10.10	4.52	44.12	53.50
1400	140 to 175 ton capacity, 19" diameter, 20' depth		130	.492		34.50	17.85	8	60.35	75.50
1500	40' depth		210	.305		32	11.05	4.95	48	58
1700	Over 30' long, L.F. cost tends to be lower									
1900	Maximum depth is about 90'									

31 63 29 – Drilled Concrete Piers and Shafts

31 63 29.13 Uncased Drilled Concrete Piers

		Crew	Daily Output	Labor-Hours	Unit	Material	Labor	Equipment	Total	Total Incl O&P
0010	**UNCASED DRILLED CONCRETE PIERS**									
0020	unless specified otherwise, not incl. pile caps or mobilization									
0050	Cast in place augered piles, no casing or reinforcing									
0060	8" diameter	B-43	540	.089	V.L.F.	3.52	2.82	6.50	12.84	15.35
0065	10" diameter		480	.100		5.60	3.18	7.30	16.08	19.05
0070	12" diameter		420	.114		7.90	3.63	8.35	19.88	23.50
0075	14" diameter		360	.133		10.65	4.24	9.70	24.59	29
0080	16" diameter		300	.160		14.35	5.10	11.65	31.10	36.50
0085	18" diameter		240	.200		17.75	6.35	14.60	38.70	45.50
0100	Cast in place, thin wall shell pile, straight sided,									
0110	not incl. reinforcing, 8" diam., 16 ga., 5.8 lb./L.F.	B-19	700	.091	V.L.F.	6.60	3.37	2.36	12.33	15.20
0200	10" diameter, 16 ga. corrugated, 7.3 lb./L.F.		650	.098		8.65	3.63	2.54	14.82	18.05
0300	12" diameter, 16 ga. corrugated, 8.7 lb./L.F.		600	.107		11.20	3.93	2.75	17.88	21.50
0400	14" diameter, 16 ga. corrugated, 10.0 lb./L.F.		550	.116		13.20	4.29	3	20.49	24.50
0500	16" diameter, 16 ga. corrugated, 11.6 lb./L.F.		500	.128		16.15	4.72	3.30	24.17	29
0800	Cast in place friction pile, 50' long, fluted,									
0810	tapered steel, 4000 psi concrete, no reinforcing									
0900	12" diameter, 7 ga.	B-19	600	.107	V.L.F.	19.80	3.93	2.75	26.48	31.50
1000	14" diameter, 7 ga.		560	.114		21.50	4.22	2.95	28.67	34
1100	16" diameter, 7 ga.		520	.123		25.50	4.54	3.17	33.21	38.50
1200	18" diameter, 7 ga.		480	.133		29.50	4.92	3.44	37.86	44
1300	End bearing, fluted, constant diameter,									
1320	4000 psi concrete, no reinforcing									
1340	12" diameter, 7 ga.	B-19	600	.107	V.L.F.	20.50	3.93	2.75	27.18	32.50
1360	14" diameter, 7 ga.		560	.114		26	4.22	2.95	33.17	38.50
1380	16" diameter, 7 ga.		520	.123		30	4.54	3.17	37.71	43.50
1400	18" diameter, 7 ga.		480	.133		33	4.92	3.44	41.36	48

31 63 Bored Piles

31 63 29 – Drilled Concrete Piers and Shafts

31 63 29.20 Cast In Place, Adds	Crew	Daily Output	Labor-Hours	Unit	Material	2007 Bare Costs Labor	Equipment	Total	Total Incl O&P
0010 **CAST IN PLACE, ADDS**									
1500 For reinforcing steel, add				Lb.	.90			.90	.98
1700 For ball or pedestal end, add	B-19	11	5.818	C.Y.	106	215	150	471	620
1900 For lengths above 60', concrete, add	"	11	5.818	"	111	215	150	476	625
2000 For steel thin shell, pipe only				Lb.	.75			.75	.83

Estimating Tips

32 01 00 Operations and Maintenance of Exterior Improvements

- Recycling of asphalt pavement is becoming very popular and is an alternative to removal and replacement of asphalt pavement. It can be a good value engineering proposal if removed pavement can be recycled either at the site or another site that is reasonably close to the project site.

32 10 00 Bases, Ballasts, and Paving

- When estimating paving, keep in mind the project schedule. If an asphaltic paving project is in a colder climate and runs through to the spring, consider placing the base course in the autumn and then topping it in the spring just prior to completion. This could save considerable costs in spring repair. Keep in mind that prices for asphalt and concrete are generally higher in the cold seasons.

32 90 00 Planting

- The timing of planting and guarantee specifications often dictate the costs for establishing tree and shrub growth and a stand of grass or ground cover. Establish the work performance schedule to coincide with the local planting season. Maintenance and growth guarantees can add from 20% to 100% to the total landscaping cost. The cost to replace trees and shrubs can be as high as 5% of the total cost depending on the planting zone, soil conditions, and time of year.

Reference Numbers

Reference numbers are shown in shaded boxes at the beginning of some major classifications. These numbers refer to related items in the Reference Section. The reference information may be an estimating procedure, an alternate pricing method, or technical information.

Note: Not all subdivisions listed here necessarily appear in this publication.

32 01 Operation and Maintenance of Exterior Improvements

32 01 13 – Flexible Paving Surface Treatment

32 01 13.61 Slurry Seal (Latex Modified)

		Crew	Daily Output	Labor-Hours	Unit	Material	2007 Bare Costs Labor	Equipment	Total	Total Incl O&P
0010	**SLURRY SEAL (LATEX MODIFIED)**									
3780	Rubberized asphalt (latex) seal	B-45	5000	.003	S.Y.	1.51	.11	.12	1.74	1.96

32 01 13.64 Sand Seal

0010	**SAND SEAL**									
2080	Sand sealing, sharp sand, asphalt emulsion, small area	B-91	10000	.006	S.Y.	.83	.22	.16	1.21	1.42
2120	Roadway or large area	"	18000	.004	"	.71	.12	.09	.92	1.07

32 01 13.66 Fog Seal

0010	**FOG SEAL**									
0012	Sealcoating, 2 coat coal tar pitch emulsion over 10,000 SY	B-45	5000	.003	S.Y.	.66	.11	.12	.89	1.03
0030	1000 to 10,000 S.Y.	"	3000	.005		.66	.18	.20	1.04	1.22
0100	Under 1000 S.Y.	B-1	1050	.023		.66	.67		1.33	1.78
0300	Petroleum resistant, over 10,000 S.Y.	B-45	5000	.003		.83	.11	.12	1.06	1.21
0320	1000 to 10,000 S.Y.	"	3000	.005		.83	.18	.20	1.21	1.40
0400	Under 1000 S.Y.	B-1	1050	.023		.83	.67		1.50	1.96
0600	Non-skid pavement renewal, over 10,000 S.Y.	B-45	5000	.003		.96	.11	.12	1.19	1.36
0620	1000 to 10,000 S.Y.	"	3000	.005		.96	.18	.20	1.34	1.55
0700	Under 1000 S.Y.	B-1	1050	.023		.96	.67		1.63	2.11
0800	Prepare and clean surface for above	A-2	8545	.003			.08	.02	.10	.15
1000	Hand seal asphalt curbing	B-1	4420	.005	L.F.	.49	.16		.65	.79
1900	Asphalt surface treatment, single course, small area									
1901	0.30 gal/S.Y. asphalt material, 20#/S.Y. aggregate	B-91	5000	.013	S.Y.	1.14	.43	.32	1.89	2.26
1910	Roadway or large area		10000	.006		1.05	.22	.16	1.43	1.66
1950	Asphalt surface treatment, dbl. course for small area		3000	.021		2.10	.72	.53	3.35	4.01
1960	Roadway or large area		6000	.011		1.89	.36	.27	2.52	2.92
1980	Asphalt surface treatment, single course, for shoulders		7500	.009		1.21	.29	.21	1.71	2

32 06 Schedules for Exterior Improvements

32 06 10 – Schedules for Bases, Ballasts, and Paving

32 06 10.10 Sidewalks, Driveways and Patios

		Crew	Daily Output	Labor-Hours	Unit	Material	Labor	Equipment	Total	Total Incl O&P
0010	**SIDEWALKS, DRIVEWAYS AND PATIOS** No base									
0020	Asphaltic concrete, 2" thick	B-37	720	.067	S.Y.	5.10	2.03	.17	7.30	8.95
0100	2-1/2" thick	"	660	.073	"	6.45	2.21	.19	8.85	10.70
0300	Concrete, 3000 psi, CIP, 6 x 6 - W1.4 x W1.4 mesh,									
0310	broomed finish, no base, 4" thick	B-24	600	.040	S.F.	1.73	1.35		3.08	3.96
0350	5" thick		545	.044		2.31	1.48		3.79	4.80
0400	6" thick		510	.047		2.70	1.58		4.28	5.40
0450	For bank run gravel base, 4" thick, add	B-18	2500	.010		.50	.28	.01	.79	1.01
0520	8" thick, add	"	1600	.015		1	.44	.02	1.46	1.81
0550	Exposed aggregate finish, add to above, minimum	B-24	1875	.013		.11	.43		.54	.78
0600	Maximum	"	455	.053		.36	1.78		2.14	3.10
1000	Crushed stone, 1" thick, white marble	2 Clab	1700	.009		.21	.27		.48	.65
1050	Bluestone	"	1700	.009		.23	.27		.50	.67
1700	Redwood, prefabricated, 4' x 4' sections	2 Carp	316	.051		8.20	1.86		10.06	11.90
1750	Redwood planks, 1" thick, on sleepers	"	240	.067		5.75	2.45		8.20	10.10
2250	Stone dust, 4" thick	B-62	900	.027	S.Y.	3.08	.84	.14	4.06	4.83

32 06 10.20 Steps

0010	**STEPS** Incl. excav., borrow & concrete base, where applicable									
0100	Brick steps	B-24	35	.686	LF Riser	11.15	23		34.15	47.50
0200	Railroad ties	2 Clab	25	.640		3.19	18.40		21.59	32
0300	Bluestone treads, 12" x 2" or 12" x 1-1/2"	B-24	30	.800		25.50	27		52.50	69

32 06 Schedules for Exterior Improvements

32 06 10 – Schedules for Bases, Ballasts, and Paving

32 06 10.20 Steps		Crew	Daily Output	Labor-Hours	Unit	Material	2007 Bare Costs Labor	Equipment	Total	Total Incl O&P
0500	Concrete, cast in place, see Division 03 30 53.40									
0600	Precast concrete, see Division 03 41 23.50									
4025	Steel edge strips, incl. stakes, 1/4" x 5"	B-1	390	.062	L.F.	3.65	1.81		5.46	6.85
4050	Edging, landscape timber or railroad ties, 6" x 8"	2 Carp	170	.094	"	2.72	3.45		6.17	8.40

32 11 Base Courses

32 11 23 – Aggregate Base Courses

32 11 23.23 Base Course Drainage Layers

		Crew	Daily Output	Labor-Hours	Unit	Material	Labor	Equipment	Total	Total Incl O&P
0010	**BASE COURSE DRAINAGE LAYERS**									
0050	Crushed 3/4" stone base, compacted, 3" deep	B-36C	5200	.008	S.Y.	3.09	.27	.56	3.92	4.42
0100	6" deep		5000	.008		6.20	.28	.58	7.06	7.85
0200	9" deep		4600	.009		9.30	.31	.63	10.24	11.35
0300	12" deep		4200	.010		12.40	.33	.69	13.42	14.85
0301	Crushed 1-1/2" stone base, compacted to 4" deep	B-36B	6000	.011		4.81	.36	.55	5.72	6.45
0302	6" deep		5400	.012		7.20	.40	.61	8.21	9.25
0303	8" deep		4500	.014		9.60	.48	.73	10.81	12.15
0304	12" deep		3800	.017		14.45	.57	.87	15.89	17.65
0350	Bank run gravel, spread and compacted									
0370	6" deep	B-32	6000	.005	S.Y.	4.15	.19	.28	4.62	5.15
0390	9" deep		4900	.007		6.25	.24	.35	6.84	7.60
0400	12" deep		4200	.008		8.30	.27	.41	8.98	10
6000	Stabilization fabric, polypropylene, 6 oz./S.Y.	B-6	10000	.002		1.07	.08	.02	1.17	1.33
6900	For small and irregular areas, add						50%	50%		
7000	Prepare and roll sub-base, small areas to 2500 S.Y.	B-32A	1500	.016	S.Y.		.56	.71	1.27	1.65
8000	Large areas over 2500 S.Y.	B-32	3700	.009	"		.31	.46	.77	.98

32 11 26 – Asphaltic Base Courses

32 11 26.19 Bituminous-Stabilized Base Courses

		Crew	Daily Output	Labor-Hours	Unit	Material	Labor	Equipment	Total	Total Incl O&P
0010	**BITUMINOUS-STABILIZED BASE COURSES**									
0020	and large paved areas									
0700	Liquid application to gravel base, asphalt emulsion	B-45	6000	.003	Gal.	3.90	.09	.10	4.09	4.54
0800	Prime and seal, cut back asphalt		6000	.003	"	4.60	.09	.10	4.79	5.30
1000	Macadam penetration crushed stone, 2 gal. per S.Y., 4" thick		6000	.003	S.Y.	7.80	.09	.10	7.99	8.85
1100	6" thick, 3 gal. per S.Y.		4000	.004		11.70	.14	.15	11.99	13.20
1200	8" thick, 4 gal. per S.Y.		3000	.005		15.60	.18	.20	15.98	17.65
8900	For small and irregular areas, add						50%	50%		

32 12 Flexible Paving

32 12 16 – Asphalt Paving

32 12 16.13 Plant-Mix Asphalt Paving

		Crew	Daily Output	Labor-Hours	Unit	Material	Labor	Equipment	Total	Total Incl O&P
0010	**PLANT-MIX ASPHALT PAVING**									
0020	and large paved areas with no hauling included									
0025	See 31 23 23.18 for hauling costs									
0080	Binder course, 1-1/2" thick	B-25	7725	.011	S.Y.	3.92	.36	.29	4.57	5.20
0120	2" thick		6345	.014		5.20	.44	.36	6	6.80
0130	2-1/2" thick		5620	.016		6.55	.49	.40	7.44	8.40
0160	3" thick		4905	.018		7.85	.57	.46	8.88	10
0170	3-1/2" thick		4520	.019		9.15	.61	.50	10.26	11.55
0200	4" thick		4140	.021		10.45	.67	.55	11.67	13.15
0300	Wearing course, 1" thick	B-25B	10575	.009		2.59	.29	.23	3.11	3.56

32 12 Flexible Paving

32 12 16 – Asphalt Paving

32 12 16.13 Plant-Mix Asphalt Paving

		Crew	Daily Output	Labor-Hours	Unit	Material	2007 Bare Costs Labor	Equipment	Total	Total Incl O&P
0340	1-1/2" thick	B-25B	7725	.012	S.Y.	3.94	.40	.32	4.66	5.30
0380	2" thick		6345	.015		5.30	.49	.39	6.18	7
0420	2-1/2" thick		5480	.018		6.55	.56	.45	7.56	8.55
0460	3" thick		4900	.020		7.80	.63	.50	8.93	10.05
0470	3-1/2" thick		4520	.021		9.15	.68	.55	10.38	11.70
0480	4" thick		4140	.023		10.45	.75	.60	11.80	13.30
0500	Open graded friction course	B-25C	5000	.010		2.29	.31	.39	2.99	3.43
0800	Alternate method of figuring paving costs									
0810	Binder course, 1-1/2" thick	B-25	630	.140	Ton	48	4.41	3.58	55.99	63.50
0811	2" thick		690	.128		48	4.03	3.27	55.30	63
0812	3" thick		800	.110		48	3.47	2.82	54.29	61.50
0813	4" thick		900	.098		48	3.09	2.51	53.60	60.50
0850	Wearing course, 1" thick	B-25B	575	.167		53	5.35	4.29	62.64	71.50
0851	1-1/2" thick		630	.152		53	4.90	3.92	61.82	70.50
0852	2" thick		690	.139		53	4.47	3.58	61.05	69.50
0853	2-1/2" thick		765	.125		53	4.03	3.23	60.26	68.50
0854	3" thick		800	.120		53	3.86	3.08	59.94	68
1000	Pavement replacement over trench, 2" thick	B-37	90	.533	S.Y.	5.40	16.25	1.36	23.01	32.50
1050	4" thick		70	.686		10.65	21	1.75	33.40	46
1080	6" thick		55	.873		16.95	26.50	2.23	45.68	62

32 12 16.14 Asphaltic Concrete Pavement, Lots and Driveways

		Crew	Daily Output	Labor-Hours	Unit	Material	Labor	Equipment	Total	Total Incl O&P
0011	**ASPHALTIC CONCRETE PAVEMENT, LOTS & DRIVEWAYS**									
0020	6" stone base, 2" binder course, 1" topping	B-25C	9000	.005	S.F.	1.73	.17	.22	2.12	2.40
0300	Binder course, 1-1/2" thick		35000	.001		.44	.04	.06	.54	.61
0400	2" thick		25000	.002		.56	.06	.08	.70	.81
0500	3" thick		15000	.003		.87	.10	.13	1.10	1.26
0600	4" thick		10800	.004		1.14	.14	.18	1.46	1.68
0800	Sand finish course, 3/4" thick		41000	.001		.25	.04	.05	.34	.39
0900	1" thick		34000	.001		.31	.05	.06	.42	.47
1000	Fill pot holes, hot mix, 2" thick	B-16	4200	.008		.59	.22	.13	.94	1.14
1100	4" thick		3500	.009		.87	.27	.15	1.29	1.54
1120	6" thick		3100	.010		1.16	.30	.17	1.63	1.94
1140	Cold patch, 2" thick	B-51	3000	.016		.66	.46	.05	1.17	1.50
1160	4" thick		2700	.018		1.26	.52	.05	1.83	2.25
1180	6" thick		1900	.025		1.96	.73	.07	2.76	3.38

32 13 Rigid Paving

32 13 13 – Concrete Paving

32 13 13.23 Concrete Paving Surface Treatment

		Crew	Daily Output	Labor-Hours	Unit	Material	Labor	Equipment	Total	Total Incl O&P
0010	**CONCRETE PAVING SURFACE TREATMENT**									
0015	Including joints, finishing and curing									
0020	Fixed form, 12' pass, unreinforced, 6" thick	B-26	3000	.029	S.Y.	22	.95	.94	23.89	26.50
0100	8" thick		2750	.032		32	1.04	1.03	34.07	38
0110	8" thick, small area		1375	.064		32	2.08	2.06	36.14	41
0200	9" thick		2500	.035		36	1.14	1.13	38.27	43
0300	10" thick		2100	.042		39.50	1.36	1.35	42.21	47
0310	10" thick, small area		1050	.084		39.50	2.72	2.69	44.91	50.50
0400	12" thick		1800	.049		45	1.59	1.57	48.16	53.50
0410	Conc pavement, w/jt,fnsh&curing,fix form,24' pass,unreinforced,6"T		6000	.015		20	.48	.47	20.95	23.50
0430	8" thick		5500	.016		29	.52	.51	30.03	33.50
0440	9" thick		5000	.018		33	.57	.57	34.14	38

32 13 Rigid Paving

32 13 13 – Concrete Paving

32 13 13.23 Concrete Paving Surface Treatment		Crew	Daily Output	Labor-Hours	Unit	Material	2007 Bare Costs Labor	Equipment	Total	Total Incl O&P
0450	10" thick	B-26	4200	.021	S.Y.	36.50	.68	.67	37.85	42
0460	12" thick		3600	.024		42	.79	.79	43.58	48
0470	15" thick		3000	.029		55.50	.95	.94	57.39	63.50
0500	Fixed form 12' pass 15" thick		1500	.059		58	1.90	1.89	61.79	68.50
0510	For small irregular areas, add				%		100%			
0520	Welded wire fabric, sheets for rigid paving 2.33 lbs/SY	2 Rodm	389	.041	S.Y.	1.15	1.70		2.85	4.05
0530	Reinforcing steel for rigid paving 12 lbs/SY		666	.024		5.65	.99		6.64	7.85
0540	Reinforcing steel for rigid paving 18 lbs/SY		444	.036		8.45	1.49		9.94	11.75
0620	Slip form, 12' pass, unreinforced, 6" thick	B-26A	5600	.016		21	.51	.53	22.04	24.50
0624	8" thick		5300	.017		31	.54	.56	32.10	35.50
0626	9" thick		4820	.018		35	.59	.61	36.20	40
0628	10" thick		4050	.022		38.50	.71	.73	39.94	44.50
0630	12" thick		3470	.025		44	.82	.85	45.67	50.50
0632	15" thick		2890	.030		56.50	.99	1.02	58.51	64.50
0640	Slip form, 24' pass, unreinforced, 6" thick		11200	.008		20.50	.26	.26	21.02	23
0644	8" thick		10600	.008		28.50	.27	.28	29.05	32
0646	9" thick		9640	.009		32.50	.30	.31	33.11	37
0648	10" thick		8100	.011		36	.35	.36	36.71	40.50
0650	12" thick		6940	.013		41.50	.41	.42	42.33	46.50
0652	15" thick		5780	.015		52.50	.49	.51	53.50	59.50
0700	Finishing, broom finish small areas	2 Cefi	120	.133			4.74		4.74	6.95
1000	Curing, with sprayed membrane by hand	2 Clab	1500	.011		.42	.31		.73	.94
1650	For integral coloring, see div. 03 31 05.35									

32 14 Unit Paving

32 14 13 – Precast Concrete Unit Paving

32 14 13.13 Interlocking Precast Concrete Unit Paving

		Crew	Daily Output	Labor-Hours	Unit	Material	Labor	Equipment	Total	Total Incl O&P
0010	**INTERLOCKING PRECAST CONCRETE UNIT PAVING**									
0020	"V" blocks for retaining soil	D-1	205	.078	S.F.	7.45	2.60		10.05	12.10

32 14 13.16 Precast Concrete Unit Paving Slabs

		Crew	Daily Output	Labor-Hours	Unit	Material	Labor	Equipment	Total	Total Incl O&P
0010	**PRECAST CONCRETE UNIT PAVING SLABS**									
0710	Precast concrete patio blocks, 2-3/8" thick, colors, 8" x 16"	D-1	265	.060	S.F.	1.30	2.01		3.31	4.49
0750	Exposed local aggregate, natural	2 Bric	250	.064		5.60	2.44		8.04	9.85
0800	Colors		250	.064		6.10	2.44		8.54	10.40
0850	Exposed granite or limestone aggregate		250	.064		6.75	2.44		9.19	11.10
0900	Exposed white tumblestone aggregate		250	.064		3.99	2.44		6.43	8.10

32 14 16 – Brick Unit Paving

32 14 16.10 Brick Paving

		Crew	Daily Output	Labor-Hours	Unit	Material	Labor	Equipment	Total	Total Incl O&P
0010	**BRICK PAVING**									
0012	4" x 8" x 1-1/2", without joints (4.5 brick/S.F.)	D-1	110	.145	S.F.	2.73	4.85		7.58	10.40
0100	Grouted, 3/8" joint (3.9 brick/S.F.)		90	.178		3.19	5.95		9.14	12.50
0200	4" x 8" x 2-1/4", without joints (4.5 bricks/S.F.)		110	.145		3.53	4.85		8.38	11.30
0300	Grouted, 3/8" joint (3.9 brick/S.F.)		90	.178		3.26	5.95		9.21	12.60
0500	Bedding, asphalt, 3/4" thick	B-25	5130	.017		.48	.54	.44	1.46	1.84
0540	Course washed sand bed, 1" thick	B-18	5000	.005		.20	.14	.01	.35	.45
0580	Mortar, 1" thick	D-1	300	.053		.59	1.78		2.37	3.36
0620	2" thick		200	.080		1.18	2.67		3.85	5.35
1500	Brick on 1" thick sand bed laid flat, 4.5 per S.F.		100	.160		2.67	5.35		8.02	11.05
2000	Brick pavers, laid on edge, 7.2 per S.F.		70	.229		2.64	7.60		10.24	14.50
2500	For 4" thick concrete bed and joints, add		595	.027		1.22	.90		2.12	2.71

32 14 Unit Paving

32 14 16 – Brick Unit Paving

32 14 16.10 Brick Paving

		Crew	Daily Output	Labor-Hours	Unit	Material	2007 Bare Costs Labor	Equipment	Total	Total Incl O&P
2800	For steam cleaning, add	A-1H	950	.008	S.F.	.06	.24	.06	.36	.52

32 14 23 – Asphalt Unit Paving

32 14 23.10 Asphalt Blocks

		Crew	Daily Output	Labor-Hours	Unit	Material	Labor	Equipment	Total	Total Incl O&P
0010	**ASPHALT BLOCKS**									
0020	Rectangular, 6" x 12" x 1-1/4", w/bed & neopr. adhesive	D-1	135	.119	S.F.	5.80	3.95		9.75	12.40
0100	3" thick		130	.123		8.10	4.10		12.20	15.20
0300	Hexagonal tile, 8" wide, 1-1/4" thick		135	.119		5.80	3.95		9.75	12.40
0400	2" thick		130	.123		8.10	4.10		12.20	15.20
0500	Square, 8" x 8", 1-1/4" thick		135	.119		5.80	3.95		9.75	12.40
0600	2" thick		130	.123		8.10	4.10		12.20	15.20
0900	For exposed aggregate (ground finish) add					.40			.40	.44
0910	For colors, add					.30			.30	.33

32 14 40 – Stone Paving

32 14 40.10 Stone Pavers

		Crew	Daily Output	Labor-Hours	Unit	Material	Labor	Equipment	Total	Total Incl O&P
0010	**STONE PAVERS**									
1100	Flagging, bluestone, irregular, 1" thick,	D-1	81	.198	S.F.	5.65	6.60		12.25	16.25
1150	Snapped random rectangular, 1" thick		92	.174		8.55	5.80		14.35	18.30
1200	1-1/2" thick		85	.188		10.30	6.30		16.60	21
1250	2" thick		83	.193		12	6.45		18.45	23
1300	Slate, natural cleft, irregular, 3/4" thick		92	.174		6.30	5.80		12.10	15.80
1350	Random rectangular, gauged, 1/2" thick		105	.152		13.65	5.10		18.75	23
1400	Random rectangular, butt joint, gauged, 1/4" thick		150	.107		14.70	3.56		18.26	21.50
1500	For interior setting, add								25%	25%
1550	Granite blocks, 3-1/2" x 3-1/2" x 3-1/2"	D-1	92	.174	S.F.	7.65	5.80		13.45	17.30
1600	4" to 12" long, 3" to 5" wide, 3" to 5" thick		98	.163		6.40	5.45		11.85	15.35
1650	6" to 15" long, 3" to 6" wide, 3" to 5" thick		105	.152		3.41	5.10		8.51	11.50

32 16 Curbs and Gutters

32 16 13 – Concrete Curbs and Gutters

32 16 13.13 Cast-in-Place Concrete Curbs and Gutters

		Crew	Daily Output	Labor-Hours	Unit	Material	Labor	Equipment	Total	Total Incl O&P
0010	**CAST-IN-PLACE CONCRETE CURBS AND GUTTERS**									
0300	Concrete, wood forms, 6" x 18", straight	C-2A	500	.096	L.F.	2.71	3.41		6.12	8.25
0400	6" x 18", radius		200	.240		2.82	8.50		11.32	16.25
0410	Steel forms, 6" x 18", straight		700	.069		4.34	2.44		6.78	8.55
0411	6" x 18", radius		400	.120		4.34	4.26		8.60	11.30
0415	Machine formed, 6" x 18", straight	B-69A	2000	.024		3.80	.76	.34	4.90	5.75
0416	6" x 18", radius	"	900	.053		3.96	1.70	.76	6.42	7.80
0421	Curb and gutter, straight									
0422	with 6" high curb and 6" thick gutter, wood forms									
0430	24" wide, .055 C.Y. per L.F.	C-2A	375	.128	L.F.	14.25	4.55		18.80	22.50
0435	30" wide, .066 C.Y. per L.F.		340	.141		15.75	5		20.75	25
0440	Steel forms, 24" wide, straight		700	.069		6.20	2.44		8.64	10.55
0441	Radius		300	.160		6.20	5.70		11.90	15.55
0442	30" wide, straight		700	.069		7.45	2.44		9.89	11.95
0443	Radius		300	.160		7.45	5.70		13.15	16.95
0445	Machine formed, 24" wide, straight	B-69A	2000	.024		6.20	.76	.34	7.30	8.35
0446	Radius		900	.053		6.20	1.70	.76	8.66	10.25
0447	30" wide, straight		2000	.024		7.45	.76	.34	8.55	9.75
0448	Radius		900	.053		7.45	1.70	.76	9.91	11.65

32 16 Curbs and Gutters

32 16 13 – Concrete Curbs and Gutters

32 16 13.26 Precast Concrete Curbs	Crew	Daily Output	Labor-Hours	Unit	Material	2007 Bare Costs Labor	Equipment	Total	Total Incl O&P
0010 **PRECAST CONCRETE CURBS**									
0550 Precast, 6" x 18", straight	B-29	700	.080	L.F.	9.05	2.51	1.29	12.85	15.25
0600 6" x 18", radius	"	325	.172	"	10.35	5.40	2.79	18.54	23

32 16 19 – Asphalt Curbs

32 16 19.10 Bituminous Concrete Curbs

	Crew	Daily Output	Labor-Hours	Unit	Material	Labor	Equipment	Total	Total Incl O&P
0010 **BITUMINOUS CONCRETE CURBS**									
0012 Curbs, asphaltic, machine formed, 8" wide, 6" high, 40 L.F./ton	B-27	1000	.032	L.F.	1.06	.94	.25	2.25	2.91
0100 8" wide, 8" high, 30 L.F. per ton	↓	900	.036		1.22	1.04	.28	2.54	3.27
0150 Asphaltic berm, 12" W, 3"-6" H, 35 L.F./ton, before pavement		700	.046		1.10	1.34	.36	2.80	3.68
0200 12" W, 1-1/2" to 4" H, 60 L.F. per ton, laid with pavement	B-2	1050	.038	↓	.67	1.11		1.78	2.47

32 16 40 – Stone Curbs

32 16 40.13 Manufactured Stone Curbs

	Crew	Daily Output	Labor-Hours	Unit	Material	Labor	Equipment	Total	Total Incl O&P
0010 **MANUFACTURED STONE CURBS**									
1000 Granite, split face, straight, 5" x 16"	D-13	500	.096	L.F.	10.65	3.39	1.15	15.19	18.10
1100 6" x 18"	"	450	.107		13.95	3.77	1.28	19	22.50
1300 Radius curbing, 6" x 18", over 10' radius	B-29	260	.215		17.10	6.75	3.48	27.33	33
1400 Corners, 2' radius		80	.700	Ea.	57.50	22	11.35	90.85	109
1600 Edging, 4-1/2" x 12", straight		300	.187	L.F.	5.30	5.85	3.02	14.17	18.15
1800 Curb inlets, (guttermouth) straight	↓	41	1.366	Ea.	128	43	22	193	231
2000 Indian granite (belgian block)									
2100 Jumbo, 10-1/2" x 7-1/2" x 4", grey	D-1	150	.107	L.F.	1.95	3.56		5.51	7.55
2150 Pink		150	.107		2.54	3.56		6.10	8.20
2200 Regular, 9" x 4-1/2" x 4-1/2", grey		160	.100		1.77	3.34		5.11	7.05
2250 Pink		160	.100		2.43	3.34		5.77	7.75
2300 Cubes, 4" x 4" x 4", grey		175	.091		1.68	3.05		4.73	6.50
2350 Pink		175	.091		1.76	3.05		4.81	6.60
2400 6" x 6" x 6", pink	↓	155	.103	↓	4.37	3.44		7.81	10.05
2500 Alternate pricing method for indian granite									
2550 Jumbo, 10-1/2" x 7-1/2" x 4" (30 lb), grey				Ton	111			111	122
2600 Pink					147			147	162
2650 Regular, 9" x 4-1/2" x 4-1/2" (20 lb), grey					125			125	137
2700 Pink					170			170	187
2750 Cubes, 4" x 4" x 4" (5 lb), grey					204			204	225
2800 Pink					227			227	249
2850 6" x 6" x 6" (25 lb), pink					170			170	187
2900 For pallets, add				↓	18			18	19.80

32 17 Paving Specialties

32 17 13 – Parking Bumpers

32 17 13.13 Metal Parking Bumpers

	Crew	Daily Output	Labor-Hours	Unit	Material	Labor	Equipment	Total	Total Incl O&P
0010 **METAL PARKING BUMPERS**									
0015 Bumper rails for garages, 12 Ga. rail, 6" wide, with steel									
0020 posts 12'-6" O.C., minimum	E-4	190	.168	L.F.	14.20	7.05	.61	21.86	29
0030 Average		165	.194		17.75	8.10	.70	26.55	35
0100 Maximum		140	.229		21.50	9.55	.82	31.87	42
0300 12" channel rail, minimum		160	.200		17.75	8.35	.72	26.82	35.50
0400 Maximum	↓	120	.267	↓	26.50	11.15	.96	38.61	50.50
1300 Pipe bollards, conc filled/paint, 8' L x 4' D hole, 6" diam.	B-6	20	1.200	Ea.	325	37.50	12.15	374.65	425
1400 8" diam.	↓	15	1.600	↓	490	50.50	16.20	556.70	635

32 17 Paving Specialties

32 17 13 – Parking Bumpers

32 17 13.13 Metal Parking Bumpers

		Crew	Daily Output	Labor-Hours	Unit	Material	2007 Bare Costs Labor	Equipment	Total	Total Incl O&P
1500	12" diam.	B-6	12	2	Ea.	640	63	20.50	723.50	825
1592	Bollards, steel, 3' H, retractable, incl hydraulic controls, min		4	6		32,300	189	61	32,550	35,900
1594	Max		2	12		36,000	375	122	36,497	40,300
2030	Folding with individual padlocks	B-2	50	.800		785	23.50		808.50	900
8000	Parking lot control, see Div. 11 12 36.10									

32 17 13.16 Plastic Parking Bumpers

		Crew	Daily Output	Labor-Hours	Unit	Material	Labor	Equipment	Total	Total Incl O&P
0010	**PLASTIC PARKING BUMPERS**									
1200	Thermoplastic, 6" x 10" x 6'-0"	B-2	120	.333	Ea.	85.50	9.70		95.20	109

32 17 13.19 Precast Concrete Parking Bumpers

		Crew	Daily Output	Labor-Hours	Unit	Material	Labor	Equipment	Total	Total Incl O&P
0010	**PRECAST CONCRETE PARKING BUMPERS**									
1000	Wheel stops, precast concrete incl. dowels, 6" x 10" x 6'-0"	B-2	120	.333	Ea.	46.50	9.70		56.20	66
1100	8" x 13" x 6'-0"	"	120	.333	"	54	9.70		63.70	74

32 17 13.26 Wood Parking Bumpers

		Crew	Daily Output	Labor-Hours	Unit	Material	Labor	Equipment	Total	Total Incl O&P
0010	**WOOD PARKING BUMPERS**									
0020	Parking barriers, timber w/saddles, treated type									
0100	4" x 4" for cars	B-2	520	.077	L.F.	2.70	2.24		4.94	6.45
0200	6" x 6" for trucks		520	.077	"	5.65	2.24		7.89	9.70
0600	Flexible fixed stanchion, 2' high, 3" diameter		100	.400	Ea.	27.50	11.65		39.15	48

32 17 23 – Pavement Markings

32 17 23.13 Painted Pavement Markings

		Crew	Daily Output	Labor-Hours	Unit	Material	Labor	Equipment	Total	Total Incl O&P
0010	**PAINTED PAVEMENT MARKINGS**									
0020	Acrylic waterborne, white or yellow, 4" wide	B-78	20000	.002	L.F.	.19	.07	.02	.28	.34
0200	6" wide		11000	.004		.18	.13	.04	.35	.44
0500	8" wide		10000	.005		.25	.14	.04	.43	.53
0600	12" wide		4000	.012		.46	.35	.10	.91	1.15
0620	Arrows or gore lines		2300	.021	S.F.	.74	.61	.18	1.53	1.95
0640	Temporary paint, white or yellow		15000	.003	L.F.	.22	.09	.03	.34	.41
0660	Removal	1 Clab	300	.027			.77		.77	1.19
0680	Temporary tape	2 Clab	1500	.011		2.13	.31		2.44	2.82
0710	Thermoplastic, white or yellow, 4" wide	B-79	15000	.003		.85	.08	.06	.99	1.13
0730	6" wide		14000	.003		1.23	.08	.07	1.38	1.56
0740	8" wide		12000	.003		1.66	.10	.08	1.84	2.06
0750	12" wide		6000	.007		2.47	.19	.16	2.82	3.18
0760	Arrows		660	.061	S.F.	2.28	1.76	1.44	5.48	6.85
0770	Gore lines		2500	.016		1.51	.47	.38	2.36	2.80
0780	Letters		660	.061		1.89	1.76	1.44	5.09	6.40

32 17 23.14 Pavement Markings

		Crew	Daily Output	Labor-Hours	Unit	Material	Labor	Equipment	Total	Total Incl O&P
0010	**PAVEMENT MARKINGS**									
0790	Layout of pavement marking	A-2	25000	.001	L.F.		.03	.01	.04	.05
0800	Parking stall, paint, white	B-78	440	.109	Stall	3.95	3.17	.93	8.05	10.30
1000	Street letters and numbers	"	1600	.030	S.F.	.76	.87	.26	1.89	2.48

32 18 Athletic and Recreational Surfacing

32 18 13 – Synthetic Grass Surfacing

32 18 13.10 Synthetic Grass Surfacing		Crew	Daily Output	Labor-Hours	Unit	Material	2007 Bare Costs Labor	Equipment	Total	Total Incl O&P
0010	**SYNTHETIC GRASS SURFACING**									
0015	Not including asphalt base or drainage,									
0020	but including cushion pad, over 50,000 S.F.									
0200	1/2" pile and 5/16" cushion pad, standard	C-17	3200	.025	S.F.	8	.96		8.96	10.30
0300	Deluxe		2560	.031		9.45	1.20		10.65	12.25
0500	1/2" pile and 5/8" cushion pad, standard		2844	.028		11.65	1.08		12.73	14.50
0600	Deluxe		2327	.034		12.75	1.32		14.07	16.05
0800	For asphaltic concrete base, 2-1/2" thick,									
0900	with 6" crushed stone sub-base, add	B-25	12000	.007	S.F.	1.56	.23	.19	1.98	2.28

32 18 23 – Athletic Surfacing

32 18 23.33 Running Track Surfacing

		Crew	Daily Output	Labor-Hours	Unit	Material	Labor	Equipment	Total	Total Incl O&P
0010	**RUNNING TRACK SURFACING**									
0020	Running track, asphalt, incl base, 3" thick	B-37	300	.160	S.Y.	16.60	4.87	.41	21.88	26.50
0100	Surface, latex rubber system, 3/8" thick, black	B-20	125	.192		7.45	6.25		13.70	17.90
0150	Colors		125	.192		13.25	6.25		19.50	24.50
0300	Urethane rubber system, 3/8" thick, black		120	.200		19.90	6.50		26.40	32
0400	Color coating		115	.209		23.50	6.80		30.30	36.50

32 18 23.53 Tennis Court Surfacing

		Crew	Daily Output	Labor-Hours	Unit	Material	Labor	Equipment	Total	Total Incl O&P
0010	**TENNIS COURT SURFACING**									
0020	Tennis court, asphalt, incl. base, 2-1/2" thick, one court	B-37	450	.107	S.Y.	20	3.25	.27	23.52	27.50
0200	Two courts		675	.071		9.10	2.16	.18	11.44	13.55
0300	Clay courts		360	.133		33	4.06	.34	37.40	42.50
0400	Pulverized natural greenstone with 4" base, fast dry		250	.192		31.50	5.85	.49	37.84	44
0800	Rubber-acrylic base resilient pavement		600	.080		39.50	2.43	.20	42.13	47.50
1000	Colored sealer, acrylic emulsion, 3 coats	2 Clab	800	.020		4.79	.58		5.37	6.15
1100	3 coat, 2 colors	"	900	.018		6.65	.51		7.16	8.10
1200	For preparing old courts, add	1 Clab	825	.010			.28		.28	.43
1400	Posts for nets, 3-1/2" diameter with eye bolts	B-1	3.40	7.059	Pr.	189	208		397	535
1500	With pulley & reel		3.40	7.059	"	269	208		477	620
1700	Net, 42' long, nylon thread with binder		50	.480	Ea.	249	14.10		263.10	296
1800	All metal		6.50	3.692	"	440	109		549	655
2000	Paint markings on asphalt, 2 coats	1 Pord	1.78	4.494	Court	80	147		227	310
2200	Complete court with fence, etc., asphaltic conc., minimum	B-37	.20	240		18,100	7,300	610	26,010	31,900
2300	Maximum		.16	300		22,500	9,125	765	32,390	39,600
2800	Clay courts, minimum		.20	240		19,600	7,300	610	27,510	33,500
2900	Maximum		.16	300		23,500	9,125	765	33,390	40,700

32 31 Fences and Gates

32 31 13 – Chain Link Fences and Gates

32 31 13.20 Fence, Chain Link Industrial

		Crew	Daily Output	Labor-Hours	Unit	Material	Labor	Equipment	Total	Total Incl O&P
0010	**FENCE, CHAIN LINK INDUSTRIAL**, schedule 40									
0020	3 strands barb wire, 2" post @ 10' O.C., set in concrete, 6' H									
0200	9 ga. wire, galv. steel	B-80C	240	.100	L.F.	14	2.87	.61	17.48	20.50
0300	Aluminized steel		240	.100		17.95	2.87	.61	21.43	25
0500	6 ga. wire, galv. steel		240	.100		22	2.87	.61	25.48	29
0600	Aluminized steel		240	.100		25	2.87	.61	28.48	32.50
0800	6 ga. wire, 6' high but omit barbed wire, galv. steel		250	.096		21	2.75	.59	24.34	28.50
0900	Aluminized steel		250	.096		29.50	2.75	.59	32.84	37.50
0920	8' H, 6 ga. wire, 2-1/2" line post, galv. steel		180	.133		34	3.82	.82	38.64	44
0940	Aluminized steel		180	.133		41.50	3.82	.82	46.14	52.50

32 31 Fences and Gates

32 31 13 – Chain Link Fences and Gates

		Crew	Daily Output	Labor-Hours	Unit	Material	2007 Bare Costs Labor	Equipment	Total	Total Incl O&P
32 31 13.20 Fence, Chain Link Industrial										
1100	Add for corner posts, 3" diam., galv. steel	B-80C	40	.600	Ea.	96.50	17.20	3.68	117.38	137
1200	Aluminized steel		40	.600		116	17.20	3.68	136.88	159
1300	Add for braces, galv. steel		80	.300		26.50	8.60	1.84	36.94	44.50
1350	Aluminized steel		80	.300		35	8.60	1.84	45.44	54
1400	Gate for 6' high fence, 1-5/8" frame, 3' wide, galv. steel		10	2.400		154	69	14.75	237.75	293
1500	Aluminized steel		10	2.400		190	69	14.75	273.75	330
2000	5'-0" high fence, 9 ga., no barbed wire, 2" line post,									
2010	10' O.C., 1-5/8" top rail									
2100	Galvanized steel	B-80C	300	.080	L.F.	12.20	2.29	.49	14.98	17.50
2200	Aluminized steel		300	.080	"	14.30	2.29	.49	17.08	19.85
2400	Gate, 4' wide, 5' high, 2" frame, galv. steel		10	2.400	Ea.	176	69	14.75	259.75	315
2500	Aluminized steel		10	2.400	"	197	69	14.75	280.75	340
3100	Overhead slide gate, chain link, 6' high, to 18' wide		38	.632	L.F.	155	18.10	3.88	176.98	202
3110	Cantilever type	B-80	48	.667		66.50	21	11.65	99.15	118
3120	8' high		24	1.333		96.50	41.50	23.50	161.50	196
3130	10' high		18	1.778		114	55.50	31	200.50	246
5000	Double swing gates, incl. posts & hardware									
5010	5' high, 12' opening	B-80C	3.40	7.059	Opng.	470	202	43.50	715.50	885
5020	20' opening		2.80	8.571		640	246	52.50	938.50	1,150
5060	6' high, 12' opening		3.20	7.500		795	215	46	1,056	1,250
5070	20' opening		2.60	9.231		1,100	265	56.50	1,421.50	1,675
5080	8' high, 12' opening	B-80	2.13	15.002		1,225	470	263	1,958	2,350
5090	20' opening		1.45	22.069		1,625	690	385	2,700	3,250
5100	10' high, 12' opening		1.31	24.427		1,400	765	430	2,595	3,200
5110	20' opening		1.03	31.068		2,100	970	545	3,615	4,425
5120	12' high, 12' opening		1.05	30.476		2,050	950	535	3,535	4,300
5130	20' opening		.85	37.647		2,625	1,175	660	4,460	5,425
5190	For aluminized steel add					20%				
32 31 13.25 Fence, Chain Link Residential										
0010	**FENCE, CHAIN LINK RESIDENTIAL**, sch. 20, 11 ga. wire, 1-5/8" post									
0020	10' O.C., 1-3/8" top rail, 2" corner post, galv. stl. 3' high	B-80C	500	.048	L.F.	4.21	1.38	.29	5.88	7.10
0050	4' high		400	.060		4.77	1.72	.37	6.86	8.35
0100	6' high		200	.120		5.80	3.44	.74	9.98	12.55
0150	Add for gate 3' wide, 1-3/8" frame, 3' high		12	2	Ea.	63	57.50	12.30	132.80	172
0170	4' high		10	2.400		72.50	69	14.75	156.25	203
0190	6' high		10	2.400		99	69	14.75	182.75	232
0200	Add for gate 4' wide, 1-3/8" frame, 3' high		9	2.667		71	76.50	16.35	163.85	215
0220	4' high		9	2.667		80	76.50	16.35	172.85	225
0240	6' high		8	3		94	86	18.40	198.40	258
0350	Aluminized steel, 11 ga. wire, 3' high		500	.048	L.F.	5.55	1.38	.29	7.22	8.55
0380	4' high		400	.060		7.15	1.72	.37	9.24	10.95
0400	6' high		200	.120		10.05	3.44	.74	14.23	17.20
0450	Add for gate 3' wide, 1-3/8" frame, 3' high		12	2	Ea.	77	57.50	12.30	146.80	187
0470	4' high		10	2.400		126	69	14.75	209.75	262
0490	6' high		10	2.400		159	69	14.75	242.75	298
0500	Add for gate 4' wide, 1-3/8" frame, 3' high		10	2.400		87	69	14.75	170.75	219
0520	4' high		9	2.667		96.50	76.50	16.35	189.35	243
0540	6' high		8	3		111	86	18.40	215.40	278
0620	Vinyl covered, 9 ga. wire, 3' high		500	.048	L.F.	4.54	1.38	.29	6.21	7.45
0640	4' high		400	.060		5.30	1.72	.37	7.39	8.95
0660	6' high		200	.120		6.75	3.44	.74	10.93	13.55
0720	Add for gate 3' wide, 1-3/8" frame, 3' high		12	2	Ea.	88	57.50	12.30	157.80	199

32 31 Fences and Gates

32 31 13 – Chain Link Fences and Gates

32 31 13.25 Fence, Chain Link Residential

		Crew	Daily Output	Labor-Hours	Unit	Material	2007 Bare Costs Labor	Equipment	Total	Total Incl O&P
0740	4' high	B-80C	10	2.400	Ea.	99	69	14.75	182.75	232
0760	6' high		10	2.400		154	69	14.75	237.75	292
0780	Add for gate 4' wide, 1-3/8" frame, 3' high		10	2.400		101	69	14.75	184.75	234
0800	4' high		9	2.667		110	76.50	16.35	202.85	258
0820	6' high		8	3		124	86	18.40	228.40	291

32 31 13.26 Tennis Court Fences and Gates

		Crew	Daily Output	Labor-Hours	Unit	Material	Labor	Equipment	Total	Total Incl O&P
0010	**TENNIS COURT FENCES AND GATES**									
0860	Tennis courts, 11 ga. wire, 2-1/2" post set									
0870	in concrete, 10' O.C., 1-5/8" top rail									
0900	10' high	B-80	190	.168	L.F.	13.85	5.25	2.95	22.05	26.50
0920	12' high		170	.188	"	16.20	5.90	3.30	25.40	30.50
1000	Add for gate 4' wide, 1-5/8" frame 7' high		10	3.200	Ea.	129	100	56	285	360
1040	Aluminized steel, 11 ga. wire 10' high		190	.168	L.F.	16.75	5.25	2.95	24.95	29.50
1100	12' high		170	.188	"	20.50	5.90	3.30	29.70	35
1140	Add for gate 4' wide, 1-5/8" frame, 7' high		10	3.200	Ea.	151	100	56	307	380
1250	Vinyl covered, 9 ga. wire, 10' high		190	.168	L.F.	16.20	5.25	2.95	24.40	29
1300	12' high		170	.188	"	27	5.90	3.30	36.20	42
1310	Fence, CL, tennis ct, transom gate, single, galv, 4' x 7' x 3'	B-80A	8.72	2.752	Ea.	279	79	21.50	379.50	450
1400	Add for gate 4' wide, 1-5/8" frame, 7' high	B-80	10	3.200	"	150	100	56	306	380

32 31 13.33 Chain Link Backstops

		Crew	Daily Output	Labor-Hours	Unit	Material	Labor	Equipment	Total	Total Incl O&P
0010	**CHAIN LINK BACKSTOPS**									
0015	Backstops, baseball, prefabricated, 30' wide, 12' high & 1 overhang	B-1	1	24	Ea.	2,925	705		3,630	4,300
0100	40' wide, 12' high & 2 overhangs	"	.75	32		3,650	940		4,590	5,500
0300	Basketball, steel, single goal	B-13	3.04	18.421		1,150	580	243	1,973	2,400
0400	Double goal	"	1.92	29.167		695	915	385	1,995	2,600
0600	Tennis, wire mesh with pair of ends	B-1	2.48	9.677	Set	1,800	285		2,085	2,425
0700	Enclosed court	"	1.30	18.462	Ea.	5,100	545		5,645	6,475
0900	Handball or squash court, outdoor, wood	2 Carp	.50	32		4,200	1,175		5,375	6,450
1000	Masonry handball/squash court	D-1	.30	53.333		23,700	1,775		25,475	28,800

32 31 13.53 High-Security Chain Link Fences and Gates

		Crew	Daily Output	Labor-Hours	Unit	Material	Labor	Equipment	Total	Total Incl O&P
0010	**HIGH-SECURITY CHAIN LINK FENCES AND GATES**									
0100	Fence, chain link, security, 7' high, no gates/signs	B-80C	480	.050	L.F.	47	1.43	.31	48.74	54
0200	Fence, barbed wire, security, 7' high, with 3 wire barbed wire arm	"	400	.060	"	6.40	1.72	.37	8.49	10.10
0300	Complete systems, including material and installation									
0310	Taunt wire fence detection system				M.L.F.				18,000	19,800
0410	Microwave fence detection system								30,000	33,000
0510	Passive magnetic fence detection system								14,400	15,840
0610	Infrared fence detection system								9,600	10,560
0710	Strain relief fence detection system								18,000	19,800
0810	Electro-shock fence detection system								26,400	29,040
0910	Photo-electric fence detection system								12,000	13,200

32 31 19 – Decorative Metal Fences and Gates

32 31 19.10 Decorative Fence

		Crew	Daily Output	Labor-Hours	Unit	Material	Labor	Equipment	Total	Total Incl O&P
0010	**DECORATIVE FENCE**									
5300	Tubular picket, steel, 6' sections, 1-9/16" posts, 4' high	B-80C	300	.080	L.F.	26.50	2.29	.49	29.28	33.50
5400	2" posts, 5' high		240	.100		37	2.87	.61	40.48	46
5600	2" posts, 6' high		200	.120		42	3.44	.74	46.18	52
5700	Staggered picket 1-9/16" posts, 4' high		300	.080		24	2.29	.49	26.78	30.50
5800	2" posts, 5' high		240	.100		39.50	2.87	.61	42.98	48.50
5900	2" posts, 6' high		200	.120		41	3.44	.74	45.18	51
6200	Gates, 4' high, 3' wide	B-1	10	2.400	Ea.	233	70.50		303.50	365

32 31 Fences and Gates

32 31 19 – Decorative Metal Fences and Gates

32 31 19.10 Decorative Fence		Crew	Daily Output	Labor-Hours	Unit	Material	2007 Bare Costs Labor	Equipment	Total	Total Incl O&P
6300	5' high, 3' wide	B-1	10	2.400	Ea.	300	70.50		370.50	440
6400	6' high, 3' wide		10	2.400		310	70.50		380.50	450
6500	4' wide		10	2.400		360	70.50		430.50	510

32 31 26 – Wire Fences and Gates

32 31 26.10 Fences, Misc. Metal

		Crew	Daily Output	Labor-Hours	Unit	Material	Labor	Equipment	Total	Total Incl O&P
0010	**FENCES, MISC. METAL**									
0012	Chicken wire, posts @ 4', 1" mesh, 4' high	B-80C	410	.059	L.F.	1.90	1.68	.36	3.94	5.10
0100	2" mesh, 6' high		350	.069		1.72	1.97	.42	4.11	5.40
0200	Galv. steel, 12 ga., 2" x 4" mesh, posts 5' O.C., 3' high		300	.080		2.58	2.29	.49	5.36	6.95
0300	5' high		300	.080		3.45	2.29	.49	6.23	7.90
0400	14 ga., 1" x 2" mesh, 3' high		300	.080		2.74	2.29	.49	5.52	7.10
0500	5' high		300	.080		3.79	2.29	.49	6.57	8.25
1000	Kennel fencing, 1-1/2" mesh, 6' long, 3'-6" wide, 6'-2" high	2 Clab	4	4	Ea.	430	115		545	655
1050	12' long		4	4		515	115		630	750
1200	Top covers, 1-1/2" mesh, 6' long		15	1.067		87.50	30.50		118	144
1250	12' long		12	1.333		140	38.50		178.50	214
1300	For kennel doors, see Division 08 31 13.40									
4500	Security fence, prison grade, set in concrete, 12' high	B-80	25	1.280	L.F.	38	40	22.50	100.50	128
4600	16' high	"	20	1.600	"	45.50	50	28	123.50	158

32 31 26.20 Wire Fencing, General

		Crew	Daily Output	Labor-Hours	Unit	Material	Labor	Equipment	Total	Total Incl O&P
0010	**WIRE FENCING, GENERAL**									
0015	Barbed wire, galvanized, domestic steel, hi-tensile 15-1/2 ga.				M.L.F.	31.50			31.50	35
0020	Standard, 12-3/4 ga.					42.50			42.50	46.50
0210	Barbless wire, 2-strand galvanized, 12-1/2 ga.					42.50			42.50	46.50
0500	Helical razor ribbon, stainless steel, 18" dia x 18" spacing				C.L.F.	123			123	135
0600	Hardware cloth galv., 1/4" mesh, 23 ga., 2' wide				C.S.F.	55			55	60.50
0700	3' wide					53.50			53.50	59
0900	1/2" mesh, 19 ga., 2' wide					48.50			48.50	53.50
1000	4' wide					48			48	52.50
1200	Chain link fabric, steel, 2" mesh, 6 ga, galvanized					182			182	201
1300	9 ga, galvanized					90			90	99
1350	Vinyl coated					75.50			75.50	83
1360	Aluminized					117			117	129
1400	2-1/4" mesh, 11.5 ga, galvanized					61			61	67
1600	1-3/4" mesh (tennis courts), 11.5 ga (core), vinyl coated					86			86	95
1700	9 ga, galvanized					77.50			77.50	85.50
2100	Welded wire fabric, galvanized, 1" x 2", 14 ga.					51.50			51.50	57
2200	2" x 4", 12-1/2 ga.					17.60			17.60	19.40

32 31 29 – Wood Fences and Gates

32 31 29.20 Fence, Wood Rail

		Crew	Daily Output	Labor-Hours	Unit	Material	Labor	Equipment	Total	Total Incl O&P
0010	**FENCE, WOOD RAIL**									
0012	Picket, No. 2 cedar, Gothic, 2 rail, 3' high	B-1	160	.150	L.F.	5.70	4.41		10.11	13.10
0050	Gate, 3'-6" wide	B-80C	9	2.667	Ea.	49	76.50	16.35	141.85	191
0400	3 rail, 4' high		150	.160	L.F.	6.55	4.59	.98	12.12	15.40
0500	Gate, 3'-6" wide		9	2.667	Ea.	59	76.50	16.35	151.85	202
0600	Open rail, rustic, No. 1 cedar, 2 rail, 3' high		160	.150	L.F.	5.10	4.30	.92	10.32	13.30
0650	Gate, 3' wide		9	2.667	Ea.	56.50	76.50	16.35	149.35	200
0700	3 rail, 4' high		150	.160	L.F.	5.80	4.59	.98	11.37	14.60
0900	Gate, 3' wide		9	2.667	Ea.	73	76.50	16.35	165.85	218
1200	Stockade, No. 2 cedar, treated wood rails, 6' high		160	.150	L.F.	7.10	4.30	.92	12.32	15.50
1250	Gate, 3' wide		9	2.667	Ea.	58	76.50	16.35	150.85	201

32 31 Fences and Gates

32 31 29 – Wood Fences and Gates

32 31 29.20 Fence, Wood Rail

		Crew	Daily Output	Labor-Hours	Unit	Material	2007 Bare Costs Labor	Equipment	Total	Total Incl O&P
1300	No. 1 cedar, 3-1/4" cedar rails, 6' high	B-80C	160	.150	L.F.	17.80	4.30	.92	23.02	27.50
1500	Gate, 3' wide		9	2.667	Ea.	148	76.50	16.35	240.85	300
1520	Open rail, split, No. 1 cedar, 2 rail, 3' high		160	.150	L.F.	5.20	4.30	.92	10.42	13.40
1540	3 rail, 4'-0" high		150	.160		6.75	4.59	.98	12.32	15.60
3300	Board, shadow box, 1" x 6", treated pine, 6' high		160	.150		10.50	4.30	.92	15.72	19.25
3400	No. 1 cedar, 6' high		150	.160		20.50	4.59	.98	26.07	31
3900	Basket weave, No. 1 cedar, 6' high		160	.150		20.50	4.30	.92	25.72	30
3950	Gate, 3'-6" wide	B-1	8	3	Ea.	140	88.50		228.50	291
4000	Treated pine, 6' high		150	.160	L.F.	12.15	4.71		16.86	20.50
4200	Gate, 3'-6" wide		9	2.667	Ea.	65.50	78.50		144	194
5000	Fence rail, redwood, 2" x 4", merch grade 8'		2400	.010	L.F.	1.11	.29		1.40	1.68
5050	Select grade, 8'		2400	.010	"	3.44	.29		3.73	4.24
6000	Fence post, select redwood, earthpacked & treated, 4" x 4" x 6'		96	.250	Ea.	11	7.35		18.35	23.50
6010	4" x 4" x 8'		96	.250		13.10	7.35		20.45	26
6020	Set in concrete, 4" x 4" x 6'		50	.480		14.20	14.10		28.30	37.50
6030	4" x 4" x 8'		50	.480		16.95	14.10		31.05	40.50
6040	Wood post, 4' high, set in concrete, incl. concrete		50	.480		8.25	14.10		22.35	31
6050	Earth packed		96	.250		5.65	7.35		13	17.65
6060	6' high, set in concrete, incl. concrete		50	.480		10.60	14.10		24.70	33.50
6070	Earth packed		96	.250		7.70	7.35		15.05	19.95

32 32 Retaining Walls

32 32 13 – Cast-In-Place Concrete Retaining Walls

32 32 13.10 Cast-In-Place Retaining Walls

		Crew	Daily Output	Labor-Hours	Unit	Material	Labor	Equipment	Total	Total Incl O&P
0010	**CAST-IN-PLACE RETAINING WALLS**									
1800	Concrete gravity wall with vertical face including excavation & backfill									
1850	No reinforcing									
1900	6' high, level embankment	C-17C	36	2.306	L.F.	76	88.50	12.65	177.15	235
2000	33° slope embankment		32	2.594		69.50	99.50	14.20	183.20	247
2200	8' high, no surcharge		27	3.074		94.50	118	16.85	229.35	305
2300	33° slope embankment		24	3.458		114	133	18.95	265.95	355
2500	10' high, level embankment		19	4.368		135	168	24	327	435
2600	33° slope embankment		18	4.611		187	177	25.50	389.50	510
2800	Reinforced concrete cantilever, incl. excavation, backfill & reinf.									
2900	6' high, 33° slope embankment	C-17C	35	2.371	L.F.	69.50	91	13	173.50	233
3000	8' high, 33° slope embankment		29	2.862		80	110	15.70	205.70	276
3100	10' high, 33° slope embankment		20	4.150		104	160	22.50	286.50	385
3200	20' high, 500 lb. per L.F. surcharge		7.50	11.067		310	425	60.50	795.50	1,075
3500	Concrete cribbing, incl. excavation and backfill									
3700	12' high, open face	B-13	210	.267	S.F.	34.50	8.35	3.52	46.37	55
3900	Closed face	"	210	.267	"	32.50	8.35	3.52	44.37	53
4100	Concrete filled slurry trench, see Div. 31 56 23.20									

32 32 23 – Segmental Retaining Walls

32 32 23.13 Segmental Conc. Unit Masonry Retaining Walls

		Crew	Daily Output	Labor-Hours	Unit	Material	Labor	Equipment	Total	Total Incl O&P
0010	**SEGMENTAL CONC. UNIT MASONRY RETAINING WALLS**									
7100	Segmental Retaining Wall system, incl pins, and void fill									
7120	base not included									
7140	Large unit, 8" high x 18" wide x 20" deep, 3 plane split	B-62	300	.080	S.F.	9.35	2.52	.41	12.28	14.60
7150	Straight split		300	.080		9.35	2.52	.41	12.28	14.60
7160	Medium, ltwt, 8" high x 18" wide x 12" deep, 3 plane split		400	.060		10.45	1.89	.30	12.64	14.75

32 32 Retaining Walls

32 32 23 – Segmental Retaining Walls

32 32 23.13 Segmental Conc. Unit Masonry Retaining Walls

		Crew	Daily Output	Labor-Hours	Unit	Material	2007 Bare Costs Labor	2007 Bare Costs Equipment	Total	Total Incl O&P
7170	Straight split	B-62	400	.060	S.F.	7.85	1.89	.30	10.04	11.90
7180	Small unit, 4" x 18" x 10" deep, 3 plane split		400	.060		11.15	1.89	.30	13.34	15.50
7190	Straight split		400	.060		11.75	1.89	.30	13.94	16.20
7200	Cap unit, 3 plane split		300	.080		13.20	2.52	.41	16.13	18.80
7210	Cap unit, st split		300	.080		13.20	2.52	.41	16.13	18.80
7260	For reinforcing, add								2.64	3.30
8000	For higher walls, add components as necessary									

32 32 26 – Metal Crib Retaining Walls

32 32 26.10 Metal Bin Retaining Walls

		Crew	Daily Output	Labor-Hours	Unit	Material	Labor	Equipment	Total	Total Incl O&P
0010	**METAL BIN RETAINING WALLS**, Aluminized steel bin, excavation									
0020	and backfill not included, 10' wide									
0100	4' high, 5.5' deep	B-13	650	.086	S.F.	19.40	2.70	1.14	23.24	27
0200	8' high, 5.5' deep		615	.091		22.50	2.85	1.20	26.55	30
0300	10' high, 7.7' deep		580	.097		23.50	3.03	1.28	27.81	32
0400	12' high, 7.7' deep		530	.106		25.50	3.31	1.40	30.21	34.50
0500	16' high, 7.7' deep		515	.109		26.50	3.41	1.44	31.35	36.50
0600	16' high, 9.9' deep		500	.112		28	3.51	1.48	32.99	38
0700	20' high, 9.9' deep		470	.119		31.50	3.74	1.57	36.81	42.50
0800	20' high, 12.1' deep		460	.122		34	3.82	1.61	39.43	45
0900	24' high, 12.1' deep		455	.123		36	3.86	1.63	41.49	47
1000	24' high, 14.3' deep		450	.124		37.50	3.90	1.64	43.04	49.50
1100	28' high, 14.3' deep		440	.127		39	3.99	1.68	44.67	51
1300	For plain galvanized bin type walls, deduct					10%				

32 32 36 – Gabion Retaining Walls

32 32 36.10 Stone Gabion Retaining Walls

		Crew	Daily Output	Labor-Hours	Unit	Material	Labor	Equipment	Total	Total Incl O&P
0010	**STONE GABION RETAINING WALLS**									
4300	Stone filled gabions, not incl. excavation,									
4310	Stone, delivered, 3' wide									
4350	Galvanized, 6' high, 33° slope embankment	B-13	49	1.143	L.F.	20.50	36	15.10	71.60	94
4500	Highway surcharge		27	2.074		43.50	65	27.50	136	178
4600	9' high, up to 33° slope embankment		24	2.333		46	73	31	150	198
4700	Highway surcharge		16	3.500		76	110	46	232	305
4900	12' high, up to 33° slope embankment		14	4		71.50	125	53	249.50	330
5000	Highway surcharge		11	5.091		108	160	67.50	335.50	440
5950	For PVC coating, add					20%				

32 32 60 – Stone Retaining Walls

32 32 60.10 Stone Retaining Walls

		Crew	Daily Output	Labor-Hours	Unit	Material	Labor	Equipment	Total	Total Incl O&P
0010	**STONE RETAINING WALLS**									
0015	Including excavation, concrete footing and									
0020	stone 3' below grade. Price is exposed face area.									
0200	Decorative random stone, to 6' high, 1'-6" thick, dry set	D-1	35	.457	S.F.	43	15.25		58.25	70
0300	Mortar set		40	.400		43	13.35		56.35	67.50
0500	Cut stone, to 6' high, 1'-6" thick, dry set		35	.457		43	15.25		58.25	70
0600	Mortar set		40	.400		43	13.35		56.35	67.50
0800	Random stone, 6' to 10' high, 2' thick, dry set		45	.356		43	11.85		54.85	65
0900	Mortar set		50	.320		43	10.65		53.65	63.50
1100	Cut stone, 6' to 10' high, 2' thick, dry set		45	.356		43	11.85		54.85	65
1200	Mortar set		50	.320		43	10.65		53.65	63.50

32 34 Fabricated Bridges

32 34 20 – Fabricated Pedestrian Bridges

32 34 20.10 Bridges, Pedestrian

		Crew	Daily Output	Labor-Hours	Unit	Material	2007 Bare Costs Labor	Equipment	Total	Total Incl O&P
0010	**BRIDGES, PEDESTRIAN** spans over streams, roadways, etc.									
0020	including erection, not including foundations									
0050	Precast concrete, complete in place, 8' wide, 60' span	E-2	215	.260	S.F.	56	10.50	7.20	73.70	87.50
0100	100' span		185	.303		61.50	12.20	8.35	82.05	97.50
0150	120' span		160	.350		66.50	14.10	9.65	90.25	109
0200	150' span		145	.386		69.50	15.60	10.65	95.75	115
0300	Steel, trussed or arch spans, compl. in place, 8' wide, 40' span		320	.175		73	7.05	4.84	84.89	97.50
0400	50' span		395	.142		65.50	5.70	3.92	75.12	86
0500	60' span		465	.120		65.50	4.86	3.33	73.69	84
0600	80' span		570	.098		78	3.96	2.71	84.67	95.50
0700	100' span		465	.120		109	4.86	3.33	117.19	132
0800	120' span		365	.153		138	6.20	4.24	148.44	167
0900	150' span		310	.181		147	7.30	4.99	159.29	179
1000	160' span		255	.220		147	8.85	6.05	161.90	183
1100	10' wide, 80' span		640	.088		81	3.53	2.42	86.95	98
1200	120' span		415	.135		105	5.45	3.73	114.18	129
1300	150' span		445	.126		118	5.10	3.48	126.58	142
1400	200' span		205	.273		125	11	7.55	143.55	165
1600	Wood, laminated type, complete in place, 80' span	C-12	203	.236		48	8.55	3.57	60.12	70
1700	130' span	"	153	.314		50	11.35	4.73	66.08	78

32 35 Screening Devices

32 35 16 – Sound Barriers

32 35 16.10 Traffic Barriers, Highway Sound Barriers

		Crew	Daily Output	Labor-Hours	Unit	Material	Labor	Equipment	Total	Total Incl O&P
0010	**TRAFFIC BARRIERS, HIGHWAY SOUND BARRIERS**									
0020	Highway sound barriers, not including footing									
0100	Precast concrete, concrete columns @ 30' OC, 8" T, 8' H	C-12	400	.120	L.F.	103	4.35	1.81	109.16	122
0110	12' H		265	.181		155	6.55	2.73	164.28	183
0120	16' H		200	.240		206	8.70	3.62	218.32	244
0130	20' H		160	.300		258	10.85	4.52	273.37	305
0400	Lt. Wt. composite panel, cementitious face, St. posts @ 12' OC, 8' H	B-80B	190	.168		118	5.20	1.43	124.63	139
0410	12' H		125	.256		176	7.90	2.17	186.07	209
0420	16' H		95	.337		235	10.35	2.86	248.21	278
0430	20' H		75	.427		294	13.15	3.62	310.77	350

32 84 Planting Irrigation

32 84 23 – Underground Sprinklers

32 84 23.10 Sprinkler Irrigation System

		Crew	Daily Output	Labor-Hours	Unit	Material	Labor	Equipment	Total	Total Incl O&P
0010	**SPRINKLER IRRIGATION SYSTEM** For lawns									
0100	Golf course with fully automatic system	C-17	.05	1600	9 holes	90,500	61,500		152,000	195,000
0200	24' diam. head at 15' O.C incl. piping, auto oper., minimum	B-20	70	.343	Head	19.90	11.15		31.05	39.50
0300	Maximum		40	.600		46	19.50		65.50	81
0500	60' diam. head at 40' O.C. incl. piping, auto oper., minimum		28	.857		60	28		88	110
0600	Maximum		23	1.043		169	34		203	238
0800	Residential system, custom, 1" supply		2000	.012	S.F.	.29	.39		.68	.93
0900	1-1/2" supply		1800	.013	"	.35	.43		.78	1.06
1020	Pop up spray head w/risers, hi-pop, full circle pattern, 4"	2 Skwk	76	.211	Ea.	3.59	8		11.59	16.40
1030	1/2 circle pattern		76	.211		3.67	8		11.67	16.50
1040	6", full circle pattern		76	.211		9.20	8		17.20	22.50

32 84 Planting Irrigation

32 84 23 – Underground Sprinklers

32 84 23.10 Sprinkler Irrigation System	Crew	Daily Output	Labor-Hours	Unit	Material	2007 Bare Costs Labor	Equipment	Total	Total Incl O&P
1050 1/2 circle pattern	2 Skwk	76	.211	Ea.	9.20	8		17.20	22.50
1060 12", full circle pattern		76	.211		13.15	8		21.15	27
1070 1/2 circle pattern		76	.211		13.20	8		21.20	27
1080 Pop up bubbler head w/risers, hi-pop bubbler head, 4"		76	.211		3.75	8		11.75	16.60
1090 6"		76	.211		9.20	8		17.20	22.50
1100 12"		76	.211		13.15	8		21.15	27
1110 Impact full/part circle sprinklers, 28'-54' 25-60 PSI		37	.432		12.25	16.45		28.70	39
1120 Spaced 37'-49' @ 25-50 PSI		37	.432		27	16.45		43.45	55
1130 Spaced 43'-61' @ 30-60 PSI		37	.432		48.50	16.45		64.95	79
1140 Spaced 54'-78' @ 40-80 PSI		37	.432		112	16.45		128.45	149
1145 Impact rotor pop-up full/part commercial circle sprinklers									
1150 Spaced 42'-65' 35-80 PSI	2 Skwk	25	.640	Ea.	24.50	24.50		49	65
1160 Spaced 48'-76' 45-85 PSI	"	25	.640	"	10.20	24.50		34.70	49.50
1165 Impact rotor pop-up part. circle comm., 53'-75', 55-100 PSI, w/ acc									
1170 Plastic case, metal cover	2 Skwk	25	.640	Ea.	171	24.50		195.50	227
1180 Rubber cover		25	.640		129	24.50		153.50	180
1190 Iron case, metal cover		22	.727		157	27.50		184.50	216
1200 Rubber cover		22	.727		170	27.50		197.50	230
1250 Plastic case, 2 nozzle, metal cover		25	.640		125	24.50		149.50	176
1260 Rubber cover		25	.640		133	24.50		157.50	184
1270 Iron case, 2 nozzle, metal cover		22	.727		182	27.50		209.50	243
1280 Rubber cover		22	.727		186	27.50		213.50	248
1282 Impact rotor pop-up full circle comm., 39'-99', 30-100 PSI									
1284 Plastic case, metal cover	2 Skwk	25	.640	Ea.	128	24.50		152.50	179
1286 Rubber cover		25	.640		142	24.50		166.50	194
1288 Iron case, metal cover		22	.727		184	27.50		211.50	246
1290 Rubber cover		22	.727		194	27.50		221.50	256
1292 Plastic case, 2 nozzle, metal cover		22	.727		138	27.50		165.50	195
1294 Rubber cover		22	.727		143	27.50		170.50	200
1296 Iron case, 2 nozzle, metal cover		20	.800		178	30.50		208.50	243
1298 Rubber cover		20	.800		190	30.50		220.50	257
1305 Electric remote control valve, plastic, 3/4"		18	.889		19.35	34		53.35	74
1310 1"		18	.889		34	34		68	90
1320 1-1/2"		18	.889		65.50	34		99.50	125
1335 Quick coupling valves, brass, locking cover									
1340 Inlet coupling valve, 3/4"	2 Skwk	18.75	.853	Ea.	47.50	32.50		80	103
1350 1"		18.75	.853		57.50	32.50		90	114
1360 Controller valve boxes, 6" round boxes		18.75	.853		4.22	32.50		36.72	55
1370 10" round boxes		14.25	1.123		9.25	42.50		51.75	76.50
1380 12" square box		9.75	1.641		18.60	62.50		81.10	118
1388 Electromech. control, 14 day 3-60 min, auto start to 23/day									
1390 4 station	2 Skwk	1.04	15.385	Ea.	201	585		786	1,125
1400 7 station		.64	25		260	950		1,210	1,750
1410 12 station		.40	40		320	1,525		1,845	2,725
1420 Dual programs, 18 station		.24	66.667		1,600	2,525		4,125	5,700
1430 23 station		.16	100		2,075	3,800		5,875	8,200
1435 Backflow preventer, bronze, 0-175 PSI, w/valves, test cocks									
1440 3/4"	2 Skwk	2	8	Ea.	131	305		436	620
1450 1"		2	8		136	305		441	625
1460 1-1/2"		2	8		270	305		575	775
1470 2"		2	8		283	305		588	785
1475 Pressure vacuum breaker, brass, 15-150 PSI									
1480 3/4"	2 Skwk	2	8	Ea.	79	305		384	560

32 84 Planting Irrigation

32 84 23 – Underground Sprinklers

32 84 23.10 Sprinkler Irrigation System		Crew	Daily Output	Labor-Hours	Unit	Material	2007 Bare Costs Labor	Equipment	Total	Total Incl O&P
1490	1"	2 Skwk	2	8	Ea.	103	305		408	590
1500	1-1/2"		2	8		204	305		509	700
1510	2"		2	8		189	305		494	685

32 91 Planting Preparation

32 91 13 – Soil Preparation

32 91 13.16 Mulching

		Crew	Daily Output	Labor-Hours	Unit	Material	Labor	Equipment	Total	Total Incl O&P
0010	**MULCHING**									
0100	Aged barks, 3" deep, hand spread	1 Clab	100	.080	S.Y.	2.46	2.30		4.76	6.30
0150	Skid steer loader	B-63	13.50	2.963	M.S.F.	273	90	9.05	372.05	450
0200	Hay, 1" deep, hand spread	1 Clab	475	.017	S.Y.	.40	.48		.88	1.19
0250	Power mulcher, small	B-64	180	.089	M.S.F.	44.50	2.55	1.45	48.50	54.50
0350	Large	B-65	530	.030	"	44.50	.86	.65	46.01	51
0400	Humus peat, 1" deep, hand spread	1 Clab	700	.011	S.Y.	4.12	.33		4.45	5.05
0450	Push spreader	"	2500	.003	"	4.12	.09		4.21	4.67
0550	Tractor spreader	B-66	700	.011	M.S.F.	460	.42	.28	460.70	505
0600	Oat straw, 1" deep, hand spread	1 Clab	475	.017	S.Y.	.43	.48		.91	1.22
0650	Power mulcher, small	B-64	180	.089	M.S.F.	48	2.55	1.45	52	58
0700	Large	B-65	530	.030	"	48	.86	.65	49.51	54.50
0750	Add for asphaltic emulsion	B-45	1770	.009	Gal.	2.21	.31	.33	2.85	3.26
0800	Peat moss, 1" deep, hand spread	1 Clab	900	.009	S.Y.	1.80	.26		2.06	2.38
0850	Push spreader	"	2500	.003	"	1.80	.09		1.89	2.12
0950	Tractor spreader	B-66	700	.011	M.S.F.	200	.42	.28	200.70	221
1000	Polyethylene film, 6 mil.	2 Clab	2000	.008	S.Y.	.19	.23		.42	.57
1100	Redwood nuggets, 3" deep, hand spread	1 Clab	150	.053	"	4.61	1.53		6.14	7.45
1150	Skid steer loader	B-63	13.50	2.963	M.S.F.	510	90	9.05	609.05	715
1200	Stone mulch, hand spread, ceramic chips, economy	1 Clab	125	.064	S.Y.	6.85	1.84		8.69	10.40
1250	Deluxe	"	95	.084	"	10.55	2.42		12.97	15.40
1300	Granite chips	B-1	10	2.400	C.Y.	34.50	70.50		105	148
1400	Marble chips		10	2.400		129	70.50		199.50	252
1600	Pea gravel		28	.857		58	25		83	103
1700	Quartz		10	2.400		166	70.50		236.50	292
1800	Tar paper, 15 Lb. felt	1 Clab	800	.010	S.Y.	.38	.29		.67	.87
1900	Wood chips, 2" deep, hand spread	"	220	.036	"	1.86	1.05		2.91	3.68
1950	Skid steer loader	B-63	20.30	1.970	M.S.F.	207	60	6	273	325

32 91 13.26 Planting Beds

		Crew	Daily Output	Labor-Hours	Unit	Material	Labor	Equipment	Total	Total Incl O&P
0010	**PLANTING BEDS**									
0100	Backfill planting pit, by hand, on site topsoil	2 Clab	18	.889	C.Y.		25.50		25.50	40
0200	Prepared planting mix	"	24	.667			19.15		19.15	30
0300	Skid steer loader, on site topsoil	B-62	340	.071			2.22	.36	2.58	3.80
0400	Prepared planting mix	"	410	.059			1.84	.30	2.14	3.16
1000	Excavate planting pit, by hand, sandy soil	2 Clab	16	1			29		29	45
1100	Heavy soil or clay	"	8	2			57.50		57.50	89.50
1200	1/2 C.Y. backhoe, sandy soil	B-11C	150	.107			3.58	1.62	5.20	7.25
1300	Heavy soil or clay	"	115	.139			4.67	2.12	6.79	9.50
2000	Mix planting soil, incl. loam, manure, peat, by hand	2 Clab	60	.267		38.50	7.65		46.15	54.50
2100	Skid steer loader	B-62	150	.160		38.50	5.05	.81	44.36	51
3000	Pile sod, skid steer loader	"	2800	.009	S.Y.		.27	.04	.31	.46
3100	By hand	2 Clab	400	.040			1.15		1.15	1.79
4000	Remove sod, F.E. loader	B-10S	2000	.006			.21	.14	.35	.47
4100	Sod cutter	B-12K	3200	.005			.17	.32	.49	.61

32 91 Planting Preparation

32 91 13 – Soil Preparation

32 91 13.26 Planting Beds		Crew	Daily Output	Labor-Hours	Unit	Material	2007 Bare Costs Labor	Equipment	Total	Total Incl O&P
4200	By hand	2 Clab	240	.067	S.Y.		1.92		1.92	2.98

32 91 19 – Landscape Grading

32 91 19.13 Topsoil Placement and Grading

		Crew	Daily Output	Labor-Hours	Unit	Material	Labor	Equipment	Total	Total Incl O&P
0010	**TOPSOIL PLACEMENT AND GRADING**									
0400	Spread from pile to rough finish grade, F.E. loader, 1.5 CY	B-10S	200	.060	E.C.Y.		2.11	1.39	3.50	4.74
0500	Up to 200' radius, by hand	1 Clab	14	.571			16.45		16.45	25.50
0600	Top dress by hand, 1 C.Y. for 600 S.F.	"	11.50	.696		21	20		41	54
0700	Furnish and place, truck dumped, screened, 4" deep	B-10S	1300	.009	S.Y.	2.60	.32	.21	3.13	3.58
0800	6" deep	"	820	.015	"	3.33	.51	.34	4.18	4.81

32 92 Turf and Grasses

32 92 19 – Seeding

32 92 19.13 Mechanical Seeding

		Crew	Daily Output	Labor-Hours	Unit	Material	Labor	Equipment	Total	Total Incl O&P
0010	**MECHANICAL SEEDING** R329219-50									
0020	Mechanical seeding, 215 lb./acre	B-66	1.50	5.333	Acre	550	197	131	878	1,050
0100	44 lb./M.S.Y.	"	2500	.003	S.Y.	.16	.12	.08	.36	.45
0300	Fine grading and seeding incl. lime, fertilizer & seed,									
0310	with equipment	B-14	1000	.048	S.Y.	.17	1.46	.24	1.87	2.72
0600	Limestone hand push spreader, 50 lbs. per M.S.F.	1 Clab	180	.044	M.S.F.	3.60	1.28		4.88	5.95
0800	Grass seed hand push spreader, 4.5 lbs. per M.S.F.	"	180	.044	"	17.80	1.28		19.08	21.50
1000	Hydro or air seeding for large areas, incl. seed and fertilizer	B-81	8900	.003	S.Y.	.16	.09	.05	.30	.37
1100	With wood fiber mulch added	"	8900	.003	"	.29	.09	.05	.43	.51
1300	Seed only, over 100 lbs., field seed, minimum				Lb.	1.19			1.19	1.31
1400	Maximum					3.96			3.96	4.36
1500	Lawn seed, minimum					1.84			1.84	2.02
1600	Maximum					4.81			4.81	5.30
1800	Aerial operations, seeding only, field seed	B-58	50	.480	Acre	385	15.10	51.50	451.60	505
1900	Lawn seed		50	.480		600	15.10	51.50	666.60	740
2100	Seed and liquid fertilizer, field seed		50	.480		465	15.10	51.50	531.60	595
2200	Lawn seed		50	.480		680	15.10	51.50	746.60	825

32 92 23 – Sodding

32 92 23.10 Sodding

		Crew	Daily Output	Labor-Hours	Unit	Material	Labor	Equipment	Total	Total Incl O&P
0010	**SODDING**									
0020	Sodding, 1" deep, bluegrass sod, on level ground, over 8 MSF	B-63	22	1.818	M.S.F.	234	55	5.55	294.55	350
0200	4 M.S.F.		17	2.353		261	71.50	7.20	339.70	405
0300	1000 S.F.		13.50	2.963		285	90	9.05	384.05	465
0500	Sloped ground, over 8 M.S.F.		6	6.667		234	202	20.50	456.50	595
0600	4 M.S.F.		5	8		261	243	24.50	528.50	690
0700	1000 S.F.		4	10		285	305	30.50	620.50	820
1000	Bent grass sod, on level ground, over 6 M.S.F.		20	2		525	60.50	6.10	591.60	675
1100	3 M.S.F.		18	2.222		580	67.50	6.80	654.30	750
1200	Sodding 1000 S.F. or less		14	2.857		665	87	8.70	760.70	875
1500	Sloped ground, over 6 M.S.F.		15	2.667		525	81	8.15	614.15	710
1600	3 M.S.F.		13.50	2.963		580	90	9.05	679.05	790
1700	1000 S.F.		12	3.333		665	101	10.15	776.15	895

32 93 Plants

32 93 10 – General Planting Costs

32 93 10.12 Travel	Crew	Daily Output	Labor-Hours	Unit	Material	2007 Bare Costs Labor	Equipment	Total	Total Incl O&P
0010 **TRAVEL** to all nursery items									
0015 10 to 20 miles, add				All					5%
0100 30 to 50 miles, add				"					10%

32 93 13 – Ground Covers

32 93 13.10 Ground Cover

	Crew	Daily Output	Labor-Hours	Unit	Material	Labor	Equipment	Total	Total Incl O&P
0010 **GROUND COVER**									
0012 Plants, pachysandra, in prepared beds	B-1	15	1.600	C	27.50	47		74.50	104
0200 Vinca minor, 1 yr, bare root		12	2	"	28.50	59		87.50	123
0600 Stone chips, in 50 lb. bags, Georgia marble		520	.046	Bag	2.56	1.36		3.92	4.93
0700 Onyx gemstone		260	.092		18.55	2.72		21.27	24.50
0800 Quartz		260	.092		6.95	2.72		9.67	11.85
0900 Pea gravel, truckload lots		28	.857	Ton	25	25		50	67

32 93 33 – Shrubs

32 93 33.10 Shrubs and Trees

	Crew	Daily Output	Labor-Hours	Unit	Material	Labor	Equipment	Total	Total Incl O&P
0010 **SHRUBS AND TREES** Evergreen, in prepared beds, B & B									
0100 Arborvitae pyramidal, 4'-5'	B-17	30	1.067	Ea.	45	33	20.50	98.50	123
0150 Globe, 12"-15"	B-1	96	.250		11.60	7.35		18.95	24
0300 Cedar, blue, 8'-10'	B-17	18	1.778		192	55	34	281	335
0500 Hemlock, canadian, 2-1/2'-3'	B-1	36	.667		24	19.60		43.60	57
0550 Holly, Savannah, 8'-10' H		9.68	2.479		545	73		618	715
0600 Juniper, andorra, 18"-24"		80	.300		16.70	8.85		25.55	32
0620 Wiltoni, 15"-18"		80	.300		16.40	8.85		25.25	32
0640 Skyrocket, 4-1/2'-5'	B-17	55	.582		50	18	11.10	79.10	95
0660 Blue pfitzer, 2'-2-1/2'	B-1	44	.545		26.50	16.05		42.55	54
0680 Ketleerie, 2-1/2'-3'		50	.480		33.50	14.10		47.60	58.50
0700 Pine, black, 2-1/2'-3'		50	.480		41	14.10		55.10	67
0720 Mugo, 18"-24"		60	.400		36	11.75		47.75	58
0740 White, 4'-5'	B-17	75	.427		53	13.20	8.15	74.35	88
0800 Spruce, blue, 18"-24"	B-1	60	.400		40	11.75		51.75	62.50
0840 Norway, 4'-5'	B-17	75	.427		90	13.20	8.15	111.35	128
0900 Yew, denisforma, 12"-15"	B-1	60	.400		24	11.75		35.75	45
1000 Capitata, 18"-24"		30	.800		20.50	23.50		44	59
1100 Hicksi, 2'-2-1/2'		30	.800		30	23.50		53.50	69.50

32 93 33.20 Shrubs

	Crew	Daily Output	Labor-Hours	Unit	Material	Labor	Equipment	Total	Total Incl O&P
0010 **SHRUBS** Broadleaf evergreen, planted in prepared beds									
0100 Andromeda, 15"-18", container	B-1	96	.250	Ea.	25.50	7.35		32.85	39.50
0200 Azalea, 15"-18", container		96	.250		28	7.35		35.35	42.50
0300 Barberry, 9"-12", container		130	.185		10.85	5.45		16.30	20.50
0400 Boxwood, 15"-18", B & B		96	.250		30	7.35		37.35	44.50
0500 Euonymus, emerald gaiety, 12" to 15", container		115	.209		18.15	6.15		24.30	29.50
0600 Holly, 15"-18", B & B		96	.250		16.70	7.35		24.05	30
0900 Mount laurel, 18"-24", B & B		80	.300		56	8.85		64.85	75.50
1000 Paxistema, 9-12" high		130	.185		18.15	5.45		23.60	28.50
1100 Rhododendron, 18"-24", container		48	.500		31	14.70		45.70	57.50
1200 Rosemary, 1 gal container		600	.040		61	1.18		62.18	69
2000 Deciduous, amelanchier, 2'-3', B & B		57	.421		90	12.40		102.40	118
2100 Azalea, 15"-18", B & B		96	.250		24	7.35		31.35	38
2300 Bayberry, 2'-3', B & B		57	.421		27.50	12.40		39.90	50
2600 Cotoneaster, 15"-18", B & B		80	.300		16.45	8.85		25.30	32
2800 Dogwood, 3'-4', B & B	B-17	40	.800		26	25	15.30	66.30	83.50
2900 Euonymus, alatus compacta, 15" to 18", container	B-1	80	.300		21.50	8.85		30.35	38

32 93 Plants

32 93 33 – Shrubs

32 93 33.20 Shrubs

		Crew	Daily Output	Labor-Hours	Unit	Material	2007 Bare Costs Labor	Equipment	Total	Total Incl O&P
3200	Forsythia, 2'-3', container	B-1	60	.400	Ea.	19	11.75		30.75	39.50
3300	Hibiscus, 3'-4', B & B	B-17	75	.427		14.85	13.20	8.15	36.20	46
3400	Honeysuckle, 3'-4', B & B	B-1	60	.400		21	11.75		32.75	42
3500	Hydrangea, 2'-3', B & B	"	57	.421		24.50	12.40		36.90	46.50
3600	Lilac, 3'-4', B & B	B-17	40	.800		25.50	25	15.30	65.80	83
3900	Privet, bare root, 18"-24"	B-1	80	.300		13.15	8.85		22	28
4100	Quince, 2'-3', B & B	"	57	.421		22	12.40		34.40	43.50
4200	Russian olive, 3'-4', B & B	B-17	75	.427		23.50	13.20	8.15	44.85	55.50
4400	Spirea, 3'-4', B & B	B-1	70	.343		27	10.10		37.10	45.50
4500	Viburnum, 3'-4', B & B	B-17	40	.800		28	25	15.30	68.30	85.50

32 93 43 – Trees

32 93 43.20 Trees

		Crew	Daily Output	Labor-Hours	Unit	Material	Labor	Equipment	Total	Total Incl O&P
0010	**TREES** Deciduous, in prep. beds, balled & burlapped (B&B)									
0100	Ash, 2" caliper	B-17	8	4	Ea.	113	124	76.50	313.50	400
0200	Beech, 5'-6'		50	.640		222	19.85	12.25	254.10	288
0300	Birch, 6'-8', 3 stems		20	1.600		122	49.50	30.50	202	244
0500	Crabapple, 6'-8'		20	1.600		159	49.50	30.50	239	285
0600	Dogwood, 4'-5'		40	.800		67	25	15.30	107.30	128
0700	Eastern redbud 4'-5'		40	.800		134	25	15.30	174.30	203
0800	Elm, 8'-10'		20	1.600		112	49.50	30.50	192	233
0900	Ginkgo, 6'-7'		24	1.333		164	41.50	25.50	231	273
1000	Hawthorn, 8'-10', 1" caliper		20	1.600		125	49.50	30.50	205	248
1100	Honeylocust, 10'-12', 1-1/2" caliper		10	3.200		149	99	61	309	385
1300	Larch, 8'		32	1		96.50	31	19.10	146.60	175
1400	Linden, 8'-10', 1" caliper		20	1.600		104	49.50	30.50	184	224
1500	Magnolia, 4'-5'		20	1.600		70	49.50	30.50	150	187
1600	Maple, red, 8'-10', 1-1/2" caliper		10	3.200		160	99	61	320	395
1700	Mountain ash, 8'-10', 1" caliper		16	2		164	62	38	264	320
1800	Oak, 2-1/2"-3" caliper		6	5.333		251	165	102	518	640
2100	Planetree, 9'-11', 1-1/4" caliper		10	3.200		102	99	61	262	335
2200	Plum, 6'-8', 1" caliper		20	1.600		90.50	49.50	30.50	170.50	210
2300	Poplar, 9'-11', 1-1/4" caliper		10	3.200		47.50	99	61	207.50	273
2500	Sumac, 2'-3'		75	.427		22.50	13.20	8.15	43.85	54.50
2700	Tulip, 5'-6'		40	.800		48.50	25	15.30	88.80	108
2800	Willow, 6'-8', 1" caliper		20	1.600		59.50	49.50	30.50	139.50	176

32 94 Planting Accessories

32 94 13 – Landscape Edging

32 94 13.20 Edging

		Crew	Daily Output	Labor-Hours	Unit	Material	Labor	Equipment	Total	Total Incl O&P
0010	**EDGING**									
0050	Aluminum alloy, including stakes, 1/8" x 4", mill finish	B-1	390	.062	L.F.	2.50	1.81		4.31	5.55
0051	Black paint		390	.062		2.90	1.81		4.71	6
0052	Black anodized		390	.062		3.35	1.81		5.16	6.50
0100	Brick, set horizontally, 1-1/2 bricks per L.F.	D-1	370	.043		1.09	1.44		2.53	3.39
0150	Set vertically, 3 bricks per L.F.	"	135	.119		3	3.95		6.95	9.30
0200	Corrugated aluminum, roll, 4" wide	1 Carp	650	.012		.40	.45		.85	1.14
0250	6" wide	"	550	.015		.50	.53		1.03	1.38
0600	Railroad ties, 6" x 8"	2 Carp	170	.094		2.72	3.45		6.17	8.40
0650	7" x 9"		136	.118		3.02	4.32		7.34	10
0750	2" x 4"		330	.048		2.27	1.78		4.05	5.25

32 94 Planting Accessories

32 94 13 – Landscape Edging

32 94 13.20 Edging	Crew	Daily Output	Labor-Hours	Unit	Material	2007 Bare Costs Labor	Equipment	Total	Total Incl O&P
0800 Steel edge strips, incl. stakes, 1/4" x 5"	B-1	390	.062	L.F.	3.65	1.81		5.46	6.85
0850 3/16" x 4"	"	390	.062	↓	2.88	1.81		4.69	6

32 94 50 – Tree Guying

32 94 50.10 Tree Guying

	Crew	Daily Output	Labor-Hours	Unit	Material	Labor	Equipment	Total	Total Incl O&P
0010 **TREE GUYING**									
0015 Tree guying Including stakes, guy wire and wrap									
0100 Less than 3" caliper, 2 stakes	2 Clab	35	.457	Ea.	16.25	13.15		29.40	38.50
0200 3" to 4" caliper, 3 stakes	"	21	.762	"	19.25	22		41.25	55
1000 Including arrowhead anchor, cable, turnbuckles and wrap									
1100 Less than 3" caliper, 3" anchors	2 Clab	20	.800	Ea.	51	23		74	92
1200 3" to 6" caliper, 4" anchors		15	1.067		73	30.50		103.50	128
1300 6" caliper, 6" anchors		12	1.333		90.50	38.50		129	159
1400 8" caliper, 8" anchors	↓	9	1.778	↓	104	51		155	194

32 96 Transplanting

32 96 23 – Plant and Bulb Transplanting

32 96 23.23 Planting

	Crew	Daily Output	Labor-Hours	Unit	Material	Labor	Equipment	Total	Total Incl O&P
0010 **PLANTING**									
0012 Moving shrubs on site, 12" ball	B-62	28	.857	Ea.	27	4.35		31.35	46.50
0100 24" ball	"	22	1.091	"	34.50	5.55		40.05	59

32 96 23.43 Moving Trees

	Crew	Daily Output	Labor-Hours	Unit	Material	Labor	Equipment	Total	Total Incl O&P
0290 **MOVING TREES**, On site									
0300 Moving trees on site, 36" ball	B-6	3.75	6.400	Ea.	201	65		266	380
0400 60" ball	"	1	24	"	755	243		998	1,425

Division Notes

	CREW	DAILY OUTPUT	LABOR-HOURS	UNIT	2007 BARE COSTS				TOTAL INCL O&P
					MAT.	LABOR	EQUIP.	TOTAL	

Estimating Tips

33 10 00 Water Utilities
33 30 00 Sanitary Sewerage Utilities
33 40 00 Storm Drainage Utilities

- Never assume that the water, sewer, and drainage lines will go in at the early stages of the project. Consider the site access needs before dividing the site in half with open trenches, loose pipe, and machinery obstructions. Always inspect the site to establish that the site drawings are complete. Check off all existing utilities on your drawings as you locate them. If you find any discrepancies, mark up the site plan for further research. Differing site conditions can be very costly if discovered later in the project.
- See also Section 33 01 00 for restoration of pipe where removal/replacement may be undesirable. Use of new types of piping materials can reduce the overall project cost. Owners/design engineers should consider the installing construction as a valuable source of current information on piping products that could lead to significant utility cost savings.

Reference Numbers

Reference numbers are shown in shaded boxes at the beginning of some major classifications. These numbers refer to related items in the Reference Section. The reference information may be an estimating procedure, an alternate pricing method, or technical information.

Note: Not all subdivisions listed here necessarily appear in this publication.

Note: **i2 Trade Service,** *in part, has been used as a reference source for some of the material prices used in Division 33.*

No part of this publication may be reproduced, stored in a retrieval system, or transmitted in any form or by any means without prior written permission of Reed Construction Data.

UTILITY TRENCH DRAINS | 33 44 16

The TRUE Cost of Storm Drainage

Example gives 3 surface drainage options for a 136,700 sf parking lot.

Cost Index provides a comparison of total installed costs against catch basins.

Catch Basins

- Site work - including excavation, - **14%** of total cost
- Concrete Surrounds - **4%**
- Pipe and fittings - **47%**
- Catch basins, grate and frame etc. - **35%**

Cost Index = 1.00

Trench Drainage

Cast-in-place
- Site work - including excavation and formwork (material & labor) - **22%** of total cost
- Concrete Surrounds - **14%**
- Pipe and fittings - **20%**
- Grate and Frame - **44%**

Cost Index = 1.09

Pre-cast Modular
- Site work - including excavation - **12%** of total cost
- Concrete Surrounds - **18%**
- Pipe and fittings - **20%**
- Pre-cast Modular Drains - **50%**

Cost Index = 0.89

ACO DRAIN - World Leader in Trench Drainage

KlassikDrain
- 4" internal width
- General purpose trench drain
- Cast-in steel edge rail
- Wide choice of grates

$ Recoup further cost savings $:
- 'QuickLok' **boltless lockings** for grates
- Choice of **installation devices** to reduce time on site

From $36.65/ft MRSP* for 4" system

FlowDrain
- 4" or 8" internal width
- General purpose fiberglass trench drain
- Choice of steel frames
- Choice of grates

$ Recoup further cost savings $:
- 'QuickLok' **boltless lockings** for grates - 4" system only
- Choice of **installation options** to reduce time & cost on site
- Long lengths - fewer parts and reduced installation time & cost

From $45.45/ft MRSP* for 4" system
From $107.56/ft MRSP* for 8" system

PowerDrain
- 4" or 12" internal width
- Heavy duty trench drain
- Cast-in ductile iron edge rail
- Choice of grates

$ Recoup further cost savings $:
- 'PowerLok' **boltless lockings** for grates
- Choice of **installation options** to reduce time & cost on site

From $64.90/ft MRSP* for 4" system
From $158.45/ft MRSP* for 12" system

*Pricing is for trench drain, grate & lockings only and varies depending upon grate (Does not included discounts)

FOR DETAILED PRICING OR FURTHER PRODUCT INFORMATION CALL (800) 543-4764

www.acousa.com

RSMEANS 2007
Job Order Contracting (JOC)

JOCWorks™ is the leading cost estimating, project management and document management software for job order contracts used by municipal, educational, healthcare, and federal government sectors. JOCWorks™ software drives the estimating and project management process with specialized features for renovation and repair projects.

The Job Order Contracting Delivery Method for renovation projects is similar to Design-Build for new construction, demonstrating a faster project delivery method over the conventional design-bid-build processes.

RSMeans Business Solutions

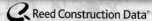

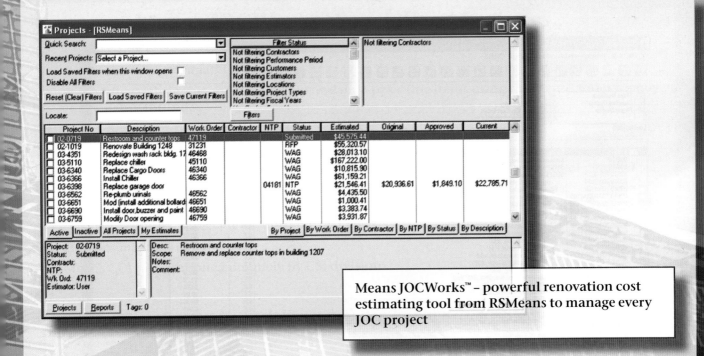

Means JOCWorks™ – powerful renovation cost estimating tool from RSMeans to manage every JOC project

Facility-Owner Benefits of Adopting JOC

- Dramatically reduces the backlog of renovation projects
- Reduces administrative costs by accomplishing more projects with the same team
- Leverages purchasing and maintenance tasks
- Dramatically reduces change orders
- Decreases up-front design and administrative costs and time
- Eliminates multiple project solicitation
- Minimizes construction claims
- Dramatically increases facility productivity

RSMeans services for developing a JOC System

A JOC-Knowledgeable Staff: Facility Management staffs must have knowledge of the JOC process, how to interact with JOC contractors, and the concept of partnering that is integral to successful JOCs. This requires familiarity with the master bidding document (i.e., Unit Price Book) and the development of JOC project specifications and contracts.

The RSMeans Unit Price Book (UPB): This customized document is the line-item listing of JOC tasks and the components of those tasks. When both the JOC client and contractor agree to this document, negotiations are reduced to project-specific peculiarities.

Training Programs: JOC specific training programs to understand how to organize a JOC program, how to effectively use Means Price Book for JOC, and how to use JOCWorks™ Software.

RSMeans Business Solutions

Reed Construction Data®

33 01 Operation and Maintenance of Utilities

33 01 10 – Operation and Maintenance of Water Utilities

33 01 10.10 Corrosion Resistance	Crew	Daily Output	Labor-Hours	Unit	Material	2007 Bare Costs Labor	Equipment	Total	Total Incl O&P
0010 **CORROSION RESISTANCE**									
0012　Wrap & coat, add to pipe, 4" dia.				L.F.	1.90			1.90	2.09
0040　　　6" diameter					1.98			1.98	2.18
0060　　　8" diameter					3.07			3.07	3.38
0100　　　12" diameter					4.58			4.58	5.05
0200　　　24" diameter					8.70			8.70	9.55
0500　Coating, bituminous, per diameter inch, 1 coat, add					.35			.35	.39
0540　　　3 coat					.54			.54	.59
0560　Coal tar epoxy, per diameter inch, 1 coat, add					.20			.20	.22
0600　　　3 coat					.44			.44	.48

33 01 30 – Operation and Maintenance of Sewer Utilities

33 01 30.72 Relining Sewers

		Crew	Daily Output	Labor-Hours	Unit	Material	Labor	Equipment	Total	Total Incl O&P
0010	**RELINING SEWERS**									
0020	Less than 10,000 L.F., urban, 6" to 10"	C-17E	130	.615	L.F.	8.10	23.50	.63	32.23	46.50
0200	24" to 36"		90	.889		12.90	34	.92	47.82	68
0300	48" to 72"		80	1		20.50	38.50	1.03	60.03	83.50

33 05 Common Work Results for Utilities

33 05 23 – Trenchless Utility Installation

33 05 23.19 Microtunneling

		Crew	Daily Output	Labor-Hours	Unit	Material	Labor	Equipment	Total	Total Incl O&P
0010	**MICROTUNNELING** Not including excavation, backfill, shoring,									
0020	or dewatering, average 50'/day, slurry method									
0100	24" to 48" outside diameter, minimum				L.F.				756	840
0110	Adverse conditions, add				%					50%
1000	Rent microtunneling machine, average monthly lease				Month				89,100	99,000
1010	Operating technician				Day				661.50	735
1100	Mobilization and demobilization, minimum				Job				44,370	49,300
1110	Maximum				"				423,000	470,000

33 05 23.20 Horizontal Boring

		Crew	Daily Output	Labor-Hours	Unit	Material	Labor	Equipment	Total	Total Incl O&P
0010	**HORIZONTAL BORING** Casing only, 100' minimum,									
0020	not incl. jacking pits or dewatering									
0100	Roadwork, 1/2" thick wall, 24" diameter casing	B-42	20	3.200	L.F.	79.50	104	64	247.50	325
0200	36" diameter		16	4		127	130	80	337	435
0300	48" diameter		15	4.267		186	139	85	410	520
0500	Railroad work, 24" diameter		15	4.267		79.50	139	85	303.50	400
0600	36" diameter		14	4.571		127	149	91	367	475
0700	48" diameter		12	5.333		186	174	106	466	595
0900	For ledge, add								186	228

33 05 26 – Utility Line Signs, Markers, and Flags

33 05 26.10 Utility Accessories

		Crew	Daily Output	Labor-Hours	Unit	Material	Labor	Equipment	Total	Total Incl O&P
0010	**UTILITY ACCESSORIES**									
0400	Underground tape, detectable, reinforced, alum. foil core, 2"	1 Clab	150	.053	C.L.F.	2	1.53		3.53	4.59
0500	6"	"	140	.057	"	5	1.64		6.64	8.05

33 11 Water Utility Distribution Piping

33 11 13 – Public Water Utility Distribution Piping

33 11 13.15 Water Supply, Ductile Iron Pipe

		Crew	Daily Output	Labor-Hours	Unit	Material	2007 Bare Costs Labor	Equipment	Total	Total Incl O&P
0010	**WATER SUPPLY, DUCTILE IRON PIPE** cement lined R331113-80									
0020	Not including excavation or backfill									
2000	Pipe, class 50 water piping, 18' lengths									
2020	Mechanical joint, 4" diameter	B-21A	200	.200	L.F.	14.15	7.20	2.88	24.23	29.50
2040	6" diameter		160	.250		16.45	9	3.60	29.05	36
2060	8" diameter		133.33	.300		18.20	10.80	4.31	33.31	41
2080	10" diameter		114.29	.350		24.50	12.60	5.05	42.15	52
2100	12" diameter		105.26	.380		30.50	13.70	5.45	49.65	60.50
2120	14" diameter		100	.400		38.50	14.40	5.75	58.65	71
2140	16" diameter		72.73	.550		42	19.80	7.90	69.70	85
2160	18" diameter		68.97	.580		53	21	8.35	82.35	99
2170	20" diameter		57.14	.700		62	25	10.05	97.05	118
2180	24" diameter		47.06	.850		79.50	30.50	12.20	122.20	147
3000	Tyton, push-on joint, 4" diameter		400	.100		8.35	3.60	1.44	13.39	16.30
3020	6" diameter		333.33	.120		9.80	4.32	1.73	15.85	19.20
3040	8" diameter		200	.200		13.25	7.20	2.88	23.33	28.50
3060	10" diameter		181.82	.220		20.50	7.90	3.16	31.56	38
3080	12" diameter		160	.250		21.50	9	3.60	34.10	41
3100	14" diameter		133.33	.300		23.50	10.80	4.31	38.61	47
3120	16" diameter		114.29	.350		33.50	12.60	5.05	51.15	61.50
3140	18" diameter		100	.400		37	14.40	5.75	57.15	69.50
3160	20" diameter		88.89	.450		41	16.20	6.45	63.65	76.50
3180	24" diameter		76.92	.520		59.50	18.70	7.50	85.70	102
8000	Fittings, mechanical joint									
8006	90° bend, 4" diameter	B-20A	16	2	Ea.	182	70		252	305
8020	6" diameter		12.80	2.500		220	87.50		307.50	375
8040	8" diameter		10.67	2.999		355	105		460	550
8060	10" diameter	B-21A	11.43	3.500		395	126	50.50	571.50	680
8080	12" diameter		10.53	3.799		645	137	54.50	836.50	975
8100	14" diameter		10	4		1,025	144	57.50	1,226.50	1,400
8120	16" diameter		7.27	5.502		1,225	198	79	1,502	1,700
8140	18" diameter		6.90	5.797		1,325	209	83.50	1,617.50	1,850
8160	20" diameter		5.71	7.005		2,500	252	101	2,853	3,250
8180	24" diameter		4.70	8.511		4,850	305	122	5,277	5,950
8200	Wye or tee, 4" diameter	B-20A	10.67	2.999		281	105		386	470
8220	6" diameter		8.53	3.751		375	131		506	615
8240	8" diameter		7.11	4.501		540	158		698	830
8260	10" diameter	B-21A	7.62	5.249		925	189	75.50	1,189.50	1,400
8280	12" diameter		7.02	5.698		1,175	205	82	1,462	1,700
8300	14" diameter		6.67	5.997		1,325	216	86	1,627	1,875
8320	16" diameter		4.85	8.247		1,375	297	119	1,791	2,075
8340	18" diameter		4.60	8.696		3,550	315	125	3,990	4,500
8360	20" diameter		3.81	10.499		5,575	380	151	6,106	6,875
8380	24" diameter		3.14	12.739		6,600	460	183	7,243	8,150
8450	Decreaser, 6" x 4" diameter	B-20A	14.22	2.250		201	79		280	340
8460	8" x 6" diameter	"	11.64	2.749		293	96.50		389.50	465
8470	10" x 6" diameter	B-21A	13.33	3.001		360	108	43	511	605
8480	12" x 6" diameter		12.70	3.150		470	113	45.50	628.50	740
8490	16" x 6" diameter		10	4		895	144	57.50	1,096.50	1,275
8500	20" x 6" diameter		8.42	4.751		1,775	171	68.50	2,014.50	2,275
8550	Butterfly valves with boxes, cast iron									
8560	4" diameter	B-20	6	4	Ea.	625	130		755	885
8570	6" diameter	"	5	4.800		935	156		1,091	1,275

33 11 Water Utility Distribution Piping

33 11 13 – Public Water Utility Distribution Piping

33 11 13.15 Water Supply, Ductile Iron Pipe		Crew	Daily Output	Labor-Hours	Unit	Material	2007 Bare Costs Labor	Equipment	Total	Total Incl O&P
8580	8" diameter	B-21	4	7	Ea.	1,100	235	41	1,376	1,600
8590	10" diameter		3.50	8		1,400	268	46.50	1,714.50	2,025
8600	12" diameter		3	9.333		1,950	315	54.50	2,319.50	2,700
8610	14" diameter		2	14		2,400	470	82	2,952	3,475
8620	16" diameter		2	14		3,725	470	82	4,277	4,900

33 11 13.25 Water Supply, Polyvinyl Chloride Pipe

		Crew	Daily Output	Labor-Hours	Unit	Material	Labor	Equipment	Total	Total Incl O&P
0010	**WATER SUPPLY, POLYVINYL CHLORIDE PIPE**									
0020	Not including excavation or backfill, unless specified									
2100	AWWA Class 160, S.D.R. 26, 1-1/2" diameter	Q-1A	750	.013	L.F.	.49	.60		1.09	1.45
2120	2" diameter		686	.015		1.17	.66		1.83	2.28
2140	2-1/2" diameter		500	.020		1.73	.90		2.63	3.26
2160	3" diameter	B-20	430	.056		2.48	1.81		4.29	5.55
2180	4" diameter		375	.064		4.05	2.08		6.13	7.70
2200	6" diameter		316	.076		8.75	2.47		11.22	13.45
2210	8" diameter		260	.092		14.80	3		17.80	21
3010	AWWA C905, PR 100, DR 41									
3030	14" diameter	B-20A	213	.150	L.F.	18.80	5.25		24.05	28.50
3040	16" diameter		200	.160		18.95	5.60		24.55	29.50
3050	18" diameter		160	.200		31	7		38	44.50
3060	20" diameter		133	.241		24	8.45		32.45	39
3070	24" diameter		107	.299		39.50	10.50		50	59
3080	30" diameter		80	.400		60.50	14		74.50	88.50
3090	36" diameter		80	.400		86	14		100	116
3100	42" diameter		60	.533		117	18.70		135.70	158
3200	48" diameter		60	.533		153	18.70		171.70	197
4520	Class 150, SDR 18, AWWA C900, 4"		380	.084		2.70	2.95		5.65	7.45
4540	8"		264	.121		9.85	4.25		14.10	17.30
4550	10"		220	.145		14.90	5.10		20	24
4560	12"		186	.172		21	6.05		27.05	32
8000	Fittings with rubber gasket									
8003	Class 150, D.R. 18									
8006	90° Bend, 4" diameter	B-20	100	.240	Ea.	48.50	7.80		56.30	65.50
8020	6" diameter		90	.267		71	8.65		79.65	91.50
8040	8" diameter		80	.300		137	9.75		146.75	165
8060	10" diameter		50	.480		241	15.60		256.60	290
8080	12" diameter		30	.800		370	26		396	445
8100	Tee, 4" diameter		90	.267		63	8.65		71.65	83
8120	6" diameter		80	.300		123	9.75		132.75	151
8140	8" diameter		70	.343		266	11.15		277.15	310
8160	10" diameter		40	.600		400	19.50		419.50	470
8180	12" diameter		20	1.200		555	39		594	670
8200	45° Bend, 4" diameter		100	.240		48.50	7.80		56.30	65.50
8220	6" diameter		90	.267		71	8.65		79.65	91.50
8240	8" diameter		50	.480		137	15.60		152.60	175
8260	10" diameter		50	.480		370	15.60		385.60	435
8280	12" diameter		30	.800		490	26		516	580
8300	Reducing tee 6"x4"		100	.240		118	7.80		125.80	142
8320	8" x 6"		90	.267		223	8.65		231.65	260
8330	10" x 6"		90	.267		212	8.65		220.65	248
8340	10" x 8"		90	.267		277	8.65		285.65	320
8350	12" x 6"		90	.267		345	8.65		353.65	395
8360	12" x 8"		90	.267		385	8.65		393.65	440

33 11 Water Utility Distribution Piping

33 11 13 – Public Water Utility Distribution Piping

33 11 13.25 Water Supply, Polyvinyl Chloride Pipe

		Crew	Daily Output	Labor-Hours	Unit	Material	2007 Bare Costs Labor	Equipment	Total	Total Incl O&P
8400	Tapped service tee (threaded type) 6" x 6" x 3/4	B-20	100	.240	Ea.	58.50	7.80		66.30	76.50
8420	6" x 6" x 3/4"		90	.267		58.50	8.65		67.15	78
8430	6" x 6" x 1"		90	.267		58.50	8.65		67.15	78
8440	6" x 6" x 1 1/2"		90	.267		58.50	8.65		67.15	78
8450	6" x 6" x 2"		90	.267		58.50	8.65		67.15	78
8460	8" x 8" x 3/4"		90	.267		207	8.65		215.65	242
8470	8" x 8" x 1"		90	.267		207	8.65		215.65	242
8480	8" x 8" x 1 1/2"		90	.267		207	8.65		215.65	242
8490	8" x 8" x 2"		90	.267		207	8.65		215.65	242
8500	Repair coupling 4"		100	.240		20.50	7.80		28.30	35
8520	6" diameter		90	.267		32.50	8.65		41.15	49
8540	8" diameter		50	.480		75.50	15.60		91.10	108
8560	10" diameter		50	.480		199	15.60		214.60	244
8580	12" diameter		50	.480		222	15.60		237.60	270
8600	Plug end 4"		100	.240		20.50	7.80		28.30	35
8620	6" diameter		90	.267		38	8.65		46.65	55
8640	8" diameter		50	.480		61	15.60		76.60	92
8660	10" diameter		50	.480		83	15.60		98.60	116
8680	12" diameter		50	.480		105	15.60		120.60	141

33 12 Water Utility Distribution Equipment

33 12 19 – Water Utility Distribution Fire Hydrants

33 12 19.10 Fire Hydrants

		Crew	Daily Output	Labor-Hours	Unit	Material	2007 Bare Costs Labor	Equipment	Total	Total Incl O&P
0010	**FIRE HYDRANTS**									
0020	Mechanical joints unless otherwise noted									
1000	Fire hydrants, two way; excavation and backfill not incl.									
1100	4-1/2" valve size, depth 2'-0"	B-21	10	2.800	Ea.	1,200	94	16.35	1,310.35	1,500
1120	2'-6"		10	2.800		1,200	94	16.35	1,310.35	1,500
1140	3'-0"		10	2.800		1,225	94	16.35	1,335.35	1,525
1300	7'-0"		6	4.667		1,375	157	27.50	1,559.50	1,775
2400	Lower barrel extensions with stems, 1'-0"	B-20	14	1.714		500	55.50		555.50	635
2480	3'-0"	"	12	2		585	65		650	745

33 12 19.40 Utility Boxes

		Crew	Daily Output	Labor-Hours	Unit	Material	2007 Bare Costs Labor	Equipment	Total	Total Incl O&P
0010	**UTILITY BOXES** Precast concrete, 6" thick									
0050	5' x 10' x 6' high, I.D.	B-13	2	28	Ea.	1,725	880	370	2,975	3,650
0100	6' x 10' x 6' high, I.D.		2	28		1,775	880	370	3,025	3,700
0150	5' x 12' x 6' high, I.D.		2	28		1,875	880	370	3,125	3,825
0200	6' x 12' x 6' high, I.D.		1.80	31.111		2,100	975	410	3,485	4,275
0250	6' x 13' x 6' high, I.D.		1.50	37.333		2,775	1,175	495	4,445	5,400
0300	8' x 14' x 7' high, I.D.		1	56		3,000	1,750	740	5,490	6,825
0350	Hand hole, precast concrete, 1-1/2" thick									
0400	1'-0" x 2'-0" x 1'-9", I.D., light duty	B-1	4	6	Ea.	292	177		469	595
0450	4'-6" x 3'-2" x 2'-0", O.D., heavy duty	B-6	3	8	"	895	252	81	1,228	1,450

33 16 Water Utility Storage Tanks

33 16 13 – Aboveground Water Utility Storage Tanks

33 16 13.13 Steel Aboveground Water Utility Storage Tanks

	33 16 13.13 Steel Aboveground Water Utility Storage Tanks	Crew	Daily Output	Labor-Hours	Unit	Material	2007 Bare Costs Labor	Equipment	Total	Total Incl O&P
0010	**STEEL ABOVEGROUND WATER UTILITY STORAGE TANKS**									
0910	Steel, ground level, ht/dia. less than 1, not incl. fdn., 100,000 gallons				Ea.				137,000	150,500
1000	250,000 gallons								145,373	159,910
1200	500,000 gallons								221,777	243,955
1250	750,000 gallons								264,223	290,645
1300	1,000,000 gallons								403,964	444,360
1500	2,000,000 gallons								652,598	717,858
1600	4,000,000 gallons								1,013,385	1,114,723
1800	6,000,000 gallons								1,406,005	1,546,606
1850	8,000,000 gallons								1,841,072	2,025,179
1910	10,000,000 gallons								2,745,000	3,020,000
2100	Steel standpipes, hgt/dia more than 1, 100' to overflow, no fdn									
2200	500,000 gallons				Ea.				291,812	320,993
2400	750,000 gallons								355,481	391,029
2500	1,000,000 gallons								432,944	476,238
2700	1,500,000 gallons								604,848	665,333
2800	2,000,000 gallons								742,795	817,075

33 16 13.16 Prestressed Conc. Aboveground Water Utility Storage

		Crew	Daily Output	Labor-Hours	Unit	Material	Labor	Equipment	Total	Total Incl O&P
0010	**PRESTRESSED CONC. ABOVEGROUND WATER UTILITY STORAGE**									
0020	Not including pipe or pumps, 250,000 gallons				Ea.				258,364	284,200
0100	500,000 gallons								350,636	385,700
0300	1,000,000 gallons								498,273	548,100
0400	2,000,000 gallons								752,945	828,240
0600	4,000,000 gallons								1,190,318	1,309,350
0700	6,000,000 gallons								1,624,000	1,786,400
0750	8,000,000 gallons								2,066,909	2,273,600
0800	10,000,000 gallons								2,491,364	2,740,500

33 16 13.19 Wood Water Storage Tanks

		Crew	Daily Output	Labor-Hours	Unit	Material	Labor	Equipment	Total	Total Incl O&P
0010	**WOOD WATER STORAGE TANKS**									
6000	Wood tanks, ground level, 2" cypress, 3,000 gallons	C-1	.19	168	Ea.	7,850	5,850		13,700	17,800
6100	2-1/2" cypress, 10,000 gallons		.12	266		20,800	9,250		30,050	37,300
6300	3" redwood or 3" fir, 20,000 gallons		.10	320		31,800	11,100		42,900	52,000
6400	30,000 gallons		.08	400		39,300	13,900		53,200	65,000
6600	45,000 gallons		.07	457		59,500	15,900		75,400	90,000
6700	Larger sizes, minimum				Gal.				.60	.66
6900	Maximum				"				.74	.81

33 16 13.23 Plastic-Coated Fabric Pillow Water Tanks

		Crew	Daily Output	Labor-Hours	Unit	Material	Labor	Equipment	Total	Total Incl O&P
0010	**PLASTIC-COATED FABRIC PILLOW WATER TANKS**									
7000	Water tanks, vinyl coated fabric pillow tanks, freestanding, 5,000 gallons	4 Clab	4	8	Ea.	9,325	230		9,555	10,700
7100	Supporting embankment not included, 25,000 gallons	6 Clab	2	24		14,400	690		15,090	17,000
7200	50,000 gallons	8 Clab	1.50	42.667		26,800	1,225		28,025	31,400
7300	100,000 gallons	9 Clab	.90	80		58,000	2,300		60,300	67,000
7400	150,000 gallons		.50	144		78,500	4,150		82,650	93,000
7500	200,000 gallons		.40	180		95,000	5,175		100,175	112,500
7600	250,000 gallons		.30	240		143,500	6,900		150,400	168,500

33 16 19 – Elevated Water Utility Storage Tanks

33 16 19.50 Elevated Water Storage Tanks

		Crew	Daily Output	Labor-Hours	Unit	Material	Labor	Equipment	Total	Total Incl O&P
0010	**ELEVATED WATER STORAGE TANKS**, not incl pipe, pumps or foundation									
3000	Elevated water tanks, 100' to bottom capacity line, incl painting									
3010	50,000 gallons				Ea.				176,149	193,764
3300	100,000 gallons								243,000	267,300

33 16 Water Utility Storage Tanks

33 16 19 – Elevated Water Utility Storage Tanks

33 16 19.50 Elevated Water Storage Tanks	Crew	Daily Output	Labor-Hours	Unit	Material	2007 Bare Costs Labor	Equipment	Total	Total Incl O&P
3400　　250,000 gallons				Ea.				335,319	368,851
3600　　500,000 gallons								541,180	595,298
3700　　750,000 gallons								745,979	820,577
3900　　1,000,000 gallons								749,255	824,180

33 21 Water Supply Wells

33 21 13 – Public Water Supply Wells

33 21 13.10 Wells and Accessories

		Crew	Daily Output	Labor-Hours	Unit	Material	Labor	Equipment	Total	Total Incl O&P
0010	**WELLS & ACCESSORIES**, domestic									
0100	Drilled, 4" to 6" diameter	B-23	120	.333	L.F.		9.70	30.50	40.20	49
0200	8" diameter	"	95.20	.420	"		12.25	38.50	50.75	61.50
0400	Gravel pack well, 40' deep, incl. gravel & casing, complete									
0500	24" diameter casing x 18" diameter screen	B-23	.13	307	Total	26,700	8,975	28,300	63,975	74,500
0600	36" diameter casing x 18" diameter screen		.12	333	"	28,800	9,725	30,700	69,225	80,500
0800	Observation wells, 1-1/4" riser pipe		163	.245	V.L.F.	14.70	7.15	22.50	44.35	52.50
0900	For flush Buffalo roadway box, add	1 Skwk	16.60	.482	Ea.	40	18.30		58.30	72.50
1200	Test well, 2-1/2" diameter, up to 50' deep (15 to 50 GPM)	B-23	1.51	26.490	"	600	770	2,450	3,820	4,525
1300	Over 50' deep, add	"	121.80	.328	L.F.	16.05	9.55	30.50	56.10	66
1500	Pumps, installed in wells to 100' deep, 4" submersible									
1510	1/2 H.P.	Q-1	3.22	4.969	Ea.	375	200		575	715
1520	3/4 H.P.		2.66	6.015		435	243		678	840
1600	1 H.P.		2.29	6.987		455	282		737	930
1700	1-1/2 H.P.	Q-22	1.60	10		1,175	405	455	2,035	2,400
1800	2 H.P.		1.33	12.030		1,375	485	545	2,405	2,850
1900	3 H.P.		1.14	14.035		1,600	565	635	2,800	3,300
2000	5 H.P.		1.14	14.035		2,175	565	635	3,375	3,925
3000	Pump, 6" submersible, 25' to 150' deep, 25 H.P., 249 to 297 GPM		.89	17.978		4,950	725	815	6,490	7,450
3100	25' to 500' deep, 30 H.P., 100 to 300 GPM		.73	21.918		5,650	885	990	7,525	8,650
8000	Steel well casing	B-23A	3020	.008	Lb.	.68	.26	1.19	2.13	2.46
9950	See Div. 31 23 19.40 for wellpoints									
9960	See Div. 31 23 19.30 for drainage wells									

33 21 13.20 Water Supply Wells, Pumps

		Crew	Daily Output	Labor-Hours	Unit	Material	Labor	Equipment	Total	Total Incl O&P
0010	**WATER SUPPLY WELLS, PUMPS** with pressure control									
1000	Deep well, jet, 42 gal. galvanized tank									
1040	3/4 HP	1 Plum	.80	10	Ea.	635	450		1,085	1,375
3000	Shallow well, jet, 30 gal. galvanized tank									
3040	1/2 HP	1 Plum	2	4	Ea.	435	179		614	745

33 31 Sanitary Utility Sewerage Piping

33 31 13 – Public Sanitary Utility Sewerage Piping

33 31 13.15 Sewage Collection, Concrete Pipe

		Crew	Daily Output	Labor-Hours	Unit	Material	Labor	Equipment	Total	Total Incl O&P
0010	**SEWAGE COLLECTION, CONCRETE PIPE**									
0020	See 33 41 13.60 for sewage/drainage collection, concrete pipe									

33 31 13.25 Sewage Collection, Polyvinyl Chloride Pipe

		Crew	Daily Output	Labor-Hours	Unit	Material	Labor	Equipment	Total	Total Incl O&P
0010	**SEWAGE COLLECTION, POLYVINYL CHLORIDE PIPE**									
0020	Not including excavation or backfill									
2000	10' lengths, S.D.R. 35, B&S, 4" diameter	B-20	375	.064	L.F.	2.33	2.08		4.41	5.80
2040	6" diameter		350	.069		4.35	2.23		6.58	8.25
2080	8" diameter		335	.072		9.05	2.33		11.38	13.60

33 31 Sanitary Utility Sewerage Piping

33 31 13 – Public Sanitary Utility Sewerage Piping

33 31 13.25 Sewage Collection, Polyvinyl Chloride Pipe		Crew	Daily Output	Labor-Hours	Unit	Material	2007 Bare Costs Labor	Equipment	Total	Total Incl O&P
2120	10" diameter	B-21	330	.085	L.F.	14.30	2.85	.50	17.65	20.50
2160	12" diameter		320	.088		16.25	2.93	.51	19.69	23
2200	15" diameter		190	.147		24.50	4.94	.86	30.30	35.50

33 36 Utility Septic Tanks

33 36 13 – Utility Septic Tank and Effluent Wet Wells

33 36 13.10 Septic Tanks

		Crew	Daily Output	Labor-Hours	Unit	Material	Labor	Equipment	Total	Total Incl O&P
0010	**SEPTIC TANKS**									
0015	Septic tanks, not incl exc or piping, precast, 1,000 gal	B-21	8	3.500	Ea.	635	117	20.50	772.50	900
0100	2,000 gallon	"	5	5.600		1,975	188	32.50	2,195.50	2,500
0200	5,000 gallon	B-13	3.50	16		6,850	500	211	7,561	8,550
0300	15,000 gallon, 4 piece	B-13B	1.70	32.941		19,000	1,025	625	20,650	23,200
0400	25,000 gallon, 4 piece		1.10	50.909		40,200	1,600	965	42,765	47,700
0500	40,000 gallon, 4 piece		.80	70		47,300	2,200	1,325	50,825	57,000
0520	50,000 gallon, 5 piece	B-13C	.60	93.333		54,500	2,925	2,850	60,275	67,500
0540	75,000 gallon, cast in place	C-14C	.25	448		66,000	15,700	85	81,785	97,500
0560	100,000 gallon	"	.15	746		82,000	26,200	142	108,342	131,500
0600	High density polyethylene, 1,000 gallon	B-21	6	4.667		1,050	157	27.50	1,234.50	1,425
0700	1,500 gallon	"	4	7		1,350	235	41	1,626	1,900
1000	Distribution boxes, concrete, 7 outlets	2 Clab	16	1		117	29		146	174
1100	9 outlets	"	8	2		435	57.50		492.50	565
1150	Leaching field chambers, 13' x 3'-7" x 1'-4", standard	B-13	16	3.500		645	110	46	801	930
1200	Heavy duty, 8' x 4' x 1'-6"		14	4		410	125	53	588	700
1300	13' x 3'-9" x 1'-6"		12	4.667		930	146	61.50	1,137.50	1,325
1350	20' x 4' x 1'-6"		5	11.200		985	350	148	1,483	1,775
1400	Leaching pit, precast concrete, 3' diameter, 3' deep	B-21	8	3.500		580	117	20.50	717.50	845
1500	6' diameter, 3' section		4.70	5.957		970	200	35	1,205	1,425
2000	Velocity reducing pit, precast conc., 6' diameter, 3' deep		4.70	5.957		1,150	200	35	1,385	1,625
2200	Excavation for septic tank, 3/4 C.Y. backhoe	B-12F	145	.110	C.Y.		3.78	3.56	7.34	9.70
2400	4' trench for disposal field, 3/4 C.Y. backhoe	"	335	.048	L.F.		1.64	1.54	3.18	4.20
2600	Gravel fill, run of bank	B-6	150	.160	C.Y.	19.35	5.05	1.62	26.02	31
2800	Crushed stone, 3/4"	"	150	.160	"	23	5.05	1.62	29.67	35

33 41 Storm Utility Drainage Piping

33 41 13 – Public Storm Utility Drainage Piping

33 41 13.40 Piping, Storm Drainage, Corrugated Metal

		Crew	Daily Output	Labor-Hours	Unit	Material	Labor	Equipment	Total	Total Incl O&P
0010	**PIPING, STORM DRAINAGE, CORRUGATED METAL**									
0020	Not including excavation or backfill									
2000	Corrugated metal pipe, galvanized and coated									
2020	Bituminous coated with paved invert, 20' lengths									
2040	8" diameter, 16 ga.	B-14	330	.145	L.F.	10.05	4.43	.74	15.22	18.70
2060	10" diameter, 16 ga.		260	.185		12.05	5.60	.94	18.59	23
2080	12" diameter, 16 ga.		210	.229		14.45	6.95	1.16	22.56	28
2100	15" diameter, 16 ga.		200	.240		17.60	7.30	1.22	26.12	32
2120	18" diameter, 16 ga.		190	.253		22.50	7.70	1.28	31.48	38
2140	24" diameter, 14 ga.		160	.300		27.50	9.15	1.52	38.17	46.50
2160	30" diameter, 14 ga.	B-13	120	.467		36.50	14.65	6.15	57.30	69.50
2180	36" diameter, 12 ga.		120	.467		53.50	14.65	6.15	74.30	88
2200	48" diameter, 12 ga.		100	.560		81	17.55	7.40	105.95	125

33 41 Storm Utility Drainage Piping

33 41 13 – Public Storm Utility Drainage Piping

33 41 13.40 Piping, Storm Drainage, Corrugated Metal

		Crew	Daily Output	Labor-Hours	Unit	Material	2007 Bare Costs Labor	Equipment	Total	Total Incl O&P
2220	60" diameter, 10 ga.	B-13B	75	.747	L.F.	105	23.50	14.15	142.65	168
2240	72" diameter, 8 ga.	"	45	1.244	↓	157	39	23.50	219.50	259
2500	Galvanized, uncoated, 20' lengths									
2520	8" diameter, 16 ga.	B-14	355	.135	L.F.	8.20	4.11	.69	13	16.15
2540	10" diameter, 16 ga.		280	.171		9	5.20	.87	15.07	18.90
2560	12" diameter, 16 ga.		220	.218		10.30	6.65	1.11	18.06	23
2580	15" diameter, 16 ga.		220	.218		13.10	6.65	1.11	20.86	26
2600	18" diameter, 16 ga.		205	.234		17.30	7.15	1.19	25.64	31.50
2620	24" diameter, 14 ga.		175	.274		25	8.35	1.39	34.74	42
2640	30" diameter, 14 ga.	B-13	130	.431		32	13.50	5.70	51.20	62.50
2660	36" diameter, 12 ga.		130	.431		52.50	13.50	5.70	71.70	85
2680	48" diameter, 12 ga.		110	.509		70	15.95	6.75	92.70	109
2690	60" diameter, 10 ga.	B-13B	78	.718	↓	110	22.50	13.60	146.10	170
2780	End sections, 8" diameter	B-14	24	2	Ea.	80	61	10.15	151.15	193
2785	10" diameter		22	2.182		82	66.50	11.05	159.55	206
2790	12" diameter		35	1.371		92.50	41.50	6.95	140.95	174
2800	18" diameter		30	1.600		110	48.50	8.10	166.60	205
2810	24" diameter	B-13	25	2.240		161	70	29.50	260.50	320
2820	30" diameter		25	2.240		305	70	29.50	404.50	475
2825	36" diameter		20	2.800		405	88	37	530	620
2830	48" diameter		10	5.600		835	176	74	1,085	1,275
2835	60" diameter	B-13B	5	11.200		1,100	350	212	1,662	1,975
2840	72" diameter	"	4	14	↓	1,975	440	265	2,680	3,125

33 41 13.60 Sewage/Drainage Collection, Concrete Pipe

		Crew	Daily Output	Labor-Hours	Unit	Material	Labor	Equipment	Total	Total Incl O&P
0010	**SEWAGE/DRAINAGE COLLECTION, CONCRETE PIPE**									
0020	Not including excavation or backfill									
1000	Non-reinforced pipe, extra strength, B&S or T&G joints									
1010	6" diameter	B-14	265.04	.181	L.F.	4.82	5.50	.92	11.24	14.85
1020	8" diameter		224	.214		5.30	6.50	1.09	12.89	17.15
1030	10" diameter		216	.222		5.90	6.75	1.13	13.78	18.15
1040	12" diameter		200	.240		7.25	7.30	1.22	15.77	20.50
1050	15" diameter		180	.267		8.45	8.10	1.35	17.90	23.50
1060	18" diameter		144	.333		10.35	10.15	1.69	22.19	29
1070	21" diameter		112	.429		12.75	13.05	2.17	27.97	36.50
1080	24" diameter	↓	100	.480	↓	15.65	14.60	2.43	32.68	42.50
2000	Reinforced culvert, class 3, no gaskets									
2010	12" diameter	B-14	150	.320	L.F.	12.80	9.75	1.62	24.17	31
2020	15" diameter		150	.320		11.90	9.75	1.62	23.27	30
2030	18" diameter		132	.364		13.85	11.05	1.84	26.74	34.50
2035	21" diameter		120	.400		17.25	12.15	2.03	31.43	40
2040	24" diameter CN	↓	100	.480		26	14.60	2.43	43.03	53.50
2045	27" diameter	B-13	92	.609		32.50	19.10	8.05	59.65	74
2050	30" diameter		88	.636		32.50	19.95	8.40	60.85	75.50
2060	36" diameter		72	.778		42.50	24.50	10.25	77.25	96
2070	42" diameter	B-13B	72	.778		61.50	24.50	14.70	100.70	122
2080	48" diameter		64	.875		89.50	27.50	16.55	133.55	159
2090	60" diameter		48	1.167		160	36.50	22	218.50	257
2100	72" diameter		40	1.400		179	44	26.50	249.50	293
2120	84" diameter		32	1.750		320	55	33	408	470
2140	96" diameter		24	2.333		385	73	44	502	580
2200	With gaskets, class 3, 12" diameter	B-21	168	.167		11.20	5.60	.97	17.77	22
2220	15" diameter	↓	160	.175	↓	13.40	5.85	1.02	20.27	25

33 41 Storm Utility Drainage Piping

33 41 13 – Public Storm Utility Drainage Piping

33 41 13.60 Sewage/Drainage Collection, Concrete Pipe

		Crew	Daily Output	Labor-Hours	Unit	Material	2007 Bare Costs Labor	Equipment	Total	Total Incl O&P
2230	18" diameter	B-21	152	.184	L.F.	16.75	6.20	1.08	24.03	29
2240	24" diameter	↓	136	.206		25	6.90	1.20	33.10	39.50
2260	30" diameter	B-13	88	.636		33.50	19.95	8.40	61.85	77
2270	36" diameter	"	72	.778		50.50	24.50	10.25	85.25	104
2290	48" diameter	B-13B	64	.875		82	27.50	16.55	126.05	151
2310	72" diameter	"	40	1.400		212	44	26.50	282.50	330
2330	Flared ends, 6'-1" long, 12" diameter	B-21	190	.147		41	4.94	.86	46.80	53.50
2340	15" diameter		155	.181		46.50	6.05	1.06	53.61	62
2400	6'-2" long, 18" diameter		122	.230		48.50	7.70	1.34	57.54	67
2420	24" diameter	↓	88	.318		56	10.65	1.86	68.51	80
2440	36" diameter	B-13	60	.933		102	29.50	12.35	143.85	171
3080	Radius pipe, add to pipe prices, 12" to 60" diameter					50%				
3090	Over 60" diameter, add				↓	20%				
3500	Reinforced elliptical, 8' lengths, C507 class 3									
3520	14" x 23" inside, round equivalent 18" diameter	B-21	82	.341	L.F.	39.50	11.45	1.99	52.94	63
3530	24" x 38" inside, round equivalent 30" diameter	B-13	58	.966		58.50	30.50	12.75	101.75	125
3540	29" x 45" inside, round equivalent 36" diameter		52	1.077		97	34	14.25	145.25	175
3550	38" x 60" inside, round equivalent 48" diameter		38	1.474		166	46	19.45	231.45	275
3560	48" x 76" inside, round equivalent 60" diameter		26	2.154		234	67.50	28.50	330	395
3570	58" x 91" inside, round equivalent 72" diameter	↓	22	2.545	↓	325	80	33.50	438.50	515
3780	Concrete slotted pipe, class 4 mortar joint									
3800	12" diameter	B-21	168	.167	L.F.	14.90	5.60	.97	21.47	26
3840	18" diameter	"	152	.184	"	23	6.20	1.08	30.28	36
3900	Class 4 O-ring									
3940	12" diameter	B-21	168	.167	L.F.	15.60	5.60	.97	22.17	27
3960	18" diameter	"	152	.184	"	21	6.20	1.08	28.28	33.50

33 42 Culverts

33 42 16 – Concrete Culverts

33 42 16.15 Oval Arch Culverts

		Crew	Daily Output	Labor-Hours	Unit	Material	Labor	Equipment	Total	Total Incl O&P
0010	**OVAL ARCH CULVERTS**									
3000	Corrugated galvanized or aluminum, coated & paved									
3020	17" x 13", 16 ga., 15" equivalent	B-14	200	.240	L.F.	31.50	7.30	1.22	40.02	47
3040	21" x 15", 16 ga., 18" equivalent		150	.320		40.50	9.75	1.62	51.87	61.50
3060	28" x 20", 14 ga., 24" equivalent		125	.384		58	11.70	1.95	71.65	84
3080	35" x 24", 14 ga., 30" equivalent	↓	100	.480		70.50	14.60	2.43	87.53	103
3100	42" x 29", 12 ga., 36" equivalent	B-13	100	.560		105	17.55	7.40	129.95	151
3120	49" x 33", 12 ga., 42" equivalent		90	.622		127	19.50	8.20	154.70	179
3140	57" x 38", 12 ga., 48" equivalent	↓	75	.747	↓	141	23.50	9.85	174.35	202
3160	Steel, plain oval arch culverts, plain									
3180	17" x 13", 16 ga., 15" equivalent	B-14	225	.213	L.F.	17	6.50	1.08	24.58	30
3200	21" x 15", 16 ga., 18" equivalent		175	.274		20	8.35	1.39	29.74	36.50
3220	28" x 20", 14 ga., 24" equivalent	↓	150	.320		32.50	9.75	1.62	43.87	52.50
3240	35" x 24", 14 ga., 30" equivalent	B-13	108	.519		40.50	16.25	6.85	63.60	77
3260	42" x 29", 12 ga., 36" equivalent		108	.519		67	16.25	6.85	90.10	106
3280	49" x 33", 12 ga., 42" equivalent		92	.609		79	19.10	8.05	106.15	125
3300	57" x 38", 12 ga., 48" equivalent		75	.747	↓	69.50	23.50	9.85	102.85	123
3320	End sections, 17" x 13"		22	2.545	Ea.	93	80	33.50	206.50	262
3340	42" x 29"	↓	17	3.294	"	380	103	43.50	526.50	625
3360	Multi-plate arch, steel	B-20	1690	.014	Lb.	1.07	.46		1.53	1.90

33 44 Storm Utility Water Drains

33 44 13 – Utility Area Drains

33 44 13.13 Catchbasins

		Crew	Daily Output	Labor-Hours	Unit	Material	2007 Bare Costs Labor	Equipment	Total	Total Incl O&P
0010	**CATCHBASINS**									
1600	Frames & covers, C.I., 24" square, 500 lb.	B-6	7.80	3.077	Ea.	310	97	31	438	525
1700	26" D shape, 600 lb.		7	3.429		535	108	35	678	790
1800	Light traffic, 18" diameter, 100 lb.		10	2.400		174	75.50	24.50	274	335
1900	24" diameter, 300 lb.		8.70	2.759		269	87	28	384	460
2000	36" diameter, 900 lb.		5.80	4.138		540	130	42	712	840
2100	Heavy traffic, 24" diameter, 400 lb.		7.80	3.077		260	97	31	388	470
2200	36" diameter, 1150 lb.		3	8		860	252	81	1,193	1,425
2300	Mass. State standard, 26" diameter, 475 lb.		7	3.429		650	108	35	793	920
2400	30" diameter, 620 lb.		7	3.429		410	108	35	553	660
2500	Watertight, 24" diameter, 350 lb.		7.80	3.077		440	97	31	568	670
2600	26" diameter, 500 lb.		7	3.429		420	108	35	563	665
2700	32" diameter, 575 lb.	↓	6	4	↓	935	126	40.50	1,101.50	1,275
2800	3 piece cover & frame, 10" deep,									
2900	1200 lbs., for heavy equipment	B-6	3	8	Ea.	1,325	252	81	1,658	1,925
3000	Raised for paving 1-1/4" to 2" high,									
3100	4 piece expansion ring									
3200	20" to 26" diameter	1 Clab	3	2.667	Ea.	146	76.50		222.50	279
3300	30" to 36" diameter	"	3	2.667	"	203	76.50		279.50	340
3320	Frames and covers, existing, raised for paving, 2", including									
3340	row of brick, concrete collar, up to 12" wide frame	B-6	18	1.333	Ea.	45.50	42	13.50	101	129
3360	20" to 26" wide frame		11	2.182		72	68.50	22	162.50	209
3380	30" to 36" wide frame	↓	9	2.667		89	84	27	200	257
3400	Inverts, single channel brick	D-1	3	5.333		100	178		278	380
3500	Concrete		5	3.200		78	107		185	248
3600	Triple channel, brick		2	8		152	267		419	570
3700	Concrete	↓	3	5.333	↓	134	178		312	420

33 46 Subdrainage

33 46 16 – Subdrainage Piping

33 46 16.20 Piping, Subdrainage, Concrete

		Crew	Daily Output	Labor-Hours	Unit	Material	Labor	Equipment	Total	Total Incl O&P
0010	**PIPING, SUBDRAINAGE, CONCRETE**									
0021	Not including excavation and backfill									
3000	Porous wall concrete underdrain, std. strength, 4" diameter	B-20	335	.072	L.F.	2.87	2.33		5.20	6.80
3020	6" diameter	"	315	.076		3.73	2.48		6.21	7.95
3040	8" diameter	B-21	310	.090		4.61	3.03	.53	8.17	10.30
3060	12" diameter		285	.098		9.75	3.30	.57	13.62	16.50
3080	15" diameter		230	.122		11.20	4.08	.71	15.99	19.40
3100	18" diameter	↓	165	.170		14.80	5.70	.99	21.49	26
4000	Extra strength, 6" diameter	B-20	315	.076		3.78	2.48		6.26	8
4020	8" diameter	B-21	310	.090		5.65	3.03	.53	9.21	11.50
4040	10" diameter		295	.095		11.35	3.18	.55	15.08	18
4060	12" diameter		285	.098		12.30	3.30	.57	16.17	19.25
4080	15" diameter		230	.122		13.60	4.08	.71	18.39	22
4100	18" diameter	↓	165	.170	↓	19.85	5.70	.99	26.54	32

33 46 16.25 Piping, Subdrainage, Corrugated Metal

		Crew	Daily Output	Labor-Hours	Unit	Material	Labor	Equipment	Total	Total Incl O&P
0010	**PIPING, SUBDRAINAGE, CORRUGATED METAL**									
0021	Not including excavation and backfill									
2010	Aluminum, perforated									
2020	6" diameter, 18 ga.	B-14	380	.126	L.F.	4.25	3.84	.64	8.73	11.35
2200	8" diameter, 16 ga.	↓	370	.130	↓	6.10	3.95	.66	10.71	13.50

33 46 Subdrainage

33 46 16 – Subdrainage Piping

33 46 16.25 Piping, Subdrainage, Corrugated Metal

		Crew	Daily Output	Labor-Hours	Unit	Material	2007 Bare Costs Labor	Equipment	Total	Total Incl O&P
2220	10" diameter, 16 ga.	B-14	360	.133	L.F.	7.60	4.06	.68	12.34	15.40
2240	12" diameter, 16 ga.		285	.168		8.50	5.15	.85	14.50	18.25
2260	18" diameter, 16 ga.	↓	205	.234	↓	12.75	7.15	1.19	21.09	26.50
3000	Uncoated galvanized, perforated									
3020	6" diameter, 18 ga.	B-20	380	.063	L.F.	6.75	2.05		8.80	10.60
3200	8" diameter, 16 ga.	"	370	.065		9.25	2.11		11.36	13.50
3220	10" diameter, 16 ga.	B-21	360	.078		13.90	2.61	.45	16.96	19.80
3240	12" diameter, 16 ga.		285	.098		14.50	3.30	.57	18.37	21.50
3260	18" diameter, 16 ga.	↓	205	.137		22	4.58	.80	27.38	32.50
4000	Steel, perforated, asphalt coated									
4020	6" diameter 18 ga.	B-20	380	.063	L.F.	5.40	2.05		7.45	9.10
4030	8" diameter 18 ga	"	370	.065		8.45	2.11		10.56	12.55
4040	10" diameter 16 ga	B-21	360	.078		9.65	2.61	.45	12.71	15.20
4050	12" diameter 16 ga		285	.098		11.10	3.30	.57	14.97	17.95
4060	18" diameter 16 ga	↓	205	.137		15.15	4.58	.80	20.53	24.50

33 46 16.40 Piping, Subdrainage, Polyvinyl Chloride

0010	**PIPING, SUBDRAINAGE, POLYVINYL CHLORIDE**									
0020	Perforated, price as solid pipe, Division 33 41 13.25									

33 49 Storm Drainage Structures

33 49 13 – Storm Drainage Manholes, Frames, and Covers

33 49 13.10 Storm Drainage Manholes, Frames and Covers

		Crew	Daily Output	Labor-Hours	Unit	Material	Labor	Equipment	Total	Total Incl O&P
0010	**STORM DRAINAGE MANHOLES, FRAMES & COVERS** not including									
0020	footing, excavation, backfill (See line items for frame & cover)									
0050	Brick, 4' inside diameter, 4' deep	D-1	1	16	Ea.	390	535		925	1,250
0100	6' deep		.70	22.857		545	760		1,305	1,750
0150	8' deep		.50	32	↓	695	1,075		1,770	2,400
0200	For depths over 8', add		4	4	V.L.F.	184	133		317	405
0400	Concrete blocks (radial), 4' I.D., 4' deep		1.50	10.667	Ea.	330	355		685	905
0500	6' deep		1	16		435	535		970	1,300
0600	8' deep		.70	22.857	↓	540	760		1,300	1,750
0700	For depths over 8', add	↓	5.50	2.909	V.L.F.	54	97		151	207
0800	Concrete, cast in place, 4' x 4', 8" thick, 4' deep	C-14H	2	24	Ea.	475	870	10.80	1,355.80	1,875
0900	6' deep		1.50	32		685	1,150	14.40	1,849.40	2,575
1000	8' deep		1	48	↓	990	1,750	21.50	2,761.50	3,850
1100	For depths over 8', add		8	6	V.L.F.	110	218	2.70	330.70	465
1110	Precast, 4' I.D., 4' deep	B-22	4.10	7.317	Ea.	850	248	60	1,158	1,375
1120	6' deep		3	10		1,050	340	81.50	1,471.50	1,800
1130	8' deep		2	15	↓	1,275	510	123	1,908	2,325
1140	For depths over 8', add	↓	16	1.875	V.L.F.	174	63.50	15.30	252.80	305
1150	5' I.D., 4' deep	B-6	3	8	Ea.	875	252	81	1,208	1,425
1160	6' deep		2	12		1,175	375	122	1,672	2,025
1170	8' deep		1.50	16	↓	1,475	505	162	2,142	2,575
1180	For depths over 8', add		12	2	V.L.F.	194	63	20.50	277.50	330
1190	6' I.D., 4' deep		2	12	Ea.	1,425	375	122	1,922	2,300
1200	6' deep		1.50	16		1,850	505	162	2,517	3,000
1210	8' deep		1	24	↓	2,300	755	243	3,298	3,950
1220	For depths over 8', add	↓	8	3	V.L.F.	300	94.50	30.50	425	510
1250	Slab tops, precast, 8" thick									
1300	4' diameter manhole	B-6	8	3	Ea.	201	94.50	30.50	326	400
1400	5' diameter manhole	↓	7.50	3.200		395	101	32.50	528.50	625

33 49 Storm Drainage Structures

33 49 13 – Storm Drainage Manholes, Frames, and Covers

33 49 13.10 Storm Drainage Manholes, Frames and Covers		Crew	Daily Output	Labor-Hours	Unit	Material	2007 Bare Costs Labor	Equipment	Total	Total Incl O&P
1500	6' diameter manhole	B-6	7	3.429	Ea.	575	108	35	718	835
3800	Steps, heavyweight cast iron, 7" x 9"	1 Bric	40	.200		16.20	7.60		23.80	29.50
3900	8" x 9"		40	.200		24.50	7.60		32.10	38
3928	12" x 10-1/2"		40	.200		23	7.60		30.60	37
4000	Standard sizes, galvanized steel		40	.200		19.50	7.60		27.10	33
4100	Aluminum		40	.200		26.50	7.60		34.10	40.50

33 51 Natural-Gas Distribution

33 51 13 – Natural-Gas Piping

33 51 13.10 Piping, Gas Service and Distribution, Polyethylene

		Crew	Daily Output	Labor-Hours	Unit	Material	Labor	Equipment	Total	Total Incl O&P
0010	**PIPING, GAS SERVICE AND DISTRIBUTION, POLYETHYLENE**									
0020	not including excavation or backfill									
1000	60 psi coils, comp cplg @ 100', 1/2" diameter, SDR 9.3	B-20A	608	.053	L.F.	1.02	1.84		2.86	3.94
1040	1-1/4" diameter, SDR 11		544	.059		1.88	2.06		3.94	5.20
1100	2" diameter, SDR 11		488	.066		2.33	2.30		4.63	6.05
1160	3" diameter, SDR 11		408	.078		4.88	2.75		7.63	9.55
1500	60 PSI 40' joints with coupling, 3" diameter, SDR 11	B-21A	408	.098		4.88	3.53	1.41	9.82	12.25
1540	4" diameter, SDR 11		352	.114		11.15	4.09	1.63	16.87	20.50
1600	6" diameter, SDR 11		328	.122		34.50	4.39	1.75	40.64	46.50
1640	8" diameter, SDR 11		272	.147		47.50	5.30	2.11	54.91	62.50

33 52 Liquid Fuel Distribution

33 52 16 – Gasoline Distribution

33 52 16.13 Gasoline Piping

		Crew	Daily Output	Labor-Hours	Unit	Material	Labor	Equipment	Total	Total Incl O&P
0010	**GASOLINE PIPING**									
0020	Primary containment pipe, fiberglass-reinforced									
0030	Plastic pipe 15' & 30' lengths									
0040	2" diameter	Q-6	425	.056	L.F.	4.05	2.38		6.43	8.05
0050	3" diameter		400	.060		5.30	2.53		7.83	9.65
0060	4" diameter		375	.064		6.85	2.70		9.55	11.55
0100	Fittings									
0110	Elbows, 90° & 45°, bell-ends, 2"	Q-6	24	1	Ea.	41	42		83	109
0120	3" diameter		22	1.091		43	46		89	117
0130	4" diameter		20	1.200		57	50.50		107.50	139
0200	Tees, bell ends, 2"		21	1.143		49.50	48		97.50	127
0210	3" diameter		18	1.333		50	56		106	140
0220	4" diameter		15	1.600		68.50	67.50		136	178
0230	Flanges bell ends, 2"		24	1		16.40	42		58.40	81.50
0240	3" diameter		22	1.091		20.50	46		66.50	92
0250	4" diameter		20	1.200		28	50.50		78.50	107
0260	Sleeve couplings, 2"		21	1.143		10.45	48		58.45	84
0270	3" diameter		18	1.333		14.80	56		70.80	101
0280	4" diameter		15	1.600		20.50	67.50		88	125
0290	Threaded adapters 2"		21	1.143		13.75	48		61.75	87.50
0300	3" diameter		18	1.333		24	56		80	111
0310	4" diameter		15	1.600		32.50	67.50		100	138
0320	Reducers, 2"		27	.889		18.25	37.50		55.75	76.50
0330	3" diameter		22	1.091		21	46		67	92.50
0340	4" diameter		20	1.200		27.50	50.50		78	106

33 52 Liquid Fuel Distribution

33 52 16 – Gasoline Distribution

33 52 16.13 Gasoline Piping

		Crew	Daily Output	Labor-Hours	Unit	Material	2007 Bare Costs Labor	2007 Bare Costs Equipment	Total	Total Incl O&P
1010	Gas station product line for secondary containment (double wall)									
1100	Fiberglass reinforced plastic pipe 25' lengths									
1120	Pipe, plain end, 3"	Q-6	375	.064	L.F.	6.95	2.70		9.65	11.70
1130	4" diameter		350	.069		11.40	2.89		14.29	16.90
1140	5" diameter		325	.074		14.80	3.12		17.92	21
1150	6" diameter		300	.080		15.25	3.37		18.62	22
1200	Fittings									
1230	Elbows, 90° & 45°, 3"	Q-6	18	1.333	Ea.	50	56		106	140
1240	4" diameter		16	1.500		87.50	63.50		151	191
1250	5" diameter		14	1.714		202	72.50		274.50	330
1260	6" diameter		12	2		205	84.50		289.50	350
1270	Tees, 3"		15	1.600		74	67.50		141.50	184
1280	4" diameter		12	2		109	84.50		193.50	247
1290	5" diameter		9	2.667		221	112		333	410
1300	6" diameter		6	4		231	169		400	510
1310	Couplings, 3"		18	1.333		35	56		91	123
1320	4" diameter		16	1.500		91	63.50		154.50	195
1330	5" diameter		14	1.714		189	72.50		261.50	315
1340	6" diameter		12	2		195	84.50		279.50	340
1350	Cross-over nipples, 3"		18	1.333		8.05	56		64.05	93.50
1360	4" diameter		16	1.500		9.40	63.50		72.90	105
1370	5" diameter		14	1.714		13.95	72.50		86.45	124
1380	6" diameter		12	2		14.65	84.50		99.15	143
1400	Telescoping, reducers, concentric 4" x 3"		18	1.333		26.50	56		82.50	114
1410	5" x 4"		17	1.412		70	59.50		129.50	167
1420	6" x 5"		16	1.500		168	63.50		231.50	279

33 71 Electrical Utility Transmission and Distribution

33 71 16 – Electrical Utility Poles

33 71 16.33 Wood Electrical Utility Poles

		Crew	Daily Output	Labor-Hours	Unit	Material	2007 Bare Costs Labor	2007 Bare Costs Equipment	Total	Total Incl O&P
0010	**WOOD ELECTRICAL UTILITY POLES**									
6200	Electric & tel sitework, ps, wd, treatment, see also 26 56 13.10, 20' hi	R-3	3.10	6.452	Ea.	251	279	52.50	582.50	750
6400	25' high		2.90	6.897		265	298	56.50	619.50	800
6600	30' high		2.60	7.692		292	335	63	690	885
6800	35' high		2.40	8.333		380	360	68	808	1,025
7000	40' high		2.30	8.696		465	375	71	911	1,150
7200	45' high		1.70	11.765		570	510	96	1,176	1,500
7400	Cross arms with hardware & insulators									
7600	4' long	1 Elec	2.50	3.200	Ea.	119	140		259	340
7800	5' long		2.40	3.333		138	146		284	370
8000	6' long		2.20	3.636		159	160		319	415

33 71 19 – Electrical Underground Ducts and Manholes

33 71 19.17 Electric and Telephone Underground

		Crew	Daily Output	Labor-Hours	Unit	Material	2007 Bare Costs Labor	2007 Bare Costs Equipment	Total	Total Incl O&P
0010	**ELECTRIC AND TELEPHONE UNDERGROUND**, Not including excavation									
0200	backfill and cast in place concrete									
0400	Hand holes, precast concrete, with concrete cover									
0600	2' x 2' x 3' deep	R-3	2.40	8.333	Ea.	295	360	68	723	940
0800	3' x 3' x 3' deep		1.90	10.526		385	455	86	926	1,200
1000	4' x 4' x 4' deep		1.40	14.286		765	620	117	1,502	1,900
1200	Manholes, precast with iron racks & pulling irons, C.I. frame									

33 71 Electrical Utility Transmission and Distribution

33 71 19 – Electrical Underground Ducts and Manholes

33 71 19.17 Electric and Telephone Underground		Crew	Daily Output	Labor-Hours	Unit	Material	2007 Bare Costs Labor	Equipment	Total	Total Incl O&P
1400	and cover, 4' x 6' x 7' deep	B-13	2	28	Ea.	1,475	880	370	2,725	3,375
1600	6' x 8' x 7' deep		1.90	29.474		1,800	925	390	3,115	3,825
1800	6' x 10' x 7' deep	↓	1.80	31.111	↓	2,025	975	410	3,410	4,175
4200	Underground duct, banks ready for concrete fill, min. of 7.5"									
4400	between conduits, ctr. to ctr.									
4580	PVC, type EB, 1 @ 2" diameter	2 Elec	480	.033	L.F.	.78	1.46		2.24	3.04
4600	2 @ 2" diameter		240	.067		1.56	2.93		4.49	6.10
4800	4 @ 2" diameter		120	.133		3.12	5.85		8.97	12.15
5000	2 @ 3" diameter		200	.080		2.12	3.51		5.63	7.60
5200	4 @ 3" diameter		100	.160		4.24	7		11.24	15.10
5400	2 @ 4" diameter		160	.100		3.22	4.39		7.61	10.10
5600	4 @ 4" diameter		80	.200		6.45	8.80		15.25	20
5800	6 @ 4" diameter		54	.296		9.65	13		22.65	30
6200	Rigid galvanized steel, 2 @ 2" diameter		180	.089		14.20	3.90		18.10	21.50
6400	4 @ 2" diameter		90	.178		28.50	7.80		36.30	43
6800	2 @ 3" diameter		100	.160		34	7		41	47.50
7000	4 @ 3" diameter		50	.320		67.50	14.05		81.55	95.50
7200	2 @ 4" diameter		70	.229		47	10.05		57.05	66.50
7400	4 @ 4" diameter		34	.471		93.50	20.50		114	134
7600	6 @ 4" diameter	↓	22	.727	↓	140	32		172	202

33 81 Communications Structures

33 81 13 – Communications Transmission Towers

33 81 13.10 Radio Towers

		Crew	Daily Output	Labor-Hours	Unit	Material	Labor	Equipment	Total	Total Incl O&P
0010	**RADIO TOWERS**									
0020	Guyed, 50'h, 40 lb. sec., 70 MPH basic wind spd.	2 Sswk	1	16	Ea.	2,600	660		3,260	4,050
0100	Wind load 90 MPH basic wind speed	"	1	16		2,600	660		3,260	4,050
0300	190' high, 40 lb. section, wind load 70 MPH basic wind speed	K-2	.33	72.727		7,000	2,750	560	10,310	13,100
0400	200' high, 70 lb. section, wind load 90 MPH basic wind speed		.33	72.727		14,100	2,750	560	17,410	20,900
0600	300' high, 70 lb. section, wind load 70 MPH basic wind speed		.20	120		20,000	4,525	925	25,450	30,900
0700	270' high, 90 lb. section, wind load 90 MPH basic wind speed		.20	120		23,100	4,525	925	28,550	34,300
0800	400' high, 100 lb. section, wind load 70 MPH basic wind speed		.14	171		33,700	6,475	1,325	41,500	49,800
0900	Self-supporting, 60' high, wind load 70 MPH basic wind speed		.80	30		5,400	1,125	232	6,757	8,175
0910	60' high, wind load 90 MPH basic wind speed		.45	53.333		9,725	2,025	410	12,160	14,700
1000	120' high, wind load 70 MPH basic wind speed		.40	60		13,300	2,275	465	16,040	19,100
1200	190' high, wind load 90 MPH basic wind speed	↓	.20	120		32,200	4,525	925	37,650	44,300
2000	For states west of Rocky Mountains, add for shipping				↓	10%				

Estimating Tips

34 11 00 Rail Tracks
This section includes items that may involve either repair of existing, or construction of new, railroad tracks. Additional preparation work, such as the roadbed earthwork, would be found in Division 31. Additional new construction siding and turnouts are found in Section 34 72. Maintenance of railroads is found under 34 01 23 Operation and Maintenance of Railways.

34 41 13 Traffic Signals
This section includes traffic signal systems. Other traffic control devices such as traffic signs are found in Section 10 14 53 Traffic Signage.

34 71 13 Vehicle Barriers
This section includes security vehicle barriers, guide and guard rails, crash barriers, and delineators. The actual maintenance and construction of concrete and asphalt pavement is found in Division 32.

Reference Numbers
Reference numbers are shown in shaded boxes at the beginning of some major classifications. These numbers refer to related items in the Reference Section. The reference information may be an estimating procedure, an alternate pricing method, or technical information.

Note: Not all subdivisions listed here necessarily appear in this publication.

34 01 Operation and Maintenance of Transportation

34 01 23 – Operation and Maintenance of Railways

34 01 23.51 Maintenance of Railroads

		Crew	Daily Output	Labor-Hours	Unit	Material	2007 Bare Costs Labor	Equipment	Total	Total Incl O&P
0010	MAINTENANCE OF RAILROADS									
0400	Resurface and realign existing track	B-14	200	.240	L.F.		7.30	1.22	8.52	12.65
0600	For crushed stone ballast, add	"	500	.096	"	14.45	2.92	.49	17.86	21

34 11 Rail Tracks

34 11 13 – Track Rails

34 11 13.23 Heavy Rail Track

0010	HEAVY RAIL TRACK R347216-10									
1000	Rail, 100 lb. prime grade				L.F.	25			25	27.50
1500	Relay rail				"	12.40			12.40	13.65

34 11 33 – Track Cross Ties

34 11 33.13 Concrete Track Cross Ties

0010	CONCRETE TRACK CROSS TIES									
1400	Ties, concrete, 8'-6" long, 30" O.C.	B-14	80	.600	Ea.	93.50	18.25	3.04	114.79	135

34 11 33.16 Timber Track Cross Ties

0010	TIMBER TRACK CROSS TIES									
1600	Wood, pressure treated, 6" x 8" x 8'-6", C.L. lots	B-14	90	.533	Ea.	39.50	16.25	2.70	58.45	71
1700	L.C.L. lots		90	.533		41	16.25	2.70	59.95	73.50
1900	Heavy duty, 7" x 9" x 8'-6", C.L. lots		70	.686		43	21	3.48	67.48	84
2000	L.C.L. lots		70	.686		43	21	3.48	67.48	84

34 11 33.17 Timber Switch Ties

0010	TIMBER SWITCH TIES									
1200	Switch timber, for a #8 switch, pressure treated	B-14	3.70	12.973	M.B.F.	2,350	395	66	2,811	3,275
1300	Complete set of timbers, 3.7 M.B.F. for #8 switch	"	1	48	Total	9,075	1,450	243	10,768	12,500

34 11 93 – Track Appurtenances and Accessories

34 11 93.50 Track Accessories

0010	TRACK ACCESSORIES									
0020	Car bumpers, test	B-14	2	24	Ea.	2,775	730	122	3,627	4,300
0100	Heavy duty R347216-20		2	24		5,250	730	122	6,102	7,025
0200	Derails hand throw (sliding)		10	4.800		875	146	24.50	1,045.50	1,225
0300	Hand throw with standard timbers, open stand & target		8	6		955	183	30.50	1,168.50	1,375
2400	Wheel stops, fixed		18	2.667	Pr.	625	81	13.50	719.50	825
2450	Hinged		14	3.429	"	840	104	17.40	961.40	1,100

34 11 93.60 Track Material

0010	TRACK MATERIAL									
0020	Track bolts				Ea.	2.19			2.19	2.41
0100	Joint bars				Pr.	66			66	73
0200	Spikes				Ea.	.58			.58	.64
0300	Tie plates				"	11.05			11.05	12.15

34 41 Roadway Signaling and Control Equipment

34 41 13 – Traffic signals

34 41 13.10 Traffic Signals

		Crew	Daily Output	Labor-Hours	Unit	Material	2007 Bare Costs Labor	Equipment	Total	Total Incl O&P
0010	**TRAFFIC SIGNALS** Mid block pedestrian crosswalk,									
0020	with pushbutton and mast arms	R-11	.30	186	Total	61,500	7,700	2,800	72,000	82,000
0600	Traffic signals, school flashing system, solar powered, remote controlled	"	1	56	Signal	15,100	2,300	840	18,240	21,000
1000	Intersection traffic signals, LED, mast, programmable, no lane control									
1010	Includes all labor, material and equip. for complete installation				Ea.	157,500			157,500	173,500
1100	Intersection traffic signals, LED, mast, programmable, R/L lane control									
1110	Includes all labor, material and equip. for complete installation				Ea.	210,000			210,000	231,000
1200	Add protective/permissive left turns to existing traffic light									
1210	Includes all labor, material and equip. for complete installation				Ea.	63,000			63,000	69,500
1300	Replace existing light heads with LED Heads wire hung									
1310	Includes all labor, material and equip. for complete installation				Ea.	42,000			42,000	46,200
1400	Replace existing light heads with LED Heads mast arm hung									
1410	Includes all labor, material and equip. for complete installation				Ea.	84,000			84,000	92,500

34 71 Roadway Construction

34 71 13 – Vehicle Barriers

34 71 13.17 Security Vehicle Barriers

		Crew	Daily Output	Labor-Hours	Unit	Material	Labor	Equipment	Total	Total Incl O&P
0010	**SECURITY VEHICLE BARRIERS**									
0020	Security planters excludes filling material									
0100	Concrete security planter, exposed aggregate 36" dia x 30" high	B-11M	8	2	Ea.	455	67	35.50	557.50	640
0200	48" dia x 36" high		8	2		740	67	35.50	842.50	955
0300	53" dia x 18" high		8	2		550	67	35.50	652.50	745
0400	72" dia x 18" high with seats		8	2		1,125	67	35.50	1,227.50	1,375
0450	84" dia x 36" high		8	2		1,100	67	35.50	1,202.50	1,375
0500	36" x 36" x 24" high square		8	2		395	67	35.50	497.50	575
0600	36" x 36" x 30" high square		8	2		455	67	35.50	557.50	640
0700	48" L x 24" W x 30" H rectangle		8	2		375	67	35.50	477.50	550
0800	72" L x 24" W x 30" H rectangle		8	2		445	67	35.50	547.50	625
0900	96" L x 24" W x 30" H rectangle		8	2		615	67	35.50	717.50	815
0950	Decorative geometric concrete barrier, 96" L x 24" W x 36" H		8	2		915	67	35.50	1,017.50	1,150
1000	Concrete security planter, filling with washed sand or gravel °1 CY		8	2		28	67	35.50	130.50	173
1050	Concrete security planter, filling with washed sand or gravel °2 CY		6	2.667		56	89.50	47.50	193	252
1200	Jersey concete barrier, 10' L x 2' by 0.5' W x 30" H	B-21B	16	2.500		268	78.50	45.50	392	465
1300	10' L x 2' by 0.5' W x 30" H, 10 or more same site		24	1.667		268	52.50	30	350.50	410
1400	Jersey concete barrier, 10' L x 2' by 0.5' W x 32" H		16	2.500		279	78.50	45.50	403	475
1500	10' L x 2' by 0.5' W x 32" H, 10 or more same site		24	1.667		279	52.50	30	361.50	420
1600	Jersey concete barrier, 20' L x 2' by 0.5' W x 30" H		12	3.333		675	105	60.50	840.50	970
1700	20' L x 2' by 0.5' W x 30" H, 10 or more same site		18	2.222		675	69.50	40	784.50	890
1800	Jersey concete barrier, 20' L x 2' by 0.5' W x 32" H		12	3.333		690	105	60.50	855.50	985
1900	20' L x 2' by 0.5' W x 32" H, 10 or more same site		18	2.222		690	69.50	40	799.50	905
2000	GFRC decorative security barrier per 10 feet section including concrete		4	10		2,000	315	181	2,496	2,875
2100	GFRC decorative security barrier per 12 feet section including concrete		4	10		2,400	315	181	2,896	3,300
2210	GFRC decorative security barrier will stop 30 MPH, 4000 lb vehicle									
2300	High security barrier base prep per 12 feet section on bare ground	B-11C	4	4	Ea.	29	134	61	224	305
2310	GFRC barrier base prep does not include haul away of excavated matl									
2400	GFRC decorative high security barrier per 12 feet section w/concrete	B-21B	3	13.333	Ea.	5,225	420	241	5,886	6,650
2410	GFRC decorative high security barrier will stop 50 MPH, 15000 lb vehl									
2500	GFRC decorative impaler security barrier per 10 feet section w/prep	B-6	4	6	Ea.	2,000	189	61	2,250	2,550
2600	GFRC decorative impaler security barrier per 12 feet section w/prep	"	4	6	"	2,400	189	61	2,650	3,000
2610	Impaler barrier should stop 50 MPH, 15000 lb vehl w/some penetr									
2700	Pipe bollards, steel, concrete filled/painted, 8' L x 4' D hole, 8" diam.	B-6	10	2.400	Ea.	335	75.50	24.50	435	515

34 71 Roadway Construction

34 71 13 – Vehicle Barriers

34 71 13.17 Security Vehicle Barriers

		Crew	Daily Output	Labor-Hours	Unit	Material	2007 Bare Costs Labor	Equipment	Total	Total Incl O&P
2710	Schedule 80 concrete bollards will stop 4000 lb vehicle @ 30 MPH									
2800	GFRC decorative jersey barrier cover per 10 feet section excludes soil	B-6	8	3	Ea.	900	94.50	30.50	1,025	1,175
2900	GFRC decorative jersey barrier cover per 12 feet section excludes soil		8	3		1,075	94.50	30.50	1,200	1,375
3000	GFRC decorative 8" diameter bollard cover		12	2		450	63	20.50	533.50	615
3100	GFRC decorative barrier face 10 foot section excludes earth backing		10	2.400		840	75.50	24.50	940	1,075
3200	Drop arm crash barrier, 20' width or less, 15000 lb vehl @ 50 MPH									
3205	Includes all material, labor for complete installation									
3210	Drop arm crash barrier, 20' width or less, 15,000lb vehl @ 50MPH				Ea.				72,600	79,750
3310	24' width								78,650	86,350
3410	Portable wedge crash barrier, 10' width								71,500	78,650
3510	12' width								78,650	86,350
3610	Sliding crash barrier, 12' width								101,750	111,650
3710	Sliding roller crash barrier, 20' width								72,600	79,750
3810	Sliding cantilever crash barrier, 20' width								78,650	86,350
3890	Note: Raised bollard crash barriers should be used w/tire shredders									
3910	Raised bollard crash barrier, 10' width				Ea.				24,200	26,620
4010	12' width								27,830	30,580
4110	Includes all labor, material and equip. for complete installation								30,800	33,880
4210	12' width								34,430	37,840
4310	In ground tire shredder, 16' width								34,800	38,280

34 71 13.26 Vehicle Guide Rails

		Crew	Daily Output	Labor-Hours	Unit	Material	Labor	Equipment	Total	Total Incl O&P
0010	**VEHICLE GUIDE RAILS**									
0012	Corrugated stl, galv. stl posts, 6'-3" O.C.	B-80	850	.038	L.F.	17.75	1.18	.66	19.59	22
0200	End sections, galvanized, flared		50	.640	Ea.	73.50	20	11.20	104.70	124
0300	Wrap around end		50	.640	"	110	20	11.20	141.20	165
0400	Timber guide rail, 4" x 8" with 6" x 8" wood posts, treated		960	.033	L.F.	23.50	1.04	.58	25.12	28
0600	Cable guide rail, 3 at 3/4" cables, steel posts, single face		900	.036		7.85	1.11	.62	9.58	11.05
0700	Wood posts		950	.034		10.90	1.05	.59	12.54	14.25
0900	Guide rail, steel box beam, 6" x 6"		120	.267		27	8.35	4.67	40.02	47.50
1100	Median barrier, steel box beam, 6" x 8"		215	.149		33	4.65	2.61	40.26	46.50
1400	Resilient guide fence and light shield, 6' high	B-2	130	.308		26	8.95		34.95	42.50
1500	Concrete posts, individual, 6'-5", triangular	B-80	110	.291	Ea.	55	9.10	5.10	69.20	80
1550	Square	"	110	.291	"	59	9.10	5.10	73.20	84.50
2000	Median, precast concrete, 3'-6" high, 2' wide, single face	B-29	380	.147	L.F.	43	4.62	2.38	50	56.50
2200	Double face	"	340	.165	"	49	5.15	2.67	56.82	65
2400	Speed bumps, thermoplastic, 10-1/2" x 2-1/4" x 48" long	B-2	120	.333	Ea.	111	9.70		120.70	137
3030	Impact barrier, UTMCD, barrel type	B-16	30	1.067	"	360	31.50	17.65	409.15	465

34 71 19 – Vehicle Delineators

34 71 19.13 Fixed Vehicle Delineators

		Crew	Daily Output	Labor-Hours	Unit	Material	Labor	Equipment	Total	Total Incl O&P
0010	**FIXED VEHICLE DELINEATORS**									
0020	Crash barriers									
0100	Traffic channelizing pavement markers, layout only	A-7	2000	.012	Ea.		.48	.03	.51	.78
0110	13" x 7-1/2" x 2-1/2" high, non-plowable install	2 Clab	96	.167		23	4.79		27.79	32.50
0200	8" x 8" x 3-1/4" high, non-plowable, install		96	.167		21	4.79		25.79	30.50
0230	4" x 4" x 3/4" high, non-plowable, install		120	.133		2.77	3.83		6.60	9
0240	9-1/4" x 5-7/8" x 1/4" high, plowable, concrete pav't	A-2A	70	.343		16.60	9.85	3.71	30.16	37.50
0250	9-1/4" x 5-7/8" x 1/4" high, plowable, asphalt pav't	"	120	.200		3.17	5.75	2.17	11.09	14.75
0300	Barrier and curb delineators, reflectorized, 2" x 4"	2 Clab	150	.107		1.81	3.07		4.88	6.75
0310	3" x 5"	"	150	.107		3.67	3.07		6.74	8.80
0500	Rumble strip, polycarbonate									
0510	24" x 3-1/2" x 1/2" high	2 Clab	50	.320	Ea.	6.90	9.20		16.10	22

34 72 Railway Construction

34 72 16 – Railway Siding

34 72 16.50 Railroad Sidings		Crew	Daily Output	Labor-Hours	Unit	Material	2007 Bare Costs Labor	2007 Bare Costs Equipment	Total	Total Incl O&P
0010	**RAILROAD SIDINGS** R347216-10									
0800	Siding, yard spur, level grade									
0810	100 lb. rail, new material on wood ties	B-14	57	.842	L.F.	98.50	25.50	4.27	128.27	152
1000	Steel ties in concrete w/100 lb new rail, fasteners & plates	"	22	2.182	"	139	66.50	11.05	216.55	268

34 72 16.60 Railroad Turnouts										
0010	**RAILROAD TURNOUTS**									
2200	Turnouts, #8, incl. 100 lb. rails, plates, bars, frog, switch pt.									
2300	Timbers and ballast 6" below bottom of tie	B-14	.50	96	Ea.	10,700	2,925	485	14,110	16,800

Division Notes

	CREW	DAILY OUTPUT	LABOR-HOURS	UNIT	2007 BARE COSTS				TOTAL INCL O&P
					MAT.	LABOR	EQUIP.	TOTAL	

Estimating Tips

35 20 16 Hydraulic Gates
This section includes various types of gates that are commonly used in waterway and canal construction. Various earthwork items and structural support is found in Division 31, and concrete work in Division 3.

35 20 23 Dredging
This section includes barge and shore dredging systems for rivers, canals, and channels.

35 31 00 Shoreline Protection
This section includes breakwaters, bulkheads, and revetments for ocean and river inlets. Additional earthwork may be required from Division 31, and concrete work from Division 3.

35 49 00 Waterway Structures
This section includes breakwaters and bulkheads for canals.

35 51 00 Floating Construction
This section includes floating piers, docks, and dock accessories. Fixed Pier Timber Construction is found in 06 13 33. Driven piles are found in Division 31, as well as sheet piling, cofferdams, and riprap.

Reference Numbers

Reference numbers are shown in shaded boxes at the beginning of some major classifications. These numbers refer to related items in the Reference Section. The reference information may be an estimating procedure, an alternate pricing method, or technical information.

Note: Not all subdivisions listed here necessarily appear in this publication.

35 20 Waterway and Marine Construction and Equipment

35 20 23 – Dredging

35 20 23.13 Mechanical Dredging

		Crew	Daily Output	Labor-Hours	Unit	Material	2007 Bare Costs Labor	Equipment	Total	Total Incl O&P
0010	**MECHANICAL DREDGING**									
0020	Dredging mobilization and demobilization., add to below, minimum	B-8	.53	120	Total		3,900	4,975	8,875	11,500
0100	Maximum	"	.10	640	"		20,600	26,300	46,900	60,500
0300	Barge mounted clamshell excavation into scows,									
0310	Dumped 20 miles at sea, minimum	B-57	310	.155	B.C.Y.		5.15	4.66	9.81	12.95
0400	Maximum	"	213	.225	"		7.45	6.80	14.25	18.85
0500	Barge mounted dragline or clamshell, hopper dumped,									
0510	pumped 1000' to shore dump, minimum	B-57	340	.141	B.C.Y.		4.68	4.25	8.93	11.80
0525	All pumping uses 2000 gallons of water per cubic yard									
0600	Maximum	B-57	243	.198	B.C.Y.		6.55	5.95	12.50	16.55

35 20 23.23 Hydraulic Dredging

		Crew	Daily Output	Labor-Hours	Unit	Material	Labor	Equipment	Total	Total Incl O&P
0010	**HYDRAULIC DREDGING**									
1000	Hydraulic method, pumped 1000' to shore dump, minimum	B-57	460	.104	B.C.Y.		3.46	3.14	6.60	8.75
1100	Maximum		310	.155			5.15	4.66	9.81	12.95
1400	Into scows dumped 20 miles, minimum		425	.113			3.74	3.40	7.14	9.45
1500	Maximum		243	.198			6.55	5.95	12.50	16.55
1600	For inland rivers and canals in South, deduct								30%	30%

35 51 Floating Construction

35 51 13 – Floating Piers

35 51 13.23 Floating Wood Piers

		Crew	Daily Output	Labor-Hours	Unit	Material	Labor	Equipment	Total	Total Incl O&P
0010	**FLOATING WOOD PIERS**									
0020	Polyethylene encased polystyrene, no pilings included	F-3	330	.121	S.F.	32	4.52	2.19	38.71	44.50
0200	Pile supported, shore constructed, bare, 3" decking		130	.308		22	11.50	5.55	39.05	48
0250	4" decking		120	.333		21	12.45	6.05	39.50	49.50
0400	Floating, small boat, prefab, no shore facilities, minimum		250	.160		10.80	5.95	2.90	19.65	24.50
0500	Maximum		150	.267		36.50	9.95	4.83	51.28	61
0700	Per slip, minimum (180 S.F. each)		1.59	25.157	Ea.	2,325	940	455	3,720	4,500
0800	Maximum		1.40	28.571	"	6,875	1,075	515	8,465	9,800

Estimating Tips

Products such as correspondence lifts, conveyors, chutes, pneumatic tube systems, material handling cranes and hoists, as well as other items specified in this division, may require trained installers. The general contractor may not have any choice as to who will perform the installation or when it will be performed. Long lead times are often required for these products, making early decisions in purchasing and scheduling necessary.

The installation of this type of equipment may require the embedment of mounting hardware during construction of floors, structural walls, or interior walls/partitions. Electrical connections will require coordination with the electrical contractor.

Reference Numbers

Reference numbers are shown in shaded boxes at the beginning of some major classifications. These numbers refer to related items in the Reference Section. The reference information may be an estimating procedure, an alternate pricing method, or technical information.

Note: Not all subdivisions listed here necessarily appear in this publication.

No part of this publication may be reproduced, stored in a retrieval system, or transmitted in any form or by any means without prior written permission of Reed Construction Data.

41 21 Conveyors

41 21 13 – Automatic Guided Vehicle Systems

41 21 13.10 Motorized Car Distribution System

		Crew	Daily Output	Labor-Hours	Unit	Material	2007 Bare Costs Labor	Equipment	Total	Total Incl O&P
0010	**MOTORIZED CAR DISTRIBUTION SYSTEM**									
0100	20 lb. per car capacity, material handling									
0200	Minimum, 50' to 100'	4 Mill	.19	168	Station	26,600	6,425		33,025	38,700
0300	Maximum, 100' to 200'	"	.15	213	"	35,900	8,150		44,050	51,500
0400	Larger system, incl. hospital transport, track type,									
0500	fully automated material handling system									
0600	Minimum, complete system	4 Mill	.05	640	Ea.	118,000	24,400		142,400	165,500
0700	Maximum, with several stations	E-6	.05	2560	"	687,500	104,500	36,200	828,200	980,000

41 21 23 – Piece Material Conveyors

41 21 23.16 Container Piece Material Conveyors

		Crew	Daily Output	Labor-Hours	Unit	Material	Labor	Equipment	Total	Total Incl O&P
0010	**CONTAINER PIECE MATERIAL CONVEYORS**									
0050	10' sections with 2 supports, 600 lb. capacity, 18" wide				Ea.	395			395	435
0100	24" wide					455			455	500
0150	1400 lb. capacity, 18" wide					365			365	400
0200	24" wide					405			405	445
0350	Horizontal belt, center drive and takeup, 60 fpm									
0400	16" belt, 26.5' length	2 Mill	.50	32	Ea.	2,775	1,225		4,000	4,850
0450	24" belt, 41.5' length		.40	40		4,800	1,525		6,325	7,525
0500	61.5' length		.30	53.333		6,000	2,025		8,025	9,600
0600	Inclined belt, 10' rise with horizontal loader and									
0620	End idler assembly, 27.5' length, 18" belt	2 Mill	.30	53.333	Ea.	8,125	2,025		10,150	12,000
0700	24" belt	"	.15	106	"	9,125	4,075		13,200	16,000
3600	Monorail, overhead, manual, channel type									
3700	125 lb. per L.F.	1 Mill	26	.308	L.F.	9.05	11.75		20.80	27.50
3900	500 lb. per L.F.	"	21	.381	"	8.20	14.55		22.75	30.50
4000	Trolleys for above, 2 wheel, 125 lb. capacity				Ea.	67.50			67.50	74
4200	4 wheel, 250 lb. capacity					171			171	188
4300	8 wheel, 1,000 lb. capacity					330			330	360

41 21 23.20 Vertical Material Handling Conveyor

		Crew	Daily Output	Labor-Hours	Unit	Material	Labor	Equipment	Total	Total Incl O&P
0010	**VERTICAL MATERIAL HANDLING CONVEYOR**									
0100	to 10 floors, base price	2 Mill	.13	120	Floor	24,300	4,600		28,900	33,500
0200	Add for electrical hook-up and testing	2 Elec	.50	32	"		1,400		1,400	2,100

41 22 Cranes and Hoists

41 22 13 – Cranes

41 22 13.10 Crane Rail

		Crew	Daily Output	Labor-Hours	Unit	Material	Labor	Equipment	Total	Total Incl O&P
0010	**CRANE RAIL**									
0020	Box beam bridge, no equipment included	E-4	3400	.009	Lb.	.96	.39	.03	1.38	1.81
0200	Running track only, 104 lb per yard	"	5600	.006	"	.50	.24	.02	.76	1

41 22 13.13 Bridge Cranes

		Crew	Daily Output	Labor-Hours	Unit	Material	Labor	Equipment	Total	Total Incl O&P
0010	**BRIDGE CRANES**									
0100	1 girder, 20' span, 3 ton	M-3	1	34	Ea.	20,100	1,450	168	21,718	24,500
0125	5 ton		1	34		22,000	1,450	168	23,618	26,600
0150	7.5 ton		1	34		26,200	1,450	168	27,818	31,200
0175	10 ton		.80	42.500		34,600	1,800	210	36,610	40,900
0200	15 ton		.80	42.500		44,400	1,800	210	46,410	52,000
0225	30' span, 3 ton		1	34		20,900	1,450	168	22,518	25,400
0250	5 ton		1	34		22,900	1,450	168	24,518	27,600
0275	7.5 ton		1	34		27,500	1,450	168	29,118	32,700
0300	10 ton		.80	42.500		35,800	1,800	210	37,810	42,200

41 22 Cranes and Hoists

41 22 13 – Cranes

41 22 13.13 Bridge Cranes

		Crew	Daily Output	Labor-Hours	Unit	Material	2007 Bare Costs Labor	Equipment	Total	Total Incl O&P
0325	15 ton	M-3	.80	42.500	Ea.	46,400	1,800	210	48,410	54,000
0350	2 girder, 40' span, 3 ton	M-4	.50	72		34,500	3,025	520	38,045	43,000
0375	5 ton		.50	72		36,100	3,025	520	39,645	44,800
0400	7.5 ton		.50	72		39,600	3,025	520	43,145	48,700
0425	10 ton		.40	90		46,300	3,775	650	50,725	57,500
0450	15 ton		.40	90		63,000	3,775	650	67,425	75,500
0475	25 ton		.30	120		74,500	5,050	870	80,420	90,500
0500	50' span, 3 ton		.50	72		39,300	3,025	520	42,845	48,300
0525	5 ton		.50	72		40,800	3,025	520	44,345	50,000
0550	7.5 ton		.50	72		43,700	3,025	520	47,245	53,000
0575	10 ton		.40	90		50,500	3,775	650	54,925	62,000
0600	15 ton		.40	90		66,000	3,775	650	70,425	79,000
0625	25 ton		.30	120		78,000	5,050	870	83,920	94,500

41 22 23 – Hoists

41 22 23.10 Material Handling

		Crew	Daily Output	Labor-Hours	Unit	Material	2007 Bare Costs Labor	Equipment	Total	Total Incl O&P
0010	**MATERIAL HANDLING**									
1500	Cranes, portable hydraulic, floor type, 2,000 lb capacity				Ea.	2,000			2,000	2,200
1600	4,000 lb capacity					3,350			3,350	3,700
1800	Movable gantry type, 12' to 15' range, 2,000 lb capacity					2,925			2,925	3,225
1900	6,000 lb capacity					4,225			4,225	4,650
2100	Hoists, electric overhead, chain, hook hung, 15' lift, 1 ton cap.					1,800			1,800	1,975
2200	3 ton capacity					2,025			2,025	2,225
2500	5 ton capacity					5,350			5,350	5,900
2600	For hand-pushed trolley, add					15%				
2700	For geared trolley, add					30%				
2800	For motor trolley, add					75%				
3000	For lifts over 15', 1 ton, add				L.F.	18.90			18.90	21
3100	5 ton, add				"	42			42	46
3300	Lifts, scissor type, portable, electric, 36" high, 2,000 lb				Ea.	3,475			3,475	3,800
3400	48" high, 4,000 lb				"	3,775			3,775	4,150

Division Notes

Estimating Tips

There are four basic types of pollution: Air, Noise, Water, and Solid.

These systems may be interrelated and care must be taken that the complete systems are estimated. For example, Air Pollution Equipment may include dust and air-entrained particles that have to be collected. The vacuum systems could be noisy, requiring silencers to reduce noise pollution, and the collected solids have to be disposed of to prevent Solid Pollution.

Water Treatment may be accomplished as comparatively small packaged plants providing Chemical, Biological or Thermal Treatment. These are usually priced by GPH or GPD.

Large plants and municipal treatment facilities are sized MGD. These facilities are usually priced in two steps. First, the large concrete structures used as settling ponds, etc.; and second, the equipment required. Storm water does not need as much treatment as sanitary wastewater. The total cost of the system is greatly affected by the level of purification required by specifications and codes before release.

An additional cost may be encountered if there is a possibility that the water being treated may contain petroleum products, which have to be isolated and disposed of in an environmentally acceptable manner.

Reference Numbers

Reference numbers are shown in shaded boxes at the beginning of some major classifications. These numbers refer to related items in the Reference Section. The reference information may be an estimating procedure, an alternate pricing method, or technical information.

Note: Not all subdivisions listed here necessarily appear in this publication.

No part of this publication may be reproduced, stored in a retrieval system, or transmitted in any form or by any means without prior written permission of Reed Construction Data.

44 11 Air Pollution Control Equipment

44 11 16 – Industrial Dust Collectors

44 11 16.10 Dust Collection Systems	Crew	Daily Output	Labor-Hours	Unit	Material	2007 Bare Costs Labor	Equipment	Total	Total Incl O&P
0010 **DUST COLLECTION SYSTEMS** Commercial / industrial									
0120 Central vacuum units									
0130 Includes stand, filters and motorized shaker									
0200 500 CFM, 10" inlet, 2 HP	Q-20	2.40	8.333	Ea.	3,650	335		3,985	4,500
0220 1000 CFM, 10" inlet, 3 HP		2.20	9.091		3,800	365		4,165	4,750
0240 1500 CFM, 10" inlet, 5 HP		2	10		3,900	400		4,300	4,900
0260 3000 CFM, 13" inlet, 10 HP		1.50	13.333		10,200	535		10,735	12,000
0280 5000 CFM, 16" inlet, 2 @ 10 HP	↓	1	20	↓	10,800	805		11,605	13,000
1000 Vacuum tubing, galvanized									
1100 2-1/8" OD, 16 ga.	Q-9	440	.036	L.F.	2.26	1.43		3.69	4.69
1110 2-1/2" OD, 16 ga.		420	.038		2.61	1.49		4.10	5.15
1120 3" OD, 16 ga.		400	.040		3.33	1.57		4.90	6.10
1130 3-1/2" OD, 16 ga.		380	.042		4.72	1.65		6.37	7.75
1140 4" OD, 16 ga.		360	.044		4.82	1.74		6.56	8
1150 5" OD, 14 ga.		320	.050		8.95	1.96		10.91	12.80
1160 6" OD, 14 ga.		280	.057		10	2.24		12.24	14.45
1170 8" OD, 14 ga.		200	.080		15.15	3.14		18.29	21.50
1180 10" OD, 12 ga.		160	.100		15.15	3.92		19.07	22.50
1190 12" OD, 12 ga.		120	.133		43.50	5.25		48.75	55.50
1200 14" OD, 12 ga.	↓	80	.200	↓	48	7.85		55.85	64.50
1940 Hose, flexible wire reinforced rubber									
1956 3" dia.	Q-9	400	.040	L.F.	5.55	1.57		7.12	8.50
1960 4" dia.		360	.044		6.30	1.74		8.04	9.60
1970 5" dia.		320	.050		7.15	1.96		9.11	10.85
1980 6" dia.	↓	280	.057	↓	7.75	2.24		9.99	11.95
2000 90° Elbow, slip fit									
2110 2-1/8" dia.	Q-9	70	.229	Ea.	9.45	8.95		18.40	24
2120 2-1/2" dia.		65	.246		13.20	9.65		22.85	29.50
2130 3" dia.		60	.267		17.90	10.45		28.35	36
2140 3-1/2" dia.		55	.291		22	11.40		33.40	42
2150 4" dia.		50	.320		27.50	12.55		40.05	49.50
2160 5" dia.		45	.356		52	13.95		65.95	78.50
2170 6" dia.		40	.400		72	15.70		87.70	103
2180 8" dia.	↓	30	.533	↓	136	21		157	182
2400 45° Elbow, slip fit									
2410 2-1/8" dia.	Q-9	70	.229	Ea.	8.25	8.95		17.20	23
2420 2-1/2" dia.		65	.246		12.20	9.65		21.85	28.50
2430 3" dia.		60	.267		14.70	10.45		25.15	32.50
2440 3-1/2" dia.		55	.291		18.50	11.40		29.90	38
2450 4" dia.		50	.320		24	12.55		36.55	46
2460 5" dia.		45	.356		40.50	13.95		54.45	66
2470 6" dia.		40	.400		54.50	15.70		70.20	84
2480 8" dia.	↓	35	.457	↓	107	17.90		124.90	146
2800 90° TY, slip fit thru 6" dia									
2810 2-1/8" dia.	Q-9	42	.381	Ea.	18.45	14.95		33.40	43.50
2820 2-1/2" dia.		39	.410		24	16.10		40.10	51.50
2830 3" dia.		36	.444		32	17.40		49.40	62
2840 3-1/2" dia.		33	.485		40.50	19		59.50	74
2850 4" dia.		30	.533		58	21		79	96
2860 5" dia.		27	.593		110	23		133	158
2870 6" dia.	↓	24	.667		150	26		176	207
2880 8" dia., butt end					320			320	350
2890 10" dia., butt end				↓	575			575	635

44 11 Air Pollution Control Equipment

44 11 16 – Industrial Dust Collectors

44 11 16.10 Dust Collection Systems		Crew	Daily Output	Labor-Hours	Unit	Material	2007 Bare Costs Labor	2007 Bare Costs Equipment	Total	Total Incl O&P
2900	12" dia., butt end				Ea.	575			575	635
2910	14" dia., butt end					1,050			1,050	1,150
2920	6" x 4" dia., butt end					92			92	101
2930	8" x 4" dia., butt end					174			174	191
2940	10" x 4" dia., butt end					228			228	250
2950	12" x 4" dia., butt end					261			261	287
3100	90° Elbow, butt end segmented									
3110	8" dia., butt end, segmented				Ea.	206			206	227
3120	10" dia., butt end, segmented					340			340	375
3130	12" dia., butt end, segmented					430			430	470
3140	14" dia., butt end, segmented					540			540	590
3200	45° Elbow, butt end segmented									
3210	8" dia., butt end, segmented				Ea.	176			176	193
3220	10" dia., butt end, segmented					245			245	270
3230	12" dia., butt end, segmented					251			251	276
3240	14" dia., butt end, segmented					310			310	340
3400	All butt end fittings require one coupling per joint.									
3410	Labor for fitting included with couplings.									
3460	Compression coupling, galvanized, neoprene gasket									
3470	2-1/8" dia.	Q-9	44	.364	Ea.	14.40	14.25		28.65	38
3480	2-1/2" dia.		44	.364		14.40	14.25		28.65	38
3490	3" dia.		38	.421		18.90	16.50		35.40	46.50
3500	3-1/2" dia.		35	.457		21	17.90		38.90	50.50
3510	4" dia.		33	.485		23	19		42	54.50
3520	5" dia.		29	.552		26	21.50		47.50	62.50
3530	6" dia.		26	.615		30.50	24		54.50	70.50
3540	8" dia.		22	.727		55.50	28.50		84	105
3550	10" dia.		20	.800		81.50	31.50		113	138
3560	12" dia.		18	.889		102	35		137	166
3570	14" dia.		16	1		144	39		183	219
3800	Air gate valves, galvanized									
3810	2-1/8" dia.	Q-9	30	.533	Ea.	109	21		130	152
3820	2-1/2" dia.		28	.571		115	22.50		137.50	161
3830	3" dia.		26	.615		133	24		157	184
3840	4" dia.		23	.696		175	27.50		202.50	234
3850	6" dia.		18	.889		241	35		276	320

44 41 Packaged Water Treatment

44 41 13 – Packaged Water Treatment Plants

44 41 13.16 Biological Packaged Water Treatment Plants

		Crew	Daily Output	Labor-Hours	Unit	Material	2007 Bare Costs Labor	2007 Bare Costs Equipment	Total	Total Incl O&P
0010	**BIOLOGICAL PACKAGED WATER TREATMENT PLANTS**									
0020	Steel packaged, blown air aeration plants									
0100	1,000 GPD				Gal.				18.60	21.40
0200	5,000 GPD								12.39	14.25
0300	15,000 GPD								6.83	7.85
0400	30,000 GPD								6.45	7.40
0500	50,000 GPD								5	5.75
0600	100,000 GPD								4.36	5
0700	200,000 GPD								3.10	3.57
0800	500,000 GPD								3.03	3.48
1000	Concrete, extended aeration, primary and secondary treatment									

44 41 Packaged Water Treatment

44 41 13 – Packaged Water Treatment Plants

44 41 13.16 Biological Packaged Water Treatment Plants

		Crew	Daily Output	Labor-Hours	Unit	Material	2007 Bare Costs Labor	Equipment	Total	Total Incl O&P
1010	10,000 GPD				Gal.				14.95	17.20
1100	30,000 GPD								6.85	7.90
1200	50,000 GPD								5.85	6.75
1400	100,000 GPD								4.35	5
1500	500,000 GPD				▼				3.10	3.57
1700	Municipal wastewater treatment facility									
1720	1.0 MGD				Gal.				5.35	6.15
1740	1.5 MGD								5.27	6.05
1760	2.0 MGD								4.52	5.20
1780	3.0 MGD								3.53	4.06
1800	5.0 MGD				▼				3.22	3.70
2000	Holding tank system, not incl. excavation or backfill									
2010	Recirculating chemical water closet	2 Plum	4	4	Ea.	875	179		1,054	1,225
2100	For voltage converter, add	"	16	1		232	45		277	325
2200	For high level alarm, add	1 Plum	7.80	1.026	▼	133	46		179	215

44 41 13.17 Wastewater Treatment System

		Crew	Daily Output	Labor-Hours	Unit	Material	Labor	Equipment	Total	Total Incl O&P
0010	**WASTEWATER TREATMENT SYSTEM**									
0020	Fiberglass, 1,000 gallon	B-21	1.29	21.705	Ea.	3,200	730	127	4,057	4,775
0100	1,500 gallon	B-21	1.03	27.184	Ea.	7,000	910	159	8,069	9,250

Reference Section

All the reference information is in one section, making it easy to find what you need to know... and easy to use the book on a daily basis. This section is visually identified by a vertical gray bar on the page edges.

In this Reference Section, we've included Equipment Rental Costs, a listing of rental and operating costs; Crew Listings, a full listing of all crews and equipment, and their costs; Historical Cost Indexes for cost comparisons over time; City Cost Indexes and Location Factors for adjusting costs to the region you are in; Reference Tables, where you will find explanations, estimating information and procedures, or technical data; Change Orders, information on pricing changes to contract documents; Square Foot Costs that allow you to make a rough estimate for the overall cost of a project; and an explanation of all the Abbreviations in the book.

Table of Contents

Construction Equipment Rental Costs	601
Crew Listings	613
Historical Cost Indexes	642
City Cost Indexes	643
Location Factors	686
Reference Tables	692
R01 General Requirements	692
R02 Existing Conditions	703
R03 Concrete	705
R04 Masonry	720
R05 Metals	722
R06 Wood, Plastics & Composites	729
R07 Thermal & Moisture Protection	729
R08 Openings	731
R09 Finishes	733
R13 Special Construction	736
R14 Conveying Equipment	738

Reference Tables (cont.)

R22 Plumbing	739
R23 Heating, Ventilating & Air Conditioning	741
R26 Electrical	744
R31 Earthwork	746
R32 Exterior Improvements	750
R33 Utilities	751
R34 Transportation	751
Change Orders	752
Square Foot Costs	757
Abbreviations	767

Equipment Rental Costs

Estimating Tips
- This section contains the average costs to rent and operate hundreds of pieces of construction equipment. This is useful information when estimating the time and material requirements of any particular operation in order to establish a unit or total cost. Equipment costs include not only rental, but also operating costs for equipment under normal use.

Rental Costs
- Equipment rental rates are obtained from industry sources throughout North America-contractors, suppliers, dealers, manufacturers, and distributors.
- Rental rates vary throughout the country, with larger cities generally having lower rates. Lease plans for new equipment are available for periods in excess of six months, with a percentage of payments applying toward purchase.
- Monthly rental rates vary from 2% to 5% of the purchase price of the equipment depending on the anticipated life of the equipment and its wearing parts.
- Weekly rental rates are about 1/3 the monthly rates, and daily rental rates are about 1/3 the weekly rate.

Operating Costs
- The operating costs include parts and labor for routine servicing, such as repair and replacement of pumps, filters and worn lines. Normal operating expendables, such as fuel, lubricants, tires and electricity (where applicable), are also included.
- Extraordinary operating expendables with highly variable wear patterns, such as diamond bits and blades, are excluded. These costs can be found as material costs in the Unit Price section.
- The hourly operating costs listed do not include the operator's wages.

Crew Equipment Cost/Day
- Any power equipment required by a crew is shown in the Crew Listings with a daily cost.
- The daily cost of crew equipment is based on dividing the weekly rental rate by 5 (number of working days in the week), and then adding the hourly operating cost times 8 (the number of hours in a day). This "Crew Equipment Cost/Day" is shown in the far right column of the Equipment Rental pages.
- If equipment is needed for only one or two days, it is best to develop your own cost by including components for daily rent and hourly operating cost. This is important when the listed Crew for a task does not contain the equipment needed, such as a crane for lifting mechanical heating/cooling equipment up onto a roof.

Mobilization/Demobilization
- The cost to move construction equipment from an equipment yard or rental company to the jobsite and back again is not included in equipment rental costs listed in the Reference section, nor in the bare equipment cost of any Unit Price line item, nor in any equipment costs shown in the Crew listings.
- Mobilization (to the site) and demobilization (from the site) costs can be found in the Unit Price section.
- If a piece of equipment is already at the jobsite, it is not appropriate to utilize mobil./demob. costs again in an estimate.

No part of this publication may be reproduced, stored in a retrieval system, or transmitted in any form or by any means without prior written permission of Reed Construction Data.

01 54 | Construction Aids

01 54 33 | Equipment Rental

		UNIT	HOURLY OPER. COST	RENT PER DAY	RENT PER WEEK	RENT PER MONTH	CREW EQUIPMENT COST/DAY	
10	0010 **CONCRETE EQUIPMENT RENTAL**, without operators							10
	0150 For batch plant, see div. 01 54 33.50							
	0200 Bucket, concrete lightweight, 1/2 C.Y.	Ea.	.60	17.35	52	156	15.20	
	0300 1 C.Y.		.65	21	63	189	17.80	
	0400 1-1/2 C.Y.		.80	28.50	85	255	23.40	
	0500 2 C.Y.		.90	33.50	100	300	27.20	
	0580 8 C.Y.		4.85	223	670	2,000	172.80	
	0600 Cart, concrete, self propelled, operator walking, 10 C.F.		2.30	58.50	175	525	53.40	
	0700 Operator riding, 18 C.F.		3.60	83.50	250	750	78.80	
	0800 Conveyer for concrete, portable, gas, 16" wide, 26' long		8.05	118	355	1,075	135.40	
	0900 46' long		8.40	143	430	1,300	153.20	
	1000 56' long		8.55	153	460	1,375	160.40	
	1100 Core drill, electric, 2-1/2 H.P., 1" to 8" bit diameter		2.20	89.50	268	805	71.20	
	1150 11 HP, 8" to 18" cores		4.75	112	335	1,000	105	
	1200 Finisher, concrete floor, gas, riding trowel, 48" diameter		5.55	91.50	275	825	99.40	
	1300 Gas, manual, 3 blade, 36" trowel	Ea.	1.15	16.65	50	150	19.20	
	1400 4 blade, 48" trowel		1.65	21	63	189	25.80	
	1500 Float, hand-operated (Bull float) 48" wide		.08	13	39	117	8.45	
	1570 Curb builder, 14 H.P., gas, single screw		10.45	228	685	2,050	220.60	
	1590 Double screw		11.10	270	810	2,425	250.80	
	1600 Grinder, concrete and terrazzo, electric, floor		2.92	127	380	1,150	99.35	
	1700 Wall grinder		1.46	63.50	190	570	49.70	
	1800 Mixer, powered, mortar and concrete, gas, 6 C.F., 18 H.P.		5.70	113	340	1,025	113.60	
	1900 10 C.F., 25 H.P.		6.95	138	415	1,250	138.60	
	2000 16 C.F.		7.30	162	485	1,450	155.40	
	2100 Concrete, stationary, tilt drum, 2 C.Y.		5.75	223	670	2,000	180	
	2120 Pump, concrete, truck mounted 4" line 80' boom		21.95	890	2,665	8,000	708.60	
	2140 5" line, 110' boom		28.70	1,175	3,545	10,600	938.60	
	2160 Mud jack, 50 C.F. per hr.		5.75	127	380	1,150	122	
	2180 225 C.F. per hr.		7.50	143	430	1,300	146	
	2190 Shotcrete pump rig, 12 CY/hr		13.70	235	705	2,125	250.60	
	2600 Saw, concrete, manual, gas, 18 H.P.		4.00	36.50	110	330	54	
	2650 Self-propelled, gas, 30 H.P.		7.80	93.50	280	840	118.40	
	2700 Vibrators, concrete, electric, 60 cycle, 2 H.P.		.42	7.65	23	69	7.95	
	2800 3 H.P.		.58	9	27	81	10.05	
	2900 Gas engine, 5 H.P.		1.05	13.65	41	123	16.60	
	3000 8 H.P.		1.55	15	45	135	21.40	
	3050 Vibrating screed, gas engine, 8 H.P.		2.71	73.50	221	665	65.90	
	3100 Concrete transit mixer, hydraulic drive							
	3120 6 x 4, 250 H.P., 8 C.Y., rear discharge		39.50	550	1,645	4,925	645	
	3200 Front discharge		46.50	680	2,040	6,125	780	
	3300 6 x 6, 285 H.P., 12 C.Y., rear discharge		45.60	640	1,920	5,750	748.80	
	3400 Front discharge		47.65	690	2,065	6,200	794.20	
20	0010 **EARTHWORK EQUIPMENT RENTAL**, without operators							20
	0040 Aggregate spreader, push type 8' to 12' wide	Ea.	2.00	24.50	73	219	30.60	
	0045 Tailgate type, 8' wide	"	1.90	32.50	97	291	34.60	
	0050 Augers for vertical drilling							
	0055 Earth auger, truck-mounted, for fence & sign posts	Ea.	10.40	485	1,460	4,375	375.20	
	0060 For borings and monitoring wells		35.95	645	1,935	5,800	674.60	
	0070 Earth auger, portable, trailer mounted		2.25	23	69	207	31.80	
	0075 Earth auger, truck-mounted, for caissons, water wells, utility poles		174.00	3,500	10,535	31,600	3,499	
	0080 Auger, horizontal boring machine, 12" to 36" diameter, 45 H.P.		19.00	192	575	1,725	267	
	0090 12" to 48" diameter, 65 H.P.		27.05	340	1,025	3,075	421.40	
	0095 Auger, for fence posts, gas engine, hand held		.40	4.67	14	42	6	
	0100 Excavator, diesel hydraulic, crawler mounted, 1/2 C.Y. cap.		17.75	360	1,080	3,250	358	
	0120 5/8 C.Y. capacity		21.40	485	1,455	4,375	462.20	
	0140 3/4 C.Y. capacity		24.80	530	1,590	4,775	516.40	

01 54 | Construction Aids

01 54 33 | Equipment Rental

Line	Description	Unit	Hourly Oper. Cost	Rent Per Day	Rent Per Week	Rent Per Month	Crew Equipment Cost/Day
0150	1 C.Y. capacity	Ea.	30.85	590	1,775	5,325	601.80
0200	1-1/2 C.Y. capacity		37.95	785	2,360	7,075	775.60
0300	2 C.Y. capacity		49.45	1,000	2,995	8,975	994.60
0320	2-1/2 C.Y. capacity		65.60	1,350	4,040	12,100	1,333
0340	3-1/2 C.Y. capacity		106.15	2,200	6,585	19,800	2,166
0341	Attachments						
0342	Bucket thumbs		2.55	212	635	1,900	147.40
0345	Grapples		2.35	182	545	1,625	127.80
0350	Gradall type, truck mounted, 3 ton @ 15' radius, 5/8 C.Y.		42.35	945	2,835	8,500	905.80
0370	1 C.Y. capacity		47.80	1,075	3,220	9,650	1,026
0400	Backhoe-loader, 40 to 45 H.P., 5/8 C.Y. capacity		9.85	197	590	1,775	196.80
0450	45 H.P. to 60 H.P., 3/4 C.Y. capacity		12.30	242	725	2,175	243.40
0460	80 H.P., 1-1/4 C.Y. capacity		15.05	275	825	2,475	285.40
0470	112 H.P., 1-1/2 C.Y. capacity		20.75	405	1,220	3,650	410
0480	Attachments						
0482	Compactor, 20,000 lb		4.45	120	360	1,075	107.60
0485	Hydraulic hammer, 750 ft-lbs		2.05	70	210	630	58.40
0486	Hydraulic hammer, 1200 ft-lbs		4.30	135	405	1,225	115.40
0500	Brush chipper, gas engine, 6" cutter head, 35 H.P.		7.25	102	305	915	119
0550	12" cutter head, 130 H.P.		11.45	153	460	1,375	183.60
0600	15" cutter head, 165 H.P.		16.30	165	495	1,475	229.40
0750	Bucket, clamshell, general purpose, 3/8 C.Y.		1.05	35	105	315	29.40
0800	1/2 C.Y.		1.15	41.50	125	375	34.20
0850	3/4 C.Y.		1.30	51.50	155	465	41.40
0900	1 C.Y.		1.35	56.50	170	510	44.80
0950	1-1/2 C.Y.		2.10	76.50	230	690	62.80
1000	2 C.Y.		2.25	86.50	260	780	70
1010	Bucket, dragline, medium duty, 1/2 C.Y.		.60	22.50	67	201	18.20
1020	3/4 C.Y.		.65	23.50	71	213	19.40
1030	1 C.Y.		.65	25.50	77	231	20.60
1040	1-1/2 C.Y.		1.00	38.50	115	345	31
1050	2 C.Y.		1.05	43.50	130	390	34.40
1070	3 C.Y.		1.60	60	180	540	48.80
1200	Compactor, manually guided 2-drum vibratory smooth roller, 7.5 H.P.		5.35	152	455	1,375	133.80
1250	Rammer compactor, gas, 1000 lb. blow		2.05	36.50	110	330	38.40
1300	Vibratory plate, gas, 18" plate, 3000 lb. blow		2.00	22	66	198	29.20
1350	21" plate, 5000 lb. blow		2.45	27.50	83	249	36.20
1370	Curb builder/extruder, 14 H.P., gas, single screw		10.45	228	685	2,050	220.60
1390	Double screw		11.10	270	810	2,425	250.80
1500	Disc harrow attachment, for tractor		.38	63	189	565	40.85
1750	Extractor, piling, see lines 2500 to 2750						
1810	Feller buncher, shearing & accumulating trees, 100 H.P.	Ea.	24.80	525	1,575	4,725	513.40
1860	Grader, self-propelled, 25,000 lb.		20.85	425	1,270	3,800	420.80
1910	30,000 lb.		25.05	510	1,525	4,575	505.40
1920	40,000 lb.		34.40	765	2,300	6,900	735.20
1930	55,000 lb.		45.75	1,100	3,275	9,825	1,021
1950	Hammer, pavement demo., hyd., gas, self-prop., 1000 to 1250 lb.		20.15	310	935	2,800	348.20
2000	Diesel 1300 to 1500 lb.		31.20	615	1,850	5,550	619.60
2050	Pile driving hammer, steam or air, 4150 ft.-lb. @ 225 BPM		6.75	275	825	2,475	219
2100	8750 ft.-lb. @ 145 BPM		8.65	445	1,340	4,025	337.20
2150	15,000 ft.-lb. @ 60 BPM		9.00	480	1,440	4,325	360
2200	24,450 ft.-lb. @ 111 BPM		11.95	530	1,590	4,775	413.60
2250	Leads, 15,000 ft.-lb. hammers	L.F.	.03	1.28	3.83	11.50	1
2300	24,450 ft.-lb. hammers and heavier	"	.05	2.33	7	21	1.80
2350	Diesel type hammer, 22,400 ft.-lb.	Ea.	28.40	615	1,840	5,525	595.20
2400	41,300 ft.-lb.		39.70	665	1,990	5,975	715.60
2450	141,000 ft.-lb.		64.00	1,150	3,415	10,200	1,195
2500	Vib. elec. hammer/extractor, 200 KW diesel generator, 34 H.P.		33.55	640	1,925	5,775	653.40

01 54 | Construction Aids

01 54 33 | Equipment Rental

			UNIT	HOURLY OPER. COST	RENT PER DAY	RENT PER WEEK	RENT PER MONTH	CREW EQUIPMENT COST/DAY
2550		80 H.P.	Ea.	59.10	935	2,810	8,425	1,035
2600		150 H.P.		110.75	1,825	5,485	16,500	1,983
2700	Extractor, steam or air, 700 ft.-lb.			15.05	470	1,405	4,225	401.40
2750		1000 ft.-lb.		17.15	570	1,715	5,150	480.20
2800	Log chipper, up to 22" diam, 600 H.P.			32.75	430	1,290	3,875	520
2850	Logger, for skidding & stacking logs, 150 H.P.			39.55	835	2,500	7,500	816.40
2900	Rake, spring tooth, with tractor			8.04	219	658	1,975	195.90
3000	Roller, vibratory, tandem, smooth drum, 20 H.P.			6.15	122	365	1,100	122.20
3050		35 H.P.		8.75	235	705	2,125	211
3100	Towed type vibratory compactor, smooth drum, 50 H.P.			20.25	315	940	2,825	350
3150	Sheepsfoot, 50 H.P.			21.25	345	1,030	3,100	376
3170	Landfill compactor, 220 HP			62.00	1,225	3,650	11,000	1,226
3200	Pneumatic tire roller, 80 H.P.			12.20	330	990	2,975	295.60
3250		120 H.P.		18.75	565	1,690	5,075	488
3300	Sheepsfoot vibratory roller, 200 H.P.			50.15	925	2,775	8,325	956.20
3320		340 H.P.		69.85	1,375	4,090	12,300	1,377
3350	Smooth drum vibratory roller, 75 H.P.			18.30	505	1,510	4,525	448.40
3400		125 H.P.		23.95	625	1,870	5,600	565.60
3410	Rotary mower, brush, 60", with tractor			11.50	238	715	2,150	235
3420	Rototiller, 5 HP, walk-behind			1.89	55.50	166	500	48.30
3450	Scrapers, towed type, 9 to 12 C.Y. capacity			5.65	163	490	1,475	143.20
3500		12 to 17 C.Y. capacity		6.25	178	535	1,600	157
3550	Scrapers, self-propelled, 4 x 4 drive, 2 engine, 14 C.Y. capacity			98.80	1,550	4,635	13,900	1,717
3600		2 engine, 24 C.Y. capacity		137.50	2,425	7,245	21,700	2,549
3640		32 - 44 C.Y. capacity		163.05	2,825	8,465	25,400	2,997
3650	Self-loading, 11 C.Y. capacity			44.85	835	2,505	7,525	859.80
3700		22 C.Y. capacity		85.05	1,775	5,350	16,100	1,750
3710	Screening plant 110 H.P. w/ 5' x 10' screen			28.65	390	1,165	3,500	462.20
3720		5' x 16' screen		30.75	490	1,465	4,400	539
3850	Shovels, see Cranes division 01590-600							
3860	Shovel/backhoe bucket, 1/2 C.Y.		Ea.	1.90	56.50	170	510	49.20
3870		3/4 C.Y.		1.95	63.50	190	570	53.60
3880		1 C.Y.		2.05	73.50	220	660	60.40
3890		1-1/2 C.Y.		2.20	86.50	260	780	69.60
3910		3 C.Y.		2.50	122	365	1,100	93
3950	Stump chipper, 18" deep, 30 H.P.			5.54	103	310	930	106.30
4110	Tractor, crawler, with bulldozer, torque converter, diesel 80 H.P.			18.60	345	1,030	3,100	354.80
4150		105 H.P.		25.60	485	1,450	4,350	494.80
4200		140 H.P.		30.55	610	1,835	5,500	611.40
4260		200 H.P.		46.30	1,025	3,090	9,275	988.40
4310		300 H.P.		59.75	1,375	4,115	12,300	1,301
4360		410 H.P.		80.30	1,725	5,175	15,500	1,677
4370		500 H.P.		106.80	2,350	7,030	21,100	2,260
4380		700 H.P.		158.50	3,575	10,690	32,100	3,406
4400	Loader, crawler, torque conv., diesel, 1-1/2 C.Y., 80 H.P.			16.75	335	1,000	3,000	334
4450		1-1/2 to 1-3/4 C.Y., 95 H.P.		19.65	405	1,210	3,625	399.20
4510		1-3/4 to 2-1/4 C.Y., 130 H.P.		26.90	625	1,870	5,600	589.20
4530		2-1/2 to 3-1/4 C.Y., 190 H.P.		39.05	870	2,610	7,825	834.40
4560		3-1/2 to 5 C.Y., 275 H.P.		53.35	1,200	3,590	10,800	1,145
4610	Tractor loader, wheel, torque conv., 4 x 4, 1 to 1-1/4 C.Y., 65 H.P.			11.70	190	570	1,700	207.60
4620		1-1/2 to 1-3/4 C.Y., 80 H.P.		15.20	260	780	2,350	277.60
4650		1-3/4 to 2 C.Y., 100 H.P.		17.20	300	900	2,700	317.60
4710		2-1/2 to 3-1/2 C.Y., 130 H.P.		18.75	330	995	2,975	349
4730		3 to 4-1/2 C.Y., 170 H.P.		24.25	490	1,470	4,400	488
4760		5-1/4 to 5-3/4 C.Y., 270 H.P.		39.60	730	2,195	6,575	755.80
4810		7 to 8 C.Y., 375 H.P.		67.35	1,325	3,990	12,000	1,337
4870		12-1/2 C.Y., 690 H.P.		93.10	2,050	6,175	18,500	1,980
4880		Wheeled, skid steer, 10 C.F., 30 H.P. gas		6.50	117	350	1,050	122

Reference codes: R015433-10, R312323-30, R312316-40, R312316-45

01 54 | Construction Aids

01 54 33 | Equipment Rental

			UNIT	HOURLY OPER. COST	RENT PER DAY	RENT PER WEEK	RENT PER MONTH	CREW EQUIPMENT COST/DAY	
20	4890	1 C.Y., 78 H.P., diesel		12.15	208	625	1,875	222.20	20
	4891	Attachments for all skid steer loaders							
	4892	Auger	Ea.	.48	80.50	241	725	52.05	
	4893	Backhoe		.69	114	343	1,025	74.10	
	4894	Broom		.63	105	314	940	67.85	
	4895	Forks		.23	37.50	113	340	24.45	
	4896	Grapple		.56	93.50	281	845	60.70	
	4897	Concrete hammer		1.01	168	504	1,500	108.90	
	4898	Tree spade		1.04	173	519	1,550	112.10	
	4899	Trencher		.70	117	352	1,050	76	
	4900	Trencher, chain, boom type, gas, operator walking, 12 H.P.		3.00	45	135	405	51	
	4910	Operator riding, 40 H.P.		9.70	248	745	2,225	226.60	
	5000	Wheel type, diesel, 4' deep, 12" wide		54.15	765	2,290	6,875	891.20	
	5100	Diesel, 6' deep, 20" wide		68.95	1,800	5,370	16,100	1,626	
	5150	Ladder type, diesel, 5' deep, 8" wide		37.95	860	2,585	7,750	820.60	
	5200	Diesel, 8' deep, 16" wide		63.00	1,825	5,510	16,500	1,606	
	5210	Tree spade, self-propelled	Ea.	12.85	267	800	2,400	262.80	
	5250	Truck, dump, tandem, 12 ton payload		24.80	283	850	2,550	368.40	
	5300	Three axle dump, 16 ton payload		33.50	435	1,310	3,925	530	
	5350	Dump trailer only, rear dump, 16-1/2 C.Y.	Ea.	4.35	122	365	1,100	107.80	
	5400	20 C.Y.		4.75	137	410	1,225	120	
	5450	Flatbed, single axle, 1-1/2 ton rating		13.30	58.50	175	525	141.40	
	5500	3 ton rating		16.90	83.50	250	750	185.20	
	5550	Off highway rear dump, 25 ton capacity		46.35	1,025	3,110	9,325	992.80	
	5600	35 ton capacity		47.25	1,050	3,180	9,550	1,014	
	5610	50 ton capacity		61.20	1,375	4,090	12,300	1,308	
	5620	65 ton capacity		65.65	1,450	4,340	13,000	1,393	
	5630	100 ton capacity		84.30	1,875	5,590	16,800	1,792	
	6000	Vibratory plow, 25 H.P., walking		5.20	58.50	175	525	76.60	
40	0010	**GENERAL EQUIPMENT RENTAL**, without operators							40
	0150	Aerial lift, scissor type, to 15' high, 1000 lb. cap., electric	Ea.	2.40	45	135	405	46.20	
	0160	To 25' high, 2000 lb. capacity		2.80	63.50	190	570	60.40	
	0170	Telescoping boom to 40' high, 500 lb. capacity, gas		14.85	285	855	2,575	289.80	
	0180	To 45' high, 500 lb. capacity		15.80	325	980	2,950	322.40	
	0190	To 60' high, 600 lb. capacity		17.80	435	1,305	3,925	403.40	
	0195	Air compressor, portable, 6.5 CFM, electric		.42	10.35	31	93	9.55	
	0196	Gasoline		.67	15.65	47	141	14.75	
	0200	Air compressor, portable, gas engine, 60 C.F.M.		10.10	46.50	140	420	108.80	
	0300	160 C.F.M.		11.70	48.50	145	435	122.60	
	0400	Diesel engine, rotary screw, 250 C.F.M.		11.80	95	285	855	151.40	
	0500	365 C.F.M.		15.85	115	345	1,025	195.80	
	0550	450 C.F.M.		20.15	147	440	1,325	249.20	
	0600	600 C.F.M.		34.90	195	585	1,750	396.20	
	0700	750 C.F.M.		35.35	212	635	1,900	409.80	
	0800	For silenced models, small sizes, add							
	0900	Large sizes, add							
	0920	Air tools and accessories							
	0930	Breaker, pavement, 60 lb.	Ea.	.40	8.65	26	78	8.40	
	0940	80 lb.		.40	9	27	81	8.60	
	0950	Drills, hand (jackhammer) 65 lb.		.50	15.35	46	138	13.20	
	0960	Track or wagon, swing boom, 4" drifter		46.05	700	2,095	6,275	787.40	
	0970	5" drifter		56.05	760	2,285	6,850	905.40	
	0975	Track mounted quarry drill, 6" diameter drill		77.10	1,150	3,445	10,300	1,306	
	0980	Dust control per drill		.91	17.35	52	156	17.70	
	0990	Hammer, chipping, 12 lb.		.40	20.50	62	186	15.60	
	1000	Hose, air with couplings, 50' long, 3/4" diameter		.04	6.65	20	60	4.30	
	1100	1" diameter		.04	6.35	19	57	4.10	

01 54 | Construction Aids

01 54 33 | Equipment Rental

			UNIT	HOURLY OPER. COST	RENT PER DAY	RENT PER WEEK	RENT PER MONTH	CREW EQUIPMENT COST/DAY	
40	1200	1-1/2" diameter	Ea.	.05	9	27	81	5.80	40
	1300	2" diameter	R312323-30	.10	17	51	153	11	
	1400	2-1/2" diameter		.12	19.35	58	174	12.55	
	1410	3" diameter	R314116-40	.18	29.50	88	264	19.05	
	1450	Drill, steel, 7/8" x 2'		.07	12.35	37	111	7.95	
	1460	7/8" x 6'	R314116-45	.08	12.65	38	114	8.25	
	1520	Moil points		.03	4.33	13	39	2.85	
	1525	Pneumatic nailer w/accessories		.43	28.50	86	258	20.65	
	1530	Sheeting driver for 60 lb. breaker		.04	6	18	54	3.90	
	1540	For 90 lb. breaker		.12	8	24	72	5.75	
	1550	Spade, 25 lb.		.35	6.65	20	60	6.80	
	1560	Tamper, single, 35 lb.		.54	36	108	325	25.90	
	1570	Triple, 140 lb.		.81	54	162	485	38.90	
	1580	Wrenches, impact, air powered, up to 3/4" bolt		.30	8	24	72	7.20	
	1590	Up to 1-1/4" bolt		.35	16.65	50	150	12.80	
	1600	Barricades, barrels, reflectorized, 1 to 50 barrels		.02	3.33	10	30	2.15	
	1610	100 to 200 barrels		.02	2.53	7.60	23	1.70	
	1620	Barrels with flashers, 1 to 50 barrels		.02	4	12	36	2.55	
	1630	100 to 200 barrels		.02	3.20	9.60	29	2.10	
	1640	Barrels with steady burn type C lights		.03	5.35	16	48	3.45	
	1650	Illuminated board, trailer mounted, with generator		.65	117	350	1,050	75.20	
	1670	Portable barricade, stock, with flashers, 1 to 6 units		.02	4	12	36	2.55	
	1680	25 to 50 units		.02	3.73	11.20	33.50	2.40	
	1690	Butt fusion machine, electric		19.15	238	715	2,150	296.20	
	1695	Electro fusion machine		14.15	102	305	915	174.20	
	1700	Carts, brick, hand powered, 1000 lb. capacity		.25	41.50	124	370	26.80	
	1800	Gas engine, 1500 lb., 7-1/2' lift		3.59	102	306	920	89.90	
	1822	Dehumidifier, medium, 6 lb/hr, 150 CFM		.79	48	144	430	35.10	
	1824	Large, 18 lb/hr, 600 CFM		1.56	95.50	286	860	69.70	
	1830	Distributor, asphalt, trailer mtd, 2000 gal., 38 H.P. diesel		8.30	300	905	2,725	247.40	
	1840	3000 gal., 38 H.P. diesel		9.50	330	985	2,950	273	
	1850	Drill, rotary hammer, electric, 1-1/2" diameter		.79	25.50	76	228	21.50	
	1860	Carbide bit for above		.04	6.65	20	60	4.30	
	1865	Rotary, crawler, 250 H.P.		115.55	1,950	5,830	17,500	2,090	
	1870	Emulsion sprayer, 65 gal., 5 H.P. gas engine		2.44	88	264	790	72.30	
	1880	200 gal., 5 H.P. engine		6.10	147	440	1,325	136.80	
	1920	Floodlight, mercury vapor, or quartz, on tripod							
	1930	1000 watt	Ea.	.31	12	36	108	9.70	
	1940	2000 watt		.54	22	66	198	17.50	
	1950	Floodlights, trailer mounted with generator, 1 - 300 watt light		2.70	66.50	200	600	61.60	
	1960	2 - 1000 watt lights		3.80	112	335	1,000	97.40	
	2000	4 - 300 watt lights		3.15	78.50	235	705	72.20	
	2020	Forklift, wheeled, for brick, 18', 3000 lb., 2 wheel drive, gas		19.30	190	570	1,700	268.40	
	2040	28', 4000 lb., 4 wheel drive, diesel		15.80	243	730	2,200	272.40	
	2050	For rough terrain, 8000 lb., 16' lift, 68 H.P.		20.35	385	1,150	3,450	392.80	
	2060	For plant, 4 T. capacity, 80 H.P., 2 wheel drive, gas		11.85	88.50	265	795	147.80	
	2080	10 T. capacity, 120 H.P., 2 wheel drive, diesel		17.35	153	460	1,375	230.80	
	2100	Generator, electric, gas engine, 1.5 KW to 3 KW		2.65	9.35	28	84	26.80	
	2200	5 KW		3.40	13.65	41	123	35.40	
	2300	10 KW		6.35	24.50	73	219	65.40	
	2400	25 KW		8.25	55	165	495	99	
	2500	Diesel engine, 20 KW		8.40	63.50	190	570	105.20	
	2600	50 KW		15.20	86.50	260	780	173.60	
	2700	100 KW		29.45	117	350	1,050	305.60	
	2800	250 KW		58.15	220	660	1,975	597.20	
	2850	Hammer, hydraulic, for mounting on boom, to 500 ft.-lb.		1.90	65	195	585	54.20	
	2860	1000 ft.-lb.		3.35	103	310	930	88.80	
	2900	Heaters, space, oil or electric, 50 MBH		1.64	7.65	23	69	17.70	

01 54 | Construction Aids

01 54 33 | Equipment Rental

			UNIT	HOURLY OPER. COST	RENT PER DAY	RENT PER WEEK	RENT PER MONTH	CREW EQUIPMENT COST/DAY	
40	3000	100 MBH	Ea.	2.96	10.65	32	96	30.10	40
	3100	300 MBH		9.48	33.50	100	300	95.85	
	3150	500 MBH		19.15	50	150	450	183.20	
	3200	Hose, water, suction with coupling, 20' long, 2" diameter		.02	3	9	27	1.95	
	3210	3" diameter		.03	4.67	14	42	3.05	
	3220	4" diameter		.03	5	15	45	3.25	
	3230	6" diameter		.11	17.65	53	159	11.50	
	3240	8" diameter		.26	43.50	131	395	28.30	
	3250	Discharge hose with coupling, 50' long, 2" diameter		.01	1.33	4	12	.90	
	3260	3" diameter		.02	2.67	8	24	1.75	
	3270	4" diameter		.02	3.67	11	33	2.35	
	3280	6" diameter		.06	9.65	29	87	6.30	
	3290	8" diameter		.31	51.50	154	460	33.30	
	3295	Insulation blower		.25	7.35	22	66	6.40	
	3300	Ladders, extension type, 16' to 36' long		.19	31.50	95	285	20.50	
	3400	40' to 60' long		.28	47.50	142	425	30.65	
	3405	Lance for cutting concrete		2.50	79.50	239	715	67.80	
	3407	Lawn mower, rotary, 22", 5HP		1.63	43	129	385	38.85	
	3408	48" self propelled		3.92	131	394	1,175	110.15	
	3410	Level, laser type, for pipe and sewer leveling		1.21	80.50	242	725	58.10	
	3430	Electronic		.76	50.50	152	455	36.50	
	3440	Laser type, rotating beam for grade control		1.26	83.50	251	755	60.30	
	3460	Builders level with tripod and rod		.08	12.65	38	114	8.25	
	3500	Light towers, towable, with diesel generator, 2000 watt		3.15	78.50	235	705	72.20	
	3600	4000 watt		3.80	112	335	1,000	97.40	
	3700	Mixer, powered, plaster and mortar, 6 C.F., 7 H.P.		1.45	17.65	53	159	22.20	
	3800	10 C.F., 9 H.P.		1.75	29	87	261	31.40	
	3850	Nailer, pneumatic		.43	28.50	86	258	20.65	
	3900	Paint sprayers complete, 8 CFM		.72	48	144	430	34.55	
	4000	17 CFM		1.13	75.50	226	680	54.25	
	4020	Pavers, bituminous, rubber tires, 8' wide, 50 H.P., diesel		33.35	965	2,900	8,700	846.80	
	4030	10' wide, 150 H.P.		70.60	1,550	4,685	14,100	1,502	
	4050	Crawler, 8' wide, 100 H.P., diesel		70.50	1,625	4,905	14,700	1,545	
	4060	10' wide, 150 H.P.		79.35	1,975	5,900	17,700	1,815	
	4070	Concrete paver, 12' to 24' wide, 250 H.P.		70.40	1,525	4,555	13,700	1,474	
	4080	Placer-spreader-trimmer, 24' wide, 300 H.P.		101.85	2,525	7,545	22,600	2,324	
	4100	Pump, centrifugal gas pump, 1-1/2", 4 MGPH		3.05	40	120	360	48.40	
	4200	2", 8 MGPH		4.05	45	135	405	59.40	
	4300	3", 15 MGPH		4.30	48.50	145	435	63.40	
	4400	6", 90 MGPH		22.40	165	495	1,475	278.20	
	4500	Submersible electric pump, 1-1/4", 55 GPM		.37	15.35	46	138	12.15	
	4600	1-1/2", 83 GPM		.41	17.65	53	159	13.90	
	4700	2", 120 GPM		1.10	22	66	198	22	
	4800	3", 300 GPM		1.80	36.50	110	330	36.40	
	4900	4", 560 GPM		8.00	158	475	1,425	159	
	5000	6", 1590 GPM		11.75	215	645	1,925	223	
	5100	Diaphragm pump, gas, single, 1-1/2" diameter	Ea.	.98	41.50	124	370	32.65	
	5200	2" diameter		3.30	51.50	155	465	57.40	
	5300	3" diameter		3.30	51.50	155	465	57.40	
	5400	Double, 4" diameter		4.40	71.50	215	645	78.20	
	5500	Trash pump, self-priming, gas, 2" diameter		3.40	20.50	62	186	39.60	
	5600	Diesel, 4" diameter		8.85	56.50	170	510	104.80	
	5650	Diesel, 6" diameter		29.10	123	370	1,100	306.80	
	5655	Grout Pump		10.15	83.50	250	750	131.20	
	5660	Rollers, see division 01590-200							
	5700	Salamanders, L.P. gas fired, 100,000 BTU	Ea.	2.97	11.35	34	102	30.55	
	5705	50,000 BTU		2.22	8	24	72	22.55	
	5720	Sandblaster, portable, open top, 3 C.F. capacity		.40	20.50	62	186	15.60	

01 54 | Construction Aids

01 54 33 | Equipment Rental

			UNIT	HOURLY OPER. COST	RENT PER DAY	RENT PER WEEK	RENT PER MONTH	CREW EQUIPMENT COST/DAY	
40	5730	6 C.F. capacity	Ea.	.70	30	90	270	23.60	40
	5740	Accessories for above		.11	19	57	171	12.30	
	5750	Sander, floor		.79	19	57	171	17.70	
	5760	Edger		.56	17.35	52	156	14.90	
	5800	Saw, chain, gas engine, 18" long		1.65	16.35	49	147	23	
	5900	36" long		.55	50	150	450	34.40	
	5950	60" long		.55	50	150	450	34.40	
	6000	Masonry, table mounted, 14" diameter, 5 H.P.		1.28	54.50	164	490	43.05	
	6050	Portable cut-off, 8 H.P.		1.80	29.50	88	264	32	
	6100	Circular, hand held, electric, 7-1/4" diameter		.20	5	15	45	4.60	
	6200	12" diameter		.28	8.65	26	78	7.45	
	6250	Wall saw, w/hydraulic power, 10 H.P		5.65	55	165	495	78.20	
	6275	Shot blaster, walk behind, 20" wide		6.65	415	1,245	3,725	302.20	
	6280	Sidewalk broom, walk-behind		2.03	62	186	560	53.45	
	6300	Steam cleaner, 100 gallons per hour		2.80	63.50	190	570	60.40	
	6310	200 gallons per hour		3.90	78.50	235	705	78.20	
	6340	Tar Kettle/Pot, 400 gallon		3.99	51.50	155	465	62.90	
	6350	Torch, cutting, acetylene-oxygen, 150' hose		.50	21.50	65	195	17	
	6360	Hourly operating cost includes tips and gas		8.10				64.80	
	6410	Toilet, portable chemical		.11	18.35	55	165	11.90	
	6420	Recycle flush type		.13	22.50	67	201	14.45	
	6430	Toilet, fresh water flush, garden hose,		.15	25	75	225	16.20	
	6440	Hoisted, non-flush, for high rise		.13	22	66	198	14.25	
	6450	Toilet, trailers, minimum		.23	38	114	340	24.65	
	6460	Maximum		.69	115	344	1,025	74.30	
	6465	Tractor, farm with attachment		10.30	223	670	2,000	216.40	
	6470	Trailer, office, see division 01520-500							
	6500	Trailers, platform, flush deck, 2 axle, 25 ton capacity	Ea.	4.35	96.50	290	870	92.80	
	6600	40 ton capacity		5.60	135	405	1,225	125.80	
	6700	3 axle, 50 ton capacity		6.05	148	445	1,325	137.40	
	6800	75 ton capacity		7.55	195	585	1,750	177.40	
	6810	Trailer mounted cable reel for H.V. line work		4.66	222	665	2,000	170.30	
	6820	Trailer mounted cable tensioning rig		9.17	435	1,310	3,925	335.35	
	6830	Cable pulling rig		62.02	2,475	7,410	22,200	1,978	
	6850	Trailer, storage, see division 01520-500							
	6900	Water tank, engine driven discharge, 5000 gallons	Ea.	5.65	127	380	1,150	121.20	
	6925	10,000 gallons		7.80	177	530	1,600	168.40	
	6950	Water truck, off highway, 6000 gallons		56.50	735	2,210	6,625	894	
	7010	Tram car for H.V. line work, powered, 2 conductor		6.10	120	361	1,075	121	
	7020	Transit (builder's level) with tripod		.08	12.65	38	114	8.25	
	7030	Trench box, 3000 lbs. 6'x8'		.60	100	300	900	64.80	
	7040	7200 lbs. 6'x20'		.73	122	367	1,100	79.25	
	7050	8000 lbs., 8' x 16'		.93	154	463	1,400	100.05	
	7060	9500 lbs., 8'x20'		1.26	210	630	1,900	136.10	
	7065	11,000 lbs., 8'x24'		1.35	225	676	2,025	146	
	7070	12,000 lbs., 10' x 20'		1.50	250	750	2,250	162	
	7100	Truck, pickup, 3/4 ton, 2 wheel drive		6.35	56.50	170	510	84.80	
	7200	4 wheel drive		6.50	65	195	585	91	
	7250	Crew carrier, 9 passenger		8.65	78.50	235	705	116.20	
	7290	Tool van, 24,000 G.V.W.		10.85	108	325	975	151.80	
	7300	Tractor, 4 x 2, 30 ton capacity, 195 H.P.	Ea.	16.70	170	510	1,525	235.60	
	7410	250 H.P.		21.65	232	695	2,075	312.20	
	7500	6 x 2, 40 ton capacity, 240 H.P.		20.90	282	845	2,525	336.20	
	7600	6 x 4, 45 ton capacity, 240 H.P.		26.80	300	905	2,725	395.40	
	7620	Vacuum truck, hazardous material, 2500 gallon		10.15	295	885	2,650	258.20	
	7625	5,000 gallon		13.03	415	1,240	3,725	352.25	
	7640	Tractor, with A frame, boom and winch, 225 H.P.		19.10	242	725	2,175	297.80	
	7650	Vacuum, H.E.P.A., 16 gal., wet/dry	Ea.	.92	22	66	198	20.55	

R312323-30
R314116-40
R314116-45

01 54 | Construction Aids

01 54 33 | Equipment Rental

			UNIT	HOURLY OPER. COST	RENT PER DAY	RENT PER WEEK	RENT PER MONTH	CREW EQUIPMENT COST/DAY	
40	7655	55 gal, wet/dry	Ea.	.89	33	99	297	26.90	40
	7660	Water tank, portable		.15	25.50	75.90	228	16.40	
	7690	Large production vacuum loader, 3150 CFM		17.37	620	1,860	5,575	510.95	
	7700	Welder, electric, 200 amp	Ea.	3.37	31.50	95	285	45.95	
	7800	300 amp		4.98	36	108	325	61.45	
	7900	Gas engine, 200 amp		10.90	23.50	71	213	101.40	
	8000	300 amp		12.50	25.50	76	228	115.20	
	8100	Wheelbarrow, any size		.07	11.65	35	105	7.55	
	8200	Wrecking ball, 4000 lb.		2.00	70	210	630	58	
50	0010	**HIGHWAY EQUIPMENT RENTAL**, without operators							50
	0050	Asphalt batch plant, portable drum mixer, 100 ton/hr.	Ea.	57.45	1,350	4,025	12,100	1,265	
	0060	200 ton/hr.		63.80	1,400	4,230	12,700	1,356	
	0070	300 ton/hr.		74.25	1,675	5,005	15,000	1,595	
	0100	Backhoe attachment, long stick, up to 185 HP, 10.5' long		.31	20.50	62	186	14.90	
	0140	Up to 250 HP, 12' long		.34	22.50	67	201	16.10	
	0180	Over 250 HP, 15' long		.45	29.50	89	267	21.40	
	0200	Special dipper arm, up to 100 HP, 32' long		.92	61	183	550	43.95	
	0240	Over 100 HP, 33' long		1.14	76	228	685	54.70	
	0300	Concrete batch plant, portable, electric, 200 CY/Hr		10.79	505	1,515	4,550	389.30	
	0500	Grader attachment, ripper/scarifier, rear mounted							
	0520	Up to 135 HP	Ea.	2.95	61.50	185	555	60.60	
	0540	Up to 180 HP		3.55	80	240	720	76.40	
	0580	Up to 250 HP		3.90	90	270	810	85.20	
	0700	Pvmt. removal bucket, for hyd. excavator, up to 90 HP		1.50	45	135	405	39	
	0740	Up to 200 HP		1.70	66.50	200	600	53.60	
	0780	Over 200 HP		1.80	80	240	720	62.40	
	0900	Aggregate spreader, self-propelled, 187 HP		42.35	800	2,400	7,200	818.80	
	1000	Chemical spreader, 3 C.Y.		2.40	41.50	125	375	44.20	
	1900	Hammermill, traveling, 250 HP		55.98	1,775	5,320	16,000	1,512	
	2000	Horizontal borer, 3" diam, 13 HP gas driven		4.90	53.50	160	480	71.20	
	2200	Hydromulchers, gas power, 3000 gal., for truck mounting		12.90	188	565	1,700	216.20	
	2400	Joint & crack cleaner, walk behind, 25 HP		2.40	48.50	145	435	48.20	
	2500	Filler, trailer mounted, 400 gal., 20 HP		6.55	183	550	1,650	162.40	
	3000	Paint striper, self propelled, double line, 30 HP		5.50	157	470	1,400	138	
	3200	Post drivers, 6" I-Beam frame, for truck mounting		10.30	420	1,255	3,775	333.40	
	3400	Road sweeper, self propelled, 8' wide, 90 HP		27.75	550	1,645	4,925	551	
	4000	Road mixer, self-propelled, 130 HP		35.40	685	2,050	6,150	693.20	
	4100	310 HP		67.30	2,125	6,405	19,200	1,819	
	4200	Cold mix paver, incl pug mill and bitumen tank,							
	4220	165 HP	Ea.	79.35	1,975	5,910	17,700	1,817	
	4250	Paver, asphalt, wheel or crawler, 130 H.P., diesel		78.05	1,875	5,630	16,900	1,750	
	4300	Paver, road widener, gas 1' to 6', 67 HP		37.40	780	2,345	7,025	768.20	
	4400	Diesel, 2' to 14', 88 HP		49.05	1,000	3,025	9,075	997.40	
	4600	Slipform pavers, curb and gutter, 2 track, 75 HP		31.95	720	2,160	6,475	687.60	
	4700	4 track, 165 HP		38.75	770	2,305	6,925	771	
	4800	Median barrier, 215 HP		39.35	800	2,395	7,175	793.80	
	4901	Trailer, low bed, 75 ton capacity		8.15	193	580	1,750	181.20	
	5000	Road planer, walk behind, 10" cutting width, 10 HP		2.30	28	84	252	35.20	
	5100	Self propelled, 12" cutting width, 64 HP		6.25	112	335	1,000	117	
	5200	Pavement profiler, 4' to 6' wide, 450 HP		197.95	3,250	9,760	29,300	3,536	
	5300	8' to 10' wide, 750 HP		313.05	4,450	13,360	40,100	5,176	
	5400	Roadway plate, steel, 1"x8'x20'		.06	10.65	32	96	6.90	
	5600	Stabilizer, self-propelled, 150 HP		36.55	580	1,735	5,200	639.40	
	5700	310 HP		62.05	1,275	3,845	11,500	1,265	
	5800	Striper, thermal, truck mounted 120 gal. paint, 150 H.P.		37.90	490	1,465	4,400	596.20	
	6000	Tar kettle, 330 gal., trailer mounted		3.67	36.50	110	330	51.35	
	7000	Tunnel locomotive, diesel, 8 to 12 ton		25.05	560	1,685	5,050	537.40	

01 54 | Construction Aids

01 54 33 | Equipment Rental

			UNIT	HOURLY OPER. COST	RENT PER DAY	RENT PER WEEK	RENT PER MONTH	CREW EQUIPMENT COST/DAY	
50	7005	Electric, 10 ton	Ea.	21.95	640	1,915	5,750	558.60	50
	7010	Muck cars, 1/2 C.Y. capacity		1.65	21.50	64	192	26	
	7020	1 C.Y. capacity		1.85	30	90	270	32.80	
	7030	2 C.Y. capacity		1.95	35	105	315	36.60	
	7040	Side dump, 2 C.Y. capacity		2.15	41.50	125	375	42.20	
	7050	3 C.Y. capacity		2.90	48.50	145	435	52.20	
	7060	5 C.Y. capacity		4.10	61.50	185	555	69.80	
	7100	Ventilating blower for tunnel, 7-1/2 H.P.		1.29	50	150	450	40.30	
	7110	10 H.P.		1.49	51.50	155	465	42.90	
	7120	20 H.P.		2.44	67.50	202	605	59.90	
	7140	40 H.P.		4.29	95	285	855	91.30	
	7160	60 H.P.		6.49	147	440	1,325	139.90	
	7175	75 H.P.		8.30	196	587	1,750	183.80	
	7180	200 H.P.		18.70	293	880	2,650	325.60	
	7800	Windrow loader, elevating		39.00	930	2,785	8,350	869	
60	0010	**LIFTING & HOISTING EQUIPMENT RENTAL**, without operators R015433-10							60
	0120	Aerial lift truck, 2 person, to 80'	Ea.	22.35	615	1,840	5,525	546.80	
	0140	Boom work platform, 40' snorkel		13.25	223	670	2,000	240	
	0150	Crane, flatbed mntd, 3 ton cap. R015433-15		17.70	217	650	1,950	271.60	
	0200	Crane, climbing, 106' jib, 6000 lb. capacity, 410 FPM R312316-45		68.15	1,400	4,230	12,700	1,391	
	0300	101' jib, 10,250 lb. capacity, 270 FPM		73.80	1,775	5,360	16,100	1,662	
	0400	Tower, static, 130' high, 106' jib,							
	0500	6200 lb. capacity at 400 FPM	Ea.	71.45	1,625	4,890	14,700	1,550	
	0600	Crawler mounted, lattice boom, 1/2 C.Y., 15 tons at 12' radius		26.58	605	1,820	5,450	576.65	
	0700	3/4 C.Y., 20 tons at 12' radius		35.44	755	2,270	6,800	737.50	
	0800	1 C.Y., 25 tons at 12' radius		47.25	1,000	3,025	9,075	983	
	0900	1-1/2 C.Y., 40 tons at 12' radius		51.20	975	2,925	8,775	994.60	
	1000	2 C.Y., 50 tons at 12' radius		54.00	1,175	3,495	10,500	1,131	
	1100	3 C.Y., 75 tons at 12' radius		66.25	1,425	4,270	12,800	1,384	
	1200	100 ton capacity, 60' boom		76.70	1,850	5,515	16,500	1,717	
	1300	165 ton capacity, 60' boom		102.30	2,300	6,890	20,700	2,196	
	1400	200 ton capacity, 70' boom		130.85	2,675	8,030	24,100	2,653	
	1500	350 ton capacity, 80' boom		182.90	3,500	10,530	31,600	3,569	
	1600	Truck mounted, lattice boom, 6 x 4, 20 tons at 10' radius		33.59	1,175	3,540	10,600	976.70	
	1700	25 tons at 10' radius		36.28	1,275	3,840	11,500	1,058	
	1800	8 x 4, 30 tons at 10' radius		49.47	1,375	4,130	12,400	1,222	
	1900	40 tons at 12' radius		50.73	1,425	4,310	12,900	1,268	
	2000	8 x 4, 60 tons at 15' radius		52.34	1,525	4,540	13,600	1,327	
	2050	82 tons at 15' radius		54.44	1,625	4,840	14,500	1,404	
	2100	90 tons at 15' radius		60.67	1,775	5,310	15,900	1,547	
	2200	115 tons at 15' radius		74.05	1,975	5,900	17,700	1,772	
	2300	150 tons at 18' radius		74.51	2,100	6,280	18,800	1,852	
	2350	165 tons at 18' radius		87.35	2,200	6,615	19,800	2,022	
	2400	Truck mounted, hydraulic, 12 ton capacity		45.25	605	1,810	5,425	724	
	2500	25 ton capacity		45.45	625	1,880	5,650	739.60	
	2550	33 ton capacity		46.30	660	1,975	5,925	765.40	
	2560	40 ton capacity		44.70	630	1,895	5,675	736.60	
	2600	55 ton capacity		66.40	880	2,645	7,925	1,060	
	2700	80 ton capacity		78.75	975	2,920	8,750	1,214	
	2720	100 ton capacity		95.75	2,500	7,525	22,600	2,271	
	2740	120 ton capacity		93.21	2,750	8,280	24,800	2,402	
	2760	150 ton capacity		114.91	3,525	10,540	31,600	3,027	
	2800	Self-propelled, 4 x 4, with telescoping boom, 5 ton		19.10	290	870	2,600	326.80	
	2900	12-1/2 ton capacity		32.75	520	1,565	4,700	575	
	3000	15 ton capacity		35.00	615	1,850	5,550	650	
	3050	20 ton capacity		35.85	640	1,920	5,750	670.80	
	3100	25 ton capacity		47.95	820	2,460	7,375	875.60	

01 54 | Construction Aids

01 54 33 | Equipment Rental

			UNIT	HOURLY OPER. COST	RENT PER DAY	RENT PER WEEK	RENT PER MONTH	CREW EQUIPMENT COST/DAY	
60	3150	40 ton capacity	Ea.	61.00	925	2,775	8,325	1,043	60
	3200	Derricks, guy, 20 ton capacity, 60' boom, 75' mast		19.00	345	1,036	3,100	359.20	
	3300	100' boom, 115' mast		30.02	590	1,770	5,300	594.15	
	3400	Stiffleg, 20 ton capacity, 70' boom, 37' mast		21.13	445	1,340	4,025	437.05	
	3500	100' boom, 47' mast		32.75	720	2,160	6,475	694	
	3550	Helicopter, small, lift to 1250 lbs. maximum, w/pilot		82.09	2,800	8,370	25,100	2,331	
	3600	Hoists, chain type, overhead, manual, 3/4 ton		.10	.67	2	6	1.20	
	3900	10 ton		.65	7.35	22	66	9.60	
	4000	Hoist and tower, 5000 lb. cap., portable electric, 40' high		4.42	199	597	1,800	154.75	
	4100	For each added 10' section, add		.09	15.65	47	141	10.10	
	4200	Hoist and single tubular tower, 5000 lb. electric, 100' high		5.95	278	833	2,500	214.20	
	4300	For each added 6'-6" section, add		.16	26.50	79	237	17.10	
	4400	Hoist and double tubular tower, 5000 lb., 100' high		6.38	305	918	2,750	234.65	
	4500	For each added 6'-6" section, add		.18	29.50	89	267	19.25	
	4550	Hoist and tower, mast type, 6000 lb., 100' high		6.91	315	952	2,850	245.70	
	4570	For each added 10' section, add		.11	19	57	171	12.30	
	4600	Hoist and tower, personnel, electric, 2000 lb., 100' @ 125 FPM		14.13	845	2,540	7,625	621.05	
	4700	3000 lb., 100' @ 200 FPM		16.14	955	2,870	8,600	703.10	
	4800	3000 lb., 150' @ 300 FPM		17.84	1,075	3,210	9,625	784.70	
	4900	4000 lb., 100' @ 300 FPM		18.50	1,100	3,270	9,800	802	
	5000	6000 lb., 100' @ 275 FPM		20.06	1,150	3,440	10,300	848.50	
	5100	For added heights up to 500', add	L.F.	.01	1.67	5	15	1.10	
	5200	Jacks, hydraulic, 20 ton	Ea.	.05	3.33	10	30	2.40	
	5500	100 ton	"	.30	9.65	29	87	8.20	
	6000	Jacks, hydraulic, climbing with 50' jackrods							
	6010	and control consoles, minimum 3 mo. rental							
	6100	30 ton capacity	Ea.	1.72	114	343	1,025	82.35	
	6150	For each added 10' jackrod section, add		.05	3.33	10	30	2.40	
	6300	50 ton capacity		2.76	184	552	1,650	132.50	
	6350	For each added 10' jackrod section, add		.06	4	12	36	2.90	
	6500	125 ton capacity		7.25	485	1,450	4,350	348	
	6550	For each added 10' jackrod section, add		.50	33	99	297	23.80	
	6600	Cable jack, 10 ton capacity with 200' cable		1.44	95.50	287	860	68.90	
	6650	For each added 50' of cable, add		.16	10.65	32	96	7.70	
70	0010	**WELLPOINT EQUIPMENT RENTAL**, without operators							70
	0020	Based on 2 months rental							
	0100	Combination jetting & wellpoint pump, 60 H.P. diesel	Ea.	13.00	283	850	2,550	274	
	0200	High pressure gas jet pump, 200 H.P., 300 psi	"	28.58	242	726	2,175	373.85	
	0300	Discharge pipe, 8" diameter	L.F.	.01	.46	1.38	4.14	.35	
	0350	12" diameter		.01	.68	2.04	6.10	.50	
	0400	Header pipe, flows up to 150 G.P.M., 4" diameter		.01	.41	1.24	3.72	.35	
	0500	400 G.P.M., 6" diameter		.01	.49	1.48	4.44	.40	
	0600	800 G.P.M., 8" diameter		.01	.68	2.04	6.10	.50	
	0700	1500 G.P.M., 10" diameter		.01	.71	2.14	6.40	.50	
	0800	2500 G.P.M., 12" diameter		.02	1.35	4.05	12.15	.95	
	0900	4500 G.P.M., 16" diameter		.03	1.73	5.18	15.55	1.30	
	0950	For quick coupling aluminum and plastic pipe, add		.03	1.79	5.36	16.10	1.30	
	1100	Wellpoint, 25' long, with fittings & riser pipe, 1-1/2" or 2" diameter	Ea.	.05	3.57	10.70	32	2.55	
	1200	Wellpoint pump, diesel powered, 4" diameter, 20 H.P.		5.78	163	490	1,475	144.25	
	1300	6" diameter, 30 H.P.		7.78	203	608	1,825	183.85	
	1400	8" suction, 40 H.P.		10.53	278	833	2,500	250.85	
	1500	10" suction, 75 H.P.		15.63	325	974	2,925	319.85	
	1600	12" suction, 100 H.P.		22.67	520	1,560	4,675	493.35	
	1700	12" suction, 175 H.P.		32.53	570	1,710	5,125	602.25	

01 54 | Construction Aids

01 54 33 | Equipment Rental

			UNIT	HOURLY OPER. COST	RENT PER DAY	RENT PER WEEK	RENT PER MONTH	CREW EQUIPMENT COST/DAY	
80	0010	**MARINE EQUIPMENT RENTAL**, without operators							80
	0200	Barge, 400 Ton, 30' wide x 90' long	Ea.	17.65	255	765	2,300	294.20	
	0240	800 Ton, 45' wide x 90' long		29.15	365	1,090	3,275	451.20	
	2000	Tugboat, diesel, 100 HP		22.25	180	540	1,625	286	
	2040	250 HP		43.65	335	1,005	3,025	550.20	
	2080	380 HP	Ea.	93.95	990	2,970	8,900	1,346	

Crews

Crew No.	Bare Costs		Incl. Subs O & P		Cost Per Labor-Hour	
Crew A-1	Hr.	Daily	Hr.	Daily	Bare Costs	Incl. O&P
1 Building Laborer	$28.75	$230.00	$44.75	$358.00	$28.75	$44.75
1 Concrete saw, gas manual		54.00		59.40	6.75	7.42
8 L.H., Daily Totals		$284.00		$417.40	$35.50	$52.17
Crew A-1A	Hr.	Daily	Hr.	Daily	Bare Costs	Incl. O&P
1 Skilled Worker	$38.00	$304.00	$59.15	$473.20	$38.00	$59.15
1 Shot Blaster, 20"		302.20		332.42	37.77	41.55
8 L.H., Daily Totals		$606.20		$805.62	$75.78	$100.70
Crew A-1B	Hr.	Daily	Hr.	Daily	Bare Costs	Incl. O&P
1 Building Laborer	$28.75	$230.00	$44.75	$358.00	$28.75	$44.75
1 Concrete Saw		118.40		130.24	14.80	16.28
8 L.H., Daily Totals		$348.40		$488.24	$43.55	$61.03
Crew A-1C	Hr.	Daily	Hr.	Daily	Bare Costs	Incl. O&P
1 Building Laborer	$28.75	$230.00	$44.75	$358.00	$28.75	$44.75
1 Chain saw, gas, 18"		23.00		25.30	2.88	3.16
8 L.H., Daily Totals		$253.00		$383.30	$31.63	$47.91
Crew A-1D	Hr.	Daily	Hr.	Daily	Bare Costs	Incl. O&P
1 Building Laborer	$28.75	$230.00	$44.75	$358.00	$28.75	$44.75
1 Vibrating plate, gas, 18"		29.20		32.12	3.65	4.01
8 L.H., Daily Totals		$259.20		$390.12	$32.40	$48.77
Crew A-1E	Hr.	Daily	Hr.	Daily	Bare Costs	Incl. O&P
1 Building Laborer	$28.75	$230.00	$44.75	$358.00	$28.75	$44.75
1 Vibratory Plate, Gas, 21"		36.20		39.82	4.53	4.98
8 L.H., Daily Totals		$266.20		$397.82	$33.27	$49.73
Crew A-1F	Hr.	Daily	Hr.	Daily	Bare Costs	Incl. O&P
1 Building Laborer	$28.75	$230.00	$44.75	$358.00	$28.75	$44.75
1 Rammer/tamper, gas, 8"		38.40		42.24	4.80	5.28
8 L.H., Daily Totals		$268.40		$400.24	$33.55	$50.03
Crew A-1G	Hr.	Daily	Hr.	Daily	Bare Costs	Incl. O&P
1 Building Laborer	$28.75	$230.00	$44.75	$358.00	$28.75	$44.75
1 Rammer/tamper, gas, 8"		38.40		42.24	4.80	5.28
8 L.H., Daily Totals		$268.40		$400.24	$33.55	$50.03
Crew A-1H	Hr.	Daily	Hr.	Daily	Bare Costs	Incl. O&P
1 Building Laborer	$28.75	$230.00	$44.75	$358.00	$28.75	$44.75
1 Pressure washer		60.40		66.44	7.55	8.30
8 L.H., Daily Totals		$290.40		$424.44	$36.30	$53.06
Crew A-1J	Hr.	Daily	Hr.	Daily	Bare Costs	Incl. O&P
1 Building Laborer	$28.75	$230.00	$44.75	$358.00	$28.75	$44.75
1 Cultivator, Walk-Behind		48.30		53.13	6.04	6.64
8 L.H., Daily Totals		$278.30		$411.13	$34.79	$51.39
Crew A-1K	Hr.	Daily	Hr.	Daily	Bare Costs	Incl. O&P
1 Building Laborer	$28.75	$230.00	$44.75	$358.00	$28.75	$44.75
1 Cultivator, Walk-Behind		48.30		53.13	6.04	6.64
8 L.H., Daily Totals		$278.30		$411.13	$34.79	$51.39

Crew No.	Bare Costs		Incl. Subs O & P		Cost Per Labor-Hour	
Crew A-1M	Hr.	Daily	Hr.	Daily	Bare Costs	Incl. O&P
1 Building Laborer	$28.75	$230.00	$44.75	$358.00	$28.75	$44.75
1 Snow Blower, Walk-Behind		53.45		58.80	6.68	7.35
8 L.H., Daily Totals		$283.45		$416.80	$35.43	$52.10
Crew A-2	Hr.	Daily	Hr.	Daily	Bare Costs	Incl. O&P
2 Laborers	$28.75	$460.00	$44.75	$716.00	$28.68	$44.53
1 Truck Driver (light)	28.55	228.40	44.10	352.80		
1 Flatbed Truck, gas, 1.5 Ton		141.40		155.54	5.89	6.48
24 L.H., Daily Totals		$829.80		$1224.34	$34.58	$51.01
Crew A-2A	Hr.	Daily	Hr.	Daily	Bare Costs	Incl. O&P
2 Laborers	$28.75	$460.00	$44.75	$716.00	$28.68	$44.53
1 Truck Driver (light)	28.55	228.40	44.10	352.80		
1 Flatbed Truck, gas, 1.5 Ton		141.40		155.54		
1 Concrete Saw		118.40		130.24	10.82	11.91
24 L.H., Daily Totals		$948.20		$1354.58	$39.51	$56.44
Crew A-3	Hr.	Daily	Hr.	Daily	Bare Costs	Incl. O&P
1 Truck Driver (heavy)	$29.55	$236.40	$45.65	$365.20	$29.55	$45.65
1 Dump Truck, 12 Ton, 8 C.Y.		368.40		405.24	46.05	50.66
8 L.H., Daily Totals		$604.80		$770.44	$75.60	$96.31
Crew A-3A	Hr.	Daily	Hr.	Daily	Bare Costs	Incl. O&P
1 Truck Driver (light)	$28.55	$228.40	$44.10	$352.80	$28.55	$44.10
1 Pickup truck, 4 x 4, 3/4 ton		91.00		100.10	11.38	12.51
8 L.H., Daily Totals		$319.40		$452.90	$39.92	$56.61
Crew A-3B	Hr.	Daily	Hr.	Daily	Bare Costs	Incl. O&P
1 Equip. Oper. (medium)	$38.40	$307.20	$57.85	$462.80	$33.98	$51.75
1 Truck Driver (heavy)	29.55	236.40	45.65	365.20		
1 Dump Truck, 16 Ton, 12 C.Y.		530.00		583.00		
1 F.E. Loader, W.M.,2.5 C.Y.		349.00		383.90	54.94	60.43
16 L.H., Daily Totals		$1422.60		$1794.90	$88.91	$112.18
Crew A-3C	Hr.	Daily	Hr.	Daily	Bare Costs	Incl. O&P
1 Equip. Oper. (light)	$36.85	$294.80	$55.55	$444.40	$36.85	$55.55
1 Loader, Skid Steer, 78 HP		222.20		244.42	27.77	30.55
8 L.H., Daily Totals		$517.00		$688.82	$64.63	$86.10
Crew A-3D	Hr.	Daily	Hr.	Daily	Bare Costs	Incl. O&P
1 Truck Driver, Light	$28.55	$228.40	$44.10	$352.80	$28.55	$44.10
1 Pickup truck, 4 x 4, 3/4 ton		91.00		100.10		
1 Flatbed Trailer, 25 Ton		92.80		102.08	22.98	25.27
8 L.H., Daily Totals		$412.20		$554.98	$51.52	$69.37
Crew A-3E	Hr.	Daily	Hr.	Daily	Bare Costs	Incl. O&P
1 Equip. Oper. (crane)	$39.80	$318.40	$60.00	$480.00	$34.67	$52.83
1 Truck Driver (heavy)	29.55	236.40	45.65	365.20		
1 Pickup truck, 4 x 4, 3/4 ton		91.00		100.10	5.69	6.26
16 L.H., Daily Totals		$645.80		$945.30	$40.36	$59.08
Crew A-3F	Hr.	Daily	Hr.	Daily	Bare Costs	Incl. O&P
1 Equip. Oper. (crane)	$39.80	$318.40	$60.00	$480.00	$34.67	$52.83
1 Truck Driver (heavy)	29.55	236.40	45.65	365.20		
1 Pickup truck, 4 x 4, 3/4 ton		91.00		100.10		
1 Truck Tractor, 240 H.P.		336.20		369.82		
1 Lowbed Trailer, 75 Ton		181.20		199.32	38.02	41.83
16 L.H., Daily Totals		$1163.20		$1514.44	$72.70	$94.65

Crews

Crew No.	Bare Costs		Incl. Subs O & P		Cost Per Labor-Hour	

Crew A-3G	Hr.	Daily	Hr.	Daily	Bare Costs	Incl. O&P
1 Equip. Oper. (crane)	$39.80	$318.40	$60.00	$480.00	$34.67	$52.83
1 Truck Driver (heavy)	29.55	236.40	45.65	365.20		
1 Pickup truck, 4 x 4, 3/4 ton		91.00		100.10		
1 Truck Tractor, 240 H.P.		395.40		434.94		
1 Lowbed Trailer, 75 Ton		181.20		199.32	41.73	45.90
16 L.H., Daily Totals		$1222.40		$1579.56	$76.40	$98.72

Crew A-4	Hr.	Daily	Hr.	Daily	Bare Costs	Incl. O&P
2 Carpenters	$36.70	$587.20	$57.15	$914.40	$35.37	$54.50
1 Painter, Ordinary	32.70	261.60	49.20	393.60		
24 L.H., Daily Totals		$848.80		$1308.00	$35.37	$54.50

Crew A-5	Hr.	Daily	Hr.	Daily	Bare Costs	Incl. O&P
2 Laborers	$28.75	$460.00	$44.75	$716.00	$28.73	$44.68
.25 Truck Driver (light)	28.55	57.10	44.10	88.20		
.25 Flatbed Truck, gas, 1.5 Ton		35.35		38.88	1.96	2.16
18 L.H., Daily Totals		$552.45		$843.09	$30.69	$46.84

Crew A-6	Hr.	Daily	Hr.	Daily	Bare Costs	Incl. O&P
1 Instrument Man	$38.00	$304.00	$59.15	$473.20	$37.02	$56.88
1 Rodman/Chainman	36.05	288.40	54.60	436.80		
1 Laser Transit/Level		58.10		63.91	3.63	3.99
16 L.H., Daily Totals		$650.50		$973.91	$40.66	$60.87

Crew A-7	Hr.	Daily	Hr.	Daily	Bare Costs	Incl. O&P
1 Chief Of Party	$47.15	$377.20	$73.10	$584.80	$40.40	$62.28
1 Instrument Man	38.00	304.00	59.15	473.20		
1 Rodman/Chainman	36.05	288.40	54.60	436.80		
1 Laser Transit/Level		58.10		63.91	2.42	2.66
24 L.H., Daily Totals		$1027.70		$1558.71	$42.82	$64.95

Crew A-8	Hr.	Daily	Hr.	Daily	Bare Costs	Incl. O&P
1 Chief of Party	$47.15	$377.20	$73.10	$584.80	$39.31	$60.36
1 Instrument Man	38.00	304.00	59.15	473.20		
2 Rodmen/Chainmen	36.05	576.80	54.60	873.60		
1 Laser Transit/Level		58.10		63.91	1.82	2.00
32 L.H., Daily Totals		$1316.10		$1995.51	$41.13	$62.36

Crew A-9	Hr.	Daily	Hr.	Daily	Bare Costs	Incl. O&P
1 Asbestos Foreman	$42.00	$336.00	$66.40	$531.20	$41.56	$65.70
7 Asbestos Workers	41.50	2324.00	65.60	3673.60		
64 L.H., Daily Totals		$2660.00		$4204.80	$41.56	$65.70

Crew A-10	Hr.	Daily	Hr.	Daily	Bare Costs	Incl. O&P
1 Asbestos Foreman	$42.00	$336.00	$66.40	$531.20	$41.56	$65.70
7 Asbestos Workers	41.50	2324.00	65.60	3673.60		
64 L.H., Daily Totals		$2660.00		$4204.80	$41.56	$65.70

Crew A-10A	Hr.	Daily	Hr.	Daily	Bare Costs	Incl. O&P
1 Asbestos Foreman	$42.00	$336.00	$66.40	$531.20	$41.67	$65.87
2 Asbestos Workers	41.50	664.00	65.60	1049.60		
24 L.H., Daily Totals		$1000.00		$1580.80	$41.67	$65.87

Crew A-10B	Hr.	Daily	Hr.	Daily	Bare Costs	Incl. O&P
1 Asbestos Foreman	$42.00	$336.00	$66.40	$531.20	$41.63	$65.80
3 Asbestos Workers	41.50	996.00	65.60	1574.40		
32 L.H., Daily Totals		$1332.00		$2105.60	$41.63	$65.80

Crew A-10C	Hr.	Daily	Hr.	Daily	Bare Costs	Incl. O&P
3 Asbestos Workers	$41.50	$996.00	$65.60	$1574.40	$41.50	$65.60
1 Flatbed Truck, gas, 1.5 Ton		141.40		155.54	5.89	6.48
24 L.H., Daily Totals		$1137.40		$1729.94	$47.39	$72.08

Crew A-10D	Hr.	Daily	Hr.	Daily	Bare Costs	Incl. O&P
2 Asbestos Workers	$41.50	$664.00	$65.60	$1049.60	$39.17	$60.58
1 Equip. Oper. (crane)	39.80	318.40	60.00	480.00		
1 Equip. Oper. Oiler	33.90	271.20	51.10	408.80		
1 Hydraulic Crane, 33 Ton		765.40		841.94	23.92	26.31
32 L.H., Daily Totals		$2019.00		$2780.34	$63.09	$86.89

Crew A-11	Hr.	Daily	Hr.	Daily	Bare Costs	Incl. O&P
1 Asbestos Foreman	$42.00	$336.00	$66.40	$531.20	$41.56	$65.70
7 Asbestos Workers	41.50	2324.00	65.60	3673.60		
2 Chipping Hammer, 12 Lb., Elec.		31.20		34.32	.49	.54
64 L.H., Daily Totals		$2691.20		$4239.12	$42.05	$66.24

Crew A-12	Hr.	Daily	Hr.	Daily	Bare Costs	Incl. O&P
1 Asbestos Foreman	$42.00	$336.00	$66.40	$531.20	$41.56	$65.70
7 Asbestos Workers	41.50	2324.00	65.60	3673.60		
1 Trk-mtd vac, 14 CY, 1500 Gal		510.95		562.04		
1 Flatbed Truck, 20,000 GVW		151.80		166.98	10.36	11.39
64 L.H., Daily Totals		$3322.75		$4933.82	$51.92	$77.09

Crew A-13	Hr.	Daily	Hr.	Daily	Bare Costs	Incl. O&P
1 Equip. Oper. (light)	$36.85	$294.80	$55.55	$444.40	$36.85	$55.55
1 Trk-mtd vac, 14 CY, 1500 Gal		510.95		562.04		
1 Flatbed Truck, 20,000 GVW		151.80		166.98	82.84	91.13
8 L.H., Daily Totals		$957.55		$1173.43	$119.69	$146.68

Crew B-1	Hr.	Daily	Hr.	Daily	Bare Costs	Incl. O&P
1 Labor Foreman (outside)	$30.75	$246.00	$47.90	$383.20	$29.42	$45.80
2 Laborers	28.75	460.00	44.75	716.00		
24 L.H., Daily Totals		$706.00		$1099.20	$29.42	$45.80

Crew B-1A	Hr.	Daily	Hr.	Daily	Bare Costs	Incl. O&P
1 Laborer Foreman	$30.75	$246.00	$47.90	$383.20	$29.42	$45.80
2 Laborers	28.75	460.00	44.75	716.00		
2 Cutting Torches		34.00		37.40		
2 Gases		129.60		142.56	6.82	7.50
24 L.H., Daily Totals		$869.60		$1279.16	$36.23	$53.30

Crew B-1B	Hr.	Daily	Hr.	Daily	Bare Costs	Incl. O&P
1 Laborer Foreman	$30.75	$246.00	$47.90	$383.20	$32.01	$49.35
2 Laborers	28.75	460.00	44.75	716.00		
1 Equip. Oper. (crane)	39.80	318.40	60.00	480.00		
2 Cutting Torches		34.00		37.40		
2 Gases		129.60		142.56		
1 Hyd. Crane, 12 Ton		724.00		796.40	27.74	30.51
32 L.H., Daily Totals		$1912.00		$2555.56	$59.75	$79.86

Crew B-2	Hr.	Daily	Hr.	Daily	Bare Costs	Incl. O&P
1 Labor Foreman (outside)	$30.75	$246.00	$47.90	$383.20	$29.15	$45.38
4 Laborers	28.75	920.00	44.75	1432.00		
40 L.H., Daily Totals		$1166.00		$1815.20	$29.15	$45.38

Crews

Crew No.	Bare Costs		Incl. Subs O & P		Cost Per Labor-Hour	
Crew B-3	Hr.	Daily	Hr.	Daily	Bare Costs	Incl. O&P
1 Labor Foreman (outside)	$30.75	$246.00	$47.90	$383.20	$30.96	$47.76
2 Laborers	28.75	460.00	44.75	716.00		
1 Equip. Oper. (med.)	38.40	307.20	57.85	462.80		
2 Truck Drivers (heavy)	29.55	472.80	45.65	730.40		
1 Crawler Loader, 3 C.Y.		834.40		917.84		
2 Dump Trucks, 16 Ton, 12 C.Y.		1060.00		1166.00	39.47	43.41
48 L.H., Daily Totals		$3380.40		$4376.24	$70.42	$91.17
Crew B-3A	Hr.	Daily	Hr.	Daily	Bare Costs	Incl. O&P
4 Laborers	$28.75	$920.00	$44.75	$1432.00	$30.68	$47.37
1 Equip. Oper. (med.)	38.40	307.20	57.85	462.80		
1 Hyd. Excavator, 1.5 C.Y.		775.60		853.16	19.39	21.33
40 L.H., Daily Totals		$2002.80		$2747.96	$50.07	$68.70
Crew B-3B	Hr.	Daily	Hr.	Daily	Bare Costs	Incl. O&P
2 Laborers	$28.75	$460.00	$44.75	$716.00	$31.36	$48.25
1 Equip. Oper. (med.)	38.40	307.20	57.85	462.80		
1 Truck Driver (heavy)	29.55	236.40	45.65	365.20		
1 Backhoe Loader, 80 H.P.		285.40		313.94		
1 Dump Truck, 16 Ton, 12 C.Y.		530.00		583.00	25.48	28.03
32 L.H., Daily Totals		$1819.00		$2440.94	$56.84	$76.28
Crew B-3C	Hr.	Daily	Hr.	Daily	Bare Costs	Incl. O&P
3 Laborers	$28.75	$690.00	$44.75	$1074.00	$31.16	$48.02
1 Equip. Oper. (med.)	38.40	307.20	57.85	462.80		
1 Crawler Loader, 4 C.Y.		1145.00		1259.50	35.78	39.36
32 L.H., Daily Totals		$2142.20		$2796.30	$66.94	$87.38
Crew B-4	Hr.	Daily	Hr.	Daily	Bare Costs	Incl. O&P
1 Labor Foreman (outside)	$30.75	$246.00	$47.90	$383.20	$29.22	$45.42
4 Laborers	28.75	920.00	44.75	1432.00		
1 Truck Driver (heavy)	29.55	236.40	45.65	365.20		
1 Truck Tractor, 195 H.P.		235.60		259.16		
1 Flatbed Trailer, 40 Ton		125.80		138.38	7.53	8.28
48 L.H., Daily Totals		$1763.80		$2577.94	$36.75	$53.71
Crew B-5	Hr.	Daily	Hr.	Daily	Bare Costs	Incl. O&P
1 Labor Foreman (outside)	$30.75	$246.00	$47.90	$383.20	$31.79	$48.94
4 Laborers	28.75	920.00	44.75	1432.00		
2 Equip. Oper. (med.)	38.40	614.40	57.85	925.60		
1 Air Compressor, 250 C.F.M.		151.40		166.54		
2 Breakers, Pavement, 60 lb.		16.80		18.48		
2 -50' Air Hoses, 1.5"		11.60		12.76		
1 Crawler Loader, 3 C.Y.		834.40		917.84	18.11	19.92
56 L.H., Daily Totals		$2794.60		$3856.42	$49.90	$68.86
Crew B-5A	Hr.	Daily	Hr.	Daily	Bare Costs	Incl. O&P
1 Foreman	$30.75	$246.00	$47.90	$383.20	$31.33	$48.25
6 Laborers	28.75	1380.00	44.75	2148.00		
2 Equip. Oper. (med.)	38.40	614.40	57.85	925.60		
1 Equip. Oper. (light)	36.85	294.80	55.55	444.40		
2 Truck Drivers (heavy)	29.55	472.80	45.65	730.40		
1 Air Compressor, 365 C.F.M.		195.80		215.38		
2 Breakers, Pavement, 60 lb.		16.80		18.48		
8 -50' Air Hoses, 1"		32.80		36.08		
2 Dump Trucks, 12 Ton, 8 C.Y.		736.80		810.48	10.23	11.25
96 L.H., Daily Totals		$3990.20		$5712.02	$41.56	$59.50

Crew No.	Bare Costs		Incl. Subs O & P		Cost Per Labor-Hour	
Crew B-5B	Hr.	Daily	Hr.	Daily	Bare Costs	Incl. O&P
1 Powderman	$38.00	$304.00	$59.15	$473.20	$33.91	$51.97
2 Equip. Oper. (med.)	38.40	614.40	57.85	925.60		
3 Truck Drivers (heavy)	29.55	709.20	45.65	1095.60		
1 F.E. Loader, W.M., 2.5 C.Y.		349.00		383.90		
3 Dump Trucks, 16 Ton, 12 C.Y.		1590.00		1749.00		
1 Air Compressor, 365 C.F.M.		195.80		215.38	44.48	48.92
48 L.H., Daily Totals		$3762.40		$4842.68	$78.38	$100.89
Crew B-5C	Hr.	Daily	Hr.	Daily	Bare Costs	Incl. O&P
3 Laborers	$28.75	$690.00	$44.75	$1074.00	$32.18	$49.31
1 Equip. Oper. (medium)	38.40	307.20	57.85	462.80		
2 Truck Drivers (heavy)	29.55	472.80	45.65	730.40		
1 Equip. Oper. (crane)	39.80	318.40	60.00	480.00		
1 Equip. Oper. Oiler	33.90	271.20	51.10	408.80		
2 Dump Trucks, 16 Ton, 12 C.Y.		1060.00		1166.00		
1 Crawler Loader, 4 C.Y.		1145.00		1259.50		
1 S.P. Crane, 4x4, 25 Ton		875.60		963.16	48.13	52.95
64 L.H., Daily Totals		$5140.20		$6544.66	$80.32	$102.26
Crew B-6	Hr.	Daily	Hr.	Daily	Bare Costs	Incl. O&P
2 Laborers	$28.75	$460.00	$44.75	$716.00	$31.45	$48.35
1 Equip. Oper. (light)	36.85	294.80	55.55	444.40		
1 Backhoe Loader, 48 H.P.		243.40		267.74	10.14	11.16
24 L.H., Daily Totals		$998.20		$1428.14	$41.59	$59.51
Crew B-6A	Hr.	Daily	Hr.	Daily	Bare Costs	Incl. O&P
.5 Labor Foreman (outside)	$30.75	$123.00	$47.90	$191.60	$33.01	$50.62
1 Laborer	28.75	230.00	44.75	358.00		
1 Equip. Oper. (med.)	38.40	307.20	57.85	462.80		
1 Vacuum Trk., 5000 Gal.		352.25		387.48	17.61	19.37
20 L.H., Daily Totals		$1012.45		$1399.88	$50.62	$69.99
Crew B-6B	Hr.	Daily	Hr.	Daily	Bare Costs	Incl. O&P
2 Labor Foremen (out)	$30.75	$492.00	$47.90	$766.40	$29.42	$45.80
4 Laborers	28.75	920.00	44.75	1432.00		
1 S.P. Crane, 4x4, 5 Ton		326.80		359.48		
1 Flatbed Truck, Gas, 1.5 Ton		141.40		155.54		
1 Butt Fusion Machine		296.20		325.82	15.93	17.52
48 L.H., Daily Totals		$2176.40		$3039.24	$45.34	$63.32
Crew B-7	Hr.	Daily	Hr.	Daily	Bare Costs	Incl. O&P
1 Labor Foreman (outside)	$30.75	$246.00	$47.90	$383.20	$30.69	$47.46
4 Laborers	28.75	920.00	44.75	1432.00		
1 Equip. Oper. (med.)	38.40	307.20	57.85	462.80		
1 Brush Chipper, 12", 130 H.P.		183.60		201.96		
1 Crawler Loader, 3 C.Y.		834.40		917.84		
2 Chainsaws, Gas, 36" Long		68.80		75.68	22.64	24.91
48 L.H., Daily Totals		$2560.00		$3473.48	$53.33	$72.36
Crew B-7A	Hr.	Daily	Hr.	Daily	Bare Costs	Incl. O&P
2 Laborers	$28.75	$460.00	$44.75	$716.00	$31.45	$48.35
1 Equip. Oper. (light)	36.85	294.80	55.55	444.40		
1 Rake w/Tractor		195.90		215.49		
2 Chain saws, gas, 18"		46.00		50.60	10.08	11.09
24 L.H., Daily Totals		$996.70		$1426.49	$41.53	$59.44

Crews

Crew No.	Bare Costs		Incl. Subs O & P		Cost Per Labor-Hour	
Crew B-8	Hr.	Daily	Hr.	Daily	Bare Costs	Incl. O&P
1 Labor Foreman (outside)	$30.75	$246.00	$47.90	$383.20	$32.26	$49.44
2 Laborers	28.75	460.00	44.75	716.00		
2 Equip. Oper. (med.)	38.40	614.40	57.85	925.60		
1 Equip. Oper. Oiler	33.90	271.20	51.10	408.80		
2 Truck Drivers (heavy)	29.55	472.80	45.65	730.40		
1 Hyd. Crane, 25 Ton		739.60		813.56		
1 Crawler Loader, 3 C.Y.		834.40		917.84		
2 Dump Trucks, 16 Ton, 12 C.Y.		1060.00		1166.00	41.16	45.27
64 L.H., Daily Totals		$4698.40		$6061.40	$73.41	$94.71
Crew B-9	Hr.	Daily	Hr.	Daily	Bare Costs	Incl. O&P
1 Labor Foreman (outside)	$30.75	$246.00	$47.90	$383.20	$29.15	$45.38
4 Laborers	28.75	920.00	44.75	1432.00		
1 Air Compressor, 250 C.F.M.		151.40		166.54		
2 Breakers, Pavement, 60 lb.		16.80		18.48		
2 -50' Air Hoses, 1.5"		11.60		12.76	4.50	4.94
40 L.H., Daily Totals		$1345.80		$2012.98	$33.65	$50.32
Crew B-9A	Hr.	Daily	Hr.	Daily	Bare Costs	Incl. O&P
2 Laborers	$28.75	$460.00	$44.75	$716.00	$29.02	$45.05
1 Truck Driver (heavy)	29.55	236.40	45.65	365.20		
1 Water Tanker, 5000 Gal.		121.20		133.32		
1 Truck Tractor, 195 H.P.		235.60		259.16		
2 -50' Discharge Hoses, 3"		3.50		3.85	15.01	16.51
24 L.H., Daily Totals		$1056.70		$1477.53	$44.03	$61.56
Crew B-9B	Hr.	Daily	Hr.	Daily	Bare Costs	Incl. O&P
2 Laborers	$28.75	$460.00	$44.75	$716.00	$29.02	$45.05
1 Truck Driver (heavy)	29.55	236.40	45.65	365.20		
2 -50' Discharge Hoses, 3"		3.50		3.85		
1 Water Tanker, 5000 Gal.		121.20		133.32		
1 Truck Tractor, 195 H.P.		235.60		259.16		
1 Pressure Washer		55.80		61.38	17.34	19.07
24 L.H., Daily Totals		$1112.50		$1538.91	$46.35	$64.12
Crew B-9C	Hr.	Daily	Hr.	Daily	Bare Costs	Incl. O&P
1 Labor Foreman (outside)	$30.75	$246.00	$47.90	$383.20	$29.15	$45.38
4 Laborers	28.75	920.00	44.75	1432.00		
1 Air Compressor, 250 C.F.M.		151.40		166.54		
2 -50' Air Hoses, 1.5"		11.60		12.76		
2 Breakers, Pavement, 60 lb.		16.80		18.48	4.50	4.94
40 L.H., Daily Totals		$1345.80		$2012.98	$33.65	$50.32
Crew B-9D	Hr.	Daily	Hr.	Daily	Bare Costs	Incl. O&P
1 Labor Foreman (Outside)	$30.75	$246.00	$47.90	$383.20	$29.15	$45.38
4 Common Laborers	28.75	920.00	44.75	1432.00		
1 Air Compressor, 250 C.F.M.		151.40		166.54		
2 -50' Air Hoses, 1.5"		11.60		12.76		
2 Air Powered Tampers		51.80		56.98	5.37	5.91
40 L.H., Daily Totals		$1380.80		$2051.48	$34.52	$51.29
Crew B-10	Hr.	Daily	Hr.	Daily	Bare Costs	Incl. O&P
1 Equip. Oper. (med.)	$38.40	$307.20	$57.85	$462.80	$35.18	$53.48
.5 Laborer	28.75	115.00	44.75	179.00		
12 L.H., Daily Totals		$422.20		$641.80	$35.18	$53.48
Crew B-10A	Hr.	Daily	Hr.	Daily	Bare Costs	Incl. O&P
1 Equip. Oper. (med.)	$38.40	$307.20	$57.85	$462.80	$35.18	$53.48
.5 Laborer	28.75	115.00	44.75	179.00		
1 Roller, 2-Drum, W.B., 7.5 HP		133.80		147.18	11.15	12.27
12 L.H., Daily Totals		$556.00		$788.98	$46.33	$65.75

Crew No.	Bare Costs		Incl. Subs O & P		Cost Per Labor-Hour	
Crew B-10B	Hr.	Daily	Hr.	Daily	Bare Costs	Incl. O&P
1 Equip. Oper. (med.)	$38.40	$307.20	$57.85	$462.80	$35.18	$53.48
.5 Laborer	28.75	115.00	44.75	179.00		
1 Dozer, 200 H.P.		988.40		1087.24	82.37	90.60
12 L.H., Daily Totals		$1410.60		$1729.04	$117.55	$144.09
Crew B-10C	Hr.	Daily	Hr.	Daily	Bare Costs	Incl. O&P
1 Equip. Oper. (med.)	$38.40	$307.20	$57.85	$462.80	$35.18	$53.48
.5 Laborer	28.75	115.00	44.75	179.00		
1 Dozer, 200 H.P.		988.40		1087.24		
1 Vibratory Roller, Towed, 23 Ton		350.00		385.00	111.53	122.69
12 L.H., Daily Totals		$1760.60		$2114.04	$146.72	$176.17
Crew B-10D	Hr.	Daily	Hr.	Daily	Bare Costs	Incl. O&P
1 Equip. Oper. (med.)	$38.40	$307.20	$57.85	$462.80	$35.18	$53.48
.5 Laborer	28.75	115.00	44.75	179.00		
1 Dozer, 200 H.P.		988.40		1087.24		
1 Sheepsft. Roller, Towed		376.00		413.60	113.70	125.07
12 L.H., Daily Totals		$1786.60		$2142.64	$148.88	$178.55
Crew B-10E	Hr.	Daily	Hr.	Daily	Bare Costs	Incl. O&P
1 Equip. Oper. (med.)	$38.40	$307.20	$57.85	$462.80	$35.18	$53.48
.5 Laborer	28.75	115.00	44.75	179.00		
1 Tandem Roller, 5 Ton		122.20		134.42	10.18	11.20
12 L.H., Daily Totals		$544.40		$776.22	$45.37	$64.69
Crew B-10F	Hr.	Daily	Hr.	Daily	Bare Costs	Incl. O&P
1 Equip. Oper. (med.)	$38.40	$307.20	$57.85	$462.80	$35.18	$53.48
.5 Laborer	28.75	115.00	44.75	179.00		
1 Tandem Roller, 10 Ton		211.00		232.10	17.58	19.34
12 L.H., Daily Totals		$633.20		$873.90	$52.77	$72.83
Crew B-10G	Hr.	Daily	Hr.	Daily	Bare Costs	Incl. O&P
1 Equip. Oper. (med.)	$38.40	$307.20	$57.85	$462.80	$35.18	$53.48
.5 Laborer	28.75	115.00	44.75	179.00		
1 Sheepsft. Roll., 130 H.P.		956.20		1051.82	79.68	87.65
12 L.H., Daily Totals		$1378.40		$1693.62	$114.87	$141.13
Crew B-10H	Hr.	Daily	Hr.	Daily	Bare Costs	Incl. O&P
1 Equip. Oper. (med.)	$38.40	$307.20	$57.85	$462.80	$35.18	$53.48
.5 Laborer	28.75	115.00	44.75	179.00		
1 Diaphragm Water Pump, 2"		57.40		63.14		
1 -20' Suction Hose, 2"		1.95		2.15		
2 -50' Discharge Hoses, 2"		1.80		1.98	5.10	5.61
12 L.H., Daily Totals		$483.35		$709.07	$40.28	$59.09
Crew B-10I	Hr.	Daily	Hr.	Daily	Bare Costs	Incl. O&P
1 Equip. Oper. (med.)	$38.40	$307.20	$57.85	$462.80	$35.18	$53.48
.5 Laborer	28.75	115.00	44.75	179.00		
1 Diaphragm Water Pump, 4"		78.20		86.02		
1 -20' Suction Hose, 4"		3.25		3.58		
2 -50' Discharge Hoses, 4"		4.70		5.17	7.18	7.90
12 L.H., Daily Totals		$508.35		$736.57	$42.36	$61.38
Crew B-10J	Hr.	Daily	Hr.	Daily	Bare Costs	Incl. O&P
1 Equip. Oper. (med.)	$38.40	$307.20	$57.85	$462.80	$35.18	$53.48
.5 Laborer	28.75	115.00	44.75	179.00		
1 Centrifugal Water Pump, 3"		63.40		69.74		
1 -20' Suction Hose, 3"		3.05		3.36		
2 -50' Discharge Hoses, 3"		3.50		3.85	5.83	6.41
12 L.H., Daily Totals		$492.15		$718.75	$41.01	$59.90

Crews

Crew No.	Bare Costs		Incl. Subs O & P		Cost Per Labor-Hour	
Crew B-10K	Hr.	Daily	Hr.	Daily	Bare Costs	Incl. O&P
1 Equip. Oper. (med.)	$38.40	$307.20	$57.85	$462.80	$35.18	$53.48
.5 Laborer	28.75	115.00	44.75	179.00		
1 Centr. Water Pump, 6"		278.20		306.02		
1 -20' Suction Hose, 6"		11.50		12.65		
2 -50' Discharge Hoses, 6"		12.60		13.86	25.19	27.71
12 L.H., Daily Totals		$724.50		$974.33	$60.38	$81.19
Crew B-10L	Hr.	Daily	Hr.	Daily	Bare Costs	Incl. O&P
1 Equip. Oper. (med.)	$38.40	$307.20	$57.85	$462.80	$35.18	$53.48
.5 Laborer	28.75	115.00	44.75	179.00		
1 Dozer, 80 H.P.		354.80		390.28	29.57	32.52
12 L.H., Daily Totals		$777.00		$1032.08	$64.75	$86.01
Crew B-10M	Hr.	Daily	Hr.	Daily	Bare Costs	Incl. O&P
1 Equip. Oper. (med.)	$38.40	$307.20	$57.85	$462.80	$35.18	$53.48
.5 Laborer	28.75	115.00	44.75	179.00		
1 Dozer, 300 H.P.		1301.00		1431.10	108.42	119.26
12 L.H., Daily Totals		$1723.20		$2072.90	$143.60	$172.74
Crew B-10N	Hr.	Daily	Hr.	Daily	Bare Costs	Incl. O&P
1 Equip. Oper. (med.)	$38.40	$307.20	$57.85	$462.80	$35.18	$53.48
.5 Laborer	28.75	115.00	44.75	179.00		
1 F.E. Loader, T.M., 1.5 C.Y		334.00		367.40	27.83	30.62
12 L.H., Daily Totals		$756.20		$1009.20	$63.02	$84.10
Crew B-10O	Hr.	Daily	Hr.	Daily	Bare Costs	Incl. O&P
1 Equip. Oper. (med.)	$38.40	$307.20	$57.85	$462.80	$35.18	$53.48
.5 Laborer	28.75	115.00	44.75	179.00		
1 F.E. Loader, T.M., 2.25 C.Y.		589.20		648.12	49.10	54.01
12 L.H., Daily Totals		$1011.40		$1289.92	$84.28	$107.49
Crew B-10P	Hr.	Daily	Hr.	Daily	Bare Costs	Incl. O&P
1 Equip. Oper. (med.)	$38.40	$307.20	$57.85	$462.80	$35.18	$53.48
.5 Laborer	28.75	115.00	44.75	179.00		
1 Crawler Loader, 3 C.Y.		834.40		917.84	69.53	76.49
12 L.H., Daily Totals		$1256.60		$1559.64	$104.72	$129.97
Crew B-10Q	Hr.	Daily	Hr.	Daily	Bare Costs	Incl. O&P
1 Equip. Oper. (med.)	$38.40	$307.20	$57.85	$462.80	$35.18	$53.48
.5 Laborer	28.75	115.00	44.75	179.00		
1 Crawler Loader, 4 C.Y.		1145.00		1259.50	95.42	104.96
12 L.H., Daily Totals		$1567.20		$1901.30	$130.60	$158.44
Crew B-10R	Hr.	Daily	Hr.	Daily	Bare Costs	Incl. O&P
1 Equip. Oper. (med.)	$38.40	$307.20	$57.85	$462.80	$35.18	$53.48
.5 Laborer	28.75	115.00	44.75	179.00		
1 F.E. Loader, W.M., 1 C.Y.		207.60		228.36	17.30	19.03
12 L.H., Daily Totals		$629.80		$870.16	$52.48	$72.51
Crew B-10S	Hr.	Daily	Hr.	Daily	Bare Costs	Incl. O&P
1 Equip. Oper. (med.)	$38.40	$307.20	$57.85	$462.80	$35.18	$53.48
.5 Laborer	28.75	115.00	44.75	179.00		
1 F.E. Loader, W.M., 1.5 C.Y.		277.60		305.36	23.13	25.45
12 L.H., Daily Totals		$699.80		$947.16	$58.32	$78.93
Crew B-10T	Hr.	Daily	Hr.	Daily	Bare Costs	Incl. O&P
1 Equip. Oper. (med.)	$38.40	$307.20	$57.85	$462.80	$35.18	$53.48
.5 Laborer	28.75	115.00	44.75	179.00		
1 F.E. Loader, W.M.,2.5 C.Y.		349.00		383.90	29.08	31.99
12 L.H., Daily Totals		$771.20		$1025.70	$64.27	$85.47

Crew No.	Bare Costs		Incl. Subs O & P		Cost Per Labor-Hour	
Crew B-10U	Hr.	Daily	Hr.	Daily	Bare Costs	Incl. O&P
1 Equip. Oper. (med.)	$38.40	$307.20	$57.85	$462.80	$35.18	$53.48
.5 Laborer	28.75	115.00	44.75	179.00		
1 F.E. Loader, W.M., 5.5 C.Y.		755.80		831.38	62.98	69.28
12 L.H., Daily Totals		$1178.00		$1473.18	$98.17	$122.77
Crew B-10V	Hr.	Daily	Hr.	Daily	Bare Costs	Incl. O&P
1 Equip. Oper. (med.)	$38.40	$307.20	$57.85	$462.80	$35.18	$53.48
.5 Laborer	28.75	115.00	44.75	179.00		
1 Dozer, 700 H.P.		3406.00		3746.60	283.83	312.22
12 L.H., Daily Totals		$3828.20		$4388.40	$319.02	$365.70
Crew B-10W	Hr.	Daily	Hr.	Daily	Bare Costs	Incl. O&P
1 Equip. Oper. (med.)	$38.40	$307.20	$57.85	$462.80	$35.18	$53.48
.5 Laborer	28.75	115.00	44.75	179.00		
1 Dozer, 105 H.P.		494.80		544.28	41.23	45.36
12 L.H., Daily Totals		$917.00		$1186.08	$76.42	$98.84
Crew B-10X	Hr.	Daily	Hr.	Daily	Bare Costs	Incl. O&P
1 Equip. Oper. (med.)	$38.40	$307.20	$57.85	$462.80	$35.18	$53.48
.5 Laborer	28.75	115.00	44.75	179.00		
1 Dozer, 410 H.P.		1677.00		1844.70	139.75	153.72
12 L.H., Daily Totals		$2099.20		$2486.50	$174.93	$207.21
Crew B-10Y	Hr.	Daily	Hr.	Daily	Bare Costs	Incl. O&P
1 Equip. Oper. (med.)	$38.40	$307.20	$57.85	$462.80	$35.18	$53.48
.5 Laborer	28.75	115.00	44.75	179.00		
1 Vibr. Roller, Towed, 12 Ton		448.40		493.24	37.37	41.10
12 L.H., Daily Totals		$870.60		$1135.04	$72.55	$94.59
Crew B-11A	Hr.	Daily	Hr.	Daily	Bare Costs	Incl. O&P
1 Equipment Oper. (med.)	$38.40	$307.20	$57.85	$462.80	$33.58	$51.30
1 Laborer	28.75	230.00	44.75	358.00		
1 Dozer, 200 H.P.		988.40		1087.24	61.77	67.95
16 L.H., Daily Totals		$1525.60		$1908.04	$95.35	$119.25
Crew B-11B	Hr.	Daily	Hr.	Daily	Bare Costs	Incl. O&P
1 Equipment Oper. (light)	$36.85	$294.80	$55.55	$444.40	$32.80	$50.15
1 Laborer	28.75	230.00	44.75	358.00		
1 Air Powered Tamper		25.90		28.49		
1 Air Compressor, 365 C.F.M.		195.80		215.38		
2 -50' Air Hoses, 1.5"		11.60		12.76	14.58	16.04
16 L.H., Daily Totals		$758.10		$1059.03	$47.38	$66.19
Crew B-11C	Hr.	Daily	Hr.	Daily	Bare Costs	Incl. O&P
1 Equipment Oper. (med.)	$38.40	$307.20	$57.85	$462.80	$33.58	$51.30
1 Laborer	28.75	230.00	44.75	358.00		
1 Backhoe Loader, 48 H.P.		243.40		267.74	15.21	16.73
16 L.H., Daily Totals		$780.60		$1088.54	$48.79	$68.03
Crew B-11J	Hr.	Daily	Hr.	Daily	Bare Costs	Incl. O&P
1 Equipment Oper. (med.)	$38.40	$307.20	$57.85	$462.80	$33.58	$51.30
1 Laborer	28.75	230.00	44.75	358.00		
1 Grader, 30,000 Lbs.		505.40		555.94		
1 Ripper, beam & 1 shank		76.40		84.04	36.36	40.00
16 L.H., Daily Totals		$1119.00		$1460.78	$69.94	$91.30

Crews

Crew No.	Bare Costs	Incl. Subs O & P	Cost Per Labor-Hour

Crew B-11K	Hr.	Daily	Hr.	Daily	Bare Costs	Incl. O&P
1 Equipment Oper. (med.)	$38.40	$307.20	$57.85	$462.80	$33.58	$51.30
1 Laborer	28.75	230.00	44.75	358.00		
1 Trencher, Chain Type, 8' D		1606.00		1766.60	100.38	110.41
16 L.H., Daily Totals		$2143.20		$2587.40	$133.95	$161.71

Crew B-11L	Hr.	Daily	Hr.	Daily	Bare Costs	Incl. O&P
1 Equipment Oper. (med.)	$38.40	$307.20	$57.85	$462.80	$33.58	$51.30
1 Laborer	28.75	230.00	44.75	358.00		
1 Grader, 30,000 Lbs.		505.40		555.94	31.59	34.75
16 L.H., Daily Totals		$1042.60		$1376.74	$65.16	$86.05

Crew B-11M	Hr.	Daily	Hr.	Daily	Bare Costs	Incl. O&P
1 Equipment Oper. (med.)	$38.40	$307.20	$57.85	$462.80	$33.58	$51.30
1 Laborer	28.75	230.00	44.75	358.00		
1 Backhoe Loader, 80 H.P.		285.40		313.94	17.84	19.62
16 L.H., Daily Totals		$822.60		$1134.74	$51.41	$70.92

Crew B-11N	Hr.	Daily	Hr.	Daily	Bare Costs	Incl. O&P
1 Labor Foreman	$30.75	$246.00	$47.90	$383.20	$31.65	$48.61
2 Equipment Operators (med.)	38.40	614.40	57.85	925.60		
6 Truck Drivers (hvy.)	29.55	1418.40	45.65	2191.20		
1 F.E. Loader, W.M., 5.5 C.Y.		755.80		831.38		
1 Dozer, 410 H.P.		1677.00		1844.70		
6 Dump Trucks, Off Hwy., 50 Ton		7848.00		8632.80	142.79	157.07
72 L.H., Daily Totals		$12559.60		$14808.88	$174.44	$205.68

Crew B-11Q	Hr.	Daily	Hr.	Daily	Bare Costs	Incl. O&P
1 Equipment Operator (med.)	$38.40	$307.20	$57.85	$462.80	$35.18	$53.48
.5 Laborer	28.75	115.00	44.75	179.00		
1 Dozer, 140 H.P.		611.40		672.54	50.95	56.05
12 L.H., Daily Totals		$1033.60		$1314.34	$86.13	$109.53

Crew B-11R	Hr.	Daily	Hr.	Daily	Bare Costs	Incl. O&P
1 Equipment Operator (med.)	$38.40	$307.20	$57.85	$462.80	$35.18	$53.48
.5 Laborer	28.75	115.00	44.75	179.00		
1 Dozer, 200 H.P.		988.40		1087.24	82.37	90.60
12 L.H., Daily Totals		$1410.60		$1729.04	$117.55	$144.09

Crew B-11S	Hr.	Daily	Hr.	Daily	Bare Costs	Incl. O&P
1 Equipment Operator (med.)	$38.40	$307.20	$57.85	$462.80	$35.18	$53.48
.5 Laborer	28.75	115.00	44.75	179.00		
1 Dozer, 300 H.P.		1301.00		1431.10		
1 Ripper, Beam & 1 Shank		76.40		84.04	114.78	126.26
12 L.H., Daily Totals		$1799.60		$2156.94	$149.97	$179.75

Crew B-11T	Hr.	Daily	Hr.	Daily	Bare Costs	Incl. O&P
1 Equipment Operator (med.)	$38.40	$307.20	$57.85	$462.80	$35.18	$53.48
.5 Laborer	28.75	115.00	44.75	179.00		
1 Dozer, 410 H.P.		1677.00		1844.70		
1 Ripper, Beam & 2 Shanks		85.20		93.72	146.85	161.54
12 L.H., Daily Totals		$2184.40		$2580.22	$182.03	$215.02

Crew B-11U	Hr.	Daily	Hr.	Daily	Bare Costs	Incl. O&P
1 Equipment Operator (med.)	$38.40	$307.20	$57.85	$462.80	$35.18	$53.48
.5 Laborer	28.75	115.00	44.75	179.00		
1 Dozer, 520 H.P.		2260.00		2486.00	188.33	207.17
12 L.H., Daily Totals		$2682.20		$3127.80	$223.52	$260.65

Crew B-11V	Hr.	Daily	Hr.	Daily	Bare Costs	Incl. O&P
3 Laborers	$28.75	$690.00	$44.75	$1074.00	$28.75	$44.75
1 Roller, 2-Drum, W.B., 7.5 HP		133.80		147.18	5.58	6.13
24 L.H., Daily Totals		$823.80		$1221.18	$34.33	$50.88

Crew B-11W	Hr.	Daily	Hr.	Daily	Bare Costs	Incl. O&P
1 Equipment Operator (med.)	$38.40	$307.20	$57.85	$462.80	$30.22	$46.59
1 Common Laborer	28.75	230.00	44.75	358.00		
10 Truck Drivers (hvy.)	29.55	2364.00	45.65	3652.00		
1 Dozer, 200 H.P.		988.40		1087.24		
1 Vibratory Roller, Towed, 23 Ton		350.00		385.00		
10 Dump Trucks, 12 Ton, 8 C.Y.		3684.00		4052.40	52.32	57.55
96 L.H., Daily Totals		$7923.60		$9997.44	$82.54	$104.14

Crew B-11Y	Hr.	Daily	Hr.	Daily	Bare Costs	Incl. O&P
1 Labor Foreman (Outside)	$30.75	$246.00	$47.90	$383.20	$32.19	$49.47
5 Common Laborers	28.75	1150.00	44.75	1790.00		
3 Equipment Operators (med.)	38.40	921.60	57.85	1388.40		
1 Dozer, 80 H.P.		354.80		390.28		
2 Roller, 2-Drum, W.B., 7.5 HP		267.60		294.36		
4 Vibratory Plates, Gas, 21"		144.80		159.28	10.66	11.72
72 L.H., Daily Totals		$3084.80		$4405.52	$42.84	$61.19

Crew B-12A	Hr.	Daily	Hr.	Daily	Bare Costs	Incl. O&P
1 Equip. Oper. (crane)	$39.80	$318.40	$60.00	$480.00	$34.27	$52.38
1 Laborer	28.75	230.00	44.75	358.00		
1 Hyd. Excavator, 1 C.Y.		601.80		661.98	37.61	41.37
16 L.H., Daily Totals		$1150.20		$1499.98	$71.89	$93.75

Crew B-12B	Hr.	Daily	Hr.	Daily	Bare Costs	Incl. O&P
1 Equip. Oper. (crane)	$39.80	$318.40	$60.00	$480.00	$34.27	$52.38
1 Laborer	28.75	230.00	44.75	358.00		
1 Hyd. Excavator, 1.5 C.Y.		775.60		853.16	48.48	53.32
16 L.H., Daily Totals		$1324.00		$1691.16	$82.75	$105.70

Crew B-12C	Hr.	Daily	Hr.	Daily	Bare Costs	Incl. O&P
1 Equip. Oper. (crane)	$39.80	$318.40	$60.00	$480.00	$34.27	$52.38
1 Laborer	28.75	230.00	44.75	358.00		
1 Hyd. Excavator, 2 C.Y.		994.60		1094.06	62.16	68.38
16 L.H., Daily Totals		$1543.00		$1932.06	$96.44	$120.75

Crew B-12D	Hr.	Daily	Hr.	Daily	Bare Costs	Incl. O&P
1 Equip. Oper. (crane)	$39.80	$318.40	$60.00	$480.00	$34.27	$52.38
1 Laborer	28.75	230.00	44.75	358.00		
1 Hyd. Excavator, 3.5 C.Y.		2166.00		2382.60	135.38	148.91
16 L.H., Daily Totals		$2714.40		$3220.60	$169.65	$201.29

Crew B-12E	Hr.	Daily	Hr.	Daily	Bare Costs	Incl. O&P
1 Equip. Oper. (crane)	$39.80	$318.40	$60.00	$480.00	$34.27	$52.38
1 Laborer	28.75	230.00	44.75	358.00		
1 Hyd. Excavator, .5 C.Y.		358.00		393.80	22.38	24.61
16 L.H., Daily Totals		$906.40		$1231.80	$56.65	$76.99

Crew B-12F	Hr.	Daily	Hr.	Daily	Bare Costs	Incl. O&P
1 Equip. Oper. (crane)	$39.80	$318.40	$60.00	$480.00	$34.27	$52.38
1 Laborer	28.75	230.00	44.75	358.00		
1 Hyd. Excavator, .75 C.Y.		516.40		568.04	32.27	35.50
16 L.H., Daily Totals		$1064.80		$1406.04	$66.55	$87.88

Crews

Crew No.	Bare Costs		Incl. Subs O & P		Cost Per Labor-Hour	
Crew B-12G	Hr.	Daily	Hr.	Daily	Bare Costs	Incl. O&P
1 Equip. Oper. (crane)	$39.80	$318.40	$60.00	$480.00	$34.27	$52.38
1 Laborer	28.75	230.00	44.75	358.00		
1 Crawler Crane, 15 Ton		576.65		634.32		
1 Clamshell Bucket, .5 C.Y.		34.20		37.62	38.18	42.00
16 L.H., Daily Totals		$1159.25		$1509.93	$72.45	$94.37
Crew B-12H	Hr.	Daily	Hr.	Daily	Bare Costs	Incl. O&P
1 Equip. Oper. (crane)	$39.80	$318.40	$60.00	$480.00	$34.27	$52.38
1 Laborer	28.75	230.00	44.75	358.00		
1 Crawler Crane, 25 Ton		983.00		1081.30		
1 Clamshell Bucket, 1 C.Y.		44.80		49.28	64.24	70.66
16 L.H., Daily Totals		$1576.20		$1968.58	$98.51	$123.04
Crew B-12I	Hr.	Daily	Hr.	Daily	Bare Costs	Incl. O&P
1 Equip. Oper. (crane)	$39.80	$318.40	$60.00	$480.00	$34.27	$52.38
1 Laborer	28.75	230.00	44.75	358.00		
1 Crawler Crane, 20 Ton		737.50		811.25		
1 Dragline Bucket, .75 C.Y.		19.40		21.34	47.31	52.04
16 L.H., Daily Totals		$1305.30		$1670.59	$81.58	$104.41
Crew B-12J	Hr.	Daily	Hr.	Daily	Bare Costs	Incl. O&P
1 Equip. Oper. (crane)	$39.80	$318.40	$60.00	$480.00	$34.27	$52.38
1 Laborer	28.75	230.00	44.75	358.00		
1 Gradall, 5/8 C.Y.		905.80		996.38	56.61	62.27
16 L.H., Daily Totals		$1454.20		$1834.38	$90.89	$114.65
Crew B-12K	Hr.	Daily	Hr.	Daily	Bare Costs	Incl. O&P
1 Equip. Oper. (crane)	$39.80	$318.40	$60.00	$480.00	$34.27	$52.38
1 Laborer	28.75	230.00	44.75	358.00		
1 Gradall, 3 Ton, 1 C.Y.		1026.00		1128.60	64.13	70.54
16 L.H., Daily Totals		$1574.40		$1966.60	$98.40	$122.91
Crew B-12L	Hr.	Daily	Hr.	Daily	Bare Costs	Incl. O&P
1 Equip. Oper. (crane)	$39.80	$318.40	$60.00	$480.00	$34.27	$52.38
1 Laborer	28.75	230.00	44.75	358.00		
1 Crawler Crane, 15 Ton		576.65		634.32		
1 F.E. Attachment, .5 C.Y.		49.20		54.12	39.12	43.03
16 L.H., Daily Totals		$1174.25		$1526.43	$73.39	$95.40
Crew B-12M	Hr.	Daily	Hr.	Daily	Bare Costs	Incl. O&P
1 Equip. Oper. (crane)	$39.80	$318.40	$60.00	$480.00	$34.27	$52.38
1 Laborer	28.75	230.00	44.75	358.00		
1 Crawler Crane, 20 Ton		737.50		811.25		
1 F.E. Attachment, .75 C.Y.		53.60		58.96	49.44	54.39
16 L.H., Daily Totals		$1339.50		$1708.21	$83.72	$106.76
Crew B-12N	Hr.	Daily	Hr.	Daily	Bare Costs	Incl. O&P
1 Equip. Oper. (crane)	$39.80	$318.40	$60.00	$480.00	$34.27	$52.38
1 Laborer	28.75	230.00	44.75	358.00		
1 Crawler Crane, 25 Ton		983.00		1081.30		
1 F.E. Attachment, 1 C.Y.		60.40		66.44	65.21	71.73
16 L.H., Daily Totals		$1591.80		$1985.74	$99.49	$124.11
Crew B-12O	Hr.	Daily	Hr.	Daily	Bare Costs	Incl. O&P
1 Equip. Oper. (crane)	$39.80	$318.40	$60.00	$480.00	$34.27	$52.38
1 Laborer	28.75	230.00	44.75	358.00		
1 Crawler Crane, 40 Ton		994.60		1094.06		
1 F.E. Attachment, 1.5 C.Y.		69.60		76.56	66.51	73.16
16 L.H., Daily Totals		$1612.60		$2008.62	$100.79	$125.54

Crew No.	Bare Costs		Incl. Subs O & P		Cost Per Labor-Hour	
Crew B-12P	Hr.	Daily	Hr.	Daily	Bare Costs	Incl. O&P
1 Equip. Oper. (crane)	$39.80	$318.40	$60.00	$480.00	$34.27	$52.38
1 Laborer	28.75	230.00	44.75	358.00		
1 Crawler Crane, 40 Ton		994.60		1094.06		
1 Dragline Bucket, 1.5 C.Y.		31.00		34.10	64.10	70.51
16 L.H., Daily Totals		$1574.00		$1966.16	$98.38	$122.89
Crew B-12Q	Hr.	Daily	Hr.	Daily	Bare Costs	Incl. O&P
1 Equip. Oper. (crane)	$39.80	$318.40	$60.00	$480.00	$34.27	$52.38
1 Laborer	28.75	230.00	44.75	358.00		
1 Hyd. Excavator, 5/8 C.Y.		462.20		508.42	28.89	31.78
16 L.H., Daily Totals		$1010.60		$1346.42	$63.16	$84.15
Crew B-12R	Hr.	Daily	Hr.	Daily	Bare Costs	Incl. O&P
1 Equip. Oper. (crane)	$39.80	$318.40	$60.00	$480.00	$34.27	$52.38
1 Laborer	28.75	230.00	44.75	358.00		
1 Hyd. Excavator, 1.5 C.Y.		775.60		853.16	48.48	53.32
16 L.H., Daily Totals		$1324.00		$1691.16	$82.75	$105.70
Crew B-12S	Hr.	Daily	Hr.	Daily	Bare Costs	Incl. O&P
1 Equip. Oper. (crane)	$39.80	$318.40	$60.00	$480.00	$34.27	$52.38
1 Laborer	28.75	230.00	44.75	358.00		
1 Hyd. Excavator, 2.5 C.Y.		1333.00		1466.30	83.31	91.64
16 L.H., Daily Totals		$1881.40		$2304.30	$117.59	$144.02
Crew B-12T	Hr.	Daily	Hr.	Daily	Bare Costs	Incl. O&P
1 Equip. Oper. (crane)	$39.80	$318.40	$60.00	$480.00	$34.27	$52.38
1 Laborer	28.75	230.00	44.75	358.00		
1 Crawler Crane, 75 Ton		1384.00		1522.40		
1 F.E. Attachment, 3 C.Y.		93.00		102.30	92.31	101.54
16 L.H., Daily Totals		$2025.40		$2462.70	$126.59	$153.92
Crew B-12V	Hr.	Daily	Hr.	Daily	Bare Costs	Incl. O&P
1 Equip. Oper. (crane)	$39.80	$318.40	$60.00	$480.00	$34.27	$52.38
1 Laborer	28.75	230.00	44.75	358.00		
1 Crawler Crane, 75 Ton		1384.00		1522.40		
1 Dragline Bucket, 3 C.Y.		48.80		53.68	89.55	98.50
16 L.H., Daily Totals		$1981.20		$2414.08	$123.83	$150.88
Crew B-13	Hr.	Daily	Hr.	Daily	Bare Costs	Incl. O&P
1 Labor Foreman (outside)	$30.75	$246.00	$47.90	$383.20	$31.35	$48.29
4 Laborers	28.75	920.00	44.75	1432.00		
1 Equip. Oper. (crane)	39.80	318.40	60.00	480.00		
1 Equip. Oper. Oiler	33.90	271.20	51.10	408.80		
1 Hyd. Crane, 25 Ton		739.60		813.56	13.21	14.53
56 L.H., Daily Totals		$2495.20		$3517.56	$44.56	$62.81
Crew B-13A	Hr.	Daily	Hr.	Daily	Bare Costs	Incl. O&P
1 Foreman	$30.75	$246.00	$47.90	$383.20	$32.02	$49.20
2 Laborers	28.75	460.00	44.75	716.00		
2 Equipment Operators	38.40	614.40	57.85	925.60		
2 Truck Drivers (heavy)	29.55	472.80	45.65	730.40		
1 Crawler Crane, 75 Ton		1384.00		1522.40		
1 Crawler Loader, 4 C.Y.		1145.00		1259.50		
2 Dump Trucks, 12 Ton, 8 C.Y.		736.80		810.48	58.32	64.15
56 L.H., Daily Totals		$5059.00		$6347.58	$90.34	$113.35

Crews

Crew No.	Bare Costs	Incl. Subs O & P	Cost Per Labor-Hour

Crew B-13B	Hr.	Daily	Hr.	Daily	Bare Costs	Incl. O&P
1 Labor Foreman (outside)	$30.75	$246.00	$47.90	$383.20	$31.35	$48.29
4 Laborers	28.75	920.00	44.75	1432.00		
1 Equip. Oper. (crane)	39.80	318.40	60.00	480.00		
1 Equip. Oper. Oiler	33.90	271.20	51.10	408.80		
1 Hyd. Crane, 55 Ton		1060.00		1166.00	18.93	20.82
56 L.H., Daily Totals		$2815.60		$3870.00	$50.28	$69.11

Crew B-13C	Hr.	Daily	Hr.	Daily	Bare Costs	Incl. O&P
1 Labor Foreman (outside)	$30.75	$246.00	$47.90	$383.20	$31.35	$48.29
4 Laborers	28.75	920.00	44.75	1432.00		
1 Equip. Oper. (crane)	39.80	318.40	60.00	480.00		
1 Equip. Oper. Oiler	33.90	271.20	51.10	408.80		
1 Crawler Crane, 100 Ton		1717.00		1888.70	30.66	33.73
56 L.H., Daily Totals		$3472.60		$4592.70	$62.01	$82.01

Crew B-13D	Hr.	Daily	Hr.	Daily	Bare Costs	Incl. O&P
1 Laborer	$28.75	$230.00	$44.75	$358.00	$34.27	$52.38
1 Equip. Oper. (crane)	39.80	318.40	60.00	480.00		
1 Hyd. Excavator, 1 C.Y.		601.80		661.98		
1 Trench Box		79.25		87.17	42.57	46.82
16 L.H., Daily Totals		$1229.45		$1587.16	$76.84	$99.20

Crew B-13E	Hr.	Daily	Hr.	Daily	Bare Costs	Incl. O&P
1 Laborer	$28.75	$230.00	$44.75	$358.00	$34.27	$52.38
1 Equip. Oper. (crane)	39.80	318.40	60.00	480.00		
1 Hyd. Excavator, 1.5 C.Y.		775.60		853.16		
1 Trench Box		79.25		87.17	53.43	58.77
16 L.H., Daily Totals		$1403.25		$1778.34	$87.70	$111.15

Crew B-13F	Hr.	Daily	Hr.	Daily	Bare Costs	Incl. O&P
1 Laborer	$28.75	$230.00	$44.75	$358.00	$34.27	$52.38
1 Equip. Oper. (crane)	39.80	318.40	60.00	480.00		
1 Hyd. Excavator, 3.5 C.Y.		2166.00		2382.60		
1 Trench Box		79.25		87.17	140.33	154.36
16 L.H., Daily Totals		$2793.65		$3307.78	$174.60	$206.74

Crew B-13G	Hr.	Daily	Hr.	Daily	Bare Costs	Incl. O&P
1 Laborer	$28.75	$230.00	$44.75	$358.00	$34.27	$52.38
1 Equip. Oper. (crane)	39.80	318.40	60.00	480.00		
1 Hyd. Excavator, .75 C.Y.		516.40		568.04		
1 Trench Box		79.25		87.17	37.23	40.95
16 L.H., Daily Totals		$1144.05		$1493.21	$71.50	$93.33

Crew B-13H	Hr.	Daily	Hr.	Daily	Bare Costs	Incl. O&P
1 Laborer	$28.75	$230.00	$44.75	$358.00	$34.27	$52.38
1 Equip. Oper. (crane)	39.80	318.40	60.00	480.00		
1 Gradall, 5/8 C.Y.		905.80		996.38		
1 Trench Box		79.25		87.17	61.57	67.72
16 L.H., Daily Totals		$1533.45		$1921.56	$95.84	$120.10

Crew B-13I	Hr.	Daily	Hr.	Daily	Bare Costs	Incl. O&P
1 Laborer	$28.75	$230.00	$44.75	$358.00	$34.27	$52.38
1 Equip. Oper. (crane)	39.80	318.40	60.00	480.00		
1 Gradall, 3 Ton, 1 C.Y.		1026.00		1128.60		
1 Trench Box		79.25		87.17	69.08	75.99
16 L.H., Daily Totals		$1653.65		$2053.78	$103.35	$128.36

Crew B-13J	Hr.	Daily	Hr.	Daily	Bare Costs	Incl. O&P
1 Laborer	$28.75	$230.00	$44.75	$358.00	$34.27	$52.38
1 Equip. Oper. (crane)	39.80	318.40	60.00	480.00		
1 Hyd. Excavator, 2.5 C.Y.		1333.00		1466.30		
1 Trench Box		79.25		87.17	88.27	97.09
16 L.H., Daily Totals		$1960.65		$2391.47	$122.54	$149.47

Crew B-14	Hr.	Daily	Hr.	Daily	Bare Costs	Incl. O&P
1 Labor Foreman (outside)	$30.75	$246.00	$47.90	$383.20	$30.43	$47.08
4 Laborers	28.75	920.00	44.75	1432.00		
1 Equip. Oper. (light)	36.85	294.80	55.55	444.40		
1 Backhoe Loader, 48 H.P.		243.40		267.74	5.07	5.58
48 L.H., Daily Totals		$1704.20		$2527.34	$35.50	$52.65

Crew B-15	Hr.	Daily	Hr.	Daily	Bare Costs	Incl. O&P
1 Equipment Oper. (med)	$38.40	$307.20	$57.85	$462.80	$31.96	$49.01
.5 Laborer	28.75	115.00	44.75	179.00		
2 Truck Drivers (heavy)	29.55	472.80	45.65	730.40		
2 Dump Trucks, 16 Ton, 12 C.Y.		1060.00		1166.00		
1 Dozer, 200 H.P.		988.40		1087.24	73.16	80.47
28 L.H., Daily Totals		$2943.40		$3625.44	$105.12	$129.48

Crew B-16	Hr.	Daily	Hr.	Daily	Bare Costs	Incl. O&P
1 Labor Foreman (outside)	$30.75	$246.00	$47.90	$383.20	$29.45	$45.76
2 Laborers	28.75	460.00	44.75	716.00		
1 Truck Driver (heavy)	29.55	236.40	45.65	365.20		
1 Dump Truck, 16 Ton, 12 C.Y.		530.00		583.00	16.56	18.22
32 L.H., Daily Totals		$1472.40		$2047.40	$46.01	$63.98

Crew B-17	Hr.	Daily	Hr.	Daily	Bare Costs	Incl. O&P
2 Laborers	$28.75	$460.00	$44.75	$716.00	$30.98	$47.67
1 Equip. Oper. (light)	36.85	294.80	55.55	444.40		
1 Truck Driver (heavy)	29.55	236.40	45.65	365.20		
1 Backhoe Loader, 48 H.P.		243.40		267.74		
1 Dump Truck, 12 Ton, 8 C.Y.		368.40		405.24	19.12	21.03
32 L.H., Daily Totals		$1603.00		$2198.58	$50.09	$68.71

Crew B-17A	Hr.	Daily	Hr.	Daily	Bare Costs	Incl. O&P
2 Laborer Foremen	$30.75	$492.00	$47.90	$766.40	$31.20	$48.57
6 Laborers	28.75	1380.00	44.75	2148.00		
1 Skilled Worker Foreman	40.00	320.00	62.25	498.00		
1 Skilled Worker	38.00	304.00	59.15	473.20		
80 L.H., Daily Totals		$2496.00		$3885.60	$31.20	$48.57

Crew B-18	Hr.	Daily	Hr.	Daily	Bare Costs	Incl. O&P
1 Labor Foreman (outside)	$30.75	$246.00	$47.90	$383.20	$29.42	$45.80
2 Laborers	28.75	460.00	44.75	716.00		
1 Vibratory Plate, gas, 21"		36.20		39.82	1.51	1.66
24 L.H., Daily Totals		$742.20		$1139.02	$30.93	$47.46

Crew B-19	Hr.	Daily	Hr.	Daily	Bare Costs	Incl. O&P
1 Pile Driver Foreman	$37.90	$303.20	$62.05	$496.40	$36.88	$58.52
4 Pile Drivers	35.90	1148.80	58.75	1880.00		
2 Equip. Oper. (crane)	39.80	636.80	60.00	960.00		
1 Equip. Oper. Oiler	33.90	271.20	51.10	408.80		
1 Crawler Crane, 40 Ton		994.60		1094.06		
60 L.F. of Leads, 15K Ft. Lbs.		60.00		66.00		
1 Hammer, Diesel, 22k Ft-Lb		595.20		654.72	25.78	28.36
64 L.H., Daily Totals		$4009.80		$5559.98	$62.65	$86.87

Crews

Crew No.	Bare Costs		Incl. Subs O & P		Cost Per Labor-Hour	
Crew B-19A	Hr.	Daily	Hr.	Daily	Bare Costs	Incl. O&P
1 Pile Driver Foreman	$37.90	$303.20	$62.05	$496.40	$36.88	$58.52
4 Pile Drivers	35.90	1148.80	58.75	1880.00		
2 Equip. Oper. (crane)	39.80	636.80	60.00	960.00		
1 Equip. Oper. Oiler	33.90	271.20	51.10	408.80		
1 Crawler Crane, 75 Ton		1384.00		1522.40		
60 L.F. Leads, 25K Ft. Lbs.		108.00		118.80		
1 Hammer, Diesel, 41k Ft-Lb		715.60		787.16	34.49	37.94
64 L.H., Daily Totals		$4567.60		$6173.56	$71.37	$96.46

Crew B-20	Hr.	Daily	Hr.	Daily	Bare Costs	Incl. O&P
1 Labor Foreman (out)	$30.75	$246.00	$47.90	$383.20	$32.50	$50.60
1 Skilled Worker	38.00	304.00	59.15	473.20		
1 Laborer	28.75	230.00	44.75	358.00		
24 L.H., Daily Totals		$780.00		$1214.40	$32.50	$50.60

Crew B-20A	Hr.	Daily	Hr.	Daily	Bare Costs	Incl. O&P
1 Labor Foreman	$30.75	$246.00	$47.90	$383.20	$35.04	$53.49
1 Laborer	28.75	230.00	44.75	358.00		
1 Plumber	44.80	358.40	67.40	539.20		
1 Plumber Apprentice	35.85	286.80	53.90	431.20		
32 L.H., Daily Totals		$1121.20		$1711.60	$35.04	$53.49

Crew B-21	Hr.	Daily	Hr.	Daily	Bare Costs	Incl. O&P
1 Labor Foreman (out)	$30.75	$246.00	$47.90	$383.20	$33.54	$51.94
1 Skilled Worker	38.00	304.00	59.15	473.20		
1 Laborer	28.75	230.00	44.75	358.00		
.5 Equip. Oper. (crane)	39.80	159.20	60.00	240.00		
.5 S.P. Crane, 4x4, 5 Ton		163.40		179.74	5.84	6.42
28 L.H., Daily Totals		$1102.60		$1634.14	$39.38	$58.36

Crew B-21A	Hr.	Daily	Hr.	Daily	Bare Costs	Incl. O&P
1 Labor Foreman	$30.75	$246.00	$47.90	$383.20	$35.99	$54.79
1 Laborer	28.75	230.00	44.75	358.00		
1 Plumber	44.80	358.40	67.40	539.20		
1 Plumber Apprentice	35.85	286.80	53.90	431.20		
1 Equip. Oper. (crane)	39.80	318.40	60.00	480.00		
1 S.P. Crane, 4x4, 12 Ton		575.00		632.50	14.38	15.81
40 L.H., Daily Totals		$2014.60		$2824.10	$50.37	$70.60

Crew B-21B	Hr.	Daily	Hr.	Daily	Bare Costs	Incl. O&P
1 Laborer Foreman	$30.75	$246.00	$47.90	$383.20	$31.36	$48.43
3 Laborers	28.75	690.00	44.75	1074.00		
1 Equip. Oper. (crane)	39.80	318.40	60.00	480.00		
1 Hyd. Crane, 12 Ton		724.00		796.40	18.10	19.91
40 L.H., Daily Totals		$1978.40		$2733.60	$49.46	$68.34

Crew B-21C	Hr.	Daily	Hr.	Daily	Bare Costs	Incl. O&P
1 Laborer Foreman	$30.75	$246.00	$47.90	$383.20	$31.35	$48.29
4 Laborers	28.75	920.00	44.75	1432.00		
1 Equip. Oper. (crane)	39.80	318.40	60.00	480.00		
1 Equip. Oper. Oiler	33.90	271.20	51.10	408.80		
2 Cutting Torches		34.00		37.40		
2 Gases		129.60		142.56		
1 Lattice Boom Crane, 90 Ton		1547.00		1701.70	30.55	33.60
56 L.H., Daily Totals		$3466.20		$4585.66	$61.90	$81.89

Crew B-22	Hr.	Daily	Hr.	Daily	Bare Costs	Incl. O&P
1 Labor Foreman (out)	$30.75	$246.00	$47.90	$383.20	$33.96	$52.48
1 Skilled Worker	38.00	304.00	59.15	473.20		
1 Laborer	28.75	230.00	44.75	358.00		
.75 Equip. Oper. (crane)	39.80	238.80	60.00	360.00		
.75 S.P. Crane, 4x4, 5 Ton		245.10		269.61	8.17	8.99
30 L.H., Daily Totals		$1263.90		$1844.01	$42.13	$61.47

Crew B-22A	Hr.	Daily	Hr.	Daily	Bare Costs	Incl. O&P
1 Labor Foreman (out)	$30.75	$246.00	$47.90	$383.20	$32.86	$50.85
1 Skilled Worker	38.00	304.00	59.15	473.20		
2 Laborers	28.75	460.00	44.75	716.00		
.75 Equipment Oper. (crane)	39.80	238.80	60.00	360.00		
.75 S.P. Crane, 4x4, 5 Ton		245.10		269.61		
1 Generator, 5 KW		35.40		38.94		
1 Butt Fusion Machine		296.20		325.82	15.18	16.69
38 L.H., Daily Totals		$1825.50		$2566.77	$48.04	$67.55

Crew B-22B	Hr.	Daily	Hr.	Daily	Bare Costs	Incl. O&P
1 Skilled Worker	$38.00	$304.00	$59.15	$473.20	$33.38	$51.95
1 Laborer	28.75	230.00	44.75	358.00		
1 Electro Fusion Machine		174.20		191.62	10.89	11.98
16 L.H., Daily Totals		$708.20		$1022.82	$44.26	$63.93

Crew B-23	Hr.	Daily	Hr.	Daily	Bare Costs	Incl. O&P
1 Labor Foreman (outside)	$30.75	$246.00	$47.90	$383.20	$29.15	$45.38
4 Laborers	28.75	920.00	44.75	1432.00		
1 Drill Rig, Truck-Mounted		3499.00		3848.90		
1 Flatbed Truck, gas, 3 Ton		185.20		203.72	92.11	101.32
40 L.H., Daily Totals		$4850.20		$5867.82	$121.26	$146.70

Crew B-23A	Hr.	Daily	Hr.	Daily	Bare Costs	Incl. O&P
1 Labor Foreman (outside)	$30.75	$246.00	$47.90	$383.20	$32.63	$50.17
1 Laborer	28.75	230.00	44.75	358.00		
1 Equip. Operator (medium)	38.40	307.20	57.85	462.80		
1 Drill Rig, Truck-Mounted		3499.00		3848.90		
1 Pickup Truck, 3/4 Ton		84.80		93.28	149.32	164.26
24 L.H., Daily Totals		$4367.00		$5146.18	$181.96	$214.42

Crew B-23B	Hr.	Daily	Hr.	Daily	Bare Costs	Incl. O&P
1 Labor Foreman (outside)	$30.75	$246.00	$47.90	$383.20	$32.63	$50.17
1 Laborer	28.75	230.00	44.75	358.00		
1 Equip. Operator (medium)	38.40	307.20	57.85	462.80		
1 Drill Rig, Truck-Mounted		3499.00		3848.90		
1 Pickup Truck, 3/4 Ton		84.80		93.28		
1 Centr. Water Pump, 6"		278.20		306.02	160.92	177.01
24 L.H., Daily Totals		$4645.20		$5452.20	$193.55	$227.18

Crew B-24	Hr.	Daily	Hr.	Daily	Bare Costs	Incl. O&P
1 Cement Finisher	$35.55	$284.40	$52.20	$417.60	$33.67	$51.37
1 Laborer	28.75	230.00	44.75	358.00		
1 Carpenter	36.70	293.60	57.15	457.20		
24 L.H., Daily Totals		$808.00		$1232.80	$33.67	$51.37

Crew B-25	Hr.	Daily	Hr.	Daily	Bare Costs	Incl. O&P
1 Labor Foreman	$30.75	$246.00	$47.90	$383.20	$31.56	$48.61
7 Laborers	28.75	1610.00	44.75	2506.00		
3 Equip. Oper. (med.)	38.40	921.60	57.85	1388.40		
1 Asphalt Paver, 130 H.P.		1750.00		1925.00		
1 Tandem Roller, 10 Ton		211.00		232.10		
1 Roller, Pneum. Whl, 12 Ton		295.60		325.16	25.64	28.21
88 L.H., Daily Totals		$5034.20		$6759.86	$57.21	$76.82

Crews

Crew No.	Bare Costs		Incl. Subs O & P		Cost Per Labor-Hour	

Crew B-25B	Hr.	Daily	Hr.	Daily	Bare Costs	Incl. O&P
1 Labor Foreman	$30.75	$246.00	$47.90	$383.20	$32.13	$49.38
7 Laborers	28.75	1610.00	44.75	2506.00		
4 Equip. Oper. (medium)	38.40	1228.80	57.85	1851.20		
1 Asphalt Paver, 130 H.P.		1750.00		1925.00		
2 Tandem Rollers, 10 Ton		422.00		464.20		
1 Roller, Pneum. Whl, 12 Ton		295.60		325.16	25.70	28.27
96 L.H., Daily Totals		$5552.40		$7454.76	$57.84	$77.65

Crew B-25C	Hr.	Daily	Hr.	Daily	Bare Costs	Incl. O&P
1 Labor Foreman	$30.75	$246.00	$47.90	$383.20	$32.30	$49.64
3 Laborers	28.75	690.00	44.75	1074.00		
2 Equip. Oper. (medium)	38.40	614.40	57.85	925.60		
1 Asphalt Paver, 130 H.P.		1750.00		1925.00		
1 Tandem Roller, 10 Ton		211.00		232.10	40.85	44.94
48 L.H., Daily Totals		$3511.40		$4539.90	$73.15	$94.58

Crew B-26	Hr.	Daily	Hr.	Daily	Bare Costs	Incl. O&P
1 Labor Foreman (outside)	$30.75	$246.00	$47.90	$383.20	$32.45	$50.19
6 Laborers	28.75	1380.00	44.75	2148.00		
2 Equip. Oper. (med.)	38.40	614.40	57.85	925.60		
1 Rodman (reinf.)	41.30	330.40	67.75	542.00		
1 Cement Finisher	35.55	284.40	52.20	417.60		
1 Grader, 30,000 Lbs.		505.40		555.94		
1 Paving Mach. & Equip.		2324.00		2556.40	32.15	35.37
88 L.H., Daily Totals		$5684.60		$7528.74	$64.60	$85.55

Crew B-26A	Hr.	Daily	Hr.	Daily	Bare Costs	Incl. O&P
1 Labor Foreman (outside)	$30.75	$246.00	$47.90	$383.20	$32.45	$50.19
6 Laborers	28.75	1380.00	44.75	2148.00		
2 Equip. Oper. (med.)	38.40	614.40	57.85	925.60		
1 Rodman (reinf.)	41.30	330.40	67.75	542.00		
1 Cement Finisher	35.55	284.40	52.20	417.60		
1 Grader, 30,000 Lbs.		505.40		555.94		
1 Paving Mach. & Equip.		2324.00		2556.40		
1 Concrete Saw		118.40		130.24	33.50	36.85
88 L.H., Daily Totals		$5803.00		$7658.98	$65.94	$87.03

Crew B-26B	Hr.	Daily	Hr.	Daily	Bare Costs	Incl. O&P
1 Labor Foreman (outside)	$30.75	$246.00	$47.90	$383.20	$32.94	$50.83
6 Laborers	28.75	1380.00	44.75	2148.00		
3 Equip. Oper. (med.)	38.40	921.60	57.85	1388.40		
1 Rodman (reinf.)	41.30	330.40	67.75	542.00		
1 Cement Finisher	35.55	284.40	52.20	417.60		
1 Grader, 30,000 Lbs.		505.40		555.94		
1 Paving Mach. & Equip.		2324.00		2556.40		
1 Concrete Pump, 110' Boom		938.60		1032.46	39.25	43.17
96 L.H., Daily Totals		$6930.40		$9024.00	$72.19	$94.00

Crew B-27	Hr.	Daily	Hr.	Daily	Bare Costs	Incl. O&P
1 Labor Foreman (outside)	$30.75	$246.00	$47.90	$383.20	$29.25	$45.54
3 Laborers	28.75	690.00	44.75	1074.00		
1 Berm Machine		250.80		275.88	7.84	8.62
32 L.H., Daily Totals		$1186.80		$1733.08	$37.09	$54.16

Crew B-28	Hr.	Daily	Hr.	Daily	Bare Costs	Incl. O&P
2 Carpenters	$36.70	$587.20	$57.15	$914.40	$34.05	$53.02
1 Laborer	28.75	230.00	44.75	358.00		
24 L.H., Daily Totals		$817.20		$1272.40	$34.05	$53.02

Crew B-29	Hr.	Daily	Hr.	Daily	Bare Costs	Incl. O&P
1 Labor Foreman (outside)	$30.75	$246.00	$47.90	$383.20	$31.35	$48.29
4 Laborers	28.75	920.00	44.75	1432.00		
1 Equip. Oper. (crane)	39.80	318.40	60.00	480.00		
1 Equip. Oper. Oiler	33.90	271.20	51.10	408.80		
1 Gradall, 5/8 C.Y.		905.80		996.38	16.18	17.79
56 L.H., Daily Totals		$2661.40		$3700.38	$47.52	$66.08

Crew B-30	Hr.	Daily	Hr.	Daily	Bare Costs	Incl. O&P
1 Equip. Oper. (med.)	$38.40	$307.20	$57.85	$462.80	$32.50	$49.72
2 Truck Drivers (heavy)	29.55	472.80	45.65	730.40		
1 Hyd. Excavator, 1.5 C.Y.		775.60		853.16		
2 Dump Trucks, 16 Ton, 12 C.Y.		1060.00		1166.00	76.48	84.13
24 L.H., Daily Totals		$2615.60		$3212.36	$108.98	$133.85

Crew B-31	Hr.	Daily	Hr.	Daily	Bare Costs	Incl. O&P
1 Labor Foreman (outside)	$30.75	$246.00	$47.90	$383.20	$30.74	$47.86
3 Laborers	28.75	690.00	44.75	1074.00		
1 Carpenter	36.70	293.60	57.15	457.20		
1 Air Compressor, 250 C.F.M.		151.40		166.54		
1 Sheeting Driver		5.75		6.33		
2 -50' Air Hoses, 1.5"		11.60		12.76	4.22	4.64
40 L.H., Daily Totals		$1398.35		$2100.03	$34.96	$52.50

Crew B-32	Hr.	Daily	Hr.	Daily	Bare Costs	Incl. O&P
1 Laborer	$28.75	$230.00	$44.75	$358.00	$35.99	$54.58
3 Equip. Oper. (med.)	38.40	921.60	57.85	1388.40		
1 Grader, 30,000 Lbs.		505.40		555.94		
1 Tandem Roller, 10 Ton		211.00		232.10		
1 Dozer, 200 H.P.		988.40		1087.24	53.27	58.60
32 L.H., Daily Totals		$2856.40		$3621.68	$89.26	$113.18

Crew B-32A	Hr.	Daily	Hr.	Daily	Bare Costs	Incl. O&P
1 Laborer	$28.75	$230.00	$44.75	$358.00	$35.18	$53.48
2 Equip. Oper. (medium)	38.40	614.40	57.85	925.60		
1 Grader, 30,000 Lbs.		505.40		555.94		
1 Roller, Vibratory, 25 Ton		565.60		622.16	44.63	49.09
24 L.H., Daily Totals		$1915.40		$2461.70	$79.81	$102.57

Crew B-32B	Hr.	Daily	Hr.	Daily	Bare Costs	Incl. O&P
1 Laborer	$28.75	$230.00	$44.75	$358.00	$35.18	$53.48
2 Equip. Oper. (medium)	38.40	614.40	57.85	925.60		
1 Dozer, 200 H.P.		988.40		1087.24		
1 Roller, Vibratory, 25 Ton		565.60		622.16	64.75	71.22
24 L.H., Daily Totals		$2398.40		$2993.00	$99.93	$124.71

Crew B-32C	Hr.	Daily	Hr.	Daily	Bare Costs	Incl. O&P
1 Labor Foreman	$30.75	$246.00	$47.90	$383.20	$33.91	$51.83
2 Laborers	28.75	460.00	44.75	716.00		
3 Equip. Oper. (medium)	38.40	921.60	57.85	1388.40		
1 Grader, 30,000 Lbs.		505.40		555.94		
1 Tandem Roller, 10 Ton		211.00		232.10		
1 Dozer, 200 H.P.		988.40		1087.24	35.52	39.07
48 L.H., Daily Totals		$3332.40		$4362.88	$69.42	$90.89

Crew B-33A	Hr.	Daily	Hr.	Daily	Bare Costs	Incl. O&P
1 Equip. Oper. (med.)	$38.40	$307.20	$57.85	$462.80	$35.64	$54.11
.5 Laborer	28.75	115.00	44.75	179.00		
.25 Equip. Oper. (med.)	38.40	76.80	57.85	115.70		
1 Scraper, Towed, 7 C.Y.		143.20		157.52		
1.25 Dozer, 300 H.P.		1626.25		1788.88	126.39	139.03
14 L.H., Daily Totals		$2268.45		$2703.90	$162.03	$193.14

Crews

Crew No.	Bare Costs		Incl. Subs O & P		Cost Per Labor-Hour	
Crew B-33B	Hr.	Daily	Hr.	Daily	Bare Costs	Incl. O&P
1 Equip. Oper. (med.)	$38.40	$307.20	$57.85	$462.80	$35.64	$54.11
.5 Laborer	28.75	115.00	44.75	179.00		
.25 Equip. Oper. (med.)	38.40	76.80	57.85	115.70		
1 Scraper, Towed, 10 C.Y.		157.00		172.70		
1.25 Dozer, 300 H.P.		1626.25		1788.88	127.38	140.11
14 L.H., Daily Totals		$2282.25		$2719.07	$163.02	$194.22
Crew B-33C	Hr.	Daily	Hr.	Daily	Bare Costs	Incl. O&P
1 Equip. Oper. (med.)	$38.40	$307.20	$57.85	$462.80	$35.64	$54.11
.5 Laborer	28.75	115.00	44.75	179.00		
.25 Equip. Oper. (med.)	38.40	76.80	57.85	115.70		
1 Scraper, Towed, 15 C.Y.		157.00		172.70		
1.25 Dozer, 300 H.P.		1626.25		1788.88	127.38	140.11
14 L.H., Daily Totals		$2282.25		$2719.07	$163.02	$194.22
Crew B-33D	Hr.	Daily	Hr.	Daily	Bare Costs	Incl. O&P
1 Equip. Oper. (med.)	$38.40	$307.20	$57.85	$462.80	$35.64	$54.11
.5 Laborer	28.75	115.00	44.75	179.00		
.25 Equip. Oper. (med.)	38.40	76.80	57.85	115.70		
1 S.P. Scraper, 14 C.Y.		1717.00		1888.70		
.25 Dozer, 300 H.P.		325.25		357.77	145.88	160.46
14 L.H., Daily Totals		$2541.25		$3003.97	$181.52	$214.57
Crew B-33E	Hr.	Daily	Hr.	Daily	Bare Costs	Incl. O&P
1 Equip. Oper. (med.)	$38.40	$307.20	$57.85	$462.80	$35.64	$54.11
.5 Laborer	28.75	115.00	44.75	179.00		
.25 Equip. Oper. (med.)	38.40	76.80	57.85	115.70		
1 S.P. Scraper, 24 C.Y.		2549.00		2803.90		
.25 Dozer, 300 H.P.		325.25		357.77	205.30	225.83
14 L.H., Daily Totals		$3373.25		$3919.18	$240.95	$279.94
Crew B-33F	Hr.	Daily	Hr.	Daily	Bare Costs	Incl. O&P
1 Equip. Oper. (med.)	$38.40	$307.20	$57.85	$462.80	$35.64	$54.11
.5 Laborer	28.75	115.00	44.75	179.00		
.25 Equip. Oper. (med.)	38.40	76.80	57.85	115.70		
1 Elev. Scraper, 11 C.Y.		859.80		945.78		
.25 Dozer, 300 H.P.		325.25		357.77	84.65	93.11
14 L.H., Daily Totals		$1684.05		$2061.05	$120.29	$147.22
Crew B-33G	Hr.	Daily	Hr.	Daily	Bare Costs	Incl. O&P
1 Equip. Oper. (med.)	$38.40	$307.20	$57.85	$462.80	$35.64	$54.11
.5 Laborer	28.75	115.00	44.75	179.00		
.25 Equip. Oper. (med.)	38.40	76.80	57.85	115.70		
1 Elev. Scraper, 20 C.Y.		1750.00		1925.00		
.25 Dozer, 300 H.P.		325.25		357.77	148.23	163.06
14 L.H., Daily Totals		$2574.25		$3040.28	$183.88	$217.16
Crew B-33H	Hr.	Daily	Hr.	Daily	Bare Costs	Incl. O&P
.25 Laborer	$28.75	$57.50	$44.75	$89.50	$36.74	$55.59
1 Equipment Operator (med.)	38.40	307.20	57.85	462.80		
.2 Equipment Operator (med.)	38.40	61.44	57.85	92.56		
1 Scraper, 32-44 C.Y.		2997.00		3296.70		
.2 Dozer, 410 H.P.		335.40		368.94	287.28	316.00
11.6 L.H., Daily Totals		$3758.54		$4310.50	$324.01	$371.59
Crew B-33J	Hr.	Daily	Hr.	Daily	Bare Costs	Incl. O&P
1 Equipment Operator (med.)	$38.40	$307.20	$57.85	$462.80	$38.40	$57.85
1 S.P. Scraper, 14 C.Y.		1717.00		1888.70	214.63	236.09
8 L.H., Daily Totals		$2024.20		$2351.50	$253.03	$293.94

Crew No.	Bare Costs		Incl. Subs O & P		Cost Per Labor-Hour	
Crew B-34A	Hr.	Daily	Hr.	Daily	Bare Costs	Incl. O&P
1 Truck Driver (heavy)	$29.55	$236.40	$45.65	$365.20	$29.55	$45.65
1 Dump Truck, 12 Ton, 8 C.Y.		368.40		405.24	46.05	50.66
8 L.H., Daily Totals		$604.80		$770.44	$75.60	$96.31
Crew B-34B	Hr.	Daily	Hr.	Daily	Bare Costs	Incl. O&P
1 Truck Driver (heavy)	$29.55	$236.40	$45.65	$365.20	$29.55	$45.65
1 Dump Truck, 16 Ton, 12 C.Y.		530.00		583.00	66.25	72.88
8 L.H., Daily Totals		$766.40		$948.20	$95.80	$118.53
Crew B-34C	Hr.	Daily	Hr.	Daily	Bare Costs	Incl. O&P
1 Truck Driver (heavy)	$29.55	$236.40	$45.65	$365.20	$29.55	$45.65
1 Truck Tractor, 240 H.P.		336.20		369.82		
1 Dump Trailer, 16.5 C.Y.		107.80		118.58	55.50	61.05
8 L.H., Daily Totals		$680.40		$853.60	$85.05	$106.70
Crew B-34D	Hr.	Daily	Hr.	Daily	Bare Costs	Incl. O&P
1 Truck Driver (heavy)	$29.55	$236.40	$45.65	$365.20	$29.55	$45.65
1 Truck Tractor, 240 H.P.		336.20		369.82		
1 Dump Trailer, 20 C.Y.		120.00		132.00	57.02	62.73
8 L.H., Daily Totals		$692.60		$867.02	$86.58	$108.38
Crew B-34E	Hr.	Daily	Hr.	Daily	Bare Costs	Incl. O&P
1 Truck Driver (heavy)	$29.55	$236.40	$45.65	$365.20	$29.55	$45.65
1 Dump Truck, Off Hwy., 25 Ton		992.80		1092.08	124.10	136.51
8 L.H., Daily Totals		$1229.20		$1457.28	$153.65	$182.16
Crew B-34F	Hr.	Daily	Hr.	Daily	Bare Costs	Incl. O&P
1 Truck Driver (heavy)	$29.55	$236.40	$45.65	$365.20	$29.55	$45.65
1 Dump Truck, Off Hwy., 35 Ton		1014.00		1115.40	126.75	139.43
8 L.H., Daily Totals		$1250.40		$1480.60	$156.30	$185.07
Crew B-34G	Hr.	Daily	Hr.	Daily	Bare Costs	Incl. O&P
1 Truck Driver (heavy)	$29.55	$236.40	$45.65	$365.20	$29.55	$45.65
1 Dump Truck, Off Hwy., 50 Ton		1308.00		1438.80	163.50	179.85
8 L.H., Daily Totals		$1544.40		$1804.00	$193.05	$225.50
Crew B-34H	Hr.	Daily	Hr.	Daily	Bare Costs	Incl. O&P
1 Truck Driver (heavy)	$29.55	$236.40	$45.65	$365.20	$29.55	$45.65
1 Dump Truck, Off Hwy., 65 Ton		1393.00		1532.30	174.13	191.54
8 L.H., Daily Totals		$1629.40		$1897.50	$203.68	$237.19
Crew B-34J	Hr.	Daily	Hr.	Daily	Bare Costs	Incl. O&P
1 Truck Driver (heavy)	$29.55	$236.40	$45.65	$365.20	$29.55	$45.65
1 Dump Truck, Off Hwy., 100 Ton		1792.00		1971.20	224.00	246.40
8 L.H., Daily Totals		$2028.40		$2336.40	$253.55	$292.05
Crew B-34K	Hr.	Daily	Hr.	Daily	Bare Costs	Incl. O&P
1 Truck Driver (heavy)	$29.55	$236.40	$45.65	$365.20	$29.55	$45.65
1 Truck Tractor, 240 H.P.		395.40		434.94		
1 Lowbed Trailer, 75 Ton		181.20		199.32	72.08	79.28
8 L.H., Daily Totals		$813.00		$999.46	$101.63	$124.93
Crew B-34N	Hr.	Daily	Hr.	Daily	Bare Costs	Incl. O&P
1 Truck Driver (heavy)	$29.55	$236.40	$45.65	$365.20	$29.55	$45.65
1 Dump Truck, 12 Ton, 8 C.Y.		368.40		405.24		
1 Flatbed Trailer, 40 Ton		125.80		138.38	61.77	67.95
8 L.H., Daily Totals		$730.60		$908.82	$91.33	$113.60

Crews

Crew No.	Bare Costs		Incl. Subs O & P		Cost Per Labor-Hour	

Crew B-34P	Hr.	Daily	Hr.	Daily	Bare Costs	Incl. O&P
1 Pipe Fitter	$45.20	$361.60	$68.00	$544.00	$37.38	$56.65
1 Truck Driver (light)	28.55	228.40	44.10	352.80		
1 Equip. Oper. (medium)	38.40	307.20	57.85	462.80		
1 Flatbed Truck, gas, 3 Ton		185.20		203.72		
1 Backhoe Loader, 48 H.P.		243.40		267.74	17.86	19.64
24 L.H., Daily Totals		$1325.80		$1831.06	$55.24	$76.29

Crew B-34Q	Hr.	Daily	Hr.	Daily	Bare Costs	Incl. O&P
1 Pipe Fitter	$45.20	$361.60	$68.00	$544.00	$37.85	$57.37
1 Truck Driver (light)	28.55	228.40	44.10	352.80		
1 Eqip. Oper. (crane)	39.80	318.40	60.00	480.00		
1 Flatbed Trailer, 25 Ton		92.80		102.08		
1 Dump Truck, 12 Ton, 8 C.Y.		368.40		405.24		
1 Hyd. Crane, 25 Ton		739.60		813.56	50.03	55.04
24 L.H., Daily Totals		$2109.20		$2697.68	$87.88	$112.40

Crew B-34R	Hr.	Daily	Hr.	Daily	Bare Costs	Incl. O&P
1 Pipe Fitter	$45.20	$361.60	$68.00	$544.00	$37.85	$57.37
1 Truck Driver (light)	28.55	228.40	44.10	352.80		
1 Eqip. Oper. (crane)	39.80	318.40	60.00	480.00		
1 Flatbed Trailer, 25 Ton		92.80		102.08		
1 Dump Truck, 12 Ton, 8 C.Y.		368.40		405.24		
1 Hyd. Crane, 25 Ton		739.60		813.56		
1 Hyd. Excavator, 1 C.Y.		601.80		661.98	75.11	82.62
24 L.H., Daily Totals		$2711.00		$3359.66	$112.96	$139.99

Crew B-34S	Hr.	Daily	Hr.	Daily	Bare Costs	Incl. O&P
2 Pipe Fitters	$45.20	$723.20	$68.00	$1088.00	$39.94	$60.41
1 Truck Driver (heavy)	29.55	236.40	45.65	365.20		
1 Eqip. Oper. (crane)	39.80	318.40	60.00	480.00		
1 Flatbed Trailer, 40 Ton		125.80		138.38		
1 Truck Tractor, 240 H.P.		336.20		369.82		
1 Hyd. Crane, 80 Ton		1214.00		1335.40		
1 Hyd. Excavator, 2 C.Y.		994.60		1094.06	83.46	91.80
32 L.H., Daily Totals		$3948.60		$4870.86	$123.39	$152.21

Crew B-34T	Hr.	Daily	Hr.	Daily	Bare Costs	Incl. O&P
2 Pipe Fitters	$45.20	$723.20	$68.00	$1088.00	$39.94	$60.41
1 Truck Driver (heavy)	29.55	236.40	45.65	365.20		
1 Eqip. Oper. (crane)	39.80	318.40	60.00	480.00		
1 Flatbed Trailer, 40 Ton		125.80		138.38		
1 Truck Tractor, 240 H.P.		336.20		369.82		
1 Hyd. Crane, 80 Ton		1214.00		1335.40	52.38	57.61
32 L.H., Daily Totals		$2954.00		$3776.80	$92.31	$118.03

Crew B-35	Hr.	Daily	Hr.	Daily	Bare Costs	Incl. O&P
1 Laborer Foreman (out)	$30.75	$246.00	$47.90	$383.20	$36.00	$55.05
1 Skilled Worker	38.00	304.00	59.15	473.20		
1 Welder (plumber)	44.80	358.40	67.40	539.20		
1 Laborer	28.75	230.00	44.75	358.00		
1 Equip. Oper. (crane)	39.80	318.40	60.00	480.00		
1 Equip. Oper. Oiler	33.90	271.20	51.10	408.80		
1 Welder, electric, 300 amp		61.45		67.59		
1 Hyd. Excavator, .75 C.Y.		516.40		568.04	12.04	13.24
48 L.H., Daily Totals		$2305.85		$3278.03	$48.04	$68.29

Crew B-35A	Hr.	Daily	Hr.	Daily	Bare Costs	Incl. O&P
1 Laborer Foreman (out)	$30.75	$246.00	$47.90	$383.20	$34.96	$53.58
2 Laborers	28.75	460.00	44.75	716.00		
1 Skilled Worker	38.00	304.00	59.15	473.20		
1 Welder (plumber)	44.80	358.40	67.40	539.20		
1 Equip. Oper. (crane)	39.80	318.40	60.00	480.00		
1 Equip. Oper. Oiler	33.90	271.20	51.10	408.80		
1 Welder, gas engine, 300 amp		115.20		126.72		
1 Crawler Crane, 75 Ton		1384.00		1522.40	26.77	29.45
56 L.H., Daily Totals		$3457.20		$4649.52	$61.74	$83.03

Crew B-36	Hr.	Daily	Hr.	Daily	Bare Costs	Incl. O&P
1 Labor Foreman (outside)	$30.75	$246.00	$47.90	$383.20	$33.01	$50.62
2 Laborers	28.75	460.00	44.75	716.00		
2 Equip. Oper. (med.)	38.40	614.40	57.85	925.60		
1 Dozer, 200 H.P.		988.40		1087.24		
1 Aggregate Spreader		30.60		33.66		
1 Tandem Roller, 10 Ton		211.00		232.10	30.75	33.83
40 L.H., Daily Totals		$2550.40		$3377.80	$63.76	$84.44

Crew B-36A	Hr.	Daily	Hr.	Daily	Bare Costs	Incl. O&P
1 Labor Foreman (outside)	$30.75	$246.00	$47.90	$383.20	$34.55	$52.69
2 Laborers	28.75	460.00	44.75	716.00		
4 Equip. Oper. (med.)	38.40	1228.80	57.85	1851.20		
1 Dozer, 200 H.P.		988.40		1087.24		
1 Aggregate Spreader		30.60		33.66		
1 Tandem Roller, 10 Ton		211.00		232.10		
1 Roller, Pneum. Whl, 12 Ton		295.60		325.16	27.24	29.97
56 L.H., Daily Totals		$3460.40		$4628.56	$61.79	$82.65

Crew B-36B	Hr.	Daily	Hr.	Daily	Bare Costs	Incl. O&P
1 Labor Foreman (outside)	$30.75	$246.00	$47.90	$383.20	$33.92	$51.81
2 Laborers	28.75	460.00	44.75	716.00		
4 Equip. Oper. (medium)	38.40	1228.80	57.85	1851.20		
1 Truck Driver, Heavy	29.55	236.40	45.65	365.20		
1 Grader, 30,000 Lbs.		505.40		555.94		
1 F.E. Loader, crl, 1.5 C.Y.		399.20		439.12		
1 Dozer, 300 H.P.		1301.00		1431.10		
1 Roller, Vibratory, 25 Ton		565.60		622.16		
1 Truck Tractor, 240 H.P.		395.40		434.94		
1 Water Tanker, 5000 Gal.		121.20		133.32	51.37	56.51
64 L.H., Daily Totals		$5459.00		$6932.18	$85.30	$108.32

Crew B-36C	Hr.	Daily	Hr.	Daily	Bare Costs	Incl. O&P
1 Labor Foreman (outside)	$30.75	$246.00	$47.90	$383.20	$35.10	$53.42
3 Equip. Oper. (medium)	38.40	921.60	57.85	1388.40		
1 Truck Driver, Heavy	29.55	236.40	45.65	365.20		
1 Grader, 30,000 Lbs.		505.40		555.94		
1 Dozer, 300 H.P.		1301.00		1431.10		
1 Roller, Vibratory, 25 Ton		565.60		622.16		
1 Truck Tractor, 240 H.P.		395.40		434.94		
1 Water Tanker, 5000 Gal.		121.20		133.32	72.22	79.44
40 L.H., Daily Totals		$4292.60		$5314.26	$107.32	$132.86

Crew B-37	Hr.	Daily	Hr.	Daily	Bare Costs	Incl. O&P
1 Labor Foreman (outside)	$30.75	$246.00	$47.90	$383.20	$30.43	$47.08
4 Laborers	28.75	920.00	44.75	1432.00		
1 Equip. Oper. (light)	36.85	294.80	55.55	444.40		
1 Tandem Roller, 5 Ton		122.20		134.42	2.55	2.80
48 L.H., Daily Totals		$1583.00		$2394.02	$32.98	$49.88

Crews

Crew No.	Bare Costs		Incl. Subs O & P		Cost Per Labor-Hour	
Crew B-38	Hr.	Daily	Hr.	Daily	Bare Costs	Incl. O&P
1 Labor Foreman (outside)	$30.75	$246.00	$47.90	$383.20	$32.70	$50.16
2 Laborers	28.75	460.00	44.75	716.00		
1 Equip. Oper. (light)	36.85	294.80	55.55	444.40		
1 Equip. Oper. (medium)	38.40	307.20	57.85	462.80		
1 Backhoe Loader, 48 H.P.		243.40		267.74		
1 Hyd. Hammer, (1200 lb)		115.40		126.94		
1 F.E. Loader, W.M., 4 C.Y.		488.00		536.80		
1 Pavt. Rem. Bucket		53.60		58.96	22.51	24.76
40 L.H., Daily Totals		$2208.40		$2996.84	$55.21	$74.92
Crew B-39	Hr.	Daily	Hr.	Daily	Bare Costs	Incl. O&P
1 Labor Foreman (outside)	$30.75	$246.00	$47.90	$383.20	$30.43	$47.08
4 Laborers	28.75	920.00	44.75	1432.00		
1 Equip. Oper. (light)	36.85	294.80	55.55	444.40		
1 Air Compressor, 250 C.F.M.		151.40		166.54		
2 Breakers, Pavement, 60 lb.		16.80		18.48		
2 -50' Air Hoses, 1.5"		11.60		12.76	3.75	4.12
48 L.H., Daily Totals		$1640.60		$2457.38	$34.18	$51.20
Crew B-40	Hr.	Daily	Hr.	Daily	Bare Costs	Incl. O&P
1 Pile Driver Foreman (out)	$37.90	$303.20	$62.05	$496.40	$36.88	$58.52
4 Pile Drivers	35.90	1148.80	58.75	1880.00		
2 Equip. Oper. (crane)	39.80	636.80	60.00	960.00		
1 Equip. Oper. Oiler	33.90	271.20	51.10	408.80		
1 Crawler Crane, 40 Ton		994.60		1094.06		
1 Vibratory Hammer & Gen.		1983.00		2181.30	46.52	51.18
64 L.H., Daily Totals		$5337.60		$7020.56	$83.40	$109.70
Crew B-40B	Hr.	Daily	Hr.	Daily	Bare Costs	Incl. O&P
1 Laborer Foreman	$30.75	$246.00	$47.90	$383.20	$31.78	$48.88
3 Laborers	28.75	690.00	44.75	1074.00		
1 Equip. Oper. (crane)	39.80	318.40	60.00	480.00		
1 Equip. Oper. Oiler	33.90	271.20	51.10	408.80		
1 Lattice Boom Crane, 40 Ton		1268.00		1394.80	26.42	29.06
48 L.H., Daily Totals		$2793.60		$3740.80	$58.20	$77.93
Crew B-41	Hr.	Daily	Hr.	Daily	Bare Costs	Incl. O&P
1 Labor Foreman (outside)	$30.75	$246.00	$47.90	$383.20	$29.85	$46.30
4 Laborers	28.75	920.00	44.75	1432.00		
.25 Equip. Oper. (crane)	39.80	79.60	60.00	120.00		
.25 Equip. Oper. Oiler	33.90	67.80	51.10	102.20		
.25 Crawler Crane, 40 Ton		248.65		273.51	5.65	6.22
44 L.H., Daily Totals		$1562.05		$2310.92	$35.50	$52.52
Crew B-42	Hr.	Daily	Hr.	Daily	Bare Costs	Incl. O&P
1 Labor Foreman (outside)	$30.75	$246.00	$47.90	$383.20	$32.60	$51.61
4 Laborers	28.75	920.00	44.75	1432.00		
1 Equip. Oper. (crane)	39.80	318.40	60.00	480.00		
1 Equip. Oper. Oiler	33.90	271.20	51.10	408.80		
1 Welder	41.35	330.80	74.90	599.20		
1 Hyd. Crane, 25 Ton		739.60		813.56		
1 Welder, gas engine, 300 amp		115.20		126.72		
1 Horz. Boring Csg. Mch.		421.40		463.54	19.94	21.93
64 L.H., Daily Totals		$3362.60		$4707.02	$52.54	$73.55

Crew No.	Bare Costs		Incl. Subs O & P		Cost Per Labor-Hour	
Crew B-43	Hr.	Daily	Hr.	Daily	Bare Costs	Incl. O&P
1 Labor Foreman (outside)	$30.75	$246.00	$47.90	$383.20	$31.78	$48.88
3 Laborers	28.75	690.00	44.75	1074.00		
1 Equip. Oper. (crane)	39.80	318.40	60.00	480.00		
1 Equip. Oper. Oiler	33.90	271.20	51.10	408.80		
1 Drill Rig, Truck-Mounted		3499.00		3848.90	72.90	80.19
48 L.H., Daily Totals		$5024.60		$6194.90	$104.68	$129.06
Crew B-44	Hr.	Daily	Hr.	Daily	Bare Costs	Incl. O&P
1 Pile Driver Foreman	$37.90	$303.20	$62.05	$496.40	$36.23	$57.73
4 Pile Drivers	35.90	1148.80	58.75	1880.00		
2 Equip. Oper. (crane)	39.80	636.80	60.00	960.00		
1 Laborer	28.75	230.00	44.75	358.00		
1 Crawler Crane, 40 Ton		994.60		1094.06		
45 L.F. of Leads, 15K Ft. Lbs.		45.00		49.50	16.24	17.87
64 L.H., Daily Totals		$3358.40		$4837.96	$52.48	$75.59
Crew B-45	Hr.	Daily	Hr.	Daily	Bare Costs	Incl. O&P
1 Equip. Oper. (med.)	$38.40	$307.20	$57.85	$462.80	$33.98	$51.75
1 Truck Driver (heavy)	29.55	236.40	45.65	365.20		
1 Dist. Tanker, 3000 Gallon		273.00		300.30		
1 Tractor, 4 x 2, 250 H.P.		312.20		343.42	36.58	40.23
16 L.H., Daily Totals		$1128.80		$1471.72	$70.55	$91.98
Crew B-46	Hr.	Daily	Hr.	Daily	Bare Costs	Incl. O&P
1 Pile Driver Foreman	$37.90	$303.20	$62.05	$496.40	$32.66	$52.30
2 Pile Drivers	35.90	574.40	58.75	940.00		
3 Laborers	28.75	690.00	44.75	1074.00		
1 Chainsaw, gas, 36" Long		34.40		37.84	.72	.79
48 L.H., Daily Totals		$1602.00		$2548.24	$33.38	$53.09
Crew B-47	Hr.	Daily	Hr.	Daily	Bare Costs	Incl. O&P
1 Blast Foreman	$30.75	$246.00	$47.90	$383.20	$32.12	$49.40
1 Driller	28.75	230.00	44.75	358.00		
1 Equip. Oper. (light)	36.85	294.80	55.55	444.40		
1 Air Track Drill, 4"		787.40		866.14		
1 Air Compressor, 600 C.F.M.		396.20		435.82		
2 -50' Air Hoses, 3"		38.10		41.91	50.90	55.99
24 L.H., Daily Totals		$1992.50		$2529.47	$83.02	$105.39
Crew B-47A	Hr.	Daily	Hr.	Daily	Bare Costs	Incl. O&P
1 Drilling Foreman	$30.75	$246.00	$47.90	$383.20	$34.82	$53.00
1 Equip. Oper. (heavy)	39.80	318.40	60.00	480.00		
1 Oiler	33.90	271.20	51.10	408.80		
1 Air Track Drill, 5"		905.40		995.94	37.73	41.50
24 L.H., Daily Totals		$1741.00		$2267.94	$72.54	$94.50
Crew B-47C	Hr.	Daily	Hr.	Daily	Bare Costs	Incl. O&P
1 Laborer	$28.75	$230.00	$44.75	$358.00	$32.80	$50.15
1 Equip. Oper. (light)	36.85	294.80	55.55	444.40		
1 Air Compressor, 750 C.F.M.		409.80		450.78		
2 -50' Air Hoses, 3"		38.10		41.91		
1 Air Track Drill, 4"		787.40		866.14	77.21	84.93
16 L.H., Daily Totals		$1760.10		$2161.23	$110.01	$135.08
Crew B-47E	Hr.	Daily	Hr.	Daily	Bare Costs	Incl. O&P
1 Laborer Foreman	$30.75	$246.00	$47.90	$383.20	$29.25	$45.54
3 Laborers	28.75	690.00	44.75	1074.00		
1 Flatbed Truck, gas, 3 Ton		185.20		203.72	5.79	6.37
32 L.H., Daily Totals		$1121.20		$1660.92	$35.04	$51.90

Crews

Crew No.	Bare Costs		Incl. Subs O & P		Cost Per Labor-Hour	
Crew B-47G	Hr.	Daily	Hr.	Daily	Bare Costs	Incl. O&P
1 Laborer Foreman	$30.75	$246.00	$47.90	$383.20	$31.27	$48.24
2 Laborers	28.75	460.00	44.75	716.00		
1 Equip. Oper. (light)	36.85	294.80	55.55	444.40		
1 Air Track Drill, 4"		787.40		866.14		
1 Air Compressor, 600 C.F.M.		396.20		435.82		
2 -50' Air Hoses, 3"		38.10		41.91		
1 Grout Pump		131.20		144.32	42.28	46.51
32 L.H., Daily Totals		$2353.70		$3031.79	$73.55	$94.74

Crew B-48	Hr.	Daily	Hr.	Daily	Bare Costs	Incl. O&P
1 Labor Foreman (outside)	$30.75	$246.00	$47.90	$383.20	$32.51	$49.83
3 Laborers	28.75	690.00	44.75	1074.00		
1 Equip. Oper. (crane)	39.80	318.40	60.00	480.00		
1 Equip. Oper. Oiler	33.90	271.20	51.10	408.80		
1 Equip. Oper. (light)	36.85	294.80	55.55	444.40		
1 Centr. Water Pump, 6"		278.20		306.02		
1 -20' Suction Hose, 6"		11.50		12.65		
1 -50' Discharge Hose, 6"		6.30		6.93		
1 Drill Rig, Truck-Mounted		3499.00		3848.90	67.77	74.54
56 L.H., Daily Totals		$5615.40		$6964.90	$100.28	$124.37

Crew B-49	Hr.	Daily	Hr.	Daily	Bare Costs	Incl. O&P
1 Labor Foreman (outside)	$30.75	$246.00	$47.90	$383.20	$33.91	$52.49
3 Laborers	28.75	690.00	44.75	1074.00		
2 Equip. Oper. (crane)	39.80	636.80	60.00	960.00		
2 Equip. Oper. Oilers	33.90	542.40	51.10	817.60		
1 Equip. Oper. (light)	36.85	294.80	55.55	444.40		
2 Pile Drivers	35.90	574.40	58.75	940.00		
1 Hyd. Crane, 25 Ton		739.60		813.56		
1 Centr. Water Pump, 6"		278.20		306.02		
1 -20' Suction Hose, 6"		11.50		12.65		
1 -50' Discharge Hose, 6"		6.30		6.93		
1 Drill Rig, Truck-Mounted		3499.00		3848.90	51.53	56.68
88 L.H., Daily Totals		$7519.00		$9607.26	$85.44	$109.17

Crew B-50	Hr.	Daily	Hr.	Daily	Bare Costs	Incl. O&P
2 Pile Driver Foremen	$37.90	$606.40	$62.05	$992.80	$35.07	$55.85
6 Pile Drivers	35.90	1723.20	58.75	2820.00		
2 Equip. Oper. (crane)	39.80	636.80	60.00	960.00		
1 Equip. Oper. Oiler	33.90	271.20	51.10	408.80		
3 Laborers	28.75	690.00	44.75	1074.00		
1 Crawler Crane, 40 Ton		994.60		1094.06		
60 L.F. of Leads, 15K Ft. Lbs.		60.00		66.00		
1 Hammer, 15K Ft. Lbs.		360.00		396.00		
1 Air Compressor, 600 C.F.M.		396.20		435.82		
2 -50' Air Hoses, 3"		38.10		41.91		
1 Chainsaw, gas, 36" Long		34.40		37.84	16.82	18.50
112 L.H., Daily Totals		$5810.90		$8327.23	$51.88	$74.35

Crew B-51	Hr.	Daily	Hr.	Daily	Bare Costs	Incl. O&P
1 Labor Foreman (outside)	$30.75	$246.00	$47.90	$383.20	$29.05	$45.17
4 Laborers	28.75	920.00	44.75	1432.00		
1 Truck Driver (light)	28.55	228.40	44.10	352.80		
1 Flatbed Truck, gas, 1.5 Ton		141.40		155.54	2.95	3.24
48 L.H., Daily Totals		$1535.80		$2323.54	$32.00	$48.41

Crew B-52	Hr.	Daily	Hr.	Daily	Bare Costs	Incl. O&P
1 Carpenter Foreman	$38.70	$309.60	$60.25	$482.00	$33.86	$52.38
1 Carpenter	36.70	293.60	57.15	457.20		
3 Laborers	28.75	690.00	44.75	1074.00		
1 Cement Finisher	35.55	284.40	52.20	417.60		
.5 Rodman (reinf.)	41.30	165.20	67.75	271.00		
.5 Equip. Oper. (med.)	38.40	153.60	57.85	231.40		
.5 Crawler Loader, 3 C.Y.		417.20		458.92	7.45	8.20
56 L.H., Daily Totals		$2313.60		$3392.12	$41.31	$60.57

Crew B-53	Hr.	Daily	Hr.	Daily	Bare Costs	Incl. O&P
1 Equip. Oper. (light)	$36.85	$294.80	$55.55	$444.40	$36.85	$55.55
1 Trencher, Chain, 12 H.P.		51.00		56.10	6.38	7.01
8 L.H., Daily Totals		$345.80		$500.50	$43.23	$62.56

Crew B-54	Hr.	Daily	Hr.	Daily	Bare Costs	Incl. O&P
1 Equip. Oper. (light)	$36.85	$294.80	$55.55	$444.40	$36.85	$55.55
1 Trencher, Chain, 40 H.P.		226.60		249.26	28.32	31.16
8 L.H., Daily Totals		$521.40		$693.66	$65.17	$86.71

Crew B-54A	Hr.	Daily	Hr.	Daily	Bare Costs	Incl. O&P
.17 Labor Foreman (outside)	$30.75	$41.82	$47.90	$65.14	$37.29	$56.40
1 Equipment Operator (med.)	38.40	307.20	57.85	462.80		
1 Wheel Trencher, 67 H.P.		891.20		980.32	95.21	104.74
9.36 L.H., Daily Totals		$1240.22		$1508.26	$132.50	$161.14

Crew B-54B	Hr.	Daily	Hr.	Daily	Bare Costs	Incl. O&P
.25 Labor Foreman (outside)	$30.75	$61.50	$47.90	$95.80	$36.87	$55.86
1 Equipment Operator (med.)	38.40	307.20	57.85	462.80		
1 Wheel Trencher, 150 H.P.		1626.00		1788.60	162.60	178.86
10 L.H., Daily Totals		$1994.70		$2347.20	$199.47	$234.72

Crew B-55	Hr.	Daily	Hr.	Daily	Bare Costs	Incl. O&P
2 Laborers	$28.75	$460.00	$44.75	$716.00	$28.68	$44.53
1 Truck Driver (light)	28.55	228.40	44.10	352.80		
1 Truck-mounted earth auger		674.60		742.06		
1 Flatbed Truck, gas, 3 Ton		185.20		203.72	35.83	39.41
24 L.H., Daily Totals		$1548.20		$2014.58	$64.51	$83.94

Crew B-56	Hr.	Daily	Hr.	Daily	Bare Costs	Incl. O&P
1 Laborer	$28.75	$230.00	$44.75	$358.00	$32.80	$50.15
1 Equip. Oper. (light)	36.85	294.80	55.55	444.40		
1 Air Track Drill, 4"		787.40		866.14		
1 Air Compressor, 600 C.F.M.		396.20		435.82		
1 -50' Air Hose, 3"		19.05		20.95	75.17	82.68
16 L.H., Daily Totals		$1727.45		$2125.32	$107.97	$132.83

Crew B-57	Hr.	Daily	Hr.	Daily	Bare Costs	Incl. O&P
1 Labor Foreman (outside)	$30.75	$246.00	$47.90	$383.20	$33.13	$50.67
2 Laborers	28.75	460.00	44.75	716.00		
1 Equip. Oper. (crane)	39.80	318.40	60.00	480.00		
1 Equip. Oper. (light)	36.85	294.80	55.55	444.40		
1 Equip. Oper. Oiler	33.90	271.20	51.10	408.80		
1 Crawler Crane, 25 Ton		983.00		1081.30		
1 Clamshell Bucket, 1 C.Y.		44.80		49.28		
1 Centr. Water Pump, 6"		278.20		306.02		
1 -20' Suction Hose, 6"		11.50		12.65		
20 -50' Discharge Hoses, 6"		126.00		138.60	30.07	33.08
48 L.H., Daily Totals		$3033.90		$4020.25	$63.21	$83.76

Crews

Crew No.	Bare Costs		Incl. Subs O & P		Cost Per Labor-Hour	
Crew B-58	Hr.	Daily	Hr.	Daily	Bare Costs	Incl. O&P
2 Laborers	$28.75	$460.00	$44.75	$716.00	$31.45	$48.35
1 Equip. Oper. (light)	36.85	294.80	55.55	444.40		
1 Backhoe Loader, 48 H.P.		243.40		267.74		
1 Small Helicopter, w/pilot		2331.00		2564.10	107.27	117.99
24 L.H., Daily Totals		$3329.20		$3992.24	$138.72	$166.34
Crew B-59	Hr.	Daily	Hr.	Daily	Bare Costs	Incl. O&P
1 Truck Driver (heavy)	$29.55	$236.40	$45.65	$365.20	$29.55	$45.65
1 Truck Tractor, 195 H.P.		235.60		259.16		
1 Water Tanker, 5000 Gal.		121.20		133.32	44.60	49.06
8 L.H., Daily Totals		$593.20		$757.68	$74.15	$94.71
Crew B-59A	Hr.	Daily	Hr.	Daily	Bare Costs	Incl. O&P
2 Laborers	$28.75	$460.00	$44.75	$716.00	$29.02	$45.05
1 Truck Driver (heavy)	29.55	236.40	45.65	365.20		
1 Water Tanker, 5000 Gal.		121.20		133.32		
1 Truck Tractor, 195 H.P.		235.60		259.16	14.87	16.35
24 L.H., Daily Totals		$1053.20		$1473.68	$43.88	$61.40
Crew B-60	Hr.	Daily	Hr.	Daily	Bare Costs	Incl. O&P
1 Labor Foreman (outside)	$30.75	$246.00	$47.90	$383.20	$33.66	$51.37
2 Laborers	28.75	460.00	44.75	716.00		
1 Equip. Oper. (crane)	39.80	318.40	60.00	480.00		
2 Equip. Oper. (light)	36.85	589.60	55.55	888.80		
1 Equip. Oper. Oiler	33.90	271.20	51.10	408.80		
1 Crawler Crane, 40 Ton		994.60		1094.06		
45 L.F. of Leads, 15K Ft. Lbs.		45.00		49.50		
1 Backhoe Loader, 48 H.P.		243.40		267.74	22.91	25.20
56 L.H., Daily Totals		$3168.20		$4288.10	$56.58	$76.57
Crew B-61	Hr.	Daily	Hr.	Daily	Bare Costs	Incl. O&P
1 Labor Foreman (outside)	$30.75	$246.00	$47.90	$383.20	$30.77	$47.54
3 Laborers	28.75	690.00	44.75	1074.00		
1 Equip. Oper. (light)	36.85	294.80	55.55	444.40		
1 Cement Mixer, 2 C.Y.		180.00		198.00		
1 Air Compressor, 160 C.F.M.		122.60		134.86	7.57	8.32
40 L.H., Daily Totals		$1533.40		$2234.46	$38.34	$55.86
Crew B-62	Hr.	Daily	Hr.	Daily	Bare Costs	Incl. O&P
2 Laborers	$28.75	$460.00	$44.75	$716.00	$31.45	$48.35
1 Equip. Oper. (light)	36.85	294.80	55.55	444.40		
1 Loader, Skid Steer, 30 HP, gas		122.00		134.20	5.08	5.59
24 L.H., Daily Totals		$876.80		$1294.60	$36.53	$53.94
Crew B-63	Hr.	Daily	Hr.	Daily	Bare Costs	Incl. O&P
4 Laborers	$28.75	$920.00	$44.75	$1432.00	$30.37	$46.91
1 Equip. Oper. (light)	36.85	294.80	55.55	444.40		
1 Loader, Skid Steer, 30 HP, gas		122.00		134.20	3.05	3.36
40 L.H., Daily Totals		$1336.80		$2010.60	$33.42	$50.27
Crew B-64	Hr.	Daily	Hr.	Daily	Bare Costs	Incl. O&P
1 Laborer	$28.75	$230.00	$44.75	$358.00	$28.65	$44.42
1 Truck Driver (light)	28.55	228.40	44.10	352.80		
1 Power Mulcher (small)		119.00		130.90		
1 Flatbed Truck, gas, 1.5 Ton		141.40		155.54	16.27	17.90
16 L.H., Daily Totals		$718.80		$997.24	$44.92	$62.33

Crew No.	Bare Costs		Incl. Subs O & P		Cost Per Labor-Hour	
Crew B-65	Hr.	Daily	Hr.	Daily	Bare Costs	Incl. O&P
1 Laborer	$28.75	$230.00	$44.75	$358.00	$28.65	$44.42
1 Truck Driver (light)	28.55	228.40	44.10	352.80		
1 Power Mulcher (large)		204.20		224.62		
1 Flatbed Truck, gas, 1.5 Ton		141.40		155.54	21.60	23.76
16 L.H., Daily Totals		$804.00		$1090.96	$50.25	$68.19
Crew B-66	Hr.	Daily	Hr.	Daily	Bare Costs	Incl. O&P
1 Equip. Oper. (light)	$36.85	$294.80	$55.55	$444.40	$36.85	$55.55
1 Loader-Backhoe		196.80		216.48	24.60	27.06
8 L.H., Daily Totals		$491.60		$660.88	$61.45	$82.61
Crew B-67	Hr.	Daily	Hr.	Daily	Bare Costs	Incl. O&P
1 Millwright	$38.20	$305.60	$56.35	$450.80	$37.52	$55.95
1 Equip. Oper. (light)	36.85	294.80	55.55	444.40		
1 Forklift, R/T, 4,000 Lb.		272.40		299.64	17.02	18.73
16 L.H., Daily Totals		$872.80		$1194.84	$54.55	$74.68
Crew B-68	Hr.	Daily	Hr.	Daily	Bare Costs	Incl. O&P
2 Millwrights	$38.20	$611.20	$56.35	$901.60	$37.75	$56.08
1 Equip. Oper. (light)	36.85	294.80	55.55	444.40		
1 Forklift, R/T, 4,000 Lb.		272.40		299.64	11.35	12.48
24 L.H., Daily Totals		$1178.40		$1645.64	$49.10	$68.57
Crew B-69	Hr.	Daily	Hr.	Daily	Bare Costs	Incl. O&P
1 Labor Foreman (outside)	$30.75	$246.00	$47.90	$383.20	$31.78	$48.88
3 Laborers	28.75	690.00	44.75	1074.00		
1 Equip Oper. (crane)	39.80	318.40	60.00	480.00		
1 Equip Oper. Oiler	33.90	271.20	51.10	408.80		
1 Hyd. Crane, 80 Ton		1214.00		1335.40	25.29	27.82
48 L.H., Daily Totals		$2739.60		$3681.40	$57.08	$76.70
Crew B-69A	Hr.	Daily	Hr.	Daily	Bare Costs	Incl. O&P
1 Labor Foreman	$30.75	$246.00	$47.90	$383.20	$31.82	$48.70
3 Laborers	28.75	690.00	44.75	1074.00		
1 Equip. Oper. (medium)	38.40	307.20	57.85	462.80		
1 Concrete Finisher	35.55	284.40	52.20	417.60		
1 Curb/Gutter Paver, 2-Track		687.60		756.36	14.32	15.76
48 L.H., Daily Totals		$2215.20		$3093.96	$46.15	$64.46
Crew B-69B	Hr.	Daily	Hr.	Daily	Bare Costs	Incl. O&P
1 Labor Foreman	$30.75	$246.00	$47.90	$383.20	$31.82	$48.70
3 Laborers	28.75	690.00	44.75	1074.00		
1 Equip. Oper. (medium)	38.40	307.20	57.85	462.80		
1 Cement Finisher	35.55	284.40	52.20	417.60		
1 Curb/Gutter Paver, 4-Track		771.00		848.10	16.06	17.67
48 L.H., Daily Totals		$2298.60		$3185.70	$47.89	$66.37
Crew B-70	Hr.	Daily	Hr.	Daily	Bare Costs	Incl. O&P
1 Labor Foreman (outside)	$30.75	$246.00	$47.90	$383.20	$33.17	$50.81
3 Laborers	28.75	690.00	44.75	1074.00		
3 Equip. Oper. (med.)	38.40	921.60	57.85	1388.40		
1 Grader, 30,000 Lbs.		505.40		555.94		
1 Ripper, beam & 1 shank		76.40		84.04		
1 Road Sweeper, SP, 8' wide		551.00		606.10		
1 F.E. Loader, W.M., 1.5 C.Y.		277.60		305.36	25.19	27.70
56 L.H., Daily Totals		$3268.00		$4397.04	$58.36	$78.52

Crews

Crew No.	Bare Costs		Incl. Subs O & P		Cost Per Labor-Hour	
Crew B-71	Hr.	Daily	Hr.	Daily	Bare Costs	Incl. O&P
1 Labor Foreman (outside)	$30.75	$246.00	$47.90	$383.20	$33.17	$50.81
3 Laborers	28.75	690.00	44.75	1074.00		
3 Equip. Oper. (med.)	38.40	921.60	57.85	1388.40		
1 Pvmt. Profiler, 750 H.P.		5176.00		5693.60		
1 Road Sweeper, SP, 8' wide		551.00		606.10		
1 F.E. Loader, W.M., 1.5 C.Y.		277.60		305.36	107.22	117.95
56 L.H., Daily Totals		$7862.20		$9450.66	$140.40	$168.76
Crew B-72	Hr.	Daily	Hr.	Daily	Bare Costs	Incl. O&P
1 Labor Foreman (outside)	$30.75	$246.00	$47.90	$383.20	$33.83	$51.69
3 Laborers	28.75	690.00	44.75	1074.00		
4 Equip. Oper. (med.)	38.40	1228.80	57.85	1851.20		
1 Pvmt. Profiler, 750 H.P.		5176.00		5693.60		
1 Hammermill, 250 H.P.		1512.00		1663.20		
1 Windrow Loader		869.00		955.90		
1 Mix Paver 165 H.P.		1817.00		1998.70		
1 Roller, Pneum. Whl, 12 Ton		295.60		325.16	151.09	166.20
64 L.H., Daily Totals		$11834.40		$13944.96	$184.91	$217.89
Crew B-73	Hr.	Daily	Hr.	Daily	Bare Costs	Incl. O&P
1 Labor Foreman (outside)	$30.75	$246.00	$47.90	$383.20	$35.03	$53.33
2 Laborers	28.75	460.00	44.75	716.00		
5 Equip. Oper. (med.)	38.40	1536.00	57.85	2314.00		
1 Road Mixer, 310 H.P.		1819.00		2000.90		
1 Tandem Roller, 10 Ton		211.00		232.10		
1 Hammermill, 250 H.P.		1512.00		1663.20		
1 Grader, 30,000 Lbs.		505.40		555.94		
.5 F.E. Loader, W.M., 1.5 C.Y.		138.80		152.68		
.5 Truck Tractor, 195 H.P.		117.80		129.58		
.5 Water Tanker, 5000 Gal.		60.60		66.66	68.20	75.02
64 L.H., Daily Totals		$6606.60		$8214.26	$103.23	$128.35
Crew B-74	Hr.	Daily	Hr.	Daily	Bare Costs	Incl. O&P
1 Labor Foreman (outside)	$30.75	$246.00	$47.90	$383.20	$34.02	$51.92
1 Laborer	28.75	230.00	44.75	358.00		
4 Equip. Oper. (med.)	38.40	1228.80	57.85	1851.20		
2 Truck Drivers (heavy)	29.55	472.80	45.65	730.40		
1 Grader, 30,000 Lbs.		505.40		555.94		
1 Ripper, beam & 1 shank		76.40		84.04		
2 Stabilizers, 310 H.P.		2530.00		2783.00		
1 Flatbed Truck, gas, 3 Ton		185.20		203.72		
1 Chem. Spreader, Towed		44.20		48.62		
1 Roller, Vibratory, 25 Ton		565.60		622.16		
1 Water Tanker, 5000 Gal.		121.20		133.32		
1 Truck Tractor, 195 H.P.		235.60		259.16	66.62	73.28
64 L.H., Daily Totals		$6441.20		$8012.76	$100.64	$125.20
Crew B-75	Hr.	Daily	Hr.	Daily	Bare Costs	Incl. O&P
1 Labor Foreman (outside)	$30.75	$246.00	$47.90	$383.20	$34.66	$52.81
1 Laborer	28.75	230.00	44.75	358.00		
4 Equip. Oper. (med.)	38.40	1228.80	57.85	1851.20		
1 Truck Driver (heavy)	29.55	236.40	45.65	365.20		
1 Grader, 30,000 Lbs.		505.40		555.94		
1 Ripper, beam & 1 shank		76.40		84.04		
2 Stabilizers, 310 H.P.		2530.00		2783.00		
1 Dist. Tanker, 3000 Gallon		273.00		300.30		
1 Truck Tractor, 240 H.P.		273.00		300.30		
1 Roller, Vibratory, 25 Ton		565.60		622.16	75.42	82.96
56 L.H., Daily Totals		$6164.60		$7603.34	$110.08	$135.77

Crew No.	Bare Costs		Incl. Subs O & P		Cost Per Labor-Hour	
Crew B-76	Hr.	Daily	Hr.	Daily	Bare Costs	Incl. O&P
1 Dock Builder Foreman	$37.90	$303.20	$62.05	$496.40	$36.77	$58.54
5 Dock Builders	35.90	1436.00	58.75	2350.00		
2 Equip. Oper. (crane)	39.80	636.80	60.00	960.00		
1 Equip. Oper. Oiler	33.90	271.20	51.10	408.80		
1 Crawler Crane, 50 Ton		1131.00		1244.10		
1 Barge, 400 Ton		294.20		323.62		
1 Hammer, 15K Ft. Lbs.		360.00		396.00		
60 L.F. of Leads, 15K Ft. Lbs.		60.00		66.00		
1 Air Compressor, 600 C.F.M.		396.20		435.82		
2 -50' Air Hoses, 3"		38.10		41.91	31.66	34.83
72 L.H., Daily Totals		$4926.70		$6722.65	$68.43	$93.37
Crew B-76A	Hr.	Daily	Hr.	Daily	Bare Costs	Incl. O&P
1 Laborer Foreman	$30.75	$246.00	$47.90	$383.20	$31.02	$47.84
5 Laborers	28.75	1150.00	44.75	1790.00		
1 Equip. Oper. (crane)	39.80	318.40	60.00	480.00		
1 Equip. Oper. Oiler	33.90	271.20	51.10	408.80		
1 Crawler Crane, 50 Ton		1131.00		1244.10		
1 Barge, 400 Ton		294.20		323.62	22.27	24.50
64 L.H., Daily Totals		$3410.80		$4629.72	$53.29	$72.34
Crew B-77	Hr.	Daily	Hr.	Daily	Bare Costs	Incl. O&P
1 Labor Foreman	$30.75	$246.00	$47.90	$383.20	$29.11	$45.25
3 Laborers	28.75	690.00	44.75	1074.00		
1 Truck Driver (light)	28.55	228.40	44.10	352.80		
1 Crack Cleaner, 25 H.P.		48.20		53.02		
1 Crack Filler, Trailer Mtd.		162.40		178.64		
1 Flatbed Truck, gas, 3 Ton		185.20		203.72	9.89	10.88
40 L.H., Daily Totals		$1560.20		$2245.38	$39.01	$56.13
Crew B-78	Hr.	Daily	Hr.	Daily	Bare Costs	Incl. O&P
1 Labor Foreman	$30.75	$246.00	$47.90	$383.20	$29.05	$45.17
4 Laborers	28.75	920.00	44.75	1432.00		
1 Truck Driver (light)	28.55	228.40	44.10	352.80		
1 Paint Striper, S.P.		138.00		151.80		
1 Flatbed Truck, gas, 3 Ton		185.20		203.72		
1 Pickup Truck, 3/4 Ton		84.80		93.28	8.50	9.35
48 L.H., Daily Totals		$1802.40		$2616.80	$37.55	$54.52
Crew B-78A	Hr.	Daily	Hr.	Daily	Bare Costs	Incl. O&P
1 Equip. Oper. (light)	$36.85	$294.80	$55.55	$444.40	$36.85	$55.55
1 Line Rem. (metal balls) 115 HP		857.60		943.36	107.20	117.92
8 L.H., Daily Totals		$1152.40		$1387.76	$144.05	$173.47
Crew B-78B	Hr.	Daily	Hr.	Daily	Bare Costs	Incl. O&P
2 Laborers	$28.75	$460.00	$44.75	$716.00	$29.65	$45.95
.25 Equip. Oper. (light)	36.85	73.70	55.55	111.10		
1 Pickup Truck, 3/4 Ton		84.80		93.28		
1 Line Rem., 11 HP, walk behind		47.80		52.58		
.25 Road Sweeper, SP, 8' wide		137.75		151.53	15.02	16.52
18 L.H., Daily Totals		$804.05		$1124.48	$44.67	$62.47
Crew B-79	Hr.	Daily	Hr.	Daily	Bare Costs	Incl. O&P
1 Labor Foreman	$30.75	$246.00	$47.90	$383.20	$29.11	$45.25
3 Laborers	28.75	690.00	44.75	1074.00		
1 Truck Driver (light)	28.55	228.40	44.10	352.80		
1 Thermo. Striper, T.M.		596.20		655.82		
1 Flatbed Truck, Gas, 3 Ton		185.20		203.72		
2 Pickup Trucks, 3/4 Ton		169.60		186.56	23.77	26.15
40 L.H., Daily Totals		$2115.40		$2856.10	$52.88	$71.40

Crews

Crew No.	Bare Costs		Incl. Subs O & P		Cost Per Labor-Hour	
Crew B-79A	Hr.	Daily	Hr.	Daily	Bare Costs	Incl. O&P
1.5 Equip. Oper. (light)	$36.85	$442.20	$55.55	$666.60	$36.85	$55.55
.5 Line Remov. (grinder) 115 HP		411.20		452.32		
1 Line Rem. (metal balls) 115 HP		857.60		943.36	105.73	116.31
12 L.H., Daily Totals		$1711.00		$2062.28	$142.58	$171.86
Crew B-80	Hr.	Daily	Hr.	Daily	Bare Costs	Incl. O&P
1 Labor Foreman	$30.75	$246.00	$47.90	$383.20	$31.23	$48.08
1 Laborer	28.75	230.00	44.75	358.00		
1 Truck Driver (light)	28.55	228.40	44.10	352.80		
1 Equip. Oper. (light)	36.85	294.80	55.55	444.40		
1 Flatbed Truck, gas, 3 Ton		185.20		203.72		
1 Fence Post Auger, T.M.		375.20		412.72	17.51	19.26
32 L.H., Daily Totals		$1559.60		$2154.84	$48.74	$67.34
Crew B-80A	Hr.	Daily	Hr.	Daily	Bare Costs	Incl. O&P
3 Laborers	$28.75	$690.00	$44.75	$1074.00	$28.75	$44.75
1 Flatbed Truck, gas, 3 Ton		185.20		203.72	7.72	8.49
24 L.H., Daily Totals		$875.20		$1277.72	$36.47	$53.24
Crew B-80B	Hr.	Daily	Hr.	Daily	Bare Costs	Incl. O&P
3 Laborers	$28.75	$690.00	$44.75	$1074.00	$30.77	$47.45
1 Equip. Oper. (light)	36.85	294.80	55.55	444.40		
1 Crane, Flatbed Mounted, 3 Ton		271.60		298.76	8.49	9.34
32 L.H., Daily Totals		$1256.40		$1817.16	$39.26	$56.79
Crew B-80C	Hr.	Daily	Hr.	Daily	Bare Costs	Incl. O&P
2 Laborers	$28.75	$460.00	$44.75	$716.00	$28.68	$44.53
1 Truck Driver (light)	28.55	228.40	44.10	352.80		
1 Flatbed Truck, gas, 1.5 Ton		141.40		155.54		
1 Manual fence post auger, gas		6.00		6.60	6.14	6.76
24 L.H., Daily Totals		$835.80		$1230.94	$34.83	$51.29
Crew B-81	Hr.	Daily	Hr.	Daily	Bare Costs	Incl. O&P
1 Laborer	$28.75	$230.00	$44.75	$358.00	$32.23	$49.42
1 Equip. Oper. (med.)	38.40	307.20	57.85	462.80		
1 Truck Driver (heavy)	29.55	236.40	45.65	365.20		
1 Hydromulcher, T.M.		216.20		237.82		
1 Truck Tractor, 195 H.P.		235.60		259.16	18.82	20.71
24 L.H., Daily Totals		$1225.40		$1682.98	$51.06	$70.12
Crew B-82	Hr.	Daily	Hr.	Daily	Bare Costs	Incl. O&P
1 Laborer	$28.75	$230.00	$44.75	$358.00	$32.80	$50.15
1 Equip. Oper. (light)	36.85	294.80	55.55	444.40		
1 Horiz. Borer, 6 H.P.		71.20		78.32	4.45	4.89
16 L.H., Daily Totals		$596.00		$880.72	$37.25	$55.05
Crew B-83	Hr.	Daily	Hr.	Daily	Bare Costs	Incl. O&P
1 Tugboat Captain	$38.40	$307.20	$57.85	$462.80	$33.58	$51.30
1 Tugboat Hand	28.75	230.00	44.75	358.00		
1 Tugboat, 250 H.P.		550.20		605.22	34.39	37.83
16 L.H., Daily Totals		$1087.40		$1426.02	$67.96	$89.13
Crew B-84	Hr.	Daily	Hr.	Daily	Bare Costs	Incl. O&P
1 Equip. Oper. (med.)	$38.40	$307.20	$57.85	$462.80	$38.40	$57.85
1 Rotary Mower/Tractor		235.00		258.50	29.38	32.31
8 L.H., Daily Totals		$542.20		$721.30	$67.78	$90.16

Crew No.	Bare Costs		Incl. Subs O & P		Cost Per Labor-Hour	
Crew B-85	Hr.	Daily	Hr.	Daily	Bare Costs	Incl. O&P
3 Laborers	$28.75	$690.00	$44.75	$1074.00	$30.84	$47.55
1 Equip. Oper. (med.)	38.40	307.20	57.85	462.80		
1 Truck Driver (heavy)	29.55	236.40	45.65	365.20		
1 Aerial Lift Truck, 80'		546.80		601.48		
1 Brush Chipper, 12", 130 H.P.		183.60		201.96		
1 Pruning Saw, Rotary		7.45		8.20	18.45	20.29
40 L.H., Daily Totals		$1971.45		$2713.64	$49.29	$67.84
Crew B-86	Hr.	Daily	Hr.	Daily	Bare Costs	Incl. O&P
1 Equip. Oper. (med.)	$38.40	$307.20	$57.85	$462.80	$38.40	$57.85
1 Stump Chipper, S.P.		106.30		116.93	13.29	14.62
8 L.H., Daily Totals		$413.50		$579.73	$51.69	$72.47
Crew B-86A	Hr.	Daily	Hr.	Daily	Bare Costs	Incl. O&P
1 Equip. Oper. (medium)	$38.40	$307.20	$57.85	$462.80	$38.40	$57.85
1 Grader, 30,000 Lbs.		505.40		555.94	63.17	69.49
8 L.H., Daily Totals		$812.60		$1018.74	$101.58	$127.34
Crew B-86B	Hr.	Daily	Hr.	Daily	Bare Costs	Incl. O&P
1 Equip. Oper. (medium)	$38.40	$307.20	$57.85	$462.80	$38.40	$57.85
1 Dozer, 200 H.P.		988.40		1087.24	123.55	135.91
8 L.H., Daily Totals		$1295.60		$1550.04	$161.95	$193.76
Crew B-87	Hr.	Daily	Hr.	Daily	Bare Costs	Incl. O&P
1 Laborer	$28.75	$230.00	$44.75	$358.00	$36.47	$55.23
4 Equip. Oper. (med.)	38.40	1228.80	57.85	1851.20		
2 Feller Bunchers, 100 H.P.		1026.80		1129.48		
1 Log Chipper, 22" Tree		520.00		572.00		
1 Dozer, 105 H.P.		494.80		544.28		
1 Chainsaw, gas, 36" Long		34.40		37.84	51.90	57.09
40 L.H., Daily Totals		$3534.80		$4492.80	$88.37	$112.32
Crew B-88	Hr.	Daily	Hr.	Daily	Bare Costs	Incl. O&P
1 Laborer	$28.75	$230.00	$44.75	$358.00	$37.02	$55.98
6 Equip. Oper. (med.)	38.40	1843.20	57.85	2776.80		
2 Feller Bunchers, 100 H.P.		1026.80		1129.48		
1 Log Chipper, 22" Tree		520.00		572.00		
2 Log Skidders, 50 H.P.		1632.80		1796.08		
1 Dozer, 105 H.P.		494.80		544.28		
1 Chainsaw, gas, 36" Long		34.40		37.84	66.23	72.85
56 L.H., Daily Totals		$5782.00		$7214.48	$103.25	$128.83
Crew B-89	Hr.	Daily	Hr.	Daily	Bare Costs	Incl. O&P
1 Equip. Oper. (light)	$36.85	$294.80	$55.55	$444.40	$32.70	$49.83
1 Truck Driver (light)	28.55	228.40	44.10	352.80		
1 Flatbed Truck, gas, 3 Ton		185.20		203.72		
1 Concrete Saw		118.40		130.24		
1 Water Tank, 65 Gal.		16.40		18.04	20.00	22.00
16 L.H., Daily Totals		$843.20		$1149.20	$52.70	$71.83
Crew B-89A	Hr.	Daily	Hr.	Daily	Bare Costs	Incl. O&P
1 Skilled Worker	$38.00	$304.00	$59.15	$473.20	$33.38	$51.95
1 Laborer	28.75	230.00	44.75	358.00		
1 Core Drill (large)		105.00		115.50	6.56	7.22
16 L.H., Daily Totals		$639.00		$946.70	$39.94	$59.17

Crews

Crew No.	Bare Costs		Incl. Subs O & P		Cost Per Labor-Hour	
Crew B-89B	Hr.	Daily	Hr.	Daily	Bare Costs	Incl. O&P
1 Equip. Oper. (light)	$36.85	$294.80	$55.55	$444.40	$32.70	$49.83
1 Truck Driver, Light	28.55	228.40	44.10	352.80		
1 Wall Saw, Hydraulic, 10 H.P.		78.20		86.02		
1 Generator, Diesel, 100 KW		305.60		336.16		
1 Water Tank, 65 Gal.		16.40		18.04		
1 Flatbed Truck, gas, 3 Ton		185.20		203.72	36.59	40.25
16 L.H., Daily Totals		$1108.60		$1441.14	$69.29	$90.07
Crew B-90	Hr.	Daily	Hr.	Daily	Bare Costs	Incl. O&P
1 Labor Foreman (outside)	$30.75	$246.00	$47.90	$383.20	$31.23	$48.07
3 Laborers	28.75	690.00	44.75	1074.00		
2 Equip. Oper. (light)	36.85	589.60	55.55	888.80		
2 Truck Drivers (heavy)	29.55	472.80	45.65	730.40		
1 Road Mixer, 310 H.P.		1819.00		2000.90		
1 Dist. Truck, 2000 Gal.		247.40		272.14	32.29	35.52
64 L.H., Daily Totals		$4064.80		$5349.44	$63.51	$83.59
Crew B-90A	Hr.	Daily	Hr.	Daily	Bare Costs	Incl. O&P
1 Labor Foreman	$30.75	$246.00	$47.90	$383.20	$34.55	$52.69
2 Laborers	28.75	460.00	44.75	716.00		
4 Equip. Oper. (medium)	38.40	1228.80	57.85	1851.20		
2 Graders, 30,000 Lbs.		1010.80		1111.88		
1 Tandem Roller, 10 Ton		211.00		232.10		
1 Roller, Pneum. Whl, 12 Ton		295.60		325.16	27.10	29.81
56 L.H., Daily Totals		$3452.20		$4619.54	$61.65	$82.49
Crew B-90B	Hr.	Daily	Hr.	Daily	Bare Costs	Incl. O&P
1 Labor Foreman	$30.75	$246.00	$47.90	$383.20	$33.91	$51.83
2 Laborers	28.75	460.00	44.75	716.00		
3 Equip. Oper. (medium)	38.40	921.60	57.85	1388.40		
1 Tandem Roller, 10 Ton		211.00		232.10		
1 Roller, Pneum. Whl, 12 Ton		295.60		325.16		
1 Road Mixer, 310 H.P.		1819.00		2000.90	48.45	53.30
48 L.H., Daily Totals		$3953.20		$5045.76	$82.36	$105.12
Crew B-91	Hr.	Daily	Hr.	Daily	Bare Costs	Incl. O&P
1 Labor Foreman (outside)	$30.75	$246.00	$47.90	$383.20	$33.92	$51.81
2 Laborers	28.75	460.00	44.75	716.00		
4 Equip. Oper. (med.)	38.40	1228.80	57.85	1851.20		
1 Truck Driver (heavy)	29.55	236.40	45.65	365.20		
1 Dist. Tanker, 3000 Gallon		273.00		300.30		
1 Truck Tractor, 240 H.P.		273.00		300.30		
1 Aggreg. Spreader, S.P.		818.80		900.68		
1 Roller, Pneum. Whl, 12 Ton		295.60		325.16		
1 Tandem Roller, 10 Ton		211.00		232.10	29.24	32.16
64 L.H., Daily Totals		$4042.60		$5374.14	$63.17	$83.97
Crew B-92	Hr.	Daily	Hr.	Daily	Bare Costs	Incl. O&P
1 Labor Foreman (outside)	$30.75	$246.00	$47.90	$383.20	$29.25	$45.54
3 Laborers	28.75	690.00	44.75	1074.00		
1 Crack Cleaner, 25 H.P.		48.20		53.02		
1 Air Compressor, 60 C.F.M.		108.80		119.68		
1 Tar Kettle, T.M.		51.35		56.48		
1 Flatbed Truck, gas, 3 Ton		185.20		203.72	12.30	13.53
32 L.H., Daily Totals		$1329.55		$1890.11	$41.55	$59.07
Crew B-93	Hr.	Daily	Hr.	Daily	Bare Costs	Incl. O&P
1 Equip. Oper. (med.)	$38.40	$307.20	$57.85	$462.80	$38.40	$57.85
1 Feller Buncher, 100 H.P.		513.40		564.74	64.17	70.59
8 L.H., Daily Totals		$820.60		$1027.54	$102.58	$128.44

Crew No.	Bare Costs		Incl. Subs O & P		Cost Per Labor-Hour	
Crew B-94A	Hr.	Daily	Hr.	Daily	Bare Costs	Incl. O&P
1 Laborer	$28.75	$230.00	$44.75	$358.00	$28.75	$44.75
1 Diaphragm Water Pump, 2"		57.40		63.14		
1 -20' Suction Hose, 2"		1.95		2.15		
2 -50' Discharge Hoses, 2"		1.80		1.98	7.64	8.41
8 L.H., Daily Totals		$291.15		$425.26	$36.39	$53.16
Crew B-94B	Hr.	Daily	Hr.	Daily	Bare Costs	Incl. O&P
1 Laborer	$28.75	$230.00	$44.75	$358.00	$28.75	$44.75
1 Diaphragm Water Pump, 4"		78.20		86.02		
1 -20' Suction Hose, 4"		3.25		3.58		
2 -50' Discharge Hoses, 4"		4.70		5.17	10.77	11.85
8 L.H., Daily Totals		$316.15		$452.76	$39.52	$56.60
Crew B-94C	Hr.	Daily	Hr.	Daily	Bare Costs	Incl. O&P
1 Laborer	$28.75	$230.00	$44.75	$358.00	$28.75	$44.75
1 Centrifugal Water Pump, 3"		63.40		69.74		
1 -20' Suction Hose, 3"		3.05		3.36		
2 -50' Discharge Hoses, 3"		3.50		3.85	8.74	9.62
8 L.H., Daily Totals		$299.95		$434.94	$37.49	$54.37
Crew B-94D	Hr.	Daily	Hr.	Daily	Bare Costs	Incl. O&P
1 Laborer	$28.75	$230.00	$44.75	$358.00	$28.75	$44.75
1 Centr. Water Pump, 6"		278.20		306.02		
1 -20' Suction Hose, 6"		11.50		12.65		
2 -50' Discharge Hoses, 6"		12.60		13.86	37.79	41.57
8 L.H., Daily Totals		$532.30		$690.53	$66.54	$86.32
Crew B-95A	Hr.	Daily	Hr.	Daily	Bare Costs	Incl. O&P
1 Equip. Oper. (crane)	$39.80	$318.40	$60.00	$480.00	$34.27	$52.38
1 Laborer	28.75	230.00	44.75	358.00		
1 Hyd. Excavator, 5/8 C.Y.		462.20		508.42	28.89	31.78
16 L.H., Daily Totals		$1010.60		$1346.42	$63.16	$84.15
Crew B-95B	Hr.	Daily	Hr.	Daily	Bare Costs	Incl. O&P
1 Equip. Oper. (crane)	$39.80	$318.40	$60.00	$480.00	$34.27	$52.38
1 Laborer	28.75	230.00	44.75	358.00		
1 Hyd. Excavator, 1.5 C.Y.		775.60		853.16	48.48	53.32
16 L.H., Daily Totals		$1324.00		$1691.16	$82.75	$105.70
Crew B-95C	Hr.	Daily	Hr.	Daily	Bare Costs	Incl. O&P
1 Equip. Oper. (crane)	$39.80	$318.40	$60.00	$480.00	$34.27	$52.38
1 Laborer	28.75	230.00	44.75	358.00		
1 Hyd. Excavator, 2.5 C.Y.		1333.00		1466.30	83.31	91.64
16 L.H., Daily Totals		$1881.40		$2304.30	$117.59	$144.02
Crew C-1	Hr.	Daily	Hr.	Daily	Bare Costs	Incl. O&P
3 Carpenters	$36.70	$880.80	$57.15	$1371.60	$34.71	$54.05
1 Laborer	28.75	230.00	44.75	358.00		
32 L.H., Daily Totals		$1110.80		$1729.60	$34.71	$54.05
Crew C-2	Hr.	Daily	Hr.	Daily	Bare Costs	Incl. O&P
1 Carpenter Foreman (out)	$38.70	$309.60	$60.25	$482.00	$35.71	$55.60
4 Carpenters	36.70	1174.40	57.15	1828.80		
1 Laborer	28.75	230.00	44.75	358.00		
48 L.H., Daily Totals		$1714.00		$2668.80	$35.71	$55.60

Crews

Crew No.	Bare Costs		Incl. Subs O & P		Cost Per Labor-Hour	
Crew C-2A	Hr.	Daily	Hr.	Daily	Bare Costs	Incl. O&P
1 Carpenter Foreman (out)	$38.70	$309.60	$60.25	$482.00	$35.52	$54.77
3 Carpenters	36.70	880.80	57.15	1371.60		
1 Cement Finisher	35.55	284.40	52.20	417.60		
1 Laborer	28.75	230.00	44.75	358.00		
48 L.H., Daily Totals		$1704.80		$2629.20	$35.52	$54.77
Crew C-3	Hr.	Daily	Hr.	Daily	Bare Costs	Incl. O&P
1 Rodman Foreman	$43.30	$346.40	$71.05	$568.40	$37.86	$60.89
4 Rodmen (reinf.)	41.30	1321.60	67.75	2168.00		
1 Equip. Oper. (light)	36.85	294.80	55.55	444.40		
2 Laborers	28.75	460.00	44.75	716.00		
3 Stressing Equipment		24.60		27.06		
.5 Grouting Equipment		73.00		80.30	1.52	1.68
64 L.H., Daily Totals		$2520.40		$4004.16	$39.38	$62.56
Crew C-4	Hr.	Daily	Hr.	Daily	Bare Costs	Incl. O&P
1 Rodman Foreman	$43.30	$346.40	$71.05	$568.40	$41.80	$68.58
3 Rodmen (reinf.)	41.30	991.20	67.75	1626.00		
3 Stressing Equipment		24.60		27.06	.77	.85
32 L.H., Daily Totals		$1362.20		$2221.46	$42.57	$69.42
Crew C-5	Hr.	Daily	Hr.	Daily	Bare Costs	Incl. O&P
1 Rodman Foreman	$43.30	$346.40	$71.05	$568.40	$40.31	$64.74
4 Rodmen (reinf.)	41.30	1321.60	67.75	2168.00		
1 Equip. Oper. (crane)	39.80	318.40	60.00	480.00		
1 Equip. Oper. Oiler	33.90	271.20	51.10	408.80		
1 Hyd. Crane, 25 Ton		739.60		813.56	13.21	14.53
56 L.H., Daily Totals		$2997.20		$4438.76	$53.52	$79.26
Crew C-6	Hr.	Daily	Hr.	Daily	Bare Costs	Incl. O&P
1 Labor Foreman (outside)	$30.75	$246.00	$47.90	$383.20	$30.22	$46.52
4 Laborers	28.75	920.00	44.75	1432.00		
1 Cement Finisher	35.55	284.40	52.20	417.60		
2 Gas Engine Vibrators		42.80		47.08	.89	.98
48 L.H., Daily Totals		$1493.20		$2279.88	$31.11	$47.50
Crew C-7	Hr.	Daily	Hr.	Daily	Bare Costs	Incl. O&P
1 Labor Foreman (outside)	$30.75	$246.00	$47.90	$383.20	$31.37	$48.09
5 Laborers	28.75	1150.00	44.75	1790.00		
1 Cement Finisher	35.55	284.40	52.20	417.60		
1 Equip. Oper. (med.)	38.40	307.20	57.85	462.80		
1 Equip. Oper. (oiler)	33.90	271.20	51.10	408.80		
2 Gas Engine Vibrators		42.80		47.08		
1 Concrete Bucket, 1 C.Y.		17.80		19.58		
1 Hyd. Crane, 55 Ton		1060.00		1166.00	15.56	17.12
72 L.H., Daily Totals		$3379.40		$4695.06	$46.94	$65.21
Crew C-7A	Hr.	Daily	Hr.	Daily	Bare Costs	Incl. O&P
1 Labor Foreman (outside)	$30.75	$246.00	$47.90	$383.20	$29.20	$45.37
5 Laborers	28.75	1150.00	44.75	1790.00		
2 Truck Drivers (Heavy)	29.55	472.80	45.65	730.40		
2 Conc. Transit Mixers		1588.40		1747.24	24.82	27.30
64 L.H., Daily Totals		$3457.20		$4650.84	$54.02	$72.67

Crew No.	Bare Costs		Incl. Subs O & P		Cost Per Labor-Hour	
Crew C-7B	Hr.	Daily	Hr.	Daily	Bare Costs	Incl. O&P
1 Labor Foreman (outside)	$30.75	$246.00	$47.90	$383.20	$31.02	$47.84
5 Laborers	28.75	1150.00	44.75	1790.00		
1 Equipment Operator (heavy)	39.80	318.40	60.00	480.00		
1 Equipment Oiler	33.90	271.20	51.10	408.80		
1 Conc. Bucket, 2 C.Y.		27.20		29.92		
1 Lattice Boom Crane, 165 Ton		2022.00		2224.20	32.02	35.22
64 L.H., Daily Totals		$4034.80		$5316.12	$63.04	$83.06
Crew C-7C	Hr.	Daily	Hr.	Daily	Bare Costs	Incl. O&P
1 Labor Foreman (outside)	$30.75	$246.00	$47.90	$383.20	$31.41	$48.42
5 Laborers	28.75	1150.00	44.75	1790.00		
2 Equipment Operators (medium)	38.40	614.40	57.85	925.60		
2 F.E. Loaders, W.M., 4 C.Y.		976.00		1073.60	15.25	16.77
64 L.H., Daily Totals		$2986.40		$4172.40	$46.66	$65.19
Crew C-7D	Hr.	Daily	Hr.	Daily	Bare Costs	Incl. O&P
1 Labor Foreman (outside)	$30.75	$246.00	$47.90	$383.20	$30.41	$47.07
5 Laborers	28.75	1150.00	44.75	1790.00		
1 Equipment Operator (med.)	38.40	307.20	57.85	462.80		
1 Concrete Conveyer		160.40		176.44	2.86	3.15
56 L.H., Daily Totals		$1863.60		$2812.44	$33.28	$50.22
Crew C-8	Hr.	Daily	Hr.	Daily	Bare Costs	Incl. O&P
1 Labor Foreman (outside)	$30.75	$246.00	$47.90	$383.20	$32.36	$49.20
3 Laborers	28.75	690.00	44.75	1074.00		
2 Cement Finishers	35.55	568.80	52.20	835.20		
1 Equip. Oper. (med.)	38.40	307.20	57.85	462.80		
1 Concrete Pump (small)		708.60		779.46	12.65	13.92
56 L.H., Daily Totals		$2520.60		$3534.66	$45.01	$63.12
Crew C-8A	Hr.	Daily	Hr.	Daily	Bare Costs	Incl. O&P
1 Labor Foreman (outside)	$30.75	$246.00	$47.90	$383.20	$31.35	$47.76
3 Laborers	28.75	690.00	44.75	1074.00		
2 Cement Finishers	35.55	568.80	52.20	835.20		
48 L.H., Daily Totals		$1504.80		$2292.40	$31.35	$47.76
Crew C-8B	Hr.	Daily	Hr.	Daily	Bare Costs	Incl. O&P
1 Labor Foreman (outside)	$30.75	$246.00	$47.90	$383.20	$31.08	$48.00
3 Laborers	28.75	690.00	44.75	1074.00		
1 Equipment Operator	38.40	307.20	57.85	462.80		
1 Vibrating Screed		65.90		72.49		
1 Roller, Vibratory, 25 Ton		565.60		622.16		
1 Dozer, 200 H.P.		988.40		1087.24	40.50	44.55
40 L.H., Daily Totals		$2863.10		$3701.89	$71.58	$92.55
Crew C-8C	Hr.	Daily	Hr.	Daily	Bare Costs	Incl. O&P
1 Labor Foreman (outside)	$30.75	$246.00	$47.90	$383.20	$31.82	$48.70
3 Laborers	28.75	690.00	44.75	1074.00		
1 Cement Finisher	35.55	284.40	52.20	417.60		
1 Equipment Operator (med.)	38.40	307.20	57.85	462.80		
1 Shotcrete Rig, 12 CY/hr		250.60		275.66	5.22	5.74
48 L.H., Daily Totals		$1778.20		$2613.26	$37.05	$54.44
Crew C-8D	Hr.	Daily	Hr.	Daily	Bare Costs	Incl. O&P
1 Labor Foreman (outside)	$30.75	$246.00	$47.90	$383.20	$32.98	$50.10
1 Laborer	28.75	230.00	44.75	358.00		
1 Cement Finisher	35.55	284.40	52.20	417.60		
1 Equipment Operator (light)	36.85	294.80	55.55	444.40		
1 Air Compressor, 250 C.F.M.		151.40		166.54		
2 -50' Air Hoses, 1"		8.20		9.02	4.99	5.49
32 L.H., Daily Totals		$1214.80		$1778.76	$37.96	$55.59

Crews

Crew No.	Bare Costs		Incl. Subs O & P		Cost Per Labor-Hour	
Crew C-8E	Hr.	Daily	Hr.	Daily	Bare Costs	Incl. O&P
1 Labor Foreman (outside)	$30.75	$246.00	$47.90	$383.20	$32.98	$50.10
1 Laborer	28.75	230.00	44.75	358.00		
1 Cement Finisher	35.55	284.40	52.20	417.60		
1 Equipment Operator (light)	36.85	294.80	55.55	444.40		
1 Air Compressor, 250 C.F.M.		151.40		166.54		
2 -50' Air Hoses, 1"		8.20		9.02		
1 Concrete Pump (small)		708.60		779.46	27.13	29.84
32 L.H., Daily Totals		$1923.40		$2558.22	$60.11	$79.94

Crew C-10	Hr.	Daily	Hr.	Daily	Bare Costs	Incl. O&P
1 Laborer	$28.75	$230.00	$44.75	$358.00	$33.28	$49.72
2 Cement Finishers	35.55	568.80	52.20	835.20		
24 L.H., Daily Totals		$798.80		$1193.20	$33.28	$49.72

Crew C-10B	Hr.	Daily	Hr.	Daily	Bare Costs	Incl. O&P
3 Laborers	$28.75	$690.00	$44.75	$1074.00	$31.47	$47.73
2 Cement Finishers	35.55	568.80	52.20	835.20		
1 Concrete Mixer, 10 CF		138.60		152.46		
2 Conc. Finishers, 46" Walk-Behind		51.60		56.76	4.75	5.23
40 L.H., Daily Totals		$1449.00		$2118.42	$36.23	$52.96

Crew C-11	Hr.	Daily	Hr.	Daily	Bare Costs	Incl. O&P
1 Struc. Steel Foreman	$43.35	$346.80	$78.50	$628.00	$40.57	$71.00
6 Struc. Steel Workers	41.35	1984.80	74.90	3595.20		
1 Equip. Oper. (crane)	39.80	318.40	60.00	480.00		
1 Equip. Oper. Oiler	33.90	271.20	51.10	408.80		
1 Lattice Boom Crane, 150 Ton		1852.00		2037.20	25.72	28.29
72 L.H., Daily Totals		$4773.20		$7149.20	$66.29	$99.29

Crew C-12	Hr.	Daily	Hr.	Daily	Bare Costs	Incl. O&P
1 Carpenter Foreman (out)	$38.70	$309.60	$60.25	$482.00	$36.23	$56.08
3 Carpenters	36.70	880.80	57.15	1371.60		
1 Laborer	28.75	230.00	44.75	358.00		
1 Equip. Oper. (crane)	39.80	318.40	60.00	480.00		
1 Hyd. Crane, 12 Ton		724.00		796.40	15.08	16.59
48 L.H., Daily Totals		$2462.80		$3488.00	$51.31	$72.67

Crew C-13	Hr.	Daily	Hr.	Daily	Bare Costs	Incl. O&P
1 Struc. Steel Worker	$41.35	$330.80	$74.90	$599.20	$39.80	$68.98
1 Welder	41.35	330.80	74.90	599.20		
1 Carpenter	36.70	293.60	57.15	457.20		
1 Welder, gas engine, 300 amp		115.20		126.72	4.80	5.28
24 L.H., Daily Totals		$1070.40		$1782.32	$44.60	$74.26

Crew C-14	Hr.	Daily	Hr.	Daily	Bare Costs	Incl. O&P
1 Carpenter Foreman (out)	$38.70	$309.60	$60.25	$482.00	$35.96	$56.19
5 Carpenters	36.70	1468.00	57.15	2286.00		
4 Laborers	28.75	920.00	44.75	1432.00		
4 Rodmen (reinf.)	41.30	1321.60	67.75	2168.00		
2 Cement Finishers	35.55	568.80	52.20	835.20		
1 Equip. Oper. (crane)	39.80	318.40	60.00	480.00		
1 Equip. Oper. Oiler	33.90	271.20	51.10	408.80		
1 Hyd. Crane, 80 Ton		1214.00		1335.40	8.43	9.27
144 L.H., Daily Totals		$6391.60		$9427.40	$44.39	$65.47

Crew No.	Bare Costs		Incl. Subs O & P		Cost Per Labor-Hour	
Crew C-14A	Hr.	Daily	Hr.	Daily	Bare Costs	Incl. O&P
1 Carpenter Foreman (out)	$38.70	$309.60	$60.25	$482.00	$36.90	$57.81
16 Carpenters	36.70	4697.60	57.15	7315.20		
4 Rodmen (reinf.)	41.30	1321.60	67.75	2168.00		
2 Laborers	28.75	460.00	44.75	716.00		
1 Cement Finisher	35.55	284.40	52.20	417.60		
1 Equip. Oper. (med.)	38.40	307.20	57.85	462.80		
1 Gas Engine Vibrator		21.40		23.54		
1 Concrete Pump (small)		708.60		779.46	3.65	4.01
200 L.H., Daily Totals		$8110.40		$12364.60	$40.55	$61.82

Crew C-14B	Hr.	Daily	Hr.	Daily	Bare Costs	Incl. O&P
1 Carpenter Foreman (out)	$38.70	$309.60	$60.25	$482.00	$36.85	$57.59
16 Carpenters	36.70	4697.60	57.15	7315.20		
4 Rodmen (reinf.)	41.30	1321.60	67.75	2168.00		
2 Laborers	28.75	460.00	44.75	716.00		
2 Cement Finishers	35.55	568.80	52.20	835.20		
1 Equip. Oper. (med.)	38.40	307.20	57.85	462.80		
1 Gas Engine Vibrator		21.40		23.54		
1 Concrete Pump (small)		708.60		779.46	3.51	3.86
208 L.H., Daily Totals		$8394.80		$12782.20	$40.36	$61.45

Crew C-14C	Hr.	Daily	Hr.	Daily	Bare Costs	Incl. O&P
1 Carpenter Foreman (out)	$38.70	$309.60	$60.25	$482.00	$35.15	$54.99
6 Carpenters	36.70	1761.60	57.15	2743.20		
2 Rodmen (reinf.)	41.30	660.80	67.75	1084.00		
4 Laborers	28.75	920.00	44.75	1432.00		
1 Cement Finisher	35.55	284.40	52.20	417.60		
1 Gas Engine Vibrator		21.40		23.54	.19	.21
112 L.H., Daily Totals		$3957.80		$6182.34	$35.34	$55.20

Crew C-14D	Hr.	Daily	Hr.	Daily	Bare Costs	Incl. O&P
1 Carpenter Foreman (out)	$38.70	$309.60	$60.25	$482.00	$36.53	$56.96
18 Carpenters	36.70	5284.80	57.15	8229.60		
2 Rodmen (reinf.)	41.30	660.80	67.75	1084.00		
2 Laborers	28.75	460.00	44.75	716.00		
1 Cement Finisher	35.55	284.40	52.20	417.60		
1 Equip. Oper. (med.)	38.40	307.20	57.85	462.80		
1 Gas Engine Vibrator		21.40		23.54		
1 Concrete Pump (small)		708.60		779.46	3.65	4.01
200 L.H., Daily Totals		$8036.80		$12195.00	$40.18	$60.98

Crew C-14E	Hr.	Daily	Hr.	Daily	Bare Costs	Incl. O&P
1 Carpenter Foreman (out)	$38.70	$309.60	$60.25	$482.00	$36.28	$57.45
2 Carpenters	36.70	587.20	57.15	914.40		
4 Rodmen (reinf.)	41.30	1321.60	67.75	2168.00		
3 Laborers	28.75	690.00	44.75	1074.00		
1 Cement Finisher	35.55	284.40	52.20	417.60		
1 Gas Engine Vibrator		21.40		23.54	.24	.27
88 L.H., Daily Totals		$3214.20		$5079.54	$36.52	$57.72

Crew C-14F	Hr.	Daily	Hr.	Daily	Bare Costs	Incl. O&P
1 Laborer Foreman (out)	$30.75	$246.00	$47.90	$383.20	$33.51	$50.07
2 Laborers	28.75	460.00	44.75	716.00		
6 Cement Finishers	35.55	1706.40	52.20	2505.60		
1 Gas Engine Vibrator		21.40		23.54	.30	.33
72 L.H., Daily Totals		$2433.80		$3628.34	$33.80	$50.39

Crews

Crew No.	Bare Costs		Incl. Subs O & P		Cost Per Labor-Hour	
Crew C-14G	Hr.	Daily	Hr.	Daily	Bare Costs	Incl. O&P
1 Laborer Foreman (out)	$30.75	$246.00	$47.90	$383.20	$32.92	$49.46
2 Laborers	28.75	460.00	44.75	716.00		
4 Cement Finishers	35.55	1137.60	52.20	1670.40		
1 Gas Engine Vibrator		21.40		23.54	.38	.42
56 L.H., Daily Totals		$1865.00		$2793.14	$33.30	$49.88

Crew C-14H	Hr.	Daily	Hr.	Daily	Bare Costs	Incl. O&P
1 Carpenter Foreman (out)	$38.70	$309.60	$60.25	$482.00	$36.28	$56.54
2 Carpenters	36.70	587.20	57.15	914.40		
1 Rodman (reinf.)	41.30	330.40	67.75	542.00		
1 Laborer	28.75	230.00	44.75	358.00		
1 Cement Finisher	35.55	284.40	52.20	417.60		
1 Gas Engine Vibrator		21.40		23.54	.45	.49
48 L.H., Daily Totals		$1763.00		$2737.54	$36.73	$57.03

Crew C-15	Hr.	Daily	Hr.	Daily	Bare Costs	Incl. O&P
1 Carpenter Foreman (out)	$38.70	$309.60	$60.25	$482.00	$34.53	$53.44
2 Carpenters	36.70	587.20	57.15	914.40		
3 Laborers	28.75	690.00	44.75	1074.00		
2 Cement Finishers	35.55	568.80	52.20	835.20		
1 Rodman (reinf.)	41.30	330.40	67.75	542.00		
72 L.H., Daily Totals		$2486.00		$3847.60	$34.53	$53.44

Crew C-16	Hr.	Daily	Hr.	Daily	Bare Costs	Incl. O&P
1 Labor Foreman (outside)	$30.75	$246.00	$47.90	$383.20	$34.34	$53.32
3 Laborers	28.75	690.00	44.75	1074.00		
2 Cement Finishers	35.55	568.80	52.20	835.20		
1 Equip. Oper. (med.)	38.40	307.20	57.85	462.80		
2 Rodmen (reinf.)	41.30	660.80	67.75	1084.00		
1 Concrete Pump (small)		708.60		779.46	9.84	10.83
72 L.H., Daily Totals		$3181.40		$4618.66	$44.19	$64.15

Crew C-17	Hr.	Daily	Hr.	Daily	Bare Costs	Incl. O&P
2 Skilled Worker Foremen	$40.00	$640.00	$62.25	$996.00	$38.40	$59.77
8 Skilled Workers	38.00	2432.00	59.15	3785.60		
80 L.H., Daily Totals		$3072.00		$4781.60	$38.40	$59.77

Crew C-17A	Hr.	Daily	Hr.	Daily	Bare Costs	Incl. O&P
2 Skilled Worker Foremen	$40.00	$640.00	$62.25	$996.00	$38.42	$59.77
8 Skilled Workers	38.00	2432.00	59.15	3785.60		
.125 Equip. Oper. (crane)	39.80	39.80	60.00	60.00		
.125 Hyd. Crane, 80 Ton		151.75		166.93	1.87	2.06
81 L.H., Daily Totals		$3263.55		$5008.52	$40.29	$61.83

Crew C-17B	Hr.	Daily	Hr.	Daily	Bare Costs	Incl. O&P
2 Skilled Worker Foremen	$40.00	$640.00	$62.25	$996.00	$38.43	$59.78
8 Skilled Workers	38.00	2432.00	59.15	3785.60		
.25 Equip. Oper. (crane)	39.80	79.60	60.00	120.00		
.25 Hyd. Crane, 80 Ton		303.50		333.85		
.25 Conc. Finish.,46" Wlk-Behind		6.45		7.09	3.78	4.16
82 L.H., Daily Totals		$3461.55		$5242.55	$42.21	$63.93

Crew C-17C	Hr.	Daily	Hr.	Daily	Bare Costs	Incl. O&P
2 Skilled Worker Foremen	$40.00	$640.00	$62.25	$996.00	$38.45	$59.78
8 Skilled Workers	38.00	2432.00	59.15	3785.60		
.375 Equip. Oper. (crane)	39.80	119.40	60.00	180.00		
.375 Hyd. Crane, 80 Ton		455.25		500.77	5.48	6.03
83 L.H., Daily Totals		$3646.65		$5462.38	$43.94	$65.81

Crew C-17D	Hr.	Daily	Hr.	Daily	Bare Costs	Incl. O&P
2 Skilled Worker Foremen	$40.00	$640.00	$62.25	$996.00	$38.47	$59.78
8 Skilled Workers	38.00	2432.00	59.15	3785.60		
.5 Equip. Oper. (crane)	39.80	159.20	60.00	240.00		
.5 Hyd. Crane, 80 Ton		607.00		667.70	7.23	7.95
84 L.H., Daily Totals		$3838.20		$5689.30	$45.69	$67.73

Crew C-17E	Hr.	Daily	Hr.	Daily	Bare Costs	Incl. O&P
2 Skilled Worker Foremen	$40.00	$640.00	$62.25	$996.00	$38.40	$59.77
8 Skilled Workers	38.00	2432.00	59.15	3785.60		
1 Hyd. Jack with Rods		82.35		90.58	1.03	1.13
80 L.H., Daily Totals		$3154.35		$4872.19	$39.43	$60.90

Crew C-18	Hr.	Daily	Hr.	Daily	Bare Costs	Incl. O&P
.125 Labor Foreman (out)	$30.75	$30.75	$47.90	$47.90	$28.97	$45.10
1 Laborer	28.75	230.00	44.75	358.00		
1 Concrete Cart, 10 C.F.		53.40		58.74	5.93	6.53
9 L.H., Daily Totals		$314.15		$464.64	$34.91	$51.63

Crew C-19	Hr.	Daily	Hr.	Daily	Bare Costs	Incl. O&P
.125 Labor Foreman (out)	$30.75	$30.75	$47.90	$47.90	$28.97	$45.10
1 Laborer	28.75	230.00	44.75	358.00		
1 Concrete Cart, 18 C.F.		78.80		86.68	8.76	9.63
9 L.H., Daily Totals		$339.55		$492.58	$37.73	$54.73

Crew C-20	Hr.	Daily	Hr.	Daily	Bare Costs	Incl. O&P
1 Labor Foreman (outside)	$30.75	$246.00	$47.90	$383.20	$31.06	$47.71
5 Laborers	28.75	1150.00	44.75	1790.00		
1 Cement Finisher	35.55	284.40	52.20	417.60		
1 Equip. Oper. (med.)	38.40	307.20	57.85	462.80		
2 Gas Engine Vibrators		42.80		47.08		
1 Concrete Pump (small)		708.60		779.46	11.74	12.91
64 L.H., Daily Totals		$2739.00		$3880.14	$42.80	$60.63

Crew C-21	Hr.	Daily	Hr.	Daily	Bare Costs	Incl. O&P
1 Labor Foreman (outside)	$30.75	$246.00	$47.90	$383.20	$31.06	$47.71
5 Laborers	28.75	1150.00	44.75	1790.00		
1 Cement Finisher	35.55	284.40	52.20	417.60		
1 Equip. Oper. (med.)	38.40	307.20	57.85	462.80		
2 Gas Engine Vibrators		42.80		47.08		
1 Concrete Conveyer		160.40		176.44	3.17	3.49
64 L.H., Daily Totals		$2190.80		$3277.12	$34.23	$51.20

Crew C-22	Hr.	Daily	Hr.	Daily	Bare Costs	Incl. O&P
1 Rodman Foreman	$43.30	$346.40	$71.05	$568.40	$41.47	$67.80
4 Rodmen (reinf.)	41.30	1321.60	67.75	2168.00		
.125 Equip. Oper. (crane)	39.80	39.80	60.00	60.00		
.125 Equip. Oper. Oiler	33.90	33.90	51.10	51.10		
.125 Hyd. Crane, 25 Ton		92.45		101.69	2.20	2.42
42 L.H., Daily Totals		$1834.15		$2949.20	$43.67	$70.22

Crew C-23	Hr.	Daily	Hr.	Daily	Bare Costs	Incl. O&P
2 Skilled Worker Foremen	$40.00	$640.00	$62.25	$996.00	$38.17	$59.05
6 Skilled Workers	38.00	1824.00	59.15	2839.20		
1 Equip. Oper. (crane)	39.80	318.40	60.00	480.00		
1 Equip. Oper. Oiler	33.90	271.20	51.10	408.80		
1 Lattice Boom Crane, 90 Ton		1547.00		1701.70	19.34	21.27
80 L.H., Daily Totals		$4600.60		$6425.70	$57.51	$80.32

Crews

Crew No.	Bare Costs		Incl. Subs O & P		Cost Per Labor-Hour	
Crew C-23A	Hr.	Daily	Hr.	Daily	Bare Costs	Incl. O&P
1 Labor Foreman (outside)	$30.75	$246.00	$47.90	$383.20	$32.39	$49.70
2 Laborers	28.75	460.00	44.75	716.00		
1 Equip. Oper. (crane)	39.80	318.40	60.00	480.00		
1 Equip. Oper. Oiler	33.90	271.20	51.10	408.80		
1 Crawler Crane, 100 Ton		1717.00		1888.70		
3 Conc. buckets, 8 C.Y.		518.40		570.24	55.88	61.47
40 L.H., Daily Totals		$3531.00		$4446.94	$88.28	$111.17
Crew C-24	Hr.	Daily	Hr.	Daily	Bare Costs	Incl. O&P
2 Skilled Worker Foremen	$40.00	$640.00	$62.25	$996.00	$38.17	$59.05
6 Skilled Workers	38.00	1824.00	59.15	2839.20		
1 Equip. Oper. (crane)	39.80	318.40	60.00	480.00		
1 Equip. Oper. Oiler	33.90	271.20	51.10	408.80		
1 Lattice Boom Crane, 150 Ton		1852.00		2037.20	23.15	25.47
80 L.H., Daily Totals		$4905.60		$6761.20	$61.32	$84.52
Crew C-25	Hr.	Daily	Hr.	Daily	Bare Costs	Incl. O&P
2 Rodmen (reinf.)	$41.30	$660.80	$67.75	$1084.00	$32.33	$53.67
2 Rodmen Helpers	23.35	373.60	39.60	633.60		
32 L.H., Daily Totals		$1034.40		$1717.60	$32.33	$53.67
Crew C-27	Hr.	Daily	Hr.	Daily	Bare Costs	Incl. O&P
2 Cement Finishers	$35.55	$568.80	$52.20	$835.20	$35.55	$52.20
1 Concrete Saw		118.40		130.24	7.40	8.14
16 L.H., Daily Totals		$687.20		$965.44	$42.95	$60.34
Crew C-28	Hr.	Daily	Hr.	Daily	Bare Costs	Incl. O&P
1 Cement Finisher	$35.55	$284.40	$52.20	$417.60	$35.55	$52.20
1 Portable Air Compressor, gas		14.75		16.23	1.84	2.03
8 L.H., Daily Totals		$299.15		$433.82	$37.39	$54.23
Crew D-1	Hr.	Daily	Hr.	Daily	Bare Costs	Incl. O&P
1 Bricklayer	$38.05	$304.40	$57.90	$463.20	$33.35	$50.75
1 Bricklayer Helper	28.65	229.20	43.60	348.80		
16 L.H., Daily Totals		$533.60		$812.00	$33.35	$50.75
Crew D-2	Hr.	Daily	Hr.	Daily	Bare Costs	Incl. O&P
3 Bricklayers	$38.05	$913.20	$57.90	$1389.60	$34.51	$52.63
2 Bricklayer Helpers	28.65	458.40	43.60	697.60		
.5 Carpenter	36.70	146.80	57.15	228.60		
44 L.H., Daily Totals		$1518.40		$2315.80	$34.51	$52.63
Crew D-3	Hr.	Daily	Hr.	Daily	Bare Costs	Incl. O&P
3 Bricklayers	$38.05	$913.20	$57.90	$1389.60	$34.40	$52.42
2 Bricklayer Helpers	28.65	458.40	43.60	697.60		
.25 Carpenter	36.70	73.40	57.15	114.30		
42 L.H., Daily Totals		$1445.00		$2201.50	$34.40	$52.42
Crew D-4	Hr.	Daily	Hr.	Daily	Bare Costs	Incl. O&P
1 Bricklayer	$38.05	$304.40	$57.90	$463.20	$33.05	$50.16
2 Bricklayer Helpers	28.65	458.40	43.60	697.60		
1 Equip. Oper. (light)	36.85	294.80	55.55	444.40		
1 Grout Pump, 50 C.F./hr		122.00		134.20	3.81	4.19
32 L.H., Daily Totals		$1179.60		$1739.40	$36.86	$54.36
Crew D-5	Hr.	Daily	Hr.	Daily	Bare Costs	Incl. O&P
1 Bricklayer	$38.05	$304.40	$57.90	$463.20	$38.05	$57.90
8 L.H., Daily Totals		$304.40		$463.20	$38.05	$57.90

Crew No.	Bare Costs		Incl. Subs O & P		Cost Per Labor-Hour	
Crew D-6	Hr.	Daily	Hr.	Daily	Bare Costs	Incl. O&P
3 Bricklayers	$38.05	$913.20	$57.90	$1389.60	$33.48	$51.01
3 Bricklayer Helpers	28.65	687.60	43.60	1046.40		
.25 Carpenter	36.70	73.40	57.15	114.30		
50 L.H., Daily Totals		$1674.20		$2550.30	$33.48	$51.01
Crew D-7	Hr.	Daily	Hr.	Daily	Bare Costs	Incl. O&P
1 Tile Layer	$35.50	$284.00	$52.10	$416.80	$31.38	$46.05
1 Tile Layer Helper	27.25	218.00	40.00	320.00		
16 L.H., Daily Totals		$502.00		$736.80	$31.38	$46.05
Crew D-8	Hr.	Daily	Hr.	Daily	Bare Costs	Incl. O&P
3 Bricklayers	$38.05	$913.20	$57.90	$1389.60	$34.29	$52.18
2 Bricklayer Helpers	28.65	458.40	43.60	697.60		
40 L.H., Daily Totals		$1371.60		$2087.20	$34.29	$52.18
Crew D-9	Hr.	Daily	Hr.	Daily	Bare Costs	Incl. O&P
3 Bricklayers	$38.05	$913.20	$57.90	$1389.60	$33.35	$50.75
3 Bricklayer Helpers	28.65	687.60	43.60	1046.40		
48 L.H., Daily Totals		$1600.80		$2436.00	$33.35	$50.75
Crew D-10	Hr.	Daily	Hr.	Daily	Bare Costs	Incl. O&P
1 Bricklayer Foreman	$40.05	$320.40	$60.95	$487.60	$36.64	$55.61
1 Bricklayer	38.05	304.40	57.90	463.20		
1 Bricklayer Helper	28.65	229.20	43.60	348.80		
1 Equip. Oper. (crane)	39.80	318.40	60.00	480.00		
1 S.P. Crane, 4x4, 12 Ton		575.00		632.50	17.97	19.77
32 L.H., Daily Totals		$1747.40		$2412.10	$54.61	$75.38
Crew D-11	Hr.	Daily	Hr.	Daily	Bare Costs	Incl. O&P
1 Bricklayer Foreman	$40.05	$320.40	$60.95	$487.60	$35.58	$54.15
1 Bricklayer	38.05	304.40	57.90	463.20		
1 Bricklayer Helper	28.65	229.20	43.60	348.80		
24 L.H., Daily Totals		$854.00		$1299.60	$35.58	$54.15
Crew D-12	Hr.	Daily	Hr.	Daily	Bare Costs	Incl. O&P
1 Bricklayer Foreman	$40.05	$320.40	$60.95	$487.60	$33.85	$51.51
1 Bricklayer	38.05	304.40	57.90	463.20		
2 Bricklayer Helpers	28.65	458.40	43.60	697.60		
32 L.H., Daily Totals		$1083.20		$1648.40	$33.85	$51.51
Crew D-13	Hr.	Daily	Hr.	Daily	Bare Costs	Incl. O&P
1 Bricklayer Foreman	$40.05	$320.40	$60.95	$487.60	$35.32	$53.87
1 Bricklayer	38.05	304.40	57.90	463.20		
2 Bricklayer Helpers	28.65	458.40	43.60	697.60		
1 Carpenter	36.70	293.60	57.15	457.20		
1 Equip. Oper. (crane)	39.80	318.40	60.00	480.00		
1 S.P. Crane, 4x4, 12 Ton		575.00		632.50	11.98	13.18
48 L.H., Daily Totals		$2270.20		$3218.10	$47.30	$67.04
Crew E-1	Hr.	Daily	Hr.	Daily	Bare Costs	Incl. O&P
1 Welder Foreman	$43.35	$346.80	$78.50	$628.00	$40.52	$69.65
1 Welder	41.35	330.80	74.90	599.20		
1 Equip. Oper. (light)	36.85	294.80	55.55	444.40		
1 Welder, gas engine, 300 amp		115.20		126.72	4.80	5.28
24 L.H., Daily Totals		$1087.60		$1798.32	$45.32	$74.93

Crews

Crew No.	Bare Costs		Incl. Subs O & P		Cost Per Labor-Hour	
Crew E-2	**Hr.**	**Daily**	**Hr.**	**Daily**	**Bare Costs**	**Incl. O&P**
1 Struc. Steel Foreman	$43.35	$346.80	$78.50	$628.00	$40.35	$69.89
4 Struc. Steel Workers	41.35	1323.20	74.90	2396.80		
1 Equip. Oper. (crane)	39.80	318.40	60.00	480.00		
1 Equip. Oper. Oiler	33.90	271.20	51.10	408.80		
1 Lattice Boom Crane, 90 Ton		1547.00		1701.70	27.63	30.39
56 L.H., Daily Totals		$3806.60		$5615.30	$67.97	$100.27
Crew E-3	**Hr.**	**Daily**	**Hr.**	**Daily**	**Bare Costs**	**Incl. O&P**
1 Struc. Steel Foreman	$43.35	$346.80	$78.50	$628.00	$42.02	$76.10
1 Struc. Steel Worker	41.35	330.80	74.90	599.20		
1 Welder	41.35	330.80	74.90	599.20		
1 Welder, gas engine, 300 amp		115.20		126.72	4.80	5.28
24 L.H., Daily Totals		$1123.60		$1953.12	$46.82	$81.38
Crew E-4	**Hr.**	**Daily**	**Hr.**	**Daily**	**Bare Costs**	**Incl. O&P**
1 Struc. Steel Foreman	$43.35	$346.80	$78.50	$628.00	$41.85	$75.80
3 Struc. Steel Workers	41.35	992.40	74.90	1797.60		
1 Welder, gas engine, 300 amp		115.20		126.72	3.60	3.96
32 L.H., Daily Totals		$1454.40		$2552.32	$45.45	$79.76
Crew E-5	**Hr.**	**Daily**	**Hr.**	**Daily**	**Bare Costs**	**Incl. O&P**
2 Struc. Steel Foremen	$43.35	$693.60	$78.50	$1256.00	$40.85	$71.75
5 Struc. Steel Workers	41.35	1654.00	74.90	2996.00		
1 Equip. Oper. (crane)	39.80	318.40	60.00	480.00		
1 Welder	41.35	330.80	74.90	599.20		
1 Equip. Oper. Oiler	33.90	271.20	51.10	408.80		
1 Lattice Boom Crane, 90 Ton		1547.00		1701.70		
1 Welder, gas engine, 300 amp		115.20		126.72	20.78	22.86
80 L.H., Daily Totals		$4930.20		$7568.42	$61.63	$94.61
Crew E-6	**Hr.**	**Daily**	**Hr.**	**Daily**	**Bare Costs**	**Incl. O&P**
3 Struc. Steel Foremen	$43.35	$1040.40	$78.50	$1884.00	$40.88	$71.95
9 Struc. Steel Workers	41.35	2977.20	74.90	5392.80		
1 Equip. Oper. (crane)	39.80	318.40	60.00	480.00		
1 Welder	41.35	330.80	74.90	599.20		
1 Equip. Oper. Oiler	33.90	271.20	51.10	408.80		
1 Equip. Oper. (light)	36.85	294.80	55.55	444.40		
1 Lattice Boom Crane, 90 Ton		1547.00		1701.70		
1 Welder, gas engine, 300 amp		115.20		126.72		
1 Air Compressor, 160 C.F.M.		122.60		134.86		
2 Impact Wrenches		25.60		28.16	14.14	15.56
128 L.H., Daily Totals		$7043.20		$11200.64	$55.03	$87.50
Crew E-7	**Hr.**	**Daily**	**Hr.**	**Daily**	**Bare Costs**	**Incl. O&P**
1 Struc. Steel Foreman	$43.35	$346.80	$78.50	$628.00	$40.85	$71.75
4 Struc. Steel Workers	41.35	1323.20	74.90	2396.80		
1 Equip. Oper. (crane)	39.80	318.40	60.00	480.00		
1 Equip. Oper. Oiler	33.90	271.20	51.10	408.80		
1 Welder Foreman	43.35	346.80	78.50	628.00		
2 Welders	41.35	661.60	74.90	1198.40		
1 Lattice Boom Crane, 90 Ton		1547.00		1701.70		
2 Welder, gas engine, 300 amp		230.40		253.44	22.22	24.44
80 L.H., Daily Totals		$5045.40		$7695.14	$63.07	$96.19

Crew No.	Bare Costs		Incl. Subs O & P		Cost Per Labor-Hour	
Crew E-8	**Hr.**	**Daily**	**Hr.**	**Daily**	**Bare Costs**	**Incl. O&P**
1 Struc. Steel Foreman	$43.35	$346.80	$78.50	$628.00	$40.62	$70.99
4 Struc. Steel Workers	41.35	1323.20	74.90	2396.80		
1 Welder Foreman	43.35	346.80	78.50	628.00		
4 Welders	41.35	1323.20	74.90	2396.80		
1 Equip. Oper. (crane)	39.80	318.40	60.00	480.00		
1 Equip. Oper. Oiler	33.90	271.20	51.10	408.80		
1 Equip. Oper. (light)	36.85	294.80	55.55	444.40		
1 Lattice Boom Crane, 90 Ton		1547.00		1701.70		
4 Welder, gas engine, 300 amp		460.80		506.88	19.31	21.24
104 L.H., Daily Totals		$6232.20		$9591.38	$59.92	$92.22
Crew E-9	**Hr.**	**Daily**	**Hr.**	**Daily**	**Bare Costs**	**Incl. O&P**
2 Struc. Steel Foremen	$43.35	$693.60	$78.50	$1256.00	$40.88	$71.95
5 Struc. Steel Workers	41.35	1654.00	74.90	2996.00		
1 Welder Foreman	43.35	346.80	78.50	628.00		
5 Welders	41.35	1654.00	74.90	2996.00		
1 Equip. Oper. (crane)	39.80	318.40	60.00	480.00		
1 Equip. Oper. Oiler	33.90	271.20	51.10	408.80		
1 Equip. Oper. (light)	36.85	294.80	55.55	444.40		
1 Lattice Boom Crane, 90 Ton		1547.00		1701.70		
5 Welder, gas engines, 300 amp		576.00		633.60	16.59	18.24
128 L.H., Daily Totals		$7355.80		$11544.50	$57.47	$90.19
Crew E-10	**Hr.**	**Daily**	**Hr.**	**Daily**	**Bare Costs**	**Incl. O&P**
1 Welder Foreman	$43.35	$346.80	$78.50	$628.00	$42.35	$76.70
1 Welder	41.35	330.80	74.90	599.20		
1 Welder, gas engine, 300 amp		115.20		126.72		
1 Flatbed Truck, gas, 3 Ton		185.20		203.72	18.77	20.65
16 L.H., Daily Totals		$978.00		$1557.64	$61.13	$97.35
Crew E-11	**Hr.**	**Daily**	**Hr.**	**Daily**	**Bare Costs**	**Incl. O&P**
2 Painters, Struc. Steel	$33.50	$536.00	$61.15	$978.40	$33.15	$55.65
1 Building Laborer	28.75	230.00	44.75	358.00		
1 Equip. Oper. (light)	36.85	294.80	55.55	444.40		
1 Air Compressor, 250 C.F.M.		151.40		166.54		
1 Sandblaster, portable, 3 C.F.		15.60		17.16		
1 Sand Blasting Accessories		12.30		13.53	5.60	6.16
32 L.H., Daily Totals		$1240.10		$1978.03	$38.75	$61.81
Crew E-12	**Hr.**	**Daily**	**Hr.**	**Daily**	**Bare Costs**	**Incl. O&P**
1 Welder Foreman	$43.35	$346.80	$78.50	$628.00	$40.10	$67.03
1 Equip. Oper. (light)	36.85	294.80	55.55	444.40		
1 Welder, gas engine, 300 amp		115.20		126.72	7.20	7.92
16 L.H., Daily Totals		$756.80		$1199.12	$47.30	$74.94
Crew E-13	**Hr.**	**Daily**	**Hr.**	**Daily**	**Bare Costs**	**Incl. O&P**
1 Welder Foreman	$43.35	$346.80	$78.50	$628.00	$41.18	$70.85
.5 Equip. Oper. (light)	36.85	147.40	55.55	222.20		
1 Welder, gas engine, 300 amp		115.20		126.72	9.60	10.56
12 L.H., Daily Totals		$609.40		$976.92	$50.78	$81.41
Crew E-14	**Hr.**	**Daily**	**Hr.**	**Daily**	**Bare Costs**	**Incl. O&P**
1 Welder Foreman	$43.35	$346.80	$78.50	$628.00	$43.35	$78.50
1 Welder, gas engine, 300 amp		115.20		126.72	14.40	15.84
8 L.H., Daily Totals		$462.00		$754.72	$57.75	$94.34
Crew E-16	**Hr.**	**Daily**	**Hr.**	**Daily**	**Bare Costs**	**Incl. O&P**
1 Welder Foreman	$43.35	$346.80	$78.50	$628.00	$42.35	$76.70
1 Welder	41.35	330.80	74.90	599.20		
1 Welder, gas engine, 300 amp		115.20		126.72	7.20	7.92
16 L.H., Daily Totals		$792.80		$1353.92	$49.55	$84.62

Crews

Crew No.	Bare Costs		Incl. Subs O & P		Cost Per Labor-Hour	
Crew E-17	Hr.	Daily	Hr.	Daily	Bare Costs	Incl. O&P
1 Structural Steel Foreman	$43.35	$346.80	$78.50	$628.00	$42.35	$76.70
1 Structural Steel Worker	41.35	330.80	74.90	599.20		
16 L.H., Daily Totals		$677.60		$1227.20	$42.35	$76.70
Crew E-18	Hr.	Daily	Hr.	Daily	Bare Costs	Incl. O&P
1 Structural Steel Foreman	$43.35	$346.80	$78.50	$628.00	$41.16	$72.21
3 Structural Steel Workers	41.35	992.40	74.90	1797.60		
1 Equipment Operator (med.)	38.40	307.20	57.85	462.80		
1 Lattice Boom Crane, 20 Ton		976.70		1074.37	24.42	26.86
40 L.H., Daily Totals		$2623.10		$3962.77	$65.58	$99.07
Crew E-19	Hr.	Daily	Hr.	Daily	Bare Costs	Incl. O&P
1 Structural Steel Worker	$41.35	$330.80	$74.90	$599.20	$40.52	$69.65
1 Structural Steel Foreman	43.35	346.80	78.50	628.00		
1 Equip. Oper. (light)	36.85	294.80	55.55	444.40		
1 Lattice Boom Crane, 20 Ton		976.70		1074.37	40.70	44.77
24 L.H., Daily Totals		$1949.10		$2745.97	$81.21	$114.42
Crew E-20	Hr.	Daily	Hr.	Daily	Bare Costs	Incl. O&P
1 Structural Steel Foreman	$43.35	$346.80	$78.50	$628.00	$40.48	$70.51
5 Structural Steel Workers	41.35	1654.00	74.90	2996.00		
1 Equip. Oper. (crane)	39.80	318.40	60.00	480.00		
1 Oiler	33.90	271.20	51.10	408.80		
1 Lattice Boom Crane, 40 Ton		1268.00		1394.80	19.81	21.79
64 L.H., Daily Totals		$3858.40		$5907.60	$60.29	$92.31
Crew E-22	Hr.	Daily	Hr.	Daily	Bare Costs	Incl. O&P
1 Skilled Worker Foreman	$40.00	$320.00	$62.25	$498.00	$38.67	$60.18
2 Skilled Workers	38.00	608.00	59.15	946.40		
24 L.H., Daily Totals		$928.00		$1444.40	$38.67	$60.18
Crew E-24	Hr.	Daily	Hr.	Daily	Bare Costs	Incl. O&P
3 Structural Steel Workers	$41.35	$992.40	$74.90	$1797.60	$40.61	$70.64
1 Equipment Operator (medium)	38.40	307.20	57.85	462.80		
1 -25 Ton Crane		739.60		813.56	23.11	25.42
32 L.H., Daily Totals		$2039.20		$3073.96	$63.73	$96.06
Crew E-25	Hr.	Daily	Hr.	Daily	Bare Costs	Incl. O&P
1 Welder Foreman	$43.35	$346.80	$78.50	$628.00	$43.35	$78.50
1 Cutting Torch		17.00		18.70		
1 Gase		64.80		71.28	10.23	11.25
8 L.H., Daily Totals		$428.60		$717.98	$53.58	$89.75
Crew F-3	Hr.	Daily	Hr.	Daily	Bare Costs	Incl. O&P
4 Carpenters	$36.70	$1174.40	$57.15	$1828.80	$37.32	$57.72
1 Equip. Oper. (crane)	39.80	318.40	60.00	480.00		
1 Hyd. Crane, 12 Ton		724.00		796.40	18.10	19.91
40 L.H., Daily Totals		$2216.80		$3105.20	$55.42	$77.63
Crew F-4	Hr.	Daily	Hr.	Daily	Bare Costs	Incl. O&P
4 Carpenters	$36.70	$1174.40	$57.15	$1828.80	$36.75	$56.62
1 Equip. Oper. (crane)	39.80	318.40	60.00	480.00		
1 Equip. Oper. Oiler	33.90	271.20	51.10	408.80		
1 Hyd. Crane, 55 Ton		1060.00		1166.00	22.08	24.29
48 L.H., Daily Totals		$2824.00		$3883.60	$58.83	$80.91
Crew F-5	Hr.	Daily	Hr.	Daily	Bare Costs	Incl. O&P
1 Carpenter Foreman	$38.70	$309.60	$60.25	$482.00	$37.20	$57.92
3 Carpenters	36.70	880.80	57.15	1371.60		
32 L.H., Daily Totals		$1190.40		$1853.60	$37.20	$57.92

Crew No.	Bare Costs		Incl. Subs O & P		Cost Per Labor-Hour	
Crew F-6	Hr.	Daily	Hr.	Daily	Bare Costs	Incl. O&P
2 Carpenters	$36.70	$587.20	$57.15	$914.40	$34.14	$52.76
2 Building Laborers	28.75	460.00	44.75	716.00		
1 Equip. Oper. (crane)	39.80	318.40	60.00	480.00		
1 Hyd. Crane, 12 Ton		724.00		796.40	18.10	19.91
40 L.H., Daily Totals		$2089.60		$2906.80	$52.24	$72.67
Crew F-7	Hr.	Daily	Hr.	Daily	Bare Costs	Incl. O&P
2 Carpenters	$36.70	$587.20	$57.15	$914.40	$32.73	$50.95
2 Building Laborers	28.75	460.00	44.75	716.00		
32 L.H., Daily Totals		$1047.20		$1630.40	$32.73	$50.95
Crew G-1	Hr.	Daily	Hr.	Daily	Bare Costs	Incl. O&P
1 Roofer Foreman	$33.80	$270.40	$57.30	$458.40	$29.67	$50.33
4 Roofers, Composition	31.80	1017.60	53.95	1726.40		
2 Roofer Helpers	23.35	373.60	39.60	633.60		
1 Application Equipment		153.20		168.52		
1 Tar Kettle/Pot		62.90		69.19		
1 Crew Truck		116.20		127.82	5.93	6.53
56 L.H., Daily Totals		$1993.90		$3183.93	$35.61	$56.86
Crew G-2	Hr.	Daily	Hr.	Daily	Bare Costs	Incl. O&P
1 Plasterer	$33.55	$268.40	$50.75	$406.00	$30.35	$46.33
1 Plasterer Helper	28.75	230.00	43.50	348.00		
1 Building Laborer	28.75	230.00	44.75	358.00		
1 Grout Pump, 50 C.F./hr		122.00		134.20	5.08	5.59
24 L.H., Daily Totals		$850.40		$1246.20	$35.43	$51.92
Crew G-2A	Hr.	Daily	Hr.	Daily	Bare Costs	Incl. O&P
1 Roofer, composition	$31.80	$254.40	$53.95	$431.60	$27.97	$46.10
1 Roofer Helper	23.35	186.80	39.60	316.80		
1 Building Laborer	28.75	230.00	44.75	358.00		
1 Grout Pump, 50 C.F./hr		122.00		134.20	5.08	5.59
24 L.H., Daily Totals		$793.20		$1240.60	$33.05	$51.69
Crew G-3	Hr.	Daily	Hr.	Daily	Bare Costs	Incl. O&P
2 Sheet Metal Workers	$43.55	$696.80	$67.15	$1074.40	$36.15	$55.95
2 Building Laborers	28.75	460.00	44.75	716.00		
32 L.H., Daily Totals		$1156.80		$1790.40	$36.15	$55.95
Crew G-4	Hr.	Daily	Hr.	Daily	Bare Costs	Incl. O&P
1 Labor Foreman (outside)	$30.75	$246.00	$47.90	$383.20	$29.42	$45.80
2 Building Laborers	28.75	460.00	44.75	716.00		
1 Flatbed Truck, gas, 1.5 Ton		141.40		155.54		
1 Air Compressor, 160 C.F.M.		122.60		134.86	11.00	12.10
24 L.H., Daily Totals		$970.00		$1389.60	$40.42	$57.90
Crew G-5	Hr.	Daily	Hr.	Daily	Bare Costs	Incl. O&P
1 Roofer Foreman	$33.80	$270.40	$57.30	$458.40	$28.82	$48.88
2 Roofers, Composition	31.80	508.80	53.95	863.20		
2 Roofer Helpers	23.35	373.60	39.60	633.60		
1 Application Equipment		153.20		168.52	3.83	4.21
40 L.H., Daily Totals		$1306.00		$2123.72	$32.65	$53.09
Crew G-6A	Hr.	Daily	Hr.	Daily	Bare Costs	Incl. O&P
2 Roofers, Composition	$31.80	$508.80	$53.95	$863.20	$31.80	$53.95
1 Small Compressor, Electric		9.55		10.51		
2 Pneumatic Nailers		41.30		45.43	3.18	3.50
16 L.H., Daily Totals		$559.65		$919.13	$34.98	$57.45

Crews

Crew No.	Bare Costs		Incl. Subs O & P		Cost Per Labor-Hour	
Crew G-7	Hr.	Daily	Hr.	Daily	Bare Costs	Incl. O&P
1 Carpenter	$36.70	$293.60	$57.15	$457.20	$36.70	$57.15
1 Small Compressor, Electric		9.55		10.51		
1 Pneumatic Nailer		20.65		22.72	3.77	4.15
8 L.H., Daily Totals		$323.80		$490.42	$40.48	$61.30
Crew H-1	Hr.	Daily	Hr.	Daily	Bare Costs	Incl. O&P
2 Glaziers	$36.05	$576.80	$54.60	$873.60	$38.70	$64.75
2 Struc. Steel Workers	41.35	661.60	74.90	1198.40		
32 L.H., Daily Totals		$1238.40		$2072.00	$38.70	$64.75
Crew H-2	Hr.	Daily	Hr.	Daily	Bare Costs	Incl. O&P
2 Glaziers	$36.05	$576.80	$54.60	$873.60	$33.62	$51.32
1 Building Laborer	28.75	230.00	44.75	358.00		
24 L.H., Daily Totals		$806.80		$1231.60	$33.62	$51.32
Crew H-3	Hr.	Daily	Hr.	Daily	Bare Costs	Incl. O&P
1 Glazier	$36.05	$288.40	$54.60	$436.80	$31.70	$48.50
1 Helper	27.35	218.80	42.40	339.20		
16 L.H., Daily Totals		$507.20		$776.00	$31.70	$48.50
Crew J-1	Hr.	Daily	Hr.	Daily	Bare Costs	Incl. O&P
3 Plasterers	$33.55	$805.20	$50.75	$1218.00	$31.63	$47.85
2 Plasterer Helpers	28.75	460.00	43.50	696.00		
1 Mixing Machine, 6 C.F.		113.60		124.96	2.84	3.12
40 L.H., Daily Totals		$1378.80		$2038.96	$34.47	$50.97
Crew J-2	Hr.	Daily	Hr.	Daily	Bare Costs	Incl. O&P
3 Plasterers	$33.55	$805.20	$50.75	$1218.00	$31.98	$48.24
2 Plasterer Helpers	28.75	460.00	43.50	696.00		
1 Lather	33.70	269.60	50.20	401.60		
1 Mixing Machine, 6 C.F.		113.60		124.96	2.37	2.60
48 L.H., Daily Totals		$1648.40		$2440.56	$34.34	$50.84
Crew J-3	Hr.	Daily	Hr.	Daily	Bare Costs	Incl. O&P
1 Terrazzo Worker	$35.15	$281.20	$51.60	$412.80	$31.73	$46.58
1 Terrazzo Helper	28.30	226.40	41.55	332.40		
1 Terrazzo Grinder, Electric		99.35		109.29		
1 Terrazzo Mixer		155.40		170.94	15.92	17.51
16 L.H., Daily Totals		$762.35		$1025.43	$47.65	$64.09
Crew J-4	Hr.	Daily	Hr.	Daily	Bare Costs	Incl. O&P
1 Tile Layer	$35.50	$284.00	$52.10	$416.80	$31.38	$46.05
1 Tile Layer Helper	27.25	218.00	40.00	320.00		
16 L.H., Daily Totals		$502.00		$736.80	$31.38	$46.05
Crew K-1	Hr.	Daily	Hr.	Daily	Bare Costs	Incl. O&P
1 Carpenter	$36.70	$293.60	$57.15	$457.20	$32.63	$50.63
1 Truck Driver (light)	28.55	228.40	44.10	352.80		
1 Flatbed Truck, gas, 3 Ton		185.20		203.72	11.57	12.73
16 L.H., Daily Totals		$707.20		$1013.72	$44.20	$63.36
Crew K-2	Hr.	Daily	Hr.	Daily	Bare Costs	Incl. O&P
1 Struc. Steel Foreman	$43.35	$346.80	$78.50	$628.00	$37.75	$65.83
1 Struc. Steel Worker	41.35	330.80	74.90	599.20		
1 Truck Driver (light)	28.55	228.40	44.10	352.80		
1 Flatbed Truck, gas, 3 Ton		185.20		203.72	7.72	8.49
24 L.H., Daily Totals		$1091.20		$1783.72	$45.47	$74.32

Crew No.	Bare Costs		Incl. Subs O & P		Cost Per Labor-Hour	
Crew L-1	Hr.	Daily	Hr.	Daily	Bare Costs	Incl. O&P
1 Electrician	$43.90	$351.20	$65.35	$522.80	$44.35	$66.38
1 Plumber	44.80	358.40	67.40	539.20		
16 L.H., Daily Totals		$709.60		$1062.00	$44.35	$66.38
Crew L-2	Hr.	Daily	Hr.	Daily	Bare Costs	Incl. O&P
1 Carpenter	$36.70	$293.60	$57.15	$457.20	$32.02	$49.77
1 Carpenter Helper	27.35	218.80	42.40	339.20		
16 L.H., Daily Totals		$512.40		$796.40	$32.02	$49.77
Crew L-3	Hr.	Daily	Hr.	Daily	Bare Costs	Incl. O&P
1 Carpenter	$36.70	$293.60	$57.15	$457.20	$40.21	$61.70
.5 Electrician	43.90	175.60	65.35	261.40		
.5 Sheet Metal Worker	43.55	174.20	67.15	268.60		
16 L.H., Daily Totals		$643.40		$987.20	$40.21	$61.70
Crew L-3A	Hr.	Daily	Hr.	Daily	Bare Costs	Incl. O&P
1 Carpenter Foreman (outside)	$38.70	$309.60	$60.25	$482.00	$40.32	$62.55
.5 Sheet Metal Worker	43.55	174.20	67.15	268.60		
12 L.H., Daily Totals		$483.80		$750.60	$40.32	$62.55
Crew L-4	Hr.	Daily	Hr.	Daily	Bare Costs	Incl. O&P
2 Skilled Workers	$38.00	$608.00	$59.15	$946.40	$34.45	$53.57
1 Helper	27.35	218.80	42.40	339.20		
24 L.H., Daily Totals		$826.80		$1285.60	$34.45	$53.57
Crew L-5	Hr.	Daily	Hr.	Daily	Bare Costs	Incl. O&P
1 Struc. Steel Foreman	$43.35	$346.80	$78.50	$628.00	$41.41	$73.29
5 Struc. Steel Workers	41.35	1654.00	74.90	2996.00		
1 Equip. Oper. (crane)	39.80	318.40	60.00	480.00		
1 Hyd. Crane, 25 Ton		739.60		813.56	13.21	14.53
56 L.H., Daily Totals		$3058.80		$4917.56	$54.62	$87.81
Crew L-5A	Hr.	Daily	Hr.	Daily	Bare Costs	Incl. O&P
1 Structural Steel Foreman	$43.35	$346.80	$78.50	$628.00	$41.46	$72.08
2 Structural Steel Workers	41.35	661.60	74.90	1198.40		
1 Equip. Oper. (crane)	39.80	318.40	60.00	480.00		
1 S.P. Crane, 4x4, 25 Ton		875.60		963.16	27.36	30.10
32 L.H., Daily Totals		$2202.40		$3269.56	$68.83	$102.17
Crew L-6	Hr.	Daily	Hr.	Daily	Bare Costs	Incl. O&P
1 Plumber	$44.80	$358.40	$67.40	$539.20	$44.50	$66.72
.5 Electrician	43.90	175.60	65.35	261.40		
12 L.H., Daily Totals		$534.00		$800.60	$44.50	$66.72
Crew L-7	Hr.	Daily	Hr.	Daily	Bare Costs	Incl. O&P
2 Carpenters	$36.70	$587.20	$57.15	$914.40	$35.46	$54.78
1 Building Laborer	28.75	230.00	44.75	358.00		
.5 Electrician	43.90	175.60	65.35	261.40		
28 L.H., Daily Totals		$992.80		$1533.80	$35.46	$54.78
Crew L-8	Hr.	Daily	Hr.	Daily	Bare Costs	Incl. O&P
2 Carpenters	$36.70	$587.20	$57.15	$914.40	$38.32	$59.20
.5 Plumber	44.80	179.20	67.40	269.60		
20 L.H., Daily Totals		$766.40		$1184.00	$38.32	$59.20

Crews

Crew No.	Bare Costs		Incl. Subs O & P		Cost Per Labor-Hour	
Crew L-9	Hr.	Daily	Hr.	Daily	Bare Costs	Incl. O&P
1 Labor Foreman (inside)	$29.25	$234.00	$45.55	$364.40	$33.34	$53.92
2 Building Laborers	28.75	460.00	44.75	716.00		
1 Struc. Steel Worker	41.35	330.80	74.90	599.20		
.5 Electrician	43.90	175.60	65.35	261.40		
36 L.H., Daily Totals		$1200.40		$1941.00	$33.34	$53.92
Crew L-10	Hr.	Daily	Hr.	Daily	Bare Costs	Incl. O&P
1 Structural Steel Foreman	$43.35	$346.80	$78.50	$628.00	$41.50	$71.13
1 Structural Steel Worker	41.35	330.80	74.90	599.20		
1 Equip. Oper. (crane)	39.80	318.40	60.00	480.00		
1 Hyd. Crane, 12 Ton		724.00		796.40	30.17	33.18
24 L.H., Daily Totals		$1720.00		$2503.60	$71.67	$104.32
Crew L-11	Hr.	Daily	Hr.	Daily	Bare Costs	Incl. O&P
2 Wreckers	$28.75	$460.00	$50.70	$811.20	$33.54	$54.24
1 Equip. Oper. (crane)	39.80	318.40	60.00	480.00		
1 Equip. Oper. (light)	36.85	294.80	55.55	444.40		
1 Hyd. Excavator, 2.5 C.Y.		1333.00		1466.30		
1 Loader, Skid Steer, 78 HP		222.20		244.42	48.60	53.46
32 L.H., Daily Totals		$2628.40		$3446.32	$82.14	$107.70
Crew M-1	Hr.	Daily	Hr.	Daily	Bare Costs	Incl. O&P
3 Elevator Constructors	$53.40	$1281.60	$79.65	$1911.60	$50.73	$75.66
1 Elevator Apprentice	42.70	341.60	63.70	509.60		
5 Hand Tools		48.00		52.80	1.50	1.65
32 L.H., Daily Totals		$1671.20		$2474.00	$52.23	$77.31
Crew M-3	Hr.	Daily	Hr.	Daily	Bare Costs	Incl. O&P
1 Electrician Foreman (out)	$45.90	$367.20	$68.35	$546.80	$42.44	$63.74
1 Common Laborer	28.75	230.00	44.75	358.00		
.25 Equipment Operator, Medium	38.40	76.80	57.85	115.70		
1 Elevator Constructor	53.40	427.20	79.65	637.20		
1 Elevator Apprentice	42.70	341.60	63.70	509.60		
.25 S.P. Crane, 4x4, 20 Ton		167.70		184.47	4.93	5.43
34 L.H., Daily Totals		$1610.50		$2351.77	$47.37	$69.17
Crew M-4	Hr.	Daily	Hr.	Daily	Bare Costs	Incl. O&P
1 Electrician Foreman (out)	$45.90	$367.20	$68.35	$546.80	$42.04	$63.16
1 Common Laborer	28.75	230.00	44.75	358.00		
.25 Equipment Operator, Crane	39.80	79.60	60.00	120.00		
.25 Equipment Operator, Oiler	33.90	67.80	51.10	102.20		
1 Elevator Constructor	53.40	427.20	79.65	637.20		
1 Elevator Apprentice	42.70	341.60	63.70	509.60		
.25 S.P. Crane, 4x4, 40 Ton		260.75		286.82	7.24	7.97
36 L.H., Daily Totals		$1774.15		$2560.63	$49.28	$71.13
Crew Q-1	Hr.	Daily	Hr.	Daily	Bare Costs	Incl. O&P
1 Plumber	$44.80	$358.40	$67.40	$539.20	$40.33	$60.65
1 Plumber Apprentice	35.85	286.80	53.90	431.20		
16 L.H., Daily Totals		$645.20		$970.40	$40.33	$60.65
Crew Q-1C	Hr.	Daily	Hr.	Daily	Bare Costs	Incl. O&P
1 Plumber	$44.80	$358.40	$67.40	$539.20	$39.68	$59.72
1 Plumber Apprentice	35.85	286.80	53.90	431.20		
1 Equip. Oper. (medium)	38.40	307.20	57.85	462.80		
1 Trencher, Chain Type, 8' D		1606.00		1766.60	66.92	73.61
24 L.H., Daily Totals		$2558.40		$3199.80	$106.60	$133.32

Crew No.	Bare Costs		Incl. Subs O & P		Cost Per Labor-Hour	
Crew Q-2	Hr.	Daily	Hr.	Daily	Bare Costs	Incl. O&P
2 Plumbers	$44.80	$716.80	$67.40	$1078.40	$41.82	$62.90
1 Plumber Apprentice	35.85	286.80	53.90	431.20		
24 L.H., Daily Totals		$1003.60		$1509.60	$41.82	$62.90
Crew Q-3	Hr.	Daily	Hr.	Daily	Bare Costs	Incl. O&P
1 Plumber Foreman (inside)	$45.30	$362.40	$68.15	$545.20	$42.69	$64.21
2 Plumbers	44.80	716.80	67.40	1078.40		
1 Plumber Apprentice	35.85	286.80	53.90	431.20		
32 L.H., Daily Totals		$1366.00		$2054.80	$42.69	$64.21
Crew Q-4	Hr.	Daily	Hr.	Daily	Bare Costs	Incl. O&P
1 Plumber Foreman (inside)	$45.30	$362.40	$68.15	$545.20	$42.69	$64.21
1 Plumber	44.80	358.40	67.40	539.20		
1 Welder (plumber)	44.80	358.40	67.40	539.20		
1 Plumber Apprentice	35.85	286.80	53.90	431.20		
1 Welder, electric, 300 amp		61.45		67.59	1.92	2.11
32 L.H., Daily Totals		$1427.45		$2122.40	$44.61	$66.32
Crew Q-5	Hr.	Daily	Hr.	Daily	Bare Costs	Incl. O&P
1 Steamfitter	$45.20	$361.60	$68.00	$544.00	$40.67	$61.17
1 Steamfitter Apprentice	36.15	289.20	54.35	434.80		
16 L.H., Daily Totals		$650.80		$978.80	$40.67	$61.17
Crew Q-6	Hr.	Daily	Hr.	Daily	Bare Costs	Incl. O&P
2 Steamfitters	$45.20	$723.20	$68.00	$1088.00	$42.18	$63.45
1 Steamfitter Apprentice	36.15	289.20	54.35	434.80		
24 L.H., Daily Totals		$1012.40		$1522.80	$42.18	$63.45
Crew Q-7	Hr.	Daily	Hr.	Daily	Bare Costs	Incl. O&P
1 Steamfitter Foreman (inside)	$45.70	$365.60	$68.75	$550.00	$43.06	$64.78
2 Steamfitters	45.20	723.20	68.00	1088.00		
1 Steamfitter Apprentice	36.15	289.20	54.35	434.80		
32 L.H., Daily Totals		$1378.00		$2072.80	$43.06	$64.78
Crew Q-8	Hr.	Daily	Hr.	Daily	Bare Costs	Incl. O&P
1 Steamfitter Foreman (inside)	$45.70	$365.60	$68.75	$550.00	$43.06	$64.78
1 Steamfitter	45.20	361.60	68.00	544.00		
1 Welder (steamfitter)	45.20	361.60	68.00	544.00		
1 Steamfitter Apprentice	36.15	289.20	54.35	434.80		
1 Welder, electric, 300 amp		61.45		67.59	1.92	2.11
32 L.H., Daily Totals		$1439.45		$2140.40	$44.98	$66.89
Crew Q-9	Hr.	Daily	Hr.	Daily	Bare Costs	Incl. O&P
1 Sheet Metal Worker	$43.55	$348.40	$67.15	$537.20	$39.20	$60.45
1 Sheet Metal Apprentice	34.85	278.80	53.75	430.00		
16 L.H., Daily Totals		$627.20		$967.20	$39.20	$60.45
Crew Q-10	Hr.	Daily	Hr.	Daily	Bare Costs	Incl. O&P
2 Sheet Metal Workers	$43.55	$696.80	$67.15	$1074.40	$40.65	$62.68
1 Sheet Metal Apprentice	34.85	278.80	53.75	430.00		
24 L.H., Daily Totals		$975.60		$1504.40	$40.65	$62.68
Crew Q-11	Hr.	Daily	Hr.	Daily	Bare Costs	Incl. O&P
1 Sheet Metal Foreman (inside)	$44.05	$352.40	$67.95	$543.60	$41.50	$64.00
2 Sheet Metal Workers	43.55	696.80	67.15	1074.40		
1 Sheet Metal Apprentice	34.85	278.80	53.75	430.00		
32 L.H., Daily Totals		$1328.00		$2048.00	$41.50	$64.00

Crews

Crew No.	Bare Costs		Incl. Subs O & P		Cost Per Labor-Hour	
Crew Q-12	Hr.	Daily	Hr.	Daily	Bare Costs	Incl. O&P
1 Sprinkler Installer	$44.30	$354.40	$66.70	$533.60	$39.88	$60.05
1 Sprinkler Apprentice	35.45	283.60	53.40	427.20		
16 L.H., Daily Totals		$638.00		$960.80	$39.88	$60.05
Crew Q-13	Hr.	Daily	Hr.	Daily	Bare Costs	Incl. O&P
1 Sprinkler Foreman (inside)	$44.80	$358.40	$67.45	$539.60	$42.21	$63.56
2 Sprinkler Installers	44.30	708.80	66.70	1067.20		
1 Sprinkler Apprentice	35.45	283.60	53.40	427.20		
32 L.H., Daily Totals		$1350.80		$2034.00	$42.21	$63.56
Crew Q-14	Hr.	Daily	Hr.	Daily	Bare Costs	Incl. O&P
1 Asbestos Worker	$41.50	$332.00	$65.60	$524.80	$37.35	$59.05
1 Asbestos Apprentice	33.20	265.60	52.50	420.00		
16 L.H., Daily Totals		$597.60		$944.80	$37.35	$59.05
Crew Q-15	Hr.	Daily	Hr.	Daily	Bare Costs	Incl. O&P
1 Plumber	$44.80	$358.40	$67.40	$539.20	$40.33	$60.65
1 Plumber Apprentice	35.85	286.80	53.90	431.20		
1 Welder, electric, 300 amp		61.45		67.59	3.84	4.22
16 L.H., Daily Totals		$706.65		$1037.99	$44.17	$64.87
Crew Q-16	Hr.	Daily	Hr.	Daily	Bare Costs	Incl. O&P
2 Plumbers	$44.80	$716.80	$67.40	$1078.40	$41.82	$62.90
1 Plumber Apprentice	35.85	286.80	53.90	431.20		
1 Welder, electric, 300 amp		61.45		67.59	2.56	2.82
24 L.H., Daily Totals		$1065.05		$1577.19	$44.38	$65.72
Crew Q-17	Hr.	Daily	Hr.	Daily	Bare Costs	Incl. O&P
1 Steamfitter	$45.20	$361.60	$68.00	$544.00	$40.67	$61.17
1 Steamfitter Apprentice	36.15	289.20	54.35	434.80		
1 Welder, electric, 300 amp		61.45		67.59	3.84	4.22
16 L.H., Daily Totals		$712.25		$1046.40	$44.52	$65.40
Crew Q-17A	Hr.	Daily	Hr.	Daily	Bare Costs	Incl. O&P
1 Steamfitter	$45.20	$361.60	$68.00	$544.00	$40.38	$60.78
1 Steamfitter Apprentice	36.15	289.20	54.35	434.80		
1 Equip. Oper. (crane)	39.80	318.40	60.00	480.00		
1 Hyd. Crane, 12 Ton		724.00		796.40		
1 Welder, electric, 300 amp		61.45		67.59	32.73	36.00
24 L.H., Daily Totals		$1754.65		$2322.80	$73.11	$96.78
Crew Q-18	Hr.	Daily	Hr.	Daily	Bare Costs	Incl. O&P
2 Steamfitters	$45.20	$723.20	$68.00	$1088.00	$42.18	$63.45
1 Steamfitter Apprentice	36.15	289.20	54.35	434.80		
1 Welder, electric, 300 amp		61.45		67.59	2.56	2.82
24 L.H., Daily Totals		$1073.85		$1590.40	$44.74	$66.27
Crew Q-19	Hr.	Daily	Hr.	Daily	Bare Costs	Incl. O&P
1 Steamfitter	$45.20	$361.60	$68.00	$544.00	$41.75	$62.57
1 Steamfitter Apprentice	36.15	289.20	54.35	434.80		
1 Electrician	43.90	351.20	65.35	522.80		
24 L.H., Daily Totals		$1002.00		$1501.60	$41.75	$62.57
Crew Q-20	Hr.	Daily	Hr.	Daily	Bare Costs	Incl. O&P
1 Sheet Metal Worker	$43.55	$348.40	$67.15	$537.20	$40.14	$61.43
1 Sheet Metal Apprentice	34.85	278.80	53.75	430.00		
.5 Electrician	43.90	175.60	65.35	261.40		
20 L.H., Daily Totals		$802.80		$1228.60	$40.14	$61.43

Crew No.	Bare Costs		Incl. Subs O & P		Cost Per Labor-Hour	
Crew Q-21	Hr.	Daily	Hr.	Daily	Bare Costs	Incl. O&P
2 Steamfitters	$45.20	$723.20	$68.00	$1088.00	$42.61	$63.92
1 Steamfitter Apprentice	36.15	289.20	54.35	434.80		
1 Electrician	43.90	351.20	65.35	522.80		
32 L.H., Daily Totals		$1363.60		$2045.60	$42.61	$63.92
Crew Q-22	Hr.	Daily	Hr.	Daily	Bare Costs	Incl. O&P
1 Plumber	$44.80	$358.40	$67.40	$539.20	$40.33	$60.65
1 Plumber Apprentice	35.85	286.80	53.90	431.20		
1 Hyd. Crane, 12 Ton		724.00		796.40	45.25	49.77
16 L.H., Daily Totals		$1369.20		$1766.80	$85.58	$110.43
Crew Q-22A	Hr.	Daily	Hr.	Daily	Bare Costs	Incl. O&P
1 Plumber	$44.80	$358.40	$67.40	$539.20	$37.30	$56.51
1 Plumber Apprentice	35.85	286.80	53.90	431.20		
1 Laborer	28.75	230.00	44.75	358.00		
1 Equip. Oper. (crane)	39.80	318.40	60.00	480.00		
1 Hyd. Crane, 12 Ton		724.00		796.40	22.63	24.89
32 L.H., Daily Totals		$1917.60		$2604.80	$59.92	$81.40
Crew Q-23	Hr.	Daily	Hr.	Daily	Bare Costs	Incl. O&P
1 Plumber Foreman	$46.80	$374.40	$70.40	$563.20	$43.33	$65.22
1 Plumber	44.80	358.40	67.40	539.20		
1 Equip. Oper. (medium)	38.40	307.20	57.85	462.80		
1 Lattice Boom Crane, 20 Ton		976.70		1074.37	40.70	44.77
24 L.H., Daily Totals		$2016.70		$2639.57	$84.03	$109.98
Crew R-1	Hr.	Daily	Hr.	Daily	Bare Costs	Incl. O&P
1 Electrician Foreman	$44.40	$355.20	$66.10	$528.80	$38.47	$57.83
3 Electricians	43.90	1053.60	65.35	1568.40		
2 Helpers	27.35	437.60	42.40	678.40		
48 L.H., Daily Totals		$1846.40		$2775.60	$38.47	$57.83
Crew R-1A	Hr.	Daily	Hr.	Daily	Bare Costs	Incl. O&P
1 Electrician	$43.90	$351.20	$65.35	$522.80	$35.63	$53.88
1 Helper	27.35	218.80	42.40	339.20		
16 L.H., Daily Totals		$570.00		$862.00	$35.63	$53.88
Crew R-2	Hr.	Daily	Hr.	Daily	Bare Costs	Incl. O&P
1 Electrician Foreman	$44.40	$355.20	$66.10	$528.80	$38.66	$58.14
3 Electricians	43.90	1053.60	65.35	1568.40		
2 Helpers	27.35	437.60	42.40	678.40		
1 Equip. Oper. (crane)	39.80	318.40	60.00	480.00		
1 S.P. Crane, 4x4, 5 Ton		326.80		359.48	5.84	6.42
56 L.H., Daily Totals		$2491.60		$3615.08	$44.49	$64.56
Crew R-3	Hr.	Daily	Hr.	Daily	Bare Costs	Incl. O&P
1 Electrician Foreman	$44.40	$355.20	$66.10	$528.80	$43.28	$64.58
1 Electrician	43.90	351.20	65.35	522.80		
.5 Equip. Oper. (crane)	39.80	159.20	60.00	240.00		
.5 S.P. Crane, 4x4, 5 Ton		163.40		179.74	8.17	8.99
20 L.H., Daily Totals		$1029.00		$1471.34	$51.45	$73.57
Crew R-4	Hr.	Daily	Hr.	Daily	Bare Costs	Incl. O&P
1 Struc. Steel Foreman	$43.35	$346.80	$78.50	$628.00	$42.26	$73.71
3 Struc. Steel Workers	41.35	992.40	74.90	1797.60		
1 Electrician	43.90	351.20	65.35	522.80		
1 Welder, gas engine, 300 amp		115.20		126.72	2.88	3.17
40 L.H., Daily Totals		$1805.60		$3075.12	$45.14	$76.88

Crews

Crew No.	Bare Costs		Incl. Subs O & P		Cost Per Labor-Hour	
Crew R-5	Hr.	Daily	Hr.	Daily	Bare Costs	Incl. O&P
1 Electrician Foreman	$44.40	$355.20	$66.10	$528.80	$37.93	$57.07
4 Electrician Linemen	43.90	1404.80	65.35	2091.20		
2 Electrician Operators	43.90	702.40	65.35	1045.60		
4 Electrician Groundmen	27.35	875.20	42.40	1356.80		
1 Crew Truck		116.20		127.82		
1 Tool Van		151.80		166.98		
1 Pickup Truck, 3/4 Ton		84.80		93.28		
.2 Hyd. Crane, 55 Ton		212.00		233.20		
.2 Hyd. Crane, 12 Ton		144.80		159.28		
.2 Drill Rig, Truck-Mounted		699.80		769.78		
1 Tractor w/Winch		297.80		327.58	19.40	21.34
88 L.H., Daily Totals		$5044.80		$6900.32	$57.33	$78.41
Crew R-6	Hr.	Daily	Hr.	Daily	Bare Costs	Incl. O&P
1 Electrician Foreman	$44.40	$355.20	$66.10	$528.80	$37.93	$57.07
4 Electrician Linemen	43.90	1404.80	65.35	2091.20		
2 Electrician Operators	43.90	702.40	65.35	1045.60		
4 Electrician Groundmen	27.35	875.20	42.40	1356.80		
1 Crew Truck		116.20		127.82		
1 Tool Van		151.80		166.98		
1 Pickup Truck, 3/4 Ton		84.80		93.28		
.2 Hyd. Crane, 55 Ton		212.00		233.20		
.2 Hyd. Crane, 12 Ton		144.80		159.28		
.2 Drill Rig, Truck-Mounted		699.80		769.78		
1 Tractor w/Winch		297.80		327.58		
3 Cable Trailers		510.90		561.99		
.5 Tensioning Rig		167.68		184.44		
.5 Cable Pulling Rig		989.00		1087.90	38.35	42.18
88 L.H., Daily Totals		$6712.38		$8734.65	$76.28	$99.26
Crew R-7	Hr.	Daily	Hr.	Daily	Bare Costs	Incl. O&P
1 Electrician Foreman	$44.40	$355.20	$66.10	$528.80	$30.19	$46.35
5 Electrician Groundmen	27.35	1094.00	42.40	1696.00		
1 Crew Truck		116.20		127.82	2.42	2.66
48 L.H., Daily Totals		$1565.40		$2352.62	$32.61	$49.01
Crew R-8	Hr.	Daily	Hr.	Daily	Bare Costs	Incl. O&P
1 Electrician Foreman	$44.40	$355.20	$66.10	$528.80	$38.47	$57.83
3 Electrician Linemen	43.90	1053.60	65.35	1568.40		
2 Electrician Groundmen	27.35	437.60	42.40	678.40		
1 Pickup Truck, 3/4 Ton		84.80		93.28		
1 Crew Truck		116.20		127.82	4.19	4.61
48 L.H., Daily Totals		$2047.40		$2996.70	$42.65	$62.43
Crew R-9	Hr.	Daily	Hr.	Daily	Bare Costs	Incl. O&P
1 Electrician Foreman	$44.40	$355.20	$66.10	$528.80	$35.69	$53.97
1 Electrician Lineman	43.90	351.20	65.35	522.80		
2 Electrician Operators	43.90	702.40	65.35	1045.60		
4 Electrician Groundmen	27.35	875.20	42.40	1356.80		
1 Pickup Truck, 3/4 Ton		84.80		93.28		
1 Crew Truck		116.20		127.82	3.14	3.45
64 L.H., Daily Totals		$2485.00		$3675.10	$38.83	$57.42
Crew R-10	Hr.	Daily	Hr.	Daily	Bare Costs	Incl. O&P
1 Electrician Foreman	$44.40	$355.20	$66.10	$528.80	$41.23	$61.65
4 Electrician Linemen	43.90	1404.80	65.35	2091.20		
1 Electrician Groundman	27.35	218.80	42.40	339.20		
1 Crew Truck		116.20		127.82		
3 Tram Cars		363.00		399.30	9.98	10.98
48 L.H., Daily Totals		$2458.00		$3486.32	$51.21	$72.63

Crew No.	Bare Costs		Incl. Subs O & P		Cost Per Labor-Hour	
Crew R-11	Hr.	Daily	Hr.	Daily	Bare Costs	Incl. O&P
1 Electrician Foreman	$44.40	$355.20	$66.10	$528.80	$41.22	$61.75
4 Electricians	43.90	1404.80	65.35	2091.20		
1 Equip. Oper. (crane)	39.80	318.40	60.00	480.00		
1 Common Laborer	28.75	230.00	44.75	358.00		
1 Crew Truck		116.20		127.82		
1 Hyd. Crane, 12 Ton		724.00		796.40	15.00	16.50
56 L.H., Daily Totals		$3148.60		$4382.22	$56.23	$78.25
Crew R-12	Hr.	Daily	Hr.	Daily	Bare Costs	Incl. O&P
1 Carpenter Foreman	$37.20	$297.60	$57.90	$463.20	$34.43	$54.39
4 Carpenters	36.70	1174.40	57.15	1828.80		
4 Common Laborers	28.75	920.00	44.75	1432.00		
1 Equip. Oper. (med.)	38.40	307.20	57.85	462.80		
1 Steel Worker	41.35	330.80	74.90	599.20		
1 Dozer, 200 H.P.		988.40		1087.24		
1 Pickup Truck, 3/4 Ton		84.80		93.28	12.20	13.41
88 L.H., Daily Totals		$4103.20		$5966.52	$46.63	$67.80
Crew R-13	Hr.	Daily	Hr.	Daily	Bare Costs	Incl. O&P
1 Electrician Foreman	$44.40	$355.20	$66.10	$528.80	$41.90	$62.52
3 Electricians	43.90	1053.60	65.35	1568.40		
.25 Equip. Oper. (crane)	39.80	79.60	60.00	120.00		
1 Equipment Oiler	33.90	271.20	51.10	408.80		
.25 Hydraulic Crane, 33 Ton		191.35		210.49	4.56	5.01
42 L.H., Daily Totals		$1950.95		$2836.49	$46.45	$67.54
Crew R-15	Hr.	Daily	Hr.	Daily	Bare Costs	Incl. O&P
1 Electrician Foreman	$44.40	$355.20	$66.10	$528.80	$42.81	$63.84
4 Electricians	43.90	1404.80	65.35	2091.20		
1 Equipment Operator	36.85	294.80	55.55	444.40		
1 Aerial Lift Truck		289.80		318.78	6.04	6.64
48 L.H., Daily Totals		$2344.60		$3383.18	$48.85	$70.48
Crew R-18	Hr.	Daily	Hr.	Daily	Bare Costs	Incl. O&P
.25 Electrician Foreman	$44.40	$88.80	$66.10	$132.20	$33.75	$51.28
1 Electrician	43.90	351.20	65.35	522.80		
2 Helpers	27.35	437.60	42.40	678.40		
26 L.H., Daily Totals		$877.60		$1333.40	$33.75	$51.28
Crew R-19	Hr.	Daily	Hr.	Daily	Bare Costs	Incl. O&P
.5 Electrician Foreman	$44.40	$177.60	$66.10	$264.40	$44.00	$65.50
2 Electricians	43.90	702.40	65.35	1045.60		
20 L.H., Daily Totals		$880.00		$1310.00	$44.00	$65.50
Crew R-21	Hr.	Daily	Hr.	Daily	Bare Costs	Incl. O&P
1 Electrician Foreman	$44.40	$355.20	$66.10	$528.80	$43.89	$65.35
3 Electricians	43.90	1053.60	65.35	1568.40		
.1 Equip. Oper. (med.)	38.40	30.72	57.85	46.28		
.1 S.P. Crane, 4x4, 25 Ton		87.56		96.32	2.67	2.94
32.8 L.H., Daily Totals		$1527.08		$2239.80	$46.56	$68.29
Crew R-22	Hr.	Daily	Hr.	Daily	Bare Costs	Incl. O&P
.66 Electrician Foreman	$44.40	$234.43	$66.10	$349.01	$36.87	$55.61
2 Helpers	27.35	437.60	42.40	678.40		
2 Electricians	43.90	702.40	65.35	1045.60		
37.28 L.H., Daily Totals		$1374.43		$2073.01	$36.87	$55.61

Crews

Crew No.	Bare Costs		Incl. Sub O & P		Cost Per Labor-Hour	
Crew R-30	Hr.	Daily	Hr.	Daily	Bare Costs	Incl. O&P
.25 Electrician Foreman (out)	$45.90	$91.80	$68.35	$136.70	$34.73	$52.90
1 Electrician	43.90	351.20	65.35	522.80		
2 Laborers, (Semi-Skilled)	28.75	460.00	44.75	716.00		
26 L.H., Daily Totals		$903.00		$1375.50	$34.73	$52.90
Crew R-31	Hr.	Daily	Hr.	Daily	Bare Costs	Incl. O&P
1 Electrician	$43.90	$351.20	$65.35	$522.80	$43.90	$65.35
1 Core Drill, Elec, 2.5 HP		71.20		78.32	8.90	9.79
8 L.H., Daily Totals		$422.40		$601.12	$52.80	$75.14
Crew W-41E	Hr.	Daily	Hr.	Daily	Bare Costs	Incl. O&P
1 Laborers, (Semi-Skilled)	$28.75	$230.00	$44.75	$358.00	$38.78	$58.94
1 Plumber	44.80	358.40	67.40	539.20		
.5 Plumber	46.80	187.20	70.40	281.60		
20 L.H., Daily Totals		$775.60		$1178.80	$38.78	$58.94

Historical Cost Indexes

The table below lists both the RSMeans Historical Cost Index based on Jan. 1, 1993 = 100 as well as the computed value of an index based on Jan. 1, 2007 costs. Since the Jan. 1, 2007 figure is estimated, space is left to write in the actual index figures as they become available through either the quarterly "RSMeans Construction Cost Indexes" or as printed in the "Engineering News-Record." To compute the actual index based on Jan. 1, 2007 = 100, divide the Historical Cost Index for a particular year by the actual Jan. 1, 2007 Construction Cost Index. Space has been left to advance the index figures as the year progresses.

Year	Historical Cost Index Jan. 1, 1993 = 100		Current Index Based on Jan. 1, 2007 = 100		Year	Historical Cost Index Jan. 1, 1993 = 100	Current Index Based on Jan. 1, 2007 = 100		Year	Historical Cost Index Jan. 1, 1993 = 100	Current Index Based on Jan. 1, 2007 = 100	
	Est.	Actual	Est.	Actual		Actual	Est.	Actual		Actual	Est.	Actual
Oct 2007					July 1992	99.4	59.8		July 1974	41.4	24.9	
July 2007					1991	96.8	58.2		1973	37.7	22.7	
April 2007					1990	94.3	56.7		1972	34.8	20.9	
Jan 2007	166.3		100.0	100.0	1989	92.1	55.4		1971	32.1	19.3	
July 2006		162.0	97.4		1988	89.9	54.0		1970	28.7	17.3	
2005		151.6	91.2		1987	87.7	52.7		1969	26.9	16.2	
2004		143.7	86.4		1986	84.2	50.6		1968	24.9	15.0	
2003		132.0	79.4		1985	82.6	49.7		1967	23.5	14.1	
2002		128.7	77.4		1984	82.0	49.3		1966	22.7	13.7	
2001		125.1	75.2		1983	80.2	48.2		1965	21.7	13.0	
2000		120.9	72.7		1982	76.1	45.8		1964	21.2	12.7	
1999		117.6	70.7		1981	70.0	42.1		1963	20.7	12.4	
1998		115.1	69.2		1980	62.9	37.8		1962	20.2	12.1	
1997		112.8	67.8		1979	57.8	34.8		1961	19.8	11.9	
1996		110.2	66.3		1978	53.5	32.2		1960	19.7	11.8	
1995		107.6	64.7		1977	49.5	29.8		1959	19.3	11.6	
1994		104.4	62.8		1976	46.9	28.2		1958	18.8	11.3	
1993		101.7	61.2		1975	44.8	26.9		1957	18.4	11.1	

Adjustments to Costs

The Historical Cost Index can be used to convert National Average building costs at a particular time to the approximate building costs for some other time.

Example:

Estimate and compare construction costs for different years in the same city.

To estimate the National Average construction cost of a building in 1970, knowing that it cost $900,000 in 2007:

INDEX in 1970 = 28.7

INDEX in 2007 = 166.3

Note: The City Cost Indexes for Canada can be used to convert U.S. National averages to local costs in Canadian dollars.

Time Adjustment using the Historical Cost Indexes:

$$\frac{\text{Index for Year A}}{\text{Index for Year B}} \times \text{Cost in Year B} = \text{Cost in Year A}$$

$$\frac{\text{INDEX 1970}}{\text{INDEX 2007}} \times \text{Cost 2007} = \text{Cost 1970}$$

$$\frac{28.7}{166.3} \times \$900{,}000 = .173 \times \$900{,}000 = \$155{,}700$$

The construction cost of the building in 1970 is $155,700.

City Cost Indexes

How to Use the City Cost Indexes

What you should know before you begin

RSMeans City Cost Indexes (CCI) are an extremely useful tool to use when you want to compare costs from city to city and region to region.

This publication contains average construction cost indexes for 731 U.S. and Canadian cities covering over 930 three-digit zip code locations, as listed directly under each city.

Keep in mind that a City Cost Index number is a percentage ratio of a specific city's cost to the national average cost of the same item at a stated time period.

In other words, these index figures represent relative construction factors (or, if you prefer, multipliers) for Material and Installation costs, as well as the weighted average for Total In Place costs for each CSI MasterFormat division. Installation costs include both labor and equipment rental costs. When estimating equipment rental rates only, for a specific location, use 01543 CONTRACTOR EQUIPMENT index.

The 30 City Average Index is the average of 30 major U.S. cities and serves as a National Average.

Index figures for both material and installation are based on the 30 major city average of 100 and represent the cost relationship as of July 1, 2006. The index for each division is computed from representative material and labor quantities for that division. The weighted average for each city is a weighted total of the components listed above it, but does not include relative productivity between trades or cities.

As changes occur in local material prices, labor rates and equipment rental rates, the impact of these changes should be accurately measured by the change in the City Cost Index for each particular city (as compared to the 30 City Average).

Therefore, if you know (or have estimated) building costs in one city today, you can easily convert those costs to expected building costs in another city.

In addition, by using the Historical Cost Index, you can easily convert National Average building costs at a particular time to the approximate building costs for some other time. The City Cost Indexes can then be applied to calculate the costs for a particular city.

Quick Calculations

Location Adjustment Using the City Cost Indexes:

$$\frac{\text{Index for City A}}{\text{Index for City B}} \times \text{Cost in City B} = \text{Cost in City A}$$

Time Adjustment for the National Average Using the Historical Cost Index:

$$\frac{\text{Index for Year A}}{\text{Index for Year B}} \times \text{Cost in Year B} = \text{Cost in Year A}$$

Adjustment from the National Average:

$$\frac{\text{Index for City A}}{100} \times \text{National Average Cost} = \text{Cost in City A}$$

Since each of the other RSMeans publications contains many different items, any *one* item multiplied by the particular city index may give incorrect results. However, the larger the number of items compiled, the closer the results should be to actual costs for that particular city.

The City Cost Indexes for Canadian cities are calculated using Canadian material and equipment prices and labor rates, in Canadian dollars. Therefore, indexes for Canadian cities can be used to convert U.S. National Average prices to local costs in Canadian dollars.

How to use this section

1. Compare costs from city to city.

In using the RSMeans Indexes, remember that an index number is not a fixed number but a ratio: It's a percentage ratio of a building component's cost at any stated time to the National Average cost of that same component at the same time period. Put in the form of an equation:

$$\frac{\text{Specific City Cost}}{\text{National Average Cost}} \times 100 = \text{City Index Number}$$

Therefore, when making cost comparisons between cities, do not subtract one city's index number from the index number of another city and read the result as a percentage difference. Instead, divide one city's index number by that of the other city. The resulting number may then be used as a multiplier to calculate cost differences from city to city.

The formula used to find cost differences between cities for the purpose of comparison is as follows:

$$\frac{\text{City A Index}}{\text{City B Index}} \times \text{City B Cost (Known)} = \text{City A Cost (Unknown)}$$

In addition, you can use RSMeans CCI to calculate and compare costs division by division between cities using the same basic formula. (Just be sure that you're comparing similar divisions.)

2. Compare a specific city's construction costs with the National Average.

When you're studying construction location feasibility, it's advisable to compare a prospective project's cost index with an index of the National Average cost.

For example, divide the weighted average index of construction costs of a specific city by that of the 30 City Average, which = 100.

$$\frac{\text{City Index}}{100} = \% \text{ of National Average}$$

As a result, you get a ratio that indicates the relative cost of construction in that city in comparison with the National Average.

3. Convert U.S. National Average to actual costs in Canadian City.

$$\frac{\text{Index for Canadian City}}{100} \times \text{National Average Cost} = \text{Cost in Canadian City in \$ CAN}$$

4. **Adjust construction cost data based on a National Average.**
When you use a source of construction cost data which is based on a National Average (such as RSMeans cost data publications), it is necessary to adjust those costs to a specific location.

$$\frac{\text{City Index}}{100} \times \text{"Book" Cost Based on National Average Costs} = \text{City Cost (Unknown)}$$

5. **When applying the City Cost Indexes to demolition projects, use the appropriate division installation index.** For example, for removal of existing doors and windows, use Division 8 (Openings) index.

What you might like to know about how we developed the Indexes

The information presented in the CCI is organized according to the Construction Specifications Institute (CSI) MasterFormat 2004.

To create a reliable index, RSMeans researched the building type most often constructed in the United States and Canada. Because it was concluded that no one type of building completely represented the building construction industry, nine different types of buildings were combined to create a composite model.

The exact material, labor and equipment quantities are based on detailed analysis of these nine building types, then each quantity is weighted in proportion to expected usage. These various material items, labor hours, and equipment rental rates are thus combined to form a composite building representing as closely as possible the actual usage of materials, labor and equipment used in the North American Building Construction Industry.

The following structures were chosen to make up that composite model:

1. Factory, 1 story
2. Office, 2–4 story
3. Store, Retail
4. Town Hall, 2–3 story
5. High School, 2–3 story
6. Hospital, 4–8 story
7. Garage, Parking
8. Apartment, 1–3 story
9. Hotel/Motel, 2–3 story

For the purposes of ensuring the timeliness of the data, the components of the index for the composite model have been streamlined. They currently consist of:

- specific quantities of 66 commonly used construction materials;
- specific labor-hours for 21 building construction trades; and
- specific days of equipment rental for 6 types of construction equipment (normally used to install the 66 material items by the 21 trades.)

A sophisticated computer program handles the updating of all costs for each city on a quarterly basis. Material and equipment price quotations are gathered quarterly from 731 cities in the United States and Canada. These prices and the latest negotiated labor wage rates for 21 different building trades are used to compile the quarterly update of the City Cost Index.

The 30 major U.S. cities used to calculate the National Average are:

Atlanta, GA
Baltimore, MD
Boston, MA
Buffalo, NY
Chicago, IL
Cincinnati, OH
Cleveland, OH
Columbus, OH
Dallas, TX
Denver, CO
Detroit, MI
Houston, TX
Indianapolis, IN
Kansas City, MO
Los Angeles, CA

Memphis, TN
Milwaukee, WI
Minneapolis, MN
Nashville, TN
New Orleans, LA
New York, NY
Philadelphia, PA
Phoenix, AZ
Pittsburgh, PA
St. Louis, MO
San Antonio, TX
San Diego, CA
San Francisco, CA
Seattle, WA
Washington, DC

What the CCI does not indicate

The weighted average for each city is a total of the divisional components weighted to reflect typical usage, but it does not include the productivity variations between trades or cities.

In addition, the CCI does not take into consideration factors such as the following:

- managerial efficiency
- competitive conditions
- automation
- restrictive union practices
- unique local requirements
- regional variations due to specific building codes

City Cost Indexes

DIVISION		UNITED STATES 30 CITY AVERAGE			ALABAMA ANNISTON 362			ALABAMA BIRMINGHAM 350 - 352			ALABAMA BUTLER 369			ALABAMA DECATUR 356			ALABAMA DOTHAN 363		
		MAT.	INST.	TOTAL	MAT.	INST.	TOTAL	MAT.	INST.	TOTAL	MAT.	INST.	TOTAL	MAT.	INST.	TOTAL	MAT.	INST.	TOTAL
01543	CONTRACTOR EQUIPMENT	.0	100.0	100.0	.0	101.4	101.4	.0	101.5	101.5	.0	97.9	97.9	.0	101.4	101.4	.0	97.9	97.9
0241, 31 - 34	SITE & INFRASTRUCTURE, DEMOLITION	100.0	100.0	100.0	87.7	91.8	90.6	82.5	93.1	90.1	101.0	85.6	90.0	80.6	91.5	88.4	98.5	85.6	89.3
0310	Concrete Forming & Accessories	100.0	100.0	100.0	91.7	33.4	41.6	94.4	71.1	74.4	87.5	43.8	50.0	96.4	46.7	53.7	97.4	43.6	51.2
0310	Concrete Reinforcing	100.0	100.0	100.0	92.3	84.3	88.4	92.3	86.0	89.2	97.6	52.8	75.7	92.3	50.6	71.9	97.6	52.8	75.7
0330	Cast-in-Place Concrete	100.0	100.0	100.0	88.7	38.3	69.7	92.7	64.1	81.9	86.5	45.8	71.2	88.2	52.8	74.9	86.5	45.8	71.2
03	CONCRETE	100.0	100.0	100.0	93.8	46.7	71.6	91.8	72.6	82.7	94.7	47.9	72.6	89.6	51.2	71.5	94.2	47.8	72.3
04	MASONRY	100.0	100.0	100.0	84.8	32.7	53.6	87.4	73.0	78.7	89.0	34.0	56.1	85.7	48.3	63.3	90.0	34.0	56.5
05	METALS	100.0	100.0	100.0	91.3	87.9	90.2	91.9	92.9	92.2	90.2	74.6	85.5	93.2	75.8	87.9	90.3	74.4	85.5
06	WOOD, PLASTICS & COMPOSITES	100.0	100.0	100.0	89.8	32.9	59.7	95.1	71.4	82.6	84.5	45.4	63.9	95.1	46.0	69.2	96.5	45.4	69.5
07	THERMAL & MOISTURE PROTECTION	100.0	100.0	100.0	96.6	38.5	73.2	98.3	82.7	92.0	96.6	51.3	78.4	96.4	56.8	80.5	96.6	46.3	76.4
08	OPENINGS	100.0	100.0	100.0	93.9	45.1	81.2	97.5	74.3	91.5	93.9	47.7	81.9	97.8	46.3	84.5	93.9	47.7	81.9
0920	Plaster & Gypsum Board	100.0	100.0	100.0	99.9	31.7	59.3	105.2	71.2	85.0	98.6	44.5	66.4	103.1	45.1	68.6	105.0	44.5	69.0
0950, 0980	Ceilings & Acoustic Treatment	100.0	100.0	100.0	96.7	31.7	57.6	102.8	71.2	83.8	96.7	44.5	65.3	100.9	45.1	67.4	96.7	44.5	65.3
0960	Flooring	100.0	100.0	100.0	102.0	33.0	83.6	105.1	46.6	89.5	108.6	26.1	86.6	105.1	48.6	90.1	116.8	31.5	94.1
0990	Wall Finishes & Painting/Coating	100.0	100.0	100.0	96.5	30.9	57.2	96.5	68.7	79.8	96.5	52.3	70.1	96.5	48.9	68.0	96.5	52.3	70.1
09	FINISHES	100.0	100.0	100.0	98.3	32.5	64.4	101.3	65.9	83.0	101.7	41.2	70.4	100.3	46.9	72.8	105.0	42.1	72.5
COVERS	DIVS. 10 - 14, 25, 28, 41, 43, 44	100.0	100.0	100.0	100.0	43.6	88.4	100.0	85.6	97.0	100.0	46.3	88.9	100.0	48.1	89.3	100.0	46.3	88.9
22, 23	PLUMBING & HVAC	100.0	100.0	100.0	99.8	35.4	73.5	99.9	64.0	85.3	97.6	37.8	73.2	99.9	45.9	77.9	97.6	37.8	73.2
26, 27, 3370	ELECTRICAL, COMMUNICATIONS & UTIL.	100.0	100.0	100.0	95.2	24.9	61.5	98.0	64.2	81.8	96.9	35.1	67.3	96.6	46.1	72.4	95.9	35.1	66.7
MF2004	WEIGHTED AVERAGE	100.0	100.0	100.0	95.3	45.1	73.6	96.2	73.1	86.2	95.7	47.4	74.9	95.7	54.1	77.7	96.0	47.4	75.0

DIVISION		ALABAMA EVERGREEN 364			ALABAMA GADSDEN 359			ALABAMA HUNTSVILLE 357 - 358			ALABAMA JASPER 355			ALABAMA MOBILE 365 - 366			ALABAMA MONTGOMERY 360 - 361		
		MAT.	INST.	TOTAL	MAT.	INST.	TOTAL	MAT.	INST.	TOTAL	MAT.	INST.	TOTAL	MAT.	INST.	TOTAL	MAT.	INST.	TOTAL
01543	CONTRACTOR EQUIPMENT	.0	97.9	97.9	.0	101.4	101.4	.0	101.4	101.4	.0	101.4	101.4	.0	97.9	97.9	.0	97.9	97.9
0241, 31 - 34	SITE & INFRASTRUCTURE, DEMOLITION	101.5	85.8	90.3	86.3	92.1	90.5	80.4	93.1	89.5	85.9	91.9	90.2	93.1	86.1	88.1	90.8	86.6	87.8
0310	Concrete Forming & Accessories	83.5	45.2	50.6	87.7	43.1	49.4	96.4	64.6	69.1	93.8	35.3	43.5	96.4	49.8	56.4	95.4	45.3	52.4
0310	Concrete Reinforcing	97.6	52.8	75.7	97.8	84.8	91.4	92.3	71.5	82.1	92.3	84.5	88.5	95.3	49.5	72.9	95.3	84.5	90.0
0330	Cast-in-Place Concrete	86.5	48.1	72.1	88.2	49.0	73.5	85.9	62.7	77.2	97.7	43.7	77.4	90.7	51.6	76.0	87.3	46.6	72.0
03	CONCRETE	94.8	49.3	73.3	93.7	54.8	75.3	88.5	66.6	78.1	97.0	49.4	74.5	91.4	51.9	72.7	89.6	54.9	73.2
04	MASONRY	89.0	38.2	58.6	84.1	40.4	58.0	86.8	67.4	75.2	81.2	35.3	53.8	87.6	48.6	64.3	86.3	33.8	54.9
05	METALS	90.3	74.4	85.5	91.3	90.2	91.0	93.2	87.9	91.6	91.3	88.7	90.5	92.1	75.6	87.1	91.9	89.5	91.2
06	WOOD, PLASTICS & COMPOSITES	79.6	45.4	61.5	85.3	41.4	62.1	95.1	64.3	78.8	92.0	33.7	61.2	95.3	49.6	71.1	93.7	45.4	68.2
07	THERMAL & MOISTURE PROTECTION	96.6	48.1	77.1	96.5	67.2	84.7	96.1	78.3	88.9	96.5	43.0	75.0	96.1	71.4	86.1	96.1	65.3	83.7
08	OPENINGS	93.9	47.7	81.9	93.8	50.9	82.7	97.8	61.0	88.3	93.8	49.8	82.4	97.8	49.6	85.3	97.8	55.3	86.8
0920	Plaster & Gypsum Board	96.0	44.5	65.4	98.3	40.4	63.8	101.0	63.9	78.9	100.0	32.5	59.8	103.1	48.7	70.7	103.1	44.5	68.2
0950, 0980	Ceilings & Acoustic Treatment	96.7	44.5	65.3	96.7	40.4	62.8	102.8	63.9	79.4	95.9	32.5	57.7	102.8	48.7	70.3	102.8	44.5	67.7
0960	Flooring	105.6	26.1	84.4	99.8	41.7	84.3	105.1	50.0	90.4	103.1	31.8	84.1	115.6	50.5	98.2	113.9	26.1	90.5
0990	Wall Finishes & Painting/Coating	96.5	52.3	70.1	96.5	59.6	74.4	96.5	60.0	74.7	96.5	37.8	61.4	100.8	53.2	72.4	96.5	52.3	70.1
09	FINISHES	100.4	42.2	70.4	97.3	43.1	69.4	100.5	60.8	80.0	98.3	33.9	65.1	105.5	49.7	76.7	104.6	41.1	71.8
COVERS	DIVS. 10 - 14, 25, 28, 41, 43, 44	100.0	47.9	89.3	100.0	77.3	95.3	100.0	83.8	96.7	100.0	44.3	88.5	100.0	61.2	92.0	100.0	77.2	95.3
22, 23	PLUMBING & HVAC	97.6	40.0	74.1	101.5	40.7	76.7	99.7	64.1	85.2	101.5	55.4	82.8	99.7	65.1	85.6	99.9	38.4	74.8
26, 27, 3370	ELECTRICAL, COMMUNICATIONS & UTIL.	94.6	35.1	66.1	96.9	64.2	81.2	97.9	68.3	83.7	96.2	25.2	62.2	97.9	47.9	73.9	97.2	64.6	81.5
MF2004	WEIGHTED AVERAGE	95.3	48.6	75.2	95.7	57.3	79.1	95.8	70.4	84.8	96.0	50.6	76.4	96.8	59.3	80.6	96.2	55.6	78.7

DIVISION		ALABAMA PHENIX CITY 368			ALABAMA SELMA 367			ALABAMA TUSCALOOSA 354			ALASKA ANCHORAGE 995 - 996			ALASKA FAIRBANKS 997			ALASKA JUNEAU 998		
		MAT.	INST.	TOTAL	MAT.	INST.	TOTAL	MAT.	INST.	TOTAL	MAT.	INST.	TOTAL	MAT.	INST.	TOTAL	MAT.	INST.	TOTAL
01543	CONTRACTOR EQUIPMENT	.0	97.9	97.9	.0	97.9	97.9	.0	101.4	101.4	.0	118.7	118.7	.0	118.7	118.7	.0	118.7	118.7
0241, 31 - 34	SITE & INFRASTRUCTURE, DEMOLITION	105.4	86.7	92.0	98.3	86.6	89.9	80.8	91.8	88.7	145.7	133.7	137.1	128.9	133.7	132.4	141.0	133.7	135.8
0310	Concrete Forming & Accessories	91.5	39.9	47.2	88.9	44.5	50.7	96.4	37.4	45.8	132.5	115.9	118.3	135.8	119.4	121.7	134.3	115.8	118.4
0310	Concrete Reinforcing	97.5	71.8	84.9	97.6	84.3	91.1	92.3	84.8	88.6	142.6	106.8	125.1	145.7	106.9	126.7	109.3	106.8	108.1
0330	Cast-in-Place Concrete	86.5	47.1	71.7	86.5	46.4	71.4	89.5	44.5	72.6	178.8	115.9	155.2	149.2	116.4	136.8	179.8	115.9	155.7
03	CONCRETE	97.3	50.2	75.1	93.5	54.4	75.0	90.3	50.7	71.6	146.2	113.6	130.8	127.9	115.4	122.0	141.5	113.6	128.3
04	MASONRY	89.0	38.6	58.8	92.8	34.1	57.7	85.8	35.9	55.9	216.0	120.7	159.0	218.1	120.7	159.8	206.3	120.7	155.1
05	METALS	90.2	84.6	88.5	90.2	88.5	89.7	92.5	90.0	91.7	125.4	102.3	118.4	125.5	102.6	118.6	125.4	102.2	118.4
06	WOOD, PLASTICS & COMPOSITES	89.4	37.8	62.2	86.3	45.4	64.7	95.1	36.0	63.9	119.4	114.2	116.7	119.7	118.5	119.1	119.4	114.2	116.7
07	THERMAL & MOISTURE PROTECTION	96.9	65.9	84.5	96.5	51.3	78.3	96.4	64.1	83.4	173.8	114.9	150.1	171.5	117.8	149.9	172.0	114.9	149.0
08	OPENINGS	93.9	48.1	82.0	93.9	55.3	83.9	97.8	55.2	86.8	125.4	110.6	121.5	122.5	113.3	120.1	122.5	110.6	119.4
0920	Plaster & Gypsum Board	100.8	36.7	62.6	99.5	44.5	66.8	103.1	34.9	62.5	140.5	114.4	125.0	151.0	118.8	131.9	140.5	114.4	125.0
0950, 0980	Ceilings & Acoustic Treatment	96.7	36.7	60.6	96.7	44.5	65.3	102.8	34.9	61.9	130.7	114.4	120.9	130.7	118.8	123.5	130.7	114.4	120.9
0960	Flooring	111.7	26.1	88.8	109.4	26.1	87.2	105.1	38.3	87.3	164.0	125.2	153.7	164.0	125.2	153.7	164.0	125.2	153.7
0990	Wall Finishes & Painting/Coating	96.5	52.3	70.1	96.5	52.3	70.1	96.5	45.0	65.7	161.6	115.4	134.0	161.6	124.9	139.7	161.6	115.4	134.0
09	FINISHES	103.3	36.9	69.0	101.8	41.2	70.5	100.8	36.7	67.7	154.3	117.9	135.5	153.9	121.5	137.1	153.0	117.9	134.9
COVERS	DIVS. 10 - 14, 25, 28, 41, 43, 44	100.0	76.6	95.2	100.0	46.3	89.0	100.0	75.0	94.8	100.0	110.8	102.2	100.0	111.3	102.3	100.0	110.8	102.2
22, 23	PLUMBING & HVAC	97.6	38.8	73.6	97.6	38.2	73.4	99.9	33.9	73.0	100.4	107.7	103.4	100.4	116.3	106.9	100.4	101.7	100.9
26, 27, 3370	ELECTRICAL, COMMUNICATIONS & UTIL.	96.5	58.7	78.4	95.6	35.1	66.6	97.4	64.2	81.5	147.2	112.7	130.7	148.4	112.9	131.4	148.4	112.8	131.4
MF2004	WEIGHTED AVERAGE	96.3	53.4	77.8	95.6	50.1	75.9	95.8	54.0	77.8	131.9	114.1	124.2	129.2	116.9	123.9	130.4	112.9	122.8

City Cost Indexes

		ALASKA			ARIZONA												
		KETCHIKAN			CHAMBERS			FLAGSTAFF			GLOBE			KINGMAN			MESA/TEMPE		
DIVISION		999			865			860			855			864			852		
		MAT.	INST.	TOTAL	MAT.	INST.	TOTAL	MAT.	INST.	TOTAL	MAT.	INST.	TOTAL	MAT.	INST.	TOTAL	MAT.	INST.	TOTAL
01543	CONTRACTOR EQUIPMENT	.0	118.7	118.7	.0	94.6	94.6	.0	94.6	94.6	.0	98.2	98.2	.0	94.6	94.6	.0	98.2	98.2
0241, 31 - 34	SITE & INFRASTRUCTURE, DEMOLITION	196.0	133.7	151.3	64.6	98.9	89.2	81.9	99.9	94.8	95.3	102.8	100.7	64.5	100.4	90.2	86.2	103.2	98.4
0310	Concrete Forming & Accessories	126.2	115.8	117.3	96.8	53.1	59.3	102.6	66.2	71.3	96.6	52.9	59.1	94.6	64.8	69.0	99.7	61.1	66.5
0310	Concrete Reinforcing	114.0	106.8	110.5	105.3	72.4	89.2	105.1	73.2	89.5	109.9	72.3	91.5	105.4	73.1	89.7	110.6	73.2	92.3
0330	Cast-in-Place Concrete	302.2	115.9	232.1	95.9	60.8	82.7	96.0	75.1	88.2	104.3	61.1	88.1	95.6	61.8	82.9	105.2	67.9	91.1
03	CONCRETE	218.7	113.5	169.1	103.2	59.5	82.6	123.2	70.5	98.3	120.7	59.6	91.8	102.8	65.3	85.1	112.0	65.8	90.2
04	MASONRY	227.3	120.7	163.5	98.6	48.5	68.6	98.3	63.9	77.7	110.4	48.5	73.4	98.6	60.7	75.9	110.5	50.3	74.5
05	METALS	125.6	102.1	118.5	97.2	65.0	87.5	97.7	69.2	89.1	100.3	65.4	89.8	97.8	69.1	89.1	100.6	69.9	91.3
06	WOOD, PLASTICS & COMPOSITES	109.6	114.2	112.0	101.0	51.3	74.7	107.3	66.1	85.5	96.0	51.4	72.5	97.1	66.1	80.7	99.6	66.3	82.0
07	THERMAL & MOISTURE PROTECTION	174.3	114.9	150.4	96.5	59.2	81.5	98.2	69.0	86.4	102.1	56.6	83.8	96.5	64.7	83.7	101.4	61.4	85.3
08	OPENINGS	122.9	110.6	119.7	101.7	55.2	89.7	101.8	66.5	92.7	98.5	55.3	87.3	101.9	66.5	92.7	98.6	66.5	90.3
0920	Plaster & Gypsum Board	136.0	114.4	123.1	90.6	50.0	66.4	93.0	65.2	76.5	92.8	50.0	67.3	87.2	65.2	74.1	94.4	65.2	77.0
0950, 0980	Ceilings & Acoustic Treatment	125.9	114.4	119.0	104.2	50.0	71.6	105.1	65.2	81.1	95.1	50.0	68.0	105.1	65.2	81.1	95.1	65.2	77.1
0960	Flooring	164.0	125.2	153.7	91.7	47.6	80.0	94.3	47.8	81.9	91.8	47.6	80.1	90.7	64.6	83.8	93.3	59.6	84.3
0990	Wall Finishes & Painting/Coating	161.6	115.4	134.0	91.3	45.6	64.0	91.3	56.0	70.2	95.2	45.6	65.6	91.3	56.0	70.2	95.2	56.0	71.8
09	FINISHES	154.4	117.9	135.6	94.2	50.4	71.6	97.1	61.1	78.5	95.2	50.5	72.1	93.6	63.2	77.9	95.0	59.7	76.8
COVERS	DIVS. 10 - 14, 25, 28, 41, 43, 44	100.0	110.8	102.2	100.0	79.9	95.9	100.0	82.9	96.5	100.0	80.3	95.9	100.0	81.6	96.2	100.0	78.1	95.5
22, 23	PLUMBING & HVAC	98.9	99.6	99.2	97.3	72.3	87.1	100.1	77.1	90.7	95.9	65.9	83.7	97.3	75.4	88.4	100.2	66.8	86.6
26, 27, 3370	ELECTRICAL, COMMUNICATIONS & UTIL.	148.4	112.8	131.4	99.5	74.6	87.6	98.6	61.4	80.7	94.6	66.3	81.0	99.5	43.2	72.5	91.6	61.4	77.1
MF2004	WEIGHTED AVERAGE	141.6	112.4	129.0	97.8	66.2	84.2	101.5	71.3	88.4	100.8	64.1	84.9	97.8	67.5	84.7	100.2	66.9	85.8

		ARIZONA											ARKANSAS					
		PHOENIX			PRESCOTT			SHOW LOW			TUCSON			BATESVILLE			CAMDEN		
DIVISION		850,853			863			859			856 - 857			725			717		
		MAT.	INST.	TOTAL	MAT.	INST.	TOTAL	MAT.	INST.	TOTAL	MAT.	INST.	TOTAL	MAT.	INST.	TOTAL	MAT.	INST.	TOTAL
01543	CONTRACTOR EQUIPMENT	.0	98.8	98.8	.0	94.6	94.6	.0	98.2	98.2	.0	98.2	98.2	.0	86.5	86.5	.0	86.5	86.5
0241, 31 - 34	SITE & INFRASTRUCTURE, DEMOLITION	86.6	104.0	99.1	70.5	98.9	90.9	97.3	102.8	101.3	82.7	103.8	97.8	73.8	84.0	81.1	74.7	83.6	81.0
0310	Concrete Forming & Accessories	100.8	70.2	74.6	98.3	52.9	59.3	104.1	53.3	60.5	100.2	69.8	74.1	83.6	55.9	59.8	83.3	36.8	43.4
0310	Concrete Reinforcing	108.9	73.3	91.5	105.1	72.7	89.3	110.6	72.5	92.0	91.3	73.2	82.4	95.8	76.3	86.3	97.1	51.6	74.9
0330	Cast-in-Place Concrete	105.3	75.7	94.1	95.9	60.7	82.7	104.3	61.3	88.1	108.3	75.5	95.9	79.2	52.3	69.1	81.2	41.7	66.3
03	CONCRETE	111.6	72.6	93.2	108.6	59.4	85.4	123.1	59.8	93.2	110.2	72.3	92.3	80.1	59.0	70.1	81.7	42.2	63.0
04	MASONRY	98.1	64.9	78.2	98.3	54.7	72.3	110.4	50.3	74.5	95.9	63.9	76.8	99.3	50.8	70.3	108.0	38.4	66.4
05	METALS	102.1	71.0	92.7	97.7	65.2	87.9	100.1	66.0	89.8	101.3	69.8	91.8	96.3	70.2	88.4	96.3	60.2	85.4
06	WOOD, PLASTICS & COMPOSITES	100.6	71.3	85.1	102.5	51.3	75.4	104.4	51.4	76.4	99.9	71.3	84.8	84.5	57.2	70.0	84.2	37.0	59.2
07	THERMAL & MOISTURE PROTECTION	101.3	69.5	88.5	97.0	58.4	81.5	102.3	59.5	85.1	102.9	66.1	88.1	98.3	53.3	80.2	98.2	38.3	74.1
08	OPENINGS	99.7	69.3	91.8	101.9	55.2	89.8	97.5	55.3	86.6	94.8	69.3	88.2	96.2	55.3	85.6	92.3	43.0	79.6
0920	Plaster & Gypsum Board	96.2	70.4	80.9	90.7	50.0	66.5	96.0	50.0	68.6	99.1	70.4	82.0	84.5	56.7	67.9	84.5	35.9	55.5
0950, 0980	Ceilings & Acoustic Treatment	101.9	70.4	82.9	103.4	50.0	71.3	95.1	50.0	68.0	98.8	70.4	81.7	90.4	56.7	70.1	90.4	35.9	57.6
0960	Flooring	93.6	62.3	85.2	92.6	47.6	80.6	95.1	54.2	84.2	94.9	47.8	82.4	104.7	71.1	95.7	104.6	47.7	89.4
0990	Wall Finishes & Painting/Coating	95.2	56.5	72.1	91.3	45.6	64.0	95.2	45.6	65.6	95.4	56.0	71.8	98.3	50.9	70.0	98.3	55.5	72.7
09	FINISHES	97.1	66.9	81.5	94.7	50.4	71.8	96.9	51.6	73.5	97.1	64.1	80.1	92.7	58.3	74.9	92.7	40.6	65.8
COVERS	DIVS. 10 - 14, 25, 28, 41, 43, 44	100.0	83.9	96.7	100.0	79.8	95.9	100.0	80.3	95.9	100.0	83.9	96.7	100.0	52.8	90.3	100.0	46.9	89.1
22, 23	PLUMBING & HVAC	100.2	77.1	90.8	100.1	71.9	88.6	95.9	72.4	86.3	100.1	68.1	87.1	95.8	44.4	74.9	95.8	36.0	71.5
26, 27, 3370	ELECTRICAL, COMMUNICATIONS & UTIL.	100.0	67.0	84.2	98.3	61.3	80.6	91.9	66.3	79.6	93.6	59.3	77.2	99.1	50.9	76.0	95.9	45.4	71.7
MF2004	WEIGHTED AVERAGE	101.0	73.9	89.3	99.2	64.9	84.4	100.9	65.9	85.8	99.4	70.2	86.8	94.1	56.5	77.9	94.1	46.0	73.3

		ARKANSAS																
		FAYETTEVILLE			FORT SMITH			HARRISON			HOT SPRINGS			JONESBORO			LITTLE ROCK		
DIVISION		727			729			726			719			724			720 - 722		
		MAT.	INST.	TOTAL	MAT.	INST.	TOTAL	MAT.	INST.	TOTAL	MAT.	INST.	TOTAL	MAT.	INST.	TOTAL	MAT.	INST.	TOTAL
01543	CONTRACTOR EQUIPMENT	.0	86.5	86.5	.0	86.5	86.5	.0	86.5	86.5	.0	86.5	86.5	.0	108.2	108.2	.0	86.5	86.5
0241, 31 - 34	SITE & INFRASTRUCTURE, DEMOLITION	73.3	84.2	81.1	78.2	84.2	82.5	78.9	84.1	82.6	78.5	83.6	82.2	101.5	99.5	100.0	77.8	84.2	82.4
0310	Concrete Forming & Accessories	78.8	46.2	50.8	100.5	43.5	51.5	88.6	56.2	60.8	80.4	43.5	48.7	87.3	60.0	63.9	94.7	73.8	76.8
0310	Concrete Reinforcing	95.9	73.5	85.0	96.9	71.2	84.3	95.4	76.2	86.0	95.3	66.5	81.2	92.7	77.2	85.1	97.1	69.5	83.6
0330	Cast-in-Place Concrete	79.2	48.9	67.8	90.6	70.9	83.2	87.9	50.0	73.6	83.1	43.1	68.1	86.3	64.5	78.1	88.6	71.2	82.1
03	CONCRETE	79.8	53.0	67.1	87.9	58.9	74.2	86.8	58.3	73.3	84.5	48.5	67.5	84.4	66.0	75.7	86.6	72.1	79.7
04	MASONRY	89.4	47.4	64.3	95.1	60.5	74.4	99.7	48.6	69.1	80.3	31.8	51.3	90.8	52.2	67.7	93.3	60.5	73.7
05	METALS	96.3	69.4	88.1	98.3	70.3	89.9	97.3	70.2	89.1	96.3	65.9	87.1	92.7	85.4	90.5	94.3	70.3	87.0
06	WOOD, PLASTICS & COMPOSITES	79.2	44.4	60.8	103.5	38.8	69.3	90.5	57.2	72.9	80.6	44.7	61.6	88.1	60.8	73.7	100.2	79.0	89.0
07	THERMAL & MOISTURE PROTECTION	99.5	53.0	80.8	99.9	55.8	82.1	98.7	51.7	79.8	98.4	42.7	76.0	102.6	59.9	85.4	98.7	60.0	83.1
08	OPENINGS	96.2	52.6	84.9	97.0	46.9	84.0	97.0	56.1	86.4	92.3	48.9	81.1	98.8	65.0	90.0	97.0	69.7	89.9
0920	Plaster & Gypsum Board	81.9	43.5	58.3	88.6	37.7	58.3	87.7	56.7	69.2	82.9	43.8	59.6	100.5	59.9	76.3	88.6	79.1	82.9
0950, 0980	Ceilings & Acoustic Treatment	90.4	43.5	62.2	95.5	37.7	60.7	94.6	56.7	71.8	90.4	43.8	62.3	94.3	59.9	73.6	95.5	79.1	85.6
0960	Flooring	101.2	71.1	93.1	113.5	71.8	102.4	107.8	71.1	98.0	103.0	71.1	94.5	75.8	64.3	72.8	115.0	71.8	103.5
0990	Wall Finishes & Painting/Coating	98.3	35.5	60.8	98.3	66.5	79.3	98.3	50.9	70.0	98.3	44.1	65.9	86.9	59.8	70.7	98.3	68.3	80.4
09	FINISHES	91.3	48.7	69.3	97.4	50.2	73.0	95.4	58.3	76.3	92.3	48.3	69.6	90.3	60.5	74.9	97.8	74.0	85.5
COVERS	DIVS. 10 - 14, 25, 28, 41, 43, 44	100.0	64.4	92.7	100.0	71.3	94.1	100.0	66.5	93.1	100.0	48.5	89.4	100.0	60.6	91.9	100.0	76.2	95.1
22, 23	PLUMBING & HVAC	95.9	49.2	76.8	100.1	52.2	80.6	95.8	51.3	77.7	95.8	41.3	73.6	100.3	52.2	80.7	100.1	72.2	88.7
26, 27, 3370	ELECTRICAL, COMMUNICATIONS & UTIL.	93.5	63.8	79.3	96.7	80.5	88.9	97.8	46.2	73.1	97.9	47.8	73.9	105.9	55.8	81.8	96.8	81.8	89.6
MF2004	WEIGHTED AVERAGE	92.8	57.0	77.3	96.7	62.3	81.9	95.5	57.4	79.0	93.2	49.6	74.4	96.4	63.8	82.3	95.8	73.2	86.0

City Cost Indexes

		ARKANSAS													CALIFORNIA					
	DIVISION	PINE BLUFF			RUSSELLVILLE			TEXARKANA			WEST MEMPHIS			ALHAMBRA			ANAHEIM		
		716			728			718			723			917 - 918			928		
		MAT.	INST.	TOTAL	MAT.	INST.	TOTAL	MAT.	INST.	TOTAL	MAT.	INST.	TOTAL	MAT.	INST.	TOTAL	MAT.	INST.	TOTAL
01543	CONTRACTOR EQUIPMENT	.0	86.5	86.5	.0	86.5	86.5	.0	87.2	87.2	.0	108.2	108.2	.0	98.1	98.1	.0	102.4	102.4
0241, 31 - 34	SITE & INFRASTRUCTURE, DEMOLITION	80.4	84.2	83.1	75.3	84.1	81.6	96.5	84.8	88.1	109.7	99.6	102.5	104.3	109.5	108.0	104.2	109.2	107.8
0310	Concrete Forming & Accessories	80.0	73.5	74.4	84.5	58.4	62.0	87.8	49.4	54.8	93.3	59.9	64.6	116.2	118.8	118.4	105.6	118.9	117.0
0310	Concrete Reinforcing	97.1	69.5	83.6	96.5	68.5	82.8	96.6	75.0	86.0	92.7	68.1	80.7	103.1	116.7	109.7	97.4	116.6	106.8
0330	Cast-in-Place Concrete	83.1	71.1	78.6	83.0	52.3	71.5	90.6	49.5	75.2	90.5	64.5	80.7	105.5	114.5	108.9	102.0	117.4	107.8
03	CONCRETE	85.0	71.9	78.9	83.0	58.2	71.3	85.0	54.9	70.8	90.5	64.3	78.1	108.3	115.9	111.9	106.3	117.1	111.4
04	MASONRY	115.2	60.5	82.5	94.4	46.5	65.8	95.1	35.7	59.5	77.6	52.2	62.4	125.2	114.2	118.6	89.5	111.2	102.5
05	METALS	97.0	70.0	88.8	96.3	59.2	85.1	89.7	69.0	83.4	91.8	82.4	89.0	88.4	99.8	91.8	107.5	102.4	106.0
06	WOOD, PLASTICS & COMPOSITES	80.1	79.0	79.5	86.0	60.6	72.6	90.2	52.7	70.4	94.8	60.6	76.8	101.2	118.0	110.1	96.2	118.2	107.9
07	THERMAL & MOISTURE PROTECTION	98.5	60.0	83.0	99.7	51.8	80.4	99.5	48.5	79.0	103.0	59.9	85.7	100.2	111.0	104.5	102.3	113.6	106.8
08	OPENINGS	92.3	69.7	86.5	96.2	53.6	85.1	97.5	55.9	86.7	98.8	63.2	89.6	95.0	116.0	100.4	104.4	116.1	107.4
0920	Plaster & Gypsum Board	82.6	79.1	80.5	84.5	60.1	70.0	87.0	52.1	66.2	102.1	59.9	77.0	107.2	118.6	114.0	102.7	118.6	112.2
0950, 0980	Ceilings & Acoustic Treatment	91.3	79.1	83.9	90.4	60.1	72.2	98.0	52.1	70.4	92.4	59.9	72.9	96.0	118.6	109.6	117.8	118.6	118.3
0960	Flooring	102.8	71.8	94.5	104.3	71.1	95.4	105.5	61.4	93.7	78.6	64.3	74.8	105.5	107.3	105.9	122.3	107.3	118.3
0990	Wall Finishes & Painting/Coating	98.3	68.3	80.4	98.3	43.2	65.4	98.3	35.1	60.5	86.9	59.8	70.7	108.3	109.8	109.2	108.7	109.8	109.4
09	FINISHES	92.5	74.0	82.9	92.8	59.4	75.5	96.1	49.9	72.3	91.6	60.5	75.6	104.8	115.6	110.4	113.7	115.7	114.8
COVERS	DIVS. 10 - 14, 25, 28, 41, 43, 44	100.0	76.2	95.1	100.0	53.2	90.4	100.0	43.1	88.3	100.0	60.6	91.9	100.0	111.9	102.4	100.0	112.4	102.5
22, 23	PLUMBING & HVAC	100.1	55.1	81.8	95.9	55.6	79.5	100.1	38.5	75.0	96.1	52.2	78.2	95.9	106.7	100.3	100.1	107.6	103.2
26, 27, 3370	ELECTRICAL, COMMUNICATIONS & UTIL.	96.1	81.8	89.2	96.7	55.7	77.0	97.8	44.8	72.4	107.3	55.8	82.6	121.9	111.7	117.0	90.2	105.0	97.3
MF2004	WEIGHTED AVERAGE	96.1	69.6	84.6	94.1	58.1	78.5	95.4	50.9	76.2	95.9	63.2	81.8	102.2	111.0	106.0	102.0	110.4	105.6

		CALIFORNIA																	
	DIVISION	BAKERSFIELD			BERKELEY			EUREKA			FRESNO			INGLEWOOD			LONG BEACH		
		932 - 933			947			955			936 - 938			903 - 905			906 - 908		
		MAT.	INST.	TOTAL	MAT.	INST.	TOTAL	MAT.	INST.	TOTAL	MAT.	INST.	TOTAL	MAT.	INST.	TOTAL	MAT.	INST.	TOTAL
01543	CONTRACTOR EQUIPMENT	.0	99.8	99.8	.0	103.0	103.0	.0	99.4	99.4	.0	99.8	99.8	.0	97.0	97.0	.0	97.0	97.0
0241, 31 - 34	SITE & INFRASTRUCTURE, DEMOLITION	109.7	106.4	107.3	144.5	103.8	115.4	120.6	104.2	108.9	110.9	106.1	107.5	97.6	105.5	103.2	105.9	105.5	105.6
0310	Concrete Forming & Accessories	98.6	118.8	115.9	120.5	139.2	136.6	116.1	124.7	123.5	101.8	125.2	121.9	108.3	118.7	117.2	102.4	118.7	116.4
0310	Concrete Reinforcing	109.7	116.5	113.0	101.7	117.4	109.4	106.4	117.3	111.7	93.0	116.1	104.6	120.0	116.3	118.4	118.9	116.7	117.8
0330	Cast-in-Place Concrete	102.0	116.4	107.4	145.0	118.0	134.8	110.7	113.7	111.8	107.5	114.0	109.9	99.6	115.3	105.5	113.5	115.3	114.2
03	CONCRETE	105.6	116.7	110.8	125.2	126.6	125.9	117.6	118.5	118.0	108.2	118.7	113.2	108.1	116.3	112.0	119.1	116.3	117.8
04	MASONRY	111.5	110.5	110.9	154.2	120.6	134.1	117.2	122.4	120.3	113.9	110.5	111.9	84.7	114.3	102.4	94.6	114.3	106.4
05	METALS	102.5	101.9	102.3	104.7	106.9	105.3	107.2	102.3	105.7	107.8	102.0	106.0	101.6	101.0	101.4	101.5	101.0	101.3
06	WOOD, PLASTICS & COMPOSITES	86.7	118.3	103.4	118.4	143.1	131.4	110.0	126.9	118.9	100.2	126.9	114.3	94.5	118.0	106.9	87.6	118.0	103.6
07	THERMAL & MOISTURE PROTECTION	99.9	108.9	103.5	113.7	127.4	119.2	106.3	113.7	109.3	96.1	110.4	101.8	101.0	112.5	105.6	101.2	112.5	105.8
08	OPENINGS	101.8	113.4	104.8	105.5	133.0	112.6	103.1	111.0	105.1	103.6	118.0	107.4	89.3	116.0	96.2	89.3	116.0	96.2
0920	Plaster & Gypsum Board	102.7	118.6	112.2	109.4	143.7	129.8	106.7	127.4	119.1	98.6	127.4	115.8	101.8	118.6	111.8	99.2	118.6	110.8
0950, 0980	Ceilings & Acoustic Treatment	119.7	118.6	119.1	119.4	143.7	134.0	121.7	127.4	125.2	120.6	127.4	124.7	112.0	118.6	116.0	112.0	118.6	116.0
0960	Flooring	117.1	94.3	111.1	123.5	124.8	123.9	122.2	119.0	121.3	129.3	138.1	131.6	112.6	107.3	111.2	108.9	107.3	108.5
0990	Wall Finishes & Painting/Coating	108.8	95.4	100.8	111.5	137.9	127.2	108.2	57.1	77.7	129.5	96.8	109.9	97.5	109.8	104.9	97.5	109.8	104.9
09	FINISHES	114.6	112.0	113.3	120.0	137.7	129.2	118.0	117.7	117.9	119.2	125.6	122.5	107.9	115.6	111.8	106.8	115.6	111.3
COVERS	DIVS. 10 - 14, 25, 28, 41, 43, 44	100.0	120.9	104.3	100.0	125.1	105.2	100.0	121.4	104.4	100.0	121.4	104.4	100.0	111.8	102.4	100.0	111.8	102.4
22, 23	PLUMBING & HVAC	100.2	104.0	101.8	96.1	136.3	112.5	95.9	112.6	102.7	100.3	115.8	106.6	95.9	106.6	100.3	95.9	106.5	100.3
26, 27, 3370	ELECTRICAL, COMMUNICATIONS & UTIL.	89.5	97.9	93.5	103.3	138.9	120.4	98.9	96.2	97.6	89.4	94.4	91.8	98.9	111.7	105.0	98.7	111.7	104.9
MF2004	WEIGHTED AVERAGE	102.1	107.7	104.5	110.0	127.9	117.7	105.7	111.5	108.2	103.9	112.0	107.4	99.0	110.8	104.1	100.8	110.8	105.1

		CALIFORNIA																	
	DIVISION	LOS ANGELES			MARYSVILLE			MODESTO			MOJAVE			OAKLAND			OXNARD		
		900 - 902			959			953			935			946			930		
		MAT.	INST.	TOTAL	MAT.	INST.	TOTAL	MAT.	INST.	TOTAL	MAT.	INST.	TOTAL	MAT.	INST.	TOTAL	MAT.	INST.	TOTAL
01543	CONTRACTOR EQUIPMENT	.0	98.9	98.9	.0	99.4	99.4	.0	99.4	99.4	.0	99.8	99.8	.0	103.0	103.0	.0	98.4	98.4
0241, 31 - 34	SITE & INFRASTRUCTURE, DEMOLITION	105.0	108.2	107.3	116.2	105.3	108.4	109.2	105.5	106.5	103.4	106.4	105.5	152.6	103.8	117.6	111.3	104.1	106.1
0310	Concrete Forming & Accessories	105.2	118.8	116.9	104.2	125.2	122.2	99.4	125.4	121.7	112.2	117.6	116.8	107.2	139.2	134.7	104.2	119.0	116.9
0310	Concrete Reinforcing	120.0	116.9	118.5	106.4	116.7	111.4	103.0	116.7	113.4	111.7	116.5	114.1	104.0	117.4	110.6	109.7	116.5	113.0
0330	Cast-in-Place Concrete	108.5	115.6	111.2	123.7	113.7	119.9	110.7	114.0	111.9	93.0	116.4	101.8	137.4	118.0	130.1	108.3	116.1	111.4
03	CONCRETE	115.3	116.5	115.8	120.0	118.6	119.3	110.1	118.8	114.2	99.8	116.1	107.5	124.0	126.6	125.3	109.1	116.9	112.8
04	MASONRY	100.2	116.7	110.1	118.2	110.3	113.5	115.4	110.5	112.5	111.9	110.5	111.1	162.8	120.6	137.6	115.9	109.8	112.3
05	METALS	108.6	101.3	106.4	106.7	101.6	105.2	103.8	102.3	103.4	104.0	101.9	103.4	99.8	106.9	101.9	102.2	102.1	102.2
06	WOOD, PLASTICS & COMPOSITES	90.6	118.0	105.1	95.6	126.9	112.1	90.5	126.9	109.7	101.5	116.9	109.5	103.1	143.1	124.2	95.1	118.3	107.4
07	THERMAL & MOISTURE PROTECTION	101.2	114.6	106.6	105.8	113.6	108.9	105.4	113.6	108.7	100.4	109.1	103.9	110.8	127.4	117.5	105.5	111.4	107.8
08	OPENINGS	95.2	116.0	100.6	102.3	120.1	106.9	101.0	120.1	106.0	97.5	112.6	101.4	105.6	133.0	112.7	100.7	116.2	104.7
0920	Plaster & Gypsum Board	102.8	118.6	112.2	100.9	127.4	116.7	103.0	127.4	117.5	109.9	117.0	114.1	103.7	143.7	127.5	102.7	118.6	112.2
0950, 0980	Ceilings & Acoustic Treatment	123.8	118.6	120.7	120.0	127.4	124.5	115.0	127.4	122.5	120.0	117.0	118.2	122.9	143.7	135.4	119.7	118.6	119.1
0960	Flooring	110.5	107.3	109.7	116.5	117.7	116.8	117.1	113.1	116.1	124.7	94.3	116.6	117.6	124.8	119.5	115.6	107.3	113.4
0990	Wall Finishes & Painting/Coating	97.5	109.8	104.9	108.2	112.6	110.8	108.2	107.3	107.7	108.2	96.7	101.3	111.5	137.9	127.2	108.2	103.7	105.5
09	FINISHES	110.5	115.6	113.1	114.5	123.7	119.3	113.3	122.4	118.0	117.7	111.1	114.3	118.7	137.7	128.6	113.9	115.1	114.5
COVERS	DIVS. 10 - 14, 25, 28, 41, 43, 44	100.0	111.8	102.4	100.0	121.5	104.4	100.0	121.5	104.4	100.0	120.8	104.3	100.0	125.1	105.2	100.0	112.7	102.6
22, 23	PLUMBING & HVAC	100.1	107.5	103.1	95.9	108.2	100.9	100.2	115.2	106.5	96.0	104.0	99.3	100.4	136.3	115.0	100.2	107.6	103.2
26, 27, 3370	ELECTRICAL, COMMUNICATIONS & UTIL.	96.3	113.8	104.7	95.7	115.0	105.0	98.2	105.6	101.8	87.9	97.7	92.7	102.6	138.9	120.0	95.7	106.3	100.8
MF2004	WEIGHTED AVERAGE	103.3	111.9	107.0	104.9	113.1	108.5	104.0	113.4	108.1	100.4	107.7	103.4	110.3	127.9	117.9	103.5	109.8	106.2

City Cost Indexes

CALIFORNIA

DIVISION		PALM SPRINGS 922			PALO ALTO 943			PASADENA 910 - 912			REDDING 960			RICHMOND 948			RIVERSIDE 925		
		MAT.	INST.	TOTAL	MAT.	INST.	TOTAL	MAT.	INST.	TOTAL	MAT.	INST.	TOTAL	MAT.	INST.	TOTAL	MAT.	INST.	TOTAL
01543	CONTRACTOR EQUIPMENT	.0	100.9	100.9	.0	103.0	103.0	.0	98.1	98.1	.0	99.4	99.4	.0	103.0	103.0	.0	100.9	100.9
0241, 31 - 34	SITE & INFRASTRUCTURE, DEMOLITION	94.9	106.9	103.5	138.9	103.8	113.8	100.7	109.5	107.0	115.9	105.4	108.4	152.2	103.8	117.5	102.3	106.9	105.6
0310	Concrete Forming & Accessories	101.5	119.0	116.5	104.9	139.1	134.3	103.5	118.8	116.6	103.9	125.2	122.2	123.8	139.1	137.0	105.9	118.9	117.1
0310	Concrete Reinforcing	111.9	116.6	114.2	101.7	117.4	109.4	103.9	116.7	110.2	106.4	116.7	111.4	101.7	117.4	109.4	108.7	116.6	112.5
0330	Cast-in-Place Concrete	97.2	117.4	104.8	122.5	118.0	120.8	100.1	114.5	105.5	122.7	113.9	119.4	141.0	118.0	132.3	105.8	117.4	110.1
03	CONCRETE	101.3	117.2	108.8	111.8	126.6	118.8	103.4	115.9	109.3	118.8	118.7	118.8	126.5	126.6	126.6	107.8	117.1	112.2
04	MASONRY	87.1	110.8	101.3	127.8	120.6	123.5	108.1	114.2	111.7	118.2	110.5	113.6	153.9	120.6	134.0	88.1	110.8	101.7
05	METALS	108.3	102.5	106.6	97.5	106.8	100.3	88.4	99.8	91.8	107.1	102.0	105.5	97.6	106.7	100.3	107.8	102.4	106.2
06	WOOD, PLASTICS & COMPOSITES	90.5	118.2	105.1	100.1	143.1	122.8	86.8	118.0	103.3	95.2	126.9	112.0	122.7	143.1	133.5	96.2	118.2	107.9
07	THERMAL & MOISTURE PROTECTION	102.5	113.1	106.7	110.4	128.8	117.8	99.9	111.0	104.3	105.9	112.2	108.5	111.1	127.4	117.6	102.6	113.5	107.0
08	OPENINGS	99.7	116.1	104.0	105.6	133.0	112.7	95.0	116.0	100.4	103.3	120.1	107.7	105.6	133.0	112.7	102.8	116.1	106.3
0920	Plaster & Gypsum Board	98.9	118.6	110.6	102.4	143.7	127.0	103.1	118.6	112.3	101.3	127.4	116.8	110.7	143.7	130.3	102.1	118.6	111.9
0950, 0980	Ceilings & Acoustic Treatment	112.5	118.6	116.2	121.1	143.7	134.6	96.0	118.6	109.6	126.4	127.4	127.0	121.1	143.7	134.6	117.2	118.6	118.1
0960	Flooring	119.2	107.3	116.1	116.6	121.6	117.9	99.5	107.3	101.6	116.4	117.7	116.7	125.6	124.8	125.4	121.5	107.3	117.7
0990	Wall Finishes & Painting/Coating	105.5	137.5	124.6	111.5	137.9	127.2	108.3	109.8	109.2	108.2	112.6	110.8	111.5	137.9	127.2	105.5	109.8	108.1
09	FINISHES	110.1	118.9	114.6	117.0	137.2	127.4	102.2	115.6	109.1	116.0	123.8	120.0	121.9	137.7	130.1	112.8	115.7	114.3
COVERS	DIVS. 10 - 14, 25, 28, 41, 43, 44	100.0	112.4	102.5	100.0	125.1	105.2	100.0	111.9	102.4	100.0	121.5	104.4	100.0	125.1	105.2	100.0	112.4	102.5
22, 23	PLUMBING & HVAC	95.9	106.7	100.3	96.1	135.3	112.1	95.9	106.7	100.3	100.2	108.3	103.5	96.1	132.3	110.9	100.1	107.6	103.1
26, 27, 3370	ELECTRICAL, COMMUNICATIONS & UTIL.	93.6	102.9	98.0	102.5	137.8	119.4	118.9	111.7	115.5	98.4	104.8	101.5	102.9	132.9	117.3	90.4	102.9	96.4
MF2004	WEIGHTED AVERAGE	99.8	110.1	104.3	105.2	127.4	114.8	99.9	111.0	104.7	106.4	111.8	108.7	109.3	126.1	116.6	101.9	109.8	105.3

CALIFORNIA

DIVISION		SACRAMENTO 942,956 - 958			SALINAS 939			SAN BERNARDINO 923 - 924			SAN DIEGO 919 - 921			SAN FRANCISCO 940 - 941			SAN JOSE 951		
		MAT.	INST.	TOTAL	MAT.	INST.	TOTAL	MAT.	INST.	TOTAL	MAT.	INST.	TOTAL	MAT.	INST.	TOTAL	MAT.	INST.	TOTAL
01543	CONTRACTOR EQUIPMENT	.0	102.6	102.6	.0	99.8	99.8	.0	100.9	100.9	.0	98.1	98.1	.0	108.3	108.3	.0	100.2	100.2
0241, 31 - 34	SITE & INFRASTRUCTURE, DEMOLITION	117.3	110.1	112.2	127.6	106.2	112.2	78.8	106.9	99.0	106.4	102.1	103.3	155.3	110.2	123.0	150.9	99.7	114.2
0310	Concrete Forming & Accessories	105.8	125.6	122.8	108.1	128.8	125.9	110.1	118.6	117.6	106.1	109.9	109.3	107.6	140.0	135.4	106.5	139.0	134.4
0310	Concrete Reinforcing	96.4	116.7	106.3	110.3	117.2	113.6	108.7	116.5	112.5	100.7	116.5	108.4	118.5	117.9	118.2	97.0	117.4	107.0
0330	Cast-in-Place Concrete	116.1	114.7	115.6	107.0	114.3	109.7	73.0	117.3	89.7	109.8	107.3	108.8	140.9	119.6	132.9	128.7	117.5	124.5
03	CONCRETE	111.9	119.1	115.3	118.6	120.5	119.5	80.6	117.1	97.8	109.4	109.6	109.5	128.1	127.6	127.9	117.6	126.4	121.8
04	MASONRY	129.8	110.5	118.3	111.5	117.2	114.9	95.2	110.8	104.5	99.1	109.3	105.2	163.4	126.8	141.5	149.5	120.7	132.3
05	METALS	95.5	101.5	97.3	106.9	103.3	105.8	107.8	102.2	106.1	105.0	101.5	104.0	105.5	108.8	106.5	101.6	108.3	103.6
06	WOOD, PLASTICS & COMPOSITES	97.9	127.1	113.3	100.4	130.8	116.4	100.7	118.2	110.0	100.5	107.1	104.0	103.1	143.3	124.3	102.0	142.8	123.6
07	THERMAL & MOISTURE PROTECTION	113.0	113.4	113.2	101.3	120.2	108.9	101.7	113.7	106.5	109.2	104.2	107.2	114.0	131.3	121.0	101.9	129.2	112.9
08	OPENINGS	118.9	120.2	119.2	102.1	126.4	108.4	99.8	116.1	104.0	103.3	108.7	104.7	110.2	133.1	116.1	93.9	132.9	104.0
0920	Plaster & Gypsum Board	99.5	127.4	116.1	103.8	131.4	120.2	103.6	118.6	112.6	107.8	107.1	107.4	106.2	143.7	128.5	98.3	143.7	125.3
0950, 0980	Ceilings & Acoustic Treatment	122.0	127.4	125.3	120.0	131.4	126.9	115.0	118.6	117.2	102.7	107.1	105.3	132.0	143.7	139.0	110.6	143.7	130.5
0960	Flooring	121.1	117.7	120.2	118.1	124.8	119.9	123.9	101.0	117.8	108.3	107.3	108.1	117.6	125.9	119.8	110.6	125.9	114.7
0990	Wall Finishes & Painting/Coating	108.8	112.6	111.0	109.2	137.9	126.3	105.5	109.8	108.1	105.4	109.8	108.1	111.5	149.7	134.3	110.1	137.9	126.7
09	FINISHES	116.9	123.9	120.5	116.6	130.1	123.6	111.7	114.5	113.2	107.9	109.0	108.5	121.5	139.6	130.8	111.4	137.8	125.0
COVERS	DIVS. 10 - 14, 25, 28, 41, 43, 44	100.0	122.1	104.5	100.0	122.0	104.5	100.0	108.1	101.7	100.0	111.0	102.3	100.0	125.7	105.3	100.0	124.5	105.0
22, 23	PLUMBING & HVAC	100.2	110.0	104.2	96.0	115.8	104.1	95.9	106.7	100.3	100.2	106.1	102.6	100.4	147.8	119.7	100.2	135.2	114.5
26, 27, 3370	ELECTRICAL, COMMUNICATIONS & UTIL.	97.3	104.8	100.9	89.4	120.7	104.4	93.6	105.5	99.3	100.0	97.8	99.0	102.7	151.3	126.0	102.4	137.8	119.4
MF2004	WEIGHTED AVERAGE	106.0	112.6	108.9	104.1	118.1	110.1	97.6	109.7	102.8	103.4	105.6	104.3	112.6	133.8	121.8	107.1	127.2	115.8

CALIFORNIA

DIVISION		SAN LUIS OBISPO 934			SAN MATEO 944			SAN RAFAEL 949			SANTA ANA 926 - 927			SANTA BARBARA 931			SANTA CRUZ 950		
		MAT.	INST.	TOTAL	MAT.	INST.	TOTAL	MAT.	INST.	TOTAL	MAT.	INST.	TOTAL	MAT.	INST.	TOTAL	MAT.	INST.	TOTAL
01543	CONTRACTOR EQUIPMENT	.0	99.8	99.8	.0	103.0	103.0	.0	103.1	103.1	.0	100.9	100.9	.0	99.8	99.8	.0	100.2	100.2
0241, 31 - 34	SITE & INFRASTRUCTURE, DEMOLITION	118.1	106.4	109.7	148.8	103.8	116.6	133.5	110.0	116.7	93.0	106.9	103.0	111.2	106.4	107.7	150.3	99.6	113.9
0310	Concrete Forming & Accessories	114.5	118.8	118.2	112.1	139.3	135.9	118.0	138.9	135.9	110.9	118.9	117.7	104.9	118.9	116.9	106.5	129.0	125.8
0310	Concrete Reinforcing	111.7	116.5	114.1	101.7	117.6	109.5	102.4	117.5	109.8	112.4	116.6	114.4	109.7	116.5	113.0	120.1	117.2	118.7
0330	Cast-in-Place Concrete	115.0	116.4	115.5	136.5	118.1	129.6	158.9	116.6	143.0	93.3	117.4	102.4	107.9	116.4	111.1	127.8	116.4	123.5
03	CONCRETE	117.4	116.7	117.1	122.4	126.7	124.4	146.2	125.8	136.6	98.8	117.1	107.4	109.0	116.8	112.7	120.8	121.6	121.2
04	MASONRY	113.6	110.4	111.7	153.5	123.9	135.8	125.7	123.8	124.6	83.4	111.2	100.0	112.0	110.4	111.0	154.2	117.3	132.1
05	METALS	104.7	101.9	103.8	97.4	107.2	100.4	98.6	103.0	100.0	107.9	102.4	106.2	102.7	102.0	102.5	109.6	107.4	108.9
06	WOOD, PLASTICS & COMPOSITES	104.0	118.3	111.6	108.9	143.1	127.0	109.4	142.8	127.0	102.5	118.2	110.8	95.1	118.3	107.4	102.0	130.9	117.3
07	THERMAL & MOISTURE PROTECTION	101.3	109.9	104.7	110.8	130.1	118.5	116.2	126.5	120.3	102.8	113.2	107.0	100.7	110.7	104.7	101.7	121.8	109.8
08	OPENINGS	99.8	113.4	103.4	105.6	133.0	112.7	106.6	132.8	112.4	98.9	116.1	103.4	101.8	116.2	105.5	95.1	126.4	103.2
0920	Plaster & Gypsum Board	110.9	118.6	115.5	105.9	143.7	128.4	108.8	143.7	129.5	103.9	118.6	112.7	102.7	118.6	112.2	107.4	131.4	121.7
0950, 0980	Ceilings & Acoustic Treatment	120.0	118.6	119.2	121.1	143.7	134.6	130.3	143.7	138.3	115.0	118.6	117.2	119.7	118.6	119.1	116.7	131.4	125.5
0960	Flooring	126.1	94.3	117.6	119.8	124.8	121.1	131.2	121.6	128.6	124.6	107.3	120.0	117.1	94.3	111.1	113.8	124.8	116.8
0990	Wall Finishes & Painting/Coating	108.2	103.7	105.5	111.5	137.9	127.2	107.5	137.9	125.6	105.5	109.8	108.1	108.2	103.7	105.5	109.4	137.9	126.4
09	FINISHES	119.2	112.4	115.7	119.1	137.7	128.7	122.6	137.0	130.0	113.1	115.7	114.5	114.6	112.9	113.7	115.3	130.2	123.0
COVERS	DIVS. 10 - 14, 25, 28, 41, 43, 44	100.0	121.0	104.3	100.0	125.1	105.2	100.0	124.3	105.0	100.0	112.4	102.5	100.0	112.7	102.6	100.0	122.4	104.6
22, 23	PLUMBING & HVAC	96.0	106.7	100.3	96.1	135.6	112.2	96.1	147.6	117.1	95.9	106.7	100.3	100.2	107.6	103.2	100.2	115.9	106.6
26, 27, 3370	ELECTRICAL, COMMUNICATIONS & UTIL.	88.0	99.3	93.4	102.5	141.3	121.1	96.8	114.9	105.5	93.6	105.0	99.1	87.0	107.2	96.7	101.5	120.7	110.7
MF2004	WEIGHTED AVERAGE	103.4	108.6	105.6	108.3	128.5	117.0	110.2	127.1	117.5	99.6	110.0	104.1	102.4	109.8	105.6	109.2	118.1	113.1

City Cost Indexes

DIVISION		CALIFORNIA															COLORADO		
		SANTA ROSA			STOCKTON			SUSANVILLE			VALLEJO			VAN NUYS			ALAMOSA		
		954			952			961			945			913 - 916			811		
		MAT.	INST.	TOTAL	MAT.	INST.	TOTAL	MAT.	INST.	TOTAL	MAT.	INST.	TOTAL	MAT.	INST.	TOTAL	MAT.	INST.	TOTAL
01543	CONTRACTOR EQUIPMENT	.0	100.1	100.1	.0	99.4	99.4	.0	99.4	99.4	.0	103.1	103.1	.0	98.1	98.1	.0	96.2	96.2
0241, 31 - 34	SITE & INFRASTRUCTURE, DEMOLITION	109.4	105.5	106.6	108.9	105.5	106.4	124.2	105.4	110.7	115.6	110.2	111.7	120.5	109.5	112.6	127.9	93.2	103.0
0310	Concrete Forming & Accessories	102.7	138.3	133.3	104.6	125.4	122.4	105.6	121.6	119.4	106.6	138.1	133.7	110.9	118.8	117.7	102.3	76.9	80.5
0310	Concrete Reinforcing	107.4	117.7	112.4	110.3	116.7	113.4	106.4	116.7	111.4	103.6	117.5	110.4	103.9	116.7	110.2	113.8	81.4	98.0
0330	Cast-in-Place Concrete	121.6	115.5	119.3	107.8	114.0	110.2	111.6	113.9	112.5	126.4	116.0	122.5	105.6	114.5	108.9	107.0	83.6	98.2
03	CONCRETE	119.5	125.3	122.2	100.0	118.8	113.7	119.4	117.1	118.3	118.4	125.3	121.6	117.2	115.9	116.6	123.1	80.2	102.8
04	MASONRY	115.9	125.4	121.5	115.5	110.5	112.5	117.2	110.5	113.2	92.2	122.6	110.3	125.2	114.2	118.6	133.0	77.2	99.6
05	METALS	108.2	104.9	107.2	104.1	102.3	103.5	106.1	102.0	104.8	98.6	103.6	100.1	87.5	99.8	91.2	99.8	83.0	94.7
06	WOOD, PLASTICS & COMPOSITES	91.8	142.6	118.7	97.0	126.9	112.8	97.4	122.0	110.4	96.3	142.8	120.9	95.6	118.0	107.5	99.3	79.1	88.6
07	THERMAL & MOISTURE PROTECTION	103.2	127.0	112.8	105.5	113.4	108.7	106.6	108.8	107.5	113.6	125.9	118.6	100.9	111.0	104.9	100.8	81.9	93.2
08	OPENINGS	100.7	132.7	109.0	101.1	120.1	106.0	103.2	117.4	106.9	118.6	132.8	122.3	94.8	116.0	100.3	93.0	83.9	90.7
0920	Plaster & Gypsum Board	99.8	143.7	125.9	103.0	127.4	117.5	101.9	122.4	114.1	104.0	143.7	127.6	105.7	118.6	113.4	85.0	78.2	81.0
0950, 0980	Ceilings & Acoustic Treatment	115.0	143.7	132.2	119.7	127.4	124.3	120.0	122.4	121.5	132.2	143.7	139.1	93.4	118.6	108.6	100.9	78.2	87.2
0960	Flooring	119.9	125.9	121.5	117.1	113.1	116.1	117.1	117.7	117.3	125.7	125.9	125.8	103.0	107.3	104.1	110.1	66.0	98.3
0990	Wall Finishes & Painting/Coating	105.5	137.9	124.8	108.2	113.0	111.1	108.2	127.1	119.5	108.3	137.9	126.0	108.3	109.8	109.2	111.6	29.2	62.3
09	FINISHES	111.9	137.4	125.1	114.5	123.0	118.9	115.6	122.6	119.2	119.1	137.5	128.6	104.2	115.6	110.1	102.7	69.7	85.7
COVERS	DIVS. 10 - 14, 25, 28, 41, 43, 44	100.0	123.2	104.8	100.0	121.6	104.4	100.0	120.9	104.3	100.0	123.8	104.9	100.0	111.9	102.4	100.0	89.7	97.9
22, 23	PLUMBING & HVAC	95.9	146.5	116.5	100.2	108.4	103.5	96.0	108.3	101.0	100.3	123.4	109.7	95.9	106.7	100.3	95.7	70.0	85.3
26, 27, 3370	ELECTRICAL, COMMUNICATIONS & UTIL.	93.9	115.0	104.0	98.2	119.4	108.4	98.7	104.8	101.7	93.2	124.5	108.2	118.9	111.7	115.5	92.0	79.2	85.8
MF2004	WEIGHTED AVERAGE	104.1	126.7	113.9	104.1	113.9	108.3	105.5	111.1	107.9	105.0	123.3	112.9	103.0	111.1	106.5	102.8	78.2	92.1

DIVISION		COLORADO																	
		BOULDER			COLORADO SPRINGS			DENVER			DURANGO			FORT COLLINS			FORT MORGAN		
		803			808 - 809			800 - 802			813			805			807		
		MAT.	INST.	TOTAL	MAT.	INST.	TOTAL	MAT.	INST.	TOTAL	MAT.	INST.	TOTAL	MAT.	INST.	TOTAL	MAT.	INST.	TOTAL
01543	CONTRACTOR EQUIPMENT	.0	95.9	95.9	.0	94.3	94.3	.0	99.5	99.5	.0	96.2	96.2	.0	95.9	95.9	.0	95.9	95.9
0241, 31 - 34	SITE & INFRASTRUCTURE, DEMOLITION	91.5	97.1	95.5	93.0	95.1	94.5	92.1	104.3	100.9	122.1	93.2	101.4	102.6	97.6	99.0	93.6	97.7	96.5
0310	Concrete Forming & Accessories	102.7	81.8	84.8	92.2	82.5	83.8	100.7	83.4	85.8	108.6	77.1	81.6	100.1	80.8	83.2	103.1	81.7	84.7
0310	Concrete Reinforcing	106.8	81.4	94.4	105.9	85.5	95.9	105.9	85.6	96.0	113.8	81.4	98.0	106.9	80.4	93.9	107.1	81.4	94.5
0330	Cast-in-Place Concrete	98.7	82.5	92.6	101.4	85.3	95.3	95.9	86.0	92.2	123.1	83.7	108.3	111.4	81.0	100.0	96.8	82.4	91.4
03	CONCRETE	105.1	82.0	94.2	108.4	84.1	96.9	102.9	84.8	94.4	126.4	80.3	104.6	115.7	80.7	99.2	103.7	81.9	93.4
04	MASONRY	97.6	77.4	85.5	101.3	77.7	87.2	103.4	81.8	90.5	119.9	77.2	94.3	113.7	56.7	79.6	113.7	77.4	92.0
05	METALS	96.8	82.9	92.6	99.6	87.8	96.1	102.0	88.1	97.8	99.8	83.2	94.8	97.9	81.4	92.9	96.6	82.9	92.4
06	WOOD, PLASTICS & COMPOSITES	102.2	85.0	93.1	91.5	85.2	88.1	100.7	85.0	92.4	107.8	79.1	92.6	100.4	85.0	92.3	102.7	85.0	93.3
07	THERMAL & MOISTURE PROTECTION	101.7	79.6	92.8	102.4	82.8	94.5	101.7	79.8	92.9	100.7	81.9	93.1	102.2	68.6	88.7	101.7	78.9	92.5
08	OPENINGS	95.1	87.1	93.0	99.3	88.4	96.5	99.1	88.3	96.3	100.3	83.9	96.1	95.0	87.1	93.0	95.0	87.1	93.0
0920	Plaster & Gypsum Board	101.6	84.8	91.6	87.5	84.8	85.9	99.7	84.8	90.9	93.0	78.2	84.2	98.4	84.8	90.3	102.6	84.8	92.0
0950, 0980	Ceilings & Acoustic Treatment	95.2	84.8	89.0	102.9	84.8	92.0	101.1	84.8	91.3	100.9	78.2	87.2	95.2	84.8	89.0	95.2	84.8	89.0
0960	Flooring	114.0	93.1	108.4	104.9	82.2	98.8	109.9	93.1	105.4	115.3	66.0	102.1	109.8	66.0	98.1	114.5	66.0	101.6
0990	Wall Finishes & Painting/Coating	108.5	72.7	87.1	108.2	50.1	73.5	108.5	75.9	89.0	111.6	29.2	62.3	108.5	49.7	73.3	108.5	66.2	83.2
09	FINISHES	101.3	82.8	91.7	98.1	79.0	88.2	101.0	84.5	92.5	104.9	69.7	86.7	99.9	74.8	86.9	101.5	77.3	89.0
COVERS	DIVS. 10 - 14, 25, 28, 41, 43, 44	100.0	89.1	97.8	100.0	89.2	97.8	100.0	89.9	97.9	100.0	89.7	97.9	100.0	88.2	97.6	100.0	89.2	97.8
22, 23	PLUMBING & HVAC	95.7	84.0	90.9	100.1	78.1	91.1	100.0	85.3	94.0	95.7	76.2	87.8	100.0	82.6	92.9	95.7	82.8	90.4
26, 27, 3370	ELECTRICAL, COMMUNICATIONS & UTIL.	96.1	90.9	93.6	99.2	84.8	92.3	100.8	90.9	96.1	91.5	76.8	84.5	96.1	90.9	93.6	96.4	90.9	93.7
MF2004	WEIGHTED AVERAGE	98.0	85.1	92.4	100.5	83.3	93.1	100.8	87.6	95.1	103.2	79.2	92.8	101.4	81.1	92.6	98.8	84.2	92.5

DIVISION		COLORADO																	
		GLENWOOD SPRINGS			GOLDEN			GRAND JUNCTION			GREELEY			MONTROSE			PUEBLO		
		816			804			815			806			814			810		
		MAT.	INST.	TOTAL	MAT.	INST.	TOTAL	MAT.	INST.	TOTAL	MAT.	INST.	TOTAL	MAT.	INST.	TOTAL	MAT.	INST.	TOTAL
01543	CONTRACTOR EQUIPMENT	.0	99.4	99.4	.0	95.9	95.9	.0	99.4	99.4	.0	95.9	95.9	.0	97.8	97.8	.0	96.2	96.2
0241, 31 - 34	SITE & INFRASTRUCTURE, DEMOLITION	136.4	101.0	111.0	103.7	98.0	99.6	121.3	100.5	106.4	90.2	96.3	94.6	130.2	96.7	106.2	113.3	94.8	100.1
0310	Concrete Forming & Accessories	99.0	81.5	84.0	95.1	81.6	83.5	107.4	80.8	84.6	97.8	50.3	57.0	98.5	81.4	83.8	104.4	82.7	85.8
0310	Concrete Reinforcing	112.7	81.4	97.4	107.1	81.4	94.5	113.0	80.3	97.0	106.8	79.8	93.6	112.5	81.3	97.3	108.9	85.5	97.5
0330	Cast-in-Place Concrete	107.0	82.9	97.9	96.9	82.5	91.5	118.5	81.9	104.7	93.1	57.4	79.7	107.0	82.8	97.9	106.2	86.5	98.8
03	CONCRETE	128.3	82.0	106.4	113.8	81.9	98.7	122.4	81.1	102.9	100.3	59.2	80.9	119.1	81.9	101.5	111.6	84.7	98.9
04	MASONRY	106.9	77.4	89.3	117.1	77.5	93.4	141.5	58.9	92.1	108.0	37.5	65.8	112.3	77.2	91.3	103.2	77.7	87.9
05	METALS	99.5	83.7	94.7	96.7	83.1	92.6	100.9	80.5	94.8	97.9	79.7	92.4	98.6	82.6	93.7	102.4	88.9	98.3
06	WOOD, PLASTICS & COMPOSITES	94.2	85.2	89.4	95.0	85.0	89.7	105.0	85.2	94.5	97.6	50.2	72.6	94.4	85.3	89.6	101.8	85.5	93.2
07	THERMAL & MOISTURE PROTECTION	100.6	78.2	91.6	102.6	76.3	92.0	99.6	65.0	85.7	101.6	57.5	83.8	100.8	82.6	93.4	99.2	82.6	92.5
08	OPENINGS	99.5	87.2	96.3	95.0	87.1	93.0	100.2	87.2	96.9	95.0	68.3	88.1	100.5	87.3	97.1	94.8	88.6	93.2
0920	Plaster & Gypsum Board	104.7	84.8	92.9	96.8	84.8	89.7	111.8	84.8	95.7	97.5	49.1	68.7	82.4	84.8	83.9	88.5	84.8	86.3
0950, 0980	Ceilings & Acoustic Treatment	99.2	84.8	90.6	95.2	84.8	89.0	99.2	84.8	90.6	95.2	49.1	67.5	100.9	84.8	91.2	111.0	84.8	95.3
0960	Flooring	109.1	60.7	96.2	107.2	66.0	96.2	114.7	66.0	101.7	108.4	66.0	97.1	112.5	54.6	97.1	111.0	93.1	106.2
0990	Wall Finishes & Painting/Coating	111.6	66.2	84.4	108.5	72.7	87.1	111.6	72.7	88.3	108.5	31.0	62.1	111.6	29.2	62.3	111.6	46.4	72.6
09	FINISHES	105.6	76.5	90.6	99.5	78.0	88.4	107.0	78.1	92.1	98.7	49.8	73.5	103.0	71.0	86.5	104.5	81.0	92.4
COVERS	DIVS. 10 - 14, 25, 28, 41, 43, 44	100.0	89.5	97.8	100.0	89.2	97.8	100.0	89.4	97.8	100.0	80.2	95.9	100.0	89.9	97.9	100.0	90.5	98.0
22, 23	PLUMBING & HVAC	95.7	76.0	87.7	95.7	83.8	90.8	99.9	69.4	87.5	100.0	77.3	90.7	95.7	76.1	87.7	99.9	70.3	87.9
26, 27, 3370	ELECTRICAL, COMMUNICATIONS & UTIL.	89.1	78.4	84.0	96.4	90.9	93.7	91.2	61.2	76.8	96.1	90.8	93.6	91.2	61.2	76.8	92.0	79.2	85.9
MF2004	WEIGHTED AVERAGE	102.7	81.3	93.4	100.2	84.4	93.4	105.1	74.9	92.1	98.9	69.9	86.4	101.7	77.8	91.4	101.2	81.3	92.6

649

City Cost Indexes

		COLORADO			CONNECTICUT													
		SALIDA			BRIDGEPORT			BRISTOL			HARTFORD			MERIDEN			NEW BRITAIN		
DIVISION		812			066			060			061			064			060		
		MAT.	INST.	TOTAL	MAT.	INST.	TOTAL	MAT.	INST.	TOTAL	MAT.	INST.	TOTAL	MAT.	INST.	TOTAL	MAT.	INST.	TOTAL
01543	CONTRACTOR EQUIPMENT	.0	97.8	97.8	.0	100.7	100.7	.0	100.7	100.7	.0	100.7	100.7	.0	101.3	101.3	.0	100.7	100.7
0241, 31 - 34	SITE & INFRASTRUCTURE, DEMOLITION	121.7	97.0	104.0	96.8	104.2	102.1	95.9	104.2	101.8	95.4	104.2	101.7	94.1	105.0	101.9	96.1	104.2	101.9
0310	Concrete Forming & Accessories	107.3	81.4	85.1	99.4	120.2	117.3	99.4	119.8	116.9	99.6	119.8	116.9	99.2	120.0	117.1	99.7	119.8	116.9
0310	Concrete Reinforcing	112.3	81.3	97.1	104.3	125.7	114.7	104.3	125.6	114.7	104.3	125.6	114.7	104.3	125.6	114.7	104.3	125.6	114.7
0330	Cast-in-Place Concrete	122.6	82.8	107.7	104.8	114.7	108.5	98.1	114.5	104.3	92.4	114.5	100.7	94.4	114.6	102.0	99.7	114.5	105.3
03	CONCRETE	121.3	81.9	102.7	109.1	119.2	113.9	105.7	118.9	112.0	102.9	118.9	110.4	103.8	119.1	111.0	106.4	118.9	112.4
04	MASONRY	143.0	77.2	103.6	107.1	127.1	119.1	99.2	127.1	115.9	98.7	127.1	115.7	98.9	127.1	115.8	101.4	127.1	116.7
05	METALS	98.3	82.6	93.6	96.9	124.1	105.2	96.9	123.7	105.0	101.7	123.7	108.3	94.4	124.0	103.3	93.6	123.7	102.7
06	WOOD, PLASTICS & COMPOSITES	103.4	85.3	93.9	99.1	118.5	109.4	99.1	118.5	109.4	100.7	118.5	110.1	99.1	118.5	109.4	99.1	118.5	109.4
07	THERMAL & MOISTURE PROTECTION	99.7	82.6	92.8	98.3	124.0	108.6	98.4	120.0	107.1	99.7	120.0	107.9	98.4	120.0	107.1	98.4	120.9	107.5
08	OPENINGS	93.1	87.3	91.6	105.6	128.2	111.5	105.6	118.5	109.0	106.3	118.5	109.4	108.3	128.2	113.5	105.6	118.5	109.0
0920	Plaster & Gypsum Board	84.7	84.8	84.8	102.1	118.2	111.7	102.1	118.2	111.7	102.1	118.2	111.7	103.6	118.2	112.3	102.1	118.2	111.7
0950, 0980	Ceilings & Acoustic Treatment	100.9	84.8	91.2	96.2	118.2	109.4	96.2	118.2	109.4	96.2	118.2	109.4	101.0	118.2	111.4	96.2	118.2	109.4
0960	Flooring	118.2	54.6	101.2	98.0	120.2	103.9	98.0	120.2	103.9	98.0	120.2	103.9	98.0	120.2	103.9	98.0	120.2	103.9
0990	Wall Finishes & Painting/Coating	111.6	29.2	62.3	91.0	114.8	105.3	91.0	114.8	105.3	91.0	114.8	105.3	91.0	118.0	107.1	91.0	114.8	105.3
09	FINISHES	104.2	71.0	87.0	100.0	119.5	110.1	100.0	119.5	110.1	100.0	119.5	110.1	101.4	119.9	111.0	100.0	119.5	110.1
COVERS	DIVS. 10 - 14, 25, 28, 41, 43, 44	100.0	90.0	97.9	100.0	110.4	102.1	100.0	110.4	102.1	100.0	110.4	102.1	100.0	110.4	102.1	100.0	110.4	102.1
22, 23	PLUMBING & HVAC	95.7	70.0	85.2	100.0	111.9	104.8	100.0	111.8	104.8	100.0	111.8	104.8	95.8	111.9	102.3	100.0	111.8	104.8
26, 27, 3370	ELECTRICAL, COMMUNICATIONS & UTIL.	91.4	79.2	85.5	98.1	109.0	103.3	98.1	108.7	103.2	98.4	109.4	103.7	98.1	108.7	103.2	98.2	108.7	103.3
MF2004	WEIGHTED AVERAGE	102.8	79.1	92.5	101.1	116.5	107.7	100.3	115.8	107.0	100.8	115.9	107.3	99.0	116.4	106.5	100.0	115.9	106.8

		CONNECTICUT																
		NEW HAVEN			NEW LONDON			NORWALK			STAMFORD			WATERBURY			WILLIMANTIC		
DIVISION		065			063			068			069			067			062		
		MAT.	INST.	TOTAL	MAT.	INST.	TOTAL	MAT.	INST.	TOTAL	MAT.	INST.	TOTAL	MAT.	INST.	TOTAL	MAT.	INST.	TOTAL
01543	CONTRACTOR EQUIPMENT	.0	101.3	101.3	.0	101.3	101.3	.0	100.7	100.7	.0	100.7	100.7	.0	100.7	100.7	.0	100.7	100.7
0241, 31 - 34	SITE & INFRASTRUCTURE, DEMOLITION	96.0	105.0	102.5	88.8	105.0	100.4	96.6	104.2	102.0	97.2	104.2	102.2	96.4	104.2	102.0	96.5	104.2	102.0
0310	Concrete Forming & Accessories	99.2	120.0	117.1	99.2	119.8	116.8	99.4	120.4	117.5	99.4	120.7	117.7	99.4	120.1	117.1	99.4	119.8	116.9
0310	Concrete Reinforcing	104.3	125.6	114.7	81.7	125.6	103.2	104.3	125.8	114.8	104.3	125.8	114.8	104.3	125.6	114.7	104.3	125.6	114.7
0330	Cast-in-Place Concrete	101.5	114.6	106.4	86.4	114.5	97.0	103.0	126.9	112.0	104.8	127.0	113.1	104.8	114.7	108.5	97.8	103.6	100.0
03	CONCRETE	121.3	119.1	120.3	93.3	118.9	105.4	108.2	123.5	115.4	109.1	123.7	116.0	109.1	119.1	113.8	105.6	115.2	110.1
04	MASONRY	99.4	127.1	116.0	97.8	127.1	115.3	98.9	128.1	116.8	99.7	128.9	117.2	99.7	127.1	116.1	99.1	127.1	115.8
05	METALS	93.8	124.0	102.9	93.5	123.7	102.7	96.9	124.5	105.3	96.9	124.9	105.4	96.9	124.0	105.1	96.7	123.7	104.9
06	WOOD, PLASTICS & COMPOSITES	99.1	118.5	109.4	99.1	118.5	109.4	99.1	118.5	109.4	99.1	118.5	109.4	99.1	118.5	109.4	99.1	118.5	109.4
07	THERMAL & MOISTURE PROTECTION	98.5	120.9	107.5	98.4	120.0	107.1	98.5	126.3	109.7	98.4	126.3	109.6	98.4	120.9	107.5	98.6	118.3	106.5
08	OPENINGS	105.6	128.2	111.5	108.9	118.0	111.2	105.6	128.2	111.5	105.6	128.2	111.5	105.6	128.2	111.5	108.9	128.2	113.9
0920	Plaster & Gypsum Board	102.1	118.2	111.7	102.1	118.2	111.7	102.1	118.2	111.7	102.1	118.2	111.7	102.1	118.2	111.7	102.1	118.2	111.7
0950, 0980	Ceilings & Acoustic Treatment	96.2	118.2	109.4	94.3	118.2	108.7	96.2	118.2	109.4	96.2	118.2	109.4	96.2	118.2	109.4	94.3	118.2	108.7
0960	Flooring	98.0	120.2	103.9	98.0	120.2	103.9	98.0	120.2	103.9	98.0	120.2	103.9	98.0	120.2	103.9	98.0	120.2	103.9
0990	Wall Finishes & Painting/Coating	91.0	114.8	105.3	91.0	114.8	105.3	91.0	114.8	105.3	91.0	114.8	105.3	91.0	118.0	107.1	91.0	118.0	107.1
09	FINISHES	100.1	119.5	110.1	99.2	119.5	109.7	100.0	119.5	110.1	100.1	119.5	110.1	99.9	119.9	110.2	99.7	119.9	110.1
COVERS	DIVS. 10 - 14, 25, 28, 41, 43, 44	100.0	110.4	102.1	100.0	110.4	102.1	100.0	110.4	102.1	100.0	110.6	102.2	100.0	110.4	102.1	100.0	110.4	102.1
22, 23	PLUMBING & HVAC	100.0	111.9	104.8	95.8	111.8	102.3	100.0	111.9	104.9	100.0	111.9	104.9	100.0	111.9	104.8	100.0	111.7	104.8
26, 27, 3370	ELECTRICAL, COMMUNICATIONS & UTIL.	98.1	108.7	103.2	95.5	108.7	101.8	98.1	104.9	101.4	98.1	150.2	123.1	97.7	109.0	103.1	98.1	109.4	103.5
MF2004	WEIGHTED AVERAGE	101.6	116.4	108.0	97.0	115.9	105.1	100.6	116.8	107.6	100.7	123.2	110.4	100.6	116.4	107.4	100.5	115.8	107.1

		D.C.			DELAWARE									FLORIDA					
		WASHINGTON			DOVER			NEWARK			WILMINGTON			DAYTONA BEACH			FORT LAUDERDALE		
DIVISION		200 - 205			199			197			198			321			333		
		MAT.	INST.	TOTAL	MAT.	INST.	TOTAL	MAT.	INST.	TOTAL	MAT.	INST.	TOTAL	MAT.	INST.	TOTAL	MAT.	INST.	TOTAL
01543	CONTRACTOR EQUIPMENT	.0	103.5	103.5	.0	118.7	118.7	.0	118.7	118.7	.0	118.9	118.9	.0	97.9	97.9	.0	89.3	89.3
0241, 31 - 34	SITE & INFRASTRUCTURE, DEMOLITION	109.6	91.5	96.6	102.8	112.3	109.6	102.8	112.3	109.6	89.0	112.6	105.9	112.4	86.7	94.0	98.6	73.9	80.9
0310	Concrete Forming & Accessories	101.4	80.4	83.4	98.5	95.7	96.1	98.5	95.7	96.1	100.0	95.7	96.3	97.4	65.6	70.1	95.2	71.5	74.8
0310	Concrete Reinforcing	110.8	88.7	100.0	97.7	91.3	94.6	98.6	91.3	95.1	98.6	91.3	95.1	95.3	82.5	89.0	95.3	66.9	81.4
0330	Cast-in-Place Concrete	122.7	84.2	108.2	93.8	89.4	92.2	93.8	89.4	92.2	88.2	89.4	88.6	88.0	67.4	80.2	92.3	66.3	82.5
03	CONCRETE	117.3	84.6	101.9	102.4	93.8	98.3	102.5	93.8	98.4	99.8	93.8	96.9	90.2	70.6	81.0	92.2	70.0	81.7
04	MASONRY	100.7	79.7	88.1	107.2	88.1	95.7	105.2	88.1	94.9	110.8	88.1	97.2	86.2	61.1	71.2	86.5	64.5	73.3
05	METALS	103.5	108.0	104.8	104.0	112.0	106.4	104.0	112.0	106.4	100.8	112.0	104.2	93.6	92.9	93.4	93.5	85.5	91.1
06	WOOD, PLASTICS & COMPOSITES	97.3	81.0	88.7	97.5	96.9	97.2	97.5	96.9	97.2	99.9	96.9	98.3	96.3	68.2	81.4	91.8	70.8	80.7
07	THERMAL & MOISTURE PROTECTION	107.9	83.7	98.1	101.7	102.1	101.9	101.8	102.1	101.9	101.5	102.1	101.7	99.2	65.7	85.7	99.2	71.7	88.1
08	OPENINGS	102.6	89.4	99.2	94.7	102.5	96.7	94.7	102.5	96.7	94.4	102.5	96.5	100.3	66.1	91.4	97.8	63.8	89.0
0920	Plaster & Gypsum Board	111.1	80.6	93.0	105.9	96.7	100.5	105.9	96.7	100.5	105.2	96.7	100.2	103.1	67.8	82.1	102.7	70.5	83.6
0950, 0980	Ceilings & Acoustic Treatment	108.6	80.6	91.7	105.6	96.7	100.3	105.6	96.7	100.3	101.3	96.7	98.6	102.8	67.8	81.8	102.8	70.5	83.4
0960	Flooring	116.2	106.1	113.5	84.0	101.5	88.7	84.0	101.5	88.7	83.9	101.5	88.6	119.4	69.3	106.0	119.4	56.2	102.6
0990	Wall Finishes & Painting/Coating	121.1	83.6	98.6	94.5	99.9	97.7	94.5	99.9	97.7	94.5	99.9	97.7	111.7	67.4	85.2	107.4	46.6	71.1
09	FINISHES	108.2	85.2	96.4	101.8	97.0	99.3	101.9	97.0	99.4	100.7	97.0	98.8	108.9	66.6	87.1	106.9	65.9	85.7
COVERS	DIVS. 10 - 14, 25, 28, 41, 43, 44	100.0	96.5	99.3	100.0	102.7	100.6	100.0	102.7	100.6	100.0	102.7	100.6	100.0	80.6	96.0	100.0	88.5	97.6
22, 23	PLUMBING & HVAC	100.2	90.7	96.3	100.3	108.4	103.6	100.3	108.4	103.6	100.3	108.4	103.6	99.9	67.1	86.5	99.9	70.9	88.1
26, 27, 3370	ELECTRICAL, COMMUNICATIONS & UTIL.	100.4	96.0	98.3	97.9	97.9	97.9	97.9	97.9	97.9	97.9	97.9	97.9	96.5	62.6	80.2	96.5	75.3	86.3
MF2004	WEIGHTED AVERAGE	104.1	90.2	98.1	100.9	101.4	101.1	100.8	101.4	101.0	99.8	101.4	100.5	97.8	70.6	86.1	97.2	71.9	86.3

City Cost Indexes

		FLORIDA																	
		FORT MYERS			GAINESVILLE			JACKSONVILLE			LAKELAND			MELBOURNE			MIAMI		
DIVISION		339,341			326,344			320,322			338			329			330 - 332,340		
		MAT.	INST.	TOTAL	MAT.	INST.	TOTAL	MAT.	INST.	TOTAL	MAT.	INST.	TOTAL	MAT.	INST.	TOTAL	MAT.	INST.	TOTAL
01543	CONTRACTOR EQUIPMENT	.0	97.9	97.9	.0	97.9	97.9	.0	97.9	97.9	.0	97.9	97.9	.0	97.9	97.9	.0	89.3	89.3
0241, 31 - 34	SITE & INFRASTRUCTURE, DEMOLITION	109.7	86.8	93.3	122.7	86.3	96.6	112.5	87.3	94.4	111.8	86.4	93.6	120.6	87.4	96.8	97.7	73.9	80.6
0310	Concrete Forming & Accessories	90.5	70.7	73.5	91.6	50.0	55.8	97.2	50.1	56.8	86.3	71.3	73.4	93.2	70.3	73.5	99.7	72.1	76.0
0310	Concrete Reinforcing	96.4	85.7	91.1	101.1	46.6	74.5	95.3	47.0	71.7	98.7	87.1	93.0	96.4	82.6	89.6	95.3	67.0	81.4
0330	Cast-in-Place Concrete	96.3	58.3	82.0	101.1	47.0	80.7	88.9	55.4	76.3	98.5	59.3	83.7	106.1	73.7	93.9	87.3	67.5	79.9
03	CONCRETE	93.1	70.4	82.4	101.1	50.0	77.0	90.6	53.0	72.9	95.1	71.2	83.8	101.6	74.9	89.0	90.0	70.6	80.9
04	MASONRY	80.3	50.7	62.6	100.1	40.6	64.5	86.1	47.8	63.2	96.5	76.0	84.3	85.4	71.9	77.3	84.9	66.6	74.0
05	METALS	95.8	91.2	94.4	92.6	75.6	87.5	92.2	76.4	87.4	95.8	92.8	94.9	101.9	93.5	99.3	98.4	85.6	94.6
06	WOOD, PLASTICS & COMPOSITES	88.5	72.7	80.2	89.7	49.0	68.2	96.3	49.0	71.3	83.8	72.7	78.0	91.6	68.2	79.2	97.2	70.8	83.2
07	THERMAL & MOISTURE PROTECTION	99.0	57.9	82.5	99.5	52.4	80.5	99.4	55.2	81.6	98.9	64.6	85.1	99.6	73.7	89.2	106.5	69.4	91.6
08	OPENINGS	98.9	69.2	91.2	98.3	45.7	84.7	100.3	46.3	86.3	98.8	67.1	90.6	99.4	71.8	92.3	97.8	64.3	89.1
0920	Plaster & Gypsum Board	99.5	72.6	83.5	99.7	48.2	69.0	103.1	48.2	70.4	97.8	72.6	82.8	99.9	67.8	80.8	102.7	70.5	83.6
0950, 0980	Ceilings & Acoustic Treatment	96.7	72.6	82.2	95.9	48.2	67.2	102.8	48.2	69.9	95.9	72.6	81.8	98.6	67.8	80.1	102.8	70.5	83.4
0960	Flooring	115.1	38.6	94.7	115.7	33.1	93.7	119.4	46.2	99.9	111.9	48.9	95.1	115.9	69.3	103.5	127.2	56.9	108.5
0990	Wall Finishes & Painting/Coating	111.7	43.5	70.9	111.7	41.9	70.0	111.7	45.0	71.8	111.7	43.5	70.9	111.7	92.9	100.4	107.4	46.6	71.1
09	FINISHES	105.4	62.3	83.2	106.4	45.2	74.8	109.0	48.2	77.6	104.1	64.4	83.6	106.9	72.2	89.0	109.2	66.5	87.1
COVERS	DIVS. 10 - 14, 25, 28, 41, 43, 44	100.0	78.9	95.7	100.0	78.6	95.6	100.0	77.3	95.3	100.0	78.9	95.7	100.0	84.7	96.9	100.0	89.1	97.8
22, 23	PLUMBING & HVAC	97.6	50.4	78.3	98.9	63.8	84.6	99.9	47.2	78.4	97.6	73.0	87.6	99.9	73.2	89.0	99.9	72.2	88.6
26, 27, 3370	ELECTRICAL, COMMUNICATIONS & UTIL.	98.5	37.8	69.4	96.8	70.0	83.9	96.2	65.3	81.4	96.7	43.8	71.3	97.9	71.9	85.4	98.1	68.5	83.9
MF2004	WEIGHTED AVERAGE	97.3	61.8	82.0	99.3	60.3	82.4	97.6	57.9	80.5	98.0	70.5	86.1	100.5	76.3	90.0	98.3	71.6	86.8

		FLORIDA																	
		ORLANDO			PANAMA CITY			PENSACOLA			SARASOTA			ST. PETERSBURG			TALLAHASSEE		
DIVISION		327 - 328,347			324			325			342			337			323		
		MAT.	INST.	TOTAL	MAT.	INST.	TOTAL	MAT.	INST.	TOTAL	MAT.	INST.	TOTAL	MAT.	INST.	TOTAL	MAT.	INST.	TOTAL
01543	CONTRACTOR EQUIPMENT	.0	97.9	97.9	.0	97.9	97.9	.0	97.9	97.9	.0	97.9	97.9	.0	97.9	97.9	.0	97.9	97.9
0241, 31 - 34	SITE & INFRASTRUCTURE, DEMOLITION	112.9	86.5	93.9	126.8	83.9	96.1	126.5	86.4	97.8	114.0	86.5	94.3	113.3	85.8	93.6	113.7	85.9	93.7
0310	Concrete Forming & Accessories	97.2	67.1	71.4	96.3	26.0	35.9	94.2	47.9	54.4	97.4	70.8	74.5	95.0	44.2	51.4	97.1	37.0	45.5
0310	Concrete Reinforcing	95.3	79.4	87.5	99.5	45.9	73.3	102.0	46.3	74.8	95.3	87.0	91.3	98.7	52.9	76.3	95.3	46.5	71.5
0330	Cast-in-Place Concrete	95.0	67.3	84.6	93.4	32.7	70.5	115.4	51.8	91.5	100.7	59.1	85.0	99.5	52.1	81.7	91.3	45.8	74.2
03	CONCRETE	95.1	70.7	83.6	98.6	33.8	68.0	109.3	50.6	81.6	96.6	70.9	84.5	96.5	50.3	74.7	91.8	43.8	69.1
04	MASONRY	89.9	61.1	72.7	90.7	25.6	51.8	109.3	46.4	71.6	87.2	76.0	80.5	134.9	45.8	81.5	88.9	36.1	57.3
05	METALS	99.1	91.4	96.7	93.4	62.2	84.0	94.4	75.6	88.7	96.4	92.3	95.1	96.6	76.4	90.5	87.1	74.9	83.4
06	WOOD, PLASTICS & COMPOSITES	96.3	70.6	82.7	95.1	26.2	58.7	93.5	48.3	69.6	96.3	72.7	83.9	93.5	44.3	67.5	94.7	35.1	63.2
07	THERMAL & MOISTURE PROTECTION	99.5	66.0	86.0	99.7	29.6	71.5	99.6	49.6	79.5	99.2	64.5	85.3	99.2	46.6	78.0	101.8	47.9	80.1
08	OPENINGS	100.3	65.9	91.4	97.8	24.9	78.9	97.8	46.7	84.6	100.2	66.6	91.5	98.8	43.3	84.4	100.7	38.2	84.5
0920	Plaster & Gypsum Board	105.2	70.4	84.5	101.6	24.7	55.8	102.5	47.5	69.7	103.1	72.6	84.9	101.1	43.4	66.7	103.1	33.9	61.9
0950, 0980	Ceilings & Acoustic Treatment	102.8	70.4	83.3	97.7	24.7	53.8	98.6	47.5	67.9	100.9	72.6	83.9	97.7	43.4	65.1	102.8	33.9	61.4
0960	Flooring	119.4	69.3	106.0	118.8	17.7	91.8	112.7	49.1	95.7	119.4	48.7	100.5	117.7	48.9	99.3	119.4	35.9	97.1
0990	Wall Finishes & Painting/Coating	111.7	51.6	75.8	111.7	23.0	58.6	111.7	52.2	76.1	111.7	43.5	70.9	111.7	43.5	70.9	111.7	37.0	67.0
09	FINISHES	109.3	66.2	87.1	108.5	23.7	64.7	106.7	48.4	76.6	108.5	64.4	85.7	106.9	44.9	74.9	109.1	35.7	71.3
COVERS	DIVS. 10 - 14, 25, 28, 41, 43, 44	100.0	80.8	96.1	100.0	40.9	87.8	100.0	45.8	88.8	100.0	78.9	95.6	100.0	47.8	89.3	100.0	69.4	93.7
22, 23	PLUMBING & HVAC	99.9	62.8	84.8	99.9	23.1	68.6	99.9	46.0	77.9	99.8	50.8	79.8	99.9	45.8	77.8	99.9	36.7	74.1
26, 27, 3370	ELECTRICAL, COMMUNICATIONS & UTIL.	98.2	41.7	71.1	95.2	31.2	64.5	99.0	60.3	80.4	96.5	34.2	66.6	96.7	45.0	71.9	98.3	38.6	69.7
MF2004	WEIGHTED AVERAGE	99.7	66.6	85.4	98.9	35.7	71.6	101.5	55.3	81.6	99.0	64.4	84.1	101.2	52.4	80.1	97.5	46.8	75.6

		FLORIDA						GEORGIA											
		TAMPA			WEST PALM BEACH			ALBANY			ATHENS			ATLANTA			AUGUSTA		
DIVISION		335 - 336,346			334,349			317,398			306			300 - 303,399			308 - 309		
		MAT.	INST.	TOTAL	MAT.	INST.	TOTAL	MAT.	INST.	TOTAL	MAT.	INST.	TOTAL	MAT.	INST.	TOTAL	MAT.	INST.	TOTAL
01543	CONTRACTOR EQUIPMENT	.0	97.9	97.9	.0	89.3	89.3	.0	90.2	90.2	.0	91.5	91.5	.0	92.2	92.2	.0	91.5	91.5
0241, 31 - 34	SITE & INFRASTRUCTURE, DEMOLITION	113.6	86.5	94.2	95.3	74.0	80.1	98.9	77.2	83.3	104.8	91.7	95.4	100.6	94.4	96.1	97.0	91.8	93.3
0310	Concrete Forming & Accessories	98.5	71.3	75.2	99.0	70.9	74.9	96.4	52.7	58.9	95.3	49.5	56.0	99.3	77.1	80.2	96.5	59.1	64.3
0310	Concrete Reinforcing	95.3	87.1	91.3	98.0	66.6	82.6	94.9	91.9	92.5	95.8	87.9	91.9	95.2	91.8	93.5	96.2	71.7	84.2
0330	Cast-in-Place Concrete	97.3	59.3	83.0	87.8	62.5	78.2	94.0	51.9	78.2	97.9	55.0	81.7	97.9	73.5	88.7	92.5	54.4	78.1
03	CONCRETE	95.0	71.2	83.8	89.4	68.3	79.5	93.0	60.8	77.8	101.9	59.1	81.7	99.4	78.6	89.6	94.9	60.2	78.6
04	MASONRY	87.4	76.0	80.6	86.1	58.8	69.8	89.4	48.8	65.1	72.8	66.6	69.1	87.3	77.0	81.1	87.5	48.5	64.1
05	METALS	95.6	92.9	94.8	92.7	84.5	90.2	92.9	89.0	91.7	87.7	73.8	83.5	88.5	81.0	86.2	87.4	69.3	81.9
06	WOOD, PLASTICS & COMPOSITES	97.8	72.7	84.5	96.6	70.8	83.0	95.1	47.6	70.0	92.5	42.9	66.3	96.9	78.8	87.3	93.8	59.7	75.8
07	THERMAL & MOISTURE PROTECTION	99.4	64.6	85.4	99.0	64.3	85.0	96.0	63.0	82.7	99.7	59.7	83.6	99.7	79.3	91.5	99.3	58.6	82.9
08	OPENINGS	100.2	67.3	91.7	97.0	63.8	88.4	97.8	53.4	86.3	93.4	50.9	82.3	99.2	75.9	93.2	93.4	58.3	84.3
0920	Plaster & Gypsum Board	103.1	72.6	84.9	104.7	70.5	84.3	100.8	46.7	68.6	118.1	41.8	72.7	119.4	78.7	95.2	118.4	59.1	83.1
0950, 0980	Ceilings & Acoustic Treatment	102.8	72.6	84.6	96.7	70.5	81.0	102.8	46.7	68.7	111.1	41.8	69.4	111.1	78.7	91.6	112.1	59.1	80.2
0960	Flooring	119.4	48.9	100.6	122.2	47.0	102.1	119.4	37.2	97.5	85.0	64.9	79.6	86.6	74.2	83.3	85.4	47.3	75.3
0990	Wall Finishes & Painting/Coating	111.7	43.5	70.9	107.4	43.1	68.9	107.4	50.0	73.1	91.4	39.6	60.4	91.4	85.9	88.1	91.4	45.2	63.8
09	FINISHES	109.0	64.4	86.0	106.4	63.6	84.3	106.3	47.9	76.2	98.9	50.1	73.7	99.3	77.4	88.0	98.8	54.9	76.1
COVERS	DIVS. 10 - 14, 25, 28, 41, 43, 44	100.0	83.8	96.7	100.0	88.3	97.6	100.0	81.0	96.1	100.0	52.0	90.1	100.0	84.3	96.8	100.0	63.7	92.5
22, 23	PLUMBING & HVAC	99.9	73.0	88.9	97.6	69.2	86.1	99.7	65.2	85.7	95.8	78.9	88.9	100.1	78.3	91.2	100.1	79.9	91.9
26, 27, 3370	ELECTRICAL, COMMUNICATIONS & UTIL.	96.3	45.0	71.7	97.9	48.6	74.3	91.9	62.2	77.7	95.5	81.6	89.0	95.8	77.2	86.9	96.4	67.7	82.6
MF2004	WEIGHTED AVERAGE	98.8	70.8	86.7	96.2	66.4	83.4	96.7	63.2	82.2	94.8	69.5	83.9	96.8	79.7	89.4	95.5	67.0	83.2

City Cost Indexes

DIVISION		GEORGIA																	
		COLUMBUS			DALTON			GAINESVILLE			MACON			SAVANNAH			STATESBORO		
		318 - 319			307			305			310 - 312			313 - 314			304		
		MAT.	INST.	TOTAL	MAT.	INST.	TOTAL	MAT.	INST.	TOTAL	MAT.	INST.	TOTAL	MAT.	INST.	TOTAL	MAT.	INST.	TOTAL
01543	CONTRACTOR EQUIPMENT	.0	90.2	90.2	.0	105.3	105.3	.0	91.5	91.5	.0	102.8	102.8	.0	91.4	91.4	.0	90.5	90.5
0241, 31 - 34	SITE & INFRASTRUCTURE, DEMOLITION	98.8	77.6	83.6	105.9	95.5	98.4	104.3	92.2	95.6	99.2	94.6	95.9	99.6	77.8	83.9	106.4	74.1	83.3
0310	Concrete Forming & Accessories	96.4	67.1	71.2	87.0	53.0	57.8	98.9	52.4	58.9	95.8	63.2	67.8	96.6	57.8	63.3	80.9	48.0	52.6
0310	Concrete Reinforcing	95.3	91.1	93.2	95.4	75.9	85.8	95.6	93.9	94.8	96.5	90.2	93.4	96.2	72.0	84.4	95.0	51.7	73.8
0330	Cast-in-Place Concrete	93.6	61.7	81.6	95.1	40.7	74.6	102.9	57.3	85.7	92.4	50.6	76.7	92.7	52.1	77.4	97.7	48.8	79.3
03	CONCRETE	92.9	70.9	82.5	101.0	54.0	78.8	103.6	62.1	84.0	92.4	65.2	79.6	92.6	59.9	77.1	101.1	50.4	77.2
04	MASONRY	88.5	68.7	76.6	76.1	39.5	54.2	82.4	41.1	57.6	101.9	46.3	68.6	87.3	57.8	69.7	75.8	48.6	59.6
05	METALS	92.5	93.0	92.6	88.7	70.5	83.2	87.0	69.5	81.7	88.5	91.4	89.3	88.8	83.2	87.1	92.0	72.1	86.0
06	WOOD, PLASTICS & COMPOSITES	95.1	65.7	79.6	75.9	56.2	65.5	96.6	47.4	70.6	102.7	64.5	82.5	108.4	56.2	80.8	68.1	49.3	58.2
07	THERMAL & MOISTURE PROTECTION	96.2	73.5	87.1	101.2	49.4	80.4	99.6	52.7	80.8	94.8	65.8	83.1	96.6	59.5	81.7	100.0	45.6	78.1
08	OPENINGS	97.8	65.5	89.4	94.7	41.9	81.0	93.3	45.3	80.9	96.2	63.7	87.8	98.9	53.8	87.2	95.5	39.4	80.9
0920	Plaster & Gypsum Board	102.9	65.3	80.5	106.0	55.6	76.0	119.4	46.5	76.0	109.8	64.1	82.6	103.1	55.6	74.8	105.8	48.5	71.7
0950, 0980	Ceilings & Acoustic Treatment	102.0	65.3	79.9	115.9	55.6	79.6	111.1	46.5	72.2	98.1	64.1	77.6	102.8	55.6	74.4	116.9	48.5	75.7
0960	Flooring	119.4	60.7	103.7	86.2	15.5	67.3	86.6	15.5	67.6	92.7	43.7	79.7	120.4	55.8	103.2	100.0	54.0	87.7
0990	Wall Finishes & Painting/Coating	107.4	56.5	76.9	82.7	47.6	61.7	91.4	41.5	61.6	109.4	55.7	77.3	109.3	56.5	77.7	89.7	40.6	60.3
09	FINISHES	106.6	63.9	84.6	104.4	45.9	74.2	99.5	42.8	70.2	95.3	58.1	76.1	107.5	57.1	81.5	107.6	48.9	77.3
COVERS	DIVS. 10 - 14, 25, 28, 41, 43, 44	100.0	83.2	96.5	100.0	26.7	84.9	100.0	52.5	90.2	100.0	80.7	96.0	100.0	64.7	92.7	100.0	48.1	89.3
22, 23	PLUMBING & HVAC	99.9	60.5	83.8	95.9	41.2	73.6	95.8	84.7	91.3	99.9	66.3	86.2	99.9	63.6	85.1	96.3	42.1	74.2
26, 27, 3370	ELECTRICAL, COMMUNICATIONS & UTIL.	91.7	65.6	79.2	108.5	70.7	90.4	95.9	81.6	89.0	90.2	64.8	78.0	94.4	59.3	77.6	96.6	38.0	68.5
MF2004	WEIGHTED AVERAGE	96.6	69.6	84.9	97.0	54.9	78.9	95.4	66.8	83.1	95.3	67.8	83.4	96.6	63.5	82.3	96.5	49.9	76.4

DIVISION		GEORGIA						HAWAII									IDAHO		
		VALDOSTA			WAYCROSS			HILO			HONOLULU			STATES & POSS., GUAM			BOISE		
		316			315			967			968			969			836 - 837		
		MAT.	INST.	TOTAL	MAT.	INST.	TOTAL	MAT.	INST.	TOTAL	MAT.	INST.	TOTAL	MAT.	INST.	TOTAL	MAT.	INST.	TOTAL
01543	CONTRACTOR EQUIPMENT	.0	90.2	90.2	.0	90.2	90.2	.0	99.6	99.6	.0	99.6	99.6	.0	176.5	176.5	.0	101.1	101.1
0241, 31 - 34	SITE & INFRASTRUCTURE, DEMOLITION	109.0	77.0	86.1	105.9	75.5	84.1	137.4	106.1	115.0	141.7	106.1	116.2	180.9	105.1	126.5	78.1	101.1	94.6
0310	Concrete Forming & Accessories	84.4	49.8	54.7	86.6	51.9	56.8	107.1	141.5	136.7	111.0	141.5	137.2	109.0	48.6	57.0	98.5	78.6	81.4
0310	Concrete Reinforcing	97.2	51.7	75.0	97.2	91.4	94.4	110.7	120.6	115.5	109.7	120.6	115.0	141.7	25.7	84.9	107.5	75.1	91.7
0330	Cast-in-Place Concrete	92.0	54.7	78.0	103.9	54.8	85.4	206.6	126.5	176.5	190.6	126.5	166.5	179.0	105.5	151.4	96.8	85.6	92.6
03	CONCRETE	96.7	53.4	76.2	100.8	61.5	82.2	159.2	131.1	146.0	151.2	131.1	141.7	150.3	64.6	109.8	107.5	80.3	94.6
04	MASONRY	94.3	52.2	69.1	95.2	48.6	67.3	136.8	126.2	130.4	130.1	126.2	127.8	190.4	34.2	96.9	130.0	69.3	93.7
05	METALS	92.1	75.3	87.0	91.3	82.9	88.7	115.3	108.6	113.3	139.6	108.6	130.2	148.8	76.4	126.9	103.7	75.1	95.1
06	WOOD, PLASTICS & COMPOSITES	80.9	46.4	62.6	84.0	53.2	67.7	99.0	146.2	123.9	103.4	146.2	126.0	109.0	48.6	77.1	94.5	78.8	86.2
07	THERMAL & MOISTURE PROTECTION	96.4	64.1	83.4	96.2	48.4	77.0	105.6	126.9	114.2	108.2	126.9	115.7	115.7	58.3	92.6	92.4	75.4	85.6
08	OPENINGS	93.2	43.7	80.4	93.2	51.4	82.4	100.6	136.5	109.9	105.9	136.5	113.8	100.4	41.0	85.0	93.6	74.4	88.6
0920	Plaster & Gypsum Board	95.8	45.5	65.9	98.1	52.5	70.9	103.0	147.2	129.3	130.1	147.2	140.3	192.8	34.3	98.4	89.0	78.1	82.5
0950, 0980	Ceilings & Acoustic Treatment	97.7	45.5	66.3	95.9	52.5	69.8	116.7	147.2	135.0	119.7	147.2	136.3	249.8	34.3	120.2	110.0	78.1	90.8
0960	Flooring	110.9	46.6	93.8	112.4	36.0	92.0	145.3	130.9	141.5	168.6	130.9	158.5	117.9	31.6	94.9	97.2	51.9	85.1
0990	Wall Finishes & Painting/Coating	107.4	39.6	67.0	107.4	49.3	72.6	101.4	145.5	127.8	102.2	145.5	128.1	108.2	38.6	66.6	99.5	48.8	69.2
09	FINISHES	102.7	47.5	74.2	102.7	48.4	74.7	123.2	141.5	132.7	136.4	141.5	139.0	189.4	44.4	114.5	98.4	70.8	84.1
COVERS	DIVS. 10 - 14, 25, 28, 41, 43, 44	100.0	63.4	92.5	100.0	48.6	89.4	100.0	122.5	104.6	100.0	122.5	104.6	100.0	93.3	98.6	100.0	76.9	95.2
22, 23	PLUMBING & HVAC	99.9	43.1	76.7	97.1	52.7	79.0	100.2	114.0	105.8	100.2	114.0	105.8	103.0	33.3	74.6	100.0	72.9	88.9
26, 27, 3370	ELECTRICAL, COMMUNICATIONS & UTIL.	90.0	38.4	65.3	93.8	70.9	82.8	107.7	120.4	113.8	111.1	120.4	115.6	146.1	36.7	93.6	95.1	73.8	84.9
MF2004	WEIGHTED AVERAGE	96.5	52.4	77.4	96.6	59.9	80.8	115.4	122.7	118.5	120.2	122.7	121.3	134.9	52.6	99.4	100.8	76.4	90.3

DIVISION		IDAHO															ILLINOIS		
		COEUR D'ALENE			IDAHO FALLS			LEWISTON			POCATELLO			TWIN FALLS			BLOOMINGTON		
		838			834			835			832			833			617		
		MAT.	INST.	TOTAL	MAT.	INST.	TOTAL	MAT.	INST.	TOTAL	MAT.	INST.	TOTAL	MAT.	INST.	TOTAL	MAT.	INST.	TOTAL
01543	CONTRACTOR EQUIPMENT	.0	94.1	94.1	.0	101.1	101.1	.0	94.1	94.1	.0	101.1	101.1	.0	101.1	101.1	.0	101.3	101.3
0241, 31 - 34	SITE & INFRASTRUCTURE, DEMOLITION	79.0	95.4	90.8	76.6	100.0	93.4	84.9	95.7	92.6	79.8	101.0	95.0	86.2	99.1	95.4	93.8	94.9	94.6
0310	Concrete Forming & Accessories	108.0	65.3	71.3	91.7	48.5	54.6	113.7	60.1	67.6	98.6	78.2	81.1	99.4	34.1	43.3	84.2	111.8	107.9
0310	Concrete Reinforcing	115.4	102.5	109.1	109.6	73.9	92.1	115.4	102.7	109.2	107.9	74.7	91.6	109.9	74.3	92.5	98.4	100.3	99.3
0330	Cast-in-Place Concrete	104.7	90.6	99.4	92.1	58.3	79.4	108.8	91.0	102.1	99.6	85.4	94.2	102.2	43.2	80.0	97.3	98.5	97.8
03	CONCRETE	114.7	81.3	98.9	99.2	57.4	79.5	118.7	81.8	101.3	106.8	79.9	94.1	114.5	45.9	82.2	95.2	105.2	99.9
04	MASONRY	131.9	82.5	102.3	125.1	37.1	72.5	132.3	82.5	102.5	127.4	61.8	88.1	130.4	36.3	74.0	110.2	109.6	109.8
05	METALS	97.0	88.3	94.3	112.0	72.2	99.9	96.4	89.2	94.2	112.0	74.0	100.5	112.1	71.7	99.8	96.4	108.0	99.9
06	WOOD, PLASTICS & COMPOSITES	96.5	59.9	77.1	87.1	47.0	65.9	102.5	59.9	80.0	94.5	78.8	86.2	95.1	33.1	62.3	82.8	110.9	97.6
07	THERMAL & MOISTURE PROTECTION	137.7	75.4	112.7	92.0	50.8	75.4	137.9	77.4	113.6	92.5	66.5	82.0	93.2	44.0	73.4	95.4	104.1	98.9
08	OPENINGS	113.8	64.5	101.0	97.6	51.0	85.5	113.8	67.7	101.9	94.3	68.1	87.5	97.6	41.0	82.9	90.3	103.0	93.6
0920	Plaster & Gypsum Board	151.4	58.9	96.3	82.9	45.4	60.6	152.7	58.9	96.8	86.9	78.1	81.7	86.7	31.1	53.6	94.8	111.0	104.5
0950, 0980	Ceilings & Acoustic Treatment	133.5	58.8	88.6	101.7	45.4	67.8	133.5	58.8	88.6	111.0	78.1	91.2	104.2	31.1	60.2	88.1	111.0	101.9
0960	Flooring	132.9	55.2	112.2	95.3	51.9	83.7	136.2	89.1	123.6	99.0	51.9	86.4	100.3	51.9	87.4	90.2	111.6	95.9
0990	Wall Finishes & Painting/Coating	123.2	72.5	92.9	99.6	40.1	64.0	123.2	72.5	92.9	99.5	50.8	70.3	99.6	29.9	57.9	91.5	108.8	101.9
09	FINISHES	156.9	62.9	108.4	94.4	47.7	70.3	158.2	69.9	112.6	98.7	71.0	84.4	98.1	36.0	66.0	90.6	112.2	101.7
COVERS	DIVS. 10 - 14, 25, 28, 41, 43, 44	100.0	63.0	92.4	100.0	48.7	89.4	100.0	78.9	95.7	100.0	76.9	95.2	100.0	43.3	88.3	100.0	100.1	100.0
22, 23	PLUMBING & HVAC	100.1	83.7	93.4	100.7	64.5	86.0	101.1	85.3	94.7	99.9	72.9	88.9	99.9	60.2	83.7	95.7	109.0	101.1
26, 27, 3370	ELECTRICAL, COMMUNICATIONS & UTIL.	83.0	81.6	82.3	82.9	75.0	79.1	81.2	81.6	81.4	91.4	75.0	83.5	84.1	67.9	76.4	94.5	83.6	89.3
MF2004	WEIGHTED AVERAGE	107.9	79.9	95.8	99.6	62.3	83.5	108.7	81.9	97.1	101.6	75.1	90.2	102.3	56.0	82.4	95.6	103.4	99.0

City Cost Indexes

ILLINOIS

DIVISION		CARBONDALE 629			CENTRALIA 628			CHAMPAIGN 618 - 619			CHICAGO 606 - 608			DECATUR 625			EAST ST. LOUIS 620 - 622		
		MAT.	INST.	TOTAL	MAT.	INST.	TOTAL	MAT.	INST.	TOTAL	MAT.	INST.	TOTAL	MAT.	INST.	TOTAL	MAT.	INST.	TOTAL
01543	CONTRACTOR EQUIPMENT	.0	108.7	108.7	.0	108.7	108.7	.0	102.2	102.2	.0	90.7	90.7	.0	102.2	102.2	.0	108.7	108.7
0241, 31 - 34	SITE & INFRASTRUCTURE, DEMOLITION	101.8	95.6	97.4	102.2	96.2	97.9	103.3	95.8	97.9	97.9	90.7	92.7	87.5	95.9	93.6	104.2	96.4	98.6
0310	Concrete Forming & Accessories	93.9	96.8	96.4	96.2	109.5	107.6	91.1	108.6	106.1	100.7	146.5	140.0	96.9	99.3	99.0	91.1	110.4	107.7
0310	Concrete Reinforcing	97.3	82.3	90.0	97.3	94.8	96.1	98.4	88.6	93.6	97.3	148.9	122.5	95.3	96.2	95.7	97.2	105.2	101.1
0330	Cast-in-Place Concrete	91.1	82.9	88.0	91.5	107.4	97.5	112.9	107.1	110.7	108.5	137.4	119.4	99.5	100.2	99.7	93.1	102.5	96.6
03	CONCRETE	86.8	90.3	88.5	87.2	106.8	96.5	107.4	104.5	106.0	106.4	142.8	123.6	96.6	99.5	98.0	88.1	107.6	97.3
04	MASONRY	73.5	96.8	87.0	73.6	107.9	94.1	133.8	89.5	107.3	96.0	138.4	121.4	70.0	103.8	90.2	73.7	109.4	95.1
05	METALS	95.7	107.3	99.2	95.7	115.4	101.7	96.4	100.4	97.6	98.3	127.7	107.2	99.0	104.6	100.7	96.7	120.4	103.9
06	WOOD, PLASTICS & COMPOSITES	98.0	96.6	97.3	100.9	111.4	106.5	90.8	108.4	100.1	102.8	146.7	126.0	99.4	95.9	97.6	95.1	111.4	103.7
07	THERMAL & MOISTURE PROTECTION	90.7	90.7	90.7	90.7	104.9	96.4	96.0	103.5	99.0	100.2	134.4	113.9	96.2	98.8	97.2	90.7	106.4	97.0
08	OPENINGS	86.5	98.5	89.6	86.5	109.8	92.5	91.0	102.3	93.9	102.6	149.7	114.8	97.0	97.4	97.1	86.5	115.7	94.1
0920	Plaster & Gypsum Board	99.0	96.4	97.4	100.0	111.6	106.9	97.7	108.5	104.1	94.1	147.9	126.1	101.5	95.7	98.0	98.4	111.6	106.2
0950, 0980	Ceilings & Acoustic Treatment	88.1	96.4	93.1	88.1	111.6	102.2	88.1	108.5	100.3	100.0	147.9	128.8	94.8	95.7	95.3	88.1	111.6	102.2
0960	Flooring	107.9	93.7	104.1	109.2	118.9	111.8	93.8	102.7	96.1	84.0	132.3	96.9	98.0	102.2	99.1	106.5	118.3	109.7
0990	Wall Finishes & Painting/Coating	99.3	90.1	93.8	99.3	98.7	98.9	91.5	102.9	98.3	83.0	141.5	118.0	91.5	98.5	95.7	99.3	98.7	98.9
09	FINISHES	94.9	95.8	95.3	95.4	110.5	103.2	92.7	107.7	100.4	90.6	144.0	118.1	95.6	99.9	97.8	94.4	109.4	102.1
COVERS	DIVS. 10 - 14, 25, 28, 41, 43, 44	100.0	98.9	99.8	100.0	101.5	100.3	100.0	86.1	97.1	100.0	124.6	105.1	100.0	84.7	96.9	100.0	102.1	100.4
22, 23	PLUMBING & HVAC	95.7	94.1	95.0	95.7	93.8	94.9	95.7	103.4	98.8	99.9	129.7	112.0	100.0	102.0	100.8	99.9	96.7	98.6
26, 27, 3370	ELECTRICAL, COMMUNICATIONS & UTIL.	93.2	99.7	96.3	94.5	97.4	95.9	97.4	95.8	96.7	96.3	131.9	113.4	97.0	88.3	92.9	94.1	99.5	96.7
MF2004	WEIGHTED AVERAGE	92.7	96.4	94.3	93.0	103.2	97.4	99.2	100.2	99.6	99.2	131.9	113.3	96.4	98.5	97.3	94.1	105.0	98.8

ILLINOIS

DIVISION		EFFINGHAM 624			GALESBURG 614			JOLIET 604			KANKAKEE 609			LA SALLE 613			NORTH SUBURBAN 600 - 603		
		MAT.	INST.	TOTAL	MAT.	INST.	TOTAL	MAT.	INST.	TOTAL	MAT.	INST.	TOTAL	MAT.	INST.	TOTAL	MAT.	INST.	TOTAL
01543	CONTRACTOR EQUIPMENT	.0	102.2	102.2	.0	101.3	101.3	.0	88.5	88.5	.0	88.5	88.5	.0	101.3	101.3	.0	88.5	88.5
0241, 31 - 34	SITE & INFRASTRUCTURE, DEMOLITION	95.0	95.6	95.4	96.2	94.9	95.2	97.7	88.8	91.4	91.6	88.5	89.4	95.6	95.3	95.4	95.7	89.1	91.3
0310	Concrete Forming & Accessories	101.5	97.2	97.8	91.0	111.8	108.9	101.5	134.2	129.5	93.8	108.4	106.3	106.3	110.7	110.0	100.6	131.3	126.9
0310	Concrete Reinforcing	98.4	91.5	95.0	97.9	100.0	98.9	97.3	129.7	113.1	98.2	107.7	102.8	98.0	100.4	99.2	97.3	140.2	118.3
0330	Cast-in-Place Concrete	99.1	74.8	90.0	100.2	100.5	100.3	108.4	124.6	114.5	101.0	119.5	108.0	100.1	106.2	102.4	108.5	135.2	118.5
03	CONCRETE	97.5	88.9	93.4	97.9	105.9	101.7	106.4	129.2	117.2	100.1	111.8	105.6	98.8	107.5	102.9	106.4	133.5	119.2
04	MASONRY	76.8	82.8	80.4	110.4	107.8	108.9	99.0	122.2	112.9	95.5	119.0	109.6	110.4	108.0	109.0	96.0	128.4	115.4
05	METALS	96.4	100.6	97.7	96.4	107.2	99.7	96.1	115.9	102.1	96.1	105.1	98.9	96.5	110.7	100.8	97.3	119.5	104.0
06	WOOD, PLASTICS & COMPOSITES	102.0	95.9	98.8	90.7	111.4	101.6	104.4	137.1	121.7	95.6	104.8	100.5	107.3	112.3	109.9	102.8	130.6	117.5
07	THERMAL & MOISTURE PROTECTION	95.7	90.2	93.5	95.6	101.3	97.9	99.6	123.8	109.3	98.8	120.5	107.5	95.8	97.5	96.5	100.2	124.5	110.0
08	OPENINGS	90.8	96.4	92.2	90.3	101.0	93.1	100.6	139.4	110.6	92.4	111.5	97.3	90.3	103.7	93.8	100.6	134.2	109.3
0920	Plaster & Gypsum Board	100.9	95.7	97.8	97.7	111.6	106.0	91.9	138.1	119.4	89.0	104.9	98.5	102.8	112.5	108.6	94.1	131.4	116.3
0950, 0980	Ceilings & Acoustic Treatment	88.1	95.7	92.6	88.1	111.6	102.2	100.0	138.1	122.9	100.0	104.9	103.0	88.1	112.5	102.8	100.0	131.4	118.9
0960	Flooring	99.4	102.7	100.3	93.7	103.7	96.4	83.5	123.4	94.2	80.3	103.0	86.4	101.9	109.8	104.0	84.0	107.2	90.2
0990	Wall Finishes & Painting/Coating	91.5	98.8	95.9	91.5	97.3	95.0	81.0	127.9	109.0	81.0	105.8	95.8	91.5	105.8	100.1	83.0	127.9	109.8
09	FINISHES	94.3	99.0	96.8	92.2	109.7	101.2	90.0	132.6	112.0	88.2	105.5	97.1	95.4	110.6	103.3	90.6	126.7	109.2
COVERS	DIVS. 10 - 14, 25, 28, 41, 43, 44	100.0	83.2	96.5	100.0	100.4	100.1	100.0	119.2	104.0	100.0	97.5	99.5	100.0	98.0	99.6	100.0	112.0	102.5
22, 23	PLUMBING & HVAC	95.7	101.4	98.0	95.7	103.3	98.8	100.0	113.2	105.4	95.8	100.9	97.8	95.7	88.1	92.6	99.9	115.4	106.2
26, 27, 3370	ELECTRICAL, COMMUNICATIONS & UTIL.	95.1	99.7	97.3	95.3	80.4	88.2	94.9	99.4	97.0	90.1	129.0	108.8	92.7	96.0	94.3	94.9	121.5	107.7
MF2004	WEIGHTED AVERAGE	94.8	95.6	95.1	96.3	101.2	98.4	98.6	116.8	106.4	95.0	109.0	101.1	96.5	100.9	98.4	98.7	120.7	108.2

ILLINOIS

DIVISION		PEORIA 615 - 616			QUINCY 623			ROCKFORD 610 - 611			ROCK ISLAND 612			SOUTH SUBURBAN 605			SPRINGFIELD 626 - 627		
		MAT.	INST.	TOTAL	MAT.	INST.	TOTAL	MAT.	INST.	TOTAL	MAT.	INST.	TOTAL	MAT.	INST.	TOTAL	MAT.	INST.	TOTAL
01543	CONTRACTOR EQUIPMENT	.0	101.3	101.3	.0	102.2	102.2	.0	101.3	101.3	.0	101.3	101.3	.0	88.5	88.5	.0	102.2	102.2
0241, 31 - 34	SITE & INFRASTRUCTURE, DEMOLITION	96.5	94.8	95.3	93.8	95.5	95.0	95.6	95.7	95.7	94.3	94.6	94.5	97.0	89.1	91.3	93.2	95.8	95.1
0310	Concrete Forming & Accessories	95.3	101.8	100.9	99.0	105.6	104.6	99.2	119.4	116.6	92.9	102.4	101.1	100.6	131.3	126.9	98.6	99.0	99.0
0310	Concrete Reinforcing	95.3	100.2	97.7	98.0	92.4	95.3	90.2	119.7	104.6	97.9	99.9	98.9	97.3	140.2	118.3	95.3	92.4	93.9
0330	Cast-in-Place Concrete	97.2	100.6	98.5	99.3	97.1	98.5	99.5	118.3	106.6	98.1	98.1	98.1	108.5	135.2	118.5	88.0	99.8	92.4
03	CONCRETE	95.4	101.5	98.3	97.1	100.4	98.7	96.0	119.2	106.9	96.1	100.9	98.4	106.4	133.5	119.2	90.9	98.5	94.5
04	MASONRY	111.0	109.7	110.2	96.1	96.3	96.2	86.2	108.7	99.7	110.2	100.7	104.6	96.0	128.4	115.4	74.0	103.1	91.4
05	METALS	99.0	107.5	101.5	96.4	102.8	98.3	99.0	118.9	105.0	96.5	106.7	99.6	97.3	119.5	104.0	101.5	102.6	101.8
06	WOOD, PLASTICS & COMPOSITES	99.4	98.0	98.7	99.2	108.4	104.1	99.4	118.3	109.4	92.8	101.7	97.5	102.8	130.6	117.5	102.0	95.9	98.8
07	THERMAL & MOISTURE PROTECTION	96.3	100.8	98.1	95.7	95.6	95.7	99.2	113.2	104.8	95.6	96.6	96.0	100.2	124.5	110.0	95.8	100.0	97.5
08	OPENINGS	97.2	97.7	97.3	91.7	103.1	94.6	97.2	118.0	102.5	90.3	96.3	91.9	100.6	134.2	109.3	98.2	96.4	97.7
0920	Plaster & Gypsum Board	101.5	97.8	99.3	100.0	108.5	105.0	101.5	118.7	111.7	98.4	101.6	100.3	94.1	131.4	116.3	101.5	95.7	98.0
0950, 0980	Ceilings & Acoustic Treatment	94.8	97.8	96.6	88.1	108.5	100.3	94.8	118.7	109.2	88.1	101.6	96.2	100.0	131.4	118.9	94.8	95.7	95.3
0960	Flooring	98.0	103.7	99.5	98.0	101.5	98.9	98.0	86.1	94.8	95.1	108.9	98.8	84.0	107.2	90.2	98.2	101.8	99.2
0990	Wall Finishes & Painting/Coating	91.5	90.1	90.7	91.5	98.5	95.7	91.5	117.1	106.9	91.5	94.9	93.5	83.0	127.9	109.8	91.5	98.5	95.7
09	FINISHES	95.7	100.9	98.4	93.7	105.3	99.7	95.7	113.0	104.6	92.6	103.6	98.3	90.6	126.7	109.2	95.7	99.7	97.8
COVERS	DIVS. 10 - 14, 25, 28, 41, 43, 44	100.0	98.5	99.7	100.0	86.7	97.3	100.0	108.8	101.8	100.0	82.1	96.3	100.0	112.0	102.5	100.0	84.5	96.8
22, 23	PLUMBING & HVAC	100.0	94.2	97.6	95.7	98.0	96.7	100.0	111.1	104.5	95.7	100.5	97.7	99.9	115.4	106.2	100.0	97.2	98.8
26, 27, 3370	ELECTRICAL, COMMUNICATIONS & UTIL.	96.1	82.9	89.8	92.8	74.5	84.0	96.6	102.8	99.6	88.1	81.3	84.9	94.9	121.5	107.7	97.7	83.2	90.8
MF2004	WEIGHTED AVERAGE	98.5	97.8	98.2	95.4	95.9	95.7	97.5	110.8	103.2	95.3	97.5	96.2	98.7	120.7	108.2	96.7	96.4	96.6

City Cost Indexes

		\multicolumn{18}{c}{INDIANA}
\multicolumn{2}{c}{DIVISION}	\multicolumn{3}{c}{ANDERSON}	\multicolumn{3}{c}{BLOOMINGTON}	\multicolumn{3}{c}{COLUMBUS}	\multicolumn{3}{c}{EVANSVILLE}	\multicolumn{3}{c}{FORT WAYNE}	\multicolumn{3}{c}{GARY}
		\multicolumn{3}{c}{460}	\multicolumn{3}{c}{474}	\multicolumn{3}{c}{472}	\multicolumn{3}{c}{476 - 477}	\multicolumn{3}{c}{467 - 468}	\multicolumn{3}{c}{463 - 464}
		MAT.	INST.	TOTAL	MAT.	INST.	TOTAL	MAT.	INST.	TOTAL	MAT.	INST.	TOTAL	MAT.	INST.	TOTAL	MAT.	INST.	TOTAL
01543	CONTRACTOR EQUIPMENT	.0	96.3	96.3	.0	85.6	85.6	.0	85.6	85.6	.0	120.9	120.9	.0	96.3	96.3	.0	96.3	96.3
0241, 31 - 34	SITE & INFRASTRUCTURE, DEMOLITION	85.3	94.7	92.0	74.5	94.7	89.0	71.7	94.5	88.0	79.1	128.8	114.8	86.1	94.7	92.3	85.9	97.4	94.1
0310	Concrete Forming & Accessories	93.8	77.5	79.8	99.0	78.1	81.1	93.1	75.3	77.8	92.2	79.6	81.4	92.5	75.5	77.9	93.9	102.9	101.6
0310	Concrete Reinforcing	96.6	77.1	87.1	86.7	82.3	84.5	87.0	76.8	82.0	95.0	78.9	87.1	96.6	78.9	88.0	96.6	95.8	96.2
0330	Cast-in-Place Concrete	94.1	80.2	88.9	96.9	77.3	89.5	96.4	73.5	87.8	92.5	89.5	91.4	100.7	78.6	92.0	98.3	108.3	102.0
03	CONCRETE	91.7	78.9	85.6	99.4	78.4	89.5	98.8	74.8	87.4	99.5	83.1	91.8	94.6	77.8	86.7	93.8	103.5	98.4
04	MASONRY	89.6	81.4	84.7	88.0	77.4	81.7	87.9	77.4	81.6	84.1	82.9	83.4	93.2	82.6	86.9	90.9	103.3	98.3
05	METALS	92.4	87.1	90.8	95.4	76.3	89.6	95.4	73.3	88.7	88.2	85.2	87.3	92.4	87.6	90.9	92.4	101.4	95.1
06	WOOD, PLASTICS & COMPOSITES	112.1	76.8	93.5	117.0	77.3	96.0	111.1	73.8	91.3	96.4	77.4	86.4	111.9	73.9	91.8	110.0	103.0	106.3
07	THERMAL & MOISTURE PROTECTION	99.0	80.2	91.5	90.9	81.6	87.1	90.5	81.4	86.9	94.9	85.6	91.2	98.8	83.6	92.7	98.1	102.7	100.0
08	OPENINGS	98.3	78.6	93.2	104.0	80.3	97.8	99.3	76.9	93.5	96.2	78.2	91.5	98.3	74.7	92.2	98.3	100.8	99.0
0920	Plaster & Gypsum Board	93.6	76.7	83.5	93.5	77.4	83.9	90.3	73.8	80.5	89.1	76.0	81.3	92.9	73.7	81.5	89.1	103.5	97.7
0950, 0980	Ceilings & Acoustic Treatment	88.3	76.7	81.3	80.3	77.4	78.6	80.3	73.8	76.4	84.8	76.0	79.5	88.3	73.7	79.5	88.3	103.5	97.4
0960	Flooring	92.8	89.4	91.9	105.3	90.3	101.3	99.6	90.3	97.1	98.7	85.5	95.2	92.8	77.9	88.8	92.8	115.6	98.9
0990	Wall Finishes & Painting/Coating	95.5	74.6	83.0	92.1	86.3	88.6	92.1	86.3	88.6	98.7	88.1	92.4	95.5	77.6	84.8	95.5	104.5	100.9
09	FINISHES	90.1	79.3	84.5	94.0	81.1	87.4	91.8	79.1	85.2	92.4	81.3	86.7	89.8	76.0	82.7	89.2	105.7	97.7
COVERS	DIVS. 10 - 14, 25, 28, 41, 43, 44	100.0	90.5	98.1	100.0	85.8	97.1	100.0	85.4	97.0	100.0	90.7	98.1	100.0	85.2	97.0	100.0	89.9	97.9
22, 23	PLUMBING & HVAC	99.9	75.2	89.8	99.7	81.9	92.4	95.4	82.9	90.3	99.9	83.7	93.3	99.9	77.6	90.8	99.9	98.5	99.4
26, 27, 3370	ELECTRICAL, COMMUNICATIONS & UTIL.	84.6	91.5	87.9	99.8	84.2	92.3	99.1	86.7	93.1	94.4	83.8	89.4	85.7	79.4	82.6	98.0	99.7	98.8
MF2004	WEIGHTED AVERAGE	94.1	82.6	89.2	97.5	81.8	90.7	95.6	81.1	89.3	94.8	87.3	91.6	94.8	80.8	88.7	95.9	100.9	98.1

		\multicolumn{18}{c}{INDIANA}
\multicolumn{2}{c}{DIVISION}	\multicolumn{3}{c}{INDIANAPOLIS}	\multicolumn{3}{c}{KOKOMO}	\multicolumn{3}{c}{LAFAYETTE}	\multicolumn{3}{c}{LAWRENCEBURG}	\multicolumn{3}{c}{MUNCIE}	\multicolumn{3}{c}{NEW ALBANY}
		\multicolumn{3}{c}{461 - 462}	\multicolumn{3}{c}{469}	\multicolumn{3}{c}{479}	\multicolumn{3}{c}{470}	\multicolumn{3}{c}{473}	\multicolumn{3}{c}{471}
		MAT.	INST.	TOTAL	MAT.	INST.	TOTAL	MAT.	INST.	TOTAL	MAT.	INST.	TOTAL	MAT.	INST.	TOTAL	MAT.	INST.	TOTAL
01543	CONTRACTOR EQUIPMENT	.0	91.3	91.3	.0	96.3	96.3	.0	85.6	85.6	.0	107.5	107.5	.0	95.7	95.7	.0	95.5	95.5
0241, 31 - 34	SITE & INFRASTRUCTURE, DEMOLITION	85.4	98.6	94.9	82.3	94.7	91.2	72.5	94.5	88.3	70.6	114.0	101.7	74.4	94.8	89.0	67.7	98.4	89.7
0310	Concrete Forming & Accessories	94.3	87.4	88.4	96.9	76.2	79.2	90.6	79.2	80.8	89.2	74.0	76.1	90.2	77.2	79.0	87.8	71.9	74.1
0310	Concrete Reinforcing	95.9	84.2	90.2	87.4	77.0	82.3	86.7	75.2	81.0	85.8	82.7	84.3	95.9	77.0	86.7	87.3	80.5	84.0
0330	Cast-in-Place Concrete	90.5	88.0	89.6	93.1	81.1	88.6	96.9	80.1	90.6	90.7	75.1	84.8	101.8	79.2	93.3	93.6	74.9	86.6
03	CONCRETE	93.3	86.6	90.1	88.7	78.6	84.0	98.9	78.4	89.2	92.3	76.6	84.9	98.8	78.4	89.2	97.4	74.9	86.8
04	MASONRY	97.7	85.1	90.2	89.3	79.5	83.5	93.4	78.7	84.6	73.5	75.6	74.8	90.0	81.3	84.8	80.7	70.3	74.4
05	METALS	94.3	79.7	89.8	89.3	86.9	88.6	94.0	72.7	87.5	90.4	87.7	89.6	97.1	86.9	94.0	92.5	82.2	89.4
06	WOOD, PLASTICS & COMPOSITES	109.1	87.2	97.5	115.6	74.9	94.1	107.4	79.1	92.6	95.2	72.8	83.4	109.5	76.6	92.1	97.2	71.7	83.7
07	THERMAL & MOISTURE PROTECTION	97.2	86.9	93.0	98.7	79.2	90.9	90.5	80.2	86.4	91.3	78.7	86.3	93.2	80.2	88.0	84.8	67.7	77.9
08	OPENINGS	106.1	86.1	100.9	92.6	77.5	88.7	97.7	79.3	93.0	97.9	76.3	92.3	98.9	78.5	93.6	95.8	75.7	90.6
0920	Plaster & Gypsum Board	92.1	87.1	89.1	97.6	74.7	84.0	87.0	79.3	82.4	76.9	72.6	74.4	88.9	76.7	81.7	88.1	71.3	78.1
0950, 0980	Ceilings & Acoustic Treatment	92.5	87.1	89.3	87.4	74.7	79.8	72.7	79.3	76.6	89.0	72.6	79.2	81.1	76.7	78.5	84.8	71.3	76.7
0960	Flooring	92.6	92.0	92.4	97.3	84.9	94.0	98.4	91.0	96.4	73.3	89.4	77.6	97.8	89.4	95.6	96.6	66.3	88.5
0990	Wall Finishes & Painting/Coating	95.5	92.0	93.4	95.5	88.7	91.4	92.1	81.5	85.8	93.8	76.6	83.5	92.1	74.6	81.7	98.7	69.8	81.4
09	FINISHES	91.1	88.9	90.0	91.7	78.9	85.1	89.1	81.8	85.3	83.8	77.3	80.4	91.0	79.1	84.9	91.8	70.6	80.9
COVERS	DIVS. 10 - 14, 25, 28, 41, 43, 44	100.0	93.8	98.7	100.0	86.2	97.2	100.0	90.3	98.0	100.0	52.6	90.3	100.0	89.9	97.9	100.0	52.6	90.3
22, 23	PLUMBING & HVAC	99.9	87.0	94.6	95.7	83.0	90.5	95.4	73.0	86.3	96.1	79.3	89.3	99.6	75.1	89.7	95.7	75.2	87.3
26, 27, 3370	ELECTRICAL, COMMUNICATIONS & UTIL.	101.6	91.4	96.7	88.7	84.5	86.7	98.6	82.1	90.7	92.2	68.8	81.0	89.7	82.2	86.1	93.2	68.6	81.4
MF2004	WEIGHTED AVERAGE	97.7	88.1	93.6	92.4	82.8	88.2	95.2	79.4	88.5	91.7	79.6	86.5	95.9	81.2	89.5	93.2	74.9	85.3

		\multicolumn{9}{c}{INDIANA}	\multicolumn{9}{c}{IOWA}
\multicolumn{2}{c}{DIVISION}	\multicolumn{3}{c}{SOUTH BEND}	\multicolumn{3}{c}{TERRE HAUTE}	\multicolumn{3}{c}{WASHINGTON}	\multicolumn{3}{c}{BURLINGTON}	\multicolumn{3}{c}{CARROLL}	\multicolumn{3}{c}{CEDAR RAPIDS}
		\multicolumn{3}{c}{465 - 466}	\multicolumn{3}{c}{478}	\multicolumn{3}{c}{475}	\multicolumn{3}{c}{526}	\multicolumn{3}{c}{514}	\multicolumn{3}{c}{522 - 524}
		MAT.	INST.	TOTAL	MAT.	INST.	TOTAL	MAT.	INST.	TOTAL	MAT.	INST.	TOTAL	MAT.	INST.	TOTAL	MAT.	INST.	TOTAL
01543	CONTRACTOR EQUIPMENT	.0	106.2	106.2	.0	120.9	120.9	.0	120.9	120.9	.0	100.1	100.1	.0	100.1	100.1	.0	96.0	96.0
0241, 31 - 34	SITE & INFRASTRUCTURE, DEMOLITION	85.1	95.1	92.3	80.6	128.6	115.0	80.8	127.2	114.0	90.7	96.2	94.6	80.1	93.9	90.0	92.8	95.8	94.9
0310	Concrete Forming & Accessories	97.9	79.0	81.7	93.6	79.6	81.6	94.2	78.6	80.8	95.4	71.6	75.0	82.0	48.3	53.0	101.3	81.1	84.0
0310	Concrete Reinforcing	96.6	73.2	85.2	95.0	82.8	89.0	87.8	60.8	74.6	94.9	77.6	86.4	95.7	49.5	73.1	95.6	84.0	89.9
0330	Cast-in-Place Concrete	89.4	83.2	87.1	89.6	88.8	89.3	97.5	94.4	96.3	103.8	52.1	84.4	103.8	56.8	86.1	104.1	77.7	94.2
03	CONCRETE	85.9	80.7	83.5	101.9	83.6	93.2	107.2	80.7	94.7	99.1	67.0	84.0	97.9	52.6	76.5	99.3	81.1	90.7
04	MASONRY	88.1	81.0	83.9	91.0	82.0	85.6	84.3	84.6	84.5	96.6	59.4	74.3	99.3	44.5	66.5	103.1	80.8	89.7
05	METALS	92.4	99.0	94.4	88.9	87.5	88.5	84.0	74.2	81.0	87.2	90.4	88.2	87.3	70.7	82.2	89.4	92.8	90.5
06	WOOD, PLASTICS & COMPOSITES	112.0	77.9	94.0	98.7	78.7	88.1	99.0	77.2	87.5	101.4	77.7	88.9	85.6	50.0	66.8	108.4	80.0	93.4
07	THERMAL & MOISTURE PROTECTION	96.7	84.5	91.8	95.0	82.4	89.9	95.1	83.6	90.4	98.3	63.4	84.3	98.6	51.7	79.7	99.3	81.5	92.1
08	OPENINGS	91.7	77.9	88.1	96.7	81.1	92.7	93.5	73.1	88.2	92.4	70.8	86.8	97.6	46.0	84.3	98.6	84.0	94.8
0920	Plaster & Gypsum Board	93.2	77.8	84.1	89.1	77.4	82.2	87.8	75.9	80.7	95.8	77.2	84.7	90.7	48.7	65.7	99.2	79.8	87.7
0950, 0980	Ceilings & Acoustic Treatment	88.3	77.8	82.0	84.8	77.4	80.4	74.7	75.9	75.4	109.0	77.2	89.9	109.0	48.7	72.7	114.0	79.8	93.4
0960	Flooring	91.4	82.3	88.9	98.7	93.7	97.3	99.8	80.5	94.7	108.6	45.5	91.8	101.4	40.0	85.0	125.6	53.6	106.4
0990	Wall Finishes & Painting/Coating	94.9	81.2	86.7	98.7	90.4	93.8	98.7	89.1	92.9	101.5	82.0	89.8	101.5	38.4	63.8	105.3	74.1	86.6
09	FINISHES	89.5	79.8	84.5	92.4	83.3	87.7	90.3	80.7	85.4	105.0	68.3	86.0	100.8	45.4	72.2	112.0	74.3	92.5
COVERS	DIVS. 10 - 14, 25, 28, 41, 43, 44	100.0	86.1	97.1	100.0	93.8	98.7	100.0	90.7	98.1	100.0	81.3	96.1	100.0	77.0	95.3	100.0	87.5	97.4
22, 23	PLUMBING & HVAC	99.9	79.0	91.4	99.9	73.4	89.1	95.7	82.2	90.2	96.1	72.8	86.6	96.1	49.3	77.0	100.3	85.6	94.3
26, 27, 3370	ELECTRICAL, COMMUNICATIONS & UTIL.	96.7	84.9	91.0	92.9	84.3	88.8	93.4	83.3	88.5	97.6	76.4	87.4	98.2	39.3	69.9	95.5	83.9	89.9
MF2004	WEIGHTED AVERAGE	94.1	83.9	89.7	95.5	85.8	91.3	93.6	85.3	90.0	96.0	74.1	86.6	95.8	53.9	77.7	98.8	84.2	92.5

City Cost Indexes

		\multicolumn{18}{c}{IOWA}															
\multicolumn{2}{c}{DIVISION}	\multicolumn{3}{c}{COUNCIL BLUFFS 515}	\multicolumn{3}{c}{CRESTON 508}	\multicolumn{3}{c}{DAVENPORT 527-528}	\multicolumn{3}{c}{DECORAH 521}	\multicolumn{3}{c}{DES MOINES 500-503,509}	\multicolumn{3}{c}{DUBUQUE 520}
		MAT.	INST.	TOTAL	MAT.	INST.	TOTAL	MAT.	INST.	TOTAL	MAT.	INST.	TOTAL	MAT.	INST.	TOTAL	MAT.	INST.	TOTAL
01543	CONTRACTOR EQUIPMENT	.0	95.7	95.7	.0	100.1	100.1	.0	100.1	100.1	.0	100.1	100.1	.0	102.0	102.0	.0	94.6	94.6
0241, 31 - 34	SITE & INFRASTRUCTURE, DEMOLITION	100.7	92.1	94.6	80.1	95.0	90.8	90.5	99.7	97.1	89.2	94.5	93.0	81.7	100.7	95.3	90.7	92.5	92.0
0310	Concrete Forming & Accessories	81.1	55.2	58.9	82.5	71.1	72.7	101.0	92.6	93.8	92.5	45.2	51.9	102.9	78.3	81.8	82.7	73.9	75.1
0310	Concrete Reinforcing	97.6	75.4	86.7	95.1	81.8	88.6	95.6	91.3	93.5	94.9	49.7	72.8	95.6	82.9	89.4	94.3	83.8	89.2
0330	Cast-in-Place Concrete	108.3	70.1	93.9	103.5	59.2	86.8	100.2	96.7	98.9	101.0	53.5	83.1	97.8	98.4	98.0	101.9	94.4	99.1
03	CONCRETE	101.4	65.4	84.4	97.7	69.8	84.5	97.3	94.2	95.9	97.0	50.2	74.9	95.1	86.8	91.2	95.9	83.6	90.1
04	MASONRY	103.4	74.2	85.9	99.3	48.3	68.8	99.0	90.3	93.8	117.3	41.4	71.9	96.5	74.7	83.4	104.0	72.3	85.0
05	METALS	94.8	87.8	92.7	87.2	86.2	86.9	89.4	100.3	92.7	87.4	70.1	82.1	89.6	94.1	90.9	88.1	92.2	89.4
06	WOOD, PLASTICS & COMPOSITES	84.3	51.2	66.8	86.2	77.9	81.8	108.4	91.3	99.3	98.1	46.4	70.7	109.8	77.9	93.0	86.4	73.8	79.7
07	THERMAL & MOISTURE PROTECTION	98.7	64.7	85.0	99.4	56.2	82.0	98.8	91.1	95.7	98.6	46.3	77.5	100.2	79.5	91.9	98.9	70.5	87.5
08	OPENINGS	97.6	58.2	87.4	110.5	74.9	101.3	98.6	92.3	97.0	95.9	48.0	83.5	98.6	83.1	94.6	97.7	81.1	93.4
0920	Plaster & Gypsum Board	90.7	50.2	66.6	90.7	77.4	82.8	99.2	91.1	94.4	94.8	45.0	65.2	96.7	77.4	85.2	91.0	73.5	80.6
0950, 0980	Ceilings & Acoustic Treatment	109.0	50.2	73.6	109.0	77.4	90.0	114.0	91.1	100.2	109.0	45.0	70.5	112.3	77.4	91.3	109.0	73.5	87.6
0960	Flooring	100.3	43.8	85.2	101.7	40.0	85.2	111.5	100.0	108.4	108.0	56.8	94.3	111.3	40.0	92.3	114.7	37.3	94.1
0990	Wall Finishes & Painting/Coating	97.6	67.4	79.5	101.5	47.4	69.2	101.5	95.1	97.7	101.5	40.7	65.1	101.5	82.1	89.9	104.2	74.1	86.2
09	FINISHES	101.3	52.8	76.3	100.8	63.7	81.7	107.6	94.2	100.7	104.6	47.1	74.9	105.7	70.7	87.6	106.0	66.2	85.4
COVERS	DIVS. 10 - 14, 25, 28, 41, 43, 44	100.0	66.0	93.0	100.0	81.9	96.3	100.0	91.1	98.2	100.0	48.7	89.4	100.0	86.5	97.2	100.0	84.5	96.8
22, 23	PLUMBING & HVAC	100.3	79.4	91.8	96.1	51.4	77.9	100.3	91.4	96.7	96.1	42.7	74.3	100.3	76.6	90.7	100.3	77.1	90.9
26, 27, 3370	ELECTRICAL, COMMUNICATIONS & UTIL.	100.0	84.9	92.8	93.7	51.2	73.3	92.2	89.1	90.7	95.5	39.1	68.4	96.1	77.8	87.3	98.8	78.7	89.2
MF2004	WEIGHTED AVERAGE	99.3	74.2	88.5	96.4	64.3	82.6	97.5	93.2	95.7	96.8	51.2	77.1	97.2	81.5	90.5	97.7	79.2	89.7

		\multicolumn{18}{c}{IOWA}
\multicolumn{2}{c}{DIVISION}	\multicolumn{3}{c}{FORT DODGE 505}	\multicolumn{3}{c}{MASON CITY 504}	\multicolumn{3}{c}{OTTUMWA 525}	\multicolumn{3}{c}{SHENANDOAH 516}	\multicolumn{3}{c}{SIBLEY 512}	\multicolumn{3}{c}{SIOUX CITY 510-511}
		MAT.	INST.	TOTAL	MAT.	INST.	TOTAL	MAT.	INST.	TOTAL	MAT.	INST.	TOTAL	MAT.	INST.	TOTAL	MAT.	INST.	TOTAL
01543	CONTRACTOR EQUIPMENT	.0	100.1	100.1	.0	100.1	100.1	.0	94.6	94.6	.0	95.7	95.7	.0	100.1	100.1	.0	100.1	100.1
0241, 31 - 34	SITE & INFRASTRUCTURE, DEMOLITION	86.4	94.2	92.0	86.4	94.5	92.2	91.4	90.6	90.8	98.7	89.5	92.1	99.8	93.9	95.6	101.0	96.2	97.5
0310	Concrete Forming & Accessories	83.2	53.0	57.3	88.4	46.3	52.2	90.4	59.8	64.1	83.0	58.9	62.3	83.5	41.3	47.3	101.3	66.8	71.6
0310	Concrete Reinforcing	95.1	49.5	72.8	94.9	63.7	79.7	94.9	82.1	88.7	97.6	73.6	85.8	97.6	63.1	80.7	95.6	63.8	80.1
0330	Cast-in-Place Concrete	97.1	48.3	78.8	97.1	53.9	80.9	104.5	60.1	87.8	104.6	49.0	83.7	102.5	49.1	82.4	103.2	55.0	85.1
03	CONCRETE	93.2	51.9	73.7	93.5	53.5	74.6	98.7	65.3	82.9	99.0	59.3	80.2	97.9	49.5	75.1	98.6	63.1	81.9
04	MASONRY	97.4	37.3	61.5	110.2	46.9	72.3	100.7	49.7	70.2	103.1	33.7	61.6	122.6	37.9	72.0	96.5	56.9	72.9
05	METALS	87.3	71.1	82.4	87.3	77.9	84.5	87.2	89.6	87.9	93.9	80.0	89.7	87.4	76.0	84.0	89.4	81.4	87.0
06	WOOD, PLASTICS & COMPOSITES	86.9	56.8	71.0	92.7	46.4	68.2	95.3	61.7	77.6	86.4	66.5	75.9	87.3	41.2	62.9	108.4	67.2	86.6
07	THERMAL & MOISTURE PROTECTION	98.7	38.3	74.4	98.3	47.6	77.9	99.1	57.7	82.4	98.0	43.8	76.2	98.3	49.0	78.5	98.8	55.4	81.3
08	OPENINGS	103.0	47.8	88.7	93.3	51.9	82.6	97.6	67.4	89.8	87.1	62.9	80.9	93.7	47.5	81.7	98.6	64.1	89.7
0920	Plaster & Gypsum Board	91.0	55.7	70.0	92.3	45.0	64.1	93.5	61.0	74.2	91.0	66.0	76.1	91.0	39.7	60.5	99.2	66.4	79.7
0950, 0980	Ceilings & Acoustic Treatment	109.0	55.7	76.9	109.0	45.0	70.5	109.0	61.0	80.1	109.0	66.0	83.1	109.0	39.7	67.3	114.0	66.4	85.4
0960	Flooring	103.4	56.8	91.0	106.2	56.8	93.0	118.5	60.4	103.0	101.1	41.3	85.2	102.5	40.9	86.0	112.0	56.5	97.2
0990	Wall Finishes & Painting/Coating	101.5	17.4	51.2	101.5	37.2	63.0	104.2	82.1	91.0	97.6	26.3	55.0	101.5	38.4	63.8	102.6	64.6	79.8
09	FINISHES	102.4	50.4	75.6	103.5	46.7	74.2	107.6	61.6	83.9	101.5	53.2	76.6	103.6	40.5	71.0	109.0	64.6	86.1
COVERS	DIVS. 10 - 14, 25, 28, 41, 43, 44	100.0	77.7	95.4	100.0	49.6	89.6	100.0	78.6	95.6	100.0	77.2	95.3	100.0	60.0	91.8	100.0	83.4	96.6
22, 23	PLUMBING & HVAC	96.1	72.3	86.4	96.1	77.4	88.5	96.1	78.5	88.9	96.1	32.6	70.2	96.1	38.9	72.8	100.3	78.6	91.5
26, 27, 3370	ELECTRICAL, COMMUNICATIONS & UTIL.	99.7	52.1	76.9	98.8	66.4	83.2	97.4	76.4	87.3	95.5	31.1	64.6	95.5	39.2	68.5	95.5	76.4	86.3
MF2004	WEIGHTED AVERAGE	96.1	60.1	80.6	96.0	64.1	82.2	96.8	72.1	86.2	96.4	51.2	76.9	97.1	49.9	76.7	98.3	72.7	87.3

		\multicolumn{6}{c}{IOWA}	\multicolumn{12}{c}{KANSAS}
\multicolumn{2}{c}{DIVISION}	\multicolumn{3}{c}{SPENCER 513}	\multicolumn{3}{c}{WATERLOO 506-507}	\multicolumn{3}{c}{BELLEVILLE 669}	\multicolumn{3}{c}{COLBY 677}	\multicolumn{3}{c}{DODGE CITY 678}	\multicolumn{3}{c}{EMPORIA 668}
		MAT.	INST.	TOTAL	MAT.	INST.	TOTAL	MAT.	INST.	TOTAL	MAT.	INST.	TOTAL	MAT.	INST.	TOTAL	MAT.	INST.	TOTAL
01543	CONTRACTOR EQUIPMENT	.0	100.1	100.1	.0	100.1	100.1	.0	103.3	103.3	.0	103.3	103.3	.0	103.3	103.3	.0	101.3	101.3
0241, 31 - 34	SITE & INFRASTRUCTURE, DEMOLITION	99.9	93.9	95.6	91.0	95.3	94.1	111.4	92.4	97.8	108.7	93.1	97.5	111.0	92.7	97.9	102.2	89.8	93.3
0310	Concrete Forming & Accessories	90.3	41.1	48.1	101.8	46.6	54.4	93.7	46.2	52.9	100.8	65.6	70.6	93.3	65.6	69.5	84.5	37.4	44.1
0310	Concrete Reinforcing	97.6	39.9	69.4	95.6	83.5	89.7	103.2	54.3	79.3	105.8	53.9	80.4	103.2	54.7	79.5	101.8	47.3	75.1
0330	Cast-in-Place Concrete	102.5	49.0	82.4	104.1	45.9	82.2	117.9	54.2	94.0	113.4	55.0	91.4	115.4	54.6	92.6	114.0	41.0	86.5
03	CONCRETE	98.4	45.0	73.2	99.3	54.8	78.3	114.8	51.9	85.1	112.1	60.9	87.9	113.2	60.8	88.5	107.7	42.1	76.7
04	MASONRY	122.6	37.9	72.0	98.1	58.3	74.2	94.5	48.5	67.0	95.6	39.5	62.0	103.4	55.1	74.5	99.9	38.7	63.3
05	METALS	87.4	65.2	80.7	89.4	90.4	89.7	94.7	75.7	89.0	95.1	78.0	89.9	96.4	77.8	90.8	94.5	70.9	87.3
06	WOOD, PLASTICS & COMPOSITES	95.2	41.2	66.6	109.0	46.4	75.9	92.7	47.1	68.6	100.6	73.1	86.1	92.0	73.1	82.0	82.9	37.3	58.8
07	THERMAL & MOISTURE PROTECTION	99.2	43.1	76.6	98.6	50.9	79.4	96.7	56.4	80.6	96.9	56.6	80.7	96.8	60.7	82.3	95.5	39.3	72.9
08	OPENINGS	107.2	40.8	90.0	94.3	59.6	85.3	96.9	47.2	84.0	97.0	61.2	87.7	96.9	61.2	87.7	94.4	39.6	80.2
0920	Plaster & Gypsum Board	93.5	39.7	61.5	99.2	45.0	66.9	99.3	45.6	67.3	102.5	72.3	84.5	98.6	72.3	82.9	95.4	35.5	59.8
0950, 0980	Ceilings & Acoustic Treatment	109.0	39.7	67.3	114.0	45.0	72.5	86.4	45.6	61.8	86.4	72.3	77.9	86.4	72.3	77.9	86.4	35.5	55.8
0960	Flooring	105.9	40.9	88.5	113.3	56.8	98.2	97.5	47.2	84.1	101.7	47.2	87.2	97.3	47.2	84.0	91.5	43.8	78.8
0990	Wall Finishes & Painting/Coating	101.5	38.4	63.8	102.6	34.4	61.8	91.5	48.1	65.6	91.5	48.1	65.6	91.5	48.1	65.6	91.5	48.1	65.6
09	FINISHES	105.0	40.5	71.7	108.2	46.3	76.3	94.2	46.0	69.3	95.8	61.2	78.0	93.9	61.2	77.1	90.9	39.0	64.2
COVERS	DIVS. 10 - 14, 25, 28, 41, 43, 44	100.0	75.9	95.0	100.0	76.7	95.2	100.0	45.4	88.8	100.0	48.6	89.4	100.0	48.6	89.4	100.0	40.7	87.8
22, 23	PLUMBING & HVAC	96.1	38.9	72.8	100.3	43.2	77.0	95.7	68.0	84.4	95.7	61.3	81.7	100.0	61.3	84.2	95.7	66.9	84.0
26, 27, 3370	ELECTRICAL, COMMUNICATIONS & UTIL.	97.0	39.2	69.3	95.5	66.7	81.7	100.7	69.2	85.5	100.7	69.1	85.6	97.8	74.3	86.5	98.1	77.0	88.0
MF2004	WEIGHTED AVERAGE	98.8	48.3	77.0	97.7	60.7	81.7	99.0	61.9	83.0	99.0	64.0	83.9	100.1	66.4	85.6	97.3	57.9	80.3

655

City Cost Indexes

		KANSAS																	
		FORT SCOTT			HAYS			HUTCHINSON			INDEPENDENCE			KANSAS CITY			LIBERAL		
	DIVISION	667			676			675			673			660 - 662			679		
		MAT.	INST.	TOTAL	MAT.	INST.	TOTAL	MAT.	INST.	TOTAL	MAT.	INST.	TOTAL	MAT.	INST.	TOTAL	MAT.	INST.	TOTAL
01543	CONTRACTOR EQUIPMENT	.0	102.4	102.4	.0	103.3	103.3	.0	103.3	103.3	.0	103.3	103.3	.0	99.6	99.6	.0	103.3	103.3
0241, 31 - 34	SITE & INFRASTRUCTURE, DEMOLITION	98.7	91.7	93.7	114.2	93.1	99.1	91.0	92.7	92.2	112.2	93.2	98.5	91.1	91.6	91.5	114.4	92.7	98.8
0310	Concrete Forming & Accessories	102.3	70.2	74.7	98.2	65.7	70.3	87.3	63.1	66.5	110.7	58.0	65.5	98.7	103.1	102.4	93.7	63.1	67.4
0310	Concrete Reinforcing	101.1	68.6	85.2	103.2	54.4	79.4	103.2	44.8	74.7	102.6	61.7	82.6	98.1	91.4	94.8	104.7	44.8	75.4
0330	Cast-in-Place Concrete	105.7	58.6	88.0	89.3	55.1	76.4	82.8	39.9	66.6	115.9	41.3	87.8	90.5	95.5	92.4	89.3	39.9	70.7
03	CONCRETE	103.0	66.8	85.9	102.4	61.0	82.9	86.8	52.7	70.7	114.5	54.2	86.1	94.5	98.5	96.4	104.3	52.7	80.0
04	MASONRY	100.7	57.3	74.7	103.7	48.7	70.8	95.2	37.6	60.7	93.0	50.5	67.6	102.8	100.2	101.2	101.4	37.6	63.2
05	METALS	94.4	83.7	91.2	94.7	78.8	89.9	94.5	71.0	87.4	94.4	80.7	90.3	101.2	99.6	100.7	94.9	70.9	87.7
06	WOOD, PLASTICS & COMPOSITES	103.2	76.0	88.8	97.5	73.1	84.6	86.1	73.1	79.2	111.5	64.1	86.4	99.1	104.4	101.9	92.7	73.1	82.4
07	THERMAL & MOISTURE PROTECTION	96.3	68.3	85.0	97.1	59.1	81.8	95.9	45.2	75.5	96.9	63.0	83.3	94.9	102.0	97.8	97.2	40.5	74.4
08	OPENINGS	94.4	70.5	88.2	96.8	61.2	87.6	96.8	58.9	87.0	94.4	58.1	85.0	95.9	96.5	96.1	96.9	58.9	87.1
0920	Plaster & Gypsum Board	101.2	75.3	85.8	101.2	72.3	84.0	97.3	72.3	82.4	107.0	63.0	80.8	95.5	104.3	100.8	99.3	72.3	83.2
0950, 0980	Ceilings & Acoustic Treatment	86.4	75.3	79.7	86.4	72.3	77.9	86.4	72.3	77.9	86.4	63.0	72.3	87.2	104.3	97.5	86.4	72.3	77.9
0960	Flooring	108.2	47.5	92.0	100.4	47.2	86.2	93.7	47.2	81.3	107.1	47.2	91.1	88.3	102.0	91.9	97.5	47.2	84.1
0990	Wall Finishes & Painting/Coating	93.7	48.1	66.4	91.5	48.1	65.6	91.5	48.1	65.6	91.5	48.1	65.6	99.3	83.7	89.9	91.5	48.1	65.6
09	FINISHES	97.0	64.2	80.1	95.6	61.2	77.9	91.4	60.2	75.3	98.3	55.2	76.0	92.0	100.1	96.2	94.7	60.2	76.9
COVERS	DIVS. 10 - 14, 25, 28, 41, 43, 44	100.0	55.6	90.9	100.0	48.6	89.4	100.0	47.0	89.1	100.0	46.3	89.0	100.0	79.5	95.8	100.0	47.0	89.1
22, 23	PLUMBING & HVAC	95.7	64.1	82.8	95.7	61.3	81.7	95.7	50.7	77.4	95.7	66.5	83.8	99.9	93.7	97.4	95.7	58.9	80.7
26, 27, 3370	ELECTRICAL, COMMUNICATIONS & UTIL.	97.4	77.1	87.7	99.7	69.1	85.0	95.0	69.1	82.6	97.1	77.1	87.5	102.6	103.7	103.1	97.8	74.2	86.5
MF2004	WEIGHTED AVERAGE	97.3	70.1	85.6	98.2	65.2	83.9	94.3	59.3	79.2	98.8	65.8	84.5	98.5	97.7	98.1	98.0	61.6	82.3

		KANSAS									KENTUCKY								
		SALINA			TOPEKA			WICHITA			ASHLAND			BOWLING GREEN			CAMPTON		
	DIVISION	674			664 - 666			670 - 672			411 - 412			421 - 422			413 - 414		
		MAT.	INST.	TOTAL	MAT.	INST.	TOTAL	MAT.	INST.	TOTAL	MAT.	INST.	TOTAL	MAT.	INST.	TOTAL	MAT.	INST.	TOTAL
01543	CONTRACTOR EQUIPMENT	.0	103.3	103.3	.0	101.3	101.3	.0	103.3	103.3	.0	98.2	98.2	.0	95.5	95.5	.0	103.1	103.1
0241, 31 - 34	SITE & INFRASTRUCTURE, DEMOLITION	100.7	93.2	95.3	93.4	90.5	91.4	94.3	93.1	93.4	98.3	85.5	89.1	67.6	99.3	90.3	75.3	100.5	93.4
0310	Concrete Forming & Accessories	89.4	46.3	52.4	98.7	49.2	56.2	96.9	54.5	60.5	85.6	123.1	117.8	83.6	84.9	84.7	86.9	56.3	60.7
0310	Concrete Reinforcing	102.6	75.9	89.6	95.3	94.6	95.0	95.3	78.5	87.1	88.8	102.5	95.5	86.1	94.7	90.3	86.9	75.7	81.4
0330	Cast-in-Place Concrete	100.1	46.8	80.0	91.6	51.7	76.6	85.6	56.1	74.5	85.2	102.2	91.6	84.5	112.9	95.2	94.3	66.6	83.9
03	CONCRETE	100.5	53.7	78.4	92.8	60.2	77.4	89.6	60.9	76.0	93.1	112.2	102.1	91.3	96.4	93.7	95.4	64.3	80.7
04	MASONRY	118.1	48.3	76.4	97.9	61.2	75.9	91.4	62.6	74.2	89.5	103.3	97.7	91.0	77.3	82.8	89.7	41.9	61.1
05	METALS	96.2	87.2	93.5	101.5	96.6	100.0	101.5	87.4	97.2	91.6	111.1	97.5	93.1	89.0	91.9	92.5	79.3	88.5
06	WOOD, PLASTICS & COMPOSITES	88.1	47.1	66.5	96.8	46.7	70.3	96.3	53.4	73.6	78.4	128.9	105.1	91.9	83.1	87.3	90.3	54.6	71.4
07	THERMAL & MOISTURE PROTECTION	96.4	55.2	79.8	97.2	68.3	85.6	96.5	62.8	82.9	86.4	103.1	93.1	84.6	90.6	87.1	95.2	56.8	79.7
08	OPENINGS	96.8	51.1	85.0	97.2	62.2	88.1	97.2	60.1	87.6	94.3	112.5	99.0	95.8	83.4	92.6	97.4	66.2	89.3
0920	Plaster & Gypsum Board	98.0	45.6	66.8	99.8	45.1	67.2	99.8	52.0	71.3	60.3	129.7	101.6	85.3	83.0	83.9	85.3	52.7	65.8
0950, 0980	Ceilings & Acoustic Treatment	86.4	45.6	61.8	87.2	45.1	61.9	87.2	52.0	66.0	80.2	129.7	110.0	84.8	83.0	83.7	84.8	52.7	65.5
0960	Flooring	95.2	34.5	79.0	99.0	43.8	84.3	98.0	75.8	92.1	78.7	100.3	84.5	94.3	91.3	93.5	95.9	49.6	83.5
0990	Wall Finishes & Painting/Coating	91.5	48.1	65.6	91.5	64.9	75.6	91.5	59.3	72.2	101.6	104.1	103.1	98.7	90.1	93.6	98.7	42.7	65.2
09	FINISHES	92.6	43.2	67.1	93.8	48.0	70.2	93.6	58.7	75.6	79.5	118.7	99.7	90.6	86.4	88.4	91.1	52.4	71.1
COVERS	DIVS. 10 - 14, 25, 28, 41, 43, 44	100.0	59.0	91.6	100.0	63.6	92.5	100.0	62.6	92.3	100.0	96.5	99.3	100.0	74.0	94.6	100.0	65.9	93.0
22, 23	PLUMBING & HVAC	100.0	61.3	84.2	100.0	69.2	87.4	100.0	62.4	84.7	95.6	93.8	94.8	99.9	84.7	93.7	95.7	78.9	88.9
26, 27, 3370	ELECTRICAL, COMMUNICATIONS & UTIL.	97.6	69.1	83.9	102.5	77.1	90.3	100.3	69.1	85.3	90.9	98.0	94.3	93.7	82.4	88.3	91.4	82.4	87.1
MF2004	WEIGHTED AVERAGE	99.0	62.0	83.0	98.5	69.2	85.8	97.5	67.4	84.5	92.2	103.4	97.0	94.1	86.9	91.0	93.9	70.4	83.7

		KENTUCKY																	
		CORBIN			COVINGTON			ELIZABETHTOWN			FRANKFORT			HAZARD			HENDERSON		
	DIVISION	407 - 409			410			427			406			417 - 418			424		
		MAT.	INST.	TOTAL	MAT.	INST.	TOTAL	MAT.	INST.	TOTAL	MAT.	INST.	TOTAL	MAT.	INST.	TOTAL	MAT.	INST.	TOTAL
01543	CONTRACTOR EQUIPMENT	.0	103.1	103.1	.0	107.5	107.5	.0	95.5	95.5	.0	103.1	103.1	.0	103.1	103.1	.0	120.9	120.9
0241, 31 - 34	SITE & INFRASTRUCTURE, DEMOLITION	72.9	100.5	92.7	71.9	117.2	104.4	62.8	98.9	88.7	73.4	101.2	93.3	73.4	102.3	94.1	70.3	128.4	112.0
0310	Concrete Forming & Accessories	83.9	55.5	59.9	82.8	115.0	110.4	78.6	83.2	82.6	97.6	68.7	72.7	83.9	48.1	53.1	90.3	82.7	83.8
0310	Concrete Reinforcing	86.5	75.8	81.3	85.5	97.9	91.5	86.5	94.9	90.6	87.7	93.4	90.5	87.3	76.5	82.0	86.2	92.7	89.4
0330	Cast-in-Place Concrete	89.2	66.5	80.6	90.2	101.0	94.3	76.4	110.6	89.3	85.5	74.4	81.3	90.7	55.1	77.3	74.7	98.3	83.6
03	CONCRETE	90.8	64.1	78.2	93.7	107.1	100.0	84.0	94.9	89.1	90.0	75.7	83.2	92.3	56.8	75.6	88.3	90.0	89.1
04	MASONRY	89.3	44.4	62.4	103.3	122.8	115.0	75.1	80.8	78.5	88.3	91.4	90.1	86.4	52.5	66.1	94.9	93.8	94.2
05	METALS	92.4	79.1	88.4	90.4	107.5	95.5	92.4	89.2	91.4	99.8	90.4	97.0	92.5	78.6	88.3	83.8	90.9	85.9
06	WOOD, PLASTICS & COMPOSITES	86.8	54.6	69.8	87.6	106.8	97.7	86.3	83.1	84.6	103.5	66.2	83.8	86.8	39.9	62.0	94.4	79.4	86.5
07	THERMAL & MOISTURE PROTECTION	95.0	58.5	80.3	91.5	106.7	97.6	84.3	83.0	83.8	95.0	81.9	89.7	95.1	56.1	79.4	94.5	96.4	95.3
08	OPENINGS	96.7	57.7	86.6	99.0	108.5	101.4	95.8	86.1	93.3	95.9	76.4	90.9	97.8	47.2	84.7	94.0	83.4	91.2
0920	Plaster & Gypsum Board	84.0	52.7	65.3	73.5	107.5	93.7	83.7	83.0	83.3	89.1	64.5	74.5	84.0	37.5	56.3	85.9	78.1	81.2
0950, 0980	Ceilings & Acoustic Treatment	84.8	52.7	65.5	86.5	107.5	99.1	84.8	83.0	83.7	84.8	64.5	72.6	84.8	37.5	56.4	74.7	78.1	76.7
0960	Flooring	93.9	49.6	82.1	70.6	95.0	77.1	91.3	72.7	86.3	99.5	65.5	90.4	93.9	44.0	80.6	97.8	91.2	96.0
0990	Wall Finishes & Painting/Coating	98.7	42.7	65.2	93.8	89.1	90.9	98.7	79.0	86.9	98.7	58.0	74.4	98.7	35.0	60.6	98.7	99.7	99.3
09	FINISHES	90.2	52.4	70.7	82.0	108.7	95.8	89.1	80.4	84.6	92.7	65.7	78.7	90.2	44.4	66.6	88.8	85.7	87.2
COVERS	DIVS. 10 - 14, 25, 28, 41, 43, 44	100.0	59.1	91.6	100.0	106.8	101.4	100.0	90.3	98.0	100.0	81.5	96.2	100.0	68.1	93.4	100.0	82.9	96.5
22, 23	PLUMBING & HVAC	95.7	78.5	88.7	96.2	103.9	99.3	95.9	82.6	90.5	99.9	80.3	91.9	95.7	51.3	77.6	95.9	81.0	89.9
26, 27, 3370	ELECTRICAL, COMMUNICATIONS & UTIL.	91.6	82.4	87.2	94.4	78.0	86.5	91.2	81.8	86.7	93.8	108.6	100.9	91.3	62.6	77.5	93.1	83.1	88.3
MF2004	WEIGHTED AVERAGE	93.0	70.0	83.1	93.6	105.1	98.5	90.7	86.1	88.7	95.6	85.4	91.2	93.2	60.1	78.9	91.6	90.0	90.9

City Cost Indexes

DIVISION		KENTUCKY																
		LEXINGTON 403 - 405			LOUISVILLE 400 - 402			OWENSBORO 423			PADUCAH 420			PIKEVILLE 415 - 416			SOMERSET 425 - 426		
		MAT.	INST.	TOTAL	MAT.	INST.	TOTAL	MAT.	INST.	TOTAL	MAT.	INST.	TOTAL	MAT.	INST.	TOTAL	MAT.	INST.	TOTAL
01543	CONTRACTOR EQUIPMENT	.0	103.1	103.1	.0	95.5	95.5	.0	120.9	120.9	.0	120.9	120.9	.0	98.2	98.2	.0	103.1	103.1
0241, 31 - 34	SITE & INFRASTRUCTURE, DEMOLITION	74.6	101.9	94.2	65.5	99.0	89.5	79.1	129.3	115.1	72.8	128.8	113.0	108.1	85.5	91.9	66.9	101.1	91.4
0310	Concrete Forming & Accessories	96.8	74.7	77.8	93.1	83.0	84.4	88.5	80.7	81.8	86.3	83.6	84.0	94.8	78.2	80.6	84.4	54.2	58.5
0310	Concrete Reinforcing	95.0	94.7	94.9	95.0	94.9	94.9	86.2	92.1	89.1	86.7	89.1	87.9	89.3	102.4	95.7	86.5	91.6	89.0
0330	Cast-in-Place Concrete	91.3	85.6	89.1	91.0	78.3	86.2	87.1	109.8	95.7	79.6	97.5	86.3	93.6	109.2	99.5	74.7	115.7	90.1
03	CONCRETE	94.0	82.5	88.6	93.7	83.7	89.0	99.7	92.9	96.5	92.5	89.4	91.1	105.7	94.6	100.4	79.4	83.1	81.2
04	MASONRY	88.0	73.3	79.2	89.7	80.8	84.4	87.8	79.8	83.0	91.0	92.9	92.1	87.9	75.2	80.3	82.1	63.5	71.0
05	METALS	94.6	90.3	93.3	101.9	89.0	98.0	85.1	89.6	86.5	82.5	88.8	84.4	91.5	105.9	95.9	92.4	86.8	90.7
06	WOOD, PLASTICS & COMPOSITES	103.0	71.9	86.6	103.5	83.1	92.7	92.3	77.0	84.2	89.9	81.6	85.5	88.8	69.4	78.5	87.9	46.2	65.8
07	THERMAL & MOISTURE PROTECTION	95.3	93.9	94.7	84.9	83.0	84.2	94.9	82.5	89.9	94.6	88.1	92.0	87.0	67.9	79.3	94.5	62.8	81.8
08	OPENINGS	96.7	79.8	92.4	96.7	86.1	94.0	94.0	82.1	90.9	93.1	83.0	90.5	95.1	79.7	91.1	96.7	65.6	88.6
0920	Plaster & Gypsum Board	91.3	70.4	78.9	92.2	83.0	86.7	84.9	75.7	79.4	84.3	80.3	81.9	63.5	68.6	66.6	84.9	44.0	60.6
0950, 0980	Ceilings & Acoustic Treatment	89.0	70.4	77.8	89.0	83.0	85.4	74.7	75.7	75.3	74.7	80.3	78.1	80.2	68.6	73.2	84.8	44.0	60.3
0960	Flooring	99.5	66.8	90.7	97.9	72.7	91.2	96.9	91.3	95.4	95.6	71.5	89.1	83.7	59.1	77.2	94.2	49.6	82.3
0990	Wall Finishes & Painting/Coating	98.7	77.6	86.1	98.7	79.0	86.9	98.7	99.7	99.3	98.7	78.9	86.8	101.6	100.4	100.9	98.7	58.9	74.9
09	FINISHES	94.1	72.3	82.9	93.5	80.4	86.7	88.8	84.2	86.4	88.1	80.4	84.1	82.1	75.8	78.9	90.0	51.7	70.2
COVERS	DIVS. 10 - 14, 25, 28, 41, 43, 44	100.0	96.4	99.3	100.0	95.7	99.1	100.0	97.4	99.5	100.0	79.2	95.7	100.0	71.4	94.1	100.0	70.6	94.0
22, 23	PLUMBING & HVAC	99.9	74.8	89.7	99.9	82.6	92.9	99.9	79.1	91.4	95.9	85.4	91.6	95.6	90.6	93.6	95.9	74.4	87.2
26, 27, 3370	ELECTRICAL, COMMUNICATIONS & UTIL.	94.1	81.8	88.2	94.7	81.8	88.5	93.2	82.5	88.1	95.3	86.4	91.0	93.5	91.5	92.5	91.7	81.8	87.0
MF2004	WEIGHTED AVERAGE	95.5	81.5	89.4	96.1	84.7	91.2	93.9	88.2	91.4	91.8	90.1	91.0	94.6	86.8	91.2	91.3	74.9	84.2

DIVISION		LOUISIANA																
		ALEXANDRIA 713 - 714			BATON ROUGE 707 - 708			HAMMOND 704			LAFAYETTE 705			LAKE CHARLES 706			MONROE 712		
		MAT.	INST.	TOTAL	MAT.	INST.	TOTAL	MAT.	INST.	TOTAL	MAT.	INST.	TOTAL	MAT.	INST.	TOTAL	MAT.	INST.	TOTAL
01543	CONTRACTOR EQUIPMENT	.0	87.2	87.2	.0	87.5	87.5	.0	88.4	88.4	.0	88.4	88.4	.0	87.5	87.5	.0	87.2	87.2
0241, 31 - 34	SITE & INFRASTRUCTURE, DEMOLITION	104.2	85.4	90.7	109.4	85.4	92.2	110.1	86.9	93.4	110.7	86.8	93.5	111.4	85.1	92.6	104.2	84.9	90.4
0310	Concrete Forming & Accessories	82.8	42.2	47.9	97.9	57.9	63.5	83.7	57.0	60.8	104.5	51.9	59.3	105.4	57.4	64.2	82.1	42.3	47.9
0310	Concrete Reinforcing	98.5	63.2	81.2	101.2	62.8	82.4	99.8	62.3	81.5	101.2	62.7	82.4	101.2	62.8	82.4	97.4	63.3	80.7
0330	Cast-in-Place Concrete	94.7	42.3	75.0	80.3	50.4	69.1	84.1	49.4	71.0	83.7	49.8	70.9	88.1	51.8	74.4	94.7	45.7	76.3
03	CONCRETE	89.6	47.3	69.7	89.6	56.8	74.2	90.5	56.0	74.2	91.8	54.0	73.9	94.1	57.1	76.6	89.4	48.5	70.1
04	MASONRY	112.3	50.2	75.2	102.9	51.6	72.2	101.8	58.2	75.7	101.8	50.7	71.2	100.5	52.6	71.8	107.4	54.0	75.4
05	METALS	88.4	73.3	83.9	95.4	70.4	87.8	91.0	71.1	85.0	90.2	70.6	84.3	90.2	70.4	84.2	88.4	71.6	83.3
06	WOOD, PLASTICS & COMPOSITES	84.0	41.3	61.4	103.8	61.3	81.3	87.6	59.2	72.6	112.9	53.9	81.7	110.7	60.6	84.2	83.2	41.7	61.3
07	THERMAL & MOISTURE PROTECTION	99.9	52.0	80.6	97.5	57.2	81.3	100.5	62.3	85.2	101.2	55.4	82.7	100.5	57.0	83.0	99.9	54.6	81.7
08	OPENINGS	99.2	47.9	85.9	102.7	58.4	91.2	98.0	61.7	88.6	102.7	54.4	90.2	102.7	59.5	91.5	99.1	52.6	87.1
0920	Plaster & Gypsum Board	83.6	40.3	57.8	102.2	60.6	77.5	94.9	58.5	73.2	102.2	53.0	72.9	102.2	59.9	77.0	83.3	40.7	58.0
0950, 0980	Ceilings & Acoustic Treatment	94.7	40.3	62.0	98.7	60.6	75.8	99.4	58.5	74.8	97.7	53.0	70.8	98.7	59.9	75.4	94.7	40.7	62.2
0960	Flooring	103.0	63.6	92.5	107.3	62.3	95.3	96.4	55.8	85.6	107.5	41.8	90.0	107.5	48.7	91.8	102.6	40.2	86.0
0990	Wall Finishes & Painting/Coating	98.3	39.4	63.1	102.9	45.1	68.3	102.9	59.7	77.1	102.9	45.1	68.3	102.9	41.6	66.2	98.3	47.0	67.6
09	FINISHES	94.5	45.0	69.0	103.3	57.7	79.8	99.0	57.1	77.4	103.1	48.8	75.1	103.3	54.4	78.1	94.4	41.5	67.1
COVERS	DIVS. 10 - 14, 25, 28, 41, 43, 44	100.0	63.7	92.5	100.0	73.4	94.5	100.0	61.1	92.0	100.0	72.1	94.3	100.0	73.4	94.5	100.0	58.2	91.4
22, 23	PLUMBING & HVAC	100.1	59.6	83.6	100.0	55.4	81.8	95.8	55.7	79.5	100.0	60.6	84.0	100.0	60.6	84.0	100.1	52.3	80.6
26, 27, 3370	ELECTRICAL, COMMUNICATIONS & UTIL.	94.1	57.8	76.7	98.0	61.0	80.3	96.6	54.5	76.4	97.7	63.3	81.2	97.2	63.3	81.0	95.8	58.7	78.0
MF2004	WEIGHTED AVERAGE	96.4	57.6	79.6	98.7	61.0	82.4	96.1	60.8	80.9	98.3	60.5	82.0	98.4	62.1	82.8	96.3	56.2	79.0

DIVISION		LOUISIANA									MAINE								
		NEW ORLEANS 700 - 701			SHREVEPORT 710 - 711			THIBODAUX 703			AUGUSTA 043			BANGOR 044			BATH 045		
		MAT.	INST.	TOTAL	MAT.	INST.	TOTAL	MAT.	INST.	TOTAL	MAT.	INST.	TOTAL	MAT.	INST.	TOTAL	MAT.	INST.	TOTAL
01543	CONTRACTOR EQUIPMENT	.0	88.7	88.7	.0	87.2	87.2	.0	88.4	88.4	.0	100.7	100.7	.0	100.7	100.7	.0	100.7	100.7
0241, 31 - 34	SITE & INFRASTRUCTURE, DEMOLITION	114.5	88.0	95.5	101.6	84.8	89.5	112.9	86.8	94.2	79.5	101.6	95.3	79.3	99.4	93.7	77.6	101.6	94.8
0310	Concrete Forming & Accessories	104.8	66.1	71.6	101.7	45.1	53.1	97.7	66.2	70.6	98.5	61.3	66.7	92.7	86.9	87.7	88.2	61.5	65.2
0310	Concrete Reinforcing	101.2	64.2	83.1	96.9	63.0	80.4	99.8	63.7	82.2	85.2	106.5	95.6	84.9	106.7	95.6	83.9	106.5	95.0
0330	Cast-in-Place Concrete	87.2	74.9	82.6	84.5	49.3	71.2	90.4	52.9	76.3	81.7	59.9	73.5	81.1	61.0	73.6	81.1	69.3	73.2
03	CONCRETE	93.6	69.3	82.1	84.9	50.9	68.8	95.5	61.6	79.5	98.7	69.6	85.0	97.5	81.1	89.8	97.5	69.7	84.4
04	MASONRY	102.0	61.1	77.5	99.0	47.5	68.2	129.8	57.1	86.3	97.2	48.8	68.3	112.6	55.9	78.7	119.2	48.4	76.8
05	METALS	102.6	74.0	93.9	84.8	71.1	80.7	91.0	71.6	85.2	84.6	83.7	84.3	84.2	80.1	83.0	82.8	83.8	83.1
06	WOOD, PLASTICS & COMPOSITES	106.3	68.5	86.3	105.8	45.4	73.9	98.1	71.1	83.8	98.7	59.5	78.0	91.8	93.0	92.4	86.2	59.5	72.1
07	THERMAL & MOISTURE PROTECTION	101.3	62.4	85.7	98.9	53.3	80.6	101.5	60.5	85.0	98.7	52.2	80.0	98.6	57.2	81.9	98.5	56.3	81.5
08	OPENINGS	103.4	68.2	94.3	97.0	50.2	84.9	103.7	68.5	94.6	102.5	58.4	91.1	102.4	78.1	96.1	102.4	60.1	91.4
0920	Plaster & Gypsum Board	101.3	68.0	81.5	88.6	44.5	62.4	99.1	70.7	82.2	100.9	57.6	75.1	97.9	92.0	94.4	95.4	57.6	72.9
0950, 0980	Ceilings & Acoustic Treatment	97.7	68.0	79.9	95.5	44.5	64.8	99.4	70.7	82.2	87.7	57.6	69.6	86.0	92.0	89.6	84.2	57.6	68.2
0960	Flooring	107.9	47.8	91.9	113.3	59.5	98.9	104.0	53.1	90.4	98.0	40.5	82.7	95.6	50.5	83.6	93.6	40.5	79.4
0990	Wall Finishes & Painting/Coating	104.5	65.1	81.0	98.3	39.4	63.1	104.5	59.7	77.7	91.0	34.0	56.9	91.0	31.4	55.3	91.0	34.0	56.9
09	FINISHES	103.6	62.6	82.4	98.6	46.5	71.7	102.2	63.7	82.3	96.7	53.4	74.3	95.1	75.3	84.9	93.6	53.4	72.8
COVERS	DIVS. 10 - 14, 25, 28, 41, 43, 44	100.0	76.4	95.1	100.0	70.7	94.0	100.0	75.3	94.9	100.0	56.6	91.1	100.0	75.4	94.9	100.0	56.6	91.1
22, 23	PLUMBING & HVAC	100.0	63.3	85.1	100.1	60.0	83.8	95.8	61.2	81.7	100.0	72.5	88.8	100.0	73.6	89.3	95.8	72.5	86.3
26, 27, 3370	ELECTRICAL, COMMUNICATIONS & UTIL.	98.2	67.8	83.7	95.0	69.0	82.6	95.2	66.1	81.2	98.1	84.1	91.4	96.6	82.4	89.8	95.0	84.1	89.7
MF2004	WEIGHTED AVERAGE	100.7	68.2	86.6	94.9	59.7	79.7	99.0	65.8	84.7	96.4	70.9	85.3	96.6	77.1	88.2	95.4	70.9	84.8

City Cost Indexes

		MAINE																	
		HOULTON			KITTERY			LEWISTON			MACHIAS			PORTLAND			ROCKLAND		
	DIVISION	047			039			042			046			040 - 041			048		
		MAT.	INST.	TOTAL	MAT.	INST.	TOTAL	MAT.	INST.	TOTAL	MAT.	INST.	TOTAL	MAT.	INST.	TOTAL	MAT.	INST.	TOTAL
01543	CONTRACTOR EQUIPMENT	.0	100.7	100.7	.0	100.7	100.7	.0	100.7	100.7	.0	100.7	100.7	.0	100.7	100.7	.0	100.7	100.7
0241, 31 - 34	SITE & INFRASTRUCTURE, DEMOLITION	79.2	101.6	95.2	74.3	101.6	93.9	77.0	99.4	93.1	78.6	101.6	95.1	80.1	99.4	94.0	75.6	101.6	94.2
0310	Concrete Forming & Accessories	96.3	74.8	77.9	88.7	61.5	65.3	98.3	86.9	88.5	93.4	61.3	65.8	98.3	86.9	88.5	94.4	61.4	66.1
0310	Concrete Reinforcing	84.9	106.5	95.5	83.4	106.5	94.7	104.3	106.7	105.5	84.9	106.5	95.4	104.3	106.7	105.5	84.9	106.6	95.5
0330	Cast-in-Place Concrete	81.1	60.0	73.2	81.1	59.9	73.1	82.8	61.0	74.6	81.1	59.9	73.1	106.9	61.0	89.6	82.8	59.9	74.2
03	CONCRETE	98.5	75.6	87.7	93.1	69.7	82.1	97.9	81.1	90.0	98.0	69.6	84.6	110.1	81.1	96.4	95.2	69.7	83.2
04	MASONRY	95.9	48.4	67.4	109.9	48.4	73.1	96.4	55.9	72.2	95.9	48.4	67.4	96.6	55.9	72.2	89.9	49.7	65.8
05	METALS	83.0	83.9	83.3	82.7	83.8	83.1	87.1	80.1	85.0	83.0	83.6	83.2	88.4	80.1	85.9	82.9	83.9	83.2
06	WOOD, PLASTICS & COMPOSITES	96.0	77.5	86.2	86.8	59.5	72.3	98.2	93.0	95.4	92.8	59.5	75.2	99.3	93.0	96.0	93.8	59.5	75.6
07	THERMAL & MOISTURE PROTECTION	98.7	58.1	82.4	98.2	56.3	81.3	98.4	57.2	81.8	98.6	56.2	81.6	101.0	57.2	83.4	98.3	56.6	81.6
08	OPENINGS	102.4	69.8	94.0	102.3	60.1	91.4	105.6	78.1	98.5	102.5	57.9	90.9	105.6	78.1	98.5	102.4	57.9	90.9
0920	Plaster & Gypsum Board	99.9	76.1	85.8	95.4	57.6	72.9	102.8	92.0	96.4	99.0	57.6	74.3	102.8	92.0	96.4	98.6	57.6	74.2
0950, 0980	Ceilings & Acoustic Treatment	85.8	76.1	80.0	84.2	57.6	68.2	96.2	92.0	93.6	85.8	57.6	68.9	96.2	92.0	93.6	85.2	57.6	68.2
0960	Flooring	97.0	40.5	81.9	93.8	40.5	79.6	98.8	50.5	85.9	95.9	19.9	75.6	98.0	50.5	85.3	96.2	40.5	81.4
0990	Wall Finishes & Painting/Coating	91.0	34.0	56.9	91.0	34.0	56.9	91.0	31.4	55.3	91.0	34.0	56.9	91.0	31.4	55.3	91.0	34.0	56.9
09	FINISHES	95.8	63.9	79.4	93.4	53.4	72.7	99.0	75.3	86.8	95.3	49.8	71.8	98.8	75.3	86.7	94.6	53.4	73.3
COVERS	DIVS. 10 - 14, 25, 28, 41, 43, 44	100.0	58.8	91.5	100.0	56.6	91.1	100.0	75.4	94.9	100.0	56.6	91.1	100.0	75.4	94.9	100.0	56.6	91.1
22, 23	PLUMBING & HVAC	95.8	72.5	86.3	95.8	72.5	86.3	100.0	73.6	89.3	95.8	72.5	86.3	100.0	73.6	89.3	95.8	72.5	86.3
26, 27, 3370	ELECTRICAL, COMMUNICATIONS & UTIL.	98.2	84.1	91.4	94.9	84.1	89.7	98.3	82.4	90.7	98.2	84.1	91.4	99.6	82.4	91.4	98.2	84.1	91.4
MF2004	WEIGHTED AVERAGE	95.0	73.7	85.8	94.2	70.9	84.1	97.1	77.1	88.5	94.9	70.3	84.2	99.0	77.1	89.6	94.1	70.9	84.1

		MAINE			MARYLAND													
		WATERVILLE			ANNAPOLIS			BALTIMORE			COLLEGE PARK			CUMBERLAND			EASTON		
	DIVISION	049			214			210 - 212			207 - 208			215			216		
		MAT.	INST.	TOTAL	MAT.	INST.	TOTAL	MAT.	INST.	TOTAL	MAT.	INST.	TOTAL	MAT.	INST.	TOTAL	MAT.	INST.	TOTAL
01543	CONTRACTOR EQUIPMENT	.0	100.7	100.7	.0	98.9	98.9	.0	103.1	103.1	.0	103.5	103.5	.0	98.9	98.9	.0	98.9	98.9
0241, 31 - 34	SITE & INFRASTRUCTURE, DEMOLITION	79.1	101.6	95.2	95.4	89.3	91.0	94.7	93.8	94.0	106.9	90.4	95.0	86.2	88.8	88.1	92.9	85.2	87.4
0310	Concrete Forming & Accessories	87.6	61.4	65.1	101.6	56.2	62.6	101.7	74.8	78.6	86.6	69.4	71.9	91.8	77.2	79.3	89.5	33.8	41.6
0310	Concrete Reinforcing	84.9	106.5	95.5	101.7	88.6	95.3	101.7	88.7	95.3	110.8	78.1	94.8	89.9	72.6	81.5	89.1	20.0	55.3
0330	Cast-in-Place Concrete	81.1	59.9	73.2	107.3	76.1	95.6	94.7	77.1	88.1	123.8	73.3	104.8	84.5	80.6	83.0	93.9	41.9	74.3
03	CONCRETE	99.1	69.7	85.2	111.7	70.6	92.3	105.3	79.3	93.0	117.1	73.7	96.6	94.0	78.5	86.7	101.9	35.8	70.7
04	MASONRY	105.8	48.8	71.7	97.3	71.2	81.7	98.9	71.2	82.3	114.4	66.9	86.0	96.8	74.9	83.7	111.2	35.8	66.1
05	METALS	83.0	83.8	83.2	96.5	98.0	96.9	97.4	99.0	97.9	89.8	95.7	91.6	93.8	91.0	93.0	94.0	59.5	83.3
06	WOOD, PLASTICS & COMPOSITES	85.5	59.5	71.7	99.1	51.4	73.9	99.1	76.1	87.0	80.5	72.1	76.0	88.1	76.5	82.0	85.2	34.1	58.2
07	THERMAL & MOISTURE PROTECTION	98.6	55.4	81.3	98.4	76.9	89.7	98.2	79.7	90.8	107.6	74.8	94.4	97.9	77.4	89.6	98.0	39.8	74.6
08	OPENINGS	102.4	58.4	91.0	91.8	69.4	86.0	93.8	84.8	91.5	97.0	75.1	91.4	93.0	80.6	89.8	91.1	31.0	75.5
0920	Plaster & Gypsum Board	95.4	57.6	72.9	108.7	50.5	74.0	108.7	75.7	89.1	103.3	71.4	84.3	105.5	76.3	88.1	104.2	32.7	61.7
0950, 0980	Ceilings & Acoustic Treatment	85.8	57.6	68.9	98.8	50.5	69.7	98.8	75.7	84.9	99.3	71.4	82.5	98.8	76.3	85.3	98.8	32.7	59.1
0960	Flooring	93.3	40.5	79.2	89.8	75.8	86.1	89.8	75.8	86.1	106.7	62.2	94.8	85.5	78.2	83.6	84.7	15.4	66.2
0990	Wall Finishes & Painting/Coating	91.0	34.0	56.9	94.2	84.7	88.8	94.9	84.7	88.8	121.1	80.5	96.8	94.9	59.9	74.0	94.9	25.1	53.2
09	FINISHES	94.1	53.4	73.1	95.2	60.8	77.5	95.2	75.3	85.0	101.8	68.8	84.8	93.0	75.2	83.8	92.9	30.3	60.6
COVERS	DIVS. 10 - 14, 25, 28, 41, 43, 44	100.0	56.6	91.1	100.0	82.6	96.4	100.0	86.1	97.1	100.0	64.0	92.6	100.0	88.7	97.7	100.0	46.6	89.0
22, 23	PLUMBING & HVAC	95.8	72.5	86.3	99.9	83.0	93.0	99.9	83.1	93.0	95.9	82.2	90.3	95.7	73.4	86.6	95.7	26.7	67.5
26, 27, 3370	ELECTRICAL, COMMUNICATIONS & UTIL.	98.2	84.1	91.4	102.2	79.0	91.0	103.7	93.6	98.9	99.7	93.4	96.7	101.0	82.3	92.0	100.5	60.7	81.4
MF2004	WEIGHTED AVERAGE	95.3	70.8	84.7	99.5	77.5	90.0	99.4	84.2	92.8	100.3	80.2	91.6	95.5	79.5	88.6	97.1	43.3	73.9

		MARYLAND															MASSACHUSETTS		
		ELKTON			HAGERSTOWN			SALISBURY			SILVER SPRING			WALDORF			BOSTON		
	DIVISION	219			217			218			209			206			020 - 022, 024		
		MAT.	INST.	TOTAL	MAT.	INST.	TOTAL	MAT.	INST.	TOTAL	MAT.	INST.	TOTAL	MAT.	INST.	TOTAL	MAT.	INST.	TOTAL
01543	CONTRACTOR EQUIPMENT	.0	98.9	98.9	.0	98.9	98.9	.0	98.9	98.9	.0	94.9	94.9	.0	94.9	94.9	.0	107.5	107.5
0241, 31 - 34	SITE & INFRASTRUCTURE, DEMOLITION	80.6	86.3	84.6	84.3	89.6	88.1	92.8	85.3	87.5	93.8	82.1	85.4	101.3	82.3	87.7	84.7	108.4	101.7
0310	Concrete Forming & Accessories	96.4	62.5	67.3	90.9	76.1	78.2	105.3	31.9	42.3	95.7	59.4	64.5	103.2	52.5	59.7	103.4	136.6	131.9
0310	Concrete Reinforcing	89.1	58.8	74.3	89.9	72.5	81.4	89.1	58.6	74.2	109.6	77.9	94.1	110.3	58.6	85.1	103.1	138.9	120.6
0330	Cast-in-Place Concrete	76.0	48.9	65.8	80.5	59.1	72.5	93.9	41.2	74.1	126.7	65.4	103.7	142.0	59.1	110.8	104.8	146.4	120.4
03	CONCRETE	86.6	58.0	73.1	90.2	70.6	80.9	102.9	41.4	73.9	115.5	66.3	92.2	127.2	57.3	94.2	111.9	139.6	125.0
04	MASONRY	97.2	39.4	62.6	103.1	74.9	86.2	110.7	33.5	64.5	112.6	67.0	85.4	96.5	63.5	76.7	115.7	151.9	137.4
05	METALS	94.1	70.8	87.0	93.9	90.3	92.8	94.1	62.2	84.4	94.2	89.1	92.6	94.2	78.3	89.4	98.9	125.5	106.9
06	WOOD, PLASTICS & COMPOSITES	93.0	69.2	80.4	86.7	75.9	81.0	103.5	30.2	64.7	87.9	59.4	72.8	95.8	53.1	73.2	102.1	136.2	120.1
07	THERMAL & MOISTURE PROTECTION	97.6	48.3	77.8	97.6	69.1	86.2	98.3	57.5	81.8	107.5	79.0	96.0	108.0	74.2	94.4	98.8	144.7	117.3
08	OPENINGS	91.1	56.5	82.1	91.1	75.8	87.1	91.3	25.9	74.4	90.6	68.2	84.8	91.4	51.7	81.1	101.6	133.5	109.9
0920	Plaster & Gypsum Board	106.8	68.8	84.2	104.5	75.7	87.4	110.3	28.7	61.7	109.2	59.3	79.5	111.4	52.9	76.6	123.7	136.4	123.7
0950, 0980	Ceilings & Acoustic Treatment	98.8	68.8	80.7	99.7	75.7	85.3	98.8	28.7	56.6	108.6	59.3	78.9	108.6	52.9	75.1	95.2	136.4	120.0
0960	Flooring	87.3	59.0	79.8	85.2	78.2	83.4	91.3	59.0	82.7	114.1	62.2	100.2	119.0	59.0	103.0	98.6	163.5	115.9
0990	Wall Finishes & Painting/Coating	94.9	80.5	86.3	94.9	38.4	61.1	94.9	28.3	55.1	127.2	80.5	99.3	127.2	66.2	90.7	93.8	155.8	130.9
09	FINISHES	93.4	64.9	78.7	92.8	72.5	82.3	95.8	35.6	64.8	103.6	61.3	81.8	105.7	54.4	79.2	99.0	144.2	122.3
COVERS	DIVS. 10 - 14, 25, 28, 41, 43, 44	100.0	52.2	90.2	100.0	88.6	97.7	100.0	45.6	88.8	100.0	60.3	91.8	100.0	48.9	89.5	100.0	120.3	104.2
22, 23	PLUMBING & HVAC	95.7	76.0	87.6	99.9	85.9	94.2	95.7	58.9	80.7	95.8	82.0	90.2	95.9	81.8	90.2	100.0	128.8	111.7
26, 27, 3370	ELECTRICAL, COMMUNICATIONS & UTIL.	102.2	61.5	82.6	100.8	82.3	91.9	99.3	60.7	80.8	95.6	93.4	94.5	93.5	93.4	93.4	97.5	134.8	115.4
MF2004	WEIGHTED AVERAGE	94.6	64.4	81.6	96.1	80.3	89.3	97.5	51.7	77.7	99.5	76.4	89.5	100.3	71.8	88.0	101.4	133.8	115.4

City Cost Indexes

DIVISION		MASSACHUSETTS																	
		BROCKTON 023			BUZZARDS BAY 025			FALL RIVER 027			FITCHBURG 014			FRAMINGHAM 017			GREENFIELD 013		
		MAT.	INST.	TOTAL	MAT.	INST.	TOTAL	MAT.	INST.	TOTAL	MAT.	INST.	TOTAL	MAT.	INST.	TOTAL	MAT.	INST.	TOTAL
01543	CONTRACTOR EQUIPMENT	.0	102.8	102.8	.0	102.8	102.8	.0	104.1	104.1	.0	100.7	100.7	.0	102.0	102.0	.0	100.7	100.7
0241, 31 - 34	SITE & INFRASTRUCTURE, DEMOLITION	82.3	104.4	98.1	73.8	104.3	95.7	81.4	104.5	97.9	75.2	104.3	96.0	72.2	104.0	95.0	78.3	102.8	95.9
0310	Concrete Forming & Accessories	103.0	123.2	120.4	100.2	122.9	119.7	103.0	123.3	120.5	94.2	133.8	128.2	101.4	123.2	120.1	92.3	101.7	100.4
0310	Concrete Reinforcing	104.3	138.5	121.0	83.6	119.0	100.9	104.3	127.7	115.7	83.6	134.1	108.3	83.5	138.5	110.4	87.4	116.8	101.8
0330	Cast-in-Place Concrete	99.7	142.7	115.9	82.8	146.6	106.8	96.4	147.3	115.6	86.8	142.2	107.7	86.9	138.5	106.3	89.3	117.8	100.0
03	CONCRETE	109.1	132.1	120.0	92.0	129.6	109.8	107.5	131.7	118.9	90.5	135.6	111.8	93.4	130.7	111.0	94.3	109.6	101.5
04	MASONRY	111.9	143.2	130.6	102.2	147.6	129.4	111.8	147.5	133.2	98.4	142.8	125.0	104.6	143.0	127.6	102.8	114.4	109.8
05	METALS	95.9	122.2	103.8	91.1	113.9	98.0	95.9	118.2	102.6	91.0	117.0	98.9	91.1	122.0	100.4	93.1	104.4	96.5
06	WOOD, PLASTICS & COMPOSITES	101.1	121.4	111.8	97.9	121.4	110.3	101.1	121.7	112.0	93.6	136.3	116.2	100.4	121.1	111.3	91.4	100.5	96.2
07	THERMAL & MOISTURE PROTECTION	98.7	138.3	114.6	98.1	136.4	113.5	98.7	135.5	113.5	98.1	134.4	112.7	98.2	139.3	114.7	98.1	110.8	103.2
08	OPENINGS	99.5	125.5	106.3	95.1	118.1	101.0	99.5	120.7	105.0	104.4	132.4	111.6	95.5	125.3	103.2	104.3	106.7	105.0
0920	Plaster & Gypsum Board	100.7	121.1	112.9	96.8	121.1	111.3	100.7	121.1	112.9	97.9	136.4	120.8	99.9	121.1	112.5	99.2	99.7	99.5
0950, 0980	Ceilings & Acoustic Treatment	97.1	121.1	111.6	85.1	121.1	106.8	97.1	121.1	111.6	84.2	136.4	115.6	84.2	121.1	106.4	94.3	99.7	97.5
0960	Flooring	99.0	163.5	116.2	97.6	163.5	115.2	98.8	163.5	116.1	95.6	163.5	113.7	97.5	163.5	115.1	94.7	128.3	103.7
0990	Wall Finishes & Painting/Coating	93.3	129.6	115.0	93.3	129.6	115.0	93.3	129.6	115.0	91.0	129.6	114.1	92.2	129.6	114.5	91.0	104.0	98.8
09	FINISHES	98.9	131.5	115.7	94.5	131.5	113.6	98.9	131.7	115.8	93.7	140.2	117.7	94.5	131.3	113.5	96.4	106.8	101.8
COVERS	DIVS. 10 - 14, 25, 28, 41, 43, 44	100.0	116.9	103.5	100.0	116.9	103.5	100.0	117.7	103.6	100.0	109.0	101.9	100.0	116.2	103.3	100.0	101.2	100.2
22, 23	PLUMBING & HVAC	100.0	107.3	103.0	95.8	107.1	100.4	100.0	107.3	103.0	96.3	105.3	100.0	96.3	123.6	107.5	96.3	93.9	95.3
26, 27, 3370	ELECTRICAL, COMMUNICATIONS & UTIL.	96.5	102.6	99.5	94.3	102.6	98.3	96.2	102.6	99.3	98.6	102.2	100.4	95.8	115.3	105.1	98.6	91.4	95.1
MF2004	WEIGHTED AVERAGE	100.0	119.9	108.6	94.4	118.8	105.0	99.7	119.7	108.3	95.5	120.6	106.3	95.1	124.8	107.9	96.8	102.4	99.2

DIVISION		MASSACHUSETTS																	
		HYANNIS 026			LAWRENCE 019			LOWELL 018			NEW BEDFORD 027			PITTSFIELD 012			SPRINGFIELD 010 - 011		
		MAT.	INST.	TOTAL	MAT.	INST.	TOTAL	MAT.	INST.	TOTAL	MAT.	INST.	TOTAL	MAT.	INST.	TOTAL	MAT.	INST.	TOTAL
01543	CONTRACTOR EQUIPMENT	.0	102.8	102.8	.0	102.8	102.8	.0	100.7	100.7	.0	104.1	104.1	.0	100.7	100.7	.0	100.7	100.7
0241, 31 - 34	SITE & INFRASTRUCTURE, DEMOLITION	79.1	104.3	97.2	83.0	104.4	98.3	81.8	104.3	97.9	79.8	104.5	97.5	82.9	102.8	97.2	82.3	103.9	97.1
0310	Concrete Forming & Accessories	94.1	122.9	118.8	102.8	123.3	120.4	99.7	123.5	120.1	103.0	123.3	120.5	99.7	96.7	97.2	99.9	107.3	106.2
0310	Concrete Reinforcing	83.6	119.0	100.9	103.4	126.8	114.8	104.3	126.8	115.3	104.3	127.7	115.7	86.7	111.7	98.9	104.3	116.8	110.4
0330	Cast-in-Place Concrete	91.0	146.6	111.9	100.5	142.7	116.4	91.4	142.8	110.7	84.9	147.3	108.4	99.7	117.4	106.4	95.1	119.5	104.3
03	CONCRETE	98.9	129.6	113.4	109.4	129.9	119.1	100.0	129.9	114.1	101.6	131.7	115.8	101.5	106.3	103.8	102.0	112.7	107.0
04	MASONRY	110.8	147.6	132.8	111.3	143.2	130.4	97.1	142.8	124.5	110.8	147.5	132.7	97.7	113.8	107.3	97.4	118.0	109.7
05	METALS	92.5	113.9	98.9	93.4	117.3	100.7	93.4	114.7	99.8	95.9	118.2	102.6	93.2	102.2	95.9	95.8	104.5	98.4
06	WOOD, PLASTICS & COMPOSITES	90.9	121.4	107.0	101.1	121.4	111.8	100.2	121.4	111.4	101.1	121.7	112.0	100.2	94.2	97.0	100.2	106.4	103.5
07	THERMAL & MOISTURE PROTECTION	98.3	136.4	113.6	98.7	138.3	114.6	98.4	138.1	114.4	98.7	135.5	113.5	98.5	109.9	103.1	98.4	112.9	104.3
08	OPENINGS	95.7	118.1	101.5	99.5	122.3	105.4	105.6	122.3	109.9	99.5	120.7	105.0	105.6	102.0	104.7	105.6	109.9	106.7
0920	Plaster & Gypsum Board	94.2	121.1	110.2	102.8	121.1	113.7	102.8	121.1	113.7	100.7	121.1	112.9	102.8	93.3	97.1	102.8	105.8	104.6
0950, 0980	Ceilings & Acoustic Treatment	87.0	121.1	107.5	96.2	121.1	111.2	96.2	121.1	111.2	97.1	121.1	111.6	96.2	93.3	94.4	96.2	105.8	101.9
0960	Flooring	95.1	163.5	113.3	98.0	163.5	115.5	98.0	163.5	115.5	98.8	163.5	116.1	98.2	128.3	106.2	97.9	129.3	106.3
0990	Wall Finishes & Painting/Coating	93.3	129.6	115.0	91.1	129.6	114.1	91.0	129.6	114.1	93.3	129.6	115.0	91.0	104.0	98.8	92.6	104.0	99.4
09	FINISHES	94.1	131.5	113.4	98.6	131.5	115.6	98.5	131.5	115.5	98.8	131.7	115.8	98.6	103.0	100.9	98.6	111.3	105.1
COVERS	DIVS. 10 - 14, 25, 28, 41, 43, 44	100.0	116.9	103.5	100.0	117.0	103.5	100.0	117.0	103.5	100.0	117.7	103.6	100.0	100.2	100.0	100.0	103.0	100.6
22, 23	PLUMBING & HVAC	100.0	107.1	102.9	100.0	113.5	105.5	100.0	123.6	109.6	100.0	107.3	103.0	100.0	93.5	97.4	100.0	95.5	98.2
26, 27, 3370	ELECTRICAL, COMMUNICATIONS & UTIL.	94.9	102.6	98.6	97.8	115.3	106.2	98.2	115.3	106.4	97.5	102.6	100.0	98.2	91.4	94.9	98.2	91.4	95.0
MF2004	WEIGHTED AVERAGE	97.1	118.8	106.5	99.7	122.1	109.4	98.5	123.9	109.4	99.1	119.7	108.0	98.7	100.8	99.6	99.1	104.4	101.4

DIVISION		MASSACHUSETTS			MICHIGAN														
		WORCESTER 015 - 016			ANN ARBOR 481			BATTLE CREEK 490			BAY CITY 487			DEARBORN 481			DETROIT 482		
		MAT.	INST.	TOTAL	MAT.	INST.	TOTAL	MAT.	INST.	TOTAL	MAT.	INST.	TOTAL	MAT.	INST.	TOTAL	MAT.	INST.	TOTAL
01543	CONTRACTOR EQUIPMENT	.0	100.7	100.7	.0	112.7	112.7	.0	104.8	104.8	.0	112.7	112.7	.0	112.7	112.7	.0	98.5	98.5
0241, 31 - 34	SITE & INFRASTRUCTURE, DEMOLITION	82.2	104.3	98.0	72.8	96.3	89.6	79.5	87.3	85.1	65.5	94.8	86.5	72.5	96.4	89.6	85.5	98.2	94.6
0310	Concrete Forming & Accessories	100.3	133.7	129.0	98.1	118.1	115.3	96.1	88.3	89.4	98.2	93.9	94.5	98.0	120.1	117.0	99.4	120.2	117.2
0310	Concrete Reinforcing	104.3	134.1	120.6	96.1	128.5	112.0	95.0	92.2	93.6	96.1	127.7	111.5	96.1	128.8	112.1	95.4	128.7	111.7
0330	Cast-in-Place Concrete	94.6	142.2	112.5	84.3	121.4	98.2	91.2	109.7	98.1	80.6	99.6	87.8	82.4	121.1	97.2	90.4	121.1	102.0
03	CONCRETE	101.7	136.2	118.0	88.9	121.4	104.3	93.9	95.7	94.8	87.0	103.0	94.6	87.9	122.3	104.1	92.0	120.9	105.7
04	MASONRY	96.8	142.8	124.4	96.6	116.1	108.3	94.9	90.2	92.1	96.3	96.2	96.3	96.5	120.5	110.9	95.6	120.5	110.5
05	METALS	95.8	118.5	102.7	97.5	124.9	105.8	93.4	87.6	91.7	98.1	121.5	105.2	97.6	125.7	106.1	101.8	104.6	102.7
06	WOOD, PLASTICS & COMPOSITES	100.6	136.3	119.5	100.2	118.9	110.1	98.7	86.1	92.0	100.2	93.3	96.5	100.2	120.2	110.8	101.1	120.2	111.2
07	THERMAL & MOISTURE PROTECTION	98.5	134.4	112.9	99.3	116.5	106.2	90.6	90.5	90.5	97.2	94.1	96.0	98.3	122.2	107.9	96.1	122.2	106.6
08	OPENINGS	105.6	133.3	112.8	96.0	115.5	101.0	91.2	83.9	89.3	96.0	97.5	96.4	96.0	116.3	101.2	97.6	117.1	102.7
0920	Plaster & Gypsum Board	102.8	136.4	122.8	92.5	118.1	107.8	90.1	81.1	84.7	92.5	91.9	92.1	92.5	119.5	108.6	92.5	119.5	108.6
0950, 0980	Ceilings & Acoustic Treatment	96.2	136.4	120.4	89.0	118.1	106.5	89.0	81.1	84.2	90.0	91.9	91.1	89.0	119.5	107.3	90.0	119.5	107.7
0960	Flooring	98.0	154.6	113.1	90.4	123.7	99.3	98.3	99.0	98.5	90.2	85.8	89.1	90.1	122.2	98.6	90.1	122.2	98.7
0990	Wall Finishes & Painting/Coating	91.0	129.6	114.1	89.2	106.4	99.5	98.7	81.0	88.1	89.2	88.9	89.0	89.2	114.2	104.1	90.8	114.2	104.8
09	FINISHES	98.5	138.4	119.1	91.5	118.4	105.4	93.5	89.0	91.2	91.4	91.5	91.4	91.4	120.6	106.3	92.4	120.2	106.7
COVERS	DIVS. 10 - 14, 25, 28, 41, 43, 44	100.0	109.0	101.9	100.0	110.5	102.2	100.0	106.4	101.3	100.0	100.0	100.0	100.0	111.4	102.3	100.0	111.4	102.3
22, 23	PLUMBING & HVAC	100.0	105.3	102.2	99.9	106.8	102.7	99.9	94.8	97.8	99.9	91.7	96.6	99.9	114.7	105.9	99.9	114.7	105.9
26, 27, 3370	ELECTRICAL, COMMUNICATIONS & UTIL.	98.3	102.2	100.2	93.8	82.1	88.2	93.7	88.6	91.3	92.5	77.8	85.4	93.8	118.5	105.6	95.2	118.4	106.3
MF2004	WEIGHTED AVERAGE	99.1	120.6	108.4	95.6	109.3	101.5	95.0	91.2	93.4	95.0	95.2	95.1	95.4	117.2	104.8	97.2	115.3	105.0

City Cost Indexes

		MICHIGAN																	
		FLINT			GAYLORD			GRAND RAPIDS			IRON MOUNTAIN			JACKSON			KALAMAZOO		
	DIVISION	484 - 485			497			493,495			498 - 499			492			491		
		MAT.	INST.	TOTAL	MAT.	INST.	TOTAL	MAT.	INST.	TOTAL	MAT.	INST.	TOTAL	MAT.	INST.	TOTAL	MAT.	INST.	TOTAL
01543	CONTRACTOR EQUIPMENT	.0	112.7	112.7	.0	107.1	107.1	.0	104.8	104.8	.0	93.8	93.8	.0	107.1	107.1	.0	104.8	104.8
0241, 31 - 34	SITE & INFRASTRUCTURE, DEMOLITION	64.1	95.2	86.4	75.9	85.2	82.6	79.4	86.8	84.7	82.5	94.7	91.2	94.5	86.8	89.0	79.8	87.3	85.2
0310	Concrete Forming & Accessories	101.4	92.5	93.8	93.3	61.1	65.7	97.2	74.6	77.8	87.5	87.2	87.2	90.9	93.9	93.4	96.1	88.0	89.1
0310	Concrete Reinforcing	96.1	127.8	111.6	88.5	109.4	98.7	95.0	90.1	92.6	88.4	97.2	92.7	85.9	128.0	106.5	95.0	92.2	93.6
0330	Cast-in-Place Concrete	84.8	98.1	89.8	91.0	74.4	84.7	92.6	94.9	93.4	107.7	92.3	101.9	90.8	98.7	93.8	93.0	109.6	99.2
03	CONCRETE	89.4	101.9	95.3	90.4	76.3	83.8	94.7	84.1	89.7	100.1	90.9	95.8	85.7	102.6	93.7	97.0	95.5	96.3
04	MASONRY	96.7	91.1	93.3	105.7	77.9	89.1	92.0	52.2	68.2	91.9	91.9	91.9	86.1	103.7	96.7	93.6	90.2	91.6
05	METALS	97.6	122.0	105.0	94.9	110.5	99.6	94.9	82.0	91.0	94.3	91.5	93.5	95.0	119.9	102.5	93.4	87.3	91.6
06	WOOD, PLASTICS & COMPOSITES	103.7	93.4	98.2	91.6	57.8	73.7	97.9	75.0	85.8	87.5	86.3	86.9	90.4	92.0	91.3	98.7	86.1	92.0
07	THERMAL & MOISTURE PROTECTION	97.2	93.1	95.6	89.3	66.5	80.1	90.8	58.6	77.9	92.2	83.8	88.8	88.7	101.9	94.0	90.6	90.5	90.5
08	OPENINGS	96.0	97.5	96.4	93.1	64.1	85.6	94.7	68.7	88.0	98.0	75.1	92.0	92.1	93.9	92.6	91.2	83.9	89.3
0920	Plaster & Gypsum Board	93.8	92.0	92.7	91.6	55.1	69.9	90.1	69.6	77.9	59.8	86.5	75.7	91.0	90.2	90.6	90.1	81.1	84.7
0950, 0980	Ceilings & Acoustic Treatment	89.0	92.0	90.8	86.5	55.1	67.6	89.0	69.6	77.3	87.1	86.5	86.7	88.2	90.2	89.4	89.0	81.1	84.2
0960	Flooring	90.2	85.8	89.1	91.4	58.0	82.5	98.3	40.8	82.9	110.8	96.8	107.1	89.6	79.8	87.0	98.3	76.2	92.4
0990	Wall Finishes & Painting/Coating	89.2	95.5	93.0	93.8	52.3	68.9	98.7	39.9	63.5	118.0	43.6	73.5	93.8	88.7	90.8	98.7	81.0	88.1
09	FINISHES	91.1	90.9	91.0	93.2	58.0	75.1	93.5	64.2	78.4	94.6	84.4	89.3	94.0	90.1	92.0	93.5	84.4	88.8
COVERS	DIVS. 10 - 14, 25, 28, 41, 43, 44	100.0	98.2	99.6	100.0	90.7	98.1	100.0	102.5	100.5	100.0	95.2	99.0	100.0	101.5	100.3	100.0	106.4	101.3
22, 23	PLUMBING & HVAC	99.9	95.4	98.1	95.9	81.9	90.2	99.9	52.6	80.6	95.8	85.0	91.4	95.9	96.9	96.3	99.9	88.8	95.4
26, 27, 3370	ELECTRICAL, COMMUNICATIONS & UTIL.	93.8	106.1	99.7	92.2	57.4	75.5	95.3	57.3	77.1	98.1	83.0	90.9	95.6	82.1	89.1	93.5	84.9	89.4
MF2004	WEIGHTED AVERAGE	95.3	99.2	97.0	94.2	75.9	86.3	95.7	67.1	83.3	96.0	87.4	92.3	93.5	96.7	94.9	95.3	88.8	92.5

		MICHIGAN																MINNESOTA		
		LANSING			MUSKEGON			ROYAL OAK			SAGINAW			TRAVERSE CITY			BEMIDJI		
	DIVISION	488 - 489			494			480,483			486			496			566		
		MAT.	INST.	TOTAL	MAT.	INST.	TOTAL	MAT.	INST.	TOTAL	MAT.	INST.	TOTAL	MAT.	INST.	TOTAL	MAT.	INST.	TOTAL
01543	CONTRACTOR EQUIPMENT	.0	112.7	112.7	.0	104.8	104.8	.0	95.5	95.5	.0	112.7	112.7	.0	93.8	93.8	.0	98.7	98.7
0241, 31 - 34	SITE & INFRASTRUCTURE, DEMOLITION	78.5	95.0	90.3	77.7	87.2	84.5	75.6	95.5	89.9	66.3	94.8	86.7	69.8	93.9	87.1	88.3	99.0	96.0
0310	Concrete Forming & Accessories	101.5	95.3	96.1	96.8	87.1	88.4	94.2	114.7	111.8	98.1	93.8	94.4	87.4	58.5	62.9	84.9	97.7	95.9
0310	Concrete Reinforcing	96.1	127.5	111.5	95.6	91.3	93.6	87.0	111.0	98.8	96.1	127.7	111.5	89.8	90.7	90.2	97.7	108.9	103.2
0330	Cast-in-Place Concrete	84.3	97.8	89.4	90.9	104.1	95.9	73.7	111.9	88.1	83.2	99.6	89.4	84.0	65.0	76.9	98.5	105.3	101.1
03	CONCRETE	89.1	103.0	95.7	92.4	93.1	92.7	77.0	111.9	93.5	88.4	103.0	95.3	82.5	67.9	75.6	94.4	103.5	98.7
04	MASONRY	90.0	95.9	93.5	92.2	78.2	83.8	91.4	119.7	108.4	98.1	96.2	97.0	90.1	72.0	79.3	97.6	108.8	104.3
05	METALS	96.0	121.3	103.7	91.1	85.7	89.5	100.9	94.4	99.0	97.6	121.4	104.8	94.3	88.5	92.5	90.5	124.8	100.9
06	WOOD, PLASTICS & COMPOSITES	103.1	93.8	98.2	95.6	86.1	90.6	95.6	114.2	105.5	96.3	93.4	94.7	87.5	55.8	70.7	76.6	95.2	86.4
07	THERMAL & MOISTURE PROTECTION	97.9	96.0	97.2	89.8	79.0	85.4	95.6	117.4	104.4	97.7	94.4	96.4	91.3	65.0	80.7	100.2	95.4	98.3
08	OPENINGS	96.0	97.8	96.4	90.5	82.2	88.3	96.0	108.5	99.3	93.8	97.5	94.8	97.9	59.0	87.9	98.4	115.1	102.8
0920	Plaster & Gypsum Board	94.6	92.4	93.3	78.6	81.1	80.1	90.5	113.4	104.1	92.5	92.0	92.2	59.8	55.1	57.0	102.8	95.4	98.4
0950, 0980	Ceilings & Acoustic Treatment	89.0	92.4	91.0	90.7	81.1	84.9	88.3	113.4	103.4	89.0	92.0	90.8	87.1	55.1	67.9	128.5	95.4	108.6
0960	Flooring	100.0	94.4	98.5	97.4	75.1	91.5	87.7	119.4	96.1	90.4	85.6	89.2	110.8	58.0	96.7	104.9	126.5	110.6
0990	Wall Finishes & Painting/Coating	101.2	88.3	93.5	98.0	57.9	74.0	90.8	106.4	100.1	89.2	88.9	89.0	118.0	53.4	79.4	97.6	99.5	98.7
09	FINISHES	96.0	94.1	95.0	91.3	81.3	86.1	90.3	115.2	103.2	91.3	91.5	91.4	93.8	56.7	74.7	109.2	103.2	106.1
COVERS	DIVS. 10 - 14, 25, 28, 41, 43, 44	100.0	101.0	100.2	100.0	106.1	101.3	100.0	106.1	101.3	100.0	100.0	100.0	100.0	74.6	94.8	100.0	97.1	99.4
22, 23	PLUMBING & HVAC	99.9	96.9	98.7	99.7	89.2	95.5	95.9	109.6	101.5	99.9	91.4	96.4	95.8	81.3	89.9	96.2	88.4	93.0
26, 27, 3370	ELECTRICAL, COMMUNICATIONS & UTIL.	92.6	82.4	87.7	94.1	79.0	86.9	95.6	106.1	100.6	91.1	77.8	84.7	94.0	56.7	76.1	101.8	94.9	98.5
MF2004	WEIGHTED AVERAGE	95.4	97.3	96.2	94.0	85.5	90.3	93.6	108.7	100.1	94.8	95.1	94.9	93.0	71.8	83.8	97.3	101.1	98.9

		MINNESOTA																	
		BRAINERD			DETROIT LAKES			DULUTH			MANKATO			MINNEAPOLIS			ROCHESTER		
	DIVISION	564			565			556 - 558			560			553 - 555			559		
		MAT.	INST.	TOTAL	MAT.	INST.	TOTAL	MAT.	INST.	TOTAL	MAT.	INST.	TOTAL	MAT.	INST.	TOTAL	MAT.	INST.	TOTAL
01543	CONTRACTOR EQUIPMENT	.0	102.0	102.0	.0	98.7	98.7	.0	101.5	101.5	.0	102.0	102.0	.0	105.9	105.9	.0	101.5	101.5
0241, 31 - 34	SITE & INFRASTRUCTURE, DEMOLITION	85.9	104.6	99.3	86.5	99.4	95.7	87.3	102.7	98.3	82.3	103.9	97.8	88.6	108.9	103.1	87.6	101.8	97.8
0310	Concrete Forming & Accessories	87.1	100.1	98.3	81.8	101.7	98.9	99.3	119.6	116.7	95.8	111.3	109.1	100.0	139.0	133.5	99.8	109.0	107.7
0310	Concrete Reinforcing	96.5	109.0	102.6	97.7	108.9	103.1	92.9	109.3	101.0	96.4	126.5	111.1	93.1	127.2	109.8	92.9	126.7	109.4
0330	Cast-in-Place Concrete	107.2	109.8	108.2	95.7	108.8	100.6	102.0	109.0	104.6	98.6	113.0	104.1	110.8	126.3	116.6	108.6	104.4	107.0
03	CONCRETE	98.6	106.1	102.1	92.0	106.5	98.8	97.9	114.7	105.8	94.0	115.6	104.2	103.9	132.7	117.5	101.3	111.8	106.2
04	MASONRY	120.5	115.5	117.5	119.3	115.3	116.9	103.1	121.0	113.8	108.4	115.4	112.6	103.9	136.6	123.5	103.2	119.5	113.0
05	METALS	91.7	124.7	101.7	90.5	124.3	100.7	93.9	127.7	104.1	91.6	133.9	104.4	96.6	140.1	109.7	93.7	137.6	107.0
06	WOOD, PLASTICS & COMPOSITES	94.3	95.0	94.6	72.9	98.1	86.2	112.6	119.3	116.2	104.8	111.5	108.4	113.1	137.8	126.1	113.1	106.8	109.7
07	THERMAL & MOISTURE PROTECTION	98.3	109.2	102.7	100.1	85.7	94.3	98.3	121.3	107.6	98.7	100.9	99.6	98.3	138.7	114.6	98.2	107.1	101.8
08	OPENINGS	86.1	115.0	93.6	98.4	116.7	103.1	94.7	121.5	101.6	91.0	130.4	101.2	97.9	144.6	110.0	94.7	127.8	103.3
0920	Plaster & Gypsum Board	86.2	95.4	91.7	101.2	98.3	99.5	98.5	120.3	111.5	90.1	112.4	103.4	98.7	139.1	122.8	98.3	107.5	103.7
0950, 0980	Ceilings & Acoustic Treatment	70.3	95.4	85.4	128.5	98.3	110.4	96.9	120.3	111.0	70.3	112.4	95.6	97.7	139.1	122.6	96.1	107.5	102.9
0960	Flooring	103.6	126.5	109.7	103.4	126.5	109.6	106.1	127.2	111.7	106.0	126.5	111.5	103.5	126.5	109.6	105.9	92.9	102.4
0990	Wall Finishes & Painting/Coating	93.1	99.5	96.9	97.6	99.5	98.7	89.4	108.5	100.8	102.0	104.9	103.7	96.1	133.6	118.5	92.0	104.9	99.7
09	FINISHES	91.9	104.6	98.4	108.4	106.5	107.4	100.9	120.4	111.0	93.6	114.1	104.2	100.7	136.7	119.3	100.7	105.9	103.4
COVERS	DIVS. 10 - 14, 25, 28, 41, 43, 44	100.0	98.9	99.8	100.0	99.7	99.9	100.0	102.7	100.6	100.0	99.6	99.9	100.0	110.5	102.2	100.0	101.0	100.2
22, 23	PLUMBING & HVAC	95.3	91.8	93.9	96.2	91.5	94.3	100.0	110.5	104.2	95.3	98.8	96.7	99.9	123.0	109.3	99.9	103.0	101.2
26, 27, 3370	ELECTRICAL, COMMUNICATIONS & UTIL.	100.0	96.4	98.2	101.5	67.1	85.0	102.5	96.4	99.5	105.6	83.5	95.0	103.9	118.7	111.0	102.5	83.5	93.4
MF2004	WEIGHTED AVERAGE	96.1	104.2	99.6	97.9	99.2	98.5	98.5	112.9	104.7	96.1	107.7	101.1	100.2	128.3	112.3	98.9	107.6	102.7

City Cost Indexes

DIVISION		MINNESOTA															MISSISSIPPI		
		SAINT PAUL			ST. CLOUD			THIEF RIVER FALLS			WILLMAR			WINDOM			BILOXI		
		550 - 551			563			567			562			561			395		
		MAT.	INST.	TOTAL	MAT.	INST.	TOTAL	MAT.	INST.	TOTAL	MAT.	INST.	TOTAL	MAT.	INST.	TOTAL	MAT.	INST.	TOTAL
01543	CONTRACTOR EQUIPMENT	.0	101.5	101.5	.0	102.0	102.0	.0	98.7	98.7	.0	102.0	102.0	.0	102.0	102.0	.0	98.5	98.5
0241, 31 - 34	SITE & INFRASTRUCTURE, DEMOLITION	90.9	103.4	99.9	80.7	106.5	99.2	87.1	99.0	95.6	80.3	100.5	94.8	74.0	99.3	92.2	100.9	86.3	90.4
0310	Concrete Forming & Accessories	94.5	131.5	126.3	84.0	130.2	123.7	85.8	97.1	95.5	83.7	57.3	61.1	88.6	53.5	58.5	95.0	39.9	47.7
0310	Concrete Reinforcing	89.7	127.1	108.0	96.6	126.7	111.3	98.0	108.9	103.3	96.2	125.9	110.8	96.2	125.0	110.3	95.3	61.7	78.9
0330	Cast-in-Place Concrete	109.7	125.5	115.7	94.5	123.8	105.5	97.7	94.5	96.5	96.0	79.0	89.6	83.0	56.6	73.1	102.3	43.7	80.2
03	CONCRETE	104.6	129.1	116.2	89.9	127.8	107.8	93.1	99.5	96.1	90.0	79.6	85.1	81.2	70.1	76.0	97.1	47.2	73.6
04	MASONRY	113.1	136.6	127.1	105.7	128.7	119.5	97.6	108.8	104.3	109.1	91.8	98.7	119.1	66.3	87.5	88.6	35.4	56.8
05	METALS	93.3	139.5	107.3	92.3	137.0	105.8	90.6	124.2	100.8	91.5	125.8	101.9	91.5	124.4	101.4	89.7	80.9	87.0
06	WOOD, PLASTICS & COMPOSITES	106.7	128.1	118.0	90.8	128.0	110.4	77.7	95.2	87.0	90.5	54.4	71.4	95.9	54.4	74.0	95.1	39.8	65.9
07	THERMAL & MOISTURE PROTECTION	98.3	136.7	113.7	98.4	119.2	106.8	101.4	81.9	93.5	98.3	60.5	83.1	98.3	50.8	79.2	96.3	44.4	75.4
08	OPENINGS	92.2	139.3	104.4	91.0	139.3	103.5	98.4	115.1	102.8	88.5	79.9	86.3	92.8	79.9	89.5	97.8	46.4	84.5
0920	Plaster & Gypsum Board	93.7	129.3	114.9	85.3	129.3	111.5	102.4	95.4	98.3	85.3	53.8	66.5	86.5	53.8	67.1	105.6	38.8	65.8
0950, 0980	Ceilings & Acoustic Treatment	94.4	129.3	115.4	70.3	129.3	105.8	128.5	95.4	108.6	70.3	53.8	60.4	70.3	53.8	60.4	102.8	38.8	64.3
0960	Flooring	99.1	126.5	106.4	99.6	126.5	106.8	104.5	126.5	110.4	101.4	126.5	108.1	104.3	126.5	110.2	119.4	38.2	97.7
0990	Wall Finishes & Painting/Coating	96.1	123.9	112.7	102.0	133.6	120.9	97.6	99.5	98.7	97.6	95.4	96.3	97.6	104.9	102.0	107.4	34.1	63.6
09	FINISHES	98.1	129.9	114.5	90.7	130.4	111.2	109.1	103.2	106.0	91.0	71.2	80.8	91.6	70.0	80.5	107.3	38.5	71.8
COVERS	DIVS. 10 - 14, 25, 28, 41, 43, 44	100.0	108.9	101.8	100.0	106.9	101.4	100.0	96.9	99.4	100.0	55.9	90.9	100.0	52.5	90.2	100.0	49.0	89.5
22, 23	PLUMBING & HVAC	99.9	121.1	108.5	99.6	121.6	108.6	96.2	88.0	92.9	95.3	95.8	95.5	95.3	81.8	89.8	99.9	51.4	80.1
26, 27, 3370	ELECTRICAL, COMMUNICATIONS & UTIL.	103.4	109.4	106.3	100.0	109.4	104.5	99.6	67.1	84.0	100.0	73.5	87.2	105.6	83.4	95.0	97.3	58.8	78.8
MF2004	WEIGHTED AVERAGE	99.4	124.3	110.1	95.5	122.9	107.4	96.9	96.2	96.6	94.4	86.7	91.1	94.9	80.6	88.7	97.4	53.7	78.5

DIVISION		MISSISSIPPI																	
		CLARKSDALE			COLUMBUS			GREENVILLE			GREENWOOD			JACKSON			LAUREL		
		386			397			387			389			390 - 392			394		
		MAT.	INST.	TOTAL	MAT.	INST.	TOTAL	MAT.	INST.	TOTAL	MAT.	INST.	TOTAL	MAT.	INST.	TOTAL	MAT.	INST.	TOTAL
01543	CONTRACTOR EQUIPMENT	.0	98.5	98.5	.0	98.5	98.5	.0	98.5	98.5	.0	98.5	98.5	.0	98.5	98.5	.0	98.5	98.5
0241, 31 - 34	SITE & INFRASTRUCTURE, DEMOLITION	98.3	85.4	89.0	99.7	85.8	89.7	104.7	86.0	91.3	101.6	85.2	89.8	96.2	86.0	88.9	106.1	84.4	90.6
0310	Concrete Forming & Accessories	84.6	17.8	27.2	82.4	28.6	36.2	80.8	32.9	39.6	94.9	19.5	30.2	96.4	37.5	45.8	82.7	22.5	31.0
0310	Concrete Reinforcing	102.6	16.1	60.3	102.0	29.2	66.4	103.3	41.2	72.9	102.3	43.8	73.9	95.3	44.7	70.6	102.3	28.6	66.5
0330	Cast-in-Place Concrete	99.7	26.1	72.0	104.3	36.0	78.6	102.7	38.3	78.5	107.2	26.0	76.7	94.8	40.6	74.4	102.0	29.1	74.6
03	CONCRETE	95.4	22.7	61.1	98.9	33.3	67.9	100.2	38.3	71.0	101.6	28.7	67.2	93.5	41.9	69.1	100.9	28.1	66.5
04	MASONRY	88.5	19.0	46.9	112.5	22.4	58.5	131.2	36.7	74.7	89.1	17.7	46.4	90.2	36.7	58.2	108.8	22.9	57.4
05	METALS	87.1	57.6	78.2	87.0	64.2	80.1	88.0	71.1	82.9	87.1	69.4	81.8	89.2	72.8	84.5	87.0	62.6	79.7
06	WOOD, PLASTICS & COMPOSITES	82.2	16.5	47.5	80.0	29.7	53.4	77.6	32.3	53.7	95.1	18.6	54.7	97.2	38.4	66.1	80.3	23.4	50.2
07	THERMAL & MOISTURE PROTECTION	96.1	26.8	68.2	96.2	26.7	68.2	96.4	39.8	73.6	96.5	22.5	66.7	96.4	41.0	74.1	96.3	25.5	67.8
08	OPENINGS	97.2	20.4	77.3	97.1	27.7	79.2	97.2	36.5	81.5	97.2	29.4	79.6	98.2	40.8	83.3	93.7	24.1	75.7
0920	Plaster & Gypsum Board	98.1	14.9	48.5	97.8	28.4	56.5	95.8	31.1	57.3	104.5	17.0	52.4	105.6	37.3	65.0	95.8	21.9	51.8
0950, 0980	Ceilings & Acoustic Treatment	95.9	14.9	47.1	95.9	28.4	55.3	97.7	31.1	57.6	95.9	17.0	48.4	102.8	37.3	63.4	95.9	21.9	51.4
0960	Flooring	112.0	22.2	88.0	110.6	19.2	86.2	109.6	32.6	89.0	119.4	20.0	92.9	119.4	37.7	97.6	108.3	18.0	84.2
0990	Wall Finishes & Painting/Coating	107.4	14.3	51.7	107.4	21.1	55.8	107.4	33.5	63.2	107.4	19.6	54.9	107.4	33.5	63.2	107.4	21.1	55.8
09	FINISHES	102.1	18.0	58.7	101.8	26.5	62.9	101.9	32.7	66.2	105.6	19.5	61.2	107.4	37.3	71.2	101.2	22.1	60.4
COVERS	DIVS. 10 - 14, 25, 28, 41, 43, 44	100.0	39.0	87.5	100.0	41.4	87.9	100.0	47.5	89.2	100.0	39.3	87.5	100.0	48.2	89.3	100.0	24.9	84.5
22, 23	PLUMBING & HVAC	98.0	17.9	65.4	97.6	24.4	67.7	99.9	32.7	72.5	98.0	18.9	65.8	99.9	32.4	72.4	97.6	20.8	66.3
26, 27, 3370	ELECTRICAL, COMMUNICATIONS & UTIL.	96.7	23.5	61.6	95.0	25.8	61.8	96.7	35.6	67.4	96.7	24.0	61.8	98.5	35.6	68.3	96.4	24.0	61.7
MF2004	WEIGHTED AVERAGE	95.6	29.9	67.2	96.9	35.5	70.4	99.1	43.1	74.9	96.9	32.3	69.0	97.2	44.6	74.5	97.0	32.4	69.1

DIVISION		MISSISSIPPI									MISSOURI								
		MCCOMB			MERIDIAN			TUPELO			BOWLING GREEN			CAPE GIRARDEAU			CHILLICOTHE		
		396			393			388			633			637			646		
		MAT.	INST.	TOTAL	MAT.	INST.	TOTAL	MAT.	INST.	TOTAL	MAT.	INST.	TOTAL	MAT.	INST.	TOTAL	MAT.	INST.	TOTAL
01543	CONTRACTOR EQUIPMENT	.0	98.5	98.5	.0	98.5	98.5	.0	98.5	98.5	.0	108.4	108.4	.0	108.4	108.4	.0	103.1	103.1
0241, 31 - 34	SITE & INFRASTRUCTURE, DEMOLITION	91.8	86.9	88.3	96.1	86.0	88.9	96.0	85.4	88.4	91.7	93.3	92.9	92.9	93.8	93.5	102.6	91.1	94.3
0310	Concrete Forming & Accessories	82.5	56.2	59.9	79.4	32.0	38.7	81.4	22.9	31.1	91.5	90.3	90.5	84.1	74.8	76.1	89.8	77.1	78.9
0310	Concrete Reinforcing	103.3	37.9	71.3	102.0	44.2	73.7	100.4	44.4	73.0	93.8	94.0	93.9	94.9	86.2	90.6	105.2	90.6	98.1
0330	Cast-in-Place Concrete	90.5	40.7	71.8	96.9	40.6	75.7	99.7	37.4	76.3	92.4	82.4	88.7	91.5	84.1	88.7	92.9	69.6	84.1
03	CONCRETE	89.0	49.0	70.1	93.1	39.3	67.7	94.8	34.2	66.2	90.0	89.6	89.8	89.0	81.8	85.6	97.1	77.8	88.0
04	MASONRY	114.3	37.4	68.3	88.3	27.4	51.8	121.3	29.2	66.2	115.5	97.0	104.4	112.4	78.3	92.0	99.9	86.8	92.1
05	METALS	87.2	72.9	82.9	88.0	71.9	83.1	87.0	71.0	82.2	95.9	113.8	101.3	96.9	109.2	100.6	93.3	94.4	93.6
06	WOOD, PLASTICS & COMPOSITES	80.0	62.9	70.9	76.3	31.0	52.4	78.2	21.8	48.4	86.2	91.5	89.0	77.9	72.2	74.9	92.9	79.0	85.6
07	THERMAL & MOISTURE PROTECTION	95.8	41.5	74.0	96.0	37.8	72.6	96.0	33.2	70.8	95.3	98.9	96.8	95.2	79.6	88.9	95.2	79.9	89.1
08	OPENINGS	97.2	51.3	85.3	97.2	35.3	81.2	97.2	27.9	79.2	96.3	86.6	93.8	96.2	74.3	90.6	91.5	73.9	87.0
0920	Plaster & Gypsum Board	97.8	62.4	76.7	95.8	29.7	56.5	95.8	20.3	50.9	100.8	91.2	95.1	97.6	71.4	82.0	93.0	78.2	84.2
0950, 0980	Ceilings & Acoustic Treatment	95.9	62.4	75.7	97.7	29.7	56.8	95.9	20.3	50.4	84.8	91.2	88.6	84.8	71.4	76.7	87.7	78.2	82.0
0960	Flooring	110.6	22.2	87.0	108.3	26.5	86.5	109.9	24.4	87.1	95.0	72.1	88.9	90.6	72.1	85.7	94.3	51.8	82.9
0990	Wall Finishes & Painting/Coating	107.4	31.6	62.1	107.4	33.1	63.0	107.4	33.5	63.2	99.5	101.1	100.5	99.5	75.8	85.3	92.1	55.4	70.2
09	FINISHES	101.3	48.6	74.1	101.0	30.5	64.6	101.1	24.0	61.3	91.9	87.9	89.9	90.0	73.2	81.4	93.6	71.1	82.0
COVERS	DIVS. 10 - 14, 25, 28, 41, 43, 44	100.0	73.3	94.5	100.0	47.4	89.2	100.0	40.2	87.7	100.0	66.6	93.1	100.0	63.5	92.5	100.0	62.3	92.2
22, 23	PLUMBING & HVAC	97.6	28.9	69.6	99.9	32.4	72.4	98.2	23.4	67.7	95.7	97.3	96.4	100.0	96.4	98.5	95.9	48.8	76.7
26, 27, 3370	ELECTRICAL, COMMUNICATIONS & UTIL.	93.7	61.4	78.2	96.4	58.7	78.3	96.4	23.5	61.5	101.4	88.6	95.2	101.3	109.8	105.4	96.8	41.0	70.0
MF2004	WEIGHTED AVERAGE	95.5	51.5	76.5	95.7	45.2	73.9	97.1	36.1	70.8	96.6	93.6	95.3	97.3	89.8	94.1	95.8	68.8	84.1

City Cost Indexes

		MISSOURI																	
		COLUMBIA			FLAT RIVER			HANNIBAL			HARRISONVILLE			JEFFERSON CITY			JOPLIN		
	DIVISION	652			636			634			647			650 - 651			648		
		MAT.	INST.	TOTAL	MAT.	INST.	TOTAL	MAT.	INST.	TOTAL	MAT.	INST.	TOTAL	MAT.	INST.	TOTAL	MAT.	INST.	TOTAL
01543	CONTRACTOR EQUIPMENT	.0	108.7	108.7	.0	108.4	108.4	.0	108.4	108.4	.0	103.1	103.1	.0	108.7	108.7	.0	107.1	107.1
0241, 31 - 34	SITE & INFRASTRUCTURE, DEMOLITION	102.2	95.0	97.0	94.5	93.9	94.1	89.0	92.8	91.7	94.2	95.2	94.9	103.9	94.4	97.1	102.4	99.0	99.9
0310	Concrete Forming & Accessories	85.5	70.5	72.6	97.8	83.7	85.7	89.7	68.6	71.6	86.5	92.4	91.6	99.1	62.4	67.6	102.4	73.0	77.2
0310	Concrete Reinforcing	99.0	112.8	105.8	94.9	104.6	99.6	93.3	79.6	86.6	104.8	98.9	101.9	97.7	112.5	105.0	108.6	72.3	90.8
0330	Cast-in-Place Concrete	89.0	86.4	88.0	95.4	82.1	90.4	87.5	76.6	83.4	95.2	96.1	95.5	90.3	70.3	82.7	100.6	70.4	89.3
03	CONCRETE	85.3	85.8	85.5	92.8	88.6	90.8	86.4	75.3	81.1	94.3	95.5	94.9	87.3	76.6	82.3	97.9	73.0	86.1
04	MASONRY	130.1	83.9	102.5	113.3	80.1	93.4	106.8	87.3	95.1	94.7	99.2	97.4	93.8	83.9	87.9	92.2	59.1	72.4
05	METALS	93.9	121.8	102.3	95.8	117.5	102.4	95.9	107.3	99.3	93.7	106.6	97.6	93.5	121.3	101.9	96.3	87.6	93.7
06	WOOD, PLASTICS & COMPOSITES	88.8	64.2	75.8	93.7	82.8	87.9	84.0	67.9	75.5	88.2	89.3	88.8	104.8	53.6	77.7	105.6	73.6	88.7
07	THERMAL & MOISTURE PROTECTION	90.8	83.7	87.9	95.5	93.5	94.7	95.2	87.5	92.1	94.5	99.8	96.6	91.1	80.3	86.8	95.7	67.3	84.3
08	OPENINGS	91.5	90.3	91.2	96.2	85.0	93.3	96.3	70.0	89.5	91.9	95.3	92.8	88.7	84.6	87.6	92.9	74.9	88.2
0920	Plaster & Gypsum Board	96.4	63.1	76.6	104.3	82.2	91.2	99.5	67.0	80.1	86.9	88.5	88.0	102.5	52.3	72.6	97.9	72.6	82.9
0950, 0980	Ceilings & Acoustic Treatment	88.1	63.1	73.1	84.8	82.2	83.3	84.8	67.0	74.1	87.7	88.8	88.4	88.1	52.3	66.5	91.1	72.6	80.0
0960	Flooring	103.2	87.8	99.1	98.6	72.1	91.5	93.8	72.1	88.0	89.9	94.8	91.2	111.8	87.8	105.4	118.9	48.5	100.1
0990	Wall Finishes & Painting/Coating	99.3	77.9	86.5	99.5	73.4	83.9	99.5	79.0	87.3	96.4	87.6	91.2	99.3	77.9	86.5	91.6	43.3	62.7
09	FINISHES	93.0	71.1	81.7	93.7	79.6	86.4	91.2	69.2	79.8	91.1	91.7	91.4	96.8	65.0	80.3	101.1	65.5	82.7
COVERS	DIVS. 10 - 14, 25, 28, 41, 43, 44	100.0	92.1	98.4	100.0	65.4	92.9	100.0	60.0	91.8	100.0	94.9	99.0	100.0	90.8	98.1	100.0	73.7	94.6
22, 23	PLUMBING & HVAC	99.9	97.9	99.1	95.7	97.2	96.3	95.7	92.1	94.2	95.8	97.4	96.4	99.9	100.7	100.2	100.1	54.8	81.7
26, 27, 3370	ELECTRICAL, COMMUNICATIONS & UTIL.	97.1	88.6	93.0	105.6	112.6	109.0	100.2	83.8	92.3	102.7	104.7	103.7	98.8	88.6	93.9	94.7	69.8	82.8
MF2004	WEIGHTED AVERAGE	96.8	90.7	94.2	97.6	94.1	96.1	95.5	84.4	90.7	95.5	98.1	96.6	95.6	88.7	92.6	97.6	69.9	85.7

		MISSOURI																	
		KANSAS CITY			KIRKSVILLE			POPLAR BLUFF			ROLLA			SEDALIA			SIKESTON		
	DIVISION	640 - 641			635			639			654 - 655			653			638		
		MAT.	INST.	TOTAL	MAT.	INST.	TOTAL	MAT.	INST.	TOTAL	MAT.	INST.	TOTAL	MAT.	INST.	TOTAL	MAT.	INST.	TOTAL
01543	CONTRACTOR EQUIPMENT	.0	104.6	104.6	.0	99.6	99.6	.0	102.2	102.2	.0	108.7	108.7	.0	99.6	99.6	.0	102.2	102.2
0241, 31 - 34	SITE & INFRASTRUCTURE, DEMOLITION	95.6	98.0	97.3	88.1	88.9	88.7	74.6	92.6	87.5	101.8	92.8	95.4	92.8	90.1	90.8	78.2	92.7	88.6
0310	Concrete Forming & Accessories	101.5	107.7	106.8	82.0	58.7	62.0	82.0	70.7	72.3	93.7	85.8	86.9	91.1	64.1	67.9	83.1	72.8	74.2
0310	Concrete Reinforcing	103.1	113.8	108.3	94.0	78.8	86.6	96.9	81.3	89.3	99.5	56.5	78.3	98.1	89.5	93.9	96.2	72.3	84.5
0330	Cast-in-Place Concrete	93.5	108.1	99.0	95.4	66.5	84.5	73.3	63.7	69.7	91.1	75.1	85.0	95.1	68.9	85.2	78.4	72.7	76.2
03	CONCRETE	93.7	109.3	101.1	101.2	66.7	84.9	79.4	71.5	75.6	87.1	77.6	82.6	97.9	71.9	85.6	82.9	73.7	78.6
04	MASONRY	97.0	106.5	102.7	118.9	72.2	91.0	111.2	60.4	80.8	106.9	48.2	71.8	112.8	55.3	78.4	111.0	65.9	84.0
05	METALS	103.7	115.1	107.2	95.5	95.9	95.6	96.1	94.7	95.7	93.4	87.9	91.8	91.9	99.3	94.1	96.4	90.2	94.5
06	WOOD, PLASTICS & COMPOSITES	104.9	107.5	106.3	72.0	57.1	64.1	71.2	72.2	71.7	97.9	97.4	97.6	91.2	57.7	73.5	72.5	74.8	73.7
07	THERMAL & MOISTURE PROTECTION	94.8	107.1	99.8	100.1	79.6	91.8	99.9	76.3	90.4	91.0	62.0	79.3	95.3	78.0	88.4	100.7	70.1	88.0
08	OPENINGS	99.2	110.0	102.0	100.0	64.4	90.8	101.1	75.7	94.5	91.5	73.9	86.9	94.8	73.7	89.4	101.1	66.9	92.2
0920	Plaster & Gypsum Board	95.0	107.5	102.4	94.1	55.8	71.3	94.7	71.4	80.8	99.0	97.2	97.9	93.1	56.4	71.3	95.7	74.0	82.7
0950, 0980	Ceilings & Acoustic Treatment	97.0	107.5	103.3	82.3	55.8	66.3	84.8	71.4	76.7	88.1	97.2	93.6	86.4	56.4	68.4	84.8	74.0	78.3
0960	Flooring	96.7	100.4	97.7	73.6	53.1	68.1	86.1	60.0	79.2	107.8	45.2	91.1	84.4	94.8	87.2	86.6	48.5	76.4
0990	Wall Finishes & Painting/Coating	96.4	115.4	107.8	95.1	43.3	64.2	94.6	84.6	88.6	99.3	87.0	91.9	99.3	87.6	92.3	94.6	62.3	75.3
09	FINISHES	96.6	107.1	102.0	88.0	54.3	70.6	89.4	70.2	79.5	94.9	80.1	87.2	90.4	70.6	80.2	89.9	67.0	78.1
COVERS	DIVS. 10 - 14, 25, 28, 41, 43, 44	100.0	97.8	99.6	100.0	55.4	90.8	100.0	58.1	91.4	100.0	60.9	92.0	100.0	87.0	97.3	100.0	59.3	91.6
22, 23	PLUMBING & HVAC	100.0	109.3	103.8	95.8	90.4	93.6	95.8	78.7	88.8	95.7	69.9	85.2	95.6	94.8	95.3	95.8	79.6	89.2
26, 27, 3370	ELECTRICAL, COMMUNICATIONS & UTIL.	105.2	106.9	106.0	99.8	83.8	92.1	100.3	109.7	104.8	95.7	88.5	92.2	96.6	104.8	100.5	99.4	112.5	105.7
MF2004	WEIGHTED AVERAGE	99.7	107.6	103.1	97.8	77.4	89.0	94.9	80.9	88.9	94.8	76.3	86.8	96.0	83.9	90.8	95.4	81.1	89.2

		MISSOURI									MONTANA								
		SPRINGFIELD			ST. JOSEPH			ST. LOUIS			BILLINGS			BUTTE			GREAT FALLS		
	DIVISION	656 - 658			644 - 645			630 - 631			590 - 591			597			594		
		MAT.	INST.	TOTAL	MAT.	INST.	TOTAL	MAT.	INST.	TOTAL	MAT.	INST.	TOTAL	MAT.	INST.	TOTAL	MAT.	INST.	TOTAL
01543	CONTRACTOR EQUIPMENT	.0	102.4	102.4	.0	103.1	103.1	.0	109.6	109.6	.0	99.0	99.0	.0	98.7	98.7	.0	98.7	98.7
0241, 31 - 34	SITE & INFRASTRUCTURE, DEMOLITION	95.1	93.6	94.0	97.1	93.5	94.5	92.6	97.5	96.1	88.7	96.6	94.4	98.4	95.2	96.1	102.2	96.0	97.7
0310	Concrete Forming & Accessories	99.8	71.3	75.3	101.4	84.3	86.7	97.8	109.9	108.2	97.5	64.3	69.0	84.6	59.3	62.9	97.6	65.1	69.7
0310	Concrete Reinforcing	95.3	112.4	103.7	101.9	98.7	100.3	86.8	113.7	99.9	95.6	72.4	84.2	103.7	72.5	88.4	95.6	72.3	84.2
0330	Cast-in-Place Concrete	96.7	64.8	84.7	93.5	103.8	97.4	91.5	114.7	100.2	107.5	68.4	92.8	118.3	68.9	99.7	125.0	55.1	98.7
03	CONCRETE	95.4	77.9	87.1	93.5	94.4	93.9	88.6	113.1	100.2	101.0	68.2	85.5	104.7	66.1	86.5	109.6	63.9	88.1
04	MASONRY	86.2	79.4	82.1	96.6	91.9	93.8	93.8	112.3	104.9	121.5	69.7	90.5	117.4	70.2	89.1	121.4	72.4	92.1
05	METALS	97.7	106.6	100.4	99.8	104.2	101.2	101.6	125.6	108.9	107.9	84.2	100.8	100.8	84.0	95.7	104.6	83.7	98.3
06	WOOD, PLASTICS & COMPOSITES	99.7	71.0	84.5	105.6	80.7	92.4	93.6	107.7	101.0	101.9	65.0	82.4	88.0	59.8	73.1	103.3	65.0	83.1
07	THERMAL & MOISTURE PROTECTION	95.4	75.9	87.5	95.2	93.1	94.3	95.2	109.4	100.9	98.8	66.6	85.9	98.7	66.8	85.9	99.3	66.0	85.9
08	OPENINGS	97.2	80.9	92.9	98.0	91.7	96.3	95.1	113.8	100.0	97.3	62.4	88.3	95.2	59.5	86.0	98.6	62.5	89.3
0920	Plaster & Gypsum Board	100.0	70.1	82.2	99.1	79.9	87.7	105.1	107.8	106.7	94.9	64.4	76.8	91.5	59.1	72.2	99.2	64.4	78.5
0950, 0980	Ceilings & Acoustic Treatment	88.1	70.1	77.2	96.2	79.9	86.4	89.9	107.8	100.6	112.1	64.4	83.4	111.5	59.1	79.9	114.0	64.4	84.2
0960	Flooring	106.7	48.5	91.2	99.4	94.8	98.2	98.4	103.2	99.6	104.9	47.2	89.5	101.3	37.6	84.3	110.1	51.0	94.3
0990	Wall Finishes & Painting/Coating	93.7	62.0	74.8	92.1	91.8	91.9	99.5	109.6	105.5	97.2	57.8	73.6	97.6	39.4	62.8	97.6	44.3	65.8
09	FINISHES	96.5	65.5	80.5	97.6	86.3	91.8	94.9	107.8	101.5	104.1	60.2	81.4	102.2	52.4	76.5	106.9	60.1	82.7
COVERS	DIVS. 10 - 14, 25, 28, 41, 43, 44	100.0	87.7	97.5	100.0	92.5	98.5	100.0	102.7	100.6	100.0	64.4	92.7	100.0	62.9	92.4	100.0	65.5	92.9
22, 23	PLUMBING & HVAC	100.0	66.6	86.4	100.1	94.2	98.0	100.0	112.5	105.1	100.2	75.1	90.0	100.3	69.7	87.9	100.3	74.6	89.9
26, 27, 3370	ELECTRICAL, COMMUNICATIONS & UTIL.	100.6	64.3	83.2	102.8	89.2	96.3	104.0	116.0	109.7	94.3	77.7	86.4	100.8	72.4	87.2	94.5	70.2	82.8
MF2004	WEIGHTED AVERAGE	97.6	76.4	88.5	98.8	92.9	96.3	97.8	112.0	103.9	101.7	73.6	89.5	101.3	70.2	87.9	102.9	72.0	89.6

City Cost Indexes

MONTANA

DIVISION		HAVRE 595 MAT.	INST.	TOTAL	HELENA 596 MAT.	INST.	TOTAL	KALISPELL 599 MAT.	INST.	TOTAL	MILES CITY 593 MAT.	INST.	TOTAL	MISSOULA 598 MAT.	INST.	TOTAL	WOLF POINT 592 MAT.	INST.	TOTAL
01543	CONTRACTOR EQUIPMENT	.0	98.7	98.7	.0	98.7	98.7	.0	98.7	98.7	.0	98.7	98.7	.0	98.7	98.7	.0	98.7	98.7
0241, 31 - 34	SITE & INFRASTRUCTURE, DEMOLITION	107.1	95.7	98.9	103.3	95.9	98.0	88.0	95.1	93.1	95.1	95.7	95.5	79.5	95.1	90.7	114.8	95.7	101.1
0310	Concrete Forming & Accessories	77.0	58.6	61.2	96.9	65.6	70.0	88.0	55.3	59.9	95.5	58.7	63.9	88.0	55.6	60.1	87.9	57.4	61.7
0310	Concrete Reinforcing	104.6	61.1	83.3	99.0	60.5	80.2	106.6	77.2	92.2	104.1	61.0	83.0	105.5	78.0	92.1	105.6	61.1	83.8
0330	Cast-in-Place Concrete	127.4	59.8	102.0	119.3	64.6	98.7	102.7	61.1	87.1	112.4	59.6	92.6	87.1	61.2	77.4	126.0	55.3	99.4
03	CONCRETE	112.5	60.5	88.0	107.2	65.1	87.3	94.7	62.6	79.5	101.7	60.4	82.2	83.5	62.9	73.8	115.9	58.4	88.8
04	MASONRY	118.1	72.1	90.6	115.7	63.9	84.7	116.2	51.0	77.2	123.1	70.5	91.6	141.0	65.7	96.0	124.2	72.1	93.0
05	METALS	97.3	77.7	91.4	103.7	77.1	95.7	97.2	84.3	93.3	96.5	77.7	90.8	97.7	85.7	94.0	96.6	77.7	90.9
06	WOOD, PLASTICS & COMPOSITES	79.1	58.0	67.9	102.5	65.0	82.7	92.3	56.2	73.2	100.0	58.0	77.8	92.3	56.2	73.2	91.2	56.2	72.7
07	THERMAL & MOISTURE PROTECTION	99.0	62.8	84.4	99.3	65.4	85.6	98.4	63.1	84.2	98.7	66.3	85.7	98.0	69.1	86.4	99.5	63.5	85.0
08	OPENINGS	95.3	55.4	84.9	98.1	59.4	88.1	95.2	59.8	86.1	94.7	55.4	84.5	95.2	59.8	86.1	94.7	54.4	84.3
0920	Plaster & Gypsum Board	88.3	57.2	69.8	98.6	64.4	78.3	92.8	55.4	70.5	98.0	57.2	73.7	92.8	55.4	70.5	94.5	55.4	71.2
0950, 0980	Ceilings & Acoustic Treatment	111.5	57.2	78.8	111.5	64.4	83.2	111.5	55.4	77.7	109.0	57.2	77.8	111.5	55.4	77.7	109.0	55.4	76.7
0960	Flooring	97.9	41.4	82.8	110.1	49.4	93.9	103.5	58.8	91.6	109.8	46.4	92.9	103.5	58.8	91.6	105.2	41.4	88.2
0990	Wall Finishes & Painting/Coating	97.6	43.8	65.4	97.6	49.2	68.7	97.6	44.3	65.8	97.6	41.3	64.0	97.6	44.3	65.8	97.6	41.3	64.0
09	FINISHES	101.2	53.5	76.5	106.3	60.8	82.8	102.6	54.4	77.7	105.0	54.2	78.8	102.1	54.4	77.5	104.2	52.2	77.4
COVERS	DIVS. 10 - 14, 25, 28, 41, 43, 44	100.0	51.8	90.1	100.0	54.1	90.6	100.0	49.1	89.5	100.0	51.7	90.1	100.0	49.1	89.5	100.0	51.6	90.0
22, 23	PLUMBING & HVAC	96.1	73.5	86.9	100.3	74.1	89.7	96.1	66.1	83.9	96.1	73.4	86.9	100.3	66.1	86.4	96.1	73.5	86.9
26, 27, 3370	ELECTRICAL, COMMUNICATIONS & UTIL.	94.5	59.0	77.5	94.5	71.4	83.5	98.1	76.7	87.8	94.5	61.1	78.5	99.0	76.7	88.3	94.5	61.1	78.5
MF2004	WEIGHTED AVERAGE	100.1	67.5	86.0	102.1	70.4	88.5	98.0	67.3	84.8	99.1	67.8	85.6	98.9	69.3	86.1	101.2	67.3	86.6

NEBRASKA

DIVISION		ALLIANCE 693 MAT.	INST.	TOTAL	COLUMBUS 686 MAT.	INST.	TOTAL	GRAND ISLAND 688 MAT.	INST.	TOTAL	HASTINGS 689 MAT.	INST.	TOTAL	LINCOLN 683 - 685 MAT.	INST.	TOTAL	MCCOOK 690 MAT.	INST.	TOTAL
01543	CONTRACTOR EQUIPMENT	.0	96.6	96.6	.0	101.3	101.3	.0	101.3	101.3	.0	101.3	101.3	.0	101.3	101.3	.0	101.3	101.3
0241, 31 - 34	SITE & INFRASTRUCTURE, DEMOLITION	94.9	95.3	95.2	94.3	88.9	90.4	98.9	91.5	93.6	97.9	89.5	91.9	89.2	91.3	90.7	98.7	88.9	91.7
0310	Concrete Forming & Accessories	87.6	24.8	33.7	96.9	24.1	34.4	96.3	48.0	54.9	100.0	44.2	52.1	101.5	44.4	52.4	93.8	30.0	39.0
0310	Concrete Reinforcing	113.6	42.5	78.8	104.4	47.5	76.5	103.8	71.4	87.9	103.8	47.5	76.2	95.3	72.2	84.0	105.8	71.0	88.8
0330	Cast-in-Place Concrete	107.1	35.2	80.0	109.7	44.7	85.2	116.2	55.7	93.4	116.2	49.2	91.0	99.9	58.1	84.2	116.2	35.3	85.7
03	CONCRETE	115.8	32.8	76.6	103.2	37.5	72.2	107.7	56.5	83.5	108.0	47.9	79.6	97.2	55.8	77.6	107.9	41.3	76.4
04	MASONRY	107.2	27.8	59.7	112.6	28.7	62.4	105.5	46.6	70.2	114.5	37.7	68.5	95.2	62.1	75.4	102.6	27.7	57.8
05	METALS	100.5	54.0	86.4	92.4	69.0	85.3	94.0	81.9	90.3	95.0	69.2	87.2	99.0	83.4	94.3	95.2	77.3	89.8
06	WOOD, PLASTICS & COMPOSITES	85.1	23.8	52.7	96.8	21.3	56.9	96.1	42.7	67.9	100.2	42.7	69.8	101.6	37.3	67.6	93.7	29.6	59.8
07	THERMAL & MOISTURE PROTECTION	103.2	33.5	75.1	95.8	30.7	69.6	95.9	55.8	79.8	96.0	44.4	75.2	95.0	59.4	80.7	95.9	33.9	71.0
08	OPENINGS	90.8	27.0	74.3	90.7	31.8	75.4	90.7	47.3	79.4	90.7	43.4	78.4	96.3	46.9	83.5	90.7	36.8	76.8
0920	Plaster & Gypsum Board	91.7	21.8	50.1	98.3	19.1	51.1	98.0	41.1	64.1	99.3	41.1	64.6	101.5	35.6	62.2	98.0	27.6	56.1
0950, 0980	Ceilings & Acoustic Treatment	88.6	21.8	48.4	86.4	19.1	45.9	86.4	41.1	59.1	86.4	41.1	59.1	94.8	35.6	59.2	86.4	27.6	51.0
0960	Flooring	95.5	27.4	77.3	96.3	34.8	79.9	96.2	33.5	79.4	97.8	34.8	81.0	98.0	40.3	82.6	95.0	27.4	77.0
0990	Wall Finishes & Painting/Coating	150.4	21.2	73.1	91.5	48.4	65.8	91.5	40.0	60.7	91.5	48.4	65.8	91.5	40.4	61.0	91.5	23.0	50.6
09	FINISHES	94.3	24.4	58.3	93.0	27.1	59.0	93.1	42.7	67.1	93.8	42.3	67.2	96.1	41.1	67.7	92.7	27.9	59.3
COVERS	DIVS. 10 - 14, 25, 28, 41, 43, 44	100.0	44.2	88.5	100.0	56.3	91.0	100.0	65.5	92.9	100.0	62.9	92.4	100.0	64.9	92.8	100.0	43.3	88.3
22, 23	PLUMBING & HVAC	95.9	25.0	67.0	95.7	67.2	84.1	100.0	77.2	90.7	95.7	72.9	86.4	100.0	77.2	90.7	95.7	64.8	83.1
26, 27, 3370	ELECTRICAL, COMMUNICATIONS & UTIL.	92.3	17.3	56.3	92.0	41.1	67.6	90.8	67.3	79.5	90.2	41.2	66.7	96.6	67.3	82.5	94.2	17.4	57.4
MF2004	WEIGHTED AVERAGE	98.9	34.8	71.2	96.1	49.2	75.8	97.5	64.6	83.3	97.2	56.0	79.4	97.7	66.0	84.0	96.9	46.6	75.2

NEBRASKA / NEVADA

DIVISION		NORFOLK 687 MAT.	INST.	TOTAL	NORTH PLATTE 691 MAT.	INST.	TOTAL	OMAHA 680 - 681 MAT.	INST.	TOTAL	VALENTINE 692 MAT.	INST.	TOTAL	CARSON CITY 897 MAT.	INST.	TOTAL	ELKO 898 MAT.	INST.	TOTAL
01543	CONTRACTOR EQUIPMENT	.0	90.7	90.7	.0	101.3	101.3	.0	90.7	90.7	.0	94.0	94.0	.0	101.1	101.1	.0	101.1	101.1
0241, 31 - 34	SITE & INFRASTRUCTURE, DEMOLITION	78.6	88.3	85.5	99.7	89.5	92.4	78.9	90.3	87.1	83.0	92.2	89.6	62.4	102.7	91.3	57.9	102.6	90.0
0310	Concrete Forming & Accessories	82.9	44.2	49.7	96.2	44.3	51.6	95.8	71.2	74.7	84.0	22.1	30.9	98.8	97.6	97.8	106.8	86.2	89.1
0310	Concrete Reinforcing	104.6	54.5	80.1	105.2	71.0	88.5	99.9	73.2	86.8	105.8	38.2	72.8	113.0	116.5	114.7	114.8	108.7	111.8
0330	Cast-in-Place Concrete	110.4	54.3	89.3	116.2	49.3	91.0	103.5	71.9	91.6	102.2	41.3	79.2	108.9	94.9	103.6	102.5	78.8	93.6
03	CONCRETE	101.9	50.4	77.6	107.9	52.5	81.7	98.5	72.0	86.0	103.9	33.1	70.5	112.4	100.2	106.6	106.4	88.1	97.8
04	MASONRY	119.1	59.0	83.1	90.3	37.7	58.8	100.5	78.3	87.2	102.3	28.8	58.3	125.3	79.4	97.8	126.1	75.7	95.9
05	METALS	95.7	63.9	86.1	94.2	78.9	89.5	98.3	75.9	91.6	106.6	54.6	90.8	97.2	103.6	99.2	96.4	98.7	97.1
06	WOOD, PLASTICS & COMPOSITES	79.9	42.1	59.9	95.9	42.7	67.8	95.6	72.6	83.4	80.5	20.7	48.9	94.3	100.0	97.3	104.2	87.2	95.2
07	THERMAL & MOISTURE PROTECTION	94.3	52.8	77.6	95.9	44.5	75.2	91.8	70.4	83.2	94.9	30.0	68.8	98.9	87.4	94.3	98.8	78.2	90.5
08	OPENINGS	92.6	44.4	80.2	89.9	49.5	79.5	99.6	65.1	90.7	92.8	25.8	75.4	94.3	107.0	97.5	96.9	90.6	95.3
0920	Plaster & Gypsum Board	97.2	41.1	63.8	98.4	41.1	64.3	103.5	72.4	84.9	100.4	19.1	52.0	85.4	99.8	94.0	93.8	86.7	89.5
0950, 0980	Ceilings & Acoustic Treatment	101.8	41.1	65.3	88.1	41.1	59.8	106.9	72.4	86.1	104.9	19.1	53.3	104.2	99.8	101.6	104.2	86.7	93.7
0960	Flooring	118.3	41.9	97.9	96.1	34.8	79.8	124.2	44.6	103.0	121.2	41.8	100.1	100.0	61.6	89.7	106.3	41.9	89.1
0990	Wall Finishes & Painting/Coating	152.3	48.4	90.2	91.5	48.4	65.8	152.3	68.8	102.3	152.3	32.5	80.6	99.6	76.3	85.6	99.6	86.8	91.9
09	FINISHES	110.4	43.3	75.8	93.5	42.3	67.1	114.4	65.7	89.2	112.9	26.3	68.3	96.5	88.3	92.3	99.5	81.8	90.4
COVERS	DIVS. 10 - 14, 25, 28, 41, 43, 44	100.0	61.4	92.0	100.0	48.9	89.5	100.0	67.7	93.4	100.0	35.7	86.8	100.0	102.6	100.5	100.0	81.2	96.1
22, 23	PLUMBING & HVAC	95.5	72.9	86.3	100.0	70.8	87.9	99.7	81.6	92.4	95.4	22.3	65.6	99.9	91.0	96.3	97.9	77.6	89.6
26, 27, 3370	ELECTRICAL, COMMUNICATIONS & UTIL.	90.9	51.7	72.1	92.5	41.2	67.9	90.4	85.9	88.2	89.6	41.1	66.4	91.2	106.2	98.4	90.7	65.5	78.6
MF2004	WEIGHTED AVERAGE	97.8	59.8	81.4	96.9	56.8	79.6	98.8	77.3	89.5	99.0	37.3	72.4	99.3	96.0	97.9	98.5	82.5	91.6

City Cost Indexes

DIVISION		NEVADA ELY 893			NEVADA LAS VEGAS 889-891			NEVADA RENO 894-895			NEW HAMPSHIRE CHARLESTON 036			NEW HAMPSHIRE CLAREMONT 037			NEW HAMPSHIRE CONCORD 032-033		
		MAT.	INST.	TOTAL	MAT.	INST.	TOTAL	MAT.	INST.	TOTAL	MAT.	INST.	TOTAL	MAT.	INST.	TOTAL	MAT.	INST.	TOTAL
01543	CONTRACTOR EQUIPMENT	.0	101.1	101.1	.0	101.1	101.1	.0	101.1	101.1	.0	100.7	100.7	.0	100.7	100.7	.0	100.7	100.7
0241, 31-34	SITE & INFRASTRUCTURE, DEMOLITION	62.8	102.1	91.0	63.5	105.3	93.5	62.9	102.7	91.4	76.7	97.3	91.4	71.2	97.3	89.9	83.1	99.0	94.5
0310	Concrete Forming & Accessories	98.8	69.6	73.7	97.6	108.1	106.6	95.5	97.7	97.4	86.3	36.2	43.3	92.9	36.2	44.2	90.7	86.2	86.9
0310	Concrete Reinforcing	113.5	100.4	107.1	107.4	123.4	115.2	107.4	122.8	114.9	83.4	55.3	69.6	83.4	55.3	69.6	83.4	88.1	85.7
0330	Cast-in-Place Concrete	110.1	76.2	97.4	105.7	105.8	105.7	116.4	94.9	108.4	97.2	51.2	79.9	89.2	51.2	74.9	97.8	95.3	96.8
03	CONCRETE	115.3	78.3	97.8	109.8	110.0	109.9	115.1	101.4	108.6	103.0	45.9	76.1	94.7	45.9	71.7	101.7	89.4	95.9
04	MASONRY	131.1	67.7	93.1	119.3	96.2	105.4	125.1	79.4	97.7	91.9	39.7	60.6	92.1	39.7	60.7	95.3	90.9	92.7
05	METALS	96.3	95.0	95.9	97.8	109.5	101.3	97.6	106.4	100.3	90.5	64.3	82.5	90.5	64.3	82.5	91.1	84.9	89.3
06	WOOD, PLASTICS & COMPOSITES	93.9	69.2	80.8	94.0	108.2	101.5	89.9	100.0	95.3	84.4	35.7	58.6	92.0	35.7	62.2	89.5	87.0	88.1
07	THERMAL & MOISTURE PROTECTION	99.3	72.9	88.7	104.8	98.8	102.4	98.8	87.4	94.2	97.8	43.6	76.0	97.7	43.6	75.9	98.3	88.6	94.4
08	OPENINGS	96.8	78.7	92.1	95.8	113.1	100.3	94.6	108.5	98.2	102.3	38.7	85.8	103.6	38.7	86.8	103.5	80.4	97.5
0920	Plaster & Gypsum Board	90.2	68.1	77.1	90.1	108.2	100.9	86.9	99.8	94.6	94.4	33.2	58.0	97.0	33.2	59.0	94.4	85.8	89.3
0950, 0980	Ceilings & Acoustic Treatment	104.2	68.1	82.5	111.2	108.2	109.4	111.0	99.8	104.3	84.2	33.2	53.5	84.2	33.2	53.5	84.2	85.8	85.2
0960	Flooring	103.7	68.9	94.4	99.6	95.7	98.5	100.0	61.6	89.7	92.1	38.6	77.8	94.9	38.6	79.9	92.6	102.1	95.2
0990	Wall Finishes & Painting/Coating	99.6	86.8	91.9	101.7	113.5	108.8	99.6	76.3	85.6	91.0	31.1	55.2	91.0	31.1	55.2	91.0	97.6	94.9
09	FINISHES	98.6	70.7	84.2	99.0	106.4	102.8	98.4	88.3	93.2	92.0	35.5	62.9	92.9	35.5	63.3	93.0	91.0	91.9
COVERS	DIVS. 10-14, 25, 28, 41, 43, 44	100.0	67.5	93.3	100.0	98.7	99.7	100.0	102.6	100.5	100.0	43.4	88.4	100.0	43.4	88.4	100.0	75.4	94.9
22, 23	PLUMBING & HVAC	97.9	73.2	87.8	99.9	102.3	100.9	99.9	91.1	96.3	95.8	37.7	72.1	95.8	37.7	72.1	100.0	86.8	94.6
26, 27, 3370	ELECTRICAL, COMMUNICATIONS & UTIL.	91.0	65.5	78.8	93.3	104.1	98.5	91.2	106.0	98.4	96.1	34.2	66.4	96.1	34.2	66.4	98.1	69.6	84.5
MF2004	WEIGHTED AVERAGE	99.8	76.4	89.7	99.6	104.7	101.8	99.9	96.5	98.4	95.7	46.2	74.3	94.8	46.2	73.8	97.5	86.0	92.5

DIVISION		NEW HAMPSHIRE KEENE 034			NEW HAMPSHIRE LITTLETON 035			NEW HAMPSHIRE MANCHESTER 031			NEW HAMPSHIRE NASHUA 030			NEW HAMPSHIRE PORTSMOUTH 038			NEW JERSEY ATLANTIC CITY 082,084		
		MAT.	INST.	TOTAL	MAT.	INST.	TOTAL	MAT.	INST.	TOTAL	MAT.	INST.	TOTAL	MAT.	INST.	TOTAL	MAT.	INST.	TOTAL
01543	CONTRACTOR EQUIPMENT	.0	100.7	100.7	.0	100.7	100.7	.0	100.7	100.7	.0	100.7	100.7	.0	100.7	100.7	.0	98.6	98.6
0241, 31-34	SITE & INFRASTRUCTURE, DEMOLITION	82.8	97.3	93.2	71.2	96.9	89.7	83.3	99.0	94.6	83.8	99.0	94.7	78.2	98.0	92.4	85.6	104.2	99.0
0310	Concrete Forming & Accessories	91.4	37.3	44.9	102.3	42.4	50.8	98.1	86.2	87.9	99.9	86.2	88.2	87.8	81.7	82.5	109.2	125.5	123.2
0310	Concrete Reinforcing	83.4	65.8	74.8	84.1	39.6	62.3	104.3	88.1	96.4	104.3	88.1	96.4	83.4	88.0	85.6	79.5	105.8	92.4
0330	Cast-in-Place Concrete	97.6	51.6	80.3	87.6	46.0	71.9	104.2	95.3	100.9	92.4	95.3	93.5	87.6	88.7	88.0	82.8	126.4	99.2
03	CONCRETE	102.8	48.5	77.2	94.1	43.8	70.4	108.7	89.4	99.6	102.8	89.4	96.5	94.0	85.0	89.8	95.8	121.0	107.7
04	MASONRY	96.2	39.6	62.3	105.0	38.3	65.1	97.8	90.9	93.7	97.9	90.9	93.7	93.4	80.1	85.4	99.9	124.1	114.4
05	METALS	91.1	69.5	84.5	91.1	57.3	80.9	95.8	84.9	92.5	95.8	84.9	92.5	92.4	83.0	89.6	90.8	99.0	93.3
06	WOOD, PLASTICS & COMPOSITES	90.2	35.7	61.4	102.1	40.9	69.7	98.5	87.0	92.4	100.2	87.0	93.2	85.8	87.0	86.4	111.6	127.0	119.8
07	THERMAL & MOISTURE PROTECTION	98.2	45.4	77.0	97.8	47.8	77.7	98.3	88.6	94.4	98.6	88.6	94.6	98.2	95.7	97.2	98.3	122.0	107.8
08	OPENINGS	100.7	47.5	87.0	104.6	38.4	87.5	105.6	80.4	99.1	105.6	80.4	99.1	106.4	74.8	98.2	103.2	118.7	107.2
0920	Plaster & Gypsum Board	96.3	33.2	58.7	107.2	38.5	66.3	102.8	85.8	92.7	102.8	85.8	92.7	94.4	85.8	89.3	105.0	126.9	118.0
0950, 0980	Ceilings & Acoustic Treatment	84.2	33.2	53.5	84.2	38.5	56.7	96.2	85.8	89.9	96.2	85.8	89.9	86.0	85.8	85.9	84.2	126.9	109.9
0960	Flooring	94.3	63.3	86.0	104.6	38.6	87.0	99.0	102.1	99.9	98.0	102.1	99.1	92.6	102.1	95.2	102.2	122.3	107.5
0990	Wall Finishes & Painting/Coating	91.0	31.1	55.2	91.0	74.1	80.9	91.0	97.6	94.9	91.0	97.6	94.9	91.0	37.7	59.1	91.0	129.6	114.1
09	FINISHES	93.9	40.6	66.4	97.5	44.5	70.1	99.0	91.0	94.8	98.9	91.0	94.8	93.2	81.5	87.1	98.0	125.9	112.4
COVERS	DIVS. 10-14, 25, 28, 41, 43, 44	100.0	61.7	92.1	100.0	45.8	88.9	100.0	75.4	94.9	100.0	75.4	94.9	100.0	71.3	94.1	100.0	114.1	102.9
22, 23	PLUMBING & HVAC	95.8	40.4	73.2	95.8	76.8	88.0	100.0	86.8	94.6	100.0	86.8	94.6	100.0	81.0	92.3	99.7	111.5	104.5
26, 27, 3370	ELECTRICAL, COMMUNICATIONS & UTIL.	96.1	34.2	66.4	97.0	69.6	83.8	99.0	69.6	84.9	98.1	69.6	84.4	96.5	69.6	83.6	94.5	127.6	110.4
MF2004	WEIGHTED AVERAGE	96.2	49.1	75.9	96.2	59.6	80.4	100.0	86.0	94.0	99.3	86.0	93.6	96.6	81.6	90.1	97.1	117.3	105.8

DIVISION		NEW JERSEY CAMDEN 081			NEW JERSEY DOVER 078			NEW JERSEY ELIZABETH 072			NEW JERSEY HACKENSACK 076			NEW JERSEY JERSEY CITY 073			NEW JERSEY LONG BRANCH 077		
		MAT.	INST.	TOTAL	MAT.	INST.	TOTAL	MAT.	INST.	TOTAL	MAT.	INST.	TOTAL	MAT.	INST.	TOTAL	MAT.	INST.	TOTAL
01543	CONTRACTOR EQUIPMENT	.0	98.6	98.6	.0	100.7	100.7	.0	100.7	100.7	.0	100.7	100.7	.0	98.6	98.6	.0	98.1	98.1
0241, 31-34	SITE & INFRASTRUCTURE, DEMOLITION	86.4	104.5	99.3	95.8	104.5	102.0	99.1	104.5	103.0	96.4	105.3	102.8	86.4	105.2	99.9	90.9	104.4	100.5
0310	Concrete Forming & Accessories	99.9	126.5	122.0	96.3	126.4	122.2	109.4	126.3	123.9	96.3	126.4	122.2	96.3	126.4	122.8	100.5	129.2	125.2
0310	Concrete Reinforcing	104.3	109.9	107.0	80.3	109.3	94.5	80.3	109.3	94.5	80.3	109.3	94.5	104.3	109.3	106.7	80.3	109.3	94.5
0330	Cast-in-Place Concrete	80.3	122.1	96.0	103.0	127.4	112.2	88.4	127.3	103.1	100.7	127.4	110.7	80.3	127.4	98.0	89.3	126.3	103.2
03	CONCRETE	96.7	120.3	107.9	104.2	122.6	112.9	99.3	122.5	110.3	102.3	122.6	111.9	96.7	122.5	108.9	101.2	123.3	111.6
04	MASONRY	91.4	124.1	111.0	94.9	122.7	111.6	111.5	122.7	118.2	99.1	122.7	113.2	89.0	122.7	109.2	103.7	124.1	115.9
05	METALS	95.7	100.7	97.2	90.8	105.5	95.3	92.2	105.3	96.2	90.9	105.5	95.3	95.7	102.8	97.9	90.9	102.3	94.4
06	WOOD, PLASTICS & COMPOSITES	100.2	127.0	114.4	99.2	127.0	113.9	115.5	127.0	121.6	99.2	127.0	113.9	100.2	127.0	114.4	101.3	131.2	117.1
07	THERMAL & MOISTURE PROTECTION	98.2	123.4	108.3	98.7	128.7	110.8	98.9	128.7	110.9	98.4	128.7	110.6	98.1	128.7	110.4	98.3	129.1	110.7
08	OPENINGS	105.6	119.8	109.3	108.8	119.6	111.6	106.7	119.6	110.1	105.9	119.6	109.5	105.6	119.6	109.2	101.6	121.9	106.9
0920	Plaster & Gypsum Board	102.8	126.9	117.1	99.5	126.9	115.8	106.0	126.9	118.4	99.5	126.9	115.8	102.8	126.9	117.1	101.1	131.3	119.1
0950, 0980	Ceilings & Acoustic Treatment	96.2	126.9	114.7	84.2	126.9	109.9	86.0	126.9	110.6	84.2	126.9	109.9	96.2	126.9	114.7	84.2	131.3	112.5
0960	Flooring	98.0	122.3	104.5	96.8	122.3	103.6	102.6	122.3	107.8	96.8	122.3	103.6	98.0	122.3	104.5	98.3	122.3	104.7
0990	Wall Finishes & Painting/Coating	91.0	129.6	114.1	90.9	129.6	114.1	90.9	129.6	114.1	90.9	129.6	114.1	91.0	129.6	114.1	91.0	129.6	114.1
09	FINISHES	99.2	125.9	113.0	95.7	127.3	112.0	99.3	127.3	113.8	95.6	127.3	112.0	99.2	127.3	113.7	96.6	128.4	113.0
COVERS	DIVS. 10-14, 25, 28, 41, 43, 44	100.0	114.1	102.9	100.0	118.9	103.9	100.0	118.9	103.9	100.0	118.9	103.9	100.0	118.9	103.9	100.0	114.5	103.0
22, 23	PLUMBING & HVAC	100.0	111.5	104.7	99.7	126.0	110.4	100.0	119.7	108.0	99.7	126.0	110.4	100.0	126.0	110.6	99.7	121.7	108.7
26, 27, 3370	ELECTRICAL, COMMUNICATIONS & UTIL.	98.6	127.6	112.5	96.1	138.1	116.2	96.6	138.1	116.5	96.1	141.9	118.1	100.0	141.9	120.1	95.8	138.1	116.1
MF2004	WEIGHTED AVERAGE	98.4	117.4	106.6	98.5	123.0	109.1	99.5	121.7	109.1	98.3	123.4	109.1	98.4	123.3	109.2	97.9	122.2	108.4

664

City Cost Indexes

DIVISION		NEW JERSEY																	
		NEWARK 070-071			NEW BRUNSWICK 088-089			PATERSON 074-075			POINT PLEASANT 087			SUMMIT 079			TRENTON 085-086		
		MAT.	INST.	TOTAL	MAT.	INST.	TOTAL	MAT.	INST.	TOTAL	MAT.	INST.	TOTAL	MAT.	INST.	TOTAL	MAT.	INST.	TOTAL
01543	CONTRACTOR EQUIPMENT	.0	100.7	100.7	.0	98.1	98.1	.0	100.7	100.7	.0	98.1	98.1	.0	100.7	100.7	.0	98.1	98.1
0241, 31-34	SITE & INFRASTRUCTURE, DEMOLITION	103.8	104.5	104.3	97.9	104.4	102.5	98.0	105.3	103.2	99.4	104.4	103.0	96.8	104.5	102.3	87.9	104.4	99.7
0310	Concrete Forming & Accessories	97.9	126.4	122.4	103.3	126.2	122.9	98.7	126.4	122.5	97.3	125.8	121.8	99.6	126.4	122.6	98.1	125.9	122.0
0310	Concrete Reinforcing	104.3	109.3	106.7	80.3	109.0	94.3	104.3	109.3	106.7	80.3	109.3	94.5	80.3	109.3	94.5	104.3	110.2	107.2
0330	Cast-in-Place Concrete	97.8	127.3	108.9	102.4	126.3	111.4	102.4	127.4	111.8	102.4	126.2	111.3	85.5	127.3	101.2	92.4	124.8	104.6
03	CONCRETE	105.5	122.6	113.6	113.6	121.9	117.5	107.9	122.6	114.8	113.2	121.8	117.2	96.0	122.6	108.5	102.8	121.4	111.6
04	MASONRY	99.9	122.7	113.5	97.9	117.2	109.4	95.4	122.7	111.8	89.3	124.1	110.1	97.6	122.7	112.6	88.5	124.1	109.8
05	METALS	95.7	105.4	98.6	90.9	101.9	94.2	90.9	105.5	95.3	90.9	102.1	94.3	90.8	105.4	95.2	90.8	101.0	93.9
06	WOOD, PLASTICS & COMPOSITES	101.6	127.0	115.0	104.7	126.9	116.5	101.9	127.0	115.2	97.5	126.9	113.1	103.2	127.0	115.8	99.2	126.9	113.9
07	THERMAL & MOISTURE PROTECTION	98.3	128.7	110.5	98.5	126.3	109.7	98.8	122.3	108.2	98.5	128.6	110.6	99.1	128.7	111.0	97.0	121.8	107.0
08	OPENINGS	111.7	119.6	113.7	97.5	119.6	103.3	111.7	119.6	113.7	99.6	120.5	105.0	114.1	119.6	115.6	106.2	118.9	109.5
0920	Plaster & Gypsum Board	102.8	126.9	117.1	102.4	126.9	117.0	102.8	126.9	117.1	99.5	126.9	115.8	101.1	126.9	116.5	102.8	126.9	117.1
0950, 0980	Ceilings & Acoustic Treatment	96.2	126.9	114.7	84.2	126.9	109.9	96.2	126.9	114.7	84.2	126.9	109.9	84.2	126.9	109.9	96.2	126.9	114.7
0960	Flooring	98.2	122.3	104.6	99.6	122.3	105.7	98.0	122.3	104.5	96.8	122.3	103.6	98.3	122.3	104.7	98.2	122.3	104.6
0990	Wall Finishes & Painting/Coating	90.9	129.6	114.1	91.0	129.6	114.1	90.9	129.6	114.1	91.0	129.6	114.1	90.9	129.6	114.1	91.0	129.6	114.1
09	FINISHES	99.7	127.3	114.0	97.8	127.3	113.0	99.3	127.3	113.8	96.5	125.9	111.7	96.6	127.3	112.5	99.2	125.9	113.0
COVERS	DIVS. 10-14, 25, 28, 41, 43, 44	100.0	118.9	103.9	100.0	118.7	103.8	100.0	118.9	103.9	100.0	109.9	102.0	100.0	118.9	103.9	100.0	114.0	102.9
22, 23	PLUMBING & HVAC	100.0	122.4	109.1	99.7	121.6	108.6	100.0	126.0	110.6	99.7	118.4	107.3	99.7	121.7	108.7	100.0	121.7	108.8
26, 27, 3370	ELECTRICAL, COMMUNICATIONS & UTIL.	99.9	141.9	120.0	95.1	138.0	115.7	100.0	138.1	118.3	94.5	138.1	115.4	96.6	138.1	116.5	98.7	142.2	119.6
MF2004	WEIGHTED AVERAGE	101.0	122.8	110.4	98.9	121.0	108.4	100.2	122.9	110.0	98.4	120.7	108.0	98.4	122.1	108.6	98.2	121.7	108.3

DIVISION		NEW JERSEY			NEW MEXICO														
		VINELAND 080,083			ALBUQUERQUE 870-872			CARRIZOZO 883			CLOVIS 881			FARMINGTON 874			GALLUP 873		
		MAT.	INST.	TOTAL	MAT.	INST.	TOTAL	MAT.	INST.	TOTAL	MAT.	INST.	TOTAL	MAT.	INST.	TOTAL	MAT.	INST.	TOTAL
01543	CONTRACTOR EQUIPMENT	.0	98.6	98.6	.0	115.6	115.6	.0	115.6	115.6	.0	115.6	115.6	.0	115.6	115.6	.0	115.6	115.6
0241, 31-34	SITE & INFRASTRUCTURE, DEMOLITION	89.8	104.5	100.3	79.4	110.2	101.5	99.2	110.2	107.1	87.5	110.2	103.8	85.2	110.2	103.1	92.8	110.2	105.3
0310	Concrete Forming & Accessories	94.6	125.7	121.3	97.0	69.1	73.1	97.1	69.1	73.1	97.1	68.8	72.8	97.1	69.1	73.1	97.1	69.1	73.1
0310	Concrete Reinforcing	79.5	105.8	92.4	108.1	66.6	87.8	115.9	66.6	91.8	117.2	54.5	86.5	118.1	66.6	92.9	113.0	66.6	90.3
0330	Cast-in-Place Concrete	89.3	126.5	103.3	108.9	74.9	96.1	101.3	74.9	91.4	101.2	74.8	91.3	109.9	74.9	96.8	103.4	74.9	92.7
03	CONCRETE	100.7	121.0	110.3	111.5	71.6	92.7	126.5	71.6	100.6	114.7	69.1	93.2	115.6	71.6	94.8	121.6	71.6	98.0
04	MASONRY	87.6	124.1	109.4	114.3	64.2	84.4	111.4	64.2	83.2	111.5	64.2	83.2	121.3	64.2	87.1	111.5	64.2	83.2
05	METALS	90.8	99.1	93.3	104.3	87.0	99.1	101.1	87.0	96.8	100.8	80.8	94.7	101.9	87.0	97.4	101.1	87.0	96.8
06	WOOD, PLASTICS & COMPOSITES	94.4	127.0	111.6	94.4	70.4	81.7	94.5	70.4	81.8	94.5	70.4	81.8	94.5	70.4	81.8	94.5	70.4	81.8
07	THERMAL & MOISTURE PROTECTION	98.1	122.0	107.7	98.5	72.5	88.0	100.1	72.5	88.9	98.9	72.5	88.2	98.7	72.5	88.1	99.7	72.5	88.8
08	OPENINGS	99.1	118.7	104.2	94.4	71.6	88.5	93.2	71.6	87.6	93.4	67.8	86.7	97.2	71.6	90.5	97.2	71.6	90.5
0920	Plaster & Gypsum Board	98.6	126.9	115.4	90.8	69.1	77.9	84.7	69.1	75.4	84.7	69.1	75.4	84.7	69.1	75.4	84.7	69.1	75.4
0950, 0980	Ceilings & Acoustic Treatment	84.2	126.9	109.9	110.2	69.1	85.5	100.9	69.1	81.7	100.9	69.1	81.7	100.9	69.1	81.7	100.9	69.1	81.7
0960	Flooring	95.9	122.3	102.9	101.6	70.0	93.2	100.0	70.0	92.0	100.0	70.0	92.0	100.0	70.0	92.0	100.0	70.0	92.0
0990	Wall Finishes & Painting/Coating	91.0	129.6	114.1	101.8	57.6	75.4	99.6	57.6	74.5	99.6	57.6	74.5	99.6	57.6	74.5	99.6	57.6	74.5
09	FINISHES	95.4	125.9	111.1	99.9	68.1	83.5	97.7	68.1	82.4	96.6	68.1	81.9	96.3	68.1	81.7	97.2	68.1	82.2
COVERS	DIVS. 10-14, 25, 28, 41, 43, 44	100.0	114.1	102.9	100.0	79.5	95.8	100.0	79.5	95.8	100.0	79.5	95.8	100.0	79.5	95.8	100.0	79.5	95.8
22, 23	PLUMBING & HVAC	99.7	117.6	107.0	100.0	69.8	87.7	97.4	69.8	86.2	97.4	69.3	86.0	99.9	69.8	87.6	97.3	69.8	86.1
26, 27, 3370	ELECTRICAL, COMMUNICATIONS & UTIL.	94.5	127.6	110.4	84.9	76.3	80.8	84.9	76.3	80.8	82.8	76.3	79.7	83.3	76.3	80.0	82.6	76.3	79.6
MF2004	WEIGHTED AVERAGE	96.4	118.6	106.0	99.8	75.7	89.4	100.6	75.7	89.8	98.5	74.5	88.1	100.2	75.7	89.6	99.9	75.7	89.4

DIVISION		NEW MEXICO																	
		LAS CRUCES 880			LAS VEGAS 877			ROSWELL 882			SANTA FE 875			SOCORRO 878			TRUTH/CONSEQUENCES 879		
		MAT.	INST.	TOTAL	MAT.	INST.	TOTAL	MAT.	INST.	TOTAL	MAT.	INST.	TOTAL	MAT.	INST.	TOTAL	MAT.	INST.	TOTAL
01543	CONTRACTOR EQUIPMENT	.0	86.2	86.2	.0	115.6	115.6	.0	115.6	115.6	.0	115.6	115.6	.0	115.6	115.6	.0	86.2	86.2
0241, 31-34	SITE & INFRASTRUCTURE, DEMOLITION	89.6	85.9	86.9	84.5	110.2	102.9	89.4	110.2	104.3	78.7	110.2	101.3	81.3	110.2	102.0	100.6	85.9	90.1
0310	Concrete Forming & Accessories	94.4	67.4	71.2	97.1	69.1	73.1	97.1	69.0	72.9	96.0	69.1	72.9	97.1	69.1	73.1	94.4	67.5	71.3
0310	Concrete Reinforcing	111.7	54.4	83.7	114.9	66.6	91.3	117.2	54.6	86.6	115.9	66.6	91.8	117.2	66.6	92.4	110.3	54.4	83.0
0330	Cast-in-Place Concrete	95.7	65.6	84.3	106.9	74.9	94.9	101.2	74.9	91.3	102.3	74.9	92.0	104.8	74.9	93.5	114.9	65.7	96.4
03	CONCRETE	92.4	64.9	79.4	112.6	71.6	93.3	115.6	69.2	93.7	109.3	71.6	91.5	111.3	71.6	92.6	105.3	65.0	86.3
04	MASONRY	107.3	63.8	81.3	111.8	64.2	83.4	122.4	64.2	87.6	115.2	64.2	84.7	111.7	64.2	83.3	108.9	63.8	81.9
05	METALS	101.8	72.8	93.0	100.9	87.0	96.7	101.9	80.9	95.5	101.9	87.0	97.4	101.1	87.0	96.8	100.7	72.9	92.3
06	WOOD, PLASTICS & COMPOSITES	86.4	69.0	77.2	94.5	70.4	81.8	94.5	70.4	81.8	93.1	70.4	81.1	94.5	70.4	81.8	86.4	69.0	77.2
07	THERMAL & MOISTURE PROTECTION	83.8	67.0	77.0	98.4	72.5	87.9	98.9	72.5	88.3	98.3	72.5	87.9	98.3	72.5	87.9	84.3	67.0	77.3
08	OPENINGS	86.9	67.0	81.8	93.4	71.6	87.7	93.2	67.8	86.6	93.3	71.6	87.7	93.2	71.6	87.6	86.9	67.0	81.7
0920	Plaster & Gypsum Board	86.6	69.1	76.1	84.7	69.1	75.4	84.7	69.1	75.4	84.7	69.1	75.4	84.7	69.1	75.4	86.6	69.1	76.1
0950, 0980	Ceilings & Acoustic Treatment	96.1	69.1	79.8	100.9	69.1	81.7	100.9	69.1	81.7	100.9	69.1	81.7	100.9	69.1	81.7	96.1	69.1	79.8
0960	Flooring	132.1	70.0	115.6	100.0	70.0	92.0	100.0	70.0	92.0	100.0	70.0	92.0	100.0	70.0	92.0	132.1	70.0	115.6
0990	Wall Finishes & Painting/Coating	92.7	57.6	71.7	99.6	57.6	74.5	99.6	57.6	74.5	99.6	57.6	74.5	99.6	57.6	74.5	92.7	57.6	71.7
09	FINISHES	109.6	67.0	87.6	96.2	68.1	81.7	96.7	68.1	81.9	96.1	68.1	81.6	96.1	68.1	81.6	110.1	67.0	87.9
COVERS	DIVS. 10-14, 25, 28, 41, 43, 44	100.0	75.8	95.0	100.0	79.5	95.8	100.0	79.5	95.8	100.0	79.5	95.8	100.0	79.5	95.8	100.0	75.9	95.0
22, 23	PLUMBING & HVAC	100.2	69.2	87.6	97.3	69.8	86.1	99.9	69.6	87.6	99.9	69.8	87.6	97.3	69.8	86.1	97.2	69.2	85.8
26, 27, 3370	ELECTRICAL, COMMUNICATIONS & UTIL.	84.8	56.4	71.2	84.9	76.3	80.8	84.0	76.3	80.3	86.1	76.3	81.4	83.1	76.3	79.8	87.3	76.3	82.0
MF2004	WEIGHTED AVERAGE	96.8	67.8	84.3	98.4	75.7	88.6	100.1	74.6	89.1	98.9	75.7	88.9	97.9	75.7	88.3	98.1	70.6	86.3

City Cost Indexes

DIVISION		NEW MEXICO TUCUMCARI 884			NEW YORK ALBANY 120 - 122			NEW YORK BINGHAMTON 137 - 139			NEW YORK BRONX 104			NEW YORK BROOKLYN 112			NEW YORK BUFFALO 140 - 142		
		MAT.	INST.	TOTAL	MAT.	INST.	TOTAL	MAT.	INST.	TOTAL	MAT.	INST.	TOTAL	MAT.	INST.	TOTAL	MAT.	INST.	TOTAL
01543	CONTRACTOR EQUIPMENT	.0	115.6	115.6	.0	115.3	115.3	.0	115.4	115.4	.0	115.7	115.7	.0	116.9	116.9	.0	94.5	94.5
0241, 31 - 34	SITE & INFRASTRUCTURE, DEMOLITION	87.2	110.2	103.7	72.7	106.4	96.9	94.3	90.2	91.3	124.5	125.9	125.5	119.8	130.4	127.4	94.6	95.9	95.5
0310	Concrete Forming & Accessories	97.1	68.8	72.8	97.6	92.2	93.0	102.4	84.7	87.2	103.7	164.1	155.6	109.3	163.7	156.0	98.4	115.1	112.8
0310	Concrete Reinforcing	114.9	54.5	85.4	97.0	94.6	95.8	96.1	91.5	93.8	97.4	189.3	142.4	97.4	189.3	142.3	95.1	103.4	99.2
0330	Cast-in-Place Concrete	101.2	74.8	91.3	83.6	99.0	89.4	104.0	92.7	99.8	96.3	155.8	118.7	105.9	153.9	124.0	104.2	113.4	108.5
03	CONCRETE	114.0	69.1	92.8	95.5	95.8	95.6	98.3	90.5	94.6	99.3	164.5	130.0	109.3	163.5	134.9	104.2	113.4	108.5
04	MASONRY	125.1	64.2	88.7	90.2	95.2	93.2	107.7	84.6	93.9	95.3	160.4	134.3	116.2	160.4	142.6	107.7	119.8	114.9
05	METALS	100.8	80.8	94.7	96.0	106.4	99.1	92.6	116.1	99.7	94.4	145.5	109.9	101.2	143.0	113.9	102.9	94.9	100.5
06	WOOD, PLASTICS & COMPOSITES	94.5	70.4	81.8	97.3	91.5	94.2	106.3	86.2	95.7	101.6	168.4	136.9	109.2	168.0	140.3	102.1	114.8	108.8
07	THERMAL & MOISTURE PROTECTION	98.8	72.5	88.2	92.3	90.4	91.5	102.9	85.0	95.7	110.3	154.0	127.9	109.1	153.5	126.9	99.0	110.1	103.5
08	OPENINGS	93.1	67.8	86.5	96.0	86.6	93.6	90.8	82.2	88.6	91.5	165.4	110.7	90.2	163.6	109.2	94.1	103.4	96.5
0920	Plaster & Gypsum Board	84.7	69.1	75.4	108.5	91.0	98.1	112.7	85.3	96.4	108.4	169.9	145.0	107.5	169.9	144.6	97.3	114.9	107.8
0950, 0980	Ceilings & Acoustic Treatment	100.9	69.1	81.7	97.3	91.0	93.5	97.3	85.3	90.1	82.8	169.9	135.2	78.9	169.9	133.6	90.4	114.9	105.2
0960	Flooring	100.0	70.0	92.0	85.9	93.1	87.8	96.4	87.9	94.2	94.0	171.8	114.7	103.1	171.8	121.4	98.7	119.2	104.2
0990	Wall Finishes & Painting/Coating	99.6	57.6	74.5	83.5	75.9	79.0	88.4	86.9	87.5	91.4	146.7	124.5	111.3	146.7	132.5	88.7	114.8	104.3
09	FINISHES	96.6	68.1	81.9	96.2	90.6	93.3	97.6	85.6	91.4	99.4	164.4	133.0	106.4	164.1	136.2	93.7	116.7	105.6
COVERS	DIVS. 10 - 14, 25, 28, 41, 43, 44	100.0	79.5	95.8	100.0	92.1	98.4	100.0	89.4	97.8	100.0	136.2	107.4	100.0	135.2	107.2	100.0	107.3	101.5
22, 23	PLUMBING & HVAC	97.4	69.3	86.0	100.3	88.8	95.6	100.5	83.1	93.4	100.3	158.0	123.8	99.7	157.8	123.4	99.7	98.5	99.2
26, 27, 3370	ELECTRICAL, COMMUNICATIONS & UTIL.	84.9	76.3	80.8	104.2	90.1	97.5	102.2	83.7	93.3	103.2	171.2	135.8	104.0	171.2	136.2	100.4	98.0	99.3
MF2004	WEIGHTED AVERAGE	99.3	74.5	88.6	97.2	94.0	95.8	98.4	88.5	94.1	99.4	157.6	124.5	103.1	157.4	126.5	100.1	105.3	102.4

DIVISION		NEW YORK ELMIRA 148 - 149			NEW YORK FAR ROCKAWAY 116			NEW YORK FLUSHING 113			NEW YORK GLENS FALLS 128			NEW YORK HICKSVILLE 115,117,118			NEW YORK JAMAICA 114		
		MAT.	INST.	TOTAL	MAT.	INST.	TOTAL	MAT.	INST.	TOTAL	MAT.	INST.	TOTAL	MAT.	INST.	TOTAL	MAT.	INST.	TOTAL
01543	CONTRACTOR EQUIPMENT	.0	116.8	116.8	.0	116.9	116.9	.0	116.9	116.9	.0	115.3	115.3	.0	116.9	116.9	.0	116.9	116.9
0241, 31 - 34	SITE & INFRASTRUCTURE, DEMOLITION	90.0	90.8	90.6	124.1	130.4	128.6	124.2	130.4	128.6	62.7	106.1	93.8	112.3	130.4	125.2	117.7	130.4	126.8
0310	Concrete Forming & Accessories	81.7	71.8	73.2	94.0	163.7	153.8	98.3	163.7	154.4	83.8	86.6	86.2	90.0	152.5	143.7	98.3	163.7	154.4
0310	Concrete Reinforcing	96.2	80.9	88.7	97.4	189.3	142.3	99.1	189.3	143.2	96.7	94.7	95.7	97.4	188.8	142.1	97.4	189.3	142.3
0330	Cast-in-Place Concrete	107.4	76.7	95.8	114.6	153.9	129.4	114.6	153.9	129.4	79.0	99.4	86.7	97.2	155.0	119.0	105.9	153.9	124.0
03	CONCRETE	98.0	77.6	88.4	115.5	163.5	138.2	116.1	163.5	138.5	87.3	93.5	90.2	100.8	158.9	128.2	108.6	163.5	134.5
04	MASONRY	103.9	68.4	82.7	121.2	160.4	144.7	114.7	160.4	142.0	93.2	94.6	94.0	111.1	158.8	139.6	118.7	160.4	143.7
05	METALS	98.2	114.7	103.2	101.3	142.9	113.9	101.3	142.9	113.9	92.5	106.2	96.6	102.6	141.1	114.3	101.3	142.9	113.9
06	WOOD, PLASTICS & COMPOSITES	84.2	70.0	76.7	91.5	168.0	132.0	96.6	168.0	134.3	81.2	85.9	83.7	87.5	153.1	122.2	96.6	168.0	134.3
07	THERMAL & MOISTURE PROTECTION	99.4	76.4	90.1	108.9	153.5	126.8	109.0	153.5	126.9	91.3	89.5	90.6	108.5	146.1	123.7	108.8	153.5	126.8
08	OPENINGS	95.2	71.2	89.0	88.6	163.6	108.0	88.6	163.6	108.0	89.5	83.6	88.0	88.6	155.5	105.9	88.6	163.6	108.0
0920	Plaster & Gypsum Board	92.6	69.0	78.6	102.1	169.9	142.4	103.7	169.9	143.1	99.7	85.3	91.1	101.4	154.5	133.0	103.7	169.9	143.1
0950, 0980	Ceilings & Acoustic Treatment	87.0	69.0	76.2	78.9	169.9	133.6	78.9	169.9	133.6	90.7	85.3	87.4	78.0	154.5	124.0	78.9	169.9	133.6
0960	Flooring	88.3	71.4	83.8	97.2	171.8	117.1	99.0	171.8	118.4	77.3	93.1	81.5	95.9	83.1	92.5	99.0	171.8	118.4
0990	Wall Finishes & Painting/Coating	87.0	85.3	86.0	111.3	146.7	132.5	111.3	146.7	132.5	83.5	75.9	79.0	111.3	146.7	132.5	111.3	146.7	132.5
09	FINISHES	88.8	72.3	80.3	104.2	164.1	135.1	104.9	164.1	135.5	89.8	86.6	88.2	102.6	137.8	120.8	104.5	164.1	135.3
COVERS	DIVS. 10 - 14, 25, 28, 41, 43, 44	100.0	86.6	97.2	100.0	135.2	107.2	100.0	135.2	107.2	100.0	63.4	92.5	100.0	134.7	107.1	100.0	135.2	107.2
22, 23	PLUMBING & HVAC	95.7	82.5	90.3	99.5	157.8	120.9	99.5	157.8	120.9	96.1	84.4	91.3	99.7	145.9	118.6	99.5	157.8	120.9
26, 27, 3370	ELECTRICAL, COMMUNICATIONS & UTIL.	99.6	94.2	97.0	110.2	171.2	139.4	110.2	171.2	139.4	98.1	90.1	94.3	103.1	153.8	127.4	102.2	171.2	135.2
MF2004	WEIGHTED AVERAGE	96.8	83.7	91.1	103.4	157.4	126.7	103.3	157.4	126.6	92.6	91.1	92.0	101.1	147.5	121.1	101.5	157.4	125.6

DIVISION		NEW YORK JAMESTOWN 147			NEW YORK KINGSTON 124			NEW YORK LONG ISLAND CITY 111			NEW YORK MONTICELLO 127			NEW YORK MOUNT VERNON 105			NEW YORK NEW ROCHELLE 108		
		MAT.	INST.	TOTAL	MAT.	INST.	TOTAL	MAT.	INST.	TOTAL	MAT.	INST.	TOTAL	MAT.	INST.	TOTAL	MAT.	INST.	TOTAL
01543	CONTRACTOR EQUIPMENT	.0	90.8	90.8	.0	116.9	116.9	.0	116.9	116.9	.0	116.9	116.9	.0	115.7	115.7	.0	115.7	115.7
0241, 31 - 34	SITE & INFRASTRUCTURE, DEMOLITION	91.4	91.1	91.2	119.6	127.7	125.4	121.3	130.4	127.8	114.6	127.9	124.1	134.2	122.7	125.9	132.7	122.7	125.5
0310	Concrete Forming & Accessories	81.8	84.5	84.1	85.5	110.0	106.5	103.2	163.7	155.1	93.5	110.2	107.9	92.7	138.7	132.2	109.3	138.7	134.5
0310	Concrete Reinforcing	96.4	86.3	91.4	97.0	126.5	111.4	97.4	189.3	142.3	96.3	126.5	111.1	96.3	187.8	141.0	96.4	187.8	141.1
0330	Cast-in-Place Concrete	111.4	107.1	109.8	106.7	124.5	113.4	109.3	153.9	126.1	99.9	124.6	109.2	107.7	131.3	116.6	107.7	131.3	116.6
03	CONCRETE	101.0	92.4	96.9	109.4	117.8	113.4	111.9	163.5	136.3	104.1	117.9	110.6	108.9	144.3	125.6	108.6	144.3	125.5
04	MASONRY	112.8	92.5	100.7	109.1	126.7	119.6	113.0	160.4	141.3	101.0	126.7	116.4	101.7	129.6	118.4	101.7	129.6	118.4
05	METALS	95.4	83.3	91.8	101.3	111.8	104.4	101.2	142.9	113.9	101.2	112.0	104.5	94.2	137.3	107.2	94.4	137.3	107.4
06	WOOD, PLASTICS & COMPOSITES	82.9	79.2	81.0	82.5	104.6	94.2	102.6	168.0	137.2	91.8	104.6	98.6	89.4	140.5	116.4	109.0	140.5	125.7
07	THERMAL & MOISTURE PROTECTION	98.9	90.3	95.5	110.8	132.1	119.4	108.9	153.5	126.8	110.5	132.1	119.2	111.2	137.4	121.8	111.4	137.4	121.8
08	OPENINGS	95.1	80.0	91.1	94.0	115.5	99.5	88.6	163.6	108.0	88.6	115.5	95.6	91.5	148.7	106.3	91.6	148.7	106.4
0920	Plaster & Gypsum Board	88.6	78.4	82.5	98.1	104.8	102.1	106.3	169.9	144.1	100.9	104.8	103.2	103.6	141.2	126.0	113.2	141.2	129.9
0950, 0980	Ceilings & Acoustic Treatment	82.8	78.4	80.2	73.8	104.8	92.5	78.9	169.9	133.6	73.8	104.8	92.5	80.2	141.2	116.9	80.2	141.2	116.9
0960	Flooring	91.1	85.5	89.6	89.8	87.3	89.2	100.6	171.8	119.6	92.9	87.3	91.4	85.8	167.0	107.4	94.0	167.0	113.5
0990	Wall Finishes & Painting/Coating	88.7	84.1	85.9	109.1	120.6	116.0	111.3	146.7	132.5	109.1	102.1	104.9	89.2	146.7	123.6	89.2	146.7	123.6
09	FINISHES	88.1	83.9	85.9	99.7	105.2	102.5	105.6	164.1	135.8	100.7	103.1	102.0	96.3	145.3	121.6	100.1	145.3	123.4
COVERS	DIVS. 10 - 14, 25, 28, 41, 43, 44	100.0	96.3	99.2	100.0	116.0	103.3	100.0	135.2	107.2	100.0	116.0	103.3	100.0	129.8	106.1	100.0	129.8	106.1
22, 23	PLUMBING & HVAC	95.5	76.9	87.9	96.0	102.8	98.8	99.7	157.8	123.4	96.0	120.7	106.1	96.2	128.6	109.4	96.2	128.6	109.4
26, 27, 3370	ELECTRICAL, COMMUNICATIONS & UTIL.	98.1	86.7	92.7	100.2	120.3	109.8	102.6	171.2	135.5	100.2	120.3	109.8	100.4	146.9	122.7	100.4	146.9	122.7
MF2004	WEIGHTED AVERAGE	96.9	85.8	92.1	101.1	114.8	107.0	102.8	157.4	126.4	99.5	118.3	107.6	99.4	137.1	115.7	99.9	137.1	116.0

City Cost Indexes

DIVISION		NEW YORK																	
		NEW YORK 100 - 102			NIAGARA FALLS 143			PLATTSBURGH 129			POUGHKEEPSIE 125 - 126			QUEENS 110			RIVERHEAD 119		
		MAT.	INST.	TOTAL	MAT.	INST.	TOTAL	MAT.	INST.	TOTAL	MAT.	INST.	TOTAL	MAT.	INST.	TOTAL	MAT.	INST.	TOTAL
01543	CONTRACTOR EQUIPMENT	.0	116.1	116.1	.0	90.8	90.8	.0	97.9	97.9	.0	116.9	116.9	.0	116.9	116.9	.0	116.9	116.9
0241, 31 - 34	SITE & INFRASTRUCTURE, DEMOLITION	134.7	127.0	129.1	93.7	92.4	92.8	85.4	102.8	97.9	115.6	127.6	124.2	115.8	130.4	126.3	113.2	130.4	125.5
0310	Concrete Forming & Accessories	108.5	180.3	170.1	81.7	114.6	109.9	88.5	81.6	82.6	85.5	115.4	111.2	90.3	163.7	153.3	95.3	152.5	144.5
0310	Concrete Reinforcing	103.1	196.2	148.6	95.1	94.2	94.6	101.1	94.1	97.7	97.0	126.6	111.5	99.1	189.3	143.2	99.2	188.8	143.0
0330	Cast-in-Place Concrete	108.5	162.3	128.7	115.3	111.8	114.0	96.6	91.5	94.7	103.3	122.7	110.6	100.6	153.9	120.6	98.9	155.0	120.0
03	CONCRETE	109.9	175.0	140.6	103.5	108.8	106.0	100.9	87.1	94.4	106.4	119.7	112.7	104.2	163.5	132.2	101.8	158.9	128.7
04	MASONRY	105.5	171.3	144.9	121.0	117.4	118.8	89.0	84.2	86.1	101.8	122.9	114.4	107.1	160.4	139.0	116.9	158.8	142.0
05	METALS	105.8	146.0	118.0	98.2	87.6	95.0	97.1	83.4	92.9	101.3	112.9	104.8	101.3	142.9	113.9	103.0	141.1	114.5
06	WOOD, PLASTICS & COMPOSITES	106.6	184.1	147.5	82.9	114.0	99.3	89.3	79.4	84.1	82.5	112.7	98.5	87.6	168.0	130.1	93.4	153.1	125.0
07	THERMAL & MOISTURE PROTECTION	110.6	163.0	131.7	99.0	106.2	101.9	101.4	83.3	94.1	110.8	131.3	119.0	108.6	153.5	126.6	109.0	146.1	124.0
08	OPENINGS	96.9	176.5	117.5	95.1	100.3	96.4	98.9	80.1	94.0	94.0	119.9	100.7	88.6	163.6	108.0	88.6	155.5	105.9
0920	Plaster & Gypsum Board	115.8	185.9	157.5	88.6	114.1	103.8	116.6	78.0	93.7	98.1	113.1	107.0	101.4	169.9	142.2	103.7	154.5	133.9
0950, 0980	Ceilings & Acoustic Treatment	107.2	185.9	154.6	82.8	114.1	101.7	98.2	78.0	86.1	73.8	113.1	97.5	78.9	169.9	133.6	82.2	154.5	125.7
0960	Flooring	95.5	171.8	115.9	91.0	112.3	96.6	99.5	93.1	97.8	89.8	128.0	100.0	95.9	171.8	116.1	97.2	83.1	93.5
0990	Wall Finishes & Painting/Coating	91.4	146.9	124.6	88.7	108.1	100.3	102.9	72.3	84.6	109.1	102.1	104.9	111.3	146.7	132.5	111.3	146.7	132.5
09	FINISHES	107.6	176.3	143.1	88.2	114.6	101.8	99.1	82.5	90.5	99.5	115.5	107.8	103.0	164.1	134.5	104.3	137.8	121.6
COVERS	DIVS. 10 - 14, 25, 28, 41, 43, 44	100.0	147.5	109.8	100.0	103.7	100.8	100.0	57.9	91.3	100.0	115.5	103.2	100.0	135.2	107.2	100.0	134.7	107.1
22, 23	PLUMBING & HVAC	100.3	163.8	126.2	95.5	97.3	96.2	96.0	84.2	91.2	96.0	102.0	98.5	99.7	157.8	123.4	99.9	145.9	118.6
26, 27, 3370	ELECTRICAL, COMMUNICATIONS & UTIL.	110.4	173.6	140.7	96.9	94.7	95.8	93.2	87.5	90.4	100.2	120.3	109.8	103.1	171.2	135.7	104.6	153.8	128.1
MF2004	WEIGHTED AVERAGE	105.3	164.5	130.9	98.0	102.2	99.8	96.7	85.4	91.8	100.2	116.1	107.1	101.2	157.4	125.4	102.0	147.5	121.6

DIVISION		NEW YORK																	
		ROCHESTER 144 - 146			SCHENECTADY 123			STATEN ISLAND 103			SUFFERN 109			SYRACUSE 130 - 132			UTICA 133 - 135		
		MAT.	INST.	TOTAL	MAT.	INST.	TOTAL	MAT.	INST.	TOTAL	MAT.	INST.	TOTAL	MAT.	INST.	TOTAL	MAT.	INST.	TOTAL
01543	CONTRACTOR EQUIPMENT	.0	118.0	118.0	.0	115.3	115.3	.0	115.7	115.7	.0	115.7	115.7	.0	115.3	115.3	.0	115.3	115.3
0241, 31 - 34	SITE & INFRASTRUCTURE, DEMOLITION	72.8	109.7	99.3	72.4	106.6	96.9	139.4	125.9	129.8	128.9	123.1	124.7	93.0	106.8	102.9	70.6	105.2	95.4
0310	Concrete Forming & Accessories	100.2	99.4	99.5	102.0	92.9	94.2	92.1	164.1	154.0	102.4	133.5	129.1	101.3	87.2	89.2	102.5	81.4	84.4
0310	Concrete Reinforcing	95.9	82.1	89.2	95.7	94.6	95.2	97.4	189.3	142.4	96.4	126.9	111.3	97.0	91.9	94.5	97.0	80.8	89.1
0330	Cast-in-Place Concrete	105.9	100.7	104.0	95.5	100.1	97.2	107.8	155.8	125.8	104.4	127.4	113.1	96.5	97.8	97.0	88.3	92.7	89.9
03	CONCRETE	107.0	97.5	102.5	101.5	96.6	99.2	110.5	164.5	136.0	105.5	129.2	116.7	100.6	92.7	96.9	98.1	86.3	92.5
04	MASONRY	104.8	98.3	100.9	91.4	97.2	94.8	108.8	160.4	139.7	101.0	124.5	115.1	97.8	97.5	97.6	90.0	89.7	89.8
05	METALS	99.3	104.7	101.0	96.1	106.4	99.2	92.4	145.5	108.4	92.4	113.1	98.6	95.9	104.9	98.6	94.1	100.3	96.0
06	WOOD, PLASTICS & COMPOSITES	101.1	100.5	100.8	102.9	91.5	96.9	88.0	168.4	130.5	100.9	139.2	121.2	102.9	84.9	93.4	102.9	81.1	91.4
07	THERMAL & MOISTURE PROTECTION	95.5	96.2	95.8	92.4	91.3	91.9	110.5	154.0	128.0	111.2	133.2	120.0	101.5	94.7	98.8	92.3	91.1	91.8
08	OPENINGS	97.8	89.8	95.8	96.0	86.6	93.6	91.5	165.4	110.7	91.6	134.2	102.6	93.9	81.3	90.6	96.0	76.7	91.0
0920	Plaster & Gypsum Board	99.0	100.6	99.9	108.5	91.0	98.1	103.9	169.9	143.1	107.5	139.8	126.7	108.5	84.2	94.1	108.5	80.3	91.7
0950, 0980	Ceilings & Acoustic Treatment	94.7	100.6	98.2	97.3	91.0	93.5	82.8	169.9	135.2	80.2	139.8	116.1	97.3	84.2	89.5	97.3	80.3	87.1
0960	Flooring	87.9	104.9	92.4	85.9	93.1	87.8	89.4	171.8	111.4	89.7	57.8	81.2	87.6	92.4	88.9	85.9	87.4	86.3
0990	Wall Finishes & Painting/Coating	86.8	100.2	94.8	83.5	75.9	79.0	91.4	146.7	124.5	89.2	106.3	99.4	89.0	85.4	86.9	83.5	83.7	83.6
09	FINISHES	94.5	101.2	98.0	95.9	91.1	93.4	98.4	164.4	132.5	97.7	116.0	107.1	97.4	87.6	92.3	96.0	82.5	89.0
COVERS	DIVS. 10 - 14, 25, 28, 41, 43, 44	100.0	96.2	99.2	100.0	92.8	98.5	100.0	136.2	107.4	100.0	127.7	105.7	100.0	95.6	99.1	100.0	88.2	97.6
22, 23	PLUMBING & HVAC	99.7	91.0	96.2	100.3	89.8	96.0	100.3	158.0	123.8	96.2	120.2	106.0	100.3	87.3	95.0	100.3	83.2	93.3
26, 27, 3370	ELECTRICAL, COMMUNICATIONS & UTIL.	104.6	92.1	98.6	102.2	90.1	96.4	103.2	171.2	135.8	107.5	129.9	118.2	102.2	93.0	97.8	99.4	87.9	93.9
MF2004	WEIGHTED AVERAGE	99.9	97.3	98.8	97.8	94.6	96.4	101.3	157.6	125.6	99.6	123.7	110.0	98.7	93.3	96.4	96.6	88.4	93.1

DIVISION		NEW YORK									NORTH CAROLINA								
		WATERTOWN 136			WHITE PLAINS 106			YONKERS 107			ASHEVILLE 287 - 288			CHARLOTTE 281 - 282			DURHAM 277		
		MAT.	INST.	TOTAL	MAT.	INST.	TOTAL	MAT.	INST.	TOTAL	MAT.	INST.	TOTAL	MAT.	INST.	TOTAL	MAT.	INST.	TOTAL
01543	CONTRACTOR EQUIPMENT	.0	115.3	115.3	.0	115.7	115.7	.0	115.7	115.7	.0	93.6	93.6	.0	93.6	93.6	.0	100.1	100.1
0241, 31 - 34	SITE & INFRASTRUCTURE, DEMOLITION	79.0	107.1	99.1	124.0	122.7	123.0	134.0	122.5	125.7	105.4	73.2	82.3	106.3	73.3	82.6	105.6	83.0	89.4
0310	Concrete Forming & Accessories	85.6	87.6	87.3	108.1	138.7	134.3	108.3	138.8	134.5	96.3	44.1	51.5	103.5	46.3	54.3	98.7	44.6	52.3
0310	Concrete Reinforcing	97.6	81.4	89.7	96.4	187.8	141.1	100.2	187.8	143.0	100.3	48.6	75.0	100.7	45.3	73.6	101.9	56.8	79.8
0330	Cast-in-Place Concrete	102.8	80.7	94.5	95.6	131.3	109.0	107.0	131.3	116.2	100.7	49.9	81.6	107.8	52.2	86.9	102.3	50.1	82.7
03	CONCRETE	110.7	85.0	98.6	98.8	144.3	120.3	108.8	144.4	125.6	106.3	48.7	79.1	109.7	49.9	81.4	106.8	50.6	80.3
04	MASONRY	91.1	90.4	90.7	100.4	129.6	117.9	105.1	129.6	119.7	85.7	45.0	61.3	92.5	49.3	66.6	91.5	40.5	61.0
05	METALS	94.2	99.8	95.9	93.7	137.3	106.9	102.2	137.4	112.9	91.1	77.5	87.0	95.4	76.4	89.6	105.1	81.2	97.9
06	WOOD, PLASTICS & COMPOSITES	83.2	89.6	86.6	107.5	140.5	124.9	107.3	140.6	124.9	94.8	43.4	67.6	104.0	45.9	73.3	97.3	45.1	69.7
07	THERMAL & MOISTURE PROTECTION	92.5	89.8	91.4	111.1	137.4	121.7	111.4	137.4	121.9	104.0	47.8	81.4	103.6	49.6	81.8	104.8	50.1	82.8
08	OPENINGS	96.0	78.2	91.4	91.6	148.7	106.4	95.0	151.4	109.6	92.2	43.4	79.5	96.3	45.1	83.0	96.3	48.0	83.8
0920	Plaster & Gypsum Board	101.8	89.1	94.2	110.0	141.2	128.6	115.4	141.2	130.7	109.8	41.8	69.3	112.9	44.3	72.0	113.5	43.5	71.8
0950, 0980	Ceilings & Acoustic Treatment	97.3	89.1	92.4	80.2	141.2	116.9	105.5	141.2	127.0	94.4	41.8	62.7	100.3	44.3	66.6	100.3	43.5	66.1
0960	Flooring	78.0	87.4	80.5	92.0	167.0	112.0	91.5	171.0	111.6	101.3	50.2	87.6	105.1	50.3	90.5	105.3	50.2	90.6
0990	Wall Finishes & Painting/Coating	83.5	68.6	74.6	89.2	146.7	123.6	89.2	146.7	123.6	112.5	39.6	68.9	112.5	45.2	72.2	112.5	41.5	70.0
09	FINISHES	93.1	85.4	89.1	98.3	145.3	122.5	105.8	145.3	126.2	99.8	44.4	71.2	102.8	46.7	73.9	103.0	45.0	73.1
COVERS	DIVS. 10 - 14, 25, 28, 41, 43, 44	100.0	74.7	94.8	100.0	129.8	106.1	100.0	135.2	107.2	100.0	77.1	95.3	100.0	77.7	95.4	100.0	75.8	95.0
22, 23	PLUMBING & HVAC	100.3	70.7	88.2	100.5	128.6	112.0	100.5	128.6	112.0	100.2	45.0	77.7	100.1	45.2	77.7	100.3	43.5	77.2
26, 27, 3370	ELECTRICAL, COMMUNICATIONS & UTIL.	102.2	87.5	95.2	100.4	146.9	122.7	107.6	146.9	126.4	98.2	42.9	71.7	100.5	55.9	79.1	99.3	54.3	77.7
MF2004	WEIGHTED AVERAGE	98.3	85.9	92.9	99.2	137.1	115.6	104.0	137.3	118.4	97.9	51.4	77.8	100.3	54.2	80.4	101.3	54.0	80.9

City Cost Indexes

DIVISION		NORTH CAROLINA																	
		ELIZABETH CITY			FAYETTEVILLE			GASTONIA			GREENSBORO			HICKORY			KINSTON		
		279			283			280			270,272 - 274			286			285		
		MAT.	INST.	TOTAL	MAT.	INST.	TOTAL	MAT.	INST.	TOTAL	MAT.	INST.	TOTAL	MAT.	INST.	TOTAL	MAT.	INST.	TOTAL
01543	CONTRACTOR EQUIPMENT	.0	105.2	105.2	.0	100.1	100.1	.0	93.6	93.6	.0	100.1	100.1	.0	100.1	100.1	.0	100.1	100.1
0241, 31 - 34	SITE & INFRASTRUCTURE, DEMOLITION	111.1	84.3	91.9	104.4	83.2	89.2	105.0	73.2	82.2	105.4	83.2	89.5	104.6	83.1	89.2	103.2	82.9	88.6
0310	Concrete Forming & Accessories	83.2	38.1	44.5	95.6	45.5	52.6	103.7	45.1	53.4	98.4	45.4	52.9	91.4	43.0	49.8	87.5	36.1	43.4
0310	Concrete Reinforcing	99.8	47.2	74.0	104.2	56.9	81.1	100.7	48.5	75.2	100.7	56.7	79.2	100.3	21.9	62.0	99.8	21.3	61.4
0330	Cast-in-Place Concrete	102.6	36.4	77.7	105.6	53.6	86.0	98.4	58.5	83.4	101.6	51.6	82.8	100.6	51.2	82.0	97.1	40.1	75.7
03	CONCRETE	107.0	41.2	75.9	108.2	52.2	81.8	105.0	52.1	80.0	106.2	51.4	80.4	106.0	43.8	76.6	102.7	36.9	71.6
04	MASONRY	102.9	29.9	59.2	88.6	46.1	63.2	90.1	42.5	61.6	88.8	41.2	60.3	74.8	46.3	57.8	81.7	32.5	52.3
05	METALS	92.6	76.3	87.7	110.0	81.3	101.3	91.6	77.5	87.3	98.5	80.8	93.1	91.1	69.4	84.6	90.0	68.0	83.4
06	WOOD, PLASTICS & COMPOSITES	79.1	40.1	58.5	93.3	45.1	67.8	104.0	45.0	72.8	97.0	45.0	69.5	88.5	43.2	64.6	84.2	36.5	59.0
07	THERMAL & MOISTURE PROTECTION	104.0	34.7	76.1	103.8	48.4	81.5	104.2	48.1	81.6	104.5	47.3	81.5	104.3	46.6	81.0	104.1	35.2	76.4
08	OPENINGS	93.1	36.7	78.5	92.3	48.0	80.8	96.3	45.2	83.0	96.3	48.0	83.7	92.2	34.5	77.3	92.3	31.3	76.5
0920	Plaster & Gypsum Board	106.9	37.5	65.6	112.7	43.5	71.5	115.0	43.4	72.4	115.0	43.4	72.4	109.8	41.5	69.1	108.5	34.6	64.5
0950, 0980	Ceilings & Acoustic Treatment	100.3	37.5	62.5	95.2	43.5	64.1	100.3	43.4	66.1	100.3	43.4	66.1	94.4	41.5	62.6	98.6	34.6	60.1
0960	Flooring	94.8	16.2	73.8	101.4	50.2	87.7	105.3	50.2	90.6	105.3	46.4	89.6	101.1	38.6	84.4	97.9	22.1	77.7
0990	Wall Finishes & Painting/Coating	112.5	51.1	75.8	112.5	39.6	68.9	112.5	45.4	72.4	112.5	38.5	68.2	112.5	38.5	68.2	112.5	35.0	66.2
09	FINISHES	99.1	35.3	66.2	100.5	45.4	72.0	103.2	45.9	73.6	103.3	44.6	73.0	99.9	41.1	69.6	99.6	33.4	65.4
COVERS	DIVS. 10 - 14, 25, 28, 41, 43, 44	100.0	75.9	95.0	100.0	76.8	95.2	100.0	77.0	95.3	100.0	77.3	95.3	100.0	77.0	95.3	100.0	74.5	94.7
22, 23	PLUMBING & HVAC	96.0	34.9	71.1	100.1	44.9	77.6	100.2	44.5	77.5	100.2	45.0	77.7	96.0	37.0	71.9	96.0	29.7	69.0
26, 27, 3370	ELECTRICAL, COMMUNICATIONS & UTIL.	98.6	34.0	67.7	98.5	49.3	74.9	97.7	39.9	70.0	98.2	43.1	71.8	96.0	36.1	67.3	95.8	56.2	76.8
MF2004	WEIGHTED AVERAGE	98.3	44.5	75.0	101.3	54.4	81.1	98.8	51.4	78.3	100.0	52.8	79.6	96.0	47.6	75.1	95.7	44.8	73.7

DIVISION		NORTH CAROLINA															NORTH DAKOTA		
		MURPHY			RALEIGH			ROCKY MOUNT			WILMINGTON			WINSTON-SALEM			BISMARCK		
		289			275 - 276			278			284			271			585		
		MAT.	INST.	TOTAL	MAT.	INST.	TOTAL	MAT.	INST.	TOTAL	MAT.	INST.	TOTAL	MAT.	INST.	TOTAL	MAT.	INST.	TOTAL
01543	CONTRACTOR EQUIPMENT	.0	93.6	93.6	.0	100.1	100.1	.0	100.1	100.1	.0	93.6	93.6	.0	100.1	100.1	.0	98.7	98.7
0241, 31 - 34	SITE & INFRASTRUCTURE, DEMOLITION	107.1	72.9	82.6	106.7	83.2	89.9	108.9	82.7	90.1	106.5	73.2	82.6	105.8	83.1	89.5	93.2	97.1	96.0
0310	Concrete Forming & Accessories	104.3	27.6	38.4	101.0	46.4	54.1	90.3	32.8	41.0	97.8	44.6	52.1	99.6	44.4	52.2	94.2	48.2	54.7
0310	Concrete Reinforcing	99.8	21.1	61.3	100.7	56.9	79.3	99.8	31.0	66.2	101.1	54.9	78.5	100.7	45.0	73.5	103.7	74.9	89.6
0330	Cast-in-Place Concrete	104.1	35.9	78.5	107.8	57.7	88.9	100.3	33.0	75.0	100.3	51.3	81.8	104.1	48.7	83.2	97.2	53.6	80.8
03	CONCRETE	109.5	31.6	72.7	109.5	54.0	83.3	107.9	34.6	73.3	106.2	50.6	80.0	107.6	47.7	79.3	96.6	56.3	77.6
04	MASONRY	77.4	29.9	49.0	91.3	47.1	64.9	82.0	29.0	50.3	75.2	42.7	55.7	89.0	42.5	61.1	101.3	62.3	78.0
05	METALS	89.1	67.3	82.5	95.9	81.3	91.5	91.9	68.1	84.7	90.6	80.1	87.5	95.9	75.8	89.8	94.3	81.1	90.3
06	WOOD, PLASTICS & COMPOSITES	104.8	25.5	62.9	100.4	45.9	71.6	87.2	33.8	59.0	96.6	44.3	69.0	97.0	44.3	69.1	87.0	43.4	64.0
07	THERMAL & MOISTURE PROTECTION	104.2	33.5	75.7	104.4	48.7	82.0	104.5	30.2	74.6	104.0	47.8	81.4	104.5	46.9	81.3	99.5	53.6	81.1
08	OPENINGS	92.1	28.8	75.7	93.0	48.1	81.4	92.3	28.9	75.9	92.3	46.4	80.4	96.3	44.4	82.8	98.7	49.3	85.9
0920	Plaster & Gypsum Board	113.6	23.4	59.9	115.0	44.2	72.9	108.4	31.8	62.8	110.6	42.7	70.2	115.0	42.6	71.9	109.7	42.2	69.5
0950, 0980	Ceilings & Acoustic Treatment	94.4	23.4	51.7	100.3	44.2	66.6	95.2	31.8	57.1	95.2	42.7	63.6	100.3	42.6	65.6	139.5	42.2	81.0
0960	Flooring	105.8	23.6	83.9	105.3	50.2	90.6	99.3	15.6	77.0	102.2	50.2	88.3	105.3	50.2	90.6	110.6	72.3	100.4
0990	Wall Finishes & Painting/Coating	112.5	18.0	56.0	112.5	43.6	71.3	112.5	37.1	67.4	112.5	38.4	68.2	112.5	40.4	69.3	97.6	36.4	61.0
09	FINISHES	101.9	25.1	62.3	103.4	46.6	74.1	99.6	30.0	63.7	100.4	44.7	71.6	103.3	44.8	73.1	115.0	50.3	81.6
COVERS	DIVS. 10 - 14, 25, 28, 41, 43, 44	100.0	73.5	94.5	100.0	77.3	95.3	100.0	72.7	94.4	100.0	76.5	95.2	100.0	76.9	95.3	100.0	65.8	93.0
22, 23	PLUMBING & HVAC	96.0	22.0	65.8	100.2	42.9	76.9	96.0	38.1	72.4	100.2	44.6	77.6	100.2	43.4	77.1	100.5	64.1	85.7
26, 27, 3370	ELECTRICAL, COMMUNICATIONS & UTIL.	99.0	30.1	66.0	98.7	42.8	71.9	100.5	41.1	72.0	98.6	41.4	71.2	98.2	47.0	73.7	93.3	70.8	82.5
MF2004	WEIGHTED AVERAGE	97.0	36.2	70.8	99.9	53.6	79.9	97.4	43.0	73.9	97.5	51.5	77.6	99.7	52.0	79.1	99.1	65.3	84.5

DIVISION		NORTH DAKOTA																	
		DEVILS LAKE			DICKINSON			FARGO			GRAND FORKS			JAMESTOWN			MINOT		
		583			586			580 - 581			582			584			587		
		MAT.	INST.	TOTAL	MAT.	INST.	TOTAL	MAT.	INST.	TOTAL	MAT.	INST.	TOTAL	MAT.	INST.	TOTAL	MAT.	INST.	TOTAL
01543	CONTRACTOR EQUIPMENT	.0	98.7	98.7	.0	98.7	98.7	.0	98.7	98.7	.0	98.7	98.7	.0	98.7	98.7	.0	98.7	98.7
0241, 31 - 34	SITE & INFRASTRUCTURE, DEMOLITION	98.8	95.6	96.5	107.7	94.6	98.3	92.1	97.1	95.7	102.8	94.6	96.9	97.8	94.6	95.5	100.1	97.1	97.9
0310	Concrete Forming & Accessories	98.8	42.7	50.6	87.5	41.9	48.4	95.3	48.6	55.2	91.7	42.0	49.1	89.2	41.7	48.4	87.1	54.8	59.4
0310	Concrete Reinforcing	103.7	75.3	89.9	104.8	49.3	77.7	95.6	74.0	85.1	102.2	75.2	89.0	104.3	58.9	82.1	105.7	75.2	90.8
0330	Cast-in-Place Concrete	114.6	52.3	91.2	103.7	51.1	83.9	99.4	55.6	83.0	103.7	51.1	83.9	113.2	51.0	89.8	103.6	52.3	84.3
03	CONCRETE	108.3	53.5	82.4	107.2	47.7	79.1	105.3	57.0	82.5	104.4	52.6	80.0	106.8	49.3	79.7	103.2	58.8	82.3
04	MASONRY	108.8	67.8	84.3	111.7	62.3	82.2	103.4	46.6	69.4	103.2	67.4	81.8	121.6	40.8	73.3	103.5	68.2	82.4
05	METALS	94.4	81.0	90.3	94.3	67.3	86.1	96.6	79.8	91.5	94.3	77.4	89.2	94.3	69.4	86.8	94.6	81.7	90.7
06	WOOD, PLASTICS & COMPOSITES	92.0	39.9	64.5	79.5	39.9	58.5	87.1	44.0	64.3	84.1	39.9	60.7	81.5	39.9	59.5	79.1	52.1	64.9
07	THERMAL & MOISTURE PROTECTION	99.9	55.2	81.9	100.4	52.8	81.3	100.2	51.2	80.5	100.1	55.2	82.0	99.8	48.4	79.1	99.9	55.9	82.2
08	OPENINGS	98.8	43.9	84.5	98.7	37.7	82.9	98.7	49.6	86.0	98.7	43.9	84.5	98.7	40.3	83.6	98.9	54.0	87.3
0920	Plaster & Gypsum Board	113.6	38.6	69.0	106.5	38.6	66.1	109.7	42.9	69.9	107.8	38.6	66.6	107.5	38.6	66.5	106.5	51.2	73.6
0950, 0980	Ceilings & Acoustic Treatment	139.5	38.6	78.8	139.5	38.6	78.8	139.5	42.9	81.4	139.5	38.6	78.8	139.5	38.6	78.8	139.5	51.2	86.4
0960	Flooring	114.9	43.2	95.8	106.3	43.2	89.5	110.4	43.2	92.5	108.8	43.2	91.3	107.5	43.2	90.3	106.1	83.2	100.0
0990	Wall Finishes & Painting/Coating	97.6	27.8	55.9	97.6	27.8	55.9	97.6	70.8	81.6	97.6	27.8	55.9	97.6	27.8	55.9	97.6	30.4	57.5
09	FINISHES	117.2	39.9	77.3	114.3	39.9	75.9	114.9	48.5	80.6	114.8	39.9	76.1	114.0	39.9	75.7	113.6	57.0	84.4
COVERS	DIVS. 10 - 14, 25, 28, 41, 43, 44	100.0	42.8	88.2	100.0	42.9	88.2	100.0	65.9	93.0	100.0	42.9	88.2	100.0	63.2	92.4	100.0	66.9	93.2
22, 23	PLUMBING & HVAC	96.2	64.7	83.4	96.2	60.5	81.7	100.5	68.0	87.3	100.5	43.1	77.1	96.2	43.0	74.5	100.5	62.4	85.0
26, 27, 3370	ELECTRICAL, COMMUNICATIONS & UTIL.	92.8	44.6	69.7	100.3	72.0	86.7	94.6	65.8	80.8	95.8	66.7	81.9	92.8	44.5	69.6	98.5	72.0	85.8
MF2004	WEIGHTED AVERAGE	100.2	59.6	82.7	100.9	59.6	83.1	100.7	63.6	84.6	100.6	57.7	82.1	100.2	50.9	78.9	100.6	67.4	86.3

City Cost Indexes

DIVISION		NORTH DAKOTA WILLISTON 588			OHIO AKRON 442 - 443			OHIO ATHENS 457			OHIO CANTON 446 - 447			OHIO CHILLICOTHE 456			OHIO CINCINNATI 451 - 452		
		MAT.	INST.	TOTAL	MAT.	INST.	TOTAL	MAT.	INST.	TOTAL	MAT.	INST.	TOTAL	MAT.	INST.	TOTAL	MAT.	INST.	TOTAL
01543	CONTRACTOR EQUIPMENT	.0	98.7	98.7	.0	96.6	96.6	.0	90.2	90.2	.0	96.6	96.6	.0	102.2	102.2	.0	101.9	101.9
0241, 31 - 34	SITE & INFRASTRUCTURE, DEMOLITION	100.9	94.6	96.4	96.0	105.2	102.6	86.6	92.8	91.1	96.1	105.3	102.7	75.3	107.2	98.2	72.8	107.3	97.5
0310	Concrete Forming & Accessories	93.6	41.9	49.2	98.0	94.1	94.7	91.5	87.3	87.9	98.0	85.4	87.2	93.7	95.7	95.4	95.9	85.9	87.3
0310	Concrete Reinforcing	106.7	49.3	78.7	96.2	94.8	95.5	91.0	86.9	89.0	96.2	82.8	89.7	88.0	85.9	87.0	93.3	85.0	89.2
0330	Cast-in-Place Concrete	103.6	51.1	83.9	96.7	104.1	99.5	99.4	99.1	99.3	97.7	100.9	98.9	90.2	95.3	92.1	83.1	86.7	84.4
03	CONCRETE	104.6	47.7	77.7	96.7	96.8	96.7	99.2	90.5	95.1	97.2	89.6	93.6	93.9	93.4	93.7	88.9	86.1	87.6
04	MASONRY	98.4	62.3	76.8	88.8	99.7	95.3	69.1	94.5	84.3	89.5	90.5	90.1	76.2	97.5	89.0	75.8	91.9	85.5
05	METALS	94.5	67.3	86.2	91.5	81.0	88.3	100.1	72.9	91.9	91.5	75.0	86.5	92.0	84.3	89.7	94.2	88.4	92.4
06	WOOD, PLASTICS & COMPOSITES	85.6	39.9	61.5	91.3	91.9	91.6	81.3	85.4	83.4	91.6	84.3	87.7	92.7	94.1	93.5	95.2	84.1	89.3
07	THERMAL & MOISTURE PROTECTION	100.1	52.8	81.1	100.5	98.3	99.6	96.9	89.7	94.0	101.2	94.2	98.4	97.1	94.9	96.2	95.3	94.2	94.9
08	OPENINGS	98.8	37.7	83.0	105.9	93.8	102.8	96.8	74.9	91.1	100.1	76.7	94.1	89.6	86.9	88.9	97.3	86.8	94.6
0920	Plaster & Gypsum Board	107.8	38.6	66.6	95.0	91.3	92.8	92.9	84.6	88.0	95.6	83.4	88.4	93.9	94.3	94.1	95.1	83.9	88.5
0950, 0980	Ceilings & Acoustic Treatment	139.5	38.6	78.8	94.5	91.3	92.6	102.0	84.6	91.5	94.5	83.4	87.8	96.6	94.3	95.2	97.4	83.9	89.3
0960	Flooring	109.9	43.2	92.1	102.3	92.3	99.6	133.4	93.2	122.6	102.5	83.7	97.5	110.1	96.7	106.5	111.3	97.9	107.7
0990	Wall Finishes & Painting/Coating	97.6	27.8	55.9	107.8	114.8	112.0	112.0	51.5	75.8	107.8	87.5	95.7	108.7	96.2	101.2	108.7	93.4	99.6
09	FINISHES	115.0	39.9	76.3	99.4	95.8	97.5	102.9	85.2	93.8	99.6	84.4	91.7	98.4	96.3	97.3	99.0	88.5	93.6
COVERS	DIVS. 10 - 14, 25, 28, 41, 43, 44	100.0	42.9	88.2	100.0	95.3	99.0	100.0	67.3	93.3	100.0	72.1	94.3	100.0	91.9	98.3	100.0	88.2	97.6
22, 23	PLUMBING & HVAC	96.2	59.7	81.4	100.1	98.0	99.2	95.5	66.6	83.8	100.1	85.1	94.0	96.1	94.6	95.5	99.8	86.7	94.5
26, 27, 3370	ELECTRICAL, COMMUNICATIONS & UTIL.	96.1	72.0	84.5	99.0	90.8	95.1	101.2	54.6	78.9	98.4	93.5	96.1	99.0	85.6	92.6	98.6	78.1	88.7
MF2004	WEIGHTED AVERAGE	99.4	59.5	82.2	97.9	95.5	96.9	96.8	77.5	88.5	97.5	87.8	93.3	93.9	93.5	93.7	95.1	88.3	92.2

DIVISION		OHIO CLEVELAND 441			OHIO COLUMBUS 430 - 432			OHIO DAYTON 453 - 454			OHIO HAMILTON 450			OHIO LIMA 458			OHIO LORAIN 440		
		MAT.	INST.	TOTAL	MAT.	INST.	TOTAL	MAT.	INST.	TOTAL	MAT.	INST.	TOTAL	MAT.	INST.	TOTAL	MAT.	INST.	TOTAL
01543	CONTRACTOR EQUIPMENT	.0	97.0	97.0	.0	95.4	95.4	.0	96.3	96.3	.0	102.2	102.2	.0	93.2	93.2	.0	96.6	96.6
0241, 31 - 34	SITE & INFRASTRUCTURE, DEMOLITION	95.9	105.4	102.7	85.1	101.2	96.7	71.7	106.7	96.8	71.9	107.3	97.3	81.6	93.3	90.0	95.3	104.5	101.9
0310	Concrete Forming & Accessories	98.1	105.4	104.3	96.9	84.5	86.3	95.8	76.7	79.4	95.8	86.2	87.6	91.4	85.0	85.9	98.0	100.9	100.5
0310	Concrete Reinforcing	96.8	95.2	96.0	92.6	86.0	89.4	93.3	80.5	87.0	93.3	85.0	89.2	91.0	80.6	85.9	96.2	94.9	95.6
0330	Cast-in-Place Concrete	94.9	112.4	101.5	90.6	89.2	90.1	77.2	86.9	80.9	82.8	87.1	84.4	91.3	102.7	95.6	92.0	104.9	96.9
03	CONCRETE	95.8	104.8	100.1	93.2	86.1	89.8	85.9	80.8	83.5	88.7	86.4	87.7	92.7	90.0	91.4	94.3	100.1	97.0
04	MASONRY	93.1	108.9	102.5	93.9	90.9	92.1	75.4	87.4	82.6	75.8	93.1	86.1	94.6	88.6	91.0	85.6	97.6	92.8
05	METALS	92.9	84.4	90.3	97.6	81.8	92.8	93.4	77.3	88.5	93.5	88.3	91.9	100.2	80.9	94.3	92.1	81.9	89.0
06	WOOD, PLASTICS & COMPOSITES	90.5	102.7	96.9	101.8	82.2	91.5	96.2	73.5	84.2	95.2	84.1	89.3	81.2	82.6	81.9	91.3	101.7	96.8
07	THERMAL & MOISTURE PROTECTION	99.3	112.6	104.7	98.0	94.2	96.5	100.6	88.5	95.7	97.2	94.9	96.3	96.5	100.0	97.9	101.1	104.8	102.6
08	OPENINGS	96.4	100.2	97.4	101.1	82.7	96.3	97.7	76.3	92.1	95.2	86.8	93.0	96.8	80.1	92.5	100.1	99.7	100.0
0920	Plaster & Gypsum Board	94.3	102.3	99.1	91.6	81.7	85.7	95.1	73.1	82.0	95.1	83.9	88.5	92.9	81.7	86.2	95.0	101.4	98.8
0950, 0980	Ceilings & Acoustic Treatment	92.8	102.3	98.5	89.6	81.7	84.8	98.3	73.1	83.1	97.4	83.9	89.3	101.1	81.7	89.4	94.5	101.4	98.6
0960	Flooring	102.1	108.2	103.7	94.1	88.1	92.5	114.1	85.5	106.5	111.3	97.2	107.5	132.5	93.0	122.0	102.5	104.2	102.9
0990	Wall Finishes & Painting/Coating	107.8	116.8	113.2	100.5	99.5	99.9	108.7	90.8	98.0	108.7	88.7	96.8	112.0	73.4	88.9	107.8	116.8	113.2
09	FINISHES	98.9	107.0	103.1	94.9	86.2	90.4	100.0	79.0	89.2	98.9	88.1	93.4	102.2	84.5	93.1	99.4	103.6	101.6
COVERS	DIVS. 10 - 14, 25, 28, 41, 43, 44	100.0	106.0	101.2	100.0	93.0	98.6	100.0	86.2	97.2	100.0	88.6	97.7	100.0	93.9	98.7	100.0	102.8	100.6
22, 23	PLUMBING & HVAC	100.1	107.8	103.2	99.8	91.0	96.2	100.8	86.7	95.0	100.3	87.3	95.0	95.5	88.5	92.7	100.1	89.3	95.7
26, 27, 3370	ELECTRICAL, COMMUNICATIONS & UTIL.	98.8	108.2	103.3	96.7	84.6	90.9	96.6	84.6	90.8	97.1	74.4	86.2	101.5	83.3	92.8	98.6	84.9	92.0
MF2004	WEIGHTED AVERAGE	97.3	104.9	100.6	97.3	88.6	93.5	94.9	85.0	90.6	94.8	88.1	91.9	97.2	87.3	92.9	97.0	94.9	96.1

DIVISION		OHIO MANSFIELD 448 - 449			OHIO MARION 433			OHIO SPRINGFIELD 455			OHIO STEUBENVILLE 439			OHIO TOLEDO 434 - 436			OHIO YOUNGSTOWN 444 - 445		
		MAT.	INST.	TOTAL	MAT.	INST.	TOTAL	MAT.	INST.	TOTAL	MAT.	INST.	TOTAL	MAT.	INST.	TOTAL	MAT.	INST.	TOTAL
01543	CONTRACTOR EQUIPMENT	.0	96.6	96.6	.0	95.0	95.0	.0	96.3	96.3	.0	100.7	100.7	.0	98.3	98.3	.0	96.6	96.6
0241, 31 - 34	SITE & INFRASTRUCTURE, DEMOLITION	91.6	105.3	101.4	80.7	100.5	94.9	72.0	105.8	96.2	120.1	110.7	113.4	84.3	101.5	96.6	95.8	105.9	103.1
0310	Concrete Forming & Accessories	87.3	99.9	98.2	93.2	82.9	84.4	95.8	82.7	84.6	94.2	89.7	90.4	96.9	94.0	94.4	98.0	90.1	91.2
0310	Concrete Reinforcing	87.5	83.3	85.4	85.4	86.2	85.8	93.3	80.5	87.0	83.1	90.3	86.7	92.6	94.3	93.4	96.2	94.9	95.6
0330	Cast-in-Place Concrete	89.5	91.4	90.2	82.7	87.7	84.6	79.3	86.7	82.1	89.9	99.0	93.3	90.6	109.3	97.6	95.8	104.9	99.2
03	CONCRETE	89.1	92.9	90.9	85.5	84.7	85.1	87.0	83.4	85.3	89.4	92.3	90.8	93.2	98.9	95.9	96.2	95.3	95.8
04	MASONRY	88.0	99.1	94.6	96.0	95.5	95.7	75.5	87.3	82.5	82.5	91.8	88.1	103.1	101.3	102.0	89.0	97.1	93.8
05	METALS	92.2	77.5	87.7	96.5	76.3	90.4	93.4	77.1	88.5	92.8	80.1	89.0	97.4	88.8	94.8	91.5	81.7	88.6
06	WOOD, PLASTICS & COMPOSITES	78.9	101.7	91.0	97.8	82.2	89.5	97.4	81.9	89.2	93.7	87.6	90.5	101.8	91.3	96.2	91.3	87.5	89.3
07	THERMAL & MOISTURE PROTECTION	100.5	96.8	99.0	97.4	89.1	94.1	100.5	89.3	96.0	109.5	94.3	103.4	99.9	106.3	102.5	101.3	99.1	100.4
08	OPENINGS	99.3	91.5	97.3	95.4	80.2	91.4	95.6	78.3	91.1	96.3	87.5	94.0	99.1	92.4	97.4	100.1	92.3	98.1
0920	Plaster & Gypsum Board	90.7	101.4	97.1	90.4	81.7	85.2	95.1	81.7	87.1	91.0	86.7	88.4	91.6	91.0	91.2	95.0	86.8	90.1
0950, 0980	Ceilings & Acoustic Treatment	97.0	101.4	99.6	89.6	81.7	84.8	98.3	81.7	88.3	86.6	86.7	86.6	89.6	91.0	90.4	94.5	86.8	89.9
0960	Flooring	97.3	107.4	100.0	92.3	84.0	90.1	114.1	85.5	106.5	120.1	95.8	113.6	93.3	93.0	93.2	102.5	93.8	100.2
0990	Wall Finishes & Painting/Coating	107.8	95.0	100.1	100.5	58.7	75.5	108.7	90.8	98.0	116.9	94.4	103.5	100.5	103.2	102.1	107.8	99.2	102.6
09	FINISHES	97.6	101.7	99.7	93.9	80.4	86.9	100.0	83.9	91.7	110.2	90.9	100.3	94.6	94.1	94.4	99.5	91.1	95.2
COVERS	DIVS. 10 - 14, 25, 28, 41, 43, 44	100.0	92.7	98.5	100.0	67.7	93.4	100.0	87.2	97.4	100.0	94.8	98.9	100.0	95.9	99.2	100.0	93.8	98.7
22, 23	PLUMBING & HVAC	95.8	88.6	92.9	95.5	89.9	93.3	100.8	86.6	95.0	95.8	91.0	93.8	99.8	102.0	100.7	100.1	92.5	97.0
26, 27, 3370	ELECTRICAL, COMMUNICATIONS & UTIL.	96.4	84.5	90.7	91.1	84.5	88.0	96.6	83.3	90.2	85.0	101.8	93.1	96.5	101.2	98.8	98.6	90.1	94.5
MF2004	WEIGHTED AVERAGE	94.9	92.4	93.8	93.9	86.4	90.7	94.8	85.9	91.0	95.4	93.5	94.6	97.5	98.6	98.0	97.3	93.2	95.6

City Cost Indexes

Table 1

DIVISION		OHIO ZANESVILLE 437 - 438			OKLAHOMA ARDMORE 734			OKLAHOMA CLINTON 736			OKLAHOMA DURANT 747			OKLAHOMA ENID 737			OKLAHOMA GUYMON 739		
		MAT.	INST.	TOTAL	MAT.	INST.	TOTAL	MAT.	INST.	TOTAL	MAT.	INST.	TOTAL	MAT.	INST.	TOTAL	MAT.	INST.	TOTAL
01543	CONTRACTOR EQUIPMENT	.0	95.0	95.0	.0	78.8	78.8	.0	77.7	77.7	.0	77.7	77.7	.0	77.7	77.7	.0	77.7	77.7
0241, 31 - 34	SITE & INFRASTRUCTURE, DEMOLITION	83.4	101.1	96.0	107.6	90.8	95.5	109.3	89.2	94.9	101.5	88.9	92.5	110.9	89.2	95.3	114.9	87.7	95.4
0310	Concrete Forming & Accessories	89.9	83.2	84.2	93.9	45.7	52.5	92.6	39.2	46.8	84.4	45.7	51.2	96.6	38.5	46.7	100.9	23.6	34.5
0310	Concrete Reinforcing	84.9	89.9	87.3	96.8	78.7	87.9	97.4	78.7	88.2	97.7	63.9	81.2	96.7	78.7	87.9	97.4	32.2	65.5
0330	Cast-in-Place Concrete	87.2	87.7	87.4	96.7	46.6	77.8	93.4	46.6	75.8	90.4	46.6	73.9	93.4	49.5	76.9	93.5	30.5	69.8
03	CONCRETE	88.9	85.6	87.4	91.4	52.4	73.0	90.5	49.6	71.2	86.3	49.6	69.0	91.0	50.2	71.7	93.3	28.1	62.5
04	MASONRY	92.8	85.6	88.5	92.4	58.9	72.3	118.8	58.9	82.9	90.9	59.5	72.1	100.5	58.9	75.6	95.3	22.3	51.6
05	METALS	98.0	80.0	92.6	90.5	67.0	83.4	90.6	67.0	83.4	90.5	60.4	81.4	91.8	67.0	84.3	91.0	38.4	75.1
06	WOOD, PLASTICS & COMPOSITES	93.8	82.2	87.7	96.9	45.0	69.5	95.7	36.4	64.4	86.1	45.0	64.4	100.2	35.4	65.9	104.7	23.1	61.6
07	THERMAL & MOISTURE PROTECTION	97.5	90.0	94.5	100.0	61.0	84.3	100.1	60.1	84.0	99.6	59.3	83.4	100.2	60.0	84.0	100.5	28.7	71.6
08	OPENINGS	95.4	83.0	92.2	95.4	54.2	84.7	95.4	49.5	83.5	95.4	50.7	83.8	95.4	48.7	83.3	95.5	25.0	77.3
0920	Plaster & Gypsum Board	88.7	81.7	84.6	85.9	44.3	61.2	85.6	35.4	55.7	82.4	44.3	59.7	86.6	34.4	55.5	87.1	21.8	48.3
0950, 0980	Ceilings & Acoustic Treatment	89.6	81.7	84.8	85.3	44.3	60.6	85.3	35.4	55.3	85.3	44.3	60.6	86.2	34.4	55.0	87.8	21.8	48.1
0960	Flooring	90.3	88.1	89.7	109.9	52.0	94.5	108.3	49.5	92.6	103.9	68.0	94.3	110.7	49.5	94.4	113.0	29.3	90.7
0990	Wall Finishes & Painting/Coating	100.5	62.9	78.0	98.3	63.4	77.4	98.3	63.4	77.4	98.3	63.4	77.4	98.3	63.4	77.4	98.3	18.8	50.8
09	FINISHES	93.2	81.7	87.2	95.3	47.3	70.5	94.9	41.9	67.5	92.6	50.8	71.0	96.1	41.2	67.8	97.6	24.0	59.6
COVERS	DIVS. 10 - 14, 25, 28, 41, 43, 44	100.0	88.0	97.5	100.0	71.1	94.1	100.0	70.1	93.8	100.0	71.3	94.1	100.0	70.0	93.8	100.0	66.0	93.0
22, 23	PLUMBING & HVAC	95.5	80.4	89.4	95.9	62.4	82.2	95.9	62.4	82.2	95.9	62.6	82.3	100.1	62.4	84.7	95.9	26.0	67.4
26, 27, 3370	ELECTRICAL, COMMUNICATIONS & UTIL.	91.7	79.8	86.0	95.5	68.5	82.5	96.5	68.5	83.0	98.0	60.6	80.0	96.5	68.5	83.0	98.0	19.2	60.2
MF2004	WEIGHTED AVERAGE	94.4	84.1	90.0	95.0	62.2	80.8	96.4	60.6	80.9	94.1	60.3	79.5	96.9	60.5	81.2	96.2	32.3	68.6

Table 2

DIVISION		OKLAHOMA LAWTON 735			OKLAHOMA MCALESTER 745			OKLAHOMA MIAMI 743			OKLAHOMA MUSKOGEE 744			OKLAHOMA OKLAHOMA CITY 730 - 731			OKLAHOMA PONCA CITY 746		
		MAT.	INST.	TOTAL	MAT.	INST.	TOTAL	MAT.	INST.	TOTAL	MAT.	INST.	TOTAL	MAT.	INST.	TOTAL	MAT.	INST.	TOTAL
01543	CONTRACTOR EQUIPMENT	.0	78.8	78.8	.0	77.7	77.7	.0	87.2	87.2	.0	87.2	87.2	.0	79.1	79.1	.0	77.7	77.7
0241, 31 - 34	SITE & INFRASTRUCTURE, DEMOLITION	105.7	90.8	95.0	93.4	89.3	90.5	95.3	86.1	88.7	95.5	84.8	87.9	106.0	91.4	95.5	102.4	89.2	92.9
0310	Concrete Forming & Accessories	100.4	52.5	59.3	82.0	47.8	52.6	96.8	56.4	62.1	101.9	34.7	44.2	101.6	45.3	53.2	91.7	46.1	52.5
0310	Concrete Reinforcing	96.9	78.7	88.0	97.4	48.1	73.2	95.9	78.7	87.5	96.7	36.8	67.4	96.9	78.7	88.0	96.7	78.6	87.9
0330	Cast-in-Place Concrete	90.4	49.5	75.0	79.2	49.7	68.1	83.0	48.6	70.1	84.0	38.0	66.7	88.1	52.6	74.8	92.8	40.9	73.3
03	CONCRETE	87.8	56.4	73.0	77.9	48.6	64.1	81.9	58.8	71.0	83.5	37.3	61.7	86.7	54.3	71.4	88.3	50.6	70.5
04	MASONRY	95.1	58.9	73.4	108.9	59.6	79.4	93.7	60.2	73.7	110.4	52.6	75.8	96.0	60.9	75.0	85.6	59.5	70.0
05	METALS	95.3	67.1	86.7	90.5	53.5	79.3	90.4	81.5	87.7	91.8	57.5	81.4	97.1	67.0	88.0	90.4	66.7	83.2
06	WOOD, PLASTICS & COMPOSITES	103.5	54.2	77.4	83.2	48.9	65.1	100.1	58.7	78.2	105.8	35.0	68.4	105.6	43.5	72.7	94.9	45.7	68.9
07	THERMAL & MOISTURE PROTECTION	100.0	62.0	84.7	99.3	60.2	83.5	99.7	62.1	84.6	99.8	46.2	78.2	98.3	61.7	83.6	99.8	60.0	83.8
08	OPENINGS	97.0	58.9	87.1	95.4	47.6	83.0	95.4	62.1	86.8	95.4	34.2	79.5	97.0	53.1	85.6	95.4	54.5	84.8
0920	Plaster & Gypsum Board	88.6	53.7	67.9	81.4	48.3	61.7	86.6	58.1	69.6	89.9	33.8	56.5	88.6	42.7	61.3	85.6	45.0	61.4
0950, 0980	Ceilings & Acoustic Treatment	95.5	53.7	70.4	85.3	48.3	63.0	85.3	58.1	69.0	95.5	33.8	58.4	95.5	42.7	63.7	85.3	45.0	61.1
0960	Flooring	113.5	49.5	96.4	102.6	49.5	88.4	111.6	68.0	100.0	114.7	39.9	94.8	113.5	49.5	96.4	107.8	49.5	92.2
0990	Wall Finishes & Painting/Coating	98.3	63.4	77.4	98.3	46.7	67.5	98.3	77.9	86.1	98.3	33.9	59.8	98.3	63.4	77.4	98.3	63.4	77.4
09	FINISHES	99.0	52.2	74.8	91.5	47.4	68.7	94.9	60.6	77.2	98.8	35.8	66.3	99.1	46.4	71.9	94.4	47.5	70.2
COVERS	DIVS. 10 - 14, 25, 28, 41, 43, 44	100.0	72.2	94.3	100.0	71.8	94.2	100.0	73.8	94.6	100.0	69.2	93.7	100.0	71.7	94.2	100.0	71.4	94.1
22, 23	PLUMBING & HVAC	100.1	62.4	84.8	95.9	36.6	71.7	95.9	62.1	82.1	100.1	27.8	70.7	100.1	63.5	85.2	95.9	61.7	81.9
26, 27, 3370	ELECTRICAL, COMMUNICATIONS & UTIL.	98.0	68.6	83.9	96.5	65.1	81.5	97.9	65.2	82.2	96.1	32.8	65.6	97.5	68.5	83.6	96.1	61.8	79.7
MF2004	WEIGHTED AVERAGE	97.2	63.7	82.8	93.5	54.4	76.6	93.9	65.8	81.8	96.3	43.0	73.3	97.3	62.8	82.4	94.1	60.8	79.7

Table 3

DIVISION		OKLAHOMA POTEAU 749			OKLAHOMA SHAWNEE 748			OKLAHOMA TULSA 740 - 741			OKLAHOMA WOODWARD 738			OREGON BEND 977			OREGON EUGENE 974		
		MAT.	INST.	TOTAL	MAT.	INST.	TOTAL	MAT.	INST.	TOTAL	MAT.	INST.	TOTAL	MAT.	INST.	TOTAL	MAT.	INST.	TOTAL
01543	CONTRACTOR EQUIPMENT	.0	86.5	86.5	.0	77.7	77.7	.0	87.2	87.2	.0	77.7	77.7	.0	99.8	99.8	.0	99.8	99.8
0241, 31 - 34	SITE & INFRASTRUCTURE, DEMOLITION	76.3	84.7	82.3	105.9	88.5	93.4	102.4	86.2	90.8	109.9	89.2	95.0	123.6	104.5	109.9	110.1	104.5	106.1
0310	Concrete Forming & Accessories	89.0	46.5	52.5	84.3	44.4	50.0	101.5	45.6	53.5	92.7	39.2	46.7	108.2	105.1	105.6	104.1	105.1	104.9
0310	Concrete Reinforcing	97.9	78.7	88.5	96.7	52.8	75.2	96.9	78.6	88.0	96.7	78.7	87.9	102.5	97.6	100.1	106.8	97.6	102.3
0330	Cast-in-Place Concrete	83.0	48.1	69.9	95.8	45.4	76.8	91.6	48.3	75.3	93.4	46.6	75.8	109.1	104.8	107.5	105.7	104.8	105.3
03	CONCRETE	83.5	54.2	69.7	90.0	46.4	69.4	88.5	53.9	72.1	90.8	49.5	71.3	116.3	103.2	110.1	107.3	103.2	105.4
04	MASONRY	94.4	59.6	73.6	110.2	57.9	78.9	94.5	61.0	74.4	89.1	58.9	71.0	118.4	102.1	108.6	115.6	102.1	107.5
05	METALS	90.5	81.3	87.7	90.4	53.3	79.2	94.6	80.6	90.4	90.6	67.0	83.5	91.6	96.1	92.9	92.2	96.0	93.4
06	WOOD, PLASTICS & COMPOSITES	91.5	45.9	67.4	85.9	44.8	64.2	104.3	44.8	72.8	95.9	36.3	64.4	99.7	105.3	102.6	94.9	105.3	100.4
07	THERMAL & MOISTURE PROTECTION	99.8	60.2	83.9	99.8	60.7	84.1	99.8	58.8	83.3	100.1	60.1	84.0	105.8	96.8	102.2	105.1	95.2	101.1
08	OPENINGS	95.4	55.2	85.0	95.4	44.3	82.1	97.0	53.4	85.7	95.4	49.4	83.5	98.5	105.6	100.3	98.9	105.6	100.6
0920	Plaster & Gypsum Board	84.9	45.0	61.2	82.4	44.0	59.6	88.6	43.9	62.0	86.2	35.3	55.9	101.6	105.2	103.8	100.0	105.2	103.1
0950, 0980	Ceilings & Acoustic Treatment	85.3	45.0	61.1	85.3	44.0	60.5	95.5	43.9	64.5	87.8	35.3	56.2	103.2	105.2	104.4	107.9	105.2	106.3
0960	Flooring	107.0	68.0	96.6	103.8	39.4	86.6	113.3	51.6	96.8	108.3	52.0	93.3	117.4	97.9	112.2	115.4	97.9	110.7
0990	Wall Finishes & Painting/Coating	98.3	63.4	77.4	98.3	42.2	64.8	98.3	50.1	69.5	98.3	63.4	77.4	113.1	72.5	88.8	113.1	72.5	88.8
09	FINISHES	92.4	51.3	71.2	92.8	42.3	66.8	98.4	46.1	71.4	95.7	42.2	68.1	111.7	100.4	105.9	110.9	100.4	105.5
COVERS	DIVS. 10 - 14, 25, 28, 41, 43, 44	100.0	71.8	94.2	100.0	70.8	94.0	100.0	72.2	94.3	100.0	70.1	93.8	100.0	102.0	100.4	100.0	102.0	100.4
22, 23	PLUMBING & HVAC	95.9	61.8	82.0	95.9	61.6	81.9	100.1	57.5	82.8	95.9	62.4	82.2	95.9	105.9	100.0	100.2	105.8	102.5
26, 27, 3370	ELECTRICAL, COMMUNICATIONS & UTIL.	96.3	65.2	81.4	98.1	68.4	83.8	98.0	42.8	71.5	97.9	68.5	83.8	99.6	96.4	98.1	98.3	96.4	97.4
MF2004	WEIGHTED AVERAGE	93.1	63.2	80.2	95.7	58.6	79.7	97.0	58.6	80.4	95.1	60.6	80.2	102.2	101.7	102.0	101.6	101.7	101.6

City Cost Indexes

DIVISION		OREGON																	
		KLAMATH FALLS 976			MEDFORD 975			PENDLETON 978			PORTLAND 970 - 972			SALEM 973			VALE 979		
		MAT.	INST.	TOTAL	MAT.	INST.	TOTAL	MAT.	INST.	TOTAL	MAT.	INST.	TOTAL	MAT.	INST.	TOTAL	MAT.	INST.	TOTAL
01543	CONTRACTOR EQUIPMENT	.0	99.8	99.8	.0	99.8	99.8	.0	96.6	96.6	.0	99.8	99.8	.0	99.8	99.8	.0	96.6	96.6
0241, 31 - 34	SITE & INFRASTRUCTURE, DEMOLITION	128.9	104.5	111.4	120.1	104.5	108.9	115.6	97.0	102.3	112.9	104.5	106.9	109.7	104.5	106.0	100.5	96.4	97.5
0310	Concrete Forming & Accessories	100.3	104.8	104.2	99.3	104.8	104.1	101.8	105.2	104.7	105.7	105.2	105.3	105.1	105.1	105.1	109.1	104.0	104.8
0310	Concrete Reinforcing	102.5	97.6	100.1	104.2	97.6	100.9	101.6	97.8	99.7	107.7	97.8	102.8	107.8	97.8	102.9	99.1	97.6	98.4
0330	Cast-in-Place Concrete	109.1	104.7	107.5	101.3	104.7	107.5	109.9	106.3	108.5	108.6	104.9	107.2	109.6	104.8	107.8	86.5	105.2	93.5
03	CONCRETE	119.0	103.1	111.5	113.6	103.1	108.6	99.9	103.8	101.7	109.1	103.3	106.4	109.6	103.3	106.6	83.6	102.8	92.6
04	MASONRY	134.9	102.1	115.3	112.7	102.1	106.4	123.2	104.6	112.1	116.7	104.5	109.4	121.0	104.5	111.1	121.4	104.4	111.2
05	METALS	91.6	95.8	92.9	91.8	95.8	93.0	98.4	96.6	97.9	93.3	96.4	94.2	92.6	96.3	93.7	98.3	93.5	96.9
06	WOOD, PLASTICS & COMPOSITES	89.8	105.3	98.0	88.5	105.3	97.4	92.6	105.4	99.4	96.3	105.3	101.0	95.7	105.3	100.8	102.1	105.4	103.8
07	THERMAL & MOISTURE PROTECTION	106.0	92.7	100.7	105.7	92.7	100.5	94.8	97.4	95.9	104.8	100.1	102.9	105.0	97.4	101.9	94.3	95.1	94.6
08	OPENINGS	98.5	105.6	100.4	101.6	105.6	102.6	94.3	105.6	97.2	96.5	105.6	98.8	98.3	105.6	100.2	94.2	98.3	95.3
0920	Plaster & Gypsum Board	98.9	105.2	102.7	98.2	105.2	102.4	80.9	105.2	95.4	102.3	105.2	104.0	98.1	105.2	102.3	85.7	105.2	97.3
0950, 0980	Ceilings & Acoustic Treatment	112.5	105.2	108.1	117.2	105.2	110.0	71.2	105.2	91.7	106.0	105.2	105.6	107.9	105.2	106.3	71.2	105.2	91.7
0960	Flooring	113.6	97.9	109.4	112.9	97.9	108.9	80.2	97.9	84.9	114.0	97.9	109.7	115.4	97.9	110.7	82.3	97.9	86.5
0990	Wall Finishes & Painting/Coating	113.1	63.5	83.4	113.1	63.5	83.4	95.7	72.5	81.9	113.6	72.5	89.1	113.1	72.5	88.8	95.7	72.5	81.9
09	FINISHES	112.8	99.4	105.9	112.8	99.4	105.9	77.4	100.5	89.3	110.6	100.4	105.3	110.5	100.4	105.3	77.9	100.5	89.6
COVERS	DIVS. 10 - 14, 25, 28, 41, 43, 44	100.0	102.0	100.4	100.0	102.0	100.4	100.0	102.2	100.4	100.0	102.0	100.4	100.0	102.0	100.4	100.0	102.3	100.5
22, 23	PLUMBING & HVAC	95.9	105.8	100.0	100.2	105.8	102.5	97.3	106.0	100.8	100.1	105.9	102.5	100.2	105.9	102.5	97.3	79.3	90.0
26, 27, 3370	ELECTRICAL, COMMUNICATIONS & UTIL.	98.4	86.3	92.6	101.5	86.3	94.2	90.2	96.2	93.1	98.8	100.9	99.8	98.6	96.4	97.5	90.2	75.5	83.1
MF2004	WEIGHTED AVERAGE	103.4	100.0	102.0	103.1	100.0	101.7	96.8	101.5	98.8	101.8	102.7	102.2	102.1	102.0	102.0	94.5	92.2	93.5

DIVISION		PENNSYLVANIA																	
		ALLENTOWN 181			ALTOONA 166			BEDFORD 155			BRADFORD 167			BUTLER 160			CHAMBERSBURG 172		
		MAT.	INST.	TOTAL	MAT.	INST.	TOTAL	MAT.	INST.	TOTAL	MAT.	INST.	TOTAL	MAT.	INST.	TOTAL	MAT.	INST.	TOTAL
01543	CONTRACTOR EQUIPMENT	.0	115.3	115.3	.0	115.3	115.3	.0	112.8	112.8	.0	115.3	115.3	.0	115.3	115.3	.0	114.5	114.5
0241, 31 - 34	SITE & INFRASTRUCTURE, DEMOLITION	91.8	106.6	102.4	96.5	106.4	103.6	99.5	103.7	102.5	91.3	106.1	101.9	87.4	107.3	101.7	86.3	104.0	99.0
0310	Concrete Forming & Accessories	100.5	114.9	112.9	83.7	83.0	83.1	85.4	84.6	84.8	86.4	84.0	84.4	85.5	95.8	94.4	85.2	83.9	84.1
0310	Concrete Reinforcing	97.0	107.7	102.3	94.1	89.5	91.9	95.1	80.4	87.9	96.1	89.6	92.9	94.7	104.7	99.6	94.5	96.0	95.3
0330	Cast-in-Place Concrete	87.4	103.4	93.4	97.4	83.0	92.0	100.6	78.9	92.4	93.0	89.7	91.8	86.0	95.1	89.4	94.5	78.6	88.5
03	CONCRETE	95.0	110.4	102.3	91.5	85.8	88.8	94.8	83.2	89.4	97.0	88.5	93.0	83.3	98.6	90.5	102.3	85.9	94.6
04	MASONRY	95.1	101.9	99.1	97.4	80.2	87.1	96.9	85.9	90.3	94.3	85.9	89.3	99.7	92.5	95.4	95.7	85.7	89.7
05	METALS	96.6	123.9	104.8	90.4	112.7	97.1	90.7	106.7	95.5	94.7	112.9	100.2	90.1	122.3	99.8	94.6	114.6	100.6
06	WOOD, PLASTICS & COMPOSITES	102.2	118.5	110.8	79.4	85.0	82.3	82.0	85.0	83.5	85.9	83.1	84.4	81.2	95.7	88.8	89.8	85.0	87.3
07	THERMAL & MOISTURE PROTECTION	101.5	117.0	107.7	100.4	91.2	96.7	102.4	89.0	97.0	101.3	93.3	98.1	100.1	98.8	99.6	100.5	76.4	90.8
08	OPENINGS	93.9	113.7	99.0	88.3	89.8	88.7	91.6	87.6	90.6	94.3	93.2	94.0	88.3	104.8	92.6	90.7	85.2	89.3
0920	Plaster & Gypsum Board	106.4	118.7	113.7	100.3	84.3	90.8	88.9	84.3	86.2	101.0	82.8	89.9	100.6	95.3	97.5	104.2	84.3	92.4
0950, 0980	Ceilings & Acoustic Treatment	88.1	118.7	106.5	92.3	84.3	87.5	89.8	84.3	86.5	90.7	82.4	85.7	93.2	95.3	94.5	90.7	84.3	86.9
0960	Flooring	87.6	93.0	89.0	81.4	54.9	74.3	85.1	55.7	77.3	80.9	101.4	86.3	82.3	97.5	86.4	85.0	55.6	77.1
0990	Wall Finishes & Painting/Coating	89.0	87.2	87.9	84.6	106.0	97.4	98.3	83.6	89.5	89.0	88.4	88.7	84.6	106.0	97.4	89.0	71.4	78.5
09	FINISHES	94.8	107.8	101.5	93.2	79.4	86.1	91.3	78.3	84.6	92.6	86.3	89.4	93.3	96.4	94.9	93.6	77.0	85.0
COVERS	DIVS. 10 - 14, 25, 28, 41, 43, 44	100.0	103.1	100.6	100.0	97.5	99.5	100.0	99.6	99.9	100.0	99.4	99.9	100.0	102.5	100.5	100.0	95.2	99.0
22, 23	PLUMBING & HVAC	100.3	110.8	104.6	99.7	86.3	94.3	95.7	89.1	93.0	96.1	89.7	93.5	95.6	92.6	94.3	96.1	91.0	94.0
26, 27, 3370	ELECTRICAL, COMMUNICATIONS & UTIL.	101.1	95.3	98.3	90.6	101.2	95.7	94.3	101.2	97.6	95.5	101.2	98.2	91.1	100.6	95.7	94.0	83.1	88.8
MF2004	WEIGHTED AVERAGE	97.7	108.3	102.2	94.3	91.5	93.1	94.5	91.4	93.1	95.6	94.3	95.0	92.3	99.9	95.6	95.7	89.5	93.0

DIVISION		PENNSYLVANIA																	
		DOYLESTOWN 189			DUBOIS 158			ERIE 164 - 165			GREENSBURG 156			HARRISBURG 170 - 171			HAZLETON 182		
		MAT.	INST.	TOTAL	MAT.	INST.	TOTAL	MAT.	INST.	TOTAL	MAT.	INST.	TOTAL	MAT.	INST.	TOTAL	MAT.	INST.	TOTAL
01543	CONTRACTOR EQUIPMENT	.0	92.4	92.4	.0	112.8	112.8	.0	115.3	115.3	.0	112.8	112.8	.0	114.5	114.5	.0	115.3	115.3
0241, 31 - 34	SITE & INFRASTRUCTURE, DEMOLITION	105.3	89.6	94.1	104.4	103.9	104.0	93.0	107.0	103.1	95.5	105.1	102.4	82.5	105.1	98.7	84.8	106.4	100.3
0310	Concrete Forming & Accessories	82.6	128.1	121.7	84.7	85.7	85.5	99.9	92.5	93.5	92.7	95.8	95.4	93.5	85.4	86.5	80.2	86.6	85.7
0310	Concrete Reinforcing	93.8	125.0	109.1	94.5	92.0	92.1	96.1	86.8	91.5	94.5	104.6	99.5	97.0	94.0	95.5	94.2	112.3	103.0
0330	Cast-in-Place Concrete	82.6	95.1	87.3	96.9	91.8	95.0	95.8	81.7	90.5	93.3	94.8	93.9	96.5	90.5	94.2	82.6	88.8	84.9
03	CONCRETE	90.4	115.8	102.4	95.3	89.9	92.8	90.6	89.0	89.8	90.2	98.4	94.0	97.6	90.3	94.1	87.8	93.7	90.6
04	MASONRY	98.8	125.7	114.9	96.9	86.2	90.5	87.1	91.8	89.9	107.2	90.1	97.0	96.0	85.7	89.8	108.1	92.3	98.7
05	METALS	94.0	119.6	101.8	90.7	112.1	97.2	90.5	111.7	96.9	90.6	121.0	99.8	98.4	116.3	103.8	96.3	124.8	104.9
06	WOOD, PLASTICS & COMPOSITES	80.9	130.4	107.1	80.9	85.0	83.0	98.7	92.2	95.2	90.2	95.6	93.1	95.8	84.4	89.8	79.2	84.4	82.0
07	THERMAL & MOISTURE PROTECTION	99.9	125.9	110.4	102.6	95.0	99.5	100.3	96.1	98.6	102.3	97.8	100.5	105.6	104.7	105.2	101.0	102.1	101.4
08	OPENINGS	95.6	132.6	105.2	91.6	94.2	92.3	88.5	89.9	88.9	91.6	104.8	95.0	93.9	90.9	93.1	94.6	96.6	95.1
0920	Plaster & Gypsum Board	99.0	130.9	118.0	88.2	84.3	85.9	106.4	91.7	97.7	90.9	95.3	93.5	106.4	83.8	92.9	99.4	83.8	90.1
0950, 0980	Ceilings & Acoustic Treatment	87.3	130.9	113.6	89.8	84.3	86.5	88.1	91.7	90.3	88.9	95.3	92.8	88.1	83.8	85.5	89.0	83.8	85.9
0960	Flooring	71.3	99.8	78.9	84.7	101.4	89.2	89.7	81.7	87.6	88.8	81.9	86.9	87.8	88.0	87.9	77.8	91.4	81.5
0990	Wall Finishes & Painting/Coating	88.4	85.3	86.5	98.3	106.0	102.9	95.6	88.4	91.3	98.3	106.0	102.9	89.0	86.6	87.5	89.0	98.8	94.9
09	FINISHES	86.2	118.6	102.9	91.5	88.8	90.1	96.3	89.8	93.0	92.2	94.4	93.4	93.7	85.6	89.5	90.6	87.1	88.8
COVERS	DIVS. 10 - 14, 25, 28, 41, 43, 44	100.0	87.0	97.3	100.0	99.6	99.9	100.0	102.4	100.5	100.0	102.5	100.5	100.0	95.2	99.0	100.0	96.6	99.3
22, 23	PLUMBING & HVAC	95.6	123.3	106.9	95.7	89.3	93.1	99.7	92.7	96.9	95.7	91.8	94.1	100.3	91.5	96.7	96.1	89.8	93.5
26, 27, 3370	ELECTRICAL, COMMUNICATIONS & UTIL.	94.3	118.5	105.9	94.8	101.2	97.9	92.4	86.8	89.7	94.8	101.2	97.9	101.1	83.7	92.5	96.2	87.8	92.2
MF2004	WEIGHTED AVERAGE	94.6	117.5	104.5	94.7	94.6	94.7	94.3	94.1	94.2	94.6	99.0	96.5	98.0	92.6	95.7	95.2	95.3	95.3

City Cost Indexes

PENNSYLVANIA

DIVISION		INDIANA 157			JOHNSTOWN 159			KITTANNING 162			LANCASTER 175 - 176			LEHIGH VALLEY 180			MONTROSE 188		
		MAT.	INST.	TOTAL	MAT.	INST.	TOTAL	MAT.	INST.	TOTAL	MAT.	INST.	TOTAL	MAT.	INST.	TOTAL	MAT.	INST.	TOTAL
01543	CONTRACTOR EQUIPMENT	.0	112.8	112.8	.0	112.8	112.8	.0	115.3	115.3	.0	114.5	114.5	.0	115.3	115.3	.0	115.3	115.3
0241, 31 - 34	SITE & INFRASTRUCTURE, DEMOLITION	93.6	104.5	101.5	100.0	104.6	103.3	90.1	107.5	102.6	77.9	105.1	97.4	89.2	106.4	101.5	88.0	105.0	100.2
0310	Concrete Forming & Accessories	86.2	87.8	87.5	84.7	85.8	85.7	85.5	95.9	94.4	87.4	84.8	85.2	93.4	111.8	109.2	81.2	86.8	86.0
0310	Concrete Reinforcing	93.8	104.8	99.2	95.1	104.5	99.7	94.7	104.9	99.7	94.2	93.9	94.1	94.2	98.5	96.3	98.7	112.1	105.3
0330	Cast-in-Place Concrete	91.5	94.8	92.7	101.5	86.3	95.8	89.4	95.2	91.6	80.3	90.3	84.1	89.4	98.8	92.9	87.6	88.1	87.8
03	CONCRETE	88.0	94.8	91.2	95.5	90.9	93.4	85.9	98.6	91.9	90.0	89.9	89.9	94.2	105.7	99.6	92.8	93.4	93.1
04	MASONRY	93.3	94.7	94.1	94.1	86.3	89.4	103.6	94.7	98.3	101.0	85.7	91.8	95.2	95.0	95.0	95.1	93.0	93.9
05	METALS	90.7	121.2	99.9	90.7	119.2	99.3	90.2	122.7	100.0	94.6	115.6	101.0	96.2	119.2	103.2	94.7	120.9	102.7
06	WOOD, PLASTICS & COMPOSITES	82.9	85.0	84.0	80.9	85.0	83.0	81.2	95.5	88.8	92.7	84.4	88.3	93.8	116.2	105.6	80.1	85.2	82.8
07	THERMAL & MOISTURE PROTECTION	102.1	96.8	100.0	102.4	93.9	99.0	100.2	99.4	99.9	100.1	99.7	100.0	101.4	106.4	103.4	101.0	94.7	98.5
08	OPENINGS	91.6	98.7	93.5	91.6	94.4	92.3	88.3	104.5	92.5	90.7	91.2	90.8	94.6	109.7	98.5	90.9	96.9	92.4
0920	Plaster & Gypsum Board	89.2	84.3	86.3	88.0	84.3	85.8	100.6	95.2	97.4	105.8	83.8	92.7	102.9	116.4	110.9	99.7	84.6	90.7
0950, 0980	Ceilings & Acoustic Treatment	89.8	84.3	86.5	88.9	84.3	86.2	93.2	95.2	94.4	90.7	83.8	86.5	89.0	116.4	105.5	90.7	84.6	87.0
0960	Flooring	85.5	101.4	89.8	84.7	69.9	80.8	82.3	101.4	87.4	86.1	88.0	86.6	83.8	97.0	87.3	78.6	69.7	76.2
0990	Wall Finishes & Painting/Coating	98.3	106.0	102.9	98.3	106.0	102.9	84.6	106.0	97.4	89.0	70.6	78.0	89.0	82.1	84.9	89.0	98.8	94.9
09	FINISHES	91.1	90.9	91.0	90.9	84.0	87.3	93.5	97.1	95.3	93.7	82.1	87.7	93.2	105.7	99.6	91.5	84.4	87.8
COVERS	DIVS. 10 - 14, 25, 28, 41, 43, 44	100.0	101.2	100.2	100.0	99.8	100.0	100.0	102.5	100.5	100.0	95.0	99.0	100.0	101.5	100.3	100.0	97.2	99.4
22, 23	PLUMBING & HVAC	95.7	91.8	94.1	95.7	89.7	93.2	95.6	95.7	95.6	96.1	91.5	94.2	96.1	108.8	101.3	96.1	94.8	95.6
26, 27, 3370	ELECTRICAL, COMMUNICATIONS & UTIL.	94.8	101.2	97.9	94.8	101.2	97.9	90.6	101.3	95.7	95.2	54.8	75.9	96.2	143.8	119.0	95.5	91.8	93.7
MF2004	WEIGHTED AVERAGE	93.4	98.1	95.4	94.4	94.9	94.6	92.8	101.1	96.4	94.5	87.9	91.7	95.8	112.1	102.8	94.6	96.0	95.2

PENNSYLVANIA

DIVISION		NEW CASTLE 161			NORRISTOWN 194			OIL CITY 163			PHILADELPHIA 190 - 191			PITTSBURGH 150 - 152			POTTSVILLE 179		
		MAT.	INST.	TOTAL	MAT.	INST.	TOTAL	MAT.	INST.	TOTAL	MAT.	INST.	TOTAL	MAT.	INST.	TOTAL	MAT.	INST.	TOTAL
01543	CONTRACTOR EQUIPMENT	.0	115.3	115.3	.0	95.9	95.9	.0	115.3	115.3	.0	95.2	95.2	.0	114.3	114.3	.0	114.5	114.5
0241, 31 - 34	SITE & INFRASTRUCTURE, DEMOLITION	87.8	107.5	101.9	95.8	96.8	96.5	86.2	106.3	100.6	103.1	96.5	98.3	98.8	107.7	105.2	80.9	104.9	98.1
0310	Concrete Forming & Accessories	85.5	97.5	95.8	83.2	130.5	123.8	85.2	84.1	84.3	101.3	133.5	129.0	101.5	98.2	98.6	78.3	85.4	84.4
0310	Concrete Reinforcing	93.5	102.1	97.7	97.2	136.8	116.6	94.7	102.0	98.3	100.3	137.3	118.4	95.6	105.1	100.3	93.4	94.9	94.2
0330	Cast-in-Place Concrete	86.8	93.2	89.2	88.6	124.9	102.3	84.3	95.1	88.4	108.8	131.4	117.3	96.9	95.0	96.2	85.6	82.6	84.5
03	CONCRETE	83.7	98.2	90.5	96.7	129.4	112.1	82.1	92.8	87.1	110.5	133.1	121.2	93.4	99.6	96.3	93.7	87.7	90.9
04	MASONRY	98.7	91.4	94.3	112.7	125.7	120.5	99.5	90.3	94.0	99.1	136.0	121.2	88.4	97.7	94.0	94.9	86.3	89.7
05	METALS	90.2	121.0	99.5	99.8	127.3	108.1	90.2	118.9	98.9	103.3	129.3	111.2	91.9	122.2	101.1	94.8	116.1	101.2
06	WOOD, PLASTICS & COMPOSITES	81.2	98.3	90.2	79.7	130.3	106.5	81.2	81.1	81.1	100.4	132.2	117.2	100.7	98.3	99.4	82.2	84.4	83.4
07	THERMAL & MOISTURE PROTECTION	100.1	98.1	99.3	101.7	130.4	113.2	100.0	95.2	98.1	102.4	137.1	116.4	102.5	100.3	101.6	100.2	101.1	100.6
08	OPENINGS	88.3	105.0	92.6	88.7	136.0	100.9	88.3	86.4	87.8	96.8	139.8	108.0	94.2	106.3	97.3	90.7	91.2	90.8
0920	Plaster & Gypsum Board	100.6	98.0	99.1	92.7	130.9	115.5	100.6	80.3	88.5	99.4	132.9	119.4	95.3	98.0	96.9	102.0	83.8	91.1
0950, 0980	Ceilings & Acoustic Treatment	93.2	98.0	96.1	99.7	130.9	118.5	93.2	80.3	85.5	99.7	132.9	119.7	89.8	98.0	94.7	90.7	83.8	86.5
0960	Flooring	82.3	69.2	78.8	77.3	133.6	92.3	82.3	101.4	87.4	85.8	140.0	100.3	93.1	104.1	96.0	81.7	88.0	83.4
0990	Wall Finishes & Painting/Coating	84.6	106.0	97.4	96.1	141.7	123.4	84.6	106.0	97.4	96.1	141.7	123.4	98.3	110.0	105.3	89.0	95.6	92.9
09	FINISHES	93.3	92.6	93.0	95.9	132.5	114.8	93.2	87.7	90.4	99.8	135.5	118.2	94.6	100.0	97.4	91.9	86.7	89.3
COVERS	DIVS. 10 - 14, 25, 28, 41, 43, 44	100.0	102.9	100.6	100.0	117.8	103.7	100.0	100.7	100.1	100.0	123.5	104.8	100.0	102.8	100.6	100.0	95.4	99.0
22, 23	PLUMBING & HVAC	95.6	92.5	94.3	95.9	129.1	109.4	95.6	92.2	94.2	100.1	135.5	114.5	99.9	99.4	99.7	96.1	92.9	94.8
26, 27, 3370	ELECTRICAL, COMMUNICATIONS & UTIL.	91.1	100.0	95.4	93.2	132.2	111.9	93.0	100.6	96.6	97.2	138.9	117.2	96.8	101.3	98.9	93.6	94.0	93.8
MF2004	WEIGHTED AVERAGE	92.3	99.1	95.3	96.9	126.7	109.7	92.3	96.3	94.0	101.2	131.6	114.3	96.0	102.7	98.9	94.3	94.2	94.3

PENNSYLVANIA

DIVISION		READING 195 - 196			SCRANTON 184 - 185			STATE COLLEGE 168			STROUDSBURG 183			SUNBURY 178			UNIONTOWN 154		
		MAT.	INST.	TOTAL	MAT.	INST.	TOTAL	MAT.	INST.	TOTAL	MAT.	INST.	TOTAL	MAT.	INST.	TOTAL	MAT.	INST.	TOTAL
01543	CONTRACTOR EQUIPMENT	.0	118.7	118.7	.0	115.3	115.3	.0	114.5	114.5	.0	115.3	115.3	.0	115.3	115.3	.0	112.8	112.8
0241, 31 - 34	SITE & INFRASTRUCTURE, DEMOLITION	100.0	112.0	108.6	92.3	106.5	102.4	83.1	105.2	98.9	86.7	104.9	99.8	91.0	106.1	101.8	94.2	105.1	102.0
0310	Concrete Forming & Accessories	100.8	85.8	87.9	100.7	86.1	88.2	84.3	83.0	83.2	87.2	88.7	88.5	91.0	83.7	84.8	78.3	95.6	93.1
0310	Concrete Reinforcing	98.6	95.1	96.9	97.0	112.3	104.5	95.3	89.6	92.5	97.3	107.2	102.1	96.1	96.3	96.2	94.5	104.3	99.3
0330	Cast-in-Place Concrete	79.4	95.8	85.6	91.3	89.4	90.6	88.0	73.1	82.4	86.0	78.5	83.2	93.6	90.1	92.3	91.5	94.7	92.7
03	CONCRETE	95.4	92.5	94.0	97.0	93.7	95.5	96.6	82.4	89.9	91.6	90.0	90.8	97.7	89.8	94.0	87.6	98.2	92.6
04	MASONRY	100.2	88.2	93.0	95.4	93.3	94.2	101.2	80.8	89.0	92.1	94.6	93.6	94.4	85.9	89.3	109.1	84.5	94.4
05	METALS	100.0	117.4	105.3	98.4	125.1	106.5	94.5	112.7	100.0	96.3	118.1	102.9	94.6	116.5	101.2	90.5	120.3	99.5
06	WOOD, PLASTICS & COMPOSITES	100.5	83.6	91.6	102.2	83.0	92.0	88.6	85.0	86.7	87.2	88.3	87.8	91.1	82.7	86.7	73.4	95.6	85.1
07	THERMAL & MOISTURE PROTECTION	101.8	108.6	104.6	101.4	99.7	100.7	100.3	94.0	97.8	101.2	82.2	93.8	101.3	99.7	100.7	102.0	96.3	99.7
08	OPENINGS	94.4	91.1	93.6	93.9	95.8	94.4	90.6	89.8	90.4	94.6	88.7	93.1	90.8	90.6	90.8	91.5	104.8	95.0
0920	Plaster & Gypsum Board	102.3	83.0	90.9	108.5	82.3	92.9	103.0	84.3	91.9	101.2	87.8	93.2	102.1	82.0	90.1	85.5	95.3	91.3
0950, 0980	Ceilings & Acoustic Treatment	88.7	83.0	85.3	97.3	82.3	88.3	88.2	84.3	85.9	87.3	87.8	87.6	88.2	82.0	84.4	88.9	95.3	92.8
0960	Flooring	84.0	98.0	87.8	87.6	91.4	88.6	84.4	59.9	77.9	81.4	61.8	76.2	82.8	51.1	74.4	81.6	77.1	80.4
0990	Wall Finishes & Painting/Coating	94.5	100.0	97.8	89.0	98.8	94.9	89.0	106.0	99.2	89.0	79.8	83.5	89.0	98.8	94.9	98.3	106.0	102.9
09	FINISHES	97.1	88.5	92.7	97.4	87.5	92.3	92.4	80.5	86.2	91.7	82.6	87.0	92.7	79.5	85.9	89.2	92.7	91.0
COVERS	DIVS. 10 - 14, 25, 28, 41, 43, 44	100.0	95.5	99.1	100.0	96.8	99.4	100.0	93.2	98.6	100.0	76.5	95.2	100.0	95.0	99.0	100.0	102.5	100.5
22, 23	PLUMBING & HVAC	100.3	106.3	102.7	100.3	95.5	98.3	96.1	88.4	93.0	96.1	95.3	95.7	96.1	91.5	94.2	95.7	91.8	94.1
26, 27, 3370	ELECTRICAL, COMMUNICATIONS & UTIL.	99.3	94.0	96.8	101.2	91.8	96.7	94.4	101.2	97.7	96.2	143.7	119.0	94.2	88.5	91.5	92.2	101.2	96.5
MF2004	WEIGHTED AVERAGE	98.8	98.9	98.8	98.4	97.1	97.9	95.2	91.6	93.6	95.1	101.4	97.8	95.5	92.5	94.1	93.6	98.1	95.5

City Cost Indexes

		PENNSYLVANIA																	
		WASHINGTON			WELLSBORO			WESTCHESTER			WILKES-BARRE			WILLIAMSPORT			YORK		
DIVISION		153			169			193			186 - 187			177			173 - 174		
		MAT.	INST.	TOTAL	MAT.	INST.	TOTAL	MAT.	INST.	TOTAL	MAT.	INST.	TOTAL	MAT.	INST.	TOTAL	MAT.	INST.	TOTAL
01543	CONTRACTOR EQUIPMENT	.0	112.8	112.8	.0	115.3	115.3	.0	95.9	95.9	.0	115.3	115.3	.0	115.3	115.3	.0	114.5	114.5
0241, 31 - 34	SITE & INFRASTRUCTURE, DEMOLITION	94.3	105.1	102.0	95.3	104.9	102.2	102.4	95.2	97.2	84.6	106.4	100.2	81.8	104.7	98.2	81.5	105.1	98.4
0310	Concrete Forming & Accessories	86.3	96.0	94.6	85.8	80.4	81.2	90.7	128.7	123.3	90.1	87.2	87.6	87.2	66.0	69.0	82.1	85.4	84.9
0310	Concrete Reinforcing	94.5	104.9	99.6	95.3	112.0	103.5	96.3	101.1	98.6	96.1	112.3	104.0	95.3	80.1	87.9	96.1	94.0	95.1
0330	Cast-in-Place Concrete	91.5	94.9	92.8	92.3	84.7	89.4	98.1	124.5	108.1	82.6	88.8	84.9	79.2	75.0	77.6	86.0	90.5	87.7
03	CONCRETE	88.1	98.5	93.0	99.1	89.4	94.5	105.2	121.8	113.0	88.7	94.0	91.2	85.5	73.7	79.9	94.9	90.3	92.7
04	MASONRY	93.5	94.6	94.2	99.9	84.7	90.8	107.0	125.8	118.3	108.8	92.3	99.0	86.4	79.0	82.0	95.4	85.7	89.6
05	METALS	90.4	121.5	99.8	94.6	121.3	102.6	99.8	113.3	103.9	94.7	125.0	103.9	94.6	108.1	98.7	96.1	116.3	102.2
06	WOOD, PLASTICS & COMPOSITES	82.9	95.6	89.7	85.2	80.0	82.5	88.3	130.3	110.5	90.1	85.1	87.4	87.2	64.2	75.1	86.4	84.4	85.3
07	THERMAL & MOISTURE PROTECTION	102.1	99.0	100.9	101.6	90.8	97.3	102.0	125.8	111.6	101.0	102.2	101.5	100.9	93.7	98.0	100.3	104.7	102.0
08	OPENINGS	91.5	104.8	95.0	94.3	94.1	94.2	88.7	110.9	94.4	90.9	97.3	92.5	90.8	67.4	84.8	90.7	90.9	90.7
0920	Plaster & Gypsum Board	89.0	95.3	92.7	100.5	79.3	87.8	95.3	130.9	116.5	102.3	84.4	91.7	102.0	63.1	78.8	103.3	83.8	91.7
0950, 0980	Ceilings & Acoustic Treatment	88.9	95.3	92.8	88.2	79.3	82.8	99.7	130.9	118.5	90.7	84.4	86.9	90.7	63.1	74.1	89.8	83.8	86.1
0960	Flooring	85.6	101.4	89.8	80.6	61.1	75.4	81.1	133.6	95.1	82.5	91.4	84.9	81.4	59.9	75.6	83.3	88.0	84.5
0990	Wall Finishes & Painting/Coating	98.3	106.0	102.9	89.0	98.8	94.9	96.1	141.7	123.4	89.0	98.8	94.9	89.0	98.8	94.9	89.0	86.6	87.5
09	FINISHES	90.9	97.3	94.2	92.1	78.0	84.8	97.3	130.8	114.8	92.9	88.5	90.6	92.3	67.1	79.3	92.4	85.6	88.9
COVERS	DIVS. 10 - 14, 25, 28, 41, 43, 44	100.0	102.5	100.5	100.0	94.1	98.8	100.0	117.6	103.6	100.0	96.7	99.3	100.0	89.7	97.9	100.0	95.2	99.0
22, 23	PLUMBING & HVAC	95.7	97.3	96.3	96.1	90.3	93.7	95.9	128.5	109.2	96.1	89.8	93.5	96.1	85.3	91.7	100.3	91.5	96.7
26, 27, 3370	ELECTRICAL, COMMUNICATIONS & UTIL.	94.2	101.2	97.6	95.5	86.5	91.2	93.1	103.6	98.1	96.2	87.8	92.2	94.6	58.0	77.1	95.2	83.1	89.4
MF2004	WEIGHTED AVERAGE	93.3	101.1	96.7	96.2	91.7	94.3	98.0	118.7	106.9	95.1	95.6	95.3	93.2	80.0	87.5	95.9	92.6	94.5

		PUERTO RICO			RHODE ISLAND						SOUTH CAROLINA					
		SAN JUAN			NEWPORT			PROVIDENCE			AIKEN			BEAUFORT			CHARLESTON		
DIVISION		009			028			029			298			299			294		
		MAT.	INST.	TOTAL	MAT.	INST.	TOTAL	MAT.	INST.	TOTAL	MAT.	INST.	TOTAL	MAT.	INST.	TOTAL	MAT.	INST.	TOTAL
01543	CONTRACTOR EQUIPMENT	.0	88.4	88.4	.0	102.5	102.5	.0	102.5	102.5	.0	99.7	99.7	.0	99.7	99.7	.0	99.7	99.7
0241, 31 - 34	SITE & INFRASTRUCTURE, DEMOLITION	115.5	90.9	97.8	77.5	102.8	95.6	77.7	103.1	95.9	125.4	84.2	95.8	119.8	81.1	92.0	101.0	81.8	87.3
0310	Concrete Forming & Accessories	92.4	18.8	29.2	102.9	115.5	113.7	103.9	115.5	113.9	100.9	69.7	74.1	99.2	29.7	39.6	98.2	38.7	47.1
0310	Concrete Reinforcing	187.3	11.5	101.3	104.3	118.7	111.3	104.3	118.7	111.3	101.9	67.9	85.3	101.0	30.4	66.4	100.7	64.4	83.0
0330	Cast-in-Place Concrete	104.5	29.3	76.2	81.1	109.2	91.6	90.3	109.2	97.4	79.4	70.9	76.2	79.4	39.4	64.4	93.2	47.3	75.9
03	CONCRETE	113.6	21.7	70.2	99.7	113.5	106.2	104.4	113.5	108.7	111.1	70.9	92.1	108.0	35.1	73.6	102.0	48.3	76.6
04	MASONRY	86.6	16.1	44.4	104.2	123.0	115.4	107.0	123.0	116.6	79.7	61.5	68.8	95.5	25.8	53.8	96.7	36.2	60.6
05	METALS	109.8	29.6	85.6	95.8	111.4	100.5	95.8	111.4	100.5	89.8	85.8	88.6	89.8	62.8	81.6	91.5	79.4	87.8
06	WOOD, PLASTICS & COMPOSITES	90.0	19.1	52.6	101.0	114.9	108.3	103.5	114.9	109.5	100.6	71.8	85.3	98.5	30.2	62.4	97.0	38.0	65.8
07	THERMAL & MOISTURE PROTECTION	140.7	22.3	93.0	98.6	112.5	104.2	97.5	112.5	103.5	105.3	68.7	90.6	105.0	33.6	76.2	104.1	43.4	79.7
08	OPENINGS	152.6	15.9	117.2	99.5	115.8	103.8	100.2	115.8	104.2	92.2	67.7	85.9	92.2	29.1	75.9	96.3	42.0	82.2
0920	Plaster & Gypsum Board	114.7	16.8	56.5	100.3	114.5	108.7	100.3	114.5	108.7	112.4	70.8	87.6	115.8	28.2	63.6	117.1	36.2	68.9
0950, 0980	Ceilings & Acoustic Treatment	199.8	16.8	89.8	93.5	114.5	106.1	95.4	114.5	106.9	94.4	70.8	80.2	100.3	28.2	56.9	100.3	36.2	61.7
0960	Flooring	201.9	15.8	152.3	98.8	127.8	106.6	98.8	127.8	106.6	104.1	17.6	81.0	105.9	32.0	86.2	105.3	40.8	88.1
0990	Wall Finishes & Painting/Coating	198.0	14.8	88.4	93.3	118.7	108.5	93.3	118.7	108.5	112.5	72.3	88.5	112.5	27.2	61.5	112.5	39.9	69.1
09	FINISHES	192.8	18.1	102.6	98.0	118.7	108.7	98.3	118.7	108.8	103.6	60.3	81.2	105.7	29.9	66.6	103.9	38.7	70.3
COVERS	DIVS. 10 - 14, 25, 28, 41, 43, 44	100.0	18.6	83.3	100.0	105.0	101.0	100.0	105.0	101.0	100.0	70.4	93.9	100.0	36.4	86.9	100.0	62.3	92.2
22, 23	PLUMBING & HVAC	95.9	14.6	62.8	100.0	104.7	101.9	100.0	104.7	101.9	96.0	65.7	83.6	96.0	26.0	67.5	100.2	44.1	77.4
26, 27, 3370	ELECTRICAL, COMMUNICATIONS & UTIL.	124.0	14.7	71.6	97.5	101.0	99.1	97.4	101.0	99.1	96.1	69.4	83.3	99.1	26.7	64.4	97.9	60.7	80.0
MF2004	WEIGHTED AVERAGE	119.0	24.6	78.3	98.4	110.3	103.5	99.1	110.3	104.0	97.7	69.6	85.6	98.5	36.6	71.8	98.7	52.3	78.7

		SOUTH CAROLINA															SOUTH DAKOTA		
		COLUMBIA			FLORENCE			GREENVILLE			ROCK HILL			SPARTANBURG			ABERDEEN		
DIVISION		290 - 292			295			296			297			293			574		
		MAT.	INST.	TOTAL	MAT.	INST.	TOTAL	MAT.	INST.	TOTAL	MAT.	INST.	TOTAL	MAT.	INST.	TOTAL	MAT.	INST.	TOTAL
01543	CONTRACTOR EQUIPMENT	.0	99.7	99.7	.0	99.7	99.7	.0	99.7	99.7	.0	99.7	99.7	.0	99.7	99.7	.0	98.7	98.7
0241, 31 - 34	SITE & INFRASTRUCTURE, DEMOLITION	100.0	81.8	87.0	112.6	81.8	90.5	107.0	81.4	88.7	104.4	81.1	87.7	106.8	81.4	88.6	88.9	94.1	92.6
0310	Concrete Forming & Accessories	102.7	40.7	49.6	84.4	40.8	47.0	98.0	40.7	48.8	96.3	28.7	38.2	101.6	40.7	49.3	95.3	37.6	45.8
0310	Concrete Reinforcing	100.7	64.4	82.9	100.4	64.3	82.8	100.3	53.1	77.2	101.2	18.8	60.9	100.3	53.1	77.2	102.3	45.5	74.5
0330	Cast-in-Place Concrete	87.0	46.9	71.9	79.4	47.1	67.3	79.4	47.1	67.2	79.4	34.9	62.6	79.4	47.1	67.2	98.8	48.1	79.7
03	CONCRETE	99.1	49.1	75.5	102.0	49.1	77.0	100.4	47.0	75.2	98.4	30.9	66.6	100.7	47.0	75.3	97.9	44.1	72.5
04	MASONRY	91.5	37.2	59.0	79.7	36.2	53.7	77.6	36.2	52.8	103.5	23.7	55.7	79.7	36.2	53.7	104.8	58.8	77.3
05	METALS	91.5	79.1	87.7	90.5	79.1	87.0	90.5	74.7	85.7	89.8	58.9	80.4	90.5	74.7	85.7	100.7	67.6	90.7
06	WOOD, PLASTICS & COMPOSITES	103.0	41.4	70.4	80.5	41.4	59.8	96.9	41.4	67.5	94.9	29.2	60.2	101.4	41.4	69.6	100.6	35.9	66.4
07	THERMAL & MOISTURE PROTECTION	103.7	43.3	79.4	104.4	43.7	80.0	104.4	43.7	80.0	104.2	29.2	74.0	104.4	43.7	80.0	98.6	50.1	79.1
08	OPENINGS	96.3	43.8	82.7	92.3	43.8	79.7	92.2	41.2	79.0	92.2	26.4	75.2	92.2	41.2	79.0	94.4	38.3	79.9
0920	Plaster & Gypsum Board	115.0	39.6	70.1	105.8	39.6	66.4	111.1	39.6	68.5	110.4	27.1	60.8	112.7	39.6	69.2	100.5	34.5	61.2
0950, 0980	Ceilings & Acoustic Treatment	100.3	39.6	63.8	95.2	39.6	61.8	94.4	39.6	61.4	94.4	27.1	53.9	94.4	39.6	61.4	111.9	34.5	65.4
0960	Flooring	103.2	40.8	86.6	95.4	40.8	80.8	102.6	41.6	86.4	101.6	31.6	82.9	104.3	41.6	87.6	110.9	54.1	95.7
0990	Wall Finishes & Painting/Coating	111.4	39.9	68.6	112.5	39.9	69.1	112.5	39.9	69.1	112.5	27.2	61.5	112.5	39.9	69.1	97.6	33.3	59.2
09	FINISHES	102.9	40.6	70.8	99.0	40.6	68.9	101.5	40.8	70.1	100.8	29.2	63.9	102.2	40.8	70.5	106.5	39.8	72.1
COVERS	DIVS. 10 - 14, 25, 28, 41, 43, 44	100.0	62.6	92.3	100.0	62.5	92.3	100.0	62.6	92.3	100.0	59.7	91.7	100.0	62.6	92.3	100.0	51.0	89.9
22, 23	PLUMBING & HVAC	100.1	37.2	74.5	100.2	37.2	74.6	100.2	37.2	74.6	96.0	18.6	64.5	100.2	37.2	74.6	100.2	36.8	74.4
26, 27, 3370	ELECTRICAL, COMMUNICATIONS & UTIL.	98.4	35.3	68.2	96.0	20.2	59.7	97.9	32.1	66.3	97.9	19.8	60.4	97.9	32.1	66.3	96.2	56.8	77.3
MF2004	WEIGHTED AVERAGE	98.0	47.8	76.4	96.8	45.6	74.7	96.9	46.5	75.2	96.8	33.3	69.4	97.2	46.5	75.3	99.5	51.8	78.9

City Cost Indexes

SOUTH DAKOTA

DIVISION		MITCHELL 573			MOBRIDGE 576			PIERRE 575			RAPID CITY 577			SIOUX FALLS 570 - 571			WATERTOWN 572		
		MAT.	INST.	TOTAL	MAT.	INST.	TOTAL	MAT.	INST.	TOTAL	MAT.	INST.	TOTAL	MAT.	INST.	TOTAL	MAT.	INST.	TOTAL
01543	CONTRACTOR EQUIPMENT	.0	98.7	98.7	.0	98.7	98.7	.0	98.7	98.7	.0	98.7	98.7	.0	99.9	99.9	.0	98.7	98.7
0241, 31 - 34	SITE & INFRASTRUCTURE, DEMOLITION	86.0	94.1	91.8	86.0	94.1	91.8	87.0	94.1	92.1	87.2	93.9	92.0	87.5	96.0	93.6	85.9	94.1	91.8
0310	Concrete Forming & Accessories	94.4	38.8	46.7	84.8	37.5	44.2	93.8	39.1	46.9	102.3	35.0	44.5	94.5	40.4	48.0	81.1	37.5	43.7
0310	Concrete Reinforcing	101.7	45.4	74.2	104.3	45.5	75.6	101.9	58.3	80.6	95.6	58.3	77.3	95.6	58.4	77.4	99.0	45.6	72.9
0330	Cast-in-Place Concrete	96.0	46.6	77.4	96.0	48.1	78.0	96.0	44.1	76.5	95.3	41.8	75.2	93.5	47.0	76.0	90.0	48.1	78.0
03	CONCRETE	95.7	44.1	71.4	95.5	44.0	71.2	95.7	45.9	72.2	94.9	43.3	70.5	92.4	47.4	71.2	94.4	44.1	70.6
04	MASONRY	93.3	57.9	72.1	101.6	58.8	76.0	101.6	57.9	75.4	101.7	51.4	71.6	98.7	57.9	74.3	126.4	61.4	87.5
05	METALS	99.8	67.3	90.0	99.9	67.5	90.1	100.7	73.3	92.4	102.4	73.6	93.7	102.6	74.1	94.0	99.8	67.7	90.1
06	WOOD, PLASTICS & COMPOSITES	99.5	37.9	66.9	88.1	35.9	60.5	98.7	37.9	66.6	104.8	33.9	67.3	99.4	38.5	67.2	83.8	35.9	58.5
07	THERMAL & MOISTURE PROTECTION	98.4	48.0	78.1	98.4	50.1	79.0	98.7	48.2	78.4	99.0	50.1	79.3	100.0	53.7	81.4	98.2	50.8	79.1
08	OPENINGS	92.8	38.9	78.9	95.8	37.5	80.7	97.6	43.1	83.5	98.6	40.9	83.7	98.9	43.4	84.5	92.8	37.5	78.5
0920	Plaster & Gypsum Board	99.5	36.6	62.0	95.1	34.5	59.0	99.2	36.6	61.9	100.5	32.5	60.0	100.5	37.2	62.8	93.1	34.5	58.2
0950, 0980	Ceilings & Acoustic Treatment	110.2	36.6	65.9	111.9	34.5	65.4	110.2	36.6	65.9	116.1	32.5	65.8	116.1	37.2	68.7	110.2	34.5	64.7
0960	Flooring	110.4	54.1	95.4	104.7	54.1	91.2	110.1	38.5	91.0	110.1	73.1	100.2	110.1	76.8	101.2	102.9	54.1	89.9
0990	Wall Finishes & Painting/Coating	97.6	37.7	61.8	97.6	33.3	59.2	97.6	38.0	62.0	97.6	38.0	62.0	97.6	38.0	62.0	97.6	33.3	59.2
09	FINISHES	105.6	41.5	72.5	103.6	39.8	70.7	105.5	38.3	70.8	107.2	42.0	73.6	107.2	46.6	75.9	102.3	39.8	70.1
COVERS	DIVS. 10 - 14, 25, 28, 41, 43, 44	100.0	44.5	88.6	100.0	50.9	89.9	100.0	60.7	91.9	100.0	58.6	91.5	100.0	60.9	92.0	100.0	50.9	89.9
22, 23	PLUMBING & HVAC	96.0	34.0	70.7	96.0	37.9	72.3	100.2	35.3	73.8	100.2	33.0	72.8	100.2	34.4	73.4	96.0	37.9	72.3
26, 27, 3370	ELECTRICAL, COMMUNICATIONS & UTIL.	94.8	42.5	69.8	96.2	37.9	68.2	92.8	52.4	73.4	93.3	52.4	73.7	92.8	73.8	83.7	94.1	37.9	67.1
MF2004	WEIGHTED AVERAGE	97.0	49.1	76.3	97.6	49.3	76.8	98.8	51.8	78.5	99.3	50.6	78.3	98.9	56.3	80.5	98.1	49.6	77.2

TENNESSEE

DIVISION		CHATTANOOGA 373 - 374			COLUMBIA 384			COOKEVILLE 385			JACKSON 383			JOHNSON CITY 376			KNOXVILLE 377 - 379		
		MAT.	INST.	TOTAL	MAT.	INST.	TOTAL	MAT.	INST.	TOTAL	MAT.	INST.	TOTAL	MAT.	INST.	TOTAL	MAT.	INST.	TOTAL
01543	CONTRACTOR EQUIPMENT	.0	105.3	105.3	.0	97.9	97.9	.0	97.9	97.9	.0	105.2	105.2	.0	98.0	98.0	.0	98.0	98.0
0241, 31 - 34	SITE & INFRASTRUCTURE, DEMOLITION	99.5	97.5	98.1	89.8	85.0	86.4	96.2	86.3	89.1	98.1	95.9	96.5	108.8	86.2	92.6	86.5	86.2	86.3
0310	Concrete Forming & Accessories	96.1	44.8	52.1	81.3	49.9	54.2	81.5	42.7	48.2	88.7	36.0	43.5	82.1	45.5	50.7	95.1	45.5	52.5
0310	Concrete Reinforcing	90.1	62.3	76.5	89.9	62.9	76.7	89.9	62.9	76.7	89.9	40.7	65.9	90.7	59.5	75.4	90.1	59.5	75.1
0330	Cast-in-Place Concrete	98.4	46.8	79.0	88.9	50.4	74.4	100.8	49.2	81.4	98.4	40.3	76.5	79.1	51.2	68.6	92.4	46.9	75.3
03	CONCRETE	92.2	50.7	72.6	92.5	54.2	74.4	102.0	50.8	77.8	94.0	40.4	68.7	95.7	51.9	75.1	89.3	50.5	71.0
04	MASONRY	98.7	41.3	64.4	111.2	51.6	75.5	106.5	52.4	74.1	111.4	32.1	64.0	111.2	41.1	69.2	75.8	41.1	55.0
05	METALS	93.8	84.6	91.1	88.2	85.1	87.2	88.2	88.2	88.2	90.2	72.1	84.7	91.2	83.3	88.8	94.4	83.6	91.1
06	WOOD, PLASTICS & COMPOSITES	99.6	45.2	70.9	70.1	52.3	60.7	70.3	42.0	55.4	85.7	36.0	59.4	74.0	46.3	59.3	89.2	46.3	66.5
07	THERMAL & MOISTURE PROTECTION	98.8	57.5	82.2	93.2	59.0	79.4	93.6	53.1	77.3	95.2	38.2	72.3	92.8	58.1	78.8	90.9	57.6	77.5
08	OPENINGS	100.7	51.4	88.0	92.1	52.5	81.9	92.2	48.8	80.9	101.4	37.8	85.0	96.5	51.9	85.0	93.2	51.9	82.5
0920	Plaster & Gypsum Board	85.0	44.3	60.8	93.6	51.6	68.6	93.6	41.0	62.3	94.1	34.4	58.8	96.0	45.4	65.9	102.4	45.4	68.5
0950, 0980	Ceilings & Acoustic Treatment	98.5	44.3	65.9	87.4	51.6	65.9	87.4	41.0	59.5	95.6	34.8	59.0	93.2	45.4	64.4	95.7	45.4	65.4
0960	Flooring	104.0	47.6	89.0	92.0	24.7	74.1	92.2	23.0	73.7	93.5	23.0	74.7	97.3	48.9	84.4	103.1	48.9	88.7
0990	Wall Finishes & Painting/Coating	110.2	44.4	70.8	96.1	41.2	63.3	96.1	45.1	65.6	97.2	30.0	57.0	107.7	52.0	74.4	107.7	52.0	74.4
09	FINISHES	98.4	44.8	70.8	95.6	44.4	69.2	96.0	37.9	66.0	96.4	33.0	63.7	99.4	46.4	72.1	95.2	46.4	70.0
COVERS	DIVS. 10 - 14, 25, 28, 41, 43, 44	100.0	49.3	89.6	100.0	55.8	90.9	100.0	54.8	90.7	100.0	53.5	90.4	100.0	56.5	91.1	100.0	56.6	91.1
22, 23	PLUMBING & HVAC	100.0	43.0	76.8	97.6	50.3	78.3	97.6	65.2	84.4	100.0	52.9	80.8	99.7	56.6	82.1	99.7	56.6	82.1
26, 27, 3370	ELECTRICAL, COMMUNICATIONS & UTIL.	104.3	73.9	89.7	93.7	66.3	80.6	95.2	66.4	81.4	102.2	48.9	76.7	92.8	55.7	75.0	97.9	62.2	80.8
MF2004	WEIGHTED AVERAGE	98.4	57.9	80.9	94.7	59.2	79.4	96.0	61.0	80.9	98.1	50.2	77.4	97.2	57.7	80.1	94.5	58.4	78.9

TENNESSEE / TEXAS

DIVISION		MCKENZIE 382			MEMPHIS 375,380 - 381			NASHVILLE 370 - 372			ABILENE 795 - 796			AMARILLO 790 - 791			AUSTIN 786 - 787		
		MAT.	INST.	TOTAL	MAT.	INST.	TOTAL	MAT.	INST.	TOTAL	MAT.	INST.	TOTAL	MAT.	INST.	TOTAL	MAT.	INST.	TOTAL
01543	CONTRACTOR EQUIPMENT	.0	97.9	97.9	.0	102.9	102.9	.0	106.4	106.4	.0	87.2	87.2	.0	87.2	87.2	.0	86.8	86.8
0241, 31 - 34	SITE & INFRASTRUCTURE, DEMOLITION	95.6	86.0	88.7	89.6	92.8	91.9	93.7	100.1	98.3	102.8	85.1	90.1	103.4	85.9	90.9	93.0	85.5	87.6
0310	Concrete Forming & Accessories	90.3	35.8	43.5	96.7	65.0	69.5	95.1	63.8	68.2	98.2	41.4	49.5	100.4	52.4	59.2	95.4	56.2	61.7
0310	Concrete Reinforcing	90.1	33.7	62.5	95.1	67.9	81.8	98.4	66.9	83.0	94.1	50.3	72.7	94.1	51.6	73.3	88.0	49.6	69.3
0330	Cast-in-Place Concrete	98.6	43.3	77.8	86.2	69.2	79.8	92.7	68.2	83.5	95.0	41.5	74.9	99.8	47.4	80.1	89.7	48.4	74.1
03	CONCRETE	100.6	40.2	72.1	85.7	68.5	77.6	92.6	67.3	80.7	89.5	44.2	68.1	92.1	51.3	72.9	79.8	52.9	67.1
04	MASONRY	109.5	51.3	74.7	89.2	73.1	79.6	83.5	64.4	72.1	98.0	53.5	71.4	99.8	49.1	69.5	88.8	52.4	67.0
05	METALS	88.2	74.7	84.1	94.7	93.0	94.2	95.8	90.5	94.2	98.3	65.3	88.3	98.3	65.8	88.5	97.4	63.7	87.2
06	WOOD, PLASTICS & COMPOSITES	80.7	33.8	55.9	94.6	65.8	79.4	96.2	64.2	79.3	99.9	40.8	68.7	101.3	55.5	77.1	96.5	60.1	77.2
07	THERMAL & MOISTURE PROTECTION	93.6	50.4	76.2	95.2	71.6	85.7	94.1	65.2	82.5	99.9	48.8	79.4	102.4	47.7	80.4	93.0	53.8	77.2
08	OPENINGS	92.2	34.1	77.1	101.3	67.4	92.5	96.9	65.6	88.8	93.0	43.5	80.2	93.0	49.6	81.7	96.8	57.3	86.5
0920	Plaster & Gypsum Board	97.5	32.5	58.8	90.3	65.7	78.9	101.1	63.7	78.9	88.6	39.8	59.6	88.6	54.9	68.5	96.6	59.6	74.6
0950, 0980	Ceilings & Acoustic Treatment	87.4	32.5	54.4	102.8	65.3	80.3	96.9	63.7	77.0	95.5	39.8	62.0	95.5	54.9	71.1	91.6	59.6	72.4
0960	Flooring	95.9	47.4	82.9	96.1	43.4	82.1	108.3	74.5	99.3	113.5	68.9	101.6	113.3	61.5	99.5	97.1	47.2	83.8
0990	Wall Finishes & Painting/Coating	96.1	52.5	70.0	101.4	58.8	75.9	111.0	66.0	84.1	97.1	52.9	70.7	97.1	36.7	60.9	93.9	41.2	62.4
09	FINISHES	97.7	38.7	67.2	94.8	59.7	76.7	105.2	65.9	84.9	98.4	47.5	72.1	98.5	52.6	74.8	93.4	52.9	72.5
COVERS	DIVS. 10 - 14, 25, 28, 41, 43, 44	100.0	32.7	86.1	100.0	80.7	96.0	100.0	80.2	95.9	100.0	74.0	94.7	100.0	71.6	94.2	100.0	72.0	94.2
22, 23	PLUMBING & HVAC	97.6	69.3	86.1	99.9	72.0	88.5	99.8	81.3	92.3	100.1	43.1	76.9	100.1	51.3	80.2	99.9	56.8	82.3
26, 27, 3370	ELECTRICAL, COMMUNICATIONS & UTIL.	95.0	64.7	80.3	99.1	73.4	86.8	102.4	66.5	85.2	98.2	46.7	73.5	98.7	59.9	80.1	101.3	66.8	84.8
MF2004	WEIGHTED AVERAGE	96.2	57.4	79.4	96.0	74.0	86.5	97.7	74.8	87.8	97.5	52.0	77.9	98.1	57.2	80.5	95.5	60.3	80.3

City Cost Indexes

		TEXAS																	
DIVISION		BEAUMONT 776 - 777			BROWNWOOD 768			BRYAN 778			CHILDRESS 792			CORPUS CHRISTI 783 - 784			DALLAS 752 - 753		
		MAT.	INST.	TOTAL	MAT.	INST.	TOTAL	MAT.	INST.	TOTAL	MAT.	INST.	TOTAL	MAT.	INST.	TOTAL	MAT.	INST.	TOTAL
01543	CONTRACTOR EQUIPMENT	.0	87.8	87.8	.0	87.2	87.2	.0	87.8	87.8	.0	87.2	87.2	.0	95.4	95.4	.0	98.2	98.2
0241, 31 - 34	SITE & INFRASTRUCTURE, DEMOLITION	98.7	85.4	89.2	110.3	83.8	91.3	87.9	86.9	87.2	116.3	84.5	93.5	131.7	80.7	95.2	127.4	86.6	98.1
0310	Concrete Forming & Accessories	104.9	51.3	58.9	96.3	27.9	37.6	81.9	42.1	47.7	96.5	50.9	57.3	98.9	37.6	46.3	94.0	57.9	63.0
0310	Concrete Reinforcing	94.3	42.2	68.8	94.5	25.3	60.6	96.6	49.6	73.7	94.3	39.4	67.5	87.3	48.0	68.1	98.6	54.8	77.2
0330	Cast-in-Place Concrete	93.0	53.3	78.0	102.5	36.9	77.8	74.5	66.6	71.5	97.5	45.7	78.0	107.1	44.9	83.7	95.8	54.9	80.4
03	CONCRETE	92.0	51.1	72.7	96.2	31.7	65.8	77.1	53.0	65.8	96.4	47.6	73.4	90.2	44.0	68.4	87.9	57.8	73.7
04	MASONRY	100.9	57.7	75.1	125.3	47.6	78.8	138.5	61.3	92.3	102.3	47.5	69.5	81.3	50.8	63.1	98.1	60.1	75.4
05	METALS	104.0	62.9	91.6	96.3	49.8	82.2	104.3	67.3	93.1	96.0	58.0	84.5	96.9	76.4	90.7	93.6	81.0	89.8
06	WOOD, PLASTICS & COMPOSITES	113.5	51.8	80.9	98.2	26.8	60.5	78.8	34.5	55.4	99.2	55.5	76.1	109.0	37.0	70.9	96.5	58.6	76.5
07	THERMAL & MOISTURE PROTECTION	105.7	57.5	86.3	100.0	40.1	75.9	99.6	58.5	83.1	100.5	44.8	78.1	95.7	46.6	75.9	96.9	61.6	82.7
08	OPENINGS	97.7	47.5	84.7	89.0	27.5	73.1	99.5	44.7	85.3	90.0	47.1	78.9	103.8	38.6	86.9	103.9	54.2	91.0
0920	Plaster & Gypsum Board	98.5	51.2	70.3	87.7	25.5	50.7	89.2	33.4	56.0	87.7	54.9	68.2	96.6	35.8	60.4	91.6	57.9	71.5
0950, 0980	Ceilings & Acoustic Treatment	104.4	51.2	72.4	90.4	25.5	51.3	97.4	33.4	58.9	90.4	54.9	69.0	91.6	35.8	58.0	96.8	57.9	73.4
0960	Flooring	112.9	71.7	101.9	112.2	39.2	92.7	84.0	63.1	78.4	111.1	41.0	92.4	109.7	46.6	92.9	101.6	57.2	89.7
0990	Wall Finishes & Painting/Coating	95.9	49.4	68.1	97.1	29.5	56.6	90.8	61.7	73.4	97.1	39.0	62.3	106.7	44.8	69.7	102.8	52.4	72.6
09	FINISHES	97.5	54.7	75.4	97.2	29.9	62.4	84.9	46.3	65.0	97.5	48.3	72.1	99.7	39.3	68.5	98.9	57.1	77.3
COVERS	DIVS. 10 - 14, 25, 28, 41, 43, 44	100.0	77.8	95.4	100.0	42.3	88.1	100.0	72.9	94.4	100.0	71.0	94.0	100.0	74.8	94.8	100.0	79.1	95.7
22, 23	PLUMBING & HVAC	100.1	63.6	85.2	95.9	28.3	68.4	95.9	70.9	85.7	95.9	50.1	77.2	99.9	44.0	77.1	99.9	66.4	86.2
26, 27, 3370	ELECTRICAL, COMMUNICATIONS & UTIL.	95.6	69.4	83.1	94.6	27.5	62.4	94.8	69.4	82.6	98.1	37.5	69.0	99.4	52.0	76.6	96.8	62.7	80.4
MF2004	WEIGHTED AVERAGE	99.1	62.3	83.2	97.7	38.3	72.0	96.6	63.4	82.3	97.2	51.6	77.5	98.4	52.0	78.3	98.0	65.6	84.0

		TEXAS																	
DIVISION		DEL RIO 788			DENTON 762			EASTLAND 764			EL PASO 798 - 799,885			FORT WORTH 760 - 761			GALVESTON 775		
		MAT.	INST.	TOTAL	MAT.	INST.	TOTAL	MAT.	INST.	TOTAL	MAT.	INST.	TOTAL	MAT.	INST.	TOTAL	MAT.	INST.	TOTAL
01543	CONTRACTOR EQUIPMENT	.0	86.8	86.8	.0	94.1	94.1	.0	87.2	87.2	.0	87.2	87.2	.0	87.2	87.2	.0	98.2	98.2
0241, 31 - 34	SITE & INFRASTRUCTURE, DEMOLITION	115.7	83.4	92.5	110.8	77.4	86.8	113.9	84.0	92.4	102.1	84.7	89.6	102.8	85.5	90.4	129.4	84.5	97.2
0310	Concrete Forming & Accessories	93.1	24.6	34.2	104.0	36.9	46.4	97.4	32.8	42.0	97.7	45.8	53.1	95.6	57.2	62.7	91.1	66.4	69.9
0310	Concrete Reinforcing	88.0	17.8	53.7	95.9	50.2	73.6	94.7	40.9	68.4	94.1	46.7	71.0	94.1	54.7	74.9	96.1	62.3	79.6
0330	Cast-in-Place Concrete	115.3	32.0	84.0	79.2	48.5	67.7	108.3	37.6	81.7	88.3	36.5	68.8	93.4	50.6	77.3	98.9	67.8	87.2
03	CONCRETE	102.9	27.1	67.1	77.1	45.4	62.1	100.6	36.9	70.5	86.1	43.6	66.0	88.6	55.1	72.8	93.6	67.6	81.3
04	MASONRY	94.9	43.0	63.8	133.4	50.7	83.9	96.5	46.7	66.7	96.1	49.2	68.1	93.4	60.1	73.5	96.5	61.3	75.5
05	METALS	95.9	45.3	80.6	95.9	76.9	90.2	96.1	52.7	82.9	98.1	61.0	86.9	95.1	67.4	86.7	106.1	88.1	100.7
06	WOOD, PLASTICS & COMPOSITES	92.2	24.9	56.6	110.2	37.1	71.6	104.7	33.6	67.1	99.9	49.0	73.0	102.2	58.5	79.1	95.6	66.8	80.4
07	THERMAL & MOISTURE PROTECTION	93.0	33.5	69.1	99.5	46.8	78.3	100.4	40.3	76.2	99.4	52.3	80.5	100.1	53.5	81.4	99.1	65.5	85.6
08	OPENINGS	97.5	21.1	77.8	104.5	41.5	88.2	77.7	30.8	65.5	93.0	44.3	80.4	87.3	54.1	78.7	104.9	65.5	94.7
0920	Plaster & Gypsum Board	93.2	23.5	51.7	91.6	35.9	58.4	87.7	32.5	54.8	88.6	48.2	64.6	88.6	57.9	70.4	95.1	66.4	78.0
0950, 0980	Ceilings & Acoustic Treatment	88.3	23.5	49.3	96.3	35.9	60.0	90.4	32.5	55.5	95.5	48.2	67.1	95.5	57.9	72.9	100.8	66.4	80.1
0960	Flooring	93.9	18.8	73.9	104.4	45.7	88.7	145.6	28.2	114.3	113.5	67.0	101.1	146.9	45.7	119.9	98.9	63.1	89.3
0990	Wall Finishes & Painting/Coating	93.9	25.9	53.3	108.1	29.8	61.3	98.3	30.8	57.9	97.1	35.2	60.1	98.3	52.3	70.8	103.8	56.7	75.6
09	FINISHES	92.5	24.1	57.2	94.8	37.3	65.2	107.4	32.0	68.5	98.4	49.1	73.0	108.5	54.6	80.7	95.3	64.7	79.5
COVERS	DIVS. 10 - 14, 25, 28, 41, 43, 44	100.0	39.8	87.6	100.0	44.1	88.5	100.0	42.8	88.2	100.0	69.8	93.8	100.0	78.8	95.6	100.0	83.1	96.5
22, 23	PLUMBING & HVAC	95.5	18.5	64.1	95.9	45.4	75.3	95.9	28.6	68.5	100.1	32.8	72.7	100.1	57.0	82.5	95.8	70.1	85.3
26, 27, 3370	ELECTRICAL, COMMUNICATIONS & UTIL.	100.6	19.5	61.7	97.7	44.5	72.2	94.4	34.2	65.5	97.2	53.8	76.4	97.0	62.6	80.5	96.0	65.7	81.4
MF2004	WEIGHTED AVERAGE	97.7	32.3	69.5	97.5	50.1	77.1	96.7	40.7	72.5	96.9	50.3	76.8	97.0	61.4	81.6	99.3	70.4	86.9

		TEXAS																	
DIVISION		GIDDINGS 789			GREENVILLE 754			HOUSTON 770 - 772			HUNTSVILLE 773			LAREDO 780			LONGVIEW 756		
		MAT.	INST.	TOTAL	MAT.	INST.	TOTAL	MAT.	INST.	TOTAL	MAT.	INST.	TOTAL	MAT.	INST.	TOTAL	MAT.	INST.	TOTAL
01543	CONTRACTOR EQUIPMENT	.0	86.8	86.8	.0	95.4	95.4	.0	98.0	98.0	.0	87.8	87.8	.0	86.8	86.8	.0	89.2	89.2
0241, 31 - 34	SITE & INFRASTRUCTURE, DEMOLITION	100.6	83.8	88.6	118.4	81.4	91.9	126.7	84.2	96.2	108.8	84.2	91.2	93.7	85.3	87.7	111.7	89.6	95.8
0310	Concrete Forming & Accessories	91.3	32.9	41.1	85.1	27.7	35.8	93.4	66.5	70.3	90.0	34.1	42.0	93.1	37.9	45.7	80.9	28.6	36.0
0310	Concrete Reinforcing	88.7	41.7	65.7	99.0	25.5	63.0	95.9	62.4	79.5	96.9	27.8	63.1	88.0	48.1	68.5	98.1	23.2	61.5
0330	Cast-in-Place Concrete	97.6	37.7	75.1	87.2	36.7	68.2	95.9	67.9	85.3	102.4	44.4	80.6	82.4	60.1	74.0	101.5	34.0	76.1
03	CONCRETE	85.6	37.1	62.7	81.5	32.4	58.3	91.5	67.7	80.2	98.9	37.6	70.0	79.2	48.4	64.7	93.8	30.6	64.0
04	MASONRY	101.5	48.6	69.8	151.2	49.5	90.3	96.3	63.1	76.5	138.0	32.6	74.9	88.9	51.5	66.5	147.3	44.6	85.8
05	METALS	95.4	52.9	82.6	91.2	63.1	82.7	108.7	88.3	102.5	104.2	52.0	88.4	98.4	62.5	87.5	84.9	48.9	74.0
06	WOOD, PLASTICS & COMPOSITES	91.4	33.6	60.8	86.9	26.9	55.2	98.4	66.8	81.6	89.1	35.8	60.9	92.2	36.9	62.9	79.8	29.1	53.0
07	THERMAL & MOISTURE PROTECTION	93.5	41.0	72.4	96.6	41.4	74.4	99.0	65.8	85.7	100.7	38.1	75.5	92.0	50.8	75.4	96.5	35.5	72.0
08	OPENINGS	96.8	30.8	79.7	102.6	27.6	83.2	107.2	65.5	96.4	99.4	30.6	81.6	97.4	39.2	82.3	92.5	26.5	75.4
0920	Plaster & Gypsum Board	92.1	32.5	56.6	88.2	25.5	50.9	97.0	66.4	78.8	93.4	34.8	58.5	94.0	35.8	59.4	85.0	27.9	51.0
0950, 0980	Ceilings & Acoustic Treatment	87.4	32.5	54.4	93.4	25.5	52.6	105.0	66.4	81.8	97.4	34.8	59.7	91.6	35.8	58.0	89.2	27.9	52.3
0960	Flooring	93.7	28.2	76.2	96.5	45.7	83.0	100.2	63.1	90.3	87.8	28.2	71.9	93.8	46.6	81.2	102.9	40.3	86.2
0990	Wall Finishes & Painting/Coating	93.9	30.8	56.1	102.8	25.4	56.5	103.8	61.7	78.6	90.8	31.6	55.4	93.9	53.8	69.9	94.8	33.8	58.3
09	FINISHES	91.2	31.9	60.6	95.5	30.5	62.0	101.7	65.2	82.9	87.8	33.1	59.6	92.1	40.2	65.4	99.4	31.2	64.2
COVERS	DIVS. 10 - 14, 25, 28, 41, 43, 44	100.0	41.2	87.9	100.0	42.1	88.1	100.0	83.2	96.5	100.0	41.2	87.9	100.0	68.1	93.4	100.0	23.2	84.2
22, 23	PLUMBING & HVAC	95.6	28.6	68.3	95.6	27.8	68.0	100.1	71.1	88.3	95.9	27.0	67.8	99.8	39.6	75.3	95.5	27.5	67.8
26, 27, 3370	ELECTRICAL, COMMUNICATIONS & UTIL.	97.7	34.2	67.3	93.5	27.5	61.9	98.0	69.5	84.3	94.8	33.5	65.4	100.8	65.0	83.6	92.8	53.5	73.9
MF2004	WEIGHTED AVERAGE	95.1	40.9	71.7	97.5	39.5	72.5	101.4	71.4	88.5	100.0	38.9	73.6	95.4	52.8	77.0	96.8	41.2	72.8

City Cost Indexes

		TEXAS																	
DIVISION		LUBBOCK			LUFKIN			MCALLEN			MCKINNEY			MIDLAND			ODESSA		
		793 - 794			759			785			750			797			797		
		MAT.	INST.	TOTAL	MAT.	INST.	TOTAL	MAT.	INST.	TOTAL	MAT.	INST.	TOTAL	MAT.	INST.	TOTAL	MAT.	INST.	TOTAL
01543	CONTRACTOR EQUIPMENT	.0	97.2	97.2	.0	89.2	89.2	.0	95.4	95.4	.0	95.4	95.4	.0	97.2	97.2	.0	87.2	87.2
0241, 31 - 34	SITE & INFRASTRUCTURE, DEMOLITION	133.8	83.2	97.5	105.0	89.7	94.0	136.8	80.7	96.6	113.3	82.2	91.0	138.1	83.2	98.7	103.0	85.4	90.4
0310	Concrete Forming & Accessories	97.8	40.7	48.7	84.3	40.1	46.4	97.8	35.4	44.3	84.0	42.3	48.2	102.1	38.9	47.8	98.1	38.7	47.1
0310	Concrete Reinforcing	95.4	50.1	73.2	99.7	31.0	66.1	87.5	48.0	68.2	99.0	50.2	75.2	96.5	50.0	73.8	94.1	50.0	72.5
0330	Cast-in-Place Concrete	95.2	47.5	77.2	90.8	39.3	71.4	116.4	43.3	88.9	81.8	49.6	69.7	101.1	44.7	79.9	95.0	43.5	75.6
03	CONCRETE	88.8	46.7	68.9	87.1	39.2	64.5	95.9	42.5	70.7	77.5	48.0	63.6	93.1	45.0	70.4	89.5	43.6	67.9
04	MASONRY	97.4	47.6	67.6	113.7	38.1	68.5	94.3	50.8	68.3	162.4	53.9	97.5	114.7	48.0	74.8	98.0	47.9	68.0
05	METALS	101.9	79.0	95.0	91.4	56.3	80.8	95.8	76.0	89.9	91.2	75.8	86.5	100.3	78.6	93.7	97.7	64.2	87.6
06	WOOD, PLASTICS & COMPOSITES	100.7	40.9	69.1	88.2	41.7	63.6	107.3	34.5	68.8	85.4	42.5	62.7	106.4	38.9	70.8	99.9	38.8	67.6
07	THERMAL & MOISTURE PROTECTION	91.4	49.5	74.6	96.3	40.6	73.9	95.7	44.8	75.2	96.5	50.2	77.9	91.7	45.8	73.2	99.9	44.9	77.8
08	OPENINGS	103.9	41.8	87.8	74.5	38.3	65.1	102.2	37.7	85.5	102.6	44.3	87.5	102.9	40.4	86.7	93.0	40.3	79.3
0920	Plaster & Gypsum Board	89.2	39.8	59.8	85.2	40.8	58.8	96.6	33.2	58.8	87.3	41.5	60.0	91.0	37.8	59.3	88.6	37.8	58.4
0950, 0980	Ceilings & Acoustic Treatment	98.0	39.8	63.0	79.9	40.8	56.4	91.6	33.2	56.5	93.4	41.5	62.2	93.7	37.8	60.1	95.5	37.8	60.8
0960	Flooring	105.1	38.8	87.4	138.0	37.6	111.2	109.1	67.3	98.0	96.1	45.7	82.7	106.8	38.3	88.6	113.5	38.3	93.4
0990	Wall Finishes & Painting/Coating	108.7	32.9	63.4	94.8	38.1	60.9	106.7	29.1	60.3	102.8	47.5	69.7	108.7	32.5	63.1	97.1	32.9	58.7
09	FINISHES	100.8	38.6	68.7	107.2	39.4	72.2	99.9	40.3	69.1	95.0	43.2	68.3	100.9	37.4	68.2	98.4	37.4	66.9
COVERS	DIVS. 10 - 14, 25, 28, 41, 43, 44	100.0	73.8	94.6	100.0	70.6	94.0	100.0	68.0	93.4	100.0	45.7	88.8	100.0	69.5	93.7	100.0	69.1	93.6
22, 23	PLUMBING & HVAC	99.6	46.1	77.8	95.5	32.0	69.7	95.6	31.0	69.3	95.6	47.1	75.8	95.4	36.1	71.2	100.1	36.0	74.0
26, 27, 3370	ELECTRICAL, COMMUNICATIONS & UTIL.	96.6	43.9	71.3	93.9	46.9	71.4	98.6	34.7	68.0	93.6	53.5	74.4	96.6	41.0	69.9	98.3	40.9	70.8
MF2004	WEIGHTED AVERAGE	99.4	51.9	78.9	94.4	45.6	73.3	98.5	46.4	76.0	97.4	53.8	78.6	99.7	48.7	77.7	97.5	47.4	75.8

		TEXAS																	
DIVISION		PALESTINE			SAN ANGELO			SAN ANTONIO			TEMPLE			TEXARKANA			TYLER		
		758			769			781 - 782			765			755			757		
		MAT.	INST.	TOTAL	MAT.	INST.	TOTAL	MAT.	INST.	TOTAL	MAT.	INST.	TOTAL	MAT.	INST.	TOTAL	MAT.	INST.	TOTAL
01543	CONTRACTOR EQUIPMENT	.0	89.2	89.2	.0	87.2	87.2	.0	89.8	89.8	.0	87.2	87.2	.0	89.2	89.2	.0	89.2	89.2
0241, 31 - 34	SITE & INFRASTRUCTURE, DEMOLITION	114.3	89.2	96.3	105.8	85.8	91.5	93.3	90.1	91.0	92.1	84.5	86.7	97.8	90.1	92.3	111.1	90.1	96.1
0310	Concrete Forming & Accessories	75.1	32.8	38.8	96.6	36.8	45.2	93.1	57.1	62.2	100.6	35.1	44.3	91.8	34.9	43.0	86.1	33.8	41.2
0310	Concrete Reinforcing	97.3	41.6	70.1	94.4	50.0	72.7	94.0	50.2	72.6	94.5	48.9	72.2	97.2	54.5	76.3	98.1	54.7	76.9
0330	Cast-in-Place Concrete	82.9	36.8	65.5	96.7	41.9	76.1	80.8	68.0	76.0	79.2	47.6	67.3	83.6	42.9	68.3	99.7	41.8	77.9
03	CONCRETE	86.0	36.7	62.7	91.8	42.1	68.3	79.3	60.2	70.3	79.1	43.1	62.1	81.5	42.5	63.1	93.4	41.7	69.0
04	MASONRY	109.0	46.0	71.3	122.1	47.5	77.5	88.8	62.5	73.1	132.6	50.5	83.4	166.5	48.0	95.6	156.3	51.2	93.4
05	METALS	91.1	51.7	79.2	96.4	62.0	86.0	99.0	66.6	89.2	96.1	61.7	85.7	84.8	65.7	79.0	91.0	66.3	83.5
06	WOOD, PLASTICS & COMPOSITES	77.4	33.5	54.2	98.5	37.1	66.0	92.2	55.6	72.9	108.1	33.6	68.7	92.3	33.1	61.0	89.9	31.2	58.9
07	THERMAL & MOISTURE PROTECTION	96.7	41.5	74.5	99.9	43.1	77.0	92.0	64.0	80.7	99.5	45.8	77.9	96.1	47.3	76.5	96.6	52.0	78.7
08	OPENINGS	74.4	31.4	63.3	89.0	39.3	76.1	99.3	54.2	87.6	75.6	36.0	65.3	92.5	39.2	78.7	74.4	38.1	65.0
0920	Plaster & Gypsum Board	81.0	32.5	52.1	87.7	36.0	56.9	94.0	55.0	70.8	89.0	32.5	55.3	88.3	32.0	54.8	85.2	30.1	52.4
0950, 0980	Ceilings & Acoustic Treatment	79.9	32.5	51.4	90.4	36.0	57.7	91.6	55.0	69.6	90.4	32.5	55.5	86.7	32.0	53.8	79.9	30.1	49.9
0960	Flooring	127.7	28.2	101.2	112.3	34.2	91.5	93.8	67.3	86.7	148.7	47.0	121.6	112.4	62.6	99.2	140.1	41.7	113.8
0990	Wall Finishes & Painting/Coating	94.8	30.8	56.5	97.1	34.0	59.3	93.9	53.8	69.9	98.3	33.9	59.8	94.8	39.7	61.8	94.8	39.7	61.8
09	FINISHES	104.2	31.6	66.8	97.0	35.3	65.1	92.1	57.8	74.4	107.3	36.3	70.6	101.5	39.7	69.6	108.2	34.3	70.0
COVERS	DIVS. 10 - 14, 25, 28, 41, 43, 44	100.0	42.6	88.2	100.0	68.0	93.4	100.0	74.6	94.8	100.0	42.8	88.2	100.0	72.3	94.3	100.0	72.0	94.2
22, 23	PLUMBING & HVAC	95.5	28.6	68.3	95.9	41.6	73.7	99.8	69.6	87.5	95.9	34.1	70.7	95.5	37.6	71.9	95.5	61.8	81.8
26, 27, 3370	ELECTRICAL, COMMUNICATIONS & UTIL.	90.6	38.8	65.7	98.1	32.6	66.7	101.3	65.0	83.9	95.5	76.7	86.5	93.8	69.4	82.1	92.8	56.0	75.1
MF2004	WEIGHTED AVERAGE	93.5	41.6	71.1	97.3	46.5	75.4	95.7	66.1	83.0	95.5	50.8	76.2	96.4	52.4	77.4	97.4	55.2	79.2

		TEXAS															UTAH		
DIVISION		VICTORIA			WACO			WAXAHACHIE			WHARTON			WICHITA FALLS			LOGAN		
		779			766 - 767			751			774			763			843		
		MAT.	INST.	TOTAL	MAT.	INST.	TOTAL	MAT.	INST.	TOTAL	MAT.	INST.	TOTAL	MAT.	INST.	TOTAL	MAT.	INST.	TOTAL
01543	CONTRACTOR EQUIPMENT	.0	96.8	96.8	.0	87.2	87.2	.0	95.4	95.4	.0	98.2	98.2	.0	87.2	87.2	.0	100.2	100.2
0241, 31 - 34	SITE & INFRASTRUCTURE, DEMOLITION	135.9	80.0	95.8	101.5	85.4	90.0	116.4	82.1	91.8	144.4	81.8	99.5	102.3	85.1	90.0	89.8	99.3	96.6
0310	Concrete Forming & Accessories	91.8	40.1	47.4	98.9	40.2	48.5	84.0	47.0	52.2	85.3	36.6	43.5	98.9	41.8	49.9	102.0	57.7	63.9
0310	Concrete Reinforcing	92.3	31.0	62.3	94.1	49.4	72.3	99.0	50.2	75.2	96.0	60.5	78.7	94.1	50.6	72.8	107.9	74.1	91.4
0330	Cast-in-Place Concrete	111.1	41.1	84.8	85.7	52.2	73.1	86.3	40.7	69.1	114.2	48.7	89.5	91.6	47.1	74.8	92.3	68.6	83.4
03	CONCRETE	101.0	40.9	72.6	84.9	47.1	67.0	80.7	47.1	64.8	104.7	47.3	77.6	87.8	46.3	68.2	114.5	65.0	91.1
04	MASONRY	112.7	38.2	68.1	94.7	57.2	72.3	151.8	51.9	92.0	97.7	33.5	59.3	95.2	57.1	72.4	113.5	57.9	80.2
05	METALS	104.3	73.5	95.0	98.2	64.5	88.0	91.2	75.6	86.5	106.1	79.8	98.1	98.2	66.0	88.5	103.5	73.5	94.4
06	WOOD, PLASTICS & COMPOSITES	99.5	41.9	69.0	105.6	36.1	68.8	85.4	50.1	66.7	88.8	37.9	61.9	105.6	40.8	71.3	87.1	56.1	70.7
07	THERMAL & MOISTURE PROTECTION	101.9	42.4	78.0	100.4	49.1	79.7	96.6	47.4	76.8	99.4	48.4	78.9	100.4	52.6	81.1	98.2	65.5	85.0
08	OPENINGS	104.7	40.0	88.0	87.3	37.4	74.4	102.6	48.5	88.6	104.9	42.0	88.6	87.3	43.4	75.9	89.1	57.6	81.0
0920	Plaster & Gypsum Board	94.0	40.8	62.3	88.6	34.9	56.7	87.5	49.3	64.7	92.5	36.7	59.3	88.6	39.8	59.6	84.9	54.8	66.9
0950, 0980	Ceilings & Acoustic Treatment	101.6	40.8	65.0	95.5	34.9	59.1	94.3	49.3	67.2	100.8	36.7	62.3	95.5	39.8	62.0	101.7	54.8	73.5
0960	Flooring	97.7	37.6	81.7	147.1	35.6	117.4	96.1	37.6	80.5	96.0	28.2	77.9	148.3	75.0	128.8	100.0	57.5	88.6
0990	Wall Finishes & Painting/Coating	103.9	38.1	64.6	98.3	33.9	59.8	102.8	38.1	64.1	103.8	31.6	60.6	102.2	53.6	73.2	99.6	46.4	67.7
09	FINISHES	93.8	39.6	65.8	108.5	37.3	71.8	95.4	44.7	69.2	94.9	34.5	63.7	109.1	48.8	78.0	97.3	55.7	75.8
COVERS	DIVS. 10 - 14, 25, 28, 41, 43, 44	100.0	45.4	88.8	100.0	76.0	95.1	100.0	46.9	89.1	100.0	43.9	88.5	100.0	69.9	93.8	100.0	67.3	93.3
22, 23	PLUMBING & HVAC	95.9	32.0	69.8	100.1	53.0	80.9	95.6	46.0	75.4	95.8	26.6	67.6	100.1	49.4	79.5	99.9	67.9	86.9
26, 27, 3370	ELECTRICAL, COMMUNICATIONS & UTIL.	100.9	47.0	75.0	98.3	76.8	88.0	93.6	60.4	77.7	99.4	33.6	67.8	100.8	64.8	83.6	91.2	71.4	81.7
MF2004	WEIGHTED AVERAGE	101.5	46.0	77.5	97.2	57.5	80.1	97.4	54.6	78.9	101.4	43.6	76.4	98.0	56.8	80.2	100.2	68.0	86.3

City Cost Indexes

DIVISION		UTAH										VERMONT					
		OGDEN			PRICE			PROVO			SALT LAKE CITY			BELLOWS FALLS			BENNINGTON		
		842,844			845			846 - 847			840 - 841			051			052		
		MAT.	INST.	TOTAL	MAT.	INST.	TOTAL	MAT.	INST.	TOTAL	MAT.	INST.	TOTAL	MAT.	INST.	TOTAL	MAT.	INST.	TOTAL
01543	CONTRACTOR EQUIPMENT	.0	100.2	100.2	.0	98.9	98.9	.0	98.9	98.9	.0	100.1	100.1	.0	100.7	100.7	.0	100.7	100.7
0241, 31 - 34	SITE & INFRASTRUCTURE, DEMOLITION	78.7	99.3	93.5	87.6	96.2	93.8	86.4	97.4	94.3	78.6	99.2	93.4	73.7	97.1	90.5	73.1	98.8	91.5
0310	Concrete Forming & Accessories	102.0	57.7	63.9	104.4	31.7	42.0	103.5	57.7	64.2	102.0	57.7	64.0	99.7	38.7	47.3	97.0	58.7	64.1
0310	Concrete Reinforcing	107.5	74.1	91.2	115.7	73.6	95.1	116.7	74.1	95.9	109.9	74.1	92.4	83.4	74.9	79.2	83.4	75.0	79.3
0330	Cast-in-Place Concrete	93.7	68.6	84.3	92.4	43.0	74.7	92.4	68.6	83.4	102.4	68.6	89.7	94.2	52.9	78.7	94.2	71.5	85.7
03	CONCRETE	104.0	65.0	85.6	115.9	45.4	82.6	114.4	65.0	91.1	124.1	65.0	96.2	99.4	51.2	76.6	99.2	66.5	83.8
04	MASONRY	107.2	57.9	77.7	119.0	39.9	71.7	119.1	57.9	82.5	121.5	57.9	83.4	95.9	36.5	60.3	105.1	55.2	75.2
05	METALS	103.9	73.5	94.7	100.9	71.6	92.0	101.8	73.6	93.3	108.2	73.6	97.7	90.7	70.1	84.5	90.7	71.8	85.0
06	WOOD, PLASTICS & COMPOSITES	87.1	56.1	70.7	90.0	29.5	58.0	88.6	56.1	71.4	89.1	56.1	71.7	100.2	36.3	66.4	96.9	53.0	73.7
07	THERMAL & MOISTURE PROTECTION	97.0	65.5	84.3	99.9	52.8	80.9	99.9	65.5	86.0	99.9	65.5	86.1	97.8	54.1	80.2	97.7	55.7	80.8
08	OPENINGS	89.1	57.6	81.0	93.4	41.9	80.0	93.4	57.6	84.1	91.0	57.6	82.4	102.3	43.9	87.2	102.3	50.5	88.9
0920	Plaster & Gypsum Board	84.9	54.8	66.9	86.1	27.5	51.2	85.2	54.8	67.1	87.0	54.8	67.8	100.5	33.8	60.8	98.9	51.0	70.4
0950, 0980	Ceilings & Acoustic Treatment	101.7	68.7	73.5	101.7	27.5	57.1	101.7	54.8	73.5	100.8	54.8	73.1	84.2	33.8	53.9	84.2	51.0	64.2
0960	Flooring	97.9	57.5	87.1	101.2	42.7	85.6	100.8	57.5	89.3	100.3	57.5	88.9	98.0	36.7	81.7	96.7	36.7	80.7
0990	Wall Finishes & Painting/Coating	99.6	46.4	67.7	99.6	46.4	67.7	99.6	59.2	75.4	102.1	59.2	76.4	91.0	33.0	56.3	91.0	33.0	56.3
09	FINISHES	95.7	55.7	75.1	98.3	33.9	65.1	98.0	57.2	76.9	97.1	57.2	76.5	94.6	36.6	64.6	93.9	50.9	71.7
COVERS	DIVS. 10 - 14, 25, 28, 41, 43, 44	100.0	67.3	93.3	100.0	50.9	89.9	100.0	67.3	93.3	100.0	67.3	93.3	100.0	45.2	88.7	100.0	54.1	90.5
22, 23	PLUMBING & HVAC	99.9	67.9	86.9	97.7	38.5	73.6	99.9	67.9	86.9	100.1	67.9	87.0	95.8	34.6	70.8	95.8	93.3	94.7
26, 27, 3370	ELECTRICAL, COMMUNICATIONS & UTIL.	91.7	71.4	81.9	96.2	42.4	70.4	91.9	74.5	83.5	95.0	74.5	85.2	97.9	88.8	93.5	97.9	67.1	83.1
MF2004	WEIGHTED AVERAGE	98.3	68.0	85.2	100.8	48.4	78.2	100.7	68.5	86.8	102.9	68.6	88.1	96.0	54.8	78.2	96.4	70.8	85.3

DIVISION		VERMONT																	
		BRATTLEBORO			BURLINGTON			GUILDHALL			MONTPELIER			RUTLAND			ST. JOHNSBURY		
		053			054			059			056			057			058		
		MAT.	INST.	TOTAL	MAT.	INST.	TOTAL	MAT.	INST.	TOTAL	MAT.	INST.	TOTAL	MAT.	INST.	TOTAL	MAT.	INST.	TOTAL
01543	CONTRACTOR EQUIPMENT	.0	100.7	100.7	.0	100.7	100.7	.0	100.7	100.7	.0	100.7	100.7	.0	100.7	100.7	.0	100.7	100.7
0241, 31 - 34	SITE & INFRASTRUCTURE, DEMOLITION	74.4	98.2	91.5	76.3	97.9	91.8	72.9	97.3	90.4	75.5	97.9	91.5	76.2	97.9	91.7	72.9	97.3	90.4
0310	Concrete Forming & Accessories	100.0	45.6	53.3	94.1	48.3	55.2	96.6	51.2	57.7	98.0	48.9	55.8	100.3	48.9	56.1	94.3	51.3	57.4
0310	Concrete Reinforcing	82.6	74.9	78.8	104.3	51.4	78.4	84.1	44.3	64.7	83.0	51.4	67.6	104.3	51.4	78.4	82.6	50.6	66.9
0330	Cast-in-Place Concrete	97.2	63.4	84.5	96.8	60.8	83.2	91.3	61.2	80.0	99.9	60.8	85.2	92.2	60.8	80.4	91.3	61.2	80.0
03	CONCRETE	101.8	57.8	81.0	104.7	54.0	80.8	96.9	53.9	76.6	103.2	54.0	80.0	102.8	54.0	79.8	96.5	55.0	76.9
04	MASONRY	105.0	54.9	75.0	105.3	57.6	76.7	105.3	42.0	67.4	89.6	57.6	70.4	86.1	57.6	69.1	131.3	42.0	77.9
05	METALS	90.7	70.0	84.4	97.4	64.3	87.4	90.7	59.3	81.2	90.7	64.4	82.7	95.8	64.4	86.3	90.7	59.4	81.2
06	WOOD, PLASTICS & COMPOSITES	100.5	36.3	66.6	92.6	47.0	68.5	95.7	53.0	73.2	96.4	47.0	70.3	100.6	47.0	72.3	91.6	53.0	71.2
07	THERMAL & MOISTURE PROTECTION	97.9	55.0	80.6	98.3	53.2	80.1	97.6	44.9	76.4	97.7	54.3	80.2	97.9	54.3	80.4	97.5	44.9	76.3
08	OPENINGS	102.3	43.9	87.2	105.6	43.6	89.6	102.3	45.1	87.5	102.3	43.6	87.1	105.6	43.6	89.6	102.3	45.1	87.5
0920	Plaster & Gypsum Board	100.5	33.8	60.8	101.4	44.8	67.7	104.7	51.0	72.7	108.2	44.8	70.5	101.4	44.8	67.7	108.2	51.0	74.1
0950, 0980	Ceilings & Acoustic Treatment	84.2	33.8	53.9	90.3	44.8	62.9	84.2	51.0	64.2	84.2	44.8	60.5	90.3	44.8	62.9	84.2	51.0	64.2
0960	Flooring	98.1	36.7	81.8	98.0	66.4	89.6	101.6	36.7	84.3	105.6	66.4	95.2	98.0	66.4	89.6	105.6	36.7	87.2
0990	Wall Finishes & Painting/Coating	91.0	33.0	56.3	91.0	38.4	59.5	91.0	25.6	51.9	91.0	38.4	59.5	91.0	38.4	59.5	91.0	26.9	52.6
09	FINISHES	94.7	41.2	67.1	96.4	50.3	72.6	96.2	45.7	70.1	98.2	50.3	73.5	96.3	50.3	72.6	98.0	45.8	71.1
COVERS	DIVS. 10 - 14, 25, 28, 41, 43, 44	100.0	52.1	90.2	100.0	87.4	97.4	100.0	47.4	89.2	100.0	87.4	97.4	100.0	87.4	97.4	100.0	47.4	89.2
22, 23	PLUMBING & HVAC	95.8	47.9	76.3	100.0	63.1	85.0	95.8	66.4	83.8	95.8	63.1	82.4	100.0	63.1	85.0	95.8	66.4	83.8
26, 27, 3370	ELECTRICAL, COMMUNICATIONS & UTIL.	97.9	88.7	93.5	98.6	67.1	83.5	97.9	67.1	83.1	97.9	67.1	83.1	97.9	67.1	83.1	97.9	67.1	83.1
MF2004	WEIGHTED AVERAGE	96.8	61.3	81.4	99.7	62.8	83.8	96.3	59.6	80.4	96.4	62.8	81.9	98.2	62.8	82.9	97.7	59.7	81.3

DIVISION		VERMONT			VIRGINIA														
		WHITE RIVER JCT.			ALEXANDRIA			ARLINGTON			BRISTOL			CHARLOTTESVILLE			CULPEPER		
		050			223			222			242			229			227		
		MAT.	INST.	TOTAL	MAT.	INST.	TOTAL	MAT.	INST.	TOTAL	MAT.	INST.	TOTAL	MAT.	INST.	TOTAL	MAT.	INST.	TOTAL
01543	CONTRACTOR EQUIPMENT	.0	100.7	100.7	.0	101.6	101.6	.0	100.1	100.1	.0	100.1	100.1	.0	105.2	105.2	.0	100.1	100.1
0241, 31 - 34	SITE & INFRASTRUCTURE, DEMOLITION	76.4	97.4	91.5	115.0	87.1	95.0	126.6	84.4	96.3	110.7	83.6	91.3	115.8	85.9	94.4	113.2	84.8	92.9
0310	Concrete Forming & Accessories	94.2	39.6	47.3	93.6	74.8	77.4	92.6	75.0	77.5	87.6	51.6	56.7	85.8	59.0	62.8	83.0	74.2	75.4
0310	Concrete Reinforcing	83.4	29.7	57.1	89.6	82.1	85.9	101.0	84.6	93.0	101.1	69.8	85.8	100.4	70.1	85.6	101.0	77.9	89.7
0330	Cast-in-Place Concrete	97.2	49.9	79.4	104.2	78.9	94.7	101.3	82.7	94.3	99.0	54.1	83.3	105.0	63.8	89.5	103.8	75.7	93.2
03	CONCRETE	103.5	42.1	74.5	108.8	78.7	94.6	113.9	80.1	97.9	109.8	57.6	85.2	110.5	64.3	88.7	107.2	76.2	92.6
04	MASONRY	117.4	44.0	73.5	92.3	71.3	79.7	105.3	73.3	86.2	93.1	48.9	66.6	119.0	56.5	81.6	107.6	72.9	86.9
05	METALS	90.7	53.2	79.4	100.0	96.4	98.9	98.8	88.0	95.5	97.7	87.8	94.7	98.0	90.3	95.7	98.0	87.7	94.9
06	WOOD, PLASTICS & COMPOSITES	93.6	39.1	64.8	96.0	75.5	85.2	92.5	75.5	83.5	84.0	50.1	66.0	81.9	57.2	68.8	80.5	75.5	77.9
07	THERMAL & MOISTURE PROTECTION	97.9	38.7	74.1	103.2	80.0	93.9	104.9	75.2	93.0	104.5	55.9	84.9	104.1	62.4	87.3	104.2	71.1	90.9
08	OPENINGS	102.3	36.6	85.3	96.3	76.8	91.2	104.2	72.7	88.6	97.6	54.6	86.5	95.5	60.2	86.4	95.9	70.2	89.3
0920	Plaster & Gypsum Board	97.9	36.7	61.5	115.0	74.7	91.0	111.9	74.7	89.7	108.2	48.5	72.7	107.2	55.1	76.2	107.7	74.7	88.1
0950, 0980	Ceilings & Acoustic Treatment	84.2	36.7	55.6	100.3	74.7	84.9	95.2	74.7	82.9	94.4	48.5	66.8	94.4	55.1	70.7	95.2	74.7	82.9
0960	Flooring	95.6	36.7	79.9	105.3	94.1	102.3	103.5	66.2	93.6	99.1	45.0	84.7	97.4	53.1	85.6	97.4	66.2	89.1
0990	Wall Finishes & Painting/Coating	91.0	20.4	48.8	124.7	83.0	99.7	124.7	83.0	99.7	112.5	51.4	75.9	112.5	72.7	88.7	124.7	64.6	88.8
09	FINISHES	93.7	36.6	64.2	104.0	79.0	91.1	102.8	74.8	88.4	99.4	49.7	73.8	98.7	58.7	78.1	99.4	72.5	85.5
COVERS	DIVS. 10 - 14, 25, 28, 41, 43, 44	100.0	45.2	88.7	100.0	86.9	97.3	100.0	86.9	97.3	100.0	73.3	94.5	100.0	82.0	96.3	100.0	73.4	94.5
22, 23	PLUMBING & HVAC	95.8	39.2	72.7	100.2	85.7	94.3	100.2	86.6	94.7	96.0	59.4	81.1	96.0	69.3	85.1	96.0	76.6	88.1
26, 27, 3370	ELECTRICAL, COMMUNICATIONS & UTIL.	97.9	67.1	83.1	98.3	92.9	95.7	96.0	91.2	93.7	97.8	44.1	72.0	97.7	71.4	85.1	100.0	65.0	83.2
MF2004	WEIGHTED AVERAGE	97.5	50.1	77.0	101.0	83.9	93.6	101.8	82.4	93.4	99.2	59.3	82.0	100.6	69.2	87.0	99.9	75.2	89.2

677

City Cost Indexes

		VIRGINIA																
	DIVISION	FAIRFAX 220 - 221			FARMVILLE 239			FREDERICKSBURG 224 - 225			GRUNDY 246			HARRISONBURG 228			LYNCHBURG 245		
		MAT.	INST.	TOTAL	MAT.	INST.	TOTAL	MAT.	INST.	TOTAL	MAT.	INST.	TOTAL	MAT.	INST.	TOTAL	MAT.	INST.	TOTAL
01543	CONTRACTOR EQUIPMENT	.0	100.1	100.1	.0	105.2	105.2	.0	100.1	100.1	.0	100.1	100.1	.0	100.1	100.1	.0	100.1	100.1
0241, 31 - 34	SITE & INFRASTRUCTURE, DEMOLITION	125.5	84.8	96.3	111.5	85.5	92.9	112.7	84.7	92.6	108.3	82.6	89.9	122.1	84.0	94.8	109.1	83.8	91.0
0310	Concrete Forming & Accessories	86.1	75.2	76.8	99.3	54.1	60.5	86.1	58.3	62.3	91.0	45.0	51.5	81.8	51.8	56.0	87.6	74.6	76.5
0310	Concrete Reinforcing	101.0	82.0	91.7	101.1	62.8	82.3	101.9	81.2	91.7	99.7	63.4	81.9	101.0	57.0	79.5	100.4	70.6	85.8
0330	Cast-in-Place Concrete	101.3	79.7	93.2	105.0	62.0	88.8	102.9	59.7	86.7	100.9	55.0	83.7	101.3	67.9	88.7	100.9	67.4	88.3
03	CONCRETE	113.5	79.0	97.2	109.6	59.9	86.1	106.9	64.7	86.9	108.5	53.3	82.4	111.0	60.2	87.1	108.4	72.5	91.4
04	MASONRY	105.3	73.4	86.2	102.7	55.7	74.6	106.2	54.1	75.0	97.1	59.0	74.3	104.0	61.4	78.5	111.5	63.8	83.0
05	METALS	98.1	94.4	97.0	97.9	83.9	93.7	98.1	94.4	97.0	97.7	76.3	91.2	98.0	91.2	95.9	97.9	89.5	95.4
06	WOOD, PLASTICS & COMPOSITES	84.0	75.5	79.5	98.3	55.2	75.5	84.0	56.4	69.4	87.7	43.8	64.5	79.2	48.8	63.1	84.0	79.1	81.4
07	THERMAL & MOISTURE PROTECTION	104.8	75.6	93.1	104.2	56.4	84.9	104.2	59.2	86.1	104.4	55.6	84.8	104.6	70.6	90.9	104.3	65.0	88.5
08	OPENINGS	94.2	76.1	89.5	95.9	54.0	85.0	95.6	62.5	87.0	97.6	43.1	83.5	95.9	61.6	87.0	95.9	70.3	89.2
0920	Plaster & Gypsum Board	108.4	74.7	88.3	113.6	53.0	77.6	108.4	55.1	76.6	109.1	42.1	69.2	107.2	47.2	71.5	108.2	78.4	90.4
0950, 0980	Ceilings & Acoustic Treatment	95.2	74.7	82.9	94.4	53.0	69.5	95.2	55.1	71.1	94.4	42.1	62.9	94.4	47.2	66.0	94.4	78.4	84.7
0960	Flooring	99.4	66.2	90.6	105.3	72.6	96.6	99.4	89.1	96.7	100.8	40.5	84.7	97.0	89.1	94.9	99.1	50.3	86.1
0990	Wall Finishes & Painting/Coating	124.7	83.0	99.7	112.5	47.9	73.8	124.7	79.4	97.6	112.5	43.4	71.2	124.7	62.7	87.6	112.5	51.4	75.9
09	FINISHES	101.0	74.8	87.5	101.9	57.3	78.9	100.1	65.7	82.3	99.9	43.5	70.8	99.6	59.1	78.7	99.2	68.3	83.2
COVERS	DIVS. 10 - 14, 25, 28, 41, 43, 44	100.0	87.0	97.3	100.0	61.2	92.0	100.0	74.3	94.7	100.0	58.5	91.5	100.0	72.8	94.4	100.0	77.7	95.4
22, 23	PLUMBING & HVAC	96.0	86.8	92.2	96.0	44.6	75.1	96.0	83.7	91.0	96.0	50.1	77.3	96.0	65.5	83.5	96.0	68.6	84.8
26, 27, 3370	ELECTRICAL, COMMUNICATIONS & UTIL.	98.8	92.9	95.9	95.3	58.5	77.7	96.1	92.9	94.6	97.8	55.6	77.5	97.9	58.2	78.9	98.7	47.2	74.0
MF2004	WEIGHTED AVERAGE	100.7	83.3	93.2	99.7	59.8	82.4	99.4	76.0	89.3	99.3	56.6	80.9	100.1	66.4	85.6	99.9	69.1	86.6

		VIRGINIA																
	DIVISION	NEWPORT NEWS 236			NORFOLK 233 - 235			PETERSBURG 238			PORTSMOUTH 237			PULASKI 243			RICHMOND 230 - 232		
		MAT.	INST.	TOTAL	MAT.	INST.	TOTAL	MAT.	INST.	TOTAL	MAT.	INST.	TOTAL	MAT.	INST.	TOTAL	MAT.	INST.	TOTAL
01543	CONTRACTOR EQUIPMENT	.0	105.2	105.3	.0	106.0	106.0	.0	105.2	105.2	.0	105.2	105.2	.0	100.1	100.1	.0	105.2	105.2
0241, 31 - 34	SITE & INFRASTRUCTURE, DEMOLITION	109.7	86.5	93.0	109.1	87.6	93.7	113.9	87.0	94.6	108.3	85.9	92.2	107.7	82.6	89.7	109.6	87.0	93.4
0310	Concrete Forming & Accessories	98.3	77.0	80.0	102.5	77.0	80.6	91.2	70.4	73.3	86.8	60.3	64.0	91.0	45.1	51.6	99.3	70.4	74.4
0310	Concrete Reinforcing	100.7	71.6	86.5	100.7	71.6	86.5	100.4	71.0	86.0	100.4	71.5	86.3	99.7	63.4	82.0	100.7	71.0	86.2
0330	Cast-in-Place Concrete	102.0	66.1	88.5	105.1	66.1	90.4	105.0	61.6	88.7	101.0	65.7	87.7	100.9	55.1	83.7	102.8	61.6	87.3
03	CONCRETE	106.5	73.3	90.8	108.3	73.3	91.7	110.5	68.7	90.7	105.1	65.6	86.5	108.5	53.4	82.5	106.9	68.7	88.9
04	MASONRY	97.9	60.0	75.2	104.3	60.0	77.8	110.3	65.4	83.4	102.8	58.2	76.1	89.9	59.0	71.4	96.2	65.4	77.8
05	METALS	100.0	91.7	97.5	99.1	91.7	96.8	98.0	92.2	96.2	99.1	89.9	96.3	97.8	76.5	91.3	102.0	92.2	99.0
06	WOOD, PLASTICS & COMPOSITES	97.0	83.6	89.9	102.6	83.6	92.5	88.2	73.8	80.6	83.4	61.4	71.7	87.7	43.8	64.5	98.2	73.8	85.3
07	THERMAL & MOISTURE PROTECTION	104.1	64.0	88.0	103.9	64.0	87.8	104.1	65.7	88.7	104.1	60.5	86.6	104.4	55.6	84.8	103.7	65.7	88.4
08	OPENINGS	96.3	73.6	90.4	96.3	73.6	90.4	95.6	69.2	88.7	96.3	61.6	87.3	97.6	43.1	83.5	96.3	69.2	89.2
0920	Plaster & Gypsum Board	115.0	82.2	95.5	115.0	82.2	95.5	109.3	72.1	87.2	108.2	59.4	79.1	109.1	42.1	69.2	115.0	72.1	89.5
0950, 0980	Ceilings & Acoustic Treatment	100.3	82.2	89.4	100.3	82.2	89.4	95.2	72.1	81.3	100.3	59.4	75.7	94.4	42.1	62.9	100.3	72.1	83.3
0960	Flooring	105.3	60.9	93.5	105.1	60.9	93.3	100.3	85.7	96.4	96.5	76.9	91.3	100.8	40.5	84.7	105.1	85.7	100.0
0990	Wall Finishes & Painting/Coating	112.5	52.0	76.3	112.5	62.8	82.8	112.5	72.7	88.7	112.5	62.8	82.8	112.5	43.4	71.2	112.5	72.7	88.7
09	FINISHES	103.4	72.2	87.3	103.3	73.4	87.9	100.1	74.0	86.6	99.7	63.2	80.9	99.9	43.5	70.8	103.2	74.0	88.2
COVERS	DIVS. 10 - 14, 25, 28, 41, 43, 44	100.0	82.7	96.4	100.0	82.7	96.4	100.0	82.3	96.4	100.0	76.3	95.1	100.0	58.5	91.5	100.0	82.3	96.4
22, 23	PLUMBING & HVAC	100.2	63.2	85.2	100.2	63.4	85.2	96.0	67.0	84.2	100.2	63.4	85.2	96.0	50.5	77.4	100.2	67.0	86.7
26, 27, 3370	ELECTRICAL, COMMUNICATIONS & UTIL.	97.8	68.1	83.6	98.0	60.5	80.0	97.9	71.4	85.2	96.3	60.5	79.1	97.8	60.0	79.7	99.7	71.4	86.1
MF2004	WEIGHTED AVERAGE	100.8	71.9	88.3	101.2	71.1	88.2	100.3	73.1	88.6	100.1	67.2	85.9	98.9	57.3	81.0	101.2	73.1	89.1

		VIRGINIA									WASHINGTON								
	DIVISION	ROANOKE 240 - 241			STAUNTON 244			WINCHESTER 226			CLARKSTON 994			EVERETT 982			OLYMPIA 985		
		MAT.	INST.	TOTAL	MAT.	INST.	TOTAL	MAT.	INST.	TOTAL	MAT.	INST.	TOTAL	MAT.	INST.	TOTAL	MAT.	INST.	TOTAL
01543	CONTRACTOR EQUIPMENT	.0	100.1	100.1	.0	105.2	105.2	.0	100.1	100.1	.0	90.2	90.2	.0	104.0	104.0	.0	104.0	104.0
0241, 31 - 34	SITE & INFRASTRUCTURE, DEMOLITION	107.2	83.8	90.4	112.2	86.1	93.5	120.7	84.1	94.5	105.2	88.3	93.1	94.4	114.6	108.9	96.3	114.7	109.5
0310	Concrete Forming & Accessories	98.1	74.4	77.7	90.7	62.0	66.1	84.5	58.0	61.7	114.1	77.8	82.9	110.0	101.0	102.3	100.5	101.4	101.2
0310	Concrete Reinforcing	100.7	70.6	85.9	100.4	61.5	81.4	100.4	84.4	92.6	108.3	92.5	100.6	108.1	93.5	101.0	108.6	93.5	101.2
0330	Cast-in-Place Concrete	114.7	67.3	96.9	105.0	59.8	88.0	101.3	63.4	87.1	111.8	83.0	100.9	97.6	105.7	100.6	91.7	107.8	97.8
03	CONCRETE	112.9	72.3	93.7	109.6	62.7	87.5	110.3	66.3	89.5	105.5	82.3	94.6	96.4	100.6	98.4	94.1	101.5	97.6
04	MASONRY	98.3	65.7	78.8	108.6	64.0	81.9	101.1	53.9	72.9	117.0	80.8	95.4	132.7	101.6	114.1	124.7	102.3	111.3
05	METALS	99.8	89.4	96.7	97.9	89.8	95.5	98.1	93.4	96.7	91.8	83.3	89.3	106.8	87.7	101.0	101.3	88.0	97.3
06	WOOD, PLASTICS & COMPOSITES	97.0	79.1	87.6	87.7	62.7	74.5	82.3	56.4	68.6	95.7	77.1	85.9	101.8	100.7	101.2	90.6	100.7	96.0
07	THERMAL & MOISTURE PROTECTION	104.1	63.8	87.9	104.1	60.9	86.7	104.7	66.4	89.3	143.1	77.0	116.5	103.0	97.4	100.8	103.2	96.3	100.4
08	OPENINGS	96.3	70.3	89.5	95.9	60.5	86.7	97.7	63.4	88.8	110.5	80.9	102.8	100.9	98.1	100.2	100.8	98.1	100.1
0920	Plaster & Gypsum Board	115.0	78.4	93.2	109.1	60.7	80.3	98.4	55.1	76.6	136.3	76.3	100.6	107.3	100.7	103.3	102.6	100.7	101.4
0950, 0980	Ceilings & Acoustic Treatment	100.3	78.4	87.1	94.4	60.7	74.1	95.2	55.1	71.1	104.9	76.3	87.7	107.9	100.7	103.6	104.3	100.7	102.1
0960	Flooring	105.3	47.1	89.8	100.0	47.5	86.0	98.7	89.1	96.1	109.7	56.7	95.6	122.9	105.2	118.2	115.8	105.2	113.0
0990	Wall Finishes & Painting/Coating	112.5	51.4	75.9	112.5	35.8	66.6	124.7	55.2	83.1	110.8	72.3	87.8	114.1	87.7	98.3	114.1	89.4	99.3
09	FINISHES	103.2	67.7	84.9	99.6	57.2	77.7	100.4	62.9	81.1	123.6	72.7	97.4	114.1	100.6	107.1	110.5	100.9	105.6
COVERS	DIVS. 10 - 14, 25, 28, 41, 43, 44	100.0	77.7	95.4	100.0	77.4	95.3	100.0	69.0	93.6	100.0	92.4	98.4	100.0	98.4	99.7	100.0	102.8	100.6
22, 23	PLUMBING & HVAC	100.2	57.7	82.9	96.0	67.5	84.4	96.0	82.1	90.3	96.2	82.5	90.6	100.1	99.6	99.9	100.2	96.7	98.7
26, 27, 3370	ELECTRICAL, COMMUNICATIONS & UTIL.	97.8	47.4	73.6	96.8	51.5	75.1	96.5	53.7	75.9	99.4	94.1	96.9	105.8	94.3	100.3	106.3	99.0	102.8
MF2004	WEIGHTED AVERAGE	101.4	67.0	86.5	99.9	66.3	85.4	100.0	69.9	87.0	103.9	83.3	95.0	104.4	99.4	102.2	102.5	99.8	101.4

City Cost Indexes

DIVISION		WASHINGTON																	
		RICHLAND 993			SEATTLE 980 - 981,987			SPOKANE 990 - 992			TACOMA 983 - 984			VANCOUVER 986			WENATCHEE 988		
		MAT.	INST.	TOTAL	MAT.	INST.	TOTAL	MAT.	INST.	TOTAL	MAT.	INST.	TOTAL	MAT.	INST.	TOTAL	MAT.	INST.	TOTAL
01543	CONTRACTOR EQUIPMENT	.0	90.2	90.2	.0	103.7	103.7	.0	90.2	90.2	.0	104.0	104.0	.0	97.2	97.2	.0	104.0	104.0
0241, 31 - 34	SITE & INFRASTRUCTURE, DEMOLITION	106.3	89.3	94.1	98.7	112.6	108.6	105.8	89.3	94.0	97.7	114.7	109.9	109.9	100.1	102.9	109.4	113.5	112.3
0310	Concrete Forming & Accessories	114.2	79.1	84.1	100.6	101.8	101.6	120.7	79.1	84.9	100.6	101.5	101.4	101.4	96.2	96.9	102.0	80.0	83.1
0310	Concrete Reinforcing	103.9	92.5	98.4	107.0	93.6	100.4	104.6	92.5	98.7	107.0	93.5	100.4	107.7	93.2	100.6	107.7	88.2	98.1
0330	Cast-in-Place Concrete	112.1	82.2	100.8	102.4	108.0	104.5	116.4	82.2	103.5	100.3	107.9	103.2	112.1	99.8	107.4	102.4	84.5	95.7
03	CONCRETE	105.1	82.6	94.5	99.2	101.9	100.5	107.9	82.6	95.9	98.2	101.6	99.8	108.0	96.6	102.6	105.4	83.0	94.8
04	MASONRY	116.8	80.8	95.3	126.5	102.3	112.0	117.5	80.8	95.6	126.3	102.3	112.0	126.9	98.8	110.1	129.5	82.7	101.5
05	METALS	92.1	83.6	89.5	108.5	90.1	102.9	94.6	83.6	91.3	108.5	88.1	102.3	105.7	88.8	100.6	106.1	83.2	99.2
06	WOOD, PLASTICS & COMPOSITES	95.9	78.6	86.8	91.6	100.7	96.4	105.7	78.6	91.4	90.8	100.7	96.0	85.1	96.4	91.1	91.7	78.5	84.7
07	THERMAL & MOISTURE PROTECTION	144.1	80.0	118.3	103.0	99.6	101.6	141.4	80.1	116.7	102.7	98.1	100.8	104.6	90.1	98.8	103.2	78.4	93.3
08	OPENINGS	111.1	76.3	102.1	103.1	98.1	101.8	111.7	76.3	102.5	101.6	98.1	100.7	99.1	94.8	98.0	101.1	75.2	94.4
0920	Plaster & Gypsum Board	136.3	77.9	101.5	102.0	100.7	101.2	129.9	77.9	99.0	104.0	100.7	102.0	102.4	96.6	98.9	104.8	77.8	88.8
0950, 0980	Ceilings & Acoustic Treatment	109.6	77.9	90.5	111.1	100.7	104.8	109.6	77.9	90.5	111.3	100.7	104.9	106.3	96.6	100.4	104.3	77.8	88.4
0960	Flooring	110.0	41.7	91.8	115.2	105.2	112.5	110.8	72.5	100.6	115.8	105.2	113.0	121.7	76.7	109.7	119.0	72.5	106.6
0990	Wall Finishes & Painting/Coating	110.8	72.3	87.8	114.1	89.4	99.3	111.7	72.3	88.1	114.1	89.4	99.3	121.3	68.6	89.8	114.1	72.3	89.1
09	FINISHES	124.9	70.6	96.8	112.0	100.9	106.3	124.2	76.9	99.8	112.5	100.9	106.5	110.8	89.5	99.8	112.8	77.2	94.5
COVERS	DIVS. 10 - 14, 25, 28, 41, 43, 44	100.0	79.5	95.8	100.0	102.9	100.6	100.0	79.4	95.8	100.0	102.8	100.6	100.0	77.4	95.3	100.0	92.9	98.5
22, 23	PLUMBING & HVAC	100.6	98.1	99.6	100.1	110.9	104.5	100.4	84.0	93.8	100.2	96.7	98.8	100.3	102.4	101.1	95.9	82.6	90.5
26, 27, 3370	ELECTRICAL, COMMUNICATIONS & UTIL.	95.3	94.1	94.7	105.7	102.7	104.2	94.7	79.9	87.6	105.7	99.0	102.5	113.1	98.7	106.2	106.4	85.2	96.3
MF2004	WEIGHTED AVERAGE	104.6	86.0	96.6	104.7	103.5	104.1	105.3	81.8	95.1	104.4	99.9	102.5	106.0	96.2	101.8	104.5	84.9	96.0

DIVISION		WASHINGTON			WEST VIRGINIA														
		YAKIMA 989			BECKLEY 258 - 259			BLUEFIELD 247 - 248			BUCKHANNON 262			CHARLESTON 250 - 253			CLARKSBURG 263 - 264		
		MAT.	INST.	TOTAL	MAT.	INST.	TOTAL	MAT.	INST.	TOTAL	MAT.	INST.	TOTAL	MAT.	INST.	TOTAL	MAT.	INST.	TOTAL
01543	CONTRACTOR EQUIPMENT	.0	104.0	104.0	.0	100.1	100.1	.0	100.1	100.1	.0	100.1	100.1	.0	100.1	100.1	.0	100.1	100.1
0241, 31 - 34	SITE & INFRASTRUCTURE, DEMOLITION	100.7	113.1	109.6	102.0	85.7	90.3	101.8	85.7	90.2	108.5	85.4	92.0	104.9	86.8	91.9	109.1	85.4	92.1
0310	Concrete Forming & Accessories	101.0	94.8	95.7	86.4	87.7	87.5	88.0	86.6	86.8	87.2	88.4	88.2	107.0	89.1	91.7	84.5	88.4	87.9
0310	Concrete Reinforcing	107.4	92.1	99.9	99.3	83.2	91.5	99.3	66.5	83.3	99.9	85.2	92.7	100.7	83.4	92.3	99.9	88.5	94.3
0330	Cast-in-Place Concrete	107.2	80.2	97.1	98.7	104.0	100.7	98.7	102.1	100.0	98.5	99.8	99.0	99.8	105.5	102.0	107.9	99.8	104.8
03	CONCRETE	103.0	88.8	96.3	103.8	93.1	98.8	103.9	88.9	96.8	107.3	92.4	100.2	105.9	94.3	100.4	111.4	93.0	102.7
04	MASONRY	118.6	68.1	88.4	96.6	89.2	92.2	91.6	89.2	90.2	103.4	86.6	93.3	95.4	91.2	92.9	104.8	86.6	93.9
05	METALS	106.7	82.2	99.3	97.9	96.7	97.5	97.9	89.6	95.4	98.1	98.1	98.1	100.0	97.7	99.3	98.1	99.4	98.5
06	WOOD, PLASTICS & COMPOSITES	91.1	100.7	96.2	83.7	88.2	86.1	85.6	88.2	86.9	84.7	87.9	86.4	107.5	87.8	97.0	81.3	87.9	84.8
07	THERMAL & MOISTURE PROTECTION	102.8	78.4	93.0	104.1	88.6	97.9	104.2	83.2	95.7	104.5	85.1	96.7	103.9	89.9	98.2	104.4	85.1	96.6
08	OPENINGS	101.0	84.2	96.7	96.9	82.0	93.0	97.8	78.2	92.7	97.8	82.3	93.8	97.5	81.8	93.4	97.8	87.3	95.1
0920	Plaster & Gypsum Board	103.5	100.7	101.8	106.0	87.7	95.1	106.3	87.7	95.2	106.7	87.4	95.2	114.6	87.3	98.3	105.4	87.4	94.7
0950, 0980	Ceilings & Acoustic Treatment	106.2	100.7	102.9	91.8	87.7	89.3	91.8	87.7	89.3	93.5	87.4	89.8	98.6	87.3	91.8	93.5	87.4	89.8
0960	Flooring	117.0	62.6	102.5	95.8	105.1	98.3	96.5	105.1	98.8	96.1	99.7	97.0	105.1	105.1	105.1	94.8	99.7	96.1
0990	Wall Finishes & Painting/Coating	114.1	72.3	89.1	112.5	43.3	71.1	112.5	43.3	71.1	112.5	98.1	103.9	112.5	92.9	100.8	112.5	98.1	103.9
09	FINISHES	111.7	87.0	98.9	96.9	86.8	91.7	97.2	86.8	91.8	98.0	91.7	94.8	102.9	92.7	97.6	97.3	91.7	94.4
COVERS	DIVS. 10 - 14, 25, 28, 41, 43, 44	100.0	98.0	99.6	100.0	67.8	93.4	100.0	67.7	93.3	100.0	90.0	97.9	100.0	90.7	98.1	100.0	90.0	97.9
22, 23	PLUMBING & HVAC	100.2	97.1	98.9	96.0	79.6	89.4	96.0	64.0	82.9	96.0	90.4	93.7	100.2	81.5	92.6	96.0	88.5	92.9
26, 27, 3370	ELECTRICAL, COMMUNICATIONS & UTIL.	108.3	94.1	101.5	94.5	83.6	89.3	96.8	53.6	76.1	98.1	92.3	95.3	97.8	83.6	91.0	98.1	92.3	95.3
MF2004	WEIGHTED AVERAGE	104.5	90.4	98.4	97.9	86.1	92.8	98.0	77.1	89.0	99.4	90.4	95.5	100.6	88.4	95.3	99.9	90.5	95.8

DIVISION		WEST VIRGINIA																	
		GASSAWAY 266			HUNTINGTON 255 - 257			LEWISBURG 249			MARTINSBURG 254			MORGANTOWN 265			PARKERSBURG 261		
		MAT.	INST.	TOTAL	MAT.	INST.	TOTAL	MAT.	INST.	TOTAL	MAT.	INST.	TOTAL	MAT.	INST.	TOTAL	MAT.	INST.	TOTAL
01543	CONTRACTOR EQUIPMENT	.0	100.1	100.1	.0	100.1	100.1	.0	100.1	100.1	.0	100.1	100.1	.0	100.1	100.1	.0	100.1	100.1
0241, 31 - 34	SITE & INFRASTRUCTURE, DEMOLITION	105.6	86.0	91.5	106.3	87.6	92.9	119.1	85.7	95.2	105.9	83.6	89.9	102.7	85.7	90.5	111.4	86.8	93.7
0310	Concrete Forming & Accessories	86.6	88.5	88.2	99.3	111.4	109.7	84.8	87.3	86.9	86.4	78.6	79.7	84.9	88.4	87.9	89.2	88.9	88.9
0310	Concrete Reinforcing	99.9	83.4	91.8	100.7	87.1	94.1	99.3	66.6	83.6	99.3	72.6	86.2	99.9	88.5	94.3	99.3	85.2	92.4
0330	Cast-in-Place Concrete	103.0	104.5	103.6	107.7	109.5	108.4	98.8	103.8	100.7	103.5	91.0	98.7	98.4	99.8	99.0	100.6	99.4	100.2
03	CONCRETE	107.5	93.7	101.0	109.4	106.3	107.9	114.4	89.8	102.8	107.6	82.4	95.7	103.6	93.0	98.6	109.3	92.5	101.4
04	MASONRY	106.2	89.4	96.1	97.7	90.4	93.3	96.5	89.2	92.2	99.0	76.8	85.6	125.7	86.6	102.3	82.1	87.4	85.3
05	METALS	98.1	97.7	97.9	100.1	99.4	99.9	98.0	90.2	95.7	98.3	86.1	94.6	98.1	99.3	98.5	98.7	98.2	98.5
06	WOOD, PLASTICS & COMPOSITES	83.9	87.9	86.0	97.0	116.6	107.4	81.6	88.2	85.1	83.7	80.5	82.1	81.6	87.9	84.9	85.8	87.9	86.9
07	THERMAL & MOISTURE PROTECTION	104.2	87.1	97.3	104.3	93.5	100.0	105.0	88.6	98.4	104.4	69.6	90.4	104.2	85.1	96.5	104.1	85.8	96.7
08	OPENINGS	95.7	81.9	92.1	96.3	98.2	96.8	97.8	78.2	92.7	99.2	69.4	91.4	99.2	87.3	96.1	96.9	81.3	92.8
0920	Plaster & Gypsum Board	106.2	87.4	95.1	113.6	116.8	115.5	105.4	87.7	94.9	106.4	79.8	90.6	105.4	87.4	94.7	108.6	87.4	96.0
0950, 0980	Ceilings & Acoustic Treatment	93.5	87.4	89.8	94.4	116.6	107.9	93.5	87.7	90.0	93.5	79.8	85.3	93.5	87.4	89.8	93.5	87.4	89.8
0960	Flooring	95.9	105.1	98.3	105.1	100.5	103.9	94.9	105.1	97.6	95.8	64.6	87.5	94.9	99.7	96.2	99.7	99.2	99.6
0990	Wall Finishes & Painting/Coating	112.5	92.9	100.8	112.5	92.3	100.4	112.5	43.3	71.1	112.5	51.1	75.8	112.5	98.1	103.9	112.5	93.6	101.2
09	FINISHES	97.6	92.3	94.8	101.5	108.9	105.3	98.3	86.8	92.3	97.7	73.8	85.3	96.9	91.7	94.3	99.4	91.4	95.3
COVERS	DIVS. 10 - 14, 25, 28, 41, 43, 44	100.0	90.0	97.9	100.0	94.8	98.9	100.0	67.7	93.3	100.0	65.3	92.9	100.0	80.5	96.0	100.0	90.5	98.0
22, 23	PLUMBING & HVAC	96.0	88.2	92.8	100.2	88.0	95.2	96.0	79.4	89.2	96.0	76.8	88.2	96.0	90.3	93.6	100.1	88.4	95.4
26, 27, 3370	ELECTRICAL, COMMUNICATIONS & UTIL.	98.1	83.6	91.1	97.8	91.9	94.9	94.5	53.6	74.9	99.7	73.9	87.1	98.2	92.3	95.4	98.2	92.8	95.6
MF2004	WEIGHTED AVERAGE	99.3	89.3	95.0	100.8	96.0	98.8	99.8	80.6	91.5	99.5	77.3	89.9	100.0	90.6	95.9	99.7	90.3	95.6

City Cost Indexes

		WEST VIRGINIA									WISCONSIN								
		PETERSBURG			ROMNEY			WHEELING			BELOIT			EAU CLAIRE			GREEN BAY		
	DIVISION	268			267			260			535			547			541 - 543		
		MAT.	INST.	TOTAL	MAT.	INST.	TOTAL	MAT.	INST.	TOTAL	MAT.	INST.	TOTAL	MAT.	INST.	TOTAL	MAT.	INST.	TOTAL
01543	CONTRACTOR EQUIPMENT	.0	100.1	100.1	.0	100.1	100.1	.0	100.1	100.1	.0	101.5	101.5	.0	101.1	101.1	.0	98.7	98.7
0241, 31 - 34	SITE & INFRASTRUCTURE, DEMOLITION	102.0	86.2	90.6	105.1	86.1	91.5	112.0	86.4	93.7	90.1	107.4	102.5	89.1	103.7	99.6	92.3	99.6	97.5
0310	Concrete Forming & Accessories	88.3	82.4	83.2	84.0	82.0	82.3	91.1	88.7	89.0	98.6	95.7	96.1	97.8	96.4	96.6	107.7	95.6	97.3
0310	Concrete Reinforcing	99.3	77.1	88.5	99.9	69.3	84.9	98.6	88.6	93.7	96.5	102.4	99.3	94.2	104.2	99.1	92.4	89.2	90.9
0330	Cast-in-Place Concrete	98.4	96.3	97.6	103.0	96.2	100.5	100.6	99.9	100.4	98.7	98.8	98.3	96.7	95.6	96.3	100.0	99.9	100.0
03	CONCRETE	103.8	87.0	95.9	107.3	85.4	97.0	109.3	93.2	101.7	95.8	98.2	96.9	96.0	97.9	96.9	97.1	96.2	96.7
04	MASONRY	99.2	86.6	91.7	96.9	72.5	82.3	106.5	86.6	94.6	101.5	99.3	100.2	89.8	99.6	95.7	120.3	98.7	107.4
05	METALS	98.2	94.6	97.1	98.2	90.5	95.9	98.8	99.6	99.1	97.4	102.4	98.9	93.7	105.0	97.1	95.7	98.1	96.4
06	WOOD, PLASTICS & COMPOSITES	85.8	80.5	83.0	80.5	80.5	80.5	87.7	87.9	87.8	108.9	95.3	101.7	111.5	95.4	103.0	117.4	95.4	105.7
07	THERMAL & MOISTURE PROTECTION	104.3	76.5	93.1	104.4	72.9	91.7	104.5	86.1	97.1	99.0	98.6	98.8	97.8	88.6	94.1	99.7	87.0	94.6
08	OPENINGS	99.2	76.5	93.4	99.1	74.9	92.9	97.8	87.3	95.1	102.6	99.6	101.8	100.1	96.1	99.1	98.4	94.4	97.4
0920	Plaster & Gypsum Board	106.7	79.8	90.7	104.8	79.8	89.9	109.0	87.4	96.1	99.8	95.6	96.9	102.3	95.6	98.3	93.8	95.6	94.8
0950, 0980	Ceilings & Acoustic Treatment	93.5	79.8	85.3	93.5	79.8	85.3	93.5	87.4	89.8	85.7	95.6	91.7	98.8	95.6	96.9	92.2	95.6	94.2
0960	Flooring	97.0	97.7	97.7	94.6	72.8	88.8	100.8	99.7	100.5	97.3	114.9	102.0	92.7	110.0	97.3	111.0	110.0	110.7
0990	Wall Finishes & Painting/Coating	112.5	92.0	100.2	112.5	92.0	100.2	112.5	98.1	103.9	94.3	97.0	95.9	88.3	91.2	90.1	97.6	77.3	85.5
09	FINISHES	97.8	86.8	92.1	97.0	81.2	88.8	99.8	91.7	95.6	95.2	99.9	97.5	97.7	98.5	98.1	101.2	96.7	98.9
COVERS	DIVS. 10 - 14, 25, 28, 41, 43, 44	100.0	79.5	95.8	100.0	89.1	97.8	100.0	97.5	99.5	100.0	92.7	98.5	100.0	94.2	98.8	100.0	93.7	98.7
22, 23	PLUMBING & HVAC	96.0	91.1	94.0	96.0	91.0	94.0	100.2	90.3	96.2	100.0	92.4	96.9	100.2	86.9	94.8	100.5	86.0	94.6
26, 27, 3370	ELECTRICAL, COMMUNICATIONS & UTIL.	101.1	81.9	91.8	100.4	81.9	91.5	95.7	92.3	94.1	94.8	91.7	93.3	99.9	87.8	94.1	95.6	87.0	91.5
MF2004	WEIGHTED AVERAGE	99.1	86.6	93.7	99.3	84.0	92.7	100.9	91.2	96.7	98.2	97.4	97.8	97.6	95.1	96.5	99.6	93.1	96.8

		WISCONSIN																
		KENOSHA			LA CROSSE			LANCASTER			MADISON			MILWAUKEE			NEW RICHMOND		
	DIVISION	531			546			538			537			530,532			540		
		MAT.	INST.	TOTAL	MAT.	INST.	TOTAL	MAT.	INST.	TOTAL	MAT.	INST.	TOTAL	MAT.	INST.	TOTAL	MAT.	INST.	TOTAL
01543	CONTRACTOR EQUIPMENT	.0	99.2	99.2	.0	101.1	101.1	.0	101.5	101.5	.0	101.5	101.5	.0	87.6	87.6	.0	102.0	102.0
0241, 31 - 34	SITE & INFRASTRUCTURE, DEMOLITION	95.7	104.1	101.8	82.7	103.7	97.7	89.2	107.2	102.1	89.0	107.6	102.3	90.7	95.8	94.4	83.1	104.7	98.6
0310	Concrete Forming & Accessories	107.4	105.3	105.6	83.4	95.8	94.0	97.8	93.8	94.4	99.3	99.6	99.6	101.6	117.3	115.1	91.9	106.3	104.2
0310	Concrete Reinforcing	96.3	97.0	96.7	93.9	88.2	91.1	97.7	88.1	93.0	96.5	88.4	92.5	96.5	97.3	96.9	91.6	104.2	97.7
0330	Cast-in-Place Concrete	106.7	102.0	104.9	86.9	96.8	90.6	97.4	97.9	97.6	93.2	100.2	95.8	95.7	112.1	101.9	100.6	92.3	97.4
03	CONCRETE	100.7	102.6	101.6	87.6	95.0	91.1	95.5	94.3	94.9	93.4	97.8	95.4	95.0	110.8	102.5	94.2	101.2	97.5
04	MASONRY	98.4	109.2	104.9	89.0	99.6	95.4	101.6	96.9	98.8	101.7	102.3	102.1	101.4	118.1	111.4	114.7	99.9	105.8
05	METALS	98.0	101.1	99.0	93.6	97.4	94.7	95.2	94.2	94.9	100.0	95.9	98.8	100.7	92.9	98.3	93.7	104.2	96.9
06	WOOD, PLASTICS & COMPOSITES	114.3	105.0	109.4	94.2	95.4	94.8	108.0	95.3	101.3	108.6	99.5	103.8	111.6	117.5	114.7	99.7	109.3	104.8
07	THERMAL & MOISTURE PROTECTION	99.2	100.7	99.8	97.2	88.0	93.5	98.7	81.8	91.9	95.4	95.5	95.4	97.5	112.4	103.5	98.5	99.7	98.9
08	OPENINGS	97.8	104.1	99.4	100.1	83.6	95.8	97.7	84.6	94.3	102.6	98.5	101.5	104.8	110.9	106.4	88.1	103.6	92.2
0920	Plaster & Gypsum Board	89.7	105.5	99.1	96.8	95.6	96.1	97.5	95.6	96.4	88.3	99.9	95.2	99.6	118.2	110.7	87.3	110.1	100.9
0950, 0980	Ceilings & Acoustic Treatment	80.2	105.5	95.4	97.2	95.6	96.2	79.8	95.6	89.3	87.6	99.9	95.0	89.1	118.2	106.6	67.8	110.1	93.3
0960	Flooring	114.1	115.7	114.5	85.2	110.0	91.8	96.5	110.5	100.3	92.0	108.0	96.3	100.0	120.2	105.4	103.6	110.0	105.3
0990	Wall Finishes & Painting/Coating	104.1	104.5	104.3	88.3	69.3	77.0	94.3	82.1	87.0	94.3	97.0	95.9	96.5	111.8	105.6	102.0	91.2	95.6
09	FINISHES	98.0	107.7	103.0	93.8	96.0	95.0	93.3	95.3	94.4	92.3	101.5	97.1	97.1	118.1	108.0	91.9	104.9	98.6
COVERS	DIVS. 10 - 14, 25, 28, 41, 43, 44	100.0	100.1	100.0	100.0	94.2	98.8	100.0	57.5	91.3	100.0	94.3	98.8	100.0	104.2	100.9	100.0	95.4	99.0
22, 23	PLUMBING & HVAC	100.2	93.8	97.6	100.2	86.5	94.6	95.7	84.7	91.2	99.8	89.0	95.4	100.0	101.8	100.7	95.3	86.5	91.7
26, 27, 3370	ELECTRICAL, COMMUNICATIONS & UTIL.	95.1	95.5	95.3	100.2	87.8	94.2	94.4	87.7	91.2	96.1	91.2	93.8	96.1	103.8	99.8	98.6	87.8	93.4
MF2004	WEIGHTED AVERAGE	98.8	101.1	99.8	95.9	93.1	94.7	96.1	91.1	93.9	98.0	96.5	97.4	99.1	106.7	102.4	95.6	97.1	96.2

		WISCONSIN																
		OSHKOSH			PORTAGE			RACINE			RHINELANDER			SUPERIOR			WAUSAU		
	DIVISION	549			539			534			545			548			544		
		MAT.	INST.	TOTAL	MAT.	INST.	TOTAL	MAT.	INST.	TOTAL	MAT.	INST.	TOTAL	MAT.	INST.	TOTAL	MAT.	INST.	TOTAL
01543	CONTRACTOR EQUIPMENT	.0	98.7	98.7	.0	101.5	101.5	.0	101.5	101.5	.0	98.7	98.7	.0	102.0	102.0	.0	98.7	98.7
0241, 31 - 34	SITE & INFRASTRUCTURE, DEMOLITION	84.1	99.4	95.1	79.8	107.4	99.6	89.9	108.1	103.0	97.0	99.3	98.7	79.8	104.8	97.8	79.7	99.5	93.9
0310	Concrete Forming & Accessories	89.6	95.3	94.5	90.2	96.0	95.1	99.0	105.3	104.4	87.0	94.9	93.8	90.2	96.5	95.6	89.0	95.2	94.4
0310	Concrete Reinforcing	92.5	89.2	90.9	97.8	88.3	93.2	96.5	97.0	96.7	92.7	92.8	92.7	91.6	93.5	92.5	92.7	88.1	90.5
0330	Cast-in-Place Concrete	92.7	99.6	95.3	83.6	102.4	90.7	96.1	101.7	98.2	105.1	99.6	103.0	94.8	100.0	96.8	86.4	99.7	91.4
03	CONCRETE	89.3	95.9	92.4	84.1	96.9	90.1	94.8	102.5	98.4	100.9	96.4	98.8	88.9	97.4	92.9	84.6	95.7	89.9
04	MASONRY	101.8	98.7	99.9	100.5	102.3	101.6	101.4	109.2	106.1	119.6	98.7	107.1	114.0	110.0	111.6	101.3	98.7	99.8
05	METALS	93.9	97.3	95.0	95.8	95.4	95.7	99.0	101.1	99.7	93.8	98.9	95.4	94.6	100.9	96.5	93.7	97.2	94.8
06	WOOD, PLASTICS & COMPOSITES	95.8	95.4	95.6	98.5	95.3	96.8	109.2	105.0	107.0	92.6	95.4	94.1	97.9	94.3	96.0	94.9	95.4	95.2
07	THERMAL & MOISTURE PROTECTION	98.8	88.6	94.7	98.2	94.6	96.8	99.0	100.3	99.5	99.6	83.3	93.0	98.2	101.1	99.4	98.6	82.2	92.0
08	OPENINGS	93.7	94.4	93.9	97.9	92.9	96.6	102.6	104.1	103.0	93.7	87.7	92.2	88.1	92.8	89.3	93.9	94.1	94.0
0920	Plaster & Gypsum Board	85.4	95.6	91.5	98.8	95.5	102.8	98.8	105.5	102.8	84.5	95.6	91.1	87.1	94.8	91.7	84.8	95.6	91.2
0950, 0980	Ceilings & Acoustic Treatment	92.2	95.6	94.2	80.7	95.6	89.6	85.7	105.5	97.6	92.2	95.6	94.2	68.6	94.8	84.4	92.2	95.6	94.2
0960	Flooring	101.4	110.0	103.7	92.6	114.9	98.5	97.3	115.7	102.2	100.5	110.0	103.0	104.9	126.8	110.8	101.2	110.0	103.5
0990	Wall Finishes & Painting/Coating	97.6	77.3	85.5	94.3	82.1	87.0	94.3	104.5	100.4	97.6	74.6	83.9	93.1	101.2	97.9	97.6	77.3	85.5
09	FINISHES	96.5	95.4	95.9	91.3	97.9	94.7	95.2	107.7	101.7	96.9	95.8	96.4	91.6	102.7	97.3	96.1	96.7	96.4
COVERS	DIVS. 10 - 14, 25, 28, 41, 43, 44	100.0	79.8	95.8	100.0	78.0	95.5	100.0	100.1	100.0	100.0	66.2	93.1	100.0	93.9	98.7	100.0	93.7	98.7
22, 23	PLUMBING & HVAC	96.2	85.9	92.0	95.7	88.8	92.9	100.0	93.8	97.5	96.2	91.2	94.2	95.3	101.8	98.0	96.2	85.6	91.9
26, 27, 3370	ELECTRICAL, COMMUNICATIONS & UTIL.	99.1	89.4	94.4	97.7	91.2	94.6	94.6	97.8	96.1	98.5	82.7	90.9	102.9	94.3	98.8	100.0	82.7	91.7
MF2004	WEIGHTED AVERAGE	95.6	92.8	94.4	94.7	95.1	94.9	98.3	101.7	99.8	98.1	92.4	95.7	95.4	100.6	97.6	94.9	92.2	93.7

City Cost Indexes

DIVISION		WYOMING																	
		CASPER 826			CHEYENNE 820			NEWCASTLE 827			RAWLINS 823			RIVERTON 825			ROCK SPRINGS 829 - 831		
		MAT.	INST.	TOTAL	MAT.	INST.	TOTAL	MAT.	INST.	TOTAL	MAT.	INST.	TOTAL	MAT.	INST.	TOTAL	MAT.	INST.	TOTAL
01543	CONTRACTOR EQUIPMENT	.0	101.1	101.1	.0	101.1	101.1	.0	101.1	101.1	.0	101.1	101.1	.0	101.1	101.1	.0	101.1	101.1
0241, 31 - 34	SITE & INFRASTRUCTURE, DEMOLITION	91.9	100.1	97.8	87.5	100.1	96.5	79.7	98.9	93.4	93.5	98.9	97.3	87.2	98.9	95.6	82.8	98.9	94.3
0310	Concrete Forming & Accessories	100.1	48.0	55.4	103.6	65.8	71.1	92.8	42.9	50.0	97.5	42.9	50.7	91.5	42.9	49.8	99.2	42.9	50.9
0310	Concrete Reinforcing	114.9	49.4	82.9	108.4	50.3	80.0	116.6	50.2	84.1	116.1	50.2	83.9	117.2	50.1	84.4	117.2	50.1	84.4
0330	Cast-in-Place Concrete	97.9	78.9	90.8	98.0	79.1	90.9	99.0	58.2	83.7	99.1	58.2	83.7	99.1	58.2	83.7	99.0	58.2	83.7
03	CONCRETE	107.2	59.6	84.7	107.0	67.7	88.4	107.6	50.3	80.5	121.4	50.3	87.8	115.8	50.3	84.9	108.1	50.3	80.8
04	MASONRY	106.9	42.2	68.2	104.3	58.3	76.8	101.0	52.2	71.8	101.0	52.2	71.8	101.0	52.2	71.8	163.7	52.2	97.0
05	METALS	100.6	62.6	89.1	101.8	64.3	90.5	97.2	61.9	86.5	97.3	61.9	86.6	97.4	61.8	86.6	98.0	61.8	87.1
06	WOOD, PLASTICS & COMPOSITES	97.4	45.3	69.9	100.9	68.7	83.8	89.1	41.6	64.0	94.3	41.6	66.4	87.8	41.6	63.4	97.1	41.6	67.8
07	THERMAL & MOISTURE PROTECTION	98.5	55.1	81.1	97.9	61.8	83.4	98.7	52.9	80.3	100.3	52.9	81.2	99.7	52.9	80.9	98.8	52.9	80.3
08	OPENINGS	93.6	44.9	81.0	95.7	57.5	85.8	99.7	43.0	85.0	99.4	43.0	84.8	99.6	43.0	84.9	100.1	43.0	85.3
0920	Plaster & Gypsum Board	96.2	43.6	64.9	89.8	67.6	76.6	84.9	39.9	58.1	86.8	39.9	58.9	84.9	39.9	58.1	91.6	39.9	60.8
0950, 0980	Ceilings & Acoustic Treatment	101.8	43.6	66.8	103.7	67.6	82.0	101.8	39.9	64.6	101.8	39.9	64.6	101.8	39.9	64.6	101.8	39.9	64.6
0960	Flooring	99.1	39.7	83.3	104.1	63.8	93.4	97.0	56.1	86.1	100.1	56.1	88.4	96.3	56.1	85.6	102.6	56.1	90.2
0990	Wall Finishes & Painting/Coating	99.3	53.2	71.7	100.9	53.2	72.4	98.7	34.1	60.0	98.7	34.1	60.0	98.7	34.1	60.0	98.7	34.1	60.0
09	FINISHES	100.1	45.9	72.2	101.3	64.6	82.4	95.5	43.2	68.5	98.1	43.2	69.8	96.1	43.2	68.8	98.3	43.2	69.9
COVERS	DIVS. 10 - 14, 25, 28, 41, 43, 44	100.0	78.2	95.5	100.0	81.1	96.1	100.0	70.5	93.9	100.0	70.5	93.9	100.0	70.4	93.9	100.0	70.4	93.9
22, 23	PLUMBING & HVAC	99.9	65.9	86.1	100.0	60.3	83.8	97.7	56.2	80.8	97.7	56.2	80.8	97.7	56.2	80.8	99.9	56.2	82.1
26, 27, 3370	ELECTRICAL, COMMUNICATIONS & UTIL.	94.4	63.7	79.5	94.4	72.2	83.7	91.6	71.2	81.8	91.6	71.2	81.8	91.6	63.3	78.0	90.1	62.9	77.1
MF2004	WEIGHTED AVERAGE	99.8	61.4	83.2	100.0	67.8	86.1	97.9	59.3	81.2	100.2	59.3	82.5	99.1	58.2	81.4	102.1	58.1	83.1

DIVISION		WYOMING												CANADA					
		SHERIDAN 828			WHEATLAND 822			WORLAND 824			YELLOWSTONE NAT'L PA 821			BARRIE, ONTARIO			BATHURST, NEW BRUNSWICK		
		MAT.	INST.	TOTAL	MAT.	INST.	TOTAL	MAT.	INST.	TOTAL	MAT.	INST.	TOTAL	MAT.	INST.	TOTAL	MAT.	INST.	TOTAL
01543	CONTRACTOR EQUIPMENT	.0	101.1	101.1	.0	101.1	101.1	.0	101.1	101.1	.0	101.1	101.1	.0	100.4	100.4	.0	98.6	98.6
0241, 31 - 34	SITE & INFRASTRUCTURE, DEMOLITION	87.5	100.1	96.5	83.8	98.9	94.6	81.3	98.9	93.9	81.3	99.3	94.2	117.4	100.4	105.2	96.8	95.5	95.8
0310	Concrete Forming & Accessories	101.8	48.2	55.8	95.0	43.4	50.7	95.1	42.9	50.3	95.2	45.6	52.6	122.8	94.6	98.6	99.3	67.1	71.7
0310	Concrete Reinforcing	117.2	50.3	84.5	116.6	50.2	84.1	117.2	50.1	84.4	119.3	50.1	85.5	179.9	91.2	136.5	148.4	61.5	105.9
0330	Cast-in-Place Concrete	102.4	78.9	93.6	103.4	58.8	86.6	99.0	58.2	83.7	99.0	62.2	85.2	167.7	91.4	139.0	138.3	62.7	109.8
03	CONCRETE	116.1	59.8	89.5	112.7	50.7	83.4	107.8	50.3	80.7	108.2	52.8	82.0	153.8	92.8	125.0	135.5	65.0	102.2
04	MASONRY	101.3	57.6	75.1	101.4	53.3	72.6	101.0	52.2	71.8	101.1	59.3	76.0	168.9	103.2	129.6	159.1	68.5	104.9
05	METALS	100.6	63.9	89.5	97.2	61.9	86.5	97.4	61.8	86.6	97.9	61.8	87.0	107.7	92.4	103.1	105.3	71.3	95.0
06	WOOD, PLASTICS & COMPOSITES	102.3	45.3	72.1	91.3	41.6	65.0	91.3	41.6	65.1	91.3	41.6	65.1	120.0	92.8	105.6	93.3	67.1	79.5
07	THERMAL & MOISTURE PROTECTION	99.8	59.2	83.4	99.0	54.0	80.9	98.7	52.9	80.3	98.3	56.1	81.6	109.8	93.6	103.3	103.8	65.4	88.3
08	OPENINGS	100.3	46.0	86.2	98.4	43.0	84.0	99.9	43.0	85.2	93.2	43.0	80.2	91.4	91.2	91.3	86.3	58.7	79.2
0920	Plaster & Gypsum Board	101.4	43.6	67.0	85.2	39.9	58.2	85.2	39.9	58.2	85.6	39.9	58.4	152.7	92.7	117.0	168.1	66.3	107.5
0950, 0980	Ceilings & Acoustic Treatment	103.7	43.6	67.6	101.8	39.9	64.6	101.8	39.9	64.6	103.5	39.9	65.2	95.2	92.7	93.7	96.1	66.3	78.1
0960	Flooring	99.1	56.1	87.7	98.6	56.1	87.2	98.6	56.1	87.2	98.6	56.1	87.2	132.6	101.8	124.3	113.6	48.6	96.2
0990	Wall Finishes & Painting/Coating	100.6	53.2	72.2	98.7	34.1	60.0	98.7	34.1	60.0	98.7	34.1	60.0	110.0	93.3	100.0	109.2	53.6	75.9
09	FINISHES	101.7	48.9	74.5	96.3	43.5	69.1	96.0	43.2	68.8	96.5	45.0	69.9	120.0	95.9	107.6	115.7	62.2	88.1
COVERS	DIVS. 10 - 14, 25, 28, 41, 43, 44	100.0	78.2	95.5	100.0	70.9	94.0	100.0	70.4	93.9	100.0	73.1	94.5	140.0	86.5	129.0	140.0	75.6	126.8
22, 23	PLUMBING & HVAC	97.7	58.9	81.9	97.7	56.8	81.0	97.7	56.2	80.8	97.7	59.9	82.3	102.8	110.0	105.7	102.7	77.2	92.3
26, 27, 3370	ELECTRICAL, COMMUNICATIONS & UTIL.	94.2	63.0	79.2	91.6	72.1	82.3	91.6	62.9	77.9	90.7	62.9	77.4	124.7	99.0	112.4	128.4	67.7	99.3
MF2004	WEIGHTED AVERAGE	100.7	62.1	84.0	98.6	59.8	81.8	98.1	58.1	80.8	97.5	60.4	81.5	119.7	99.1	110.8	115.3	71.2	96.3

DIVISION		CANADA																	
		BRANDON, MANITOBA			BRANTFORD, ONTARIO			BRIDGEWATER, NS			CALGARY, ALBERTA			CAP-DE-LA-MADELEINE, PQ			CHARLESBOURG, QUEBEC		
		MAT.	INST.	TOTAL	MAT.	INST.	TOTAL	MAT.	INST.	TOTAL	MAT.	INST.	TOTAL	MAT.	INST.	TOTAL	MAT.	INST.	TOTAL
01543	CONTRACTOR EQUIPMENT	.0	102.9	102.9	.0	100.4	100.4	.0	98.6	98.6	.0	107.0	107.0	.0	99.2	99.2	.0	99.2	99.2
0241, 31 - 34	SITE & INFRASTRUCTURE, DEMOLITION	112.5	99.5	103.2	115.0	101.0	104.9	99.7	97.2	98.0	117.3	105.7	109.0	94.6	98.6	97.5	94.6	98.6	97.5
0310	Concrete Forming & Accessories	122.0	76.6	83.0	122.7	102.6	105.4	92.5	75.6	78.0	126.8	94.8	99.3	129.6	93.0	98.2	129.6	93.0	98.2
0310	Concrete Reinforcing	161.5	57.8	110.8	174.8	89.8	133.2	148.4	50.0	100.6	161.5	66.1	114.9	148.4	85.7	117.7	148.4	85.7	117.7
0330	Cast-in-Place Concrete	151.2	78.8	123.9	163.7	114.1	145.0	168.2	73.7	132.6	184.0	104.6	154.1	132.0	100.5	120.1	132.0	100.5	120.1
03	CONCRETE	136.2	74.3	107.0	150.9	103.9	128.7	148.3	70.7	111.7	159.4	92.9	128.0	131.1	94.2	113.7	131.1	94.2	113.7
04	MASONRY	163.5	71.3	108.3	165.5	108.1	131.1	160.1	75.8	109.6	179.0	88.4	124.8	160.3	91.8	119.3	160.3	91.8	119.3
05	METALS	107.8	78.0	98.8	107.1	93.8	103.1	106.3	74.4	96.6	134.9	86.5	120.3	105.5	89.0	100.5	105.5	89.0	100.5
06	WOOD, PLASTICS & COMPOSITES	115.9	77.2	95.5	118.2	100.8	109.0	84.3	74.6	79.2	116.0	94.2	104.5	131.2	92.5	110.8	131.2	92.5	110.8
07	THERMAL & MOISTURE PROTECTION	104.1	75.4	92.6	109.7	99.3	105.5	104.7	73.4	92.1	124.8	90.7	111.1	104.3	94.2	100.2	104.3	94.2	100.2
08	OPENINGS	92.3	68.2	86.0	91.2	97.6	92.9	82.9	68.1	79.1	92.3	84.2	90.2	92.3	85.4	90.5	92.3	85.4	90.5
0920	Plaster & Gypsum Board	124.8	76.3	95.9	152.7	100.9	121.9	162.5	74.0	109.8	140.8	93.6	112.7	188.5	92.3	131.2	188.5	92.3	131.2
0950, 0980	Ceilings & Acoustic Treatment	95.2	76.3	83.8	95.2	100.9	98.6	95.2	72.0	82.4	108.7	93.6	99.6	95.2	92.3	93.5	95.2	92.3	93.5
0960	Flooring	132.5	72.3	116.4	132.5	101.8	124.3	109.5	69.3	98.8	134.9	95.4	124.3	132.5	102.5	124.5	132.5	102.5	124.5
0990	Wall Finishes & Painting/Coating	109.3	61.0	80.4	109.2	102.5	105.2	109.2	66.6	83.7	109.2	100.8	104.2	109.2	95.2	100.8	109.2	95.2	100.8
09	FINISHES	114.5	74.3	93.7	118.9	102.6	110.5	113.2	73.4	92.7	121.2	95.8	108.1	124.2	95.2	109.2	124.2	95.2	109.2
COVERS	DIVS. 10 - 14, 25, 28, 41, 43, 44	140.0	78.5	127.4	140.0	89.0	129.5	140.0	77.1	127.1	140.0	111.1	134.1	140.0	99.2	131.6	140.0	99.2	131.6
22, 23	PLUMBING & HVAC	102.7	93.2	98.8	102.7	113.3	107.0	102.7	93.5	99.0	101.1	92.7	97.7	103.1	100.8	102.2	103.1	100.8	102.2
26, 27, 3370	ELECTRICAL, COMMUNICATIONS & UTIL.	127.1	76.2	103.0	124.7	98.5	112.2	130.5	70.5	101.7	119.9	90.5	105.8	123.7	79.5	102.5	123.7	79.5	102.5
MF2004	WEIGHTED AVERAGE	117.0	80.6	101.3	118.9	103.3	112.2	116.7	79.1	100.5	124.8	93.1	111.1	116.0	93.0	106.1	116.0	93.0	106.1

City Cost Indexes

		CANADA																
	DIVISION	CHARLOTTETOWN, PEI			CHICOUTIMI, QUEBEC			CORNER BROOK, NFLD			CORNWALL, ONTARIO			DALHOUSIE, NB			DARTMOUTH, NOVA SCOTIA		
		MAT.	INST.	TOTAL	MAT.	INST.	TOTAL	MAT.	INST.	TOTAL	MAT.	INST.	TOTAL	MAT.	INST.	TOTAL	MAT.	INST.	TOTAL
01543	CONTRACTOR EQUIPMENT	.0	98.6	98.6	.0	99.2	99.2	.0	98.6	98.6	.0	100.4	100.4	.0	98.6	98.6	.0	98.6	98.6
0241, 31 - 34	SITE & INFRASTRUCTURE, DEMOLITION	107.7	94.8	98.5	94.0	97.9	96.8	116.0	95.2	101.1	112.8	100.5	103.9	96.7	95.5	95.8	108.0	97.2	100.3
0310	Concrete Forming & Accessories	92.8	61.3	65.7	129.6	98.8	103.2	101.1	64.9	70.0	120.2	94.7	98.3	99.3	67.1	71.7	92.6	75.6	78.0
0310	Concrete Reinforcing	158.5	50.8	105.8	112.0	101.0	106.6	148.4	52.8	101.6	174.8	89.5	133.1	148.4	61.5	105.9	155.0	50.8	104.0
0330	Cast-in-Place Concrete	159.4	62.0	122.8	130.2	103.9	120.3	172.4	71.6	134.5	147.1	103.8	130.8	138.3	62.7	109.8	170.0	73.7	133.8
03	CONCRETE	153.4	60.1	109.4	124.4	100.7	113.2	161.8	65.6	116.4	142.4	96.8	120.9	135.5	65.0	102.2	150.2	70.7	112.7
04	MASONRY	160.5	66.0	104.0	160.6	98.4	123.4	160.5	67.6	104.9	164.2	99.1	125.3	157.1	68.5	104.0	171.8	75.8	114.3
05	METALS	106.3	67.0	94.4	105.1	94.0	101.7	108.2	72.1	97.3	107.1	92.2	102.6	105.3	71.3	95.0	108.3	74.4	98.1
06	WOOD, PLASTICS & COMPOSITES	84.6	60.5	71.9	131.2	99.0	114.2	94.5	63.6	78.2	116.7	93.7	104.6	93.3	67.1	79.5	84.3	74.6	79.2
07	THERMAL & MOISTURE PROTECTION	104.7	62.8	87.9	104.3	100.9	102.9	106.7	64.9	89.9	109.5	92.4	102.6	106.6	65.4	90.0	104.7	73.4	92.1
08	OPENINGS	94.5	52.4	83.6	91.8	84.8	90.0	98.6	60.0	88.6	92.3	90.8	91.9	86.3	58.7	79.2	82.9	68.1	79.1
0920	Plaster & Gypsum Board	166.0	59.5	102.6	188.5	98.9	135.2	168.3	62.7	105.5	227.7	93.6	147.9	168.1	66.3	107.5	165.1	74.0	110.9
0950, 0980	Ceilings & Acoustic Treatment	109.5	59.5	79.5	95.2	98.9	97.4	96.9	62.7	76.3	98.6	93.6	95.6	96.1	66.3	78.1	107.0	74.0	87.1
0960	Flooring	109.6	64.8	97.7	132.5	102.5	124.5	114.4	58.9	99.6	132.5	100.3	123.9	113.6	48.6	96.2	109.5	69.3	98.8
0990	Wall Finishes & Painting/Coating	109.2	44.6	70.6	109.2	109.5	109.4	109.2	64.2	82.3	109.2	95.5	101.0	109.2	53.6	75.9	109.2	66.6	83.7
09	FINISHES	118.0	60.3	88.2	124.2	101.3	112.4	117.0	63.5	89.4	132.2	96.0	113.5	115.7	62.2	88.1	116.4	73.4	94.2
COVERS	DIVS. 10 - 14, 25, 28, 41, 43, 44	140.0	75.1	126.6	140.0	100.8	131.9	140.0	75.8	126.8	140.0	86.1	128.9	140.0	75.6	126.8	140.0	77.1	127.1
22, 23	PLUMBING & HVAC	102.7	70.6	89.6	102.7	102.1	102.5	102.7	79.5	93.3	103.1	111.0	106.3	102.7	77.2	92.3	102.7	93.5	99.0
26, 27, 3370	ELECTRICAL, COMMUNICATIONS & UTIL.	124.7	57.3	92.4	123.7	89.8	107.4	123.7	63.8	95.0	125.7	99.6	113.2	127.7	67.7	98.9	128.0	70.5	100.4
MF2004	WEIGHTED AVERAGE	118.4	66.4	95.9	115.1	97.8	107.6	120.2	71.4	99.1	119.3	99.5	110.7	115.2	71.2	96.2	118.1	79.1	101.3

		CANADA																
	DIVISION	EDMONTON, ALBERTA			FORT MCMURRAY, ALBERTA			FREDERICTON, NB			GATINEAU, QUEBEC			GRANBY, QUEBEC			HALIFAX, NOVA SCOTIA		
		MAT.	INST.	TOTAL	MAT.	INST.	TOTAL	MAT.	INST.	TOTAL	MAT.	INST.	TOTAL	MAT.	INST.	TOTAL	MAT.	INST.	TOTAL
01543	CONTRACTOR EQUIPMENT	.0	107.0	107.0	.0	103.7	103.7	.0	98.6	98.6	.0	99.2	99.2	.0	99.2	99.2	.0	98.6	98.6
0241, 31 - 34	SITE & INFRASTRUCTURE, DEMOLITION	126.1	105.7	111.5	112.1	101.8	104.7	96.1	95.5	95.7	94.4	98.6	97.4	95.0	98.6	97.6	100.1	96.6	97.6
0310	Concrete Forming & Accessories	124.7	94.8	99.0	119.6	86.9	91.5	120.5	67.4	74.9	129.6	92.9	98.1	129.6	92.8	98.0	92.7	78.5	80.6
0310	Concrete Reinforcing	161.5	66.1	114.9	161.5	66.1	114.9	149.3	61.6	106.4	157.1	85.7	122.2	157.1	85.7	122.2	155.0	62.5	109.7
0330	Cast-in-Place Concrete	198.7	104.6	163.3	151.2	100.4	132.1	131.6	62.8	105.7	130.2	100.5	119.0	134.5	100.5	121.7	171.3	76.1	135.5
03	CONCRETE	166.7	92.9	131.8	142.3	88.0	116.6	133.8	65.2	101.4	131.6	94.1	113.9	133.8	94.0	115.0	150.9	75.1	115.1
04	MASONRY	175.5	88.4	123.4	160.5	82.6	113.9	156.6	70.1	104.8	160.1	91.8	119.2	160.5	91.8	119.4	173.4	83.8	119.8
05	METALS	135.7	86.5	120.8	107.8	86.5	101.4	105.2	71.8	95.1	105.5	88.9	100.5	105.5	88.8	100.5	120.4	79.4	108.0
06	WOOD, PLASTICS & COMPOSITES	112.5	94.2	102.8	112.5	86.4	98.7	119.1	67.1	91.6	131.2	92.5	110.8	131.2	92.5	110.8	84.3	78.3	81.1
07	THERMAL & MOISTURE PROTECTION	123.4	90.7	110.3	114.9	86.3	103.4	106.9	66.4	90.6	104.3	94.2	100.2	104.3	92.6	99.6	104.9	77.8	94.0
08	OPENINGS	92.3	84.2	90.2	92.3	79.9	89.1	86.2	57.5	78.8	92.3	80.4	89.2	92.3	80.4	89.2	82.9	72.5	80.2
0920	Plaster & Gypsum Board	146.9	93.6	115.1	129.9	85.6	103.5	183.5	66.3	113.7	137.8	92.3	110.7	139.9	92.3	111.6	165.1	77.7	113.1
0950, 0980	Ceilings & Acoustic Treatment	123.9	93.6	105.6	95.2	85.6	89.4	104.5	66.3	81.5	95.2	92.3	93.5	95.2	92.3	93.5	107.0	77.7	89.4
0960	Flooring	134.9	95.4	124.3	132.5	95.4	122.6	127.6	48.6	106.5	132.5	102.5	124.5	132.5	102.5	124.5	109.5	69.3	98.8
0990	Wall Finishes & Painting/Coating	109.3	99.2	103.2	109.3	84.3	94.4	109.2	68.3	84.8	109.2	95.2	100.8	109.2	95.2	100.8	109.2	80.0	91.7
09	FINISHES	126.7	95.6	110.7	115.6	87.9	101.3	124.5	63.8	93.2	115.7	95.2	105.1	116.1	95.2	105.3	116.4	77.2	96.2
COVERS	DIVS. 10 - 14, 25, 28, 41, 43, 44	140.0	111.1	134.1	140.0	107.7	133.4	140.0	75.6	126.8	140.0	99.2	131.6	140.0	99.2	131.6	140.0	77.7	127.2
22, 23	PLUMBING & HVAC	101.1	92.7	97.7	102.7	105.8	104.0	102.7	87.9	96.7	103.1	100.8	102.2	102.7	100.8	101.9	100.7	80.6	92.5
26, 27, 3370	ELECTRICAL, COMMUNICATIONS & UTIL.	120.0	90.5	105.9	115.9	93.5	105.2	135.9	84.4	111.2	123.7	79.5	102.5	124.7	79.5	103.0	130.5	79.6	106.1
MF2004	WEIGHTED AVERAGE	126.3	93.1	111.9	116.6	93.1	106.4	116.9	76.2	99.3	115.3	92.8	105.6	115.7	92.7	105.8	119.8	80.4	102.8

		CANADA																
	DIVISION	HAMILTON, ONTARIO			HULL, QUEBEC			JOLIETTE, QUEBEC			KAMLOOPS, BC			KINGSTON, ONTARIO			KITCHENER, ONTARIO		
		MAT.	INST.	TOTAL	MAT.	INST.	TOTAL	MAT.	INST.	TOTAL	MAT.	INST.	TOTAL	MAT.	INST.	TOTAL	MAT.	INST.	TOTAL
01543	CONTRACTOR EQUIPMENT	.0	103.1	103.1	.0	99.2	99.2	.0	99.2	99.2	.0	103.5	103.5	.0	103.1	103.1	.0	103.0	103.0
0241, 31 - 34	SITE & INFRASTRUCTURE, DEMOLITION	115.6	105.0	108.0	94.4	98.6	97.4	95.1	98.6	97.6	117.8	102.7	107.0	112.8	104.7	107.0	102.1	104.6	103.9
0310	Concrete Forming & Accessories	121.7	99.5	102.6	129.6	92.9	98.1	129.6	93.0	98.2	118.5	96.1	99.2	120.3	94.8	98.4	116.3	91.4	94.9
0310	Concrete Reinforcing	175.4	87.0	132.2	157.1	85.7	122.2	148.4	85.7	117.7	116.5	83.5	100.4	174.8	89.5	133.1	110.4	86.9	98.9
0330	Cast-in-Place Concrete	154.0	99.7	133.6	130.2	100.5	119.0	135.6	100.5	122.4	117.3	104.9	112.6	147.1	103.8	130.8	140.8	84.1	119.5
03	CONCRETE	146.1	97.0	122.9	131.6	94.1	113.9	133.0	94.2	114.6	136.6	96.6	117.7	142.4	96.8	120.9	122.9	88.1	106.5
04	MASONRY	166.7	100.2	126.9	160.1	91.8	119.2	160.5	91.8	119.4	167.3	100.0	127.1	170.7	99.2	127.9	162.1	96.3	122.7
05	METALS	138.4	90.8	124.0	105.5	88.9	100.5	105.5	89.0	100.5	107.7	89.2	102.1	108.6	92.1	103.6	116.1	90.4	108.3
06	WOOD, PLASTICS & COMPOSITES	117.2	99.8	108.0	131.2	92.5	110.8	131.2	92.5	110.8	98.0	94.1	95.9	116.7	93.8	104.6	110.7	90.7	100.1
07	THERMAL & MOISTURE PROTECTION	110.1	95.3	104.1	104.3	94.2	100.2	104.3	94.2	100.2	120.9	91.1	109.0	109.5	93.6	103.1	108.8	91.8	102.0
08	OPENINGS	92.3	95.6	93.1	92.3	80.4	89.2	92.3	85.4	90.5	88.1	90.2	88.6	92.3	90.5	91.8	83.5	88.9	84.9
0920	Plaster & Gypsum Board	184.9	99.9	134.3	137.8	92.3	110.7	188.5	92.3	131.2	137.4	93.5	111.3	231.1	93.7	149.4	151.8	90.5	115.3
0950, 0980	Ceilings & Acoustic Treatment	112.1	99.9	104.7	95.2	92.3	93.5	95.2	92.3	93.5	95.2	93.5	94.2	113.7	93.7	101.7	107.0	90.5	97.1
0960	Flooring	132.5	101.8	124.3	132.5	102.5	124.5	132.5	102.5	124.5	130.1	57.3	110.7	132.5	100.3	123.9	128.1	101.8	121.1
0990	Wall Finishes & Painting/Coating	109.2	104.9	106.7	109.2	95.2	100.8	109.2	95.2	100.8	109.2	87.2	96.0	109.2	88.2	96.6	109.2	94.9	100.7
09	FINISHES	128.5	100.9	114.3	115.7	95.2	105.1	124.2	95.2	109.2	117.4	88.0	102.2	136.4	95.2	115.2	119.6	93.8	106.3
COVERS	DIVS. 10 - 14, 25, 28, 41, 43, 44	140.0	115.0	134.9	140.0	99.2	131.6	140.0	99.2	131.6	140.0	111.7	134.2	140.0	86.1	128.9	140.0	112.9	134.4
22, 23	PLUMBING & HVAC	101.1	96.2	99.1	102.7	100.8	101.9	102.7	100.8	101.9	102.7	104.6	103.5	103.1	111.1	106.4	100.7	93.1	97.6
26, 27, 3370	ELECTRICAL, COMMUNICATIONS & UTIL.	132.2	100.0	116.8	126.7	79.5	104.0	124.7	79.5	103.0	129.3	89.7	110.3	125.7	98.1	112.5	126.0	97.2	112.2
MF2004	WEIGHTED AVERAGE	124.8	98.7	113.5	115.6	92.8	105.7	116.3	93.0	106.3	117.8	96.4	108.6	120.2	99.6	111.3	115.5	94.5	106.4

City Cost Indexes

		\multicolumn{18}{c}{CANADA}															
DIVISION		LAVAL, QUEBEC			LETHBRIDGE, ALBERTA			LLOYDMINSTER, ALBERTA			LONDON, ONTARIO			MEDICINE HAT, ALBERTA			MONCTON, NEW BRUNSWICK		
		MAT.	INST.	TOTAL	MAT.	INST.	TOTAL	MAT.	INST.	TOTAL	MAT.	INST.	TOTAL	MAT.	INST.	TOTAL	MAT.	INST.	TOTAL
01543	CONTRACTOR EQUIPMENT	.0	99.2	99.2	.0	103.7	103.7	.0	103.7	103.7	.0	103.2	103.2	.0	103.7	103.7	.0	98.6	98.6
0241, 31 - 34	SITE & INFRASTRUCTURE, DEMOLITION	95.0	98.6	97.6	112.8	101.8	104.9	112.1	101.8	104.7	116.3	104.9	108.2	110.8	101.8	104.4	96.1	95.5	95.6
0310	Concrete Forming & Accessories	129.7	92.9	98.1	121.8	86.9	91.8	119.6	86.9	91.5	127.5	94.0	98.7	121.8	86.9	91.8	99.3	67.4	71.9
0310	Concrete Reinforcing	157.1	85.7	122.2	161.5	66.1	114.9	161.5	66.1	114.9	135.3	85.5	111.0	161.5	66.1	114.9	148.4	64.7	107.5
0330	Cast-in-Place Concrete	134.5	100.5	121.7	165.5	100.4	141.0	151.2	100.4	132.1	150.8	98.4	131.1	151.2	100.4	132.1	133.1	62.9	106.7
03	CONCRETE	133.8	94.1	115.1	149.7	87.9	120.5	142.3	88.0	116.6	138.4	93.9	117.4	142.4	87.9	116.7	132.9	65.8	101.3
04	MASONRY	160.4	91.8	119.4	161.6	82.6	114.3	160.5	82.6	113.9	166.3	98.7	125.9	160.5	82.6	113.9	158.7	68.5	104.7
05	METALS	105.6	88.9	100.5	107.8	86.4	101.4	107.8	86.5	101.4	124.0	90.4	113.8	107.8	86.4	101.4	105.3	74.4	95.9
06	WOOD, PLASTICS & COMPOSITES	131.2	92.5	110.8	116.0	86.4	100.3	112.5	86.4	98.7	117.2	92.9	104.4	116.0	86.4	100.3	93.3	67.1	79.5
07	THERMAL & MOISTURE PROTECTION	104.5	94.2	100.1	114.9	86.3	103.4	113.5	86.3	102.5	112.7	93.2	104.9	117.7	86.3	105.1	106.6	67.3	90.8
08	OPENINGS	92.3	80.4	89.2	92.3	79.9	89.1	92.3	79.9	89.1	93.2	90.7	92.6	92.3	79.9	89.1	86.3	60.5	79.6
0920	Plaster & Gypsum Board	142.1	92.3	112.4	131.5	85.6	104.2	129.9	85.6	103.5	185.2	92.8	130.2	131.5	85.6	104.2	168.1	66.3	107.5
0950, 0980	Ceilings & Acoustic Treatment	95.2	92.3	93.5	95.2	85.6	89.4	95.2	85.6	89.4	113.7	92.8	101.1	95.2	85.6	89.4	96.1	66.3	78.1
0960	Flooring	132.5	102.5	124.5	132.5	95.4	122.6	132.5	95.4	122.6	135.0	101.8	126.2	132.5	95.4	122.6	113.6	59.8	99.2
0990	Wall Finishes & Painting/Coating	109.2	95.2	100.8	109.2	84.3	94.3	109.2	84.3	94.4	109.2	102.7	105.3	109.2	84.3	94.3	109.2	53.6	75.9
09	FINISHES	116.4	95.2	105.5	115.7	87.9	101.4	115.6	87.9	101.3	129.9	96.5	112.7	115.7	87.9	101.4	115.5	64.5	89.3
COVERS	DIVS. 10 - 14, 25, 28, 41, 43, 44	140.0	99.2	131.6	140.0	107.7	133.4	140.0	107.7	133.4	140.0	114.0	134.7	140.0	107.7	133.4	140.0	75.6	126.8
22, 23	PLUMBING & HVAC	100.7	100.8	100.7	100.7	102.0	102.4	103.1	105.8	104.2	101.2	93.7	98.1	102.7	102.0	102.4	102.7	77.3	92.4
26, 27, 3370	ELECTRICAL, COMMUNICATIONS & UTIL.	125.8	79.5	103.6	115.9	93.5	105.2	115.9	93.5	105.2	132.8	97.2	115.7	115.9	93.5	105.2	132.6	67.7	101.5
MF2004	WEIGHTED AVERAGE	115.4	92.8	105.6	117.5	92.3	106.7	116.6	93.1	106.5	122.0	96.2	110.9	116.7	92.3	106.2	115.6	72.0	96.8

		\multicolumn{18}{c}{CANADA}															
DIVISION		MONTREAL, QUEBEC			MOOSE JAW, SASKATCHEWAN			NEWCASTLE, NB			NEW GLASGOW, NOVA SCOTIA			NORTH BAY, ONTARIO			OSHAWA, ONTARIO		
		MAT.	INST.	TOTAL	MAT.	INST.	TOTAL	MAT.	INST.	TOTAL	MAT.	INST.	TOTAL	MAT.	INST.	TOTAL	MAT.	INST.	TOTAL
01543	CONTRACTOR EQUIPMENT	.0	101.0	101.0	.0	98.6	98.6	.0	98.6	98.6	.0	98.6	98.6	.0	100.4	100.4	.0	103.0	103.0
0241, 31 - 34	SITE & INFRASTRUCTURE, DEMOLITION	96.7	98.2	97.7	112.1	95.8	100.4	96.8	95.5	95.8	100.0	97.2	98.0	115.0	100.1	104.3	114.9	104.4	107.4
0310	Concrete Forming & Accessories	130.0	99.1	103.5	103.4	63.9	69.4	99.3	67.1	71.7	92.5	75.6	78.0	122.7	92.1	96.4	122.9	94.6	98.6
0310	Concrete Reinforcing	159.7	101.0	131.0	114.0	67.3	91.2	148.4	61.5	105.9	148.4	50.8	100.6	174.8	89.0	132.8	174.8	91.8	134.2
0330	Cast-in-Place Concrete	132.0	105.5	122.0	146.7	72.8	118.8	138.3	62.7	109.8	170.0	73.7	133.8	163.7	89.6	135.8	162.9	90.8	135.8
03	CONCRETE	133.0	101.4	118.1	125.4	68.1	98.3	135.5	65.0	102.2	149.2	70.7	112.2	150.9	90.7	122.5	150.6	92.7	123.3
04	MASONRY	165.0	98.4	125.2	159.1	67.9	104.5	159.1	68.5	104.9	160.3	75.5	109.7	165.5	94.9	123.2	165.1	100.7	126.6
05	METALS	124.5	94.5	115.4	104.5	74.9	95.6	105.3	71.3	95.0	106.3	74.4	96.6	107.1	91.9	102.5	107.1	92.8	102.8
06	WOOD, PLASTICS & COMPOSITES	131.2	99.3	114.3	115.1	62.0	77.6	93.3	67.1	79.5	84.3	74.6	79.2	118.2	92.2	104.5	118.2	92.8	104.8
07	THERMAL & MOISTURE PROTECTION	105.1	101.5	103.6	103.6	67.4	89.1	106.6	65.4	90.0	104.7	73.4	92.1	109.7	89.9	101.7	109.8	92.3	102.7
08	OPENINGS	92.3	86.8	90.8	87.1	59.7	80.0	86.3	58.7	79.2	82.9	68.1	79.1	91.2	88.2	90.5	91.2	92.6	91.6
0920	Plaster & Gypsum Board	144.7	98.9	117.4	132.1	61.0	89.8	168.1	66.3	107.5	162.5	74.0	109.8	152.7	92.1	116.6	153.8	92.7	117.4
0950, 0980	Ceilings & Acoustic Treatment	107.0	98.9	102.1	95.2	61.0	74.7	96.1	66.3	78.1	95.2	74.0	82.4	95.2	92.1	93.3	100.3	92.7	95.7
0960	Flooring	132.5	102.5	124.5	119.8	65.1	105.2	113.6	48.6	96.2	109.5	69.3	98.8	132.5	100.3	123.9	132.5	104.4	125.0
0990	Wall Finishes & Painting/Coating	109.2	109.5	109.4	109.2	69.4	85.4	109.2	53.6	75.9	109.2	66.6	83.7	109.2	94.7	100.5	109.2	109.6	109.5
09	FINISHES	120.0	101.5	110.4	112.0	64.2	87.4	115.7	62.2	88.1	113.2	73.4	92.7	118.9	94.2	106.1	120.3	98.0	108.8
COVERS	DIVS. 10 - 14, 25, 28, 41, 43, 44	140.0	101.5	132.1	140.0	74.9	126.6	140.0	75.6	126.8	140.0	77.1	127.1	140.0	84.6	128.6	140.0	114.4	134.7
22, 23	PLUMBING & HVAC	101.2	102.2	101.6	103.1	84.3	95.5	102.7	77.2	92.3	102.7	93.5	99.0	102.7	108.8	105.2	100.7	112.2	105.4
26, 27, 3370	ELECTRICAL, COMMUNICATIONS & UTIL.	133.2	89.8	112.4	128.2	68.7	99.7	127.7	67.7	98.9	124.2	70.5	98.4	124.7	99.4	112.6	127.1	99.0	113.6
MF2004	WEIGHTED AVERAGE	119.8	98.1	110.5	114.3	73.8	96.8	115.3	71.2	96.3	116.1	79.1	100.1	118.9	97.2	109.5	118.8	100.7	111.0

		\multicolumn{18}{c}{CANADA}															
DIVISION		OTTAWA, ONTARIO			OWEN SOUND, ONTARIO			PETERBOROUGH, ONTARIO			PORTAGE LA PRAIRIE, MB			PRINCE ALBERT, SS			PRINCE GEORGE, BC		
		MAT.	INST.	TOTAL	MAT.	INST.	TOTAL	MAT.	INST.	TOTAL	MAT.	INST.	TOTAL	MAT.	INST.	TOTAL	MAT.	INST.	TOTAL
01543	CONTRACTOR EQUIPMENT	.0	103.0	103.0	.0	100.4	100.4	.0	100.4	100.4	.0	102.9	102.9	.0	98.6	98.6	.0	103.5	103.5
0241, 31 - 34	SITE & INFRASTRUCTURE, DEMOLITION	113.8	104.9	107.4	117.4	100.3	105.1	115.0	100.4	104.5	112.9	99.5	103.3	106.8	96.0	99.0	121.8	102.7	108.1
0310	Concrete Forming & Accessories	120.4	96.7	100.1	122.8	90.4	95.0	122.7	93.2	97.4	122.0	76.5	82.9	103.4	63.6	69.3	109.1	96.1	97.9
0310	Concrete Reinforcing	173.8	92.2	133.9	179.9	91.1	136.5	174.8	89.5	133.1	161.5	57.7	110.8	119.1	67.3	93.7	116.5	83.5	100.4
0330	Cast-in-Place Concrete	155.1	103.0	135.5	167.7	85.1	136.7	163.7	91.4	136.5	151.2	78.8	123.9	132.8	72.7	110.2	147.2	104.9	131.2
03	CONCRETE	146.3	97.9	123.4	153.8	88.8	123.1	150.9	91.9	123.1	136.2	74.2	106.9	119.2	68.0	95.0	151.1	96.6	125.4
04	MASONRY	159.3	101.4	124.7	168.9	100.7	128.1	165.5	101.9	127.4	163.5	71.3	108.3	158.1	67.9	104.1	169.6	100.0	128.0
05	METALS	116.1	92.6	109.0	107.7	92.2	103.0	107.1	92.4	102.7	107.8	77.9	98.8	104.6	74.7	95.5	107.7	89.2	102.1
06	WOOD, PLASTICS & COMPOSITES	116.7	96.0	105.8	120.8	88.7	103.8	118.2	90.8	103.7	115.9	77.2	95.5	95.1	62.0	77.6	98.0	94.1	95.9
07	THERMAL & MOISTURE PROTECTION	110.2	96.1	104.6	109.9	90.8	102.1	109.7	95.7	104.0	104.1	75.4	92.6	103.5	66.2	88.5	116.0	91.1	106.0
08	OPENINGS	92.3	93.5	92.6	91.4	87.5	90.4	91.2	90.1	90.9	92.3	68.2	86.0	85.7	59.7	79.0	88.1	90.2	88.6
0920	Plaster & Gypsum Board	229.4	96.0	150.0	152.7	88.5	114.5	152.7	90.6	115.7	124.8	76.3	95.9	132.1	61.0	89.8	135.3	93.5	110.4
0950, 0980	Ceilings & Acoustic Treatment	106.2	96.0	100.0	95.2	88.5	91.2	95.2	90.6	92.4	95.2	76.3	83.8	95.2	61.0	74.7	95.2	93.5	94.2
0960	Flooring	132.5	100.3	123.9	132.6	101.8	124.3	132.5	100.3	123.9	132.5	72.3	116.4	119.8	65.1	105.2	126.8	57.3	108.3
0990	Wall Finishes & Painting/Coating	109.2	96.7	101.8	110.0	93.3	100.0	109.2	97.1	102.0	109.3	61.0	80.4	109.2	59.2	79.3	109.2	87.2	96.0
09	FINISHES	134.3	97.5	115.3	120.0	92.9	106.0	118.9	94.9	106.5	114.6	74.3	93.8	112.0	63.1	86.8	116.1	88.0	101.6
COVERS	DIVS. 10 - 14, 25, 28, 41, 43, 44	140.0	111.9	134.2	140.0	85.1	128.7	140.0	86.3	129.0	140.0	78.5	127.3	140.0	74.9	126.6	140.0	111.7	134.2
22, 23	PLUMBING & HVAC	101.2	94.7	98.5	102.8	108.7	105.2	102.7	112.7	106.8	102.7	93.2	98.8	103.1	76.0	92.1	102.7	104.6	103.5
26, 27, 3370	ELECTRICAL, COMMUNICATIONS & UTIL.	125.8	98.2	112.6	126.4	98.1	112.8	124.7	99.0	112.4	127.7	65.7	98.0	128.2	68.7	99.7	124.9	89.7	108.0
MF2004	WEIGHTED AVERAGE	120.7	97.8	110.8	119.9	97.2	110.1	118.9	99.3	110.4	117.0	79.2	100.7	113.2	71.9	95.4	118.9	96.4	109.2

683

City Cost Indexes

		\multicolumn{18}{c	}{CANADA}

	DIVISION	QUEBEC, QUEBEC			RED DEER, ALBERTA			REGINA, SASKATCHEWAN			RIMOUSKI, QUEBEC			ROUYN-NORANDA, QUEBEC			SAINT HYACINTHE, QUEBEC		
		MAT.	INST.	TOTAL	MAT.	INST.	TOTAL	MAT.	INST.	TOTAL	MAT.	INST.	TOTAL	MAT.	INST.	TOTAL	MAT.	INST.	TOTAL
01543	CONTRACTOR EQUIPMENT	.0	101.5	101.5	.0	103.7	103.7	.0	98.6	98.6	.0	99.2	99.2	.0	99.2	99.2	.0	99.2	99.2
0241, 31 - 34	SITE & INFRASTRUCTURE, DEMOLITION	96.5	98.3	97.8	110.8	101.8	104.4	113.5	95.8	100.8	94.9	97.9	97.0	94.4	98.6	97.4	95.0	98.6	97.6
0310	Concrete Forming & Accessories	129.9	99.4	103.7	135.5	86.9	93.7	103.5	63.9	69.5	129.6	98.8	103.2	129.6	92.9	98.1	129.6	92.9	98.1
0310	Concrete Reinforcing	148.4	101.0	125.2	161.5	66.1	114.9	126.2	67.3	97.4	112.0	101.0	106.6	157.1	85.7	122.2	157.1	85.7	122.2
0330	Cast-in-Place Concrete	145.5	105.9	130.6	151.2	100.4	132.1	156.3	72.8	124.9	136.9	103.9	124.5	130.2	100.5	119.0	134.5	100.5	121.7
03	CONCRETE	138.0	101.7	120.8	143.3	87.9	117.2	132.2	68.1	102.0	127.9	100.7	115.0	131.6	94.1	113.9	133.8	94.1	115.0
04	MASONRY	167.8	98.4	126.3	160.5	82.6	113.9	166.3	68.3	107.7	159.8	98.4	123.0	160.1	91.8	119.2	160.4	91.8	119.3
05	METALS	120.3	94.8	112.6	107.8	86.4	101.4	105.5	74.9	96.2	105.1	94.0	101.7	105.5	88.9	100.5	105.5	88.9	100.5
06	WOOD, PLASTICS & COMPOSITES	131.9	99.3	114.7	116.0	86.4	100.3	95.1	62.0	77.6	131.2	99.0	114.2	131.2	92.5	110.8	131.2	92.5	110.8
07	THERMAL & MOISTURE PROTECTION	104.7	101.7	103.5	125.5	86.3	109.7	105.3	67.5	90.1	104.3	100.9	102.9	104.3	94.2	100.2	104.5	94.2	100.4
08	OPENINGS	92.3	94.8	92.9	92.3	79.9	89.1	87.1	59.7	80.0	91.8	84.8	90.0	92.3	80.4	89.2	92.3	80.4	89.2
0920	Plaster & Gypsum Board	190.1	98.9	135.8	131.5	85.6	104.2	165.2	61.0	103.2	187.9	98.9	134.9	137.3	92.3	110.5	141.5	92.3	112.2
0950, 0980	Ceilings & Acoustic Treatment	96.9	98.9	98.1	95.2	85.6	89.4	120.5	61.0	84.7	92.7	98.9	96.4	92.7	92.3	92.5	92.7	92.3	92.5
0960	Flooring	132.5	102.5	124.5	135.0	95.4	124.5	119.6	65.1	105.2	132.5	102.5	124.5	132.5	102.5	124.5	132.5	102.5	124.5
0990	Wall Finishes & Painting/Coating	109.7	99.5	109.6	109.2	84.3	94.3	109.2	69.4	85.4	109.2	109.5	109.4	109.2	95.2	100.8	109.2	95.2	100.8
09	FINISHES	124.9	101.6	112.9	116.5	87.9	101.7	123.9	64.2	93.1	123.5	101.3	112.0	115.0	95.2	104.8	115.7	95.2	105.1
COVERS	DIVS. 10 - 14, 25, 28, 41, 43, 44	140.0	101.7	132.1	140.0	107.7	133.4	140.0	74.9	126.6	140.0	100.8	131.9	140.0	99.2	131.6	140.0	99.2	131.6
22, 23	PLUMBING & HVAC	101.1	102.2	101.6	102.7	102.0	102.4	100.9	84.3	94.2	102.7	102.1	102.5	102.7	100.8	101.9	99.3	100.8	99.9
26, 27, 3370	ELECTRICAL, COMMUNICATIONS & UTIL.	129.9	89.5	110.6	115.9	93.5	105.2	129.3	68.7	100.2	124.7	89.6	108.0	124.7	79.5	103.0	125.8	79.5	103.6
MF2004	WEIGHTED AVERAGE	119.9	98.5	110.7	117.1	92.3	106.4	116.3	73.9	98.0	115.5	97.8	107.8	115.3	92.8	105.6	115.0	92.8	105.4

		\multicolumn{18}{c	}{CANADA}

	DIVISION	SAINT JOHN, N B			SARNIA, ONTARIO			SASKATOON, SASKATCHEWAN			SAULT STE MARIE, ON			SHERBROOKE, QUEBEC			SOREL, QUEBEC		
		MAT.	INST.	TOTAL	MAT.	INST.	TOTAL	MAT.	INST.	TOTAL	MAT.	INST.	TOTAL	MAT.	INST.	TOTAL	MAT.	INST.	TOTAL
01543	CONTRACTOR EQUIPMENT	.0	98.6	98.6	.0	100.4	100.4	.0	98.6	98.6	.0	100.4	100.4	.0	99.2	99.2	.0	99.2	99.2
0241, 31 - 34	SITE & INFRASTRUCTURE, DEMOLITION	96.8	96.3	96.4	113.3	100.5	104.1	108.1	96.0	99.4	102.8	100.1	100.8	95.0	98.6	97.6	95.1	98.6	97.6
0310	Concrete Forming & Accessories	120.5	66.6	74.2	121.3	100.9	103.7	103.5	63.8	69.4	109.6	92.4	94.8	129.6	92.9	98.1	129.6	93.0	98.2
0310	Concrete Reinforcing	148.4	64.9	107.6	123.7	91.0	107.7	119.1	67.3	93.8	112.3	89.0	100.9	157.1	85.7	122.2	148.4	85.7	117.7
0330	Cast-in-Place Concrete	136.2	62.8	108.6	151.0	105.5	133.8	142.1	72.7	116.0	135.6	89.7	118.3	134.5	100.5	121.7	135.6	100.5	122.4
03	CONCRETE	136.0	65.5	102.7	136.3	100.4	119.3	123.9	68.0	97.5	120.1	90.8	106.3	133.8	94.1	115.0	133.0	94.2	114.6
04	MASONRY	177.3	70.3	113.3	176.2	104.8	133.5	166.0	68.3	107.6	161.9	95.1	121.9	160.4	91.8	119.4	160.5	91.8	119.4
05	METALS	105.2	74.9	96.0	107.0	92.8	102.7	105.5	74.8	96.2	106.3	90.4	101.5	105.5	88.9	100.5	105.5	89.0	100.5
06	WOOD, PLASTICS & COMPOSITES	119.1	65.7	90.8	117.2	99.7	108.0	93.7	62.0	76.9	103.0	92.2	97.3	131.2	92.5	110.8	131.2	92.5	110.8
07	THERMAL & MOISTURE PROTECTION	106.9	67.1	90.9	109.7	99.4	105.6	104.1	66.4	88.9	108.5	91.6	101.7	104.3	94.2	100.2	104.4	94.2	100.2
08	OPENINGS	86.2	58.5	79.0	93.2	94.2	93.5	86.1	59.7	79.3	84.0	88.5	85.2	92.3	80.4	89.2	92.3	85.4	90.5
0920	Plaster & Gypsum Board	183.5	64.8	112.9	182.2	99.8	133.1	144.1	61.0	94.7	141.1	92.2	111.9	139.4	92.3	111.4	187.9	92.3	131.0
0950, 0980	Ceilings & Acoustic Treatment	104.5	64.8	80.6	100.3	99.8	100.0	120.5	61.0	84.7	95.2	92.1	93.3	92.7	92.3	92.5	92.7	92.3	92.5
0960	Flooring	127.6	59.8	109.5	132.5	106.0	125.4	119.8	65.1	105.2	123.3	100.3	117.2	132.5	102.5	124.5	132.5	102.5	124.5
0990	Wall Finishes & Painting/Coating	109.2	74.3	88.3	109.2	109.8	109.6	109.2	59.2	79.3	109.2	101.7	104.7	109.2	95.2	100.8	109.2	95.2	100.8
09	FINISHES	124.5	65.9	94.3	125.1	103.1	113.7	120.1	63.1	90.7	113.6	95.0	104.0	115.4	95.2	105.0	123.5	95.2	108.9
COVERS	DIVS. 10 - 14, 25, 28, 41, 43, 44	140.0	75.4	126.7	140.0	88.0	129.3	140.0	74.9	126.6	140.0	112.8	134.4	140.0	99.2	131.6	140.0	99.2	131.6
22, 23	PLUMBING & HVAC	102.7	82.9	94.6	102.7	119.5	109.5	100.8	84.3	94.1	102.7	104.9	103.6	103.1	100.8	102.2	102.7	100.8	101.9
26, 27, 3370	ELECTRICAL, COMMUNICATIONS & UTIL.	135.4	84.2	110.8	129.2	101.5	115.9	129.6	68.7	100.4	127.2	99.5	113.9	124.7	79.5	103.0	124.7	79.5	103.0
MF2004	WEIGHTED AVERAGE	118.2	75.9	99.9	119.0	104.0	112.5	114.8	73.7	97.0	113.7	97.3	106.6	115.7	92.8	105.8	116.2	93.0	106.2

		\multicolumn{18}{c	}{CANADA}

	DIVISION	ST CATHARINES, ONTARIO			ST JEROME, QUEBEC			ST JOHNS, NEWFOUNDLAND			SUDBURY, ONTARIO			SUMMERSIDE, PEI			SYDNEY, NOVA SCOTIA		
		MAT.	INST.	TOTAL	MAT.	INST.	TOTAL	MAT.	INST.	TOTAL	MAT.	INST.	TOTAL	MAT.	INST.	TOTAL	MAT.	INST.	TOTAL
01543	CONTRACTOR EQUIPMENT	.0	100.4	100.4	.0	99.2	99.2	.0	98.6	98.6	.0	100.4	100.4	.0	98.6	98.6	.0	98.6	98.6
0241, 31 - 34	SITE & INFRASTRUCTURE, DEMOLITION	102.7	100.7	101.3	94.4	98.6	97.4	116.5	96.2	102.0	102.8	100.3	101.0	108.0	94.8	98.6	94.7	97.2	96.5
0310	Concrete Forming & Accessories	114.0	96.6	99.1	129.6	92.9	98.1	101.1	68.9	73.4	109.6	92.0	94.4	92.8	61.3	65.7	92.5	75.6	78.0
0310	Concrete Reinforcing	111.3	86.9	99.4	157.1	85.7	122.2	165.0	62.0	114.6	112.3	85.0	99.0	148.4	50.8	100.7	148.4	50.8	100.6
0330	Cast-in-Place Concrete	134.5	99.5	121.3	130.2	100.5	119.0	172.4	79.8	137.6	135.6	95.8	120.6	161.3	62.0	123.9	130.9	73.7	109.4
03	CONCRETE	119.7	95.7	108.3	131.6	94.1	113.9	164.5	71.9	120.7	120.1	92.0	106.8	152.7	60.1	109.0	129.4	70.7	101.7
04	MASONRY	161.5	100.2	124.8	160.1	91.8	119.2	163.5	70.0	107.5	161.6	94.3	121.3	159.6	66.0	103.6	157.2	75.8	108.5
05	METALS	106.3	90.4	101.5	105.5	88.9	100.5	108.2	75.6	98.4	106.3	90.3	101.5	106.3	67.0	94.4	106.3	74.4	96.6
06	WOOD, PLASTICS & COMPOSITES	108.0	96.5	102.0	131.2	92.5	110.8	94.5	69.0	81.0	103.0	92.2	97.3	84.6	60.5	71.9	84.3	74.6	79.2
07	THERMAL & MOISTURE PROTECTION	108.8	94.6	103.1	104.3	94.2	100.2	108.3	67.2	91.8	108.5	90.6	101.3	104.7	62.8	87.9	104.7	73.4	92.1
08	OPENINGS	83.0	92.4	85.5	92.3	80.4	89.2	98.6	64.6	89.8	84.0	88.5	85.2	94.5	52.4	83.6	82.9	68.1	79.1
0920	Plaster & Gypsum Board	136.4	96.5	112.6	137.3	92.3	110.5	170.4	68.2	109.6	141.1	92.2	111.9	162.8	59.5	101.3	162.8	74.0	109.8
0950, 0980	Ceilings & Acoustic Treatment	100.3	96.5	98.0	92.7	92.3	92.5	106.2	68.2	83.3	95.2	92.1	93.3	95.2	59.5	73.8	95.2	74.0	82.4
0960	Flooring	126.2	101.8	119.7	132.5	102.5	124.5	114.4	58.9	99.6	123.3	100.3	117.2	109.6	64.8	97.7	109.5	69.3	98.8
0990	Wall Finishes & Painting/Coating	109.2	104.9	106.7	109.2	95.2	100.8	109.2	69.4	85.4	109.2	94.7	100.5	109.2	44.6	70.6	109.2	66.6	83.7
09	FINISHES	114.9	99.0	106.7	115.0	95.2	104.8	119.6	67.0	92.5	113.6	94.0	103.5	114.0	60.3	86.3	113.2	73.4	92.7
COVERS	DIVS. 10 - 14, 25, 28, 41, 43, 44	140.0	86.5	129.0	140.0	99.2	131.6	140.0	76.1	126.9	140.0	112.5	134.4	140.0	75.1	126.6	140.0	77.1	127.1
22, 23	PLUMBING & HVAC	100.7	94.8	98.3	102.7	100.8	101.9	101.1	69.9	88.3	102.8	94.4	99.4	102.7	70.6	89.6	102.7	93.5	99.0
26, 27, 3370	ELECTRICAL, COMMUNICATIONS & UTIL.	128.5	98.3	114.0	125.7	79.5	103.5	125.1	70.3	98.8	124.7	98.2	112.0	122.7	57.3	91.3	124.2	70.5	98.4
MF2004	WEIGHTED AVERAGE	113.4	96.3	106.0	115.4	92.8	105.6	120.7	72.5	99.9	113.3	94.8	105.4	117.7	66.4	95.5	113.5	79.1	98.7

City Cost Indexes

CANADA

DIVISION		THUNDER BAY, ONTARIO			TIMMINS, ONTARIO			TORONTO, ONTARIO			TROIS RIVIERES, QUEBEC			TRURO, NOVA SCOTIA			VANCOUVER, B C		
		MAT.	INST.	TOTAL	MAT.	INST.	TOTAL	MAT.	INST.	TOTAL	MAT.	INST.	TOTAL	MAT.	INST.	TOTAL	MAT.	INST.	TOTAL
01543	CONTRACTOR EQUIPMENT	.0	100.4	100.4	.0	100.4	100.4	.0	103.2	103.2	.0	99.2	99.2	.0	98.6	98.6	.0	111.3	111.3
0241, 31 - 34	SITE & INFRASTRUCTURE, DEMOLITION	108.4	100.7	102.9	115.0	100.1	104.3	116.2	105.6	108.6	95.1	98.6	97.6	100.0	97.3	98.0	119.8	107.4	110.9
0310	Concrete Forming & Accessories	122.8	97.1	100.7	122.7	92.1	96.4	123.2	105.3	107.8	129.6	93.0	98.2	92.5	75.6	78.0	122.7	84.0	89.5
0310	Concrete Reinforcing	99.5	86.2	93.0	174.8	89.0	132.8	170.9	87.7	130.2	148.4	85.7	117.7	148.4	50.8	100.6	173.4	77.8	126.6
0330	Cast-in-Place Concrete	148.1	98.4	129.4	163.7	89.6	135.8	149.4	107.9	133.8	135.6	100.5	122.4	170.0	73.7	133.8	151.9	96.6	131.1
03	CONCRETE	128.3	95.4	112.7	150.9	90.7	122.5	143.1	102.6	124.0	133.0	94.2	114.6	149.2	70.8	112.2	150.2	87.5	120.6
04	MASONRY	162.3	100.2	125.1	165.5	94.9	123.2	175.1	105.2	133.3	160.5	91.8	119.4	160.3	75.8	109.7	167.8	87.2	119.6
05	METALS	106.2	89.4	101.1	107.1	91.9	102.5	128.6	92.5	117.7	105.5	89.0	100.5	106.3	74.5	96.6	143.7	88.5	127.0
06	WOOD, PLASTICS & COMPOSITES	118.2	96.8	106.9	118.2	92.2	104.5	118.2	105.4	111.4	131.2	92.5	110.8	84.3	74.6	79.2	113.5	80.9	96.3
07	THERMAL & MOISTURE PROTECTION	109.0	94.9	103.4	109.7	89.9	101.7	110.7	100.1	106.5	104.3	94.2	100.2	104.7	73.4	92.1	133.3	87.5	114.9
08	OPENINGS	82.0	91.8	84.6	91.2	88.2	90.5	91.2	100.6	93.7	92.3	85.4	90.5	82.9	68.1	79.1	94.0	81.6	90.8
0920	Plaster & Gypsum Board	167.5	96.8	125.4	152.7	92.1	116.6	166.6	105.6	130.3	187.9	92.3	131.0	162.5	74.0	109.8	144.0	79.2	105.4
0950, 0980	Ceilings & Acoustic Treatment	95.2	96.8	96.2	95.2	92.1	93.3	119.6	105.6	111.2	92.7	92.3	92.5	95.2	74.0	82.4	109.5	79.9	91.7
0960	Flooring	132.5	58.6	112.8	132.5	100.3	123.9	132.5	107.9	125.9	132.5	102.5	124.5	109.5	69.3	98.8	134.9	98.8	125.2
0990	Wall Finishes & Painting/Coating	109.2	98.7	102.9	109.2	94.7	100.5	111.0	109.6	110.2	109.2	95.2	100.8	109.2	66.6	83.7	109.2	95.4	100.9
09	FINISHES	121.0	91.3	105.7	118.9	94.2	106.1	127.7	106.8	116.9	123.5	95.2	108.9	113.2	73.4	92.7	122.6	87.5	104.5
COVERS	DIVS. 10 - 14, 25, 28, 41, 43, 44	140.0	87.0	129.1	140.0	84.6	128.6	140.0	117.0	135.3	140.0	99.2	131.6	140.0	77.1	127.1	140.0	108.7	133.6
22, 23	PLUMBING & HVAC	100.7	95.0	98.4	102.7	108.8	105.2	101.1	99.9	100.6	102.7	100.8	101.9	102.7	93.5	99.0	101.1	79.6	92.4
26, 27, 3370	ELECTRICAL, COMMUNICATIONS & UTIL.	126.0	96.5	111.8	127.2	99.4	113.9	131.4	100.8	116.7	124.7	79.5	103.0	124.2	70.5	98.4	134.2	80.1	108.3
MF2004	WEIGHTED AVERAGE	114.8	95.0	106.3	119.2	97.2	109.7	123.1	102.2	114.1	116.2	93.0	106.2	116.1	79.1	100.1	126.9	86.9	109.6

CANADA

DIVISION		VICTORIA, B C			WHITEHORSE, YUKON			WINDSOR, ONTARIO			WINNIPEG, MANITOBA			YARMOUTH, NOVA SCOTIA			YELLOWKNIFE, NWT		
		MAT.	INST.	TOTAL	MAT.	INST.	TOTAL	MAT.	INST.	TOTAL	MAT.	INST.	TOTAL	MAT.	INST.	TOTAL	MAT.	INST.	TOTAL
01543	CONTRACTOR EQUIPMENT	.0	107.3	107.3	.0	99.2	99.2	.0	100.4	100.4	.0	105.3	105.3	.0	98.6	98.6	.0	98.7	98.7
0241, 31 - 34	SITE & INFRASTRUCTURE, DEMOLITION	121.8	106.8	111.1	108.7	96.0	99.6	98.0	100.7	100.0	116.1	101.1	105.3	99.7	97.2	98.0	121.9	100.3	106.4
0310	Concrete Forming & Accessories	109.1	83.1	86.7	103.5	62.8	68.5	122.8	94.0	98.0	122.5	70.4	77.8	92.5	75.6	78.0	103.5	86.2	88.6
0310	Concrete Reinforcing	116.5	77.6	97.5	119.1	65.2	92.8	109.1	85.5	97.5	161.5	60.2	112.0	148.4	50.8	100.6	119.1	65.3	92.8
0330	Cast-in-Place Concrete	147.2	95.7	127.8	146.7	71.7	118.5	137.7	99.5	123.4	174.4	74.8	136.9	168.2	73.7	132.6	243.6	91.6	186.4
03	CONCRETE	151.1	86.5	120.6	126.2	66.9	98.2	121.5	94.3	108.7	154.0	70.6	114.6	148.3	70.7	111.7	175.3	84.9	132.6
04	MASONRY	168.8	87.2	119.9	158.5	66.2	103.2	161.7	98.6	123.9	178.2	66.8	111.5	160.1	75.8	109.6	166.7	80.3	115.0
05	METALS	107.7	83.4	100.3	105.6	73.9	96.1	106.2	90.3	101.4	138.4	77.3	119.9	106.3	74.4	96.6	106.4	80.6	98.6
06	WOOD, PLASTICS & COMPOSITES	98.0	80.8	88.9	95.1	61.1	77.1	118.2	93.3	105.0	115.9	71.0	92.2	84.3	74.6	79.2	95.1	87.5	91.1
07	THERMAL & MOISTURE PROTECTION	116.0	85.7	103.8	103.5	65.2	88.1	108.9	93.1	102.5	105.1	71.6	91.6	104.7	73.4	92.1	104.9	83.4	96.3
08	OPENINGS	88.1	77.7	85.4	85.7	58.3	78.6	81.7	90.5	84.0	92.3	65.3	85.3	82.9	68.1	79.1	85.7	73.9	82.6
0920	Plaster & Gypsum Board	135.3	79.9	102.4	115.2	60.0	82.4	156.9	93.1	119.0	148.6	69.7	101.6	162.5	74.0	109.8	146.8	87.2	111.3
0950, 0980	Ceilings & Acoustic Treatment	95.2	79.9	86.0	95.2	60.0	74.1	95.2	93.1	94.0	107.0	69.7	84.6	95.2	74.0	82.4	95.2	87.2	90.4
0960	Flooring	126.8	47.7	105.7	119.8	63.2	104.7	132.5	102.5	124.5	132.5	72.3	116.4	109.5	69.3	98.8	119.8	96.9	113.7
0990	Wall Finishes & Painting/Coating	109.2	81.3	92.5	109.2	58.2	78.7	109.2	98.1	102.6	109.3	51.7	74.8	109.2	66.6	83.7	109.2	84.8	94.6
09	FINISHES	116.1	76.9	95.8	109.2	61.9	84.8	118.9	96.3	107.3	121.5	69.0	94.4	113.2	73.4	92.7	114.5	87.2	100.4
COVERS	DIVS. 10 - 14, 25, 28, 41, 43, 44	140.0	80.7	127.8	140.0	74.0	126.4	140.0	85.9	128.9	140.0	77.1	127.1	140.0	77.1	127.1	140.0	78.9	127.4
22, 23	PLUMBING & HVAC	102.7	77.6	92.5	102.7	82.9	94.6	100.7	94.5	98.2	101.1	71.3	88.9	102.7	93.5	99.0	102.7	106.0	104.0
26, 27, 3370	ELECTRICAL, COMMUNICATIONS & UTIL.	125.5	79.8	103.6	128.2	68.1	99.4	132.7	98.2	116.2	134.5	71.6	104.3	124.2	70.5	98.4	129.2	94.9	112.7
MF2004	WEIGHTED AVERAGE	118.9	83.4	103.6	113.9	72.6	96.1	114.3	95.4	106.1	125.8	73.5	103.2	116.0	79.1	100.1	121.1	90.8	108.0

Location Factors

Costs shown in *RSMeans cost data publications* are based on National Averages for materials and installation. To adjust these costs to a specific location, simply multiply the base cost by the factor and divide by 100 for that city. The data is arranged alphabetically by state and postal zip code numbers. For a city not listed, use the factor for a nearby city with similar economic characteristics.

STATE/ZIP	CITY	MAT.	INST.	TOTAL
ALABAMA				
350-352	Birmingham	96.2	73.1	86.2
354	Tuscaloosa	95.8	54.0	77.8
355	Jasper	96.0	50.6	76.4
356	Decatur	95.7	54.1	77.7
357-358	Huntsville	95.8	70.4	84.8
359	Gadsden	95.7	57.3	79.1
360-361	Montgomery	96.2	55.6	78.7
362	Anniston	95.3	45.1	73.6
363	Dothan	96.0	47.4	75.0
364	Evergreen	95.3	48.6	75.2
365-366	Mobile	96.8	59.3	80.6
367	Selma	95.6	50.1	75.9
368	Phenix City	96.3	53.4	77.8
369	Butler	95.7	47.4	74.9
ALASKA				
995-996	Anchorage	131.9	114.1	124.2
997	Fairbanks	129.2	116.9	123.9
998	Juneau	130.4	112.9	122.8
999	Ketchikan	141.6	112.4	129.0
ARIZONA				
850,853	Phoenix	101.0	73.9	89.3
852	Mesa/Tempe	100.2	66.9	85.8
855	Globe	100.8	64.1	84.9
856-857	Tucson	99.4	70.2	86.8
859	Show Low	100.9	65.9	85.8
860	Flagstaff	101.5	71.3	88.4
863	Prescott	99.2	64.9	84.4
864	Kingman	97.8	67.5	84.7
865	Chambers	97.8	66.2	84.2
ARKANSAS				
716	Pine Bluff	96.1	69.6	84.6
717	Camden	94.1	46.0	73.3
718	Texarkana	95.4	50.9	76.2
719	Hot Springs	93.2	49.6	74.4
720-722	Little Rock	95.8	73.2	86.0
723	West Memphis	95.9	63.2	81.8
724	Jonesboro	96.4	63.8	82.3
725	Batesville	94.1	56.5	77.9
726	Harrison	95.5	57.4	79.0
727	Fayetteville	92.8	57.0	77.3
728	Russellville	94.1	58.1	78.5
729	Fort Smith	96.7	62.3	81.9
CALIFORNIA				
900-902	Los Angeles	103.3	111.9	107.0
903-905	Inglewood	99.0	110.8	104.1
906-908	Long Beach	100.8	110.8	105.1
910-912	Pasadena	99.9	111.0	104.7
913-916	Van Nuys	103.0	111.0	106.5
917-918	Alhambra	102.2	111.0	106.0
919-921	San Diego	103.4	105.6	104.3
922	Palm Springs	99.8	110.1	104.3
923-924	San Bernardino	97.6	109.7	102.8
925	Riverside	101.9	109.8	105.3
926-927	Santa Ana	99.6	110.0	104.1
928	Anaheim	102.0	110.4	105.6
930	Oxnard	103.5	109.8	106.2
931	Santa Barbara	102.4	109.8	105.6
932-933	Bakersfield	102.1	107.7	104.5
934	San Luis Obispo	103.4	108.6	105.6
935	Mojave	100.4	107.5	103.4
936-938	Fresno	103.9	112.0	107.4
939	Salinas	104.1	118.1	110.1
940-941	San Francisco	112.6	133.8	121.8
942,956-958	Sacramento	106.0	112.6	108.9
943	Palo Alto	105.2	127.4	114.8
944	San Mateo	108.3	128.5	117.0
945	Vallejo	105.0	123.3	112.9
946	Oakland	110.3	127.9	117.9
947	Berkeley	110.0	127.9	117.7
948	Richmond	109.3	126.1	116.6
949	San Rafael	110.2	127.1	117.5
950	Santa Cruz	109.2	118.1	113.1

STATE/ZIP	CITY	MAT.	INST.	TOTAL
CALIFORNIA (CONT'D)				
951	San Jose	107.1	127.2	115.8
952	Stockton	104.1	113.9	108.3
953	Modesto	104.0	113.4	108.1
954	Santa Rosa	104.1	126.7	113.9
955	Eureka	105.7	111.5	108.2
959	Marysville	104.9	113.1	108.5
960	Redding	106.4	111.8	108.7
961	Susanville	105.5	111.1	107.9
COLORADO				
800-802	Denver	100.8	87.6	95.1
803	Boulder	98.0	85.1	92.4
804	Golden	100.2	84.4	93.4
805	Fort Collins	101.4	81.1	92.6
806	Greeley	98.9	69.9	86.4
807	Fort Morgan	98.8	84.2	92.5
808-809	Colorado Springs	100.5	83.3	93.1
810	Pueblo	101.2	81.3	92.6
811	Alamosa	102.8	78.2	92.1
812	Salida	102.8	79.1	92.5
813	Durango	103.2	79.2	92.8
814	Montrose	101.7	77.8	91.4
815	Grand Junction	105.1	74.9	92.1
816	Glenwood Springs	102.7	81.3	93.4
CONNECTICUT				
060	New Britain	100.0	115.9	106.8
061	Hartford	100.8	115.9	107.3
062	Willimantic	100.5	115.8	107.1
063	New London	97.0	115.9	105.1
064	Meriden	99.0	116.4	106.5
065	New Haven	101.6	116.4	108.0
066	Bridgeport	101.1	116.5	107.7
067	Waterbury	100.6	116.4	107.4
068	Norwalk	100.6	116.8	107.6
069	Stamford	100.7	123.2	110.4
D.C.				
200-205	Washington	104.1	90.2	98.1
DELAWARE				
197	Newark	100.8	101.4	101.0
198	Wilmington	99.8	101.4	100.5
199	Dover	100.9	101.4	101.1
FLORIDA				
320,322	Jacksonville	97.6	57.9	80.5
321	Daytona Beach	97.8	70.6	86.1
323	Tallahassee	97.5	46.8	75.6
324	Panama City	98.9	35.7	71.6
325	Pensacola	101.5	55.3	81.6
326,344	Gainesville	99.3	60.3	82.4
327-328,347	Orlando	99.7	66.6	85.4
329	Melbourne	100.5	76.3	90.0
330-332,340	Miami	98.3	71.6	86.8
333	Fort Lauderdale	97.2	71.9	86.3
334,349	West Palm Beach	96.2	66.4	83.4
335-336,346	Tampa	98.8	70.8	86.7
337	St. Petersburg	101.2	52.4	80.1
338	Lakeland	98.0	70.5	86.1
339,341	Fort Myers	97.3	61.8	82.0
342	Sarasota	99.0	64.4	84.1
GEORGIA				
300-303,399	Atlanta	96.8	79.7	89.4
304	Statesboro	96.5	49.9	76.4
305	Gainesville	95.4	66.8	83.1
306	Athens	94.8	69.5	83.9
307	Dalton	97.0	54.9	78.9
308-309	Augusta	95.5	67.0	83.2
310-312	Macon	95.3	67.8	83.4
313-314	Savannah	96.6	63.5	82.3
315	Waycross	96.6	59.9	80.8
316	Valdosta	96.5	52.4	77.4
317,398	Albany	96.7	63.2	82.2
318-319	Columbus	96.6	69.6	84.9

Location Factors

STATE/ZIP	CITY	MAT.	INST.	TOTAL
HAWAII				
967	Hilo	115.4	122.7	118.5
968	Honolulu	120.2	122.7	121.3
STATES & POSS.				
969	Guam	134.9	52.6	99.4
IDAHO				
832	Pocatello	101.6	75.1	90.2
833	Twin Falls	102.3	56.2	82.4
834	Idaho Falls	99.6	62.3	83.5
835	Lewiston	108.7	81.9	97.1
836-837	Boise	100.8	76.4	90.3
838	Coeur d'Alene	107.9	79.9	95.8
ILLINOIS				
600-603	North Suburban	98.7	120.7	108.2
604	Joliet	98.6	116.8	106.4
605	South Suburban	98.7	120.7	108.2
606-608	Chicago	99.2	131.9	113.3
609	Kankakee	95.0	109.0	101.1
610-611	Rockford	97.5	110.8	103.2
612	Rock Island	95.3	97.5	96.2
613	La Salle	96.5	100.9	98.4
614	Galesburg	96.3	101.2	98.4
615-616	Peoria	98.5	97.8	98.2
617	Bloomington	95.6	103.4	99.0
618-619	Champaign	99.2	100.2	99.6
620-622	East St. Louis	94.1	105.0	98.8
623	Quincy	95.4	95.9	95.7
624	Effingham	94.8	95.6	95.1
625	Decatur	96.4	98.5	97.3
626-627	Springfield	96.7	96.4	96.6
628	Centralia	93.0	103.2	97.4
629	Carbondale	92.7	96.4	94.3
INDIANA				
460	Anderson	94.1	82.6	89.2
461-462	Indianapolis	97.7	88.1	93.6
463-464	Gary	95.9	100.9	98.1
465-466	South Bend	94.1	83.9	89.7
467-468	Fort Wayne	94.8	80.8	88.7
469	Kokomo	92.4	82.8	88.2
470	Lawrenceburg	91.7	79.6	86.5
471	New Albany	93.2	74.9	85.3
472	Columbus	95.6	81.1	89.3
473	Muncie	95.9	81.2	89.5
474	Bloomington	97.5	81.8	90.7
475	Washington	93.6	85.3	90.0
476-477	Evansville	94.8	87.3	91.6
478	Terre Haute	95.5	85.8	91.3
479	Lafayette	95.2	79.6	88.5
IOWA				
500-503,509	Des Moines	97.2	81.5	90.5
504	Mason City	96.0	64.1	82.2
505	Fort Dodge	96.1	60.1	80.6
506-507	Waterloo	97.7	60.7	81.7
508	Creston	96.4	64.3	82.6
510-511	Sioux City	98.3	72.7	87.3
512	Sibley	97.1	49.9	76.7
513	Spencer	98.8	48.3	77.0
514	Carroll	95.8	53.9	77.7
515	Council Bluffs	99.3	74.2	88.5
516	Shenandoah	96.4	51.2	76.9
520	Dubuque	97.7	79.2	89.7
521	Decorah	96.8	51.2	77.1
522-524	Cedar Rapids	98.8	84.2	92.5
525	Ottumwa	96.8	72.1	86.2
526	Burlington	96.0	74.1	86.6
527-528	Davenport	97.5	93.2	95.7
KANSAS				
660-662	Kansas City	98.5	97.7	98.1
664-666	Topeka	98.5	69.2	85.8
667	Fort Scott	97.3	70.1	85.6
668	Emporia	97.3	57.9	80.3
669	Belleville	99.0	61.9	83.0
670-672	Wichita	97.5	67.4	84.5
673	Independence	98.8	65.8	84.5
674	Salina	99.0	62.0	83.0
675	Hutchinson	94.3	59.3	79.2
676	Hays	98.2	65.2	83.9
677	Colby	99.0	64.0	83.9

STATE/ZIP	CITY	MAT.	INST.	TOTAL
KANSAS (CONT'D)				
678	Dodge City	100.1	66.4	85.6
679	Liberal	98.0	61.6	82.3
KENTUCKY				
400-402	Louisville	96.1	84.7	91.2
403-405	Lexington	95.5	81.5	89.4
406	Frankfort	95.6	85.4	91.2
407-409	Corbin	93.0	70.0	83.1
410	Covington	93.6	105.1	98.5
411-412	Ashland	92.2	103.4	97.0
413-414	Campton	93.9	70.4	83.7
415-416	Pikeville	94.6	86.8	91.2
417-418	Hazard	93.2	60.1	78.9
420	Paducah	91.8	90.1	91.0
421-422	Bowling Green	94.1	86.9	91.0
423	Owensboro	93.9	88.2	91.4
424	Henderson	91.6	90.0	90.9
425-426	Somerset	91.3	74.9	84.2
427	Elizabethtown	90.7	86.1	88.7
LOUISIANA				
700-701	New Orleans	100.7	68.2	86.6
703	Thibodaux	99.0	65.8	84.7
704	Hammond	96.1	60.8	80.9
705	Lafayette	98.3	60.5	82.0
706	Lake Charles	98.4	62.1	82.8
707-708	Baton Rouge	98.7	61.0	82.4
710-711	Shreveport	94.9	59.7	79.7
712	Monroe	96.3	56.2	79.0
713-714	Alexandria	96.4	57.6	79.6
MAINE				
039	Kittery	94.2	70.9	84.1
040-041	Portland	99.0	77.1	89.6
042	Lewiston	97.1	77.1	88.5
043	Augusta	96.4	70.7	85.3
044	Bangor	96.6	77.1	88.2
045	Bath	95.4	70.9	84.8
046	Machias	94.9	70.3	84.2
047	Houlton	95.0	73.7	85.8
048	Rockland	94.1	70.9	84.1
049	Waterville	95.3	70.8	84.7
MARYLAND				
206	Waldorf	100.3	71.8	88.0
207-208	College Park	100.3	80.2	91.6
209	Silver Spring	99.5	76.4	89.5
210-212	Baltimore	99.4	84.2	92.8
214	Annapolis	99.5	77.5	90.0
215	Cumberland	95.5	79.5	88.6
216	Easton	97.1	43.3	73.9
217	Hagerstown	96.1	80.3	89.3
218	Salisbury	97.5	51.7	77.7
219	Elkton	94.6	64.4	81.6
MASSACHUSETTS				
010-011	Springfield	99.1	104.4	101.4
012	Pittsfield	98.7	100.8	99.6
013	Greenfield	96.8	102.4	99.2
014	Fitchburg	95.5	120.6	106.3
015-016	Worcester	99.1	120.6	108.4
017	Framingham	95.1	124.8	107.9
018	Lowell	98.5	123.9	109.4
019	Lawrence	99.7	122.1	109.4
020-022, 024	Boston	101.4	133.8	115.4
023	Brockton	100.0	119.9	108.6
025	Buzzards Bay	94.4	118.8	105.0
026	Hyannis	97.1	118.8	106.5
027	New Bedford	99.1	119.7	108.0
MICHIGAN				
480,483	Royal Oak	93.6	108.7	100.1
481	Ann Arbor	95.6	109.3	101.5
482	Detroit	97.2	115.3	105.0
484-485	Flint	95.3	99.2	97.0
486	Saginaw	94.8	95.1	94.9
487	Bay City	95.0	95.2	95.1
488-489	Lansing	95.4	97.3	96.2
490	Battle Creek	95.0	91.2	93.4
491	Kalamazoo	95.3	88.8	92.5
492	Jackson	93.5	96.7	94.9
493,495	Grand Rapids	95.7	67.1	83.3
494	Muskegon	94.0	85.5	90.3

Location Factors

STATE/ZIP	CITY	MAT.	INST.	TOTAL
MICHIGAN (CONT'D)				
496	Traverse City	93.0	71.8	83.8
497	Gaylord	94.2	75.9	86.3
498-499	Iron Mountain	96.0	87.4	92.3
MINNESOTA				
550-551	Saint Paul	99.4	124.3	110.1
553-555	Minneapolis	100.2	128.3	112.3
556-558	Duluth	98.5	112.9	104.7
559	Rochester	98.9	107.6	102.7
560	Mankato	96.1	107.7	101.1
561	Windom	94.9	80.6	88.7
562	Willmar	94.4	86.7	91.1
563	St. Cloud	95.5	122.9	107.4
564	Brainerd	96.1	104.2	99.6
565	Detroit Lakes	97.9	99.2	98.5
566	Bemidji	97.3	101.1	98.9
567	Thief River Falls	96.9	96.2	96.6
MISSISSIPPI				
386	Clarksdale	95.6	29.9	67.2
387	Greenville	99.1	43.1	74.9
388	Tupelo	97.1	36.1	70.8
389	Greenwood	96.9	32.3	69.0
390-392	Jackson	97.2	44.6	74.5
393	Meridian	95.7	45.2	73.9
394	Laurel	97.0	32.4	69.1
395	Biloxi	97.4	53.7	78.5
396	McComb	95.5	51.5	76.5
397	Columbus	96.9	35.5	70.4
MISSOURI				
630-631	St. Louis	97.8	112.0	103.9
633	Bowling Green	96.6	93.6	95.3
634	Hannibal	95.5	84.4	90.7
635	Kirksville	97.8	77.4	89.0
636	Flat River	97.6	94.1	96.1
637	Cape Girardeau	97.3	89.8	94.1
638	Sikeston	95.4	81.1	89.2
639	Poplar Bluff	94.9	80.9	88.9
640-641	Kansas City	99.7	107.6	103.1
644-645	St. Joseph	98.8	92.9	96.3
646	Chillicothe	95.8	68.8	84.1
647	Harrisonville	95.5	98.1	96.6
648	Joplin	97.6	69.9	85.7
650-651	Jefferson City	95.6	88.7	92.6
652	Columbia	96.8	90.7	94.2
653	Sedalia	96.0	83.9	90.8
654-655	Rolla	94.8	76.3	86.8
656-658	Springfield	97.6	76.4	88.5
MONTANA				
590-591	Billings	101.7	73.6	89.5
592	Wolf Point	101.2	67.3	86.6
593	Miles City	99.1	67.8	85.6
594	Great Falls	102.9	72.0	89.6
595	Havre	100.1	67.5	86.0
596	Helena	102.1	70.4	88.5
597	Butte	101.3	70.2	87.9
598	Missoula	98.9	69.3	86.1
599	Kalispell	98.0	67.3	84.8
NEBRASKA				
680-681	Omaha	98.8	77.3	89.5
683-685	Lincoln	97.7	66.0	84.0
686	Columbus	96.1	49.2	75.8
687	Norfolk	97.8	59.8	81.4
688	Grand Island	97.5	64.6	83.3
689	Hastings	97.2	56.0	79.4
690	Mccook	96.9	46.6	75.2
691	North Platte	96.9	56.8	79.6
692	Valentine	99.0	37.3	72.4
693	Alliance	98.9	34.8	71.2
NEVADA				
889-891	Las Vegas	99.6	104.7	101.8
893	Ely	99.8	76.4	89.7
894-895	Reno	99.9	96.5	98.4
897	Carson City	99.3	96.0	97.9
898	Elko	98.5	82.5	91.6
NEW HAMPSHIRE				
030	Nashua	99.3	86.0	93.6
031	Manchester	100.0	86.0	94.0

STATE/ZIP	CITY	MAT.	INST.	TOTAL
NEW HAMPSHIRE (CONT'D)				
032-033	Concord	97.5	86.0	92.5
034	Keene	96.2	49.1	75.9
035	Littleton	96.2	59.6	80.4
036	Charleston	95.7	46.2	74.3
037	Claremont	94.8	46.2	73.8
038	Portsmouth	96.6	81.6	90.1
NEW JERSEY				
070-071	Newark	101.0	122.8	110.4
072	Elizabeth	99.5	121.7	109.1
073	Jersey City	98.4	123.3	109.2
074-075	Paterson	100.2	122.9	110.0
076	Hackensack	98.3	123.4	109.1
077	Long Branch	97.9	122.2	108.4
078	Dover	98.5	123.0	109.1
079	Summit	98.4	122.1	108.6
080,083	Vineland	96.4	118.6	106.0
081	Camden	98.4	117.4	106.6
082,084	Atlantic City	97.1	117.3	105.8
085-086	Trenton	98.2	121.7	108.3
087	Point Pleasant	98.4	120.7	108.0
088-089	New Brunswick	98.9	121.0	108.4
NEW MEXICO				
870-872	Albuquerque	99.8	75.7	89.4
873	Gallup	99.9	75.7	89.4
874	Farmington	100.2	75.7	89.6
875	Santa Fe	98.9	75.7	88.9
877	Las Vegas	98.4	75.7	88.6
878	Socorro	97.9	75.7	88.3
879	Truth/Consequences	98.1	70.6	86.3
880	Las Cruces	96.8	67.8	84.3
881	Clovis	98.5	74.5	88.1
882	Roswell	100.1	74.6	89.1
883	Carrizozo	100.6	75.7	89.8
884	Tucumcari	99.3	74.5	88.6
NEW YORK				
100-102	New York	105.3	164.5	130.9
103	Staten Island	101.3	157.6	125.6
104	Bronx	99.4	157.6	124.5
105	Mount Vernon	99.4	137.1	115.7
106	White Plains	99.2	137.1	115.6
107	Yonkers	104.0	137.3	118.4
108	New Rochelle	99.9	137.1	116.0
109	Suffern	99.6	123.7	110.0
110	Queens	101.2	157.4	125.4
111	Long Island City	102.8	157.4	126.4
112	Brooklyn	103.1	157.4	126.5
113	Flushing	103.3	157.4	126.6
114	Jamaica	101.5	157.4	125.6
115,117,118	Hicksville	101.1	147.5	121.1
116	Far Rockaway	103.4	157.4	126.7
119	Riverhead	102.0	147.5	121.6
120-122	Albany	97.2	94.0	95.8
123	Schenectady	97.8	94.6	96.4
124	Kingston	101.0	114.8	107.0
125-126	Poughkeepsie	100.2	116.1	107.1
127	Monticello	99.5	118.3	107.6
128	Glens Falls	92.6	91.1	92.0
129	Plattsburgh	96.7	85.4	91.8
130-132	Syracuse	98.7	93.3	96.4
133-135	Utica	96.6	88.4	93.1
136	Watertown	98.3	85.9	92.9
137-139	Binghamton	98.4	88.5	94.1
140-142	Buffalo	100.1	105.3	102.4
143	Niagara Falls	98.0	102.2	99.8
144-146	Rochester	99.9	97.3	98.8
147	Jamestown	96.9	85.8	92.1
148-149	Elmira	96.8	83.7	91.1
NORTH CAROLINA				
270,272-274	Greensboro	100.0	52.8	79.6
271	Winston-Salem	99.7	52.0	79.1
275-276	Raleigh	99.9	53.6	79.9
277	Durham	101.3	54.0	80.9
278	Rocky Mount	97.4	43.0	73.9
279	Elizabeth City	98.3	44.5	75.0
280	Gastonia	98.8	51.4	78.3
281-282	Charlotte	100.3	54.2	80.4
283	Fayetteville	101.3	54.4	81.1
284	Wilmington	97.5	51.5	77.6
285	Kinston	95.7	44.8	73.7

Location Factors

STATE/ZIP	CITY	MAT.	INST.	TOTAL
NORTH CAROLINA (CONT'D)				
286	Hickory	96.0	47.6	75.1
287-288	Asheville	97.9	51.4	77.8
289	Murphy	97.0	36.2	70.8
NORTH DAKOTA				
580-581	Fargo	100.7	63.6	84.6
582	Grand Forks	100.6	57.7	82.1
583	Devils Lake	100.2	59.6	82.7
584	Jamestown	100.2	50.9	78.9
585	Bismarck	99.1	65.3	84.5
586	Dickinson	100.9	59.6	83.1
587	Minot	100.6	67.4	86.3
588	Williston	99.4	59.5	82.2
OHIO				
430-432	Columbus	97.3	88.6	93.5
433	Marion	93.9	86.4	90.7
434-436	Toledo	97.5	98.6	98.0
437-438	Zanesville	94.4	84.1	90.0
439	Steubenville	95.4	93.5	94.6
440	Lorain	97.0	94.9	96.1
441	Cleveland	97.3	104.9	100.6
442-443	Akron	97.9	95.5	96.9
444-445	Youngstown	97.3	93.2	95.6
446-447	Canton	97.5	87.8	93.3
448-449	Mansfield	94.9	92.4	93.8
450	Hamilton	94.8	88.1	91.9
451-452	Cincinnati	95.1	88.3	92.2
453-454	Dayton	94.9	85.0	90.6
455	Springfield	94.8	85.9	91.0
456	Chillicothe	93.9	93.5	93.7
457	Athens	96.8	77.5	88.5
458	Lima	97.2	87.3	92.9
OKLAHOMA				
730-731	Oklahoma City	97.3	62.8	82.4
734	Ardmore	95.0	62.2	80.8
735	Lawton	97.2	63.7	82.8
736	Clinton	96.4	60.6	80.9
737	Enid	96.9	60.5	81.2
738	Woodward	95.1	60.6	80.2
739	Guymon	96.2	32.3	68.6
740-741	Tulsa	97.0	58.6	80.4
743	Miami	93.9	65.8	81.8
744	Muskogee	96.3	43.0	73.3
745	Mcalester	93.5	54.4	76.6
746	Ponca City	94.1	60.8	79.7
747	Durant	94.1	60.3	79.5
748	Shawnee	95.7	58.6	79.7
749	Poteau	93.1	63.2	80.2
OREGON				
970-972	Portland	101.8	102.7	102.2
973	Salem	102.1	102.0	102.0
974	Eugene	101.6	101.7	101.6
975	Medford	103.1	100.0	101.7
976	Klamath Falls	103.4	100.0	102.0
977	Bend	102.2	101.7	102.0
978	Pendleton	96.8	101.5	98.8
979	Vale	94.5	92.2	93.5
PENNSYLVANIA				
150-152	Pittsburgh	96.0	102.7	98.9
153	Washington	93.3	101.1	96.7
154	Uniontown	93.6	98.1	95.5
155	Bedford	94.5	91.4	93.1
156	Greensburg	94.6	99.0	96.5
157	Indiana	93.4	98.1	95.4
158	Dubois	94.7	94.6	94.7
159	Johnstown	94.4	94.9	94.6
160	Butler	92.3	99.9	95.6
161	New Castle	92.3	99.1	95.3
162	Kittanning	92.8	101.1	96.4
163	Oil City	92.3	96.3	94.0
164-165	Erie	94.3	94.1	94.2
166	Altoona	94.3	91.5	93.1
167	Bradford	95.6	94.3	95.0
168	State College	95.2	91.6	93.6
169	Wellsboro	96.2	91.7	94.3
170-171	Harrisburg	98.0	92.6	95.7
172	Chambersburg	95.7	89.5	93.0
173-174	York	95.9	92.6	94.5
175-176	Lancaster	94.5	87.9	91.7

STATE/ZIP	CITY	MAT.	INST.	TOTAL
PENNSYLVANIA (CONT'D)				
177	Williamsport	93.2	80.0	87.5
178	Sunbury	95.3	92.5	94.1
179	Pottsville	94.3	94.2	94.3
180	Lehigh Valley	95.8	112.1	102.8
181	Allentown	97.7	108.3	102.2
182	Hazleton	95.2	95.3	95.3
183	Stroudsburg	95.1	101.4	97.8
184-185	Scranton	98.4	97.1	97.9
186-187	Wilkes-Barre	95.1	95.6	95.3
188	Montrose	94.6	96.0	95.2
189	Doylestown	94.6	117.5	104.5
190-191	Philadelphia	101.2	131.6	114.3
193	Westchester	98.0	118.7	106.9
194	Norristown	96.9	126.7	109.7
195-196	Reading	98.8	98.9	98.8
PUERTO RICO				
009	San Juan	119.0	24.6	78.3
RHODE ISLAND				
028	Newport	98.4	110.3	103.5
029	Providence	99.1	110.3	104.0
SOUTH CAROLINA				
290-292	Columbia	98.0	47.8	76.4
293	Spartanburg	97.2	46.5	75.3
294	Charleston	98.7	52.3	78.7
295	Florence	96.8	45.6	74.7
296	Greenville	96.9	46.5	75.2
297	Rock Hill	96.8	33.3	69.4
298	Aiken	97.7	69.6	85.6
299	Beaufort	98.5	36.6	71.8
SOUTH DAKOTA				
570-571	Sioux Falls	98.9	56.3	80.5
572	Watertown	98.1	49.6	77.2
573	Mitchell	97.0	49.1	76.3
574	Aberdeen	99.5	51.8	78.9
575	Pierre	98.8	51.8	78.5
576	Mobridge	97.6	49.3	76.8
577	Rapid City	99.3	50.6	78.3
TENNESSEE				
370-372	Nashville	97.7	74.8	87.8
373-374	Chattanooga	98.4	57.9	80.9
375,380-381	Memphis	96.0	74.0	86.5
376	Johnson City	97.2	57.7	80.1
377-379	Knoxville	94.5	58.4	78.9
382	Mckenzie	96.2	57.4	79.4
383	Jackson	98.1	50.2	77.4
384	Columbia	94.7	59.2	79.4
385	Cookeville	96.0	61.0	80.9
TEXAS				
750	Mckinney	97.4	53.8	78.6
751	Waxahackie	97.4	54.6	78.9
752-753	Dallas	98.0	65.6	84.0
754	Greenville	97.5	39.5	72.5
755	Texarkana	96.4	52.4	77.4
756	Longview	96.8	41.2	72.8
757	Tyler	97.4	55.2	79.2
758	Palestine	93.5	41.6	71.1
759	Lufkin	94.4	45.6	73.3
760-761	Fort Worth	97.0	61.4	81.6
762	Denton	97.5	50.1	77.1
763	Wichita Falls	98.0	56.8	80.2
764	Eastland	96.7	40.7	72.5
765	Temple	95.5	50.8	76.2
766-767	Waco	97.2	57.5	80.1
768	Brownwood	97.7	38.3	72.0
769	San Angelo	97.3	46.5	75.4
770-772	Houston	101.4	71.4	88.5
773	Huntsville	100.0	38.9	73.6
774	Wharton	101.4	43.6	76.4
775	Galveston	99.3	70.4	86.9
776-777	Beaumont	99.1	62.3	83.2
778	Bryan	96.6	63.4	82.3
779	Victoria	101.5	46.0	77.5
780	Laredo	95.4	52.8	77.0
781-782	San Antonio	95.7	66.1	83.0
783-784	Corpus Christi	98.4	52.0	78.3
785	Mc Allen	98.5	46.4	76.0
786-787	Austin	95.5	60.3	80.3

Location Factors

STATE/ZIP	CITY	MAT.	INST.	TOTAL
TEXAS (CONT'D)				
788	Del Rio	97.7	32.3	69.5
789	Giddings	95.1	40.9	71.7
790-791	Amarillo	98.1	57.2	80.5
792	Childress	97.2	51.6	77.5
793-794	Lubbock	99.4	51.9	78.9
795-796	Abilene	97.5	52.0	77.9
797	Midland	99.7	48.7	77.7
798-799,885	El Paso	96.9	50.3	76.8
UTAH				
840-841	Salt Lake City	102.9	68.6	88.1
842,844	Ogden	98.3	68.0	85.2
843	Logan	100.2	68.0	86.3
845	Price	100.8	48.4	78.2
846-847	Provo	100.7	68.5	86.8
VERMONT				
050	White River Jct.	97.5	50.1	77.0
051	Bellows Falls	96.0	54.8	78.2
052	Bennington	96.4	70.8	85.3
053	Brattleboro	96.8	61.3	81.4
054	Burlington	99.7	62.8	83.8
056	Montpelier	96.4	62.8	81.9
057	Rutland	98.2	62.8	82.9
058	St. Johnsbury	97.7	59.7	81.3
059	Guildhall	96.3	59.6	80.4
VIRGINIA				
220-221	Fairfax	100.7	83.3	93.2
222	Arlington	101.8	82.4	93.4
223	Alexandria	101.0	83.9	93.6
224-225	Fredericksburg	99.4	76.0	89.3
226	Winchester	100.0	69.9	87.0
227	Culpeper	99.9	75.2	89.2
228	Harrisonburg	100.1	66.4	85.6
229	Charlottesville	100.6	69.2	87.0
230-232	Richmond	101.2	73.1	89.1
233-235	Norfolk	101.2	71.1	88.2
236	Newport News	100.8	71.9	88.3
237	Portsmouth	100.1	67.2	85.9
238	Petersburg	100.3	73.1	88.6
239	Farmville	99.7	59.8	82.4
240-241	Roanoke	101.4	67.0	86.5
242	Bristol	99.2	59.3	82.0
243	Pulaski	98.9	57.3	81.0
244	Staunton	99.9	66.3	85.4
245	Lynchburg	99.9	69.1	86.6
246	Grundy	99.3	56.6	80.9
WASHINGTON				
980-981,987	Seattle	104.7	103.5	104.1
982	Everett	104.4	99.4	102.2
983-984	Tacoma	104.4	99.9	102.5
985	Olympia	102.5	99.8	101.4
986	Vancouver	106.0	96.2	101.8
988	Wenatchee	104.5	84.9	96.0
989	Yakima	104.5	90.4	98.4
990-992	Spokane	105.3	81.8	95.1
993	Richland	104.6	86.0	96.6
994	Clarkston	103.9	83.3	95.0
WEST VIRGINIA				
247-248	Bluefield	98.0	77.1	89.0
249	Lewisburg	99.8	80.6	91.5
250-253	Charleston	100.6	88.4	95.3
254	Martinsburg	99.5	77.3	89.9
255-257	Huntington	100.8	96.0	98.8
258-259	Beckley	97.9	86.1	92.8
260	Wheeling	100.9	91.2	96.7
261	Parkersburg	99.7	90.3	95.6
262	Buckhannon	99.4	90.4	95.5
263-264	Clarksburg	99.9	90.5	95.8
265	Morgantown	100.0	90.6	95.9
266	Gassaway	99.3	89.3	95.0
267	Romney	99.3	84.0	92.7
268	Petersburg	99.1	86.6	93.7
WISCONSIN				
530,532	Milwaukee	99.1	106.7	102.4
531	Kenosha	98.8	101.1	99.8
534	Racine	98.3	101.7	99.8
535	Beloit	98.2	97.4	97.8
537	Madison	98.0	96.5	97.4

STATE/ZIP	CITY	MAT.	INST.	TOTAL
WISCONSIN (CONT'D)				
538	Lancaster	96.1	91.1	93.9
539	Portage	94.7	95.1	94.9
540	New Richmond	95.6	97.1	96.2
541-543	Green Bay	99.6	93.1	96.8
544	Wausau	94.9	92.2	93.7
545	Rhinelander	98.1	92.4	95.7
546	La Crosse	95.9	93.1	94.7
547	Eau Claire	97.6	95.1	96.5
548	Superior	95.4	100.6	97.6
549	Oshkosh	95.6	92.8	94.4
WYOMING				
820	Cheyenne	100.0	67.8	86.1
821	Yellowstone Nat'l Park	97.5	60.4	81.5
822	Wheatland	98.6	59.8	81.8
823	Rawlins	100.2	59.3	82.5
824	Worland	98.1	58.1	80.8
825	Riverton	99.1	58.2	81.4
826	Casper	99.8	61.4	83.2
827	Newcastle	97.9	59.3	81.2
828	Sheridan	100.7	62.1	84.0
829-831	Rock Springs	102.1	58.1	83.1
CANADIAN FACTORS (reflect Canadian currency)				
ALBERTA				
	Calgary	124.8	93.1	111.1
	Edmonton	126.3	93.1	111.9
	Fort McMurray	116.6	93.1	106.4
	Lethbridge	117.5	92.3	106.7
	Lloydminster	116.6	93.1	106.5
	Medicine Hat	116.7	92.3	106.2
	Red Deer	117.1	92.3	106.4
BRITISH COLUMBIA				
	Kamloops	117.8	96.4	108.6
	Prince George	118.9	96.4	109.2
	Vancouver	126.9	86.9	109.6
	Victoria	118.9	83.4	103.6
MANITOBA				
	Brandon	117.0	80.6	101.3
	Portage la Prairie	117.0	79.2	100.7
	Winnipeg	125.8	73.5	103.2
NEW BRUNSWICK				
	Bathurst	115.3	71.2	96.3
	Dalhousie	115.2	71.2	96.2
	Fredericton	116.9	76.2	99.3
	Moncton	115.6	72.0	96.8
	Newcastle	115.3	71.2	96.3
	Saint John	118.2	75.9	99.9
NEWFOUNDLAND				
	Corner Brook	120.2	71.4	99.1
	St. John's	120.7	72.5	99.9
NORTHWEST TERRITORIES				
	Yellowknife	121.1	90.8	108.0
NOVA SCOTIA				
	Bridgewater	116.7	79.1	100.5
	Dartmouth	118.1	79.1	101.3
	Halifax	119.8	80.4	102.8
	New Glasgow	116.1	79.1	100.1
	Sydney	113.5	79.1	98.7
	Truro	116.1	79.1	100.1
	Yarmouth	116.0	79.1	100.1
ONTARIO				
	Barrie	119.7	99.1	110.8
	Brantford	118.9	103.3	112.2
	Cornwall	119.3	99.5	110.7
	Hamilton	124.8	98.7	113.5
	Kingston	120.2	99.6	111.3
	Kitchener	115.5	94.5	106.4
	London	122.0	96.2	110.9
	North Bay	118.9	97.2	109.5
	Oshawa	118.8	100.7	111.0
	Ottawa	120.7	97.8	110.8
	Owen Sound	119.9	97.2	110.1
	Peterborough	118.9	99.3	110.4
	Sarnia	119.0	104.0	112.5

Location Factors

STATE/ZIP	CITY	MAT.	INST.	TOTAL
ONTARIO (CONT'D)				
	Sault Ste Marie	113.7	97.3	106.6
	St. Catharines	113.4	96.3	106.0
	Sudbury	113.4	94.8	105.4
	Thunder Bay	114.8	95.0	106.3
	Timmins	119.2	97.2	109.7
	Toronto	123.1	102.2	114.1
	Windsor	114.3	95.4	106.1
PRINCE EDWARD ISLAND				
	Charlottetown	118.4	66.4	95.9
	Summerside	117.7	66.4	95.5
QUEBEC				
	Cap-de-la-Madeleine	116.0	93.0	106.1
	Charlesbourg	116.0	93.0	106.1
	Chicoutimi	115.1	97.8	107.6
	Gatineau	115.3	92.8	105.6
	Granby	115.7	92.7	105.8
	Hull	115.6	92.8	105.7
	Joliette	116.3	93.0	106.3
	Laval	115.4	92.8	105.6
	Montreal	119.8	98.1	110.5
	Quebec	119.9	98.5	110.7
	Rimouski	115.5	97.8	107.8
	Rouyn-Noranda	115.3	92.8	105.6
	Saint Hyacinthe	115.0	92.8	105.4
	Sherbrooke	115.7	92.8	105.8
	Sorel	116.2	93.0	106.2
	St Jerome	115.4	92.8	105.6
	Trois Rivieres	116.2	93.0	106.2
SASKATCHEWAN				
	Moose Jaw	114.3	73.8	96.8
	Prince Albert	113.2	71.9	95.4
	Regina	116.3	73.9	98.0
	Saskatoon	114.8	73.7	97.0
YUKON				
	Whitehorse	113.9	72.6	96.1

General Requirements — R0111 Summary of Work

R011105-05 Tips for Accurate Estimating

1. Use pre-printed or columnar forms for orderly sequence of dimensions and locations and for recording telephone quotations.
2. Use only the front side of each paper or form except for certain pre-printed summary forms.
3. Be consistent in listing dimensions: For example, length x width x height. This helps in rechecking to ensure that, the total length of partitions is appropriate for the building area.
4. Use printed (rather than measured) dimensions where given.
5. Add up multiple printed dimensions for a single entry where possible.
6. Measure all other dimensions carefully.
7. Use each set of dimensions to calculate multiple related quantities.
8. Convert foot and inch measurements to decimal feet when listing. Memorize decimal equivalents to .01 parts of a foot (1/8" equals approximately .01').
9. Do not "round off" quantities until the final summary.
10. Mark drawings with different colors as items are taken off.
11. Keep similar items together, different items separate.
12. Identify location and drawing numbers to aid in future checking for completeness.
13. Measure or list everything on the drawings or mentioned in the specifications.
14. It may be necessary to list items not called for to make the job complete.
15. Be alert for: Notes on plans such as N.T.S. (not to scale); changes in scale throughout the drawings; reduced size drawings; discrepancies between the specifications and the drawings.
16. Develop a consistent pattern of performing an estimate. For example:
    a. Start the quantity takeoff at the lower floor and move to the next higher floor.
    b. Proceed from the main section of the building to the wings.
    c. Proceed from south to north or vice versa, clockwise or counterclockwise.
    d. Take off floor plan quantities first, elevations next, then detail drawings.
17. List all gross dimensions that can be either used again for different quantities, or used as a rough check of other quantities for verification (exterior perimeter, gross floor area, individual floor areas, etc.).
18. Utilize design symmetry or repetition (repetitive floors, repetitive wings, symmetrical design around a center line, similar room layouts, etc.). Note: Extreme caution is needed here so as not to omit or duplicate an area.
19. Do not convert units until the final total is obtained. For instance, when estimating concrete work, keep all units to the nearest cubic foot, then summarize and convert to cubic yards.
20. When figuring alternatives, it is best to total all items involved in the basic system, then total all items involved in the alternates. Therefore you work with positive numbers in all cases. When adds and deducts are used, it is often confusing whether to add or subtract a portion of an item; especially on a complicated or involved alternate.

R011110-10 Architectural Fees

Tabulated below are typical percentage fees by project size, for good professional architectural service. Fees may vary from those listed depending upon degree of design difficulty and economic conditions in any particular area.

Rates can be interpolated horizontally and vertically. Various portions of the same project requiring different rates should be adjusted proportionately. For alterations, add 50% to the fee for the first $500,000 of project cost and add 25% to the fee for project cost over $500,000.

Architectural fees tabulated below include Structural, Mechanical and Electrical Engineering Fees. They do not include the fees for special consultants such as kitchen planning, security, acoustical, interior design, etc.

Civil Engineering fees are included in the Architectural fee for project sites requiring minimal design such as city sites. However, separate Civil Engineering fees must be added when utility connections require design, drainage calculations are needed, stepped foundations are required, or provisions are required to protect adjacent wetlands.

Building Types	Total Project Size in Thousands of Dollars						
	100	250	500	1,000	5,000	10,000	50,000
Factories, garages, warehouses, repetitive housing	9.0%	8.0%	7.0%	6.2%	5.3%	4.9%	4.5%
Apartments, banks, schools, libraries, offices, municipal buildings	12.2	12.3	9.2	8.0	7.0	6.6	6.2
Churches, hospitals, homes, laboratories, museums, research	15.0	13.6	12.7	11.9	9.5	8.8	8.0
Memorials, monumental work, decorative furnishings	—	16.0	14.5	13.1	10.0	9.0	8.3

General Requirements — R0111 Summary of Work

R011110-30 Engineering Fees

Typical **Structural Engineering Fees** based on type of construction and total project size. These fees are included in Architectural Fees.

Type of Construction	Total Project Size (in thousands of dollars)			
	$500	$500-$1,000	$1,000-$5,000	Over $5000
Industrial buildings, factories & warehouses	Technical payroll times 2.0 to 2.5	1.60%	1.25%	1.00%
Hotels, apartments, offices, dormitories, hospitals, public buildings, food stores		2.00%	1.70%	1.20%
Museums, banks, churches and cathedrals		2.00%	1.75%	1.25%
Thin shells, prestressed concrete, earthquake resistive		2.00%	1.75%	1.50%
Parking ramps, auditoriums, stadiums, convention halls, hangars & boiler houses		2.50%	2.00%	1.75%
Special buildings, major alterations, underpinning & future expansion		Add to above 0.5%	Add to above 0.5%	Add to above 0.5%

For complex reinforced concrete or unusually complicated structures, add 20% to 50%.

Typical **Mechanical and Electrical Engineering Fees** are based on the size of the subcontract. The fee structure for both are shown below. These fees are included in Architectural Fees.

Type of Construction	Subcontract Size							
	$25,000	$50,000	$100,000	$225,000	$350,000	$500,000	$750,000	$1,000,000
Simple structures	6.4%	5.7%	4.8%	4.5%	4.4%	4.3%	4.2%	4.1%
Intermediate structures	8.0	7.3	6.5	5.6	5.1	5.0	4.9	4.8
Complex structures	10.1	9.0	9.0	8.0	7.5	7.5	7.0	7.0

For renovations, add 15% to 25% to applicable fee.

General Requirements — R0121 Allowances

R012157-20 Construction Time Requirements

Table at left is average construction time in months for different types of building projects. Table at right is the construction time in months for different size projects. Design time runs 25% to 40% of construction time.

Type Building	Construction Time	Project Value	Construction Time
Industrial Buildings	12 Months	Under $1,400,000	10 Months
Commercial Buildings	15 Months	Up to $3,800,000	15 Months
Research & Development	18 Months	Up to $19,000,000	21 Months
Institutional Buildings	20 Months	over $19,000,000	28 Months

General Requirements — R0129 Payment Procedures

R012909-80 Sales Tax by State

State sales tax on materials is tabulated below (5 states have no sales tax). Many states allow local jurisdictions, such as a county or city, to levy additional sales tax.

Some projects may be sales tax exempt, particularly those constructed with public funds.

State	Tax (%)	State	Tax (%)	State	Tax (%)	State	Tax (%)
Alabama	4	Illinois	6.25	Montana	0	Rhode Island	7
Alaska	0	Indiana	6	Nebraska	5.5	South Carolina	5
Arizona	5.6	Iowa	5	Nevada	6.5	South Dakota	4
Arkansas	6	Kansas	5.3	New Hampshire	0	Tennessee	7
California	7.25	Kentucky	6	New Jersey	6	Texas	6.25
Colorado	2.9	Louisiana	4	New Mexico	5	Utah	4.75
Connecticut	6	Maine	5	New York	4	Vermont	6
Delaware	0	Maryland	5	North Carolina	4.5	Virginia	5
District of Columbia	5.75	Massachusetts	5	North Dakota	5	Washington	6.5
Florida	6	Michigan	6	Ohio	5.5	West Virginia	6
Georgia	4	Minnesota	6.5	Oklahoma	4.5	Wisconsin	5
Hawaii	4	Mississippi	7	Oregon	0	Wyoming	4
Idaho	5	Missouri	4.225	Pennsylvania	6	Average	4.84 %

Sales Tax by Province (Canada)

GST - a value-added tax, which the government imposes on most goods and services provided in or imported into Canada. PST - a retail sales tax, which five of the provinces impose on the price of most goods and some services. QST - a value-added tax, similar to the federal GST, which Quebec imposes. HST - Three provinces have combined their retail sales tax with the federal GST into one harmonized tax.

Province	PST (%)	QST (%)	GST (%)	HST (%)
Alberta	0	0	6	0
British Columbia	7	0	6	0
Manitoba	7	0	6	0
New Brunswick	0	0	0	14
Newfoundland	0	0	0	14
Northwest Territories	0	0	6	0
Nova Scotia	0	0	0	14
Ontario	8	0	6	0
Prince Edward Island	10	0	6	0
Quebec	0	7.5	6	0
Saskatchewan	7	0	6	0
Yukon	0	0	6	0

General Requirements — R0129 Payment Procedures

R012909-85 Unemployment Taxes and Social Security Taxes

State Unemployment Tax rates vary not only from state to state, but also with the experience rating of the contractor. The Federal Unemployment Tax rate is 6.2% of the first $7,000 of wages. This is reduced by a credit of up to 5.4% for timely payment to the state. The minimum Federal Unemployment Tax is 0.8% after all credits.

Social Security (FICA) for 2007 is estimated at time of publication to be 7.65% of wages up to $94,200.

R012909-90 Overtime

One way to improve the completion date of a project or eliminate negative float from a schedule is to compress activity duration times. This can be achieved by increasing the crew size or working overtime with the proposed crew.

To determine the costs of working overtime to compress activity duration times, consider the following examples. Below is an overtime efficiency and cost chart based on a five, six, or seven day week with an eight through twelve hour day. Payroll percentage increases for time and one half and double time are shown for the various working days.

Days per Week	Hours per Day	Production Efficiency					Payroll Cost Factors	
		1st Week	2nd Week	3rd Week	4th Week	Average 4 Weeks	@ 1-1/2 Times	@ 2 Times
5	8	100%	100%	100%	100%	100 %	100 %	100 %
	9	100	100	95	90	96.25	105.6	111.1
	10	100	95	90	85	91.25	110.0	120.0
	11	95	90	75	65	81.25	113.6	127.3
	12	90	85	70	60	76.25	116.7	133.3
6	8	100	100	95	90	96.25	108.3	116.7
	9	100	95	90	85	92.50	113.0	125.9
	10	95	90	85	80	87.50	116.7	133.3
	11	95	85	70	65	78.75	119.7	139.4
	12	90	80	65	60	73.75	122.2	144.4
7	8	100	95	85	75	88.75	114.3	128.6
	9	95	90	80	70	83.75	118.3	136.5
	10	90	85	75	65	78.75	121.4	142.9
	11	85	80	65	60	72.50	124.0	148.1
	12	85	75	60	55	68.75	126.2	152.4

General Requirements — R0131 Project Management & Coordination

R013113-40 Builder's Risk Insurance

Builder's Risk Insurance is insurance on a building during construction. Premiums are paid by the owner or the contractor. Blasting, collapse and underground insurance would raise total insurance costs above those listed. Floater policy for materials delivered to the job runs $.75 to $1.25 per $100 value. Contractor equipment insurance runs $.50 to $1.50 per $100 value. Insurance for miscellaneous tools to $1,500 value runs from $3.00 to $7.50 per $100 value.

Tabulated below are New England Builder's Risk insurance rates in dollars per $100 value for $1,000 deductible. For $25,000 deductible, rates can be reduced 13% to 34%. On contracts over $1,000,000, rates may be lower than those tabulated. Policies are written annually for the total completed value in place. For "all risk" insurance (excluding flood, earthquake and certain other perils) add $.025 to total rates below.

Coverage	Frame Construction (Class 1)		Brick Construction (Class 4)		Fire Resistive (Class 6)	
	Range	Average	Range	Average	Range	Average
Fire Insurance	$.350 to $.850	$.600	$.158 to $.189	$.174	$.052 to $.080	$.070
Extended Coverage	.115 to .200	.158	.080 to .105	.101	.081 to .105	.100
Vandalism	.012 to .016	.014	.008 to .011	.011	.008 to .011	.010
Total Annual Rate	$.477 to $1.066	$.772	$.246 to $.305	$.286	$.141 to $.196	$.180

General Requirements — R0131 Project Management & Coordination

R013113-50 General Contractor's Overhead

There are two distinct types of overhead on a construction project: Project Overhead and Main Office Overhead. Project Overhead includes those costs at a construction site not directly associated with the installation of construction materials. Examples of Project Overhead costs include the following:

1. Superintendent
2. Construction office and storage trailers
3. Temporary sanitary facilities
4. Temporary utilities
5. Security fencing
6. Photographs
7. Clean up
8. Performance and payment bonds

The above Project Overhead items are also referred to as General Requirements and therefore are estimated in Division 1. Division 1 is the first division listed in the CSI MasterFormat but it is usually the last division estimated. The sum of the costs in Divisions 1 through 49 is referred to as the sum of the direct costs.

All construction projects also include indirect costs. The primary components of indirect costs are the contractor's Main Office Overhead and profit. The amount of the Main Office Overhead expense varies depending on the the following:

1. Owner's compensation
2. Project managers and estimator's wages
3. Clerical support wages
4. Office rent and utilities
5. Corporate legal and accounting costs
6. Advertising
7. Automobile expenses
8. Association dues
9. Travel and entertainment expenses

These costs are usually calculated as a percentage of annual sales volume. This percentage can range from 35% for a small contractor doing less than $500,000 to 5% for a large contractor with sales in excess of $100 million.

General Requirements — R0131 Project Management & Coordination

R013113-60 Workers' Compensation Insurance Rates by Trade

The table below tabulates the national averages for Workers' Compensation insurance rates by trade and type of building. The average "Insurance Rate" is multiplied by the "% of Building Cost" for each trade. This produces the "Workers' Compensation Cost" by % of total labor cost, to be added for each trade by building type to determine the weighted average Workers' Compensation rate for the building types analyzed.

Trade	Insurance Rate (% Labor Cost) Range	Insurance Rate (% Labor Cost) Average	% of Building Cost Office Bldgs.	% of Building Cost Schools & Apts.	% of Building Cost Mfg.	Workers' Compensation Office Bldgs.	Workers' Compensation Schools & Apts.	Workers' Compensation Mfg.
Excavation, Grading, etc.	4.5% to 10.5%	10.4%	4.8%	4.9%	4.5%	.50%	.51%	.47%
Piles & Foundations	8.9 to 42.9	21.4	7.1	5.2	8.7	1.52	1.11	1.86
Concrete	5.1 to 29.8	15.5	5.0	14.8	3.7	.78	2.29	.57
Masonry	5.4 to 34.5	14.9	6.9	7.5	1.9	1.03	1.12	.28
Structural Steel	8.6 to 104.1	40.8	10.7	3.9	17.6	4.37	1.59	7.18
Miscellaneous & Ornamental Metals	5.2 to 22.6	11.9	2.8	4.0	3.6	.33	.48	.43
Carpentry & Millwork	7.6 to 53.2	18.4	3.7	4.0	0.5	.68	.74	.09
Metal or Composition Siding	6.8 to 34.1	16.7	2.3	0.3	4.3	.38	.05	.72
Roofing	8.4 to 77.1	32.3	2.3	2.6	3.1	.74	.84	1.00
Doors & Hardware	5 to 24.9	11.1	0.9	1.4	0.4	.10	.16	.04
Sash & Glazing	5.6 to 28.9	14.0	3.5	4.0	1.0	.49	.56	.14
Lath & Plaster	3.1 to 35.4	14.0	3.3	6.9	0.8	.46	.97	.11
Tile, Marble & Floors	2 to 29.4	9.5	2.6	3.0	0.5	.25	.29	.05
Acoustical Ceilings	2.3 to 50.2	11.6	2.4	0.2	0.3	.28	.02	.03
Painting	4.9 to 29.6	13.2	1.5	1.6	1.6	.20	.21	.21
Interior Partitions	7.6 to 53.2	18.4	3.9	4.3	4.4	.72	.79	.81
Miscellaneous Items	2.3 to 197.9	16.8	5.2	3.7	9.7	.88	.62	1.63
Elevators	2.9 to 13.7	6.9	2.1	1.1	2.2	.14	.08	.15
Sprinklers	2.8 to 15.3	8.3	0.5	—	2.0	.04	—	.17
Plumbing	2.9 to 14.2	8.1	4.9	7.2	5.2	.40	.58	.42
Heat., Vent., Air Conditioning	4.6 to 24.8	11.9	13.5	11.0	12.9	1.61	1.31	1.54
Electrical	2.7 to 13.6	6.6	10.1	8.4	11.1	.67	.55	.73
Total	2% to 197.9%	—	100.0%	100.0%	100.0%	16.57%	14.87%	18.63%

Overall Weighted Average 16.69%

Workers' Compensation Insurance Rates by States

The table below lists the weighted average Workers' Compensation base rate for each state with a factor comparing this with the national average of 16.3%.

State	Weighted Average	Factor	State	Weighted Average	Factor	State	Weighted Average	Factor
Alabama	26.2%	161	Kentucky	19.6%	120	North Dakota	13.7%	84
Alaska	25.0	153	Louisiana	28.4	174	Ohio	14.1	87
Arizona	7.5	46	Maine	21.1	129	Oklahoma	14.6	90
Arkansas	14.0	86	Maryland	19.2	118	Oregon	13.3	82
California	18.9	116	Massachusetts	15.3	94	Pennsylvania	13.0	80
Colorado	13.3	82	Michigan	17.4	107	Rhode Island	21.3	131
Connecticut	21.3	131	Minnesota	25.7	158	South Carolina	17.0	104
Delaware	19.0	117	Mississippi	18.5	113	South Dakota	18.6	114
District of Columbia	15.9	98	Missouri	18.1	111	Tennessee	15.6	96
Florida	20.4	125	Montana	17.9	110	Texas	13.2	81
Georgia	22.8	140	Nebraska	24.4	150	Utah	12.4	76
Hawaii	17.0	104	Nevada	12.5	77	Vermont	24.3	149
Idaho	12.1	74	New Hampshire	21.5	132	Virginia	12.6	77
Illinois	18.8	115	New Jersey	12.9	79	Washington	11.0	67
Indiana	6.6	40	New Mexico	19.4	119	West Virginia	9.5	58
Iowa	12.0	74	New York	13.8	85	Wisconsin	14.8	91
Kansas	8.7	53	North Carolina	17.3	106	Wyoming	9.9	61

Weighted Average for U.S. is 16.7% of payroll = 100%

Rates in the following table are the base or manual costs per $100 of payroll for Workers' Compensation in each state. Rates are usually applied to straight time wages only and not to premium time wages and bonuses.

The weighted average skilled worker rate for 35 trades is 16.3%. For bidding purposes, apply the full value of Workers' Compensation directly to total labor costs, or if labor is 38%, materials 42% and overhead and profit 20% of total cost, carry 38/80 x 16.3% =7.7% of cost (before overhead and profit) into overhead. Rates vary not only from state to state but also with the experience rating of the contractor.

Rates are the most current available at the time of publication.

General Requirements — R0131 Project Management & Coordination

R013113-60 Workers' Compensation Insurance Rates by Trade and State (cont.)

State	Carpentry — 3 stories or less 5651	Carpentry — interior cab. work 5437	Carpentry — general 5403	Concrete Work — NOC 5213	Concrete Work — flat (flr., sdwk.) 5221	Electrical Wiring — inside 5190	Excavation — earth NOC 6217	Excavation — rock 6217	Glaziers 5462	Insulation Work 5479	Lathing 5443	Masonry 5022	Painting & Decorating 5474	Pile Driving 6003	Plastering 5480	Plumbing 5183	Roofing 5551	Sheet Metal Work (HVAC) 5538	Steel Erection — door & sash 5102	Steel Erection — inter., ornam. 5102	Steel Erection — structure 5040	Steel Erection — NOC 5057	Tile Work — (interior ceramic) 5348	Waterproofing 9014	Wrecking 5701
AL	34.13	19.34	33.95	12.96	10.74	9.34	14.00	14.00	23.27	16.95	13.45	26.80	27.85	42.90	27.83	10.23	57.63	24.81	16.77	16.77	64.80	28.72	12.05	7.74	64.80
AK	20.75	14.96	16.46	15.76	14.11	13.57	20.48	20.48	28.87	34.86	12.30	34.35	24.83	41.80	33.32	11.83	46.92	12.89	13.20	13.20	53.01	25.10	9.07	8.50	53.01
AZ	6.88	5.05	11.62	5.56	3.49	3.85	4.74	4.74	7.40	7.69	4.27	5.37	4.87	9.83	5.20	3.90	12.19	5.65	9.50	9.50	18.98	8.11	2.04	2.34	50.31
AR	15.09	8.06	16.49	13.30	6.63	5.14	8.49	8.49	11.79	23.72	6.63	9.32	9.64	16.15	13.39	5.21	28.01	13.09	7.75	7.75	36.16	21.58	6.29	4.74	36.16
CA	28.28	9.07	28.28	13.18	13.18	9.93	7.88	7.88	16.37	23.34	10.90	14.55	19.45	21.18	18.12	11.59	39.79	15.33	13.55	13.55	23.93	21.91	7.74	19.45	21.91
CO	16.28	8.03	12.76	13.19	7.46	5.18	10.35	10.35	9.79	13.91	6.03	14.30	10.29	15.08	9.00	8.45	23.95	11.46	8.38	8.38	35.52	15.85	8.60	5.74	15.85
CT	18.32	16.06	25.93	24.48	12.29	8.04	11.27	11.27	19.79	16.24	22.62	25.73	16.24	24.89	18.63	10.52	37.59	15.83	16.42	16.42	50.14	26.51	9.01	6.09	56.71
DE	19.38	19.38	15.10	15.16	12.03	8.22	12.07	12.07	14.07	19.38	16.67	15.69	20.18	25.15	16.67	10.12	34.39	12.62	16.11	16.11	35.78	16.11	12.50	15.69	35.78
DC	10.42	10.52	11.61	12.06	13.54	6.02	10.60	10.60	20.19	9.97	11.19	17.94	9.87	13.76	12.24	14.19	18.73	8.23	13.03	13.03	45.04	17.90	29.37	4.20	45.04
FL	24.38	15.34	24.74	21.86	10.37	8.91	11.12	11.12	15.53	14.31	10.58	17.48	16.78	38.24	28.82	10.04	35.40	14.90	11.74	11.74	44.71	25.87	10.03	7.76	44.71
GA	32.85	16.50	24.47	14.19	10.63	9.63	16.68	16.68	15.64	22.01	14.48	19.28	18.57	26.27	18.16	10.69	43.72	20.35	15.97	15.97	51.69	40.46	9.92	8.47	51.69
HI	17.38	11.62	28.20	13.74	12.27	7.10	7.73	7.73	20.55	21.19	10.94	17.87	11.33	20.15	15.61	6.06	33.24	8.02	11.11	11.11	31.71	21.70	9.78	11.54	31.71
ID	11.92	6.31	13.47	10.64	5.99	5.25	6.90	6.90	9.58	8.14	6.31	9.91	8.88	15.41	9.45	5.99	29.54	9.11	7.94	7.94	33.81	14.13	13.67	4.09	33.81
IL	17.33	15.90	18.72	26.98	10.79	8.21	9.42	9.42	19.93	15.96	10.26	18.26	10.08	25.32	13.09	9.80	29.37	14.49	15.54	15.54	50.39	21.95	14.63	4.66	50.39
IN	6.77	5.04	7.58	5.08	3.27	2.66	4.64	4.64	5.63	5.55	2.25	5.64	5.54	9.05	4.04	2.94	11.82	4.60	5.20	5.20	21.57	8.02	3.19	2.68	21.57
IA	11.41	6.24	9.43	11.35	6.29	4.24	5.17	5.17	11.28	8.38	5.51	8.91	7.85	8.93	7.26	5.08	16.80	7.56	7.75	7.75	46.05	40.59	5.48	3.96	27.19
KS	8.44	8.42	10.36	7.95	5.18	3.44	4.51	4.51	8.63	8.03	4.28	7.83	6.15	11.55	6.35	5.42	8.44	7.02	7.09	7.09	25.33	12.78	5.36	3.20	27.57
KY	22.12	12.10	19.05	18.45	7.85	6.61	11.25	11.25	20.44	20.69	13.02	10.03	13.56	25.37	15.33	7.48	50.97	19.58	14.49	14.49	46.98	23.88	11.89	6.30	46.98
LA	23.14	24.93	53.17	26.43	15.61	9.88	17.49	17.49	20.14	21.43	24.51	28.33	29.58	31.25	22.37	8.64	77.12	21.20	18.98	18.98	51.76	24.21	13.77	13.78	66.41
ME	13.43	10.42	43.59	24.29	11.12	3.72	12.56	12.56	13.35	15.39	14.03	17.26	16.07	31.44	17.95	9.07	32.07	10.07	15.08	15.08	37.84	61.81	11.66	5.92	37.84
MD	13.71	5.95	10.55	18.82	8.70	5.15	12.91	12.91	25.52	20.07	11.37	13.31	9.37	26.22	14.21	8.58	48.72	12.74	13.66	13.66	56.25	37.81	8.03	7.14	26.80
MA	9.03	6.88	16.48	22.32	9.18	4.18	6.08	6.08	9.25	14.59	5.89	15.67	6.85	15.92	6.62	4.80	47.57	6.72	13.48	13.48	43.69	34.06	8.78	2.65	24.03
MI	18.40	11.39	20.64	19.45	8.95	4.72	10.20	10.20	12.41	14.75	13.59	15.54	13.64	40.27	15.18	6.84	30.67	10.00	10.16	10.16	40.27	22.21	11.11	4.87	39.06
MN	19.53	19.80	41.48	14.65	15.73	7.53	14.95	14.95	15.68	11.90	22.05	19.18	16.95	24.63	22.05	10.88	70.18	14.23	10.58	10.58	104.13	33.30	14.55	6.68	36.03
MS	15.56	12.84	23.21	11.49	8.98	6.57	12.44	12.44	15.13	18.03	7.40	9.64	14.52	24.87	35.36	8.27	35.09	19.67	13.57	13.57	35.55	29.83	10.22	5.63	35.55
MO	23.87	12.38	16.85	16.75	11.59	7.61	9.87	9.87	11.40	18.47	10.12	14.58	15.09	18.46	14.96	10.16	33.26	15.34	11.85	11.85	47.95	33.09	9.40	6.12	47.95
MT	16.07	12.88	21.13	12.84	11.28	7.97	14.85	14.85	11.94	30.09	15.86	13.71	9.70	33.39	11.78	10.03	39.71	11.20	9.70	9.70	36.35	18.79	7.76	6.04	36.35
NE	27.13	14.63	21.57	29.20	14.30	10.22	17.88	17.88	25.50	35.50	12.00	25.82	22.27	24.63	20.60	11.52	39.10	18.35	14.60	14.60	65.40	26.27	12.02	7.63	58.22
NV	16.61	7.10	11.57	9.28	6.51	5.39	9.50	9.50	12.80	10.21	8.27	8.90	9.60	14.19	8.05	6.78	14.14	19.04	10.96	10.96	23.34	26.59	6.18	5.18	33.18
NH	26.55	11.21	21.62	32.68	12.33	7.21	17.79	17.79	11.16	29.06	9.32	22.91	15.04	21.80	13.65	10.84	58.14	14.88	12.51	12.51	49.35	18.42	11.43	6.01	49.35
NJ	13.50	9.36	13.50	12.14	8.95	4.45	8.62	8.62	7.20	13.18	11.54	13.14	11.44	13.54	11.54	6.26	34.09	6.04	11.03	11.03	25.41	13.62	4.87	4.83	23.62
NM	24.50	8.22	21.85	20.69	11.35	7.41	11.51	11.51	19.30	14.97	10.20	16.38	16.09	24.75	12.34	9.96	33.44	12.58	18.13	18.13	52.69	29.16	7.79	7.63	52.69
NY	14.21	12.39	12.39	17.19	11.26	6.69	8.37	8.37	10.18	9.91	11.27	15.70	10.41	15.68	9.04	7.33	33.93	12.79	9.59	9.59	21.71	15.20	8.72	6.92	24.30
NC	15.87	13.33	16.07	15.61	7.64	10.21	10.51	10.51	11.41	13.79	14.12	11.11	12.95	17.63	13.75	9.20	26.70	13.53	9.22	9.22	79.26	20.69	6.50	5.68	79.26
ND	10.92	10.92	10.92	6.26	6.26	3.90	6.46	6.46	10.92	10.92	8.84	7.94	6.41	22.61	8.84	5.47	22.68	5.47	22.61	22.61	22.61	22.61	10.92	22.68	12.19
OH	7.70	7.61	9.64	11.57	10.29	5.45	8.10	8.10	7.09	17.31	50.15	11.49	11.69	28.24	3.08	6.58	24.34	9.30	8.14	8.14	26.41	10.66	9.21	5.70	26.41
OK	13.42	9.17	11.76	11.76	6.34	5.73	10.83	10.83	16.07	19.47	8.45	10.02	8.47	20.25	12.19	7.09	21.13	9.05	13.76	13.76	39.72	23.97	6.50	5.71	39.78
OR	16.00	8.81	15.78	13.64	8.79	4.26	9.88	9.88	14.41	10.15	7.41	14.86	10.88	15.81	11.82	5.87	23.65	9.60	8.95	8.95	28.75	14.60	11.37	4.49	28.75
PA	12.67	6.29	11.41	13.57	9.77	5.98	7.88	7.88	9.15	11.41	10.98	11.40	13.00	15.49	10.98	6.94	26.17	7.49	13.91	13.91	21.88	13.91	7.85	11.40	21.88
RI	19.53	11.65	18.07	18.23	16.24	4.43	10.38	10.38	12.85	22.78	11.97	25.11	24.13	37.66	17.25	8.52	33.92	10.29	14.07	14.07	59.49	37.50	14.36	7.70	78.79
SC	22.26	17.21	19.93	13.51	7.81	9.42	11.95	11.95	14.32	11.08	8.53	13.27	16.47	18.32	18.15	8.83	43.19	10.97	11.47	11.47	24.55	27.29	8.90	5.98	47.05
SD	22.94	7.74	24.42	28.24	8.33	5.60	12.27	12.27	11.43	12.46	8.11	9.46	10.34	21.36	12.74	12.97	19.88	12.48	8.97	8.97	84.39	33.82	8.80	4.25	84.39
TN	17.21	13.36	22.83	16.35	9.27	7.36	12.04	12.04	13.17	11.21	11.71	14.74	10.49	18.13	12.60	9.19	24.86	13.88	9.95	9.95	32.82	20.02	8.84	5.72	32.82
TX	13.44	9.34	13.44	11.97	7.27	7.70	8.65	8.65	9.83	12.10	7.83	12.38	9.08	20.20	9.08	6.92	19.68	16.57	12.11	12.11	30.40	12.18	6.82	6.89	13.00
UT	9.45	6.23	10.52	7.68	7.99	7.05	6.38	6.38	9.55	10.95	12.32	13.75	15.29	17.24	10.60	6.41	26.27	6.30	8.99	8.99	23.51	23.51	6.06	5.83	26.53
VT	23.50	10.31	19.77	29.84	12.44	8.34	14.86	14.86	24.74	25.93	10.12	20.66	14.15	21.89	15.99	10.47	45.95	14.75	14.56	14.56	74.50	62.00	10.22	10.24	74.50
VA	11.41	7.88	10.02	11.54	5.24	5.47	7.04	7.04	8.77	10.82	18.78	8.94	9.54	12.50	9.73	5.46	21.49	8.23	9.60	9.60	43.89	22.16	5.03	2.89	43.89
WA	9.30	9.30	9.30	8.12	8.12	3.35	8.04	8.04	12.92	10.56	9.30	10.21	9.88	17.82	11.50	4.64	18.57	4.61	8.65	8.65	8.65	8.65	10.40	18.57	7.87
WV	9.77	7.20	10.47	8.58	4.65	4.82	5.92	5.92	7.80	7.09	7.11	8.14	8.24	11.16	8.73	4.78	19.23	7.63	6.45	6.45	26.02	12.50	4.53	2.83	23.67
WI	12.40	11.36	16.86	12.29	9.90	4.24	7.44	7.44	13.31	11.62	8.09	18.99	14.32	13.64	12.11	5.42	34.72	7.16	9.18	9.18	37.88	24.06	13.99	5.52	37.88
WY	8.93	8.93	8.93	8.93	8.93	8.93	8.93	8.93	8.93	8.93	8.93	8.93	8.93	8.93	8.93	8.93	8.93	8.93	8.93	8.93	8.93	8.93	8.93	8.93	8.93
AVG.	16.75	11.08	18.39	15.53	9.55	6.59	10.39	10.39	14.05	15.81	11.61	14.92	13.19	21.39	14.04	8.10	32.30	11.89	11.86	11.86	40.80	23.82	9.53	7.16	39.10

General Requirements — R0131 Project Management & Coordination

R013113-60 Workers' Compensation (cont.) (Canada in Canadian dollars)

Province		Alberta	British Columbia	Manitoba	Ontario	New Brunswick	Newfndld. & Labrador	Northwest Territories	Nova Scotia	Prince Edward Island	Quebec	Saskatchewan	Yukon
Carpentry—3 stories or less	Rate	8.17	5.91	4.48	4.58	3.94	8.84	4.16	7.67	5.73	15.41	7.33	4.79
	Code	42143	721028	40102	723	4226	4226	4-41	4226	401	80110	B13-17	202
Carpentry—interior cab. work	Rate	2.56	6.13	4.48	4.58	4.67	4.92	4.16	5.74	3.98	15.41	3.68	4.79
	Code	42133	721021	40102	723	4279	4270	4-41	4274	402	80110	B11-27	202
CARPENTRY—general	Rate	8.17	5.91	4.48	4.58	3.94	4.92	4.16	7.67	5.73	15.41	7.33	4.79
	Code	42143	721028	40102	723	4226	4299	4-41	4226	401	80110	B13-17	202
CONCRETE WORK—NOC	Rate	5.12	6.02	7.42	15.40	3.94	8.84	4.16	4.83	5.73	16.51	7.33	3.86
	Code	42104	721010	40110	748	4224	4224	4-41	4224	401	80100	B13-14	203
CONCRETE WORK—flat (flr. sidewalk)	Rate	5.12	6.02	7.42	15.40	3.94	8.84	4.16	4.83	5.73	16.51	7.33	3.86
	Code	42104	721010	40110	748	4224	4224	4-41	4224	401	80100	B13-14	203
ELECTRICAL Wiring—inside	Rate	2.22	2.18	2.56	3.25	2.45	2.92	3.91	2.23	3.98	7.64	3.68	3.86
	Code	42124	721019	40203	704	4261	4261	4-46	4261	402	80170	B11-05	206
EXCAVATION—earth NOC	Rate	2.95	4.20	3.78	4.55	2.88	4.29	4.04	4.11	4.23	8.36	4.37	3.86
	Code	40604	721031	40706	711	4214	4214	4-43	4214	404	80030	R11-06	207
EXCAVATION—rock	Rate	2.95	4.20	3.78	4.55	2.88	4.29	4.04	4.11	4.23	8.36	4.37	3.86
	Code	40604	721031	40706	711	4214	4214	4-43	4214	404	80030	R11-06	207
GLAZIERS	Rate	3.66	4.07	4.48	8.90	5.20	6.52	4.16	7.67	3.98	14.36	7.33	3.86
	Code	42121	715020	40109	751	4233	4233	4-41	4233	402	80150	B13-04	212
INSULATION WORK	Rate	3.03	7.02	4.48	8.90	5.20	6.52	4.16	7.67	5.73	15.41	6.15	4.79
	Code	42184	721029	40102	751	4234	4234	4-41	4234	401	80110	B12-07	202
LATHING	Rate	6.18	8.48	4.48	4.58	4.67	4.92	4.16	5.74	3.98	15.41	7.33	4.79
	Code	42135	721033	40102	723	4273	4279	4-41	4271	402	80110	B13-16	202
MASONRY	Rate	5.12	8.48	4.48	11.79	5.20	6.52	4.16	7.67	5.73	16.51	7.33	4.79
	Code	42102	721037	40102	741	4231	4231	4-41	4231	401	80100	B13-18	202
PAINTING & DECORATING	Rate	5.08	5.21	36.17	6.75	4.67	4.92	4.16	5.74	3.98	15.41	6.15	4.79
	Code	42111	721041	40105	719	4275	4275	4-41	4275	402	80110	B12-01	202
PILE DRIVING	Rate	5.12	5.52	3.78	6.26	3.94	9.78	4.04	4.83	5.73	8.36	7.33	4.79
	Code	42159	722004	40706	732	4221	4221	4-43	4221	401	80030	B13-10	202
PLASTERING	Rate	6.18	8.48	5.38	6.75	4.67	4.92	4.16	5.74	3.98	15.41	6.15	4.79
	Code	42135	721042	40108	719	4271	4271	4-41	4271	402	80110	B12-21	202
PLUMBING	Rate	2.22	3.89	2.92	4.02	2.89	3.24	3.91	2.23	3.98	7.61	3.68	3.27
	Code	42122	721043	40204	707	4241	4241	4-46	4241	402	80160	B11-01	214
ROOFING	Rate	9.20	10.86	7.12	12.53	8.10	8.84	4.16	9.49	5.73	22.49	7.33	4.79
	Code	42118	721036	40403	728	4236	4236	4-41	4236	401	80130	B13-20	202
SHEET METAL WORK (HVAC)	Rate	2.22	3.89	7.12	4.02	2.89	3.24	3.91	3.23	3.98	7.61	3.68	4.35
	Code	42117	721043	40402	707	4244	4244	4-46	4244	402	80160	B11-07	208
STEEL ERECTION—door & sash	Rate	3.03	16.01	12.96	15.40	3.94	8.84	4.16	7.67	5.73	29.92	7.33	4.79
	Code	42106	722005	40502	748	4227	4227	4-41	4227	401	80080	B13-22	202
STEEL ERECTION—inter., ornam.	Rate	3.03	16.01	12.96	15.40	3.94	8.84	4.16	7.67	5.73	29.92	7.33	4.79
	Code	42106	722005	40502	748	4227	4227	4-41	4227	401	80080	B13-22	202
STEEL ERECTION—structure	Rate	3.03	16.01	12.96	15.47	3.94	8.84	4.16	7.67	5.73	29.92	7.33	4.79
	Code	42106	722005	40502	748	4227	4227	4-41	4227	401	80080	B13-22	202
STEEL ERECTION—NOC	Rate	3.03	16.01	12.96	15.40	3.94	8.84	4.16	7.67	5.73	29.92	7.33	4.79
	Code	42106	722005	40502	748	4227	4227	4-41	4227	401	80080	B13-22	202
TILE WORK—inter. (ceramic)	Rate	4.22	6.58	2.05	6.75	4.67	4.92	4.16	5.74	3.98	15.41	7.33	4.79
	Code	42113	721054	40103	719	4276	4276	4-41	4276	402	80110	B13-01	202
WATERPROOFING	Rate	5.08	5.19	4.48	4.58	5.20	4.92	4.16	7.67	3.98	22.49	6.15	4.79
	Code	42139	721016	40102	723	4239	4299	4-41	4239	402	80130	B12-17	202
WRECKING	Rate	2.95	5.74	7.07	15.40	2.88	4.29	4.04	4.11	5.73	15.41	7.33	4.79
	Code	40604	721005	40106	748	4211	4211	4-43	4211	401	80110	B13-09	202

General Requirements — R0131 Project Management & Coordination

R013113-80 Performance Bond

This table shows the cost of a Performance Bond for a construction job scheduled to be completed in 12 months. Add 1% of the premium cost per month for jobs requiring more than 12 months to complete. The rates are "standard" rates offered to contractors that the bonding company considers financially sound and capable of doing the work. Preferred rates are offered by some bonding companies based upon financial strength of the contractor. Actual rates vary from contractor to contractor and from bonding company to bonding company. Contractors should prequalify through a bonding agency before submitting a bid on a contract that requires a bond.

Contract Amount	Building Construction Class B Projects	Highways & Bridges Class A New Construction	Highways & Bridges Class A-1 Highway Resurfacing
First $100,000 bid	$25.00 per M	$15.00 per M	$9.40 per M
Next 400,000 bid	$2,500 plus $15.00 per M	$1,500 plus $10.00 per M	$940 plus $7.20 per M
Next 2,000,000 bid	8,500 plus 10.00 per M	5,500 plus 7.00 per M	3,820 plus 5.00 per M
Next 2,500,000 bid	28,500 plus 7.50 per M	19,500 plus 5.50 per M	15,820 plus 4.50 per M
Next 2,500,000 bid	47,250 plus 7.00 per M	33,250 plus 5.00 per M	28,320 plus 4.50 per M
Over 7,500,000 bid	64,750 plus 6.00 per M	45,750 plus 4.50 per M	39,570 plus 4.00 per M

General Requirements — R0151 Temporary Utilities

R015113-65 Temporary Power Equipment

Cost data for the temporary equipment was developed utilizing the following information.

1) Re-usable material-services, transformers, equipment and cords are based on new purchase and prorated to three projects.
2) PVC feeder includes trench and backfill.
3) Connections include disconnects and fuses.
4) Labor units include an allowance for removal.
5) No utility company charges or fees are included.
6) Concrete pads or vaults are not included.
7) Utility company conduits not included.

General Requirements   R0154 Construction Aids

R015423-10 Steel Tubular Scaffolding

On new construction, tubular scaffolding is efficient up to 60' high or five stories. Above this it is usually better to use a hung scaffolding if construction permits. Swing scaffolding operations may interfere with tenants. In this case, the tubular is more practical at all heights.

In repairing or cleaning the front of an existing building the cost of tubular scaffolding per S.F. of building front increases as the height increases above the first tier. The first tier cost is relatively high due to leveling and alignment.

The minimum efficient crew for erecting and dismantling is three workers. They can set up and remove 18 frame sections per day up to 5 stories high. For 6 to 12 stories high, a crew of four is most efficient. Use two or more on top and two on the bottom for handing up or hoisting. They can also set up and remove 18 frame sections per day. At 7' horizontal spacing, this will run about 800 S.F. per day of erecting and dismantling. Time for placing and removing planks must be added to the above. A crew of three can place and remove 72 planks per day up to 5 stories. For over 5 stories, a crew of four can place and remove 80 planks per day.

The table below shows the number of pieces required to erect tubular steel scaffolding for 1000 S.F. of building frontage. This area is made up of a scaffolding system that is 12 frames (11 bays) long by 2 frames high.

For jobs under twenty-five frames, add 50% to rental cost. Rental rates will be lower for jobs over three months duration. Large quantities for long periods can reduce rental rates by 20%.

Description of Component	Number of Pieces for 1000 S.F. of Building Front	Unit
5' Wide Standard Frame, 6'-4" High	24	Ea.
Leveling Jack & Plate	24	
Cross Brace	44	
Side Arm Bracket, 21"	12	
Guardrail Post	12	
Guardrail, 7' section	22	
Stairway Section	2	
Stairway Starter Bar	1	
Stairway Inside Handrail	2	
Stairway Outside Handrail	2	
Walk-Thru Frame Guardrail	2	

Scaffolding is often used as falsework over 15' high during construction of cast-in-place concrete beams and slabs. Two foot wide scaffolding is generally used for heavy beam construction. The span between frames depends upon the load to be carried with a maximum span of 5'.

Heavy duty shoring frames with a capacity of 10,000#/leg can be spaced up to 10' O.C. depending upon form support design and loading.

Scaffolding used as horizontal shoring requires less than half the material required with conventional shoring.

On new construction, erection is done by carpenters.

Rolling towers supporting horizontal shores can reduce labor and speed the job. For maintenance work, catwalks with spans up to 70' can be supported by the rolling towers.

R015423-20 Pump Staging

Pump staging is generally not available for rent. The table below shows the number of pieces required to erect pump staging for 2400 S.F. of building frontage. This area is made up of a pump jack system that is 3 poles (2 bays) wide by 2 poles high.

Item	Number of Pieces for 2400 S.F. of Building Front	Unit
Aluminum pole section, 24' long	6	Ea.
Aluminum splice joint, 6' long	3	
Aluminum foldable brace	3	
Aluminum pump jack	3	
Aluminum support for workbench/back safety rail	3	
Aluminum scaffold plank/workbench, 14" wide x 24' long	4	
Safety net, 22' long	2	
Aluminum plank end safety rail	2	

The cost in place for this 2400 S.F. will depend on how many uses are realized during the life of the equipment.

General Requirements — R0154 Construction Aids

R015433-10 Contractor Equipment

Rental Rates shown elsewhere in the book pertain to late model high quality machines in excellent working condition, rented from equipment dealers. Rental rates from contractors may be substantially lower than the rental rates from equipment dealers depending upon economic conditions; for older, less productive machines, reduce rates by a maximum of 15%. Any overtime must be added to the base rates. For shift work, rates are lower. Usual rule of thumb is 150% of one shift rate for two shifts; 200% for three shifts.

For periods of less than one week, operated equipment is usually more economical to rent than renting bare equipment and hiring an operator.

Costs to move equipment to a job site (mobilization) or from a job site (demobilization) are not included in rental rates, nor in any Equipment costs on any Unit Price line items or crew listings. These costs can be found elsewhere. If a piece of equipment is already at a job site, it is not appropriate to utilize mob/demob costs in an estimate again.

Rental rates vary throughout the country with larger cities generally having lower rates. Lease plans for new equipment are available for periods in excess of six months with a percentage of payments applying toward purchase.

Monthly rental rates vary from 2% to 5% of the cost of the equipment depending on the anticipated life of the equipment and its wearing parts. Weekly rates are about 1/3 the monthly rates and daily rental rates about 1/3 the weekly rate.

The hourly operating costs for each piece of equipment include costs to the user such as fuel, oil, lubrication, normal expendables for the equipment, and a percentage of mechanic's wages chargeable to maintenance. The hourly operating costs listed do not include the operator's wages.

The daily cost for equipment used in the standard crews is figured by dividing the weekly rate by five, then adding eight times the hourly operating cost to give the total daily equipment cost, not including the operator. This figure is in the right hand column of the Equipment listings under Crew Equipment Cost/Day.

Pile Driving rates shown for pile hammer and extractor do not include leads, crane, boiler or compressor. Vibratory pile driving requires an added field specialist during set-up and pile driving operation for the electric model. The hydraulic model requires a field specialist for set-up only. Up to 125 reuses of sheet piling are possible using vibratory drivers. For normal conditions, crane capacity for hammer type and size are as follows.

Crane Capacity	Hammer Type and Size		
	Air or Steam	Diesel	Vibratory
25 ton	to 8,750 ft.-lb.		70 H.P.
40 ton	15,000 ft.-lb.	to 32,000 ft.-lb.	170 H.P.
60 ton	25,000 ft.-lb.		300 H.P.
100 ton		112,000 ft.-lb.	

Cranes should be specified for the job by size, building and site characteristics, availability, performance characteristics, and duration of time required.

Backhoes & Shovels rent for about the same as equivalent size cranes but maintenance and operating expense is higher. Crane operators rate must be adjusted for high boom heights. Average adjustments: for 150' boom add 2% per hour; over 185', add 4% per hour; over 210', add 6% per hour; over 250', add 8% per hour and over 295', add 12% per hour.

Tower Cranes of the climbing or static type have jibs from 50' to 200' and capacities at maximum reach range from 4,000 to 14,000 pounds. Lifting capacities increase up to maximum load as the hook radius decreases.

Typical rental rates, based on purchase price are about 2% to 3% per month.

Erection and dismantling runs between 500 and 2000 labor hours. Climbing operation takes 10 labor hours per 20' climb. Crane dead time is about 5 hours per 40' climb. If crane is bolted to side of the building add cost of ties and extra mast sections. Climbing cranes have from 80' to 180' of mast while static cranes have 80' to 800' of mast.

Truck Cranes can be converted to tower cranes by using tower attachments. Mast heights over 400' have been used.

A single 100' high material **Hoist and Tower** can be erected and dismantled in about 400 labor hours; a double 100' high hoist and tower in about 600 labor hours. Erection times for additional heights are 3 and 4 labor hours per vertical foot respectively up to 150', and 4 to 5 labor hours per vertical foot over 150' high. A 40' high portable Buck hoist takes about 160 labor hours to erect and dismantle. Additional heights take 2 labor hours per vertical foot to 80' and 3 labor hours per vertical foot for the next 100'. Most material hoists do not meet local code requirements for carrying personnel.

A 150' high **Personnel Hoist** requires about 500 to 800 labor hours to erect and dismantle. Budget erection time at 5 labor hours per vertical foot for all trades. Local code requirements or labor scarcity requiring overtime can add up to 50% to any of the above erection costs.

Earthmoving Equipment: The selection of earthmoving equipment depends upon the type and quantity of material, moisture content, haul distance, haul road, time available, and equipment available. Short haul cut and fill operations may require dozers only, while another operation may require excavators, a fleet of trucks, and spreading and compaction equipment. Stockpiled material and granular material are easily excavated with front end loaders. Scrapers are most economically used with hauls between 300' and 1-1/2 miles if adequate haul roads can be maintained. Shovels are often used for blasted rock and any material where a vertical face of 8' or more can be excavated. Special conditions may dictate the use of draglines, clamshells, or backhoes. Spreading and compaction equipment must be matched to the soil characteristics, the compaction required and the rate the fill is being supplied.

R015433-15 Heavy Lifting

Hydraulic Climbing Jacks

The use of hydraulic heavy lift systems is an alternative to conventional type crane equipment. The lifting, lowering, pushing, or pulling mechanism is a hydraulic climbing jack moving on a square steel jackrod from 1-5/8" to 4" square, or a steel cable. The jackrod or cable can be vertical or horizontal, stationary or movable, depending on the individual application. When the jackrod is stationary, the climbing jack will climb the rod and push or pull the load along with itself. When the climbing jack is stationary, the jackrod is movable with the load attached to the end and the climbing jack will lift or lower the jackrod with the attached load. The heavy lift system is normally operated by a single control lever located at the hydraulic pump.

The system is flexible in that one or more climbing jacks can be applied wherever a load support point is required, and the rate of lift synchronized.

Economic benefits have been demonstrated on projects such as: erection of ground assembled roofs and floors, complete bridge spans, girders and trusses, towers, chimney liners and steel vessels, storage tanks, and heavy machinery. Other uses are raising and lowering offshore work platforms, caissons, tunnel sections and pipelines.

Existing Conditions — R0241 Demolition

R024119-10 Demolition Defined

Whole Building Demolition - Demolition of the whole building with no concern for any particular building element, component, or material type being demolished. This type of demolition is accomplished with large pieces of construction equipment that break up the structure, load it into trucks and haul it to a disposal site, but disposal or dump fees are not included. Demolition of below-grade foundation elements, such as footings, foundation walls, grade beams, slabs on grade, etc., is not included. Certain mechanical equipment containing flammable liquids or ozone-depleting refrigerants, electric lighting elements, communication equipment components, and other building elements may contain hazardous waste, and must be removed, either selectively or carefully, as hazardous waste before the building can be demolished.

Foundation Demolition - Demolition of below-grade foundation footings, foundation walls, grade beams, and slabs on grade. This type of demolition is accomplished by hand or pneumatic hand tools, and does not include saw cutting, or handling, loading, hauling, or disposal of the debris.

Gutting - Removal of building interior finishes and electrical/mechanical systems down to the load-bearing and sub-floor elements of the rough building frame, with no concern for any particular building element, component, or material type being demolished. This type of demolition is accomplished by hand or pneumatic hand tools, and includes loading into trucks, but not hauling, disposal or dump fees, scaffolding, or shoring. Certain mechanical equipment containing flammable liquids or ozone-depleting refrigerants, electric lighting elements, communication equipment components, and other building elements may contain hazardous waste, and must be removed, either selectively or carefully, as hazardous waste, before the building is gutted.

Selective Demolition - Demolition of a selected building element, component, or finish, with some concern for surrounding or adjacent elements, components, or finishes (see the first Subdivision (s) at the beginning of appropriate Divisions). This type of demolition is accomplished by hand or pneumatic hand tools, and does not include handling, loading, storing, hauling, or disposal of the debris, scaffolding, or shoring. "Gutting" methods may be used in order to save time, but damage that is caused to surrounding or adjacent elements, components, or finishes may have to be repaired at a later time.

Careful Removal - Removal of a piece of service equipment, building element or component, or material type, with great concern for both the removed item and surrounding or adjacent elements, components or finishes. The purpose of careful removal may be to protect the removed item for later re-use, preserve a higher salvage value of the removed item, or replace an item while taking care to protect surrounding or adjacent elements, components, connections, or finishes from cosmetic and/or structural damage. An approximation of the time required to perform this type of removal is 1/3 to 1/2 the time it would take to install a new item of like kind (see Reference Number R220105-10). This type of removal is accomplished by hand or pneumatic hand tools, and does not include loading, hauling, or storing the removed item, scaffolding, shoring, or lifting equipment.

Cutout Demolition - Demolition of a small quantity of floor, wall, roof, or other assembly, with concern for the appearance and structural integrity of the surrounding materials. This type of demolition is accomplished by hand or pneumatic hand tools, and does not include saw cutting, handling, loading, hauling, or disposal of debris, scaffolding, or shoring.

Rubbish Handling - Work activities that involve handling, loading or hauling of debris. Generally, the cost of rubbish handling must be added to the cost of all types of demolition, with the exception of whole building demolition.

Minor Site Demolition - Demolition of site elements outside the footprint of a building. This type of demolition is accomplished by hand or pneumatic hand tools, or with larger pieces of construction equipment, and may include loading a removed item onto a truck (check the Crew for equipment used). It does not include saw cutting, hauling or disposal of debris, and, sometimes, handling or loading.

Existing Conditions — R0265 Underground Storage Tank Removal

R026510-20 Underground Storage Tank Removal

Underground Storage Tank Removal can be divided into two categories: Non-Leaking and Leaking. Prior to removing an underground storage tank, tests should be made, with the proper authorities present, to determine whether a tank has been leaking or the surrounding soil has been contaminated.

To safely remove Liquid Underground Storage Tanks:
1. Excavate to the top of the tank.
2. Disconnect all piping.
3. Open all tank vents and access ports.
4. Remove all liquids and/or sludge.
5. Purge the tank with an inert gas.
6. Provide access to the inside of the tank and clean out the interior using proper personal protective equipment (PPE).
7. Excavate soil surrounding the tank using proper PPE for on-site personnel.
8. Pull and properly dispose of the tank.
9. Clean up the site of all contaminated material.
10. Install new tanks or close the excavation.

Existing Conditions — R0282 Asbestos Remediation

R028213-20 Asbestos Removal Process

Asbestos removal is accomplished by a specialty contractor who understands the federal and state regulations regarding the handling and disposal of the material. The process of asbestos removal is divided into many individual steps. An accurate estimate can be calculated only after all the steps have been priced.

The steps are generally as follows:
1. Obtain an asbestos abatement plan from an industrial hygienist.
2. Monitor the air quality in and around the removal area and along the path of travel between the removal area and transport area. This establishes the background contamination.
3. Construct a two part decontamination chamber at entrance to removal area.
4. Install a HEPA filter to create a negative pressure in the removal area.
5. Install wall, floor and ceiling protection as required by the plan, usually 2 layers of fireproof 6 mil polyethylene.
6. Industrial hygienist visually inspects work area to verify compliance with plan.
7. Provide temporary supports for conduit and piping affected by the removal process.
8. Proceed with asbestos removal and bagging process. Monitor air quality as described in Step #2. Discontinue operations when contaminate levels exceed applicable standards.
9. Document the legal disposal of materials in accordance with EPA standards.
10. Thoroughly clean removal area including all ledges, crevices and surfaces.
11. Post abatement inspection by industrial hygienist to verify plan compliance.
12. Provide a certificate from a licensed industrial hygienist attesting that contaminate levels are within acceptable standards before returning area to regular use.

Existing Conditions — R0283 Lead Remediation

R028319-60 Lead Paint Remediation Methods

Lead paint remediation can be accomplished by the following methods.
1. Abrasive blast
2. Chemical stripping
3. Power tool cleaning with vacuum collection system
4. Encapsulation
5. Remove and replace
6. Enclosure

Each of these methods has strengths and weakness depending on the specific circumstances of the project. The following is an overview of each method.

1. **Abrasive blasting** is usually accomplished with sand or recyclable metallic blast. Before work can begin, the area must be contained to ensure the blast material with lead does not escape to the atmosphere. The use of vacuum blast greatly reduces the containment requirements. Lead abatement equipment that may be associated with this work includes a negative air machine. In addition, it is necessary to have an industrial hygienist monitor the project on a continual basis. When the work is complete, the spent blast sand with lead must be disposed of as a hazardous material. If metallic shot was used, the lead is separated from the shot and disposed of as hazardous material. Worker protection includes disposable clothing and respiratory protection.

2. **Chemical stripping** requires strong chemicals be applied to the surface to remove the lead paint. Before the work can begin, the area under/adjacent to the work area must be covered to catch the chemical and removed lead. After the chemical is applied to the painted surface it is usually covered with paper. The chemical is left in place for the specified period, then the paper with lead paint is pulled or scraped off. The process may require several chemical applications. The paper with chemicals and lead paint adhered to it, plus the containment and loose scrapings collected by a HEPA (High Efficiency Particulate Air Filter) vac, must be disposed of as a hazardous material. The chemical stripping process usually requires a neutralizing agent and several wash downs after the paint is removed. Worker protection includes a neoprene or other compatible protective clothing and respiratory protection with face shield. An industrial hygienist is required intermittently during the process.

3. **Power tool cleaning** is accomplished using shrouded needle blasting guns. The shrouding with different end configurations is held up against the surface to be cleaned. The area is blasted with hardened needles and the shroud captures the lead with a HEPA vac and deposits it in a holding tank. An industrial hygienist monitors the project, protective clothing and a respirator is required until air samples prove otherwise. When the work is complete the lead must be disposed of as a hazardous material.

4. **Encapsulation** is a method that leaves the well bonded lead paint in place after the peeling paint has been removed. Before the work can begin, the area under/adjacent to the work must be covered to catch the scrapings. The scraped surface is then washed with a detergent and rinsed. The prepared surface is covered with approximately 10 mils of paint. A reinforcing fabric can also be embedded in the paint covering. The scraped paint and containment must be disposed of as a hazardous material. Workers must wear protective clothing and respirators.

5. **Remove and replace** is an effective way to remove lead paint from windows, gypsum walls and concrete masonry surfaces. The painted materials are removed and new materials are installed. Workers should wear a respirator and tyvek suit. The demolished materials must be disposed of as hazardous waste if it fails the TCLP (Toxicity Characteristic Leachate Process) test.

6. **Enclosure** is the process that permanently seals lead painted materials in place. This process has many applications such as covering lead painted drywall with new drywall, covering exterior construction with tyvek paper then residing, or covering lead painted structural members with aluminum or plastic. The seams on all enclosing materials must be securely sealed. An industrial hygienist monitors the project, and protective clothing and a respirator is required until air samples prove otherwise.

All the processes require clearance monitoring and wipe testing as required by the hygienist.

Concrete — R0311 Concrete Forming

R031113-10 Wall Form Materials

Aluminum Forms
Approximate weight is 3 lbs. per S.F.C.A. Standard widths are available from 4" to 36" with 36" most common. Standard lengths of 2', 4', 6' to 8' are available. Forms are lightweight and fewer ties are needed with the wider widths. The form face is either smooth or textured.

Metal Framed Plywood Forms
Manufacturers claim over 75 reuses of plywood and over 300 reuses of steel frames. Many specials such as corners, fillers, pilasters, etc. are available. Monthly rental is generally about 15% of purchase price for first month and 9% per month thereafter with 90% of rental applied to purchase for the first month and decreasing percentages thereafter. Aluminum framed forms cost 25% to 30% more than steel framed.

After the first month, extra days may be prorated from the monthly charge. Rental rates do not include ties, accessories, cleaning, loss of hardware or freight in and out. Approximate weight is 5 lbs. per S.F. for steel; 3 lbs. per S.F. for aluminum.

Forms can be rented with option to buy.

Plywood Forms, Job Fabricated
There are two types of plywood used for concrete forms.
1. Exterior plyform which is completely waterproof. This is face oiled to facilitate stripping. Ten reuses can be expected with this type with 25 reuses possible.
2. An overlaid type consists of a resin fiber fused to exterior plyform. No oiling is required except to facilitate cleaning. This is available in both high density (HDO) and medium density overlaid (MDO). Using HDO, 50 reuses can be expected with 200 possible.

Plyform is available in 5/8" and 3/4" thickness. High density overlaid is available in 3/8", 1/2", 5/8" and 3/4" thickness.

5/8" thick is sufficient for most building forms, while 3/4" is best on heavy construction.

Plywood Forms, Modular, Prefabricated
There are many plywood forming systems without frames. Most of these are manufactured from 1-1/8" (HDO) plywood and have some hardware attached. These are used principally for foundation walls 8' or less high. With care and maintenance, 100 reuses can be attained with decreasing quality of surface finish.

Steel Forms
Approximate weight is 6-1/2 lbs. per S.F.C.A. including accessories. Standard widths are available from 2" to 24", with 24" most common. Standard lengths are from 2' to 8', with 4' the most common. Forms are easily ganged into modular units.

Forms are usually leased for 15% of the purchase price per month prorated daily over 30 days.

Rental may be applied to sale price, and usually rental forms are bought. With careful handling and cleaning 200 to 400 reuses are possible.

Straight wall gang forms up to 12' x 20' or 8' x 30' can be fabricated. These crane handled forms usually lease for approx. 9% per month.

Individual job analysis is available from the manufacturer at no charge.

R031113-30 Slipforms

The slipform method of forming may be used for forming circular silo and multi-celled storage bin type structures over 30' high, and building core shear walls over eight stories high. The shear walls, usually enclose elevator shafts, stairwells, mechanical spaces, and toilet rooms. Reuse of the form on duplicate structures will reduce the height necessary and spread the cost of building the form. Slipform systems can be used to cast chimneys, towers, piers, dams, underground shafts or other structures capable of being extruded.

Slipforms are usually 4' high and are raised semi-continuously by jacks climbing on rods which are embedded in the concrete. The jacks are powered by a hydraulic, pneumatic, or electric source and are available in 3, 6, and 22 ton capacities. Interior work decks and exterior scaffolds must be provided for placing inserts, embedded items, reinforcing steel, and concrete. Scaffolds below the form for finishers may be required. The interior work decks are often used as roof slab forms on silos and bin work. Form raising rates will range from 6" to 20" per hour for silos; 6" to 30" per hour for buildings; and 6" to 48" per hour for shaft work.

Reinforcing bars and stressing strands are usually hoisted by crane or gin pole, and the concrete material can be hoisted by crane, winch-powered skip, or pumps. The slipform system is operated on a continuous 24-hour day when a monolithic structure is desired. For least cost, the system is operated only during normal working hours.

Placing concrete will range from 0.5 to 1.5 labor-hours per C.Y. Bucks, blockouts, keyways, weldplates, etc. are extra.

Concrete — R0311 Concrete Forming

R031113-40 Forms for Reinforced Concrete

Design Economy

Avoid many sizes in proportioning beams and columns.

From story to story avoid changing column dimensions. Gain strength by adding steel or using a richer mix. If a change in size of column is necessary, vary one dimension only to minimize form alterations. Keep beams and columns the same width.

From floor to floor in a multi-story building vary beam depth, not width, as that will leave slab panel form unchanged. It is cheaper to vary the strength of a beam from floor to floor by means of steel area than by 2" changes in either width or depth.

Cost Factors

Material includes the cost of lumber, cost of rent for metal pans or forms if used, nails, form ties, form oil, bolts and accessories.

Labor includes the cost of carpenters to make up, erect, remove and repair, plus common labor to clean and move. Having carpenters remove forms minimizes repairs.

Improper alignment and condition of forms will increase finishing cost. When forms are heavily oiled, concrete surfaces must be neutralized before finishing. Special curing compounds will cause spillages to spall off in first frost. Gang forming methods will reduce costs on large projects.

Materials Used

Boards are seldom used unless their architectural finish is required. Generally, steel, fiberglass and plywood are used for contact surfaces. Labor on plywood is 10% less than with boards. The plywood is backed up with 2 x 4's at 12" to 32" O.C. Walers are generally 2 - 2 x 4's. Column forms are held together with steel yokes or bands. Shoring is with adjustable shoring or scaffolding for high ceilings.

Reuse

Floor and column forms can be reused four or possibly five times without excessive repair. Remember to allow for 10% waste on each reuse.

When modular sized wall forms are made, up to twenty uses can be expected with exterior plyform.

When forms are reused, the cost to erect, strip, clean and move will not be affected. 10% replacement of lumber should be included and about one hour of carpenter time for repairs on each reuse per 100 S.F.

The reuse cost for certain accessory items normally rented on a monthly basis will be lower than the cost for the first use.

After fifth use, new material required plus time needed for repair prevent form cost from dropping further and it may go up. Much depends on care in stripping, the number of special bays, changes in beam or column sizes and other factors.

Costs for multiple use of formwork may be developed as follows:

2 Uses
$$\frac{(\text{1st Use} + \text{Reuse})}{2} = \text{avg. cost/2 uses}$$

3 Uses
$$\frac{(\text{1st Use} + \text{2 Reuse})}{3} = \text{avg. cost/3 uses}$$

4 Uses
$$\frac{(\text{1st use} + \text{3 Reuse})}{4} = \text{avg. cost/4 uses}$$

Concrete — R0311 Concrete Forming

R031113-60 Formwork Labor-Hours

Item	Unit	Fabricate	Erect & Strip	Clean & Move	Total Hours 1 Use	2 Use	3 Use	4 Use
Beam and Girder, interior beams, 12" wide	100 S.F.	6.4	8.3	1.3	16.0	13.3	12.4	12.0
Hung from steel beams		5.8	7.7	1.3	14.8	12.4	11.6	11.2
Beam sides only, 36" high		5.8	7.2	1.3	14.3	11.9	11.1	10.7
Beam bottoms only, 24" wide		6.6	13.0	1.3	20.9	18.1	17.2	16.7
Box out for openings		9.9	10.0	1.1	21.0	16.6	15.1	14.3
Buttress forms, to 8' high		6.0	6.5	1.2	13.7	11.2	10.4	10.0
Centering, steel, 3/4" rib lath			1.0		1.0			
3/8" rib lath or slab form			0.9		0.9			
Chamfer strip or keyway	100 L.F.		1.5		1.5	1.5	1.5	1.5
Columns, fiber tube 8" diameter			20.6		20.6			
12"			21.3		21.3			
16"			22.9		22.9			
20"			23.7		23.7			
24"			24.6		24.6			
30"			25.6		25.6			
Columns, round steel, 12" diameter			22.0		22.0	22.0	22.0	22.0
16"			25.6		25.6	25.6	25.6	25.6
20"			30.5		30.5	30.5	30.5	30.5
24"			37.7		37.7	37.7	37.7	37.7
Columns, plywood 8" x 8"	100 S.F.	7.0	11.0	1.2	19.2	16.2	15.2	14.7
12" x 12"		6.0	10.5	1.2	17.7	15.2	14.4	14.0
16" x 16"		5.9	10.0	1.2	17.1	14.7	13.8	13.4
24" x 24"		5.8	9.8	1.2	16.8	14.4	13.6	13.2
Columns, steel framed plywood 8" x 8"			10.0	1.0	11.0	11.0	11.0	11.0
12" x 12"			9.3	1.0	10.3	10.3	10.3	10.3
16" x 16"			8.5	1.0	9.5	9.5	9.5	9.5
24" x 24"			7.8	1.0	8.8	8.8	8.8	8.8
Drop head forms, plywood		9.0	12.5	1.5	23.0	19.0	17.7	17.0
Coping forms		8.5	15.0	1.5	25.0	21.3	20.0	19.4
Culvert, box			14.5	4.3	18.8	18.8	18.8	18.8
Curb forms, 6" to 12" high, on grade		5.0	8.5	1.2	14.7	12.7	12.1	11.7
On elevated slabs		6.0	10.8	1.2	18.0	15.5	14.7	14.3
Edge forms to 6" high, on grade	100 L.F.	2.0	3.5	0.6	6.1	5.6	5.4	5.3
7" to 12" high	100 S.F.	2.5	5.0	1.0	8.5	7.8	7.5	7.4
Equipment foundations		10.0	18.0	2.0	30.0	25.5	24.0	23.3
Flat slabs, including drops		3.5	6.0	1.2	10.7	9.5	9.0	8.8
Hung from steel		3.0	5.5	1.2	9.7	8.7	8.4	8.2
Closed deck for domes		3.0	5.8	1.2	10.0	9.0	8.7	8.5
Open deck for pans		2.2	5.3	1.0	8.5	7.9	7.7	7.6
Footings, continuous, 12" high		3.5	3.5	1.5	8.5	7.3	6.8	6.6
Spread, 12" high		4.7	4.2	1.6	10.5	8.7	8.0	7.7
Pile caps, square or rectangular		4.5	5.0	1.5	11.0	9.3	8.7	8.4
Grade beams, 24" deep		2.5	5.3	1.2	9.0	8.3	8.0	7.9
Lintel or Sill forms		8.0	17.0	2.0	27.0	23.5	22.3	21.8
Spandrel beams, 12" wide		9.0	11.2	1.3	21.5	17.5	16.2	15.5
Stairs			25.0	4.0	29.0	29.0	29.0	29.0
Trench forms in floor		4.5	14.0	1.5	20.0	18.3	17.7	17.4
Walls, Plywood, at grade, to 8' high		5.0	6.5	1.5	13.0	11.0	9.7	9.5
8' to 16'		7.5	8.0	1.5	17.0	13.8	12.7	12.1
16' to 20'		9.0	10.0	1.5	20.5	16.5	15.2	14.5
Foundation walls, to 8' high		4.5	6.5	1.0	12.0	10.3	9.7	9.4
8' to 16' high		5.5	7.5	1.0	14.0	11.8	11.0	10.6
Retaining wall to 12' high, battered		6.0	8.5	1.5	16.0	13.5	12.7	12.3
Radial walls to 12' high, smooth		8.0	9.5	2.0	19.5	16.0	14.8	14.3
2' chords		7.0	8.0	1.5	16.5	13.5	12.5	12.0
Prefabricated modular, to 8' high		—	4.3	1.0	5.3	5.3	5.3	5.3
Steel, to 8' high		—	6.8	1.2	8.0	8.0	8.0	8.0
8' to 16' high		—	9.1	1.5	10.6	10.3	10.2	10.2
Steel framed plywood to 8' high		—	6.8	1.2	8.0	7.5	7.3	7.2
8' to 16' high		—	9.3	1.2	10.5	9.5	9.2	9.0

Concrete — R0321 Reinforcing Steel

R032110-10 Reinforcing Steel Weights and Measures

Bar Designation No.**	Nominal Weight Lb./Ft.	U.S. Customary Units Nominal Dimensions*			SI Units Nominal Dimensions*			
		Diameter in.	Cross Sectional Area, in.^{2}	Perimeter in.	Nominal Weight kg/m	Diameter mm	Cross Sectional Area, cm^2	Perimeter mm
3	.376	.375	.11	1.178	.560	9.52	.71	29.9
4	.668	.500	.20	1.571	.994	12.70	1.29	39.9
5	1.043	.625	.31	1.963	1.552	15.88	2.00	49.9
6	1.502	.750	.44	2.356	2.235	19.05	2.84	59.8
7	2.044	.875	.60	2.749	3.042	22.22	3.87	69.8
8	2.670	1.000	.79	3.142	3.973	25.40	5.10	79.8
9	3.400	1.128	1.00	3.544	5.059	28.65	6.45	90.0
10	4.303	1.270	1.27	3.990	6.403	32.26	8.19	101.4
11	5.313	1.410	1.56	4.430	7.906	35.81	10.06	112.5
14	7.650	1.693	2.25	5.320	11.384	43.00	14.52	135.1
18	13.600	2.257	4.00	7.090	20.238	57.33	25.81	180.1

* The nominal dimensions of a deformed bar are equivalent to those of a plain round bar having the same weight per foot as the deformed bar.
** Bar numbers are based on the number of eighths of an inch included in the nominal diameter of the bars.

R032110-20 Metric Rebar Specification - ASTM A615-81

	Grade 300 (300 MPa* = 43,560 psi; +8.7% vs. Grade 40)			
	Grade 400 (400 MPa* = 58,000 psi; –3.4% vs. Grade 60)			
Bar No.	Diameter mm	Area mm^2	Equivalent in.^{2}	Comparison with U.S. Customary Bars
10M	11.3	100	.16	Between #3 & #4
15M	16.0	200	.31	#5 (.31 in.^{2})
20M	19.5	300	.47	#6 (.44 in.^{2})
25M	25.2	500	.78	#8 (.79 in.^{2})
30M	29.9	700	1.09	#9 (1.00 in.^{2})
35M	35.7	1000	1.55	#11 (1.56 in.^{2})
45M	43.7	1500	2.33	#14 (2.25 in.^{2})
55M	56.4	2500	3.88	#18 (4.00 in.^{2})

* MPa = megapascals

Concrete — R0321 Reinforcing Steel

R032110-40 Weight of Steel Reinforcing Per Square Foot of Wall (PSF)

Reinforced Weights: The table below suggests the weights per square foot for reinforcing steel in walls. Weights are approximate and will be the same for all grades of steel bars. For bars in two directions, add weights for each size and spacing.

C/C Spacing in Inches	#3 Wt. (PSF)	#4 Wt. (PSF)	#5 Wt. (PSF)	#6 Wt. (PSF)	#7 Wt. (PSF)	#8 Wt. (PSF)	#9 Wt. (PSF)	#10 Wt. (PSF)	#11 Wt. (PSF)
2"	2.26	4.01	6.26	9.01	12.27				
3"	1.50	2.67	4.17	6.01	8.18	10.68	13.60	17.21	21.25
4"	1.13	2.01	3.13	4.51	6.13	8.10	10.20	12.91	15.94
5"	.90	1.60	2.50	3.60	4.91	6.41	8.16	10.33	12.75
6"	.752	1.34	2.09	3.00	4.09	5.34	6.80	8.61	10.63
8"	.564	1.00	1.57	2.25	3.07	4.01	5.10	6.46	7.97
10"	.451	.802	1.25	1.80	2.45	3.20	4.08	5.16	6.38
12"	.376	.668	1.04	1.50	2.04	2.67	3.40	4.30	5.31
18"	.251	.445	.695	1.00	1.32	1.78	2.27	2.86	3.54
24"	.188	.334	.522	.751	1.02	1.34	1.70	2.15	2.66
30"	.150	.267	.417	.600	.817	1.07	1.36	1.72	2.13
36"	.125	.223	.348	.501	.681	.890	1.13	1.43	1.77
42"	.107	.191	.298	.429	.584	.753	.97	1.17	1.52
48"	.094	.167	.261	.376	.511	.668	.85	1.08	1.33

R032110-50 Minimum Wall Reinforcement Weight (PSF)

This table lists the approximate minimum wall reinforcement weights per S.F. according to the specification of .12% of gross area for vertical bars and .20% of gross area for horizontal bars.

Location	Wall Thickness	Bar Size	Horizontal Steel Spacing C/C	Sq. In. Req'd per S.F.	Total Wt. per S.F.	Bar Size	Vertical Steel Spacing C/C	Sq. In. Req'd per S.F.	Total Wt. per S.F.	Horizontal & Vertical Steel Total Weight per S.F.
Both Faces	10"	#4	18"	.24	.89#	#3	18"	.14	.50#	1.39#
	12"	#4	16"	.29	1.00	#3	16"	.17	.60	1.60
	14"	#4	14"	.34	1.14	#3	13"	.20	.69	1.84
	16"	#4	12"	.38	1.34	#3	11"	.23	.82	2.16
	18"	#5	17"	.43	1.47	#4	18"	.26	.89	2.36
One Face	6"	#3	9"	.15	.50	#3	18"	.09	.25	.75
	8"	#4	12"	.19	.67	#3	11"	.12	.41	1.08
	10"	#5	15"	.24	.83	#4	16"	.14	.50	1.34

R032110-70 Bend, Place and Tie Reinforcing

Placing and tying by rodmen for footings and slabs runs from nine hrs. per ton for heavy bars to fifteen hrs. per ton for light bars. For beams, columns, and walls, production runs from eight hrs. per ton for heavy bars to twenty hrs. per ton for light bars. Overall average for typical reinforced concrete buildings is about fourteen hrs. per ton. These production figures include the time for placing of accessories and usual inserts, but not their material cost (allow 15% of the cost of delivered bent rods). Equipment handling is necessary for the larger-sized bars so that installation costs for the very heavy bars will not decrease proportionately.

Installation costs for splicing reinforcing bars include allowance for equipment to hold the bars in place while splicing as well as necessary scaffolding for iron workers.

R032110-80 Shop-Fabricated Reinforcing Steel

The material prices for reinforcing, shown in the unit cost sections of the book, are for 50 tons or more of shop-fabricated reinforcing steel and include:

1. Mill base price of reinforcing steel
2. Mill grade/size/length extras
3. Mill delivery to the fabrication shop
4. Shop storage and handling
5. Shop drafting/detailing
6. Shop shearing and bending
7. Shop listing
8. Shop delivery to the job site

Both material and installation costs can be considerably higher for small jobs consisting primarily of smaller bars, while material costs may be slightly lower for larger jobs.

Concrete — R0322 Welded Wire Fabric Reinforcing

R032205-30 Common Stock Styles of Welded Wire Fabric

This table provides some of the basic specifications, sizes, and weights of welded wire fabric used for reinforcing concrete.

	New Designation Spacing — Cross Sectional Area (in.) — (Sq. in. 100)	Old Designation Spacing — Wire Gauge (in.) — (AS & W)		Steel Area per Foot Longitudinal		Transverse		Approximate Weight per 100 S.F.	
				in.	cm	in.	cm	lbs	kg
Rolls	6 x 6 — W1.4 x W1.4	6 x 6 — 10 x 10		.028	.071	.028	.071	21	9.53
	6 x 6 — W2.0 x W2.0	6 x 6 — 8 x 8	1	.040	.102	.040	.102	29	13.15
	6 x 6 — W2.9 x W2.9	6 x 6 — 6 x 6		.058	.147	.058	.147	42	19.05
	6 x 6 — W4.0 x W4.0	6 x 6 — 4 x 4		.080	.203	.080	.203	58	26.91
	4 x 4 — W1.4 x W1.4	4 x 4 — 10 x 10		.042	.107	.042	.107	31	14.06
	4 x 4 — W2.0 x W2.0	4 x 4 — 8 x 8	1	.060	.152	.060	.152	43	19.50
	4 x 4 — W2.9 x W2.9	4 x 4 — 6 x 6		.087	.227	.087	.227	62	28.12
	4 x 4 — W4.0 x W4.0	4 x 4 — 4 x 4		.120	.305	.120	.305	85	38.56
Sheets	6 x 6 — W2.9 x W2.9	6 x 6 — 6 x 6		.058	.147	.058	.147	42	19.05
	6 x 6 — W4.0 x W4.0	6 x 6 — 4 x 4		.080	.203	.080	.203	58	26.31
	6 x 6 — W5.5 x W5.5	6 x 6 — 2 x 2	2	.110	.279	.110	.279	80	36.29
	4 x 4 — W1.4 x W1.4	4 x 4 — 4 x 4		.120	.305	.120	.305	85	38.56

NOTES: 1. Exact W—number size for 8 gauge is W2.1
2. Exact W—number size for 2 gauge is W5.4

Concrete — R0330 Cast-In-Place Concrete

R033053-50 Industrial Chimneys

Foundation requirements in C.Y. of concrete for various sized chimneys.

Size Chimney	2 Ton Soil	3 Ton Soil	Size Chimney	2 Ton Soil	3 Ton Soil	Size Chimney	2 Ton Soil	3 Ton Soil
75' x 3'-0"	13 C.Y.	11 C.Y.	160' x 6'-6"	86 C.Y.	76 C.Y.	300' x 10'-0"	325 C.Y.	245 C.Y.
85' x 5'-6"	19	16	175' x 7'-0"	108	95	350' x 12'-0"	422	320
100' x 5'-0"	24	20	200' x 6'-0"	125	105	400' x 14'-0"	520	400
125' x 5'-6"	43	36	250' x 8'-0"	230	175	500' x 18'-0"	725	575

Concrete R0331 Structural Concrete

R033105-10 Proportionate Quantities

The tables below show both quantities per S.F. of floor areas as well as form and reinforcing quantities per C.Y. Unusual structural requirements would increase the ratios below. High strength reinforcing would reduce the steel weights. Figures are for 3000 psi concrete and 60,000 psi reinforcing unless specified otherwise.

Type of Construction	Live Load	Span	Per S.F. of Floor Area				Per C.Y. of Concrete		
			Concrete	Forms	Reinf.	Pans	Forms	Reinf.	Pans
Flat Plate	50 psf	15 Ft.	.46 C.F.	1.06 S.F.	1.71 lb.		62 S.F.	101 lb.	
		20	.63	1.02	2.40		44	104	
		25	.79	1.02	3.03		35	104	
	100	15	.46	1.04	2.14		61	126	
		20	.71	1.02	2.72		39	104	
		25	.83	1.01	3.47		33	113	
Flat Plate (waffle construction) 20" domes	50	20	.43	1.00	2.10	.84 S.F.	63	135	53 S.F.
		25	.52	1.00	2.90	.89	52	150	46
		30	.64	1.00	3.70	.87	42	155	37
	100	20	.51	1.00	2.30	.84	53	125	45
		25	.64	1.00	3.20	.83	42	135	35
		30	.76	1.00	4.40	.81	36	160	29
Waffle Construction 30" domes	50	25	.69	1.06	1.83	.68	42	72	40
		30	.74	1.06	2.39	.69	39	87	39
		35	.86	1.05	2.71	.69	33	85	39
		40	.78	1.00	4.80	.68	35	165	40
Flat Slab (two way with drop panels)	50	20	.62	1.03	2.34		45	102	
		25	.77	1.03	2.99		36	105	
		30	.95	1.03	4.09		29	116	
	100	20	.64	1.03	2.83		43	119	
		25	.79	1.03	3.88		35	133	
		30	.96	1.03	4.66		29	131	
	200	20	.73	1.03	3.03		38	112	
		25	.86	1.03	4.23		32	133	
		30	1.06	1.03	5.30		26	135	
One Way Joists 20" Pans	50	15	.36	1.04	1.40	.93	78	105	70
		20	.42	1.05	1.80	.94	67	120	60
		25	.47	1.05	2.60	.94	60	150	54
	100	15	.38	1.07	1.90	.93	77	140	66
		20	.44	1.08	2.40	.94	67	150	58
		25	.52	1.07	3.50	.94	55	185	49
One Way Joists 8" x 16" filler blocks	50	15	.34	1.06	1.80	.81 Ea.	84	145	64 Ea.
		20	.40	1.08	2.20	.82	73	145	55
		25	.46	1.07	3.20	.83	63	190	49
	100	15	.39	1.07	1.90	.81	74	130	56
		20	.46	1.09	2.80	.82	64	160	48
		25	.53	1.10	3.60	.83	56	190	42
One Way Beam & Slab	50	15	.42	1.30	1.73		84	111	
		20	.51	1.28	2.61		68	138	
		25	.64	1.25	2.78		53	117	
	100	15	.42	1.30	1.90		84	122	
		20	.54	1.35	2.69		68	154	
		25	.69	1.37	3.93		54	145	
	200	15	.44	1.31	2.24		80	137	
		20	.58	1.40	3.30		65	163	
		25	.69	1.42	4.89		53	183	
Two Way Beam & Slab	100	15	.47	1.20	2.26		69	130	
		20	.63	1.29	3.06		55	131	
		25	.83	1.33	3.79		43	123	
	200	15	.49	1.25	2.70		41	149	
		20	.66	1.32	4.04		54	165	
		25	.88	1.32	6.08		41	187	

Concrete — R0331 Structural Concrete

R033105-10 Proportionate Quantities (cont.)

4000 psi Concrete and 60,000 psi Reinforcing—Form and Reinforcing Quantities per C.Y.

Item	Size	Forms	Reinforcing	Minimum	Maximum
Columns (square tied)	10" x 10"	130 S.F.C.A.	#5 to #11	220 lbs.	875 lbs.
	12" x 12"	108	#6 to #14	200	955
	14" x 14"	92	#7 to #14	190	900
	16" x 16"	81	#6 to #14	187	1082
	18" x 18"	72	#6 to #14	170	906
	20" x 20"	65	#7 to #18	150	1080
	22" x 22"	59	#8 to #18	153	902
	24" x 24"	54	#8 to #18	164	884
	26" x 26"	50	#9 to #18	169	994
	28" x 28"	46	#9 to #18	147	864
	30" x 30"	43	#10 to #18	146	983
	32" x 32"	40	#10 to #18	175	866
	34" x 34"	38	#10 to #18	157	772
	36" x 36"	36	#10 to #18	175	852
	38" x 38"	34	#10 to #18	158	765
	40" x 40"	32	#10 to #18	143	692

Item	Size	Form	Spiral	Reinforcing	Minimum	Maximum
Columns (spirally reinforced)	12" diameter	34.5 L.F.	190 lbs.	#4 to #11	165 lbs.	1505 lb.
		34.5	190	#14 & #18	—	1100
	14"	25	170	#4 to #11	150	970
		25	170	#14 & #18	800	1000
	16"	19	160	#4 to #11	160	950
		19	160	#14 & #18	605	1080
	18"	15	150	#4 to #11	160	915
		15	150	#14 & #18	480	1075
	20"	12	130	#4 to #11	155	865
		12	130	#14 & #18	385	1020
	22"	10	125	#4 to #11	165	775
		10	125	#14 & #18	320	995
	24"	9	120	#4 to #11	195	800
		9	120	#14 & #18	290	1150
	26"	7.3	100	#4 to #11	200	729
		7.3	100	#14 & #18	235	1035
	28"	6.3	95	#4 to #11	175	700
		6.3	95	#14 & #18	200	1075
	30"	5.5	90	#4 to #11	180	670
		5.5	90	#14 & #18	175	1015
	32"	4.8	85	#4 to #11	185	615
		4.8	85	#14 & #18	155	955
	34"	4.3	80	#4 to #11	180	600
		4.3	80	#14 & #18	170	855
	36"	3.8	75	#4 to #11	165	570
		3.8	75	#14 & #18	155	865
	40"	3.0	70	#4 to #11	165	500
		3.0	70	#14 & #18	145	765

Concrete — R0331 Structural Concrete

R033105-10 Proportionate Quantities (cont.)

		3000 psi Concrete and 60,000 psi Reinforcing—Form and Reinforcing Quantities per C.Y.				
Item	Type	Loading	Height	C.Y./L.F.	Forms/C.Y.	Reinf./C.Y.
Retaining Walls	Cantilever	Level Backfill	4 Ft.	0.2 C.Y.	49 S.F.	35 lbs.
			8	0.5	42	45
			12	0.8	35	70
			16	1.1	32	85
			20	1.6	28	105
		Highway Surcharge	4	0.3	41	35
			8	0.5	36	55
			12	0.8	33	90
			16	1.2	30	120
			20	1.7	27	155
		Railroad Surcharge	4	0.4	28	45
			8	0.8	25	65
			12	1.3	22	90
			16	1.9	20	100
			20	2.6	18	120
	Gravity, with Vertical Face	Level Backfill	4	0.4	37	None
			7	0.6	27	↓
			10	1.2	20	
		Sloping Backfill	4	0.3	31	
			7	0.8	21	
			10	1.6	15	↓

		Live Load in Kips per Linear Foot						
	Span	Under 1 Kip		2 to 3 Kips		4 to 5 Kips		6 to 7 Kips	
		Forms	Reinf.	Forms	Reinf.	Forms	Reinf.	Forms	Reinf.
Beams	10 Ft.	—	—	90 S.F.	170 #	85 S.F.	175 #	75 S.F.	185 #
	16	130 S.F.	165 #	85	180	75	180	65	225
	20	110	170	75	185	62	200	51	200
	26	90	170	65	215	62	215	—	—
	30	85	175	60	200	—	—	—	—

Item	Size	Type	Forms per C.Y.	Reinforcing per C.Y.
Spread Footings	Under 1 C.Y.	1,000 psf soil	24 S.F.	44 lbs.
		5,000	24	42
		10,000	24	52
	1 C.Y. to 5 C.Y.	1,000	14	49
		5,000	14	50
		10,000	14	50
	Over 5 C.Y.	1,000	9	54
		5,000	9	52
		10,000	9	56
Pile Caps (30 Ton Concrete Piles)	Under 5 C.Y.	shallow caps	20	65
		medium	20	50
		deep	20	40
	5 C.Y. to 10 C.Y.	shallow	14	55
		medium	15	45
		deep	15	40
	10 C.Y. to 20 C.Y.	shallow	11	60
		medium	11	45
		deep	12	35
	Over 20 C.Y.	shallow	9	60
		medium	9	45
		deep	10	40

Concrete

R0331 Structural Concrete

R033105-10 Proportionate Quantities (cont.)

		3000 psi Concrete and 60,000 psi Reinforcing — Form and Reinforcing Quantities per C.Y.					
Item	Size	Pile Spacing	50 T Pile	100 T Pile	50 T Pile	100 T Pile	
Pile Caps (Steel H Piles)	Under 5 C.Y.	24" O.C.	24 S.F.	24 S.F.	75 lbs.	90 lbs.	
		30"	25	25	80	100	
		36"	24	24	80	110	
	5 C.Y. to 10 C.Y.	24"	15	15	80	110	
		30"	15	15	85	110	
		36"	15	15	75	90	
	Over 10 C.Y.	24"	13	13	85	90	
		30"	11	11	85	95	
		36"	10	10	85	90	

		8" Thick		10" Thick		12" Thick		15" Thick	
	Height	Forms	Reinf.	Forms	Reinf.	Forms	Reinf.	Forms	Reinf.
Basement Walls	7 Ft.	81 S.F.	44 lbs.	65 S.F.	45 lbs.	54 S.F.	44 lbs.	41 S.F.	43 lbs.
	8		44		45		44		43
	9		46		45		44		43
	10		57		45		44		43
	12		83		50		52		43
	14		116		65		64		51
	16				86		90		65
	18						106		70

R033105-20 Materials for One C.Y. of Concrete

This is an approximate method of figuring quantities of cement, sand and coarse aggregate for a field mix with waste allowance included.

With crushed gravel as coarse aggregate, to determine barrels of cement required, divide 10 by total mix; that is, for 1:2:4 mix, 10 divided by 7 = 1-3/7 barrels.

If the coarse aggregate is crushed stone, use 10-1/2 instead of 10 as given for gravel.

To determine tons of sand required, multiply barrels of cement by parts of sand and then by 0.2; that is, for the 1:2:4 mix, as above, 1-3/7 x 2 x .2 = .57 tons.

Tons of crushed gravel are in the same ratio to tons of sand as parts in the mix, or 4/2 x .57 = 1.14 tons.

1 bag cement = 94#
4 bags = 1 barrel

1 C.Y. sand or crushed gravel = 2700#
1 ton sand or crushed gravel = 20 C.F.

1 C.Y. crushed stone = 2575#
1 ton crushed stone = 21 C.F.

Average carload of cement is 692 bags; of sand or gravel is 56 tons.

Do not stack stored cement over 10 bags high.

R033105-30 Metric Equivalents of Cement Content for Concrete Mixes

94 Pound Bags per Cubic Yard	Kilograms per Cubic Meter	94 Pound Bags per Cubic Yard	Kilograms per Cubic Meter
1.0	55.77	7.0	390.4
1.5	83.65	7.5	418.3
2.0	111.5	8.0	446.2
2.5	139.4	8.5	474.0
3.0	167.3	9.0	501.9
3.5	195.2	9.5	529.8
4.0	223.1	10.0	557.7
4.5	251.0	10.5	585.6
5.0	278.8	11.0	613.5
5.5	306.7	11.5	641.3
6.0	334.6	12.0	669.2
6.5	362.5	12.5	697.1

a. If you know the cement content in pounds per cubic yard, multiply by .5933 to obtain kilograms per cubic meter.

b. If you know the cement content in 94 pound bags per cubic yard, multiply by 55.77 to obtain kilograms per cubic meter.

Concrete — R0331 Structural Concrete

R033105-40 Metric Equivalents of Common Concrete Strengths
(to convert other psi values to megapascals, multiply by 0.006895)

U.S. Values psi	SI Value Megapascals	Non-SI Metric Value kgf/cm^2*
2000	14	140
2500	17	175
3000	21	210
3500	24	245
4000	28	280
4500	31	315
5000	34	350
6000	41	420
7000	48	490
8000	55	560
9000	62	630
10,000	69	705

* kilograms force per square centimeter

R033105-50 Quantities of Cement, Sand and Stone for One C.Y. of Concrete per Various Mixes

This table can be used to determine the quantities of the ingredients for smaller quantities of site mixed concrete.

Concrete (C.Y.)	Mix = 1:1:1-3/4			Mix = 1:2:2.25			Mix = 1:2.25:3			Mix = 1:3:4		
	Cement (sacks)	Sand (C.Y.)	Stone (C.Y.)	Cement (sacks)	Sand (C.Y.)	Stone (C.Y.)	Cement (sacks)	Sand (C.Y.)	Stone (C.Y.)	Cement (sacks)	Sand (C.Y.)	Stone (C.Y.)
1	10	.37	.63	7.75	.56	.65	6.25	.52	.70	5	.56	.74
2	20	.74	1.26	15.50	1.12	1.30	12.50	1.04	1.40	10	1.12	1.48
3	30	1.11	1.89	23.25	1.68	1.95	18.75	1.56	2.10	15	1.68	2.22
4	40	1.48	2.52	31.00	2.24	2.60	25.00	2.08	2.80	20	2.24	2.96
5	50	1.85	3.15	38.75	2.80	3.25	31.25	2.60	3.50	25	2.80	3.70
6	60	2.22	3.78	46.50	3.36	3.90	37.50	3.12	4.20	30	3.36	4.44
7	70	2.59	4.41	54.25	3.92	4.55	43.75	3.64	4.90	35	3.92	5.18
8	80	2.96	5.04	62.00	4.48	5.20	50.00	4.16	5.60	40	4.48	5.92
9	90	3.33	5.67	69.75	5.04	5.85	56.25	4.68	6.30	45	5.04	6.66
10	100	3.70	6.30	77.50	5.60	6.50	62.50	5.20	7.00	50	5.60	7.40
11	110	4.07	6.93	85.25	6.16	7.15	68.75	5.72	7.70	55	6.16	8.14
12	120	4.44	7.56	93.00	6.72	7.80	75.00	6.24	8.40	60	6.72	8.88
13	130	4.82	8.20	100.76	7.28	8.46	81.26	6.76	9.10	65	7.28	9.62
14	140	5.18	8.82	108.50	7.84	9.10	87.50	7.28	9.80	70	7.84	10.36
15	150	5.56	9.46	116.26	8.40	9.76	93.76	7.80	10.50	75	8.40	11.10
16	160	5.92	10.08	124.00	8.96	10.40	100.00	8.32	11.20	80	8.96	11.84
17	170	6.30	10.72	131.76	9.52	11.06	106.26	8.84	11.90	85	9.52	12.58
18	180	6.66	11.34	139.50	10.08	11.70	112.50	9.36	12.60	90	10.08	13.32
19	190	7.04	11.98	147.26	10.64	12.36	118.76	9.84	13.30	95	10.64	14.06
20	200	7.40	12.60	155.00	11.20	13.00	125.00	10.40	14.00	100	11.20	14.80
21	210	7.77	13.23	162.75	11.76	13.65	131.25	10.92	14.70	105	11.76	15.54
22	220	8.14	13.86	170.05	12.32	14.30	137.50	11.44	15.40	110	12.32	16.28
23	230	8.51	14.49	178.25	12.88	14.95	143.75	11.96	16.10	115	12.88	17.02
24	240	8.88	15.12	186.00	13.44	15.60	150.00	12.48	16.80	120	13.44	17.76
25	250	9.25	15.75	193.75	14.00	16.25	156.25	13.00	17.50	125	14.00	18.50
26	260	9.64	16.40	201.52	14.56	16.92	162.52	13.52	18.20	130	14.56	19.24
27	270	10.00	17.00	209.26	15.12	17.56	168.76	14.04	18.90	135	15.02	20.00
28	280	10.36	17.64	217.00	15.68	18.20	175.00	14.56	19.60	140	15.68	20.72
29	290	10.74	18.28	224.76	16.24	18.86	181.26	15.08	20.30	145	16.24	21.46

Concrete — R0331 Structural Concrete

R033105-65 Field-Mix Concrete

Presently most building jobs are built with ready-mixed concrete except at isolated locations and some larger jobs requiring over 10,000 C.Y. where land is readily available for setting up a temporary batch plant.

The most economical mix is a controlled mix using local aggregate proportioned by trial to give the required strength with the least cost of material.

R033105-70 Placing Ready-Mixed Concrete

For ground pours allow for 5% waste when figuring quantities.

Prices in the front of the book assume normal deliveries. If deliveries are made before 8 A.M. or after 5 P.M. or on Saturday afternoons add 30%. Negotiated discounts for large volumes are not included in prices in front of book.

For the lower floors without truck access, concrete may be wheeled in rubber-tired buggies, conveyer handled, crane handled or pumped. Pumping is economical if there is top steel. Conveyers are more efficient for thick slabs.

At higher floors the rubber-tired buggies may be hoisted by a hoisting tower and wheeled to location. Placement by a conveyer is limited to three floors and is best for high-volume pours. Pumped concrete is best when building has no crane access. Concrete may be pumped directly as high as thirty-six stories using special pumping techniques. Normal maximum height is about fifteen stories.

Best pumping aggregate is screened and graded bank gravel rather than crushed stone.

Pumping downward is more difficult than pumping upward. Horizontal distance from pump to pour may increase preparation time prior to pour. Placing by cranes, either mobile, climbing or tower types, continues as the most efficient method for high-rise concrete buildings.

R033105-85 Lift Slabs

The cost advantage of the lift slab method is due to placing all concrete, reinforcing steel, inserts and electrical conduit at ground level and in reduction of formwork. Minimum economical project size is about 30,000 S.F. Slabs may be tilted for parking garage ramps.

It is now used in all types of buildings and has gone up to 22 stories high in apartment buildings. Current trend is to use post-tensioned flat plate slabs with spans from 22′ to 35′. Cylindrical void forms are used when deep slabs are required. One pound of prestressing steel is about equal to seven pounds of conventional reinforcing.

To be considered cured for stressing and lifting, a slab must have attained 75% of design strength. Seven days are usually sufficient with four to five days possible if high early strength cement is used. Slabs can be stacked using two coats of a non-bonding agent to insure that slabs do not stick to each other. Lifting is done by companies specializing in this work. Lift rate is 5′ to 15′ per hour with an average of 10′ per hour. Total areas up to 33,000 S.F. have been lifted at one time. 24 to 36 jacking columns are common. Most economical bay sizes are 24′ to 28′ with four to fourteen stories most efficient. Continuous design reduces reinforcing steel cost. Use of post-tensioned slabs allows larger bay sizes.

Concrete — R0341 Precast Structural Concrete

R034105-30 Prestressed Precast Concrete Structural Units

Type	Location	Depth	Span in Ft.		Live Load Lb. per S.F.
Double Tee (8' to 10')	Floor	28" to 34"	60 to 80		50 to 80
	Roof	12" to 24"	30 to 50		40
	Wall	Width 8'	Up to 55' high		Wind
Multiple Tee (8')	Roof	8" to 12"	15 to 40		40
	Floor	8" to 12"	15 to 30		100
Plank	Roof or Floor		Roof	Floor	
		4"	13	12	40 for Roof
		6"	22	18	
		8"	26	25	
		10"	33	29	100 for Floor
		12"	42	32	
Single Tee (8' to 10')	Roof	28"	40		40
		32"	80		
		36"	100		
		48"	120		
AASHO Girder	Bridges	Type 4	100		Highway
		5	110		
		6	125		
Box Beam (4')	Bridges	15"	40 to 100		Highway
		27"			
		33"			

The majority of precast projects today utilize double tees rather than single tees because of speed and ease of installation. As a result casting beds at manufacturing plants are normally formed for double tees. Single tee projects will therefore require an initial set up charge to be spread over the individual single tee costs.

For floors, a 2" to 3" topping is field cast over the shapes. For roofs, insulating concrete or rigid insulation is placed over the shapes.

Member lengths up to 40' are standard haul, 40' to 60' require special permits and lengths over 60' must be escorted. Over width and/or over length can add up to 100% on hauling costs.

Large heavy members may require two cranes for lifting which would increase erection costs by about 45%. An eight man crew can install 12 to 20 double tees, or 45 to 70 quad tees or planks per day.

Grouting of connections must also be included.

Several system buildings utilizing precast members are available. Heights can go up to 22 stories for apartment buildings. Optimum design ratio is 3 S.F. of surface to 1 S.F. of floor area.

Concrete — R0341 Precast Structural Concrete

R034136-90 Prestressed Concrete, Post-Tensioned

In post-tensioned concrete the steel tendons are tensioned after the concrete has reached about 3/4 of its ultimate strength. The cableways are grouted after tensioning to provide bond between the steel and concrete. If bond is to be prevented, the tendons are coated with a corrosion-preventative grease and wrapped with waterproofed paper or plastic. Bonded tendons are usually used when ultimate strength (beams & girders) are controlling factors.

High strength concrete is used to fully utilize the steel, thereby reducing the size and weight of the member. A plasticizing agent may be added to reduce water content. Maximum size aggregate ranges from 1/2" to 1-1/2" depending on the spacing of the tendons.

The types of steel commonly used are bars and strands. Job conditions determine which is best suited. Bars are best for vertical prestresses since they are easy to support. The trend is for steel manufacturers to supply a finished package, cut to length, which reduces field preparation to a minimum.

Bars vary from 3/4" to 1-3/8" diameter. Table below gives time in labor-hours per tendon for placing, tensioning and grouting (if required) a 75' beam. Tendons used in buildings are not usually grouted; tendons for bridges usually are grouted. For strands the table indicates the labor-hours per pound for typical prestressed units 100' long. Simple span beams usually require one-end stressing regardless of lengths. Continuous beams are usually stressed from two ends. Long slabs are poured from the center outward and stressed in 75' increments after the initial 150' center pour.

Labor Hours per Tendon and per Pound of Prestressed Steel						
Length	100' Beam		75' Beam		100' Slab	
Type Steel	Strand		Bars		Strand	
Diameter	0.5"		3/4"	1-3/8"	0.5"	0.6"
Number	4	12	1	1	1	1
Force in Kips	100	300	42	143	25	35
Preparation & Placing Cables	3.6	7.4	0.9	2.9	0.9	1.1
Stressing Cables	2.0	2.4	0.8	1.6	0.5	0.5
Grouting, if required	2.5	3.0	0.6	1.3		
Total Labor Hours	8.1	12.8	2.3	5.8	1.4	1.6
Prestressing Steel Weights (Lbs.)	215	640	115	380	53	74
Labor-hours per Lb. Bonded	0.038	0.020	0.020	0.015		
Non-bonded					0.026	0.022

Flat Slab construction — 4000 psi concrete with span-to-depth ratio between 36 and 44. Two way post-tensioned steel averages 1.0 lb. per S.F. for 24' to 28' bays (usually strand) and additional reinforcing steel averages .5 lb. per S.F.

Pan and Joist construction — 4000 psi concrete with span-to-depth ratio 28 to 30. Post-tensioned steel averages .8 lb. per S.F. and reinforcing steel about 1.0 lb. per S.F. Placing and stressing averages 40 hours per ton of total material.

Beam construction — 4000 to 5000 psi concrete. Steel weights vary greatly.

Labor cost per pound goes down as the size and length of the tendon increase. The primary economic consideration is the cost per kip for the member.

Post-tensioning becomes feasible for beams and girders over 30' long; for continuous two-way slabs over 20' clear; also in transferring upper building loads over longer spans at lower levels. Post-tension suppliers will provide engineering services at no cost to the user. Substantial economies are possible by using post-tensioned Lift Slabs.

Concrete — R0345 Precast Architectural Concrete

R034513-10 Precast Concrete Wall Panels

Panels are either solid or insulated with plain, colored or textured finishes. Transportation is an important cost factor. Prices shown in the unit cost section of the book are based on delivery within 50 miles of a plant including fabricators' overhead and profit. Engineering data is available from fabricators to assist with construction details. Usual minimum job size for economical use of panels is about 5000 S.F. Small jobs can double the prices shown. For large, highly repetitive jobs, deduct up to 15% from the prices shown.

2" thick panels cost about the same as 3" thick panels, and maximum panel size is less. For building panels faced with granite, marble or stone, add the material prices from those unit cost sections to the plain panel price shown. There is a growing trend toward aggregate facings and broken rib finish rather than plain gray concrete panels.

No allowance has been made in the unit cost section for supporting steel framework. On one story buildings, panels may rest on grade beams and require only wind bracing and fasteners. On multi-story buildings panels can span from column to column and floor to floor. Plastic-designed steel-framed structures may have large deflections which slow down erection and raise costs.

Large panels are more economical than small panels on a S.F. basis. When figuring areas include all protrusions, returns, etc. Overhangs can triple erection costs. Panels over 45'- have been produced. Larger flat units should be prestressed. Vacuum lifting of smooth finish panels eliminates inserts and can speed erection.

Concrete — R0347 Site-Cast Concrete

R034713-20 Tilt Up Concrete Panels

The advantage of tilt up construction is in the low cost of forms and the placing of concrete and reinforcing. Panels up to 75' high and 5-1/2" thick have been tilted using strongbacks. Tilt up has been used for one to five story buildings and is well-suited for warehouses, stores, offices, schools and residences.

The panels are cast in forms on the floor slab. Most jobs use 5-1/2" thick solid reinforced concrete panels. Sandwich panels with a layer of insulating materials are also used. Where dampness is a factor, lightweight aggregate is used. Optimum panel size is 300 to 500 S.F.

Slabs are usually poured with 3000 psi concrete which permits tilting seven days after pouring. Slabs may be stacked on top of each other and are separated from each other by either two coats of bond breaker or a film of polyethylene. Use of high early-strength cement allows tilting two days after a pour. Tilting up is done with a roller outrigger crane with a capacity of at least 1-1/2 times the weight of the panel at the required reach. Exterior precast columns can be set at the same time as the panels; interior precast columns can be set first and the panels clipped directly to them. The use of cast-in-place concrete columns is diminishing due to shrinkage problems. Structural steel columns are sometimes used if crane rails are planned. Panels can be clipped to the columns or lowered between the flanges. Steel channels with anchors may be used as edge forms for the slab. When the panels are lifted the channels form an integral steel column to take structural loads. Roof loads can be carried directly by the panels for wall heights to 14'.

Requirements of local building codes may be a limiting factor and should be checked. Building floor slabs should be poured first and should be a minimum of 5" thick with 100% compaction of soil or 6" thick with less than 100% compaction.

Setting times as fast as nine minutes per panel have been observed, but a safer expectation would be four panels per hour with a crane and a four-man setting crew. If crane erects from inside building, some provision must be made to get crane out after walls are erected. Good yarding procedure is important to minimize delays. Equalizing three-point lifting beams and self-releasing pick-up hooks speed erection. If panels must be carried to their final location, setting time per panel will be increased and erection costs may approach the erection cost range of architectural precast wall panels. Placing panels into slots formed in continuous footers will speed erection.

Reinforcing should be with #5 bars with vertical bars on the bottom. If surface is to be sandblasted, stainless steel chairs should be used to prevent rust staining.

Use of a broom finish is popular since the unavoidable surface blemishes are concealed.

Precast columns run from three to five times the C.Y. price of the panels only.

Concrete — R0352 Lightweight Concrete Roof Insulation

R035216-10 Lightweight Concrete

Lightweight aggregate concrete is usually purchased ready mixed, but it can also be field mixed.

Vermiculite or Perlite comes in bags of 4 C.F. under various trade names. Weight is about 8 lbs. per C.F. For insulating roof fill use 1:6 mix. For structural deck use 1:4 mix over gypsum boards, steeltex, steel centering, etc., supported by closely spaced joists or bulb trees. For structural slabs use 1:3:2 vermiculite sand concrete over steeltex, metal lath, steel centering, etc., on joists spaced 2' 0" O.C. for maximum L.L. of 80 P.S.F. Use same mix for slab base fill over steel flooring or regular reinforced concrete slab when tile, terrazzo or other finish is to be laid over.

For slabs on grade use 1:3:2 mix when tile, etc., finish is to be laid over. If radiant heating units are installed use a 1:6 mix for a base. After coils are in place, cover with a regular granolithic finish (mix 1:3:2) to a minimum depth of 1-1/2" over top of units.

Reinforce all slabs with 6 x 6 or 10 x 10 welded wire mesh.

Masonry — R0401 Maintenance of Masonry

R040130-10 Cleaning Face Brick

On smooth brick a person can clean 70 S.F. an hour; on rough brick 50 S.F. per hour. Use one gallon muriatic acid to 20 gallons of water for 1000 S.F. Do not use acid solution until wall is at least seven days old, but a mild soap solution may be used after two days.

Time has been allowed for clean-up in brick prices.

Masonry — R0405 Common Work Results for Masonry

R040513-10 Cement Mortar (material only)

Type N - 1:1:6 mix by volume. Use everywhere above grade except as noted below. - 1:3 mix using conventional masonry cement which saves handling two separate bagged materials.

Type M - 1:1/4:3 mix by volume, or 1 part cement, 1/4 (10% by wt.) lime, 3 parts sand. Use for heavy loads and where earthquakes or hurricanes may occur. Also for reinforced brick, sewers, manholes and everywhere below grade.

Mix Proportions by Volume and Compressive Strength of Mortar

Where Used	Mortar Type	Allowable Proportions by Volume				Compressive Strength @ 28 days
		Portland Cement	Masonry Cement	Hydrated Lime	Masonry Sand	
Plain Masonry	M	1	1	—	6	2500 psi
		1	—	1/4	3	
	S	1/2	1	—	4	1800 psi
		1	—	1/4 to 1/2	4	
	N	—	1	—	3	750 psi
		1	—	1/2 to 1-1/4	6	
	O	—	1	—	3	350 psi
		1	—	1-1/4 to 2-1/2	9	
	K	1	—	2-1/2 to 4	12	75 psi
Reinforced Masonry	PM	1	1	—	6	2500 psi
	PL	1	—	1/4 to 1/2	4	2500 psi

Note: The total aggregate should be between 2.25 to 3 times the sum of the cement and lime used.

The labor cost to mix the mortar is included in the productivity and labor cost of unit price lines in unit cost sections for brickwork, blockwork and stonework.

The material cost of mixed mortar is included in the material cost of those same unit price lines and includes the cost of renting and operating a 10 C.F. mixer at the rate of 200 C.F. per day.

There are two types of mortar color used. One type is the inert additive type with about 100 lbs. per M brick as the typical quantity required. These colors are also available in smaller-batch-sized bags (1 lb. to 15 lb.) which can be placed directly into the mixer without measuring. The other type is premixed and replaces the masonry cement. Dark green color has the highest cost.

R040519-50 Masonry Reinforcing

Horizontal joint reinforcing helps prevent wall cracks where wall movement may occur and in many locations is required by code. Horizontal joint reinforcing is generally not considered to be structural reinforcing and an unreinforced wall may still contain joint reinforcing.

Reinforcing strips come in 10' and 12' lengths and in truss and ladder shapes, with and without drips. Field labor runs between 2.7 to 5.3 hours per 1000 L.F. for wall thicknesses up to 12".

The wire meets ASTM A82 for cold drawn steel wire and the typical size is 9 ga. sides and ties with 3/16" diameter also available. Typical finish is mill galvanized with zinc coating at .10 oz. per S.F. Class I (.40 oz. per S.F.) and Class III (.80 oz per S.F.) are also available, as is hot dipped galvanizing at 1.50 oz. per S.F.

Masonry — R0421 Clay Unit Masonry

R042110-10 Economy in Bricklaying

Have adequate supervision. Be sure bricklayers are always supplied with materials so there is no waiting. Place best bricklayers at corners and openings.

Use only screened sand for mortar. Otherwise, labor time will be wasted picking out pebbles. Use seamless metal tubs for mortar as they do not leak or catch the trowel. Locate stack and mortar for easy wheeling.

Have brick delivered for stacking. This makes for faster handling, reduces chipping and breakage, and requires less storage space. Many dealers will deliver select common in 2' x 3' x 4' pallets or face brick packaged. This affords quick handling with a crane or forklift and easy tonging in units of ten, which reduces waste.

Use wider bricks for one wythe wall construction. Keep scaffolding away from wall to allow mortar to fall clear and not stain wall.

On large jobs develop specialized crews for each type of masonry unit.

Consider designing for prefabricated panel construction on high rise projects.

Avoid excessive corners or openings. Each opening adds about 50% to labor cost for area of opening.

Bolting stone panels and using window frames as stops reduces labor costs and speeds up erection.

R042110-20 Common and Face Brick

Common building brick manufactured according to ASTM C62 and facing brick manufactured according to ASTM C216 are the two standard bricks available for general building use.

Building brick is made in three grades; SW, where high resistance to damage caused by cyclic freezing is required; MW, where moderate resistance to cyclic freezing is needed; and NW, where little resistance to cyclic freezing is needed. Facing brick is made in only the two grades SW and MW. Additionally, facing brick is available in three types; FBS, for general use; FBX, for general use where a higher degree of precision and lower permissible variation in size than FBS is needed; and FBA, for general use to produce characteristic architectural effects resulting from non-uniformity in size and texture of the units.

In figuring the material cost of brickwork, an allowance of 25% mortar waste and 3% brick breakage was included. If bricks are delivered palletized with 280 to 300 per pallet, or packaged, allow only 1-1/2% for breakage. Packaged or palletized delivery is practical when a job is big enough to have a crane or other equipment available to handle a package of brick. This is so on all industrial work but not always true on small commercial buildings.

The use of buff and gray face is increasing, and there is a continuing trend to the Norman, Roman, Jumbo and SCR brick.

Common red clay brick for backup is not used that often. Concrete block is the most usual backup material with occasional use of sand lime or cement brick. Building brick is commonly used in solid walls for strength and as a fire stop.

Brick panels built on the ground and then crane erected to the upper floors have proven to be economical. This allows the work to be done under cover and without scaffolding.

R042110-50 Brick, Block & Mortar Quantities

Type Brick	Nominal Size (incl. mortar) L H W	Modular Coursing	Number of Brick per S.F.	C.F. of Mortar per M Bricks, Waste Included 3/8" Joint	1/2" Joint	Bond Type	Description	Factor
Standard	8 x 2-2/3 x 4	3C=8"	6.75	10.3	12.9	Common	full header every fifth course	+20%
Economy	8 x 4 x 4	1C=4"	4.50	11.4	14.6		full header every sixth course	+16.7%
Engineer	8 x 3-1/5 x 4	5C=16"	5.63	10.6	13.6	English	full header every second course	+50%
Fire	9 x 2-1/2 x 4-1/2	2C=5"	6.40	550 # Fireclay	—	Flemish	alternate headers every course	+33.3%
Jumbo	12 x 4 x 6 or 8	1C=4"	3.00	23.8	30.8		every sixth course	+5.6%
Norman	12 x 2-2/3 x 4	3C=8"	4.50	14.0	17.9	Header = W x H exposed		+100%
Norwegian	12 x 3-1/5 x 4	5C=16"	3.75	14.6	18.6	Rowlock = H x W exposed		+100%
Roman	12 x 2 x 4	2C=4"	6.00	13.4	17.0	Rowlock stretcher = L x W exposed		+33.3%
SCR	12 x 2-2/3 x 6	3C=8"	4.50	21.8	28.0	Soldier = H x L exposed		—
Utility	12 x 4 x 4	1C=4"	3.00	15.4	19.6	Sailor = W x L exposed		-33.3%

Concrete Blocks Nominal Size	Approximate Weight per S.F. Standard	Lightweight	Blocks per 100 S.F.	Mortar per M block, waste included Partitions	Back up
2" x 8" x 16"	20 PSF	15 PSF	113	27 C.F.	36 C.F.
4"	30	20		41	51
6"	42	30		56	66
8"	55	38		72	82
10"	70	47		87	97
12"	85	55		102	112

Masonry

R0422 Concrete Unit Masonry

R042210-20 Concrete Block

The material cost of special block such as corner, jamb and head block can be figured at the same price as ordinary block of same size. Labor on specials is about the same as equal-sized regular block.

Bond beam and 16" high lintel blocks are more expensive than regular units of equal size. Lintel blocks are 8" long and either 8" or 16" high.

Use of motorized mortar spreader box will speed construction of continuous walls.

Hollow non-load-bearing units are made according to ASTM C129 and hollow load-bearing units according to ASTM C90.

Metals

R0505 Common Work Results for Metals

R050516-30 Coating Structural Steel

On field-welded jobs, the shop-applied primer coat is necessarily omitted. All painting must be done in the field and usually consists of red oxide rust inhibitive paint or an aluminum paint. The table below shows paint coverage and daily production for field painting.

See Division 09 97 13.23 for hot-dipped galvanizing and for field-applied cold galvanizing and other paints and protective coatings.

See Division 05 01 10.51 for steel surface preparation treatments such as wire brushing, pressure washing and sand blasting.

Type Construction	Surface Area per Ton	Coat	One Gallon Covers		In 8 Hrs. Person Covers		Average per Ton Spray	
			Brush	Spray	Brush	Spray	Gallons	Labor-hours
Light Structural	300 S.F. to 500 S.F.	1st	500 S.F.	455 S.F.	640 S.F.	2000 S.F.	0.9 gals.	1.6 L.H.
		2nd	450	410	800	2400	1.0	1.3
		3rd	450	410	960	3200	1.0	1.0
Medium	150 S.F. to 300 S.F.	All	400	365	1600	3200	0.6	0.6
Heavy Structural	50 S.F. to 150 S.F.	1st	400	365	1920	4000	0.2	0.2
		2nd	400	365	2000	4000	0.2	0.2
		3rd	400	365	2000	4000	0.2	0.2
Weighted Average	225 S.F.	All	400	365	1350	3000	0.6	0.6

R050521-20 Welded Structural Steel

Usual weight reductions with welded design run 10% to 20% compared with bolted or riveted connections. This amounts to about the same total cost compared with bolted structures since field welding is more expensive than bolts. For normal spans of 18' to 24' figure 6 to 7 connections per ton.

Trusses — For welded trusses add 4% to weight of main members for connections. Up to 15% less steel can be expected in a welded truss compared to one that is shop bolted. Cost of erection is the same whether shop bolted or welded.

General — Typical electrodes for structural steel welding are E6010, E6011, E60T and E70T. Typical buildings vary between 2# to 8# of weld rod per ton of steel. Buildings utilizing continuous design require about three times as much welding as conventional welded structures. In estimating field erection by welding, it is best to use the average linear feet of weld per ton to arrive at the welding cost per ton. The type, size and position of the weld will have a direct bearing on the cost per linear foot. A typical field welder will deposit 1.8# to 2# of weld rod per hour manually. Using semiautomatic methods can increase production by as much as 50% to 75%.

R050523-10 High Strength Bolts

Common bolts (A307) are usually used in secondary connections (see Division 05 05 23.10).

High strength bolts (A325 and A490) are usually specified for primary connections such as column splices, beam and girder connections to columns, column bracing, connections for supports of operating equipment or of other live loads which produce impact or reversal of stress, and in structures carrying cranes of over 5-ton capacity.

Allow 20 field bolts per ton of steel for a 6 story office building, apartment house or light industrial building. For 6 to 12 stories allow 25 bolts per ton, and above 12 stories, 30 bolts per ton. On power stations, 20 to 25 bolts per ton are needed.

Metals

R0512 Structural Steel Framing

R051223-10 Structural Steel

The bare material prices for structural steel, shown in the unit cost sections of the book, are for 100 tons of shop-fabricated structural steel and include:
1. Mill base price of structural steel
2. Mill scrap/grade/size/length extras
3. Mill delivery to a metals service center (warehouse)
4. Service center storage and handling
5. Service center delivery to a fabrication shop
6. Shop storage and handling
7. Shop drafting/detailing
8. Shop fabrication
9. Shop coat of primer paint
10. Shop listing
11. Shop delivery to the job site

In unit cost sections of the book that contain items for field fabrication of steel components, the bare material cost of steel includes:
1. Mill base price of structural steel
2. Mill scrap/grade/size/length extras
3. Mill delivery to a metals service center (warehouse)
4. Service center storage and handling
5. Service center delivery to the job site

R051223-20 Steel Estimating Quantities

One estimate on erection is that a crane can handle 35 to 60 pieces per day. Say the average is 45. With usual sizes of beams, girders, and columns, this would amount to about 20 tons per day. The type of connection greatly affects the speed of erection. Moment connections for continuous design slow down production and increase erection costs.

Short open web bar joists can be set at the rate of 75 to 80 per day, with 50 per day being the average for setting long span joists.

After main members are calculated, add the following for usual allowances: base plates 2% to 3%; column splices 4% to 5%; and miscellaneous details 4% to 5%, for a total of 10% to 13% in addition to main members.

The ratio of column to beam tonnage varies depending on type of steels used, typical spans, story heights and live loads.

It is more economical to keep the column size constant and to vary the strength of the column by using high strength steels. This also saves floor space. Buildings have recently gone as high as ten stories with 8" high strength columns. For light columns under W8X31 lb. sections, concrete filled steel columns are economical.

High strength steels may be used in columns and beams to save floor space and to meet head room requirements. High strength steels in some sizes sometimes require long lead times.

Round, square and rectangular columns, both plain and concrete filled, are readily available and save floor area, but are higher in cost per pound than rolled columns. For high unbraced columns, tube columns may be less expensive.

Below are average minimum figures for the weights of the structural steel frame for different types of buildings using A36 steel, rolled shapes and simple joints. For economy in domes, rise to span ratio = .13. Open web joist framing systems will reduce weights by 10% to 40%. Composite design can reduce steel weight by up to 25% but additional concrete floor slab thickness may be required. Continuous design can reduce the weights up to 20%. There are many building codes with different live load requirements and different structural requirements, such as hurricane and earthquake loadings which can alter the figures.

Structural Steel Weights per S.F. of Floor Area

Type of Building	No. of Stories	Avg. Spans	L.L. #/S.F.	Lbs. Per S.F.	Type of Building	No. of Stories	Avg. Spans	L.L. #/S.F.	Lbs. Per S.F.
Steel Frame Mfg.	1	20'x20'	40	8	Apartments	2-8	20'x20'	40	8
		30'x30'		13		9-25			14
		40'x40'		18	Office	to 10	Various	80	10
Parking garage	4	Various	80	8.5		20			18
Domes (Schwedler)*	1	200'	30	10		30			26
		300'		15		over 50			35

Metals — R0512 Structural Steel Framing

R051223-25 Common Structural Steel Specifications

ASTM A992 (formerly A36, then A572 Grade 50) is the all-purpose carbon grade steel widely used in building and bridge construction.

The other high-strength steels listed below may each have certain advantages over ASTM A992 stuctural carbon steel, depending on the application. They have proven to be economical choices where, due to lighter members, the reduction of dead load and the associated savings in shipping cost can be significant.

ASTM A588 atmospheric weathering, high-strength low-alloy steels can be used in the bare (uncoated) condition, where exposure to normal atmosphere causes a tightly adherant oxide to form on the surface protecting the steel from further oxidation. ASTM A242 corrosion-resistant, high-strength low-alloy steels have enhanced atmospheric corrosion resistance of at least two times that of carbon structural steels with copper, or four times that of carbon structural steels without copper. The reduction or elimination of maintenance resulting from the use of these steels often offsets their higher initial cost.

Steel Type	ASTM Designation	Minimum Yield Stress in KSI	Shapes Available
Carbon	A36	36	All structural shape groups, and plates & bars up thru 8" thick
	A529	42	Structural shape group 1, and plates & bars up thru 1/2" thick
High-Strength Low-Alloy Manganese-Vanadium	A441	40	Plates & bars over 4" up thru 8" thick
		42	Structural shape groups 4 & 5, and plates & bars over 1-1/2" up thru 4" thick
		46	Structural shape group 3, and plates & bars over 3/4" up thru 1-1/2" thick
		50	Structural shape groups 1 & 2, and plates & bars up thru 3/4" thick
High-Strength Low-Alloy Columbium-Vanadium	A572	42	All structural shape groups, and plates & bars up thru 6" thick
		50	All structural shape groups, and plates & bars up thru 4" thick
		60	Structural shape groups 1 & 2, and plates & bars up thru 1-1/4" thick
		65	Structural shape group 1, and plates & bars up thru 1-1/4" thick
High-Strength Low-Alloy Columbium-Vanadium	A992	50	All structural shape groups
Corrosion-Resistant High-Strength Low-Alloy	A242	42	Structural shape groups 4 & 5, and plates & bars over 1-1/2" up thru 4" thick
		46	Structural shape group 3, and plates & bars over 3/4" up thru 1-1/2" thick
		50	Structural shape groups 1 & 2, and plates & bars up thru 3/4" thick
Weathering High-Strength Low-Alloy	A588	42	Plates & bars over 5" up thru 8" thick
		46	Plates & bars over 4" up thru 5" thick
		50	All structural shape groups, and plates & bars up thru 4" thick
Quenched and Tempered Low-Alloy	A852	70	Plates & bars up thru 4" thick
Quenched and Tempered Alloy	A514	90	Plates & bars over 2-1/2" up thru 6" thick
		100	Plates & bars up thru 2-1/2" thick

R051223-30 High Strength Steels

The mill price of high strength steels may be higher than A992 carbon steel but their proper use can achieve overall savings thru total reduced weights. For columns with L/r over 100, A992 steel is best; under 100, high strength steels are economical. For heavy columns, high strength steels are economical when cover plates are eliminated. There is no economy using high strength steels for clip angles or supports or for beams where deflection governs. Thinner members are more economical than thick.

The per ton erection and fabricating costs of the high strength steels will be higher than for A992 since the same number of pieces, but less weight, will be installed.

Metals — R0512 Structural Steel Framing

R051223-35 Common Steel Sections

The upper portion of this table shows the name, shape, common designation and basic characteristics of commonly used steel sections. The lower portion explains how to read the designations used for the above illustrated common sections.

Shape & Designation	Name & Characteristics	Shape & Designation	Name & Characteristics
W	W Shape — Parallel flange surfaces	MC	Miscellaneous Channel — Infrequently rolled by some producers
S	American Standard Beam (I Beam) — Sloped inner flange	L	Angle — Equal or unequal legs, constant thickness
M	Miscellaneous Beams — Cannot be classified as W, HP or S; infrequently rolled by some producers	T	Structural Tee — Cut from W, M or S on center of web
C	American Standard Channel — Sloped inner flange	HP	Bearing Pile — Parallel flanges and equal flange and web thickness

Common drawing designations follow:

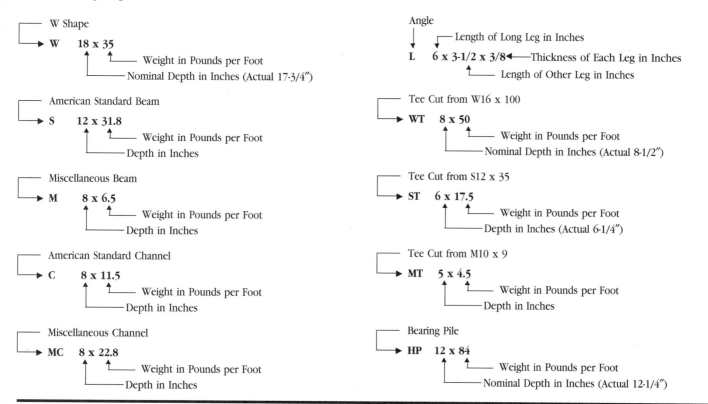

725

Metals — R0512 Structural Steel Framing

R051223-45 Installation Time for Structural Steel Building Components

The following tables show the expected average installation times for various structural steel shapes. Table A presents installation times for columns, Table B for beams, Table C for light framing and bolts, and Table D for structural steel for various project types.

Table A		
Description	Labor-Hours	Unit
Columns		
Steel, Concrete Filled		
3-1/2" Diameter	.933	Ea.
6-5/8" Diameter	1.120	Ea.
Steel Pipe		
3" Diameter	.933	Ea.
8" Diameter	1.120	Ea.
12" Diameter	1.244	Ea.
Structural Tubing		
4" x 4"	.966	Ea.
8" x 8"	1.120	Ea.
12" x 8"	1.167	Ea.
W Shape 2 Tier		
W8 x 31	.052	L.F.
W8 x 67	.057	L.F.
W10 x 45	.054	L.F.
W10 x 112	.058	L.F.
W12 x 50	.054	L.F.
W12 x 190	.061	L.F.
W14 x 74	.057	L.F.
W14 x 176	.061	L.F.

Table B				
Description	Labor-Hours	Unit	Labor-Hours	Unit
Beams, W Shape				
W6 x 9	.949	Ea.	.093	L.F.
W10 x 22	1.037	Ea.	.085	L.F.
W12 x 26	1.037	Ea.	.064	L.F.
W14 x 34	1.333	Ea.	.069	L.F.
W16 x 31	1.333	Ea.	.062	L.F.
W18 x 50	2.162	Ea.	.088	L.F.
W21 x 62	2.222	Ea.	.077	L.F.
W24 x 76	2.353	Ea.	.072	L.F.
W27 x 94	2.581	Ea.	.067	L.F.
W30 x 108	2.857	Ea.	.067	L.F.
W33 x 130	3.200	Ea.	.071	L.F.
W36 x 300	3.810	Ea.	.077	L.F.

Table C		
Description	Labor-Hours	Unit
Light Framing		
Angles 4" and Larger	.055	lbs.
Less than 4"	.091	lbs.
Channels 8" and Larger	.048	lbs.
Less than 8"	.072	lbs.
Cross Bracing Angles	.055	lbs.
Rods	.034	lbs.
Hanging Lintels	.069	lbs.
High Strength Bolts in Place		
3/4" Bolts	.070	Ea.
7/8" Bolts	.076	Ea.

Table D				
Description	Labor-Hours	Unit	Labor-Hours	Unit
Apartments, Nursing Homes, etc.				
1-2 Stories	4.211	Piece	7.767	Ton
3-6 Stories	4.444	Piece	7.921	Ton
7-15 Stories	4.923	Piece	9.014	Ton
Over 15 Stories	5.333	Piece	9.209	Ton
Offices, Hospitals, etc.				
1-2 Stories	4.211	Piece	7.767	Ton
3-6 Stories	4.741	Piece	8.889	Ton
7-15 Stories	4.923	Piece	9.014	Ton
Over 15 Stories	5.120	Piece	9.209	Ton
Industrial Buildings				
1 Story	3.478	Piece	6.202	Ton

R051223-50 Subpurlins

Bulb tee subpurlins are structural members designed to support and reinforce a variety of roof deck systems such as precast cement fiber roof deck tiles, monolithic roof deck systems, and gypsum or lightweight concrete over formboard. Other uses include interstitial service ceiling systems, wall panel systems, and joist anchoring in bond beams. See Unit Price section for pricing on a square foot basis at 32-5/8" O.C. Maximum span is based on a 3-span condition with a total allowable vertical load of 40 psf.

Metals — R0512 Structural Steel Framing

R051223-80 Dimensions and Weights of Sheet Steel

Gauge No.	Approximate Thickness				Weight		
	Inches (in fractions) Wrought Iron	Inches (in decimal parts) Wrought Iron	Steel	Millimeters Steel	per S.F. in Ounces	per S.F. in Lbs.	per Square Meter in Kg.
0000000	1/2"	.5	.4782	12.146	320	20.000	97.650
000000	15/32"	.46875	.4484	11.389	300	18.750	91.550
00000	7/16"	.4375	.4185	10.630	280	17.500	85.440
0000	13/32"	.40625	.3886	9.870	260	16.250	79.330
000	3/8"	.375	.3587	9.111	240	15.000	73.240
00	11/32"	.34375	.3288	8.352	220	13.750	67.130
0	5/16"	.3125	.2989	7.592	200	12.500	61.030
1	9/32"	.28125	.2690	6.833	180	11.250	54.930
2	17/64"	.265625	.2541	6.454	170	10.625	51.880
3	1/4"	.25	.2391	6.073	160	10.000	48.820
4	15/64"	.234375	.2242	5.695	150	9.375	45.770
5	7/32"	.21875	.2092	5.314	140	8.750	42.720
6	13/64"	.203125	.1943	4.935	130	8.125	39.670
7	3/16"	.1875	.1793	4.554	120	7.500	36.320
8	11/64"	.171875	.1644	4.176	110	6.875	33.570
9	5/32"	.15625	.1495	3.797	100	6.250	30.520
10	9/64"	.140625	.1345	3.416	90	5.625	27.460
11	1/8"	.125	.1196	3.038	80	5.000	24.410
12	7/64"	.109375	.1046	2.657	70	4.375	21.360
13	3/32"	.09375	.0897	2.278	60	3.750	18.310
14	5/64"	.078125	.0747	1.897	50	3.125	15.260
15	9/128"	.0713125	.0673	1.709	45	2.813	13.730
16	1/16"	.0625	.0598	1.519	40	2.500	12.210
17	9/160"	.05625	.0538	1.367	36	2.250	10.990
18	1/20"	.05	.0478	1.214	32	2.000	9.765
19	7/160"	.04375	.0418	1.062	28	1.750	8.544
20	3/80"	.0375	.0359	.912	24	1.500	7.324
21	11/320"	.034375	.0329	.836	22	1.375	6.713
22	1/32"	.03125	.0299	.759	20	1.250	6.103
23	9/320"	.028125	.0269	.683	18	1.125	5.490
24	1/40"	.025	.0239	.607	16	1.000	4.882
25	7/320"	.021875	.0209	.531	14	.875	4.272
26	3/160"	.01875	.0179	.455	12	.750	3.662
27	11/640"	.0171875	.0164	.417	11	.688	3.357
28	1/64"	.015625	.0149	.378	10	.625	3.052

Metals — R0531 Steel Decking

R053100-10 Decking Descriptions

General - All Deck Products

Steel deck is made by cold forming structural grade sheet steel into a repeating pattern of parallel ribs. The strength and stiffness of the panels are the result of the ribs and the material properties of the steel. Deck lengths can be varied to suit job conditions, but because of shipping considerations, are usually less than 40 feet. Standard deck width varies with the product used but full sheets are usually 12″, 18″, 24″, 30″, or 36″. Deck is typically furnished in a standard width with the ends cut square. Any cutting for width, such as at openings or for angular fit, is done at the job site.

Deck is typically attached to the building frame with arc puddle welds, self-drilling screws, or powder or pneumatically driven pins. Sheet to sheet fastening is done with screws, button punching (crimping), or welds.

Composite Floor Deck

After installation and adequate fastening, floor deck serves several purposes. It (a) acts as a working platform, (b) stabilizes the frame, (c) serves as a concrete form for the slab, and (d) reinforces the slab to carry the design loads applied during the life of the building. Composite decks are distinguished by the presence of shear connector devices as part of the deck. These devices are designed to mechanically lock the concrete and deck together so that the concrete and the deck work together to carry subsequent floor loads. These shear connector devices can be rolled-in embossments, lugs, holes, or wires welded to the panels. The deck profile can also be used to interlock concrete and steel.

Composite deck finishes are either galvanized (zinc coated) or phosphatized/painted. Galvanized deck has a zinc coating on both the top and bottom surfaces. The phosphatized/painted deck has a bare (phosphatized) top surface that will come into contact with the concrete. This bare top surface can be expected to develop rust before the concrete is placed. The bottom side of the deck has a primer coat of paint.

Composite floor deck is normally installed so the panel ends do not overlap on the supporting beams. Shear lugs or panel profile shape often prevent a tight metal to metal fit if the panel ends overlap; the air gap caused by overlapping will prevent proper fusion with the structural steel supports when the panel end laps are shear stud welded.

Adequate end bearing of the deck must be obtained as shown on the drawings. If bearing is actually less in the field than shown on the drawings, further investigation is required.

Roof Deck

Roof deck is not designed to act compositely with other materials. Roof deck acts alone in transferring horizontal and vertical loads into the building frame. Roof deck rib openings are usually narrower than floor deck rib openings. This provides adequate support of rigid thermal insulation board.

Roof deck is typically installed to endlap approximately 2″ over supports. However, it can be butted (or lapped more than 2″) to solve field fit problems. Since designers frequently use the installed deck system as part of the horizontal bracing system (the deck as a diaphragm), any fastening substitution or change should be approved by the designer. Continuous perimeter support of the deck is necessary to limit edge deflection in the finished roof and may be required for diaphragm shear transfer.

Standard roof deck finishes are galvanized or primer painted. The standard factory applied paint for roof deck is a primer paint and is not intended to weather for extended periods of time. Field painting or touching up of abrasions and deterioration of the primer coat or other protective finishes is the responsibility of the contractor.

Cellular Deck

Cellular deck is made by attaching a bottom steel sheet to a roof deck or composite floor deck panel. Cellular deck can be used in the same manner as floor deck. Electrical, telephone, and data wires are easily run through the chase created between the deck panel and the bottom sheet.

When used as part of the electrical distribution system, the cellular deck must be installed so that the ribs line up and create a smooth cell transition at abutting ends. The joint that occurs at butting cell ends must be taped or otherwise sealed to prevent wet concrete from seeping into the cell. Cell interiors must be free of welding burrs, or other sharp intrusions, to prevent damage to wires.

When used as a roof deck, the bottom flat plate is usually left exposed to view. Care must be maintained during erection to keep good alignment and prevent damage.

Cellular deck is sometimes used with the flat plate on the top side to provide a flat working surface. Installation of the deck for this purpose requires special methods for attachment to the frame because the flat plate, now on the top, can prevent direct access to the deck material that is bearing on the structural steel. It may be advisable to treat the flat top surface to prevent slipping.

Cellular deck is always furnished galvanized or painted over galvanized.

Form Deck

Form deck can be any floor or roof deck product used as a concrete form. Connections to the frame are by the same methods used to anchor floor and roof deck. Welding washers are recommended when welding deck that is less than 20 gauge thickness.

Form deck is furnished galvanized, prime painted, or uncoated. Galvanized deck must be used for those roof deck systems where form deck is used to carry a lightweight insulating concrete fill.

Wood, Plastics & Comp. — R0611 Wood Framing

R061110-30 Lumber Product Material Prices

The price of forest products fluctuates widely from location to location and from season to season depending upon economic conditions. The bare material prices in the unit cost sections of the book show the National Average material prices in effect Jan. 1 of this book year. It must be noted that lumber prices in general may change significantly during the year.

Availability of certain items depends upon geographic location and must be checked prior to firm-price bidding.

Wood, Plastics & Comp. — R0616 Sheathing

R061636-20 Plywood

There are two types of plywood used in construction: interior, which is moisture-resistant but not waterproofed, and exterior, which is waterproofed.

The grade of the exterior surface of the plywood sheets is designated by the first letter: A, for smooth surface with patches allowed; B, for solid surface with patches and plugs allowed; C, which may be surface plugged or may have knot holes up to 1" wide; and D, which is used only for interior type plywood and may have knot holes up to 2-1/2" wide. "Structural Grade" is specifically designed for engineered applications such as box beams. All CC & DD grades have roof and floor spans marked on them.

Underlayment-grade plywood runs from 1/4" to 1-1/4" thick. Thicknesses 5/8" and over have optional tongue and groove joints which eliminate the need for blocking the edges. Underlayment 19/32" and over may be referred to as Sturd-i-Floor.

The price of plywood can fluctuate widely due to geographic and economic conditions.

Typical uses for various plywood grades are as follows:

AA-AD Interior — cupboards, shelving, paneling, furniture
BB Plyform — concrete form plywood
CDX — wall and roof sheathing
Structural — box beams, girders, stressed skin panels
AA-AC Exterior — fences, signs, siding, soffits, etc.
Underlayment — base for resilient floor coverings
Overlaid HDO — high density for concrete forms & highway signs
Overlaid MDO — medium density for painting, siding, soffits & signs
303 Siding — exterior siding, textured, striated, embossed, etc.

Thermal & Moist. Protec. — R0731 Shingles & Shakes

R073126-20 Roof Slate

16", 18" and 20" are standard lengths, and slate usually comes in random widths. For standard 3/16" thickness use 1-1/2" copper nails. Allow for 3% breakage.

Thermal & Moist. Protec. — R0751 Built-Up Bituminous Roofing

R075113-20 Built-Up Roofing

Asphalt is available in kegs of 100 lbs. each; coal tar pitch in 560 lb. kegs. Prepared roofing felts are available in a wide range of sizes, weights and characteristics. However, the most commonly used are #15 (432 S.F. per roll, 13 lbs. per square) and #30 (216 S.F. per roll, 27 lbs. per square).

Inter-ply bitumen varies from 24 lbs. per sq. (asphalt) to 30 lbs. per sq. (coal tar) per ply, MF4@ 25%. Flood coat bitumen also varies from 60 lbs. per sq. (asphalt) to 75 lbs. per sq. (coal tar), MF4@ 25%. Expendable equipment (mops, brooms, screeds, etc.) runs about 16% of the bitumen cost. For new, inexperienced crews this factor may be much higher.

Rigid insulation board is typically applied in two layers. The first is mechanically attached to nailable decks or spot or solid mopped to non-nailable decks; the second layer is then spot or solid mopped to the first layer. Membrane application follows the insulation, except in protected membrane roofs, where the membrane goes down first and the insulation on top, followed with ballast (stone or concrete pavers). Insulation and related labor costs are NOT included in prices for built-up roofing.

Thermal & Moist. Protec. — R0752 Modified Bituminous Membrane Roofing

R075213-30 Modified Bitumen Roofing

The cost of modified bitumen roofing is highly dependent on the type of installation that is planned. Installation is based on the type of modifier used in the bitumen. The two most popular modifiers are atactic polypropylene (APP) and styrene butadiene styrene (SBS). The modifiers are added to heated bitumen during the manufacturing process to change its characteristics. A polyethylene, polyester or fiberglass reinforcing sheet is then sandwiched between layers of this bitumen. When completed, the result is a pre-assembled, built-up roof that has increased elasticity and weatherablility. Some manufacturers include a surfacing material such as ceramic or mineral granules, metal particles or sand.

The preferred method of adhering SBS-modified bitumen roofing to the substrate is with hot-mopped asphalt (much the same as built-up roofing). This installation method requires a tar kettle/pot to heat the asphalt, as well as the labor, tools and equipment necessary to distribute and spread the hot asphalt.

The alternative method for applying APP and SBS modified bitumen is as follows. A skilled installer uses a torch to melt a small pool of bitumen off the membrane. This pool must form across the entire roll for proper adhesion. The installer must unroll the roofing at a pace slow enough to melt the bitumen, but fast enough to prevent damage to the rest of the membrane.

Modified bitumen roofing provides the advantages of both built-up and single-ply roofing. Labor costs are reduced over those of built-up roofing because only a single ply is necessary. The elasticity of single-ply roofing is attained with the reinforcing sheet and polymer modifiers. Modifieds have some self-healing characteristics and because of their multi-layer construction, they offer the reliability and safety of built-up roofing.

Thermal & Moist. Protec. — R0784 Firestopping

R078413-30 Firestopping

Firestopping is the sealing of structural, mechanical, electrical and other penetrations through fire-rated assemblies. The basic components of firestop systems are safing insulation and firestop sealant on both sides of wall penetrations and the top side of floor penetrations.

Pipe penetrations are assumed to be through concrete, grout, or joint compound and can be sleeved or unsleeved. Costs for the penetrations and sleeves are not included. An annular space of 1" is assumed. Escutcheons are not included.

Metallic pipe is assumed to be copper, aluminum, cast iron or similar metallic material. Insulated metallic pipe is assumed to be covered with a thermal insulating jacket of varying thickness and materials.

Non-metallic pipe is assumed to be PVC, CPVC, FR Polypropylene or similar plastic piping material. Intumescent firestop sealant or wrap strips are included. Collars on both sides of wall penetrations and a sheet metal plate on the underside of floor penetrations are included.

Ductwork is assumed to be sheet metal, stainless steel or similar metallic material. Duct penetrations are assumed to be through concrete, grout or joint compound. Costs for penetrations and sleeves are not included. An annular space of 1/2" is assumed.

Multi-trade openings include costs for sheet metal forms, firestop mortar, wrap strips, collars and sealants as necessary.

Structural penetrations joints are assumed to be 1/2" or less. CMU walls are assumed to be within 1-1/2" of metal deck. Drywall walls are assumed to be tight to the underside of metal decking.

Metal panel, glass or curtain wall systems include a spandrel area of 5' filled with mineral wool foil-faced insulation. Fasteners and stiffeners are included.

Openings — R0813 Metal Doors

R081313-20 Steel Door Selection Guide

Standard steel doors are classified into four levels, as recommended by the Steel Door Institute in the chart below. Each of the four levels offers a range of construction models and designs, to meet architectural requirements for preference and appearance, including full flush, seamless, and stile & rail. Recommended minimum gauge requirements are also included.

For complete standard steel door construction specifications and available sizes, refer to the Steel Door Institute Technical Data Series, ANSI A250.8-98 (SDI-100), and ANSI A250.4-94 Test Procedure and Acceptance Criteria for Physical Endurance of Steel Door and Hardware Reinforcements.

Level		Model	Construction	For Full Flush or Seamless		
				Min. Gauge	Thickness (in)	Thickness (mm)
I	Standard Duty	1	Full Flush	20	0.032	0.8
		2	Seamless			
II	Heavy Duty	1	Full Flush	18	0.042	1.0
		2	Seamless			
III	Extra Heavy Duty	1	Full Flush	16	0.053	1.3
		2	Seamless			
		3	*Stile & Rail			
IV	Maximum Duty	1	Full Flush	14	0.067	1.6
		2	Seamless			

*Stiles & rails are 16 gauge; flush panels, when specified, are 18 gauge

Openings — R0851 Metal Windows

R085123-10 Steel Sash

Ironworker crew will erect 25 S.F. or 1.3 sash unit per hour, whichever is less.

Mechanic will point 30 L.F. per hour.

Painter will paint 90 S.F. per coat per hour.

Glazier production depends on light size.

Allow 1 lb. special steel sash putty per 16" x 20" light.

Openings — R0852 Wood Windows

R085216-10 Window Estimates

To ensure a complete window estimate, be sure to include the material and labor costs for each window, as well as the material and labor costs for an interior wood trim set.

Openings — R0871 Door Hardware

R087120-10 Hinges

All closer equipped doors should have ball bearing hinges. Lead lined or extremely heavy doors require special strength hinges. Usually 1-1/2 pair of hinges are used per door up to 7'-6" high openings. Table below shows typical hinge requirements.

Use Frequency	Type Hinge Required	Type of Opening	Type of Structure
High	Heavy weight ball bearing	Entrances	Banks, Office buildings, Schools, Stores & Theaters
		Toilet Rooms	Office buildings and Schools
Average	Standard weight ball bearing	Entrances	Dwellings
		Corridors	Office buildings and Schools
		Toilet Rooms	Stores
Low	Plain bearing	Interior	Dwellings

Door Thickness	Weight of Doors in Pounds per Square Foot				
	White Pine	Oak	Hollow Core	Solid Core	Hollow Metal
1-3/8"	3 psf	6 psf	1-1/2 psf	3-1/2 — 4 psf	6-1/2 psf
1-3/4"	3-1/2	7	2-1/2	4-1/2 — 5-1/4	6-1/2
2-1/4"	4-1/2	9	—	5-1/2 — 6-3/4	6-1/2

Openings — R0881 Glass Glazing

R088110-10 Glazing Productivity

Some glass sizes are estimated by the "united inch" (height + width). The table below shows the number of lights glazed in an eight-hour period by the crew size indicated, for glass up to 1/4" thick. Square or nearly square lights are more economical on a S.F. basis. Long slender lights will have a high S.F. installation cost. For insulated glass reduce production by 33%. For 1/2" float glass reduce production by 50%. Production time for glazing with two glaziers per day averages: 1/4" float glass 120 S.F.; 1/2" float glass 55 S.F.; 1/2" insulated glass 95 S.F.; 3/4" insulated glass 75 S.F.

Glazing Method	United Inches per Light							
	40"	60"	80"	100"	135"	165"	200"	240"
Number of Men in Crew	1	1	1	1	2	3	3	4
Industrial sash, putty	60	45	24	15	18	—	—	—
With stops, putty bed	50	36	21	12	16	8	4	3
Wood stops, rubber	40	27	15	9	11	6	3	2
Metal stops, rubber	30	24	14	9	9	6	3	2
Structural glass	10	7	4	3	—	—	—	—
Corrugated glass	12	9	7	4	4	4	3	—
Storefronts	16	15	13	11	7	6	4	4
Skylights, putty glass	60	36	21	12	16	—	—	—
Thiokol set	15	15	11	9	9	6	3	2
Vinyl set, snap on	18	18	13	12	12	7	5	4
Maximum area per light	2.8 S.F.	6.3 S.F.	11.1 S.F.	17.4 S.F.	31.6 S.F.	47 S.F.	69 S.F.	100 S.F.

Finishes — R0920 Plaster & Gypsum Board

R092000-50 Lath, Plaster and Gypsum Board

Gypsum board lath is available in 3/8" thick x 16" wide x 4' long sheets as a base material for multi-layer plaster applications. It is also available as a base for either multi-layer or veneer plaster applications in 1/2" and 5/8" thick—4' wide x 8', 10' or 12' long sheets. Fasteners are screws or blued ring shank nails for wood framing and screws for metal framing.

Metal lath is available in diamond mesh pattern with flat or self-furring profiles. Paper backing is available for applications where excessive plaster waste needs to be avoided. A slotted mesh ribbed lath should be used in areas where the span between structural supports is greater than normal. Most metal lath comes in 27" x 96" sheets. Diamond mesh weighs 1.75, 2.5 or 3.4 pounds per square yard, slotted mesh lath weighs 2.75 or 3.4 pounds per square yard. Metal lath can be nailed, screwed or tied in place.

Many **accessories** are available. Corner beads, flat reinforcing strips, casing beads, control and expansion joints, furring brackets and channels are some examples. Note that accessories are not included in plaster or stucco line items.

Plaster is defined as a material or combination of materials that when mixed with a suitable amount of water, forms a plastic mass or paste. When applied to a surface, the paste adheres to it and subsequently hardens, preserving in a rigid state the form or texture imposed during the period of elasticity.

Gypsum plaster is made from ground calcined gypsum. It is mixed with aggregates and water for use as a base coat plaster.

Vermiculite plaster is a fire-retardant plaster covering used on steel beams, concrete slabs and other heavy construction materials. Vermiculite is a group name for certain clay minerals, hydrous silicates or aluminum, magnesium and iron that have been expanded by heat.

Perlite plaster is a plaster using perlite as an aggregate instead of sand. Perlite is a volcanic glass that has been expanded by heat.

Gauging plaster is a mix of gypsum plaster and lime putty that when applied produces a quick drying finish coat.

Veneer plaster is a one or two component gypsum plaster used as a thin finish coat over special gypsum board.

Keenes cement is a white cementitious material manufactured from gypsum that has been burned at a high temperature and ground to a fine powder. Alum is added to accelerate the set. The resulting plaster is hard and strong and accepts and maintains a high polish, hence it is used as a finishing plaster.

Stucco is a Portland cement based plaster used primarily as an exterior finish.

Plaster is used on both interior and exterior surfaces. Generally it is applied in multiple-coat systems. A three-coat system uses the terms scratch, brown and finish to identify each coat. A two-coat system uses base and finish to describe each coat. Each type of plaster and application system has attributes that are chosen by the designer to best fit the intended use.

Gypsum Plaster Quantities for 100 S.Y.	2 Coat, 5/8" Thick		3 Coat, 3/4" Thick		
	Base	Finish	Scratch	Brown	Finish
	1:3 Mix	2:1 Mix	1:2 Mix	1:3 Mix	2:1 Mix
Gypsum plaster	1,300 lb.		1,350 lb.	650 lb.	
Sand	1.75 C.Y.		1.85 C.Y.	1.35 C.Y.	
Finish hydrated lime		340 lb.			340 lb.
Gauging plaster		170 lb.			170 lb.

Vermiculite or Perlite Plaster Quantities for 100 S.Y.	2 Coat, 5/8" Thick		3 Coat, 3/4" Thick		
	Base	Finish	Scratch	Brown	Finish
Gypsum plaster	1,250 lb.		1,450 lb.	800 lb.	
Vermiculite or perlite	7.8 bags		8.0 bags	3.3 bags	
Finish hydrated lime		340 lb.			340 lb.
Gauging plaster		170 lb.			170 lb.

Stucco–Three-Coat System Quantities for 100 S.Y.	On Wood Frame	On Masonry
Portland cement	29 bags	21 bags
Sand	2.6 C.Y.	2.0 C.Y.
Hydrated lime	180 lb.	120 lb.

Finishes — R0929 Gypsum Board

R092910-10 Levels of Gypsum Drywall Finish

In the past, contract documents often used phrases such as "industry standard" and "workmanlike finish" to specify the expected quality of gypsum board wall and ceiling installations. The vagueness of these descriptions led to unacceptable work and disputes.

In order to resolve this problem, four major trade associations concerned with the manufacture, erection, finish and decoration of gypsum board wall and ceiling systems have developed an industry-wide *Recommended Levels of Gypsum Board Finish*.

The finish of gypsum board walls and ceilings for specific final decoration is dependent on a number of factors. A primary consideration is the location of the surface and the degree of decorative treatment desired. Painted and unpainted surfaces in warehouses and other areas where appearance is normally not critical may simply require the taping of wallboard joints and 'spotting' of fastener heads. Blemish-free, smooth, monolithic surfaces often intended for painted and decorated walls and ceilings in habitated structures, ranging from single-family dwellings through monumental buildings, require additional finishing prior to the application of the final decoration.

Other factors to be considered in determining the level of finish of the gypsum board surface are (1) the type of angle of surface illumination (both natural and artificial lighting), and (2) the paint and method of application or the type and finish of wallcovering specified as the final decoration. Critical lighting conditions, gloss paints, and thin wallcoverings require a higher level of gypsum board finish than do heavily textured surfaces which are subsequently painted or surfaces which are to be decorated with heavy grade wallcoverings.

The following descriptions were developed jointly by the Association of the Wall and Ceiling Industries-International (AWCI), Ceiling & Interior Systems Construction Association (CISCA), Gypsum Association (GA), and Painting and Decorating Contractors of America (PDCA) as a guide.

Level 0: No taping, finishing, or accessories required. This level of finish may be useful in temporary construction or whenever the final decoration has not been determined.

Level 1: All joints and interior angles shall have tape set in joint compound. Surface shall be free of excess joint compound. Tool marks and ridges are acceptable. Frequently specified in plenum areas above ceilings, in attics, in areas where the assembly would generally be concealed or in building service corridors, and other areas not normally open to public view.

Level 2: All joints and interior angles shall have tape embedded in joint compound and wiped with a joint knife leaving a thin coating of joint compound over all joints and interior angles. Fastener heads and accessories shall be covered with a coat of joint compound. Surface shall be free of excess joint compound. Tool marks and ridges are acceptable. Joint compound applied over the body of the tape at the time of tape embedment shall be considered a separate coat of joint compound and shall satisfy the conditions of this level. Specified where water-resistant gypsum backing board is used as a substrate for tile; may be specified in garages, warehouse storage, or other similar areas where surface appearance is not of primary concern.

Level 3: All joints and interior angles shall have tape embedded in joint compound and one additional coat of joint compound applied over all joints and interior angles. Fastener heads and accessories shall be covered with two separate coats of joint compound. All joint compound shall be smooth and free of tool marks and ridges. Typically specified in appearance areas which are to receive heavy- or medium-texture (spray or hand applied) finishes before final painting, or where heavy-grade wallcoverings are to be applied as the final decoration. This level of finish is not recommended where smooth painted surfaces or light to medium wallcoverings are specified.

Level 4: All joints and interior angles shall have tape embedded in joint compound and two separate coats of joint compound applied over all flat joints and one separate coat of joint compound applied over interior angles. Fastener heads and accessories shall be covered with three separate coats of joint compound. All joint compound shall be smooth and free of tool marks and ridges. This level should be specified where flat paints, light textures, or wallcoverings are to be applied. In critical lighting areas, flat paints applied over light textures tend to reduce joint photographing. Gloss, semi-gloss, and enamel paints are not recommended over this level of finish. The weight, texture, and sheen level of wallcoverings applied over this level of finish should be carefully evaluated. Joints and fasteners must be adequately concealed if the wallcovering material is lightweight, contains limited pattern, has a gloss finish, or any combination of these finishes is present. Unbacked vinyl wallcoverings are not recommended over this level of finish.

Level 5: All joints and interior angles shall have tape embedded in joint compound and two separate coats of joint compound applied over all flat joints and one separate coat of joint compound applied over interior angles. Fastener heads and accessories shall be covered with three separate coats of joint compound. A thin skim coat of joint compound or a material manufactured especially for this purpose, shall be applied to the entire surface. The surface shall be smooth and free of tool marks and ridges. This level of finish is highly recommended where gloss, semi-gloss, enamel, or nontextured flat paints are specified or where severe lighting conditions occur. This highest quality finish is the most effective method to provide a uniform surface and minimize the possibility of joint photographing and of fasteners showing through the final decoration.

Finishes — R0966 Terrazzo Flooring

R096613-10   Terrazzo Floor

The table below lists quantities required for 100 S.F. of 5/8" terrazzo topping, either bonded or not bonded.

Description	Bonded to Concrete 1-1/8" Bed, 1:4 Mix	Not Bonded 2-1/8" Bed and 1/4" Sand
Portland cement, 94 lb. Bag	6 bags	8 bags
Sand	10 C.F.	20 C.F.
Divider strips, 4' squares	50 L.F.	50 L.F.
Terrazzo fill, 50 lb. Bag	12 bags	12 bags
15 Lb. tarred felt		1 C.S.F.
Mesh 2 x 2 #14 galvanized		1 C.S.F.
Crew J-3	0.77 days	0.87 days

2' x 2' panels require 1.00 L.F. divider strip per S.F.
3' x 3' panels require 0.67 L.F. divider strip per S.F.
4' x 4' panels require 0.50 L.F. divider strip per S.F.
5' x 5' panels require 0.40 L.F. divider strip per S.F.
6' x 6' panels require 0.33 L.F. divider strip per S.F.

Finishes — R0972 Wall Coverings

R097223-10   Wall Covering

The table below lists the quantities required for 100 S.F. of wall covering.

Description	Medium-Priced Paper	Expensive Paper
Paper	1.6 dbl. rolls	1.6 dbl. rolls
Wall sizing	0.25 gallon	0.25 gallon
Vinyl wall paste	0.6 gallon	0.6 gallon
Apply sizing	0.3 hour	0.3 hour
Apply paper	1.2 hours	1.5 hours

Most wallpapers now come in double rolls only.
To remove old paper, allow 1.3 hours per 100 S.F.

Finishes — R0991 Painting

R099100-20 Painting

Item	Coat	One Gallon Covers			In 8 Hours a Laborer Covers			Labor-Hours per 100 S.F.		
		Brush	Roller	Spray	Brush	Roller	Spray	Brush	Roller	Spray
Paint wood siding	prime	250 S.F.	225 S.F.	290 S.F.	1150 S.F.	1300 S.F.	2275 S.F.	.695	.615	.351
	others	270	250	290	1300	1625	2600	.615	.492	.307
Paint exterior trim	prime	400	—	—	650	—	—	1.230	—	—
	1st	475	—	—	800	—	—	1.000	—	—
	2nd	520	—	—	975	—	—	.820	—	—
Paint shingle siding	prime	270	255	300	650	975	1950	1.230	.820	.410
	others	360	340	380	800	1150	2275	1.000	.695	.351
Stain shingle siding	1st	180	170	200	750	1125	2250	1.068	.711	.355
	2nd	270	250	290	900	1325	2600	.888	.603	.307
Paint brick masonry	prime	180	135	160	750	800	1800	1.066	1.000	.444
	1st	270	225	290	815	975	2275	.981	.820	.351
	2nd	340	305	360	815	1150	2925	.981	.695	.273
Paint interior plaster or drywall	prime	400	380	495	1150	2000	3250	.695	.400	.246
	others	450	425	495	1300	2300	4000	.615	.347	.200
Paint interior doors and windows	prime	400	—	—	650	—	—	1.230	—	—
	1st	425	—	—	800	—	—	1.000	—	—
	2nd	450	—	—	975	—	—	.820	—	—

Special Construction — R1311 Swimming Pools

R131113-20 Swimming Pools

Pool prices given per square foot of surface area include pool structure, filter and chlorination equipment, pumps, related piping, ladders/steps, maintenance kit, skimmer and vacuum system. Decks and electrical service to equipment are not included.

Residential in-ground pool construction can be divided into two categories: vinyl lined and gunite. Vinyl lined pool walls are constructed of different materials including wood, concrete, plastic or metal. The bottom is often graded with sand over which the vinyl liner is installed. Vermiculite or soil cement bottoms may be substituted for an added cost.

Gunite pool construction is used both in residential and municipal installations. These structures are steel reinforced for strength and finished with a white cement limestone plaster.

Municipal pools will have a higher cost because plumbing codes require more expensive materials, chlorination equipment and higher filtration rates.

Municipal pools greater than 1,800 S.F. require gutter systems to control waves. This gutter may be formed into the concrete wall. Often a vinyl/stainless steel gutter or gutter/wall system is specified, which will raise the pool cost.

Competition pools usually require tile bottoms and sides with contrasting lane striping, which will also raise the pool cost.

Special Construction   R1331 Fabric Structures

R133113-10   Air Supported Structures

Air supported structures are made from fabrics that can be classified into two groups: temporary and permanent. Temporary fabrics include nylon, woven polyethylene, vinyl film, and vinyl coated dacron. These have lifespans that range from five to fifteen plus years. The cost per square foot includes a fabric shell, tension cables, primary and back-up inflation systems and doors. The lower cost structures are used for construction shelters, bulk storage and pond covers. The more expensive are used for recreational structures and warehouses.

Permanent fabrics are teflon coated fiberglass. The life of this structure is twenty plus years. The high cost limits its application to architectural designed structures which call for a clear span covered area, such as stadiums and convention centers. Both temporary and permanent structures are available in translucent fabrics which eliminates the need for daytime lighting.

Areas to be covered vary from 10,000 S.F. to any area up to 1000 foot wide by any length. Height restrictions range from a maximum of 1/2 of width to a minimum of 1/6 of the width. Erection of even the largest of the temporary structures requires no more than a week.

Centrifugal fans provide the inflation necessary to support the structure during application of live loads. Airlocks are usually used at large entrances to prevent loss of static pressure. Some manufacturers employ propeller fans which generate sufficient airflow (30,000 CFM) to eliminate the need for airlocks. These fans may also be automatically controlled to resist high wind conditions, regulate humidity (air changes), and provide cooling and heat.

Insulation can be provided with the addition of a second or even third interior liner, creating a dead air space with an "R" value of four to nine. Some structures allow for the liner to be collapsed into the outer shell to enable the internal heat to melt accumulated snow. For cooling or air conditioning, the exterior face of the liner can be aluminized to reflect the sun's heat.

R133113-90   Seismic Bracing

Sometimes referred to as anti-sway bracing, this support system is required in earthquake areas. The individual components must be assembled to make a required system.

Example			Additionally, height factors must be taken into account. Add the following percentages to labor for elevated installations:	
	C-Clamp 3/8" rod	2 ea.	15' to 20' high	10%
	Rod, continuous thread 3/8"	10 L.F.	21' to 25' high	20%
	Field, weld 1"	2 ea.	26' to 30' high	30%
			31' to 35' high	40%
			36' to 40' high	50%
			41' to 50' high	60%

Special Construction   R1334 Fabricated Engineered Structures

R133419-10   Pre-Engineered Steel Buildings

These buildings are manufactured by many companies and normally erected by franchised dealers throughout the U.S. The four basic types are: Rigid Frames, Truss type, Post and Beam and the Sloped Beam type. Most popular roof slope is low pitch of 1" in 12". The minimum economical area of these buildings is about 3000 S.F. of floor area. Bay sizes are usually 20' to 24' but can go as high as 30' with heavier girts and purlins. Eave heights are usually 12' to 24' with 18' to 20' most typical.

Material prices shown in the Unit Price section are bare costs for the building shell only and do not include floors, foundations, anchor bolts, interior finishes or utilities. Costs assume at least three bays of 24' each, a 1" in 12" roof slope, and they are based on 30 psf roof load and 20 psf wind load (wind load is a function of wind speed, building height, and terrain characteristics; this should be determined by a Registered Structural Engineer) and no unusual requirements. Costs include the structural frame, 26 ga. non-insulated colored corrugated or ribbed roofing and siding panels, fasteners, closures, trim and flashing but no allowance for insulation, doors, windows, skylights, gutters or downspouts. Very large projects would generally cost less for materials than the prices shown. For roof panel substitutions and wall panel substitutions, see appropriate Unit Price sections.

Conditions at the site, weather, shape and size of the building, and labor availability will affect the erection cost of the building.

R133423-30   Dome Structures

Steel — The four types are Lamella, Schwedler, Arch and Geodesic. For maximum economy, rise should be about 15 to 20% of diameter. Most common diameters are in the 200' to 300' range. Lamella domes weigh about 5 P.S.F. of floor area less than Schwedler domes. Schwedler dome weight in lbs. per S.F. approaches .046 times the diameter. Domes below 125' diameter weigh .07 times diameter and the cost per ton of steel is higher. See R051223-20 for estimating weight.

Wood — Small domes are of sawn lumber, larger ones are laminated. In larger sizes, triaxial and triangular cost about the same; radial domes cost more. Radial domes are economical in the 60' to 70' diameter range. Most economical range of all types is 80' to 200' diameters. Diameters can run over 400'. All costs are quoted above the foundation. Prices include 2" decking and a tension tie ring in place.

Plywood — Stock prefab geodesic domes are available with diameters from 24' to 60'.

Fiberglass — Aluminum framed translucent sandwich panels with spans from 5' to 45' are commercially available.

Aluminum — Stressed skin aluminum panels form geodesic domes with spans ranging from 82' to 232'. An aluminum space truss, triangulated or nontriangulated, with aluminum or clear acrylic closure panels can be used for clear spans of 40' to 415'.

Conveying Equipment — R1420 Elevators

R142000-10 Freight Elevators

Capacities run from 2,000 lbs. to over 100,000 lbs. with 3,000 lbs. to 10,000 lbs. most common. Travel speeds are generally lower and control less intricate than on passenger elevators. Costs in the Unit Price sections are for hydraulic and geared elevators.

R142000-20 Elevator Selective Costs See R142000-40 for cost development.

A. Base Unit	Passenger		Freight		Hospital	
	Hydraulic	Electric	Hydraulic	Electric	Hydraulic	Electric
Capacity	1,500 lb.	2,000 lb.	2,000 lb.	4,000 lb.	4,000 lb.	4,000 lb.
Speed	100 F.P.M.	200 F.P.M.	100 F.P.M.	200 F.P.M.	100 F.P.M.	200 F.P.M.
#Stops/Travel Ft.	2/12	4/40	2/20	4/40	2/20	4/40
Push Button Oper.	Yes	Yes	Yes	Yes	Yes	Yes
Telephone Box & Wire	"	"	"	"	"	"
Emergency Lighting	"	"	No	No	"	"
Cab	Plastic Lam. Walls	Plastic Lam. Walls	Painted Steel	Painted Steel	Plastic Lam. Walls	Plastic Lam. Walls
Cove Lighting	Yes	Yes	No	No	Yes	Yes
Floor	V.C.T.	V.C.T.	Wood w/Safety Treads	Wood w/Safety Treads	V.C.T.	V.C.T.
Doors, & Speedside Slide	Yes	Yes	Yes	Yes	Yes	Yes
Gates, Manual	No	No	No	No	No	No
Signals, Lighted Buttons	Car and Hall	Car and Hall	Car and Hall	Car and Hall	Car and Hall	Car and Hall
O.H. Geared Machine	N.A.	Yes	N.A.	Yes	N.A.	Yes
Variable Voltage Contr.	"	"	N.A.	"	"	"
Emergency Alarm	Yes	"	Yes	"	Yes	"
Class "A" Loading	N.A.	N.A.	"	"	N.A.	N.A.

R142000-30 Passenger Elevators

Electric elevators are used generally but hydraulic elevators can be used for lifts up to 70' and where large capacities are required. Hydraulic speeds are limited to 200 F.P.M. but cars are self leveling at the stops. On low rises, hydraulic installation runs about 15% less than standard electric types but on higher rises this installation cost advantage is reduced. Maintenance of hydraulic elevators is about the same as electric type but underground portion is not included in the maintenance contract.

In electric elevators there are several control systems available, the choice of which will be based upon elevator use, size, speed and cost criteria. The two types of drives are geared for low speeds and gearless for 450 F.P.M. and over.

The tables on the preceding pages illustrate typical installed costs of the various types of elevators available.

R142000-40 Elevator Cost Development

To price a new car or truck from the factory, you must start with the manufacturer's basic model, then add or exchange optional equipment and features. The same is true for pricing elevators.

Requirement: One-passenger elevator, five-story hydraulic, 2,500 lb. capacity, 12' floor to floor, speed 150 F.P.M., emergency power switching and maintenance contract.

Example:

Description	Adjustment
A. Base Elevator: Hydraulic Passenger, 1500 lb. Capacity, 100 fpm, 2 Stops, Standard Finish	1 Ea.
B. Capacity Adjustment (2,500 lb.)	1 Ea.
C. Excess Travel Adjustment: 48' Total Travel (4 x 12') minus 12' Base Unit Travel =	36 V.L.F.
D. Stops Adjustment: 5 Total Stops minus 2 Stops (Base Unit) =	3 Stops
E. Speed Adjustment (150 F.P.M.)	1 Ea.
F. Options:	
1. Intercom Service	1 Ea.
2. Emergency Power Switching, Automatic	1 Ea.
3. Stainless Steel Entrance Doors	5 Ea.
4. Maintenance Contract (12 Months)	1 Ea.
5. Position Indicator for main floor level (none indicated in Base Unit)	1 Ea.

Conveying Equipment — R1431 Escalators

R143110-10 Escalators

Moving stairs can be used for buildings where 600 or more people are to be carried to the second floor or beyond. Freight cannot be carried on escalators and at least one elevator must be available for this function.

Carrying capacity is 5,000 to 8,000 people per hour. Power requirement is 2 to 3 KW per hour and incline angle is 30°.

Conveying Equipment — R1432 Moving Walks

R143210-20 Moving Ramps and Walks

These are a specialized form of conveyor 3' to 6' wide with capacities of 3,600 to 18,000 persons per hour. Maximum speed is 140 F.P.M. and normal incline is 0° to 15°.

Local codes will determine the maximum angle. Outdoor units would require additional weather protection.

Plumbing — R2201 Operation & Maintenance of Plumbing

R220105-10 Demolition (Selective vs. Removal for Replacement)

Demolition can be divided into two basic categories.

One type of demolition involves the removal of material with no concern for its replacement. The labor-hours to estimate this work are found under "Selective Demolition" in the Fire Protection, Plumbing and HVAC Divisions. It is selective in that individual items or all the material installed as a system or trade grouping such as plumbing or heating systems are removed. This may be accomplished by the easiest way possible, such as sawing, torch cutting, or sledge hammer as well as simple unbolting.

The second type of demolition is the removal of some item for repair or replacement. This removal may involve careful draining, opening of unions, disconnecting and tagging of electrical connections, capping of pipes/ducts to prevent entry of debris or leakage of the material contained as well as transport of the item away from its in-place location to a truck/dumpster. An approximation of the time required to accomplish this type of demolition is to use half of the time indicated as necessary to install a new unit. For example; installation of a new pump might be listed as requiring 6 labor-hours so if we had to estimate the removal of the old pump we would allow an additional 3 hours for a total of 9 hours. That is, the complete replacement of a defective pump with a new pump would be estimated to take 9 labor-hours.

Plumbing — R2211 Facility Water Distribution

R221113-50 Pipe Material Considerations

1. Malleable fittings should be used for gas service.
2. Malleable fittings are used where there are stresses/strains due to expansion and vibration.
3. Cast fittings may be broken as an aid to disassembling of heating lines frozen by long use, temperature and minerals.
4. Cast iron pipe is extensively used for underground and submerged service.
5. Type M (light wall) copper tubing is available in hard temper only and is used for nonpressure and less severe applications than K and L.
6. Type L (medium wall) copper tubing, available hard or soft for interior service.
7. Type K (heavy wall) copper tubing, available in hard or soft temper for use where conditions are severe. For underground and interior service.
8. Hard drawn tubing requires fewer hangers or supports but should not be bent. Silver brazed fittings are recommended, however soft solder is normally used.
9. Type DMV (very light wall) copper tubing designed for drainage, waste and vent plus other non-critical pressure services.

Domestic/Imported Pipe and Fittings Cost

The prices shown in this publication for steel/cast iron pipe and steel, cast iron, malleable iron fittings are based on domestic production sold at the normal trade discounts. The above listed items of foreign manufacture may be available at prices of 1/3 to 1/2 those shown. Some imported items after minor machining or finishing operations are being sold as domestic to further complicate the system.

Caution: Most pipe prices in this book also include a coupling and pipe hangers which for the larger sizes can add significantly to the per foot cost and should be taken into account when comparing "book cost" with quoted supplier's cost.

Plumbing — R2240 Plumbing Fixtures

R224000-40 Plumbing Fixture Installation Time

Item	Rough-In	Set	Total Hours	Item	Rough-In	Set	Total Hours
Bathtub	5	5	10	Shower head only	2	1	3
Bathtub and shower, cast iron	6	6	12	Shower drain	3	1	4
Fire hose reel and cabinet	4	2	6	Shower stall, slate		15	15
Floor drain to 4 inch diameter	3	1	4	Slop sink	5	3	8
Grease trap, single, cast iron	5	3	8	Test 6 fixtures			14
Kitchen gas range		4	4	Urinal, wall	6	2	8
Kitchen sink, single	4	4	8	Urinal, pedestal or floor	6	4	10
Kitchen sink, double	6	6	12	Water closet and tank	4	3	7
Laundry tubs	4	2	6	Water closet and tank, wall hung	5	3	8
Lavatory wall hung	5	3	8	Water heater, 45 gals. gas, automatic	5	2	7
Lavatory pedestal	5	3	8	Water heaters, 65 gals. gas, automatic	5	2	7
Shower and stall	6	4	10	Water heaters, electric, plumbing only	4	2	6

Fixture prices in front of book are based on the cost per fixture set in place. The rough-in cost, which must be added for each fixture, includes carrier, if required, some supply, waste and vent pipe connecting fittings and stops. The lengths of rough-in pipe are nominal runs which would connect to the larger runs and stacks. The supply runs and DWV runs and stacks must be accounted for in separate entries. In the eastern half of the United States it is common for the plumber to carry these to a point 5' outside the building.

R224000-50 Water Cooler Application

Type of Service	Requirement
Office, School or Hospital	12 persons per gallon per hour
Office, Lobby or Department Store	4 or 5 gallons per hour per fountain
Light manufacturing	7 persons per gallon per hour
Heavy manufacturing	5 persons per gallon per hour
Hot heavy manufacturing	4 persons per gallon per hour
Hotel	.08 gallons per hour per room
Theatre	1 gallon per hour per 100 seats

Heating, Ventilating & A.C. R2331 HVAC Ducts & Casings

R233100-20 Ductwork

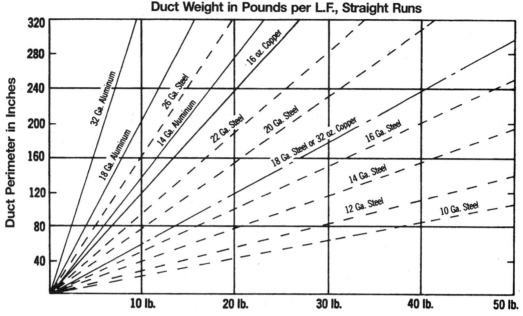

Add to the above for fittings; 90° elbow is 3 L.F.; 45° elbow is 2.5 L.F.; offset is 4 L.F.; transition offset is 6 L.F.; square-to-round transition is 4 L.F.; 90° reducing elbow is 5 L.F. For bracing and waste, add 20% to aluminum and copper, 15% to steel.

Heating, Ventilating & A.C. R2334 HVAC Fans

R233400-10 Recommended Ventilation Air Changes

Table below lists range of time in minutes per change for various types of facilities.

Facility	Min.	Facility	Min.	Facility	Min.
Assembly Halls	2-10	Dance Halls	2-10	Laundries	1-3
Auditoriums	2-10	Dining Rooms	3-10	Markets	2-10
Bakeries	2-3	Dry Cleaners	1-5	Offices	2-10
Banks	3-10	Factories	2-5	Pool Rooms	2-5
Bars	2-5	Garages	2-10	Recreation Rooms	2-10
Beauty Parlors	2-5	Generator Rooms	2-5	Sales Rooms	2-10
Boiler Rooms	1-5	Gymnasiums	2-10	Theaters	2-8
Bowling Alleys	2-10	Kitchens-Hospitals	2-5	Toilets	2-5
Churches	5-10	Kitchens-Restaurant	1-3	Transformer Rooms	1-5

CFM air required for changes = Volume of room in cubic feet ÷ Minutes per change.

Heating, Ventilating & A.C. R2350 Central Heating Equipment

R235000-35 Heating (42° degrees latitude)

$$\text{Approximate S.F. radiation} = \frac{\text{S.F. sash}}{2} + \frac{\text{S.F. wall + roof} - \text{sash}}{20} + \frac{\text{C.F. building}}{200}$$

Heating, Ventilating & A.C. R2350 Central Heating Equipment

R235000-50 Factor for Determining Heat Loss for Various Types of Buildings

General: While the most accurate estimates of heating requirements would naturally be based on detailed information about the building being considered, it is possible to arrive at a reasonable approximation using the following procedure:

1. Calculate the cubic volume of the room or building.
2. Select the appropriate factor from Table 1 below. Note that the factors apply only to inside temperatures listed in the first column and to 0° F outside temperature.
3. If the building has bad north and west exposures, multiply the heat loss factor by 1.1.
4. If the outside design temperature is other than 0° F, multiply the factor from Table 1 by the factor from Table 2.
5. Multiply the cubic volume by the factor selected from Table 1. This will give the estimated BTUH heat loss which must be made up to maintain inside temperature.

Table 1 — Building Type	Conditions	Qualifications	Loss Factor*
Factories & Industrial Plants General Office Areas at 70° F	One Story	Skylight in Roof	6.2
		No Skylight in Roof	5.7
	Multiple Story	Two Story	4.6
		Three Story	4.3
		Four Story	4.1
		Five Story	3.9
		Six Story	3.6
	All Walls Exposed	Flat Roof	6.9
		Heated Space Above	5.2
	One Long Warm Common Wall	Flat Roof	6.3
		Heated Space Above	4.7
	Warm Common Walls on Both Long Sides	Flat Roof	5.8
		Heated Space Above	4.1
Warehouses at 60° F	All Walls Exposed	Skylights in Roof	5.5
		No Skylight in Roof	5.1
		Heated Space Above	4.0
	One Long Warm Common Wall	Skylight in Roof	5.0
		No Skylight in Roof	4.9
		Heated Space Above	3.4
	Warm Common Walls on Both Long Sides	Skylight in Roof	4.7
		No Skylight in Roof	4.4
		Heated Space Above	3.0

*Note: This table tends to be conservative particularly for new buildings designed for minimum energy consumption.

Table 2 — Outside Design Temperature Correction Factor (for Degrees Fahrenheit)									
Outside Design Temperature	50	40	30	20	10	0	-10	-20	-30
Correction Factor	.29	.43	.57	.72	.86	1.00	1.14	1.28	1.43

Heating, Ventilating & A.C. R2356 Solar Energy Heating Equipment

R235616-60 Solar Heating (Space and Hot Water)

Collectors should face as close to due South as possible, however, variations of up to 20 degrees on either side of true South are acceptable. Local climate and collector type may influence the choice between east or west deviations. Obviously they should be located so they are not shaded from the sun's rays. Incline collectors at a slope of latitude minus 5 degrees for domestic hot water and latitude plus 15 degrees for space heating.

Flat plate collectors consist of a number of components as follows: Insulation to reduce heat loss through the bottom and sides of the collector. The enclosure which contains all the components in this assembly is usually weatherproof and prevents dust, wind and water from coming in contact with the absorber plate. The cover plate usually consists of one or more layers of a variety of glass or plastic and reduces the reradiation by creating an air space which traps the heat between the cover and the absorber plates.

The absorber plate must have a good thermal bond with the fluid passages. The absorber plate is usually metallic and treated with a surface coating which improves absorptivity. Black or dark paints or selective coatings are used for this purpose, and the design of this passage and plate combination helps determine a solar system's effectiveness.

Heat transfer fluid passage tubes are attached above and below or integral with an absorber plate for the purpose of transferring thermal energy from the absorber plate to a heat transfer medium. The heat exchanger is a device for transferring thermal energy from one fluid to another.

Piping and storage tanks should be well insulated to minimize heat losses.

Size domestic water heating storage tanks to hold 20 gallons of water per user, minimum, plus 10 gallons per dishwasher or washing machine. For domestic water heating an optimum collector size is approximately 3/4 square foot of area per gallon of water storage. For space heating of residences and small commercial applications the collector is commonly sized between 30% and 50% of the internal floor area. For space heating of large commercial applications, collector areas less than 30% of the internal floor area can still provide significant heat reductions.

A supplementary heat source is recommended for Northern states for December through February.

The solar energy transmission per square foot of collector surface varies greatly with the material used. Initial cost, heat transmittance and useful life are obviously interrelated.

Heating, Ventilating & A.C. R2360 Central Cooling Equipment

R236000-20 Air Conditioning Requirements

BTU's per hour per S.F. of floor area and S.F. per ton of air conditioning.

Type of Building	BTU per S.F.	S.F. per Ton	Type of Building	BTU per S.F.	S.F. per Ton	Type of Building	BTU per S.F.	S.F. per Ton
Apartments, Individual	26	450	Dormitory, Rooms	40	300	Libraries	50	240
Corridors	22	550	Corridors	30	400	Low Rise Office, Exterior	38	320
Auditoriums & Theaters	40	300/18*	Dress Shops	43	280	Interior	33	360
Banks	50	240	Drug Stores	80	150	Medical Centers	28	425
Barber Shops	48	250	Factories	40	300	Motels	28	425
Bars & Taverns	133	90	High Rise Office—Ext. Rms.	46	263	Office (small suite)	43	280
Beauty Parlors	66	180	Interior Rooms	37	325	Post Office, Individual Office	42	285
Bowling Alleys	68	175	Hospitals, Core	43	280	Central Area	46	260
Churches	36	330/20*	Perimeter	46	260	Residences	20	600
Cocktail Lounges	68	175	Hotel, Guest Rooms	44	275	Restaurants	60	200
Computer Rooms	141	85	Corridors	30	400	Schools & Colleges	46	260
Dental Offices	52	230	Public Spaces	55	220	Shoe Stores	55	220
Dept. Stores, Basement	34	350	Industrial Plants, Offices	38	320	Shop'g. Ctrs., Supermarkets	34	350
Main Floor	40	300	General Offices	34	350	Retail Stores	48	250
Upper Floor	30	400	Plant Areas	40	300	Specialty	60	200

*Persons per ton
12,000 BTU = 1 ton of air conditioning

Electrical — R2605 Common Work Results for Electrical

R260519-92 Minimum Copper and Aluminum Wire Size Allowed for Various Types of Insulation

Minimum Wire Sizes

Amperes	Copper THW THWN or XHHW	Copper THHN XHHW *	Aluminum THW XHHW	Aluminum THHN XHHW *	Amperes	Copper THW THWN or XHHW	Copper THHN XHHW *	Aluminum THW XHHW	Aluminum THHN XHHW *
15A	#14	#14	#12	#12	195	3/0	2/0	250kcmil	4/0
20	#12	#12	#10	#10	200	3/0	3/0	250kcmil	4/0
25	#10	#10	#10	#10	205	4/0	3/0	250kcmil	4/0
30	#10	#10	#8	#8	225	4/0	3/0	300kcmil	250kcmil
40	#8	#8	#8	#8	230	4/0	4/0	300kcmil	250kcmil
45	#8	#8	#6	#8	250	250kcmil	4/0	350kcmil	300kcmil
50	#8	#8	#6	#6	255	250kcmil	4/0	400kcmil	300kcmil
55	#6	#8	#4	#6	260	300kcmil	4/0	400kcmil	350kcmil
60	#6	#6	#4	#6	270	300kcmil	250kcmil	400kcmil	350kcmil
65	#6	#6	#4	#4	280	300kcmil	250kcmil	500kcmil	350kcmil
75	#4	#6	#3	#4	285	300kcmil	250kcmil	500kcmil	400kcmil
85	#4	#4	#2	#3	290	350kcmil	250kcmil	500kcmil	400kcmil
90	#3	#4	#2	#2	305	350kcmil	300kcmil	500kcmil	400kcmil
95	#3	#4	#1	#2	310	350kcmil	300kcmil	500kcmil	500kcmil
100	#3	#3	#1	#2	320	400kcmil	300kcmil	600kcmil	500kcmil
110	#2	#3	1/0	#1	335	400kcmil	350kcmil	600kcmil	500kcmil
115	#2	#2	1/0	#1	340	500kcmil	350kcmil	600kcmil	500kcmil
120	#1	#2	1/0	1/0	350	500kcmil	350kcmil	700kcmil	500kcmil
130	#1	#2	2/0	1/0	375	500kcmil	400kcmil	700kcmil	600kcmil
135	1/0	#1	2/0	1/0	380	500kcmil	400kcmil	750kcmil	600kcmil
150	1/0	#1	3/0	2/0	385	600kcmil	500kcmil	750kcmil	600kcmil
155	2/0	1/0	3/0	3/0	420	600kcmil	500kcmil		700kcmil
170	2/0	1/0	4/0	3/0	430		500kcmil		750kcmil
175	2/0	2/0	4/0	3/0	435		600kcmil		750kcmil
180	3/0	2/0	4/0	4/0	475		600kcmil		

*Dry Locations Only

Notes:
1. Size #14 to 4/0 is in AWG units (American Wire Gauge).
2. Size 250 to 750 is in kcmil units (Thousand Circular Mils).
3. Use next higher ampere value if exact value is not listed in table.
4. For loads that operate continuously increase ampere value by 25% to obtain proper wire size.
5. Refer to Table R260519-91 for the maximum circuit length for the various size wires.
6. Table R260519-92 has been written for estimating purpose only, based on ambient temperature of 30°C (86° F); for ambient temperature other than 30°C (86° F), ampacity correction factors will be applied.

Electrical

R2605 Common Work Results for Electrical

R260533-22 Conductors in Conduit

Table below lists maximum number of conductors for various sized conduit using THW, TW or THWN insulations.

Copper Wire Size	1/2"			3/4"			1"			1-1/4"			1-1/2"			2"			2-1/2"			3"			3-1/2"			4"		
	TW	THW	THWN	TW	THW	THWN	TW	THW	THWN	TW	THW	THWN	TW	THW	THWN	TW	THW	THWN	TW	THW	THWN	TW	THW	THWN	TW	THW	THWN	TW	THW	THWN
#14	9	6	13	15	10	24	25	16	39	44	29	69	60	40	94	99	65	154	142	93			143			192				
#12	7	4	10	12	8	18	19	13	29	35	24	51	47	32	70	78	53	114	111	76	164		117			157				
#10	5	4	6	9	6	11	15	11	18	26	19	32	36	26	44	60	43	73	85	61	104		95	160		127			163	
#8	2	1	3	4	3	5	7	5	9	12	10	16	17	13	22	28	22	36	40	32	51		49	79		66	106		85	136
#6		1	1	2		4	4		6	7		11	10		15	16		26	23		37	36		57	48		76	62		98
#4		1	1	1		2	3		4	5		7	7		9	12		16	17		22	27		35	36		47	47		60
#3		1	1	1		1	2		3	4		6	6		8	10		13	15		19	23		29	31		39	40		51
#2		1	1	1		1	2		3	4		5	5		7	9		11	13		16	20		25	27		33	34		43
#1				1		1	1		1	3		3	4		5	6		8	9		12	14		18	19		25	25		32
1/0				1		1	1		1	2		3	3		4	5		7	8		10	12		15	16		21	21		27
2/0				1		1	1		1	1		2	3		3	5		6	7		8	10		13	14		17	18		22
3/0				1		1	1		1	1		1	2		3	4		5	6		7	9		11	12		14	15		18
4/0						1	1		1	1		1	1		2	3		4	5		6	7		9	10		12	13		15
250 kcmil							1		1	1		1	1		1	2		3	4		4	6		7	8		10	10		12
300							1		1	1		1	1		1	2		3	3		4	5		6	7		8	9		11
350									1	1		1	1		1	1		2	3		3	4		5	6		7	8		9
400							1		1	1		1	1		1	2		3	4		5	5		6	7		8			
500							1		1	1		1	1		1	1		2	3		4	4		5	6		7			
600									1	1		1	1		1	1		1	3		3	4		4	5		5			
700										1		1	1		1	1		1	2		3	3		4	4		5			
750										1		1	1		1	1		1	2		2	3		3	4		4			

EARTHWORK — R3123 Excavation & Fill

R312316-40 Excavating

The selection of equipment used for structural excavation and bulk excavation or for grading is determined by the following factors.

1. Quantity of material
2. Type of material
3. Depth or height of cut
4. Length of haul
5. Condition of haul road
6. Accessibility of site
7. Moisture content and dewatering requirements
8. Availability of excavating and hauling equipment

Some additional costs must be allowed for hand trimming the sides and bottom of concrete pours and other excavation below the general excavation.

Number of B.C.Y. per truck = 1.5 C.Y. bucket x 8 passes = 12 loose C.Y.

$$= 12 \times \frac{100}{118} = 10.2 \text{ B.C.Y. per truck}$$

Truck Haul Cycle:

Load truck, 8 passes	=	4 minutes
Haul distance, 1 mile	=	9 minutes
Dump time	=	2 minutes
Return, 1 mile	=	7 minutes
Spot under machine	=	1 minute
		23 minute cycle

Add the mobilization and demobilization costs to the total excavation costs. When equipment is rented for more than three days, there is often no mobilization charge by the equipment dealer. On larger jobs outside of urban areas, scrapers can move earth economically provided a dump site or fill area and adequate haul roads are available. Excavation within sheeting bracing or cofferdam bracing is usually done with a clamshell and production is low, since the clamshell may have to be guided by hand between the bracing. When excavating or filling an area enclosed with a wellpoint system, add 10% to 15% to the cost to allow for restricted access. When estimating earth excavation quantities for structures, allow work space outside the building footprint for construction of the foundation and a slope of 1:1 unless sheeting is used.

When planning excavation and fill, the following should also be considered.

1. Swell factor
2. Compaction factor
3. Moisture content
4. Density requirements

A typical example for scheduling and estimating the cost of excavation of a 15′ deep basement on a dry site when the material must be hauled off the site is outlined below.

Assumptions:

1. Swell factor, 18%
2. No mobilization or demobilization
3. Allowance included for idle time and moving on job
4. No dewatering, sheeting, or bracing
5. No truck spotter or hand trimming

Fleet Haul Production per day in B.C.Y.

$$4 \text{ trucks} \times \frac{50 \text{ min. hour}}{23 \text{ min. haul cycle}} \times 8 \text{ hrs.} \times 10.2 \text{ B.C.Y.}$$

$$= 4 \times 2.2 \times 8 \times 10.2 = 718 \text{ B.C.Y./day}$$

EARTHWORK — R3123 Excavation & Fill

R312316-45 Excavating Equipment

The table below lists THEORETICAL hourly production in C.Y./hr. bank measure for some typical excavation equipment. Figures assume 50 minute hours, 83% job efficiency, 100% operator efficiency, 90° swing and properly sized hauling units, which must be modified for adverse digging and loading conditions. Actual production costs in the front of the book average about 50% of the theoretical values listed here.

Equipment	Soil Type	B.C.Y. Weight	% Swell	1 C.Y.	1-1/2 C.Y.	2 C.Y.	2-1/2 C.Y.	3 C.Y.	3-1/2 C.Y.	4 C.Y.
Hydraulic Excavator "Backhoe" 15' Deep Cut	Moist loam, sandy clay	3400 lb.	40%	85	125	175	220	275	330	380
	Sand and gravel	3100	18	80	120	160	205	260	310	365
	Common earth	2800	30	70	105	150	190	240	280	330
	Clay, hard, dense	3000	33	65	100	130	170	210	255	300
Power Shovel Optimum Cut (Ft.)	Moist loam, sandy clay	3400	40	170 (6.0)	245 (7.0)	295 (7.8)	335 (8.4)	385 (8.8)	435 (9.1)	475 (9.4)
	Sand and gravel	3100	18	165 (6.0)	225 (7.0)	275 (7.8)	325 (8.4)	375 (8.8)	420 (9.1)	460 (9.4)
	Common earth	2800	30	145 (7.8)	200 (9.2)	250 (10.2)	295 (11.2)	335 (12.1)	375 (13.0)	425 (13.8)
	Clay, hard, dense	3000	33	120 (9.0)	175 (10.7)	220 (12.2)	255 (13.3)	300 (14.2)	335 (15.1)	375 (16.0)
Drag Line Optimum Cut (Ft.)	Moist loam, sandy clay	3400	40	130 (6.6)	180 (7.4)	220 (8.0)	250 (8.5)	290 (9.0)	325 (9.5)	385 (10.0)
	Sand and gravel	3100	18	130 (6.6)	175 (7.4)	210 (8.0)	245 (8.5)	280 (9.0)	315 (9.5)	375 (10.0)
	Common earth	2800	30	110 (8.0)	160 (9.0)	190 (9.9)	220 (10.5)	250 (11.0)	280 (11.5)	310 (12.0)
	Clay, hard, dense	3000	33	90 (9.3)	130 (10.7)	160 (11.8)	190 (12.3)	225 (12.8)	250 (13.3)	280 (12.0)

Equipment	Soil Type	B.C.Y. Weight	% Swell	Wheel Loaders				Track Loaders		
				3 C.Y.	4 C.Y.	6 C.Y.	8 C.Y.	2-1/4 C.Y.	3 C.Y.	4 C.Y.
Loading Tractors	Moist loam, sandy clay	3400	40	260	340	510	690	135	180	250
	Sand and gravel	3100	18	245	320	480	650	130	170	235
	Common earth	2800	30	230	300	460	620	120	155	220
	Clay, hard, dense	3000	33	200	270	415	560	110	145	200
	Rock, well-blasted	4000	50	180	245	380	520	100	130	180

EARTHWORK — R3123 Excavation & Fill

R312319-90 Wellpoints

A single stage wellpoint system is usually limited to dewatering an average 15' depth below normal ground water level. Multi-stage systems are employed for greater depth with the pumping equipment installed only at the lowest header level. Ejectors with unlimited lift capacity can be economical when two or more stages of wellpoints can be replaced or when horizontal clearance is restricted, such as in deep trenches or tunneling projects, and where low water flows are expected. Wellpoints are usually spaced on 2-1/2' to 10' centers along a header pipe. Wellpoint spacing, header size, and pump size are all determined by the expected flow as dictated by soil conditions.

In almost all soils encountered in wellpoint dewatering, the wellpoints may be jetted into place. Cemented soils and stiff clays may require sand wicks about 12" in diameter around each wellpoint to increase efficiency and eliminate weeping into the excavation. These sand wicks require 1/2 to 3 C.Y. of washed filter sand and are installed by using a 12" diameter steel casing and hole puncher jetted into the ground 2' deeper than the wellpoint. Rock may require predrilled holes.

Labor required for the complete installation and removal of a single stage wellpoint system is in the range of 3/4 to 2 labor-hours per linear foot of header, depending upon jetting conditions, wellpoint spacing, etc.

Continuous pumping is necessary except in some free draining soil where temporary flooding is permissible (as in trenches which are backfilled after each day's work). Good practice requires provision of a stand-by pump during the continuous pumping operation.

Systems for continuous trenching below the water table should be installed three to four times the length of expected daily progress to ensure uninterrupted digging, and header pipe size should not be changed during the job.

For pervious free draining soils, deep wells in place of wellpoints may be economical because of lower installation and maintenance costs. Daily production ranges between two to three wells per day, for 25' to 40' depths, to one well per day for depths over 50'.

Detailed analysis and estimating for any dewatering problem is available at no cost from wellpoint manufacturers. Major firms will quote "sufficient equipment" quotes or their affiliates offer lump sum proposals to cover complete dewatering responsibility.

	Description for 200' System with 8" Header	Quantities
Equipment & Material	Wellpoints 25' long, 2" diameter @ 5' O.C.	40 Each
	Header pipe, 8" diameter	200 L.F.
	Discharge pipe, 8" diameter	100 L.F.
	8" valves	3 Each
	Combination jetting & wellpoint pump (standby)	1 Each
	Wellpoint pump, 8" diameter	1 Each
	Transportation to and from site	1 Day
	Fuel for 30 days x 60 gal./day	1800 Gallons
	Lubricants for 30 days x 16 lbs./day	480 Lbs.
	Sand for points	40 C.Y.
Labor	Technician to supervise installation	1 Week
	Labor for installation and removal of system	300 Labor-hours
	4 Operators straight time 40 hrs./wk. for 4.33 wks.	693 Hrs.
	4 Operators overtime 2 hrs./wk. for 4.33 wks.	35 Hrs.

R312323-30 Compacting Backfill

Compaction of fill in embankments, around structures, in trenches, and under slabs is important to control settlement. Factors affecting compaction are:

1. Soil gradation
2. Moisture content
3. Equipment used
4. Depth of fill per lift
5. Density required

Production Rate:

$$\frac{1.75' \text{ plate width} \times 50 \text{ F.P.M.} \times 50 \text{ min./hr.} \times .67' \text{ lift}}{27 \text{ C.F. per C.Y.}} = 108.5 \text{ C.Y./hr.}$$

Production Rate for 4 Passes:

$$\frac{108.5 \text{ C.Y.}}{4 \text{ passes}} = 27.125 \text{ C.Y./hr.} \times 8 \text{ hrs.} = 217 \text{ C.Y./day}$$

Example:

Compact granular fill around a building foundation using a 21" wide x 24" vibratory plate in 8" lifts. Operator moves at 50 F.P.M. working a 50 minute hour to develop 95% Modified Proctor Density with 4 passes.

Earthwork — R3141 Shoring

R314116-40 Wood Sheet Piling

Wood sheet piling may be used for depths to 20' where there is no ground water. If moderate ground water is encountered Tongue & Groove sheeting will help to keep it out. When considerable ground water is present, steel sheeting must be used.

For estimating purposes on trench excavation, sizes are as follows:

Depth	Sheeting	Wales	Braces	B.F. per S.F.
To 8'	3 x 12's	6 x 8's, 2 line	6 x 8's, @ 10'	4.0 @ 8'
8' x 12'	3 x 12's	10 x 10's, 2 line	10 x 10's, @ 9'	5.0 average
12' to 20'	3 x 12's	12 x 12's, 3 line	12 x 12's, @ 8'	7.0 average

Sheeting to be toed in at least 2' depending upon soil conditions. A five person crew with an air compressor and sheeting driver can drive and brace 440 SF/day at 8' deep, 360 SF/day at 12' deep, and 320 SF/day at 16' deep.

For normal soils, piling can be pulled in 1/3 the time to install. Pulling difficulty increases with the time in the ground. Production can be increased by high pressure jetting.

R314116-45 Steel Sheet Piling

Limiting weights are 22 to 38#/S.F. of wall surface with 27#/S.F. average for usual types and sizes. (Weights of piles themselves are from 30.7#/L.F. to 57#/L.F. but they are 15" to 21" wide.) Lightweight sections 12" to 28" wide from 3 ga. to 12 ga. thick are also available for shallow excavations. Piles may be driven two at a time with an impact or vibratory hammer (use vibratory to pull) hung from a crane without leads. A reasonable estimate of the life of steel sheet piling is 10 uses with up to 125 uses possible if a vibratory hammer is used. Used piling costs from 50% to 80% of new piling depending on location and market conditions. Sheet piling and H piles can be rented for about 30% of the delivered mill price for the first month and 5% per month thereafter. Allow 1 labor-hour per pile for cleaning and trimming after driving. These costs increase with depth and hydrostatic head. Vibratory drivers are faster in wet granular soils and are excellent for pile extraction. Pulling difficulty increases with the time in the ground and may cost more than driving. It is often economical to abandon the sheet piling, especially if it can be used as the outer wall form. Allow about 1/3 additional length or more for toeing into ground. Add bracing, waler and strut costs. Waler costs can equal the cost per ton of sheeting.

Earthwork — R3145 Vibroflotation & Densification

R314513-90 Vibroflotation and Vibro Replacement Soil Compaction

Vibroflotation is a proprietary system of compacting sandy soils in place to increase relative density to about 70%. Typical bearing capacities attained will be 6000 psf for saturated sand and 12,000 psf for dry sand. Usual range is 4000 to 8000 psf capacity. Costs in the front of the book are for a vertical foot of compacted cylinder 6' to 10' in diameter.

Vibro replacement is a proprietary system of improving cohesive soils in place to increase bearing capacity. Most silts and clays above or below the water table can be strengthened by installation of stone columns.

The process consists of radial displacement of the soil by vibration. The created hole is then backfilled in stages with coarse granular fill which is thoroughly compacted and displaced into the surrounding soil in the form of a column.

The total project cost would depend on the number and depth of the compacted cylinders. The installing company guarantees relative soil density of the sand cylinders after compaction and the bearing capacity of the soil after the replacement process. Detailed estimating information is available from the installer at no cost.

Earthwork — R3163 Drilled Caissons

R316326-60 Caissons

The three principal types of cassions are:

(1) Belled Caissons, which except for shallow depths and poor soil conditions, are generally recommended. They provide more bearing than shaft area. Because of its conical shape, no horizontal reinforcement of the bell is required.

(2) Straight Shaft Caissons are used where relatively light loads are to be supported by caissons that rest on high value bearing strata. While the shaft is larger in diameter than for belled types this is more than offset by the saving in time and labor.

(3) Keyed Caissons are used when extremely heavy loads are to be carried. A keyed or socketed caisson transfers its load into rock by a combination of end-bearing and shear reinforcing of the shaft. The most economical shaft often consists of a steel casing, a steel wide flange core and concrete. Allowable compressive stresses of .225 f'c for concrete, 16,000 psi for the wide flange core, and 9,000 psi for the steel casing are commonly used. The usual range of shaft diameter is 18" to 84". The number of sizes specified for any one project should be limited due to the problems of casing and auger storage. When hand work is to be performed, shaft diameters should not be less than 32". When inspection of borings is required a minimum shaft diameter of 30" is recommended. Concrete caissons are intended to be poured against earth excavation so permanent forms which add to cost should not be used if the excavation is clean and the earth sufficiently impervious to prevent excessive loss of concrete.

Soil Conditions for Belling		
Good	Requires Handwork	Not Recommended
Clay	Hard Shale	Silt
Sandy Clay	Limestone	Sand
Silty Clay	Sandstone	Gravel
Clayey Silt	Weathered Mica	Igneous Rock
Hard-pan		
Soft Shale		
Decomposed Rock		

Exterior Improvements — R3292 Turf & Grasses

R329219-50 Seeding

The type of grass is determined by light, shade and moisture content of soil plus intended use. Fertilizer should be disked 4" before seeding. For steep slopes disk five tons of mulch and lay two tons of hay or straw on surface per acre after seeding. Surface mulch can be staked, lightly disked or tar emulsion sprayed. Material for mulch can be wood chips, peat moss, partially rotted hay or straw, wood fibers and sprayed emulsions. Hemp seed blankets with fertilizer are also available. For spring seeding, watering is necessary. Late fall seeding may have to be reseeded in the spring. Hydraulic seeding, power mulching, and aerial seeding can be used on large areas.

Utilities — R3311 Water Utility Distribution Piping

R331113-80 Piping Designations

There are several systems currently in use to describe pipe and fittings. The following paragraphs will help to identify and clarify classifications of piping systems used for water distribution.

Piping may be classified by schedule. Piping schedules include 5S, 10S, 10, 20, 30, Standard, 40, 60, Extra Strong, 80, 100, 120, 140, 160 and Double Extra Strong. These schedules are dependent upon the pipe wall thickness. The wall thickness of a particular schedule may vary with pipe size.

Ductile iron pipe for water distribution is classified by Pressure Classes such as Class 150, 200, 250, 300 and 350. These classes are actually the rated water working pressure of the pipe in pounds per square inch (psi). The pipe in these pressure classes is designed to withstand the rated water working pressure plus a surge allowance of 100 psi.

The American Water Works Association (AWWA) provides standards for various types of **plastic pipe**. C-900 is the specification for polyvinyl chloride (PVC) piping used for water distribution in sizes ranging from 4" through 12". C-901 is the specification for polyethylene (PE) pressure pipe, tubing and fittings used for water distribution in sizes ranging from 1/2" through 3". C-905 is the specification for PVC piping sizes 14" and greater.

PVC pressure-rated pipe is identified using the standard dimensional ratio (SDR) method. This method is defined by the American Society for Testing and Materials (ASTM) Standard D 2241. This pipe is available in SDR numbers 64, 41, 32.5, 26, 21, 17, and 13.5. Pipe with an SDR of 64 will have the thinnest wall while pipe with an SDR of 13.5 will have the thickest wall. When the pressure rating (PR) of a pipe is given in psi, it is based on a line supplying water at 73 degrees F.

The National Sanitation Foundation (NSF) seal of approval is applied to products that can be used with potable water. These products have been tested to ANSI/NSF Standard 14.

Valves and strainers are classified by American National Standards Institute (ANSI) Classes. These Classes are 125, 150, 200, 250, 300, 400, 600, 900, 1500 and 2500. Within each class there is an operating pressure range dependent upon temperature. Design parameters should be compared to the appropriate material dependent, pressure-temperature rating chart for accurate valve selection.

Transportation — R3472 Railway Construction

R347216-10 Single Track R.R. Siding

The costs for a single track RR siding in the Unit Price section include the components shown in the table below.

Description of Component	Qty. per L.F. of Track	Unit
Ballast, 1-1/2" crushed stone	.667	C.Y.
6" x 8" x 8'-6" Treated timber ties, 22" O.C.	.545	Ea.
Tie plates, 2 per tie	1.091	Ea.
Track rail	2.000	L.F.
Spikes, 6", 4 per tie	2.182	Ea.
Splice bars w/ bolts, lock washers & nuts, @ 33' O.C.	.061	Pair
Crew B-14 @ 57 L.F./Day	.018	Day

R347216-20 Single Track, Steel Ties, Concrete Bed

The costs for a R.R. siding with steel ties and a concrete bed in the Unit Price section include the components shown in the table below.

Description of Component	Qty. per L.F. of Track	Unit
Concrete bed, 9' wide, 10" thick	.278	C.Y.
Ties, W6x16 x 6'-6" long, @ 30" O.C.	.400	Ea.
Tie plates, 4 per tie	1.600	Ea.
Track rail	2.000	L.F.
Tie plate bolts, 1", 8 per tie	3.200	Ea.
Splice bars w/bolts, lock washers & nuts, @ 33' O.C.	.061	Pair
Crew B-14 @ 22 L.F./Day	.045	Day

Change Orders

Change Order Considerations

A Change Order is a written document, usually prepared by the design professional, and signed by the owner, the architect/engineer and the contractor. A change order states the agreement of the parties to: an addition, deletion, or revision in the work; an adjustment in the contract sum, if any; or an adjustment in the contract time, if any. Change orders, or "extras" in the construction process occur after execution of the construction contract and impact architects/engineers, contractors and owners.

Change orders that are properly recognized and managed can ensure orderly, professional and profitable progress for all who are involved in the project. There are many causes for change orders and change order requests. In all cases, change orders or change order requests should be addressed promptly and in a precise and prescribed manner. The following paragraphs include information regarding change order pricing and procedures.

The Causes of Change Orders

Reasons for issuing change orders include:

- Unforeseen field conditions that require a change in the work
- Correction of design discrepancies, errors or omissions in the contract documents
- Owner-requested changes, either by design criteria, scope of work, or project objectives
- Completion date changes for reasons unrelated to the construction process
- Changes in building code interpretations, or other public authority requirements that require a change in the work
- Changes in availability of existing or new materials and products

Procedures

Properly written contract documents must include the correct change order procedures for all parties—owners, design professionals and contractors—to follow in order to avoid costly delays and litigation.

Being "in the right" is not always a sufficient or acceptable defense. The contract provisions requiring notification and documentation must be adhered to within a defined or reasonable time frame.

The appropriate method of handling change orders is by a written proposal and acceptance by all parties involved. Prior to starting work on a project, all parties should identify their authorized agents who may sign and accept change orders, as well as any limits placed on their authority.

Time may be a critical factor when the need for a change arises. For such cases, the contractor might be directed to proceed on a "time and materials" basis, rather than wait for all paperwork to be processed—a delay that could impede progress. In this situation, the contractor must still follow the prescribed change order procedures, including but not limited to, notification and documentation.

All forms used for change orders should be dated and signed by the proper authority. Lack of documentation can be very costly, especially if legal judgments are to be made and if certain field personnel are no longer available. For time and material change orders, the contractor should keep accurate daily records of all labor and material allocated to the change. Forms that can be used to document change order work are available in *Means Forms for Building Construction Professionals*.

Owners or awarding authorities who do considerable and continual building construction (such as the federal government) realize the inevitability of change orders for numerous reasons, both predictable and unpredictable. As a result, the federal government, the American Institute of Architects (AIA), the Engineers Joint Contract Documents Committee (EJCDC) and other contractor, legal and technical organizations have developed standards and procedures to be followed by all parties to achieve contract continuance and timely completion, while being financially fair to all concerned.

In addition to the change order standards put forth by industry associations, there are also many books available on the subject.

Pricing Change Orders

When pricing change orders, regardless of their cause, the most significant factor is when the change occurs. The need for a change may be perceived in the field or requested by the architect/engineer *before* any of the actual installation has begun, or may evolve or appear *during* construction when the item of work in question is partially installed. In the latter cases, the original sequence of construction is disrupted, along with all contiguous and supporting systems. Change orders cause the greatest impact when they occur *after* the installation has been completed and must be uncovered, or even replaced. Post-completion changes may be caused by necessary design changes, product failure, or changes in the owner's requirements that are not discovered until the building or the systems begin to function.

Specified procedures of notification and record keeping must be adhered to and enforced regardless of the stage of construction: *before*, *during*, or *after* installation. Some bidding documents anticipate change orders by requiring that unit prices including overhead and profit percentages—for additional as well as deductible changes—be listed. Generally these unit prices do not fully take into account the ripple effect, or impact on other trades, and should be used for general guidance only.

When pricing change orders, it is important to classify the time frame in which the change occurs. There are two basic time frames for change orders: *pre-installation change orders*, which occur before the start of construction, and *post-installation change orders*, which involve reworking after the original installation. Change orders that occur between these stages may be priced according to the extent of work completed using a combination of techniques developed for pricing *pre-* and *post-installation* changes.

The following factors are the basis for a check list to use when preparing a change order estimate.

Factors To Consider When Pricing Change Orders

As an estimator begins to prepare a change order, the following questions should be reviewed to determine their impact on the final price.

General

- Is the change order work *pre-installation* or *post-installation*?

Change order work costs vary according to how much of the installation has been completed. Once workers have the project scoped in their mind, even though they have not started, it can be difficult to refocus.

Consequently they may spend more than the normal amount of time understanding the change. Also, modifications to work in place such as trimming or refitting usually take more time than was initially estimated. The greater the amount of work in place, the more reluctant workers are to change it. Psychologically they may resent the change and as a result the rework takes longer than normal. Post-installation change order estimates must include demolition of existing work as required to accomplish the change. If the work is performed at a later time, additional obstacles such as building finishes may be present which must be protected. Regardless of whether the change occurs pre-installation or post-installation, attempt to isolate the identifiable factors and price them separately. For example, add shipping costs that may be required pre-installation or any demolition required post-installation. Then analyze the potential impact on productivity of psychological and/or learning curve factors and adjust the output rates accordingly. One approach is to break down the typical workday into segments and quantify the impact on each segment. The following chart may be useful as a guide:

	Activities (Productivity) Expressed as Percentages of a Workday		
Task	Means Mechanical Cost Data (for New Construction)	Pre-Installation Change Orders	Post-Installation Change Orders
1. Study plans	3%	6%	6%
2. Material procurement	3%	3%	3%
3. Receiving and storing	3%	3%	3%
4. Mobilization	5%	5%	5%
5. Site movement	5%	5%	8%
6. Layout and marking	8%	10%	12%
7. Actual installation	64%	59%	54%
8. Clean-up	3%	3%	3%
9. Breaks—non-productive	6%	6%	6%
Total	100%	100%	100%

Change Order Installation Efficiency

The labor-hours expressed (for new construction) are based on average installation time, using an efficiency level of approximately 60-65%. For change order situations, adjustments to this efficiency level should reflect the daily labor-hour allocation for that particular occurrence.

If any of the specific percentages expressed in the above chart do not apply to a particular project situation, then those percentage points should be reallocated to the appropriate task(s). Example: Using data for new construction, assume there is no new material being utilized. The percentages for Tasks 2 and 3 would therefore be reallocated to other tasks. If the time required for Tasks 2 and 3 can now be applied to installation, we can add the time allocated for *Material Procurement* and *Receiving and Storing* to the *Actual Installation* time for new construction, thereby increasing the Actual Installation percentage.

This chart shows that, due to reduced productivity, labor costs will be higher than those for new construction by 5% to 15% for pre-installation change orders and by 15% to 25% for post-installation change orders. Each job and change order is unique and must be examined individually. Many factors, covered elsewhere in this section, can each have a significant impact on productivity and change order costs. All such factors should be considered in every case.

- Will the change substantially delay the original completion date?

  A significant change in the project may cause the original completion date to be extended. The extended schedule may subject the contractor to new wage rates dictated by relevant labor contracts. Project supervision and other project overhead must also be extended beyond the original completion date. The schedule extension may also put installation into a new weather season. For example, underground piping scheduled for October installation was delayed until January. As a result, frost penetrated the trench area, thereby changing the degree of difficulty of the task. Changes and delays may have a ripple effect throughout the project. This effect must be analyzed and negotiated with the owner.

- What is the net effect of a deduct change order?

  In most cases, change orders resulting in a deduction or credit reflect only bare costs. The contractor may retain the overhead and profit based on the original bid.

Materials

- Will you have to pay more or less for the new material, required by the change order, than you paid for the original purchase?

  The same material prices or discounts will usually apply to materials purchased for change orders as new construction. In some instances, however, the contractor may forfeit the advantages of competitive pricing for change orders. Consider the following example:

  A contractor purchased over $20,000 worth of fan coil units for an installation, and obtained the maximum discount. Some time later it was determined the project required an additional matching unit. The contractor has to purchase this unit from the original supplier to ensure a match. The supplier at this time may not discount the unit because of the small quantity, and the fact that he is no longer in a competitive situation. The impact of quantity on purchase can add between 0% and 25% to material prices and/or subcontractor quotes.

- If materials have been ordered or delivered to the job site, will they be subject to a cancellation charge or restocking fee?

  Check with the supplier to determine if ordered materials are subject to a cancellation charge. Delivered materials not used as result of a change order may be subject to a restocking fee if returned to the supplier. Common restocking charges run between 20% and 40%. Also, delivery charges to return the goods to the supplier must be added.

Labor

- How efficient is the existing crew at the actual installation?

  Is the same crew that performed the initial work going to do the change order? Possibly the change consists of the installation of a unit identical to one already installed; therefore the change should take less time. Be sure to consider this potential productivity increase and modify the productivity rates accordingly.

- If the crew size is increased, what impact will that have on supervision requirements?

  Under most bargaining agreements or management practices, there is a point at which a working foreman is replaced by a nonworking foreman. This replacement increases project overhead by adding a nonproductive worker. If additional workers are added to accelerate the project or to perform changes while maintaining the schedule, be sure to add additional supervision time if warranted. Calculate the hours involved and the additional cost directly if possible.

- What are the other impacts of increased crew size?

  The larger the crew, the greater the potential for productivity to decrease. Some of the factors that cause this productivity loss are: overcrowding (producing restrictive conditions in the working space), and possibly a shortage of any special tools and equipment required. Such factors affect not only the crew working on the elements directly involved in the change order, but other crews whose movement may also be hampered.

  As the crew increases, check its basic composition for changes by the addition or deletion of apprentices or nonworking foreman and quantify the potential effects of equipment shortages or other logistical factors.

- As new crews, unfamiliar with the project, are brought onto the site, how long will it take them to become oriented to the project requirements?

  The orientation time for a new crew to become 100% effective varies with the site and type of project. Orientation is easiest at a new construction site, and most difficult at existing, very restrictive renovation sites. The type of work also affects orientation time. When all elements of the work are exposed, such as concrete or masonry work, orientation is decreased. When the work is concealed or less visible, such as existing electrical systems, orientation takes longer. Usually orientation can be accomplished in one day or less. Costs for added orientation should be itemized and added to the total estimated cost.

- How much actual production can be gained by working overtime?

  Short term overtime can be used effectively to accomplish more work in a day. However, as overtime is scheduled to run beyond several weeks, studies have shown marked decreases in output. The following chart shows the effect of long term overtime on worker efficiency. If the anticipated change requires extended overtime to keep the job on schedule, these factors can be used as a guide to predict the impact on time and cost. Add project overhead, particularly supervision, that may also be incurred.

Days per Week	Hours per Day	Production Efficiency					Payroll Cost Factors	
		1 Week	2 Weeks	3 Weeks	4 Weeks	Average 4 Weeks	@ 1-1/2 Times	@ 2 Times
5	8	100%	100%	100%	100%	100%	100%	100%
	9	100	100	95	90	96.25	105.6	111.1
	10	100	95	90	85	91.25	110.0	120.0
	11	95	90	75	65	81.25	113.6	127.3
	12	90	85	70	60	76.25	116.7	133.3
6	8	100	100	95	90	96.25	108.3	116.7
	9	100	95	90	85	92.50	113.0	125.9
	10	95	90	85	80	87.50	116.7	133.3
	11	95	85	70	65	78.75	119.7	139.4
	12	90	80	65	60	73.75	122.2	144.4
7	8	100	95	85	75	88.75	114.3	128.6
	9	95	90	80	70	83.75	118.3	136.5
	10	90	85	75	65	78.75	121.4	142.9
	11	85	80	65	60	72.50	124.0	148.1
	12	85	75	60	55	68.75	126.2	152.4

Effects of Overtime

Caution: Under many labor agreements, Sundays and holidays are paid at a higher premium than the normal overtime rate.

The use of long-term overtime is counterproductive on almost any construction job; that is, the longer the period of overtime, the lower the actual production rate. Numerous studies have been conducted, and while they have resulted in slightly different numbers, all reach the same conclusion. The figure above tabulates the effects of overtime work on efficiency.

As illustrated, there can be a difference between the *actual* payroll cost per hour and the *effective* cost per hour for overtime work. This is due to the reduced production efficiency with the increase in weekly hours beyond 40. This difference between actual and effective cost results from overtime work over a prolonged period. Short-term overtime work does not result in as great a reduction in efficiency, and in such cases, effective cost may not vary significantly from the actual payroll cost. As the total hours per week are increased on a regular basis, more time is lost because of fatigue, lowered morale, and an increased accident rate.

As an example, assume a project where workers are working 6 days a week, 10 hours per day. From the figure above (based on productivity studies), the average effective productive hours over a four-week period are:

$$0.875 \times 60 = 52.5$$

Depending upon the locale and day of week, overtime hours may be paid at time and a half or double time. For time and a half, the overall (average) *actual* payroll cost (including regular and overtime hours) is determined as follows:

$$\frac{40 \text{ reg. hrs.} + (20 \text{ overtime hrs.} \times 1.5)}{60 \text{ hrs.}} = 1.167$$

Based on 60 hours, the payroll cost per hour will be 116.7% of the normal rate at 40 hours per week. However, because the effective production (efficiency) for 60 hours is reduced to the equivalent of 52.5 hours, the effective cost of overtime is calculated as follows:

For time and a half:

$$\frac{40 \text{ reg. hrs.} + (20 \text{ overtime hrs.} \times 1.5)}{52.5 \text{ hrs.}} = 1.33$$

Installed cost will be 133% of the normal rate (for labor).

Thus, when figuring overtime, the actual cost per unit of work will be higher than the apparent overtime payroll dollar increase, due to the reduced productivity of the longer workweek. These efficiency calculations are true only for those cost factors determined by hours worked. Costs that are applied weekly or monthly, such as equipment rentals, will not be similarly affected.

Equipment

- What equipment is required to complete the change order?

Change orders may require extending the rental period of equipment already on the job site, or the addition of special equipment brought in to accomplish the change work. In either case, the additional rental charges and operator labor charges must be added.

Summary

The preceding considerations and others you deem appropriate should be analyzed and applied to a change order estimate. The impact of each should be quantified and listed on the estimate to form an audit trail.

Change orders that are properly identified, documented, and managed help to ensure the orderly, professional and profitable progress of the work. They also minimize potential claims or disputes at the end of the project.

Square Foot Costs

Estimating Tips

- The cost figures in this Square Foot Cost section were derived from approximately 11,200 projects contained in the RSMeans database of completed construction projects. They include the contractor's overhead and profit, but do not generally include architectural fees or land costs. The figures have been adjusted to January of the current year. New projects are added to our files each year, and outdated projects are discarded. For this reason, certain costs may not show a uniform annual progression. In no case are all subdivisions of a project listed.

- These projects were located throughout the U.S. and reflect a tremendous variation in square foot (S.F.) and cubic foot (C.F.) costs. This is due to differences, not only in labor and material costs, but also in individual owners' requirements. For instance, a bank in a large city would have different features than one in a rural area. This is true of all the different types of buildings analyzed. Therefore, caution should be exercised when using these Square Foot costs. For example, for court houses, costs in the database are local court house costs and will not apply to the larger, more elaborate federal court houses. As a general rule, the projects in the 1/4 column do not include any site work or equipment, while the projects in the 3/4 column may include both equipment and site work. The median figures do not generally include site work.

- None of the figures "go with" any others. All individual cost items were computed and tabulated separately. Thus, the sum of the median figures for Plumbing, HVAC and Electrical will not normally total up to the total Mechanical and Electrical costs arrived at by separate analysis and tabulation of the projects.

- Each building was analyzed as to total and component costs and percentages. The figures were arranged in ascending order with the results tabulated as shown. The 1/4 column shows that 25% of the projects had lower costs and 75% had higher. The 3/4 column shows that 75% of the projects had lower costs and 25% had higher. The median column shows that 50% of the projects had lower costs and 50% had higher.

- There are two times when square foot costs are useful. The first is in the conceptual stage when no details are available. Then square foot costs make a useful starting point. The second is after the bids are in and the costs can be worked back into their appropriate units for information purposes. As soon as details become available in the project design, the square foot approach should be discontinued and the project priced as to its particular components. When more precision is required, or for estimating the replacement cost of specific buildings, the current edition of *RSMeans Square Foot Costs* should be used.

- In using the figures in this section, it is recommended that the median column be used for preliminary figures if no additional information is available. The median figures, when multiplied by the total city construction cost index figures (see City Cost Indexes) and then multiplied by the project size modifier at the end of this section, should present a fairly accurate base figure, which would then have to be adjusted in view of the estimator's experience, local economic conditions, code requirements, and the owner's particular requirements. There is no need to factor the percentage figures, as these should remain constant from city to city. All tabulations mentioning air conditioning had at least partial air conditioning.

- The editors of this book would greatly appreciate receiving cost figures on one or more of your recent projects, which would then be included in the averages for next year. All cost figures received will be kept confidential, except that they will be averaged with other similar projects to arrive at Square Foot cost figures for next year's book. See the last page of the book for details and the discount available for submitting one or more of your projects.

50 17 | Square Foot Costs

		50 17 00	S.F. Costs	UNIT	UNIT COSTS 1/4	UNIT COSTS MEDIAN	UNIT COSTS 3/4	% OF TOTAL 1/4	% OF TOTAL MEDIAN	% OF TOTAL 3/4	
01	0010	APARTMENTS Low Rise (1 to 3 story)		S.F.	60	75.50	101				01
	0020	Total project cost		C.F.	5.40	7.15	8.80				
	0100	Site work		S.F.	5.15	7	12.30	6.05%	10.55%	14.05%	
	0500	Masonry			1.18	2.75	4.75	1.54%	3.67%	6.35%	
	1500	Finishes			6.35	8.75	10.80	9.05%	10.75%	12.85%	
	1800	Equipment			1.96	2.97	4.42	2.73%	4.03%	5.95%	
	2720	Plumbing			4.68	6	7.65	6.65%	8.95%	10.05%	
	2770	Heating, ventilating, air conditioning			2.98	3.67	5.40	4.20%	5.60%	7.60%	
	2900	Electrical			3.47	4.61	6.20	5.20%	6.65%	8.40%	
	3100	Total: Mechanical & Electrical			12.05	15.30	19.10	15.90%	18.05%	23%	
	9000	Per apartment unit, total cost		Apt.	56,000	85,500	126,000				
	9500	Total: Mechanical & Electrical		"	10,600	16,600	21,700				
02	0010	APARTMENTS Mid Rise (4 to 7 story)		S.F.	79.50	96	119				02
	0020	Total project costs		C.F.	6.20	8.55	11.70				
	0100	Site work		S.F.	3.18	6.30	11.35	5.25%	6.70%	9.15%	
	0500	Masonry			5.30	7.30	10.40	5.10%	7.25%	10.50%	
	1500	Finishes			10	13.95	16.45	10.55%	13.45%	17.70%	
	1800	Equipment			2.43	3.73	4.75	2.54%	3.48%	4.31%	
	2500	Conveying equipment			1.80	2.22	2.65	1.94%	2.27%	2.69%	
	2720	Plumbing			4.67	7.45	8.30	5.70%	7.20%	8.95%	
	2900	Electrical			5.25	7.50	8.75	6.65%	7.20%	8.95%	
	3100	Total: Mechanical & Electrical			16.80	21	25.50	18.50%	21%	23%	
	9000	Per apartment unit, total cost		Apt.	90,000	106,000	176,000				
	9500	Total: Mechanical & Electrical		"	17,000	19,700	24,800				
03	0010	APARTMENTS High Rise (8 to 24 story)		S.F.	90	109	133				03
	0020	Total project costs		C.F.	8.75	10.70	13				
	0100	Site work		S.F.	3.27	5.30	7.40	2.58%	4.84%	6.15%	
	0500	Masonry			5.20	9.50	11.80	4.74%	9.65%	11.05%	
	1500	Finishes			10	12.50	14.80	9.75%	11.80%	13.70%	
	1800	Equipment			2.90	3.57	4.73	2.78%	3.49%	4.35%	
	2500	Conveying equipment			2.05	3.11	4.23	2.23%	2.78%	3.37%	
	2720	Plumbing			6.65	7.85	11	6.80%	7.20%	10.45%	
	2900	Electrical			6.20	7.85	10.60	6.45%	7.65%	8.80%	
	3100	Total: Mechanical & Electrical			18.55	23.50	28.50	17.95%	22.50%	24.50%	
	9000	Per apartment unit, total cost		Apt.	94,000	103,500	143,500				
	9500	Total: Mechanical & Electrical		"	20,300	23,200	24,500				
04	0010	AUDITORIUMS		S.F.	93	126	182				04
	0020	Total project costs		C.F.	5.85	8.15	11.70				
	2720	Plumbing		S.F.	5.95	8.20	10.40	5.85%	7.20%	8.70%	
	2900	Electrical			7.25	10.50	14.10	6.80%	8.95%	11.30%	
	3100	Total: Mechanical & Electrical			14.40	20.50	41.50	24.50%	30.50%	31.50%	
05	0010	AUTOMOTIVE SALES		S.F.	67.50	94	115				05
	0020	Total project costs		C.F.	4.57	5.50	7.10				
	2720	Plumbing		S.F.	3.16	5.50	6	2.89%	6.05%	6.50%	
	2770	Heating, ventilating, air conditioning			4.88	7.45	8.05	4.61%	10%	10.35%	
	2900	Electrical			5.60	8.60	11.90	7.40%	9.95%	12.40%	
	3100	Total: Mechanical & Electrical			15.60	22	28	19.15%	20.50%	26%	
06	0010	BANKS		S.F.	135	169	213				06
	0020	Total project costs		C.F.	9.70	13.20	17.45				
	0100	Site work		S.F.	15.55	23.50	34	7.85%	12.95%	17%	
	0500	Masonry			7.05	13.25	24.50	3.36%	6.95%	10.35%	
	1500	Finishes			12.05	16.45	21	5.85%	8.45%	11.25%	
	1800	Equipment			5.45	11.25	24	1.34%	5.95%	10.65%	
	2720	Plumbing			4.26	6.10	8.90	2.82%	3.90%	4.93%	
	2770	Heating, ventilating, air conditioning			8.10	10.80	14.40	4.86%	7.15%	8.50%	
	2900	Electrical			12.85	17.15	22.50	8.20%	10.20%	12.20%	
	3100	Total: Mechanical & Electrical			30	40.50	49	16.55%	19.45%	23%	
	3500	See also division 11020 & 11030 (MF2004 11 16 00 & 11 17 00)									

50 17 | Square Foot Costs

50 17 00 | S.F. Costs

				UNIT COSTS			% OF TOTAL		
			UNIT	1/4	MEDIAN	3/4	1/4	MEDIAN	3/4
13	0010	**CHURCHES**	S.F.	91.50	116	150			
	0020	Total project costs	C.F.	5.70	7.15	9.45			
	1800	Equipment	S.F.	1.10	2.62	5.55	.95%	2.24%	4.50%
	2720	Plumbing		3.57	4.99	7.35	3.51%	4.96%	6.25%
	2770	Heating, ventilating, air conditioning		8.35	10.85	15.40	7.50%	10%	12%
	2900	Electrical		7.70	10.55	14.15	7.30%	8.75%	10.95%
	3100	Total: Mechanical & Electrical		23.50	30.50	41.50	18.25%	22%	24.50%
	3500	See also division 11040 (MF2004 11 91 00)							
15	0010	**CLUBS, COUNTRY**	S.F.	98.50	118	149			
	0020	Total project costs	C.F.	7.90	9.65	13.35			
	2720	Plumbing	S.F.	6.30	8.80	20	5.60%	7.90%	10%
	2900	Electrical		7.75	10.60	13.80	7%	8.95%	11%
	3100	Total: Mechanical & Electrical		23.50	41	51.50	19%	26.50%	29.50%
17	0010	**CLUBS, SOCIAL Fraternal**	S.F.	78.50	113	151			
	0020	Total project costs	C.F.	4.90	7.45	8.85			
	2720	Plumbing	S.F.	4.93	6.15	9.30	5.60%	6.90%	8.55%
	2770	Heating, ventilating, air conditioning		7.10	8.60	11.05	8.20%	9.25%	14.40%
	2900	Electrical		6.25	9.70	11.75	6.50%	9.50%	10.55%
	3100	Total: Mechanical & Electrical		17.40	33	42	21%	23%	23.50%
18	0010	**CLUBS, Y.M.C.A.**	S.F.	98.50	133	163			
	0020	Total project costs	C.F.	4.54	7.60	11.30			
	2720	Plumbing	S.F.	6.20	12.40	13.90	5.65%	7.60%	10.85%
	2900	Electrical		7.90	10.20	14.50	6.25%	7.80%	10.20%
	3100	Total: Mechanical & Electrical		28	33	37.50	18.40%	22.50%	28.50%
19	0010	**COLLEGES Classrooms & Administration**	S.F.	109	144	196			
	0020	Total project costs	C.F.	7.95	11.30	17.80			
	0500	Masonry	S.F.	7.30	13.75	16.90	5.65%	8.25%	10.50%
	2720	Plumbing		5.40	10.55	19.35	5.10%	6.60%	8.95%
	2900	Electrical		8.95	13.65	17.15	7.70%	9.85%	12%
	3100	Total: Mechanical & Electrical		33	46	55	24%	28%	31.50%
21	0010	**COLLEGES Science, Engineering, Laboratories**	S.F.	185	216	263			
	0020	Total project costs	C.F.	10.60	15.50	17.55			
	1800	Equipment	S.F.	10.30	23.50	25.50	2%	6.45%	12.65%
	2900	Electrical		15.25	21	33.50	7.10%	9.40%	12.10%
	3100	Total: Mechanical & Electrical		56.50	67	104	28.50%	31.50%	41%
	3500	See also division 11600 (MF2004 11 53 00)							
23	0010	**COLLEGES Student Unions**	S.F.	118	165	194			
	0020	Total project costs	C.F.	6.60	8.65	11.10			
	3100	Total: Mechanical & Electrical	S.F.	31	48	57	23.50%	26%	29%
25	0010	**COMMUNITY CENTERS**	"	96.50	120	160			
	0020	Total project costs	C.F.	6.35	9.10	11.75			
	1800	Equipment	S.F.	2.45	4.17	6.55	1.69%	3.12%	5.60%
	2720	Plumbing		4.73	8.05	11.75	4.85%	7%	8.95%
	2770	Heating, ventilating, air conditioning		7.80	11.20	15.40	6.80%	10.35%	12.90%
	2900	Electrical		8.25	10.65	15.45	7.35%	9.10%	10.85%
	3100	Total: Mechanical & Electrical		29	34.50	49.50	20.50%	25.50%	32.50%
28	0010	**COURT HOUSES**	S.F.	140	161	188			
	0020	Total project costs	C.F.	10.75	12.90	16.25			
	2720	Plumbing	S.F.	6.70	9.35	13.40	5.95%	7.45%	8.20%
	2900	Electrical		13.75	16.25	19.60	8.55%	9.95%	11.55%
	3100	Total: Mechanical & Electrical		38	52	57	22.50%	29.50%	30.50%
30	0010	**DEPARTMENT STORES**	S.F.	52	70.50	88.50			
	0020	Total project costs	C.F.	2.78	3.61	4.90			
	2720	Plumbing	S.F.	1.62	2.04	3.10	1.82%	4.21%	5.90%
	2770	Heating, ventilating, air conditioning		4.72	7.30	10.95	8.20%	9.10%	14.80%

50 17 | Square Foot Costs

	50 17 00	S.F. Costs	UNIT	UNIT COSTS 1/4	UNIT COSTS MEDIAN	UNIT COSTS 3/4	% OF TOTAL 1/4	% OF TOTAL MEDIAN	% OF TOTAL 3/4	
30	2900	Electrical	S.F.	5.95	8.20	9.65	9.05%	12.15%	14.95%	30
	3100	Total: Mechanical & Electrical		10.50	13.40	23.50	13.20%	21.50%	50%	
31	0010	DORMITORIES Low Rise (1 to 3 story)	S.F.	97.50	125	156				31
	0020	Total project costs	C.F.	5.90	9	13.50				
	2720	Plumbing	S.F.	5.95	7.95	10	8.05%	9%	9.65%	
	2770	Heating, ventilating, air conditioning		6.25	7.50	10	4.61%	8.05%	10%	
	2900	Electrical		6.40	9.55	13.30	6.55%	8.90%	9.55%	
	3100	Total: Mechanical & Electrical		33	36	56.50	22.50%	26%	29%	
	9000	Per bed, total cost	Bed	41,800	46,300	99,500				
32	0010	DORMITORIES Mid Rise (4 to 8 story)	S.F.	121	158	193				32
	0020	Total project costs	C.F.	13.35	14.65	17.55				
	2900	Electrical	S.F.	12.85	14.60	19	8.20%	10.20%	11.10%	
	3100	Total: Mechanical & Electrical	"	33.50	37.50	72.50	19.50%	30.50%	37.50%	
	9000	Per bed, total cost	Bed	17,200	39,300	81,200				
34	0010	FACTORIES	S.F.	45.50	68	105				34
	0020	Total project costs	C.F.	2.94	4.38	7.25				
	0100	Site work	S.F.	5.25	9.55	15.10	6.95%	11.45%	17.95%	
	2720	Plumbing		2.47	4.59	7.60	3.73%	6.05%	8.10%	
	2770	Heating, ventilating, air conditioning		4.81	6.90	9.30	5.25%	8.45%	11.35%	
	2900	Electrical		5.70	9	13.75	8.10%	10.50%	14.20%	
	3100	Total: Mechanical & Electrical		16.25	21.50	33	21%	28.50%	35.50%	
36	0010	FIRE STATIONS	S.F.	90.50	124	162				36
	0020	Total project costs	C.F.	5.30	7.30	9.70				
	0500	Masonry	S.F.	12.80	24	31.50	8.60%	12.40%	16.45%	
	1140	Roofing		2.96	8	9.10	1.90%	4.94%	5.05%	
	1580	Painting		2.43	3.43	3.51	1.37%	1.57%	2.07%	
	1800	Equipment		1.92	2.72	6.55	.81%	2.25%	3.69%	
	2720	Plumbing		5.75	8.50	12.45	5.85%	7.35%	9.50%	
	2770	Heating, ventilating, air conditioning		4.95	7.85	12.30	4.86%	7.25%	9.25%	
	2900	Electrical		6.45	11.10	14.80	6.80%	8.75%	10.60%	
	3100	Total: Mechanical & Electrical		32	38.50	44.50	19.60%	23%	26%	
37	0010	FRATERNITY HOUSES and Sorority Houses	S.F.	91	117	160				37
	0020	Total project costs	C.F.	9.05	9.45	12.10				
	2720	Plumbing	S.F.	6.85	7.85	14.40	6.80%	8%	10.85%	
	2900	Electrical		6	12.95	15.85	6.60%	9.90%	10.65%	
	3100	Total: Mechanical & Electrical		16	23	27.75		15.10%	15.90%	
38	0010	FUNERAL HOMES	S.F.	96	131	237				38
	0020	Total project costs	C.F.	9.80	10.90	21				
	2900	Electrical	S.F.	4.23	7.75	8.50	3.58%	4.44%	5.95%	
	3100	Total: Mechanical & Electrical	"	15	22.50	30.50	12.90%	12.90%	12.90%	
39	0010	GARAGES, COMMERCIAL (Service)	S.F.	56	84	116				39
	0020	Total project costs	C.F.	3.65	5.25	7.65				
	1800	Equipment	S.F.	3.06	6.90	10.70	2.69%	4.62%	6.80%	
	2720	Plumbing		3.76	5.80	10.55	5.45%	7.85%	10.65%	
	2730	Heating & ventilating		4.94	6.75	9.30	5.25%	6.85%	8.20%	
	2900	Electrical		5.25	8	11.40	7.15%	9.25%	10.85%	
	3100	Total: Mechanical & Electrical		11.70	21.50	32.50	13.60%	17.40%	27%	
40	0010	GARAGES, MUNICIPAL (Repair)	S.F.	79.50	106	150				40
	0020	Total project costs	C.F.	4.98	6.30	10.85				
	0500	Masonry	S.F.	7.50	14.65	22.50	5.60%	9.15%	12.50%	
	2720	Plumbing		3.58	6.85	12.95	3.59%	6.70%	7.95%	
	2730	Heating & ventilating		6.10	8.85	17.10	6.15%	7.45%	13.50%	
	2900	Electrical		5.90	9.25	13.35	6.65%	8.15%	11.15%	
	3100	Total: Mechanical & Electrical		18.95	38	55	21.50%	25.50%	28.50%	

50 17 | Square Foot Costs

50 17 00 | S.F. Costs

				UNIT COSTS			% OF TOTAL		
			UNIT	1/4	MEDIAN	3/4	1/4	MEDIAN	3/4
41	0010	**GARAGES, PARKING**	S.F.	31	45.50	78			
	0020	Total project costs	C.F.	2.92	3.96	5.75			
	2720	Plumbing	S.F.	.88	1.36	2.10	1.72%	2.70%	3.85%
	2900	Electrical		1.70	2.09	3.28	4.33%	5.20%	6.30%
	3100	Total: Mechanical & Electrical	↓	3.48	4.85	6.05	7%	8.90%	11.05%
	3200								
	9000	Per car, total cost	Car	13,100	16,400	21,000			
43	0010	**GYMNASIUMS**	S.F.	86.50	115	147			
	0020	Total project costs	C.F.	4.30	5.85	7.15			
	1800	Equipment	S.F.	2.05	3.85	7.40	2.03%	3.30%	5.20%
	2720	Plumbing		5.45	6.75	8.35	4.95%	6.75%	7.75%
	2770	Heating, ventilating, air conditioning		5.85	8.95	17.95	5.80%	9.80%	11.10%
	2900	Electrical		6.55	8.85	11.05	6.60%	8.50%	10.30%
	3100	Total: Mechanical & Electrical	↓	23.50	33	39	20.50%	24%	29%
	3500	See also division 11480 (MF2004 11 67 00)							
46	0010	**HOSPITALS**	S.F.	164	203	300			
	0020	Total project costs	C.F.	12.55	15.60	22.50			
	1800	Equipment	S.F.	4.20	8.10	13.95	1.10%	2.68%	5%
	2720	Plumbing		14.25	19.95	25.50	7.60%	9.10%	10.85%
	2770	Heating, ventilating, air conditioning		21	27	36	7.80%	12.95%	16.65%
	2900	Electrical		18.05	23.50	36.50	9.85%	11.55%	13.90%
	3100	Total: Mechanical & Electrical	↓	51	68.50	110	27%	33.50%	37%
	9000	Per bed or person, total cost	Bed	135,500	217,000	290,000			
	9900	See also division 11700 (MF2004 11 71 00)							
48	0010	**HOUSING For the Elderly**	S.F.	81.50	103	127			
	0020	Total project costs	C.F.	5.80	8.05	10.30			
	0100	Site work	S.F.	6.05	8.95	12.90	5.05%	7.90%	12.10%
	0500	Masonry		2.48	9.25	13.55	1.30%	6.05%	11%
	1800	Equipment		1.97	2.71	4.32	1.88%	3.23%	4.43%
	2510	Conveying systems		1.98	2.66	3.61	1.78%	2.20%	2.81%
	2720	Plumbing		6.05	7.70	9.70	8.15%	9.55%	10.50%
	2730	Heating, ventilating, air conditioning		3.10	4.40	6.55	3.30%	5.60%	7.25%
	2900	Electrical		6.05	8.25	10.55	7.30%	8.50%	10.25%
	3100	Total: Mechanical & Electrical	↓	21	25	33	18.10%	22.50%	29%
	9000	Per rental unit, total cost	Unit	75,500	88,500	98,500			
	9500	Total: Mechanical & Electrical	"	16,900	19,400	22,600			
50	0010	**HOUSING Public (Low Rise)**	S.F.	68.50	95	124			
	0020	Total project costs	C.F.	6.10	7.60	9.45			
	0100	Site work	S.F.	8.70	12.55	20.50	8.35%	11.75%	16.50%
	1800	Equipment		1.86	3.04	4.63	2.26%	3.03%	4.24%
	2720	Plumbing		4.94	6.50	8.25	7.15%	9.05%	11.60%
	2730	Heating, ventilating, air conditioning		2.48	4.81	5.25	4.26%	6.05%	6.45%
	2900	Electrical		4.14	6.15	8.55	5.10%	6.55%	8.25%
	3100	Total: Mechanical & Electrical	↓	19.65	25.50	28.50	14.50%	17.55%	26.50%
	9000	Per apartment, total cost	Apt.	75,000	85,500	107,500			
	9500	Total: Mechanical & Electrical	"	16,000	19,800	21,900			
51	0010	**ICE SKATING RINKS**	S.F.	58.50	137	150			
	0020	Total project costs	C.F.	4.30	4.40	5.05			
	2720	Plumbing	S.F.	2.19	4.10	4.19	3.12%	3.23%	5.65%
	2900	Electrical		6.25	9.60	10.15	6.30%	10.15%	15.05%
	3100	Total: Mechanical & Electrical	↓	10.45	14.75	18.45	18.95%	18.95%	18.95%
52	0010	**JAILS**	S.F.	178	230	297			
	0020	Total project costs	C.F.	16.05	22.50	27.50			
	1800	Equipment	S.F.	6.95	20.50	35	2.80%	5.55%	11.90%
	2720	Plumbing		18.15	23	30.50	7%	8.90%	13.35%
	2770	Heating, ventilating, air conditioning		16.05	21.50	41.50	7.50%	9.45%	17.75%
	2900	Electrical	↓	18.55	24.50	30.50	8.20%	11.55%	14.70%

50 17 | Square Foot Costs

50 17 00 | S.F. Costs

			UNIT	UNIT COSTS 1/4	MEDIAN	3/4	% OF TOTAL 1/4	MEDIAN	3/4	
52	3100	Total: Mechanical & Electrical	S.F.	47.50	89	105	27.50%	30%	34%	52
53	0010	**LIBRARIES**	S.F.	113	141	186				53
	0020	Total project costs	C.F.	7.70	9.65	12.30				
	0500	Masonry	S.F.	8.80	15.60	26	5.80%	7.80%	12.35%	
	1800	Equipment		1.54	4.14	6.25	.41%	1.50%	4.16%	
	2720	Plumbing		4.16	6.05	8.20	3.38%	4.60%	5.70%	
	2770	Heating, ventilating, air conditioning		9.20	15.60	20.50	7.80%	10.95%	12.80%	
	2900	Electrical		11.30	14.75	18.65	8.30%	10.25%	11.95%	
	3100	Total: Mechanical & Electrical		34	43	53.50	19.65%	22.50%	26.50%	
54	0010	**LIVING, ASSISTED**	S.F.	104	123	145				54
	0020	Total project costs	C.F.	8.75	10.25	11.65				
	0500	Masonry	S.F.	3.06	3.65	4.29	2.37%	3.16%	3.86%	
	1800	Equipment		2.37	2.75	3.53	2.12%	2.45%	2.66%	
	2720	Plumbing		8.75	11.70	12.10	6.05%	8.15%	10.60%	
	2770	Heating, ventilating, air conditioning		10.35	10.85	11.85	7.95%	9.35%	9.70%	
	2900	Electrical		10.20	11.25	13.05	9%	10%	10.70%	
	3100	Total: Mechanical & Electrical		28.50	33.50	38.50	26%	29%	31.50%	
55	0010	**MEDICAL CLINICS**	S.F.	106	131	166				55
	0020	Total project costs	C.F.	7.75	10.05	13.35				
	1800	Equipment	S.F.	2.87	6	9.35	1.05%	2.94%	6.35%	
	2720	Plumbing		7.05	9.90	13.25	6.15%	8.40%	10.10%	
	2770	Heating, ventilating, air conditioning		8.40	11	16.20	6.65%	8.85%	11.35%	
	2900	Electrical		9.10	12.95	16.90	8.10%	10%	12.25%	
	3100	Total: Mechanical & Electrical		29	39.50	54	22.50%	27%	33.50%	
	3500	See also division 11700 (MF2004 11 71 00)								
57	0010	**MEDICAL OFFICES**	S.F.	100	123	151				57
	0020	Total project costs	C.F.	7.45	10.10	13.65				
	1800	Equipment	S.F.	3.46	6.50	9.25	.98%	5.10%	7.05%	
	2720	Plumbing		5.50	8.50	11.45	5.60%	6.80%	8.50%	
	2770	Heating, ventilating, air conditioning		6.65	9.80	12.70	6.15%	8.05%	9.70%	
	2900	Electrical		8	11.60	16.20	7.60%	9.80%	11.70%	
	3100	Total: Mechanical & Electrical		21.50	31	47	19.35%	23%	30.50%	
59	0010	**MOTELS**	S.F.	63	91	119				59
	0020	Total project costs	C.F.	5.60	7.50	12.25				
	2720	Plumbing	S.F.	6.40	8.10	9.70	9.45%	10.60%	12.55%	
	2770	Heating, ventilating, air conditioning		3.89	5.80	10.40	5.60%	5.60%	10%	
	2900	Electrical		5.95	7.50	9.35	7.45%	9.05%	10.45%	
	3100	Total: Mechanical & Electrical		20	25.50	43.50	18.50%	24%	25.50%	
	5000									
	9000	Per rental unit, total cost	Unit	32,000	61,000	66,000				
	9500	Total: Mechanical & Electrical	"	6,250	9,450	11,000				
60	0010	**NURSING HOMES**	S.F.	98.50	127	156				60
	0020	Total project costs	C.F.	7.75	9.70	13.25				
	1800	Equipment	S.F.	3.11	4.12	6.85	2.02%	3.62%	4.99%	
	2720	Plumbing		8.45	12.80	15.45	8.75%	10.10%	12.70%	
	2770	Heating, ventilating, air conditioning		8.90	13.50	17.90	9.70%	11.45%	11.80%	
	2900	Electrical		9.75	12.20	16.60	9.40%	10.55%	12.45%	
	3100	Total: Mechanical & Electrical		23.50	32.50	54.50	26%	29.50%	30.50%	
	9000	Per bed or person, total cost	Bed	43,800	55,000	70,500				
61	0010	**OFFICES Low Rise (1 to 4 story)**	S.F.	83	107	139				61
	0020	Total project costs	C.F.	5.95	8.20	10.75				
	0100	Site work	S.F.	6.45	11.05	16.40	5.90%	9.70%	13.55%	
	0500	Masonry	"	2.86	6.35	11.75	2.62%	5.45%	8.20%	
	1800	Equipment	S.F.	.88	1.73	4.71	.73%	1.50%	3.66%	
	2720	Plumbing		2.95	4.59	6.70	3.66%	4.50%	6.10%	
	2770	Heating, ventilating, air conditioning		6.60	9.15	13.40	7.20%	10.30%	11.70%	
	2900	Electrical		6.75	9.65	13.60	7.45%	9.65%	11.40%	

50 17 | Square Foot Costs

50 17 00 | S.F. Costs

				UNIT COSTS			% OF TOTAL		
			UNIT	1/4	MEDIAN	3/4	1/4	MEDIAN	3/4
61	3100	Total: Mechanical & Electrical	S.F.	17.60	25	37	18%	22.50%	27%
62	0010	**OFFICES Mid Rise (5 to 10 story)**	S.F.	88	107	141			
	0020	Total project costs	C.F.	6.25	7.95	11.30			
	2720	Plumbing	S.F.	2.66	4.12	5.90	2.83%	3.74%	4.50%
	2770	Heating, ventilating, air conditioning		6.70	9.55	15.25	7.65%	9.40%	11%
	2900	Electrical		6.55	8.35	11.60	6.35%	7.80%	10%
	3100	Total: Mechanical & Electrical		16.70	21.50	41.50	18.95%	21%	27.50%
63	0010	**OFFICES High Rise (11 to 20 story)**	S.F.	108	136	168			
	0020	Total project costs	C.F.	7.55	9.45	13.55			
	2900	Electrical	S.F.	6.55	8	11.90	5.80%	7.85%	10.50%
	3100	Total: Mechanical & Electrical	"	21	28.50	48	16.90%	23.50%	34%
64	0010	**POLICE STATIONS**	S.F.	130	166	212			
	0020	Total project costs	C.F.	10.30	12.65	17.30			
	0500	Masonry	S.F.	12.20	21.50	27	6.70%	10.55%	11.35%
	1800	Equipment		2.06	9.05	14.35	1.43%	4.07%	6.70%
	2720	Plumbing		7.25	14.45	18	5.65%	6.90%	10.75%
	2770	Heating, ventilating, air conditioning		11.30	15.05	20.50	5.85%	10.55%	11.70%
	2900	Electrical		14.10	20.50	27	9.80%	11.85%	14.80%
	3100	Total: Mechanical & Electrical		46.50	56	75.50	28.50%	32%	32.50%
65	0010	**POST OFFICES**	S.F.	102	126	161			
	0020	Total project costs	C.F.	6.15	7.80	8.85			
	2720	Plumbing	S.F.	4.61	5.70	7.20	4.24%	5.30%	5.60%
	2770	Heating, ventilating, air conditioning		7.20	8.90	9.90	6.65%	7.15%	9.35%
	2900	Electrical		8.45	11.90	14.10	7.25%	9%	11%
	3100	Total: Mechanical & Electrical		24.50	32	36	16.25%	18.80%	22%
66	0010	**POWER PLANTS**	S.F.	710	940	1,725			
	0020	Total project costs	C.F.	19.55	42.50	91			
	2900	Electrical	S.F.	50	106	159	9.30%	12.75%	21.50%
	8100	Total: Mechanical & Electrical	"	125	405	910	32.50%	32.50%	52.50%
67	0010	**RELIGIOUS EDUCATION**	S.F.	82.50	108	132			
	0020	Total project costs	C.F.	4.58	6.55	8.20			
	2720	Plumbing	S.F.	3.45	4.89	6.85	4.40%	5.30%	7.10%
	2770	Heating, ventilating, air conditioning		8.70	9.85	13.95	10.05%	11.45%	12.35%
	2900	Electrical		6.55	9.25	12.20	7.60%	9.05%	10.35%
	3100	Total: Mechanical & Electrical		26.50	35	42.50	22%	23%	27%
69	0010	**RESEARCH Laboratories and Facilities**	S.F.	123	177	258			
	0020	Total project costs	C.F.	9.65	18.55	22			
	1800	Equipment	S.F.	5.50	10.80	26.50	.90%	4.58%	8.80%
	2720	Plumbing		12.40	15.85	25.50	6.15%	8.30%	10.80%
	2770	Heating, ventilating, air conditioning		11.10	37.50	44	7.25%	16.50%	17.50%
	2900	Electrical		14.20	24	40.50	9.45%	11.15%	15.40%
	3100	Total: Mechanical & Electrical		43.50	83.50	118	29.50%	37%	45.50%
70	0010	**RESTAURANTS**	S.F.	119	153	200			
	0020	Total project costs	C.F.	10.05	13.20	17.30			
	1800	Equipment	S.F.	7.75	19	28.50	6.10%	13%	15.65%
	2720	Plumbing		9.50	11.50	15.10	6.10%	8.15%	9%
	2770	Heating, ventilating, air conditioning		12.05	16.65	22	9.20%	12%	12.40%
	2900	Electrical		12.65	15.60	20.50	8.35%	10.55%	11.55%
	3100	Total: Mechanical & Electrical		39	41.50	53	19.25%	24%	29.50%
	9000	Per seat unit, total cost	Seat	4,400	5,850	6,925			
	9500	Total: Mechanical & Electrical	"	1,100	1,450	1,725			
72	0010	**RETAIL STORES**	S.F.	55.50	75	99			
	0020	Total project costs	C.F.	3.79	5.40	7.50			
	2720	Plumbing	S.F.	2.03	3.38	5.75	3.26%	4.60%	6.80%
	2770	Heating, ventilating, air conditioning		4.38	6	9	6.75%	8.75%	10.15%
	2900	Electrical		5	6.85	9.95	7.25%	9.90%	11.65%
	3100	Total: Mechanical & Electrical		12.85	17.15	23	17.05%	21%	23.50%

50 17 | Square Foot Costs

50 17 00 | S.F. Costs

			UNIT	UNIT COSTS			% OF TOTAL		
				1/4	MEDIAN	3/4	1/4	MEDIAN	3/4
74	0010	**SCHOOLS Elementary**	S.F.	90	111	137			
	0020	Total project costs	C.F.	5.95	7.65	9.85			
	0500	Masonry	S.F.	8.15	13.50	20.50	5.80%	10.65%	14.95%
	1800	Equipment		2.65	4.41	8.30	1.90%	3.38%	4.98%
	2720	Plumbing		5.25	7.40	9.90	5.70%	7.15%	9.35%
	2730	Heating, ventilating, air conditioning		7.85	12.50	17.45	8.15%	10.80%	14.90%
	2900	Electrical		8.50	11.15	14	8.40%	10%	11.70%
	3100	Total: Mechanical & Electrical	↓	30	38	45.50	24.50%	27.50%	30%
	9000	Per pupil, total cost	Ea.	10,500	15,600	46,200			
	9500	Total: Mechanical & Electrical	"	2,950	3,750	13,400			
76	0010	**SCHOOLS Junior High & Middle**	S.F.	92.50	115	137			
	0020	Total project costs	C.F.	5.95	7.70	8.65			
	0500	Masonry	S.F.	11.70	15.10	17.65	8%	11.60%	14.30%
	1800	Equipment		3	4.84	7.45	1.81%	3.26%	4.96%
	2720	Plumbing		5.95	6.75	8.75	5.40%	6.80%	7.25%
	2770	Heating, ventilating, air conditioning		10.95	13.30	23.50	9%	11.80%	17.45%
	2900	Electrical		9.20	11.10	14.25	7.90%	9.30%	10.60%
	3100	Total: Mechanical & Electrical	↓	28.50	38	47.50	23%	26.50%	29.50%
	9000	Per pupil, total cost	Ea.	11,900	15,600	21,000			
78	0010	**SCHOOLS Senior High**	S.F.	96	118	149			
	0020	Total project costs	C.F.	6.10	8.70	14.40			
	1800	Equipment	S.F.	2.57	6.25	8.95	1.86%	3.22%	4.80%
	2720	Plumbing		5.60	8.30	15.35	5.70%	7%	8.35%
	2770	Heating, ventilating, air conditioning		11.20	12.85	24.50	8.95%	11.60%	15%
	2900	Electrical		9.80	12.50	19.30	8.45%	10.05%	11.95%
	3100	Total: Mechanical & Electrical	↓	32.50	37.50	64	23%	26.50%	28.50%
	9000	Per pupil, total cost	Ea.	9,225	18,800	23,500			
80	0010	**SCHOOLS Vocational**	S.F.	79	112	140			
	0020	Total project costs	C.F.	4.92	7.05	9.75			
	0500	Masonry	S.F.	4.65	11.50	17.55	3.53%	4.61%	10.95%
	1800	Equipment	"	2.37	3.26	8.50	1.24%	3.13%	4.68%
	2720	Plumbing	S.F.	5.05	7.55	11.10	5.40%	6.90%	8.55%
	2770	Heating, ventilating, air conditioning		7.10	13.20	22	8.60%	11.90%	14.65%
	2900	Electrical		8.25	10.80	15.40	8.45%	10.95%	13.20%
	3100	Total: Mechanical & Electrical	↓	28.50	31.50	54.50	23.50%	29.50%	31%
	9000	Per pupil, total cost	Ea.	11,000	29,500	44,000			
83	0010	**SPORTS ARENAS**	S.F.	69	92.50	142			
	0020	Total project costs	C.F.	3.76	6.75	8.70			
	2720	Plumbing	S.F.	4.01	6.10	12.85	4.35%	6.35%	9.40%
	2770	Heating, ventilating, air conditioning		8.65	10.20	14.20	8.80%	10.20%	13.55%
	2900	Electrical		7.20	9.80	12.65	8.60%	9.90%	12.25%
	3100	Total: Mechanical & Electrical	↓	17.95	32	42	21.50%	25%	27.50%
85	0010	**SUPERMARKETS**	S.F.	64	74	87			
	0020	Total project costs	C.F.	3.56	4.30	6.50			
	2720	Plumbing	S.F.	3.57	4.50	5.25	5.40%	6%	7.45%
	2770	Heating, ventilating, air conditioning		5.25	7	8.50	8.60%	8.65%	9.60%
	2900	Electrical		8	9.20	10.90	10.40%	12.45%	13.60%
	3100	Total: Mechanical & Electrical	↓	20.50	22.50	31	20.50%	26.50%	31%
86	0010	**SWIMMING POOLS**	S.F.	103	173	370			
	0020	Total project costs	C.F.	8.30	10.35	11.30			
	2720	Plumbing	S.F.	9.55	10.95	15	4.80%	9.70%	20.50%
	2900	Electrical		7.80	12.60	18.35	6.50%	7.25%	7.60%
	3100	Total: Mechanical & Electrical	↓	18.95	48	66	11.15%	14.10%	23.50%
87	0010	**TELEPHONE EXCHANGES**	S.F.	137	201	255			
	0020	Total project costs	C.F.	8.55	13.75	18.85			
	2720	Plumbing	S.F.	5.80	9.20	13.10	4.52%	5.80%	6.90%
	2770	Heating, ventilating, air conditioning	↓	13.45	27	33.50	11.80%	16.05%	18.40%

50 17 | Square Foot Costs

		50 17 00	S.F. Costs	UNIT	UNIT COSTS 1/4	UNIT COSTS MEDIAN	UNIT COSTS 3/4	% OF TOTAL 1/4	% OF TOTAL MEDIAN	% OF TOTAL 3/4	
87	2900		Electrical	S.F.	14	22	39	10.90%	14%	17.85%	87
	3100		Total: Mechanical & Electrical	↓	41.50	78.50	111	29.50%	33.50%	44.50%	
91	0010	**THEATERS**		S.F.	86.50	107	163				91
	0020		Total project costs	C.F.	3.99	5.90	8.70				
	2720		Plumbing	S.F.	2.88	3.12	12.75	2.92%	4.70%	6.80%	
	2770		Heating, ventilating, air conditioning		8.40	10.15	12.60	8%	12.25%	13.40%	
	2900		Electrical		7.55	10.20	21	8.05%	9.95%	12.25%	
	3100		Total: Mechanical & Electrical	↓	19.45	29	59.50	23%	26.50%	27.50%	
94	0010	**TOWN HALLS City Halls & Municipal Buildings**		S.F.	100	127	159				94
	0020		Total project costs	C.F.	8.80	10.75	14.85				
	2720		Plumbing	S.F.	4.03	7.55	13.85	4.31%	5.95%	7.95%	
	2770		Heating, ventilating, air conditioning		7.30	14.45	21	7.05%	9.05%	13.45%	
	2900		Electrical		9.15	13.40	17.85	8.05%	9.50%	12.05%	
	3100		Total: Mechanical & Electrical	↓	31	36	48	22%	26.50%	31%	
97	0010	**WAREHOUSES & Storage Buildings**		S.F.	37.50	54	77				97
	0020		Total project costs	C.F.	1.95	2.95	4.88				
	0100		Site work	S.F.	3.72	7.40	11.15	6.05%	12.95%	19.85%	
	0500		Masonry		2.25	5.10	11.05	3.73%	7.40%	12.30%	
	1800		Equipment		.58	1.25	7	.91%	1.82%	5.55%	
	2720		Plumbing		1.20	2.15	4.03	2.90%	4.80%	6.55%	
	2730		Heating, ventilating, air conditioning		1.37	3.86	5.20	2.41%	5%	8.90%	
	2900		Electrical		2.13	4.01	6.65	5.15%	7.20%	10.10%	
	3100		Total: Mechanical & Electrical	↓	5.95	9.10	19.95	12.75%	18.90%	26%	
99	0010	**WAREHOUSE & OFFICES Combination**		S.F.	44	59	79.50				99
	0020		Total project costs	C.F.	2.27	3.29	4.86				
	1800		Equipment	S.F.	.77	1.48	2.21	.52%	1.21%	2.40%	
	2720		Plumbing		1.71	3.03	4.44	3.74%	4.76%	6.30%	
	2770		Heating, ventilating, air conditioning		2.70	4.22	5.90	5%	5.65%	10.05%	
	2900		Electrical		2.97	4.40	6.95	5.85%	8%	10%	
	3100		Total: Mechanical & Electrical	↓	8.25	12.60	20	14.40%	19.95%	24.50%	

Square Foot Project Size Modifier

One factor that affects the S.F. cost of a particular building is the size. In general, for buildings built to the same specifications in the same locality, the larger building will have the lower S.F. cost. This is due mainly to the decreasing contribution of the exterior walls plus the economy of scale usually achievable in larger buildings. The Area Conversion Scale shown below will give a factor to convert costs for the typical size building to an adjusted cost for the particular project.

The Square Foot Base Size lists the median costs, most typical project size in our accumulated data, and the range in size of the projects.

The Size Factor for your project is determined by dividing your project area in S.F. by the typical project size for the particular Building Type. With this factor, enter the Area Conversion Scale at the appropriate Size Factor and determine the appropriate cost multiplier for your building size.

Example: Determine the cost per S.F. for a 100,000 S.F. Mid-rise apartment building.

$$\frac{\text{Proposed building area} = 100,000 \text{ S.F.}}{\text{Typical size from below} = 50,000 \text{ S.F.}} = 2.00$$

Enter Area Conversion scale at 2.0, intersect curve, read horizontally the appropriate cost multiplier of .94. Size adjusted cost becomes .94 x $96.00 = $90.25 based on national average costs.

Note: For Size Factors less than .50, the Cost Multiplier is 1.1
For Size Factors greater than 3.5, the Cost Multiplier is .90

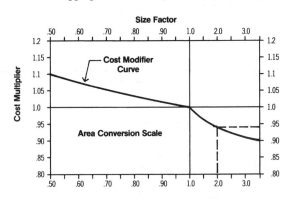

Square Foot Base Size

Building Type	Median Cost per S.F.	Typical Size Gross S.F.	Typical Range Gross S.F.	Building Type	Median Cost per S.F.	Typical Size Gross S.F.	Typical Range Gross S.F.
Apartments, Low Rise	$75.50	21,000	9,700 - 37,200	Jails	$230.00	40,000	5,500 - 145,000
Apartments, Mid Rise	96.00	50,000	32,000 - 100,000	Libraries	141.00	12,000	7,000 - 31,000
Apartments, High Rise	109.00	145,000	95,000 - 600,000	Living, Assisted	123.00	32,300	23,500 - 50,300
Auditoriums	126.00	25,000	7,600 - 39,000	Medical Clinics	131.00	7,200	4,200 - 15,700
Auto Sales	94.00	20,000	10,800 - 28,600	Medical Offices	123.00	6,000	4,000 - 15,000
Banks	169.00	4,200	2,500 - 7,500	Motels	91.00	40,000	15,800 - 120,000
Churches	116.00	17,000	2,000 - 42,000	Nursing Homes	127.00	23,000	15,000 - 37,000
Clubs, Country	118.00	6,500	4,500 - 15,000	Offices, Low Rise	107.00	20,000	5,000 - 80,000
Clubs, Social	113.00	10,000	6,000 - 13,500	Offices, Mid Rise	107.00	120,000	20,000 - 300,000
Clubs, YMCA	133.00	28,300	12,800 - 39,400	Offices, High Rise	136.00	260,000	120,000 - 800,000
Colleges (Class)	144.00	50,000	15,000 - 150,000	Police Stations	166.00	10,500	4,000 - 19,000
Colleges (Science Lab)	216.00	45,600	16,600 - 80,000	Post Offices	126.00	12,400	6,800 - 30,000
College (Student Union)	165.00	33,400	16,000 - 85,000	Power Plants	940.00	7,500	1,000 - 20,000
Community Center	120.00	9,400	5,300 - 16,700	Religious Education	108.00	9,000	6,000 - 12,000
Court Houses	161.00	32,400	17,800 - 106,000	Research	177.00	19,000	6,300 - 45,000
Dept. Stores	70.50	90,000	44,000 - 122,000	Restaurants	153.00	4,400	2,800 - 6,000
Dormitories, Low Rise	125.00	25,000	10,000 - 95,000	Retail Stores	75.00	7,200	4,000 - 17,600
Dormitories, Mid Rise	158.00	85,000	20,000 - 200,000	Schools, Elementary	111.00	41,000	24,500 - 55,000
Factories	68.00	26,400	12,900 - 50,000	Schools, Jr. High	115.00	92,000	52,000 - 119,000
Fire Stations	124.00	5,800	4,000 - 8,700	Schools, Sr. High	118.00	101,000	50,500 - 175,000
Fraternity Houses	117.00	12,500	8,200 - 14,800	Schools, Vocational	112.00	37,000	20,500 - 82,000
Funeral Homes	131.00	10,000	4,000 - 20,000	Sports Arenas	92.50	15,000	5,000 - 40,000
Garages, Commercial	84.00	9,300	5,000 - 13,600	Supermarkets	74.00	44,000	12,000 - 60,000
Garages, Municipal	106.00	8,300	4,500 - 12,600	Swimming Pools	173.00	20,000	10,000 - 32,000
Garages, Parking	45.50	163,000	76,400 - 225,300	Telephone Exchange	201.00	4,500	1,200 - 10,600
Gymnasiums	115.00	19,200	11,600 - 41,000	Theaters	107.00	10,500	8,800 - 17,500
Hospitals	203.00	55,000	27,200 - 125,000	Town Halls	127.00	10,800	4,800 - 23,400
House (Elderly)	103.00	37,000	21,000 - 66,000	Warehouses	54.00	25,000	8,000 - 72,000
Housing (Public)	95.00	36,000	14,400 - 74,400	Warehouse & Office	59.00	25,000	8,000 - 72,000
Ice Rinks	137.00	29,000	27,200 - 33,600				

Abbreviations

A	Area Square Feet; Ampere	Cab.	Cabinet	Dcmob.	Demobilization
ABS	Acrylonitrile Butadiene Stryrene; Asbestos Bonded Steel	Cair.	Air Tool Laborer	d.f.u.	Drainage Fixture Units
A.C.	Alternating Current; Air-Conditioning; Asbestos Cement; Plywood Grade A & C	Calc	Calculated	D.H.	Double Hung
		Cap.	Capacity	DHW	Domestic Hot Water
		Carp.	Carpenter	Diag.	Diagonal
		C.B.	Circuit Breaker	Diam.	Diameter
A.C.I.	American Concrete Institute	C.C.A.	Chromate Copper Arsenate	Distrib.	Distribution
AD	Plywood, Grade A & D	C.C.F.	Hundred Cubic Feet	Dk.	Deck
Addit.	Additional	cd	Candela	D.L.	Dead Load; Diesel
Adj.	Adjustable	cd/sf	Candela per Square Foot	DLH	Deep Long Span Bar Joist
af	Audio-frequency	CD	Grade of Plywood Face & Back	Do.	Ditto
A.G.A.	American Gas Association	CDX	Plywood, Grade C & D, exterior glue	Dp.	Depth
Agg.	Aggregate			D.P.S.T.	Double Pole, Single Throw
A.H.	Ampere Hours	Cefi.	Cement Finisher	Dr.	Driver
A hr.	Ampere-hour	Cem.	Cement	Drink.	Drinking
A.H.U.	Air Handling Unit	CF	Hundred Feet	D.S.	Double Strength
A.I.A.	American Institute of Architects	C.F.	Cubic Feet	D.S.A.	Double Strength A Grade
AIC	Ampere Interrupting Capacity	CFM	Cubic Feet per Minute	D.S.B.	Double Strength B Grade
Allow.	Allowance	c.g.	Center of Gravity	Dty.	Duty
alt.	Altitude	CHW	Chilled Water; Commercial Hot Water	DWV	Drain Waste Vent
Alum.	Aluminum			DX	Deluxe White, Direct Expansion
a.m.	Ante Meridiem	C.I.	Cast Iron	dyn	Dyne
Amp.	Ampere	C.I.P.	Cast in Place	e	Eccentricity
Anod.	Anodized	Circ.	Circuit	E	Equipment Only; East
Approx.	Approximate	C.L.	Carload Lot	Ea.	Each
Apt.	Apartment	Clab.	Common Laborer	E.B.	Encased Burial
Asb.	Asbestos	Clam	Common maintenance laborer	Econ.	Economy
A.S.B.C.	American Standard Building Code	C.L.F.	Hundred Linear Feet	E.C.Y	Embankment Cubic Yards
Asbe.	Asbestos Worker	CLF	Current Limiting Fuse	EDP	Electronic Data Processing
A.S.H.R.A.E.	American Society of Heating, Refrig. & AC Engineers	CLP	Cross Linked Polyethylene	EIFS	Exterior Insulation Finish System
		cm	Centimeter	E.D.R.	Equiv. Direct Radiation
A.S.M.E.	American Society of Mechanical Engineers	CMP	Corr. Metal Pipe	Eq.	Equation
		C.M.U.	Concrete Masonry Unit	Elec.	Electrician; Electrical
A.S.T.M.	American Society for Testing and Materials	CN	Change Notice	Elev.	Elevator; Elevating
		Col.	Column	EMT	Electrical Metallic Conduit; Thin Wall Conduit
Attchmt.	Attachment	CO_2	Carbon Dioxide		
Avg.	Average	Comb.	Combination	Eng.	Engine, Engineered
A.W.G.	American Wire Gauge	Compr.	Compressor	EPDM	Ethylene Propylene Diene Monomer
AWWA	American Water Works Assoc.	Conc.	Concrete		
Bbl.	Barrel	Cont.	Continuous; Continued	EPS	Expanded Polystyrene
B&B	Grade B and Better; Balled & Burlapped	Corr.	Corrugated	Eqhv.	Equip. Oper., Heavy
		Cos	Cosine	Eqlt.	Equip. Oper., Light
B.&S.	Bell and Spigot	Cot	Cotangent	Eqmd.	Equip. Oper., Medium
B.&W.	Black and White	Cov.	Cover	Eqmm.	Equip. Oper., Master Mechanic
b.c.c.	Body-centered Cubic	C/P	Cedar on Paneling	Eqol.	Equip. Oper., Oilers
B.C.Y.	Bank Cubic Yards	CPA	Control Point Adjustment	Equip.	Equipment
BE	Bevel End	Cplg.	Coupling	ERW	Electric Resistance Welded
B.F.	Board Feet	C.P.M.	Critical Path Method	E.S.	Energy Saver
Bg. cem.	Bag of Cement	CPVC	Chlorinated Polyvinyl Chloride	Est.	Estimated
BHP	Boiler Horsepower; Brake Horsepower	C.Pr.	Hundred Pair	esu	Electrostatic Units
		CRC	Cold Rolled Channel	E.W.	Each Way
B.I.	Black Iron	Creos.	Creosote	EWT	Entering Water Temperature
Bit.; Bitum.	Bituminous	Crpt.	Carpet & Linoleum Layer	Excav.	Excavation
Bk.	Backed	CRT	Cathode-ray Tube	Exp.	Expansion, Exposure
Bkrs.	Breakers	CS	Carbon Steel, Constant Shear Bar Joist	Ext.	Exterior
Bldg.	Building			Extru.	Extrusion
Blk.	Block	Csc	Cosecant	f.	Fiber stress
Bm.	Beam	C.S.F.	Hundred Square Feet	F	Fahrenheit; Female; Fill
Boil.	Boilermaker	CSI	Construction Specifications Institute	Fab.	Fabricated
B.P.M.	Blows per Minute			FBGS	Fiberglass
BR	Bedroom	C.T.	Current Transformer	F.C.	Footcandles
Brg.	Bearing	CTS	Copper Tube Size	f.c.c.	Face-centered Cubic
Brhe.	Bricklayer Helper	Cu	Copper, Cubic	f'c.	Compressive Stress in Concrete; Extreme Compressive Stress
Bric.	Bricklayer	Cu. Ft.	Cubic Foot		
Brk.	Brick	cw	Continuous Wave	F.E.	Front End
Brng.	Bearing	C.W.	Cool White; Cold Water	FEP	Fluorinated Ethylene Propylene (Teflon)
Brs.	Brass	Cwt.	100 Pounds		
Brz.	Bronze	C.W.X.	Cool White Deluxe	F.G.	Flat Grain
Bsn.	Basin	C.Y.	Cubic Yard (27 cubic feet)	F.H.A.	Federal Housing Administration
Btr.	Better	C.Y./Hr.	Cubic Yard per Hour	Fig.	Figure
BTU	British Thermal Unit	Cyl.	Cylinder	Fin.	Finished
BTUH	BTU per Hour	d	Penny (nail size)	Fixt.	Fixture
B.U.R.	Built-up Roofing	D	Deep; Depth; Discharge	Fl. Oz.	Fluid Ounces
BX	Interlocked Armored Cable	Dis.;Disch.	Discharge	Flr.	Floor
c	Conductivity, Copper Sweat	Db.	Decibel	F.M.	Frequency Modulation; Factory Mutual
C	Hundred; Centigrade	Dbl.	Double		
		DC	Direct Current	Fmg.	Framing
C/C	Center to Center, Cedar on Cedar	DDC	Direct Digital Control	Fndtn.	Foundation

Abbreviations

Fori.	Foreman, Inside	I.W.	Indirect Waste	M.C.F.	Thousand Cubic Feet
Foro.	Foreman, Outside	J	Joule	M.C.F.M.	Thousand Cubic Feet per Minute
Fount.	Fountain	J.I.C.	Joint Industrial Council	M.C.M.	Thousand Circular Mils
FPM	Feet per Minute	K	Thousand; Thousand Pounds; Heavy Wall Copper Tubing, Kelvin	M.C.P.	Motor Circuit Protector
FPT	Female Pipe Thread			MD	Medium Duty
Fr.	Frame	K.A.H.	Thousand Amp. Hours	M.D.O.	Medium Density Overlaid
F.R.	Fire Rating	KCMIL	Thousand Circular Mils	Med.	Medium
FRK	Foil Reinforced Kraft	KD	Knock Down	MF	Thousand Feet
FRP	Fiberglass Reinforced Plastic	K.D.A.T.	Kiln Dried After Treatment	M.F.B.M.	Thousand Feet Board Measure
FS	Forged Steel	kg	Kilogram	Mfg.	Manufacturing
FSC	Cast Body; Cast Switch Box	kG	Kilogauss	Mfrs.	Manufacturers
Ft.	Foot; Feet	kgf	Kilogram Force	mg	Milligram
Ftng.	Fitting	kHz	Kilohertz	MGD	Million Gallons per Day
Ftg.	Footing	Kip.	1000 Pounds	MGPH	Thousand Gallons per Hour
Ft. Lb.	Foot Pound	KJ	Kiljoule	MH, M.H.	Manhole; Metal Halide; Man-Hour
Furn.	Furniture	K.L.	Effective Length Factor	MHz	Megahertz
FVNR	Full Voltage Non-Reversing	K.L.F.	Kips per Linear Foot	Mi.	Mile
FXM	Female by Male	Km	Kilometer	MI	Malleable Iron; Mineral Insulated
Fy.	Minimum Yield Stress of Steel	K.S.F.	Kips per Square Foot	mm	Millimeter
g	Gram	K.S.I.	Kips per Square Inch	Mill.	Millwright
G	Gauss	kV	Kilovolt	Min., min.	Minimum, minute
Ga.	Gauge	kVA	Kilovolt Ampere	Misc.	Miscellaneous
Gal.	Gallon	K.V.A.R.	Kilovar (Reactance)	ml	Milliliter, Mainline
Gal./Min.	Gallon per Minute	KW	Kilowatt	M.L.F.	Thousand Linear Feet
Galv.	Galvanized	KWh	Kilowatt-hour	Mo.	Month
Gen.	General	L	Labor Only; Length; Long; Medium Wall Copper Tubing	Mobil.	Mobilization
G.F.I.	Ground Fault Interrupter			Mog.	Mogul Base
Glaz.	Glazier	Lab.	Labor	MPH	Miles per Hour
GPD	Gallons per Day	lat	Latitude	MPT	Male Pipe Thread
GPH	Gallons per Hour	Lath.	Lather	MRT	Mile Round Trip
GPM	Gallons per Minute	Lav.	Lavatory	ms	Millisecond
GR	Grade	lb.; #	Pound	M.S.F.	Thousand Square Feet
Gran.	Granular	L.B.	Load Bearing; L Conduit Body	Mstz.	Mosaic & Terrazzo Worker
Grnd.	Ground	L. & E.	Labor & Equipment	M.S.Y.	Thousand Square Yards
H	High; High Strength Bar Joist; Henry	lb./hr.	Pounds per Hour	Mtd.	Mounted
		lb./L.F.	Pounds per Linear Foot	Mthe.	Mosaic & Terrazzo Helper
H.C.	High Capacity	lbf/sq.in.	Pound-force per Square Inch	Mtng.	Mounting
H.D.	Heavy Duty; High Density	L.C.L.	Less than Carload Lot	Mult.	Multi; Multiply
H.D.O.	High Density Overlaid	L.C.Y.	Loose Cubic Yard	M.V.A.	Million Volt Amperes
Hdr.	Header	Ld.	Load	M.V.A.R.	Million Volt Amperes Reactance
Hdwe.	Hardware	LE	Lead Equivalent	MV	Megavolt
Help.	Helper Average	LED	Light Emitting Diode	MW	Megawatt
HEPA	High Efficiency Particulate Air Filter	L.F.	Linear Foot	MXM	Male by Male
		Lg.	Long; Length; Large	MYD	Thousand Yards
Hg	Mercury	L & H	Light and Heat	N	Natural; North
HIC	High Interrupting Capacity	LH	Long Span Bar Joist	nA	Nanoampere
HM	Hollow Metal	L.H.	Labor Hours	NA	Not Available; Not Applicable
H.O.	High Output	L.L.	Live Load	N.B.C.	National Building Code
Horiz.	Horizontal	L.L.D.	Lamp Lumen Depreciation	NC	Normally Closed
H.P.	Horsepower; High Pressure	lm	Lumen	N.E.M.A.	National Electrical Manufacturers Assoc.
H.P.F.	High Power Factor	lm/sf	Lumen per Square Foot		
Hr.	Hour	lm/W	Lumen per Watt	NEHB	Bolted Circuit Breaker to 600V.
Hrs./Day	Hours per Day	L.O.A.	Length Over All	N.L.B.	Non-Load-Bearing
HSC	High Short Circuit	log	Logarithm	NM	Non-Metallic Cable
Ht.	Height	L-O-L	Lateralolet	nm	Nanometer
Htg.	Heating	L.P.	Liquefied Petroleum; Low Pressure	No.	Number
Htrs.	Heaters	L.P.F.	Low Power Factor	NO	Normally Open
HVAC	Heating, Ventilation & Air-Conditioning	LR	Long Radius	N.O.C.	Not Otherwise Classified
		L.S.	Lump Sum	Nose.	Nosing
Hvy.	Heavy	Lt.	Light	N.P.T.	National Pipe Thread
HW	Hot Water	Lt. Ga.	Light Gauge	NQOD	Combination Plug-on/Bolt on Circuit Breaker to 240V.
Hyd.; Hydr.	Hydraulic	L.T.L.	Less than Truckload Lot		
Hz.	Hertz (cycles)	Lt. Wt.	Lightweight	N.R.C.	Noise Reduction Coefficient
I.	Moment of Inertia	L.V.	Low Voltage	N.R.S.	Non Rising Stem
I.C.	Interrupting Capacity	M	Thousand; Material; Male; Light Wall Copper Tubing	ns	Nanosecond
ID	Inside Diameter			nW	Nanowatt
I.D.	Inside Dimension; Identification	M^2CA	Meters Squared Contact Area	OB	Opposing Blade
I.F.	Inside Frosted	m/hr; M.H.	Man-hour	OC	On Center
I.M.C.	Intermediate Metal Conduit	mA	Milliampere	OD	Outside Diameter
In.	Inch	Mach.	Machine	O.D.	Outside Dimension
Incan.	Incandescent	Mag. Str.	Magnetic Starter	ODS	Overhead Distribution System
Incl.	Included; Including	Maint.	Maintenance	O.G.	Ogee
Int.	Interior	Marb.	Marble Setter	O.H.	Overhead
Inst.	Installation	Mat; Mat'l.	Material	O&P	Overhead and Profit
Insul.	Insulation/Insulated	Max.	Maximum	Oper.	Operator
I.P.	Iron Pipe	MBF	Thousand Board Feet	Opng.	Opening
I.P.S.	Iron Pipe Size	MBH	Thousand BTU's per hr.	Orna.	Ornamental
I.P.T.	Iron Pipe Threaded	MC	Metal Clad Cable	OSB	Oriented Strand Board

Abbreviations

O.S.&Y.	Outside Screw and Yoke	Rsr	Riser	Th.;Thk.	Thick
Ovhd.	Overhead	RT	Round Trip	Thn.	Thin
OWG	Oil, Water or Gas	S.	Suction; Single Entrance; South	Thrded	Threaded
Oz.	Ounce	SC	Screw Cover	Tilf.	Tile Layer, Floor
P.	Pole; Applied Load; Projection	SCFM	Standard Cubic Feet per Minute	Tilh.	Tile Layer, Helper
p.	Page	Scaf.	Scaffold	THHN	Nylon Jacketed Wire
Pape.	Paperhanger	Sch.; Sched.	Schedule	THW.	Insulated Strand Wire
P.A.P.R.	Powered Air Purifying Respirator	S.C.R.	Modular Brick	THWN;	Nylon Jacketed Wire
PAR	Parabolic Reflector	S.D.	Sound Deadening	T.L.	Truckload
Pc., Pcs.	Piece, Pieces	S.D.R.	Standard Dimension Ratio	T.M.	Track Mounted
P.C.	Portland Cement; Power Connector	S.E.	Surfaced Edge	Tot.	Total
P.C.F.	Pounds per Cubic Foot	Sel.	Select	T-O-L	Threadolet
P.C.M.	Phase Contrast Microscopy	S.E.R.; S.E.U.	Service Entrance Cable	T.S.	Trigger Start
P.E.	Professional Engineer; Porcelain Enamel; Polyethylene; Plain End	S.F.	Square Foot	Tr.	Trade
		S.F.C.A.	Square Foot Contact Area	Transf.	Transformer
		S.F. Flr.	Square Foot of Floor	Trhv.	Truck Driver, Heavy
Perf.	Perforated	S.F.G.	Square Foot of Ground	Trlr	Trailer
Ph.	Phase	S.F. Hor.	Square Foot Horizontal	Trlt.	Truck Driver, Light
P.I.	Pressure Injected	S.F.R.	Square Feet of Radiation	TTY	Teletypewriter
Pile.	Pile Driver	S.F. Shlf.	Square Foot of Shelf	TV	Television
Pkg.	Package	S4S	Surface 4 Sides	T.W.	Thermoplastic Water Resistant Wire
Pl.	Plate	Shee.	Sheet Metal Worker		
Plah.	Plasterer Helper	Sin.	Sine	UCI	Uniform Construction Index
Plas.	Plasterer	Skwk.	Skilled Worker	UF	Underground Feeder
Pluh.	Plumbers Helper	SL	Saran Lined	UGND	Underground Feeder
Plum.	Plumber	S.L.	Slimline	U.H.F.	Ultra High Frequency
Ply.	Plywood	Sldr.	Solder	U.L.	Underwriters Laboratory
p.m.	Post Meridiem	SLH	Super Long Span Bar Joist	Unfin.	Unfinished
Pntd.	Painted	S.N.	Solid Neutral	URD	Underground Residential Distribution
Pord.	Painter, Ordinary	S-O-L	Socketolet		
pp	Pages	sp	Standpipe	US	United States
PP; PPL	Polypropylene	S.P.	Static Pressure; Single Pole; Self-Propelled	USP	United States Primed
P.P.M.	Parts per Million			UTP	Unshielded Twisted Pair
Pr.	Pair	Spri.	Sprinkler Installer	V	Volt
P.E.S.B.	Pre-engineered Steel Building	spwg	Static Pressure Water Gauge	V.A.	Volt Amperes
Prefab.	Prefabricated	S.P.D.T.	Single Pole, Double Throw	V.C.T.	Vinyl Composition Tile
Prefin.	Prefinished	SPF	Spruce Pine Fir	VAV	Variable Air Volume
Prop.	Propelled	S.P.S.T.	Single Pole, Single Throw	VC	Veneer Core
PSF; psf	Pounds per Square Foot	SPT	Standard Pipe Thread	Vent.	Ventilation
PSI; psi	Pounds per Square Inch	Sq.	Square; 100 Square Feet	Vert.	Vertical
PSIG	Pounds per Square Inch Gauge	Sq. Hd.	Square Head	V.F.	Vinyl Faced
PSP	Plastic Sewer Pipe	Sq. In.	Square Inch	V.G.	Vertical Grain
Pspr.	Painter, Spray	S.S.	Single Strength; Stainless Steel	V.H.F.	Very High Frequency
Psst.	Painter, Structural Steel	S.S.B.	Single Strength B Grade	VHO	Very High Output
P.T.	Potential Transformer	sst	Stainless Steel	Vib.	Vibrating
P. & T.	Pressure & Temperature	Sswk.	Structural Steel Worker	V.L.F.	Vertical Linear Foot
Ptd.	Painted	Sswl.	Structural Steel Welder	Vol.	Volume
Ptns.	Partitions	St.;Stl.	Steel	VRP	Vinyl Reinforced Polyester
Pu	Ultimate Load	S.T.C.	Sound Transmission Coefficient	W	Wire; Watt; Wide; West
PVC	Polyvinyl Chloride	Std.	Standard	w/	With
Pvmt.	Pavement	STK	Select Tight Knot	W.C.	Water Column; Water Closet
Pwr.	Power	STP	Standard Temperature & Pressure	W.F.	Wide Flange
Q	Quantity Heat Flow	Stpi.	Steamfitter, Pipefitter	W.G.	Water Gauge
Quan.;Qty.	Quantity	Str.	Strength; Starter; Straight	Wldg.	Welding
Q.C.	Quick Coupling	Strd	Stranded	W. Mile	Wire Mile
r	Radius of Gyration	Struct.	Structural	W-O-L	Weldolet
R	Resistance	Sty.	Story	W.R.	Water Resistant
R.C.P.	Reinforced Concrete Pipe	Subj.	Subject	Wrck.	Wrecker
Rect.	Rectangle	Subs.	Subcontractors	W.S.P.	Water, Steam, Petroleum
Reg.	Regular	Surf.	Surface	WT., Wt.	Weight
Reinf.	Reinforced	Sw.	Switch	WWF	Welded Wire Fabric
Req'd.	Required	Swbd.	Switchboard	XFER	Transfer
Res.	Resistant	S.Y.	Square Yard	XFMR	Transformer
Resi.	Residential	Syn.	Synthetic	XHD	Extra Heavy Duty
Rgh.	Rough	S.Y.P.	Southern Yellow Pine	XHHW; XLPE	Cross-Linked Polyethylene Wire Insulation
RGS	Rigid Galvanized Steel	Sys.	System		
R.H.W.	Rubber, Heat & Water Resistant; Residential Hot Water	t.	Thickness	XLP	Cross-linked Polyethylene
		T	Temperature; Ton	Y	Wye
rms	Root Mean Square	Tan	Tangent	yd	Yard
Rnd.	Round	T.C.	Terra Cotta	yr	Year
Rodm.	Rodman	T & C	Threaded and Coupled	Δ	Delta
Rofc.	Roofer, Composition	T.D.	Temperature Difference	%	Percent
Rofp.	Roofer, Precast	Tdd	Telecommunications Device for the Deaf	~	Approximately
Rohe.	Roofer Helpers (Composition)			Ø	Phase
Rots.	Roofer, Tile & Slate	T.E.M.	Transmission Electron Microscopy	@	At
R.O.W.	Right of Way	TFE	Tetrafluoroethylene (Teflon)	#	Pound; Number
RPM	Revolutions per Minute	T. & G.	Tongue & Groove; Tar & Gravel	<	Less Than
R.S.	Rapid Start			>	Greater Than

Index

A

Abatement asbestos 31, 32
ABC extinguisher 321, 322
Abrasive floor 66
   floor tile 276
   silicon carbide 66
   terrazzo 283
   tile 275
   tread 276
ABS DWV pipe 423
Absorber shock 425
Absorption testing 13
   water chillers 470
A/C packaged terminal 473
Accelerator sprinkler system .. 410
Access control 253, 517
   control card type 253
   control explosive detector ... 516
   control facial 254
   control hand scan 253
   control metal detector 516
   control X-ray machine 516
   door and panels 225
   door basement 69
   door duct 453, 455
   door fire rated 225
   door floor 225
   door metal 225
   door roof 206
   door stainless steel 225
   floor 287
   road and parking area 20
Accessories bathroom .. 275, 317, 318
   column formwork 50
   door 254
   drywall 274
   duct 454
   formwork hanger 52
   plaster 269
   reinforcing 55
   reinforcing steel 57
   roof 206
   toilet 317
   wall formwork 54
Accessory drainage 203
   fireplace 319
   formwork 50, 51, 53
   masonry 80
Accordion door 222
Acid proof floor 278, 284
   resistant pipe 442
Acoustic ceiling board 277
Acoustical batt 289
   block 89
   booth 391
   ceiling 277, 278
   door 229, 316
   enclosure 391, 393
   folding partition 316
   glass 259
   metal deck 123
   panel 289, 315
   partition 315, 316
   phone booth 310
   room 382, 383
   sealant 209, 273
   spray 207
   underlayment 289
   wall tile 278
   wallboard 273
   window wall 233
Acrylic ceiling 277
   floor 284
   rubber roofing 198
   sign 309

wall coating 302
wallcovering 287
wood block 279
Adhesive cement 282
   EPDM 180
   neoprene 180
   PVC 180
   roof 194
   wallpaper 287
Adjustable astragal 252
   jack post 110
Adjustment factor 8
Admixture cement 66
Admixtures concrete 40
Adobe brick 91
Aeration sewage 597
Aerator 441
Aerial photography 12
   seeding 562
   survey 24
Agent form release 40
Aggregate base course 547
   concrete 41
   exposed 66
   masonry 79
   panels exposed 192
   stone 563
   testing 13
Aids staging 18
Air balancing 445
   cleaner electronic 463
   compressor control system . 447
   compressor dental 352
   conditioner cooling & heating . 473
   conditioner direct expansion . 474
   conditioner fan coil 474
   conditioner gas heat 472
   conditioner packaged 473
   conditioner packaged terminal . 473
   conditioner receptacle ... 493
   conditioner removal 444
   conditioner rooftop 472
   conditioner thru-wall 473
   conditioner ventilating .. 453
   conditioner wiring 495
   conditioning 743
   conditioning computer 473
   conditioning fan 456
   conditioning ventilating .. 447, 454,
        456, 457, 461, 470, 471, 473
   control 450
   cooled belt drive condenser . 469
   cooled condensing unit ... 469
   cooled direct drive condenser . 470
   curtain 458, 459
   curtain door 458
   entraining agent 40
   extractor 453
   filter 462
   filter roll type 462
   filters washable 462
   filtration 31
   handling fan 456
   handling troffer 506
   handling unit 474
   lock 385
   make-up unit 472
   register 460
   return grille 460
   sampling 31
   supply pneumatic control .. 447
   supply register 460
   supported building ... 15, 384
   supported enclosure 387
   supp. storage tank cov demo . 376
   supported storage tank cover . 384

   supported structures 737
   tube system 405
   unit make-up 472
   vent automatic 450
   vent roof 207
   wall 316
Air-compressor mobilization .. 521
Airfoil fan centrifugal 457
Airless sprayer 31
Airplane hangar 392
   hangar door 228
Air-source heat pumps 473
Alarm burglar 516
   exit control 516
   residential 495
   sprinkler 517
   standpipe 517
   valve sprinkler 408
Alley bowling 349
Altar 355
Alteration fee 8
Alternating tread ladder 134
Aluminum astragal 252
   bench 373
   blind 358
   cable tray 488
   cast tread 138
   ceiling tile 278
   column 139, 171
   column base 139, 140
   column cap 140
   conduit 486
   coping 97
   cross 355
   curtain wall 235
   diffuser perforated 459
   directory board 308
   dome 391
   door 213, 217, 230, 231
   door frame 234
   downspout 202
   drip edge 205
   ductwork 453
   edging 564
   entrance 234
   expansion joint 204
   extrusion 117
   fence 553, 554
   flagpole 327
   flashing 191, 200, 428
   foil 183, 186, 290
   framing 117
   grandstand 387
   grating tread 138
   grating 136
   gravel stop 202
   grille 460
   gutter 203
   handrail 287
   joist shoring 53
   louver 259, 461
   mansard 191
   mesh grating 136
   nail 148
   operating louver 461
   pipe 576, 579
   plank grating 136
   plank scaffold 18
   register 460
   reglet 204
   rivet 108
   roof 189
   salvage 414
   sash 236
   screw 191
   service entrance cable .. 484

   shade 360
   sheet metal 202
   shelter 391
   shingle 187
   siding 191
   siding paint 294
   sign 309
   steeple 326
   storefront 234
   storm door 213
   structural 117
   tile 189, 276
   transom 234
   trench cover 138
   tube frame 232
   weatherstrip 252
   weld rod 109
   window 236
   window demolition 212
Ammeter 496
Analysis petrographic 13
   sieve 13
Anchor bolt 50, 79, 102
   brick 80
   buck 79
   bumper 331
   chemical 73, 104
   epoxy 73, 104
   expansion 104
   framing 150
   hollow wall 105
   jute fiber 105
   lead screw 106
   machinery 107
   masonry 79
   metal nailing 105
   nailing 105
   nylon nailing 105
   partition 80
   plastic screw 106
   rafter 150
   rigid 80
   roof safety 18
   screw 105
   self-drilling 105
   sill 150
   steel 80
   stone 80
   super high-tensile 535
   toggle-bolt 105
   wedge 106
Anechoic chamber 382
Angle corner 316
   curb edging 111
   fiberglass 173
   valve bronze 415
   valve plastic 417
Antenna system 512
Anti-siphon water 425
Apartment call system 513
Appliance 336
   plumbing 431, 432
   residential 334, 336, 337, 495
Approach ramp 287
Apron wood 165
Arch culverts oval 577
   laminated 164
   radial 164
Architectural equipment 336
   fee 8, 692
Area clean-up 34
   wall 326
Ark 355
Armored cable 482
Arrow 552

Index

Artificial turf 553
Asbestos abatement 31, 32
   demolition 33
   disposal 34
   felt 195
   removal 32, 33
   removal process 704
As-built drawings 22
Ash conveyor 354
   receiver 318, 366
   urn 366
Ashlar stone 94
   veneer 93
Asphalt base sheet 195
   block 550
   block floor 550
   coating 179
   concrete 520
   curb 551
   cutting 29
   felt 194, 196
   flashing 200
   flood coat 195
   mix design 13
   pavement 553
   primer 178, 282
   sealcoat rubberized 546
   sheathing 163
   shingle 186
   sidewalk 546
   testing 12
Asphaltic binder 547
   concrete 548
   emulsion 561
   pavement 547
   wearing course 548
Aspirator dental 352
Astragal adjustable 252
   aluminum 252
   magnetic 252
   molding 166
   one piece 252
   overlapping 252
   rubber 252
   split 252
   steel 252
Astronomy observatory 390
Athletic equipment 348, 555
   or recreational screening . 555
   paving 553
   pole 349
   post 350
Atomizer water 32
Attic stair 338
Audio masking 393
Audiometric room 382
Auditorium chair 371
Auger boring 25
   hole 24
Augering 520
Auto park drain 428
Autoclaved concrete block ... 85
Automatic fire suppression .. 410
   flush 435
   opener 245
   operator 245
   teller 333
   washing machine 337
Automotive equipment 330, 404, 448
   lift 404
Autopsy equipment 351, 353
Awning canvas 325
   window 236, 237, 325
Axial flow fan 456

B

Backer rod 209
   rod polyethylene 52
Backerboard 179, 271
Backfill 532, 533
   compaction 748
   dozer 532, 533
   general 532
   planting pit 561
   structural 532
   trench 524, 526
Backflow preventer 425
Backhoe 528
   excavation 523
Backsplash 363
   countertop 364
Backstop baseball 555
   basketball 348, 555
   electric 348
   handball court 555
   squash court 555
   tennis court 555
Backup block 86
Backwater valve 430
Baffle roof 207
Bag disposable 32
   glove 33
Baked enamel door 216
   enamel frame 214
Balanced door 234
Balancing air 445
   water 446
Balcony fire escape 134
Bale hay 534
Baler 354
Ball valve 417
   valve bronze 415
   valve plastic 417
Ballast fixture 507
   railroad 584
Baluster 169
Balustrade painting 300
Band joist framing 128
   molding 165
   riser 346
Banding iron 52
Bank counter 333
   equipment 333
   window 333
Bankrun gravel 520
Banquet booth 369
Baptistry 354
Bar bell 347
   chair reinforcing 56
   front 393
   grab 317
   grouted 61
   joist painting 121
   panic 248
   parallel 351
   restaurant 368
   spacer reinforcing 57
   tie reinforcing 55
   touch 247
   towel 318
   tub 318
   ungrouted 61
   Zee 274
Barbed wire fence 553
Barber equipment 334
Bark mulch 561
   mulch redwood 561
Barrel bolts 245
Barricade 20, 586
   tape 20

Barrier delineator 586
   impact 586
   median 586
   moisture 180
   parking 552
   security 332
   separation 32, 36
   waterstop 51
   X-ray 394
Barriers and enclosures 20
   crash 586
   highway sound 559
Base cabinet 360, 361
   carpet 286
   ceramic tile 275
   column 150
   course 547
   course aggregate 547
   cove 281
   gravel 547
   masonry 90
   molding 165
   plate column 116
   quarry tile 276
   resilient 281
   road 547
   sheet 195, 196
   sink 360
   stone 94, 547
   terrazzo 283
   vanity 362
   wood 165
Baseball backstop 555
   scoreboard 348
Baseboard demolition 148
   heat 475
   heat electric 476
   heater electric 476
Basement stair 69
Baseplate scaffold 17
   shoring 16
Basic finish materials 262
Basketball backstop 348, 555
   scoreboard 348
Batch trial 13
Bath paraffin 351
   steam 383
   whirlpool 352
Bathroom 434
   accessories 275, 317, 318
   exhaust fan 458
   faucet 435
   fixture 433, 434
   heater & fan 458
   soaking 434
Bathtub 434
   bar 318
   enclosure 313
   removal 414
Batt acoustical 289
   insulation 182
   thermal 289
Battery light 507
Bead blast demo 263
   board insulation 382
   casing 274
   corner 269, 274
   parting 167
Beam & girder framing 153
   and girder formwork 42
   bolster reinforcing 55
   bond 79, 86
   bondcrete 270
   bottom formwork 43
   castellated 116
   ceiling 170

   concrete 61, 62, 64
   drywall 272
   fiberglass wide flange 173
   fireproofing 207
   grade 65
   hanger 150
   laminated 164
   mantel 170
   plaster 269
   precast 68
   precast concrete 68
   precast concrete tee 68
   reinforcing 59
   side 43
   soldier 538
   test 13
   wood 153, 161
Bearing pad 108
Bed molding 165
Bedding brick 549
   pipe 533
Beech tread 170
Belgian block 551
Bell 347
   & spigot pipe 427
   bar 347
   caisson 542
   church 347
   system 503
Bellows expansion joint 450
Belt material handling 592
Bench aluminum 373
   fiberglass 373
   folding 371
   greenhouse 386
   park 373
   planter 372
   players 373
   wood 373
   work 346
Benches 373
Bentonite 180, 539
Berm pavement 551
   road 551
Bevel glass 255
   siding 193
Bicycle rack 349
   trainer 347, 348
Bi-fold door 217, 222
Bin parts 323
   retaining walls metal 558
Binder asphaltic 547
Bi-passing closet door 222
Birch door 222-224
   molding 166
   paneling 168
   stair 169
   wood frame 172
Bit drill 74
Bituminous block 550
   coating 179, 569
   concrete curbs 551
Blackboard 315
Blanket curing 68
   insulation 182, 446
   sound attenuation 289
Blast curtains 359
   demo bead 263
   floor shot 263
Blasting 528
   cap 528
   mat 528
   water 76
Bleachers 387
   outdoor 387
   stadium 387

771

Index

telescoping	371
Blind exterior	172
venetian	358
window	358
Block acoustical	89
asphalt	550
backup	86
belgian	551
bituminous	550
Block, brick and mortar	721
Block cap	90
chimney	85
column	86
concrete	85-89, 722
concrete bond beam	86
concrete exterior	88
corner	90
decorative concrete	86, 87
filler	300
floor	280
glass	91
glazed	90
grooved	86
ground face	87
hexagonal face	87
high strength	88
insulation	86
interlocking	88
lightweight	89
lintel	89
manhole	579
partition	89, 90
patio	94
planter	549
profile	87
reflective	91
removal	27
scored	87
scored split face	87
sill	90
slotted	89
slump	87
solar screen	90
split rib	87
wall removal	27
Blocking carpentry	153
steel	153
wood	153, 156
Blood pressure unit	351
Bloodbank refrigeration	345
Blower pneumatic tube	405
Blown in cellulose	183
in fiberglass	183
in insulation	183
Blueboard	270
partition	265
Bluegrass sod	562
Bluestone	93
sidewalk	546
sill	95
step	546
Board & batten siding	193
batten fence	557
bulletin	307
ceiling	277
control	308, 347
directory	308, 309
dock	331
drain	180, 181
gypsum	272-274
insulation	181
paneling	169
partition gypsum	264, 265
ridge	158
sheathing	163
valance	361

verge	166
Body harness	18
Boiler control system	447
demolition	444
electric	464
electric steam	464
gas fired	465
gas/oil combination	465
general	414
hot water	465, 466
insulation	447
mobilization	521
oil fired	465
removal	444
steam	465
Bollard	110
Bollards pipe	551
Bolt anchor	50, 79, 102
flush	245
high strength	106
steel	150
Bolt-on circuit-breaker	498
Bolts barrel	245
remove	101
Bond beam	79, 86
performance	11, 700
Bondcrete	270
Bonding surface	78
Bookcase	324, 361
Bookshelf	370
Booster fan	456
Boot pile	541
Booth acoustical	391
banquet	369
fixed	369
mounted	369
painting	330
portable	391
telephone	310
ticket	391
Border light	508
Bored pile	541
Boring and exploratory drilling	24
auger	25
cased	24
horizontal	569
Borosilicate pipe	442
Borrow	520, 533
Bosuns chair	19
Bottle storage	340
Boulder excavation	528
Bow window	240
Bowling alley	349
Bowstring truss	164
Box buffalo	574
culvert formwork	44
distribution	575
electrical	488
locker	322
mail	323
pull	487
safety deposit	333
stair	169
steel mesh	535
storage	15
termination	513
vent	80
Boxed beam headers	124, 128
Boxes & wiring device	487, 488
utility	572
Brace cross	112
formwork	54
Bracing	153
let-in	152
metal joist	127
roof rafter	130

shoring	16
stud wall	123
Bracket roof	18
scaffold	17
sidewall	18
Braided bronze hose	424
Branch circuit fusible switch	498
Brass hinge	250, 251
pipe	420
salvage	414
screw	149
valve	415
Brazed connection	486
Break glass station	517
Breaker circuit	503
vacuum	425
Breeching draft inducer	463
insulation	447
Brick	84
adobe	91
anchor	80
bedding	549
Brick, block and mortar	721
Brick catch basin	579
chimney simulated	319
cleaning	76
concrete	90
demolition	77, 262
economy	83
edging	564
engineer	83
face	83, 721
floor	550
molding	166
oversized	82
paving	549
removal	26
saw	76
sidewalk	549
sill	95
simulated	97
stair	92
step	546
structural	83
testing	13
veneer	81, 82
veneer demolition	78
vent	461
vent box	80
wall	92
wall panel	92
wash	76, 77
Bricklaying	721
Bridge	559
concrete	559
pedestrian	559
sidewalk	17
Bridging	153
joist	120
metal joist	127
roof rafter	130
steel	153
stud wall	123
wood	153
Broiler	341
Bronze angle valve	415
ball valve	415
body strainer	451
cast tread	138
cross	355
globe valve	416
letter	309
plaque	309
push-pull plate	249
swing check valve	415
valve	415

Broom cabinet	361
finish concrete	66
Brown coat	284
Brownstone	95
Brush clearing	522
mowing	522
Bubbler	440
drinking	440
Buck anchor	79
rough	155
Buffalo box	574
Buggy concrete	65
Builder's risk insurance	695
Building air supported	384
directory	308, 513
disposal	28
greenhouse	386
hangar	392
hardware	245
insulation	183
model	8
moving	30
paper	186
permit	12
portable	15
prefabricated	386
sprinkler	410
temporary	15
tension	385
Built-in range	336
shower	435
Built-up roof	194, 195
roofing	729
Bulb tee subpurlin	71, 116
Bulk bank measure excavating	528
storage dome	390
Bulkhead door	225
Bull winch	51
Bulldozer	532
Bullet resistant window	333
Bulletin board	307
board cork	307
Bulletproof glass	259
Bullnose block	88
Bumper car	316, 552
dock	331
door	246, 249
parking	552
railroad	584
wall	249, 317
Burglar alarm	516
alarm indicating panel	516
Burlap curing	68
rubbing	67
Bus-duct cable tap box	501
copper	500
fitting	500
plug in	500, 501
switch	501
tap box	501
Bush hammer	67
hammer finish concrete	67
Butterfly valve	444, 570
Butyl caulking	209
expansion joint	204
flashing	201
waterproofing	180

C

Cabinet base	360, 361
broom	361
casework	361
corner base	360
corner wall	360

772

Index

current transformer	502
demolition	148
door	361
electrical	487
electrical hinged	488
fan coil	474
fire equipment	321
hardboard	360
hardware	362
hinge	362
hose rack	321
hotel	319
key	254, 362
kitchen	360, 363
laboratory	362
medicine	319
metal	363
oven	361
portable	340
school	362
shower	312
stain	297
storage	362
strip	513
transformer	502
valve	321
varnish	297
wall	360
wardrobe	324
Cable armored	482
control	485
copper	485
copper shielded	481
crimp termination	483
electric	482-484
guide rail	586
mineral insulated	484
PVC jacket	485
railing	141
safety railing	118
sheathed nonmetallic	484
sheathed romex	484
shielded	481
tap box bus-duct	501
termination	483
tray	488
tray aluminum	488
tray ladder type	488
Cafe door	223
Caissons	541, 750
bell	542
concrete	541
displacement	543
foundation	541
Calcium chloride	535
Call system apartment	513
system emergency	513
system nurse	513
Camera TV closed circuit	516
Canopy	325
entrance	325, 388
entry	325
framing	110
patio/deck	325
wood	159
Cant roof	158
Cantilever retaining	557
Canvas awning	325
Cap blasting	528
block	90
dowel	60
pile	63, 65
post	150
Capacitor indoor	504
Car bumper	316, 551, 552
motorized	592
Carbon dioxide exting.	321, 410
Carborundum rub finish	67
Card catalog file	370
Carillon	347
Carousel compactor	354
telephone	310
Carpentry finish	360
Carpet	285, 286
base	286
computer room	287
felt pad	285
floor	286
maintenance	262
nylon	286
pad	285
padding	285
removal	262
stair	286
tile	285
urethane pad	285
wool	286
Carport	326
Carrel	370
Carrier ceiling	267
channel	278
fixture	437
fixture support	437
Cart concrete	65
mounted extinguisher	322
Carving stone	94
Case display	363
exhibit	363
refrigerated	339
work	361
Cased boring	24
Casement window	236-238, 240
window vinyl	243
Casework	363
cabinet	361
custom	360
demolition	148
ground	160
metal	254
painting	297
varnish	297
Cash register	334
Casing bead	274
wood	165
Cast in place concrete	61, 62
in place concrete piles	543
in place pile	543
in place retaining walls	557
in place terrazzo	283
iron bench	373
iron damper	320
iron grate cover	138
iron manhole cover	578
iron pipe	427
iron pull box	488
iron radiator	475
iron trap	427
iron tread	138
trim lock	249
Castellated beam	116
Caster scaffold	17
Castings construction	139
fiberglass	173
Cast-iron column base	139
weld rod	109
Catch basin	578
basin brick	579
basin frame and cover	578
basin masonry	579
basin precast	579
basin removal	25
door	362
Cathodic protection	505
Catwalk scaffold	18
Caulking	210
lead	427
oakum	427
polyurethane	210
sealant	209
Cavity wall	92
wall grout	79
wall insulation	183
wall reinforcing	80
Cedar closet	168
fence	556
paneling	169
post	171
roof deck	162
roof plank	161
shingle	187
siding	193
Ceiling acoustical	277, 278
beam	170
board	277
board acoustic	277
bondcrete	270
carrier	267
demolition	262
diffuser	459
drilling	74
drywall	271, 272
eggcrate	277
fan	456
framing	154
furring	160, 265, 267
heater	338
insulation	183
lath	267, 268
luminous	277, 509
molding	165
painting	300
plaster	269
polystyrene	382
register	460
stair	338
support	110
suspended	267, 271, 277
tile	277, 278
woven wire	314
Cell prison	393
Cellar door	225
wine	340
Cellular concrete	62
decking	123
Cellulose blown in	183
insulation	183
Cement adhesive	282
admixture	66
color	79
concrete curbs	550
content	714
flashing	178
grout	79, 537
gunite	67
keene	270
liner	569
masonry	78
masonry unit	86-89
mortar	276, 720
parging	179
plaster	381
Portland	41
stair pan fill	54
terrazzo portland	283
testing	13
underlayment self	72
vermiculite	207
Cementitious waterproofing	180
wood fiber plank	71
Center bulb waterstop	54
Central vacuum	335
Centrifugal airfoil fan	457
fan	456
liquid chiller	470
Ceramic coating	302
mulch	561
tile	275
tile countertop	364
tile demolition	262
tile floor	275
Certification welding	14
Chain core rope	332
hoist	593
hoist door	229
link fence	21, 553, 554, 556
link fence paint	292
link fence removal	26
link gate	554
steel	139
Chair	369
barber	334
bosuns	19
dental	352
hydraulic	352
library	370
life guard	380
lifts	403
locker bench	322
molding	166
movie	344, 371
portable	346
rail demolition	148
seating	371
Chalkboard	306
electric	306
freestanding	307
liquid chalk	306
metal	315
sliding	306
swing leaf	307
wall hung	306
Chamber anechoic	382
decontamination	32, 36
echo	393
Chamfer strip	50
strip galvanized	51
Channel carrier	278
curb edging	111
frame	215
furring	265, 274
siding	193
steel	269
Channelizing traffic	586
Charge powder	108
Charges disposal	34
dump	28
Charging desk	370
Check floor	250
swing valve	415, 444
valve	408, 445
Checkered plate	138
plate cover	138
plate platform	138
Checkout counter	334
scanner	334
supermarket	334
Chemical anchor	73, 104
cleaning masonry	76
dry extinguisher	321
water closet	598
Chest	369
Chilled water coil	474
Chiller water	470
Chimney	82

773

Index

accessories	319
all fuel vent	463
block	85
brick	82
demolition	77
flue	96
foundation	61
high temperature vent	464
metal	82, 319, 463, 464
screen	319
simulated brick	319
vent	463
Chipping stump	521
Chips quartz	41
silica	41
white marble	41
wood	561
Chloride calcium	535
Chlorination system	441
Church bell	347
equipment	354
pew	371
Chute linen	404
mail	323
package	405
refuse	404
rubbish	29
Chutes	404
Cinema equipment	344
Circuit-breaker	503
bolt-on	498
feeder section	498
Circular slipforms	705
Circulating pump	452
Clamp water pipe ground	486
Clamshell	528, 590
Clapboard painting	293
Classroom heating & cooling	475
seating	371
Clay fire	96
roofing tile	188
tennis court	553
tile	188
tile coping	97
Clean control joint	66
room	381
steam	76
tank	30
Cleaning & disposal equipment	343
brick	76
face brick	720
masonry	76
metal	101
steam	76
up	22
Cleanout door	320
floor	418
pipe	418
PVC	418
tee	418
Clean-up area	34
Clear and grub	521
Clearing brush	522
selective	522
site	521
Clevis hooks	139
Clip plywood	150
sleeper	57
tie reinforcing	55
wire rope	117
Clock	366
& bell system	513
system	513
timer	494
Closed circuit T.V.	333
deck grandstand	387
Closer concealed	246
door	247
electronic	246
floor	246
holder	246
Closers door	246
Closet cedar	168
door	217, 222
pole	166
rod	324
water	433
Closets water	436
Closing coordinator door	250
Closure trash	372
Cloth hardware	556
Clothes dryer commercial	335
dryer residential	338
CMU	85
Coal tar pitch	178, 195
Coat brown	284
glaze	300
hook	318
rack	324, 361
scratch	284
Coated roofing foamed	198
Coating bituminous	179, 569
ceramic	302
epoxy	66, 302, 569
flood	196
polyethylene	569
reinforcing	60
roof	178
rubber	181
silicone	181
spray	179
structural steel	722
trowel	179
wall	302
water repellent	181
waterproofing	179
Cocks gas	415
Coffee urn	341
Cofferdam excavation	529
Cofferdams	538
Cohesion testing	13
Coil cooling	474
flanged	474
Coiling door and grilles	226
Coin operated gate	332
CO_2 extinguisher	322
Cold roofing	194
storage door	228
Collection box lobby	323
systems dust	596
Collector solar energy sys.	467, 468
Colonial door	222, 224
wood frame	171, 172
Color floor	66
wheel	508
Coloring concrete	41
Column	171
aluminum	139
base	150
base aluminum	139, 140
base cast-iron	139
base plate	116
block	86
bondcrete	270
brick	82
cap aluminum	140
capital formwork	43
concrete	61, 64
demolition	77
drywall	272
fireproof	207
form	51
formwork	43
formwork accessories	50
lally	110
laminated wood	165
lath	267
plaster	269
plywood formwork	43
precast	68
precast concrete	68
reinforcing	59
removal	77
round fiber tube formwork	43
round fiberglass formwork	43
round steel formwork	43
steel framed formwork	44
stone	94
structural shape	111
wood	154, 161, 171
Columns framing	154
lightweight	110
ornamental	142
structural	110
Combination device	492
storm door	221, 224
Comfort station	390
Command dog	21
Commercial door	215, 216, 221, 225
folding partition	316
gas water heater	432
greenhouse	386
gutting	29
oil water heater	432
refrigeration	381
water heater	431, 432
water heater electric	432
Commissioning	22
Common brick	83
nail	148, 149
Communicating lockset	248
Communication system	513
Compact fill	533
Compaction	534
backfill	748
soil	532
structural	534
test Proctor	14
vibroflotation	749
Compactor	354
residential	337
waste	354
Compartments shower	312
toilet	310
Compensation workers'	10
Component sound system	512
Composite decking	157
door	216
insulation	185
metal deck	122
rafter	158
Composition flooring	279, 284
flooring removal	262
Compressive strength	13
Compressor tank mounted	447
Computer air conditioning	473
floor	287
room carpet	287
Concealed closer	246
Concrete	711-715
admixtures	40
aggregate	41
asphalt	520
asphaltic	548
beam	61, 62, 64
block	85-89, 722
block autoclaved	85
block back-up	85
block bond beam	86
block decorative	86, 87
block demolition	28
block exterior	88
block foundation	88
block grout	79
block insulation	183
block paver	549
block planter	549
brick	90
bridge	559
broom finish	66
buggy	65
bush hammer finish	67
caisson	541
cart	65
cast in place	61, 64
catch basin	579
cellular	62
chimney	82
coloring	41
column	61, 64
conversion	715
coping	97
core	73
crane and bucket	64
cribbing	557
curb	550
curing	68, 549
curing compound	68
cutout	28
cutting	29
cylinder	13
demolition	28
direct chute	65
drill	73, 74
drilling	74
elevated slab	64
field mix	64, 716
finish	66, 549
flagpole	327
float finish	66, 67
footing	65
forms	705, 706
formwork insert	52
formwork shoring	53
furring	160
granolithic finish	66
grout	79
hand hole	581
hand trowel finish	66
hardener	66
hydrodemolition	25
insulating lightweight agg	72
insulating lightweight cell	72
integral finish	66
joist	62
lift slab	716
lightweight	62-64, 719
machine trowel finish	66
manhole	579, 581
materials	714
membrane curing	68
mix design	13
mixes	714, 715
monolithic finish	66
paper curing	68
patio block	549
paving	548
pier	62
piers and shafts drilled	543
piles prestressed	539
pipe	574, 576-578
pipe removal	26
placing	64, 716
planter	372

774

Index

post 586
post tensioned 70
prestressed 718
prestressed precast 717
pump 64
pumped 64
ready mix 64
removal 25
retaining 557
sandblast finish 67
saw 29
scarify 262
septic tank 575
shingle 189
sidewalk 546
silo 392
slab 62, 65
slabs X-ray 14
stair 64
stair tread insert 54
stamping 67
strengths 715
tank 573
testing 12
ties railroad 584
tile 189
tilt-up 719
topping 66
utility vault 572
vertical shoring 53
wall 65
wall finish 67
water curing 68
wheeling 65
Condensate piping 452
  steam meter 452
Condenser 469
  air cooled belt drive . 469
  air cooled direct drive 470
Condensing unit air cooled . 469
Conductive floor ... 282, 284, 285
  terrazzo 284
Conductor 482, 745
  & grounding 481-484
  wire 484
Conductors 745
Conduit aluminum 486
  electrical 486
  flexible 490
  high installation 487
  in slab 489
  in slab PVC 489
  in trench electrical .. 489
  in trench steel 489
  intermediate 486
  intermediate steel 486
  raceway 486
  rigid in slab 489
  rigid steel 486
Cone traffic 20
Confessional 355
Connection brazed 486
  motor 490
  standpipe 408
Connector flexible 424
  gas 424
  joist 150
  screw-on 483
  timber 150
  water copper tubing ... 424
  wire 483
Conrete wall panel tilt-up . 71
Consolidation test 13
Construction aids 15
  castings 139
  cost index 9, 643

management fee 8
photography 12
sump hole 531
temporary 20
time 8
Contaminated soil 30
Contingencies 9
Continuous footing formwork 45
  hinge 251
Contractor overhead 22
  scale 328
Control & motor starter .. 500
  air 450
  board 308, 347
  cable 485
  damper volume 454
  hand-off-automatic 504
  joint 66, 80
  joint clean 66
  joint PVC 80
  joint rubber 80
  joint sawn 66
  joint sealant 66
  radiator supply 450
  station 504
  station stop-start 504
  system air compressor . 447
  system boiler 447
  system cooling 447
  system electronic 447
  system split system ... 447
  system ventilator 447
  systems pneumatic 447
  tower 392
  valve heating 450
Controller time system ... 513
Convection oven 342
Convector heating unit ... 475
Conveyor 592
  ash 354
  door 459
  material handling 592
  system 404
  vertical 592
Cooking equipment ... 336, 341
  range 336
Cooler 382
  beverage 339
  door 459
  water 441
Cooling & heating A/C 473
  coil 474
  control system 447
  tower 471
  tower louver 461
  towers fiberglass 471
  towers stainless 471
Coping 94, 97
  aluminum 97
  clay tile 97
  concrete 97
  removal 78
  terra cotta 84, 97
Copper bus-duct 500
  cable 482, 483, 485
  downspout 202
  drum trap 427
  DWV tubing 420
  flashing 200, 201, 428
  gutter 204
  pipe 420
  reglet 204
  rivet 108
  roof 199
  salvage 414
  shielded cable 481

wall covering 287
wire 484, 485
Core drill 13, 73, 74
  testing 13
Coreboard 273
Cork floor 282
  insulation 382
  tile 282
  wall tile 287
Corner base cabinet 360
  bead 269, 274
  block 90
  guard 316, 317
  protection 317
  wall cabinet 360
Cornerboards PVC 174
Cornice drain 430
  molding 165, 166
  painting 300
  stone 94
Correspondence lift 404
Corrosion resistant fan .. 457
  resistant pipe 442
Corrugated metal pipe . 575, 578
  pipe 579
  roof tile 188
  siding 189-191
Cost control 12
  mark-up 11
Cot prison 369
Counter bank 333
  checkout 334
  door 226
  flashing 204
  hospital 363
  top 364
  top demolition 148
  window 323
Countertop backsplash 364
  sink 434
Course wearing 547
Court air supported 385
  handball 384
  paddle tennis 350
  racquetball 384
  squash 384
Cove base 281
  base ceramic tile 275
  base terrazzo 283
  molding 165
  molding scotia 165
Cover aluminum trench 138
  cast iron grate 138
  checkered plate 138
  ground 563
  manhole 578
  pool 380
  stadium 385
  stair tread 21
  tank 384
  trench 138
  walkway 326
Covering wall 735
CPVC pipe 424
  valve 417
Crane 593
  and bucket concrete ... 64
  material handling 593
  mobilization 521
  rail 592
  scale 328
Cranes overhead bridge ... 592
Crash barriers 586
Crematory 353
Crew survey 21
Cribbing concrete 557

Crimp termination cable .. 483
Critical path schedule ... 12
Cross brace 112
  wall 355
Crown molding 166
Crushed stone sidewalk ... 546
Cubicle 313
  curtain 313
  detention 244
  shower 434
  toilet 312
Cultured stone 97
Culvert end 577
  reinforced 576
Cupola 326
Cupric oxychloride floor . 284
Curb 550
  and gutter 550
  asphalt 551
  concrete 550
  edging 111, 551
  edging angle 111
  edging channel 111
  granite 93, 551
  inlet 551
  precast 551
  removal 25
  roof 158
  seal 546
  terrazzo 283
Curing blanket 68
  compound concrete 68
  concrete 68, 549
  paper 186
Current transformer 496
  transformer cabinet ... 502
Curtain 235
  air 458, 459
  blast 359
  cubicle 313
  damper fire 454
  divider 349
  gymnasium divider 349
  rod 317
  sound 393
  track 347
  wall glass 235
Custom casework 360
Cut stone trim 95
Cutoff pile 520
Cutout counter 364
  demolition 28
  slab 28
Cutting asphalt 29
  block 364
  concrete 29
  saw 73
  steel 29, 102
  torch 102
  tree 522
Cylinder concrete 13
  lockset 248
  recore 248, 249

D

Dairy case 339
Damper 454
  fire curtain 454
  fireplace 320
  foundation vent 80
  multi-blade 454
Dampproofing 180
Darkroom 382
  door 228

775

Index

equipment ... 336
revolving ... 383
sink ... 336
Dasher hockey ... 381
Data safe ... 332
Day gate ... 229
Deadbolt ... 248, 250
Deadlocking latch ... 248
Deciduous shrub ... 563, 564
tree ... 564
Deck drain ... 428
edge form ... 122
framing ... 156, 157
metal floor ... 123
roof ... 161, 163
slab form ... 122
wood ... 161, 162
Decking ... 728
cellular ... 123
composite ... 157
floor ... 121, 123
form ... 122
roof ... 122
Decontamination chamber ... 32, 36
enclosure ... 32, 33, 36
equipment ... 31
Decorative block ... 86
fence ... 555
Decorator device ... 491
switch ... 491
Deep freeze ... 337
seal trap ... 427
therapy room ... 393
Deep-longspan joist ... 119
Dehumidifier ... 338
Delicatessen case ... 339
Delineator barrier ... 586
Delivery charge ... 19
Deluge valve assembly sprinkler ... 410
Demo air supp. storage tank cov. ... 376
concrete ... 25
concrete selective ... 40
garden house ... 376
geodesic domes ... 376
greenhouses ... 376
hangars ... 376
lightning protection ... 376
pre-engineered steel buildings ... 377
silos ... 377
sound control ... 377
special purpose rooms ... 377
steel selective ... 101
storage tank ... 378
swimming pool ... 378
tension structures ... 379
thermal & moisture protection ... 178
X-ray/radio freq protection ... 379
Demolition ... 26, 27, 30, 78, 263, 414, 703
asbestos ... 33
baseboard ... 148
boiler ... 444
brick ... 77, 262
cabinet ... 148
casework ... 148
ceiling ... 262
ceramic tile ... 262
chimney ... 77
column ... 77
concrete ... 28
concrete block ... 28
cutout ... 28
door ... 212
drywall ... 262
ductwork ... 444
electric ... 480, 481

enclosure ... 34
explosive ... 27
fireplace ... 78
flooring ... 262
framing ... 145
glass ... 212
granite ... 78
gutter ... 178
house ... 27
HVAC ... 444
implosive ... 27
joist ... 145
lath ... 262
masonry ... 28, 77, 263
metal deck ... 178
metal stud ... 263
millwork ... 148
mold contaminated area ... 37
paneling ... 148
partition ... 263
pavement ... 25
plaster ... 262, 263
plenum ... 263
plumbing ... 414
plywood ... 178, 262, 263
post ... 146
rafter ... 146
railing ... 148
roofing ... 33, 178
selective ... 144, 212
siding ... 178
site ... 25
steel window ... 212
stucco ... 263
terra cotta ... 263
terrazzo ... 263
tile ... 262
trim ... 148
wall ... 263
walls and partitions ... 263
window ... 212
wood ... 262
Demountable partition ... 314
Dental equipment ... 352
metal interceptor ... 429
Depository night ... 333
Derail railroad ... 584
Desk ... 369, 370
Detection leak ... 518
probe leak ... 518
system ... 516
tape ... 569
Detector infra-red ... 516
motion ... 516
smoke ... 518
temperature rise ... 517
ultrasonic motion ... 516
Detention cubicle ... 244
doors and frames ... 229
equipment ... 334
Developing tank ... 336
Device & box wiring ... 488
combination ... 492
decorator ... 491
exit ... 247
GFI ... 493
panic ... 247
receptacle ... 493
residential ... 491
wiring ... 502
Dewater ... 531
pumping ... 531
Dewatering ... 531, 748
Diamond lath ... 268
Diaper station ... 317
Diesel generator ... 504

Diffuser ceiling ... 459
opposed blade damper ... 459
perforated aluminum ... 459
rectangular ... 459
steel ... 460
T-bar mount ... 460
Dimensions and weights of sheet steel ... 727
Dimmer switch ... 491, 503
Direct expansion A/C ... 475
shute concrete ... 65
Directional sign ... 310
Directory ... 308
board ... 308, 309
building ... 308, 513
shelf ... 310
Disappearing stairs ... 338
Dishwasher ... 337, 343
Diskette safe ... 332
Dispenser hot water ... 431
napkin ... 318
product ... 448
soap ... 317, 318
toilet tissue ... 317, 318
towel ... 317, 318
Dispensing equipment food ... 342
Dispersion nozzle ... 410
Displacement caisson ... 543
Display case ... 363
Disposable bag ... 32
Disposal ... 27, 28
asbestos ... 34
building ... 28
charges ... 34
field ... 575
garbage ... 337
waste ... 354
Distiller medical ... 351
water ... 351
Distribution box ... 575
pipe water ... 570
section electric ... 498
Ditching ... 531, 539
Divider curtain ... 349
strip terrazzo ... 283
Diving board ... 380
Dock board ... 331
board magnesium ... 331
boat jetties ... 590
bumper ... 331
leveler ... 331
loading ... 331
shelter ... 331
truck ... 331
Dog command ... 21
Dome drain ... 430
geodesic ... 386, 391
roof ... 235
structures ... 737
Domes ... 390
Door accessories ... 254
accordion ... 222
acoustical ... 229, 316
air curtain ... 458
air lock ... 385
aluminum ... 213, 217
and grilles coiling ... 226
and panels access ... 225
and window matls ... 212
automatic entrance ... 233
balanced ... 234
bell residential ... 494
bell system ... 503
bi-fold ... 217, 222
bi-passing closet ... 222
birch ... 223

blind ... 173
bulkhead ... 225
bullet resistant ... 333
bumper ... 246, 249
cabinet ... 361
cafe ... 223
casing PVC ... 174
catch ... 362
chain hoist ... 229
cleanout ... 320
closer ... 247
Door closers ... 246
Door closet ... 217, 222
closing coordinator ... 250
cold storage ... 228
combination storm ... 221
commercial ... 215, 216, 221
composite ... 216
conveyor ... 459
cooler ... 459
counter ... 226
darkroom ... 228
demolition ... 212
double ... 216
duct access ... 455
dutch ... 222
dutch oven ... 320
entrance ... 213, 222, 233
fiberglass ... 223
fire ... 216, 220, 227, 517
fire rated access ... 225
flexible ... 232
floor ... 225, 226
flush ... 219, 220, 222
flush wood ... 218
folding ... 316
folding accordion ... 222
frame ... 171, 214, 215
frame grout ... 79
frame interior ... 172
frame lead lined ... 393
freezer ... 459
garage ... 229, 230
glass ... 226, 233
handle ... 362
hangar ... 228, 392
hardware ... 245
industrial ... 227
jamb molding ... 167
kennel ... 226
kick plate ... 251
knob ... 249
labeled ... 216, 220
metal ... 230, 389
metal access ... 225
mirror ... 257
molding ... 167
moulded ... 221
opener ... 212, 227, 231, 245
operator ... 212, 245
overhead ... 229-231
overhead commercial ... 229
panel ... 221
paneled ... 216
partition ... 315
passage ... 222-224
prefinished ... 219
pre-hung ... 223, 224
prison ... 229
pull ... 250
refrigerator ... 228
release ... 513
removal ... 212
residential ... 172, 216, 221
residential garage ... 230
revolving ... 228, 233, 385

776

Index

revolving entrance 233
rolling 227
roof 206
sauna 383
seal 331
sectional 229
sectional overhead 230
security vault 229
shower 313
side light 217, 223, 224
sill 167, 171, 250
sliding 217
special 229
stain 297
stainless steel (access) 225
steel 215, 216, 230, 389, 731
stop 232, 249
storm 213
swing 224
switch burglar alarm 516
telescoping 231
threshold 172
traffic 231
varnish 297
vault 229
vertical lift 231
weatherstrip 254
weatherstrip garage 253
wire partition 314
wood 218, 219, 221
Doors & windows exterior paint . 295
& windows interior paint 298
and frames detention 229
Dormer gable 160
Dormitory furniture 369
Double acting door 231
brick 83
hung window 237, 238, 240
precast concrete tee beam .. 69
wall pipe 580, 581
weight hinge 251
Dovetail anchor system 51
Dowel cap 60
reinforcing 59
sleeve 59
Downspout 202, 203, 326
aluminum 202
copper 202
elbow 203
lead coated copper 203
steel 203
strainer 202
Dozer backfill 532, 533
excavating bulk 529
excavation 529, 530
Draft inducer breeching 463
Dragline 528, 590
Drain 427, 430
board 180, 181
deck 428
dome 430
floor 428
main 430
pipe 531
scupper 430
sediment bucket 428
shower 427
trench 430
Drainage accessory 203
mat 367
pipe 442, 574-576
site 579
trap 427
Drains roof 430
Drapery hardware 358
Drawer kitchen 360

track 362
type cabinet 363
wood 361
Drawings as-built 22
Dredge 590
hydraulic 590
Dredging 590
Dressing unit 370
Drill auger 25
bit 74
concrete 73, 74
console dental 352
core 13, 73, 74
earth 24, 541
hole 535
rig 24
rig mobilization 521
rock 24, 528
shop 346
wood 150
Drilled concrete piers and shafts 543
pier 542
Drilling ceiling 74
concrete 74
horizontal 569
only rock 527
plaster 265
quarry 528
rock bolt 527
steel 102
Drinking bubbler 440
fountain 440, 441
fountain deck rough-in .. 440
fountain floor rough-in .. 440
fountain handicap 440
fountain support 437
fountain wall rough-in .. 440
Drip edge 205
edge aluminum 205
Dripproof motor 510
Driven pile 539
Drive-up window 333
Driveway 546
removal 25
Drop pan ceiling 277
Drum trap copper 427
Dry chemical extinguisher .. 321
fall painting 301
pipe sprinkler head 410
transformer 497
wall leaded 394
Drycleaner 334
Dryer clothes 338
commercial clothes 335
darkroom 336
hand 318
receptacle 493
residential 338
vent 338
Drywall 272
accessories 274
ceiling 271
column 272
cutout 28
demolition 262, 263
finish 734
frame 214
gypsum 272-274
high abuse 273
nail 148
painting 300
partition 264
prefinished 273
removal 263
screw 274
Duck tarpaulin 20

Duckboard 367
Duct access door 453
accessories 454
connection fabric flexible . 453
electric 489, 500
electrical underfloor 490
fitting trench 490
fitting underfloor 490
flexible insulated 455
flexible noninsulated ... 455
furnace 467
grille 287
heater electric 474
humidifier 477
HVAC 453
insulation 446
liner 455
mechanical 453
silencer 455
steel trench 490
trench 489
underfloor 490
underground 582
utility 582
Ductile iron fitting 570
iron fitting mech joint ... 570
iron grooved join 422
iron pipe 570
Ducts flexible 455
Ductwork 453, 741
aluminum 453
demolition 444
fabric flexible 455
fabricated 453
fiberglass 454
galvanized 453
laboratory 345
metal 453
rectangular 453
rigid 453
Dumbbell 347
waterstop 54
Dumbwaiter 398
electric 398
Dump charges 28, 29
truck 533
Dumpster 29
Duplex pump 452
receptacle 503
Dust collection systems .. 596
collector shop 346
partition 29
Dustproofing 66
Dutch door 222
oven door 320
DWV ABS pipe 423
pipe ABS 423
PVC pipe 423
tubing copper 420

E

Earth drill 24, 541
rolling 534
vibrator 524, 532, 534
Earthwork equipment 532
haul 533
Eave overhang 388
Ecclesiastical equipment . 354
Echo chamber 393
Economy brick 83
Edge drip 205
Edging 564
aluminum 564
angle curb 111

curb 111, 551
Educational TV studio 517
Efflorescence testing 13
Eggcrate ceiling 277
EIFS 185
Ejector pump 429
Elastomeric roof 198
Elbow downspout 203
pipe 422
Electric appliance 336
backstop 348
ballast 507
baseboard heater 476
bed 370
boiler 464
cable 482-484
capacitor 504
chalkboard 306
commercial water heater . 432
demolition 480, 481
duct 489, 500
duct heater 474
dumbwaiter 398
feeder 500
fixture 506, 509
furnace 466
generator set 504
heater 338
heating 476
hinge 251
hoist 593
lamp 509
log 320
metallic tubing 486
meter 496
motor 510
panelboard 499
pole 581
pool heater 466
service 503
stair 338
switch 502, 503
utility 582
water heaters 431
wire 484
Electrical box 488
cabinet 487
conduit 486
fee 8
laboratory 346
telephone 581
telephone underground . 581
tubing 486
Electronic air cleaner 463
closer 246
control system 447
Elevated conduit 487
floor 62
installation add pipe 414
pipe add 414
slab 62
slab concrete 64
slab formwork 44
water tank 573
Elevating scraper 530
Elevator 398, 399, 738
cab 401
controls 401
fee 8
options 401
shaft wall 264
Embossed print door 221
Emergency call system ... 513
lighting 507
Employer liability 10
EMT 486

777

Index

Emulsion adhesive 282
   asphaltic 561
   pavement 546
   penetration 547
Encapsulation 34
   pipe 34
Encasement pile 541
Encasing steel beam formwork ... 42
Enclosure acoustical 393
   bathtub 313
   decontamination 32, 33, 36
   demolition 34
   shower 313
   swimming pool 387
   telephone 310
End culvert 577
Endwall overhang 389
Energy circulator air solar 468
   system controller solar 468
   system heat exchanger solar .. 468
Engineer brick 83
Engineered stone countertops ... 365
Engineering fees 8, 693
Entrance aluminum 234
   and storefront 233
   canopy 325, 388
   door 213, 222, 224, 233
   frame 171, 172
   lock 250
   screen 311
   sliding 233
   strip 231
Entry canopy 325
EPDM adhesive 180
   roof 197
Epoxy 537
   anchor 73, 104
   coated reinforcing 60
   coating 66, 302, 569
   fiberglass wound pipe 442
   floor 284, 285
   grating 174
   grout 275, 276, 537
   lined silo 392
   terrazzo 284
   wall coating 302
Equipment athletic 555
   bank 333
   barber 334
   cinema 344
   cleaning & disposal 343
   darkroom 336
   detention 334
   earthwork 532
   ecclesiastical 354
   fire 409
   foundation formwork 45
   general 11
   gymnasium 348
   health club 347
   hospital 351
   installation 330
   insulation 447
   insurance 10
   laboratory 345, 362
   laundry 334, 335
   loading dock 331
   medical 351, 353
   mobilization 19, 543
   pad 62
   parking ... 140, 330, 331, 369, 371, 393
   playfield 349
   playground 349, 350, 555
   rental 702
   security and vault 332

shop 346
stage 346
swimming pool 380
waste handling 354
Erosion control synthetic 534
Escalators 402, 739
Escape fire 134
Estimate electrical heating 476
Estimates window 731
Estimating 692
   reinforcing 59
Evergreen shrub 563
   tree 563
Excavating bulk bank measure .. 528
   bulk dozer 529
   equipment 747
Excavation 523, 528, 534, 746
   backhoe 523
   bulk scraper 530
   cofferdam 529
   dozer 529, 530
   hand 524, 527, 532
   machine 532
   planting pit 561
   rock 528
   scraper 530
   septic tank 575
   structural 527
   trench 523-527
Exchanger heat 469
Exercise equipment 348
   ladder 348
   rope 348
   weight 347
Exhaust hood 337, 345
   vent 461
Exhauster roof 457
Exhibit case 363
Exit and emergency lighting 507
   control alarm 516
   device 247
   light 507
Expansion anchor 104
   joint 51, 204, 269, 450
   joint assembly 210
   joint bellows 450
   joint floor 210
   joint roof 204
   shield 104
   tank 450
   tank steel 450
Expense office 11
Explosionproof fixture 508
   lighting 508
Explosive demolition 27
Explosives 528
Exposed aggregate 66, 546
   aggregate coating 302
   aggregate panels 192
Exterior blind 172
   concrete block 88
   door frame 171
   floodlamp 509
   insulation 185
   insulation finish system 185
   lighting fixture 508, 509
   molding 166
   plaster 270
   pre-hung door 224
   PVC molding 174
   residential door 221
   signage 309
   sprinkler 559
   tile 275
   wood frame 172
Extinguisher ABC 322

carbon dioxide 321
chemical dry 321
CO_2 322
fire 321
installation 322
pressurized fire 322
standard 321
Extra work 11
Extractor air 453
   industrial 334
Extrusion aluminum 117
Eye bolt screw 57
   wash fountain 439
   wash portable 439

F

Fabric flashing 200
   stabilization 547
   stile 252
   structure 384
   waterproofing 179
   welded wire 60
   wire 556
Fabricated ductwork 453
Face brick 83, 721
   brick cleaning 720
   wash fountain 439
Faceted glass 255
Facing panel 95
   stone 94
   tile structural 84
Fall painting dry 301
Fan 456
   air conditioning 456
   air handling 456
   axial flow 456
   bathroom exhaust 458
   booster 456
   ceiling 456
   centrifugal 456
   coil air conditioning 474
   coil cabinet 474
   corrosion resistant 457
   induced draft 463
   in-line 456
   kitchen exhaust 458
   low sound 456
   paddle 495
   propeller exhaust 458
   residential 495
   roof 457
   utility set 456, 457
   vane-axial 456
   ventilation 495
   ventilator 456
   wall 458
   wall exhaust 457
   wiring 495
Farm type siding 191
Fascia board demolition 145
   metal 202
   PVC 174, 175
   wood 158, 166
Fast food equipment 341
Fastener timber 148
   wood 149
Faucet & fitting 435
   bathroom 435
   laundry 435
   lavatory 435, 436
Fee architectural 8, 692
   engineering 8, 693
Feeder electric 500
   section 498

section branch circuit 498
section circuit-breaker 498
section frame 498
Felt 196
   asphalt 194, 196
   carpet pad 285
   tarred 196
   waterproofing 179
Fence aluminum 554
   and gate 553
   board batten 557
   cedar 556
   chain link 21, 553, 556
   decorative 555
   helical topping 556
   mesh 553
   metal 553
   picket paint 292
   plywood 21
   removal 26
   residential 554
   security 555, 556
   snow 30
   steel 555
   temporary 21
   tennis court 555
   treated pine 557
   wire 21, 556
Fences 292
   plastic 556
Fencing wire 556
Fenders 551
Fiber cement formboard 49
   cement shingle 187
   cement siding 194
   optic cable 512
   optics 512
Fiberboard insulation 184
Fibergalss tank 448
Fiberglass angle 173
   area wall cap 326
   bench 373
   blown in 183
   castings 173
   ceiling board 277
   cooling towers 471
   cross 355
   door 223, 230
   ductwork 454
   flagpole 327
   flat sheet 173
   formboard 49
   grating 174
   handrail 173
   insulation 181-183, 389, 446
   panel 20, 190, 289
   panel skylight 389
   planter 372
   reinforced ceiling 382
   reinforced plastic panel ... 288
   round bar 173
   round tube 173
   shade 360
   shower stall 312
   square bar 173
   square tube 173
   stair tread 173
   steeple 326
   tank 448
   threaded rod 173
   trash receptacle 372
   tread 138
   trench drain 430
   wall lamination 302

Index

waterproofing	179
wide flange beam	173
wool	183
Field disposal	575
mix concrete	64, 716
office	15
office expense	15
personnel	10
seeding	562
Fieldstone	93
Fill	532, 533
by borrow	533
floor	62
gravel	520, 533
Filler block	300
strip	252
Fillet welding	102
Film equipment	336, 344
polyethylene	561
security	258
Filter air	462
grille	460
mechanical media	462
stone	536
swimming pool	441
Filtration air	31
equipment	380
Fine grade	523, 562
Finish carpentry	360
concrete	66, 549
floor	280, 302
grading	523
hardware	246
keene cement	270
lime	78
nail	148
refrigeration	382
Fir column	171
floor	280
molding	166
roof deck	162
roof plank	161
Fire alarm	517
brick	96
call pullbox	517
clay	96
damper curtain	454
door	216, 220, 227, 517
door frame	214
equipment	409
equipment cabinet	321
escape	134
escape balcony	134
escape disappearing	338
escape stairs	134
extinguisher	321
extinguishing system	321, 410
horn	517
hose	409
hose adapter	409
hose nozzle	409
hose storage cabinet	321
hose valve	409
hydrant	408, 572
protection	321
protection kitchen	462
pump	411
rated closer	246
rated tile	84
resistant drywall	272, 273
resistant wall	264
retardant lumber	152
retardant pipe	442
retardant plywood	152
signal bell	517
suppression automatic	410
Fireplace accessory	319
box	96
built-in	320
damper	320
demolition	78
form	319
free standing	319
mantel	170
masonry	96
prefabricated	319
Fireproofing	207
plaster	207
plastic	207
spray	207
Firestop gypsum	267
wood	155
Firestopping	208, 730
Fitting bus-duct	500
ductile iron	570
grooved joint	422
grooved joint pipe	422
poke-thru	487
PVC	571, 572
underfloor duct	490
vent chimney	464
Fixed blade louver	461
booth	369
roof tank	449
Fixture ballast	507
bathroom	433, 434
carrier	437
electric	506
explosionproof	508
fluorescent	506
incandescent	506
incandescent vaportight	507
interior light	506
plumbing	429, 433, 440, 452, 574
removal	414
residential	494
sodium high pressure	509
support handicap	437
vandalproof	507
Flagging	278, 550
slate	550
Flagpole	327
aluminum	327
fiberglass	327
foundation	327
wall	327
Flanged coil	474
Flashing	200-202
aluminum	191, 200, 428
asphalt	200
butyl	201
cement	178
copper	200, 201, 428
counter	204
fabric	200
lead coated	201
masonry	200
membrane	196
metal	389
PVC	201
stainless	200
valley	187
vent	428
Flat sheet fiberglass	173
Flexible conduit	490
connector	424
door	232
duct connection fabric	453
ducts	455
ductwork fabric	455
insulated duct	455
metal hose	424
noninsulated duct	455
plastic netting	328
sign	309
Float finish concrete	66, 67
glass	255
Floater equipment	10
Floating dock	590
floor	280
pin	250
roof tank	449
Flood coating	196
Floodlamp exterior	509
Floodlight pole mounted	509
Floor	278
abrasive	66
access	287
acid proof	278
asphalt block	278
brick	278, 550
carpet	286
ceramic tile	275
check	250
cleaning	22
cleanout	418
closer	246
color	66
conductive	282, 284, 285
cork	282
decking	121, 123
door	225, 226
drain	428
elevated	62
epoxy	284, 285
expansion joint	210
fill	62
finish	302
flagging	550
framing removal	28
grating	136, 174
hatch	226
marble	94
mastic	284
mat	366
nail	149
neoprene	284
oak	280
paint	299
parquet	279
patching concrete	40
pedestal	287
plank	161
plywood	162
polyacrylate	284
polyester	285
polyethylene	286
polyurethane	286
portable	346
quarry tile	276
refrigeration	382
register	460
removal	27
rubber	281
scale	328
scupper	430
slate	95
sleeper	159
stain	299
subfloor	162
terrazzo	283, 735
tile terrazzo	283
topping	66
underlayment	162
varnish	299
vinyl	282
wood	280
wood athletic	280
wood composition	279
Flooring	278
composition	284
demolition	262
Flow meter	452
Flue chimney	96
chimney metal	463
liner	82, 96
prefab metal	463
screen	319
tile	96
Fluid heat transfer	468
Fluorescent fixture	506
Fluoroscopy room	393
Flush automatic	435
bolt	245
door	219, 220, 222
tube framing	232
valve	435
wood door	218
Flying truss shoring	53
FM200 fire extinguisher	411
Foam glass insulation	181
insulation	183, 184
pipe covering	419
urethane	210
Foamed coated roofing	198
Fog seal	546
Foil aluminum	183, 186, 290
back insulation	182
barrier	290
Folder laundry	334
Folding accordion door	222
accordion partition	316
bench	371
door	316
door shower	313
gate	314
partition	316
table	368
Food dispensing equipment	342
mixer	341
service equipment	341
warmer	342
Foot valve	417
Football goalpost	351
scoreboard	348
Footing concrete	65
keyway	46
keyway formwork	46
reinforcing	59
removal	27
spread	63
Forged valves	445
Form deck edge	122
decking	122
fireplace	319
material	705
release agent	40
Formblock	86
Formboard fiber cement	49
fiberglass	49
Formwork wall plywood	49
Forms	706
concrete	706
Formwork	43, 47
accessory	50, 51, 53
beam and girder	42
beam bottom	43
box culvert	44
brace	54
column	43
column capital	43
column plywood	43
column round fiber tube	43
column round fiberglass	43

779

Index

column round steel 43
column steel framed 44
continuous footing 45
elevated slab 44
encasing steel beam 42
equipment foundation 45
footing keyway 46
gang wall 48
gas station 46
gas station island 46
girder 42
grade beam 46
hanger accessories 52
insert 52
insulating concrete 49
interior beam 42
labor hours 707
light base 46
mat foundation 46
oil 54
patch 54
pile cap 46
plywood 42, 44
radial wall 48
retaining wall 48
sign base 46
slab blockout 47
slab box out opening 45
slab bulkhead 45, 46
slab curb 45, 47
slab edge 45, 47
slab fiberglass domes 45
slab flat plate 44
slab metal domes 45
slab metal pan 44
slab on grade 46
slab screed 47
slab trench 47
slab voids 45, 47
slab with drop panel 44
sleeve 53
slipform building 42
slipform silo 42
snap tie 53
spandrel beam 42
spread footing 46
stairs 50
stake 54
steel framed plywood 49
upstanding beam 43
wall 47
wall boxout 47
wall brick shelf 47
wall bulkhead 47
wall buttress 47
wall corbel 47
wall liner 48
wall lintel 48
wall pilaster 49
wall plywood 47
wall rustication strip 48
wall sill 49
wall steel framed 49
Fossil fuel boiler 465
Foundation caisson 541
chimney 61
concrete block 88
flagpole 327
mat 63, 65
pile 543
scale pit 328
underpin 538
ven 80
wall 88
Fountain drinking 440, 441
eye wash 439

face wash 439
lighting 442
pump 441
wash 437
Frame baked enamel 214
door 171, 214, 215
drywall 214
entrance 171, 172
fire door 214
grating 136
labeled 214
metal 214
metal butt 214
scaffold 16
steel 214
trench grating 138
welded 214
window 239
wood 171, 361
Frames shoring 53
Framing aluminum 117
anchor 150
band joist 128
beam & girder 153
canopy 110
ceiling 154
columns 154
deck 156, 157
demolition 145
heavy 161
joist 154
laminated 164
lightweight 111
lightweight angle 112
lightweight channel 112
lightweight junior beam ... 112
lightweight tee 112
lightweight zee 112
metal joist 128
metal partition 124
metal stud 124
opening 389
partition 266, 267
pipe support 113
porch 156
removal 144
roof metal rafter 131
roof metal truss 132
roof rafters 158
roof soffits 132
roof truss 164
sill 158
sleepers 159
slotted channel 112
stair stringer 155
steel 115
suspended ceiling 154
timber 161
treated lumber 159
tube 232
wall 159
window 389
window wall 232
wood 153-155, 158-161
wood plate 159
Freestanding chalkboard ... 307
Freeze deep 337
Freezer 339, 382
door 459
Freight elevators 738
Friction collar 51
pile 541, 543
Frieze PVC 175
Front end loader 528
vault 229
FRP panel 288

Fryer 341
Fuel storage tank 449
tank 448
Full vision door 213
vision glass 255
Fume hood 345
Furnace duct 467
electric 466
gas 467
gas fired 466
hot air 466
oil fired 466
wall 467
Furniture dormitory 369
hospital 370
hotel 368
library 370
office 367, 368
restaurant 368
school 369
Furring and lathing 265
ceiling 160, 265, 267
channel 265, 274
metal 265
steel 265
wall 160, 266
Fusible link closer 246
switch branch circuit 498

G

Gabion 535, 558
retaining walls stone 558
Gable dormer 160
Galvanized chamfer strip ... 51
ductwork 453
plank grating 136
reinforcing 60
roof 189
steel reglet 204
wire gabion 558
Galvanizing 113, 303
metal in field 302
metal in shop 101
Gang wall formwork 48
Gantry crane 593
Garage 390
door 229, 230
door weatherstrip 253
residential 390
Garbage disposal 337, 343
Garden house 391
house demo 376
Gas cocks 415
connector 424
fired boiler 465
fired furnace 466
fired infra-red heater 475
fired space heater 467
furnace 467
generator 504
heat air conditioner 472
incinerator 353
log 320
pipe 580
station formwork 46
station island formwork 46
station piping 580
station tank 448
stop valve 415
vent 463
water heater 432
water heater commercial ... 432
Gasket neoprene 210
Gas/oil combination boiler 465

Gasoline generator 504
pump 448
tank 449
Gate chain link 554
day 229
fence 554
folding 314
security 314
swing 554
valve 444, 445
valve soldered 415
Gauging plaster 270
General boiler 414
contractor's overhead 696
equipment 11
Generator diesel 504
emergency 504
gasoline 504
set 504
steam 351
Geodesic dome 386, 391
domes demo 376
GFI receptacle 493
Girder formwork 42
joist 121
reinforcing 59
wood 153, 161
Glass 255, 257, 258
acoustical 259
and glazing 259
bead 232
bevel 255
block 91
block skylight 245
bulletin board 308
bulletproof 259
curtain wall 235
demolition 212
door 213, 226, 233
door astragal 252
door shower 313
faceted 255
float 255
full vision 255
heat reflective 256
insulating 256
laminated 259
lead 394
lined water heater 338
low emissivity 255
mirror 257, 318
mosaic 276
obscure 256
patterned 256
pipe 442
reflective 257
safety 256
sandblast 255
sheet 256
shower stall 312
spandrel 256
tempered 255
tile 257
tinted 255
vinyl 258
window 257
window wall 233
wire 257
Glassware washer 345
Glaze coat 300
Glazed aluminum curtain wall 235
Glazed block 90
brick 83
ceramic tile 275
wall coating 302

780

Index

Glazing 255
   plastic 257
   polycarbonate 258
   productivity 732
Globe valve 445
   valve bronze 416
Glove bag 33
   box 345
Glued laminated 164
Goalpost 350
   football 351
   soccer 351
Golf shelter 350
   tee surface 286
Gore line 552
Grab bar 317
Gradall 529
Grade beam 65
   beam formwork 46
   fine 523, 562
   wall 63
Grading 523, 534
Grandstand 387
Granite 93
   building 93
   chips 561
   conductive floor 285
   curb 93, 551
   demolition 78
   indian 551
   paver 93
   paving block 550
   sidewalk 550
Granolithic finish concrete 66
Grass cloth wallpaper 288
   lawn 562
   seed 562
   sprinkler 559
Grating aluminum 136
   aluminum mesh 136
   aluminum plank 136
   area wall 326
   area way 326
   fiberglass 174
   floor 136, 174
   frame 136
   galvanized plank 136
   plank 136
   stainless 137
   stainless plank 136
   stair 134
   steel 137
   steel mesh 137
Gravel bankrun 520
   base 547
   fill 520, 533
   pack well 574
   pea 561
   roof 41
   stop 202
Gravity retaining wall 557
Grease interceptor 428
   trap 428
Greenhouse 386
   air supported 385
   cooling 386
Greenhouses demo 376
Grid spike 150
Griddle 341
Grille air return 460
   aluminum 460
   decorative wood 170
   duct 287
   filter 460
   painting 299
   plastic 460

side coiling 227
top coiling 226
window 241
Grinder pump system 429
   shop 346
Grooved block 86
   joint ductile iron 422
   joint fitting 421, 422
   joint pipe 421
Ground 160
   box hydrant 426
   clamp water pipe 486
   cover 563
   face block 87
   fault protection 496
   rod 485, 486
   socket 349
   water monitoring 14
Grounding 485
   & conductor 481-484
   wire brazed 486
Group shower 437
   wash fountain 437
Grout 79
   cavity wall 79
   cement 79
   concrete 79
   concrete block 79
   door frame 79
   epoxy 275, 276, 537
   metallic non-shrink 72
   non-metallic non-shrink . 72
   tile 275
   topping 72
   wall 79
Grouted bar 61
   strand 61
Grouting pressure 537
Guard corner 316, 317
   gutter 204
   house 390
   lamp 509
   service 21
   snow 206
   wall 316, 317
   window 140
Guardrail scaffold 17
   temporary 20
Guards snow 206
   wall & corner 316
Guidance inertial 24
Guide rail 586
   rail cable 586
   rail removal 26
   rail timber 586
Guide/guard rail 586
Gunite 67
   mesh 67
Gutter 204, 389
   aluminum 203
   copper 204
   demolition 178
   guard 204
   lead coated copper 204
   monolithic 550
   stainless 204
   steel 204
   strainer 204
   swimming pool 380
   valley 389
   vinyl 204
   wood 204
Gutting 29
Guyed tower 582
Gym floor underlayment 280
   mat 348

Gymnasium divider curtain .. 349
   equipment 347, 348
   floor 280, 286
   squash court 384
Gypsum block demolition 28
   board 272-274
   board accessories 274
   board leaded 394
   board partition 264, 265
   board removal 263
   board system 264
   drywall 272-274
   fabric wallcovering 287
   firestop 267
   lath 267
   lath nail 149
   partition 265, 314
   plaster 269
   restoration 262
   roof plank demolition ... 178
   shaft wall 264
   sheathing 163
   underlayment 72
   weatherproof 163

H

H pile 540
Hair interceptor 428
Half round molding 166
Hammer bush 67
   pile mobilization 521
Hand carved door 218
   clearing 522
   dryer 318
   excavation 524, 527, 532
   hole 572
   hole concrete 581
   split shake 188
   trowel finish concrete .. 66
Handball court 384
   court backstop 555
Handicap drinking fountain . 440
   fixture support 437
   lever 250
   opener 245
   ramp 63
   water cooler 441
Handle door 362
Handling material 593
   rubbish 29
   unit air 474
   waste 353, 354
Hand-off-automatic control . 504
Handrail 317
   aluminum 287
   fiberglass 173
   wood 166, 170
Handwasher-dryer module 436
Hangar 392
   door 228
Hangars demo 376
Hanger beam 150
   joist 150
Hanging lintel 112
   wire 278
Hardboard cabinet 360
   overhead door 230
   paneling 167
   siding 193
   tempered 167
   underlayment 162
Hardener concrete 66
Hardware 217, 245
   cabinet 362

cloth 556
door 245
drapery 358
finish 246
panic 247
rough 152
window 245
Hardwood carrel 370
   floor stage 346
   grille 170
Harness body 18
Hasp 250
Hat and coat strip 318
   rack 324
Hatch floor 226
   roof 206
   smoke 206
Haul earthwork 533
Hauling 533
   truck 534
Hay 561
   bale 534
Hazardous waste cleanup 31
   waste disposal 31
   waste handling 354
Hea space 467
Head sprinkler 409
Header wood 159
Headers boxed beam 124, 128
Health club equipment 347
Hearth 96
Heat baseboard 475
   electric baseboard 476
   exchanger 469
   exchanger plate 469
   greenhouse 386
   loss 742
   pump 473
   pump residential 496
   pumps air-source 473
   pumps water-source 474
   recovery 354
   reflective glass 256
   temporary 14, 42
   therapy 351
   transfer fluid 468
   transfer package 433
Heated doorway curtain 459
   pads 42
   ready-mix concrete 42
Heater & fan bathroom 458
   electric 338
   floor mounted space 467
   gas fired infra-red 475
   infrared 475
   sauna 383
   swimming pool 466
   terminal 475
   unit 467, 475
   warm air 467
   water 338, 431, 432
Heating . 419, 446, 449, 464-469, 741
   & cooling classroom 475
   electric 476
   estimate electrical 476
   hot air 466, 467, 475
   hydronic 465, 466, 475
   insulation 447
   subsoil 381
   unit convector 475
Heavy construction 559
   duty shoring 16, 17
   framing 161
   lifting 702
   timber 161
Hemlock column 171

Index

Hex bolt steel 104
Hexagonal face block 87
High abuse drywall 273
  build coating 302
  installation conduit 487
  rib lath 268
  rise glazing 255
  strength block 86, 88
  strength bolts 106, 722
  strength steels 116, 724
  temperature vent chimney 464
Highway paver 550
  sign 310
Hinge brass 250, 251
  cabinet 362
  continuous 251
  electric 251
  hospital 251
  paumelle 251
  prison 251
  residential 251
  school 251
  security 251
  special 251
  steel 250
Hinges 732
Hip rafter 158
Historical Cost Index 9
Hockey dasher 381
  scoreboard 348
Hoist automotive 404
  chain 593
  electric 593
  overhead 593
Holder closer 246
Holding tank 598
Holdown 150
Hole drill 535
Hollow core door 218, 223
  metal 214
  metal door 215
  metal stud partition 268
  precast concrete plank 69
  wall anchor 105
Honed block 88
Hood exhaust 345
  fire protection 462
  fume 345
  range 337
Hook coat 318
  robe 318
Hooks clevis 139
Hopper refuse 405
Horizontal boring 569
  drilling 569
Horn fire 517
Hose adapter fire 409
  bibb sillcock 436
  braided bronze 424
  equipment 409
  fire 409
  metal flexible 424
  nozzle 409
  rack 409
  rack cabinet 321
  valve cabinet 321
Hospital cabinet 363
  door hardware 246
  equipment 351
  furniture 370
  hinge 251
  kitchen equipment 342
  partition 313
  tip pin 250, 251
Hot air furnace 466
  air heating 466, 467, 475

water boiler 465, 466
water dispenser 431
water heating 464, 475
water-steam exchange ... 433, 469
water-water exchange 469
Hotel cabinet 319
  furniture 368
  lockset 248, 249
House demolition 27
  garden 391
  guard 390
  telephone 513
Housewrap 186
Hubbard tank 351
Humidification equipment 386
Humidifier 338, 477
  duct 477
  room 477
Humus peat 561
HVAC demolition 444
  duct 453
  piping specialties 450, 451
Hydrant fire 408, 572
  ground box 426
  removal 26
  wall 426
  water 426
Hydrated lime 78
Hydraulic chair 352
  crane 593
  dredge 590
  jacking 702
  seeding 562
Hydrodemolition concrete 25
Hydronic heating 465, 466, 475
Hypalon neoprene roofing 198

I

Ice machine 343
  skating equipment 381
Icemaker 337, 343
Impact barrier 586
Implosive demolition 27
Improvement site 555
Incandescent fixture 506
Incinerator gas 353
  municipal 354
  waste 353
Inclined ramp 403
Increaser vent chimney 464
Incubator 345
Index construction cost 9, 643
Indian granite 551
Indicating panel burglar alarm . 516
Induced draft fan 463
Industrial address system 512
  chimneys 710
  door 225, 227
  equipment installation 330
  folding partition 316
  lighting 506
  railing 135
  safety fixture 439
  window 237
Inert gas 30
Inertial guidance 24
Infra-red broiler 341
  detector 516
Infrared heater 475
Infra-red heater gas fired 475
Inlet curb 551
In-line fan 456
Insecticide 535
Insert concrete formwork 52

formwork 52
  slab lifting 57
In slab conduit 14
Inspection technician 14
Installation extinguisher 322
Instrument switchboard 496
Insulated glass spandrel 257
  panels 185
  precast concrete wall panel .. 70
Insulating concrete formwork .. 49
  glass 256
  lightweight agg concrete 72
  lightweight cel concrete 72
Insulation 181, 183, 419, 447
  batt 182
  blanket 182, 446
  board 181
  boiler 447
  breeching 447
  building 183
  cavity wall 183
  cellulose 183
  composite 185
  duct 446
  equipment 447
  exterior 185
  fiberglass 181-183, 389, 446
  finish system exterior 185
  foam 183, 184
  foam glass 181
  insert 86
  isocyanurate 181
  masonry 183
  mineral fiber 182
  pipe 419
  piping 419
  polystyrene 182, 183
  refrigeration 382
  removal 33, 178
  roof 184, 389
  roof deck 184
  sandwich panel 191
  shingle 187
  spray 184
  sub 381
  urethane 184
  vapor barrier 186
  vermiculite 183
  wall 86, 182, 389
Insurance 10
  builder risk 10
  equipment 10
  public liability 10
Intake-exhaust louver 461
  vent 461
Integral finish concrete 66
  waterproofing 42
Interceptor 428
  grease 428
  hair 428
  metal recovery 429
  oil 428, 429
Intercom 513
Interior beam formwork 42
  door frame 172
  light fixture 506
  pre-hung door 224
  residential door 222
  shutter 359
  wood frame 172
Interlocking block 88
  retaining walls 557
Intermediate conduit 486
Interval timer 491
Intrusion system 516
Invert manhole 578

Inverted bucket steam trap ... 451
  tee beam 68
Iron alloy mechanical joint pipe . 442
  body valve 444, 445
Ironer laundry 335
Ironing center 319
Ironspot brick 279
Irrigation system 559
Isocyanurate insulation 181
I.V. track system 313

J

Jack ladder 18
  post adjustable 110
  pump 16
  roof 158
  screw 536
Jackhammer 528
Jacking 569
Jail equipment 140
Jet water system 574
Jetties dock boat 590
Jeweler safe 333
Job condition 9
Jockey pump fire 411
Joint 209
  assembly expansion 210
  control 66, 80
  expansion 51, 204, 269, 450
  push-on 570
  reinforcing 80
  roof 205
  tyton 570
Jointer shop 346
Joist bridging 120
  chair reinforcing 56
  concrete 62
  connector 150
  deep-longspan 119
  demolition 145
  framing 154
  girder 121
  hanger 150
  longspan 119
  metal framing 128
  open-web 120
  precast concrete 69
  removal 145
  truss 121
  web stiffeners 129
  wood 154, 164
Jumbo brick 83
Jute fiber anchor 105
  mesh 534

K

Kalamein door 221
Keene cement 270
Kennel door 226
  fence 556
Kettle 342
Key cabinet 254, 362
  keeper 323
Keyless lock 249
Keyway footing 46
Kick plate 251
  plate door 251
Kiln dried lumber 152
  vocational 346
King brick 83
Kiosk 391
Kitchen appliance 336

782

Index

cabinet . . . 360, 363
equipment . . . 339, 341
equipment fee . . . 8
exhaust fan . . . 458
sink . . . 434
sink faucet . . . 435
unit . . . 336
K-lath . . . 268
Knob door . . . 249
Kraft paper . . . 186

L

Labeled door . . . 216, 220
frame . . . 214
Labor formwork . . . 707
index . . . 9
Laboratory cabinet . . . 362
equipment . . . 345, 362
sink . . . 345
table . . . 345
Ladder alternating tread . . . 134
exercise . . . 348
jack . . . 18
monkey . . . 349
rolling . . . 18
ship . . . 134
swimming pool . . . 380
towel . . . 318
type cable tray . . . 488
vertical metal . . . 134
Lag screw . . . 107
screw shield . . . 105
Lagging . . . 538
Lally column . . . 110
Laminated beam . . . 164
countertop . . . 364
countertop plastic . . . 364
epoxy & fiberglass . . . 302
framing . . . 164
glass . . . 259
lead . . . 393
roof deck . . . 162
veneer members . . . 165
wood . . . 164
Lamp guard . . . 509
metal halide . . . 509
post . . . 140, 508
Lampholder . . . 503
Lamphouse . . . 344
Landfill fees . . . 28
Landing metal pan . . . 133
newel . . . 169
stair . . . 284
Landscape fee . . . 8
surface . . . 286
Latch deadlocking . . . 248
set . . . 248, 249
Lateral arm awning retractable . . . 325
Latex caulking . . . 209
modified emulsion . . . 546
underlayment . . . 281
Lath demolition . . . 262
gypsum . . . 267
metal . . . 264, 268
Lath, plaster and gypsum board . 733
Lath rib . . . 268
Lathe shop . . . 346
Lattice molding . . . 166
Lauan door . . . 222, 223
Laundry equipment . . . 334, 335
faucet . . . 435
sink . . . 435
tray . . . 435
Lava stone . . . 94
Lavatory . . . 433
faucet . . . 435, 436
removal . . . 414
sink . . . 433
support . . . 437
vanity top . . . 433
wall hung . . . 434
Lawn grass . . . 562
seed . . . 562
Lazy susan . . . 360, 361
Leaching field chamber . . . 575
pit . . . 575
Lead barrier . . . 290
caulking . . . 427
coated copper downspout . . . 203
coated copper gutter . . . 204
coated downspout . . . 203
coated flashing . . . 201
flashing . . . 200
glass . . . 394
gypsum board . . . 394
lined darkroom . . . 383
lined door frame . . . 393
paint encapsulation . . . 35
paint remediation methods . . . 704
paint removal . . . 35
plastic . . . 394
roof . . . 199
salvage . . . 414
screw anchor . . . 106
sheets . . . 393
shielding . . . 393
Leak detection . . . 518
detection probe . . . 518
detection tank . . . 518
Lean-to type greenhouse . . . 386
Lectern . . . 355
Lecture hall seating . . . 371
Ledger wood . . . 158
Lens movie . . . 344
Let-in bracing . . . 152
Letter sign . . . 309
slot . . . 323
Leveler dock . . . 331
truck . . . 331
Leveling jack shoring . . . 16
Lever handicap . . . 250
Lexan . . . 258
Liability employer . . . 10
insurance . . . 696
Library chair . . . 370
equipment . . . 370
furniture . . . 370
shelf . . . 370
Life guard chair . . . 380
Lift . . . 404
automotive . . . 404
correspondence . . . 404
parcel . . . 404
scissor . . . 593
slab . . . 63, 716
wheelchair . . . 403
Lifter platform . . . 331
Lifting insert slab . . . 57
Light base formwork . . . 46
border . . . 508
dental . . . 352
exit . . . 507
fixture interior . . . 506, 508
loading dock . . . 331
nurse call . . . 513
pole . . . 508, 581
pole aluminum . . . 508
pole steel . . . 508
shield . . . 586
stand . . . 31
strobe . . . 508
support . . . 110
temporary . . . 14
underwater . . . 380
Lighting . . . 506, 507
darkroom . . . 336
emergency . . . 507
exit and emergency . . . 507
explosionproof . . . 508
fixture exterior . . . 508, 509
fountain . . . 442
incandescent . . . 507
industrial . . . 506
outlet . . . 494
residential . . . 494
roadway . . . 508
stage . . . 508
strip . . . 506
surgical . . . 351
Lightning protection demo . . . 376
suppressor . . . 491
Lightweight angle framing . . . 112
block . . . 89
channel framing . . . 112
columns . . . 110
concrete . . . 62, 64, 719
floor fill . . . 62
framing . . . 111
junior beam framing . . . 112
natural stone . . . 94
tee framing . . . 112
Zee framing . . . 112
Lime finish . . . 78
hydrated . . . 78
Limestone . . . 94, 562
coping . . . 97
Line gore . . . 552
Linen chute . . . 404
collector . . . 405
wallcovering . . . 288
Liner cement . . . 569
duct . . . 455
flue . . . 96
pipe . . . 569
Link mat . . . 366
Lint collector . . . 334
Lintel . . . 94
block . . . 89
hanging . . . 112
precast concrete . . . 71
steel . . . 112, 113
Liquid chiller centrifugal . . . 470
Load center residential . . . 491
Load test pile . . . 520
Loader front end . . . 528
vacuum . . . 32
Loading dock . . . 331
dock equipment . . . 331
dock light . . . 331
Loam . . . 520, 523
Lobby collection box . . . 323
Lock electric release . . . 249
entrance . . . 250
keyless . . . 249
tile . . . 367
time . . . 229
tubular . . . 248
Locker metal . . . 322
steel . . . 322
wall mount . . . 322
wire mesh . . . 322
Locking receptacle . . . 493
Lockset communicating . . . 248
cylinder . . . 248
hotel . . . 248, 249
mortise . . . 248
Log electric . . . 320
gas . . . 320
Longspan joist . . . 119
Louver . . . 259, 460
aluminum . . . 259, 461
aluminum operating . . . 461
coating . . . 461
cooling tower . . . 461
fixed blade . . . 461
intake-exhaust . . . 461
midget . . . 259
mullion type . . . 461
redwood . . . 260
ventilation . . . 260
wall . . . 260
wood . . . 170
Louvered door . . . 223, 224
Low sound fan . . . 456
Low-voltage silicon rectifier . . . 502
switching . . . 502
switchplate . . . 502
transformer . . . 502
Lube equipment . . . 330
Lug terminal . . . 483
Lumber core paneling . . . 168
kiln dried . . . 152
product prices . . . 729
treated . . . 158
treatment . . . 152
Luminous ceiling . . . 277, 509
panel . . . 277

M

Macadam . . . 547
penetration . . . 547
Machine excavation . . . 532
screw . . . 107
trowel finish concrete . . . 66
Machinery anchor . . . 107
Magazine shelving . . . 370
Magnesium dock board . . . 331
oxychloride . . . 207
Magnetic astragal . . . 252
motor starter . . . 500
particle test . . . 14
Mahogany door . . . 218
Mail box . . . 323
box call system . . . 513
chute . . . 323
slot . . . 323
Main drain . . . 430
office expense . . . 11
Maintenance . . . 22
carpet . . . 262
Make-up air unit . . . 472
Mall front . . . 234
Management fee construction . . . 8
Manhole . . . 579
concrete . . . 579, 581
cover . . . 578
electric service . . . 582
invert . . . 578
precast . . . 579
raise . . . 578
removal . . . 25
step . . . 580
Mansard aluminum . . . 191
Mantel beam . . . 170
fireplace . . . 170
Map rail . . . 307
Maple countertop . . . 364
floor . . . 280
wall . . . 384
Marble . . . 94

Index

chips ... 561	Mercury vapor lamp ... 509	stud ... 264, 266, 267	metal ... 269
coping ... 97	Mesh fence ... 553	stud demolition ... 263	pine ... 166
countertop ... 364	gunite ... 67	stud framing ... 124	trim ... 166
floor ... 94	partition ... 314	studs ... 124	window and door ... 167
screen ... 312	security ... 269	support assemblies ... 265	wood ... 169
shower stall ... 312	stucco ... 268	threshold ... 250	wood transition ... 280
sill ... 95	wire ... 556	tile ... 276	Monel rivet ... 108
soffit ... 94	Metal aluminum ... 213	toilet partition ... 311, 312	Money safe ... 333
stair ... 94	bin retaining walls ... 558	trash receptacle ... 366	Monitor support ... 110
synthetic ... 279	bookshelf ... 370	truss framing ... 132	Monkey ladder ... 349
tile ... 279	butt frame ... 214	water blasting ... 101	Monolithic finish concrete ... 66
Marina small boat ... 590	cabinet ... 362, 363	window ... 236, 237, 389	gutter ... 550
Markings pavement ... 552	casework ... 254	wire brushing ... 101	terrazzo ... 283
Mark-up cost ... 11	chalkboard ... 306, 315	Metallic foil ... 183	Monorail ... 592
Masking ... 76	chimney ... 82, 319, 463, 464	hardener ... 66	Monument survey ... 24
Mason scaffold ... 16	cleaning ... 101	non-shrink grout ... 72	Mop holder strip ... 318
Masonry ... 77	deck ... 121	Meter electric ... 496	roof ... 196
accessory ... 80	deck acoustical ... 123	steam condensate ... 452	sink ... 437
aggregate ... 79	deck composite ... 122	venturi flow ... 452	Mortar ... 78
anchor ... 79	deck demolition ... 178	water supply ... 424	Mortar, brick and block ... 721
base ... 90	deck ventilated ... 123	water supply domestic ... 424	Mortar cement ... 276, 720
brick ... 82, 278	door ... 213, 230, 389	Metric conversion ... 715	masonry cement ... 78
catch basin ... 579	door frame ... 389	rebar specs ... 708	pigments ... 79
cement ... 78	door residential ... 216	Microphone ... 512	Portland cement ... 78
cleaning ... 76	ductwork ... 453	Microtunneling ... 569	restoration ... 78
color ... 79	facing panel ... 192	Microwave oven ... 337	sand ... 79
demolition ... 28, 77, 263	fascia ... 202	Mill construction ... 161	testing ... 13
fireplace ... 96	fence ... 553	extras steel ... 116	thinset ... 276
flashing ... 200	flashing ... 389	Millwork ... 165, 361	Mortise lockset ... 248
furring ... 160	flexible hose ... 424	demolition ... 148	Mortuary equipment ... 353
insulation ... 183	floor deck ... 123	Mineral fiber ceiling ... 277, 278	refrigeration ... 353
manhole ... 579	flue chimney ... 463	fiber insulation ... 182	Mosaic glass ... 276
nail ... 149	frame ... 214	insulated cable ... 484	Moss peat ... 561
needle ... 538	framing parapet ... 131	roof ... 199	Motion detector ... 516
panel ... 92	furring ... 265	wool blown in ... 183	Motor connection ... 490
pointing ... 76	gutter ... 389	Minor site demolition ... 25	dripproof ... 510
reinforcing ... 80, 720	halide lamp ... 509	Mirror ... 257, 318	electric ... 510
removal ... 26, 77	hollow ... 214	ceiling board ... 277	starter ... 500
sill ... 95	in field galvanizing ... 302	door ... 257	starter & control ... 500
step ... 546	in field painting ... 302	glass ... 257	starter magnetic ... 500
testing ... 13	in shop galvanizing ... 101	plexiglass ... 258	starter w/circuit protector ... 500
toothing ... 28	interceptor dental ... 429	wall ... 257, 258	starter w/fused switch ... 500
ventilator ... 80	joist bracing ... 127	Miscellaneous metal ... 326	support ... 110
wall ... 86, 558	joist bridging ... 127	painting ... 292, 297	Motorized car ... 592
wall tie ... 79	joist framing ... 128	Mix design asphalt ... 13	roof ... 387
waterproofing ... 79	lath ... 264, 268	planting pit ... 561	Motors ... 510
Massage table ... 348	locker ... 322	Mixed bituminous concrete plant ... 520	Moulded door ... 221
Master clock system ... 513	miscellaneous ... 326	Mixer food ... 341	Mounted booth ... 369
Mastic floor ... 284	molding ... 269	Mixes concrete ... 714, 715	Mounting board plywood ... 163
Mat blasting ... 528	nailing anchor ... 105	Mixing valve ... 436	Movable louver blind ... 359
floor ... 366	ornamental ... 326	Mobile shelving ... 370	office partition ... 314
foundation ... 63, 65	overhead door ... 230	X-ray ... 353	Movie equipment ... 344
foundation formwork ... 46	pan ceiling ... 277	Mobilization ... 521, 542	lens ... 344
gym ... 348	pan landing ... 133	air-compressor ... 521	projector ... 344
wall ... 348	pan stair ... 133	equipment ... 537, 543	screen ... 344
Material handling ... 592, 593	panel ... 191	or demob. ... 19	Moving building ... 30
handling belt ... 592	partition framing ... 124	Model building ... 8	ramp ... 403
handling system ... 592	pipe ... 575	Modification to cost ... 9	ramps and walks ... 739
index ... 9	pipe removal ... 414	Modified bitumen roofing ... 730	shrub ... 565
Materials concrete ... 714	plate stair ... 133	Modular office system ... 314, 315	tree ... 565
Meat case ... 339	pressure washing ... 101	playground ... 350	walk ... 403
Mechanical ... 739	rafter framing ... 131	Module handwasher-dryer ... 436	Mowing brush ... 522
duct ... 453	recovery interceptor ... 429	Modulus of elasticity ... 13	Mud sill ... 158
fee ... 8	roof ... 189	Moisture barrier ... 180	Mulch bark ... 561
media filter ... 462	roof parapet framing ... 130	content test ... 13, 14	ceramic ... 561
Median barrier ... 586	roof truss ... 116	Mold abatement ... 36	stone ... 561
precast ... 586	sandblasting ... 101	abatement work area ... 36	Mulching ... 561
Medical distiller ... 351	sash ... 237	contaminated area demolition ... 37	Mullion type louver ... 461
equipment ... 351, 353	screen ... 237	Molding ... 165	vertical ... 232
exam equipment ... 351	sheet ... 199	base ... 165	Multi-blade damper ... 454
X-ray ... 353	shelf ... 323	brick ... 166	Multi-channel rack enclosure ... 512
Medicine cabinet ... 319	shingle ... 187	ceiling ... 165	Multizone air cond. rooftop ... 472
Membrane ... 194	siding ... 189, 191	chair ... 166	Municipal incinerator ... 354
curing concrete ... 68	sign ... 309	cornice ... 166	Muntin window ... 241
flashing ... 196	soffit ... 194	exterior ... 166	Mushroom ventilator stationary ... 461
waterproofing ... 179	steam cleaning ... 101	hardboard ... 168	Music room ... 383

Index

Mylar tarpaulin 20

N

Nail 148, 149
   lead head 393
   stake 54
Nailer wood 155
Nailing anchor 105
Napkin dispenser 318
Needle masonry 538
Neoprene adhesive 180
   expansion joint 204
   flashing 201
   floor 284
   gasket 210
   roof 198
   waterproofing 180
Net safety 15
   tennis court 553
Netting flexible plastic 328
Newel 169
Newspaper rack 370
Night depository 333
No hub pipe 427
Non-destructive testing 14
Non-metallic non-shrink grout .. 72
Non-removable pin 250
Norwegian brick 83
Nosing mat 367
   rubber 281
   safety 281
   stair 281, 283
Nozzle dispersion 410
   fire hose 409
   fog 409
Nurse call light 513
   call system 513
   speaker station 513
   station cabinet 363
Nursing home bed 370
Nuts remove 101
Nylon carpet 168, 286, 288
   nailing anchor 105

O

Oak door frame 171
   floor 280
   molding 166
   paneling 168
   stair tread 170
   threshold 172
Oakum caulking 427
Obscure glass 256
Observation dome 390
   well 574
Office expense 11
   field 15
   floor 287
   furniture 367, 368
   partition 314
   partition movable 314
   safe 332
   system modular 315
   trailer 15
Oil filled transformer 496
   fired boiler 465
   fired furnace 466
   fired space heater 467
   formwork 54
   interceptor 428, 429
   storage tank 449
   water heater 432

water heater commercial 432
Olive knuckle hinge 251
Omitted work 11
One piece astragal 252
One-way vent 207
Onyx 283
Opener automatic 245
   door 212, 227, 231, 245
   handicap 245
Opening framing 389
   roof frame 112
Open-web joist 120
Operable partition 316
Operating room equipment 353
Operator automatic 245
Options elevator 401
Ornamental aluminum rail ... 141
   columns 142
   glass rail 141
   metal 326
   railing 141
   steel rail 141
   wrought iron rail 141
OSHA testing 33
Outdoor bleacher 387
Outlet box steel 487
   lighting 494
Outrigger wall pole 327
Oval arch culverts 577
Oven 336, 341
   cabinet 361
   convection 342
   microwave 337
Overbed table 371
Overhang eave 388
   endwall 389
Overhaul 29, 534
Overhead & profit 11
   bridge cranes 592
   commercial door 229
   contractor 22, 696
   door 229-231
   hoist 593
   support 214
Overlapping astragal 252
Overlay face door 219, 220
Overpass 559
Oversized brick 82
Overtime 9, 11, 695
Oxygen lance cutting 29

P

P condensate 452
   trap 427
   trap running 427
Package chute 405
   receiver 333
Packaging waste 34
Pad bearing 108
   equipment 62
   vibration 108
Padding carpet 285
Paddle blade air circulator ... 458
   fan 495
   tennis court 350
Pads heated 42
Paint aluminum siding 294
   chain link fence 292
   doors & windows exterior ... 295
   doors & windows interior ... 298
   encapsulation lead 35
   fence picket 292
   floor 299
   floor concrete 299

floor wood 299
   removal 35, 36
   siding 294
   trim exterior 296
   walls masonry, exterior ... 296
Painted traffic lines and markings ....
   552
Painting 736
   balustrade 300
   bar joist 121
   booth 330
   casework 297
   ceiling 300
   clapboard 293
   cornice 300
   decking 293
   drywall 300
   grille 299
   metal in field 302
   miscellaneous 292, 297
   parking stall 552
   pavement 552
   pipe 299
   plaster 300
   railing 293
   reflective 552
   shutters 293
   siding 293
   stair stringers 293
   steel 302
   steel siding 293
   stucco 293
   swimming pool 380
   temporary road 552
   tennis court 553
   thermoplastic 552
   trellis/lattice 293
   trim 299
   truss 300
   wall 294, 300
   window 297
Paints & coatings 290
Palladian windows 240
Pallet rack 323
Pan shower 201
   slab 62
Panel acoustical 289, 315
   brick wall 92
   ceramic tile 276
   divider 315
   door 221, 222
   facing 95
   fiberglass 20, 190
   fire 517
   FRP 288
   luminous 277
   masonry 92
   metal 191
   metal facing 192
   portable 315
   precast concrete double wall ... 69
   sandwich 190
   sound 393
   spandrel 257
   structural 160
Panel vision 244
Panel wall 63, 97
   woven wire 314
Panelboard 499
   electric 499
   w/circuit-breaker .. 498, 499
Paneled door 216
   pine door 223
Paneling 167, 168
   board 169
   cutout 28

demolition 148
hardboard 167
plywood 168
wood 167
Panelized shingle 188
Panels insulated 185
   prefabricated 185
Panic bar 248
   device 247
   device door hardware 246
Paper building 186
   curing concrete 68
   sheathing 186
   tubing reinforcing 57
Paperhanging 287
Paperholder 318
Paraffin bath 351
Parallel bar 348, 351
Parapet metal frami 131
Parcel lift 404
Parging cement 179
Park bench 373
Parking barrier 552
   barrier precast 552
   bumper 552
   equipment 140, 330, 331, 369, 371,
   393
   garage 390
   lots paving 548
   stall painting 552
Parquet floor 279
Particle board siding 193
   board underlayment ... 162
   core door 219
Parting bead 167
Partition 315
   acoustical 315, 316
   anchor 80
   block 89, 90
   blueboard 265
   bulletproof 333
   demolition 263
   demountable 314
   door 315
   drywall 264
   dust 29
   folding 316
   folding leaf 316
   framing 266, 267
   gypsum 265, 314
   hospital 313
   mesh 314
   movable office 314
   office 314
   operable 316
   plaster 264
   portable 316
   refrigeration 382
   shower 94, 312
   steel 315, 316
   stud 268
   support 110
   thin plaster 265
   tile 85
   toilet 94, 310-312
   wall 264, 265
   wire 314
   wire mesh 314
   wood frame 155
   woven wire 314
Parts bin 323
Passage door 222-224
Passenger elevators 738
Patch core hole 13
   formwork 54
   roof 178

Index

Patching asphalt 520
   concrete floor 40
   concrete wall 40
Patient nurse call 513
Patio 547
   block 94, 196
   block concrete 549
   door 226
Patio/deck canopy 325
Patterned glass 256
Paumelle hinge 251
Pavement 549
   asphalt 553
   asphaltic 547
   berm 551
   demolition 25
   emulsion 546
   markings 551, 552
   painting 552
   replacement 548
   sealer 553
   slate 95
Paver concrete block 549
   floor 278
   highway 550
Paving athletic 553
   block granite 550
   brick 549
   concrete 548
   parking lots 548
Pea gravel 561
   stone 41
Peastone 563
Peat humus 561
   moss 561
Pedestal floor 287
   type seating 371
Pedestrian bridge 559
   traffic control 332
Peephole 249
Pegboard 167
Penetration macadam 547
   test 13
Penthouse roof louver 461
Perforated aluminum pipe 578
   ceiling 277
   pipe 579
   PVC pipe 579
Performance bond 11, 700
Perlite insulation 182
   plaster 269
   sprayed 302
Permit building 12
Personal respirator 31
Personnel field 10
   protection 15
Petrographic analysis 13
Pew church 371
Phone booth 310
Photography 12
   aerial 12
   construction 12
   time lapse 12
Physician's scale 351
PIB roof 197
Picket railing 135
Picture window 236, 237, 239
Pier brick 82
   concrete 62
Pigments mortar 79
Pilaster toilet partition 312
   wood column 171
Pile 543
   boot 541
   bored 541
   cap 63, 65

cap formwork 46
cutoff 520
driven 539
driving 702
driving mobilization 521
encasement 541
foundation 543
friction 541
H 538, 540
high strength 536
lightweight 536
load test 520
mobilization hammer 521
pipe 540, 543
point 540
point heavy duty 540
precast 539
prestressed 539, 543
sod 561
splice 540, 541
steel 540
steel sheet 536
step tapered 540
testing 520
timber 540
treated 541
wood sheet 536
Piling sheet 536, 749
Pillow T 573
Pin floating 250
   non-removable 250
   powder 108
Pine door 222
   door frame 171
   fireplace mantel 170
   floor 280
   molding 166
   roof deck 162
   shelving 324
   siding 193
   stair 169
   stair tread 169
Pipe 579
   & fittings .. 416, 417, 420-423, 425,
   426, 428, 430, 438, 439, 442, 445, 739
   acid resistant 442
   add elevated 414
   aluminum 576, 579
   bedding 533
   bedding trench 527
   bollards 551
   brass 420
   bumper 317
   cast iron 427
   cleanout 418
   concrete 574, 576-578
   copper 420
   corrosion resistant 442
   corrugated 579
   corrugated metal 575, 578
   covering 419
   covering fiberglass 419
   CPVC 424
   double wall 580, 581
   drain 531
   drainage 442, 574-576
   ductile iron 570
   DWV ABS 423
   DWV PVC 423
   elbow 422
   elevated installation add 414
   encapsulation 34
   epoxy fiberglass wound 442
   fire retardant 442
   fitting grooved joint 422
   gas 580

glass 442
grooved joint 421
insulation 419
insulation removal 33
iron alloy mechanical joint 442
liner 569
metal 575
no hub 427
painting 299
perforated aluminum 578
pile 540, 543
plastic 423, 424
plastic, FRP 423
polyethylene 580
polypropylene 442
proxylene 442
PVC 571, 574, 575
rail aluminum 134
rail galvanized 135
rail stainless 135
rail steel 135
rail wall 135
railing 134
reinforced concrete 576
relay 531
removal 26
removal metal 414
sewage 574-576
shock absorber 426
single hub 427
sleeve 53
soil 427
stainless steel 423
steel 420, 421, 576, 579
subdrainage 578, 579
support framing 113
tee 422
water 570
weld joint 421
wrapping 569
Piping designations 751
Piping gas station 580
   HVAC specialties 450
   insulation 419
   specialties HVAC 450, 451
   subdrainage 578
Pit excavation 527
   leaching 575
   scale 328
   sump 531
Pitch coal tar 195
   emulsion tar 546
   pocket 207
Pits test 25
Pivoted window 237
Placing concrete 64, 716
   reinforcment 709
Plain tube framing 232
Planer shop 346
Plank floor 161
   grating 136
   hollow precast concrete 69
   precast concrete nailable 69
   precast concrete roof 69
   precast concrete slab 69
   roof 161
   scaffolding 17
   sheathing 163
Plant bed preparation 561
   mixed bituminous concrete 520
Planter 372
   bench 372
   concrete 372
   fiberglass 372
Planting 565
Plaque 309

Plaster 264
   accessories 269
   beam 269
   ceiling 269
   cement 381
   column 269
   cutout 28
   demolition 262, 263
   drilling 265
   gauging 270
   ground 160
   gypsum 269
   painting 300
   partition 264
   partition thin 265
   perlite 269
   thincoat 271
   vermiculite 269
   wall 269
Plasterboard 272
Plastic angle valve 417
   ball valve 417
   faced hardboard 167
   fences 556
   fireproofing 207
Plastic, FRP pipe 423
Plastic glazing 257
   grating 174
   grille 460
   laminate door 219, 220
   laminated countertop 364
   lead 394
   pipe 423, 424
   roof ventilator 457
   screw anchor 106
   sign 309
   skylight 244
   valve 417
   window 242
Plate checkered 138
   glass greenhouse 386
   heat exchanger 469
   push-pull 250
   shear 150
   steel 113
   stiffener 113
   wall switch 503
   wood 159
Platform checkered plate 138
   lifter 331
   telescoping 346
   tennis 350
Plating zinc 149
Players bench 373
Playfield equipment 349
Playground equipment . 349, 350, 555
   modular 350
   slide 350
   surface 282
   whirler 350
Plenum demolition 263
Plexiglass 257
   mirror 258
Plug in circuit-breaker 501
   in switch 501
   wall 81
Plugmold raceway 502
Plumbing 425
   appliance 431, 432
   demolition 414
   fixture 429, 433, 440, 452, 574, 740
   fixtures removal 414
   laboratory 346
Plywood 729
   clip 150
   demolition 178, 262, 263

Index

fence 21	cabinet 340	fireplace 319	operator 532
floor 162	eye wash 439	panels 185	sewage ejector 429
formwork 42, 44	fire extinguisher 321	stair 169	shallow well 574
formwork steel framed 49	floor 281	Prefinished door 219	staging 16
joist 164	panel 315	drywall 273	submersible 431
mounting board 163	partition 316	floor 280	sump 338, 430
paneling 168	post 332	shelving 324	water 574
sheathing roof & walls ... 163	scale 328	Preformed roof panel 189	Pumped concrete 64
shelving 324	stage 346	Pre-hung door 223, 224	Pumping 531
sidewalk 21	Portland cement 41	Preparation plant bed 561	dewater 531
siding 193	cement terrazzo 283	surface 290, 291	station 429
sign 309	Post 332	Pressure grouting cement ... 537	Purlin roof 161
soffit 167	athletic 350	reducing valve water 416	Push button lock 248
subfloor 162	cap 150	regulator oil 424	Push-on joint 570
treatment 152	cedar 171	regulator steam 424	Push-pull plate 249
underlayment 162	concrete 586	relief valve 416	Putlog scaffold 18
Pneumatic control systems 447	demolition 146	switch switchboard 498	Putting surface 286
stud driver 108	fence 554	valve relief 416	Puttying 290
tube 333	lamp 140, 508	wash 291	PVC adhesive 180
tube system 405	portable 332	washing metal 101	blind 358
Pocket door frame 172	recreational 349	Pressurized fire extinguisher 322	cleanout 418
pitch 207	shores 53	Prestressed concrete 718	conduit in slab 489
Point heavy duty pile ... 540	tennis court 553	concrete piles 539	control joint 80
pile 540, 541	tensioned concrete 70	pile 539, 543	cornerboards 174
Pointing masonry 76	Postal specialty 323	precast concrete 717	door casing 174
Poisoning soil 535	Posts sign 310	Prestressing steel 61	DWV pipe 423
Poke-thru fitting 487	Post-tensioned concrete ... 718	Pretreatment termite 535	fascia 174, 175
Pole aluminum light 508	Potable water softener ... 431	Prices lumber products ... 729	fitting 571, 572
athletic 349	water treatment 431	Prime coat 547	flashing 201
closet 166	Potters wheel 346	Primer asphalt 178, 282	frieze 175
cross arm 581	Poured insulation 183	steel 302	gravel stop 202
electric 581	Powder actuated tool 107	Prison door 229	molding exterior 174
light 508	charge 108	equipment 229	pipe 423, 571, 574, 575
portable decorative 332	pin 108	fence 556	rake 175
steel light 508	Power equipment 700	hinge 251	sheet 180, 286
utility 508, 581	system & capacitor 504	toilet 334	siding 193
Polyacrylate floor 284	temporary 14	Process air handling fan ... 456	soffit 175
Polycarbonate glazing 258	wiring 503	Proctor compaction test ... 14	underground duct 582
Polyester floor 285	Preblast survey 528	Produce case 339	valve 417
Polyethylene backer rod ... 52	Precast beam 68	Product dispenser 448	waterstop 54
coating 569	bridge 559	piping 581	Pyrex pipe 442
film 561	catch basin 579	Productivity glazing 732	
floor 286	column 68	Profile block 87	
pipe 580	concrete beam 68	Progress schedule 12	**Q**
pool cover 380	concrete channel slab 69	Project overhead 11	
septic tank 575	concrete column 68	sign 21	Quad precast concrete tee beam . 69
tarpaulin 20	concrete joist 69	Projected window 236, 237	Quarry drilling 528
waterproofing 179, 180	concrete lintel 71	Projection screen 307, 344	tile 276
Polymer trench drain 430	concrete nailable plank ... 69	Projector movie 344	Quarter round molding ... 166
Polyolefin roofing thermoplastic . 198	concrete roof plank 69	Propeller exhaust fan 458	Quartz 561
Polypropylene pipe 442	concrete slab plank 69	unit heaters 476	chips 41
shower 434	concrete stairs 69	Property line survey 24	Quoins 94
valve 417	concrete tee beam 68	Protection corner 317	
Polystyrene blind 173	concrete tee beam double ... 69	fire 321	
ceiling 277, 382	concrete tee beam quad ... 69	slope 534, 536	**R**
ceiling panel 277	concrete tee beam single ... 69	stile 252	
insulation 182, 183, 382	concrete wall pan insulated 70	termite 535	Raceway 487, 488, 490
Polysulfide caulking 209	concrete wall panel tilt-up ... 71	winter 20, 42	conduit 486
Polyurethane caulking ... 210	concrete wall wall 70	worker 32	plugmold 502
floor 284, 286	concrete window sill 71	Proxylene pipe 442	surface 502
roof 198	coping 97	P&T relief valve 416	wiremold 502
varnish 302	curb 551	Public address system ... 512	wireway 487
Polyvinyl chloride roof ... 197	manhole 579	Pull box 487	Rack bicycle 349
soffit 167, 194	median 586	door 362	coat 324, 361
tarpaulin 20	members 717	Pulpit church 355	hat 324
Pool accessory 380	parking barrier 552	Pump 426, 452, 574	hose 409
cover 380	pile 539	circulating 452	pallet 323
cover polyethylene 380	receptor 312	concrete 64	Racquetball court 384
heater electric 466	septic tank 575	condensate return 452	Radial arch 164
swimming 379	terrazzo 283	contractor 531	wall 48
therapeutic 381	wall 719	duplex 452	wall formwork 48
Porcelain tile 275	Pre-engineered steel bldg demo ... 377	fire 411	Radiator cast iron 475
Porch framing 156	steel buildings 387, 737	fountain 441	supply control 450
molding 166	Prefab metal flue 463	gasoline 448	thermostat co 447
Portable booth 391	Prefabricated building 386	heat 473	Radio frequency shielding ... 395
building 15	comfort station 390	jack 16	

787

Index

tower	582
Radiography test	14
Radiology equipment	353
Rafter anchor	150
composite	158
demolition	146
framing metal	131
hip	158
metal bracing	130
metal bridging	130
tie	158
valley	158
wood	158
Rail aluminum pipe	134
crane	592
crash	317
dock shelter	332
galvanized pipe	135
guide	586
guide/guard	586
map	307
ornamental aluminum	141
ornamental glass	141
ornamental steel	141
ornamental wrought iron	141
railroad	587
stainless pipe	135
steel pipe	135
trolley	317
wall pipe	135
Railing cable	141
church	355
demolition	148
industrial	135
ornamental	141
picket	135
pipe	134
wood	166, 169, 170
Railroad ballast	584
bumper	584
concrete ties	584
derail	584
rail	587
siding	587, 751
tie	547, 564
tie step	546
timber switch ties	584
timber ties	584
track accessories	584
track heavy rail	584
track maintenance	584
track removal	26
turnout	587
wheel stop	584
Raise manhole	578
manhole frame	578
Raised floor	287
Rake PVC	175
Ramp approach	287
handicap	63
moving	403
temporary	20
Ranch plank floor	280
Range cooking	336
hood	337
receptacle	493, 503
restaurant	342
shooting	349
Ratio water cement	13
Razor wire	556
Reading table	370
Ready mix concrete	64, 716
Ready-mix concrete heated	42
Receiver ash	366
trash	366
Receptacle air conditioner	493
device	493
dryer	493
duplex	503
GFI	493
locking	493
range	493, 503
telephone	494
television	494
trash	372
waste	318
weatherproof	493
Receptor precast	312
shower	312, 437
terrazzo	312
Recessed mat	366
Reciprocating water chiller	470
Recirculating chemical toilet	598
Recorder videotape	517
Recore cylinder	248, 249
Recreational post	349
Rectangular diffuser	459
ductwork	453
Rectifier low-voltage silicon	502
Redwood bark mulch	561
cupola	326
louver	260
paneling	169
siding	193
tank	573
trim	166
wine cellar	340
Refinish floor	280
Reflective block	91
glass	257
insulation	183
painting	552
sign	309
Refrigerant removal	34
Refrigerated case	339
wine cellar	340
Refrigeration	381
bloodbank	345
commercial	339
equipment	381
floor	382
insulation	382
mortuary	353
panel fiberglass	382
partition	382
residential	337
walk-in	382
Refrigerator door	228
Refuse chute	404
hopper	405
Register air supply	460
cash	334
return	460
steel	460
wall	460
Reglet	204
aluminum	204
galvanized steel	204
Regulator oil pressure	424
steam pressure	424
Reinforced concrete pipe	576
culvert	576
plastic panel fiberglass	288
PVC roof	197
Reinforcement	709
welded wire	710
Reinforcing	59, 709
accessories	55
bar chair	56
bar spacer	57
bar tie	55
beam	59
beam bolster	55
clip tie	55
coating	60
column	59
dowel	59
epoxy coated	60
estimating	59
flange clip	55
footing	59
galvanized	60
girder	59
high chair	56
joint	80
joist chair	56
masonry	80
metric	708
paper tubing	57
slab	59
slab bolster	55
sorting	59
spiral	59
splice	58
splicing	58
steel	708, 709
steel accessory	56, 57
steel fiber	61
subgrade chair	57
synthetic fiber	61
testing	13
tie wire	58
wall	59
Relay pipe	531
Release door	513
Relief valve temperature	416
vent ventilator	461
water vacuum	425
Removal air conditioner	444
asbestos	32, 33
bathtub	414
block wall	27
boiler	444
catch basin	25
chain link fence	26
concrete	25
concrete pipe	26
curb	25
driveway	25
fence	26
fixture	414
floor	27
guide rail	26
hydrant	26
insulation	33
lavatory	414
masonry	26
paint	35
pipe	26
pipe insulation	33
plumbing fixtures	414
railroad track	26
refrigerant	34
shingle	178
sidewalk	26
sink	414
sod	561
steel pipe	26
stone	26
stump	522
tank	30
tree	521, 522
urinal	414
VAT	33
water closet	414
water fountain	414
water heater	414
window	212
Remove bolts	101
nuts	101
Rendering	8
Repellent water	300
Replacement pavement	548
sash	238
Restaurant roof	178
Residential alarm	495
appliance	334, 336, 337, 495
closet door	217
device	491
door	172, 216, 221
door bell	494
dryer	338
elevator	399
fan	495
fence	554
fixture	494
folding partition	316
garage	390
garage door	230
greenhouse	386
gutting	29
heat pump	496
hinge	251
lighting	494
load center	491
lock	248
overhead door	230
refrigeration	337
roof jack	458
service	491
smoke detector	495
stair	169
storm door	213
switch	491
transition	458
wall cap	458
washer	337
water heater	431, 495
wiring	491, 495
Resilient base	281
floor	281, 282
pavement	553
Respirator	32
personal	31
Resquared shingle	187, 188
Restaurant furniture	368
range	342
Restoration gypsum	262
mortar	78
window	36
Retaining wall	63, 558
wall formwork	48
wall segmental	557
walls interlocking	557
walls stone	558
Retarder vapor	186
Retractable lateral arm awning	325
Return register	460
Revolving darkroom	383
dome	390
door	228, 233, 385
entrance door	233
Rewind table	344
Rib lath	268
Ribbed siding	191
waterstop	54
Ridge board	158
cap	187, 189
flashing	389
roll	190
vent	206
Rig drill	24
Rigid anchor	80
conduit in-trench	489

Index

in slab conduit 489
insulation 182
Ring split 150
toothed 151
Rip rap 536
Riser rubber 281
stair 283
terrazzo 283
wood stair 169
Rivet 108
aluminum 108
copper 108
monel 108
stainless 108
steel 108
tool 108
Road base 547
berm 551
sign 310
temporary 20
Roadway lighting 508
Robe hook 318
Rock bolt drilling 527
bolting 535
drill 24, 528
excavation 528
Rod backer 209
closet 324
curtain 317
ground 485, 486
tie 112, 536
weld 109
Roll ridge 190
roof 194
roofing 199
type air filter 462
Roller compaction 532
sheepsfoot 532
Rolling door 227, 234
earth 534
ladder 18
roof 235
service door 227
tower scaffold 18
Roman brick 83
Romex copper 484
Roof accessories 206
adhesive 194
aluminum 189
baffle 207
beam 164
bracket 18
built-up 194, 195
cant 158, 195
clay tile 188
coating 178
copper 199
CSPE 197
curb 158
deck 161
deck insulation 184
deck laminated 162
deck wood 162
decking 122
dome 235
drains 430
elastomeric 198
EPDM 197
exhauster 457
expansion joint 205
fan 457
fiberglass 190
fill 719
frame opening 112
framing removal 28
gravel 41
hatch 206
insulation 184, 389
jack 158
jack residential 458
joint 205
lead 199
metal 189
metal rafter framing 131
metal truss framing 132
mineral 199
modified bitumen 196
mop 196
nail 149
panel preformed 189
patch 178
PIB 197
polyvinyl chloride 197
purlin 161
rafter bracing 130
rafter bridging 130
rafters framing 158
reinforced PVC 197
resaturant 178
roll 194
rolling 235
safety anchor 18
sheathing 163
sheet metal 199
shingle 187
single-ply 197
skylight 244
slate 187, 729
soffits framing 132
specialties, prefab 202
stainless steel 199
steel 189
stressed skin 113
thermoplastic polyolefin 197
tile 189
TPO 198
truss 161, 164
vent 206, 390
ventilator 207
ventilator plastic 457
walkway 196
zinc 199
Roofing built-up 729
cold 194
demolition 33, 178
modified bitumen 196
roll 199
single ply 197
Rooftop air conditioner 472
multizone air cond. 472
Room acoustical 383
audiometric 382
clean 381
humidifier 477
Rope decorative 332
exercise 348
safety line 18
steel wire 118
Rosewood door 218
Rosin paper 186
Rough buck 155
hardware 152
stone wall 93
Rough-in drinking fountain deck 440
drinking fountain floor 440
drinking fountain wall 440
sink countertop 434
sink raised deck 434
sink service floor 437
sink service wall 437
tub 434
Round bar fiberglass 173
diffuser 459
rail fence 556
table 368
tube fiberglass 173
Rowing machine 347
Rubber astragal 252
base 281
coating 181
control joint 80
dock bumper 331
floor 281
floor tile 282
mat 366
nosing 281
pavement 553
pipe insulation 419
riser 281
sheet 281
stair 281
threshold 250
tile 282
waterproofing 180
waterstop 54
Rubberized asphalt sealcoat 546
Rubbing wall 67
Rubbish chute 29
handling 29
Rumble strip 586
Running track 280, 281
trap 427
Rupture testing 13

S

Safe data 332
diskette 332
office 332
Safety deposit box 333
equipment laboratory 345
fixture industrial 439
glass 256
line rope 18
net 15
nosing 281
railing cable 118
shower 439
switch 503
Sales tax 9, 694
Salt treatment lumber 152
Sampling air 31
Sand 79
fill 520
screened 79
seal 546
Sandblast finish concrete 67
glass 255
Sandblasting metal 101
Sanding 290
floor 280
Sandstone 95
flagging 550
Sandwich panel 190
panel aluminum 191
panel skylight 244
wall panel 192
Sanitary base cove 275
Sash 389
aluminum 236
metal 237
replacement 238
security 237
steel 236, 237
wood 239
Sauna 383
door 383
Saw brick 76
concrete 29
cutting 73
shop 346
table 346
Sawing 76
Sawn control joint 66
Scaffold aluminum plank 18
baseplate 17
bracket 17
caster 17
catwalk 18
frame 16
guardrail 17
mason 16
putlog 18
rolling tower 18
specialties 17
stairway 17
wood plank 17, 18
Scaffolding 701
plank 17
tubular 16
Scale 328
contractor 328
crane 328
floor 328
physician's 351
pit 328
portable 328
truck 328
warehouse 328
Scanner checkout 334
Scarify concrete 262
Schedule 12
board 308
critical path 12
progress 12
School cabinet 362
crosswalk 585
door hardware 246
equipment 348
furniture 369
hinge 251
T.V. 517
Scissor gate 314
lift 593
Scoreboard baseball 348
Scored block 87
split face block 87
SCR brick 83
Scraper elevating 530
excavation 530
excavation bulk 530
mobilization 19
self propelled 530
Scratch coat 284
Screed base 57
holder 57
Screen chimney 319
entrance 311
fence 556
metal 237
molding 166
projection 307, 344
security 140, 237
sight 315
squirrel and bird 319
urinal 312
window 237, 241
wood 241
Screened sand 79
Screw aluminum 191
anchor 57, 105
anchor bolt 57
brass 149

Index

drywall 274
eye bolt 57
jack 536
lag 107
machine 107
sheet metal 149
steel 149
wood 149
Screw-on connector 483
Scrub station 353
Scupper drain 430
  floor 430
Seal curb 546
  door 331
  fog 546
  pavement 546
  security 252
Sealant 179, 186
  acoustical 273
  caulking 209
  control joint 66
  tape 210
Sealcoat 546
Sealer pavement 553
Seamless floor 284
Seating church 355
  movie 344
Secondary treatment plant .. 598
Sectional door 229, 230
  overhead door 230
Security and vault equipment ... 332
  barrier 332
  fence 555, 556
  film 258
  gate 314
  hinge 251
  mesh 268, 269
Security planter 585
Security sash 237
  screen 140, 237
  seal 252
  turnstile 332
  vault door 229
Security vehicle barriers .. 585
Sediment bucket drain 428
  strainer Y valve 417
Seeding 562, 750
  general 562
See-saw 349
Segmental retaining wall ... 557
Seismic bracing 737
Selective clearing 522
  demo concrete 40
  demo steel 101
  demolition 27, 77, 144, 212
Self-closing relief valve .. 416
Self-contained air conditioner ... 473
Self-drilling anchor 105
Self-propelled scraper 530
Self-supporting tower 582
Sentry dog 21
Separation barrier 32, 36
Septic system 575
  tanks 575
Service door 227
  electric 503
  entrance cable aluminum . 484
  residential 491
  sink 436
  sink faucet 436
  station equipment 330
Sewage aeration 597
  ejector pump 429
  holding tank 598
  municipal waste water .. 598
  pipe 574-576

pumping station 429
treatment plants 597
Shade 359
Shaft wall 264
Shake wood 188
Shear connector welded 108
  plate 150
  test 14
  wall 163
Sheathed nonmetallic cable . 484
  romex cable 484
Sheathing 162, 163
  asphalt 163
  gypsum 163
  paper 186
  roof 163
  roof & walls plywood ... 163
  wall 163
Sheepsfoot roller 532, 534
Sheet base 196
  floor 285
  glass 256
  metal 199
  metal aluminum 202
  metal screw 149, 191
  piling 536, 749
Sheeting 536
  open 538
  tie back 539
  wale 536
  wood 531, 536, 749
Sheets lead 393
Shelf bathroom 318
  bin 324
  directory 310
  library 370
  metal 323
Shellac door 297
Shelter aluminum 391
  dock 331
  golf 350
  rail dock 332
  temporary 42
Shelving 324
  mobile 370
  pine 324
  plywood 324
  prefinished 324
  refrigeration 382
  steel 323
  storage 323
  wood 324
Shield expansion 104
  lag screw 105
  light 586
Shielded cable 481
  copper cable 481
Shielding lead 393
  radio frequency 395
Shift work 9
Shingle 186, 187
  aluminum 187
  asphalt 186
  concrete 189
  metal 187
  panelized 188
  removal 178
  roof 187
  strip 186
  wood 187
Ship ladder 134
Shipping door 458
Shock absorber 425
  absorber pipe 426
  absorbing door 212, 232
Shooting range 349

Shop drill 346
  equipment 346
Shoring 536, 537
  aluminum joist 53
  baseplate 16
  bracing 16
  concrete formwork 53
  concrete vertical 53
  flying truss 53
  frames 53
  heavy duty 16, 17
  leveling jack 16
  post shores 53
  slab 17
  steel beam 53
  temporary 537
Shot blast floor 263
Shovel 529
Shower arm 435
  built-in 435
  by-pass valve 436
  compartments 312
  cubicle 434
  door 313
  drain 427
  enclosure 313
  glass door 313
  group 437
  pan 201
  partition 94, 312
  polypropylene 434
  receptor 312, 437
  safety 439
  stall 434
  surround 313
Shower/tub control set 436
Shower-tub valve spout set . 436
Shredder 354
  compactor 354
Shrinkage test 13
Shrub broadleaf evergreen .. 563
  deciduous 563, 564
  evergreen 563
  moving 565
Shutter 359
  interior 359
  wood 359
Siamese 408
Side beam 43
  coiling grille 227
  light door 223, 224
Sidelight 172
  door 217
Sidewalk 278, 546, 549
  asphalt 546
  brick 549
  bridge 17
  concrete 546
  removal 26
  temporary 21
Sidewall bracket 18
Siding aluminum 191
  bevel 193
  cedar 193
  demolition 178
  fiber cement 194
  fiberglass 190
  hardboard 193
  metal 189, 191
  nail 149
  paint 294
  painting 293
  plywood 193
  railroad 587, 751
  redwood 193
  removal 178

ribbed 191
steel 191
vinyl 193
wood 193
wood product 193
Sieve analysis 13
Sign 21, 309
  aluminum 309
  base formwork 46
  directional 310
  flexible 309
  letter 309
  posts 310
  project 21
  reflective 309
  road 310
  street 309
  traffic 310
Signage exterior 309
Signal bell fire 517
  traffic 585
Silencer duct 455
Silica chips 41
Silicon carbide abrasive ... 66
Silicone caulking 210
  coating 181
  water repellent 181
Sill 94
  anchor 150
  block 90
  door 167, 171, 250
  framing 158
  masonry 95
  mud 158
  precast concrete window . 71
  quarry tile 276
  stone 93, 95
  window 239
  wood 158
Sillcock hose bibb 436
Silo 392
Silos demo 377
Silt fence 534
Simulated brick 97
  stone 98
Single hub pipe 427
  hung window 236
  ply roofing 197
  precast concrete tee beam . 69
  zone rooftop unit 472
Single-ply roof 197
Sink 434
  barber 334
  base 360
  countertop 434
  countertop rough-in 434
  darkroom 336
  kitchen 434
  laboratory 345
  laundry 435
  lavatory 433
  mop 437
  raised deck rough-in ... 434
  removal 414
  service 436
  service floor rough-in . 437
  service wall rough-in .. 437
  support 438
Site clearing 521
  demolition 25
  demolition minor 25
  drainage 579
  improvement 373, 555
  plan 8
  preparation 24
Skating mat 367

Index

Skirtboard ... 170
Skylight ... 244
   fiberglass panel ... 389
   removal ... 178
   roof ... 244
Skyroof ... 235
Slab blockout formwork ... 47
   bolster reinforcing ... 55
   box out opening formwork ... 45
   bulkhead formwork ... 45, 46
   concrete ... 62, 65
   curb formwork ... 45, 47
   cutout ... 28
   edge form ... 122
   edge formwork ... 45, 47
   elevated ... 62
   fiberglass domes formwork ... 45
   flat plate formwork ... 44
   lift ... 63, 716
   lifting insert ... 57
   metal domes formwork ... 45
   metal pan formwork ... 44
   on grade ... 63
   on grade formwork ... 46
   on grade removal ... 26
   pan ... 62
   precast concrete channel ... 69
   reinforcing ... 59
   screed formwork ... 47
   shoring ... 17
   textured ... 63
   trench formwork ... 47
   void formwork ... 45, 47
   waffle ... 62
   with drop panel formwork ... 44
Slate ... 95
   flagging ... 550
   pavement ... 95
   removal ... 178
   roof ... 187, 729
   shingle ... 187
   sidewalk ... 550
   sill ... 96
   stair ... 95
   tile ... 279
Slatwall ... 168, 288
Sleeper clip ... 57
   floor ... 159
   wood ... 159
Sleepers framing ... 159
Sleeve dowel ... 59
   formwork ... 53
   pipe ... 53
Slide gate ... 554
   playground ... 350
   swimming pool ... 380
Sliding chalkboard ... 306
   door ... 217
   door shower ... 313
   entrance ... 233
   glass door ... 226
   mirror ... 319
   panel door ... 234
   window ... 236, 239
Slipform building formwork ... 42
   silo formwork ... 42
Slipforms ... 705
Slop sink ... 435
Slope protection ... 534, 536
Slot letter ... 323
Slotted block ... 89
   channel framing ... 112
   pipe ... 577
Slump block ... 87
Slurry trench ... 539
Small tools ... 19

Smoke detector ... 518
   hatch ... 206
   vent ... 206
   vent chimney ... 463
Smokestack ... 82
Snap tie formwork ... 53
Snow fence ... 30
   guards ... 206
Soaking bathroom ... 434
   tub ... 434
Soap dispenser ... 317, 318
   holder ... 318
   tank ... 318
Soccer goalpost ... 351
Socket ground ... 349
Sod ... 562
   tennis court ... 553
Sodding ... 562
Sodium low pressure fixture ... 509
Soffit drywall ... 272
   marble ... 94
   metal ... 194
   plaster ... 269
   plywood ... 167
   PVC ... 175
   steel ... 388
   stucco ... 270
   vent ... 260
   vinyl ... 194
   wood ... 159, 167
Softener water ... 431
Soil compaction ... 532
   decontamination ... 30
   pipe ... 427
   poisoning ... 535
   sample ... 25
   stabilization ... 535
   tamping ... 532
   test ... 12-14
Solar energy ... 467, 468
   energy circulator air ... 468
   energy system collector ... 468
   energy system controller ... 468
   energy system heat exchanger ... 468
   film glass ... 258
   heating ... 743
   screen block ... 90
Soldier beam ... 538
Solid core door ... 219
   surface countertops ... 364
   wood door ... 221
Sorting reinforcing ... 59
Sound attenuation ... 289
   control demo ... 377
   curtain ... 393
   movie ... 344
   panel ... 393
   proof enclosure ... 382
   system component ... 512
   system speaker ... 512
Space heater ... 467
   heater floor mounted ... 467
Spandrel beam formwork ... 42
   glass ... 256
Spanish roof tile ... 188
Speaker movie ... 344
   sound system ... 512
   station nurse ... 513
Special construction ... 393
   door ... 229
   hinge ... 251
   purpose rooms demo ... 377
   systems ... 513, 516
Specialties ... 318
   piping HVAC ... 450, 451
   scaffold ... 17

Specific gravity ... 13
   gravity testing ... 13
Speed bump ... 586
Spike grid ... 150
Spinner ventilator ... 461
Spiral reinforcing ... 59
   stair ... 141, 169
Spire church ... 326
Splice pile ... 540, 541
   reinforcing ... 58
Splicer movie ... 344
Splicing reinforcing ... 58
Split astragal ... 252
   rail fence ... 557
   rib block ... 87
   ring ... 150
   system control system ... 447
Splitter damper assembly ... 454
Spotlight ... 508
Spotter ... 534
Spray acoustical ... 207
   coating ... 179
   fireproofing ... 207
   insulation ... 184
   substrate ... 34
Sprayer airless ... 31
Spread footing ... 63
   footing formwork ... 46
Spring bolt astragal ... 252
   bronze weatherstrip ... 252
   hinge ... 251
Sprinkler alarm ... 517
   grass ... 559
   head ... 409
   system accelerator ... 410
   systems ... 409, 559
Square bar fiberglass ... 173
   tube fiberglass ... 173
Squash court ... 384
   court backstop ... 555
Stabilization fabric ... 547
   soil ... 535
Stacked bond block ... 88
Stadium ... 387
   bleacher ... 387
   cover ... 385
Stage equipment ... 346
   lighting ... 508
   portable ... 346
Staging aids ... 18
   pump ... 16
   swing ... 19
Stain cabinet ... 297
   door ... 297
   floor ... 299
   lumber ... 165
   siding ... 193
   truss ... 300
Stainless cooling towers ... 471
   duct ... 453
   flashing ... 200
   grating ... 137
   gutter ... 204
   plank grating ... 136
   reglet ... 204
   rivet ... 108
   screen ... 311
   sign ... 309
   steel corner guard ... 317
   steel cot ... 369
   steel cross ... 355
   steel downspout ... 203
   steel gravel stop ... 202
   steel hinge ... 250, 251
   steel pipe ... 423
   steel roof ... 199

   steel shelf ... 318
   steel storefront ... 234
   weld rod ... 110
Stair ... 169
   basement ... 69, 169
   brick ... 92
   carpet ... 286
   ceiling ... 338
   climber ... 403
   concrete ... 64
   electric ... 338
   finish ... 66
   grating ... 134
   landing ... 284
   marble ... 94
   metal pan ... 133
   metal plate ... 133
   nosing ... 281, 283
   pan fill cement ... 54
   prefabricated ... 169
   railroad tie ... 547
   removal ... 147
   residential ... 169
   riser ... 283
   riser vinyl ... 281
   rubber ... 281
   slate ... 95
   spiral ... 141, 169
   stone ... 585
   stringer ... 284
   stringer framing ... 155
   stringer wood ... 155
   temporary protection ... 21
   terrazzo ... 283
   tread fiberglass ... 173
   tread insert concrete ... 54
   tread terrazzo ... 283
   tread tile ... 276
   tread wood ... 170
   treads ... 93, 138, 281
   wood ... 169
Stairlift wheelchair ... 403
Stairs disappearing ... 338
   fire escape ... 134
   formwork ... 50
   precast concrete ... 69
Stairway door hardware ... 246
   scaffold ... 17
Stairwork and handrails ... 170
Stake formwork ... 54
   nail ... 54
   subgrade ... 58
Stall shower ... 434
   toilet ... 310-312
   type urinal ... 436
   urinal ... 436
Stamp time ... 513
Stamping concrete ... 67
   texture ... 67
Standard extinguisher ... 321
Standpipe alarm ... 517
   connection ... 408
   steel ... 573
Starter board & switch ... 498, 499
   motor ... 500
Starting newel ... 169
Station control ... 504
   diaper ... 317
   hospital ... 352
   transfer ... 354
   weather ... 395
Stationary ventilator ... 461
   ventilator mushroom ... 461
Steam bath ... 383
   bath residential ... 384
   boiler ... 465

Index

boiler electric 464
  clean 76
  clean masonry 76
  cleaning 76
  cleaning metal 101
  condensate meter 452
  humidifier 477
  jacketed kettle 342
  pressure valve 451
  regulator pressure 424
  trap 451
Steamer 342
Steel anchor 80
  astragal 252
  beam hanger 52
  beam shoring 53
  beam W-shape 113
  bin wall 558
  blind 358
  blocking 153
  bolt 150
  bridge 559
  bridging 153
  building components 726
  building pre-engineered 387
  chain 139
  channel 269
  conduit in slab 489
  conduit in trench 489
  conduit intermediate 486
  conduit rigid 486
  corner guard 316
  cutting 29, 102
  diffuser 460
  dome 390
  door . 215, 216, 224, 226, 229, 230,
                  389, 731
  downspout 203
  drilling 102
  edging 547, 565
  estimating 723
  expansion tank 450
  fence 553, 555
  fiber reinforcing 61
  frame 214
  framing 115
  furring 265
  grating 137
  grating tread 138
  gravel stop 202
  gutter 204
  hex bolt 104
  high strength 116
  hinge 250
  lath 268
  lintel 112, 113
  locker 322
  members structural 113
  mesh box 535
  mesh grating 137
  mesh tread 139
  mill extras 116
  painting 302
  partition 315, 316
  pile 540
  pipe 420, 421, 576, 579
  pipe downspout 203
  pipe removal 26
  plate 113
  prestressing 61
  primer 302
  projects structural 115
  register 460
  reinforcing 708, 709
  rivet 108
  roof 189
  salvage 414
  sash 236, 237, 731
  screw 149
  sections 725
  sheet pile 536
  shelving 323
  shingle 187
  siding 191
  silo 392
  standpipe 573
  structural 723
  stud 266, 267
  tank 450, 573
  testing 12
  tower 582
  underground duct 582
  valve 445
  weld rod 109
  well casing 574
  window 236, 237, 389
  window demolition 212
  wire rope 118
Steeple 326
  aluminum 326
  tip pin 250
Step 310
  bluestone 546
  brick 546
  manhole 580
  masonry 546
  railroad tie 546
  stone 94
  tapered pile 540
Sterilizer barber 334
  dental 352
  medical 351
Stiffener plate 113
Stiffeners joist web 129
Stile fabric 252
  protection 252
Stockpiling of soil 523
Stone aggregate 563
  anchor 80
  ashlar 94
  base 94, 547
  cast 97
  column 94
  countertops engineered 365
  cultured 97
  curbs 551
  dust 546
  fill 520
  filter 536
  floor 94
  gabion retaining walls 558
  ground cover 563
  mulch 561
  paver 93, 550
  pea 41
  removal 26
  retaining walls 558
  sill 93, 95
  simulated 98
  stair 585
  step 94
  stool 95
  tread 93
  trim cut 95
  wall 93, 558
Stool cap 167
  doctor 352
  stone 95
  window 95, 96
Stop door 167
  gravel 202
Stop-start control station .. 504
Storage bottle 340
  box 15
  cabinet 362
  dome 390
  dome bulk 390
  door cold 228
  shelving 323
  tank 448, 573
  tank cover 384
  tank demo 378
  tanks underground 448
Storefront 233
  aluminum 234
Storm door 213
  drainage manholes frames .. 579
  window 241, 389
Stove 342
  wood burning 321
Strainer bronze body 451
  downspout 202
  gutter 204
  roof 202
  wire 203
  Y type 451
  Y type iron body 451
Strand grouted 61
  ungrouted 61
Stranded wire 484
Strap tie 151
Straw 561
Street sign 309
Streetlight 508
Strength compressive 13
Stressed skin roof 113
Stringer stair 284
  stair terrazzo 284
Strip cabinet 513
  chamfer 50
  entrance 231
  filler 252
  floor 280
  footing 63
  lighting 506
  rumble 586
  shingle 186
  soil 523
Stripping topsoil 523
Strobe light 508
Structural aluminum 117
  backfill 532
  brick 83
  columns 110
  excavation 527
  fabrication 116
  face tile 84
  facing tile 84
  fee 8
  framing 724
  insulated panel 160
  panel 160
  shape column 111
  steel 723
  steel members 113
  steel projects 115
  tile 84
  welding 102, 116
Structure fabric 384
  tension 385, 386
Stub switchboard 501
Stucco 270
  demolition 263
  mesh 268
  painting 293
Stud demolition 147
  driver pneumatic 108
  metal 264, 266, 267
  partition 156, 268
  steel 266, 267
  wall 155, 156, 159, 264-267
  wall bracing 123
  wall bridging 123
  welded 109
  wood 159, 264
Studs metal 124
Stump chipping 521
  removal 522
Subcontractor O & P 11
Subdrainage pipe 578, 579
  piping 578
  system 578
Subfloor 162
  plywood 162
  wood 162
Subgrade chair reinforcing ... 57
  stake 58
Submersible pump 431, 574
  sump pump 431
Subpurlin bulb tee 71, 116
Subpurlins 726
Subsoil heating 381
Subsurface exploration 25
Sump hole construction 531
  pit 531
  pump 338, 430
  pump submersible 431
Super high-tensile anchor ... 535
Supermarket checkout 334
  scanner 334
Supply wells pumps water 574
Support carrier fixture 437
  ceiling 110
  drinking fountain 437
  framing pipe 113
  lavatory 437
  light 110
  monitor 110
  motor 110
  partition 110
  sink 438
  urinal 438
  water closet 438
  water cooler 439
  X-ray 110
Suppressor lightning 491
Surface bonding 78
  countertops solid 364
  landscape 286
  playground 282
  preparation 290, 291
  raceway 502
  treatment 546
Surfacing 546
Surfactant 34
Surgery equipment 353
  table 351
Surgical lighting 351
Surround shower 313
  tub 313
Surveillance system TV 516
Survey aerial 24
  crew 21
  monument 24
  preblast 528
  property line 24
  stake 21
  topographic 24
Suspended ceiling ... 267, 271, 277
  ceiling framing 154
Suspension mounted heater ... 467
  system ceiling 278
Swell testing 13
Swimming pools 379, 380, 736

Index

pool blanket 380
pool demo 378
pool enclosure 387
pool equipment 380
pool heater 466
pool hydraulic lift 380
pool ladder 380
pool painting 380
pool ramp 380
Swing 350
  check valve 415
  check valve bronze 415
  clear hinge 251
  door 224
  gate 554
  staging 19
Swing-up overhead door 230
Switch box 487
  bus-duct 501
  decorator 491
  dimmer 491, 503
  electric 502, 503
  general duty 503
  residential 491
  safety 503
  toggle 502
Switchboard electric 504
  instrument 496
  pressure switch 498
  stub 501
  w/bus bar 497
  w/CT compartment 497
Switching low-voltage 502
Switchplate low-voltage 502
Synthetic erosion control ... 534
  fiber reinforcing 61
  marble 279
  turf 282
System antenna 512
  ceiling suspension 278
  clock 513
  control 447
  dovetail anchor 51
  fire extinguishing .. 321, 410
  grinder pump 429
  irrigation 559
  septic 575
  sprinkler 410, 559
  subdrainage 578
  tube 405
  T.V. 512
  T.V. surveillance 516
  UHF 512
  VHF 512
Systems sprinkler 409

T

T pillow 573
Table folding 368
  laboratory 345
  massage 348
  overbed 371
  physical therapy 352
  reading 370
  rewind 344
  round 368
  saw 346
  surgery 351
Tamper 524
Tamping soil 532
Tank clean 30
  cover 384
  darkroom 336
  developing 336

disposal 30
expansion 450
fibergalss 448
fixed roof 449
gasoline 449
holding 598
hubbard 351
leak detection 518
removal 30
septic 575
soap 318
steel 450, 573
storage 448, 573
testing 14
water 573
water storage solar 468
Tanks septic 575
Tap box bus-duct 501
Tape barricade 20
  detection 569
  sealant 210
  temporary 552
  underground 569
Tar paper 561
  pitch emulsion 546
  roof 178
Target range 349
Tarpaulin 20
  duck 20
  mylar 20
  polyethylene 20
  polyvinyl 20
Tarred felt 196
Taxes 9, 10, 695
  sales 9, 694
  social security 10, 695
  unemployment 10, 695
T-bar mount diffuser 460
Teak floor 279
  molding 166
  paneling 168
Technician inspection 14
Tee beam 68
  cleanout 418
  pipe 422
  precast concrete beam .. 68
Telephone booth 310
  electrical 581
  enclosure 310
  house 513
  manhole 582
  pole 581
  receptacle 494
  underground electrical . 581
Telescoping bleacher 371
  door 231
  platform 346
Television equipment 512
  receptacle 494
Teller automatic 333
  window 333
Temperature relief valve ... 416
  rise detector 517
Tempered glass 255
  glass greenhouse 386
  hardboard 167
Tempering valve 416
  valve water 416
Temporary building 15
  construction 14, 20
  facility 20
  fence 21
  guardrail 20
  heat 14, 42
  light 14
  road painting 552

shelter 42
shoring 537
tape 552
utilities 14
utility 14
Tennis court air supported .. 385
  court backstop 555
  court clay 553
  court fence 555, 556
  court net 553
  court painting 553
  court post 553
  court sod 553
  court surface 286
Tensile test 13
Tension structure 385
  structures demo 379
Terminal A/C packaged ... 473
  heater 475
  lug 483
Termination box 513
  cable 483
Termite pretreatment 535
  protection 535
Terne coated flashing 200
Terra cotta 84, 85
  cotta coping 97
  cotta demolition ... 28, 263
Terrazzo 283
  abrasive 283
  base 283
  conductive 284
  demolition 263
  epoxy 284
  floor 283, 735
  monolithic 283
  precast 283
  receptor 312
  stair 283
  venetian 283
  wainscot 283, 284
Test beam 13
  load pile 520
  moisture content 14
  pile load 520
  pits 25
  soil 13, 14
  ultrasonic 14
  well 574
Testing 12
  OSHA 33
  pile 520
  sulfate soundness 13
  tank 14
Texture stamping 67
Textured slab 63
Therapeutic pool 381
Thermal & moisture prot. demo . 178
  batt 289
Thermometer 345
Thermoplastic painting ... 552
  polyolefin roofing .. 197, 198
Thermostat control radiator .. 447
  integral 476
  wire 496
Thimble wire rope 117
Thin plaster partition 265
Thincoat plaster 271
Thinning tree 521
Thinset ceramic tile 275
  mortar 276
Threaded rod fiberglass ... 173
Threshold 250
  door 172
  stone 94
  wood 167

Thru-wall air conditioner 473
Ticket booth 391
Tie back sheeting 539
  formwork snap 53
  rafter 158
  railroad 547, 564
  rod 112, 536
  strap 151
  wall 80
  wire reinforcing 58
Tier locker 322
Tile 275, 281
  abrasive 275
  aluminum 189
  carpet 285
  ceiling 277, 278
  ceramic 275
  clay 188
  concrete 189
  cork 282
  cork wall 287
  demolition 262
  exterior 275
  fire rated 84
  flue 96
  glass 257, 276
  grout 275
  marble 279
  metal 276
  partition 85
  porcelain 275
  quarry 276
  roof 189
  rubber 282
  slate 279
  stainless steel 276
  stair tread 276
  structural 84
  terra cotta 85
  vinyl 282
  wall 275, 582
  window sill 276
Tilt-up concrete 719
  conrete wall panel 71
  precast concrete wall panel 71
Timber connector 150
  fastener 148
  framing 161
  guide rail 586
  heavy 161
  laminated 164
  pile 540
  roof deck 161
  switch ties railroad ... 584
  ties railroad 584
Time lapse photography 12
  lock 229
  stamp 513
  system controller 513
Timer clock 494
  interval 491
Tin clad door 221
Tinted glass 255
Toaster 342
Toggle switch 502
Toggle-bolt anchor 105
Toilet 598
  accessories 317
  bowl 433
  compartments 310
  door hardware 246
  partition 94, 310-312
  partition removal 263
  prison 334
  stall 310-312
  tissue dispenser 317, 318

Index

Tool powder actuated 107
   rivet 108
Tools small 19
Toothed ring 151
Toothing masonry 28
Top coiling grille 226
   demolition counter 148
   dressing 562
   vanity 365
Topographic survey 24
Topping concrete 66
   epoxy 285
   floor 66
   grout 72
Topsoil 520, 523, 562
   stripping 523
Torch cutting 29, 102
Touch bar 247
Tour station watchman 514
Towel bar 318
   dispenser 317, 318
Tower control 392
   cooling 471
   crane 702
   radio 582
TPO roof 198
Track accessories railroad 584
   curtain 347
   drawer 362
   heavy rail railroad 584
   maintenance railroad 584
   running 280, 281
   surface 286
   traverse 358
Tractor 524, 532
Traffic sound barriers highway .. 559
   channelizing 586
   cone 20
   control pedestrian 332
   door 231
   line 552
   lines and markings painted .. 552
   sign 310
   signal 585
Trailer office 15
Trainer bicycle 347
Training 22
Transceiver 512
Transfer station 354
Transformer 496
   & bus-duct 497, 501
   cabinet 502
   current 496
   dry type 497
   fused potential 496
   low-voltage 502
   oil filled 496
Transition molding wood 280
   residential 458
Transom aluminum 234
   lite frame 215
   windows 241
Trap cast iron 427
   deep seal 427
   drainage 427
   grease 428
   inverted bucket steam 451
   P 427
   rock surface 66
   steam 451
Trash closure 372
   compactor 343
   receiver 366
   receptacle 372
   receptacle fiberglass 372
Traverse 358

track 358
Travertine 94
Tray cable 488
   laundry 435
Tread abrasive 276
   aluminum cast 138
   aluminum grating 138
   bronze cast 138
   cast iron 138
   fiberglass 138
   rubber 281
   stair 93
   steel grating 138
   steel mesh 139
   stone 94, 95
   vinyl 281
   wood 170
Treadmill 348
Treads stair 138
Treated lumber 158
   lumber framing 159
   pile 541
   pine fence 557
Treatment equipment water 597
   lumber 152
   plant secondary 598
   plants sewage 597
   plywood 152
   potable water 431
   wood 152
Tree 564
   cutting 522
   deciduous 564
   evergreen 563
   guying 565
   moving 565
   removal 521, 522
   thinning 521
Trench backfill 524, 526
   cover 138
   drain 430
   drain polymer 430
   duct 489
   duct fitting 490
   duct steel 490
   excavation 523-527
   grating frame 138
   slurry 539
   utility 526
Trenching 528, 531
Trial batch 13
Trim demolition 148
   exterior 166
   painting 299
   redwood 166
   tile 275
   wood 166
Triple brick 83
   weight hinge 251
Troffer air handling 506
Trolley rail 317
Trowel coating 179
Truck dock 331
   hauling 534
   leveler 331
   loading 527
   scale 328
Trucking 534
Truss bowstring 164
   framing metal 132
   joist 121
   metal roof 116
   painting 300
   plate 151
   roof 161, 164
   stain 300

varnish 300
Tub bar 318
   rough-in 434
   soaking 434
   surround 313
Tube framing 232
   pneumatic 333
   system 405
   system air 405
   system pneumatic 405
Tubing copper 420
   electric metallic 486
   electrical 486
Tubular fence 555
   lock 248
   scaffolding 16
   steel door 224
   steel joist 164
Tumbler holder 318
Turf artificial 553
   synthetic 282
Turnbuckle wire rope 117
Turned column 171
Turnout railroad 587
Turnstile 332
   security 332
TV closed circuit camera 516
   system 512
Two piece astragal 252
Tyton joint 570

U

UHF system 512
Ultrasonic cleaner 352
   motion detector 516
   test 14
Underdrain 579
Undereave vent 260
Underfloor duct 490
   duct electrical 490
   duct fitting 490
Underground duct 582
   storage tank removal 703
   storage tanks 448
   tape 569
Underlayment 162
   acoustical 289
   gym floor 280
   gypsum 72
   hardboard 162
   latex 281
   self cement 72
Underpin foundation 538
Underwater light 380
Undisturbed soil 13
Unemployment tax 10, 695
Ungrouted bar 61
   strand 61
Unit heater 467, 475
   heaters propeller 476
   kitchen 336
Upstanding beam formwork 43
Urethane foam 210, 289
   insulation 184, 382
   wall coating 302
Urinal 436
   removal 414
   screen 312
   stall 436
   stall type 436
   support 438
   wall hung 436
Urn ash 366
Utensil washer medical 351

Utilities temporary 14
Utility accessories 569
   boxes 572
   brick 83, 84
   duct 582
   electric 582
   pole 508, 581
   set fan 456, 457
   sitework 581
   temporary 14
   trench 526

V

Vacuum breaker 425
   central 335
   cleaning 335
   loader 32
   relief water 425
Valance board 361
Valley flashing 187
   gutter 389
   rafter 158
Valve 415-417, 444, 445
   assembly sprinkler deluge . 410
   backwater 430
   ball 417
   brass 415
   bronze 415
   butterfly 444
   cabinet 321
   check 408, 445
   check swing 444
   CPVC 417
   fire hose 409
   flush 435
   foot 417
   gas stop 415
   gate 444, 445
   globe 445
   hot water radiator 450
   iron body 444, 445
   mixing 436
   plastic 417
   polypropylene 417
   PVC 417
   relief pressure 416
   shower by-pass 436
   soldered gate 415
   spout set shower-tub 436
   sprinkler alarm 408
   steam pressure 451
   steel 445
   swing check 415
   tempering 416
   water pressure 416
   Y sediment strainer 417
Valves forged 445
Vandalproof fixture 507
Vane-axial fan 456
Vanity base 362
   top 365
   top lavatory 433
Vapor barrier 186, 382
   barrier sheathing 163
   retarder 186
Vaportight fixture incandescent . 507
Varnish cabinet 297
   casework 297
   door 297
   floor 299, 302
   polyurethane 302
   truss 300
VAT removal 33
Vault door 229

Index

front 229
Vaulting side horse 348
VCT removal 262
Vehicle barriers security 585
Veneer ashlar 93
  brick 81, 82
  core paneling 168
  granite 93
  members laminated 165
  removal 78
Venetian blind 358
  terrazzo 283
Vent automatic air 450
  box 80
  brick 461
  chimney 463
  chimney all fuel 463
  chimney fitting 464
  chimney high temperature .. 464
  chimney increaser 464
  dryer 338
  exhaust 461
  flashing 428
  foundation 80
  gas 463
  intake-exhaust 461
  metal chimney 463
  one-way 207
  ridge 206
  ridge strip 260
  roof 206, 390
  smoke 206
  soffit 260
Ventilated metal deck 123
Ventilating air conditioning . 447, 453,
    454, 456, 457, 461, 470, 471, 473
Ventilation 741
  fan 495
  louver 260
Ventilator control system 447
  fan 456
  masonry 80
  mushroom stationary 461
  relief vent 461
  roof 207
  spinner 461
  stationary mushroom 461
Venturi flow meter 452
Verge board 166
Vermiculite cement 207
  insulation 183
  plaster 269
Vertical blinds 358
  conveyor 592
  lift door 231
  metal ladder 134
VHF system 512
Vibration pad 108
Vibrator earth 524, 532, 534
  plate 532
Vibroflotation 537
  compaction 749
Videotape recorder 517
Vinyl blind 173
  casement window 243
  coated fence 554
  composition floor 282
  downspout 203
  faced wallboard 272
  floor 282
  glass 258
  gutter 204
  lead barrier 290
  mat 367
  plastic waterproofing 180
  roof 198

siding 193
soffit 194
stair riser 281
stair tread 281
tile 282
tread 281
wall coating 302
wallpaper 287
window 242, 244
Vision panel 244
Vocational kiln 346
  shop equipment 346
Voltmeter 496
Volume control damper 454

W

Waffle slab 62
Wainscot ceramic T 275
  molding 166
  quarry tile 276
  terrazzo 283, 284
Wale sheeting 536
Wales 536
Walk 546
  moving 403
Walk-in refrigeration 382
Walkway cover 326
  roof 196
Wall & corner guards 316
  aluminum 191
  area 326
  boxout formwork 47
  brick 92
  brick shelf formwork 47
  bulkhead formwork 47
  bumper 249, 317
  buttress formwork 47
  cabinet 360, 362, 363
  canopy 325
  cap residential 458
  cavity 92
  ceramic tile 275
  coating 302
  concrete 65
  corbel formwork 47
  covering 735
  cross 355
  cutout 28
  demolition 263
  drywall 272
  exhaust fan 457
  fan 458
  finish concrete 67
  flagpole 327
  forms 705
  formwork 47
  formwork accessories 54
  foundation 88
  framing 159
  framing removal 28
  furnace 467
  furring 160, 266
  grade 63
  grout 79
  guard 316, 317
  heater 338
  hung lavatory 434
  hung urinal 436
  hydrant 408, 426
  insulation 86, 181-183, 389
  lath 267, 268
  liner formwork 48
  lintel formwork 48
  louver 260

masonry 86, 558
mat 348
mirror 257, 258
painting 294, 300
panel 63, 97
panel brick 92
panel ceramic tile 276
panel precast concrete double . 69
panel woven wire 314
paneling 168
partition 264, 265
patching concrete 40
pilaster formwork 49
plaster 269
plug 81
plywood formwork 47, 49
precast concrete 719
precast concrete wall 70
radial 48
register 460
reinforcing 59, 709
removal 27
retaining 63, 558
rubbing 67
rustication strip formwork 48
shaft 264
shear 163
sheathing 163
siding 193
sill formwork 49
steel bin 558
steel framed formwork 49
stone 93
stucco 270
stud 155, 156, 159, 264-267
switch plate 503
tie 80
tie masonry 79
tile 275, 582
tile cork 287
urn ash receiver 318
window 233, 235
Wallboard acoustical 273
Wallcovering 287
  acrylic 287
  gypsum fabric 287
Wallguard 317
Wallpaper 287, 288, 735
  grass cloth 288
  vinyl 287
Walls and partitions demolition .. 263
Walnut door frame 171
  floor 279
Wardrobe cabinet 324
  wood 361
Warehouse 384
  scale 328
Warm air heater 467
Wash bowl 433
  brick 76, 77
  fountain 437
  fountain group 437
Washable air filters 462
Washer 151
  commercial 335
  darkroom 336
  residential 337, 338
Washing machine automatic .. 337
Waste compactor 354
  disposal 354
  handling 353, 354
  handling equipment 354
  incinerator 353
  packaging 34
  receptacle 318
Wastewater treatment system 598

Watchdog 21
Watchman service 21
  tour station 514
Wate balancing 446
Water anti-siphon 425
  atomizer 32
  balancing 446
  blasting 76
  blasting metal 101
  cement ratio 13
  chiller 470
  chiller reciprocating 470
  chillers absorption 470
  closet 433
  closet chemical 598
  closet removal 414
  closet support 438, 439
  closets 436
  coil chilled 474
  cooler 441, 740
  cooler handicap 441
  cooler support 439
  copper tubing connector ... 424
  curing concrete 68
  dispenser hot 431
  distiller 351
  distribution pipe 570, 571
  effect testing 13
  fountain removal 414
  hammer arrester 425
  heater 338, 432, 495
  heater commercial 431, 432
  heater electric 431
  heater gas 432
  heater oil 432
  heater removal 414
  heater residential 431
  heating hot 464, 475
  hydrant 426
  pipe 570
  pipe ground clamp 486
  pressure reducing valve 416
  pressure relief valve 416
  pressure valve 416
  pump 338, 452, 574
  pump fire 411
  pumping 531
  repellent 300
  repellent coating 181
  repellent silicone 181
  softener 431
  softener potable 431
  storage solar tank 468
  supply domestic meter 424
  supply meter 424
  supply wells pumps 574
  tank 573
  tank elevated 573
  tempering valve 416
  treatment 598
  treatment equipment 597
  vacuum relief 425
  well 574
Waterproofing 178, 180
  butyl 180
  cementitious 180
  coating 179
  integral 42
  masonry 79
  membrane 179
  neoprene 180
  rubber 180
Water-source heat pumps 474
Waterstop 54
  barrier 51
  center bulb 54

Index

dumbbell ... 54
    fitting ... 55
    PVC ... 54
    ribbed ... 54
    rubber ... 54
Water-water exchange hot ... 469
Watt meter ... 496
Wearing course ... 547
Weather station ... 395
Weatherproof receptacle ... 493
Weatherstrip ... 227, 253
    aluminum ... 252
    door ... 254
Weatherstripping ... 253
Wedge anchor ... 106
Weight exercise ... 347
    lifting multi station ... 348
Weld joint pipe ... 421
    rod ... 109
    rod aluminum ... 109
    rod cast-iron ... 109
    rod stainless ... 110
    rod steel ... 109
Welded frame ... 214
    shear connector ... 108
    structural steel ... 722
    stud ... 109
    wire fabric ... 60, 710
Welder arc ... 346
Welding certification ... 14
    fillet ... 102
    structural ... 102, 116
Well ... 531, 574
    casing steel ... 574
    gravel pack ... 574
    pump shallow ... 574
    water ... 574
Wellpoints ... 531, 748
Wheel color ... 508
    guard ... 316
    potters ... 346
    stop railroad ... 584
Wheelchair lift ... 403
    stairlift ... 403
Whirler playground ... 350
Whirlpool bath ... 352
Whitemarble chips ... 41
Wide flange beam fiberglass ... 173
    throw hinge ... 251
Winch bull ... 51
Window ... 236
    aluminum ... 236
    and door molding ... 167
    awning ... 237, 325
    bank ... 333
    blind ... 172, 358
    bullet resistant ... 333
    casement ... 236, 238, 240
    casing ... 165
    counter ... 323
    demolition ... 212
    double hung ... 237, 238, 240
    drive-up ... 333
    estimates ... 731
    frame ... 239
    framing ... 389
    glass ... 257
    grille ... 241
    guard ... 140
    hardware ... 245
    industrial ... 237
    metal ... 236, 389
    muntin ... 241
    painting ... 297
    picture ... 237, 239
    pivoted ... 237
    plastic ... 242
    precast concrete sill ... 71
    projected ... 237
    removal ... 212
    restoration ... 36
    screen ... 237, 241
    sill ... 95
    sill marble ... 94
    sill tile ... 276
    sliding ... 239
    steel ... 236, 237, 389
    stool ... 95, 96
    storm ... 241, 389
    teller ... 333
    trim set ... 167
    vinyl ... 242, 244
    wall ... 233
    wall framing ... 232
    wood ... 237-240
Windows transom ... 241
Wine cellar ... 340
Winter protection ... 20, 42
Wire ... 744
    brushing metal ... 101
    connector ... 483
    copper ... 484, 485
    electric ... 484
    fabric welded ... 60
    fence ... 21, 556
    fences misc. metal ... 556
    fencing ... 556
    glass ... 257
    ground ... 486
    hanging ... 278
    mesh ... 270, 556
    mesh locker ... 322
    mesh partition ... 314
    partition ... 314
    razor ... 556
    reinforcement ... 710
    rope clip ... 117
    rope thimble ... 117
    rope turnbuckle ... 117
    strainer ... 203
    thermostat ... 496
    THW ... 484
    THWN-THHN ... 485
    tying ... 52
Wiremold raceway ... 502
Wireway raceway ... 487
Wiring air conditioner ... 495
    device ... 488, 502
    device & box ... 488
    fan ... 495
    power ... 503
    residential ... 491, 495
Wood base ... 165
    beam ... 153, 161
    bench ... 373
    blind ... 172, 359
    block floor ... 279
    block floor demolition ... 263
    blocking ... 153, 156
    bridge ... 559
    bridging ... 153
    cabinet ... 362
    canopy ... 159
    casing ... 165
    chips ... 561
    column ... 154, 161, 171
    cupola ... 326
    deck ... 161, 162
    demolition ... 262
    dome ... 390
    door ... 218, 219, 221
    drawer ... 361
    fascia ... 158, 166
    fastener ... 149
    fiber insulation ... 183
    fiber plank cementitious ... 71
    fiber sheathing ... 163
    fiber underlayment ... 162
    firestop ... 155
    flagpole ... 327
    floor ... 279, 280
    floor demolition ... 263
    folding partition ... 316
    frame ... 171, 361
    framing ... 153, 154, 160
    framing plate ... 159
    furring ... 160
    girder ... 153
    gutter ... 204
    handrail ... 166
    header ... 159
    joist ... 154, 164
    laminated ... 164
    ledger ... 158
    louver ... 170
    molding ... 169
    nailer ... 155
    overhead door ... 230
    panel door ... 221
    paneling ... 167
    parquet ... 279
    partition ... 156
    plank scaffold ... 17, 18
    planter ... 372
    plate ... 159
    pole ... 581
    product siding ... 193
    rafter ... 158
    railing ... 169, 170
    roof deck ... 161
    roof deck demolition ... 178
    roof truss ... 164
    sash ... 239
    screen ... 241
    screw ... 149
    shade ... 359
    shake ... 188
    sheathing ... 163
    sheet piling ... 749
    sheeting ... 531, 536, 749
    shelving ... 324
    shingle ... 187
    sidewalk ... 546
    siding ... 193
    siding demolition ... 179
    sill ... 158
    sleeper ... 159
    soffit ... 159, 167
    stair ... 169
    stair stringer ... 155
    storm door ... 224
    stud ... 159, 264
    subfloor ... 162
    tank ... 573
    threshold ... 167
    tread ... 170
    treatment ... 152
    trim ... 166
    veneer wallpaper ... 287
    wardrobe ... 361
    window ... 237-240
    window demolition ... 212
Wool carpet ... 286
    fiberglass ... 183
Work extra ... 11
Worker protection ... 32
Workers' compensation ... 10, 697-699
Woven wire partition ... 314
Wrapping pipe ... 569
Wrestling mat ... 348

X

X-ray barrier ... 394
    concrete slabs ... 14
    dental ... 352
    medical ... 353
    mobile ... 353
    protection ... 393
    support ... 110
X-ray/radio freq protection demo ... 379

Y

Y sediment strainer valve ... 417
    type iron body strainer ... 451
    type strainer ... 451
Yellow pine floor ... 280

Z

Z bar suspension ... 278
Zee bar ... 274
Zinc divider strip ... 283
    plating ... 149
    roof ... 199
    terrazzo strip ... 283
    weatherstrip ... 253

Notes

Notes

Notes

Notes

Notes

Notes

Notes

Notes

Notes

Notes

Reed Construction Data/RSMeans... a tradition of excellence in Construction Cost Information and Services since 1942.

Table of Contents
Annual Cost Guides
Reference Books
Seminars
Electronic Data
New Titles
Order Form

For more information visit RSMeans Web site at www.rsmeans.com

Book Selection Guide

The following table provides definitive information on the content of each cost data publication. The number of lines of data provided in each unit price or assemblies division, as well as the number of reference tables and crews, is listed for each book. The presence of other elements such as equipment rental costs, historical cost indexes, city cost indexes, square foot models, or cross-referenced index is also indicated. You can use the table to help select the RSMeans book that has the quantity and type of information you most need in your work.

Unit Cost Divisions	Building Construction Costs	Mechanical	Electrical	Repair & Remodel	Square Foot	Site Work Landsc.	Assemblies	Interior	Concrete Masonry	Open Shop	Heavy Construc.	Light Commercial	Facil. Construc.	Plumbing	Western Construction Costs	Residential
1	548	283	368	477		490		304	455	547	499	209	976	317	546	160
2	540	237	58	528		768		331	189	539	665	410	1088	260	540	218
3	1469	179	165	797		1336		261	1852	1463	1478	256	1392	148	1464	288
4	847	18	0	649		660		561	1060	823	582	426	1098	0	833	354
5	1802	205	159	917		749		954	686	1770	989	801	1781	282	1742	725
6	1398	78	78	1358		487		1127	317	1376	645	1437	1356	22	1735	1591
7	1275	167	71	1284		476		493	430	1270	0	969	1329	176	1274	759
8	1906	61	10	1936		299		1670	671	1868	0	1325	2107	0	1865	1270
9	1650	70	23	1474		237		1743	370	1614	0	1373	1866	55	1641	1253
10	877	17	10	512		223		777	156	879	36	396	998	233	877	212
11	944	213	169	488		124		796	28	929	0	216	1053	176	925	104
12	529	0	2	273		267		1681	31	520	19	339	1734	27	520	312
13	700	119	111	233		385		245	69	692	287	71	707	62	610	68
14	287	36	0	233		29		264	0	286	0	12	304	21	285	6
21	78	0	16	25		0		244	0	78	0	59	362	367	77	0
22	1024	6968	135	1039		1450		598	20	1032	1655	728	6407	8663	1057	531
23	1093	7172	606	765		159		661	38	1094	110	568	5021	1851	1071	326
26	1226	493	9755	985		715		1040	55	1221	527	1046	9702	438	1162	556
27	75	0	195	18		9		67	0	75	35	48	197	0	66	4
28	87	59	93	65		0		69	0	72	0	42	108	45	87	27
31	853	199	108	311		1375		7	657	826	1470	90	973	127	854	97
32	706	76	0	575		3802		360	251	680	1242	344	1331	188	692	412
33	517	1030	430	199		1893		0	225	520	1852	95	1561	1191	514	93
34	99	0	20	4		134		0	31	55	110	0	119	0	99	0
35	18	0	0	0		166		0	0	18	166	0	83	0	18	0
41	58	0	0	24		7		28	0	58	30	0	65	14	58	0
44	98	82	0	0		29		0	0	23	29	0	98	105	23	0
Totals	20704	17711	12582	15169		16269		14281	7591	20328	12426	11260	43735	14768	20635	9366

Assembly Divisions	Building Construction Costs	Mechanical	Electrical	Repair & Remodel	Square Foot	Site Work Landsc.	Assemblies	Interior	Concrete Masonry	Open Shop	Heavy Construc.	Light Commercial	Facil. Construc.	Plumbing	Western Construction Costs	Asm Div	Residential
A		19	0	192	150	540	612	0	550		580	153	24	0		1	374
B		0	0	809	2479	0	5588	332	1914		368	2022	144	0		2	217
C		0	0	635	850	0	1206	1550	131		0	753	238	0		3	588
D		1014	789	693	1783	0	2401	753	0		0	1262	1011	890		4	867
E		0	0	85	257	0	294	5	0		0	257	5	0		5	393
F		0	0	0	123	0	124	0	0		0	123	3	0		6	358
G		465	172	331	111	1857	584	0	482		910	110	113	559		7	300
																8	760
																9	80
																10	0
																11	0
																12	0
Totals		1498	961	2745	5753	2397	10809	2640	3077		1858	4680	1538	1449			3937

Reference Section	Building Construction Costs	Mechanical	Electrical	Repair & Remodel	Square Foot	Site Work Landsc.	Assemblies	Interior	Concrete Masonry	Open Shop	Heavy Construc.	Light Commercial	Facil. Construc.	Plumbing	Western Construction Costs	Residential
Reference Tables	yes	yes	yes	yes	no	yes	yes	yes	yes	yes	yes	yes	yes	yes	yes	yes
Models					105							46				32
Crews	461	461	461	443		461		461	461	440	461	440	443	461	461	461
Equipment Rental Costs	yes	yes	yes	yes		yes		yes	yes	yes	yes	yes	yes	yes	yes	yes
Historical Cost Indexes	yes	yes	yes	yes	yes	yes	yes	yes	yes	yes	yes	yes	yes	yes	yes	no
City Cost Indexes	yes	yes	yes	yes	yes	yes	yes	yes	yes	yes	yes	yes	yes	yes	yes	yes

Annual Cost Guides

For more information visit RSMeans Web site at www.rsmeans.com

RSMeans Building Construction Cost Data 2007

Available in Both Softbound and Looseleaf Editions

Many customers enjoy the convenience and flexibility of the looseleaf binder, which increases the usefulness of *RSMeans Building Construction Cost Data 2007* by making it easy to add and remove pages. You can insert your own cost information pages, so everything is in one place. Copying pages for faxing is easier also. Whichever edition you prefer, softbound or the convenient looseleaf edition, you'll be eligible to receive *RSMeans Quarterly Update Service* FREE. Current subscribers can receive *RSMeans Quarterly Update Service* via e-mail.

Unit Prices Now Updated to MasterFormat 2004!

Now Available in Spanish!

$169.95 per copy, Looseleaf
Available Oct. 2006
Catalog No. 61017

RSMeans Building Construction Cost Data 2007

Offers you unchallenged unit price reliability in an easy-to-use arrangement. Whether used for complete, finished estimates or for periodic checks, it supplies more cost facts better and faster than any comparable source. Over 20,000 unit prices for 2007. The City Cost Indexes and Location Factors cover over 930 areas, for indexing to any project location in North America. Order and get *RSMeans Quarterly Update Service* FREE. You'll have year-long access to the RSMeans Estimating **HOTLINE** FREE with your subscription. Expert assistance when using RSMeans data is just a phone call away.

$136.95 per copy (English or Spanish)
Catalog No. 60017 (English) Available Oct. 2006
Catalog No. 60717 (Spanish) Available Jan. 2007

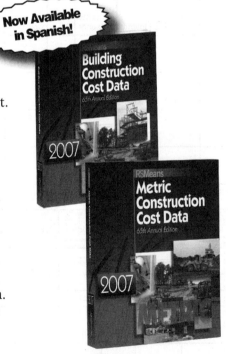

Now Available in Spanish!

Unit Prices Now Updated to MasterFormat 2004!

RSMeans Metric Construction Cost Data 2007

A massive compendium of all the data from both the 2007 *RSMeans Building Construction Cost Data* AND *Heavy Construction Cost Data*, in **metric** format! Access all of this vital information from one complete source. It contains more than 600 pages of unit costs and 40 pages of assemblies costs. The Reference Section contains over 200 pages of tables, charts and other estimating aids. A great way to stay in step with today's construction trends and rapidly changing costs.

$162.95 per copy
Available Dec. 2006
Catalog No. 63017

For more information visit RSMeans Web site at www.rsmeans.com

Annual Cost Guides

RSMeans Mechanical Cost Data 2007

- **HVAC**
- **Controls**

Total unit and systems price guidance for mechanical construction... materials, parts, fittings, and complete labor cost information. Includes prices for piping, heating, air conditioning, ventilation, and all related construction.

Plus new 2007 unit costs for:

- Over 2500 installed HVAC/controls assemblies
- "On Site" Location Factors for over 930 cities and towns in the U.S. and Canada
- Crews, labor, and equipment

$136.95 per copy
Available Oct. 2006
Catalog No. 60027

Unit Prices Now Updated to MasterFormat 2004!

RSMeans Plumbing Cost Data 2007

Comprehensive unit prices and assemblies for plumbing, irrigation systems, commercial and residential fire protection, point-of-use water heaters, and the latest approved materials. This publication and its companion, *RSMeans Mechanical Cost Data*, provide full-range cost estimating coverage for all the mechanical trades.

New for '07: More lines of no-hub CI soil pipe fittings, more flange-type escutcheons, fiberglass pipe insulation in a full range of sizes for 2-1/2" and 3" wall thicknesses, 220 lines of grease duct, and much more.

$136.95 per copy
Available Oct. 2006
Catalog No. 60217

RSMeans Electrical Cost Data 2007

Pricing information for every part of electrical cost planning. More than 13,000 unit and systems costs with design tables; clear specifications and drawings; engineering guides; illustrated estimating procedures; complete labor-hour and materials costs for better scheduling and procurement; and the latest electrical products and construction methods.

- A variety of special electrical systems including cathodic protection
- Costs for maintenance, demolition, HVAC/mechanical, specialties, equipment, and more

$136.95 per copy
Available Oct. 2006
Catalog No. 60037

Unit Prices Now Updated to MasterFormat 2004!

RSMeans Electrical Change Order Cost Data 2007

RSMeans Electrical Change Order Cost Data 2007 provides you with electrical unit prices exclusively for pricing change orders—based on the recent, direct experience of contractors and suppliers. Analyze and check your own change order estimates against the experience others have had doing the same work. It also covers productivity analysis and change order cost justifications. With useful information for calculating the effects of change orders and dealing with their administration.

$136.95 per copy
Available Nov. 2006
Catalog No. 60237

RSMeans Square Foot Costs 2007

It's Accurate and Easy To Use!

- **Updated 2007 price information**, based on nationwide figures from suppliers, estimators, labor experts, and contractors
- "How-to-Use" sections, with **clear examples** of commercial, residential, industrial, and institutional structures
- Realistic graphics, offering true-to-life illustrations of building projects
- Extensive information on using square foot cost data, including sample estimates and alternate pricing methods

$148.95 per copy
Over 450 pages, illustrated, available Nov. 2006
Catalog No. 60057

RSMeans Repair & Remodeling Cost Data 2007

Commercial/Residential

Use this valuable tool to estimate commercial and residential renovation and remodeling.

Includes: New costs for hundreds of unique methods, materials, and conditions that only come up in repair and remodeling, PLUS:

- Unit costs for over 15,000 construction components
- Installed costs for over 90 assemblies
- Over 930 "On-Site" localization factors for the U.S. and Canada

Unit Prices Now Updated to MasterFormat 2004!

$116.95 per copy
Available Nov. 2006
Catalog No. 60047

Annual Cost Guides

For more information visit RSMeans Web site at www.rsmeans.com

RSMeans Facilities Construction Cost Data 2007

For the maintenance and construction of commercial, industrial, municipal, and institutional properties. Costs are shown for new and remodeling construction and are broken down into materials, labor, equipment, overhead, and profit. Special emphasis is given to sections on mechanical, electrical, furnishings, site work, building maintenance, finish work, and demolition.

More than 43,000 unit costs, plus assemblies costs and a comprehensive Reference Section are included.

$323.95 per copy
Available Dec. 2006
Catalog No. 60207

Unit Prices Now Updated to MasterFormat 2004!

RSMeans Light Commercial Cost Data 2007

Specifically addresses the light commercial market, which is a specialized niche in the construction industry. Aids you, the owner/designer/contractor, in preparing all types of estimates—from budgets to detailed bids. Includes new advances in methods and materials.

Assemblies Section allows you to evaluate alternatives in the early stages of design/planning.

Over 11,000 unit costs for 2007 ensure you have the prices you need... when you need them.

$116.95 per copy
Available Dec. 2006
Catalog No. 60187

RSMeans Residential Cost Data 2007

Contains square foot costs for 30 basic home models with the look of today, plus hundreds of custom additions and modifications you can quote right off the page. With costs for the 100 residential systems you're most likely to use in the year ahead. Complete with blank estimating forms, sample estimates, and step-by-step instructions.

Now contains line items for cultured stone and brick, PVC trim lumber, and TPO roofing.

$116.95 per copy
Available Oct. 2006
Catalog No. 60177

Unit Prices Now Updated to MasterFormat 2004!

RSMeans Site Work & Landscape Cost Data 2007

Includes unit and assemblies costs for earthwork, sewerage, piped utilities, site improvements, drainage, paving, trees & shrubs, street openings/repairs, underground tanks, and more. Contains 57 tables of Assemblies Costs for accurate conceptual estimates.

2007 update includes:
- Estimating for infrastructure improvements
- Environmentally-oriented construction
- ADA-mandated handicapped access
- Hazardous waste line items

$136.95 per copy
Available Nov. 2006
Catalog No. 60287

RSMeans Assemblies Cost Data 2007

RSMeans Assemblies Cost Data 2007 takes the guesswork out of preliminary or conceptual estimates. Now you don't have to try to calculate the assembled cost by working up individual component costs. We've done all the work for you.

Presents detailed illustrations, descriptions, specifications, and costs for every conceivable building assembly—240 types in all—arranged in the easy-to-use UNIFORMAT II system. Each illustrated "assembled" cost includes a complete grouping of materials and associated installation costs, including the installing contractor's overhead and profit.

$223.95 per copy
Available Oct. 2006
Catalog No. 60067

RSMeans Open Shop Building Construction Cost Data 2007

The latest costs for accurate budgeting and estimating of new commercial and residential construction... renovation work... change orders... cost engineering.

RSMeans Open Shop "BCCD" will assist you to:
- Develop benchmark prices for change orders
- Plug gaps in preliminary estimates and budgets
- Estimate complex projects
- Substantiate invoices on contracts
- Price ADA-related renovations

Unit Prices Now Updated to MasterFormat 2004!

$136.95 per copy
Available Dec. 2006
Catalog No. 60157

Annual Cost Guides

For more information visit RSMeans Web site at www.rsmeans.com

RSMeans Building Construction Cost Data 2007
Western Edition

This regional edition provides more precise cost information for western North America. Labor rates are based on union rates from 13 western states and western Canada. Included are western practices and materials not found in our national edition: tilt-up concrete walls, glu-lam structural systems, specialized timber construction, seismic restraints, and landscape and irrigation systems.

$136.95 per copy
Available Dec. 2006
Catalog No. 60227

Unit Prices Now Updated to MasterFormat 2004!

RSMeans Heavy Construction Cost Data 2007

A comprehensive guide to heavy construction costs. Includes costs for highly specialized projects such as tunnels, dams, highways, airports, and waterways. Information on labor rates, equipment, and material costs is included. Features unit price costs, systems costs, and numerous reference tables for costs and design.

$136.95 per copy
Available Dec. 2006
Catalog No. 60167

RSMeans Construction Cost Indexes 2007

Who knows what 2007 holds? What materials and labor costs will change unexpectedly? By how much?

- Breakdowns for 316 major cities
- National averages for 30 key cities
- Expanded five major city indexes
- Historical construction cost indexes

$294.00 per year (subscription)
$73.50 individual quarters
Catalog No. 60147 A,B,C,D

RSMeans Interior Cost Data 2007

Provides you with prices and guidance needed to make accurate interior work estimates. Contains costs on materials, equipment, hardware, custom installations, furnishings, and labor costs... for new and remodel commercial and industrial interior construction, including updated information on office furnishings, and reference information.

Unit Prices Now Updated to MasterFormat 2004!

$136.95 per copy
Available Nov. 2006
Catalog No. 60097

RSMeans Concrete & Masonry Cost Data 2007

Provides you with cost facts for virtually all concrete/masonry estimating needs, from complicated formwork to various sizes and face finishes of brick and block—all in great detail. The comprehensive unit cost section contains more than 7,500 selected entries. Also contains an Assemblies Cost section, and a detailed Reference section that supplements the cost data.

$124.95 per copy
Available Dec. 2006
Catalog No. 60117

Unit Prices Now Updated to MasterFormat 2004!

RSMeans Labor Rates for the Construction Industry 2007

Complete information for estimating labor costs, making comparisons, and negotiating wage rates by trade for over 300 U.S. and Canadian cities. With 46 construction trades listed by local union number in each city, and historical wage rates included for comparison. Each city chart lists the county and is alphabetically arranged with handy visual flip tabs for quick reference.

$296.95 per copy
Available Dec. 2006
Catalog No. 60127

RSMeans Facilities Maintenance & Repair Cost Data 2007

RSMeans Facilities Maintenance & Repair Cost Data gives you a complete system to manage and plan your facility repair and maintenance costs and budget efficiently. Guidelines for auditing a facility and developing an annual maintenance plan. Budgeting is included, along with reference tables on cost and management, and information on frequency and productivity of maintenance operations.

The only nationally recognized source of maintenance and repair costs. Developed in cooperation with the Civil Engineering Research Laboratory (CERL) of the Army Corps of Engineers.

$296.95 per copy
Available Dec. 2006
Catalog No. 60307

Reference Books

For more information visit RSMeans Web site at www.rsmeans.com

Home Addition & Renovation Project Costs

This essential home remodeling reference gives you 35 project estimates, and guidance for some of the most popular home renovation and addition projects... from opening up a simple interior wall to adding an entire second story. Each estimate includes a floor plan, color photos, and detailed costs. Use the project estimates as backup for pricing, to check your own estimates, or as a cost reference for preliminary discussion with homeowners.

Includes:
- Case studies—with creative solutions and design ideas.
- Alternate materials costs—so you can match the estimates to the particulars of your projects.
- Location Factors—easy multipliers to adjust the book's costs to your own location.

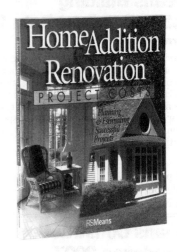

$29.95 per copy
Over 200 pages, illustrated, Softcover
Catalog No. 67349

Kitchen & Bath Project Costs:
Planning & Estimating Successful Projects

Project estimates for 35 of the most popular kitchen and bath renovations... from replacing a single fixture to whole-room remodels. Each estimate includes:
- All materials needed for the project
- Labor-hours to install (and demolish/remove) each item
- Subcontractor costs for certain trades and services
- An allocation for overhead and profit

PLUS! Takeoff and pricing worksheets—forms you can photocopy or access electronically from the book's Web site; alternate materials—unit costs for different finishes and fixtures; location factors—easy multipliers to adjust the costs to your location; and expert guidance on estimating methods, project design, contracts, marketing, working with homeowners, and tips for each of the estimated projects.

$29.95 per copy
Over 175 pages
Catalog No. 67347

Residential & Light Commercial Construction Standards 2nd Edition
By RSMeans and Contributing Authors

For contractors, subcontractors, owners, developers, architects, engineers, attorneys, and insurance personnel, this book provides authoritative requirements and recommendations compiled from the nation's leading professional associations, industry publications, and building code organizations.

It's an all-in-one reference for establishing a standard for workmanship, quickly resolving disputes, and avoiding defect claims. Includes practical guidance from professionals who are well-known in their respective fields for quality design and construction.

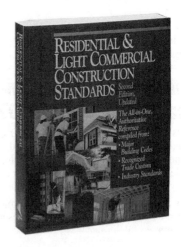

$59.95 per copy
600 pages, illustrated, Softcover
Catalog No. 67322A

For more information visit RSMeans Web site at www.rsmeans.com

Reference Books

Value Engineering: Practical Applications
By Alphonse Dell'Isola, PE
For Design, Construction, Maintenance & Operations

A tool for immediate application—for engineers, architects, facility managers, owners, and contractors. Includes: making the case for VE—the management briefing, integrating VE into planning, budgeting, and design, conducting life cycle costing, using VE methodology in design review and consultant selection, case studies, VE workbook, and a life cycle costing program on disk.

$79.95 per copy
Over 450 pages, illustrated, Softcover
Catalog No. 67319

Facilities Operations & Engineering Reference
By the Association for Facilities Engineering and RSMeans

An all-in-one technical reference for planning and managing facility projects and solving day-to-day operations problems. Selected as the official Certified Plant Engineer reference, this handbook covers financial analysis, maintenance, HVAC and energy efficiency, and more.

$54.98 per copy
Over 700 pages, illustrated, Hardcover
Catalog No. 67318

The Building Professional's Guide to Contract Documents
3rd Edition
By Waller S. Poage, AIA, CSI, CVS

A comprehensive reference for owners, design professionals, contractors, and students.

- Structure your documents for maximum efficiency.
- Effectively communicate construction requirements.
- Understand the roles and responsibilities of construction professionals.
- Improve methods of project delivery.

$32.48 per copy, 400 pages
Diagrams and construction forms, Hardcover
Catalog No. 67261A

Building Security: Strategies & Costs
By David Owen

This comprehensive resource will help you evaluate your facility's security needs, and design and budget for the materials and devices needed to fulfill them.

Includes over 130 pages of RSMeans cost data for installation of security systems and materials, plus a review of more than 50 security devices and construction solutions.

$44.98 per copy
350 pages, illustrated, Hardcover
Catalog No. 67339

Cost Planning & Estimating for Facilities Maintenance

In this unique book, a team of facilities management authorities shares their expertise on:

- Evaluating and budgeting maintenance operations
- Maintaining and repairing key building components
- Applying *RSMeans Facilities Maintenance & Repair Cost Data* to your estimating

Covers special maintenance requirements of the ten major building types.

$89.95 per copy
Over 475 pages, Hardcover
Catalog No. 67314

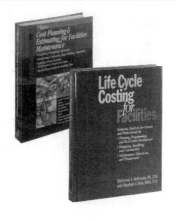

Life Cycle Costing for Facilities
By Alphonse Dell'Isola and Dr. Steven Kirk

Guidance for achieving higher quality design and construction projects at lower costs! Cost-cutting efforts often sacrifice quality to yield the cheapest product. Life cycle costing enables building designers and owners to achieve both. The authors of this book show how LCC can work for a variety of projects — from roads to HVAC upgrades to different types of buildings.

$99.95 per copy
450 pages, Hardcover
Catalog No. 67341

Planning & Managing Interior Projects 2nd Edition
By Carol E. Farren, CFM
Expert guidance on managing renovation & relocation projects.

This book guides you through every step in relocating to a new space or renovating an old one. From initial meeting through design and construction, to post-project administration, it helps you get the most for your company or client. Includes sample forms, spec lists, agreements, drawings, and much more!

$69.95 per copy
200 pages, Softcover
Catalog No. 67245A

Builder's Essentials: Best Business Practices for Builders & Remodelers
An Easy-to-Use Checklist System
By Thomas N. Frisby

A comprehensive guide covering all aspects of running a construction business, with more than 40 user-friendly checklists. Provides expert guidance on: increasing your revenue and keeping more of your profit, planning for long-term growth, keeping good employees, and managing subcontractors.

$29.95 per copy
Over 220 pages, Softcover
Catalog No. 67329

Reference Books

For more information visit RSMeans Web site at www.rsmeans.com

[Interior] Home Improvement [Costs] New 9th Edition

Updated estimates for the most popular remodeling and repair projects—from small, do-it-yourself jobs to major renovations and new construction. Includes: Kitchens & Baths; New Living Space from your Attic, Basement, or Garage; New Floors, Paint, and Wallpaper; Tearing Out or Building New Walls; Closets, Stairs, and Fireplaces; New Energy-Saving Improvements, Home Theaters, and More!

$24.95 per copy
250 pages, illustrated, Softcover
Catalog No. 67308E

Exterior Home Improvement Costs New 9th Edition

Updated estimates for the most popular remodeling and repair projects—from small, do-it-yourself jobs, to major renovations and new construction. Includes: Curb Appeal Projects—Landscaping, Patios, Porches, Driveways, and Walkways; New Windows and Doors; Decks, Greenhouses, and Sunrooms; Room Additions and Garages; Roofing, Siding, and Painting; "Green" Improvements to Save Energy & Water.

$24.95 per copy
Over 275 pages, illustrated, Softcover
Catalog No. 67309E

Builder's Essentials: Plan Reading & Material Takeoff
By Wayne J. DelPico
For Residential and Light Commercial Construction

A valuable tool for understanding plans and specs, and accurately calculating material quantities. Step-by-step instructions and takeoff procedures based on a full set of working drawings.

$35.95 per copy
Over 420 pages, Softcover
Catalog No. 67307

Builder's Essentials: Framing & Rough Carpentry 2nd Edition
By Scot Simpson

Develop and improve your skills with easy-to-follow instructions and illustrations. Learn proven techniques for framing walls, floors, roofs, stairs, doors, and windows. Updated guidance on standards, building codes, safety requirements, and more. Also available in Spanish!

$24.95 per copy
Over 150 pages, Softcover
Catalog No. 67298A
Spanish Catalog No. 67298AS

Concrete Repair and Maintenance Illustrated
By Peter Emmons

Hundreds of illustrations show users how to analyze, repair, clean, and maintain concrete structures for optimal performance and cost effectiveness. From parking garages to roads and bridges to structural concrete, this comprehensive book describes the causes, effects, and remedies for concrete wear and failure. Invaluable for planning jobs, selecting materials, and training employees, this book is a must-have for concrete specialists, general contractors, facility managers, civil and structural engineers, and architects.

$34.98 per copy
300 pages, illustrated, Softcover
Catalog No. 67146

Means Unit Price Estimating Methods
New 3rd Edition

This new edition includes up-to-date cost data and estimating examples, updated to reflect changes to the CSI numbering system and new features of RSMeans cost data. It describes the most productive, universally accepted ways to estimate, and uses checklists and forms to illustrate shortcuts and timesavers. A model estimate demonstrates procedures. A new chapter explores computer estimating alternatives.

$29.98 per copy
Over 350 pages, illustrated, Hardcover
Catalog No. 67303A

Total Productive Facilities Management
By Richard W. Sievert, Jr.

Today, facilities are viewed as strategic resources... elevating the facility manager to the role of asset manager supporting the organization's overall business goals. Now, Richard Sievert Jr., in this well-articulated guidebook, sets forth a new operational standard for the facility manager's emerging role... a comprehensive program for managing facilities as a true profit center.

$29.98 per copy
275 pages, Softcover
Catalog No. 67321

Means Environmental Remediation Estimating Methods 2nd Edition
By Richard R. Rast

Guidelines for estimating 50 standard remediation technologies. Use it to prepare preliminary budgets, develop estimates, compare costs and solutions, estimate liability, review quotes, negotiate settlements.

$49.98 per copy
Over 750 pages, illustrated, Hardcover
Catalog No. 64777A

For more information
visit RSMeans Web site
at www.rsmeans.com

Reference Books

Means Illustrated Construction Dictionary Condensed, 2nd Edition
Recognized in the industry as the best resource of its kind.

This essential tool has been further enhanced with updates to existing terms and the addition of hundreds of new terms and illustrations—in keeping with recent developments. For contractors, architects, insurance and real estate personnel, homeowners, and anyone who needs quick, clear definitions for construction terms.

$59.95 per copy
Over 500 pages, Softcover
Catalog No. 67282A

Means Repair & Remodeling Estimating New 4th Edition
By Edward B. Wetherill & RSMeans

This important reference focuses on the unique problems of estimating renovations of existing structures, and helps you determine the true costs of remodeling through careful evaluation of architectural details and a site visit.

New section on disaster restoration costs.

$69.95 per copy
Over 450 pages, illustrated, Hardcover
Catalog No. 67265B

Facilities Planning & Relocation
New, lower price and user-friendly format.
By David D. Owen

A complete system for planning space needs and managing relocations. Includes step-by-step manual, over 50 forms, and extensive reference section on materials and furnishings.

$89.95 per copy
Over 450 pages, Softcover
Catalog No. 67301

Means Square Foot & Assemblies Estimating Methods 3rd Edition

Develop realistic square foot and assemblies costs for budgeting and construction funding. The new edition features updated guidance on square foot and assemblies estimating using UNIFORMAT II. An essential reference for anyone who performs conceptual estimates.

$34.98 per copy
Over 300 pages, illustrated, Hardcover
Catalog No. 67145B

Means Electrical Estimating Methods 3rd Edition

Expanded edition includes sample estimates and cost information in keeping with the latest version of the CSI MasterFormat and UNIFORMAT II. Complete coverage of fiber optic and uninterruptible power supply electrical systems, broken down by components, and explained in detail. Includes a new chapter on computerized estimating methods. A practical companion to *RSMeans Electrical Cost Data*.

$64.95 per copy
Over 325 pages, Hardcover
Catalog No. 67230B

Means Mechanical Estimating Methods 3rd Edition

This guide assists you in making a review of plans, specs, and bid packages, with suggestions for takeoff procedures, listings, substitutions, and pre-bid scheduling. Includes suggestions for budgeting labor and equipment usage. Compares materials and construction methods to allow you to select the best option.

$64.95 per copy
Over 350 pages, illustrated, Hardcover
Catalog No. 67294A

Means ADA Compliance Pricing Guide New Second Edition
By Adaptive Environments and RSMeans

Completely updated and revised to the new 2004 *Americans with Disabilities Act Accessibility Guidelines*, this book features more than 70 of the most commonly needed modifications for ADA compliance. Projects range from installing ramps and walkways, widening doorways and entryways, and installing and refitting elevators, to relocating light switches and signage.

$79.95 per copy
Over 350 pages, illustrated, Softcover
Catalog No. 67310A

Project Scheduling & Management for Construction
New 3rd Edition
By David R. Pierce, Jr.

A comprehensive yet easy-to-follow guide to construction project scheduling and control—from vital project management principles through the latest scheduling, tracking, and controlling techniques. The author is a leading authority on scheduling, with years of field and teaching experience at leading academic institutions. Spend a few hours with this book and come away with a solid understanding of this essential management topic.

$64.95 per copy
Over 300 pages, illustrated, Hardcover
Catalog No. 67247B

Reference Books

For more information visit RSMeans Web site at www.rsmeans.com

The Practice of Cost Segregation Analysis
by Bruce A. Desrosiers and Wayne J. DelPico

This expert guide walks you through the practice of cost segregation analysis, which enables property owners to defer taxes and benefit from "accelerated cost recovery" through depreciation deductions on assets that are properly identified and classified.

With a glossary of terms, sample cost segregation estimates for various building types, key information resources, and updates via a dedicated Web site, this book is a critical resource for anyone involved in cost segregation analysis.

$99.95 per copy
Over 225 pages
Catalog No. 67345

Preventive Maintenance for Multi-Family Housing
by John C. Maciha

Prepared by one of the nation's leading experts on multi-family housing.

This complete PM system for apartment and condominium communities features expert guidance, checklists for buildings and grounds maintenance tasks and their frequencies, a reusable wall chart to track maintenance, and a dedicated Web site featuring customizable electronic forms. A must-have for anyone involved with multi-family housing maintenance and upkeep.

$89.95 per copy
225 pages
Catalog No. 67346

Means Landscape Estimating Methods
4th Edition
By Sylvia H. Fee

This revised edition offers expert guidance for preparing accurate estimates for new landscape construction and grounds maintenance. Includes a complete project estimate featuring the latest equipment and methods, and chapters on Life Cycle Costing and Landscape Maintenance Estimating.

$62.95 per copy
Over 300 pages, illustrated, Hardcover
Catalog No. 67295B

Job Order Contracting
Expediting Construction Project Delivery
by Allen Henderson

Expert guidance to help you implement JOC—fast becoming the preferred project delivery method for repair and renovation, minor new construction, and maintenance projects in the public sector and in many states and municipalities. The author, a leading JOC expert and practitioner, shows how to:

- Establish a JOC program
- Evaluate proposals and award contracts
- Handle general requirements and estimating
- Partner for maximum benefits

$89.95 per copy
192 pages, illustrated, Hardcover
Catalog No. 67348

Builder's Essentials: Estimating Building Costs
For the Residential & Light Commercial Contractor
By Wayne J. DelPico

Step-by-step estimating methods for residential and light commercial contractors. Includes a detailed look at every construction specialty—explaining all the components, takeoff units, and labor needed for well-organized, complete estimates. Covers correctly interpreting plans and specifications, and developing accurate and complete labor and material costs.

$29.95 per copy
Over 400 pages, illustrated, Softcover
Catalog No. 67343

Building & Renovating Schools

This all-inclusive guide covers every step of the school construction process—from initial planning, needs assessment, and design, right through moving into the new facility. A must-have resource for anyone concerned with new school construction or renovation. With square foot cost models for elementary, middle, and high school facilities, and real-life case studies of recently completed school projects.

The contributors to this book—architects, construction project managers, contractors, and estimators who specialize in school construction—provide start-to-finish, expert guidance on the process.

$99.95 per copy
Over 425 pages, Hardcover
Catalog No. 67342

For more information visit RSMeans Web site at www.rsmeans.com

Reference Books

Historic Preservation: Project Planning & Estimating
By Swanke Hayden Connell Architects

Expert guidance on managing historic restoration, rehabilitation, and preservation building projects and determining and controlling their costs. Includes:

- How to determine whether a structure qualifies as historic
- Where to obtain funding and other assistance
- How to evaluate and repair more than 75 historic building materials

$49.98 per copy
Over 675 pages, Hardcover
Catalog No. 67323

Means Illustrated Construction Dictionary
Unabridged 3rd Edition, with CD-ROM

Long regarded as the industry's finest, *Means Illustrated Construction Dictionary* is now even better. With the addition of over 1,000 new terms and hundreds of new illustrations, it is the clear choice for the most comprehensive and current information. The companion CD-ROM that comes with this new edition adds many extra features: larger graphics, expanded definitions, and links to both CSI MasterFormat numbers and product information.

$99.95 per copy
Over 790 pages, illustrated, Hardcover
Catalog No. 67292A

Designing & Building with the IBC 2nd Edition
By Rolf Jensen & Associates, Inc.

This updated, comprehensive guide helps building professionals make the transition to the 2003 International Building Code®. Includes a side-by-side code comparison of the IBC 2003 to the IBC 2000 and the three primary model codes, a quick-find index, and professional code commentary. With illustrations, abbreviations key, and an extensive Resource section.

$99.95 per copy
Over 875 pages, Softcover
Catalog No. 67328A

Means Plumbing Estimating Methods 3rd Edition
By Joseph Galeno and Sheldon Greene

Updated and revised! This practical guide walks you through a plumbing estimate, from basic materials and installation methods through change order analysis. *Plumbing Estimating Methods* covers residential, commercial, industrial, and medical systems, and features sample takeoff and estimate forms and detailed illustrations of systems and components.

$29.98 per copy
330+ pages, Softcover
Catalog No. 67283B

Builder's Essentials: Advanced Framing Methods
By Scot Simpson

A highly illustrated, "framer-friendly" approach to advanced framing elements. Provides expert, but easy to interpret, instruction for laying out and framing complex walls, roofs, and stairs, and special requirements for earthquake and hurricane protection. Also helps bring framers up to date on the latest building code changes, and provides tips on the lead framer's role and responsibilities, how to prepare for a job, and how to get the crew started.

$24.95 per copy
250 pages, illustrated, Softcover
Catalog No. 67330

Means Estimating Handbook
2nd Edition

Updated Second Edition answers virtually any estimating technical question—all organized by CSI MasterFormat. This comprehensive reference covers the full spectrum of technical data required to estimate construction costs. The book includes information on sizing, productivity, equipment requirements, code-mandated specifications, design standards, and engineering factors.

$99.95 per copy
Over 900 pages, Hardcover
Catalog No. 67276A

Preventive Maintenance Guidelines for School Facilities
By John C. Maciha

A complete PM program for K-12 schools that ensures sustained security, safety, property integrity, user satisfaction, and reasonable ongoing expenditures.

Includes schedules for weekly, monthly, semiannual, and annual maintenance in hard copy and electronic format.

$149.95 per copy
Over 225 pages, Hardcover
Catalog No. 67326

Preventive Maintenance for Higher Education Facilities
By Applied Management Engineering, Inc.

An easy-to-use system to help facilities professionals establish the value of PM, and to develop and budget for an appropriate PM program for their college or university. Features interactive campus building models typical of those found in different-sized higher education facilities, and PM checklists linked to each piece of equipment or system in hard copy and electronic format.

$149.95 per copy
150 pages, Hardcover
Catalog No. 67337

For more information visit RSMeans Web site at www.rsmeans.com

Seminars

Means CostWorks® Training

This one-day seminar has been designed with the intention of assisting both new and existing users to become more familiar with the *Means CostWorks* program. The class is broken into two unique sections: (1) A one-half day presentation on the function of each icon; and each student will be shown how to use the software to develop a cost estimate. (2) Hands-on estimating exercises that will ensure that each student thoroughly understands how to use *CostWorks*. You must bring your own laptop computer to this course.

Means CostWorks Benefits/Features:
- Estimate in your own spreadsheet format
- Power of RSMeans National Database
- Database automatically regionalized
- Save time with keyword searches
- Save time by establishing common estimate items in "Bookmark" files
- Customize your spreadsheet template
- Hot Key to Product Manufacturers' listings and specs
- Merge capability for networking environments
- View crews and assembly components
- AutoSave capability
- Enhanced sorting capability

Unit Price Estimating

This interactive two-day seminar teaches attendees how to interpret project information and process it into final, detailed estimates with the greatest accuracy level.

The single most important credential an estimator can take to the job is the ability to visualize construction in the mind's eye, and thereby estimate accurately.

Some Of What You'll Learn:
- Interpreting the design in terms of cost
- The most detailed, time-tested methodology for accurate "pricing"
- Key cost drivers—material, labor, equipment, staging, and subcontracts
- Understanding direct and indirect costs for accurate job cost accounting and change order management

Who Should Attend: Corporate and government estimators and purchasers, architects, engineers... and others needing to produce accurate project estimates.

Square Foot and Assemblies Estimating

This two-day course teaches attendees how to quickly deliver accurate square foot estimates using limited budget and design information.

Some Of What You'll Learn:
- How square foot costing gets the estimate done faster
- Taking advantage of a "systems" or "assemblies" format
- The RSMeans "building assemblies/square foot cost approach"
- How to create a very reliable preliminary and systems estimate using bare-bones design information

Who Should Attend: Facilities managers, facilities engineers, estimators, planners, developers, construction finance professionals... and others needing to make quick, accurate construction cost estimates at commercial, government, educational, and medical facilities.

Repair and Remodeling Estimating

This two-day seminar emphasizes all the underlying considerations unique to repair/remodeling estimating and presents the correct methods for generating accurate, reliable R&R project costs using the unit price and assemblies methods.

Some Of What You'll Learn:
- Estimating considerations—like labor-hours, building code compliance, working within existing structures, purchasing materials in smaller quantities, unforeseen deficiencies
- Identifying problems and providing solutions to estimating building alterations
- Rules for factoring in minimum labor costs, accurate productivity estimates, and allowances for project contingencies
- R&R estimating examples calculated using unit price and assemblies data

Who Should Attend: Facilities managers, plant engineers, architects, contractors, estimators, builders... and others who are concerned with the proper preparation and/or evaluation of repair and remodeling estimates.

Mechanical and Electrical Estimating

This two-day course teaches attendees how to prepare more accurate and complete mechanical/electrical estimates, avoiding the pitfalls of omission and double-counting, while understanding the composition and rationale within the RSMeans Mechanical/Electrical database.

Some Of What You'll Learn:
- The unique way mechanical and electrical systems are interrelated
- M&E estimates–conceptual, planning, budgeting, and bidding stages
- Order of magnitude, square foot, assemblies, and unit price estimating
- Comparative cost analysis of equipment and design alternatives

Who Should Attend: Architects, engineers, facilities managers, mechanical and electrical contractors... and others needing a highly reliable method for developing, understanding, and evaluating mechanical and electrical contracts.

Plan Reading and Material Takeoff

This two-day program teaches attendees to read and understand construction documents and to use them in the preparation of material takeoffs.

Some of What You'll Learn:
- Skills necessary to read and understand typical contract documents—blueprints and specifications
- Details and symbols used by architects and engineers
- Construction specifications' importance in conjunction with blueprints
- Accurate takeoff of construction materials and industry-accepted takeoff methods

Who Should Attend: Facilities managers, construction supervisors, office managers... and others responsible for the execution and administration of a construction project, including government, medical, commercial, educational, or retail facilities.

Facilities Maintenance and Repair Estimating

This two-day course teaches attendees how to plan, budget, and estimate the cost of ongoing and preventive maintenance and repair for existing buildings and grounds.

Some Of What You'll Learn:
- The most financially favorable maintenance, repair, and replacement scheduling and estimating
- Auditing and value engineering facilities
- Preventive planning and facilities upgrading
- Determining both in-house and contract-out service costs
- Annual, asset-protecting M&R plan

Who Should Attend: Facility managers, maintenance supervisors, buildings and grounds superintendents, plant managers, planners, estimators... and others involved in facilities planning and budgeting.

Scheduling and Project Management

This two-day course teaches attendees the most current and proven scheduling and management techniques needed to bring projects in on time and on budget.

Some Of What You'll Learn:
- Crucial phases of planning and scheduling
- How to establish project priorities and develop realistic schedules and management techniques
- Critical Path and Precedence Methods
- Special emphasis on cost control

Who Should Attend: Construction project managers, supervisors, engineers, estimators, contractors... and others who want to improve their project planning, scheduling, and management skills.

Assessing Scope of Work for Facility Construction Estimating

This two-day course is a practical training program that addresses the vital importance of understanding the SCOPE of projects in order to produce accurate cost estimates in a facility repair and remodeling environment.

Some Of What You'll Learn:
- Discussions on site visits, plans/specs, record drawings of facilities, and site-specific lists
- Review of CSI divisions, including means, methods, materials, and the challenges of scoping each topic
- Exercises in SCOPE identification and SCOPE writing for accurate estimating of projects
- Hands-on exercises that require SCOPE, take-off, and pricing

Who Should Attend: Corporate and government estimators, planners, facility managers... and others needing to produce accurate project estimates.

Seminars

2007 RSMeans Seminar Schedule

For more information visit RSMeans Web site at www.rsmeans.com

Location	Dates
Las Vegas, NV	March
Washington, DC	April
Phoenix, AZ	April
Denver, CO	May
San Francisco, CA	June
Philadelphia, PA	June
Washington, DC	September
Dallas, TX	September
Las Vegas, NV	October
Orlando, FL	November
Atlantic City, NJ	November
San Diego, CA	December

Note: Call for exact dates and details.

Registration Information

Register Early... Save up to $100! Register 30 days before the start date of a seminar and save $100 off your total fee. *Note: This discount can be applied only once per order. It cannot be applied to team discount registrations or any other special offer.*

How to Register Register by phone today! RSMeans' toll-free number for making reservations is: **1-800-334-3509.**

Individual Seminar Registration Fee $935. *Means CostWorks*® Training Registration Fee $375. To register by mail, complete the registration form and return with your full fee to: Seminar Division, Reed Construction Data, RSMeans Seminars, 63 Smiths Lane, Kingston, MA 02364.

Federal Government Pricing All federal government employees save 25% off regular seminar price. Other promotional discounts cannot be combined with Federal Government discount.

Team Discount Program Two to four seminar registrations, call for pricing: 1-800-334-3509, Ext. 5115

Multiple Course Discounts When signing up for two or more courses, call for pricing.

Refund Policy Cancellations will be accepted up to ten days prior to the seminar start. There are no refunds for cancellations received later than ten working days prior to the first day of the seminar. A $150 processing fee will be applied for all cancellations. Written notice of cancellation is required. Substitutions can be made at any time before the session starts. **No-shows are subject to the full seminar fee.**

AACE Approved Courses Many seminars described and offered here have been approved for 14 hours (1.4 recertification credits) of credit by the AACE International Certification Board toward meeting the continuing education requirements for recertification as a Certified Cost Engineer/Certified Cost Consultant.

AIA Continuing Education We are registered with the AIA Continuing Education System (AIA/CES) and are committed to developing quality learning activities in accordance with the CES criteria. Many seminars meet the AIA/CES criteria for Quality Level 2. AIA members may receive (14) learning units (LUs) for each two-day RSMeans course.

NASBA CPE Sponsor Credits We are part of the National Registry of CPE Sponsors. Attendees may be eligible for (16) CPE credits.

Daily Course Schedule The first day of each seminar session begins at 8:30 A.M. and ends at 4:30 P.M. The second day is 8:00 A.M.–4:00 P.M. Participants are urged to bring a hand-held calculator since many actual problems will be worked out in each session.

Continental Breakfast Your registration includes the cost of a continental breakfast, a morning coffee break, and an afternoon break. These informal segments will allow you to discuss topics of mutual interest with other members of the seminar. (You are free to make your own lunch and dinner arrangements.)

Hotel/Transportation Arrangements RSMeans has arranged to hold a block of rooms at most host hotels. To take advantage of special group rates when making your reservation, be sure to mention that you are attending the RSMeans Seminar. You are, of course, free to stay at the lodging place of your choice. (**Hotel reservations and transportation arrangements should be made directly by seminar attendees.**)

Important Class sizes are limited, so please register as soon as possible.

Note: Pricing subject to change.

Registration Form

Call **1-800-334-3509** to register or FAX **1-800-632-6732**. Visit our Web site: www.rsmeans.com

Please register the following people for the RSMeans Construction Seminars as shown here. We understand that we must make our own hotel reservations if overnight stays are necessary.

☐ Full payment of $_____ enclosed.

☐ Bill me

Name of Registrant(s)
(To appear on certificate of completion)

P.O. #:
GOVERNMENT AGENCIES MUST SUPPLY PURCHASE ORDER NUMBER OR TRAINING FORM.

Firm Name
Address
City/State/Zip
Telephone No.   Fax No.
E-mail Address
Charge our registration(s) to: ☐ MasterCard ☐ VISA ☐ American Express ☐ Discover
Account No.   Exp. Date
Cardholder's Signature
Seminar Name   City   Dates

Please mail check to: Seminar Division, Reed Construction Data, RSMeans Seminars, 63 Smiths Lane, P.O. Box 800, Kingston, MA 02364 USA

MeansData™

CONSTRUCTION COSTS FOR SOFTWARE APPLICATIONS
Your construction estimating software is only as good as your cost data.

A proven construction cost database is a mandatory part of any estimating package. The following list of software providers can offer you MeansData™ as an added feature for their estimating systems. See the table below for what types of products and services they offer (match their numbers). Visit online at www.rsmeans.com/demosource/ for more information and free demos. Or call their numbers listed below.

1. **3D International**
   713-871-7000
   venegas@3di.com

2. **4Clicks-Solutions, LLC**
   719-574-7721
   mbrown@4clicks-solutions.com

3. **Aepco, Inc.**
   301-670-4642
   blueworks@aepco.com

4. **Applied Flow Technology**
   800-589-4943
   info@aft.com

5. **ArenaSoft Estimating**
   888-370-8806
   info@arenasoft.com

6. **Ares Corporation**
   925-299-6700
   sales@arescorporation.com

7. **Beck Technology**
   214-303-6293
   stewartcarroll@beckgroup.com

8. **BSD - Building Systems Design, Inc.**
   888-273-7638
   bsd@bsdsoftlink.com

9. **CMS - Computerized Micro Solutions**
   800-255-7407
   cms@proest.com

10. **Corecon Technologies, Inc.**
    714-895-7222
    sales@corecon.com

11. **CorVet Systems**
    301-622-9069
    sales@corvetsys.com

12. **Earth Tech**
    303-771-3103
    kyle.knudson@earthtech.com

13. **Estimating Systems, Inc.**
    800-967-8572
    esipulsar@adelphia.net

14. **HCSS**
    800-683-3196
    info@hcss.com

15. **MC^2 - Management Computer**
    800-225-5622
    vkeys@mc2-ice.com

16. **Maximus Asset Solutions**
    800-659-9001
    assetsolutions@maximus.com

17. **Sage Timberline Office**
    800-628-6583
    productinfo.timberline@sage.com

18. **Shaw Beneco Enterprises, Inc.**
    877-719-4748
    inquire@beneco.com

19. **US Cost, Inc.**
    800-372-4003
    sales@uscost.com

20. **Vanderweil Facility Advisors**
    617-451-5100
    info@VFA.com

21. **WinEstimator, Inc.**
    800-950-2374
    sales@winest.com

TYPE	1	2	3	4	5	6	7	8	9	10	11	12	13	14	15	16	17	18	19	20	21
BID					•			•	•		•			•	•		•				•
Estimating		•			•	•	•	•	•	•	•	•	•	•	•		•	•	•		•
DOC/JOC/SABER		•			•			•			•		•			•	•	•			•
ID/IQ		•									•		•			•		•			•
Asset Mgmt.																•				•	•
Facility Mgmt.	•		•													•				•	
Project Mgmt.	•	•					•			•	•			•			•	•			
TAKE-OFF					•				•	•	•	•	•	•	•		•		•		
EARTHWORK											•		•		•						
Pipe Flow				•										•							
HVAC/Plumbing						•			•			•									
Roofing						•			•			•									
Design	•					•		•									•	•			•
Other Offers/Links:																					
Accounting/HR		•			•											•	•	•			
Scheduling						•	•										•		•		•
CAD							•										•				
PDA																					
Lt. Versions		•							•								•				
Consulting	•	•			•		•		•		•	•				•	•	•	•	•	
Training		•			•		•	•	•	•	•	•	•	•	•	•	•	•	•		•

Qualified re-seller applications now being accepted. Call Carol Polio, Ext. 5107.

FOR MORE INFORMATION
CALL 1-800-448-8182, EXT. 5107 OR FAX 1-800-632-6732

For more information visit RSMeans Web site at www.rsmeans.com

New Titles

Understanding & Negotiating Construction Contracts

By Kit Werremeyer

Take advantage of the author's 30 years' experience in small-to-large (including international) construction projects. Learn how to identify, understand, and evaluate high risk terms and conditions typically found in all construction contracts—then negotiate to lower or eliminate the risk, improve terms of payment, and reduce exposure to claims and disputes. The author avoids "legalese" and gives real-life examples from actual projects.

$69.95 per copy
300 pages, Softcover
Catalog No. 67350

Green Building: Project Planning & Cost Estimating, 2nd Edition

This new edition has been completely updated with the latest in green building technologies, design concepts, standards, and costs. Now includes a 2007 Green Building *CostWorks* CD with more than 300 green building assemblies and over 5,000 unit price line items for sustainable building. The new edition is also full-color with all new case studies—plus a new chapter on deconstruction, a key aspect of green building.

$129.95 per copy
350 pages, Softcover
Catalog No. 67338A

How to Estimate with Means Data & CostWorks
New 3rd Edition

By RSMeans and Saleh A. Mubarak, Ph.D.

New 3rd Edition—fully updated with new chapters, plus new CD with updated *CostWorks* cost data and MasterFormat organization. Includes all major construction items—with more than 300 exercises and two sets of plans that show how to estimate for a broad range of construction items and systems—including general conditions and equipment costs.

$59.95 per copy
272 pages, Softcover
Includes CostWorks CD
Catalog No. 67324B

Means Spanish/English Construction Dictionary
2nd Edition

By RSMeans and the International Code Council

This expanded edition features thousands of the most common words and useful phrases in the construction industry with easy-to-follow pronunciations in both Spanish and English. Over 800 new terms, phrases, and illustrations have been added. It also features a new stand-alone "Safety & Emergencies" section, with colored pages for quick access. Unique to this dictionary are the systems illustrations showing the relationship of components in the most common building systems for all major trades.

$23.95 per copy
Over 400 pages
Catalog No. 67327A

Construction Business Management

By Nick Ganaway

Only 43% of construction firms stay in business after four years. Make sure your company thrives with valuable guidance from a pro with 25 years of success as a commercial contractor. Find out what it takes to build all aspects of a business that is profitable, enjoyable, and enduring. With a bonus chapter on retail construction.

$49.95 per copy
200 pages, Softcover
Catalog No. 67352

The Homeowner's Guide to Mold

Expert guidance to protect your health and your home.

Mold, whether caused by leaks, humidity or flooding, is a real health and financial issue—for homeowners and contractors. This full-color book explains:

- Construction and maintenance practices to prevent mold
- How to inspect for and remove mold
- Mold remediation procedures and costs
- What to do after a flood
- How to deal with insurance companies if you're thinking of submitting a mold damages claim

$21.95 per copy
144 pages, Softcover
Catalog No. 67344

ORDER TOLL FREE 1-800-334-3509
OR FAX 1-800-632-6732

2007 Order Form

Qty.	Book No.	COST ESTIMATING BOOKS	Unit Price	Total
	60067	Assemblies Cost Data 2007	$223.95	
	60017	Building Construction Cost Data 2007	136.95	
	61017	Building Const. Cost Data–Looseleaf Ed. 2007	169.95	
	60717	Building Const. Cost Data–Spanish 2007	136.95	
	60227	Building Const. Cost Data–Western Ed. 2007	136.95	
	60117	Concrete & Masonry Cost Data 2007	124.95	
	50147	Construction Cost Indexes 2007 (subscription)	294.00	
	60147A	Construction Cost Index–January 2007	73.50	
	60147B	Construction Cost Index–April 2007	73.50	
	60147C	Construction Cost Index–July 2007	73.50	
	60147D	Construction Cost Index–October 2007	73.50	
	60347	Contr. Pricing Guide: Resid. R & R Costs 2007	39.95	
	60337	Contr. Pricing Guide: Resid. Detailed 2007	39.95	
	60327	Contr. Pricing Guide: Resid. Sq. Ft. 2007	39.95	
	60237	Electrical Change Order Cost Data 2007	136.95	
	60037	Electrical Cost Data 2007	136.95	
	60207	Facilities Construction Cost Data 2007	323.95	
	60307	Facilities Maintenance & Repair Cost Data 2007	296.95	
	60167	Heavy Construction Cost Data 2007	136.95	
	60097	Interior Cost Data 2007	136.95	
	60127	Labor Rates for the Const. Industry 2007	296.95	
	60187	Light Commercial Cost Data 2007	116.95	
	60027	Mechanical Cost Data 2007	136.95	
	63017	Metric Construction Cost Data 2007	162.95	
	60157	Open Shop Building Const. Cost Data 2007	136.95	
	60217	Plumbing Cost Data 2007	136.95	
	60047	Repair and Remodeling Cost Data 2007	116.95	
	60177	Residential Cost Data 2007	116.95	
	60287	Site Work & Landscape Cost Data 2007	136.95	
	60057	Square Foot Costs 2007	148.95	
	62017	Yardsticks for Costing (2007)	136.95	
	62016	Yardsticks for Costing (2006)	126.95	
		REFERENCE BOOKS		
	67310A	ADA Compliance Pricing Guide, 2nd Ed.	79.95	
	67330	Bldrs Essentials: Adv. Framing Methods	24.95	
	67329	Bldrs Essentials: Best Bus. Practices for Bldrs	29.95	
	67298A	Bldrs Essentials: Framing/Carpentry 2nd Ed.	24.95	
	67298AS	Bldrs Essentials: Framing/Carpentry Spanish	24.95	
	67307	Bldrs Essentials: Plan Reading & Takeoff	35.95	
	67342	Building & Renovating Schools	99.95	
	67261A	Bldg. Prof. Guide to Contract Documents, 3rd Ed.	32.48	
	67339	Building Security: Strategies & Costs	44.98	
	67146	Concrete Repair & Maintenance Illustrated	34.98	
	67352	Construction Business Management	49.95	
	67314	Cost Planning & Est. for Facil. Maint.	89.95	
	67328A	Designing & Building with the IBC, 2nd Ed.	99.95	
	67230B	Electrical Estimating Methods, 3rd Ed.	64.95	
	64777A	Environmental Remediation Est. Methods, 2nd Ed.	49.98	
	67343	Estimating Bldg. Costs for Resi. & Lt. Comm.	29.95	

Qty.	Book No.	REFERENCE BOOKS (Cont.)	Unit Price	Total
	67276A	Estimating Handbook, 2nd Ed.	$99.95	
	67318	Facilities Operations & Engineering Reference	54.98	
	67301	Facilities Planning & Relocation	89.95	
	67338A	Green Building: Proj. Planning & Cost Est., 2nd Ed.	129.95	
	67323	Historic Preservation: Proj. Planning & Est.	49.98	
	67349	Home Addition & Renovation Project Costs	29.95	
	67308E	Home Improvement Costs–Int. Projects, 9th Ed.	24.95	
	67309E	Home Improvement Costs–Ext. Projects, 9th Ed.	24.95	
	67344	Homeowner's Guide to Mold	21.95	
	67324B	How to Est.w/Means Data & CostWorks, 3rd Ed.	59.95	
	67282A	Illustrated Const. Dictionary, Condensed, 2nd Ed.	59.95	
	67292A	Illustrated Const. Dictionary, w/CD-ROM, 3rd Ed.	99.95	
	67348	Job Order Contracting	89.95	
	67347	Kitchen & Bath Project Costs	29.95	
	67295B	Landscape Estimating Methods, 4th Ed.	62.95	
	67341	Life Cycle Costing for Facilities	99.95	
	67294A	Mechanical Estimating Methods, 3rd Ed.	64.95	
	67245A	Planning & Managing Interior Projects, 2nd Ed.	69.95	
	67283B	Plumbing Estimating Methods, 3rd Ed.	29.98	
	67345	Practice of Cost Segregation Analysis	99.95	
	67337	Preventive Maint. for Higher Education Facilities	149.95	
	67346	Preventive Maint. for Multi-Family Housing	89.95	
	67326	Preventive Maint. Guidelines for School Facil.	149.95	
	67247B	Project Scheduling & Management for Constr. 3rd Ed.	64.95	
	67265B	Repair & Remodeling Estimating Methods, 4th Ed.	69.95	
	67322A	Resi. & Light Commercial Const. Stds., 2nd Ed.	59.95	
	67327A	Spanish/English Construction Dictionary, 2nd Ed.	23.95	
	67145B	Sq. Ft. & Assem. Estimating Methods, 3rd Ed.	34.98	
	67321	Total Productive Facilities Management	29.98	
	67350	Understanding and Negotiating Const. Contracts	69.95	
	67303A	Unit Price Estimating Methods, 3rd Ed.	29.98	
	67319	Value Engineering: Practical Applications	79.95	

MA residents add 5% state sales tax		
Shipping & Handling**		
Total (U.S. Funds)*		

Prices are subject to change and are for U.S. delivery only. *Canadian customers may call for current prices. **Shipping & handling charges: Add 7% of total order for check and credit card payments. Add 9% of total order for invoiced orders.

Send Order To: ADDV-1000

Name (Please Print) _____

Company _____

☐ Company
☐ Home Address _____

_____

City/State/Zip _____

Phone # _____ P.O. # _____

(Must accompany all orders being billed)

Mail To: RSMeans, P.O. Box 800, Kingston, MA 02364-0800

Reed Construction Data, Inc.

Reed Construction Data, Inc. is a leading worldwide provider of total construction information solutions. The company's portfolio of information products and services is designed specifically to help construction industry professionals advance their businesses with timely, accurate, and actionable project, product, and cost data. Each of these groups offers a variety of innovative products and services created for the full spectrum of design, construction, distribution, and manufacturing professionals. Reed Construction Data is a division of Reed Business Information, a member of the Reed Elsevier plc group of companies.

Cost Information
RSMeans, the undisputed market leader and authority on construction costs, publishes current cost and estimating information in annual cost books, on *Means CostWorks*® CD, and on the Web. RSMeans furnishes the construction industry with a rich library of corresponding reference books and a series of professional seminars that are designed to sharpen personal skills and maximize the effective use of cost estimating and management tools. RSMeans also provides construction cost consulting for owners, manufacturers, designers, and contractors.

Project Data
Reed Construction Data maintains a significant role in the overall construction process by facilitating the assembly of public and private project data and delivering this information to contractors, distributors, and building product manufacturers in a secure, accurate, and timely manner. Acting on behalf of architects, engineers, owner/developers, and government agencies, Reed Construction Data is the construction community's premier resource for project leads and bid documents. Our wide variety of products and services provides the most up-to-date, enhanced details on many of the country's largest public sector, commercial, industrial, and multi-family residential projects.

Reed Bulletin and Reed CONNECT™ provide project leads and project data through all stages of construction. Customers are supplied industry data through leads, project reports, contact lists, plans, specifications, and addenda—either online, via e-mail, or in paper format.

Building Product Information
The Reed First Source suite of products is the only integrated building product information system offered to the commercial construction industry for searching, selecting, and specifying building products. These online and print resources include the *First Source* catalog, SPEC-DATA™, MANU-SPEC ™, First Source CAD, and manufacturer's catalogs. Written by industry professionals and organized using CSI MasterFormat 2004 criteria, the design and construction community uses this information to make better, more informed design decisions.

Research & Analytics
Reed Construction Data's forecasting tools cover most aspects of the construction business. Our vast network of resources makes us uniquely qualified to give you the information you need to keep your business profitable.

Associated Construction Publications (ACP)
Reed Construction Data's regional construction magazines cover the nation through a network of 14 regional magazines. Serving the construction market for more than 100 years, our magazines are a trusted source of news and information in the local and national construction communities.

International
Reed Construction Data Canada is the "voice of construction" in Canada, and leading provider of project information and construction market intelligence. Products and services include: *Building Reports*, Canadian ICI project leads tracking construction projects through life cycles—from concept to construction. *Building Reports* can be customized and delivered through CONNECT™, RCD's web-based sales and contact management system, or by fax or email; *Daily Commercial News*, Ontario's leading construction industry daily newspaper, contains ICI project leads, Building Reports, Bidders' Register, news and trends, construction tenders, certificates of substantial performance, and CanaData Economic Snapshot; *Journal of Commerce*, Alberta and British Columbia's twice-weekly construction industry newspaper, is a one-stop source for private and public projects, tenders and expressions of interest, top industry news and construction trends, and more; *CanaData*, a variety of products tracking market performance, economic analysis, forecasts, and trends in the construction market; *Buildcore*, organized by MasterFormat™, is Canada's most comprehensive listing of commercial building products and manufacturers active and available in Canada—in print and online at www.buildcore.com.

Reed Business Information – Scandinavia, with offices in Denmark, Norway, Finland, and Sweden, is the construction industry's source for project information in the Nordic region.

Reed Construction Data—Australia (formerly Cordell Building Information Services) is Australia's leading provider and respected industry authority on building and construction information.

For more information, please visit our website at www. reedconstructiondata.com

Reed Construction Data, Inc.
30 Technology Parkway South
Norcross, GA 30092-2912
(800) 322-6996
(800) 895-8661 (fax)
Email: info@reedbusiness.com

RSMeans Project Cost Report

By filling out this report, your project data will contribute to the database that supports the RSMeans Project Cost Square Foot Data. When you fill out this form, RSMeans will provide a $30 discount off one of the RSMeans products advertised in the preceding pages. Please complete the form including all items where you have cost data, and all the items marked (✔).

$30.00 Discount per product for each report you submit.

Project Description (No remodeling projects, please.)

✔ Building Use (Office, School...) _____

✔ Address (City, State) _____

_____

✔ Frame (Wood, Steel...) _____

✔ Exterior Wall (Brick, Tilt-up...) _____

✔ Basement: (check one) ☐ Full ☐ Partial ☐ None

✔ Number Stories _____

✔ Floor-to-Floor Height _____

% Air Conditioned _____ Tons _____

Comments _____

Total Project Cost $ _____

Owner _____

Architect _____

General Contractor _____

✔ Bid Date _____

Typical Bay Size _____

✔ Labor Force: _____ % Union   _____ % Non-Union

✔ Project Description (Circle one number in each line)

   1. Economy 2. Average   3. Custom   4. Luxury
   1. Square   2. Rectangular 3. Irregular 4. Very Irregular

			$				$
A	✔	General Conditions	$	K	✔	Specialties	$
B	✔	Site Work	$	L	✔	Equipment	$
C	✔	Concrete	$	M	✔	Furnishings	$
D	✔	Masonry	$	N	✔	Special Construction	$
E	✔	Metals	$	P	✔	Conveying Systems	$
F	✔	Wood & Plastics	$	Q	✔	Mechanical	$
G	✔	Thermal & Moisture Protection		QP		Plumbing	$
GR		Roofing & Flashing	$	QB		HVAC	$
H	✔	Doors and Windows	$	R	✔	Electrical	$
J	✔	Finishes	$	S	✔	Mech./Elec. Combined	$
JP		Painting & Wall Covering	$				

Please specify the RSMeans product you wish to receive.
Complete the address information.

Product Name _____

Product Number _____

Your Name _____

Title _____

Company _____

  ☐ Company
  ☐ Home   Street Address _____

City, State, Zip _____

Method of Payment:

Credit Card # _____

Expiration Date _____

Check _____

Purchase Order _____

Reed Construction Data/RSMeans
Square Foot Costs Department
P.O. Box 800
Kingston, MA 02364-9988

Return by mail or Fax 888-492-6770.